Code of Federal Regulations

Interpharm Master Keyword Guide: 21 CFR Regulations of the U.S. Food and Drug Administration

2002-2003 Edition

Revised as of April 1, 2002

Interpharm/CRC

Boca Raton London New York Washington, D.C.

Library of Congress Cataloging-in-Publication Data

Catalog record is available from the Library of Congress

Direct all inquiries to CRC Press LLC, 2000 N.W. Corporate Blvd., Boca Raton, Florida 33431.

Visit the CRC Press Web site at www.crcpress.com

International Standard Book Number 0-8493-1851-3
Printed in the United States of America 1 2 3 4 5 6 7 8 9 0
Printed on acid-free paper

Interpharm Master Keyword Guide: 21 CFR
2002–2003 Edition
USER'S GUIDE

Each entry in the Master Keyword Guide is arranged as follows:

KEYWORD	SECTION NUMBER	SECTION TITLE (contains the keyword in context) Subpart Title and/or Part Title

The KEYWORD and SECTION TITLE are shown in upper case.
The Subpart Title and/or Part Title are shown in lower case with initial capitals.

Scope, Purpose, Definitions, and Applicability appear at the start of each Part for which they are included, and are not listed separately in the Master Keyword Guide.

USING THE MASTER KEYWORD GUIDE

1. Look up the keyword of interest.
2. Note the context in which the keyword is mentioned in the section title and the details of the subpart or part title.
3. If the reference is of interest, look up the details of the regulations in the Code of Federal Regulations Title 21 (edition revised as of April 2002) under the section number provided.

Readers should note that where a concept or keyword is part of the text of regulations but the word is not included by FDA/DEA in the section title, then the concept-keyword does not appear in this Keyword Guide. Thus, for example, "Good Clinical Practices" is a concept incorporating many different regulations, but the phrase is never used as part of a section title, and hence does not appear in the Keyword Guide.

Table of Contents

A

A2 ASSAY, HEMOGLOBIN	864.7400	HEMOGLOBIN A2 ASSAY Hematology Kits And Packages; Hematology And Pathology Devices
A, FACTOR	866.5330	FACTOR XIII, A, S, IMMUNOLOGICAL TEST SYSTEM Immunological Test Systems; Immunology And Microbiology Devices
A, FIBRINOPEPTIDE	866.5350	FIBRINOPEPTIDE A IMMUNOLOGICAL TEST SYSTEM Immunological Test Systems; Immunology And Microbiology Devices
A, IMMUNOGLOBULIN	866.5510	IMMUNOGLOBULINS A, G, M, D, AND E IMMUNOLOGICAL TEST SYSTEM Immunological Test Systems; Immunology And Microbiology Devices
A, VITAMIN	131.127	NONFAT DRY MILK FORTIFIED WITH VITAMINS A AND D Requirements For Specific Standardized Milk And Cream; Milk And Cream
A, VITAMIN	184.1930	VITAMIN A Listing Of Specific Substances Affirmed As GRAS; Direct Food Substances Affirmed As Generally Recognized As Safe
A, VITAMIN	582.5930	VITAMIN A Nutrients And/or Dietary Supplements; Substances Generally Recognized As Safe
A, VITAMIN	582.5933	VITAMIN A ACETATE Nutrients And/or Dietary Supplements; Substances Generally Recognized As Safe
A, VITAMIN	582.5936	VITAMIN A PALMITATE Nutrients And/or Dietary Supplements; Substances Generally Recognized As Safe
A, VITAMIN	862.1805	VITAMIN A TEST SYSTEM Clinical Chemistry Test Systems; Clinical Chemistry And Clinical Toxicology Devices
ABANDONMENT	1303.27	ABANDONMENT OF QUOTA Individual Manufacturing Quotas; Quotas
ABBREVIATED APPLICATION	314.92	DRUG PRODUCTS FOR WHICH ABBREVIATED APPLICATIONS MAY BE SUBMITTED Applications For FDA Approval To Market A New Drug
ABBREVIATED APPLICATION	314.94	CONTENT AND FORMAT OF AN ABBREVIATED APPLICATION Applications For FDA Approval To Market A New Drug
ABBREVIATED APPLICATION	314.96	AMENDMENTS TO AN UNAPPROVED ABBREVIATED APPLICATION Applications For FDA Approval To Market A New Drug
ABBREVIATED APPLICATION	314.97	SUPPLEMENTS AND OTHER CHANGES TO AN APPROVED ABBREVIATED APPLICATION Applications For FDA Approval To Market A New Drug
ABBREVIATED APPLICATION	314.99	OTHER RESPONSIBILITIES OF AN APPLICANT OF AN ABBREVIATED APPLICATION Applications For FDA Approval To Market A New Drug
ABBREVIATED APPLICATION	314.105	APPROVAL OF AN APPLICATION AND AN ABBREVIATED APPLICATION Applications For FDA Approval To Market A New Drug
ABBREVIATED APPLICATION	314.122	SUBMITTING AN ABBREVIATED APPLICATION FOR, OR A 505(J)(2)(C) PETITION Applications For FDA Approval To Market A New Drug
ABBREVIATED APPLICATION	314.150	WITHDRAWAL OF APPROVAL OF AN APPLICATION OR ABBREVIATED APPLICATION Applications For FDA Approval To Market A New Drug
ABBREVIATED APPLICATION	314.152	NOTICE OF WITHDRAWAL OF APPROVAL OF AN APPLICATION OR ABBREVIATED APPLICATION FOR A NEW DRUG Applications For FDA Approval To Market A New Drug
ABBREVIATED APPLICATION	314.430	AVAILABILITY FOR PUBLIC DISCLOSURE OF DATA AND INFORMATION IN AN APPLICATION OR ABBREVIATED APPLICATION Applications For FDA Approval To Market A New Drug
ABBREVIATED APPLICATIONS	314.100	TIME FRAMES FOR REVIEWING APPLICATIONS AND ABBREVIATED APPLICATIONS Applications For FDA Approval To Market A New Drug
ABBREVIATED APPLICATIONS	314.440	ADDRESSES FOR APPLICATIONS AND ABBREVIATED APPLICATIONS Applications For FDA Approval To Market A New Drug
ABBREVIATED NEW DRUG APPLICATION	314.107	EFFECTIVE DATE OF APPROVAL OF A 505(B)(2) APPLICATION OR ABBREVIATED NEW DRUG APPLICATION Applications For FDA Approval To Market A New Drug
ABBREVIATED NEW DRUG APPLICATION	314.127	REFUSAL TO APPROVE AN ABBREVIATED NEW DRUG APPLICATION Applications For FDA Approval To Market A New Drug
ABBREVIATED NEW DRUG APPLICATION	314.151	WITHDRAWAL OF APPROVAL OF AN ABBREVIATED NEW DRUG APPLICATION UNDER SECTION 505(J)(5) Applications For FDA Approval To Market A New Drug
ABBREVIATED NEW DRUG APPLICATION	314.153	SUSPENSION OF APPROVAL OF AN ABBREVIATED NEW DRUG APPLICATION Applications For FDA Approval To Market A New Drug
ABBREVIATED NEW DRUG APPLICATIONS	5.106	SUBMISSION OF AND EFFECTIVE APPROVAL DATES FOR ABBREVIATED NEW DRUG APPLICATIONS AND CERTAIN NEW DRUG APPLICATIONS Human Drugs; Redelegations Of Authority; Delegations Of Authority And Organization

ABBREVIATED NEW DRUG APPLICATIONS	5.503	SUBMISSION OF AND EFFECTIVE APPROVAL DATES FOR ABBREVIATED NEW ANIMAL DRUG APPLICATIONS AND CERTAIN NEW ANIMAL DRUG APPLICATIONS Animal Drugs; Redelegations Of Authority; Delegations Of Authority And Organization
ABBREVIATED REPORTS	1002.12	ABBREVIATED REPORTS Required Manufacturers' Reports For Listed Electronic Products; Records And Reports
ABDUCTION	890.3665	CONGENITAL HIP DISLOCATION ABDUCTION SPLINT Physical Medicine Prosthetic Devices; Physical Medicine Devices
ABDOMINAL	884.5225	ABDOMINAL DECOMPRESSION CHAMBER Obstetrical And Gynecological Therapeutic Devices; Obstetrical And Gynecological Devices
ABNORMAL	864.7415	ABNORMAL HEMOGLOBIN ASSAY Hematology Kits And Packages; Hematology And Pathology Devices
ABORTION	884.5050	METREURYNTER-BALLOON ABORTION SYSTEM Obstetrical And Gynecological Therapeutic Devices; Obstetrical And Gynecological Devices
ABORTION	884.5070	VACUUM ABORTION SYSTEM Obstetrical And Gynecological Therapeutic Devices; Obstetrical And Gynecological Devices
ABRASIVE AGENT	872.6030	ORAL CAVITY ABRASIVE POLISHING AGENT Miscellaneous Devices; Dental Devices
ABRASIVE DEVICE	872.6010	ABRASIVE DEVICE AND ACCESSORIES Miscellaneous Devices; Dental Devices
ABROAD, ENVIRONMENTAL IMPACT	25.60	ENVIRONMENTAL EFFECTS ABROAD OF MAJOR AGENCY ACTIONS Other Requirements; Environmental Impact Considerations
ABSORBABLE	878.4490	ABSORBABLE HEMOSTATIC AGENT AND DRESSING Surgical Devices; General And Plastic Surgery Devices
ABSORBABLE IMPLANT	886.3300	ABSORBABLE IMPLANT (SCLERAL BUCKLING METHOD) Prosthetic Devices; Ophthalmic Devices
ABSORBABLE POWDER	878.4480	ABSORBABLE POWDER FOR LUBRICATING A SURGEON'S GLOVE Surgical Devices; General And Plastic Surgery Devices
ABSORBABLE SUTURE	878.4493	ABSORBABLE POLY(GLYCOLIDE/L-LACTIDE) SURGICAL SUTURE Surgical Devices; General And Plastic Surgery Devices
ABSORBENT	868.5300	CARBON DIOXIDE ABSORBENT Therapeutic Devices; Anesthesiology Devices
ABSORBENT	880.5300	MEDICAL ABSORBENT FIBER General Hospital And Personal Use Therapeutic Devices; General Hospital And Personal Use Devices
ABSORBENT	880.6025	ABSORBENT TIPPED APPLICATOR General Hospital And Personal Use Miscellaneous Devices; General Hospital And Personal Use Devices
ABSORBER	868.5310	CARBON DIOXIDE ABSORBER Therapeutic Devices; Anesthesiology Devices
ABSORBER	872.6050	SALIVA ABSORBER Miscellaneous Devices; Dental Devices
ABSORPTION	862.2850	ATOMIC ABSORPTION SPECTROPHOTOMETER FOR CLINICAL USE Clinical Laboratory Instruments; Clinical Chemistry And Clinical Toxicology Devices
ABUSE, DRUGS OF	809.40	RESTRICTIONS ON THE SALE, DISTRIBUTION, AND USE OF OTC TEST SAMPLE COLLECTION SYSTEMS FOR DRUGS OF ABUSE TESTING Requirements For Manufacturers And Producers; In Vitro Diagnostic Products For Human Use
ABUSE, DRUGS OF	864.3260	OTC TEST SAMPLE COLLECTION SYSTEMS FOR DRUGS OF ABUSE TESTING Pathology Instrumentation And Accessories; Hematology And Pathology Devices
ABUSE, POTENTIAL	314.104	DRUGS WITH POTENTIAL FOR ABUSE FDA Action On Applications; Applications For FDA Approval To Market A New Drug
ACACIA	184.1330	ACACIA (GUM ARABIC) Listing Of Specific Substances Affirmed As GRAS; Direct Food Substances Affirmed As Generally Recognized As Safe
ACACIA	872.3400	KARAYA AND SODIUM BORATE WITH OR WITHOUT ACACIA DENTURE ADHESIVE Prosthetic Devices; Dental Devices
ACCEPTABLE LEVEL, OZONE	801.415	MAXIMUM ACCEPTABLE LEVEL OF OZONE Special Requirements For Specific Devices; Labeling
ACCEPTANCE	5.606	ACCEPTANCE OF ASSISTANCE FROM STATE AND LOCAL AUTHORITIES FOR ENFORCEMENT OF RADIATION CONTROL LEGISLATION AND REGULATIONS Radiation Control; Redelegations Of Authority; Delegations Of Authority And Organization
ACCEPTANCE	820.80	RECEIVING, IN-PROCESS, AND FINISHED DEVICE ACCEPTANCE Acceptance Activities; Quality System Regulation
ACCEPTANCE	820.86	ACCEPTANCE STATUS Acceptance Activities; Quality System Regulation

ACCEPTANCE	1301.14	FILING OF APPLICATION; ACCEPTANCE FOR FILING; DEFECTIVE APPLICATIONS	Registration Of Manufacturers, Distributors, And Dispensers Of Controlled Substances
ACCEPTANCE	1309.34	ACCEPTANCE FOR FILING; DEFECTIVE APPLICATIONS	Registration Of Manufacturers, Distributors, Importers And Exporters Of List I Chemicals
ACCESS	20.21	UNIFORM ACCESS TO RECORDS	General Policy; Public Information
ACCESS	21.40	PROCEDURES FOR SUBMITTING REQUESTS FOR NOTIFICATION AND ACCESS	Procedures For Notification Of And Access To Records In Privacy Act Record Systems; Protection Of Privacy
ACCESS	21.43	ACCESS TO REQUESTED RECORDS	Procedures For Notification Of And Access To Records In Privacy Act Record Systems; Protection Of Privacy
ACCESS	21.65	ACCESS TO RECORDS IN EXEMPT SYSTEMS	Exemptions; Protection Of Privacy
ACCESS	806.30	FDA ACCESS TO RECORDS	Medical Devices; Reports Of Corrections And Removals
ACCESS	876.5540	BLOOD ACCESS DEVICE AND ACCESSORIES	Therapeutic Devices; Gastroenterology-urology Devices
ACCESS	1401.4	ACCESS TO INFORMATION	Public Availability Of Information
ACCESS	1403.42	RETENTION AND ACCESS REQUIREMENTS FOR RECORDS	Reports, Records, Retention, And Enforcement, Uniform Administrative Requirements For Grants And Coop. Agreements To State And Local Gov.
ACCESSORIES	864.3600	MICROSCOPES AND ACCESSORIES	Pathology Instrumentation And Accessories; Hematology And Pathology Devices
ACCESSORIES	868.5860	PRESSURE TUBING AND ACCESSORIES	Therapeutic Devices; Anesthesiology Devices
ACCESSORIES	872.3980	ENDOSSEOUS DENTAL IMPLANT ACCESSORIES	Prosthetic Devices; Dental Devices
ACCESSORIES	872.4120	BONE CUTTING INSTRUMENT AND ACCESSORIES	Surgical Devices; Dental Devices
ACCESSORIES	872.4200	DENTAL HANDPIECE AND ACCESSORIES	Surgical Devices; Dental Devices
ACCESSORIES	872.4920	DENTAL ELECTROSURGICAL UNIT AND ACCESSORIES	Surgical Devices; Dental Devices
ACCESSORIES	872.5410	ORTHODONTIC APPLIANCE AND ACCESSORIES	Therapeutic Devices; Dental Devices
ACCESSORIES	872.6010	ABRASIVE DEVICE AND ACCESSORIES	Miscellaneous Devices; Dental Devices
ACCESSORIES	872.6250	DENTAL CHAIR AND ACCESSORIES	Miscellaneous Devices; Dental Devices
ACCESSORIES	872.6300	RUBBER DAM AND ACCESSORIES	Miscellaneous Devices; Dental Devices
ACCESSORIES	872.6640	DENTAL OPERATIVE UNIT AND ACCESSORIES	Miscellaneous Devices; Dental Devices
ACCESSORIES	874.4680	BRONCHOSCOPE (FLEXIBLE OR RIGID) AND ACCESSORIES	Surgical Devices; Ear, Nose, And Throat Devices
ACCESSORIES	874.4710	ESOPHAGOSCOPE (FLEXIBLE OR RIGID) AND ACCESSORIES	Surgical Devices; Ear, Nose, And Throat Devices
ACCESSORIES	874.4720	MEDIASTINOSCOPE AND ACCESSORIES	Surgical Devices; Ear, Nose, And Throat Devices
ACCESSORIES	874.4760	NASOPHARYNGOSCOPE (FLEXIBLE OR RIGID) AND ACCESSORIES	Surgical Devices; Ear, Nose, And Throat Devices
ACCESSORIES	876.1500	ENDOSCOPE AND ACCESSORIES	Diagnostic Devices; Gastroenterology-urology Devices
ACCESSORIES	876.4300	ENDOSCOPIC ELECTROSURGICAL UNIT AND ACCESSORIES	Surgical Devices; Gastroenterology-urology Devices
ACCESSORIES	876.4730	MANUAL GASTROENTEROLOGY-UROLOGY SURGICAL INSTRUMENT AND ACCESSORIES	Surgical Devices; Gastroenterology-urology Devices
ACCESSORIES	876.4890	UROLOGICAL TABLE AND ACCESSORIES	Surgical Devices; Gastroenterology-urology Devices
ACCESSORIES	876.5010	BILIARY CATHETER AND ACCESSORIES	Therapeutic Devices; Gastroenterology-urology Devices
ACCESSORIES	876.5090	SUPRAPUBIC UROLOGICAL CATHETER AND ACCESSORIES	Therapeutic Devices; Gastroenterology-urology Devices
ACCESSORIES	876.5130	UROLOGICAL CATHETER AND ACCESSORIES	Therapeutic Devices; Gastroenterology-urology Devices
ACCESSORIES	876.5250	URINE COLLECTOR AND ACCESSORIES	Therapeutic Devices; Gastroenterology-urology Devices

ACCESSORIES	876.5540	BLOOD ACCESS DEVICE AND ACCESSORIES Therapeutic Devices; Gastroenterology-urology Devices
ACCESSORIES	876.5630	PERITONEAL DIALYSIS SYSTEM AND ACCESSORIES Therapeutic Devices; Gastroenterology-urology Devices
ACCESSORIES	876.5820	HEMODIALYSIS SYSTEM AND ACCESSORIES Therapeutic Devices; Gastroenterology-urology Devices
ACCESSORIES	876.5880	ISOLATED KIDNEY PERFUSION AND TRANSPORT SYSTEM AND ACCESSORIES Therapeutic Devices; Gastroenterology-urology Devices
ACCESSORIES	876.5900	OSTOMY POUCH AND ACCESSORIES Therapeutic Devices; Gastroenterology-urology Devices
ACCESSORIES	876.5980	GASTROINTESTINAL TUBE AND ACCESSORIES Therapeutic Devices; Gastroenterology-urology Devices
ACCESSORIES	878.1800	SPECULUM AND ACCESSORIES Diagnostic Devices; General And Plastic Surgery Devices
ACCESSORIES	878.4160	SURGICAL CAMERA AND ACCESSORIES Surgical Devices; General And Plastic Surgery Devices
ACCESSORIES	878.4200	INTRODUCTION/DRAINAGE CATHETER AND ACCESSORIES Surgical Devices; General And Plastic Surgery Devices
ACCESSORIES	878.4400	ELECTROSURGICAL CUTTING AND COAGULATION DEVICE AND ACCESSORIES Surgical Devices; General And Plastic Surgery Devices
ACCESSORIES	878.4820	SURGICAL INSTRUMENT MOTORS AND ACCESSORIES/ATTACHMENTS Surgical Devices; General And Plastic Surgery Devices
ACCESSORIES	878.4950	MANUAL OPERATING TABLE/CHAIR AND ACCESSORIES Surgical Devices; General And Plastic Surgery Devices
ACCESSORIES	878.4960	OPERATING TABLES/CHAIRS AND ACCESSORIES Surgical Devices; General And Plastic Surgery Devices
ACCESSORIES	882.4300	MANUAL CRANIAL DRILLS, BURRS, TREPHINES, AND THEIR ACCESSORIES Neurological Surgical Devices; Neurological Devices
ACCESSORIES	882.4305	POWERED COMPOUND CRANIAL DRILLS, BURRS, TREPHINES, AND THEIR ACCESSORIES Neurological Surgical Devices; Neurological Devices
ACCESSORIES	882.4310	POWERED SIMPLE CRANIAL DRILLS, BURRS, TREPHINES, AND THEIR ACCESSORIES Neurological Surgical Devices; Neurological Devices
ACCESSORIES	884.1175	ENDOMETRIAL SUCTION CURETTE AND ACCESSORIES Obstetrical And Gynecological Diagnostic Devices; Obstetrical And Gynecological Devices
ACCESSORIES	884.1300	UTEROTUBAL CARBON DIOXIDE INSUFFLATOR AND ACCESSORIES Obstetrical And Gynecological Diagnostic Devices; Obstetrical And Gynecological Devices
ACCESSORIES	884.1600	TRANSABDOMINAL AMNIOSCOPE (FETOSCOPE) AND ACCESSORIES Obstetrical And Gynecological Diagnostic Devices; Obstetrical And Gynecological Devices
ACCESSORIES	884.1640	CULDOSCOPE AND ACCESSORIES Obstetrical And Gynecological Diagnostic Devices; Obstetrical And Gynecological Devices
ACCESSORIES	884.1660	TRANSCERVICAL ENDOSCOPE (AMNIOSCOPE) AND ACCESSORIES Obstetrical And Gynecological Diagnostic Devices; Obstetrical And Gynecological Devices
ACCESSORIES	884.1690	HYSTEROSCOPE AND ACCESSORIES Obstetrical And Gynecological Diagnostic Devices; Obstetrical And Gynecological Devices
ACCESSORIES	884.1720	GYNECOLOGIC LAPAROSCOPE AND ACCESSORIES Obstetrical And Gynecological Diagnostic Devices; Obstetrical And Gynecological Devices
ACCESSORIES	884.2640	FETAL PHONOCARDIOGRAPH MONITOR AND ACCESSORIES Obstetrical And Gynecological Monitoring Devices; Obstetrical And Gynecological Devices
ACCESSORIES	884.2660	FETAL ULTRASONIC MONITOR AND ACCESSORIES Obstetrical And Gynecological Monitoring Devices; Obstetrical And Gynecological Devices
ACCESSORIES	884.2700	INTRAUTERINE PRESSURE MONITOR AND ACCESSORIES Obstetrical And Gynecological Monitoring Devices; Obstetrical And Gynecological Devices
ACCESSORIES	884.2720	EXTERNAL UTERINE CONTRACTION MONITOR AND ACCESSORIES Obstetrical And Gynecological Monitoring Devices; Obstetrical And Gynecological Devices
ACCESSORIES	884.2740	PERINATAL MONITORING SYSTEM AND ACCESSORIES Obstetrical And Gynecological Monitoring Devices; Obstetrical And Gynecological Devices

ACCESSORIES	884.2960	OBSTETRIC ULTRASONIC TRANSDUCER AND ACCESSORIES Obstetrical And Gynecological Monitoring Devices; Obstetrical And Gynecological Devices
ACCESSORIES	884.4100	ENDOSCOPIC ELECTROCAUTERY AND ACCESSORIES Obstetrical And Gynecological Surgical Devices; Obstetrical And Gynecological Devices
ACCESSORIES	884.4120	GYNECOLOGIC ELECTROCAUTERY AND ACCESSORIES Obstetrical And Gynecological Surgical Devices; Obstetrical And Gynecological Devices
ACCESSORIES	884.4150	BIPOLAR ENDOSCOPIC COAGULATOR-CUTTER AND ACCESSORIES Obstetrical And Gynecological Surgical Devices; Obstetrical And Gynecological Devices
ACCESSORIES	884.4160	UNIPOLAR ENDOSCOPIC COAGULATOR-CUTTER AND ACCESSORIES Obstetrical And Gynecological Surgical Devices; Obstetrical And Gynecological Devices
ACCESSORIES	884.4900	OBSTETRIC TABLE AND ACCESSORIES Obstetrical And Gynecological Surgical Devices; Obstetrical And Gynecological Devices
ACCESSORIES	884.5350	CONTRACEPTIVE DIAPHRAGM AND ACCESSORIES Obstetrical And Gynecological Therapeutic Devices; Obstetrical And Gynecological Devices
ACCESSORIES	884.6120	ASSISTED REPRODUCTION ACCESSORIES Assisted Reproduction Devices; Obstetrical And Gynecological Devices
ACCESSORIES	884.6190	ASSISTED REPRODUCTIVE MICROSCOPES AND MICROSCOPE ACCESSORIES Assisted Reproduction Devices; Obstetrical And Gynecological Devices
ACCESSORIES	886.1930	TONOMETER AND ACCESSORIES Diagnostic Devices; Ophthalmic Devices
ACCESSORIES	888.4580	SONIC SURGICAL INSTRUMENT AND ACCESSORIES/ATTACHMENTS Surgical Devices; Orthopedic Devices
ACCESSORIES	888.5850	NONPOWERED ORTHOPEDIC TRACTION APPARATUS AND ACCESSORIES Surgical Devices; Orthopedic Devices
ACCESSORY	870.4200	CARDIOPULMONARY BYPASS ACCESSORY EQUIPMENT Cardiovascular Surgical Devices; Cardiovascular Devices
ACCESSORY	890.3910	WHEELCHAIR ACCESSORY Physical Medicine Prosthetic Devices; Physical Medicine Devices
ACCESSORY	890.5160	TRACTION ACCESSORY Physical Medicine Therapeutic Devices; Physical Medicine Devices
ACCIDENTAL INGESTION	369.9	GENERAL WARNINGS RE ACCIDENTAL INGESTION BY CHILDREN Definitions And Interpretations; Interpretative Statements Re Warnings On Drugs And Devices For OTC Sale
ACCIDENTAL RADIATION	1002.20	REPORTING OF ACCIDENTAL RADIATION OCCURRENCES Manufacturers' Reports On Accidental Radiation Occurrences; Records And Reports
ACCOUNTING	21.70	DISCLOSURE AND INTRA-AGENCY USE OF RECORDS IN PRIVACY ACT RECORD SYSTEMS; NO ACCOUNTING REQUIRED Disclosure Of Records In Privacy Act Record Systems To Persons Other Than The Subject Individual; Protection Of Privacy
ACCOUNTING	21.71	DISCLOSURE OF RECORDS IN PRIVACY ACT RECORD SYSTEMS; ACCOUNTING REQUIRED Disclosure Of Records In Privacy Act Record Systems To Persons Other Than The Subject Individual; Protection Of Privacy
ACCREDITATION	5.417	AUTHORITY PERTAINING TO ACCREDITATION FUNCTIONS FOR MEDICAL DEVICES Medical Devices And Radiological Health; Redelegations Of Authority; Delegations Of Authority And Organization
ACCREDITATION	900.3	APPLICATION FOR APPROVAL AS AN ACCREDITATION BODY Accreditation; Mammography
ACCREDITATION	900.4	STANDARDS FOR ACCREDITATION BODIES Accreditation; Mammography
ACCREDITATION	900.13	REVOCATION OF ACCREDITATION AND OF ACCREDITATION BODY APPROVAL Quality Standards And Certification; Mammography
ACCREDITATION	900.15	APPEALS OF ADVERSE ACCREDITATION OR REACCREDITATION DECISIONS Quality Standards And Certification; Mammography
ACCREDITING BODIES	5.303	ESTABLISHING STANDARDS AND APPROVING ACCREDITING BODIES UNDER THE NATIONAL LABORATORY ACCREDITATION PROGRAM Foods And Cosmetics; Redelegations Of Authority; Delegations Of Authority And Organization
ACCURACY	21.73	ACCURACY, COMPLETENESS, TIMELINESS, AND RELEVANCE OF RECORDS DISCLOSED FROM PRIVACY ACT RECORD SYSTEMS Disclosure Of Records In Privacy Act Record Systems To Persons Other ThanThe Subject Individual; Protection Of Privacy

A.C.E.	862.1090	ANGIOTENSIN CONVERTING ENZYME (A.C.E.) TEST SYSTEM Clinical Chemistry Test Systems; Clinical Chemistry And Clinical Toxicology Devices
ACEPROMAZINE MALEATE	520.23	ACEPROMAZINE MALEATE TABLETS Oral Dosage Form New Animal Drugs
ACEPROMAZINE MALEATE	522.23	ACEPROMAZINE MALEATE INJECTION Implantation Or Injectable Dosage Form New Animal Drugs
ACESULFAME	172.800	ACESULFAME POTASSIUM Multipurpose Additives; Food Additives Permitted For Direct Addition To Food For Human Consumption
ACETABULAR	888.3320	HIP JOINT METAL/METAL SEMI-CONSTRAINED, WITH A CEMENTED ACETABULAR COMPONENT, PROSTHESIS Prosthetic Devices; Orthopedic Devices
ACETABULAR	888.3330	HIP JOINT METAL/METAL SEMI-CONSTRAINED, WITH AN UNCEMENTED ACETABULAR COMPONENT, PROSTHESIS Prosthetic Devices; Orthopedic Devices
ACETABULAR	888.3370	HIP JOINT (HEMI-HIP) ACETABULAR METAL CEMENTED PROSTHESIS Prosthetic Devices; Orthopedic Devices
ACETAMINOPHEN	862.3030	ACETAMINOPHEN TEST SYSTEM Clinical Toxicology Test Systems; Clinical Chemistry And Clinical Toxicology Devices
ACETATE, CALCIUM	184.1185	CALCIUM ACETATE Listing Of Specific Substances Affirmed As GRAS; Direct Food Substances Affirmed As Generally Recognized As Safe
ACETATE, CALCIUM	582.6185	CALCIUM ACETATE Sequestrants; Substances Generally Recognized As Safe
ACETATE, ETHYL	173.228	ETHYL ACETATE Solvents, Lubricants, Release Agents And Related Substances; Secondary Direct Food Additives Permitted In Food For Human Consumption
ACETATE, ETHYL	584.200	ETHYL ALCOHOL CONTAINING ETHYL ACETATE Listing Of Substances Affirmed As GRAS; Food Substances Affirmed As GRAS In Feed And Drinking Water Of Animals
ACETATE, HYDROCORTISONE	524.155	BACITRACIN ZINC-POLYMIXIN B SULFATE-NEOMYCIN SULFATE-HYDROCORTISONE OR HYDROCORTISONE ACETATE OPHTHALMIC OINTMENT Ophthalmic And Topical Dosage Form New Animal Drugs
ACETATE, HYDROCORTISONE	524.1204	KANAMYCIN SULFATE, CALCIUM AMPHOMYCIN, AND HYDROCORTISONE ACETATE Ophthalmic And Topical Dosage Form New Animal Drugs
ACETATE, HYDROCORTISONE	524.1484d	NEOMYCIN SULFATE, HYDROCORTISONE ACETATE, TETRACAINE HYDROCHLORIDE EAR OINTMENT Ophthalmic And Topical Dosage Form New Animal Drug
ACETATE, HYDROCORTISONE	524.1484i	NEOMYCIN SULFATE, HYDROCORTISONE ACETATE, STERILE OINTMENT Ophthalmic And Topical Dosage Form New Animal Drugs
ACETATE, LEAD	73.2396	LEAD ACETATE Cosmetics; Listing Of Color Additives Exempt From Certification
ACETATE, MELENGESTROL	556.380	MELENGESTROL ACETATE Tolerances For Residues Of New Animal Drugs In Food
ACETATE, MELENGESTROL	558.342	MELENGESTROL ACETATE New Animal Drugs For Use In Animal Feeds
ACETATE, SODIUM	184.1721	SODIUM ACETATE Listing Of Specific Substances Affirmed As GRAS; Direct Food Substances Affirmed As Generally Recognized As Safe
ACETATE, SODIUM	582.1721	SODIUM ACETATE General Purpose Food Additives; Substances Generally Recognized As Safe
ACETATE, TOCOPHEROL	182.8892	α-TOCOPHEROL ACETATE Nutrients; Substances Generally Recognized As Safe
ACETATE, TOCOPHEROL	582.5892	α-TOCOPHEROL ACETATE Nutrients And/or Dietary Supplements; Substances Generally Recognized As Safe
ACETATE, VINYL	175.350	VINYL ACETATE/CROTONIC ACID COPOLYMER Substances For Use As Components Of Coatings; Indirect Food Additives
ACETATE, VINYL	177.1350	ETHYLENE-VINYL ACETATE-VINYL ALCOHOL COPOLYMERS Substances For Use As Basic Components Of Single And Repeated Use Food Contact Surfaces; Indirect Food Additives: Polymers
ACETATE, VINYL	177.1360	ETHYLENE-VINYL ACETATE-VINYL ALCOHOL COPOLYMERS Substances For Use As Basic Components Of Single And Repeated Use Food Contact Surfaces; Indirect Food Additives: Polymers
ACETATE, VITAMIN A	582.5933	VITAMIN A ACETATE Nutrients And/or Dietary Supplements; Substances Generally Recognized As Safe
ACETAZOLAMIDE SODIUM	520.44	ACETAZOLAMIDE SODIUM SOLUBLE POWDER Oral Dosage Form New Animal Drugs
ACETAZOLAMIDE, SODIUM	522.44	STERILE SODIUM ACETAZOLAMIDE Implantation Or Injectable Dosage Form New Animal Drugs

ACETIC ACID	184.1005	ACETIC ACID Listing Of Specific Substances Affirmed As GRAS; Direct Food Substances Affirmed As Generally Recognized As Safe
ACETIC ACID	582.1005	ACETIC ACID General Purpose Food Additives; Substances Generally Recognized As Safe
ACETIC ACID	862.1390	5-HYDROXYINDOLE ACETIC ACID/SEROTONIN TEST SYSTEM Clinical Chemistry Test Systems; Clinical Chemistry And Toxicology Devices
ACETOLACTATE	173.115	ALPHA-ACETOLACTATE DECARBOXYLASE (ALPHA-ALDC) ENZYME PREPARATION DERIVED FROM A RECOMBINANT BACILLUS SUBTILIS Enzyme Preparations And Microorganisms; Secondary Direct Food Additives Permitted In Food For Human Consumption
ACETONE	73.1150	DIHYDROXYACETONE Drugs; Listing Of Color Additives Exempt From Certification
ACETONE	73.2150	DIHYDROXYACETONE Cosmetics; Listing Of Color Additives Exempt From Certification
ACETONE	172.802	ACETONE PEROXIDES Multipurpose Additives; Food Additives Permitted For Direct Addition To Food For Human Consumption
ACETONE	173.210	ACETONE Solvents, Lubricants, Release Agents And Related Substances; Secondary Direct Food Additives Permitted In Food For Human Consumption
ACETONIDE	524.2481	TRIAMCINOLONE ACETONIDE CREAM Ophthalmic And Topical Dosage Form New Animal Drugs
ACETOPHENETIDIN	201.309	ACETOPHENETIDIN (PHENACETIN)-CONTAINING PREPARATIONS; NECESSARY WARNING STATEMENT Specific Labeling Requirements For Specific Drug Products; Labeling
ACETYL	172.372	N-ACETYL-L-METHIONINE Food Additives Permitted For Direct Addition To Food For Human Consumption
ACETYLAMINO	556.20	2-ACETYLAMINO-5-NITROTHIAZOLE Tolerances For Residues Of New Animal Drugs In Food
ACETYLATED MONOGLYCERIDES	172.828	ACETYLATED MONOGLYCERIDES Multipurpose Additives; Food Additives Permitted For Direct Addition To Food For Human Consumption
ACID, ACETIC	582.1005	ACETIC ACID General Purpose Food Additives; Substances Generally Recognized As Safe
ACID, ACETIC	862.1390	5-HYDROXYINDOLE ACETIC ACID/SEROTONIN TEST SYSTEM Clinical Chemistry Test Systems; Clinical Chemistry And Toxicology Devices
ACID, ACRYLIC	176.110	ACRYLAMIDE-ACRYLIC ACID RESINS Substances For Use Only As Components Of Paper And Paperboard; Indirect Food Additives
ACID, ACRYLIC	177.1310	ETHYLENE-ACRYLIC ACID COPOLYMERS Substances For Use As Basic Components Of Single And Repeated Use Food Contact Surfaces; Indirect Food Additives: Polymers
ACID, ACRYLIC	573.120	ACRYLAMIDE-ACRYLIC ACID RESIN Food Additive Listing; Food Additives Permitted In Feed And Drinking Water Of Animals
ACID, ADIPIC	184.1009	ADIPIC ACID Listing Of Specific Substances Affirmed As GRAS; Direct Food Substances Affirmed As Generally Recognized As Safe
ACID, ADIPIC	582.1009	ADIPIC ACID General Purpose Food Additives; Substances Generally Recognized As Safe
ACID, ALGINIC	184.1011	ALGINIC ACID Listing Of Specific Substances Affirmed As GRAS; Direct Food Substances Affirmed As Generally Recognized As Safe
ACID, ALGINIC	201.319	WATER SOLUBLE GUMS, AND HYDROPHILIC MUCILLOIDS (INCLUDING ALGINIC ACID) AS ACTIVE INGREDIENTS; WARNINGS AND DIRECTIONS Specific Labeling Requirements For Specific Drug Products; Labeling Recognized As Safe
ACID, AMINOACETIC	170.50	GLYCINE (AMINOACETIC ACID) IN FOOD FOR HUMAN CONSUMPTION Specific Administrative Rulings And Decisions; Food Additives
ACID, AMINOACETIC	582.5049	AMINOACETIC ACID Nutrients And/or Dietary Supplements; Substances Generally Recognized As Safe
ACID, AMINOLEVULINIC	862.1060	DELTA-AMINOLEVULINIC ACID TEST SYSTEM Clinical Chemistry Test Systems; Clinical Chemistry And Clinical Toxicology Devices
ACID, ARSANILIC	558.62	ARSANILIC ACID New Animal Drugs For Use In Animal Feeds
ACID, ASCORBIC	172.315	NICOTINAMIDE-ASCORBIC ACID COMPLEX Special Dietary And Nutritional Additives; Food Additives Permitted For Direct Addition To Food For Human Consumption
ACID, ASCORBIC	182.3013	ASCORBIC ACID Chemical Preservatives; Substances Generally Recognized As Safe
ACID, ASCORBIC	182.8013	ASCORBIC ACID Nutrients; Substances Generally Recognized As Safe

ACID, ASCORBIC	582.3013	ASCORBIC ACID Chemical Preservatives; Substances Generally Recognized As Safe
ACID, ASCORBIC	582.5013	ASCORBIC ACID Nutrients And/or Dietary Supplements; Substances Generally Recognized As Safe
ACID, ASCORBIC	862.1095	ASCORBIC ACID TEST SYSTEM Clinical Chemistry Test Systems; Clinical Chemistry And Clinical Toxicology Devices
ACID, ASPARTIC	582.5017	ASPARTIC ACID Nutrients And/or Dietary Supplements; Substances Generally Recognized As Safe
ACID, BENZOIC	184.1021	BENZOIC ACID Listing Of Specific Substances Affirmed As GRAS; Direct Food Substances Affirmed As Generally Recognized As Safe
ACID, BENZOIC	582.3021	BENZOIC ACID Chemical Preservatives; Substances Generally Recognized As Safe
ACID, CAPRYLIC	184.1025	CAPRYLIC ACID Listing Of Specific Substances Affirmed As GRAS; Direct Food Substances Affirmed As Generally Recognized As Safe
ACID, CITRIC	173.280	SOLVENT EXTRACTION PROCESS FOR CITRIC ACID Solvents, Lubricants, Release Agents And Related Substances; Secondary Direct Food Additives Permitted In Food For Human Consumption
ACID, CITRIC	184.1033	CITRIC ACID Listing Of Specific Substances Affirmed As GRAS; Direct Food Substances Affirmed As Generally Recognized As Safe
ACID, CITRIC	582.1033	CITRIC ACID General Purpose Food Additives; Substances Generally Recognized As Safe
ACID, CITRIC	582.6033	CITRIC ACID Sequestrants; Substances Generally Recognized As Safe
ACID, CROTONIC	175.350	VINYL ACETATE/CROTONIC ACID COPOLYMER Substances For Use As Components Of Coatings; Indirect Food Additives
ACID, DEHYDROACETIC	172.130	DEHYDROACETIC ACID Food Preservatives; Food Additives Permitted For Direct Addition To Food For Human Consumption
ACID, DIPHOSPHOGLYCERIC	862.1255	2,3 DIPHOSPHOGLYCERIC ACID TEST SYSTEM Clinical Chemistry Test Systems; Clinical Chemistry And Clinical Toxicology Devices
ACID, ERYTHORBIC	182.3041	ERYTHORBIC ACID Chemical Preservatives; Substances Generally Recognized As Safe
ACID, ERYTHORBIC	582.3041	ERYTHORBIC ACID Chemical Preservatives; Substances Generally Recognized As Safe
ACID, FATTY	101.62	NUTRIENT CONTENT CLAIMS FOR FAT, FATTY ACID, AND CHOLESTEROL CONTENT OF FOODS Specific Requirements For Nutrient Content Claims; Food Labeling
ACID, FATTY	172.850	LACTYLATED FATTY ACID ESTERS OF GLYCEROL AND PROPYLENE GLYCOL Multipurpose Additives; Food Additives Permitted For Direct Addition To Food For Human Consumption
ACID FATTY	172.859	SUCROSE FATTY ACID ESTERS Multipurpose Additives; Food Additives Permitted For Direct Addition To Food For Human Consumption
ACID, FOLIC	172.345	FOLIC ACID (FOLACIN) Special Dietary And Nutritional Additives; Food Additives Permitted For Direct Addition To Food For Human Consumption
ACID, FOLIC	862.1295	FOLIC ACID TEST SYSTEM Clinical Chemistry Test Systems; Clinical Chemistry And Clinical Toxicology Devices
ACID, FORMIC	186.1316	FORMIC ACID Listing Of Specific Substances Affirmed As GRAS; Indirect Food Substances Affirmed As Generally Recognized As Safe
ACID, FORMIC	573.480	FORMIC ACID Food Additive Listing; Food Additives Permitted In Feed and Drinking Water Of Animals
ACID, FORMIMINOGLUTAMIC	862.1305	FORMIMINOGLUTAMIC ACID (FIGLU) TEST SYSTEM Clinical Chemistry Test Systems; Clinical Chemistry And Clinical Toxicology Devices
ACID, FUMARIC	172.350	FUMARIC ACID AND SALTS OF FUMARIC ACID Special Dietary And Nutritional Additives; Food Additives Permitted For Direct Addition To Food For Human Consumption
ACID, GIBBERELLIC	172.725	GIBBERELLIC ACID AND ITS POTASSIUM SALT Other Specific Usage Additives; Food Additives Permitted For Direct Addition To Food For Human Consumption
ACID, GLUTAMIC	182.1047	GLUTAMIC ACID HYDROCHLORIDE Multiple Purpose GRAS Food Substances; Substances Generally Recognized As Safe

ACID, GLUTAMIC	573.500	CONDENSED, EXTRACTED GLUTAMIC ACID FERMENTATION PRODUCT Food Additive Listing; Food Additives Permitted In Feed and Drinking Water Of Animals
ACID, HYDROCHLORIC	582.1057	HYDROCHLORIC ACID General Purpose Food Additives; Substances Generally Recognized As Safe
ACID, LACTIC	184.1061	LACTIC ACID Listing Of Specific Substances Affirmed As GRAS; Direct Food Substances Affirmed As Generally Recognized As Safe
ACID, LACTIC	582.1061	LACTIC ACID General Purpose Food Additives; Substances Generally Recognized As Safe
ACID, LACTIC	862.1450	LACTIC ACID TEST SYSTEM Clinical Chemistry Test Systems; Clinical Chemistry And Clinical Toxicology Devices
ACID, LINOLEIC	184.1065	LINOLEIC ACID Listing Of Specific Substances Affirmed As GRAS; Direct Food Substances Affirmed As Generally Recognized As Safe
ACID, LINOLEIC	582.5065	LINOLEIC ACID Nutrients And/or Dietary Supplements; Substances Generally Recognized As Safe
ACID, LOW	108.35	THERMALLY PROCESSED LOW-ACID FOODS PACKAGED IN HERMETICALLY SEALED CONTAINERS Specific Requirements And Conditions For Exemption From Or Compliance With An Emergency Permit; Emergency Permit Control
ACID, LOW	500.23	THERMALLY PROCESSED LOW-ACID FOODS PACKAGED IN HERMETICALLY SEALED CONTAINERS Specific Administrative Rulings And Decisions; General
ACID, MALIC	184.1069	MALIC ACID Listing Of Specific Substances Affirmed As GRAS; Direct Food Substances Affirmed As Generally Recognized As Safe
ACID, MALIC	582.1069	MALIC ACID General Purpose Food Additives; Substances Generally Recognized As Safe
ACID, MECLOFENAMIC	520.1330	MECLOFENAMIC ACID GRANULES Oral Dosage Form New Animal Drugs
ACID, MECLOFENAMIC	520.1331	MECLOFENAMIC ACID TABLETS Oral Dosage Form New Animal Drugs
ACID, METHACRYLIC	172.775	METHACRYLIC ACID-DIVINYLBENZENE COPOLYMER Other Specific Usage Additives; Food Additives Permitted For Direct Addition To Food For Human Consumption
ACID, METHYLMALONIC	862.1509	METHYLMALONIC ACID (NONQUANTITATIVE) TEST SYSTEM Clinical Chemistry Test Systems; Clinical Chemistry And Clinical Toxicology Devices
ACID, MONOCHLOROACETIC	189.144	MONOCHLOROACETIC ACID Substances Generally Prohibited From Direct Addition Or Use As Human Food; Substances Prohibited From Use In Human Food
ACID, NORDIHYDROGUAIARETIC	189.165	NORDIHYDROGUAIARETIC ACID (NDGA) Substances Generally Prohibited From Direct Addition Or Use As Human Food; Substances Prohibited From Use In Human Food
ACID, OLEIC	172.862	OLEIC ACID DERIVED FROM TALL OIL FATTY ACIDS Multipurpose Additives; Food Additives Permitted For Direct Addition To Food For Human Consumption
ACID PHOSPHATASE	862.1020	ACID PHOSPHATASE (TOTAL OR PROSTATIC) TEST SYSTEM Clinical Chemistry Test Systems; Clinical Chemistry And Clinical Toxicology Devices
ACID PHOSPHATE	182.6085	SODIUM ACID PHOSPHATE Sequestrants; Substances Generally Recognized As Safe
ACID PHOSPHATE	582.6085	SODIUM ACID PHOSPHATE Sequestrants; Substances Generally Recognized As Safe
ACID, PHOSPHORIC	175.260	PARTIAL PHOSPHORIC ACID ESTERS OF POLYESTER RESINS Substances For Use As Components Of Coatings; Indirect Food Additives
ACID, PHOSPHORIC	582.1073	PHOSPHORIC ACID General Purpose Food Additives; Substances Generally Recognized As Safe
ACID, POLYMALEIC	173.45	POLYMALEIC ACID AND ITS SODIUM SALT Polymer Substances For Food Treatment; Secondary Direct Food Additives Permitted In Food For Human Consumption
ACID, PROPIONIC	184.1081	PROPIONIC ACID Listing Of Specific Substances Affirmed As GRAS; Direct Food Substances Affirmed As Generally Recognized As Safe
ACID, PROPIONIC	582.3081	PROPIONIC ACID Chemical Preservatives; Substances Generally Recognized As Safe
ACID PYROPHOSPHATE	582.1087	SODIUM ACID PYROPHOSPHATE General Purpose Food Additives; Substances Generally Recognized As Safe
ACID, PYRUVIC	862.1655	PYRUVIC ACID TEST SYSTEM Clinical Chemistry Test Systems; Clinical Chemistry And Clinical Toxicology Devices

ACID, REDUCED	146.148	REDUCED ACID FROZEN CONCENTRATED ORANGE JUICE Requirements For Specific Standardized Canned Fruit Juices And Beverages; Canned Fruit Juices
ACID, SALICYLIC	529.2090	SALICYLIC ACID Certain Other Dosage Form New Animal Drugs
ACID, SALICYLIC	556.590	SALICYLIC ACID Tolerances For Residues Of New Animal Drugs In Food
ACID, SORBIC	518.3089	SORBIC ACID Chemical Preservatives; Substances Generally Recognized As Safe
ACID, SORBIC	582.3089	SORBIC ACID Chemical Preservatives; Substances Generally Recognized As Safe
ACID, STEARIC	184.1090	STEARIC ACID Listing Of Specific Substances Affirmed As GRAS; Direct Food Substances Affirmed As Generally Recognized As Safe
ACID, SUCCINIC	184.1091	SUCCINIC ACID Listing Of Specific Substances Affirmed As GRAS; Direct Food Substances Affirmed As Generally Recognized As Safe
ACID, SUCCINIC	582.1091	SUCCINIC ACID General Purpose Food Additives; Substances Generally Recognized As Safe
ACID, SULFAMIC	186.1093	SULFAMIC ACID Listing Of Specific Substances Affirmed As GRAS; Indirect Food Substances Affirmed As Generally Recognized As Safe
ACID, SULFOLITHOCHOLIC	862.1187	CONJUGATED SULFOLITHOCHOLIC ACID (SLCG) TEST SYSTEM Clinical Chemistry Test Systems; Clinical Chemistry And Clinical Toxicology Devices
ACID, SULFONIC	173.395	TRIFLUOROMETHANE SULFONIC ACID Specific Use Additives; Secondary Direct Food Additives Permitted In Food For Human Consumption
ACID, SULFURIC	184.1095	SULFURIC ACID Listing Of Specific Substances Affirmed As GRAS; Direct Food Substances Affirmed As Generally Recognized As Safe
ACID, SULFURIC	582.1095	SULFURIC ACID General Purpose Food Additives; Substances Generally Recognized As Safe
ACID, TANNIC	184.1097	TANNIC ACID Listing Of Specific Substances Affirmed As GRAS; Direct Food Substances Affirmed As Generally Recognized As Safe
ACID, TANNIC	201.304	TANNIC ACID AND BARIUM ENEMA PREPARATIONS Labeling
ACID TARTARATE	184.1077	POTASSIUM ACID TARTARATE Listing Of Specific Substances Affirmed As GRAS; Direct Food Substances Affirmed As Generally Recognized As Safe
ACID TARTARATE	582.1077	POTASSIUM ACID TARTARATE General Purpose Food Additives; Substances Generally Recognized As Safe
ACID, TARTARIC	184.1099	TARTARIC ACID Listing Of Specific Substances Affirmed As GRAS; Direct Food Substances Affirmed As Generally Recognized As Safe
ACID, TARTARIC	184.1101	DIACETYL TARTARIC ACID ESTERS OF MONO AND DIGLYCERIDES Listing Of Specific Substances Affirmed As GRAS; Direct Food Substances Affirmed As Generally Recognized As Safe
ACID, TARTARIC	582.1099	TARTARIC ACID General Purpose Food Additives; Substances Generally Recognized As Safe
ACID, TARTARIC	582.4101	DIACETYL TARTARIC ACID ESTERS OF MONO- AND DIGLYCERIDES OF EDIBLE FATS Emulsifying Agents; Substances Generally Recognized As Safe
ACID, TARTARIC	582.6099	TARTARIC ACID Sequestrants; Substances Generally Recognized As Safe
ACID, THIODIPROPIONIC	182.3109	THIODIPROPIONIC ACID Chemical Preservatives; Substances Generally Recognized As Safe
ACID, THIODIPROPIONIC	582.3109	THIODIPROPIONIC ACID Chemical Preservatives; Substances Generally Recognized As Safe
ACID, URIC	862.1775	URIC ACID Clinical Chemistry Test Systems; Clinical Chemistry And Toxicology Devices
ACID, VANILMANDELIC	862.1795	VANILMANDELIC ACID TEST SYSTEM Clinical Chemistry Test Systems; Clinical Chemistry And Clinical Toxicology Devices
ACIDIFIED	173.325	ACIDIFIED SODIUM CHLORITE SOLUTIONS Specific Use Additives; Secondary Direct Food Additives Permitted In Food For Human Consumption
ACIDIFIED FOODS	108.25	ACIDIFIED FOODS Specific Requirements And Conditions For Exemption From Or Compliance With An Emergency Permit; Emergency Permit Control
ACIDIFIED MILK	131.111	MILK Requirements For Specific Standardized Milk And Cream; Milk And Cream

ACIDIFIED SOUR CREAM	131.162	ACIDIFIED SOUR CREAM Requirements For Specific Standardized Milk And Cream; Milk And Cream
ACIDIFIERS	310.540	DRUG PRODUCTS CONTAINING ACTIVE INGREDIENTS OFFERED OVER-THE-COUNTER FOR USE AS STOMACH ACIDIFIERS Requirements For Specific New Drugs Or Devices; New Drugs
ACIDITY	862.1320	GASTRIC ACIDITY TEST SYSTEM Clinical Chemistry Test Systems; Clinical Chemistry And Clinical Toxicology Devices
ACIDS, AMINO	172.320	AMINO ACIDS Special Dietary And Nutritional Additives; Food Additives Permitted For Direct Addition To Food For Human Consumption
ACIDS, FATTY	172.225	METHYL AND ETHYL ESTERS OF FATTY ACIDS PRODUCED FROM EDIBLE FATS AND OILS Coatings, Films And Related Substances; Food Additives Permitted For Direct Addition To Food For Human Consumption
ACIDS, FATTY	172.848	LACTYLIC ESTERS OF FATTY ACIDS Multipurpose Additives; Food Additives Permitted For Direct Addition To Food For Human Consumption
ACIDS, FATTY	172.852	GLYCERYL-LACTO ESTERS OF FATTY ACIDS Multipurpose Additives; Food Additives Permitted For Direct Addition To Food For Human Consumption
ACIDS, FATTY	172.854	POLYGLYCEROL ESTERS OF FATTY ACIDS Multipurpose Additives; Food Additives Permitted For Direct Addition To Food For Human Consumption
ACIDS, FATTY	172.856	PROPYLENE GLYCOL MONO- AND DIESTERS OF FATS AND FATTY ACIDS Multipurpose Additives; Food Additives Permitted For Direct Addition To Food For Human Consumption
ACIDS, FATTY	172.860	FATTY ACIDS Multipurpose Additives; Food Additives Permitted For Direct Addition To Food For Human Consumption
ACIDS, FATTY	172.862	OLEIC ACID DERIVED FROM TALL OIL FATTY ACIDS Multipurpose Additives; Food Additives Permitted For Direct Addition To Food For Human Consumption
ACIDS, FATTY	172.863	SALTS OF FATTY ACIDS Multipurpose Additives; Food Additives Permitted For Direct Addition To Food For Human Consumption
ACIDS, FATTY	573.640	METHYL ESTERS OF HIGHER FATTY ACIDS Food Additive Listing; Food Additives Permitted In Feed And Drinking Water Of Animals
ACIDS, FATTY	573.914	SALTS OF VOLATILE FATTY ACIDS Food Additive Listing; Food Additives Permitted In Feed And Drinking Water Of Animals
ACIDS, FATTY	582.4101	DIACETYL TARTARTIC ACID ESTERS OF MONO- AND DIGLYCERIDES OF EDIBLE FAT-FORMING FATTY ACIDS Emulsifying Agents; Substances Generally Recognized As Safe
ACIDS, FATTY	582.4521	MONOSODIUM PHOSPHATE DERIVATIVES OF EDIBLE FAT-FORMING FATTY ACIDS Emulsifying Agents; Substances Generally Recognized As Safe
ACIDS, FATTY	862.1290	FATTY ACIDS TEST SYSTEM Clinical Chemistry Test Systems; Clinical Chemistry And Clinical Toxicology Devices
ACIDS, MONOBASIC	178.3780	POLYHYDRIC ALCOHOL ESTERS OF LONG CHAIN MONOBASIC ACIDS Certain Adjuvants And Production Aids; Indirect Food Additives: Adjuvants, Production Aids, And Sanitizers
ACIDS, MONTAN WAX	178.3770	POLYHYDRIC ALCOHOL ESTERS OF OXIDATIVELY REFINED (GERSTHOFEN PROCESS) MONTAN WAX ACIDS Certain Adjuvants And Production Aids; Indirect Food Additives: Adjuvants, Production Aids, And Sanitizers
ACIDS, PALMITIC	178.3450	ESTERS OF STEARIC AND PALMITIC ACIDS Certain Adjuvants And Production Aids; Indirect Food Additives: Adjuvants, Production Aids, And Sanitizers
ACIDS, PEROXY	173.370	PEROXYACIDS Secondary Direct Food Additives Permitted In Food For Human Consumption
ACIDS, STEARIC	178.3450	ESTERS OF STEARIC AND PALMITIC ACIDS Certain Adjuvants And Production Aids; Indirect Food Additives: Adjuvants, Production Aids, And Sanitizers
ACINETOBACTER CALCOACETICUS	866.3010	ACINETOBACTER CALCOACETICUS SEROLOGICAL REAGENTS Serological Reagents; Immunology And Microbiology Devices
ACNE	333.310	ACNE ACTIVE INGREDIENTS Topical Acne Drug Products; Topical Antimicrobial Drug Products For Over-the-counter Human Use
ACNE	333.350	LABELING OF ACNE DRUG PRODUCTS Topical Acne Drug Products; Topical Antimicrobial Drug Products For Over-the-counter Human Use

ACONITIC ACID	184.1007	ACONITIC ACID Listing Of Specific Substances Affirmed As GRAS; Direct Food Substances Affirmed As Generally Recognized As Safe
ACOUSTIC	874.1060	ACOUSTIC CHAMBER FOR AUDIOMETRIC TESTING Diagnostic Devices; Ear, Nose, And Throat Devices
ACRYLAMIDE	173.5	ACRYLATE-ACRYLAMIDE RESINS Polymer Substances For Food Treatment; Secondary Direct Food Additives Permitted In Food For Human Consumption
ACRYLAMIDE-ACRYLIC ACID	176.110	ACRYLAMIDE-ACRYLIC ACID RESINS Substances For Use Only As Components Of Paper And Paperboard; Indirect Food Additives
ACRYLAMIDE-ACRYLIC ACID	573.120	ACRYLAMIDE-ACRYLIC ACID RESIN Food Additive Listing; Food Additives Permitted In Feed And Drinking Water Of Animals
ACRYLATE	173.5	ACRYLATE-ACRYLAMIDE RESINS Polymer Substances For Food Treatment; Secondary Direct Food Additives Permitted In Food For Human Consumption
ACRYLATE	177.1320	ETHYLENE-ETHYL ACRYLATE COPOLYMERS Substances For Use As Basic Components Of Single And Repeated Use Food Contact Surfaces; Indirect Food Additives: Polymers
ACRYLATE	177.1340	ETHYLENE-METHYL ACRYLATE COPOLYMER RESINS Substances For Use As Basic Components Of Single And Repeated Use Food Contact Surfaces; Indirect Food Additives: Polymers
ACRYLATE	177.1480	NITRILE RUBBER MODIFIED ACRYLONITRILE-METHYL ACRYLATE COPOLYMERS Substances For Use As Basic Components Of Single And Repeated Use Food Contact Surfaces; Indirect Food Additives: Polymers
ACRYLATE	177.1990	VINYLIDENE CHLORIDE/METHYL ACRYLATE COPOLYMERS Substances For Use As Basic Components Of Single And Repeated Use Food Contact Surfaces; Indirect Food Additives: Polymers
ACRYLATE	177.2000	VINYLIDENE CHLORIDE/METHYL ACRYLATE/METHYL METHACRYLATE POLYMERS Substances For Use As Basic Components Of Single And Repeated Use Food Contact Surfaces; Indirect Food Additives: Polymers
ACRYLATE ESTER	175.210	ACRYLATE ESTER COPOLYMER COATINGS Substances For Use As Components Of Coatings; Indirect Food Additives
ACRYLIC ACID	176.110	ACRYLAMIDE-ACRYLIC ACID RESINS Substances For Use Only As Components Of Paper And Paperboard; Indirect Food Additives
ACRYLIC ACID	177.1310	ETHYLENE-ACRYLIC ACID COPOLYMERS Substances For Use As Basic Components Of Single And Repeated Use Food Contact Surfaces; Indirect Food Additives: Polymers
ACRYLIC ACID	573.120	ACRYLAMIDE-ACRYLIC ACID RESIN Food Additive Listing; Food Additives Permitted In Feed And Drinking Water Of Animals
ACRYLIC COPOLYMERS	177.1060	*N*-ALKYLGLUTARIMIDE/ACRYLIC COPOLYMERS Substances For Use As Basic Components Of Single And Repeated Use Food Contact Surfaces; Indirect Food Additives: Polymers
ACRYLIC PLASTICS	177.1010	ACRYLIC AND MODIFIED ACRYLIC PLASTICS, SEMIRIGID AND RIGID Substances For Use As Basic Components Of Single And Repeated Use Food Contact Surfaces; Indirect Food Additives: Polymers
ACRYLONITRILE	177.1020	ACRYLONITRILE/BUTADIENE/STYRENE COPOLYMER Substances For Use As Basic Components Of Single And Repeated Use Food Contact Surfaces; Indirect Food Additives: Polymers
ACRYLONITRILE	177.1030	ACRYLONITRILE/BUTADIENE/STYRENE/METHYL METHACRYLATE COPOLYMER Substances For Use As Basic Components Of Single And Repeated Use Food Contact Surfaces; Indirect Food Additives: Polymers
ACRYLONITRILE	177.1040	ACRYLONITRILE/STYRENE COPOLYMER Substances For Use As Basic Components Of Single And Repeated Use Food Contact Surfaces; Indirect Food Additives: Polymers
ACRYLONITRILE	177.1050	ACRYLONITRILE/STYRENE COPOLYMER MODIFIED WITH BUTADIENE/STYRENE ELASTOMER Substances For Use As Basic Components Of Single And Repeated Use Food Contact Surfaces; Indirect Food Additives: Polymers
ACRYLONITRILE	177.1480	NITRILE RUBBER MODIFIED ACRYLONITRILE-METHYL ACRYLATE COPOLYMERS Substances For Use As Basic Components Of Single And Repeated Use Food Contact Surfaces; Indirect Food Additives: Polymers
ACRYLONITRILE	180.22	ACRYLONITRILE COPOLYMERS Specific Requirements For Certain Food Additives; Food Additives Permitted In Food Or In Contact With Food On Interim Basis Pending Study
ACRYLONITRILE	181.32	ACRYLONITRILE COPOLYMERS AND RESINS Specific Prior-sanctioned Food Ingredients; Prior-sanctioned Food Ingredients

ACTH	862.1025	ADRENOCORTICOTROPIC HORMONE (ACTH) TEST SYSTEM Clinical Chemistry Test Systems; Clinical Chemistry And Clinical Toxicology Devices
ACTION	10.45	COURT REVIEW OF FINAL ADMINISTRATIVE ACTION; EXHAUSTION OF ADMINISTRATIVE REMEDIES General Administrative Procedures; Administrative Practices And Procedures
ACTION	12.35	NOTICE OF HEARING; STAY OF ACTION Initiation Of Proceedings; Formal Evidentiary Public Hearing
ACTION	12.139	RECONSIDERATION AND STAY OF ACTION Initial And Final Decisions; Formal Evidentiary Public Hearing
ACTION	16.119	RECONSIDERATION AND STAY OF ACTION Reconsideration And Stay; Regulatory Hearing Before The Food And Drug Administration
ACTION	60.20	FDA ACTION ON REGULATORY REVIEW PERIOD DETERMINATIONS Regulatory Review Period Determinations; Patent Term Restoration
ACTION	60.26	FINAL ACTION ON REGULATORY REVIEW PERIOD DETERMINATIONS Regulatory Review Period Determinations; Patent Term Restoration
ACTION	60.34	FDA ACTION ON PETITIONS Due Diligence Petitions; Patent Term Restoration
ACTION	99.301	AGENCY ACTION ON A SUBMISSION FDA Action On Submissions, Requests, and Applications; Dissemination Of Information On Unapproved/New Uses For Marketed Drugs, Biologics, and Devices
ACTION	807.100	FDA ACTION ON A PREMARKET NOTIFICATION Establishment Registration And Device Listing For Manufacturers And Initial Importers Of Devices
ACTION	812.30	FDA ACTION ON APPLICATIONS Application And Administrative Action; Investigational Device Exemptions
ACTION	820.100	CORRECTIVE AND PREVENTIVE ACTION Corrective And Preventive Action; Quality System Regulation
ACTION	1301.32	ACTION ON APPLICATIONS FOR RESEARCH IN SCHEDULE I SUBSTANCES Action On Application For Registration: Revocation Or Suspension Of Registration Registration Of Manufacturers, Distributors, And Dispensers Of Controlled Substances
ACTION, ADMINISTRATIVE	10.45	COURT REVIEW OF FINAL ADMINISTRATIVE ACTION; EXHAUSTION OF ADMINISTRATIVE REMEDIES General Administrative Procedures; Administrative Practices And Procedures
ACTION, ENFORCEMENT	20.91	USE OF DATA OR INFORMATION FOR ADMINISTRATIVE OR COURT ENFORCEMENT ACTION Limitations On Exemptions; Public Information
ACTION LEVELS	109.4	ESTABLISHMENT OF TOLERANCES, REGULATORY LIMITS, AND ACTION LEVELS Unavoidable Contaminants In Food For Human Consumption & Food-packaging Material
ACTION LEVELS	509.4	ESTABLISHMENT OF TOLERANCES, REGULATORY LIMITS, AND ACTION LEVELS General Provisions; Unavoidable Contaminants In Animal Food And Food-packaging Material
ACTION, RECONSIDERATION	10.33	ADMINISTRATIVE RECONSIDERATION OF ACTION General Administrative Procedures; Administrative Practices And Procedures
ACTION, STAY OF	10.35	ADMINISTRATIVE STAY OF ACTION General Administrative Procedures; Administrative Practices And Procedures
ACTION, STAY OF	16.119	RECONSIDERATION AND STAY OF ACTION Regulatory Hearing Before The FDA
ACTIONS	25.20	ACTIONS REQUIRING PREPARATION OF AN ENVIRONMENTAL ASSESSMENT Agency Actions Requiring Environmental Consideration; Environmental Impact Considerations
ACTIONS	25.22	ACTIONS REQUIRING PREPARATION OF AN ENVIRONMENTAL IMPACT STATEMENT Agency Actions Requiring Environmental Consideration; Environmental Impact Considerations
ACTIONS	56.124	ACTIONS ALTERNATIVE OR ADDITIONAL TO DISQUALIFICATION Administrative Actions For Noncompliance; Institutional Review Boards
ACTIONS	58.210	ACTIONS UPON DISQUALIFICATION Disqualification Of Testing Facilities; Good Laboratory Practice For Nonclinical Laboratory Studies
ACTIONS	58.215	ALTERNATIVE OR ADDITIONAL ACTIONS TO DISQUALIFICATION Disqualification Of Testing Facilities; Good Laboratory Practice For Nonclinical Laboratory Studies
ACTIONS, CORRECTIVE	99.401	CORRECTIVE ACTIONS AND CESSATION OF DISSEMINATION Dissemination Of Information On Unapproved/New Uses For Marketed Drugs, Biologics, and Devices
ACTIONS, CORRECTIVE	120.10	CORRECTIVE ACTIONS General Provisions; Hazard Analysis And Critical Control Point (HACCP) Plan

ACTIONS, CORRECTIVE	123.7	CORRECTIVE ACTIONS General Provisions; Fish And Fishery Products
ACTIONS, LESSER	56.120	LESSER ADMINISTRATIVE ACTIONS Administrative Actions For Noncompliance; Institutional Review Boards
ACTIONS, MAJOR	25.60	ENVIRONMENTAL EFFECTS ABROAD OF MAJOR AGENCY ACTIONS Other Requirements; Environmental Impact Considerations
ACTIVATED ATTAPULGITE	520.1204	KANAMYCIN SULFATE, AMINOPENTAMIDE HYDROGEN SULFATE, PECTIN, BISMUTH SUBCARBONATE, ACTIVATED ATTAPULGITE SUSPENSION Oral Dosage Form New Animal Drugs
ACTIVATED ATTAPULGITE	520.1205	KANAMYCIN SULFATE, PECTIN, BISMUTH SUBCARBONATE, ACTIVATED ATTAPULGITE TABLETS Oral Dosage Form New Animal Drugs
ACTIVATOR	872.6070	ULTRAVIOLET ACTIVATOR FOR POLYMERIZATION Miscellaneous Devices; Dental Devices
ACTIVE INGREDIENT	340.10	STIMULANT ACTIVE INGREDIENT Active Ingredient; Stimulant Drug Products For Over-the-counter Human Use
ACTIVE INGREDIENT	341.18	EXPECTORANT ACTIVE INGREDIENT Cold, Cough, Allergy, Bronchodilator, And Antiasthmatic Drug Products For OTC Human Use
ACTIVE INGREDIENT	344.10	EARWAX REMOVAL AID ACTIVE INGREDIENT Active Ingredients; Topical Otic Drug Products For Over-the-counter Human Use
ACTIVE INGREDIENT	344.12	EAR DRYING AID ACTIVE INGREDIENT Active Ingredients; Topical Otic Drug Products For Over-the-counter Human Use
ACTIVE INGREDIENT	357.110	ANTHELMINTIC ACTIVE INGREDIENT Miscellaneous Internal Drug Products For Over-the-counter Human Use
ACTIVE INGREDIENT	1310.14	EXEMPTION OF DRUG PRODUCTS CONTAINING EPHEDRINE OR ANOTHER ACTIVE MEDICINAL INGREDIENT Records And Reports Of Listed Chemicals And Certain Machines
ACTIVE INGREDIENT	1310.15	EXEMPT DRUG PRODUCTS CONTAINING EPHEDRINE AND ANOTHER ACTIVE MEDICINAL INGREDIENT Records And Reports Of Listed Chemicals And Certain Machines
ACTIVE INGREDIENTS	201.319	WATER-SOLUBLE GUMS, HYDROPHILIC GUMS AND HYDROPHILIC MUCILLOIDS (INCLUDING AGAR, ALGINIC ACID) AS ACTIVE INGREDIENTS; WARNINGS AND DIRECTIONS Specific Labeling Requirements For Specific Drug Products; Labeling
ACTIVE INGREDIENTS	201.322	OVER-THE COUNTER DRUG PRODUCTS CONTAINING INTERNAL ANALGESIC/ANTIPYRETIC ACTIVE INGREDIENTS Specific Labeling Requirements For Specific Drug Products; Labeling
ACTIVE INGREDIENTS	310.527	DRUG PRODUCTS CONTAINING ACTIVE INGREDIENTS OFFERED OTC FOR EXTERNAL USE AS HAIR GROWERS OR FOR HAIR LOSS PREVENTION Requirements For Specific New Drugs Or Devices; New Drugs
ACTIVE INGREDIENTS	310.528	DRUG PRODUCTS CONTAINING ACTIVE INGREDIENTS OFFERED OTC FOR USE AS AN APHRODISIAC Requirements For Specific New Drugs Or Devices; New Drugs
ACTIVE INGREDIENTS	310.529	DRUG PRODUCTS CONTAINING ACTIVE INGREDIENTS OFFERED OTC FOR ORAL USE AS INSECT REPELLENTS Requirements For Specific New Drugs Or Devices; New Drugs
ACTIVE INGREDIENTS	310.531	DRUG PRODUCTS CONTAINING ACTIVE INGREDIENTS OFFERED OTC FOR TREATMENT OF BOILS Requirements For Specific New Drugs Or Devices; New Drugs
ACTIVE INGREDIENTS	310.532	DRUG PRODUCTS CONTAINING ACTIVE INGREDIENTS OFFERED OTC TO RELIEVE SYMPTOMS OF BENIGN PROSTATIC HYPERTROPHY Requirements For Specific New Drugs Or Devices; New Drugs
ACTIVE INGREDIENTS	310.533	DRUG PRODUCTS CONTAINING ACTIVE INGREDIENTS OFFERED OTC FOR HUMAN USE AS AN ANTICHOLINERGIC IN COUGH-COLD DRUG PRODUCTS Requirements For Specific New Drugs Or Devices; New Drugs
ACTIVE INGREDIENTS	310.534	DRUG PRODUCTS CONTAINING ACTIVE INGREDIENTS OFFERED OTC FOR HUMAN USE AS ORAL WOUND HEALING AGENTS Requirements For Specific New Drugs Or Devices; New Drugs
ACTIVE INGREDIENTS	310.536	DRUG PRODUCTS CONTAINING ACTIVE INGREDIENTS OFFERED OTC FOR USE AS A NAILBITING OR THUMBSUCKING DETERRENT Requirements For Specific New Drugs Or Devices; New Drugs
ACTIVE INGREDIENTS	310.537	DRUG PRODUCTS CONTAINING ACTIVE INGREDIENTS OFFERED OTC FOR TREATMENT OF FEVER BLISTERS AND COLD SORES Requirements For Specific New Drugs Or Devices; New Drugs
ACTIVE INGREDIENTS	310.538	DRUG PRODUCTS CONTAINING ACTIVE INGREDIENTS OFFERED OTC FOR USE FOR INGROWN TOENAIL RELIEF Requirements For Specific New Drugs Or Devices; New Drugs
ACTIVE INGREDIENTS	310.540	DRUG PRODUCTS CONTAINING ACTIVE INGREDIENTS OFFERED OTC FOR USE AS STOMACH ACIDIFIERS Requirements For Specific New Drugs Or Devices; New Drugs
ACTIVE INGREDIENTS	310.541	DRUG PRODUCTS CONTAINING ACTIVE INGREDIENTS OFFERED FOR USE IN TREATMENT OF HYPOPHOSPHATEMIA Requirements For Specific New Drugs Or Devices; New Drugs

ACTIVE INGREDIENTS	310.542	DRUG PRODUCTS CONTAINING ACTIVE INGREDIENTS OFFERED FOR USE IN TREATMENT OF HYPERPHOSPHATEMIA Requirements For Specific New Drugs Or Devices; New Drugs
ACTIVE INGREDIENTS	310.543	DRUG PRODUCTS CONTAINING ACTIVE INGREDIENTS OFFERED FOR HUMAN USE IN EXOCRINE PANCREATIC INSUFFICIENCY Requirements For Specific New Drugs Or Devices; New Drugs
ACTIVE INGREDIENTS	310.544	DRUG PRODUCTS CONTAINING CERTAIN ACTIVE INGREDIENTS OFFERED OTC FOR USE AS A SMOKING DETERRENT Requirements For Specific New Drugs Or Devices; New Drugs
ACTIVE INGREDIENTS	310.545	DRUG PRODUCTS CONTAINING CERTAIN ACTIVE INGREDIENTS OFFERED OTC FOR CERTAIN USES Requirements For Specific New Drugs Or Devices; New Drugs
ACTIVE INGREDIENTS	310.546	DRUG PRODUCTS CONTAINING ACTIVE INGREDIENTS OFFERED OTC FOR TREATMENT AND/OR PREVENTION OF NOCTURNAL LEG MUSCLE CRAMPS Requirements For Specific New Drugs Or Devices; New Drugs
ACTIVE INGREDIENTS	331.10	ANTACID ACTIVE INGREDIENTS Active Ingredients; Antacid Products For Over-the-counter (OTC) Human Use
ACTIVE INGREDIENTS	331.11	LISTING OF SPECIFIC ACTIVE INGREDIENTS Active Ingredients; Antacid Products For Over-the-counter (OTC) Human Use
ACTIVE INGREDIENTS	331.15	COMBINATION WITH NONANTACID ACTIVE INGREDIENTS Active Ingredients; Antacid Products For Over-the-counter (OTC) Human Use
ACTIVE INGREDIENTS	331.20	DETERMINATION OF PERCENT CONTRIBUTION OF ACTIVE INGREDIENTS Testing Procedures; Antacid Products For Over-the-counter (OTC) Human Use
ACTIVE INGREDIENTS	332.10	ANTIFLATULENT ACTIVE INGREDIENTS Active Ingredients; Antiflatulent Products For Over-the-counter Human Use
ACTIVE INGREDIENTS	332.15	COMBINATION WITH NON-ANTIFLATULENT ACTIVE INGREDIENTS Active Ingredients; Antiflatulent Products For Over-the-counter Human Use
ACTIVE INGREDIENTS	333.110	FIRST AID ANTIBIOTIC ACTIVE INGREDIENTS First Aid Antibiotic Drug Products; Topical Antimicrobial Drug Products For Over-the-counter Human Use
ACTIVE INGREDIENTS	333.120	PERMITTED COMBINATIONS OF ACTIVE INGREDIENTS First Aid Antibiotic Drug Products; Topical Antimicrobial Drug Products For Over-the-counter Human Use
ACTIVE INGREDIENTS	333.160	LABELING OF PERMITTED COMBINATIONS OF ACTIVE INGREDIENTS First Aid Antibiotic Drug Products; Topical Antimicrobial Drug Products For Over-the-counter Human Use
ACTIVE INGREDIENTS	333.210	ANTIFUNGAL ACTIVE INGREDIENTS Topical Antifungal Drug Products; Topical Antimicrobial Drug Products For Over-the-counter Human Use
ACTIVE INGREDIENTS	333.310	ACNE ACTIVE INGREDIENTS Topical Acne Drug Products; Topical Antimicrobial Drug Products For Over-the-counter Human Use
ACTIVE INGREDIENTS	333.320	PERMITTED COMBINATIONS OF ACTIVE INGREDIENTS Topical Acne Drug Products; Topical Antimicrobial Drug Products For Over-the-counter Human Use
ACTIVE INGREDIENTS	336.10	ANTIEMETIC ACTIVE INGREDIENTS Antiemetic Drug Products For Over-the-counter Human Use
ACTIVE INGREDIENTS	338.10	NIGHTTIME SLEEP-AID ACTIVE INGREDIENTS Nighttime Sleep-aid Drug Products For Over-the-counter Human Use
ACTIVE INGREDIENTS	341.12	ANTIHISTAMINE ACTIVE INGREDIENTS Cold, Cough, Allergy, Bronchodilator, And Antiasthmatic Drug Products For OTC Human Use
ACTIVE INGREDIENTS	341.14	ANTITUSSIVE ACTIVE INGREDIENTS Cold, Cough, Allergy, Bronchodilator, And Antiasthmatic Drug Products For OTC Human Use
ACTIVE INGREDIENTS	341.16	BRONCHODILATOR ACTIVE INGREDIENTS Cold, Cough, Allergy, Bronchodilator, And Antiasthmatic Drug Products OTC Human Use
ACTIVE INGREDIENTS	341.20	NASAL DECONGESTANT ACTIVE INGREDIENTS Cold, Cough, Allergy, Bronchodilator, And Antiasthmatic Drug Products For OTC Human Use
ACTIVE INGREDIENTS	343.12	CARDIOVASCULAR ACTIVE INGREDIENTS Internal Analgesic, Antipyretic, And Antirheumatic Drug Products For OTC Human Use
ACTIVE INGREDIENTS	343.13	RHEUMATOLOGIC ACTIVE INGREDIENTS Internal Analgesic, Antipyretic, And Antirheumatic Drug Products For OTC Human Use
ACTIVE INGREDIENTS	343.22	PERMITTED COMBINATIONS OF ACTIVE INGREDIENTS FOR CARDIOVASCULAR-RHEUMATOLOGIC USE Internal Analgesic, Antipyretic, And Antirheumatic Drug Products For OTC Human Use
ACTIVE INGREDIENTS	346.10	LOCAL ANESTHETIC ACTIVE INGREDIENTS Anorectal Drug Products For Over-the-counter Human Use

ACTIVE INGREDIENTS	346.12	VASOCONSTRICTOR ACTIVE INGREDIENTS Anorectal Drug Products For Over-the-counter Human Use
ACTIVE INGREDIENTS	346.14	PROTECTANT ACTIVE INGREDIENTS Anorectal Drug Products For Over-the-counter Human Use
ACTIVE INGREDIENTS	346.16	ANALGESIC, ANESTHETIC, AND ANTIPRURITIC ACTIVE INGREDIENTS Anorectal Drug Products For Over-the-counter Human Use
ACTIVE INGREDIENTS	346.18	ASTRINGENT ACTIVE INGREDIENTS Anorectal Drug Products For Over-the-counter Human Use
ACTIVE INGREDIENTS	346.20	KERATOLYTIC ACTIVE INGREDIENTS Anorectal Drug Products For Over-the-counter Human Use
ACTIVE INGREDIENTS	346.22	PERMITTED COMBINATIONS OF ANORECTAL ACTIVE INGREDIENTS Anorectal Drug Products For Over-the-counter Human Use
ACTIVE INGREDIENTS	346.52	LABELING OF PERMITTED COMBINATIONS OF ANORECTAL ACTIVE INGREDIENTS Anorectal Drug Products For Over-the-counter Human Use
ACTIVE INGREDIENTS	347.10	ASTRINGENT ACTIVE INGREDIENTS Skin Protectant Drug Products For Over-the-counter Human Use
ACTIVE INGREDIENTS	348.10	ANALGESIC, ANESTHETIC, AND ANTIPRURITIC ACTIVE INGREDIENTS External Analgesic Drug Products For Over-the-counter Human Use
ACTIVE INGREDIENTS	349.30	PERMITTED COMBINATIONS OF ACTIVE INGREDIENTS Ophthalmic Drug Products For Over-the-counter Human Use
ACTIVE INGREDIENTS	349.79	LABELING OF PERMITTED COMBINATIONS OF ACTIVE INGREDIENTS Ophthalmic Drug Products For Over-the-counter Human Use
ACTIVE INGREDIENTS	352.10	SUNSCREEN ACTIVE INGREDIENTS Sunscreen Drug Products For Over-the-counter Human Use
ACTIVE INGREDIENTS	352.20	PERMITTED COMBINATIONS OF ACTIVE INGREDIENTS Sunscreen Drug Products For Over-the-counter Human Use
ACTIVE INGREDIENTS	352.60	LABELING OF PERMITTED COMBINATIONS OF ACTIVE INGREDIENTS Sunscreen Drug Products For Over-the-counter Human Use
ACTIVE INGREDIENTS	355.10	ANTICARIES ACTIVE INGREDIENTS Anticaries Drug Products For Over-the-counter Human Use
ACTIVE INGREDIENTS	357.210	CHOLECYSTOKINETIC ACTIVE INGREDIENTS Cholecystokinetic Drug Products; Miscellaneous Internal Drug Products For Over-the-counter Human Use
ACTIVE INGREDIENTS	358.110	WART REMOVER ACTIVE INGREDIENTS Miscellaneous External Drug Products For Over-the-counter Human Use
ACTIVE INGREDIENTS	358.510	CORN AND CALLUS REMOVER ACTIVE INGREDIENTS Miscellaneous External Drug Products For Over-the-counter Human Use
ACTIVE INGREDIENTS	358.610	PEDICULICIDE ACTIVE INGREDIENTS Miscellaneous External Drug Products For Over-the-counter Human Use
ACTIVE INGREDIENTS	358.710	ACTIVE INGREDIENTS FOR THE CONTROL OF DANDRUFF, SEBORRHEIC DERMATITIS, OR PSORIASIS Miscellaneous External Drug Products For Over-the-counter Human Use
ACTIVE INGREDIENTS	358.720	PERMITTED COMBINATIONS OF ACTIVE INGREDIENTS Miscellaneous External Drug Products For Over-the-counter Human Use
ACTIVITIES	5.31	ENFORCEMENT ACTIVITIES General Redelegations Of Authority; Delegations Of Authority And Organization
ACTIVITIES	10.95	PARTICIPATION IN OUTSIDE STANDARD-SETTING ACTIVITIES General Administrative Procedures; Administrative Practices And Procedures
ACTIVITIES, ILLICIT	1301.92	ILLICIT ACTIVITIES BY EMPLOYEES Employee Screening: Non-practitioners; Registration Of Manufacturers, Distributors, And Dispensers Of Controlled Substances
ACTIVITIES, INDEPENDENT	1309.22	SEPARATE REGISTRATION FOR INDEPENDENT ACTIVITIES Registration Of Manufacturers, Distributors, Importers And Exporters Of List I Chemicals
ACTIVITIES, TRANSITION	26.8	OTHER TRANSITION ACTIVITIES Specific Sector Provisions For Pharmaceutical GMPs; Mutual Recognition Of Pharmaceutical GMP Reports, Medical Device Quality System Audit Reports
ACTIVITY	201.316	DRUGS WITH THYROID HORMONE ACTIVITY FOR HUMAN USE; REQUIRED WARNING Specific Labeling Requirements For Specific Drug Products; Labeling
ACTIVITY ASSIST	890.5050	DAILY ACTIVITY ASSIST DEVICE Physical Medicine Therapeutic Devices; Physical Medicine Devices
ACTIVITY, UTERINE	884.2730	HOME UTERINE ACTIVITY MONITOR Obstetrical and Gynecological Devices
ACTS, PROHIBITED	5.601	EXEMPTION OF ELECTRONIC PRODUCTS FROM PERFORMANCE STANDARDS AND PROHIBITED ACTS Radiation Control; Redelegations Of Authority; Delegations Of Authority And Organization
ACTUATOR	870.1670	SYRINGE ACTUATOR FOR AN INJECTOR Cardiovascular Diagnostic Devices; Cardiovascular Devices
ACUITY	886.1150	VISUAL ACUITY CHART Diagnostic Devices; Ophthalmic Devices

ACUPUNCTURE NEEDLE	880.5580	ACUPUNCTURE NEEDLE General Hospital And Personal Use Devices
ACUTE EFFECT	320.28	CORRELATION OF BIOAVAILABILITY WITH AN ACUTE PHARMACOLOGICAL EFFECT OR CLINICAL EVIDENCE Procedures For Determining The Bioavailability Of Drug Products; Bioavailability And Bioequivalence Requirements
ADAPTER, SISI	874.1070	SHORT INCREMENT SENSITIVITY INDEX (SISI) ADAPTER Diagnostic Devices; Ear, Nose, And Throat Devices
ADAPTOMETER	886.1050	ADAPTOMETER (BIOPHOTOMETER) Diagnostic Devices; Ophthalmic Devices
ADAPTOR	870.2350	ELECTROCARDIOGRAPH LEAD SWITCHING ADAPTOR Cardiovascular Monitoring Devices; Cardiovascular Devices
ADAPTOR	870.3620	PACEMAKER LEAD ADAPTOR Cardiovascular Prosthetic Devices; Cardiovascular Devices
ADAPTOR	870.4290	CARDIOPULMONARY BYPASS ADAPTOR, STOPCOCK, MANIFOLD, OR FITTING Cardiovascular Surgical Devices; Cardiovascular Devices
ADDED DIRECTLY	184.1	SUBSTANCES ADDED DIRECTLY TO HUMAN FOOD AFFIRMED AS GENERALLY RECOGNIZED AS SAFE (GRAS) General Provisions; Direct Food Substances Affirmed As Generally Recognized As Safe
ADDED INDIRECTLY	186.1	SUBSTANCES ADDED INDIRECTLY TO HUMAN FOOD AFFIRMED AS GENERALLY RECOGNIZED AS SAFE (GRAS) Indirect Food Substances Affirmed As Generally Recognized As Safe
ADDITIONAL INFORMATION	1301.15	ADDITIONAL INFORMATION Registration Of Manufacturers, Distributors, And Dispensers Of Controlled Substances
ADDITIONS	1403.6	ADDITIONS AND EXCEPTIONS Uniform Administrative Requirements For Grants And Coop. Agreements To State & Local Gov.
ADDITIVE, COLOR	14.140	ESTABLISHMENT OF A COLOR ADDITIVE ADVISORY COMMITTEE Color Additive Advisory Committees; Public Hearing Before A Public Advisory Committee
ADDITIVE, COLOR	14.142	FUNCTIONS OF A COLOR ADDITIVE ADVISORY COMMITTEE Color Additive Advisory Committees; Public Hearing Before A Public Advisory Committee
ADDITIVE, COLOR	14.145	PROCEDURES OF A COLOR ADDITIVE ADVISORY COMMITTEE Color Additive Advisory Committees; Public Hearing Before A Public Advisory Committee
ADDITIVE, COLOR	14.147	MEMBERSHIP OF A COLOR ADDITIVE ADVISORY COMMITTEE Color Additive Advisory Committees; Public Hearing Before A Public Advisory Committee
ADDITIVE, COLOR	14.155	FEES AND COMPENSATION PERTAINING TO A COLOR ADDITIVE ADVISORY COMMITTEE Color Additive Advisory Committees; Public Hearing Before A Public Advisory Committee
ADDITIVE, COLOR	71.15	CONFIDENTIALITY OF DATA AND INFORMATION IN COLOR ADDITIVE PETITIONS General Provisions; Color Additive Petitions
ADDITIVE, COLOR	71.22	DECEPTION AS A BASIS FOR REFUSING TO ISSUE REGULATIONS; DECEPTIVE USE OF A COLOR ADDITIVE FOR WHICH A REGULATION HAS ISSUED Administrative Action On Petitions; Color Additive Petitions
ADDITIVE, COLOR	73.1	DILUENTS IN COLOR ADDITIVE MIXTURES FOR FOOD USE EXEMPT FROM CERTIFICATION Foods; Listing Of Color Additives Exempt From Certification
ADDITIVE, COLOR	73.1001	DILUENTS IN COLOR ADDITIVE MIXTURES FOR DRUG USE EXEMPT FROM CERTIFICATION Drugs; Listing Of Color Additives Exempt From Certification
ADDITIVE, COLOR	80.35	COLOR ADDITIVE MIXTURES, CERTIFICATION AND EXEMPTION FROM CERTIFICATION Certification Procedures; Color Additive Certification
ADDITIVE, FOOD	170.6	OPINION LETTERS ON FOOD ADDITIVE STATUS General Provisions; Food Additives
ADDITIVE, FOOD	170.38	DETERMINATION OF FOOD ADDITIVE STATUS Food Additive Safety; Food Additives
ADDITIVE, FOOD	570.6	OPINION LETTERS ON FOOD ADDITIVE STATUS General Provisions; Food Additives
ADDITIVE, FOOD	570.38	DETERMINATION OF FOOD ADDITIVE STATUS Food Additive Safety; Food Additives
ADDITIVES	173.310	BOILER WATER ADDITIVES Specific Usage Additives; Secondary Direct Food Additives Permitted In Food For Human Consumption

ADDITIVES, COLOR	5.27	PATENT TERM EXTENSIONS FOR HUMAN DRUG PRODUCTS, MEDICAL DEVICES, AND FOOD AND COLOR ADDITIVES General Redelegations Of Authority; Delegations Of Authority And Organization
ADDITIVES, COLOR	5.300	FOOD STANDARDS, FOOD ADDITIVES, GRAS SUBSTANCES, COLOR ADDITIVES, NUTRIENT CONTENT CLAIMS, AND HEALTH CLAIMS Foods And Cosmetics; Redelegations Of Authority; Delegations Of Authority And Organization
ADDITIVES, COLOR	25.32	FOODS, FOOD ADDITIVES, AND COLOR ADDITIVES Environmental Impact Considerations
ADDITIVES, COLOR	70.5	GENERAL RESTRICTIONS ON USE OF COLOR ADDITIVES General Provisions; Color Additives
ADDITIVES, COLOR	70.10	COLOR ADDITIVES IN STANDARDIZED FOODS AND NEW DRUGS General Provisions; Color Additives
ADDITIVES, COLOR	70.25	LABELING REQUIREMENTS FOR COLOR ADDITIVES Packaging And Labeling; Color Additives
ADDITIVES, COLOR	70.42	CRITERIA FOR EVALUATING THE SAFETY OF COLOR ADDITIVES Safety Evaluation; Color Additives
ADDITIVES, COLOR	70.45	ALLOCATION OF COLOR ADDITIVES Safety Evaluation; Color Additives
ADDITIVES, COLOR	71.37	EXEMPTION OF COLOR ADDITIVES FOR INVESTIGATIONAL USE Administrative Action On Petitions; Color Additive Petitions
ADDITIVES, COLOR	81.1	PROVISIONAL LISTS OF COLOR ADDITIVES General Specs+restrictns Provisional Color Additives For Foods, Drugs, And Cosmetics
ADDITIVES, COLOR	81.10	TERMINATION OF PROVISIONAL LISTINGS OF COLOR ADDITIVES General Specs+restrictns Provisional Color Additives For Foods, Drugs, And Cosmetics
ADDITIVES, FOOD	5.27	PATENT TERM EXTENSIONS FOR HUMAN DRUG PRODUCTS, MEDICAL DEVICES, AND FOOD AND COLOR ADDITIVES General Redelegations Of Authority; Delegations Of Authority And Organization
ADDITIVES, FOOD	5.300	FOOD STANDARDS, FOOD ADDITIVES, GRAS SUBSTANCES, COLOR ADDITIVES, NUTRIENT CONTENT CLAIMS, AND HEALTH CLAIMS Foods And Cosmetics; Redelegations Of Authority; Delegations Of Authority And Organization
ADDITIVES, FOOD	25.32	FOODS, FOOD ADDITIVES, AND COLOR ADDITIVES Environmental Impact Considerations
ADDITIVES, FOOD	130.20	FOOD ADDITIVES PROPOSED FOR USE IN FOODS FOR WHICH DEFINITIONS AND STANDARDS OF IDENTITY ARE ESTABLISHED Food Additives In Standardized Foods; Food Standards: General
ADDITIVES, FOOD	170.10	FOOD ADDITIVES IN STANDARDIZED FOODS General Provisions; Food Additives
ADDITIVES, FOOD	170.18	TOLERANCES FOR RELATED FOOD ADDITIVES General Provisions; Food Additives
ADDITIVES, FOOD	170.20	GENERAL PRINCIPLES FOR EVALUATING THE SAFETY OF FOOD ADDITIVES Food Additive Safety; Food Additives
ADDITIVES, FOOD	172.5	GENERAL PROVISIONS FOR DIRECT FOOD ADDITIVES General Provisions; Food Additives Permitted For Direct Addition To Food For Human Consumption
ADDITIVES, FOOD	174.5	GENERAL PROVISIONS APPLICABLE TO INDIRECT FOOD ADDITIVES Indirect Food Additives: General
ADDITIVES, FOOD	570.13	INDIRECT FOOD ADDITIVES RESULTING FROM PACKAGING MATERIALS PRIOR SANCTIONED FOR ANIMAL FEED AND PET FOOD General Provisions; Food Additives
ADDITIVES, FOOD	570.14	INDIRECT FOOD ADDITIVES RESULTING FROM PACKAGING MATERIALS FOR ANIMAL FEED AND PET FOOD General Provisions; Food Additives
ADDITIVES, FOOD	570.18	TOLERANCES FOR RELATED FOOD ADDITIVES General Provisions; Food Additives
ADDITIVES, FOOD	570.20	GENERAL PRINCIPLES FOR EVALUATING THE SAFETY OF FOOD ADDITIVES Food Additive Safety; Food Additives
ADDRESS	312.140	ADDRESS FOR CORRESPONDENCE Miscellaneous; Investigational New Drug Application
ADDRESS	316.4	ADDRESS FOR SUBMISSIONS General Provisions; Orphan Drugs
ADDRESS	610.64	NAME AND ADDRESS OF DISTRIBUTOR Labeling Standards; General Biological Products Standards
ADDRESS	812.19	ADDRESS FOR IDE CORRESPONDENCE General Provisions; Investigational Device Exemptions
ADDRESS	1316.45	FILINGS; ADDRESS; HOURS Administrative Hearings; Administrative Functions, Practices, And Procedures
ADDRESSES	20.119	LISTS OF NAMES AND ADDRESSES Availability Of Specific Categories Of Records; Public Information
ADDRESSES	314.440	ADDRESSES FOR APPLICATIONS Miscellaneous Provisions; Applications For FDA Approval To Market A New Drug

ADDRESSES	510.600	NAMES, ADDRESSES, AND DRUG LABELER CODES OF SPONSORS OF APPROVED APPLICATIONS Sponsors Of Approved Applications; New Animal Drugs
ADENOSINE TRIPHOSPHATE	864.7040	ADENOSINE TRIPHOSPHATE RELEASE ASSAY Hematology Kits And Packages; Hematology And Pathology Devices
ADENOVIRUS	866.3020	ADENOVIRUS SEROLOGICAL REAGENTS Serological Reagents; Immunology And Microbiology Devices
ADEQUATE STUDIES	314.126	ADEQUATE AND WELL-CONTROLLED STUDIES FDA Action On Applications; Applications For FDA Approval To Market A New Drug
ADEQUATE STUDIES	514.117	ADEQUATE AND WELL-CONTROLLED STUDIES Administrative Actions On Applications; New Animal Drug Applications
ADHESION	864.6650	PLATELET ADHESION TEST Manual Hematology Devices; Hematology And Pathology Devices
ADHESIVE	872.3750	BRACKET ADHESIVE RESIN AND TOOTH CONDITIONER Prosthetic Devices; Dental Devices
ADHESIVE, DENTURE	872.3400	KARAYA AND SODIUM BORATE WITH OR WITHOUT ACACIA DENTURE ADHESIVE Prosthetic Devices; Dental Devices
ADHESIVE, DENTURE	872.3410	ETHYLENE OXIDE HOMOPOLYMER AND/OR CARBOXYMETHYLCELLULOSE SODIUM DENTURE ADHESIVE Prosthetic Devices; Dental Devices
ADHESIVE, DENTURE	872.3420	CARBOXYMETHYLCELLULOSE SODIUM AND CATIONIC POLYACRYLAMIDE POLYMER DENTURE ADHESIVE Prosthetic Devices; Dental Devices
ADHESIVE, DENTURE	872.3450	ETHYLENE OXIDE HOMOPOLYMER AND/OR KARAYA DENTURE ADHESIVE Prosthetic Devices; Dental Devices
ADHESIVE, DENTURE	872.3480	POLYACRYLAMIDE POLYMER DENTURE ADHESIVE Prosthetic Devices; Dental Devices
ADHESIVE, DENTURE	872.3490	CARBOXYMETHYLCELLULOSE SODIUM AND/OR POLYVINYLMETHYLETHER MALEIC ACID CALCIUM-SODIUM DOUBLE SALT DENTURE ADHESIVE Prosthetic Devices; Dental Devices
ADHESIVE, DENTURE	872.3500	POLYVINYLMETHYLETHER MALEIC ANHYDRIDE (PVM-MA), ACID COPOLYMER, AND CARBOXYMETHYLCELLULOSE SODIUM (NACMC) DENTURE ADHESIVE Prosthetic Devices; Dental Devices
ADHESIVE, DRAPE	878.4380	DRAPE ADHESIVE Surgical Devices; General And Plastic Surgery Devices
ADHESIVE, PROSTHESIS	878.3750	EXTERNAL PROSTHESIS ADHESIVE Prosthetic Devices; General And Plastic Surgery Devices
ADHESIVE TAPE	878.4730	SURGICAL SKIN DEGREASER OR ADHESIVE TAPE SOLVENT Surgical Devices; General And Plastic Surgery Devices
ADHESIVE TAPE	880.5240	MEDICAL ADHESIVE TAPE AND ADHESIVE BANDAGE General Hospital And Personal Use Therapeutic Devices; General Hospital And Personal Use Devices
ADHESIVES	175.105	ADHESIVES Substances For Use Only As Components Of Adhesives; Indirect Food Additives
ADHESIVES	175.125	PRESSURE-SENSITIVE ADHESIVES Substances For Use Only As Components Of Adhesives; Indirect Food Additives
ADIPATE-STEARATE	178.3690	PENTAERYTHRITOL ADIPATE-STEARATE Certain Adjuvants And Production Aids; Indirect Food Additives: Adjuvants, Production Aids, And Sanitizers
ADIPIC ACID	184.1009	ADIPIC ACID Listing Of Specific Substances Affirmed As GRAS; Direct Food Substances Affirmed As Generally Recognized As Safe
ADIPIC ACID	582.1009	ADIPIC ACID General Purpose Food Additives; Substances Generally Recognized As Safe
ADJUSTABLE	880.5100	AC-POWERED ADJUSTABLE HOSPITAL BED General Hospital And Personal Use Therapeutic Devices; General Hospital And Personal Use Devices
ADJUSTABLE	880.5110	HYDRAULIC ADJUSTABLE HOSPITAL BED General Hospital And Personal Use Therapeutic Devices; General Hospital And Personal Use Devices
ADJUSTABLE	880.5120	MANUAL ADJUSTABLE HOSPITAL BED General Hospital And Personal Use Therapeutic Devices; General Hospital And Personal Use Devices
ADJUSTMENTS	1303.13	ADJUSTMENTS TO AGGREGATE PRODUCTION QUOTAS Aggregate Production And Procurement Quotas; Quotas
ADJUSTMENTS	1403.51	LATER DISALLOWANCES AND ADJUSTMENTS Post-award Requirements; Uniform Administrative Requirements For Grants And Coop. Agreements To State & Local Gov.
ADJUVANT	178.3010	ADJUVANT SUBSTANCES USED IN THE MANUFACTURE OF FOAMED PLASTICS Certain Adjuvants And Production Aids; Indirect Food Additives: Adjuvants, Production Aids, And Sanitizers

ADJUVANTS	172.515	SYNTHETIC FLAVORING SUBSTANCES AND ADJUVANTS Flavoring Agents And Related Substances; Food Additives Permitted For Direct Addition To Food For Human Consumption
ADJUVANTS	172.710	ADJUVANTS FOR PESTICIDE USE DILUTIONS Other Specific Usage Additives; Food Additives Permitted For Direct Addition To Food For Human Consumption
ADJUVANTS	182.60	SYNTHETIC FLAVORING SUBSTANCES AND ADJUVANTS General Provisions; Substances Generally Recognized As Safe
ADJUVANTS	182.99	ADJUVANTS FOR PESTICIDE CHEMICALS General Provisions; Substances Generally Recognized As Safe
ADJUVANTS	582.60	SYNTHETIC FLAVORING SUBSTANCES AND ADJUVANTS General Provisions; Substances Generally Recognized As Safe
ADJUVANTS	582.99	ADJUVANTS FOR PESTICIDE CHEMICALS General Provisions; Substances Generally Recognized As Safe
ADMIN. APPARATUS, N. O.	868.5165	NITRIC OXIDE ADMINISTRATION APPARATUS Therapeutic Devices; Anesthesiology Devices
ADMINISTERING	1306.07	ADMINISTERING OR DISPENSING OF NARCOTIC DRUGS General Information; Prescriptions
ADMINISTRATION	19.10	FDA CONFLICT OF INTEREST REVIEW BOARD General Provisions; Standards Of Conduct And Conflicts Of Interest
ADMINISTRATION	20.1	TESTIMONY BY FDA EMPLOYEES Official Testimony And Information; Public Information
ADMINISTRATION	20.2	PRODUCTION OF RECORDS BY FDA EMPLOYEES Official Testimony And Information; Public Information
ADMINISTRATION	20.3	CERTIFICATION AND AUTHENTICATION OF FDA RECORDS Official Testimony And Information; Public Information
ADMINISTRATION	20.30	FDA FREEDOM OF INFORMATION STAFF General Policy; Public Information
ADMINISTRATION	20.31	RETENTION SCHEDULE OF REQUESTS FOR FDA RECORDS General Policy; Public Information
ADMINISTRATION	20.32	DISCLOSURE OF FDA EMPLOYEE NAMES General Policy; Public Information
ADMINISTRATION	20.105	TESTING AND RESEARCH CONDUCTED BY OR WITH FUNDS PROVIDED BY THE FDA Availability Of Specific Categories Of Records; Public Information
ADMINISTRATION	21.20	PROCEDURES FOR NOTICE OF FDA PRIVACY ACT RECORD SYSTEMS FDA Privacy Act Record Systems; Protection Of Privacy
ADMINISTRATION	21.31	RECORDS STORED BY THE NATIONAL ARCHIVES AND RECORDS ADMINISTRATION Requirements For Specific Categories Of Records; Protection Of Privacy
ADMINISTRATION	1307.22	DISPOSAL OF CONTROLLED SUBSTANCES BY THE ADMINISTRATION Disposal Of Controlled Substances; Miscellaneous
ADMINISTRATION	1308.03	ADMINISTRATION CONTROLLED SUBSTANCES CODE NUMBER General Information; Schedules Of Controlled Substances
ADMINISTRATION, DRUG	874.5220	EAR, NOSE, AND THROAT DRUG ADMINISTRATION DEVICE Therapeutic Devices; Ear, Nose, and Throat Devices
ADMINISTRATION, INTRAVASCULAR	880.5440	INTRAVASCULAR ADMINISTRATION SET General Hospital And Personal Use Therapeutic Devices; General Hospital And Personal Use Devices
ADMINISTRATION-REQUESTED	7.45	FDA-REQUESTED RECALL Recalls - Guidance on Policy, Procedures & Industry Responsibilities; Enforcement Policy
ADMINISTRATIVE ACTION	20.91	USE OF DATA OR INFORMATION FOR ADMINISTRATIVE OR COURT ENFORCEMENT ACTION Limitations On Exemptions; Public Information
ADMINISTRATIVE ACTIONS	56.120	LESSER ADMINISTRATIVE ACTIONS Administrative Actions For Noncompliance; Institutional Review Boards
ADMINISTRATIVE APPEALS	21.52	ADMINISTRATIVE APPEALS OF REFUSALS TO AMEND RECORDS Procedures For Requests For Amendment Of Records; Protection Of Privacy
ADMINISTRATIVE CAUSE	1316.10	ADMINISTRATIVE PROBABLE CAUSE Administrative Inspections; Administrative Functions, Practices, And Procedures
ADMINISTRATIVE DECISION	13.50	RECORD FOR ADMINISTRATIVE DECISION Records Of A Hearing Before A Board; Public Hearing Before A Public Board Of Inquiry
ADMINISTRATIVE DECISION	16.95	ADMINISTRATIVE DECISION AND RECORD FOR DECISION Administrative Record And Decision; Regulatory Hearing Before The FDA
ADMINISTRATIVE DECISION	60.46	ADMINISTRATIVE DECISION Due Diligence Hearings; Patent Term Restoration
ADMINISTRATIVE DETENTION	800.55	ADMINISTRATIVE DETENTION Requirements For Specific Medical Devices; General
ADMINISTRATIVE FILE	10.70	DOCUMENTATION OF SIGNIFICANT DECISIONS IN ADMINISTRATIVE FILE General Administrative Procedures; Administrative Practices And Procedures

ADMINISTRATIVE FORFEITURE	1316.77	ADMINISTRATIVE FORFEITURE Seizure, Forfeiture, And Disposition Of Property; Administrative Functions, Practices, And Procedures
ADMINISTRATIVE FORFEITURE	1316.92	PETITION FOR EXPEDITED RELEASE IN AN ADMINISTRATIVE FORFEITURE ACTION Expedited Forfeiture Proceedings For Certain Property; Administrative Functions, Practices, And Procedures
ADMINISTRATIVE FORFEITURE	1316.93	RULING ON PETITION FOR EXPEDITED RELEASE IN AN ADMINISTRATIVE FORFEITURE ACTION Expedited Forfeiture Proceedings For Certain Property; Administrative Functions, Practices, And Procedures
ADMINISTRATIVE FORFEITURE	1316.94	POSTING OF SUBSTITUTE RES IN AN ADMINISTRATIVE FORFEITURE ACTION Expedited Forfeiture Proceedings For Certain Property; Administrative Functions, Practices, And Procedures
ADMINISTRATIVE INSPECTION	1316.07	REQUIREMENT FOR ADMINISTRATIVE INSPECTION WARRANT; EXCEPTIONS Administrative Inspections; Administrative Functions, Practices, And Procedures
ADMINISTRATIVE INSPECTION	1316.09	APPLICATION FOR ADMINISTRATIVE INSPECTION WARRANT Administrative Inspections; Administrative Functions, Practices, And Procedures
ADMINISTRATIVE INSPECTIONS	1316.13	FREQUENCY OF ADMINISTRATIVE INSPECTIONS Administrative Inspections; Administrative Functions, Practices, And Procedures
ADMINISTRATIVE PRACTICES	10.10	SUMMARIES OF ADMINISTRATIVE PRACTICES AND PROCEDURES General Provisions; Administrative Practices And Procedures
ADMINISTRATIVE PROCEEDINGS	10.25	INITIATION OF ADMINISTRATIVE PROCEEDINGS General Administrative Procedures; Administrative Practices And Procedures
ADMINISTRATIVE PROCEEDINGS	10.205	ELECTRONIC MEDIA COVERAGE OF PUBLIC ADMINISTRATIVE PROCEEDINGS Electronic Media Coverage Of Public Administrative Proceedings; Guidelines On Policy And Procedures; Administrative Practices And Procedures
ADMINISTRATIVE PROCEEDINGS	10.206	PROCEDURES FOR ELECTRONIC MEDIA COVERAGE OF AGENCY PUBLIC ADMINISTRATIVE PROCEEDINGS Electronic Media Coverage Of Public Administrative Proceedings; Guidelines On Policy And Procedures; Administrative Practices And Procedures
ADMINISTRATIVE PROCEEDINGS	20.86	DISCLOSURE IN ADMINISTRATIVE OR COURT PROCEEDINGS Limitations On Exemptions; Public Information
ADMINISTRATIVE RECONSIDERATION	10.33	ADMINISTRATIVE RECONSIDERATION OF ACTION General Administrative Procedures; Administrative Practices And Procedures
ADMINISTRATIVE RECORD	12.100	ADMINISTRATIVE RECORD OF A HEARING Administrative Record; Formal Evidentiary Public Hearing
ADMINISTRATIVE RECORD	13.40	ADMINISTRATIVE RECORD OF A BOARD Records Of A Hearing Before A Board; Public Hearing Before A Public Board Of Inquiry
ADMINISTRATIVE RECORD	13.45	EXAMINATION OF ADMINISTRATIVE RECORD Records Of A Hearing Before A Board; Public Hearing Before A Public Board Of Inquiry
ADMINISTRATIVE RECORD	14.70	ADMINISTRATIVE RECORD OF A PUBLIC HEARING BEFORE AN ADVISORY COMMITTEE Records Of Meetings And Hearings Before Advisory Committees; Public Hearings Before A Public Advisory Committee
ADMINISTRATIVE RECORD	14.75	EXAMINATION OF ADMINISTRATIVE RECORD AND OTHER ADVISORY COMMITTEE RECORDS Records Of Meetings And Hearings Before Advisory Committees; Public Hearings Before A Public Advisory Committee
ADMINISTRATIVE RECORD	15.40	ADMINISTRATIVE RECORD Records Of A Public Hearing Before The Commissioner; Public Hearing Before The Commissioner
ADMINISTRATIVE RECORD	15.45	EXAMINATION OF ADMINISTRATIVE RECORD Records Of A Public Hearing Before The Commissioner; Public Hearing Before The Commissioner
ADMINISTRATIVE RECORD	16.80	ADMINISTRATIVE RECORD OF A REGULATORY HEARING Administrative Record And Decision; Regulatory Hearing Before The FDA
ADMINISTRATIVE RECORD	16.85	EXAMINATION OF ADMINISTRATIVE RECORD Administrative Record And Decision; Regulatory Hearing Before The FDA
ADMINISTRATIVE RECORD	17.41	THE ADMINISTRATIVE RECORD Civil Money Penalties Hearings
ADMINISTRATIVE RECORDS	20.101	ADMINISTRATIVE ENFORCEMENT RECORDS Availability Of Specific Categories Of Records; Public Information
ADMINISTRATIVE REMEDIES	10.45	COURT REVIEW OF FINAL ADMINISTRATIVE ACTION; EXHAUSTION OF ADMINISTRATIVE REMEDIES General Administrative Procedures; Administrative Practices And Procedures
ADMINISTRATIVE REMEDIES	14.7	ADMINISTRATIVE REMEDIES General Provisions; Public Hearing Before A Public Advisory Committee

ADMINISTRATIVE REVIEW	1301.31	ADMINISTRATIVE REVIEW GENERALLY Registration Of Manufacturers, Distributors, And Dispensers Of Controlled Substances
ADMINISTRATIVE REVIEW	1309.41	ADMINISTRATIVE REVIEW GENERALLY Registration Of Manufacturers, Distributors, Importers And Exporters Of List I Chemicals
ADMINISTRATIVE STAY	10.35	ADMINISTRATIVE STAY OF ACTION General Administrative Procedures; Administrative Practices And Procedures
ADMINISTRATIVE SUPPORT	13.15	SEPARATION OF FUNCTIONS; EX PARTE COMMUNICATIONS; ADMINISTRATIVE SUPPORT General Provisions; Public Hearing Before a Public Board Of Inquiry
ADMINISTRATIVE SYSTEMS	203.34	POLICIES AND PROCEDURES; ADMINISTRATIVE SYSTEMS Samples; Prescription Drug Marketing
ADMINISTRATIVE WARRANT	1316.12	REFUSAL TO ALLOW INSPECTION WITH AN ADMINISTRATIVE WARRANT Administrative Inspections; Administrative Functions, Practices, And Procedures
ADMISSION	1.94	HEARING ON REFUSAL OF ADMISSION Imports And Exports; General Enforcement Regulations
ADOPTION	170.15	ADOPTION OF REGULATION ON INITIATIVE OF COMMISSIONER General Provisions; Food Additives
ADOPTION	570.15	ADOPTION OF REGULATION ON INITIATIVE OF COMMISSIONER General Provisions; Food Additives
ADRENOCORTICOTROPIC	862.1025	ADRENOCORTICOTROPIC HORMONE (ACTH) TEST SYSTEM Clinical Chemistry Test Systems; Clinical Chemistry And Clinical Toxicology Devices
ADULTERATED	5.402	DETENTION OF ADULTERATED OR MISBRANDED MEDICAL DEVICES Medical Devices And Radiological Health; Redelegations Of Authority; Delegations Of Authority And Organization
ADULTERATION	2.25	GRAIN SEED TREATED WITH POISONOUS SUBSTANCES; COLOR IDENTIFICATION TO PREVENT ADULTERATION OF HUMAN AND ANIMAL FOOD Human And Animal Foods; General Administrative Rulings And Decisions
ADULTERATION	99.405	APPLICABILITY OF LABELING ADULTERATION, AND MISBRANDING AUTHORITY Dissemination Of Information On Unapproved/New Uses For Marketed Drugs, Biologics and Devices
ADULTERATION	314.170	ADULTERATION AND MISBRANDING OF AN APPROVED DRUG FDA Action On Applications; Applications For FDA Approval To Market A New Drug
ADULTERATION	800.20	PATIENT EXAMINATION GLOVES AND SURGEON'S GLOVES; SAMPLE PLANS AND TEST METHOD FOR LEAKAGE DEFECTS; ADULTERATION Requirements For Specific Medical Devices; General
ADVANCE NOTICE	1312.32	SCHEDULES II, III, IV: ADVANCE NOTICE Transshipment And In-Transit Shipment Of Controlled Substances; Importation And Exportation Of Controlled Substances
ADVANCE NOTICE	1313.15	WAIVER OF 15-DAY ADVANCE NOTICE FOR REGULAR IMPORTERS Importation Of Listed Chemicals; Importation And Exportation Of Precursors And Essential Chemicals
ADVANCE NOTICE	1313.24	WAIVER OF 15-DAY ADVANCE NOTICE FOR CHEMICAL EXPORTERS Exportation Of Listed Chemicals; Importation And Exportation Of Precursors And Essential Chemicals
ADVANCE NOTICE	1313.31	ADVANCE NOTICE OF IMPORTATION FOR TRANSSHIPMENT OR TRANSFER Transshipment, In-Transit Shipments And International Transactions Involving Listed Chemicals; Importation And Exportation Of Precursors And Essential Chemicals
ADVERSE DECISION	203.12	APPEAL FROM AN ADVERSE DECISION BY DISTRICT OFFICE Reimportation; Prescription Drug Marketing
ADVERSE DECISIONS	900.15	APPEALS OF ADVERSE ACCREDITATION OR REACCREDITATION DECISIONS Quality Standards And Certification; Mammography
ADVERSE DRUG EXPERIENCES	310.305	RECORDS AND REPORTS CONCERNING ADVERSE DRUG EXPERIENCES ON MARKETED PRESCRIPTION DRUGS Records And Reports; New Drugs
ADVERSE DRUG EXPERIENCES	314.80	POSTMARKETING REPORTING OF ADVERSE DRUG EXPERIENCES Applications; Applications For FDA Approval To Market A New Drug
ADVERSE EVENT	803.30	INDIVIDUAL ADVERSE EVENT REPORTS; USER FACILITIES User Facility Reporting Requirements; Medical Device Reporting
ADVERSE EVENT	803.32	INDIVIDUAL ADVERSE EVENT REPORT DATA ELEMENTS User Facility Reporting Requirements; Medical Device Reporting
ADVERSE EVENT	803.40	INDIVIDUAL ADVERSE EVENT REPORTING REQUIREMENTS; IMPORTERS Importer Reporting Requirement; Medical Device Reporting
ADVERSE EVENT	803.42	INDIVIDUAL ADVERSE EVENT REPORT DATA ELEMENTS Importer Reporting Requirement; Medical Device Reporting
ADVERSE EVENT	803.50	INDIVIDUAL ADVERSE EVENT REPORTS; MANUFACTURERS Manufacturer Reporting Requirements; Medical Device Reporting
ADVERSE EVENT	803.52	INDIVIDUAL ADVERSE EVENT REPORT DATA ELEMENTS Manufacturer Reporting Requirements; Medical Device Reporting

ADVERSE EXPERIENCES	600.80	POSTMARKETING REPORTING OF ADVERSE EXPERIENCES Biological Products: General
ADVERSE REACTION	606.170	ADVERSE REACTION FILE Records And Reports; Current Good Manufacturing Practice For Blood And Blood Components
ADVERTISEMENT	1316.75	ADVERTISEMENT Seizure, Forfeiture, And Disposition Of Property; Administrative Functions, Practices, And Procedures
ADVERTISEMENTS	200.200	PRESCRIPTION DRUGS; REMINDER ADVERTISEMENTS AND REMINDER LABELING TO PROVIDE PRICE INFORMATION TO CONSUMERS Prescription Drug Consumer Price Listing; General
ADVERTISEMENTS	202.1	PRESCRIPTION DRUG ADVERTISEMENTS Prescription Drug Advertising
ADVERTISING	201.200	DISCLOSURE OF DRUG EFFICACY STUDY EVALUATIONS IN LABELING AND ADVERTISING Labeling Claims For Drugs In Drug Efficacy Study; Labeling
ADVERTISING	530.4	ADVERTISING AND PROMOTION General Provisions; Extralabel Drug Use In Animals
ADVICE	12.50	ADVICE ON PUBLIC PARTICIPATION IN HEARINGS Appearance And Participation; Formal Evidentiary Public Hearing
ADVICE	14.174	ADVICE AND RECOMMENDATIONS IN WRITING Advisory Committees For Human Prescription Drugs; Public Hearing Before A Public Advisory Committee
ADVISORY COMMITTEE	14.5	PURPOSE OF PROCEEDINGS BEFORE AN ADVISORY COMMITTEE General Provisions; Public Hearing Before A Public Advisory Committee
ADVISORY COMMITTEE	14.22	MEETINGS OF AN ADVISORY COMMITTEE Meeting Procedures; Public Hearings Before A Public Advisory Committee
ADVISORY COMMITTEE	14.27	DETERMINATION TO CLOSE PORTIONS OF ADVISORY COMMITTEE MEETINGS Meeting Procedures; Public Hearing Before A Public Advisory Committee
ADVISORY COMMITTEE	14.60	MINUTES AND REPORTS OF ADVISORY COMMITTEE MEETINGS Records Of Meetings And Hearings Before Advisory Committees; Public Hearings Before A Public Advisory Committee
ADVISORY COMMITTEE	14.70	ADMINISTRATIVE RECORD OF A PUBLIC HEARING BEFORE AN ADVISORY COMMITTEE Records Of Meetings And Hearings Before Advisory Committees; Public Hearings Before A Public Advisory Committee
ADVISORY COMMITTEE	14.140	ESTABLISHMENT OF A COLOR ADDITIVE ADVISORY COMMITTEE Color Additive Advisory Committees; Public Hearings Before A Public Advisory Committee
ADVISORY COMMITTEE	20.118	ADVISORY COMMITTEE RECORDS Availability Of Specific Categories Of Records; Public Information
ADVISORY COMMITTEE	70.51	ADVISORY COMMITTEE ON THE APPLICATION OF THE ANTICANCER CLAUSE Safety Evaluation; Color Additives
ADVISORY COMMITTEES	5.30	AUTHORITY TO SELECT TEMPORARY VOTING MEMBERS FOR ADVISORY COMMITTEES AND AUTHORITY TO SIGN CONFLICT OF INTEREST WAIVERS General Redelegations Of Authority; Delegations Of Authority And Organization
ADVISORY COMMITTEES	14.40	ESTABLISHMENT AND RENEWAL OF ADVISORY COMMITTEES Establishment Of Advisory Committees; Public Hearings Before A Public Advisory Committee
ADVISORY COMMITTEES	14.55	TERMINATION OF ADVISORY COMMITTEES Establishment Of Advisory Committees; Public Hearings Before A Public Advisory Committee
ADVISORY COMMITTEES	14.80	QUALIFICATIONS FOR MEMBERS OF STANDING POLICY AND TECHNICAL ADVISORY COMMITTEES Members Of Advisory Committees; Public Hearings Before A Public Advisory Committee
ADVISORY COMMITTEES	14.100	LIST OF STANDING ADVISORY COMMITTEES Standing Advisory Committees; Public Hearings Before A Public Advisory Committee
ADVISORY COMMITTEES	14.160	ESTABLISHMENT OF STANDING TECHNICAL ADVISORY COMMITTEES FOR HUMAN PRESCRIPTION DRUGS Advisory Committees For Human Prescription Drugs; Public Hearings Before A Public Advisory Committee
ADVISORY COMMITTEES	20.84	DISCLOSURE TO CONSULTANTS, ADVISORY COMMITTEES, AND OTHER SPECIAL GOVERNMENT EMPLOYEES Limitations On Exemptions; Public Information
ADVISORY COMMITTEES	861.38	STANDARDS ADVISORY COMMITTEES Procedures For Performance Standards Development
ADVISORY OPINIONS	10.85	ADVISORY OPINIONS General Administrative Procedures; Administrative Practices And Procedures
ADVISORY OPINIONS	12.83	ADVISORY OPINIONS Hearing Procedures; Formal Evidentiary Public Hearing

ADVISORY OPINIONS	808.5	ADVISORY OPINIONS Exemption Procedures; Exemptions From Federal Preemption Of State And Local Medical Device Requirements
AERATOR	880.6100	ETHYLENE OXIDE GAS AERATOR CABINET General Hospital And Personal Use Miscellaneous Devices; General Hospital And Personal Use Devices
AEROGEL	182.1711	SILICA AEROGEL Multiple Purpose GRAS Food Substances; Substances Generally Recognized As Safe
AEROGEL	582.1711	SILICA AEROGEL General Purpose Food Additives; Substances Generally Recognized As Safe
AEROSOL	524.1301	MAFENIDE ACETATE AND NITORFURAZONE AEROSOL POWDER Ophthalmic And Topical Dosage Form New Animal Drugs
AEROSOL	700.14	USE OF VINYL CHLORIDE AS AN INGREDIENT, INCLUDING PROPELLANT OF COSMETIC AEROSOL PRODUCTS Requirements For Specific Cosmetic Products; General
AEROSOL	700.16	USE OF AEROSOL COSMETIC PRODUCTS CONTAINING ZIRCONIUM Requirements For Specific Cosmetic Products; General
AEROSOL POWDER	524.1005	FURAZOLIDONE AEROSOL POWDER Ophthalmic And Topical Dosage Form New Animal Drugs
AEROSOLS	201.305	SOPROTERENOL INHALATION PREPARATIONS (PRESSURIZED AEROSOLS, NEBULIZERS, POWDERS) FOR HUMAN USE; WARNINGS Specific Labeling Requirements For Specific Drug Products; Labeling
AESTHETIC RESTORATION	878.3800	EXTERNAL AESTHETIC RESTORATION PROSTHESIS Prosthetic Devices; General And Plastic Surgery Devices
AFFIDAVITS	1316.57	SUBMISSION OF DOCUMENTARY EVIDENCE AND AFFIDAVITS SUBSEQUENT TO PREHEARING CONFERENCE Administrative Hearings; Administrative Functions, Practices, And Procedures
AFFIDAVITS	1316.58	SUMMARY OF TESTIMONY; AFFIDAVITS Administrative Hearings; Administrative Functions, Practices, And Procedures
AFFIRMATION	170.35	AFFIRMATION OF GENERALLY RECOGNIZED AS SAFE (GRAS) STATUS Food Additive Safety; Food Additives
AFFIRMATION	570.35	AFFIRMATION OF GENERALLY RECOGNIZED AS SAFE (GRAS) STATUS Food Additive Safety; Food Additives
AFFIRMED	184.1	SUBSTANCES ADDED DIRECTLY TO HUMAN FOOD AFFIRMED AS GENERALLY RECOGNIZED AS SAFE (GRAS) General Provisions; Direct Food Substances Affirmed As Generally Recognized As Safe
AFTERIMAGE	886.1300	AFTERIMAGE FLASHER Diagnostic Devices; Ophthalmic Devices
AGAR	201.319	WATER-SOLUBLE GUMS, HYDROPHILIC GUMS and HYDROPHILIC MUCILLOIDS (INCLUDING AGAR) AS ACTIVE INGREDIENTS; WARNINGS AND DIRECTIONS Specific Labeling Requirements For Specific Drug Products; Labeling
AGAR	866.4600	OUCHTERLONY AGAR PLATE Immunology Laboratory Equipment And Reagents; Immunology And Microbiology Devices
AGAR-AGAR	582.7115	AGAR-AGAR Stabilizers; Substances Generally Recognized As Safe
AGENCIES, GOVERNMENT	20.85	DISCLOSURE TO OTHER FEDERAL GOVERNMENT DEPARTMENTS AND AGENCIES Limitations On Exemptions; Public Information
AGENCIES	20.108	AGREEMENTS BETWEEN THE FDA AND OTHER DEPARTMENTS, AGENCIES, AND ORGANIZATIONS Availability Of Specific Categories Of Records; Public Information
AGENCIES	21.54	AMENDED OR DISPUTED RECORDS RECEIVED FROM OTHER AGENCIES Procedures For Requests For Amendment Of Records; Protection Of Privacy
AGENCIES	25.44	LEAD AND COOPERATING AGENCIES Preparation of Environmental Documents; Environmental Impact Considerations
AGENCIES, CERTIFICATION	900.22	STANDARDS FOR CERTIFICATION AGENCIES States As Certifiers; Mammography
AGENCY	25.45	RESPONSIBLE AGENCY OFFICIAL Preparation of Environmental Documents; Environmental Impact Considerations
AGENCY	54.5	AGENCY EVALUATION OF FINANCIAL INTERESTS Financial Disclosure By Clinical Investigators
AGENCY	99.201	MANUFACTURER'S SUBMISSION TO THE AGENCY Dissemination Of Information On Unapproved/New Uses For Marketed Drugs, Biologics, and Devices
AGENCY ACTION	99.301	AGENCY ACTION ON A SUBMISSION Dissemination Of Information On Unapproved/New Uses For Marketed Drugs, Biologics, and Devices
AGENCY ACTIONS	25.60	ENVIRONMENTAL EFFECTS ABROAD OF MAJOR AGENCY ACTIONS Other Requirements; Environmental Impact Considerations

AGENCY, CERTIFICATION	900.21	APPLICATION FOR APPROVAL AS A CERTIFICATION AGENCY States As Certifiers; Mammography
AGENCY COMPONENT	3.4	DESIGNATED AGENCY COMPONENT Assignment of Agency Component for Review of Premarket Applications; Product Jurisdiction
AGENCY COMPONENT	3.5	PROCEDURES FOR IDENTIFYING THE DESIGNATED AGENCY COMPONENT Assignment of Agency Component for Review of Premarket Applications; Product Jurisdiction
AGENCY PROCEEDINGS	10.206	PROCEDURES FOR ELECTRONIC MEDIA COVERAGE OF AGENCY PUBLIC ADMINISTRATIVE PROCEEDINGS Electronic Media Coverage Of Public Admin Proceedings; Administrative Practices And Procedures
AGENCY REVIEW	10.75	INTERNAL AGENCY REVIEW OF DECISIONS General Administrative Procedures; Administrative Practices And Procedures
AGENT	316.22	PERMANENT-RESIDENT AGENT FOR FOREIGN SPONSOR Orphan Drugs
AGENT, HEMOSTATIC	878.4490	ABSORBABLE HEMOSTATIC AGENT AND DRESSING Surgical Devices; General And Plastic Surgery Devices
AGENT, HYPERTONICITY	349.16	OPHTHALMIC HYPERTONICITY AGENT Ophthalmic Drug Products For Over-the-counter Human Use
AGENT, POLISHING	872.6030	ORAL CAVITY ABRASIVE POLISHING AGENT Miscellaneous Devices; Dental Devices
AGENT, TOOTH-BONDING	872.3200	RESIN TOOTH BONDING AGENT Prosthetic Devices; Dental Devices
AGENTS	5.802	MANUFACTURER'S RESIDENT IMPORT AGENTS Imports And Exports; Redelegations Of Authority; Delegations Of Authority And Organization
AGENTS	173.340	DEFOAMING AGENTS Specific Usage Additives; Secondary Direct Food Additives Permitted In Food For Human Consumption
AGENTS	173.357	MATERIALS USED AS FIXING AGENTS IN THE IMMOBILIZATION OF ENZYME PREPARATIONS Specific Usage Additives; Secondary Direct Food Additives Permitted In Food For Human Consumption
AGENTS	176.150	CHELATING AGENTS USED IN THE MANUFACTURE OF PAPER AND PAPERBOARD Substances For Use Only As Components Of Paper And Paperboard; Indirect Food Additives
AGENTS	176.200	DEFOAMING AGENTS USED IN COATINGS Substances For Use Only As Components Of Paper And Paperboard; Indirect Food Additives
AGENTS	176.210	DEFOAMING AGENTS USED IN THE MANUFACTURE OF PAPER AND PAPERBOARD Substances For Use Only As Components Of Paper And Paperboard; Indirect Food Additives
AGENTS	178.3125	ANTICORROSIVE AGENTS Certain Adjuvants And Production Aids; Indirect Food Additives: Adjuvants, Production Aids, And Sanitizers
AGENTS	178.3130	ANTISTATIC AND/OR ANTIFOGGING AGENTS IN FOOD-PACKAGING MATERIALS Certain Adjuvants And Production Aids; Indirect Food Additives: Adjuvants, Production Aids, And Sanitizers
AGENTS	178.3295	CLARIFYING AGENTS FOR POLYMERS Certain Adjuvants And Production Aids; Indirect Food Additives: Adjuvants, Production Aids, And Sanitizers
AGENTS	178.3400	EMULSIFIERS AND/OR SURFACE-ACTIVE AGENTS Certain Adjuvants And Production Aids; Indirect Food Additives: Adjuvants, Production Aids, And Sanitizers
AGENTS	178.3860	RELEASE AGENTS Certain Adjuvants And Production Aids; Indirect Food Additives: Adjuvants, Production Aids, And Sanitizers
AGENTS	181.28	RELEASE AGENTS Specific Prior-sanctioned Food Ingredients; Prior-sanctioned Food Ingredients
AGENTS	310.534	DRUG PRODUCTS CONTAINING ACTIVE INGREDIENTS OFFERED OTC FOR HUMAN USE AS ORAL WOUND HEALING AGENTS Requirements For Specific New Drugs Or Devices; New Drugs
AGENTS	700.13	USE OF MERCURY COMPOUNDS IN COSMETICS INCLUDING USE AS SKINBLEACHING AGENTS IN COSMETIC PREPARATIONS ALSO REGARDED AS DRUGS Requirements For Specific Cosmetic Products; General
AGENTS	1301.22	EXEMPTION OF AGENTS AND EMPLOYEES; AFFILIATED PRACTITIONERS Registration Of Manufacturers, Distributors, And Dispensers Of Controlled Substances

AGENTS	1309.24	EXEMPTION OF AGENTS AND EMPLOYEES Registration Of Manufacturers, Distributors, Importers And Exporters Of List I Chemicals
AGGREGATION	864.5700	AUTOMATED PLATELET AGGREGATION SYSTEM Automated And Semi-automated Hematology Devices; Hematology And Pathology Devices
AGGREGOMETER	864.6675	PLATELET AGGREGOMETER Manual Hematology Devices; Hematology And Pathology Devices
AGREEMENTS	5.413	APPROVAL, DISAPPROVAL, OR WITHDRAWAL OF APPROVAL OF APPLICATIONS AND ENTERING INTO AGREEMENTS FOR INVESTIGATIONAL DEVICE EXEMPTIONS Medical Devices And Radiological Health; Redelegations Of Authority; Delegations Of Authority And Organization
AGREEMENTS	10.90	FDA REGULATIONS, GUIDELINES, RECOMMENDATIONS, AND AGREEMENTS General Administrative Procedures; Administrative Practices And Procedures
AGREEMENTS	20.108	AGREEMENTS BETWEEN THE FDA AND OTHER DEPARTMENTS, AGENCIES, AND ORGANIZATIONS Availability Of Specific Categories Of Records; Public Information
AGREEMENTS	20.114	DATA AND INFORMATION SUBMITTED PURSUANT TO COOPERATIVE QUALITY ASSURANCE AGREEMENTS Availability Of Specific Categories Of Records; Public Information
AGREEMENTS	26.78	AGREEMENTS WITH OTHER COUNTRIES Framework Provisions; Mutual Recognition Of Pharmaceutical GMP Reports, Medical Device Quality System Audit Reports
AGREEMENTS	203.36	FULFILLMENT HOUSES, SHIPPING AND MAILING SERVICES, COMARKETING AGREEMENTS, AND THIRD-PARTY RECORDKEEPING Samples; Prescription Drug Marketing
AGGREGATE	1303.11	AGGREGATE PRODUCTION QUOTAS Aggregate Production And Procurement Quotas; Quotas
AGGREGATE	1303.13	ADJUSTMENTS TO AGGREGATE PRODUCTION QUOTAS Aggregate Production And Procurement Quotas; Quotas
AHF	640.50	CRYOPRECIPITATE AHF Cryoprecipitate; Additional Standards For Human Blood And Blood Products
AID, EAR DRYING	344.12	EAR DRYING AID ACTIVE INGREDIENT Active Ingredients; Topical Otic Drug Products For Over-the-counter Human Use
AID, EAR DRYING	344.52	LABELING OF EAR DRYING AID DRUG PRODUCTS Labeling; Topical Otic Drug Products For Over-the-counter Human Use
AID, EARWAX REMOVAL	344.10	EARWAX REMOVAL AID ACTIVE INGREDIENT Active Ingredients; Topical Otic Drug Products For Over-the-counter Human Use
AID, EARWAX REMOVAL	344.50	LABELING OF EARWAX REMOVAL AID DRUG PRODUCTS Labeling; Topical Otic Drug Products For Over-the-counter Human Use
AID, HEARING	801.420	HEARING AID DEVICES; PROFESSIONAL AND PATIENT LABELING Special Requirements For Specific Devices; Labeling
AID, HEARING	801.421	HEARING AID DEVICES; CONDITIONS FOR SALE Special Requirements For Specific Devices; Labeling
AID, HEARING	874.3300	HEARING AID Prosthetic Devices; Ear, Nose, And Throat Devices
AID, HEARING	874.3310	HEARING AID CALIBRATOR AND ANALYSIS SYSTEM Prosthetic Devices; Ear, Nose, And Throat Devices
AID, HEARING	874.3320	GROUP HEARING AID OR GROUP AUDITORY TRAINER Prosthetic Devices; Ear, Nose, And Throat Devices
AID, HEARING	874.3330	MASTER HEARING AID Prosthetic Devices; Ear, Nose, And Throat Devices
AID, SLEEP	338.10	NIGHTTIME SLEEP-AID ACTIVE INGREDIENTS Active Ingredients; Nighttime Sleep-aid Drug Products For OTC Human Use
AID, SLEEP	338.50	LABELING OF NIGHTTIME SLEEP-AID DRUG PRODUCTS Labeling; Nighttime Sleep-aid Drug Products For OTC Human Use
AID, VISION	886.5900	ELECTRONIC VISION AID Therapeutic Devices; Ophthalmic Devices
AID, VISION	886.5910	IMAGE INTENSIFICATION VISION AID Therapeutic Devices; Ophthalmic Devices
AID, VISION	886.5915	OPTICAL VISION AID Therapeutic Devices; Ophthalmic Devices
AIR	868.2025	ULTRASONIC AIR EMBOLISM MONITOR Monitoring Devices; Anesthesiology Devices
AIR	868.6250	PORTABLE AIR COMPRESSOR Miscellaneous; Anesthesiology Devices
AIR	880.5045	MEDICAL RECIRCULATING AIR CLEANER General Hospital And Personal Use Therapeutic Devices; General Hospital And Personal Use Devices
AIR	880.5550	ALTERNATING PRESSURE AIR FLOTATION MATTRESS General Hospital And Personal Use Therapeutic Devices; General Hospital And Personal Use Devices

AIR CALORIC STIMULATOR	874.1800	AIR OR WATER CALORIC STIMULATOR Diagnostic Devices; Ear, Nose, And Throat Devices
AIR CONVEYANCES	1250.53	DISCHARGE OF WASTES ON AIR CONVEYANCES Equipment And Operation Of Land And Air Conveyances; Interstate Conveyance Sanitation
AIR FILTRATION	211.46	VENTILATION, AIR FILTRATION, AIR HEATING AND COOLING Buildings And Facilities; Current Good Manufacturing Practice For Finished Pharmaceuticals
AIR-FLUIDIZED	890.5160	AIR-FLUIDIZED BED Physical Medicine Therapeutic Devices; Physical Medicine Devices
AIR-HANDLING APPARATUS	878.5070	AIR-HANDLING APPARATUS FOR A SURGICAL OPERATING ROOM Therapeutic Devices; General And Plastic Surgery Devices
AIR PURIFIER	880.6500	MEDICAL ULTRAVIOLET AIR PURIFIER General Hospital And Personal Use Miscellaneous Devices; General Hospital And Personal Use Devices
AIRBRUSH	872.6080	AIRBRUSH Miscellaneous Devices; Dental Devices
AIRCRAFT	1301.25	REGISTRATION REGARDING OCEAN VESSELS, AIRCRAFT, AND OTHER ENTITIES Registration; Registration Of Manufacturers, Distributors, And Dispensers Of Controlled Substances
AIRWAY	868.1780	INSPIRATORY AIRWAY PRESSURE METER Diagnostic Devices; Anesthesiology Devices
AIRWAY	868.2600	AIRWAY PRESSURE MONITOR Monitoring Devices; Anesthesiology Devices
AIRWAY	868.5090	EMERGENCY AIRWAY NEEDLE Therapeutic Devices; Anesthesiology Devices
AIRWAY	868.5100	NASOPHARYNGEAL AIRWAY Therapeutic Devices; Anesthesiology Devices
AIRWAY	868.5110	OROPHARYNGEAL AIRWAY Therapeutic Devices; Anesthesiology Devices
AIRWAY	868.5115	DEVICE TO RELIEVE ACUTE UPPER AIRWAY OBSTRUCTION Therapeutic Devices; Anesthesiology Devices
AIRWAY	868.5810	AIRWAY CONNECTOR Therapeutic Devices; Anesthesiology Devices
AKLOMIDE	520.2320	SULFANITRAN AND AKLOMIDE IN COMBINATION Oral Dosage Form New Animal Drugs
AKLOMIDE	556.30	AKLOMIDE Specific Tolerances For Residues Of New Animal Drugs; Tolerances For Residues Of New Animal Drugs In Food
AKLOMIDE	558.35	AKLOMIDE Specific New Animal Drugs For Use In Animal Feeds; New Animal Drugs For Use In Animal Feeds
ALANINE	172.540	DL-ALANINE Flavoring Agents And Related Substances; Food Additives Permitted For Direct Addition To Food For Human Consumption
ALANINE	582.5118	ALANINE Nutrients And/or Dietary Supplements; Substances Generally Recognized As Safe
ALANINE	862.1030	ALANINE AMINO TRANSFERASE (ALT/SGPT) TEST SYSTEM Clinical Chemistry Test Systems; Clinical Chemistry And Clinical Toxicology Devices
ALARM	870.1025	ARRHYTHMIA DETECTOR AND ALARM Cardiovascular Diagnostic Devices; Cardiovascular Devices
ALARM	870.1100	BLOOD PRESSURE ALARM Cardiovascular Diagnostic Devices; Cardiovascular Devices
ALARM	870.2640	PORTABLE LEAKAGE CURRENT ALARM Cardiovascular Monitoring Devices; Cardiovascular Devices
ALARM	870.2300	CARDIAC MONITOR (INCLUDING CARDIOTACHOMETER AND RATE ALARM) Cardiovascular Monitoring Devices; Cardiovascular Devices
ALARM	876.2040	ENURESIS ALARM Monitoring Devices; Gastroenterology-urology Devices
ALBENDAZOLE	520.45	ALBENDAZOLE ORAL DOSAGE FORMS Oral Dosage Form New Animal Drugs
ALBENDAZOLE	556.34	ALBENDAZOLE Specific Tolerances For Residues Of New Animal Drugs; Tolerances For Residues Of New Animal Drugs In Food
ALBUMIN	862.1035	ALBUMIN TEST SYSTEM Clinical Chemistry Test Systems; Clinical Chemistry And Clinical Toxicology Devices
ALBUMIN	862.1645	URINARY PROTEIN OR ALBUMIN TEST (NON-QUANTITATIVE) SYSTEM Clinical Chemistry Test Systems; Clinical Chemistry And Clinical Toxicology Devices
ALBUMIN	866.5040	ALBUMIN IMMUNOLOGICAL TEST SYSTEM Immunological Test Systems; Immunology And Microbiology Devices

ALBUMIN, HUMAN	640.80	ALBUMIN (HUMAN) Albumin (Human); Additional Standards For Human Blood And Blood Products
ALBUTEROL	529.40	ALBUTEROL Certain Other Dosage Form New Animal Drugs
ALCOHOL	201.322	OVER-THE COUNTER DRUG PRODUCTS CONTAINING INTERNAL ANALGESIC/ANTIPYRETIC ACTIVE INGREDIENTS; REQUIRED ALCOHOL WARNING Specific Labeling Requirements For Specific Drug Products; Labeling
ALCOHOL	328.10	ALCOHOL Over-the-counter Drug Products Intended For Oral Ingestion That Contain Alcohol
ALCOHOL	328.50	PRINCIPAL DISPLAY PANEL OF ALL OTC DRUG PRODUCTS INTENDED FOR ORAL INGESTION THAT CONTAIN ALCOHOL Labeling; Over-the-counter Drug Products Intended For Oral Ingestion That Contain Alcohol
ALCOHOL	862.3040	ALCOHOL TEST SYSTEM Clinical Toxicology Test Systems; Clinical Chemistry And Clinical Toxicology Devices
ALCOHOL, BREATH	862.3050	BREATH-ALCOHOL TEST SYSTEM Clinical Toxicology Test Systems; Clinical Chemistry And Clinical Toxicology Devices
ALCOHOL ESTERS	178.3770	POLYHYDRIC ALCOHOL ESTERS OF OXIDATIVELY REFINED (GERSTHOFEN PROCESS) MONTAN WAX ACIDS Certain Adjuvants And Production Aids; Indirect Food Additives: Adjuvants, Production Aids, And Sanitizers
ALCOHOL ESTERS	178.3780	POLYHYDRIC ALCOHOL ESTERS OF LONG CHAIN MONOBASIC ACIDS Certain Adjuvants And Production Aids; Indirect Food Additives: Adjuvants, Production Aids, And Sanitizers
ALCOHOL, ETHYL	184.1293	ETHYL ALCOHOL Listing Of Specific Substances Affirmed As GRAS; Direct Food Substances Affirmed As Generally Recognized As Safe
ALCOHOL, ETHYL	584.200	ETHYL ALCOHOL CONTAINING ETHYL ACETATE Listing Of Substances Affirmed As GRAS; Food Substances Affirmed As GRAS In Feed And Drinking Water Of Animals
ALCOHOL, ISOPROPYL	173.240	ISOPROPYL ALCOHOL Solvents, Lubricants, Release Agents And Related Substances; Secondary Direct Food Additives Permitted In Food For Human Consumption
ALCOHOL, PANTOTHENYL	582.5580	D-PANTOTHENYL ALCOHOL Nutrients And/or Dietary Supplements; Substances Generally Recognized As Safe
ALCOHOL, POLYVINYL	177.1670	POLYVINYL ALCOHOL FILM Substances For Use As Basic Components Of Single And Repeated Use Food Contact Surfaces; Indirect Food Additives: Polymers
ALCOHOL, PROPYL	573.880	NORMAL PROPYL ALCOHOL Food Additive Listing; Food Additives Permitted In Feed And Drinking Water Of Animals
ALCOHOL RESIDUES	173.250	METHYL ALCOHOL RESIDUES Solvents, Lubricants, Release Agents And Related Substances; Secondary Direct Food Additives Permitted In Food For Human Consumption
ALCOHOL, VINYL	73.3127	VINYL ALCOHOL/METHYL METHACRYLATE-DYE REACTION PRODUCTS Listing Of Color Additives Exempt From Certification
ALCOHOL, VINYL	177.1360	ETHYLENE-VINYL ACETATE-VINYL ALCOHOL COPOLYMERS Substances For Use As Basic Components Of Single And Repeated Use Food Contact Surfaces; Indirect Food Additives: Polymers
ALCOHOLS	101.80	HEALTH CLAIMS: DIETARY SUGAR ALCOHOLS AND DENTAL CARIES Specific Requirements For Health Claims; Food Labeling
ALCOHOLS, FATTY	172.864	SYNTHETIC FATTY ALCOHOLS Multipurpose Additives; Food Additives Permitted For Direct Addition To Food For Human Consumption
ALCOHOLS, FATTY	178.3480	FATTY ALCOHOLS, SYNTHETIC Certain Adjuvants And Production Aids; Indirect Food Additives: Adjuvants, Production Aids, And Sanitizers
ALDOLASE	862.1040	ALDOLASE TEST SYSTEM Clinical Chemistry Test Systems; Clinical Chemistry And Clinical Toxicology Devices
ALDOSTERONE	862.1045	ALDOSTERONE TEST SYSTEM Clinical Chemistry Test Systems; Clinical Chemistry And Clinical Toxicology Devices
ALERT SYSTEM	26.20	ALERT SYSTEM Specific Sector Provisions For Pharmaceutical GMPs; Mutual Recognition Of Pharmaceutical GMP Reports, Medical Device Quality System Audit Reports
ALERT SYSTEM	26.50	ALERT SYSTEM AND EXCHANGE OF POSTMARKET VIGILANCE REPORTS Specific Sector Provisions For Medical Devices; Mutual Recognition Of Pharmaceutical GMP Reports, Medical Device Quality System Audit Reports
ALFAPROSTOL	522.46	ALFAPROSTOL Implantation Or Injectable Dosage Form New Animal Drugs

ALGAE	184.1120	BROWN ALGAE Listing Of Specific Substances Affirmed As GRAS; Direct Food Substances Affirmed As Generally Recognized As Safe
ALGAE	184.1121	RED ALGAE Listing Of Specific Substances Affirmed As GRAS; Direct Food Substances Affirmed As Generally Recognized As Safe
ALGAE MEAL	73.185	HAEMATOCOCCUS ALGAE MEAL Foods; Listing Of Color Additives Exempt From Certification
ALGAE MEAL	73.275	DRIED ALGAE MEAL Foods; Listing Of Color Additives Exempt From Certification
ALGESIMETER	868.1030	MANUAL ALGESIMETER Diagnostic Devices; Anesthesiology Devices
ALGESIMETER	868.1040	POWERED ALGESIMETER Diagnostic Devices; Anesthesiology Devices
ALGINATE, AMMONIUM	184.1133	AMMONIUM ALGINATE Listing Of Specific Substances Affirmed As GRAS; Direct Food Substances Affirmed As Generally Recognized As Safe
ALGINATE, AMMONIUM	582.7133	AMMONIUM ALGINATE Stabilizers; Substances Generally Affirmed As Safe
ALGINATE, CALCIUM	184.1187	CALCIUM ALGINATE Listing Of Specific Substances Affirmed As GRAS; Direct Food Substances Affirmed As Generally Recognized As Safe
ALGINATE, CALCIUM	582.7187	CALCIUM ALGINATE Stabilizers; Substances Generally Affirmed As Safe
ALGINATE, POTASSIUM	184.1610	POTASSIUM ALGINATE Listing Of Specific Substances Affirmed As GRAS; Direct Food Substances Affirmed As Generally Recognized As Safe
ALGINATE, POTASSIUM	582.7610	POTASSIUM ALGINATE Stabilizers; Substances Generally Affirmed As Safe
ALGINATE, PROPYLENE GLYCOL	172.858	PROPYLENE GLYCOL ALGINATE Multipurpose Additives; Food Additives Permitted For Direct Addition To Food For Human Consumption
ALGINATE, SODIUM	184.1724	SODIUM ALGINATE Listing Of Specific Substances Affirmed As GRAS; Direct Food Substances Affirmed As Generally Recognized As Safe
ALGINATE, SODIUM	582.7724	SODIUM ALGINATE Stabilizers; Substances Generally Affirmed As Safe
ALGINIC ACID	184.1011	ALGINIC ACID Listing Of Specific Substances Affirmed As GRAS; Direct Food Substances Affirmed As Generally Recognized As Safe
ALGINIC ACID	201.319	WATER-SOLUBLE GUMS, HYDROPHILIC GUMS AND HYDROPHILIC MUCILLOIDS (INCLUDING AGAR, ALGINIC ACID) AS ACTIVE INGREDIENTS; WARNINGS AND DIRECTIONS Specific Labeling Requirements For Specific Drug Products; Labeling Recognized As Safe
ALIGNMENT	872.1820	DENTAL X-RAY EXPOSURE ALIGNMENT DEVICE Diagnostic Devices; Dental Devices
ALKALINE PHOSPHATASE	862.1050	ALKALINE PHOSPHATASE OR ISOENZYMES TEST SYSTEM Clinical Chemistry Test Systems; Clinical Chemistry And Clinical Toxicology Devices
ALKALINE PHOSPHATASE	864.7660	LEUKOCYTE ALKALINE PHOSPHATASE TEST Hematology Kits And Packages; Hematology And Pathology Devices
ALKYLGLUTARIMIDE	177.1060	N-ALKYLGLUTARIMIDE/ACRYLIC COPOLYMERS Substances For Use As Basic Components Of Single And Repeated Use Food Contact Surfaces; Indirect Food Additives: Polymers
ALKYL KETENE DIMERS	176.120	ALKYL KETENE DIMERS Substances For Use Only As Components Of Paper And Paperboard; Indirect Food Additives
ALLERGENIC	680.1	ALLERGENIC PRODUCTS Additional Standards For Miscellaneous Products
ALLERGENIC	680.2	MANUFACTURE OF ALLERGENIC PRODUCTS Additional Standards For Miscellaneous Products
ALLOCATION	70.45	ALLOCATION OF COLOR ADDITIVES Safety Evaluation; Color Additives
ALLOTYPIC MARKER	866.5065	HUMAN ALLOTYPIC MARKER IMMUNOLOGICAL TEST SYSTEM Immunological Test Systems; Immunology And Microbiology Devices
ALLOWABLE COSTS	1403.22	ALLOWABLE COSTS Post-award Requirements; Uniform Administrative Requirements For Grants And Coop. Agreements To State & Local Gov.
ALLOWANCE, INVENTORY	1303.24	INVENTORY ALLOWANCE Individual Manufacturing Quotas; Quotas
ALLOY	872.3050	AMALGAM ALLOY Prosthetic Devices; Dental Devices

ALLOY	872.3080	MERCURY ALLOY DISPENSER Prosthetic Devices; Dental Devices
ALLOY	872.3710	BASE METAL ALLOY Prosthetic Devices; Dental Devices
ALLOYS	872.3060	GOLD BASED ALLOYS AND PRECIOUS METAL ALLOYS FOR CLINICAL USE Prosthetic Devices; Dental Devices
ALPHA-1-ANTICHYMOTRYPSIN	866.5080	ALPHA-1-ANTICHYMOTRYPSIN IMMUNOLOGICAL TEST SYSTEM Immunological Test Systems; Immunology And Microbiology Devices
ALPHA-1-ANTITRYPSIN	866.5130	ALPHA-1-ANTITRYPSIN IMMUNOLOGICAL TEST SYSTEM Immunological Test Systems; Immunology And Microbiology Devices
ALPHA-1-GLYCOPROTEINS	866.5420	ALPHA-1-GLYCOPROTEINS IMMUNOLOGICAL TEST SYSTEM Immunological Test Systems; Immunology And Microbiology Devices
ALPHA-1-LIPOPROTEIN	866.5580	ALPHA-1-LIPOPROTEIN IMMUNOLOGICAL TEST SYSTEM Immunological Test Systems; Immunology And Microbiology Devices
ALPHA-2-GLYCOPROTEINS	866.5425	ALPHA-2-GLYCOPROTEINS IMMUNOLOGICAL TEST SYSTEM Immunological Test Systems; Immunology And Microbiology Devices
ALPHA-2-MACROGLOBULIN	866.5620	ALPHA-2-MACROGLOBULIN IMMUNOLOGICAL TEST SYSTEM Immunological Test Systems; Immunology And Microbiology Devices
ALPHA-ACETOLACTATE	173.115	ALPHA-ACETOLACTATE DECARBOXYLASE (ALPHA-ALDC) ENZYME PREPARATION DERIVED FROM A RECOMBINANT BACILLUS SUBTILIS Enzyme Preparations And Microorganisms; Secondary Direct Food Additives Permitted In Food For Human Consumption
ALPHA-AMYLASE	184.1012	α-AMYLASE ENZYME PREPARATION FROM BACILLUS STEAROTHERMOPHILUS Listing Of Specific Substances Affirmed As GRAS; Direct Food Substances Affirmed As Generally Recognized As Safe
ALPHA-GALACTOSIDASE	173.145	ALPHA-GALACTOSIDASE DERIVED FROM MORTEIRELLA VINACEAE Enzyme Preparations And Microorganisms; Secondary Direct Food Additives Permitted In Food For Human Consumption
ALPHA-GLOBULIN	866.5400	ALPHA-GLOBULIN IMMUNOLOGICAL TEST SYSTEM Immunological Test Systems; Immunology And Microbiology Devices
ALPHA-METHYLSTYRENE	178.3610	α-METHYLSTYRENE-VINYLTOLUENE RESINS, HYDROGENATED Certain Adjuvants And Production Aids; Indirect Food Additives: Adjuvants, Production Aids, And Sanitizers
ALPHA MONITOR	882.1610	ALPHA MONITOR Neurological Diagnostic Devices; Neurological Devices
ALPHA-TOCOPHEROL ACETATE	182.8892	α-TOCOPHEROL ACETATE Nutrients; Substances Generally Recognized As Safe
ALPHA-TOCOPHEROL ACETATE	582.5892	α-TOCOPHEROL ACETATE Nutrients And/or Dietary Supplements; Substances Generally Recognized As Safe
ALPHA-TOCOPHEROLS	184.1890	α-TOCOPHEROLS Listing Of Specific Substances Affirmed As GRAS; Direct Food Substances Affirmed As Generally Recognized As Safe
ALTERABLE	882.5320	PREFORMED ALTERABLE CRANIOPLASTY PLATE Neurological Therapeutic Devices; Neurological Devices
ALTERNATING	880.5550	ALTERNATING PRESSURE AIR FLOTATION MATTRESS General Hospital And Personal Use Therapeutic Devices; General Hospital And Personal Use Devices
ALTERNATIVE	5.403	AUTHORIZATION TO USE ALTERNATIVE EVIDENCE FOR DETERMINATION OF THE EFFECTIVENESS OF MEDICAL DEVICES Medical Devices And Radiological Health; Redelegations Of Authority; Delegations Of Authority And Organization
ALTERNATIVE	12.32	REQUEST FOR ALTERNATIVE FORM OF HEARING Initiation Of Proceedings; Formal Evidentiary Public Hearing
ALTERNATIVE	56.124	ACTIONS ALTERNATIVE OR ADDITIONAL TO DISQUALIFICATION Administrative Actions For Noncompliance; Institutional Review Boards
ALTERNATIVE	58.215	ALTERNATIVE OR ADDITIONAL ACTIONS TO DISQUALIFICATION Disqualification Of Testing Facilities; Good Laboratory Practice For Nonclinical Laboratory Studies
ALTERNATIVE PROCEDURES	640.120	ALTERNATIVE PROCEDURES Alternative Procedures; Additional Standards For Human Blood And Blood Products
ALTRENOGEST	520.48	ALTRENOGEST SOLUTION Oral Dosage Form New Animal Drugs
ALUMINA	73.1010	ALUMINA (DRIED ALUMINUM HYDROXIDE) Drugs; Listing Of Color Additives Exempt From Certification
ALUMINOSILICATE	182.2727	SODIUM ALUMINOSILICATE Anticaking Agents; Substances Generally Recognized As Safe
ALUMINOSILICATE	182.2729	SODIUM CALCIUM ALUMINOSILICATE, HYDRATED Anticaking Agents; Substances Generally Recognized As Safe
ALUMINOSILICATE	582.2727	SODIUM ALUMINOSILICATE Anticaking Agents; Substances Generally Recognized As Safe
ALUMINOSILICATE	582.2729	HYDRATED SODIUM CALCIUM ALUMINOSILICATE Anticaking Agents; Substances Generally Recognized As Safe

ALUMINUM	73.1010	ALUMINA (DRIED ALUMINUM HYDROXIDE) Drugs; Listing Of Color Additives Exempt From Certification
ALUMINUM	182.1127	ALUMINUM AMMONIUM SULFATE Multiple Purpose GRAS Food Substances; Substances Generally Recognized As Safe
ALUMINUM	182.1129	ALUMINUM POTASSIUM SULFATE Multiple Purpose GRAS Food Substances; Substances Generally Recognized As Safe
ALUMINUM	182.1131	ALUMINUM SODIUM SULFATE Multiple Purpose GRAS Food Substances; Substances Generally Recognized As Safe
ALUMINUM	182.2122	ALUMINUM CALCIUM SILICATE Anticaking Agents; Substances Generally Recognized As Safe
ALUMINUM	201.323	ALUMINUM IN LARGE AND SMALL VOLUME PARENTERALS USED IN TOTAL PARENTERAL NUTRITION Specific Labeling Requirements For Specific Drug Products; Labeling
ALUMINUM	582.1125	ALUMINUM SULFATE General Purpose Food Additives; Substances Generally Recognized As Safe
ALUMINUM	582.1127	ALUMINUM AMMONIUM SULFATE General Purpose Food Additives; Substances Generally Recognized As Safe
ALUMINUM	582.1129	ALUMINUM POTASSIUM SULFATE General Purpose Food Additives; Substances Generally Recognized As Safe
ALUMINUM	582.1131	ALUMINUM SODIUM SULFATE General Purpose Food Additives; Substances Generally Recognized As Safe
ALUMINUM	582.2122	ALUMINUM CALCIUM SILICATE Anticaking Agents; Substances Generally Recognized As Safe
ALUMINUM HYDROXIDE, DRIED	73.1010	ALUMINA (DRIED ALUMINUM HYDROXIDE) Drugs; Listing Of Color Additives Exempt From Certification
ALUMINUM NICOTINATE	172.310	ALUMINUM NICOTINATE Special Dietary And Nutritional Additive; Food Additives Permitted For Direct Addition To Food For Human Consumption
ALUMINUM OXIDE	73.1015	CHROMIUM-COBALT-ALUMINUM OXIDE Drugs; Listing Of Color Additives Exempt From Certification
ALUMINUM OXIDE	73.3110a	CHROMIUM-COBALT-ALUMINUM OXIDE Medical Devices; Listing Of Color Additives Exempt From Certification
ALUMINUM PHOSPHATE	182.1781	SODIUM ALUMINUM PHOSPHATE Multiple Purpose GRAS Food Substances; Substances Generally Recognized As Safe
ALUMINUM PHOSPHATE	582.1781	SODIUM ALUMINUM PHOSPHATE General Purpose Food Additives; Substances Generally Recognized As Safe
ALUMINUM POWDER	73.1645	ALUMINUM POWDER Drugs; Listing Of Color Additives Exempt From Certification
ALUMINUM POWDER	73.2645	ALUMINUM POWDER Cosmetics; Listing Of Color Additives Exempt From Certification
ALUMINUM SULFATE	182.1125	ALUMINUM SULFATE Multiple Purpose GRAS Food Substances; Substances Generally Recognized As Safe
ALUMINUM SULFATE	582.1125	ALUMINUM SULFATE General Purpose Food Additives; Substances Generally Recognized As Safe
AMALGAM	872.3050	AMALGAM ALLOY Prosthetic Devices; Dental Devices
AMALGAM	872.3110	DENTAL AMALGAM CAPSULE Prosthetic Devices; Dental Devices
AMALGAMATOR, DENTAL	872.3100	DENTAL AMALGAMATOR Prosthetic Devices; Dental Devices
AMEND	21.52	ADMINISTRATIVE APPEALS OF REFUSALS TO AMEND RECORDS Procedures For Requests For Amendment Of Records; Protection Of Privacy
AMEND	1271.26	WHEN MUST I AMEND MY ESTABLISHMENT REGISTRATION? Procedures For Registration And Listing; Human Cells, Tissues, And Cellular And Tissue-Based Products
AMENDED	21.54	AMENDED OR DISPUTED RECORDS RECEIVED FROM OTHER AGENCIES Procedures For Requests For Amendment Of Records; Protection Of Privacy
AMENDED	514.6	AMENDED APPLICATIONS General Provisions; New Animal Drug Applications
AMENDING	171.130	PROCEDURES FOR AMENDING AND REPEALING TOLERANCES OR EXEMPTIONS FROM TOLERANCES Food Additive Petitions
AMENDING	320.32	PROCEDURES FOR ESTABLISHING OR AMENDING A BIOEQUIVALENCE REQUIREMENT Bioavailability And Bioequivalence Requirements
AMENDING	571.130	PROCEDURE FOR AMENDING AND REPEALING TOLERANCES OR EXEMPTIONS FROM TOLERANCES Administrative Actions On Applications; Food Additive Petitions

AMENDMENT	12.20	INITIATION OF A HEARING INVOLVING THE ISSUANCE, AMENDMENT, OR REVOCATION OF A REGULATION Initiation Of Proceedings; Formal Evidentiary Public Hearing
AMENDMENT	12.21	INITIATION OF A HEARING INVOLVING THE ISSUANCE, AMENDMENT, OR REVOCATION OF AN ORDER Initiation Of Proceedings; Formal Evidentiary Public Hearing
AMENDMENT	21.50	PROCEDURES FOR SUBMITTING REQUESTS FOR AMENDMENT OF RECORDS Procedures For Requests For Amendment Of Records; Protection Of Privacy
AMENDMENT	21.51	RESPONSES TO REQUESTS FOR AMENDMENT OF RECORDS Procedures For Requests For Amendment Of Records; Protection Of Privacy
AMENDMENT	26.80	ENTRY INTO FORCE, AMENDMENT, AND TERMINATION Framework Provisions; Mutual Recognition Of Pharmaceutical GMP Reports, Medical Device Quality System Audit Reports
AMENDMENT	171.6	AMENDMENT OF PETITION General Provisions; Food Additive Petitions
AMENDMENT	316.26	AMENDMENT TO ORPHAN-DRUG DESIGNATION Orphan Drugs
AMENDMENT	571.6	AMENDMENT OF PETITION General Provisions; Food Additive Petitions
AMENDMENT	861.34	AMENDMENT OR REVOCATION OF A STANDARD Procedures For Performance Standards Development And Publication; Procedures For Performance Standards Development
AMENDMENTS	5.100	ISSUANCE OF NOTICE IMPLEMENTING THE PROVISIONS OF THE DRUG AMENDMENTS OF 1962 Human Drugs; Redelegations Of Authority; Delegations Of Authority And Organization
AMENDMENTS	71.6	EXTENSION OF TIME FOR STUDYING PETITIONS; SUBSTANTIVE AMENDMENTS General Provisions; Color Additive Petitions
AMENDMENTS	207.26	AMENDMENTS TO REGISTRATION Procedures For Domestic Drug Establishments; Registration Of Drug Producers And Listing Of Drugs In Commercial Distribution
AMENDMENTS	312.30	PROTOCOL AMENDMENTS Investigational New Drug Application
AMENDMENTS	312.31	INFORMATION AMENDMENTS Investigational New Drug Application
AMENDMENTS	314.60	AMENDMENTS TO AN UNAPPROVED APPLICATION Applications; Applications For FDA Approval To Market A New Drug
AMENDMENTS	314.96	AMENDMENTS TO AN UNAPPROVED ABBREVIATED APPLICATION Abbreviated Applications; Applications For FDA Approval To Market A New Drug
AMENDMENTS	607.26	AMENDMENTS TO ESTABLISHMENT REGISTRATION Procedures For Domestic Blood Product Establishments; Establishment Registratn+product Listing, Manufacturers Human Blood/b.products
AMENDMENTS	710.5	AMENDMENTS TO REGISTRATION Voluntary Registration Of Cosmetic Product Establishments
AMENDMENTS	720.6	AMENDMENTS TO STATEMENT Voluntary Filing Of Cosmetic Product Ingredient Composition Statements
AMENDMENTS	807.26	AMENDMENTS TO ESTABLISHMENT REGISTRATION Procedures For Device Establishments; Establishment Registration And Device Listing For Manufacturers And Initial Importers Of Devices
AMENDMENTS	814.37	PMA AMENDMENTS AND RESUBMITTED PMA'S Premarket Approval Application; Premarket Approval Of Medical Devices
AMENDMENTS	814.106	HDE AMENDMENTS AND RESUBMITTED HDE'S Humanitarian Use Devices; Premarket Approval Of Medical Devices
AMENDMENTS	1301.16	AMENDMENTS TO AND WITHDRAWAL OF APPLICATIONS Registration Of Manufacturers, Distributors, And Dispensers Of Controlled Substances
AMENDMENTS	1309.36	AMENDMENTS TO AND WITHDRAWALS OF APPLICATIONS Registration Of Manufacturers, Distributors, Importers And Exporters Of List I Chemicals
AMERICAN CHEESE	133.147	GRATED AMERICAN CHEESE FOOD Requirements For Specific Standardized Cheese And Related Products; Cheeses And Related Cheese Products
AMERICAN, NATIVE	1307.31	NATIVE AMERICAN CHURCH Special Exempt Persons; Miscellaneous
AMIKACIN	862.3035	AMIKACIN TEST SYSTEM Clinical Toxicology Test Systems; Clinical Chemistry And Clinical Toxicology Devices
AMIKACIN SULFATE	522.56	AMIKACIN SULFATE INJECTION Implantation Or Injectable Dosage Form New Animal Drugs
AMIKACIN SULFATE	529.50	AMIKACIN SULFATE INTRAUTERINE SOLUTION Certain Other Dosage Form New Animal Drugs

AMINATED RESIN	173.70	CHLOROMETHYLATED AMINATED STYRENE-DIVINYLBENZENE RESIN Polymer Substances For Food Treatment; Secondary Direct Food Additives Permitted In Food For Human Consumption
AMINO ACIDS	172.320	AMINO ACIDS Special Dietary And Nutritional Additives; Food Additives Permitted For Direct Addition To Food For Human Consumption
AMINO TRANSFERASE	862.1030	ALANINE AMINO TRANSFERASE (ALT/SGPT) TEST SYSTEM Clinical Chemistry Test Systems; Clinical Chemistry And Clinical Toxicology Devices
AMINO TRANSFERASE	862.1100	ASPARTATE AMINO TRANSFERASE (AST/SGOT) TEST SYSTEM Clinical Chemistry Test Systems; Clinical Chemistry And Clinical Toxicology Devices
AMINOACETIC ACID	170.50	GLYCINE (AMINOACETIC ACID) IN FOOD FOR HUMAN CONSUMPTION Specific Administrative Rulings And Decisions; Food Additives
AMINOACETIC ACID	582.5049	AMINOACETIC ACID Nutrients And/or Dietary Supplements; Substances Generally Recognized As Safe
AMINOGLYCOSIDE	173.170	AMINOGLYCOSIDE PHOSPHOTRANSFERASE II Secondary Direct Food Additives Permitted In Food For Human Consumption
AMINOGLYCOSIDE	573.130	AMINOGLYCOSIDE 3'-PHOPHOTRANSFERASE II Food Additives Permitted In Feed And Drinking Water Of Animals
AMINOLEVULINIC ACID	862.1060	DELTA-AMINOLEVULINIC ACID TEST SYSTEM Clinical Chemistry Test Systems; Clinical Chemistry And Clinical Toxicology Devices
AMINOPENTAMIDE	520.62	AMINOPENTAMIDE HYDROGEN SULPHATE TABLETS Oral Dosage Form New Animal Drugs
AMINOPENTAMIDE	520.1204	KANAMYCIN SULFATE, AMINOPENTAMIDE HYDROGEN SULFATE, PECTIN, BISMUTH SUBCARBONATE, ACTIVATED ATTAPULGITE SUSPENSION Oral Dosage Form New Animal Drugs
AMINOPENTAMIDE	522.62	AMINOPENTAMIDE HYDROGEN SULFATE INJECTION Implantation Or Injectable Dosage Form New Animal Drugs
AMINOPENTAMIDE HYDROGEN-S	522.1222b	KETAMINE HYDROCHLORIDE WITH PROMAZINE HYDROCHLORIDE AND AMINOPENTAMIDE HYDROGEN SULFATE INJECTION Implantation Or Injectable Dosage Form New Animal Drugs
AMINOPEPTIDASE	184.1985	AMINOPEPTIDASE ENZYME PREPARATION DERIVED FROM LACTOCOCCUS LACTIS Listing Of Specific Substances Affirmed As GRAS; Direct Food Substances Affirmed As Generally Recognized As Safe
AMINOPEPTIDASE	862.1460	LEUCINE AMINOPEPTIDASE TEST SYSTEM Clinical Chemistry Test Systems; Clinical Chemistry And Clinical Toxicology Devices
AMINOPROPAZINE FUMARATE	520.82	AMINOPROPAZINE FUMARATE ORAL DOSAGE FORMS Oral Dosage Form New Animal Drugs
AMINOPROPAZINE FUMARATE	522.82	AMINOPROPAZINE FUMARATE STERILE SOLUTION INJECTION Implantation Or Injectable Dosage Form New Animal Drugs
AMINOPROPRIONITRILE	522.84	BETA-AMINOPROPRIONITRILE FUMARATE Implantation Or Injectable Dosage Form New Animal Drugs
AMITRAZ	524.86	AMITRAZ LIQUID Ophthalmic And Topical Dosage Form New Animal Drugs
AMMONIA	2.110	DEFINITION OF AMMONIA UNDER FEDERAL CAUSTIC POISON ACT Caustic Poisons; General Administrative Rulings And Decisions
AMMONIA	862.1065	AMMONIA TEST SYSTEM Clinical Chemistry Test Systems; Clinical Chemistry And Clinical Toxicology Devices
AMMONIA, ANHYDROUS	573.180	ANHYDROUS AMMONIA Food Additive Listing; Food Additives Permitted In Feed And Drinking Water Of Animals
AMMONIATED COTTONSEED	573.140	AMMONIATED COTTONSEED MEAL Food Additive Listing; Food Additives Permitted In Feed And Drinking Water Of Animals
AMMONIATED RICE	573.100	AMMONIATED RICE HULLS Food Additive Listing; Food Additives Permitted In Feed And Drinking Water Of Animals
AMMONIATED WHEY	573.450	FERMENTED AMMONIATED CONDENSED WHEY Food Additive Listing; Food Additives Permitted In Feed And Drinking Water Of Animals
AMMONIUM ALGINATE	184.1133	AMMONIUM ALGINATE Listing Of Specific Substances Affirmed As GRAS; Direct Food Substances Affirmed As Generally Recognized As Safe
AMMONIUM ALGINATE	582.7133	AMMONIUM ALGINATE Stabilizers; Substances Generally Recognized As Safe
AMMONIUM BICARBONATE	184.1135	AMMONIUM BICARBONATE Listing Of Specific Substances Affirmed As GRAS; Direct Food Substances Affirmed As Generally Recognized As Safe
AMMONIUM BICARBONATE	582.1135	AMMONIUM BICARBONATE General Purpose Food Additives; Substances Generally Recognized As Safe

AMMONIUM CARBONATE	184.1137	AMMONIUM CARBONATE Listing Of Specific Substances Affirmed As GRAS; Direct Food Substances Affirmed As Generally Recognized As Safe
AMMONIUM CARBONATE	582.1137	AMMONIUM CARBONATE General Purpose Food Additives; Substances Generally Recognized As Safe
AMMONIUM CHLORIDE	184.1138	AMMONIUM CHLORIDE Listing Of Specific Substances Affirmed As GRAS; Direct Food Substances Affirmed As Generally Recognized As Safe
AMMONIUM CHLORIDE	520.310	CARAMIPHEN ETHANEDISULFONATE AND AMMONIUM CHLORIDE TABLETS Oral Dosage Form New Animal Drugs
AMMONIUM CHLORIDE (QUAT)	172.165	QUATERNARY AMMONIUM CHLORIDE COMBINATION Food Preservatives; Food Additives Permitted For Direct Addition To Food For Human Consumption
AMMONIUM CITRATE	184.1140	AMMONIUM CITRATE, DIBASIC Listing Of Specific Substances Affirmed As GRAS; Direct Food Substances Affirmed As Generally Recognized As Safe
AMMONIUM CITRATE	184.1296	FERRIC AMMONIUM CITRATE Listing Of Specific Substances Affirmed As GRAS; Direct Food Substances Affirmed As Generally Recognized As Safe
AMMONIUM CITRATE, IRON	172.430	IRON AMMONIUM CITRATE Anticaking Agents; Food Additives Permitted For Direct Addition To Food For Human Consumption
AMMONIUM CITRATE, IRON	573.560	IRON AMMONIUM CITRATE Food Additive Listing; Food Additives Permitted In Feed And Drinking Water Of Animals
AMMONIUM FERROCYANIDE	73.1298	FERRIC AMMONIUM FERROCYANIDE Drugs; Listing Of Color Additives Exempt From Certification
AMMONIUM FERROCYANIDE	73.2298	FERRIC AMMONIUM FERROCYANIDE Cosmetics; Listing Of Color Additives Exempt From Certification
AMMONIUM HYDROXIDE	184.1139	AMMONIUM HYDROXIDE Listing Of Specific Substances Affirmed As GRAS; Direct Food Substances Affirmed As Generally Recognized As Safe
AMMONIUM HYDROXIDE	582.1139	AMMONIUM HYDROXIDE General Purpose Food Additives; Substances Generally Recognized As Safe
AMMONIUM, MADURAMICIN	556.375	MADURAMICIN AMMONIUM Tolerances For Residues Of New Animal Drugs In Food
AMMONIUM, MADURAMICIN	558.340	MADURAMICIN AMMONIUM New Animal Drugs For Use In Animal Feeds
AMMONIUM PHOSPHATE	184.1141a	AMMONIUM PHOSPHATE, MONOBASIC Listing Of Specific Substances Affirmed As GRAS; Direct Food Substances Affirmed As Generally Recognized As Safe
AMMONIUM PHOSPHATE	184.1141b	AMMONIUM PHOSPHATE, DIBASIC Listing Of Specific Substances Affirmed As GRAS; Direct Food Substances Affirmed As Generally Recognized As Safe
AMMONIUM PHOSPHATE	582.1141	AMMONIUM PHOSPHATE General Purpose Food Additives; Substances Generally Recognized As Safe
AMMONIUM SACCHARIN	180.37	SACCHARIN, AMMONIUM SACCHARIN, CALCIUM SACCHARIN, AND SODIUM SACCHARIN Specific Requirements For Certain Food Additives; Food Additives Permitted On Interim Basis Or In Contact With Food Pending Study
AMMONIUM SULFATE	182.1127	ALUMINUM AMMONIUM SULFATE Multiple Purpose GRAS Food Substances; Substances Generally Recognized As Safe
AMMONIUM SULFATE	184.1143	AMMONIUM SULFATE Listing Of Specific Substances Affirmed As GRAS; Direct Food Substances Affirmed As Generally Recognized As Safe
AMMONIUM SULFATE	582.1127	ALUMINUM AMMONIUM SULFATE General Purpose Food Additives; Substances Generally Recognized As Safe
AMMONIUM SULFATE	582.1143	AMMONIUM SULFATE General Purpose Food Additives; Substances Generally Recognized As Safe
AMNIOCENTESIS TRAY	884.1550	AMNIOTIC FLUID SAMPLER (AMNIOCENTESIS TRAY) Obstetrical And Gynecological Diagnostic Devices; Obstetrical And Gynecological Devices
AMNIOSCOPE	884.1600	TRANSABDOMINAL AMNIOSCOPE (FETOSCOPE) AND ACCESSORIES Obstetrical And Gynecological Diagnostic Devices; Obstetrical And Gynecological Devices
AMNIOSCOPE	884.1660	TRANSCERVICAL ENDOSCOPE (AMNIOSCOPE) AND ACCESSORIES Obstetrical And Gynecological Diagnostic Devices; Obstetrical And Gynecological Devices
AMNIOTIC	862.1455	LECITHIN/SPHINGOMYELIN RATIO IN AMNIOTIC FLUID TEST SYSTEM Clinical Chemistry Test Systems; Clinical Chemistry And Clinical Toxicology Devices
AMNIOTIC	884.1550	AMNIOTIC FLUID SAMPLER (AMNIOCENTESIS TRAY) Obstetrical And Gynecological Diagnostic Devices; Obstetrical And Gynecological Devices

AMOUNT	17.34	DETERMINING THE AMOUNT OF PENALTIES AND ASSESSMENTS Civil Money Penalties Hearings
AMOUNT	1312.15	SHIPMENTS IN GREATER OR LESS AMOUNT THAN AUTHORIZED Importation Of Controlled Substances; Importation And Exportation Of Controlled Substances
AMOUNTS	101.12	REFERENCE AMOUNTS CUSTOMARILY CONSUMED PER EATING OCCASION General Provisions; Food Labeling
AMOUNTS	1309.11	FEE AMOUNTS Registration Of Manufacturers, Distributors, Importers And Exporters Of List I Chemicals
AMOUNTS DUE	1403.52	COLLECTION OF AMOUNTS DUE After-The-Grant Requirements; Uniform Administrative Requirements For Grants And Coop. Agreements To State And Local Gov.
AMOXICILLIN	520.88	AMOXICILLIN ORAL DOSAGE FORMS Oral Dosage Form New Animal Drugs
AMOXICILLIN	556.38	AMOXICILLIN Specific Tolerances For Residues Of New Animal Drugs; Tolerances For Residues Of New Animal Drugs In Food
AMOXICILLIN TRIHYDRATE	522.88	STERILE AMOXICILLIN TRIHYDRATE FOR SUSPENSION Implantation Or Injectable Dosage Form New Animal Drugs
AMOXICILLIN TRIHYDRATE	526.88	AMOXICILLIN TRIHYDRATE FOR INTRAMAMMARY INFUSION Intramammary Dosage Forms
AMPHETAMINE	250.101	AMPHETAMINE AND METHAMPHETAMINE INHALERS REGARDED AS PRESCRIPTION DRUGS New Drug Or Prescription Status Of Specific Drugs; Specific Requirements For Specific Human Drugs
AMPHETAMINE	862.3100	AMPHETAMINE TEST SYSTEM Clinical Toxicology Test Systems; Clinical Chemistry And Clinical Toxicology Devices
AMPHOMYCIN	524.1204	KANAMYCIN SULFATE, CALCIUM AMPHOMYCIN, AND HYDROCORTISONE ACETATE Ophthalmic And Topical Dosage Form New Animal Drugs
AMPICILLIN	520.90	AMPICILLIN ORAL DOSAGE FORMS Oral Dosage Form New Animal Drug Applications
AMPICILLIN	522.90	AMPICILLIN IMPLANTATION AND INJECTABLE DOSAGE FORMS Implantation Or Injectable Dosage Form New Animal Drugs
AMPICILLIN	556.40	AMPICILLIN Specific Tolerances For Residues Of New Animal Drugs; Tolerances For Residues Of New Animal Drugs In Food
AMPLIFIER	870.2050	BIOPOTENTIAL AMPLIFIER AND SIGNAL CONDITIONER Cardiovascular Monitoring Devices; Cardiovascular Devices
AMPLIFIER	870.2060	TRANSDUCER SIGNAL AMPLIFIER AND SIGNAL CONDITIONER Cardiovascular Monitoring Devices; Cardiovascular Devices
AMPLIFIER	882.1835	PHYSIOLOGICAL SIGNAL AMPLIFIER Neurological Diagnostic Devices; Neurological Devices
AMPROLIUM	520.100	AMPROLIUM ORAL DOSAGE FORMS Oral Dosage Form New Animal Drugs
AMPROLIUM	556.50	AMPROLIUM Specific Tolerances For Residues Of New Animal Drugs; Tolerances For Residues Of New Animal Drugs In Food
AMPROLIUM	558.55	AMPROLIUM Specific New Animal Drugs For Use In Animal Feeds; New Animal Drugs For Use In Animal Feeds
AMPROLIUM	558.58	AMPROLIUM AND ETHOPABATE Specific New Animal Drugs For Use In Animal Feeds; New Animal Drugs For Use In Animal Feeds
AMSLER GRID	886.1330	AMSLER GRID Diagnostic Devices; Ophthalmic Devices
AMYL NITRITE	250.100	AMYL NITRITE INHALANT AS A PRESCRIPTION DRUG FOR HUMAN USE New Drug Or Prescription Status Of Specific Drugs; Specific Requirements For Specific Human Drugs
AMYLASE	184.1012	α-AMYLASE ENZYME PREPARATION FROM BACILLUS STEAROTHERMOPHILUS Listing Of Specific Substances Affirmed As GRAS; Direct Food Substances Affirmed As Generally Recognized As Safe
AMYLASE	862.1070	AMYLASE TEST SYSTEM Clinical Chemistry Test Systems; Clinical Chemistry And Clinical Toxicology Devices
AMYLOGLUCOSIDASE	173.110	AMYLOGLUCOSIDASE DERIVED FROM RHIZOPUS NIVEUS Enzyme Preparations And Microorganisms; Secondary Direct Food Additives Permitted In Food For Human Consumption
ANABOLIC STEROID	1308.25	EXCLUSION OF A VETERINARY ANABOLIC STEROID IMPLANT PRODUCT Schedules Of Controlled Substances

ANABOLIC STEROID	1308.26	EXCLUDED VETERINARY ANABOLIC STEROID IMPLANT PRODUCTS Schedules Of Controlled Substances
ANABOLIC STEROID	1308.33	EXEMPTION OF CERTAIN ANABOLIC STEROID PRODUCTS Schedules Of Controlled Substances
ANABOLIC STEROID	1308.34	EXEMPT ANABOLIC STEROID PRODUCTS Schedules Of Controlled Substances
ANAEROBIC	866.2120	ANAEROBIC CHAMBER Microbiology Devices; Immunology And Microbiology Devices
ANALGESIA	868.5160	GAS MACHINE FOR ANESTHESIA OR ANALGESIA Therapeutic Devices; Anesthesiology Devices
ANALGESIC	201.322	OVER-THE-COUNTER DRUG PRODUCTS CONTAINING INTERNAL ANALGESIC/ANTIPYRETIC ACTIVE INGREDIENTS Specific Labeling Requirements For Specific Drug Products; Labeling
ANALGESIC	346.16	ANALGESIC, ANESTHETIC, AND ANTIPRURITIC ACTIVE INGREDIENTS Anorectal Drug Products For Over-the-counter Human Use
ANALGESIC	348.10	ANALGESIC, ANESTHETIC, AND ANTIPRURITIC ACTIVE INGREDIENTS External Analgesic Drug Products For Over-the-counter Human Use
ANALGESIC, EXTERNAL	348.50	LABELING OF EXTERNAL ANALGESIC DRUG PRODUCTS External Analgesic Drug Products For Over-the-counter Human Use
ANALOG, METHIONINE	582.5477	METHIONINE HYDROXY ANALOG AND ITS CALCIUM SALTS Nutrients And/or Dietary Supplements; Substances Generally Recognized As Safe
ANALYSIS	801.125	MEDICAL DEVICES FOR USE IN TEACHING, LAW ENFORCEMENT, RESEARCH, AND ANALYSIS Exemptions From Adequate Directions For Use; Labeling
ANALYSIS	864.7440	ELECTROPHORETIC HEMOGLOBIN ANALYSIS SYSTEM Hematology Kits And Packages; Hematology And Pathology Devices
ANALYSIS	874.3310	HEARING AID CALIBRATOR AND ANALYSIS SYSTEM Prosthetic Devices; Ear, Nose, And Throat Devices
ANALYSIS	1230.34	ANALYSIS Administrative Procedures; Regulations Under The Federal Caustic Poison Act
ANALYSIS, HAZARD	120.7	HAZARD ANALYSIS General Provisions; Hazard Analysis And Critical Control Point (HACCP) Systems
ANALYSIS, HAZARD	120.8	HAZARD ANALYSIS AND CRITICAL CONTROL POINT (HACCP) PLAN General Provisions; Hazard Analysis And Critical Control Point (HACCP) Systems
ANALYSIS, HAZARD	123.6	HAZARD ANALYSIS AND HAZARD ANALYSIS CRITICAL CONTROL POINT (HACCP) PLAN Fish And Fishery Products
ANALYSIS, METHODS	2.19	METHODS OF ANALYSIS General Provisions; General Administrative Rulings And Decisions
ANALYSIS, METHODS	133.5	METHODS OF ANALYSIS General Provisions; Cheeses And Related Cheese Products
ANALYSIS, METHODS	163.5	METHODS OF ANALYSIS General Provisions; Cacao Products
ANALYSIS, RESEARCH	201.125	DRUGS FOR USE IN TEACHING, LAW ENFORCEMENT, RESEARCH, AND ANALYSIS Exemptions From Adequate Directions For Use; Labeling
ANALYSIS, RISK-BENEFIT	312.84	RISK-BENEFIT ANALYSIS IN REVIEW OF MARKETING APPLICATIONS FOR DRUGS TO TREAT LIFE-THREATENING AND SEVERELY-DEBILITATING ILLNESSES Drugs Intended To Treat Life-threatening And Severely Debilitating Illnesses; Investigational New Drug Application
ANALYSTS	1304.23	RECORDS FOR CHEMICAL ANALYSTS Continuing Records; Records And Reports Of Registrants
ANALYTE	809.30	RESTRICTIONS ON THE SALE, DISTRIBUTION AND USE OF ANALYTE SPECIFIC REAGENTS Requirements For Manufacturers And Producers; In Vitro Diagnostic Products For Human Use
ANALYTE	864.4020	ANALYTE SPECIFIC REAGENTS Specimen Preparation Reagents; Hematology And Pathology Devices
ANALYTICAL METHOD	5.500	ISSUANCE OF FEDERAL REGISTER DOCUMENTS PERTAINING TO NOTICE OF NEED FOR DEVELOPMENT OF AN ANALYTICAL METHOD, NOTICE OF AVAILABILITY OF A DEVELOPED ANALYTICAL METHOD Animal Drugs; Redelegations Of Authority; Delegations Of Authority And Organization
ANALYTICAL METHODS	320.29	ANALYTICAL METHODS FOR AN IN VIVO BIOAVAILABILITY STUDY Procedures For Determining The Bioavailability Of Drug Products; Bioavailability And Bioequivalence Requirements
ANALYTICAL METHODS	530.22	SAFE LEVELS AND ANALYTICAL METHODS FOR FOOD-PRODUCING ANIMALS Extralabel Drug Use In Animals
ANALYTICAL METHODS	530.24	PROCEDURE FOR ANNOUNCING ANALYTICAL METHODS FOR DRUG RESIDUE QUANTIFICATION Extralabel Drug Use In Animals

ANALYTICAL METHODS	530.40	SAFE LEVELS AND AVAILABILITY OF ANALYTICAL METHODS Extralabel Drug Use In Animals
ANALYZER	862.2140	CENTRIFUGAL CHEMISTRY ANALYZER FOR CLINICAL USE Clinical Laboratory Instruments; Clinical Chemistry And Clinical Toxicology Devices
ANALYZER	862.2150	CONTINUOUS FLOW SEQUENTIAL MULTIPLE CHEMISTRY ANALYZER FOR CLINICAL USE Clinical Laboratory Instruments; Clinical Chemistry And Clinical Toxicology Devices
ANALYZER	862.2160	DISCRETE PHOTOMETRIC CHEMISTRY ANALYZER FOR CLINICAL USE Clinical Laboratory Instruments; Clinical Chemistry And Clinical Toxicology Devices
ANALYZER	862.2170	MICRO CHEMISTRY ANALYZER FOR CLINICAL USE Clinical Laboratory Instruments; Clinical Chemistry And Clinical Toxicology Devices
ANALYZER	862.2500	ENZYME ANALYZER FOR CLINICAL USE Clinical Laboratory Instruments; Clinical Chemistry And Clinical Toxicology Devices
ANALYZER	864.5680	AUTOMATED HEPARIN ANALYZER Automated And Semi-automated Hematology Devices; Hematology And Pathology Devices
ANALYZER	868.1075	ARGON GAS ANALYZER Diagnostic Devices; Anesthesiology Devices
ANALYZER	868.1120	INDWELLING BLOOD OXYHEMOGLOBIN CONCENTRATION ANALYZER Diagnostic Devices; Anesthesiology Devices
ANALYZER	868.1150	INDWELLING BLOOD CARBON DIOXIDE PARTIAL PRESSURE (PCO2) ANALYZER Diagnostic Devices; Anesthesiology Devices
ANALYZER	868.1170	INDWELLING BLOOD HYDROGEN ION CONCENTRATION (PH) ANALYZER Diagnostic Devices; Anesthesiology Devices
ANALYZER	868.1200	INDWELLING BLOOD OXYGEN PARTIAL PRESSURE (PO2) ANALYZER Diagnostic Devices; Anesthesiology Devices
ANALYZER	868.1400	CARBON DIOXIDE GAS ANALYZER Diagnostic Devices; Anesthesiology Devices
ANALYZER	868.1430	CARBON MONOXIDE GAS ANALYZER Diagnostic Devices; Anesthesiology Devices
ANALYZER	868.1500	ENFLURANE GAS ANALYZER Diagnostic Devices; Anesthesiology Devices
ANALYZER	868.1620	HALOTHANE GAS ANALYZER Diagnostic Devices; Anesthesiology Devices
ANALYZER	868.1640	HELIUM GAS ANALYZER Diagnostic Devices; Anesthesiology Devices
ANALYZER	868.1670	NEON GAS ANALYZER Diagnostic Devices; Anesthesiology Devices
ANALYZER	868.1690	NITROGEN GAS ANALYZER Diagnostic Devices; Anesthesiology Devices
ANALYZER	868.1700	NITROUS OXIDE GAS ANALYZER Diagnostic Devices; Anesthesiology Devices
ANALYZER	868.1720	OXYGEN GAS ANALYZER Diagnostic Devices; Anesthesiology Devices
ANALYZER	868.1975	WATER VAPOR ANALYZER Diagnostic Devices; Anesthesiology Devices
ANALYZER	868.2380	NITRIC OXIDE ANALYZER Monitoring Devices; Anesthesiology Devices
ANALYZER	868.2385	NITROGEN DIOXIDE ANALYZER Monitoring Devices; Anesthesiology Devices
ANALYZER	870.3630	PACEMAKER GENERATOR FUNCTION ANALYZER Cardiovascular Prosthetic Devices; Cardiovascular Devices
ANALYZER	870.3640	INDIRECT PACEMAKER GENERATOR FUNCTION ANALYZER Cardiovascular Prosthetic Devices; Cardiovascular Devices
ANALYZER	882.1020	RIGIDITY ANALYZER Neurological Diagnostic Devices; Neurological Devices
ANALYZER	882.1420	ELECTROENCEPHALOGRAM (EEG) SIGNAL SPECTRUM ANALYZER Neurological Diagnostic Devices; Neurological Devices
ANALYZER	884.2050	OBSTETRIC DATA ANALYZER Obstetrical And Gynecological Monitoring Devices; Obstetrical And Gynecological Devices
ANALYZING	1230.32	ANALYZING OF SAMPLES Administrative Procedures; Regulations Under The Federal Caustic Poison Act
ANAPHYLACTOID SHOCK	500.65	EPINEPHRINE INJECTION FOR EMERGENCY TREATMENT OF ANAPHYLACTOID SHOCK IN CATTLE, HORSES, SHEEP, AND SWINE Requirements For Specific Animal Drugs; General
ANCHOR	872.3130	PREFORMED ANCHOR Prosthetic Devices; Dental Devices
ANDA APPLICATION	314.101	FILING AN APPLICATION AND RECEIVING AN ABBREVIATED NEW DRUG APPLICATION Applications For FDA Approval To Market A New Drug

ANDA APPLICATION	314.107	EFFECTIVE DATE OF APPROVAL OF A 505(B)(2) APPLICATION OR ABBREVIATED NEW DRUG APPLICATION Applications For FDA Approval To Market A New Drug
ANDA APPLICATION	314.127	REFUSAL TO APPROVE AN ABBREVIATED NEW DRUG APPLICATION Applications For FDA Approval To Market A New Drug
ANDA APPLICATION	314.151	WITHDRAWAL OF APPROVAL OF AN ABBREVIATED NEW DRUG APPLICATION Applications For FDA Approval To Market A New Drug
ANDA APPLICATION	314.153	SUSPENSION OF APPROVAL OF AN ABBREVIATED NEW DRUG APPLICATION Applications For FDA Approval To Market A New Drug
ANDROSTENEDIONE	862.1075	ANDROSTENEDIONE TEST SYSTEM Clinical Chemistry Test Systems; Clinical Chemistry And Clinical Toxicology Devices
ANDROSTERONE	862.1080	ANDROSTERONE TEST SYSTEM Clinical Chemistry Test Systems; Clinical Chemistry And Clinical Toxicology Devices
ANEMIA, PERNICIOUS	250.201	PREPARATIONS FOR THE TREATMENT OF PERNICIOUS ANEMIA Requirements For Drugs And Foods; Specific Requirements For Specific Human Drugs
ANESTHESIA	868.5120	ANESTHESIA CONDUCTION CATHETER Therapeutic Devices; Anesthesiology Devices
ANESTHESIA	868.5130	ANESTHESIA CONDUCTION FILTER Therapeutic Devices; Anesthesiology Devices
ANESTHESIA	868.5140	ANESTHESIA CONDUCTION KIT Therapeutic Devices; Anesthesiology Devices
ANESTHESIA	868.5150	ANESTHESIA CONDUCTION NEEDLE Therapeutic Devices; Anesthesiology Devices
ANESTHESIA	868.5160	GAS MACHINE FOR ANESTHESIA OR ANALGESIA Therapeutic Devices; Anesthesiology Devices
ANESTHESIA	868.5170	LARYNGOTRACHEAL TOPICAL ANESTHESIA APPLICATOR Therapeutic Devices; Anesthesiology Devices
ANESTHESIA	868.5240	ANESTHESIA BREATHING CIRCUIT Therapeutic Devices; Anesthesiology Devices
ANESTHESIA	868.6700	ANESTHESIA STOOL Miscellaneous; Anesthesiology Devices
ANESTHESIA	884.5100	OBSTETRIC ANESTHESIA SET Obstetrical And Gynecological Therapeutic Devices; Obstetrical And Gynecological Devices
ANESTHETIC	346.16	ANALGESIC, ANESTHETIC, AND ANTIPRURITIC ACTIVE INGREDIENTS Anorectal Drug Products For Over-the-counter Human Use
ANESTHETIC	348.10	ANALGESIC, ANESTHETIC, AND ANTIPRURITIC ACTIVE INGREDIENTS External Analgesic Drug Products For Over-the-counter Human Use
ANESTHETIC	868.5550	ANESTHETIC GAS MASK Therapeutic Devices; Anesthesiology Devices
ANESTHETIC	868.5880	ANESTHETIC VAPORIZER Therapeutic Devices; Anesthesiology Devices
ANESTHETIC	868.6100	ANESTHETIC CABINET, TABLE, OR TRAY Miscellaneous; Anesthesiology Devices
ANESTHETIC	872.6100	ANESTHETIC WARMER Miscellaneous Devices; Dental Devices
ANEURYSM	882.4175	ANEURYSM CLIP APPLIER Neurological Surgical Devices; Neurological Devices
ANEURYSM	882.5200	ANEURYSM CLIP Neurological Therapeutic Devices; Neurological Devices
ANEURYSMORRHAPHY	882.5030	METHYL METHACRYLATE FOR ANEURYSMORRHAPHY Neurological Therapeutic Devices; Neurological Devices
ANGIOGRAPHIC	870.1650	ANGIOGRAPHIC INJECTOR AND SYRINGE Cardiovascular Diagnostic Devices; Cardiovascular Devices
ANGIOGRAPHIC	892.1600	ANGIOGRAPHIC X-RAY SYSTEM Diagnostic Devices; Radiology Devices
ANGIOTENSIN	862.1085	ANGIOTENSIN I AND RENIN TEST SYSTEM Clinical Chemistry Test Systems; Clinical Chemistry And Clinical Toxicology Devices
ANGIOTENSIN	862.1090	ANGIOTENSIN CONVERTING ENZYME (A.C.E.) TEST SYSTEM Clinical Chemistry Test Systems; Clinical Chemistry And Clinical Toxicology Devices
ANHYDRASE	866.5200	CARBONIC ANHYDRASE B AND C IMMUNOLOGICAL TEST SYSTEM Immunological Test Systems; Immunology And Microbiology Devices
ANHYDRIDE, STYRENE-MALEIC	177.1820	STYRENE-MALEIC ANHYDRIDE COPOLYMERS Substances For Use As Basic Components Of Single And Repeated Use Food Contact Surfaces; Indirect Food Additives: Polymers
ANHYDROUS AMMONIA	573.180	ANHYDROUS AMMONIA Food Additive Listing; Food Additives Permitted In Feed And Drinking Water Of Animals

ANHYDROUS DEXTROSE	168.110	DEXTROSE ANHYDROUS Requirements For Specific Standardized Sweeteners And Table Sirups; Sweeteners And Table Sirups
ANIMAL	5.301	ISSUANCE OF INITIAL EMERGENCY PERMIT ORDERS AND NOTICES OF CONFIRMATION OF EFFECTIVE DATE OF FINAL REGULATIONS ON FOOD FOR HUMAN AND ANIMAL CONSUMPTION Foods And Cosmetics; Redelegations Of Authority; Delegations Of Authority And Organization
ANIMAL	589.2000	ANIMAL PROTEINS PROHIBITED IN RUMINANT FEED Substances Prohibited From Use In Animal Food Or Feed
ANIMAL	864.2280	CULTURED ANIMAL AND HUMAN CELLS Cell And Tissue Culture Products; Hematology And Pathology Devices
ANIMAL	864.2800	ANIMAL AND HUMAN SERA Cell And Tissue Culture Products; Hematology And Pathology Devices
ANIMAL CARE	58.90	ANIMAL CARE Testing Facilities Operation; Good Laboratory Practice For Nonclinical Laboratory Studies
ANIMAL CARE FACILITIES	58.43	ANIMAL CARE FACILITIES Facilities; Good Laboratory Practice For Nonclinical Laboratory Studies
ANIMAL DIETS	579.22	IONIZING RADIATION FOR TREATMENT OF ANIMAL DIETS Radiation And Radiation Sources; Irradiation In The Production, Process & Handling Of Animal Feed And Pet Food
ANIMAL DRUG	500.30	GENTIAN VIOLET FOR ANIMAL DRUG USE General
ANIMAL DRUG	530.20	CONDITIONS FOR PERMITTED EXTRALABEL ANIMAL AND HUMAN DRUG USE IN FOOD-PRODUCING ANIMALS Extralabel Drug Use In Animals
ANIMAL DRUG APPLICATION	514.11	CONFIDENTIALITY OF DATA AND INFORMATION IN A NEW ANIMAL DRUG APPLICATION FILE General Provisions; New Animal Drug Applications
ANIMAL DRUG APPLICATIONS	5.501	APPROVAL OF NEW ANIMAL DRUG APPLICATIONS, MEDICATED FEED MILL LICENSE APPLICATIONS AND THEIR SUPPLEMENTS Animal Drugs; Redelegations Of Authority; Delegations Of Authority And Organization
ANIMAL DRUG APPLICATIONS	5.503	SUBMISSION OF AND EFFECTIVE APPROVAL DATES FOR ABBREVIATED NEW ANIMAL DRUG APPLICATIONS Animal Drugs; Redelegations Of Authority; Delegations Of Authority And Organization
ANIMAL DRUG APPLICATIONS	514.8	SUPPLEMENTAL NEW ANIMAL DRUG APPLICATIONS General Provisions; New Animal Drug Applications
ANIMAL DRUG, NEW	510.455	NEW ANIMAL DRUG REQUIREMENTS REGARDING FREE-CHOICE ADMINISTRATION IN FEEDS Requirements For Specific New Animal Drugs; New Animal Drugs
ANIMAL DRUG, NEW	558.5	NEW ANIMAL DRUG REQUIREMENTS FOR LIQUID TYPE-B FEEDS General Provisions; New Animal Drugs For Use In Animal Feeds
ANIMAL DRUGS	25.33	ANIMAL DRUGS Categorical Exclusions; Environmental Impact Considerations
ANIMAL DRUGS	500.46	HEXACHLOROPHENE IN ANIMAL DRUGS Specific Administrative Rulings And Decisions; General
ANIMAL DRUGS	500.51	LABELING OF ANIMAL DRUGS; MISBRANDING Animal Drug Labeling Requirements; General
ANIMAL DRUGS	530.10	PROVISION PERMITTING EXTRALABEL USE OF ANIMAL DRUGS Extralabel Drug Use In Animals
ANIMAL DRUGS	530.13	EXTRALABEL USE FROM COMPOUNDING OF APPROVED NEW ANIMAL AND HUMAN DRUGS Extralabel Drug Use In Animals
ANIMAL DRUGS, NEW	5.502	ISSUANCE OF NOTICES, PROPOSALS, AND ORDERS RELATING TO NEW ANIMAL DRUGS AND MEDICATED FEED MILL LICENSE APPLICATIONS Animal Drugs; Redelegations Of Authority; Delegations Of Authority And Organization
ANIMAL DRUGS, NEW	5.504	ISSUANCE OF WRITTEN NOTICES CONCERNING PATENT INFORMATION, CURRENT GOOD MANUFACTURING PRACTICES, AND FALSE OR MISLEADING LABELING OF NEW ANIMAL DRUGS AND FEEDS BEARING OR CONTAINING NEW ANIMAL DRUGS Animal Drugs; Redelegations Of Authority; Delegations Of Authority And Organization
ANIMAL DRUGS, NEW	201.115	NEW DRUGS OR NEW ANIMAL DRUGS Exemptions From Adequate Directions For Use; Labeling
ANIMAL DRUGS, NEW	510.7	CONSIGNEES OF NEW ANIMAL DRUGS FOR USE IN THE MANUFACTURE OF ANIMAL FEED General Provisions; New Animal Drugs
ANIMAL DRUGS, NEW	510.300	RECORDS AND REPORTS CONCERNING EXPERIENCE WITH NEW ANIMAL DRUGS FOR WHICH AN APPROVED APPLICATION IS IN EFFECT Records And Reports; New Animal Drugs

ANIMAL DRUGS, NEW	510.301	RECORDS AND REPORTS CONCERNING EXPERIENCE WITH ANIMAL FEEDS BEARING OR CONTAINING NEW ANIMAL DRUGS FOR WHICH AN APPROVED APPLICATION IS IN EFFECT Records And Reports; New Animal Drugs
ANIMAL DRUGS, NEW	510.305	MAINTENANCE OF COPIES OF APPROVED MEDICATED FEED MILL LICENSES TO MANUFACTURE ANIMAL FEED BEARING OR CONTAINING NEW ANIMAL DRUGS Records And Reports; New Animal Drugs
ANIMAL DRUGS, NEW	510.515	ANIMAL FEEDS BEARING OR CONTAINING NEW ANIMAL DRUGS SUBJECT TO THE PROVISIONS OF SECTION 512(N) Animal Use Exemptions From Certification And Labeling Requirements; New Animal Drugs
ANIMAL DRUGS, NEW	511.1	NEW ANIMAL DRUGS FOR INVESTIGATIONAL USE EXEMPT FROM SECTION 512(A) New Animal Drugs For Investigational Use
ANIMAL DRUGS, NEW	514.12	CONFIDENTIALITY OF DATA AND INFORMATION IN AN INVESTIGATIONAL NEW ANIMAL DRUG NOTICE General Provisions; New Animal Drug Applications
ANIMAL DRUGS, NEW	514.80	RECORDS AND REPORTS CONCERNING EXPERIENCE WITH APPROVED NEW ANIMAL DRUGS Administrative Actions On Applications; New Animal Drug Applications
ANIMAL DRUGS, NEW	530.13	EXTRALABEL USE FROM COMPOUNDING FOR APPROVED NEW ANIMAL AND APPROVED HUMAN DRUGS Extralabel Drug Use In Animals
ANIMAL DRUGS, NEW	556.1	GENERAL CONSIDERATIONS; TOLERANCES FOR RESIDUES OF NEW ANIMAL DRUGS IN FOOD General Provisions; Tolerances For Residues Of New Animal Drugs In Food
ANIMAL FEED	2.35	USE OF SECONDHAND CONTAINERS FOR THE SHIPMENT AND STORAGE OF FOOD AND ANIMAL FEED Human And Animal Foods; General Administrative Rulings And Decisions
ANIMAL FEED	500.45	USE OF POLYCHLORINATED BIPHENYLS (PCB'S) IN THE PRODUCTION, HANDLING AND STORAGE OF ANIMAL FEED Specific Administrative Rulings And Decisions; General
ANIMAL FEED	501.110	ANIMAL FEED LABELING; COLLECTIVE NAMES FOR FEED INGREDIENTS Exemptions From Animal Food Labeling Requirements; Animal Food Labeling
ANIMAL FEED	510.7	CONSIGNEES OF NEW ANIMAL DRUGS FOR USE IN MANUFACTURE OF ANIMAL FEED General Provisions; New Animal Drugs
ANIMAL FEED	510.305	MAINTENANCE OF COPIES OF APPROVED MEDICATED FEED MILL LICENSES TO MANUFACTURE ANIMAL FEED BEARING OR CONTAINING NEW ANIMAL DRUGS Records And Reports; New Animal Drugs
ANIMAL FEED	558.15	ANTIBIOTIC, NITROFURAN, AND SULFONAMIDE DRUGS IN THE FEED OF ANIMALS General Provisions; New Animal Drugs For Use In Animal Feeds
ANIMAL FEED	570.13	INDIRECT FOOD ADDITIVES RESULTING FROM PACKAGING MATERIALS PRIOR SANCTIONED FOR ANIMAL FEED AND PET FOOD General Provisions; Food Additives
ANIMAL FEED	570.14	INDIRECT FOOD ADDITIVES RESULTING FROM PACKAGING MATERIALS FOR ANIMAL FEED AND PET FOOD General Provisions; Food Additives
ANIMAL FEEDS	500.29	GENTIAN VIOLET FOR USE IN ANIMAL FEEDS Specific Administrative Rulings And Decisions; General
ANIMAL FEEDS	500.35	ANIMAL FEEDS CONTAMINATED WITH SALMONELLA MICROORGANISMS Specific Administrative Rulings And Decisions; General
ANIMAL FEEDS	510.301	RECORDS AND REPORTS CONCERNING EXPERIENCE WITH ANIMAL FEEDS BEARING OR CONTAINING NEW ANIMAL DRUGS Records And Reports; New Animal Drugs
ANIMAL FEEDS	510.515	ANIMAL FEEDS BEARING OR CONTAINING NEW ANIMAL DRUGS SUBJECT TO THE PROVISIONS OF SECTION 512(N) Animal Use Exemptions From Certification And Labeling Requirements; New Animal Drugs
ANIMAL FEEDS	573.380	ETHOXYQUIN IN ANIMAL FEEDS Food Additive Listing; Food Additives Permitted In Feed And Drinking Water Of Animals
ANIMAL FEEDS	582.80	TRACE MINERALS ADDED TO ANIMAL FEEDS General Provisions; Substances Generally Recognized As Safe
ANIMAL FOOD	2.25	GRAIN SEED TREATED WITH POISONOUS SUBSTANCES; COLOR IDENTIFICATION TO PREVENT ADULTERATION OF HUMAN AND ANIMAL FOOD Human And Animal Foods; General Administrative Rulings And Decisions
ANIMAL FOOD	501.1	PRINCIPAL DISPLAY PANEL OF PACKAGE FORM ANIMAL FOOD General Provisions; Animal Food Labeling

ANIMAL FOOD	501.2	INFORMATION PANEL OF PACKAGE FOR ANIMAL FOOD General Provisions; Animal Food Labeling
ANIMAL FOOD	501.3	IDENTITY LABELING OF ANIMAL FOOD IN PACKAGE FORM General Provisions; Animal Food Labeling
ANIMAL FOOD	501.4	ANIMAL FOOD; DESIGNATION OF INGREDIENTS General Provisions; Animal Food Labeling
ANIMAL FOOD	501.5	ANIMAL FOOD; NAME AND PLACE OF BUSINESS OF MANUFACTURER, PACKER, OR DISTRIBUTOR General Provisions; Animal Food Labeling
ANIMAL FOOD	501.8	LABELING OF ANIMAL FOOD WITH NUMBER OF SERVINGS General Provisions; Animal Food Labeling
ANIMAL FOOD	501.15	ANIMAL FOOD; PROMINENCE OF REQUIRED STATEMENTS General Provisions; Animal Food Labeling
ANIMAL FOOD	501.17	ANIMAL FOOD LABELING WARNING STATEMENTS General Provisions; Animal Food Labeling
ANIMAL FOOD	501.18	MISBRANDING OF ANIMAL FOOD General Provisions; Animal Food Labeling
ANIMAL FOOD	501.100	ANIMAL FOOD; EXEMPTIONS FROM LABELING Exemptions From Animal Food Labeling Requirements; Animal Food Labeling
ANIMAL FOOD	589.1	SUBSTANCES PROHIBITED FROM USE IN ANIMAL FOOD OR FEED General Provisions; Substances Prohibited From Use In Animal Food Or Feed
ANIMAL FOODS	501.22	ANIMAL FOODS; LABELING OF SPICES, FLAVORINGS, COLORINGS, AND CHEMICAL PRESERVATIVES Specific Animal Food Labeling Requirements; Animal Food Labeling
ANIMAL GLUE	178.3120	ANIMAL GLUE Certain Adjuvants And Production Aids; Indirect Food Additives: Adjuvants, Production Aids, And Sanitizers
ANIMAL LIPASE	184.1415	ANIMAL LIPASE Listing Of Specific Substances Affirmed As GRAS; Direct Food Substances Affirmed As Generally Recognized As Safe
ANIMAL PROTEIN HYDROLYSATE	573.200	CONDENSED ANIMAL PROTEIN HYDROLYSATE Food Additive Listing; Food Additives Permitted In Feed And Drinking Water Of Animals
ANIMAL SUPPLY	58.45	ANIMAL SUPPLY FACILITIES Facilities; Good Laboratory Practice For Nonclinical Laboratory Studies
ANIMALS	5.505	TERMINATION OF EXEMPTIONS FOR NEW DRUGS FOR INVESTIGATIONAL USE IN ANIMALS Animal Drugs; Redelegations Of Authority; Delegations Of Authority And Organization
ANIMALS	312.160	DRUGS FOR INVESTIGATIONAL USE IN LABORATORY RESEARCH ANIMALS OR IN VITRO TESTS Drugs For Investigational Use In Laboratory Research Animals Or In Vitro Tests; Investigational New Drug Application
ANIMALS	500.25	ANTHELMINTIC DRUGS FOR USE IN ANIMALS Specific Administrative Rulings And Decisions; General
ANIMALS	500.27	METHYLENE BLUE-CONTAINING DRUGS FOR USE IN ANIMALS Specific Administrative Rulings And Decisions; General
ANIMALS	500.52	USE OF TERMS SUCH AS "TONIC," "TONE" "TONER," OR "CONDITIONER" IN THE LABELING OF PREPARATIONS INTENDED FOR USE IN OR ON ANIMALS Animal Drug Labeling Requirements; General
ANIMALS	510.410	CORTICOSTEROIDS FOR ORAL, INJECTABLE, AND OPHTHALMIC USE IN ANIMALS; WARNINGS AND LABELING REQUIREMENTS Requirements For Specific New Animal Drugs; New Animal Drugs
ANIMALS	530.21	PROHIBITIONS FOR FOOD-PRODUCING ANIMALS Extralabel Drug Use In Animals
ANIMALS	530.41	DRUGS PROHIBITED FOR EXTRALABEL USE IN ANIMALS Extralabel Drug Use In Animals
ANIMALS	570.17	EXEMPTION FOR INVESTIGATIONAL USE AND PROCEDURE FOR OBTAINING AUTHORIZATION TO MARKET EDIBLE PRODUCTS FROM EXPERIMENTAL ANIMALS General Provisions; Food Additives
ANIMALS	600.11	PHYSICAL ESTABLISHMENT, EQUIPMENT, ANIMALS, AND CARE Establishment Standards; Biological Products: General
ANIMALS, EXPERIMENTAL	170.17	EXEMPTION FOR INVESTIGATIONAL USE AND PROCEDURE FOR OBTAINING AUTHORIZATION TO MARKET EDIBLE PRODUCTS FROM EXPERIMENTAL ANIMALS General Provisions; Food Additives
ANIMALS, FOOD-PRODUCING	510.110	ANTIBIOTICS USED IN FOOD-PRODUCING ANIMALS Specific Administrative Rulings And Decisions; New Animal Drugs
ANIMALS, FOOD-PRODUCING	530.20	CONDITIONS FOR PERMITTED EXTRALABEL ANIMAL AND HUMAN DRUG USE IN FOOD-PRODUCING ANIMALS Extralabel Drug Use In Animals
ANIMALS, FOOD-PRODUCING	530.21	PROHIBITIONS FOR FOOD-PRODUCING ANIMALS Extralabel Drug Use In Animals

ANIMALS, FOOD-PRODUCING	530.22	SAFE LEVELS AND ANALYTICAL METHODS FOR FOOD-PRODUCING ANIMALS Extralabel Drug Use In Animals
ANIMALS, FOOD-PRODUCING	530.25	ORDERS PROHIBITING EXTRALABEL USES FOR DRUGS IN FOOD-PRODUCING ANIMALS Extralabel Drug Use In Animals
ANIMALS, LABORATORY	211.173	LABORATORY ANIMALS Laboratory Controls; Current Good Manufacturing Practice For Finished Pharmaceuticals
ANIMALS, MILK-PRODUCING	510.105	LABELING OF DRUGS FOR USE IN MILK-PRODUCING ANIMALS Specific Administrative Rulings And Decisions; New Animal Drugs
ANIMALS, MILK-PRODUCING	510.106	LABELING OF ANTIBIOTIC AND ANTIBIOTIC-CONTAINING DRUGS INTENDED FOR USE IN MILK-PRODUCING ANIMALS Specific Administrative Rulings And Decisions; New Animal Drugs
ANIMALS, NONFOOD	530.30	EXTRALABEL DRUG USE IN NONFOOD ANIMALS Extralabel Drug Use In Animals
ANKLE JOINT	888.3100	ANKLE JOINT METAL/COMPOSITE SEMI-CONSTRAINED CEMENTED PROSTHESIS Prosthetic Devices; Orthopedic Devices
ANKLE JOINT	888.3110	ANKLE JOINT METAL/POLYMER SEMI-CONSTRAINED CEMENTED PROSTHESIS Prosthetic Devices; Orthopedic Devices
ANKLE JOINT	888.3120	ANKLE JOINT METAL/POLYMER NON-CONSTRAINED CEMENTED PROSTHESIS Prosthetic Devices; Orthopedic Devices
ANNATTO	73.2030	ANNATTO Cosmetics; Listing Of Color Additives Exempt From Certification
ANNATTO EXTRACT	73.30	ANNATTO EXTRACT Foods; Listing Of Color Additives Exempt From Certification
ANNATTO EXTRACT	73.1030	ANNATTO EXTRACT Drugs; Listing Of Color Additives Exempt From Certification
ANNUAL REPORTS	312.33	ANNUAL REPORTS Investigational New Drug Application (IND)
ANNUAL REPORTS	316.30	ANNUAL REPORTS OF HOLDER OF ORPHAN-DRUG DESIGNATION Orphan Drugs
ANNUAL REPORTS	601.28	ANNUAL REPORTS OF POSTMARKETING PEDIATRIC STUDIES Biologics Licensing; Licensing
ANNUAL REPORTS	601.70	ANNUAL PROGRESS REPORTS OF POSTMARKETING STUDIES Postmarketing Studies; Licensing
ANNUAL REPORTS	803.33	ANNUAL REPORTS User Facility Reporting Requirements; Medical Device Reporting
ANNUAL REPORTS	1002.13	ANNUAL REPORTS Required Manufacturers' Reports For Listed Electronic Products; Records And Reports
ANNULOPLASTY	870.3800	ANNULOPLASTY RING Cardiovascular Prosthetic Devices; Cardiovascular Devices
ANOMALOSCOPE	886.1070	ANOMALOSCOPE Diagnostic Devices; Ophthalmic Devices
ANORECTAL	346.22	PERMITTED COMBINATIONS OF ANORECTAL ACTIVE INGREDIENTS Active Ingredients; Anorectal Drug Products For Over-the-counter Human Use
ANORECTAL	346.50	LABELING OF ANORECTAL DRUG PRODUCTS Labeling; Anorectal Drug Products For Over-the-counter Human Use
ANORECTAL	346.52	LABELING OF PERMITTED COMBINATIONS OF ANORECTAL ACTIVE INGREDIENTS Labeling; Anorectal Drug Products For Over-the-counter Human Use
ANOXOMER	172.105	ANOXOMER Food Preservatives; Food Additives Permitted For Direct Addition To Food For Human Consumption
ANSWER	17.9	ANSWER Civil Money Penalties Hearings
ANSWER	17.11	DEFAULT UPON FAILURE TO FILE AN ANSWER Civil Money Penalties Hearings
ANTACID	331.10	ANTACID ACTIVE INGREDIENTS Active Ingredients; Antacid Products For Over-the-counter (OTC) Human Use
ANTACID	331.30	LABELING OF ANTACID PRODUCTS Labeling; Antacid Products For Over-the-counter (OTC) Human Use
ANTHELMINTHIC	357.110	ANTHELMINTHIC ACTIVE INGREDIENT Anthelmintic Drug Products; Miscellaneous Internal Drug Products For Over-the-counter Human Use
ANTHELMINTHIC DRUG	357.150	LABELING OF ANTHELMINTHIC DRUG PRODUCTS Anthelmintic Drug Products; Miscellaneous Internal Drug Products For Over-the-counter Human Use

ANTHELMINTIC DRUG	357.152	PACKAGE INSERTS FOR ANTHELMINTIC DRUG PRODUCTS Anthelmintic Drug Products; Miscellaneous Internal Drug Products For Over-the-counter Human Use
ANTHELMINTIC DRUGS	500.25	ANTHELMINTIC DRUGS FOR USE IN ANIMALS Specific Administrative Rulings And Decisions; General
ANTHRACENEDIONE	73.3100	1,4-BIS[(2-HYDROXYETHYL)AMINO]-9,10-ANTHRACENEDIONE BIS(2-PROPENOIC)ESTER COPOLYMERS Medical Devices; Listing Of Color Additives Exempt From Certification
ANTHRACENEDIONE	73.3105	1,4-BIS[(2-METHYLPHENYL)AMINO]-9,10-ANTHRACENEDIONE Medical Devices; Listing Of Color Additives Exempt From Certification
ANTHRANILATE, CINNAMYL	189.113	CINNAMYL ANTHRANILATE Substances Generally Prohibited From Direct Addition Or Use As Human Food; Substances Prohibited From Use In Human Food
ANTHRAQUINONE	73.3106	1,4-BIS[4-(2-METHACRYLOXYETHYL)PHENYLAMINO]ANTHRAQUINONE COPOLYMERS Medical Devices; Listing Of Color Additives Exempt From Certification
ANTHRAZINETETRONE	73.3119	7,16-DICHLORO-6,15-DIHYDRO-5,9,14,18-ANTHRAZINETETRONE Medical Devices; Listing Of Color Additives Exempt From Certification
ANTHROPOMORPHIC	892.1370	NUCLEAR ANTHROPOMORPHIC PHANTOM Diagnostic Devices; Radiology Devices
ANTHROPOMORPHIC	892.1950	RADIOGRAPHIC ANTHROPOMORPHIC PHANTOM Diagnostic Devices; Radiology Devices
ANTIBIOTIC	333.110	FIRST AID ANTIBIOTIC ACTIVE INGREDIENTS First Aid Antibiotic Drug Products; Topical Antimicrobial Drug Products For Over-the-counter Human Use
ANTIBIOTIC	333.150	LABELING OF FIRST AID ANTIBIOTIC DRUG PRODUCTS First Aid Antibiotic Drug Products; Topical Antimicrobial Drug Products For Over-the-counter Human Use
ANTIBIOTIC	510.106	LABELING OF ANTIBIOTIC AND ANTIBIOTIC-CONTAINING DRUGS INTENDED FOR USE IN MILK-PRODUCING ANIMALS Specific Administrative Rulings And Decisions; New Animal Drugs
ANTIBIOTIC	558.15	ANTIBIOTIC, NITROFURAN, AND SULFONAMIDE DRUGS IN THE FEED OF ANIMALS General Provisions; New Animal Drugs For Use In Animal Feeds
ANTIBIOTICS	510.110	ANTIBIOTICS USED IN FOOD-PRODUCING ANIMALS Specific Administrative Rulings And Decisions; New Animal Drugs
ANTIBIOTICS	510.112	ANTIBIOTICS USED IN VETERINARY MEDICINE AND FOR NONMEDICAL PURPOSES; REQUIRED DATA Specific Administrative Rulings And Decisions; New Animal Drugs
ANTIBODIES	660.54	POTENCY TESTS, SPECIFICITY TESTS, TESTS FOR HETEROSPECIFIC ANTIBODIES, AND ADDITIONAL TESTS FOR NONSPECIFIC PROPERTIES Anti-Human Globulin; Additional Standards For Diagnostic Substances For Laboratory Tests
ANTIBODY	660.1	ANTIBODY TO HEPATITIS B SURFACE ANTIGEN Antibody To Hepatitis B Surface Antigen; Additional Standards For Diagnostic Substances For Laboratory Tests
ANTIBODY	864.9175	AUTOMATED BLOOD GROUPING AND ANTIBODY TEST SYSTEM Products Used In Establishments That Manufacture Blood And Blood Products; Hematology And Pathology Devices
ANTIBODY	866.5090	ANTIMITOCHONDRIAL ANTIBODY IMMUNOLOGICAL TEST SYSTEM Immunological Test Systems; Immunology And Microbiology Devices
ANTIBODY	866.5100	ANTINUCLEAR ANTIBODY IMMUNOLOGICAL TEST SYSTEM Immunological Test Systems; Immunology And Microbiology Devices
ANTIBODY	866.5110	ANTIPARIETAL ANTIBODY IMMUNOLOGICAL TEST SYSTEM Immunological Test Systems; Immunology And Microbiology Devices
ANTIBODY	866.5120	ANTISMOOTH MUSCLE ANTIBODY IMMUNOLOGICAL TEST SYSTEM Immunological Test Systems; Immunology And Microbiology Devices
ANTIBODY	866.5785	ANTI-SACCHAROMYCES CEREVISIAE ANTIBODY (ASCA) TEST SYSTEMS Immunological Test Systems; Immunology And Microbiology Devices
ANTIBODY TEST, GONOCOCCAL	866.3290	GONOCOCCAL ANTIBODY TEST (GAT) Serological Reagents; Immunology And Microbiology Devices
ANTICANCER	70.51	ADVISORY COMMITTEE ON THE APPLICATION OF THE ANTICANCER CLAUSE Safety Evaluation; Color Additives
ANTICARIES	355.10	ANTICARIES ACTIVE INGREDIENTS Anticaries Drug Products For Over-the-counter Human Use
ANTICARIES	355.50	LABELING OF ANTICARIES PRODUCTS Anticaries Drug Products For Over-the-counter Human Use
ANTICHOKE	874.5350	SUCTION ANTICHOKE DEVICE Therapeutic Devices; Ear, Nose And Throat Devices
ANTICHOKE	874.5370	TONGS ANTICHOKE DEVICE Therapeutic Devices; Ear, Nose And Throat Devices

ANTICHOLINERGIC DRUGS	310.533	DRUG PRODUCTS CONTAINING ACTIVE INGREDIENTS OFFERED OVER-THE-COUNTER (OTC) FOR HUMAN USE AS AN ANTICHOLINERGIC IN COUGH-COLD DRUG PRODUCTS Requirements For Specific New Drugs Or Devices; New Drugs
ANTICHYMOTRYPSIN	866.5080	ALPHA-1-ANTICHYMOTRYPSIN IMMUNOLOGICAL TEST SYSTEM Immunological Test Systems; Immunology And Microbiology Devices
ANTICORROSIVE	178.3125	ANTICORROSIVE AGENTS Certain Adjuvants And Production Aids; Indirect Food Additives: Adjuvants, Production Aids, And Sanitizers
ANTIDEPRESSANT	862.3910	TRICYCLIC ANTIDEPRESSANT DRUGS TEST SYSTEM Clinical Toxicology Text Systems; Clinical Chemistry And Toxicology Test Systems
ANTIEMETIC	336.10	ANTIEMETIC ACTIVE INGREDIENTS Active Ingredients; Antiemetic Drug Products For Over-the-counter Human Use
ANTIEMETIC	336.50	LABELING OF ANTIEMETIC DRUG PRODUCTS Labeling; Antiemetic Drug Products For Over-the-counter Human Use
ANTIFLATULENT	332.10	ANTIFLATULENT ACTIVE INGREDIENTS Active Ingredients; Antiflatulent Products For Over-the-counter Human Use
ANTIFLATULENT	332.30	LABELING OF ANTIFLATULENT PRODUCTS Labeling; Antiflatulent Products For Over-the-counter Human Use
ANTIFOGGING	178.3130	ANTISTATIC AND/OR ANTIFOGGING AGENTS IN FOOD-PACKAGING MATERIALS Certain Adjuvants And Production Aids; Indirect Food Additives: Adjuvants, Production Aids, And Sanitizers
ANTIFUNGAL	333.210	ANTIFUNGAL ACTIVE INGREDIENTS Topical Antifungal Drug Products; Topical Antimicrobial Drug Products For Over-the-counter Human Use
ANTIFUNGAL	333.250	LABELING OF ANTIFUNGAL DRUG PRODUCTS Topical Antifungal Drug Products; Topical Antimicrobial Drug Products For Over-the-counter Human Use
ANTIGEN, HEPATITIS	660.1	ANTIBODY TO HEPATITIS B SURFACE ANTIGEN Antibody To Hepatitis B Surface Antigen; Additional Standards For Diagnostic Substances For Laboratory Tests
ANTIGEN, HEPATITIS	660.40	HEPATITIS B SURFACE ANTIGEN Hepatitis B Surface Antigen; Additional Standards For Diagnostic Substances For Laboratory Tests
ANTIGEN, TUMOR ASSOC.	866.6010	TUMOR ASSOCIATED ANTIGEN IMMUNOLOGICAL TEST SYSTEM Tumor Associated Antigen Immunological Test Systems; Immunology And Microbiology Devices
ANTIHISTAMINE	341.12	ANTIHISTAMINE ACTIVE INGREDIENTS Cold, Cough, Allergy, Bronchodilator, And Antiasthmatic Drug Products For Over-the-counter Human Use
ANTIHISTAMINE	341.72	LABELING OF ANTIHISTAMINE DRUG PRODUCTS Cold, Cough, Allergy, Bronchodilator, And Antiasthmatic Drug Products For Over-the-counter Human Use
ANTI-HUMAN GLOBULIN	660.50	ANTI-HUMAN GLOBULIN Anti-Human Globulin; Additional Standards For Diagnostic Substances For Laboratory Tests
ANTIMICROBIAL	866.1620	ANTIMICROBIAL SUSCEPTIBILITY TEST DISC Diagnostic Devices; Immunology And Microbiology Devices
ANTIMICROBIAL	866.1640	ANTIMICROBIAL SUSCEPTIBILITY TEST POWDER Diagnostic Devices; Immunology And Microbiology Devices
ANTIMICROBIAL	866.1700	CULTURE MEDIUM FOR ANTIMICROBIAL SUSCEPTIBILITY TESTS Diagnostic Devices; Immunology And Microbiology Devices
ANTIMITOCHONDRIAL	866.5090	ANTIMITOCHONDRIAL ANTIBODY IMMUNOLOGICAL TEST SYSTEM Immunological Test Systems; Immunology And Microbiology Devices
ANTIMONY	862.3110	ANTIMONY TEST SYSTEM Clinical Toxicology Test Systems; Clinical Chemistry And Clinical Toxicology Devices
ANTIMYCOTICS	181.23	ANTIMYCOTICS Specific Prior-sanctioned Food Ingredients; Prior-sanctioned Food Ingredients
ANTINUCLEAR	866.5100	ANTINUCLEAR ANTIBODY IMMUNOLOGICAL TEST SYSTEM Immunological Test Systems; Immunology And Microbiology Devices
ANTI-OFFSET SUBSTANCES	176.130	ANTI-OFFSET SUBSTANCES Substances For Use Only As Components Of Paper And Paperboard; Indirect Food Additives
ANTIOXIDANTS	178.2010	ANTIOXIDANTS AND/OR STABILIZERS FOR POLYMERS Antioxidants And Stabilizers; Indirect Food Additives: Adjuvants, Production Aids, And Sanitizers
ANTIOXIDANTS	181.24	ANTIOXIDANTS Specific Prior-sanctioned Food Ingredients; Prior-sanctioned Food Ingredients
ANTIPARIETAL	866.5110	ANTIPARIETAL ANTIBODY IMMUNOLOGICAL TEST SYSTEM Immunological Test Systems; Immunology And Microbiology Devices
ANTIPRURITIC	346.16	ANALGESIC, ANESTHETIC, AND ANTIPRURITIC ACTIVE INGREDIENTS Active Ingredients; Anorectal Drug Products For Over-the-counter Human Use

ANTIPRURITIC	348.10	ANALGESIC, ANESTHETIC, AND ANTIPRURITIC ACTIVE INGREDIENTS Active Ingredients; External Analgesic Drug Products For Over-the-counter Human Use
ANTIPYRETIC	201.322	OVER THE COUNTER OTC DRUG PRODUCTS CONTAINING INTERNAL ANALGESIC ANTIPYRETIC Labeling
ANTI-SACCHAROMYCES	866.5785	ANTI-SACCHAROMYCES CEREVISIAE ANTIBODY (ASCA) TEST SYSTEMS Immunological Test Systems; Immunology And Microbiology Devices
ANTISMOOTH MUSCLE	866.5120	ANTISMOOTH MUSCLE ANTIBODY IMMUNOLOGICAL TEST SYSTEM Immunological Test Systems; Immunology And Microbiology Devices
ANTISTAMMERING	874.5840	ANTISTAMMERING DEVICE Therapeutic Devices; Ear, Nose And Throat Devices
ANTISTATIC	178.3130	ANTISTATIC AND/OR ANTIFOGGING AGENTS IN FOOD-PACKAGING MATERIALS Certain Adjuvants And Production Aids; Indirect Food Additives: Adjuvants, Production Aids, And Sanitizers
ANTITHROMBIN	864.7060	ANTITHROMBIN III ASSAY Hematology Kits And Packages; Hematology And Pathology Devices
ANTITRYPSIN	866.5130	ALPHA-1-ANTITRYPSIN IMMUNOLOGICAL TEST SYSTEM Immunological Test Systems; Immunology And Microbiology Devices
ANTITUSSIVE	341.14	ANTITUSSIVE ACTIVE INGREDIENTS Cold, Cough, Allergy, Bronchodilator, And Antiasthmatic Drug Products For OTC Human Use
ANTITUSSIVE	341.74	LABELING OF ANTITUSSIVE DRUG PRODUCTS Cold, Cough, Allergy, Bronchodilator, And Antiasthmatic Drug Products For OTC Human Use
ANVIL	882.4030	SKULL PLATE ANVIL Neurological Surgical Devices; Neurological Devices
AORTIC	870.3535	INTRA-AORTIC BALLOON AND CONTROL SYSTEM Cardiovascular Prosthetic Devices; Cardiovascular Devices
APEX	870.2310	APEX CARDIOGRAPH (VIBROCARDIOGRAPH) Cardiovascular Monitoring Devices; Cardiovascular Devices
APEX	870.2840	APEX CARDIOGRAPHIC TRANSDUCER Cardiovascular Monitoring Devices; Cardiovascular Devices
APGAR	880.2930	APGAR TIMER General Hospital And Personal Use Devices
APHRODISIAC	310.528	DRUGS PRODUCTS CONTAINING ACTIVE INGREDIENTS OFFERED OVER-THE-COUNTER FOR USE AS IN APHRODISIAC Requirements For Specific New Drugs Or Devices; New Drugs
APPARATUS	864.5240	AUTOMATED BLOOD CELL DILUTING APPARATUS Automated And Semi-automated Hematology Devices; Hematology And Pathology Devices
APPARATUS	864.9225	CELL-FREEZING APPARATUS AND REAGENTS FOR IN VITRO DIAGNOSTIC USE Products Used In Establishments That Manufacture Blood And Blood Products; Hematology And Pathology Devices
APPARATUS	868.5400	ELECTROANESTHESIA APPARATUS Therapeutic Devices; Anesthesiology Devices
APPARATUS	868.5430	GAS-SCAVENGING APPARATUS Therapeutic Devices; Anesthesiology Devices
APPARATUS	868.5830	AUTOTRANSFUSION APPARATUS Therapeutic Devices; Anesthesiology Devices
APPARATUS	870.5050	PATIENT CARE SUCTION APPARATUS Therapeutic Devices; Cardiovascular Devices
APPARATUS	878.4680	NONPOWERED, SINGLE PATIENT, PORTABLE SUCTION APPARATUS Surgical Devices; General And Plastic Surgery Devices
APPARATUS	880.6740	VACUUM-POWERED BODY FLUID SUCTION APPARATUS General Hospital And Personal Use Miscellaneous Devices; General Hospital And Personal Use Devices
APPARATUS	884.5900	THERAPEUTIC VAGINAL DOUCHE APPARATUS Obstetrical And Gynecological Therapeutic Devices; Obstetrical And Gynecological Devices
APPARATUS	886.4100	RADIOFREQUENCY ELECTROSURGICAL CAUTERY APPARATUS Surgical Devices; Ophthalmic Devices
APPARATUS	888.5850	NONPOWERED ORTHOPEDIC TRACTION APPARATUS AND ACCESSORIES Surgical Devices; Orthopedic Devices
APPAREL, SURGICAL	878.4040	SURGICAL APPAREL Surgical Devices; General And Plastic Surgery Devices
APPEAL	12.97	INTERLOCUTORY APPEAL FROM RULING OF PRESIDING OFFICER Hearing Procedures; Formal Evidentiary Public Hearing
APPEAL	12.125	APPEAL FROM OR REVIEW OF INITIAL DECISION Initial And Final Decisions; Formal Evidentiary Public Hearing
APPEAL	12.130	DECISION BY COMMISSIONER ON APPEAL OR REVIEW OF INITIAL DECISION Initial And Final Decisions; Formal Evidentiary Public Hearing

APPEAL	17.18	INTERLOCUTORY APPEAL FROM RULING OF PRESIDING OFFICER Civil Money Penalties Hearings
APPEAL	203.12	APPEAL FROM AN ADVERSE DECISION BY THE DISTRICT OFFICE Reimportation; Prescription Drug Marketing
APPEAL	1316.62	APPEAL FROM RULING OF PRESIDING OFFICER Administrative Hearings; Administrative Functions, Practices, And Procedures
APPEAL	1402.5	APPEAL PROCEDURE Mandatory Declassification Review
APPEALS	17.47	APPEALS Civil Money Penalties Hearings
APPEALS	21.52	ADMINISTRATIVE APPEALS OF REFUSALS TO AMEND RECORDS Procedures For Requests For Amendment Of Records; Protection Of Privacy
APPEALS	900.15	APPEALS OF ADVERSE ACCREDITATION OR REACCREDITATION DECISIONS Quality Standards And Certification; Mammography
APPEALS	900.16	APPEALS OF DENIALS OF CERTIFICATION Quality Standards And Certification; Mammography
APPEALS	900.25	HEARINGS AND APPEALS States As Certifiers; Mammography
APPEALS	1401.9	APPEALS Public Availability Of Information
APPEARANCE	12.40	APPEARANCE Appearance And Participation; Formal Evidentiary Public Hearing
APPEARANCE	1301.43	REQUEST FOR HEARING OR APPEARANCE; WAIVER Hearings; Registration Of Manufacturers, Distributors, And Dispensers Of Controlled Substances
APPEARANCE	1303.34	REQUEST FOR HEARING OR APPEARANCE; WAIVER Hearings; Quotas
APPEARANCE	1308.44	REQUEST FOR HEARING OR APPEARANCE; WAIVER Hearings; Schedules Of Controlled Substances
APPEARANCE	1309.53	REQUEST FOR HEARING OR APPEARANCE; WAIVER Registration Of Manufacturers, Distributors, Importers And Exporters Of List I Chemicals
APPEARANCE	1312.44	REQUEST FOR HEARING OR APPEARANCE; WAIVER Hearings; Importation And Exportation Of Controlled Substances
APPEARANCE	1316.48	NOTICE OF APPEARANCE Administrative Hearings; Administrative Functions, Practices, And Procedures
APPEARANCE	1316.50	APPEARANCE; REPRESENTATION; AUTHORIZATION Administrative Hearings; Administrative Functions, Practices, And Procedures
APPLESAUCE	145.110	CANNED APPLESAUCE Requirements For Specific Standardized Canned Fruits; Canned Fruits
APPLIANCE, FRACTURE	878.3250	EXTERNAL FACIAL FRACTURE FIXATION APPLIANCE Prosthetic Devices; General And Plastic Surgery Devices
APPLICANT	60.32	APPLICANT RESPONSE TO PETITION Due Diligence Petitions; Patent Term Restoration
APPLICANT	312.99	OTHER RESPONSIBILITIES OF AN APPLICANT OF AN ABBREVIATED APPLICATION Abbreviated Applications; Applications For FDA Approval To Market A New Drug
APPLICANT	314.65	WITHDRAWAL BY THE APPLICANT OF AN UNAPPROVED APPLICATION Applications; Applications For FDA Approval To Market A New Drug
APPLICANT	314.99	OTHER RESPONSIBILITIES OF AN APPLICANT OF AN ABBREVIATED APPLICATION Abbreviated Applications; Applications For FDA Approval To Market A New Drug
APPLICANT	314.110	APPROVABLE LETTER TO THE APPLICANT FDA Action On Applications; Applications For FDA Approval To Market A New Drug
APPLICANT	314.120	NOT APPROVABLE LETTER TO THE APPLICANT FDA Action On Applications; Applications For FDA Approval To Market A New Drug
APPLICANTS	314.102	COMMUNICATIONS BETWEEN FDA AND APPLICANTS FDA Action On Applications; Applications For FDA Approval To Market A New Drug
APPLICATION	1.95	APPLICATION FOR AUTHORIZATION TO RELABEL AND RECONDITION Imports And Exports; General Enforcement Regulations
APPLICATION	5.34	ISSUANCE OF NOTICES RELATING TO PROPOSALS AND ORDERS FOR DEBARMENT AND DENIAL OF AN APPLICATION TO TERMINATE DEBARMENT General Redelegations Of Authority; Delegations Of Authority And Organization
APPLICATION	99.205	APPLICATION FOR EXEMPTION FROM THE REQUIREMENT TO FILE A SUPPLEMENTAL APPLIC Dissemination Of Information On Unapproved/New Uses For Marketed Drugs, Biologics And Devices
APPLICATION	203.70	APPLICATION FOR A REWARD Prescription Drug Marketing
APPLICATION	314.50	CONTENT AND FORMAT OF AN APPLICATION General Provisions; Applications For FDA Approval To Market A New Drug

APPLICATION, PMA	814.9	CONFIDENTIALITY OF DATA AND INFORMATION IN A PREMARKET APPROVAL APPLICATION (PMA) General; Premarket Approval Of Medical Devices
APPLICATION OF REGULATIONS	20.25	RETROACTIVE APPLICATION OF REGULATIONS General Policy; Public Information
APPLICATION, SUPPLEMENTAL	99.305	EXEMPTION FROM THE REQUIREMENT TO FILE A SUPPLEMENTAL APPLICATION Dissemination Of Information On Unapproved/New Uses For Marketed Drugs, Biologics and Devices
APPLICATION, TERRITORIAL	26.79	TERRITORIAL APPLICATION Framework Provisions; Mutual Recognition Of Pharmaceutical GMP Reports, Medical Device Quality System Audit Reports
APPLICATION, TIME FOR	1309.31	TIME FOR APPLICATION FOR REGISTRATION; EXPIRATION DATE Registration Of Manufacturers, Distributors, Importers And Exporters Of List I Chemicals
APPLICATION, VETERINARY	1308.25	EXCLUSION OF A VETERINARY ANABOLIC STEROID IMPLANT PRODUCT; APPLICATION Schedules Of Controlled Substances
APPLICATIONS	5.202	ISSUANCE OF NOTICES OF OPPORTUNITY FOR A HEARING ON PROPOSALS FOR DENIAL OF APPROVAL OF APPLICATIONS FOR LICENSES Biologics; Redelegations Of Authority; Delegations Of Authority And Organization
APPLICATIONS	5.413	APPROVAL, DISAPPROVAL, OR WITHDRAWAL OF APPROVAL OF APPLICATIONS AND ENTERING INTO AGREEMENTS FOR INVESTIGATIONAL DEVICE EXEMPTIONS Medical Devices And Radiological Health; Redelegations Of Authority; Delegations Of Authority And Organization
APPLICATIONS	5.501	APPROVAL OF NEW ANIMAL DRUG APPLICATIONS, MEDICATED FEED MILL LICENSE APPLICATIONS AND THEIR SUPPLEMENTS Animal Drugs; Redelegations Of Authority; Delegations Of Authority And Organization
APPLICATIONS	5.502	ISSUANCE OF NOTICES, PROPOSALS AND ORDERS RELATING TO NEW ANIMAL DRUGS AND MEDICATED FEED MILL LICENSE APPLICATIONS Animal Drugs; Redelegations Of Authority; Delegations Of Authority And Organization
APPLICATIONS	203.11	APPLICATIONS FOR REIMPORTATION TO PROVIDE EMERGENCY MEDICAL CARE Prescription Drug Marketing
APPLICATIONS	312.84	RISK-BENEFIT ANALYSIS IN REVIEW OF MARKETING APPLICATIONS FOR DRUGS TO TREAT LIFE-THREATENING AND SEVERELY-DEBILITATING ILLNESSES Drugs Intended To Treat Life-threatening And Severely Debilitating Illnesses; Investigational New Drug Application
APPLICATIONS	314.92	DRUG PRODUCTS FOR WHICH ABBREVIATED APPLICATIONS MAY BE SUBMITTED Abbreviated Applications; Applications For FDA Approval To Market A New Drug
APPLICATIONS	314.94	CONTENT AND FORMAT OF AN ABBREVIATED APPLICATION Abbreviated Applications; Applications For FDA Approval To Market A New Drug
APPLICATIONS	314.96	AMENDMENTS TO AN UNAPPROVED ABBREVIATED APPLICATION Abbreviated Applications; Applications For FDA Approval To Market A New Drug
APPLICATIONS	314.97	SUPPLEMENTS AND OTHER CHANGES TO AN APPROVED ABBREVIATED APPLICATION Abbreviated Applications; Applications For FDA Approval To Market A New Drug
APPLICATIONS	314.98	OTHER RESPONSIBILITIES OF AN APPLICANT OF AN ABBREVIATED APPLICATION Abbreviated Applications; Applications For FDA Approval To Market A New Drug
APPLICATIONS	314.100	TIME FRAMES FOR REVIEWING APPLICATIONS AND ABBREVIATED APPLICATIONS FDA Action On Applications; Applications For FDA Approval To Market A New Drug
APPLICATIONS	314.440	ADDRESSES FOR APPLICATIONS AND ABBREVIATED APPLICATIONS Miscellaneous Provisions; Applications For FDA Approval To Market A New Drug
APPLICATIONS	316.52	AVAILABILITY FOR PUBLIC DISCLOSURE OF DATA AND INFORMATION IN REQUESTS AND APPLICATIONS Availability Of Information; Orphan Drugs
APPLICATIONS	510.600	NAMES, ADDRESSES, AND DRUG LABELER CODES OF SPONSORS OF APPROVED APPLICATIONS Sponsors Of Approved Applications; New Animal Drugs
APPLICATIONS	514.1	APPLICATIONS General Provisions; New Animal Drug Applications
APPLICATIONS	514.6	AMENDED APPLICATIONS General Provisions; New Animal Drug Applications
APPLICATIONS	514.15	UNTRUE STATEMENTS IN APPLICATIONS General Provisions; New Animal Drug Applications
APPLICATIONS	514.100	EVALUATION AND COMMENT ON APPLICATIONS Administrative Actions On Applications; New Animal Drug Applications

APPLICATIONS	514.105	APPROVAL OF APPLICATIONS Administrative Actions On Applications; New Animal Drug Applications
APPLICATIONS	514.106	APPROVAL OF SUPPLEMENTAL APPLICATIONS Administrative Actions On Applications; New Animal Drug Applications
APPLICATIONS	514.110	REASONS FOR REFUSING TO FILE APPLICATIONS Administrative Actions On Applications; New Animal Drug Applications
APPLICATIONS	514.115	WITHDRAWAL OF APPROVAL OF APPLICATIONS Administrative Actions On Applications; New Animal Drug Applications
APPLICATIONS	515.10	MEDICATED FEED MILL LICENSE APPLICATIONS Applications; Medicated Feed Mill License
APPLICATIONS	515.11	SUPPLEMENTAL MEDICATED FEED MILL LICENSE APPLICATIONS Applications; Medicated Feed Mill License
APPLICATIONS	515.20	APPROVAL OF MEDICATED FEED MILL LICENSE APPLICATIONS Administrative Actions On Licenses; Medicated Feed Mill License
APPLICATIONS	601.2	APPLICATIONS FOR BIOLOGICS LICENSES; PROCEDURE FOR FILING General Provisions; Licensing
APPLICATIONS	601.51	CONFIDENTIALITY OF DATA AND INFORMATION IN APPLICATIONS FOR ESTABLISHMENT AND PRODUCT LICENSES Confidentiality Of Information; Licensing
APPLICATIONS	812.30	FDA ACTION ON APPLICATIONS Application And Administrative Action; Investigational Device Exemptions
APPLICATIONS	812.35	SUPPLEMENTAL APPLICATIONS Application And Administrative Action; Investigational Device Exemptions
APPLICATIONS	814.104	ORIGINAL APPLICATIONS Humanitarian Use Devices; Premarket Approval Of Medical Devices
APPLICATIONS	814.108	SUPPLEMENTAL APPLICATIONS Humanitarian Use Devices; Premarket Approval Of Medical Devices
APPLICATIONS	1301.16	AMENDMENTS TO AND WITHDRAWAL OF APPLICATIONS Registration Of Manufacturers, Distributors, And Dispensers Of Controlled Substances
APPLICATIONS	1301.17	SPECIAL PROCEDURES FOR CERTAIN APPLICATIONS Registration Of Manufacturers, Distributors, And Dispensers Of Controlled Substances
APPLICATIONS	1301.32	ACTION ON APPLICATIONS FOR RESEARCH IN SCHEDULE I SUBSTANCES Action On Application For Registration: Revocation Or Suspension Of Registration; Registration Of Manufacturers, Distributors, And Dispensers Of Controlled Substances
APPLICATIONS, ANDA	5.503	SUBMISSION OF AND EFFECTIVE APPROVAL DATES FOR ABBREVIATED NEW ANIMAL DRUG APPLICATIONS Animal Drugs; Redelegations Of Authority; Delegations Of Authority And Organization
APPLICATIONS, DEFECTIVE	1301.14	FILING OF APPLICATION; ACCEPTANCE FOR FILING; DEFECTIVE APPLICATIONS Registration Of Manufacturers, Distributors, And Dispensers Of Controlled Substances
APPLICATIONS, DEFECTIVE	1309.34	ACCEPTANCE FOR FILING; DEFECTIVE APPLICATIONS Registration Of Manufacturers, Distributors, Importers And Exporters Of List I Chemicals
APPLICATIONS, EXEMPTION	99.403	TERMINATION OF APPROVALS OF APPLICATIONS FOR EXEMPTION Dissemination Of Information On Unapproved/New Uses For Marketed Drugs, Biologics and Devices
APPLICATIONS, FILING	1301.14	FILING OF APPLICATION; ACCEPTANCE FOR FILING; DEFECTIVE APPLICATIONS Registration Of Manufacturers, Distributors, And Dispensers Of Controlled Substances
APPLICATIONS, FILING	1309.33	FILING OF APPLICATIONS; JOINT FILINGS Registration Of Manufacturers, Distributors, Importers And Exporters Of List I Chemicals
APPLICATIONS, NADA	5.501	APPROVAL OF NEW ANIMAL DRUG APPLICATIONS AND THEIR SUPPLEMENTS Animal Drugs; Redelegations Of Authority; Delegations Of Authority And Organization
APPLICATIONS, NDA	5.103	APPROVAL OF NEW DRUG APPLICATIONS AND THEIR SUPPLEMENTS Human Drugs; Redelegations Of Authority; Delegations Of Authority And Organization
APPLICATIONS, NDA	5.105	ISSUANCE OF NOTICES RELATING TO PROPOSALS TO REFUSE APPROVAL OR TO WITHDRAW APPROVAL OF NEW DRUG APPLICATIONS AND THEIR SUPPLEMENTS Human Drugs; Redelegations Of Authority; Delegations Of Authority And Organization
APPLICATIONS, NDA	5.106	SUBMISSION OF AND EFFECTIVE APPROVAL DATES FOR ABBREVIATED NEW DRUG APPLICATIONS AND CERTAIN NEW DRUG APPLICATIONS Human Drugs; Redelegations Of Authority; Delegations Of Authority And Organization

APPLICATIONS, NDA	310.201	EXEMPTION FOR CERTAIN DRUGS LIMITED BY NEW DRUG APPLICATIONS TO PRESCRIPTION SALE New Drugs Exempted From Prescription-Dispensing Requirements; New Drugs
APPLICATIONS, NDA	310.303	CONTINUATION OF LONG-TERM STUDIES, RECORDS, AND REPORTS ON CERTAIN DRUGS FOR WHICH NEW DRUG APPLICATIONS HAVE BEEN APPROVED Records And Reports; New Drugs
APPLICATIONS, NDA	310.305	RECORDS AND REPORTS CONTAINING ADVERSE DRUG EXPERIENCES ON MARKETED PRESCRIPTION DRUGS FOR HUMAN USE WITHOUT APPROVED NEW DRUG APPLICATIONS Records And Reports; New Drugs
APPLICATIONS, NDA	514.8	SUPPLEMENTAL NEW ANIMAL DRUG APPLICATIONS General Provisions; New Animal Drug Applications
APPLICATIONS, PREMARKET	5.407	APPROVAL, DISAPPROVAL, OR WITHDRAWAL OF APPROVAL OF PRODUCT DEVELOPMENT PROTOCOLS AND APPLICATIONS FOR PREMARKET APPROVAL FOR MEDICAL DEVICES Medical Devices And Radiological Health; Redelegations Of Authority; Delegations Of Authority And Organization
APPLICATIONS, SUPPLEMENTAL	514.106	APPROVAL OF SUPPLEMENTAL APPLICATIONS Administrative Actions On Applications; New Animal Drug Applications
APPLICATIONS, WITHDRAWAL	514.7	WITHDRAWAL OF APPLICATIONS WITHOUT PREJUDICE General Provisions; New Animal Drug Applications
APPLICATIONS, WITHDRAWAL	1301.16	AMENDMENTS TO AND WITHDRAWAL OF APPLICATIONS Registration Of Manufacturers, Distributors, And Dispensers Of Controlled Substances
APPLICATIONS, WITHDRAWALS	1309.36	AMENDMENTS TO AND WITHDRAWAL OF APPLICATIONS Registration Of Manufacturers, Distributors, Importers And Exporters Of List I Chemicals
APPLICATOR	868.5170	LARYNGOTRACHEAL TOPICAL ANESTHESIA APPLICATOR Therapeutic Devices; Anesthesiology Devices
APPLICATOR	872.3140	RESIN APPLICATOR Prosthetic Devices; Dental Devices
APPLICATOR	880.6025	ABSORBENT TIPPED APPLICATOR General Hospital And Personal Use Miscellaneous Devices; General Hospital And Personal Use Devices
APPLICATOR	892.5650	MANUAL RADIONUCLIDE APPLICATOR SYSTEM Therapeutic Devices; Radiology Devices
APPLICATOR	892.5700	REMOTE CONTROLLED RADIONUCLIDE APPLICATOR SYSTEM Therapeutic Devices; Radiology Devices
APPLIER	882.4175	ANEURYSM CLIP APPLIER Neurological Surgical Devices; Neurological Devices
APPRAISEMENT	1316.74	APPRAISEMENT Seizure, Forfeiture, And Disposition Of Property; Administrative Functions, Practices, And Procedures
APPROVABLE	50.54	CLINICAL INVESTIGATIONS NOT OTHERWISE APPROVABLE THAT PRESENT AN OPPORTUNITY TO UNDERSTAND, PREVENT, OR ALLEVIATE A SERIOUS PROBLEM AFFECTING THE HEALTH OR WELFARE OF CHILDREN Additional Safeguards For Children In Clinical Investigations; Protection Of Human Subjects
APPROVABLE LETTER	314.110	APPROVABLE LETTER TO THE APPLICANT FDA Action On Applications; Applications For FDA Approval To Market A New Drug
APPROVABLE LETTER	314.120	NOT APPROVABLE LETTER TO THE APPLICANT FDA Action On Applications; Applications For FDA Approval To Market A New Drug
APPROVAL	5.501	APPROVAL OF NEW ANIMAL DRUG APPLICATIONS, MEDICATED FEED MILL LICENSE APPLICATIONS AND THEIR SUPPLEMENTS Animal Drugs; Redelegations Of Authority; Delegations Of Authority And Organization
APPROVAL	5.102	AUTHORITY TO APPROVE AND WITHDRAW APPROVAL OF A CHARGE FOR INVESTIGATIONAL NEW DRUGS Human Drugs; Redelegations Of Authority; Delegations Of Authority And Organization
APPROVAL	5.103	APPROVAL OF NEW DRUG APPLICATIONS AND THEIR SUPPLEMENTS Human Drugs; Redelegations Of Authority; Delegations Of Authority And Organization
APPROVAL	5.105	ISSUANCE OF NOTICES RELATING TO PROPOSALS TO REFUSE APPROVAL OR TO WITHDRAW APPROVAL OF NEW DRUG APPLICATIONS AND THEIR SUPPLEMENTS Human Drugs; Redelegations Of Authority; Delegations Of Authority And Organization
APPROVAL	5.202	ISSUANCE OF NOTICES OF OPPORTUNITY FOR HEARING ON PROPOSALS FOR DENIAL OF APPROVAL OF APPLICATIONS FOR LICENSES OR REVOCATION OF LICENSES AND CERTAIN NOTICES Biologics; Redelegations Of Authority; Delegations Of Authority And Organization

APPROVAL	5.304	APPROVAL OF SCHOOLS PROVIDING FOOD-PROCESSING INSTRUCTION Foods And Cosmetics; Redelegations Of Authority; Delegations Of Authority And Organization
APPROVAL	5.407	APPROVAL, DISAPPROVAL, OR WITHDRAWAL OF APPROVAL OF PRODUCT DEVELOPMENT PROTOCOLS AND APPLICATIONS FOR PREMARKET APPROVAL FOR MEDICAL DEVICES Medical Devices And Radiological Health; Redelegations Of Authority; Delegations Of Authority And Organization
APPROVAL	5.413	APPROVAL, DISAPPROVAL, OR WITHDRAWAL OF APPROVAL OF APPLICATIONS AND ENTERING INTO AGREEMENTS FOR INVESTIGATIONAL DEVICE EXEMPTIONS Medical Devices And Radiological Health; Redelegations Of Authority; Delegations Of Authority And Organization
APPROVAL	56.111	CRITERIA FOR IRB APPROVAL OF RESEARCH IRB Functions And Operations; Institutional Review Boards
APPROVAL	56.113	SUSPENSION OR TERMINATION OF IRB APPROVAL OF RESEARCH IRB Functions And Operations; Institutional Review Boards
APPROVAL	211.84	TESTING AND APPROVAL OR REJECTION OF COMPONENTS, DRUG PRODUCT CONTAINERS, AND CLOSURES Current Good Manufacturing Practice For Finished Pharmaceuticals
APPROVAL	314.105	APPROVAL OF AN APPLICATION OR ABBREVIATED APPLICATION FDA Action On Applications; Applications For FDA Approval To Market A New Drug
APPROVAL	314.150	WITHDRAWAL OF APPROVAL OF AN APPLICATION OR ABBREVIATED APPLICATION FDA Action On Applications; Applications For FDA Approval To Market A New Drug
APPROVAL	314.151	WITHDRAWAL OF APPROVAL OF AN ABBREVIATED NEW DRUG APPLICATION UNDER SECTION 505(J)(5) FDA Action On Applications; Applications For FDA Approval To Market A New Drug
APPROVAL	314.152	NOTICE OF WITHDRAWAL OF APPROVAL OF AN APPLICATION OR ABBREVIATED APPLICATION FOR A NEW DRUG FDA Action On Applications; Applications For FDA Approval To Market A New Drug
APPROVAL	314.153	SUSPENSION OF APPROVAL OF AN ABBREVIATED NEW DRUG APPLICATION FDA Action On Applications; Applications For FDA Approval To Market A New Drug
APPROVAL	314.160	APPROVAL OF AN APPLICATION OR ABBREVIATED APPLICATION FOR WHICH APPROVAL WAS PREVIOUSLY REFUSED, SUSPENDED, OR WITHDRAWN FDA Action On Applications; Applications For FDA Approval To Market A New Drug
APPROVAL	314.510	APPROVAL BASED ON A SURROGATE ENDPOINT OR ON AN EFFECT ON A CLINICAL ENDPOINT Accelerated Approval of New Drugs; Applications For FDA Approval To Market A New Drug
APPROVAL	314.520	APPROVAL WITH RESTRICTIONS TO ASSURE SAFE USE Accelerated Approval of New Drugs; Applications For FDA Approval To Market A New Drug
APPROVAL	514.105	APPROVAL OF APPLICATIONS Administrative Actions On Applications; New Animal Drug Applications
APPROVAL	514.106	APPROVAL OF SUPPLEMENTAL APPLICATIONS Administrative Actions On Applications; New Animal Drug Applications
APPROVAL	514.115	WITHDRAWAL OF APPROVAL OF APPLICATIONS Administrative Actions On Applications; New Animal Drug Applications
APPROVAL	514.116	NOTICE OF WITHDRAWAL OF APPROVAL OF APPLICATION Administrative Actions On Applications; New Animal Drug Applications
APPROVAL	514.120	REVOCATION OF ORDER REFUSING TO APPROVE AN APPLICATION OR SUSPENDING OR WITHDRAWING APPROVAL OF AN APPLICATION Administrative Actions On Applications; New Animal Drug Applications
APPROVAL	515.20	APPROVAL OF MEDICATED FEED MILL LICENSE APPLICATIONS Administrative Actions On Licenses; Medicated Feed Mill License
APPROVAL	515.22	SUSPENSION AND/OR REVOCATION OF APPROVAL OF A MEDICATED FEED MILL LICENSE Administrative Actions On Licenses; Medicated Feed Mill License
APPROVAL	601.41	APPROVAL BASED ON A SURROGATE ENDPOINT OR EFFECT ON A CLINICAL ENDPOINT Accelerated Approval Of Biological Products; Licensing
APPROVAL	601.42	APPROVAL WITH RESTRICTIONS TO ASSURE SAFE USE Accelerated Approval Of Biological Products; Licensing
APPROVAL	812.42	FDA AND IRB APPROVAL Responsibilities Of Sponsors; Investigational Device Exemptions
APPROVAL	812.62	IRB APPROVAL IRB Review And Approval; Investigational Device Exemptions
APPROVAL	900.3	APPLICATION FOR APPROVAL AS AN ACCREDITATION BODY Accreditation; Mammography
APPROVAL	900.6	WITHDRAWAL OF APPROVAL Accreditation; Mammography

APPROVAL	900.13	REVOCATION OF ACCREDITATION AND OF ACCREDITATION BODY APPROVAL Quality Standards And Certification; Mammography
APPROVAL	900.21	APPLICATION FOR APPROVAL AS A CERTIFICATION AGENCY States As Certifiers; Mammography
APPROVAL	900.24	WITHDRAWAL OF APPROVAL States As Certifiers; Mammography
APPROVAL	1004.6	APPROVAL OF PLANS Repurchase, Repairs, Or Replacement Of Electronic Products
APPROVAL	1240.83	APPROVAL OF WATERING POINTS Source And Use Of Potable Water; Control Of Communicable Diseases
APPROVAL	1240.90	APPROVAL OF TREATMENT ABOARD CONVEYANCES Source And Use Of Potable Water; Control Of Communicable Diseases
APPROVAL	1250.6	INSPECTION AND APPROVAL Servicing Areas For Land And Air Conveyances; Interstate Conveyance Sanitation
APPROVAL	1312.31	SCHEDULE I: APPLICATION FOR PRIOR WRITTEN APPROVAL Transshipment And In-transit Shipment Of Controlled Substances; Importation And Exportation Of Controlled Substances
APPROVAL, DATE	314.107	EFFECTIVE DATE OF APPROVAL OF A 505(B)(2) APPLICATION OR ANDA Applications For FDA Approval To Market A New Drug
APPROVAL DATES	5.106	SUBMISSION OF AND EFFECTIVE APPROVAL DATES FOR ABBREVIATED NEW DRUG APPLICATIONS Human Drugs; Redelegations Of Authority; Delegations Of Authority And Organization
APPROVAL DATES	5.503	SUBMISSION OF AND EFFECTIVE APPROVAL DATES FOR ABBREVIATED NEW ANIMAL DRUG APPLICATIONS Animal Drugs; Redelegations Of Authority; Delegations Of Authority And Organization
APPROVAL, EXCLUSIVE	316.31	SCOPE OF ORPHAN-DRUG EXCLUSIVE APPROVAL Orphan Drugs
APPROVAL, EXCLUSIVE	316.34	FDA RECOGNITION OF EXCLUSIVE APPROVAL Orphan Drugs
APPROVAL, HDE	814.118	DENIAL OF APPROVAL OR WITHDRAWAL OF APPROVAL OF AN HDE Humanitarian Use Devices; Premarket Approval Of Medical Devices
APPROVAL, HDE	814.120	TEMPORARY SUSPENSION OF APPROVAL OF A HDE Humanitarian Use Devices; Premarket Approval Of Medical Devices
APPROVAL, PMA	814.45	DENIAL OF APPROVAL OF A PMA FDA Action On A PMA; Premarket Approval Of Medical Devices
APPROVAL, PMA	814.46	WITHDRAWAL OF APPROVAL OF A PMA FDA Action On A PMA; Premarket Approval Of Medical Devices
APPROVAL, PMA	814.47	TEMPORARY SUSPENSION OF APPROVAL OF A PMA FDA Action On A PMA; Premarket Approval Of Medical Devices
APPROVAL, PREMARKET	5.407	APPROVAL, DISAPPROVAL, OR WITHDRAWAL OF APPROVAL OF PRODUCT DEVELOPMENT PROTOCOLS AND APPLICATIONS FOR PREMARKET APPROVAL FOR MEDICAL DEVICES Medical Devices And Radiological Health; Redelegations Of Authority; Delegations Of Authority And Organization
APPROVAL, PREMARKET	5.408	DETERMINATIONS CONCERNING THE TYPE OF VALID SCIENTIFIC EVIDENCE SUBMITTED IN A PREMARKET APPROVAL APPLICATION Medical Devices And Radiological Health; Redelegations Of Authority; Delegations Of Authority And Organization
APPROVAL, PREMARKET	5.701	PREMARKET APPROVAL OF A PRODUCT THAT IS OR CONTAINS A BIOLOGIC, DEVICE, OR DRUG Product Designation; Redelegations Of Authority; Delegations Of Authority And Organization
APPROVAL, PREMARKET	814.9	CONFIDENTIALITY OF DATA AND INFORMATION IN A PREMARKET APPROVAL APPLICATION (PMA) FILE General; Premarket Approval Of Medical Devices
APPROVAL, PREMARKET	860.132	PROCEDURES WHEN THE COMMISSIONER INITIATES A PERFORMANCE STANDARD OR PREMARKET APPROVAL PROCEEDING UNDER SECTION 514(B) OR 515(B) Reclassification; Medical Device Classification Procedures
APPROVAL, PREMARKET	862.3	EFFECTIVE DATES OF REQUIREMENT FOR PREMARKET APPROVAL Clinical Chemistry And Clinical Toxicology Devices
APPROVAL, PREMARKET	864.3	EFFECTIVE DATES OF REQUIREMENT FOR PREMARKET APPROVAL Hematology And Pathology Devices
APPROVAL, PREMARKET	866.3	EFFECTIVE DATES OF REQUIREMENT FOR PREMARKET APPROVAL Immunology And Microbiology Devices
APPROVAL, PREMARKET	868.3	EFFECTIVE DATES OF REQUIREMENT FOR PREMARKET APPROVAL Anesthesiology Devices
APPROVAL, PREMARKET	870.3	EFFECTIVE DATES OF REQUIREMENT FOR PREMARKET APPROVAL Cardiovascular Devices
APPROVAL, PREMARKET	872.3	EFFECTIVE DATES OF REQUIREMENT FOR PREMARKET APPROVAL Dental Devices

APPROVAL, PREMARKET	874.3	EFFECTIVE DATES OF REQUIREMENT FOR PREMARKET APPROVAL Ear, Nose, And Throat Devices
APPROVAL, PREMARKET	876.3	EFFECTIVE DATES OF REQUIREMENT FOR PREMARKET APPROVAL Gastroenterology-urology Devices
APPROVAL, PREMARKET	878.3	EFFECTIVE DATES OF REQUIREMENT FOR PREMARKET APPROVAL General And Plastic Surgery Devices
APPROVAL, PREMARKET	880.3	EFFECTIVE DATES OF REQUIREMENT FOR PREMARKET APPROVAL General Hospital And Personal Use Devices
APPROVAL, PREMARKET	882.3	EFFECTIVE DATES OF REQUIREMENT FOR PREMARKET APPROVAL Neurological Devices
APPROVAL, PREMARKET	884.3	EFFECTIVE DATES OF REQUIREMENT FOR PREMARKET APPROVAL Obstetrical And Gynecological Devices
APPROVAL, PREMARKET	886.3	EFFECTIVE DATES OF REQUIREMENT FOR PREMARKET APPROVAL Ophthalmic Devices
APPROVAL, PREMARKET	888.3	EFFECTIVE DATES OF REQUIREMENT FOR PREMARKET APPROVAL Orthopedic Devices
APPROVAL, PREMARKET	890.3	EFFECTIVE DATES OF REQUIREMENT FOR PREMARKET APPROVAL Physical Medicine Devices
APPROVAL, PREMARKET	892.3	EFFECTIVE DATES OF REQUIREMENTS FOR PREMARKET APPROVAL Radiology Devices
APPROVALS, TERMINATION OF	99.403	TERMINATION OF APPROVALS OF APPLICATIONS FOR EXEMPTION Dissemination Of Information On Unapproved/New Uses For Marketed Drugs, Biologics, And Devices
APPROVE	314.125	REFUSAL TO APPROVE AN APPLICATION FDA Action On Applications; Applications For FDA Approval To Market A New Drug
APPROVE	314.127	REFUSAL TO APPROVE AN ABBREVIATED NEW DRUG APPLICATION Applications For FDA Approval To Market A New Drug
APPROVE	514.111	REFUSAL TO APPROVE AN APPLICATION Administrative Actions On Applications; New Animal Drug Applications
APPROVE	515.21	REFUSAL TO APPROVE A MEDICATED FEED MILL LICENSE APPLICATION Administrative Actions On Licenses; Medicated Feed Mill License
APPROVE	515.25	REVOCATION OF ORDER REFUSING TO APPROVE A MEDICATED FEEL MILL LICENSE APPLICATION OR SUSPENDING OR REVOKING A LICENSE Administrative Actions On Licenses; Medicated Feed Mill License
APPROVED APPLICATION	314.70	SUPPLEMENTS AND OTHER CHANGES TO AN APPROVED APPLICATION Applications; Applications For FDA Approval To Market A New Drug
APPROVED APPLICATION	314.71	PROCEDURES FOR SUBMISSION OF A SUPPLEMENT TO AN APPROVED APPLICATION Applications; Applications For FDA Approval To Market A New Drug
APPROVED APPLICATION	510.300	RECORDS AND REPORTS CONCERNING EXPERIENCE WITH NEW ANIMAL DRUGS FOR WHICH AN APPROVED APPLICATION IS IN EFFECT Records And Reports; New Animal Drugs
APPROVED APPLICATION	510.301	RECORDS AND REPORTS CONCERNING EXPERIENCE WITH ANIMAL FEEDS BEARING OR CONTAINING NEW ANIMAL DRUGS FOR WHICH AN APPROVED MEDICATED FEED MILL LICENSE APPLICATION IS IN EFFECT Records And Reports; New Animal Drugs
APPROVED APPLICATION	601.12	CHANGES TO AN APPROVED APPLICATION Biologics Licensing; Licensing
APPROVED APPLICATIONS	510.600	NAMES, ADDRESSES, AND DRUG LABELER CODES OF SPONSORS OF APPROVED APPLICATIONS Sponsors Of Approved Applications; New Animal Drugs
APPROVED COMPONENTS	211.86	USE OF APPROVED COMPONENTS, DRUG PRODUCT CONTAINERS, AND CLOSURES Current Good Manufacturing Practice For Finished Pharmaceuticals
APPROVED COMPONENTS	211.87	RETESTING OF APPROVED COMPONENTS, DRUG PRODUCT CONTAINERS, AND CLOSURES Current Good Manufacturing Practice For Finished Pharmaceuticals
APPROVED DRUG	314.170	ADULTERATION AND MISBRANDING OF AN APPROVED DRUG FDA Action On Applications; Applications For FDA Approval To Market A New Drug
APPROVED DRUGS	316.23	TIMING OF REQUESTS FOR ORPHAN-DRUG DESIGNATION, DESIGNATION OF ALREADY APPROVED DRUGS Orphan Drugs
APPROVED DRUGS	514.80	RECORDS AND REPORTS CONCERNING EXPERIENCE WITH APPROVED NEW ANIMAL DRUGS Administrative Actions On Applications; New Animal Drug Applications
APPROVED DRUGS	530.13	EXTRALABEL USE FROM COMPOUNDING OF APPROVED NEW ANIMAL AND HUMAN DRUGS Extralabel Drug Use In Animals
APPROVED LICENSE	510.305	MAINTENANCE OF COPIES OF APPROVED MEDICATED FEED MILL LICENSES TO MANUFACTURE ANIMAL FEED BEARING OR CONTAINING NEW ANIMAL DRUGS Records And Reports; New Animal Drugs

APPROVED NDA	310.305	RECORDS AND REPORTS CONTAINING ADVERSE DRUG EXPERIENCES ON MARKETED PRESCRIPTION DRUGS FOR HUMAN USE WITHOUT APPROVED NEW DRUG APPLICATIONS Records And Reports; New Drugs
APPROVED RESEARCH	56.110	EXPEDITED REVIEW PROCEDURES FOR CERTAIN KINDS OF RESEARCH INVOLVING NO MORE THAN MINIMAL RISK, AND FOR MINOR CHANGES IN APPROVED RESEARCH IRB Functions And Operations; Institutional Review Boards
APRAMYCIN	520.110	APRAMYCIN SULFATE SOLUBLE POWDER Oral Dosage Form New Animal Drugs
APRAMYCIN	556.52	APRAMYCIN Tolerances For Residues Of New Animal Drugs In Food
APRAMYCIN	558.59	APRAMYCIN New Animal Drugs For Use In Animal Feeds
APRICOTS	145.115	CANNED APRICOTS Requirements For Specific Standardized Canned Fruits; Canned Fruits
APRICOTS	145.116	ARTIFICIALLY SWEETENED CANNED APRICOTS Requirements For Specific Standardized Canned Fruits; Canned Fruits
AQUEOUS	522.90c	AMPICILLIN SODIUM FOR AQUEOUS INJECTION Implantation Or Injectable Dosage Form New Animal Drugs
AQUEOUS	522.144	ARSENAMIDE SODIUM AQUEOUS INJECTION Implantation Or Injectable Dosage Form New Animal Drugs
AQUEOUS	522.161	BETAMETHASONE ACETATE AND BETAMETHASONE DISODIUM PHOSPHATE AQUEOUS SUSPENSION Implantation Or Injectable Dosage Form New Animal Drugs
AQUEOUS	522.163	BETAMETHASONE DIPROPIONATE AND BETAMETHASONE SODIUM PHOSPHATE AQUEOUS SUSPENSION Implantation Or Injectable Dosage Form New Animal Drugs
AQUEOUS	522.380	CHLORAL HYDRATE, PENTOBARBITAL, AND MAGNESIUM SULFATE STERILE AQUEOUS SOLUTION Implantation Or Injectable Dosage Form New Animal Drugs
AQUEOUS	522.1696b	PENICILLIN G PROCAINE AQUEOUS SUSPENSION Implantation Or Injectable Dosage Form New Animal Drugs
AQUEOUS	524.1200b	KANAMYCIN OPHTHALMIC AQUEOUS SOLUTION Ophthalmic And Topical Dosage Form New Animal Drugs
AQUEOUS FOODS	176.170	COMPONENTS OF PAPER AND PAPERBOARD IN CONTACT WITH AQUEOUS AND FATTY FOODS Substances For Use Only As Components Of Paper And Paperboard; Indirect Food Additives
AQUEOUS SHUNT	886.3920	AQUEOUS SHUNT Prosthetic Devices; Ophthalmic Devices
AQUEOUS-BASED	200.51	AQUEOUS-BASED DRUG PRODUCTS FOR ORAL INHALATION General Provisions; General
ARABIC, GUM	184.1330	ACACIA (GUM ARABIC) Listing Of Specific Substances Affirmed As GRAS; Direct Food Substances Affirmed As Generally Recognized As Safe
ARABIC, GUM	582.7330	GUM ARABIC Stabilizers; Substances Generally Recognized As Safe
ARABINOGALACTAN	172.610	ARABINOGALACTAN Gums, Chewing Gum Bases And Related Substances; Food Additives Permitted For Direct Addition To Food For Human Consumption
ARCHIVES, NATIONAL	21.31	RECORDS STORED BY THE NATIONAL ARCHIVES AND RECORDS ADMINISTRATION Requirements For Specific Categories Of Records; Protection Of Privacy
ARCHIVING	892.2050	PICTURE ARCHIVING AND COMMUNICATIONS SYSTEM Diagnostic Devices; Radiology Devices
ARCOS	1304.33	REPORTS TO ARCOS Reports, Records And Reports Of Registrants
ARGININE	582.5145	ARGININE Nutrients And/or Dietary Supplements; Substances Generally Recognized As Safe
ARGON GAS	868.1075	ARGON GAS ANALYZER Diagnostic Devices; Anesthesiology Devices
ARGON LASER	874.4490	ARGON LASER FOR OTOLOGY, RHINOLOGY AND LARYNGOLOGY Surgical Devices; Ear, Nose And Throat Devices
ARGUMENT	12.96	BRIEFS AND ARGUMENT Hearing Procedures; Formal Evidentiary Public Hearing
ARIZONA	808.53	ARIZONA Listing Of Specific State And Local Exemptions; Exemption From Federal Preemption Of State & Local Medical Device Requirements
ARIZONA SPP.	866.3035	ARIZONA SPP. SEROLOGICAL REAGENTS Serological Reagents; Immunology And Microbiology Devices
ARM SLING	890.3640	ARM SLING Physical Medicine Prosthetic Devices; Physical Medicine Devices

ARRHYTHMIA	870.1025	ARRHYTHMIA DETECTOR AND ALARM Cardiovascular Diagnostic Devices; Cardiovascular Devices
ARSANILATE SODIUM	558.60	ARSANILATE SODIUM Specific New Animal Drugs For Use In Animal Feeds; New Animal Drugs For Use In Animal Feeds
ARSANILIC ACID	558.62	ARSANILIC ACID Specific New Animal Drugs For Use In Animal Feeds; New Animal Drugs For Use In Animal Feeds
ARSENAMIDE SODIUM	522.144	ARSENAMIDE SODIUM AQUEOUS INJECTION Implantation Or Injectable Dosage Form New Animal Drugs
ARSENIC	556.60	ARSENIC Tolerances For Residues Of New Animal Drugs In Food
ARSENIC	862.3120	ARSENIC TEST SYSTEM Clinical Toxicology Test Systems; Clinical Chemistry And Clinical Toxicology Devices
ARTERIAL	868.1100	ARTERIAL BLOOD SAMPLING KIT Diagnostic Devices; Anesthesiology Devices
ARTERIAL	870.3300	ARTERIAL EMBOLIZATION DEVICE Cardiovascular Prosthetic Devices; Cardiovascular Devices
ARTERIAL	870.4260	CARDIOPULMONARY BYPASS ARTERIAL LINE BLOOD FILTER Cardiovascular Surgical Devices; Cardiovascular Devices
ARTERY	870.4875	INTRALUMINAL ARTERY STRIPPER Cardiovascular Surgical Devices; Cardiovascular Devices
ARTERY	882.5175	CAROTID ARTERY CLAMP Neurological Therapeutic Devices; Neurological Devices
ARTHROSCOPE	888.1100	ARTHROSCOPE Diagnostic Devices; Orthopedic Devices
ARTICLE	58.105	TEST AND CONTROL ARTICLE CHARACTERIZATION Test And Control Articles; Good Laboratory Practice For Nonclinical Laboratory Studies
ARTICLE	58.107	TEST AND CONTROL ARTICLE HANDLING Test And Control Articles; Good Laboratory Practice For Nonclinical Laboratory Studies
ARTICLES	58.47	FACILITIES FOR HANDLING TEST AND CONTROL ARTICLES Facilities; Good Laboratory Practice For Nonclinical Laboratory Studies
ARTICLES	58.113	MIXTURE OF ARTICLES WITH CARRIERS Test And Control Articles; Good Laboratory Practice For Nonclinical Laboratory Studies
ARTICLES	177.1460	MELAMINE-FORMALDEHYDE RESINS IN MOLDED ARTICLES Substances For Use As Basic Components Of Single And Repeated Use Food Contact Surfaces; Indirect Food Additives: Polymers
ARTICLES	177.1900	UREA-FORMALDEHYDE RESINS IN MOLDED ARTICLES Substances For Use As Basic Components Of Single And Repeated Use Food Contact Surfaces; Indirect Food Additives: Polymers
ARTICLES	177.2410	PHENOLIC RESINS IN MOLDED ARTICLES Substances For Use Only As Components Of Articles Intended For Repeated Use; Indirect Food Additives: Polymers
ARTICLES	177.2600	RUBBER ARTICLES INTENDED FOR REPEATED USE Substances For Use Only As Components Of Articles Intended For Repeated Use; Indirect Food Additives: Polymers
ARTICLES	178.3910	SURFACE LUBRICANTS USED IN THE MANUFACTURE OF METALLIC ARTICLES Certain Adjuvants And Production Aids; Indirect Food Additives: Adjuvants, Production Aids, And Sanitizers
ARTICLES	801.405	LABELING OF ARTICLES INTENDED FOR LAY USE IN THE REPAIRING AND/OR REFITTING OF DENTURES Special Requirements For Specific Devices; Labeling
ARTICLES, FOOD-CONTACT	170.39	THRESHOLD OF REGULATION FOR SUBSTANCES USED IN FOOD-CONTACT ARTICLES Food Additives
ARTICLES, FOOD-CONTACT	171.8	THRESHOLD OF REGULATIONS FOR SUBSTANCES USED IN FOOD-CONTACT ARTICLES Food Additive Petitions
ARTICLES FOOD-CONTACT	174.6	THRESHOLD OF REGULATIONS FOR SUBSTANCES USED IN FOOD-CONTACT ARTICLES Indirect Food Additives: General
ARTICULATION	872.6140	ARTICULATION PAPER Miscellaneous Devices; Dental Devices
ARTICULATOR	872.3150	ARTICULATOR Prosthetic Devices; Dental Devices
ARTIFICIAL EYE	886.3200	ARTIFICIAL EYE Prosthetic Devices; Ophthalmic Devices
ARTIFICIAL NOSE	868.5375	HEAT AND MOISTURE CONDENSER (ARTIFICIAL NOSE) Therapeutic Devices; Anesthesiology Devices

ARTIFICIAL LARYNX	874.3375	BATTERY-POWERED ARTIFICIAL LARYNX Prosthetic Devices; Ear, Nose, And Throat Devices
ARTIFICIAL TOOTH	872.3900	POSTERIOR ARTIFICIAL TOOTH WITH A METAL INSERT Prosthetic Devices; Dental Devices
ARTIFICIAL TOOTH	872.3910	BACKING AND FACING FOR AN ARTIFICIAL TOOTH Prosthetic Devices; Dental Devices
ARTIFICIALLY SWEETENED	145.116	ARTIFICIALLY SWEETENED CANNED APRICOTS Requirements For Specific Standardized Canned Fruits; Canned Fruits
ARTIFICIALLY SWEETENED	145.126	ARTIFICIALLY SWEETENED CANNED CHERRIES Requirements For Specific Standardized Canned Fruits; Canned Fruits
ARTIFICIALLY SWEETENED	145.131	ARTIFICIALLY SWEETENED CANNED FIGS Requirements For Specific Standardized Canned Fruits; Canned Fruits
ARTIFICIALLY SWEETENED	145.136	ARTIFICIALLY SWEETENED CANNED FRUIT COCKTAIL Requirements For Specific Standardized Canned Fruits; Canned Fruits
ARTIFICIALLY SWEETENED	145.171	ARTIFICIALLY SWEETENED CANNED PEACHES Requirements For Specific Standardized Canned Fruits; Canned Fruits
ARTIFICIALLY SWEETENED	145.176	ARTIFICIALLY SWEETENED CANNED PEARS Requirements For Specific Standardized Canned Fruits; Canned Fruits
ARTIFICIALLY SWEETENED	145.181	ARTIFICIALLY SWEETENED CANNED PINEAPPLE Requirements For Specific Standardized Canned Fruits; Canned Fruits
ARTIFICIALLY SWEETENED	146.121	FROZEN CONCENTRATE FOR ARTIFICIALLY SWEETENED LEMONADE Canned Fruit Juices
ARTIFICIALLY SWEETENED	150.141	ARTIFICIALLY SWEETENED FRUIT JELLY Fruit Butters, Jellies, Preserves, And Related Products
ARTIFICIALLY SWEETENED	150.161	ARTIFICIALLY SWEETENED FRUIT PRESERVES AND JAMS Fruit Butters, Jellies, Preserves, And Related Products
ARYLETHERKETONE	177.2415	POLY(ARYLETHERKETONE) RESINS Substances For Use Only As Components Of Articles Intended For Repeated Use; Indirect Food Additives: Polymers
ASCA	866.5785	ANTI-SACCHAROMYCES CEREVISIAE ANTIBODY (ASCA) TEST SYSTEMS Immunological Test Systems; Immunology And Microbiology Devices
ASCORBATE	182.3189	CALCIUM ASCORBATE Chemical Preservatives; Substances Generally Recognized As Safe
ASCORBATE	182.3731	SODIUM ASCORBATE Chemical Preservatives; Substances Generally Recognized As Safe
ASCORBATE	184.1307a	FERROUS ASCORBATE Listing Of Specific Substances Affirmed As GRAS; Direct Food Substances Affirmed As Generally Recognized as Safe
ASCORBATE	582.3189	CALCIUM ASCORBATE Chemical Preservatives; Substances Generally Recognized As Safe
ASCORBATE	582.3731	SODIUM ASCORBATE Chemical Preservatives; Substances Generally Recognized As Safe
ASCORBIC ACID	172.315	NICOTINAMIDE-ASCORBIC ACID COMPLEX Special Dietary And Nutritional Additives; Food Additives Permitted For Direct Addition To Food For Human Consumption
ASCORBIC ACID	182.3013	ASCORBIC ACID Chemical Preservatives; Substance Generally Recognized As Safe
ASCORBIC ACID	182.8013	ASCORBIC ACID Nutrients; Substances Generally Recognized As Safe
ASCORBIC ACID	582.3013	ASCORBIC ACID Chemical Preservatives; Substances Generally Recognized As Safe
ASCORBIC ACID	582.5013	ASCORBIC ACID Nutrients And/or Dietary Supplements; Substances Generally Recognized As Safe
ASCORBIC ACID	862.1095	ASCORBIC ACID TEST SYSTEM Clinical Chemistry Test Systems; Clinical Chemistry And Clinical Toxicology Devices
ASCORBYL PALMITATE	182.3149	ASCORBYL PALMITATE Chemical Preservatives; Substances Generally Recognized As Safe
ASCORBYL PALMITATE	582.3149	ASCORBYL PALMITATE Chemical Preservatives; Substances Generally Recognized As Safe
ASIAGO CHEESE	133.102	ASIAGO FRESH AND ASIAGO SOFT CHEESE Requirements For Specific Standardized Cheese And Related Products; Cheeses And Related Cheese Products
ASIAGO MEDIUM CHEESE	133.103	ASIAGO MEDIUM CHEESE Requirements For Specific Standardized Cheese And Related Products; Cheeses And Related Cheese Products
ASIAGO OLD CHEESE	133.104	ASIAGO OLD CHEESE Requirements For Specific Standardized Cheese And Related Products; Cheeses And Related Cheese Products
ASPARTAME	172.804	ASPARTAME Multipurpose Additives; Food Additives Permitted For Direct Addition To Food For Human Consumption

ASPARTAME	201.21	DECLARATION OF PRESENCE OF PHENYLALANINE AS A COMPONENT OF ASPARTAME IN OVER-THE-COUNTER AND PRESCRIPTION DRUGS FOR HUMAN USE General Labeling Provisions; Labeling
ASPARTATE	862.1100	ASPARTATE AMINO TRANSFERASE (AST/SGOT) TEST SYSTEM Clinical Chemistry Test Systems; Clinical Chemistry And Clinical Toxicology Devices
ASPARTIC ACID	582.5017	ASPARTIC ACID Nutrients And/or Dietary Supplements; Substances Generally Recognized As Safe
ASPERGILLUS NIGER	173.120	CARBOHYDRASE AND CELLULASE DERIVED FROM ASPERGILLUS NIGER Enzyme Preparations And Microorganisms; Secondary Direct Food Additives Permitted In Food For Human Consumption
ASPERGILLUS SPP.	866.3040	ASPERGILLUS SPP. SEROLOGICAL REAGENTS Serological Reagents; Immunology And Microbiology Devices
ASPIRATION	886.4150	VITREOUS ASPIRATION AND CUTTING INSTRUMENT Surgical Devices; Ophthalmic Devices
ASPIRATOR	884.1050	ENDOCERVICAL ASPIRATOR Obstetrical And Gynecological Diagnostic Devices; Obstetrical And Gynecological Devices
ASPIRATOR	884.1060	ENDOMETRIAL ASPIRATOR Obstetrical And Gynecological Diagnostic Devices; Obstetrical And Gynecological Devices
ASPIRIN	520.1409	METHYLPREDNISOLONE, ASPIRIN TABLETS Oral Dosage Form New Animal Drugs
ASSAY	864.7040	ADENOSINE TRIPHOSPHATE RELEASE ASSAY Hematology Kits And Packages; Hematology And Pathology Devices
ASSAY	864.7060	ANTITHROMBIN III ASSAY Hematology Kits And Packages; Hematology And Pathology Devices
ASSAY	864.7100	RED BLOOD CELL ENZYME ASSAY Hematology Kits And Packages; Hematology And Pathology Devices
ASSAY	864.7250	ERYTHROPOIETIN ASSAY Hematology Kits And Packages; Hematology And Pathology Devices
ASSAY	864.7320	FIBRINOGEN/FIBRIN DEGRADATION PRODUCTS ASSAY Hematology Kits And Packages; Hematology And Pathology Devices
ASSAY	864.7360	ERYTHROCYTIC GLUCOSE-6-PHOSPHATE DEHYDROGENASE ASSAY Hematology Kits And Packages; Hematology And Pathology Devices
ASSAY	864.7375	GLUTATHIONE REDUCTASE ASSAY Hematology Kits And Packages; Hematology And Pathology Devices
ASSAY	864.7415	ABNORMAL HEMOGLOBIN ASSAY Hematology Kits And Packages; Hematology And Pathology Devices
ASSAY	864.7425	CARBOXYHEMOGLOBIN ASSAY Hematology Kits And Packages; Hematology And Pathology Devices
ASSAY	864.7455	FETAL HEMOGLOBIN ASSAY Hematology Kits And Packages; Hematology And Pathology Devices
ASSAY	864.7470	GLYCOSYLATED HEMOGLOBIN ASSAY Hematology Kits And Packages; Hematology And Pathology Devices
ASSAY	864.7490	SULFHEMOGLOBIN ASSAY Hematology Kits And Packages; Hematology And Pathology Devices
ASSAY	864.7525	HEPARIN: ASSAY Hematology Kits And Packages; Hematology And Pathology Devices
ASSAY	866.2350	MICROBIOLOGICAL ASSAY CULTURE MEDIUM Microbiology Devices; Immunology And Microbiology Devices
ASSAYS	864.7500	WHOLE BLOOD HEMOGLOBIN ASSAYS Hematology Kits And Packages; Hematology And Pathology Devices
ASSAYS, LABORATORY	225.158	LABORATORY ASSAYS Product Quality Assurance; Current Good Manufacturing Practice For Medicated Feeds
ASSEMBLY	868.6885	MEDICAL GAS YOKE ASSEMBLY Miscellaneous; Anesthesiology Devices
ASSENT	50.55	REQUIREMENTS FOR PERMISSION BY PARENTS OR GUARDIANS AND FOR ASSENT BY CHILDREN Additional Safeguards For Children In Clinical Investigations; Protection Of Human Subjects
ASSESSMENT, CONFORMITY	26.67	SUSPENSION OF LISTED CONFORMITY ASSESSMENT BODIES Framework Provisions; Mutual Recognition Of Pharmaceutical GMP Reports, Medical Device Quality System Audit Reports
ASSESSMENT, CONFORMITY	26.68	WITHDRAWAL OF LISTED CONFORMITY ASSESSMENT BODIES Framework Provisions; Mutual Recognition Of Pharmaceutical GMP Reports, Medical Device Quality System Audit Reports
ASSESSMENT, CONFORMITY	26.69	MONITORING OF CONFORMITY ASSESSMENT BODIES Framework Provisions; Mutual Recognition Of Pharmaceutical GMP Reports, Medical Device Quality System Audit Reports

ASSESSMENT, CONFORMITY	26.70	CONFORMITY ASSESSMENT BODIES Framework Provisions; Mutual Recognition Of Pharmaceutical GMP Reports, Medical Device Quality System Audit Reports
ASSESSMENT, ENVIRONMENTAL	25.20	ACTIONS REQUIRING PREPARATION OF AN ENVIRONMENTAL ASSESSMENT Agency Actions Requiring Environmental Consideration; Environmental Impact Considerations
ASSESSMENT, EQUIVALENCE	26.6	EQUIVALENCE ASSESSMENT Specific Sector Provisions For Pharmaceutical GMPs; Mutual Recognition Of Pharmaceutical GMP Reports, Medical Device Quality System Audit Reports
ASSESSMENT, EQUIVALENCE	26.7	PARTICIPATION IN THE EQUIVALENCE ASSESSMENT AND DETERMINATION Specific Sector Provisions For Pharmaceutical GMPs; Mutual Recognition Of Pharmaceutical GMP Reports, Medical Device Quality System Audit Reports
ASSESSMENT, EQUIVALENCE	26.39	EQUIVALENCE ASSESSMENT Specific Sector Provisions For Medical Devices; Mutual Recognition Of Pharmaceutical GMP Reports, Medical Device Quality System Audit Reports
ASSESSMENTS	17.34	DETERMINING THE AMOUNT OF PENALTIES AND ASSESSMENTS Civil Money Penalties Hearings
ASSESSMENTS, ENVIRONMENTAL	25.31	ENVIRONMENTAL ASSESSMENTS Preparation Of Environmental Documents; Environmental Impact Considerations
ASSESSMENTS, ENVIRONMENTAL	25.51	ENVIRONMENTAL ASSESSMENTS AND FINDINGS OF NO SIGNIFICANT IMPACT Public Participation And Notification Of Environmental Documents; Environmental Impact Considerations
ASSISTANCE	5.606	ACCEPTANCE OF ASSISTANCE FROM STATE AND LOCAL AUTHORITIES FOR ENFORCEMENT OF RADIATION CONTROL LEGISLATION AND REGULATIONS Radiation Control; Redelegations Of Authority; Delegations Of Authority And Organization
ASSISTANCE	60.10	FDA ASSISTANCE ON ELIGIBILITY Eligibility Assistance; Patent Term Restoration
ASSISTED REPRODUCTION	884.6100	ASSISTED REPRODUCTION NEEDLES Assisted Reproduction Devices; Obstetrical And Gynecological Devices
ASSISTED REPRODUCTION	884.6110	ASSISTED REPRODUCTION CATHETERS Assisted Reproduction Devices; Obstetrical And Gynecological Devices
ASSISTED REPRODUCTION	84.6120	ASSISTED REPRODUCTION ACCESSORIES Assisted Reproduction Devices; Obstetrical And Gynecological Devices
ASSISTED REPRODUCTION	884.6130	ASSISTED REPRODUCTION MICROTOOLS Assisted Reproduction Devices; Obstetrical And Gynecological Devices
ASSISTED REPRODUCTION	884.6140	ASSISTED REPRODUCTION MICROPIPETTE FABRICATION INSTRUMENTS Assisted Reproduction Devices; Obstetrical And Gynecological Devices
ASSISTED REPRODUCTION	884.6150	ASSISTED REPRODUCTION MICROMANIPULATORS AND MICROINJECTORS Assisted Reproduction Devices; Obstetrical And Gynecological Devices
ASSISTED REPRODUCTION	884.6160	ASSISTED REPRODUCTION LABWARE Assisted Reproduction Devices; Obstetrical And Gynecological Devices
ASSISTED REPRODUCTION	884.6170	ASSISTED REPRODUCTION WATER AND WATER PURIFICATION SYSTEMS Assisted Reproduction Devices; Obstetrical And Gynecological Devices
ASSISTED REPRODUCTIVE	884.6190	ASSISTED REPRODUCTIVE MICROSCOPES AND MICROSCOPE ACCESSORIES Assisted Reproduction Devices; Obstetrical And Gynecological Devices
ASSURANCE, QUALITY	20.114	DATA AND INFORMATION SUBMITTED PURSUANT TO COOPERATIVE QUALITY ASSURANCE AGREEMENTS Availability Of Specific Categories Of Records; Public Information
ASSURANCE, QUALITY	58.35	QUALITY ASSURANCE UNIT Organization And Personnel; Good Laboratory Practice For Nonclinical Laboratory Studies
ASSURANCE, QUALITY	1000.55	RECOMMENDATIONS FOR QUALITY ASSURANCE PROGRAMS IN DIAGNOSTIC RADIOLOGY FACILITIES Radiation Protection Recommendations; General
ASTAXANTHIN	73.35	ASTAXANTHIN Foods; Listing of Color Additives Exempt From Certification
ASTRINGENT	346.18	ASTRINGENT ACTIVE INGREDIENTS Active Ingredients; Anorectal Drug Products For Over-the-counter Human Use
ASTRINGENT	347.10	ASTRINGENT ACTIVE INGREDIENTS Skin Protectant Drug Products For Over-the-counter Human Use
ASTRINGENT DRUG	347.50	LABELING OF ASTRINGENT DRUG PRODUCTS Skin Protectant Drug Products For Over-the-counter Human Use
ASTRINGENT DRUG	349.55	LABELING OF OPHTHALMIC ASTRINGENT DRUG PRODUCTS Labeling; Ophthalmic Drug Products For Over-the-counter Human Use
ASTRINGENT, OPHTHALMIC	349.10	OPHTHALMIC ASTRINGENT Active Ingredients; Ophthalmic Drug Products For Over-the-counter Human Use
AST/SGOT	862.1100	ASPARTATE AMINO TRANSFERASE (AST/SGOT) TEST SYSTEM Clinical Chemistry Test Systems; Clinical Chemistry And Clinical Toxicology Devices
ATAXIAGRAPH	882.1030	ATAXIAGRAPH Neurological Diagnostic Devices; Neurological Devices

ATIPAMEZOLE HYDROCHLORIDE	522.147	ATIPAMEZOLE HYDROCHLORIDE Implantation Or Injectable Dosage Form New Animal Drugs
ATOMIC ABSORPTION	862.2850	ATOMIC ABSORPTION SPECTROPHOTOMETER FOR CLINICAL USE Clinical Laboratory Instruments; Clinical Chemistry And Clinical Toxicology Devices
ATOMIZER	868.5640	MEDICINAL NONVENTILATORY NEBULIZER (ATOMIZER) Therapeutic Devices; Anesthesiology Devices
ATROPINE	520.2520b	TRICHLORFON AND ATROPINE Oral Dosage Form New Animal Drugs
ATROX REAGENT	864.8100	BOTHROPS ATROX REAGENT Hematology Reagents; Hematology And Pathology Devices
ATTACHMENT	868.5955	INTERMITTENT MANDATORY VENTILATION ATTACHMENT Therapeutic Devices; Anesthesiology Devices
ATTACHMENT	868.5965	POSITIVE END EXPIRATORY PRESSURE BREATHING ATTACHMENT Therapeutic Devices; Anesthesiology Devices
ATTACHMENT	872.3165	PRECISION ATTACHMENT Prosthetic Devices; Dental Devices
ATTACHMENTS	878.4820	SURGICAL INSTRUMENT MOTORS AND ACCESSORIES/ATTACHMENTS Surgical Devices; General And Plastic Surgery Devices
ATTAPULGITE	520.1204	KANAMYCIN SULFATE, AMINOPENTAMIDE HYDROGEN SULFATE, PECTIN, BISMUTH SUBCARBONATE, ACTIVATED ATTAPULGITE SUSPENSION Oral Dosage Form New Animal Drugs
ATTAPULGITE	520.1205	KANAMYCIN SULFATE, PECTIN, BISMUTH SUBCARBONATE, ACTIVATED ATTAPULGITE TABLETS Oral Dosage Form New Animal Drugs
ATTORNEY, POWER OF	1305.07	POWER OF ATTORNEY Order Forms
AUDIOMETER	874.1050	AUDIOMETER Diagnostic Devices; Ear, Nose, And Throat Devices
AUDIOMETER	874.1080	AUDIOMETER CALIBRATION SET Diagnostic Devices; Ear, Nose, And Throat Devices
AUDIOMETRIC	874.1060	ACOUSTIC CHAMBER FOR AUDIOMETRIC TESTING Diagnostic Devices; Ear, Nose, And Throat Devices
AUDIOMETRIC	874.1100	EARPHONE CUSHION FOR AUDIOMETRIC TESTING Diagnostic Devices; Ear, Nose, And Throat Devices
AUDIOMETRIC	874.1120	ELECTRONIC NOISE GENERATOR FOR AUDIOMETRIC TESTING Diagnostic Devices; Ear, Nose, And Throat Devices
AUDIT, NON-FEDERAL	1403.26	NON-FEDERAL AUDIT Post-Award Requirements; Uniform Administrative Requirements For Grants And Coop. Agreements To State & Local Gov.
AUDIT, QUALITY	820.22	QUALITY AUDIT Quality System Requirements; Quality System Regulation
AUDITORY	874.1090	AUDITORY IMPEDANCE TESTER Diagnostic Devices; Ear, Nose, And Throat Devices
AUDITORY	874.3320	GROUP HEARING AID OR GROUP AUDITORY TRAINER Prosthetic Devices; Ear, Nose, And Throat Devices
AUDITORY	882.1900	EVOKED RESPONSE AUDITORY STIMULATOR Neurological Diagnostic Devices; Neurological Devices
AUTHENTICATION	20.3	CERTIFICATION AND AUTHENTICATION OF FOOD AND DRUG ADMINISTRATION RECORDS Official Testimony And Information; Public Information
AUTHORITIES, DESIGNATING	26.65	DESIGNATING AUTHORITIES Framework Provisions; Mutual Recognition Of Pharmaceutical GMP Reports, Medical Device Quality System Audit Reports
AUTHORITIES, REGULATORY	26.10	REGULATORY AUTHORITIES NOT LISTED AS CURRENTLY EQUIVALENT Specific Sector Provisions For Pharmaceutical GMPs; Mutual Recognition Of Pharmaceutical GMP Reports, Medical Device Quality System Audit Reports
AUTHORITIES, REGULATORY	26.34	REGULATORY AUTHORITIES Specific Sector Provisions For Medical Devices; Mutual Recognition Of Pharmaceutical GMP Reports, Medical Device Quality System Audit Reports
AUTHORITIES, STATE/LOCAL	5.606	ACCEPTANCE OF ASSISTANCE FROM STATE AND LOCAL AUTHORITIES FOR ENFORCEMENT OF RADIATION CONTROL LEGISLATION AND REGULATIONS Radiation Control; Redelegations Of Authority; Delegations Of Authority And Organization
AUTHORITY	5.11	RESERVATION OF AUTHORITY Delegations Of Authority To The Commissioner; Delegations Of Authority And Organization
AUTHORITY	5.20	GENERAL REDELEGATIONS OF AUTHORITY FROM THE COMMISSIONER TO OTHER OFFICERS OF THE FOOD AND DRUG ADMINISTRATION General Redelegations Of Authority; Delegations Of Authority And Organization
AUTHORITY	5.24	AUTHORITY RELATING TO TECHNOLOGY TRANSFER General Redelegations Of Authority; Delegations Of Authority And Organization

AUTHORITY	5.27	PATENT TERM EXTENSIONS FOR HUMAN DRUG PRODUCTS, MEDICAL DEVICES, AND FOOD AND COLOR ADDITIVES; AND AUTHORITY TO PERFORM DUE DILIGENCE DETERMINATIONS AND INFORMAL HEARINGS General Redelegations Of Authority; Delegations Of Authority And Organization
AUTHORITY	5.30	AUTHORITY TO SELECT TEMPORARY VOTING MEMBERS FOR ADVISORY COMMITTEES AND TO SIGN CONFLICT OF INTEREST WAIVERS General Redelegations Of Authority; Delegations Of Authority And Organization
AUTHORITY	5.102	AUTHORITY TO APPROVE AND TO WITHDRAW APPROVAL OF A CHARGE FOR INVESTIGATIONAL NEW DRUGS Human Drugs; Redelegations Of Authority; Delegations Of Authority And Organization
AUTHORITY	5.108	AUTHORITY RELATING TO WAIVERS OR REDUCTIONS OF PRESCRIPTION DRUG USER FEES Human Drugs; Redelegations Of Authority; Delegations Of Authority And Organization
AUTHORITY	5.415	AUTHORITY RELATING TO MEDICAL DEVICE REPORTING PROCEDURES Medical Devices And Radiological Health; Redelegations Of Authority; Delegations Of Authority And Organization
AUTHORITY	5.417	AUTHORITY PERTAINING TO ACCREDITATION FUNCTIONS FOR MEDICAL DEVICES Medical Devices And Radiological Health; Redelegations Of Authority; Delegations Of Authority And Organization
AUTHORITY	5.700	AUTHORITY RELATING TO DETERMINATION OF PRODUCT PRIMARY JURISDICTION Product Designation; Redelegations Of Authority; Delegations Of Authority And Organization
AUTHORITY	5.1000	AUTHORITY TO ENSURE THAT MAMMOGRAPHY FACILITIES MEET QUALITY STANDARDS Mammography Facilities; Redelegations Of Authority; Delegations Of Authority And Organization
AUTHORITY	12.70	AUTHORITY OF PRESIDING OFFICER Presiding Officer; Formal Evidentiary Public Hearing
AUTHORITY	17.19	AUTHORITY OF THE PRESIDING OFFICER Civil Money Penalties Hearings
AUTHORITY	80.34	AUTHORITY TO REFUSE CERTIFICATION SERVICE Certification Procedures; Color Additive Certification
AUTHORITY	1220.17	AUTHORITY TO SAMPLE AND INSPECT Inspection And Testing; Regulations Under The Federal Import Milk Act
AUTHORITY	1316.03	AUTHORITY TO MAKE INSPECTIONS Administrative Inspections; Administrative Functions, Practices, And Procedures
AUTHORITY	1316.31	AUTHORITY FOR ENFORCEMENT PROCEEDING Enforcement Proceedings; Administrative Functions, Practices, And Procedures
AUTHORITY CITATIONS	1.4	AUTHORITY CITATIONS General Provisions; General Regulations For The Enforcement Of Federal Food, Drug, Cosmetic Act
AUTHORITY, MISBRANDING	99.405	APPLICABILITY OF LABELING ADULTERATION, AND MISBRANDING AUTHORITY Dissemination Of Information On Unapproved/New Uses For Marketed Drugs, Biologics and Devices
AUTHORITY, RECALL	5.411	MEDICAL DEVICE RECALL AUTHORITY Medical Devices And Radiological Health; Redelegations Of Authority; Delegations Of Authority And Organization
AUTHORITY, REGULATORY	26.74	PRESERVATION OF REGULATORY AUTHORITY Framework Provisions; Mutual Recognition Of Pharmaceutical GMP Reports, Medical Device Quality System Audit Reports
AUTHORIZATION	1.95	APPLICATION FOR AUTHORIZATION TO RELABEL AND RECONDITION Imports And Exports; General Enforcement Regulations
AUTHORIZATION	1.96	GRANTING OF AUTHORIZATION TO RELABEL AND RECONDITION Imports And Exports; General Enforcement Regulations
AUTHORIZATION	5.23	DISCLOSURE OF OFFICIAL RECORDS AND AUTHORIZATION OF TESTIMONY General Redelegations Of Authority; Delegations Of Authority And Organization
AUTHORIZATION	5.403	AUTHORIZATION TO USE ALTERNATIVE EVIDENCE FOR DETERMINATION OF THE EFFECTIVENESS OF MEDICAL DEVICES Medical Devices And Radiological Health; Redelegations Of Authority; Delegations Of Authority And Organization
AUTHORIZATION	1312.11	REQUIREMENT OF AUTHORIZATION TO IMPORT Importation Of Controlled Substances; Importation And Exportation Of Controlled Substances
AUTHORIZATION	1312.21	REQUIREMENT OF AUTHORIZATION TO EXPORT Exportation Of Controlled Substances; Importation And Exportation Of Controlled Substances
AUTHORIZATION	1313.12	REQUIREMENT OF AUTHORIZATION TO IMPORT Importation Of Listed Chemicals; Importation And Exportation Of Precursors And Essential Chemicals

AUTHORIZATION	1313.21	REQUIREMENT OF AUTHORIZATION TO EXPORT Exportation Of Listed Chemicals; Importation And Exportation Of Precursors And Essential Chemicals
AUTHORIZATION	1313.32	REQUIREMENT OF AUTHORIZATION FOR INTERNATIONAL TRANSACTIONS Importation And Exportation Of Precursors And Essential Chemicals
AUTHORIZATION	1316.50	APPEARANCE; REPRESENTATION; AUTHORIZATION Administrative Hearings; Administrative Functions, Practices, And Procedures
AUTHORIZATION, MARKET	170.17	EXEMPTION FOR INVESTIGATIONAL USE AND PROCEDURE FOR OBTAINING AUTHORIZATION TO MARKET EDIBLE PRODUCTS FROM EXPERIMENTAL ANIMALS General Provisions; Food Additives
AUTHORIZATION, MARKET	570.17	EXEMPTION FOR INVESTIGATIONAL USE AND PROCEDURE FOR OBTAINING AUTHORIZATION TO MARKET EDIBLE PRODUCTS FROM EXPERIMENTAL ANIMALS General Provisions; Food Additives
AUTHORIZED	1312.15	SHIPMENTS IN GREATER OR LESS AMOUNT THAN AUTHORIZED Importation Of Controlled Substances; Importation And Exportation Of Controlled Substances
AUTOANTIBODIES	866.5660	MULTIPLE AUTOANTIBODIES IMMUNOLOGICAL TEST SYSTEM Immunological Test Systems; Immunology And Microbiology Devices
AUTOANTIBODY, THYROID	866.5870	THYROID AUTOANTIBODY IMMUNOLOGICAL TEST SYSTEM Immunological Test Systems; Immunology And Microbiology Devices
AUTOMATED	862.2900	AUTOMATED URINALYSIS SYSTEM Clinical Laboratory Instruments; Clinical Chemistry And Clinical Toxicology Devices
AUTOMATED	864.3800	AUTOMATED SLIDE STAINER Pathology Instrumentation And Accessories; Hematology And Pathology Devices
AUTOMATED	864.3875	AUTOMATED TISSUE PROCESSOR Pathology Instrumentation And Accessories; Hematology And Pathology Devices
AUTOMATED	864.5200	AUTOMATED CELL COUNTER Automated And Semi-automated Hematology Devices; Hematology And Pathology Devices
AUTOMATED	864.5220	AUTOMATED DIFFERENTIAL CELL COUNTER Automated And Semi-automated Hematology Devices; Hematology And Pathology Devices
AUTOMATED	864.5240	AUTOMATED BLOOD CELL DILUTING APPARATUS Automated And Semi-automated Hematology Devices; Hematology And Pathology Devices
AUTOMATED	864.5260	AUTOMATED CELL-LOCATING DEVICE Automated And Semi-automated Hematology Devices; Hematology And Pathology Devices
AUTOMATED	864.5600	AUTOMATED HEMATOCRIT INSTRUMENT Automated And Semi-automated Hematology Devices; Hematology And Pathology Devices
AUTOMATED	864.5620	AUTOMATED HEMOGLOBIN SYSTEM Automated And Semi-automated Hematology Devices; Hematology And Pathology Devices
AUTOMATED	864.5680	AUTOMATED HEPARIN ANALYZER Automated And Semi-automated Hematology Devices; Hematology And Pathology Devices
AUTOMATED	864.5700	AUTOMATED PLATELET AGGREGATION SYSTEM Automated And Semi-automated Hematology Devices; Hematology And Pathology Devices
AUTOMATED	864.5800	AUTOMATED SEDIMENTATION RATE DEVICE Automated And Semi-automated Hematology Devices; Hematology And Pathology Devices
AUTOMATED	864.5850	AUTOMATED SLIDE SPINNER Automated And Semi-automated Hematology Devices; Hematology And Pathology Devices
AUTOMATED	864.9175	AUTOMATED BLOOD GROUPING AND ANTIBODY TEST SYSTEM Products Used In Establishments That Manufacture Blood And Blood Products; Hematology And Pathology Devices
AUTOMATED	864.9245	AUTOMATED BLOOD CELL SEPARATOR Products Used In Establishments That Manufacture Blood And Blood Products; Hematology And Pathology Devices
AUTOMATED	864.9285	AUTOMATED CELL-WASHING CENTRIFUGE FOR IMMUNO-HEMATOLOGY Products Used In Establishments That Manufacture Blood And Blood Products; Hematology And Pathology Devices
AUTOMATED	864.9300	AUTOMATED COOMBS TEST SYSTEMS Products Used In Establishments That Manufacture Blood And Blood Products; Hematology And Pathology Devices
AUTOMATED	866.2170	AUTOMATED COLONY COUNTER Microbiology Devices; Immunology And Microbiology Devices
AUTOMATED	866.2440	AUTOMATED MEDIUM DISPENSING AND STACKING DEVICE Microbiology Devices; Immunology And Microbiology Devices

AUTOMATED	866.2850	AUTOMATED ZONE READER Microbiology Devices; Immunology And Microbiology Devices
AUTOMATIC	870.5925	AUTOMATIC ROTATING TOURNIQUET Cardiovascular Therapeutic Devices; Cardiovascular Devices
AUTOMATIC EQUIPMENT	211.68	AUTOMATIC, MECHANICAL, AND ELECTRONIC EQUIPMENT Equipment And Facilities; Current Good Manufacturing Practice For Finished Pharmaceuticals
AUTOTRANSFUSION	868.5830	AUTOTRANSFUSION APPARATUS Therapeutic Devices; Anesthesiology Devices
AVAILABILITY	10.20	SUBMISSION OF DOCUMENTS TO DOCKETS MANAGEMENT BRANCH; AVAILABILITY FOR PUBLIC DISCLOSURE General Administrative Procedures; Administrative Practices And Procedures
AVAILABILITY	316.52	AVAILABILITY FOR PUBLIC DISCLOSURE OF DATA AND INFORMATION IN REQUESTS AND APPLICATIONS Availability Of Information; Orphan Drugs
AVAILABILITY	1210.10	AVAILABILITY FOR EXAMINATION AND INSPECTION Inspection And Testing; Regulations Under The Federal Import Milk Act
AVAILABILITY OF FUNDS	1403.23	PERIOD OF AVAILABILITY OF FUNDS Post-Award Requirements; Uniform Administrative Requirements For Grants And Coop. Agreements To State & Local Gov.
AVAILABILITY, PUBLIC	803.9	PUBLIC AVAILABILITY OF REPORTS General Provisions; Medical Device Reporting
AVAILABILITY, PUBLIC	806.40	PUBLIC AVAILABILITY OF REPORTS Medical Device Corrections And Removals
AVERSIVE	882.5235	AVERSIVE CONDITIONING DEVICE Neurological Therapeutic Devices; Neurological Devices
AVIDITY	660.26	SPECIFICITY TESTS AND AVIDITY TESTS Blood Grouping Reagent; Additional Standards For Diagnostic Substances For Laboratory Tests
AZAPERONE	522.150	AZAPERONE INJECTION Implantation Or Injectable Dosage Form New Animal Drugs
AZODICARBONAMIDE	172.806	AZODICARBONAMIDE Multipurpose Additives; Food Additives Permitted For Direct Addition To Food For Human Consumption
AZTEC MARIGOLD	73.295	TAGETES (AZTEC MARIGOLD) MEAL AND EXTRACT Foods; Listing Of Color Additives Exempt From Certification

B

B12, VITAMIN	184.1945	VITAMIN B12 Listing Of Specific Substances Affirmed As GRAS; Direct Food Substances Affirmed As Generally Recognized As Safe
B12, VITAMIN	582.5945	VITAMIN B12 Nutrients And/or Dietary Supplements; Substances Generally Recognized As Safe
B12, VITAMIN	862.1810	VITAMIN B12 TEST SYSTEM Clinical Chemistry Test Systems; Clinical Chemistry And Clinical Toxicology Devices
B, FACTOR	866.5320	PROPERIDIN FACTOR B IMMUNOLOGICAL TEST SYSTEM Immunological Test Systems; Immunology And Microbiology Devices
B, TYPE	558.5	NEW ANIMAL DRUG REQUIREMENTS FOR LIQUID TYPE B FEEDS New Animal Drugs For Use In Animal Feeds
B-TYPE	862.1117	B-TYPE NATRIURETIC PEPTIDE TEST SYSTEM Clinical Chemistry Test Systems; Clinical Chemistry And Clinical Toxicology Devices
BACILLUS SUBTILIS	173.115	ALPHA-ACETOLACTATE DECARBOXYLASE (ALPHA-ALDC) ENZYME PREPARATION DERIVED FROM A RECOMBINANT BACILLUS SUBTILIS Enzyme Preparations And Microorganisms; Secondary Direct Food Additives Permitted In Food For Human Consumption
BACILLUS STEAROTHERMOPHILUS	184.1012	α-AMYLASE ENZYME PREPARATION FROM BACILLUS STEAROTHERMOPHILUS Listing Of Specific Substances Affirmed As GRAS; Direct Food Substances Affirmed As Generally Recognized As Safe
BACITRACIN	520.154	BACITRACIN ORAL DOSAGE FORMS Oral Dosage Form New Animal Drugs
BACITRACIN	524.154	BACITRACIN OR BACITRACIN ZINC-NEOMYCIN SULFATE-POLYMIXIN B SULFATE OPHTHALMIC OINTMENT Ophthalmic And Topical Dosage Form New Animal Drugs
BACITRACIN	524.155	BACITRACIN ZINC-POLYMIXIN B SULFATE-NEOMYCIN SULFATE-HYDROCORTISONE OR HYDROCORTISONE ACETATE OPHTHALMIC OINTMENT Ophthalmic And Topical Dosage Form New Animal Drugs
BACITRACIN	556.70	BACITRACIN Specific Tolerances For Residues Of New Animal Drugs; Tolerances For Residues Of New Animal Drugs In Food
BACITRACIN	558.76	BACITRACIN METHYLENE DISALICYLATE Specific New Animal Drugs For Use In Animal Feeds; New Animal Drugs For Use In Animal Feeds
BACITRACIN ZINC	558.78	BACITRACIN ZINC Specific New Animal Drugs For Use In Animal Feeds; New Animal Drugs For Use In Animal Feeds
BACKING	872.3910	BACKING AND FACING FOR AN ARTIFICIAL TOOTH Prosthetic Devices; Dental Devices
BACTERIAL	868.5260	BREATHING CIRCUIT BACTERIAL FILTER Therapeutic Devices; Anesthesiology Devices
BACTERIAL	1210.16	METHOD OF BACTERIAL COUNT Inspection And Testing; Regulations Under The Federal Import Milk Act
BACTERICIDAL	1210.10	EFFECTIVE BACTERICIDAL TREATMENT General Provisions; Control Of Communicable Diseases
BACTERIOPHAGE	866.2050	STAPHYLOCOCCAL TYPING BACTERIOPHAGE Microbiology Devices; Immunology And Microbiology Devices
BAG	868.5320	RESERVOIR BAG Therapeutic Devices; Anesthesiology Devices
BAG	870.3650	PACEMAKER POLYMERIC MESH BAG Cardiovascular Prosthetic Devices; Cardiovascular Devices
BAG	878.4100	ORGAN BAG Surgical Devices; General And Plastic Surgery Devices
BAG	880.5420	PRESSURE INFUSOR FOR AN I.V. BAG General Hospital And Personal Use Therapeutic Devices; General Hospital And Personal Use Devices
BAG	880.6050	ICE BAG General Hospital And Personal Use Miscellaneous Devices; General Hospital And Personal Use Devices
BAGOLINI LENS	886.1375	BAGOLINI LENS Diagnostic Devices; Ophthalmic Devices
BAGS	178.3730	PIPERONYL BUTOXIDE AND PYRETHRINS AS COMPONENTS OF BAGS Certain Adjuvants And Production Aids; Indirect Food Additives: Adjuvants, Production Aids, And Sanitizers

BAKERS YEAST	172.325	BAKERS YEAST PROTEIN Special Dietary And Nutritional Additives; Food Additives Permitted For Direct Addition To Food For Human Consumption
BAKERS YEAST	172.898	BAKERS YEAST GLYCAN Multipurpose Additives; Food Additives Permitted For Direct Addition To Food For Human Consumption
BAKERS YEAST	184.1983	BAKERS YEAST EXTRACT Listing Of Specific Substances Affirmed As GRAS; Direct Food Substances Affirmed As Generally Recognized As Safe
BALANCED SALT	864.2875	BALANCED SALT SOLUTIONS OR FORMULATIONS Cell And Tissue Culture Products; Hematology And Pathology Devices
BALLISTOCARDIOGRAPH	870.2320	BALLISTOCARDIOGRAPH Cardiovascular Monitoring Devices; Cardiovascular Devices
BALLOON	870.1350	CATHETER BALLOON REPAIR KIT Cardiovascular Diagnostic Devices; Cardiovascular Devices
BALLOON	870.3535	INTRA-AORTIC BALLOON AND CONTROL SYSTEM Cardiovascular Prosthetic Devices; Cardiovascular Devices
BALLOON	874.4100	EPISTAXIS BALLOON Surgical Devices; Ear, Nose And Throat Devices
BALLOON	884.5050	METREURYNTER-BALLOON ABORTION SYSTEM Obstetrical And Gynecological Therapeutic Devices; Obstetrical And Gynecological Devices
BALSAM, PERU	524.2620	LIQUID CRYSTALLINE TRYPSIN, PERU BALSAM, CASTOR OIL Ophthalmic And Topical Dosage Form New Animal Drugs
BAMBERMYCINS	558.95	BAMBERMYCINS Specific New Animal Drugs For Use In Animal Feeds; New Animal Drugs For Use In Animal Feeds
BANDAGE	880.5075	ELASTIC BANDAGE General Hospital And Personal Use Therapeutic Devices; General Hospital And Personal Use Devices
BANDAGE	880.5090	LIQUID BANDAGE General Hospital And Personal Use Therapeutic Devices; General Hospital And Personal Use Devices
BANDAGE	880.5240	MEDICAL ADHESIVE TAPE AND ADHESIVE BANDAGE General Hospital And Personal Use Therapeutic Devices; General Hospital And Personal Use Devices
BANK, BLOOD	864.9050	BLOOD BANK SUPPLIES Products Used In Establishments That Manufacture Blood And Blood Products; Hematology And Pathology Devices
BANK, BLOOD	864.9275	BLOOD BANK CENTRIFUGE FOR IN VITRO DIAGNOSTIC USE Products Used In Establishments That Manufacture Blood And Blood Products; Hematology And Pathology Devices
BANKING, BLOOD	864.9650	QUALITY CONTROL KIT FOR BLOOD BANKING REAGENTS Products Used In Establishments That Manufacture Blood And Blood Products; Hematology And Pathology Devices
BANKS, BLOOD	607.7	ESTABLISHMENT REGISTRATION AND PRODUCT LISTING OF BLOOD BANKS AND OTHER FIRMS MANUFACTURING HUMAN BLOOD AND BLOOD PRODUCTS General Provisions; Establishment Registration+product Listing, Manufacturers Human Blood/b.products
BANNING	895.21	PROCEDURES FOR BANNING A DEVICE General Provisions; Banned Devices
BAR PRISM	886.1650	OPHTHALMIC BAR PRISM Diagnostic Devices; Ophthalmic Devices
BAR READER	886.5800	OPHTHALMIC BAR READER Therapeutic Devices; Ophthalmic Devices
BARBITURATE	862.3150	BARBITURATE TEST SYSTEM Clinical Toxicology Test Systems; Clinical Chemistry And Clinical Toxicology Devices
BARIUM ENEMA	201.304	TANNIC ACID AND BARIUM ENEMA PREPARATIONS Specific Labeling Requirements For Specific Drug Products; Labeling
BARR, EPSTEIN-	866.3235	EPSTEIN-BARR VIRUS SEROLOGICAL REAGENTS Serological Reagents; Immunology And Microbiology Devices
BASE	172.585	SUGAR BEET EXTRACT FLAVOR BASE Flavoring Agents And Related Substances; Food Additives Permitted For Direct Addition To Food For Human Consumption
BASE	172.615	CHEWING GUM BASE Gums, Chewing Gum Bases And Related Substances; Food Additives Permitted For Direct Addition To Food For Human Consumption
BASE METAL	872.3710	BASE METAL ALLOY Prosthetic Devices; Dental Devices
BASE PLATE	872.6200	BASE PLATE SHELLAC Miscellaneous Devices; Dental Devices

BASELINE	803.55	BASELINE REPORTS Manufacturing Reporting Requirements; Medical Device Reporting
BASIS, LEGAL	120.9	LEGAL BASIS Hazard Analysis And Critical Control Point (HACCP) Systems
BATCH	80.37	TREATMENT OF BATCH PENDING CERTIFICATION Certification Procedures; Color Additive Certification
BATCH	80.38	TREATMENT OF BATCH AFTER CERTIFICATION Certification Procedures; Color Additive Certification
BATCH PRODUCTION	211.188	BATCH PRODUCTION AND CONTROL RECORDS Records And Reports; Current Good Manufacturing Practice For Finished Pharmaceuticals
BATCH PRODUCTION	226.102	MASTER-FORMULA AND BATCH-PRODUCTION RECORDS Records And Reports; Current Good Manufacturing Practice For Type A Medicated Articles
BATCH TESTING	320.34	REQUIREMENTS FOR BATCH TESTING AND CERTIFICATION BY THE FDA Bioavailability And Bioequivalence Requirements
BATCH TESTING	320.35	REQUIREMENTS FOR IN VITRO TESTING OF EACH BATCH Bioavailability And Bioequivalence Requirements
BATH PRODUCTS	740.17	FOAMING DETERGENT BATH PRODUCTS Warning Statements; Cosmetic Product Warning Statements
BATH, PARAFFIN	890.5110	PARAFFIN BATH Physical Medicine Therapeutic Devices; Physical Medicine Devices
BATH, SITZ	890.5125	NONPOWERED SITZ BATH Physical Medicine Therapeutic Devices; Physical Medicine Devices
BEAM, LIGHT	892.5780	LIGHT BEAM PATIENT POSITION INDICATOR Therapeutic Devices; Radiology Devices
BEAM-LIMITING	892.1610	DIAGNOSTIC X-RAY BEAM-LIMITING DEVICE Diagnostic Devices; Radiology Devices
BEAM-SHAPING	892.5710	RADIATION THERAPY BEAM-SHAPING BLOCK Therapeutic Devices; Radiology Devices
BEAN, LOCUST	582.7343	LOCUST BEAN GUM Stabilizers; Substances Generally Recognized As Safe
BEAN, LOCUST (CAROB)	184.1343	LOCUST (CAROB) BEAN GUM Listing Of Specific Substances Affirmed As GRAS; Direct Food Substances Affirmed As Generally Recognized As Safe
BEANS	155.120	CANNED GREEN BEANS AND CANNED WAX BEANS Requirements For Specific Standardized Canned Vegetables; Canned Vegetables
BED	890.5160	AIR-FLUIDIZED BED Physical Medicine Therapeutic Devices; Physical Medicine Devices
BED BOARD	880.6070	BED BOARD General Hospital And Personal Use Miscellaneous Devices; General Hospital And Personal Use Devices
BED, HOSPITAL	880.5100	AC-POWERED ADJUSTABLE HOSPITAL BED General Hospital And Personal Use Therapeutic Devices; General Hospital And Personal Use Devices
BED, HOSPITAL	880.5110	HYDRAULIC ADJUSTABLE HOSPITAL BED General Hospital And Personal Use Therapeutic Devices; General Hospital And Personal Use Devices
BED, HOSPITAL	880.5120	MANUAL ADJUSTABLE HOSPITAL BED General Hospital And Personal Use Therapeutic Devices; General Hospital And Personal Use Devices
BED, HOSPITAL	880.5140	PEDIATRIC HOSPITAL BED General Hospital And Personal Use Therapeutic Devices; General Hospital And Personal Use Devices
BED, NUCLEAR SCANNING	892.1350	NUCLEAR SCANNING BED Diagnostic Devices; Radiology Devices
BED-PATIENT	880.2400	BED-PATIENT MONITOR General Hospital And Personal Use Monitoring Devices; General Hospital And Personal Use Devices
BED, ROCKING	868.5180	ROCKING BED Therapeutic Devices; Anesthesiology Devices
BED, ROTATION	890.5180	MANUAL PATIENT ROTATION BED Physical Medicine Therapeutic Devices; Physical Medicine Devices
BED, ROTATION	890.5225	POWERED PATIENT ROTATION BED Physical Medicine Therapeutic Devices; Physical Medicine Devices
BED, THERAPY	890.5170	POWERED FLOTATION THERAPY BED Physical Medicine Therapeutic Devices; Physical Medicine Devices
BEDDING	880.6060	MEDICAL DISPOSABLE BEDDING General Hospital And Personal Use Miscellaneous Devices; General Hospital And Personal Use Devices
BEESWAX	184.1973	BEESWAX (YELLOW AND WHITE) Listing Of Specific Substances Affirmed As GRAS; Direct Food Substances Affirmed As Generally Recognized As Safe

BEESWAX	582.1973	BEESWAX General Purpose Food Additives; Substances Generally Recognized As Safe
BEESWAX	582.1975	BLEACHED BEESWAX General Purpose Food Additives; Substances Generally Recognized As Safe
BEET EXTRACT	172.585	SUGAR BEET EXTRACT FLAVOR BASE Flavoring Agents And Related Substances; Food Additives Permitted For Direct Addition To Food For Human Consumption
BEET POWDER	73.40	DEHYDRATED BEETS (BEET POWDER) Foods; Listing Of Color Additives Exempt From Certification
BEET-SUGAR	173.320	CHEMICALS FOR CONTROLLING MICROORGANISMS IN CANE-SUGAR AND BEET-SUGAR MILLS Specific Usage Additives; Secondary Direct Food Additives Permitted In Food For Human Consumption
BEETS, DEHYDRATED	73.40	DEHYDRATED BEETS (BEET POWDER) Foods; Listing Of Color Additives Exempt From Certification
BEHENATE, GLYCERAL	184.1328	GLYCERAL BEHENATE Listing Of Specific Substances Affirmed As GRAS; Direct Food Substances Affirmed As Generally Recognized as Safe
BENCE-JONES PROTEINS	866.5150	BENCE-JONES PROTEINS IMMUNOLOGICAL TEST SYSTEM Immunological Test Systems; Immunology And Microbiology Devices
BENEFIT	50.52	CLINICAL INVESTIGATIONS INVOLVING GREATER THAN MINIMAL RISK BUT PRESENTING THE PROSPECT OF DIRECT BENEFIT TO INDIVIDUAL SUBJECTS Additional Safeguards For Children In Clinical Investigations; Protection Of Human Subjects
BENEFIT	50.53	CLINICAL INVESTIGATIONS INVOLVING GREATER THAN MINIMAL RISK AND NO PROSPECT OF DIRECT BENEFIT TO INDIVIDUAL SUBJECTS, BUT LIKELY TO YIELD GENERALIZABLE KNOWLEDGE ABOUT THE SUBJECTS' DISORDER OR CONDITION Additional Safeguards For Children In Clinical Investigations; Protection Of Human Subjects
BENIGN PROSTATIC HYPERTROPHY	310.532	DRUG PRODUCTS CONTAINING ACTIVE INGREDIENTS OFFERED OTC TO RELIEVE THE SYMPTOMS OF BENIGN PROSTATIC HYPERTROPHY Requirements For Specific New Drugs Or Devices; New Drugs
BENTONITE	184.1155	BENTONITE Listing Of Specific Substances Affirmed As GRAS; Direct Food Substances Affirmed As Generally Recognized As Safe
BENTONITE	582.115	BENTONITE General Purpose Food Additives; Substances Generally Recognized As Safe
BENZATHINE	522.1696a	PENICILLIN G BENZATHINE AND PENICILLIN G PROCAINE STERILE SUSPENSION Implantation Or Injectable Dosage Form New Animal Drugs
BENZATHINE	526.363	CEPHAPIRAN BENZATHINE Intramammary Dosage Forms
BENZATHINE	526.464a	CLOXACILLIN BENZATHINE FOR INTRAMAMMARY INFUSION Intramammary Dosage Forms
BENZATHINE	526.464b	CLOXACILLIN BENZATHINE FOR INTRAMAMMARY INFUSION, STERILE Intramammary Dosage Forms
BENZENE, DIVINYL	172.775	METHACRYLIC ACID-DIVINYLBENZENE COPOLYMER Multipurpose Additives; Food Additives Permitted For Direct Addition To Food For Human Consumption
BENZENE, DIVINYL	177.2710	STYRENE/DIVINYLBENZENE RESINS, CROSS-LINKED Substances For Use Only As Components Of Articles Intended For Repeated Use; Indirect Food Additives: Polymers
BENZENETRIOL	73.3115	2-[[2,5-DIETHOXY-4-[(4-METHYLPHENYL)THIOL]PHENYL]AZO]-1,3,5-BENZENETRIOL Listing Of Color Additives Exempt From Certification
BENZOATE	184.1736	SODIUM BENZOATE Listing Of Specific Substances Affirmed As GRAS; Direct Food Substances Affirmed As Generally Recognized As Safe
BENZOATE	582.3733	SODIUM BENZOATE Chemical Preservatives; Substances Affirmed As Generally Recognized As Safe
BENZO[B]THIOPHEN	73.3123	6-ETHOXY-2-(6-ETHOXY-3-OXOBENZO[B]THIEN-2(3H)-YLIDENE) BENZO[B]THIOPEN-3 (2H)-ONE Medical Devices; Listing Of Color Additives Exempt From Certification
BENZODIAZEPAM	862.3170	BENZODIAZEPAM TEST SYSTEM Clinical Toxicology Test Systems; Clinical Chemistry And Clinical Toxicology Devices
BENZOIC ACID	184.1021	BENZOIC ACID Listing Of Specific Substances Affirmed As GRAS; Direct Food Substances Affirmed As Generally Recognized As Safe
BENZOIC ACID	582.3021	BENZOIC ACID Chemical Preservatives; Substances Generally Recognized As Safe

BENZOYL	184.1157	BENZOYL PEROXIDE Listing Of Specific Substances Affirmed As GRAS; Direct Food Substances Affirmed As Generally Recognized As Safe
BERRIES	145.120	CANNED BERRIES Requirements For Specific Standardized Canned Fruits; Canned Fruits
BETA-2-GLYCOPROTEIN	866.5430	BETA-2-GLYCOPROTEIN I IMMUNOLOGICAL TEST SYSTEM Immunological Test Systems; Immunology And Microbiology Devices
BETA-2-GLYCOPROTEIN	866.5440	BETA-2-GLYCOPROTEIN III IMMUNOLOGICAL TEST SYSTEM Immunological Test Systems; Immunology And Microbiology Devices
BETA-2-MICROGLOBULIN	866.5630	BETA-2-MICROGLOBULIN IMMUNOLOGICAL TEST SYSTEM Immunological Test Systems; Immunology And Microbiology Devices
BETA-AMINOPROPIONITRILE	522.84	BETA-AMINOPROPIONITRILE FUMARATE Implantation Or Injectable Dosage Form New Animal Drugs
BETA-APO-CAROTENAL	73.90	BETA-APO-8'-CAROTENAL Foods; Listing Of Color Additives Exempt From Certification
BETA-CAROTENE	73.95	BETA-CAROTENE Foods; Listing Of Color Additives Exempt From Certification
BETA-CAROTENE	73.1095	BETA-CAROTENE Drugs; Listing Of Color Additives Exempt From Certification
BETA-CAROTENE	73.2095	BETA-CAROTENE Cosmetics; Listing Of Color Additives Exempt From Certification
BETA-CAROTENE	184.1245	BETA-CAROTENE Listing Of Specific Substances Affirmed As GRAS; Direct Food Substances Affirmed As Generally Recognized As Safe
BETA COUNTER	862.2320	BETA OR GAMMA COUNTER FOR CLINICAL USE Clinical Laboratory Instruments; Clinical Chemistry And Clinical Toxicology Devices
BETA-GLOBULIN	866.5160	BETA-GLOBULIN IMMUNOLOGICAL TEST SYSTEM Immunological Test Systems; Immunology And Microbiology Devices
BETA RADIATION	886.5100	OPHTHALMIC BETA RADIATION SOURCE Therapeutic Devices; Ophthalmic Devices
BETAMETHASONE ACETATE	522.161	BETAMETHASONE ACETATE AND BETAMETHASONE DISODIUM PHOSPHATE AQUEOUS SUSPENSION Implantation Or Injectable Dosage Form New Animal Drugs
BETAMETHASONE DIPROPIONATE	522.163	BETAMETHASONE DIPROPIONATE AND BETAMETHASONE SODIUM PHOSPHATE AQUEOUS SUSPENSION Implantation Or Injectable Dosage Form New Animal Drugs
BETAMETHASONE VALERATE	524.1044b	GENTAMICIN SULFATE, BETAMETHASONE VALERATE OTIC SOLUTION Ophthalmic And Topical Dosage Form New Animal Drugs
BETAMETHASONE VALERATE	524.1044d	GENTAMICIN SULFATE, BETAMETHASONE VALERATE OINTMENT Ophthalmic And Topical Dosage Form New Animal Drugs
BETAMETHASONE VALERATE	524.1044f	GENTAMICIN SULFATE, BETAMETHASONE VALERATE TOPICAL SPRAY Ophthalmic And Topical Dosage Form New Animal Drugs
BETAMETHASONE VALERATE	524.1044g	GENTAMICIN SULFATE, BETAMETHASONE VALERATE, CLOTRIMAZOLE OINTMENT Ophthalmic And Topical Dosage Form New Animal Drugs
BEVERAGES	101.30	PERCENTAGE JUICE DECLARATION FOR FOODS PURPORTING TO BE BEVERAGES THAT CONTAIN FRUIT OR VEGETABLE JUICE Food Labeling
BEVERAGES	102.33	BEVERAGES THAT CONTAIN FRUIT OR VEGETABLE JUICE Requirements For Specific Nonstandardized Foods; Common Or Usual Name For Nonstandardized Foods
BHA	172.110	BHA Food Preservatives; Food Additives Permitted For Direct Addition To Food For Human Consumption
BHT	172.115	BHT Food Preservatives; Food Additives Permitted For Direct Addition To Food For Human Consumption
BICARBONATE	184.1135	AMMONIUM BICARBONATE Listing Of Specific Substances Affirmed As GRAS; Direct Food Substances Affirmed As Generally Recognized As Safe
BICARBONATE	184.1613	POTASSIUM BICARBONATE Listing Of Specific Substances Affirmed As GRAS; Direct Food Substances Affirmed As Generally Recognized As Safe
BICARBONATE	184.1736	SODIUM BICARBONATE Listing Of Specific Substances Affirmed As GRAS; Direct Food Substances Affirmed As Generally Recognized As Safe
BICARBONATE	582.1135	AMMONIUM BICARBONATE General Purpose Food Additives; Substances Generally Recognized As Safe
BICARBONATE	582.1613	POTASSIUM BICARBONATE Substances Generally Recognized As Safe
BICARBONATE	582.1736	SODIUM BICARBONATE General Purpose Food Additives; Substances Affirmed As Generally Recognized As Safe

BICARBONATE	862.1160	BICARBONATE/CARBON DIOXIDE TEST SYSTEM Clinical Chemistry Test Systems; Clinical Chemistry And Toxicology Devices
BICYCLOHEXYLAMMONIUM	520.182	BICYCLOHEXYLAMMONIUM FUMAGILLIN Oral Dosage Form New Animal Drugs
BILE EXTRACT	184.1560	OX BILE EXTRACT Listing Of Specific Substances Affirmed As GRAS; Direct Food Substances Affirmed As Generally Recognized As Safe
BILIARY	876.5010	BILIARY CATHETER AND ACCESSORIES Therapeutic Devices; Gastroenterology-urology Devices
BILIRUBIN	862.1110	BILIRUBIN (TOTAL OR DIRECT) TEST SYSTEM Clinical Chemistry Test Systems; Clinical Chemistry And Clinical Toxicology Devices
BILIRUBIN	862.1113	BILIRUBIN (TOTAL AND UNBOUND) IN THE NEONATE TEST SYSTEM Clinical Chemistry Test Systems; Clinical Chemistry And Clinical Toxicology Devices
BILIRUBIN	862.1115	URINARY BILIRUBIN AND ITS CONJUGATES (NONQUANTITATIVE) TEST SYSTEM Clinical Chemistry Test Systems; Clinical Chemistry And Clinical Toxicology Devices
BINDER	880.5160	THERAPEUTIC MEDICAL BINDER General Hospital And Personal Use Therapeutic Devices; General Hospital And Personal Use Devices
BINOCULAR	886.5120	LOW-POWER BINOCULAR LOUPE Therapeutic Devices; Ophthalmic Devices
BIOAVAILABILITY	320.24	TYPES OF EVIDENCE TO ESTABLISH BIOAVAILABILITY OR BIOEQUIVALENCE Procedures For Determining The Bioavailability Of Drug Products; Bioavailability And Bioequivalence Requirements
BIOAVAILABILITY	320.28	CORRELATION OF BIOAVAILABILITY WITH AN ACUTE PHARMACOLOGICAL EFFECT OR CLINICAL EVIDENCE Procedures For Determining The Bioavailability Of Drug Products; Bioavailability And Bioequivalence Requirements
BIOAVAILABILITY	320.30	INQUIRIES REGARDING BIOAVAILABILITY AND BIOEQUIVALENCE REQUIREMENTS AND REVIEW OF PROTOCOLS BY THE FDA Procedures For Determining The Bioavailability Of Drug Products; Bioavailability And Bioequivalence Requirements
BIOAVAILABILITY	320.38	RETENTION OF BIOAVAILABILITY SAMPLES Procedures For Determining The Bioavailability Of Drug Products; Bioavailability And Bioequivalence Requirements
BIOAVAILABILITY, IN VIVO	320.21	REQUIREMENTS FOR SUBMISSION OF IN VIVO BIOAVAILABILITY AND BIOEQUIVALENCE DATA Procedures For Determining The Bioavailability Of Drug Products; Bioavailability And Bioequivalence Requirements
BIOAVAILABILITY, IN VIVO	320.22	CRITERIA FOR WAIVER OF EVIDENCE OF IN VIVO BIOAVAILABILITY OR BIOEQUIVALENCE Procedures For Determining The Bioavailability Of Drug Products; Bioavailability And Bioequivalence Requirements
BIOAVAILABILITY, IN VIVO	320.23	BASIS FOR DEMONSTRATING IN VIVO BIOAVAILABILITY OR BIOEQUIVALENCE Procedures For Determining The Bioavailability Of Drug Products; Bioavailability And Bioequivalence Requirements
BIOAVAILABILITY, IN VIVO	320.25	GUIDELINES FOR THE CONDUCT OF AN IN VIVO BIOAVAILABILITY STUDY Procedures For Determining The Bioavailability Of Drug Products; Bioavailability And Bioequivalence Requirements
BIOAVAILABILITY, IN VIVO	320.26	GUIDELINES ON THE DESIGN OF A SINGLE-DOSE IN VIVO BIOAVAILABILITY STUDY Procedures For Determining The Bioavailability Of Drug Products; Bioavailability And Bioequivalence Requirements
BIOAVAILABILITY, IN VIVO	320.27	GUIDELINES ON THE DESIGN OF A MULTIPLE-DOSE IN VIVO BIOAVAILABILITY STUDY Procedures For Determining The Bioavailability Of Drug Products; Bioavailability And Bioequivalence Requirements
BIOAVAILABILITY, IN VIVO	320.29	ANALYTICAL METHODS FOR AN IN VIVO BIOAVAILABILITY STUDY Procedures For Determining The Bioavailability Of Drug Products; Bioavailability And Bioequivalence Requirements
BIOEQUIVALENCE	320.21	REQUIREMENTS FOR SUBMISSION OF IN VIVO BIOAVAILABILITY AND BIOEQUIVALENCE DATA Procedures For Determining The Bioavailability Of Drug Products; Bioavailability And Bioequivalence Requirements
BIOEQUIVALENCE	320.22	CRITERIA FOR WAIVER OF EVIDENCE OF IN VIVO BIOAVAILABILITY AND BIOEQUIVALENCE Procedures For Determining The Bioavailability Of Drug Products; Bioavailability And Bioequivalence Requirements

BIOEQUIVALENCE	320.23	BASIS FOR DEMONSTRATING IN VIVO BIOAVAILABILITY OR BIOEQUIVALENCE Procedures For Determining The Bioavailability Of Drug Products; Bioavailability And Bioequivalence Requirements
BIOEQUIVALENCE	320.24	TYPES OF EVIDENCE TO ESTABLISH BIOAVAILABILITY OR BIOEQUIVALENCE Procedures For Determining The Bioavailability Of Drug Products; Bioavailability And Bioequivalence Requirements
BIOEQUIVALENCE	320.30	INQUIRIES REGARDING BIOAVAILABILITY AND BIOEQUIVALENCE REQUIREMENTS AND REVIEW OF PROTOCOLS BY THE FDA Procedures For Determining The Bioavailability Of Drug Products; Bioavailability And Bioequivalence Requirements
BIOEQUIVALENCE	320.32	PROCEDURES FOR ESTABLISHING OR AMENDING A BIOEQUIVALENCE REQUIREMENT Bioavailability And Bioequivalence Requirements
BIOEQUIVALENCE	320.33	CRITERIA AND EVIDENCE TO ASSESS ACTUAL OR POTENTIAL BIOEQUIVALENCE PROBLEMS Bioavailability And Bioequivalence Requirements
BIOEQUIVALENCE	320.36	REQUIREMENTS FOR MAINTENANCE OF RECORDS OF BIOEQUIVALENCE TESTING Bioavailability And Bioequivalence Requirements
BIOEQUIVALENCE	320.63	RETENTION OF BIOEQUIVALENCE SAMPLES Bioavailability And Bioequivalence Requirements
BIOFEEDBACK	882.5050	BIOFEEDBACK DEVICE Neurological Therapeutic Devices; Neurological Devices
BIOLOGIC	5.701	PREMARKET APPROVAL OF A PRODUCT THAT IS OR CONTAINS A BIOLOGIC, DEVICE, OR DRUG Product Designation; Redelegations Of Authority; Delegations Of Authority And Organization
BIOLOGICAL PRODUCT	600.14	REPORTING OF BIOLOGICAL PRODUCT DEVIATIONS BY LICENSED MANUFACTURERS Establishment Standards; Biological Products: General
BIOLOGICAL PRODUCT	601.50	CONFIDENTIALITY OF DATA AND INFORMATION IN AN INVESTIGATIONAL NEW DRUG NOTICE FOR A BIOLOGICAL PRODUCT Confidentiality Of Information; Licensing
BIOLOGICAL PRODUCTS	5.203	ISSUANCE AND REVOCATION OF LICENSES FOR THE PROPAGATION OR MANUFACTURE AND PREPARATION OF BIOLOGICAL PRODUCTS Biologics; Redelegations Of Authority; Delegations Of Authority And Organization
BIOLOGICAL PRODUCTS	5.204	NOTIFICATION OF RELEASE FOR DISTRIBUTION OF BIOLOGICAL PRODUCTS Biologics; Redelegations Of Authority; Delegations Of Authority And Organization
BIOLOGICAL PRODUCTS	601.25	REVIEW PROCEDURES TO DETERMINE LICENSED BIOLOGICAL PRODUCTS ARE SAFE, EFFECTIVE AND NOT MISBRANDED UNDER USE CONDITIONS Biologics Licensing; Licensing
BIOLOGICAL PRODUCTS	601.26	RECLASSIFICATION PROCEDURES TO DETERMINE LICENSED BIOLOGICAL PRODUCTS ARE SAFE, EFFECTIVE AND NOT MISBRANDED UNDER USE CONDITIONS Biologics Licensing; Licensing
BIOLOGICAL PRODUCTS	610.53	DATING PERIODS FOR LICENSED BIOLOGICAL PRODUCTS Dating Period Limitations; General Biological Products Standards
BIOLOGICS	5.201	REDELEGATION OF THE CENTER FOR BIOLOGICS EVALUATION AND RESEARCH DIRECTOR'S PROGRAM AUTHORITIES Biologics; Redelegations Of Authority; Delegations Of Authority And Organization
BIOLOGICS	25.31	HUMAN DRUGS AND BIOLOGICS Categorical Exclusions; Environmental Impact Considerations
BIOLOGICS	310.4	BIOLOGICS; PRODUCTS SUBJECT TO LICENSE CONTROL General Provisions; New Drugs
BIOLOGICS	510.4	BIOLOGICS; PRODUCTS SUBJECT TO LICENSE CONTROL General Provisions; New Animal Drugs
BIOLOGICS	601.2	APPLICATIONS FOR BIOLOGICS LICENSES; PROCEDURES FOR FILING General Provisions; Licensing
BIOLOGICS	601.20	BIOLOGICS LICENSES; ISSUANCE AND CONDITIONS Biologics Licensing; Licensing
BIOLOGICS	601.51	CONFIDENTIALITY OF DATA AND INFORMATION IN APPLICATIONS FOR BIOLOGICS LICENSES Confidentiality Of Information; Licensing
BIOMICROSCOPE	886.1850	AC-POWERED SLITLAMP BIOMICROSCOPE Diagnostic Devices; Ophthalmic Devices
BIOPHOTOMETER	866.1050	ADAPTOMETER (BIOPHOTOMETER) Diagnostic Devices; Ophthalmic Devices
BIOPOTENTIAL	870.2050	BIOPOTENTIAL AMPLIFIER AND SIGNAL CONDITIONER Cardiovascular Monitoring Devices; Cardiovascular Devices
BIOPSY	870.4075	ENDOMYOCARDIAL BIOPSY DEVICE Cardiovascular Surgical Devices; Cardiovascular Devices
BIOPSY	876.1075	GASTROENTEROLOGY-UROLOGY BIOPSY INSTRUMENT Diagnostic Devices; Gastroenterology-urology Devices

BIOTIN	182.8159	BIOTIN Nutrients; Substances Generally Recognized As Safe
BIOTIN	582.5159	BIOTIN Nutrients; Substances Generally Recognized As Safe
BIOTINIDASE	862.1118	BIOTINIDASE TEST SYSTEM Clinical Chemistry Test Systems; Clinical Chemistry And Clinical Toxicology Devices
BIPHENYLS (PCB'S)	109.15	USE OF POLYCHLORINATED BIPHENYLS (PCB'S) IN ESTABLISHMENTS MANUFACTURING FOOD-PACKAGING MATERIALS General Provisions; Unavoidable Contaminants In Food For Human Consumption & Food-packaging Material
BIPHENYLS (PCB'S)	109.30	TOLERANCES FOR POLYCHLORINATED BIPHENYLS (PCB'S) Tolerances For Unavoidable Poisonous Or Deleterious Substances; Unavoidable Contaminants In Food For Human Consumption & Food-packaging Material
BIPHENYLS (PCB'S)	500.45	USE OF POLYCHLORINATED BIPHENYLS (PCB'S) IN THE PRODUCTION, HANDLING, AND STORAGE OF ANIMAL FOOD Specific Administrative Rulings And Decisions; General
BIPHENYLS (PCB'S)	509.15	USE OF POLYCHLORINATED BIPHENYLS (PCB'S) IN ESTABLISHMENTS MANUFACTURING FOOD-PACKAGING MATERIALS General Provisions; Unavoidable Contaminants In Animal Food And Food-packaging Materials
BIPHENYLS (PCB'S)	509.30	TEMPORARY TOLERANCES FOR POLYCHLORINATED BIPHENYLS (PCB'S) Tolerances For Unavoidable Poisonous Or Deleterious Substances; Unavoidable Contaminants In Animal Food And Food-packaging Materials
BIRDS	1210.65	PSITTACINE BIRDS Specific Administrative Decisions Regarding Interstate Shipments; Control Of Communicable Diseases
BISBENZAMIDE	73.3118	N,N'-(9,10-DIHYDRO-9,10-DIOXO-1,5-ANTHRACENEDIYL) BISBENZAMIDE Medical Devices; Listing Of Color Additives Exempt From Certification
BISMUTH CITRATE	73.2110	BISMUTH CITRATE Cosmetics; Listing Of Color Additives Exempt From Certification
BISMUTH OXYCHLORIDE	73.1162	BISMUTH OXYCHLORIDE Drugs; Listing Of Color Additives Exempt From Certification
BISMUTH OXYCHLORIDE	73.2162	BISMUTH OXYCHLORIDE Cosmetics; Listing Of Color Additives Exempt From Certification
BISMUTH SUBCARBONATE	520.1204	KANAMYCIN SULFATE, AMINO-PENTAMIDE HYDROGEN SULFATE, PECTIN, BISMUTH SUBCARBONATE, ACTIVATED ATTAPULGITE SUSPENSION Oral Dosage Form New Animal Drugs
BISMUTH SUBCARBONATE	520.1205	KANAMYCIN SULFATE, PECTIN, BISMUTH SUBCARBONATE, ACTIVATED ATTAPULGITE TABLETS Oral Dosage Form New Animal Drugs
BISULFATE	520.445a	CHLORTETRACYCLINE BISULFATE/SULFAMETHAZINE BISULFATE SOLUBLE POWDER Oral Dosage Form New Animal Drugs
BISULFATE	520.445b	CHLORTETRACYCLINE POWDER (CHLORTETRACYCLINE HYDROCHLORIDE OR BISULFATE) Oral Dosage Form New Animal Drugs
BISULFITE	182.3616	POTASSIUM BISULFITE Chemical Preservatives; Substances Generally Recognized As Safe
BISULFITE	182.3739	SODIUM BISULFITE Chemical Preservatives; Substances Generally Recognized As Safe
BISULFITE	573.620	MENADIONE DIMETHYLPYRIMIDINOL BISULFITE Food Additives Permitted In Feed And Drinking Water Of Animals
BISULFITE	573.625	MENADIONE NICOTINAMIDE BISULFITE Food Additives Permitted In Feed And Drinking Water Of Animals
BISULFITE	582.3616	POTASSIUM BISULFITE Chemical Preservatives; Substances Generally Recognized As Safe
BISULFITE	582.3739	SODIUM BISULFITE Chemical Preservatives; Substances Generally Recognized As Safe
BITARTRATE	182.8250	CHOLINE BITARTRATE Nutrients; Substances Generally Recognized As Safe
BITARTRATE	582.5250	CHOLINE BITARTRATE Nutrients And/or Dietary Supplements; Substances Generally Recognized As Safe
BITE BLOCK	882.5070	BITE BLOCK Neurological Therapeutic Devices; Neurological Devices
BITHIONOL	700.11	COSMETICS CONTAINING BITHIONOL Requirements For Specific Cosmetic Products; General
BIURET	573.220	FEED-GRADE BIURET Food Additive Listing; Food Additives Permitted In Feed And Drinking Water Of Animals
BLADDER	882.5850	IMPLANTED SPINAL CORD STIMULATOR FOR BLADDER EVACUATION Neurological Therapeutic Devices; Neurological Devices
BLASTOMYCES	866.3060	BLASTOMYCES DERMATITIDIS SEROLOGICAL REAGENTS Serological Reagents; Immunology And Microbiology Devices

BLEACHING TEETH	872.6475	HEAT SOURCE FOR BLEACHING TEETH Miscellaneous Devices; Dental Devices
BLEEDING	864.6100	BLEEDING TIME DEVICE Manual Hematology Devices; Hematology And Pathology Devices
BLENDED CHEESE	133.10	NOTICE TO MANUFACTURERS, PACKERS, AND DISTRIBUTORS OF PASTEURIZED BLENDED CHEESE AND RELATED FOODS General Provisions; Cheeses And Related Cheese Products
BLENDED CHEESE	133.167	PASTEURIZED BLENDED CHEESE Requirements For Specific Standardized Cheese And Related Products; Cheeses And Related Cheese Products
BLENDED CHEESE	133.168	PASTEURIZED BLENDED CHEESE WITH FRUITS, VEGETABLES, OR MEATS Requirements For Specific Standardized Cheese And Related Products; Cheeses And Related Cheese Products
BLISTERS	310.537	DRUG PRODUCTS CONTAINING ACTIVE INGREDIENTS OFFERED OTC FOR TREATMENT OF FEVER BLISTERS AND COLD SORES Requirements For Specific New Drugs Or Devices; New Drugs
BLOCK	177.1810	STYRENE BLOCK POLYMERS Substances For Use As Basic Components Of Single And Repeated Use Food Contact Surfaces; Indirect Food Additives: Polymers
BLOCK	882.1925	ULTRASONIC SCANNER CALIBRATION TEST BLOCK Neurological Diagnostic Devices; Neurological Devices
BLOCK	892.5710	RADIATION THERAPY BEAM-SHAPING BLOCK Therapeutic Devices; Radiology Devices
BLOCK, BITE	882.5070	BITE BLOCK Neurological Therapeutic Devices; Neurological Devices
BLOCK, FEED	520.2380	THIABENDAZOLE TOP DRESSING AND MINERAL PROTEIN FEED BLOCK Oral Dosage Form New Animal Drugs
BLOCKS	520.905e	FENBENDAZOLE BLOCKS Oral Dosage Form New Animal Drugs
BLOCKS	520.1448a	MONENSIN BLOCKS Oral Dosage Form New Animal Drugs
BLOCKS	520.1846	POLYOXYETHYLENE (23) LAURYL ETHER BLOCKS Oral Dosage Form New Animal Drugs
BLOOD	207.7	ESTABLISHMENT REGISTRATION AND PRODUCT LISTING FOR HUMAN BLOOD AND BLOOD PRODUCTS AND FOR MEDICAL DEVICES General; Registration Of Drug Producers And Listing Of Drugs In Commercial Distribution
BLOOD	606.171	REPORTING OF PRODUCT DEVIATIONS BY MANUFACTURERS, UNLICENSED BLOOD ESTABLISHMENTS, AND TRANSFUSION SERVICES Records And Reports; Current Good Manufacturing Practices For Blood And Blood Components
BLOOD	607.7	ESTABLISHMENT REGISTRATION AND PRODUCT LISTING OF BLOOD BANKS AND OTHER FIRMS MANUFACTURING HUMAN BLOOD AND BLOOD PRODUCTS General Provisions; Establishment Registratn+product Listing, Manufacturers Human Blood/b.products
BLOOD	640.1	WHOLE BLOOD Whole Blood; Additional Standards For Human Blood And Blood Products
BLOOD	640.4	COLLECTION OF THE BLOOD Whole Blood; Additional Standards For Human Blood And Blood Products
BLOOD	640.5	TESTING THE BLOOD Whole Blood; Additional Standards For Human Blood And Blood Products
BLOOD	640.6	MODIFICATIONS OF WHOLE BLOOD Whole Blood; Additional Standards For Human Blood And Blood Products
BLOOD	640.13	COLLECTION OF THE BLOOD Red Blood Cells; Additional Standards For Human Blood And Blood Products
BLOOD	640.14	TESTING THE BLOOD Red Blood Cells; Additional Standards For Human Blood And Blood Products
BLOOD	640.23	TESTING THE BLOOD Platelets; Additional Standards For Human Blood And Blood Products
BLOOD	640.33	TESTING THE BLOOD Plasma; Additional Standards For Human Blood And Blood Products
BLOOD	640.53	TESTING THE BLOOD Cryoprecipitate; Additional Standards For Human Blood And Blood Products
BLOOD	640.64	COLLECTION OF BLOOD FOR SOURCE PLASMA Source Plasma; Additional Standards For Human Blood And Blood Products
BLOOD	862.1120	BLOOD GASES AND BLOOD PH TEST SYSTEM Clinical Chemistry Test Systems; Clinical Chemistry And Clinical Toxicology Devices
BLOOD	862.1130	BLOOD VOLUME TEST SYSTEM Clinical Chemistry Test Systems; Clinical Chemistry And Clinical Toxicology Devices

BLOOD	862.1675	BLOOD SPECIMEN COLLECTION DEVICE Clinical Chemistry Test Systems; Clinical Chemistry And Clinical Toxicology Devices
BLOOD	864.5950	BLOOD VOLUME MEASURING DEVICE Automated And Semi-automated Hematology Devices; Hematology And Pathology Devices
BLOOD	864.6150	CAPILLARY BLOOD COLLECTION TUBE Manual Hematology Devices; Hematology And Pathology Devices
BLOOD	864.6550	OCCULT BLOOD TEST Manual Hematology Devices; Hematology And Pathology Devices
BLOOD	864.7140	ACTIVATED WHOLE BLOOD CLOTTING TIME TESTS Hematology Kits And Packages; Hematology And Pathology Devices
BLOOD	864.7500	WHOLE BLOOD HEMOGLOBIN ASSAYS Hematology Kits And Packages; Hematology And Pathology Devices
BLOOD	864.9100	EMPTY CONTAINER FOR THE COLLECTION AND PROCESSING OF BLOOD AND BLOOD COMPONENTS Products Used In Establishments That Manufacture Blood And Blood Products; Hematology And Pathology Devices
BLOOD	864.9125	VACUUM-ASSISTED BLOOD COLLECTION SYSTEM Products Used In Establishments That Manufacture Blood And Blood Products; Hematology And Pathology Devices
BLOOD	864.9145	PROCESSING SYSTEM FOR FROZEN BLOOD Products Used In Establishments That Manufacture Blood And Blood Products; Hematology And Pathology Devices
BLOOD	864.9195	BLOOD MIXING AND WEIGHING DEVICES Products Used In Establishments That Manufacture Blood And Blood Products; Hematology And Pathology Devices
BLOOD	864.9205	BLOOD AND PLASMA WARMING DEVICE Products Used In Establishments That Manufacture Blood And Blood Products; Hematology And Pathology Devices
BLOOD	864.9700	BLOOD STORAGE REFRIGERATOR AND BLOOD STORAGE FREEZER Products Used In Establishments That Manufacture Blood And Blood Products; Hematology And Pathology Devices
BLOOD	868.1100	ARTERIAL BLOOD SAMPLING KIT Diagnostic Devices; Anesthesiology Devices
BLOOD	868.1120	INDWELLING BLOOD OXYHEMOGLOBIN CONCENTRATION ANALYZER Diagnostic Devices; Anesthesiology Devices
BLOOD	868.1150	INDWELLING BLOOD CARBON DIOXIDE PARTIAL PRESSURE (PCO2) ANALYZER Diagnostic Devices; Anesthesiology Devices
BLOOD	868.1170	INDWELLING BLOOD HYDROGEN ION CONCENTRATION (PH) ANALYZER Diagnostic Devices; Anesthesiology Devices
BLOOD	868.1200	INDWELLING BLOOD OXYGEN PARTIAL PRESSURE (PO2) ANALYZER Diagnostic Devices; Anesthesiology Devices
BLOOD	870.2100	CARDIOVASCULAR BLOOD FLOWMETER Cardiovascular Monitoring Devices; Cardiovascular Devices
BLOOD	870.2120	EXTRAVASCULAR BLOOD FLOW PROBE Cardiovascular Monitoring Devices; Cardiovascular Devices
BLOOD	870.4260	CARDIOPULMONARY BYPASS ARTERIAL LINE BLOOD FILTER Cardiovascular Surgical Devices; Cardiovascular Devices
BLOOD	870.4270	CARDIOPULMONARY BYPASS CARDIOTOMY SUCTION LINE BLOOD FILTER Cardiovascular Surgical Devices; Cardiovascular Devices
BLOOD	870.4330	CARDIOPULMONARY BYPASS ON-LINE BLOOD GAS MONITOR Cardiovascular Surgical Devices; Cardiovascular Devices
BLOOD	870.4360	NONROLLER-TYPE CARDIOPULMONARY BYPASS BLOOD PUMP Cardiovascular Surgical Devices; Cardiovascular Devices
BLOOD	870.4370	ROLLER-TYPE CARDIOPULMONARY BYPASS BLOOD PUMP Cardiovascular Surgical Devices; Cardiovascular Devices
BLOOD	870.4400	CARDIOPULMONARY BYPASS BLOOD RESERVOIR Cardiovascular Surgical Devices; Cardiovascular Devices
BLOOD	870.4410	CARDIOPULMONARY BYPASS IN-LINE BLOOD GAS SENSOR Cardiovascular Surgical Devices; Cardiovascular Devices
BLOOD	876.5540	BLOOD ACCESS DEVICE AND ACCESSORIES Therapeutic Devices; Gastroenterology-urology Devices
BLOOD	884.1560	FETAL BLOOD SAMPLER Obstetrical And Gynecological Diagnostic Devices; Obstetrical And Gynecological Devices
BLOOD BANK	864.9050	BLOOD BANK SUPPLIES Products Used In Establishments That Manufacture Blood And Blood Products; Hematology And Pathology Devices
BLOOD BANK	864.9275	BLOOD BANK CENTRIFUGE FOR IN VITRO DIAGNOSTIC USE Products Used In Establishments That Manufacture Blood And Blood Products; Hematology And Pathology Device

BLOOD BANKING	864.9650	QUALITY CONTROL KIT FOR BLOOD BANKING REAGENTS Products Used In Establishments That Manufacture Blood And Blood Products; Hematology And Pathology Devices
BLOOD BANKS	607.7	ESTABLISHMENT REGISTRATION AND PRODUCT LISTING OF BLOOD BANKS AND OTHER FIRMS MANUFACTURING HUMAN BLOOD AND BLOOD PRODUCTS General Provisions; Establishment Registratn+product Listing, Manufacturers Human Blood/b.products
BLOOD CELL	660.23	RED BLOOD CELL PREPARATIONS Blood Grouping Reagent; Additional Standards For Diagnostic Substances For Laboratory Tests
BLOOD CELL	864.5240	AUTOMATED BLOOD CELL DILUTING APPARATUS Automated And Semi-automated Hematology Devices; Hematology And Pathology Devices
BLOOD CELL	864.6160	MANUAL BLOOD CELL COUNTING DEVICE Manual Hematology Devices; Hematology And Pathology Devices
BLOOD CELL	864.7100	RED BLOOD CELL ENZYME ASSAY Hematology Kits And Packages; Hematology And Pathology Devices
BLOOD CELL	864.8200	BLOOD CELL DILUENT Hematology Reagents; Hematology And Pathology Devices
BLOOD CELL	864.9245	AUTOMATED BLOOD CELL SEPARATOR Products Used In Establishments That Manufacture Blood And Blood Products; Hematology And Pathology Devices
BLOOD CELLS	640.10	RED BLOOD CELLS Red Blood Cells; Additional Standards For Human Blood And Blood Products
BLOOD CELLS	660.30	REAGENT RED BLOOD CELLS Reagent Red Blood Cells; Additional Standards For Diagnostic Substances For Laboratory Tests
BLOOD GROUP	864.9160	BLOOD GROUP SUBSTANCES OF NONHUMAN ORIGIN FOR IN VITRO DIAGNOSTIC USE Products Used In Establishments That Manufacture Blood And Blood Products; Hematology And Pathology Devices
BLOOD GROUPING	660.20	BLOOD GROUPING SERUM Blood Grouping Reagent; Additional Standards For Diagnostic Substances For Laboratory Tests
BLOOD GROUPING	864.9175	AUTOMATED BLOOD GROUPING AND ANTIBODY TEST SYSTEM Products Used In Establishments That Manufacture Blood And Blood Products; Hematology And Pathology Devices
BLOOD GROUPING	864.9185	BLOOD GROUPING VIEW BOX Products Used In Establishments That Manufacture Blood And Blood Products; Hematology And Pathology Devices
BLOOD PRESSURE	870.1100	BLOOD PRESSURE ALARM Cardiovascular Diagnostic Devices; Cardiovascular Devices
BLOOD PRESSURE	870.1110	BLOOD PRESSURE COMPUTER Cardiovascular Diagnostic Devices; Cardiovascular Devices
BLOOD PRESSURE	870.1120	BLOOD PRESSURE CUFF Cardiovascular Diagnostic Devices; Cardiovascular Devices
BLOOD PRESSURE	870.1130	NONINVASIVE BLOOD PRESSURE MEASUREMENT SYSTEM Cardiovascular Diagnostic Devices; Cardiovascular Devices
BLOOD PRESSURE	870.1140	VENOUS BLOOD PRESSURE MANOMETER Cardiovascular Diagnostic Devices; Cardiovascular Devices
BLOOD PRESSURE	870.2850	EXTRAVASCULAR BLOOD PRESSURE TRANSDUCER Cardiovascular Monitoring Devices; Cardiovascular Devices
BLOOD PRODUCT	607.20	WHO MUST REGISTER AND SUBMIT A BLOOD PRODUCT LIST Procedures For Domestic Blood Product Establishments; Establishment Registratn+product Listing, Manufacturers Human Blood/b.products
BLOOD PRODUCT	607.21	TIMES FOR ESTABLISHMENT REGISTRATION AND BLOOD PRODUCT LISTING Procedures For Domestic Blood Product Establishments; Establishment Registratn+product Licting, Manufacturers Human Blood/b.products
BLOOD PRODUCT	607.25	INFORMATION REQUIRED FOR ESTABLISHMENT REGISTRATION AND BLOOD PRODUCT LISTING Procedures For Domestic Blood Product Establishments; Establishment Registratn+product Listing, Manufacturers Human Blood/b.products
BLOOD PRODUCT	607.30	UPDATING BLOOD PRODUCT LISTING INFORMATION Procedures For Domestic Blood Product Establishments; Establishment Registratn+product Listing, Manufacturers Human Blood/b.products
BLOOD PRODUCT	607.31	ADDITIONAL BLOOD PRODUCT LISTING INFORMATION Procedures For Domestic Blood Product Establishments; Establishment Registratn+product Listing, Manufacturers Human Blood/b.products
BLOOD PRODUCT	607.35	NOTIFICATION OF REGISTRANT; BLOOD PRODUCT ESTABLISHMENT REGISTRATION NUMBER AND NDC LABELER CODE Procedures For Domestic Blood Product Establishments; Establishment Registratn+product Listing, Manufacturers Human Blood/b.products

BLOOD PRODUCT	607.37	INSPECTION OF ESTABLISHMENT REGISTRATIONS AND BLOOD PRODUCT LISTINGS Procedures For Domestic Blood Product Establishments; Establishment Registratn+product Listing, Manufacturers Human Blood/b.products
BLOOD PRODUCT	607.40	ESTABLISHMENT REGISTRATION AND BLOOD PRODUCT LISTING REQUIREMENTS FOR FOREIGN BLOOD PRODUCT ESTABLISHMENTS Procedures For Foreign Blood Product Establishments; Establishment Registratn+product Listing, Manufacturers Human Blood/b.products
BLOOD PRODUCT	607.65	EXEMPTIONS FOR BLOOD PRODUCT ESTABLISHMENTS Exemptions; Establishment Registratn+product Listing, Manufacturers Human Blood/ b.products
BLOOD PRODUCTS	207.7	ESTABLISHMENT REGISTRATION AND PRODUCT LISTING FOR HUMAN BLOOD AND BLOOD PRODUCTS AND FOR MEDICAL DEVICES General; Registration Of Drug Producers And Listing Of Drugs In Commercial Distribution
BLOOD PRODUCTS	607.7	ESTABLISHMENT REGISTRATION AND PRODUCT LISTING OF BLOOD BANKS AND OTHER FIRMS MANUFACTURING HUMAN BLOOD AND BLOOD PRODUCTS General Provisions; Establishment Registratn+product Listing, Manufacturers Human Blood/b.products
BLOOD PRODUCTS	607.22	HOW AND WHERE TO REGISTER ESTABLISHMENTS AND LIST BLOOD PRODUCTS Procedures For Domestic Blood Product Establishments; Establishment Registratn+product Listing, Manufacturers Human Blood/b.products
BLOW BOTTLE	868.5220	BLOW BOTTLE Therapeutic Devices; Anesthesiology Devices
BLUE CHEESE	133.106	BLUE CHEESE Requirements For Specific Standardized Cheese And Related Products; Cheeses And Related Cheese Products
BLUE, METHYLENE	500.27	METHYLENE BLUE-CONTAINING DRUGS FOR USE IN ANIMALS Specific Administrative Rulings and Decisions; General
BLUE NO. 1	74.101	FD&C BLUE NO. 1 Foods; Listing Of Color Additives Subject To Certification
BLUE NO. 1	74.1101	FD&C BLUE NO. 1 Drugs; Listing Of Color Additives Subject To Certification
BLUE NO. 1	74.2101	FD&C BLUE NO. 1 Cosmetics; Listing Of Color Additives Subject To Certification
BLUE NO. 1	82.101	FD&C BLUE NO. 1 Foods, Drugs, And Cosmetics; Listing Of Certified Provisionally Listed Colors And Specifications
BLUE NO. 2	74.102	FD&C BLUE NO. 2 Foods; Listing Of Color Additives Subject To Certification
BLUE NO. 2	74.1102	FD&C BLUE NO. 2 Drugs; Listing Of Color Additives Subject To Certification
BLUE NO. 2	74.3102	FD&C BLUE NO. 2 Medical Devices; Listings Of Color Additives Subject To Certification
BLUE NO. 2	82.102	FD&C BLUE NO. 2 Foods, Drugs, And Cosmetics; Listing Of Certified Provisionally Listed Colors And Specifications
BLUE NO. 4	74.1104	D&C BLUE NO. 4 Drugs; Listing Of Color Additives Subject To Certification
BLUE NO. 4	74.2104	D&C BLUE NO. 4 Cosmetics; Listing Of Color Additives Subject To Certification
BLUE NO. 4	82.1104	D&C BLUE NO. 4 Drugs And Cosmetics; Listing Of Certified Provisionally Listed Colors And Specifications
BLUE NO. 6	74.3106	D&C BLUE NO. 6 Medical Devices; Listing Of Color Additives Subject To Certification
BLUE NO. 9	74.1109	D&C BLUE NO. 9 Drugs; Listing Of Color Additives Subject To Certification
BLUE, ULTRAMARINE.	73.50	ULTRAMARINE BLUE Foods; Listing Of Color Additives Exempt From Certification
BLUE-MOLD CHEESE	133.184	ROQUEFORT, SHEEP'S MILK BLUE-MOLD, AND BLUE-MOLD CHEESE FROM SHEEP'S MILK Requirements For Specific Standardized Cheese And Related Products; Cheeses And Related Cheese Products
BOARD	13.5	NOTICE OF A HEARING BEFORE A BOARD General Provisions; Public Hearing Before A Public Board Of Inquiry
BOARD	13.10	MEMBERS OF A BOARD General Provisions; Public Hearing Before A Public Board Of Inquiry
BOARD	13.20	SUBMISSIONS TO A BOARD Hearing Procedures; Public Hearing Before A Public Board Of Inquiry
BOARD	13.30	PROCEEDINGS OF A BOARD Hearing Procedures; Public Hearing Before A Public Board Of Inquiry

BOARD	13.40	ADMINISTRATIVE RECORD OF A BOARD Records Of A Hearing Before A Board; Public Hearing Before A Public Board Of Inquiry
BOARD, BED	880.6070	BED BOARD General Hospital And Personal Use Miscellaneous Devices; General Hospital And Personal Use Devices
BOARD, RESUSCITATION	880.6080	CARDIOPULMONARY RESUSCITATION BOARD General Hospital And Personal Use Miscellaneous Devices; General Hospital And Personal Use Devices
BOATS	1240.95	SANITATION OF WATER BOATS Source And Use Of Potable Water; Control Of Communicable Diseases
BODY	701.20	DETERGENT SUBSTANCES, OTHER THAN SOAP, INTENDED FOR USE IN CLEANSING THE BODY Labeling Of Specific Ingredients; Cosmetic Labeling
BODY COUNTER, NUCLEAR	892.1130	NUCLEAR WHOLE BODY COUNTER Diagnostic Devices; Radiology Devices
BODY FLUID	880.6740	VACUUM-POWERED BODY FLUID SUCTION APPARATUS General Hospital And Personal Use Miscellaneous Devices; General Hospital And Personal Use Devices
BODY SCANNER	892.1330	NUCLEAR WHOLE BODY SCANNER Diagnostic Devices; Radiology Devices
BODY WASTE	880.6730	BODY WASTE RECEPTACLE General Hospital And Personal Use Miscellaneous Devices; General Hospital And Personal Use Devices
BODY WASTE	880.6800	WASHER FOR BODY WASTE RECEPTACLES General Hospital And Personal Use Miscellaneous Devices; General Hospital And Personal Use Devices
BODY WEIGHT	105.66	LABEL STATEMENTS RELATING TO USEFULNESS IN REDUCING OR MAINTAINING CALORIC INTAKE OR BODY WEIGHT Label Statements; Foods For Special Dietary Use
BOILER WATER	173.310	BOILER WATER ADDITIVES Specific Usage Additives; Secondary Direct Food Additives Permitted In Food For Human Consumption
BOILING WATER STERILIZER	872.6710	BOILING WATER STERILIZER Miscellaneous Devices; Dental Devices
BOILS	310.531	DRUG PRODUCTS CONTAINING ACTIVE INGREDIENTS OFFERED OTC FOR TREATMENT OF BOILS Requirements For Specific New Drugs Or Devices; New Drugs
BOLDENONE UNDECYLENATE	522.204	BOLDENONE UNDECYLENATE INJECTION Implantation Or Injectable Dosage Form New Animal Drugs
BOLTED CORN MEAL	137.255	BOLTED WHITE CORN MEAL Requirements For Specific Standardized Cereal Flours And Related Products; Cereal Flours And Related Products
BOLTED CORN MEAL	137.280	BOLTED YELLOW CORN MEAL Requirements For Specific Standardized Cereal Flours And Related Products; Cereal Flours And Related Products
BOLUS	520.1197	IVERMECTIN SUSTAINED-RELEASE BOLUS Oral Dosage Form New Animal Drugs
BOLUS	520.1242b	LEVAMISOLE HYDROCHLORIDE TABLET OR OBLET (BOLUS) Oral Dosage Form New Animal Drugs
BOLUS	520.1450a	MORANTEL TARTRATE BOLUS Oral Dosage Form New Animal Drugs
BOLUS	520.2200a	SULFACHLORPYRIDAZINE BOLUS Oral Dosage Form New Animal Drugs
BOLUS	520.2260a	SULFAMETHAZINE OBLET, TABLET AND BOLUS Oral Dosage Form New Animal Drugs
BOLUS	520.2380c	THIABENDAZOLE BOLUS Oral Dosage Form New Animal Drugs
BOLUSES	520.88e	AMOXICILLIN TRIHYDRATE BOLUSES Oral Dosage Form New Animal Drugs
BOLUSES	520.90f	AMPICILLIN TRIHYDRATE BOLUSES Oral Dosage Form New Animal Drugs
BOLUSES	520.420	CHLOROTHIAZIDE TABLETS AND BOLUSES Oral Dosage Form New Animal Drugs
BOLUSES	520.445c	CHLORTETRACYCLINE TABLETS AND BOLUSES Oral Dosage Form New Animal Drugs
BOLUSES	520.540	DEXAMETHASONE TABLETS AND BOLUSES Oral Dosage Form New Animal Drugs
BOLUSES	520.1120b	HALOXON BOLUSES Oral Dosage Form New Animal Drugs
BOLUSES	520.1158	IODOCHLORHYDROXYQUIN BOLUSES Oral Dosage Form New Animal Drugs
BOLUSES	520.1660c	OXYTETRACYCLINE HYDROCHLORIDE TABLETS/BOLUSES Oral Dosage Form New Animal Drugs

BOLUSES	520.1720a	PHENYLBUTAZONE TABLETS AND BOLUSES Oral Dosage Form New Animal Drugs
BOLUSES	520.1802	PIPERAZINE-CARBON DISULFIDE COMPLEX BOLUSES Oral Dosage Form New Animal Drugs
BOLUSES	520.2170	SULFABROMOMETHAZINE SODIUM BOLUSES Oral Dosage Form New Animal Drugs
BOLUSES	520.2220b	SULFADIMETHOXINE TABLETS AND BOLUSES Oral Dosage Form New Animal Drugs
BOLUSES	520.2260b	SULFAMETHAZINE SUSTAINED-RELEASE BOLUSES Oral Dosage Form New Animal Drugs
BOLUSES	520.2345c	TETRACYCLINE BOLUSES Oral Dosage Form New Animal Drugs
BOLUSES	520.2520e	TRICHLOROFON BOLUSES Oral Dosage Form New Animal Drugs
BOND	1316.76	REQUIREMENTS AS TO CLAIM AND BOND Seizure, Forfeiture, And Disposition Of Property; Administrative Functions, Practices, And Procedures
BOND	1316.97	INITIATING JUDICIAL FORFEITURE PROCEEDING AGAINST A CONVEYANCE WITHIN 60 DAYS OF FILING A CLAIM AND COST BOND Expedited Forfeiture Proceedings For Certain Property; Administrative Functions, Practices, And Procedures
BOND	1316.98	SUBSTITUTE RES BOND IN A JUDICIAL FORFEITURE ACTION AGAINST A CONVEYANCE Expedited Forfeiture Proceedings For Certain Property; Administrative Functions, Practices, And Procedures
BONDING	872.3200	RESIN TOOTH BONDING AGENT Prosthetic Devices; Dental Devices
BONDS	1.97	BONDS Imports And Exports; General Enforcement Regulations
BONDS	1005.23	BONDS Bonding And Compliance Procedures; Importation Of Electronic Products
BONE CAP	888.3000	BONE CAP Prosthetic Devices; Orthopedic Devices
BONE CEMENT	888.3027	POLYMETHYLMETHACRYLATE (PMMA) BONE CEMENT Prosthetic Devices; Orthopedic Devices
BONE CUTTING	872.4120	BONE CUTTING INSTRUMENT AND ACCESSORIES Surgical Devices; Dental Devices
BONE DENSITOMETER	892.1170	BONE DENSITOMETER Diagnostic Devices; Radiology Devices
BONE FIXATION	888.3030	SINGLE/MULTIPLE COMPONENT METALLIC BONE FIXATION APPLIANCES AND ACCESSORIES Prosthetic Devices; Orthopedic Devices
BONE FIXATION	888.3040	SMOOTH OR THREADED METALLIC BONE FIXATION FASTENER Prosthetic Devices; Orthopedic Devices
BONE HETEROGRAFT	888.3000	BONE HETEROGRAFT Prosthetic Devices; Orthopedic Devices
BONE PARTICLE COLLECTOR	874.4800	BONE PARTICLE COLLECTOR Surgical Devices; Ear, Nose And Throat Devices
BONE PLATE	872.4760	BONE PLATE Surgical Devices; Dental Devices
BONE REPAIR	872.3930	TRICALCIUM PHOSPHATE GRANULES FOR DENTAL BONE REPAIR Prosthetic Devices; Dental Devices
BONITO	102.47	BONITO Requirements For Specific Nonstandardized Foods; Common Or Usual Name For Nonstandardized Foods
BORATE	872.3400	KARAYA AND SODIUM BORATE WITH OR WITHOUT ACACIA DENTURE ADHESIVE Prosthetic Devices; Dental Devices
BORDETELLA SPP.	866.3065	BORDETELLA SPP. SEROLOGICAL REAGENTS Serological Reagents; Immunology And Microbiology Devices
BOTHROPS ATROX	864.8100	BOTHROPS ATROX REAGENT Hematology Reagents; Hematology And Pathology Devices
BOTTLE	868.5220	BLOW BOTTLE Therapeutic Devices; Anesthesiology Devices
BOTTLE	880.6085	HOT/COLD WATER BOTTLE General Hospital And Personal Use Miscellaneous Devices; General Hospital And Personal Use Devices
BOTTLED WATER	165.110	BOTTLED WATER Beverages
BOTTLES, CONSTANT TEMP	1250.42	WATER SYSTEMS; CONSTANT TEMPERATURE BOTTLES Equipment And Operation Of Land And Air Conveyances; Interstate Conveyance Sanitation

BOTTLES, CONSTANT TEMP	1250.85	DRINKING FOUNTAINS AND COOLERS; ICE; CONSTANT TEMPERATURE BOTTLES Sanitation Facilities And Conditions On Vessels; Interstate Conveyance Sanitation
BOTTLES, WINE	189.301	TIN-COATED LEAD FOIL CAPSULES FOR WINE BOTTLES Substances Prohibited From Use In Human Foods
BOURDON GAUGE	868.2300	BOURDON GAUGE FLOWMETER Monitoring Devices; Anesthesiology Devices
BOX	864.9185	BLOOD GROUPING VIEW BOX Products Used In Establishments That Manufacture Blood And Blood Products; Hematology And Pathology Devices
BRACE	882.4325	CRANIAL DRILL HANDPIECE (BRACE) Neurological Surgical Devices; Neurological Devices
BRACHYTHERAPY	892.5730	RADIONUCLIDE BRACHYTHERAPY SOURCE Therapeutic Devices; Radiology Devices
BRACKET	872.5470	ORTHODONTIC PLASTIC BRACKET Therapeutic Devices; Dental Devices
BRACKET ADHESIVE	872.3750	BRACKET ADHESIVE RESIN AND TOOTH CONDITIONER Prosthetic Devices; Dental Devices
BRAN WAX	172.890	RICE BRAN WAX Multipurpose Additives; Food Additives Permitted For Direct Addition To Food For Human Consumption
BREAD	136.110	BREAD, ROLLS, AND BUNS Requirements For Specific Standardized Bakery Products; Bakery Products
BREAD	136.115	ENRICHED BREAD, ROLLS, AND BUNS Requirements For Specific Standardized Bakery Products; Bakery Products
BREAD	136.130	MILK BREAD, ROLLS, AND BUNS Requirements For Specific Standardized Bakery Products; Bakery Products
BREAD	136.160	RAISIN BREAD, ROLLS, AND BUNS Requirements For Specific Standardized Bakery Products; Bakery Products
BREAD	136.180	WHOLE WHEAT BREAD, ROLLS, AND BUNS Requirements For Specific Standardized Bakery Products; Bakery Products
BREADED SHRIMP	102.55	NONSTANDARDIZED BREADED COMPOSITE SHRIMP UNITS Requirements For Specific Nonstandardized Foods; Common Or Usual Name For Nonstandardized Foods
BREADED SHRIMP	161.175	FROZEN RAW BREADED SHRIMP Requirements For Specific Standardized Fish And Shellfish; Fish And Shellfish
BREADED SHRIMP	161.176	FROZEN RAW LIGHTLY BREADED SHRIMP Requirements For Specific Standardized Fish And Shellfish; Fish And Shellfish
BREAKFAST COCOA	163.112	BREAKFAST COCOA Requirements For Specific Standardized Cacao Products; Cacao Products
BREAST EVALUATION	892.1990	TRANSILLUMINATOR FOR BREAST EVALUATION Diagnostic Devices; Radiology Devices
BREAST-FEEDING	201.63	PREGNANCY/BREAST-FEEDING WARNING Labeling Requirements For Over-the-counter Drugs; Labeling
BREAST MILK	866.5170	BREAST MILK IMMUNOLOGICAL TEST SYSTEM Immunological Test Systems; Immunology And Microbiology Devices
BREAST PUMP	884.5150	NONPOWERED BREAST PUMP Obstetrical And Gynecological Devices
BREAST PUMP	884.5160	POWERED BREAST PUMP Obstetrical And Gynecological Devices
BREAST PROSTHESIS	878.3530	SILICONE INFLATABLE BREAST PROSTHESIS Prosthetic Devices; General And Plastic Surgery Devices
BREAST PROSTHESIS	878.3540	SILICONE GEL-FILLED BREAST PROSTHESIS Prosthetic Devices; General And Plastic Surgery Devices
BREATH-ALCOHOL	862.3050	BREATH-ALCOHOL TEST SYSTEM Clinical Toxicology Test Systems; Clinical Chemistry And Clinical Toxicology Devices
BREATHING	868.2375	BREATHING FREQUENCY MONITOR Monitoring Devices; Anesthesiology Devices
BREATHING	868.5240	ANESTHESIA BREATHING CIRCUIT Therapeutic Devices; Anesthesiology Devices
BREATHING	868.5250	BREATHING CIRCUIT CIRCULATOR Therapeutic Devices; Anesthesiology Devices
BREATHING	868.5260	BREATHING CIRCUIT BACTERIAL FILTER Therapeutic Devices; Anesthesiology Devices
BREATHING	868.5270	BREATHING SYSTEM HEATER Therapeutic Devices; Anesthesiology Devices
BREATHING	868.5280	BREATHING TUBE SUPPORT Therapeutic Devices; Anesthesiology Devices
BREATHING	868.5330	BREATHING GAS MIXER Therapeutic Devices; Anesthesiology Devices
BREATHING	868.5620	BREATHING MOUTHPIECE Therapeutic Devices; Anesthesiology Devices

BREATHING	868.5965	POSITIVE END EXPIRATORY PRESSURE BREATHING ATTACHMENT Therapeutic Devices; Anesthesiology Devices
BRICK CHEESE	133.108	BRICK CHEESE Requirements For Specific Standardized Cheese And Related Products; Cheeses And Related Cheese Products
BRICK CHEESE	133.109	BRICK CHEESE FOR MANUFACTURING Requirements For Specific Standardized Cheese And Related Products; Cheeses And Related Cheese Products
BRIDGE	872.3770	TEMPORARY CROWN AND BRIDGE RESIN Prosthetic Devices; Dental Devices
BRIEFS	12.96	BRIEFS AND ARGUMENT Hearing Procedures; Formal Evidentiary Public Hearing
BRIEFS	17.43	POSTHEARING BRIEFS Civil Money Penalties Hearings
BROMATE, POTASSIUM	172.730	POTASSIUM BROMATE Other Specific Usage Additives; Food Additives Permitted For Direct Addition To Food For Human Consumption
BROMATED FLOUR	137.155	BROMATED FLOUR Requirements For Specific Standardized Cereal Flours And Related Products; Cereal Flours And Related Products
BROMATED FLOUR	137.160	ENRICHED BROMATED FLOUR Requirements For Specific Standardized Cereal Flours And Related Products; Cereal Flours And Related Products
BROMATED WHEAT	137.205	BROMATED WHOLE WHEAT FLOUR Requirements For Specific Standardized Cereal Flours And Related Products; Cereal Flours And Related Products
BROMELAIN	184.1024	BROMELAIN Listing Of Specific Substances Affirmed As GRAS; Direct Food Substances Affirmed As Generally Recognized As Safe
BROMINATED	180.30	BROMINATED VEGETABLE OIL Specific Requirements For Certain Food Additives; Food Additives Permitted On Interim Basis Or In Contact With Food Pending Study
BRONCHIAL	868.5720	BRONCHIAL TUBE Therapeutic Devices; Anesthesiology Devices
BRONCHIAL	868.5740	TRACHEAL/BRONCHIAL DIFFERENTIAL VENTILATION TUBE Therapeutic Devices; Anesthesiology Devices
BRONCHODILATOR	341.16	BRONCHODILATOR ACTIVE INGREDIENTS Active Ingredients; Cold, Cough, Allergy, Bronchodilator, Antiasthmatic Drug Products OTC Human Use
BRONCHODILATOR	341.76	LABELING OF BRONCHODILATOR DRUG PRODUCTS Labeling; Cold, Cough, Allergy, Bronchodilator, Antiasthmatic Drug Products OTC Human Use
BRONCHOSCOPE	874.4680	BRONCHOSCOPE (FLEXIBLE OR RIGID) AND ACCESSORIES Surgical Devices; Ear, Nose And Throat Devices
BRONZE POWDER	73.1646	BRONZE POWDER Drugs; Listing Of Color Additives Exempt From Certification
BRONZE POWDER	73.2646	BRONZE POWDER Cosmetics; Listing Of Color Additives Exempt From Certification
BROWN	74.2151	D&C BROWN NO. 1 Cosmetics; Listing Of Color Additives Subject To Certification
BROWN	184.1120	BROWN ALGAE Listing Of Specific Substances Affirmed As GRAS; Direct Food Substances Affirmed As Generally Recognized As Safe
BRUCELLA SPP.	866.3085	BRUCELLA SPP. SEROLOGICAL REAGENTS Serological Reagents; Immunology And Microbiology Devices
BRUSH	868.5795	TRACHEAL TUBE CLEANING BRUSH Therapeutic Devices; Anesthesiology Devices
BRUSH, ENDOMETRIAL	884.1100	ENDOMETRIAL BRUSH Obstetrical And Gynecological Diagnostic Devices; Obstetrical And Gynecological Devices
BRUSH, HAIDLINGER	886.1090	HAIDLINGER BRUSH Diagnostic Devices; Ophthalmic Devices
BUBBLE	870.4205	CARDIOPULMONARY BYPASS BUBBLE DETECTOR Cardiovascular Surgical Devices; Cardiovascular Devices
BUCKLING	886.3300	ABSORBABLE IMPLANT (SCLERAL BUCKLING METHOD) Prosthetic Devices; Ophthalmic Devices
BUILDINGS	225.20	BUILDINGS Construction And Maintenance Of Facilities And Equipment; Current Good Manufacturing Practice For Medicated Feeds
BUILDINGS	225.120	BUILDINGS AND GROUNDS Facilities And Equipment; Current Good Manufacturing Practice For Medicated Feeds

BUILDINGS	226.20	BUILDINGS Construction And Maintenance Of Facilities And Equipment; Current Good Manufacturing Practice For Type A Medicated Articles
BULK MANUFACTURE	1301.33	APPLICATION FOR BULK MANUFACTURE OF SCHEDULE I AND II SUBSTANCES Action On Application For Registration: Revocation Or Suspension Of Registration; Registration Of Manufacturers, Distributors, And Dispensers Of Controlled Substances
BUNAMIDINE HYDROCHLORIDE	520.222	BUNAMIDINE HYDROCHLORIDE Oral Dosage Form New Animal Drugs
BUNS	136.110	BREAD, ROLLS, AND BUNS Requirements For Specific Standardized Bakery Products; Bakery Products
BUNS	136.115	ENRICHED BREAD, ROLLS, AND BUNS Requirements For Specific Standardized Bakery Products; Bakery Products
BUNS	136.130	MILK BREAD, ROLLS, AND BUNS Requirements For Specific Standardized Bakery Products; Bakery Products
BUNS	136.160	RAISIN BREAD, ROLLS, AND BUNS Requirements For Specific Standardized Bakery Products; Bakery Products
BUNS	136.180	WHOLE WHEAT BREAD, ROLLS, AND BUNS Requirements For Specific Standardized Bakery Products; Bakery Products
BUQUINOLATE	556.90	BUQUINOLATE Specific Tolerances For Residues Of New Animal Drugs; Tolerances For Residues Of New Animal Drugs In Food
BUR, DENTAL	872.3240	DENTAL BUR Prosthetic Devices; Dental Devices
BUR, ENT	874.4140	EAR, NOSE AND THROAT BUR Surgical Devices; Ear, Nose And Throat Devices
BURDEN OF PROOF	12.87	PURPOSE; ORAL AND WRITTEN TESTIMONY; BURDEN OF PROOF Hearing Procedures; Formal Evidentiary Hearing
BURDEN OF PROOF	17.33	THE HEARING AND BURDEN OF PROOF Civil Money Penalties Hearings
BURDEN OF PROOF	1301.44	BURDEN OF PROOF Hearings; Registration Of Manufacturers, Distributors, And Dispensers Of Controlled Substances
BURDEN OF PROOF	1303.35	BURDEN OF PROOF Hearings; Quotas
BURDEN OF PROOF	1309.54	BURDEN OF PROOF Hearings; Registration Of Manufacturers, Distributors, Importers And Exporters Of List I Chemicals
BURDEN OF PROOF	1312.45	BURDEN OF PROOF Hearings; Importation And Exportation Of Controlled Substances
BURDEN OF PROOF	1313.55	BURDEN OF PROOF Hearings; Importation And Exportation Of Precursors And Essential Chemicals
BURDEN OF PROOF	1316.56	BURDEN OF PROOF Administrative Hearings; Administrative Functions, Practices, And Procedures
BURDENSOME REQUESTS	20.48	NONSPECIFIC AND OVERLY BURDENSOME REQUESTS Procedures And Fees; Public Information
BURN DRESSING	878.4022	HYDROGEL WOUND DRESSING AND BURN DRESSING Surgical Devices; General And Plastic Surgery Devices
BURN SHEET	880.5180	BURN SHEET General Hospital And Personal Use Therapeutic Devices; General Hospital And Personal Use Devices
BURR	882.5250	BURR HOLE COVER Neurological Therapeutic Devices; Neurological Devices
BURR	886.4070	POWERED CORNEAL BURR Surgical Devices; Ophthalmic Devices
BURRS	882.4300	MANUAL CRANIAL DRILLS, BURRS, TREPHINES, AND THEIR ACCESSORIES Neurological Surgical Devices; Neurological Devices
BURRS	882.4305	POWERED COMPOUND CRANIAL DRILLS, BURRS, TREPHINES, AND THEIR ACCESSORIES Neurological Surgical Devices; Neurological Devices
BURRS	882.4310	POWERED SIMPLE CRANIAL DRILLS, BURRS, TREPHINES, AND THEIR ACCESSORIES Neurological Surgical Devices; Neurological Devices
BUSINESS	101.5	FOOD; NAME AND PLACE OF BUSINESS OF MANUFACTURER, PACKER, OR DISTRIBUTOR General Provisions; Food Labeling
BUSINESS	201.1	DRUGS; NAME AND PLACE OF BUSINESS OF MANUFACTURER, PACKER OR DISTRIBUTOR General Labeling Provisions; Labeling
BUSINESS	501.5	ANIMAL FOOD; NAME AND PLACE OF BUSINESS OF MANUFACTURER, PACKER, OR DISTRIBUTOR General Provisions; Animal Food Labeling

BUSINESS	701.12	NAME AND PLACE OF BUSINESS OF MANUFACTURER, PACKER, OR DISTRIBUTOR Package Form; Cosmetic Labeling
BUSINESS	801.1	MEDICAL DEVICES; NAME AND PLACE OF BUSINESS OF MANUFACTURER, PACKER, OR DISTRIBUTOR General Labeling Provisions; Labeling
BUSINESS, DISCONTINUANCE	1301.52	TERMINATION OF REGISTRATION; TRANSFER OF REGISTRATION; DISTRIBUTION UPON DISCONTINUANCE OF BUSINESS Registration Of Manufacturers, Distributors, And Dispensers Of Controlled Substances
BUTACAINE	524.1580e	NITROFURAZONE OINTMENT WITH BUTACAINE SULFATE Ophthalmic And Topical Dosage Form New Animal Drugs
BUTADIENE/STYRENE	177.1020	ACRYLONITRILE/BUTADIENE/STYRENE COPOLYMER Substances For Use As Basic Components Of Single And Repeated Use Food Contact Surfaces; Indirect Food Additives: Polymers
BUTADIENE/STYRENE	177.1030	ACRYLONITRILE/BUTADIENE/STYRENE/METHYL METHACRYLATE COPOLYMER Substances For Use As Basic Components Of Single And Repeated Use Food Contact Surfaces; Indirect Food Additives: Polymers
BUTADIENE/STYRENE	177.1050	ACRYLONITRILE/STYRENE COPOLYMER MODIFIED WITH BUTADIENE/STYRENE ELASTOMER Substances For Use As Basic Components Of Single And Repeated Use Food Contact Surfaces; Indirect Food Additives: Polymers
BUTAMISOLE HYDROCHLORIDE	522.234	BUTAMISOLE HYDROCHLORIDE Implantation Or Injectable Dosage Form New Animal Drugs
BUTANE	184.1165	N-BUTANE AND ISO-BUTANE Listing Of Specific Substances Affirmed As GRAS; Direct Food Substances Affirmed As Generally Recognized As Safe
BUTANE	582.1165	BUTANE General Purpose Food Additives; Substances Generally Recognized As Safe
BUTENE	177.1430	ISOBUTYLENE-BUTENE COPOLYMERS Substances For Use As Basic Components Of Single And Repeated Use Food Contact Surfaces; Indirect Food Additives: Polymers
BUTENE	177.1570	POLY-1-BUTENE RESINS AND BUTENE/ETHYLENE COPOLYMERS Substances For Use As Basic Components Of Single And Repeated Use Food Contact Surfaces; Indirect Food Additives: Polymers
BUTORPHANOL TARTRATE	520.246	BUTORPHANOL TARTRATE TABLETS Oral Dosage Form New Animal Drugs
BUTORPHANOL TARTRATE	522.246	BUTORPHANOL TARTRATE INJECTION Implantation Or Injectable Dosage Form New Animal Drugs
BUTOXIDE	178.3730	PIPERONYL BUTOXIDE AND PYRETHRINS AS COMPONENTS OF BAGS Certain Adjuvants And Production Aids; Indirect Food Additives: Adjuvants, Production Aids, And Sanitizers
BUTTER	101.67	USE OF NUTRITION CONTENT CLAIMS FOR BUTTER Specific Requirements For Nutrient Content Claims; Food Labeling
BUTTER, COCOA	172.861	COCOA BUTTER SUBSTITUTE FROM COCONUT OIL, PALM KERNEL OIL, OR BOTH OILS Food Additives Permitted For Direct Addition To Food For Human Consumption
BUTTER, COCOA	184.1259	COCOA BUTTER SUBSTITUTE Listing Of Specific Substances Affirmed As GRAS; Direct Food Substances Affirmed As Generally Recognized As Safe
BUTTER, FRUIT	150.110	FRUIT BUTTER Requirements For Specific Standardized Fruit Butters, Jellies, Preserves; Fruit Butters, Jellies, Preserves, And Related Products
BUTTER, PEANUT	164.150	PEANUT BUTTER Requirements For Specific Standardized Tree Nut And Peanut Products; Tree Nut And Peanut Products
BUTTERMILK	163.135	BUTTERMILK CHOCOLATE Requirements For Specific Standardized Cacao Products; Cacao Products
BUTYL CHLORIDE	520.260	N-BUTYL CHLORIDE CAPSULES Oral Dosage Form New Animal Drugs
BUTYL OLEATE	172.270	SULFATED BUTYL OLEATE Coatings, Films And Related Substances; Food Additives Permitted For Direct Addition To Food For Human Consumption
BUTYL-PHENOL	172.150	4-HYDROXYMETHYL-2,6-DI-TERT-BUTYL-PHENOL Food Preservatives; Food Additives Permitted For Direct Addition To Food For Human Consumption
BUTYL-PHENOL	178.2550	4-HYDROXYMETHYL-2,6-DI-TERT. BUTYL-PHENOL Antioxidants And Stabilizers; Indirect Food Additives: Adjuvants, Production Aids, And Sanitizers
BUTYLATED HYDROXYANISOLE	182.3169	BUTYLATED HYDROXYANISOLE Chemical Preservatives; Substances Generally Recognized As Safe
BUTYLATED HYDROXYANISOLE	582.3169	BUTYLATED HYDROXYANISOLE Chemical Preservatives; Substances Generally Recognized As Safe

BUTYLATED HYDROXYTOLUENE	182.3173	BUTYLATED HYDROXYTOLUENE Chemical Preservatives; Substances Generally Recognized As Safe
BUTYLATED HYDROXYTOLUENE	582.3173	BUTYLATED HYDROXYTOLUENE Chemical Preservatives; Substances Generally Recognized As Safe
BUTYLENE GLYCOL	573.225	1,3-BUTYLENE GLYCOL Food Additive Listing; Food Additives Permitted In Feed And Drinking Water Of Animals
BUTYLENE GLYCOL	173.220	1,3-BUTYLENE GLYCOL Solvents, Lubricants, Release Agents And Related Substances; Secondary Direct Food Additives Permitted In Food For Human Consumption
BYPASS	870.3545	VENTRICULAR BYPASS (ASSIST) DEVICE Cardiovascular Prosthetic Devices; Cardiovascular Devices
BYPASS	870.4200	CARDIOPULMONARY BYPASS ACCESSORY EQUIPMENT Cardiovascular Surgical Devices; Cardiovascular Devices
BYPASS	870.4205	CARDIOPULMONARY BYPASS BUBBLE DETECTOR Cardiovascular Surgical Devices; Cardiovascular Devices
BYPASS	870.4210	CARDIOPULMONARY BYPASS VASCULAR CATHETER, CANNULA, OR TUBING Cardiovascular Surgical Devices; Cardiovascular Devices
BYPASS	870.4220	CARDIOPULMONARY BYPASS HEART-LUNG MACHINE CONSOLE Cardiovascular Surgical Devices; Cardiovascular Devices
BYPASS	870.4230	CARDIOPULMONARY BYPASS DEFOAMER Cardiovascular Surgical Devices; Cardiovascular Devices
BYPASS	870.4240	CARDIOPULMONARY BYPASS HEAT EXCHANGER Cardiovascular Surgical Devices; Cardiovascular Devices
BYPASS	870.4250	CARDIOPULMONARY BYPASS TEMPERATURE CONTROLLER Cardiovascular Surgical Devices; Cardiovascular Devices
BYPASS	870.4260	CARDIOPULMONARY BYPASS ARTERIAL LINE BLOOD FILTER Cardiovascular Surgical Devices; Cardiovascular Devices
BYPASS	870.4270	CARDIOPULMONARY BYPASS CARDIOTOMY SUCTION LINE BLOOD FILTER Cardiovascular Surgical Devices; Cardiovascular Devices
BYPASS	870.4290	CARDIOPULMONARY BYPASS ADAPTOR, STOPCOCK, MANIFOLD, OR FITTING Cardiovascular Surgical Devices; Cardiovascular Devices
BYPASS	870.4300	CARDIOPULMONARY BYPASS GAS CONTROL UNIT Cardiovascular Surgical Devices; Cardiovascular Devices
BYPASS	870.4310	CARDIOPULMONARY BYPASS CORONARY PRESSURE GAUGE Cardiovascular Surgical Devices; Cardiovascular Devices
BYPASS	870.4320	CARDIOPULMONARY BYPASS PULSATILE FLOW GENERATOR Cardiovascular Surgical Devices; Cardiovascular Devices
BYPASS	870.4330	CARDIOPULMONARY BYPASS ON-LINE BLOOD GAS MONITOR Cardiovascular Surgical Devices; Cardiovascular Devices
BYPASS	870.4340	CARDIOPULMONARY BYPASS LEVEL SENSING MONITOR AND/OR CONTROL Cardiovascular Surgical Devices; Cardiovascular Devices
BYPASS	870.4350	CARDIOPULMONARY BYPASS OXYGENATOR Cardiovascular Surgical Devices; Cardiovascular Devices
BYPASS	870.4360	NONROLLER-TYPE CARDIOPULMONARY BYPASS BLOOD PUMP Cardiovascular Surgical Devices; Cardiovascular Devices
BYPASS	870.4370	ROLLER-TYPE CARDIOPULMONARY BYPASS BLOOD PUMP Cardiovascular Surgical Devices; Cardiovascular Devices
BYPASS	870.4380	CARDIOPULMONARY BYPASS PUMP SPEED CONTROL Cardiovascular Surgical Devices; Cardiovascular Devices
BYPASS	870.4390	CARDIOPULMONARY BYPASS PUMP TUBING Cardiovascular Surgical Devices; Cardiovascular Devices
BYPASS	870.4400	CARDIOPULMONARY BYPASS BLOOD RESERVOIR Cardiovascular Surgical Devices; Cardiovascular Devices
BYPASS	870.4410	CARDIOPULMONARY BYPASS IN-LINE BLOOD GAS SENSOR Cardiovascular Surgical Devices; Cardiovascular Devices
BYPASS	870.4420	CARDIOPULMONARY BYPASS CARDIOTOMY RETURN SUCKER Cardiovascular Surgical Devices; Cardiovascular Devices
BYPASS	870.4430	CARDIOPULMONARY BYPASS INTRACARDIAC SUCTION CONTROL Cardiovascular Surgical Devices; Cardiovascular Devices

C

C1 INHIBITOR	866.5250	COMPLEMENT C1 INHIBITOR (INACTIVATOR) IMMUNOLOGICAL TEST SYSTEM Immunological Test Systems; Immunology And Microbiology Devices
C3B COMPLEMENT, INACTIVATOR	866.5260	COMPLEMENT C3B INACTIVATOR IMMUNOLOGICAL TEST SYSTEM Immunological Test Systems; Immunology And Microbiology Devices
C. I. VAT ORANGE 1	73.3112	C. I. VAT ORANGE 1 Medical Devices; Listing Of Color Additives Exempt From Certification
C PEPTIDES	862.1135	C-PEPTIDES OF PROINSULIN TEST SYSTEM Clinical Chemistry Test Systems; Clinical Chemistry And Clinical Toxicology Devices
C REACTIVE PROTEIN	866.5270	C-REACTIVE PROTEIN IMMUNOLOGICAL TEST SYSTEM Immunological Test Systems; Immunology And Microbiology Devices
C, TYPE	558.465	POLOXALENE FREE-CHOICE LIQUID TYPE C FEED New Animal Drugs For Use In Animal Feeds
CABINET	868.6100	ANESTHETIC CABINET, TABLE, OR TRAY Miscellaneous; Anesthesiology Devices
CABINET	880.6100	ETHYLENE OXIDE GAS AERATOR CABINET General Hospital And Personal Use Miscellaneous Devices; General Hospital And Personal Use Devices
CABINET	890.5250	MOIST STEAM CABINET Physical Medicine Therapeutic Devices; Physical Medicine Devices
CABINET	1020.40	CABINET X-RAY SYSTEM Performance Standards For Ionizing Radiation Emitting Products
CABLE	870.2900	PATIENT TRANSDUCER AND ELECTRODE CABLE Cardiovascular Monitoring Devices; Cardiovascular Devices
CABLE	890.1175	ELECTRODE CABLE Physical Medicine Diagnostic Devices; Physical Medicine Devices
CABS	26.36	LISTING OF CABS Specific Sector Provisions For Medical Devices; Mutual Recognition Of Pharmaceutical GMP Reports, Medical Device Quality System Audit Reports
CABS	26.46	LISTING OF ADDITIONAL CABS Specific Sector Provisions For Medical Devices; Mutual Recognition Of Pharmaceutical GMP Reports, Medical Device Quality System Audit Reports
CACAO NIBS	163.110	CACAO NIBS Requirements For Specific Standardized Cacao Products; Cacao Products
CACIOCAVALLO CHEESE	133.111	CACIOCAVALLO SICILIANO CHEESE Requirements For Specific Standardized Cheese And Related Products; Cheeses And Related Cheese Products
CAFFEINE	182.1180	CAFFEINE Multiple Purpose GRAS Food Substances; Substances Generally Recognized As Safe
CALAMUS DERIVATIVES	189.110	CALAMUS AND ITS DERIVATIVES Substances Generally Prohibited From Direct Addition Or Use As Human Food; Substances Prohibited From Use In Human Food
CALCITONIN	862.1140	CALCITONIN TEST SYSTEM Clinical Chemistry Test Systems; Clinical Chemistry And Clinical Toxicology Devices
CALCIUM	101.72	HEALTH CLAIMS: CALCIUM AND OSTEOPOROSIS Specific Requirements For Health Claims; Food Labeling
CALCIUM	862.1145	CALCIUM TEST SYSTEM Clinical Chemistry Test Systems; Clinical Chemistry And Clinical Toxicology Devices
CALCIUM ACETATE	184.1185	CALCIUM ACETATE Listing Of Specific Substances Affirmed As GRAS; Direct Food Substances Affirmed As Generally Recognized As Safe
CALCIUM ACETATE	582.6185	CALCIUM ACETATE Sequestrants; Substances Generally Recognized As Safe
CALCIUM ALGINATE	184.1187	CALCIUM ALGINATE Listing Of Specific Substances Affirmed As GRAS; Direct Food Substances Affirmed As Generally Recognized As Safe
CALCIUM ALGINATE	582.7187	CALCIUM ALGINATE Stabilizers; Substances Generally Recognized As Safe
CALCIUM ALUMINOSILICATE	182.2729	SODIUM CALCIUM ALUMINOSILICATE, HYDRATED Anticaking Agents; Substances Generally Recognized As Safe
CALCIUM ALUMINOSILICATE	582.2729	HYDRATED SODIUM CALCIUM ALUMINOSILICATE Anticaking Agents; Substances Generally Recognized As Safe
CALCIUM AMPHOMYCIN	524.1204	KANAMYCIN SULFATE, CALCIUM AMPHOMYCIN, AND HYDROCORTISONE ACETATE Ophthalmic And Topical Dosage Form New Animal Drugs

CALCIUM ASCORBATE	182.3189	CALCIUM ASCORBATE Chemical Preservatives; Substances Generally Recognized As Safe
CALCIUM ASCORBATE	582.3189	CALCIUM ASCORBATE Chemical Preservatives; Substances Generally Recognized As Safe
CALCIUM CARBONATE	73.1070	CALCIUM CARBONATE Drugs; Listing Of Color Additives Exempt From Certification
CALCIUM CARBONATE	184.1191	CALCIUM CARBONATE Listing Of Specific Substances Affirmed As GRAS; Direct Food Substances Affirmed As Generally Recognized As Safe
CALCIUM CARBONATE	582.1191	CALCIUM CARBONATE General Purpose Food Additives; Substances Generally Recognized As Safe
CALCIUM CARBONATE	582.5191	CALCIUM CARBONATE Nutrients And/or Dietary Supplements; Substances Generally Recognized As Safe
CALCIUM CHLORIDE	172.330	CALCIUM PANTOTHENATE, CALCIUM CHLORIDE DOUBLE SALT Special Dietary And Nutritional Additives; Food Additives Permitted For Direct Addition To Food For Human Consumption
CALCIUM CHLORIDE	184.1193	CALCIUM CHLORIDE Listing Of Specific Substances Affirmed As GRAS; Direct Food Substances Affirmed As Generally Recognized As Safe
CALCIUM CHLORIDE	582.1193	CALCIUM CHLORIDE General Purpose Food Additives; Substances Generally Recognized As Safe
CALCIUM CHLORIDE	582.6193	CALCIUM CHLORIDE Sequestrants; Substances Generally Recognized As Safe
CALCIUM CITRATE	182.8195	CALCIUM CITRATE Nutrients; Substances Generally Recognized As Safe
CALCIUM CITRATE	184.1195	CALCIUM CITRATE Listing Of Specific Substances Affirmed As GRAS; Direct Food Substances Affirmed As Generally Recognized As Safe
CALCIUM CITRATE	582.1195	CALCIUM CITRATE General Purpose Food Additives; Substances Generally Recognized As Safe
CALCIUM CITRATE	582.5195	CALCIUM CITRATE Nutrients And/or Dietary Supplements; Substances Generally Recognized As Safe
CALCIUM CITRATE	582.6195	CALCIUM CITRATE Sequestrants; Substances Generally Recognized As Safe
CALCIUM DIACETATE	182.6197	CALCIUM DIACETATE Sequestrants; Substances Generally Recognized As Safe
CALCIUM DIACETATE	582.6197	CALCIUM DIACETATE Sequestrants; Substances Generally Recognized As Safe
CALCIUM EDTA	172.120	CALCIUM DISODIUM EDTA Food Preservatives; Food Additives Permitted For Direct Addition To Food For Human Consumption
CALCIUM GLUCONATE	184.1199	CALCIUM GLUCONATE Listing Of Specific Substances Affirmed As GRAS; Direct Food Substances Affirmed As Generally Recognized As Safe
CALCIUM GLUCONATE	582.1199	CALCIUM GLUCONATE General Purpose Food Additives; Substances Generally Recognized As Safe
CALCIUM GLUCONATE	582.6199	CALCIUM GLUCONATE Sequestrants; Substances Generally Recognized As Safe
CALCIUM GLYCEROPHOSPHATE	184.1201	CALCIUM GLYCEROPHOSPHATE Listing Of Specific Substances Affirmed As GRAS; Direct Food Substances Affirmed As Generally Recognized As Safe
CALCIUM GLYCEROPHOSPHATE	582.5201	CALCIUM GLYCEROPHOSPHATE Nutrients And/or Dietary Supplements; Substances Generally Recognized As Safe
CALCIUM HEXAMETAPHOSPHATE	182.6203	CALCIUM HEXAMETAPHOSPHATE Sequestrants; Substances Generally Recognized As Safe
CALCIUM HEXAMETAPHOSPHATE	582.6203	CALCIUM HEXAMETAPHOSPHATE Sequestrants; Substances Generally Recognized As Safe
CALCIUM HYDROXIDE	184.1205	CALCIUM HYDROXIDE Listing Of Specific Substances Affirmed As GRAS; Direct Food Substances Affirmed As Generally Recognized As Safe
CALCIUM HYDROXIDE	582.1205	CALCIUM HYDROXIDE General Purpose Food Additives; Substances Generally Recognized As Safe
CALCIUM HYDROXIDE	872.3250	CALCIUM HYDROXIDE CAVITY LINER Prosthetic Devices; Dental Devices
CALCIUM IODATE	184.1206	CALCIUM IODATE Listing Of Specific Substances Affirmed As GRAS; Direct Food Substances Affirmed As Generally Recognized As Safe
CALCIUM LACTATE	184.1207	CALCIUM LACTATE Listing Of Specific Substances Affirmed As GRAS; Direct Food Substances Affirmed As Generally Recognized As Safe
CALCIUM LACTATE	582.1207	CALCIUM LACTATE General Purpose Food Additives; Substances Generally Recognized As Safe

CALCIUM LACTOBIONATE	172.720	CALCIUM LACTOBIONATE Other Specific Usage Additives; Food Additives Permitted For Direct Addition To Food For Human Consumption
CALCIUM LACTYLATE	172.844	CALCIUM STEAROYL-2-LACTYLATE Multipurpose Additives; Food Additives Permitted For Direct Addition To Food For Human Consumption
CALCIUM LIGNOSULFONATE	172.715	CALCIUM LIGNOSULFONATE Other Specific Usage Additives; Food Additives Permitted For Direct Addition To Food For Human Consumption
CALCIUM OXIDE	184.1210	CALCIUM OXIDE Listing Of Specific Substances Affirmed As GRAS; Direct Food Substances Affirmed As Generally Recognized As Safe
CALCIUM OXIDE	582.1210	CALCIUM OXIDE General Purpose Food Additives; Substances Generally Recognized As Safe
CALCIUM OXIDE	582.5210	CALCIUM OXIDE Nutrients And/or Dietary Supplements; Substances Generally Recognized As Safe
CALCIUM PANTOTHENATE	172.330	CALCIUM PANTOTHENATE, CALCIUM CHLORIDE DOUBLE SALT Special Dietary And Nutritional Additives; Food Additives Permitted For Direct Addition To Food For Human Consumption
CALCIUM PANTOTHENATE	184.1212	CALCIUM PANTOTHENATE Listing Of Specific Substances Affirmed As GRAS; Direct Food Substances Affirmed As Generally Recognized As Safe
CALCIUM PANTOTHENATE	582.5212	CALCIUM PANTOTHENATE Nutrients And/or Dietary Supplements; Substances Generally Recognized As Safe
CALCIUM PERIODATE	573.240	CALCIUM PERIODATE Food Additive Listing; Food Additives Permitted In Feed And Drinking Water Of Animals
CALCIUM PHOSPHATE	182.1217	CALCIUM PHOSPHATE Multiple Purpose GRAS Food Substances; Substances Generally Recognized As Safe
CALCIUM PHOSPHATE	182.6215	MONOBASIC CALCIUM PHOSPHATE Sequestrants; Substances Generally Recognized As Safe
CALCIUM PHOSPHATE	182.8217	CALCIUM PHOSPHATE Nutrients; Substances Generally Recognized As Safe
CALCIUM PHOSPHATE	582.1217	CALCIUM PHOSPHATE General Purpose Food Additives; Substances Generally Recognized As Safe
CALCIUM PHOSPHATE	582.5217	CALCIUM PHOSPHATE Nutrients And/or Dietary Supplements; Substances Generally Recognized As Safe
CALCIUM PHOSPHATE	582.6215	MONOBASIC CALCIUM PHOSPHATE Sequestrants; Substances Generally Recognized As Safe
CALCIUM PHYTATE	582.6219	CALCIUM PHYTATE Sequestrants; Substances Generally Recognized As Safe
CALCIUM POLYCARBOPHIL	201.319	WATER-SOLUBLE GUMS, AND HYDROPHILIC MUCILLOIDS (INCLUDING CALCIUM POLYCARBOPHIL) AS ACTIVE INGREDIENTS; WARNINGS AND DIRECTIONS Specific Labeling Requirements For Specific Drug Products; Labeling
CALCIUM PROPIONATE	184.1221	CALCIUM PROPIONATE Listing Of Specific Substances Affirmed As GRAS; Direct Food Substances Affirmed As Generally Recognized As Safe
CALCIUM PROPIONATE	582.3221	CALCIUM PROPIONATE Chemical Preservatives; Substances Generally Recognized As Safe
CALCIUM PYROPHOSPHATE	182.8223	CALCIUM PYROPHOSPHATE Nutrients; Substances Generally Recognized As Safe
CALCIUM PYROPHOSPHATE	582.5223	CALCIUM PYROPHOSPHATE Nutrients And/or Dietary Supplements; Substances Generally Recognized As Safe
CALCIUM SACCHARIN	180.37	SACCHARIN, AMMONIUM SACCHARIN, CALCIUM SACCHARIN, AND SODIUM SACCHARIN Specific Requirements For Certain Food Additives; Food Additives Permitted In Food Or In Contact With Food On Interim Basis Pending Study
CALCIUM SALTS	582.5477	METHIONINE HYDROXY ANALOG AND ITS CALCIUM SALTS Nutrients And/or Dietary Supplements; Substances Generally Recognized As Safe
CALCIUM SILICATE	172.410	CALCIUM SILICATE Anticaking Agents; Food Additives Permitted For Direct Addition To Food For Human Consumption
CALCIUM SILICATE	182.2122	ALUMINUM CALCIUM SILICATE Anticaking Agents; Substances Generally Recognized As Safe
CALCIUM SILICATE	182.2227	CALCIUM SILICATE Anticaking Agents; Substances Generally Recognized As Safe
CALCIUM SILICATE	573.260	CALCIUM SILICATE Food Additive Listing; Food Additives Permitted In Feed And Drinking Water Of Animals
CALCIUM SILICATE	582.2122	ALUMINUM CALCIUM SILICATE Anticaking Agents; Substances Generally Recognized As Safe

CALCIUM SILICATE	582.2227	CALCIUM SILICATE Anticaking Agents; Substances Generally Recognized As Safe
CALCIUM SORBATE	182.3225	CALCIUM SORBATE Chemical Preservatives; Substances Generally Recognized As Safe
CALCIUM SORBATE	582.3225	CALCIUM SORBATE Chemical Preservatives; Substances Generally Recognized As Safe
CALCIUM STEARATE	184.1229	CALCIUM STEARATE Listing Of Specific Substances Affirmed As GRAS; Direct Food Substances Affirmed As Generally Recognized As Safe
CALCIUM STEARATE	573.280	FEED-GRADE CALCIUM STEARATE AND SODIUM STEARATE Food Additive Listing; Food Additives Permitted In Feed And Drinking Water Of Animals
CALCIUM STEAROLY-LACTYLATE	172.844	CALCIUM STEAROYL-2-LACTYLATE Multipurpose Additives; Food Additives Permitted For Direct Addition To Food For Human Consumption
CALCIUM SULFATE	184.1230	CALCIUM SULFATE Listing Of Specific Substances Affirmed As GRAS; Direct Food Substances Affirmed As Generally Recognized As Safe
CALCIUM SULFATE	582.5230	CALCIUM SULFATE Nutrients And/or Dietary Supplements; Substances Generally Recognized As Safe
CALCOACETICUS, ACINETOBACTER	866.3010	ACINETOBACTER CALCOACETICUS SEROLOGICAL REAGENTS Serological Reagents; Immunology And Microbiology Devices
CALCULATION, YIELD	211.103	CALCULATION OF YIELD Production And Process Controls; Current Good Manufacturing Practice For Finished Pharmaceuticals
CALCULATOR	862.2100	CALCULATOR/DATA PROCESSING MODULE FOR CLINICAL USE Clinical Laboratory Instruments; Clinical Chemistry And Clinical Toxicology Devices
CALCULATOR	868.1880	PULMONARY-FUNCTION DATA CALCULATOR Diagnostic Devices; Anesthesiology Devices
CALCULATOR	868.1890	PREDICTIVE PULMONARY-FUNCTION VALUE CALCULATOR Diagnostic Devices; Anesthesiology Devices
CALCULATOR	868.1900	DIAGNOSTIC PULMONARY-FUNCTION INTERPRETATION CALCULATOR Diagnostic Devices; Anesthesiology Devices
CALCULI	862.1780	URINARY CALCULI (STONES) TEST SYSTEM Clinical Chemistry Test Systems; Clinical Chemistry And Clinical Toxicology Devices
CALENDARS, PUBLIC	10.100	PUBLIC CALENDARS General Administrative Procedures; Administrative Practices And Procedures
CALIBRATION	58.63	MAINTENANCE AND CALIBRATION OF EQUIPMENT Equipment; Good Laboratory Practice For Nonclinical Laboratory Studies
CALIBRATION	868.2350	GAS CALIBRATION FLOWMETER Monitoring Devices; Anesthesiology Devices
CALIBRATION	868.6400	CALIBRATION GAS Miscellaneous; Anesthesiology Devices
CALIBRATION	874.1080	AUDIOMETER CALIBRATION SET Diagnostic Devices; Ear, Nose, And Throat Devices
CALIBRATION	882.1925	ULTRASONIC SCANNER CALIBRATION TEST BLOCK Neurological Diagnostic Devices; Neurological Devices
CALIBRATION	892.1400	NUCLEAR SEALED CALIBRATION SOURCE Diagnostic Devices; Radiology Devices
CALIBRATOR	862.1150	CALIBRATOR Clinical Chemistry Test Systems; Clinical Chemistry And Clinical Toxicology Devices
CALIBRATOR	862.3200	CLINICAL TOXICOLOGY CALIBRATOR Clinical Toxicology Test Systems; Clinical Chemistry And Clinical Toxicology Devices
CALIBRATOR	864.8150	CALIBRATOR FOR CELL INDICES Hematology Reagents; Hematology And Pathology Devices
CALIBRATOR	864.8165	CALIBRATOR FOR HEMOGLOBIN OR HEMATOCRIT MEASUREMENT Hematology Reagents; Hematology And Pathology Devices
CALIBRATOR	864.8175	CALIBRATOR FOR PLATELET COUNTING Hematology Reagents; Hematology And Pathology Devices
CALIBRATOR	864.8185	CALIBRATOR FOR RED CELL AND WHITE CELL COUNTING Hematology Reagents; Hematology And Pathology Devices
CALIBRATOR	868.1870	GAS VOLUME CALIBRATOR Diagnostic Devices; Anesthesiology Devices
CALIBRATOR	868.2620	GAS PRESSURE CALIBRATOR Monitoring Devices; Anesthesiology Devices
CALIBRATOR	874.3310	HEARING AID CALIBRATOR AND ANALYSIS SYSTEM Prosthetic Devices; Ear, Nose, And Throat Devices
CALIBRATOR	892.1360	RADIONUCLIDE DOSE CALIBRATOR Diagnostic Devices; Radiology Devices
CALIBRATORS	862.2	REGULATION OF CALIBRATORS General Provisions; Clinical Chemistry And Clinical Toxicology Devices

CALIFORNIA	808.55	CALIFORNIA Listing Of Specific State And Local Exemptions; Exemption From Federal Preemption Of State & Local Medical Device Requirements
CALIPERS	888.4150	CALIPERS FOR CLINICAL USE Surgical Devices; Orthopedic Devices
CALLUS REMOVER	358.510	CORN AND CALLUS REMOVER ACTIVE INGREDIENTS Corn And Callus Remover Drug Products; Miscellaneous External Drug Products For Over-the-counter Human Use
CALLUS REMOVER	358.550	LABELING OF CORN AND CALLUS REMOVER DRUG PRODUCTS Corn And Callus Remover Drug Products; Miscellaneous External Drug Products For Over-the-counter Human Use
CALORIC STIMULATOR	874.4120	AIR OR WATER CALORIC STIMULATOR Diagnostic Devices; Ear, Nose, And Throat Devices
CALORIE CONTENT	101.60	NUTRIENT CONTENT CLAIMS FOR THE CALORIE CONTENT OF FOODS Specific Requirements For Nutrient Content Claims; Food Labeling
CAMBENDAZOLE	520.300	CAMBENDAZOLE ORAL DOSAGE FORMS Oral Dosage Form New Animal Drugs
CAMERA, OPHTHALMIC	886.1120	OPHTHALMIC CAMERA Diagnostic Devices; Ophthalmic Devices
CAMERA, POSITRON	892.1110	POSITRON CAMERA Diagnostic Devices; Radiology Devices
CAMERA, SCINTILLATION	892.1100	SCINTILLATION (GAMMA) CAMERA Diagnostic Devices; Radiology Devices
CAMERA, SURGICAL	878.4160	SURGICAL CAMERA AND ACCESSORIES Surgical Devices; General And Plastic Surgery Devices
CAMERA, X-RAY	892.1620	CINE OR SPOT FLUOROGRAPHIC X-RAY CAMERA Diagnostic Devices; Radiology Devices
CAMPIMETER	886.1810	TANGENT SCREEN (CAMPIMETER) Diagnostic Devices; Ophthalmic Devices
CAMPYLOBACTER	866.3110	CAMPYLOBACTER FETUS SEROLOGICAL REAGENTS Serological Reagents; Immunology And Microbiology Devices
CANCELLATION	81.30	CANCELLATION OF CERTIFICATES General Specs+restrictns For Provisional Color Additives For Foods, Drugs, And Cosmetics
CANCELLATION	1305.15	CANCELLATION AND VOIDING OF ORDER FORMS Order Forms
CANCELLATION	1312.16	CANCELLATION OF PERMIT; EXPIRATION DATE Importation Of Controlled Substances; Importation And Exportation Of Controlled Substances
CANCER	70.50	APPLICATION OF THE CANCER CLAUSE OF SECTION 721 OF THE ACT Safety Evaluation; Color Additives
CANCER	101.73	HEALTH CLAIMS: DIETARY LIPIDS AND CANCER Specific Requirements For Health Claims; Food Labeling
CANCER	101.76	HEALTH CLAIMS: FIBER-CONTAINING GRAIN PRODUCTS, FRUITS, AND VEGETABLES AND CANCER Specific Requirements For Health Claims; Food Labeling
CANCER	101.78	HEALTH CLAIMS: FRUITS AND VEGETABLES AND CANCER Specific Requirements For Health Claims; Food Labeling
CANCER	571.115	APPLICATION OF THE CANCER CLAUSE OF SECTION 409 OF THE ACT Administrative Actions On Applications; Food Additive Petitions
CANCER	740.18	COAL TAR HAIR DYES POSING A RISK OF CANCER Warning Statements; Cosmetic Product Warning Statements
CANDELILLA WAX	184.1976	CANDELILLA WAX Listing Of Specific Substances Affirmed As GRAS; Direct Food Substances Affirmed As Generally Recognized As Safe
CANDIDA GUILLIERMONDII	173.160	CANDIDA GUILLIERMONDII Enzyme Preparations And Microorganisms; Secondary Direct Food Additives Permitted In Food For Human Consumption
CANDIDA LIPOLYTICA	173.165	CANDIDA LIPOLYTICA Enzyme Preparations And Microorganisms; Secondary Direct Food Additives Permitted In Food For Human Consumption
CANDIDA PSEUDOTROPICALIS	184.1387	LACTASE ENZYME PREPARATION FROM CANDIDA PSEUDOTROPICALIS Listing Of Specific Substances Affirmed As GRAS; Direct Food Substances Affirmed As Generally Recognized As Safe
CANE	890.3075	CANE Physical Medicine Prosthetic Devices; Physical Medicine Devices
CANE	890.3790	CANE, CRUTCH, AND WALKER TIPS AND PADS Physical Medicine Prosthetic Devices; Physical Medicine Devices
CANE SIRUP	168.130	CANE SIRUP Requirements For Specific Standardized Sweeteners And Table Sirups; Sweeteners And Table Sirups

CANE-SUGAR	173.320	CHEMICALS FOR CONTROLLING MICROORGANISMS IN CANE-SUGAR AND BEET-SUGAR MILLS Specific Usage Additives; Secondary Direct Food Additives Permitted In Food For Human Consumption
CANNABIS	1308.35	EXEMPTION OF CERTAIN CANNABIS PLANT MATERIAL, AND PRODUCTS MADE THEREFROM, THAT CONTAIN TETRAHYDROCANNABINOIDS Schedules Of Controlled Substances
CANNED APPLESAUCE	145.110	CANNED APPLESAUCE Requirements For Specific Standardized Canned Fruits
CANNED APRICOTS	145.115	CANNED APRICOTS Requirements For Specific Standardized Canned Fruits; Canned Fruits
CANNED APRICOTS	145.116	ARTIFICIALLY SWEETENED CANNED APRICOTS Requirements For Specific Standardized Canned Fruits; Canned Fruits
CANNED BEANS	155.120	CANNED GREEN BEANS AND WAX BEANS Requirements For Specific Standardized Canned Vegetables; Canned Vegetables
CANNED BERRIES	145.120	CANNED BERRIES Requirements For Specific Standardized Canned Fruits; Canned Fruits
CANNED CHERRIES	145.125	CANNED CHERRIES Requirements For Specific Standardized Canned Fruits; Canned Fruits
CANNED CHERRIES	145.126	ARTIFICIALLY SWEETENED CANNED CHERRIES Requirements For Specific Standardized Canned Fruits; Canned Fruits
CANNED CORN	155.130	CANNED CORN Requirements For Specific Standardized Canned Vegetables; Canned Vegetables
CANNED CORN	155.131	CANNED FIELD CORN Requirements For Specific Standardized Canned Vegetables; Canned Vegetables
CANNED FIGS	145.130	CANNED FIGS Requirements For Specific Standardized Canned Fruits; Canned Fruits
CANNED FIGS	145.131	ARTIFICIALLY SWEETENED CANNED FIGS Requirements For Specific Standardized Canned Fruits; Canned Fruits
CANNED FIGS	145.134	CANNED PRESERVED FIGS Requirements For Specific Standardized Canned Fruits; Canned Fruits
CANNED FRUIT COCKTAIL	145.135	CANNED FRUIT COCKTAIL Requirements For Specific Standardized Canned Fruits; Canned Fruits
CANNED FRUIT COCKTAIL	145.136	ARTIFICIALLY SWEETENED CANNED FRUIT COCKTAIL Requirements For Specific Standardized Canned Fruits; Canned Fruits
CANNED GRAPES	145.140	CANNED SEEDLESS GRAPES Requirements For Specific Standardized Canned Fruits; Canned Fruits
CANNED GRAPEFRUIT	145.145	CANNED GRAPEFRUIT Requirements For Specific Standardized Canned Fruits; Canned Fruits
CANNED MUSHROOMS	155.201	CANNED MUSHROOMS Requirements For Specific Standardized Canned Vegetables; Canned Vegetables
CANNED O.J.	146.141	CANNED ORANGE JUICE Canned Fruit Juices
CANNED O.J.	146.150	CANNED CONCENTRATED ORANGE JUICE Canned Fruit Juices
CANNED OYSTERS	161.30	DECLARATION OF QUANTITY OF CONTENTS ON LABELS FOR CANNED OYSTERS General Provisions; Fish And Shellfish
CANNED OYSTERS	161.145	CANNED OYSTERS Requirements For Specific Standardized Fish and Shellfish; Fish And Shellfish
CANNED PEACHES	145.170	CANNED PEACHES Requirements For Specific Standardized Canned Fruits; Canned Fruits
CANNED PEACHES	145.171	ARTIFICIALLY SWEETENED CANNED PEACHES Requirements For Specific Standardized Canned Fruits; Canned Fruits
CANNED PEARS	145.175	CANNED PEARS Requirements For Specific Standardized Canned Fruits; Canned Fruits
CANNED PEARS	145.176	ARTIFICIALLY SWEETENED CANNED PEARS Requirements For Specific Standardized Canned Fruits; Canned Fruits
CANNED PEAS	155.170	CANNED PEAS Requirements For Specific Standardized Canned Vegetables; Canned Vegetables
CANNED PEAS	155.172	CANNED DRY PEAS Requirements For Specific Standardized Canned Vegetables; Canned Vegetables
CANNED PINEAPPLE	145.180	CANNED PINEAPPLE Requirements For Specific Standardized Canned Fruits; Canned Fruits
CANNED PINEAPPLE	145.181	ARTIFICIALLY SWEETENED CANNED PINEAPPLE Requirements For Specific Standardized Canned Fruits; Canned Fruits
CANNED PLUMS	145.185	CANNED PLUMS Requirements For Specific Standardized Canned Fruits; Canned Fruits
CANNED PRUNE JUICE	146.187	CANNED PRUNE JUICE Requirements For Specific Standardized Canned Fruits Juices And Beverages; Canned Fruit Juices
CANNED PRUNES	145.190	CANNED PRUNES Requirements For Specific Standardized Canned Fruits; Canned Fruits

CANNED SALMON	161.170	CANNED PACIFIC SALMON Requirements For Specific Standardized Fish and Shellfish; Fish And Shellfish
CANNED SHRIMP	161.173	CANNED WET PACK SHRIMP Requirements For Specific Standardized Fish and Shellfish; Fish And Shellfish
CANNED TOMATOES	155.190	CANNED TOMATOES Requirements For Specific Standardized Canned Vegetables; Canned Vegetables
CANNED TUNA	161.190	CANNED TUNA Requirements For Specific Standardized Fish and Shellfish; Fish And Shellfish
CANNED VEGETABLES	155.100	CERTAIN OTHER CANNED VEGETABLES Requirements For Specific Standardized Canned Vegetables; Canned Vegetables
CANNULA	868.5340	NASAL OXYGEN CANNULA Therapeutic Devices; Anesthesiology Devices
CANNULA	870.1300	CATHETER CANNULA Cardiovascular Diagnostic Devices; Cardiovascular Devices
CANNULA	870.4210	CARDIOPULMONARY BYPASS VASCULAR CATHETER, CANNULA, OR TUBING Cardiovascular Surgical Devices; Cardiovascular Devices
CANNULA	882.4060	VENTRICULAR CANNULA Neurological Surgical Devices; Neurological Devices
CANTHAXANTHIN	73.75	CANTHAXANTHIN Foods; Listing Of Color Additives Exempt From Certification
CANTHAXANTHIN	73.1075	CANTHAXANTHIN Drugs; Listing Of Color Additives Exempt From Certification
CAP	884.5250	CERVICAL CAP Obstetrical And Gynecological Therapeutic Devices; Obstetrical And Gynecological Devices
CAP	888.3000	BONE CAP Prosthetic Devices; Orthopedic Devices
CAPACITY	130.12	GENERAL METHODS FOR WATER CAPACITY AND FILL OF CONTAINERS General Provisions; Food Standards: General
CAPILLARY	864.6150	CAPILLARY BLOOD COLLECTION TUBE Manual Hematology Devices; Hematology And Pathology Devices
CAPROFEN	520.309	CAPROFEN CAPSULES Oral Dosage Form New Animal Drugs
CAPRYLIC ACID	184.1025	CAPRYLIC ACID Listing Of Specific Substances Affirmed As GRAS; Direct Food Substances Affirmed As Generally Recognized As Safe
CAPSULE	872.3110	DENTAL AMALGAM CAPSULE Prosthetic Devices; Dental Devices
CAPSULE	876.1300	INGESTIBLE TELEMETRIC GASTROINTESTINAL CAPSULE IMAGING SYSTEM Diagnostic Devices; Gastroenterology-urology Devices
CAPSULES	189.301	TIN-COATED LEAD FOIL CAPSULES FOR WINE BOTTLES Substances Prohibited From Use In Human Foods
CAPSULES	520.90	AMPICILLIN CAPSULES Oral Dosage Form New Animal Drugs
CAPSULES	520.90c	AMPICILLIN TRIHYDRATE CAPSULES Oral Dosage Form New Animal Drugs
CAPSULES	520.260	N-BUTYL CHLORIDE CAPSULES Oral Dosage Form New Animal Drugs
CAPSULES	520.390b	CHORAMPHENICOL CAPSULES Oral Dosage Form New Animal Drugs
CAPSULES	520.446	CLINDAMYCIN HYDROCHLORIDE CAPSULES Oral Dosage Form New Animal Drugs
CAPSULES	520.580	DICHLOROPHENE AND TOLUENE CAPSULES Oral Dosage Form New Animal Drugs
CAPSULES	520.608	DICLOXACILLIN SODIUM MONOHYDRATE CAPSULES Oral Dosage Form New Animal Drug Applications
CAPSULES	520.622d	DIETHYLCARBAMAZINE CITRATE CAPSULES Oral Dosage Form New Animal Drugs
CAPSULES	520.1130a	HETACILLIN POTASSIUM CAPSULES Oral Dosage Form New Animal Drugs
CAPSULES	520.1660b	OXYTETRACYCLINE HYDROCHLORIDE CAPSULES Oral Dosage Form New Animal Drugs
CAPSULES	520.1803	PIPERAZINE CITRATE CAPSULES Oral Dosage Form New Animal Drugs
CAPSULES	520.1804	PIPERAZINE PHOSPHATE CAPSULES Oral Dosage Form New Animal Drugs
CAPSULES	520.1920	PROCHLORPERAZINE, ISOPROPAMIDE SUSTAINED-RELEASE CAPSULES Oral Dosage Form New Animal Drugs
CAPSULES	520.1921	PROCHLORPERAZINE, ISOPROPAMIDE, WITH NEOMYCIN SUSTAINED-RELEASE CAPSULES Oral Dosage Form New Animal Drugs
CAPSULES	520.2100	SELENIUM, VITAMIN E CAPSULES Oral Dosage Form New Animal Drugs

CAPSULES	520.2345a	TETRACYCLINE HYDROCHLORIDE CAPSULES Oral Dosage Form New Animal Drugs
CAPSULES	520.2345f	TETRACYCLINE PHOSPHATE COMPLEX AND SODIUM NOVOBIOCIN CAPSULES Oral Dosage Form New Animal Drugs
CAPSULES	520.2605	TRIMEPRAZINE TARTRATE AND PREDNISOLONE CAPSULES Oral Dosage Form New Animal Drugs
CAPSULES	529.1526	NIFURPIRINOL CAPSULES Certain Other Dosage Form New Animal Drugs
CAPSULOTOMY	886.4392	ND:YAG LASER FOR POSTERIOR CAPSULOTOMY Surgical Devices; Ophthalmic Devices
CARAMEL	73.85	CARAMEL Foods; Listing Of Color Additives Exempt From Certification
CARAMEL	73.1085	CARAMEL Drugs; Listing Of Color Additives Exempt From Certification
CARAMEL	73.2085	CARAMEL Cosmetics; Listing Of Color Additives Exempt From Certification
CARAMEL	182.1235	CARAMEL Multiple Purpose GRAS Food Substances; Substances Generally Recognized As Safe
CARAMEL	582.1235	CARAMEL General Purpose Food Additives; Substances Generally Recognized As Safe
CARAMIPHEN	520.310	CARAMIPHEN ETHANEDISULFONATE AND AMMONIUM CHLORIDE TABLETS Oral Dosage Form New Animal Drugs
CARBADOX	556.100	CARBADOX Specific Tolerances For Residues Of New Animal Drugs; Tolerances For Residues Of New Animal Drugs In Food
CARBADOX	558.115	CARBADOX Specific New Animal Drugs For Use In Animal Feeds; New Animal Drugs For Use In Animal Feeds
CARBAMYL TRANSFERASE	862.1535	ORNITHINE CARBAMYL TRANSFERASE TEST SYSTEM Clinical Chemistry Test Systems; Clinical Chemistry And Clinical Toxicology Devices
CARBARSONE	558.120	CARBARSONE (NOT U.S.P.) Specific New Animal Drugs For Use In Animal Feeds; New Animal Drugs For Use In Animal Feeds
CARBAZOLE-HEXONE	73.3117	16,23-DIHYDRODINAPHTHO[2,3-A:2',3'-I] NAPHTH [2',3':6,7] INDOLO [2,3-C] CARBAZOLE-5,10,15,17,22,24-HEXONE Medical Devices; Listing Of Color Additives Exempt From Certification
CARBAZOLE VIOLET	73.3107	CARBAZOLE VIOLET Medical Devices; Listing Of Color Additives Exempt From Certification
CARBOHYDRASE	173.120	CARBOHYDRASE AND CELLULASE DERIVED FROM ASPERGILLUS NIGER Enzyme Preparations And Microorganisms; Secondary Direct Food Additives Permitted In Food For Human Consumption
CARBOHYDRASE	173.130	CARBOHYDRASE DERIVED FROM RHIZOPUS ORYZAE Enzyme Preparations And Microorganisms; Secondary Direct Food Additives Permitted In Food For Human Consumption
CARBOHYDRASE	184.1027	MIXED CARBOHYDRASE AND PROTEASE ENZYME PRODUCT Listing Of Specific Substances Affirmed As GRAS; Direct Food Substances Affirmed As Generally Recognized Safe
CARBOHYDRASE	184.1148	BACTERIALLY DERIVED CARBOHYDRASE ENZYME Listing Of Specific Substances Affirmed As GRAS; Direct Food Substances Affirmed As Generally Recognized As Safe
CARBOHYDRATES	172.866	SYNTHETIC GLYCERIN PRODUCED BY THE HYDROGENOLYSIS OF CARBOHYDRATES Multipurpose Additives; Food Additives Permitted For Direct Addition To Food For Human Consumption
CARBOMYCIN	520.1660a	OXYTETRACYCLINE AND CARBOMYCIN IN COMBINATION Oral Dosage Form New Animal Drugs
CARBOMYCIN	556.110	CARBOMYCIN Specific Tolerances For Residues Of New Animal Drugs; Tolerances For Residues Of New Animal Drugs In Food
CARBON DIOXIDE	184.1240	CARBON DIOXIDE Listing Of Specific Substances Affirmed As GRAS; Direct Food Substances Affirmed As Generally Recognized As Safe
CARBON DIOXIDE	201.161	CARBON DIOXIDE AND CERTAIN OTHER GASES Other Exemptions; Labeling
CARBON DIOXIDE	582.1240	CARBON DIOXIDE General Purpose Food Additives; Substances Generally Recognized As Safe
CARBON DIOXIDE	862.1160	BICARBONATE/CARBON DIOXIDE TEST SYSTEM Clinical Chemistry Test Systems; Clinical Chemistry And Clinical Toxicology Devices

CARBON DIOXIDE	868.1150	INDWELLING BLOOD CARBON DIOXIDE PARTIAL PRESSURE (PCO2) ANALYZER Diagnostic Devices; Anesthesiology Devices
CARBON DIOXIDE	868.1400	CARBON DIOXIDE GAS ANALYZER Diagnostic Devices; Anesthesiology Devices
CARBON DIOXIDE	868.2480	CUTANEOUS CARBON DIOXIDE (PCCO2) MONITOR Diagnostic Devices; Anesthesiology Devices
CARBON DIOXIDE	868.5300	CARBON DIOXIDE ABSORBENT Therapeutic Devices; Anesthesiology Devices
CARBON DIOXIDE	868.5310	CARBON DIOXIDE ABSORBER Therapeutic Devices; Anesthesiology Devices
CARBON DIOXIDE	874.4500	EAR, NOSE AND THROAT MICROSURGICAL CARBON-DIOXIDE LASER Surgical Devices; Ear, Nose, And Throat Devices
CARBON DIOXIDE	884.1300	UTEROTUBAL CARBON DIOXIDE INSUFFLATOR AND ACCESSORIES Obstetrical And Gynecological Diagnostic Devices; Obstetrical And Gynecological Devices
CARBON DISULFIDE	520.1802	PIPERAZINE-CARBON DISULFIDE COMPLEX ORAL DOSAGE FORMS Oral Dosage Form New Animal Drugs
CARBON FIBERS	878.3500	POLYTETRAFLUOROETHYLENE WITH CARBON FIBERS COMPOSITE IMPLANT MATERIAL Prosthetic Devices; General And Plastic Surgery Devices
CARBON MATERIALS	872.3680	POLYTETRAFLUOROETHYLENE (PTFE) VITREOUS CARBON MATERIALS Prosthetic Devices; Dental Devices
CARBON MONOXIDE	177.1312	ETHYLENE-CARBON MONOXIDE COPOLYMERS Substances For Use As Basic Components Of Single And Repeated Use Food Contact Surfaces; Indirect Food Additives: Polymers
CARBON MONOXIDE	862.3220	CARBON MONOXIDE TEST SYSTEM Clinical Toxicology Test Systems; Clinical Chemistry And Clinical Toxicology Devices
CARBON MONOXIDE	868.1430	CARBON MONOXIDE ANALYZER Diagnostic Devices; Anesthesiology Devices
CARBONATE	73.1070	CALCIUM CARBONATE Drugs; Listing Of Color Additives Exempt From Certification
CARBONATE	184.1137	AMMONIUM CARBONATE Listing Of Specific Substances Affirmed As GRAS; Direct Food Substances Affirmed As Generally Recognized Safe
CARBONATE	184.1191	CALCIUM CARBONATE Listing Of Specific Substances Affirmed As GRAS; Direct Food Substances Affirmed As Generally Recognized Safe
CARBONATE	184.1307b	FERROUS CARBONATE Listing Of Specific Substances Affirmed As GRAS; Direct Food Substances Affirmed As Generally Recognized Safe
CARBONATE	184.1425	MAGNESIUM CARBONATE Listing Of Specific Substances Affirmed As GRAS; Direct Food Substances Affirmed As Generally Recognized Safe
CARBONATE	184.1619	POTASSIUM CARBONATE Listing Of Specific Substances Affirmed As GRAS; Direct Food Substances Affirmed As Generally Recognized Safe
CARBONATE	184.1742	SODIUM CARBONATE Listing Of Specific Substances Affirmed As GRAS; Direct Food Substances Affirmed As Generally Recognized Safe
CARBONATE	582.1137	AMMONIUM CARBONATE General Purpose Food Additives; Substances Generally Recognized As Safe
CARBONATE	582.1181	CALCIUM CARBONATE General Purpose Food Additives; Substances Generally Recognized As Safe
CARBONATE	582.1425	MAGNESIUM CARBONATE General Purpose Food Additives; Substances Generally Recognized As Safe
CARBONATE	582.1619	POTASSIUM CARBONATE General Purpose Food Additives; Substances Generally Recognized As Safe
CARBONATE	582.1742	SODIUM CARBONATE General Purpose Food Additives; Substances Generally Recognized As Safe
CARBONATE	582.5191	CALCIUM CARBONATE Nutrients And/or Dietary Supplements; Substances Generally Recognized As Safe
CARBONIC ANHYDRASE	866.5200	CARBONIC ANHYDRASE B AND C IMMUNOLOGICAL TEST SYSTEM Immunological Test Systems; Immunology And Microbiology Devices
CARBOXYHEMOGLOBIN	864.7425	CARBOXYHEMOGLOBIN ASSAY Hematology Kits And Packages; Hematology And Pathology Devices
CARBOXYMETHYLCELLULOSE	182.1745	SODIUM CARBOXYMETHYLCELLULOSE Multiple Purpose GRAS Food Substances; Substances Generally Recognized As Safe
CARBOXYMETHYLCELLULOSE	201.319	WATER-SOLUBLE GUMS, HYDROPHILIC GUMS AND HYDROPHILIC MUCILLOIDS (INCLUDING CARBOXYMETHYLCELLULOSE SODIUM) AS ACTIVE INGREDIENTS; WARNINGS AND DIRECTIONS Specific Labeling Requirements For Specific Drug Products; Labeling

CARBOXYMETHYLCELLULOSE	582.1745	SODIUM CARBOXYMETHYLCELLULOSE General Purpose Food Additives; Substances Generally Recognized As Safe
CARBOXYMETHYLCELLULOSE	872.3410	ETHYLENE OXIDE HOMOPOLYMER AND/OR CARBOXYMETHYLCELLULOSE SODIUM DENTURE ADHESIVE Prosthetic Devices; Dental Devices
CARBOXYMETHYLCELLULOSE	872.3420	CARBOXYMETHYLCELLULOSE SODIUM AND CATIONIC POLYACRYLAMIDE POLYMER DENTURE ADHESIVE Prosthetic Devices; Dental Devices
CARBOXYMETHYLCELLULOSE	872.3490	CARBOXYMETHYLCELLULOSE SODIUM AND/OR POLYVINYLMETHYLETHER MALEIC ACID CALCIUM-SODIUM DOUBLE SALT DENTURE ADHESIVE Prosthetic Devices; Dental Devices
CARBOXYMETHYLCELLULOSE	872.3500	POLYVINYLMETHYLETHER MALEIC ANHYDRIDE (PVM-MA), ACID COPOLYMER, AND CARBOXYMETHYLCELLULOSE SODIUM (NACMC) DENTURE ADHESIVE Prosthetic Devices; Dental Devices
CARDIAC COMPRESSOR	870.5200	EXTERNAL CARDIAC COMPRESSOR Cardiovascular Therapeutic Devices; Cardiovascular Devices
CARDIAC MONITOR	870.2300	CARDIAC MONITOR (INCLUDING CARDIOTACHOMETER AND RATE ALARM) Cardiovascular Monitoring Devices; Cardiovascular Devices
CARDIAC MONITOR	884.2600	FETAL CARDIAC MONITOR Obstetrical And Gynecological Monitoring Devices; Obstetrical And Gynecological Devices
CARDIAC PACEMAKER	870.5550	EXTERNAL TRANSCUTANEOUS CARDIAC PACEMAKER (NONINVASIVE) Cardiovascular Therapeutic Devices; Cardiovascular Devices
CARDIAC TREATMENT	868.5365	POSTURE CHAIR FOR CARDIAC OR PULMONARY TREATMENT Therapeutic Devices; Anesthesiology Devices
CARDIOGRAPH	870.2310	APEX CARDIOGRAPH (VIBROCARDIOGRAPH) Cardiovascular Monitoring Devices; Cardiovascular Devices
CARDIOGRAPHIC	870.2840	APEX CARDIOGRAPHIC TRANSDUCER Cardiovascular Monitoring Devices; Cardiovascular Devices
CARDIOPULMONARY	868.6175	CARDIOPULMONARY EMERGENCY CART Miscellaneous; Anesthesiology Devices
CARDIOPULMONARY	870.4200	CARDIOPULMONARY BYPASS ACCESSORY EQUIPMENT Cardiovascular Surgical Devices; Cardiovascular Devices
CARDIOPULMONARY	870.4205	CARDIOPULMONARY BYPASS BUBBLE DETECTOR Cardiovascular Surgical Devices; Cardiovascular Devices
CARDIOPULMONARY	870.4210	CARDIOPULMONARY BYPASS VASCULAR CATHETER, CANNULA, OR TUBING Cardiovascular Surgical Devices; Cardiovascular Devices
CARDIOPULMONARY	870.4220	CARDIOPULMONARY BYPASS HEART-LUNG MACHINE CONSOLE Cardiovascular Surgical Devices; Cardiovascular Devices
CARDIOPULMONARY	870.4230	CARDIOPULMONARY BYPASS DEFOAMER Cardiovascular Surgical Devices; Cardiovascular Devices
CARDIOPULMONARY	870.4240	CARDIOPULMONARY BYPASS HEAT EXCHANGER Cardiovascular Surgical Devices; Cardiovascular Devices
CARDIOPULMONARY	870.4250	CARDIOPULMONARY BYPASS TEMPERATURE CONTROLLER Cardiovascular Surgical Devices; Cardiovascular Devices
CARDIOPULMONARY	870.4260	CARDIOPULMONARY BYPASS ARTERIAL LINE BLOOD FILTER Cardiovascular Surgical Devices; Cardiovascular Devices
CARDIOPULMONARY	870.4270	CARDIOPULMONARY BYPASS CARDIOTOMY SUCTION LINE BLOOD FILTER Cardiovascular Surgical Devices; Cardiovascular Devices
CARDIOPULMONARY	870.4280	CARDIOPULMONARY PREBYPASS FILTER Cardiovascular Surgical Devices; Cardiovascular Devices
CARDIOPULMONARY	870.4290	CARDIOPULMONARY BYPASS ADAPTOR, STOPCOCK, MANIFOLD, OR FITTING Cardiovascular Surgical Devices; Cardiovascular Devices
CARDIOPULMONARY	870.4300	CARDIOPULMONARY BYPASS GAS CONTROL UNIT Cardiovascular Surgical Devices; Cardiovascular Devices
CARDIOPULMONARY	870.4310	CARDIOPULMONARY BYPASS CORONARY PRESSURE GAUGE Cardiovascular Surgical Devices; Cardiovascular Devices
CARDIOPULMONARY	870.4320	CARDIOPULMONARY BYPASS PULSATILE FLOW GENERATOR Cardiovascular Surgical Devices; Cardiovascular Devices
CARDIOPULMONARY	870.4330	CARDIOPULMONARY BYPASS ON-LINE BLOOD GAS MONITOR Cardiovascular Surgical Devices; Cardiovascular Devices
CARDIOPULMONARY	870.4340	CARDIOPULMONARY BYPASS LEVEL SENSING MONITOR AND/OR CONTROL Cardiovascular Surgical Devices; Cardiovascular Devices
CARDIOPULMONARY	870.4350	CARDIOPULMONARY BYPASS OXYGENATOR Cardiovascular Surgical Devices; Cardiovascular Devices
CARDIOPULMONARY	870.4360	NONROLLER- TYPE CARDIOPULMONARY BYPASS BLOOD PUMP Cardiovascular Surgical Devices; Cardiovascular Devices
CARDIOPULMONARY	870.4370	ROLLER-TYPE CARDIOPULMONARY BYPASS BLOOD PUMP Cardiovascular Surgical Devices; Cardiovascular Devices
CARDIOPULMONARY	870.4380	CARDIOPULMONARY BYPASS PUMP SPEED CONTROL Cardiovascular Surgical Devices; Cardiovascular Devices

CARDIOPULMONARY	870.4390	CARDIOPULMONARY BYPASS PUMP TUBING Cardiovascular Surgical Devices; Cardiovascular Devices
CARDIOPULMONARY	870.4400	CARDIOPULMONARY BYPASS BLOOD RESERVOIR Cardiovascular Surgical Devices; Cardiovascular Devices
CARDIOPULMONARY	870.4410	CARDIOPULMONARY BYPASS IN-LINE BLOOD GAS SENSOR Cardiovascular Surgical Devices; Cardiovascular Devices
CARDIOPULMONARY	870.4420	CARDIOPULMONARY BYPASS CARDIOTOMY RETURN SUCKER Cardiovascular Surgical Devices; Cardiovascular Devices
CARDIOPULMONARY	870.4430	CARDIOPULMONARY BYPASS INTRACARDIAC SUCTION CONTROL Cardiovascular Surgical Devices; Cardiovascular Devices
CARDIOPULMONARY	880.6080	CARDIOPULMONARY RESUSCITATION BOARD General Hospital And Personal Use Miscellaneous Devices; General Hospital And Personal Use Devices
CARDIOTACHOMETER	870.2300	CARDIAC MONITOR (INCLUDING CARDIOTACHOMETER AND RATE ALARM) Cardiovascular Monitoring Devices; Cardiovascular Devices
CARDIOTOMY	870.4270	CARDIOPULMONARY BYPASS CARDIOTOMY SUCTION LINE BLOOD FILTER Cardiovascular Surgical Devices; Cardiovascular Devices
CARDIOTOMY	870.4420	CARDIOPULMONARY BYPASS CARDIOTOMY RETURN SUCKER Cardiovascular Surgical Devices; Cardiovascular Devices
CARDIOTONIC DRUGS	201.317	DIGITALIS AND RELATED CARDIOTONIC DRUGS FOR HUMAN USE IN ORAL DOSAGE FORMS; REQUIRED WARNING Specific Labeling Requirements For Specific Drug Products; Labeling
CARDIOVASCULAR	343.12	CARDIOVASCULAR ACTIVE INGREDIENTS Internal Analgesic, Antipyretic And Antirheumatic Drug Products For OTC Human Use
CARDIOVASCULAR	343.22	PERMITTED COMBINATIONS OF ACTIVE INGREDIENTS FOR CARDIOVASCULAR-RHEUMATIC USE Internal Analgesic, Antipyretic And Antirheumatic Drug Products For OTC Human Use
CARDIOVASCULAR	870.2100	CARDIOVASCULAR BLOOD FLOWMETER Cardiovascular Monitoring Devices; Cardiovascular Devices
CARDIOVASCULAR	870.3375	CARDIOVASCULAR INTRAVASCULAR FILTER Cardiovascular Prosthetic Devices; Cardiovascular Devices
CARDIOVASCULAR	870.3680	CARDIOVASCULAR PERMANENT OR TEMPORARY PACEMAKER ELECTRODE Cardiovascular Prosthetic Devices; Cardiovascular Devices
CARDIOVASCULAR	870.4500	CARDIOVASCULAR SURGICAL INSTRUMENTS Cardiovascular Surgical Devices; Cardiovascular Devices
CARE	600.11	PHYSICAL ESTABLISHMENT, EQUIPMENT, ANIMALS, AND CARE Establishment Standards; Biological Products: General
CARE, ANIMAL	58.43	ANIMAL CARE FACILITIES Facilities; Good Laboratory Practice For Nonclinical Laboratory Studies
CARE, ANIMAL	58.90	ANIMAL CARE Testing Facilities Operation; Good Laboratory Practice For Nonclinical Laboratory Studies
CARE, PATIENT	870.5050	PATIENT CARE SUCTION APPARATUS Cardiovascular Therapeutic Devices; Cardiovascular Devices
CARE, PATIENT	880.5450	PATIENT CARE REVERSE ISOLATION CHAMBER General Hospital And Personal Use Therapeutic Devices; General Hospital And Personal Use Devices
CARFENTANIL CITRATE	522.311	CARFENTANIL CITRATE INJECTION Implantation Or Injectable Dosage Form New Animal Drugs
CARIES	101.80	HEALTH CLAIMS: DIETARY SUGAR ALCOHOLS AND DENTAL CARIES Food Labeling
CARIES	872.1740	CARIES DETECTION DEVICE Diagnostic Devices; Dental Devices
CARIES	872.1745	LASER FLUORESCENCE CARIES DETECTION DEVICE Diagnostic Devices; Dental Devices
CARMINE	73.100	COCHINEAL EXTRACT; CARMINE Foods; Listing Of Color Additives Exempt From Certification
CARMINE	73.1100	COCHINEAL EXTRACT; CARMINE Drugs; Listing Of Color Additives Exempt From Certification
CARMINE	73.2087	CARMINE Cosmetics; Listing Of Color Additives Exempt From Certification
CARNAUBA WAX	184.1978	CARNAUBA WAX Listing Of Specific Substances Affirmed As GRAS; Direct Food Substances Affirmed As Generally Recognized As Safe
CARNAUBA WAX	582.1978	CARNAUBA WAX General Purpose Food Additives; Substances Generally Recognized As Safe
CARNIDAZOLE	520.316	CARNIDAZOLE TABLETS Oral Dosage Form New Animal Drugs
CAROB BEAN	184.1343	LOCUST (CAROB) BEAN GUM Listing Of Specific Substances Affirmed As GRAS; Direct Food Substances Affirmed As Generally Recognized As Safe

CAROTENAL, BETA-APO	73.90	BETA-APO-8'-CAROTENAL Foods; Listing Of Color Additives Exempt From Certification
CAROTENE	582.5245	CAROTENE Nutrients And/or Dietary Supplements; Substances Generally Recognized As Safe
CAROTENE (BETA)	73.95	BETA-CAROTENE Foods; Listing Of Color Additives Exempt From Certification
CAROTENE (BETA)	73.1095	BETA-CAROTENE Drugs; Listing Of Color Additives Exempt From Certification
CAROTENE (BETA)	73.2095	BETA-CAROTENE Cosmetics; Listing Of Color Additives Exempt From Certification
CAROTENE (BETA)	184.1245	BETA-CAROTENE Listing Of Specific Substances Affirmed As GRAS; Direct Food Substances Affirmed As Generally Recognized As Safe
CAROTID ARTERY	882.5175	CAROTID ARTERY CLAMP Neurological Therapeutic Devices; Neurological Devices
CAROTID SINUS	870.3850	CAROTID SINUS NERVE STIMULATOR Cardiovascular Prosthetic Devices; Cardiovascular Devices
CARPAL	888.3750	WRIST JOINT CARPAL LUNATE POLYMER PROSTHESIS Prosthetic Devices; Orthopedic Devices
CARPAL	888.3760	WRIST JOINT CARPAL SCAPHOID POLYMER PROSTHESIS Prosthetic Devices; Orthopedic Devices
CARPAL	888.3770	WRIST JOINT CARPAL TRAPEZIUM POLYMER PROSTHESIS Prosthetic Devices; Orthopedic Devices
CARRAGEENAN	172.620	CARRAGEENAN Gums, Chewing Gum Bases And Related Substances; Food Additives Permitted For Direct Addition To Food For Human Consumption
CARRAGEENAN	172.623	CARRAGEENAN WITH POLYSORBATE 80 Gums, Chewing Gum Bases And Related Substances; Food Additives Permitted For Direct Addition To Food For Human Consumption
CARRAGEENAN	172.626	SALTS OF CARRAGEENAN Gums, Chewing Gum Bases And Related Substances; Food Additives Permitted For Direct Addition To Food For Human Consumption
CARRAGEENAN	201.319	WATER-SOLUBLE GUMS, HYDROPHILIC GUMS AND HYDROPHILIC MUCILLOIDS (INCLUDING CARRAGEENAN) AS ACTIVE INGREDIENTS; WARNINGS AND DIRECTIONS Specific Labeling Requirements For Specific Drug Products; Labeling
CARRIER	874.4350	EAR, NOSE AND THROAT FIBEROPTIC LIGHT SOURCE CARRIER Surgical Devices; Ear, Nose, And Throat Devices
CARRIER, COMMON	203.3	SAMPLE DISTRIBUTION BY MAIL OR COMMON CARRIER Samples; Prescription Drug Marketing
CARRIER, COMMON	203.31	SAMPLE DISTRIBUTION BY MEANS OTHER THAN MAIL OR COMMON CARRIER Samples; Prescription Drug Marketing
CARRIERS	58.113	MIXTURE OF ARTICLES WITH CARRIERS Test And Control Articles; Good Laboratory Practice For Nonclinical Laboratory Studies
CARROT OIL	73.300	CARROT OIL Foods; Listing Of Color Additives Exempt From Certification
CART	868.6175	CARDIOPULMONARY EMERGENCY CART Miscellaneous; Anesthesiology Devices
CARTRIDGE	520.1450	MORANTEL TARTRATE CARTRIDGE Oral Dosage Form New Animal Drugs
CARTRIDGE SYRINGE	872.6770	CARTRIDGE SYRINGE Miscellaneous Devices; Dental Devices
CASEIN, IODINATED	558.295	IODINATED CASEIN Specific New Animal Drugs For Use In Animal Feeds; New Animal Drugs For Use In Animal Feeds
CASEIN, IODINATED	520.1157	IODINATED CASEIN TABLETS Oral Dosage Form New Animal Drugs
CASEINATE	182.1748	SODIUM CASEINATE Multiple Purpose GRAS Food Substances; Substances Generally Recognized As Safe
CASEINATE	582.1748	SODIUM CASEINATE General Purpose Food Additives; Substances Generally Recognized As Safe
CASSETTE	892.1850	RADIOGRAPHIC FILM CASSETTE Diagnostic Devices; Radiology Devices
CASSETTE	892.1860	RADIOGRAPHIC FILM/CASSETTE CHANGER Diagnostic Devices; Radiology Devices
CASSETTE	892.1870	RADIOGRAPHIC FILM/CASSETTE CHANGER PROGRAMMER Diagnostic Devices; Radiology Devices
CASSETTE	892.1880	WALL-MOUNTED RADIOGRAPHIC CASSETTE HOLDER Diagnostic Devices; Radiology Devices

CAST	880.6185	CAST COVER General Hospital And Personal Use Miscellaneous Devices; General Hospital And Personal Use Devices
CAST	888.5940	CAST COMPONENT INSTRUMENT Surgical Devices; Orthopedic Devices
CAST	888.5960	CAST REMOVAL INSTRUMENT Surgical Devices; Orthopedic Devices
CAST	888.5940	MANUAL CAST APPLICATION AND REMOVAL INSTRUMENT Surgical Devices; Orthopedic Devices
CASTOR OIL	172.876	CASTOR OIL Multipurpose Additives; Food Additives Permitted For Direct Addition To Food For Human Consumption
CASTOR OIL	178.3280	CASTOR OIL, HYDROGENATED Certain Adjuvants And Production Aids; Indirect Food Additives: Adjuvants, Production Aids, And Sanitizers
CASTOR OIL	524.2620	LIQUID CRYSTALLINE TRYPSIN, PERU BALSAM, CASTOR OIL Ophthalmic And Topical Dosage Form New Animal Drugs
CAT FOOD	500.50	PROPYLENE GLYCOL IN OR ON CAT FOOD Specific Administrative Rulings and Decisions; General
CAT FOOD	589.1001	PROPYLENE GLYCOL IN OR ON CAT FOOD Substances Prohibited From Use In Animal Food Or Feed
CATALASE	173.135	CATALASE DERIVED FROM MICROCCOCUS LYSODEIKTICUS Enzyme Preparations And Microorganisms; Secondary Direct Food Additives Permitted In Food For Human Consumption
CATALASE	184.1034	CATALASE Listing Of Specific Substances Affirmed As GRAS; Direct Food Substances Affirmed As Generally Recognized As Safe
CATECHOLAMINES	862.1165	CATECHOLAMINES (TOTAL) TEST SYSTEM Clinical Chemistry Test Systems; Clinical Chemistry And Clinical Toxicology Devices
CATEGORIES	1401.12	FEES TO BE CHARGED - CATEGORIES OF REQUESTERS Public Availability Of Information
CATEGORIES, DRUG	330.5	DRUG CATEGORIES General Provisions; Over-the-counter Human Drugs Generally Recognized As Safe And Effective & Not Misbranded
CATHETER	868.5120	ANESTHESIA CONDUCTION CATHETER Therapeutic Devices; Anesthesiology Devices
CATHETER	868.5350	NASAL OXYGEN CATHETER Therapeutic Devices; Anesthesiology Devices
CATHETER	868.6810	TRACHEOBRONCHIAL SUCTION CATHETER Miscellaneous; Anesthesiology Devices
CATHETER	870.1200	DIAGNOSTIC INTRAVASCULAR CATHETER Cardiovascular Diagnostic Devices; Cardiovascular Devices
CATHETER	870.1210	CONTINUOUS FLUSH CATHETER Cardiovascular Diagnostic Devices; Cardiovascular Devices
CATHETER	870.1220	ELECTRODE RECORDING CATHETER OR ELECTRODE RECORDING PROBE Cardiovascular Diagnostic Devices; Cardiovascular Devices
CATHETER	870.1230	FIBEROPTIC OXIMETER CATHETER Cardiovascular Diagnostic Devices; Cardiovascular Devices
CATHETER	870.1240	FLOW-DIRECTED CATHETER Cardiovascular Diagnostic Devices; Cardiovascular Devices
CATHETER	870.1250	PERCUTANEOUS CATHETER Cardiovascular Diagnostic Devices; Cardiovascular Devices
CATHETER	870.1280	STEERABLE CATHETER Cardiovascular Diagnostic Devices; Cardiovascular Devices
CATHETER	870.1290	STEERABLE CATHETER CONTROL SYSTEM Cardiovascular Diagnostic Devices; Cardiovascular Devices
CATHETER	870.1300	CATHETER CANNULA Cardiovascular Diagnostic Devices; Cardiovascular Devices
CATHETER	870.1330	CATHETER GUIDE WIRE Cardiovascular Diagnostic Devices; Cardiovascular Devices
CATHETER	870.1340	CATHETER INTRODUCER Cardiovascular Diagnostic Devices; Cardiovascular Devices
CATHETER	870.1350	CATHETER BALLOON REPAIR KIT Cardiovascular Diagnostic Devices; Cardiovascular Devices
CATHETER	870.1370	CATHETER TIP OCCLUDER Cardiovascular Diagnostic Devices; Cardiovascular Devices
CATHETER	870.1380	CATHETER STYLET Cardiovascular Diagnostic Devices; Cardiovascular Devices
CATHETER	870.2870	CATHETER TIP PRESSURE TRANSDUCER Cardiovascular Monitoring Devices; Cardiovascular Devices
CATHETER	870.4210	CARDIOPULMONARY BYPASS VASCULAR CATHETER Cardiovascular Surgical Devices; Cardiovascular Devices

CATHETER	870.5150	EMBOLECTOMY CATHETER Cardiovascular Therapeutic Devices; Cardiovascular Devices
CATHETER	870.5175	SEPTOSTOMY CATHETER Cardiovascular Therapeutic Devices; Cardiovascular Devices
CATHETER	874.4175	NASOPHARYNGEAL CATHETER Surgical Devices; Ear, Nose And Throat Devices
CATHETER	876.4020	FIBEROPTIC LIGHT URETERAL CATHETER Surgical Devices; Gastroenterology-urology Devices
CATHETER	876.5010	BILIARY CATHETER AND ACCESSORIES Therapeutic Devices; Gastroenterology-urology Devices
CATHETER	876.5030	CONTINENT ILEOSTOMY CATHETER Therapeutic Devices; Gastroenterology-urology Devices
CATHETER	876.5090	SUPRAPUBIC UROLOGICAL CATHETER AND ACCESSORIES Therapeutic Devices; Gastroenterology-urology Devices
CATHETER	876.5130	UROLOGICAL CATHETER AND ACCESSORIES Therapeutic Devices; Gastroenterology-urology Devices
CATHETER	878.4200	INTRODUCTION/DRAINAGE CATHETER AND ACCESSORIES Surgical Devices; General And Plastic Surgery Devices
CATHETER	880.5200	INTRAVASCULAR CATHETER General Hospital And Personal Use Therapeutic Devices; General Hospital And Personal Use Devices
CATHETER	880.5210	INTRAVASCULAR CATHETER SECUREMENT DEVICE General Hospital And Personal Use Therapeutic Devices; General Hospital And Personal Use Devices
CATHETER	880.5965	SUBCUTANEOUS, IMPLANTED, INTRAVASCULAR INFUSION PORT AND CATHETER General Hospital And Personal Use Therapeutic Devices; General Hospital And Personal Use Devices
CATHETER	880.5970	PERCUTANEOUS, IMPLANTED, LONG-TERM INTRAVASCULAR CATHETER General Hospital And Personal Use Therapeutic Devices; General Hospital And Personal Use Devices
CATHETER	882.4100	VENTRICULAR CATHETER Neurological Surgical Devices; Neurological Devices
CATHETER	882.5150	INTRAVASCULAR OCCLUDING CATHETER Neurological Therapeutic Devices; Neurological Devices
CATHETERIZATION	870.1310	VESSEL DILATOR FOR PERCUTANEOUS CATHETERIZATION Cardiovascular Diagnostic Devices; Cardiovascular Devices
CATHETERS	884.6110	ASSISTED REPRODUCTION CATHETERS Obstetrical And Gynecological Devices
CATHODE	1020.20	COLD-CATHODE GAS DISCHARGE TUBES Performance Standards For Ionizing Radiation Emitting Products
CATHODE-RAY	870.2450	MEDICAL CATHODE-RAY TUBE DISPLAY Cardiovascular Monitoring Devices; Cardiovascular Devices
CATSUP	155.194	CATSUP Requirements For Specific Standardized Canned Vegetables; Canned Vegetables
CATTLE	500.65	EPINEPHRINE INJECTION FOR EMERGENCY TREATMENT OF ANAPHYLACTOID SHOCK IN CATTLE, HORSES, SHEEP, AND SWINE Requirements For Specific Animal Drugs; General
CAUSE, PROBABLE	1316.10	ADMINISTRATIVE PROBABLE CAUSE Administrative Inspections; Administrative Functions, Practices, And Procedures
CAUSE, SHOW	1301.37	ORDER TO SHOW CAUSE Action On Application For Registration: Revocation Or Suspension Of Registration; Registration Of Manufacturers, Distributors, And Dispensers Of Controlled Substances
CAUSE, SHOW	1309.46	ORDER TO SHOW CAUSE Registration Of Manufacturers, Distributors, Importers And Exporters Of List I Chemicals
CAUSES	1404.305	CAUSES FOR DEBARMENT Debarment; Governmentwide Debarment And Suspension And Requirements For Drug-Free Workplace
CAUSES	1404.405	CAUSES FOR SUSPENSION Suspension; Governmentwide Debarment And Suspension And Requirements For Drug-Free Workplace
CAUSTIC POISON ACT	2.110	DEFINITION OF AMMONIA UNDER FEDERAL CAUSTIC POISON ACT Caustic Poisons; General Administrative Rulings And Decisions
CAUTERY	886.4100	RADIOFREQUENCY ELECTROSURGICAL CAUTERY APPARATUS Surgical Devices; Ophthalmic Devices
CAUTERY	886.4115	THERMAL CAUTERY APPARATUS Surgical Devices; Ophthalmic Devices
CAUTION	369.20	DRUGS; RECOMMENDED WARNING AND CAUTION STATEMENTS Warning And Caution Statement For Drugs; Interpretative Statements Re: Warnings On Drugs And Devices For OTC Sale

CAUTION	369.21	DRUGS; WARNING AND CAUTION STATEMENTS REQUIRED BY REGULATIONS Warning And Caution Statement For Drugs; Interpretative Statements Re: Warnings On Drugs And Devices For OTC Sale
CAUTION	369.22	DRUGS; WARNING AND CAUTION STATEMENTS SPECIFICALLY REQUIRED BY LAW Warning And Caution Statement For Drugs; Interpretative Statements Re: Warnings On Drugs And Devices For OTC Sale
CAVITY LINER	872.3250	CALCIUM HYDROXIDE CAVITY LINER Prosthetic Devices; Dental Devices
CAVITY VARNISH	872.3260	CAVITY VARNISH Prosthetic Devices; Dental Devices
CEASE DISTRIBUTION	810.10	CEASE DISTRIBUTION AND NOTIFICATION ORDER Mandatory Medical Device Recall Procedures; Medical Device Recall Authority
CEASE DISTRIBUTION	810.12	WRITTEN REQUEST FOR REVIEW OF CEASE DISTRIBUTION AND NOTIFICATION ORDER Mandatory Medical Device Recall Procedures; Medical Device Recall Authority
CEASE DISTRIBUTION	810.14	CEASE DISTRIBUTION AND NOTIFICATION OR MANDATORY RECALL STRATEGY Mandatory Medical Device Recall Procedures; Medical Device Recall Authority
CEASE DISTRIBUTION	810.15	COMMUNICATIONS CONCERNING A CEASE DISTRIBUTION AND NOTIFICATION OR MANDATORY RECALL ORDER Mandatory Medical Device Recall Procedures; Medical Device Recall Authority
CEASE DISTRIBUTION	810.16	CEASE DISTRIBUTION AND NOTIFICATION OR MANDATORY RECALL STATUS REPORTS Mandatory Medical Device Recall Procedures; Medical Device Recall Authority
CEASE DISTRIBUTION	810.17	TERMINATION OF A CEASE DISTRIBUTION AND NOTIFICATION OR MANDATORY RECALL ORDER Mandatory Medical Device Recall Procedures; Medical Device Recall Authority
CEFADROXIL	520.314	CEFADROXIL TABLETS Oral Dosage Form New Animal Drugs
CEFADROXIL	520.315	CEFADROXIL POWDER FOR ORAL SUSPENSION Oral Dosage Form New Animal Drugs
CEFTIOFUR	522.313	CEFTIOFUR SODIUM POWDER FOR INJECTION Implantation Or Injectable Dosage Form New Animal Drugs
CEFTIOFUR	522.314	CEFTIOFUR HYDROCHLORIDE STERILE SUSPENSION Implantation Or Injectable Dosage Form New Animal Drugs
CEFTIOFUR	556.113	CEFTIOFUR Specific Tolerances For Residues Of New Animal Drugs; Tolerances For Residues Of New Animal Drugs In Food
CELL	864.8150	CALIBRATOR FOR CELL INDICES Hematology Reagents; Hematology And Pathology Devices
CELL, BLOOD	864.5240	AUTOMATED BLOOD CELL DILUTING APPARATUS Automated And Semi-automated Hematology Devices; Hematology And Pathology Devices
CELL, BLOOD	864.6160	MANUAL BLOOD CELL COUNTING DEVICE Manual Hematology Devices; Hematology And Pathology Devices
CELL, BLOOD	864.7100	RED BLOOD CELL ENZYME ASSAY Hematology Kits And Packages; Hematology And Pathology Devices
CELL, BLOOD	864.8200	BLOOD CELL DILUENT Hematology Reagents; Hematology And Pathology Devices
CELL, BLOOD	864.9245	AUTOMATED BLOOD CELL SEPARATOR Products Used In Establishments That Manufacture Blood And Blood Products; Hematology And Pathology Devices
CELL COUNTER	864.5200	AUTOMATED CELL COUNTER Automated And Semi-automated Hematology Devices; Hematology And Pathology Devices
CELL COUNTER	864.5220	AUTOMATED DIFFERENTIAL CELL COUNTER Automated And Semi-automated Hematology Devices; Hematology And Pathology Devices
CELL CULTURE	864.2220	SYNTHETIC CELL AND TISSUE CULTURE MEDIA AND COMPONENTS Cell And Tissue Culture Products; Hematology And Pathology Devices
CELL CULTURE	864.2240	CELL AND TISSUE CULTURE SUPPLIES AND EQUIPMENT Cell And Tissue Culture Products; Hematology And Pathology Devices
CELL CULTURE	876.5885	TISSUE CULTURE MEDIA FOR HUMAN EX VIVO TISSUE AND CELL CULTURE PROCESSING APPLICATIONS Therapeutic Devices; Gastroenterology-urology Devices
CELL-FREEZING	864.9225	CELL-FREEZING APPARATUS AND REAGENTS FOR IN VITRO DIAGNOSTIC USE Products Used In Establishments That Manufacture Blood And Blood Products; Hematology And Pathology Devices
CELL-LOCATING	864.5260	AUTOMATED CELL-LOCATING DEVICE Automated And Semi-automated Hematology Devices; Hematology And Pathology Devices

CELL, RED	864.5300	RED CELL INDICES DEVICE Automated And Semi-automated Hematology Devices; Hematology And Pathology Devices
CELL, RED	864.8540	RED CELL LYSING REAGENT Hematology Reagents; Hematology And Pathology Devices
CELL, RED/WHITE	864.8185	CALIBRATOR FOR RED CELL AND WHITE CELL COUNTING Hematology Reagents; Hematology And Pathology Devices
CELL-WASHING	864.9285	AUTOMATED CELL-WASHING CENTRIFUGE FOR IMMUNO-HEMATOLOGY Products Used In Establishments That Manufacture Blood And Blood Products; Hematology And Pathology Devices
CELLOPHANE	177.1200	CELLOPHANE Substances For Use As Basic Components Of Single And Repeated Use Food Contact Surfaces; Indirect Food Additives: Polymers
CELLS	864.2280	CULTURED ANIMAL AND HUMAN CELLS Cell And Tissue Culture Products; Hematology And Pathology Devices
CELLS	1271.22	HOW AND WHERE DO I REGISTER AND SUBMIT AN HCT/P (HUMAN CELLS/TISSUES/PRODUCTS) LIST? Procedures For Registration And Listing; Human Cells, Tissues, And Cellular And Tissue-Based Products
CELLS, RED BLOOD	640.10	RED BLOOD CELLS Red Blood Cells; Additional Standards For Human Blood And Blood Products
CELLS, RED BLOOD	660.30	REAGENT RED BLOOD CELLS Reagent Red Blood Cells; Additional Standards For Diagnostic Substances For Laboratory Tests
CELLULAR PRODUCTS	1271.22	HOW AND WHERE DO I REGISTER AND SUBMIT AN HCT/P (HUMAN CELLS/TISSUES/PRODUCTS) LIST? Procedures For Registration And Listing; Human Cells, Tissues, And Cellular And Tissue-Based Products
CELLULASE	173.120	CARBOHYDRASE AND CELLULASE DERIVED FROM ASPERGILLUS NIGER Enzyme Preparations And Microorganisms; Secondary Direct Food Additives Permitted In Food For Human Consumption
CELLULASE	184.1250	CELLULASE ENZYME PREPARATION DERIVED FROM TRICHODERMA LONGIBRANCHIATUM Listing Of Specific Substances Affirmed As GRAS; Direct Food Substances Affirmed As Generally Recognized As Safe
CELLULOSE	172.868	ETHYL CELLULOSE Multipurpose Additives; Food Additives Permitted For Direct Addition To Food For Human Consumption
CELLULOSE	172.870	HYDROXYPROPYL CELLULOSE Multipurpose Additives; Food Additives Permitted For Direct Addition To Food For Human Consumption
CELLULOSE	172.872	METHYL ETHYL CELLULOSE Multipurpose Additives; Food Additives Permitted For Direct Addition To Food For Human Consumption
CELLULOSE	177.1400	HYDROXYETHYL CELLULOSE FILM, WATER-INSOLUBLE Substances For Use As Basic Components Of Single And Repeated Use Food Contact Surfaces; Indirect Food Additives: Polymers
CELLULOSE	573.420	ETHYL CELLULOSE Food Additive Listing; Food Additives Permitted In Feed And Drinking Water Of Animals
CEMENT	888.3027	POLYMETHYLMETHACRYLATE (PMMA) BONE CEMENT Prosthetic Devices; Orthopedic Devices
CEMENT	872.3275	DENTAL CEMENT Prosthetic Devices; Dental Devices
CEMENT	888.4200	CEMENT DISPENSER Surgical Devices; Orthopedic Devices
CEMENT	888.4210	CEMENT MIXER FOR CLINICAL USE Surgical Devices; Orthopedic Devices
CEMENT	888.4220	CEMENT MONOMER VAPOR EVACUATOR Surgical Devices; Orthopedic Devices
CEMENT	888.4230	CEMENT VENTILATION TUBE Surgical Devices; Orthopedic Devices
CENTER, BIOLOGICS	5.201	REDELEGATION OF THE CENTER FOR BIOLOGICS EVALUATION AND RESEARCH DIRECTOR'S PROGRAM AUTHORITIES Biologics; Redelegations Of Authority; Delegations Of Authority And Organization
CENTRAL NERVOUS SYSTEM	882.5550	CENTRAL NERVOUS SYSTEM FLUID SHUNT AND COMPONENTS Neurological Therapeutic Devices; Neurological Devices
CENTRIFUGAL	862.2140	CENTRIFUGAL CHEMISTRY ANALYZER FOR CLINICAL USE Clinical Laboratory Instruments; Clinical Chemistry And Clinical Toxicology Devices
CENTRIFUGE	864.5350	MICROSEDIMENTATION CENTRIFUGE Automated And Semi-automated Hematology Devices; Hematology And Pathology Devices

CENTRIFUGE	864.9275	BLOOD BANK CENTRIFUGE FOR IN VITRO DIAGNOSTIC USE Products Used In Establishments That Manufacture Blood And Blood Products; Hematology And Pathology Devices
CENTRIFUGE	864.9285	AUTOMATED CELL-WASHING CENTRIFUGE FOR IMMUNO-HEMATOLOGY Products Used In Establishments That Manufacture Blood And Blood Products; Hematology And Pathology Devices
CEPHALOMETER	872.1830	CEPHALOMETER Diagnostic Devices; Dental Devices
CEPHAPIRIN	526.363	CEPHAPIRIN BENZATHINE Intramammary Dosage Forms
CEPHAPIRIN	529.365	CEPHAPIRIN SODIUM FOR INTRAMAMMARY INFUSION Certain Other Dosage Form New Animal Drugs
CEPHAPIRIN	556.115	CEPHAPIRIN Specific Tolerances For Residues Of New Animal Drugs; Tolerances For Residues Of New Animal Drugs In Food
CERAMIC	888.3353	HIP JOINT METAL/CERAMIC/POLYMER SEMI-CONSTRAINED CEMENTED OR NONPOROUS UNCEMENTED PROSTHESIS Prosthetic Devices; Orthopedic Devices
CERAMICWARE	109.16	ORNAMENTAL AND DECORATIVE CERAMICWARE Unavoidable Contaminants In Food For Human Consumption + Food Packaging Materials
CERCLAGE	888.3010	BONE FIXATION CERCLAGE Prosthetic Devices; Orthopedic Devices
CEREBELLAR	882.5820	IMPLANTED CEREBELLAR STIMULATOR Neurological Therapeutic Devices; Neurological Devices
CERTIFIABLE MIXTURES	82.6	CERTIFIABLE MIXTURES General Provisions; Listing Of Certified Provisionally Listed Colors And Specifications
CERTIFICATE OF REGISTRATION	1301.35	CERTIFICATE OF REGISTRATION; DENIAL OF REGISTRATION Action On Application For Registration: Revocation Or Suspension Of Registration; Registration Of Manufacturers, Distributors, And Dispensers Of Controlled Substances
CERTIFICATE OF REGISTRATION	1309.42	CERTIFICATE OF REGISTRATION; DENIAL OF REGISTRATION Registration Of Manufacturers, Distributors, Importers And Exporters Of List I Chemicals
CERTIFICATES	80.32	LIMITATIONS OF CERTIFICATES Certification Procedures; Color Additive Certification
CERTIFICATES	81.30	CANCELLATION OF CERTIFICATES General Specs+restrictns Provisional Color Additives For Foods, Drugs, And Cosmetics
CERTIFICATES	81.32	LIMITATION OF CERTIFICATES General Specs+restrictns Provisional Color Additives For Foods, Drugs, And Cosmetics
CERTIFICATES	900.14	SUSPENSION OR REVOCATION OF CERTIFICATES Quality Standards And Certification; Mammography
CERTIFICATES	1210.23	PERMITS GRANTED ON CERTIFICATES Permit Control; Regulations Under The Federal Import Milk Act
CERTIFICATES	1240.20	ISSUANCE AND POSTING OF CERTIFICATES FOLLOWING INSPECTION Control Of Communicable Diseases
CERTIFICATION	5.22	CERTIFICATION OF TRUE COPIES AND USE OF DEPARTMENT SEAL General Redelegations Of Authority; Delegations Of Authority And Organization
CERTIFICATION	5.32	CERTIFICATION FOLLOWING INSPECTIONS General Redelegations Of Authority; Delegations Of Authority And Organization
CERTIFICATION	5.35	OFFICIALS AUTHORIZED TO MAKE CERTIFICATION UNDER 5 U.S.C. 605(B) FOR ANY PROPOSED AND FINAL RULES General Redelegations Of Authority; Delegations Of Authority And Organization
CERTIFICATION	5.602	TESTING PROGRAMS AND METHODS OF CERTIFICATION AND IDENTIFICATION FOR ELECTRONIC PRODUCTS Radiation Control; Redelegations Of Authority; Delegations Of Authority And Organization
CERTIFICATION	20.3	CERTIFICATION AND AUTHENTICATION OF FDA RECORDS Official Testimony And Information; Public Information
CERTIFICATION	54.4	CERTIFICATION AND DISCLOSURE REQUIREMENTS Financial Disclosure By Clinical Investigators
CERTIFICATION	71.10	PETITION FOR EXEMPTION FROM CERTIFICATION General Provisions; Color Additive Petitions
CERTIFICATION	71.25	CONDITION FOR CERTIFICATION Administrative Action On Petitions; Color Additive Petitions
CERTIFICATION	71.26	REVOCATION OF EXEMPTION FROM CERTIFICATION Administrative Action On Petitions; Color Additive Petitions
CERTIFICATION	71.27	LISTING AND EXEMPTION FROM CERTIFICATION ON THE COMMISSIONER'S INITIATIVE Administrative Action On Petitions; Color Additive Petitions

CERTIFICATION	73.1	DILUENTS IN COLOR ADDITIVE MIXTURES FOR FOOD USE EXEMPT FROM CERTIFICATION Foods; Listing Of Color Additives Exempt From Certification
CERTIFICATION	73.1001	DILUENTS IN COLOR ADDITIVE MIXTURES FOR DRUG USE EXEMPT FROM CERTIFICATION Drugs; Listing Of Color Additives Exempt From Certification
CERTIFICATION	80.10	FEES FOR CERTIFICATION SERVICES General Provisions; Color Additive Certification
CERTIFICATION	80.21	REQUEST FOR CERTIFICATION Certification Procedures; Color Additive Certification
CERTIFICATION	80.22	SAMPLES TO ACCOMPANY REQUESTS FOR CERTIFICATION Certification Procedures; Color Additive Certification
CERTIFICATION	80.31	CERTIFICATION Certification Procedures; Color Additive Certification
CERTIFICATION	80.34	AUTHORITY TO REFUSE CERTIFICATION SERVICE Certification Procedures; Color Additive Certification
CERTIFICATION	80.35	COLOR ADDITIVE MIXTURES; CERTIFICATION AND EXEMPTION FROM CERTIFICATION Certification Procedures; Color Additive Certification
CERTIFICATION	80.37	TREATMENT OF BATCH PENDING CERTIFICATION Certification Procedures; Color Additive Certification
CERTIFICATION	80.38	TREATMENT OF BATCH AFTER CERTIFICATION Certification Procedures; Color Additive Certification
CERTIFICATION	320.34	REQUIREMENTS FOR BATCH TESTING AND CERTIFICATION BY THE FDA Bioavailability And Bioequivalence Requirements
CERTIFICATION	900.11	REQUIREMENTS FOR CERTIFICATION Quality Standards And Certification; Mammography
CERTIFICATION	900.15	APPEALS OF ADVERSE ACCREDITATION OR REACCREDITATION DECISIONS THAT PRECLUDE CERTIFICATION OR RECERTIFICATION Quality Standards And Certification; Mammography
CERTIFICATION	900.16	APPEALS OF DENIALS OF CERTIFICATION Quality Standards And Certification; Mammography
CERTIFICATION	900.21	APPLICATION FOR APPROVAL AS A CERTIFICATION AGENCY States As Certifiers; Mammography
CERTIFICATION	900.22	STANDARDS FOR CERTIFICATION AGENCIES States As Certifiers; Mammography
CERTIFICATION	1010.2	CERTIFICATION Performance Standards For Electronic Products: General
CERTIFICATION	1404.610	CERTIFICATION REQUIREMENTS AND PROCEDURES Governmentwide Debarment And Suspension And Requirements For Drug-Free Workplace
CERTIFICATION, CLASS III	807.94	FORMAT OF A CLASS III CERTIFICATION Establishment Registration And Device Listing For Manufacturers And Initial Importers Of Devices
CERTIFICATION, PATENT	314.52	NOTICE OF CERTIFICATION OF INVALIDITY OR NONINFRINGEMENT OF A PATENT Applications; Applications For FDA Approval To Market A New Drug
CERTIFICATION, PATENT	314.95	NOTICE OF CERTIFICATION OF INVALIDITY OR NONINFRINGEMENT OF A PATENT Abbreviated Applications; Applications For FDA Approval To Market A New Drug
CERULOPLASMIN	866.5210	CERULOPLASMIN IMMUNOLOGICAL TEST SYSTEM Immunological Test Systems; Immunology And Microbiology Devices
CERVICAL	884.1040	VISCOMETER FOR CERVICAL MUCUS Obstetrical And Gynecological Diagnostic Devices; Obstetrical And Gynecological Devices
CERVICAL	884.3200	CERVICAL DRAIN Obstetrical And Gynecological Prosthetic Devices; Obstetrical And Gynecological Devices
CERVICAL	884.4250	EXPANDABLE CERVICAL DILATOR Obstetrical And Gynecological Surgical Devices; Obstetrical And Gynecological Devices
CERVICAL	884.4260	HYGROSCOPIC LAMINARIA CERVICAL DILATOR Obstetrical And Gynecological Surgical Devices; Obstetrical And Gynecological Devices
CERVICAL	884.4270	VIBRATORY CERVICAL DILATORS Obstetrical And Gynecological Surgical Devices; Obstetrical And Gynecological Devices
CERVICAL	884.5250	CERVICAL CAP Obstetrical And Gynecological Therapeutic Devices; Obstetrical And Gynecological Devices
CESSATION	99.401	CORRECTIVE ACTIONS AND CESSATION OF DISSEMINATION OF INFORMATION Dissemination Of Information On Unapproved/New Uses For Marketed Drugs, Biologics And Devices

CHAIN, LONG	178.3780	POLYHYDRIC ALCOHOL ESTERS OF LONG CHAIN MONOBASIC ACIDS Certain Adjuvants And Production Aids; Indirect Food Additives: Adjuvants, Production Aids, And Sanitizers
CHAIN SPECIFIC	866.5550	IMMUNOGLOBULIN (LIGHT CHAIN SPECIFIC) IMMUNOLOGICAL TEST SYSTEM Immunological Test Systems; Immunology And Microbiology Devices
CHAIR	868.5365	POSTURE CHAIR FOR CARDIAC OR PULMONARY TREATMENT Therapeutic Devices; Anesthesiology Devices
CHAIR	872.6250	DENTAL CHAIR AND ACCESSORIES Miscellaneous Devices; Dental Devices
CHAIR	878. 4950	MANUAL OPERATING TABLE AND MANUAL OPERATING CHAIR AND ACCESSORIES Surgical Devices; General And Plastic Surgery Devices
CHAIR	880.6140	MEDICAL CHAIR AND TABLE General Hospital And Personal Use Miscellaneous Devices; General Hospital And Personal Use Devices
CHAIR	882.4125	NEUROSURGICAL CHAIR Neurological Surgical Devices; Neurological Devices
CHAIR	886.1140	OPHTHALMIC CHAIR Diagnostic Devices; Ophthalmic Devices
CHAIR	890.3100	MECHANICAL CHAIR Physical Medicine Prosthetic Devices; Physical Medicine Devices
CHAIR	890.3110	ELECTRIC POSITIONING CHAIR Physical Medicine Prosthetic Devices; Physical Medicine Devices
CHAIR	892.1820	PNEUMOENCEPHALOGRAPHIC CHAIR Diagnostic Devices; Radiology Devices
CHAIRMAN	14.30	CHAIRMAN OF AN ADVISORY COMMITTEE Meeting Procedures; Public Hearing Before A Public Advisory Committee
CHAIRS	878.4960	OPERATING TABLES AND ACCESSORIES AND OPERATING CHAIRS AND ACCESSORIES Surgical Devices; General And Plastic Surgery Devices
CHAMBER, ABDOMINAL	884.5225	ABDOMINAL DECOMPRESSION CHAMBER Obstetrical And Gynecological Therapeutic Devices; Obstetrical And Gynecological Devices
CHAMBER, ACOUSTIC	874.1060	ACOUSTIC CHAMBER FOR AUDIOMETRIC TESTING Diagnostic Devices; Ear, Nose, And Throat Devices
CHAMBER, ANAEROBIC	866.2120	ANAEROBIC CHAMBER Microbiology Devices; Immunology And Microbiology Devices
CHAMBER, ENVIRO.	864.9575	ENVIRONMENTAL CHAMBER FOR STORAGE OF PLATELET CONCENTRATE Products Used In Establishments That Manufacture Blood And Blood Products; Hematology And Pathology Devices
CHAMBER, HYPERBARIC	868.5470	HYPERBARIC CHAMBER Therapeutic Devices; Anesthesiology Devices
CHAMBER, ISOLATION	880.5450	PATIENT CARE REVERSE ISOLATION CHAMBER General Hospital And Personal Use Therapeutic Devices; General Hospital And Personal Use Devices
CHAMBER, OXYGEN	878.5650	TOPICAL OXYGEN CHAMBER FOR EXTREMITIES Therapeutic Devices; General And Plastic Surgery Devices
CHANGE, COLOR	880.2900	CLINICAL COLOR CHANGE THERMOMETER General Hospital And Personal Use Monitoring Devices; General Hospital And Personal Use Devices
CHANGE, DRUG	314.54	PROCEDURE FOR SUBMISSION OF AN APPLICATION REQUIRING INVESTIGATIONS FOR APPROVAL OF A NEW INDICATION FOR, OR OTHER CHANGE FROM, A LISTED DRUG Applications; Applications For FDA Approval To Market A New Drug
CHANGE, DRUG	314.93	PETITION TO REQUEST A CHANGE FROM A LISTED DRUG Abbreviated Applications; Applications For FDA Approval To Market A New Drug
CHANGE, OWNERSHIP	314.72	CHANGE IN OWNERSHIP OF AN APPLICATION Applications; Applications For FDA Approval To Market A New Drug
CHANGE, OWNERSHIP	316.27	CHANGE IN OWNERSHIP OF ORPHAN DRUG DESIGNATION Orphan Drugs
CHANGER, CASSETTE	892.1860	RADIOGRAPHIC FILM/CASSETTE CHANGER Diagnostic Devices; Radiology Devices
CHANGER, CASSETTE	892.1870	RADIOGRAPHIC FILM/CASSETTE CHANGER PROGRAMMER Diagnostic Devices; Radiology Devices
CHANGES	1403.30	CHANGES Post-Award Requirements; Uniform Administrative Requirements For Grants And Coop. Agreements To State & Local Gov.
CHANGES, APPLICATION	314.70	SUPPLEMENTS AND OTHER CHANGES TO AN APPROVED APPLICATION Applications; Applications For FDA Approval To Market A New Drug
CHANGES, APPLICATION	314.97	SUPPLEMENTS AND OTHER CHANGES TO AN APPROVED ABBREVIATED APPLICATION Abbreviated Applications; Applications For FDA Approval To Market A New Drug

CHANGES, APPLICATION	601.12	CHANGES TO AN APPROVED APPLICATION Biologics Licensing; Licensing
CHANGES, RESEARCH	56.110	EXPEDITED REVIEW PROCEDURES FOR CERTAIN KINDS OF RESEARCH INVOLVING NO MORE THAN MINIMAL RISK, AND FOR MINOR CHANGES IN APPROVED RESEARCH IRB Functions And Operations; Institutional Review Boards
CHANGES, SYSTEMS	21.21	CHANGES IN SYSTEMS AND NEW SYSTEMS FDA Privacy Act Record Systems; Protection Of Privacy
CHARACTERIZATION	58.105	TEST AND CONTROL ARTICLE CHARACTERIZATION Test And Control Articles; Good Laboratory Practice For Nonclinical Laboratory Studies
CHARGE	5.102	AUTHORITY TO APPROVE AND TO WITHDRAW APPROVAL OF A CHARGE FOR INVESTIGATIONAL NEW DRUGS Human Drugs; Redelegations Of Authority; Delegations Of Authority And Organization
CHARGE-IN	211.101	CHARGE-IN OF COMPONENTS Production And Process Controls; Current Good Manufacturing Practice For Finished Pharmaceuticals
CHARGEABLE COSTS	1.99	COSTS CHARGEABLE IN CONNECTION WITH RELABELING AND RECONDITIONING INADMISSIBLE IMPORTS Imports And Exports; General Enforcement Regulations
CHARGED-PARTICLE	892.5050	MEDICAL CHARGED-PARTICLE RADIATION THERAPY SYSTEM Therapeutic Devices; Radiology Devices
CHARGER	870.3670	PACEMAKER CHARGER Cardiovascular Prosthetic Devices; Cardiovascular Devices
CHARGING	312.7	PROMOTION AND CHARGING FOR INVESTIGATIONAL DRUGS General Provisions; Investigational New Drug Application
CHARITABLE INST.	203.39	DONATION OF DRUG SAMPLES TO CHARITABLE INSTITUTIONS Prescription Drug Marketing
CHART	866.1150	VISUAL ACUITY CHART Diagnostic Devices; Ophthalmic Devices
CHART RECORDER	870.2810	PAPER CHART RECORDER Cardiovascular Monitoring Devices; Cardiovascular Devices
CHECKS, EMPLOYEE	1301.93	SOURCES OF INFORMATION FOR EMPLOYEE CHECKS Employee Screening: Non-practitioners; Registration Of Manufacturers, Distributors, And Dispensers Of Controlled Substances
CHEDDAR CHEESE	133.113	CHEDDAR CHEESE Requirements For Specific Standardized Cheese And Related Products; Cheeses And Related Cheese Products
CHEDDAR CHEESE	133.114	CHEDDAR CHEESE FOR MANUFACTURING Requirements For Specific Standardized Cheese And Related Products; Cheeses And Related Cheese Products
CHEDDAR CHEESE	133.116	LOW SODIUM CHEDDAR CHEESE Requirements For Specific Standardized Cheese And Related Products; Cheeses And Related Cheese Products
CHEESE	133.10	NOTICE TO MANUFACTURERS PACKERS DISTRIBUTORS OF PASTEURIZED BLENDED CHEESE, PASTEURIZED PROCESS CHEESE, CHEESE FOOD, CHEESE SPREAD, AND RELATED FOODS General Provisions; Cheeses And Related Cheese Products
CHEESE, AMERICAN	133.147	GRATED AMERICAN CHEESE FOOD Requirements For Specific Standardized Cheese And Related Products; Cheeses And Related Cheese Products
CHEESE, ASIAGO	133.102	ASIAGO FRESH AND ASIAGO SOFT CHEESE Requirements For Specific Standardized Cheese And Related Products; Cheeses And Related Cheese
CHEESE, ASIAGO	133.103	ASIAGO MEDIUM CHEESE Requirements For Specific Standardized Cheese And Related Products; Cheeses And Related Cheese Products
CHEESE, ASIAGO	133.104	ASIAGO OLD CHEESE Requirements For Specific Standardized Cheese And Related Products; Cheeses And Related Cheese Products
CHEESE, BLENDED	133.10	NOTICE TO MANUFACTURERS, PACKERS, AND DISTRIBUTORS OF PASTEURIZED BLENDED CHEESE, PASTEURIZED PROCESS CHEESE, CHEESE FOOD, CHEESE SPREAD, AND RELATED FOODS General Provisions; Cheeses And Related Cheese Products
CHEESE, BLENDED	133.167	PASTEURIZED BLENDED CHEESE Requirements For Specific Standardized Cheese And Related Products; Cheeses And Related Cheese Products
CHEESE, BLENDED	133.168	PASTEURIZED BLENDED CHEESE WITH FRUITS, VEGETABLES, OR MEATS Requirements For Specific Standardized Cheese And Related Products; Cheeses And Related Cheese Products
CHEESE, BLUE	133.106	BLUE CHEESE Requirements For Specific Standardized Cheese And Related Products; Cheeses And Related Cheese Products

CHEESE, BLUE-MOLD	133.184	ROQUEFORT, SHEEP'S MILK BLUE-MOLD, AND BLUE-MOLD CHEESE FROM SHEEP'S MILK Requirements For Specific Standardized Cheese And Related Products; Cheeses And Related Cheese Products
CHEESE, BRICK	133.108	BRICK CHEESE Requirements For Specific Standardized Cheese And Related Products; Cheeses And Related Cheese Products
CHEESE, BRICK	133.109	BRICK CHEESE FOR MANUFACTURING Requirements For Specific Standardized Cheese And Related Products; Cheeses And Related Cheese Products
CHEESE, CACIOCAVALLO	133.111	CACIOCAVALLO SICILIANO CHEESE Requirements For Specific Standardized Cheese And Related Products; Cheeses And Related Cheese Products
CHEESE, CHEDDAR	133.113	CHEDDAR CHEESE Requirements For Specific Standardized Cheese And Related Products; Cheeses And Related Cheese Products
CHEESE, CHEDDAR	133.114	CHEDDAR CHEESE FOR MANUFACTURING Requirements For Specific Standardized Cheese And Related Products; Cheeses And Related Cheese Products
CHEESE, CHEDDAR	133.116	LOW SODIUM CHEDDAR CHEESE Requirements For Specific Standardized Cheese And Related Products; Cheeses And Related Cheese Products
CHEESE, CLUB	133.123	COLD-PACK AND CLUB CHEESE Requirements For Specific Standardized Cheese And Related Products; Cheeses And Related Cheese Products
CHEESE, COLBY	133.118	COLBY CHEESE Requirements For Specific Standardized Cheese And Related Products; Cheeses And Related Cheese Products
CHEESE, COLBY	133.119	COLBY CHEESE FOR MANUFACTURING Requirements For Specific Standardized Cheese And Related Products; Cheeses And Related Cheese Products
CHEESE, COLBY	133.121	LOW SODIUM COLBY CHEESE Requirements For Specific Standardized Cheese And Related Products; Cheeses And Related Cheese Products
CHEESE, COLD-PACK	133.123	COLD-PACK AND CLUB CHEESE Requirements For Specific Standardized Cheese And Related Products; Cheeses And Related Cheese Products
CHEESE, COLD-PACK	133.124	COLD-PACK CHEESE FOOD Requirements For Specific Standardized Cheese And Related Products; Cheeses And Related Cheese Products
CHEESE, COLD-PACK	133.125	COLD-PACK CHEESE FOOD WITH FRUITS, VEGETABLES, OR MEATS Requirements For Specific Standardized Cheese And Related Products; Cheeses And Related Cheese Products
CHEESE, COOK	133.127	COOK CHEESE, KOCH KAESE Requirements For Specific Standardized Cheese And Related Products; Cheeses And Related Cheese Products
CHEESE, COTTAGE	133.128	COTTAGE CHEESE Requirements For Specific Standardized Cheese And Related Products; Cheeses And Related Cheese Products
CHEESE, COTTAGE	133.129	DRY CURD COTTAGE CHEESE Requirements For Specific Standardized Cheese And Related Products; Cheeses And Related Cheese Products
CHEESE, CREAM	133.133	CREAM CHEESE Requirements For Specific Standardized Cheese And Related Products; Cheeses And Related Cheese Products
CHEESE, CREAM	133.134	CREAM CHEESE WITH OTHER FOODS Requirements For Specific Standardized Cheese And Related Products; Cheeses And Related Cheese Products
CHEESE, CURD	133.136	WASHED CURD AND SOAKED CURD CHEESE Requirements For Specific Standardized Cheese And Related Products; Cheeses And Related Cheese Products
CHEESE, CURD	133.137	WASHED CURD CHEESE FOR MANUFACTURING Requirements For Specific Standardized Cheese And Related Products; Cheeses And Related Cheese Products
CHEESE, CURD	133.144	GRANULAR AND STIRRED CURD CHEESE Requirements For Specific Standardized Cheese And Related Products; Cheeses And Related Cheese Products
CHEESE, EDAM	133.138	EDAM CHEESE Requirements For Specific Standardized Cheese And Related Products; Cheeses And Related Cheese Products
CHEESE, EMMENTALER	133.195	SWISS AND EMMENTALER CHEESE Requirements For Specific Standardized Cheese And Related Products; Cheeses And Related Cheese Products

CHEESE FOOD	133.10	NOTICE TO MANUFACTURERS, PACKERS, AND DISTRIBUTORS OF PASTEURIZED BLENDED CHEESE, PASTEURIZED PROCESS CHEESE, CHEESE FOOD, CHEESE SPREAD, AND RELATED FOODS General Provisions; Cheeses And Related Cheese Products
CHEESE FOOD	133.124	COLD-PACK CHEESE FOOD Requirements For Specific Standardized Cheese And Related Products; Cheeses And Related Cheese Products
CHEESE FOOD	133.125	COLD-PACK CHEESE FOOD WITH FRUITS, VEGETABLES, OR MEATS Requirements For Specific Standardized Cheese And Related Products; Cheeses And Related Cheese
CHEESE FOOD	133.147	GRATED AMERICAN CHEESE FOOD Requirements For Specific Standardized Cheese And Related Products; Cheeses And Related Cheese Products
CHEESE, GAMMELOST	133.140	GAMMELOST CHEESE Requirements For Specific Standardized Cheese And Related Products; Cheeses And Related Cheese Products
CHEESE, GORGONZOLA	133.141	GORGONZOLA CHEESE Requirements For Specific Standardized Cheese And Related Products; Cheeses And Related Cheese Products
CHEESE, GOUDA	133.142	GOUDA CHEESE Requirements For Specific Standardized Cheese And Related Products; Cheeses And Related Cheese Products
CHEESE, GRANULAR	133.145	GRANULAR CHEESE FOR MANUFACTURING Requirements For Specific Standardized Cheese And Related Products; Cheeses And Related Cheese Products
CHEESE, GRUYERE	133.149	GRUYERE CHEESE Requirements For Specific Standardized Cheese And Related Products; Cheeses And Related Cheese Products
CHEESE, HIGH-MOISTURE	133.154	HIGH-MOISTURE JACK CHEESE Requirements For Specific Standardized Cheese And Related Products; Cheeses And Related Cheese Products
CHEESE, LIMBURGER	133.152	LIMBURGER CHEESE Requirements For Specific Standardized Cheese And Related Products; Cheeses And Related Cheese Products
CHEESE, MONTEREY JACK	133.153	MONTEREY CHEESE AND MONTEREY JACK CHEESE Requirements For Specific Standardized Cheese And Related Products; Cheeses And Related Cheese Products
CHEESE, MOZZARELLA	133.155	MOZZARELLA CHEESE AND SCAMORZA CHEESE Requirements For Specific Standardized Cheese And Related Products; Cheeses And Related Cheese Products
CHEESE, MOZZARELLA	133.157	PART-SKIM MOZZARELLA AND SCAMORZA CHEESE Requirements For Specific Standardized Cheese And Related Products; Cheeses And Related Cheese Products
CHEESE, MOZZARELLA	133.158	LOW-MOISTURE PART-SKIM MOZZARELLA AND SCAMORZA CHEESE Requirements For Specific Standardized Cheese And Related Products; Cheeses And Related Cheese Products
CHEESE, MUENSTER/MUNSTER	133.160	MUENSTER AND MUNSTER CHEESE Requirements For Specific Standardized Cheese And Related Products; Cheeses And Related Cheese Products
CHEESE, MUENSTER/MUNSTER	133.161	MUENSTER AND MUNSTER CHEESE FOR MANUFACTURING Requirements For Specific Standardized Cheese And Related Products; Cheeses And Related Cheese Products
CHEESE, NEUFCHATEL	133.162	NEUFCHATEL CHEESE Requirements For Specific Standardized Cheese And Related Products; Cheeses And Related Cheese Products
CHEESE, NEUFCHATEL	133.178	PASTEURIZED NEUFCHATEL CHEESE SPREAD WITH OTHER FOODS Requirements For Specific Standardized Cheese And Related Products; Cheeses And Related Cheese Products
CHEESE, NUWORLD	133.164	NUWORLD CHEESE Requirements For Specific Standardized Cheese And Related Products; Cheeses And Related Cheese Products
CHEESE, PARMESAN/REGGIANO	133.165	PARMESAN AND REGGIANO CHEESE Requirements For Specific Standardized Cheese And Related Products; Cheeses And Related Cheese Products
CHEESE, PIMENTO	133.171	PASTEURIZED PROCESS PIMENTO CHEESE Requirements For Specific Standardized Cheese And Related Products; Cheeses And Related Cheese Products
CHEESE FOOD	133.10	NOTICE TO MANUFACTURERS, PACKERS, AND DISTRIBUTORS OF PASTEURIZED BLENDED CHEESE, PASTEURIZED PROCESS CHEESE, CHEESE FOOD, CHEESE SPREAD, AND RELATED FOODS General Provisions; Cheeses And Related Cheese Products
CHEESE, PROCESS	133.169	PASTEURIZED PROCESS CHEESE Requirements For Specific Standardized Cheese And Related Products; Cheeses And Related Cheese Products

CHEESE, PROCESS	133.170	PASTEURIZED PROCESS CHEESE WITH FRUITS, VEGETABLES, OR MEATS Requirements For Specific Standardized Cheese And Related Products; Cheeses And Related Cheese Products
CHEESE, PROCESS	133.173	PASTEURIZED PROCESS CHEESE FOOD Requirements For Specific Standardized Cheese And Related Products; Cheeses And Related Cheese Products
CHEESE, PROCESS	133.174	PASTEURIZED PROCESS CHEESE FOOD WITH FRUITS, VEGETABLES, OR MEATS Requirements For Specific Standardized Cheese And Related Products; Cheeses And Related Cheese Products
CHEESE, PROVOLONE	133.181	PROVOLONE CHEESE Requirements For Specific Standardized Cheese And Related Products; Cheeses And Related Cheese Products
CHEESE, ROMANO	133.183	ROMANO CHEESE Requirements For Specific Standardized Cheese And Related Products; Cheeses And Related Cheese Products
CHEESE, ROQUEFORT	133.184	ROQUEFORT, SHEEP'S MILK BLUE-MOLD, AND BLUE-MOLD CHEESE FROM SHEEP'S MILK Requirements For Specific Standardized Cheese And Related Products; Cheeses And Related Cheese Products
CHEESE, SAMSOE	133.185	SAMSOE CHEESE Requirements For Specific Standardized Cheese And Related Products; Cheeses And Related Cheese Products
CHEESE, SAP SAGO	133.186	SAP SAGO CHEESE Requirements For Specific Standardized Cheese And Related Products; Cheeses And Related Cheese Products
CHEESE, SCAMORZA	133.155	MOZZARELLA CHEESE AND SCAMORZA CHEESE Requirements For Specific Standardized Cheese And Related Products; Cheeses And Related Cheese Products
CHEESE, SCAMORZA	133.156	LOW-MOISTURE MOZZARELLA AND SCAMORZA CHEESE Requirements For Specific Standardized Cheese And Related Products; Cheeses And Related Cheese Products
CHEESE, SCAMORZA	133.157	PART-SKIM MOZZARELLA AND SCAMORZA CHEESE Requirements For Specific Standardized Cheese And Related Products; Cheeses And Related Cheese Products
CHEESE, SKIM MILK	133.189	SKIM MILK CHEESE FOR MANUFACTURING Requirements For Specific Standardized Cheese And Related Products; Cheeses And Related Cheese Products
CHEESE SPREAD	133.10	NOTICE TO MANUFACTURERS, PACKERS, AND DISTRIBUTORS OF PASTEURIZED BLENDED CHEESE, PASTEURIZED PROCESS CHEESE, CHEESE FOOD, CHEESE SPREAD, AND RELATED FOODS General Provisions; Cheeses And Related Cheese Products
CHEESE SPREAD	133.175	PASTEURIZED CHEESE SPREAD Requirements For Specific Standardized Cheese And Related Products; Cheeses And Related Cheese Products
CHEESE SPREAD	133.176	PASTEURIZED CHEESE SPREAD WITH FRUITS, VEGETABLES OR MEATS Requirements For Specific Standardized Cheese And Related Products; Cheeses And Related Cheese Products
CHEESE SPREAD	133.179	PASTEURIZED PROCESS CHEESE SPREAD Requirements For Specific Standardized Cheese And Related Products; Cheeses And Related Cheese Products
CHEESE SPREAD	133.180	PASTEURIZED PROCESS CHEESE SPREAD WITH FRUITS, VEGETABLES, OR MEATS Requirements For Specific Standardized Cheese And Related Products; Cheeses And Related Cheese Products
CHEESE, SWISS	133.195	SWISS AND EMMENTALER CHEESE Requirements For Specific Standardized Cheese And Related Products; Cheeses And Related Cheese Products
CHEESE, SWISS	133.196	SWISS CHEESE FOR MANUFACTURING Requirements For Specific Standardized Cheese And Related Products; Cheeses And Related Cheese Products
CHEESES, GRATED	133.146	GRATED CHEESES Requirements For Specific Standardized Cheese And Related Products; Cheeses And Related Cheese Products
CHEESES, HARD	133.148	HARD GRATING CHEESES Requirements For Specific Standardized Cheese And Related Products; Cheeses And Related Cheese Products
CHEESES, HARD	133.150	HARD CHEESES Requirements For Specific Standardized Cheese And Related Products; Cheeses And Related Cheese Products
CHEESES, SEMISOFT	133.187	SEMISOFT CHEESES Requirements For Specific Standardized Cheese And Related Products; Cheeses And Related Cheese Products

CHEESES, SEMISOFT	133.188	SEMISOFT PART-SKIM CHEESES Requirements For Specific Standardized Cheese And Related Products; Cheeses And Related Cheese Products
CHEESES, SOFT RIPENED	133.182	SOFT RIPENED CHEESES Requirements For Specific Standardized Cheese And Related Products; Cheeses And Related Cheese Products
CHEESES, SPICED	133.190	SPICED CHEESES Requirements For Specific Standardized Cheese And Related Products; Cheeses And Related Cheese Products
CHEESES, SPICED	133.191	PART-SKIM SPICED CHEESES Requirements For Specific Standardized Cheese And Related Products; Cheeses And Related Cheese Products
CHEESES, STANDARDIZED	133.193	SPICED, FLAVORED STANDARDIZED CHEESES Requirements For Specific Standardized Cheese And Related Products; Cheeses And Related Cheese Products
CHELATING AGENTS	176.150	CHELATING AGENTS USED IN THE MANUFACTURE OF PAPER AND PAPERBOARD Substances For Use Only As Components Of Paper And Paperboard; Indirect Food Additives
CHEMICAL ANALYSTS	1304.23	RECORDS FOR CHEMICAL ANALYSTS Continuing Records; Records And Reports Of Registrants
CHEMICAL COLD PACK	880.5760	CHEMICAL COLD PACK SNAKEBITE KIT General Hospital And Personal Use Therapeutic Devices; General Hospital And Personal Use Devices
CHEMICAL DERIVATIVES	329.1	HABIT-FORMING DRUGS WHICH ARE CHEMICAL DERIVATIVES OF SUBSTANCES SPECIFIED IN SECTION 502(D) OF THE FEDERAL FOOD, DRUG, AND COSMETIC ACT Derivatives Designated As Habit Forming; Habit-forming Drugs
CHEMICAL EXPORTERS	1313.24	WAIVER OF 15-DAY ADVANCE NOTICE FOR CHEMICAL EXPORTERS Exportation Of Listed Chemicals; Importation And Exportation Of Precursors And Essential Chemicals
CHEMICAL PREPARATIONS	1308.23	EXEMPTION OF CERTAIN CHEMICAL PREPARATIONS; APPLICATION Exempt Chemical Preparations; Schedules Of Controlled Substances
CHEMICAL PREPARATIONS	1308.24	EXEMPTION CHEMICAL PREPARATIONS Exempt Chemical Preparations; Schedules Of Controlled Substances
CHEMICAL PRESERVATIVES	101.22	FOODS; LABELING OF SPICES, FLAVORINGS, COLORINGS AND CHEMICAL PRESERVATIVES Specific Food Labeling Requirements; Food Labeling
CHEMICAL PRESERVATIVES	501.22	ANIMAL FOODS; LABELING OF SPICES, FLAVORINGS, COLORINGS, AND CHEMICAL PRESERVATIVES Specific Animal Food Labeling Requirements; Animal Food Labeling
CHEMICAL SOLUTION STAINS	864.1850	DYE AND CHEMICAL SOLUTION STAINS Biological Stains; Hematology And Pathology Devices
CHEMICAL STERILANTS	880.6885	LIQUID CHEMICAL STERILANTS/HIGH LEVEL DISINFECTANTS General Hospital And Personal Use Miscellaneous Devices; General Hospital And Personal Use Devices
CHEMICALS	170.19	PESTICIDE CHEMICALS IN PROCESSED FOODS General Provisions; Food Additives
CHEMICALS	173.315	CHEMICALS USED IN WASHING OR TO ASSIST IN THE PEELING OF FRUITS AND VEGETABLES Specific Usage Additives; Secondary Direct Food Additives Permitted In Food For Human Consumption
CHEMICALS	173.320	CHEMICALS FOR CONTROLLING MICROORGANISMS IN CANE-SUGAR AND BEET-SUGAR MILLS Specific Usage Additives; Secondary Direct Food Additives Permitted In Food For Human Consumption
CHEMICALS	173.322	CHEMICALS USED IN DELINTING COTTONSEED Specific Usage Additives; Secondary Direct Food Additives Permitted In Food For Human Consumption
CHEMICALS	182.9	ADJUVANTS FOR PESTICIDE CHEMICALS General Provisions; Substances Generally Recognized As Safe
CHEMICALS	570.19	PESTICIDE CHEMICALS IN PROCESSED FOODS General Provisions; Food Additives
CHEMICALS	582.99	ADJUVANTS FOR PESTICIDE CHEMICALS General Provisions; Substances Generally Recognized As Safe
CHEMICALS, LISTED	1300.2	DEFINITIONS RELATING TO LISTED CHEMICALS Definitions
CHEMICALS, PRESCRIPTION	201.120	PRESCRIPTION CHEMICALS AND OTHER PRESCRIPTION COMPONENTS Exemptions From Adequate Directions For Use; Labeling
CHEMISTRY	862.2140	CENTRIFUGAL CHEMISTRY ANALYZER FOR CLINICAL USE Clinical Laboratory Instruments; Clinical Chemistry And Clinical Toxicology Devices
CHEMISTRY	862.2150	CONTINUOUS FLOW SEQUENTIAL MULTIPLE CHEMISTRY ANALYZER FOR CLINICAL USE Clinical Laboratory Instruments; Clinical Chemistry And Clinical Toxicology Devices

CHEMISTRY	862.2160	DISCRETE PHOTOMETRIC CHEMISTRY ANALYZER FOR CLINICAL USE Clinical Laboratory Instruments; Clinical Chemistry And Clinical Toxicology Devices
CHEMISTRY	862.2170	MICRO CHEMISTRY ANALYZER FOR CLINICAL USE Clinical Laboratory Instruments; Clinical Chemistry And Clinical Toxicology Devices
CHERRIES	145.125	CANNED CHERRIES Requirements For Specific Standardized Canned Fruits; Canned Fruits
CHERRIES	145.126	ARTIFICIALLY SWEETENED CANNED CHERRIES Requirements For Specific Standardized Canned Fruits; Canned Fruits
CHERRY	152.126	FROZEN CHERRY PIE Requirements For Specific Standardized Fruit Pies; Fruit Pies
CHEWABLE	520.540c	DEXAMETHASONE CHEWABLE TABLETS Oral Dosage Form New Animal Drugs
CHEWABLE	520.622c	DIETHYLCARBAMAZINE CITRATE CHEWABLE TABLETS Oral Dosage Form New Animal Drugs
CHEWABLE	520.623	DIETHYLCARBAMAZINE CITRATE, OXIBENDAZOLE CHEWABLE TABLETS Oral Dosage Form New Animal Drugs
CHEWABLE	520.1196	IVERMECTIN AND PYRANTEL PAMOATE CHEWABLE TABLET Oral Dosage Form New Animal Drugs
CHEWABLE	520.2041	PYRANTEL PAMOATE CHEWABLE TABLETS Oral Dosage Form New Animal Drugs
CHEWABLE	520.2150b	STANOZOLOL CHEWABLE TABLETS Oral Dosage Form New Animal Drugs
CHEWABLES	520.1193	IVERMECTIN TABLETS AND CHEWABLES Oral Dosage Form New Animal Drugs
CHEWING GUM	172.615	CHEWING GUM BASE Gums, Chewing Gum Bases And Related Substances; Food Additives Permitted For Direct Addition To Food For Human Consumption
CHIEF COUNSEL	5.1105	CHIEF COUNSEL, FOOD AND DRUG ADMINISTRATION Organization; Delegations Of Authority And Organization
CHILDREN	50.54	CLINICAL INVESTIGATIONS NOT OTHERWISE APPROVABLE THAT PRESENT AN OPPORTUNITY TO UNDERSTAND, PREVENT, OR ALLEVIATE A SERIOUS PROBLEM AFFECTING THE HEALTH OR WELFARE OF CHILDREN Additional Safeguards For Children In Clinical Investigations; Protection Of Human Subjects
CHILDREN	50.55	REQUIREMENTS FOR PERMISSION BY PARENTS OR GUARDIANS AND FOR ASSENT BY CHILDREN Additional Safeguards For Children In Clinical Investigations; Protection Of Human Subjects
CHILDREN	369.9	GENERAL WARNINGS RE ACCIDENTAL INGESTION BY CHILDREN Definitions And Interpretations; Interpretative Statements Re Warnings On Drugs And Devices For OTC Sale
CHILLING UNIT	890.5940	CHILLING UNIT Physical Medicine Therapeutic Devices; Physical Medicine Devices
CHIN PROSTHESIS	878.3550	CHIN PROSTHESIS Prosthetic Devices; General And Plastic Surgery Devices
CHIPS, POTATO	102.41	POTATO CHIPS MADE FROM DRIED POTATOES Requirements For Specific Nonstandardized Foods; Common Or Usual Name For Nonstandardized Foods
CHLAMYDIA	866.3120	CHLAMYDIA SEROLOGICAL REAGENTS Serological Reagents; Immunology And Microbiology Devices
CHLOMIPRAMINE HYDROCHLORIDE	520.455	CHLOMIPRAMINE HYDROCHLORIDE TABLETS Oral Dosage Form New Animal Drugs
CLEANLINESS	1250.49	CLEANLINESS OF CONVEYANCES Equipment And Operation Of Land And Air Conveyances; Interstate Conveyance Sanitation
CLOPIDAL	558.175	CLOPIDAL New Animal Drugs For Use In Animal Feeds
CHLORAL HYDRATE	522.380	CHLORAL HYDRATE, PENTOBARBITAL, AND MAGNESIUM SULFATE STERILE AQUEOUS SOLUTION Implantation Or Injectable Dosage Form New Animal Drugs
CHLORAMPHENICOL	520.390	CHLORAMPHENICOL ORAL DOSAGE FORMS Oral Dosage Form New Animal Drug Applications
CHLORAMPHENICOL	522.390	CHLORAMPHENICOL INJECTION Implantation Or Injectable Dosage Form New Animal Drugs
CHLORAMPHENICOL	524.390	CHLORAMPHENICOL OPHTHALMIC AND TOPICAL DOSAGE FORMS Ophthalmic And Topical Dosage Form New Animal Drugs
CHLORHEXIDINE	529.400	CHLORHEXIDINE TABLETS AND SUSPENSION Certain Other Dosage Form New Animal Drugs
CHLORHEXIDINE	556.120	CHLORHEXIDINE Specific Tolerances For Residues Of New Animal Drugs; Tolerances For Residues Of New Animal Drugs In Food
CHLORHEXIDINE DIACETATE	524.402	CHLORHEXIDINE DIACETATE OINTMENT Ophthalmic And Topical Dosage Form New Animal Drugs

CHLORIDE	862.1170	CHLORIDE TEST SYSTEM Clinical Chemistry Test Systems; Clinical Chemistry And Clinical Toxicology Devices
CHLORIDE, AMMONIUM	172.165	QUATERNARY AMMONIUM CHLORIDE COMBINATION Food Preservatives; Food Additives Permitted For Direct Addition To Food For Human Consumption
CHLORIDE, AMMONIUM	184.1138	AMMONIUM CHLORIDE Listing Of Specific Substances Affirmed As GRAS; Direct Food Substances Affirmed As Generally Recognized As Safe
CHLORIDE, AMMONIUM	520.310	CARAMIPHEN ETHANEDISULFONATE AND AMMONIUM CHLORIDE TABLETS Oral Dosage Form New Animal Drugs
CHLORIDE, BUTYL	520.260	N-BUTYL CHLORIDE CAPSULES Oral Dosage Form New Animal Drugs
CHLORIDE, CALCIUM	172.330	CALCIUM PANTOTHENATE, CALCIUM CHLORIDE DOUBLE SALT Special Dietary And Nutritional Additives; Food Additives Permitted For Direct Addition To Food For Human Consumption
CHLORIDE, CALCIUM	184.1193	CALCIUM CHLORIDE Listing Of Specific Substances Affirmed As GRAS; Direct Food Substances Affirmed As Generally Recognized As Safe
CHLORIDE, CALCIUM	582.1193	CALCIUM CHLORIDE General Purpose Food Additives; Substances Generally Recognized As Safe
CHLORIDE, CALCIUM	582.6193	CALCIUM CHLORIDE Sequestrants; Substances Generally Recognized As Safe
CHLORIDE, CHOLINE	182.8252	CHOLINE CHLORIDE Nutrients; Substances Generally Recognized As Safe
CHLORIDE, CHOLINE	582.5252	CHOLINE CHLORIDE Nutrients And/or Dietary Supplements; Substances Generally Recognized As Safe
CHLORIDE, CHROMIC	178.3290	CHROMIC CHLORIDE COMPLEXES Certain Adjuvants And Production Aids; Indirect Food Additives: Adjuvants, Production Aids, And Sanitizers
CHLORIDE, DIMETHYLDIALKYL-AMMONIUM	173.400	DIMETHYLDIALKYLAMMONIUM CHLORIDE Specific Usage Additives; Secondary Direct Food Additives Permitted In Food For Human Consumption
CHLORIDE, FERRIC	184.1297	FERRIC CHLORIDE Listing Of Specific Substances Affirmed As GRAS; Direct Food Substances Affirmed As Generally Recognized As Safe
CHLORIDE, MAGNESIUM	184.1426	MAGNESIUM CHLORIDE Listing Of Specific Substances Affirmed As GRAS; Direct Food Substances Affirmed As Generally Recognized As Safe
CHLORIDE, MANGANESE	184.1446	MANGANESE CHLORIDE Listing Of Specific Substances Affirmed As GRAS; Direct Food Substances Affirmed As Generally Recognized As Safe
CHLORIDE, MANGANESE	582.5446	MANGANESE CHLORIDE Nutrients And/or Dietary Supplements; Substances Generally Recognized As Safe
CHLORIDE, METHYLENE	173.255	METHYLENE CHLORIDE Solvents, Lubricants, Release Agents And Related Substances; Secondary Direct Food Additives Permitted In Food For Human Consumption
CHLORIDE, METHYLENE	700.19	USE OF METHYLENE CHLORIDE AS AN INGREDIENT IN COSMETIC PRODUCTS Requirements For Specific Cosmetic Products; General
CHLORIDE, POTASSIUM	184.1622	POTASSIUM CHLORIDE Listing Of Specific Substances Affirmed As GRAS; Direct Food Substances Affirmed As Generally Recognized As Safe
CHLORIDE, POTASSIUM	582.1193	POTASSIUM CHLORIDE General Purpose Food Additives; Substances Generally Recognized As Safe
CHLORIDE, STANNOUS	172.180	STANNOUS CHLORIDE Food Preservatives; Food Additives Permitted For Direct Addition To Food For Human Consumption
CHLORIDE, STANNOUS	184.1845	STANNOUS CHLORIDE Listing Of Specific Substances Affirmed As GRAS; Direct Food Substances Affirmed As Generally Recognized As Safe
CHLORIDE, STANNOUS	582.3845	STANNOUS CHLORIDE Chemical Preservatives; Substances Generally Recognized As Safe
CHLORIDE, VINYL	177.1950	VINYL CHLORIDE-ETHYLENE COPOLYMERS Substances For Use As Basic Components Of Single And Repeated Use Food Contact Surfaces; Indirect Food Additives: Polymers
CHLORIDE, VINYL	177.1960	VINYL CHLORIDE-HEXENE-1 COPOLYMERS Substances For Use As Basic Components Of Single And Repeated Use Food Contact Surfaces; Indirect Food Additives: Polymers
CHLORIDE, VINYL	177.1970	VINYL CHLORIDE-LAURYL VINYL ETHER COPOLYMERS Substances For Use As Basic Components Of Single And Repeated Use Food Contact Surfaces; Indirect Food Additives: Polymers

CHLORIDE, VINYL	177.1980	VINYL CHLORIDE-PROPYLENE COPOLYMERS Substances For Use As Basic Components Of Single And Repeated Use Food Contact Surfaces; Indirect Food Additives: Polymers
CHLORIDE, VINYL	178.2650	ORGANOTIN STABILIZERS IN VINYL CHLORIDE PLASTICS Antioxidants And Stabilizers; Indirect Food Additives: Adjuvants, Production Aids, And Sanitizers
CHLORIDE, VINYL	178.3790	POLYMER MODIFIERS IN SEMIRIGID AND RIGID VINYL CHLORIDE PLASTICS Certain Adjuvants And Production Aids; Indirect Food Additives: Adjuvants, Production Aids, And Sanitizers
CHLORIDE, VINYL	700.14	USE OF VINYL CHLORIDE AS AN INGREDIENT, INCLUDING PROPELLANT OF COSMETIC AEROSOL PRODUCTS Requirements For Specific Cosmetic Products; General
CHLORIDE, VINYLIDENE	175.360	VINYLIDENE CHLORIDE COPOLYMER COATINGS FOR NYLON FILM Substances For Use As Components Of Coatings; Indirect Food Additives: Adhesives And Components Of Coatings
CHLORIDE, VINYLIDENE	175.365	VINYLIDENE CHLORIDE COPOLYMER COATINGS FOR POLYCARBONATE FILM Substances For Use As Components Of Coatings; Indirect Food Additives: Adhesives And Components Of Coatings
CHLORIDE, VINYLIDENE	177.1990	VINYLIDENE CHLORIDE/METHYL ACRYLATE COPOLYMERS Substances For Use As Basic Components Of Single And Repeated Use Food Contact Surfaces; Indirect Food Additives: Polymers
CHLORIDE, VINYLIDENE	177.2000	VINYLIDENE CHLORIDE/METHYL ACRYLATE/METHYL METHACRYLATE POLYMERS Substances For Use As Basic Components Of Single And Repeated Use Food Contact Surfaces; Indirect Food Additives: Polymers
CHLORIDE, ZINC	182.8985	ZINC CHLORIDE Nutrients; Substances Generally Recognized As Safe
CHLORIDE, ZINC	582.5985	ZINC CHLORIDE Nutrients And/or Dietary Supplements; Substances Generally Recognized As Safe
CHLORINATED	177.1610	POLYETHYLENE, CHLORINATED Substances For Use As Basic Components Of Single And Repeated Use Food Contact Surfaces; Indirect Food Additives: Polymers
CHLORINATED	177.2430	POLYETHER RESINS, CHLORINATED Substances For Use Only As Components Of Articles Intended For Repeated Use; Indirect Food Additives: Polymers
CHLORINE DIOXIDE	173.300	CHLORINE DIOXIDE Specific Usage Additives; Secondary Direct Food Additives Permitted In Food For Human Consumption
CHLORITE, SODIUM	173.325	ACIDIFIED SODIUM CHLORITE SOLUTIONS Specific Usage Additives; Secondary Direct Food Additives Permitted In Food For Human Consumption
CHLORITE, SODIUM	186.1750	SODIUM CHLORITE Listing Of Specific Substances Affirmed As GRAS; Indirect Food Substances Affirmed As Generally Recognized As Safe
CHLOROANALINE	189.280	4,4'-METHYLENEBIS (2-CHLOROANALINE) Substances Prohibited From Indirect Addition To Human Food Thru Food Contact Surfaces; Substances Prohibited From Use In Human Food
CHLOROBUTANOL	556.140	CHLOROBUTANOL Specific Tolerances For Residues Of New Animal Drugs; Tolerances For Residues Of New Animal Drugs In Food
CHLOROFLUOROCARBON	2.125	USE OF CHLOROFLUOROCARBON PROPELLANTS IN SELF-PRESSURIZED CONTAINERS General Administrative Rulings And Decisions
CHLOROFLUOROCARBON	189.191	CHLOROFLUOROCARBON PROPELLANTS Substances Generally Prohibited From Direct Addition Or Use As Human Food; Substances Prohibited From Use In Human Food
CHLOROFLUOROCARBON	300.100	CHLOROFLUOROCARBON PROPELLANTS Substances Generally Prohibited From Drugs; General
CHLOROFLUOROCARBON	700.23	CHLOROFLUOROCARBON PROPELLANTS Requirements For Specific Cosmetic Products; General
CHLOROFLUOROCARBON	801.417	CHLOROFLUOROCARBON PROPELLANTS Special Requirements For Specific Devices; Labeling
CHLOROFLUOROCARBON 113	173.342	CHLOROFLUOROCARBON 113 AND PERFLUOROHEXANE Specific Usage Additives; Secondary Direct Food Additives Permitted In Food For Human Consumption
CHLOROFLUROCARBONS	201.320	WARNING STATEMENTS FOR DRUG PRODUCTS CONTAINING OR MANUFACTURED WITH CHLOROFLUOROCARBONS Labeling
CHLOROFLUOROCARBONS	801.63	MEDICAL DEVICES; WARNING STATEMENT FOR DEVICES CONTAINING OR MANUFACTURED WITH CHLOROFLUOROCARBONS Labeling Requirements For Over-the-counter Devices; Labeling
CHLOROFORM	700.18	USE OF CHLOROFORM AS AN INGREDIENT IN COSMETIC PRODUCTS Requirements For Specific Cosmetic Products; General

CHLOROMETHYLATED	173.70	CHLOROMETHYLATED AMINATED STYRENE-DIVINYLBENZENE RESIN Polymer Substances For Food Treatment; Secondary Direct Food Additives Permitted In Food For Human Consumption
CHLOROPENTAFLUOROETHANE	173.345	CHLOROPENTAFLUOROETHANE Specific Usage Additives; Secondary Direct Food Additives Permitted In Food For Human Consumption
CHLOROPHYLLIN	3.1125	POTASSIUM SODIUM COPPER CHLOROPHYLLIN (CHLOROPHYLLIN-COPPER COMPLEX) Drugs; Listing Of Color Additives Exempt From Certification
CHLOROPHYLLIN	73.2125	POTASSIUM SODIUM COPPER CHLOROPHYLLIN (CHLOROPHYLLIN-COPPER COMPLEX) Cosmetics; Listing Of Color Additives Exempt From Certification
CHLOROPHYLLIN	73.3110	CHLOROPHYLLIN-COPPER COMPLEX, OIL SOLUBLE Medical Devices; Listing Of Color Additives Exempt From Certification
CHLOROSULFONATED	177.2210	ETHYLENE POLYMER, CHLOROSULFONATED Substances For Use Only As Components Of Articles Intended For Repeated Use; Indirect Food Additives: Polymers
CHLOROTHIAZIDE	520.420	CHLOROTHIAZIDE TABLETS AND BOLUSES Oral Dosage Form New Animal Drugs
CHLORPHENESIN CARBAMATE	520.434	CHLORPHENESIN CARBAMATE TABLETS Oral Dosage Form New Animal Drugs
CHLORTETRACYCLINE	520.445	CHLORTETRACYCLINE ORAL DOSAGE FORMS Oral Dosage Form New Animal Drugs
CHLORTETRACYCLINE	556.150	CHLORTETRACYCLINE Specific Tolerances For Residues Of New Animal Drugs; Tolerances For Residues Of New Animal Drugs In Food
CHLORTETRACYCLINE	558.128	CHLORTETRACYCLINE New Animal Drugs For Use In Animal Feeds
CHLORTETRACYCLINE	558.140	CHLORTETRACYCLINE AND SULFAMETHAZINE New Animal Drugs For Use In Animal Feeds
CHLORTETRACYCLINE	558.145	CHLORTETRACYCLINE, PROCAINE PENICILLIN, AND SULFAMETHAZINE New Animal Drugs For Use In Animal Feeds
CHLORTETRACYCLINE	558.155	CHLORTETRACYCLINE, SULFATHIAZOLE, PENICILLIN New Animal Drugs For Use In Animal Feeds
CHOCOLATE	163.111	CHOCOLATE LIQUOR Requirements For Specific Standardized Cacao Products; Cacao Products
CHOCOLATE	163.123	SWEET CHOCOLATE Requirements For Specific Standardized Cacao Products; Cacao Products
CHOCOLATE	163.130	MILK CHOCOLATE Requirements For Specific Standardized Cacao Products; Cacao Products
CHOCOLATE	163.135	BUTTERMILK CHOCOLATE Requirements For Specific Standardized Cacao Products; Cacao Products
CHOCOLATE	163.140	SKIM MILK CHOCOLATE Requirements For Specific Standardized Cacao Products; Cacao Products
CHOCOLATE	163.153	SWEET CHOCOLATE AND VEGETABLE FAT COATING Requirements For Specific Standardized Cacao Products; Cacao Products
CHOCOLATE	163.155	MILK CHOCOLATE AND VEGETABLE FAT COATING Requirements For Specific Standardized Cacao Products; Cacao Products
CHOCOLATES	163.145	MIXED DAIRY PRODUCT CHOCOLATES Requirements For Specific Standardized Cacao Products; Cacao Products
CHOLECYSTOKINETIC	357.210	CHOLECYSTOKINETIC ACTIVE INGREDIENTS Cholecystokinetic Drug Products; Miscellaneous Internal Drug Products For Over-the-counter Human Use
CHOLECYSTOKINETIC	357.250	LABELING OF CHOLECYSTOKINETIC DRUG PRODUCTS Cholecystokinetic Drug Products; Miscellaneous Internal Drug Products For Over-the-counter Human Use
CHOLERAE, VIBRIO	866.3930	VIBRIO CHOLERAE SEROLOGICAL REAGENTS Serological Reagents; Immunology And Microbiology Devices
CHOLESTEROL	101.62	NUTRIENT CONTENT CLAIMS FOR FAT, FATTY ACID, AND CHOLESTEROL CONTENT OF FOODS Specific Requirements For Nutrient Content Claims; Food Labeling
CHOLESTEROL	101.75	HEALTH CLAIMS: DIETARY SATURATED FAT AND CHOLESTEROL AND RISK OF CORONARY HEART DISEASE Specific Requirements For Health Claims; Food Labeling
CHOLESTEROL	862.1175	CHOLESTEROL (TOTAL) TEST SYSTEM Clinical Chemistry Test Systems; Clinical Chemistry And Clinical Toxicology Devices
CHOLINE	172.370	IRON-CHOLINE CITRATE COMPLEX Special Dietary And Nutritional Additives; Food Additives Permitted For Direct Addition To Food For Human Consumption
CHOLINE	182.5250	CHOLINE BITARTRATE Dietary Supplements; Substances Generally Recognized As Safe
CHOLINE	182.5252	CHOLINE CHLORIDE Dietary Supplements; Substances Generally Recognized As Safe

CHOLINE	182.8250	CHOLINE BITARTRATE Nutrients; Substances Generally Recognized As Safe
CHOLINE	182.8252	CHOLINE CHLORIDE Nutrients; Substances Generally Recognized As Safe
CHOLINE	573.300	CHOLINE XANTHATE Food Additive Listing; Food Additives Permitted In Feed And Drinking Water Of Animals
CHOLINE	573.580	IRON-CHOLINE CITRATE Food Additive Listing; Food Additives Permitted In Feed And Drinking Water Of Animals
CHOLINE	582.5250	CHOLINE BITARTRATE Nutrients And/or Dietary Supplements; Substances Generally Recognized As Safe
CHOLINE	582.5252	CHOLINE CHLORIDE Nutrients And/or Dietary Supplements; Substances Generally Recognized As Safe
CHOLINESTERASE	862.3240	CHOLINESTERASE TEST SYSTEM Clinical Toxicology Test Systems; Clinical Chemistry And Clinical Toxicology Devices
CHOLYLGLYCINE	862.1177	CHOLYLGLYCINE TEST SYSTEM Clinical Chemistry Test Systems; Clinical Chemistry And Clinical Toxicology Devices
CHONDRUS	182.7255	CHONDRUS EXTRACT Stabilizers; Substances Generally Recognized As Safe
CHONDRUS	201.319	WATER-SOLUBLE GUMS, HYDROPHILIC GUM AND HYDROPHILIC MUCILLOIDS (INCLUDING CHONDRUS) AS ACTIVE INGREDIENTS; WARNINGS AND DIRECTIONS Specific Labeling Requirements For Specific Drug Products; Labeling
CHONDRUS	582.7255	CHONDRUS EXTRACT Stabilizers; Substances Generally Recognized As Safe
CHORIOMENINGITIS	866.3360	LYMPHOCYTIC CHORIOMENINGITIS VIRUS SEROLOGICAL REAGENTS Serological Reagents; Immunology And Microbiology Devices
CHORIONIC GONADOTROPIN	522.1079	SERUM GONADOTROPIN AND CHORIONIC GONADOTROPIN Implantation Or Injectable Dosage Form New Animal Drugs
CHORIONIC GONADOTROPIN	522.1081	CHORIONIC GONADOTROPIN FOR INJECTION; CHORIONIC GONADOTROPIN SUSPENSION Implantation Or Injectable Dosage Form New Animal Drugs
CHORIONIC GONADOTROPIN	862.1155	HUMAN CHORIONIC GONADOTROPIN (HCG) TEST SYSTEM Clinical Chemistry Test Systems; Clinical Chemistry And Clinical Toxicology Devices
CHROMATOGRAPHIC	862.2230	CHROMATOGRAPHIC SEPARATION MATERIAL FOR CLINICAL USE Clinical Laboratory Instruments; Clinical Chemistry And Clinical Toxicology Devices
CHROMATOGRAPHY	862.2250	GAS LIQUID CHROMATOGRAPHY SYSTEM FOR CLINICAL USE Clinical Laboratory Instruments; Clinical Chemistry And Clinical Toxicology Devices
CHROMATOGRAPHY	862.2260	HIGH PRESSURE LIQUID CHROMATOGRAPHY SYSTEM FOR CLINICAL USE Clinical Laboratory Instruments; Clinical Chemistry And Clinical Toxicology Devices
CHROMATOGRAPHY	862.2270	THIN LAYER LIQUID CHROMATOGRAPHY SYSTEM FOR CLINICAL USE Clinical Laboratory Instruments; Clinical Chemistry And Clinical Toxicology Devices
CHROMIC CHLORIDE	178.3290	CHROMIC CHLORIDE COMPLEXES Certain Adjuvants And Production Aids; Indirect Food Additives: Adjuvants, Production Aids, And Sanitizers
CHROMIUM COMPLEX	176.160	CHROMIUM (CR III) COMPLEX N-ETHYL-N-HEPTADECYLFLUORO-OCTANE SULFONYL GLYCINE Substances For Use Only As Components Of Paper And Paperboard; Indirect Food Additives
CHROMIUM-COBALT-ALUMINUM	73.1015	CHROMIUM-COBALT-ALUMINUM OXIDE Drugs; Listing Of Color Additives Exempt From Certification
CHROMIUM-COBALT-ALUMINUM	73.3110a	CHROMIUM-COBALT-ALUMINUM OXIDE Medical Devices; Listing Of Color Additives Exempt From Certification
CHROMIUM HYDROXIDE GREEN	73.1326	CHROMIUM HYDROXIDE GREEN Drugs; Listing Of Color Additives Exempt From Certification
CHROMIUM HYDROXIDE GREEN	73.2326	CHROMIUM HYDROXIDE GREEN Cosmetics; Listing Of Color Additives Exempt From Certification
CHROMIUM OXIDE GREENS	73.1327	CHROMIUM OXIDE GREENS Drugs; Listing Of Color Additives Exempt From Certification
CHROMIUM OXIDE GREENS	73.2327	CHROMIUM OXIDE GREENS Cosmetics; Listing Of Color Additives Exempt From Certification
CHROMIUM OXIDE GREENS	73.3111	CHROMIUM OXIDE GREENS Medical Devices; Listing Of Color Additives Exempt From Certification
CHROMOSOME	864.2260	CHROMOSOME CULTURE KIT Cell And Tissue Culture Products; Hematology And Pathology Devices
CHRONAXIMETER	890.1225	CHRONAXIMETER Physical Medicine Diagnostic Devices; Physical Medicine Devices
CHUB, SMOKED	172.177	SODIUM NITRITE USED IN PROCESSING SMOKED CHUB Food Preservatives; Food Additives Permitted For Direct Addition To Food For Human Consumption

CHURCH, NATIVE AMERICAN	1307.31	NATIVE AMERICAN CHURCH Special Exempt Persons; Miscellaneous
CHYMOSIN	184.1685	RENNET (ANIMAL-DERIVED) AND CHYMOSIN PREPARATION (FERMENTATION-DERIVED) Listing Of Specific Substances Affirmed As GRAS; Direct Food Substances Affirmed As Generally Recognized As Safe
CHYMOTRYPSIN	862.1180	CHYMOTRYPSIN TEST SYSTEM Clinical Chemistry Test Systems; Clinical Chemistry And Clinical Toxicology Devices
CINE	892.1620	CINE OR SPOT FLUOROGRAPHIC X-RAY CAMERA Diagnostic Devices; Radiology Devices
CINNAMYL ANTHRANILATE	189.113	CINNAMYL ANTHRANILATE Substances Generally Prohibited From Direct Addition Or Use As Human Food; Substances Prohibited From Use In Human Food
CIRCUIT	868.5240	ANESTHESIA BREATHING CIRCUIT Therapeutic Devices; Anesthesiology Devices
CIRCUIT	868.5250	BREATHING CIRCUIT CIRCULATOR Therapeutic Devices; Anesthesiology Devices
CIRCUIT	868.5260	BREATHING CIRCUIT BACTERIAL FILTER Therapeutic Devices; Anesthesiology Devices
CIRCULAR	606.122	INSTRUCTION CIRCULAR Finished Product Control; Current Good Manufacturing Practice For Blood And Blood Components
CIRCULAR ELECTRODE	884.2675	FETAL SCALP CIRCULAR (SPIRAL) ELECTRODE AND APPLICATOR Obstetrical And Gynecological Monitoring Devices; Obstetrical And Gynecological Devices
CIRCULATOR	868.5250	BREATHING CIRCUIT CIRCULATOR Therapeutic Devices; Anesthesiology Devices
CITATIONS	1.4	AUTHORITY CITATIONS General Provisions; General Enforcement Regulations
CITIZEN PETITION	10.30	CITIZEN PETITION General Administrative Procedures; Administrative Practices And Procedures
CITRATE, AMMONIUM	184.1140	AMMONIUM CITRATE, DIBASIC Listing Of Specific Substances Affirmed As GRAS; Direct Food Substances Affirmed As Generally Recognized As Safe
CITRATE, BISMUTH	73.2110	BISMUTH CITRATE Cosmetics; Listing Of Color Additives Exempt From Certification
CITRATE, CALCIUM	184.1195	CALCIUM CITRATE Listing Of Specific Substances Affirmed As GRAS; Direct Food Substances Affirmed As Generally Recognized As Safe
CITRATE, CALCIUM	582.1195	CALCIUM CITRATE General Purpose Food Additives; Substances Generally Recognized As Safe
CITRATE, CALCIUM	582.5195	CALCIUM CITRATE Nutrients And/or Dietary Supplements; Substances Generally Recognized As Safe
CITRATE, CALCIUM	582.6195	CALCIUM CITRATE Sequestrants; Substances Generally Recognized As Safe
CITRATE, DIETHYLCARBAMAZINE	520.622	DIETHYLCARBAMAZINE CITRATE SYRUP Oral Dosage Form New Animal Drugs
CITRATE, FERRIC	184.1298	FERRIC CITRATE Listing Of Specific Substances Affirmed As GRAS; Direct Food Substances Affirmed As Generally Recognized As Safe
CITRATE, FERRIC AMMONIUM	73.1025	FERRIC AMMONIUM CITRATE Drugs; Listing Of Color Additives Exempt From Certification
CITRATE, FERRIC AMMONIUM	184.1296	FERRIC AMMONIUM CITRATE Listing Of Specific Substances Affirmed As GRAS; Direct Food Substances Affirmed As Generally Recognized Safe
CITRATE, FERROUS	184.1307c	FERROUS CITRATE Listing Of Specific Substances Affirmed As GRAS; Direct Food Substances Affirmed As Generally Recognized As Safe
CITRATE, IRON AMMONIUM	172.430	IRON AMMONIUM CITRATE Anticaking Agents; Food Additives Permitted For Direct Addition To Food For Human Consumption
CITRATE, IRON AMMONIUM	573.560	IRON AMMONIUM CITRATE Food Additive Listing; Food Additives Permitted In Feed And Drinking Water Of Animals
CITRATE, IRON-CHOLINE	172.370	IRON-CHOLINE CITRATE COMPLEX Special Dietary And Nutritional Additives; Food Additives Permitted For Direct Addition To Food For Human Consumption
CITRATE, IRON-CHOLINE	573.580	IRON-CHOLINE CITRATE COMPLEX Food Additive Listing; Food Additives Permitted In Feed And Drinking Water Of Animals
CITRATE, ISOPROPYL	184.1386	ISOPROPYL CITRATE Listing Of Specific Substances Affirmed As GRAS; Direct Food Substances Affirmed As Generally Recognized Safe

CITRATE, ISOPROPYL	582.6386	ISOPROPYL CITRATE Sequestrants; Substances Generally Recognized As Safe
CITRATE, MANGANESE	184.1449	MANGANESE CITRATE Listing Of Specific Substances Affirmed As GRAS; Direct Food Substances Affirmed As Generally Recognized As Safe
CITRATE, MANGANESE	582.5449	MANGANESE CITRATE Nutrients And/or Dietary Supplements; Substances Generally Recognized As Safe
CITRATE, MONOGLYCERIDE	172.832	MONOGLYCERIDE CITRATE Multipurpose Additives; Food Additives Permitted For Direct Addition To Food For Human Consumption
CITRATE, MONOISOPROPYL	582.6115	MONOISOPROPYL CITRATE Sequestrants; Substances Generally Recognized As Safe
CITRATE, POTASSIUM	184.1625	POTASSIUM CITRATE Listing Of Specific Substances Affirmed As GRAS; Direct Food Substances Affirmed As Generally Recognized As Safe
CITRATE, POTASSIUM	582.1625	POTASSIUM CITRATE General Purpose Food Additives; Substances Generally Recognized As Safe
CITRATE, POTASSIUM	582.6625	POTASSIUM CITRATE Sequestrants; Substances Generally Recognized As Safe
CITRATE, SODIUM	184.1751	SODIUM CITRATE Listing Of Specific Substances Affirmed As GRAS; Direct Food Substances Affirmed As Generally Recognized As Safe
CITRATE, SODIUM	582.1751	SODIUM CITRATE General Purpose Food Additives; Substances Generally Recognized As Safe
CITRATE, SODIUM	582.6751	SODIUM CITRATE Sequestrants; Substances Generally Recognized As Safe
CITRATE, STEARYL	172.755	STEARYL MONOGLYCERIDYL CITRATE Other Specific Usage Additives; Food Additives Permitted For Direct Addition To Food For Human Consumption
CITRATE, STEARYL	184.1851	STEARYL CITRATE Listing Of Specific Substances Affirmed As GRAS; Direct Food Substances Affirmed As Generally Recognized As Safe
CITRATE, STEARYL	582.6851	STEARYL CITRATE Sequestrants; Substances Generally Recognized As Safe
CITRATE, TRIETHYL	184.1911	TRIETHYL CITRATE Listing Of Specific Substances Affirmed As GRAS; Direct Food Substances Affirmed As Generally Recognized As Safe
CITRIC ACID	173.280	SOLVENT EXTRACTION PROCESS FOR CITRIC ACID Solvents, Lubricants, Release Agents And Related Substances; Secondary Direct Food Additives Permitted In Food For Human Consumption
CITRIC ACID	184.1033	CITRIC ACID Listing Of Specific Substances Affirmed As GRAS; Direct Food Substances Affirmed As Generally Recognized As Safe
CITRIC ACID	582.1033	CITRIC ACID General Purpose Food Additives; Substances Generally Recognized As Safe
CITRIC ACID	582.6033	CITRIC ACID Sequestrants; Substances Generally Recognized As Safe
CITROBACTER SPP	866.3125	CITROBACTER SPP. SEROLOGICAL REAGENTS Serological Reagents; Immunology And Microbiology Devices
CITRUS	172.210	COATINGS ON FRESH CITRUS FRUIT Coatings, Films And Related Substances; Food Additives Permitted For Direct Addition To Food For Human Consumption
CITRUS RED NO 2	74.302	CITRUS RED NO. 2 Foods; Listing Of Color Additives Subject To Certification
CLAIM	1316.70	REQUIREMENTS AS TO CLAIM AND BOND Seizure, Forfeiture, And Disposition Of Property; Administrative Functions, Practices, And Procedures
CLAIM	1316.97	INITIATING JUDICIAL FORFEITURE PROCEEDING AGAINST A CONVEYANCE WITHIN 60 DAYS OF FILING A CLAIM AND COST BOND Expedited Forfeiture Proceedings For Certain Property; Administrative Functions, Practices, And Procedures
CLAIM, NUTRIENT CONTENT	130.10	REQUIREMENTS FOR FOODS NAMED BY USE OF A NUTRIENT CONTENT CLAIM AND A STANDARDIZED TERM Food Standards: General
CLAIMS, HEALTH	5.300	FOOD STANDARDS, FOOD ADDITIVES, GRAS SUBSTANCES, COLOR ADDITIVES, NUTRIENT CONTENT CLAIMS, AND HEALTH CLAIMS Foods And Cosmetics; Redelegations Of Authority; Delegations Of Authority And Organization
CLAIMS, HEALTH	101.14	HEALTH CLAIMS: GENERAL REQUIREMENTS General Provisions; Food Labeling
CLAIMS, HEALTH	101.70	PETITIONS FOR HEALTH CLAIMS Specific Requirements For Health Claims; Food Labeling
CLAIMS, HEALTH	101.71	HEALTH CLAIMS: CLAIMS NOT AUTHORIZED Specific Requirements For Health Claims; Food Labeling

CLAIMS, HEALTH	101.72	HEALTH CLAIMS: CALCIUM AND OSTEOPOROSIS Specific Requirements For Health Claims; Food Labeling
CLAIMS, HEALTH	101.73	HEALTH CLAIMS: DIETARY LIPIDS AND CANCER Specific Requirements For Health Claims; Food Labeling
CLAIMS, HEALTH	101.74	HEALTH CLAIMS: SODIUM AND HYPERTENSION Specific Requirements For Health Claims; Food Labeling
CLAIMS, HEALTH	101.75	HEALTH CLAIMS: DIETARY SATURATED FAT AND CHOLESTEROL AND RISK OF CORONARY HEART DISEASE Specific Requirements For Health Claims; Food Labeling
CLAIMS, HEALTH	101.76	HEALTH CLAIMS: FIBER-CONTAINING GRAIN PRODUCTS, FRUITS, AND VEGETABLES AND CANCER Specific Requirements For Health Claims; Food Labeling
CLAIMS, HEALTH	101.77	HEALTH CLAIMS: FRUITS, VEGETABLES, AND GRAIN PRODUCTS THAT CONTAIN FIBER, AND RISK OF CORONARY HEART DISEASE Specific Requirements For Health Claims; Food Labeling
CLAIMS, HEALTH	101.78	HEALTH CLAIMS: FRUITS AND VEGETABLES AND CANCER Specific Requirements For Health Claims; Food Labeling
CLAIMS, HEALTH	101.79	HEALTH CLAIMS: FOLATE AND NEURAL TUBE DEFECTS Specific Requirements For Health Claims; Food Labeling
CLAIMS, HEALTH	101.80	HEALTH CLAIMS: DIETARY SUGAR ALCOHOLS AND DENTAL CARIES Specific Requirements For Health Claims; Food Labeling
CLAIMS, HEALTH	101.81	HEALTH CLAIMS: SOLUBLE FIBER FROM CERTAIN FOODS AND RISK OF CORONARY HEART DISEASE Specific Requirements For Health Claims; Food Labeling
CLAIMS, HEALTH	101.82	HEALTH CLAIMS: SOY PROTEIN AND RISK OF CORONARY HEART DISEASE Specific Requirements For Health Claims; Food Labeling
CLAIMS, HEALTH	101.83	HEALTH CLAIMS: PLANT STEROL/STANOL ESTERS AND RISK OF CORONARY HEART DISEASE Specific Requirements For Health Claims; Food Labeling
CLAIMS, NUTRIENT CONTENT	5.300	FOOD STANDARDS, FOOD ADDITIVES, GRAS SUBSTANCES, COLOR ADDITIVES, NUTRIENT CONTENT CLAIMS, AND HEALTH CLAIMS Foods And Cosmetics; Redelegations Of Authority; Delegations Of Authority And Organization
CLAIMS, NUTRIENT CONTENT	101.13	NUTRIENT CONTENT CLAIMS - GENERAL PRINCIPLES General Provisions; Food Labeling
CLAIMS, NUTRIENT CONTENT	101.54	NUTRIENT CONTENT CLAIMS FOR "GOOD SOURCE," "HIGH," "MORE," AND "HIGH POTENCY" Specific Requirements For Nutrient Content Claims; Food Labeling
CLAIMS, NUTRIENT CONTENT	101.56	NUTRIENT CONTENT CLAIMS FOR "LIGHT" OR "LITE" Specific Requirements For Nutrient Content Claims; Food Labeling
CLAIMS, NUTRIENT CONTENT	101.60	NUTRIENT CONTENT CLAIMS FOR THE CALORIE CONTENT OF FOODS Specific Requirements For Nutrient Content Claims; Food Labeling
CLAIMS, NUTRIENT CONTENT	101.61	NUTRIENT CONTENT CLAIMS FOR THE SODIUM CONTENT OF FOODS Specific Requirements For Nutrient Content Claims; Food Labeling
CLAIMS, NUTRIENT CONTENT	101.62	NUTRIENT CONTENT CLAIMS FOR FAT, FATTY ACID, AND CHOLESTEROL CONTENT OF FOODS Specific Requirements For Nutrient Content Claims; Food Labeling
CLAIMS, NUTRIENT CONTENT	101.65	IMPLIED NUTRIENT CONTENT CLAIMS AND RELATED LABEL STATEMENTS Specific Requirements For Nutrient Content Claims; Food Labeling
CLAIMS, NUTRIENT CONTENT	101.67	USE OF NUTRIENT CONTENT CLAIMS FOR BUTTER Specific Requirements For Nutrient Content Claims; Food Labeling
CLAIMS, NUTRIENT CONTENT	101.69	PETITIONS FOR NUTRIENT CONTENT CLAIMS Specific Requirements For Nutrient Content Claims; Food Labeling
CLAMP	870.4450	VASCULAR CLAMP Cardiovascular Surgical Devices; Cardiovascular Devices
CLAMP	876.5160	UROLOGICAL CLAMP FOR MALES Therapeutic Devices; Gastroenterology-urology Devices
CLAMP	882.4460	NEUROSURGICAL HEAD HOLDER (SKULL CLAMP) Neurological Surgical Devices; Neurological Devices
CLAMP	882.5175	CAROTID ARTERY CLAMP Neurological Therapeutic Devices; Neurological Devices
CLAMS	102.49	FRIED CLAMS MADE FROM MINCED CLAMS Requirements For Specific Nonstandardized Foods; Common Or Usual Name For Nonstandardized Foods
CLARIFYING AGENTS	178.3295	CLARIFYING AGENTS FOR POLYMERS Certain Adjuvants And Production Aids; Indirect Food Additives: Adjuvants, Production Aids, And Sanitizers
CLASP	872.3285	PREFORMED CLASP Prosthetic Devices; Dental Devices
CLASS I	801.63	MEDICAL DEVICES; WARNING STATEMENT FOR DEVICES CONTAINING OR MANUFACTURED WITH CHLOROFLUOROCARBONS AND OTHER CLASS I OZONE-DEPLETING SUBSTANCES Labeling Requirements For Over-the-counter Devices; Labeling

CLASS III CERTIFICATION	807.94	FORMAT OF A CLASS III CERTIFICATION Establishment Registration And Device Listing For Manufacturers And Initial Importers Of Devices
CLASSIFICATION	5.405	DETERMINATION OF CLASSIFICATION OF DEVICES Medical Devices And Radiological Health; Redelegations Of Authority; Delegations Of Authority And Organization
CLASSIFICATION	7.41	HEALTH HAZARD EVALUATION AND RECALL CLASSIFICATION Recalls - Guidance On Policy, Procedures & Industry Responsibilities; Enforcement Policy
CLASSIFICATION	170.30	ELIGIBILITY FOR CLASSIFICATION AS GENERALLY RECOGNIZED AS SAFE (GRAS) Food Additive Safety; Food Additives
CLASSIFICATION	570.30	ELIGIBILITY FOR CLASSIFICATION AS GENERALLY RECOGNIZED AS SAFE (GRAS) Food Additive Safety; Food Additives
CLASSIFICATION	860.5	CONFIDENTIALITY AND USE OF DATA AND INFORMATION SUBMITTED IN CONNECTION WITH CLASSIFICATION AND RECLASSIFICATION General; Medical Device Classification Procedures
CLASSIFICATION	860.84	CLASSIFICATION PROCEDURES FOR "OLD DEVICES" Classification; Medical Device Classification Procedures
CLASSIFICATION	860.93	CLASSIFICATION OF IMPLANTS, LIFE-SUPPORTING OR LIFE-SUSTAINING DEVICES Classification; Medical Device Classification Procedures
CLASSIFIED	1402.4	INFORMATION CLASSIFIED BY ANOTHER AGENCY Mandatory Declassification Review
CLAUSE, ANTICANCER	70.51	ADVISORY COMMITTEE ON THE APPLICATION OF THE ANTICANCER CLAUSE Safely Evaluation; Color Additives
CLAUSE, CANCER	70.50	APPLICATION OF THE CANCER CLAUSE OF SECTION 721 OF THE ACT Safely Evaluation; Color Additives
CLAUSE, CANCER	571.115	APPLICATION OF THE CANCER CLAUSE OF SECTION 409 OF THE ACT Administrative Actions On Applications; Food Additive Petitions
CLAUSE, SAFEGUARD	26.21	SAFEGUARD CLAUSE Specific Sector Provisions For Pharmaceutical GMPs; Mutual Recognition Of Pharmaceutical GMP Reports, Medical Device Quality System Audit Reports
CLAVULANAT	520.88g	AMOXICILLIN TRIHYDRATE AND CLAVULANATE POTASSIUM FILM-COATED TABLETS Oral Dosage Form New Animal Drugs
CLAVULANATE	520.88h	AMOXICILLIN TRIHYDRATE AND CLAVULANATE POTASSIUM FOR ORAL SUSPENSION Oral Dosage Form New Animal Drugs
CLAY	186.1256	CLAY (KAOLIN) Listing Of Specific Substances Affirmed As GRAS; Indirect Food Substances Affirmed As Generally Recognized As Safe
CLEANER, AIR	880.5045	MEDICAL RECIRCULATING AIR CLEANER General Hospital And Personal Use Therapeutic Devices; General Hospital And Personal Use Devices
CLEANER, DENTURE	872.3530	MECHANICAL DENTURE CLEANER Prosthetic Devices; Dental Devices
CLEANER, ULTRASONIC	880.6150	ULTRASONIC CLEANER FOR MEDICAL INSTRUMENTS General Hospital And Personal Use Miscellaneous Devices; General Hospital And Personal Use Devices
CLEANING BRUSH	868.5795	TRACHEAL TUBE CLEANING BRUSH Therapeutic Devices; Anesthesiology Devices
CLEANING, EQUIPMENT	211.67	EQUIPMENT CLEANING AND MAINTENANCE Equipment; Current Good Manufacturing Practice For Finished Pharmaceuticals
CLEANING, EQUIPMENT	211.182	EQUIPMENT CLEANING AND USE LOG Records And Reports; Current Good Manufacturing Practice For Finished Pharmaceuticals
CLEANOUT, EQUIPMENT	225.165	EQUIPMENT CLEANOUT PROCEDURES Product Quality Assurance; Current Good Manufacturing Practice For Medicated Feeds
CLEANOUT, EQUIPMENT	225.65	EQUIPMENT CLEANOUT PROCEDURES Product Quality Control; Current Good Manufacturing Practice For Medicated Feeds
CLEANSER, DENTURE	872.3520	OTC DENTURE CLEANSER Prosthetic Devices; Dental Devices
CLEANSING	701.20	DETERGENT SUBSTANCES, OTHER THAN SOAP, INTENDED FOR USE IN CLEANSING THE BODY Labeling Of Specific Ingredients; Cosmetic Labeling
CLENBUTEROL SYRUP	520.452	CLENBUTEROL SYRUP Oral Dosage Form New Animal Drugs
CLINDAMYCIN	520.446	CLINDAMYCIN HYDROCHLORIDE CAPSULES Oral Dosage Form New Animal Drugs

CLINDAMYCIN	520.447	CLINDAMYCIN HYDROCHLORIDE LIQUID Oral Dosage Form New Animal Drugs
CLINICAL	862.2310	CLINICAL SAMPLE CONCENTRATOR Clinical Laboratory Instruments; Clinical Chemistry And Clinical Toxicology Devices
CLINICAL	862.3200	CLINICAL TOXICOLOGY CALIBRATOR Clinical Toxicology Test Systems; Clinical Chemistry And Clinical Toxicology Devices
CLINICAL	862.3280	CLINICAL TOXICOLOGY CONTROL MATERIAL Clinical Toxicology Test Systems; Clinical Chemistry And Clinical Toxicology Devices
CLINICA	880.2900	CLINICAL COLOR CHANGE THERMOMETER General Hospital And Personal Use Monitoring Devices; General Hospital And Personal Use Devices
CLINICAL	880.2910	CLINICAL ELECTRONIC THERMOMETER General Hospital And Personal Use Monitoring Devices; General Hospital And Personal Use Devices
CLINICAL	880.2920	CLINICAL MERCURY THERMOMETER General Hospital And Personal Use Monitoring Devices; General Hospital And Personal Use Devices
CLINICAL ENDPOINT	314.510	APPROVAL BASED ON A SURROGATE ENDPOINT OR ON AN EFFECT ON A CLINICAL ENDPOINT Applications For FDA Approval To Market A New Drug
CLINICAL ENDPOINT	601.41	APPROVAL BASED ON A SURROGATE ENDPOINT OR ON AN EFFECT ON A CLINICAL ENDPOINT Accelerated Approval Of Biological Products; Licensing
CLINICAL EVIDENCE	320.28	CORRELATION OF BIOAVAILABILITY WITH AN ACUTE PHARMACOLOGICAL EFFECT OR CLINICAL EVIDENCE Procedures For Determining The Bioavailability Of Drug Products; Bioavailability And Bioequivalence Requirements
CLINICAL HOLDS	312.42	CLINICAL HOLDS AND REQUESTS FOR MODIFICATION Investigational New Drug Application
CLINICAL INVESTIGATION	312.40	GENERAL REQUIREMENTS FOR USE OF AND INVESTIGATIONAL NEW DRUG IN A CLINICAL INVESTIGATION Investigational New Drug Application
CLINICAL INVESTIGATIONS	50.51	CLINICAL INVESTIGATIONS NOT INVOLVING GREATER THAN MINIMAL RISK Additional Safeguards For Children In Clinical Investigations; Protection Of Human Subjects
CLINICAL INVESTIGATIONS	50.52	CLINICAL INVESTIGATIONS INVOLVING GREATER THAN MINIMAL RISK BUT PRESENTING THE PROSPECT OF DIRECT BENEFIT TO INDIVIDUAL SUBJECTS Additional Safeguards For Children In Clinical Investigations; Protection Of Human Subjects
CLINICAL INVESTIGATIONS	50.53	CLINICAL INVESTIGATIONS INVOLVING GREATER THAN MINIMAL RISK AND NO PROSPECT OF DIRECT BENEFIT TO INDIVIDUAL SUBJECTS, BUT LIKELY TO YIELD GENERALIZABLE KNOWLEDGE ABOUT THE SUBJECTS' DISORDER OR CONDITION Additional Safeguards For Children In Clinical Investigations; Protection Of Human Subjects
CLINICAL INVESTIGATIONS	50.54	CLINICAL INVESTIGATIONS NOT OTHERWISE APPROVABLE THAT PRESENT AN OPPORTUNITY TO UNDERSTAND, PREVENT, OR ALLEVIATE A SERIOUS PROBLEM AFFECTING THE HEALTH OR WELFARE OF CHILDREN Additional Safeguards For Children In Clinical Investigations; Protection Of Human Subjects
CLINICAL INVESTIGATOR	312.70	DISQUALIFICATION OF A CLINICAL INVESTIGATOR Investigational New Drug Application
CLINICAL INVESTIGATOR	812.119	DISQUALIFICATION OF A CLINICAL INVESTIGATOR Responsibilities Of Investigators; Investigational Device Exemptions
CLINICAL STUDIES	312.120	FOREIGN CLINICAL STUDIES NOT CONDUCTED UNDER AN IND Investigational New Drug Application
CLINICAL TRIALS	312.87	ACTIVE MONITORING OF CONDUCT AND EVALUATION OF CLINICAL TRIALS Investigational New Drug Application
CLIP	868.6225	NOSE CLIP Miscellaneous; Anesthesiology Devices
CLIP	870.3250	VASCULAR CLIP Cardiovascular Prosthetic Devices; Cardiovascular Devices
CLIP	870.3260	VENA CAVA CLIP Cardiovascular Prosthetic Devices; Cardiovascular Devices
CLIP	878.4300	IMPLANTABLE CLIP Surgical Devices; General And Plastic Surgery Devices
CLIP	878.4320	REMOVABLE SKIN CLIP Surgical Devices; General And Plastic Surgery Devices
CLIP	882.4150	SCALP CLIP Neurological Surgical Devices; Neurological Devices

CLIP	882.4175	ANEURYSM CLIP APPLIER Neurological Surgical Devices; Neurological Devices
CLIP	882.4190	CLIP FORMING/CUTTING INSTRUMENT Neurological Surgical Devices; Neurological Devices
CLIP	882.4200	CLIP REMOVAL INSTRUMENT Neurological Surgical Devices; Neurological Devices
CLIP	882.4215	CLIP RACK Neurological Surgical Devices; Neurological Devices
CLIP	882.5200	ANEURYSM CLIP Neurological Therapeutic Devices; Neurological Devices
CLIP	882.5225	IMPLANTED MALLEABLE CLIP Neurological Therapeutic Devices; Neurological Devices
CLIP	886.1410	OPHTHALMIC TRIAL LENS CLIP Diagnostic Devices; Ophthalmic Devices
CLIP	886.3100	OPHTHALMIC TANTALUM CLIP Prosthetic Devices; Ophthalmic Devices
CLITORAL	884.5970	CLITORAL ENGORGEMENT DEVICE Obstetrical And Gynecological Therapeutic Devices; Obstetrical And Gynecological Devices
CLOPIDOL	556.160	CLOPIDOL Specific Tolerances For Residues Of New Animal Drugs; Tolerances For Residues Of New Animal Drugs In Food
CLOPIDOL	558.175	CLOPIDOL Specific New Animal Drugs For Use In Animal Feeds; New Animal Drugs For Use In Animal Feeds
CLOPROSTENOL SODIUM	522.460	CLOPROSTENOL SODIUM Implantation Or Injectable Dosage Form New Animal Drugs
CLORSULON	520.462	CLORSULON DRENCH Oral Dosage Form New Animal Drugs
CLORSULON	522.1193	IVERMECTIN AND CLORSULON INJECTION Implantation Or Injectable Dosage Form New Animal Drugs
CLORSULON	556.163	CLORSULON Specific Tolerances For Residues Of New Animal Drugs; Tolerances For Residues Of New Animal Drugs In Food
CLOSE MEETINGS	14.27	DETERMINATION TO CLOSE PORTIONS OF ADVISORY COMMITTEE MEETINGS Meeting Procedures; Public Hearing Before A Public Advisory Committee
CLOSED-CIRCUIT	886.5820	CLOSED-CIRCUIT TELEVISION READING SYSTEM Therapeutic Devices; Ophthalmic Devices
CLOSED SYSTEMS	11.10	CONTROLS FOR CLOSED SYSTEMS Electronic Records; Electronic Signatures
CLOSEOUT	1403.50	CLOSEOUT Uniform Administrative Requirements For Grants And Coop. Agreements To State & Local Gov.
CLOSURE	211.184	COMPONENT, DRUG PRODUCT CONTAINER, CLOSURE, AND LABELING RECORDS Records And Reports; Current Good Manufacturing Practice For Finished Pharmaceuticals
CLOSURES	177.1210	CLOSURES WITH SEALING GASKETS FOR FOOD CONTAINERS Substances For Use As Basic Components Of Single And Repeated Use Food Contact Surfaces; Indirect Food Additives: Polymers
CLOSURES	211.82	RECEIPT AND STORAGE OF UNTESTED COMPONENTS, DRUG PRODUCT CONTAINERS, AND CLOSURES Control Of Components And Drug Product Containers And Closures; Current Good Manufacturing Practice For Finished Pharmaceuticals
CLOSURES	211.84	TESTING AND APPROVAL OR REJECTION OF COMPONENTS, DRUG PRODUCT CONTAINERS, AND CLOSURES Control Of Components And Drug Product Containers And Closures; Current Good Manufacturing Practice For Finished Pharmaceuticals
CLOSURES	211.86	USE OF APPROVED COMPONENTS, DRUG PRODUCT CONTAINERS, AND CLOSURES Control Of Components And Drug Product Containers And Closures; Current Good Manufacturing Practice For Finished Pharmaceuticals
CLOSURES	211.87	RETESTING OF APPROVED COMPONENTS, DRUG PRODUCT CONTAINERS, AND CLOSURES Control Of Components And Drug Product Containers And Closures; Current Good Manufacturing Practice For Finished Pharmaceuticals
CLOSURES	211.89	REJECTED COMPONENTS, DRUG PRODUCT CONTAINERS, AND CLOSURES Control Of Components And Drug Product Containers And Closures; Current Good Manufacturing Practice For Finished Pharmaceuticals
CLOSURES	211.94	DRUG PRODUCT CONTAINERS AND CLOSURES Control Of Components And Drug Product Containers And Closures; Current Good Manufacturing Practice For Finished Pharmaceuticals

COATINGS	175.360	VINYLIDENE CHLORIDE COPOLYMER COATINGS FOR NYLON FILM Substances For Use As Components Of Coatings; Indirect Food Additives: Adhesives And Components Of Coatings
COATINGS	175.365	VINYLIDENE CHLORIDE COPOLYMER COATINGS FOR POLYCARBONATE FILM Substances For Use As Components Of Coatings; Indirect Food Additives: Adhesives And Components Of Coatings
COATINGS	175.390	ZINC-SILICON DIOXIDE MATRIX COATINGS Substances For Use As Components Of Coatings; Indirect Food Additives: Adhesives And Components Of Coatings
COATINGS	176.200	DEFOAMING AGENTS USED IN COATINGS Substances For Use Only As Components Of Paper And Paperboard; Indirect Food Additives: Paper And Paperboard Components
COBALT	73.1015	CHROMIUM-COBALT-ALUMINUM OXIDE Drugs; Listing Of Color Additives Exempt From Certification
COBALT	73.3110a	CHROMIUM-COBALT-ALUMINUM OXIDE Medical Devices; Listing Of Color Additives Exempt From Certification
COBALTOUS SALTS	189.120	COBALTOUS SALTS AND ITS DERIVATIVES Substances Generally Prohibited From Direct Addition Or Use As Human Food; Substances Prohibited From Use In Human Food
COCA LEAVES	1304.32	REPORTS OF MANUFACTURERS IMPORTING COCA LEAVES Reports; Records And Reports Of Registrants
COCAINE	862.3250	COCAINE AND COCAINE METABOLITE TEST SYSTEM Clinical Toxicology Test Systems; Clinical Chemistry And Clinical Toxicology Devices
COCCIDIOIDES IMMITIS	866.3135	COCCIDIOIDES IMMITIS SEROLOGICAL REAGENTS Serological Reagents; Immunology And Microbiology Devices
COCHINEAL	73.100	COCHINEAL EXTRACT; CARMINE Foods; Listing Of Color Additives Exempt From Certification
COCHINEAL	73.1100	COCHINEAL EXTRACT; CARMINE Drugs; Listing Of Color Additives Exempt From Certification
COCKTAIL, FRUIT	145.135	CANNED FRUIT COCKTAIL Requirements For Specific Standardized Canned Fruits; Canned Fruits
COCKTAIL, FRUIT	145.136	ARTIFICIALLY SWEETENED CANNED FRUIT COCKTAIL Requirements For Specific Standardized Canned Fruits; Canned Fruits
COCKTAILS, SEAFOOD	102.54	SEAFOOD COCKTAILS Requirements For Specific Nonstandardized Foods; Common Or Usual Name For Nonstandardized Foods
COCOA	163.112	BREAKFAST COCOA Requirements For Specific Standardized Cacao Products; Cacao Products
COCOA	163.113	COCOA Requirements For Specific Standardized Cacao Products; Cacao Products
COCOA	163.114	LOW-FAT COCOA Requirements For Specific Standardized Cacao Products; Cacao Products
COCOA	163.117	COCOA WITH DIOCTYL SODIUM SULFOSUCCINATE FOR MANUFACTURING Requirements For Specific Standardized Cacao Products; Cacao Products
COCOA	163.150	SWEET COCOA AND VEGETABLE FAT COATING Requirements For Specific Standardized Cacao Products; Cacao Products
COCOA	172.520	COCOA WITH DIOCTYL SODIUM SULFOSUCCINATE FOR MANUFACTURING Flavoring Agents And Related Substances; Food Additives Permitted For Direct Addition To Food For Human Consumption
COCOA BUTTER	172.861	COCOA BUTTER SUBSTITUTE FROM COCONUT OIL, PALM KERNEL OIL, OR BOTH OILS Food Additives Permitted For Direct Addition To Food For Human Consumption
COCOA BUTTER	184.1259	COCOA BUTTER SUBSTITUTE Listing Of Specific Substances Affirmed As GRAS; Direct Food Substances Affirmed As Generally Recognized As Safe
COCONUT OIL	172.816	METHYL GLUCOSIDE COCONUT OIL ESTER Multipurpose Additives; Food Additives Permitted For Direct Addition To Food For Human Consumption
COCONUT OIL	172.861	COCOA BUTTER SUBSTITUTE FROM COCONUT OIL, PALM KERNEL OIL, OR BOTH OILS Food Additives Permitted For Direct Addition To Food For Human Consumption
COCONUT OIL	178.3600	METHYL GLUCOSIDE-COCONUT OIL ESTER Indirect Food Additives: Adjuvants, Production Aids, And Sanitizers
COCONUT OIL	573.660	METHYL GLUCOSIDE-COCONUT OIL ESTER Food Additive Listing; Food Additives Permitted In Feed And Drinking Water Of Animals
CODE OF ETHICS	19.6	CODE OF ETHICS FOR GOVERNMENT SERVICE General Provisions; Standards Of Conduct And Conflicts Of Interest
CODE IMPRINT	206.10	CODE IMPRINT REQUIRED Imprinting Of Solid Oral Dosage Form Drug Products For Human Use

CODE, LABELER	607.35	NOTIFICATION OF REGISTRANT; BLOOD PRODUCT ESTABLISHMENT REGISTRATION NUMBER AND NDC LABELER CODE Procedures For Domestic Blood Product Establishments; Establishment Registratn+product Listing, Manufacturers Human Blood/b.products
CODE NUMBER	1308.03	ADMINISTRATION CONTROLLED SUBSTANCES CODE NUMBER General Information; Schedules Of Controlled Substances
CODE NUMBERS	201.2	DRUGS AND DEVICES; NATIONAL DRUG CODE NUMBERS General Labeling Provisions; Labeling
CODEINE	862.3270	CODEINE TEST SYSTEM Clinical Toxicology Test Systems; Clinical Chemistry And Clinical Toxicology Devices
CODES, LABELER	510.600	NAMES, ADDRESSES, AND DRUG LABELER CODES OF SPONSORS OF APPROVED APPLICATIONS Sponsors Of Approved Applications; New Animal Drugs
CODES/PASSWORDS	11.300	CONTROLS FOR IDENTIFICATION CODES/PASSWORDS Electronic Records; Electronic Signatures
CODES, PRODUCT	20.115	PRODUCT CODES FOR MANUFACTURING OR SALES DATES Availability Of Specific Categories Of Records; Public Information
CODES, REPORTING	803.21	REPORTING CODES Generally Applicable Requirements For Individual Adverse Event Reports; Medical Device Reporting
CODEX ALIMENTARIUS	130.6	REVIEW OF CODEX ALIMENTARIUS FOOD STANDARDS General Provisions; Food Standards: General
CODING	106.90	CODING Quality Control Procedures For Assuring Nutrient Content Of Infant Formulas; Infant Formula Quality Control Procedures
CODY TACK	874.3760	SACCULOTOMY TACK (CODY TACK) Prosthetic Devices; Ear, Nose And Throat Devices
COHN FRACTION II	866.5220	COHN FRACTION II IMMUNOLOGICAL TEST SYSTEM Immunological Test Systems; Immunology And Microbiology Devices
COHN FRACTION IV	866.5360	COHN FRACTION IV IMMUNOLOGICAL TEST SYSTEM Immunological Test Systems; Immunology And Microbiology Devices
COHN FRACTION V	866.5370	COHN FRACTION V IMMUNOLOGICAL TEST SYSTEM Immunological Test Systems; Immunology And Microbiology Devices
COLBY CHEESE	133.118	COLBY CHEESE Requirements For Specific Standardized Cheese And Related Products; Cheeses And Related Cheese Products
COLBY CHEESE	133.119	COLBY CHEESE FOR MANUFACTURING Requirements For Specific Standardized Cheese And Related Products; Cheeses And Related Cheese Products
COLBY CHEESE	133.121	LOW SODIUM COLBY CHEESE Requirements For Specific Standardized Cheese And Related Products; Cheeses And Related Cheese Products
COLD	310.533	DRUG PRODUCTS CONTAINING ACTIVE INGREDIENTS OFFERED OTC FOR HUMAN USE AS AN ANTICHOLINERGIC IN COUGH-COLD DRUG PRODUCTS Requirements For Specific New Drugs Or Devices; New Drugs
COLD-CATHODE	1020.20	COLD-CATHODE GAS DISCHARGE TUBES Performance Standards For Ionizing Radiation Emitting Products
COLD PACK	880.5760	CHEMICAL COLD PACK SNAKEBITE KIT General Hospital And Personal Use Therapeutic Devices; General Hospital And Personal Use Devices
COLD PACK	890.5700	COLD PACK Physical Medicine Therapeutic Devices; Physical Medicine Devices
COLD PACK	890.5710	HOT OR COLD DISPOSABLE PACK Physical Medicine Therapeutic Devices; Physical Medicine Devices
COLD PACK	890.5720	WATER CIRCULATING HOT OR COLD PACK Physical Medicine Therapeutic Devices; Physical Medicine Devices
COLD-PACK CHEESE	133.123	COLD-PACK AND CLUB CHEESE Requirements For Specific Standardized Cheese And Related Products; Cheeses And Related Cheese Products
COLD-PACK CHEESE	133.124	COLD-PACK CHEESE FOOD Requirements For Specific Standardized Cheese And Related Products; Cheeses And Related Cheese Products
COLD-PACK CHEESE	133.125	COLD-PACK CHEESE FOOD WITH FRUITS, VEGETABLES, OR MEATS Requirements For Specific Standardized Cheese And Related Products; Cheeses And Related Cheese Products
COLD SORES	310.537	DRUG PRODUCTS CONTAINING ACTIVE INGREDIENTS OFFERED OTC FOR TREATMENT OF FEVER BLISTERS AND COLD SORES Requirements For Specific New Drugs Or Devices; New Drugs
COLD WATER BOTTLE	880.6085	HOT/COLD WATER BOTTLE General Hospital And Personal Use Miscellaneous Devices; General Hospital And Personal Use Devices
COLI, ESCHERICHIA	866.3255	ESCHERICHIA COLI SEROLOGICAL REAGENTS Serological Reagents; Immunology And Microbiology Devices

COLISTIMETHATE	522.468	COLISTIMETHATE SODIUM POWDER FOR INJECTION Implantation Or Injectable Dosage Form New Animal Drugs
COLISTIMETHATE	556.167	COLISTIMETHATE Specific Tolerances For Residues Of New Animal Drugs; Tolerances For Residues Of New Animals Drugs In Food
COLLABORATION, REGULATORY	26.18	REGULATORY COLLABORATION Specific Sector Provisions For Pharmaceutical GMPs; Mutual Recognition Of Pharmaceutical GMP Reports, Medical Device Quality System Audit Reports
COLLECTED	1230.31	WHERE SAMPLES MAY BE COLLECTED Regulations Under The Federal Caustic Poison Act
COLLECTION	640.4	COLLECTION OF THE BLOOD Whole Blood; Additional Standards For Human Blood And Blood Products
COLLECTION	640.13	COLLECTION OF THE BLOOD Red Blood Cells; Additional Standards For Human Blood And Blood Products
COLLECTION	640.22	COLLECTION OF SOURCE MATERIAL Platelets; Additional Standards For Human Blood And Blood Products
COLLECTION	640.32	COLLECTION OF SOURCE MATERIAL Plasma; Additional Standards For Human Blood And Blood Products
COLLECTION	640.52	COLLECTION OF SOURCE MATERIAL Cryoprecipitate; Additional Standards For Human Blood And Blood Products
COLLECTION	640.64	COLLECTION OF BLOOD FOR SOURCE PLASMA Source Plasma; Additional Standards For Human Blood And Blood Products
COLLECTION	660.32	COLLECTION OF SOURCE MATERIAL Reagent Red Blood Cells; Additional Standards For Diagnostic Substances For Laboratory Tests
COLLECTION	803.55	RESTRICTIONS ON THE SALE, DISTRIBUTION, AND USE OF OTC TEST SAMPLE COLLECTION SYSTEMS FOR DRUGS OF ABUSE TESTING Requirements For Manufacturers And Producers; In Vitro Diagnostic Products For Human Use
COLLECTION	862.1675	BLOOD SPECIMEN COLLECTION DEVICE Clinical Chemistry Test Systems; Clinical Chemistry And Clinical Toxicology Devices
COLLECTION	864.3260	OTC TEST SAMPLE COLLECTION SYSTEMS FOR DRUGS OF ABUSE TESTING Pathology Instrumentation And Accessories; Hematology And Pathology Devices
COLLECTION	864.6150	CAPILLARY BLOOD COLLECTION TUBE Manual Hematology Devices; Hematology And Pathology Devices
COLLECTION	864.9100	EMPTY CONTAINER FOR THE COLLECTION AND PROCESSING OF BLOOD AND BLOOD COMPONENTS Products Used In Establishments That Manufacture Blood And Blood Products; Hematology And Pathology Devices
COLLECTION	864.9125	VACUUM-ASSISTED BLOOD COLLECTION SYSTEM Products Used In Establishments That Manufacture Blood And Blood Products; Hematology And Pathology Devices
COLLECTION	866.2900	MICROBIOLOGICAL SPECIMEN COLLECTION AND TRANSPORT DEVICE Microbiology Devices; Immunology And Microbiology Devices
COLLECTION	868.1575	GAS COLLECTION VESSEL Diagnostic Devices; Anesthesiology Devices
COLLECTION	1230.30	COLLECTION OF SAMPLES Regulations Under The Federal Caustic Poison Act
COLLECTION	1403.52	COLLECTION OF AMOUNTS DUE Uniform Administrative Requirements For Grants And Coop. Agreements To State & Local Gov.
COLLECTIVE NAMES	501.110	ANIMAL FEED LABELING; COLLECTIVE NAMES FOR FEED INGREDIENTS Exemptions From Animal Food Labeling Requirements; Animal Food Labeling
COLLECTOR, BONE PARTICLE	874.4800	BONE PARTICLE COLLECTOR Surgical Devices, Ear, Nose And Throat Devices
COLLECTOR, URINE	876.5250	URINE COLLECTOR AND ACCESSORIES Therapeutic Devices; Gastroenterology-urology Devices
COLLOIDAL FERRIC OXIDE	522.940	COLLOIDAL FERRIC OXIDE INJECTION Implantation Or Injectable Dosage Form New Animal Drugs
COLLOIDAL SILVER	310.548	DRUG PRODUCTS CONTAINING COLLOIDAL SILVER INGREDIENTS Requirements For Specific New Drugs Or Devices; New Drugs
COLONIC	876.5220	COLONIC IRRIGATION SYSTEM Therapeutic Devices; Gastroenterology-urology Devices
COLONY	866.2170	AUTOMATED COLONY COUNTER Microbiology Devices; Immunology And Microbiology Devices
COLONY	866.2180	MANUAL COLONY COUNTER Microbiology Devices; Immunology And Microbiology Devices
COLOR	2.25	GRAIN SEED TREATED WITH POISONOUS SUBSTANCES; COLOR IDENTIFICATION TO PREVENT ADULTERATION OF HUMAN AND ANIMAL FOOD Human And Animal Foods; General Administrative Rulings And Decisions
COLOR	73.169	GRAPE COLOR EXTRACT Listing Of Color Additives Exempt From Certification

COLOR	880.2900	CLINICAL COLOR CHANGE THERMOMETER General Hospital And Personal Use Monitoring Devices; General Hospital And Personal Use Devices
COLOR ADDITIVE	14.140	ESTABLISHMENT OF A COLOR ADDITIVE ADVISORY COMMITTEE Color Additive Advisory Committees; Public Hearing Before A Public Advisory Committee
COLOR ADDITIVE	14.142	FUNCTIONS OF A COLOR ADDITIVE ADVISORY COMMITTEE Color Additive Advisory Committees; Public Hearing Before A Public Advisory Committee
COLOR ADDITIVE	14.145	PROCEDURES OF A COLOR ADDITIVE ADVISORY COMMITTEE Color Additive Advisory Committees; Public Hearing Before A Public Advisory Committee
COLOR ADDITIVE	14.147	MEMBERSHIP OF A COLOR ADDITIVE ADVISORY COMMITTEE Coir Additive Advisory Committees; Public Hearing Before A Public Advisory Committee
COLOR ADDITIVE	14.155	FEES AND COMPENSATION PERTAINING TO A COLOR ADDITIVE ADVISORY COMMITTEE Color Additive Advisory Committees; Public Hearing Before A Public Advisory Committee
COLOR ADDITIVE	71.15	CONFIDENTIALITY OF DATA AND INFORMATION IN COLOR ADDITIVE PETITIONS General Provisions; Color Additive Petitions
COLOR ADDITIVE	71.22	DECEPTION AS A BASIS FOR REFUSING TO ISSUE REGULATIONS; DECEPTIVE USE OF A COLOR ADDITIVE FOR WHICH A REGULATION HAS ISSUED Administrative Action On Petitions; Color Additive Petitions
COLOR ADDITIVE	73.1	DILUENTS IN COLOR ADDITIVE MIXTURES FOR FOOD USE EXEMPT FROM CERTIFICATION Foods; Listing Of Color Additives Exempt From Certification
COLOR ADDITIVE	73.1001	DILUENTS IN COLOR ADDITIVE MIXTURES FOR DRUG USE EXEMPT FROM CERTIFICATION Drugs; Listing Of Color Additives Exempt From Certification
COLOR ADDITIVE	80.35	COLOR ADDITIVE MIXTURES; CERTIFICATION AND EXEMPTION FROM CERTIFICATION Certification Procedures; Color Additive Certification
COLOR ADDITIVES	5.27	PATENT TERM EXTENSIONS FOR HUMAN DRUG PRODUCTS, MEDICAL DEVICES, AND FOOD AND COLOR ADDITIVES General Redelegations Of Authority; Delegations Of Authority And Organization
COLOR ADDITIVES	5.300	FOOD STANDARDS, FOOD ADDITIVES, GRAS SUBSTANCES, COLOR ADDITIVES, NUTRIENT CONTENT CLAIMS, AND HEALTH CLAIMS Foods And Cosmetics; Redelegations Of Authority; Delegations Of Authority And Organization
COLOR ADDITIVES	25.32	FOODS, FOOD ADDITIVES, AND COLOR ADDITIVES Categorical Exclusions; Environmental Impact Considerations
COLOR ADDITIVES	70.5	GENERAL RESTRICTIONS ON USE OF COLOR ADDITIVES General Provisions; Color Additives
COLOR ADDITIVES	70.10	COLOR ADDITIVES IN STANDARDIZED FOODS, NEW DRUGS, AND ANTIBIOTICS General Provisions; Color Additives
COLOR ADDITIVES	70.25	LABELING REQUIREMENTS FOR COLOR ADDITIVES (OTHER THAN HAIR DYES) Packaging And Labeling; Color Additives
COLOR ADDITIVES	71.37	EXEMPTION OF COLOR ADDITIVES FOR INVESTIGATIONAL USE Administrative Action On Petitions; Color Additive Petitions
COLOR ADDITIVES	70.42	CRITERIA FOR EVALUATING THE SAFETY OF COLOR ADDITIVES Safety Evaluation; Color Additives
COLOR ADDITIVES	70.45	ALLOCATION OF COLOR ADDITIVES Safely Evaluation; Color Additives
COLOR ADDITIVES	81.1	PROVISIONAL LISTS OF COLOR ADDITIVES General Specs+restrictns Provisional Color Additives For Foods, Drugs, Cosmetics
COLOR ADDITIVES	81.10	TERMINATION OF PROVISIONAL LISTINGS OF COLOR ADDITIVES General Specs+restrictns Provisional Color Additives For Foods, Drugs, Cosmetics
COLOR VISION	866.1070	COLOR VISION TESTER Diagnostic Devices; Ophthalmic Devices
COLOR VISION	886.1160	COLOR VISION PLATE ILLUMINATOR Diagnostic Devices; Ophthalmic Devices
COLORANTS	178.3297	COLORANTS FOR POLYMERS Certain Adjuvants And Production Aids; Indirect Food Additives: Adjuvants, Production Aids, And Sanitizers
COLORED LEMONADE	146.126	FROZEN CONCENTRATE FOR COLORED LEMONADE Requirements For Specific Standardized Canned Fruit Juices And Beverages; Canned Fruit Juices

COLORIMETER	862.2300	COLORIMETER, PHOTOMETER, OR SPECTROPHOTOMETER FOR CLINICAL USE Clinical Laboratory Instruments; Clinical Chemistry And Clinical Toxicology Devices
COLORINGS	101.22	FOODS; LABELING OF SPICES, FLAVORINGS, COLORINGS AND CHEMICAL PRESERVATIVES Specific Food Labeling Requirements; Food Labeling
COLORINGS	501.22	ANIMAL FOODS; LABELING OF SPICES, FLAVORINGS, COLORINGS, AND CHEMICAL PRESERVATIVES Specific Animal Food Labeling Requirements; Animal Food Labeling
COLORS, STRAIGHT	70.20	PACKAGING REQUIREMENTS FOR STRAIGHT COLORS (OTHER THAN HAIR DYES) Packaging And Labeling; Color Additives
COLORS, STRAIGHT	82.5	GENERAL SPECIFICATIONS FOR STRAIGHT COLORS General Provisions; Listing Of Certified Provisionally Listed Colors And Specifications
COLOSTOMY	876.4270	COLOSTOMY ROD Surgical Devices; Gastroenterology-urology Devices
COLOSTRUM	866.5230	COLOSTRUM IMMUNOLOGICAL TEST SYSTEM Immunological Test Systems; Immunology And Microbiology Devices
COLPOSCOPE	884.1630	COLPOSCOPE Obstetrical And Gynecological Diagnostic Devices; Obstetrical And Gynecological Devices
COLUMBIA, DISTRICT OF	808.101	DISTRICT OF COLUMBIA Listing Of Specific State And Local Exemptions; Exemption From Federal Preemption Of State & Local Medical Device Requirements
COMARKETING	203.36	FULFILLMENT HOUSES, SHIPPING AND MAILING SERVICES, COMARKETING AGREEMENTS, AND THIRD-PARTY RECORDKEEPING Samples; Prescription Drug Marketing
COMBINATION	172.165	QUATERNARY AMMONIUM CHLORIDE COMBINATION Food Preservatives; Food Additives Permitted For Direct Addition To Food For Human Consumption
COMBINATION	300.50	FIXED-COMBINATION PRESCRIPTION DRUGS FOR HUMANS Combination Drugs; General
COMBINATION	331.15	COMBINATION WITH NONANTACID ACTIVE INGREDIENTS Active Ingredients; Antacid Products For Over-the-counter Human Use
COMBINATION	332.15	COMBINATION WITH NON-ANTIFLATULENT ACTIVE INGREDIENTS Active Ingredients; Antiflatulent Products For Over-the-counter Human Use
COMBINATION	341.70	LABELING OF OTC DRUG PRODUCTS CONTAINING INGREDIENTS USED FOR TREATING CONCURRENT SYMPTOMS (IN EITHER SINGLE-INGREDIENT OR COMBINATION DRUG PRODUCT) Labeling; Cold, Cough, Allergy, Bronchodilator, And Antiasthmatic Drug Products For OTC Human Use
COMBINATION	520.1660a	OXYTETRACYCLINE AND CARBOMYCIN IN COMBINATION Oral Dosage Form New Animal Drugs
COMBINATION	520.2320	SULFANITRAN AND AKLOMIDE IN COMBINATION Oral Dosage Form New Animal Drugs
COMBINATION	522.1940	PROGESTERONE AND ESTRADIOL BENZOATE IN COMBINATION Implantation Or Injectable Dosage Form New Animal Drugs
COMBINATION	522.842	ESTRADIOL BENZOATE AND TESTOSTERONE PROPIONATE IN COMBINATION Implantation Or Injectable Dosage Form New Animal Drugs
COMBINATION	522.850	ESTRADIOL VALERATE AND NORGESTOMET IN COMBINATION Implantation Or Injectable Dosage Form New Animal Drugs
COMBINATIONS	333.120	PERMITTED COMBINATIONS OF ACTIVE INGREDIENTS First Aid Antibiotic Drug Products; Topical Antimicrobial Drug Products For Over-the-counter Human Use
COMBINATIONS	333.160	LABELING OF PERMITTED COMBINATIONS OF ACTIVE INGREDIENTS First Aid Antibiotic Drug Products, Topical Antimicrobial Drug Products For Over-the-counter Human Use
COMBINATIONS	333.320	PERMITTED COMBINATIONS OF ACTIVE INGREDIENTS Topical Acne Drugs; Topical Antimicrobial Drug Products For Over-the-counter Human Use
COMBINATIONS	343.22	PERMITTED COMBINATIONS OF ACTIVE INGREDIENTS FOR CARDIOVASCULAR-RHEUMATOLOGIC USE Internal Analgesic, Antipyretic And Antirheumatic Drug Products For OTC Human Use
COMBINATIONS	346.22	PERMITTED COMBINATIONS OF ANORECTAL ACTIVE INGREDIENTS Anorectal Drug Products For Over-the-counter Human Use
COMBINATIONS	346.52	LABELING OF PERMITTED COMBINATIONS OF ANORECTAL ACTIVE INGREDIENTS Labeling; Anorectal Drug Products For Over-the-counter Human Use
COMBINATIONS	349.30	PERMITTED COMBINATIONS OF ACTIVE INGREDIENTS Ophthalmic Drug Products For Over-the-counter Human Use

COMBINATIONS	349.79	LABELING OF PERMITTED COMBINATIONS OF ACTIVE INGREDIENTS Ophthalmic Drug Products For Over-the-counter Human Use
COMBINATIONS	352.20	PERMITTED COMBINATIONS OF ACTIVE INGREDIENTS Sunscreen Drug Products For Over-the-counter Human Use
COMBINATIONS	352.60	LABELING OF PERMITTED COMBINATIONS OF ACTIVE INGREDIENTS Sunscreen Drug Products For Over-the-counter Human Use
COMBINATIONS	358.720	PERMITTED COMBINATIONS OF ACTIVE INGREDIENTS Drug Products For Control Of Dandruff, Seborrheic Dermatitis, And Psoriasis; Miscellaneous External Drug Products For OTC Human Use
COMBINATIONS	610.17	PERMISSIBLE COMBINATIONS General Provisions; General Biological Products Standards
COMBUSTION GAS	173.350	COMBUSTION PRODUCT GAS Specific Usage Additives; Secondary Direct Food Additives Permitted In Food For Human Consumption
COMMENCEMENT	12.62	COMMENCEMENT OF FUNCTIONS Presiding Officer; Formal Evidentiary Public Hearing
COMMENT	312.41	COMMENT AND ADVICE ON AN IND Investigational New Drug Application
COMMENT	514.100	EVALUATION AND COMMENT ON APPLICATIONS Administrative Actions On Applications; New Animal Drug Applications
COMMERCIAL INFORMATION	20.53	INDEXING TRADE SECRETS AND CONFIDENTIAL COMMERCIAL OR FINANCIAL INFORMATION Procedures And Fees; Public Information
COMMERCIAL INFORMATION	20.61	TRADE SECRETS AND COMMERCIAL OR FINANCIAL INFORMATION WHICH IS PRIVILEGED OR CONFIDENTIAL Exemptions; Public Information
COMMISSIONER	5.10	DELEGATIONS FROM THE SECRETARY OF HEALTH AND HUMAN SERVICES TO THE COMMISSIONER OF FOOD AND DRUGS Delegations Of Authority To The Commissioner Of Food And Drugs; Delegations Of Authority And Organization
COMMISSIONER	5.20	GENERAL REDELEGATIONS OF AUTHORITY FROM THE COMMISSIONER TO OTHER OFFICERS OF THE FDA General Redelegations Of Authority; Delegations Of Authority And Organization
COMMISSIONER	12.130	DECISION BY COMMISSIONER ON APPEAL OR REVIEW OF INITIAL DECISION Initial And Final Decisions; Formal Evidentiary Public Hearing
COMMISSIONER	15.20	NOTICE OF A PUBLIC HEARING BEFORE THE COMMISSIONER Procedures For Public Hearing Before The Commissioner; Public Hearing Before The Commissioner
COMMISSIONER	15.30	CONDUCT OF A PUBLIC HEARING BEFORE THE COMMISSIONER Procedures For Public Hearing Before The Commissioner; Public Hearing Before The Commissioner
COMMISSIONER	16.40	COMMISSIONER Commissioner And Presiding Officer; Regulatory Hearing Before The FDA
COMMISSIONER	16.44	COMMUNICATION TO PRESIDING OFFICER AND COMMISSIONER Commissioner And Presiding Officer; Regulatory Hearing Before The FDA
COMMISSIONER	20.82	DISCRETIONARY DISCLOSURE BY THE COMMISSIONER Limitations On Exemptions; Public Information
COMMISSIONER	170.15	ADOPTION OF REGULATION ON INITIATIVE OF COMMISSIONER General Provisions; Food Additives
COMMISSIONER	570.15	ADOPTION OF REGULATION ON INITIATIVE OF COMMISSIONER General Provisions; Food Additives
COMMISSIONER	860.132	PROCEDURES WHEN THE COMMISSIONER INITIATES A PERFORMANCE STANDARD OR PREMARKET APPROVAL PROCEEDING UNDER SECTION 514(B) OR 515(B) Reclassification; Medical Device Classification Procedures
COMMISSIONER'S	71.27	LISTING AND EXEMPTION FROM CERTIFICATION ON THE COMMISSIONER'S INITIATIVE Administrative Action On Petitions; Color Additive Petitions
COMMITTEE	14.5	PURPOSE OF PROCEEDINGS BEFORE AN ADVISORY COMMITTEE General Provisions; Public Hearing Before A Public Advisory Committee
COMMITTEE	14.20	NOTICE OF HEARING BEFORE AN ADVISORY COMMITTEE Meeting Procedures; Public Hearing Before A Public Advisory Committee
COMMITTEE	14.22	MEETINGS OF AN ADVISORY COMMITTEE Meeting Procedures; Public Hearing Before A Public Advisory Committee
COMMITTEE	14.25	PORTIONS OF ADVISORY COMMITTEE MEETINGS Meeting Procedures; Public Hearing Before A Public Advisory Committee
COMMITTEE	14.27	DETERMINATION TO CLOSE PORTIONS OF ADVISORY COMMITTEE MEETINGS Meeting Procedures; Public Hearing Before A Public Advisory Committee
COMMITTEE	14.29	CONDUCT OF A HEARING BEFORE AN ADVISORY COMMITTEE Meeting Procedures; Public Hearing Before A Public Advisory Committee
COMMITTEE	14.30	CHAIRMAN OF AN ADVISORY COMMITTEE Meeting Procedures; Public Hearing Before A Public Advisory Committee

COMMITTEE	14.31	CONSULTATION BY AN ADVISORY COMMITTEE WITH OTHER PERSONS Meeting Procedures; Public Hearing Before A Public Advisory Committee
COMMITTEE	14.33	COMPILATION OF MATERIALS FOR MEMBERS OF AN ADVISORY COMMITTEE Meeting Procedures; Public Hearing Before A Public Advisory Committee
COMMITTEE	14.35	WRITTEN SUBMISSIONS TO AN ADVISORY COMMITTEE Meeting Procedures; Public Hearing Before A Public Advisory Committee
COMMITTEE	14.39	ADDITIONAL RULES FOR A PARTICULAR ADVISORY COMMITTEE Meeting Procedures; Public Hearing Before A Public Advisory Committee
COMMITTEE	14.60	MINUTES AND REPORTS OF ADVISORY COMMITTEE MEETINGS Records Of Meetings And Hearings Before Advisory Committees; Public Hearing Before A Public Advisory Committee
COMMITTEE	14.61	TRANSCRIPTS OF ADVISORY COMMITTEE MEETINGS Records Of Meetings And Hearings Before Advisory Committees; Public Hearing Before A Public Advisory Committee
COMMITTEE	14.65	PUBLIC INQUIRIES AND REQUESTS FOR ADVISORY COMMITTEE RECORDS Records Of Meetings And Hearings Before Advisory Committees; Public Hearing Before A Public Advisory Committee
COMMITTEE	14.70	ADMINISTRATIVE RECORD OF A PUBLIC HEARING BEFORE AN ADVISORY COMMITTEE Records Of Meetings And Hearings Before Advisory Committees; Public Hearing Before A Public Advisory Committee
COMMITTEE	14.75	EXAMINATION OF ADMINISTRATIVE RECORD AND OTHER ADVISORY COMMITTEE RECORDS Records Of Meetings And Hearings Before Advisory Committees; Public Hearing Before A Public Advisory Committee
COMMITTEE	14.90	AD HOC ADVISORY COMMITTEE MEMBERS Members Of Advisory Committees; Public Hearing Before A Public Advisory Committee
COMMITTEE	14.95	COMPENSATION OF ADVISORY COMMITTEE MEMBERS Members Of Advisory Committees; Public Hearing Before A Public Advisory Committee
COMMITTEE	14.12	ESTABLISHMENT OF THE TECHNICAL ELECTRONIC PRODUCT RADIATION SAFETY STANDARD COMMITTEE (TEPRSSC) Technical Electronic Products Radiation Safety Standards Committee; Public Hearing Before A Public Advisory Committee
COMMITTEE	14.140	ESTABLISHMENT OF A COLOR ADDITIVE ADVISORY COMMITTEE Color Additive Advisory Committees; Public Hearing Before A Public Advisory Committee
COMMITTEE	14.142	FUNCTIONS OF A COLOR ADDITIVE ADVISORY COMMITTEE Color Additive Advisory Committees; Public Hearing Before A Public Advisory Committee
COMMITTEE	14.145	PROCEDURES OF A COLOR ADDITIVE ADVISORY COMMITTEE Color Additive Advisory Committees; Public Hearing Before A Public Advisory Committee
COMMITTEE	14.147	MEMBERSHIP OF A COLOR ADDITIVE ADVISORY COMMITTEE Color Additive Advisory Committees; Public Hearing Before A Public Advisory Committee
COMMITTEE	14.155	FEES AND COMPENSATION PERTAINING TO A COLOR ADDITIVE ADVISORY COMMITTEE Color Additive Advisory Committees; Public Hearing Before A Public Advisory Committee
COMMITTEE	14.17	UTILIZATION OF AN ADVISORY COMMITTEE ON THE INITIATIVE OF FDA Advisory Committees For Human Prescription Drugs; Public Hearing Before A Public Advisory Committee
COMMITTEE	14.172	UTILIZATION OF AN ADVISORY COMMITTEE AT THE REQUEST OF AN INTERESTED PERSON Advisory Committees For Human Prescription Drugs; Public Hearing Before A Public Advisory Committee
COMMITTEE	20.11	ADVISORY COMMITTEE RECORDS Availability Of Specific Categories Of Records; Public Information
COMMITTEE	70.51	ADVISORY COMMITTEE ON THE APPLICATION OF THE ANTICANCER CLAUSE Safety Evaluation; Color Additives
COMMITTEE, JOINT	26.73	JOINT COMMITTEE Framework Provisions; Mutual Recognition Of Pharmaceutical GMP Reports, Medical Device Quality System Audit Reports
COMMITTEE, JOINT SECTORAL	26.17	ROLE AND COMPOSITION OF THE JOINT SECTORAL COMMITTEE Specific Sector Provisions For Pharmaceutical GMPs; Mutual Recognition Of Pharmaceutical GMP Reports, Medical Device Quality System Audit Reports
COMMITTEE, JOINT SECTORAL	26.47	ROLE AND COMPOSITION OF THE JOINT SECTORAL COMMITTEE Specific Sector Provisions For Medical Devices; Mutual Recognition Of Pharmaceutical GMP Reports, Medical Device Quality System Audit Reports

COMMITTEES	5.30	AUTHORITY TO SELECT TEMPORARY VOTING MEMBERS FOR ADVISORY COMMITTEES AND AUTHORITY TO SIGN CONFLICT OF INTEREST WAIVERS General Redelegations Of Authority; Delegations Of Authority And Organization
COMMITTEES	14.15	COMMITTEES WORKING UNDER A CONTRACT WITH FDA General Provisions; Public Hearing Before A Public Advisory Committee
COMMITTEES	14.40	ESTABLISHMENT AND RENEWAL OF ADVISORY COMMITTEES Establishment Of Advisory Committees; Public Hearing Before A Public Advisory Committee
COMMITTEES	14.55	TERMINATION OF ADVISORY COMMITTEES Establishment Of Advisory Committees; Public Hearing Before A Public Advisory Committee
COMMITTEES	14.80	QUALIFICATIONS FOR MEMBERS OF STANDING POLICY AND TECHNICAL ADVISORY COMMITTEES Members Of Advisory Committees; Public Hearing Before A Public Advisory Committee
COMMITTEES	14.82	NOMINATIONS OF VOTING MEMBERS OF STANDING ADVISORY COMMITTEES Members Of Advisory Committees; Public Hearing Before A Public Advisory Committee
COMMITTEES	14.84	NOMINATIONS AND SELECTION OF NONVOTING MEMBERS OF STANDING TECHNICAL ADVISORY COMMITTEES Members Of Advisory Committees; Public Hearing Before A Public Advisory Committee
COMMITTEES	14.86	RIGHTS AND RESPONSIBILITIES OF NONVOTING MEMBERS OF ADVISORY COMMITTEES Members Of Advisory Committees; Public Hearing Before A Public Advisory Committee
COMMITTEES	14.100	LIST OF STANDING ADVISORY COMMITTEES Standing Advisory Committees; Public Hearing Before A Public Advisory Committee
COMMITTEES	14.160	ESTABLISHMENT OF STANDING TECHNICAL ADVISORY COMMITTEES FOR HUMAN PRESCRIPTION DRUGS Advisory Committees For Human Prescription Drugs; Public Hearing Before A Public Advisory Committee
COMMITTEES	20.84	DISCLOSURE TO CONSULTANTS, ADVISORY COMMITTEES, STATE AND LOCAL GOVT OFFICIALS AND OTHER SPECIAL GOVERNMENT EMPLOYEES Limitations On Exemptions; Public Information
COMMITTEES	861.38	STANDARDS ADVISORY COMMITTEES Procedures For Performance Standards Development And Publication; Procedures For Performance Standards Development
COMMON CARRIER	203.30	SAMPLE DISTRIBUTION BY MAIL OR COMMON CARRIER Samples; Prescription Drug Marketing
COMMON CARRIER	203.31	SAMPLE DISTRIBUTION BY MEANS OTHER THAN MAIL OR COMMON CARRIER Samples; Prescription Drug Marketing
COMMUNICATION	16.44	COMMUNICATION TO PRESIDING OFFICER AND COMMISSIONER Commissioner And Presiding Officer; Regulatory Hearing Before The Food And Drug Administration
COMMUNICATION SYSTEM	890.3700	NONPOWERED COMMUNICATION SYSTEM Physical Medicine Prosthetic Devices; Physical Medicine Devices
COMMUNICATION SYSTEM	890.3710	POWERED COMMUNICATION SYSTEM Physical Medicine Prosthetic Devices; Physical Medicine Devices
COMMUNICATIONS	7.49	RECALL COMMUNICATIONS Recalls - Guidance On Policy, Procedures & Industry Responsibilities; Enforcement Policy
COMMUNICATIONS	20.88	COMMUNICATIONS WITH STATE AND LOCAL GOVERNMENT OFFICIALS Limitations On Exemptions; Public Information
COMMUNICATIONS	20.89	COMMUNICATIONS WITH FOREIGN GOVERNMENT OFFICIALS Limitations On Exemptions; Public Information
COMMUNICATIONS	314.102	COMMUNICATIONS BETWEEN FDA AND APPLICANTS FDA Action On Applications; Applications For FDA Approval To Market A New Drug
COMMUNICATIONS	810.15	COMMUNICATIONS CONCERNING A CEASE DISTRIBUTION AND NOTIFICATION OR MANDATORY RECALL ORDER Mandatory Medical Device Recall Procedures; Medical Device Recall Authority
COMMUNICATIONS	1003.22	COPIES OF COMMUNICATIONS SENT TO PURCHASERS, DEALERS, OR DISTRIBUTORS Notification Of Defects Or Failure To Comply
COMMUNICATIONS DEVICE	892.2020	MEDICAL IMAGE COMMUNICATIONS DEVICE Diagnostic Devices; Radiology Devices
COMMUNICATIONS, EX PARTE	10.55	SEPARATION OF FUNCTIONS; EX PARTE COMMUNICATIONS General Administrative Procedures; Administrative Practices And Procedures
COMMUNICATIONS, EX PARTE	13.15	SEPARATION OF FUNCTIONS; EX PARTE COMMUNICATIONS; ADMINISTRATIVE SUPPORT General Provisions; Public Hearing Before A Public Board Of Inquiry

COMMUNICATIONS, EX PARTE	1316.51	CONDUCT OF HEARING AND PARTIES; EX PARTE COMMUNICATIONS Administrative Hearings; Administrative Functions, Practices, And Procedures
COMMUNICATIONS SYSTEM	892.2050	PICTURE ARCHIVING AND COMMUNICATIONS SYSTEM Diagnostic Devices; Radiology Devices
COMPATIBILITY	606.151	COMPATIBILITY TESTING Laboratory Controls; Current Good Manufacturing Practice For Blood And Blood Components
COMPENDIA, OFFICIAL	369.7	WARNINGS REQUIRED BY OFFICIAL COMPENDIA Definitions And Interpretations; Interpretative Statements Re Warnings On Drugs And Devices For OTC Sale
COMPENDIAL NAME	299.5	DRUGS; COMPENDIAL NAME General Provisions; Drugs; Official Names And Established Names
COMPENSATED	868.2340	COMPENSATED THORPE TUBE FLOWMETER Monitoring Devices; Anesthesiology Devices
COMPENSATION	14.95	COMPENSATION OF ADVISORY COMMITTEE MEMBERS Members Of Advisory Committees; Public Hearing Before A Public Advisory Committee
COMPENSATION	14.155	FEES AND COMPENSATION PERTAINING TO A COLOR ADDITIVE ADVISORY COMMITTEE Color Additive Advisory Committees; Public Hearing Before A Public Advisory Committee
COMPETITIVE	529.469	COMPETITIVE EXCLUSION CULTURE Certain Other Dosage Form New Animal Drugs
COMPILATION	14.33	COMPILATION OF MATERIALS FOR MEMBERS OF AN ADVISORY COMMITTEE Meeting Procedures; Public Hearing Before A Public Advisory Committee
COMPLAINT	17.5	COMPLAINT Civil Money Penalties Hearings
COMPLAINT	17.7	SERVICE OF COMPLAINT Civil Money Penalties Hearings
COMPLAINT FILES	211.198	COMPLAINT FILES Records And Reports; Current Good Manufacturing Practice For Finished Pharmaceuticals
COMPLAINT FILES	225.115	COMPLAINT FILES Records And Reports; Current Good Manufacturing Practice For Medicated Feeds
COMPLAINT FILES	226.115	COMPLAINT FILES Records And Reports; Current Good Manufacturing Practice For Type A Medicated Articles
COMPLAINT FILES	803.31	COMPLAINT FILES Reports And Records; Medical Device Reporting
COMPLAINT FILES	820.198	COMPLAINT FILES Records; Quality System Regulation
COMPLAINTS	1402.7	SUGGESTIONS AND COMPLAINTS Mandatory Declassification Review
COMPLEMENT	866.4100	COMPLEMENT REAGENT Immunology Laboratory Equipment And Reagents; Immunology And Microbiology Devices
COMPLEMENT	866.5240	COMPLEMENT COMPONENTS IMMUNOLOGICAL TEST SYSTEM Immunological Test Systems; Immunology And Microbiology Devices
COMPLEMENT INACTIVATOR	866.5260	COMPLEMENT C3B INACTIVATOR IMMUNOLOGICAL TEST SYSTEM Immunological Test Systems; Immunology And Microbiology Devices
COMPLEMENT INHIBITOR	866.5250	COMPLEMENT C1 INHIBITOR (INACTIVATOR) IMMUNOLOGICAL TEST SYSTEM Immunological Test Systems; Immunology And Microbiology Devices
COMPLETENESS	21.73	ACCURACY, COMPLETENESS, TIMELINESS, AND RELEVANCE OF RECORDS DISCLOSED FROM PRIVACY ACT RECORD SYSTEMS Disclosure Of Records In Privacy Act Record Systems To Persons Other ThanThe Subject Individual; Protection Of Privacy
COMPLETING STUDIES	99.203	REQUEST TO EXTEND THE TIME FOR COMPLETING PLANNED STUDIES Dissemination Of Information On Unapproved/New Uses For Marketed Drugs, Biologics, and Devices
COMPLETING STUDIES	99.303	REQUEST TO EXTEND THE TIME FOR COMPLETING PLANNED STUDIES Dissemination Of Information On Unapproved/New Uses For Marketed Drugs, Biologics, and Devices
COMPLEX	73.1125	POTASSIUM SODIUM COPPER CHLOROPHYLLIN (CHLOROPHYLLIN-COPPER COMPLEX) Drugs; Listing Of Color Additives Exempt From Certification
COMPLEX	73.2125	POTASSIUM SODIUM COPPER CHLOROPHYLLIN (CHLOROPHYLLIN-COPPER COMPLEX) Cosmetics; Listing Of Color Additives Exempt From Certification
COMPLEX	73.3110	CLOROPHYLLIN-COPPER COMPLEX, OIL SOLUBLE Medical Devices; Listing Of Color Additives Exempt From Certification
COMPLEX	172.315	NICOTINAMIDE-ASCORBIC ACID COMPLEX Special Dietary And Nutritional Additives; Food Additives Permitted For Direct Addition To Food For Human Consumption

COMPLEX	172.370	IRON-CHOLINE CITRATE COMPLEX Special Dietary And Nutritional Additives; Food Additives Permitted For Direct Addition To Food For Human Consumption
COMPLEX	176.160	CHROMIUM (CR III) COMPLEX N-ETHYL-N-HEPTADECYLFLUORO-OCTANE SULFONYL GLYCINE Substances For Use Only As Components Of Paper And Paperboard; Indirect Food Additives: Paper And Paperboard Components
COMPLEX	176.320	SODIUM NITRATE-UREA COMPLEX Substances For Use Only As Components Of Paper And Paperboard; Indirect Food Additives: Paper And Paperboard Components
COMPLEX	520.1802	PIPERAZINE-CARBON DISULFIDE COMPLEX ORAL DOSAGE FORMS Oral Dosage Form New Animal Drugs
COMPLEX	520.2345f	TETRACYCLINE PHOSPHATE COMPLEX AND SODIUM NOVOBIOCIN CAPSULES Oral Dosage Form New Animal Drugs
COMPLEX	522.1182	IRON DEXTRAN COMPLEX INJECTION Implantation Or Injectable Dosage Form New Animal Drugs
COMPLEX	573.580	IRON-CHOLINE-CITRATE COMPLEX Food Additive Listing; Food Additives Permitted In Feed And Drinking Water Of Animals
COMPLEXES	178.3290	CHROMIC CHLORIDE COMPLEXES Certain Adjuvants And Production Aids; Indirect Food Additives: Adjuvants, Production Aids, And Sanitizers
COMPLIANCE	5.107	EXTENSIONS OR STAYS OF EFFECTIVE DATES FOR COMPLIANCE WITH CERTAIN LABELING REQUIREMENTS FOR HUMAN PRESCRIPTION DRUGS Human Drugs; Redelegations Of Authority; Delegations Of Authority And Organization
COMPLIANCE	101.43	SUBSTANTIAL COMPLIANCE OF FOOD RETAILERS WITH THE GUIDELINES FOR VOLUNTARY NUTRITION LABELING Food Labeling
COMPLIANCE	107.270	COMPLIANCE WITH THIS SUBPART Infant Formula Recalls; Infant Formula
COMPLIANCE	1003.10	DISCOVER OF DEFECT OF FAILURE OF COMPLIANCE BY MANUFACTURERS' NOTICE REQUIREMENTS Notification Of Defects Or Failure To Comply
COMPLIANCE	1005.21	APPLICATION FOR PERMISSION TO BRING PRODUCT INTO COMPLIANCE Bonding And Compliance Procedures; Importation Of Electronic Products
COMPLIANCE	1005.22	GRANTING PERMISSION TO BRING PRODUCT INTO COMPLIANCE Bonding And Compliance Procedures; Importation Of Electronic Products
COMPLIANCE	1005.24	COSTS OF BRINGING PRODUCT INTO COMPLIANCE Bonding And Compliance Procedures; Importation Of Electronic Products
COMPLIANCE DATES	898.13	COMPLIANCE DATES Performance Standards For Electrode Lead Wires And Patient Cables
COMPLY	1003.11	DETERMINATION BY SECRETARY THAT PRODUCT FAILS TO COMPLY OR HAS A DEFECT Notification Of Defects Or Failure To Comply
COMPONENT	3.4	DESIGNATED AGENCY COMPONENT Assignment Of Agency Component For Review Of Premarket Applications; Product Jurisdiction
COMPONENT	3.5	PROCEDURES FOR IDENTIFYING THE DESIGNATED AGENCY COMPONENT Assignment Of Agency Component For Review Of Premarket Applications; Product Jurisdiction
COMPONENT	201.21	DECLARATION OF PRESENCE OF PHENYLALANINE AS A COMPONENT OF ASPARTAME IN OVER-THE-COUNTER AND PRESCRIPTION DRUGS FOR HUMAN USE General Labeling Provisions; Labeling
COMPONENT	211.184	COMPONENT, DRUG PRODUCT CONTAINER, CLOSURE, AND LABELING RECORDS Records And Reports; Current Good Manufacturing Practice For Finished Pharmaceuticals
COMPONENT	250.250	HEXACHLOROPHENE, AS A COMPONENT OF DRUG AND COSMETIC PRODUCTS Requirements For Drugs And Cosmetics; Specific Requirements For Specific Human Drugs
COMPONENT	866.5380	FREE SECRETORY COMPONENT IMMUNOLOGICAL TEST SYSTEM Immunological Test Systems; Immunology And Microbiology Devices
COMPONENT	888.5890	NONINVASIVE TRACTION COMPONENT Surgical Devices; Orthopedic Devices
COMPONENT	888.5940	CAST COMPONENT Surgical Devices; Orthopedic Devices
COMPONENT	890.3410	EXTERNAL LIMB ORTHOTIC COMPONENT Physical Medicine Prosthetic Devices; Physical Medicine Devices
COMPONENT	890.3420	EXTERNAL LIMB PROSTHETIC COMPONENT Physical Medicine Prosthetic Devices; Physical Medicine Devices

COMPONENT	890.3920	WHEELCHAIR COMPONENT Physical Medicine Prosthetic Devices; Physical Medicine Devices
COMPONENT	890.5350	EXERCISE COMPONENT Physical Medicine Therapeutic Devices; Physical Medicine Devices
COMPONENTS	11.200	ELECTRONIC SIGNATURE COMPONENTS AND CONTROLS Electronic Records; Electronic Signatures
COMPONENTS	176.170	COMPONENTS OF PAPER AND PAPERBOARD IN CONTACT WITH AQUEOUS AND FATTY FOODS Substances For Use Only As Components Of Paper And Paperboard; Indirect Food Additives: Paper And Paperboard Components
COMPONENTS	176.180	COMPONENTS OF PAPER AND PAPERBOARD IN CONTACT WITH DRY FOOD Substances For Use Only As Components Of Paper And Paperboard; Indirect Food Additives: Paper And Paperboard Components
COMPONENTS	178.3730	PIPERONYL BUTOXIDE AND PYRETHRINS AS COMPONENTS OF BAGS Certain Adjuvants And Production Aids; Indirect Food Additives: Adjuvants, Production Aids, And Sanitizers
COMPONENTS	181.26	DRYING OILS AS COMPONENTS OF FINISHED RESINS Specific Prior-sanctioned Food Ingredients; Prior-sanctioned Food Ingredients
COMPONENTS	201.120	PRESCRIPTION CHEMICALS AND OTHER PRESCRIPTION COMPONENTS Exemptions From Adequate Directions For Use; Labeling
COMPONENTS	211.82	RECEIPT AND STORAGE OF UNTESTED COMPONENTS, DRUG PRODUCT CONTAINERS, AND CLOSURES Control Of Components And Drug Product Containers And Closures; Current Good Manufacturing Practice For Finished Pharmaceuticals
COMPONENTS	211.84	TESTING AND APPROVAL OR REJECTION OF COMPONENTS, DRUG PRODUCT CONTAINERS, AND CLOSURES Control Of Components And Drug Product Containers And Closures; Current Good Manufacturing Practice For Finished Pharmaceuticals
COMPONENTS	211.86	USE OF APPROVED COMPONENTS, DRUG PRODUCT CONTAINERS, AND CLOSURES Control Of Components And Drug Product Containers And Closures; Current Good Manufacturing Practice For Finished Pharmaceuticals
COMPONENTS	211.87	RETESTING OF APPROVED COMPONENTS, DRUG PRODUCT CONTAINERS, AND CLOSURES Control Of Components And Drug Product Containers And Closures; Current Good Manufacturing Practice For Finished Pharmaceuticals
COMPONENTS	211.89	REJECTED COMPONENTS, DRUG PRODUCT CONTAINERS, AND CLOSURES Control Of Components And Drug Product Containers And Closures; Current Good Manufacturing Practice For Finished Pharmaceuticals
COMPONENTS	211.101	CHARGE-IN OF COMPONENTS Production And Process Controls; Current Good Manufacturing Practice For Finished Pharmaceuticals
COMPONENTS	225.42	COMPONENTS Product Quality Control; Current Good Manufacturing Practice For Medicated Feeds
COMPONENTS	225.142	COMPONENTS Product Quality Assurance; Current Good Manufacturing Practice For Medicated Feeds
COMPONENTS	226.42	COMPONENTS Product Quality Control; Current Good Manufacturing Practice For Type A Medicated Articles
COMPONENTS	864.2220	SYNTHETIC CELL AND TISSUE CULTURE MEDIA AND COMPONENTS Cell And Tissue Culture Products; Hematology And Pathology Devices
COMPONENTS	864.2360	MYCOPLASMA DETECTION MEDIA AND COMPONENTS Cell And Tissue Culture Products; Hematology And Pathology Devices
COMPONENTS	864.9100	EMPTY CONTAINER FOR THE COLLECTION AND PROCESSING OF BLOOD AND BLOOD COMPONENTS Products Used In Establishments That Manufacture Blood And Blood Products; Hematology And Pathology Devices
COMPONENTS	866.5240	COMPLEMENT COMPONENTS IMMUNOLOGICAL TEST SYSTEM Immunological Test Systems; Immunology And Microbiology Devices
COMPONENTS	882.5550	CENTRAL NERVOUS SYSTEM FLUID SHUNT AND COMPONENTS Neurological Therapeutic Devices; Neurological Devices
COMPOSITE IMPLANT	878.3500	POLYTETRAFLUOROETHYLENE WITH CARBON FIBERS COMPOSITE IMPLANT MATERIAL Prosthetic Devices; General And Plastic Surgery Devices
COMPOSITE SHRIMP	102.55	NONSTANDARDIZED BREADED COMPOSITE SHRIMP UNITS Requirements For Specific Nonstandardized Foods; Common Or Usual Name For Nonstandardized Foods
COMPOSITION, IRB	812.60	IRB COMPOSITION, DUTIES, AND FUNCTIONS IRB Review And Approval; Investigational Device Exemptions
COMPOSITION, JSC	26.17	ROLE AND COMPOSITION OF THE JOINT SECTORAL COMMITTEE Specific Sector Provisions For Pharmaceutical GMPs; Mutual Recognition Of Pharmaceutical GMP Reports, Medical Device Quality System Audit Reports

COMPOSITION, JSC	26.47	ROLE AND COMPOSITION OF THE JOINT SECTORAL COMMITTEE Specific Sector Provisions For Medical Devices; Mutual Recognition Of Pharmaceutical GMP Reports, Medical Device Quality System Audit Reports
COMPOUND	1304.25	RECORDS FOR TREATMENT PROGRAMS WHICH COMPOUND NARCOTICS Continuing Records; Records And Reports Of Registrants
COMPOUND DRILLS	882.4305	POWERED COMPOUND CRANIAL DRILLS, BURRS, TREPHINES, AND THEIR ACCESSORIES Neurological Surgical Devices; Neurological Devices
COMPOUND S	862.1185	COMPOUND S (11-DEOXYCORTISOL) TEST SYSTEM Clinical Chemistry Test Systems; Clinical Chemistry And Clinical Toxicology Devices
COMPOUNDERS	1301.72	PHYSICAL SECURITY CONTROLS FOR NON-PRACTITIONERS; NARCOTIC TREATMENT PROGRAMS AND COMPOUNDERS FOR NARCOTIC TREATMENT PROGRAMS; STORAGE AREAS Security Requirements; Registration Of Manufacturers, Distributors, And Dispensers Of Controlled Substances
COMPOUNDERS	1301.73	PHYSICAL SECURITY CONTROLS FOR NON-PRACTITIONERS; COMPOUNDERS FOR NARCOTIC TREATMENT PROGRAMS; MANUFACTURING AND COMPOUNDING AREAS Security Requirements; Registration Of Manufacturers, Distributors, And Dispensers Of Controlled Substances
COMPOUNDERS	1301.74	OTHER SECURITY CONTROLS FOR NON-PRACTITIONERS; NARCOTIC TREATMENT PROGRAMS AND COMPOUNDERS FOR NARCOTIC TREATMENT PROGRAMS Security Requirements; Registration Of Manufacturers, Distributors, And Dispensers Of Controlled Substances
COMPOUNDING	530.13	EXTRALABEL USE FROM COMPOUNDING OF APPROVED NEW ANIMAL AND APPROVED HUMAN DRUGS Extralabel Drug Use In Animals
COMPOUNDING AREAS	1301.73	PHYSICAL SECURITY CONTROLS FOR NON-PRACTITIONERS; COMPOUNDERS FOR NARCOTIC TREATMENT PROGRAMS; MANUFACTURING AND COMPOUNDING AREAS Security Requirement; Registration Of Manufacturers, Distributors, And Dispensers Of Controlled Substances
COMPRESSIBLE	870.5800	COMPRESSIBLE LIMB SLEEVE Cardiovascular Therapeutic Devices; Cardiovascular Devices
COMPRESSOR	868.6250	PORTABLE AIR COMPRESSOR Miscellaneous; Anesthesiology Devices
COMPRESSOR	870.5200	EXTERNAL CARDIAC COMPRESSOR Cardiovascular Therapeutic Devices; Cardiovascular Devices
COMPUTATION OF TIME	10.20	SUBMISSION OF DOCUMENTS TO DOCKETS MANAGEMENT BRANCH; COMPUTATION OF TIME; AVAILABILITY FOR PUBLIC DISCLOSURE General Administrative Procedures; Administrative Practices And Procedures
COMPUTATION OF TIME	17.30	COMPUTATION OF TIME Civil Money Penalties Hearings
COMPUTATION OF TIME	810.3	COMPUTATION OF TIME Medical Device Recall Authority
COMPUTED TOMOGRAPHY	892.1200	EMISSION COMPUTED TOMOGRAPHY SYSTEM Diagnostic Devices; Radiology Devices
COMPUTED TOMOGRAPHY	892.1750	COMPUTED TOMOGRAPHY X-RAY SYSTEM Diagnostic Devices; Radiology Devices
COMPUTED TOMOGRAPHY	1020.33	COMPUTED TOMOGRAPHY (CT) EQUIPMENT Performance Standards For Ionizing Radiation Emitting Products
COMPUTER	868.1730	OXYGEN UPTAKE COMPUTER Diagnostic Devices; Anesthesiology Devices
COMPUTER	870.1110	BLOOD PRESSURE COMPUTER Cardiovascular Diagnostic Devices; Cardiovascular Devices
COMPUTER	870.1425	PROGRAMMABLE DIAGNOSTIC COMPUTER Cardiovascular Diagnostic Devices; Cardiovascular Devices
COMPUTER	870.1435	SINGLE-FUNCTION, PREPROGRAMMED DIAGNOSTIC COMPUTER Cardiovascular Diagnostic Devices; Cardiovascular Devices
CONCENTRATE	146.120	FROZEN CONCENTRATE FOR LEMONADE Canned Fruit Juices
CONCENTRATE	146.121	FROZEN CONCENTRATE FOR ARTIFICIALLY SWEETENED LEMONADE Canned Fruit Juices
CONCENTRATE	146.126	FROZEN CONCENTRATE FOR COLORED LEMONADE Canned Fruit Juices
CONCENTRATE	146.145	ORANGE JUICE FROM CONCENTRATE Canned Fruit Juices
CONCENTRATE	172.385	WHOLE FISH PROTEIN CONCENTRATE Special Dietary And Nutritional Additives; Food Additives Permitted For Direct Addition To Food For Human Consumption

CONCENTRATE	184.1979c	WHEY PROTEIN CONCENTRATE Listing Of Specific Substances Affirmed As GRAS; Direct Food Substances Affirmed As Generally Recognized As Safe
CONCENTRATE	520.2456	TIAMULIN LIQUID CONCENTRATE Oral Dosage Form New Animal Drugs
CONCENTRATE	864.9575	ENVIRONMENTAL CHAMBER FOR STORAGE OF PLATELET CONCENTRATE Products Used In Establishments That Manufacture Blood and Blood Products; Hematology And Pathology Devices
CONCENTRATED	169.176	CONCENTRATED VANILLA EXTRACT Food Dressings And Flavorings
CONCENTRATED	169.178	CONCENTRATED VANILLA FLAVORING Food Dressings And Flavorings
CONCENTRATED O.J.	146.146	FROZEN CONCENTRATED ORANGE JUICE Canned Fruit Juices
CONCENTRATED O.J.	146.148	REDUCED ACID FROZEN CONCENTRATED ORANGE JUICE Canned Fruit Juices
CONCENTRATED O.J.	146.150	CANNED CONCENTRATED ORANGE JUICE Canned Fruit Juices
CONCENTRATED O.J.	146.153	CONCENTRATED ORANGE JUICE FOR MANUFACTURING Canned Fruit Juices
CONCENTRATED O.J.	146.154	CONCENTRATED ORANGE JUICE WITH PRESERVATIVE Canned Fruit Juices
CONCENTRATED MILK	131.120	CONCENTRATED MILK Requirements For Specific Standardized Milk And Cream; Milk And Cream
CONCENTRATES	155.191	TOMATO CONCENTRATES Requirements For Specific Standardized Canned Vegetables; Canned Vegetables
CONCENTRATOR	862.2310	CLINICAL SAMPLE CONCENTRATOR Clinical Laboratory Instruments; Clinical Chemistry And Clinical Toxicology Devices
CONCLUSIONS	1316.64	PROPOSED FINDINGS OF FACT AND CONCLUSIONS OF LAW Administrative Hearings; Administrative Functions, Practices, And Procedures
CONCURRENT SYMPTOMS	41.70	LABELING OF OTC DRUG PRODUCTS CONTAINING INGREDIENTS USED FOR TREATING CONCURRENT SYMPTOMS Labeling; Cold, Cough, Allergy, Bronchodilator, And Antiasthmatic Drug Products For OTC Human Use
CONDENSATES	172.808	COPOLYMER CONDENSATES OF ETHYLENE OXIDE AND PROPYLENE OXIDE Multipurpose Additives; Food Additives Permitted For Direct Addition To Food For Human Consumption
CONDENSED ANIMAL PROTEIN	573.200	CONDENSED ANIMAL PROTEIN HYDROLYSATE Food Additive Listing; Food Additives Permitted In Feed And Drinking Water Of Animals
CONDENSED GLUTAMIC ACID	573.500	CONDENSED, EXTRACTED GLUTAMIC ACID FERMENTATION PRODUCT Food Additive Listing; Food Additives Permitted In Feed And Drinking Water Of Animals
CONDENSED WHEY	573.450	FERMENTED AMMONIATED CONDENSED WHEY Food Additive Listing; Food Additives Permitted In Feed And Drinking Water Of Animals
CONDENSED MILK	131.120	SWEETENED CONDENSED MILK Requirements For Specific Standardized Milk And Cream; Milk And Cream
CONDENSER	868.5375	HEAT AND MOISTURE CONDENSER (ARTIFICIAL NOSE) Therapeutic Devices; Anesthesiology Devices
CONDENSING LENS	886.1380	DIAGNOSTIC CONDENSING LENS Diagnostic Devices; Ophthalmic Devices
CONDITION	71.25	CONDITION FOR CERTIFICATION Administrative Action On Petitions; Color Additive Petitions
CONDITIONER	500.52	USE OF TERMS SUCH AS "TONIC," "TONE," "TONER," OR "CONDITIONER" IN THE LABELING OF PREPARATIONS INTENDED FOR USE IN OR ON ANIMALS Animal Drug Labeling Requirements; General
CONDITIONER	870.2050	BIOPOTENTIAL AMPLIFIER AND SIGNAL CONDITIONER Cardiovascular Monitoring Devices; Cardiovascular Devices
CONDITIONER	870.2060	TRANSDUCER SIGNAL AMPLIFIER AND SIGNAL CONDITIONER Cardiovascular Monitoring Devices; Cardiovascular Devices
CONDITIONER	872.3750	BRACKET ADHESIVE RESIN AND TOOTH CONDITIONER Prosthetic Devices; Dental Devices
CONDITIONER	872.3765	PIT AND FISSURE SEALANT AND CONDITIONER Prosthetic Devices; Dental Devices
CONDITIONER	882.1845	PHYSIOLOGICAL SIGNAL CONDITIONER Neurological Diagnostic Devices; Neurological Devices
CONDITIONING	882.5235	AVERSIVE CONDITIONING DEVICE Neurological Therapeutic Devices; Neurological Devices
CONDITIONS	107.50	TERMS AND CONDITIONS Exempt Infant Formulas; Infant Formula
CONDITIONS	310.500	DIGOXIN PRODUCTS FOR ORAL USE; CONDITIONS FOR MARKETING Requirements For Specific New Drugs Or Devices; New Drugs

CONDUCTION	868.5150	ANESTHESIA CONDUCTION NEEDLE Therapeutic Devices; Anesthesiology Devices
CONDUCTION	882.1550	NERVE CONDUCTION VELOCITY MEASUREMENT DEVICE Neurological Diagnostic Devices; Neurological Devices
CONDUCTORS	868.1920	ESOPHAGEAL STETHOSCOPE WITH ELECTRICAL CONDUCTORS Diagnostic Devices; Anesthesiology Devices
CONDYLE	872.3960	MANDIBULAR CONDYLE PROSTHESIS Prosthetic Devices; Dental Devices
CONFERENCE	12.91	TIME AND PLACE OF PREHEARING CONFERENCE Hearing Procedures; Formal Evidentiary Public Hearing
CONFERENCE	12.92	PREHEARING CONFERENCE PROCEDURE Hearing Procedures; Formal Evidentiary Public Hearing
CONFERENCE	1316.54	PREHEARING CONFERENCE Administrative Hearings; Administrative Functions, Practices, And Procedures
CONFERENCE	1316.57	SUBMISSION OF DOCUMENTARY EVIDENCE AND AFFIDAVITS AND IDENTIFICATION OF WITNESS SUBSEQUENT TO PREHEARING CONFERENCE Administrative Hearings; Administrative Functions, Practices, And Procedures
CONFERENCES	12.90	CONDUCT AT ORAL HEARINGS OR CONFERENCES Hearing Procedures; Formal Evidentiary Public Hearing
CONFERENCES	17.21	PREHEARING CONFERENCES Civil Money Penalties Hearings
CONFIDENCE BUILDING	26.37	CONFIDENCE BUILDING ACTIVITIES Specific Sector Provisions For Medical Devices; Mutual Recognition Of Pharmaceutical GMP Reports, Medical Device Quality System Audit Reports
CONFIDENTIAL INFORMATION	20.53	INDEXING TRADE SECRETS AND CONFIDENTIAL COMMERCIAL OR FINANCIAL INFORMATION Procedures And Fees; Public Information
CONFIDENTIAL INFORMATION	20.61	TRADE SECRETS AND COMMERCIAL OR FINANCIAL INFORMATION WHICH IS PRIVILEGED OR CONFIDENTIAL Exemptions; Public Information
CONFIDENTIAL RECORDS	20.27	SUBMISSION OF RECORDS MARKED AS CONFIDENTIAL General Policy; Public Information
CONFIDENTIALITY	20.28	FDA DETERMINATIONS OF CONFIDENTIALITY General Policy; Public Information
CONFIDENTIALITY	20.44	PRESUBMISSION REVIEW OF REQUEST FOR CONFIDENTIALITY OF VOLUNTARILY SUBMITTED DATA OR INFORMATION Procedures And Fees; Public Information
CONFIDENTIALITY	20.45	SITUATIONS IN WHICH CONFIDENTIALITY IS UNCERTAIN Procedures And Fees; Public Information
CONFIDENTIALITY	26.76	CONFIDENTIALITY Framework Provisions; Mutual Recognition Of Pharmaceutical GMP Reports, Medical Device Quality System Audit Reports
CONFIDENTIALITY	71.15	CONFIDENTIALITY OF DATA AND INFORMATION IN COLOR ADDITIVE PETITIONS General Provisions; Color Additive Petitions
CONFIDENTIALITY	514.11	CONFIDENTIALITY OF DATA AND INFORMATION IN A NEW ANIMAL DRUG APPLICATION FILE General Provisions; New Animal Drug Applications
CONFIDENTIALITY	514.12	CONFIDENTIALITY OF DATA AND INFORMATION IN AN INVESTIGATIONAL NEW ANIMAL DRUG NOTICE General Provisions; New Animal Drug Applications
CONFIDENTIALITY	601.50	CONFIDENTIALITY OF DATA AND INFORMATION IN AN INVESTIGATIONAL NEW DRUG NOTICE FOR A BIOLOGICAL PRODUCT Confidentiality Of Information; Licensing
CONFIDENTIALITY	601.51	CONFIDENTIALITY OF DATA AND INFORMATION IN APPLICATIONS FOR BIOLOGICS LICENSES Confidentiality Of Information; Licensing
CONFIDENTIALITY	720.8	CONFIDENTIALITY OF STATEMENTS Voluntary Filing Of Cosmetic Product Ingredient Composition Statements
CONFIDENTIALITY	807.95	CONFIDENTIALITY OF INFORMATION Premarket Notification Procedures; Establishment Registration And Device Listing For Manufacturers And Initial Importers Of Devices
CONFIDENTIALITY	809.4	CONFIDENTIALITY OF SUBMITTED INFORMATION General Provisions; In Vitro Diagnostic Products For Human Use
CONFIDENTIALITY	812.38	CONFIDENTIALITY OF DATA AND INFORMATION Application And Administrative Action; Investigational Device Exemptions
CONFIDENTIALITY	814.9	CONFIDENTIALITY OF DATA AND INFORMATION IN A PREMARKET APPROVAL APPLICATION (PMA) FILE General; Premarket Approval Of Medical Devices
CONFIDENTIALITY	814.122	CONFIDENTIALITY OF DATA AND INFORMATION Humanitarian Use Devices; Premarket Approval Of Medical Devices
CONFIDENTIALITY	821.55	CONFIDENTIALITY Records And Inspections; Medical Device Tracking Requirements

CONFIDENTIALITY	860.5	CONFIDENTIALITY AND USE OF DATA AND INFORMATION SUBMITTED IN CONNECTION WITH CLASSIFICATION AND RECLASSIFICATION General; Medical Device Classification Procedures
CONFIDENTIALITY	1002.4	CONFIDENTIALITY OF INFORMATION General Provisions; Records and Reports
CONFIDENTIALITY	1002.42	CONFIDENTIALITY OF RECORDS FURNISHED BY DEALERS AND DISTRIBUTORS Dealer And Distributor Records; Records and Reports
CONFIDENTIALITY	1316.23	CONFIDENTIALITY OF IDENTITY OF RESEARCH SUBJECTS Protection Of Researchers And Research Subjects; Administrative Functions, Practices, And Procedures
CONFIRMATION	5.301	ISSUANCE OF INITIAL EMERGENCY PERMIT ORDERS AND NOTICES OF CONFIRMATION OF EFFECTIVE DATE OF FINAL REGULATIONS ON FOOD FOR HUMAN AND ANIMAL CONSUMPTION Foods And Cosmetics; Redelegations Of Authority; Delegations Of Authority And Organization
CONFLICT OF INTEREST	5.30	AUTHORITY TO SELECT TEMPORARY VOTING MEMBERS FOR ADVISORY COMMITTEES AND SIGN CONFLICT OF INTEREST WAIVERS General Redelegations Of Authority; Delegations Of Authority And Organization
CONFLICT OF INTEREST	19.10	FDA CONFLICT OF INTEREST REVIEW BOARD General Provisions; Standards Of Conduct And Conflicts Of Interest
CONFORMER	886.3130	OPHTHALMIC CONFORMER Prosthetic Devices; Ophthalmic Devices
CONFORMITY	130.8	CONFORMITY TO DEFINITIONS AND STANDARDS OF IDENTITY General Provisions; Food Standards: General
CONFORMITY ASSESSMENT	26.67	SUSPENSION OF LISTED CONFORMITY ASSESSMENT BODIES Framework Provisions; Mutual Recognition Of Pharmaceutical GMP Reports, Medical Device Quality System Audit Reports
CONFORMITY ASSESSMENT	26.68	WITHDRAWAL OF LISTED CONFORMITY ASSESSMENT BODIES Framework Provisions; Mutual Recognition Of Pharmaceutical GMP Reports, Medical Device Quality System Audit Reports
CONFORMITY ASSESSMENT	26.69	MONITORING AND CONFORMITY ASSESSMENT BODIES Framework Provisions; Mutual Recognition Of Pharmaceutical GMP Reports, Medical Device Quality System Audit Reports
CONFORMITY ASSESSMENT	26.70	CONFORMITY ASSESSMENT BODIES Framework Provisions; Mutual Recognition Of Pharmaceutical GMP Reports, Medical Device Quality System Audit Reports
CONGENITAL	890.3665	CONGENITAL HIP DISLOCATION ABDUCTION SPLINT Physical Medicine Prosthetic Devices; Physical Medicine Devices
CONGRESS	14.10	APPLICABILITY TO CONGRESS General Provisions; Public Hearing Before A Public Advisory Committee
CONGRESS	20.87	DISCLOSURE TO CONGRESS Limitations On Exemptions; Public Information
CONJUGATED	862.1187	CONJUGATED SULFOLITHOCHOLIC ACID (SLCG) TEST SYSTEM Clinical Chemistry Test Systems; Clinical Chemistry And Clinical Toxicology Devices
CONJUGATES	862.1115	URINARY BILIRUBIN AND ITS CONJUGATES TEST SYSTEM Clinical Chemistry Test Systems; Clinical Chemistry And Clinical Toxicology Devices
CONNECTICUT	808.57	CONNECTICUT Listing Of Specific State And Local Exemptions; Exemption From Federal Preemption Of State & Local Medical Device Requirements
CONNECTOR	868.5810	AIRWAY CONNECTOR Therapeutic Devices; Anesthesiology Devices
CONNECTOR	870.2900	PATIENT TRANSDUCER AND ELECTRODE CABLE (INCLUDING CONNECTOR) Cardiovascular Monitoring Devices; Cardiovascular Devices
CONSENT	21.72	INDIVIDUAL CONSENT TO DISCLOSURE OF RECORDS TO OTHER PERSONS Disclosure Of Records In Privacy Act Record Systems To Persons Other ThanThe Subject Individual; Protection Of Privacy
CONSENT	1316.08	CONSENT TO INSPECTION Administrative Inspections; Administrative Functions, Practices, And Procedures
CONSENT, INFORMED	50.20	GENERAL REQUIREMENTS FOR INFORMED CONSENT Informed Consent Of Human Subjects; Protection Of Human Subjects
CONSENT, INFORMED	50.24	EXEMPTION FROM INFORMED CONSENT REQUIREMENTS FOR EMERGENCY RESEARCH Informed Consent Of Human Subjects; Protection Of Human Subjects
CONSENT, INFORMED	50.25	ELEMENTS OF INFORMED CONSENT Informed Consent Of Human Subjects; Protection Of Human Subjects
CONSENT, INFORMED	50.27	DOCUMENTATION OF INFORMED CONSENT Informed Consent Of Human Subjects; Protection Of Human Subjects
CONSENT, INFORMED	640.61	INFORMED CONSENT Source Plasma; Additional Standards For Human Blood And Blood Products

CONSIDERATIONS	556.1	GENERAL CONSIDERATIONS; TOLERANCES FOR RESIDUES OF NEW ANIMAL DRUGS IN FOOD General Provisions; Tolerances For Residues Of New Animal Drugs In Food
CONSIDERATIONS	558.3	DEFINITIONS AND GENERAL CONSIDERATIONS APPLICABLE TO THIS PART General Provisions; New Animal Drugs For Use In Animal Feeds
CONSIGNEES	510.7	CONSIGNEES OF NEW ANIMAL DRUGS FOR USE IN THE MANUFACTURE OF ANIMAL FEED General Provisions; New Animal Drugs
CONSOLE	870.4220	CARDIOPULMONARY BYPASS HEART-LUNG MACHINE CONSOLE Cardiovascular Surgical Devices; Cardiovascular Devices
CONSPICUOUSNESS, WARNING	369.10	CONSPICUOUSNESS OF WARNING STATEMENTS Definitions And Interpretations; Interpretative Statements Re Warnings On Drugs And Devices For OTC Sale
CONSPICUOUSNESS, WARNING	740.2	CONSPICUOUSNESS OF WARNING STATEMENTS General; Cosmetic Product Warning Statements
CONSTANT TEMPERATURE	1250.42	WATER SYSTEMS; CONSTANT TEMPERATURE BOTTLES Equipment And Operation Of Land And Air Conveyances; Interstate Conveyance Sanitation
CONSTANT TEMPERATURE	1250.85	DRINKING FOUNTAINS AND COOLERS; ICE; CONSTANT TEMPERATURE BOTTLES Sanitation Facilities And Conditions On Vessels; Interstate Conveyance Sanitation
CONSTITUENT	610.15	CONSTITUENT MATERIALS General Provisions; General Biological Products Standards
CONSTRAINT	888.6	DEGREE OF CONSTRAIN General Provisions; Orthopedic Devices
CONSTRUCTION	129.20	PLANT CONSTRUCTION AND DESIGN Buildings And Facilities; Processing And Bottling Of Bottled Drinking Water
CONSTRUCTION	211.42	DESIGN AND CONSTRUCTION FEATURES Buildings And Facilities; Current Good Manufacturing Practice For Finished Pharmaceuticals
CONSTRUCTION	211.65	EQUIPMENT CONSTRUCTION Equipment And Facilities; Current Good Manufacturing Practice For Finished Pharmaceuticals
CONSTRUCTION	1250.30	CONSTRUCTION, MAINTENANCE AND USE OF PLACES WHERE FOOD IS PREPARED, SERVED, OR STORED Food Service Sanitation On Land And Air Conveyances, And Vessels; Interstate Conveyance Sanitation
CONSTRUCTION	1250.41	SUBMITTAL OF CONSTRUCTION PLANS Equipment And Operation Of Land And Air Conveyances; Interstate Conveyance Sanitation
CONSTRUCTION	1250.62	SUBMITTAL OF CONSTRUCTION PLANS Servicing Areas For Land And Air Conveyances; Interstate Conveyance Sanitation
CONSULTANTS	20.84	DISCLOSURE TO CONSULTANTS, ADVISORY COMMITTEES, STATE AND LOCAL GOVT OFFICIALS AND OTHER SPECIAL GOVERNMENT EMPLOYEES Limitations On Exemptions; Public Information
CONSULTANTS	211.34	CONSULTANTS Organization And Personnel; Current Good Manufacturing Practice For Finished Pharmaceuticals
CONSULTATION	14.31	CONSULTATION BY AN ADVISORY COMMITTEE WITH OTHER PERSONS Meeting Procedures; Public Hearing Before A Public Advisory Committee
CONSULTATION	312.82	EARLY CONSULTATION Drugs Intended To Treat Life-threatening and Severely-debilitating Illnesses; Investigational New Drug Application
CONSULTATION	860.125	CONSULTATION WITH PANELS Reclassification; Medical Device Classification Procedures
CONSULTING LABORATORIES	200.10	CONTRACT FACILITIES (INCLUDING CONSULTING LABORATORIES) UTILIZED AS EXTRAMURAL FACILITIES BY PHARMACEUTICAL MANUFACTURERS General Provisions; General
CONSUMERS	200.200	PRESCRIPTION DRUGS; REMINDER ADVERTISEMENTS AND REMINDER LABELING TO PROVIDE PRICE INFORMATION TO CONSUMERS Prescription Drug Consumer Price Listing; General
CONSUMPTION	864.7720	PROTHROMBIN CONSUMPTION TEST Hematology Kits And Packages; Hematology And Pathology Devices
CONSUMPTION, HUMAN	170.50	GLYCINE (AMINOACETIC ACID) IN FOOD FOR HUMAN CONSUMPTION Specific Administrative Rulings And Decisions; Food Additives
CONSUMPTION, HUMAN	172.894	MODIFIED COTTONSEED PRODUCTS INTENDED FOR HUMAN CONSUMPTION Multipurpose Additives; Food Additives Permitted For Direct Addition To Food For Human Consumption
CONSUMPTION, HUMAN	1240.61	MANDATORY PASTEURIZATION FOR ALL MILK AND MILK PRODUCTS IN FINAL PACKAGE FORM INTENDED FOR DIRECT HUMAN CONSUMPTION Specific Administrative Decisions Regarding Interstate Shipments; Control Of Communicable Diseases

CONSUMPTION, HUMAN/ANIMAL	5.301	ISSUANCE OF INITIAL EMERGENCY PERMIT ORDERS AND NOTICES OF CONFIRMATION OF EFFECTIVE DATE OF FINAL REGULATIONS ON FOOD FOR HUMAN AND ANIMAL CONSUMPTION Foods And Cosmetics; Redelegations Of Authority; Delegations Of Authority And Organization
CONTACT	176.170	COMPONENTS OF PAPER AND PAPERBOARD IN CONTACT WITH AQUEOUS AND FATTY FOODS Indirect Food Additives: Paper And Paperboard Components
CONTACT	176.180	COMPONENTS OF PAPER AND PAPERBOARD IN CONTACT WITH DRY FOOD Indirect Food Additives: Paper And Paperboard Components
CONTACT, FOOD	170.39	THRESHOLD OF REGULATION FOR SUBSTANCES USED IN FOOD-CONTACT ARTICLES Food Additives
CONTACT, FOOD	171.8	THRESHOLD OF REGULATIONS FOR SUBSTANCES USED IN FOOD-CONTACT ARTICLES Food Additive Petitions
CONTACT, FOOD	174.6	THRESHOLD OF REGULATION FOR SUBSTANCES USED IN FOOD-CONTACT ARTICLES Indirect Food Additives: General
CONTACT, FOOD	178.3570	LUBRICANTS WITH INCIDENTAL FOOD CONTACT Indirect Food Additives: Adjuvants, Production Aids, And Sanitizers
CONTACT LENS	800.10	CONTACT LENS SOLUTIONS; STERILITY Requirements For Specific Medical Devices; General
CONTACT LENS	800.12	CONTACT LENS SOLUTIONS AND TABLETS; TAMPER-RESISTANT PACKAGING Requirements For Specific Medical Devices; General
CONTACT LENS	886.1385	POLYMETHYLMETHACRYLATE (PMMA) DIAGNOSTIC CONTACT LENS Diagnostic Devices; Ophthalmic Devices
CONTACT LENS	886.1430	OPHTHALMIC CONTACT LENS RADIUS MEASURING DEVICE Diagnostic Devices; Ophthalmic Devices
CONTACT LENS	886.5420	CONTACT LENS INSERTER/REMOVER Therapeutic Devices; Ophthalmic Devices
CONTACT LENS	886.5916	RIGID GAS PERMEABLE CONTACT LENS Therapeutic Devices; Ophthalmic Devices
CONTACT LENS	886.5918	RIGID GAS PERMEABLE CONTACT LENS CARE PRODUCTS Therapeutic Devices; Ophthalmic Devices
CONTACT LENS	886.5925	SOFT (HYDROPHILIC) CONTACT LENS Therapeutic Devices; Ophthalmic Devices
CONTACT LENS	886.5928	SOFT (HYDROPHILIC) CONTACT LENS CARE PRODUCTS Therapeutic Devices; Ophthalmic Devices
CONTACT POINTS, SECTORAL	26.72	SECTORAL CONTACT POINTS Framework Provisions; Mutual Recognition Of Pharmaceutical GMP Reports, Medical Device Quality System Audit Reports
CONTACTS, EX PARTE	17.20	EX PARTE CONTACTS Civil Money Penalties Hearings
CONTAINER	130.14	GENERAL STATEMENTS OF SUBSTANDARD QUALITY AND SUBSTANDARD FILL OF CONTAINER General Provisions; Food Standards: General
CONTAINER	606.121	CONTAINER LABEL Finished Product Control; Current Good Manufacturing Practice For Blood And Blood Components
CONTAINER	610.60	CONTAINER LABEL Labeling Standards; General Biological Products Standards
CONTAINER	864.3250	SPECIMEN TRANSPORT AND STORAGE CONTAINER Pathology Instrumentation And Accessories; Hematology And Pathology Devices
CONTAINER	864.9100	EMPTY CONTAINER FOR THE COLLECTION AND PROCESSING OF BLOOD AND BLOOD COMPONENTS Products Used in Establishments That Manufacture Blood And Blood Products; Hematology And Pathology Devices
CONTAINER	880.5025	I.V. CONTAINER General Hospital And Personal Use Therapeutic Devices; General Hospital And Personal Use Devices
CONTAINER	880.6175	SPECIMEN CONTAINER General Hospital And Personal Use Miscellaneous Devices; General Hospital And Personal Use Devices
CONTAINER, DRUG	211.184	COMPONENT, DRUG PRODUCT CONTAINER, CLOSURE, AND LABELING RECORDS Records And Reports; Current Good Manufacturing Practice For Finished Pharmaceuticals
CONTAINERS	100.100	MISLEADING CONTAINERS Misbranding For Reasons Other Than Labeling; General
CONTAINERS	113.60	CONTAINERS Control Of Components, Food Product Containers, Closures & In-process Material; Thermally Processed Low-acid Foods Packaged In Hermetically Sealed Containers

CONTAINERS	130.12	GENERAL METHODS FOR WATER CAPACITY AND FILL OF CONTAINERS General Provisions; Food Standards: General
CONTAINERS	161.173	CANNED WET PACKED SHRIMP IN TRANSPARENT OR NONTRANSPARENT CONTAINERS Requirements For Specific Standardized Fish And Shellfish; Fish And Shellfish
CONTAINERS	164.120	SHELLED NUTS IN RIGID OR SEMIRIGID CONTAINERS Requirements For Specific Standardized Tree Nut And Peanut Products; Tree Nut And Peanut Products
CONTAINERS	177.1210	CLOSURES WITH SEALING GASKETS FOR FOOD CONTAINERS Substances For Use As Basic Components Of Single And Repeated Use Food Contact Surfaces; Indirect Food Additives: Polymers
CONTAINERS	310.509	PARENTERAL DRUG PRODUCTS IN PLASTIC CONTAINERS Requirements For Specific New Drugs Or Devices; New Drugs
CONTAINERS	500.65	EPINEPHRINE INJECTION 1:1,000 IN 10-MILLILITER CONTAINERS FOR EMERGENCY TREATMENT OF ANAPHYLACTOID SHOCK IN CATTLE, HORSES, SHEEP, AND SWINE Requirements For Specific Animal Drugs; General
CONTAINERS	1230.41	DELIVERY OF CONTAINERS Imports; Regulations Under The Federal Caustic Poison Act
CONTAINERS	1230.47	REJECTED CONTAINERS Imports; Regulations Under The Federal Caustic Poison Act
CONTAINERS	1230.48	RELABELING OF CONTAINERS Imports; Regulations Under The Federal Caustic Poison Act
CONTAINERS, DRUG	211.82	RECEIPT AND STORAGE OF UNTESTED COMPONENTS, DRUG PRODUCT CONTAINERS, AND CLOSURES Control Of Components And Drug Product Containers And Closures; Current Good Manufacturing Practice For Finished Pharmaceuticals
CONTAINERS, DRUG	211.84	TESTING AND APPROVAL OR REJECTION OF COMPONENTS, DRUG PRODUCT CONTAINERS, AND CLOSURES Control Of Components And Drug Product Containers And Closures; Current Good Manufacturing Practice For Finished Pharmaceuticals
CONTAINERS, DRUG	211.86	USE OF APPROVED COMPONENTS, DRUG PRODUCT CONTAINERS, AND CLOSURES Control Of Components And Drug Product Containers And Closures; Current Good Manufacturing Practice For Finished Pharmaceuticals
CONTAINERS, DRUG	211.8	RETESTING OF APPROVED COMPONENTS, DRUG PRODUCT CONTAINERS, AND CLOSURES Control Of Components And Drug Product Containers And Closures; Current Good Manufacturing Practice For Finished Pharmaceuticals
CONTAINERS, DRUG	211.89	REJECTED COMPONENTS, DRUG PRODUCT CONTAINERS, AND CLOSURES Control Of Components And Drug Product Containers And Closures; Current Good Manufacturing Practice For Finished Pharmaceuticals
CONTAINERS, DRUG	211.94	DRUG PRODUCT CONTAINERS AND CLOSURES Control Of Components And Drug Product Containers And Closures; Current Good Manufacturing Practice For Finished Pharmaceuticals
CONTAINERS, SEALED	108.35	THERMALLY PROCESSED LOW-ACID FOODS PACKAGED IN HERMETICALLY SEALED CONTAINERS Specific Requirements And Conditions For Exemption From Or Compliance With An Emergency Permit; Emergency Permit Control
CONTAINERS, SEALED	500.23	THERMALLY PROCESSED LOW-ACID FOODS PACKAGED IN HERMETICALLY SEALED CONTAINERS Specific Administrative Rulings And Decisions; General
CONTAINERS, SECONDHAND	2.35	USE OF SECONDHAND CONTAINERS FOR THE SHIPMENT AND STORAGE OF FOOD AND ANIMAL FEED Human And Animal Foods; General Administrative Rulings And Decisions
CONTAINERS, SELF-PRESSURIZED	2.125	USE OF CHLOROFLUOROCARBON PROPELLANTS IN SELF-PRESSURIZED CONTAINERS Specific Provisions; General Administrative Rulings And Decisions
CONTAINERS, SELF-PRESSURIZED	740.11	COSMETICS IN SELF-PRESSURIZED CONTAINERS Warning Statements, Cosmetic Product Warning Statements
CONTAMINATED	500.35	ANIMAL FEEDS CONTAMINATED WITH SALMONELLA MICROORGANISMS Specific Administrative Rulings And Decisions; General
CONTAMINATION, MICROBIO.	211.113	CONTROL OF MICROBIOLOGICAL CONTAMINATION Production And Process Controls; Current Good Manufacturing Practice For Finished Pharmaceuticals
CONTAMINATION, PENICILLIN	211.176	PENICILLIN CONTAMINATION Laboratory Controls; Current Good Manufacturing Practice For Finished Pharmaceuticals
CONTENT	60.30	FILING, FORMAT, AND CONTENT OF PETITIONS Due Diligence Petitions; Patent Term Restoration
CONTENT	201.56	GENERAL REQUIREMENTS ON CONTENT AND FORMAT OF LABELING FOR HUMAN PRESCRIPTION DRUGS Labeling Requirements For Prescription Drugs And/or Insulin; Labeling

CONTENT	201.57	SPECIFIC REQUIREMENTS ON CONTENT AND FORMAT OF LABELING FOR HUMAN PRESCRIPTION DRUGS Labeling Requirements For Prescription Drugs And/or Insulin; Labeling
CONTENT	201.66	FORMAT AND CONTENT REQUIREMENTS FOR OTC DRUG PRODUCT LABELING Labeling Requirements For Over-the-counter Drugs; Labeling
CONTENT	208.20	CONTENT AND FORMAT OF A MEDICATION GUIDE Medication Guides For Prescription Drug Products
CONTENT	312.23	IND CONTENT AND FORMAT Investigational New Drug Application
CONTENT	314.50	CONTENT AND FORMAT OF AN APPLICATION Applications; Applications For FDA Approval To Market A New Drug
CONTENT	314.94	CONTENT AND FORMAT OF AN ABBREVIATED APPLICATION Abbreviated Applications; Applications For FDA Approval To Market A New Drug
CONTENT	316.10	CONTENT AND FORMAT OF A REQUEST FOR WRITTEN RECOMMENDATIONS Orphan Drugs
CONTENT	316.20	CONTENT AND FORMAT OF A REQUEST FOR ORPHAN-DRUG DESIGNATION Orphan Drugs
CONTENT	821.25	DEVICE TRACKING SYSTEM AND CONTENT REQUIREMENTS: MANUFACTURER REQUIREMENTS Medical Device Tracking Requirements
CONTENT	860.123	RECLASSIFICATION PETITION: CONTENT AND FORM Reclassification; Medical Device Classification Procedures
CONTENT	1310.06	CONTENT OF RECORDS AND REPORTS Records and Reports Of Listed Chemicals And Certain Machines
CONTENT, 510(K)	807.92	CONTENT AND FORMAT OF A 510(K) SUMMARY Establishment Registration And Device Listing For Manufacturers And Initial Importers Of Devices
CONTENT, 510(K)	807.93	CONTENT AND FORMAT OF A 510(K) STATEMENT Establishment Registration And Device Listing For Manufacturers And Initial Importers Of Devices
CONTENT, CALORIE	101.60	NUTRIENT CONTENT CLAIMS FOR THE CALORIE CONTENT OF FOODS Specific Requirements For Nutrient Content Claims; Food Labeling
CONTENT, CHOLESTEROL	101.62	NUTRIENT CONTENT CLAIMS FOR FAT, FATTY ACID, AND CHOLESTEROL CONTENT OF FOODS Specific Requirements For Nutrient Content Claims; Food Labeling
CONTENT, NUTRIENT	5.300	FOOD STANDARDS, FOOD ADDITIVES, GRAS SUBSTANCES, COLOR ADDITIVES, NUTRIENT CONTENT CLAIMS, AND HEALTH CLAIMS Foods And Cosmetics; Redelegations Of Authority; Delegations Of Authority And Organization
CONTENT, NUTRIENT	101.13	NUTRIENT CONTENT CLAIMS - GENERAL PRINCIPLES General Provisions; Food Labeling
CONTENT, NUTRIENT	101.54	NUTRIENT CONTENT CLAIMS FOR "GOOD SOURCE," "HIGH," "MORE," AND "HIGH POTENCY" Specific Requirements For Nutrient Content Claims; Food Labeling
CONTENT, NUTRIENT	101.56	NUTRIENT CONTENT CLAIMS FOR "LIGHT" OR "LITE" Specific Requirements For Nutrient Content Claims; Food Labeling
CONTENT, NUTRIENT	101.60	NUTRIENT CONTENT CLAIMS FOR THE CALORIE CONTENT OF FOODS Specific Requirements For Nutrient Content Claims; Food Labeling
CONTENT, NUTRIENT	101.61	NUTRIENT CONTENT CLAIMS FOR THE SODIUM CONTENT OF FOODS Specific Requirements For Nutrient Content Claims; Food Labeling
CONTENT, NUTRIENT	101.62	NUTRIENT CONTENT CLAIMS FOR FAT, FATTY ACID, AND CHOLESTEROL CONTENT OF FOODS Specific Requirements For Nutrient Content Claims; Food Labeling
CONTENT, NUTRIENT	101.65	IMPLIED NUTRIENT CONTENT CLAIMS AND RELATED LABEL STATEMENTS Specific Requirements For Nutrient Content Claims; Food Labeling
CONTENT, NUTRIENT	101.67	USE OF NUTRIENT CONTENT CLAIMS FOR BUTTER Specific Requirements For Nutrient Content Claims; Food Labeling
CONTENT, NUTRIENT	101.69	PETITIONS FOR NUTRIENT CONTENT CLAIMS Specific Requirements For Nutrient Content Claims; Food Labeling
CONTENT, NUTRIENT	130.10	REQUIREMENTS FOR FOODS NAMED BY USE OF A NUTRIENT CONTENT CLAIM AND A STANDARDIZED TERM Food Standards: General
CONTENT, SODIUM	101.61	NUTRIENT CONTENT CLAIMS FOR THE SODIUM CONTENT OF FOODS Specific Requirements For Nutrient Content Claims; Food Labeling
CONTENTS	101.105	DECLARATION OF NET QUANTITY OF CONTENTS WHEN EXEMPT Exemptions From Food Labeling Requirements; Food Labeling
CONTENTS	161.30	DECLARATION OF QUANTITY OF CONTENTS ON LABELS FOR CANNED OYSTERS General Provisions; Fish And Shellfish
CONTENTS	201.51	DECLARATION OF NET QUANTITY OF CONTENTS Labeling Requirements For Prescription Drugs And/or Insulin; Labeling

CONTENTS	201.62	DECLARATION OF NET QUANTITY OF CONTENTS Labeling Requirements For Over-the-counter Drugs; Labeling
CONTENTS	501.105	DECLARATION OF NET QUANTITY OF CONTENTS WHEN EXEMPT Exemptions From Animal Food Labeling Requirements; Animal Food Labeling
CONTENTS	514.200	CONTENTS OF NOTICE OF OPPORTUNITY FOR A HEARING Hearing Procedures; New Animal Drug Applications
CONTENTS	515.30	CONTENTS OF NOTICE OF OPPORTUNITY FOR A HEARING Hearing Procedures; Medicated Feed Mill License
CONTENTS	701.13	DECLARATION OF NET QUANTITY OF CONTENTS Package Form; Cosmetic Labeling
CONTENTS	801.62	DECLARATION OF NET QUANTITY OF CONTENTS Labeling Requirements For Over-the-counter Devices; Labeling
CONTENTS	861.7	CONTENTS OF STANDARDS General; Procedures For Performance Standards Development
CONTENTS	1301.13	APPLICATION FOR REGISTRATION; APPLICATION FORMS, FEES, CONTENTS, AND SIGNATURE Registration Of Manufacturers, Distributors, And Dispensers Of Controlled Substances
CONTENTS	1309.32	APPLICATION FORMS; CONTENTS; SIGNATURE Application For Registration; Registration Of Manufacturers, Distributors, Importers And Exporters Of List I Chemicals
CONTENTS	1312.18	CONTENTS OF IMPORT DECLARATION Importation Of Controlled Substances; Importation And Exportation Of Controlled Substances
CONTENTS	1312.27	CONTENTS OF SPECIAL CONTROLLED SUBSTANCES INVOICE Exportation Of Controlled Substances; Importation And Exportation Of Controlled Substances
CONTENTS	1313.13	CONTENTS OF IMPORT DECLARATION Importation Of Listed Chemicals; Importation And Exportation Of Precursors And Essential Chemicals
CONTENTS	1313.22	CONTENTS OF EXPORT DECLARATION Exportation Of Listed Chemicals; Importation And Exportation Of Precursors And Essential Chemicals
CONTENTS	1313.33	CONTENTS OF AN INTERNATIONAL TRANSACTION DECLARATION Shipments Involving Listed Chemicals; Exportation Of Listed Chemicals; Importation And Exportation Of Precursors And Essential Chemicals
CONTEST	1404.313	OPPORTUNITY TO CONTEST PROPOSED DEBARMENT Debarment; Governmentwide Debarment And Suspension And Requirements For Drug-Free Workplace
CONTEST	1404.412	OPPORTUNITY TO CONTEST SUSPENSION Suspension; Governmentwide Debarment And Suspension And Requirements For Drug-Free Workplace
CONTINENCE	876.5270	IMPLANTED ELECTRICAL URINARY CONTINENCE DEVICE Therapeutic Devices; Gastroenterology-urology Devices
CONTINENCE	876.5280	IMPLANTED MECHANICAL/HYDRAULIC URINARY CONTINENCE DEVICE Therapeutic Devices; Gastroenterology-urology Devices
CONTINENCE	876.5310	NONIMPLANTED, PERIPHERAL ELECTRICAL CONTINENCE DEVICE Therapeutic Devices; Gastroenterology-urology Devices
CONTINENCE	876.5320	NONIMPLANTED ELECTRICAL CONTINENCE DEVICE Therapeutic Devices; Gastroenterology-urology Devices
CONTINENT	876.5030	CONTINENT ILEOSTOMY CATHETER Therapeutic Devices; Gastroenterology-urology Devices
CONTINUATION	310.303	CONTINUATION OF LONG-TERM STUDIES, RECORDS, AND REPORTS ON CERTAIN DRUGS FOR WHICH NEW DRUG APPLICATIONS HAVE BEEN APPROVED Records And Reports; New Drugs
CONTINUATION	1404.220	CONTINUATION OF COVERED TRANSACTIONS Governmentwide Debarment And Suspension And Requirements For Drug-Free Workplace
CONTINUED EQUIVALENCE	26.15	MONITORING CONTINUED EQUIVALENCE Specific Sector Provisions For Pharmaceutical GMPs; Mutual Recognition Of Pharmaceutical GMP Reports, Medical Device Quality System Audit Reports
CONTINUED EQUIVALENCE	26.45	MONITORING CONTINUED EQUIVALENCE Specific Sector Provisions For Medical Devices; Mutual Recognition Of Pharmaceutical GMP Reports, Medical Device Quality System Audit Reports
CONTINUING RECORDS	1304.21	GENERAL REQUIREMENTS FOR CONTINUING RECORDS Continuing Records; Records And Reports Of Registrants
CONTINUING REVIEW	812.64	IRB'S CONTINUING REVIEW IRB Review And Approval; Investigational Device Exemptions
CONTINUOUS FLOW	862.2150	CONTINUOUS FLOW SEQUENTIAL MULTIPLE CHEMISTRY ANALYZER FOR CLINICAL USE Clinical Laboratory Instruments; Clinical Chemistry And Clinical Toxicology Devices
CONTINUOUS FLUSH	870.1210	CONTINUOUS FLUSH CATHETER Cardiovascular Diagnostic Devices; Cardiovascular Devices

CONTINUOUS VENTILATOR	868.5895	CONTINUOUS VENTILATOR Therapeutic Devices; Anesthesiology Devices
CONTRACEPTIVE	884.5350	CONTRACEPTIVE DIAPHRAGM AND ACCESSORIES Obstetrical And Gynecological Therapeutic Devices; Obstetrical And Gynecological Devices
CONTRACEPTIVE	884.5360	CONTRACEPTIVE INTRAUTERINE DEVICE (IUD) AND INTRODUCER Obstetrical And Gynecological Therapeutic Devices; Obstetrical And Gynecological Devices
CONTRACEPTIVE	884.5380	CONTRACEPTIVE TUBAL OCCLUSION DEVICE (TOD) AND INTRODUCER Obstetrical And Gynecological Therapeutic Devices; Obstetrical And Gynecological Devices
CONTRACEPTIVES	310.501	PATIENT PACKAGE INSERTS FOR ORAL CONTRACEPTIVES Requirements For Specific New Drugs Or Devices; New Drugs
CONTRACT	14.15	COMMITTEES WORKING UNDER A CONTRACT WITH FDA General Provisions; Public Hearing Before A Public Advisory Committee
CONTRACT	20.109	DATA AND INFORMATION OBTAINED BY CONTRACT Availability Of Specific Categories Of Records; Public Information
CONTRACT FACILITIES	200.10	CONTRACT FACILITIES (INCLUDING CONSULTING LABORATORIES) UTILIZED AS EXTRAMURAL FACILITIES BY PHARMACEUTICAL MANUFACTURERS General Provisions; General
CONTRACT RESEARCH ORG.	312.52	TRANSFER OF OBLIGATIONS TO A CONTRACT RESEARCH ORGANIZATION Investigational New Drug Application
CONTRACTION	884.2720	EXTERNAL UTERINE CONTRACTION MONITOR AND ACCESSORIES Obstetrical And Gynecological Monitoring Devices; Obstetrical And Gynecological Devices
CONTRACTOR, PRIVATE	20.51	USE OF PRIVATE CONTRACTOR FOR COPYING Procedures And Fees; Public Information
CONTRACTORS	20.90	DISCLOSURE TO CONTRACTORS Limitations On Exemptions; Public Information
CONTRACTORS	21.30	RECORDS OF CONTRACTORS Requirements For Specific Categories Of Records; Protection Of Privacy
CONTRACTS	58.10	APPLICABILITY TO STUDIES PERFORMED UNDER GRANTS AND CONTRACTS General Provisions; Good Laboratory Practice For Nonclinical Laboratory Studies
CONTRIBUTION	331.20	DETERMINATION OF PERCENT CONTRIBUTION OF ACTIVE INGREDIENTS Testing Procedures; Antacid Products For Over-the-counter (OTC) Human Use
CONTROL	113.89	DEVIATIONS IN PROCESSING, VENTING, OR CONTROL OF CRITICAL FACTORS Production And Process Controls; Thermally Processed Low-acid Foods Packaged In Hermetically Sealed Containers
CONTROL	211.113	CONTROL OF MICROBIOLOGICAL CONTAMINATION Current Good Manufacturing Practice For Finished Pharmaceuticals
CONTROL	312.61	CONTROL OF THE INVESTIGATIONAL DRUG Investigational New Drug Application
CONTROL	870.1290	STEERABLE CATHETER CONTROL SYSTEM Cardiovascular Diagnostic Devices; Cardiovascular Devices
CONTROL	870.3535	INTRA-AORTIC BALLOON AND CONTROL SYSTEM Cardiovascular Prosthetic Devices; Cardiovascular Devices
CONTROL	870.4300	CARDIOPULMONARY BYPASS GAS CONTROL UNIT Cardiovascular Surgical Devices; Cardiovascular Devices
CONTROL	870.4340	CARDIOPULMONARY BYPASS LEVEL SENSING MONITOR AND/OR CONTROL Cardiovascular Surgical Devices; Cardiovascular Devices
CONTROL	870.4380	CARDIOPULMONARY BYPASS PUMP SPEED CONTROL Cardiovascular Surgical Devices; Cardiovascular Devices
CONTROL	870.4430	CARDIOPULMONARY BYPASS INTRACARDIAC SUCTION CONTROL Cardiovascular Surgical Devices; Cardiovascular Devices
CONTROL	1308.46	CONTROL REQUIRED UNDER INTERNATIONAL TREATY Hearings; Schedules Of Controlled Substances
CONTROL	1308.47	CONTROL OF IMMEDIATE PRECURSORS Hearings; Schedules Of Controlled Substances
CONTROL ARTICLE	58.105	TEST AND CONTROL ARTICLE CHARACTERIZATION Test And Control Articles; Good Laboratory Practice For Nonclinical Laboratory Studies
CONTROL ARTICLE	58.107	TEST AND CONTROL ARTICLE HANDLING Test And Control Articles; Good Laboratory Practice For Nonclinical Laboratory Studies
CONTROL ARTICLES	58.47	FACILITIES FOR HANDLING TEST AND CONTROL ARTICLES Facilities; Good Laboratory Practice For Nonclinical Laboratory Studies
CONTROL, DANDRUFF	358.710	ACTIVE INGREDIENTS FOR THE CONTROL OF DANDRUFF, SEBORRHEIC DERMATITIS, OR PSORIASIS Miscellaneous External Drug Products For Over-the-counter Human Use
CONTROL, EMERGENCY PERMIT	500.24	EMERGENCY PERMIT CONTROL Specific Administrative Rulings And Decisions; General

CONTROL, IN-PROCESS	106.25	IN-PROCESS CONTROL Quality Control Procedures For Assuring Nutrient Content Of Infant Formulas; Infant Formula Quality Control Procedures
CONTROL, INGREDIENT	106.20	INGREDIENT CONTROL Quality Control Procedures For Assuring Nutrient Content Of Infant Formulas; Infant Formula Quality Control Procedures
CONTROL, INSECT	1250.95	INSECT CONTROL Sanitation Facilities And Conditions On Vessels; Interstate Conveyance Sanitation
CONTROL, LICENSE	310.	BIOLOGICS; PRODUCTS SUBJECT TO LICENSE CONTROL General Provisions; New Drugs
CONTROL, LICENSE	510.4	BIOLOGICS; PRODUCTS SUBJECT TO LICENSE CONTROL General Provisions; New Animal Drugs
CONTROL, LOCAL	1240.30	MEASURES IN THE EVENT OF INADEQUATE LOCAL CONTROL Control Of Communicable Diseases
CONTROL NUMBERS	201.18	DRUGS; SIGNIFICANCE OF CONTROL NUMBERS General Labeling Provisions; Labeling
CONTROL NUMBERS	203.38	SAMPLE LOT OR CONTROL NUMBERS Samples; Prescription Drug Marketing
CONTROL MATERIAL	862.3280	CLINICAL TOXICOLOGY CONTROL MATERIAL Clinical Toxicology Test Systems; Clinical Chemistry And Clinical Toxicology Devices
CONTROL POINT	120.8	HAZARD ANALYSIS AND CRITICAL CONTROL POINT (HACCP) PLAN Hazard Analysis And Critical Control Point (HACCP) Systems
CONTROL POINT	123.6	HAZARD ANALYSIS AND HAZARD ANALYSIS CRITICAL CONTROL POINT (HACCP) PLAN Fish And Fishery Products
CONTROL PROCEDURES	123.11	SANITATION CONTROL PROCEDURES Fish And Fishery Products
CONTROL PROCEDURES	226.40	PRODUCTION AND CONTROL PROCEDURES Product Quality Control; Current Good Manufacturing Practice For Type A Medicated Articles
CONTROL, QUALITY	106.1	STATUS AND APPLICABILITY OF THE QUALITY CONTROL PROCEDURES REGULATION General Provisions; Infant Formula Quality Control Procedures
CONTROL, QUALITY	211.22	RESPONSIBILITIES OF QUALITY CONTROL UNIT Organization And Personnel; Current Good Manufacturing Practice For Finished Pharmaceuticals
CONTROL, QUALITY	640.56	QUALITY CONTROL TEST FOR POTENCY Cryoprecipitate; Additional Standards For Human Blood And Blood Products
CONTROL, QUALITY	862.1660	QUALITY CONTROL MATERIAL (ASSAYED AND UNASSAYED) Clinical Chemistry Test Systems; Clinical Chemistry And Clinical Toxicology Devices
CONTROL, QUALITY	864.8625	HEMATOLOGY QUALITY CONTROL MIXTURE Hematology Reagents; Hematology And Pathology Devices
CONTROL, QUALITY	864.9650	QUALITY CONTROL KIT FOR BLOOD BANKING REAGENTS Products Used In Establishments That Manufacture Blood And Blood Products; Hematology And Pathology Devices
CONTROL, QUALITY	866.2480	QUALITY CONTROL KIT FOR CULTURE MEDIA Microbiology Devices; Immunology And Microbiology Devices
CONTROL, RADIATION	5.606	ACCEPTANCE OF ASSISTANCE FROM STATE AND LOCAL AUTHORITIES FOR ENFORCEMENT OF RADIATION CONTROL LEGISLATION AND REGULATIONS Radiation Control; Redelegations Of Authority; Delegations Of Authority And Organization
CONTROL RECORDS	211.186	MASTER PRODUCTION AND CONTROL RECORDS Records And Reports; Current Good Manufacturing Practice For Finished Pharmaceuticals
CONTROL RECORDS	211.188	BATCH PRODUCTION AND CONTROL RECORDS Records And Reports; Current Good Manufacturing Practice For Finished Pharmaceuticals
CONTROL, RODENT	1250.96	RODENT CONTROL Sanitation Facilities And Conditions On Vessels; Interstate Conveyance Sanitation
CONTROL SYSTEM	890.3725	POWERED ENVIRONMENTAL CONTROL SYSTEM Physical Medicine Prosthetic Devices; Physical Medicine Devices
CONTROLLED STUDIES	201.58	REQUESTS FOR WAIVER OF REQUIREMENT FOR ADEQUATE AND WELL-CONTROLLED STUDIES TO SUBSTANTIATE CERTAIN LABELING STATEMENTS Labeling Requirements For Prescription Drugs And/or Insulin; Labeling
CONTROLLED STUDIES	314.126	ADEQUATE AND WELL-CONTROLLED STUDIES FDA Action On Applications; Applications For FDA Approval To Market A New Drug
CONTROLLED STUDIES	514.117	ADEQUATE AND WELL-CONTROLLED STUDIES Administrative Actions On Applications; New Animal Drug Applications
CONTROLLED SUBSTANCE	1309.25	EXEMPTION OF CERTAIN CONTROLLED SUBSTANCE REGISTRANTS Registration Of Manufacturers, Distributors, Importers And Exporters Of List I Chemicals

CONTROLLED SUBSTANCES	290.1	CONTROLLED SUBSTANCES General Provisions; Controlled Drugs
CONTROLLED SUBSTANCES	312.69	HANDLING OF CONTROLLED SUBSTANCES Investigational New Drug Application
CONTROLLED SUBSTANCES	1300.01	DEFINITIONS RELATING TO CONTROLLED SUBSTANCES Definitions
CONTROLLED SUBSTANCES	1302.06	SEALING OF CONTROLLED SUBSTANCES Labeling And Packaging Requirements For Controlled Substances
CONTROLLED SUBSTANCES	1306.25	TRANSFER BETWEEN PHARMACIES OF PRESCRIPTION INFORMATION FOR SCHEDULES III, IV, AND V CONTROLLED SUBSTANCES FOR REFILL PURPOSES Controlled Substances Listed In Schedules III, IV, And V; Prescriptions
CONTROLLED SUBSTANCES	1307.13	INCIDENTAL MANUFACTURE OF CONTROLLED SUBSTANCES Special Exemptions For Manufacture And Distribution Of Controlled Substances; Miscellaneous
CONTROLLED SUBSTANCES	1307.21	PROCEDURE FOR DISPOSING OF CONTROLLED SUBSTANCES Disposal Of Controlled Substances; Miscellaneous
CONTROLLED SUBSTANCES	1307.22	DISPOSAL OF CONTROLLED SUBSTANCES BY THE ADMINISTRATION Disposal Of Controlled Substances; Miscellaneous
CONTROLLED SUBSTANCES	1308.03	ADMINISTRATION CONTROLLED SUBSTANCES CODE NUMBER General Information; Schedules Of Controlled Substances
CONTROLLED SUBSTANCES	1312.27	CONTENTS OF SPECIAL CONTROLLED SUBSTANCES INVOICE Exportation Of Controlled Substances; Importation And Exportation Of Controlled Substances
CONTROLLED SUBSTANCES	1312.28	DISTRIBUTION OF SPECIAL CONTROLLED SUBSTANCES INVOICE Exportation Of Controlled Substances; Importation And Exportation Of Controlled Substances
CONTROLLED SUBSTANCES	1312.30	SCHEDULE III, IV, AND V NON-NARCOTIC CONTROLLED SUBSTANCES REQUIRING AN IMPORT AND EXPORT PERMIT Exportation Of Controlled Substances; Importation And Exportation Of Controlled Substances
CONTROLLER	870.4250	CARDIOPULMONARY BYPASS TEMPERATURE CONTROLLER Cardiovascular Surgical Devices; Cardiovascular Devices
CONTROLLING	179.21	SOURCES OF RADIATION USED FOR INSPECTION OF FOOD, FOR INSPECTION OF PACKAGED FOOD, AND FOR CONTROLLING FOOD PROCESSING Radiation And Radiation Sources; Irradiation In The Production, Processing And Handling Of Food
CONTROLS	11.10	CONTROLS FOR CLOSED SYSTEMS Electronic Records; Electronic Signatures
CONTROLS	11.30	CONTROLS FOR OPEN SYSTEMS Electronic Records; Electronic Signatures
CONTROLS	11.200	ELECTRONIC SIGNATURE COMPONENTS AND CONTROLS Electronic Records; Electronic Signatures
CONTROLS	11.300	CONTROLS FOR IDENTIFICATION CODES/PASSWORDS Electronic Records; Electronic Signatures
CONTROLS	110.37	SANITARY FACILITIES AND CONTROLS Buildings And Facilities; Current Good Manufacturing Practice In Manufacturing, Packing, Or Holding Human Food
CONTROLS	110.80	PROCESSES AND CONTROLS Production And Process Controls; Current Good Manufacturing Practice In Manufacturing, Packing, Or Holding Human Food
CONTROLS	114.80	PROCESSES AND CONTROLS Production And Process Controls; Acidified Foods
CONTROLS	129.80	PROCESSES AND CONTROLS Production And Process Controls; Processing And Bottling Of Bottled Drinking Water
CONTROLS	660.53	CONTROLS FOR SEROLOGICAL PROCEDURES Anti-human Globulin; Additional Standards For Diagnostic Substances For Laboratory Tests
CONTROLS, DESIGN	820.30	DESIGN CONTROLS Quality System Regulation
CONTROLS, DOCUMENT	820.40	DOCUMENT CONTROLS Quality System Regulation
CONTROLS, LABORATORY	225.58	LABORATORY CONTROLS Product Quality Control; Current Good Manufacturing Practice For Medicated Feeds
CONTROLS, LABORATORY	226.58	LABORATORY CONTROLS Product Quality Control; Current Good Manufacturing Practice For Type A Medicated Articles
CONTROLS, LABORATORY	606.140	LABORATORY CONTROLS Laboratory Controls; Current Good Manufacturing Practice For Blood And Blood Components
CONTROLS, PROCESS	120.24	PROCESS CONTROLS Pathogen Reduction; Hazard Analysis And Critical Control Point (HACCP) Systems

CONTROLS, PROCESS	123.16	PROCESS CONTROLS Smoked And Smoke-flavored Fishery Products; Fish And Fishery Products
CONTROLS, PROCESS	820.70	PRODUCTION AND PROCESS CONTROLS Quality System Regulation
CONTROLS, PURCHASING	820.50	PURCHASING CONTROLS Quality System Regulation
CONTROLS, SECURITY	1301.72	PHYSICAL SECURITY CONTROLS FOR NONPRACTITIONERS; NARCOTIC TREATMENT PROGRAMS AND COMPOUNDERS FOR NARCOTIC TREATMENT PROGRAMS; STORAGE AREAS Security Requirements; Registration Of Manufacturers, Distributors, And Dispensers Of Controlled Substances
CONTROLS, SECURITY	1301.73	PHYSICAL SECURITY CONTROLS FOR NON-PRACTITIONERS; COMPOUNDERS FOR NARCOTIC TREATMENT PROGRAMS; MANUFACTURING AND COMPOUNDING AREAS Security Requirements; Registration Of Manufacturers, Distributors, And Dispensers Of Controlled Substances
CONTROLS, SECURITY	1301.74	OTHER SECURITY CONTROLS FOR NON-PRACTITIONERS; NARCOTIC TREATMENT PROGRAMS AND COMPOUNDERS FOR NARCOTIC TREATMENT PROGRAMS Security Requirements; Registration Of Manufacturers, Distributors, And Dispensers Of Controlled Substances
CONTROLS, SECURITY	1301.75	PHYSICAL SECURITY CONTROLS FOR PRACTITIONERS Security Requirements; Registration Of Manufacturers, Distributors, And Dispensers Of Controlled Substances
CONTROLS, SECURITY	1301.76	OTHER SECURITY CONTROLS FOR PRACTITIONERS Security Requirements; Registration Of Manufacturers, Distributors, And Dispensers Of Controlled Substances
CONTROLS, SECURITY	1301.77	SECURITY CONTROLS FOR FREIGHT FORWARDING FACILITIES Security Requirements; Registration Of Manufacturers, Distributors, And Dispensers Of Controlled Substances
CONTROLS, SOURCE	123.28	SOURCE CONTROLS Raw Molluscan Shellfish; Fish And Fishery Products
CONVENIENCE	1403.44	TERMINATION FOR CONVENIENCE Reports, Records, Retention, and Enforcement, Uniform Administrative Requirements For Grants And Coop. Agreements To State And Local Gov.
CONVENIENCES	1240.70	EMPLOYEE CONVENIENCES Servicing Areas For Land And Air Conveyances; Interstate Conveyance Sanitation
CONVEYANCE	1316.95	PETITION FOR EXPEDITED RELEASE OF A CONVEYANCE IN A JUDICIAL FORFEITURE ACTION Expedited Forfeiture Proceedings For Certain Property; Administrative Functions, Practices, And Procedures
CONVEYANCE	1316.96	RULING ON A PETITION FOR EXPEDITED RELEASE OF A CONVEYANCE IN A JUDICIAL FORFEITURE ACTION Expedited Forfeiture Proceedings For Certain Property; Administrative Functions, Practices, And Procedures
CONVEYANCE	1316.97	INITIATING JUDICIAL FORFEITURE PROCEEDING AGAINST A CONVEYANCE WITHIN 60 DAYS OF FILING A CLAIM Expedited Forfeiture Proceedings For Certain Property; Administrative Functions, Practices, And Procedures
CONVEYANCE	1316.98	SUBSTITUTE RES BOND IN A JUDICIAL FORFEITURE ACTION AGAINST A CONVEYANCE Expedited Forfeiture Proceedings For Certain Property; Administrative Functions, Practices, And Procedures
CONVEYANCES	1240.95	APPROVAL OF TREATMENT ABOARD CONVEYANCES Source And Use Of Potable Water; Control Of Communicable Diseases
CONVEYANCES	1250.45	FOOD HANDLING FACILITIES ON RAILROAD CONVEYANCES Equipment And Operation Of Land And Air Conveyances; Interstate Conveyance Sanitation
CONVEYANCES	1250.49	CLEANLINESS OF CONVEYANCES Equipment And Operation Of Land And Air Conveyances; Interstate Conveyance Sanitation
CONVEYANCES	1250.51	RAILROAD CONVEYANCES; DISCHARGE OF WASTES Equipment And Operation Of Land And Air Conveyances; Interstate Conveyance Sanitation
CONVEYANCES	1250.52	DISCHARGE OF WASTES ON HIGHWAY CONVEYANCES Equipment And Operation Of Land And Air Conveyances; Interstate Conveyance Sanitation
CONVEYANCES	1250.53	DISCHARGE OF WASTES ON AIR CONVEYANCES Equipment And Operation Of Land And Air Conveyances; Interstate Conveyance Sanitation
CONVICTION, FELONY	1309.72	FELONY CONVICTION; EMPLOYER RESPONSIBILITIES Security Requirements; Registration Of Manufacturers, Distributors, Importers And Exporters Of List I Chemicals

CONVICTIONS	1404.635	REPORTING OF AND EMPLOYEE SANCTIONS FOR CONVICTIONS OF CRIMINAL DRUG OFFENSES Governmentwide Debarment And Suspension And Requirements For Drug-Free Workplace
COOK, CHEESE	133.127	COOK CHEESE, KOCH KAESE Requirements For Specific Standardized Cheese And Related Products; Cheeses And Related Cheese Products
COOKED COTTONSEED FLOUR	73.140	TOASTED PARTIALLY DEFATTED COOKED COTTONSEED FLOUR Foods; Listing Of Color Additives Exempt From Certification
COOLERS	1250.85	DRINKING FOUNTAINS AND COOLERS; ICE; CONSTANT TEMPERATURE BOTTLES Sanitation Facilities And Conditions On Vessels; Interstate Conveyance Sanitation
COOLING	211.46	VENTILATION, AIR FILTRATION, AIR HEATING AND COOLING Buildings And Facilities; Current Good Manufacturing Practice For Finished Pharmaceuticals
COOMBS TEST	864.9300	AUTOMATED COOMBS TEST SYSTEMS Products Used In Establishments That Manufacture Blood And Blood Products; Hematology And Pathology Devices
COOPERATING	25.44	LEAD AND COOPERATING AGENCIES Preparation Of Environmental Documents; Environmental Impact Considerations
COOPERATION, REGULATORY	26.49	REGULATORY COOPERATION Specific Sector Provisions For Medical Devices; Mutual Recognition Of Pharmaceutical GMP Reports, Medical Device Quality System Audit Reports
COOPERATIVE	20.114	DATA AND INFORMATION SUBMITTED PURSUANT TO COOPERATIVE QUALITY ASSURANCE AGREEMENTS Availability Of Specific Categories Of Records; Public Information
COOPERATIVE	56.114	COOPERATIVE RESEARCH IRB Functions And Operations; Institutional Review Boards
COPIES	5.22	CERTIFICATION OF TRUE COPIES AND USE OF DEPARTMENT SEAL General Redelegations Of Authority; Delegations Of Authority And Organization
COPIES	12.159	COPIES OF PETITIONS FOR JUDICIAL REVIEW Judicial Review; Formal Evidentiary Public Hearing
COPIES	510.305	MAINTENANCE OF COPIES OF APPROVED MEDICATED FEEL MILL LICENSES TO MANUFACTURE ANIMAL FEED BEARING OR CONTAINING NEW ANIMAL DRUGS Records And Reports; New Animal Drugs
COPIES	1003.22	COPIES OF COMMUNICATIONS SENT TO PURCHASERS, DEALERS, OR DISTRIBUTORS Notification Of Defects Or Failure To Comply
COPIES	1312.14	DISTRIBUTION OF COPIES OF IMPORT PERMIT Importation Of Controlled Substances; Importation And Exportation Of Controlled Substances
COPIES	1312.24	DISTRIBUTION OF COPIES OF EXPORT PERMIT Exportation Of Controlled Substances; Importation And Exportation Of Controlled Substances
COPIES	1316.68	COPIES OF PETITIONS FOR JUDICIAL REVIEW Administrative Hearings; Administrative Functions, Practices, And Procedures
COPOLYMER	172.775	METHACRYLIC ACID-DIVINYLBENZENE COPOLYMER Other Specific Usage Additives; Food Additives Permitted For Direct Addition To Food For Human Consumption
COPOLYMER	172.808	COPOLYMER CONDENSATES OF ETHYLENE OXIDE AND PROPYLENE OXIDE Multipurpose Additives; Food Additives Permitted For Direct Addition To Food For Human Consumption
COPOLYMER	173.60	DIMETHYLAMINE-EPICHLOROHYDRIN COPOLYMER Polymer Substances For Food Treatment; Secondary Direct Food Additives Permitted In Food For Human Consumption
COPOLYMER	173.65	DIVINYLBENZENE COPOLYMER Polymer Substances For Food Treatment; Secondary Direct Food Additives Permitted In Food For Human Consumption
COPOLYMER	175.210	ACRYLATE-ESTER COPOLYMER COATING Substances For Use As Components Of Coatings; Indirect Food Additives
COPOLYMER	175.350	VINYL ACETATE/CROTONIC ACID COPOLYMER Substances For Use As Components Of Coatings; Indirect Food Additives
COPOLYMER	175.360	VINYLIDENE CHLORIDE COPOLYMER COATINGS FOR NYLON FILM Substances For Use As Components Of Coatings; Indirect Food Additives
COPOLYMER	175.365	VINYLIDENE CHLORIDE COPOLYMER COATINGS FOR POLYCARBONATE FILM Substances For Use As Components Of Coatings; Indirect Food Additives
COPOLYMER	177.102	ACRYLONITRILE/BUTADIENE/STYRENE COPOLYMER Substances For Use As Basic Components Of Single And Repeated Use Food Contact Surfaces; Indirect Food Additives: Polymers

COPOLYMER	177.1030	ACRYLONITRILE/BUTADIENE/STYRENE/METHYL METHACRYLATE COPOLYMER Substances For Use As Basic Components Of Single And Repeated Use Food Contact Surfaces; Indirect Food Additives: Polymers
COPOLYMER	177.1040	ACRYLONITRILE/STYRENE COPOLYMER Substances For Use As Basic Components Of Single And Repeated Use Food Contact Surfaces; Indirect Food Additives: Polymers
COPOLYMER	177.1050	ACRYLONITRILE/STYRENE COPOLYMER MODIFIED WITH BUTADIENE/STYRENE ELASTOMER Substances For Use As Basic Components Of Single And Repeated Use Food Contact Surfaces; Indirect Food Additives: Polymers
COPOLYMER	177.1240	1,4-CYCLOHEXYLENE DIMETHYLENE TEREPHTHALATE AND 1,4-CYCLOHEXYLENE DIMETHYLENE ISOPHTHALATE COPOLYMER Substances For Use As Basic Components Of Single And Repeated Use Food Contact Surfaces; Indirect Food Additives: Polymers
COPOLYMER	177.1340	ETHYLENE-METHYL ACRYLATE COPOLYMER RESINS Substances For Use As Basic Components Of Single And Repeated Use Food Contact Surfaces; Indirect Food Additives: Polymers
COPOLYMER	177.1345	ETHYLENE/1,3-PHENYLENE OXYETHYLENE ISOPHTHALATE/TEREPHTHALATE COPOLYMER Substances For Use As Basic Components Of Single And Repeated Use Food Contact Surfaces; Indirect Food Additives: Polymers
COPOLYMER	177.2470	POLYOXYMETHYLENE COPOLYMER Substances For Use Only As Components Of Articles Intended For Repeated Use; Indirect Food Additives: Polymers
COPOLYMER	872.3500	POLYVINYLMETHYLETHER MALEIC ANHYDRIDE (PVM MA), ACID COPOLYMER, AND CARBOXYMETHYLCELLULOSE SODIUM (NACMC) DENTURE ADHESIVE Prosthetic Devices; Dental Devices
COPOLYMERS	73.3100	1,4-BIS[(2-HYDROXYETHYL)AMINO]-9,10-ANTHRACENEDIONE BIS(2-PROPENOIC)ESTER COPOLYMERS Medical Devices; Listing Of Color Additives Exempt From Certification
COPOLYMERS	73.3106	1,4-BIS[4-(2-METHACRYLOXYETHYL)PHENYLAMINO]ANTHRAQUINONE COPOLYMERS Medical Devices; Listing Of Color Additives Exempt From Certification
COPOLYMERS	73.3121	POLY(HYDROXYETHYL METHACRYLATE)-DYE COPOLYMERS Medical Devices; Listing Of Color Additives Exempt From Certification
COPOLYMERS	177.1060	N-ALKYLGLUTARIMIDE/ACRYLIC COPOLYMERS Substances For Use As Basic Components Of Single And Repeated Use Food Contact Surfaces; Indirect Food Additives: Polymers
COPOLYMERS	177.1211	CROSS-LINKED POLYACRYLATE COPOLYMERS Substances For Use As Basic Components Of Single And Repeated Use Food Contact Surfaces; Indirect Food Additives: Polymers
COPOLYMERS	177.1310	ETHYLENE-ACRYLIC ACID COPOLYMERS Substances For Use As Basic Components Of Single And Repeated Use Food Contact Surfaces; Indirect Food Additives: Polymers
COPOLYMERS	177.1312	ETHYLENE-CARBON MONOXIDE COPOLYMERS Substances For Use As Basic Components Of Single And Repeated Use Food Contact Surfaces; Indirect Food Additives: Polymers
COPOLYMERS	177.1315	ETHYLENE-1,4-CYCLOHEXYLENE DIMETHYLENE TEREPHTHALATE COPOLYMERS Substances For Use As Basic Components Of Single And Repeated Use Food Contact Surfaces; Indirect Food Additives: Polymers
COPOLYMERS	177.1320	ETHYLENE-ETHYL ACRYLATE COPOLYMERS Substances For Use As Basic Components Of Single And Repeated Use Food Contact Surfaces; Indirect Food Additives: Polymers
COPOLYMERS	177.1350	ETHYLENE-VINYL ACETATE COPOLYMERS Substances For Use As Basic Components Of Single And Repeated Use Food Contact Surfaces; Indirect Food Additives: Polymers
COPOLYMERS	177.1360	ETHYLENE-VINYL ACETATE-VINYL ALCOHOL COPOLYMERS Substances For Use As Basic Components Of Single And Repeated Use Food Contact Surfaces; Indirect Food Additives: Polymers
COPOLYMERS	177.1430	ISOBUTYLENE-BUTENE COPOLYMERS Substances For Use As Basic Components Of Single And Repeated Use Food Contact Surfaces; Indirect Food Additives: Polymers
COPOLYMERS	177.1480	NITRILE RUBBER MODIFIED ACRYLONITRILE-METHYL ACRYLATE COPOLYMERS Substances For Use As Basic Components Of Single And Repeated Use Food Contact Surfaces; Indirect Food Additives: Polymers
COPOLYMERS	177.1570	POLY-1-BUTENE RESINS AND BUTENE/ETHYLENE COPOLYMERS Substances For Use As Basic Components Of Single And Repeated Use Food Contact Surfaces; Indirect Food Additives: Polymers

COPOLYMERS	177.1820	STYRENE-MALEIC ANHYDRIDE COPOLYMERS Substances For Use As Basic Components Of Single And Repeated Use Food Contact Surfaces; Indirect Food Additives: Polymers
COPOLYMERS	177.1830	STYRENE-METHYL METHACRYLATE COPOLYMERS Substances For Use As Basic Components Of Single And Repeated Use Food Contact Surfaces; Indirect Food Additives: Polymers
COPOLYMERS	177.1950	VINYL CHLORIDE-ETHYLENE COPOLYMERS Substances For Use As Basic Components Of Single And Repeated Use Food Contact Surfaces; Indirect Food Additives: Polymers
COPOLYMERS	177.1960	VINYL CHLORIDE-HEXENE-1 COPOLYMERS Substances For Use As Basic Components Of Single And Repeated Use Food Contact Surfaces; Indirect Food Additives: Polymers
COPOLYMERS	177.1970	VINYL CHLORIDE-LAURYL VINYL ETHER COPOLYMERS Substances For Use As Basic Components Of Single And Repeated Use Food Contact Surfaces; Indirect Food Additives: Polymers
COPOLYMERS	177.1980	VINYL CHLORIDE-PROPYLENE COPOLYMERS Substances For Use As Basic Components Of Single And Repeated Use Food Contact Surfaces; Indirect Food Additives: Polymers
COPOLYMERS	177.1990	VINYLIDENE CHLORIDE/METHYL ACRYLATE COPOLYMERS Substances For Use As Basic Components Of Single And Repeated Use Food Contact Surfaces; Indirect Food Additives: Polymers
COPOLYMERS	177.2465	POLYMETHYLMETHACRYLATE/POLY(TRIMETHOXYSILYLPROPYL) METHACRYLATE COPOLYMERS Substances For Use Only As Components Of Articles Intended For Repeated Use; Indirect Food Additives: Polymers
COPOLYMERS	180.22	ACRYLONITRILE COPOLYMERS Specific Requirements For Certain Food Additives; Food Additives Permitted In Food Or In Contact With Food On Interim Basis Pending Study
COPOLYMERS	181.32	ACRYLONITIRLE COPOLYMERS AND RESINS Prior-sanctioned Food Ingredients
COPPER	73.2120	DISODIUM EDTA-COPPER Cosmetics; Listing Of Color Additives Exempt From Certification
COPPER	73.3110	CLOROPHYLLIN-COPPER COMPLEX, OIL SOLUBLE Medical Devices; Listing Of Color Additives Exempt From Certification
COPPER	74.3045	[PHTHALOCYANINATO(2-)] COPPER Listing Of Color Additives Subject To Certification
COPPER	862.1190	COPPER TEST SYSTEM Clinical Chemistry Test Systems; Clinical Chemistry And Clinical Toxicology Devices
COPPER CHLOROPHYLLIN	73.1125	POTASSIUM SODIUM COPPER CHLOROPHYLLIN (CHLOROPHYLLIN-COPPER COMPLEX) Drugs; Listing Of Color Additives Exempt From Certification
COPPER CHLOROPHYLLIN	73.2125	POTASSIUM SODIUM COPPER CHLOROPHYLLIN (CHLOROPHYLLIN-COPPER COMPLEX) Cosmetics; Listing Of Color Additives Exempt From Certification
COPPER GLUCONATE	184.1260	COPPER GLUCONATE Listing Of Specific Substances Affirmed As GRAS; Direct Food Substances Affirmed As Generally Recognized As Safe
COPPER GLUCONATE	582.5260	COPPER GLUCONATE Nutrients And/or Dietary Supplements; Substances Generally Recognized As Safe
COPPER NAPHTHENATE	524.463	COPPER NAPHTHENATE SOLUTION Ophthalmic And Topical Dosage Form New Animal Drugs
COPPER POWDER	73.1647	COPPER POWDER Drugs; Listing Of Color Additives Exempt From Certification
COPPER POWDER	73.2647	COPPER POWDER Cosmetics; Listing Of Color Additives Exempt From Certification
COPPER SULFATE	184.1261	COPPER SULFATE Listing Of Specific Substances Affirmed As GRAS; Direct Food Substances Affirmed As Generally Recognized As Safe
COPPER SULFATE	864.9320	COPPER SULFATE SOLUTION FOR SPECIFIC GRAVITY DETERMINATIONS Products Used In Establishments That Manufacture Blood And Blood Products; Hematology And Pathology Devices
COPYING	20.51	USE OF PRIVATE CONTRACTOR FOR COPYING Procedures And Fees; Public Information
COPYING	20.52	REQUEST FOR REVIEW WITHOUT COPYING Procedures And Fees; Public Information
COPYRIGHTS	1403.34	COPYRIGHTS Post-Award Requirements; Uniform Administrative Requirements For Grants And Coop. Agreements To State & Local Gov.
CORD, SPINAL	882.5850	IMPLANTED SPINAL CORD STIMULATOR FOR BLADDER EVACUATION Neurological Therapeutic Devices; Neurological Devices
CORD, SPINAL	882.5880	IMPLANTED SPINAL CORD STIMULATOR FOR PAIN RELIEF Neurological Therapeutic Devices; Neurological Devices

CORN	155.130	CANNED CORN Requirements For Specific Standardized Canned Vegetables; Canned Vegetables
CORN	155.131	CANNED FIELD CORN Requirements For Specific Standardized Canned Vegetables; Canned Vegetables
CORN FLOUR	137.211	WHITE CORN FLOUR Requirements For Specific Standardized Cereal Flours And Related Products; Cereal Flours And Related Products
CORN FLOUR	137.215	YELLOW CORN FLOUR Requirements For Specific Standardized Cereal Flours And Related Products; Cereal Flours And Related Products
CORN GLUTEN	184.1321	CORN GLUTEN Listing Of Specific Substances Affirmed As GRAS; Direct Food Substances Affirmed As Generally Recognized As Safe
CORN MEAL	137.250	WHITE CORN MEAL Requirements For Specific Standardized Cereal Flours And Related Products; Cereal Flours And Related Products
CORN MEAL	137.255	BOLTED WHITE CORN MEAL Requirements For Specific Standardized Cereal Flours And Related Products; Cereal Flours And Related Products
CORN MEAL	137.265	DEGERMINATED WHITE CORN MEAL Requirements For Specific Standardized Cereal Flours And Related Products; Cereal Flours And Related Products
CORN MEAL	137.270	SELF-RISING WHITE CORN MEAL Requirements For Specific Standardized Cereal Flours And Related Products; Cereal Flours And Related Products
CORN MEAL	137.275	YELLOW CORN MEAL Requirements For Specific Standardized Cereal Flours And Related Products; Cereal Flours And Related Products
CORN MEAL	137.280	BOLTED YELLOW CORN MEAL Requirements For Specific Standardized Cereal Flours And Related Products; Cereal Flours And Related Products
CORN MEAL	137.285	DEGERMINATED YELLOW CORN MEAL Requirements For Specific Standardized Cereal Flours And Related Products; Cereal Flours And Related Products
CORN MEAL	137.290	SELF-RISING YELLOW CORN MEAL Requirements For Specific Standardized Cereal Flours And Related Products; Cereal Flours And Related Products
CORN MEALS	137.260	ENRICHED CORN MEALS Requirements For Specific Standardized Cereal Flours And Related Products; Cereal Flours And Related Products
CORN OIL	73.315	CORN ENDOSPERM OIL Foods; Listing Of Color Additives Exempt From Certification
CORN REMOVER	358.510	CORN AND CALLUS REMOVER ACTIVE INGREDIENTS Corn And Callus Remover Drug Products; Miscellaneous External Drug Products For Over-the-counter Human Use
CORN REMOVER	358.550	LABELING OF CORN AND CALLUS REMOVER DRUG PRODUCTS Corn And Callus Remover Drug Products; Miscellaneous External Drug Products For Over-the-counter Human Use
CORN SILK	184.1262	CORN SILK AND CORN SILK EXTRACT Listing Of Specific Substances Affirmed As GRAS; Direct Food Substances Affirmed As Generally Recognized As Safe
CORN SUGAR	184.1857	CORN SUGAR Listing Of Specific Substances Affirmed As GRAS; Direct Food Substances Affirmed As Generally Recognized As Safe
CORN SYRUP	182.1866	HIGH FRUCTOSE CORN SYRUP Multiple Purpose GRAS Food Substances; Substances Generally Recognized As Safe
CORN SYRUP	184.1865	CORN SYRUP Listing Of Specific Substances Affirmed As GRAS; Direct Food Substances Affirmed As Generally Recognized As Safe
CORN SYRUP	573.530	HYDROGENATED CORN SYRUP Food Additive Listing; Food Additives Permitted In Feed And Drinking Water Of Animals
CORNEAL BURR	886.4070	POWERED CORNEAL BURR Surgical Devices; Ophthalmic Devices
CORNEAL ELECTRODE	886.1220	CORNEAL ELECTRODE Diagnostic Devices; Ophthalmic Devices
CORNEAL RADIUS	886.1450	CORNEAL RADIUS MEASURING DEVICE Diagnostic Devices; Ophthalmic Devices
CORONARY	101.75	HEALTH CLAIMS: DIETARY SATURATED FAT AND CHOLESTEROL AND RISK OF CORONARY HEART DISEASE Food Labeling

CORONARY	101.77	HEALTH CLAIMS: FRUITS, VEGETABLES, AND GRAIN PRODUCTS THAT CONTAIN FIBER, PARTICULARLY SOLUBLE FIBER, AND RISK OF CORONARY HEART DISEASE Food Labeling
CORONARY	101.81	HEALTH CLAIMS: SOLUBLE FIBER FROM CERTAIN FOODS AND RISK OF CORONARY HEART DISEASE Food Labeling
CORONARY	101.82	HEALTH CLAIMS: SOY PROTEIN AND RISK OF CORONARY HEART DISEASE Food Labeling
CORONARY	101.83	HEALTH CLAIMS: PLANT STEROL/STANOL ESTERS AND RISK OF CORONARY HEART DISEASE Food Labeling
CORONARY	250.102	DRUG PREPARATIONS INTENDED FOR HUMAN USE CONTAINING CERTAIN "CORONARY VASODILATORS" New Drug Or Prescription Status Of Specific Drugs; Specific Requirements For Specific Human Drugs
CORONARY	870.4310	CARDIOPULMONARY BYPASS CORONARY PRESSURE GAUGE Cardiovascular Surgical Devices; Cardiovascular Devices
CORRECTIONS	806.10	REPORTS OF CORRECTIONS AND REMOVALS Reports And Records; Medical Device Corrections And Removals
CORRECTIONS	806.20	RECORDS OF CORRECTIONS AND REMOVALS NOT REQUIRED TO BE REPORTED Reports And Records; Medical Device Corrections And Removals
CORRECTIONS	1316.63	OFFICIAL TRANSCRIPT; INDEX; CORRECTIONS Administrative Hearings; Administrative Functions, Practices, And Procedures
CORRECTIVE ACTION	820.100	CORRECTIVE AND PREVENTIVE ACTION Quality System Regulation
CORRECTIVE ACTIONS	99.401	CORRECTIVE ACTIONS AND CESSATION OF DISSEMINATION OF INFORMATION Dissemination Of Information On Unapproved/New Uses For Marketed Drugs, Biologics And Devices
CORRECTIVE ACTIONS	120.10	CORRECTIVE ACTIONS Hazard Analysis And Critical Control Point (HACCP) Systems
CORRECTIVE ACTIONS	123.7	CORRECTIVE ACTIONS Fish And Fishery Products
CORRELATION	320.28	CORRELATION OF BIOAVAILABILITY WITH AN ACUTE PHARMACOLOGICAL EFFECT OR CLINICAL EVIDENCE Procedures For Determining The Bioavailability Of Drug Products; Bioavailability And Bioequivalence Requirements
CORRESPONDENCE	10.65	MEETINGS AND CORRESPONDENCE General Administrative Procedures; Administrative Practices And Procedures
CORRESPONDENCE	20.103	CORRESPONDENCE Availability Of Specific Categories Of Records; Public Information
CORRESPONDENCE	312.140	ADDRESS FOR CORRESPONDENCE Investigational New Drug Application
CORRESPONDENCE	812.19	ADDRESS FOR IDE CORRESPONDENCE General Provisions; Investigational Device Exemptions
CORROSION	178.3300	CORROSION INHIBITORS USED FOR STEEL OR TINPLATE Certain Adjuvants And Production Aids; Indirect Food Additives: Adjuvants, Production Aids, And Sanitizers
CORTICAL	882.1310	CORTICAL ELECTRODE Neurological Diagnostic Devices; Neurological Devices
CORTICOIDS	862.1195	CORTICOIDS TEST SYSTEM Clinical Chemistry Test Systems; Clinical Chemistry And Clinical Toxicology Devices
CORTICOSTEROIDS	510.410	CORTICOSTEROIDS FOR ORAL, INJECTABLE, AND OPHTHALMIC USE IN ANIMALS; WARNINGS AND LABELING REQUIREMENTS Requirements For Specific New Animal Drugs; New Animal Drugs
CORTICOSTERONE	862.1200	CORTICOSTERONE TEST SYSTEM Clinical Chemistry Test Systems; Clinical Chemistry And Clinical Toxicology Devices
CORTICOTROPIN	522.480	REPOSITORY CORTICOTROPIN INJECTION Implantation Or Injectable Dosage Form New Animal Drugs
CORTISOL	862.1205	CORTISOL (HYDROCORTISONE AND HYDROXYCORTICOSTERONE) TEST SYSTEM Clinical Chemistry Test Systems; Clinical Chemistry And Clinical Toxicology Devices
CORYNEBACTERIUM SPP.	866.3140	CORYNEBACTERIUM SPP. SEROLOGICAL REAGENTS Serological Reagents; Immunology And Microbiology Devices
COSMETIC	250.250	HEXACHLOROPHENE, AS A COMPONENT OF DRUG AND COSMETIC PRODUCTS Requirements For Drugs And Cosmetics; Specific Requirements For Specific Human Drugs

COSMETIC	700.13	USE OF MERCURY COMPOUNDS IN COSMETICS INCLUDING USE AS SKINBLEACHING AGENTS IN COSMETIC PREPARATIONS ALSO REGARDED AS DRUGS Requirements For Specific Cosmetic Products; General
COSMETIC	700.14	USE OF VINYL CHLORIDE AS AN INGREDIENT, INCLUDING PROPELLANT OF COSMETIC AEROSOL PRODUCTS Requirements For Specific Cosmetic Products; General
COSMETIC	700.15	USE OF CERTAIN HALOGENATED SALICYLANILIDES AS INGREDIENTS IN COSMETIC PRODUCTS Requirements For Specific Cosmetic Products; General
COSMETIC	700.16	USE OF AEROSOL COSMETIC PRODUCTS CONTAINING ZIRCONIUM Requirements For Specific Cosmetic Products; General
COSMETIC	700.18	USE OF CHLOROFORM AS AN INGREDIENT IN COSMETIC PRODUCTS Requirements For Specific Cosmetic Products; General
COSMETIC	700.19	USE OF METHYLENE CHLORIDE AS AN INGREDIENT IN COSMETIC PRODUCTS Requirements For Specific Cosmetic Products; General
COSMETIC	700.25	TAMPER-RESISTANT PACKAGING REQUIREMENTS FOR COSMETIC PRODUCTS Requirements For Specific Cosmetic Products; General
COSMETIC	701.30	INGREDIENT NAMES ESTABLISHED FOR COSMETIC INGREDIENT LABELING Labeling Of Specific Ingredients; Cosmetic Labeling
COSMETIC	710.6	NOTIFICATION OF REGISTRANT; COSMETIC PRODUCT ESTABLISHMENT REGISTRATION NUMBER Voluntary Registration Of Cosmetic Product Establishments
COSMETIC	720.4	INFORMATION REQUESTED ABOUT COSMETIC PRODUCTS Voluntary Filing Of Cosmetic Product Ingredient Composition Statements
COSMETIC	720.7	NOTIFICATION OF PERSON SUBMITTING COSMETIC PRODUCT INGREDIENT STATEMENT Voluntary Filing Of Cosmetic Product Ingredient Composition Statements
COSMETIC	740.10	LABELING OF COSMETIC PRODUCTS FOR WHICH ADEQUATE SUBSTANTIATION OF SAFETY HAS NOT BEEN OBTAINED Warning Statements; Cosmetic Product Warning Statements
COSMETICS	700.11	COSMETICS CONTAINING BITHIONOL Requirements For Specific Cosmetic Products; General
COSMETICS	700.35	COSMETICS CONTAINING SUNSCREEN INGREDIENTS Requirements For Specific Cosmetic Products; General
COSMETICS	740.11	COSMETICS IN SELF-PRESSURIZED CONTAINERS Warning Statements; Cosmetic Product Warning Statements
COST	1316.97	INITIATING JUDICIAL FORFEITURE PROCEEDING AGAINST A CONVEYANCE WITHIN 60 DAYS OF FILING A CLAIM AND COST BOND Expedited Forfeiture Proceedings For Certain Property; Administrative Functions, Practices, And Procedures
COST, REFUND	1004.1	MANUFACTURER'S OBLIGATION TO REPAIR, REPLACE OR REFUND COST OF ELECTRONIC PRODUCTS Repurchase, Repairs, Or Replacement of Electronic Products
COST, REFUNDING	1004.4	PLANS FOR REFUNDING THE COST OF ELECTRONIC PRODUCTS Repurchase, Repairs, Or Replacement of Electronic Products
COST SHARING	1403.24	MATCHING OR COST SHARING Post-Award Requirements; Uniform Administrative Requirements For Grants And Coop. Agreements To State & Local Gov.
COSTS	1.99	COSTS CHARGEABLE IN CONNECTION WITH RELABELING AND RECONDITIONING INADMISSABLE IMPORTS Imports And Exports, General Enforcement Regulations
COSTS	1005.24	COSTS OF BRINGING PRODUCT INTO COMPLIANCE Bonding And Compliance Procedures; Importation Of Electronic Products
COSTS, ALLOWABLE	1403.22	ALLOWABLE COSTS Uniform Administrative Requirements For Grants And Coop. Agreements To State & Local Gov.
COTTAGE CHEESE	133.128	COTTAGE CHEESE Requirements For Specific Standardized Cheese And Related Products; Cheeses And Related Cheese Products
COTTAGE CHEESE	133.129	DRY CURD COTTAGE CHEESE Requirements For Specific Standardized Cheese And Related Products; Cheeses And Related Cheese Products
COTTON	182.70	SUBSTANCES MIGRATING FROM COTTON AND COTTON FABRICS USED IN DRY FOOD PACKAGING General Provisions; Substances Generally Recognized As Safe
COTTONOID	882.4700	COTTONOID PADDIE Neurological Surgical Devices; Neurological Devices
COTTONSEED	172.894	MODIFIED COTTONSEED PRODUCTS INTENDED FOR HUMAN CONSUMPTION Multipurpose Additives; Food Additives Permitted For Direct Addition To Food For Human Consumption

COTTONSEED	173.322	CHEMICALS USED IN DELINTING COTTONSEED Specific Usage Additives; Secondary Direct Food Additives Permitted In Food For Human Consumption
COTTONSEED FLOUR	73.140	TOASTED PARTIALLY DEFATTED COOKED COTTONSEED FLOUR Foods; Listing Of Color Additives Exempt From Certification
COTTONSEED MEAL	573.140	AMMONIATED COTTONSEED MEAL Food Additive Listing; Food Additives Permitted In Feed And Drinking Water Of Animals
COUGH-COLD	310.533	DRUG PRODUCTS CONTAINING ACTIVE INGREDIENTS OFFERED OVER-THE-COUNTER (OTC) FOR HUMAN USE AS AN ANTICHOLINERGIC IN COUGH-COLD DRUG PRODUCTS Requirements For Specific New Drugs Or Devices; New Drugs
COUMAPHOS	558.185	COUMAPHOS Specific New Animal Drugs For Use In Animal Feeds; New Animal Drugs For Use In Animal Feeds
COUMARIN	189.130	COUMARIN Substances Generally Prohibited From Direct Addition Or Use As Human Food; Substances Prohibited From Use In Human Food
COUMARONE-INDENE	172.215	COUMARONE-INDENE RESIN Coatings, Films And Related Substances; Food Additives Permitted For Direct Addition To Food For Human Consumption
COUNSEL	16.62	RIGHT TO COUNSEL Procedures Before A Regulatory Hearing; Regulatory Hearing Before The FDA
COUNSEL, CHIEF	5.1105	CHIEF COUNSEL, FOOD AND DRUG ADMINISTRATION Organization; Delegations Of Authority And Organization
COUNT, BACTERIAL	1210.16	METHOD OF BACTERIAL COUNT Inspection And Testing; Regulations Under The Federal Import Milk Act
COUNTER	862.2320	BETA OR GAMMA COUNTER FOR CLINICAL USE Clinical Laboratory Instruments; Clinical Chemistry And Clinical Toxicology Devices
COUNTER	864.5200	AUTOMATED CELL COUNTER Automated And Semi-automated Hematology Devices; Hematology And Pathology Devices
COUNTER	864.5220	AUTOMATED DIFFERENTIAL CELL COUNTER Automated And Semi-automated Hematology Devices; Hematology And Pathology Devices
COUNTER	866.2170	AUTOMATED COLONY COUNTER Microbiology Devices; Immunology And Microbiology Devices
COUNTER	866.2180	MANUAL COLONY COUNTER Microbiology Devices; Immunology And Microbiology Devices
COUNTER	892.1130	NUCLEAR WHOLE BODY COUNTER Diagnostic Devices; Radiology Devices
COUNTER-PULSATING	870.5225	EXTERNAL COUNTER-PULSATING DEVICE Cardiovascular Therapeutic Devices; Cardiovascular Devices
COUNTING	864.6160	MANUAL BLOOD CELL COUNTING DEVICE Manual Hematology Devices; Hematology And Pathology Devices
COUNTING	864.8175	CALIBRATOR FOR PLATELET COUNTING Hematology Reagents; Hematology And Pathology Devices
COUNTING	864.8185	CALIBRATOR FOR RED CELL AND WHITE CELL COUNTING Hematology Reagents; Hematology And Pathology Devices
COUNTRIES, OTHER	26.78	AGREEMENTS WITH OTHER COUNTRIES Framework Provisions; Mutual Recognition Of Pharmaceutical GMP Reports, Medical Device Quality System Audit Reports
COURT	10.60	REFERRAL BY COURT General Administrative Procedures; Administrative Practices And Procedures
COURT	20.86	DISCLOSURE IN ADMINISTRATIVE OR COURT PROCEEDINGS Limitations On Exemptions; Public Information
COURT	20.91	USE OF DATA OR INFORMATION FOR ADMINISTRATIVE OR COURT ENFORCEMENT ACTION Limitations On Exemptions; Public Information
COURT	20.102	COURT ENFORCEMENT RECORDS Availability Of Specific Categories Of Records; Public Information
COURT ORDER	20.83	DISCLOSURE REQUIRED BY COURT ORDER Limitations On Exemptions; Public Information
COURT REVIEW	10.45	COURT REVIEW OF FINAL ADMINISTRATIVE ACTION; EXHAUSTION OF ADMINISTRATIVE REMEDIES General Administrative Procedures; Administrative Practices And Procedures
COURTS	12.140	REVIEW BY THE COURTS Judicial Review; Formal Evidentiary Public Hearing
COVER	880.6185	CAST COVER General Hospital And Personal Use Miscellaneous Devices; General Hospital And Personal Use Devices
COVER	880.6190	MATTRESS COVER FOR MEDICAL PURPOSES General Hospital And Personal Use Miscellaneous Devices; General Hospital And Personal Use Devices

COVER	882.5250	BURR HOLE COVER Neurological Therapeutic Devices; Neurological Devices
COVERAGE	1404.110	COVERAGE Governmentwide Debarment And Suspension And Requirements For Drug-Free Workplace
COVERAGE	1404. 610	COVERAGE Governmentwide Debarment And Suspension And Requirements For Drug-Free Workplace
COVERAGE, MEDIA	10.205	ELECTRONIC MEDIA COVERAGE OF PUBLIC ADMINISTRATIVE PROCEEDINGS Electronic Media Coverage Public Admin Proceedings; Guideline Policy Procedures; Administrative Practices And Procedures
COVERAGE, MEDIA	10.206	PROCEDURES FOR ELECTRONIC MEDIA COVERAGE OF AGENCY PUBLIC ADMINISTRATIVE PROCEEDINGS Electronic Media Coverage Public Admin Proceedings; Guideline Policy Procedures; Administrative Practices And Procedures
COVERAGE, PRODUCT	26.4	PRODUCT COVERAGE Specific Sector Provisions For Pharmaceutical GMPs; Mutual Recognition Of Pharmaceutical GMP Reports, Medical Device Quality System Audit Reports
COVERAGE, PRODUCT	26.33	PRODUCT COVERAGE Specific Sector Provisions For Medical Devices; Mutual Recognition Of Pharmaceutical GMP Reports, Medical Device Quality System Audit Reports
COVERED TRANSACTIONS	404.220	CONTINUATION OF COVERED TRANSACTIONS Governmentwide Debarment And Suspension And Requirements For Drug-Free Workplace
COWS	526.1696b	PENICILLIN G PROCAINE-DIHYDROSTREPTOMYCIN IN SOYBEAN OIL FOR INTRAMAMMARY INFUSION (DRY COWS) Intramammary Dosage Forms
COWS	526.1696c	PENICILLIN G PROCAINE-DIHYDROSTREPTOMYCIN SULFATE FOR INTRAMAMMARY INFUSION (DRY COWS) Intramammary Dosage Forms
COWS	1210.12	PHYSICAL INSPECTION OF COWS Inspection And Testing; Regulations Under The Federal Import Milk Act
COXSACKIEVIRUS	866.3145	COXSACKIEVIRUS SEROLOGICAL REAGENTS Serological Reagents; Immunology And Microbiology Devices
CRABMEAT	102.50	CRABMEAT Requirements For Specific Nonstandardized Foods; Common Or Usual Name For Nonstandardized Foods
CRACKED WHEAT	137.190	CRACKED WHEAT Requirements For Specific Standardized Cereal Flours And Related Products; Cereal Flours And Related Products
CRADLE	892.1830	RADIOLOGIC PATIENT CRADLE Diagnostic Devices; Radiology Devices
CRAMBE MEAL	573.310	CRAMBE MEAL, HEAT TOASTED Food Additive Listing; Food Additives Permitted In Feed And Drinking Water Of Animals
CRAMPS	310.546	DRUG PRODUCTS CONTAINING ACTIVE INGREDIENTS OFFERED OTC FOR THE TREATMENT AND/OR PREVENTION OF NOCTURNAL LEG MUSCLE CRAMPS Requirements For Specific New Drugs Or Devices; New Drugs
CRANIAL	882.4300	MANUAL CRANIAL DRILLS, BURRS, TREPHINES, AND THEIR ACCESSORIES Neurological Surgical Devices; Neurological Devices
CRANIAL	882.4305	POWERED COMPOUND CRANIAL DRILLS, BURRS, TREPHINES, AND THEIR ACCESSORIES Neurological Surgical Devices; Neurological Devices
CRANIAL	882.4310	POWERED SIMPLE CRANIAL DRILLS, BURRS, TREPHINES, AND THEIR ACCESSORIES Neurological Surgical Devices; Neurological Devices
CRANIAL	882.4325	CRANIAL DRILL HANDPIECE (BRACE) Neurological Surgical Devices; Neurological Devices
CRANIAL	882.4360	ELECTRIC CRANIAL DRILL MOTOR Neurological Surgical Devices; Neurological Devices
CRANIAL	882.4370	PNEUMATIC CRANIAL DRILL MOTOR Neurological Surgical Devices; Neurological Devices
CRANIAL	882.5800	CRANIAL ELECTROTHERAPHY STIMULATOR Neurological Therapeutic Devices; Neurological Devices
CRANIAL	882.5970	CRANIAL ORTHOSIS Neurological Therapeutic Devices; Neurological Devices
CRANIOPLASTY	882.4500	CRANIOPLASTY MATERIAL FORMING INSTRUMENT Neurological Surgical Devices; Neurological Devices
CRANIOPLASTY	882.5300	METHYL METHACRYLATE FOR CRANIOPLASTY Neurological Therapeutic Devices; Neurological Devices
CRANIOPLASTY	882.5320	PREFORMED ALTERABLE CRANIOPLASTY PLATE Neurological Therapeutic Devices; Neurological Devices

CRANIOPLASTY	882.5330	PREFORMED NONALTERABLE CRANIOPLASTY PLATE Neurological Therapeutic Devices; Neurological Devices
CRANIOPLASTY	882.5360	CRANIOPLASTY PLATE FASTENER Neurological Therapeutic Devices; Neurological Devices
CRANIOSYNOSTOSIS	882.5900	PREFORMED CRANIOSYNOSTOSIS STRIP Neurological Therapeutic Devices; Neurological Devices
CREAM	524.450	CLOTRIMAZOLE CREAM Ophthalmic And Topical Dosage Form New Animal Drugs
CREAM	524.520	CUPRIMYXIN CREAM Ophthalmic And Topical Dosage Form New Animal Drugs
CREAM	524.981a	FLUOCINOLONE ACETONIDE CREAM Ophthalmic And Topical Dosage Form New Animal Drugs
CREAM	524.1443	MICONAZOLE NITRATE CREAM; MICONAZOLE NITRATE LOTION; MICONAZOLE NITRATE SPRAY Ophthalmic And Topical Dosage Form New Animal Drugs
CREAM	524.2350	TOLNAFTATE CREAM Ophthalmic And Topical Dosage Form New Animal Drugs
CREAM	524.2481	TRIAMCINOLONE ACETONIDE CREAM Ophthalmic And Topical Dosage Form New Animal Drugs
CREAM	878.4470	SURGEON'S GLOVING CREAM Surgical Devices; General And Plastic Surgery Devices
CREAM	1210.25	PERMITS FOR PASTEURIZED MILK OR CREAM Permit Control; Regulations Under The Federal Import Milk Act
CREAM	1210.26	PERMITS FOR RAW MILK OR CREAM Permit Control; Regulations Under The Federal Import Milk Act
CREAM CHEESE	133.133	CREAM CHEESE Requirements For Specific Standardized Cheese And Related Products; Cheeses And Related Cheese Products
CREAM CHEESE	133.134	CREAM CHEESE WITH OTHER FOODS Requirements For Specific Standardized Cheese And Related Products; Cheeses And Related Cheese Products
CREAM, DRY	131.149	DRY CREAM Requirements For Specific Standardized Milk And Cream; Milk And Cream
CREAM, HEAVY	131.150	HEAVY CREAM Requirements For Specific Standardized Milk And Cream; Milk And Cream
CREAM, ICE	135.110	ICE CREAM AND FROZEN CUSTARD Requirements For Specific Standardized Frozen Desserts; Frozen Desserts
CREAM, ICE	135.115	GOAT'S MILK ICE CREAM Requirements For Specific Standardized Frozen Desserts; Frozen Desserts
CREAM, LIGHT	131.155	LIGHT CREAM Requirements For Specific Standardized Milk And Cream; Milk And Cream
CREAM, LIGHT WHIPPING	131.157	LIGHT WHIPPING CREAM Requirements For Specific Standardized Milk And Cream; Milk And Cream
CREAM, SOUR	131.160	SOUR CREAM Requirements For Specific Standardized Milk And Cream; Milk And Cream
CREAM, SOUR	131.162	ACIDIFIED SOUR CREAM Requirements For Specific Standardized Milk And Cream; Milk And Cream
CREAM, WHIPPED	131.25	WHIPPED CREAM PRODUCTS CONTAINING FLAVORING OR SWEETENING General Provisions; Milk And Cream
CREATINE	862.1210	CREATINE TEST SYSTEM Clinical Chemistry Test Systems; Clinical Chemistry And Clinical Toxicology Devices
CREATINE	862.1215	CREATINE PHOSPHOKINASE/CREATINE KINASE OR ISOENZYMES TEST SYSTEM Clinical Chemistry Test Systems; Clinical Chemistry And Clinical Toxicology Devices
CREATININE	862.1225	CREATININE TEST SYSTEM Clinical Chemistry Test Systems; Clinical Chemistry And Clinical Toxicology Devices
CRIMINAL	7.84	OPPORTUNITY FOR PRESENTATION OF VIEWS BEFORE REPORT OF CRIMINAL VIOLATION Criminal Violations; Enforcement Policy
CRIMINAL	7.85	CONDUCT OF A PRESENTATION OF VIEWS BEFORE REPORT OF CRIMINAL VIOLATION Criminal Violations; Enforcement Policy
CRIMINAL	7.87	RECORDS RELATED TO OPPORTUNITIES FOR PRESENTATION OF VIEWS CONDUCTED BEFORE REPORT OF CRIMINAL VIOLATION Criminal Violations; Enforcement Policy
CRIMINAL	1404.635	REPORTING OF AND EMPLOYEE SANCTIONS FOR CONVICTIONS OF CRIMINAL DRUG OFFENSES Governmentwide Debarment And Suspension And Requirements For Drug-Free Workplace
CRITERIA	56.111	CRITERIA FOR IRB APPROVAL OF RESEARCH IRB Functions And Operations; Institutional Review Boards

CRITERIA	70.42	CRITERIA FOR EVALUATING THE SAFETY OF COLOR ADDITIVES Safety Evaluation; Color Additives
CRITERIA	211.122	MATERIALS EXAMINATION AND USAGE CRITERIA Packaging And Labeling Control; Current Good Manufacturing Practice For Finished Pharmaceuticals
CRITERIA	320.22	CRITERIA FOR WAIVER OF EVIDENCE OF IN VIVO BIOAVAILABILITY Procedures For Determining The Bioavailability Of Drug Products; Bioavailability And Bioequivalence Requirements
CRITERIA	320.33	CRITERIA AND EVIDENCE TO ASSESS ACTUAL OR POTENTIAL BIOEQUIVALENCE PROBLEMS Bioequivalence Requirements; Bioavailability And Bioequivalence Requirements
CRITERIA	330.14	ADDITIONAL CRITERIA AND PROCEDURES FOR CLASSIFYING OTC DRUGS AS GENERALLY RECOGNIZED AS SAFE AND EFFECTIVE AND NOT MISBRANDED Administrative Procedures; Over-the-counter Human Drugs Generally Recognized As Safe And Effective & Not Misbranded
CRITERIA	1271.20	IF MY HCT/P'S DO NOT MEET THE CRITERIA IN SECTION 1271.10, AND I DO NOT QUALIFY FOR ANY OF THE EXCEPTIONS IN SECTION 1271.15, WHAT REGULATIONS APPLY? General Provisions; Human Cells, Tissues, And Cellular And Tissue-Based Products
CRITICAL CONTROL POINT	120.8	HAZARD ANALYSIS AND CRITICAL CONTROL POINT (HACCP) PLAN Hazard Analysis And Critical Control Point (HACCP) Systems
CRITICAL CONTROL POINT	123.6	HAZARD ANALYSIS AND HAZARD ANALYSIS CRITICAL CONTROL POINT (HACCP) PLAN Fish And Fishery Products
CRITICAL FACTORS	113.89	DEVIATIONS IN PROCESSING, VENTING, OR CONTROL OF CRITICAL FACTORS Production And Process Controls; Thermally Processed Low-acid Foods Packaged In Hermetically Sealed Containers
CROPS	573.400	ETHOXYQUIN IN CERTAIN DEHYDRATED FORAGE CROPS Food Additive Listing; Food Additives Permitted In Feed And Drinking Water Of Animals
CROSS-LINKED	177.1211	CROSS LINKED POLYACRYLATE COPOLYMERS Substances For Use As Basic Components Of Single And Repeated Use Food Contact Surfaces; Indirect Food Additives: Polymers
CROSS-LINKED	177.2420	POLYESTER RESINS, CROSS-LINKED Substances For Use Only As Components Of Articles Intended For Repeated Use; Indirect Food Additives: Polymers
CROSS-LINKED	177.2710	STYRENE-DIVINYLBENZENE RESINS, CROSS-LINKED Substances For Use Only As Components Of Articles Intended For Repeated Use; Indirect Food Additives: Polymers
CROSS-REFERENCE	20.100	APPLICABILITY; CROSS-REFERENCE TO OTHER REGULATIONS Availability Of Specific Categories Of Records; Public Information
CROTONIC ACID	175.350	VINYL ACETATE/CROTONIC ACID COPOLYMER Substances For Use As Components Of Coatings; Indirect Food Additives
CROWN	872.3330	PREFORMED CROWN Prosthetic Devices; Dental Devices
CROWN	872.3770	TEMPORARY CROWN AND BRIDGE RESIN Prosthetic Devices; Dental Devices
CRUMBLES	520.100c	AMPROLIUM CRUMBLES Oral Dosage Form New Animal Drugs
CRUSHED WHEAT	137.195	CRUSHED WHEAT Requirements For Specific Standardized Cereal Flours And Related Products; Cereal Flours And Related Products
CRUTCH	886.5600	PTOSIS CRUTCH Therapeutic Devices; Ophthalmic Devices
CRUTCH	890.3150	CRUTCH Physical Medicine Prosthetic Devices; Physical Medicine Devices
CRUTCH	890.3790	CANE, CRUTCH, AND WALKER TIPS AND PADS Physical Medicine Prosthetic Devices; Physical Medicine Devices
CRYOGENIC	882.4250	CRYOGENIC SURGICAL DEVICE Neurological Surgical Devices; Neurological Devices
CRYOPHTHALMIC	886.4170	CRYOPHTHALMIC UNIT Surgical Devices; Ophthalmic Devices
CRYOPRECIPITATE	640.50	CRYOPRECIPITATE AHF Cryoprecipitate; Additional Standards For Human Blood And Blood Products
CRYOSURGICAL	878.4350	CRYOSURGICAL UNIT AND ACCESSORIES Surgical Devices; General And Plastic Surgery Devices
CRYPTOCOCCUS NEOFORMANS	866.3165	CRYPTOCOCCUS NEOFORMANS SEROLOGICAL REAGENTS Serological Reagents; Immunology And Microbiology Devices
CRYSTAL	880.2200	LIQUID CRYSTAL FOREHEAD TEMPERATURE STRIP General Hospital And Personal Use Monitoring Devices; General Hospital And Personal Use Devices

CRYSTAL	880.6970	LIQUID CRYSTAL VEIN LOCATOR General Hospital And Personal Use Miscellaneous Devices; General Hospital And Personal Use Devices
CRYSTAL	884.2982	LIQUID CRYSTAL THERMOGRAPHIC SYSTEM Obstetrical And Gynecological Monitoring Devices; Obstetrical And Gynecological Devices
CRYSTALLINE TRYPSIN	524.2620	LIQUID CRYSTALLINE TRYPSIN, PERU BALSAM, CASTOR OIL Ophthalmic And Topical Dosage Form New Animal Drugs
CT	1020.33	COMPUTED TOMOGRAPHY (CT) EQUIPMENT Performance Standards For Ionizing Radiation Emitting Products
CUFF	868.5750	INFLATABLE TRACHEAL TUBE CUFF Therapeutic Devices; Anesthesiology Devices
CUFF	868.5760	CUFF SPREADER Therapeutic Devices; Anesthesiology Devices
CUFF	868.5800	THRACHEOSTOMY TUBE AND TUBE CUFF Therapeutic Devices; Anesthesiology Devices
CUFF	870.1120	BLOOD PRESSURE CUFF Cardiovascular Diagnostic Devices; Cardiovascular Devices
CUFF	882.5275	NERVE CUFF Neurological Therapeutic Devices; Neurological Devices
CULDOSCOPE	884.1640	CULDOSCOPE AND ACCESSORIES Obstetrical And Gynecological Diagnostic Devices; Obstetrical And Gynecological Devices
CULINARY	1240.80	GENERAL REQUIREMENTS FOR WATER FOR DRINKING AND CULINARY PURPOSES Source And Use Of Potable Water; Control Of Communicable Diseases
CULTURE	529.469	COMPETITIVE EXCLUSION CULTURE Certain Other Dosage Form New Animal Drugs
CULTURE	864.2240	CELL AND TISSUE CULTURE SUPPLIES AND EQUIPMENT Cell And Tissue Culture Products; Hematology And Pathology Devices
CULTURE	864.2260	CHROMOSOME CULTURE KIT Cell And Tissue Culture Products; Hematology And Pathology Devices
CULTURE MEDIA	864.2220	SYNTHETIC CELL AND TISSUE CULTURE MEDIA AND COMPONENTS Cell And Tissue Culture Products; Hematology And Pathology Devices
CULTURE MEDIA	866.2450	SUPPLEMENT FOR CULTURE MEDIA Microbiology Devices; Immunology And Microbiology Devices
CULTURE MEDIA	866.2480	QUALITY CONTROL KIT FOR CULTURE MEDIA Microbiology Devices; Immunology And Microbiology Devices
CULTURE MEDIA	876.5885	TISSUE CULTURE MEDIA FOR HUMAN EX VIVO TISSUE AND CELL CULTURE PROCESSING APPLICATIONS Therapeutic Devices; Gastroenterology-urology Devices
CULTURE MEDIUM	866.1700	CULTURE MEDIUM FOR ANTIMICROBIAL SUSCEPTIBILITY TESTS Diagnostic Devices; Immunology And Microbiology Devices
CULTURE MEDIUM	866.2300	MULTIPURPOSE CULTURE MEDIUM Microbiology Devices; Immunology And Microbiology Devices
CULTURE MEDIUM	866.2320	DIFFERENTIAL CULTURE MEDIUM Microbiology Devices; Immunology And Microbiology Devices
CULTURE MEDIUM	866.2330	ENRICHED CULTURE MEDIUM Microbiology Devices; Immunology And Microbiology Devices
CULTURE MEDIUM	866.2350	MICROBIOLOGICAL ASSAY CULTURE MEDIUM Microbiology Devices; Immunology And Microbiology Devices
CULTURE MEDIUM	866.2360	SELECTIVE CULTURE MEDIUM Microbiology Devices; Immunology And Microbiology Devices
CULTURE MEDIUM	866.2390	TRANSPORT CULTURE MEDIUM Microbiology Devices; Immunology And Microbiology Devices
CULTURE MEDIUM	866.2410	CULTURE MEDIUM FOR PATHOGENIC NEISSERIA SPP. Microbiology Devices; Immunology And Microbiology Devices
CULTURED CELLS	864.2280	CULTURED ANIMAL AND HUMAN CELLS Cell And Tissue Culture Products; Hematology And Pathology Devices
CULTURED MILK	131.112	CULTURED MILK Requirements For Specific Standardized Milk And Cream; Milk And Cream
CULTURES	610.18	CULTURES General Provisions; General Biological Products Standards
CUP	872.6290	PROPHYLAXIS CUP Miscellaneous Devices; Dental Devices
CUP	884.5400	MENSTRUAL CUP Obstetrical And Gynecological Therapeutic Devices; Obstetrical And Gynecological Devices
CUPRIC GLYCINATE	522.518	CUPRIC GLYCINATE INJECTION Implantation Or Injectable Dosage Form New Animal Drugs
CUPRIMYXIN	524.520	CUPRIMYXIN CREAM Ophthalmic And Topical Dosage Form New Animal Drugs

CUPROUS IODIDE	184.1265	CUPROUS IODIDE Listing Of Specific Substances Affirmed As GRAS; Direct Food Substances Affirmed As Generally Recognized As Safe
CURD CHEESE	133.136	WASHED CURD AND SOAKED CURD CHEESE Requirements For Specific Standardized Cheese And Related Products; Cheeses And Related Cheese Products
CURD CHEESE	133.137	WASHED CURD CHEESE FOR MANUFACTURING Requirements For Specific Standardized Cheese And Related Products; Cheeses And Related Cheese Products
CURD CHEESE	133.144	GRANULAR AND STIRRED CURD CHEESE Requirements For Specific Standardized Cheese And Related Products; Cheeses And Related Cheese Products
CURD, DRY	133.129	DRY CURD COTTAGE CHEESE Requirements For Specific Standardized Cheese And Related Products; Cheeses And Related Cheese Products
CURDLAN	172.809	CURDLAN Multipurpose Additives; Food Additives Permitted For Direct Addition To Food For Human Consumption
CURED ELASTOMERS	177.2400	PERFLUOROCARBON CURED ELASTOMERS Substances For Use Only As Components Of Articles Intended For Repeated Use; Indirect Food Additives: Polymers
CURETTE	884.1175	ENDOMETRIAL SUCTION CURETTE AND ACCESSORIES Obstetrical And Gynecological Diagnostic Devices; Obstetrical And Gynecological Devices
CURING PREMIXES	170.60	NITRITES AND/OR NITRATES IN CURING PREMIXES Specific Administrative Rulings And Decisions; Food Additives
CURRENT	870.2640	PORTABLE LEAKAGE CURRENT ALARM Cardiovascular Monitoring Devices; Cardiovascular Devices
CURRENT GMP	5.109	ISSUANCE OF WRITTEN NOTICES CONCERNING PATENT INFORMATION, CURRENT GOOD MANUFACTURING PRACTICES AND FALSE OR MISLEADING LABELING OF NEW DRUGS Human Drugs; Redelegations Of Authority; Delegations Of Authority And Organization
CURRENT GMP	5.504	ISSUANCE OF WRITTEN NOTICES CONCERNING PATENT INFORMATION, CURRENT GOOD MANUFACTURING PRACTICES AND FALSE OR MISLEADING LABELING OF NEW ANIMAL DRUGS AND FEEDS BEARING OR CONTAINING NEW ANIMAL DRUGS Animal Drugs; Redelegations Of Authority; Delegations Of Authority And Organization
CURRENT GMP	110.5	CURRENT GOOD MANUFACTURING PRACTICE Current Good Manufacturing Practice In Manufacturing, Packing, Or Holding Human Food
CURRENT GMP	113.5	CURRENT GOOD MANUFACTURING PRACTICE Thermally Processed Low-acid Foods Packaged In Hermetically Sealed Containers
CURRENT GMP	114.5	CURRENT GOOD MANUFACTURING PRACTICE Acidified Foods
CURRENT GMP	120.5	CURRENT GOOD MANUFACTURING PRACTICE Hazard Analysis And Critical Control Point (HACCP) Systems
CURRENT GMP	123.5	CURRENT GOOD MANUFACTURING PRACTICE Fish and Fishery Products
CURRENT GMP	129.1	CURRENT GOOD MANUFACTURING PRACTICE Processing And Bottling Of Bottled Drinking Water
CURRENT GMP	210.1	STATUS OF CURRENT GOOD MANUFACTURING PRACTICE REGULATIONS Current Good Manufacturing Practice In Manufacturing, Processing, Or Holding Of Drugs; General
CURRENT GMP	210.2	APPLICABILITY OF CURRENT GOOD MANUFACTURING PRACTICE REGULATIONS Current Good Manufacturing Practice In Manufacturing, Processing, Or Holding Of Drugs; General
CURRENT GMP	225.1	CURRENT GOOD MANUFACTURING PRACTICE Current Good Manufacturing Practice For Medicated Feeds
CURRENT GMP	226.1	CURRENT GOOD MANUFACTURING PRACTICE Current Good Manufacturing Practice For Type A Medicated Articles
CUSHION	874.1100	EARPHONE CUSHION FOR AUDIOMETRIC TESTING Diagnostic Devices; Ear, Nose, And Throat Devices
CUSHION	874.3540	OTC DENTURE CUSHION OR PAD Prosthetic Devices; Dental Devices
CUSHION	890.3175	FLOTATION CUSHION Physical Medicine Prosthetic Devices; Physical Medicine Devices
CUSP	872.3350	GOLD OR STAINLESS STEEL CUSP Prosthetic Devices; Dental Devices
CUSP	872.3360	PREFORMED CUSP Prosthetic Devices; Dental Devices

CUSTARD, FROZEN	135.110	ICE CREAM AND FROZEN CUSTARD Requirements For Specific Standardized Frozen Desserts; Frozen Desserts
CUSTODY	1316.73	CUSTODY AND OTHER DUTIES Seizure, Forfeiture, And Disposition Of Property; Administrative Functions, Practices, And Procedures
CUSTODY	1402.3	INFORMATION IN THE CUSTODY OF ONDCP Mandatory Declassification Review
CUTANEOUS	868.2480	CUTANEOUS CARBON DIOXIDE (PCCO2) MONITOR Monitoring Devices; Anesthesiology Devices
CUTANEOUS	868.2500	CUTANEOUS OXYGEN MONITOR Monitoring Devices; Anesthesiology Devices
CUTANEOUS	882.1320	CUTANEOUS ELECTRODE Neurological Diagnostic Devices; Neurological Devices
CUTTER	880.6200	RING CUTTER General Hospital And Personal Use Miscellaneous Devices; General Hospital And Personal Use Devices
CUTTER	884.4150	BIPOLAR ENDOSCOPIC COAGULATOR-CUTTER AND ACCESSORIES Obstetrical And Gynecological Surgical Devices; Obstetrical And Gynecological Devices
CUTTER	884.4160	UNIPOLAR ENDOSCOPIC COAGULATOR-CUTTER AND ACCESSORIES Obstetrical And Gynecological Surgical Devices; Obstetrical And Gynecological Devices
CUTTING	872.4120	BONE CUTTING INSTRUMENT AND ACCESSORIES Surgical Devices; Dental Devices
CUTTING	878.4400	ELECTROSURGICAL CUTTING AND COAGULATION DEVICE AND ACCESSORIES Surgical Devices; General And Plastic Surgery Devices
CUTTING	880.4190	CLIP FORMING/CUTTING INSTRUMENT Surgical Devices; Neurological Devices
CUTTING	882.4275	DOWEL CUTTING INSTRUMENT Neurological Surgical Devices; Neurological Devices
CUTTING	886.4150	VITREOUS ASPIRATION UNIT AND CUTTING INSTRUMENT Surgical Devices; Ophthalmic Devices
CYCLAMATE	189.135	CYCLAMATE AND ITS DERIVATIVES Substances Generally Prohibited From Direct Addition Or Use As Human Food; Substances Prohibited From Use In Human Food
CYCLIC AMP	862.1230	CYCLIC AMP TEST SYSTEM Clinical Chemistry Test Systems; Clinical Chemistry And Clinical Toxicology Devices
CYCLOHEXYLENE	177.1240	1,4-CYCLOHEXYLENE DIMETHYLENE TEREPHTHALATE AND 1,4-CYCLOHEXYLENE DIMETHYLENE ISOPHTHALATE COPOLYMER Substances For Use As Basic Components Of Single And Repeated Use Food Contact Surfaces; Indirect Food Additives: Polymers
CYCLOHEXYLENE	177.1315	ETHYLENE-1,4-CYCLOHEXYLENE DIMETHYLENE TEREPHTHALATE COPOLYMERS Substances For Use As Basic Components Of Single And Repeated Use Food Contact Surfaces; Indirect Food Additives: Polymers
CYCLOSPORINE	524.575	CYCLOSPORINE OPHTHALMIC OINTMENT Ophthalmic And Topical Dosage Form New Animal Drugs
CYLINDER	886.1840	SIMULATAN (INCLUDING CROSSED CYLINDER) Diagnostic Devices; Ophthalmic Devices
CYLINDER/SHEET	520.1450c	MORANTEL TARTRATE SUSTAINED-RELEASE TRILAMINATE CYLINDER/SHEET Oral Dosage Form New Animal Drugs
CYSTEINE	184.1271	L-CYSTEINE Listing Of Specific Substances Affirmed As GRAS; Direct Food Substances Affirmed As Generally Recognized As Safe
CYSTEINE	184.1272	L-CYSTEINE MONOHYDROCHLORIDE Listing Of Specific Substances Affirmed As GRAS; Direct Food Substances Affirmed As Generally Recognized As Safe
CYSTEINE	582.5271	CYSTEINE Nutrients And/or Dietary Supplements; Substances Generally Recognized As Safe
CYSTINE	582.5273	CYSTINE Nutrients And/or Dietary Supplements; Substances Generally Recognized As Safe
CYSTINE	862.1240	CYSTINE TEST SYSTEM Clinical Chemistry Test Systems; Clinical Chemistry And Clinical Toxicology Devices
CYTHIOATE	520.530	CYTHIOATE ORAL LIQUID Oral Dosage Form New Animal Drugs Not Subject To Certification
CYTHIOATE	520.531	CYTHIOATE TABLETS Oral Dosage Form New Animal Drugs Not Subject To Certification
CYTOCENTRIFUGE	864.3300	CYTOCENTRIFUGE Pathology Instrumentation And Accessories; Hematology And Pathology Devices
CYTOMEGALOVIRUS	866.3175	CYTOMEGALOVIRUS SEROLOGICAL REAGENTS Serological Reagents; Immunology And Microbiology Devices

D

D, IMMUNOGLOBULIN	866.5510	IMMUNOGLOBULINS A, G, M, D, AND E IMMUNOLOGICAL TEST SYSTEM Immunological Test Systems; Immunology And Microbiology Devices
D, VITAMIN	131.127	NONFAT DRY MILK FORTIFIED WITH VITAMINS A AND D Requirements For Specific Standardized Milk And Cream; Milk And Cream
D, VITAMIN	184.1950	VITAMIN D Listing Of Specific Substances Affirmed As GRAS; Direct Food Substances Affirmed As Generally Recognized As Safe
D, VITAMIN	862.1825	VITAMIN D TEST SYSTEM Clinical Chemistry Test Systems; Clinical Chemistry And Clinical Toxicology Devices
D2, VITAMIN	582.5950	VITAMIN D2 Nutrients And/or Dietary Supplements; Substances Generally Recognized As Safe
D3, VITAMIN	582.5953	VITAMIN D3 Nutrients And/or Dietary Supplements; Substances Generally Recognized As Safe
D&C BLUE NO. 4	74.1104	D&C BLUE NO. 4 Drugs; Listing Of Color Additives Subject To Certification
D&C BLUE NO. 4	74.2104	D&C BLUE NO. 4 Cosmetics; Listing Of Color Additives Subject To Certification
D&C BLUE NO. 4	82.1104	D&C BLUE NO. 4 Drugs And Cosmetics; Listing Of Certified Provisionally Listed Colors And Specifications
D&C BLUE NO. 6	74.3106	D&C BLUE NO. 6 Medical Devices; Listing Of Color Additives Subject To Certification
D&C BLUE NO. 9	74.1109	D&C BLUE NO. 9 Drugs; Listing Of Color Additives Subject To Certification
D&C BROWN NO. 1	74.2151	D&C BROWN NO. 1 Cosmetics; Listing Of Color Additives Subject To Certification
D&C GREEN NO. 5	74.1205	D&C GREEN NO. 5 Drugs; Listing Of Color Additives Subject To Certification
D&C GREEN NO. 5	74.2205	D&C GREEN NO. 5 Cosmetics; Listing Of Color Additives Subject To Certification
D&C GREEN NO. 5	82.1205	D&C GREEN NO. 5 Drugs And Cosmetics; Listing Of Certified Provisionally Listed Colors And Specifications
D&C GREEN NO. 6	74.1206	D&C GREEN NO. 6 Drugs; Listing Of Color Additives Subject To Certification
D&C GREEN NO. 6	74.2206	D&C GREEN NO. 6 Cosmetics; Listing Of Color Additives Subject To Certification
D&C GREEN NO. 6	74.3206	D&C GREEN NO. 6 Medical Devices; Listing Of Color Additives Subject To Certification
D&C GREEN NO. 6	82.1206	D&C GREEN NO. 6 Drugs And Cosmetics; Listing Of Certified Provisionally Listed Colors And Specifications
D&C GREEN NO. 8	74.1208	D&C GREEN NO. 8 Drugs; Listing Of Color Additives Subject To Certification
D&C GREEN NO. 8	74.2208	D&C GREEN NO. 8 Cosmetics; Listing Of Color Additives Subject To Certification
D&C ORANGE NO. 4	74.1254	D&C ORANGE NO. 4 Drugs; Listing Of Color Additives Subject To Certification
D&C ORANGE NO. 4	74.2254	D&C ORANGE NO. 4 Cosmetics; Listing Of Color Additives Subject To Certification
D&C ORANGE NO. 4	82.1254	D&C ORANGE NO. 4 Drugs And Cosmetics; Listing Of Certified Provisionally Listed Colors And Specifications
D&C ORANGE NO. 5	74.1255	D&C ORANGE NO. 5 Drugs; Listing Of Color Additives Subject To Certification
D&C ORANGE NO. 5	74.2255	D&C ORANGE NO. 5 Cosmetics; Listing Of Color Additives Subject To Certification
D&C ORANGE NO. 5	82.1255	D&C ORANGE NO. 5 Drugs And Cosmetics; Listing Of Certified Provisionally Listed Colors And Specifications
D&C ORANGE NO. 10	74.1260	D&C ORANGE NO. 10 Drugs; Listing Of Color Additives Subject To Certification
D&C ORANGE NO. 10	74.2260	D&C ORANGE NO. 10 Cosmetics; Listing Of Color Additives Subject To Certification
D&C ORANGE NO. 10	82.1260	D&C ORANGE NO. 10 Drugs And Cosmetics; Listing Of Certified Provisionally Listed Colors And Specifications
D&C ORANGE NO. 11	74.1261	D&C ORANGE NO. 11 Drugs; Listing Of Color Additives Subject To Certification

D&C ORANGE NO. 11	74.2261	D&C ORANGE NO. 11 Cosmetics; Listing Of Color Additives Subject To Certification
D&C ORANGE NO. 11	82.1261	D&C ORANGE NO. 11 Drugs And Cosmetics; Listing Of Certified Provisionally Listed Colors And Specifications
D&C RED NO. 6	74.1306	D&C RED NO. 6 Drugs; Listing Of Color Additives Subject To Certification
D&C RED NO. 6	74.2306	D&C RED NO. 6 Cosmetics; Listing Of Color Additives Subject To Certification
D&C RED NO. 6	82.1306	D&C RED NO. 6 Drugs And Cosmetics; Listing Of Certified Provisionally Listed Colors And Specifications
D&C RED NO. 7	74.1307	D&C RED NO. 7 Drugs; Listing Of Color Additives Subject To Certification
D&C RED NO. 7	74.2307	D&C RED NO. 7 Cosmetics; Listing Of Color Additives Subject To Certification
D&C RED NO. 7	82.1307	D&C RED NO. 7 Drugs And Cosmetics; Listing Of Certified Provisionally Listed Colors And Specifications
D&C RED NO. 17	74.1317	D&C RED NO. 17 Drugs; Listing Of Color Additives Subject To Certification
D&C RED NO. 17	74.2317	D&C RED NO. 17 Cosmetics; Listing Of Color Additives Subject To Certification
D&C RED NO. 17	74.3230	D&C RED NO. 17 Medical Devices; Listing Of Color Additives Subject To Certification
D&C RED NO. 17	82.1317	D&C RED NO. 17 Drugs And Cosmetics; Listing Of Certified Provisionally Listed Colors And Specifications
D&C RED NO. 21	74.1321	D&C RED NO. 21 Drugs; Listing Of Color Additives Subject To Certification
D&C RED NO. 21	74.2321	D&C RED NO. 21 Cosmetics; Listing Of Color Additives Subject To Certification
D&C RED NO. 21	82.1321	D&C RED NO.21 Drugs And Cosmetics; Listing Of Certified Provisionally Listed Colors And Specifications
D&C RED NO. 22	74.1322	D&C RED NO. 22 Drugs; Listing Of Color Additives Subject To Certification
D&C RED NO. 22	74.2322	D&C RED NO. 22 Cosmetics; Listing Of Color Additives Subject To Certification
D&C RED NO. 22	82.1322	D&C RED NO. 22 Drugs And Cosmetics; Listing Of Certified Provisionally Listed Colors And Specifications
D&C RED NO. 27	74.1327	D&C RED NO. 27 Drugs; Listing Of Color Additives Subject To Certification
D&C RED NO. 27	74.2327	D&C RED NO. 27 Cosmetics; Listing Of Color Additives Subject To Certification
D&C RED NO. 27	82.1327	D&C RED NO. 27 Drugs And Cosmetics; Listing Of Certified Provisionally Listed Colors And Specifications
D&C RED NO. 28	74.1328	D&C RED NO. 28 Drugs; Listing Of Color Additives Subject To Certification
D&C RED NO. 28	74.2328	D&C RED NO. 28 Cosmetics; Listing Of Color Additives Subject To Certification
D&C RED NO. 28	82.1328	D&C RED NO. 28 Drugs And Cosmetics; Listing Of Certified Provisionally Listed Colors And Specifications
D&C RED NO. 30	74.1330	D&C RED NO. 30 Drugs; Listing Of Color Additives Subject To Certification
D&C RED NO. 30	74.2330	D&C RED NO. 30 Cosmetics; Listing Of Color Additives Subject To Certification
D&C RED NO. 30	82.1330	D&C RED NO. 30 Drugs And Cosmetics; Listing Of Certified Provisionally Listed Colors And Specifications
D&C RED NO. 31	74.1331	D&C RED NO. 31 Drugs; Listing Of Color Additives Subject To Certification
D&C RED NO. 31	74.2331	D&C RED NO. 31 Cosmetics; Listing Of Color Additives Subject To Certification
D&C RED NO. 31	82.1331	D&C RED NO. 31 Drugs And Cosmetics; Listing Of Certified Provisionally Listed Colors And Specifications
D&C RED NO. 33	74.1333	D&C RED NO. 33 Drugs; Listing Of Color Additives Subject To Certification
D&C RED NO. 33	74.2333	D&C RED NO. 33 Cosmetics; Listing Of Color Additives Subject To Certification

D&C RED NO. 33	82.1333	D&C RED NO. 33 Drugs And Cosmetics; Listing Of Certified Provisionally Listed Colors And Specifications
D&C RED NO. 34	74.1334	D&C RED NO. 34 Drugs; Listing Of Color Additives Subject To Certification
D&C RED NO. 34	74.2334	D&C RED NO. 34 Cosmetics; Listing Of Color Additives Subject To Certification
D&C RED NO. 34	82.1334	D&C RED NO. 34 Drugs And Cosmetics; Listing Of Certified Provisionally Listed Colors And Specifications
D&C RED NO. 36	74.1336	D&C RED NO. 36 Drugs; Listing Of Color Additives Subject To Certification
D&C RED NO. 36	74.2336	D&C RED NO. 36 Cosmetics; Listing Of Color Additives Subject To Certification
D&C RED NO. 36	82.1336	D&C RED NO. 36 Drugs And Cosmetics; Listing Of Certified Provisionally Listed Colors And Specifications
D&C RED NO. 39	74.1339	D&C RED NO. 39 Drugs; Listing Of Coir Additives Subject To Certification
D&C VIOLET NO. 2	74.1602	D&C VIOLET NO. 2 Drugs; Listing Of Color Additives Subject To Certification
D&C VIOLET NO. 2	74.2602	D&C VIOLET NO. 2 Cosmetics; Listing Of Color Additives Subject To Certification
D&C VIOLET NO. 2	74.2602a	EXT. D&C VIOLET NO. 2 Cosmetics; Listing Of Color Additives Subject To Certification
D&C VIOLET NO. 2	74.3602	D&C VIOLET NO. 2 Medical Devices; Listing Of Color Additives Subject To Certification
D&C VIOLET NO. 2	82.1602	D&C VIOLET NO. 2 Drugs And Cosmetics; Listing Of Certified Provisionally Listed Colors And Specifications
D&C YELLOW NO. 7	74.1707	D&C YELLOW NO. 7 Drugs; Listing Of Color Additives Subject To Certification
D&C YELLOW NO. 7	74.1707a	EXT. D&C YELLOW NO. 7 Drugs; Listing Of Color Additives Subject To Certification
D&C YELLOW NO. 7	74.2707	D&C YELLOW NO. 7 Cosmetics; Listing Of Color Additives Subject To Certification
D&C YELLOW NO. 7	74.2707a	EXT. D&C YELLOW NO. 7 Cosmetics; Listing Of Color Additives Subject To Certification
D&C YELLOW NO. 7	82.1707	D&C YELLOW NO. 7 Drugs And Cosmetics; Listing Of Certified Provisionally Listed Colors And Specifications
D&C YELLOW NO. 7	82.2707a	EXT. D&C YELLOW NO. 7 Externally Applied Drugs And Cosmetics; Listing Of Certified Provisionally Listed Colors And Specifications
D&C YELLOW NO. 8	74.1708	D&C YELLOW NO. 8 Drugs; Listing Of Color Additives Subject To Certification
D&C YELLOW NO. 8	74.2708	D&C YELLOW NO. 8 Cosmetics; Listing Of Color Additives Subject To Certification
D&C YELLOW NO. 8	82.1708	D&C YELLOW NO. 8 Drugs And Cosmetics; Listing Of Certified Provisionally Listed Colors And Specifications
D&C YELLOW NO. 10	74.1710	D&C YELLOW NO. 10 Drugs; Listing Of Color Additives Subject To Certification
D&C YELLOW NO. 10	74.2710	D&C YELLOW NO. 10 Cosmetics; Listing Of Color Additives Subject To Certification
D&C YELLOW NO. 10	74.3710	D&C YELLOW NO. 10 Medical Devices; Listing Of Color Additives Subject To Certification
D&C YELLOW NO. 10	82.1710	D&C YELLOW NO. 10 Drugs And Cosmetics; Listing Of Certified Provisionally Listed Colors And Specifications
D&C YELLOW NO. 11	74.1711	D&C YELLOW NO. 11 Drugs; Listing Of Color Additives Subject To Certification
D&C YELLOW NO. 11	74.2711	D&C YELLOW NO. 11 Cosmetics; Listing Of Color Additives Subject To Certification
D&C, LAKES	82.1051	LAKES (D&C) Drugs And Cosmetics; Listing Of Certified Provisionally Listed Colors And Specifications
D&C, LAKES	82.2051	LAKES (EXT. D&C) Externally Applied Drugs And Cosmetics; Listing Of Certified Provisionally Listed Colors And Specifications
DAILY ACTIVITY ASSIST	890.5050	DAILY ACTIVITY ASSIST DEVICE Physical Medicine Therapeutic Devices; Physical Medicine Devices
DAIRY CHOCOLATES	163.145	MIXED DAIRY PRODUCT CHOCOLATES Requirements For Specific Standardized Cacao Products; Cacao Products

DAIRY FARMS	1210.11	SANITARY INSPECTION OF DAIRY FARMS Regulations Under The Federal Import Milk Act
DAM	872.6300	RUBBER DAM AND ACCESSORIES Miscellaneous Devices; Dental Devices
DANDRUFF	358.710	ACTIVE INGREDIENTS FOR THE CONTROL OF DANDRUFF, SEBORRHEIC DERMATITIS, OR PSORIASIS Miscellaneous External Drug Products For Over-the-counter Human Use
DANDRUFF	358.750	LABELING OF DRUG PRODUCTS FOR THE CONTROL OF DANDRUFF, SEBORRHEIC DERMATITIS OR PSORIASIS Miscellaneous External Drug Products For Over-the-counter Human Use
DATA	5.604	MANUFACTURERS REQUIREMENT TO PROVIDE DATA TO ULTIMATE PURCHASERS OF ELECTRONIC PRODUCTS Radiation Control; Redelegations Of Authority; Delegations Of Authority And Organization
DATA	5.605	DEALER AND DISTRIBUTOR DIRECTION TO PROVIDE DATA TO MANUFACTURERS OF ELECTRONIC PRODUCTS Radiation Control; Redelegations Of Authority; Delegations Of Authority And Organization
DATA	58.51	SPECIMEN AND DATA STORAGE FACILITIES Facilities; Good Laboratory Practice For Nonclinical Laboratory Studies
DATA	58.190	STORAGE AND RETRIEVAL OF RECORDS AND DATA Records And Reports; Good Laboratory Practice For Nonclinical Laboratory Studies
DATA	320.21	REQUIREMENTS FOR SUBMISSION OF IN VIVO BIOAVAILABILITY AND BIOEQUIVALENCE DATA Procedures For Determining The Bioavailability Of Drug Products; Bioavailability And Bioequivalence Requirements
DATA	510.112	ANTIBIOTICS USED IN VETERINARY MEDICINE AND FOR NONMEDICAL PURPOSES; REQUIRED DATA Specific Administrative Rulings And Decisions; New Animal Drugs
DATA ANALYZER	884.2050	OBSTETRIC DATA ANALYZER Obstetrical And Gynecological Monitoring Devices; Obstetrical And Gynecological Devices
DATA CALCULATOR	868.1880	PULMONARY-FUNCTION DATA CALCULATOR Diagnostic Devices; Anesthesiology Devices
DATA, FOREIGN	314.106	FOREIGN DATA FDA Action On Applications; Applications For FDA Approval To Market A New Drug
DATA & INFORMATION	12.85	DISCLOSURE OF DATA AND INFORMATION BY THE PARTICIPANTS Hearing Procedures; Formal Evidentiary Public Hearing
DATA & INFORMATION	13.25	DISCLOSURE OF DATA AND INFORMATION BY THE PARTICIPANTS Hearing Procedures; Public Hearing Before A Public Board Of Inquiry
DATA & INFORMATION	20.81	DATA AND INFORMATION PREVIOUSLY DISCLOSED TO THE PUBLIC Limitations On Exemptions; Public Information
DATA & INFORMATION	20.109	DATA AND INFORMATION OBTAINED BY CONTRACT Availability Of Specific Categories Of Records; Public Information
DATA & INFORMATION	20.110	DATA AND INFORMATION ABOUT FDA EMPLOYEES Availability Of Specific Categories Of Records; Public Information
DATA & INFORMATION	20.111	DATA AND INFORMATION SUBMITTED VOLUNTARILY TO THE FDA Availability Of Specific Categories Of Records; Public Information
DATA & INFORMATION	20.114	DATA AND INFORMATION SUBMITTED PURSUANT TO COOPERATIVE QUALITY ASSURANCE AGREEMENTS Availability Of Specific Categories Of Records; Public Information
DATA & INFORMATION	71.15	CONFIDENTIALITY OF DATA AND INFORMATION IN COLOR ADDITIVE PETITIONS General Provisions; Color Additive Petitions
DATA & INFORMATION	312.130	AVAILABILITY FOR PUBLIC DISCLOSURE OF DATA AND INFORMATION IN AN IND Miscellaneous; New Drugs For Investigational Use
DATA & INFORMATION	314.430	AVAILABILITY FOR PUBLIC DISCLOSURE OF DATA AND INFORMATION IN AN APPLICATION Miscellaneous Provisions; Applications For FDA Approval To Market A New Drug
DATA & INFORMATION	316.52	AVAILABILITY FOR PUBLIC DISCLOSURE OF DATA AND INFORMATION IN REQUESTS AND APPLICATIONS Orphan Drugs
DATA & INFORMATION	514.11	CONFIDENTIALITY OF DATA AND INFORMATION IN A NEW ANIMAL DRUG APPLICATION FILE General Provisions; New Animal Drug Applications
DATA & INFORMATION	514.12	CONFIDENTIALITY OF DATA AND INFORMATION IN AN INVESTIGATIONAL NEW ANIMAL DRUG NOTICE General Provisions; New Animal Drug Applications
DATA & INFORMATION	601.50	CONFIDENTIALITY OF DATA AND INFORMATION IN AN INVESTIGATIONAL NEW DRUG NOTICE FOR A BIOLOGICAL PRODUCT Confidentiality Of Information; Licensing

DATA & INFORMATION	601.51	CONFIDENTIALITY OF DATA AND INFORMATION IN APPLICATIONS FOR BIOLOGICS LICENSES Confidentiality Of Information; Licensing
DATA & INFORMATION	812.38	CONFIDENTIALITY OF DATA AND INFORMATION Application And Administrative Action; Investigational Device Exemptions
DATA & INFORMATION	814.9	CONFIDENTIALITY OF DATA AND INFORMATION IN A PREMARKET APPROVAL APPLICATION (PMA) FILE General; Premarket Approval Of Medical Devices
DATA & INFORMATION	814.122	CONFIDENTIALITY OF DATA AND INFORMATION Humanitarian Use Devices; Premarket Approval Of Medical Devices
DATA & INFORMATION	860.5	CONFIDENTIALITY AND USE OF DATA AND INFORMATION SUBMITTED IN CONNECTION WITH CLASSIFICATION AND RECLASSIFICATION General; Medical Device Classification Procedures
DATA & INFORMATION	895.22	SUBMISSION OF DATA AND INFORMATION BY THE MANUFACTURER, DISTRIBUTOR, OR IMPORTER General Provisions; Banned Devices
DATA OR INFORMATION	20.44	PRESUBMISSION REVIEW OF REQUEST FOR CONFIDENTIALITY OF VOLUNTARILY SUBMITTED DATA OR INFORMATION Procedures And Fees; Public Information
DATA OR INFORMATION	20.91	USE OF DATA OR INFORMATION FOR ADMINISTRATIVE OR COURT ENFORCEMENT ACTION Limitations On Exemptions; Public Information
DATA PROCESSING	862.2100	CALCULATOR/DATA PROCESSING MODULE FOR CLINICAL USE Clinical Laboratory Instruments; Clinical Chemistry And Clinical Toxicology Devices
DATA AND REPORTS	1002.7	SUBMISSION OF DATA AND REPORTS General Provisions; Records and Reports
DATA, TECHNICAL	1002.3	NOTIFICATION TO USER OF PERFORMANCE AND TECHNICAL DATA General Provisions; Records and Reports
DATE, EFFECTIVE	5.301	ISSUANCE OF INITIAL EMERGENCY PERMIT ORDERS AND NOTICES OF CONFIRMATION OF EFFECTIVE DATE OF FINAL REGULATIONS ON FOOD FOR HUMAN AND ANIMAL CONSUMPTION Foods And Cosmetics; Redelegations Of Authority; Delegations Of Authority And Organization
DATE, EFFECTIVE	12.37	EFFECTIVE DATE OF A REGULATION Initiation Of Proceedings; Formal Evidentiary Public Hearing
DATE, EFFECTIVE	12.38	DATE OF AN ORDER Initiation Of Proceedings; Formal Evidentiary Public Hearing
DATE, EFFECTIVE	171.102	EFFECTIVE DATE OF REGULATIONS Administrative Actions On Applications; Food Additive Petitions
DATE, EFFECTIVE	314.107	EFFECTIVE DATE OF APPROVAL OF A 505(B)(2) APPLICATION OR ABBREVIATED NEW DRUG APPLICATION UNDER SECTION 505(J) Applications For FDA Approval To Market A New Drug
DATE, EFFECTIVE	571.102	EFFECTIVE DATE OF REGULATION Food Additive Petitions
DATE, EFFECTIVE	895.30	SPECIAL EFFECTIVE DATE General Provisions; Banned Devices
DATE, EXPIRATION	201.17	DRUGS; LOCATION OF EXPIRATION DATE General Labeling Provisions; Labeling
DATE, EXPIRATION	1301.13	APPLICATION FOR REGISTRATION; TIME FOR APPLICATION; EXPIRATION DATE; APPLICATION FORMS Registration; Registration Of Manufacturers, Distributors, And Dispensers Of Controlled Substances
DATE, EXPIRATION	1309.31	TIME FOR APPLICATION FOR REGISTRATION; EXPIRATION DATE Registration Of Manufacturers, Distributors, Importers And Exporters Of List I Chemicals
DATE, EXPIRATION	1312.25	EXPIRATION DATE Importation And Exportation Of Controlled Substances
DATE, MANUFACTURE	610.50	DATE OF MANUFACTURE Dating Period Limitations; General Biological Products Standards
DATES	20.115	PRODUCT CODES FOR MANUFACTURING OR SALES DATES Availability Of Specific Categories Of Records; Public Information
DATES, APPROVAL	5.106	SUBMISSION OF AND EFFECTIVE APPROVAL DATES FOR ABBREVIATED NEW DRUG APPLICATIONS AND CERTAIN NEW DRUG APPLICATIONS Human Drugs; Redelegations Of Authority; Delegations Of Authority And Organization
DATES, APPROVAL	5.503	SUBMISSION OF AND EFFECTIVE APPROVAL DATES FOR ABBREVIATED NEW ANIMAL DRUG APPLICATIONS AND CERTAIN NEW ANIMAL DRUG APPLICATIONS Animal Drugs; Redelegations Of Authority; Delegations Of Authority And Organization
DATES, COMPLIANCE	898.13	COMPLIANCE DATES Performance Standards For Electrode Lead Wires And Patient Cables

DATES, EFFECTIVE	5.107	EXTENSIONS OR STAYS OF EFFECTIVE DATES FOR COMPLIANCE WITH CERTAIN LABELING REQUIREMENTS FOR HUMAN PRESCRIPTION DRUGS Human Drugs; Redelegations Of Authority; Delegations Of Authority And Organization
DATES, EFFECTIVE	61.36	EFFECTIVE DATES Procedures For Performance Standards Development And Publication; Procedures For Performance Standards Development
DATES, EFFECTIVE	861.36	EFFECTIVE DATES Procedures For Performance Standards Development And Publication; Procedures For Performance Standards Development
DATES, EFFECTIVE	862.3	EFFECTIVE DATES OF REQUIREMENT FOR PREMARKET APPROVAL Clinical Chemistry And Clinical Toxicology Devices
DATES, EFFECTIVE	864.3	EFFECTIVE DATES OF REQUIREMENT FOR PREMARKET APPROVAL Hematology And Pathology Devices
DATES, EFFECTIVE	866.3	EFFECTIVE DATES OF REQUIREMENT FOR PREMARKET APPROVAL Immunology And Microbiology Devices
DATES, EFFECTIVE	868.3	EFFECTIVE DATES OF REQUIREMENT FOR PREMARKET APPROVAL Anesthesiology Devices
DATES, EFFECTIVE	870.3	EFFECTIVE DATES OF REQUIREMENT FOR PREMARKET APPROVAL Cardiovascular Devices
DATES, EFFECTIVE	872.3	EFFECTIVE DATES OF REQUIREMENT FOR PREMARKET APPROVAL Dental Devices
DATES, EFFECTIVE	874.3	EFFECTIVE DATES OF REQUIREMENT FOR PREMARKET APPROVAL Ear, Nose, And Throat Devices
DATES, EFFECTIVE	876.3	EFFECTIVE DATES OF REQUIREMENT FOR PREMARKET APPROVAL Gastroenterology-urology Devices
DATES, EFFECTIVE	878.3	EFFECTIVE DATES OF REQUIREMENT FOR PREMARKET APPROVAL General And Plastic Surgery Devices
DATES, EFFECTIVE	880.3	EFFECTIVE DATES OF REQUIREMENT FOR PREMARKET APPROVAL General Hospital And Personal Use Devices
DATES, EFFECTIVE	882.3	EFFECTIVE DATES OF REQUIREMENT FOR PREMARKET APPROVAL Neurological Devices
DATES, EFFECTIVE	884.3	EFFECTIVE DATES OF REQUIREMENT FOR PREMARKET APPROVAL Obstetrical And Gynecological Devices
DATES, EFFECTIVE	886.3	EFFECTIVE DATES OF REQUIREMENT FOR PREMARKET APPROVAL Ophthalmic Devices
DATES, EFFECTIVE	888.3	EFFECTIVE DATES OF REQUIREMENT FOR PREMARKET APPROVAL Orthopedic Devices
DATES, EFFECTIVE	890.3	EFFECTIVE DATES OF REQUIREMENT FOR PREMARKET APPROVAL Physical Medicine Devices
DATES, EFFECTIVE	892.3	EFFECTIVE DATES OF REQUIREMENTS FOR PREMARKET APPROVAL Radiology Devices
DATES, EFFECTIVE	1302.05	EFFECTIVE DATES OF LABELING REQUIREMENTS Labeling And Packaging Requirements For Controlled Substances
DATING, EXPIRATION	211.137	EXPIRATION DATING Packaging And Labeling Control; Current Good Manufacturing Practice For Finished Pharmaceuticals
DATING PERIODS	610.53	DATING PERIODS FOR LICENSED BIOLOGICAL PRODUCTS Dating Period Limitations; General Biological Products Standards
DAYTIME SEDATIVES	310.519	DRUG PRODUCTS MARKETED AS OVER-THE-COUNTER DAYTIME SEDATIVES Requirements For Specific New Drugs Or Devices; New Drugs
DC-DEFIBRILLATOR	870.5300	DC-DEFIBRILLATOR (INCLUDING PADDLES) Cardiovascular Therapeutic Devices; Cardiovascular Devices
DEA	5.104	RESPONSES TO DRUG ENFORCEMENT ADMINISTRATION TEMPORARY SCHEDULING NOTICES Human Drugs; Redelegations Of Authority; Delegations Of Authority And Organization
DEALER	5.605	DEALER AND DISTRIBUTOR DIRECTION TO PROVIDE DATA TO MANUFACTURERS OF ELECTRONIC PRODUCTS Radiation Control; Redelegations Of Authority; Delegations Of Authority And Organization
DEALERS	1002.40	RECORDS TO BE OBTAINED BY DEALERS AND DISTRIBUTORS Dealer And Distributor Records; Records and Reports
DEALERS	1002.41	DISPOSITION OF RECORDS OBTAINED BY DEALERS AND DISTRIBUTORS Dealer And Distributor Records; Records and Reports
DEALERS	1002.42	CONFIDENTIALITY OF RECORDS FURNISHED BY DEALERS AND DISTRIBUTORS Dealer And Distributor Records; Records and Reports
DEALERS	1003.22	COPIES OF COMMUNICATIONS SENT TO PURCHASERS, DEALERS OR DISTRIBUTORS Notification; Notification Of Defects Or Failure To Comply

DECISIONS, SIGNIFICANT	10.70	DOCUMENTATION OF SIGNIFICANT DECISIONS IN ADMINISTRATIVE FILE General Administrative Procedures; Administrative Practices And Procedures
DECISIONS, SUMMARY	12.93	SUMMARY DECISIONS Hearing Procedures; Formal Evidentiary Public Hearing
DECISIONS, SUMMARY	17.17	SUMMARY DECISIONS Civil Money Penalties Hearings
DECLARATION	101.30	PERCENTAGE JUICE DECLARATION FOR FOODS PURPORTING TO BE BEVERAGES THAT CONTAIN FRUIT OR VEGETABLE JUICE Food Labeling
DECLARATION	161.30	DECLARATION OF QUANTITY OF CONTENTS ON LABELS FOR CANNED OYSTERS General Provisions; Fish And Shellfish
DECLARATION	201.20	DECLARATION OF PRESENCE OF FD&C YELLOW NO. 5 AND/OR NO. 6 IN CERTAIN DRUGS FOR HUMAN USE General Labeling Provisions; Labeling
DECLARATION	201.21	DECLARATION OF PRESENCE OF PHENYLALANINE AS A COMPONENT OF ASPARTAME IN OVER-THE-COUNTER AND PRESCRIPTION DRUGS FOR HUMAN USE General Labeling Provisions; Labeling
DECLARATION, CONTENTS	101.105	DECLARATION OF NET QUANTITY OF CONTENTS WHEN EXEMPT Exemptions From Food Labeling Requirements; Food Labeling
DECLARATION, CONTENTS	201.51	DECLARATION OF NET QUANTITY OF CONTENTS Labeling Requirements For Prescription Drugs And/or Insulin; Labeling
DECLARATION, CONTENTS	201.62	DECLARATION OF NET QUANTITY OF CONTENTS Labeling Requirements For Over-the-counter Drugs; Labeling
DECLARATION, CONTENTS	501.105	DECLARATION OF NET QUANTITY OF CONTENTS WHEN EXEMPT Exemptions From Animal Food Labeling Requirements; Animal Food Labeling
DECLARATION, CONTENTS	701.13	DECLARATION OF NET QUANTITY OF CONTENTS Package Form; Cosmetic Labeling
DECLARATION, CONTENTS	801.62	DECLARATION OF NET QUANTITY OF CONTENTS Labeling Requirements For Over-the-counter Devices; Labeling
DECLARATION, EXPORT	1313.22	CONTENTS OF EXPORT DECLARATION Importation And Exportation Of Precursors And Essential Chemicals
DECLARATION, EXPORT	1313.23	DISTRIBUTION OF EXPORT DECLARATION Importation And Exportation Of Precursors And Essential Chemicals
DECLARATION, IMPORT	1312.18	CONTENTS OF IMPORT DECLARATION Importation Of Controlled Substances; Importation And Exportation Of Controlled Substances
DECLARATION, IMPORT	1312.19	DISTRIBUTION OF IMPORT DECLARATION Importation Of Controlled Substances; Importation And Exportation Of Controlled Substances
DECLARATION, IMPORT	1313.13	CONTENTS OF IMPORT DECLARATION Importation And Exportation Of Precursors And Essential Chemicals
DECLARATION, IMPORT	1313.14	DISTRIBUTION OF IMPORT DECLARATION Importation And Exportation Of Precursors And Essential Chemicals
DECLARATION, INTERNATIONAL	1313.33	CONTENTS OF AN INTERNATIONAL TRANSACTION DECLARATION Importation And Exportation Of Precursors And Essential Chemicals
DECLARATION, INTERNATIONAL	1313.34	DISTRIBUTION OF THE INTERNATIONAL TRANSACTION DECLARATION Importation And Exportation Of Precursors And Essential Chemicals
DECLARATION, JUICE	101.30	PERCENTAGE JUICE DECLARATION FOR FOOD PURPORTING TO BE BEVERAGES THAT CONTAIN FRUIT OR VEGETABLE JUICE Specific Food Labeling Requirements; Food Labeling
DECLARATION, LABEL	201.312	MAGNESIUM SULFATE HEPTAHYDRATE; LABEL DECLARATION ON DRUG PRODUCTS Specific Labeling Requirements For Specific Drug Products; Labeling
DECLARATION, LABEL	501.103	PETITIONS REQUESTING EXEMPTIONS FROM OR SPECIAL REQUIREMENTS FOR LABEL DECLARATION OF INGREDIENTS Exemptions From Animal Food Labeling Requirements; Animal Food Labeling
DECOMPRESSION	884.5225	ABDOMINAL DECOMPRESSION CHAMBER Obstetrical And Gynecological Therapeutic Devices; Obstetrical And Gynecological Devices
DECONGESTANT, NASAL	341.20	NASAL DECONGESTANT ACTIVE INGREDIENTS Cold, Cough, Allergy, Bronchodilator, And Antiasthmatic Drug Products For OTC Human Use
DECONGESTANT, NASAL	341.80	LABELING NASAL DECONGESTANT DRUG PRODUCTS Cold, Cough, Allergy, Bronchodilator, And Antiasthmatic Drug Products For OTC Human Use
DECOQUINATE	520.534	DECOQUINATE Oral Dosage Form New Animal Drugs
DECOQUINATE	556.170	DECOQUINATE Specific Tolerances For Residues Of New Animal Drugs; Tolerances For Residues Of New Animal Drugs In Food

DECOQUINATE	558.195	DECOQUINATE Specific New Animal Drugs For Use In Animal Feeds; New Animal Drugs For Use In Animal Feeds
DECORATIVE CERAMICWARE	109.16	ORNAMENTAL AND DECORATIVE CERAMICWARE Unavoidable Contaminants In Food For Human Consumption And Food Packaging Materials
DEFATTED COTTONSEED FLOUR	73.140	TOASTED PARTIALLY DEFATTED COOKED COTTONSEED FLOUR Foods; Listing Of Color Additives Exempt From Certification
DEFAULT	17.11	DEFAULT UPON FAILURE TO FILE AN ANSWER Civil Money Penalties Hearings
DEFECT	1003.2	DEFECT IN AN ELECTRONIC PRODUCT General Provisions; Notification Of Defects Or Failure To Comply
DEFECT	1003.10	DISCOVERY OF DEFECT OR FAILURE OF COMPLIANCE BY MANUFACTURER; NOTICE REQUIREMENTS Discovery Of Defect Or Failure To Comply; Notification Of Defects Or Failure To Comply
DEFECT	1003.11	DETERMINATION BY SECRETARY THAT PRODUCT FAILS TO COMPLY OR HAS A DEFECT Discovery Of Defect Or Failure To Comply; Notification Of Defects Or Failure To Comply
DEFECT REPORTS	20.113	VOLUNTARY PRODUCT DEFECT REPORTS Availability Of Specific Categories Of Records; Public Information
DEFECTIVE APPLICATIONS	1301.14	FILING OF APPLICATION; ACCEPTANCE FOR FILING; DEFECTIVE APPLICATIONS Registration Of Manufacturers, Distributors, And Dispensers Of Controlled Substances
DEFECTIVE APPLICATIONS	1309.34	ACCEPTANCE FOR FILING; DEFECTIVE APPLICATIONS Registration Of Manufacturers, Distributors, Importers And Exporters Of List I Chemicals
DEFECTIVE ORDER FORMS	1305.11	UNACCEPTED AND DEFECTIVE ORDER FORMS Order Forms
DEFECTS	5.603	NOTIFICATION OF DEFECTS IN, AND REPAIR OR REPLACEMENT OF, ELECTRONIC PRODUCTS Radiation Control; Redelegations Of Authority; Delegations Of Authority And Organization
DEFECTS	110.110	NATURAL OR UNAVOIDABLE DEFECTS IN FOOD FOR HUMAN USE THAT PRESENT NO HEALTH HAZARD Defect Action Levels; Current Good Manufacturing Practice In Manufacturing, Packing, Or Holding Human Food
DEFECTS, LEAKAGE	800.20	PATIENT EXAMINATION GLOVES AND SURGEON'S GLOVES; SAMPLE PLANS AND TEST METHOD FOR LEAKAGE DEFECTS; ADULTERATION Requirements For Specific Medical Devices; General
DEFECTS, NEURAL TUBE	101.79	HEALTH CLAIMS: FOLATE AND NEURAL TUBE DEFECTS Food Labeling
DEFERRAL, DONOR	610.41	DONOR DEFERRAL Testing Requirements For Communicable Disease Agents; General Biological Products Standards
DEFERRALS	208.26	EXEMPTIONS AND DEFERRALS Medication Guides For Prescription Drug Products
DEFIBRILLATOR	870.5300	DC-DEFIBRILLATOR (INCLUDING PADDLES) Cardiovascular Therapeutic Devices; Cardiovascular Devices
DEFIBRILLATOR	870.5325	DEFIBRILLATOR TESTER Cardiovascular Therapeutic Devices; Cardiovascular Devices
DEFICIENCIES	5.406	NOTIFICATION TO SPONSORS OF DEFICIENCIES IN PETITIONS FOR RECLASSIFICATION OF MEDICAL DEVICES Medical Devices And Radiological Health; Redelegations Of Authority; Delegations Of Authority And Organization
DEFICIENCY	864.7290	FACTOR DEFICIENCY TEST Hematology Kits And Packages; Hematology And Pathology Devices
DEFINITIONS	130.17	TEMPORARY PERMITS FOR INTERSTATE SHIPMENT OF EXPERIMENTAL PACKS OF FOOD VARYING FROM THE REQUIREMENTS OF DEFINITIONS AND STANDARDS OF IDENTITY Food Standards: General
DEFINITIONS	130.20	FOOD ADDITIVES PROPOSED FOR USE IN FOODS FOR WHICH DEFINITIONS AND STANDARDS OF IDENTITY ARE ESTABLISHED Food Standards: General
DEFOAMER	870.4230	CARDIOPULMONARY BYPASS DEFOAMER Cardiovascular Surgical Devices; Cardiovascular Devices
DEFOAMING AGENTS	173.340	DEFOAMING AGENTS Specific Usage Additives; Secondary Direct Food Additives Permitted In Food For Human Consumption
DEFOAMING AGENTS	176.200	DEFOAMING AGENTS USED IN COATINGS Substances For Use Only As Components Of Paper And Paperboard; Indirect Food Additives

DEFOAMING AGENTS	176.210	DEFOAMING AGENTS USED IN THE MANUFACTURE OF PAPER AND PAPERBOARD Substances For Use Only As Components Of Paper And Paperboard; Indirect Food Additives
DEGERMINATED CORN	137.265	DEGERMINATED WHITE CORN MEAL Requirements For Specific Standardized Cereal Flours And Related Products; Cereal Flours And Related Products
DEGERMINATED CORN	137.285	DEGERMINATED YELLOW CORN MEAL Requirements For Specific Standardized Cereal Flours And Related Products; Cereal Flours And Related Products
DEGRADATION	864.7320	FIBRINOGEN/FIBRIN DEGRADATION PRODUCTS ASSAY Hematology Kits And Packages; Hematology And Pathology Devices
DEGREASER, SKIN	878.4730	SURGICAL SKIN DEGREASER OR ADHESIVE TAPE SOLVENT Surgical Devices; General And Plastic Surgery Devices
DEGREE OF CONSTRAINT	888.6	DEGREE OF CONSTRAINT General Provisions; Orthopedic Devices
DEHYDRATED BEETS	73.40	DEHYDRATED BEETS (BEET POWDER) Foods; Listing Of Color Additives Exempt From Certification
DEHYDRATED FORAGE CROPS	573.400	ETHOXYQUIN IN CERTAIN DEHYDRATED FORAGE CROPS Food Additive Listing; Food Additives Permitted In Feed And Drinking Water Of Animals
DEHYDROACETIC ACID	172.130	DEHYDROACETIC ACID Food Preservatives; Food Additives Permitted For Direct Addition To Food For Human Consumption
DEHYDROEPIANDROSTERONE	862.1245	DEHYDROEPIANDROSTERONE (FREE AND SULFATE) TEST SYSTEM Clinical Chemistry Test Systems; Clinical Chemistry And Clinical Toxicology Devices
DEHYDROGENASE	862.1380	HYDROXYBUTYRIC DEHYDROGENASE TEST SYSTEM Clinical Chemistry Test Systems; Clinical Chemistry And Clinical Toxicology Devices
DEHYDROGENASE	862.1420	ISOCITRIC DEHYDROGENASE TEST SYSTEM Clinical Chemistry Test Systems; Clinical Chemistry And Clinical Toxicology Devices
DEHYDROGENASE	862.1440	LACTATE DEHYDROGENASE TEST SYSTEM Clinical Chemistry Test Systems; Clinical Chemistry And Clinical Toxicology Devices
DEHYDROGENASE	862.1445	LACTATE DEHYDROGENASE ISOENZYMES TEST SYSTEM Clinical Chemistry Test Systems; Clinical Chemistry And Clinical Toxicology Devices
DEHYDROGENASE	862.1500	MALIC DEHYDROGENASE TEST SYSTEM Clinical Chemistry Test Systems; Clinical Chemistry And Clinical Toxicology Devices
DEHYDROGENASE	862.1565	6-PHOSPHOGLUCONATE DEHYDROGENASE TEST SYSTEM Clinical Chemistry Test Systems; Clinical Chemistry And Clinical Toxicology Devices
DEHYDROGENASE	862.1670	SORBITOL DEHYDROGENASE TEST SYSTEM Clinical Chemistry Test Systems; Clinical Chemistry And Clinical Toxicology Devices
DEHYDROGENASE	864.7360	ERYTHROCYTIC GLUCOSE-6-PHOSPHATE DEHYDROGENASE ASSAY Hematology Kits And Packages; Hematology And Pathology Devices
DEHYDROGENASE	866.5560	LACTIC DEHYDROGENASE IMMUNOLOGICAL TEST SYSTEM Immunological Test Systems; Immunology And Microbiology Devices
DELEGATIONS	5.10	DELEGATIONS FROM THE SECRETARY OF HEALTH AND HUMAN SERVICES TO THE COMMISSIONER OF FOOD AND DRUGS Delegations Of Authority To The Commissioner Of Food And Drugs; Delegations Of Authority And Organization
DELETERIOUS SUBSTANCES	109.6	ADDED POISONOUS OR DELETERIOUS SUBSTANCES General Provisions; Unavoidable Contaminants In Food For Human Consumption And Food-packaging Material
DELETERIOUS SUBSTANCES	509.6	ADDED POISONOUS OR DELETERIOUS SUBSTANCES General Provisions; Unavoidable Contaminants In Animal Food And Food-packaging Material
DELINTING	173.322	CHEMICALS USED IN DELINTING COTTONSEED Specific Usage Additives; Secondary Direct Food Additives Permitted In Food For Human Consumption
DELIVERY	1230.41	DELIVERY OF CONTAINERS Regulations Under The Federal Caustic Poison Act
DELIVERY, DIRECT	203.31	SAMPLE DISTRIBUTION BY MEANS OTHER THAN MAIL OR COMMON CARRIER (DIRECT DELIVERY BY A REPRESENTATIVE OR DETAILER) Samples; Prescription Drug Marketing
DELIVERY SYSTEM	876.5600	SORBENT REGENERATED DIALYSATE DELIVERY SYSTEM FOR HEMODIALYSIS Therapeutic Devices; Gastroenterology-urology Devices

DELTA-AMINOLEVULINIC	862.1060	DELTA-AMINOLEVULINIC ACID TEST SYSTEM Clinical Chemistry Test Systems; Clinical Chemistry And Clinical Toxicology Devices
DELTA-LACTONE	184.1318	GLUCONO DELTA-LACTONE Listing Of Specific Substances Affirmed As GRAS; Direct Food Substances Affirmed As Generally Recognized As Safe
DEMULCENT	349.60	LABELING OF OPHTHALMIC DEMULCENT DRUG PRODUCTS Ophthalmic Drug Products For Over-the-counter Human Use
DEMULCENTS, OPHTHALMIC	349.12	OPHTHALMIC DEMULCENTS Active Ingredients; Ophthalmic Drug Products For Over-the-counter Human Use
DENIAL OF APPLICATION	5.34	ISSUANCE OF NOTICES RELATING TO PROPOSALS AND ORDERS FOR DEBARMENT AND DENIAL OF AN APPLICATION TO TERMINATE DEBARMENT General Redelegations Of Authority; Delegations Of Authority And Organization
DENIAL OF APPROVAL	5.202	ISSUANCE OF NOTICES OF OPPORTUNITY FOR HEARING ON PROPOSALS FOR DENIAL OF APPROVAL OF APPLICATIONS FOR LICENSES Biologics; Redelegations Of Authority; Delegations Of Authority And Organization
DENIAL OF APPROVAL	814.45	DENIAL OF APPROVAL OF A PMA FDA Action On A PMA; Premarket Approval Of Medical Devices
DENIAL OF APPROVAL	814.118	DENIAL OF APPROVAL OR WITHDRAWAL OF APPROVAL OF AN HDE Humanitarian Use Devices; Premarket Approval Of Medical Devices
DENIAL OF HEARING	12.28	DENIAL OF HEARING IN WHOLE OR IN PART Initiation Of Proceedings; Formal Evidentiary Public Hearing
DENIAL OF HEARING	16.26	DENIAL OF HEARING AND SUMMARY DECISION Initiation Of Proceedings; Regulatory Hearing Before The Food And Drug Administration
DENIAL OF HEARING	314.200	NOTICE OF OPPORTUNITY FOR HEARING; NOTICE OF PARTICIPATION AND REQUEST FOR HEARING; GRANT OR DENIAL OF HEARING Hearing Procedures For New Drugs; Applications For FDA Approval To Market A New Drug
DENIAL OF LICENSE	601.4	ISSUANCE AND DENIAL OF LICENSE General Provisions; Licensing
DENIAL OF PERMIT	108.7	ISSUANCE OR DENIAL OF PERMIT General Provisions; Emergency Permit Control
DENIAL, PERMIT	1210.30	HEARING PROCEDURES FOR PERMIT DENIAL, SUSPENSION, AND REVOCATION Regulations Under The Federal Import Milk Act
DENIAL OF REGISTRATION	1301.35	CERTIFICATE OF REGISTRATION; DENIAL OF REGISTRATION Action On Application For Registration: Revocation Or Suspension Of Registration; Registration Of Manufacturers, Distributors, And Dispensers Of Controlled Substances
DENIAL OF REGISTRATION	1309.42	CERTIFICATE OF REGISTRATION; DENIAL OF REGISTRATION Registration Of Manufacturers, Distributors, Importers And Exporters Of List I Chemicals
DENIAL OF REQUEST	20.47	DENIAL OF A REQUEST FOR RECORDS Procedures And Fees; Public Information
DENIALS	900.16	APPEALS OF DENIALS OF CERTIFICATION Quality Standards And Certification; Mammography
DENIS BROWN	890.3675	DENIS BROWN SPLINT Physical Medicine Prosthetic Devices; Physical Medicine Devices
DENSITOMETER	862.2400	DENSITOMETER/SCANNER (INTEGRATING, REFLECTANCE, TLC, OR RADIOCHROMATOGRAM) FOR CLINICAL USE Clinical Laboratory Instruments; Clinical Chemistry And Clinical Toxicology Devices
DENSITOMETER	870.1450	DENSITOMETER Cardiovascular Diagnostic Devices; Cardiovascular Devices
DENSITOMETER	892.1170	BONE DENSITOMETER Diagnostic Devices; Radiology Devices
DENTAL AMALGAM	872.3110	DENTAL AMALGAM CAPSULE Prosthetic Devices; Dental Devices
DENTAL AMALGAMATOR	872.3100	DENTAL AMALGAMATOR Prosthetic Devices; Dental Devices
DENTAL BONE REPAIR	872.3930	TRICALCIUM PHOSPHATE GRANULES FOR DENTAL BONE REPAIR Prosthetic Devices; Dental Devices
DENTAL BUR	872.3240	DENTAL BUR Prosthetic Devices; Dental Devices
DENTAL CARIES	101.80	HEALTH CLAIMS: DIETARY SUGAR ALCOHOLS AND DENTAL CARIES Food Labeling
DENTAL CEMENT	872.3275	DENTAL CEMENT Prosthetic Devices; Dental Devices
DENTAL CHAIR	872.6250	DENTAL CHAIR AND ACCESSORIES Miscellaneous Devices; Dental Devices
DENTAL DRILL	872.4130	INTRAORAL DENTAL DRILL Surgical Devices; Dental Devices
DENTAL ELECTROSURGICAL	872.4920	DENTAL ELECTROSURGICAL UNIT AND ACCESSORIES Surgical Devices; Dental Devices

DENTAL FLOSS	872.6390	DENTAL FLOSS Miscellaneous Devices; Dental Devices
DENTAL HANDPIECE	872.4200	DENTAL HANDPIECE AND ACCESSORIES Surgical Devices; Dental Devices
DENTAL IMPLANT	872.3980	ENDOSSEOUS DENTAL IMPLANT ACCESSORIES Prosthetic Devices; Dental Devices
DENTAL INSTRUMENT	872.4535	DENTAL DIAMOND INSTRUMENT Surgical Devices; Dental Devices
DENTAL INSTRUMENT	872.4565	DENTAL HAND INSTRUMENT Surgical Devices; Dental Devices
DENTAL LIGHT	872.4620	FIBER OPTIC DENTAL LIGHT Surgical Devices; Dental Devices
DENTAL LIGHT	872.4630	DENTAL OPERATING LIGHT Surgical Devices; Dental Devices
DENTAL MERCURY	872.3700	DENTAL MERCURY Prosthetic Devices; Dental Devices
DENTAL NEEDLE	872.4730	DENTAL INJECTING NEEDLE Surgical Devices; Dental Devices
DENTAL OPERATIVE UNIT	872.6640	DENTAL OPERATIVE UNIT AND ACCESSORIES Miscellaneous Devices; Dental Devices
DENTAL PROTECTOR	868.5820	DENTAL PROTECTOR Therapeutic Devices; Anesthesiology Devices
DENTAL WAX	872.6890	INTRAORAL DENTAL WAX Miscellaneous Devices; Dental Devices
DENTAL X-RAY	872.1820	DENTAL X-RAY EXPOSURE ALIGNMENT DEVICE Diagnostic Devices; Dental Devices
DENTAL X-RAY	872.1840	DENTAL X-RAY POSITION INDICATING DEVICE Diagnostic Devices; Dental Devices
DENTAL X-RAY	872.1905	DENTAL X-RAY FILM HOLDER Diagnostic Devices; Dental Devices
DENTAL X-RAY	1000.60	RECOMMENDATIONS ON ADMINISTRATIVELY REQUIRED DENTAL X-RAY EXAMINATIONS Radiation Protection Recommendations; General
DENTIFRICE, FLUORIDE	355.70	TESTING PROCEDURES FOR FLUORIDE DENTIFRICE PRODUCTS Anticaries Drug Products For Over-the-counter Human Use
DENTURE ADHESIVE	872.3400	KARAYA AND SODIUM BORATE WITH OR WITHOUT ACACIA DENTURE ADHESIVE Prosthetic Devices; Dental Devices
DENTURE ADHESIVE	872.3410	ETHYLENE OXIDE HOMOPOLYMER AND/OR CARBOXYMETHYLCELLULOSE SODIUM DENTURE ADHESIVE Prosthetic Devices; Dental Devices
DENTURE ADHESIVE	872.3420	CARBOXYMETHYLCELLULOSE SODIUM AND CATIONIC POLYACRYLAMIDE POLYMER DENTURE ADHESIVE Prosthetic Devices; Dental Devices
DENTURE ADHESIVE	872.3450	ETHYLENE OXIDE HOMOPOLYMER AND/OR KARAYA DENTURE ADHESIVE Prosthetic Devices; Dental Devices
DENTURE ADHESIVE	872.3480	POLYACRYLAMIDE POLYMER DENTURE ADHESIVE Prosthetic Devices; Dental Devices
DENTURE ADHESIVE	872.3490	CARBOXYMETHYLCELLULOSE SODIUM AND/OR POLYVINYLMETHYLETHER MALEIC ACID CALCIUM-SODIUM DOUBLE SALT DENTURE ADHESIVE Prosthetic Devices; Dental Devices
DENTURE ADHESIVE	872.3500	POLYVINYLMETHYLETHER MALEIC ANHYDRIDE (PVM-MA), ACID COPOLYMER, AND CARBOXYMETHYLCELLULOSE SODIUM (NACMC) DENTURE ADHESIVE Prosthetic Devices; Dental Devices
DENTURE CLEANER	872.3530	MECHANICAL DENTURE CLEANER Prosthetic Devices; Dental Devices
DENTURE CLEANSER	872.3520	OTC DENTURE CLEANSER Prosthetic Devices; Dental Devices
DENTURE CUSHION	872.3540	OTC DENTURE CUSHION OR PAD Prosthetic Devices; Dental Devices
DENTURE KIT	872.3600	PARTIALLY FABRICATED DENTURE KIT Prosthetic Devices; Dental Devices
DENTURE RELINER	872.3560	OTC DENTURE RELINER Prosthetic Devices; Dental Devices
DENTURE RELINING	872.3760	DENTURE RELINING, REPAIRING, OR REBASING RESIN Prosthetic Devices; Dental Devices
DENTURE REPAIR	872.3570	OTC DENTURE REPAIR KIT Prosthetic Devices; Dental Devices
DENTURE TOOTH	872.3580	PREFORMED GOLD DENTURE TOOTH Prosthetic Devices; Dental Devices
DENTURE TOOTH	872.3590	PREFORMED PLASTIC DENTURE TOOTH Prosthetic Devices; Dental Devices

DENTURES	801.405	LABELING OF ARTICLES INTENDED FOR LAY USE IN THE REPAIRING AND/OR REFITTING OF DENTURES Special Requirements For Specific Devices; Labeling
DENTURES	872.3300	HYDROPHILIC RESIN COATING FOR DENTURES Prosthetic Devices; Dental Devices
DEODORANT	357.810	ACTIVE INGREDIENTS FOR DEODORANT DRUG PRODUCTS FOR INTERNAL USE Miscellaneous Internal Drug Products For Over-the-counter Human Use
DEODORANT	357.850	LABELING OF DEODORANT DRUG PRODUCTS FOR INTERNAL USE Miscellaneous Internal Drug Products For Over-the-counter Human Use
DEODORANT	740.12	FEMININE DEODORANT SPRAYS Warning Statements; Cosmetic Product Warning Statements
DEODORIZED	884.5425	SCENTED OR SCENTED DEODORIZED MENSTRUAL PAD Obstetrical And Gynecological Therapeutic Devices; Obstetrical And Gynecological Devices
DEODORIZED	884.5460	SCENTED OR SCENTED DEODORIZED MENSTRUAL TAMPON Obstetrical And Gynecological Therapeutic Devices; Obstetrical And Gynecological Devices
DEOXYCORTISOL	862.1185	COMPOUND S (11-DEOXYCORTISOL) TEST SYSTEM Clinical Chemistry Test Systems; Clinical Chemistry And Clinical Toxicology Devices
DEPARTMENT REGULATIONS	19.5	REFERENCE TO DEPARTMENT REGULATIONS General Provisions; Standards Of Conduct And Conflicts Of Interest
DEPARTMENT SEAL	5.22	CERTIFICATION OF TRUE COPIES AND USE OF DEPARTMENT SEAL General Redelegations Of Authority; Delegations Of Authority And Organization
DEPARTMENTS	20.85	DISCLOSURE TO OTHER FEDERAL GOVERNMENT DEPARTMENTS AND AGENCIES Limitations On Exemptions; Public Information
DEPARTMENTS	20.108	AGREEMENTS BETWEEN THE FDA AND OTHER DEPARTMENTS, AGENCIES, AND ORGANIZATIONS Availability of Specific Categories Of Records; Public Information
DEPC	189.140	DIETHYLPYROCARBONATE (DEPC) Substances Generally Prohibited From Direct Addition Or Use As Human Food; Substances Prohibited From Use In Human Food
DEPOSIT	17.54	DEPOSIT IN THE TREASURY OF THE UNITED STATES Civil Money Penalties Hearings
DEPRESSOR	880.6230	TONGUE DEPRESSOR General Hospital And Personal Use Miscellaneous Devices; General Hospital And Personal Use Devices
DEPTH ELECTRODE	882.1330	DEPTH ELECTRODE Neurological Diagnostic Devices; Neurological Devices
DEPTH GAUGE	888.4300	DEPTH GAUGE FOR CLINICAL USE Surgical Devices; Orthopedic Devices
DERMATITIS	358.710	ACTIVE INGREDIENTS FOR THE CONTROL OF DANDRUFF, SEBORRHEIC DERMATITIS, OR PSORIASIS Miscellaneous External Drug Products For Over-the-counter Human Use
DERMATITIS	358.750	LABELING OF DRUG PRODUCTS FOR THE CONTROL OF DANDRUFF, SEBORRHEIC DERMATITIS Miscellaneous External Drug Products For Over-the-counter Human Use
DERMATOLOGIC	878.4630	ULTRAVIOLET LAMP FOR DERMATOLOGIC DISORDERS Surgical Devices; General And Plastic Surgery Devices
DERMATOLOGY	878.4810	LASER SURGICAL INSTRUMENT FOR USE IN GENERAL AND PLASTIC SURGERY AND IN DERMATOLOGY Surgical Devices; General And Plastic Surgery Devices
DESI	330.12	STATUS OF OVER-THE-COUNTER DRUGS PREVIOUSLY REVIEWED UNDER THE DRUG EFFICACY STUDY (DESI) Administrative Procedures; Over-the-counter Human Drugs Generally Recognized As Safe And Effective & Not Misbranded
DESIGN	58.61	EQUIPMENT DESIGN Equipment; Good Laboratory Practice For Nonclinical Laboratory Studies
DESIGN	129.20	PLANT CONSTRUCTION AND DESIGN Buildings And Facilities; Processing And Bottling Of Bottled Drinking Water
DESIGN	211.42	DESIGN AND CONSTRUCTION FEATURES Buildings And Facilities; Current Good Manufacturing Practice For Finished Pharmaceuticals
DESIGN	211.63	EQUIPMENT DESIGN, SIZE, AND LOCATION Equipment And Facilities; Current Good Manufacturing Practice For Finished Pharmaceuticals
DESIGN	320.26	GUIDELINES ON THE DESIGN OF A SINGLE-DOSE IN VIVO BIOAVAILABILITY STUDY Procedures For Determining The Bioavailability Of Drug Products; Bioavailability And Bioequivalence Requirements

DESIGN	320.27	GUIDELINES ON THE DESIGN OF A MULTIPLE-DOSE IN VIVO BIOAVAILABILITY STUDY Procedures For Determining The Bioavailability Of Drug Products; Bioavailability And Bioequivalence Requirements
DESIGN	820.30	DESIGN CONTROLS Quality System Regulation
DESIGNATED AGENCIES	3.5	PROCEDURES FOR IDENTIFYING THE DESIGNATED AGENCY COMPONENT Product Jurisdiction
DESIGNATED AGENCY	3.4	DESIGNATED AGENCY COMPONENT Product Jurisdiction
DESIGNATING AUTHORITIES	26.65	DESIGNATING AUTHORITIES Framework Provisions; Mutual Recognition Of Pharmaceutical GMP Reports, Medical Device Quality System Audit Reports
DESIGNATION	3.7	REQUEST FOR DESIGNATION Product Jurisdiction
DESIGNATION	3.8	LETTER OF DESIGNATION Product Jurisdiction
DESIGNATION	3.9	EFFECT OF LETTER OF DESIGNATION Product Jurisdiction
DESIGNATION	26.66	DESIGNATION AND LISTING PROCEDURES Framework Provisions; Mutual Recognition Of Pharmaceutical GMP Reports, Medical Device Quality System Audit Reports
DESIGNATION	814.102	DESIGNATION OF HUD STATUS Humanitarian Use Devices; Premarket Approval Of Medical Devices
DESIGNATION, INGREDIENTS	101.4	FOOD; DESIGNATION OF INGREDIENTS General Provisions; Food Labeling
DESIGNATION, INGREDIENTS	501.4	ANIMAL FOOD; DESIGNATION OF INGREDIENTS General Provisions; Animal Food Labeling
DESIGNATION, INGREDIENTS	701.3	DESIGNATION OF INGREDIENTS General Provisions; Cosmetic Labeling
DESIGNATION, ORPHAN-DRUG	316.20	CONTENT AND FORMAT OF A REQUEST FOR ORPHAN-DRUG DESIGNATION Orphan Drugs
DESIGNATION, ORPHAN-DRUG	316.23	TIMING OF REQUESTS FOR ORPHAN-DRUG DESIGNATION; DESIGNATION OF ALREADY APPROVED DRUGS Orphan Drugs
DESIGNATION, ORPHAN-DRUG	316.24	GRANTING ORPHAN-DRUG DESIGNATION Orphan Drugs
DESIGNATION, ORPHAN-DRUG	316.25	REFUSAL TO GRANT ORPHAN-DRUG DESIGNATION Orphan Drugs
DESIGNATION, ORPHAN-DRUG	316.26	AMENDMENT TO ORPHAN-DRUG DESIGNATION Orphan Drugs
DESIGNATION, ORPHAN-DRUG	316.27	CHANGE IN OWNERSHIP OF ORPHAN DRUG-DESIGNATION Orphan Drugs
DESIGNATION, ORPHAN-DRUG	316.29	REVOCATION OF ORPHAN-DRUG DESIGNATION Orphan Drugs
DESIGNATION, ORPHAN-DRUG	316.30	ANNUAL REPORTS OF HOLDER OF ORPHAN-DRUG DESIGNATION Orphan Drugs
DESIGNATIONS, LABEL	130.11	LABEL DESIGNATIONS OF INGREDIENTS FOR STANDARDIZED FOOD Food Standards: General
DESIGNATIONS, ORPHAN-DRUG	316.28	PUBLICATION OF ORPHAN-DRUG DESIGNATIONS Orphan Drugs
DESLORELIN	522.533	DESLORELIN ACETATE Implantation Or Injectable Dosage Form New Animal Drugs
DESOXYCORTICOSTERONE	522.535	DESOXYCORTICOSTERONE PIVALATE Implantation Or Injectable Dosage Form New Animal Drugs
DESOXYCORTICOSTERONE	862.1250	DESOXYCORTICOSTERONE TEST SYSTEM Clinical Chemistry Test Systems; Clinical Chemistry And Clinical Toxicology Devices
DESTRUCTION	1270.43	RETENTION, RECALL, AND DESTRUCTION OF HUMAN TISSUE Human Tissue Intended For Transplantation
DETECTION	864.2360	MYCOPLASMA DETECTION MEDIA AND COMPONENTS Cell And Tissue Culture Products; Hematology And Pathology Devices
DETECTION	872.1745	LASER FLUORESCENCE CARIES DETECTION DEVICE Diagnostic Devices; Dental Devices
DETECTION	872.1870	SULFIDE DETECTION DEVICE Diagnostic Devices ;Dental Devices
DETECTOR	870.1025	ARRHYTHMIA DETECTOR AND ALARM Cardiovascular Diagnostic Devices; Cardiovascular Devices
DETECTOR	870.4205	CARDIOPULMONARY BYPASS BUBBLE DETECTOR Cardiovascular Surgical Devices; Cardiovascular Devices
DETECTOR	872.6350	ULTRAVIOLET DETECTOR Miscellaneous Devices; Dental Devices

DETENTION	5.302	DETENTION OF MEAT, POULTRY, EGGS AND RELATED PRODUCTS Foods And Cosmetics; Redelegations Of Authority; Delegations Of Authority And Organization
DETENTION	5.402	DETENTION OF ADULTERATED OR MISBRANDED MEDICAL DEVICES Medical Devices And Radiological Health; Redelegations Of Authority; Delegations Of Authority And Organization
DETENTION	800.55	ADMINISTRATIVE DETENTION Administrative Practices And Procedures; General
DETERGENT	701.20	DETERGENT SUBSTANCES, OTHER THAN SOAP, INTENDED FOR USE IN CLEANSING THE BODY Labeling Of Specific Ingredients; Cosmetic Labeling
DETERGENT	740.17	FOAMING DETERGENT BATH PRODUCTS Warning Statements; Cosmetic Product Warning Statements
DETERMINATION	5.403	AUTHORIZATION TO USE ALTERNATIVE EVIDENCE FOR DETERMINATION OF THE EFFECTIVENESS OF MEDICAL DEVICES Medical Devices And Radiological Health; Redelegations Of Authority; Delegations Of Authority And Organization
DETERMINATION	5.405	DETERMINATION OF CLASSIFICATION OF DEVICES Medical Devices And Radiological Health; Redelegations Of Authority; Delegations Of Authority And Organization
DETERMINATION	5.500	ISSUANCE OF FEDERAL REGISTER DOCUMENTS PERTAINING TO THE DETERMINATION OF SAFE LEVELS, NOTICE OF NEED FOR DEVELOPMENT OF AN ANALYTICAL METHOD, AND PROHIBITION OF CERTAIN EXTRALABEL DRUG USE Animal Drugs; Redelegations Of Authority; Delegations Of Authority And Organization
DETERMINATION	5.700	AUTHORITY RELATING TO DETERMINATION OF PRODUCT PRIMARY JURISDICTION Product Designation; Redelegations Of Authority; Delegations Of Authority And Organization
DETERMINATION	14.27	DETERMINATION TO CLOSE PORTIONS OF ADVISORY COMMITTEE MEETINGS Meeting Procedures; Public Hearing Before A Public Advisory Committee
DETERMINATION	26.7	PARTICIPATION IN THE EQUIVALENCE ASSESSMENT AND DETERMINATION Specific Sector Provisions For Pharmaceutical GMPs; Mutual Recognition Of Pharmaceutical GMP Reports, Medical Device Quality System Audit Reports
DETERMINATION	26.9	EQUIVALENCE DETERMINATION Specific Sector Provisions For Pharmaceutical GMPs; Mutual Recognition Of Pharmaceutical GMP Reports, Medical Device Quality System Audit Reports
DETERMINATION	108.5	DETERMINATION OF THE NEED FOR A PERMIT General Provisions; Emergency Permit Control
DETERMINATION	108.6	REVOCATION OF DETERMINATION OF NEED FOR PERMIT General Provisions; Emergency Permit Control
DETERMINATION	170.38	DETERMINATION OF FOOD ADDITIVE STATUS Food Additive Safety; Food Additives
DETERMINATION	314.161	DETERMINATION OF REASONS FOR VOLUNTARY WITHDRAWAL OF A LISTED DRUG Applications For FDA Approval To Market A New Drug
DETERMINATION	331.20	DETERMINATION OF PERCENT CONTRIBUTION OF ACTIVE INGREDIENTS Testing Procedures; Antacid Products For Over-the-counter (OTC) Human Use
DETERMINATION	352.73	DETERMINATION OF SPF VALUE Sunscreen Drug Products For OTC Human Use
DETERMINATION	352.76	DETERMINATION IF A PRODUCT IS WATER RESISTANT OR VERY WATER RESISTANT Sunscreen Drug Products For OTC Human Use
DETERMINATION	570.38	DETERMINATION OF FOOD ADDITIVE STATUS Food Additive Safety; Food Additives
DETERMINATION	860.7	DETERMINATION OF SAFETY AND EFFECTIVENESS General; Medical Device Classification Procedures
DETERMINATION	864.7340	FIBRINOGEN DETERMINATION SYSTEM Hematology Kits And Packages; Hematology And Pathology Devices
DETERMINATION	1003.11	DETERMINATION BY SECRETARY THAT PRODUCT FAILS TO COMPLY OR HAS A DEFECT Discovery Of Defect Or Failure To Comply, Notification Of Defects Or Failure To Comply
DETERMINATION	1270.21	DETERMINATION OF DONOR SUITABILITY FOR HUMAN TISSUE INTENDED FOR TRANSPLANTATION Human Tissue Intended For Transplantation
DETERMINATIONS	5.404	NOTIFICATION TO PETITIONERS OF DETERMINATIONS MADE ON PETITIONS FOR RECLASSIFICATION OF MEDICAL DEVICES Medical Devices And Radiological Health; Redelegations Of Authority; Delegations Of Authority And Organization

DETERMINATIONS	5.408	DETERMINATIONS CONCERNING THE TYPE OF VALID SCIENTIFIC EVIDENCE SUBMITTED IN A PREMARKET APPROVAL APPLICATION Medical Devices And Radiological Health; Redelegations Of Authority; Delegations Of Authority And Organization
DETERMINATIONS	5.409	DETERMINATIONS THAT MEDICAL DEVICES PRESENT UNREASONABLE RISK OF SUBSTANTIAL HARM Medical Devices And Radiological Health; Redelegations Of Authority; Delegations Of Authority And Organization
DETERMINATIONS	20.28	FDA DETERMINATIONS OF CONFIDENTIALITY General Policy; Public Information
DETERMINATIONS	60.20	FDA ACTION ON REGULATORY REVIEW PERIOD DETERMINATIONS Regulatory Review Period Determinations; Patent Term Restoration
DETERMINATIONS	60.22	REGULATORY REVIEW PERIOD DETERMINATIONS Regulatory Review Period Determinations; Patent Term Restoration
DETERMINATIONS	60.24	REVISION OF REGULATORY REVIEW PERIOD DETERMINATIONS Regulatory Review Period Determinations; Patent Term Restoration
DETERMINATIONS	60.26	FINAL ACTION ON REGULATORY REVIEW PERIOD DETERMINATIONS Regulatory Review Period Determinations; Patent Term Restoration
DETERMINATIONS	812.66	SIGNIFICANT RISK DEVICE DETERMINATIONS IRB Review And Approval; Investigational Device Exemptions
DETERMINATIONS	864.9320	COPPER SULFATE SOLUTION FOR SPECIFIC GRAVITY DETERMINATIONS Products Used In Establishments That Manufacture Blood And Blood Products; Hematology And Pathology Devices
DETERMINE	601.25	REVIEW PROCEDURES TO DETERMINE LICENSED BIOLOGICAL PRODUCTS ARE SAFE, EFFECTIVE AND NOT MISBRANDED UNDER PRESCRIBED, RECOMMENDED, OR SUGGESTED USE CONDITIONS Biologics Licensing; Licensing
DETERMINE	601.26	RECLASSIFICATION PROCEDURES DETERMINE LICENSED BIOLOGICAL PRODUCTS ARE SAFE, EFFECTIVE, AND NOT MISBRANDED UNDER PRESCRIBED, RECOMMENDED, OR SUGGESTED USE CONDITIONS Biologics Licensing; Licensing
DETERMINING	60.28	TIME FRAME FOR DETERMINING REGULATORY REVIEW PERIODS Regulatory Review Period Determinations; Patent Term Restoration
DETERRENT, NAILBITING	310.536	DRUG PRODUCTS CONTAINING ACTIVE INGREDIENTS OFFERED OTC FOR USE AS A NAILBITING OR THUMBSUCKING DETERRENT Requirements For Specific New Drugs Or Devices; New Drugs
DETERRENT, SMOKING	310.544	DRUG PRODUCTS CONTAINING ACTIVE INGREDIENTS OFFERED OTC FOR USE AS A SMOKING DETERRENT Requirements For Specific New Drugs Or Devices; New Drugs
DETOMIDINE	522.536	DETOMIDINE HYDROCHLORIDE INJECTION Implantation Or Injectable Dosage Form New Animal Drugs
DETOXIFICATION	1304.24	RECORDS FOR MAINTENANCE TREATMENT PROGRAMS AND DETOXIFICATION TREATMENT PROGRAMS Continuing Records; Records And Reports Of Registrants
DEVELOPMENT	5.407	APPROVAL, DISAPPROVAL, OR WITHDRAWAL OF APPROVAL OF PRODUCT DEVELOPMENT PROTOCOLS AND APPLICATIONS FOR PREMARKET APPROVAL FOR MEDICAL DEVICES Medical Devices And Radiological Health; Redelegations Of Authority; Delegations Of Authority And Organization
DEVELOPMENT	601.21	PRODUCTS UNDER DEVELOPMENT Biologics Licensing; Licensing
DEVELOPMENT	814.19	PRODUCT DEVELOPMENT PROTOCOL (PDP) General; Premarket Approval Of Medical Devices
DEVELOPMENT	861.20	SUMMARY OF STANDARDS DEVELOPMENT PROCESS Procedures For Performance Standards Development And Publication; Procedures For Performance Standards Development
DEVELOPMENT	861.30	DEVELOPMENT OF STANDARDS Procedures For Performance Standards Development And Publication; Procedures For Performance Standards Development
DEVIATIONS	113.89	DEVIATIONS IN PROCESSING, VENTING, OR CONTROL OF CRITICAL FACTORS Production And Process Controls; Thermally Processed Low-acid Foods Packaged In Hermetically Sealed Containers
DEVIATIONS	114.89	DEVIATIONS FROM SCHEDULED PROCESSING Production And Process Controls; Acidified Foods
DEVIATIONS	211.100	WRITTEN PROCEDURES; DEVIATIONS Production And Process Controls; Current Good Manufacturing Practice For Finished Pharmaceuticals
DEVIATIONS	330.11	NDA DEVIATIONS FROM APPLICABLE MONOGRAPH Administrative Procedures; Over-the-counter Human Drugs Generally Recognized As Safe And Effective & Not Misbranded
DEVIATIONS	600.14	REPORTING OF BIOLOGICAL PRODUCT DEVIATIONS BY LICENSED MANUFACTURERS Establishment Standards; Biological Products: General

DEVIATIONS	606.171	REPORTING OF PRODUCT DEVIATIONS BY MANUFACTURERS, UNLICENSED BLOOD ESTABLISHMENTS, AND TRANSFUSION SERVICES Records And Reports; Current Good Manufacturing Practices For Blood And Blood Components
DEVICE	5.413	APPROVAL, DISAPPROVAL, OR WITHDRAWAL OF APPROVAL OF APPLICATIONS AND ENTERING INTO AGREEMENTS FOR INVESTIGATIONAL DEVICE EXEMPTIONS Medical Devices And Radiological Health; Redelegations Of Authority; Delegations Of Authority And Organization
DEVICE	5.415	AUTHORITY RELATING TO MEDICAL DEVICE REPORTING PROCEDURES Medical Devices And Radiological Health; Redelegations Of Authority; Delegations Of Authority And Organization
DEVICE	5.701	PREMARKET APPROVAL OF A PRODUCT THAT IS, OR CONTAINS A BIOLOGIC, A DEVICE, OR A DRUG Product Designation; Redelegations Of Authority; Delegations Of Authority And Organization
DEVICE	801.433	WARNING STATEMENTS FOR PRESCRIPTION AND RESTRICTED DEVICE PRODUCTS CONTAINING OR MANUFACTURED WITH CHLOROFLUOROCARBONS Labeling
DEVICE	807.65	EXEMPTIONS FOR DEVICE ESTABLISHMENTS Exemptions; Establishment Registration And Device Listing For Manufacturers And Initial Importers Of Devices
DEVICE	812.66	SIGNIFICANT RISK DEVICE DETERMINATIONS IRB Review And Approval; Investigational Device Exemptions
DEVICE	820.120	DEVICE LABELING Packaging And Labeling Control; Quality System Regulation
DEVICE	820.121	CRITICAL DEVICES, DEVICE LABELING Packaging And Labeling Control; Quality System Regulation
DEVICE	820.130	DEVICE PACKAGING Packaging And Labeling Control; Quality System Regulation
DEVICE	820.160	FINISHED DEVICE INSPECTION Device Evaluation; Quality System Regulation
DEVICE	820.161	CRITICAL DEVICES, FINISHED DEVICE INSPECTION Device Evaluation; Quality System Regulation
DEVICE	821.25	DEVICE TRACKING SYSTEM AND CONTENT REQUIREMENTS: MANUFACTURER REQUIREMENTS Medical Device Tracking Requirements
DEVICE	821.30	TRACKING OBLIGATIONS OF PERSONS OTHER THAN DEVICE MANUFACTURERS: DISTRIBUTOR REQUIREMENTS Medical Device Tracking Requirements
DEVICE	862.1675	BLOOD SPECIMEN COLLECTION DEVICE Clinical Chemistry Test Systems; Clinical Chemistry And Clinical Toxicology Devices
DEVICE	864.3400	DEVICE FOR SEALING MICROSECTIONS Pathology Instrumentation And Accessories; Hematology And Pathology Devices
DEVICE	864.5260	AUTOMATED CELL-LOCATING DEVICE Automated And Semi-automated Hematology Devices; Hematology And Pathology Devices
DEVICE	864.5300	RED CELL INDICES DEVICE Automated And Semi-automated Hematology Devices; Hematology And Pathology Devices
DEVICE	864.5800	AUTOMATED SEDIMENTATION RATE DEVICE Automated And Semi-automated Hematology Devices; Hematology And Pathology Devices
DEVICE	864.5950	BLOOD VOLUME MEASURING DEVICE Automated And Semi-automated Hematology Devices; Hematology And Pathology Devices
DEVICE	864.6100	BLEEDING TIME DEVICE Manual Hematology Devices; Hematology And Pathology Devices
DEVICE	864.6160	MANUAL BLOOD CELL COUNTING DEVICE Manual Hematology Devices; Hematology And Pathology Devices
DEVICE	864.6400	HEMATOCRIT MEASURING DEVICE Manual Hematology Devices; Hematology And Pathology Devices
DEVICE	864.9205	BLOOD AND PLASMA WARMING DEVICE Products Used In Establishments That Manufacture Blood And Blood Products; Hematology And Pathology Devices
DEVICE	864.9750	HEAT-SEALING DEVICE Products Used In Establishments That Manufacture Blood And Blood Products; Hematology And Pathology Devices
DEVICE	866.2440	AUTOMATED MEDIUM DISPENSING AND STACKING DEVICE Microbiology Devices; Immunology And Microbiology Devices
DEVICE	866.2500	MICROTITER DILUTING AND DISPENSING DEVICE Microbiology Devices; Immunology And Microbiology Devices

DEVICE	866.2580	GAS-GENERATING DEVICE Microbiology Devices; Immunology And Microbiology Devices
DEVICE	866.2660	MICROORGANISM DIFFERENTIATION AND IDENTIFICATION DEVICE Microbiology Devices; Immunology And Microbiology Devices
DEVICE	866.2900	MICROBIOLOGICAL SPECIMEN COLLECTION AND TRANSPORT DEVICE Microbiology Devices; Immunology And Microbiology Devices
DEVICE	868.5115	DEVICE TO RELIEVE ACUTE UPPER AIRWAY OBSTRUCTION Therapeutic Devices; Anesthesiology Devices
DEVICE	868.5675	REBREATHING DEVICE Therapeutic Devices; Anesthesiology Devices
DEVICE	868.5770	TRACHEAL TUBE FIXATION DEVICE Therapeutic Devices; Anesthesiology Devices
DEVICE	870.3300	ARTERIAL EMBOLIZATION DEVICE Cardiovascular Prosthetic Devices; Cardiovascular Devices
DEVICE	870.3545	VENTRICULAR BYPASS (ASSIST) DEVICE Cardiovascular Prosthetic Devices; Cardiovascular Devices
DEVICE	870.4075	ENDOMYOCARDIAL BIOPSY DEVICE Cardiovascular Surgical Devices; Cardiovascular Devices
DEVICE	870.5225	EXTERNAL COUNTER-PULSATING DEVICE Cardiovascular Therapeutic Devices; Cardiovascular Devices
DEVICE	872.1740	CARIES DETECTION DEVICE Diagnostic Devices; Dental Devices
DEVICE	872.1745	LASER FLUORESCENCE CARIES DETECTION DEVICE Diagnostic Devices; Dental Devices
DEVICE	872.1820	DENTAL X-RAY EXPOSURE ALIGNMENT DEVICE Diagnostic Devices; Dental Devices
DEVICE	872.1870	SULFIDE DETECTION DEVICE Diagnostic Devices; Dental Devices
DEVICE	872.1840	DENTAL X-RAY POSITION INDICATING DEVICE Diagnostic Devices; Dental Devices
DEVICE	872.6010	ABRASIVE DEVICE AND ACCESSORIES Miscellaneous Devices; Dental Devices
DEVICE	874.5220	EAR, NOSE AND THROAT DRUG ADMINISTRATION DEVICE Therapeutic Devices; Ear, Nose And Throat Devices
DEVICE	874.5350	SUCTION ANTICHOKE DEVICE Therapeutic Devices; Ear, Nose And Throat Devices
DEVICE	874.5370	TONGS ANTICHOKE DEVICE Therapeutic Devices; Ear, Nose And Throat Devices
DEVICE	874.5840	ANT\STAMMERING DEVICE Therapeutic Devices; Ear, Nose And Throat Devices
DEVICE	876.5270	IMPLANTED ELECTRICAL URINARY CONTINENCE DEVICE Therapeutic Devices; Gastroenterology-urology Devices
DEVICE	876.5280	IMPLANTED MECHANICAL/HYDRAULIC URINARY CONTINENCE DEVICE Therapeutic Devices; Gastroenterology-urology Devices
DEVICE	876.5310	NONIMPLANTED, PERIPHERAL ELECTRICAL CONTINENCE DEVICE Therapeutic Devices; Gastroenterology-urology Devices
DEVICE	876.5320	NONIMPLANTED ELECTRICAL CONTINENCE DEVICE Therapeutic Devices; Gastroenterology-urology Devices
DEVICE	876.5540	BLOOD ACCESS DEVICE AND ACCESSORIES Therapeutic Devices; Gastroenterology-urology Devices
DEVICE	878.4400	ELECTROSURGICAL CUTTING AND COAGULATION DEVICE AND ACCESSORIES Surgical Devices; General And Plastic Surgery Devices
DEVICE	878.4930	SUTURE RETENTION DEVICE Surgical Devices; General And Plastic Surgery Devices
DEVICE	880.5210	INTRAVASCULAR CATHETER SECUREMENT DEVICE General Hospital And Personal Use Therapeutic Devices; General Hospital And Personal Use Devices
DEVICE	880.5950	UMBILICAL OCCLUSION DEVICE General Hospital And Personal Use Therapeutic Devices; General Hospital And Personal Use Devices
DEVICE	880.6775	POWERED PATIENT TRANSFER DEVICE General Hospital And Personal Use Miscellaneous Devices; General Hospital And Personal Use Devices
DEVICE	880.6785	MANUAL PATIENT TRANSFER DEVICE General Hospital And Personal Use Miscellaneous Devices; General Hospital And Personal Use Devices
DEVICE	882.1540	GALVANIC SKIN RESPONSE MEASUREMENT DEVICE Neurological Diagnostic Devices; Neurological Devices
DEVICE	882.1550	NERVE CONDUCTION VELOCITY MEASUREMENT DEVICE Neurological Diagnostic Devices; Neurological Devices
DEVICE	882.1560	SKIN POTENTIAL MEASUREMENT DEVICE Neurological Diagnostic Devices; Neurological Devices

DEVICE	882.1570	POWERED DIRECT-CONTACT TEMPERATURE MEASUREMENT DEVICE Neurological Diagnostic Devices; Neurological Devices
DEVICE	882.1620	INTRACRANIAL PRESSURE MONITORING DEVICE Neurological Diagnostic Devices; Neurological Devices
DEVICE	882.4250	CRYOGENIC SURGICAL DEVICE Neurological Surgical Devices; Neurological Devices
DEVICE	882.5050	BIOFEEDBACK DEVICE Neurological Therapeutic Devices; Neurological Devices
DEVICE	882.5235	AVERSIVE CONDITIONING DEVICE Neurological Therapeutic Devices; Neurological Devices
DEVICE	882.5940	ELECTROCONVULSIVE THERAPY DEVICE Neurological Therapeutic Devices; Neurological Devices
DEVICE	882.5950	ARTIFICIAL EMBOLIZATION DEVICE Neurological Therapeutic Devices; Neurological Devices
DEVICE	884.5360	CONTRACEPTIVE INTRAUTERINE DEVICE (IUD) AND INTRODUCER Obstetrical And Gynecological Therapeutic Devices; Obstetrical And Gynecological Devices
DEVICE	884.5380	CONTRACEPTIVE TUBAL OCCLUSION DEVICE (TOD) AND INTRODUCER Obstetrical And Gynecological Therapeutic Devices; Obstetrical And Gynecological Devices
DEVICE	884.5970	CLITORAL ENGORGEMENT DEVICE Obstetrical And Gynecological Therapeutic Devices; Obstetrical And Gynecological Devices
DEVICE	886.1290	FIXATION DEVICE Diagnostic Devices; Ophthalmic Devices
DEVICE	886.1430	OPHTHALMIC CONTACT LENS MEASURING DEVICE Diagnostic Devices; Ophthalmic Devices
DEVICE	886.1450	CORNEAL RADIUS MEASURING DEVICE Diagnostic Devices; Ophthalmic Devices
DEVICE	886.4280	INTRAOCULAR PRESSURE MEASURING DEVICE Surgical Devices; Ophthalmic Devices
DEVICE	886.4360	OCULAR SURGERY IRRIGATION DEVICE Surgical Devices; Ophthalmic Devices
DEVICE	890.5050	DAILY ACTIVITY ASSIST DEVICE Physical Medicine Therapeutic Devices; Physical Medicine Devices
DEVICE	890.5525	IONTOPHORESIS DEVICE Physical Medicine Therapeutic Devices; Physical Medicine Devices
DEVICE	890.5545	POWERED EXTERNAL LIMB OVERLOAD WARNING DEVICE Physical Medicine Therapeutic Devices; Physical Medicine Devices
DEVICE	890.5765	PRESSURE-APPLYING DEVICE Physical Medicine Therapeutic Devices; Physical Medicine Devices
DEVICE	892.1000	MAGNETIC RESONANCE DIAGNOSTIC DEVICE Diagnostic Devices; Radiology Devices
DEVICE	892.1610	DIAGNOSTIC X-RAY BEAM-LIMITING DEVICE Diagnostic Devices; Radiology Devices
DEVICE	892.1670	SPOT-FILM DEVICE Diagnostic Devices; Radiology Devices
DEVICE	892.2010	MEDICAL IMAGE STORAGE DEVICE Diagnostic Devices; Radiology Devices
DEVICE	892.2020	MEDICAL IMAGE COMMUNICATIONS DEVICE Diagnostic Devices; Radiology Devices
DEVICE	892.2050	MEDICAL IMAGE HARDCOPY DEVICE Diagnostic Devices; Radiology Devices
DEVICE	895.21	PROCEDURES FOR BANNING A DEVICE General Provisions; Banned Devices
DEVICE, FINISHED	820.80	RECEIVING, IN-PROCESS, AND FINISHED DEVICE ACCEPTANCE Quality System Regulation
DEVICE HISTORY RECORD	820.184	DEVICE HISTORY RECORD Records; Quality System Regulation
DEVICE HISTORY RECORD	820.185	CRITICAL DEVICES, DEVICE HISTORY RECORD Records; Quality System Regulation
DEVICE LIST	807.20	WHO MUST REGISTER AND SUBMIT A DEVICE LIST Procedures For Device Establishments; Establishment Registration And Device Listing For Manufacturers And Initial Importers Of Devices
DEVICE LISTING	20.116	DRUG AND DEVICE LISTING INFORMATION Availability Of Specific Categories Of Records; Public Information
DEVICE LISTING	807.21	TIMES FOR ESTABLISHMENT REGISTRATION AND DEVICE LISTING Procedures For Device Establishments; Establishment Registration And Device Listing For Manufacturers And Initial Importers Of Devices
DEVICE LISTING	807.25	INFORMATION REQUIRED OR REQUESTED FOR ESTABLISHMENT REGISTRATION AND DEVICE LISTING Procedures For Device Establishments; Establishment Registration And Device Listing For Manufacturers And Initial Importers Of Devices

DEVICE LISTING	807.30	UPDATING DEVICE LISTING INFORMATION Procedures For Device Establishments; Establishment Registration And Device Listing For Manufacturers And Initial Importers Of Devices
DEVICE LISTING	807.40	ESTABLISHMENT REGISTRATION AND DEVICE LISTING FOR FOREIGN ESTABLISHMENTS IMPORTING OR OFFERING FOR IMPORT DEVICES INTO THE UNITED STATES Registration Procedures For Foreign Device Establishments; Establishment Registration And Device Listing For Manufacturers And Initial Importers Of Devices
DEVICE LISTINGS	807.37	INSPECTION OF ESTABLISHMENT REGISTRATION AND DEVICE LISTINGS Procedures For Device Establishments; Establishment Registration And Device Listing For Manufacturers And Initial Importers Of Devices
DEVICE MASTER RECORD	820.181	DEVICE MASTER RECORD Records; Quality System Regulation
DEVICE, MEDICAL	5.411	MEDICAL DEVICE RECALL AUTHORITY Medical Devices And Radiological Health; Redelegations Of Authority; Delegations Of Authority And Organization
DEVICE, MEDICAL	5.412	TEMPORARY SUSPENSION OF A MEDICAL DEVICE APPLICATION Medical Devices And Radiological Health; Redelegations Of Authority; Delegations Of Authority And Organization
DEVICE, MEDICAL	5.415	AUTHORITY RELATING TO MEDICAL DEVICE REPORTING PROCEDURES Medical Devices And Radiological Health; Redelegations Of Authority; Delegations Of Authority And Organization
DEVICE, MEDICAL	5.416	MEDICAL DEVICE TRACKING Medical Devices And Radiological Health; Redelegations Of Authority; Delegations Of Authority And Organization
DEVICES	5.405	DETERMINATION OF CLASSIFICATION OF DEVICES Medical Devices And Radiological Health; Redelegations Of Authority; Delegations Of Authority And Organization
DEVICES	25.34	DEVICES AND ELECTRONIC PRODUCTS Environmental Impact Considerations
DEVICES	201.2	DRUGS AND DEVICES; NATIONAL DRUG CODE NUMBERS General Labeling Provisions; Labeling
DEVICES	201.2	DRUGS AND DEVICES; NATIONAL DRUG CODE NUMBERS General Labeling Provisions; Labeling
DEVICES	801.109	PRESCRIPTION DEVICES Exemptions From Adequate Directions For Use; Labeling
DEVICES	801.437	USER LABELING FOR DEVICES THAT CONTAIN NATURAL RUBBER Labeling
DEVICES	807.22	HOW AND WHERE TO REGISTER ESTABLISHMENTS AND LIST DEVICES Procedures For Device Establishments; Establishment Registration And Device Listing For Manufacturers And Initial Importers Of Devices
DEVICES	807.40	ESTABLISHMENT REGISTRATION AND DEVICE LISTING FOR FOREIGN ESTABLISHMENTS IMPORTING OR OFFERING FOR IMPORT DEVICES INTO THE UNITED STATES Registration Procedures For Foreign Device Establishments; Establishment Registration And Device Listing For Manufacturers And Initial Importers Of Devices
DEVICES	812.5	LABELING OF INVESTIGATIONAL DEVICES General Provisions; Investigational Device Exemptions
DEVICES	821.20	DEVICES SUBJECT TO TRACKING Medical Device Tracking Requirements
DEVICES	864.9195	BLOOD MIXING DEVICES AND BLOOD WEIGHING DEVICES Products Used In Establishments That Manufacture Blood And Blood Products; Hematology And Pathology Devices
DEVICES, HEARING AID	801.420	HEARING AID DEVICES; PROFESSIONAL AND PATIENT LABELING Special Requirements For Specific Devices; Labeling
DEVICES, HEARING AID	801.421	HEARING AID DEVICES; CONDITIONS FOR SALE Special Requirements For Specific Devices; Labeling
DEVICES, IMPORTED	821.4	IMPORTED DEVICES Medical Device Tracking Requirements
DEVICES, LIFE-SUPPORTING	860.93	CLASSIFICATION OF IMPLANTS, LIFE-SUPPORTING OR LIFE-SUSTAINING DEVICES Classification; Medical Device Classification Procedures
DEVICES, MEDICAL	5.27	PATENT TERM EXTENSIONS FOR HUMAN DRUG PRODUCTS, MEDICAL DEVICES, AND FOOD AND COLOR ADDITIVES; AND AUTHORITY TO PERFORM DUE DILIGENCE DETERMINATIONS AND INFORMAL HEARINGS General Redelegations Of Authority; Delegations Of Authority And Organization
DEVICES, MEDICAL	5.402	DETENTION OF ADULTERATED OR MISBRANDED MEDICAL DEVICES Medical Devices And Radiological Health; Redelegations Of Authority; Delegations Of Authority And Organization
DEVICES, MEDICAL	5.403	AUTHORIZATION TO USE ALTERNATIVE EVIDENCE FOR DETERMINATION OF THE EFFECTIVENESS OF MEDICAL DEVICES Medical Devices And Radiological Health; Redelegations Of Authority; Delegations Of Authority And Organization

DEVICES, MEDICAL	5.404	NOTIFICATION TO PETITIONERS OF DETERMINATIONS MADE ON PETITIONS FOR RECLASSIFICATION OF MEDICAL DEVICES Medical Devices And Radiological Health; Redelegations Of Authority; Delegations Of Authority And Organization
DEVICES, MEDICAL	5.406	NOTIFICATION TO SPONSORS OF DEFICIENCIES IN PETITIONS FOR RECLASSIFICATION OF MEDICAL DEVICES Medical Devices And Radiological Health; Redelegations Of Authority; Delegations Of Authority And Organization
DEVICES, MEDICAL	5.407	APPROVAL, DISAPPROVAL, OR WITHDRAWAL OF APPROVAL OF PRODUCT DEVELOPMENT PROTOCOLS AND APPLICATIONS FOR PREMARKET APPROVAL FOR MEDICAL DEVICES Medical Devices And Radiological Health; Redelegations Of Authority; Delegations Of Authority And Organization
DEVICES, MEDICAL	5.409	DETERMINATIONS THAT MEDICAL DEVICES PRESENT UNREASONABLE RISK OF SUBSTANTIAL HARM Medical Devices And Radiological Health; Redelegations Of Authority; Delegations Of Authority And Organization
DEVICES, MEDICAL	5.410	ORDERS TO REPAIR OR REPLACE, OR MAKE REFUNDS FOR, MEDICAL DEVICES Medical Devices And Radiological Health; Redelegations Of Authority; Delegations Of Authority And Organization
DEVICES, MEDICAL	5.417	AUTHORITY PERTAINING TO ACCREDITATION FUNCTIONS FOR MEDICAL DEVICES Medical Devices And Radiological Health; Redelegations Of Authority; Delegations Of Authority And Organization
DEVICES, MEDICAL	207.7	ESTABLISHMENT REGISTRATION AND PRODUCT LISTING FOR HUMAN BLOOD AND BLOOD PRODUCTS AND FOR MEDICAL DEVICES General; Registration Of Drug Producers And Listing Of Drugs In Commercial Distribution
DEVICES, MEDICAL	801.1	MEDICAL DEVICES; NAME AND PLACE OF BUSINESS OF MANUFACTURER, PACKER OR DISTRIBUTOR General Labeling Provisions; Labeling
DEVICES, MEDICAL	801.5	MEDICAL DEVICES; ADEQUATE DIRECTIONS FOR USE General Labeling Provisions; Labeling
DEVICES, MEDICAL	801.6	MEDICAL DEVICES; MISLEADING STATEMENTS General Labeling Provisions; Labeling
DEVICES, MEDICAL	801.15	MEDICAL DEVICES; PROMINENCE OF REQUIRED LABEL STATEMENTS General Labeling Provisions; Labeling
DEVICES, MEDICAL	801.16	MEDICAL DEVICES; SPANISH-LANGUAGE VERSION OF CERTAIN REQUIRED STATEMENTS General Labeling Provisions; Labeling
DEVICES, MEDICAL	801.63	MEDICAL DEVICES; WARNING STATEMENTS FOR DEVICES CONTAINING OR MANUFACTURED WITH CHLOROFLUOROCARBONS Labeling Requirements For Over-the-counter Devices; Labeling
DEVICES, MEDICAL	801.116	MEDICAL DEVICES HAVING COMMONLY KNOWN DIRECTIONS Exemptions From Adequate Directions For Use; Labeling
DEVICES, MEDICAL	801.122	MEDICAL DEVICES FOR PROCESSING, REPACKING, OR MANUFACTURING Exemptions From Adequate Directions For Use; Labeling
DEVICES, MEDICAL	801.125	MEDICAL DEVICES FOR USE IN TEACHING, LAW ENFORCEMENT, RESEARCH, AND ANALYSIS Exemptions From Adequate Directions For Use; Labeling
DEVICES, MEDICAL	801.127	MEDICAL DEVICES; EXPIRATION OF EXEMPTIONS Exemptions From Adequate Directions For Use; Labeling
DEVICES, MEDICAL	801.150	MEDICAL DEVICES; PROCESSING, LABELING, OR REPACKING Other Exemptions; Labeling
DEVICES, NEW	860.134	PROCEDURES FOR "NEW DEVICES" UNDER SECTION 513(F) OF THE ACT Reclassification; Medical Device Classification Procedures
DEVICES, OLD	860.84	CLASSIFICATION PROCEDURES FOR "OLD DEVICES" Classification; Medical Device Classification Procedures
DEVICES, PRESCRIPTION	801.110	RETAIL EXEMPTIONS FOR PRESCRIPTION DEVICES Exemptions From Adequate Directions For Use; Labeling
DEXAMETHASONE	520.540	DEXAMETHASONE ORAL DOSAGE FORMS Oral Dosage Form New Animal Drugs
DEXAMETHASONE	522.540	DEXAMETHASONE INJECTION Implantation Or Injectable Dosage Form New Animal Drugs
DEXAMETHASONE	522.542	DEXAMETHASONE-21-ISONICOTINATE SUSPENSION Implantation Or Injectable Dosage Form New Animal Drugs
DEXAMETHASONE	524.1484g	NEOMYCIN SULFATE-THIABENDAZOLE-DEXAMETHASONE SOLUTION Ophthalmic And Topical Dosage Form New Animal Drugs
DEXTRAN	520.1182	IRON DEXTRAN ORAL SUSPENSION Oral Dosage Form New Animal Drugs
DEXTRAN	522.1182	IRON DEXTRAN COMPLEX INJECTION Implantation Or Injectable Dosage Form New Animal Drugs

DEXTRAN	522.1183	IRON HYDROGENATED DEXTRAN INJECTION Implantation Or Injectable Dosage Form New Animal Drugs
DEXTRANS	186.1275	DEXTRANS Listing Of Specific Substances Affirmed As GRAS; Indirect Food Substances Affirmed As Generally Recognized As Safe
DEXTRANS	582.1275	DEXTRANS General Purpose Food Additives; Substances Generally Recognized As Safe
DEXTRIN	184.1277	DEXTRIN Listing Of Specific Substances Affirmed As GRAS; Direct Food Substances Affirmed As Generally Recognized As Safe
DEXTROSE	168.110	DEXTROSE ANHYDROUS Requirements For Specific Standardized Sweeteners And Table Sirups; Sweeteners And Table Sirups
DEXTROSE	168.111	DEXTROSE MONOHYDRATE Requirements For Specific Standardized Sweeteners And Table Sirups; Sweeteners And Table Sirups
DEXTROSE	520.550	DEXTROSE/GLYCINE/ELECTROLYTE Oral Dosage Form New Animal Drugs
DI-(2-ETHYLHEXOATE)	178.3940	GLYCOL DI-(2-ETHYLHEXOATE) Certain Adjuvants And Production Aids; Indirect Food Additives: Adjuvants, Production Aids, And Sanitizers
DIACETATE	182.6197	CALCIUM DIACETATE Sequestrants; Substances Generally Recognized As Safe
DIACETATE	184.1754	SODIUM DIACETATE Listing Of Specific Substances Affirmed As GRAS; Direct Food Substances Affirmed As Generally Recognized As Safe
DIACETATE	582.6197	CALCIUM DIACETATE Sequestrants; Substances Generally Recognized As Safe
DIACETATE	582.6754	SODIUM DIACETATE Sequestrants; Substances Generally Recognized As Safe
DIACETYL	184.1278	DIACETYL Listing Of Specific Substances Affirmed As GRAS; Direct Food Substances Affirmed As Generally Recognized As Safe
DIACETYL TARTARIC ACID	184.1101	DIACETYL TARTARIC ACID ESTERS OF MONO- AND DIGLYCERIDES Listing Of Specific Substances Affirmed As GRAS; Direct Food Substances Affirmed As Generally Recognized As Safe
DIACETYL TARTARIC ACID	582.4101	DIACETYL TARTARIC ACID ESTERS OF MONO- AND DIGLYCERIDES OF EDIBLE FATS OR OILS, OR EDIBLE FAT-FORMING FATTY ACIDS Emulsifying Agents; Substances Generally Recognized As Safe
DIAGNOSTIC	868.1840	DIAGNOSTIC SPIROMETER Diagnostic Devices; Anesthesiology Devices
DIAGNOSTIC	868.1900	DIAGNOSTIC PULMONARY-FUNCTION VALUE CALCULATOR Diagnostic Devices; Anesthesiology Devices
DIAGNOSTIC	870.1200	DIAGNOSTIC INTRAVASCULAR CATHETER Cardiovascular Diagnostic Devices; Cardiovascular Devices
DIAGNOSTIC	870.1425	PROGRAMMABLE DIAGNOSTIC COMPUTER Cardiovascular Diagnostic Devices; Cardiovascular Devices
DIAGNOSTIC	870.1435	SINGLE-FUNCTION, PREPROGRAMMED DIAGNOSTIC COMPUTER Cardiovascular Diagnostic Devices; Cardiovascular Devices
DIAGNOSTIC	874.1925	TOYNBEE DIAGNOSTIC TUBE Diagnostic Devices; Ear, Nose, And Throat Devices
DIAGNOSTIC	886.1380	DIAGNOSTIC CONDENSING LENS Diagnostic Devices; Ophthalmic Devices
DIAGNOSTIC	886.1385	POLYMETHYLMETHACRYLATE (PMMA) DIAGNOSTIC CONTACT LENS Diagnostic Devices; Ophthalmic Devices
DIAGNOSTIC	886.1390	FLEXIBLE DIAGNOSTIC FRESNEL LENS Diagnostic Devices; Ophthalmic Devices
DIAGNOSTIC	886.1395	DIAGNOSTIC HRUBY FUNDUS LENS Diagnostic Devices; Ophthalmic Devices
DIAGNOSTIC	890.1375	DIAGNOSTIC ELECTROMYOGRAPH Physical Medicine Diagnostic Devices; Physical Medicine Devices
DIAGNOSTIC	890.1385	DIAGNOSTIC ELECTROMYOGRAPH NEEDLE ELECTRODE Physical Medicine Diagnostic Devices; Physical Medicine Devices
DIAGNOSTIC	890.1850	DIAGNOSTIC MUSCLE STIMULATOR Physical Medicine Diagnostic Devices; Physical Medicine Devices
DIAGNOSTIC	892.1000	MAGNETIC RESONANCE DIAGNOSTIC DEVICE Diagnostic Devices; Radiology Devices
DIAGNOSTIC	892.1570	DIAGNOSTIC ULTRASONIC TRANSDUCER Diagnostic Devices; Radiology Devices
DIAGNOSTIC, IN-VITRO	201.119	IN VITRO DIAGNOSTIC PRODUCTS Exemptions From Adequate Directions For Use; Labeling
DIAGNOSTIC, IN-VITRO	801.119	IN VITRO DIAGNOSTIC PRODUCTS Exemptions From Adequate Directions For Use; Labeling

DIAGNOSTIC, IN-VITRO	809.10	LABELING FOR IN VITRO DIAGNOSTIC PRODUCTS Labeling; In Vitro Diagnostic Products For Human Use
DIAGNOSTIC, IN-VITRO	809.20	GENERAL REQUIREMENTS FOR MANUFACTURERS AND PRODUCERS OF IN VITRO DIAGNOSTIC PRODUCTS Requirements For Manufacturers And Producers; In Vitro Diagnostic Products For Human Use
DIAGNOSTIC, IN VITRO	864.9160	BLOOD GROUP SUBSTANCES OF NONHUMAN ORIGIN FOR IN VITRO DIAGNOSTIC USE Products Used In Establishments That Manufacture Blood And Blood Products; Hematology And Pathology Devices
DIAGNOSTIC, IN-VITRO	864.9225	CELL-FREEZING APPARATUS AND REAGENTS FOR IN VITRO DIAGNOSTIC USE Products Used In Establishments That Manufacture Blood And Blood Products; Hematology And Pathology Devices
DIAGNOSTIC, IN-VITRO	864.9275	BLOOD BANK CENTRIFUGE FOR IN VITRO DIAGNOSTIC USE Products Used In Establishments That Manufacture Blood And Blood Products; Hematology And Pathology Devices
DIAGNOSTIC, IN-VITRO	864.9600	POTENTIATING MEDIA FOR IN VITRO DIAGNOSTIC USE Products Used In Establishments That Manufacture Blood And Blood Products; Hematology And Pathology Devices
DIAGNOSTIC RADIOLOGY	1000.55	RECOMMENDATION FOR QUALITY ASSURANCE PROGRAMS IN DIAGNOSTIC RADIOLOGY FACILITIES Radiation Protection Recommendations; General
DIAGNOSTIC X-RAY	892.1610	DIAGNOSTIC X-RAY BEAM-LIMITING DEVICE Diagnostic Devices; Radiology Devices
DIAGNOSTIC X-RAY	892.1700	DIAGNOSTIC X-RAY HIGH VOLTAGE GENERATOR Diagnostic Devices; Radiology Devices
DIAGNOSTIC X-RAY	892.1760	DIAGNOSTIC X-RAY TUBE HOUSING ASSEMBLY Diagnostic Devices; Radiology Devices
DIAGNOSTIC X-RAY	892.1770	DIAGNOSTIC X-RAY TUBE MOUNT Diagnostic Devices; Radiology Devices
DIAGNOSTIC X-RAY	1000.50	RECOMMENDATIONS FOR THE USE OF SPECIFIC AREA GONAD SHIELDING ON PATIENTS DURING MEDICAL DIAGNOSTIC X-RAY PROCEDURES Radiation Protection Recommendations; General
DIAGNOSTIC X-RAY	1020.30	DIAGNOSTIC X-RAY SYSTEMS AND THEIR MAJOR COMPONENTS Performance Standards For Ionizing Radiation Emitting Products
DIALYSATE	876.5600	SORBENT REGENERATED DIALYSATE DELIVERY SYSTEM FOR HEMODIALYSIS Therapeutic Devices; Gastroenterology-urology Devices
DIALYSIS	876.5630	PERITONEAL DIALYSIS SYSTEM AND ACCESSORIES Therapeutic Devices; Gastroenterology-urology Devices
DIAMMONIUM PHOSPHATE	573.320	DIAMMONIUM PHOSPHATE Food Additive Listing; Food Additives Permitted In Feed And Drinking Water Of Animals
DIAMOND	872.4535	DENTAL DIAMOND INSTRUMENT Surgical Devices; Dental Devices
DIAPHRAGM	884.5350	CONTRACEPTIVE DIAPHRAGM AND ACCESSORIES Obstetrical And Gynecological Therapeutic Devices; Obstetrical And Gynecological Devices
DIAPHRAGMATIC/PHRENIC	882.5830	IMPLANTED DIAPHRAGMATIC/PHRENIC NERVE STIMULATOR Neurological Therapeutic Devices; Neurological Devices
DIATHERMY	890.5275	MICROWAVE DIATHERMY Physical Medicine Therapeutic Devices; Physical Medicine Devices
DIATHERMY	890.5290	SHORTWAVE DIATHERMY Physical Medicine Therapeutic Devices; Physical Medicine Devices
DIATHERMY	890.5300	ULTRASONIC DIATHERMY Physical Medicine Therapeutic Devices; Physical Medicine Devices
DIATOMACEOUS EARTH	573.340	DIATOMACEOUS EARTH Food Additive listing; Food Additives Permitted In Feed And Drinking Water Of Animals
DIATRIZOATE	520.563	DIATRIZOATE MEGLUMINE AND DIATRIZOATE SODIUM ORAL SOLUTION Oral Dosage Form New Animal Drugs
DIATRIZOATE	522.563	DIATRIZOATE MEGLUMINE AND DIATRIZOATE SODIUM INJECTION Implantation Or Injectable Dosage Form New Animal Drugs
DICARBONATE	172.133	DIMETHYL DICARBONATE Food Preservatives; Food Additives Permitted For Direct Addition To Food For Human Consumption
DICHLORO-DIHYDRO-ANTHRAZINETETRONE	73.3119	7,16-DICHLORO-6,15-DIHYDRO-5,9,14,18-ANTHRAZINETETRONE Medical Devices; Listing Of Color Additives Exempt From Certification
DICHLORIDE, ETHYLENE	173.230	ETHYLENE DICHLORIDE Solvents, Lubricants, Release Agents And Related Substances; Secondary Direct Food Additives Permitted In Food For Human Consumption

DICHLORIDE, ETHYLENE	573.440	ETHYLENE DICHLORIDE Food Additive Listing; Food Additives Permitted In Feed And Drinking Water Of Animals
DICHLORODIFLUOROMETHANE	173.355	DICHLORODIFLUOROMETHANE Specific Usage Additives; Secondary Direct Food Additives Permitted In Food For Human Consumption
DICHLOROPHENE	520.580	DICHLOROPHENE AND TOLUENE CAPSULES Oral Dosage Form New Animal Drugs
DICHLOROPHENE	520.581	DICHLOROPHENE TABLETS Oral Dosage Form New Animal Drugs
DICHLORVOS	520.600	DICHLORVOS Oral Dosage Form New Animal Drugs
DICHLORVOS	556.180	DICHLORVOS Specific Tolerances For Residues Of New Animal Drugs; Tolerances For Residues Of New Animal Drugs In Food
DICHLORVOS	558.205	DICHLORVOS Specific New Animal Drugs For Use In Animal Feeds; New Animal Drugs For Use In Animal Feeds
DICLAZURIL	556.185	DICLAZURIL Specific Tolerances For Residues Of New Animal Drugs; Tolerances For Residues Of New Animal Drugs In Food
DICLAZURIL	558.198	DICLAZURIL New Animal Drugs For Use In Animal Feeds
DICLOXACILLIN SODIUM	520.608	DICLOXACILLIN SODIUM MONOHYDRATE CAPSULES Oral Dosage Form New Animal Drugs
DIESTERS	172.856	PROPYLENE GLYCOL MONO- AND DIESTERS OF FATS AND FATTY ACIDS Multipurpose Additives; Food Additives Permitted For Direct Addition To Food For Human Consumption
DIETARY FAT	101.75	HEALTH CLAIMS: DIETARY SATURATED FAT AND CHOLESTEROL AND RISK OF CORONARY HEART DISEASE Food Labeling
DIETARY LIPIDS	101.73	HEALTH CLAIMS: DIETARY LIPIDS AND CANCER Food Labeling
DIETARY SUGAR ALCOHOLS	101.80	HEALTH CLAIMS: DIETARY SUGAR ALCOHOLS AND DENTAL CARIES Food Labeling
DIETARY SUPPLEMENTS	101.36	NUTRITION LABELING OF DIETARY SUPPLEMENTS OF VITAMINS AND MINERALS Food Labeling
DIETARY SUPPLEMENT	101.93	CERTAIN TYPES OF STATEMENTS FOR DIETARY SUPPLEMENTS Specific Requirements For Descriptive Claims That Are Neither Nutrient Content Nor Health Claims; Food Labeling
DIETARY SUPPLEMENTS	111.50	PACKAGING OF IRON-CONTAINING DIETARY SUPPLEMENTS Current Good Manufacturing Practice For Dietary Supplements
DIETHYLCARBAMAZINE	520.620	DIETHYLCARBAMAZINE ORAL DOSAGE FORMS Oral Dosage Form New Animal Drugs
DIETHYLCARBAMAZINE	520.2160	STYRYLPYRIDINIUM, DIETHYLCARBAMAZINE ORAL DOSAGE FORMS Oral Dosage Form New Animal Drugs
DIETHYLCARBAMAZINE CITRATE	520.622	DIETHYLCARBAMAZINE CITRATE SYRUP Oral Dosage Form New Animal Drugs
DIETHYLPYROCARBONATE	189.140	DIETHYLPYROCARBONATE (DEPC) Substances Generally Prohibited From Direct Addition Or Use As Human Food; Substances Prohibited From Use In Human Food
DIETS	579.22	IONIZING RADIATION FOR TREATMENT OF ANIMAL DIETS Radiation And Radiation Sources; Irradiation In The Production, Process & Handling Of Animal Feed And Pet Food
DIFFERENTIAL	864.5220	AUTOMATED DIFFERENTIAL CELL COUNTER Automated And Semi-automated Hematology Devices; Hematology And Pathology Devices
DIFFERENTIAL	866.2320	DIFFERENTIAL CULTURE MEDIUM Microbiology Devices; Immunology And Microbiology Devices
DIFFERENTIAL	868.2875	DIFFERENTIAL PRESSURE TRANSDUCER Monitoring Devices; Anesthesiology Devices
DIFFERENTIAL	868.5740	TRACHEAL/BRONCHIAL DIFFERENTIAL VENTILATION TUBE Therapeutic Devices; Anesthesiology Devices
DIFFERENTIATION	866.2660	MICROORGANISM DIFFERENTIATION AND IDENTIFICATION DEVICE Microbiology Devices; Immunology And Microbiology Devices
DIFLOXACIN	520.645	DIFLOXACIN Oral Dosage Form New Animal Drugs
DIGITALIS	201.317	DIGITALIS AND RELATED CARDIOTONIC DRUGS FOR HUMAN USE IN ORAL DOSAGE FORMS; REQUIRED WARNING Specific Labeling Requirements For Specific Drug Products; Labeling
DIGITIZER, MEDICAL IMAGE	892.2030	MEDICAL IMAGE DIGITIZER Diagnostic Devices; Radiology Devices

DIGITOXIN	862.3300	DIGITOXIN TEST SYSTEM Clinical Toxicology Test Systems; Clinical Chemistry And Clinical Toxicology Devices
DIGLYCERIDES	172.834	ETHOXYLATED MONO- AND DIGLYCERIDES Multipurpose Additives; Food Additives Permitted For Direct Addition To Food For Human Consumption
DIGLYCERIDES	184.1101	DIACETYL TARTARIC ACID ESTERS OF MONO- AND DIGLYCERIDES Listing Of Specific Substances Affirmed As GRAS; Direct Food Substances Affirmed As Generally Recognized As Safe
DIGLYCERIDES	184.1505	MONO- AND DIGLYCERIDES Listing Of Specific Substances Affirmed As GRAS; Direct Food Substances Affirmed As Generally Recognized As Safe
DIGLYCERIDES	184.1521	MONOSODIUM PHOSPHATE DERIVATIVES OF MONO- AND DIGLYCERIDES Listing Of Specific Substances Affirmed As GRAS; Direct Food Substances Affirmed As Generally Recognized As Safe
DIGLYCERIDES	582.4101	DIACETYL TARTARIC ACID ESTERS OF MONO- AND DIGLYCERIDES OF EDIBLE FATS OR OILS, OR EDIBLE FAT-FORMING FATTY ACIDS Emulsifying Agents; Substances Generally Recognized As Safe
DIGLYCERIDES	582.4505	MONO- AND DIGLYCERIDES OF EDIBLE FATS OR OILS, OR EDIBLE FAT-FORMING-ACIDS Emulsifying Agents; Substances Generally Recognized As Safe
DIGLYCERIDES	582.4521	MONOSODIUM PHOSPHATE DERIVATIVES OF MONO- AND DIGLYCERIDES OF EDIBLE FATS OR OILS, OR EDIBLE FAT-FORMING FATTY ACIDS Emulsifying Agents; Substances Generally Recognized As Safe
DIGOXIN	310.500	DIGOXIN PRODUCTS FOR ORAL USE; CONDITIONS FOR MARKETING Requirements For Specific New Drugs Or Devices; New Drugs
DIGOXIN	862.3320	DIGOXIN TEST SYSTEM Clinical Toxicology Test Systems; Clinical Chemistry And Clinical Toxicology Devices
DIHYDRO-	73.3122	4-[(2,4-DIMETHYLPHENYL)AZO]-2,4-DIHYDRO-5-METHYL-2-PHENYL-3H-PYRAZOL-3-ONE Drugs; Listing Of Color Additives Exempt From Certification
DIHYDRO-DIOXO-ANTHRACENEDIYL	73.3118	N,N'-(9,10-DIHYDRO-9,10-DIOXO-1,5-ANTHRACENEDIYL) BISBENZAMIDE Medical Devices; Listing Of Color Additives Exempt From Certification
DIHYDROCHLORIDE	520.2122	SPECTINOMYCIN DIHYDROCHLORIDE ORAL SOLUTION Oral Dosage Form New Animal Drugs
DIHYDROCHLORIDE	520.2123	SPECTINOMYCIN DIHYDROCHLORIDE PENTAHYDRATE ORAL DOSAGE FORMS Oral Dosage Form New Animal Drugs
DIHYDRODINAPHTHO-CARBAZOLE	73.3117	16,23-DIHYDRODINAPHTHO[2,3-A:2',3'-I]NAPHTH [2',3':6,7] INDOLO [2,3-C] CARBAZOLE-5,10,15,17,22,24-HEXONE Medical Devices; Listing Of Color Additives Exempt From Certification
DIHYDROSTREPTOMYCIN	520.2158	STREPTOMYCIN/DIHYDROSTREPTOMYCIN ORAL DOSAGE FORMS Oral Dosage Form New Animal Drugs
DIHYDROSTREPTOMYCIN	522.650	DIHYDROSTREPTOMYCIN SULFATE INJECTION Implantation Or Injectable Dosage Form New Animal Drugs
DIHYDROSTREPTOMYCIN	526.1696b	PENICILLIN G PROCAINE-DIHYDROSTREPTOMYCIN IN SOYBEAN OIL FOR INTRAMAMMARY INFUSION (DRY COWS) Intramammary Dosage Forms
DIHYDROSTREPTOMYCIN	526.1696c	PENICILLIN G PROCAINE-DIHYDROSTREPTOMYCIN SULFATE FOR INTRAMAMMARY INFUSION (DRY COWS) Intramammary Dosage Forms
DIHYDROSTREPTOMYCIN	556.200	DIHYDROSTREPTOMYCIN Specific Tolerances For Residues Of New Animal Drugs; Tolerances For Residues Of New Animal Drugs In Food
DIHYDROXYACETONE	73.1150	DIHYDROXYACETONE Drugs; Listing Of Color Additives Exempt From Certification
DIHYDROXYACETONE	73.2150	DIHYDROXYACETONE Cosmetics; Listing Of Color Additives Exempt From Certification
DILATOR	870.1310	VESSEL DILATOR FOR PERCUTANEOUS CATHETERIZATION Cardiovascular Diagnostic Devices; Cardiovascular Devices
DILATOR	870.4475	SURGICAL VESSEL DILATOR Cardiovascular Surgical Devices; Cardiovascular Devices
DILATOR	874.3900	NASAL DILATOR Ear, Nose And Throat Devices
DILATOR	876.5365	ESOPHAGEAL DILATOR Therapeutic Devices; Gastroenterology-urology Devices
DILATOR	876.5450	RECTAL DILATOR Therapeutic Devices; Gastroenterology-urology Devices
DILATOR	876.5470	URETERAL DILATOR Therapeutic Devices; Gastroenterology-urology Devices
DILATOR	876.5520	URETHRAL DILATOR Therapeutic Devices; Gastroenterology-urology Devices

DILATOR	884.4250	EXPANDABLE CERVICAL DILATOR Obstetrical And Gynecological Surgical Devices; Obstetrical And Gynecological Devices
DILATOR	884.4260	HYGROSCOPIC LAMINARIA CERVICAL DILATOR Obstetrical And Gynecological Surgical Devices; Obstetrical And Gynecological Devices
DILATORS	884.4270	VIBRATORY CERVICAL DILATORS Obstetrical And Gynecological Surgical Devices; Obstetrical And Gynecological Devices
DILAURYL THIODIPROPIONATE	182.3280	DILAURYL THIODIPROPIONATE Chemical Preservatives; Substances Generally Recognized As Safe
DILAURYL THIODIPROPIONATE	582.3280	DILAURYL THIODIPROPIONATE Chemical Preservatives; Substances Generally Recognized As Safe
DILIGENCE, DUE	5.27	PATENT TERM EXTENSIONS FOR HUMAN DRUG PRODUCTS, MEDICAL DEVICES, AND FOOD AND COLOR ADDITIVES; AND AUTHORITY TO PERFORM DUE DILIGENCE DETERMINATIONS AND INFORMAL HEARINGS General Redelegations Of Authority; Delegations Of Authority And Organization
DILIGENCE, DUE	60.36	STANDARD OF DUE DILIGENCE Due Diligence Petitions; Patent Term Restoration
DILL	184.1282	DILL AND ITS DERIVATIVES Listing Of Specific Substances Affirmed As GRAS; Direct Food Substances Affirmed As Generally Recognized As Safe
DILUENT	864.8200	BLOOD CELL DILUENT Hematology Reagents; Hematology And Pathology Devices
DILUENTS	73.1	DILUENTS IN COLOR ADDITIVE MIXTURES FOR FOOD USE EXEMPT FROM CERTIFICATION Foods; Listing Of Color Additives Exempt From Certification
DILUENTS	73.1001	DILUENTS IN COLOR ADDITIVE MIXTURES FOR DRUG USE EXEMPT FROM CERTIFICATION Drugs; Listing Of Color Additives Exempt From Certification
DILUTING	862.2540	PIPETTING AND DILUTING SYSTEM FOR CLINICAL USE Clinical Laboratory Instruments; Clinical Chemistry And Clinical Toxicology Devices
DILUTING	864.5240	AUTOMATED BLOOD CELL DILUTING APPARATUS Hematology And Pathology Devices
DILUTING	866.2500	MICROTITER DILUTING AND DISPENSING DEVICE Microbiology Devices; Immunology And Microbiology Devices
DILUTIONS	172.710	ADJUVANTS FOR PESTICIDE USE DILUTIONS Other Specific Usage Additives; Food Additives Permitted For Direct Addition To Food For Human Consumption
DIMERS, ALKYL KETENE	176.120	ALKYL KETENE DIMERS Substances For Use Only As Components Of Paper And Paperboard; Indirect Food Additives
DIMETHOXYDINAPHTHO PERYLENE	73.3120	16,17-DIMETHOXYDINAPHTHO [1,2,3-CD:3',2',1 '-1 M] PERYLENE-5,1O-DIONE Medical Devices; Listing Of Color Additives Exempt From Certification
DIMETHYL	176.230	3,5-DIMETHYL-1,3,5,*2H*-TETRAHYDROTHIADIAZINE-2-THIONE Substances For Use Only As Components Of Paper And Paperboard; Indirect Food Additives
DIMETHYL DICARBONATE	172.133	DIMETHYL DICARBONATE Food Preservatives; Food Additives Permitted For Direct Addition To Food For Human Consumption
DIMETHYL NAPHTHALENE	172.824	SODIUM MONO- AND DIMETHYL NAPHTHALENE SULFONATES Multipurpose Additives; Food Additives Permitted For Direct Addition To Food For Human Consumption
DIMETHYL-PHENYLENE OXIDE	177.2460	POLY(2,6-DIMETHYL-1,4-PHENYLENE) OXIDE RESINS Substances For Use Only As Components Of Articles Intended For Repeated Use; Indirect Food Additives: Polymers
DIMETHYL PHOSPHORODITHIOATE	524.1742	N-(MERCAPTOMETHYL) PHTHALIMIDE S-(*O,O*-DIMETHYL PHOSPHORODITHIOATE) EMULSIFIABLE LIQUID Ophthalmic And Topical Dosage Form New Animal Drugs
DIMETHYL SULFOXIDE	524.660	DIMETHYL SULFOXIDE OPHTHALMIC AND TOPICAL DOSAGE FORMS Ophthalmic And Topical Dosage Form New Animal Drugs
DIMETHYL SULFOXIDE	524.981d	FLUOCINOLONE ACETONIDE, DIMETHYL SULFOXIDE SOLUTION Ophthalmic And Topical Dosage Form New Animal Drugs
DIMETHYL SULFOXIDE	524.981e	FLUOCINOLONE ACETONIDE, DIMETHYL SULFOXIDE OTIC SOLUTION Ophthalmic And Topical Dosage Form New Animal Drugs
DIMETHYLAMINE	173.60	DIMETHYLAMINE-EPICHLOROHYDRIN COPOLYMER Polymer Substances For Food Treatment; Secondary Direct Food Additives Permitted In Food For Human Consumption
DIMETHYLDIALKYLAMMONIUM	173.400	DIMETHYLDIALKYLAMMONIUM CHLORIDE Secondary Direct Food Additives Permitted In Food For Human Consumption
DIMETHYLENE	177.1240	1,4-CYCLOHEXYLENE DIMETHYLENE TEREPHTHALATE AND 1,4-CYCLOHEXYLENE DIMETHYLENE ISOPHTHALATE COPOLYMER Substances For Use As Basic Components Of Single And Repeated Use Food Contact Surfaces; Indirect Food Additives: Polymers

DIMETHYLENE	177.1315	ETHYLENE-1,4-CYCLOHEXYLENE DIMETHYLENE TEREPHTHALATE COPOLYMERS Substances For Use As Basic Components Of Single And Repeated Use Food Contact Surfaces; Indirect Food Additives: Polymers
DIMETHYLPHENYL	73.3122	4-[(2,4-DIMETHYLPHENYL)AZO]-2,4-DIHYDRO-5-METHYL-2-PHENYL-3H-PYRAZOL-3-ONE Drugs; Listing Of Color Additives Exempt From Certification
DIMETHYLPYRIMIDINOL	573.620	MENADIONE DIMETHYLPYRIMIDINOL BISULFITE Food Additive Listing; Food Additives Permitted In Feed And Drinking Water Of Animals
DINITROBENZAMINE	556.220	3,5-DINITROBENZAMINE Tolerances For Residues Of New Animal Drugs In Food
DINNER, FROZEN	104.47	FROZEN "HEAT AND SERVE" DINNER Nutritional Quality Guidelines For Foods
DINNERS	102.28	FOODS PACKAGED FOR USE IN THE PREPARATION OF "MAIN DISHES" OR "DINNERS" Requirements For Specific Nonstandardized Foods; Common Or Usual Name For Nonstandardized Foods
DINNERS, FROZEN	102.26	FROZEN "HEAT AND SERVE" DINNERS Requirements For Specific Nonstandardized Foods; Common Or Usual Name For Nonstandardized Foods
DINOPROST	522.690	DINOPROST TROMETHAMINE STERILE SOLUTION Implantation Or Injectable Dosage Form New Animal Drugs
DIOCTYL	163.117	COCOA WITH DIOCTYL SODIUM SULFOSUCCINATE FOR MANUFACTURING Requirements For Specific Standardized Cacao Products; Cacao Products
DIOCTYL	172.520	COCOA WITH DIOCTYL SODIUM SULFOSUCCINATE FOR MANUFACTURING Flavoring Agents And Related Substances; Food Additives Permitted For Direct Addition To Food For Human Consumption
DIOCTYL	172.810	DIOCTYL SODIUM SULFOSUCCINATE Multipurpose Additives; Food Additives Permitted For Direct Addition To Food For Human Consumption
DIOLEATE	573.800	POLYOXYETHYLENE GLYCOL (400) MONO- AND DIOLEATE Food Additive Listing; Food Additives Permitted In Feed And Drinking Water Of Animals
DIOLEATES	573.820	POLYOXYETHYLENE GLYCOL (400) MONO- AND DIOLEATES Food Additive Listing; Food Additives Permitted In Feed And Drinking Water Of Animals
DIOXIDE, CARBON	184.1240	CARBON DIOXIDE Listing Of Specific Substances Affirmed As GRAS; Direct Food Substances Affirmed As Generally Recognized As Safe
DIOXIDE, CARBON	201.161	CARBON DIOXIDE AND CERTAIN OTHER GASES Other Exemptions; Labeling
DIOXIDE, CARBON	582.1240	CARBON DIOXIDE General Purpose Food Additives; Substances Generally Recognized As Safe
DIOXIDE, CARBON	862.1160	BICARBONATE/CARBON DIOXIDE TEST SYSTEM Clinical Chemistry Test Systems; Clinical Chemistry And Clinical Toxicology Devices
DIOXIDE, CARBON	868.1150	INDWELLING BLOOD CARBON DIOXIDE PARTIAL PRESSURE (PCO2) ANALYZER Diagnostic Devices; Anesthesiology Devices
DIOXIDE, CARBON	868.1400	CARBON DIOXIDE GAS ANALYZER Diagnostic Devices; Anesthesiology Devices
DIOXIDE, CARBON	868.2480	CUTANEOUS CARBON DIOXIDE (PCCO2) MONITOR Diagnostic Devices; Anesthesiology Devices
DIOXIDE, CARBON	868.5300	CARBON DIOXIDE ABSORBENT Therapeutic Devices; Anesthesiology Devices
DIOXIDE, CARBON	868.5310	CARBON DIOXIDE ABSORBER Therapeutic Devices; Anesthesiology Devices
DIOXIDE, CARBON	874.4500	EAR, NOSE, AND THROAT MICROSURGICAL CARBON DIOXIDE LASER Surgical Devices; Ear, Nose And Throat Devices
DIOXIDE, CARBON	884.1300	UTEROTUBAL CARBON DIOXIDE INSUFFLATOR AND ACCESSORIES Obstetrical And Gynecological Diagnostic Devices; Obstetrical And Gynecological Devices
DIOXIDE, CHLORINE	173.300	CHLORINE DIOXIDE Specific Usage Additives; Secondary Direct Food Additives Permitted In Food For Human Consumption
DIOXIDE, NITROGEN	868.2385	NITROGEN DIOXIDE ANALYZER Monitoring Devices; Anesthesiology Devices
DIOXIDE, SILICON	172.480	SILICON DIOXIDE Anticaking Agents; Food Additives Permitted For Direct Addition To Food For Human Consumption
DIOXIDE, SILICON	175.390	ZINC-SILICON DIOXIDE MATRIX COATINGS Substances For Use As Components Of Coatings; Indirect Food Additives

DIOXIDE, SILICON	573.940	SILICON DIOXIDE Food Additive Listing; Food Additives Permitted In Feed And Drinking Water Of Animals
DIOXIDE, SULFUR	182.3862	SULFUR DIOXIDE Chemical Preservatives; Substances Generally Recognized As Safe
DIOXIDE, SULFUR	582.3862	SULFUR DIOXIDE Chemical Preservatives; Substances Generally Recognized As Safe
DIOXIDE, TITANIUM	73.575	TITANIUM DIOXIDE Foods; Listing Of Color Additives Exempt From Certification
DIOXIDE, TITANIUM	73.1575	TITANIUM DIOXIDE Drugs; Listing Of Color Additives Exempt From Certification
DIOXIDE, TITANIUM	73.2575	TITANIUM DIOXIDE Cosmetics; Listing Of Color Additives Exempt From Certification
DIOXIDE, TITANIUM	73.3126	TITANIUM DIOXIDE Drugs; Listing Of Color Additives Exempt From Certification
DIPHENOLPHOSPHITE	189.300	HYDROGENATED 4,4'-ISOPROPYLIDENE-DIPHENOLPHOSPHITE ESTER RESINS Substances Prohibited From Indirect Addition To Human Food Through Food-contact Surfaces; Substances Prohibited From Use In Human Food
DIPHENYLHYDANTOIN	862.3350	DIPHENYLHYDANTOIN TEST SYSTEM Clinical Toxicology Test Systems; Clinical Chemistry And Clinical Toxicology Devices
DIPHOSPHOGLYCERIC ACID	862.1250	2,3-DIPHOSPHOGLYCERIC ACID TEST SYSTEM Clinical Chemistry Test Systems; Clinical Chemistry And Clinical Toxicology Devices
DIPOTASSIUM PHOSPHATE	182.6285	DIPOTASSIUM PHOSPHATE Sequestrants; Substances Generally Recognized As Safe
DIPOTASSIUM PHOSPHATE	582.6285	DIPOTASSIUM PHOSPHATE Sequestrants; Substances Generally Recognized As Safe
DIPRENORPHINE HCL	522.723	DIPRENORPHINE HYDROCHLORIDE INJECTION Implantation Or Injectable Dosage Form New Animal Drugs
DIRECT-CONTACT	882.1570	POWERED DIRECT-CONTACT TEMPERATURE MEASUREMENT DEVICE Neurological Diagnostic Devices; Neurological Devices
DIRECT DELIVERY	203.31	SAMPLE DISTRIBUTION BY MEANS OTHER THAN MAIL OR COMMON CARRIER (DIRECT DELIVERY BY A REPRESENTATIVE OR DETAILER) Samples; Prescription Drug Marketing
DIRECT FOOD ADDITIVES	172.5	GENERAL PROVISIONS FOR DIRECT FOOD ADDITIVES Food Additives Permitted For Direct Addition To Food For Human Consumption
DIRECTION	5.605	DEALER AND DISTRIBUTOR DIRECTION TO PROVIDE DATA TO MANUFACTURERS OF ELECTRONIC PRODUCTS Radiation Control; Redelegations Of Authority; Delegations Of Authority And Organization
DIRECTIONS	107.20	DIRECTIONS FOR USE Labeling; Infant Formula
DIRECTIONS	201.5	DRUGS; ADEQUATE DIRECTIONS FOR USE Labeling
DIRECTIONS	201.116	DRUGS HAVING COMMONLY KNOWN DIRECTIONS Exemptions From Adequate Directions For Use; Labeling
DIRECTIONS	201.307	SODIUM PHOSPHATES, PACKAGE SIZE LIMITATION, WARNINGS AND DIRECTIONS FOR OVER-THE-COUNTER SALE Specific Labeling Requirements For Specific Drug Products; Labeling
DIRECTIONS	201.308	IPECAC SYRUP; WARNINGS AND DIRECTIONS FOR USE FOR OVER-THE-COUNTER SALE Specific Labeling Requirements For Specific Drug Products; Labeling
DIRECTIONS	201.319	WATER-SOLUBLE GUMS, HYDROPHILIC GUMS, AND HYDROPHILIC MUCILLOIDS AS ACTIVE INGREDIENTS; REQUIRED WARNINGS AND DIRECTIONS Specific Labeling Requirements For Specific Drug Products; Labeling
DIRECTIONS	250.12	STRAMONIUM PREPARATIONS LABELED WITH DIRECTIONS FOR USE IN SELF-MEDICATION REGARDED AS MISBRANDED Drugs Regarded As Misbranded; Specific Requirements For Specific Human Drugs
DIRECTIONS	801.5	MEDICAL DEVICES; ADEQUATE DIRECTIONS FOR USE General Labeling Provisions; Labeling
DIRECTIONS	801.116	MEDICAL DEVICES HAVING COMMONLY KNOWN DIRECTIONS Exemptions From Adequate Directions For Use; Labeling
DIRECTIONS	1230.14	DIRECTIONS FOR TREATMENT Labeling; Regulations Under The Federal Caustic Poison Act
DIRECTIONS	1230.15	RESPONSIBILITY FOR LABELING DIRECTIONS FOR TREATMENT Labeling; Regulations Under The Federal Caustic Poison Act
DIRECTIONS	1230.16	EXEMPTION FROM LABELING DIRECTIONS FOR TREATMENT Labeling; Regulations Under The Federal Caustic Poison Act
DIRECTIVE DRUGS	558.6	VETERINARY FEED DIRECTIVE DRUGS General Provisions; New Animal Drugs For Use In Animal Feeds

DIRECTOR, RESEARCH	5.201	REDELEGATION OF THE CENTER FOR BIOLOGICS EVALUATION AND RESEARCH DIRECTOR'S PROGRAM AUTHORITIES Biologics; Redelegations Of Authority; Delegations Of Authority And Organization
DIRECTOR, STUDY	58.33	STUDY DIRECTOR Organization And Personnel; Good Laboratory Practice For Nonclinical Laboratory Studies
DISALICYLATE	558.76	BACITRACIN METHYLENE DISALICYLATE New Animal Drugs For Use In Animal Feeds
DISALLOWANCES	1403.51	LATER DISALLOWANCES AND ADJUSTMENTS Post-award Requirements; Uniform Administrative Requirements For Grants And Coop. Agreements To State & Local Gov.
DISAPPROVAL	5.407	APPROVAL, DISAPPROVAL, OR WITHDRAWAL OF APPROVAL OF PRODUCT DEVELOPMENT PROTOCOLS AND APPLICATIONS FOR PREMARKET APPROVAL FOR MEDICAL DEVICES Medical Devices And Radiological Health; Redelegations Of Authority; Delegations Of Authority And Organization
DISAPPROVAL	5.413	APPROVAL, DISAPPROVAL, WITHDRAWAL OF APPLICATIONS AND ENTERING INTO AGREEMENTS FOR INVESTIGATIONAL DEVICE EXEMPTIONS Medical Devices And Radiological Health; Redelegations Of Authority; Delegations Of Authority And Organization
DISC	866.1620	ANTIMICROBIAL SUSCEPTIBILITY TEST DISC Diagnostic Devices; Immunology And Microbiology Devices
DISC	872.3970	INTERARTICULAR DISC PROSTHESIS (INTERPOSITIONAL IMPLANT) Dental Devices
DISCHARGE	1020.20	COLD-CATHODE GAS DISCHARGE TUBES Performance Standards For Ionizing Radiation Emitting Products
DISCHARGE	1040.30	HIGH-INTENSITY MERCURY VAPOR DISCHARGE LAMPS Performance Standards For Light-emitting Products
DISCHARGE	1250.51	RAILROAD CONVEYANCES; DISCHARGE OF WASTES Equipment And Operation Of Land And Air Conveyances; Interstate Conveyance Sanitation
DISCHARGE	1250.52	DISCHARGE OF WASTES ON HIGHWAY CONVEYANCES Equipment And Operation Of Land And Air Conveyances; Interstate Conveyance Sanitation
DISCHARGE	1250.53	DISCHARGE OF WASTES ON AIR CONVEYANCES Equipment And Operation Of Land And Air Conveyances; Interstate Conveyance Sanitation
DISCHARGE	1250.93	DISCHARGE OF WASTES Sanitation Facilities And Conditions On Vessels; Interstate Conveyance Sanitation
DISCLAIMERS	803.16	DISCLAIMERS Medical Device Reporting
DISCLOSED	20.81	DATA AND INFORMATION PREVIOUSLY DISCLOSED TO THE PUBLIC Limitations On Exemptions; Public Information
DISCLOSED	21.73	ACCURACY, COMPLETENESS, TIMELINESS, AND RELEVANCE OF RECORDS DISCLOSED FROM PRIVACY ACT RECORD SYSTEMS Disclosure Of Records In Privacy Act Record Systems To Persons Other ThanThe Subject Individual; Protection Of Privacy
DISCLOSURE	5.23	DISCLOSURE OF OFFICIAL RECORDS AND AUTHORIZATION OF TESTIMONY General Redelegations Of Authority; Delegations Of Authority And Organization
DISCLOSURE	12.85	DISCLOSURE OF DATA AND INFORMATION BY THE PARTICIPANTS Hearing Procedures; Formal Evidentiary Public Hearing
DISCLOSURE	13.25	DISCLOSURE OF DATA AND INFORMATION BY THE PARTICIPANTS Hearing Procedures; Public Hearing Before A Public Board Of Inquiry
DISCLOSURE	20.20	POLICY ON DISCLOSURE OF FDA RECORDS General Policy; Public Information
DISCLOSURE	20.22	PARTIAL DISCLOSURE OF RECORDS General Policy; Public Information
DISCLOSURE	20.32	DISCLOSURE OF FDA EMPLOYEE NAMES General Policy; Public Information
DISCLOSURE	20.46	JUDICIAL REVIEW OF PROPOSED DISCLOSURE Procedures And Fees; Public Information
DISCLOSURE	20.63	PERSONNEL, MEDICAL, AND SIMILAR FILES, DISCLOSURE OF WHICH CONSTITUTES A CLEARLY UNWARRANTED INVASION OF PERSONAL PRIVACY Exemptions; Public Information
DISCLOSURE	20.82	DISCRETIONARY DISCLOSURE BY THE COMMISSIONER Limitations On Exemptions; Public Information
DISCLOSURE	20.83	DISCLOSURE REQUIRED BY COURT ORDER Limitations On Exemptions; Public Information
DISCLOSURE	20.84	DISCLOSURE TO CONSULTANTS, ADVISORY COMMITTEES, STATE AND LOCAL GOVT. OFFICIALS AND OTHER SPECIAL GOVERNMENT EMPLOYEES Limitations On Exemptions; Public Information

DISCLOSURE	20.85	DISCLOSURE TO OTHER FEDERAL GOVERNMENT DEPARTMENTS AND AGENCIES Limitations On Exemptions; Public Information
DISCLOSURE	20.86	DISCLOSURE IN ADMINISTRATIVE OR COURT PROCEEDINGS Limitations On Exemptions; Public Information
DISCLOSURE	20.87	DISCLOSURE TO CONGRESS Limitations On Exemptions; Public Information
DISCLOSURE	20.90	DISCLOSURE TO CONTRACTORS Limitations On Exemptions; Public Information
DISCLOSURE	21.53	NOTATION AND DISCLOSURE OF DISPUTED RECORDS Procedures For Requests For Amendment Of Records; Protection Of Privacy
DISCLOSURE	21.70	DISCLOSURE AND INTRA-AGENCY USE OF RECORDS IN PRIVACY ACT RECORD SYSTEMS; NO ACCOUNTING REQUIRED Disclosure Of Records In Privacy Act Record Systems To Persons Other ThanThe Subject Individual; Protection Of Privacy
DISCLOSURE	21.71	DISCLOSURE OF RECORDS IN PRIVACY ACT RECORD SYSTEMS; ACCOUNTING REQUIRED Disclosure Of Records In Privacy Act Record Systems To Persons Other ThanThe Subject Individual; Protection Of Privacy
DISCLOSURE	21.72	INDIVIDUAL CONSENT TO DISCLOSURE OF RECORDS TO OTHER PERSONS Disclosure Of Records In Privacy Act Record Systems To Persons Other ThanThe Subject Individual; Protection Of Privacy
DISCLOSURE	54.4	CERTIFICATION AND DISCLOSURE REQUIREMENTS Financial Disclosure By Clinical Investigators
DISCLOSURE	201.200	DISCLOSURE OF DRUG EFFICACY STUDY EVALUATIONS IN LABELING AND ADVERTISING Labeling Claims For Drugs In Drug Efficacy Studies; Labeling
DISCLOSURE, PUBLIC	10.20	SUBMISSION OF DOCUMENTS TO DOCKETS MANAGEMENT BRANCH; COMPUTATION OF TIME; AVAILABILITY FOR PUBLIC DISCLOSURE General Administrative Procedures; Administrative Practices And Procedures
DISCLOSURE, PUBLIC	56.122	PUBLIC DISCLOSURE OF INFORMATION REGARDING REVOCATION Administrative Action For Noncompliance; Institutional Review Boards
DISCLOSURE, PUBLIC	58.213	PUBLIC DISCLOSURE OF INFORMATION REGARDING DISQUALIFICATION Disqualification Of Testing Facilities; Good Laboratory Practice For Nonclinical Laboratory Studies
DISCLOSURE, PUBLIC	312.130	AVAILABILITY FOR PUBLIC DISCLOSURE OF DATA AND INFORMATION IN AN IND Investigational New Drug Application
DISCLOSURE, PUBLIC	314.430	AVAILABILITY FOR PUBLIC DISCLOSURE OF DATA AND INFORMATION IN AN APPLICATION Miscellaneous Provisions; Applications For FDA Approval To Market A New Drug
DISCLOSURE, PUBLIC	316.52	AVAILABILITY FOR PUBLIC DISCLOSURE OF DATA AND INFORMATION IN REQUESTS AND APPLICATIONS Orphan Drugs
DISCONTINUANCE	1301.52	TERMINATION OF REGISTRATION; TRANSFER OF REGISTRATION; DISTRIBUTION UPON DISCONTINUANCE OF BUSINESS Registration Of Manufacturers, Distributors, And Dispensers Of Controlled Substances
DISCOVERY	17.23	DISCOVERY Civil Money Penalties Hearings
DISCOVERY	1003.10	DISCOVERY OF DEFECT OR FAILURE OF COMPLIANCE BY MANUFACTURER; NOTICE REQUIREMENTS Discovery Of Defect Or Failure To Comply; Notification Of Defects Or Failure To Comply
DISCRETE	862.2160	DISCRETE PHOTOMETRIC CHEMISTRY ANALYZER FOR CLINICAL USE Clinical Laboratory Instruments; Clinical Chemistry And Toxicology Devices
DISCRETIONARY	20.82	DISCRETIONARY DISCLOSURE BY THE COMMISSIONER Limitations On Exemptions; Public Information
DISCRIMINATOR	882.1200	TWO-POINT DISCRIMINATOR Neurological Diagnostic Devices; Neurological Devices
DISCUSSIONS, ORAL	20.104	SUMMARIES OF ORAL DISCUSSIONS Availability Of Specific Categories Of Records; Public Information
DISEASE	310.548	DRUG PRODUCTS CONTAINING COLLOIDAL SILVER INGREDIENTS OFFERED OTC FOR THE TREATMENT OF DISEASE Requirements For Specific New Drugs Or Devices; New Drugs
DISEASE	1240.45	REPORT OF DISEASE Control Of Communicable Diseases
DISEASE, CORONARY HEART	101.75	HEALTH CLAIMS: DIETARY SATURATED FAT AND CHOLESTEROL AND RISK OF CORONARY HEART DISEASE Food Labeling
DISEASE, CORONARY HEART	101.77	HEALTH CLAIMS: FRUITS, VEGETABLES, AND GRAIN PRODUCTS THAT CONTAIN FIBER, PARTICULARLY SOLUBLE FIBER, AND RISK OF CORONARY HEART DISEASE Food Labeling

DISEASE, CORONARY HEART	101.81	HEALTH CLAIMS: SOLUBLE FIBER FROM CERTAIN FOODS AND RISK OF CORONARY HEART DISEASE Food Labeling
DISEASE, CORONARY HEART	101.82	HEALTH CLAIMS: SOY PROTEIN AND RISK OF CORONARY HEART DISEASE Food Labeling
DISEASE, CORONARY HEART	101.83	HEALTH CLAIMS: PLANT STEROL/STANOL ESTERS AND RISK OF CORONARY HEART DISEASE Food Labeling
DISINFECTANTS	880.6885	LIQUID CHEMICAL STERILANTS/HIGH LEVEL DISINFECTANTS General Hospital And Personal Use Miscellaneous Devices; General Hospital And Personal Use Devices
DISINFECTANTS	880.6890	GENERAL PURPOSE DISINFECTANTS General Hospital And Personal Use Miscellaneous Devices; General Hospital And Personal Use Devices
DISLOCATION	890.3665	CONGENITAL HIP DISLOCATION ABDUCTION SPLINT Physical Medicine Prosthetic Devices; Physical Medicine Devices
DISLODGER	876.4650	WATER JET RENAL STONE DISLODGER SYSTEM Surgical Devices; Gastroenterology-urology Devices
DISLODGER	876.4680	URETERAL STONE DISLODGER Surgical Devices; Gastroenterology-urology Devices
DISODIUM EDTA	73.2120	DISODIUM EDTA-COPPER Cosmetics; Listing Of Color Additives Exempt From Certification
DISODIUM EDTA	172.120	CALCIUM DISODIUM EDTA Food Preservatives; Food Additives Permitted For Direct Addition To Food For Human Consumption
DISODIUM EDTA	172.135	DISODIUM EDTA Food Preservatives; Food Additives Permitted For Direct Addition To Food For Human Consumption
DISODIUM EDTA	573.360	DISODIUM EDTA Food Additive Listing; Food Additives Permitted In Feed And Drinking Water Of Animals
DISODIUM GUANYLATE	172.530	DISODIUM GUANYLATE Flavoring Agents And Related Substances; Food Additives Permitted For Direct Addition To Food For Human Consumption
DISODIUM INOSINATE	172.535	DISODIUM INOSINATE Flavoring Agents And Related Substances; Food Additives Permitted For Direct Addition To Food For Human Consumption
DISODIUM PHOSPHATE	182.6290	DISODIUM PHOSPHATE Sequestrants; Substances Generally Recognized As Safe
DISODIUM PHOSPHATE	582.6290	DISODIUM PHOSPHATE Sequestrants; Substances Generally Recognized As Safe
DISPENSER	872.3080	MERCURY AND ALLOY DISPENSER Prosthetic Devices; Dental Devices
DISPENSER	880.6430	LIQUID MEDICATION DISPENSER General Hospital And Personal Use Miscellaneous Devices; General Hospital And Personal Use Devices
DISPENSER	888.4200	CEMENT DISPENSER Diagnostic Devices; Orthopedic Devices
DISPENSER	1307.11	DISTRIBUTION BY DISPENSER TO ANOTHER PRACTITIONER Special Exemptions For Manufacture And Distribution Of Controlled Substances; Miscellaneous
DISPENSERS	200.50	OPHTHALMIC PREPARATIONS AND DISPENSERS Requirements For Specific Classes Of Drugs; General
DISPENSERS	1304.22	RECORDS FOR MANUFACTURERS, DISTRIBUTORS, DISPENSERS, RESEARCHERS, IMPORTERS AND EXPORTERS Continuing Records; Records And Reports Of Registrants
DISPENSING	208.24	DISTRIBUTING AND DISPENSING A MEDICATION GUIDE Medication Guides For Prescription Drug Products
DISPENSING	866.2440	AUTOMATED MEDIUM DISPENSING AND STACKING DEVICE Microbiology Devices; Immunology And Microbiology Devices
DISPENSING	866.2500	MICROTITER DILUTING AND DISPENSING DEVICE Microbiology Devices; Immunology And Microbiology Devices
DISPENSING	1306.07	ADMINISTERING OR DISPENSING OF NARCOTIC DRUGS General Information; Prescriptions
DISPENSING	1306.26	DISPENSING WITHOUT PRESCRIPTION Controlled Substances Listed In Schedules III, IV, AND V; Prescriptions
DISPENSING REQUIREMENTS	369.3	WARNINGS REQUIRED ON DRUGS EXEMPTED FROM PRESCRIPTION-DISPENSING REQUIREMENTS OF SECTION 503(B)(1)(C) Definitions And Interpretations; Interpretative Statements Re: Warnings On Drugs And Devices For OTC Sale
DISPERSANTS	178.3725	PIGMENT DISPERSANTS Indirect Food Additives: Adjuvants, Production Aids, And Sanitizers
DISPLAY	870.2450	MEDICAL CATHODE-RAY TUBE DISPLAY Cardiovascular Monitoring Devices; Cardiovascular Devices

DISPLAY PANEL	101.1	PRINCIPAL DISPLAY PANEL OF PACKAGE FORM FOOD General Provisions; Food Labeling
DISPLAY PANEL	201.60	PRINCIPAL DISPLAY PANEL Labeling Requirements For Over-the-counter Drugs; Labeling
DISPLAY PANEL	328.50	PRINCIPAL DISPLAY PANEL OF ALL OTC DRUG PRODUCTS INTENDED FOR ORAL INGESTION THAT CONTAIN ALCOHOL Over-the-counter Drug Products Intended For Oral Ingestion That Contain Alcohol
DISPLAY PANEL	352.50	PRINCIPAL DISPLAY PANEL OF ALL SUNSCREEN DRUG PRODUCTS Sunscreen Drug Products For Over-the-counter Human Use
DISPLAY PANEL	355.55	PRINCIPAL DISPLAY PANEL OF ALL FLUORIDE RINSE DRUG PRODUCTS Anticaries Drug Products For Over-the-counter Human Use
DISPLAY PANEL	501.1	PRINCIPAL DISPLAY PANEL OF PACKAGE FORM ANIMAL FOOD General Provisions; Animal Food Labeling
DISPLAY PANEL	701.10	PRINCIPAL DISPLAY PANEL Package Form; Cosmetic Labeling
DISPLAY PANEL	801.60	PRINCIPAL DISPLAY PANEL Labeling Requirements For Over-the-counter Devices; Labeling
DISPOSABLE BEDDING	880.6060	MEDICAL DISPOSABLE BEDDING General Hospital And Personal Use Miscellaneous Devices; General Hospital And Personal Use Devices
DISPOSABLE INSERT	876.5830	HEMODIALYZER WITH DISPOSABLE INSERT (KIIL TYPE) Therapeutic Devices; Gastroenterology-urology Devices
DISPOSABLE PACK	890.5710	HOT OR COLD DISPOSABLE PACK Physical Medicine Therapeutic Devices; Physical Medicine Devices
DISPOSABLE SCISSORS	880.6820	MEDICAL DISPOSABLE SCISSORS General Hospital And Personal Use Miscellaneous Devices; General Hospital And Personal Use Devices
DISPOSABLE TRAY	872.6870	DISPOSABLE FLUORIDE TRAY Miscellaneous Devices; Dental Devices
DISPOSAL	1250.70	DISPOSAL OF HUMAN WASTES Servicing Areas For Land And Air Conveyances; Interstate Conveyance Sanitation
DISPOSAL	1250.79	GARBAGE DISPOSAL Servicing Areas For Land And Air Conveyances; Interstate Conveyance Sanitation
DISPOSAL	1307.22	DISPOSAL OF CONTROLLED SUBSTANCES BY THE ADMINISTRATION Disposal Of Controlled Substances; Miscellaneous
DISPOSING	1307.21	PROCEDURE FOR DISPOSING OF CONTROLLED SUBSTANCES Disposal Of Controlled Substances; Miscellaneous
DISPOSITION	312.59	DISPOSITION OF UNUSED SUPPLY OF INVESTIGATIONAL DRUG Responsibilities Of Sponsors And Investigators; Investigational New Drug Application
DISPOSITION	1002.41	DISPOSITION OF RECORDS OBTAINED BY DEALERS AND DISTRIBUTORS Dealer And Distributor Records; Records and Reports
DISPOSITION	1250.39	GARBAGE EQUIPMENT AND DISPOSITION Food Service Sanitation On Land And Air Conveyances; Interstate Conveyance Sanitation
DISPUTE	312.48	DISPUTE RESOLUTION Responsibilities Of Sponsors And Investigators; Investigational New Drug Application
DISPUTE	314.103	DISPUTE RESOLUTION FDA Action On Applications; Applications For FDA Approval To Market A New Drug
DISPUTED	21.53	NOTATION AND DISCLOSURE OF DISPUTED RECORDS Procedures For Requests For Amendment Of Records; Protection Of Privacy
DISPUTED	21.54	AMENDED OR DISPUTED RECORDS RECEIVED FROM OTHER AGENCIES Procedures For Requests For Amendment Of Records; Protection Of Privacy
DISPUTED	21.74	PROVIDING NOTICE THAT A RECORD IS DISPUTED Disclosure Of Records In Privacy Act Record Systems To Persons Other Than The Subject Individual; Protection Of Privacy
DISQUALIFICATION	12.75	DISQUALIFICATION OF PRESIDING OFFICER Presiding Officer; Formal Evidentiary Public Hearing
DISQUALIFICATION	19.45	TEMPORARY DISQUALIFICATION OF FORMER EMPLOYEES Disqualification Conditions; Standards Of Conduct And Conflicts Of Interest
DISQUALIFICATION	19.55	PERMANENT DISQUALIFICATION OF FORMER EMPLOYEES Disqualification Conditions; Standards Of Conduct And Conflicts Of Interest
DISQUALIFICATION	56.121	DISQUALIFICATION OF AN IRB OR AN INSTITUTION Administrative Action For Noncompliance; Institutional Review Boards
DISQUALIFICATION	56.124	ACTIONS ALTERNATIVE OR ADDITIONAL TO DISQUALIFICATION Administrative Action For Noncompliance; Institutional Review Boards
DISQUALIFICATION	58.202	GROUNDS FOR DISQUALIFICATION Disqualification Of Testing Facilities; Good Laboratory Practice For Nonclinical Laboratory Studies
DISQUALIFICATION	58.204	NOTICE OF AND OPPORTUNITY FOR HEARING ON PROPOSED DISQUALIFICATION Disqualification Of Testing Facilities; Good Laboratory Practice For Nonclinical Laboratory Studies

DISQUALIFICATION	58.206	FINAL ORDER ON DISQUALIFICATION Disqualification Of Testing Facilities; Good Laboratory Practice For Nonclinical Laboratory Studies
DISQUALIFICATION	58.210	ACTIONS UPON DISQUALIFICATION Disqualification Of Testing Facilities; Good Laboratory Practice For Nonclinical Laboratory Studies
DISQUALIFICATION	58.213	PUBLIC DISCLOSURE OF INFORMATION REGARDING DISQUALIFICATION Disqualification Of Testing Facilities; Good Laboratory Practice For Nonclinical Laboratory Studies
DISQUALIFICATION	58.215	ALTERNATIVE OR ADDITIONAL ACTIONS TO DISQUALIFICATION Disqualification Of Testing Facilities; Good Laboratory Practice For Nonclinical Laboratory Studies
DISQUALIFICATION	312.70	DISQUALIFICATION OF A CLINICAL INVESTIGATOR Responsibilities Of Sponsors And Investigators; Investigational New Drug Application
DISQUALIFICATION	812.119	DISQUALIFICATION OF A CLINICAL INVESTIGATOR Investigational Device Exemptions
DISQUALIFIED	58.219	REINSTATEMENT OF A DISQUALIFIED TESTING FACILITY Disqualification Of Testing Facilities; Good Laboratory Practice For Nonclinical Laboratory Studies
DISSEMINATED	99.101	INFORMATION THAT MAY BE DISSEMINATED Dissemination Of Information On Unapproved/New Uses For Marketed Drugs, Biologics, and Devices
DISSEMINATION	10.80	DISSEMINATION OF DRAFT FEDERAL REGISTER NOTICES AND REGULATIONS General Administrative Procedures; Administrative Practices And Procedures
DISSEMINATION	99.401	CORRECTIVE ACTIONS AND CESSATION OF DISSEMINATION OF INFORMATION Dissemination Of Information On Unapproved/New Uses For Marketed Drugs, Biologics And Devices
DISSOCIATION TEST	886.1910	SPECTACLE DISSOCIATION TEST SYSTEM Diagnostic Devices; Ophthalmic Devices
DISSOLUTION	343.90	DISSOLUTION AND DRUG RELEASE TESTING Internal Analgesic, Antipyretic And Antirheumatic Drug Products For OTC Human Use
DISTILLATE	184.1848	STARTER DISTILLATE Listing Of Specific Substances Affirmed As GRAS; Direct Food Substances Affirmed As Generally Recognized As Safe
DISTILLATES	182.20	ESSENTIAL OILS, OLEORESINS (SOLVENT-FREE), AND NATURAL EXTRACTIVES (INCLUDING DISTILLATES) General Provisions; Substances Generally Recognized As Safe
DISTILLATES	582.20	ESSENTIAL OILS, OLEORESINS (SOLVENT-FREE), AND NATURAL EXTRACTIVES (INCLUDING DISTILLATES) General Provisions; Substances Generally Recognized As Safe
DISTOMETER	886.1190	DISTOMETER Diagnostic Devices; Ophthalmic Devices
DISTRIBUTING	208.24	DISTRIBUTING AND DISPENSING A MEDICATION GUIDE Medication Guides For Prescription Drug Products
DISTRIBUTION	5.204	NOTIFICATION OF RELEASE FOR DISTRIBUTION OF BIOLOGICAL PRODUCTS Biologics; Redelegations Of Authority; Delegations Of Authority And Organization
DISTRIBUTION	80.39	RECORDS OF DISTRIBUTION Certification Procedures; Color Additive Certification
DISTRIBUTION	110.93	WAREHOUSING AND DISTRIBUTION Production And Process Controls; Current GMP In Manufacturing, Packing, Or Holding Human Food
DISTRIBUTION	205.50	MINIMUM REQUIREMENTS FOR THE STORAGE AND HANDLING OF PRESCRIPTION DRUGS AND FOR THE ESTABLISHMENT AND MAINTENANCE OF PRESCRIPTION DRUG DISTRIBUTION RECORDS Guidelines For State Licensing Of Wholesale Prescription Drug Distributors
DISTRIBUTION	211.150	DISTRIBUTION PROCEDURES Holding And Distribution; Current Good Manufacturing Practice For Finished Pharmaceuticals
DISTRIBUTION	211.165	TESTING AND RELEASE FOR DISTRIBUTION Laboratory Controls; Current Good Manufacturing Practice For Finished Pharmaceuticals
DISTRIBUTION	225.110	DISTRIBUTION RECORDS Records And Reports; Current Good Manufacturing Practice For Medicated Feeds
DISTRIBUTION	606.165	DISTRIBUTION AND RECEIPT; PROCEDURES AND RECORDS Records And Reports; Current Good Manufacturing Practice For Blood And Blood Components
DISTRIBUTION	809.30	RESTRICTIONS ON THE SALE, DISTRIBUTION AND USE OF ANALYTE SPECIFIC REAGENTS Requirements For Manufacturers And Producers; In Vitro Diagnostic Products For Human Use

DISTRIBUTION	809.40	RESTRICTIONS ON THE SALE, DISTRIBUTION AND USE OF OTC TEST SAMPLE COLLECTION SYSTEMS FOR DRUGS OF ABUSE TESTING Requirements For Manufacturers And Producers; In Vitro Diagnostic Products For Human Use
DISTRIBUTION	820.160	DISTRIBUTION Handling, Storage, Distribution, And Installation; Quality System Regulation
DISTRIBUTION	1301.52	TERMINATION OF REGISTRATION; TRANSFER OF REGISTRATION; DISTRIBUTION UPON DISCONTINUANCE OF BUSINESS Registration Of Manufacturers, Distributors, And Dispensers Of Controlled Substances
DISTRIBUTION	1307.11	DISTRIBUTION BY DISPENSER TO ANOTHER PRACTITIONER Special Exemptions For Manufacture And Distribution Of Controlled Substances; Miscellaneous
DISTRIBUTION	1307.12	DISTRIBUTION TO SUPPLIER Special Exemptions For Manufacture And Distribution Of Controlled Substances; Miscellaneous
DISTRIBUTION	1312.14	DISTRIBUTION OF COPIES OF IMPORT PERMIT Importation Of Controlled Substances; Importation And Exportation Of Controlled Substances
DISTRIBUTION	1312.19	DISTRIBUTION OF IMPORT DECLARATION Importation Of Controlled Substances; Importation And Exportation Of Controlled Substances
DISTRIBUTION	1312.24	DISTRIBUTION OF COPIES OF EXPORT PERMIT Exportation Of Controlled Substances; Importation And Exportation Of Controlled Substances
DISTRIBUTION	1312.28	DISTRIBUTION OF SPECIAL CONTROLLED SUBSTANCES INVOICE Exportation Of Controlled Substances; Importation And Exportation Of Controlled Substances
DISTRIBUTION	1313.14	DISTRIBUTION OF IMPORT DECLARATION Importation And Exportation Of Precursors And Essential Chemicals
DISTRIBUTION	1313.23	DISTRIBUTION OF EXPORT DECLARATION Importation And Exportation Of Precursors And Essential Chemicals
DISTRIBUTION	1313.34	DISTRIBUTION OF THE INTERNATIONAL TRANSACTION DECLARATION Importation And Exportation Of Precursors And Essential Chemicals
DISTRIBUTION, CEASE	810.10	CEASE DISTRIBUTION AND NOTIFICATION ORDER Mandatory Medical Device Recall Procedures; Medical Device Recall Authority
DISTRIBUTION, CEASE	810.12	WRITTEN REQUEST FOR REVIEW OF CEASE DISTRIBUTION AND NOTIFICATION ORDER Mandatory Medical Device Recall Procedures; Medical Device Recall Authority
DISTRIBUTION, CEASE	810.14	CEASE DISTRIBUTION AND NOTIFICATION OR MANDATORY RECALL STRATEGY Mandatory Medical Device Recall Procedures; Medical Device Recall Authority
DISTRIBUTION, CEASE	810.15	COMMUNICATIONS CONCERNING A CEASE DISTRIBUTION AND NOTIFICATION OR MANDATORY RECALL ORDER Mandatory Medical Device Recall Procedures; Medical Device Recall Authority
DISTRIBUTION, CEASE	810.16	CEASE DISTRIBUTION AND NOTIFICATION OR MANDATORY RECALL STATUS REPORTS Mandatory Medical Device Recall Procedures; Medical Device Recall Authority
DISTRIBUTION, CEASE	810.17	TERMINATION OF A CEASE DISTRIBUTION AND NOTIFICATION RO MANDATORY RECALL ORDER Mandatory Medical Device Recall Procedures; Medical Device Recall Authority
DISTRIBUTION RECORDS	205.50	MINIMUM REQUIREMENTS FOR STORAGE AND HANDLING OF PRESCRIPTION DRUGS AND ESTABLISHMENT/MAINTENANCE OF PRESCRIPTION DRUG DISTRIBUTION RECORDS Guidelines For State Licensing Of Wholesale Prescription Drug Distributors
DISTRIBUTION RECORDS	211.196	DISTRIBUTION RECORDS Records And Reports; Current Good Manufacturing Practice For Finished Pharmaceuticals
DISTRIBUTION RECORDS	225.202	FORMULA, PRODUCTION, AND DISTRIBUTION RECORDS Records; Current Good Manufacturing Practice For Medicated Feeds
DISTRIBUTION RECORDS	226.110	DISTRIBUTION RECORDS Records And Reports; Current Good Manufacturing Practice For Type A Medicated Articles
DISTRIBUTION REPORTS	600.81	DISTRIBUTION REPORTS Reporting Of Adverse Experiences; Biological Products: General
DISTRIBUTION, RETAIL	115.50	REFRIGERATION OF SHELL EGGS HELD FOR RETAIL DISTRIBUTION Shell Eggs
DISTRIBUTION, SAMPLE	203.30	SAMPLE DISTRIBUTION BY MAIL OR COMMON CARRIER Prescription Drug Marketing
DISTRIBUTION, SAMPLE	203.31	SAMPLE DISTRIBUTION BY MEANS OTHER THAN MAIL OR COMMON CARRIER Prescription Drug Marketing

DISTRIBUTION, WHOLESALE	203.50	REQUIREMENTS FOR WHOLESALE DISTRIBUTION OF PRESCRIPTION DRUGS Prescription Drug Marketing
DISTRIBUTIONS	1305.03	DISTRIBUTIONS REQUIRING ORDER FORMS Order Forms
DISTRIBUTOR	5.605	DEALER AND DISTRIBUTOR DIRECTION TO PROVIDE DATA TO MANUFACTURERS OF ELECTRONIC PRODUCTS Radiation Control; Redelegations Of Authority; Delegations Of Authority And Organization
DISTRIBUTOR	101.5	FOOD; NAME AND PLACE OF BUSINESS OF MANUFACTURER, PACKER, OR DISTRIBUTOR General Provisions; Food Labeling
DISTRIBUTOR	201.1	DRUGS; NAME AND PLACE OF BUSINESS OF MANUFACTURER, PACKER OR DISTRIBUTOR General Labeling Provisions; Labeling
DISTRIBUTOR	205.4	WHOLESALE DRUG DISTRIBUTOR LICENSING REQUIREMENT Guidelines For State Licensing Of Wholesale Prescription Drug Distributors
DISTRIBUTOR	501.5	ANIMAL FOOD; NAME AND PLACE OF BUSINESS OF MANUFACTURER, PACKER, OR DISTRIBUTER General Provisions; Animal Food Labeling
DISTRIBUTOR	610.64	NAME AND ADDRESS OF DISTRIBUTOR Labeling Standards; General Biological Products Standards
DISTRIBUTOR	701.12	NAME AND PLACE OF BUSINESS OF MANUFACTURER, PACKER, OR DISTRIBUTOR Package Form; Cosmetic Labeling
DISTRIBUTOR	801.1	MEDICAL DEVICES; NAME AND PLACE OF BUSINESS OF MANUFACTURER, PACKER OR DISTRIBUTOR General Labeling Provisions, Labeling
DISTRIBUTOR	803.18	FILES AND DISTRIBUTOR RECORDS Medical Device Reporting
DISTRIBUTOR	821.30	TRACKING OBLIGATIONS OF PERSONS OTHER THAN DEVICE MANUFACTURERS: DISTRIBUTOR REQUIREMENTS Medical Device Tracking Requirements
DISTRIBUTOR	895.22	SUBMISSION OF DATA AND INFORMATION BY THE MANUFACTURER, DISTRIBUTOR, OR IMPORTER Banned Devices
DISTRIBUTOR	1230.12	MANUFACTURER; DISTRIBUTOR Labeling; Regulations Under The Federal Caustic Poison Act
DISTRIBUTORS	133.10	NOTICE TO MANUFACTURERS PACKERS AND DISTRIBUTORS OF PASTEURIZED BLENDED CHEESE, PASTEURIZED PROCESS CHEESE, CHEESE FOOD, CHEESE SPREAD, AND RELATED FOODS General Provisions; Cheeses And Related Cheese Products
DISTRIBUTORS	201.300	NOTICE TO MANUFACTURERS, PACKERS, AND DISTRIBUTORS OF GLANDULAR PREPARATIONS Specific Labeling Requirements For Specific Drug Products; Labeling
DISTRIBUTORS	201.301	NOTICE TO MANUFACTURERS, PACKERS, AND DISTRIBUTORS OF ESTROGENIC HORMONE PREPARATIONS Specific Labeling Requirements For Specific Drug Products; Labeling
DISTRIBUTORS	201.302	NOTICE TO MANUFACTURERS, PACKERS, AND DISTRIBUTORS OF DRUGS FOR INTERNAL USE WHICH CONTAIN MINERAL OIL Specific Labeling Requirements For Specific Drug Products; Labeling
DISTRIBUTORS	1002.40	RECORDS TO BE OBTAINED BY DEALERS AND DISTRIBUTORS Dealer And Distributor Records; Records and Reports
DISTRIBUTORS	1002.41	DISPOSITION OF RECORDS OBTAINED BY DEALERS AND DISTRIBUTORS Dealer And Distributor Records; Records and Reports
DISTRIBUTORS	1002.42	CONFIDENTIALITY OF RECORDS FURNISHED BY DEALERS AND DISTRIBUTORS Dealer And Distributor Records; Records and Reports
DISTRIBUTORS	1003.22	COPIES OF COMMUNICATIONS SENT TO PURCHASERS, DEALERS OR DISTRIBUTORS Notification; Notification Of Defects Or Failure To Comply
DISTRIBUTORS	1304.22	RECORDS FOR MANUFACTURERS, DISTRIBUTORS, DISPENSERS, RESEARCHERS, IMPORTERS AND EXPORTERS Continuing Records; Records And Reports Of Registrants
DISTRIBUTORS	1309.28	EXEMPTION OF DISTRIBUTORS OF REGULATED PRESCRIPTION DRUG PRODUCTS Registration Of Manufacturers, Distributors, Importers And Exporters Of List I Chemicals
DISTRICT OF COLUMBIA	808.101	DISTRICT OF COLUMBIA Listing Of Specific State And Local Exemptions; Exemption From Federal Preemption Of State & Local Medical Device Requirements
DISTRICT OFFICE	203.12	AN APPEAL FROM AN ADVERSE DECISION BY THE DISTRICT OFFICE Reimportation; Prescription Drug Marketing

DISULFIDE, CARBON	520.1802	PIPERAZINE-CARBON DISULFIDE COMPLEX ORAL DOSAGE FORMS Oral Dosage Form New Animal Drugs
DISULFIDE, SELENIUM	524.2101	SELENIUM DISULFIDE SUSPENSION Ophthalmic And Topical Dosage Form New Animal Drugs
DITHIAZANINE IODIDE	520.763	DITHIAZANINE IODIDE ORAL DOSAGE FORMS Oral Dosage Form New Animal Drugs
DIVERSION	1309.73	EMPLOYEE RESPONSIBILITY TO REPORT DIVERSION Registration Of Manufacturers, Distributors, Importers And Exporters Of List I Chemicals
DIVERSION, DRUG	1301.91	EMPLOYEE RESPONSIBILITY TO REPORT DRUG DIVERSION Employee Screening: Non-practitioners; Registration Of Manufacturers, Distributors, And Dispensers Of Controlled Substances
DIVIDED RESPONSIBILITY	610.63	DIVIDED MANUFACTURING RESPONSIBILITY TO BE SHOWN Labeling Standards; General Biological Products Standards
DIVINYLBENZENE	172.775	METHACRYLIC ACID-DIVINYLBENZENE COPOLYMER Other Specific Usage Additives; Food Additives Permitted For Direct Addition To Food For Human Consumption
DIVINYLBENZENE	173.65	DIVINYLBENZENE COPOLYMER Polymer Substances For Food Treatment; Secondary Direct Food Additives Permitted In Food For Human Consumption
DIVINYLBENZENE	173.70	CHLOROMETHYLATED AMINATED STYRENE-DIVINYLBENZENE RESIN Polymer Substances For Food Treatment; Secondary Direct Food Additives Permitted In Food For Human Consumption
DIVINYLBENZENE	177.2710	STYRENE/DIVINYLBENZENE RESINS, CROSS-LINKED Substances For Use Only As Components Of Articles Intended For Repeated Use; Indirect Food Additives: Polymers
DMF	314.420	DRUG MASTER FILES Miscellaneous Provisions; Applications For FDA Approval To Market A New Drug
DOCKETS MANAGEMENT BRANCH	10.20	SUBMISSION OF DOCUMENTS TO DOCKETS MANAGEMENT BRANCH; COMPUTATION OF TIME; AVAILABILITY FOR PUBLIC DISCLOSURE General Administrative Procedures; Administrative Practices And Procedures
DOCUMENT CONTROLS	820.40	DOCUMENT CONTROLS Quality System Regulation
DOCUMENTARY EVIDENCE	1316.57	SUBMISSION OF DOCUMENTARY EVIDENCE AND AFFIDAVITS AND IDENTIFICATION OF WITNESSES SUBSEQUENT TO PREHEARING CONFERENCE Administrative Hearings; Administrative Functions, Practices, And Procedures
DOCUMENTATION	10.70	DOCUMENTATION OF SIGNIFICANT DECISIONS IN ADMINISTRATIVE FILE General Administrative Procedures; Administrative Practices And Procedures
DOCUMENTATION	50.27	DOCUMENTATION OF INFORMED CONSENT Informed Consent Of Human Subjects; Protection Of Human Subjects
DOCUMENTS, FED. REGISTER	5.400	ISSUANCE OF FEDERAL REGISTER DOCUMENTS TO RECOGNIZE OR TO WITHDRAW RECOGNITION OF A STANDARD TO MEET PREMARKET SUBMISSION REQUIREMENTS Medical Devices And Radiological Health; Redelegations Of Authority; Delegations Of Authority And Organization
DOCUMENTS, FED. REGISTER	5.401	ISSUANCE OF FEDERAL REGISTER DOCUMENTS PERTAINING TO EXEMPTIONS FROM PREMARKET NOTIFICATION Medical Devices And Radiological Health; Redelegations Of Authority; Delegations Of Authority And Organization
DOCUMENTS, FED. REGISTER	5.500	ISSUANCE OF FEDERAL REGISTER DOCUMENTS PERTAINING TO THE DETERMINATION OF SAFE LEVELS, DEVELOPMENT OF AN ANALYTICAL METHOD AND PROHIBITION OF CERTAIN EXTRALABEL USE Animal Drugs; Redelegations Of Authority; Delegations Of Authority And Organization
DOCUMENTS, GUIDANCE	312.145	GUIDANCE DOCUMENTS Investigational New Drug Application
DOCUMENTS, GUIDANCE	314.445	GUIDANCE DOCUMENTS Applications For FDA Approval To Market A New Drug
DOCUMENTS, GUIDANCE	316.52	GUIDANCE DOCUMENTS Orphan Drugs
DOCUMENTS, GUIDANCE	601.29	GUIDANCE DOCUMENTS Biologics Licensing; Licensing
DOCUMENTS, SUBMISSION OF	10.20	SUBMISSION OF DOCUMENTS TO DOCKETS MANAGEMENT BRANCH; COMPUTATION OF TIME; AVAILABILITY FOR PUBLIC DISCLOSURE General Administrative Procedures; Administrative Practices And Procedures
DOMESTIC RELEASE	1312.29	DOMESTIC RELEASE PROHIBITED Exportation Of Controlled Substances; Importation And Exportation Of Controlled Substances
DONATION	203.39	DONATION OF DRUG SAMPLES TO CHARITABLE INSTITUTIONS Prescription Drug Marketing
DONOR	610.41	DONOR DEFERRAL Testing Requirements For Communicable Disease Agents; General Biological Products Standards

DONOR	640.3	SUITABILITY OF DONOR Whole Blood; Additional Standards For Human Blood And Blood Products
DONOR	640.12	SUITABILITY OF DONOR Red Blood Cells; Additional Standards For Human Blood And Blood Products
DONOR	640.63	SUITABILITY OF DONOR Source Plasma; Additional Standards For Human Blood And Blood Products
DONOR	640.73	REPORTING OF FATAL DONOR REACTIONS Source Plasma; Additional Standards For Human Blood And Blood Products
DONOR	660.31	SUITABILITY OF THE DONOR Reagent Red Blood Cells; Additional Standards For Diagnostic Substances For Laboratory Tests
DONOR	1270.21	DETERMINATION OF DONOR SUITABILITY FOR HUMAN TISSUE INTENDED FOR TRANSPLANTATION Human Tissue Intended For Transplantation
DONORS	640.21	SUITABILITY OF DONORS Platelets; Additional Standards For Human Blood And Blood Products
DONORS	640.31	SUITABILITY OF DONORS Plasma; Additional Standards For Human Blood And Blood Products
DONORS	640.51	SUITABILITY OF DONORS Cryoprecipitate; Additional Standards For Human Blood And Blood Products
DONORS	640.66	IMMUNIZATION OF DONORS Source Plasma; Additional Standards For Human Blood And Blood Products
DOPPLER	892.1550	ULTRASONIC PULSED DOPPLER IMAGING SYSTEM Diagnostic Devices; Radiology Devices
DORAMECTIN	522.770	DORAMECTIN Implantation Or Injectable Dosage Form New Animal Drugs
DORAMECTIN	524.770	DORAMECTIN Ophthalmic And Topical Dosage Form New Animal Drugs
DORAMECTIN	556.225	DORAMECTIN Tolerances For Residues Of New Animal Drugs In Food
DOSAGE	201.55	STATEMENT OF DOSAGE Labeling Requirements For Prescription Drugs And/or Insulin; Labeling
DOSAGE FORM	330.3	IMPRINTING OF SOLID ORAL DOSAGE FORM DRUG PRODUCTS Over-the-counter Human Drugs Generally Recognized As Safe And Effective & Not Misbranded
DOSAGE FORM	500.26	TIME-RELEASE DOSAGE FORM DRUGS Specific Administrative Rulings And Decisions; General
DOSAGE INFORMATION	200.7	SUPPLYING PHARMACISTS WITH INDICATIONS AND DOSAGE INFORMATION General Provisions; General
DOSE	892.1360	RADIONUCLIDE DOSE CALIBRATOR Diagnostic Devices; Radiology Devices
DOSE, MULTIPLE	320.27	GUIDELINES ON THE DESIGN OF A MULTIPLE-DOSE IN VIVO BIOAVAILABILITY STUDY Procedures For Determining The Bioavailability Of Drug Products; Bioavailability And Bioequivalence Requirements
DOSE, SINGLE	320.26	GUIDELINES ON THE DESIGN OF A SINGLE-DOSE IN VIVO BIOAVAILABILITY STUDY Procedures For Determining The Bioavailability Of Drug Products; Bioavailability And Bioequivalence Requirements
DOUBLE SALT	172.330	CALCIUM PANTOTHENATE, CALCIUM CHLORIDE DOUBLE SALT Special Dietary And Nutritional Additives; Food Additives Permitted For Direct Addition To Food For Human Consumption
DOUBLE SALT	872.3490	CARBOXYMETHYLCELLULOSE SODIUM AND/OR POLYVINYLMETHYLETHER MALEIC ACID CALCIUM-SODIUM DOUBLE SALT DENTURE ADHESIVE Prosthetic Devices; Dental Devices
DOUCHE	884.5900	THERAPEUTIC VAGINAL DOUCHE APPARATUS Obstetrical And Gynecological Therapeutic Devices; Obstetrical And Gynecological Devices
DOWEL	882.4275	DOWEL CUTTING INSTRUMENT Neurological Surgical Devices; Neurological Devices
DOXAPRAM HYDROCHLORIDE	522.775	DOXAPRAM HYDROCHLORIDE INJECTION Implantation Or Injectable Dosage Form New Animal Drugs
DOXYCYCLINE HYCLATE	522.778	DOXYCYCLINE HYCLATE Implantation Or Injectable Dosage Form New Animal Drugs
DOXYLAMINE SUCCINATE	520.784	DOXYLAMINE SUCCINATE TABLETS Oral Dosage Form New Animal Drugs
DOXYLAMINE SUCCINATE	522.784	DOXYLAMINE SUCCINATE INJECTION Implantation Or Injectable Dosage Form New Animal Drugs
DRAFT NOTICES, FED. REG.	10.80	DISSEMINATION OF DRAFT FEDERAL REGISTER NOTICES AND REGULATIONS General Administrative Procedures; Administrative Practices And Procedures
DRAIN	868.5995	TEE DRAIN (WATER TRAP) Therapeutic Devices; Anesthesiology Devices

DRAIN	884.3200	CERVICAL DRAIN Obstetrical And Gynecological Prosthetic Devices; Obstetrical And Gynecological Devices
DRAINAGE	1250.65	DRAINAGE Servicing Areas For Land And Air Conveyances; Interstate Conveyance Sanitation
DRAINAGE CATHETER	878.4200	INTRODUCTION/DRAINAGE CATHETER AND ACCESSORIES Surgical Devices; General And Plastic Surgery Devices
DRAPE ADHESIVE	878.4380	DRAPE ADHESIVE Surgical Devices; General And Plastic Surgery Devices
DRAPE, SURGICAL	878.4370	SURGICAL DRAPE AND DRAPE ACCESSORIES Surgical Devices; General And Plastic Surgery Devices
DRENCH	520.100b	AMPROLIUM DRENCH Oral Dosage Form New Animal Drugs
DRENCH	520.462	CLORSULON DRENCH Oral Dosage Form New Animal Drugs
DRENCH	520.1120b	HALOXON DRENCH Oral Dosage Form New Animal Drugs
DRENCH	520.1194	IVERMECTIN DRENCH Oral Dosage Form New Animal Drugs
DRENCH	520.1242a	LEVAMISOLE HYDROCHLORIDE DRENCH AND DRINKING WATER Oral Dosage Form New Animal Drugs
DRENCH	520.2325b	SULFAQUINOXALINE DRENCH Oral Dosage Form New Animal Drugs
DRENCH	520.2380b	THIABENDAZOLE DRENCH OR ORAL PASTE Oral Dosage Form New Animal Drugs
DRESSING	878.4490	ABSORBABLE HEMOSTATIC AGENT AND DRESSING Surgical Devices; General And Plastic Surgery Devices
DRESSING, BURN	878.4022	HYDROGEL WOUND DRESSING AND BURN DRESSING Surgical Devices; General And Plastic Surgery Devices
DRESSING, FRENCH	169.115	FRENCH DRESSING Requirements For Specific Standardized Food Dressings And Flavorings; Food Dressings And Flavorings
DRESSING, SALAD	169.150	SALAD DRESSING Requirements For Specific Standardized Food Dressings And Flavorings; Food Dressings And Flavorings
DRESSING, TOP	520.2380a	THIABENDAZOLE TOP DRESSING AND MINERAL PROTEIN FEED BLOCK Oral Dosage Form New Animal Drugs
DRESSING, WOUND	878.4018	HYDROPHILIC WOUND DRESSING Surgical Devices; General And Plastic Surgery Devices
DRESSING, WOUND	878.4020	OCCLUSIVE WOUND DRESSING Surgical Devices; General And Plastic Surgery Devices
DRESSING, WOUND,	878.4022	HYDROGEL WOUND DRESSING AND BURN DRESSING Surgical Devices; General And Plastic Surgery Devices
DRIED ALGAE MEAL	73.275	DRIED ALGAE MEAL Foods; Listing Of Color Additives Exempt From Certification
DRIED ALUMINUM HYDROXIDE	73.1010	ALUMINA (DRIED ALUMINUM HYDROXIDE) Drugs; Listing Of Color Additives Exempt From Certification
DRIED EGG	160.145	DRIED EGG WHITES Requirements For Specific Standardized Eggs And Egg Products; Eggs And Egg Products
DRIED EGG	160.185	DRIED EGG YOLKS Requirements For Specific Standardized Eggs And Egg Products; Eggs And Egg Products
DRIED EGGS	160.105	DRIED EGGS Requirements For Specific Standardized Eggs And Egg Products; Eggs And Egg Products
DRIED GLUCOSE	168.121	DRIED GLUCOSE SIRUP Requirements For Specific Standardized Sweeteners And Table Sirups; Sweeteners And Table Sirups
DRIED POTATOES	102.41	POTATO CHIPS MADE FROM DRIED POTATOES Requirements For Specific Nonstandardized Foods; Common Or Usual Name For Nonstandardized Foods
DRIED YEAST	573.750	PICHIA PASTORIS DRIED YEAST Food Additives Permitted In Feed And Drinking Water Of Animals
DRIED YEASTS	172.896	DRIED YEASTS Multipurpose Additives; Food Additives Permitted For Direct Addition To Food For Human Consumption
DRIERS	181.25	DRIERS Specific Prior-sanctioned Food Ingredients; Prior-sanctioned Food Ingredients
DRILL	872.4130	INTRAORAL DENTAL DRILL Surgical Devices; Dental Devices
DRILL	874.4250	EAR, NOSE AND THROAT ELECTRIC OR PNEUMATIC SURGICAL DRILL Surgical Devices; Ear, Nose And Throat Devices

DRILL	882.4325	CRANIAL DRILL HANDPIECE (BRACE) Neurological Surgical Devices; Neurological Devices
DRILL	882.4360	ELECTRIC CRANIAL DRILL MOTOR Neurological Surgical Devices; Neurological Devices
DRILL	882.4370	PNEUMATIC CRANIAL DRILL MOTOR Neurological Surgical Devices; Neurological Devices
DRILLS	882.4300	MANUAL CRANIAL DRILLS, BURRS, TREPHINES, AND THEIR ACCESSORIES Neurological Surgical Devices; Neurological Devices
DRILLS	882.4305	POWERED COMPOUND CRANIAL DRILLS, BURRS, TREPHINES, AND THEIR ACCESSORIES Neurological Surgical Devices; Neurological Devices
DRILLS	882.4310	POWERED SIMPLE CRANIAL DRILLS, BURRS, TREPHINES, AND THEIR ACCESSORIES Neurological Surgical Devices; Neurological Devices
DRINK	1250.25	SOURCE IDENTIFICATION AND INSPECTION OF FOOD AND DRINK Food Service Sanitation On Land And Air Conveyances, And Vessels; Interstate Conveyance Sanitation
DRINKING FOUNTAINS	1250.85	DRINKING FOUNTAINS AND COOLERS Sanitation Facilities And Conditions On Vessels; Interstate Conveyance Sanitation
DRINKING WATER	520.100b	AMPROLIUM DRINKING WATER Oral Dosage Form New Animal Drugs
DRINKING WATER	520.1242a	LEVAMISOLE HYDROCHLORIDE DRENCH AND DRINKING WATER Oral Dosage Form New Animal Drugs
DRINKING WATER	520.1696b	PENICILLIN G POTASSIUM IN DRINKING WATER Oral Dosage Form New Animal Drugs
DRINKING WATER	520.2200b	SULFACHLORPYRIDAZINE MEDICATED MILK AND DRINKING WATER Oral Dosage Form New Animal Drugs
DRINKING WATER	520.2240a	SULFAETHOXYPYRIDAZINE DRINKING WATER Oral Dosage Form New Animal Drugs
DRINKING WATER	520.2261a	SULFAMETHAZINE SODIUM DRINKING WATER SOLUTION Oral Dosage Form New Animal Drugs
DRINKING WATER	520.2325	SULFAQUINOXALINE DRINKING WATER Oral Dosage Form New Animal Drugs
DRINKING WATER	1240.80	GENERAL REQUIREMENTS FOR WATER FOR DRINKING AND CULINARY PURPOSES Source And Use Of Potable Water; Control Of Communicable Diseases
DRINKING UTENSILS	1250.44	DRINKING UTENSILS AND TOILET ARTICLES Equipment And Operation Of Land And Air Conveyances; Interstate Conveyance Sanitation
DROPERIDOL	522.800	DROPERIDOL AND FENTANYL CITRATE INJECTION Implantation Or Injectable Dosage Form New Animal Drugs
DRUG	5.701	PREMARKET APPROVAL OF A PRODUCT THAT IS OR CONTAINS A BIOLOGIC, DEVICE, OR DRUG Product Designation; Redelegations Of Authority; Delegations Of Authority And Organization
DRUG	73.1001	DILUENTS IN COLOR ADDITIVE MIXTURES FOR DRUG USE EXEMPT FROM CERTIFICATION Drugs; Listing Of Color Additives Exempt From Certification
DRUG ADMINISTRATION	874.5220	EAR, NOSE, AND THROAT DRUG ADMINISTRATION DEVICE Therapeutic Devices; Ear, Nose And Throat Devices
DRUG AMENDMENTS	5.100	ISSUANCE OF NOTICE IMPLEMENTING THE PROVISIONS OF THE DRUG AMENDMENTS OF 1962 Human Drugs; Redelegations Of Authority; Delegations Of Authority And Organization
DRUG, ANIMAL	500.30	GENTIAN VIOLET FOR ANIMAL DRUG USE General
DRUG, ANIMAL	530.20	CONDITIONS FOR PERMITTED EXTRALABEL ANIMAL AND HUMAN DRUG USE IN FOOD-PRODUCING ANIMALS Extralabel Drug Use In Animals
DRUG APPLICATION	314.101	FILING AN APPLICATION AND RECEIVING AN ABBREVIATED NEW DRUG APPLICATION Applications For FDA Approval To Market A New Drug
DRUG APPLICATION	314.127	REFUSAL TO APPROVE AN ABBREVIATED NEW DRUG APPLICATION Applications For FDA Approval To Market A New Drug
DRUG APPLICATION	314.151	WITHDRAWAL OF APPROVAL OF AN ABBREVIATED NEW DRUG APPLICATION UNDER SECTION 505(J)(5) Applications For FDA Approval To Market A New Drug
DRUG APPLICATION	314.152	NOTICE OF WITHDRAWAL OF APPROVAL OF AN ABBREVIATED NEW DRUG APPLICATION UNDER SECTION 505(J)(5) FDA Action On Applications; Applications For FDA Approval To Market A New Drug
DRUG APPLICATION	314.153	SUSPENSION OF APPROVAL OF AN ABBREVIATED NEW DRUG APPLICATION Applications For FDA Approval To Market A New Drug

DRUG APPLICATION	320.31	APPLICABILITY OF REQUIREMENTS REGARDING AN "INVESTIGATIONAL NEW DRUG APPLICATION" Procedures For Determining The Bioavailability Of Drug Products; Bioavailability And Bioequivalence Requirements
DRUG APPLICATION	514.11	CONFIDENTIALITY OF DATA AND INFORMATION IN A NEW ANIMAL DRUG APPLICATION FILE General Provisions; New Animal Drug Applications
DRUG APPLICATIONS	5.103	APPROVAL OF NEW DRUG APPLICATIONS AND THEIR SUPPLEMENTS Human Drugs; Redelegations Of Authority; Delegations Of Authority And Organization
DRUG APPLICATIONS	5.105	ISSUANCE OF NOTICES RELATING TO PROPOSALS TO REFUSE APPROVAL OR TO WITHDRAW APPROVAL OF NEW DRUG APPLICATIONS AND THEIR SUPPLEMENTS Human Drugs; Redelegations Of Authority; Delegations Of Authority And Organization
DRUG APPLICATIONS	5.106	SUBMISSION OF AND EFFECTIVE APPROVAL DATES FOR ABBREVIATED NEW DRUG APPLICATIONS AND CERTAIN NEW DRUG APPLICATIONS Human Drugs; Redelegations Of Authority; Delegations Of Authority And Organization
DRUG APPLICATIONS	5.501	APPROVAL OF NEW ANIMAL DRUG APPLICATIONS AND THEIR SUPPLEMENTS Animal Drugs; Redelegations Of Authority; Delegations Of Authority And Organization
DRUG APPLICATIONS	5.503	SUBMISSION OF AND EFFECTIVE APPROVAL DATES FOR ABBREVIATED NEW ANIMAL DRUG APPLICATIONS Animal Drugs; Redelegations Of Authority; Delegations Of Authority And Organization
DRUG APPLICATIONS	310.201	EXEMPTION FOR CERTAIN DRUGS LIMITED BY NEW DRUG APPLICATIONS TO PRESCRIPTION SALE New Drugs Exempted From Prescription-dispensing Requirements; New Drugs
DRUG APPLICATIONS	310.303	CONTINUATION OF LONG-TERM STUDIES, RECORDS, AND REPORTS ON CERTAIN DRUGS FOR WHICH NEW DRUG APPLICATIONS HAVE BEEN APPROVED Records And Reports; New Drugs
DRUG APPLICATIONS	310.305	RECORDS AND REPORTS CONTAINING ADVERSE DRUG EXPERIENCES ON MARKETED PRESCRIPTION DRUGS FOR HUMAN USE WITHOUT APPROVED NEW DRUG APPLICATIONS Records And Reports; New Drugs
DRUG APPLICATIONS	514.8	SUPPLEMENTAL NEW ANIMAL DRUG APPLICATIONS General Provisions; New Animal Drug Applications
DRUG, APPROVED	314.170	ADULTERATION AND MISBRANDING OF AN APPROVED DRUG FDA Action On Applications; Applications For FDA Approval To Market A New Drug
DRUG CATEGORIES	330.5	DRUG CATEGORIES General Provisions; Over-the-counter Human Drugs Generally Recognized As Safe And Effective & Not Misbranded
DRUG CODE	201.2	DRUGS AND DEVICES; NATIONAL DRUG CODE NUMBERS Labeling
DRUG CODES	510.600	NAMES, ADDRESSES, AND DRUG LABELER CODES OF SPONSORS OF APPROVED APPLICATIONS Sponsors Of Approved Applications; New Animal Drugs
DRUG CONTAINER	211.184	COMPONENT, DRUG PRODUCT CONTAINER, CLOSURE, AND LABELING RECORDS Records And Reports; Current Good Manufacturing Practice For Finished Pharmaceuticals
DRUG CONTAINERS	211.82	RECEIPT AND STORAGE OF UNTESTED COMPONENTS, DRUG PRODUCT CONTAINERS, AND CLOSURES Control Of Components And Drug Product Containers And Closures; Current Good Manufacturing Practice For Finished Pharmaceuticals
DRUG CONTAINERS	211.84	TESTING AND APPROVAL OR REJECTION OF COMPONENTS, DRUG PRODUCT CONTAINERS, AND CLOSURES Control Of Components And Drug Product Containers And Closures; Current Good Manufacturing Practice For Finished Pharmaceuticals
DRUG CONTAINERS	211.86	USE OF APPROVED COMPONENTS, DRUG PRODUCT CONTAINERS, AND CLOSURES Control Of Components And Drug Product Containers And Closures; Current Good Manufacturing Practice For Finished Pharmaceuticals
DRUG CONTAINERS	211.87	RETESTING OF APPROVED COMPONENTS, DRUG PRODUCT CONTAINERS, AND CLOSURES Control Of Components And Drug Product Containers And Closures; Current Good Manufacturing Practice For Finished Pharmaceuticals
DRUG CONTAINERS	211.89	REJECTED COMPONENTS, DRUG PRODUCT CONTAINERS AND CLOSURES Control Of Components And Drug Product Containers And Closures; Current Good Manufacturing Practice For Finished Pharmaceuticals

DRUG CONTAINERS	211.94	DRUG PRODUCT CONTAINERS AND CLOSURES Control Of Components And Drug Product Containers And Closures; Current Good Manufacturing Practice For Finished Pharmaceuticals
DRUG CONTROL, OFFICE	1401.2	THE OFFICE OF NATIONAL DRUG CONTROL POLICY - ORGANIZATION AND FUNCTIONS Public Availability Of Information
DRUG DISPOSITION	312.59	DISPOSITION OF UNUSED SUPPLY OF INVESTIGATIONAL DRUG Responsibilities Of Sponsors And Investigators; Investigational New Drug Application
DRUG DISTRIBUTOR	205.4	WHOLESALE DRUG DISTRIBUTOR LICENSING REQUIREMENT Guidelines For State Licensing Of Wholesale Prescription Drug Distributors
DRUG DIVERSION	1301.91	EMPLOYEE RESPONSIBILITY TO REPORT DRUG DIVERSION Employee Screening: Non-practitioners; Registration Of Manufacturers, Distributors, And Dispensers Of Controlled Substances
DRUG EFFICACY STUDY	201.200	DISCLOSURE OF DRUG EFFICACY STUDY EVALUATIONS IN LABELING AND ADVERTISING Labeling Claims For Drugs In Drug Efficacy Studies; Labeling
DRUG EFFICACY STUDY	310.6	APPLICABILITY OF "NEW DRUG" OR SAFETY/EFFECTIVENESS FINDINGS IN DRUG EFFICACY STUDY IMPLEMENTATION NOTICES AND NOTICES OF OPPORTUNITY FOR HEARING TO IDENTICAL, RELATED, AND SIMILAR DRUG PRODUCTS General Provisions; New Drugs
DRUG EFFICACY STUDY	330.12	STATUS OF OVER-THE-COUNTER (OTC) DRUGS PREVIOUSLY REVIEWED UNDER THE DRUG EFFICACY STUDY (DESI) Administrative Procedures; Over-the-counter Human Drugs Generally Recognized As Safe And Effective & Not Misbranded
DRUG ESTABLISHMENT	207.35	NOTIFICATION OF REGISTRANT; DRUG ESTABLISHMENT REGISTRATION NUMBER AND DRUG LISTING NUMBER Procedures For Domestic Drug Establishments; Registration Of Drug Producers And Listing Of Drugs In Commercial Distribution
DRUG ESTABLISHMENTS	200.11	USE OF OCTADECYLAMINE IN STEAM LINES OF DRUG ESTABLISHMENTS General Provisions; General
DRUG EXPERIENCE REPORTS	20.112	VOLUNTARY DRUG EXPERIENCE REPORTS SUBMITTED BY PHYSICIANS AND HOSPITALS Availability Of Specific Categories Of Records; Public Information
DRUG EXPERIENCES, ADVERSE	310.305	RECORDS AND REPORTS CONTAINING ADVERSE DRUG EXPERIENCES ON MARKETED PRESCRIPTION DRUGS FOR HUMAN USE WITHOUT APPROVED NEW DRUG APPLICATIONS Records And Reports; New Drugs
DRUG EXPERIENCES, ADVERSE	314.80	POSTMARKETING REPORTING OF ADVERSE DRUG EXPERIENCES Applications; Applications For FDA Approval To Market A New Drug
DRUG, EXTRALABEL	5.500	ISSUANCE OF FEDERAL REGISTER DOCUMENTS PERTAINING TO THE DETERMINATION OF SAFE LEVELS, NOTICE OF NEED FOR DEVELOPMENT OF AN ANALYTICAL METHOD, AND PROHIBITION OF CERTAIN EXTRALABEL DRUG USE Animal Drugs; Redelegations Of Authority; Delegations Of Authority And Organization
DRUG, EXTRALABEL	530.30	EXTRALABEL DRUG USE IN NONFOOD ANIMALS Extralabel Drug Use In Animals
DRUG, HUMAN	211.132	TAMPER-EVIDENT PACKAGING REQUIREMENTS FOR OTC HUMAN DRUG PRODUCTS Current Good Manufacturing Practice For Finished Pharmaceuticals
DRUG, HUMAN	530.20	CONDITIONS FOR PERMITTED EXTRALABEL ANIMAL AND HUMAN DRUG USE IN FOOD-PRODUCING ANIMALS Extralabel Drug Use In Animals
DRUG, INVESTIGATIONAL	312.61	CONTROL OF THE INVESTIGATIONAL DRUG Investigational New Drug Application
DRUG, INVESTIGATIONAL NEW	312.6	LABELING OF AN INVESTIGATIONAL NEW DRUG Investigational New Drug Application
DRUG, INVESTIGATIONAL NEW	312.34	TREATMENT USE OF AN INVESTIGATIONAL NEW DRUG Investigational New Drug Application
DRUG, INVESTIGATIONAL NEW	312.36	EMERGENCY USE OF AN INVESTIGATIONAL NEW DRUG Investigational New Drug Application
DRUG, INVESTIGATIONAL NEW	312.40	GENERAL REQUIREMENTS FOR USE OF AN INVESTIGATIONAL NEW DRUG IN A CLINICAL INVESTIGATION Investigational New Drug Application
DRUG, INVESTIGATIONAL NEW	320.31	APPLICABILITY OF REQUIREMENTS REGARDING AN "INVESTIGATIONAL NEW DRUG APPLICATION" Procedures For Determining The Bioavailability Of Drug Products; Bioavailability And Bioequivalence Requirements
DRUG-LABELING	500.55	EXEMPTION FROM CERTAIN DRUG-LABELING REQUIREMENTS Animal Drug Labeling Requirements; General

DRUG LIST	207.20	WHO MUST REGISTER AND SUBMIT A DRUG LIST Procedures For Domestic Drug Establishments; Registration Of Drug Producers And Listing Of Drugs In Commercial Distribution
DRUG, LISTED	314.54	PROCEDURE FOR SUBMISSION OF AN APPLICATION REQUIRING INVESTIGATIONS FOR APPROVAL OF A NEW INDICATION FOR, OR OTHER CHANGE FROM, A LISTED DRUG Applications For FDA Approval To Market A New Drug
DRUG, LISTED	314.93	PETITION TO REQUEST A CHANGE FROM A LISTED DRUG Applications For FDA Approval To Market A New Drug
DRUG, LISTED	314.122	SUBMITTING AN ABBREVIATED APPLICATION FOR A LISTED DRUG THAT IS NO LONGER MARKETED Applications For FDA Approval To Market A New Drug
DRUG, LISTED	314.161	DETERMINATION OF REASONS FOR VOLUNTARY WITHDRAWAL OF A LISTED DRUG Applications For FDA Approval To Market A New Drug
DRUG LISTING	20.116	DRUG AND DEVICE LISTING INFORMATION Availability Of Specific Categories Of Records; Public Information
DRUG LISTING	207.21	TIMES FOR REGISTRATION AND DRUG LISTING Procedures For Domestic Drug Establishments; Registration Of Drug Producers And Listing Of Drugs In Commercial Distribution
DRUG LISTING	207.25	INFORMATION REQUIRED IN REGISTRATION AND DRUG LISTING Procedures For Domestic Drug Establishments; Registration Of Drug Producers And Listing Of Drugs In Commercial Distribution
DRUG LISTING	207.30	UPDATING DRUG LISTING INFORMATION Procedures For Domestic Drug Establishments; Registration Of Drug Producers And Listing Of Drugs In Commercial Distribution
DRUG LISTING	207.31	ADDITIONAL DRUG LISTING INFORMATION Procedures For Domestic Drug Establishments; Registration Of Drug Producers And Listing Of Drugs In Commercial Distribution
DRUG LISTING	207.35	NOTIFICATION OF REGISTRANT; DRUG ESTABLISHMENT REGISTRATION NUMBER AND DRUG LISTING NUMBER Procedures For Domestic Drug Establishments; Registration Of Drug Producers And Listing Of Drugs In Commercial Distribution
DRUG LISTING	207.40	ESTABLISHMENT REGISTRATION AND DRUG LISTING REQUIREMENTS FOR FOREIGN ESTABLISHMENTS Procedure For Foreign Drug Establishments; Registration Of Drug Producers And Listing Of Drugs In Commercial Distribution
DRUG LISTINGS	207.37	INSPECTION OF REGISTRATIONS AND DRUG LISTINGS Procedures For Domestic Drug Establishments; Registration Of Drug Producers And Listing Of Drugs In Commercial Distribution
DRUG MASTER FILES	314.420	DRUG MASTER FILES Miscellaneous Provisions; Applications For FDA Approval To Market A New Drug
DRUG, NEW	5.103	APPROVAL OF NEW DRUG APPLICATIONS AND THEIR SUPPLEMENTS Human Drugs; Redelegations Of Authority; Delegations Of Authority And Organization
DRUG, NEW	5.105	ISSUANCE OF NOTICES RELATING TO PROPOSALS TO REFUSE APPROVAL OR TO WITHDRAW APPROVAL OF NEW DRUG APPLICATIONS AND THEIR SUPPLEMENTS Human Drugs; Redelegations Of Authority; Delegations Of Authority And Organization
DRUG, NEW	5.106	SUBMISSION OF AND EFFECTIVE APPROVAL DATES FOR ABBREVIATED NEW DRUG APPLICATIONS AND CERTAIN NEW DRUG APPLICATIONS Human Drugs; Redelegations Of Authority; Delegations Of Authority And Organization
DRUG, NEW	20.117	NEW DRUG INFORMATION Availability Of Specific Categories Of Records; Public Information
DRUG, NEW	310.6	APPLICABILITY OF "NEW DRUG" OR SAFETY/EFFECTIVENESS FINDINGS IN DRUG EFFICACY STUDY IMPLEMENTATION NOTICES AND NOTICES OF OPPORTUNITY FOR HEARING TO IDENTICAL, RELATED, AND SIMILAR DRUG PRODUCTS General Provisions; New Drugs
DRUG, NEW	310.100	NEW DRUG STATUS OPINIONS; STATEMENT OF POLICY Specific Administrative Rulings And Decisions; New Drugs
DRUG, NEW	310.103	NEW DRUG SUBSTANCES INTENDED FOR HYPERSENSITIVITY TESTING Specific Administrative Rulings And Decisions; New Drugs
DRUG, NEW	310.201	EXEMPTION FOR CERTAIN DRUGS LIMITED BY NEW DRUG APPLICATIONS TO PRESCRIPTION SALE New Drugs Exempted From Prescription-dispensing Requirements; New Drugs
DRUG, NEW	310.303	CONTINUATION OF LONG-TERM STUDIES, RECORDS, AND REPORTS ON CERTAIN DRUGS FOR WHICH NEW DRUG APPLICATIONS HAVE BEEN APPROVED Records And Reports; New Drugs

DRUG, NEW	310.305	RECORDS AND REPORTS CONTAINING ADVERSE DRUG EXPERIENCES ON MARKETED PRESCRIPTION DRUGS FOR HUMAN USE WITHOUT APPROVED NEW DRUG APPLICATIONS Records And Reports; New Drugs
DRUG, NEW	310.502	CERTAIN DRUGS ACCORDED NEW DRUG STATUS THROUGH RULEMAKING PROCEDURES Requirements For Specific New Drugs Or Devices; New Drugs
DRUG, NEW	312.6	LABELING OF AN INVESTIGATIONAL NEW DRUG Investigational New Drug Application
DRUG, NEW	312.34	TREATMENT USE OF AN INVESTIGATIONAL NEW DRUG Investigational New Drug Application
DRUG, NEW	312.36	EMERGENCY USE OF AN INVESTIGATIONAL NEW DRUG Investigational New Drug Application
DRUG, NEW	312.40	GENERAL REQUIREMENTS FOR USE OF AN INVESTIGATIONAL NEW DRUG IN A CLINICAL INVESTIGATION Investigational New Drug Application
DRUG, NEW	314.101	FILING AN APPLICATION AND RECEIVING AN ABBREVIATED NEW DRUG APPLICATION Applications For FDA Approval To Market A New Drug
DRUG, NEW	314.108	NEW DRUG PRODUCT EXCLUSIVITY Applications For FDA Approval To Market A New Drug
DRUG, NEW	314.127	REFUSAL TO APPROVE AN ABBREVIATED NEW DRUG APPLICATION Applications For FDA Approval To Market A New Drug
DRUG, NEW	314.152	NOTICE OF WITHDRAWAL OF APPROVAL OF AN ABBREVIATED NEW DRUG APPLICATION UNDER SECTION 505(J)(5) FDA Action On Applications; Applications For FDA Approval To Market A New Drug
DRUG, NEW	314.153	SUSPENSION OF APPROVAL OF AN ABBREVIATED NEW DRUG APPLICATION Applications For FDA Approval To Market A New Drug
DRUG, NEW	320.31	APPLICABILITY OF REQUIREMENTS REGARDING AN "INVESTIGATIONAL NEW DRUG APPLICATION" Procedures For Determining The Bioavailability Of Drug Products; Bioavailability And Bioequivalence Requirements
DRUG, NEW	601.50	CONFIDENTIALITY OF DATA AND INFORMATION IN AN INVESTIGATIONAL NEW DRUG NOTICE FOR A BIOLOGICAL PRODUCT Confidentiality Of Information; Licensing
DRUG, NEW ANIMAL	5.501	APPROVAL OF NEW ANIMAL DRUG APPLICATIONS AND THEIR SUPPLEMENTS Animal Drugs; Redelegations Of Authority; Delegations Of Authority And Organization
DRUG, NEW ANIMAL	5.503	SUBMISSION OF AND EFFECTIVE APPROVAL DATES FOR ABBREVIATED NEW ANIMAL DRUG APPLICATIONS Animal Drugs; Redelegations Of Authority; Delegations Of Authority And Organization
DRUG, NEW ANIMAL	510.455	NEW ANIMAL DRUG REQUIREMENTS REGARDING FREE-CHOICE ADMINISTRATION IN FEEDS Requirements For Specific New Animal Drugs; New Animal Drugs
DRUG, NEW ANIMAL	514.8	SUPPLEMENTAL NEW ANIMAL DRUG APPLICATIONS General Provisions; New Animal Drug Applications
DRUG, NEW ANIMAL	514.11	CONFIDENTIALITY OF DATA AND INFORMATION IN A NEW ANIMAL DRUG APPLICATION FILE General Provisions; New Animal Drug Applications
DRUG, NEW ANIMAL	514.12	CONFIDENTIALITY OF DATA AND INFORMATION IN AN INVESTIGATIONAL NEW ANIMAL DRUG NOTICE General Provisions; New Animal Drug Applications
DRUG, NEW ANIMAL	558.5	NEW ANIMAL DRUG REQUIREMENTS FOR LIQUID TYPE B FEEDS General Provisions; New Animal Drugs For Use In Animal Feeds
DRUG OFFENSES	1404.635	REPORTING OF AND EMPLOYEE SANCTIONS FOR CONVICTIONS OF CRIMINAL DRUG OFFENSES Governmentwide Debarment And Suspension And Requirements For Drug-Free Workplace (Grants)
DRUG, ORPHAN	316.20	CONTENT AND FORMAT OF A REQUEST FOR ORPHAN-DRUG DESIGNATION Orphan Drugs
DRUG, ORPHAN	316.21	VERIFICATION OF ORPHAN-DRUG STATUS Orphan Drugs
DRUG, ORPHAN	316.23	TIMING OF REQUESTS FOR ORPHAN-DRUG DESIGNATION; DESIGNATION OF ALREADY APPROVED DRUGS Orphan Drugs
DRUG, ORPHAN	316.24	GRANTING ORPHAN-DRUG STATUS Orphan Drugs
DRUG, ORPHAN	316.25	REFUSAL TO GRANT ORPHAN-DRUG DESIGNATION Orphan Drugs
DRUG, ORPHAN	316.26	AMENDMENT TO ORPHAN-DRUG DESIGNATION Orphan Drugs

DRUG, ORPHAN	316.27	CHANGE IN OWNERSHIP OF ORPHAN-DRUG DESIGNATION Orphan Drugs
DRUG, ORPHAN	316.28	PUBLICATION OF ORPHAN-DRUG DESIGNATIONS Orphan Drugs
DRUG, ORPHAN	316.29	REVOCATION OF ORPHAN-DRUG DESIGNATION Orphan Drugs
DRUG, ORPHAN	316.30	ANNUAL REPORTS OF HOLDER OF ORPHAN-DRUG DESIGNATION Orphan Drugs
DRUG, ORPHAN	316.31	SCOPE OF ORPHAN-DRUG EXCLUSIVE APPROVAL Orphan Drugs
DRUG, ORPHAN	316.40	TREATMENT USE OF A DESIGNATED ORPHAN-DRUG Orphan Drugs
DRUG, OTC	201.66	FORMAT AND CONTENT REQUIREMENTS FOR OTC DRUG PRODUCT LABELING Labeling Requirements For Prescription Drugs And/or Insulin; Labeling
DRUG, OTC	328.50	PRINCIPAL DISPLAY PANEL OF ALL OTC DRUG PRODUCTS INTENDED FOR ORAL INGESTION THAT CONTAIN ALCOHOL Over-the-counter Drug Products Intended For Oral Ingestion That Contain Alcohol
DRUG, OTC	330.13	CONDITIONS FOR MARKETING INGREDIENTS RECOMMENDED FOR OVER-THE-COUNTER (OTC) USE UNDER THE OTC DRUG REVIEW Administrative Procedures; Over-the-counter Human Drugs Generally Recognized As Safe And Effective & Not Misbranded
DRUG PREPARATIONS	201.303	LABELING OF DRUG PREPARATIONS CONTAINING SIGNIFICANT PROPORTIONS OF WINTERGREEN OIL Specific Labeling Requirements For Specific Drug Products; Labeling
DRUG PREPARATIONS	201.310	PHENINDIONE; LABELING OF DRUG PREPARATIONS INTENDED FOR USE BY MAN Specific Labeling Requirements For Specific Drug Products; Labeling
DRUG PREPARATIONS	201.314	LABELING OF DRUG PREPARATIONS CONTAINING SALICYLATES Specific Labeling Requirements For Specific Drug Products; Labeling
DRUG PREPARATIONS	250.11	THYROID-CONTAINING DRUG PREPARATIONS INTENDED FOR TREATMENT OF OBESITY IN HUMANS Drugs Regarded As Misbranded; Specific Requirements For Specific Human Drugs
DRUG PREPARATIONS	250.102	DRUG PREPARATIONS INTENDED FOR HUMAN USE CONTAINING CERTAIN "CORONARY VASODILATORS" New Drug Or Prescription Status Of Specific Drugs; Specific Requirements For Specific Human Drugs
DRUG, PRESCRIPTION	5.108	AUTHORITY RELATING TO WAIVERS OR REDUCTIONS OF PRESCRIPTION DRUG USER FEES Human Drugs; Redelegations Of Authority; Delegations Of Authority And Organization
DRUG, PRESCRIPTION	202.1	PRESCRIPTION DRUG ADVERTISEMENTS Prescription Drug Advertising
DRUG, PRESCRIPTION	250.100	AMYL NITRITE INHALANT AS A PRESCRIPTION DRUG FOR HUMAN USE New Drug Or Prescription Status Of Specific Drugs; Specific Requirements For Specific Human Drugs
DRUG PRODUCT	201.66	FORMAT AND CONTENT REQUIREMENTS FOR OTC DRUG PRODUCT LABELING Labeling Requirements For Over-the-counter Drugs; Labeling
DRUG PRODUCT	211.134	DRUG PRODUCT INSPECTION Packaging And Labeling Control; Current Good Manufacturing Practice For Finished Pharmaceuticals
DRUG PRODUCT	211.208	DRUG PRODUCT SALVAGING Returned And Salvaged Drug Products; Current Good Manufacturing Practice For Finished Pharmaceuticals
DRUG PRODUCT	314.108	NEW DRUG PRODUCT EXCLUSIVITY Applications For FDA Approval To Market A New Drug
DRUG PRODUCT	314.162	REMOVAL OF A DRUG PRODUCT FROM THE LIST Applications For FDA Approval To Market A New Drug
DRUG PRODUCTS	5.27	PATENT TERM EXTENSIONS FOR HUMAN DRUG PRODUCTS, MEDICAL DEVICES, AND FOOD AND COLOR ADDITIVES General Redelegations Of Authority; Delegations Of Authority And Organization
DRUG PRODUCTS	200.51	AQUEOUS-BASED DRUG PRODUCTS FOR ORAL INHALATION Requirements For Specific Classes Of Drugs; General
DRUG PRODUCTS	201.312	MAGNESIUM SULFATE HEPTAHYDRATE; LABEL DECLARATION ON DRUG PRODUCTS Specific Labeling Requirements For Specific Drug Products; Labeling
DRUG PRODUCTS	201.320	WARNING STATEMENTS FOR DRUG PRODUCTS CONTAINING OR MANUFACTURED WITH CHLOROFLUOROCARBONS Specific Labeling Requirements For Specific Drug Products; Labeling
DRUG PRODUCTS	201.322	OVER-THE-COUNTER DRUG PRODUCTS CONTAINING INTERNAL ANALGESIC/ANTIPYRETIC ACTIVE INGREDIENTS; REQUIRED ALCOHOL WARNING Specific Labeling Requirements For Specific Drug Products; Labeling

DRUG PRODUCTS	211.110	SAMPLING AND TESTING OF IN-PROCESS MATERIALS AND DRUG PRODUCTS Equipment; Current Good Manufacturing Practice For Finished Pharmaceuticals
DRUG PRODUCTS	211.132	TAMPER-EVIDENT PACKAGING REQUIREMENTS FOR OTC HUMAN DRUG PRODUCTS Current Good Manufacturing Practice For Finished Pharmaceuticals
DRUG PRODUCTS	211.204	RETURNED DRUG PRODUCTS Returned And Salvaged Drug Products; Current Good Manufacturing Practice For Finished Pharmaceuticals
DRUG PRODUCTS	216.24	DRUG PRODUCTS WITHDRAWN OR REMOVED FROM THE MARKET FOR REASONS OF SAFETY OR EFFECTIVENESS Pharmacy Compounding
DRUG PRODUCTS	250.250	HEXACHLOROPHENE, AS A COMPONENT OF DRUG AND COSMETIC PRODUCTS Requirements For Drugs And Cosmetics; Specific Requirements For Specific Human Drugs
DRUG PRODUCTS	310.6	APPLICABILITY OF "NEW DRUG" OR SAFETY/EFFECTIVENESS FINDINGS IN DRUG EFFICACY STUDY IMPLEMENTATION NOTICES AND NOTICES OF OPPORTUNITY FOR HEARING TO IDENTICAL, RELATED, AND SIMILAR DRUG PRODUCTS General Provisions; New Drugs
DRUG PRODUCTS	310.509	PARENTERAL DRUG PRODUCTS IN PLASTIC CONTAINERS Requirements For Specific New Drugs Or Devices; New Drugs
DRUG PRODUCTS	310.518	DRUG PRODUCTS CONTAINING IRON OR IRON SALTS Requirements For Specific New Drugs Or Devices; New Drugs
DRUG PRODUCTS	310.519	DRUG PRODUCTS MARKETED AS OTC DAYTIME SEDATIVES Requirements For Specific New Drugs Or Devices; New Drugs
DRUG PRODUCTS	310.527	DRUG PRODUCTS CONTAINING ACTIVE INGREDIENTS OFFERED OTC FOR EXTERNAL USE A HAIR GROWERS OR FOR HAIR LOSS PREVENTION Requirements For Specific New Drugs Or Devices; New Drugs
DRUG PRODUCTS	310.528	DRUG PRODUCTS CONTAINING ACTIVE INGREDIENTS OFFERED OTC FOR USE AS AN APHRODISIAC Requirements For Specific New Drugs Or Devices; New Drugs
DRUG PRODUCTS	310.529	DRUG PRODUCTS CONTAINING ACTIVE INGREDIENTS OFFERED OTC FOR ORAL USE AS INSECT REPELLENTS Requirements For Specific New Drugs Or Devices; New Drugs
DRUG PRODUCTS	310.530	TOPICALLY-APPLIED HORMONE-CONTAINING DRUG PRODUCTS FOR OTC HUMAN USE Requirements For Specific New Drugs Or Devices; New Drugs
DRUG PRODUCTS	310.531	DRUG PRODUCTS CONTAINING ACTIVE INGREDIENTS OFFERED OTC FOR THE TREATMENT OF BOILS Requirements For Specific New Drugs Or Devices; New Drugs
DRUG PRODUCTS	310.532	DRUG PRODUCTS CONTAINING ACTIVE INGREDIENTS OFFERED OTC TO RELIEVE THE SYMPTOMS OF BENIGN PROSTATIC HYPERTROPHY Requirements For Specific New Drugs Or Devices; New Drugs
DRUG PRODUCTS	310.533	DRUG PRODUCTS CONTAINING ACTIVE INGREDIENTS OFFERED OTC FOR HUMAN USE AS AN ANTICHOLINERGIC IN COUGH-COLD DRUG PRODUCTS Requirements For Specific New Drugs Or Devices; New Drugs
DRUG PRODUCTS	310.534	DRUG PRODUCTS CONTAINING ACTIVE INGREDIENTS OFFERED OTC FOR HUMAN USE AS ORAL WOUND HEALING AGENTS Requirements For Specific New Drugs Or Devices; New Drugs
DRUG PRODUCTS	310.536	DRUG PRODUCTS CONTAINING ACTIVE INGREDIENTS OFFERED OTC FOR USE AS A NAILBITING OR THUMBSUCKING DETERRENT Requirements For Specific New Drugs Or Devices; New Drugs
DRUG PRODUCTS	310.537	DRUG PRODUCTS CONTAINING ACTIVE INGREDIENTS OFFERED OTC FOR TREATMENT OF FEVER BLISTERS AND COLD SORES Requirements For Specific New Drugs Or Devices; New Drugs
DRUG PRODUCTS	310.538	DRUG PRODUCTS CONTAINING ACTIVE INGREDIENTS OFFERED OTC FOR USE FOR INGROWN TOENAIL RELIEF Requirements For Specific New Drugs Or Devices; New Drugs
DRUG PRODUCTS	310.540	DRUG PRODUCTS CONTAINING ACTIVE INGREDIENTS OFFERED OTC FOR USE AS STOMACH ACIDIFIERS Requirements For Specific New Drugs Or Devices; New Drugs
DRUG PRODUCTS	310.541	OTC DRUG PRODUCTS CONTAINING ACTIVE INGREDIENTS OFFERED FOR USE IN THE TREATMENT OF HYPOPHOSPHATEMIA Requirements For Specific New Drugs Or Devices; New Drugs
DRUG PRODUCTS	310.542	OTC DRUG PRODUCTS CONTAINING ACTIVE INGREDIENTS OFFERED FOR USE IN THE TREATMENT OF HYPERPHOSPHATEMIA Requirements For Specific New Drugs Or Devices; New Drugs
DRUG PRODUCTS	310.543	DRUG PRODUCTS CONTAINING ACTIVE INGREDIENTS OFFERED OTC FOR HUMAN USE IN EXOCRINE PANCREATIC INSUFFICIENCY Requirements For Specific New Drugs Or Devices; New Drugs

DRUG PRODUCTS	310.544	DRUG PRODUCTS CONTAINING ACTIVE INGREDIENTS OFFERED OTC FOR USE AS A SMOKING DETERRENT Requirements For Specific New Drugs Or Devices; New Drugs
DRUG PRODUCTS	310.545	DRUG PRODUCTS CONTAINING CERTAIN ACTIVE INGREDIENTS OFFERED OTC FOR CERTAIN USES Requirements For Specific New Drugs Or Devices; New Drugs
DRUG PRODUCTS	310.546	DRUG PRODUCTS CONTAINING ACTIVE INGREDIENTS OFFERED OTC FOR THE TREATMENT AND/OR PREVENTION OF NOCTURNAL LEG MUSCLE CRAMPS Requirements For Specific New Drugs Or Devices; New Drugs
DRUG PRODUCTS	310.547	DRUG PRODUCTS CONTAINING QUININE OFFERED OTC FOR THE TREATMENT AND/OR PREVENTION OF MALARIA Requirements For Specific New Drugs Or Devices; New Drugs
DRUG PRODUCTS	310.548	DRUG PRODUCTS CONTAINING COLLOIDAL SILVER INGREDIENTS OFFERED OTC FOR THE TREATMENT OF DISEASE Requirements For Specific New Drugs Or Devices; New Drugs
DRUG PRODUCTS	314.92	DRUG PRODUCTS FOR WHICH ABBREVIATED APPLICATIONS MAY BE SUBMITTED Applications For FDA Approval To Market A New Drug
DRUG PRODUCTS	328.50	PRINCIPAL DISPLAY PANEL OF ALL OTC DRUG PRODUCTS INTENDED FOR ORAL INGESTION THAT CONTAIN ALCOHOL Over-the-counter Drug Products Intended For Oral Ingestion That Contain Alcohol
DRUG PRODUCTS	330.3	IMPRINTING OF SOLID ORAL DOSAGE FORM DRUG PRODUCTS Over-the-counter Human Drugs Generally Recognized As Safe And Effective & Not Misbranded
DRUG PRODUCTS	333.150	LABELING OF FIRST AID ANTIBIOTIC DRUG PRODUCTS Topical Antimicrobial Drug Products For Over-the-counter Human Use
DRUG PRODUCTS	333.250	LABELING OF ANTIFUNGAL DRUG PRODUCTS Topical Antimicrobial Drug Products For Over-the-counter Human Use
DRUG PRODUCTS	333.350	LABELING OF ACNE DRUG PRODUCTS Topical Antimicrobial Drug Products For Over-the-counter Human Use
DRUG PRODUCTS	336.50	LABELING OF ANTIEMETIC DRUG PRODUCTS Antiemetic Drug Products For Over-the-counter Human Use
DRUG PRODUCTS	338.50	LABELING OF NIGHTTIME SLEEP-AID DRUG PRODUCTS Nighttime Sleep-aid Drug Products For Over-the-counter Human Use
DRUG PRODUCTS	340.50	LABELING OF STIMULANT DRUG PRODUCTS Stimulant Drug Products For Over-the-counter Human Use
DRUG PRODUCTS	341.70	LABELING OF OTC DRUG PRODUCTS CONTAINING INGREDIENTS USED FOR TREATING CONCURRENT SYMPTOMS Cold, Cough, Allergy, Bronchodilator, And Antiasthmatic Drug Products For OTC Human Use
DRUG PRODUCTS	341.72	LABELING OF ANTIHISTAMINE DRUG PRODUCTS Cold, Cough, Allergy, Bronchodilator, And Antiasthmatic Drug Products For OTC Human Use
DRUG PRODUCTS	341.74	LABELING OF ANTITUSSIVE DRUG PRODUCTS Cold, Cough, Allergy, Bronchodilator, And Antiasthmatic Drug Products For OTC Human Use
DRUG PRODUCTS	341.76	LABELING OF BRONCHODILATOR DRUG PRODUCTS Cold, Cough, Allergy, Bronchodilator, And Antiasthmatic Drug Products For OTC Human Use
DRUG PRODUCTS	341.78	LABELING OF EXPECTORANT DRUG PRODUCTS Cold, Cough, Allergy, Bronchodilator, And Antiasthmatic Drug Products For OTC Human Use
DRUG PRODUCTS	341.80	LABELING OF NASAL DECONGESTANT DRUG PRODUCTS Cold, Cough, Allergy, Bronchodilator, And Antiasthmatic Drug Products For OTC Human Use
DRUG PRODUCTS	344.50	LABELING OF EARWAX REMOVAL AID DRUG PRODUCTS Topical Otic Over-the-counter Products For Human Use
DRUG PRODUCTS	344.52	LABELING OF EAR DRYING AID DRUG PRODUCTS Topical Otic Over-the-counter Products For Human Use
DRUG PRODUCTS	346.52	LABELING OF ANORECTAL DRUG PRODUCTS Anorectal Drug Products For Over-the-counter Human Use
DRUG PRODUCTS	347.50	LABELING OF ASTRINGENT DRUG PRODUCTS Skin Protectant Drug Products For Over-the-counter Human Use
DRUG PRODUCTS	348.50	LABELING OF EXTERNAL ANALGESIC DRUG PRODUCTS External Analgesic Drug Products For Over-the-counter Human Use
DRUG PRODUCTS	349.50	LABELING OF OPHTHALMIC DRUG PRODUCTS Ophthalmic Drug Products For Over-the-counter Human Use
DRUG PRODUCTS	349.55	LABELING OF OPHTHALMIC ASTRINGENT DRUG PRODUCTS Ophthalmic Drug Products For Over-the-counter Human Use
DRUG PRODUCTS	349.60	LABELING OF OPHTHALMIC DEMULCENT DRUG PRODUCTS Ophthalmic Drug Products For Over-the-counter Human Use
DRUG PRODUCTS	349.65	LABELING OF OPHTHALMIC EMOLLIENT DRUG PRODUCTS Ophthalmic Drug Products For Over-the-counter Human Use

DRUG PRODUCTS	349.70	LABELING OF OPHTHALMIC HYPERTONICITY DRUG PRODUCTS Ophthalmic Drug Products For Over-the-counter Human Use
DRUG PRODUCTS	349.75	LABELING OF OPHTHALMIC VASOCONSTRICTOR DRUG PRODUCTS Ophthalmic Drug Products For Over-the-counter Human Use
DRUG PRODUCTS	349.78	LABELING OF OPHTHALMIC EYEWASH DRUG PRODUCTS Ophthalmic Drug Products For Over-the-counter Human Use
DRUG PRODUCTS	352.50	PRINCIPAL DISPLAY PANEL OF ALL SUNSCREEN DRUG PRODUCTS Sunscreen Drug Products For Over-the-counter Human Use
DRUG PRODUCTS	352.52	LABELING OF SUNSCREEN DRUG PRODUCTS Sunscreen Drug Products For Over-the-counter Human Use
DRUG PRODUCTS	355.50	LABELING OF ANTICARIES DRUG PRODUCTS Anticaries Drug Products For Over-the-counter Human Use
DRUG PRODUCTS	355.55	PRINCIPAL DISPLAY PANEL OF ALL FLUORIDE RINSE DRUG PRODUCTS Anticaries Drug Products For Over-the-counter Human Use
DRUG PRODUCTS	355.70	TESTING PROCEDURES FOR FLUORIDE DENTIFRICE DRUG PRODUCTS Anticaries Drug Products For Over-the-counter Human Use
DRUG PRODUCTS	357.110	LABELING OF ANTHELMINTIC DRUG PRODUCTS Anthelmintic Drug Products; Miscellaneous Internal Drug Products For Over-the-counter Human Use
DRUG PRODUCTS	357.152	PACKAGE INSERTS FOR ANTHELMINTIC DRUG PRODUCTS Anthelmintic Drug Products; Miscellaneous Internal Drug Products For Over-the-counter Human Use
DRUG PRODUCTS	357.250	LABELING OF CHOLECYSTOKINETIC DRUG PRODUCTS Cholecystokinetic Drug Products; Miscellaneous Internal Drug Products For Over-the-counter Human Use
DRUG PRODUCTS	357.810	ACTIVE INGREDIENTS FOR DEODORANT DRUG PRODUCTS FOR INTERNAL USE Deodorant Drug Products For Internal Use; Miscellaneous Internal Drug Products For Over-the-counter Human Use
DRUG PRODUCTS	357.850	LABELING OF DEODORANT DRUG PRODUCTS FOR INTERNAL USE Deodorant Drug Products For Internal Use; Miscellaneous Internal Drug Products For Over-the-counter Human Use
DRUG PRODUCTS	358.150	LABELING OF WART REMOVER DRUG PRODUCTS Miscellaneous External Drug Products For Over-the-counter Human Use
DRUG PRODUCTS	358.550	LABELING OF CORN AND CALLUS REMOVER DRUG PRODUCTS Miscellaneous External Drug Products For Over-the-counter Human Use
DRUG PRODUCTS	358.650	LABELING OF PEDICULICIDE DRUG PRODUCTS Miscellaneous External Drug Products For Over-the-counter Human Use
DRUG PRODUCTS	358.750	LABELING OF DRUG PRODUCTS FOR THE CONTROL OF DANDRUFF, SEBORRHEIC DERMATITIS Miscellaneous External Drug Products For Over-the-counter Human Use
DRUG PRODUCTS	1309.28	EXEMPTION OF DISTRIBUTORS OF REGULATED PRESCRIPTION DRUG PRODUCTS Registration Of Manufacturers, Distributors, Importers And Exporters Of List I Chemicals
DRUG PRODUCTS	1310.11	REINSTATEMENT OF THE EXEMPTION OF DRUG PRODUCTS DISTRIBUTED UNDER THE FOOD, DRUG AND COSMETIC ACT Records And Reports Of Listed Chemicals And Certain Machines
DRUG PRODUCTS	1310.14	EXEMPTION OF DRUG PRODUCTS CONTAINING EPHEDRINE AND ANOTHER ACTIVE MEDICINAL INGREDIENT Records And Reports Of Listed Chemicals And Certain Machines
DRUG PRODUCTS	1310.15	EXEMPT DRUG PRODUCTS CONTAINING EPHEDRINE AND ANOTHER ACTIVE MEDICINAL INGREDIENT Records And Reports Of Listed Chemicals And Certain Machines
DRUG RELEASE	343.90	DISSOLUTION AND DRUG RELEASE TESTING Internal Analgesic, Antipyretic And Antirheumatic Drug Products For OTC Human Use
DRUG RESIDUE	530.24	PROCEDURE FOR ANNOUNCING ANALYTICAL METHODS FOR DRUG RESIDUE QUANTIFICATION Extralabel Drug Use In Animals
DRUG SAMPLE	203.32	DRUG SAMPLE STORAGE AND HANDLING REQUIREMENTS Prescription Drug Marketing
DRUG SAMPLES	203.33	DRUG SAMPLES FORMS Prescription Drug Marketing
DRUG SAMPLES	203.39	DONATION OF DRUG SAMPLES TO CHARITABLE INSTITUTIONS Prescription Drug Marketing
DRUG USE, EXTRALABEL	5.500	ISSUANCE OF FEDERAL REGISTER DOCUMENTS PERTAINING TO THE DETERMINATION OF SAFE LEVELS, NOTICE OF NEED FOR DEVELOPMENT OF AN ANALYTICAL METHOD, AND PROHIBITION OF CERTAIN EXTRALABEL DRUG USE Animal Drugs; Redelegations Of Authority; Delegations Of Authority And Organization
DRUG USE, EXTRALABEL	530.30	EXTRALABEL DRUG USE IN NONFOOD ANIMALS Extralabel Drug Use In Animals

DRUGS	314.104	DRUGS WITH POTENTIAL FOR ABUSE FDA Action On Applications; Applications For FDA Approval To Market A New Drug
DRUGS	369.3	WARNINGS REQUIRED ON DRUGS EXEMPTED FROM PRESCRIPTION-DISPENSING REQUIREMENTS OF SECTION 503(B)(1)(C) Definitions And Interpretations; Interpretative Statements Re: Warnings On Drugs And Devices For OTC Sale
DRUGS	369.4	WARNINGS SUGGESTED FOR DRUGS BY FORMAL OR INFORMAL STATEMENTS OF POLICY Definitions And Interpretations; Interpretative Statements Re: Warnings On Drugs And Devices For OTC Sale
DRUGS	369.20	DRUGS; RECOMMENDED WARNING AND CAUTION STATEMENTS Warning And Caution Statement For Drugs; Interpretative Statements Re: Warnings On Drugs And Devices For OTC Sale
DRUGS	369.21	DRUGS; WARNING AND CAUTION STATEMENTS REQUIRED BY REGULATIONS Warning And Caution Statement For Drugs; Interpretative Statements Re: Warnings On Drugs And Devices For OTC Sale
DRUGS	369.22	DRUGS; WARNING AND CAUTION STATEMENTS SPECIFICALLY REQUIRED BY LAW Warning And Caution Statement For Drugs; Interpretative Statements Re: Warnings On Drugs And Devices For OTC Sale
DRUGS	510.105	LABELING OF DRUGS FOR USE IN MILK-PRODUCING ANIMALS Specific Administrative Rulings And Decisions; New Animal Drugs
DRUGS	530.25	ORDERS PROHIBITING EXTRALABEL USES FOR DRUGS IN FOOD-PRODUCING ANIMALS Extralabel Drug Use In Animals
DRUGS	530.41	DRUGS PROHIBITED FOR EXTRALABEL USE IN ANIMALS Extralabel Drug Use In Animals
DRUGS	700.13	USE OF MERCURY COMPOUNDS IN COSMETICS INCLUDING USE AS SKINBLEACHING AGENTS IN COSMETIC PREPARATIONS ALSO REGARDED AS DRUGS Requirements For Specific Cosmetic Products; General
DRUGS	1310.10	REMOVAL OF THE EXEMPTION FOR DRUGS DISTRIBUTED UNDER THE FOOD, DRUG AND COSMETIC ACT Records And Reports Of Listed Chemicals And Certain Machines
DRUGS, ABUSE	809.40	RESTRICTIONS ON THE SALE, DISTRIBUTION AND USE OF OTC TEST SAMPLE COLLECTION SYSTEMS FOR DRUGS OF ABUSE TESTING Requirements For Manufacturers And Producers; In Vitro Diagnostic Products For Human Use
DRUGS, ABUSE	864.3260	OTC TEST SAMPLE COLLECTION SYSTEMS FOR DRUGS OF ABUSE TESTING Pathology Instrumentation And Accessories; Hematology And Pathology Devices
DRUGS, ANIMAL	25.33	ANIMAL DRUGS Categorical Exclusions; Environmental Impact Considerations
DRUGS, ANIMAL	500.46	HEXACHLOROPHENE IN ANIMAL DRUGS Specific Administrative Rulings And Decisions; General
DRUGS, ANIMAL	500.51	LABELING OF ANIMAL DRUGS; MISBRANDING Animal Drug Labeling Requirements; General
DRUGS, ANIMAL	530.10	PROVISION PERMITTING EXTRALABEL USE OF ANIMAL DRUGS Extralabel Drug Use In Animals
DRUGS, ANTHELMINTIC	500.25	ANTHELMINTIC DRUGS FOR USE IN ANIMALS Specific Administrative Rulings And Decisions; General
DRUGS, ANTIBIOTIC	510.106	LABELING OF ANTIBIOTIC AND ANTIBIOTIC CONTAINING DRUGS INTENDED FOR USE IN MILK-PRODUCING ANIMALS Specific Administrative Rulings And Decisions; New Animal Drugs
DRUGS, ANTIBIOTIC	558.15	ANTIBIOTIC, NITROFURAN, AND SULFONAMIDE DRUGS IN THE FEED OF ANIMALS General Provisions; New Animal Drugs For Use In Animal Feeds
DRUGS, HABIT-FORMING	329.1	HABIT FORMING DRUGS WHICH ARE CHEMICAL DERIVATIVES OF SUBSTANCES SPECIFIED IN SECTION 502(D) OF THE FEDERAL FOOD, DRUG, AND COSMETIC ACT Derivatives Designated As Habit Forming; Habit-forming Drugs
DRUGS, HABIT-FORMING	329.10	LABELING REQUIREMENTS FOR HABIT-FORMING DRUGS Labeling; Habit-forming Drugs
DRUGS, HABIT-FORMING	329.20	EXEMPTION OF CERTAIN HABIT-FORMING DRUGS FROM PRESCRIPTION REQUIREMENTS Exemptions; Habit-forming Drugs
DRUGS, HUMAN	25.31	HUMAN DRUGS AND BIOLOGICS Categorical Exclusions; Environmental Impact Considerations
DRUGS, HUMAN	530.13	EXTRALABEL USE FROM COMPOUNDING OF APPROVED NEW ANIMAL AND APPROVED HUMAN DRUGS Extralabel Drug Use In Animals
DRUGS, HYPOGLYCEMIC	310.517	LABELING FOR ORAL HYPOGLYCEMIC DRUGS OF THE SULFONYLUREA CLASS Requirements For Specific New Drugs Or Devices; New Drugs

DRUGS, INVESTIGATIONAL	312.7	PROMOTION AND CHARGING FOR INVESTIGATIONAL DRUGS Investigational New Drug Application
DRUGS, INVESTIGATIONAL NEW	5.102	AUTHORITY TO APPROVE AND TO WITHDRAW APPROVAL OF A CHARGE FOR INVESTIGATIONAL NEW DRUGS Human Drugs; Redelegations Of Authority; Delegations Of Authority And Organization
DRUGS, METHYLENE-BLUE	500.27	METHYLENE BLUE-CONTAINING DRUGS FOR USE IN ANIMALS Specific Administrative Rulings And Decisions; General
DRUGS, NARCOTIC	1306.07	ADMINISTERING OR DISPENSING OF NARCOTIC DRUGS General Information; Prescriptions
DRUGS, NEW	5.101	TERMINATION OF EXEMPTIONS FOR NEW DRUGS FOR INVESTIGATIONAL USE IN HUMAN BEINGS Human Drugs; Redelegations Of Authority; Delegations Of Authority And Organization
DRUGS, NEW	5.109	ISSUANCE OF WRITTEN NOTICES CONCERNING PATENT INFORMATION, CURRENT GOOD MANUFACTURING PRACTICES AND FALSE OR MISLEADING LABELING OF NEW DRUGS Human Drugs; Redelegations Of Authority; Delegations Of Authority And Organization
DRUGS, NEW	5.505	TERMINATION OF EXEMPTIONS FOR NEW DRUGS FOR INVESTIGATIONAL USE IN ANIMALS Animal Drugs; Redelegations Of Authority; Delegations Of Authority And Organization
DRUGS, NEW	70.10	COLOR ADDITIVES IN STANDARDIZED FOODS AND NEW DRUGS General Provisions; Color Additives
DRUGS, NEW	201.115	NEW DRUGS OR NEW ANIMAL DRUGS Exemptions From Adequate Directions For Use; Labeling
DRUGS, NEW	314.410	IMPORTS AND EXPORTS OF NEW DRUGS Miscellaneous Provisions; Applications For FDA Approval To Market A New Drug
DRUGS, NEW ANIMAL	5.502	ISSUANCE OF NOTICES, PROPOSALS, AND ORDERS RELATING TO NEW ANIMAL DRUGS AND MEDICATED FEED MILL LICENSE APPLICATIONS Animal Drugs; Redelegations Of Authority; Delegations Of Authority And Organization
DRUGS, NEW ANIMAL	5.504	ISSUANCE OF WRITTEN NOTICES CONCERNING PATENT INFORMATION, CURRENT GOOD MANUFACTURING PRACTICES AND FALSE OR MISLEADING LABELING OF NEW ANIMAL DRUGS AND FEEDS BEARING OR CONTAINING NEW ANIMAL DRUGS Animal Drugs; Redelegations Of Authority; Delegations Of Authority And Organization
DRUGS, NEW ANIMAL	201.115	NEW DRUGS OR NEW ANIMAL DRUGS Exemptions From Adequate Directions For Use; Labeling
DRUGS, NEW ANIMAL	510.7	CONSIGNEES OF NEW ANIMAL DRUGS FOR USE IN THE MANUFACTURE OF ANIMAL FEED General Provisions; New Animal Drugs
DRUGS, NEW ANIMAL	510.300	RECORDS AND REPORTS CONCERNING EXPERIENCE WITH NEW ANIMAL DRUGS FOR WHICH AN APPROVED APPLICATION IS IN EFFECT Records And Reports; New Animal Drugs
DRUGS, NEW ANIMAL	510.301	RECORDS AND REPORTS CONCERNING EXPERIENCE WITH ANIMAL FEEDS BEARING OR CONTAINING NEW ANIMAL DRUGS FOR WHICH AN APPROVED MEDICATED FEED MILL LICENSE IS IN EFFECT Records And Reports; New Animal Drugs
DRUGS, NEW ANIMAL	510.305	MAINTENANCE OF COPIES OF APPROVED MEDICATED FEED MILL LICENSES TO MANUFACTURE ANIMAL FEED BEARING OR CONTAINING NEW ANIMAL DRUGS Records And Reports; New Animal Drugs
DRUGS, NEW ANIMAL	510.515	ANIMAL FEEDS BEARING OR CONTAINING NEW ANIMAL DRUGS SUBJECT TO THE PROVISIONS OF SECTION 512(N) Animal Use Exemptions From Certification And Labeling Requirements; New Animal Drugs
DRUGS, NEW ANIMAL	511.1	NEW ANIMAL DRUGS FOR INVESTIGATIONAL USE EXEMPT FROM SECTION 512(A) New Animal Drugs For Investigational Use
DRUGS, NEW ANIMAL	514.2	APPLICATIONS FOR ANIMAL FEEDS BEARING OR CONTAINING NEW ANIMAL DRUGS General Provisions; New Animal Drug Applications
DRUGS, NEW ANIMAL	514.80	RECORDS AND REPORTS CONCERNING EXPERIENCE WITH APPROVED NEW ANIMAL DRUGS Administrative Actions On Applications; New Animal Drug Applications
DRUGS, NEW ANIMAL	530.13	EXTRALABEL USE FROM COMPOUNDING OF APPROVED NEW ANIMAL AND APPROVED HUMAN DRUGS Extralabel Drug Use In Animals
DRUGS, NITROFURAN	558.15	ANTIBIOTIC, NITROFURAN, AND SULFONAMIDE DRUGS IN THE FEED OF ANIMALS General Provisions; New Animal Drugs For Use In Animal Feeds

DRUGS, ORPHAN	316.36	INSUFFICIENT QUANTITIES OF ORPHAN DRUGS Orphan Drugs
DRUGS, OTC	201.21	DECLARATION OF PRESENCE OF PHENYLALANINE AS A COMPONENT OF ASPARTAME IN OVER-THE-COUNTER AND PRESCRIPTION DRUGS FOR HUMAN USE General Labeling Provisions; Labeling
DRUGS, OTC	201.315	OVER-THE-COUNTER DRUGS FOR MINOR SORE THROATS; SUGGESTED WARNING Specific Labeling Requirements For Specific Drug Products; Labeling
DRUGS, OTC	330.10	PROCEDURES FOR CLASSIFYING OTC DRUGS AS GENERALLY RECOGNIZED AS SAFE AND EFFECTIVE AND NOT MISBRANDED Administrative Procedures; Over-the-counter Human Drugs Generally Recognized As Safe And Effective & Not Misbranded
DRUGS, OTC	330.12	STATUS OF OVER-THE-COUNTER (OTC) DRUGS PREVIOUSLY REVIEWED UNDER THE DRUG EFFICACY STUDY (DESI) Administrative Procedures; Over-the-counter Human Drugs Generally Recognized As Safe And Effective & Not Misbranded
DRUGS, OTC	330.14	ADDITIONAL CRITERIA AND PROCEDURES FOR CLASSIFYING OTC DRUGS AS GENERALLY RECOGNIZED AS SAFE AND EFFECTIVE AND NOT MISBRANDED Administrative Procedures; Over-the-counter Human Drugs Generally Recognized As Safe And Effective & Not Misbranded
DRUGS, PRESCRIPTION	5.107	EXTENSIONS OR STAYS OF EFFECTIVE DATES FOR COMPLIANCE WITH CERTAIN LABELING REQUIREMENTS FOR HUMAN PRESCRIPTION DRUGS Human Drugs; Redelegations Of Authority; Delegations Of Authority And Organization
DRUGS, PRESCRIPTION	14.160	ESTABLISHMENT OF STANDING TECHNICAL ADVISORY COMMITTEES FOR HUMAN PRESCRIPTION DRUGS Advisory Committees For Human Prescription Drugs; Public Hearing Before A Public Advisory Committee
DRUGS, PRESCRIPTION	200.200	PRESCRIPTION DRUGS; REMINDER ADVERTISEMENTS AND REMINDER LABELING TO PROVIDE PRICE INFORMATION TO CONSUMERS Prescription Drug Consumer Price Listing; General
DRUGS, PRESCRIPTION	201.21	DECLARATION OF PRESENCE OF PHENYLALANINE AS A COMPONENT OF ASPARTAME IN OVER-THE-COUNTER AND PRESCRIPTION DRUGS FOR HUMAN USE General Labeling Provisions; Labeling
DRUGS, PRESCRIPTION	201.22	PRESCRIPTION DRUGS CONTAINING SULFITES; REQUIRED WARNING STATEMENTS General Labeling Provisions; Labeling
DRUGS, PRESCRIPTION	201.56	GENERAL REQUIREMENTS ON CONTENT AND FORMAT OF LABELING FOR HUMAN PRESCRIPTION DRUGS Labeling Requirements For Prescription Drugs And/or Insulin; Labeling
DRUGS, PRESCRIPTION	201.57	SPECIFIC REQUIREMENTS ON CONTENT AND FORMAT OF LABELING FOR HUMAN PRESCRIPTION DRUGS Labeling Requirements For Prescription Drugs And/or Insulin; Labeling
DRUGS, PRESCRIPTION	201.100	PRESCRIPTION DRUGS FOR HUMAN USE Exemptions From Adequate Directions For Use; Labeling
DRUGS, PRESCRIPTION	203.50	REQUIREMENTS FOR WHOLESALE DISTRIBUTION OF PRESCRIPTION DRUGS Wholesale Distribution; Prescription Drug Marketing
DRUGS, PRESCRIPTION	205.50	MINIMUM REQUIREMENTS FOR THE STORAGE AND HANDLING OF PRESCRIPTION DRUGS Guidelines For State Licensing Of Wholesale Prescription Drug Distributors
DRUGS, PRESCRIPTION	250.101	AMPHETAMINE AND METHAMPHETAMINE INHALERS REGARDED AS PRESCRIPTION DRUGS New Drug Or Prescription Status Of Specific Drugs; Specific Requirements For Specific Human Drugs
DRUGS, PRESCRIPTION	250.105	GELSEMIUM-CONTAINING PREPARATIONS REGARDED AS PRESCRIPTION DRUGS New Drug Or Prescription Status Of Specific Drugs; Specific Requirements For Specific Human Drugs
DRUGS, PRESCRIPTION	250.108	POTASSIUM PERMANGANATE PREPARATIONS AS PRESCRIPTION DRUGS New Drug Or Prescription Status Of Specific Drugs; Specific Requirements For Specific Human Drugs
DRUGS, PRESCRIPTION	300.50	FIXED-COMBINATION PRESCRIPTION DRUGS FOR HUMANS Combination Drugs; General
DRUGS, PRESCRIPTION	310.305	RECORDS AND REPORTS CONTAINING ADVERSE DRUG EXPERIENCES ON MARKETED PRESCRIPTION DRUGS FOR HUMAN USE WITHOUT APPROVED NEW DRUG APPLICATIONS Records And Reports; New Drugs

DRUGS, PRESCRIPTION	1309.28	EXEMPTION OF DISTRIBUTORS OF REGULATED PRESCRIPTION DRUG PRODUCTS Registration Of Manufacturers, Distributors, Importers And Exporters Of List I Chemicals
DRUGS, RADIOACTIVE	201.129	DRUGS; EXEMPTION FOR RADIOACTIVE DRUGS FOR RESEARCH USE Exemptions From Adequate Directions For Use; Labeling
DRUGS, RADIOACTIVE	310.503	REQUIREMENTS REGARDING CERTAIN RADIOACTIVE DRUGS Requirements For Specific New Drugs Or Devices; New Drugs
DRUGS, SULFONAMIDE	558.15	ANTIBIOTIC, NITROFURAN, AND SULFONAMIDE DRUGS IN THE FEED OF ANIMALS General Provisions; New Animal Drugs For Use In Animal Feeds
DRUGS, TIMED-RELEASE	500.26	TIMED-RELEASE DOSAGE FORM DRUGS Specific Administrative Rulings And Decisions; General
DRUGS, VETERINARY	201.105	VETERINARY DRUGS Exemptions From Adequate Directions For Use; Labeling
DRUGS, VETERINARY	558.6	VETERINARY FEED DIRECTIVE DRUGS General Provisions; New Animal Drugs For Use In Animal Feeds
DRUM	886.1200	OPTOKINETIC DRUM Diagnostic Devices; Ophthalmic Devices
DRUM	886.4230	OPHTHALMIC KNIFE TEST DRUM Surgical Devices; Ophthalmic Devices
DRY COWS	526.1696b	PENICILLIN G PROCAINE-DIHYDROSTREPTOMYCIN IN SOYBEAN OIL FOR INTRAMAMMARY INFUSION (DRY COWS) Intramammary Dosage Forms
DRY COWS	526.1696c	PENICILLIN G PROCAINE-DIHYDROSTREPTOMYCIN SULFATE FOR INTRAMAMMARY INFUSION (DRY COWS) Intramammary Dosage Forms
DRY CREAM	131.149	DRY CREAM Requirements For Specific Standardized Milk And Cream; Milk And Cream
DRY CURD	133.129	DRY CURD COTTAGE CHEESE Requirements For Specific Standardized Cheese And Related Products; Cheeses And Related Cheese Products
DRY FOOD	176.180	COMPONENTS OF PAPER AND PAPERBOARD IN CONTACT WITH DRY FOOD Substances For Use Only As Components Of Paper And Paperboard; Indirect Food Additives
DRY FOOD	182.70	SUBSTANCES MIGRATING FROM COTTON AND COTTON FABRICS USED IN DRY FOOD PACKAGING General Provisions; Substances Generally Recognized As Safe
DRY HEAT	872.6370	ENDODONTIC DRY HEAT STERILIZER Miscellaneous Devices; Dental Devices
DRY MILK	131.125	NONFAT DRY MILK Requirements For Specific Standardized Milk And Cream; Milk And Cream
DRY MILK	131.127	NONFAT DRY MILK FORTIFIED WITH VITAMINS A AND D Requirements For Specific Standardized Milk And Cream; Milk And Cream
DRY MILK	131.147	DRY WHOLE MILK Requirements For Specific Standardized Milk And Cream; Milk And Cream
DRY PEAS	155.172	CANNED DRY PEAS Requirements For Specific Standardized Canned Vegetables; Canned Vegetables
DRY-HEAT STERILIZER	880.6870	DRY-HEAT STERILIZER General Hospital And Personal Use Miscellaneous Devices; General Hospital And Personal Use Devices
DRYING OILS	181.26	DRYING OILS AS COMPONENTS OF FINISHED RESINS Specific Prior-sanctioned Food Ingredients; Prior-sanctioned Food Ingredients
DUE, AMOUNTS	1403.53	COLLECTION OF AMOUNTS DUE Post-award Requirements; Uniform Administrative Requirements For Grants And Coop. Agreements To State & Local Gov.
DUE DILIGENCE	5.27	PATENT TERM EXTENSIONS FOR HUMAN DRUG PRODUCTS, MEDICAL DEVICES, AND FOOD AND COLOR ADDITIVES; AND AUTHORITY TO PERFORM DUE DILIGENCE DETERMINATIONS AND INFORMAL HEARINGS General Redelegations Of Authority; Delegations Of Authority And Organization
DUE DILIGENCE	60.36	STANDARD OF DUE DILIGENCE Due Diligence Petitions; Patent Term Restoration
DULCIN	189.145	DULCIN Substances Generally Prohibited From Direct Addition Or Use As Human Food; Substances Prohibited From Use In Human Food
DURA	882.5910	DURA SUBSTITUTE Neurological Therapeutic Devices; Neurological Devices
DURUM	137.220	DURUM FLOUR Requirements For Specific Standardized Cereal Flours And Related Products; Cereal Flours And Related Products
DURUM	137.225	WHOLE DURUM FLOUR Requirements For Specific Standardized Cereal Flours And Related Products; Cereal Flours And Related Products

DUTIES	1316.73	CUSTODY AND OTHER DUTIES Seizure, Forfeiture, And Disposition Of Property; Administrative Functions, Practices, And Procedures
DUTIES, INSPECTOR	600.22	DUTIES OF INSPECTOR Establishment Inspection; Biological Products: General
DUTIES, IRB	50.50	IRB DUTIES Additional Safeguards For Children In Clinical Investigations; Protection Of Human Subjects
DUTIES, IRB	812.60	IRB COMPOSITION, DUTIES, AND FUNCTIONS IRB Review And Approval; Investigational Device Exemptions
DUTY	19.21	DUTY TO REPORT VIOLATIONS Reporting Of Violations; Standards Of Conduct And Conflicts Of Interest
DYE	73.3121	POLY(HYDROXYETHYL METHACRYLATE)-DYE COPOLYMERS Medical Devices; Listing Of Color Additives Exempt From Certification
DYE	73.3127	VINYL ALCOHOL/METHYL METHACRYLATE-DYE REACTION PRODUCTS Listing Of Color Additives Exempt From Certification
DYE	864.1850	DYE AND CHEMICAL SOLUTION STAINS Biological Stains; Hematology And Pathology Devices
DYES, HAIR	70.20	PACKAGING REQUIREMENTS FOR STRAIGHT COLORS (OTHER THAN HAIR DYES) Packaging And Labeling; Color Additives
DYES, HAIR	70.25	LABELING REQUIREMENTS FOR COLOR ADDITIVES (OTHER THAN HAIR DYES) Packaging And Labeling; Color Additives
DYES, HAIR	740.18	COAL TAR HAIR DYES POSING A RISK OF CANCER Warning Statements; Cosmetic Product Warning Statements
DYNAMOMETER	888.1240	AC-POWERED DYNAMOMETER Diagnostic Devices; Orthopedic Devices
DYNAMOMETER	888.1250	NONPOWERED DYNAMOMETER Diagnostic Devices; Orthopedic Devices

E

E, IMMUNOGLOBULIN	866.5510	IMMUNOGLOBULINS A, G, M, D, AND E IMMUNOLOGICAL TEST SYSTEM Immunological Test Systems; Immunology And Microbiology Devices
E, VITAMIN	522.2100	SELENIUM, VITAMIN E INJECTION Implantation Or Injectable Dosage Form New Animal Drugs
E, VITAMIN	862.1815	VITAMIN E TEST SYSTEM Clinical Chemistry Test Systems; Clinical Chemistry And Clinical Toxicology Devices
EAR	524.1484	NEOMYCIN SULFATE, HYDROCORTISONE ACETATE, TETRACAINE HYDROCHLORIDE EAR OINTMENT Ophthalmic And Topical Dosage Form New Animal Drugs
EAR	870.2710	EAR OXIMETER Cardiovascular Monitoring Devices; Cardiovascular Devices
EAR	874.3620	EAR, NOSE, AND THROAT SYNTHETIC POLYMER MATERIAL Prosthetic Devices; Ear, Nose And Throat Devices
EAR	874.4140	EAR, NOSE, AND THROAT BURR Surgical Devices; Ear, Nose And Throat Devices
EAR	874.4250	EAR, NOSE, AND THROAT ELECTRIC OR PNEUMATIC SURGICAL DRILL Surgical Devices; Ear, Nose And Throat Devices
EAR	874.4350	EAR, NOSE, AND THROAT FIBEROPTIC LIGHT SOURCE AND CARRIER Surgical Devices; Ear, Nose And Throat Devices
EAR	874.4420	EAR, NOSE, AND THROAT MANUAL SURGICAL INSTRUMENT Surgical Devices; Ear, Nose And Throat Devices
EAR	4500	EAR, NOSE, AND THROAT MICROSURGICAL CARBON DIOXIDE LASER Surgical Devices; Ear, Nose And Throat Devices
EAR	874.5220	EAR, NOSE, AND THROAT DRUG ADMINISTRATION DEVICE Therapeutic Devices; Ear, Nose And Throat Devices
EAR	874.5300	EAR, NOSE AND THROAT EXAMINATION AND TREATMENT UNIT Therapeutic Devices; Ear, Nose, And Throat Devices
EAR DRYING	344.12	EAR DRYING AID ACTIVE INGREDIENT Active Ingredients; Topical Otic Drug Products For Over-the-counter Human Use
EAR DRYING	344.52	LABELING OF EAR DRYING AID DRUG PRODUCTS Labeling; Topical Otic Drug Products For Over-the-counter Human Use
EAR, MIDDLE	874.3430	MIDDLE EAR MOLD Prosthetic Devices; Ear, Nose And Throat Devices
EAR PROSTHESIS	878.3590	EAR PROSTHESIS Prosthetic Devices; General And Plastic Surgery Devices
EARDROPS	524.1484f	NEOMYCIN SULFATE, PREDNISOLONE ACETATE, TETRACAINE HYDROCHLORIDE EARDROPS Ophthalmic And Topical Dosage Form New Animal Drugs
EARLY CONSULTATION	312.82	EARLY CONSULTATION Drugs Intended To Treat Life-threatening And Severely-debilitating Illnesses; Investigational New Drug Application
EARPHONE	874.1100	EARPHONE CUSHION FOR AUDIOMETRIC TESTING Diagnostic Devices; Ear, Nose, And Throat Devices
EARTH, DIATOMACEOUS	573.340	DIATOMACEOUS EARTH Food Additive Listing; Food Additives Permitted In Feed And Drinking Water Of Animals
EARWAX	344.10	EARWAX REMOVAL AID ACTIVE INGREDIENT Active Ingredients; Topical Otic Drug Products For Over-the-counter Human Use
EARWAX	344.50	LABELING OF EARWAX REMOVAL AID DRUG PRODUCTS Labeling; Topical Otic Drug Products For Over-the-counter Human Use
EATING OCCASION	101.12	REFERENCE AMOUNTS CUSTOMARILY CONSUMED PER EATING OCCASION General Provisions; Food Labeling
ECG	892.1970	RADIOGRAPHIC ECG/RESPIRATOR SYNCHRONIZER Diagnostic Devices; Radiology Devices
ECHINOCOCCUS SPP.	866.3200	ECHINOCOCCUS SPP. SEROLOGICAL REAGENTS Serological Reagents; Immunology And Microbiology Devices
ECHO	892.1560	ULTRASONIC PULSED ECHO IMAGING SYSTEM Diagnostic Devices; Radiology Devices
ECHOCARDIOGRAPH	870.2330	ECHOCARDIOGRAPH Cardiovascular Monitoring Devices; Cardiovascular Devices
ECHOENCEPHALOGRAPH	882.1240	ECHOENCEPHALOGRAPH Neurological Diagnostic Devices; Neurological Devices
ECHOVIRUS	866.3205	ECHOVIRUS SEROLOGICAL REAGENTS Serological Reagents; Immunology And Microbiology Devices
EDAM CHEESE	133.138	EDAM CHEESE. Requirements For Specific Standardized Cheese And Related Products; Cheeses And Related Cheese Products

EDIBLE FAT	102.37	MIXTURE OF EDIBLE FAT OR OIL AND OLIVE OIL Requirements For Specific Nonstandardized Foods; Common Or Usual Name For Nonstandardized Foods
EDIBLE FATS	172.225	METHYL AND ETHYL ESTERS OF FATTY ACIDS PRODUCED FROM EDIBLE FATS AND OILS Coatings, Films And Related Substances; Food Additives Permitted For Direct Addition To Food For Human Consumption
EDIBLE FATS	582.4101	DIACETYL TARTARIC ACID ESTERS OF MONO- AND DIGLYCERIDES OF EDIBLE FATS OR OILS, OR EDIBLE FAT-FORMING FATTY ACIDS Emulsifying Agents; Substances Generally Recognized As Safe
EDIBLE FATS	582.4505	MONO- AND DIGLYCERIDES OF EDIBLE FATS OR OILS, OR EDIBLE FAT-FORMING ACIDS Emulsifying Agents; Substances Generally Recognized As Safe
EDIBLE FATS	582.452	MONOSODIUM PHOSPHATE DERIVATIVES OF MONO- AND DIGLYCERIDES OF EDIBLE FATS OR OILS, OR EDIBLE FAT-FORMING FATTY ACIDS Emulsifying Agents; Substances Generally Recognized As Safe
EDIBLE PRODUCTS	170.17	EXEMPTION FOR INVESTIGATIONAL USE AND PROCEDURE FOR OBTAINING AUTHORIZATION TO MARKET EDIBLE PRODUCTS FROM EXPERIMENTAL ANIMALS General Provisions; Food Additives
EDIBLE PRODUCTS	570.17	EXEMPTION FOR INVESTIGATIONAL USE AND PROCEDURE FOR OBTAINING AUTHORIZATION TO MARKET EDIBLE PRODUCTS FROM EXPERIMENTAL ANIMALS General Provisions; Food Additives
EDTA, DISODIUM	73.2120	DISODIUM EDTA-COPPER Cosmetics; Listing Of Color Additives Exempt From Certification
EDTA, DISODIUM	172.120	CALCIUM DISODIUM EDTA Food Preservatives; Food Additives Permitted For Direct Addition To Food For Human Consumption
EDTA, DISODIUM	172.135	DISODIUM EDTA Food Preservatives; Food Additives Permitted For Direct Addition To Food For Human Consumption
EDTA, DISODIUM	573.360	DISODIUM EDTA Food Additive Listing; Food Additives Permitted In Feed And Drinking Water Of Animals
EEG	882.1420	ELECTROENCEPHALOGRAM (EEG) SIGNAL SPECTRUM ANALYZER Neurological Diagnostic Devices; Neurological Devices
EEG	882.1855	ELECTROENCEPHALOGRAM (EEG) TELEMETRY SYSTEM Neurological Diagnostic Devices; Neurological Devices
EFFECT	3.9	EFFECT OF LETTER OF DESIGNATION Product J
EFFECT	107.220	SCOPE AND EFFECT OF INFANT FORMULA RECALLS Infant Formula Recalls; Infant Formula
EFFECT	314.510	APPROVAL BASED ON A SURROGATE ENDPOINT OR ON AN EFFECT ON A CLINICAL ENDPOINT OTHER THAN SURVIVAL OR IRREVERSIBLE MORBIDITY Applications For FDA Approval To Market A New Drug
EFFECT	320.28	CORRELATION OF BIOAVAILABILITY WITH AN ACUTE PHARMACOLOGICAL EFFECT OR CLINICAL EVIDENCE Procedures For Determining The Bioavailability Of Drug Products; Bioavailability And Bioequivalence
EFFECT	601.41	APPROVAL BASED ON A SURROGATE ENDPOINT OR ON AN EFFECT ON A CLINICAL ENDPOINT OTHER THAN SURVIVAL OR IRREVERSIBLE MORBIDITY Accelerated Approval Of Biological Products; Licensing
EFFECT	1003.5	EFFECT OF REGULATIONS ON OTHER LAWS General Provisions; Notification Of Defects Or Failure To Comply
EFFECT	1404.620	EFFECT OF VIOLATION Governmentwide Debarment And Suspension And Requirements For Drug-Free Workplace (Grants)
EFFECTIVE	330.1	GENERAL CONDITIONS FOR GENERAL RECOGNITION AS SAFE, EFFECTIVE AND NOT MISBRANDED General Provisions; Over-the-counter Human Drugs Generally Recognized As Safe And Effective & Not Misbranded
EFFECTIVE	330.10	PROCEDURES FOR CLASSIFYING OTC DRUGS AS GENERALLY RECOGNIZED AS SAFE AND EFFECTIVE AND NOT MISBRANDED, AND ESTABLISHING MONOGRAPHS Administrative Procedures; Over-the-counter Human Drugs Generally Recognized As Safe And Effective & Not Misbranded
EFFECTIVE	330.14	ADDITIONAL CRITERIA AND PROCEDURES FOR CLASSIFYING OTC DRUGS AS GENERALLY RECOGNIZED AS SAFE AND EFFECTIVE AND NOT MISBRANDED Administrative Procedures; Over-the-counter Human Drugs Generally Recognized As Safe And Effective & Not Misbranded

EFFECTIVE	601.25	REVIEW PROCEDURES TO DETERMINE THAT LICENSED BIOLOGICAL PRODUCTS ARE SAFE, EFFECTIVE NOT MISBRANDED UNDER PRESCRIBED, RECOMMENDED, OR SUGGESTED CONDITIONS OF USE Biologics Licensing; Licensing
EFFECTIVE	601.26	RECLASSIFICATION PROCEDURES TO DETERMINE THAT LICENSED BIOLOGICAL PRODUCTS ARE SAFE, EFFECTIVE, AND NOT MISBRANDED UNDER PRESCRIBED, RECOMMENDED, OR SUGGESTED CONDITIONS OF USE Biologics Licensing; Licensing
EFFECTIVE	1240.10	EFFECTIVE BACTERICIDAL TREATMENT Control Of Communicable Diseases
EFFECTIVE APPROVAL DATES	5.106	SUBMISSION OF AND EFFECTIVE APPROVAL DATES FOR ABBREVIATED NEW DRUG APPLICATIONS AND CERTAIN NEW DRUG APPLICATIONS Human Drugs; Redelegations Of Authority; Delegations Of Authority And Organization
EFFECTIVE APPROVAL DATES	5.503	SUBMISSION OF AND EFFECTIVE APPROVAL DATES FOR ABBREVIATED NEW ANIMAL DRUG APPLICATIONS AND CERTAIN NEW ANIMAL DRUG APPLICATIONS Animal Drugs; Redelegations Of Authority; Delegations Of Authority And Organization
EFFECTIVE DATE	5.301	ISSUANCE OF INITIAL EMERGENCY PERMIT ORDERS AND NOTICES OF CONFIRMATION OF EFFECTIVE DATE OF FINAL REGULATIONS ON FOOD FOR HUMAN AND ANIMAL CONSUMPTION Foods And Cosmetics; Redelegations Of Authority; Delegations Of Authority And Organization
EFFECTIVE DATE	12.37	EFFECTIVE DATE OF A REGULATION Initiation Of Proceedings; Formal Evidentiary Public Hearing
EFFECTIVE DATE	12.38	EFFECTIVE DATE OF AN ORDER Initiation Of Proceedings; Formal Evidentiary Public Hearing
EFFECTIVE DATE	50.21	EFFECTIVE DATE Informed Consent Of Human Subjects; Protection Of Human Subjects
EFFECTIVE DATE	171.102	EFFECTIVE DATE OF REGULATIONS Administrative Actions On Applications; Food Additive Petitions
EFFECTIVE DATE	201.59	EFFECTIVE DATE OF SECTIONS 201.56, 201.57, 201.100(D)(3), AND 201.100(E) Labeling Requirements For Prescription Drugs And/or Insulin; Labeling
EFFECTIVE DATE	314.107	EFFECTIVE DATE OF APPROVAL OF A 505(B)(2) APPLICATION OR ABBREVIATED NEW DRUG APPLICATION UNDER SECTION 505(J) Applications For FDA Approval To Market A New Drug
EFFECTIVE DATE	571.102	EFFECTIVE DATE OF REGULATION Administrative Actions On Applications; Food Additive Petitions
EFFECTIVE DATE	895.30	SPECIAL EFFECTIVE DATE General Provisions; Banned Devices
EFFECTIVE DATES	5.107	EXTENSIONS OR STAYS OF EFFECTIVE DATES FOR COMPLIANCE WITH CERTAIN LABELING REQUIREMENTS FOR HUMAN PRESCRIPTION DRUGS Human Drugs; Redelegations Of Authority; Delegations Of Authority And Organization
EFFECTIVE DATES	861.36	EFFECTIVE DATES Procedures For Performance Standards Development And Publication; Procedures For Performance Standards Development
EFFECTIVE DATES	862.3	EFFECTIVE DATES OF REQUIREMENT FOR PREMARKET APPROVAL Clinical Chemistry And Clinical Toxicology Devices
EFFECTIVE DATES	864.3	EFFECTIVE DATES OF REQUIREMENT FOR PREMARKET APPROVAL Hematology And Pathology Devices
EFFECTIVE DATES	866.3	EFFECTIVE DATES OF REQUIREMENT FOR PREMARKET APPROVAL Immunology And Microbiology Devices
EFFECTIVE DATES	868.3	EFFECTIVE DATES OF REQUIREMENT FOR PREMARKET APPROVAL Anesthesiology Devices
EFFECTIVE DATES	870.3	EFFECTIVE DATES OF REQUIREMENT FOR PREMARKET APPROVAL Cardiovascular Devices
EFFECTIVE DATES	872.3	EFFECTIVE DATES OF REQUIREMENT FOR PREMARKET APPROVAL Dental Devices
EFFECTIVE DATES	874.3	EFFECTIVE DATES OF REQUIREMENT FOR PREMARKET APPROVAL Ear, Nose, And Throat Devices
EFFECTIVE DATES	876.3	EFFECTIVE DATES OF REQUIREMENT FOR PREMARKET APPROVAL Gastroenterology-urology Devices
EFFECTIVE DATES	878.3	EFFECTIVE DATES OF REQUIREMENT FOR PREMARKET APPROVAL General And Plastic Surgery Devices
EFFECTIVE DATES	880.3	EFFECTIVE DATES OF REQUIREMENT FOR PREMARKET APPROVAL General Hospital And Personal Use Devices
EFFECTIVE DATES	882.3	EFFECTIVE DATES OF REQUIREMENT FOR PREMARKET APPROVAL Neurological Devices
EFFECTIVE DATES	884.3	EFFECTIVE DATES OF REQUIREMENT FOR PREMARKET APPROVAL Obstetrical And Gynecological Devices

EFFECTIVE DATES	886.3	EFFECTIVE DATES OF REQUIREMENT FOR PREMARKET APPROVAL Ophthalmic Devices
EFFECTIVE DATES	888.3	EFFECTIVE DATES OF REQUIREMENT FOR PREMARKET APPROVAL Orthopedic Devices
EFFECTIVE DATES	890.3	EFFECTIVE DATES OF REQUIREMENT FOR PREMARKET APPROVAL Physical Medicine Devices
EFFECTIVE DATES	892.3	EFFECTIVE DATES OF REQUIREMENT FOR PREMARKET APPROVAL Radiology Devices
EFFECTIVE DATES	1302.05	EFFECTIVE DATES OF LABELING REQUIREMENTS Labeling And Packaging Requirements For Controlled Substances
EFFECTIVENESS	5.403	AUTHORIZATION TO USE ALTERNATIVE EVIDENCE FOR DETERMINATION OF THE EFFECTIVENESS OF MEDICAL DEVICES Medical Devices And Radiological Health; Redelegations Of Authority; Delegations Of Authority And Organization
EFFECTIVENESS	216.24	DRUG PRODUCTS WITHDRAWN OR REMOVED FROM THE MARKET FOR REASONS OF SAFETY OR EFFECTIVENESS Pharmacy Compounding
EFFECTIVENESS	310.6	APPLICABILITY OF "NEW DRUG" OR SAFETY/EFFECTIVENESS FINDINGS IN DRUG EFFICACY STUDY IMPLEMENTATION NOTICES AND NOTICES OF OPPORTUNITY FOR HEARING TO IDENTICAL, RELATED, AND SIMILAR DRUG PRODUCTS General Provisions; New Drugs
EFFECTIVENESS	315.3	GENERAL FACTORS RELEVANT TO SAFETY AND EFFECTIVENESS Diagnostic Radiopharmaceuticals
EFFECTIVENESS	315.5	EVALUATION OF EFFECTIVENESS Diagnostic Radiopharmaceuticals
EFFECTIVENESS	601.32	GENERAL FACTORS RELEVANT TO SAFETY AND EFFECTIVENESS Diagnostic Radiopharmaceuticals; Licensing
EFFECTIVENESS	601.34	EVALUATION OF EFFECTIVENESS Diagnostic Radiopharmaceuticals; Licensing
EFFECTIVENESS	860.7	DETERMINATION OF SAFETY AND EFFECTIVENESS General; Medical Device Classification Procedures
EFFERVESCENT	520.1242e	LEVAMISOLE HYDROCHLORIDE EFFERVESCENT TABLETS Oral Dosage Form New Animal Drugs
EFFICACY STUDY	201.200	DISCLOSURE OF DRUG EFFICACY STUDY EVALUATIONS IN LABELING AND ADVERTISING Labeling Claims For Drugs In Drug Efficacy Studies; Labeling
EFFICACY STUDY	310.6	APPLICABILITY OF "NEW DRUG" OR SAFETY/EFFECTIVENESS FINDINGS IN DRUG EFFICACY STUDY IMPLEMENTATION NOTICES AND NOTICES OF OPPORTUNITY FOR HEARING TO IDENTICAL, RELATED, AND SIMILAR DRUG PRODUCTS General Provisions; New Drugs
EFFICACY STUDY	330.12	STATUS OF OVER-THE-COUNTER (OTC) DRUGS PREVIOUSLY REVIEWED UNDER THE DRUG EFFICACY STUDY (DESI) Administrative Procedures; Over-the-counter Human Drugs Generally Recognized As Safe And Effective & Not Misbranded
EFROTOMYCIN	558.235	EFROTOMYCIN New Animal Drugs For Use In Animal Feeds
EGG	160.180	EGG YOLKS Requirements For Specific Standardized Eggs And Egg Products; Eggs And Egg Products
EGG	160.185	DRIED EGG YOLKS Requirements For Specific Standardized Eggs And Egg Products; Eggs And Egg Products
EGG	160.190	FROZEN EGG YOLKS Requirements For Specific Standardized Eggs And Egg Products; Eggs And Egg Products
EGG WHITES	160.140	EGG WHITES Requirements For Specific Standardized Eggs And Egg Products; Eggs And Egg Products
EGG WHITES	160.145	DRIED EGG WHITES Requirements For Specific Standardized Eggs And Egg Products; Eggs And Egg Products
EGG WHITES	160.150	FROZEN EGG WHITES Requirements For Specific Standardized Eggs And Egg Products; Eggs And Egg Products
EGGNOG	131.170	EGGNOG Requirements For Specific Standardized Milk And Cream; Milk And Cream
EGGS	5.302	DETENTION OF MEAT, POULTRY, EGGS AND RELATED PRODUCTS Foods And Cosmetics; Redelegations Of Authority; Delegations Of Authority And Organization
EGGS	115.50	REFRIGERATION OF SHELL EGGS HELD FOR RETAIL DISTRIBUTION Shell Eggs

EGGS	160.100	EGGS Requirements For Specific Standardized Eggs And Egg Products; Eggs And Egg Products
EGGS	160.105	DRIED EGGS Requirements For Specific Standardized Eggs And Egg Products; Eggs And Egg Products
EGGS	160.110	FROZEN EGGS Requirements For Specific Standardized Eggs And Egg Products; Eggs And Egg Products
EGGS	160.115	LIQUID EGGS Requirements For Specific Standardized Eggs And Egg Products; Eggs And Egg Products
ELASTIC	880.5075	ELASTIC BANDAGE General Hospital And Personal Use Therapeutic Devices; General Hospital And Personal Use Devices
ELASTOMER	177.1050	ACRYLONITRILE/STYRENE COPOLYMER MODIFIED WITH BUTADIENE/STYRENE ELASTOMER Substances For Use As Basic Components Of Single And Repeated Use Food Contact Surfaces; Indirect Food Additives: Polymers
ELASTOMERS	177.1590	POLYESTER ELASTOMERS Substances For Use As Basic Components Of Single And Repeated Use Food Contact Surfaces; Indirect Food Additives: Polymers
ELASTOMERS	177.2400	PERFLUOROCARBON CURED ELASTOMERS Substances For Use Only As Components Of Articles Intended For Repeated Use; Indirect Food Additives: Polymers
ELBOW JOINT	888.3150	ELBOW JOINT METAL/POLYMER CONSTRAINED CEMENTED PROSTHESIS Prosthetic Devices; Orthopedic Devices
ELBOW JOINT	888.3160	ELBOW JOINT METAL/POLYMER SEMI-CONSTRAINED CEMENTED PROSTHESIS Prosthetic Devices; Orthopedic Devices
ELBOW JOINT	888.3170	ELBOW JOINT RADIAL (HEMI-ELBOW) POLYMER PROSTHESIS Prosthetic Devices; Orthopedic Devices
ELBOW JOINT	888.3180	ELBOW JOINT HUMERAL (HEMI-ELBOW) METALLIC UNCEMENTED PROSTHESIS Prosthetic Devices; Orthopedic Devices
ELECTRIC CHAIR	890.3110	ELECTRIC POSITIONING CHAIR Physical Medicine Prosthetic Devices; Physical Medicine Devices
ELECTRIC DRILL	874.4250	EAR, NOSE AND THROAT ELECTRIC OR PNEUMATIC SURGICAL DRILL Surgical Devices; Ear, Nose And Throat Devices
ELECTRIC DRILL	882.4360	ELECTRIC CRANIAL DRILL MOTOR Neurological Surgical Devices; Neurological Devices
ELECTRICAL	868.1920	ESOPHAGEAL STETHOSCOPE WITH ELECTRICAL CONDUCTORS Diagnostic Devices; Anaesthesiology Devices
ELECTRICAL	868.2775	ELECTRICAL PERIPHERAL NERVE STIMULATOR Monitoring Devices; Anesthesiology Devices
ELECTRICAL	876.5270	IMPLANTED ELECTRICAL URINARY CONTINENCE DEVICE Therapeutic Devices; Gastroenterology-urology Devices
ELECTRICAL	876.5310	NONIMPLANTED, PERIPHERAL ELECTRICAL CONTINENCE DEVICE Therapeutic Devices; Gastroenterology-urology Devices
ELECTRICAL	876.5320	NONIMPLANTED ELECTRICAL CONTINENCE DEVICE Therapeutic Devices; Gastroenterology-urology Devices
ELECTRICAL	882.1870	EVOKED RESPONSE ELECTRICAL STIMULATOR Neurological Diagnostic Devices; Neurological Devices
ELECTRICAL	882.5890	TRANSCUTANEOUS ELECTRICAL NERVE STIMULATOR FOR PAIN RELIEF Neurological Therapeutic Devices; Neurological Devices
ELECTRICALLY POWERED	880.2400	ELECTRICALLY POWERED SPINAL FLUID PRESSURE MONITOR General Hospital And Personal Use Monitoring Devices; General Hospital And Personal Use Devices
ELECTRICALLY POWERED	868.5710	ELECTRICALLY POWERED OXYGEN TENT Therapeutic Devices; Anesthesiology Devices
ELECTROANESTHESIA	868.5400	ELECTROANESTHESIA APPARATUS Therapeutic Devices; Anesthesiology Devices
ELECTROCARDIOGRAPH	870.2340	ELECTROCARDIOGRAPH Cardiovascular Monitoring Devices; Cardiovascular Devices
ELECTROCARDIOGRAPH	870.2350	ELECTROCARDIOGRAPH LEAD SWITCHING ADAPTOR Cardiovascular Monitoring Devices; Cardiovascular Devices
ELECTROCARDIOGRAPH	870.2360	ELECTROCARDIOGRAPH ELECTRODE Cardiovascular Monitoring Devices; Cardiovascular Devices
ELECTROCARDIOGRAPH	870.2370	ELECTROCARDIOGRAPH SURFACE ELECTRODE TESTER Cardiovascular Monitoring Devices; Cardiovascular Devices
ELECTROCARDIOGRAPH	870.2920	TELEPHONE ELECTROCARDIOGRAPH TRANSMITTER AND RECEIVER Cardiovascular Monitoring Devices; Cardiovascular Devices
ELECTROCARDIOGRAPH	892.1410	NUCLEAR ELECTROCARDIOGRAPH SYNCHRONIZER Diagnostic Devices; Radiology Devices

ELECTROCAUTERY	884.4100	ENDOSCOPIC ELECTROCAUTERY AND ACCESSORIES Obstetrical And Gynecological Surgical Devices; Obstetrical And Gynecological Devices
ELECTROCAUTERY	884.4120	GYNECOLOGIC ELECTROCAUTERY AND ACCESSORIES Obstetrical And Gynecological Surgical Devices; Obstetrical And Gynecological Devices
ELECTROCONDUCTIVE	882.1275	ELECTROCONDUCTIVE MEDIA Neurological Diagnostic Devices; Neurological Devices
ELECTROCONVULSIVE	882.5940	ELECTROCONVULSIVE THERAPY DEVICE Neurological Therapeutic Devices; Neurological Devices
ELECTRODE	870.1220	ELECTRODE RECORDING CATHETER OR ELECTRODE RECORDING PROBE Cardiovascular Diagnostic Devices; Cardiovascular Devices
ELECTRODE	870.2360	ELECTROCARDIOGRAPH ELECTRODE Cardiovascular Monitoring Devices; Cardiovascular Devices
ELECTRODE	870.2370	ELECTROCARDIOGRAPH SURFACE ELECTRODE TESTER Cardiovascular Monitoring Devices; Cardiovascular Devices
ELECTRODE	870.2900	PATIENT TRANSDUCER AND ELECTRODE CABLE (INCLUDING CONNECTOR) Cardiovascular Monitoring Devices; Cardiovascular Devices
ELECTRODE	870.3680	CARDIOVASCULAR PERMANENT OR TEMPORARY PACEMAKER ELECTRODE Cardiovascular Prosthetic Devices; Cardiovascular Devices
ELECTRODE	870.3720	PACEMAKER ELECTRODE FUNCTION TESTER Cardiovascular Prosthetic Devices; Cardiovascular Devices
ELECTRODE	872.1730	ELECTRODE GEL FOR PULP TESTERS Diagnostic Devices; Dental Devices
ELECTRODE	876.1400	STOMACH PH ELECTRODE Diagnostic Devices; Gastroenterology-urology Devices
ELECTRODE	882.1310	CORTICAL ELECTRODE Neurological Diagnostic Devices; Neurological Devices
ELECTRODE	882.1320	CUTANEOUS ELECTRODE Neurological Diagnostic Devices; Neurological Devices
ELECTRODE	882.1330	DEPTH ELECTRODE Neurological Diagnostic Devices; Neurological Devices
ELECTRODE	882.1340	NASOPHARYNGEAL ELECTRODE Neurological Diagnostic Devices; Neurological Devices
ELECTRODE	882.1350	NEEDLE ELECTRODE Neurological Diagnostic Devices; Neurological Devices
ELECTRODE	882.1410	ELECTROENCEPHALOGRAPH ELECTRODE/LEAD TESTER Neurological Diagnostic Devices; Neurological Devices
ELECTRODE	884.2675	FETAL SCALP CIRCULAR (SPIRAL) ELECTRODE AND APPLICATOR Obstetrical And Gynecological Monitoring Devices; Obstetrical And Gynecological Devices
ELECTRODE	884.2685	FETAL SCALP CLIP ELECTRODE AND APPLICATOR Obstetrical And Gynecological Monitoring Devices; Obstetrical And Gynecological Devices
ELECTRODE	886.1220	CORNEAL ELECTRODE Diagnostic Devices; Ophthalmic Devices
ELECTRODE	890.1175	ELECTRODE CABLE Physical Medicine Diagnostic Devices; Physical Medicine Devices
ELECTROENCEPHALOGRAM	882.1420	ELECTROENCEPHALOGRAM (EEG) SIGNAL SPECTRUM ANALYZER Neurological Diagnostic Devices; Neurological Devices
ELECTROENCEPHALOGRAM	882.1855	ELECTROENCEPHALOGRAM (EEG) TELEMETRY SYSTEM Neurological Diagnostic Devices; Neurological Devices
ELECTROENCEPHALOGRAPH	882.1400	ELECTROENCEPHALOGRAPH Neurological Diagnostic Devices; Neurological Devices
ELECTROENCEPHALOGRAPH	882.1410	ELECTROENCEPHALOGRAPH ELECTRODE/LEAD TESTER Neurological Diagnostic Devices; Neurological Devices
ELECTROENCEPHALOGRAPH	882.1430	ELECTROENCEPHALOGRAPH TEST SIGNAL GENERATOR Neurological Diagnostic Devices; Neurological Devices
ELECTROENCEPHALOGRAPHIC	884.2620	FETAL ELECTROENCEPHALOGRAPHIC MONITOR Obstetrical And Gynecological Monitoring Devices; Obstetrical And Gynecological Devices
ELECTROGASTROGRAPHY	876.1735	ELECTROGASTROGRAPHY SYSTEM Diagnostic Devices; Gastroenterology-urology Devices
ELECTROGLOTTOGRAPH	874.1325	ELECTROGLOTTOGRAPH Diagnostic Devices; Ear, Nose, And Throat Devices
ELECTROHYDRAULIC	876.4480	ELECTROHYDRAULIC LITHOTRIPTOR Surgical Devices; Gastroenterology-urology Devices
ELECTROLYSIS	886.4250	OPHTHALMIC ELECTROLYSIS UNIT Surgical Devices; Ophthalmic Devices
ELECTROLYTE	520.550	DEXTROSE/GLYCINE/ELECTROLYTE Oral Dosage Form New Animal Drugs
ELECTROMYOGRAPH	890.1375	DIAGNOSTIC ELECTROMYOGRAPH Physical Medicine Diagnostic Devices; Physical Medicine Devices

ELECTROMYOGRAPH	890.1385	DIAGNOSTIC ELECTROMYOGRAPH NEEDLE ELECTRODE Physical Medicine Diagnostic Devices; Physical Medicine Devices
ELECTRONIC	874.1120	ELECTRONIC NOISE GENERATOR FOR AUDIOMETRIC TESTING Diagnostic Devices; Ear, Nose, And Throat Devices
ELECTRONIC	880.2420	ELECTRONIC MONITOR FOR GRAVITY FLOW INFUSION SYSTEMS General Hospital And Personal Use Monitoring Devices; General Hospital And Personal Use Devices
ELECTRONIC	880.2910	CLINICAL ELECTRONIC THERMOMETER General Hospital And Personal Use Monitoring Devices; General Hospital And Personal Use Devices
ELECTRONIC	886.4400	ELECTRONIC METAL LOCATOR Surgical Devices; Ophthalmic Devices
ELECTRONIC	886.5900	ELECTRONIC VISION AID Therapeutic Devices; Ophthalmic Devices
ELECTRONIC EQUIPMENT	211.68	AUTOMATIC, MECHANICAL, AND ELECTRONIC EQUIPMENT Equipment; Current Good Manufacturing Practice For Finished Pharmaceuticals
ELECTRONIC MEDIA	10.205	ELECTRONIC MEDIA COVERAGE OF PUBLIC ADMINISTRATIVE PROCEEDINGS Electronic Media Coverage Public Admin. Proceedings; Administrative Practices And Procedures
ELECTRONIC MEDIA	10.206	PROCEDURES FOR ELECTRONIC MEDIA COVERAGE OF AGENCY PUBLIC ADMINISTRATIVE PROCEEDINGS Electronic Media Coverage Public Admin. Proceedings; Administrative Practices And Procedures
ELECTRONIC PRODUCT	14.120	ESTABLISHMENT OF THE TECHNICAL ELECTRONIC PRODUCT RADIATION SAFETY STANDARD COMMITTEE (TEPRSSC) Technical Electronic Products Radiation Safety Standards Committee; Public Hearing Before A Public Advisory Committee
ELECTRONIC PRODUCT	1003.2	DEFECT IN AN ELECTRONIC PRODUCT General Provisions; Notification Of Defects Or Failure To Comply
ELECTRONIC PRODUCTS	5.600	VARIANCES FROM PERFORMANCE STANDARDS FOR ELECTRONIC PRODUCTS Radiation Control; Redelegations Of Authority; Delegations Of Authority And Organization
ELECTRONIC PRODUCTS	5.601	EXEMPTION OF ELECTRONIC PRODUCTS FROM PERFORMANCE STANDARDS AND PROHIBITED ACTS Radiation Control; Redelegations Of Authority; Delegations Of Authority And Organization
ELECTRONIC PRODUCTS	5.602	TESTING PROGRAMS AND METHODS OF CERTIFICATION AND IDENTIFICATION FOR ELECTRONIC PRODUCTS Radiation Control; Redelegations Of Authority; Delegations Of Authority And Organization
ELECTRONIC PRODUCTS	5.603	NOTIFICATION OF DEFECTS IN, AND REPAIR OR REPLACEMENT OF, ELECTRONIC PRODUCTS Radiation Control; Redelegations Of Authority; Delegations Of Authority And Organization
ELECTRONIC PRODUCTS	5.604	MANUFACTURERS REQUIREMENT TO PROVIDE DATA TO ULTIMATE PURCHASERS OF ELECTRONIC PRODUCTS Radiation Control; Redelegations Of Authority; Delegations Of Authority And Organization
ELECTRONIC PRODUCTS	5.605	DEALER AND DISTRIBUTOR DIRECTION TO PROVIDE DATA TO MANUFACTURERS OF ELECTRONIC PRODUCTS Radiation Control; Redelegations Of Authority; Delegations Of Authority And Organization
ELECTRONIC PRODUCTS	25.34	DEVICES AND ELECTRONIC PRODUCTS Environmental Impact Considerations
ELECTRONIC PRODUCTS	1000.15	EXAMPLES OF ELECTRONIC PRODUCTS SUBJECT TO THE RADIATION CONTROL FOR HEALTH AND SAFETY ACT OF 1968 Statements Of Policy And Interpretation; General
ELECTRONIC PRODUCTS	1004.1	MANUFACTURER'S OBLIGATION TO REPAIR, REPLACE, OR REFUND COST OF ELECTRONIC PRODUCTS Repurchase, Repairs, Or Replacement Of Electronic Products
ELECTRONIC PRODUCTS	1004.2	PLANS FOR THE REPAIR OF ELECTRONIC PRODUCTS Repurchase, Repairs, Or Replacement Of Electronic Products
ELECTRONIC PRODUCTS	1004.3	PLANS FOR THE REPLACEMENT OF ELECTRONIC PRODUCTS Repurchase, Repairs, Or Replacement Of Electronic Products
ELECTRONIC PRODUCTS	1004.4	PLANS FOR REFUNDING THE COST OF ELECTRONIC PRODUCTS Repurchase, Repairs, Or Replacement Of Electronic Products
ELECTRONIC PRODUCTS	1010.20	ELECTRONIC PRODUCTS INTENDED FOR EXPORT Performance Standards For Electronic Products: General
ELECTRONIC REPORTING	803.13	ELECTRONIC REPORTING Medical Device Reporting
ELECTRONIC SIGNATURE	11.200	ELECTRONIC SIGNATURE COMPONENTS AND CONTROLS Electronic Records; Electronic Signatures

ELECTROPHORESIS	862.2485	ELECTROPHORESIS APPARATUS FOR CLINICAL USE Clinical Laboratory Instruments; Clinical Chemistry And Clinical Toxicology Devices
ELECTROPHORETIC	864.7440	ELECTROPHORETIC HEMOGLOBIN ANALYSIS SYSTEM Hematology Kits And Packages; Hematology And Pathology Devices
ELECTROSTATIC	892.1630	ELECTROSTATIC X-RAY FLUOROSCOPIC X-RAY SYSTEM Diagnostic Devices; Radiology Devices
ELECTROSURGICAL	872.4920	DENTAL ELECTROSURGICAL UNIT AND ACCESSORIES Surgical Devices; Dental Devices
ELECTROSURGICAL	876.4300	ENDOSCOPIC ELECTROSURGICAL UNIT AND ACCESSORIES Surgical Devices; Gastroenterology-urology Devices
ELECTROSURGICAL	878.4400	ELECTROSURGICAL CUTTING AND COAGULATION DEVICE AND ACCESSORIES Surgical Devices; General And Plastic Surgery Devices
ELECTROSURGICAL	886.1400	RADIOFREQUENCY ELECTROSURGICAL CAUTERY APPARATUS Surgical Devices; Ophthalmic Devices
ELECTROTHERAPHY	882.5800	CRANIAL ELECTROTHERAPHY STIMULATOR Neurological Therapeutic Devices; Neurological Devices
ELEMENTS	50.25	ELEMENTS OF INFORMED CONSENT Informed Consent Of Human Subjects; Protection Of Human Subjects
ELEVATOR	890.3930	WHEELCHAIR ELEVATOR Physical Medicine Prosthetic Devices; Physical Medicine Devices
ELIGIBILITY	60.10	FDA ASSISTANCE ON ELIGIBILITY Eligibility Assistance; Patent Term Restoration
ELIGIBILITY	170.30	ELIGIBILITY FOR CLASSIFICATION AS GENERALLY RECOGNIZED AS SAFE (GRAS) Food Additive Safety; Food Additives
ELIGIBILITY	570.30	ELIGIBILITY FOR CLASSIFICATION AS GENERALLY RECOGNIZED AS SAFE (GRAS) Food Additive Safety; Food Additives
EMBOLECTOMY	870.5150	EMBOLECTOMY CATHETER Cardiovascular Therapeutic Devices; Cardiovascular Devices
EMBOLISM	868.2025	ULTRASONIC AIR EMBOLISM MONITOR Monitoring Devices; Anesthesiology Devices
EMBOLIZATION	870.3300	ARTERIAL EMBOLIZATION DEVICE Cardiovascular Prosthetic Devices; Cardiovascular Devices
EMBOLIZATION	882.5950	ARTIFICIAL EMBOLIZATION DEVICE Neurological Therapeutic Devices; Neurological Devices
EMERGENCIES	25.16	PUBLIC HEALTH AND SAFETY EMERGENCIES Environmental Impact Considerations
EMERGENCY AIRWAY	869.5090	EMERGENCY AIRWAY NEEDLE Therapeutic Devices; Anesthesiology Devices
EMERGENCY CART	868.6175	CARDIOPULMONARY EMERGENCY CART Miscellaneous; Anesthesiology Devices
EMERGENCY CARE	203.11	APPLICATIONS FOR REIMPORTATION TO PROVIDE EMERGENCY MEDICAL CARE Reimportation; Prescription Drug Marketing
EMERGENCY FUNCTIONS	5.21	EMERGENCY FUNCTIONS General Redelegations Of Authority; Delegations Of Authority And Organization
EMERGENCY PERMIT	5.301	ISSUANCE OF INITIAL EMERGENCY PERMIT ORDERS AND NOTICES OF CONFIRMATION OF EFFECTIVE DATE OF FINAL REGULATIONS ON FOOD FOR HUMAN AND ANIMAL CONSUMPTION Foods And Cosmetics; Redelegations Of Authority; Delegations Of Authority And Organization
EMERGENCY PERMIT	500.24	EMERGENCY PERMIT CONTROL Animal Drugs, Feeds, And Related Products
EMERGENCY PROVISIONS	640.27	EMERGENCY PROVISIONS Platelets; Additional Standards For Human Blood And Blood Products
EMERGENCY RESEARCH	50.24	EXEMPTION FROM INFORMED CONSENT REQUIREMENTS FOR EMERGENCY RESEARCH Protection Of Human Subjects
EMERGENCY RESEARCH	312.54	EMERGENCY RESEARCH UNDER SECTION 50.24 OF THIS CHAPTER Investigational New Drug Application
EMERGENCY RESEARCH	812.47	EMERGENCY RESEARCH UNDER SECTION 50.24 OF THIS CHAPTER Investigational Drug Exemptions
EMERGENCY SCHEDULING	1308.49	EMERGENCY SCHEDULING Hearings; Schedule Of Controlled Substances
EMERGENCY SITUATION	290.10	DEFINITION OF EMERGENCY SITUATION General Provisions; Controlled Drugs
EMERGENCY TREATMENT	500.65	EPINEPHRINE INJECTION FOR EMERGENCY TREATMENT OF ANAPHYLACTOID SHOCK IN CATTLE, HORSES, SHEEP, AND SWINE Requirements For Specific Animal Drugs; General
EMERGENCY USE	314.36	EMERGENCY USE OF AN INVESTIGATIONAL NEW DRUG Investigational New Drug Application

EMERGENCY VENTILATOR	868.5915	MANUAL EMERGENCY VENTILATOR Therapeutic Devices; Anesthesiology Devices
EMERGENCY VENTILATOR	868.5925	POWERED EMERGENCY VENTILATOR Therapeutic Devices; Anesthesiology Devices
EMISSION	862.2540	FLAME EMISSION PHOTOMETER FOR CLINICAL USE Clinical Laboratory Instruments; Clinical Chemistry And Clinical Toxicology Devices
EMISSION	892.1200	EMISSION COMPUTED TOMOGRAPHY SYSTEM Diagnostic Devices; Radiology Devices
EMMENTALER CHEESE	133.195	SWISS AND EMMENTALER CHEESE Requirements For Specific Standardized Cheese And Related Products; Cheeses And Related Cheese Products
EMOLLIENT	349.65	LABELING OF OPHTHALMIC EMOLLIENT DRUG PRODUCTS Ophthalmic Drug Products For Over-the-counter Human Use
EMOLLIENTS	349.14	OPHTHALMIC EMOLLIENTS Active Ingredients; Ophthalmic Drug Products For Over-the-counter Human Use
EMPLOYEE CHECKS	1301.93	SOURCES OF INFORMATION FOR EMPLOYEE CHECKS Employee Screening: Non-practitioners; Registration Of Manufacturers, Distributors, And Dispensers Of Controlled Substances
EMPLOYEE CONVENIENCES	1250.70	EMPLOYEE CONVENIENCES Servicing Areas For Land And Air Conveyances; Interstate Conveyance Sanitation
EMPLOYEE NAMES, FDA	20.32	DISCLOSURE OF FDA EMPLOYEE NAMES General Policy; Public Information
EMPLOYEE RESPONSIBILITY	1301.91	EMPLOYEE RESPONSIBILITY TO REPORT DRUG DIVERSION Employee Screening: Non-practitioners; Registration Of Manufacturers, Distributors, And Dispensers Of Controlled Substances
EMPLOYEE SCREENING	1301.90	EMPLOYEE SCREENING PROCEDURES Employee Screening: Non-practitioners; Registration Of Manufacturers, Distributors, And Dispensers Of Controlled Substances
EMPLOYEE, RESPONSIBILITY	1309.73	EMPLOYEE RESPONSIBILITY TO REPORT DIVERSION Registration Of Manufacturers, Distributors, Importers And Exporters Of List I Chemicals
EMPLOYEE SANCTIONS	1404.635	REPORTING OF AND EMPLOYEE SANCTIONS FOR CONVICTIONS OF CRIMINAL DRUG OFFENSES Governmentwide Debarment And Suspension And Requirements For Drug-Free Workplace (Grants)
EMPLOYEES	1250.38	TOILET AND LAVATORY FACILITIES FOR USE OF FOOD-HANDLING EMPLOYEES Food Service Sanitation On Land And Air Conveyances, And Vessels; Interstate Conveyance Sanitation
EMPLOYEES	1301.22	EXEMPTION OF AGENTS AND EMPLOYEES; AFFILIATED PRACTITIONERS Exceptions To Registration And Fees; Registration Of Manufacturers, Distributors, And Dispensers Of Controlled Substances
EMPLOYEES	1301.92	ILLICIT ACTIVITIES BY EMPLOYEES Employee Screening: Non-practitioners; Registration Of Manufacturers, Distributors, And Dispensers Of Controlled Substances
EMPLOYEES	1309.24	EXEMPTION OF AGENTS AND EMPLOYEES Registration Of Manufacturers, Distributors, Importers And Exporters Of List I Chemicals
EMPLOYEES, FDA	20.1	TESTIMONY BY FDA EMPLOYEES Official Testimony And Information; Public Information
EMPLOYEES, FDA	20.2	PRODUCTION OF RECORDS BY FDA EMPLOYEES Official Testimony And Information; Public Information
EMPLOYEES, FDA	20.110	DATA AND INFORMATION ABOUT FDA EMPLOYEES Availability Of Specific Categories Of Records; Public Information
EMPLOYEES, FORMER	19.45	TEMPORARY DISQUALIFICATION OF FORMER EMPLOYEES Disqualification Conditions; Standards Of Conduct And Conflicts Of Interest
EMPLOYEES, FORMER	19.55	PERMANENT DISQUALIFICATION OF FORMER EMPLOYEES Disqualification Conditions; Standards Of Conduct And Conflicts Of Interest
EMPLOYEES, GOVERNMENT	20.84	DISCLOSURE TO CONSULTANTS, ADVISORY COMMITTEES, STATE AND LOCAL GOVT. OFFICIALS AND OTHER SPECIAL GOVERNMENT EMPLOYEES Limitations On Exemptions; Public Information
EMPLOYER RESPONSIBILITIES	1309.72	FELONY CONVICTION; EMPLOYER RESPONSIBILITIES Registration Of Manufacturers, Distributors, Importers And Exporters Of List I Chemicals
EMPTY CONTAINER	864.9100	EMPTY CONTAINER FOR THE COLLECTION AND PROCESSING OF BLOOD AND BLOOD COMPONENTS Products Used In Establishments That Manufacture Blood And Blood Products; Hematology And Pathology Devices
EMULSIFIABLE LIQUID	524.1742	N-(MERCAPTOMETHYL) PHTHALIMIDE S-(*O,O*-DIMETHYL PHOSPHORODITHIOATE) EMULSIFIABLE LIQUID Ophthalmic And Topical Dosage Form New Animal Drugs
EMULSIFIERS	178.3400	EMULSIFIERS AND/OR SURFACE-ACTIVE AGENTS Certain Adjuvants And Production Aids; Indirect Food Additives: Adjuvants, Production Aids, And Sanitizers

ENFLURANE GAS	868.1500	ENFLURANE GAS ANALYZER Diagnostic Devices; Anesthesiology Devices
ENFORCEMENT	5.31	ENFORCEMENT ACTIVITIES General Redelegations Of Authority; Delegations Of Authority And Organization
ENFORCEMENT	5.606	ACCEPTANCE OF ASSISTANCE FROM STATE AND LOCAL AUTHORITIES FOR ENFORCEMENT OF RADIATION CONTROL LEGISLATION AND REGULATIONS Radiation Control; Redelegations Of Authority; Delegations Of Authority And Organization
ENFORCEMENT	10.40	PROMULGATION OF REGULATIONS FOR THE EFFICIENT ENFORCEMENT OF THE LAW General Administrative Procedures; Administrative Practices And Procedures
ENFORCEMENT	20.91	USE OF DATA OR INFORMATION FOR ADMINISTRATIVE OR COURT ENFORCEMENT ACTION Limitations On Exemptions; Public Information
ENFORCEMENT	100.2	STATE ENFORCEMENT OF FEDERAL REGULATIONS State And Local Requirements; General
ENFORCEMENT	1230.43	ENFORCEMENT Administrative Procedures; Regulations Under The Federal Caustic Poison Act
ENFORCEMENT	1403.43	ENFORCEMENT Post-award Requirements; Uniform Administrative Requirements For Grants And Coop. Agreements To State & Local Gov.
ENFORCEMENT, LAW	20.64	RECORDS OR INFORMATION COMPILED FOR LAW ENFORCEMENT PURPOSES Exemptions; Public Information
ENFORCEMENT, LAW	201.125	DRUGS FOR USE IN TEACHING, LAW ENFORCEMENT, RESEARCH, AND ANALYSIS Exemptions From Adequate Directions For Use; Labeling
ENFORCEMENT, LAW	801.125	MEDICAL DEVICES FOR USE IN TEACHING, LAW ENFORCEMENT, RESEARCH, AND ANALYSIS Exemptions From Adequate Directions For Use; Labeling
ENFORCEMENT, LAW	1301.24	EXEMPTION OF LAW ENFORCEMENT OFFICIALS Exceptions To Registration And Fees; Registration Of Manufacturers, Distributors, And Dispensers Of Controlled Substances
ENFORCEMENT, LAW	1309.26	EXEMPTION OF LAW ENFORCEMENT OFFICIALS Requirements For Registration; Registration Of Manufacturers, Distributors, Importers, And Exporters Of List I Chemicals
ENFORCEMENT PROCEEDING	1316.31	AUTHORITY FOR ENFORCEMENT PROCEEDING Enforcement Proceedings; Administrative Functions, Practices, And Procedures
ENFORCEMENT RECORDS	20.101	ADMINISTRATIVE ENFORCEMENT RECORDS Availability Of Specific Categories Of Records; Public Information
ENFORCEMENT RECORDS	20.102	COURT ENFORCEMENT RECORDS Availability Of Specific Categories Of Records; Public Information
ENGLISH	803.13	ENGLISH REPORTING REQUIREMENT Medical Device Reporting
ENGORGEMENT	884.5970	CLITORAL ENGORGEMENT DEVICE Obstetrical And Gynecological Therapeutic Devices; Obstetrical And Gynecological Devices
ENOCIANINA	73.170	GRAPE SKIN EXTRACT (ENOCIANINA) Foods; Listing Of Color Additives Exempt From Certification
ENRICHED	866.2330	ENRICHED CULTURE MEDIUM Microbiology Devices; Immunology And Microbiology Devices
ENRICHED BREAD	136.115	ENRICHED BREAD, ROLLS, AND BUNS Requirements For Specific Standardized Bakery Products; Bakery Products
ENRICHED CORN	137.260	ENRICHED CORN MEALS Requirements For Specific Standardized Cereal Flours And Related Products; Cereal Flours And Related Products
ENRICHED FARINA	137.305	ENRICHED FARINA Requirements For Specific Standardized Cereal Flours And Related Products; Cereal Flours And Related Products
ENRICHED FLOUR	137.160	ENRICHED BROMATED FLOUR Requirements For Specific Standardized Cereal Flours And Related Products; Cereal Flours And Related Products
ENRICHED FLOUR	137.165	ENRICHED FLOUR Requirements For Specific Standardized Cereal Flours And Related Products; Cereal Flours And Related Products
ENRICHED FLOUR	137.185	ENRICHED SELF-RISING FLOUR Requirements For Specific Standardized Cereal Flours And Related Products; Cereal Flours And Related Products
ENRICHED MACARONI	139.115	ENRICHED MACARONI PRODUCTS Requirements For Specific Standardized Macaroni And Noodle Products; Macaroni And Noodle Products
ENRICHED MACARONI	139.117	ENRICHED MACARONI PRODUCTS WITH FORTIFIED PROTEIN Requirements For Specific Standardized Macaroni And Noodle Products; Macaroni And Noodle Products

ENRICHED MACARONI	139.122	ENRICHED NONFAT MILK MACARONI PRODUCTS Requirements For Specific Standardized Macaroni And Noodle Products; Macaroni And Noodle Products
ENRICHED MACARONI	139.135	ENRICHED VEGETABLE MACARONI PRODUCTS Requirements For Specific Standardized Macaroni And Noodle Products; Macaroni And Noodle Products
ENRICHED NOODLE	139.155	ENRICHED NOODLE PRODUCTS Requirements For Specific Standardized Macaroni And Noodle Products; Macaroni And Noodle Products
ENRICHED NOODLE	139.165	ENRICHED VEGETABLE NOODLE PRODUCTS Requirements For Specific Standardized Macaroni And Noodle Products; Macaroni And Noodle Products
ENRICHED RICE	137.350	ENRICHED RICE Requirements For Specific Standardized Cereal Flours And Related Products; Cereal Flours And Related Products
ENROFLOXACIN	520.812	ENROFLOXACIN TABLETS Oral Dosage Form New Animal Drugs
ENROFLOXACIN	520.813	ENROFLOXACIN ORAL SOLUTION Oral Dosage Form New Animal Drugs
ENROFLOXACIN	522.812	ENROFLOXACIN SOLUTION Implantation Or Injectable Dosage Form New Animal Drugs
ENROFLOXACIN	524.802	ENROFLOXACIN, SILVER SULFADIAZINE EMULSION Ophthalmic And Topical Dosage Form New Animal Drugs
ENROFLOXACIN	556.228	ENROFLOXACIN Tolerances For Residues Of New Animal Drugs In Food
ENTAMOEBA HISTOLYTICA	866.3220	ENTAMOEBA HISTOLYTICA SEROLOGICAL REAGENTS Serological Reagents; Immunology And Microbiology Devices
ENTRY	1316.05	ENTRY Administrative Inspections; Administrative Functions, Practices, And Procedures
ENTRY INTO FORCE	26.80	ENTRY INTO FORCE, AMENDMENT, AND TERMINATION Framework Provisions; Mutual Recognition Of Pharmaceutical GMP Reports, Medical Device Quality System Audit Reports
ENURESIS	876.2040	ENURESIS ALARM Monitoring Devices; Gastroenterology-urology Devices
ENVIRONMENTAL ASSESSMENT	25.20	ACTIONS REQUIRING PREPARATION OF AN ENVIRONMENTAL ASSESSMENT Agency Actions Requiring Environmental Consideration; Environmental Impact Considerations
ENVIRONMENTAL ASSESSMENTS	25.40	ENVIRONMENTAL ASSESSMENTS Preparation Of Environmental Documents; Environmental Impact Considerations
ENVIRONMENTAL ASSESSMENTS	25.51	ENVIRONMENTAL ASSESSMENTS AND FINDINGS OF NO SIGNIFICANT IMPACT Public Participation And Notification Of Environmental Documents; Environmental Impact Considerations
ENVIRONMENTAL CHAMBER	864.9575	ENVIRONMENTAL CHAMBER FOR STORAGE OF PLATELET CONCENTRATE Products Used In Establishments That Manufacture Blood And Blood Products; Hematology And Pathology Devices
ENVIRONMENTAL CONTROL	820.46	ENVIRONMENTAL CONTROL Buildings; Good Manufacturing Practice For Medical Devices
ENVIRONMENTAL CONTROL	890.3725	POWERED ENVIRONMENTAL CONTROL SYSTEM Physical Medicine Prosthetic Devices; Physical Medicine Devices
ENVIRONMENTAL EFFECTS	25.60	ENVIRONMENTAL EFFECTS ABROAD OF MAJOR AGENCY ACTIONS Other Requirements; Environmental Impact Considerations
ENVIRONMENTAL IMPACT	25.22	ACTIONS REQUIRING PREPARATION OF AN ENVIRONMENTAL IMPACT STATEMENT Agency Actions Requiring Environmental Consideration; Environmental Impact Considerations
ENVIRONMENTAL IMPACT	25.42	ENVIRONMENTAL IMPACT STATEMENTS Preparation Of Environmental Documents; Environmental Impact Considerations
ENVIRONMENTAL IMPACT	25.52	ENVIRONMENTAL IMPACT STATEMENTS Public Participation And Notification Of Environmental Documents; Environmental Impact Considerations
ENZYME, AMINOPEPTIDASE	184.1985	AMINOPEPTIDASE ENZYME PREPARATION DERIVED FROM LACTOCOCCUS LACTIS Listing Of Specific Substances Affirmed As GRAS; Direct Food Substances Affirmed As Generally Recognized As Safe
ENZYME, AMYLASE	184.1012	ALPHA-AMYLASE ENZYME PREPARATION FROM BACILLUS STEAROTHERMOPHILUS Listing Of Specific Substances Affirmed As GRAS; Direct Food Substances Affirmed As Generally Recognized As Safe
ENZYME ANALYZER	862.2500	ENZYME ANALYZER FOR CLINICAL USE Clinical Laboratory Instruments; Clinical Chemistry And Toxicology Devices

ENZYME, ACETOLACTATE. DECARB	173.115	ALPHA-ACETOLACTATE DECARBOXYLASE (ALPHA-ALDC) ENZYME PREPARATION DERIVED FROM A RECOMBINANT BACILLUS SUBTILIS Enzyme Preparations And Microorganisms; Secondary Direct Food Additives Permitted In Food For Human Consumption
ENZYME, ANGIOTENSIN-CONVERTING	862.1090	ANGIOTENSIN CONVERTING ENZYME (A.C.E.) TEST SYSTEM Clinical Chemistry Test Systems; Clinical Chemistry And Clinical Toxicology Devices
ENZYME, CARBOHYDRASE	184.1148	BACTERIALLY DERIVED CARBOHYDRASE ENZYME PREPARATION Listing Of Specific Substances Affirmed As GRAS; Direct Food Substances Affirmed As Generally Recognized As Safe
ENZYME, CELLULASE	184.1250	CELLULASE ENZYME PREPARATION DERIVED FROM TRICHODERMA LONGIBRACHIATUM Listing Of Specific Substances Affirmed As GRAS; Direct Food Substances Affirmed As Generally Recognized As Safe
ENZYME, GLUCOSE ISOMERASE	184.1372	INSOLUBLE GLUCOSE ISOMERASE ENZYME PREPARATIONS Listing Of Specific Substances Affirmed As GRAS; Direct Food Substances Affirmed As Generally Recognized As Safe
ENZYME, LACTASE	184.1387	LACTASE ENZYME PREPARATION FROM CANDIDA PSEUDOTROPICALIS Listing Of Specific Substances Affirmed As GRAS; Direct Food Substances Affirmed As Generally Recognized As Safe
ENZYME, LACTASE	184.1388	LACTASE ENZYME PREPARATION FROM KLUYVEROMYCES LACTIS Listing Of Specific Substances Affirmed As GRAS; Direct Food Substances Affirmed As Generally Recognize As Safe
ENZYME, LIPASE	184.1420	LIPASE ENZYME PREPARATION DERIVED FROM RHIZOPUS NIVEUS Listing Of Specific Substances Affirmed As GRAS; Direct Food Substances Affirmed As Generally Recognized As Safe
ENZYME-MODIFIED	184.1063	ENZYME-MODIFIED LECITHIN Listing Of Specific Substances Affirmed As GRAS; Direct Food Substances Affirmed As Generally Recognized As Safe
ENZYME-MODIFIED	184.1287	ENZYME-MODIFIED FATS Listing Of Specific Substances Affirmed As GRAS; Direct Food Substances Affirmed As Generally Recognized As Safe
ENZYME PREPARATIONS	173.357	MATERIALS USED AS FIXING AGENTS IN THE IMMOBILIZATION OF ENZYME PREPARATIONS Specific Usage Additives; Secondary Direct Food Additives Permitted In Food For Human Consumption
ENZYME PREPARATIONS	864.4400	ENZYME PREPARATIONS Specimen Preparations Reagents; Hematology And Pathology Devices
ENZYME PRODUCT	184.1027	MIXED CARBOHYDRASE AND PROTEASE ENZYME PRODUCT Listing Of Specific Substances Affirmed As GRAS; Direct Food Substances Affirmed As Generally Recognized As Safe
ENZYME, PROTEASE	184.1150	BACTERIALLY DERIVED PROTEASE ENZYME PREPARATION Listing Of Specific Substances Affirmed As GRAS; Direct Food Substances Affirmed As Generally Recognized As Safe
ENZYME, RBC	864.7100	RED BLOOD CELL ENZYME ASSAY Hematology Kits And Packages; Hematology And Pathology Devices
ENZYME SOLUTION	4.9400	STABILIZED ENZYME SOLUTION Products Used In Establishments That Manufacture Blood And Blood Products; Hematology And Pathology Devices
ENZYME, UREASE	184.1924	UREASE ENZYME PREPARATION FROM LACTOBACILLUS FERMENTUM Listing Of Specific Substances Affirmed As GRAS; Direct Food Substances Affirmed As Generally Recognized As Safe
ENZYMES	173.150	MILK-CLOTTING ENZYMES, MICROBIAL Enzyme Preparations And Microorganisms; Secondary Direct Food Additives Permitted In Food For Human Consumption
EPHEDRINE	1310.14	EXEMPTION OF DRUGS PRODUCTS CONTAINING EPHEDRINE AND ANOTHER MEDICALLY ACTIVE INGREDIENT Records And Reports Of Listed Chemicals And Certain Machines
EPHEDRINE	1310.15	EXEMPT DRUG PRODUCTS CONTAINING EPHEDRINE AND ANOTHER MEDICALLY ACTIVE INGREDIENT Records And Reports Of Listed Chemicals And Certain Machines
EPICHLOROHYDRIN	173.60	DIMETHYLAMINE EPICHLOROHYDRIN COPOLYMER Polymer Substances For Food Treatment; Secondary Direct Food Additives Permitted In Food For Human Consumption
EPICHLOROHYDRIN	175.380	XYLENE-FORMALDEHYDE RESINS CONDENSED WITH 4,4'-ISOPROPYLIDENEDIPHENOL- EPICHLOROHYDRIN EPOXY RESINS Substances For Use As Components Of Coatings; Indirect Food Additives
EPICHLOROHYDRIN	177.1440	4,4'-ISOPROPYLIDENEDIPHENOL-EPICHLOROHYDRIN RESINS MINIMUM MOLECULAR WEIGHT 10,000 Substances For Use As Basic Components Of Single And Repeated Use Food Contact Surfaces; Indirect Food Additives: Polymers
EPICHLOROHYDRIN	177.1460	MELAMINE-FORMALDEHYDE RESINS IN MOLDED ARTICLES Substances For Use As Basic Components Of Single And Repeated Use Food Contact Surfaces; Indirect Food Additives: Polymers

EPICHLOROHYDRIN	177.2280	4,4'-ISOPROPYLIDENEDIPHENOL-EPICHLOROHYDRIN THERMOSETTING EPOXY RESINS Substances For Use Only As Components Of Articles Intended For Repeated Use; Indirect Food Additives: Polymers
EPILATOR	878.5350	NEEDLE-TYPE EPILATOR Therapeutic Devices; General And Plastic Surgery Devices
EPILATOR	878.5360	TWEEZER-TYPE EPILATOR Therapeutic Devices; General And Plastic Surgery Devices
EPINEPHRINE	500.65	EPINEPHRINE INJECTION FOR EMERGENCY TREATMENT OF ANAPHYLACTOID SHOCK IN CATTLE, HORSES, SHEEP, AND SWINE Requirements For Specific Animal Drugs; General
EPISTAXIS	874.4100	EPISTAXIS BALLOON Surgical Devices; Ear, Nose And Throat Devices
EPOXIDIZED SOYBEAN OIL	172.723	EPOXIDIZED SOYBEAN OIL Food Additive Permitted For Direct Addition To Food For Human Consumption
EPOXY	175.380	XYLENE-FORMALDEHYDE RESINS CONDENSED WITH 4,4'-ISOPROPYLIDENEDIPHENOL-EPICHLOROHYDRIN EPOXY RESINS Substances For Use As Components Of Coatings; Indirect Food Additives
EPOXY	177.2280	4,4'-ISOPROPYLIDENEDIPHENOL-EPICHLOROHYDRIN THERMOSETTING EPOXY RESINS Substances For Use Only As Components Of Articles Intended For Repeated Use; Indirect Food Additives: Polymers
EPRINOMECTIN	524.814	EPRINOMECTIN Ophthalmic And Topical Dosage Form New Animal Drugs
EPRINOMECTIN	556.227	EPRINOMECTIN Tolerances For Residues Of New Animal Drugs In Food
EPSIPRANTEL	520.816	EPSIPRANTEL TABLETS Oral Dosage Form New Animal Drugs
EPSTEIN-BARR VIRUS	866.3235	EPSTEIN-BARR VIRUS SEROLOGICAL REAGENTS Serological Reagents; Immunology And Microbiology Devices
EQUINE ENCEPHALOMYELITIS	866.3240	EQUINE ENCEPHALOMYELITIS VIRUS SEROLOGICAL REAGENTS Serological Reagents; Immunology And Microbiology Devices
EQUIPMENT	58.63	MAINTENANCE AND CALIBRATION OF EQUIPMENT Equipment; Good Laboratory Practice For Nonclinical Laboratory Studies
EQUIPMENT	110.40	EQUIPMENT AND UTENSILS Equipment; Current Good Manufacturing Practice In Manufacturing, Packing, Or Holding Human Food
EQUIPMENT	113.40	EQUIPMENT AND PROCEDURES Equipment; Thermally Processed Low-acid Foods Packaged In Hermetically Sealed Containers
EQUIPMENT	129.40	EQUIPMENT AND PROCEDURES Equipment And Facilities; Processing And Bottling Of Bottled Drinking Water
EQUIPMENT	211.68	AUTOMATIC, MECHANICAL, AND ELECTRONIC EQUIPMENT Equipment; Current Good Manufacturing Practice For Finished Pharmaceuticals
EQUIPMENT	225.30	EQUIPMENT Construction And Maintenance Of Facilities And Equipment; Current Good Manufacturing Practice For Medicated Feeds
EQUIPMENT	25.35	USE OF WORK AREAS, EQUIPMENT, AND STORAGE AREAS FOR OTHER MANUFACTURING AND STORAGE PURPOSES Construction And Maintenance Of Facilities And Equipment; Current Good Manufacturing Practice For Medicated Feeds
EQUIPMENT	225.130	EQUIPMENT Facilities And Equipment; Current Good Manufacturing Practice For Medicated Feeds
EQUIPMENT	226.30	EQUIPMENT Construction And Maintenance Of Facilities And Equipment; Current Good Manufacturing Practice For Type A Medicated Articles
EQUIPMENT	600.11	PHYSICAL ESTABLISHMENT, EQUIPMENT, ANIMALS, AND CARE Establishment Standards; Biological Products: General
EQUIPMENT	606.60	EQUIPMENT Equipment; Current Good Manufacturing Practice For Blood And Blood Components
EQUIPMENT	820.72	INSPECTION, MEASURING, AND TEST EQUIPMENT Production And Process Controls; Quality System Regulation
EQUIPMENT	862.2050	GENERAL PURPOSE LABORATORY EQUIPMENT LABELED OR PROMOTED FOR A SPECIFIC MEDICINAL USE Clinical Laboratory Instruments; Clinical Chemistry And Clinical Toxicology Devices
EQUIPMENT	864.2240	CELL AND TISSUE CULTURE SUPPLIES AND EQUIPMENT Cell And Tissue Culture Products; Hematology And Pathology Devices
EQUIPMENT	864.3010	TISSUE PROCESSING EQUIPMENT Pathology Instrumentation And Accessories; Hematology And Pathology Devices
EQUIPMENT	866.4500	IMMUNOELECTROPHORESIS EQUIPMENT Immunology Laboratory Equipment And Reagents; Immunology And Microbiology Devices

EQUIPMENT	866.4520	IMMUNOFLUOROMETER EQUIPMENT Immunology Laboratory Equipment And Reagents; Immunology And Microbiology Devices
EQUIPMENT	866.4540	IMMUNONEPHELOMETER EQUIPMENT Immunology Laboratory Equipment And Reagents; Immunology And Microbiology Devices
EQUIPMENT	866.4830	ROCKET IMMUNOELECTROPHORESIS EQUIPMENT Immunology Laboratory Equipment And Reagents; Immunology And Microbiology Devices
EQUIPMENT	870.4200	CARDIOPULMONARY BYPASS ACCESSORY EQUIPMENT Cardiovascular Surgical Devices; Cardiovascular Devices
EQUIPMENT	890.5360	MEASURING EXERCISE EQUIPMENT Physical Medicine Therapeutic Devices; Physical Medicine Devices
EQUIPMENT	890.5370	NONMEASURING EXERCISE EQUIPMENT Physical Medicine Therapeutic Devices; Physical Medicine Devices
EQUIPMENT	890.5380	POWERED EXERCISE EQUIPMENT Physical Medicine Therapeutic Devices; Physical Medicine Devices
EQUIPMENT	890.5900	POWERED TRACTION EQUIPMENT Physical Medicine Therapeutic Devices; Physical Medicine Devices
EQUIPMENT	1020.31	RADIOGRAPHIC EQUIPMENT Performance Standards For Ionizing Radiation Emitting Products
EQUIPMENT	1020.32	FLUOROSCOPIC EQUIPMENT Performance Standards For Ionizing Radiation Emitting Products
EQUIPMENT	1020.33	COMPUTED TOMOGRAPHY (CT) EQUIPMENT Performance Standards For Ionizing Radiation Emitting Products
EQUIPMENT	1210.15	PASTEURIZATION; EQUIPMENT AND METHODS Inspection And Testing; Regulations Under The Federal Import Milk Act
EQUIPMENT	1250.33	UTENSILS AND EQUIPMENT Food Service Sanitation On Land And Air Conveyances, And Vessels; Interstate Conveyance Sanitation
EQUIPMENT	1250.34	REFRIGERATION EQUIPMENT Food Service Sanitation On Land And Air Conveyances, And Vessels; Interstate Conveyance Sanitation
EQUIPMENT	1250.39	GARBAGE EQUIPMENT AND DISPOSITION Food Service Sanitation On Land And Air Conveyances, And Vessels; Interstate Conveyance Sanitation
EQUIPMENT	1250.67	WATERING EQUIPMENT Servicing Areas For Land And Air Conveyances; Interstate Conveyance Sanitation
EQUIPMENT	1403.32	EQUIPMENT Post-award Requirements; Uniform Administrative Requirements For Grants And Coop. Agreements To State & Local Gov.
EQUIPMENT CLEANING	211.67	EQUIPMENT CLEANING AND MAINTENANCE Equipment; Current Good Manufacturing Practice For Finished Pharmaceuticals
EQUIPMENT CLEANING/USE	211.182	EQUIPMENT CLEANING AND USE LOG Records And Reports; Current Good Manufacturing Practice For Finished Pharmaceuticals
EQUIPMENT CLEANOUT	225.65	EQUIPMENT CLEANOUT PROCEDURES Product Quality Control; Current Good Manufacturing Practice For Medicated Feeds
EQUIPMENT CLEANOUT	225.165	EQUIPMENT CLEANOUT PROCEDURES Product Quality Assurance; Current Good Manufacturing Practice For Medicated Feeds
EQUIPMENT CONSTRUCTION	211.65	EQUIPMENT CONSTRUCTION Equipment; Current Good Manufacturing Practice For Finished Pharmaceuticals
EQUIPMENT DESIGN	58.61	EQUIPMENT DESIGN Equipment; Good Laboratory Practice For Nonclinical Laboratory Studies
EQUIPMENT DESIGN	211.63	EQUIPMENT DESIGN, SIZE, AND LOCATION Equipment; Current Good Manufacturing Practice For Finished Pharmaceuticals
EQUIPMENT IDENTIFICATION	211.105	EQUIPMENT IDENTIFICATION Production And Process Controls; Current Good Manufacturing Practice For Finished Pharmaceuticals
EQUIVALENCE ASSESSMENT	26.6	EQUIVALENCE ASSESSMENT Specific Sector Provisions For Pharmaceutical GMPs; Mutual Recognition Of Pharmaceutical GMP Reports, Medical Device Quality System Audit Reports
EQUIVALENCE ASSESSMENT	26.7	PARTICIPATION IN THE EQUIVALENCE ASSESSMENT AND DETERMINATION Specific Sector Provisions For Pharmaceutical GMPs; Mutual Recognition Of Pharmaceutical GMP Reports, Medical Device Quality System Audit Reports
EQUIVALENCE ASSESSMENT	26.39	EQUIVALENCE ASSESSMENT Specific Sector Provisions For Medical Devices; Mutual Recognition Of Pharmaceutical GMP Reports, Medical Device Quality System Audit Reports
EQUIVALENCE, CONTINUED	26.15	MONITORING CONTINUED EQUIVALENCE Specific Sector Provisions For Pharmaceutical GMPs; Mutual Recognition Of Pharmaceutical GMP Reports, Medical Device Quality System Audit Reports

EQUIVALENCE, CONTINUED	26.45	MONITORING CONTINUED EQUIVALENCE Specific Sector Provisions For Medical Devices; Mutual Recognition Of Pharmaceutical GMP Reports, Medical Device Quality System Audit Reports
EQUIVALENCE DETERMINATION	26.9	EQUIVALENCE DETERMINATION Specific Sector Provisions For Pharmaceutical GMPs; Mutual Recognition Of Pharmaceutical GMP Reports, Medical Device Quality System Audit Reports
EQUIVALENT	26.10	REGULATORY AUTHORITIES NOT LISTED AS CURRENTLY EQUIVALENT Specific Sector Provisions For Pharmaceutical GMPs; Mutual Recognition Of Pharmaceutical GMP Reports, Medical Device Quality System Audit Reports
EQUIVALENT METHODS	610.9	EQUIVALENT METHODS AND PROCESSES General Provisions; General Biological Products Standards
ERROR, HARMLESS	17.48	HARMLESS ERROR Civil Money Penalties Hearings
ERYSIPELOTHRIX RHUSIOPATHIAE	866.3250	ERYSIPELOTHRIX RHUSIOPATHIAE SEROLOGICAL REAGENTS Serological Reagents; Immunology And Microbiology Devices
ERYTHEMATOSUS, LUPUS	866.5820	SYSTEMIC LUPUS ERYTHEMATOSUS IMMUNOLOGICAL TEST SYSTEM Immunological Test Systems; Immunology And Microbiology Devices
ERYTHROBIC ACID	182.3041	ERYTHROBIC ACID Chemical Preservatives; Substances Generally Recognized As Safe
ERYTHORBIC ACID	582.3041	ERYTHORBIC ACID Chemical Preservatives; Substances Generally Recognized As Safe
ERYTHROCYTE	864.6700	ERYTHROCYTE SEDIMENTATION RATE TEST Manual Hematology Devices; Hematology And Pathology Devices
ERYTHROCYTIC	864.7360	ERYTHROCYTIC GLUCOSE-6-PHOSPHATE DEHYDROGENASE ASSAY Hematology Kits And Packages; Hematology And Pathology Devices
ERYTHROMYCIN	522.820	ERYTHROMYCIN INJECTION Implantation Or Injectable Dosage Form New Animal Drugs
ERYTHROMYCIN	526.820	ERYTHROMYCIN Intramammary Dosage Forms
ERYTHROMYCIN	556.230	ERYTHROMYCIN Specific Tolerances For Residues Of New Animal Drugs; Tolerances For Residues Of New Animal Drugs In Food
ERYTHROMYCIN PHOSPHATE	520.823	ERYTHROMYCIN PHOSPHATE Oral Dosage Form New Animal Drugs
ERYTHROMYCIN THIOCYANATE	558.248	ERYTHROMYCIN THIOCYANATE Specific New Animal Drugs For Use In Animal Feeds; New Animal Drugs For Use In Animal Feeds
ERYTHROPOIETIN	864.7250	ERYTHROPOIETIN ASSAY Hematology Kits And Packages; Hematology And Pathology Devices
ESCHERICHIA COLI	866.3255	ESCHERICHIA COLI SEROLOGICAL REAGENTS Serological Reagents; Immunology And Microbiology Devices
ESOPHAGEAL	868.1910	ESOPHAGEAL STETHOSCOPE Diagnostic Devices; Anesthesiology Devices
ESOPHAGEAL	868.1920	ESOPHAGEAL STETHOSCOPE WITH ELECTRICAL CONDUCTORS Diagnostic Devices; Anesthesiology Devices
ESOPHAGEAL	868.5650	ESOPHAGEAL OBTURATOR Therapeutic Devices; Anesthesiology Devices
ESOPHAGEAL	876.5365	ESOPHAGEAL DILATOR Therapeutic Devices; Gastroenterology-urology Devices
ESOPHAGEAL	878.3610	ESOPHAGEAL PROSTHESIS Prosthetic Devices; General And Plastic Surgery Devices
ESOPHAGOSCOPE	874.4710	ESOPHAGOSCOPE (FLEXIBLE OR RIGID) AND ACCESSORIES Surgical Devices; Ear, Nose And Throat Devices
ESSENTIAL OILS	182.20	ESSENTIAL OILS, OLEORESINS, AND NATURAL EXTRACTIVES (INCLUDING DISTILLATES) General Provisions; Substances Generally Recognized As Safe
ESSENTIAL OILS	582.20	ESSENTIAL OILS, OLEORESINS, AND NATURAL EXTRACTIVES (INCLUDING DISTILLATES) General Provisions; Substances Generally Recognized As Safe
ESSENTIAL OILS	582.50	CERTAIN OTHER SPICES, SEASONINGS, ESSENTIAL OILS, OLEORESINS, AND NATURAL EXTRACTS General Provisions; Substances Generally Recognized As Safe
ESTABLISH	320.24	TYPES OF EVIDENCE TO ESTABLISH BIOAVAILABILITY OR BIOEQUIVALENCE Procedures For Determining The Bioavailability Of Drug Products; Bioavailability And Bioequivalence Requirements
ESTABLISHED NAMES	299.4	ESTABLISHED NAMES FOR DRUGS Drugs; Official Names And Established Names
ESTABLISHED NAMES	701.30	INGREDIENT NAMES ESTABLISHED FOR COSMETIC INGREDIENT LABELING Labeling Of Specific Ingredients; Cosmetic Labeling
ESTABLISHING	320.32	PROCEDURES FOR ESTABLISHING OR AMENDING A BIOEQUIVALENCE REQUIREMENT Procedures For Determining The Bioavailability Of Drug Products; Bioavailability And Bioequivalence Requirements

ESTABLISHMENT	14.40	ESTABLISHMENT AND RENEWAL OF ADVISORY COMMITTEES Establishment Of Advisory Committees; Public Hearing Before A Public Advisory Committee
ESTABLISHMENT	14.120	ESTABLISHMENT OF THE TECHNICAL ELECTRONIC PRODUCT RADIATION SAFETY STANDARD COMMITTEE (TEPRSSC) Technical Electronic Products Radiation Safety Standards Committee; Public Hearing Before A Public Advisory Committee
ESTABLISHMENT	14.140	ESTABLISHMENT OF A COLOR ADDITIVE ADVISORY COMMITTEE Color Additive Advisory Committees; Public Hearing Before A Public Advisory Committee
ESTABLISHMENT	14.160	ESTABLISHMENT OF STANDING TECHNICAL ADVISORY COMMITTEES FOR HUMAN PRESCRIPTION DRUGS Advisory Committees For Human Prescription Drugs; Public Hearing Before A Public Advisory Committee
ESTABLISHMENT	108.19	ESTABLISHMENT OF REQUIREMENTS FOR EXEMPTION FROM SECTION 404 General Provisions; Emergency Permit Control
ESTABLISHMENT	109.4	ESTABLISHMENT OF TOLERANCES, REGULATORY LIMITS AND ACTION LEVELS General Provisions; Unavoidable Contaminants In Food For Human Consumption & Food-packaging Material
ESTABLISHMENT	509.4	ESTABLISHMENT OF TOLERANCES, REGULATORY LIMITS AND ACTION LEVELS General Provisions; Unavoidable Contaminants In Animal Food And Food-packaging Material
ESTABLISHMENT	600.11	PHYSICAL ESTABLISHMENT, EQUIPMENT, ANIMALS, AND CARE Establishment Standards; Biological Products: General
ESTABLISHMENT	740.1	ESTABLISHMENT OF WARNING STATEMENTS General; Cosmetic Product Warning Statements
ESTABLISHMENT REGISTRATION	207.7	ESTABLISHMENT REGISTRATION AND PRODUCT LISTING FOR HUMAN BLOOD AND BLOOD PRODUCTS AND FOR MEDICAL DEVICES General; Registration Of Drug Producers And Listing Of Drugs In Commercial Distribution
ESTABLISHMENT REGISTRATION	207.35	NOTIFICATION OF REGISTRANT; DRUG ESTABLISHMENT REGISTRATION NUMBER AND DRUG LISTING NUMBER Procedures For Domestic Drug Establishments; Registration Of Drug Producers And Listing Of Drugs In Commercial Distribution
ESTABLISHMENT REGISTRATION	207.40	ESTABLISHMENT REGISTRATION AND DRUG LISTING REQUIREMENTS FOR FOREIGN ESTABLISHMENTS Procedure For Foreign Drug Establishments; Registration Of Drug Producers And Listing Of Drugs In Commercial Distribution
ESTABLISHMENT REGISTRATION	607.7	ESTABLISHMENT REGISTRATION AND PRODUCT LISTING OF BLOOD BANKS AND OTHER FIRMS MANUFACTURING HUMAN BLOOD AND BLOOD PRODUCTS General Provisions; Establishment Registratn+product Listing, Manufacturers Human Blood/b.products
ESTABLISHMENT REGISTRATION	607.21	TIMES FOR ESTABLISHMENT REGISTRATION AND BLOOD PRODUCT LISTING Procedures For Domestic Blood Product Establishments; Establishment Registratn+product Listing, Manufacturers Human Blood/b.products
ESTABLISHMENT REGISTRATION	607.25	INFORMATION REQUIRED FOR ESTABLISHMENT REGISTRATION AND BLOOD PRODUCT LISTING Procedures For Domestic Blood Product Establishments; Establishment Registratn+product Listing, Manufacturers Human Blood/b.products
ESTABLISHMENT REGISTRATION	607.26	AMENDMENTS TO ESTABLISHMENT REGISTRATION Procedures For Domestic Blood Product Establishments; Establishment Registratn+product Listing, Manufacturers Human Blood/b.products
ESTABLISHMENT REGISTRATION	607.35	NOTIFICATION OF REGISTRANT; BLOOD PRODUCT ESTABLISHMENT REGISTRATION NUMBER AND NDC LABELER CODE Procedures For Domestic Blood Product Establishments; Establishment Registratn+product Listing, Manufacturers Human Blood/b.products
ESTABLISHMENT REGISTRATION	607.39	MISBRANDING BY REFERENCE TO ESTABLISHMENT REGISTRATION OR TO REGISTRATION NUMBER Procedures For Domestic Blood Product Establishments; Establishment Registratn+product Listing, Manufacturers Human Blood/b.products
ESTABLISHMENT REGISTRATION	607.40	ESTABLISHMENT REGISTRATION AND BLOOD PRODUCT LISTING REQUIREMENTS FOR FOREIGN BLOOD PRODUCT ESTABLISHMENTS Procedures For Foreign Blood Product Establishments; Establishment Registratn+product Listing, Manufacturers Human Blood/b.products
ESTABLISHMENT REGISTRATION	710.6	NOTIFICATION OF REGISTRANT; COSMETIC PRODUCT ESTABLISHMENT REGISTRATION NUMBER Voluntary Registration Of Cosmetic Product Establishments
ESTABLISHMENT REGISTRATION	807.21	TIMES FOR ESTABLISHMENT REGISTRATION AND DEVICE LISTING Procedures For Device Establishments; Establishment Registration And Device Listing For Manufacturers And Initial Importers Of Devices

ESTABLISHMENT REGISTRATION	807.25	INFORMATION REQUIRED OR REQUESTED FOR ESTABLISHMENT REGISTRATION AND DEVICE LISTING Procedures For Device Establishments; Establishment Registration And Device Listing For Manufacturers And Initial Importers Of Devices
ESTABLISHMENT REGISTRATION	807.26	AMENDMENTS TO ESTABLISHMENT REGISTRATION Procedures For Device Establishments; Establishment Registration And Device Listing For Manufacturers And Initial Importers Of Devices
ESTABLISHMENT REGISTRATION	807.37	INSPECTION OF ESTABLISHMENT REGISTRATION AND DEVICE LISTINGS Procedures For Device Establishments; Establishment Registration And Device Listing For Manufacturers And Initial Importers Of Devices
ESTABLISHMENT REGISTRATION	807.39	MISBRANDING BY REFERENCE TO ESTABLISHMENT REGISTRATION OR TO REGISTRATION NUMBER Procedures For Device Establishments; Establishment Registration And Device Listing For Manufacturers And Initial Importers Of Devices
ESTABLISHMENT REGISTRATION	807.40	ESTABLISHMENT REGISTRATION AND DEVICE LISTING FOR FOREIGN ESTABLISHMENTS IMPORTING OR OFFERING FOR IMPORT DEVICES INTO THE UNITED STATES Registration Procedures For Foreign Device Establishments; Establishment Registration And Device Listing For Manufacturers And Initial Importers Of Devices
ESTABLISHMENT REGISTRATION	1271.25	WHAT INFORMATION IS REQUIRED FOR ESTABLISHMENT REGISTRATION AND HCT/P LISTING? Procedures For Registration And Listing; Human Cells, Tissues, And Cellular And Tissue-Based Products
ESTABLISHMENT REGISTRATION	1271.26	WHEN MUST I AMEND MY ESTABLISHMENT REGISTRATION? Procedures For Registration And Listing; Human Cells, Tissues, And Cellular And Tissue-Based Products
ESTABLISHMENT REGISTRATIONS	607.37	INSPECTION OF ESTABLISHMENT REGISTRATIONS AND BLOOD PRODUCT LISTINGS Procedures For Domestic Blood Product Establishments; Establishment Registratn+product Listing, Manufacturers Human Blood/b.products
ESTABLISHMENT REGISTRATIONS	1271.37	WILL ESTABLISHMENT REGISTRATIONS AND HCT/P LISTINGS BE AVAILABLE FOR INSPECTION, AND HOW DO I REQUEST INFORMATION? Procedures For Registration And Listing; Human Cells, Tissues, And Cellular And Tissue-Based Products
ESTABLISHMENTS	109.15	USE OF POLYCHLORINATED BIPHENYLS (PCB'S) IN ESTABLISHMENTS MANUFACTURING FOOD-PACKAGING MATERIALS General Provisions; Unavoidable Contaminants In Food For Human Consumption & Food-packaging Material
ESTABLISHMENTS	200.11	USE OF OCTADECYLAMINE IN STEAM LINES OF DRUG ESTABLISHMENTS General Provisions; General
ESTABLISHMENTS	207.10	EXEMPTIONS FOR ESTABLISHMENTS Exemptions; Registration Of Drug Producers And Listing Of Drugs In Commercial Distribution
ESTABLISHMENTS	509.15	USE OF POLYCHLORINATED BIPHENYLS (PCB'S) IN ESTABLISHMENTS MANUFACTURED FOOD-PACKAGING MATERIALS General Provisions; Unavoidable Contaminants In Animal Food And Food-packaging Material
ESTABLISHMENTS	606.171	REPORTING OF PRODUCT DEVIATIONS BY LICENSED MANUFACTURERS, UNLICENSED REGISTERED BLOOD ESTABLISHMENTS, AND TRANSFUSION SERVICES Current Good Manufacturing Practice For Blood And Blood Components
ESTABLISHMENTS	607.22	HOW AND WHERE TO REGISTER ESTABLISHMENTS AND LIST BLOOD PRODUCTS Procedures For Domestic Blood Product Establishments; Establishment Registratn+product Listing, Manufacturers Human Blood/b.products
ESTABLISHMENTS	607.65	EXEMPTIONS FOR BLOOD PRODUCT ESTABLISHMENTS Exemptions; Establishment Registratn+product Listing, Manufacturers Human Blood/ b.products
ESTABLISHMENTS	807.22	HOW AND WHERE TO REGISTER ESTABLISHMENTS AND LIST DEVICES Procedures For Device Establishments; Establishment Registration And Device Listing For Manufacturers And Initial Importers Of Devices
ESTABLISHMENTS	807.65	EXEMPTIONS FOR DEVICE ESTABLISHMENTS Exemptions; Establishment Registration And Device Listing For Manufacturers And Initial Importers Of Devices
ESTABLISHMENTS, FOREIGN	207.40	ESTABLISHMENT REGISTRATION AND DRUG LISTING REQUIREMENTS FOR FOREIGN ESTABLISHMENTS Procedure For Foreign Drug Establishments; Registration Of Drug Producers And Listing Of Drugs In Commercial Distribution
ESTABLISHMENTS, FOREIGN	601.15	FOREIGN ESTABLISHMENTS AND PRODUCTS: SAMPLES FOR EACH IMPORTATION Biologics Licensing; Licensing

ESTABLISHMENTS, FOREIGN	607.40	ESTABLISHMENT REGISTRATION AND BLOOD PRODUCT LISTING REQUIREMENTS FOR FOREIGN BLOOD PRODUCT ESTABLISHMENTS Procedures For Foreign Blood Product Establishments; Establishment Registratn+product Listing, Manufacturers Human Blood/b.products
ESTABLISHMENTS, FOREIGN	807.40	ESTABLISHMENT REGISTRATION AND DEVICE LISTING FOR FOREIGN ESTABLISHMENTS IMPORTING OR OFFERING FOR IMPORT DEVICES INTO THE UNITED STATES Registration Procedures For Foreign Device Establishments; Establishment Registration And Device Listing For Manufacturers And Initial Importers Of Devices
ESTER, ACRYLATE	175.210	ACRYLATE ESTER COPOLYMER COATING Substances For Use As Components Of Coatings; Indirect Food Additives
ESTER, GLUCOSIDE-COCONUT	172.816	METHYL GLUCOSIDE-COCONUT OIL ESTER Multipurpose Additives; Food Additives Permitted For Direct Addition To Food For Human Consumption
ESTER, GLUCOSIDE-COCONUT	178.3600	METHYL GLUCOSIDE-COCONUT OIL ESTER Certain Adjuvants And Production Aids; Indirect Food Additives: Adjuvants, Production Aids, And Sanitizers
ESTER, GLUCOSIDE-COCONUT	573.660	METHYL GLUCOSIDE-COCONUT OIL ESTER Food Additive Listing; Food Additives Permitted In Feed And Drinking Water Of Animals
ESTER COPOLYMERS	73.31001	,4-BIS[(2-HYDROXYETHYL)AMINO]-9,10-ANTHRACENEDIONE BIS(2-PROPENOIC) ESTER COPOLYMERS Medical Devices; Listing Of Color Additives Exempt From Certification
ESTER, GLYCEROL	172.735	GLYCEROL ESTER OF WOOD ROSIN Other Specific Usage Additives; Food Additives Permitted For Direct Addition To Food For Human Consumption
ESTER, METHYL GLUCOSIDE	172.816	METHYL GLUCOSIDE-COCONUT OIL ESTER Multipurpose Additives; Food Additives Permitted For Direct Addition To Food For Human Consumption
ESTER, METHYL GLUCOSIDE	178.3600	METHYL GLUCOSIDE-COCONUT OIL ESTER Certain Adjuvants And Production Aids; Indirect Food Additives: Adjuvants, Production Aids, And Sanitizers
ESTER RESINS	189.300	HYDROGENATED 4,4'-ISOPROPYLIDENE-DIPHENOLPHOSPHITE ESTER RESINS Substances Prohibited From Indirect Addition To Human Food Through Food-contact Surfaces; Substances Prohibited From Use In Human Food
ESTERASE-LIPASE	173.140	ESTERASE-LIPASE DERIVED FROM MUCOR MIEHEI Enzyme Preparations And Microorganisms; Secondary Direct Food Additives Permitted In Food For Human Consumption
ESTERS	101.83	HEALTH CLAIMS: PLANT STEROL/STANOL ESTERS AND RISK OF CORONARY HEART DISEASE Specific Requirements For Health Claims; Food Labeling
ESTERS	107.83	HEALTH CLAIMS: PLANT STEROL/STANOL ESTERS AND RISK OF CORONARY HEART DISEASE Food Labeling
ESTERS	172.225	METHYL AND ETHYL ESTERS OF FATTY ACIDS PRODUCED FROM EDIBLE FATS AND OILS Coatings, Films And Related Substances; Food Additives Permitted For Direct Addition To Food For Human Consumption
ESTERS	172.848	LACTYLIC ESTERS OF FATTY ACIDS Multipurpose Additives; Food Additives Permitted For Direct Addition To Food For Human Consumption
ESTERS	172.850	LACTYLATED FATTY ACID ESTERS OF GLYCEROL AND PROPYLENE GLYCOL Multipurpose Additives; Food Additives Permitted For Direct Addition To Food For Human Consumption
ESTERS	172.852	GLYCERYL-LACTO ESTERS OF FATTY ACIDS Multipurpose Additives, Food Additives Permitted For Direct Addition To Food For Human Consumption
ESTERS	172.854	POLYGLYCEROL ESTERS OF FATTY ACIDS Multipurpose Additives; Food Additives Permitted For Direct Addition To Food For Human Consumption
ESTERS	172.859	SUCROSE FATTY ACID ESTERS Multipurpose Additives; Food Additives Permitted For Direct Addition To Food For Human Consumption
ESTERS	175.260	PARTIAL PHOSPHORIC ACID ESTERS OF POLYESTER RESINS Substances For Use As Components Of Coatings; Indirect Food Additives
ESTERS	178.3450	ESTERS OF STEARIC AND PALMITIC ACIDS Certain Adjuvants And Production Aids; Indirect Food Additives: Adjuvants, Production Aids, And Sanitizers
ESTERS	178.3770	POLYHYDRIC ALCOHOL ESTERS OF OXIDATIVELY REFINED (GERSTHOFEN PROCESS) MONTAN WAX ACIDS Certain Adjuvants And Production Aids; Indirect Food Additives: Adjuvants, Production Aids, And Sanitizers

ESTERS	178.3780	POLYHYDRIC ALCOHOL ESTERS OF LONG CHAIN MONOBASIC ACIDS Certain Adjuvants And Production Aids; Indirect Food Additives: Adjuvants, Production Aids, And Sanitizers
ESTERS	184.110	DIACETYL TARTARIC ACID ESTERS OF MONO- AND DIGLYCERIDES Listing Of Specific Substances Affirmed As GRAS; Direct Food Substances Affirmed As Generally Recognized As Safe
ESTERS	556.240	ESTRADIOL BENZOATE AND RELATED ESTERS Specific Tolerances For Residues Of New Animal Drugs; Tolerances For Residues Of New Animal Drugs In Food
ESTERS	573.640	METHYL ESTERS OF HIGHER FATTY ACIDS Food Additive Listing; Food Additives Permitted In Feed And Drinking Water Of Animals
ESTERS	582.4101	DIACETYL TARTARIC ACID ESTERS OF MONO- AND DIGLYCERIDES OF EDIBLE FATS OR OILS, OR EDIBLE FAT-FORMING FATTY ACIDS Emulsifying Agents; Substances Generally Recognized As Safe
ESTHESIOMETER	882.1500	ESTHESIOMETER Neurological Diagnostic Devices; Neurological Devices
ESTRADIOL	201.313	ESTRADIOL LABELING Specific Labeling Requirements For Specific Drug Products; Labeling
ESTRADIOL	522.840	ESTRADIOL Implantation Or Injectable Dosage Form New Animal Drugs
ESTRADIOL	522.2477	TRENBOLONE ACETATE AND ESTRADIOL IN COMBINATION Implantation Or Injectable Dosage Form New Animal Drugs
ESTRADIOL	862.1260	ESTRADIOL TEST SYSTEM Clinical Chemistry Test Systems; Clinical Chemistry And Clinical Toxicology Devices
ESTRADIOL BENZOATE	522.842	ESTRADIOL BENZOATE AND TESTOSTERONE PROPIONATE IN COMBINATION Implantation Or Injectable Dosage Form New Animal Drugs
ESTRADIOL BENZOATE	522.1940	PROGESTERONE AND ESTRADIOL BENZOATE IN COMBINATION Implantation Or Injectable Dosage Form New Animal Drugs
ESTRADIOL BENZOATE	522.2478	TRENBOLONE ACETATE AND ESTRADIOL BENZOATE Implantation Or Injectable Dosage Form New Animal Drugs
ESTRADIOL BENZOATE	556.240	ESTRADIOL BENZOATE AND RELATED ESTERS Specific Tolerances For Residues Of New Animal Drugs; Tolerances For Residues Of New Animal Drugs In Food
ESTRADIOL VALERATE	522.850	ESTRADIOL VALERATE AND NORGESTOMET IN COMBINATION Implantation Or Injectable Dosage Form New Animal Drugs
ESTRIOL	862.1265	ESTRIOL TEST SYSTEM Clinical Chemistry Test Systems; Clinical Chemistry And Clinical Toxicology Devices
ESTROGENIC HORMONE PREPARATIONS	201.301	NOTICE TO MANUFACTURERS, PACKERS, AND DISTRIBUTORS OF ESTROGENIC HORMONE PREPARATIONS Specific Labeling Requirements For Specific Drug Products; Labeling
ESTROGENS	310.515	PATIENT PACKAGE INSERTS FOR ESTROGENS Requirements For Specific New Drugs Or Devices; New Drugs
ESTROGENS	862.1270	ESTROGENS (TOTAL, IN PREGNANCY) TEST SYSTEM Clinical Chemistry Test Systems; Clinical Chemistry And Clinical Toxicology Devices
ESTROGENS	862.1275	ESTROGENS (TOTAL, NONPREGNANCY) TEST SYSTEM Clinical Chemistry Test Systems; Clinical Chemistry And Clinical Toxicology Devices
ESTRONE	862.1280	ESTRONE TEST SYSTEM Clinical Chemistry Test Systems; Clinical Chemistry And Clinical Toxicology Devices
ETHANEDISULFONATE	520.310	CARAMIPHEN ETHANEDISULFONATE AND AMMONIUM CHLORIDE TABLETS Oral Dosage Form New Animal Drugs
ETHANEDIYLOXYCARBONYL	177.1637	POLY(OXY-1,2-ETHANEDIYLOXYCARBONYL-2,6-NAPHTHALENEDIYLCARBONYL) RESINS Substances For Use As Basic Components Of Single And Repeated Use Food Contact Surfaces; Indirect Food Additives: Polymers
ETHER	177.1970	VINYL CHLORIDE-LAURYL VINYL ETHER COPOLYMERS Substances For Use As Basic Components Of Single And Repeated Use Food Contact Surfaces; Indirect Food Additives: Polymers
ETHER BLOCKS	520.1846	POLYOXYETHYLENE (23) LAURYL ETHER BLOCKS Oral Dosage Form New Animal Drugs
ETHER HOOK	868.5420	ETHER HOOK Therapeutic Devices; Anesthesiology Devices
ETHICS	19.6	CODE OF ETHICS FOR GOVERNMENT SERVICE General Provisions; Standards Of Conduct And Conflicts Of Interest
ETHOPABATE	556.260	ETHOPABATE Tolerances For Residues Of New Animal Drugs In Food
ETHOPABATE	558.58	AMPROLIUM AND ETHOPABATE General Provisions; New Animal Drugs For Use In Animal Feeds

ETHOSUXIMIDE	862.3380	ETHOSUXIMIDE TEST SYSTEM Clinical Toxicology Test Systems; Clinical Chemistry And Toxicology Devices
ETHOXY-OXOBENZO	73.3123	6-ETHOXY-2-(6-ETHOXY-3-OXOBENZO[B]THIEN-2(3H)-YLIDENE) BENZO[B]THIOPEN-3 (2H)-ONE Medical Devices; Listing Of Color Additives Exempt From Certification
ETHOXYLATED MONO-DIGLYCERIDES	172.834	ETHOXYLATED MONO- AND DIGLYCERIDES Multipurpose Additives; Food Additives Permitted For Direct Addition To Food For Human Consumption
ETHOXYQUIN	172.140	ETHOXYQUIN Food Preservatives; Food Additives Permitted For Direct Addition To Food For Human Consumption
ETHOXYQUIN	573.380	ETHOXYQUIN IN ANIMAL FEEDS Food Additive Listing; Food Additives Permitted In Feed And Drinking Water Of Animals
ETHOXYQUIN	573.400	ETHOXYQUIN IN CERTAIN DEHYDRATED FORAGE CROPS Food Additive Listing; Food Additives Permitted In Feed And Drinking Water Of Animals
ETHYL ACETATE	173.228	ETHYL ACETATE Solvents, Lubricants, Release Agents And Related Substances; Secondary Direct Food Additives Permitted In Food For Human Consumption
ETHYL ACRYLATE	177.1320	ETHYLENE-ETHYL ACRYLATE COPOLYMERS Substances For Use As Basic Components Of Single And Repeated Use Food Contact Surfaces; Indirect Food Additives: Polymers
ETHYL ALCOHOL	184.1293	ETHYL ALCOHOL Listing Of Specific Substances Affirmed As GRAS; Direct Food Substances Affirmed As Generally Recognized As Safe
ETHYL ALCOHOL	584.200	ETHYL ALCOHOL CONTAINING ETHYL ACETATE Listing Of Substances Affirmed As GRAS; Food Substances Affirmed As GRAS In Feed And Drinking Water Of Animals
ETHYL CELLULOS	172.868	ETHYL CELLULOSE Multipurpose Additives; Food Additives Permitted For Direct Addition To Food For Human Consumption
ETHYL CELLULOSE	172.872	METHYL ETHYL CELLULOSE Multipurpose Additives; Food Additives Permitted For Direct Addition To Food For Human Consumption
ETHYL CELLULOSE	573.420	ETHYL CELLULOSE Food Additive Listing; Food Additives Permitted In Feed And Drinking Water Of Animals
ETHYL ESTERS	172.225	METHYL AND ETHYL ESTERS OF FATTY ACIDS PRODUCED FROM EDIBLE FATS AND OILS Coatings, Films And Related Substances; Food Additives Permitted For Direct Addition To Food For Human Consumption
ETHYL FORMATE	184.1295	ETHYL FORMATE Listing Of Specific Substances Affirmed As GRAS; Direct Food Substances Affirmed As Generally Recognized As Safe
ETHYLENE	177.1310	ETHYLENE-ACRYLIC ACID COPOLYMERS Substances For Use As Basic Components Of Single And Repeated Use Food Contact Surfaces; Indirect Food Additives: Polymers
ETHYLENE	177.1312	ETHYLENE-CARBON MONOXIDE COPOLYMERS Substances For Use As Basic Components Of Single And Repeated Use Food Contact Surfaces; Indirect Food Additives: Polymers
ETHYLENE	177.1315	ETHYLENE-1,4 CYCLOHEXYLENE DIMETHYLENE TEREPHTHALATE COPOLYMERS Substances For Use As Basic Components Of Single And Repeated Use Food Contact Surfaces; Indirect Food Additives: Polymers
ETHYLENE	177.1320	ETHYLENE-ETHYL ACRYLATE COPOLYMERS Substances For Use As Basic Components Of Single And Repeated Use Food Contact Surfaces; Indirect Food Additives: Polymers
ETHYLENE	177.1340	ETHYLENE-METHYL ACRYLATE COPOLYMER RESINS Substances For Use As Basic Components Of Single And Repeated Use Food Contact Surfaces; Indirect Food Additives: Polymers
ETHYLENE	177.1345	ETHYLENE/1,3-PHENYLENE OXYETHYLENE ISOPHTHALATE/TEREPHTHALATE COPOLYMER Substances For Use As Basic Components Of Single And Repeated Use Food Contact Surfaces; Indirect Food Additives: Polymers
ETHYLENE	177.1350	ETHYLENE-VINYL ACETATE COPOLYMERS Substances For Use As Basic Components Of Single And Repeated Use Food Contact Surfaces; Indirect Food Additives: Polymers
ETHYLENE	177.1360	ETHYLENE-VINYL ACETATE-VINYL ALCOHOL COPOLYMERS Substances For Use As Basic Components Of Single And Repeated Use Food Contact Surfaces; Indirect Food Additives: Polymers
ETHYLEN	177.1570	POLY-1-BUTENE RESINS AND BUTENE/ETHYLENE COPOLYMERS Substances For Use As Basic Components Of Single And Repeated Use Food Contact Surfaces; Indirect Food Additives: Polymers

ETHYLENE	177.1950	VINYL CHLORIDE-ETHYLENE COPOLYMERS Substances For Use As Basic Components Of Single And Repeated Use Food Contact Surfaces; Indirect Food Additives: Polymers
ETHYLENE	177.2210	ETHYLENE POLYMER, CHLOROSULFONATED Substances For Use Only As Components Of Articles Intended For Repeated Use; Indirect Food Additives: Polymers
ETHYLENE DICHLORIDE	173.230	ETHYLENE DICHLORIDE Solvents, Lubricants, Release Agents And Related Substances; Secondary Direct Food Additives Permitted In Food For Human Consumption
ETHYLENE DICHLORIDE	573.440	ETHYLENE DICHLORIDE Food Additive Listing; Food Additives Permitted In Feed And Drinking Water Of Animals
ETHYLENE OXIDE	172.770	ETHYLENE OXIDE POLYMER Other Specific Usage Additives; Food Additives Permitted For Direct Addition To Food For Human Consumption
ETHYLENE OXIDE	172.808	COPOLYMER CONDENSATES OF ETHYLENE OXIDE AND PROPYLENE OXIDE Multipurpose Additives; Food Additives Permitted For Direct Addition To Food For Human Consumption
ETHYLENE OXIDE	872.3410	ETHYLENE OXIDE HOMOPOLYMER AND/OR CARBOXYMETHYLCELLULOSE SODIUM DENTURE ADHESIVE Prosthetic Devices; Dental Devices
ETHYLENE OXIDE	872.3450	ETHYLENE OXIDE HOMOPOLYMER AND/OR KARAYA DENTURE ADHESIVE Prosthetic Devices; Dental Devices
ETHYLENE OXIDE	880.6100	ETHYLENE OXIDE GAS AERATOR CABINET General Hospital And Personal Use Miscellaneous Devices; General Hospital And Personal Use Devices
ETHYLENE OXIDE	880.6860	ETHYLENE OXIDE GAS STERILIZER General Hospital And Personal Use Miscellaneous Devices; General Hospital And Personal Use Devices
ETHYLENE TEREPHTHALATE	878.5000	NONABSORABLE POLY(ETHYLENE TEREPHTHALATE) SURGICAL SUTURE Surgical Devices; General And Plastic Surgery Devices
ETHYLENEDIAMINE	556.270	ETHYLENEDIAMINE Specific Tolerances For Residues Of New Animal Drugs; Tolerances For Residues Of New Animal Drugs In Food
ETHYLHEXOATE	178.3940	TETRAETHYLENE GLYCOL DI-(2-ETHYLHEXOATE) Certain Adjuvants And Production Aids; Indirect Food Additives: Adjuvants, Production Aids, And Sanitizers
ETHYLISOBUTRAZINE HCL	520.863	ETHYLISOBUTRAZINE HYDROCHLORIDE TABLETS Oral Dosage Form New Animal Drugs
ETHYLISOBUTRAZINE HCL	522.863	ETHYLISOBUTRAZINE HYDROCHLORIDE INJECTION Implantation Or Injectable Dosage Form New Animal Drugs
ETIOCHOLANOLONE	862.1285	ETIOCHOLANOLONE TEST SYSTEM Clinical Chemistry Test Systems; Clinical Chemistry And Clinical Toxicology Devices
ETODOLAC	520.870	ETODOLAC Oral Dosage Form New Animal Drugs
ETORPHINE HYDROCHLORIDE	522.883	ETORPHINE HYDROCHLORIDE INJECTION Implantation Or Injectable Dosage Form New Animal Drugs
EUGLOBULIN	864.7275	EUGLOBULIN LYSIS TIME TESTS Hematology Kits And Packages; Hematology And Pathology Devices
EUTHANASIA	522.900	EUTHANASIA SOLUTION Implantation Or Injectable Dosage Form New Animal Drugs
EUTHYSCOPE	886.1250	EUTHYSCOPE Diagnostic Devices; Ophthalmic Devices
EVACUATION	882.5850	IMPLANTED SPINAL CORD STIMULATOR FOR BLADDER EVACUATION Neurological Therapeutic Devices; Neurological Devices
EVACUATOR	876.4370	GASTROENTEROLOGY-UROLOGY EVACUATOR Surgical Devices; Gastroenterology-urology Devices
EVACUATOR	888.4220	CEMENT MONOMER VAPOR EVACUATOR Surgical Devices; Orthopedic Devices
EVALUATING	70.42	CRITERIA FOR EVALUATING THE SAFETY OF COLOR ADDITIVES Safely Evaluation; Color Additives
EVALUATING	170.20	GENERAL PRINCIPLES FOR EVALUATING THE SAFETY OF FOOD ADDITIVES Food Additive Safety; Food Additives
EVALUATING	570.20	GENERAL PRINCIPLES FOR EVALUATING THE SAFETY OF FOOD ADDITIVES Food Additive Safely; Food Additives
EVALUATION	54.5	AGENCY EVALUATION OF FINANCIAL INTERESTS Financial Disclosure By Clinical Investigators
EVALUATION	312.87	ACTIVE MONITORING OF CONDUCT AND EVALUATION OF CLINICAL TRIALS Drugs Intended To Treat Life-threatening And Severely-debilitating Illnesses; Investigational New Drug Application
EVALUATION	315.5	EVALUATION OF EFFECTIVENESS Diagnostic Radiopharmaceuticals

EVALUATION	315.6	EVALUATION OF SAFETY Diagnostic Radiopharmaceuticals
EVALUATION	514.100	EVALUATION AND COMMENT ON APPLICATIONS Administrative Actions On Applications; New Animal Drug Applications
EVALUATION	601.34	EVALUATON OF EFFECTIVENESS Diagnostic Radiopharmaceuticals; Licensing
EVALUATION	900.5	EVALUATION Accreditation; Mammography
EVALUATION	900.23	EVALUATION States As Certifiers; Mammography
EVALUATION, BIOLOGICS	5.201	REDELEGATION OF THE CENTER FOR BIOLOGICS EVALUATION AND RESEARCH DIRECTOR'S PROGRAM AUTHORITIES Biologics; Redelegations Of Authority; Delegations Of Authority And Organization
EVALUATION, BREAST	892.1990	TRANSILLUMINATOR FOR BREAST EVALUATION Diagnostic Devices; Radiology Devices
EVALUATION, HAZARD	7.41	HEALTH HAZARD EVALUATION AND RECALL CLASSIFICATION Recalls - Guidance On Policy, Procedures & Industry Responsibilities; Enforcement Policy
EVALUATION, PRODUCT	106.30	FINISHED PRODUCT EVALUATION Quality Control Procedures For Assuring Nutrient Content Of Infant Formulas; Infant Formula Quality Control Procedures
EVALUATION REPORTS	26.41	EXCHANGE AND ENDORSEMENT OF QUALITY SYSTEM EVALUATION REPORTS Specific Sector Provisions For Medical Devices; Mutual Recognition Of Pharmaceutical GMP Reports, Medical Device Quality System Audit Reports
EVALUATION REPORTS	26.42	EXCHANGE AND ENDORSEMENT OF PRODUCT EVALUATION REPORTS Specific Sector Provisions For Medical Devices; Mutual Recognition Of Pharmaceutical GMP Reports, Medical Device Quality System Audit Reports
EVALUATION REPORTS	26.43	TRANSMISSION OF QUALITY SYSTEM EVALUATION REPORTS Specific Sector Provisions For Medical Devices; Mutual Recognition Of Pharmaceutical GMP Reports, Medical Device Quality System Reports
EVALUATION REPORTS	26.44	TRANSMISSION OF PRODUCT EVALUATION REPORTS Specific Sector Provisions For Medical Devices; Mutual Recognition Of Pharmaceutical GMP Reports, Medical Device Quality System Audit Reports
EVALUATION SYSTEM	890.1925	ISOKINETIC TESTING AND EVALUATION SYSTEM Physical Medicine Diagnostic Devices; Physical Medicine Devices
EVALUATIONS	201.200	DISCLOSURE OF DRUG EFFICACY STUDY EVALUATIONS IN LABELING AND ADVERTISING Labeling Claims For Drugs In Drug Efficacy Studies; Labeling
EVAPORATED MILK	131.130	EVAPORATED MILK Requirements For Specific Standardized Milk And Cream; Milk And Cream
EVENT, ADVERSE	803.30	INDIVIDUAL ADVERSE EVENT REPORTS; USER FACILITIES User Facility Reporting Requirements; Medical Device Reporting
EVENT, ADVERSE	803.32	INDIVIDUAL ADVERSE EVENT REPORT DATA ELEMENTS User Facility Reporting Requirements; Medical Device Reporting
EVENT, ADVERSE	803.40	INDIVIDUAL ADVERSE EVENT REPORTING REQUIREMENTS; IMPORTERS Importer Reporting Requirement; Medical Device Reporting
EVENT, ADVERSE	803.42	INDIVIDUAL ADVERSE EVENT REPORT DATA ELEMENTS Importer Reporting Requirement; Medical Device Reporting
EVENT, ADVERSE	803.50	INDIVIDUAL ADVERSE EVENT REPORTS; MANUFACTURERS Manufacturer Reporting Requirements; Medical Device Reporting
EVENT, ADVERSE	803.52	INDIVIDUAL ADVERSE EVENT REPORT DATA ELEMENTS Manufacturer Reporting Requirements; Medical Device Reporting
EVIDENCE	5.403	AUTHORIZATION TO USE ALTERNATIVE EVIDENCE FOR DETERMINATION OF THE EFFECTIVENESS OF MEDICAL DEVICES Medical Devices And Radiological Health; Redelegations Of Authority; Delegations Of Authority And Organization
EVIDENCE	5.408	DETERMINATIONS CONCERNING THE TYPE OF VALID SCIENTIFIC EVIDENCE SUBMITTED IN A PREMARKET APPROVAL APPLICATION Medical Devices And Radiological Health; Redelegations Of Authority; Delegations Of Authority And Organization
EVIDENCE	12.94	RECEIPT OF EVIDENCE Hearing Procedures; Formal Evidentiary Public Hearing
EVIDENCE	17.39	EVIDENCE Civil Money Penalties Hearings
EVIDENCE	320.22	CRITERIA FOR WAIVER OF EVIDENCE OF IN VIVO BIOAVAILABILITY Procedures For Determining The Bioavailability Of Drug Products; Bioavailability And Bioequivalence Requirements
EVIDENCE	320.24	TYPES OF EVIDENCE TO ESTABLISH BIOAVAILABILITY OR BIOEQUIVALENCE Procedures For Determining The Bioavailability Of Drug Products; Bioavailability And Bioequivalence Requirements

EVIDENCE	320.33	CRITERIA AND EVIDENCE TO ASSESS ACTUAL OR POTENTIAL BIOEQUIVALENCE PROBLEMS Procedures For Determining The Bioavailability Of Drug Products; Bioavailability And Bioequivalence Requirements
EVIDENCE	1316.59	SUBMISSION AND RECEIPT OF EVIDENCE Administrative Hearings; Administrative Functions, Practices, And Procedures
EVIDENCE, CLINICAL	320.28	CORRELATION OF BIOAVAILABILITY WITH AN ACUTE PHARMACOLOGICAL EFFECT OR CLINICAL EVIDENCE Procedures For Determining The Bioavailability Of Drug Products; Bioavailability And Bioequivalence Requirements
EVIDENCE, DOCUMENTARY	1316.57	SUBMISSION OF DOCUMENTARY EVIDENCE AND AFFIDAVITS AND IDENTIFICATION OF WITNESS SUBSEQUENT TO PREHEARING CONFERENCE Administrative Hearings; Administrative Functions, Practices, And Procedures
EVIDENCE, SUBSTANTIAL	514.4	SUBSTANTIAL EVIDENCE New Animal Drug Applications
EVIDENTIARY	10.50	PROMULGATION OF REGULATIONS AND ORDERS AFTER AN OPPORTUNITY FOR A FORMAL EVIDENTIARY PUBLIC HEARING General Administrative Procedures; Administrative Practices And Procedures
EVOKED RESPONSE	882.1870	EVOKED RESPONSE ELECTRICAL STIMULATOR Neurological Diagnostic Devices; Neurological Devices
EVOKED RESPONSE	882.1880	EVOKED RESPONSE MECHANICAL STIMULATOR Neurological Diagnostic Devices; Neurological Devices
EVOKED RESPONSE	882.1890	EVOKED RESPONSE PHOTIC STIMULATOR Neurological Diagnostic Devices; Neurological Devices
EVOKED RESPONSE	882.1900	EVOKED RESPONSE AUDITORY STIMULATOR Neurological Diagnostic Devices; Neurological Devices
EX PARTE COMMUNICATIONS	10.55	SEPARATION OF FUNCTIONS; EX PARTE COMMUNICATIONS General Administrative Procedures; Administrative Practices And Procedures
EX PARTE COMMUNICATIONS	13.15	SEPARATION OF FUNCTIONS; EX PARTE COMMUNICATIONS; ADMINISTRATIVE SUPPORT General Provisions; Public Hearing Before A Public Board Of Inquiry
EX PARTE COMMUNICATIONS	1316.51	CONDUCT OF HEARING AND PARTIES; EX PARTE COMMUNICATIONS Administrative Hearings; Administrative Functions, Practices, And Procedures
EX PARTE CONTACTS	17.20	EX PARTE CONTACTS Civil Money Penalties Hearings
EX VIVO	876.5885	TISSUE CULTURE MEDIA FOR HUMAN EX VIVO TISSUE AND CELL CULTURE PROCESSING APPLICATIONS Therapeutic Devices; Gastroenterology-urology Devices
EXAMINATION	2.10	EXAMINATION AND INVESTIGATION SAMPLES General Provisions; General Administrative Rulings And Decisions
EXAMINATION	12.105	EXAMINATION OF RECORD Administrative Record; Formal Evidentiary Public Hearing
EXAMINATION	13.45	EXAMINATION OF ADMINISTRATIVE RECORD Records Of A Hearing Before A Board; Public Hearing Before A Public Board Of Inquiry
EXAMINATION	14.75	EXAMINATION OF ADMINISTRATIVE RECORD AND OTHER ADVISORY COMMITTEE RECORDS Records Of Meetings And Hearings Before Advisory Committees; Public Hearing Before A Public Advisory Committee
EXAMINATION	15.45	EXAMINATION OF ADMINISTRATIVE RECORD Records Of A Public Hearing Before The Commissioner; Public Hearing Before The Commissioner
EXAMINATION	16.85	EXAMINATION OF ADMINISTRATIVE RECORD Administrative Record And Decision; Regulatory Hearing Before The Food And Drug Administration
EXAMINATION	211.122	MATERIALS EXAMINATION AND USAGE CRITERIA Packaging And Labeling Control; Current Good Manufacturing Practice For Finished Pharmaceuticals
EXAMINATION	800.20	PATIENT EXAMINATION GLOVES AND SURGEON'S GLOVES; SAMPLE PLANS AND TEST METHOD FOR LEAKAGE DEFECTS; ADULTERATION Requirements For Specific Medical Devices; General
EXAMINATION	874.5300	EAR, NOSE AND THROAT EXAMINATION AND TREATMENT UNIT Therapeutic Devices; Ear, Nose, And Throat Devices
EXAMINATION	880.6250	PATIENT EXAMINATION GLOVE General Hospital And Personal Use Miscellaneous Devices; General Hospital And Personal Use Devices
EXAMINATIO	880.6265	EXAMINATION GOWN General Hospital And Personal Use Miscellaneous Devices; General Hospital And Personal Use Devices
EXAMINATION	880.6320	AC-POWERED MEDICAL EXAMINATION LIGHT General Hospital And Personal Use Miscellaneous Devices; General Hospital And Personal Use Devices

EXAMINATION	880.6350	BATTERY-POWERED MEDICAL EXAMINATION LIGHT General Hospital And Personal Use Miscellaneous Devices; General Hospital And Personal Use Devices
EXAMINATION	1210.10	AVAILABILITY FOR EXAMINATION AND INSPECTION Inspection And Testing; Regulations Under The Federal Import Milk Act
EXAMINATION	1210.12	PHYSICAL EXAMINATION OF COWS Inspection And Testing; Regulations Under The Federal Import Milk Act
EXAMINATIONS	1000.60	RECOMMENDATION ON ADMINISTRATIVELY REQUIRED DENTAL X-RAY EXAMINATIONS Radiation Protection Recommendations; General
EXCEPTION	50.23	EXCEPTION FROM GENERAL REQUIREMENTS Informed Consent Of Human Subjects; Protection Of Human Subjects
EXCEPTION	50.24	EXCEPTION FROM INFORMED CONSENT REQUIREMENTS FOR EMERGENCY RESEARCH Informed Consent Of Human Subjects; Protection Of Human Subjects
EXCEPTION	1301.21	EXCEPTION FROM FEES Exceptions To Registration And Fees; Registration Of Manufacturers, Distributors, And Dispensers Of Controlled Substances
EXCEPTION	1404.215	EXCEPTION PROVISION Governmentwide Debarment And Suspension And Requirements For Drug-Free Workplace
EXCEPTION	1404.625	EXCEPTION PROVISION Governmentwide Debarment And Suspension And Requirements For Drug-Free Workplace
EXCEPTIONS	1271.15	ARE THERE ANY EXCEPTIONS FROM THE REQUIREMENTS OF THIS PART? General Provisions; Human Cells, Tissues, And Cellular And Tissue-Based Products
EXCEPTIONS	1271.20	IF MY HCT/P'S DO NOT MEET THE CRITERIA IN SECTION 1271.10, AND I DO NOT QUALIFY FOR ANY OF THE EXCEPTIONS IN SECTION 1271.15, WHAT REGULATIONS APPLY? General Provisions; Human Cells, Tissues, And Cellular And Tissue-Based Products
EXCEPTIONS	1302.03	SYMBOL REQUIRED; EXCEPTIONS Labeling And Packaging Requirements For Controlled Substances
EXCEPTIONS	1307.03	EXCEPTIONS TO REGULATIONS General Information; Miscellaneous
EXCEPTIONS	1316.07	REQUIREMENT FOR ADMINISTRATIVE INSPECTION WARRANT; EXCEPTIONS Administrative Inspections; Administrative Functions, Practices, And Procedures
EXCEPTIONS	1316.61	EXCEPTIONS TO RULINGS Administrative Hearings; Administrative Functions, Practices, And Procedures
EXCEPTIONS	1316.66	EXCEPTIONS Administrative Hearings; Administrative Functions, Practices, And Procedures
EXCEPTIONS	1403.6	ADDITIONS AND EXCEPTIONS Post-award Requirements; Uniform Administrative Requirements For Grants And Coop. Agreements To State & Local Gov.
EXCHANGE	17.25	EXCHANGE OF WITNESS LISTS, WITNESS STATEMENTS, AND EXHIBITS Civil Money Penalties Hearings
EXCHANGE	26.41	EXCHANGE AND ENDORSEMENT OF QUALITY SYSTEM EVALUATION REPORTS Specific Sector Provisions For Medical Devices; Mutual Recognition Of Pharmaceutical GMP Reports, Medical Device Quality System Audit Reports
EXCHANGE	26.42	EXCHANGE AND ENDORSEMENT OF PRODUCT EVALUATION REPORTS Specific Sector Provisions For Medical Devices; Mutual Recognition Of Pharmaceutical GMP Reports, Medical Device Quality System Audit Reports
EXCHANGE	26.71	EXCHANGE OF INFORMATION Framework Provisions; Mutual Recognition Of Pharmaceutical GMP Reports, Medical Device Quality System Audit Reports
EXCHANGER	870.4240	CARDIOPULMONARY BYPASS HEAT EXCHANGER Cardiovascular Surgical Devices; Cardiovascular Devices
EXCLUDED	1308.22	EXCLUDED SUBSTANCES Excluded Nonnarcotic Substances; Schedules Of Controlled Substances
EXCLUDED	1308.26	EXCLUDED VETERINARY ANABOLIC STEROID IMPLANT PRODUCTS Schedules Of Controlled Substances
EXCLUDED	1310.08	EXCLUDED TRANSACTIONS Records and Reports Of Listed Chemicals and Certain machines
EXCLUSION	1308.21	APPLICATION FOR EXCLUSION OF A NONNARCOTIC SUBSTANCE Schedules Of Controlled Substances
EXCLUSION	1308.25	EXCLUSION OF A VETERINARY ANABOLIC STEROID IMPLANT PRODUCT; APPLICATION Schedules Of Controlled Substances
EXCLUSION	1316.04	EXCLUSION FROM INSPECTION Administrative Functions, Practices, And Procedures

EXCLUSION	1404.210	VOLUNTARY EXCLUSION Governmentwide Debarment And Suspension And Requirements For Drug-Free Workplace (Grants)
EXCLUSION	1404.315	SETTLEMENT AND VOLUNTARY EXCLUSION Governmentwide Debarment And Suspension And Requirements For Drug-Free Workplace (Grants)
EXCLUSION, COMPETITIVE	529.469	COMPETITIVE EXCLUSION CULTURE Certain Other Dosage Form New Animal Drugs
EXCLUSIONS	110.19	EXCLUSIONS General Provisions; Current Good Manufacturing Practice In Manufacturing, Packing, Or Holding Human Food
EXCLUSIONS	203.22	EXCLUSIONS Prescription Drug Marketing
EXCLUSIVE APPROVAL	316.31	SCOPE OF ORPHAN DRUG EXCLUSIVE APPROVAL Orphan Drugs
EXCLUSIVE APPROVAL	316.34	FDA RECOGNITION OF EXCLUSIVE APPROVAL Orphan Drugs
EXCLUSIVITY	314.108	NEW DRUG PRODUCT EXCLUSIVITY Applications For FDA Approval To Market A New Drug
EXECUTE	1305.04	PERSONS ENTITLED TO OBTAIN AND EXECUTE ORDER FORMS Order Forms
EXECUTING	1305.06	PROCEDURE FOR EXECUTING ORDER FORMS Order Forms
EXECUTION	1316.11	EXECUTION OF WARRANTS Administrative Inspections; Administrative Functions, Practices, And Procedures
EXEMPT	21.61	EXEMPT SYSTEMS Exemptions; Protection Of Privacy
EXEMPT	21.65	ACCESS TO RECORDS IN EXEMPT SYSTEMS Exemptions; Protection Of Privacy
EXEMPT	73.1	DILUENTS IN COLOR ADDITIVE MIXTURES FOR FOOD USE EXEMPT FROM CERTIFICATION Foods; Listing Of Color Additives Exempt From Certification
EXEMPT	73.1001	DILUENTS IN COLOR ADDITIVE MIXTURES FOR DRUG USE EXEMPT FROM CERTIFICATION Drugs; Listing Of Color Additives Exempt From Certification
EXEMPT	501.105	DECLARATION OF NET QUANTITY OF CONTENTS WHEN EXEMPT Exemptions From Animal Food Labeling Requirements; Animal Food Labeling
EXEMPT	511.1	NEW ANIMAL DRUGS FOR INVESTIGATIONAL USE EXEMPT FROM SECTION 512(A) OF THE ACT New Animal Drugs For Investigational Use
EXEMPT	1308.34	EXEMPT ANABOLIC STEROID PRODUCTS Schedules Of Controlled Substances
EXEMPT	1310.15	EXEMPT DRUG PRODUCTS CONTAINING EPHEDRINE AND ANOTHER ACTIVE MEDICINAL INGREDIENT Records And Reports Of Listed Chemicals And Certain Machines
EXEMPTED	369.3	WARNINGS REQUIRED ON DRUGS EXEMPTED FROM PRESCRIPTION-DISPENSING REQUIREMENTS OF SECTION 503(B)(1)(C) Definitions And Interpretations; Interpretative Statements Re: Warnings On Drugs And Devices For OTC Sale
EXEMPTED	1308.32	EXEMPTED PRESCRIPTION PRODUCTS Schedules Of Controlled Substances
EXEMPTION	5.601	EXEMPTION OF ELECTRONIC PRODUCTS FROM PERFORMANCE STANDARDS AND PROHIBITED ACTS Radiation Control; Redelegations Of Authority; Delegations Of Authority And Organization
EXEMPTION	71.27	LISTING AND EXEMPTION FROM CERTIFICATION ON THE COMMISSIONER'S INITIATIVE Administrative Action On Petitions; Color Additive Petitions
EXEMPTION	71.37	EXEMPTION OF COLOR ADDITIVES FOR INVESTIGATIONAL USE Administrative Action On Petitions; Color Additive Petitions
EXEMPTION	80.35	COLOR ADDITIVE MIXTURES; CERTIFICATION AND EXEMPTION FROM CERTIFICATION Certification Procedures; Color Additive Certification
EXEMPTION	99.205	APPLICATION FOR EXEMPTION FROM THE REQUIREMENT TO FILE A SUPPLEMENTAL APPLICATION Dissemination Of Information On Unapproved/New Uses For Marketed Drugs, Biologics, and Devices
EXEMPTION	99.305	EXEMPTION FROM THE REQUIREMENT TO FILE A SUPPLEMENTAL APPLICATION Dissemination Of Information On Unapproved/New Uses For Marketed Drugs, Biologics, and Devices
EXEMPTION	99.403	TERMINATION OF APPROVALS OF APPLICATIONS FOR EXEMPTION Dissemination Of Information On Unapproved/New Uses For Marketed Drugs, Biologics, and Devices

EXEMPTION	100.1	PETITIONS REQUESTING EXEMPTION FROM PREEMPTION FOR STATE OR LOCAL REQUIREMENTS State And Local Requirements; General
EXEMPTION	108.19	ESTABLISHMENT OF REQUIREMENTS FOR EXEMPTION FROM SECTION 404 OF THE ACT General Provisions; Emergency Permit Control
EXEMPTION	170.17	EXEMPTION FOR INVESTIGATIONAL USE AND PROCEDURE FOR OBTAINING AUTHORIZATION TO MARKET EDIBLE PRODUCTS FORM EXPERIMENTAL ANIMALS General Provisions; Food Additives
EXEMPTION	201.129	DRUGS; EXEMPTION FOR RADIOACTIVE DRUGS FOR RESEARCH USE Exemptions From Adequate Directions For Use; Labeling
EXEMPTION	290.2	EXEMPTION FROM PRESCRIPTION REQUIREMENTS General Provisions; Controlled Drugs
EXEMPTION	310.200	PRESCRIPTION-EXEMPTION PROCEDURE New Drugs Exempted From Prescription-dispensing Requirements; New Drugs
EXEMPTION	310.201	EXEMPTION FOR CERTAIN DRUGS LIMITED BY NEW DRUG APPLICATIONS TO PRESCRIPTION SALE New Drugs Exempted From Prescription-dispensing Requirements; New Drugs
EXEMPTION	329.20	EXEMPTION OF CERTAIN HABIT-FORMING DRUGS FROM PRESCRIPTION REQUIREMENTS Exemptions; Habit-forming Drugs
EXEMPTION	500.55	EXEMPTION FROM CERTAIN DRUG-LABELING REQUIREMENTS Animal Drug Labeling Requirements; General
EXEMPTION	570.17	EXEMPTION FOR INVESTIGATIONAL USE AND PROCEDURE FOR OBTAINING AUTHORIZATION TO MARKET EDIBLE PRODUCTS FROM EXPERIMENTAL ANIMALS General Provisions; Food Additives
EXEMPTION	801.110	RETAIL EXEMPTION FOR PRESCRIPTION DEVICES Exemptions From Adequate Directions For Use; Labeling
EXEMPTION	807.85	EXEMPTION FROM PREMARKET NOTIFICATION Premarket Notification Procedures; Establishment Registration And Device Listing For Manufacturers And Initial Importers Of Devices
EXEMPTION	808.35	REVOCATION OF AN EXEMPTION Exemption Procedures; Exemption From Federal Preemption Of State & Local Medical Device Requirements
EXEMPTION	1003.30	APPLICATION FOR EXEMPTION FROM NOTIFICATION REQUIREMENTS Exemptions From Notification Requirements; Notification Of Defects Or Failure To Comply
EXEMPTION	1003.31	GRANTING THE EXEMPTION Exemptions From Notification Requirements; Notification Of Defects Or Failure To Comply
EXEMPTION	1230.16	EXEMPTION FROM LABELING DIRECTIONS FOR TREATMENT Labeling; Regulations Under The Federal Caustic Poison Act
EXEMPTION	1301.22	EXEMPTION OF AGENTS AND EMPLOYEES; AFFILIATED PRACTITIONERS Exceptions To Registration And Fees; Registration Of Manufacturers, Distributors, And Dispensers Of Controlled Substances
EXEMPTION	1301.23	EXEMPTION OF CERTAIN MILITARY AND OTHER PERSONNEL Exceptions To Registration And Fees; Registration Of Manufacturers, Distributors, And Dispensers Of Controlled Substances
EXEMPTION	1301.26	EXEMPTION FROM IMPORT OR EXPORT REQUIREMENTS FOR PERSONAL MEDICAL USE Exceptions To Registration And Fees; Registration Of Manufacturers, Distributors, And Dispensers Of Controlled Substances
EXEMPTION	1308.23	EXEMPTION OF CERTAIN CHEMICAL PREPARATIONS; APPLICATION Exempt Chemical Preparations; Schedules Of Controlled Substances
EXEMPTION	1308.24	EXEMPTION CHEMICAL PREPARATIONS Exempt Chemical Preparations; Schedules Of Controlled Substances
EXEMPTION	1308.31	APPLICATION FOR EXEMPTION OF A NONNARCOTIC PRESCRIPTION DRUG Exempted Prescription Products; Schedules Of Controlled Substances
EXEMPTION	1308.33	EXEMPTION OF CERTAIN ANABOLIC STEROID PRODUCTS; APPLICATION Schedules Of Controlled Substances
EXEMPTION	1308.35	EXEMPTION OF CERTAIN CANNABIS PLANT MATERIAL, AND PRODUCTS MADE THEREFROM, THAT CONTAIN TETRAHYDROCANNABINOIDS Schedules Of Controlled Substances
EXEMPTION	1309.24	EXEMPTION OF AGENTS AND EMPLOYEES Registration Of Manufacturers, Distributors, Importers And Exporters Of List I Chemicals
EXEMPTION	1309.25	EXEMPTION OF CERTAIN CONTROLLED SUBSTANCE REGISTRANTS Registration Of Manufacturers, Distributors, Importers And Exporters Of List I Chemicals
EXEMPTION	1309.26	EXEMPTION OF LAW ENFORCEMENT OFFICIALS Registration Of Manufacturers, Distributors, Importers And Exporters Of List I Chemicals

EXEMPTION	1309.27	EXEMPTION OF CERTAIN MANUFACTURERS Registration Of Manufacturers, Distributors, Importers And Exporters Of List I Chemicals
EXEMPTION	1309.28	EXEMPTION OF DISTRIBUTORS OF REGULATED PRESCRIPTION DRUG PRODUCTS Registration Of Manufacturers, Distributors, Importers And Exporters Of List I Chemicals
EXEMPTION	1310.09	TEMPORARY EXEMPTION FROM REGISTRATION Records And Reports Of Listed Chemicals And Certain Machines
EXEMPTION	1310.10	REMOVAL OF THE EXEMPTION OF DRUGS DISTRIBUTED UNDER THE FOOD, DRUG AND COSMETIC ACT Records And Reports Of Listed Chemicals And Certain Machines
EXEMPTION	1310.11	REINSTATEMENT OF THE EXEMPTION FOR DRUG PRODUCTS DISTRIBUTED UNDER THE FOOD, DRUG AND COSMETIC ACT Records And Reports Of Listed Chemicals And Certain Machines
EXEMPTION	1310.14	EXEMPTION OF DRUG PRODUCTS CONTAINING EPHEDRINE AND ANOTHER ACTIVE MEDICINAL INGREDIENT Records And Reports Of Listed Chemicals And Certain Machines
EXEMPTION	1316.22	EXEMPTION Protection Of Researchers And Research Subjects; Administrative Functions, Practices, And Procedures
EXEMPTION	1316.24	EXEMPTION FROM PROSECUTION FOR RESEARCHERS Protection Of Researchers And Research Subjects; Administrative Functions, Practices, And Procedures
EXEMPTIONS	1.23	PROCEDURES FOR REQUESTING VARIATIONS AND EXEMPTIONS FROM REQUIRED LABEL STATEMENTS General Labeling Requirements; General Enforcement Regulations
EXEMPTIONS	1.24	EXEMPTIONS FROM REQUIRED LABEL STATEMENTS General Labeling Requirements; General Enforcement Regulations
EXEMPTIONS	5.101	TERMINATION OF EXEMPTIONS FOR NEW DRUGS FOR INVESTIGATIONAL USE OF HUMAN BEINGS Human Drugs; Redelegations Of Authority; Delegations Of Authority And Organization
EXEMPTIONS	5.401	ISSUANCE OF FEDERAL REGISTER DOCUMENTS PERTAINING TO EXEMPTIONS FROM PREMARKET NOTIFICATION Medical Devices And Radiological Health; Redelegations Of Authority; Delegations Of Authority And Organization
EXEMPTIONS	5.413	APPROVAL, DISAPPROVAL, OR WITHDRAWAL OF APPROVAL OF APPLICATIONS AND ENTERING INTO AGREEMENTS FOR INVESTIGATIONAL DEVICE EXEMPTIONS Medical Devices And Radiological Health; Redelegations Of Authority; Delegations Of Authority And Organization
EXEMPTIONS	5.505	TERMINATION OF EXEMPTIONS FOR NEW DRUGS FOR INVESTIGATIONAL USE IN ANIMALS Animal Drugs; Redelegations Of Authority; Delegations Of Authority And Organization
EXEMPTIONS	20.60	APPLICABILITY OF EXEMPTIONS Exemptions; Public Information
EXEMPTIONS	20.80	APPLICABILITY OF LIMITATIONS ON EXEMPTIONS Limitations On Exemptions; Public Information
EXEMPTIONS	56.104	EXEMPTIONS FROM IRB REQUIREMENT General Provisions; Institutional Review Boards
EXEMPTIONS	101.100	FOOD; EXEMPTIONS FROM LABELING Exemptions From Food Labeling Requirements; Food Labeling
EXEMPTIONS	101.108	TEMPORARY EXEMPTIONS FOR PURPOSES OF CONDUCTING AUTHORIZED FOOD LABELING EXPERIMENTS Exemptions From Food Labeling Requirements; Food Labeling
EXEMPTIONS	107.30	EXEMPTIONS Labeling; Infant Formula
EXEMPTIONS	171.130	PROCEDURE FOR AMENDING AND REPEALING TOLERANCES OR EXEMPTIONS FROM TOLERANCES Administrative Actions On Applications; Food Additive Petitions
EXEMPTIONS	201.127	DRUGS; EXPIRATION OF EXEMPTIONS Exemptions From Adequate Directions For Use; Labeling
EXEMPTIONS	206.7	EXEMPTIONS Imprinting Of Solid Oral Dosage Form Drug Products For Human Use
EXEMPTIONS	207.10	EXEMPTIONS FOR ESTABLISHMENTS Exemptions; Registration Of Drug Producers And Listing Of Drugs In Commercial Distribution
EXEMPTIONS	208.26	EXEMPTIONS AND DEFERRALS Medication Guides For Prescription Drug Products
EXEMPTIONS	501.100	ANIMAL FOOD; EXEMPTIONS FROM LABELING Exemptions From Animal Food Labeling Requirements; Animal Food Labeling

EXEMPTIONS	501.103	PETITIONS REQUESTING EXEMPTIONS FROM OR SPECIAL REQUIREMENTS FOR LABEL DECLARATION OF INGREDIENTS Exemptions From Animal Food Labeling Requirements; Animal Food Labeling
EXEMPTIONS	571.130	PROCEDURE FOR AMENDING AND REPEALING TOLERANCES OR EXEMPTIONS FROM TOLERANCES Administrative Actions On Applications; Food Additive Petitions
EXEMPTIONS	607.65	EXEMPTIONS FOR BLOOD PRODUCT ESTABLISHMENTS Exemptions; Establishment Registratn+product Listing, Manufacturers Human Blood/ b.products
EXEMPTIONS	701.9	EXEMPTIONS FROM LABELING REQUIREMENTS General Provisions; Cosmetic Labeling
EXEMPTIONS	710.9	EXEMPTIONS Voluntary Registration Of Cosmetic Product Establishments
EXEMPTIONS	801.127	MEDICAL DEVICES; EXPIRATION OF EXEMPTIONS Exemptions From Adequate Directions For Use; Labeling
EXEMPTIONS	803.19	EXEMPTIONS OF VARIANCES, AND ALTERNATIVE REPORTING REQUIREMENTS Reports And Records; Medical Device Reporting
EXEMPTIONS	807.65	EXEMPTIONS FOR DEVICE ESTABLISHMENTS Exemptions; Establishment Registration And Device Listing For Manufacturers And Initial Importers Of Devices
EXEMPTIONS	821.2	EXEMPTIONS AND VARIANCES Medical Device Tracking Requirements
EXEMPTIONS	860.95	EXEMPTIONS FROM SECTIONS 510, 519, AND 520(F) OF THE ACT Classification; Medical Device Classification Procedures
EXEMPTIONS	898.14	EXEMPTIONS AND VARIANCES Performance Standards For Electrode Lead Wires And Patient Cables
EXEMPTIONS	1002.50	SPECIAL EXEMPTIONS Exemptions From Records And Reports Requirements; Records and Reports
EXEMPTIONS	1002.51	EXEMPTIONS FOR MANUFACTURERS OF PRODUCTS INTENDED FOR THE U.S. GOVERNMENT Exemptions From Records And Reports Requirements; Records and Reports
EXEMPTIONS	1010.5	EXEMPTIONS FOR PRODUCTS INTENDED FOR UNITED STATES GOVERNMENT USE Performance Standards For Electronic Products: General
EXEMPTIONS, LIMITATIONS	862.9	LIMITATIONS OF EXEMPTIONS FROM SECTION 510(K) OF THE FEDERAL FOOD, DRUG, AND COSMETIC ACT Clinical Chemistry And Clinical Toxicology Devices
EXEMPTIONS, LIMITATIONS	864.9	LIMITATIONS OF EXEMPTIONS FROM SECTION 510(K) OF THE FEDERAL FOOD, DRUG, AND COSMETIC ACT Hematology And Pathology Devices
EXEMPTIONS, LIMITATIONS	866.9	LIMITATIONS OF EXEMPTIONS FROM SECTION 510(K) OF THE FEDERAL FOOD, DRUG, AND COSMETIC ACT Immunology And Microbiology Devices
EXEMPTIONS, LIMITATIONS	868.9	LIMITATIONS OF EXEMPTIONS FROM SECTION 510(K) OF THE FEDERAL FOOD, DRUG, AND COSMETIC ACT Anesthesiology Devices
EXEMPTIONS, LIMITATIONS	870.9	LIMITATIONS OF EXEMPTIONS FROM SECTION 510(K) OF THE FEDERAL FOOD, DRUG, AND COSMETIC ACT Cardiovascular Devices
EXEMPTIONS, LIMITATIONS	872.9	LIMITATIONS OF EXEMPTIONS FROM SECTION 510(K) OF THE FEDERAL FOOD, DRUG, AND COSMETIC ACT Dental Devices
EXEMPTIONS, LIMITATIONS	874.9	LIMITATIONS OF EXEMPTIONS FROM SECTION 510(K) OF THE FEDERAL FOOD, DRUG, AND COSMETIC ACT Ear, Nose, And Throat Devices
EXEMPTIONS, LIMITATIONS	876.9	LIMITATIONS OF EXEMPTIONS FROM SECTION 510(K) OF THE FEDERAL FOOD, DRUG, AND COSMETIC ACT Gastroenterology-urology Devices
EXEMPTIONS, LIMITATIONS	878.9	LIMITATIONS OF EXEMPTIONS FROM SECTION 510(K) OF THE FEDERAL FOOD, DRUG, AND COSMETIC ACT General And Plastic Surgery Devices
EXEMPTIONS, LIMITATIONS	880.9	LIMITATIONS OF EXEMPTIONS FROM SECTION 510(K) OF THE FEDERAL FOOD, DRUG, AND COSMETIC ACT General Hospital And Personal Use Devices
EXEMPTIONS, LIMITATIONS	882.9	LIMITATIONS OF EXEMPTIONS FROM SECTION 510(K) OF THE FEDERAL FOOD, DRUG, AND COSMETIC ACT Neurological Devices
EXEMPTIONS, LIMITATIONS	884.9	LIMITATIONS OF EXEMPTIONS FROM SECTION 510(K) OF THE FEDERAL FOOD, DRUG, AND COSMETIC ACT Obstetrical And Gynecological Devices
EXEMPTIONS, LIMITATIONS	886.9	LIMITATIONS OF EXEMPTIONS FROM SECTION 510(K) OF THE FEDERAL FOOD, DRUG, AND COSMETIC ACT Ophthalmic Devices

EXEMPTIONS, LIMITATIONS	888.9	LIMITATIONS OF EXEMPTIONS FROM SECTION 510(K) OF THE FEDERAL FOOD, DRUG, AND COSMETIC ACT Orthopedic Devices
EXEMPTIONS, LIMITATIONS	890.9	LIMITATIONS OF EXEMPTIONS FROM SECTION 510(K) OF THE FEDERAL FOOD, DRUG, AND COSMETIC ACT Physical Medicine Devices
EXEMPTIONS, LIMITATIONS	892.9	LIMITATIONS OF EXEMPTIONS FROM SECTION 510(K) OF THE FEDERAL FOOD, DRUG, AND COSMETIC ACT Radiology Devices
EXERCISE	890.5350	EXERCISE COMPONENT Physical Medicine Therapeutic Devices; Physical Medicine Devices
EXERCISE	890.5360	MEASURING EXERCISE EQUIPMENT Physical Medicine Therapeutic Devices; Physical Medicine Devices
EXERCISE	890.5370	NONMEASURING EXERCISE EQUIPMENT Physical Medicine Therapeutic Devices; Physical Medicine Devices
EXERCISE	890.5380	POWERED EXERCISE EQUIPMENT Physical Medicine Therapeutic Devices; Physical Medicine Devices
EXERCISER	890.5410	POWERED FINGER EXERCISER Physical Medicine Therapeutic Devices; Physical Medicine Devices
EXHAUSTION OF REMEDIES	10.45	COURT REVIEW OF FINAL ADMINISTRATIVE ACTION; EXHAUSTION OF ADMINISTRATIVE REMEDIES General Administrative Procedures; Administrative Practices And Procedures
EXHIBITS	17.25	EXCHANGES OF WITNESS LISTS, WITNESS STATEMENTS, AND EXHIBITS Civil Money Penalties Hearings
EXISTING RECORDS	20.23	REQUEST FOR EXISTING RECORDS General Policy; Public Information
EXISTING STANDARD	861.24	EXISTING STANDARD AS A PROPOSED STANDARD Procedures For Performance Standards Development
EXOCRINE	310.543	DRUG PRODUCTS CONTAINING ACTIVE INGREDIENTS OFFERED OTC FOR HUMAN USE IN EXOCRINE PANCREATIC INSUFFICIENCY Requirements For Specific New Drugs Or Devices; New Drugs
EXOENZYME	866.3720	STREPTOCOCCUS SPP. EXOENZYME REAGENTS Serological Reagents; Immunology And Microbiology Devices
EXOPHTHALMOMETER	886.1270	EXOPHTHALMOMETER Diagnostic Devices; Ophthalmic Devices
EXPANDABLE	884.4250	EXPANDABLE CERVICAL DILATOR Obstetrical And Gynecological Surgical Devices; Obstetrical And Gynecological Devices
EXPECTORANT	341.18	EXPECTORANT ACTIVE INGREDIENT Active Ingredients; Cold, Cough, Allergy, Bronchodilator And Antiasthmatic Drugs Products (OTC)
EXPECTORANT	341.78	LABELING OF EXPECTORANT DRUG PRODUCTS Labeling; Cold, Cough, Allergy, Bronchodilator And Antiasthmatic Drugs Products (OTC)
EXPEDITED PROCESS	1401.6	EXPEDITED PROCESS Public Availability Of Information
EXPEDITED RELEASE	1316.92	PETITION FOR EXPEDITED RELEASE IN AN ADMINISTRATIVE FORFEITURE ACTION Administrative Functions, Practices, And Procedures
EXPEDITED RELEASE	1316.93	RULING ON PETITION FOR EXPEDITED RELEASE IN AN ADMINISTRATIVE FORFEITURE ACTION Administrative Functions, Practices, And Procedures
EXPEDITED RELEASE	1316.95	PETITION FOR EXPEDITED RELEASE OF A CONVEYANCE IN A JUDICIAL FORFEITURE ACTION Administrative Functions, Practices, And Procedures
EXPEDITED RELEASE	1316.96	RULING ON A PETITION FOR EXPEDITED RELEASE OF A CONVEYANCE IN A JUDICIAL FORFEITURE ACTION Expedited Forfeiture Proceedings For Certain Property; Administrative Functions, Practices, And Procedures
EXPEDITED REVIEW	56.110	EXPEDITED REVIEW PROCEDURES FOR CERTAIN KINDS OF RESEARCH INVOLVING NO MORE THAN MINIMAL RISK, AND FOR MINOR CHANGES IN APPROVED RESEARCH IRB Functions And Operations; Institutional Review Boards
EXPERIENCE	510.300	RECORDS AND REPORTS CONCERNING EXPERIENCE WITH NEW ANIMAL DRUGS FOR WHICH AN APPROVED APPLICATION IS IN EFFECT Records And Reports; New Animal Drugs
EXPERIENCE	510.301	RECORDS AND REPORTS CONCERNING EXPERIENCE WITH ANIMAL FEEDS BEARING OR CONTAINING NEW ANIMAL DRUGS FOR WHICH AN APPROVED MEDICATED FEED MILL LICENSE APPLICATION IS IN EFFECT Records And Reports; New Animal Drugs
EXPERIENCE	514.80	RECORDS AND REPORTS CONCERNING EXPERIENCE WITH APPROVED NEW ANIMAL DRUGS Administrative Actions On Applications; New Animal Drug Applications

EXPERIENCE REPORTS	20.112	VOLUNTARY DRUG EXPERIENCE REPORTS SUBMITTED BY PHYSICIANS AND HOSPITALS Availability Of Specific Categories Of Records; Public Information
EXPERIENCES, ADVERSE	310.305	RECORDS AND REPORTS CONTAINING ADVERSE DRUG EXPERIENCES ON MARKETED PRESCRIPTION DRUGS FOR HUMAN USE WITHOUT APPROVED NEW DRUG APPLICATIONS Records And Reports; New Drugs
EXPERIENCES, ADVERSE	314.80	POSTMARKETING REPORTING OF ADVERSE DRUG EXPERIENCES Applications; Applications For FDA Approval To Market A New Drug
EXPERIENCES, ADVERSE	600.80	POSTMARKETING REPORTING OF ADVERSE EXPERIENCES Biological Products: General
EXPERIMENTAL	130.17	TEMPORARY PERMITS FOR INTERSTATE SHIPMENT OF EXPERIMENTAL PACKS OF FOOD VARYING FROM THE REQUIREMENTS OF DEFINITIONS AND STANDARDS OF IDENTITY General Provisions; Food Standards: General
EXPERIMENTAL ANIMALS	170.17	EXEMPTION FOR INVESTIGATIONAL USE AND PROCEDURE FOR OBTAINING AUTHORIZATION TO MARKET EDIBLE PRODUCTS FORM EXPERIMENTAL ANIMALS General Provisions; Food Additives
EXPERIMENTAL ANIMALS	570.17	EXEMPTION FOR INVESTIGATIONAL USE AND PROCEDURE FOR OBTAINING AUTHORIZATION TO MARKET EDIBLE PRODUCTS FROM EXPERIMENTAL ANIMALS General Provisions; Food Additives
EXPERIMENTS	101.108	TEMPORARY EXEMPTIONS FOR PURPOSES OF CONDUCTING AUTHORIZED FOOD LABELING EXPERIMENTS Exemptions From Food Labeling Requirements; Food Labeling
EXPIRATION	201.127	DRUGS; EXPIRATION OF EXEMPTIONS Exemptions From Adequate Directions For Use; Labeling
EXPIRATION	801.127	MEDICAL DEVICES; EXPIRATION OF EXEMPTIONS Exemptions From Adequate Directions For Use; Labeling
EXPIRATION DATE	201.17	DRUGS; LOCATION OF EXPIRATION DATE General Labeling Provisions; Labeling
EXPIRATION DATE	1301.13	TIME FOR APPLICATION FOR REGISTRATION; EXPIRATION DATE Applications For Registration; Registration Of Manufacturers, Distributors, And Dispensers Of Controlled Substances
EXPIRATION DATE	1309.31	TIME FOR APPLICATION FOR REGISTRATION; EXPIRATION DATE Registration Of Manufacturers, Distributors, Importers And Exporters Of List I Chemicals
EXPIRATION DATE	1312.16	CANCELLATION OF PERMIT; EXPIRATION DATE Importation Of Controlled Substances; Importation And Exportation Of Controlled Substances
EXPIRATION DATE	1312.25	EXPIRATION DATE Exportation Of Controlled Substances; Importation And Exportation Of Controlled Substances
EXPIRATION DATING	211.137	EXPIRATION DATING Packaging And Labeling Control; Current Good Manufacturing Practice For Finished Pharmaceuticals
EXPIRATORY	868.5965	POSITIVE END EXPIRATORY PRESSURE BREATHING ATTACHMENT Therapeutic Devices; Anesthesiology Devices
EXPORT	5.801	EXPORT OF UNAPPROVED DRUGS Imports And Exports; Redelegations Of Authority; Delegations Of Authority And Organization
EXPORT	610.65	PRODUCTS FOR EXPORT Labeling Standards; General Biological Products Standards
EXPORT	1010.20	ELECTRONIC PRODUCTS INTENDED FOR EXPORT Performance Standards For Electronic Products: General
EXPORT	1312.21	REQUIREMENT OF AUTHORIZATION TO EXPORT Exportation Of Controlled Substances; Importation And Exportation Of Controlled Substances
EXPORT DECLARATION	1313.22	CONTENTS OF EXPORT DECLARATION Importation And Exportation Of Precursors And Essential Chemicals
EXPORT DECLARATION	1313.23	DISTRIBUTION OF EXPORT DECLARATION Importation And Exportation Of Precursors And Essential Chemicals
EXPORT PERMIT	1312.22	APPLICATION FOR EXPORT PERMIT Exportation Of Controlled Substances; Importation And Exportation Of Controlled Substances
EXPORT PERMIT	1312.23	ISSUANCE OF EXPORT PERMIT Exportation Of Controlled Substances; Importation And Exportation Of Controlled Substances
EXPORT PERMIT	1312.24	DISTRIBUTION OF COPIES OF EXPORT PERMIT Exportation Of Controlled Substances; Importation And Exportation Of Controlled Substances

EXPORT PERMIT	1312.30	SCHEDULE III, IV, AND V NON-NARCOTIC CONTROLLED SUBSTANCES REQUIRING AN IMPORT AND EXPORT PERMIT Exportation Of Controlled Substances; Importation And Exportation Of Controlled Substances
EXPORT REQUIREMENTS	312.110	IMPORT AND EXPORT REQUIREMENTS Miscellaneous; New Drugs For Investigational Use
EXPORT REQUIREMENTS	812.18	IMPORT AND EXPORT REQUIREMENTS General Provisions; Investigational Device Exemptions
EXPORT REQUIREMENTS	1301.26	EXEMPTIONS FROM IMPORT OR EXPORT REQUIREMENTS FOR PERSONAL MEDICAL USE Registration Of Manufacturers, Distributors, And Dispensers Of Controlled Substances
EXPORTED SUBSTANCES	1302.07	LABELING AND PACKAGING REQUIREMENTS FOR IMPORTED AND EXPORTED SUBSTANCES Labeling And Packaging Requirements For Controlled Substances
EXPORTER	1312.26	RECORDS REQUIRED OF EXPORTER Exportation Of Controlled Substances; Importation And Exportation Of Controlled Substances
EXPORTERS	1304.22	RECORDS FOR MANUFACTURERS, DISTRIBUTORS, DISPENSERS, RESEARCHERS, IMPORTERS AND EXPORTERS Records and Reports Of Registrants
EXPORTS	5.800	IMPORTS AND EXPORTS Imports And Exports; Redelegations Of Authority; Delegations Of Authority And Organization
EXPORTS	314.410	IMPORTS AND EXPORTS OF NEW DRUGS Miscellaneous Provisions; Applications For FDA Approval To Market A New Drug
EXPOSURE ALIGNMENT	872.1820	DENTAL X-RAY EXPOSURE ALIGNMENT DEVICE Diagnostic devices; Dental Devices
EXT. D&C, LAKES	82.2051	LAKES (EXT. D&C) Externally Applied Drugs And Cosmetics; Listing Of Certified Provisionally Listed Colors And Specifications
EXT. D&C VIOLET NO. 2	74.2602a	EXT. D&C VIOLET NO. 2 Cosmetics; Listing Of Color Additives Subject To Certification
EXT. D&C YELLOW NO. 7	74.1707a	EXT. D&C YELLOW NO. 7 Drugs; Listing Of Color Additives Subject To Certification
EXT. D&C YELLOW NO. 7	74.2707a	EXT. D&C YELLOW NO. 7 Cosmetics; Listing Of Color Additives Subject To Certification
EXT. D&C YELLOW NO. 7	82.2707a	EXT. D&C YELLOW NO. 7 Externally Applied Drugs And Cosmetics; Listing Of Certified Provisionally Listed Colors And Specifications
EXTEND TIME	99.203	REQUEST TO EXTEND THE TIME FOR COMPLETING PLANNED STUDIES Dissemination Of Information On Unapproved/New Uses For Marketed Drugs, Biologics, And Devices
EXTENSION OF REGISTRATION	1301.36	EXTENSION OF REGISTRATION PENDING FINAL ORDER Action On Application For Registration: Revocation Or Suspension Of Registration; Registration Of Manufacturers, Distributors, And Dispensers Of Controlled Substances
EXTENSION OF REGISTRATION	1309.45	EXTENSION OF REGISTRATION PENDING FINAL ORDER Registration Of Manufacturers, Distributors, Importers And Exporters Of List I Chemicals
EXTENSION OF TIME	71.6	EXTENSION OF TIME FOR STUDYING PETITIONS; SUBSTANTIVE AMENDMENTS; WITHDRAWAL OF PETITIONS General Provisions; Color Additive Petitions
EXTENSION OF TIME	99.303	EXTENSION OF TIME FOR COMPLETING PLANNED STUDIES Dissemination Of Information On Unapproved/New Uses For Marketed Drugs, Biologics, And Devices
EXTENSION OF TIME	1401.8	EXTENSION OF TIME Public Availability Of Information
EXTENSIONS	5.27	PATENT TERM EXTENSIONS FOR HUMAN DRUG PRODUCTS, MEDICAL DEVICES, AND FOOD AND COLOR ADDITIVES General Redelegations Of Authority; Delegations Of Authority And Organization
EXTENSIONS	5.107	EXTENSIONS OR STAYS OF EFFECTIVE DATES FOR COMPLIANCE WITH CERTAIN LABELING REQUIREMENTS FOR HUMAN PRESCRIPTION DRUGS Human Drugs; Redelegations Of Authority; Delegations Of Authority And Organization
EXTERNAL	310.527	DRUG PRODUCTS CONTAINING ACTIVE INGREDIENTS OFFERED OTC FOR EXTERNAL USE A HAIR GROWERS OR FOR HAIR LOSS PREVENTION Requirements For Specific New Drugs Or Devices; New Drugs
EXTERNAL	348.50	LABELING OF EXTERNAL ANALGESIC DRUG PRODUCTS External Analgesic Drug Products For Over-the-counter Human Use
EXTERNAL	868.5935	EXTERNAL NEGATIVE PRESSURE VENTILATOR Therapeutic Devices; Anesthesiology Devices
EXTERNAL	870.1750	EXTERNAL PROGRAMMABLE PACEMAKER PULSE GENERATOR Cardiovascular Diagnostic Devices; Cardiovascular Devices

EXTERNAL	870.3600	EXTERNAL PACEMAKER PULSE GENERATOR Cardiovascular Prosthetic Devices; Cardiovascular Devices
EXTERNAL	870.4885	EXTERNAL VEIN STRIPPER Cardiovascular Surgical Devices; Cardiovascular Devices
EXTERNAL	870.5200	EXTERNAL CARDIAC COMPRESSOR Cardiovascular Therapeutic Devices; Cardiovascular Devices
EXTERNAL	870.5225	EXTERNAL COUNTER-PULSATING DEVICE Cardiovascular Therapeutic Devices; Cardiovascular Devices
EXTERNAL	870.5550	EXTERNAL TRANSCUTANEOUS CARDIAC PACEMAKER (NONINVASIVE) Cardiovascular Therapeutic Devices; Cardiovascular Devices
EXTERNAL	874.5800	EXTERNAL NASAL SPLINT Therapeutic Devices; Ear, Nose And Throat Devices
EXTERNAL	878.3250	EXTERNAL FACIAL FRACTURE FIXATION APPLIANCE Prosthetic Devices; General And Plastic Surgery Devices
EXTERNAL	878.3750	EXTERNAL PROSTHESIS ADHESIVE Prosthetic Devices; General And Plastic Surgery Devices
EXTERNAL	878.3800	EXTERNAL AESTHETIC RESTORATION PROSTHESIS Prosthetic Devices; General And Plastic Surgery Devices
EXTERNAL	878.4014	NONRESORBABLE GAUZE SPONGE FOR EXTERNAL USE Surgical Devices; General And Plastic Surgery Devices
EXTERNAL	882.5810	EXTERNAL FUNCTIONAL NEUROMUSCULAR STIMULATOR Neurological Therapeutic Devices; Neurological Devices
EXTERNAL	884.2720	EXTERNAL UTERINE CONTRACTION MONITOR AND ACCESSORIES Obstetrical And Gynecological Monitoring Devices; Obstetrical And Gynecological Devices
EXTERNAL	890.3410	EXTERNAL LIMB ORTHOTIC COMPONENT Physical Medicine Prosthetic Devices; Physical Medicine Devices
EXTERNAL	890.3420	EXTERNAL LIMB PROSTHETIC COMPONENT Physical Medicine Prosthetic Devices; Physical Medicine Devices
EXTERNAL	890.3500	EXTERNAL ASSEMBLED LOWER LIMB PROSTHESIS Physical Medicine Prosthetic Devices; Physical Medicine Devices
EXTERNAL	890.5575	POWERED EXTERNAL LIMB OVERLOAD WARNING DEVICE Physical Medicine Therapeutic Devices; Physical Medicine Devices
EXTRACORPOREAL	876.5990	EXTRACORPOREAL SHOCK WAVE LITHOTRIPTER Therapeutic Devices; Gastroenterology-urology Devices
EXTRACT, ANNATTO	73.30	ANNATTO EXTRACT Foods; Listing Of Color Additives Exempt From Certification
EXTRACT, ANNATTO	73.1030	ANNATTO EXTRACT Drugs; Listing Of Color Additives Exempt From Certification
EXTRACT, BAKERS YEAST	184.1983	BAKERS YEAST EXTRACT Listing Of Specific Substances Affirmed As GRAS; Direct Food Substances Affirmed As Generally Recognized As Safe
EXTRACT, CHRONDRUS	182.7255	CHRONDRUS EXTRACT Stabilizers; Substances Generally Recognized As Safe
EXTRACT, CHRONDRUS	582.7255	CHRONDRUS EXTRACT Stabilizers; Substances Generally Recognized As Safe
EXTRACT, COCHINEAL	73.100	COCHINEAL EXTRACT; CARMINE Foods; Listing Of Color Additives Exempt From Certification
EXTRACT, COCHINEAL	73.1100	COCHINEAL EXTRACT; CARMINE Drugs; Listing Of Color Additives Exempt From Certification
EXTRACT, CORN SILK	184.1262	CORN SILK AND CORN SILK EXTRACT Listing Of Specific Substances Affirmed As GRAS; Direct Food Substances Affirmed As Generally Recognized As Safe
EXTRACT, GRAPE COLOR	73.169	GRAPE COLOR EXTRACT Foods; Listing Of Color Additives Exempt From Certification
EXTRACT, GRAPE SKIN	73.170	GRAPE SKIN EXTRACT (ENOCIANINA) Foods; Listing Of Color Additives Exempt From Certification
EXTRACT, HEMICELLULOSE	573.520	HEMICELLULOSE EXTRACT Food Additive Listing; Food Additives Permitted In Feed And Drinking Water Of Animals
EXTRACT, HOP	172.560	MODIFIED HOP EXTRACT Flavoring Agents And Related Substances; Food Additives Permitted For Direct Addition To Food For Human Consumption
EXTRACT, LOGWOOD	73.1410	LOGWOOD EXTRACT Drugs; Listing Of Color Additives Exempt From Certification
EXTRACT, MALT	172.590	YEAST-MALT SPROUT EXTRACT Flavoring Agents And Related Substances; Food Additives Permitted For Direct Addition To Food For Human Consumption
EXTRACT, MALT	184.1445	MALT SYRUP (MALT EXTRACT) Listing Of Specific Substances Affirmed As GRAS; Direct Food Substances Affirmed As Generally Recognized As Safe
EXTRACT, OX BILE	184.1560	OX BILE EXTRACT Listing Of Specific Substances Affirmed As GRAS; Direct Food Substances Affirmed As Generally Recognized As Safe

EXTRACT, SASSAFRAS	172.580	SAFROLE-FREE EXTRACT OF SASSAFRAS Flavoring Agents And Related Substances; Food Additives Permitted For Direct Addition To Food For Human Consumption
EXTRACT, SUGAR BEET	172.585	SUGAR BEET EXTRACT FLAVOR BASE Flavoring Agents And Related Substances; Food Additives Permitted For Direct Addition To Food For Human Consumption
EXTRACT, TAGETES	73.295	TAGETES (AZTEC MARIGOLD) MEAL AND EXTRACT Foods; Listing Of Color Additives Exempt From Certification
EXTRACT, VANILLA	169.175	VANILLA EXTRACT Requirements For Specific Standardized Food Dressings And Flavorings; Food Dressings And Flavorings
EXTRACT, VANILLA	169.176	CONCENTRATED VANILLA EXTRACT Requirements For Specific Standardized Food Dressings And Flavorings; Food Dressings And Flavorings
EXTRACT, VANILLIN	169.180	VANILLA-VANILLIN EXTRACT Requirements For Specific Standardized Food Dressings And Flavorings; Food Dressings And Flavorings
EXTRACTED GLUTAMIC ACID	573.500	CONDENSED, EXTRACTED GLUTAMIC ACID FERMENTATION PRODUCT Food Additive Listing; Food Additives Permitted In Feed And Drinking Water Of Animals
EXTRACTION, SOLVENT	173.280	SOLVENT EXTRACTION PROCESS FOR CITRIC ACID Solvents, Lubricants, Release Agents And Related Substances; Secondary Direct Food Additives Permitted In Food For Human Consumption
EXTRACTIVES	182.20	ESSENTIAL OILS, OLEORESINS, AND NATURAL EXTRACTIVES (INCLUDING DISTILLATES) General Provisions; Substances Generally Recognized As Safe
EXTRACTIVES	182.40	NATURAL EXTRACTIVES USED IN CONJUNCTION WITH SPICES, SEASONINGS, AND FLAVORINGS General Provisions; Substances Generally Recognized As Safe
EXTRACTIVES	582.20	ESSENTIAL OILS, OLEORESINS, AND NATURAL EXTRACTIVES (INCLUDING DISTILLATES) General Provisions; Substances Generally Recognized As Safe
EXTRACTIVES	582.40	NATURAL EXTRACTIVES (SOLVENT-FREE) USED IN CONJUNCTION WITH SPICES, SEASONINGS, AND FLAVORINGS General Provisions; Substances Generally Recognized As Safe
EXTRACTOR, FETAL	884.4340	FETAL VACUUM EXTRACTOR Obstetrical And Gynecological Surgical Devices; Obstetrical And Gynecological Devices
EXTRACTS	182.50	CERTAIN OTHER SPICES, SEASONINGS, ESSENTIAL OILS, OLEORESINS, AND NATURAL EXTRACTS General Provisions; Substances Generally Recognized As Safe
EXTRACTS	582.50	CERTAIN OTHER SPICES, SEASONINGS, ESSENTIAL OILS, OLEORESINS, AND NATURAL EXTRACTS General Provisions; Substances Generally Recognized As Safe
EXTRALABEL USE	5.500	ISSUANCE OF FEDERAL REGISTER DOCUMENTS PERTAINING TO THE DETERMINATION OF SAFE LEVELS, NOTICE OF NEED FOR DEVELOPMENT OF AN ANALYTICAL METHOD, PROHIBITION OF CERTAIN EXTRALABEL DRUG USE Animal Drugs; Redelegations Of Authority; Delegations Of Authority And Organization
EXTRALABEL USE	530.10	PROVISION PERMITTING EXTRALABEL USE OF ANIMAL DRUGS Extralabel Drug Use In Animals
EXTRALABEL USE	530.13	EXTRALABEL USE FROM COMPOUNDING OF APPROVED NEW ANIMAL AND HUMAN DRUGS Extralabel Drug Use In Animals
EXTRALABEL USE	530.20	CONDITIONS FOR PERMITTED EXTRALABEL ANIMAL AND HUMAN DRUG USE IN FOOD-PRODUCTS Extralabel Drug Use In Animals
EXTRALABEL USE	530.30	EXTRALABEL DRUG USE IN NONFOOD ANIMALS Extralabel Drug Use In Animals
EXTRALABEL USE	530.41	DRUGS PROHIBITED FOR EXTRALABEL USE IN ANIMALS Extralabel Drug Use In Animals
EXTRALABEL USES	530.25	ORDERS PROHIBITING EXTRALABEL USES FOR DRUGS IN FOOD-PRODUCING ANIMALS Extralabel Drug Use In Animals
EXTRAMURAL FACILITIES	200.10	CONTRACT FACILITIES (INCLUDING CONSULTING LABORATORIES) UTILIZED AS EXTRAMURAL FACILITIES BY PHARMACEUTICAL MANUFACTURERS General Provisions; General
EXTRAOCULAR	886.3340	EXTRAOCULAR ORBITAL IMPLANT Prosthetic Devices; Ophthalmic Devices
EXTRAORAL	872.1800	EXTRAORAL SOURCE X-RAY SYSTEM Diagnostic devices; Dental Devices
EXTRAORAL	872.5500	EXTRAORAL ORTHODONTIC HEADGEAR Therapeutic Devices; Dental Devices

EXTRAORDINARY CIRCUMSTANCES	25.21	EXTRAORDINARY CIRCUMSTANCES Agency Actions Requiring Environmental Consideration; Environmental Impact Considerations
EXTRAVASCULAR	870.2120	EXTRAVASCULAR BLOOD FLOW PROBE Cardiovascular Monitoring Devices; Cardiovascular Devices
EXTRAVASCULAR	870.2850	EXTRAVASCULAR BLOOD PRESSURE TRANSDUCER Cardiovascular Monitoring Devices; Cardiovascular Devices
EXTREMITIES	878.5650	TOPICAL OXYGEN CHAMBER FOR EXTREMITIES Therapeutic Devices; General And Plastic Surgery Devices
EXTREMITY	878.3900	INFLATABLE EXTREMITY SPLINT Prosthetic Devices; General And Plastic Surgery Devices
EXTREMITY	878.3910	NONINFLATABLE EXTREMITY SPLINT Prosthetic Devices; General And Plastic Surgery Devices
EYE, ARTIFICIAL	886.3200	ARTIFICAL EYE Prosthetic Devices; Ophthalmic Devices
EYE IMPLANT	886.3320	EYE SPHERE IMPLANT Prosthetic Devices; Ophthalmic Devices
EYE MOVEMENT	886.1510	EYE MOVEMENT MONITOR Diagnostic Devices; Ophthalmic Devices
EYE PAD	878.4440	EYE PAD Surgical Devices; General And Plastic Surgery Devices
EYE PAD	880.5270	NEONATAL EYE PAD General Hospital And Personal Use Therapeutic Devices; General Hospital And Personal Use Devices
EYE SHIELD	886.4750	OPHTHALMIC EYE SHIELD Surgical Devices; Ophthalmic Devices
EYEGLASSES	801.410	USE OF IMPACT-RESISTANT LENSES IN EYEGLASSES AND SUNGLASSES Special Requirements For Specific Devices; Labeling
EYEWASH	349.78	LABELING OF EYEWASH DRUG PRODUCTS Labeling; Ophthalmic Drug Products For Over-the-counter Human Use
EYEWASHES	349.20	EYEWASHES Active Ingredients; Ophthalmic Drug Products For Over-the-counter Human Use

F

FAB FRAGMENT	866.5520	IMMUNOGLOBULIN G (FAB FRAGMENT SPECIFIC) IMMUNOLOGICAL TEST SYSTEM Immunological Test Systems; Immunology And Microbiology Devices
FABRICATED	872.3600	PARTIALLY FABRICATED DENTURE KIT Prosthetic Devices; Dental Devices
FABRICATION INSTRUMENTS	884.6140	ASSISTED REPRODUCTION MICROPIPETTE FABRICATION INSTRUMENTS Obstetrical And Gynecological Devices
FABRICS	182.70	SUBSTANCES MIGRATING FROM COTTON AND COTTON FABRICS USED IN DRY FOOD PACKAGING General Provisions; Substances Generally Recognized As Safe
FACEBOW	872.3220	FACEBOW Prosthetic Devices; Dental Devices
FACIAL FRACTURE	878.3250	EXTERNAL FACIAL FRACTURE FIXATION APPLIANCE Prosthetic Devices; General And Plastic Surgery Devices
FACIAL PROSTHESIS	874.3695	MANDIBULAR IMPLANT FACIAL PROSTHESIS Prosthetic Devices; Ear, Nose And Throat Devices
FACILITIES	58.43	ANIMAL CARE FACILITIES Facilities; Good Laboratory Practice For Nonclinical Laboratory Studies
FACILITIES	58.45	ANIMAL SUPPLY FACILITIES Facilities; Good Laboratory Practice For Nonclinical Laboratory Studies
FACILITIES	58.47	FACILITIES FOR HANDLING TEST AND CONTROL ARTICLES Facilities; Good Laboratory Practice For Nonclinical Laboratory Studies
FACILITIES	58.51	SPECIMEN AND DATA STORAGE FACILITIES Facilities; Good Laboratory Practice For Nonclinical Laboratory Studies
FACILITIES	606.40	FACILITIES Plant And Facilities; Current Good Manufacturing Practice For Blood And Blood Components
FACILITIES	1000.55	RECOMMENDATION FOR QUALITY ASSURANCE PROGRAMS IN DIAGNOSTIC RADIOLOGY FACILITIES Radiation Protection Recommendations; General
FACILITIES	1250.45	FOOD HANDLING FACILITIES ON RAILROAD CONVEYANCES Equipment And Operation Of Land And Air Conveyances; Interstate Conveyance Sanitation
FACILITIES	1301.77	SECURITY CONTROLS FOR FREIGHT FORWARDING FACILITIES Registration Of Manufacturers, Distributors, And Dispensers Of Controlled Substances
FACILITIES, CONTRACT	200.10	CONTRACT FACILITIES (INCLUDING CONSULTING LABORATORIES) UTILIZED AS EXTRAMURAL FACILITIES BY PHARMACEUTICAL MANUFACTURERS General Provisions; General
FACILITIES, MAMMOGRAPHY	5.1000	AUTHORITY TO ENSURE THAT MAMMOGRAPHY FACILITIES MEET QUALITY STANDARDS Mammography Facilities; Redelegations Of Authority; Delegations Of Authority And Organization
FACILITIES, SANITARY	110.35	SANITARY FACILITIES AND CONTROLS Buildings And Facilities; Current Good Manufacturing Practice In Manufacturing, Packing, Or Holding Human Food
FACILITIES, SANITARY	129.35	SANITARY FACILITIES Buildings And Facilities; Processing And Bottling Of Bottled Drinking Water
FACILITIES, TOILET	211.52	WASHING AND TOILET FACILITIES Buildings And Facilities; Current Good Manufacturing Practice For Finished Pharmaceuticals
FACILITIES, TOILET	1250.38	TOILET AND LAVATORY FACILITIES FOR USE OF FOOD-HANDLING EMPLOYEES Food Service Sanitation On Land And Air Conveyances, And Vessels; Interstate Conveyance Sanitation
FACILITIES, TOILET	1250.50	TOILET AND LAVATORY FACILITIES Equipment And Operation Of Land And Air Conveyances; Interstate Conveyance Sanitation
FACILITIES, USER	803.10	GENERAL DESCRIPTION OF REPORTS REQUIRED FROM USER FACILITIES, IMPORTERS, AND MANUFACTURERS Medical Device Reporting
FACILITIES, USER	803.30	INDIVIDUAL ADVERSE EVENT REPORTS; USER FACILITIES Medical Device Reporting
FACILITY	58.15	INSPECTION OF A TESTING FACILITY General Provisions; Good Laboratory Practice For Nonclinical Laboratory Studies
FACILITY	58.31	TESTING FACILITY MANAGEMENT Organization And Personnel; Good Laboratory Practice For Nonclinical Laboratory Studies

FACILITY	58.217	SUSPENSION OR TERMINATION OF A TESTING FACILITY BY A SPONSOR Disqualification Of Testing Facilities; Good Laboratory Practice For Nonclinical Laboratory Studies
FACILITY	58.219	REINSTATEMENT OF A DISQUALIFIED TESTING FACILITY Disqualification Of Testing Facilities; Good Laboratory Practice For Nonclinical Laboratory Studies
FACING	872.3910	BACKING AND FACING FOR AN ARTIFICIAL TOOTH Prosthetic Devices; Dental Devices
FACT, FINDINGS OF	1316.64	PROPOSED FINDINGS OF FACT AND CONCLUSIONS OF LAW Administrative Hearings; Administrative Functions, Practices, And Procedures
FACTOR 4	864.7695	PLATELET FACTOR 4 RADIOIMMUNOASSAY Hematology Kits And Packages; Hematology And Pathology Devices
FACTOR XIII	866.5330	FACTOR XIII, A, S, IMMUNOLOGICAL TEST SYSTEM Immunological Test Systems; Immunology And Microbiology Devices
FACTOR B	866.5320	PROPERIDIN FACTOR B IMMUNOLOGICAL TEST SYSTEM Immunological Test Systems; Immunology And Microbiology Devices
FACTOR DEFICIENCY	864.7290	FACTOR DEFICIENCY TEST Hematology Kits And Packages; Hematology And Pathology Devices
FACTOR, RHEUMATOID	866.5775	RHEUMATOID FACTOR IMMUNOLOGICAL TEST SYSTEM Immunological Test Systems; Immunology And Microbiology Devices
FACTORS, CRITICAL	113.89	DEVIATIONS IN PROCESSING, VENTING, OR CONTROL OF CRITICAL FACTORS Production And Process Controls; Thermally Processed Low-acid Foods Packaged In Hermetically Sealed Containers
FACTORS, GENERAL	315.3	GENERAL FACTORS RELEVANT TO SAFETY AND EFFECTIVENESS Diagnostic Radiopharmaceuticals
FACTORS, SAFETY	70.40	SAFETY FACTORS TO BE CONSIDERED Safety Evaluation; Color Additives
FACTORS, SAFETY	170.22	SAFETY FACTORS TO BE CONSIDERED Food Additive Safety; Food Additives
FACTS, MATERIAL	1.21	FAILURE TO REVEAL MATERIAL FACTS General Labeling Requirements; General Enforcement Regulations
FAILS TO COMPLY	1003.11	DETERMINATION BY SECRETARY THAT PRODUCT FAILS TO COMPLY OR HAS A DEFECT Discovery Of Defect Or Failure To Comply; Notification Of Defects Or Failure To Comply
FAILURE TO ADHERE	1404.225	FAILURE TO ADHERE TO RESTRICTIONS Governmentwide Debarment And Suspension And Requirements For Drug-Free Workplace
FAILURE TO COMPLY	1003.10	DISCOVERY OF DEFECT OR FAILURE OF COMPLIANCE BY MANUFACTURER; NOTICE REQUIREMENTS Discovery Of Defect Or Failure To Comply; Notification Of Defects Or Failure To Comply
FAILURE TO FILE	17.11	DEFAULT UPON FAILURE TO FILE AN ANSWER Civil Money Penalties Hearings
FAILURE TO REVEAL	1.21	FAILURE TO REVEAL MATERIAL FACTS General Labeling Requirements; General Enforcement Regulations
FALLOPIAN TUBE	884.3650	FALLOPIAN TUBE PROSTHESIS Obstetrical And Gynecological Prosthetic Devices; Obstetrical And Gynecological Devices
FALSE LABELING	5.109	ISSUANCE OF WRITTEN NOTICES CONCERNING PATENT INFORMATION, CURRENT GOOD MANUFACTURING PRACTICES AND FALSE OR MISLEADING LABELING OF NEW DRUGS Human Drugs; Redelegations Of Authority; Delegations Of Authority And Organization
FALSE LABELING	5.504	ISSUANCE OF WRITTEN NOTICES CONCERNING PATENT INFORMATION, CURRENT GOOD MANUFACTURING PRACTICES AND FALSE OR MISLEADING LABELING OF NEW ANIMAL DRUGS AND FEEDS BEARING OR CONTAINING NEW ANIMAL DRUGS Animal Drugs; Redelegations Of Authority; Delegations Of Authority And Organization
FAMPHUR	520.1242g	LEVAMISOLE RESINATE AND FAMPHUR PASTE Oral Dosage Form New Animal Drugs
FAMPHUR	524.900	FAMPHUR Ophthalmic And Topical Dosage Form New Animal Drugs
FAMPHUR	556.273	FAMPHUR Tolerances For Residues Of New Animal Drugs In Food
FAMPHUR	558.254	FAMPHUR New Animal Drugs For Use In Animal Feeds
FARINA	137.300	FARINA Requirements For Specific Standardized Cereal Flours And Related Products; Cereal Flours And Related Products

FARINA	137.305	ENRICHED FARINA Requirements For Specific Standardized Cereal Flours And Related Products; Cereal Flours And Related Products
FARMS	1210.11	SANITARY INSPECTION OF DIARY FARMS Regulations Under The Federal Import Milk Act
FASTENER	882.5360	CRANIOPLASTY PLATE FASTENER Neurological Therapeutic Devices; Neurological Devices
FASTENER	888.3040	SMOOTH OR THREADED METALLIC BONE FIXATION FASTENER Prosthetic Devices; Orthopedic Devices
FAT	101.62	NUTRIENT CONTENT CLAIMS FOR FAT, FATTY ACID, AND CHOLESTEROL CONTENT OF FOODS Specific Requirements For Nutrient Content Claims; Food Labeling
FAT, EDIBLE	102.37	MIXTURE OF EDIBLE FAT OR OIL AND OLIVE OIL Requirements For Specific Nonstandardized Foods; Common Or Usual Name For Nonstandardized Foods
FAT, VEGETABLE	163.150	SWEET COCOA AND VEGETABLE FAT COATING Requirements For Specific Standardized Cacao Products; Cacao Products
FAT, VEGETABLE	163.153	SWEET CHOCOLATE AND VEGETABLE FAT COATING Requirements For Specific Standardized Cacao Products; Cacao Products
FAT, VEGETABLE	163.155	MILK CHOCOLATE AND VEGETABLE FAT COATING Requirements For Specific Standardized Cacao Products; Cacao Products
FAT-FORMING	582.4101	DIACETYL TARTARIC ACID ESTERS OF MONO- AND DIGLYCERIDES OF EDIBLE FATS OR OILS, OR EDIBLE FAT-FORMING FATTY ACIDS Emulsifying Agents; Substances Generally Recognized As Safe
FAT-FORMING	582.4505	MONO- AND DIGLYCERIDES OF EDIBLE FATS OR OILS, OR EDIBLE FAT-FORMING ACIDS Emulsifying Agents; Substances Generally Recognized As Safe
FAT-FORMING	582.4521	MONOSODIUM PHOSPHATE DERIVATIVES OF MONO- AND DIGLYCERIDES OF EDIBLE FATS OR OILS, OR EDIBLE FAT-FORMING FATTY ACIDS Emulsifying Agents; Substances Generally Recognized As Safe
FAT, SATURATED	101.75	HEALTH CLAIMS: DIETARY SATURATED FAT AND CHOLESTEROL AND RISK OF CORONARY HEART DISEASE Food Labeling
FATAL	640.73	REPORTING OF FATAL DONOR REACTIONS Source Plasma; Additional Standards For Human Blood And Blood Products
FATS	172.856	PROPYLENE GLYCOL MONO- AND DIESTERS OF FATS AND FATTY ACIDS Multipurpose Additives; Food Additives Permitted For Direct Addition To Food For Human Consumption
FATS	184.1287	ENZYME-MODIFIED FATS Listing Of Specific Substances Affirmed As GRAS; Direct Food Substances Affirmed As Generally Recognized As Safe
FATS, EDIBLE	172.225	METHYL AND ETHYL ESTERS OF FATTY ACIDS PRODUCED FROM EDIBLE FATS AND OILS Coatings, Films And Related Substances; Food Additives Permitted For Direct Addition To Food For Human Consumption
FATS, EDIBLE	582.4101	DIACETYL TARTARIC ACID ESTERS OF MONO- AND DIGLYCERIDES OF EDIBLE FATS OR OILS, OR EDIBLE FAT-FORMING FATTY ACIDS Emulsifying Agents; Substances Generally Recognized As Safe
FATS, EDIBLE	582.4505	MONO- AND DIGLYCERIDES OF EDIBLE FATS OR OILS, OR EDIBLE FAT-FORMING ACIDS Emulsifying Agents; Substances Generally Recognized As Safe
FATS, EDIBLE	582.4521	MONOSODIUM PHOSPHATE DERIVATIVES OF MONO- AND DIGLYCERIDES OF EDIBLE FATS OR OILS, OR EDIBLE FAT-FORMING FATTY ACIDS Emulsifying Agents; Substances Generally Recognized As Safe
FATTY ACID	101.62	NUTRIENT CONTENT CLAIMS FOR FAT, FATTY ACID, AND CHOLESTEROL CONTENT OF FOODS Specific Requirements For Nutrient Content Claims; Food Labeling
FATTY ACID	172.859	SUCROSE FATTY ACID ESTERS Multipurpose Additives; Food Additives Permitted For Direct Addition To Food For Human Consumption
FATTY ACIDS	172.225	METHYL AND ETHYL ESTERS OF FATTY ACIDS PRODUCED FROM EDIBLE FATS AND OILS Coatings, Films And Related Substances; Food Additives Permitted For Direct Addition To Food For Human Consumption
FATTY ACIDS	72.848	LACTYLIC ESTERS OF FATTY ACIDS Multipurpose Additives; Food Additives Permitted For Direct Addition To Food For Human Consumption
FATTY ACIDS	172.850	LACTYLATED FATTY ACID ESTERS OF GLYCEROL AND PROPYLENE GLYCOL Multipurpose Additives; Food Additives Permitted For Direct Addition To Food For Human Consumption
FATTY ACIDS	172.852	GLYCERYL-LACTO ESTERS OF FATTY ACIDS Multipurpose Additives; Food Additives Permitted For Direct Addition To Food For Human Consumption

FATTY ACIDS	172.854	POLYGLYCEROL ESTERS OF FATTY ACIDS Multipurpose Additives; Food Additives Permitted For Direct Addition To Food For Human Consumption
FATTY ACIDS	172.856	PROPYLENE GLYCOL MONO- AND DIESTERS OF FATS AND FATTY ACIDS Multipurpose Additives; Food Additives Permitted For Direct Addition To Food For Human Consumption
FATTY ACIDS	172.860	FATTY ACIDS Multipurpose Additives; Food Additives Permitted For Direct Addition To Food For Human Consumption
FATTY ACIDS	172.862	OLEIC ACID DERIVED FROM TALL OIL FATTY ACIDS Multipurpose Additives; Food Additives Permitted For Direct Addition To Food For Human Consumption
FATTY ACIDS	172.863	SALTS OF FATTY ACIDS Multipurpose Additives; Food Additives Permitted For Direct Addition To Food For Human Consumption
FATTY ACIDS	573.640	METHYL ESTERS OF HIGHER FATTY ACIDS Food Additive Listing; Food Additives Permitted In Feed And Drinking Water Of Animals
FATTY ACIDS	573.914	SALTS OF VOLATILE FATTY ACIDS Food Additive Listing; Food Additives Permitted In Feed And Drinking Water Of Animals
FATTY ACIDS	582.4101	DIACETYL TARTARIC ACID ESTERS OF MONO- AND DIGLYCERIDES OF EDIBLE FATS OR OILS, OR EDIBLE FAT-FORMING FATTY ACIDS Emulsifying Agents; Substances Generally Recognized As Safe
FATTY ACIDS	582.4521	MONOSODIUM PHOSPHATE DERIVATIVES OF MONO- AND DIGLYCERIDES OF EDIBLE FATS OR OILS, OR EDIBLE FAT-FORMING FATTY ACIDS Emulsifying Agents; Substances Generally Recognized As Safe
FATTY ACIDS	862.1290	FATTY ACIDS TEST SYSTEM Clinical Chemistry Test Systems; Clinical Chemistry And Clinical Toxicology Devices
FATTY ALCOHOLS	172.864	SYNTHETIC FATTY ALCOHOLS Multipurpose Additives; Food Additives Permitted For Direct Addition To Food For Human Consumption
FATTY ALCOHOLS	178.3480	FATTY ALCOHOLS, SYNTHETIC Certain Adjuvants And Production Aids; Indirect Food Additives: Adjuvants, Production Aids, And Sanitizers
FATTY FOODS	176.170	COMPONENTS OF PAPER AND PAPERBOARD IN CONTACT WITH AQUEOUS AND FATTY FOODS Substances For Use Only As Components Of Paper And Paperboard; Indirect Food Additives
FC FRAGMENT	866.5530	IMMUNOGLOBULIN G (FC FRAGMENT SPECIFIC) IMMUNOLOGICAL TEST SYSTEM Immunological Test Systems; Immunology And Microbiology Devices
FD&C BLUE NO. 1	74.101	FD&C BLUE NO. 1 Foods; Listing Of Color Additives Subject To Certification
FD&C BLUE NO. 1	74.1101	FD&C BLUE NO. 1 Drugs; Listing Of Color Additives Subject To Certification
FD&C BLUE NO. 1	74.2101	FD&C BLUE NO. 1 Cosmetics; Listing Of Color Additives Subject To Certification
FD&C BLUE NO. 1	82.101	FD&C BLUE NO. 1 Foods, Drugs, And Cosmetics; Listing Of Certified Provisionally Listed Colors And Specifications
FD&C BLUE NO. 2	74.102	FD&C BLUE NO. 2 Foods; Listing Of Color Additives Subject To Certification
FD&C BLUE NO. 2	74.1102	FD&C BLUE NO. 2 Drugs; Listing Of Color Additives Subject To Certification
FD&C BLUE NO. 2	74.3102	FD&C BLUE NO. 2 Medical Devices; Listing Of Color Additives Subject To Certification
FD&C BLUE NO. 2	82.102	FD&C BLUE NO. 2 Foods, Drugs, And Cosmetics; Listing Of Certified Provisionally Listed Colors And Specifications
FD&C GREEN NO. 3	74.203	FD&C GREEN NO. 3 Foods; Listing Of Color Additives Subject To Certification
FD&C GREEN NO. 3	74.1203	FD&C GREEN NO. 3 Drugs; Listing Of Color Additives Subject To Certification
FD&C GREEN NO. 3	74.2203	FD&C GREEN NO. 3 Cosmetics; Listing Of Color Additives Subject To Certification
FD&C GREEN NO. 3	82.203	FD&C GREEN NO. 3 Foods, Drugs, And Cosmetics; Listing Of Certified Provisionally Listed Colors And Specifications
FD&C LAKES	82.51	LAKES (FD&C) Foods, Drugs, And Cosmetics; Listing Of Certified Provisionally Listed Colors And Specifications

FD&C RED NO. 3	74.303	FD&C RED NO. 3 Foods; Listing Of Color Additives Subject To Certification
FD&C RED NO. 3	74.1303	FD&C RED NO. 3 Drugs; Listing Of Color Additives Subject To Certification
FD&C RED NO. 4	74.1304	FD&C RED NO. 4 Drugs; Listing Of Color Additives Subject To Certification
FD&C RED NO. 4	74.2304	FD&C RED NO. 4 Cosmetics; Listing Of Color Additives Subject To Certification
FD&C RED NO. 4	82.304	FD&C RED NO. 4 Foods, Drugs, And Cosmetics; Listing Of Certified Provisionally Listed Colors And Specifications
FD&C RED NO. 40	74.340	FD&C RED NO. 40 Foods; Listing Of Color Additives Subject To Certification
FD&C RED NO. 40	74.1340	FD&C RED NO. 40 Drugs; Listing Of Color Additives Subject To Certification
FD&C RED NO. 40	74.2340	FD&C RED NO. 40 Cosmetics; Listing Of Color Additives Subject To Certification
FD&C YELLOW NO. 5	74.705	FD&C YELLOW NO. 5 Foods; Listing Of Color Additives Subject To Certification
FD&C YELLOW NO. 5	74.1705	FD&C YELLOW NO. 5 Drugs; Listing Of Color Additives Subject To Certification
FD&C YELLOW NO. 5	74.2705	FD&C YELLOW NO. 5 Cosmetics; Listing Of Color Additives Subject To Certification
FD&C YELLOW NO. 5	82.705	FD&C YELLOW NO. 5 Foods, Drugs, And Cosmetics; Listing Of Certified Provisionally Listed Colors And Specifications
FD&C YELLOW NO. 5/6	201.20	DECLARATION OF PRESENCE OF FD&C YELLOW NO. 5 AND/OR FD&C YELLOW NO. 6 IN CERTAIN DRUGS FOR HUMAN USE General Labeling Provisions; Labeling
FD&C YELLOW NO. 6	74.706	FD&C YELLOW NO. 6 Foods; Listing Of Color Additives Subject To Certification
FD&C YELLOW NO. 6	74.1706	FD&C YELLOW NO. 6 Drugs; Listing Of Color Additives Requiring Certification
FD&C YELLOW NO. 6	74.2706	FD&C YELLOW NO. 6 Cosmetics; Listing Of Color Additives Requiring Certification
FD&C YELLOW NO. 6	82.706	FD&C YELLOW NO. 6 Foods, Drugs, And Cosmetics; Listing Of Certified Provisionally Listed Colors And Specifications
FD FRAGMENT	866.5540	IMMUNOGLOBULIN G (FD FRAGMENT SPECIFIC) IMMUNOLOGICAL TEST SYSTEM Immunological Test Systems; Immunology And Microbiology Devices
FDA	5.20	GENERAL REDELEGATIONS OF AUTHORITY FROM THE COMMISSIONER TO OTHER OFFICERS OF THE FOOD AND DRUG ADMINISTRATION General Redelegations Of Authority; Delegations Of Authority And Organization
FDA	5.1105	CHIEF COUNSEL, FOOD AND DRUG ADMINISTRATION Organization; Delegations Of Authority And Organization
FDA	5.1110	FDA PUBLIC INFORMATION OFFICES Organization; Delegations Of Authority And Organization
FDA	10.90	FOOD AND DRUG ADMINISTRATION REGULATIONS, GUIDELINES, RECOMMENDATIONS, AND AGREEMENTS General Administrative Procedures; Administrative Practices And Procedures
FDA	14.15	COMMITTEES WORKING UNDER A CONTRACT WITH FDA General Provisions; Public Hearing Before A Public Advisory Committee
FDA	14.171	UTILIZATION OF AN ADVISORY COMMITTEE ON THE INITIATIVE OF FDA Advisory Committees For Human Prescription Drugs; Public Hearing Before A Public Advisory Committee
FDA	19.10	FOOD AND DRUG ADMINISTRATION CONFLICT OF INTEREST REVIEW BOARD Standards Of Conduct And Conflicts Of Interest
FDA	20.108	AGREEMENTS BETWEEN THE FOOD AND DRUG ADMINISTRATION AND OTHER DEPARTMENTS, AGENCIES, AND ORGANIZATIONS Availability Of Specific Categories Of Records; Public Information
FDA	20.111	DATA AND INFORMATION SUBMITTED VOLUNTARILY TO THE FOOD AND DRUG ADMINISTRATION Availability Of Specific Categories Of Records; Public Information
FDA	314.102	COMMUNICATIONS BETWEEN FDA AND APPLICANTS FDA Action On Applications; Applications For FDA Approval To Market A New Drug
FDA ACCESS	806.30	FDA ACCESS TO RECORDS Medical Device Corrections And Removals
FDA ACTION	60.20	FDA ACTION ON REGULATORY REVIEW PERIOD DETERMINATIONS Regulatory Review Period Determinations; Patent Term Restoration
FDA ACTION	60.34	FDA ACTION ON PETITIONS Due Diligence Petitions; Patent Term Restoration

FDA ACTION	807.100	FDA ACTION ON A PREMARKET NOTIFICATION Establishment Registration And Device Listing For Manufacturers And Initial Importers Of Devices
FDA ACTION	812.30	FDA ACTION ON APPLICATIONS Application And Administrative Action; Investigational Device Exemptions
FDA APPROVAL	812.42	FDA AND IRB APPROVAL Responsibilities Of Sponsors; Investigational Device Exemptions
FDA ASSISTANCE	60.10	FDA ASSISTANCE ON ELIGIBILITY Eligibility Assistance; Patent Term Restoration
FDA DETERMINATIONS	20.28	FOOD AND DRUG ADMINISTRATION DETERMINATIONS OF CONFIDENTIALITY General Policy; Public Information
FDA EMPLOYEE NAMES	20.32	DISCLOSURE OF FOOD AND DRUG ADMINISTRATION EMPLOYEE NAMES General Policy; Public Information
FDA EMPLOYEES	20.1	TESTIMONY BY FOOD AND DRUG ADMINISTRATION EMPLOYEES Official Testimony And Information; Public Information
FDA EMPLOYEES	20.2	PRODUCTION OF RECORDS BY FOOD AND DRUG ADMINISTRATION EMPLOYEES Official Testimony And Information; Public Information
FDA EMPLOYEES	20.110	DATA AND INFORMATION ABOUT FOOD AND DRUG ADMINISTRATION EMPLOYEES Availability Of Specific Categories Of Records; Public Information
FDA FILES	20.29	PROHIBITION ON WITHDRAWAL OF RECORDS FROM FOOD AND DRUG ADMINISTRATION FILES General Policy; Public Information
FDA FREEDOM OF INFORMATION	20.30	FOOD AND DRUG ADMINISTRATION FREEDOM OF INFORMATION STAFF General Policy; Public Information
FDA FUNDS	20.105	TESTING AND RESEARCH CONDUCTED BY OR WITH FUNDS PROVIDED BY THE FOOD AND DRUG ADMINISTRATION Availability Of Specific Categories Of Records; Public Information
FDA FUNDS	20.106	STUDIES AND REPORTS PREPARED BY OR WITH FUNDS PROVIDED BY THE FOOD AND DRUG ADMINISTRATION Availability Of Specific Categories Of Records; Public Information
FDA MANUALS	20.107	FOOD AND DRUG ADMINISTRATION MANUALS Availability Of Specific Categories Of Records; Public Information
FDA PRIVACY ACT	21.20	PROCEDURES FOR NOTICE OF FOOD AND DRUG ADMINISTRATION PRIVACY ACT RECORD SYSTEMS FDA Privacy Act Record Systems; Protection Of Privacy
FDA RECALL	7.45	FDA-REQUESTED RECALL Recalls - Guidance On Policy, Procedures & Industry Responsibilities; Enforcement Policy
FDA RECALL	107.200	FDA-REQUIRED RECALL Infant Formula Recalls; Infant Formula
FDA RECOGNITION	316.34	FDA RECOGNITION OF EXCLUSIVE APPROVAL Orphan Drugs
FDA RECORDS	20.3	CERTIFICATION AND AUTHENTICATION OF FOOD AND DRUG ADMINISTRATION RECORDS Official Testimony And Information; Public Information
FDA RECORDS	20.20	POLICY ON DISCLOSURE OF FOOD AND DRUG ADMINISTRATION RECORDS General Policy; Public Information
FDA RECORDS	20.31	RETENTION SCHEDULE OF REQUESTS FOR FOOD AND DRUG ADMINISTRATION RECORDS General Policy; Public Information
FDA REGULATORY RESEARCH	312.86	FOCUSED FDA REGULATORY RESEARCH Drugs Intended To Treat Life-threatening And Severely-debilitating Illnesses; Investigational New Drug Application
FDA REQUIREMENTS	320.30	INQUIRIES REGARDING BIOAVAILABILITY REQUIREMENTS AND REVIEW OF PROTOCOLS BY THE FOOD AND DRUG ADMINISTRATION Procedures For Determining The Bioavailability Of Drug Products; Bioavailability And Bioequivalence Requirements
FDA REQUIREMENTS	320.35	REQUIREMENTS FOR BATCH TESTING AND CERTIFICATION BY THE FOOD AND DRUG ADMINISTRATION Procedures For Determining The Bioavailability Of Drug Products; Bioavailability And Bioequivalence Requirements
FEBANTEL	520.903	FEBANTEL ORAL DOSAGE FORMS Oral Dosage Form New Animal Drugs
FEBANTEL-PRAZIQUANTEL	520.903d	FEBANTEL-PRAZIQUANTEL PASTE Oral Dosage Form New Animal Drugs
FEBANTEL	520.1872	PRAZIQUANTEL, PYRANTEL PAMOATE AND FEBANTEL TABLETS Oral Dosage Form New Animal Drugs
FEDERAL CAUSTIC POISON ACT	2.110	DEFINITION OF AMMONIA UNDER FEDERAL CAUSTIC POISON ACT Caustic Poisons; General Administrative Rulings And Decisions

FEDERAL DEPTS/AGENCIES	20.85	DISCLOSURE TO OTHER FEDERAL GOVERNMENT DEPARTMENTS AND AGENCIES Limitations On Exemptions; Public Information
FEDERAL IMPORT MILK ACT	1210.27	PERMITS WAIVING CLAUSES 2 AND 5, SECTION 2 OF THE FEDERAL IMPORT MILK ACT Permit Control; Regulations Under The Federal Import Milk Act
FEDERAL IMPORT MILK ACT	1210.28	PERMITS WAIVING CLAUSE 4, SECTION 2 OF THE FEDERAL IMPORT MILK ACT Permit Control; Regulations Under The Federal Import Milk Act
FEDERAL LAW	1307.02	APPLICATION OF STATE LAW AND OTHER FEDERAL LAW General Information; Miscellaneous
FEDERAL REGISTER	5.400	ISSUANCE OF FEDERAL REGISTER DOCUMENTS TO RECOGNIZE OR TO WITHDRAW RECOGNITION OF A STANDARD TO MEET PREMARKET SUBMISSION REQUIREMENTS Medical Devices And Radiological Health; Redelegations Of Authority; Delegations Of Authority And Organization
FEDERAL REGISTER	5.401	ISSUANCE OF FEDERAL REGISTER DOCUMENTS PERTAINING TO EXEMPTIONS FROM PREMARKET NOTIFICATION Medical Devices And Radiological Health; Redelegations Of Authority; Delegations Of Authority And Organization
FEDERAL REGISTER	5.500	ISSUANCE OF FEDERAL REGISTER DOCUMENTS PERTAINING TO THE DETERMINATION OF SAFE LEVELS, NOTICE OF NEED FOR DEVELOPMENT OF AN ANALYTICAL METHOD, PROHIBITION OF CERTAIN EXTRALABEL DRUG USE Animal Drugs; Redelegations Of Authority; Delegations Of Authority And Organization
FEDERAL REGISTER	10.80	DISSEMINATION OF DRAFT FEDERAL REGISTER NOTICES AND REGULATIONS General Administrative Procedures; Administrative Practices And Procedures
FEDERAL REGULATIONS	100.2	STATE ENFORCEMENT OF FEDERAL REGULATIONS General
FEED	558.465	POLOXALENE FREE-CHOICE LIQUID TYPE C FEED Specific New Animal Drugs For Use In Animal Feeds; New Animal Drugs For Use In Animal Feeds
FEED, ANIMAL	2.35	USE OF SECONDHAND CONTAINERS FOR THE SHIPMENT AND STORAGE OF FOOD AND ANIMAL FEED Human And Animal Foods; General Administrative Rulings And Decisions
FEED, ANIMAL	500.45	USE OF POLYCHLORINATED BIPHENYLS (PCB'S) IN THE PRODUCTION, HANDLING AND STORAGE OF ANIMAL FEED Specific Administrative Rulings And Decisions; General
FEED, ANIMAL	501.110	ANIMAL FEED LABELING; COLLECTIVE NAMES FOR FEED INGREDIENTS Exemptions From Animal Food Labeling Requirements; Animal Food Labeling
FEED, ANIMAL	510.7	CONSIGNEES OF NEW ANIMAL DRUGS FOR USE IN THE MANUFACTURE OF ANIMAL FEED General Provisions; New Animal Drugs
FEED, ANIMAL	510.305	MAINTENANCE OF COPIES OF APPROVED APPLICATIONS FOR ANIMAL FEED BEARING OR CONTAINING NEW ANIMAL DRUGS Records And Reports; New Animal Drugs
FEED, ANIMAL	570.13	INDIRECT FOOD ADDITIVES RESULTING FROM PACKAGING MATERIALS PRIOR SANCTIONED FOR ANIMAL FEED AND PET FOOD General Provisions; Food Additives
FEED, ANIMAL	570.14	INDIRECT FOOD ADDITIVES RESULTING FROM PACKAGING MATERIALS FOR ANIMAL FEED AND PET FOOD General Provisions; Food Additives
FEED, ANIMAL	589.1	SUBSTANCES PROHIBITED FROM USE IN ANIMAL FOOD OR FEED General Provisions; Substances Prohibited From Use In Animal Food Or Feed
FEED, ANIMALS	558.15	ANTIBIOTIC, NITROFURAN, AND SULFONAMIDE DRUGS IN THE FEED OF ANIMALS General Provisions; New Animal Drugs For Use In Animal Feeds
FEED BLOCKS	520.2380a	THIABENDAZOLE TOP DRESSING AND MINERAL PROTEIN FEED BLOCKS Oral Dosage Form New Animal Drugs
FEED-GRADE	573.280	FEED-GRADE CALCIUM STEARATE AND SODIUM STEARATE Food Additive Listing; Food Additives Permitted In Feed And Drinking Water Of Animals
FEED-GRADE	573.220	FEED-GRADE BIURET Food Additive Listing; Food Additives Permitted In Feed And Drinking Water Of Animals
FEED MILL	5.501	APPROVAL OF NEW ANIMAL DRUG APPLICATIONS, MEDICATED FEED MILL LICENSE APPLICATIONS AND THEIR SUPPLEMENTS Animal Drugs; Redelegations Of Authority; Delegations Of Authority And Organization

FEED MILL	5.502	ISSUANCE OF NOTICES, PROPOSALS, AND ORDERS RELATING TO NEW ANIMAL DRUGS AND MEDICATED FEED MILL LICENSE APPLICATIONS Animal Drugs; Redelegations Of Authority; Delegations Of Authority And Organization
FEED MILL	510.301	RECORDS AND REPORTS CONCERNING EXPERIENCE WITH ANIMAL FEEDS BEARING OR CONTAINING NEW ANIMAL DRUGS FOR WHICH AN APPROVED MEDICATED FEED MILL LICENSE APPLICATION IS IN EFFECT Records And Reports; New Animal Drugs
FEED MILL	510.305	MAINTENANCE OF COPIES OF APPROVED MEDICATED FEED MILL LICENSES TO MANUFACTURE ANIMAL FEED BEARING OR CONTAINING NEW ANIMAL DRUGS Records And Reports; New Animal Drugs
FEED MILL	515.10	MEDICATED FEED MILL LICENSE APPLICATIONS Applications; Medicated Feed Mill License
FEED MILL	515.11	SUPPLEMENTAL MEDICATED FEED MILL LICENSE APPLICATIONS Applications; Medicated Feed Mill License
FEED MILL	515.20	APPROVAL OF MEDICATED FEED MILL LICENSE APPLICATIONS Administrative Actions On Licenses; Medicated Feed Mill License
FEED MILL	515.21	REFUSAL TO APPROVE A MEDICATED FEED MILL LICENSE APPLICATION Administrative Actions On Licenses; Medicated Feed Mill License
FEED MILL	515.22	SUSPENSION AND/OR REVOCATION OF APPROVAL OF A MEDICATED FEED MILL LICENSE Administrative Actions On Licenses; Medicated Feed Mill License
FEED MILL	515.23	VOLUNTARY REVOCATION OF A MEDICATED FEED MILL LICENSE Administrative Actions On Licenses; Medicated Feed Mill License
FEED MILL	515.24	NOTICE OF REVOCATION OF A MEDICATED FEED MILL LICENSE Administrative Actions On Licenses; Medicated Feed Mill License
FEED MILL	515.25	REVOCATION OF ORDER REFUSING TO APPROVE A MEDICATED FEED MILL LICENSE APPLICATION OR SUSPENDING OR REVOKING A LICENSE Administrative Actions On Licenses; Medicated Feed Mill License
FEED MILL	558.4	REQUIREMENT OF A MEDICATED FEED MILL LICENSE General Provisions; New Animal Drugs For Use In Animal Feeds
FEED, POULTRY	579.40	IONIZING RADIATION FOR THE TREATMENT OF POULTRY FEED AND POULTRY FEED INGREDIENTS Irradiation In The Production, Processing And Handling Of Animal Feed And Pet Food
FEED, RUMINANT	589.2000	ANIMAL PROTEINS PROHIBITED IN RUMINANT FEED Substances Prohibited From Use In Animal Food Or Feed
FEED, VETERINARY	558.6	VETERINARY FEED DIRECTIVE DRUGS New Animal Drugs For Use In Animal Feeds
FEEDING	880.5640	LAMB FEEDING NIPPLE General Hospital And Personal Use Therapeutic Devices; General Hospital And Personal Use Devices
FEEDS	5.504	ISSUANCE OF WRITTEN NOTICES CONCERNING PATENT INFORMATION, CURRENT GOOD MANUFACTURING PRACTICES AND FALSE OR MISLEADING LABELING OF NEW ANIMAL DRUGS AND FEEDS BEARING OR CONTAINING NEW ANIMAL DRUGS Animal Drugs; Redelegations Of Authority; Delegations Of Authority And Organization
FEEDS	510.455	NEW ANIMAL DRUG REQUIREMENTS REGARDING FREE-CHOICE ADMINISTRATION IN FEEDS Requirements For Specific New Animal Drugs; New Animal Drugs
FEEDS	558.5	NEW ANIMAL DRUG REQUIREMENTS FOR LIQUID TYPE B FEED General Provisions; New Animal Drugs For Use In Animal Feeds
FEEDS, ANIMAL	500.29	GENTIAN VIOLET FOR USE IN ANIMAL FEEDS Specific Administrative Rulings And Decisions; General
FEEDS, ANIMAL	500.35	ANIMAL FEEDS CONTAMINATED WITH SALMONELLA MICROORGANISMS Specific Administrative Rulings And Decisions; General
FEEDS, ANIMAL	510.301	RECORDS AND REPORTS CONCERNING EXPERIENCE WITH ANIMAL FEEDS BEARING OR CONTAINING NEW ANIMAL DRUGS FOR WHICH AN APPROVED MEDICATED FEED MILL LICENSE APPLICATION IS IN EFFECT Records And Reports; New Animal Drugs
FEEDS, ANIMAL	510.515	ANIMAL FEEDS BEARING OR CONTAINING NEW ANIMAL DRUGS SUBJECT TO THE PROVISIONS OF SECTION 512(N) OF THE ACT Animal Use Exemptions From Certification And Labeling Requirements; New Animal Drugs
FEEDS, ANIMAL	573.380	ETHOXYQUIN IN ANIMAL FEEDS Food Additive Listing; Food Additives Permitted In Feed And Drinking Water Of Animals
FEEDS, ANIMAL	582.80	TRACE MINERALS ADDED TO ANIMAL FEED General Provisions; Substances Generally Recognized As Safe

FEES	5.108	AUTHORITY RELATING TO WAIVERS OR REDUCTIONS OF PRESCRIPTION DRUG USER FEES Human Drugs; Redelegations Of Authority; Delegations Of Authority And Organization
FEES	14.155	FEES AND COMPENSATION PERTAINING TO A COLOR ADDITIVE ADVISORY COMMITTEE Color Additive Advisory Committees; Public Hearing Before A Public Advisory Committee
FEES	17.29	FEES Civil Money Penalties Hearings
FEES	20.42	FEES TO BE CHARGED Procedures And Fees; Public Information
FEES	20.43	WAIVER OF FEES Procedures And Fees; Public Information
FEES	21.45	FEES Procedures For Notification Of And Access To Records In Privacy Act Record Systems; Protection Of Privacy
FEES	26.77	FEES Framework Provisions; Mutual Recognition Of Pharmaceutical GMP Reports, Medical Device Quality System Audit Reports
FEES	70.19	FEES FOR LISTING General Provisions; Color Additives
FEES	80.10	FEES FOR CERTIFICATION SERVICES General Procedures; Color Additive Certification
FEES	1301.13	APPLICATION FOR REGISTRATION; APPLICATION FORMS, FEES, CONTENTS, AND SIGNATURE; APPLICATION FORMS Registration Of Manufacturers, Distributors, And Dispensers Of Controlled Substances
FEES	1301.21	EXCEPTION FROM FEES Registration Of Manufacturers, Distributors, And Dispensers Of Controlled Substances
FEES	1309.11	FEE AMOUNTS Registration Of Manufacturers, Distributors, Importers And Exporters Of List I Chemicals
FEES	1401.10	FEES TO BE CHARGED - GENERAL Public Availability Of Information
FEES	1401.11	FEES TO BE CHARGED - MISCELLANEOUS PROVISIONS Public Availability Of Information
FEES	1401.12	FEES TO BE CHARGED - CATEGORIES OF REQUESTERS Public Availability Of Information
FEES	1401.13	WAIVER OR REDUCTION OF FEES Public Availability Of Information
FEES	1402.6	FEES Mandatory Declassification Review
FELLOWSHIPS	5.26	SERVICE FELLOWSHIPS General Redelegations Of Authority; Delegations Of Authority And Organization
FELONY CONVICTION	1309.72	FELONY CONVICTION; EMPLOYER RESPONSIBILITIES Registration Of Manufacturers, Distributors, Importers And Exporters Of List I Chemicals
FEMALE CONDOM	884.5330	FEMALE CONDOM Obstetrical And Gynecological Therapeutic Devices; Obstetrical And Gynecological Devices
FEMININE	740.12	FEMININE DEODORANT SPRAYS Warning Statements; Cosmetic Product Warning Statements
FEMORAL	888.3360	HIP JOINT FEMORAL (HEMI-HIP) METALLIC CEMENTED OR UNCEMENTED PROSTHESIS Prosthetic Devices; Orthopedic Devices
FEMORAL	888.3370	HIP JOINT (HEMI-HIP) ACETABULAR METAL CEMENTED PROSTHESIS Prosthetic Devices; Orthopedic Devices
FEMORAL	888.3380	HIP JOINT FEMORAL (HEMI-HIP) TRUNION-BEARING METAL/POLYACETAL CEMENTED PROSTHESIS Prosthetic Devices; Orthopedic Devices
FEMORAL	888.3390	HIP JOINT FEMORAL (HEMI-HIP) METAL/POLYMER CEMENTED OR UNCEMENTED PROSTHESIS Prosthetic Devices; Orthopedic Devices
FEMORAL	888.3400	HIP JOINT FEMORAL (HEMI-HIP) METALLIC RESURFACING PROSTHESIS Prosthetic Devices; Orthopedic Devices
FEMORAL	888.3570	KNEE JOINT FEMORAL (HEMI-KNEE) METALLIC UNCEMENTED PROSTHESIS Prosthetic Devices; Orthopedic Devices
FEMOROTIBIAL	888.3480	KNEE JOINT FEMOROTIBIAL METALLIC CONSTRAINED CEMENTED PROSTHESIS Prosthetic Devices; Orthopedic Devices

FEMOROTIBIAL	888.3490	KNEE JOINT FEMOROTIBIAL METAL/COMPOSITE NON-CONSTRAINED CEMENTED PROSTHESIS Prosthetic Devices; Orthopedic Devices
FEMOROTIBIAL	888.3500	KNEE JOINT FEMOROTIBIAL METAL/COMPOSITE SEMI-CONSTRAINED CEMENTED PROSTHESIS Prosthetic Devices; Orthopedic Devices
FEMOROTIBIAL	888.3510	KNEE JOINT FEMOROTIBIAL METAL/POLYMER CONSTRAINED CEMENTED PROSTHESIS Prosthetic Devices; Orthopedic Devices
FEMOROTIBIAL	888.3520	KNEE JOINT FEMOROTIBIAL METAL/POLYMER NON-CONSTRAINED CEMENTED PROSTHESIS Prosthetic Devices; Orthopedic Devices
FEMOROTIBIAL	888.3530	KNEE JOINT FEMOROTIBIAL METAL/POLYMER SEMI-CONSTRAINED CEMENTED PROSTHESIS Prosthetic Devices; Orthopedic Devices
FENBENDAZOLE	520.905	FENBENDAZOLE ORAL DOSAGE FORMS Oral Dosage Form New Animal Drugs
FENBENDAZOLE	556.275	FENBENDAZOLE Specific Tolerances For Residues Of New Animal Drugs; Tolerances For Residues Of New Animal Drugs In Food
FENBENDAZOLE	558.258	FENBENDAZOLE Specific New Animal Drugs For Use In Animal Feeds; New Animal Drugs For Use In Animal Feeds
FENPROSTALENE	522.914	FENPROSTALENE SOLUTION Implantation Or Injectable Dosage Form New Animal Drugs
FENPROSTALENE	556.277	FENPROSTALENE Specific Tolerances For Residues Of New Animal Drugs; Tolerances For Residues Of New Animal Drugs In Food
FENTANYL CITRATE	522.800	DROPERIDOL AND FENTANYL CITRATE INJECTION Implantation Or Injectable Dosage Form New Animal Drugs
FENTHION	524.920	FENTHION Ophthalmic And Topical Dosage Form New Animal Drugs
FERMENTATION	573.500	CONDENSED, EXTRACTED GLUTAMIC ACID FERMENTATION PRODUCT Food Additive Listing; Food Additives Permitted In Feed And Drinking Water Of Animals
FERMENTED WHEY	573.450	FERMENTED AMMONIATED CONDENSED WHEY Food Additive Listing; Food Additives Permitted In Feed And Drinking Water Of Animals
FERRIC AMMONIUM CITRATE	73.1025	FERRIC AMMONIUM CITRATE Drugs; Listing Of Color Additives Exempt From Certification
FERRIC AMMONIUM CITRATE	184.1296	FERRIC AMMONIUM CITRATE Listing Of Specific Substances Affirmed As GRAS; Direct Food Substances Affirmed As Generally Recognized As Safe
FERRIC AMMONIUM FERROCYAN	73.1298	FERRIC AMMONIUM FERROCYANIDE Drugs; Listing Of Color Additives Exempt From Certification
FERRIC AMMONIUM FERROCYAN	73.2298	FERRIC AMMONIUM FERROCYANIDE Cosmetics; Listing Of Color Additives Exempt From Certification
FERRIC CHLORIDE	184.1297	FERRIC CHLORIDE Listing Of Specific Substances Affirmed As GRAS; Direct Food Substances Affirmed As Generally Recognized As Safe
FERRIC CITRATE	184.1298	FERRIC CITRATE Listing Of Specific Substances Affirmed As GRAS; Direct Food Substances Affirmed As Generally Recognized As Safe
FERRIC FERROCYANIDE	73.1299	FERRIC FERROCYANIDE Drugs; Listing Of Color Additives Exempt From Certification
FERRIC FERROCYANIDE	73.2299	FERRIC FERROCYANIDE Cosmetics; Listing Of Color Additives Exempt From Certification
FERRIC OXIDE	186.1300	FERRIC OXIDE Listing Of Specific Substances Affirmed As GRAS; Indirect Food Substances Affirmed As Generally Recognized As Safe
FERRIC OXIDE	522.940	COLLOIDAL FERRIC OXIDE INJECTION Implantation Or Injectable Dosage Form New Animal Drugs
FERRIC PHOSPHATE	184.1301	FERRIC PHOSPHATE Listing Of Specific Substances Affirmed As GRAS; Direct Food Substances Affirmed As Generally Recognized As Safe
FERRIC PHOSPHATE	582.5301	FERRIC PHOSPHATE Nutrients And/or Dietary Supplements; Substances Generally Recognized As Safe
FERRIC PYROPHOSPHATE	184.1304	FERRIC PYROPHOSPHATE Listing Of Specific Substances Affirmed As GRAS; Direct Food Substances Affirmed As Generally Recognized As Safe
FERRIC PYROPHOSPHATE	582.5304	FERRIC PYROPHOSPHATE Nutrients And/or Dietary Supplements; Substances Generally Recognized As Safe
FERRIC PYROPHOSPHATE	582.5306	FERRIC SODIUM PYROPHOSPHATE Nutrients And/or Dietary Supplements; Substances Generally Recognized As Safe

FERRIC SULFATE	184.1307	FERRIC SULFATE Listing Of Specific Substances Affirmed As GRAS; Direct Food Substances Affirmed As Generally Recognized As Safe
FERRITIN	866.5340	FERRITIN IMMUNOLOGICAL TEST SYSTEM Immunological Test Systems; Immunology And Microbiology Devices
FERROCYANIDE	73.1298	FERRIC AMMONIUM FERROCYANIDE Drugs; Listing Of Color Additives Exempt From Certification
FERROCYANIDE	73.1299	FERRIC FERROCYANIDE Drugs; Listing Of Color Additives Exempt From Certification
FERROCYANIDE	73.2298	FERRIC AMMONIUM FERROCYANIDE Cosmetics; Listing Of Color Additives Exempt From Certification
FERROCYANIDE	73.2299	FERRIC FERROCYANIDE Cosmetics; Listing Of Color Additives Exempt From Certification
FERROUS ASCORBATE	184.1307a	FERROUS ASCORBATE Listing Of Specific Substances Affirmed As GRAS; Direct Food Substances Affirmed As Generally Recognized As Safe
FERROUS CARBONATE	184.1307b	FERROUS CARBONATE Listing Of Specific Substances Affirmed As GRAS; Direct Food Substances Affirmed As Generally Recognized As Safe
FERROUS CITRATE	184.1307c	FERROUS CITRATE Listing Of Specific Substances Affirmed As GRAS; Direct Food Substances Affirmed As Generally Recognized As Safe
FERROUS FUMARATE	184.1307d	FERROUS FUMARATE Listing Of Specific Substances Affirmed As GRAS; Direct Food Substances Affirmed As Generally Recognized As Safe
FERROUS GLUCONATE	73.160	FERROUS GLUCONATE Foods; Listing Of Color Additives Exempt From Certification
FERROUS GLUCONATE	184.1308	FERROUS GLUCONATE Listing Of Specific Substances Affirmed As GRAS; Direct Food Substances Affirmed As Generally Recognized As Safe
FERROUS GLUCONATE	582.5308	FERROUS GLUCONATE Nutrients And/or Dietary Supplements; Substances Generally Recognized As Safe
FERROUS LACTATE	73.165	FERROUS LACTATE Listing Of Color Additives Exempt From Certification
FERROUS LACTATE	184.1311	FERROUS LACTATE Listing Of Specific Substances Affirmed As GRAS; Direct Food Substances Affirmed As Generally Recognized As Safe
FERROUS LACTATE	582.5311	FERROUS LACTATE Nutrients And/or Dietary Supplements; Substances Generally Recognized As Safe
FERROUS SULFATE	184.1315	FERROUS SULFATE Listing Of Specific Substances Affirmed As GRAS; Direct Food Substances Affirmed As Generally Recognized As Safe
FERROUS SULFATE	582.5315	FERROUS SULFATE Nutrients And/or Dietary Supplements; Substances Generally Recognized As Safe
FETAL BLOOD	884.1560	FETAL BLOOD SAMPLER Obstetrical And Gynecological Diagnostic Devices; Obstetrical And Gynecological Devices
FETAL EXTRACTOR	884.4340	FETAL VACUUM EXTRACTOR Obstetrical And Gynecological Surgical Devices; Obstetrical And Gynecological Devices
FETAL DESTRUCTIVE	884.4500	OBSTETRIC FETAL DESTRUCTIVE INSTRUMENT Obstetrical And Gynecological Surgical Devices; Obstetrical And Gynecological Devices
FETAL HEMOGLOBIN	864.7455	FETAL HEMOGLOBIN ASSAY Hematology Kits And Packages; Hematology And Pathology Devices
FETAL MONITOR	884.2600	FETAL CARDIAC MONITOR Obstetrical And Gynecological Monitoring Devices; Obstetrical And Gynecological Devices
FETAL MONITOR	884.2620	FETAL ENCEPHALOGRAPHIC MONITOR Obstetrical And Gynecological Monitoring Devices; Obstetrical And Gynecological Devices
FETAL MONITOR	884.2640	FETAL PHONOCARDIOGRAPHIC MONITOR Obstetrical And Gynecological Monitoring Devices; Obstetrical And Gynecological Devices
FETAL MONITOR	884.2660	FETAL ULTRASONIC MONITOR AND ACCESSORIES Obstetrical And Gynecological Monitoring Devices; Obstetrical And Gynecological Devices
FETAL SCALP	884.2675	FETAL SCALP CIRCULAR (SPIRAL) ELECTRODE AND APPLICATOR Obstetrical And Gynecological Monitoring Devices; Obstetrical And Gynecological Devices
FETAL SCALP	884.2685	FETAL SCALP CLIP ELECTRODE AND APPLICATOR Obstetrical And Gynecological Monitoring Devices; Obstetrical And Gynecological Devices

FETAL STETHOSCOPE	884.2900	FETAL STETHOSCOPE Obstetrical And Gynecological Monitoring Devices; Obstetrical And Gynecological Devices
FETOSCOPE	884.1600	TRANSABDOMINAL AMNIOSCOPE (FETOSCOPE) AND ACCESSORIES Obstetrical And Gynecological Diagnostic Devices; Obstetrical And Gynecological Devices
FEVER BLISTERS	310.537	DRUG PRODUCTS CONTAINING ACTIVE INGREDIENTS OFFERED OTC FOR TREATMENT OF FEVER BLISTERS AND COLD SORES Requirements For Specific New Drugs Or Devices; New Drugs
FIBER	101.76	HEALTH CLAIMS: FIBER-CONTAINING GRAIN PRODUCT, FRUITS, AND VEGETABLES AND CANCER Specific Requirements For Health Claims; Food Labeling
FIBER	101.77	HEALTH CLAIMS: FRUITS, VEGETABLES, AND GRAIN PRODUCTS THAT CONTAIN FIBER, PARTICULARLY SOLUBLE FIBER, AND RISK OF CORONARY HEART DISEASE Specific Requirements For Health Claims; Food Labeling
FIBER	101.81	HEALTH CLAIMS: SOLUBLE FIBER FROM CERTAIN FOODS AND RISK OF CORONARY HEART DISEASE Specific Requirements For Health Claims; Food Labeling
FIBER	176.260	PULP FROM RECLAIMED FIBER Substances For Use Only As Components Of Paper And Paperboard; Indirect Food Additives
FIBER	880.5300	MEDICAL ABSORBENT FIBER General Hospital And Personal Use Therapeutic Devices; General Hospital And Personal Use Devices
FIBER OPTIC	872.4620	FIBER OPTIC DENTAL LIGHT Surgical Devices; Dental Devices
FIBEROPTIC	870.1230	FIBEROPTIC OXIMETER CATHETER Cardiovascular Diagnostic Devices; Cardiovascular Devices
FIBEROPTIC	874.4350	EAR, NOSE AND THROAT FIBEROPTIC LIGHT SOURCE AND CARRIER Surgical Devices; Ear, Nose And Throat Devices
FIBEROPTIC	876.4020	FIBEROPTIC LIGHT URETERAL CATHETER Surgical Devices; Gastroenterology-urology Devices
FIBEROPTIC	876.4530	GASTROENTEROLOGY-UROLOGY FIBEROPTIC RETRACTOR Surgical Devices; Gastroenterology-urology Devices
FIBERS, CARBON	878.3500	POLYTETRAFLUOROETHYLENE WITH CARBON FIBERS COMPOSITE IMPLANT MATERIAL Prosthetic Devices; General And Plastic Surgery Devices
FIBERS, HAIR	895.101	PROSTHETIC HAIR FIBERS Listing Of Banned Devices; Banned Devices
FIBERS, TEXTILE	177.2800	TEXTILES AND TEXTILE FIBERS Substances For Use Only As Components Of Articles Intended For Repeated Use; Indirect Food Additives: Polymers
FIBRIN	864.7300	FIBRIN MONOMER PARACOAGULATION TEST Hematology Kits And Packages; Hematology And Pathology Devices
FIBRINOGEN	864.7340	FIBRINOGEN DETERMINATION SYSTEM Hematology Kits And Packages; Hematology And Pathology Devices
FIBRINOGEN/FIBRIN	864.7320	FIBRINOGEN/FIBRIN DEGRADATION PRODUCTS ASSAY Hematology Kits And Packages; Hematology And Pathology Devices
FIBRINOPEPTIDE A	866.5350	FIBRINOPEPTIDE A IMMUNOLOGICAL TEST SYSTEM Immunological Test Systems; Immunology And Microbiology Devices
FICIN	184.1316	FICIN Listing Of Specific Substances Affirmed As GRAS; Direct Food Substances Affirmed As Generally Recognized As Safe
FIELD CORN	155.131	CANNED FIELD CORN Requirements For Specific Standardized Canned Vegetables; Canned Vegetables
FIELD STRUCTURE	5.1115	FIELD STRUCTURE Organization; Delegations Of Authority And Organization
FIGLU	862.1305	FORMIMINOGLUTAMIC ACID (FIGLU) TEST SYSTEM Clinical Chemistry Test Systems; Clinical Chemistry And Clinical Toxicology Devices
FIGS	145.130	CANNED FIGS Requirements For Specific Standardized Canned Fruits; Canned Fruits
FIGS	145.131	ARTIFICIALLY SWEETENED CANNED FIGS Requirements For Specific Standardized Canned Fruits; Canned Fruits
FIGS	145.134	CANNED PRESERVED FIGS Requirements For Specific Standardized Canned Fruits; Canned Fruits
FILE	99.205	APPLICATION FOR EXEMPTION FROM THE REQUIREMENT TO FILE A SUPPLEMENTAL APPLICATION Dissemination Of Information On Unapproved/New Uses For Marketed Drugs, Biologics And Devices

FILE	99.305	EXEMPTION FROM THE REQUIREMENT TO FILE A SUPPLEMENTAL APPLICATION Dissemination Of Information On Unapproved/New Uses For Marketed Drugs, Biologics, and Devices
FILE	720.1	WHO SHOULD FILE Voluntary Filing Of Cosmetic Product Ingredient Composition Statements
FILE	720.3	HOW AND WHERE TO FILE Voluntary Filing Of Cosmetic Product Ingredient Composition Statements
FILE	803.22	WHEN NOT TO FILE Medical Device Reporting
FILE, ADMINISTRATIVE	10.70	DOCUMENTATION OF SIGNIFICANT DECISIONS IN ADMINISTRATIVE FILE General Administrative Procedures; Administrative Practices And Procedures
FILE, ADVERSE REACTION	606.170	ADVERSE REACTION FILE Records And Reports; Current Good Manufacturing Practice For Blood And Blood Components
FILE APPLICATIONS	514.101	REASONS FOR REFUSING TO FILE APPLICATIONS Administrative Actions On Applications; New Animal Drug Applications
FILE, FAILURE TO	17.11	DEFAULT UPON FAILURE TO FILE AN ANSWER Civil Money Penalties Hearings
FILE, MASTER RECORD	225.102	MASTER RECORD FILE AND PRODUCTION RECORDS Records And Reports; Current Good Manufacturing Practice For Medicated Feeds
FILE, PMA	814.9	CONFIDENTIALITY OF DATA AND INFORMATION IN A PREMARKET APPROVAL APPLICATION (PMA) FILE General; Premarket Approval Of Medical Devices
FILE REPORTS	1304.03	PERSONS REQUIRED TO KEEP RECORDS AND FILE REPORTS Records and Reports Of Registrants
FILE REPORTS	1310.03	PERSONS REQUIRED TO KEEP RECORDS AND FILE REPORTS Records and Reports Of Listed Chemicals and Certain Machines
FILES	20.63	PERSONNEL, MEDICAL, AND SIMILAR FILES, DISCLOSURE OF WHICH CONSTITUTES A CLEARLY UNWARRANTED INVASION OF PERSONAL PRIVACY Exemptions; Public Information
FILES	803.18	FILES AND DISTRIBUTOR RECORDS General Provisions; Medical Device Reporting
FILES, COMPLAINT	211.198	COMPLAINT FILES Records And Reports; Current Good Manufacturing Practice For Finished Pharmaceuticals
FILES, COMPLAINT	225.115	COMPLAINT FILES Records And Reports; Current Good Manufacturing Practice For Medicated Feeds
FILES, COMPLAINT	226.115	COMPLAINT FILES Records And Reports; Current Good Manufacturing Practice For Type A Medicated Articles
FILES, COMPLAINT	820.198	COMPLAINT FILES Records; Good Manufacturing Practice For Medical Devices
FILES, DRUG MASTER	314.420	DRUG MASTER FILES Applications For FDA Approval To Market A New Drug
FILES, FDA	20.29	PROHIBITION ON WITHDRAWAL OF RECORDS FROM FOOD AND DRUG ADMINISTRATION FILES General Policy; Public Information
FILING	12.22	FILING OBJECTIONS AND REQUESTS FOR A HEARING ON A REGULATION OR ORDER Initiation Of Proceedings; Formal Evidentiary Public Hearing
FILING	12.23	NOTICE OF FILING OF OBJECTIONS Initiation Of Proceedings; Formal Evidentiary Public Hearing
FILING	12.80	FILING AND SERVICE OF SUBMISSIONS Hearing Procedures; Formal Evidentiary Public Hearing
FILING	17.31	FORM, FILING, AND SERVICE OF PAPERS Civil Money Penalties Hearings
FILING	20.40	FILING A REQUEST FOR RECORDS Procedures And Fees; Public Information
FILING	60.30	FILING, FORMAT, AND CONTENT OF PETITIONS Due Diligence Petitions; Patent Term Restoration
FILING	71.2	NOTICE OF FILING OF PETITION General Provisions; Color Additive Petitions
FILING	71.30	PROCEDURE FOR FILING OBJECTIONS TO REGULATIONS Administrative Action On Petitions; Color Additive Petitions
FILING	314.101	FILING AN APPLICATION AND RECEIVING AN ABBREVIATED NEW DRUG APPLICATION FDA Action On Applications; Applications For FDA Approval To Market A New Drug
FILING	601.2	APPLICATIONS FOR BIOLOGICS LICENSES; PROCEDURE FOR FILING General Provisions; Licensing
FILING	720.2	TIMES FOR FILING Voluntary Filing Of Cosmetic Product Ingredient Composition Statements

FILING	720.9	MISBRANDING BY REFERENCE TO FILING OR TO STATEMENT NUMBER Voluntary Filing Of Cosmetic Product Ingredient Composition Statements
FILING	814.42	FILING A PMA FDA Action On A PMA; Premarket Approval Of Medical Devices
FILING	814.112	FILING AN HDE Humanitarian Use Devices; Premarket Approval Of Medical Devices
FILING	1301.14	FILING OF APPLICATION; ACCEPTANCE FOR FILING; DEFECTIVE APPLICATIONS Registration Of Manufacturers, Distributors, And Dispensers Of Controlled Substances
FILING	1309.33	FILING OF APPLICATION; JOINT FILINGS Registration Of Manufacturers, Distributors, Importers And Exporters Of List I Chemicals
FILING	1309.34	ACCEPTANCE FOR FILING; DEFECTIVE APPLICATIONS Registration Of Manufacturers, Distributors, Importers And Exporters Of List I Chemicals
FILING	1316.80	TIME FOR FILING PETITIONS Seizure, Forfeiture, And Disposition Of Property; Administrative Functions, Practices, And Procedures
FILING	1316.97	INITIATING JUDICIAL FORFEITURE PROCEEDING AGAINST A CONVEYANCE WITHIN 60 DAYS OF FILING A CLAIM AND COST BOND Expedited Forfeiture Proceedings For Certain Property; Administrative Functions, Practices, And Procedures
FILINGS	1316.45	FILINGS; ADDRESS; HOURS Administrative Hearings; Administrative Functions, Practices, And Procedures
FILINGS, JOINT	1309.33	FILING OF APPLICATIONS; JOINT FILINGS Registration Of Manufacturers, Distributors, Importers And Exporters Of List I Chemicals
FILL	130.12	GENERAL METHODS FOR WATER CAPACITY AND FILL OF CONTAINERS General Provisions; Food Standards: General
FILL	130.14	GENERAL STATEMENTS OF SUBSTANDARD QUALITY AND SUBSTANDARD FILL OF CONTAINER General Provisions; Food Standards: General
FILL ORDER FORMS	1305.08	PERSONS ENTITLED TO FILL ORDER FORMS Order Forms
FILL PRESCRIPTIONS	1306.06	PERSONS ENTITLED TO FILL PRESCRIPTIONS General Information; Prescriptions
FILLING	872.3820	ROOT CANAL FILLING RESIN Prosthetic Devices; Dental Devices
FILLING ORDER FORMS	1305.09	PROCEDURE FOR FILLING ORDER FORMS Order Forms
FILLING ORDER FORMS	1305.16	SPECIAL PROCEDURE FOR FILLING CERTAIN ORDER FORMS Order Forms
FILLING PRESCRIPTIONS	1306.13	PARTIAL FILLING OF PRESCRIPTIONS Controlled Substances Listed In Schedule II; Prescriptions
FILLING PRESCRIPTIONS	1306.14	LABELING OF SUBSTANCES AND FILLING OF PRESCRIPTIONS Controlled Substances Listed In Schedule II; Prescriptions
FILLING PRESCRIPTIONS	1306.23	PARTIAL FILLING OF PRESCRIPTIONS Controlled Substances Listed In Schedules III, IV, AND V; Prescriptions
FILLINGS	872.3310	COATING MATERIAL FOR RESIN FILLINGS Prosthetic Devices; Dental Devices
FILM	177.1400	HYDROXYETHYL CELLULOSE FILM, WATER-INSOLUBLE Substances For Use As Basic Components Of Single And Repeated Use Food Contact Surfaces; Indirect Food Additives: Polymers
FILM	177.1670	POLYVINYL ALCOHOL FILM Substances For Use As Basic Components Of Single And Repeated Use Food Contact Surfaces; Indirect Food Additives: Polymers
FILM	892.1840	RADIOGRAPHIC FILM Diagnostic Devices; Radiology Devices
FILM	892.1850	RADIOGRAPHIC FILM CASSETTE Diagnostic Devices; Radiology Devices
FILM/CASSETTE	892.1860	RADIOGRAPHIC FILM/CASSETTE CHANGER Diagnostic Devices; Radiology Devices
FILM/CASSETTE	892.1870	RADIOGRAPHIC FILM/CASSETTE CHANGER PROGRAMMER Diagnostic Devices; Radiology Devices
FILM HOLDER	872.1905	DENTAL X-RAY FILM HOLDER Diagnostic Devices; Dental Devices
FILM ILLUMINATOR	892.1890	RADIOGRAPHIC FILM ILLUMINATOR Diagnostic Devices; Radiology Devices
FILM MARKING	892.1640	RADIOGRAPHIC FILM MARKING SYSTEM Diagnostic Devices; Radiology Devices
FILM, NYLON	175.360	VINYLIDENE CHLORIDE COPOLYMER COATINGS FOR NYLON FILM Substances For Use As Components Of Coatings; Indirect Food Additives

FILM, POLYCARBONATE	175.365	VINYLIDENE CHLORIDE COPOLYMER COATINGS FOR POLYCARBONATE FILM Substances For Use As Components Of Coatings; Indirect Food Additives:
FILM PROCESSOR	892.1900	AUTOMATIC RADIOGRAPHIC FILM PROCESSOR Diagnostic Devices; Radiology Devices
FILM, SPOT	892.1670	SPOT-FILM DEVICE Diagnostic Devices; Radiology Devices
FILMS, POLYOLEFIN	175.320	RESINOUS AND POLYMERIC COATINGS FOR POLYOLEFIN FILMS Substances For Use As Components Of Coatings; Indirect Food Additives
FILTER	868.5130	ANESTHESIA CONDUCTION FILTER Therapeutic Devices; Anesthesiology Devices
FILTER	868.5260	BREATHING CIRCUIT BACTERIAL FILTER Therapeutic Devices; Anesthesiology Devices
FILTER	870.3375	CARDIOVASCULAR INTRAVASCULAR FILTER Cardiovascular Prosthetic Devices; Cardiovascular Devices
FILTER	870.4260	CARDIOPULMONARY BYPASS ARTERIAL LINE BLOOD FILTER Cardiovascular Surgical Devices; Cardiovascular Devices
FILTER	870.4270	CARDIOPULMONARY BYPASS CARDIOTOMY SUCTION LINE BLOOD FILTER Cardiovascular Surgical Devices; Cardiovascular Devices
FILTER	870.4280	CARDIOPULMONARY PREBYPASS FILTER Cardiovascular Surgical Devices; Cardiovascular Devices
FILTERS	177.2250	FILTERS, MICROPOROUS POLYMERIC Substances For Use Only As Components Of Articles Intended For Repeated Use; Indirect Food Additives: Polymers
FILTERS	177.2260	FILTERS, RESIN-BONDED Substances For Use Only As Components Of Articles Intended For Repeated Use; Indirect Food Additives: Polymers
FILTERS	211.72	FILTERS Equipment; Current Good Manufacturing Practice For Finished Pharmaceuticals
FILTRATION, AIR	211.46	VENTILATION, AIR FILTRATION, AIR HEATING AND COOLING Buildings And Facilities; Current Good Manufacturing Practice For Finished Pharmaceuticals
FILTRATION, ULTRA	177.2910	ULTRA-FILTRATION MEMBRANES Substances For Use Only As Components Of Articles Intended For Repeated Use; Indirect Food Additives: Polymers
FINAL ACTION	10.45	COURT REVIEW OF FINAL ADMINISTRATIVE ACTION; EXHAUSTION OF ADMINISTRATIVE REMEDIES General Administrative Procedures; Administrative Practices And Procedures
FINAL ACTION	60.26	FINAL ACTION ON REGULATORY REVIEW PERIOD DETERMINATIONS Regulatory Review Period Determinations; Patent Term Restoration
FINAL ORDER	58.206	FINAL ORDER ON DISQUALIFICATION Disqualification Of Testing Facilities; Good Laboratory Practice For Nonclinical Laboratory Studies
FINAL ORDER	1301.36	SUSPENSION OR REVOCATION OF REGISTRATION; SUSPENSION PENDING FINAL ORDER; EXTENSION OF REGISTRATION PENDING FINAL ORDER Action On Application For Registration: Revocation Or Suspension Of Registration; Registration Of Manufacturers, Distributors, And Dispensers Of Controlled Substances
FINAL ORDER	1301.46	FINAL ORDER Hearings; Registration Manufacturers, Distributors, And Dispensers Of Controlled Substances
FINAL ORDER	1303.37	FINAL ORDER Hearings; Quotas
FINAL ORDER	1308.45	FINAL ORDER Hearings; Schedules Of Controlled Substances
FINAL ORDER	1309.44	SUSPENSION OF REGISTRATION PENDING FINAL ORDER Registration Of Manufacturers, Distributors, Importers And Exporters Of List I Chemicals
FINAL ORDER	1309.45	EXTENSION OF REGISTRATION PENDING FINAL ORDER Registration Of Manufacturers, Distributors, Importers And Exporters Of List I Chemicals
FINAL ORDER	1312.47	FINAL ORDER Hearings; Importation And Exportation Of Controlled Substances
FINAL ORDER	1313.56	FINAL ORDER Importation and Exportation Of Precursors And Essential Chemicals
FINAL ORDER	1316.67	FINAL ORDER Administrative Hearings; Administrative Functions, Practices, And Procedures
FINAL PACKAGE	1240.61	MANDATORY PASTEURIZATION FOR ALL MILK AND MILK PRODUCTS IN FINAL PACKAGE FORM INTENDED FOR DIRECT HUMAN CONSUMPTION Specific Administrative Decisions Regarding Interstate Shipments; Control Of Communicable Diseases
FINAL PRODUCT	640.82	TESTS ON FINAL PRODUCT Albumin (Human); Additional Standards For Human Blood And Blood Products

FINAL PRODUCT	640.92	TESTS ON FINAL PRODUCT Plasma Protein Fraction (Human); Additional Standards For Human Blood And Blood Products
FINAL PRODUCT	640.103	THE FINAL PRODUCT Immune Globulin (Human); Additional Standards For Human Blood And Blood Products
FINAL PROVISIONS	26.81	FINAL PROVISIONS Framework Provisions; Mutual Recognition Of Pharmaceutical GMP Reports, Medical Device Quality System Audit Reports
FINAL REGULATIONS	5.301	ISSUANCE OF INITIAL EMERGENCY PERMIT ORDERS AND NOTICES OF CONFIRMATION OF EFFECTIVE DATE OF FINAL REGULATIONS ON FOOD FOR HUMAN AND ANIMAL CONSUMPTION Foods And Cosmetics; Redelegations Of Authority; Delegations Of Authority And Organization
FINAL RULES	5.35	OFFICIALS AUTHORIZED TO MAKE CERTIFICATION UNDER 5 U.S.C. 605(B) FOR ANY PROPOSED AND FINAL RULES General Redelegations Of Authority; Delegations Of Authority And Organization
FINANCIAL INFORMATION	20.53	INDEXING TRADE SECRETS AND CONFIDENTIAL COMMERCIAL OR FINANCIAL INFORMATION Procedures And Fees; Public Information
FINANCIAL INFORMATION	20.61	TRADE SECRETS AND COMMERCIAL OR FINANCIAL INFORMATION WHICH IS PRIVILEGED OR CONFIDENTIAL Exemptions; Public Information
FINANCIAL INTERESTS	54.5	AGENCY EVALUATION OF FINANCIAL INTERESTS Financial Disclosure By Clinical Investigators
FINANCIAL MANAGEMENT	1403.20	STANDARDS FOR FINANCIAL MANAGEMENT Reports, Records, Retention, and Enforcement; Uniform Administrative Requirements For Grants And Coop. Agreements To State And Local Gov.
FINANCIAL REPORTING	1403.41	FINANCIAL REPORTING Post-award Requirements; Uniform Administrative Requirements For Grants And Coop. Agreements To State & Local Gov.
FINDINGS	25.41	ENVIRONMENTAL ASSESSMENTS AND FINDINGS OF NO SIGNIFICANT IMPACT Preparation Of Environmental Documents; Environmental Impact Considerations
FINDINGS	25.51	FINDINGS OF NO SIGNIFICANT IMPACT Public Participation And Notification Of Environmental Documents; Environmental Impact Considerations
FINDINGS	310.6	APPLICABILITY OF "NEW DRUG" OR SAFETY/EFFECTIVENESS FINDINGS IN DRUG EFFICACY STUDY IMPLEMENTATION NOTICES AND NOTICES OF OPPORTUNITY FOR HEARING TO IDENTICAL, RELATED, AND SIMILAR DRUG PRODUCTS General Provisions; New Drugs
FINDINGS	1316.64	PROPOSED FINDINGS OF FACT AND CONCLUSIONS OF LAW Administrative Hearings; Administrative Functions, Practices, And Procedures
FINGER EXERCISER	890.5410	POWERED FINGER EXERCISER Physical Medicine Therapeutic Devices; Physical Medicine Devices
FINGER JOINT	888.3200	FINGER JOINT METAL/METAL CONSTRAINED UNCEMENTED PROSTHESIS Prosthetic Devices; Orthopedic Devices
FINGER JOINT	888.3210	FINGER JOINT METAL/METAL CONSTRAINED CEMENTED PROSTHESIS Prosthetic Devices; Orthopedic Devices
FINGER JOINT	888.3220	FINGER JOINT METAL/POLYMER CONSTRAINED CEMENTED PROSTHESIS Prosthetic Devices; Orthopedic Devices
FINGER JOINT	888.3230	FINGER JOINT POLYMER CONSTRAINED CEMENTED PROSTHESIS Prosthetic Devices; Orthopedic Devices
FINISHED DEVICE	820.80	RECEIVING, IN-PROCESS, AND FINISHED DEVICE ACCEPTANCE Acceptance Activities; Quality System Regulation
FINISHED PRODUCT	106.30	FINISHED PRODUCT EVALUATION Quality Control Procedures For Assuring Nutrient Content Of Infant Formulas; Infant Formula Quality Control Procedures
FINISHED RESINS	181.26	DRYING OILS AS COMPONENTS OF FINISHED RESINS Specific Prior-sanctioned Food Ingredients; Prior-sanctioned Food Ingredients
FIRM-INITIATED RECALL	7.46	FIRM-INITIATED RECALL Recalls - Guidance On Policy, Procedures & Industry Responsibilities; Enforcement Policy
FIRM-INITIATED REMOVALS	107.210	FIRM-INITIATED RECALLS Infant Formula Recalls; Infant Formula
FIRMS	607.7	ESTABLISHMENT REGISTRATION AND PRODUCT LISTING OF BLOOD BANKS AND OTHER FIRMS MANUFACTURING HUMAN BLOOD AND BLOOD PRODUCTS General Provisions; Establishment Registratn+product Listing, Manufacturers Human Blood/b.products
FIRST AID	333.110	FIRST AID ANTIBIOTIC ACTIVE INGREDIENTS First Aid Antibiotic Drug Products; Topical Antimicrobial Drug Products For Over-the-counter Human Use

FIRST AID	333.150	LABELING OF FIRST AID ANTIBIOTIC DRUG PRODUCTS First Aid Antibiotic Drug Products; Topical Antimicrobial Drug Products For Over-the-counter Human Use
FISH	101.42	NUTRITION LABELING OF RAW FRUIT, VEGETABLES AND FISH Food Labeling
FISH	101.43	SUBSTANTIAL COMPLIANCE OF FOOD RETAILERS WITH THE GUIDELINES FOR THE VOLUNTARY NUTRITIONAL LABELING OF RAW FRUIT, VEGETABLES AND FISH Food Labeling
FISH	101.44	IDENTIFICATION OF THE 20 MOST FREQUENTLY CONSUMED RAW FRUIT, VEGETABLES AND FISH IN THE UNITED STATES Food Labeling
FISH	101.45	GUIDELINES FOR THE VOLUNTARY NUTRITIONAL LABELING OF RAW FRUIT, VEGETABLES AND FISH Food Labeling
FISH OIL	186.1551	HYDROGENATED FISH OIL Listing Of Specific Substances Affirmed As GRAS; Indirect Food Substances Affirmed As Generally Recognized As Safe
FISH PROTEIN	172.340	FISH PROTEIN ISOLATE Special Dietary And Nutritional Additives; Food Additives Permitted For Direct Addition To Food For Human Consumption
FISH PROTEIN	172.385	WHOLE FISH PROTEIN CONCENTRATE Special Dietary And Nutritional Additives; Food Additives Permitted For Direct Addition To Food For Human Consumption
FISH STICKS	102.45	FISH STICKS OR PORTIONS MADE FROM MINCED FISH Requirements For Specific Nonstandardized Foods; Common Or Usual Name For Nonstandardized Foods
FISSURE SEALANT	872.3765	PIT AND FISSURE SEALANT AND CONDITIONER Prosthetic Devices; Dental Devices
FITTING	870.4290	CARDIOPULMONARY BYPASS ADAPTOR, STOPCOCK, MANIFOLD, OR FITTING Cardiovascular Surgical Devices; Cardiovascular Devices
FIVE-DAY REPORTS	803.53	FIVE-DAY REPORTS Medical Device Reporting
FIVE HUNDRED TEN (K)	807.92	CONTENT AND FORMAT OF A 510(K) SUMMARY Establishment Registration And Device Listing For Manufacturers And Initial Importers Of Devices
FIVE HUNDRED TEN (K)	807.93	CONTENT AND FORMAT OF A 510(K) STATEMENT Establishment Registration And Device Listing For Manufacturers And Initial Importers Of Devices
FIXATION, BONE	888.3010	BONE FIXATION CERCLAGE Prosthetic Devices; Orthopedic Devices
FIXATION, BONE	888.3030	SINGLE/MULTIPLE COMPONENT METALLIC BONE FIXATION APPLIANCES AND ACCESSORIES Prosthetic Devices; Orthopedic Devices
FIXATION, BONE	888.3040	SMOOTH OR THREADED METALLIC BONE FIXATION FASTENER Prosthetic Devices; Orthopedic Devices
FIXATION DEVICE	868.5770	TRACHEAL TUBE FIXATION DEVICE Therapeutic Devices; Anesthesiology Devices
FIXATION DEVICE	886.1290	FIXATION DEVICE Diagnostic Devices; Ophthalmic Devices
FIXATION ORTHOSIS	888.3050	SPINAL INTERLAMINAL FIXATION ORTHOSIS Prosthetic Devices; Orthopedic Devices
FIXATION ORTHOSIS	888.3060	SPINAL INTERVERTEBRAL BODY FIXATION ORTHOSIS Prosthetic Devices; Orthopedic Devices
FIXATION ROD	888.3020	INTRAMEDULLARY FIXATION ROD Prosthetic Devices; Orthopedic Devices
FIXATION SCREW/WIRE	872.4880	INTRAOSSEOUS FIXATION SCREW OR WIRE Surgical Devices; Dental Devices
FIXED-COMBINATION DRUGS	300.5	FIXED-COMBINATION PRESCRIPTION DRUGS FOR HUMANS Combination Drugs; General
FIXING AGENTS	173.357	MATERIALS USED AS FIXING AGENTS IN THE IMMOBILIZATION OF ENZYME PREPARATIONS Specific Usage Additives; Secondary Direct Food Additives Permitted In Food For Human Consumption
FIXING QUOTAS	1303.23	PROCEDURE FOR FIXING INDIVIDUAL MANUFACTURING QUOTAS Individual Manufacturing Quotas; Quotas
FLAME EMISSION	862.2540	FLAME EMISSION PHOTOMETER FOR CLINICAL USE Clinical Laboratory Instruments; Clinical Chemistry And Toxicology Devices
FLASHER	886.1300	AFTERIMAGE FLASHER Diagnostic Devices; Ophthalmic Devices
FLAVOBACTERIUM SPP.	866.3270	FLAVOBACTERIUM SPP. SEROLOGICAL REAGENTS Serological Reagents; Immunology And Microbiology Devices

FLAVOR	172.585	SUGAR BEET EXTRACT FLAVOR BASE Flavoring Agents And Related Substances; Food Additives Permitted For Direct Addition To Food For Human Consumption
FLAVORED CHEESES	133.193	SPICED, FLAVORED STANDARDIZED CHEESES Requirements For Specific Standardized Cheese And Related Products; Cheeses And Related Cheese Products
FLAVORING	172.230	MICROCAPSULES FOR FLAVORING SUBSTANCES Coatings, Films And Related Substances; Food Additives Permitted For Direct Addition To Food For Human Consumption
FLAVORING, NATURAL	172.510	NATURAL FLAVORING SUBSTANCES AND NATURAL SUBSTANCES USED IN CONJUNCTION WITH FLAVORS Flavoring Agents And Related Substances; Food Additives Permitted For Direct Addition To Food For Human Consumption
FLAVORING, SYNTHETIC	172.515	SYNTHETIC FLAVORING SUBSTANCES AND ADJUVANTS Favoring Agents And Related Substances; Food Additives Permitted For Direct Addition To Food For Human Consumption
FLAVORING, SYNTHETIC	182.60	SYNTHETIC FLAVORING SUBSTANCES AND ADJUVANTS General Provisions; Substances Generally Recognized As Safe
FLAVORING, SYNTHETIC	582.60	SYNTHETIC FLAVORING SUBSTANCES AND ADJUVANTS General Provisions; Substances Generally Recognized As Safe
FLAVORING, VANILLA	169.177	VANILLA FLAVORING Requirements For Specific Standardized Food Dressings And Flavorings; Food Dressings And Flavorings
FLAVORING, VANILLA	169.178	CONCENTRATED VANILLA FLAVORING Requirements For Specific Standardized Food Dressings And Flavorings; Food Dressings And Flavorings
FLAVORING, VANILLA-VANILLIN	169.181	VANILLA-VANILLIN FLAVORING Requirements For Specific Standardized Food Dressings And Flavorings; Food Dressings And Flavorings
FLAVORINGS	101.22	FOODS; LABELING OF SPICES, FLAVORINGS, COLORINGS AND CHEMICAL PRESERVATIVES Specific Food Labeling Requirements; Food Labeling
FLAVORINGS	131.25	WHIPPED CREAM PRODUCTS CONTAINING FLAVORING OR SWEETENING General Provisions; Milk And Cream
FLAVORINGS	182.10	SPICES AND OTHER NATURAL SEASONINGS AND FLAVORINGS General Provisions; Substances Generally Recognized As Safe
FLAVORINGS	182.40	NATURAL EXTRACTIVES (SOLVENT-FREE) USED IN CONJUNCTION WITH SPICES, SEASONINGS, AND FLAVORINGS General Provisions; Substances Generally Recognized As Safe
FLAVORINGS	501.22	ANIMAL FOODS; LABELING OF SPICES; FLAVORINGS, COLORINGS, AND CHEMICAL PRESERVATIVES Specific Animal Food Labeling Requirements; Animal Food Labeling
FLAVORINGS	582.10	SPICES AND OTHER NATURAL SEASONINGS AND FLAVORINGS General Provisions; Substances Generally Recognized As Safe
FLAVORINGS	582.30	NATURAL SUBSTANCES USED IN CONJUNCTION WITH SPICES AND OTHER NATURAL SEASONINGS AND FLAVORINGS General Provisions; Substances Generally Recognized As Safe
FLAVORINGS	582.40	NATURAL EXTRACTIVES (SOLVENT-FREE) USED IN CONJUNCTION WITH SPICES, SEASONINGS, AND FLAVORINGS General Provisions; Substances Generally Recognized As Safe
FLAVORS	172.510	NATURAL FLAVORING SUBSTANCES AND NATURAL SUBSTANCES USED IN CONJUNCTION WITH FLAVORS Flavoring Agents And Related Substances; Food Additives Permitted For Direct Addition To Food For Human Consumption
FLECTOL H	189.220	FLECTOL H Substances Prohibited From Indirect Addition To Human Food Thru Contact Surface; Substances Prohibited From Use In Human Food
FLEXIBLE	868.5530	FLEXIBLE LARYNGOSCOPE Therapeutic Devices; Anesthesiology Devices
FLOOD SOURCE	892.1380	NUCLEAR FLOOD SOURCE PHANTOM Diagnostic Devices; Radiology Devices
FLORFENICOL	522.955	FLORFENICOL Implantation Or Injectable Dosage Form New Animal Drugs
FLORFENICOL	556.283	FLORFENICOL Tolerances For Residues Of New Animal Drugs In Food
FLORIDA	808.59	FLORIDA Listing Of Specific State And Local Exemptions; Exemption From Federal Preemption Of State & Local Medical Device Requirements
FLOSS, DENTAL	872.6390	DENTAL FLOSS Miscellaneous Devices; Dental Devices
FLOTATION	880.5150	NONPOWERED FLOTATION THERAPY MATTRESS General Hospital And Personal Use Therapeutic Devices; General Hospital And Personal Use Devices

FLOTATION	880.5550	ALTERNATING PRESSURE AIR FLOTATION MATTRESS General Hospital And Personal Use Therapeutic Devices; General Hospital And Personal Use Devices
FLOTATION	890.3175	FLOTATION CUSHION Physical Medicine Prosthetic Devices; Physical Medicine Devices
FLOUR	137.105	FLOUR Requirements For Specific Standardized Cereal Flours And Related Products; Cereal Flours And Related Products
FLOUR	137.155	BROMATED FLOUR Requirements For Specific Standardized Cereal Flours And Related Products; Cereal Flours And Related Products
FLOUR	137.160	ENRICHED BROMATED FLOUR Requirements For Specific Standardized Cereal Flours And Related Products; Cereal Flours And Related Products
FLOUR	137.165	ENRICHED FLOUR Requirements For Specific Standardized Cereal Flours And Related Products; Cereal Flours And Related Products
FLOUR	137.175	PHOSPHATED FLOUR Requirements For Specific Standardized Cereal Flours And Related Products; Cereal Flours And Related Products
FLOUR	137.180	SELF-RISING FLOUR Requirements For Specific Standardized Cereal Flours And Related Products; Cereal Flours And Related Products
FLOUR	137.185	ENRICHED SELF-RISING FLOUR Requirements For Specific Standardized Cereal Flours And Related Products; Cereal Flours And Related Products
FLOUR	137.200	WHOLE WHEAT FLOUR Requirements For Specific Standardized Cereal Flours And Related Products; Cereal Flours And Related Products
FLOUR	137.205	BROMATED WHOLE WHEAT FLOUR Requirements For Specific Standardized Cereal Flours And Related Products; Cereal Flours And Related Products
FLOUR	137.211	WHITE CORN FLOUR Requirements For Specific Standardized Cereal Flours And Related Products; Cereal Flours And Related Products
FLOUR	137.215	YELLOW CORN FLOUR Requirements For Specific Standardized Cereal Flours And Related Products; Cereal Flours And Related Products
FLOUR	137.220	DURUM FLOUR Requirements For Specific Standardized Cereal Flours And Related Products; Cereal Flours And Related Products
FLOUR	137.225	WHOLE DURUM FLOUR Requirements For Specific Standardized Cereal Flours And Related Products; Cereal Flours And Related Products
FLOUR, COTTONSEED	73.140	TOASTED PARTIALLY DEFATTED COOKED COTTONSEED FLOUR Foods; Listing Of Color Additives Exempt From Certification
FLOURS	137.170	INSTANTIZED FLOURS Requirements For Specific Standardized Cereal Flours And Related Products; Cereal Flours And Related Products
FLOTATION	890.5170	POWERED FLOTATION THERAPY BED Physical Medicine Therapeutic Devices; Physical Medicine Devices
FLOW, BLOOD	870.2120	EXTRAVASCULAR BLOOD FLOW PROBE Cardiovascular Monitoring Devices; Cardiovascular Devices
FLOW, CONTINUOUS	862.2150	CONTINUOUS FLOW SEQUENTIAL MULTIPLE CHEMISTRY ANALYZER FOR CLINICAL USE Clinical Laboratory Instruments; Clinical Chemistry And Toxicology Devices
FLOW-DIRECTED	870.1240	FLOW-DIRECTED CATHETER Cardiovascular Diagnostic Devices; Cardiovascular Devices
FLOW, GAS	868.2885	GAS FLOW TRANSDUCER Monitoring Devices; Anesthesiology Devices
FLOW, GRAVITY	880.2420	ELECTRONIC MONITOR FOR GRAVITY FLOW INFUSION SYSTEMS General Hospital And Personal Use Monitoring Devices; General Hospital And Personal Use Devices
FLOW, PEAK	868.1860	PEAK-FLOW METER FOR SPIROMETRY Diagnostic Devices; Anesthesiology Devices
FLOW, PULSATILE	870.4320	CARDIOPULMONARY BYPASS PULSATILE FLOW GENERATOR Cardiovascular Surgical Devices; Cardiovascular Devices
FLOW, URINE	876.1800	URINE FLOW OR VOLUME MEASURING SYSTEM Diagnostic Devices; Gastroenterology-urology Devices
FLOWMETER	868.2300	BOURDON GAUGE FLOWMETER Monitoring Devices; Anesthesiology Devices
FLOWMETER	868.2320	UNCOMPENSATED THORPE TUBE FLOWMETER Monitoring Devices; Anesthesiology Devices

FLOWMETER	868.2340	COMPENSATED THORPE TUBE FLOWMETER Monitoring Devices; Anesthesiology Devices
FLOWMETER	868.2350	GAS CALIBRATION FLOWMETER Monitoring Devices; Anesthesiology Devices
FLOWMETER	870.2100	CARDIOVASCULAR BLOOD FLOWMETER Cardiovascular Monitoring Devices; Cardiovascular Devices
FLUID	872.1500	GINGIVAL FLUID MEASURER Diagnostic Devices; Dental Devices
FLUID	880.2460	ELECTRICALLY POWERED SPINAL FLUID PRESSURE MONITOR General Hospital And Personal Use Monitoring Devices; General Hospital And Personal Use Devices
FLUID	880.2500	SPINAL FLUID MANOMETER General Hospital And Personal Use Monitoring Devices; General Hospital And Personal Use Devices
FLUID	880.5430	NONELECTRICALLY POWERED FLUID INJECTOR General Hospital And Personal Use Therapeutic Devices; General Hospital And Personal Use Devices
FLUID	880.6740	VACUUM-POWERED BODY FLUID SUCTION APPARATUS General Hospital And Personal Use Miscellaneous Devices; General Hospital And Personal Use Devices
FLUID	882.5550	CENTRAL NERVOUS SYSTEM FLUID SHUNT AND COMPONENTS Neurological Therapeutic Devices; Neurological Devices
FLUID	884.1550	AMNIOTIC FLUID SAMPLER (AMNIOCENTESIS TRAY) Obstetrical And Gynecological Diagnostic Devices; Obstetrical And Gynecological Devices
FLUID	866.5800	SEMINAL FLUID (SPERM) IMMUNOLOGICAL TEST SYSTEM Immunological Test Systems; Immunology And Microbiology Devices
FLUID	866.5860	TOTAL SPINAL FLUID (SPERM) IMMUNOLOGICAL TEST SYSTEM Immunological Test Systems; Immunology And Microbiology Devices
FLUID	886.4275	INTRAOCULAR FLUID Surgical Devices; Ophthalmic Devices
FLUMETHASONE	520.960	FLUMETHASONE TABLETS Oral Dosage Form New Animal Drugs
FLUMETHASONE	522.960	FLUMETHASONE IMPLANTATION OF INJECTABLE DOSAGE FORMS Implantation Or Injectable Dosage Form New Animal Drugs
FLUMETHASONE	524.960	FLUMETHASONE, NEOMYCIN SULFATE, AND POLYMYXIN B SULFATE OPHTHALMIC SOLUTION Ophthalmic And Topical Dosage Form New Animal Drugs
FLUNIXIN	520.970	FLUNIXIN ORAL DOSAGE FORMS Oral Dosage Form New Animal Drugs
FLUNIXIN	522.970	FLUNIXIN MEGLUMINE SOLUTION Implantation Or Injectable Dosage Form New Animal Drugs
FLUNIXIN	556.286	FLUNIXIN MEGLUMINE Tolerances For Residues Of New Animal Drugs In Food
FLUOCINOLONE ACETONIDE	524.981	FLUOCINOLONE ACETONIDE OPHTHALMIC AND TOPICAL DOSAGE FORMS Ophthalmic And Topical Dosage Form New Animal Drugs
FLUORESCENCE	872.1745	LASER FLUORESCENCE CARIES DETECTION DEVICE Diagnostic Devices; Dental Devices
FLUORESCENT LAMP, WOOD'S	866.2600	WOOD'S FLUORESCENT LAMP Microbiology Devices; Immunology And Microbiology Devices
FLUORESCENT SCANNER	892.1220	FLUORESCENT SCANNER Diagnostic Devices; Radiology Devices
FLUORIDE DENTIFRICE	355.70	TESTING PROCEDURES FOR FLUORIDE DENTIFRICE PRODUCTS Anticaries Drug Products For Over-the-counter Human Use
FLUORIDE RINSE	355.55	PRINCIPAL DISPLAY PANEL OF ALL FLUORIDE RINSE DRUG PRODUCTS Anticaries Drug Products For Over-the-counter Human Use
FLUORIDE RESINS	175.270	POLY(VINYL FLUORIDE) RESINS Substances For Use As Components Of Coatings; Indirect Food Additives; Adhesives And Components Of Coatings
FLUORIDE RESINS	177.2510	POLYVINYLIDENE FLUORIDE RESINS Substances For Use Only As Components Of Articles Intended For Repeated Use; Indirect Food Additives: Polymers
FLUORIDE TRAY	872.6870	DISPOSABLE FLUORIDE TRAY Miscellaneous Devices; Dental Devices
FLUORINATED	177.1615	POLYETHYLENE, FLUORINATED Substances For Use As Basic Components Of Single And Repeated Use Food Contact Surfaces; Indirect Food Additives: Polymers
FLUORINE	170.45	FLUORINE-CONTAINING COMPOUNDS Specific Administrative Rulings And Decisions; Food Additives
FLUOROCARBON RESINS	177.1380	FLUOROCARBON RESINS Substances For Use As Basic Components Of Single And Repeated Use Food Contact Surfaces; Indirect Food Additives: Polymers
FLUOROMETER	862.2560	FLUOROMETER FOR CLINICAL USE Clinical Laboratory Instruments; Clinical Chemistry And Toxicology Devices

FLUOROSCOPIC	892.1650	IMAGE-INTENSIFIED FLUOROSCOPIC X-RAY SYSTEM Diagnostic Devices; Radiology Devices
FLUOROSCOPIC	892.1660	NON-IMAGE-INTENSIFIED FLUOROSCOPIC X-RAY SYSTEM Diagnostic Devices; Radiology Devices
FLUOROSCOPIC	1020.32	FLUOROSCOPIC EQUIPMENT Performance Standards For Ionizing Radiation Emitting Products
FLUPROSTENOL SODIUM	522.995	FLUPROSTENOL SODIUM INJECTION Implantation Or Injectable Dosage Form New Animal Drugs
FLUROGESTONE ACETATE	529.1003	FLUROGESTONE ACETATE-IMPREGNATED VAGINAL SPONGE Certain Other Dosage Form New Animal Drugs
FLUOROGRAPHIC	892.1620	CINE OR SPOT FLUOROGRAPHIC X-RAY CAMERA Diagnostic Devices; Radiology Devices
FLUSH	870.1210	CONTINUOUS FLUSH CATHETER Cardiovascular Diagnostic Devices; Cardiovascular Devices
FOAMED	178.3010	ADJUVANT SUBSTANCES USED IN THE MANUFACTURE OF FOAMED PLASTICS Certain Adjutants And Production Aids; Indirect Food Additives: Adjuvants, Production Aids, And Sanitizers
FOAMING	740.17	FOAMING DETERGENT BATH PRODUCTS Warning Statements; Cosmetic Product Warning Statements
FOIL, TIN-COATED LEAD	189.301	TIN-COATED LEAD FOIL CAPSULES FOR WINE BOTTLES Substances Prohibited From Use In Human Foods
FOLACIN	172.345	FOLIC ACID (FOLACIN) Special Dietary And Nutritional Additives; Food Additives Permitted For Direct Addition To Food For Human Consumption
FOLATE	101.79	HEALTH CLAIMS: FOLATE AND NEURAL TUBE DEFECTS Food Labeling
FOLIC ACID	172.345	FOLIC ACID (FOLACIN) Special Dietary And Nutritional Additives; Food Additives Permitted For Direct Addition To Food For Human Consumption
FOLIC ACID	862.1295	FOLIC ACID TEST SYSTEM Clinical Chemistry Test Systems; Clinical Chemistry And Clinical Toxicology Devices
FOLLICLE STIMULATING HORMONE	522.1002	FOLLICLE STIMULATING HORMONE-PITUITARY FOR INJECTION Implantation Or Injectable Dosage Form New Animal Drugs
FOLLICLE-STIMULATING HORMONE	862.1300	FOLLICLE-STIMULATING HORMONE TEST SYSTEM Clinical Chemistry Test Systems; Clinical Chemistry And Clinical Toxicology Devices
FOMEPIZOLE	522.1004	FOMEPIZOLE Implantation Or Injectable Dosage Form New Animal Drugs
FOOD	5.301	ISSUANCE OF INITIAL EMERGENCY PERMIT ORDERS AND NOTICES OF CONFIRMATION OF EFFECTIVE DATE OF FINAL REGULATIONS ON FOOD FOR HUMAN AND ANIMAL CONSUMPTION Foods And Cosmetics; Redelegations Of Authority; Delegations Of Authority And Organization
FOOD	101.3	IDENTITY LABELING OF FOOD IN PACKAGED FORM Food Labeling
FOOD	101.5	FOOD; NAME AND PLACE OF BUSINESS OF MANUFACTURER, PACKER, OR DISTRIBUTOR General Provisions; Food Labeling
FOOD	101.15	FOOD; PROMINENCE OF REQUIRED STATEMENTS General Provisions; Food Labeling
FOOD	101.100	FOOD; EXEMPTIONS FROM LABELING Exemptions From Food Labeling Requirements; Food Labeling
FOOD	130.17	TEMPORARY PERMITS FOR INTERSTATE SHIPMENT OF EXPERIMENTAL PACKS OF FOOD VARYING FROM THE REQUIREMENTS OF DEFINITIONS AND STANDARDS OF IDENTITY General Provisions; Food Standards: General
FOOD	179.21	SOURCES OF RADIATION USED FOR INSPECTION OF FOOD OR PACKAGED FOOD, AND FOR CONTROLLING FOOD PROCESSING Radiation And Radiation Sources; Irradiation In The Production, Processing And Handling Of Food
FOOD	179.25	GENERAL PROVISIONS FOR FOOD IRRADIATION Radiation And Radiation Sources; Irradiation In The Production, Processing And Handling Of Food
FOOD	179.30	RADIOFREQUENCY RADIATION FOR THE HEATING OF FOOD, INCLUDING MICROWAVE FREQUENCIES Radiation And Radiation Sources; Irradiation In The Production, Processing And Handling Of Food
FOOD	179.39	ULTRAVIOLET RADIATION FOR THE PROCESSING AND TREATMENT OF FOOD Radiation And Radiation Sources; Irradiation In The Production, Processing And Handling Of Food

FOOD	179.41	PULSED LIGHT FOR THE TREATMENT OF FOOD Radiation And Radiation Sources; Irradiation In The Production, Processing And Handling Of Food
FOOD	182.70	SUBSTANCES MIGRATING FROM COTTON AND COTTON FABRICS USED IN DRY FOOD PACKAGING General Provisions; Substances Generally Regarded As Safe
FOOD	182.90	SUBSTANCES MIGRATING TO FOOD FROM PAPER AND PAPERBOARD PRODUCTS General Provisions; Substances Generally Regarded As Safe
FOOD	189.1	SUBSTANCES PROHIBITED FROM USE IN HUMAN FOOD General Provisions; Substances Prohibited From Use In Human Food
FOOD	556.1	GENERAL CONSIDERATIONS; TOLERANCES FOR RESIDUES OF NEW ANIMAL DRUGS IN FOOD General Provisions; Tolerances For Residues Of New Animal Drugs In Food
FOOD	1250.25	SOURCE IDENTIFICATION AND INSPECTION OF FOOD AND DRINK Food Service Sanitation On Land And Air Conveyances, And Vessels; Interstate Conveyance Sanitation
FOOD	1250.26	SPECIAL FOOD REQUIREMENTS Food Service Sanitation On Land And Air Conveyances, And Vessels; Interstate Conveyance Sanitation
FOOD	1250.30	CONSTRUCTION, MAINTENANCE AND USE OF PLACES WHERE FOOD IS PREPARED, SERVED, OR STORED Food Service Sanitation On Land And Air Conveyances, And Vessels; Interstate Conveyance Sanitation
FOOD ADDITIVE	170.6	OPINION LETTERS ON FOOD ADDITIVE STATUS General Provisions; Food Additives
FOOD ADDITIVE	170.38	DETERMINATION OF FOOD ADDITIVE STATUS Food Additive Safety; Food Additives
FOOD ADDITIVE	570.6	OPINION LETTERS ON FOOD ADDITIVE STATUS General Provisions; Food Additives
FOOD ADDITIVE	570.38	DETERMINATION OF FOOD ADDITIVE STATUS Food Additive Safety; Food Additives
FOOD ADDITIVES	5.27	PATENT TERM EXTENSIONS FOR HUMAN DRUG PRODUCTS, MEDICAL DEVICES, AND FOOD AND COLOR ADDITIVES General Redelegations Of Authority; Delegations Of Authority And Organization
FOOD ADDITIVES	5.300	FOOD STANDARDS, FOOD ADDITIVES, GRAS SUBSTANCES, AND COLOR ADDITIVES Foods And Cosmetics; Redelegations Of Authority; Delegations Of Authority And Organization
FOOD ADDITIVES	25.32	FOODS, FOOD ADDITIVES AND COLOR ADDITIVES Categorical Exclusions; Environmental Impact Considerations
FOOD ADDITIVES	130.20	FOOD ADDITIVES PROPOSED FOR USE IN FOODS FOR WHICH DEFINITIONS AND STANDARDS OF IDENTITY ARE ESTABLISHED Food Standards: General
FOOD ADDITIVES	170.10	FOOD ADDITIVES IN STANDARDIZED FOODS General Provisions; Food Additives
FOOD ADDITIVES	170.18	TOLERANCES FOR RELATED FOOD ADDITIVES General Provisions; Food Additives
FOOD ADDITIVES	170.20	GENERAL PRINCIPLES FOR EVALUATING THE SAFETY OF FOOD ADDITIVES Food Additive Safety; Food Additives
FOOD ADDITIVES	172.5	GENERAL PROVISIONS FOR DIRECT FOOD ADDITIVES General Provisions; Food Additives Permitted For Direct Addition To Food For Human Consumption
FOOD ADDITIVES	174.5	GENERAL PROVISIONS APPLICABLE TO INDIRECT FOOD ADDITIVES Indirect Food Additives: General
FOOD ADDITIVES	570.13	INDIRECT FOOD ADDITIVES RESULTING FROM PACKAGING MATERIALS PRIOR SANCTIONED FOR ANIMAL FEED AND PET FOOD General Provisions; Food Additives
FOOD ADDITIVES	570.14	INDIRECT FOOD ADDITIVES RESULTING FROM PACKAGING MATERIALS FOR ANIMAL FEED AND PET FOOD General Provisions; Food Additives
FOOD ADDITIVES	570.18	TOLERANCES FOR RELATED FOOD ADDITIVES General Provisions; Food Additives
FOOD, ANIMAL	2.25	GRAIN SEED TREATED WITH POISONOUS SUBSTANCES; COLOR IDENTIFICATION TO PREVENT ADULTERATION OF HUMAN AND ANIMAL FOOD Human And Animal Foods; General Administrative Rulings And Decisions
FOOD, ANIMAL	2.35	USE OF SECONDHAND CONTAINERS FOR THE SHIPMENT AND STORAGE OF FOOD AND ANIMAL FEED Human And Animal Foods; General Administrative Rulings And Decisions
FOOD, ANIMAL	501.1	PRINCIPAL DISPLAY PANEL OF PACKAGE FORM ANIMAL FOOD General Provisions; Animal Food Labeling
FOOD, ANIMAL	501.2	INFORMATION PANEL OF PACKAGE FOR ANIMAL FOOD General Provisions; Animal Food Labeling

FOOD, HANDLING	1250.35	HEALTH OF PERSONS HANDLING FOOD Food Service Sanitation On Land And Air Conveyances, And Vessels; Interstate Conveyance Sanitation
FOOD-HANDLING	1250.38	TOILET AND LAVATORY FACILITIES FOR USE OF FOOD-HANDLING EMPLOYEES Food Service Sanitation On Land And Air Conveyances, And Vessels; Interstate Conveyance Sanitation
FOOD HANDLING	1250.45	FOOD HANDLING FACILITIES ON RAILROAD CONVEYANCES Equipment And Operation Of Land And Air Conveyances; Interstate Conveyance Sanitation
FOOD, HUMAN	170.50	GLYCINE (AMINOACETIC ACID) IN FOOD FOR HUMAN CONSUMPTION Specific Administrative Rulings And Decisions; Food Additives
FOOD, HUMAN	2.25	GRAIN SEED TREATED WITH POISONOUS SUBSTANCES; COLOR IDENTIFICATION TO PREVENT ADULTERATION OF HUMAN AND ANIMAL FOOD Human And Animal Foods; General Administrative Rulings And Decisions
FOOD, HUMAN	184.1	SUBSTANCES ADDED DIRECTLY TO HUMAN FOOD AFFIRMED AS GENERALLY RECOGNIZED AS SAFE (GRAS) General Provisions; Direct Food Substances Affirmed As Generally Recognized As Safe
FOOD, HUMAN	186.1	SUBSTANCES ADDED INDIRECTLY TO HUMAN FOOD AFFIRMED AS GENERALLY RECOGNIZED AS SAFE (GRAS) General Provisions; Indirect Food Substances Affirmed As Generally Recognized As Safe
FOOD, HUMAN	189.1	SUBSTANCES PROHIBITED FROM USE IN HUMAN FOOD General Provisions; Substances Prohibited From Use In Human Food
FOOD LABELING	101.9	NUTRITION LABELING OF FOOD General Provisions; Food Labeling
FOOD LABELING	101.17	FOOD LABELING WARNING, NOTICE, AND SAFE HANDLING STATEMENTS General Provisions; Food Labeling
FOOD RETAILERS	101.43	SUBSTANTIAL COMPLIANCE OF FOOD RETAILERS WITH THE GUIDELINES FOR THE VOLUNTARY NUTRITIONAL LABELING OF RAW FRUIT, VEGETABLES AND FISH Food Labeling
FOOD, MISBRANDING	101.18	MISBRANDING OF FOOD General Provisions; Food Labeling
FOOD, PACKAGE	101.1	PRINCIPAL DISPLAY PANEL OF PACKAGE FORM FOOD General Provisions; Food Labeling
FOOD, PACKAGE	101.2	INFORMATION PANEL OF PACKAGE FORM FOOD General Provisions; Food Labeling
FOOD-PACKAGING	109.15	USE OF POLYCHLORINATED BIPHENYLS (PCB'S) IN ESTABLISHMENTS MANUFACTURING FOOD-PACKAGING MATERIALS General Provisions; Unavoidable Contaminants In Food For Human Consumption & Food-packaging Material
FOOD-PACKAGING	178.3130	ANTISTATIC AND/OR ANTIFOGGING AGENTS IN FOOD-PACKAGING MATERIALS Certain Adjuvants And Production Aids; Indirect Food Additives: Adjuvants, Production Aids, And Sanitizers
FOOD-PACKAGING	181.22	CERTAIN SUBSTANCES EMPLOYED IN THE MANUFACTURE OF FOOD-PACKAGING MATERIALS Specific Prior-sanctioned Food Ingredients; Prior-sanctioned Food Ingredients
FOOD-PACKAGING	181.30	SUBSTANCES USED IN THE MANUFACTURE OF PAPER AND PAPERBOARD PRODUCTS USED IN FOOD-PACKAGING Specific Prior-sanctioned Food Ingredients; Prior-sanctioned Food Ingredients
FOOD-PACKAGING	509.15	USE OF POLYCHLORINATED BIPHENYLS (PCB'S) IN ESTABLISHMENTS MANUFACTURED FOOD-PACKAGING MATERIALS General Provisions; Unavoidable Contaminants In Animal Food And Food-packaging Material
FOOD, PET	570.13	INDIRECT FOOD ADDITIVES RESULTING FROM PACKAGING MATERIALS PRIOR SANCTIONED FOR ANIMAL FEED AND PET FOOD General Provisions; Food Additives
FOOD, PET	570.14	INDIRECT FOOD ADDITIVES RESULTING FROM PACKAGING MATERIALS FOR ANIMAL FEED AND PET FOOD General Provisions; Food Additives
FOOD-PROCESSING	5.304	APPROVAL OF SCHOOLS PROVIDING FOOD-PROCESSING INSTRUCTION Foods And Cosmetics; Redelegations Of Authority; Delegations Of Authority And Organization
FOOD-PRODUCING ANIMALS	510.110	ANTIBIOTICS USED IN FOOD-PRODUCING ANIMALS Specific Administrative Rulings And Decisions; New Animal Drugs
FOOD-PRODUCING ANIMALS	530.20	CONDITIONS FOR PERMITTED EXTRALABEL ANIMAL AND HUMAN DRUG USE IN FOOD-PRODUCING ANIMALS Extralabel Drug Use In Animals
FOOD-PRODUCING ANIMALS	530.21	PROHIBITIONS FOR FOOD-PRODUCING ANIMALS Extralabel Drug Use In Animals

FOOD-PRODUCING ANIMALS	530.22	SAFE LEVELS AND ANALYTICAL METHODS FOR FOOD-PRODUCING ANIMALS Extralabel Drug Use In Animals
FOOD-PRODUCING ANIMALS	530.25	ORDERS PROHIBITING EXTRALABEL USES FOR DRUGS IN FOOD-PRODUCING ANIMALS Extralabel Drug Use In Animals
FOOD RETAILERS	101.43	SUBSTANTIAL COMPLIANCE OF FOOD RETAILERS WITH THE GUIDELINES FOR THE VOLUNTARY NUTRITION LABELING OF RAW FRUIT, VEGETABLES, AND FISH Food Labeling
FOOD STANDARD	130.5	PROCEDURE FOR ESTABLISHING A FOOD STANDARD General Provisions; Food Standards: General
FOOD STANDARDS	5.300	FOOD STANDARDS, FOOD ADDITIVES, GRAS SUBSTANCES, COLOR ADDITIVES, NUTRIENT CONTENT CLAIMS, AND HEALTH CLAIMS Foods And Cosmetics; Redelegations Of Authority; Delegations Of Authority And Organization
FOOD STANDARDS	130.6	REVIEW OF CODEX ALIMENTARIUS FOOD STANDARDS General Provisions; Food Standards: General
FOOD, STANDARDIZED	130.9	SULFITES IN STANDARDIZED FOOD Food Standards: General
FOOD STARCH	172.892	FOOD STARCH-MODIFIED Multipurpose Additives; Food Additives Permitted For Direct Addition To Food For Human Consumption
FOOD USE	73.1	DILUENTS IN COLOR ADDITIVE MIXTURES FOR FOOD USE EXEMPT FROM CERTIFICATION Listing Of Color Additives Exempt From Certification
FOODS	25.32	FOODS, FOOD ADDITIVES AND COLOR ADDITIVES Categorical Exclusions; Environmental Impact Considerations
FOODS	101.81	HEALTH CLAIMS: SOLUBLE FIBER FROM CERTAIN FOODS AND RISK OF CORONARY HEART DISEASE Food Labeling
FOODS	102.28	FOODS PACKAGED FOR USE IN THE PREPARATION OF "MAIN DISHES" OR "DINNERS" Requirements For Specific Nonstandardized Foods; Common Or Usual Name For Nonstandardized Foods
FOODS	130.10	REQUIREMENTS FOR FOODS NAMED BY USE OF A NUTRIENT CONTENT CLAIM AND A STANDARDIZED TERM Food Standards: General
FOODS	130.20	FOOD ADDITIVES PROPOSED FOR USE IN FOODS FOR WHICH DEFINITIONS AND STANDARDS OF IDENTITY ARE ESTABLISHED Food Standards: General
FOODS	133.134	CREAM CHEESE WITH OTHER FOODS Requirements For Specific Standardized Cheese And Related Products; Cheeses And Related Cheese Products
FOODS	133.178	PASTEURIZED NEUFCHATEL CHEESE SPREAD WITH OTHER FOODS Requirements For Specific Standardized Cheese And Related Products; Cheeses And Related Cheese Products
FOODS, ACIDIFIED	108.25	ACIDIFIED FOODS Specific Requirements And Conditions For Exemption From Or Compliance With An Emergency Permit; Emergency Permit Control
FOODS, ANIMAL	501.22	ANIMAL FOODS; LABELING OF SPICES, FLAVORINGS, COLORINGS, AND CHEMICAL PRESERVATIVES Specific Animal Food Labeling Requirements; Animal Food Labeling
FOODS, AQUEOUS/FATTY	176.170	COMPONENTS OF PAPER AND PAPERBOARD IN CONTACT WITH AQUEOUS AND FATTY FOODS Substances For Use Only As Components Of Paper And Paperboard; Indirect Food Additives
FOODS, BEVERAGES	101.30	PERCENTAGE JUICE DECLARATION FOR FOODS PURPORTING TO BE BEVERAGES THAT CONTAIN FRUIT OR VEGETABLE JUICE Food Labeling
FOODS, CONTENT	101.60	NUTRIENT CONTENT CLAIMS FOR THE CALORIE CONTENT OF FOODS Specific Requirements For Nutrient Content Claims; Food Labeling
FOODS, CONTENT	101.61	NUTRIENT CONTENT CLAIMS FOR THE SODIUM CONTENT OF FOODS Specific Requirements For Nutrient Content Claims; Food Labeling
FOODS, CONTENT	101.62	NUTRIENT CONTENT CLAIMS FOR FAT, FATTY ACID, AND CHOLESTEROL CONTENT OF FOODS Specific Requirements For Nutrient Content Claims; Food Labeling
FOODS, HYPOALLERGENIC	105.6	HYPOALLERGENIC FOODS Label Statements; Foods For Special Dietary Use
FOODS, INFANT	105.65	INFANT FOODS Label Statements; Foods For Special Dietary Use
FOODS, LABELING	101.22	FOODS; LABELING OF SPICES, FLAVORINGS, COLORINGS AND CHEMICAL PRESERVATIVES Specific Food Labeling Requirements; Food Labeling

FOODS, LOW-ACID	108.35	THERMAL PROCESSING OF LOW-ACID FOODS PACKAGED IN HERMETICALLY SEALED CONTAINERS Specific Requirements And Conditions For Exemption From Or Compliance With An Emergency Permit; Emergency Permit Control
FOODS, LOW-ACID	500.23	THERMALLY PROCESSED LOW-ACID FOODS PACKAGED IN HERMETICALLY SEALED CONTAINERS Specific Administrative Rulings And Decisions; General
FOODS "MAIN DISHES/DINNERS"	102.28	FOODS PACKAGED FOR USE IN THE PREPARATION OF "MAIN DISHES" OR "DINNERS" Requirements For Specific Nonstandardized Foods; Common Or Usual Name For Nonstandardized Foods
FOODS, PREPACKAGED	179.45	PACKAGING MATERIALS FOR USE DURING THE IRRADIATION OF PREPACKAGED FOODS Packaging Materials For Irradiated Foods; Irradiation In The Production, Processing And Handling Of Food
FOODS, PROCESSED	170.19	PESTICIDE CHEMICALS IN PROCESSED FOODS General Provisions; Food Additives
FOODS, PROCESSED	570.19	PESTICIDE CHEMICALS IN PROCESSED FOODS General Provisions; Food Additives
FOOD, RESTAURANT	101.10	NUTRITION LABELING OF RESTAURANT FOODS General Provisions; Food Labeling
FOODS, STANDARDIZED	70.10	COLOR ADDITIVES IN STANDARDIZED FOODS AND NEW DRUGS General Provisions; Color Additives
FOODS, STANDARDIZED	130.11	LABEL DESIGNATIONS OF INGREDIENTS FOR STANDARDIZED FOODS Food Standards: General
FOODS, STANDARDIZED	170.10	FOOD ADDITIVES IN STANDARDIZED FOODS General Provisions; Food Additives
FORAGE CROPS	573.400	ETHOXYQUIN IN CERTAIN DEHYDRATED FORAGE CROPS Food Additive Listing; Food Additives Permitted In Feed And Drinking Water Of Animals
FORCE, ENTRY INTO	26.80	ENTRY INTO FORCE, AMENDMENT, AND TERMINATION Framework Provisions; Mutual Recognition Of Pharmaceutical GMP Reports, Medical Device Quality System Audit Reports
FORCE-MEASURING	890.1575	FORCE-MEASURING PLATFORM Physical Medicine Diagnostic Devices; Physical Medicine Devices
FORCEPS	868.5780	TUBE INTRODUCTION FORCEPS Therapeutic Devices; Anaesthesiology Devices
FORCEPS	884.4400	OBSTETRIC FORCEPS Obstetrical And Gynecological Surgical Devices; Obstetrical And Gynecological Devices
FOREHEAD	880.2200	LIQUID CRYSTAL FOREHEAD TEMPERATURE STRIP General Hospital And Personal Use Monitoring Devices; General Hospital And Personal Use Devices
FOREIGN CLINICAL STUDIES	312.110	FOREIGN CLINICAL STUDIES NOT CONDUCTED UNDER AN IND Miscellaneous; New Drugs For Investigational Use
FOREIGN DATA	314.106	FOREIGN DATA FDA Action On Applications; Applications For FDA Approval To Market A New Drug
FOREIGN ESTABLISHMENTS	207.40	ESTABLISHMENT REGISTRATION AND DRUG LISTING REQUIREMENTS FOR FOREIGN ESTABLISHMENTS Procedure For Foreign Drug Establishments; Registration Of Drug Producers And Listing Of Drugs In Commercial Distribution
FOREIGN ESTABLISHMENTS	601.15	FOREIGN ESTABLISHMENTS AND PRODUCTS: SAMPLES FOR EACH IMPORTATION Biologics Licensing; Licensing
FOREIGN ESTABLISHMENTS	607.40	ESTABLISHMENT REGISTRATION AND BLOOD PRODUCT LISTING REQUIREMENTS FOR FOREIGN BLOOD PRODUCT ESTABLISHMENTS Procedures For Foreign Blood Product Establishments; Establishment Registratn+product Listing, Manufacturers Human Blood/b.products
FOREIGN ESTABLISHMENTS	807.40	ESTABLISHMENT REGISTRATION AND DEVICE LISTING FOR FOREIGN ESTABLISHMENTS IMPORTING OR OFFERING FOR IMPORT DEVICES INTO THE UNITED STATES Registration Procedures For Foreign Device Establishments; Establishment Registration And Device Listing For Manufacturers And Initial Importers Of Devices
FOREIGN GOVERNMENT	20.89	COMMUNICATIONS WITH FOREIGN GOVERNMENT OFFICIALS Limitations On Exemptions; Public Information
FOREIGN IMPORT RESTRICTIONS	1313.25	FOREIGN IMPORT RESTRICTIONS Importation And Exportation Of Precursors and Essential Chemicals
FOREIGN MANUFACTURERS	803.58	FOREIGN MANUFACTURERS Medical Device Reporting
FOREIGN SPONSOR	316.22	PERMANENT RESIDENT AGENT FOR FOREIGN SPONSOR Orphan Drugs
FORFEITURE	1316.79	PETITIONS FOR REMISSION OR MITIGATION OF FORFEITURE Seizure, Forfeiture, And Disposition Of Property; Administrative Functions, Practices, And Procedures

FORFEITURE, ADMINISTRATIVE	1316.77	ADMINISTRATIVE FORFEITURE Seizure, Forfeiture, And Disposition Of Property; Administrative Functions, Practices, And Procedures
FORFEITURE, ADMINISTRATIVE	1316.92	PETITION FOR EXPEDITED RELEASE IN AN ADMINISTRATIVE FORFEITURE ACTION Expedited Forfeiture Proceedings For Certain Property; Administrative Functions, Practices, And Procedures
FORFEITURE, ADMINISTRATIVE	1316.93	RULING ON PETITION FOR EXPEDITED RELEASE IN AN ADMINISTRATIVE FORFEITURE ACTION Expedited Forfeiture Proceedings For Certain Property; Administrative Functions, Practices, And Procedures
FORFEITURE, ADMINISTRATIVE	1316.94	POSTING OF SUBSTITUTE RES IN AN ADMINISTRATIVE FORFEITURE ACTION Expedited Forfeiture Proceedings For Certain Property; Administrative Functions, Practices, And Procedures
FORFEITURE, JUDICIAL	1316.78	JUDICIAL FORFEITURE Seizure, Forfeiture, And Disposition Of Property; Administrative Functions, Practices, And Procedures
FORFEITURE, JUDICIAL	1316.95	PETITION FOR EXPEDITED RELEASE OF A CONVEYANCE IN A JUDICIAL FORFEITURE ACTION Expedited Forfeiture Proceedings For Certain Property; Administrative Functions, Practices, And Procedures
FORFEITURE, JUDICIAL	1316.96	RULING ON A PETITION FOR EXPEDITED RELEASE OF A CONVEYANCE IN A JUDICIAL FORFEITURE ACTION Expedited Forfeiture Proceedings For Certain Property; Administrative Functions, Practices, And Procedures
FORFEITURE, JUDICIAL	1316.97	INITIATING JUDICIAL FORFEITURE PROCEEDING AGAINST A CONVEYANCE WITHIN 60 DAYS OF FILING A CLAIM AND COST BOND Expedited Forfeiture Proceedings For Certain Property; Administrative Functions, Practices, And Procedures
FORFEITURE, JUDICIAL	1316.98	SUBSTITUTE RES BOND IN A JUDICIAL FORFEITURE ACTION AGAINST A CONVEYANCE Expedited Forfeiture Proceedings For Certain Property; Administrative Functions, Practices, And Procedures
FORK	882.1525	TUNING FORK Neurological Diagnostic Devices; Neurological Devices
FORM	12.32	REQUEST FOR ALTERNATIVE FORM OF HEARING Initiation Of Proceedings; Formal Evidentiary Public Hearing; Civil Money Penalties Hearing
FORM	501.1	PRINCIPAL DISPLAY PANEL OF PACKAGE FORM ANIMAL FOOD General Provisions; Animal Food Labeling
FORM	501.3	IDENTITY LABELING OF ANIMAL FOOD IN PACKAGE FORM General Provisions; Animal Food Labeling
FORM	701.2	FORM OF STATING LABELING REQUIREMENTS General Provisions; Cosmetic Labeling
FORMA PAUPERIS	12.82	PETITION TO PARTICIPATE IN FORMA PAUPERIS Hearing Procedures; Formal Evidentiary Public Hearing
FORMAL HEARING	10.50	PROMULGATION OF REGULATIONS AND ORDERS AFTER AN OPPORTUNITY FOR A FORMAL EVIDENTIARY PUBLIC HEARING General Administrative Procedures; Administrative Practices And Procedures
FORMAL STATEMENTS	369.4	WARNINGS SUGGESTED FOR DRUGS BY FORMAL OR INFORMAL STATEMENTS OF POLICY Definitions And Interpretations; Interpretative Statements Re. Warnings On Drugs And Devices For OTC Sale
FORMALDEHYDE	175.380	XYLENE FORMALDEHYDE RESINS CONDENSED WITH 4,4'-ISOPROPYLIDENEDIPHENOL-EPICHLOROHYDRIN EPOXY RESINS Substances For Use As Components Of Coatings; Indirect Food Additives
FORMALDEHYDE	177.1460	MELAMINE-FORMALDEHYDE RESINS IN MOLDED ARTICLES Substances For Use As Basic Components Of Single And Repeated Use Food Contact Surfaces; Indirect Food Additives: Polymers
FORMALDEHYDE	177.1900	UREA-FORMALDEHYDE RESINS IN MOLDED ARTICLES Substances For Use As Basic Components Of Single And Repeated Use Food Contact Surfaces; Indirect Food Additives: Polymers
FORMALDEHYDE	573.460	FORMALDEHYDE Food Additive Listing; Food Additives Permitted In Feed And Drinking Water Of Animals
FORMALIN	529.1030	FORMALIN SOLUTION Certain Other Dosage Form New Animal Drugs
FORMAT	60.30	FILING, FORMAT, AND CONTENT OF PETITIONS Due Diligence Petitions; Patent Term Restoration
FORMAT	201.56	GENERAL REQUIREMENTS ON CONTENT AND FORMAT OF LABELING FOR HUMAN PRESCRIPTION DRUGS Labeling Requirements For Prescription Drugs And/or Insulin; Labeling

FORMAT	201.57	SPECIFIC REQUIREMENTS ON CONTENT AND FORMAT OF LABELING FOR HUMAN PRESCRIPTION DRUGS Labeling Requirements For Prescription Drugs And/or Insulin; Labeling
FORMAT	201.66	FORMAT AND CONTENT REQUIREMENTS FOR OTC DRUG PRODUCT LABELING Labeling Requirements For Over-the-counter Drugs; Labeling
FORMAT	208.20	CONTENT AND FORMAT OF A MEDICATION GUIDE Medication Guides For Prescription Drug Products
FORMAT	312.23	IND CONTENT AND FORMAT Investigational New Drug Application
FORMAT	314.50	CONTENT AND FORMAT OF AN APPLICATION Applications; Applications For FDA Approval To Market A New Drug
FORMAT	314.94	CONTENT AND FORMAT OF AN ABBREVIATED APPLICATION Abbreviated Applications; Applications For FDA Approval To Market A New Drug
FORMAT	316.10	CONTENT AND FORMAT OF A REQUEST FOR WRITTEN RECOMMENDATIONS Orphan Drugs
FORMAT	316.20	CONTENT AND FORMAT OF A REQUEST FOR ORPHAN DRUG DESIGNATION Orphan Drugs
FORMAT	807.90	FORMAT OF A PREMARKET NOTIFICATION SUBMISSION Premarket Notification Procedures; Establishment Registration And Device Listing For Manufacturers And Initial Importers Of Devices
FORMAT, 510(K)	807.92	CONTENT AND FORMAT OF A 510(K) SUMMARY Establishment Registration And Device Listing For Manufacturers And Initial Importers Of Devices
FORMAT, 510(K)	807.93	CONTENT AND FORMAT OF A 510(K) STATEMENT Establishment Registration And Device Listing For Manufacturers And Initial Importers Of Devices
FORMAT, CLASS III CERTIF.	807.94	FORMAT OF A CLASS III CERTIFICATION Establishment Registration And Device Listing For Manufacturers And Initial Importers Of Devices
FORMATE	184.1295	ETHYL FORMATE Listing Of Specific Substances Affirmed As GRAS; Direct Food Substances Affirmed As Generally Recognized As Safe
FORMATE	186.1756	SODIUM FORMATE Listing Of Specific Substances Affirmed As GRAS; Indirect Food Substances Affirmed As Generally Recognized As Safe
FORMER EMPLOYEES	19.45	TEMPORARY DISQUALIFICATION OF FORMER EMPLOYEES Disqualification Conditions; Standards Of Conduct And Conflicts Of Interest
FORMER EMPLOYEES	19.55	PERMANENT DISQUALIFICATION OF FORMER EMPLOYEES Disqualification Conditions; Standards Of Conduct And Conflicts Of Interest
FORMIMINOGLUTAMIC	862.1305	FORMIMINOGLUTAMIC ACID (FIGLU) TEST SYSTEM Clinical Chemistry Test Systems; Clinical Chemistry And Clinical Toxicology Devices
FORMIC ACID	186.1316	FORMIC ACID Listing Of Specific Substances Affirmed As GRAS; Indirect Food Substances Affirmed As Generally Recognized As Safe
FORMIC ACID	573.480	FORMIC ACID Food Additive Listing; Food Additives Permitted In Feed And Drinking Water Of Animals
FORMING, CRANIOPLASTY	882.4500	CRANIOPLASTY MATERIAL FORMING INSTRUMENT Neurological Surgical Devices; Neurological Devices
FORMING, CLIP	882.4190	CLIP FORMING/CUTTING INSTRUMENT Neurological Surgical Devices; Neurological Devices
FORMS	203.33	DRUG SAMPLES FORMS Prescription Drug Marketing
FORMS	203.60	REQUEST AND RECEIPT FORMS, REPORTS AND RECORDS Prescription Drug Marketing
FORMS	803.11	OBTAINING THE FORMS Medical Device Reporting
FORMS	1403.12	FORMS FOR APPLYING FOR GRANTS Pre-Award Requirements; Uniform Administrative Requirements For Grants And Coop. Agreements To State & Local Gov.
FORMS, APPLICATION	1301.13	APPLICATION FOR REGISTRATION; APPLICATION FORMS, FEES, CONTENTS, AND SIGNATURE; APPLICATION FORMS Registration Of Manufacturers, Distributors, And Dispensers Of Controlled Substances
FORMS, APPLICATION	1309.32	APPLICATION FORMS; CONTENTS, SIGNATURE Registration Of Manufacturers, Distributors, Importers And Exporters Of List I Chemicals
FORMS OF GUARANTY	7.13	SUGGESTED FORMS OF GUARANTY General Provisions; Enforcement Policy
FORMS, ORDER	1305.03	DISTRIBUTIONS REQUIRING ORDER FORMS Order Forms

FORMS, ORDER	1305.04	PERSONS ENTITLED TO OBTAIN AND EXECUTE ORDER FORMS Order Forms
FORMS, ORDER	1305.05	PROCEDURES FOR OBTAINING ORDER FORMS Order Forms
FORMS, ORDER	1305.06	PROCEDURE FOR EXECUTING ORDER FORMS Order Forms
FORMS, ORDER	1305.08	PERSONS ENTITLED TO FILL ORDER FORMS Order Forms
FORMS, ORDER	1305.09	PROCEDURE FOR FILLING ORDER FORMS Order Forms
FORMS, ORDER	1305.10	PROCEDURE FOR ENDORSING ORDER FORMS Order Forms
FORMS, ORDER	1305.11	UNACCEPTED AND DEFECTIVE ORDER FORMS Order Forms
FORMS, ORDER	1305.12	LOST AND STOLEN ORDER FORMS Order Forms
FORMS, ORDER	1305.13	PRESERVATION OF ORDER FORMS Order Forms
FORMS, ORDER	1305.14	RETURN OF UNUSED ORDER FORMS Order Forms
FORMS, ORDER	1305.15	CANCELLATION AND VOIDING OF ORDER FORMS Order Forms
FORMS, ORDER	1305.16	SPECIAL PROCEDURE FOR FILLING CERTAIN ORDER FORMS Order Forms
FORMS, REPORTING	510.302	REPORTING FORMS Records And Reports; New Animal Drugs
FORMULA	225.202	FORMULA, PRODUCTION, AND DISTRIBUTION RECORDS Records; Current Good Manufacturing Practice For Medicated Feeds
FORMULA	226.302	MASTER-FORMULA AND BATCH PRODUCTION RECORDS Records And Reports; Current Good Manufacturing Practice For Type A Medicated Articles
FORMULA, INFANT	107.220	SCOPE AND EFFECT OF INFANT FORMULA RECALL Infant Formula Recalls; Infant Formula
FORMULA, INFANT	107.230	ELEMENTS OF AN INFANT FORMULA RECALL Infant Formula Recalls; Infant Formula
FORMULA, INFANT	107.250	TERMINATION OF AN INFANT FORMULA RECALL Infant Formula Recalls; Infant Formula
FORMULA, INFANT	107.260	REVISION OF AN INFANT FORMULA RECALL Infant Formula Recalls; Infant Formula
FORMULATIONS	864.2875	BALANCED SALT SOLUTIONS OR FORMULATIONS Cell And Tissue Culture Products; Hematology And Pathology Devices
FORMULATIONS, NEW	106.120	NEW FORMULATIONS AND REFORMULATIONS Notification Requirements; Infant Formula Quality Control Procedures
FORNIXSCOPE	886.1320	FORNIXSCOPE Diagnostic Devices; Ophthalmic Devices
FORTIFIED MILK	131.127	NONFAT DRY MILK FORTIFIED WITH VITAMINS A AND D Requirements For Specific Standardized Milk And Cream; Milk And Cream
FORTIFIED PROTEIN	139.11	ENRICHED MACARONI PRODUCTS WITH FORTIFIED PROTEIN Requirements For Specific Standardized Macaroni and Noodle Products; Macaroni And Noodle Products
FORWARDING FACILITIES	1301.77	SECURITY CONTROLS FOR FREIGHT FORWARDING FACILITIES Registration Of Manufacturers, Distributors, And Dispensers Of Controlled Substances
FOSSA	872.3950	GLENOID FOSSA PROSTHESIS Dental Devices
FOUNTAINS, DRINKING	1250.85	DRINKING FOUNTAINS AND COOLERS; ICE; CONSTANT TEMPERATURE BOTTLES Sanitation Facilities And Conditions On Vessels; Interstate Conveyance Sanitation
FRACTION II, COHN	866.5220	COHN FRACTION II IMMUNOLOGICAL TEST SYSTEM Immunological Test Systems; Immunology And Microbiology Devices
FRACTION IV, COHN	866.5360	COHN FRACTION IV IMMUNOLOGICAL TEST SYSTEM Immunological Test Systems; Immunology And Microbiology Devices
FRACTION V, COHN	866.5370	COHN FRACTION V IMMUNOLOGICAL TEST SYSTEM Immunological Test Systems; Immunology And Microbiology Devices
FRACTION, PLASMA PROTEIN	640.90	PLASMA PROTEIN FRACTION (HUMAN) Plasma Protein Fraction (Human); Additional Standards For Human Blood And Blood Products
FRACTIONATION	862.1630	PROTEIN (FRACTIONATION) TEST SYSTEM Clinical Chemistry Test Systems; Clinical Chemistry And Toxicology Devices
FRACTURE, FACIAL	878.3250	EXTERNAL FACIAL FRACTURE FIXATION APPLIANCE Prosthetic Devices; General And Plastic Surgery Devices
FRAGILITY	864.6600	OSMOTIC FRAGILITY TEST Manual Hematology Devices; Hematology And Pathology Devices

FRAGMENT SPECIFIC, FAB	866.5520	IMMUNOGLOBULIN G (FAB FRAGMENT SPECIFIC) IMMUNOLOGICAL TEST SYSTEM Immunological Test Systems; Immunology And Microbiology Devices
FRAGMENT SPECIFIC, FC	866.5530	IMMUNOGLOBULIN G (FC FRAGMENT SPECIFIC) IMMUNOLOGICAL TEST SYSTEM Immunological Test Systems; Immunology And Microbiology Devices
FRAGMENT SPECIFIC, FD	866.5540	IMMUNOGLOBULIN G (FD FRAGMENT SPECIFIC) IMMUNOLOGICAL TEST SYSTEM Immunological Test Systems; Immunology And Microbiology Devices
FRANCISELLA TULARENSIS	866.3280	FRANCISELLA TULARENSIS SEROLOGICAL REAGENTS Serological Reagents; Immunology And Microbiology Devices
FRAME	886.1415	OPHTHALMIC TRIAL LENS FRAME Diagnostic Devices; Ophthalmic Devices
FRAME	886.5842	SPECTACLE FRAME Therapeutic Devices; Ophthalmic Devices
FREE-CHOICE	510.455	NEW ANIMAL DRUG REQUIREMENTS REGARDING FREE-CHOICE ADMINISTRATION IN FEEDS Requirements For Specific New Animal Drugs; New Animal Drugs
FREE-CHOICE	558.465	POLOXALENE FREE-CHOICE LIQUID TYPE C FEED Specific New Animal Drugs For Use In Animal Feeds; New Animal Drugs For Use In Animal Feeds
FREE SECRETORY	866.5380	FREE SECRETORY COMPONENT IMMUNOLOGICAL TEST SYSTEM Immunological Test Systems; Immunology And Microbiology Devices
FREEDOM OF INFORMATION	20.30	FOOD AND DRUG ADMINISTRATION FREEDOM OF INFORMATION STAFF General Policy; Public Information
FREEZER	864.9700	BLOOD STORAGE REFRIGERATOR AND BLOOD STORAGE FREEZER Products Used In Establishments That Manufacture Blood And Blood Products; Hematology And Pathology Devices
FREEZING	864.9225	CELL-FREEZING APPARATUS AND REAGENTS FOR IN VITRO DIAGNOSTIC USE Products Used In Establishments That Manufacture Blood And Blood Products; Hematology And Pathology Devices
FREIGHT	1301.77	SECURITY CONTROLS FOR FREIGHT FORWARDING FACILITIES Registration Of Manufacturers, Distributors, And Dispensers Of Controlled Substances
FRENCH DRESSING	169.115	FRENCH DRESSING Requirements For Specific Standardized Food Dressings And Flavorings; Food Dressings And Flavorings
FREQUENCIES	179.30	RADIOFREQUENCY RADIATION FOR THE HEATING OF FOOD, INCLUDING MICROWAVE FREQUENCIES Radiation And Radiation Sources; Irradiation In The Production, Processing And Handling Of Food
FREQUENCY	1316.13	FREQUENCY OF ADMINISTRATIVE INSPECTIONS Administrative Inspections; Administrative Functions, Practices, And Procedures
FREQUENCY, BREATHING	868.2375	BREATHING FREQUENCY MONITOR Monitoring Devices; Anesthesiology Devices
FRESH	101.95	"FRESH," "FRESHLY FROZEN," "FRESH FROZEN," "FROZEN FRESH" Specific Requirements For Descriptive Claims That Are Neither Nutrient Content Nor Health Claims; Food Labeling
FRESH ASIAGO CHEESE	133.102	ASIAGO FRESH AND ASIAGO SOFT CHEESE Requirements For Specific Standardized Cheese And Related Products; Cheeses And Related Cheese Products
FRESH CITRUS	172.210	COATINGS ON FRESH CITRUS FRUIT Coatings, Films And Related Substances; Food Additives Permitted For Direct Addition To Food For Human Consumption
FRESHLY FROZEN	101.95	"FRESH," "FRESHLY FROZEN," "FRESH FROZEN," "FROZEN FRESH" Specific Requirements For Descriptive Claims That Are Neither Nutrient Content Nor Health Claims; Food Labeling
FRESNEL	886.1390	FLEXIBLE DIAGNOSTIC FRESNEL LENS Diagnostic Devices; Ophthalmic Devices
FRESNEL	886.1655	OPHTHALMIC FRESNEL PRISM Diagnostic Devices; Ophthalmic Devices
FRIED CLAMS	102.49	FRIED CLAMS MADE FROM MINCED CLAMS Requirements For Specific Nonstandardized Foods; Common Or Usual Name For Nonstandardized Foods
FROZEN BLOOD	864.9145	PROCESSING SYSTEM FOR FROZEN BLOOD Products Used In Establishments That Manufacture Blood And Blood Products; Hematology And Pathology Devices
FROZEN CHERRY PIE	152.126	FROZEN CHERRY PIE Requirements For Specific Standardized Fruit Butters, Jellies, Preserves; Fruit Pie
FROZEN CUSTARD	135.110	ICE CREAM AND FROZEN CUSTARD Requirements For Specific Standardized Frozen Desserts; Frozen Desserts

FROZEN DINNER	104.47	FROZEN "HEAT AND SERVE" DINNER Nutritional Quality Guidelines For Food; Common Or Usual Name For Nonstandardized Foods
FROZEN DINNERS	102.26	FROZEN "HEAT AND SERVE" DINNERS Requirements For Specific Nonstandardized Foods; Common Or Usual Name For Nonstandardized Foods
FROZEN EGG	160.190	FROZEN EGG YOLKS Requirements For Specific Standardized Eggs And Egg Products; Eggs And Egg Products
FROZEN EGG WHITES	160.150	FROZEN EGG WHITES Requirements For Specific Standardized Eggs And Egg Products; Eggs And Egg Products
FROZEN EGGS	160.110	FROZEN EGGS Requirements For Specific Standardized Eggs And Egg Products; Eggs And Egg Products
FROZEN FRESH	101.9	"FRESH," "FRESHLY FROZEN," "FRESH FROZEN," "FROZEN FRESH" Specific Requirements For Descriptive Claims That Are Neither Nutrient Content Nor Health Claims; Food Labeling
FROZEN LEMONADE	146.120	FROZEN CONCENTRATE FOR LEMONADE Requirements For Specific Standardized Canned Fruit Juices And Beverages; Canned Fruit Juices
FROZEN LEMONADE	146.121	FROZEN CONCENTRATE FOR ARTIFICIALLY SWEETENED LEMONADE Requirements For Specific Standardized Canned Fruit Juices And Beverages; Canned Fruit Juices
FROZEN LEMONADE	146.126	FROZEN CONCENTRATE FOR COLORED LEMONADE Requirements For Specific Standardized Canned Fruit Juices And Beverages; Canned Fruit Juices
FROZEN ORANGE JUICE	146.137	FROZEN ORANGE JUICE Requirements For Specific Standardized Canned Fruit Juices And Beverages; Canned Fruit Juices
FROZEN ORANGE JUICE	146.146	FROZEN CONCENTRATED ORANGE JUICE Requirements For Specific Standardized Canned Fruit Juices And Beverages; Canned Fruit Juices
FROZEN ORANGE JUICE	146.148	REDUCED ACID FROZEN CONCENTRATED ORANGE JUICE Requirements For Specific Standardized Canned Fruit Juices And Beverages; Canned Fruit Juices
FROZEN PEAS	158.170	FROZEN PEAS Requirements For Specific Standardized Frozen Vegetables; Frozen Vegetables
FROZEN SHRIMP	161.175	FROZEN RAW BREADED SHRIMP Requirements For Specific Standardized Fish And Shellfish; Fish And Shellfish
FROZEN SHRIMP	161.176	FROZEN RAW LIGHTLY BREADED SHRIMP Requirements For Specific Standardized Fish And Shellfish; Fish And Shellfish
FRUCTOSE	184.1560	HIGH FRUCTOSE CORN SYRUP Listing Of Specific Substances Affirmed As GRAS; Direct Food Substances Affirmed As Generally Recognized As Safe
FRUIT	172.210	COATINGS ON FRESH CITRUS FRUIT Coatings, Films And Related Substances; Food Additives Permitted For Direct Addition To Food For Human Consumption
FRUIT BUTTER	150.110	FRUIT BUTTER Requirements For Specific Standardized Fruit Butters, Jellies, Preserves; Fruit Butters, Jellies, Preserves, And Related Products
FRUIT COCKTAIL	145.135	CANNED FRUIT COCKTAIL Requirements For Specific Standardized Canned Fruits; Canned Fruits
FRUIT COCKTAIL	145.136	ARTIFICIALLY SWEETENED CANNED FRUIT COCKTAIL Requirements For Specific Standardized Canned Fruits; Canned Fruits
FRUIT JELLY	150.140	FRUIT JELLY Requirements For Specific Standardized Fruit Butters, Jellies, Preserves; Fruit Butters, Jellies, Preserves, And Related Products
FRUIT JELLY	150.141	ARTIFICIALLY SWEETENED FRUIT JELLY Requirements For Specific Standardized Fruit Butters, Jellies, Preserves; Fruit Butters, Jellies, Preserves, And Related Products
FRUIT JUICE	73.250	FRUIT JUICE Foods; Listing Of Color Additives Exempt From Certification
FRUIT JUICE	101.30	PERCENTAGE JUICE DECLARATION FOR FOOD PURPORTING TO BE BEVERAGES THAT CONTAIN FRUIT OR VEGETABLE JUICE Specific Food Labeling Requirements; Food Labeling
FRUIT JUICE	102.33	BEVERAGES THAT CONTAIN FRUIT OR VEGETABLE JUICE Requirements For Specific Nonstandardized Foods; Common Or Usual Name For Nonstandardized Foods
FRUIT LABELING	101.42	NUTRITION LABELING OF RAW FRUIT, VEGETABLES AND FISH Food Labeling

FRUIT LABELING	101.43	SUBSTANTIAL COMPLIANCE OF FOOD RETAILERS WITH THE GUIDELINES FOR THE VOLUNTARY NUTRITION LABELING OF RAW FRUIT, VEGETABLES, AND FISH Food Labeling
FRUIT LABELING	101.45	GUIDELINES FOR THE VOLUNTARY NUTRITION LABELING OF RAW FRUIT, VEGETABLES, AND FISH Food Labeling
FRUIT PRESERVES	150.160	FRUIT PRESERVES AND JAMS Requirements For Specific Standardized Fruit Butters, Jellies, Preserves; Fruit Butters, Jellies, Preserves, And Related Products
FRUIT PRESERVES	150.161	ARTIFICIALLY SWEETENED FRUIT PRESERVES AND JAMS Requirements For Specific Standardized Fruit Butters, Jellies, Preserves; Fruit Butters, Jellies, Preserves, And Related Products
FRUIT, RAW	101.44	IDENTIFICATION OF THE 20 MOST FREQUENTLY CONSUMED RAW FRUIT, VEGETABLES AND FISH IN THE UNITED STATES Food Labeling
FRUITS	101.76	HEALTH CLAIMS: FIBER-CONTAINING GRAIN PRODUCTS, FRUITS AND VEGETABLES AND CANCER Food Labeling
FRUITS	101.77	HEALTH CLAIMS: FRUITS, VEGETABLES, AND GRAIN PRODUCTS THAT CONTAIN FIBER, PARTICULARLY SOLUBLE FIBER, AND THE RISK OF CORONARY HEART DISEASE Food Labeling
FRUITS	107.78	HEALTH CLAIMS: FRUITS AND VEGETABLES AND CANCER Food Labeling
FRUITS	133.125	COLD-PACK CHEESE FOOD WITH FRUITS, VEGETABLES, OR MEATS Requirements For Specific Standardized Cheese And Related Products; Cheeses And Related Cheese Products
FRUITS	133.168	PASTEURIZED BLENDED CHEESE WITH FRUITS, VEGETABLES, OR MEATS Requirements For Specific Standardized Cheese And Related Products; Cheeses And Related Cheese Products
FRUITS	133.170	PASTEURIZED PROCESS CHEESE WITH FRUITS, VEGETABLES, OR MEATS Requirements For Specific Standardized Cheese And Related Products; Cheeses And Related Cheese Products
FRUITS	133.174	PASTEURIZED PROCESS CHEESE FOOD WITH FRUITS, VEGETABLES, OR MEATS Requirements For Specific Standardized Cheese And Related Products; Cheeses And Related Cheese Products
FRUITS	133.176	PASTEURIZED CHEESE SPREAD WITH FRUITS, VEGETABLES OR MEATS Requirements For Specific Standardized Cheese And Related Products; Cheeses And Related Cheese Products
FRUITS	133.180	PASTEURIZED PROCESS CHEESE SPREAD WITH FRUITS, VEGETABLES, OR MEATS Requirements For Specific Standardized Cheese And Related Products; Cheeses And Related Cheese Products
FRUITS	173.315	CHEMICALS USED IN WASHING OR TO ASSIST IN THE PEELING OF FRUITS AND VEGETABLES Specific Usage Additives; Secondary Direct Food Additives Permitted In Food For Human Consumption
FULFILLMENT HOUSES	203.36	FULFILLMENT HOUSES, SHIPPING AND MAILING SERVICES Prescription Drug Marketing
FUMAGILLIN	520.182	BICYCLOHEXYLAMMONIUM FUMAGILLIN Oral Dosage Form New Animal Drugs
FUMARATE	172.826	SODIUM STEARYL FUMARATE Multipurpose Additives; Food Additives Permitted For Direct Addition To Food For Human Consumption
FUMARATE	184.1307d	FERROUS FUMARATE Listing Of Specific Substances Affirmed As GRAS; Direct Food Substances Affirmed As Generally Recognized As Safe
FUMARATE	520.82	AMINOPROPAZINE FUMARATE ORAL DOSAGE FORMS Oral Dosage Form New Animal Drugs
FUMARATE	522.82	BETA-AMINOPROPRIONITRILE FUMARATE Implantation Or Injectable Dosage Form New Animal Drugs
FUMARATE	522.84	AMINOPROMAZINE FUMARATE STERILE SOLUTION INJECTION Implantation Or Injectable Dosage Form New Animal Drugs
FUMARIC ACID	172.350	FUMARIC ACID AND SALTS OF FUMARIC ACIDS Special Dietary And Nutritional Additives; Food Additives Permitted For Direct Addition To Food For Human Consumption
FUNCTION ANALYZER	870.3630	PACEMAKER GENERATOR FUNCTION ANALYZER Cardiovascular Prosthetic Devices; Cardiovascular Devices
FUNCTION ANALYZER	870.3640	INDIRECT PACEMAKER GENERATOR FUNCTION ANALYZER Cardiovascular Prosthetic Devices; Cardiovascular Devices
FUNCTION, PULMONARY	868.1880	PULMONARY-FUNCTION DATA CALCULATOR Diagnostic Devices; Anesthesiology Devices

FUNCTION, PULMONARY	868.1890	PREDICTIVE PULMONARY-FUNCTION VALUE CALCULATOR Diagnostic Devices; Anesthesiology Devices
FUNCTION, PULMONARY	868.1900	DIAGNOSTIC PULMONARY-FUNCTION INTERPRETATION CALCULATOR Diagnostic Devices; Anesthesiology Devices
FUNCTION TESTER	870.3720	PACEMAKER ELECTRODE FUNCTION TESTER Cardiovascular Prosthetic Devices; Cardiovascular Devices
FUNCTIONAL	882.5810	EXTERNAL FUNCTIONAL NEUROMUSCULAR STIMULATOR Neurological Therapeutic Devices; Neurological Devices
FUNCTIONS	5.21	EMERGENCY FUNCTIONS General Redelegations Of Authority; Delegations Of Authority And Organization
FUNCTIONS	5.417	AUTHORITY PERTAINING TO ACCREDITATION FUNCTIONS FOR MEDICAL DEVICES Medical Devices And Radiological Health; Redelegations Of Authority; Delegations Of Authority And Organization
FUNCTIONS	12.62	COMMENCEMENT OF FUNCTIONS Presiding Officer; Formal Evidentiary Public Hearing
FUNCTIONS	14.122	FUNCTIONS OF TEPRSSC Technical Electronic Products Radiation Safety Standards Committee; Public Hearing Before A Public Advisory Committee
FUNCTIONS	14.142	FUNCTIONS OF A COLOR ADDITIVE ADVISORY COMMITTEE Color Additive Advisory Committees; Public Hearing Before A Public Advisory Committee
FUNCTIONS	56.108	IRB FUNCTIONS AND OPERATIONS IRB Functions And Operations; Institutional Review Boards
FUNCTIONS	812.60	IRB COMPOSITION, DUTIES, AND FUNCTIONS IRB Review And Approval; Investigational Device Exemptions
FUNCTIONS	1401.2	THE OFFICE OF NATIONAL DRUG CONTROL POLICY - ORGANIZATION AND FUNCTIONS Public Availability Of Information
FUNCTIONS, SEPARATION OF	10.55	SEPARATION OF FUNCTIONS; EX PARTE COMMUNICATIONS General Administrative Procedures; Administrative Practices And Procedures
FUNCTIONS, SEPARATION OF	13.15	SEPARATION OF FUNCTIONS; EX PARTE COMMUNICATIONS; ADMINISTRATIVE SUPPORT General Provisions; Public Hearing Before A Public Board Of Inquiry
FUNDS	1403.23	PERIOD OF AVAILABILITY OF FUNDS Post-award Requirements; Uniform Administrative Requirements For Grants And Coop. Agreements To State & Local Gov.
FUNDS, FDA	20.105	TESTING AND RESEARCH CONDUCTED BY OR WITH FUNDS PROVIDED BY THE FOOD AND DRUG ADMINISTRATION Availability Of Specific Categories Of Records; Public Information
FUNDS, FDA	20.106	STUDIES AND REPORTS PREPARED BY OR WITH FUNDS PROVIDED BY THE FOOD AND DRUG ADMINISTRATION Availability Of Specific Categories Of Records; Public Information
FUNDUS	886.1395	DIAGNOSTIC HRUBY FUNDUS LENS Diagnostic Devices; Ophthalmic Devices
FURAZOLIDONE	524.1005	FURAZOLIDONE AEROSOL POWDER Ophthalmic & Topical Dosage Form New Animal Drugs
FURAZOLIDONE	556.290	FURAZOLIDONE Specific Tolerances For Residues Of New Animal Drugs; Tolerances For Residues Of New Animal Drugs In Food
FURCELLERAN	172.655	FURCELLERAN Gums, Chewing Gum Bases And Related Substances; Food Additives Permitted For Direct Addition To Food For Human Consumption
FURCELLERAN	172.660	SALTS OF FURCELLERAN Gums, Chewing Gum Bases And Related Substances; Food Additives Permitted For Direct Addition To Food For Human Consumption
FUROSEMIDE	520.1010	FUROSEMIDE Oral Dosage Form New Animal Drugs
FUROSEMIDE	522.1010	FUROSEMIDE Implantation Or Injectable Dosage Form New Animal Drugs
FUSION	886.1880	FUSION AND STEREOSCOPIC TARGET Diagnostic Devices; Ophthalmic Devices

G

G, IMMUNOGLOBULIN	866.5510	IMMUNOGLOBULINS A, G, M, D, AND E IMMUNOLOGICAL TEST SYSTEM Immunological Test Systems; Immunology And Microbiology Devices
G, IMMUNOGLOBULIN	866.5520	IMMUNOGLOBULIN G (FAB FRAGMENT SPECIFIC) IMMUNOLOGICAL TEST SYSTEM Immunological Test Systems; Immunology And Microbiology Devices
G, IMMUNOGLOBULIN	866.5530	IMMUNOGLOBULIN G (FC FRAGMENT SPECIFIC) IMMUNOLOGICAL TEST SYSTEM Immunological Test Systems; Immunology And Microbiology Devices
G, IMMUNOGLOBULIN	866.5540	IMMUNOGLOBULIN G (FD FRAGMENT SPECIFIC) IMMUNOLOGICAL TEST SYSTEM Immunological Test Systems; Immunology And Microbiology Devices
GALACTOSE	862.1310	GALACTOSE TEST SYSTEM Clinical Chemistry Test Systems; Clinical Chemistry And Clinical Toxicology Devices
GALACTOSE-1-PHOSPHATE	862.1315	GALACTOSE-1-PHOSPHATE URIDYL TRANSFERASE TEST SYSTEM Clinical Chemistry Test Systems; Clinical Chemistry And Clinical Toxicology Devices
GALACTOSIDASE	173.145	ALPHA-GALACTOSIDASE DERIVED FROM MORTEIRELLA VINACEAE Enzyme Preparations And Microorganisms; Secondary Direct Food Additives Permitted In Food For Human Consumption
GALLATE	184.1660	PROPYL GALLATE Listing Of Specific Substances Affirmed As GRAS; Direct Food Substances Affirmed As Generally Recognized As Safe
GALLATE	582.3660	PROPYL GALLATE Chemical Preservatives; Substances Generally Recognized As Safe
GALLEYS	1250.84	WATER IN GALLEYS AND MEDICAL CARE SPACES Sanitation Facilities And Conditions On Vessels; Interstate Conveyance Sanitation
GALVANIC	882.1540	GALVANIC SKIN RESPONSE MEASUREMENT DEVICE Neurological Diagnostic Devices; Neurological Devices
GAMMA CAMERA	892.1100	SCINTILLATION (GAMMA) CAMERA Diagnostic Devices; Radiology Devices
GAMMA COUNTER	862.2320	BETA OR GAMMA COUNTER FOR CLINICAL USE Clinical Laboratory Instruments; Clinical Chemistry And Toxicology Devices
GAMMA-GLUTAMYL TRANSPEPTIDASE	862.1360	GAMMA-GLUTAMYL TRANSPEPTIDASE AND ISOENZYMES TEST SYSTEM Clinical Chemistry Test Systems; Clinical Chemistry And Clinical Toxicology Devices
GAMMELOST CHEESE	133.140	GAMMELOST CHEESE Requirements For Specific Standardized Cheese And Related Products; Cheeses And Related Cheese Products
GARBAGE	1240.75	GARBAGE Specific Administrative Decisions Regarding Interstate Shipments; Control Of Communicable Diseases
GARBAGE	1250.39	GARBAGE EQUIPMENT AND DISPOSITION Food Service Sanitation On Land And Air Conveyances; Interstate Conveyance Sanitation
GARBAGE	1250.79	GARBAGE DISPOSAL Servicing Areas For Land And Air Conveyances; Interstate Conveyance Sanitation
GARLIC	184.1317	GARLIC AND ITS DERIVATIVES Listing Of Specific Substances Affirmed As GRAS; Direct Food Substances Affirmed As Generally Recognized As Safe
GARMENT	876.5920	PROTECTIVE GARMENT FOR INCONTINENCE Therapeutic Devices; Gastroenterology-urology Devices
GAS	862.2250	GAS LIQUID CHROMATOGRAPHY SYSTEM FOR CLINICAL USE Clinical Laboratory Instruments; Clinical Chemistry And Toxicology Devices
GAS	868.1075	ARGON GAS ANALYZER Diagnostic Devices; Anesthesiology Devices
GAS	868.1400	CARBON DIOXIDE GAS ANALYZER Diagnostic Devices; Anesthesiology Devices
GAS	868.1430	CARBON MONOXIDE GAS ANALYZER Diagnostic Devices; Anesthesiology Devices
GAS	868.1500	ENFLURANE GAS ANALYZER Diagnostic Devices; Anesthesiology Devices
GAS	868.1575	GAS COLLECTION VESSEL Diagnostic Devices; Anesthesiology Devices
GAS	868.1620	HALOTHANE GAS ANALYZER Diagnostic Devices; Anesthesiology Devices
GAS	868.1640	HELIUM GAS ANALYZER Diagnostic Devices; Anesthesiology Devices
GAS	868.1670	NEON GAS ANALYZER Diagnostic Devices; Anesthesiology Devices

GAS	868.1690	NITROGEN GAS ANALYZER Diagnostic Devices; Anesthesiology Devices
GAS	868.1700	NITROUS OXIDE GAS ANALYZER Diagnostic Devices; Anesthesiology Devices
GAS	868.1720	OXYGEN GAS ANALYZER Diagnostic Devices; Anesthesiology Devices
GAS	868.1870	GAS VOLUME CALIBRATOR Diagnostic Devices; Anesthesiology Devices
GAS	868.2350	GAS CALIBRATION FLOWMETER Monitoring Devices; Anesthesiology Devices
GAS	868.2610	GAS PRESSURE GAUGE Monitoring Devices; Anesthesiology Devices
GAS	868.2620	GAS PRESSURE CALIBRATOR Monitoring Devices; Anesthesiology Devices
GAS	868.2885	GAS FLOW TRANSDUCER Monitoring Devices; Anesthesiology Devices
GAS	868.2900	GAS PRESSURE TRANSDUCER Monitoring Devices; Anesthesiology Devices
GAS	868.5160	GAS MACHINE FOR ANESTHESIA OR ANALGESIA Therapeutic Devices; Anesthesiology Devices
GAS	868.5330	BREATHING GAS MIXER Therapeutic Devices; Anesthesiology Devices
GAS	868.5450	RESPIRATORY GAS HUMIDIFIER Therapeutic Devices; Anesthesiology Devices
GAS	868.5550	ANESTHETIC GAS MASK Therapeutic Devices; Anesthesiology Devices
GAS	868.5560	GAS MASK HEAD STRAP Therapeutic Devices; Anesthesiology Devices
GAS	868.6400	CALIBRATION GAS Miscellaneous; Anesthesiology Devices
GAS	868.6885	MEDICAL GAS YOKE ASSEMBLY Miscellaneous; Anesthesiology Devices
GAS	870.4300	CARDIOPULMONARY BYPASS GAS CONTROL UNIT Cardiovascular Surgical Devices; Cardiovascular Devices
GAS	870.4330	CARDIOPULMONARY BYPASS ON-LINE BLOOD GAS MONITOR Cardiovascular Surgical Devices; Cardiovascular Devices
GAS	870.4410	CARDIOPULMONARY BYPASS IN-LINE BLOOD GAS SENSOR Cardiovascular Surgical Devices; Cardiovascular Devices
GAS	880.6100	ETHYLENE OXIDE GAS AERATOR CABINET General Hospital And Personal Use Miscellaneous Devices; General Hospital And Personal Use Devices
GAS	886.4270	INTRAOCULAR GAS Surgical Devices; Ophthalmic Devices
GAS STERILIZER	880.6860	ETHYLENE OXIDE GAS STERILIZER General Hospital And Personal Use Miscellaneous Devices; General Hospital And Personal Use Devices
GAS, COMBUSTION	173.350	COMBUSTION PRODUCT GAS Specific Usage Additives; Secondary Direct Food Additives Permitted In Food For Human Consumption
GAS DISCHARGE	1020.20	COLD-CATHODE GAS DISCHARGE TUBES Performance Standards For Ionizing Radiation Emitting Products
GAS-GENERATING DEVICE	866.2580	GAS-GENERATING DEVICE Microbiology Devices; Immunology And Microbiology Devices
GAS PERMEABLE	886.5916	RIGID GAS PERMEABLE CONTACT LENS Therapeutic Devices; Ophthalmic Devices
GAS PERMEABLE	886.5918	RIGID GAS PERMEABLE CONTACT LENS CARE PRODUCTS Therapeutic Devices; Ophthalmic Devices
GAS-POWERED	872.4465	GAS-POWERED JET INJECTOR Surgical Devices; Dental Devices
GAS-SCAVENGING	868.5430	GAS-SCAVENGING APPARATUS Therapeutic Devices; Anesthesiology Devices
GASES, BLOOD	862.1120	BLOOD GASES (PCO2PO2) AND BLOOD PH TEST SYSTEM Clinical Chemistry Test Systems; Clinical Chemistry And Clinical Toxicology Devices
GASES, CARBON DIOXIDE	201.161	CARBON DIOXIDE AND CERTAIN OTHER GASES Other Exemptions; Labeling
GASKETS	177.1210	CLOSURES WITH SEALING GASKETS FOR FOOD CONTAINERS Substances For Use As Basic Components Of Single And Repeated Use Food Contact Surfaces; Indirect Food Additives: Polymers
GASTRIC ACIDITY	862.1320	GASTRIC ACIDITY TEST SYSTEM Clinical Chemistry Test Systems; Clinical Chemistry And Clinical Toxicology Devices

GASTRIN	862.1325	GASTRIN TEST SYSTEM Clinical Chemistry Test Systems; Clinical Chemistry And Clinical Toxicology Devices
GASTROENTEROLOGY-UROLOGY	876.1075	GASTROENTEROLOGY-UROLOGY BIOPSY INSTRUMENT Diagnostic Devices; Gastroenterology-urology Devices
GASTROENTEROLOGY-UROLOGY	876.4370	GASTROENTEROLOGY-UROLOGY EVACUATOR Surgical Devices; Gastroenterology-urology Devices
GASTROENTEROLOGY-UROLOGY	876.4530	GASTROENTEROLOGY-UROLOGY FIBEROPTIC RETRACTOR Surgical Devices; Gastroenterology-urology Devices
GASTROENTEROLOGY-UROLOGY	876.4730	MANUAL GASTROENTEROLOGY-UROLOGY SURGICAL INSTRUMENT AND ACCESSORIES Surgical Devices; Gastroenterology-urology Devices
GASTROINTESTINAL	876.1300	INGESTIBLE TELEMETRIC GASTROINTESTINAL CAPSULE IMAGING SYSTEM Diagnostic Devices; Gastroenterology-urology Devices
GASTROINTESTINAL	876.1725	GASTROINTESTINAL MOTILITY MONITORING SYSTEM Diagnostic Devices; Gastroenterology-urology Devices
GASTROINTESTINAL	876.5980	GASTROINTESTINAL TUBE AND ACCESSORIES Therapeutic Devices; Gastroenterology-urology Devices
GAT	866.3290	GONOCOCCAL ANTIBODY TEST (GAT) Serological Reagents; Immunology And Microbiology Devices
GAUGE, BOURDON	868.2300	BOURDON GAUGE FLOWMETER Monitoring Devices; Anesthesiology Devices
GAUGE, DEPTH	888.4300	DEPTH GAUGE FOR CLINICAL USE Surgical Devices; Orthopedic Devices
GAUGE, LENS	886.1420	OPHTHALMIC LENS GUAGE Diagnostic Devices; Ophthalmic Devices
GAUGE, PRESSURE	868.2610	GAS PRESSURE GAUGE Monitoring Devices; Anesthesiology Devices
GAUGE, PRESSURE	870.4310	CARDIOPULMONARY BYPASS CORONARY PRESSURE GAUGE Cardiovascular Surgical Devices; Cardiovascular Devices
GAUZE	878.4014	NONRESORBABLE GAUZE SPONGE FOR EXTERNAL USE Surgical Devices; General And Plastic Surgery Devices
GAUZE	878.4450	NONABSORBABLE GAUZE FOR INTERNAL USE Surgical Devices; General And Plastic Surgery Devices
GEL	520.1242f	LEVAMISOLE HYDROCHLORIDE GEL Oral Dosage Form New Animal Drugs
GEL	520.1452	MOXIDECTIN GEL Oral Dosage Form New Animal Drugs
GEL	520.1720d	PHENYLBUTAZONE GEL Oral Dosage Form New Animal Drugs
GEL	524.660b	DIMETHYL SULFOXIDE GEL Ophthalmic And Topical Dosage Form New Animal Drugs
GEL, ELECTRODE	872.1730	ELECTRODE GEL FOR PULP TESTERS Diagnostic devices; Dental Devices
GEL-FILLED	878.3540	SILICONE GEL-FILLED BREAST PROSTHESIS Prosthetic Devices; General And Plastic Surgery Devices
GEL, SUPPORT	866.4900	SUPPORT GEL Immunology Laboratory Equipment And Reagents; Immunology And Microbiology Devices
GELATIN	522.1020	GELATIN SOLUTION Implantation Or Injectable Dosage Form New Animal Drugs
GELLAN GUM	172.665	GELLAN GUM Gums, Chewing Gum Bases And Related Substances; Food Additives Permitted For Direct Addition To Food For Human Consumption
GELSEMIUM PREPARATIONS	250.105	GELSEMIUM-CONTAINING PREPARATIONS REGARDED AS PRESCRIPTION DRUGS New Drug Or Prescription Status Of Specific Drugs; Specific Requirements For Specific Human Drugs
GENERAL	82.5	GENERAL SPECIFICATIONS FOR STRAIGHT COLORS Foods, Drugs And Cosmetics; Listing Of Certified Provisionally Listed Colors And Specifications
GENERAL	82.50	GENERAL Foods, Drugs And Cosmetics; Listing Of Certified Provisionally Listed Colors And Specifications
GENERAL	82.1050	GENERAL Drugs And Cosmetics; Listing Of Certified Provisionally Listed Colors And Specifications
GENERAL	82.2050	GENERAL Externally Applied Drugs And Cosmetics; Listing Of Certified Provisionally Listed Colors And Specifications
GENERAL CONSIDERATIONS	556.1	GENERAL CONSIDERATIONS; TOLERANCES FOR RESIDUES OF NEW ANIMAL DRUGS IN FOOD Tolerances For Residues Of New Animal Drugs In Food

GENERAL FACTORS	315.3	GENERAL FACTORS RELEVANT TO SAFETY AND EFFECTIVENESS Diagnostic Radiopharmaceuticals
GENERAL FACTORS	601.32	GENERAL FACTORS RELEVANT TO SAFETY AND EFFECTIVENESS Diagnostic Radiopharmaceuticals; Licensing
GENERAL GUARANTY	1230.20	GENERAL GUARANTY Regulations Under The Federal Caustic Poison Act
GENERAL GUIDANCE	7.59	GENERAL INDUSTRY GUIDANCE Recalls - Guidance on Policy, Procedures & Industry Responsibilities; Enforcement Policy
GENERAL INSTRUMENT	884.4520	OBSTETRIC-GYNECOLOGIC GENERAL MANUAL INSTRUMENT Obstetrical And Gynecological Surgical Devices; Obstetrical And Gynecological Devices
GENERAL OBLIGATIONS	26.62	GENERAL OBLIGATIONS Framework Provisions; Mutual Recognition Of Pharmaceutical GMP Reports, Medical Device Quality System Audit Reports
GENERAL PRINCIPLES	101.13	NUTRIENT CONTENT CLAIMS - GENERAL PRINCIPLES Food Labeling
GENERAL PRINCIPLES	102.5	GENERAL PRINCIPLES Common Or Usual Name For Nonstandardized Foods
GENERAL PRINCIPLES	104.5	GENERAL PRINCIPLES Nutritional Quality Guidelines For Foods
GENERAL PRINCIPLES	170.20	GENERAL PRINCIPLES FOR EVALUATING THE SAFETY OF FOOD ADDITIVES Food Additives
GENERAL PRINCIPLES	312.22	GENERAL PRINCIPLES OF THE IND SUBMISSION Investigational New Drug Application
GENERAL PRINCIPLES	502.5	GENERAL PRINCIPLES Common Or Usual Names For Nonstandardized Animal Foods
GENERAL PRINCIPLES	570.20	GENERAL PRINCIPLES FOR EVALUATING THE SAFETY OF FOOD ADDITIVES Food Additive Safety; Food Additives
GENERAL PROCEDURES	25.15	GENERAL PROCEDURES Agency Actions Requiring Environmental Consideration; Environmental Impact Considerations
GENERAL PROVISIONS	172.5	GENERAL PROVISIONS FOR DIRECT FOOD ADDITIVES Food Additives Permitted For Direct Addition To Food For Human Consumption
GENERAL PROVISIONS	174.5	GENERAL PROVISIONS APPLICABLE TO INDIRECT FOOD ADDITIVES Indirect Food Additives
GENERAL PROVISIONS	179.25	GENERAL PROVISIONS FOR FOOD IRRADIATION Radiation And Radiation Sources; Irradiation In The Production, Processing And Handling Of Food
GENERAL PURPOSE	862.2050	GENERAL PURPOSE LABORATORY EQUIPMENT LABELED OR PROMOTED FOR A SPECIFIC MEDICINAL USE Clinical Laboratory Instruments; Clinical Chemistry And Clinical Toxicology Devices
GENERAL PURPOSE	864.4010	GENERAL PURPOSE REAGENT Specimen Preparation Reagents; Hematology and Pathology Devices
GENERAL PURPOSE	880.6890	GENERAL PURPOSE DISINFECTANTS General Hospital And Personal Use Miscellaneous Devices; General Hospital And Personal Use Devices
GENERAL RECOGNITION	330.1	GENERAL CONDITIONS FOR GENERAL RECOGNITION AS SAFE, EFFECTIVE, AND NOT MISBRANDED Over-the-counter Human Drugs Generally Recognized As Safe And Effective & Not Misbranded
GENERAL REDELEGATIONS	5.20	GENERAL REDELEGATIONS OF AUTHORITY FROM THE COMMISSIONER TO OTHER OFFICERS OF THE FDA General Redelegations Of Authority; Delegations Of Authority And Organization
GENERAL REQUIREMENTS	11.100	GENERAL REQUIREMENTS Electronic Records; Electronic Signatures
GENERAL REQUIREMENTS	50.20	GENERAL REQUIREMENTS FOR INFORMED CONSENT Informed Consent Of Human Subjects; Protection Of Human Subjects
GENERAL REQUIREMENTS	50.23	EXCEPTION FROM GENERAL REQUIREMENTS Informed Consent Of Human Subjects; Protection Of Human Subjects
GENERAL REQUIREMENTS	101.14	HEALTH CONTENT CLAIMS - GENERAL REQUIREMENTS Food Labeling
GENERAL REQUIREMENTS	201.56	GENERAL REQUIREMENTS ON CONTENT AND FORMAT OF LABELING FOR HUMAN PRESCRIPTION DRUGS Labeling Requirements For Prescription Drugs and/or Insulin; Labeling
GENERAL REQUIREMENTS	211.80	GENERAL REQUIREMENTS Control Of Components And Drug Product Containers And Closures; Current Good Manufacturing Practice For Finished Pharmaceuticals
GENERAL REQUIREMENTS	211.160	GENERAL REQUIREMENTS Laboratory Controls; Current Good Manufacturing Practice For Finished Pharmaceuticals
GENERAL REQUIREMENTS	211.180	GENERAL REQUIREMENTS Records and Reports; Current Good Manufacturing Practice For Finished Pharmaceuticals

GENERAL REQUIREMENTS	312.40	GENERAL REQUIREMENTS FOR USE OF AN INVESTIGATIONAL NEW DRUG IN A CLINICAL INVESTIGATION Investigational New Drug Application
GENERAL REQUIREMENTS	606.120	LABELING; GENERAL REQUIREMENTS Finished Product Control; Current Good Manufacturing Practice For Blood And Blood Components
GENERAL REQUIREMENTS	640.2	GENERAL REQUIREMENTS Whole Blood; Additional Standards For Human Blood And Blood Products
GENERAL REQUIREMENTS	640.11	GENERAL REQUIREMENTS Red Blood Cells; Additional Standards For Human Blood And Blood Products
GENERAL REQUIREMENTS	640.25	GENERAL REQUIREMENTS Platelets; Additional Standards For Human Blood And Blood Products
GENERAL REQUIREMENTS	640.69	GENERAL REQUIREMENTS Source Plasma; Additional Standards For Human Blood And Blood Products
GENERAL REQUIREMENTS	640.83	GENERAL REQUIREMENTS Albumin (Human); Additional Standards For Human Blood And Blood Products
GENERAL REQUIREMENTS	640.93	GENERAL REQUIREMENTS Plasma Protein Fraction (Human); Additional Standards For Human Blood And Blood Products
GENERAL REQUIREMENTS	640.101	GENERAL REQUIREMENTS Additional Standards For Diagnostic Substances For Laboratory Tests
GENERAL REQUIREMENTS	660.2	GENERAL REQUIREMENTS Plasma Protein Fraction (Human); Additional Standards For Human Blood And Blood Products
GENERAL REQUIREMENTS	809.20	GENERAL REQUIREMENTS FOR MANUFACTURERS AND PRODUCERS OF IN VITRO DIAGNOSTIC PRODUCTS Requirements For Manufacturers And Producers; In Vitro Diagnostic Products For Human Use
GENERAL REQUIREMENTS	820.180	GENERAL REQUIREMENTS Records; Quality System Regulation
GENERAL REQUIREMENTS	1240.80	GENERAL REQUIREMENTS FOR WATER FOR DRINKING AND CULINARY PURPOSES Source And Use Of Potable Water; Control Of Communicable Diseases
GENERAL REQUIREMENTS	1250.22	GENERAL REQUIREMENTS Food Service Sanitation On Land And Air Conveyances, And Vessels; Interstate Conveyance Sanitation
GENERAL REQUIREMENTS	1250.63	GENERAL REQUIREMENTS Servicing Areas For Land And Air Conveyances; Interstate Conveyance Sanitation
GENERAL REQUIREMENTS	1270.33	RECORDS, GENERAL REQUIREMENTS Procedures and Records; Human Tissue Intended For Transplantation
GENERAL REQUIREMENTS	1304.21	GENERAL REQUIREMENTS FOR CONTINUING RECORDS Records and Reports Of Registrants
GENERAL REQUIREMENTS	1309.71	GENERAL SECURITY REQUIREMENTS Registration Of Manufacturers, Distributors, Importers And Exporters Of List I Chemicals
GENERAL RESPONSIBILITIES	312.50	GENERAL RESPONSIBILITIES OF SPONSORS Responsibilities Of Sponsors And Investigators; Investigational New Drug Application
GENERAL RESPONSIBILITIES	312.60	GENERAL RESPONSIBILITIES OF INVESTIGATORS Responsibilities Of Sponsors And Investigators; Investigational New Drug Application
GENERAL RESPONSIBILITIES	812.40	GENERAL RESPONSIBILITIES OF SPONSORS Responsibilities Of Sponsors; Investigational Device Exemptions
GENERAL RESPONSIBILITIES	812.100	GENERAL RESPONSIBILITIES OF INVESTIGATORS Responsibilities Of Investigators; Investigational Device Exemptions
GENERAL RESTRICTIONS	70.5	GENERAL RESTRICTIONS ON USE OF COLOR ADDITIVES Color Additives
GENERAL SAFETY	610.11	GENERAL SAFETY General Biological Products Standards
GENERAL SAFETY	610.11a	INACTIVATED INFLUENZA VACCINE, GENERAL SAFETY TEST General Biological Products Standards
GENERAL WARNINGS	369.9	GENERAL WARNINGS RE ACCIDENTAL INGESTION BY CHILDREN Definitions And Interpretations; Interpretative Statements Re Warnings On Drugs And Devices For OTC Sale
GENERALLY RECOGNIZED SAFE	5.300	FOOD STANDARDS, FOOD ADDITIVES, GENERALLY RECOGNIZED AS SAFE (GRAS) SUBSTANCES, COLOR ADDITIVES, NUTRIENT CONTENT CLAIMS, AND HEALTH CLAIMS Foods And Cosmetics; Redelegations Of Authority; Delegations Of Authority And Organization
GENERALLY RECOGNIZED SAFE	170.30	ELIGIBILITY FOR CLASSIFICATION AS GENERALLY RECOGNIZED AS SAFE (GRAS) Food Additive Safety; Food Additives
GENERALLY RECOGNIZED SAFE	170.35	AFFIRMATION OF GENERALLY RECOGNIZED AS SAFE (GRAS) STATUS Food Additive Safety; Food Additives

GENERALLY RECOGNIZED SAFE	182.1	SUBSTANCES THAT ARE GENERALLY RECOGNIZED AS SAFE General Provisions; Substances Generally Recognized As Safe
GENERALLY RECOGNIZED SAFE	184.1	SUBSTANCES ADDED DIRECTLY TO HUMAN FOOD AFFIRMED AS GENERALLY RECOGNIZED AS SAFE (GRAS) General Provisions; Direct Food Substances Affirmed As Generally Recognized As Safe
GENERALLY RECOGNIZED SAFE	186.1	SUBSTANCES ADDED INDIRECTLY TO HUMAN FOOD AFFIRMED AS GENERALLY RECOGNIZED AS SAFE (GRAS) General Provisions; Indirect Food Substances Affirmed As Generally Recognized As Safe
GENERALLY RECOGNIZED SAFE	330.10	PROCEDURES FOR CLASSIFYING OTC DRUGS AS GENERALLY RECOGNIZED AS SAFE AND EFFECTIVE AND NOT MISBRANDED Administrative Procedures; Over-the-counter Human Drugs Generally Recognized As Safe And Effective & Not Misbranded
GENERALLY RECOGNIZED SAFE	330.14	ADDITIONAL CRITERIA AND PROCEDURES FOR CLASSIFYING OTC DRUGS AS GENERALLY RECOGNIZED AS SAFE AND EFFECTIVE AND NOT MISBRANDED Administrative Procedures; Over-the-counter Human Drugs Generally Recognized As Safe And Effective & Not Misbranded
GENERALLY RECOGNIZED SAFE	570.30	ELIGIBILITY FOR CLASSIFICATION AS GENERALLY RECOGNIZED AS SAFE (GRAS) Food Additive Safety; Food Additives
GENERALLY RECOGNIZED SAFE	570.35	AFFIRMATION OF GENERALLY RECOGNIZED AS SAFE (GRAS) STATUS Food Additive Safety; Food Additives
GENERALLY RECOGNIZED SAFE	582.1	SUBSTANCES THAT ARE GENERALLY RECOGNIZED AS SAFE General Provisions; Substances Generally Recognized As Safe
GENERATION	864.7900	THROMBOPLASTIN GENERATION TEST Hematology Kits And Packages; Hematology And Pathology Devices
GENERATOR	868.5440	PORTABLE OXYGEN GENERATOR Therapeutic Devices; Anesthesiology Devices
GENERATOR	870.1750	EXTERNAL PROGRAMMABLE PACEMAKER PULSE GENERATOR Cardiovascular Diagnostic Devices; Cardiovascular Devices
GENERATOR	870.3600	EXTERNAL PACEMAKER PULSE GENERATOR Cardiovascular Prosthetic Devices; Cardiovascular Devices
GENERATOR	870.3610	IMPLANTABLE PACEMAKER PULSE GENERATOR Cardiovascular Prosthetic Devices; Cardiovascular Devices
GENERATOR	870.3630	PACEMAKER GENERATOR FUNCTION ANALYZER Cardiovascular Prosthetic Devices; Cardiovascular Devices
GENERATOR	870.3640	INDIRECT PACEMAKER GENERATOR FUNCTION ANALYZER Cardiovascular Prosthetic Devices; Cardiovascular Devices
GENERATOR	870.4320	CARDIOPULMONARY BYPASS PULSATILE FLOW GENERATOR Cardiovascular Surgical Devices; Cardiovascular Devices
GENERATOR	874.1120	ELECTRONIC NOISE GENERATOR FOR AUDIOMETRIC TESTING Diagnostic Devices; Ear, Nose, And Throat Devices
GENERATOR	882.1430	ELECTROENCEPHALOGRAPH TEST SIGNAL GENERATOR Neurological Diagnostic Devices; Neurological Devices
GENERATOR	882.4400	RADIOFREQUENCY LESION GENERATOR Neurological Surgical Devices; Neurological Devices
GENERATOR	892.1700	DIAGNOSTIC X-RAY HIGH VOLTAGE GENERATOR Diagnostic Devices; Radiology Devices
GENITAL VIBRATOR	884.5960	GENITAL VIBRATOR FOR THERAPEUTIC USE Obstetrical And Gynecological Therapeutic Devices; Obstetrical And Gynecological Devices
GENTAMICIN	862.3450	GENTAMICIN TEST SYSTEM Clinical Toxicology Test Systems; Clinical Chemistry And Toxicology Devices
GENTAMICIN SULFATE	520.1044	GENTAMICIN SULFATE ORAL DOSAGE FORMS Oral Dosage Form New Animal Drugs
GENTAMICIN SULFATE	522.1044	GENTAMICIN SULFATE INJECTION Implantation Or Injectable Dosage Form New Animal Drugs
GENTAMICIN SULFATE	524.1044	GENTAMICIN SULFATE OPHTHALMIC AND TOPICAL DOSAGE FORMS Ophthalmic And Topical Dosage Form New Animal Drugs
GENTAMICIN SULFATE	529.1044	GENTAMICIN SULFATE IN CERTAIN OTHER DOSAGE FORMS Certain Other Dosage Form New Animal Drugs
GENTAMICIN SULFATE	556.300	GENTAMICIN SULFATE Specific Tolerances For Residues Of New Animal Drugs; Tolerances For Residues Of New Animal Drugs In Food
GENTIAN VIOLET	500.29	GENTIAN VIOLET FOR USE IN ANIMAL FEED General
GENTIAN VIOLET	500.30	GENTIAN VIOLET FOR ANIMAL DRUG USE General
GENTIAN VIOLET	589.1000	GENTIAN VIOLET Substances Prohibited From Use In Animal Food Or Feed

GERSTHOFEN PROCESS	178.3770	POLYHYDRIC ALCOHOL ESTERS OF OXIDATIVELY REFINED (GERSTHOFEN PROCESS) MONTAN WAX ACIDS Certain Adjuvants And Production Aids; Indirect Food Additives: Adjuvants, Production Aids, And Sanitizers
GHATTI, GUM	184.1333	GUM GHATTI Listing Of Specific Substances Affirmed As GRAS; Direct Food Substances Affirmed As Generally Recognized As Safe
GHATTI, GUM	582.7333	GUM GHATTI Stabilizers; Substances Generally Recognized As Safe
GIBBERELLIC ACID	172.725	GIBBERELLIC ACID AND ITS POTASSIUM SALT Other Specific Usage Additives; Food Additives Permitted For Direct Addition To Food For Human Consumption
GINGIVAL FLUID	872.1500	GINGIVAL FLUID MEASURER Diagnostic Devices; Dental Devices
GLANDULAR PREPARATIONS	201.300	NOTICE TO MANUFACTURERS, PACKERS, AND DISTRIBUTORS OF GLANDULAR PREPARATIONS Specific Labeling Requirements For Specific Drug Products; Labeling
GLANS	884.5320	GLANS SHEATH Obstetrical And Gynecological Therapeutic Devices; Obstetrical And Gynecological Devices
GLENOID	872.3950	GLENOID FOSSA PROSTHESIS Prosthetic Devices; Dental Devices
GLENOID	888.3680	SHOULDER JOINT GLENOID (HEMI-SHOULDER) METALLIC CEMENTED PROSTHESIS Prosthetic Devices; Orthopedic Devices
GLEPTOFERRON	522.1055	GLEPTOFERRON INJECTION Implantation Or Injectable Dosage Form New Animal Drugs
GLOBULIN	640.100	IMMUNE GLOBULIN (HUMAN) Immune Globulin (Human); Additional Standards For Human Blood And Blood Products
GLOBULIN	640.102	MANUFACTURE OF IMMUNE GLOBULIN (HUMAN) Immune Globulin (Human); Additional Standards For Human Blood And Blood Products
GLOBULIN	660.50	ANTI-HUMAN GLOBULIN Anti-human Globulin; Additional Standards For Diagnostic Substances For Laboratory Tests
GLOBULIN	862.1330	GLOBULIN TEST SYSTEM Clinical Chemistry Test Systems; Clinical Chemistry And Clinical Toxicology Devices
GLOBULIN	862.1685	THYROXINE-BINDING GLOBULIN TEST SYSTEM Clinical Chemistry Test Systems; Clinical Chemistry And Clinical Toxicology Devices
GLOBULIN	866.5160	BETA-GLOBULIN IMMUNOLOGICAL TEST SYSTEM Immunological Test Systems; Immunology And Microbiology Devices
GLOBULIN	866.5400	ALPHA-GLOBULIN IMMUNOLOGICAL TEST SYSTEM Immunological Test Systems; Immunology And Microbiology Devices
GLOVE	878.4460	SURGEON'S GLOVE Surgical Devices; General And Plastic Surgery Devices
GLOVE	880.6250	PATIENT EXAMINATION GLOVE General Hospital And Personal Use Miscellaneous Devices; General Hospital And Personal Use Devices
GLOVE POWDER	878.4480	ABSORBABLE POWDER FOR LUBRICATING A SURGEON'S GLOVE Surgical Devices; General And Plastic Surgery Devices
GLOVES	800.20	PATIENT EXAMINATION GLOVES AND SURGEON'S GLOVES; SAMPLE PLANS AND TEST METHODS FOR LEAKAGE DEFECTS; ADULTERATION Requirements For Specific Medical Device; General
GLOVING CREAM	878.4470	SURGEON'S GLOVING CREAM Surgical Devices; General And Plastic Surgery Devices
GLUCAGON	862.1335	GLUCAGON TEST SYSTEM Clinical Chemistry Test Systems; Clinical Chemistry And Clinical Toxicology Devices
GLUCOMANNAN	201.310	WATER-SOLUBLE GUMS, HYDROPHILIC GUMS AND HYDROPHILIC MUCILLOIDS (INCLUDING GLUCOMANNAN (B-1,4 LINKED) POLYMANNOSE ACETATE)) AS ACTIVE INGREDIENTS; WARNINGS AND DIRECTIONS Specific Labeling Requirements For Specific Drug Products; Labeling
GLUCONATE, CALCIUM	184.1199	CALCIUM GLUCONATE Listing Of Specific Substances Affirmed As GRAS; Direct Food Substances Affirmed As Generally Recognized As Safe
GLUCONATE, CALCIUM	582.1205	CALCIUM GLUCONATE General Purpose Food Additives; Substances Generally Recognized As Safe
GLUCONATE, CALCIUM	582.6199	CALCIUM GLUCONATE Sequestrants; Substances Affirmed As Generally Recognized As Safe

GLUCONATE, COPPER	184.1260	COPPER GLUCONATE Listing Of Specific Substances Affirmed As GRAS; Direct Food Substances Affirmed As Generally Recognized As Safe
GLUCONATE, COPPER	582.5260	COPPER GLUCONATE Nutrients And/or Dietary Supplements; Substances Affirmed As Generally Recognized As Safe
GLUCONATE, FERROUS	73.160	FERROUS GLUCONATE Foods; Listing Of Color Additives Exempt From Certification
GLUCONATE, FERROUS	184.1308	FERROUS GLUCONATE Listing Of Specific Substances Affirmed As GRAS; Direct Food Substances Affirmed As Generally Recognized As Safe
GLUCONATE, FERROUS	582.5308	FERROUS GLUCONATE Nutrients And/or Dietary Supplements; Substances Affirmed As Generally Recognized As Safe
GLUCONATE, MANGANESE	184.1452	MANGANESE GLUCONATE Listing Of Specific Substances Affirmed As GRAS; Direct Food Substances Affirmed As Generally Recognized As Safe
GLUCONATE, MANGANESE	582.5452	MANGANESE GLUCONATE Nutrients And/or Dietary Supplements; Substances Generally Recognized As Safe
GLUCONATE, SODIUM	182.6757	SODIUM GLUCONATE Sequestrants; Substances Affirmed As Generally Recognized As Safe
GLUCONATE, SODIUM	582.6757	SODIUM GLUCONATE Sequestrants; Substances Affirmed As Generally Recognized As Safe
GLUCONATE, ZINC	182.5988	ZINC GLUCONATE Nutrients; Substances Affirmed As Generally Recognized As Safe
GLUCONATE, ZINC	582.5988	ZINC GLUCONATE Nutrients And/or Dietary Supplements; Substances Affirmed As Generally Recognized As Safe
GLUCONO	184.1318	GLUCONO DELTA-LACTONE Listing Of Specific Substances Affirmed As GRAS; Direct Food Substances Affirmed As Generally Recognized As Safe
GLUCOSE	168.120	GLUCOSE SIRUP Requirements For Specific Standardized Sweeteners And Table Sirups; Sweeteners And Table Sirups
GLUCOSE	168.121	DRIED GLUCOSE SIRUP Requirements For Specific Standardized Sweeteners And Table Sirups; Sweeteners And Table Sirups
GLUCOSE	862.1340	URINARY GLUCOSE (NONQUANTITATIVE) TEST SYSTEM Clinical Chemistry Test Systems; Clinical Chemistry And Clinical Toxicology Devices
GLUCOSE	862.1345	GLUCOSE TEST SYSTEM Clinical Chemistry Test Systems; Clinical Chemistry And Clinical Toxicology Devices
GLUCOSE-6-PHOSPHATE	864.7360	ERYTHROCYTIC GLUCOSE-6-PHOSPHATE DEHYDROGENASE ASSAY Hematology Kits And Packages; Hematology And Pathology Devices
GLUCOSE ISOMERASE	184.1372	INSOLUBLE GLUCOSE ISOMERASE ENZYME PREPARATIONS Listing Of Specific Substances Affirmed As GRAS; Direct Food Substances Affirmed As Generally Recognized As Safe
GLUCOSIDE	172.816	METHYL GLUCOSIDE-COCONUT OIL ESTER Multipurpose Additives; Food Additives Permitted For Direct Addition To Food For Human Consumption
GLUCOSIDE	178.3600	METHYL GLUCOSIDE-COCONUT OIL ESTER Certain Adjuvants And Production Aids; Indirect Food Additives: Adjuvants, Production Aids, And Sanitizers
GLUCOSIDE	573.660	METHYL GLUCOSIDE-COCONUT OIL ESTER Food Additive Listing; Food Additives Permitted In Feed And Drinking Water Of Animals
GLUE	178.3120	ANIMAL GLUE Certain Adjuvants And Production Aids; Indirect Food Additives: Adjuvants, Production Aids, And Sanitizers
GLUTAMATE	182.1500	MONOAMMONIUM GLUTAMATE Multiple Purpose GRAS Food Substances; Substances Generally Recognized As Safe
GLUTAMATE	182.1516	MONOPOTASSIUM GLUTAMATE Multiple Purpose GRAS Food Substances; Substances Generally Recognized As Safe
GLUTAMATE	582.1500	MONOAMMONIUM GLUTAMATE General Purpose Food Additives; Substances Generally Recognized As Safe
GLUTAMATE	582.1516	MONOPOTASSIUM GLUTAMATE General Purpose Food Additives; Substances Generally Recognized As Safe
GLUTAMER-200, HEMOGLOBIN	522.1125	HEMOGLOBIN GLUTAMER-200 (BOVINE) Implantation Or Injectable Dosage Form New Animal Drugs

GLUTAMIC ACID	182.1045	GLUTAMIC ACID Multiple Purpose GRAS Food Substances; Substances Generally Recognized As Safe
GLUTAMIC ACID	182.1047	GLUTAMIC ACID HYDROCHLORIDE Multiple Purpose GRAS Food Substances; Substances Generally Recognized As Safe
GLUTAMIC ACID	573.500	CONDENSED, EXTRACTED GLUTAMIC ACID FERMENTATION PRODUCT Food Additive Listing; Food Additives Permitted In Feed And Drinking Water Of Animals
GLUTAMYL TRANSPEPTIDASE	862.1360	GAMMA-GLUTAMYL TRANSPEPTIDASE AND ISOENZYMES TEST SYSTEM Clinical Chemistry Test Systems; Clinical Chemistry And Clinical Toxicology Devices
GLUTATHIONE	862.1365	GLUTATHIONE TEST SYSTEM Clinical Chemistry Test Systems; Clinical Chemistry And Clinical Toxicology Devices
GLUTATHIONE REDUCTASE	864.7375	GLUTATHIONE REDUCTASE ASSAY Hematology Kits And Packages; Hematology And Pathology Devices
GLUTEN	184.1321	CORN GLUTEN Listing Of Specific Substances Affirmed As GRAS; Direct Food Substances Affirmed As Generally Recognized As Safe
GLUTEN	184.1322	WHEAT GLUTEN Listing Of Specific Substances Affirmed As GRAS; Direct Food Substances Affirmed As Generally Recognized As Safe
GLYCAN	172.898	BAKERS YEAST GLYCAN Multipurpose Additives; Food Additives Permitted For Direct Addition To Food For Human Consumption
GLYCERIN	172.866	SYNTHETIC GLYCERIN PRODUCED BY THE HYDROGENOLYSIS OF CARBOHYDRATES Multipurpose Additives; Food Additives Permitted For Direct Addition To Food For Human Consumption
GLYCERIN	178.3500	GLYCERIN, SYNTHETIC Certain Adjuvants And Production Aids; Indirect Food Additives: Adjuvants, Production Aids, And Sanitizers
GLYCERIN	182.1320	GLYCERIN Multiple Purpose GRAS Food Substances; Substances Generally Recognized As Safe
GLYCERIN	582.1320	GLYCERIN General Purpose Food Additives; Substances Generally Recognized As Safe
GLYCEROL	172.735	GLYCEROL ESTER OF WOOD ROSIN Other Specific Usage Additives; Food Additives Permitted For Direct Addition To Food For Human Consumption
GLYCEROL	172.850	LACTYLATED FATTY ACID ESTERS OF GLYCEROL AND PROPYLENE GLYCOL Multipurpose Additives; Food Additives Permitted For Direct Addition To Food For Human Consumption
GLYCEROPHOSPHATE	184.1201	CALCIUM GLYCEROPHOSPHATE Listing Of Specific Substances Affirmed As GRAS; Direct Food Substances Affirmed As Generally Recognized As Safe
GLYCEROPHOSPHATE	582.5201	CALCIUM GLYCEROPHOSPHATE Nutrients And/or Dietary Supplements; Substances Generally Recognized As Safe
GLYCEROPHOSPHATE	582.5455	MANGANESE GLYCEROPHOSPHATE Nutrients And/or Dietary Supplements; Substances Generally Recognized As Safe
GLYCEROPHOSPHATE	582.5628	POTASSIUM GLYCEROPHOSPHATE Nutrients And/or Dietary Supplements; Substances Generally Recognized As Safe
GLYCERYL BEHENATE	184.1328	GLYCERYL BEHENATE Listing Of Specific Substances Affirmed As GRAS; Direct Food Substances Affirmed As Generally Recognized As Safe
GLYCERYL-LACTO ESTERS	172.852	GLYCERYL-LACTO ESTERS OF FATTY ACIDS Multipurpose Additives; Food Additives Permitted For Direct Addition To Food For Human Consumption
GLYCERYL MONOOLEATE	184.1323	GLYCERYL MONOOLEATE Listing Of Specific Substances Affirmed As GRAS; Direct Food Substances Affirmed As Generally Recognized As Safe
GLYCERYL MONOSTEARATE	184.1324	GLYCERYL MONOSTEARATE Listing Of Specific Substances Affirmed As GRAS; Direct Food Substances Affirmed As Generally Recognized As Safe
GLYCERYL MONOSTEARATE	582.1324	GLYCERYL MONOSTEARATE General Purpose Food Additives; Substances Generally Recognized As Safe
GLYCERYL PALMITOSTEARATE	184.1329	GLYCERYL PALMITOSTEARATE Listing Of Specific Substances Affirmed As GRAS; Direct Food Substances Affirmed As Generally Recognized As Safe
GLYCERYL TRI-ACETOXYSTEARATE	178.3505	GLYCERYL TRI-(12-ACETOXYSTEARATE) Certain Adjuvants And Production Aids; Indirect Food Additives: Adjuvants, Production Aids, And Sanitizers

GLYCERYL TRISTEARATE	172.811	GLYCERYL TRISTEARATE Multipurpose Additives; Food Additives Permitted For Direct Addition To Food For Human Consumption
GLYCINATE	522.518	CUPRIC GLYCINATE INJECTION Implantation Or Injectable Dosage Form New Animal Drugs
GLYCINE	170.50	GLYCINE (AMINOACETIC ACID) IN FOOD FOR HUMAN CONSUMPTION Specific Administrative Rulings And Decisions; Food Additives
GLYCINE	172.812	GLYCINE Multipurpose Additives; Food Additives Permitted For Direct Addition To Food For Human Consumption
GLYCINE	176.160	CHROMIUM (CR III) COMPLEX N-ETHYL-N-HEPTADECYLFLUORO-OCTANE SULFONYL GLYCINE Substances For Use Only As Components Of Paper And Paperboard; Indirect Food Additives
GLYCINE	520.550	DEXTROSE/GLYCINE/ELECTROLYTE Oral Dosage Form New Animal Drugs
GLYCOL, BUTYLENE	172.712	1,3-BUTYLENE GLYCOL Other Specific Usage Additives; Food Additives Permitted For Direct Addition To Food For Human Consumption
GLYCOL, BUTYLENE	173.220	1,3-BUTYLENE GLYCOL Solvents, Lubricants, Release Agents And Related Substances; Secondary Direct Food Additives Permitted In Food For Human Consumption
GLYCOL, BUTYLENE	573.225	1,3-BUTYLENE GLYCOL Food Additive Listing; Food Additives Permitted In Feed And Drinking Water Of Animals
GLYCOL, POLYETHYLENE	172.820	POLYETHYLENE GLYCOL (MEAN MOLECULAR WEIGHT 200-9,500) Multipurpose Additives; Food Additives Permitted For Direct Addition To Food For Human Consumption
GLYCOL, POLYETHYLENE	178.3750	POLYETHYLENE GLYCOL (MEAN MOLECULAR WEIGHT 200-9,500) Certain Adjuvants And Production Aids; Indirect Food Additives: Adjuvants, Production Aids, And Sanitizers
GLYCOL, POLYETHYLENE	178.3760	POLYETHYLENE GLYCOL (400) MONOLAURATE Certain Adjuvants And Production Aids; Indirect Food Additives: Adjuvants, Production Aids, And Sanitizers
GLYCOL, POLYETHYLENE	573.800	POLYETHYLENE GLYCOL (400) MONO- AND DIOLEATE Food Additive Listing; Food Additives Permitted In Feed And Drinking Water Of Animals
GLYCOL, POLYETHYLENE	573.820	POLYETHYLENE GLYCOL (400) MONO- AND DIOLEATES Food Additive Listing; Food Additives Permitted In Feed And Drinking Water Of Animals
GLYCOL, PROPYLENE	172.765	SUCCISTEARIN (STEAROYL PROPYLENE GLYCOL HYDROGEN SUCCINATE) Other Specific Usage Additives; Food Additives Permitted For Direct Addition To Food For Human Consumption
GLYCOL, PROPYLENE	172.850	LACTYLATED FATTY ACID ESTERS OF GLYCEROL AND PROPYLENE GLYCOL Multipurpose Additives; Food Additives Permitted For Direct Addition To Food For Human Consumption
GLYCOL, PROPYLENE	172.856	PROPYLENE GLYCOL MONO- AND DIESTERS OF FATS AND FATTY ACIDS Multipurpose Additives; Food Additives Permitted For Direct Addition To Food For Human Consumption
GLYCOL, PROPYLENE	172.858	PROPYLENE GLYCOL ALGINATE Multipurpose Additives; Food Additives Permitted For Direct Addition To Food For Human Consumption
GLYCOL, PROPYLENE	184.1666	PROPYLENE GLYCOL Listing Of Specific Substances Affirmed As GRAS; Direct Food Substances Affirmed As Generally Recognized As Safe
GLYCOL, PROPYLENE	500.50	PROPYLENE GLYCOL IN OR ON CAT FOOD Specific Administrative Rulings And Decisions; General
GLYCOL, PROPYLENE	582.1666	PROPYLENE GLYCOL General Purpose Food Additives; Substances Generally Recognized As Safe
GLYCOL, PROPYLENE	582.4666	PROPYLENE GLYCOL Emulsifying Agents; Substances Generally Recognized As Safe
GLYCOL, PROPYLENE	589.1001	PROPYLENE GLYCOL IN OR ON CAT FOOD Substances Prohibited From Use In Animal Food Or Feed
GLYCOL, TETRAETHYLENE	178.3940	TETRAETHYLENE GLYCOL DI-(2-ETHYLHEXOATE) Certain Adjuvants And Production Aids; Indirect Food Additives: Adjuvants, Production Aids, And Sanitizers
GLYCOLIDE	878.4493	ABSORBABLE POLY(GLYCOLIDE/L-LACTIDE) SURGICAL SUTURE Surgical Devices; General And Plastic Surgery Devices
GLYCOPROTEIN	866.5430	BETA-2-GLYCOPROTEIN I IMMUNOLOGICAL TEST SYSTEM Immunological Test Systems; Immunology And Microbiology Devices
GLYCOPROTEIN	866.5440	BETA-2-GLYCOPROTEIN III IMMUNOLOGICAL TEST SYSTEM Immunological Test Systems; Immunology And Microbiology Devices

GLYCOPROTEINS	866.5420	ALPHA-1-GLYCOPROTEINS IMMUNOLOGICAL TEST SYSTEM Immunological Test Systems; Immunology And Microbiology Devices
GLYCOPROTEINS	866.5425	ALPHA-2-GLYCOPROTEINS IMMUNOLOGICAL TEST SYSTEM Immunological Test Systems; Immunology And Microbiology Devices
GLYCOPYRROLATE	522.1066	GLYCOPYRROLATE INJECTION Implantation Or Injectable Dosage Form New Animal Drugs
GLYCOSAMINOGLYCAN	522.1850	POLYSULFATED GLYCOSAMINOGLYCAN Implantation Or Injectable Dosage Form New Animal Drugs
GLYCOSYLATED	864.7470	GLYCOSYLATED HEMOGLOBIN ASSAY Hematology Kits And Packages; Hematology And Pathology Devices
GMP	5.109	ISSUANCE OF WRITTEN NOTICES CONCERNING PATENT INFORMATION, CURRENT GOOD MANUFACTURING PRACTICES AND FALSE OR MISLEADING LABELING OF NEW DRUGS Human Drugs; Redelegations Of Authority; Delegations Of Authority And Organization
GMP	5.504	ISSUANCE OF WRITTEN NOTICES CONCERNING PATENT INFORMATION, CURRENT GOOD MANUFACTURING PRACTICES AND FALSE OR MISLEADING LABELING OF NEW ANIMAL DRUGS AND FEEDS BEARING OR CONTAINING NEW ANIMAL DRUGS Animal Drugs; Redelegations Of Authority; Delegations Of Authority And Organization
GMP	110.5	CURRENT GOOD MANUFACTURING PRACTICE Current Good Manufacturing Practice In Manufacturing, Packing, Or Holding Human Food
GMP	113.5	CURRENT GOOD MANUFACTURING PRACTICE Thermally Processed Low-acid Foods Packaged In Hermetically Sealed Containers
GMP	114.5	CURRENT GOOD MANUFACTURING PRACTICES Acidified Foods
GMP	120.5	CURRENT GOOD MANUFACTURING PRACTICE Hazard Analysis And Critical Control Point (HACCP) Systems
GMP	123.5	CURRENT GOOD MANUFACTURING PRACTICE Fish And Fishery Products
GMP	129.1	CURRENT GOOD MANUFACTURING PRACTICE Processing And Bottling Of Bottled Drinking Water
GMP	210.1	STATUS OF CURRENT GOOD MANUFACTURING PRACTICE REGULATIONS Current Good Manufacturing Practice In Manufacturing, Processing, Packing, Or Holding Of Drugs; General
GMP	210.2	APPLICABILITY OF CURRENT GOOD MANUFACTURING PRACTICE REGULATIONS Current Good Manufacturing Practice In Manufacturing, Processing, Packing, Or Holding Of Drugs; General
GMP	225.1	CURRENT GOOD MANUFACTURING PRACTICE Current Good Manufacturing Practice For Medicated Feeds
GMP	226.1	CURRENT GOOD MANUFACTURING PRACTICE Current Good Manufacturing Practice For Type A Medicated Articles
GMP REGULATIONS	210.1	STATUS OF CURRENT GOOD MANUFACTURING PRACTICE REGULATIONS Current Good Manufacturing Practice In Manufacturing, Processing, Packing, Or Holding Of Drugs; General
GMP REGULATIONS	210.2	APPLICABILITY OF CURRENT GOOD MANUFACTURING PRACTICE REGULATIONS Current Good Manufacturing Practice In Manufacturing, Processing, Packing, Or Holding Of Drugs; General
GOAT'S MILK	135.115	GOAT'S MILK ICE CREAM Requirements For Specific Standardized Frozen Desserts; Frozen Desserts
GOLD	872.3060	GOLD BASED ALLOYS AND PRECIOUS METAL ALLOYS FOR CLINICAL USE Prosthetic Devices; Dental Devices
GOLD	872.3350	GOLD OR STAINLESS STEEL CUSP Prosthetic Devices; Dental Devices
GOLD	872.3380	PREFORMED GOLD DENTURE TOOTH Prosthetic Devices; Dental Devices
GONADORELIN	522.1077	GONADORELIN DIACETATE TETRAHYDRATE INJECTABLE Implantation Or Injectable Dosage Form New Animal Drugs
GONADORELIN	522.1078	GONADORELIN DIACETATE TETRAHYDRATE INJECTION Implantation Or Injectable Dosage Form New Animal Drugs
GONAD SHIELDING	1000.50	RECOMMENDATIONS FOR THE USE OF SPECIFIC AREA GONAD SHIELDING ON PATIENTS DURING MEDICAL DIAGNOSTIC X-RAY PROCEDURES Radiation Protection Recommendations; General
GONADOTROPIN	522.1079	SERUM GONADOTROPIN AND CHORIONIC GONADOTROPIN Implantation Or Injectable Dosage Form New Animal Drugs
GONADOTROPIN	522.1081	CHORIONIC GONADOTROPIN FOR INJECTION; CHORIONIC GONADOTROPIN SUSPENSION Implantation Or Injectable Dosage Form New Animal Drugs
GONADOTROPIN	556.304	GONADOTROPIN Tolerances For Residues Of New Animal Drugs In Food

GONADOTROPIN	862.1155	HUMAN CHORIONIC GONADOTROPIN (HCG) TEST SYSTEM Clinical Chemistry Test Systems; Clinical Chemistry And Clinical Toxicology Devices
GONIMETER	888.1500	GONIMETER Diagnostic Devices; Orthopedic Devices
GONIMETER	888.1520	NONPOWERED GONIMETER Diagnostic Devices; Orthopedic Devices
GONIOSCOPIC	886.1660	GONIOSCOPIC PRISM Diagnostic Devices; Ophthalmic Devices
GONOCOCCAL ANTIBODY TEST	866.3290	GONOCOCCAL ANTIBODY TEST (GAT) Serological Reagents; Immunology And Microbiology Devices
GONORRHEA	866.2420	OXIDASE SCREENING TEST FOR GONORRHEA Microbiology Devices; Immunology And Microbiology Devices
GOOD GUIDANCE	10.115	GOOD GUIDANCE PRACTICES General Administrative Procedures; Administrative Practices And Procedures
GOOD MANUFACTURING PRACTICE	5.109	ISSUANCE OF WRITTEN NOTICES CONCERNING PATENT INFORMATION, CURRENT GOOD MANUFACTURING PRACTICES AND FALSE OR MISLEADING LABELING OF NEW DRUGS Human Drugs; Redelegations Of Authority; Delegations Of Authority And Organization
GOOD MANUFACTURING PRACTICE	5.504	ISSUANCE OF WRITTEN NOTICES CONCERNING PATENT INFORMATION, CURRENT GOOD MANUFACTURING PRACTICES AND FALSE OR MISLEADING LABELING OF NEW ANIMAL DRUGS AND FEEDS BEARING OR CONTAINING NEW ANIMAL DRUGS Animal Drugs; Redelegations Of Authority; Delegations Of Authority And Organization
GOOD MANUFACTURING PRACTICE	110.5	CURRENT GOOD MANUFACTURING PRACTICE Current Good Manufacturing Practice In Manufacturing, Packing, Or Holding Human Food
GOOD MANUFACTURING PRACTICE	113.5	CURRENT GOOD MANUFACTURING PRACTICE Thermally Processed Low-acid Foods Packaged In Hermetically Sealed Containers
GOOD MANUFACTURING PRACTICE	114.5	CURRENT GOOD MANUFACTURING PRACTICES Acidified Foods
GOOD MANUFACTURING PRACTICE	120.5	CURRENT GOOD MANUFACTURING PRACTICE Hazard Analysis And Critical Control Point (HACCP) Systems
GOOD MANUFACTURING PRACTICE	123.5	CURRENT GOOD MANUFACTURING PRACTICE Fish And Fishery Products
GOOD MANUFACTURING PRACTICE	129.1	CURRENT GOOD MANUFACTURING PRACTICE Processing And Bottling Of Bottled Drinking Water
GOOD MANUFACTURING PRACTICE	210.1	STATUS OF CURRENT GOOD MANUFACTURING PRACTICE REGULATIONS Current Good Manufacturing Practice In Manufacturing, Processing, Packing, Or Holding Of Drugs; General
GOOD MANUFACTURING PRACTICE	210.2	APPLICABILITY OF CURRENT GOOD MANUFACTURING PRACTICE REGULATIONS Current Good Manufacturing Practice In Manufacturing, Processing, Packing, Or Holding Of Drugs; General
GOOD MANUFACTURING PRACTICE	225.1	CURRENT GOOD MANUFACTURING PRACTICE Current Good Manufacturing Practice For Medicated Feeds
GOOD MANUFACTURING PRACTICE	226.1	CURRENT GOOD MANUFACTURING PRACTICE Current Good Manufacturing Practice For Type A Medicated Articles
GOOD SOURCE	101.54	NUTRIENT CONTENT CLAIMS FOR "GOOD SOURCE," "HIGH," "MORE," AND "HIGH POTENCY" Specific Requirements For Nutrient Content Claims; Food Labeling
GORGONZOLA CHEESE	133.141	GORGONZOLA CHEESE Requirements For Specific Standardized Cheese And Related Products; Cheeses And Related Cheese Products
GOUDA CHEESE	133.142	GOUDA CHEESE Requirements For Specific Standardized Cheese And Related Products; Cheeses And Related Cheese Products
GOVERNMENT	19.6	CODE OF ETHICS FOR GOVERNMENT SERVICE General Provisions; Standards Of Conduct And Conflicts Of Interest
GOVERNMENT DEPARTMENTS	20.85	DISCLOSURE TO OTHER FEDERAL GOVERNMENT DEPARTMENTS AND AGENCIES Limitations On Exemptions; Public Information
GOVERNMENT EMPLOYEES	20.84	DISCLOSURE TO CONSULTANTS, ADVISORY COMMITTEES, STATE AND LOCAL GOVT. OFFICIALS AND OTHER SPECIAL GOVERNMENT EMPLOYEES Limitations On Exemptions; Public Information
GOVERNMENT, FOREIGN	20.89	COMMUNICATIONS WITH FOREIGN GOVERNMENT OFFICIALS Limitations On Exemptions; Public Information
GOVERNMENT OFFICIALS	20.84	DISCLOSURE TO CONSULTANTS, ADVISORY COMMITTEES, STATE AND LOCAL GOVT. OFFICIALS AND OTHER SPECIAL GOVERNMENT EMPLOYEES Limitations On Exemptions; Public Information
GOVERNMENT OFFICIALS	20.88	COMMUNICATIONS WITH STATE AND LOCAL GOVERNMENT OFFICIALS Limitations On Exemptions; Public Information

GOVERNMENT SERVICE	19.6	CODE OF ETHICS FOR GOVERNMENT SERVICE General Provisions; Standards Of Conduct And Conflicts Of Interest
GOVERNMENT, UNITED STATES	1010.5	EXEMPTIONS FOR PRODUCTS INTENDED FOR UNITED STATES GOVERNMENT USE Performance Standards For Electronic Products: General
GOVERNMENT, U.S.	1002.51	EXEMPTIONS FOR MANUFACTURERS OF PRODUCTS INTENDED FOR THE U.S. GOVERNMENT Exemptions From Records And Reports Requirements; Records and Reports
GOVERNMENTS, FOREIGN	20.89	COMMUNICATIONS WITH FOREIGN GOVERNMENT OFFICIALS Limitations On Exemptions; Public Information
GOWN	880.6265	EXAMINATION GOWN General Hospital And Personal Use Miscellaneous Devices; General Hospital And Personal Use Devices
GRAFT	870.3450	VASCULAR GRAFT PROSTHESIS Cardiovascular Prosthetic Devices; Cardiovascular Devices
GRAIN PRODUCTS	101.76	HEALTH CLAIMS: FIBER-CONTAINING GRAIN PRODUCTS, FRUITS, VEGETABLES AND CANCER Food Labeling
GRAIN PRODUCTS	101.77	HEALTH CLAIMS: FRUITS, VEGETABLES, AND GRAIN PRODUCTS THAT CONTAIN FIBER, PARTICULARLY SOLUBLE FIBER, AND THE RISK OF CORONARY HEAT DISEASE Food Labeling
GRAIN SEED	2.25	GRAIN SEED TREATED WITH POISONOUS SUBSTANCES; COLOR IDENTIFICATION TO PREVENT ADULTERATION OF HUMAN AND ANIMAL FOOD Human And Animal Foods; General Administrative Rulings And Decisions
GRANT/GRANTEES	1403.12	SPECIAL GRANT OR SUBGRANT CONDITIONS FOR "HIGH-RISK" GRANTEES Pre-Award Requirements; Uniform Administrative Requirements For Grants And Coop. Agreements To State & Local Gov.
GRANT OF HEARING	314.200	NOTICE OF OPPORTUNITY FOR HEARING; NOTICE OF PARTICIPATION AND REQUEST FOR HEARING; GRANT OR DENIAL OF HEARING Hearing Procedures For New Drugs; Applications For FDA Approval To Market A New Drug
GRANTS	58.10	APPLICABILITY TO STUDIES PERFORMED UNDER GRANTS AND CONTRACTS General Provisions; Good Laboratory Practice For Nonclinical Laboratory Studies
GRANTS	1403.12	FORMS FOR APPLYING FOR GRANTS Pre-Award Requirements; Uniform Administrative Requirements For Grants And Coop. Agreements To State & Local Gov.
GRANTS	1404.615	GROUNDS FOR SUSPENSION OF PAYMENTS, SUSPENSION OR TERMINATION OF GRANTS, OR SUSPENSION OR DEBARMENT Governmentwide Debarment And Suspension And Requirements For Drug-Free Workplace (Grants)
GRANULAR CHEESE	133.145	GRANULAR CHEESE FOR MANUFACTURING Requirements For Specific Standardized Cheese And Related Products; Cheeses And Related Cheese Products
GRANULAR CURD CHEESE	133.144	GRANULAR AND STIRRED CURD CHEESE Requirements For Specific Standardized Cheese And Related Products; Cheeses And Related Cheese Products
GRANULES	520.905b	FENBENDAZOLE GRANULES Oral Dosage Form New Animal Drugs
GRANULES	520.970a	FLUNIXIN MEGLUMINE GRANULES Oral Dosage Form New Animal Drug
GRANULES	520.1330	MECLOFENAMIC ACID GRANULES Oral Dosage Form New Animal Drugs
GRANULES	520.1448	MONENSIN-MINERAL GRANULES Oral Dosage Form New Animal Drugs
GRANULES	520.1468	NAPROXEN GRANULES Oral Dosage Form New Animal Drugs
GRANULES	520.1720b	PHENYLBUTAZONE GRANULES Oral Dosage Form New Animal Drugs
GRANULES	520.2473a	TIOXIDAZOLE GRANULES Oral Dosage Form New Animal Drugs
GRANULES	520.2520f	TRICHLOROFON GRANULE Oral Dosage Form New Animal Drugs
GRANULES	872.3930	TRICALCIUM PHOSPHATE GRANULES FOR DENTAL BONE REPAIR Prosthetic Devices; Dental Devices
GRAPE COLOR EXTRACT	73.169	GRAPE COLOR EXTRACT Foods; Listing Of Color Additives Exempt From Certification
GRAPE SKIN EXTRACT	73.170	GRAPE SKIN EXTRACT (ENOCIANINA) Foods; Listing Of Color Additives Exempt From Certification
GRAPEFRUIT	145.145	CANNED GRAPEFRUIT Requirements For Specific Standardized Canned Fruits; Canned Fruits

GRAPEFRUIT	146.132	GRAPEFRUIT JUICE Requirements For Specific Standardized Canned Fruit Juices And Beverages; Canned Fruit Juices
GRAPES	145.140	CANNED SEEDLESS GRAPES Requirements For Specific Standardized Canned Fruits; Canned Fruits
GRAS	5.300	FOOD STANDARDS, FOOD ADDITIVES, GENERALLY RECOGNIZED AS SAFE (GRAS) SUBSTANCES, COLOR ADDITIVES, NUTRIENT CONTENT CLAIMS, AND HEALTH CLAIMS Foods And Cosmetics; Redelegations Of Authority; Delegations Of Authority And Organization
GRAS	170.30	ELIGIBILITY FOR CLASSIFICATION AS GENERALLY RECOGNIZED AS SAFE (GRAS) Food Additive Safety; Food Additives
GRAS	170.35	AFFIRMATION OF GENERALLY RECOGNIZED AS SAFE (GRAS) STATUS Food Additive Safety; Food Additives
GRAS	184.1	SUBSTANCES ADDED DIRECTLY TO HUMAN FOOD AFFIRMED AS GENERALLY RECOGNIZED AS SAFE (GRAS) General Provisions; Direct Food Substances Affirmed As Generally Recognized As Safe
GRAS	186.1	SUBSTANCES ADDED INDIRECTLY TO HUMAN FOOD AFFIRMED AS GENERALLY RECOGNIZED AS SAFE (GRAS) General Provisions; Indirect Food Substances Affirmed As Generally Recognized As Safe
GRAS	570.30	ELIGIBILITY FOR CLASSIFICATION AS GENERALLY RECOGNIZED AS SAFE (GRAS) Food Additive Safety; Food Additive
GRAS	570.35	AFFIRMATION OF GENERALLY RECOGNIZED AS SAFE (GRAS) STATUS Food Additive Safety; Food Additives
GRAS	582.1	SUBSTANCES THAT ARE GENERALLY RECOGNIZED AS SAFE General Provisions; Substances Generally Recognized As Safe
GRATED CHEESE	133.147	GRATED AMERICAN CHEESE FOOD Requirements For Specific Standardized Cheese And Related Products; Cheeses And Related Cheese Products
GRATED CHEESES	133.146	GRATED CHEESES Requirements For Specific Standardized Cheese And Related Products; Cheeses And Related Cheese Products
GRATING, CHEESES	133.148	HARD GRATING CHEESES Requirements For Specific Standardized Cheese And Related Products; Cheeses And Related Cheese Products
GRAVITY	880.2420	ELECTRONIC MONITOR FOR GRAVITY FLOW INFUSION SYSTEMS General Hospital And Personal Use Monitoring Devices; General Hospital And Personal Use Devices
GRAVITY, SPECIFIC	864.9320	COPPER SULFATE SOLUTION FOR SPECIFIC GRAVITY DETERMINATIONS Products Used In Establishments That Manufacture Blood And Blood Products; Hematology And Pathology Devices
GREEN BEANS	155.120	CANNED GREEN BEANS AND CANNED WAX BEANS Requirements For Specific Standardized Canned Vegetables; Canned Vegetables
GREEN NO. 3	74.203	FD&C GREEN NO. 3 Foods; Listing Of Color Additives Subject To Certification
GREEN NO. 3	74.1203	FD&C GREEN NO. 3 Drugs; Listing Of Color Additives Subject To Certification
GREEN NO. 3	74.2203	FD&C GREEN NO. 3 Cosmetics; Drugs; Listing Of Color Additives Subject To Certification
GREEN NO. 3	82.203	FD&C GREEN NO. 3 Foods, Drugs, And Cosmetics; Listing Of Certified Provisionally Listed Colors And Specifications
GREEN NO. 5	74.1205	D&C GREEN NO. 5 Drugs; Listing Of Color Additives Subject To Certification
GREEN NO. 5	74.2205	D&C GREEN NO. 5 Cosmetics; Listing Of Color Additives Subject To Certification
GREEN NO. 5	82.1205	D&C GREEN NO. 5 Drugs And Cosmetics; Listing Of Certified Provisionally Listed Colors And Specifications
GREEN NO. 6	74.1206	D&C GREEN NO. 6 Drugs; Listing Of Color Additives Subject To Certification
GREEN NO. 6	74.2206	D&C GREEN NO. 6 Cosmetics; Listing Of Color Additives Subject To Certification
GREEN NO. 6	74.3206	D&C GREEN NO. 6 Medical Devices; Listing Of Color Additives Subject To Certification
GREEN NO. 6	82.1206	D&C GREEN NO. 6 Drugs And Cosmetics; Listing Of Certified Provisionally Listed Colors And Specifications
GREEN NO. 8	74.1208	D&C GREEN NO. 8 Drugs; Listing Of Color Additives Subject To Certification

GREEN NO. 8	74.2208	D&C GREEN NO. 8 Cosmetics; Listing Of Color Additives Subject To Certification
GREEN, CHROMIUM HYDROXIDE	73.1326	CHROMIUM HYDROXIDE GREEN Drugs; Listing Of Color Additives Exempt From Certification
GREEN, CHROMIUM HYDROXIDE	73.2326	CHROMIUM HYDROXIDE GREEN Cosmetics; Listing Of Color Additives Exempt From Certification
GREEN, PHTHALOCYANINE	73.3124	PHTHALOCYANINE GREEN Medical Devices; Listing Of Color Additives Exempt From Certification
GREENLAND TURBOT	102.57	GREENLAND TURBOT (REINHARDITUS HIPPOGLOSSOIDES) Requirements For Specific Nonstandardized Foods; Common Or Usual Name For Nonstandardized Foods
GREENS, CHROMIUM OXIDE	73.1327	CHROMIUM OXIDE GREENS Drugs; Listing Of Color Additives Exempt From Certification
GREENS, CHROMIUM OXIDE	73.2327	CHROMIUM OXIDE GREENS Cosmetics; Listing Of Color Additives Exempt From Certification
GREENS, CHROMIUM OXIDE	73.3111	CHROMIUM OXIDE GREENS Medical Devices; Listing Of Color Additives Exempt From Certification
GRID, AMSLER	886.1330	AMSLER GRID Diagnostic Devices; Ophthalmic Devices
GRID, RADIOGRAPHIC	892.1910	RADIOGRAPHIC GRID Diagnostic Devices; Radiology Devices
GRISEOFULVIN	520.1100	GRISEOFULVIN Oral Dosage Form New Animal Drugs
GROUNDS	58.202	GROUNDS FOR DISQUALIFICATION Disqualification Of Testing Facilities; Good Laboratory Practice For Nonclinical Laboratory Studies
GROUNDS	110.20	PLANT AND GROUNDS Buildings And Facilities; Current Good Manufacturing Practice In Manufacturing, Packing, Or Holding Human Food
GROUNDS	225.120	BUILDINGS AND GROUNDS Facilities And Equipment; Current Good Manufacturing Practice For Medicated Feeds
GROUNDS	1404.615	GROUNDS FOR SUSPENSION OF PAYMENTS, SUSPENSION OR TERMINATION OF GRANTS, OR SUSPENSION OR DEBARMENT Governmentwide Debarment And Suspension And Requirements For Drug-Free Workplace (Grants)
GROUP	874.3320	GROUP HEARING AID OR GROUP AUDITORY TRAINER Prosthetic Devices; Ear, Nose, And Throat Devices
GROUP A STREP.	610.19	STATUS OF SPECIFIC PRODUCTS; GROUP A STREPTOCOCCUS General Provisions; General Biological Products Standards
GROUP, BLOOD	864.9160	BLOOD GROUP SUBSTANCES OF NONHUMAN ORIGIN FOR IN VITRO DIAGNOSTIC USE Products Used In Establishments That Manufacture Blood And Blood Products; Hematology And Pathology Devices
GROUPING, BLOOD	660.20	BLOOD GROUPING REAGENT Blood Grouping Reagent; Additional Standards For Diagnostic Substances For Laboratory Tests
GROUPING, BLOOD	864.9175	AUTOMATED BLOOD GROUPING AND ANTIBODY TEST SYSTEM Products Used In Establishments That Manufacture Blood And Blood Products; Hematology And Pathology Devices
GROUPING, BLOOD	864.9185	BLOOD GROUPING VIEW BOX Products Used In Establishments That Manufacture Blood And Blood Products; Hematology And Pathology Devices
GROWTH HORMONE	862.1370	HUMAN GROWTH HORMONE TEST SYSTEM Clinical Chemistry Test Systems; Clinical Chemistry And Clinical Toxicology Devices
GROWTH, MICROBIAL	866.2560	MICROBIAL GROWTH MONITOR Microbiology Devices; Immunology And Microbiology Devices
GRUYERE CHEESE	133.149	GRUYERE CHEESE Requirements For Specific Standardized Cheese And Related Products; Cheeses And Related Cheese Products
GSA	1404.500	GSA RESPONSIBILITIES Governmentwide Debarment And Suspension And Requirements For Drug-Free Workplace (Grants)
GUAIAC, GUM	582.3336	GUM GUAIAC Chemical Preservatives; Substances Generally Recognized As Safe
GUAIAZULENE	73.2180	GUAIAZULENE Cosmetics; Listing Of Color Additives Exempt From Certification
GUAIFENESIN	522.1085	GUAIFENESIN STERILE POWDER Implantation Or Injectable Dosage Form New Animal Drugs
GUAIFENESIN	522.1086	GUAIFENESIN INJECTION Implantation Or Injectable Dosage Form New Animal Drugs
GUANINE	73.1329	GUANINE Drugs; Listing Of Color Additives Exempt From Certification

GUANINE	73.2329	GUANINE Cosmetics; Listing Of Color Additives Exempt From Certification
GUANYLATE, DISODIUM	172.530	DISODIUM GUANYLATE Flavoring Agents And Related Substances; Food Additives Permitted For Direct Addition To Food For Human Consumption
GUAR GUM	184.1339	GUAR GUM Listing Of Specific Substances Affirmed As GRAS; Direct Food Substances Affirmed As Generally Recognized As Safe
GUAR GUM	201.319	WATER-SOLUBLE GUMS, AND HYDROPHILIC MUCILLOIDS (INCLUDING GUAR GUM) AS ACTIVE INGREDIENTS; WARNINGS AND DIRECTIONS Specific Labeling Requirements For Specific Drug Products; Labeling
GUAR GUM	582.7339	GUAR GUM Stabilizers; Substances Generally Recognized As Safe
GUARANTY	7.12	GUARANTY General Provisions; Enforcement Policy
GUARANTY	7.13	SUGGESTED FORMS OF GUARANTY General Provisions; Enforcement Policy
GUARANTY	1230.20	GENERAL GUARANTY Regulations Under The Federal Caustic Poison Act
GUARANTY	1230.21	SPECIFIC GUARANTY Regulations Under The Federal Caustic Poison Act
GUARDIANS	50.55	REQUIREMENTS FOR PERMISSION BY PARENTS OR GUARDIANS AND FOR ASSENT BY CHILDREN Additional Safeguards For Children In Clinical Investigations; Protection Of Human Subjects
GUARDIANS	21.75	RIGHTS OF LEGAL GUARDIANS Disclosure Of Records In Privacy Act Record Systems To Persons Other ThanThe Subject Individual; Protection Of Privacy
GUIDANCE	7.59	GENERAL INDUSTRY GUIDANCE Recalls - Guidance On Policy, Procedures & Industry Responsibilities; Enforcement Policy
GUIDANCE	10.115	GOOD GUIDANCE PRACTICES General Administrative Procedures; Administrative Practices And Procedures
GUIDANCE DOCUMENTS	312.145	GUIDANCE DOCUMENTS Investigational New Drug Application
GUIDANCE DOCUMENTS	314.445	GUIDANCE DOCUMENTS Applications For FDA Approval To Market A New Drug
GUIDANCE DOCUMENTS	316.52	GUIDANCE DOCUMENTS Orphan Drugs
GUIDANCE DOCUMENTS	601.29	GUIDANCE DOCUMENTS Biologics Licensing; Licensing
GUIDE, LENS	886.4300	INTRAOCULAR LENS GUIDE Surgical Devices; Ophthalmic Devices
GUIDE, MEDICATION	208.20	CONTENT AND FORMAT OF A MEDICATION GUIDE Medication Guides For Prescription Drug Products
GUIDE, MEDICATION	208.24	DISTRIBUTION AND DISPENSING A MEDICATION GUIDE Medication Guides For Prescription Drug Products
GUIDE WIRE	870.1330	CATHETER GUIDE WIRE Cardiovascular Diagnostic Devices; Cardiovascular Devices
GUIDELINES	320.25	GUIDELINES FOR THE CONDUCT OF AN IN VIVO BIOAVAILABILITY STUDY Bioavailability And Bioequivalence Requirements
GUIDELINES	320.2	GUIDELINES ON THE DESIGN OF A SINGLE-DOSE IN VIVO BIOAVAILABILITY STUDY Bioavailability And Bioequivalence Requirements
GUIDELINES	320.27	GUIDELINES ON THE DESIGN OF A MULTIPLE-DOSE IN VIVO BIOAVAILABILITY STUDY Bioavailability And Bioequivalence Requirements
GUIDELINES, LABELING	101.43	SUBSTANTIAL COMPLIANCE OF FOOD RETAILERS WITH THE GUIDELINES FOR THE VOLUNTARY NUTRITION LABELING OF RAW FRUIT, VEGETABLES, AND FISH Food Labeling
GUIDELINES, LABELING	101.45	GUIDELINES FOR THE VOLUNTARY NUTRITION LABELING OF RAW FRUIT, VEGETABLES, AND FISH Food Labeling
GUM ARABIC	184.1330	ACACIA (GUM ARABIC) Listing Of Specific Substances Affirmed As GRAS; Direct Food Substances Affirmed As Generally Recognized As Safe
GUM ARABIC	582.7330	GUM ARABIC Stabilizers; Substances Generally Recognized As Safe
GUM, CHEWING	172.615	CHEWING GUM BASE Gums, Chewing Gum Bases And Related Substances; Food Additives Permitted For Direct Addition To Food For Human Consumption

GUM, GELLAN	172.665	GELLAN GUM Gums, Chewing Gum Bases And Related Substances; Food Additives Permitted For Direct Addition To Food For Human Consumption
GUM GHATTI	184.1333	GUM GHATTI Listing Of Specific Substances Affirmed As GRAS; Direct Food Substances Affirmed As Generally Recognized As Safe
GUM GHATTI	582.7333	GUM GHATTI Stabilizers; Substances Generally Recognized As Safe
GUM, GUAR	184.1339	GUAR GUM Listing Of Specific Substances Affirmed As GRAS; Direct Food Substances Affirmed As Generally Recognized As Safe
GUM, GUAR	201.319	WATER-SOLUBLE GUMS, AND HYDROPHILIC MUCILLOIDS (INCLUDING GUAR GUM, KARAYA GUM, XANTHAN GUM) AS ACTIVE INGREDIENTS; WARNINGS AND DIRECTIONS Specific Labeling Requirements For Specific Drug Products; Labeling
GUM, GUAR	582.7339	GUAR GUM Stabilizers; Substances Generally Recognized As Safe
GUM GUAIC	582.3336	GUM GUAIAC Chemical Preservatives; Substances Generally Recognized As Safe
GUM, KARAYA	184.1349	KARAYA GUM (STERCULIA GUM) Listing Of Specific Substances Affirmed As GRAS; Direct Food Substances Affirmed As Generally Recognized As Safe
GUM, KARAYA	201.319	WATER-SOLUBLE GUMS, AND HYDROPHILIC MUCILLOIDS (INCLUDING GUAR GUM, KARAYA GUM, XANTHAN GUM) AS ACTIVE INGREDIENTS; WARNINGS AND DIRECTIONS Specific Labeling Requirements For Specific Drug Products; Labeling
GUM, LOCUST BEAN	184.1343	LOCUST (CAROB) BEAN GUM Listing Of Specific Substances Affirmed As GRAS; Direct Food Substances Affirmed As Generally Recognized As Safe
GUM, LOCUST BEAN	582.7343	LOCUST BEAN GUM Stabilizers; Substances Generally Recognized As Safe
GUM, STERCULIA	184.1349	KARAYA GUM (STERCULIA GUM) Listing Of Specific Substances Affirmed As GRAS; Direct Food Substances Affirmed As Generally Recognized As Safe
GUM, STERCULIA	582.7349	STERCULIA GUM Stabilizers; Substances Generally Recognized As Safe
GUM TRAGACANTH	184.1351	GUM TRAGACANTH Listing Of Specific Substances Affirmed As GRAS; Direct Food Substances Affirmed As Generally Recognized As Safe
GUM TRAGACANTH	582.7351	GUM TRAGACANTH Stabilizers; Substances Generally Recognized As Safe
GUM, XANTHAN	172.695	XANTHAN GUM Gums, Chewing Gum Bases And Related Substances; Food Additives Permitted For Direct Addition To Food For Human Consumption
GUM, XANTHAN	201.319	WATER-SOLUBLE GUMS, AND HYDROPHILIC MUCILLOIDS (INCLUDING GUAR GUM, KARAYA GUM, XANTHAN GUM) AS ACTIVE INGREDIENTS; WARNINGS AND DIRECTIONS Specific Labeling Requirements For Specific Drug Products; Labeling
GUM, XANTHAN	573.1010	XANTHAN GUM Food Additive Listing; Food Additives Permitted In Feed And Drinking Water Of Animals
GUMS	201.319	WATER SOLUBLE GUMS, HYDROPHILIC GUMS AND HYDROPHILIC MUCILLOIDS (INCLUDING GUAR GUM, KARAYA GUM, XANTHAN GUM) AS ACTIVE INGREDIENTS; WARNINGS AND DIRECTIONS Specific Labeling Requirements For Specific Drug Products; Labeling
GUSTOMETER	874.1500	GUSTOMETER Diagnostic Devices; Ear, Nose, And Throat Devices
GUT SUTURE	878.4830	ABSORBABLE SURGICAL GUT SUTURE Surgical Devices; General And Plastic Surgery Devices
GUTTA PERCHA	872.3850	GUTTA PERCHA Prosthetic Devices; Dental Devices
GYNECOLOGIC	884.1720	GYNECOLOGIC LAPAROSCOPE AND ACCESSORIES Obstetrical And Gynecological Diagnostic Devices; Obstetrical And Gynecological Devices
GYNECOLOGIC	884.2225	OBSTETRIC-GYNECOLOGIC ULTRASONIC IMAGER Obstetrical And Gynecological Monitoring Devices; Obstetrical And Gynecological Devices
GYNECOLOGIC	884.4120	GYNECOLOGIC ELECTROCAUTERY AND ACCESSORIES Obstetrical And Gynecological Surgical Devices; Obstetrical And Gynecological Devices
GYNECOLOGIC	884.4520	OBSTETRIC-GYNECOLOGIC GENERAL MANUAL INSTRUMENT Obstetrical And Gynecological Surgical Devices; Obstetrical And Gynecological Devices

GYNECOLOGIC	884.4530	OBSTETRIC-GYNECOLOGIC SPECIALIZED MANUAL INSTRUMENT Obstetrical And Gynecological Surgical Devices; Obstetrical And Gynecological Devices
GYNECOLOGIC	884.4550	GYNECOLOGIC SURGICAL LASER Obstetrical And Gynecological Surgical Devices; Obstetrical And Gynecological Devices

H

HABIT-FORMING DRUGS	329.1	HABIT-FORMING DRUGS WHICH ARE CHEMICAL DERIVATIVES OF SUBSTANCES SPECIFIED IN SECTION 502(D) OF THE FEDERAL FOOD, DRUG, AND COSMETIC ACT Derivatives Designated As Habit Forming; Habit-forming Drugs
HABIT-FORMING DRUGS	329.10	LABELING REQUIREMENTS FOR HABIT-FORMING DRUGS Labeling; Habit-forming Drugs
HABIT-FORMING DRUGS	329.20	EXEMPTION OF CERTAIN HABIT-FORMING DRUGS FROM PRESCRIPTION REQUIREMENTS Exemptions; Habit-forming Drugs
HACCP	120.8	HAZARD ANALYSIS AND CRITICAL CONTROL POINT (HACCP) PLAN Hazard Analysis And Critical Control Point (HACCP) Systems
HACCP	123.6	HAZARD ANALYSIS AND HAZARD ANALYSIS CRITICAL CONTROL POINT (HACCP) PLAN Fish And Fishery Products
HAEMATOCOCCUS	73.185	HAEMATOCOCCUS ALGAE MEAL Foods; Listing Of Color Additives Exempt From Certification
HAEMOPHILUS SPP.	866.3300	HAEMOPHILUS SPP. SEROLOGICAL REAGENTS Serological Reagents; Immunology And Microbiology Devices
HAIDLINGER BRUSH	886.1090	HAIDLINGER BRUSH Diagnostic Devices; Ophthalmic Devices
HAIR DYES	70.20	PACKAGING REQUIREMENTS FOR STRAIGHT COLORS (OTHER THAN HAIR DYES) Packaging And Labeling; Color Additives
HAIR DYES	70.25	LABELING REQUIREMENTS FOR COLOR ADDITIVES (OTHER THAN HAIR DYES) Packaging And Labeling; Color Additives
HAIR DYES	740.18	COAL TAR HAIR DYES POSING A RISK OF CANCER Warning Statements; Cosmetic Product Warning Statements
HAIR GROWERS/HAIR LOSS	310.527	DRUG PRODUCTS CONTAINING ACTIVE INGREDIENTS OFFERED OTC FOR EXTERNAL USE A HAIR GROWERS OR FOR HAIR LOSS PREVENTION Requirements For Specific New Drugs Or Devices; New Drugs
HALF-AND-HALF	131.180	HALF-AND-HALF Requirements For Specific Standardized Milk And Cream; Milk And Cream
HALOFUGINONE HYDROBROMIDE	556.308	HALOFUGINONE HYDROBROMIDE Specific Tolerances For Residues Of New Animal Drugs; Tolerances For Residues Of New Animal Drugs In Food
HALOFUGINONE HYDROBROMIDE	558.265	HALOFUGINONE HYDROBROMIDE Specific New Animal Drugs For Use In Animal Feeds; New Animal Drugs For Use In Animal Feeds
HALOGENATED SALICYLANILIDES	700.15	USE OF CERTAIN HALOGENATED SALICYLANILIDES AS INGREDIENTS IN COSMETIC PRODUCTS Requirements For Specific Cosmetic Products; General
HALOTHANE	529.1115	HALOTHANE Certain Other Dosage Form New Animal Drugs
HALOTHANE	868.1620	HALOTHANE GAS ANALYZER Diagnostic Devices; Anesthesiology Devices
HALOXON	520.1120	HALOXON DRENCH Oral Dosage Form New Animal Drugs
HALOXON	556.310	HALOXON Specific Tolerances For Residues Of New Animal Drugs; Tolerances For Residues Of New Animal Drugs In Food
HAMMER	890.1450	POWERED REFLEX HAMMER Physical Medicine Diagnostic Devices; Physical Medicine Devices
HAND-CARRIED	880.6900	HAND CARRIED STRETCHER General Hospital And Personal Use Miscellaneous Devices; General Hospital And Personal Use Devices
HANDLING	58.47	FACILITIES FOR HANDLING TEST AND CONTROL ARTICLES Facilities; Good Laboratory Practice For Nonclinical Laboratory Studies
HANDLING	58.107	TEST AND CONTROL ARTICLE HANDLING Test And Control Articles; Good Laboratory Practice For Nonclinical Laboratory Studies
HANDLING	203.32	DRUG SAMPLE STORAGE AND HANDLING REQUIREMENTS Prescription Drug Marketing
HANDLING	205.50	MINIMUM REQUIREMENTS FOR THE STORAGE AND HANDLING OF PRESCRIPTION DRUGS AND FOR ESTABLISHMENT AND MAINTENANCE OF PRESCRIPTION DRUG DISTRIBUTION RECORDS Guidelines For State Licensing Of Wholesale Prescription Drug Distributors
HANDLING	312.69	HANDLING OF CONTROLLED SUBSTANCES Responsibilities Of Sponsors And Investigators; Investigational New Drug Application

HCT/P LIST	1271.21	WHEN DO I REGISTER, SUBMIT AN HCT/P LIST, AND SUBMIT UPDATES? Procedures For Registration And Listing; Human Cells, Tissues, And Cellular And Tissue-Based Products
HCT/P LIST	1271.22	HOW AND WHERE DO I REGISTER AND SUBMIT AN HCT/P LIST? Procedures For Registration And Listing; Human Cells, Tissues, And Cellular And Tissue-Based Products
HCT/P LISTING,	1271.25	WHAT INFORMATION IS REQUIRED FOR ESTABLISHMENT REGISTRATION AND HCT/P LISTING? Procedures For Registration And Listing; Human Cells, Tissues, And Cellular And Tissue-Based Products
HCT/P LISTINGS	1271.37	WILL ESTABLISHMENT REGISTRATIONS AND HCT/P LISTINGS BE AVAILABLE FOR INSPECTION, AND HOW DO I REQUEST INFORMATION? Procedures For Registration And Listing; Human Cells, Tissues, And Cellular And Tissue-Based Products
HCT/P'S	1271.10	ARE MY HCT/P'S REGULATED SOLELY UNDER SECTION 361 OF THE PHS ACT AND REGULATIONS IN THIS PART, AND IF SO WHAT MUST I DO? General Provisions; Human Cells, Tissues, And Cellular And Tissue-Based Products
HCT/P'S	1271.20	IF MY HCT/P'S DO NOT MEET THE CRITERIA IN SECTION 1271.10, AND I DO NOT QUALIFY FOR ANY OF THE EXCEPTIONS IN SECTION 1271.15, WHAT REGULATIONS APPLY? General Provisions; Human Cells, Tissues, And Cellular And Tissue-Based Products
HDE	814.106	HDE AMENDMENTS AND RESUBMITTED HDE'S Humanitarian Use Devices; Premarket Approval Of Medical Devices
HDE	814.112	FILING AN HDE Humanitarian Use Devices; Premarket Approval Of Medical Devices
HDE	814.114	TIMEFRAMES FOR REVIEWING AN HDE Humanitarian Use Devices; Premarket Approval Of Medical Devices
HDE	814.116	PROCEDURES FOR REVIEW OF AN HDE Humanitarian Use Devices; Premarket Approval Of Medical Devices
HDE	814.118	DENIAL OF APPROVAL OR WITHDRAWAL OF APPROVAL OF AN HDE Humanitarian Use Devices; Premarket Approval Of Medical Devices
HDE	814.120	TEMPORARY SUSPENSION OF APPROVAL OF AN HDE Humanitarian Use Devices; Premarket Approval Of Medical Devices
HEAD	868.1930	STETHOSCOPE HEAD Diagnostic Devices; Anesthesiology Devices
HEAD HOLDER	882.4460	NEUROSURGICAL HEAD HOLDER (SKULL CLAMP) Neurological Surgical Devices; Neurological Devices
HEAD HOLDER	892.1920	RADIOGRAPHIC HEAD HOLDER Diagnostic Devices; Radiology Devices
HEAD STRAP	868.5560	GAS MASK HEAD STRAP Therapeutic Devices; Anesthesiology Devices
HEADBAND	886.1500	HEADBAND MIRROR Diagnostic Devices; Ophthalmic Devices
HEADGEAR	872.5500	EXTRAORAL ORTHODONTIC HEADGEAR Therapeutic Devices; Dental Devices
HEADLAMP	886.4335	OPERATING HEADLAMP Surgical Devices; Ophthalmic Devices
HEADQUARTERS	5.1100	HEADQUARTERS Organization; Delegations Of Authority And Organization
HEADRESTS	882.4440	NEUROSURGICAL HEADRESTS Neurological Surgical Devices; Neurological Devices
HEALING	310.534	DRUG PRODUCTS CONTAINING ACTIVE INGREDIENTS OFFERED OTC FOR HUMAN USE AS ORAL WOUND HEALING AGENTS Requirements For Specific New Drugs Or Devices; New Drugs
HEALTH	5.10	DELEGATIONS FROM THE SECRETARY OF HEALTH AND HUMAN SERVICES TO THE COMMISSIONER OF FOOD AND DRUGS Delegations Of Authority To The Commissioner Of Food And Drugs; Delegations Of Authority And Organization
HEALTH	5.25	RESEARCH, INVESTIGATION, AND TESTING PROGRAMS AND HEALTH INFORMATION AND HEALTH PROMOTION PROGRAMS General Redelegations Of Authority; Delegations Of Authority And Organization
HEALTH	50.54	CLINICAL INVESTIGATIONS NOT OTHERWISE APPROVABLE THAT PRESENT AN OPPORTUNITY TO UNDERSTAND, PREVENT, OR ALLEVIATE A SERIOUS PROBLEM AFFECTING THE HEALTH OR WELFARE OF CHILDREN Additional Safeguards For Children In Clinical Investigations; Protection Of Human Subjects
HEALTH	1250.35	HEALTH OF PERSONS HANDLING FOOD Food Service Sanitation On Land And Air Conveyances, And Vessels; Interstate Conveyance Sanitation

HEALTH CLAIMS	5.300	FOOD STANDARDS, FOOD ADDITIVES, GRAS SUBSTANCES, COLOR ADDITIVES, NUTRIENT CONTENT CLAIMS, AND HEALTH CLAIMS Foods And Cosmetics; Redelegations Of Authority; Delegations Of Authority And Organization
HEALTH CLAIMS	101.14	HEALTH CLAIMS: GENERAL REQUIREMENTS Food Labeling
HEALTH CLAIMS	101.70	HEALTH CLAIMS: PETITIONS FOR HEALTH CLAIMS Food Labeling
HEALTH CLAIMS	101.71	HEALTH CLAIMS: CLAIMS NOT AUTHORIZED Food Labeling
HEALTH CLAIMS	101.72	HEALTH CLAIMS: CALCIUM AND OSTEOPOROSIS Food Labeling
HEALTH CLAIMS	101.73	HEALTH CLAIMS: DIETARY LIPIDS AND CANCER Food Labeling
HEALTH CLAIMS	101.74	HEALTH CLAIMS: SODIUM AND HYPERTENSION Food Labeling
HEALTH CLAIMS	101.75	HEALTH CLAIMS: DIETARY SATURATED FAT AND CHOLESTEROL AND RISK OF CORONARY HEART DISEASE Food Labeling
HEALTH CLAIMS	101.76	HEALTH CLAIMS: FIBER-CONTAINING GRAIN PRODUCTS, FRUITS, VEGETABLES AND CANCER Food Labeling
HEALTH CLAIMS	101.77	HEALTH CLAIMS: FRUITS, VEGETABLES, AND GRAIN PRODUCTS THAT CONTAIN FIBER, PARTICULARLY SOLUBLE FIBER, AND RISK OF CORONARY HEART DISEASE Food Labeling
HEALTH CLAIMS	101.78	HEALTH CLAIMS: FRUITS AND VEGETABLES AND CANCER Food Labeling
HEALTH CLAIMS	101.79	HEALTH CLAIMS: FOLATE AND NEURAL TUBE DEFECTS Food Labeling
HEALTH CLAIMS	101.80	HEALTH CLAIMS: DIETARY SUGAR ALCOHOLS AND DENTAL CARIES Food Labeling
HEALTH CLAIMS	101.81	HEALTH CLAIMS: SOLUBLE FIBER FROM CERTAIN FOODS AND RISK OF CORONARY HEART DISEASE Food Labeling
HEALTH CLAIMS	107.82	HEALTH CLAIMS: SOY PROTEIN AND RISK OF CORONARY HEART DISEASE Food Labeling
HEALTH CLAIMS	107.83	HEALTH CLAIMS: PLANT STEROL/STANOL ESTERS AND RISK OF CORONARY HEART DISEASE Food Labeling
HEALTH HAZARD	7.41	HEALTH HAZARD EVALUATION AND RECALL CLASSIFICATION Recalls - Guidance On Policy, Procedures & Industry Responsibilities; Enforcement Policy
HEALTH HAZARD	110.110	NATURAL OR UNAVOIDABLE DEFECTS IN FOOD FOR HUMAN USE THAT PRESENT NO HEALTH HAZARD Defect Action Levels; Current Good Manufacturing Practice In Manufacturing, Packing, Or Holding Human Food
HEALTH, PUBLIC	2.5	IMMINENT HAZARD TO THE PUBLIC HEALTH General Provisions; General Administrative Rulings And Decisions
HEALTH, PUBLIC	25.16	PUBLIC HEALTH AND SAFETY EMERGENCIES Environmental Impact Considerations
HEARING	1.94	HEARING ON REFUSAL OF ADMISSION Imports And Exports; General Enforcement
HEARING	5.202	ISSUANCE OF NOTICES OF OPPORTUNITY FOR HEARING ON PROPOSALS FOR DENIAL OF APPROVAL OF APPLICATIONS FOR LICENSES, SUSPENSION OR REVOCATION OF LICENSES AND CERTAIN NOTICES Biologics; Redelegations Of Authority; Delegations Of Authority And Organization
HEARING	10.50	PROMULGATION OF REGULATIONS AND ORDERS AFTER AN OPPORTUNITY FOR A FORMAL EVIDENTIARY PUBLIC HEARING General Administrative Procedures; Administrative Practices And Procedures
HEARING	12.20	INITIATION OF A HEARING INVOLVING THE ISSUANCE, AMENDMENT, OR REVOCATION OF A REGULATION Initiation Of Proceedings; Formal Evidentiary Public Hearing
HEARING	12.21	INITIATION OF A HEARING INVOLVING THE ISSUANCE, AMENDMENT, OR REVOCATION OF AN ORDER Initiation Of Proceedings; Formal Evidentiary Public Hearing
HEARING	12.22	FILING OBJECTIONS AND REQUESTS FOR A HEARING ON A REGULATION OR ORDER Initiation Of Proceedings; Formal Evidentiary Public Hearing
HEARING	12.24	RULING ON OBJECTIONS AND REQUESTS FOR HEARING Initiation Of Proceedings; Formal Evidentiary Public Hearing
HEARING	12.28	DENIAL OF HEARING IN WHOLE OR IN PART Initiation Of Proceedings; Formal Evidentiary Public Hearing

HEARING	12.30	JUDICIAL REVIEW AFTER WAIVER OF HEARING ON A REGULATION Initiation Of Proceedings; Formal Evidentiary Public Hearing
HEARING	12.32	REQUEST FOR ALTERNATIVE FORM OF HEARING Initiation Of Proceedings; Formal Evidentiary Public Hearing
HEARING	12.35	NOTICE OF HEARING; STAY OF ACTION Initiation Of Proceedings; Formal Evidentiary Public Hearing
HEARING	12.100	ADMINISTRATIVE RECORD OF A HEARING Administrative Record; Formal Evidentiary Public Hearing
HEARING	13.5	NOTICE OF A HEARING BEFORE A BOARD General Provisions; Public Hearing Before A Public Board Of Inquiry
HEARING	14.20	NOTICE OF HEARING BEFORE AN ADVISORY COMMITTEE Meeting Procedures; Public Hearing Before A Public Advisory Committee
HEARING	14.29	CONDUCT OF A HEARING BEFORE AN ADVISORY COMMITTEE Meeting Procedures; Public Hearing Before A Public Advisory Committee
HEARING	14.70	ADMINISTRATIVE RECORD OF A PUBLIC HEARING BEFORE AN ADVISORY COMMITTEE Records Of Meetings And Hearings Before Advisory Committees; Public Hearing Before A Public Advisory Committee
HEARING	15.20	NOTICE OF A PUBLIC HEARING BEFORE THE COMMISSIONER Procedures For Public Hearing Before The Commissioner; Public Hearing Before The Commissioner
HEARING	15.21	NOTICE OF PARTICIPATION; SCHEDULE FOR HEARING Procedures For Public Hearing Before The Commissioner; Public Hearing Before The Commissioner
HEARING	15.30	CONDUCT OF A PUBLIC HEARING BEFORE THE COMMISSIONER Procedures For Public Hearing Before The Commissioner; Public Hearing Before The Commissioner
HEARING	16.22	INITIATION OF REGULATORY HEARING Initiation Of Proceedings; Regulatory Hearing Before The Food And Drug Administration
HEARING	16.24	REGULATORY HEARING REQUIRED BY THE ACT OR A REGULATION Initiation Of Proceedings; Regulatory Hearing Before The Food And Drug Administration
HEARING	16.26	DENIAL OF HEARING AND SUMMARY DECISION Initial Proceedings; Regulatory Hearing Before The Food And Drug Administration
HEARING	16.60	HEARING PROCEDURE Procedure For Regulatory Hearing; Regulatory Hearing Before The Food And Drug Administration
HEARING	16.80	ADMINISTRATIVE RECORD OF A REGULATORY HEARING Administrative Record And Decision; Regulatory Hearing Before The Food And Drug Administration
HEARING	17.13	NOTICE OF HEARING Civil Money Penalties Hearings
HEARING	17.15	PARTIES TO THE HEARING Civil Money Penalties Hearings
HEARING	17.27	HEARING SUBPOENAS Civil Money Penalties Hearings
HEARING	17.33	THE HEARING AND BURDEN OF PROOF Civil Money Penalties Hearings
HEARING	58.204	NOTICE OF AND OPPORTUNITY FOR HEARING ON PROPOSED DISQUALIFICATION Disqualification Of Testing Facilities; Good Laboratory Practice For Nonclinical Laboratory Studies
HEARING	60.40	REQUEST FOR HEARING Due Diligence Hearings; Patent Term Restoration
HEARING	60.42	NOTICE OF HEARING Due Diligence Hearings; Patent Term Restoration
HEARING	60.44	HEARING PROCEDURES Due Diligence Hearings; Patent Term Restoration
HEARING	310.6	APPLICABILITY OF "NEW DRUG" OR SAFETY/EFFECTIVENESS FINDINGS IN DRUG EFFICACY STUDY IMPLEMENTATION NOTICES AND NOTICES OF OPPORTUNITY FOR HEARING TO IDENTICAL, RELATED, AND SIMILAR DRUG PRODUCTS General Provisions; New Drugs
HEARING	314.200	NOTICE OF OPPORTUNITY FOR HEARING; NOTICE OF PARTICIPATION AND REQUEST FOR HEARING; GRANT OR DENIAL OF HEARING Hearing Procedures For New Drugs; Applications For FDA Approval To Market A New Drug
HEARING	514.200	CONTENTS OF NOTICE OF OPPORTUNITY FOR A HEARING Hearing Procedures; New Animal Drug Applications
HEARING	515.30	CONTENTS OF NOTICE OF OPPORTUNITY FOR A HEARING Hearing Procedures; Medicated Feed Mill License
HEARING	810.11	REGULATORY HEARING Mandatory Medical Device Recall Procedures; Medical Device Recall Authority

HEARING	1005.20	HEARING Importation Of Electronic Products
HEARING	1210.30	HEARING PROCEDURE FOR PERMIT DENIAL, SUSPENSION, AND REVOCATION Regulations Under The Federal Import Milk Act
HEARING	1210.31	HEARING BEFORE PROSECUTION Regulations Under The Federal Import Milk Act
HEARING	1301.42	PURPOSE OF HEARING Hearings; Registration Of Manufacturers, Distributors, And Dispensers Of Controlled Substances
HEARING	1301.43	REQUEST FOR HEARING OR APPEARANCE; WAIVER Hearings; Registration Of Manufacturers, Distributors, And Dispensers Of Controlled Substances
HEARING	1301.45	TIME AND PLACE OF HEARING Hearings; Registration Of Manufacturers, Distributors, And Dispensers Of Controlled Substances
HEARING	1303.32	PURPOSE OF HEARING Hearings; Quotas
HEARING	1303.34	REQUEST FOR HEARING OR APPEARANCE; WAIVER Hearings; Quotas
HEARING	1303.36	TIME AND PLACE OF HEARING Hearings; Quotas
HEARING	1308.42	PURPOSE OF HEARING Hearings; Schedules Of Controlled Substances
HEARING	1308.44	REQUEST FOR HEARING OR APPEARANCE; WAIVER Hearings; Schedules Of Controlled Substances
HEARING	1309.52	PURPOSE OF HEARING Registration Of Manufacturers, Distributors, Importers And Exporters Of List I Chemicals
HEARING	1309.53	REQUEST FOR HEARING OR APPEARANCE; WAIVER Registration Of Manufacturers, Distributors, Importers And Exporters Of List I Chemicals
HEARING	1309.55	TIME AND PLACE OF HEARING Registration Of Manufacturers, Distributors, Importers And Exporters Of List I Chemicals
HEARING	1312.42	PURPOSE OF HEARING Hearings; Importation And Exportation Of Controlled Substances
HEARING	1312.44	REQUEST FOR HEARING OR APPEARANCE; WAIVER Hearings; Importation And Exportation Of Controlled Substances
HEARING	1312.46	TIME AND PLACE OF HEARING Hearings; Importation And Exportation Of Controlled Substances
HEARING	1313.52	PURPOSE OF HEARING Hearings; Importation and Exportation Of Precursors And Essential Chemicals
HEARING	1313.54	REQUEST FOR HEARING Hearings; Importation and Exportation Of Precursors And Essential Chemicals
HEARING	1313.56	TIME AND PLACE OF HEARING Hearings; Importation and Exportation Of Precursors And Essential Chemicals
HEARING	1316.47	REQUEST FOR HEARING Administrative Hearings; Administrative Functions, Practices, And Procedures
HEARING	1316.49	WAIVER OF HEARING Administrative Hearings; Administrative Functions, Practices, And Procedures
HEARING	1316.51	CONDUCT OF HEARING AND PARTIES; EX PARTE COMMUNICATIONS Administrative Hearings; Administrative Functions, Practices, And Procedures
HEARING	1316.53	TIME AND PLACE OF HEARING Administrative Hearings; Administrative Functions, Practices, And Procedures
HEARING AID	801.420	HEARING AID DEVICES; PROFESSIONAL AND PATIENT LABELING Special Requirements For Specific Devices; Labeling
HEARING AID	801.421	HEARING AID DEVICES; CONDITIONS FOR SALE Special Requirements For Specific Devices; Labeling
HEARING AID	874.3300	HEARING AID Prosthetic Devices; Ear, Nose, And Throat Devices
HEARING AID	874.3310	HEARING AID CALIBRATOR AND ANALYSIS SYSTEM Prosthetic Devices; Ear, Nose, And Throat Devices
HEARING AID	874.3320	GROUP HEARING AID OR GROUP AUDITORY TRAINER Prosthetic Devices; Ear, Nose, And Throat Devices
HEARING AID	874.3330	MASTER HEARING AID Prosthetic Devices; Ear, Nose, And Throat Devices
HEARINGS	5.27	PATENT TERM EXTENSIONS FOR HUMAN DRUG PRODUCTS, MEDICAL DEVICES, AND FOOD AND COLOR ADDITIVES; AND AUTHORITY TO PERFORM DUE DILIGENCE DETERMINATIONS AND INFORMAL HEARINGS General Redelegations Of Authority; Delegations Of Authority And Organization
HEARINGS	5.28	HEARINGS General Redelegations Of Authority; Delegations Of Authority And Organization

HEARINGS	12.50	ADVICE ON PUBLIC PARTICIPATION IN HEARINGS Appearance And Participation; Formal Evidentiary Public Hearing
HEARINGS	12.90	CONDUCT AT ORAL HEARINGS OR CONFERENCES Hearing Procedures; Formal Evidentiary Public Hearing
HEARINGS	171.110	PROCEDURE FOR OBJECTIONS AND HEARINGS Administrative Actions On Applications; Food Additive Petitions
HEARINGS	314.201	PROCEDURE FOR HEARINGS Hearing Procedures For New Drugs; Applications For FDA Approval To Market A New Drug
HEARINGS	514.201	PROCEDURES FOR HEARINGS Hearing Procedures; New Animal Drug Applications
HEARINGS	571.110	PROCEDURE FOR OBJECTIONS AND HEARINGS Administrative Actions On Applications; Food Additive Petitions
HEARINGS	601.7	PROCEDURE FOR HEARINGS General Provisions; Licensing
HEARINGS	900.7	HEARINGS Accreditation; Mammography
HEARINGS	900.25	HEARINGS AND APPEALS States As Certifiers; Mammography
HEARINGS	1230.35	HEARINGS Regulations Under The Federal Caustic Poison Act
HEARINGS	1230.36	HEARINGS; WHEN NOT PROVIDED FOR Regulations Under The Federal Caustic Poison Act
HEARINGS	1301.51	HEARINGS GENERALLY Hearings; Registration Of Manufacturers, Distributors, And Dispensers Of Controlled Substances
HEARINGS	1303.31	HEARINGS GENERALLY Hearings; Quotas
HEARINGS	1308.41	HEARINGS GENERALLY Hearings; Schedules Of Controlled Substances
HEARINGS	1309.51	HEARINGS GENERALLY Registration Of Manufacturers, Distributors, Importers And Exporters Of List I Chemicals
HEARINGS	1312.41	HEARING GENERALLY Hearings; Importation And Exportation Of Controlled Substances
HEARINGS	1313.51	HEARINGS GENERALLY Hearings; Importation and Exportation Of Precursors And Essential Chemicals
HEART	870.2860	HEART SOUND TRANSDUCER Cardiovascular Monitoring Devices; Cardiovascular Devices
HEART DISEASE	101.75	HEALTH CLAIMS: DIETARY SATURATED FAT AND CHOLESTEROL AND RISK OF CORONARY HEART DISEASE Food Labeling
HEART DISEASE	101.77	HEALTH CLAIMS: FRUITS, VEGETABLES, AND GRAIN PRODUCTS THAT CONTAIN FIBER, PARTICULARLY SOLUBLE FIBER, AND RISK OF CORONARY HEART DISEASE Food Labeling
HEART DISEASE	101.81	HEALTH CLAIMS: SOLUBLE FIBER FROM CERTAIN FOODS AND RISK OF CORONARY HEART DISEASE Food Labeling
HEART DISEASE	107.82	HEALTH CLAIMS: SOY PROTEIN AND RISK OF CORONARY HEART DISEASE Food Labeling
HEART DISEASE	107.83	HEALTH CLAIMS: PLANT STEROL/STANOL ESTERS AND RISK OF CORONARY HEART DISEASE Food Labeling
HEART-LUNG	870.4220	CARDIOPULMONARY BYPASS HEART-LUNG MACHINE CONSOLE Cardiovascular Surgical Devices; Cardiovascular Devices
HEART VALVE	870.3925	REPLACEMENT HEART VALVE Cardiovascular Prosthetic Devices; Cardiovascular Devices
HEART VALVE	870.3935	PROSTHETIC HEART VALVE HOLDER Cardiovascular Prosthetic Devices; Cardiovascular Devices
HEART VALVE	870.3945	PROSTHETIC HEART VALVE SIZER Cardiovascular Prosthetic Devices; Cardiovascular Devices
HEAT	868.5375	HEAT AND MOISTURE CONDENSER (ARTIFICIAL NOSE) Therapeutic Devices; Anesthesiology Devices
HEAT AND SERVE	102.26	FROZEN "HEAT AND SERVE" DINNERS Requirements For Specific Nonstandardized Foods; Common Or Usual Name For Nonstandardized Foods
HEAT AND SERVE	104.47	FROZEN "HEAT AND SERVE" DINNER Nutritional Quality Guidelines For Foods
HEAT, DRY	880.5420	DRY-HEAT STERILIZER General Hospital And Personal Use Miscellaneous Devices; General Hospital And Personal Use Devices
HEAT EXCHANGER	870.4240	CARDIOPULMONARY BYPASS HEAT EXCHANGER Cardiovascular Surgical Devices; Cardiovascular Devices

HEAT PACK	890.5730	MOIST HEAT PACK Physical Medicine Therapeutic Devices; Physical Medicine Devices
HEAT-SEALING	864.9750	HEAT-SEALING DEVICE Products Used In Establishments That Manufacture Blood And Blood Products; Hematology And Pathology Devices
HEAT SOURCE	872.6475	HEAT SOURCE FOR BLEACHING TEETH Miscellaneous Devices; Dental Devices
HEAT STERILIZER	872.6730	ENDODONTIC DRY HEAT STERILIZER Miscellaneous Devices; Dental Devices
HEAT TOASTED	573.310	CRAMBE MEAL, HEAT TOASTED Food Additive Listing; Food Additives Permitted In Feed And Drinking Water Of Animals
HEATER	868.5270	BREATHING SYSTEM HEATER Therapeutic Devices; Anesthesiology Devices
HEATER	884.5390	PERINEAL HEATER Obstetrical And Gynecological Therapeutic Devices; Obstetrical And Gynecological Devices
HEATING	179.30	RADIOFREQUENCY RADIATION FOR THE HEATING OF FOOD, INCLUDING MICROWAVE FREQUENCIES Radiation And Radiation Sources; Irradiation In The Production, Processing And Handling Of Food
HEATING	211.46	VENTILATION, AIR FILTRATION, AIR HEATING AND COOLING Buildings And Facilities; Current Good Manufacturing Practice For Finished Pharmaceuticals
HEATING PAD	890.5740	POWERED HEATING PAD Physical Medicine Therapeutic Devices; Physical Medicine Devices
HEATING UNIT	890.5950	POWERED HEATING UNIT Physical Medicine Therapeutic Devices; Physical Medicine Devices
HEAVY CREAM	131.150	HEAVY CREAM Requirements For Specific Standardized Milk And Cream; Milk And Cream
HELIUM	184.1355	HELIUM Listing Of Specific Substances Affirmed As GRAS; Direct Food Substances Affirmed As Generally Recognized As Safe
HELIUM	582.1355	HELIUM General Purpose Food Additives; Substances Generally Recognized As Safe
HELIUM	868.1640	HELIUM GAS ANALYZER Diagnostic Devices; Anesthesiology Devices
HEMATOCRIT	864.5600	AUTOMATED HEMATOCRIT INSTRUMENT Automated And Semi-automated Hematology Devices; Hematology And Pathology Devices
HEMATOCRIT	864.6400	HEMATOCRIT MEASURING DEVICE Manual Hematology Devices; Hematology And Pathology Devices
HEMATOCRIT	864.8165	CALIBRATOR FOR HEMOGLOBIN OR HEMATOCRIT MEASUREMENT Hematology Reagents; Hematology And Pathology Devices
HEMATOLOGY	864.8625	HEMATOLOGY QUALITY CONTROL MIXTURE Hematology Reagents; Hematology And Pathology Devices
HEMATOLOGY	864.9285	AUTOMATED CELL-WASHING CENTRIFUGE FOR IMMUNO-HEMATOLOGY Products Used In Establishments That Manufacture Blood And Blood Products; Hematology And Pathology Devices
HEMICELLULOSE	573.520	HEMICELLULOSE EXTRACT Food Additive Listing; Food Additives Permitted In Feed And Drinking Water Of Animals
HEMI-ELBOW	888.3170	ELBOW JOINT RADIAL (HEMI-ELBOW) POLYMER PROSTHESIS Prosthetic Devices; Orthopedic Devices
HEMI-ELBOW	888.3180	ELBOW JOINT HUMERAL (HEMI-ELBOW) METALLIC UNCEMENTED PROSTHESIS Prosthetic Devices; Orthopedic Devices
HEMI-HIP	888.3360	HIP JOINT FEMORAL (HEMI-HIP) METALLIC CEMENTED OR UNCEMENTED PROSTHESIS Prosthetic Devices; Orthopedic Devices
HEMI-HIP	888.3370	HIP JOINT (HEMI-HIP) ACETABULAR METAL CEMENTED PROSTHESIS Prosthetic Devices; Orthopedic Devices
HEMI-HIP	888.3380	HIP JOINT FEMORAL (HEMI-HIP) TRUNION-BEARING METAL/POLYACETAL CEMENTED PROSTHESIS Prosthetic Devices; Orthopedic Devices
HEMI-HIP	888.3390	HIP JOINT FEMORAL (HEMI-HIP) METAL/POLYMER CEMENTED OR UNCEMENTED PROSTHESIS Prosthetic Devices; Orthopedic Devices
HEMI-HIP	888.3400	HIP JOINT FEMORAL (HEMI-HIP) METALLIC RESURFACING PROSTHESIS Prosthetic Devices; Orthopedic Devices
HEMI-KNEE	888.3570	KNEE JOINT FEMORAL (HEMI-KNEE) METALLIC UNCEMENTED PROSTHESIS Prosthetic Devices; Orthopedic Devices

HEMI-KNEE	888.3580	KNEE JOINT PATELLAR (HEMI-KNEE) METALLIC RESURFACING UNCEMENTED PROSTHESIS Prosthetic Devices; Orthopedic Devices
HEMI-KNEE	888.3590	KNEE JOINT TIBIAL (HEMI-KNEE) METALLIC RESURFACING UNCEMENTED PROSTHESIS Prosthetic Devices; Orthopedic Devices
HEMI-SHOULDER	888.3680	SHOULDER JOINT GLENOID (HEMI-SHOULDER) METALLIC CEMENTED PROSTHESIS Prosthetic Devices; Orthopedic Devices
HEMI-SHOULDER	888.3690	SHOULDER JOINT HUMERAL (HEMI-SHOULDER) METALLIC UNCEMENTED PROSTHESIS Prosthetic Devices; Orthopedic Devices
HEMI-TOE	888.3730	TOE JOINT PHALANGEAL (HEMI-TOE) POLYMER PROSTHESIS Prosthetic Devices; Orthopedic Devices
HEMI-WRIST	888.3810	WRIST JOINT ULNAR (HEMI-WRIST) POLYMER PROSTHESIS Prosthetic Devices; Orthopedic Devices
HEMODIALYSIS	876.5600	SORBENT REGENERATED DIALYSATE DELIVERY SYSTEM FOR HEMODIALYSIS Therapeutic Devices; Gastroenterology-urology Devices
HEMODIALYSIS	876.5665	WATER PURIFICATION SYSTEM FOR HEMODIALYSIS Therapeutic Devices; Gastroenterology-urology Devices
HEMODIALYSIS	876.5820	HEMODIALYSIS SYSTEM AND ACCESSORIES Therapeutic Devices; Gastroenterology-urology Devices
HEMODIALYSIS	876.5860	HIGH PERMEABILITY HEMODIALYSIS SYSTEM Therapeutic Devices; Gastroenterology-urology Devices
HEMODIALYZER	876.5830	HEMODIALYZER WITH DISPOSABLE INSERT (KIIL TYPE) Therapeutic Devices; Gastroenterology-urology Devices
HEMOGLOBIN	864.5620	AUTOMATED HEMOGLOBIN SYSTEM Automated And Semi-automated Hematology Devices; Hematology And Pathology Devices
HEMOGLOBIN	864.7400	HEMOGLOBIN A2 ASSAY Hematology Kits And Packages; Hematology And Pathology Devices
HEMOGLOBIN	864.7415	ABNORMAL HEMOGLOBIN ASSAY Hematology Kits And Packages; Hematology And Pathology Devices
HEMOGLOBIN	864.7440	ELECTROPHORETIC HEMOGLOBIN ANALYSIS SYSTEM Hematology Kits And Packages; Hematology And Pathology Devices
HEMOGLOBIN	864.7455	FETAL HEMOGLOBIN ASSAY Hematology Kits And Packages; Hematology And Pathology Devices
HEMOGLOBIN	864.7470	GLYCOSYLATED HEMOGLOBIN ASSAY Hematology Kits And Packages; Hematology And Pathology Devices
HEMOGLOBIN	864.7500	WHOLE BLOOD HEMOGLOBIN ASSAYS Hematology Kits And Packages; Hematology And Pathology Devices
HEMOGLOBIN	864.8165	CALIBRATOR FOR HEMOGLOBIN OR HEMATOCRIT MEASUREMENT Hematology Reagents; Hematology And Pathology Devices
HEMOGLOBIN	866.5470	HEMOGLOBIN IMMUNOLOGICAL TEST SYSTEM Immunological Test Systems; Immunology And Microbiology Devices
HEMOGLOBIN GLUTAMER	522.1125	HEMOGLOBIN GLUTAMER-200 (BOVINE) Implantation Or Injectable Dosage Form New Animal Drugs
HEMOPERFUSION	876.5870	SORBENT HEMOPERFUSION SYSTEM Therapeutic Devices; Gastroenterology-urology Devices
HEMOPEXIN	866.5490	HEMOPEXIN IMMUNOLOGICAL TEST SYSTEM Immunological Test Systems; Immunology And Microbiology Devices
HEMORRHOIDAL	876.4400	HEMORRHOIDAL LIGATOR Surgical Devices; Gastroenterology-urology Devices
HEMOSTATIC AGENT	878.4490	ABSORBABLE HEMOSTATIC AGENT AND DRESSING Surgical Devices; General And Plastic Surgery Devices
HENNA	73.2190	HENNA Cosmetics; Listing Of Color Additives Exempt From Certification
HEPARIN	864.5680	AUTOMATED HEPARIN ANALYZER Automated And Semi-automated Hematology Devices; Hematology And Pathology Devices
HEPARIN	864.7525	HEPARIN ASSAY Hematology Kits And Packages; Hematology And Pathology Devices
HEPATITIS B	660.1	ANTIBODY TO HEPATITIS B SURFACE ANTIGEN Antibody To Hepatitis B Surface Antigen; Additional Standards For Diagnostic Substances For Laboratory Tests
HEPATITIS B	660.40	HEPATITIS B SURFACE ANTIGEN Hepatitis B Surface Antigen; Additional Standards For Diagnostic Substances For Laboratory Tests
HEPTYLPARABEN	172.145	HEPTYLPARABEN Food Preservatives; Food Additives Permitted For Direct Addition To Food For Human Consumption

HERMETICALLY SEALED	108.35	THERMAL PROCESSING OF LOW-ACID FOODS PACKAGED IN HERMETICALLY SEALED CONTAINERS Specific Requirements And Conditions For Exemption From Or Compliance With An Emergency Permit; Emergency Permit Control
HERMETICALLY SEALED	500.23	THERMALLY PROCESSED LOW-ACID FOODS PACKAGED IN HERMETICALLY SEALED CONTAINERS Specific Administrative Rulings And Decisions; General
HERNIA	876.5970	HERNIA SUPPORT Therapeutic Devices; Gastroenterology-urology Devices
HERPES SIMPLEX VIRUS	866.330	HERPES SIMPLEX VIRUS SEROLOGICAL REAGENTS Serological Reagents; Immunology And Microbiology Devices
HETACILLIN	520.1130	HETACILLIN ORAL DOSAGE FORMS Oral Dosage Form New Animal Drug Applications
HETACILLIN	526.1130	HETACILLIN POTASSIUM FOR INTRAMAMMARY INFUSION Intramammary Dosage Forms
HETEROGRAFT	888.3010	BONE HETEROGRAFT Prosthetic Devices; Orthopedic Devices
HETEROSPECIFIC	660.54	POTENCY TESTS, SPECIFICITY TESTS, TESTS FOR HETEROSPECIFIC ANTIBODIES, AND ADDITIONAL TESTS FOR NONSPECIFIC PROPERTIES Anti-human Globulin; Additional Standards For Diagnostic Substances For Laboratory Tests
HEXACHLOROPHENE	250.250	HEXACHLOROPHENE, AS A COMPONENT OF DRUG AND COSMETIC PRODUCTS Requirements For Drugs And Cosmetics; Specific Requirements For Specific Human Drugs
HEXACHLOROPHENE	500.46	HEXACHLOROPHENE IN ANIMAL DRUGS Specific Administrative Rulings And Decisions; General
HEXAMETAPHOSPHATE	182.6203	CALCIUM HEXAMETAPHOSPHATE Sequestrants; Substances Generally Recognized As Safe
HEXAMETAPHOSPHATE	182.6760	SODIUM HEXAMETAPHOSPHATE Sequestrants; Substances Generally Recognized As Safe
HEXAMETAPHOSPHATE	582.6203	CALCIUM HEXAMETAPHOSPHATE Sequestrants; Substances Generally Recognized As Safe
HEXAMETAPHOSPHATE	582.6760	SODIUM HEXAMETAPHOSPHATE Sequestrants; Substances Generally Recognized As Safe
HEXANE	173.270	HEXANE Solvents, Lubricants, Release Agents And Related Substances; Secondary Direct Food Additives Permitted In Food For Human Consumption
HEXENE	177.1960	VINYL CHLORIDE-HEXENE-1 COPOLYMERS Substances For Use As Basic Components Of Single And Repeated Use Food Contact Surfaces; Indirect Food Additives: Polymers
HIGH	101.54	NUTRIENT CONTENT CLAIMS FOR "GOOD SOURCE," "HIGH," "MORE," AND "HIGH POTENCY" Specific Requirements For Nutrient Content Claims; Food Labeling
HIGH-MOISTURE	133.154	HIGH-MOISTURE JACK CHEESE Requirements For Specific Standardized Cheese And Related Products; Cheeses And Related Cheese Products
HIGH PERMEABILITY	876.5860	HIGH PERMEABILITY HEMODIALYSIS SYSTEM Therapeutic Devices; Gastroenterology-urology Devices
HIGH POTENCY	101.54	NUTRIENT CONTENT CLAIMS FOR "GOOD SOURCE," "HIGH," "MORE," AND "HIGH POTENCY" Specific Requirements For Nutrient Content Claims; Food Labeling
HIGH PRESSURE	862.2260	HIGH PRESSURE LIQUID CHROMATOGRAPHY SYSTEM FOR CLINICAL USE Clinical Laboratory Instruments; Clinical Chemistry And Clinical Toxicology Devices
HIGH-RISK	1403.12	SPECIAL GRANT OR SUBGRANT CONDITIONS FOR "HIGH-RISK" GRANTEES Pre-Award Requirements; Uniform Administrative Requirements For Grants And Coop. Agreements To State & Local Gov.
HIGH VOLTAGE	892.1700	DIAGNOSTIC X-RAY HIGH VOLTAGE GENERATOR Diagnostic Devices; Radiology Devices
HIGHER FATTY ACIDS	573.640	METHYL ESTERS OF HIGHER FATTY ACIDS Food Additive Listing; Food Additives Permitted In Feed And Drinking Water Of Animals
HIGHWAY CONVEYANCES	1250.52	DISCHARGE OF WASTES ON HIGHWAY CONVEYANCES Equipment And Operation Of Land And Air Conveyances; Interstate Conveyance Sanitation
HIP DISLOCATION	890.3665	CONGENITAL HIP DISLOCATION ABDUCTION SPLINT Physical Medicine Prosthetic Devices; Physical Medicine Devices
HIP JOINT	888.3300	HIP JOINT METAL CONSTRAINED CEMENTED OR UNCEMENTED PROSTHESIS Prosthetic Devices; Orthopedic Devices
HIP JOINT	888.3310	HIP JOINT METAL/POLYMER CONSTRAINED CEMENTED OR UNCEMENTED PROSTHESIS Prosthetic Devices; Orthopedic Devices

HIP JOINT	888.3320	HIP JOINT METAL/METAL SEMI-CONSTRAINED, WITH A CEMENTED ACETABULAR COMPONENT, PROSTHESIS Prosthetic Devices; Orthopedic Devices
HIP JOINT	888.3330	HIP JOINT METAL/METAL SEMI-CONSTRAINED, WITH AN UNCEMENTED ACETABULAR COMPONENT, PROSTHESIS Prosthetic Devices; Orthopedic Devices
HIP JOINT	888.3340	HIP JOINT METAL/COMPOSITE SEMI-CONSTRAINED CEMENTED PROSTHESIS Prosthetic Devices; Orthopedic Devices
HIP JOINT	888.3350	HIP JOINT METAL/POLYMER SEMI-CONSTRAINED CEMENTED PROSTHESIS Prosthetic Devices; Orthopedic Devices
HIP JOINT	888.3353	HIP JOINT METAL/CERAMIC/POLYMER SEMI-CONSTRAINED CEMENTED OR NONPOROUS UNCEMENTED PROSTHESIS Prosthetic Devices; Orthopedic Devices
HIP JOINT	888.3358	HIP JOINT METAL/POLYMER/METAL SEMI-CONSTRAINED POROUS-COATED UNCEMENTED PROSTHESIS Prosthetic Devices; Orthopedic Devices
HIP JOINT	888.3360	HIP JOINT FEMORAL (HEMI-HIP) METALLIC CEMENTED OR UNCEMENTED PROSTHESIS Prosthetic Devices; Orthopedic Devices
HIP JOINT	888.3370	HIP JOINT (HEMI-HIP) ACETABULAR METAL CEMENTED PROSTHESIS Prosthetic Devices; Orthopedic Devices
HIP JOINT	888.3380	HIP JOINT FEMORAL (HEMI-HIP) TRUNION-BEARING METAL/POLYACETAL CEMENTED PROSTHESIS Prosthetic Devices; Orthopedic Devices
HIP JOINT	888.3390	HIP JOINT FEMORAL (HEMI-HIP) METAL/POLYMER CEMENTED OR UNCEMENTED PROSTHESIS Prosthetic Devices; Orthopedic Devices
HIP JOINT	888.3400	HIP JOINT FEMORAL (HEMI-HIP) METALLIC RESURFACING PROSTHESIS Prosthetic Devices; Orthopedic Devices
HIP JOINT	888.3410	HIP JOINT METAL/POLYMER SEMI-CONSTRAINED RESURFACING CEMENTED PROSTHESIS Prosthetic Devices; Orthopedic Devices
HISTIDINE	582.5361	HISTIDINE Nutrients And/or Dietary Supplements; Substances Generally Recognized As Safe
HISTIDINE	862.1375	HISTIDINE TEST SYSTEM Clinical Chemistry Test Systems; Clinical Chemistry And Clinical Toxicology Devices
HISTOPLASMA CAPSULATUM	866.3320	HISTOPLASMA CAPSULATUM SEROLOGICAL REAGENTS Serological Reagents; Immunology And Microbiology Devices
HISTORY RECORD, DEVICE	820.184	DEVICE HISTORY RECORD Records; Quality System Regulation
HOLDER, FILM	872.1905	DENTAL X-RAY FILM HOLDER Diagnostic Devices; Dental Devices
HOLDER, HEAD	882.4460	NEUROSURGICAL HEAD HOLDER (SKULL CLAMP) Neurological Surgical Devices; Neurological Devices
HOLDER, HEAD	892.1920	RADIOGRAPHIC HEAD HOLDER Diagnostic Devices; Radiology Devices
HOLDER, HEART VALVE	870.3935	PROSTHETIC HEART VALVE HOLDER Cardiovascular Prosthetic Devices; Cardiovascular Devices
HOLDER, ORPHAN DRUG	316.30	ANNUAL REPORTS OF HOLDER OF ORPHAN DRUG DESIGNATION Orphan Drugs
HOLDER, POSITION	880.5680	PEDIATRIC POSITION HOLDER General Hospital And Personal Use Therapeutic Devices; General Hospital And Personal Use Devices
HOLDER, RADIOGRAPHIC CASSETTE	892.1880	WALL-MOUNTED RADIOGRAPHIC CASSETTE HOLDER Diagnostic Devices; Radiology Devices
HOLDS, CLINICAL	312.42	CLINICAL HOLDS AND REQUESTS FOR MODIFICATION Investigational New Drug Application
HOLE	882.5250	BURR HOLE COVER Neurological Therapeutic Devices; Neurological Devices
HOME MONITOR	884.2730	HOME UTERINE ACTIVITY MONITOR Obstetrical And Gynecological Monitoring Devices; Obstetrical And Gynecological Devices
HOME USE	868.5460	THERAPEUTIC HUMIDIFIER FOR HOME USE Therapeutic Devices; Anesthesiology Devices
HOMOCYSTINE	862.1377	URINARY HOMOCYSTINE (NONQUANTITATIVE) TEST SYSTEM Clinical Chemistry Test Systems; Clinical Chemistry And Clinical Toxicology Devices
HOMOPOLYMER	177.2480	POLYOXYMETHYLENE HOMOPOLYMER Substances For Use Only As Components Of Articles Intended For Repeated Use; Indirect Food Additives: Polymers

HOMOPOLYMER	872.3410	ETHYLENE OXIDE HOMOPOLYMER AND/OR CARBOXYMETHYLCELLULOSE SODIUM DENTURE ADHESIVE Prosthetic Devices; Dental Devices
HOOK	868.5420	ETHER HOOK Therapeutic Devices; Anesthesiology Devices
HOP EXTRACT	172.560	MODIFIED HOP EXTRACT Flavoring Agents And Related Substances; Food Additives Permitted For Direct Addition To Food For Human Consumption
HORMONE, ADRENOCORTICOTROPIC	862.1025	ADRENOCORTICOTROPIC HORMONE (ACTH) TEST SYSTEM Clinical Chemistry Test Systems; Clinical Chemistry And Clinical Toxicology Devices
HORMONE-CONTAINING	310.530	TOPICALLY APPLIED HORMONE-CONTAINING DRUG PRODUCTS FOR OVER-THE-COUNTER HUMAN USE Requirements For Specific New Drugs Or Devices; New Drugs
HORMONE, ESTROGENIC PREPS	201.301	NOTICE TO MANUFACTURERS, PACKERS, AND DISTRIBUTORS OF ESTROGENIC HORMONE PREPARATIONS Specific Labeling Requirements For Specific Drug Products; Labeling
HORMONE, FOLLICLE STIMULATING	522.1002	FOLLICLE STIMULATING HORMONE Implantation Or Injectable Dosage Form New Animal Drugs
HORMONE, FOLLICLE-STIMULATING	862.1300	FOLLICLE-STIMLUATING HORMONE TEST SYSTEM Clinical Chemistry Test Systems; Clinical Chemistry And Clinical Toxicology Devices
HORMONE, GROWTH	862.1370	HUMAN GROWTH HORMONE TEST SYSTEM Clinical Chemistry Test Systems; Clinical Chemistry And Clinical Toxicology Devices
HORMONE, LUTEINIZING	522.1820	PITUITARY LUTEINIZING HORMONE FOR INJECTION Implantation Or Injectable Dosage Form New Animal Drugs
HORMONE, LUTEINIZING	862.1485	LUTEINIZING HORMONE TEST SYSTEM Clinical Chemistry Test Systems; Clinical Chemistry And Clinical Toxicology Devices
HORMONE, PARATHYROID	862.1485	PARATHYROID HORMONE TEST SYSTEM Clinical Chemistry Test Systems; Clinical Chemistry And Clinical Toxicology Devices
HORMONE, THYROID	201.316	DRUGS WITH THYROID HORMONE ACTIVITY FOR HUMAN USE; REQUIRED WARNING Specific Labeling Requirements For Specific Drug Products; Labeling
HORMONE, THYROID-STIM.	862.1690	THYROID-STIMULATING HORMONE TEST SYSTEM Clinical Chemistry Test Systems; Clinical Chemistry And Clinical Toxicology Devices
HORSES	500.65	EPINEPHRINE INJECTION FOR EMERGENCY TREATMENT OF ANAPHYLACTOID SHOCK IN CATTLE, HORSES, SHEEP, AND SWINE Requirements For Specific Animal Drugs; General
HOSPITAL BED	880.5100	AC-POWERED ADJUSTABLE HOSPITAL BED General Hospital And Personal Use Therapeutic Devices; General Hospital And Personal Use Devices
HOSPITAL BED	880.5110	HYDRAULIC ADJUSTABLE HOSPITAL BED General Hospital And Personal Use Therapeutic Devices; General Hospital And Personal Use Devices
HOSPITAL BED	880.5120	MANUAL ADJUSTABLE HOSPITAL BED General Hospital And Personal Use Therapeutic Devices; General Hospital And Personal Use Devices
HOSPITAL BED	880.5140	PEDIATRIC HOSPITAL BED General Hospital And Personal Use Therapeutic Devices; General Hospital And Personal Use Devices
HOSPITALS	20.112	VOLUNTARY DRUG EXPERIENCE REPORTS SUBMITTED BY PHYSICIANS AND HOSPITALS Availability Of Specific Categories Of Records; Public Information
HOT/COLD WATER	880.6085	HOT/COLD WATER BOTTLE General Hospital And Personal Use Miscellaneous Devices; General Hospital And Personal Use Devices
HOT OR COLD PACK	890.5710	HOT OR COLD DISPOSABLE PACK Physical Medicine Therapeutic Devices; Physical Medicine Devices
HOT OR COLD PACK	890.5720	WATER CIRCULATING HOT OR COLD PACK Physical Medicine Therapeutic Devices; Physical Medicine Devices
HOT-MELT	175.230	HOT-MELT STRIPPABLE FOOD COATINGS Substances For Use As Components Of Coatings; Indirect Food Additives
HOURS	1316.45	FILINGS; ADDRESS; HOURS Administrative Hearings; Administrative Functions, Practices, And Procedures
HOUSING ASSEMBLY	892.1760	DIAGNOSTIC X-RAY TUBE HOUSING ASSEMBLY Diagnostic Devices; Radiology Devices
HOUSING ASSEMBLY	892.5930	THERAPEUTIC X-RAY TUBE HOUSING ASSEMBLY Therapeutic Devices; Radiology Devices
HRUBY	886.1395	DIAGNOSTIC HRUBY FUNDUS LENS Diagnostic Devices; Ophthalmic Devices

HUD	814.102	DESIGNATION OF HUD STATUS Humanitarian Use Devices; Premarket Approval Of Medical Devices
HULLS, RICE	573.160	AMMONIATED RICE HULLS Food Additive Listing; Food Additives Permitted In Feed And Drinking Water Of Animals
HUMAN ALBUMIN	640.80	ALBUMIN (HUMAN) Albumin (Human); Additional Standards For Human Blood And Blood Products
HUMAN, ANTI	660.50	ANTI-HUMAN GLOBULIN Anti-human Globulin; Additional Standards For Diagnostic Substances For Laboratory Tests
HUMAN BLOOD	207.7	ESTABLISHMENT REGISTRATION AND PRODUCT LISTING FOR HUMAN BLOOD AND BLOOD PRODUCTS AND FOR MEDICAL DEVICES General; Registration Of Drug Producers And Listing Of Drugs In Commercial Distribution
HUMAN BLOOD	607.7	ESTABLISHMENT REGISTRATION AND PRODUCT LISTING OF BLOOD BANKS AND OTHER FIRMS MANUFACTURING HUMAN BLOOD AND BLOOD PRODUCTS General Provisions; Establishment Registratn+product Listing, Manufacturers Human Blood/b.products
HUMAN CELLS	864.2280	CULTURED ANIMAL AND HUMAN CELLS Cell And Tissue Culture Products; Hematology And Pathology Devices
HUMAN CELLS	1271.22	HOW AND WHERE DO I REGISTER AND SUBMIT AN HCT/P (HUMAN CELLS/TISSUES/PRODUCTS) LIST? Procedures For Registration And Listing; Human Cells, Tissues, And Cellular And Tissue-Based Products
HUMAN CHORIONIC	862.1155	HUMAN CHORIONIC GONADOTROPIN (HCG) TEST SYSTEM Clinical Chemistry Test Systems; Clinical Chemistry And Clinical Toxicology Devices
HUMAN CONSUMPTION	5.301	ISSUANCE OF INITIAL EMERGENCY PERMIT ORDERS AND NOTICES OF CONFIRMATION OF EFFECTIVE DATE OF FINAL REGULATIONS ON FOOD FOR HUMAN AND ANIMAL CONSUMPTION Foods And Cosmetics; Redelegations Of Authority; Delegations Of Authority And Organization
HUMAN CONSUMPTION	170.50	GLYCINE (AMINOACETIC ACID) IN FOOD FOR HUMAN CONSUMPTION Specific Administrative Rulings And Decisions; Food Additives
HUMAN CONSUMPTION	172.894	MODIFIED COTTONSEED PRODUCTS INTENDED FOR HUMAN CONSUMPTION Multipurpose Additives; Food Additives Permitted For Direct Addition To Food For Human Consumption
HUMAN CONSUMPTION	1240.61	MANDATORY PASTEURIZATION FOR ALL MILK AND MILK PRODUCTS IN FINAL PACKAGE FORM INTENDED FOR DIRECT HUMAN CONSUMPTION Specific Administrative Decisions Regarding Interstate Shipments; Control Of Communicable Diseases
HUMAN DRUG	5.27	PATENT TERM EXTENSIONS FOR HUMAN DRUG PRODUCTS, MEDICAL DEVICES, AND FOOD AND COLOR ADDITIVES General Redelegations Of Authority; Delegations Of Authority And Organization
HUMAN DRUG	211.132	TAMPER-EVIDENT PACKAGING REQUIREMENTS FOR OTC HUMAN DRUG PRODUCTS Current Good Manufacturing Practice For Finished Pharmaceuticals
HUMAN DRUG	530.20	CONDITIONS FOR PERMITTED EXTRALABEL ANIMAL AND HUMAN DRUG USE IN FOOD-PRODUCING ANIMALS Extralabel Drug Use In Animals
HUMAN DRUGS	5.101	TERMINATION OF EXEMPTIONS FOR NEW DRUGS FOR INVESTIGATIONAL USE IN HUMAN BEINGS Human Drugs; Redelegations Of Authority; Delegations Of Authority And Organization
HUMAN DRUGS	5.107	EXTENSIONS OR STAYS OF EFFECTIVE DATES FOR COMPLIANCE WITH CERTAIN LABELING REQUIREMENTS FOR HUMAN PRESCRIPTION DRUGS Human Drugs; Redelegations Of Authority; Delegations Of Authority And Organization
HUMAN DRUGS	14.160	ESTABLISHMENT OF STANDING TECHNICAL ADVISORY COMMITTEES FOR HUMAN PRESCRIPTION DRUGS Advisory Committees For Human Prescription Drugs; Public Hearing Before A Public Advisory Committee
HUMAN DRUGS	25.31	HUMAN DRUGS AND BIOLOGICS Categorical Exclusions; Environmental Impact Considerations
HUMAN DRUGS	201.21	DECLARATION OF PRESENCE OF PHENYLALANINE AS A COMPONENT OF ASPARTAME IN OVER-THE-COUNTER AND PRESCRIPTION DRUGS FOR HUMAN USE General Labeling Provisions; Labeling
HUMAN DRUGS	201.56	GENERAL REQUIREMENTS ON CONTENT AND FORMAT OF LABELING FOR HUMAN PRESCRIPTION DRUGS Labeling Requirements For Prescription Drugs And/or Insulin; Labeling

HUMAN DRUGS	201.57	SPECIFIC REQUIREMENTS ON CONTENT AND FORMAT OF LABELING FOR HUMAN PRESCRIPTION DRUGS Labeling Requirements For Prescription Drugs And/or Insulin; Labeling
HUMAN DRUGS	201.100	PRESCRIPTION DRUGS FOR HUMAN USE Exemptions From Adequate Directions For Use; Labeling
HUMAN DRUGS	530.13	EXTRALABEL USE FROM COMPOUNDING OF APPROVED NEW ANIMAL AND HUMAN DRUGS Extralabel Drug Use In Animals
HUMAN FOOD	2.25	GRAIN SEED TREATED WITH POISONOUS SUBSTANCES; COLOR IDENTIFICATION TO PREVENT ADULTERATION OF HUMAN AND ANIMAL FOOD Human And Animal Foods; General Administrative Rulings And Decisions
HUMAN FOOD	184.1	SUBSTANCES ADDED DIRECTLY TO HUMAN FOOD AFFIRMED AS GENERALLY RECOGNIZED AS SAFE (GRAS) General Provisions; Direct Food Substances Affirmed As Generally Recognized As Safe
HUMAN FOOD	186.1	SUBSTANCES ADDED INDIRECTLY TO HUMAN FOOD AFFIRMED AS GENERALLY RECOGNIZED AS SAFE (GRAS) General Provisions; Indirect Food Substances Affirmed As Generally Recognized As Safe
HUMAN FOOD	189.1	SUBSTANCES PROHIBITED FROM USE IN HUMAN FOOD General Provisions; Substances Prohibited From Use In Human Food
HUMAN GLOBULIN	640.100	IMMUNE GLOBULIN (HUMAN) Immune Globulin (Human); Additional Standards For Human Blood And Blood Products
HUMAN GLOBULIN	640.102	MANUFACTURE OF IMMUNE GLOBULIN (HUMAN) Immune Globulin (Human); Additional Standards For Human Blood And Blood Products
HUMAN GLOBULIN	660.50	ANTI-HUMAN GLOBULIN Anti-human Globulin; Additional Standards For Diagnostic Substances For Laboratory Tests
HUMAN GROWTH HORMONE	862.1370	HUMAN GROWTH HORMONE TEST SYSTEM Clinical Chemistry Test Systems; Clinical Chemistry And Clinical Toxicology Devices
HUMAN MARKER	866.5065	HUMAN ALLOTYPIC MARKER IMMUNOLOGICAL TEST SYSTEM Immunological Test Systems; Immunology And Microbiology Devices
HUMAN PLACENTAL	862.1585	HUMAN PLACENTAL LACTOGEN TEST SYSTEM Clinical Chemistry Test Systems; Clinical Chemistry And Clinical Toxicology Devices
HUMAN PLASMA	866.5700	WHOLE HUMAN PLASMA OR SERUM IMMUNOLOGICAL TEST SYSTEM Immunological Test Systems; Immunology And Microbiology Devices
HUMAN PLASMA PROTEIN	640.90	PLASMA PROTEIN FRACTION (HUMAN) Plasma Protein Fraction (Human); Additional Standards For Human Blood And Blood Products
HUMAN SERA	864.2800	ANIMAL AND HUMAN SERA Cell And Tissue Culture Products; Hematology And Pathology Devices
HUMAN SERVICES	5.10	DELEGATIONS FROM THE SECRETARY OF HEALTH AND HUMAN SERVICES TO THE COMMISSIONER OF FOOD AND DRUGS Delegations Of Authority To The Commissioner Of Food And Drugs; Delegations Of Authority And Organization
HUMAN TISSUE	1270.21	DETERMINATION OF DONOR SUITABILITY FOR HUMAN TISSUE INTENDED FOR TRANSPLANTATION Human Tissue Intended For Transplantation
HUMAN TISSUE	1270.42	HUMAN TISSUE OFFERED FOR IMPORT Inspection Of Tissue Establishments; Human Tissue Intended For Transplantation
HUMAN TISSUE	1270.43	RETENTION, RECALL, AND DESTRUCTION OF HUMAN TISSUE Human Tissue Intended For Transplantation
HUMAN TISSUE CULTURE	876.5885	TISSUE CULTURE MEDIA FOR HUMAN EX VIVO TISSUE AND CELL CULTURE PROCESSING APPLICATIONS Therapeutic Devices; Gastroenterology-urology Devices
HUMAN TISSUES	1271.22	HOW AND WHERE DO I REGISTER AND SUBMIT AN HCT/P (HUMAN CELLS/TISSUES/PRODUCTS) LIST? Procedures For Registration And Listing; Human Cells, Tissues, And Cellular And Tissue-Based Products
HUMAN USE	110.110	NATURAL OR UNAVOIDABLE DEFECTS IN FOOD FOR HUMAN USE THAT PRESENT NO HEALTH HAZARD Defect Action Levels; Current Good Manufacturing Practice In Manufacturing, Packing, Or Holding Human Food
HUMAN USE	201.20	DECLARATION OF PRESENCE OF FD&C YELLOW NO. 5 AND/OR FD&C YELLOW NO. 6 IN CERTAIN DRUGS FOR HUMAN USE General Labeling Provisions; Labeling

HUMAN USE	201.21	DECLARATION OF PRESENCE OF PHENYLALANINE AS A COMPONENT OF ASPARTAME IN OVER-THE-COUNTER AND PRESCRIPTION DRUGS FOR HUMAN USE General Labeling Provisions; Labeling
HUMAN USE	201.100	PRESCRIPTION DRUGS FOR HUMAN USE Exemptions From Adequate Directions For Use; Labeling
HUMAN USE	201.305	ISOPROTERENOL INHALATION PREPARATIONS (PRESSURIZED AEROSOLS, NEBULIZERS, POWDERS) FOR HUMAN USE; WARNINGS
HUMAN USE	201.316	DRUGS WITH THYROID HORMONE ACTIVITY FOR HUMAN USE; REQUIRED WARNING Specific Labeling Requirements For Specific Drug Products; Labeling
HUMAN USE	201.317	DIGITALIS AND RELATED CARDIOTONIC DRUGS FOR HUMAN USE IN ORAL DOSAGE FORMS; REQUIRED WARNING Specific Labeling Requirements For Specific Drug Products; Labeling
HUMAN USE	250.100	AMYL NITRITE INHALANT AS A PRESCRIPTION DRUG FOR HUMAN USE New Drug Or Prescription Status Of Specific Drugs; Specific Requirements For Specific Human Drugs
HUMAN USE	250.102	DRUG PREPARATIONS INTENDED FOR HUMAN USE CONTAINING CERTAIN "CORONARY VASODILATORS" New Drug Or Prescription Status Of Specific Drugs; Specific Requirements For Specific Human Drugs
HUMAN USE	310.305	RECORDS AND REPORTS CONTAINING ADVERSE DRUG EXPERIENCES ON MARKETED PRESCRIPTION DRUGS FOR HUMAN USE WITHOUT APPROVED NEW DRUG APPLICATIONS Records And Reports; New Drugs
HUMAN USE	310.530	TOPICALLY APPLIED HORMONE-CONTAINING DRUG PRODUCTS FOR OVER-THE-COUNTER HUMAN USE Requirements For Specific New Drugs Or Devices; New Drugs
HUMAN USE	310.533	DRUG PRODUCTS CONTAINING ACTIVE INGREDIENTS OFFERED OTC FOR HUMAN USE AS AN ANTICHOLINERGIC IN COUGH-COLD DRUG PRODUCTS Requirements For Specific New Drugs Or Devices; New Drugs
HUMAN USE	310.534	DRUG PRODUCTS CONTAINING ACTIVE INGREDIENTS OFFERED OTC FOR HUMAN USE AS ORAL WOUND HEALING AGENTS Requirements For Specific New Drugs Or Devices; New Drugs
HUMAN USE	310.543	DRUG PRODUCTS CONTAINING ACTIVE INGREDIENTS OFFERED OTC FOR HUMAN USE IN EXOCRINE PANCREATIC INSUFFICIENCY Requirements For Specific New Drugs Or Devices; New Drugs
HUMAN WASTES	1250.70	DISPOSAL OF HUMAN WASTES Servicing Areas For Land And Air Conveyances; Interstate Conveyance Sanitation
HUMANS	250.11	THYROID-CONTAINING DRUG PREPARATIONS INTENDED FOR TREATMENT OF OBESITY IN HUMANS Drugs Regarded As Misbranded; Specific Requirements For Specific Human Drugs
HUMANS	300.50	FIXED-COMBINATION PRESCRIPTION DRUGS FOR HUMANS Combination Drugs; General
HUMERAL	888.3180	ELBOW JOINT HUMERAL (HEMI-ELBOW) METALLIC UNCEMENTED PROSTHESIS Prosthetic Devices; Orthopedic Devices
HUMERAL	888.3690	SHOULDER JOINT HUMERAL (HEMI-SHOULDER) METALLIC UNCEMENTED PROSTHESIS Prosthetic Devices; Orthopedic Devices
HUMIDIFIER	868.5450	RESPIRATORY GAS HUMIDIFIER Therapeutic Devices; Anesthesiology Devices
HUMIDIFIER	868.5460	THERAPEUTIC HUMIDIFIER FOR HOME USE Therapeutic Devices; Anesthesiology Devices
HYALURONATE SODIUM	522.1145	HYALURONATE SODIUM INJECTION Implantation Or Injectable Dosage Form New Animal Drugs
HYCLATE	522.778	DOXYCYCLINE HYCLATE Implantation Or Injectable Dosage Form New Animal Drugs
HYDRATE, CHLORAL	522.380	CHLORAL HYDRATE Implantation Or Injectable Dosage Form New Animal Drugs
HYDRATED	182.2729	SODIUM CALCIUM ALUMINOSILICATE, HYDRATED Anticaking Agents; Substances Generally Recognized As Safe
HYDRATED	582.2729	HYDRATED SODIUM CALCIUM ALUMINOSILICATE Anticaking Agents; Substances Generally Recognized As Safe
HYDRAULIC	870.2780	HYDRAULIC, PNEUMATIC, OR PHOTOELECTRIC PLETHYSMOGRAPHS Cardiovascular Monitoring Devices; Cardiovascular Devices
HYDRAULIC	876.5280	IMPLANTED MECHANICAL/HYDRAULIC URINARY CONTINENCE DEVICE Therapeutic Devices; Gastroenterology-urology Devices
HYDRAULIC	880.5110	HYDRAULIC ADJUSTABLE HOSPITAL BED General Hospital And Personal Use Therapeutic Devices; General Hospital And Personal Use Devices
HYDROBATH	890.5100	IMMERSION HYDROBATH Physical Medicine Therapeutic Devices; Physical Medicine Devices

HYDROBROMIDE	556.308	HALOFUGINONE HYDROBROMIDE Specific Tolerances For Residues Of New Animal Drugs; Tolerances For Residues Of New Animal Drugs In Food
HYDROCARBONS	172.882	SYNTHETIC ISOPARAFFINIC PETROLEUM HYDROCARBONS Multipurpose Additives; Food Additives Permitted For Direct Addition To Food For Human Consumption
HYDROCARBONS	172.884	ODORLESS LIGHT PETROLEUM HYDROCARBONS Multipurpose Additives; Food Additives Permitted For Direct Addition To Food For Human Consumption
HYDROCARBONS	178.3530	ISOPARAFFINIC PETROLEUM HYDROCARBONS, SYNTHETIC Certain Adjuvants And Production Aids; Indirect Food Additives: Adjuvants, Production Aids, And Sanitizers
HYDROCARBONS	178.3650	ODORLESS LIGHT PETROLEUM HYDROCARBONS Certain Adjuvants And Production Aids; Indirect Food Additives: Adjuvants, Production Aids, And Sanitizers
HYDROCARBONS	573.740	ODORLESS LIGHT PETROLEUM HYDROCARBONS Food Additive Listing; Food Additives Permitted In Feed And Drinking Water Of Animals
HYDROCHLORIC ACID	182.1057	HYDROCHLORIC ACID Multiple Purpose GRAS Food Substances; Substances Generally Recognized As Safe
HYDROCHLORIC ACID	582.1057	HYDROCHLORIC ACID General Purpose Food Additives; Substances Generally Recognized As Safe
HYDROCHLORIDE, GLUTAMIC ACID	182.1047	GLUTAMIC ACID HYDROCHLORIDE Multiple Purpose GRAS Food Substances; Substances Generally Recognized As Safe
HYDROCHLORIDE, LEVAMISOLE	556.350	LEVAMISOLE HYDROCHLORIDE Specific Tolerances For Residues Of New Animal Drugs; Tolerances For Residues Of New Animal Drugs In Food
HYDROCHLORIDE, PYRIDOXINE	184.1676	PYRIDOXINE HYDROCHLORIDE Listing Of Specific Substances Affirmed As GRAS; Direct Food Substances Affirmed As Generally Recognized As Safe
HYDROCHLORIDE, PYRIDOXINE	582.5676	PYRIDOXINE HYDROCHLORIDE Nutrients And/or Dietary Supplements; Substances Generally Recognized As Safe
HYDROCHLORIDE, ROBENIDINE	556.580	ROBENIDINE HYDROCHLORIDE Specific Tolerances For Residues Of New Animal Drugs; Tolerances For Residues Of New Animal Drugs In Food
HYDROCHLORIDE, THIAMINE	184.1875	THIAMINE HYDROCHLORIDE Listing Of Specific Substances Affirmed As GRAS; Direct Food Substances Affirmed As Generally Recognized As Safe
HYDROCHLORIDE, THIAMINE	582.5875	THIAMINE HYDROCHLORIDE Nutrients And/or Dietary Supplements; Substances Generally Recognized As Safe
HYDROCHLOROTHIAZIDE	522.1150	HYDROCHLOROTHIAZIDE INJECTION Implantation Or Injectable Dosage Form New Animal Drugs
HYDROCORTISONE	524.155	BACITRACIN ZINC-POLYMYXIN B SULFATE, NEOMYCIN SULFATE-HYDROCORTISONE OR HYDROCORTISONE ACETATE OPHTHALMIC OINTMENT Ophthalmic And Topical Dosage Form New Animal Drugs
HYDROCORTISONE	524.1204	KANAMYCIN SULFATE, CALCIUM AMPHOMYCIN, AND HYDROCORTISONE ACETATE Ophthalmic And Topical Dosage Form New Animal Drugs
HYDROCORTISONE	524.1484h	NEOMYCIN, PENICILLIN, POLYMYXIN, HYDROCORTISONE SUSPENSION Ophthalmic And Topical Dosage Form New Animal Drugs
HYDROCORTISONE	524.1662a	OXYTETRACYCLINE HYDROCHLORIDE AND HYDROCORTISONE SPRAY Ophthalmic And Topical Dosage Form New Animal Drugs
HYDROCORTISONE	862.1205	CORTISOL (HYDROCORTISONE AND HYDROXYCORTICOSTERONE) TEST SYSTEM Clinical Chemistry Test Systems; Clinical Chemistry And Clinical Toxicology Devices
HYDROCORTISONE	556.320	HYDROCORTISONE Specific Tolerances For Residues Of New Animal Drugs; Tolerances For Residues Of New Animal Drugs In Food
HYDROCORTISONE ACETATE	524.155	BACITRACIN ZINC-POLYMIXIN B SULFATE-NEOMYCIN SULFATE-HYDROCORTISONE OR HYDROCORTISONE ACETATE OPHTHALMIC OINTMENT Ophthalmic And Topical Dosage Form New Animal Drugs
HYDROCORTISONE ACETATE	524.1204	KANAMYCIN SULFATE, CALCIUM AMPHOMYCIN, AND HYDROCORTISONE ACETATE Ophthalmic And Topical Dosage Form New Animal Drugs
HYDROCORTISONE ACETATE	524.1484d	NEOMYCIN SULFATE, HYDROCORTISONE ACETATE, TETRACAINE HYDROCHLORIDE EAR OINTMENT Ophthalmic And Topical Dosage Form New Animal Drug
HYDROCORTISONE ACETATE	524.1484i	NEOMYCIN SULFATE, HYDROCORTISONE ACETATE, STERILE OINTMENT Ophthalmic And Topical Dosage Form New Animal Drugs

HYDROGEL WOUND DRESSING	878.4022	HYDROGEL WOUND DRESSING AND BURN DRESSING Surgical Devices; General And Plastic Surgery Devices
HYDROGEN ION (pH)	868.1170	INDWELLING BLOOD HYDROGEN ION CONCENTRATION (pH) ANALYZER Diagnostic Devices; Anesthesiology Devices
HYDROGEN PEROXIDE	178.1005	HYDROGEN PEROXIDE SOLUTION Substances Utilized To Control The Growth Of Microorganisms; Indirect Food Additives: Adjuvants, Production Aids, And Sanitizers
HYDROGEN PEROXIDE	184.1366	HYDROGEN PEROXIDE Listing Of Specific Substances Affirmed As GRAS; Direct Food Substances Affirmed As Generally Recognized As Safe
HYDROGEN PEROXIDE	582.1366	HYDROGEN PEROXIDE General Purpose Food Additives; Substances Generally Recognized As Safe
HYDROGEN SUCCINATE	172.765	SUCCISTEARIN (STEAROYL PROPYLENE GLYCOL HYDROGEN SUCCINATE) Other Specific Usage Additives; Food Additives Permitted For Direct Addition To Food For Human Consumption
HYDROGEN SULPHATE	520.62	AMINOPENTAMIDE HYDROGEN SULPHATE TABLETS Oral Dosage Form New Animal Drugs
HYDROGEN SULFATE	520.1204	KANAMYCIN SULFATE, AMINOPENTAMIDE, HYDROGEN SULFATE, PECTIN, BISMUTH SUBCARBONATE, ACTIVATED ATTAPULGITE ORAL SUSPENSION Oral Dosage Form New Animal Drugs
HYDROGEN SULPHATE	522.62	AMINOPENTAMIDE HYDROGEN SULPHATE INJECTION Implantation Or Injectable Dosage Form New Animal Drugs
HYDROGENATED	178.3280	CASTOR OIL, HYDROGENATED Certain Adjuvants And Production Aids; Indirect Food Additives: Adjuvants, Production Aids, And Sanitizers
HYDROGENATED	178.3610	α-METHYLSTYRENE-VINYLTOLUENE RESINS, HYDROGENATED Certain Adjuvants And Production Aids; Indirect Food Additives: Adjuvants, Production Aids, And Sanitizers
HYDROGENATED CORN SYRUP	573.530	HYDROGENATED CORN SYRUP Food Additive Listing; Food Additives Permitted In Feed And Drinking Water Of Animals
HYDROGENATED DEXTRAN	522.1183	IRON HYDROGENATED DEXTRAN INJECTION Implantation Or Injectable Dosage Form New Animal Drugs
HYDROGENATED ESTER	189.300	HYDROGENATED 4,4'-ISOPROPYLIDENE-DIPHENOLPHOSPHITE ESTER RESINS Substances Prohibited From Indirect Addition To Human Food Through Food-contact Surfaces; Substances Prohibited From Use In Human Food
HYDROGENATED FISH OIL	186.1551	HYDROGENATED FISH OIL Listing Of Specific Substances Affirmed As GRAS; Indirect Food Substances Affirmed As Generally Recognized As Safe
HYDROGENATED SPERM OIL	173.275	HYDROGENATED SPERM OIL Solvents, Lubricants, Release Agents And Related Substances; Secondary Direct Food Additives Permitted In Food For Human Consumption
HYDROGENOLYSIS	172.866	SYNTHETIC GLYCERIN PRODUCED BY THE HYDROGENOLYSIS OF CARBOHYDRATES Multipurpose Additives; Food Additives Permitted For Direct Addition To Food For Human Consumption
HYDROLYSATES, PROTEIN	102.22	PROTEIN HYDROLYSATES Requirements For Specific Nonstandardized Foods; Common Or Usual Name For Nonstandardized Foods
HYDROLYSATES, PROTEIN	573.200	CONDENSED ANIMAL PROTEIN HYDROLYSATES Food Additive Listing; Food Additives Permitted In Feed And Drinking Water Of Animals
HYDROLYZED LEATHER MEAL	573.540	HYDROLYZED LEATHER MEAL Food Additive Listing; Food Additives Permitted In Feed And Drinking Water Of Animals
HYDROPHILIC CONTACT LENS	886.5925	SOFT (HYDROPHILIC) CONTACT LENS Therapeutic Devices; Ophthalmic Devices
HYDROPHILIC CONTACT LENS	886.5928	SOFT (HYDROPHILIC) CONTACT LENS CARE PRODUCTS Therapeutic Devices; Ophthalmic Devices
HYDROPHILIC GUMS	201.319	WATER-SOLUBLE GUMS, HYDROPHILIC GUMS AND HYDROPHILIC MUCILLOIDS (INCLUDING GUAR GUM, KARAYA GUM, XANTHAN GUM) AS ACTIVE INGREDIENTS; WARNINGS AND DIRECTIONS Specific Labeling Requirements For Specific Drug Products; Labeling
HYDROPHILIC RESIN	872.3300	HYDROPHILIC RESIN COATING FOR DENTURES Prosthetic Devices; Dental Devices
HYDROPHILIC WOUND DRESSING	878.4018	HYDROPHILIC WOUND DRESSING Surgical Devices; General And Plastic Surgery Devices
HYDROPHOBIC SILICAS	584.700	HYDROPHOBIC SILICAS Food Substances Affirmed As Generally Recognized As Safe In Feed And Drinking Water
HYDROXIDE, ALUMINUM	73.1010	ALUMINUM HYDROXIDE Drugs; Listing Of Color Additives Exempt From Certification

HYDROXIDE, AMMONIUM	184.1139	AMMONIUM HYDROXIDE Listing Of Specific Substances Affirmed As GRAS; Direct Food Substances Affirmed As Generally Recognized As Safe
HYDROXIDE, AMMONIUM	582.1139	AMMONIUM HYDROXIDE General Purpose Food Additives; Substances Generally Recognized As Safe
HYDROXIDE, CALCIUM	184.1205	CALCIUM HYDROXIDE Listing Of Specific Substances Affirmed As GRAS; Direct Food Substances Affirmed As Generally Recognized As Safe
HYDROXIDE, CALCIUM	582.1205	CALCIUM HYDROXIDE General Purpose Food Additives; Substances Generally Recognized As Safe
HYDROXIDE, CALCIUM	872.3250	CALCIUM HYDROXIDE CAVITY LINER Prosthetic Devices; Dental Devices
HYDROXIDE, CHROMIUM	73.1326	CHROMIUM HYDROXIDE GREEN Drugs; Listing Of Color Additives Exempt From Certification
HYDROXIDE, CHROMIUM	73.2326	CHROMIUM HYDROXIDE GREEN Cosmetics; Listing Of Color Additives Exempt From Certification
HYDROXIDE, MAGNESIUM	184.1428	MAGNESIUM HYDROXIDE Listing Of Specific Substances Affirmed As GRAS; Direct Food Substances Affirmed As Generally Recognized As Safe
HYDROXIDE, MAGNESIUM	582.1428	MAGNESIUM HYDROXIDE General Purpose Food Additives; Substances Generally Recognized As Safe
HYDROXIDE, POTASSIUM	184.1631	POTASSIUM HYDROXIDE Listing Of Specific Substances Affirmed As GRAS; Direct Food Substances Affirmed As Generally Recognized As Safe
HYDROXIDE, POTASSIUM	582.1631	POTASSIUM HYDROXIDE General Purpose Food Additives; Substances Generally Recognized As Safe
HYDROXIDE, SODIUM	184.1763	SODIUM HYDROXIDE Listing Of Specific Substances Affirmed As GRAS; Direct Food Substances Affirmed As Generally Recognized As Safe
HYDROXIDE, SODIUM	582.1763	SODIUM HYDROXIDE General Purpose Food Additives; Substances Generally Recognized As Safe
HYDROXY ANALOG	582.5477	METHIONINE HYDROXY ANALOG AND ITS CALCIUM SALTS Nutrients And/or Dietary Supplements; Substances Generally Recognized As Safe
HYDROXYANISOLE	182.3169	BUTYLATED HYDROXYANISOLE Chemical Preservatives; Substances Generally Recognized As Safe
HYDROXYANISOLE	582.3169	BUTYLATED HYDROXYANISOLE Chemical Preservatives; Substances Generally Recognized As Safe
HYDROXYBUTYRIC	862.1380	HYDROXYBUTYRIC DEHYDROGENASE TEST SYSTEM Clinical Chemistry Test Systems; Clinical Chemistry And Clinical Toxicology Devices
HYDROXYCORTICOSTEROIDS	862.1385	17-HYDROXYCORTICOSTEROIDS (17-KETOGENIC STEROIDS) TEST SYSTEM Clinical Chemistry Test Systems; Clinical Chemistry And Clinical Toxicology Devices
HYDROXYCORTICOSTERONE	862.1205	CORTISOL (HYDROCORTISONE AND HYDROXYCORTICOSTERONE) TEST SYSTEM Clinical Chemistry Test Systems; Clinical Chemistry And Clinical Toxicology Devices
HYDROXYETHYL	73.3121	POLY(HYDROXYETHYL METHACRYLATE)-DYE COPOLYMERS Medical Devices; Listing Of Color Additives Exempt From Certification
HYDROXYETHYL	177.1400	HYDROXYETHYL CELLULOSE FILM, WATER-INSOLUBLE Substances For Use As Basic Components Of Single And Repeated Use Food Contact Surfaces; Indirect Food Additives: Polymers
HYDROXYHEXADECANE	176.250	POLY-1,4,7,10,13-PENTAAZA-15-HYDROXYHEXADECANE Substances For Use Only As Components Of Paper And Paperboard; Indirect Food Additives
HYDROXYINDOLE	862.1390	5-HYDROXYINDOLE ACETIC ACID/SEROTONIN TEST SYSTEM Clinical Chemistry Test Systems; Clinical Chemistry And Clinical Toxicology Devices
HYDROXYLATED LECITHIN	172.814	HYDROXYLATED LECITHIN Multipurpose Additives; Food Additives Permitted For Direct Addition To Food For Human Consumption
HYDROXYMETHYL-BUTYL-PHENOL	172.150	4-HYDROXYMETHYL-2,6-DI-TERT-BUTYL-PHENOL Food Preservatives; Food Additives Permitted For Direct Addition To Food For Human Consumption
HYDROXYMETHYL-BUTYL-PHENOL	178.2550	4-HYDROXYMETHYL-2,6-DI-TERT-BUTYL-PHENOL Antioxidants And Stabilizers; Indirect Food Additives: Adjuvants, Production Aids, And Sanitizers
HYDROXYPROGESTERONE	862.1395	17-HYDROXYPROGESTERONE TEST SYSTEM Clinical Chemistry Test Systems; Clinical Chemistry And Clinical Toxicology Devices
HYDROXYPROLINE	862.1400	HYDROXPROLINE TEST SYSTEM Clinical Chemistry Test Systems; Clinical Chemistry And Clinical Toxicology Devices

HYDROXYPROPYL CELLULOSE	172.870	HYDROXYPROPYL CELLULOSE Multipurpose Additives; Food Additives Permitted For Direct Addition To Food For Human Consumption
HYDROXYPROPYL METHYLCELLULOSE	172.874	HYDROXYPROPYL METHYLCELLULOSE Multipurpose Additives; Food Additives Permitted For Direct Addition To Food For Human Consumption
HYDROXYTOLUENE	182.3173	BUTYLATED HYDROXYTOLUENE Chemical Preservatives; Substances Generally Recognized As Safe
HYDROXYTOLUENE	582.3173	BUTYLATED HYDROXYTOLUENE Chemical Preservatives; Substances Generally Recognized As Safe
HYGROMYCIN	556.330	HYGROMYCIN B Specific Tolerances For Residues Of New Animal Drugs; Tolerances For Residues Of New Animal Drugs In Food
HYGROMYCIN	558.274	HYGROMYCIN B Specific New Animal Drugs For Use In Animal Feeds; New Animal Drugs For Use In Animal Feeds
HYGROSCOPIC	884.4250	HYGROSCOPIC LAMINARIA CERVICAL DILATOR Obstetrical And Gynecological Surgical Devices; Obstetrical And Gynecological Devices
HYPERBARIC CHAMBER	868.5470	HYPERBARIC CHAMBER Therapeutic Devices; Anesthesiology Devices
HYPERPHOSPHATEMIA	310.542	OVER-THE-COUNTER (OTC) DRUG PRODUCTS CONTAINING ACTIVE INGREDIENTS OFFERED FOR USE IN TREATMENT OF HYPERPHOSPHATEMIA Requirements For Specific New Drugs Or Devices; New Drugs
HYPERSENSITIVITY	866.5500	HYPERSENSITIVITY PNEUMONITIS IMMUNOLOGICAL TEST SYSTEM Immunological Test Systems; Immunology And Microbiology Devices
HYPERSENSITIVITY TESTING	310.103	NEW DRUG SUBSTANCES INTENDED FOR HYPERSENSITIVITY TESTING Specific Administrative Rulings And Decisions; New Drugs
HYPERTENSION	101.74	HEALTH CLAIMS: SODIUM AND HYPERTENSION Food Labeling
HYPERTONIC	349.70	LABELING OF OPHTHALMIC HYPERTONIC DRUG PRODUCTS Labeling; Ophthalmic Drug Products For Over-the-counter Human Use
HYPERTONICITY	349.16	OPHTHALMIC HYPERTONICITY AGENT Active Ingredients; Ophthalmic Drug Products For Over-the-counter Human Use
HYPERTROPHY, PROSTATIC	310.532	DRUG PRODUCTS CONTAINING ACTIVE INGREDIENTS OFFERED OVER-THE-COUNTER (OTC) TO RELIEVE THE SYMPTOMS OF BENIGN PROSTATIC HYPERTROPHY Requirements For Specific New Drugs Or Devices; New Drugs
HYPOALLERGENIC FOODS	105.62	HYPOALLERGENIC FOODS Label Statements; Foods For Special Dietary Use
HYPODERMIC	880.5570	HYPODERMIC SINGLE LUMEN NEEDLE General Hospital And Personal Use Therapeutic Devices; General Hospital And Personal Use Devices
HYPOGLYCEMIC	310.517	LABELING FOR ORAL HYPOGLYCEMIC DRUGS OF THE SULFONYLUREA CLASS Requirements For Specific New Drugs Or Devices; New Drugs
HYPOPHOSPHATEMIA	310.541	OVER-THE-COUNTER (OTC) DRUG PRODUCTS CONTAINING ACTIVE INGREDIENTS OFFERED FOR USE IN THE TREATMENT OF HYPOPHOSPHATEMIA Requirements For Specific New Drugs Or Devices; New Drugs
HYPOPHOSPHITE	184.1764	SODIUM HYPOPHOSPHITE Listing Of Specific Substances Affirmed As GRAS; Direct Food Substances Affirmed As Generally Recognized As Safe
HYPOPHOSPHITE	582.5458	MANGANESE HYPOPHOSPHITE Nutrients And/or Dietary Supplements; Substances Generally Recognized As Safe
HYSTEROSCOPE	884.1690	HYSTEROSCOPE AND ACCESSORIES Obstetrical And Gynecological Diagnostic Devices; Obstetrical And Gynecological Devices
HYSTEROSCOPIC	884.1700	HYSTEROSCOPIC INSUFFLATOR Obstetrical And Gynecological Diagnostic Devices; Obstetrical And Gynecological Devices

I. V. BAG	880.5420	PRESSURE INFUSOR FOR AN I. V. BAG General Hospital And Personal Use Therapeutic Devices; General Hospital And Personal Use Devices
I. V. CONTAINER	880.5025	I. V. CONTAINER General Hospital And Personal Use Therapeutic Devices; General Hospital And Personal Use Devices
ICE	1250.28	SOURCE AND HANDLING OF ICE Food Service Sanitation On Land And Air Conveyances; Interstate Conveyance Sanitation
ICE	1250.43	ICE Equipment And Operation Of Land And Air Conveyances; Interstate Conveyance Sanitation
ICE	1250.85	DRINKING FOUNTAINS AND COOLERS; ICE; CONSTANT TEMPERATURE BOTTLES Sanitation Facilities And Conditions On Vessels; Interstate Conveyance Sanitation
ICE	1250.86	WATER FOR MAKING ICE Sanitation Facilities And Conditions On Vessels; Interstate Conveyance Sanitation
ICE BAG	880.6050	ICE BAG General Hospital And Personal Use Miscellaneous Devices; General Hospital And Personal Use Devices
ICE CREAM	135.110	ICE CREAM AND FROZEN CUSTARD Requirements For Specific Standardized Frozen Desserts; Frozen Desserts
ICE CREAM	135.115	GOATS MILK ICE CREAM Requirements For Specific Standardized Frozen Desserts; Frozen Desserts
ICES, WATER	135.160	WATER ICES Requirements For Specific Standardized Frozen Desserts; Frozen Desserts
IDE CORRESPONDENCE	812.19	ADDRESS FOR IDE CORRESPONDENCE General Provisions; Investigational Device Exemptions
IDENTICAL DRUG PRODUCTS	310.6	APPLICABILITY OF "NEW DRUG" OR SAFETY OR EFFECTIVENESS FINDINGS IN DRUG EFFICACY STUDY IMPLEMENTATION NOTICES AND NOTICES OF OPPORTUNITY FOR HEARING IDENTICAL, RELATED, AND SIMILAR DRUG PRODUCTS General Provisions; New Drugs
IDENTIFICATION	2.25	GRAIN SEED TREATED WITH POISONOUS SUBSTANCES; COLOR IDENTIFICATION TO PREVENT ADULTERATION OF HUMAN AND ANIMAL FOOD Human And Animal Foods; General Administrative Rulings And Decisions
IDENTIFICATION	5.602	TESTING PROGRAMS AND METHODS OF CERTIFICATION AND IDENTIFICATION FOR ELECTRONIC PRODUCTS Radiation Control; Redelegations Of Authority; Delegations Of Authority And Organization
IDENTIFICATION	211.105	EQUIPMENT IDENTIFICATION Production And Process Controls; Current Good Manufacturing Practice For Finished Pharmaceuticals
IDENTIFICATION	820.60	IDENTIFICATION Identification And Traceability; Quality System Regulation
IDENTIFICATION	866.2660	MICROORGANISM DIFFERENTIATION AND IDENTIFICATION DEVICE Microbiology Devices; Immunology And Microbiology Devices
IDENTIFICATION	1010.3	IDENTIFICATION Performance Standards For Electronic Products: General
IDENTIFICATION	1250.25	SOURCE IDENTIFICATION AND INSPECTION OF FOOD AND DRINK Food Service Sanitation On Land And Air Conveyances; Interstate Conveyance Sanitation
IDENTIFICATION	1316.57	SUBMISSION OF DOCUMENTARY EVIDENCE AND AFFIDAVITS AND IDENTIFICATION OF WITNESSES SUBSEQUENT TO PREHEARING CONFERENCE Administrative Hearings; Administrative Functions, Practices, And Procedures
IDENTIFICATION CODES	11.300	CONTROLS FOR IDENTIFICATION CODES/PASSWORDS Electronic Records; Electronic Signatures
IDENTIFICATION, TOP 20	101.44	IDENTIFICATION OF THE 20 MOST FREQUENTLY CONSUMED RAW FRUIT, VEGETABLES AND FISH IN THE UNITED STATES Food Labeling
IDENTIFYING	3.5	PROCEDURES FOR IDENTIFYING THE DESIGNATED AGENCY COMPONENT Product Jurisdiction
IDENTITY	21.44	VERIFICATION OF IDENTITY Procedures For Notification Of And Access To Records In Privacy Act Record Systems; Protection Of Privacy
IDENTITY	201.50	STATEMENT OF IDENTITY Labeling Requirements For Prescription Drugs And/or Insulin; Labeling

IDENTITY	201.61	STATEMENT OF IDENTITY Labeling Requirements For Over-the-counter Drugs; Labeling
IDENTITY	610.14	IDENTITY General Provisions; General Biological Products Standards
IDENTITY	801.61	STATEMENT OF IDENTITY Labeling Requirements For Over-the-counter Devices; Labeling
IDENTITY	1310.07	PROOF OF IDENTITY Records and Reports Of Listed Chemicals And Certain Machines
IDENTITY	1316.23	CONFIDENTIALITY OF IDENTITY OF RESEARCH SUBJECTS Protection Of Researchers And Research Subjects; Administrative Functions, Practices, And Procedures
IDENTITY LABELING	101.3	IDENTITY LABELING OF FOOD IN PACKAGED FORM General Provisions; Food Labeling
IDENTITY LABELING	501.3	IDENTITY LABELING OF ANIMAL FOOD IN PACKAGE FORM General Provisions; Animal Food Labeling
IDENTITY LABELING	701.11	IDENTITY LABELING Package Form; Cosmetic Labeling
IDENTITY, STANDARDS	130.8	CONFORMITY TO DEFINITIONS AND STANDARDS OF IDENTITY General Provisions; Food Standards: General
IDENTITY, STANDARDS	130.17	TEMPORARY PERMITS FOR INTERSTATE SHIPMENT OF EXPERIMENTAL PACKS OF FOOD VARYING FROM THE REQUIREMENTS OF DEFINITIONS AND STANDARDS OF IDENTITY General Provisions; Food Standards: General
IDENTITY, STANDARDS	130.20	FOOD ADDITIVES PROPOSED FOR USE IN FOODS FOR WHICH DEFINITIONS AND STANDARDS OF IDENTITY ARE ESTABLISHED Food Additives In Standardized Foods; Food Standards: General
ILEOSTOMY	876.5030	CONTINENT ILEOSTOMY CATHETER Therapeutic Devices; Gastroenterology-urology Devices
ILLICIT ACTIVITIES	1301.92	ILLICIT ACTIVITIES BY EMPLOYEES Employee Screening: Non-practitioners; Registration Of Manufacturers, Distributors, And Dispensers Of Controlled Substances
ILLNESSES	312.84	RISK-BENEFIT ANALYSIS IN REVIEW OF MARKETING APPLICATIONS FOR DRUGS TO TREAT LIFE-THREATENING AND SEVERELY-DEBILITATING ILLNESSES Investigational New Drug Application
ILLUMINATOR, COLOR VISION	886.1160	COLOR VISION PLATE ILLUMINATOR Diagnostic Devices; Ophthalmic Devices
ILLUMINATOR, RADIOGRAPHIC	892.1890	RADIOGRAPHIC FILM ILLUMINATOR Diagnostic Devices; Radiology Devices
IMAGE INTENSIFICATION	886.5910	IMAGE INTENSIFICATION VISION AID Therapeutic Devices; Ophthalmic Devices
IMAGE-INTENSIFIED	892.1650	IMAGE-INTENSIFIED FLUOROSCOPIC X-RAY SYSTEM Diagnostic Devices; Radiology Devices
IMAGE-INTENSIFIED	892.1660	NON-IMAGE-INTENSIFIED FLUOROSCOPIC X-RAY Diagnostic Devices; Radiology Devices
IMAGE, MEDICAL	892.2010	MEDICAL IMAGE STORAGE DEVICE Diagnostic Devices; Radiology Devices
IMAGE, MEDICAL	892.2020	MEDICAL IMAGE COMMUNICATIONS DEVICE Diagnostic Devices; Radiology Devices
IMAGE, MEDICAL	892.2030	MEDICAL IMAGE DIGITIZER Diagnostic Devices; Radiology Devices
IMAGE, MEDICAL	892.2040	MEDICAL IMAGE HARD COPY DEVICE Diagnostic Devices; Radiology Devices
IMAGER	884.2225	OBSTETRIC-GYNECOLOGIC ULTRASONIC IMAGER Obstetrical And Gynecological Monitoring Devices; Obstetrical And Gynecological Devices
IMAGING SYSTEM	876.1300	INGESTIBLE TELEMETRIC GASTROINTESTINAL CAPSULE IMAGING SYSTEM Diagnostic Devices; Gastroenterology-urology Devices
IMAGING SYSTEM	892.1550	ULTRASONIC PULSED DOPPLER IMAGING SYSTEM Diagnostic Devices; Radiology Devices
IMAGING SYSTEM	892.1560	ULTRASONIC PULSED ECHO IMAGING SYSTEM Diagnostic Devices; Radiology Devices
IMAGING SYSTEM	892.1630	ELECTROSTATIC X-RAY IMAGING SYSTEM Diagnostic Devices; Radiology Devices
IMIDOCARB DIPROPIONATE	522.1155	IMIDOCARB DIPROPIONATE STERILE POWDER Implantation Or Injectable Dosage Form New Animal Drugs
IMIDOCARB DIPROPIONATE	522.1156	IMIDOCARB DIPROPIONATE SOLUTION Implantation Or Injectable Dosage Form New Animal Drugs
IMMEDIATE PRECURSORS	1308.47	CONTROL OF IMMEDIATE PRECURSORS Hearings; Schedules Of Controlled Substances
IMMERSION	890.5100	IMMERSION HYDROBATH Physical Medicine Therapeutic Devices; Physical Medicine Devices
IMMINENT HAZARD	2.5	IMMINENT HAZARD TO THE PUBLIC HEALTH General Provisions; General Administrative Rulings And Decisions

IMMOBILIZATION	173.357	MATERIALS USED AS FIXING AGENTS IN THE IMMOBILIZATION OF ENZYME PREPARATIONS Specific Usage Additives; Secondary Direct Food Additives Permitted In Food For Human Consumption
IMMUNE GLOBULIN	640.100	IMMUNE GLOBULIN (HUMAN) Immune Globulin (Human); Additional Standards For Human Blood And Blood Products
IMMUNE GLOBULIN	640.102	MANUFACTURE OF IMMUNE GLOBULIN (HUMAN) Immune Globulin (Human); Additional Standards For Human Blood And Blood Products
IMMUNIZATION	640.66	IMMUNIZATION OF DONORS Source Plasma; Additional Standards For Human Blood And Blood Products
IMMUNO-HEMATOLOGY	864.9285	AUTOMATED CELL-WASHING CENTRIFUGE FOR IMMUNO-HEMATOLOGY Products Used In Establishments That Manufacture Blood And Blood Products; Hematology And Pathology Devices
IMMUNODIFFUSION	866.4800	RADIAL IMMUNODIFFUSION PLATE Immunology Laboratory Equipment And Reagents; Immunology And Microbiology Devices
IMMUNOELECTROPHORESIS	866.4500	IMMUNOELECTROPHORESIS EQUIPMENT Immunology Laboratory Equipment And Reagents; Immunology And Microbiology Devices
IMMUNOELECTROPHORESIS	866.4830	ROCKET IMMUNOELECTROPHORESIS EQUIPMENT Immunology Laboratory Equipment And Reagents; Immunology And Microbiology Devices
IMMUNOFLUORESCENT REAGENTS	866.3370	MYCOBACTERIUM TUBERCULOSIS IMMUNOFLUORESCENT REAGENTS Serological Reagents; Immunology And Microbiology Devices
IMMUNOFLUORESCENT REAGENTS	866.3460	RABIESVIRUS IMMUNOFLUORESCENT REAGENTS Serological Reagents; Immunology And Microbiology Devices
IMMUNOFLUOROMETER	866.4520	IMMUNOFLUOROMETER EQUIPMENT Immunology Laboratory Equipment And Reagents; Immunology And Microbiology Devices
IMMUNOGLOBULIN	866.5520	IMMUNOGLOBULIN G (FAB FRAGMENT SPECIFIC) IMMUNOLOGICAL TEST SYSTEM Immunological Test Systems; Immunology And Microbiology Devices
IMMUNOGLOBULIN	866.5530	IMMUNOGLOBULIN G (FC FRAGMENT SPECIFIC) IMMUNOLOGICAL TEST SYSTEM Immunological Test Systems; Immunology And Microbiology Devices
IMMUNOGLOBULIN	866.5540	IMMUNOGLOBULIN G (FD FRAGMENT SPECIFIC) IMMUNOLOGICAL TEST SYSTEM Immunological Test Systems; Immunology And Microbiology Devices
IMMUNOGLOBULIN	866.5550	IMMUNOGLOBULIN (LIGHT CHAIN SPECIFIC) IMMUNOLOGICAL TEST SYSTEM Immunological Test Systems; Immunology And Microbiology Devices
IMMUNOGLOBULINS	866.5510	IMMUNOGLOBULINS A, G, M, D, AND E IMMUNOLOGICAL TEST SYSTEM Immunological Test Systems; Immunology And Microbiology Devices
IMMUNOHISTOCHEMISTRY	864.1860	IMMUNOHISTOCHEMISTRY REAGENTS AND KITS Hematology And Pathology Devices
IMMUNOLOGICAL	866.5040	ALBUMIN IMMUNOLOGICAL TEST SYSTEM Immunological Test Systems; Immunology And Microbiology Device
IMMUNOLOGICAL	866.5060	PREALBUMIN IMMUNOLOGICAL TEST SYSTEM Immunological Test Systems; Immunology And Microbiology Devices
IMMUNOLOGICAL	866.5065	HUMAN ALLOTYPIC MARKER IMMUNOLOGICAL TEST SYSTEM Immunological Test Systems; Immunology And Microbiology Devices
IMMUNOLOGICAL	866.5080	ALPHA-1-ANTICHYMOTRYPSIN IMMUNOLOGICAL TEST SYSTEM Immunological Test Systems; Immunology And Microbiology Devices
IMMUNOLOGICAL	866.5090	ANTIMITOCHONDRIAL ANTIBODY IMMUNOLOGICAL TEST SYSTEM Immunological Test Systems; Immunology And Microbiology Devices
IMMUNOLOGICAL	866.5100	ANTINUCLEAR ANTIBODY IMMUNOLOGICAL TEST SYSTEM Immunological Test Systems; Immunology And Microbiology Devices
IMMUNOLOGICAL	866.5110	ANTIPARIETAL ANTIBODY IMMUNOLOGICAL TEST SYSTEM Immunological Test Systems; Immunology And Microbiology Devices
IMMUNOLOGICAL	866.5120	ANTISMOOTH MUSCLE ANTIBODY IMMUNOLOGICAL TEST SYSTEM Immunological Test Systems; Immunology And Microbiology Devices
IMMUNOLOGICAL	866.5130	ALPHA-1-ANTITRYPSIN IMMUNOLOGICAL TEST SYSTEM Immunological Test Systems; Immunology And Microbiology Devices
IMMUNOLOGICAL	866.5150	BENCE-JONES PROTEINS IMMUNOLOGICAL TEST SYSTEM Immunological Test Systems; Immunology And Microbiology Devices
IMMUNOLOGICAL	866.5160	BETA-GLOBULIN IMMUNOLOGICAL TEST SYSTEM Immunological Test Systems; Immunology And Microbiology Devices
IMMUNOLOGICAL	866.5170	BREAST MILK IMMUNOLOGICAL TEST SYSTEM Immunological Test Systems; Immunology And Microbiology Devices
IMMUNOLOGICAL	866.5200	CARBONIC ANHYDRASE B AND C IMMUNOLOGICAL TEST SYSTEM Immunological Test Systems; Immunology And Microbiology Devices

IMMUNOLOGICAL	866.5210	CERULOPLASMIN IMMUNOLOGICAL TEST SYSTEM Immunological Test Systems; Immunology And Microbiology Devices
IMMUNOLOGICAL	866.5220	COHN FRACTION II IMMUNOLOGICAL TEST SYSTEM Immunological Test Systems; Immunology And Microbiology Devices
IMMUNOLOGICAL	866.5230	COLOSTRUM IMMUNOLOGICAL TEST SYSTEM Immunological Test Systems; Immunology And Microbiology Devices
IMMUNOLOGICAL	866.5240	COMPLEMENT COMPONENTS IMMUNOLOGICAL TEST SYSTEM Immunological Test Systems; Immunology And Microbiology Devices
IMMUNOLOGICAL	866.5250	COMPLEMENT C1 INHIBITOR (INACTIVATOR) IMMUNOLOGICAL TEST SYSTEM Immunological Test Systems; Immunology And Microbiology Devices
IMMUNOLOGICAL	866.5260	COMPLEMENT C3B INACTIVATOR IMMUNOLOGICAL TEST SYSTEM Immunological Test Systems; Immunology And Microbiology Devices
IMMUNOLOGICAL	866.5270	C-REACTIVE PROTEIN IMMUNOLOGICAL TEST SYSTEM Immunological Test Systems; Immunology And Microbiology Devices
IMMUNOLOGICAL	866.5320	PROPERIDIN FACTOR B IMMUNOLOGICAL TEST SYSTEM Immunological Test Systems; Immunology And Microbiology Devices
IMMUNOLOGICAL	866.5330	FACTOR XIII, A, S, IMMUNOLOGICAL TEST SYSTEM Immunological Test Systems; Immunology And Microbiology Devices
IMMUNOLOGICAL	866.5340	FERRITIN IMMUNOLOGICAL TEST SYSTEM Immunological Test Systems; Immunology And Microbiology Devices
IMMUNOLOGICAL	866.5350	FIBRINOPEPTIDE A IMMUNOLOGICAL TEST SYSTEM Immunological Test Systems; Immunology And Microbiology Devices
IMMUNOLOGICAL	866.5360	COHN FRACTION IV IMMUNOLOGICAL TEST SYSTEM Immunological Test Systems; Immunology And Microbiology Devices
IMMUNOLOGICAL	866.5370	COHN FRACTION V IMMUNOLOGICAL TEST SYSTEM Immunological Test Systems; Immunology And Microbiology Devices
IMMUNOLOGICAL	866.5380	FREE SECRETORY COMPONENT IMMUNOLOGICAL TEST SYSTEM Immunological Test Systems; Immunology And Microbiology Devices
IMMUNOLOGICAL	866.5400	ALPHA-GLOBULIN IMMUNOLOGICAL TEST SYSTEM Immunological Test Systems; Immunology And Microbiology Devices
IMMUNOLOGICAL	866.5420	ALPHA-1-GLYCOPROTEINS IMMUNOLOGICAL TEST SYSTEM Immunological Test Systems; Immunology And Microbiology Devices
IMMUNOLOGICAL	866.5425	ALPHA-2-GLYCOPROTEINS IMMUNOLOGICAL TEST SYSTEM Immunological Test Systems; Immunology And Microbiology Devices
IMMUNOLOGICAL	866.5430	BETA-2-GLYCOPROTEIN I IMMUNOLOGICAL TEST SYSTEM Immunological Test Systems; Immunology And Microbiology Devices
IMMUNOLOGICAL	866.5440	BETA-2-GLYCOPROTEIN III IMMUNOLOGICAL TEST SYSTEM Immunological Test Systems; Immunology And Microbiology Devices
IMMUNOLOGICAL	866.5460	HAPTOGLOBIN IMMUNOLOGICAL TEST SYSTEM Immunological Test Systems; Immunology And Microbiology Devices
IMMUNOLOGICAL	866.5470	HEMOGLOBIN IMMUNOLOGICAL TEST SYSTEM Immunological Test Systems; Immunology And Microbiology Devices
IMMUNOLOGICAL	866.5490	HEMOPEXIN IMMUNOLOGICAL TEST SYSTEM Immunological Test Systems; Immunology And Microbiology Devices
IMMUNOLOGICAL	866.5500	HYPERSENSITIVITY PNEUMONITIS IMMUNOLOGICAL TEST SYSTEM Immunological Test Systems; Immunology And Microbiology Devices
IMMUNOLOGICAL	866.5510	IMMUNOGLOBULINS A, G, M, D, AND E IMMUNOLOGICAL TEST SYSTEM Immunological Test Systems; Immunology And Microbiology Devices
IMMUNOLOGICAL	866.5520	IMMUNOGLOBULIN G (FAB FRAGMENT SPECIFIC) IMMUNOLOGICAL TEST SYSTEM Immunological Test Systems; Immunology And Microbiology Devices
IMMUNOLOGICAL	866.5530	IMMUNOGLOBULIN G (FC FRAGMENT SPECIFIC) IMMUNOLOGICAL TEST SYSTEM Immunological Test Systems; Immunology And Microbiology Devices
IMMUNOLOGICAL	866.5540	IMMUNOGLOBULIN G (FD FRAGMENT SPECIFIC) IMMUNOLOGICAL TEST SYSTEM Immunological Test Systems; Immunology And Microbiology Devices
IMMUNOLOGICAL	866.5550	IMMUNOGLOBULIN (LIGHT CHAIN SPECIFIC) IMMUNOLOGICAL TEST SYSTEM Immunological Test Systems; Immunology And Microbiology Devices
IMMUNOLOGICAL	866.5560	LACTIC DEHYDROGENASE IMMUNOLOGICAL TEST SYSTEM Immunological Test Systems; Immunology And Microbiology Devices
IMMUNOLOGICAL	866.5570	LACTOFERRIN IMMUNOLOGICAL TEST SYSTEM Immunological Test Systems; Immunology And Microbiology Devices
IMMUNOLOGICAL	866.5580	ALPHA-1-LIPOPROTEIN IMMUNOLOGICAL TEST SYSTEM Immunological Test Systems; Immunology And Microbiology Devices
IMMUNOLOGICAL	866.5590	LIPOPROTEIN X IMMUNOLOGICAL TEST SYSTEM Immunological Test Systems; Immunology And Microbiology Devices
IMMUNOLOGICAL	866.5600	LOW-DENSITY LIPOPROTEIN IMMUNOLOGICAL TEST SYSTEM Immunological Test Systems; Immunology And Microbiology Devices
IMMUNOLOGICAL	866.5620	ALPHA-2-MACROGOLOBULIN IMMUNOLOGICAL TEST SYSTEM Immunological Test Systems; Immunology And Microbiology Devices

IMMUNOLOGICAL	866.5630	BETA-2-MICROGLOBULIN IMMUNOLOGICAL TEST SYSTEM Immunological Test Systems; Immunology And Microbiology Devices
IMMUNOLOGICAL	866.5640	INFECTIOUS MONONUCLEOSIS IMMUNOLOGICAL TEST SYSTEM Immunological Test Systems; Immunology And Microbiology Devices
IMMUNOLOGICAL	866.5660	MULTIPLE AUTOANTIBODIES IMMUNOLOGICAL TEST SYSTEM Immunological Test Systems; Immunology And Microbiology Devices
IMMUNOLOGICAL	866.5680	MYOGLOBIN IMMUNOLOGICAL TEST SYSTEM Immunological Test Systems; Immunology And Microbiology Devices
IMMUNOLOGICAL	866.5700	WHOLE HUMAN PLASMA OR SERUM IMMUNOLOGICAL TEST SYSTEM Immunological Test Systems; Immunology And Microbiology Devices
IMMUNOLOGICAL	866.5715	PLASMINOGEN IMMUNOLOGICAL TEST SYSTEM Immunological Test Systems; Immunology And Microbiology Devices
IMMUNOLOGICAL	866.5735	PROTHROMBIN IMMUNOLOGICAL TEST SYSTEM Immunological Test Systems; Immunology And Microbiology Devices
IMMUNOLOGICAL	866.5750	RADIOALLERGOSORBENT (RAST) IMMUNOLOGICAL TEST SYSTEM Immunological Test Systems; Immunology And Microbiology Devices
IMMUNOLOGICAL	866.5765	RETINOL-BINDING PROTEIN IMMUNOLOGICAL TEST SYSTEM Immunological Test Systems; Immunology And Microbiology Devices
IMMUNOLOGICAL	866.5775	RHEUMATOID FACTOR IMMUNOLOGICAL TEST SYSTEM Immunological Test Systems; Immunology And Microbiology Devices
IMMUNOLOGICAL	866.5800	SEMINAL FLUID (SPERM) IMMUNOLOGICAL TEST SYSTEM Immunological Test Systems; Immunology And Microbiology Devices
IMMUNOLOGICAL	866.5820	SYSTEMIC LUPUS ERYTHEMATOSUS IMMUNOLOGICAL TEST SYSTEM Immunological Test Systems; Immunology And Microbiology Devices
IMMUNOLOGICAL	866.5860	TOTAL SPINAL FLUID IMMUNOLOGICAL TEST SYSTEM Immunological Test Systems; Immunology And Microbiology Devices
IMMUNOLOGICAL	866.5870	THYROID AUTOANTIBODY IMMUNOLOGICAL TEST SYSTEM Immunological Test Systems; Immunology And Microbiology Devices
IMMUNOLOGICAL	866.5880	TRANSFERRIN IMMUNOLOGICAL TEST SYSTEM Immunological Test Systems; Immunology And Microbiology Devices
IMMUNOLOGICAL	866.5890	INTER-ALPHA TRYPSIN INHIBITOR IMMUNOLOGICAL TEST SYSTEM Immunological Test Systems; Immunology And Microbiology Devices
IMMUNOLOGICAL	866.6010	TUMOR ASSOCIATED ANTIGEN IMMUNOLOGICAL TEST SYSTEM Tumor Associated Antigen Immunological Test Systems; Immunology And Microbiology Devices
IMMUNONEPHELOMETER	866.4540	IMMUNONEPHELOMETER EQUIPMENT Immunology Laboratory Equipment And Reagents; Immunology And Microbiology Devices
IMMUNOREACTIVE INSULIN	862.1405	IMMUNOREACTIVE INSULIN TEST SYSTEM Clinical Chemistry Test Systems; Clinical Chemistry And Clinical Toxicology Devices
IMPACT	25.41	FINDINGS OF NO SIGNIFICANT IMPACT Preparation Of Environmental Documents; Environmental Impact Considerations
IMPACT	25.51	ENVIRONMENTAL ASSESSMENTS AND FINDINGS OF NO SIGNIFICANT IMPACT Public Participation And Notification Of Environmental Documents; Environmental Impact Considerations
IMPACT STATEMENTS	25.22	ACTIONS REQUIRING PREPARATION OF AN ENVIRONMENTAL IMPACT STATEMENT Agency Actions Requiring Environmental Consideration; Environmental Impact Considerations
IMPACT STATEMENTS	25.42	ENVIRONMENTAL IMPACT STATEMENTS Agency Decisionmaking; Environmental Impact Considerations
IMPACT STATEMENTS	25.52	ENVIRONMENTAL IMPACT STATEMENTS Public Participation And Notification Of Environmental Documents; Environmental Impact Considerations
IMPACT-RESISTANT LENSES	801.410	USE OF IMPACT-RESISTANT LENSES IN EYEGLASSES AND SUNGLASSES Special Requirements For Specific Devices; Labeling
IMPEDANCE	870.2750	IMPEDANCE PHLEBOGRAPH Cardiovascular Monitoring Devices; Cardiovascular Devices
IMPEDANCE	870.2770	IMPEDANCE PLETHYSMOGRAPH Cardiovascular Monitoring Devices; Cardiovascular Devices
IMPEDANCE	874.1090	AUDITORY IMPEDANCE TESTER Diagnostic Devices; Ear, Nose, And Throat Devices
IMPLANT	522.1350	MELATONIN IMPLANT Implantation Or Injectable Dosage Form New Animal Drugs
IMPLANT	872.3640	ENDOSSEOUS IMPLANT Prosthetic Devices; Dental Devices
IMPLANT	872.3970	INTERARTICULAR DISC PROSTHESIS (INTERPOSITIONAL IMPLANT) Prosthetic Devices; Dental Devices
IMPLANT	872.3980	ENDOSSEOUS DENTAL IMPLANT ACCESSORIES Prosthetic Devices; Dental Devices
IMPLANT	874.3695	MANDIBULAR IMPLANT FACIAL PROSTHESIS Prosthetic Devices; Ear, Nose And Throat Devices

IMPLANT	876.3350	PENILE INFLATABLE IMPLANT Prosthetic Devices; Gastroenterology-urology Devices
IMPLANT	876.3630	PENILE RIGIDITY IMPLANT Prosthetic Devices; Gastroenterology-urology Devices
IMPLANT	886.3300	ABSORBABLE IMPLANT (SCLERAL BUCKLING METHOD) Prosthetic Devices; Ophthalmic Devices
IMPLANT	886.3320	EYE SPHERE IMPLANT Prosthetic Devices; Ophthalmic Devices
IMPLANT	886.3340	EXTRAOCULAR ORBITAL IMPLANT Prosthetic Devices; Ophthalmic Devices
IMPLANT	1308.25	EXCLUSION OF A VETERINARY ANABOLIC STEROID IMPLANT PRODUCT; APPLICATION Schedules Of Controlled Substances
IMPLANT	1308.26	EXCLUDED VETERINARY ANABOLIC STEROID IMPLANT PRODUCTS Schedules Of Controlled Substances
IMPLANT MATERIAL	872.3645	SUBPERIOSTEAL IMPLANT MATERIAL Prosthetic Devices; Dental Devices
IMPLANT MATERIAL	878.3500	POLYTETRAFLUOROETHYLENE WITH CARBON FIBERS COMPOSITE IMPLANT MATERIAL Prosthetic Devices; General And Plastic Surgery Devices
IMPLANTABLE	870.3610	IMPLANTABLE PACEMAKER PULSE GENERATOR Cardiovascular Prosthetic Devices; Cardiovascular Devices
IMPLANTABLE CLIP	878.4300	IMPLANTABLE CLIP Surgical Devices; General And Plastic Surgery Devices
IMPLANTABLE STAPLE	878.4750	IMPLANTABLE STAPLE Surgical Devices; General And Plastic Surgery Devices
IMPLANTATION	522.90	AMPICILLIN IMPLANTATION AND INJECTABLE DOSAGE FORM Implantation Or Injectable Dosage Form New Animal Drugs
IMPLANTATION	522.960	FLUMETHASONE IMPLANTATION OR INJECTABLE DOSAGE FORMS Implantation Or Injectable Dosage Form New Animal Drug
IMPLANTATION	522.1662	OXYTETRACYCLINE HYDROCHLORIDE IMPLANTATION OR INJECTABLE DOSAGE FORMS Implantation Or Injectable Dosage Form New Animal Drugs
IMPLANTATION	522.1696	PENICILLIN G PROCAINE IMPLANTATION AND INJECTABLE DOSAGE FORM Implantation Or Injectable Dosage Form New Animal Drugs
IMPLANTATION	522.2444	SODIUM THIOPENTAL IMPLANTATION OR INJECTABLE DOSAGE FORMS Implantation Or Injectable Dosage Form New Animal Drug
IMPLANTATION	882.4545	SHUNT SYSTEM IMPLANTATION INSTRUMENT Neurological Surgical Devices; Neurological Devices
IMPLANTED	876.5270	IMPLANTED ELECTRICAL URINARY CONTINENCE DEVICE Therapeutic Devices; Gastroenterology-urology Devices
IMPLANTED	876.5280	IMPLANTED MECHANICAL/HYDRAULIC URINARY CONTINENCE DEVICE Therapeutic Devices; Gastroenterology-urology Devices
IMPLANTED	880.5965	SUBCUTANEOUS, IMPLANTED, INTRAVASCULAR INFUSION PORT AND CATHETER General Hospital And Personal Use Therapeutic Devices; General Hospital And Personal Use Devices
IMPLANTED	880.5970	PERCUTANEOUS, IMPLANTED, LONG-TERM INTRAVASCULAR CATHETER General Hospital And Personal Use Therapeutic Devices; General Hospital And Personal Use Devices
IMPLANTED	882.5225	IMPLANTED MALLEABLE CLIP Neurological Therapeutic Devices; Neurological Devices
IMPLANTED	882.5820	IMPLANTED CEREBELLAR STIMULATOR Neurological Therapeutic Devices; Neurological Devices
IMPLANTED	882.5830	IMPLANTED DIAPHRAGMATIC/PHRENIC NERVE STIMULATOR Neurological Therapeutic Devices; Neurological Devices
IMPLANTED	882.5840	IMPLANTED INTRACEREBRAL/SUBCORTICAL STIMULATOR FOR PAIN RELIEF Neurological Therapeutic Devices; Neurological Devices
IMPLANTED	882.5850	IMPLANTED SPINAL CORD STIMULATOR FOR BLADDER EVACUATION Neurological Therapeutic Devices; Neurological Devices
IMPLANTED	882.5860	IMPLANTED NEUROMUSCULAR STIMULATOR Neurological Therapeutic Devices; Neurological Devices
IMPLANTED	882.5870	IMPLANTED PERIPHERAL NERVE STIMULATOR FOR PAIN RELIEF Neurological Therapeutic Devices; Neurological Devices
IMPLANTED	882.5880	IMPLANTED SPINAL CORD STIMULATOR FOR PAIN RELIEF Neurological Therapeutic Devices; Neurological Devices
IMPLANTS	860.93	CLASSIFICATION OF IMPLANTS, LIFE-SUPPORTING OR LIFE-SUSTAINING DEVICES Classification; Medical Device Classification Procedures
IMPLEMENTATION NOTICES	310.6	APPLICABILITY OF “NEW DRUG” OR SAFETY/EFFECTIVENESS FINDINGS IN DRUG EFFICACY STUDY IMPLEMENTATION NOTICES AND NOTICES OF OPPORTUNITY FOR HEARING TO IDENTICAL, RELATED, AND SIMILAR DRUG PRODUCTS General Provisions; New Drugs

IMPORT	5.802	MANUFACTURER'S RESIDENT IMPORT AGENTS Imports And Exports; Redelegations Of Authority; Delegations Of Authority And Organization
IMPORT	312.110	IMPORT AND EXPORT REQUIREMENTS Miscellaneous; New Drugs For Investigational Use
IMPORT	807.40	ESTABLISHMENT REGISTRATION AND DEVICE LISTING FOR FOREIGN ESTABLISHMENTS IMPORTING OR OFFERING FOR IMPORT DEVICES INTO THE UNITED STATES Registration Procedures For Foreign Device Establishments; Establishment Registration And Device Listing For Manufacturers And Initial Importers Of Devices
IMPORT	812.18	IMPORT AND EXPORT REQUIREMENTS General Provisions; Investigational Device Exemptions
IMPORT	1270.42	HUMAN TISSUE OFFERED FOR IMPORT Inspection Of Tissue Establishments; Human Tissue Intended For Transplantation
IMPORT	1301.26	EXEMPTIONS FOR IMPORT OR EXPORT REQUIREMENTS FOR PERSONAL MEDICAL USE Registration Of Manufacturers, Distributors, And Dispensers Of Controlled Substances
IMPORT	1312.11	REQUIREMENT OF AUTHORIZATION TO IMPORT Importation Of Controlled Substances; Importation And Exportation Of Controlled Substances
IMPORT DECLARATION	1312.18	CONTENTS OF IMPORT DECLARATION Importation Of Controlled Substances; Importation And Exportation Of Controlled Substances
IMPORT DECLARATION	1312.19	DISTRIBUTION OF IMPORT DECLARATION Importation Of Controlled Substances; Importation And Exportation Of Controlled Substances
IMPORT, FOREIGN	1313.25	FOREIGN IMPORT RESTRICTIONS Importation Of Listed Chemicals; Importation And Exportation Of Precursors And Essential Chemicals
IMPORT MILK ACT	1210.27	PERMITS WAIVING CLAUSES 2 AND 5, SECTION 2 OF THE FEDERAL IMPORT MILK ACT Permit Control; Regulations Under The Federal Import Milk Act
IMPORT MILK ACT	1210.28	PERMITS WAIVING CLAUSE 4, SECTION 2 OF THE FEDERAL IMPORT MILK ACT Permit Control; Regulations Under The Federal Import Milk Act
IMPORT PERMIT	1312.12	APPLICATION FOR IMPORT PERMIT Importation Of Controlled Substances; Importation And Exportation Of Controlled Substances
IMPORT PERMIT	1312.13	ISSUANCE OF IMPORT PERMIT Importation Of Controlled Substances; Importation And Exportation Of Controlled Substances
IMPORT PERMIT	1312.14	DISTRIBUTION OF COPIES OF IMPORT PERMIT Importation Of Controlled Substances; Importation And Exportation Of Controlled Substances
IMPORT PERMIT	1312.30	SCHEDULE III, IV, AND V NON-NARCOTIC CONTROLLED SUBSTANCES REQUIRING AN IMPORT AND EXPORT PERMIT Exportation Of Controlled Substances; Importation And Exportation Of Controlled Substances
IMPORTANT INFORMATION	200.5	MAILING OF IMPORTANT INFORMATION ABOUT DRUGS General Provisions; General
IMPORTATION	601.15	FOREIGN ESTABLISHMENTS AND PRODUCTS: SAMPLES FOR EACH IMPORTATION Biologics Licensing; Licensing
IMPORTATION	1005.3	IMPORTATION OF NONCOMPLYING GOODS PROHIBITED Importation Of Electronic Products
IMPORTATION	1301.34	APPLICATION FOR IMPORTATION OF SCHEDULE I AND II SUBSTANCES Action On Application For Registration; Registration Of Manufacturers, Distributors, And Dispensers Of Controlled Substances
IMPORTATION	1313.31	ADVANCE NOTICE OF IMPORTATION FOR TRANSSHIPMENT OR TRANSFER Transshipment, In-Transit Shipments And International Transactions Involving Listed Chemicals; Importation And Exportation Of Precursors And Essential Chemicals
IMPORTED DEVICES	821.4	IMPORTED DEVICES Medical Device Tracking Requirements
IMPORTED PRODUCTS	120.14	APPLICATION OF REQUIREMENTS TO IMPORTED PRODUCTS Hazard Analysis And Critical Control Point (HACCP) Systems
IMPORTED PRODUCTS	123.12	SPECIAL REQUIREMENTS FOR IMPORTED PRODUCTS Fish And Fishery Products
IMPORTED SUBSTANCES	1302.07	LABELING AND PACKAGING REQUIREMENTS FOR IMPORTED AND EXPORTED SUBSTANCES Labeling And Packaging Requirements For Controlled Substances

IMPORTER	895.22	SUBMISSION OF DATA AND INFORMATION BY THE MANUFACTURER, DISTRIBUTOR, OR IMPORTER Banned Devices
IMPORTERS	803.10	GENERAL DESCRIPTION OF REPORTS REQUIRED FROM USER FACILITIES, IMPORTERS, AND MANUFACTURERS Medical Device Reporting
IMPORTERS	803.40	INDIVIDUAL ADVERSE EVENT REPORTING REQUIREMENTS; IMPORTERS Importer Reporting Requirement; Medical Device Reporting
IMPORTERS	1304.22	RECORDS FOR MANUFACTURERS, DISTRIBUTORS, DISPENSERS, RESEARCHERS, IMPORTERS AND EXPORTERS Continuing Records; Records And Reports Of Registrants
IMPORTERS	1312.17	SPECIAL REPORT FROM IMPORTERS Importation Of Controlled Substances; Importation And Exportation Of Controlled Substances
IMPORTERS	1313.15	WAIVER OF 15-DAY ADVANCE NOTICE FOR REGULAR IMPORTERS Importation Of Listed Chemicals; Importation And Exportation Of Precursors And Essential Chemicals
IMPORTING	807.40	ESTABLISHMENT REGISTRATION AND DEVICE LISTING FOR FOREIGN ESTABLISHMENTS IMPORTING OR OFFERING FOR IMPORT DEVICES INTO THE UNITED STATES Registration Procedures For Foreign Device Establishments; Establishment Registration And Device Listing For Manufacturers And Initial Importers Of Devices
IMPORTING	1304.31	REPORTS OF MANUFACTURERS IMPORTING NARCOTIC RAW MATERIAL Reports; Records And Reports Of Registrants
IMPORTING	1304.32	REPORTS OF MANUFACTURERS IMPORTING COCA LEAVES Reports; Records And Reports Of Registrants
IMPORTS	5.800	IMPORTS AND EXPORTS Imports And Exports; Redelegations Of Authority; Delegations Of Authority And Organization
IMPORTS	314.410	IMPORTS AND EXPORTS OF NEW DRUGS Miscellaneous Provisions; Applications For FDA Approval To Market A New Drug
IMPORTS, INADMISSABLE	1.99	COSTS CHARGEABLE IN CONNECTION WITH RELABELING AND RECONDITIONING INADMISSABLE IMPORTS Imports And Exports; General Enforcement Regulation
IMPREGNATED SPONGE	529.1003	FLUROGESTONE ACETATE-IMPREGNATED VAGINAL SPONGE Certain Other Dosage Form New Animal Drugs
IMPRESSION	872.3660	IMPRESSION MATERIAL Prosthetic Devices; Dental Devices
IMPRESSION	872.3670	RESIN IMPRESSION TRAY MATERIAL Prosthetic Devices; Dental Devices
IMPRESSION	872.6570	IMPRESSION TUBE Miscellaneous Devices; Dental Devices
IMPRESSION	872.6880	PREFORMED IMPRESSION TRAY Miscellaneous Devices; Dental Devices
IMPRINT, CODE	206.10	CODE IMPRINT REQUIRED Imprinting Of Solid Oral Dosage Form Drug Products For Human Use
IMPRINTING	330.3	IMPRINTING OF SOLID ORAL DOSAGE FORM DRUG PRODUCTS Over-the-counter Human Drugs Generally Recognized As Safe And Effective & Not Misbranded
IN-LINE SENSOR	870.4410	CARDIOPULMONARY BYPASS IN-LINE BLOOD GAS SENSOR Cardiovascular Surgical Devices; Cardiovascular Devices
IN-PROCESS ACCEPTANCE	820.80	RECEIVING, IN-PROCESS, AND FINISHED DEVICE ACCEPTANCE Acceptance Activities; Quality System Regulation
IN-PROCESS CONTROL	106.25	IN-PROCESS CONTROL Quality Control Procedures For Assuring Nutrient Content Of Infant Formulas; Infant Formula Quality Control Procedures
IN-PROCESS MATERIALS	211.110	SAMPLING AND TESTING OF IN-PROCESS MATERIALS AND DRUG PRODUCTS Production And Process Controls; Current Good Manufacturing Practice For Finished Pharmaceuticals
IN VITRO DIAGNOSTIC	201.119	IN VITRO DIAGNOSTIC PRODUCTS Exemptions From Adequate Directions For Use; Labeling
IN VITRO DIAGNOSTIC	801.119	IN VITRO DIAGNOSTIC PRODUCTS Exemptions From Adequate Directions For Use; Labeling
IN VITRO DIAGNOSTIC	809.10	LABELING FOR IN VITRO DIAGNOSTIC PRODUCTS Labeling; In Vitro Diagnostic Products For Human Use
IN VITRO DIAGNOSTIC	809.20	GENERAL REQUIREMENTS FOR MANUFACTURERS AND PRODUCERS OF IN VITRO DIAGNOSTIC PRODUCTS Requirements For Manufacturers And Producers; In Vitro Diagnostic Products For Human Use
IN VITRO DIAGNOSTIC	864.9160	BLOOD GROUP SUBSTANCES OF NONHUMAN ORIGIN FOR IN VITRO DIAGNOSTIC USE Products Used In Establishments That Manufacture Blood And Blood Products; Hematology And Pathology Devices

IN VITRO DIAGNOSTIC	864.9225	CELL-FREEZING APPARATUS AND REAGENTS FOR IN VITRO DIAGNOSTIC USE Products Used In Establishments That Manufacture Blood And Blood Products; Hematology And Pathology Devices
IN VITRO DIAGNOSTIC	864.9275	BLOOD BANK CENTRIFUGE FOR IN VITRO DIAGNOSTIC USE Products Used In Establishments That Manufacture Blood And Blood Products; Hematology And Pathology Devices
IN VITRO DIAGNOSTIC	864.9600	POTENTIATING MEDIA FOR IN VITRO DIAGNOSTIC USE Products Used In Establishments That Manufacture Blood And Blood Products; Hematology And Pathology Devices
IN VITRO STUDIES	864.5425	MULTIPURPOSE SYSTEM FOR IN VITRO COAGULATION STUDIES Automated And Semi-automated Hematology Devices; Hematology And Pathology Devices
IN VITRO TESTS	312.160	DRUGS FOR INVESTIGATIONAL USE IN LABORATORY RESEARCH ANIMALS OR IN VITRO TESTS Investigational New Drug Application
IN VITRO TESTING	320.35	REQUIREMENTS FOR IN VITRO TESTING OF EACH BATCH Bioequivalence Requirements; Bioavailability And Bioequivalence Requirements
IN VIVO BIOAVAILABILITY	320.21	REQUIREMENTS FOR SUBMISSION OF IN VIVO BIOAVAILABILITY AND BIOEQUIVALENCE DATA Procedures For Determining The Bioavailability Of Drug Products; Bioavailability And Bioequivalence Requirements
IN VIVO BIOAVAILABILITY	320.22	CRITERIA FOR WAIVER OF EVIDENCE OF IN VIVO BIOAVAILABILITY OR BIOEQUIVALENCE Procedures For Determining The Bioavailability Of Drug Products; Bioavailability And Bioequivalence Requirements
IN VIVO BIOAVAILABILITY	320.23	BASIS FOR DEMONSTRATING IN VIVO BIOAVAILABILITY OR BIOEQUIVALENCE Procedures For Determining The Bioavailability Of Drug Products; Bioavailability And Bioequivalence Requirements
IN VIVO BIOAVAILABILITY	320.25	GUIDELINES FOR THE CONDUCT OF AN IN VIVO BIOAVAILABILITY STUDY Procedures For Determining The Bioavailability Of Drug Products; Bioavailability And Bioequivalence Requirements
IN VIVO BIOAVAILABILITY	320.26	GUIDELINES ON THE DESIGN OF A SINGLE-DOSE IN VIVO BIOAVAILABILITY STUDY Procedures For Determining The Bioavailability Of Drug Products; Bioavailability And Bioequivalence Requirements
IN VIVO BIOAVAILABILITY	320.27	GUIDELINES ON THE DESIGN OF A MULTIPLE-DOSE IN VIVO BIOAVAILABILITY STUDY Procedures For Determining The Bioavailability Of Drug Products; Bioavailability And Bioequivalence Requirements
IN VIVO BIOAVAILABILITY	320.29	ANALYTICAL METHODS FOR AN IN VIVO BIOAVAILABILITY STUDY Procedures For Determining The Bioavailability Of Drug Products; Bioavailability And Bioequivalence Requirements
INACTIVATED	610.11a	INACTIVATED INFLUENZA VACCINE, GENERAL SAFETY TEST General Provisions; General Biological Products Standards
INACTIVATOR	866.5250	COMPLEMENT C1 INHIBITOR (INACTIVATOR) IMMUNOLOGICAL TEST SYSTEM Immunological Test Systems; Immunology And Microbiology Devices
INACTIVATOR	866.5260	COMPLEMENT C3B INACTIVATOR IMMUNOLOGICAL TEST SYSTEM Immunological Test Systems; Immunology And Microbiology Devices
INACTIVE INGREDIENTS	201.117	INACTIVE INGREDIENTS Exemptions From Adequate Directions For Use; Labeling
INACTIVE STATUS	312.45	INACTIVE STATUS Investigational New Drug Application
INADEQUATE CONTROL	1240.30	MEASURES IN THE EVENT OF INADEQUATE LOCAL CONTROL Control Of Communicable Diseases
INADMISSABLE IMPORTS	1.99	COSTS CHARGEABLE IN CONNECTION WITH RELABELING AND RECONDITIONING INADMISSABLE IMPORTS Imports And Exports; General Enforcement Regulations
INCENTIVE	868.5690	INCENTIVE SPIROMETER Therapeutic Devices; Anesthesiology Devices
INCOME	1403.25	PROGRAM INCOME Financial Administration; Uniform Administrative Requirements For Grants And Coop. Agreements To State & Local Gov.
INCONTINENCE	876.5920	PROTECTIVE GARMENT FOR INCONTINENCE Therapeutic Devices; Gastroenterology-urology Devices
INCREMENT	874.1070	SHORT INCREMENT SENSITIVITY INDEX (SISI) ADAPTER Diagnostic Devices; Ear, Nose, And Throat Devices
INCUBATOR	866.2540	MICROBIOLOGICAL INCUBATOR Microbiology Devices; Immunology And Microbiology Devices
INCUBATOR	880.5400	NEONATAL INCUBATOR General Hospital And Personal Use Therapeutic Devices; General Hospital And Personal Use Devices

INDOLO-CARBAZOLE-HEXONE	73.3117	16,23-DIHYDRODINAPHTHOL[2,3-A:2',3'-I]NAPHTH [2',3':6,7] INDOLO [2,3-C] CARBAZOLE-5,10,15,17,22,24-HEXONE Medical Devices; Listing Of Color Additives Exempt From Certification
INDUSTRIAL	178.3520	INDUSTRIAL STARCH-MODIFIED Certain Adjuvants And Production Aids; Indirect Food Additives: Adjuvants, Production Aids, And Sanitizers
INDUSTRY	7.59	GENERAL INDUSTRY GUIDANCE Recalls - Guidance On Policy, Procedures & Industry Responsibilities; Enforcement Policy
INDWELLING	868.1120	INDWELLING BLOOD OXYHEMOGLOBIN CONCENTRATION ANALYZER Diagnostic Devices; Anesthesiology Devices
INDWELLING	868.1150	INDWELLING BLOOD CARBON DIOXIDE PARTIAL PRESSURE (PCO2) ANALYZER Diagnostic Devices; Anesthesiology Devices
INDWELLING	868.1170	INDWELLING BLOOD HYDROGEN ION CONCENTRATION (pH) ANALYZER Diagnostic Devices; Anesthesiology Devices
INDWELLING	868.1200	INDWELLING BLOOD OXYGEN PARTIAL PRESSURE (PO2) ANALYZER Diagnostic Devices; Anesthesiology Devices
INELIGIBLE	1404.205	INELIGIBLE PERSONS Governmentwide Debarment And Suspension And Requirements For Drug-Free Workplace (Grants)
INFANT	201.19	DRUGS; USE OF TERM "INFANT" General Labeling Provisions; Labeling
INFANT	880.5130	INFANT RADIANT WARMER General Hospital And Personal Use Therapeutic Devices; General Hospital And Personal Use Devices
INFANT FOODS	105.65	INFANT FOODS Label Statements; Foods For Special Dietary Use
INFANT FORMULA	107.220	SCOPE AND EFFECT OF INFANT FORMULA RECALLS Infant Formula Recalls; Infant Formula
INFANT FORMULA	107.230	ELEMENTS OF AN INFANT FORMULA RECALL Infant Formula Recalls; Infant Formula
INFANT FORMULA	107.250	TERMINATION OF AN INFANT FORMULA RECALL Infant Formula Recalls; Infant Formula
INFANT FORMULA	107.260	REVISION OF AN INFANT FORMULA RECALL Infant Formula Recalls; Infant Formula
INFECTIOUS	866.5640	INFECTIOUS MONONUCLEOSIS IMMUNOLOGICAL TEST SYSTEM Immunological Test Systems; Immunology And Microbiology Devices
INFLATABLE CUFF	868.5750	INFLATABLE TRACHEAL TUBE CUFF Therapeutic Devices; Anesthesiology Devices
INFLATABLE IMPLANT	876.3350	PENILE INFLATABLE IMPLANT Prosthetic Devices; Gastroenterology-urology Devices
INFLATABLE MASSAGER	890.5650	POWERED INFLATABLE TUBE MASSAGER Physical Medicine Therapeutic Devices; Physical Medicine Devices
INFLATABLE PROSTHESIS	878.3540	SILICONE INFLATABLE BREAST PROSTHESIS Prosthetic Devices; General And Plastic Surgery Devices
INFLATABLE SPLINT	878.3900	INFLATABLE EXTREMITY SPLINT Prosthetic Devices; General And Plastic Surgery Devices
INFLUENZA VACCINE	610.11	INACTIVATED INFLUENZA VACCINE, GENERAL SAFETY TEST General Provisions; General Biological Products Standards
INFLUENZA VIRUS	866.3330	INFLUENZA VIRUS SEROLOGICAL REAGENTS Serological Reagents; Immunology And Microbiology Devices
INFORMAL HEARINGS	5.27	PATENT TERM EXTENSIONS FOR HUMAN DRUG PRODUCTS, MEDICAL DEVICES, AND FOOD AND COLOR ADDITIVES; AND AUTHORITY TO PERFORM DUE DILIGENCE DETERMINATIONS AND INFORMAL HEARINGS General Redelegations Of Authority; Delegations Of Authority And Organization
INFORMAL STATEMENTS	369.4	WARNINGS SUGGESTED FOR DRUGS BY FORMAL OR INFORMAL STATEMENTS OF POLICY Definitions And Interpretations; Interpretative Statements Re. Warnings On Drugs And Devices For OTC Sale
INFORMATION	1.20	PRESENCE OF MANDATORY LABEL INFORMATION General Labeling Requirements; General Enforcement Regulations
INFORMATION	20.44	PRESUBMISSION REVIEW OF REQUEST FOR CONFIDENTIALITY OF VOLUNTARILY SUBMITTED DATA OR INFORMATION Procedures And Fees; Public Information
INFORMATION	20.50	AVAILABILITY OF RECORDS AT NATIONAL TECHNICAL INFORMATION SERVICE Procedures And Fees; Public Information
INFORMATION	20.53	INDEXING TRADE SECRETS AND CONFIDENTIAL COMMERCIAL OR FINANCIAL INFORMATION Procedures And Fees; Public Information
INFORMATION	20.64	RECORDS OR INFORMATION COMPILED FOR LAW ENFORCEMENT PURPOSES Exemptions; Public Information

INFORMATION	1301.93	SOURCES OF INFORMATION FOR EMPLOYEE CHECKS Registration Of Manufacturers, Distributors, And Dispensers Of Controlled Substances
INFORMATION	1309.03	INFORMATION: SPECIAL INSTRUCTIONS Registration Of Manufacturers, Distributors, Importers And Exporters Of List I Chemicals
INFORMATION	1309.35	ADDITIONAL INFORMATION Registration Of Manufacturers, Distributors, Importers And Exporters Of List I Chemicals
INFORMATION	1316.43	INFORMATION; SPECIAL INSTRUCTIONS Administrative Hearings; Administrative Functions, Practices, And Procedures
INFORMATION	1401.4	ACCESS TO INFORMATION Public Availability Of Information
INFORMATION	1402.3	INFORMATION IN THE CUSTODY OF ONDCP Mandatory Declassification Review
INFORMATION	1402.4	INFORMATION CLASSIFIED BY ANOTHER AGENCY Mandatory Declassification Review
INFORMATION, BLOOD PRODUCT	607.25	INFORMATION REQUIRED FOR ESTABLISHMENT REGISTRATION AND BLOOD PRODUCT LISTING Procedures For Domestic Blood Product Establishments; Establishment Registratn+product Listing, Manufacturers Human Blood/b.products
INFORMATION, BLOOD PRODUCT	607.30	UPDATING BLOOD PRODUCT LISTING INFORMATION Procedures For Domestic Blood Product Establishments; Establishment Registratn+product Listing, Manufacturers Human Blood/b.products
INFORMATION, BLOOD PRODUCT	607.31	ADDITIONAL BLOOD PRODUCT LISTING INFORMATION Procedures For Domestic Blood Product Establishments; Establishment Registratn+product Listing, Manufacturers Human Blood/b.products
INFORMATION, COMMERCIAL/FINANCIAL	20.53	INDEXING TRADE SECRETS AND CONFIDENTIAL COMMERCIAL OR FINANCIAL INFORMATION Procedures And Fees; Public Information
INFORMATION, COMMERCIAL/FINANCIAL	20.61	TRADE SECRETS AND COMMERCIAL OR FINANCIAL INFORMATION WHICH IS PRIVILEGED OR CONFIDENTIAL Exemptions; Public Information
INFORMATION, CONFIDENTIALITY	20.44	PRESUBMISSION REVIEW OF REQUEST FOR CONFIDENTIALITY OF VOLUNTARILY SUBMITTED DATA OR INFORMATION Procedures And Fees; Public Information
INFORMATION, CONFIDENTIALITY	71.15	CONFIDENTIALITY OF DATA AND INFORMATION IN COLOR ADDITIVE PETITIONS General Provisions; Color Additive Petitions
INFORMATION, CONFIDENTIALITY	514.11	CONFIDENTIALITY OF DATA AND INFORMATION IN A NEW ANIMAL DRUG APPLICATION FILE General Provisions; New Animal Drug Applications
INFORMATION, CONFIDENTIALITY	514.12	CONFIDENTIALITY OF DATA AND INFORMATION IN AN INVESTIGATIONAL NEW ANIMAL DRUG NOTICE General Provisions; New Animal Drug Applications
INFORMATION, CONFIDENTIALITY	601.50	CONFIDENTIALITY OF DATA AND INFORMATION IN AN INVESTIGATIONAL NEW DRUG NOTICE FOR A BIOLOGICAL PRODUCT Confidentiality Of Information; Licensing
INFORMATION, CONFIDENTIALITY	601.51	CONFIDENTIALITY OF DATA AND INFORMATION IN APPLICATIONS FOR ESTABLISHMENT AND PRODUCT LICENSES Confidentiality Of Information; Licensing
INFORMATION, CONFIDENTIALITY	807.05	CONFIDENTIALITY OF INFORMATION Premarket Notification Procedures; Establishment Registration And Device Listing For Manufacturers And Initial Importers Of Devices
INFORMATION, CONFIDENTIALITY	809.4	CONFIDENTIALITY OF SUBMITTED INFORMATION General Provisions; In Vitro Diagnostic Products For Human Use
INFORMATION, CONFIDENTIALITY	814.9	CONFIDENTIALITY OF DATA AND INFORMATION IN A PREMARKET APPROVAL APPLICATION (PMA) FILE General; Premarket Approval Of Medical Devices
INFORMATION, CONFIDENTIALITY	812.38	CONFIDENTIALITY OF DATA AND INFORMATION Application And Administrative Action; Investigational Device Exemptions
INFORMATION, CONFIDENTIALITY	860.5	CONFIDENTIALITY AND USE OF DATA AND INFORMATION SUBMITTED IN CONNECTION WITH CLASSIFICATION AND RECLASSIFICATION General; Medical Device Classification Procedures
INFORMATION, CONFIDENTIALITY	1002.4	CONFIDENTIALITY OF INFORMATION General Provisions; Records and Reports
INFORMATION, COSMETIC PRODUCTS	720.4	INFORMATION REQUESTED ABOUT COSMETIC PRODUCTS Voluntary Filing Of Cosmetic Product Ingredient Composition Statements
INFORMATION, DEVICE LISTING	20.116	DRUG AND DEVICE LISTING INFORMATION Availability Of Specific Categories Of Records; Public Information
INFORMATION, DEVICE LISTING	807.25	INFORMATION REQUIRED OR REQUESTED FOR ESTABLISHMENT REGISTRATION AND DEVICE LISTING Procedures For Device Establishments; Establishment Registration And Device Listing For Manufacturers And Initial Importers Of Devices

INFORMATION, DEVICE LISTING	807.30	UPDATING DEVICE LISTING INFORMATION Procedures For Device Establishments; Establishment Registration And Device Listing For Manufacturers And Initial Importers Of Devices
INFORMATION, DISCLOSED	20.81	DATA AND INFORMATION PREVIOUSLY DISCLOSED TO THE PUBLIC Limitations On Exemptions; Public Information
INFORMATION, DISCLOSURE	12.85	DISCLOSURE OF DATA AND INFORMATION BY THE PARTICIPANTS Hearing Procedures; Formal Evidentiary Public Hearing
INFORMATION, DISCLOSURE	13.25	DISCLOSURE OF DATA AND INFORMATION BY THE PARTICIPANTS Hearing Procedures; Public Hearing Before A Public Board Of Inquiry
INFORMATION, DISCLOSURE	56.122	PUBLIC DISCLOSURE OF INFORMATION REGARDING REVOCATION Administrative Action For Noncompliance; Institutional Review Boards
INFORMATION, DISCLOSURE	58.213	PUBLIC DISCLOSURE OF INFORMATION REGARDING DISQUALIFICATION Disqualification Of Testing Facilities; Good Laboratory Practice For Nonclinical Laboratory Studies
INFORMATION, DISCLOSURE	312.130	AVAILABILITY FOR PUBLIC DISCLOSURE OF DATA AND INFORMATION IN AN IND Investigational New Drug Application
INFORMATION, DISCLOSURE	314.430	AVAILABILITY FOR PUBLIC DISCLOSURE OF DATA AND INFORMATION IN AN APPLICATION Miscellaneous Provisions; Applications For FDA Approval To Market A New Drug
INFORMATION, DISCLOSURE	316.52	AVAILABILITY FOR PUBLIC DISCLOSURE OF DATA AND INFORMATION IN REQUESTS AND APPLICATIONS Orphan Drugs
INFORMATION, DOSAGE	200.7	SUPPLYING PHARMACISTS WITH INDICATIONS AND DOSAGE INFORMATION General Provisions; General
INFORMATION, DRUG LISTING	20.116	DRUG AND DEVICE LISTING INFORMATION Availability Of Specific Categories Of Records; Public Information
INFORMATION, DRUG LISTING	207.25	INFORMATION REQUIRED IN REGISTRATION AND DRUG LISTING Procedures For Domestic Drug Establishments; Registration Of Drug Producers And Listing Of Drugs In Commercial Distribution
INFORMATION, DRUG LISTING	207.30	UPDATING DRUG LISTING INFORMATION Procedures For Domestic Drug Establishments; Registration Of Drug Producers And Listing Of Drugs In Commercial Distribution
INFORMATION, DRUG LISTING	207.31	ADDITIONAL DRUG LISTING INFORMATION Procedures For Domestic Drug Establishments; Registration Of Drug Producers And Listing Of Drugs In Commercial Distribution
INFORMATION, FREEDOM OF	20.30	FOOD AND DRUG ADMINISTRATION FREEDOM OF INFORMATION STAFF General Policy; Public Information
INFORMATION, HEALTH	5.25	RESEARCH, INVESTIGATION, AND TESTING PROGRAMS AND HEALTH INFORMATION AND PROMOTION PROGRAMS General Redelegations Of Authority; Delegations Of Authority And Organization
INFORMATION, IMPORTANT	200.5	MAILING OF IMPORTANT INFORMATION ABOUT DRUGS General Provisions; General
INFORMATION, LABEL	1.20	PRESENCE OF MANDATORY LABEL INFORMATION General Labeling Requirements; General Enforcement Regulations
INFORMATION, NUTRIENT	107.10	NUTRIENT INFORMATION Labeling; Infant Formula
INFORMATION OFFICES	5.1110	FDA PUBLIC INFORMATION OFFICES Organization; Delegations Of Authority And Organization
INFORMATION PANEL	101.2	INFORMATION PANEL OF PACKAGE FORM FOOD General Provisions; Food Labeling
INFORMATION PANEL	501.2	INFORMATION PANEL OF PACKAGE FOR ANIMAL FOOD General Provisions; Animal Food Labeling
INFORMATION, PATENT	5.109	ISSUANCE OF WRITTEN NOTICES CONCERNING PATENT INFORMATION, CURRENT GOOD MANUFACTURING PRACTICES AND FALSE OR MISLEADING LABELING OF NEW DRUGS Human Drugs; Redelegations Of Authority; Delegations Of Authority And Organization
INFORMATION, PATENT	5.504	ISSUANCE OF WRITTEN NOTICES CONCERNING PATENT INFORMATION, CURRENT GOOD MANUFACTURING PRACTICES AND FALSE OR MISLEADING LABELING OF NEW ANIMAL DRUGS AND FEEDS BEARING OR CONTAINING NEW ANIMAL DRUGS Animal Drugs; Redelegations Of Authority; Delegations Of Authority And Organization
INFORMATION, PATENT	314.53	SUBMISSION OF PATENT INFORMATION Applications For FDA Approval To Market A New Drug
INFORMATION, PEDIATRIC	314.55	PEDIATRIC USE INFORMATION Applications For FDA Approval To Market A New Drug
INFORMATION, PREMARKET	807.87	INFORMATION REQUIRED IN A PREMARKET NOTIFICATION SUBMISSION Premarket Notification Procedures; Establishment Registration And Device Listing For Manufacturers And Initial Importers Of Devices
INFORMATION, PREMARKET	814.9	CONFIDENTIALITY OF DATA AND INFORMATION IN A PREMARKET APPROVAL APPLICATION (PMA) FILE General; Premarket Approval Of Medical Devices

INFORMATION, PRESCRIPTION	1306.25	TRANSFER BETWEEN PHARMACIES OF PRESCRIPTION INFORMATION FOR SCHEDULES III, IV, AND V CONTROLLED SUBSTANCES FOR REFILL PURPOSES Controlled Substances Listed In Schedules III, IV, AND V; Prescriptions
INFORMATION, PRICE	200.200	PRESCRIPTION DRUGS; REMINDER ADVERTISEMENTS AND REMINDER LABELING TO PROVIDE PRICE INFORMATION TO CONSUMERS Prescription Drug Consumer Price Listing; General
INFORMATION, QUALITY	26.19	INFORMATION RELATING TO QUALITY ASPECTS Specific Sector Provisions For Pharmaceutical GMPs; Mutual Recognition Of Pharmaceutical GMP Reports, Medical Device Quality System Audit Reports
INFORMATION, REGISTRATION	207.25	INFORMATION REQUIRED IN REGISTRATION AND DRUG LISTING Procedures For Domestic Drug Establishments; Registration Of Drug Producers And Listing Of Drugs In Commercial Distribution
INFORMATION, REGISTRATION	807.25	INFORMATION REQUIRED OR REQUESTED FOR ESTABLISHMENT REGISTRATION AND DEVICE LISTING Procedures For Device Establishments; Establishment Registration And Device Listing For Manufacturers And Initial Importers Of Devices
INFORMED CONSENT	50.20	GENERAL REQUIREMENTS FOR INFORMED CONSENT Informed Consent Of Human Subjects; Protection Of Human Subjects
INFORMED CONSENT	50.24	EXCEPTION FROM INFORMED CONSENT REQUIREMENTS FOR EMERGENCY RESEARCH Informed Consent Of Human Subjects; Protection Of Human Subjects
INFORMED CONSENT	50.25	ELEMENTS OF INFORMED CONSENT Informed Consent Of Human Subjects; Protection Of Human Subjects
INFORMED CONSENT	50.27	DOCUMENTATION OF INFORMED CONSENT Informed Consent Of Human Subjects; Protection Of Human Subjects
INFORMED CONSENT	640.61	INFORMED CONSENT Source Plasma; Additional Standards For Human Blood And Blood Products
INFORMING	312.55	INFORMING INVESTIGATORS Investigational New Drug Application
INFORMING	812.45	INFORMING INVESTIGATORS Responsibilities Of Sponsors; Investigational Device Exemptions
INFRARED LAMP	890.5500	INFRARED LAMP Physical Medicine Therapeutic Devices; Physical Medicine Devices
INFUSION	526.88	AMOXICILLIN TRIHYDRATE FOR INTRAMAMMARY INFUSION Intramammary Dosage Forms
INFUSION	526.365	CEPHAPIRIN SODIUM FOR INTRAMAMMARY INFUSION Certain Other Dosage Form New Animal Drugs
INFUSION	526.464a	CLOXACILLIN BENZATHINE FOR INTRAMAMMARY INFUSION Intramammary Dosage Forms
INFUSION	526.464b	CLOXACILLIN BENZATHINE FOR INTRAMAMMARY INFUSION, STERILE Intramammary Dosage Forms
INFUSION	526.464c	CLOXACILLIN SODIUM FOR INTRAMAMMARY INFUSION, STERILE Intramammary Dosage Forms
INFUSION	526.464d	CLOXACILLIN SODIUM FOR INTRAMAMMARY INFUSION Intramammary Dosage Forms
INFUSION	526.1130	HETACILLIN POTASSIUM FOR INTRAMAMMARY INFUSION Intramammary Dosage Forms
INFUSION	526.1696b	PENICILLIN G PROCAINE-DIHYDROSTREPTOMYCIN IN SOYBEAN OIL FOR INTRAMAMMARY INFUSION (DRY COWS) Intramammary Dosage Forms
INFUSION	526.1696c	PENICILLIN G PROCAINE-DIHYDROSTREPTOMYCIN SULFATE FOR INTRAMAMMARY INFUSION (DRY COWS) Intramammary Dosage Forms
INFUSION	526.1696d	PENICILLIN G PROCAINE-NOVOBIOCIN FOR INTRAMAMMARY INFUSION Intramammary Dosage Forms
INFUSION	880.2420	ELECTRONIC MONITOR FOR GRAVITY FLOW INFUSION SYSTEMS General Hospital And Personal Use Monitoring Devices; General Hospital And Personal Use Devices
INFUSION PORT	880.5965	SUBCUTANEOUS, IMPLANTED, INTRAVASCULAR INFUSION PORT AND CATHETER General Hospital And Personal Use Therapeutic Devices; General Hospital And Personal Use Devices
INFUSION PUMP	870.1800	WITHDRAWAL-INFUSION PUMP Cardiovascular Diagnostic Devices; Cardiovascular Devices
INFUSION PUMP	880.5725	INFUSION PUMP General Hospital And Personal Use Therapeutic Devices; General Hospital And Personal Use Devices
INFUSION STAND	880.6990	INFUSION STAND General Hospital And Personal Use Miscellaneous Devices; General Hospital And Personal Use Devices
INFUSOR	880.5420	PRESSURE INFUSOR FOR AN I. V. BAG General Hospital And Personal Use Therapeutic Devices; General Hospital And Personal Use Devices

INGESTIBLE	876.1300	INGESTIBLE TELEMETRIC GASTROINTESTINAL CAPSULE IMAGING SYSTEM Diagnostic Devices; Gastroenterology-urology Devices
INGESTION, ACCIDENTAL	369.9	GENERAL WARNINGS RE ACCIDENTAL INGESTION BY CHILDREN Definitions And Interpretations; Interpretative Statements Re Warnings On Drugs And Devices For OTC Sale
INGESTION, ORAL	201.306	POTASSIUM SALT PREPARATIONS INTENDED FOR ORAL INGESTION BY MAN Specific Labeling Requirements For Specific Drug Products; Labeling
INGESTION, ORAL	328.50	PRINCIPAL DISPLAY PANEL OF ALL OTC DRUG PRODUCTS INTENDED FOR ORAL INGESTION THAT CONTAIN ALCOHOL Over-the-counter Drug Products Intended For Oral Ingestion That Contain Alcohol
INGREDIENT	106.20	INGREDIENT CONTROL Quality Control Procedures For Assuring Nutrient Content Of Infant Formulas; Infant Formula Quality Control Procedures
INGREDIENT	700.14	USE OF VINYL CHLORIDE AS AN INGREDIENT, INCLUDING PROPELLANT OF COSMETIC AEROSOL PRODUCTS Requirements For Specific Cosmetic Products; General
INGREDIENT	700.18	USE OF CHLOROFORM AS AN INGREDIENT IN COSMETIC PRODUCTS Requirements For Specific Cosmetic Products; General
INGREDIENT	700.19	USE OF METHYLENE CHLORIDE AS AN INGREDIENT OF COSMETIC PRODUCTS Requirements For Specific Cosmetic Products; General
INGREDIENT	701.30	INGREDIENT NAMES ESTABLISHED FOR COSMETIC INGREDIENT LABELING Labeling Of Specific Ingredients; Cosmetic Labeling
INGREDIENT	720.7	NOTIFICATION OF PERSON SUBMITTING COSMETIC PRODUCT INGREDIENT STATEMENT Voluntary Filing Of Cosmetic Product Ingredient Composition Statements
INGREDIENT, ACTIVE	340.10	STIMULANT ACTIVE INGREDIENT Stimulant Drug Products For Over-the-counter Human Use
INGREDIENT, ACTIVE	341.18	EXPECTORANT ACTIVE INGREDIENT Cold, Cough, Allergy, Bronchodilator, And Antiasthmatic Drug Products For OTC Human Use
INGREDIENT, ACTIVE	344.10	EARWAX REMOVAL AID ACTIVE INGREDIENT Topical Otic Drug Products For Over-the-counter Human Use
INGREDIENT, ACTIVE	344.12	EAR DRYING AID ACTIVE INGREDIENT Topical Otic Drug Products For Over-the-counter Human Use
INGREDIENT, ACTIVE	357.110	ANTHELMINTIC ACTIVE INGREDIENT Miscellaneous Internal Drug Products For Over-the-counter Human Use
INGREDIENT, ACTIVE	1310.14	EXEMPTION OF DRUG PRODUCTS CONTAINING EPHEDRINE OR ANOTHER ACTIVE MEDICINAL INGREDIENT Records And Reports Of Listed Chemicals And Certain Machines
INGREDIENT, ACTIVE	1310.15	EXEMPT DRUG PRODUCTS CONTAINING EPHEDRINE AND ANOTHER ACTIVE MEDICINAL INGREDIENT Records And Reports Of Listed Chemicals And Certain Machines
INGREDIENTS	101.4	FOOD; DESIGNATION OF INGREDIENTS General Provisions; Food Labeling
INGREDIENTS	130.11	LABEL DESIGNATIONS OF INGREDIENTS FOR STANDARDIZED FOOD Food Standards: General
INGREDIENTS	330.13	CONDITIONS FOR MARKETING INGREDIENTS RECOMMENDED FOR OVER-THE-COUNTER (OTC) USE UNDER THE OTC DRUG REVIEW Administrative Procedures; Over-the-counter Human Drugs Generally Recognized As Safe And Effective & Not Misbranded
INGREDIENTS	341.70	LABELING OF OTC DRUG PRODUCTS CONTAINING INGREDIENTS THAT ARE USED FOR TREATING CONCURRENT SYMPTOMS Cold, Cough, Allergy, Bronchodilator, And Antiasthmatic Drug Products For OTC Human Use
INGREDIENTS	501.4	ANIMAL FOOD; DESIGNATION OF INGREDIENTS General Provisions; Animal Food Labeling
INGREDIENTS	501.103	PETITIONS REQUESTING EXEMPTIONS FROM OR SPECIAL REQUIREMENTS FOR LABEL DECLARATION OF INGREDIENTS Exemptions From Animal Food Labeling Requirements; Animal Food Labeling
INGREDIENTS	501.110	ANIMAL FEED LABELING; COLLECTIVE NAMES FOR FEED INGREDIENTS Exemptions From Animal Food Labeling Requirements; Animal Food Labeling
INGREDIENTS	700.15	USE OF CERTAIN HALOGENATED SALICYLANILIDES AS INGREDIENTS IN COSMETIC PRODUCTS Requirements For Specific Cosmetic Products; General
INGREDIENTS	701.3	DESIGNATION OF INGREDIENTS General Provisions; Cosmetic Labeling
INGREDIENTS, ACTIVE	201.319	WATER-SOLUBLE GUMS, AND HYDROPHILIC MUCILLOIDS AS ACTIVE INGREDIENTS; WARNINGS AND DIRECTIONS Specific Labeling Requirements For Specific Drug Products; Labeling
INGREDIENTS, ACTIVE	201.322	OVER-THE COUNTER DRUG PRODUCTS CONTAINING INTERNAL ANALGESIC/ANTIPYRETIC ACTIVE INGREDIENTS Specific Labeling Requirements For Specific Drug Products; Labeling

INGREDIENTS, ACTIVE	310.527	DRUG PRODUCTS CONTAINING ACTIVE INGREDIENTS OFFERED OTC FOR EXTERNAL USE AS HAIR GROWERS OR FOR HAIR LOSS PREVENTION Requirements For Specific New Drugs Or Devices; New Drugs
INGREDIENTS, ACTIVE	310.528	DRUG PRODUCTS CONTAINING ACTIVE INGREDIENTS OFFERED OTC FOR USE AS AN APHRODISIAC Requirements For Specific New Drugs Or Devices; New Drugs
INGREDIENTS, ACTIVE	310.529	DRUG PRODUCTS CONTAINING ACTIVE INGREDIENTS OFFERED OTC FOR ORAL USE AS INSECT REPELLENTS Requirements For Specific New Drugs Or Devices; New Drugs
INGREDIENTS, ACTIVE	310.531	DRUG PRODUCTS CONTAINING ACTIVE INGREDIENTS OFFERED OTC FOR TREATMENT OF BOILS Requirements For Specific New Drugs Or Devices; New Drugs
INGREDIENTS, ACTIVE	310.532	DRUG PRODUCTS CONTAINING ACTIVE INGREDIENTS OFFERED OTC TO RELIEVE SYMPTOMS OF BENIGN PROSTATIC HYPERTROPHY Requirements For Specific New Drugs Or Devices; New Drugs
INGREDIENTS, ACTIVE	310.533	DRUG PRODUCTS CONTAINING ACTIVE INGREDIENTS OFFERED OTC FOR HUMAN USE AS AN ANTICHOLINERGIC IN COUGH-COLD DRUG PRODUCTS Requirements For Specific New Drugs Or Devices; New Drugs
INGREDIENTS, ACTIVE	310.534	DRUG PRODUCTS CONTAINING ACTIVE INGREDIENTS OFFERED OTC FOR HUMAN USE AS ORAL WOUND HEALING AGENTS Requirements For Specific New Drugs Or Devices; New Drugs
INGREDIENTS, ACTIVE	310.536	DRUG PRODUCTS CONTAINING ACTIVE INGREDIENTS OFFERED OTC FOR USE AS A NAILBITING OR THUMBSUCKING DETERRENT Requirements For Specific New Drugs Or Devices; New Drugs
INGREDIENTS, ACTIVE	310.537	DRUG PRODUCTS CONTAINING ACTIVE INGREDIENTS OFFERED OTC FOR TREATMENT OF FEVER BLISTERS AND COLD SORES Requirements For Specific New Drugs Or Devices; New Drugs
INGREDIENTS, ACTIVE	310.538	DRUG PRODUCTS CONTAINING ACTIVE INGREDIENTS OFFERED OTC FOR USE FOR INGROWN TOENAIL RELIEF Requirements For Specific New Drugs Or Devices; New Drugs
INGREDIENTS, ACTIVE	310.540	DRUG PRODUCTS CONTAINING ACTIVE INGREDIENTS OFFERED OTC FOR USE AS STOMACH ACIDIFIERS Requirements For Specific New Drugs Or Devices; New Drugs
INGREDIENTS, ACTIVE	310.541	DRUG PRODUCTS CONTAINING ACTIVE INGREDIENTS OFFERED FOR USE IN TREATMENT OF HYPOPHOSPHATEMIA Requirements For Specific New Drugs Or Devices; New Drugs
INGREDIENTS, ACTIVE	310.542	DRUG PRODUCTS CONTAINING ACTIVE INGREDIENTS OFFERED FOR USE IN TREATMENT OF HYPERPHOSPHATEMIA Requirements For Specific New Drugs Or Devices; New Drugs
INGREDIENTS, ACTIVE	310.543	DRUG PRODUCTS CONTAINING ACTIVE INGREDIENTS OFFERED FOR HUMAN USE IN EXOCRINE PANCREATIC INSUFFICIENCY Requirements For Specific New Drugs Or Devices; New Drugs
INGREDIENTS, ACTIVE	310.544	DRUG PRODUCTS CONTAINING CERTAIN ACTIVE INGREDIENTS OFFERED OTC FOR USE AS A SMOKING DETERRENT Requirements For Specific New Drugs Or Devices; New Drugs
INGREDIENTS, ACTIVE	310.545	DRUG PRODUCTS CONTAINING CERTAIN ACTIVE INGREDIENTS OFFERED OTC FOR CERTAIN USES Requirements For Specific New Drugs Or Devices; New Drugs
INGREDIENTS, ACTIVE	310.546	DRUG PRODUCTS CONTAINING ACTIVE INGREDIENTS OFFERED OTC FOR TREATMENT AND/OR PREVENTION OF NOCTURNAL LEG MUSCLE CRAMPS Requirements For Specific New Drugs Or Devices; New Drugs
INGREDIENTS, ACTIVE	331.10	ANTACID ACTIVE INGREDIENTS Antacid Products For Over-the-counter (OTC) Human Use
INGREDIENTS, ACTIVE	331.11	LISTING OF SPECIFIC ACTIVE INGREDIENTS Antacid Products For Over-the-counter (OTC) Human Use
INGREDIENTS, ACTIVE	331.15	COMBINATION WITH NONANTACID ACTIVE INGREDIENTS Antacid Products For Over-the-counter (OTC) Human Use
INGREDIENTS, ACTIVE	331.20	DETERMINATION OF PERCENT CONTRIBUTION OF ACTIVE INGREDIENTS Testing Procedures; Antacid Products For Over-the-counter (OTC) Human Use
INGREDIENTS, ACTIVE	332.10	ANTIFLATULENT ACTIVE INGREDIENTS Antiflatulent Products For Over-the-counter Human Use
INGREDIENTS, ACTIVE	332.15	COMBINATION WITH NON-ANTIFLATULENT ACTIVE INGREDIENTS Antiflatulent Products For Over-the-counter Human Use
INGREDIENTS, ACTIVE	333.110	FIRST AID ANTIBIOTIC ACTIVE INGREDIENTS First Aid Antibiotic Drug Products; Topical Antimicrobial Drug Products For Over-the-counter Human Use
INGREDIENTS, ACTIVE	333.120	PERMITTED COMBINATIONS OF ACTIVE INGREDIENTS First Aid Antibiotic Drug Products; Topical Antimicrobial Drug Products For Over-the-counter Human Use
INGREDIENTS, ACTIVE	333.160	LABELING OF PERMITTED COMBINATIONS OF ACTIVE INGREDIENTS First Aid Antibiotic Drug Products; Topical Antimicrobial Drug Products For Over-the-counter Human Use

INGREDIENTS, ACTIVE	333.210	ANTIFUNGAL ACTIVE INGREDIENTS Topical Antifungal Drug Products; Topical Antimicrobial Drug Products For Over-the-counter Human Use
INGREDIENTS, ACTIVE	333.310	ACNE ACTIVE INGREDIENTS Topical Acne Drug Products; Topical Antimicrobial Drug Products For Over-the-counter Human Use
INGREDIENTS, ACTIVE	333.320	PERMITTED COMBINATIONS OF ACTIVE INGREDIENTS Topical Acne Drug Products; Topical Antimicrobial Drug Products For Over-the-counter Human Use
INGREDIENTS, ACTIVE	336.10	ANTIEMETIC ACTIVE INGREDIENTS Antiemetic Drug Products For Over-the-counter Human Use
INGREDIENTS, ACTIVE	338.10	NIGHTTIME SLEEP-AID ACTIVE INGREDIENTS Nighttime Sleep-aid Drug Products For Over-the-counter Human Use
INGREDIENTS, ACTIVE	341.12	ANTIHISTAMINE ACTIVE INGREDIENTS Cold, Cough, Allergy, Bronchodilator, And Antiasthmatic Drug Products For OTC Human Use
INGREDIENTS, ACTIVE	341.14	ANTITUSSIVE ACTIVE INGREDIENTS Cold, Cough, Allergy, Bronchodilator, And Antiasthmatic Drug Products For OTC Human Use
INGREDIENTS, ACTIVE	341.16	BRONCHODILATOR ACTIVE INGREDIENTS Cold, Cough, Allergy, Bronchodilator, And Antiasthmatic Drug Products OTC Human Use
INGREDIENTS, ACTIVE	341.20	NASAL DECONGESTANT ACTIVE INGREDIENTS Cold, Cough, Allergy, Bronchodilator, And Antiasthmatic Drug Products For OTC Human Use
INGREDIENTS, ACTIVE	343.12	CARDIOVASCULAR ACTIVE INGREDIENTS Internal Analgesic, Antipyretic, And Antirheumatic Drug Products For OTC Human Use
INGREDIENTS, ACTIVE	343.13	RHEUMATOLOGIC ACTIVE INGREDIENTS Internal Analgesic, Antipyretic, And Antirheumatic Drug Products For OTC Human Use
INGREDIENTS, ACTIVE	343.22	PERMITTED COMBINATIONS OF ACTIVE INGREDIENTS FOR CARDIOVASCULAR-RHEUMATOLOGIC USE Internal Analgesic, Antipyretic, And Antirheumatic Drug Products For OTC Human Use
INGREDIENTS, ACTIVE	346.10	LOCAL ANESTHETIC ACTIVE INGREDIENTS Anorectal Drug Products For Over-the-counter Human Use
INGREDIENTS, ACTIVE	346.12	VASOCONSTRICTOR ACTIVE INGREDIENTS Anorectal Drug Products For Over-the-counter Human Use
INGREDIENTS, ACTIVE	346.14	PROTECTANT ACTIVE INGREDIENTS Anorectal Drug Products For Over-the-counter Human Use
INGREDIENTS, ACTIVE	346.16	ANALGESIC, ANESTHETIC, AND ANTIPRURITIC ACTIVE INGREDIENTS Anorectal Drug Products For Over-the-counter Human Use
INGREDIENTS, ACTIVE	346.18	ASTRINGENT ACTIVE INGREDIENTS Anorectal Drug Products For Over-the-counter Human Use
INGREDIENTS, ACTIVE	346.20	KERATOLYTIC ACTIVE INGREDIENTS Anorectal Drug Products For Over-the-counter Human Use
INGREDIENTS, ACTIVE	346.22	PERMITTED COMBINATIONS OF ANORECTAL ACTIVE INGREDIENTS Anorectal Drug Products For Over-the-counter Human Use
INGREDIENTS, ACTIVE	346.52	LABELING OF PERMITTED COMBINATIONS OF ANORECTAL ACTIVE INGREDIENTS Anorectal Drug Products For Over-the-counter Human Use
INGREDIENTS, ACTIVE	347.10	ASTRINGENT ACTIVE INGREDIENTS Skin Protectant Drug Products For Over-the-counter Human Use
INGREDIENTS, ACTIVE	348.10	ANALGESIC, ANESTHETIC, AND ANTIPRURITIC ACTIVE INGREDIENTS External Analgesic Drug Products For Over-the-counter Human Use
INGREDIENTS, ACTIVE	349.30	PERMITTED COMBINATIONS OF ACTIVE INGREDIENTS Ophthalmic Drug Products For Over-the-counter Human Use
INGREDIENTS, ACTIVE	349.79	LABELING OF PERMITTED COMBINATIONS OF ACTIVE INGREDIENTS Ophthalmic Drug Products For Over-the-counter Human Use
INGREDIENTS, ACTIVE	352.10	SUNSCREEN ACTIVE INGREDIENTS Sunscreen Drug Products For Over-the-counter Human Use
INGREDIENTS, ACTIVE	352.20	PERMITTED COMBINATIONS OF ACTIVE INGREDIENTS Sunscreen Drug Products For Over-the-counter Human Use
INGREDIENTS, ACTIVE	352.60	LABELING OF PERMITTED COMBINATIONS OF ACTIVE INGREDIENTS Sunscreen Drug Products For Over-the-counter Human Use
INGREDIENTS, ACTIVE	355.10	ANTICARIES ACTIVE INGREDIENTS Anticaries Drug Products For Over-the-counter Human Use
INGREDIENTS, ACTIVE	357.210	CHOLECYSTOKINETIC ACTIVE INGREDIENTS Cholecystokinetic Drug Products; Miscellaneous Internal Drug Products For Over-the-counter Human Use
INGREDIENTS, ACTIVE	357.810	ACTIVE INGREDIENTS FOR DEODORANT DRUG PRODUCTS FOR INTERNAL USE Miscellaneous Internal Drug Products For Over-the-counter Human Use

INGREDIENTS, ACTIVE	358.110	WART REMOVER ACTIVE INGREDIENTS Miscellaneous External Drug Products For Over-the-counter Human Use
INGREDIENTS, ACTIVE	358.510	CORN AND CALLUS REMOVER ACTIVE INGREDIENTS Miscellaneous External Drug Products For Over-the-counter Human Use
INGREDIENTS, ACTIVE	358.610	PEDICULICIDE ACTIVE INGREDIENTS Miscellaneous External Drug Products For Over-the-counter Human Use
INGREDIENTS, ACTIVE	358.710	ACTIVE INGREDIENTS FOR THE CONTROL OF DANDRUFF, SEBORRHEIC DERMATITIS, OR PSORIASIS Miscellaneous External Drug Products For Over-the-counter Human Use
INGREDIENTS, ACTIVE	358.720	PERMITTED COMBINATIONS OF ACTIVE INGREDIENTS Miscellaneous External Drug Products For Over-the-counter Human Use
INGREDIENTS, INACTIVE	201.117	INACTIVE INGREDIENTS Exemptions From Adequate Directions For Use; Labeling
INGREDIENTS, POULTRY FEED	579.40	IONIZING RADIATION FOR THE TREATMENT OF POULTRY FEED AND POULTRY FEED INGREDIENTS Irradiation In The Production, Processing And Handling Of Animal Feed And Pet Food
INGREDIENTS, SILVER	310.548	DRUG PRODUCTS CONTAINING COLLOIDAL SILVER INGREDIENTS Requirements For Specific New Drugs Or Devices; New Drugs
INGREDIENTS, STATEMENT	201.10	DRUGS; STATEMENT OF INGREDIENTS General Labeling Provisions; Labeling
INGREDIENTS, SUNSCREEN	700.35	COSMETICS CONTAINING SUNSCREEN INGREDIENTS Requirements For Specific Cosmetic Products; General
INGROWN TOENAIL	310.538	DRUG PRODUCTS CONTAINING ACTIVE INGREDIENTS OFFERED OTC FOR USE FOR INGROWN TOENAIL RELIEF Requirements For Specific New Drugs Or Devices; New Drugs
INHALANT	250.100	AMYL NITRITE INHALANT AS A PRESCRIPTION DRUG FOR HUMAN USE New Drug Or Prescription Status Of Specific Drugs; Specific Requirements For Specific Human Drugs
INHALATION	200.51	AQUEOUS-BASED DRUG PRODUCTS FOR ORAL INHALATION Requirements For Specific Classes Of Drugs; General
INHALATION	201.305	ISOPROTERENOL INHALATION PREPARATIONS (PRESSURIZED AEROSOLS, NEBULIZERS, POWDERS) FOR HUMAN USE; WARNINGS Specific Labeling Requirements For Specific Drug Products; Labeling
INHALERS	250.101	AMPHETAMINE AND METHAMPHETAMINE INHALERS REGARDED AS PRESCRIPTION DRUGS New Drug Or Prescription Status Of Specific Drugs; Specific Requirements For Specific Human Drugs
INHIBITOR, COMPLEMENT C1	866.5250	COMPLEMENT C1 INHIBITOR (INACTIVATOR) IMMUNOLOGICAL TEST SYSTEM Immunological Test Systems; Immunology And Microbiology Devices
INHIBITOR, TRYPSIN	866.5890	INTER-ALPHA TRYPSIN INHIBITOR IMMUNOLOGICAL TEST SYSTEM Immunological Test Systems; Immunology And Microbiology Devices
INHIBITORS	178.3300	CORROSION INHIBITORS USED FOR STEEL OR TINPLATE Certain Adjuvants And Production Aids; Indirect Food Additives: Adjuvants, Production Aids, And Sanitizers
INITIATES	860.132	PROCEDURES WHEN THE COMMISSIONER INITIATES A PERFORMANCE STANDARD OR PREMARKET APPROVAL PROCEEDING UNDER SECTION 514(B) OR 515(B) OF THE ACT Reclassification; Medical Device Classification Procedures
INITIAL DECISION	12.120	INITIAL DECISION Initial And Final Decisions; Formal Evidentiary Public Hearing
INITIAL DECISION	12.125	APPEAL FROM OR REVIEW OF INITIAL DECISION Initial And Final Decisions; Formal Evidentiary Public Hearing
INITIAL DECISION	12.130	DECISION BY COMMISSIONER ON APPEAL OR REVIEW OF INITIAL DECISION Initial And Final Decisions; Formal Evidentiary Public Hearing
INITIAL DECISION	17.45	INITIAL DECISION Civil Money Penalties Hearings
INITIAL MANUFACTURING	601.22	PRODUCTS IN SHORT SUPPLY; INITIAL MANUFACTURING AT OTHER THAN LICENSED ESTABLISHMENT Biologics Licensing; Licensing
INITIATION	10.25	INITIATION OF ADMINISTRATIVE PROCEEDINGS General Administrative Procedures; Administrative Practices And Procedures
INITIATION	12.20	INITIATION OF A HEARING INVOLVING THE ISSUANCE, AMENDMENT, OR REVOCATION OF A REGULATION Initiation Of Proceedings; Formal Evidentiary Public Hearing
INITIATION	12.21	INITIATION OF A HEARING INVOLVING THE ISSUANCE, AMENDMENT, OR REVOCATION OF AN ORDER Initiation Of Proceedings; Formal Evidentiary Public Hearing
INITIATION	16.22	INITIATION OF REGULATORY HEARING Initiation Of Proceedings; Regulatory Hearing Before The Food And Drug Administration
INITIATION	1308.43	INITIATION OF PROCEEDINGS FOR RULEMAKING Hearings; Schedules Of Controlled Substances

INITIATING	1316.97	INITIATING JUDICIAL FORFEITURE PROCEEDING AGAINST A CONVEYANCE WITHIN 60 DAYS OF FILING A CLAIM AND COST BOND Expedited Forfeiture Proceedings For Certain Property; Administrative Functions, Practices, And Procedures
INITIATIVE	14.171	UTILIZATION OF AN ADVISORY COMMITTEE ON THE INITIATIVE OF FDA Advisory Committees For Human Prescription Drugs; Public Hearing Before A Public Advisory Committee
INITIATIVE	71.27	LISTING AND EXEMPTION FROM CERTIFICATION ON THE COMMISSIONER'S INITIATIVE Administrative Action On Petitions; Color Additive Petitions
INITIATIVE	170.15	ADOPTION OF REGULATION ON INITIATIVE OF COMMISSIONER General Provisions; Food Additives
INITIATIVE	570.15	ADOPTION OF REGULATION ON INITIATIVE OF COMMISSIONER General Provisions; Food Additives
INJECTABLE	510.410	CORTICOSTEROIDS FOR ORAL, INJECTABLE, AND OPHTHALMIC USE IN ANIMALS; WARNINGS AND LABELING REQUIREMENTS Requirements For Specific New Animal Drugs; New Animal Drugs
INJECTABLE	510.440	INJECTABLE IRON PREPARATIONS Requirements For Specific New Animal Drugs; New Animal Drugs
INJECTABLE	522.90	AMPICILLIN IMPLANTATION AND INJECTABLE DOSAGE FORMS Implantation Or Injectable Dosage Form New Animal Drugs
INJECTABLE	522.960	FLUMETHASONE IMPLANTATION OR INJECTABLE DOSAGE FORMS Implantation Or Injectable Dosage Form New Animal Drugs
INJECTABLE	522.1077	GONADORELIN INJECTABLE Implantation Or Injectable Dosage Form New Animal Drugs
INJECTABLE	522.1222	KETAMINE HYDROCHLORIDE INJECTABLE DOSAGE FORMS Implantation Or Injectable Dosage Form New Animal Drugs
INJECTABLE	522.1662	OXYTETRACYCLINE HYDROCHLORIDE IMPLANTATION OR INJECTABLE DOSAGE FORMS Implantation Or Injectable Dosage Form New Animal Drugs
INJECTABLE	522.1696	PENICILLIN G PROCAINE IMPLANTATION AND INJECTABLE DOSAGE FORMS Implantation Or Injectable Dosage Form New Animal Drugs
INJECTABLE	522.1870	PRAZIQUANTEL INJECTABLE SOLUTION Implantation Or Injectable Dosage Form New Animal Drugs
INJECTABLE	522.2260	SULFAMETHAZINE INJECTABLE SOLUTION Implantation Or Injectable Dosage Form New Animal Drugs
INJECTABLE	522.2444	SODIUM THIOPENTAL IMPLANTATION OR INJECTABLE DOSAGE FORMS Implantation Or Injectable Dosage Form New Animal Drugs
INJECTABLE	522.2640	TYLOSIN INJECTABLE DOSAGE FORMS Implantation Or Injectable Dosage Form New Animal Drugs
INJECTABLE	522.2670	YOHIMBINE INJECTABLE Implantation Or Injectable Dosage Form New Animal Drugs
INJECTABLE	878.4420	POLYTETRAFLUOROETHYLENE INJECTABLE Surgical Devices; General And Plastic Surgery Devices
INJECTING	872.4730	DENTAL INJECTING NEEDLE Surgical Devices; Dental Devices
INJECTION	500.65	EPINEPHRINE INJECTION FOR EMERGENCY TREATMENT OF ANAPHYLACTOID SHOCK IN CATTLE, HORSES, SHEEP, AND SWINE Requirements For Specific Animal Drugs; General
INJECTION	522.23	ACEPROMAZINE MALEATE INJECTION Implantation Or Injectable Dosage Form New Animal Drugs
INJECTION	522.62	AMINOPENTAMIDE HYDROGEN SULFATE INJECTION Implantation Or Injectable Dosage Form New Animal Drugs
INJECTION	522.82	AMINOPROPAZINE FUMARATE STERILE SOLUTION INJECTION Implantation Or Injectable Dosage Form New Animal Drugs
INJECTION	522.90c	AMPICILLIN SODIUM FOR AQUEOUS INJECTION Implantation Or Injectable Dosage Form New Animal Drugs
INJECTION	522.144	ARSENAMIDE SODIUM AQUEOUS INJECTION Implantation Or Injectable Dosage Form New Animal Drugs
INJECTION	522.150	AZAPERONE INJECTION Implantation Or Injectable Dosage Form New Animal Drugs
INJECTION	522.204	BOLDENONE UNDECYLENATE INJECTION Implantation Or Injectable Dosage Form New Animal Drugs
INJECTION	522.246	BUTORPHANOL TARTRATE INJECTION Implantation Or Injectable Dosage Form New Animal Drugs
INJECTION	522.311	CARFENTANIL CITRATE INJECTION Implantation Or Injectable Dosage Form New Animal Drugs
INJECTION	522.313	CEFTIOFUR SODIUM POWDER FOR INJECTION Implantation Or Injectable Dosage Form New Animal Drugs
INJECTION	522.390	CHLORAMPHENOL INJECTION Implantation Or Injectable Dosage Form New Animal Drugs
INJECTION	522.468	COLISTIMETHATE SODIUM POWDER FOR INJECTION Implantation Or Injectable Dosage Form New Animal Drugs

INJECTION	522.480	REPOSITORY CORTICOTROPIN INJECTION Implantation Or Injectable Dosage Form New Animal Drugs
INJECTION	522.518	CUPRIC GLYCINATE INJECTION Implantation Or Injectable Dosage Form New Animal Drugs
INJECTION	522.536	DETOMIDINE HYDROCHLORIDE INJECTION Implantation Or Injectable Dosage Form New Animal Drugs
INJECTION	522.540	DEXAMETHASONE INJECTION Implantation Or Injectable Dosage Form New Animal Drugs
INJECTION	522.563	DIATRIZOATE MEGLUMINE AND DIATRIZOATE SODIUM INJECTION Implantation Or Injectable Dosage Form New Animal Drugs
INJECTION	522.575	DIAZEPAM INJECTION Implantation Or Injectable Dosage Form New Animal Drugs
INJECTION	522.650	DIHYDROSTREPTOMYCIN SULFATE INJECTION Implantation Or Injectable Dosage Form New Animal Drugs
INJECTION	522.723	DIPRENORPHINE HYDROCHLORIDE INJECTION Implantation Or Injectable Dosage Form New Animal Drugs
INJECTION	522.775	DOXAPRAM HYDROCHLORIDE INJECTION Implantation Or Injectable Dosage Form New Animal Drugs
INJECTION	522.784	DOXYLAMINE SUCCINATE INJECTION Implantation Or Injectable Dosage Form New Animal Drugs
INJECTION	522.800	DROPERIDOL AND FENTANYL CITRATE INJECTION Implantation Or Injectable Dosage Form New Animal Drugs
INJECTION	522.820	ERYTHROMYCIN INJECTION Implantation Or Injectable Dosage Form New Animal Drugs
INJECTION	522.863	ETHYLISOBUTRAZINE HYDROCHLORIDE INJECTION Implantation Or Injectable Dosage Form New Animal Drugs
INJECTION	522.883	ETORPHINE HYDROCHLORIDE INJECTION Implantation Or Injectable Dosage Form New Animal Drugs
INJECTION	522.940	COLLOIDAL FERRIC OXIDE INJECTION Implantation Or Injectable Dosage Form New Animal Drugs
INJECTION	522.960b	FLUMETHASONE ACETATE INJECTION Implantation Or Injectable Dosage Form New Animal Drugs
INJECTION	522.995	FLUPROSTENOL SODIUM INJECTION Implantation Or Injectable Dosage Form New Animal Drugs
INJECTION	522.1010	FUROSEMIDE Implantation Or Injectable Dosage Form New Animal Drugs
INJECTION	522.1044	GENTAMICIN SULFATE INJECTION Implantation Or Injectable Dosage Form New Animal Drugs
INJECTION	522.1055	GLEPTOFERRON INJECTION Implantation Or Injectable Dosage Form New Animal Drugs
INJECTION	522.1066	GLYCOPYRROLATE INJECTION Implantation Or Injectable Dosage Form New Animal Drugs
INJECTION	522.1078	GONADORELIN DIACETATE TETRAHYDRATE INJECTION Implantation Or Injectable Dosage Form New Animal Drugs
INJECTION	522.1081	CHORIONIC GONADOTROPIN FOR INJECTION; CHORIONIC GONADOTROPIN SUSPENSION Implantation Or Injectable Dosage Form New Animal Drugs
INJECTION	522.1086	GUAIFENESIN INJECTION Implantation Or Injectable Dosage Form New Animal Drugs
INJECTION	522.1145	HYALURONATE SODIUM INJECTION Implantation Or Injectable Dosage Form New Animal Drugs
INJECTION	522.1150	HYDROCHLOROTHIAZIDE INJECTION Implantation Or Injectable Dosage Form New Animal Drugs
INJECTION	522.1182	IRON DEXTRAN COMPLEX INJECTION Implantation Or Injectable Dosage Form New Animal Drugs
INJECTION	522.1183	IRON HYDROGENATED DEXTRAN INJECTION Implantation Or Injectable Dosage Form New Animal Drugs
INJECTION	522.1192	IVERMECTIN INJECTION Implantation Or Injectable Dosage Form New Animal Drugs
INJECTION	522.1193	IVERMECTIN AND CLORSULON INJECTION Implantation Or Injectable Dosage Form New Animal Drugs
INJECTION	522.1204	KANAMYCIN SULFATE INJECTION Implantation Or Injectable Dosage Form New Animal Drugs
INJECTION	522.1222a	KETAMINE HYDROCHLORIDE INJECTION Implantation Or Injectable Dosage Form New Animal Drugs
INJECTION	522.1222b	KETAMINE HYDROCHLORIDE WITH PROMAZINE HYDROCHLORIDE AND AMINOPENTAMIDE HYDROGEN SULFATE INJECTION Implantation Or Injectable Dosage Form New Animal Drugs
INJECTION	522.1244	LEVAMISOLE PHOSPHATE INJECTION Implantation Or Injectable Dosage Form New Animal Drugs
INJECTION	522.1260	LINCOMYCIN INJECTION Implantation Or Injectable Dosage Form New Animal Drugs
INJECTION	522.1335	MEDETOMIDINE HYDROCHLORIDE INJECTION Implantation Or Injectable Dosage Form New Animal Drugs

INJECTION	522.1362	MELARSOMINE DIHYDROCHLORIDE FOR INJECTION Implantation Or Injectable Dosage Form New Animal Drugs
INJECTION	522.1372	MEPIVACAINE HYDROCHLORIDE INJECTION Implantation Or Injectable Dosage Form New Animal Drugs
INJECTION	522.1380	METHOCARBAMOL INJECTION Implantation Or Injectable Dosage Form New Animal Drugs
INJECTION	522.1452	NALORPHINE HYDROCHLORIDE INJECTION Implantation Or Injectable Dosage Form New Animal Drugs
INJECTION	522.1462	NALOXONE HYDROCHLORIDE INJECTION Implantation Or Injectable Dosage Form New Animal Drugs
INJECTION	522.1465	NALTREXONE HYDROCHLORIDE INJECTION Implantation Or Injectable Dosage Form New Animal Drugs
INJECTION	522.1468	NAPROXEN FOR INJECTION Implantation Or Injectable Dosage Form New Animal Drugs
INJECTION	522.1503	NEOSTIGMINE METHYLSULFATE INJECTION Implantation Or Injectable Dosage Form New Animal Drugs
INJECTION	522.1620	ORGOTEIN FOR INJECTION Implantation Or Injectable Dosage Form New Animal Drugs
INJECTION	522.1642	OXYMORPHONE HYDROCHLORIDE INJECTION Implantation Or Injectable Dosage Form New Animal Drugs
INJECTION	522.1660	OXYTETRACYCLINE INJECTION Implantation Or Injectable Dosage Form New Animal Drugs
INJECTION	522.1662a	OXYTETRACYCLINE HYDROCHLORIDE INJECTION Implantation Or Injectable Dosage Form New Animal Drugs
INJECTION	522.1662b	OXYTETRACYCLINE HYDROCHLORIDE WITH LIDOCAINE INJECTION Implantation Or Injectable Dosage Form New Animal Drugs
INJECTION	522.1680	OXYTOCIN INJECTION Implantation Or Injectable Dosage Form New Animal Drugs
INJECTION	522.1698	PENTAZOCINE LACTATE INJECTION Implantation Or Injectable Dosage Form New Animal Drugs
INJECTION	522.1704	SODIUM PENTOBARBITAL INJECTION Implantation Or Injectable Dosage Form New Animal Drugs
INJECTION	522.1720	PHENYLBUTAZONE INJECTION Implantation Or Injectable Dosage Form New Animal Drugs
INJECTION	522.1820	PITUITARY LUTEINIZING HORMONE FOR INJECTION Implantation Or Injectable Dosage Form New Animal Drugs
INJECTION	522.1883	PREDNISOLONE SODIUM PHOSPHATE INJECTION, STERILE Implantation Or Injectable Dosage Form New Animal Drugs
INJECTION	522.1884	PREDNISOLONE SODIUM SUCCINATE INJECTION Implantation Or Injectable Dosage Form New Animal Drugs
INJECTION	522.1920	PROCHLORPERAZINE, ISOPROPAMIDE FOR INJECTION Implantation Or Injectable Dosage Form New Animal Drugs
INJECTION	522.1962	PROMAZINE HYDROCHLORIDE INJECTION Implantation Or Injectable Dosage Form New Animal Drugs
INJECTION	522.2002	PROPIOPROMAZINE HYDROCHLORIDE INJECTION Implantation Or Injectable Dosage Form New Animal Drugs
INJECTION	522.2005	PROPOFOL INJECTION Implantation Or Injectable Dosage Form New Animal Drugs
INJECTION	522.2063	PYRILAMINE MALEATE INJECTION Implantation Or Injectable Dosage Form New Animal Drugs
INJECTION	522.2100	SELENIUM, VITAMIN E INJECTION Implantation Or Injectable Dosage Form New Animal Drugs
INJECTION	522.2120	SPECTINOMYCIN DIHYDROCHLORIDE INJECTION Implantation Or Injectable Dosage Form New Animal Drugs
INJECTION	522.2220	SULFADIMETHOXINE INJECTION Implantation Or Injectable Dosage Form New Animal Drugs
INJECTION	522.2404	THIALBARBITONE SODIUM FOR INJECTION Implantation Or Injectable Dosage Form New Animal Drugs
INJECTION	522.2424	SODIUM THIAMYLAL FOR INJECTION Implantation Or Injectable Dosage Form New Animal Drugs
INJECTION	522.2444a	SODIUM THIOPENTAL FOR INJECTION Implantation Or Injectable Dosage Form New Animal Drugs
INJECTION	522.2444b	SODIUM THIOPENTAL, SODIUM PENTOBARBITAL FOR INJECTION Implantation Or Injectable Dosage Form New Animal Drugs
INJECTION	522.2470	TILETAMINE HYDROCHLORIDE AND ZOLAZEPAM HYDROCHLORIDE FOR INJECTION Implantation Or Injectable Dosage Form New Animal Drugs
INJECTION	522.2471	TILMICOSIN PHOSPHATE INJECTION Implantation Or Injectable Dosage Form New Animal Drugs
INJECTION	522.2474	TOLAZOLINE HYDROCHLORIDE INJECTION Implantation Or Injectable Dosage Form New Animal Drugs
INJECTION	522.2582	TRIFLUPROMAZINE HYDROCHLORIDE INJECTION Implantation Or Injectable Dosage Form New Animal Drugs

INJECTION	522.2615	TRIPELENNAMINE HYDROCHLORIDE INJECTION Implantation Or Injectable Dosage Form New Animal Drugs
INJECTION	522.2640a	TYLOSIN INJECTION Implantation Or Injectable Dosage Form New Animal Drugs
INJECTION	522.2662	XYLAZINE HYDROCHLORIDE INJECTION Implantation Or Injectable Dosage Form New Animal Drugs
INJECTOR	870.1650	ANGIOGRAPHIC INJECTOR AND SYRINGE Cardiovascular Diagnostic Devices; Cardiovascular Devices
INJECTOR	870.1660	INDICATOR INJECTOR Cardiovascular Diagnostic Devices; Cardiovascular Devices
INJECTOR	870.1670	SYRINGE ACTUATOR FOR AN INJECTOR Cardiovascular Diagnostic Devices; Cardiovascular Devices
INJECTOR	872.4465	GAS-POWERED JET INJECTOR Surgical Devices; Dental Devices
INJECTOR	872.4475	SPRING-POWERED JET INJECTOR Surgical Devices; Dental Devices
INJECTOR	880.5430	NONELECTRICALLY POWERED FLUID INJECTOR General Hospital And Personal Use Therapeutic Devices; General Hospital And Personal Use Devices
INOSINATE	172.535	DISODIUM INOSINATE Flavoring Agents And Related Substances; Food Additives Permitted For Direct Addition To Food For Human Consumption
INOSITOL	184.1370	INOSITOL Listing Of Specific Substances Affirmed As GRAS; Direct Food Substances Affirmed As Generally Recognized As Safe
INOSITOL	582.5370	INOSITOL Nutrients And/or Dietary Supplements; Substances Generally Recognized As Safe
INQUIRIES	14.65	PUBLIC INQUIRIES AND REQUESTS FOR ADVISORY COMMITTEE RECORDS Records Of Meetings And Hearings Before Advisory Committees; Public Hearing Before A Public Advisory Committee
INSECT CONTROL	1250.95	INSECT CONTROL Sanitation Facilities And Conditions On Vessels; Interstate Conveyance Sanitation
INSECT REPELLANTS, ORAL	310.529	DRUG PRODUCTS CONTAINING ACTIVE INGREDIENTS OFFERED OVER-THE-COUNTER (OTC) FOR ORAL USE AS INSECT REPELLENTS Requirements For Specific New Drugs Or Devices; New Drugs
INSERT	872.3900	POSTERIOR ARTIFICIAL TOOTH WITH A METAL INSERT Prosthetic Devices; Dental Devices
INSERT (KIIL TYPE)	876.5830	HEMODIALYZER WITH DISPOSABLE INSERT (KIIL TYPE) Therapeutic Devices; Gastroenterology-urology Devices
INSERTER	886.5540	CONTACT LENS INSERTER/REMOVER Therapeutic Devices; Ophthalmic Devices
INSERTS, PACKAGE	310.501	PATIENT PACKAGE INSERTS FOR ORAL CONTRACEPTIVES Requirements For Specific New Drugs Or Devices; New Drugs
INSERTS, PACKAGE	310.515	PATIENT PACKAGE INSERTS FOR ESTROGENS Requirements For Specific New Drugs Or Devices; New Drugs
INSERTS, PACKAGE	357.152	PACKAGE INSERTS FOR ANTHELMINTIC DRUG PRODUCTS Anthelmintic Drug Products; Miscellaneous Internal Drug Products For Over-the-counter Human Use
INSOLE	880.6280	MEDICAL INSOLE General Hospital And Personal Use Miscellaneous Devices; General Hospital And Personal Use Devices
INSOLUBLE	177.1400	HYDROXYETHYL CELLULOSE FILM, WATER-INSOLUBLE Substances For Use As Basic Components Of Single And Repeated Use Food Contact Surfaces; Indirect Food Additives: Polymers
INSOLUBLE	184.1372	INSOLUBLE GLUCOSE ISOMERASE ENZYME PREPARATIONS Listing Of Specific Substances Affirmed As GRAS; Direct Food Substances Affirmed As Generally Recognized As Safe
INSPECTION	58.15	INSPECTION OF A TESTING FACILITY General Provisions; Good Laboratory Practice For Nonclinical Laboratory Studies
INSPECTION	179.21	SOURCES OF RADIATION USED FOR INSPECTION OF FOOD OR PACKAGED FOOD, AND FOR CONTROLLING FOOD PROCESSING Radiation And Radiation Sources; Irradiation In The Production, Processing And Handling Of Food
INSPECTION	207.37	INSPECTION OF REGISTRATIONS AND DRUG LISTINGS Procedures For Domestic Drug Establishments; Registration Of Drug Producers And Listing Of Drugs In Commercial Distribution
INSPECTION	211.134	DRUG PRODUCT INSPECTION Packaging And Labeling Control; Current Good Manufacturing Practice For Finished Pharmaceuticals
INSPECTION	312.58	INSPECTION OF SPONSOR'S RECORDS AND REPORTS Investigational New Drug Application
INSPECTION	312.68	INSPECTION OF INVESTIGATOR'S RECORDS AND REPORTS Investigational New Drug Application

INSPECTION	600.21	TIMES FOR INSPECTION Establishment Inspection; Biological Products: General
INSPECTION	607.37	INSPECTION OF ESTABLISHMENT REGISTRATIONS AND BLOOD PRODUCT LISTINGS Procedures For Domestic Blood Product Establishments; Establishment Registratn+product Listing, Manufacturers Human Blood/b.products
INSPECTION	710.7	INSPECTION OF REGISTRATIONS Voluntary Registration Of Cosmetic Product Establishments
INSPECTION	807.37	INSPECTION OF ESTABLISHMENT REGISTRATION AND DEVICE LISTINGS Procedures For Device Establishments; Establishment Registration And Device Listing For Manufacturers And Initial Importers Of Devices
INSPECTION	820.72	INSPECTION, MEASURING, AND TEST EQUIPMENT Production And Process Controls; Quality System Regulation
INSPECTION	1002.31	PRESERVATION AND INSPECTION OF RECORDS Manufacturers' Records; Records and Reports
INSPECTION	1210.10	AVAILABILITY FOR EXAMINATION AND INSPECTION Regulations Under The Federal Import Milk Act
INSPECTION	1210.11	SANITARY INSPECTION OF DIARY FARMS Regulations Under The Federal Import Milk Act
INSPECTION	1210.14	SANITARY INSPECTION OF PLANTS Regulations Under The Federal Import Milk Act
INSPECTION	1210.17	AUTHORITY TO SAMPLE AND INSPECT Regulations Under The Federal Import Milk Act
INSPECTION	1250.21	INSPECTION Food Service Sanitation On Land And Air Conveyances; Interstate Conveyance Sanitation
INSPECTION	1250.25	SOURCE IDENTIFICATION AND INSPECTION OF FOOD AND DRINK Food Service Sanitation On Land And Air Conveyances; Interstate Conveyance Sanitation
INSPECTION	1250.71	INSPECTION Sanitation Facilities And Conditions On Vessels; Interstate Conveyance Sanitation
INSPECTION	1271.37	WILL ESTABLISHMENT REGISTRATIONS AND HCT/P LISTINGS BE AVAILABLE FOR INSPECTION, AND HOW DO I REQUEST INFORMATION? Procedures For Registration And Listing; Human Cells, Tissues, And Cellular And Tissue-Based Products
INSPECTION	1316.04	EXCLUSION FROM INSPECTION Administrative Inspections; Administrative Functions, Practices, And Procedures
INSPECTION	1316.06	NOTICE OF INSPECTION Administrative Inspections; Administrative Functions, Practices, And Procedures
INSPECTION	1316.07	REQUIREMENT FOR ADMINISTRATIVE INSPECTION WARRANT Administrative Inspections; Administrative Functions, Practices, And Procedures
INSPECTION	1316.08	CONSENT TO INSPECTION Administrative Inspections; Administrative Functions, Practices, And Procedures
INSPECTION	1316.09	APPLICATION FOR ADMINISTRATIVE INSPECTION WARRANT Administrative Inspections; Administrative Functions, Practices, And Procedures
INSPECTION	1316.12	REFUSAL TO ALLOW INSPECTION WITH AN ADMINISTRATIVE WARRANT Administrative Inspections; Administrative Functions, Practices, And Procedures
INSPECTION	1316.46	INSPECTION OF RECORD Administrative Hearings; Administrative Functions, Practices, And Procedures
INSPECTION REPORTS	26.12	NATURE OF RECOGNITION OF INSPECTION REPORTS Specific Sector Provisions For Pharmaceutical GMPs; Mutual Recognition Of Pharmaceutical GMP Reports, Medical Device Quality System Audit Reports
INSPECTION REPORTS	26.13	TRANSMISSION OF POSTAPPROVAL INSPECTION REPORTS Specific Sector Provisions For Pharmaceutical GMPs; Mutual Recognition Of Pharmaceutical GMP Reports, Medical Device Quality System Audit Reports
INSPECTION REPORTS	26.14	TRANSMISSION OF PREAPPROVAL INSPECTION REPORTS Specific Sector Provisions For Pharmaceutical GMPs; Mutual Recognition Of Pharmaceutical GMP Reports, Medical Device Quality System Audit Reports
INSPECTIONS	5.32	CERTIFICATION FOLLOWING INSPECTIONS General Redelegations Of Authority; Delegations Of Authority And Organization
INSPECTIONS	1270.41	INSPECTIONS Human Tissue Intended For Transplantation
INSPECTIONS	1316.03	AUTHORITY TO MAKE INSPECTIONS Administrative Inspections; Administrative Functions, Practices, And Procedures
INSPECTIONS	1316.13	FREQUENCY OF ADMINISTRATIVE INSPECTIONS Administrative Inspections; Administrative Functions, Practices, And Procedures
INSPECTIONS	812.145	INSPECTIONS Records And Reports; Investigational Device Exemptions
INSPECTIONS	1270.41	INSPECTIONS Inspection Of Tissue Establishments; Human Tissue Intended For Transplantation
INSPECTION	1240.20	ISSUANCE AND POSTING OF CERTIFICATES FOLLOWING INSPECTION Control Of Communicable Diseases
INSPECTOR	600.22	DUTIES OF INSPECTOR Establishment Inspection; Biological Products: General

INSPECTORS	600.20	INSPECTORS Establishment Inspection; Biological Products: General
INSPIRATORY	868.1780	INSPIRATORY AIRWAY PRESSURE METER Diagnostic Devices; Anesthesiology Devices
INSTALLATION	820.170	INSTALLATION Holding, Storage, Distribution, And Installation; Quality System Regulation
INSTANTIZED FLOURS	137.170	INSTANTIZED FLOURS Requirements For Specific Standardized Cereal Flours And Related Products; Cereal Flours And Related Products
INSTITUTION	56.112	REVIEW BY INSTITUTION IRB Functions And Operations; Institutional Review Boards
INSTITUTION	56.121	DISQUALIFICATION OF AN IRB OR AN INSTITUTION Administrative Action For Noncompliance; Institutional Review Boards
INSTITUTION	56.123	REINSTATEMENT OF AN IRB OR AN INSTITUTION Administrative Action For Noncompliance; Institutional Review Boards
INSTITUTIONAL REVIEW BOARD	814.124	INSTITUTIONAL REVIEW BOARD REQUIREMENTS Humanitarian Use Devices; Premarket Approval Of Medical Devices
INSTRUCTION	5.304	APPROVAL OF SCHOOLS PROVIDING FOOD-PROCESSING INSTRUCTION Foods And Cosmetics; Redelegations Of Authority; Delegations Of Authority And Organization
INSTRUCTION	606.122	INSTRUCTION CIRCULAR Finished Product Control; Current Good Manufacturing Practice For Blood And Blood Components
INSTRUCTIONS	1301.03	INFORMATION; SPECIAL INSTRUCTIONS General Information; Registration Of Manufacturers, Distributors, And Dispensers Of Controlled Substances
INSTRUCTIONS	1309.3	INFORMATION: SPECIAL INSTRUCTIONS Registration Of Manufacturers, Distributors, Importers And Exporters Of List I Chemicals
INSTRUCTIONS	1316.43	INFORMATION; SPECIAL INSTRUCTIONS Administrative Hearings; Administrative Functions, Practices, And Procedures
INSTRUMENT	864.5400	COAGULATION INSTRUMENT Automated And Semi-automated Hematology Devices; Hematology And Pathology Devices
INSTRUMENT	864.5600	AUTOMATED HEMATOCRIT INSTRUMENT Automated And Semi-automated Hematology Devices; Hematology And Pathology Devices
INSTRUMENT	872.4120	BONE CUTTING INSTRUMENT AND ACCESSORIES Surgical Devices; Dental Devices
INSTRUMENT	872.4535	DENTAL DIAMOND INSTRUMENT Surgical Devices; Dental Devices
INSTRUMENT	872.4565	DENTAL HAND INSTRUMENT Surgical Devices; Dental Devices
INSTRUMENT	874.3540	PROSTHESIS MODIFICATION INSTRUMENT FOR OSSICULAR REPLACEMENT SURGERY Prosthetic Devices; Ear, Nose And Throat Devices
INSTRUMENT	874.4420	EAR, NOSE AND THROAT MANUAL SURGICAL INSTRUMENT Surgical Devices; Ear, Nose And Throat Devices
INSTRUMENT	876.1075	GASTROENTEROLOGY-UROLOGY BIOPSY INSTRUMENT Diagnostic Devices; Gastroenterology-urology Devices
INSTRUMENT	876.4730	MANUAL GASTROENTEROLOGY-UROLOGY SURGICAL INSTRUMENT AND ACCESSORIES Surgical Devices; Gastroenterology-urology Devices
INSTRUMENT	878.4800	MANUAL SURGICAL INSTRUMENT FOR GENERAL USE Surgical Devices; General And Plastic Surgery Devices
INSTRUMENT	878.4810	LASER SURGICAL INSTRUMENT FOR USE IN PLASTIC SURGERY AND IN DERMATOLOGY Surgical Devices; General And Plastic Surgery Devices
INSTRUMENT	882.4190	CLIP FORMING/CUTTING INSTRUMENT Neurological Surgical Devices; Neurological Devices
INSTRUMENT	882.4200	CLIP REMOVAL INSTRUMENT Neurological Surgical Devices; Neurological Devices
INSTRUMENT	882.4275	DOWEL CUTTING INSTRUMENT Neurological Surgical Devices; Neurological Devices
INSTRUMENT	882.4500	CRANIOPLASTY MATERIAL FORMING INSTRUMENT Neurological Surgical Devices; Neurological Devices
INSTRUMENT	882.4525	MICROSURGICAL INSTRUMENT Neurological Surgical Devices; Neurological Devices
INSTRUMENT	882.4535	NONPOWERED NEUROSURGICAL INSTRUMENT Neurological Surgical Devices; Neurological Devices
INSTRUMENT	882.4545	SHUNT SYSTEM IMPLANTATION INSTRUMENT Neurological Surgical Devices; Neurological Devices
INSTRUMENT	882.4560	STEREOTAXIC INSTRUMENT Neurological Surgical Devices; Neurological Devices

INSTRUMENT	884.4500	OBSTETRIC FETAL DESTRUCTIVE INSTRUMENT Obstetrical And Gynecological Surgical Devices; Obstetrical And Gynecological Devices
INSTRUMENT	884.4520	OBSTETRIC-GYNECOLOGIC GENERAL MANUAL INSTRUMENT Obstetrical And Gynecological Surgical Devices; Obstetrical And Gynecological Devices
INSTRUMENT	884.4530	OBSTETRIC-GYNECOLOGIC SPECIALIZED MANUAL INSTRUMENT Obstetrical And Gynecological Surgical Devices; Obstetrical And Gynecological Devices
INSTRUMENT	886.1360	VISUAL FIELD LASER INSTRUMENT Diagnostic Devices; Ophthalmic Devices
INSTRUMENT	886.1425	LENS MEASURING INSTRUMENT Diagnostic Devices; Ophthalmic Devices
INSTRUMENT	886.1460	STEREOPSIS MEASURING INSTRUMENT Diagnostic Devices; Ophthalmic Devices
INSTRUMENT	886.4350	MANUAL OPHTHALMIC SURGICAL INSTRUMENT Surgical Devices; Ophthalmic Devices
INSTRUMENT	888.4540	ORTHOPEDIC MANUAL SURGICAL INSTRUMENT Surgical Devices; Orthopedic Devices
INSTRUMENT	888.4580	SONIC SURGICAL INSTRUMENT AND ACCESSORIES/ATTACHMENTS Surgical Devices; Orthopedic Devices
INSTRUMENT	888.5960	CAST REMOVAL INSTRUMENT Surgical Devices; Orthopedic Devices
INSTRUMENT	888.5980	MANUAL CAST APPLICATION AND REMOVAL INSTRUMENT Surgical Devices; Orthopedic Devices
INSTRUMENT	892.1940	RADIOLOGIC QUALITY ASSURANCE INSTRUMENT Diagnostic Devices; Radiology Devices
INSTRUMENT MOTORS	878.4820	SURGICAL INSTRUMENT MOTORS AND ACCESSORIES/ATTACHMENTS Surgical Devices; General And Plastic Surgery Devices
INSTRUMENT STAND	886.1860	OPHTHALMIC INSTRUMENT STAND Diagnostic Devices; Ophthalmic Devices
INSTRUMENT TABLE	886.4855	OPHTHALMIC INSTRUMENT TABLE Surgical Devices; Ophthalmic Devices
INSTRUMENTS, FABRICATION	884.6140	ASSISTED REPRODUCTION MICROPIPETTE FABRICATION INSTRUMENTS Obstetrical And Gynecological Devices
INSTRUMENTS, MEDICAL	880.6150	ULTRASONIC CLEANER FOR MEDICAL INSTRUMENTS General Hospital And Personal Use Miscellaneous Devices; General Hospital And Personal Use Devices
INSTRUMENTS, SURGICAL	870.4500	CARDIOVASCULAR SURGICAL INSTRUMENTS Cardiovascular Surgical Devices; Cardiovascular Devices
INSUFFICIENCY, PANCREATIC	310.543	DRUG PRODUCTS CONTAINING ACTIVE INGREDIENTS OFFERED OTC FOR HUMAN USE IN EXOCRINE PANCREATIC INSUFFICIENCY Requirements For Specific New Drugs Or Devices; New Drugs
INSUFFICIENT QUANTITIES	316.36	INSUFFICIENT QUANTITIES OF ORPHAN DRUGS Orphan Drugs
INSUFFLATOR	884.1300	UTEROTUBAL CARBON DIOXIDE INSUFFLATOR AND ACCESSORIES Obstetrical And Gynecological Diagnostic Devices; Obstetrical And Gynecological Devices
INSUFFLATOR	884.1700	HYSTEROSCOPIC INSUFFLATOR Obstetrical And Gynecological Diagnostic Devices; Obstetrical And Gynecological Devices
INSUFFLATOR	884.1730	LAPAROSCOPIC INSUFFLATOR Obstetrical And Gynecological Diagnostic Devices; Obstetrical And Gynecological Devices
INSULIN	200.15	DEFINITION OF TERM "INSULIN" General Provisions; General
INSULIN	862.1405	IMMUNOREACTIVE INSULIN TEST SYSTEM Clinical Chemistry Test Systems; Clinical Chemistry And Clinical Toxicology Devices
INTEGRATING DENSITOMETER	862.2400	DENSITOMETER/SCANNER (INTEGRATING, REFLECTANCE, TLC, OR RADIOCHROMATOGRAM) FOR CLINICAL USE Clinical Laboratory Instruments; Clinical Chemistry And Clinical Toxicology Devices
INTENDED USES	201.128	MEANING OF "INTENDED USES" Exemptions From Adequate Directions For Use; Labeling
INTENDED USES	801.4	MEANING OF "INTENDED USES" General Labeling Provisions; Labeling
INTENSIFIED	892.1650	IMAGE-INTENSIFIED FLUOROSCOPIC X-RAY SYSTEM Diagnostic Devices; Radiology Devices
INTENSIFIED	892.1660	NON-IMAGE-INTENSIFIED FLUOROSCOPIC X-RAY Diagnostic Devices; Radiology Devices
INTENSIFYING SCREEN	892.1960	RADIOGRAPHIC INTENSIFYING SCREEN Diagnostic Devices; Radiology Devices
INTER-ALPHA	866.5890	INTER-ALPHA TRYPSIN INHIBITOR IMMUNOLOGICAL TEST SYSTEM Immunological Test Systems; Immunology And Microbiology Devices

INTERARTICULAR	872.3970	INTERARTICULAR DISC PROSTHESIS (INTERPOSITIONAL IMPLANT) Prosthetic Devices; Dental Devices
INTERLAMINAL	888.3050	SPINAL INTERLAMINAL FIXATION ORTHOSIS Prosthetic Devices; Orthopedic Devices
INTEREST, CONFLICT OF	5.30	AUTHORITY TO SELECT TEMPORARY VOTING MEMBERS FOR ADVISORY COMMITTEES AND TO SIGN CONFLICT OF INTEREST WAIVERS General Redelegations Of Authority; Delegations Of Authority And Organization
INTEREST, CONFLICT OF	19.10	FOOD AND DRUG ADMINISTRATION CONFLICT OF INTEREST REVIEW BOARD Standards Of Conduct And Conflicts Of Interest
INTERESTS, FINANCIAL	54.5	AGENCY EVALUATION OF FINANCIAL INTERESTS Financial Disclosure By Clinical Investigators
INTERLOCKING	876.4590	INTERLOCKING URETHRAL SOUND Surgical Devices; Gastroenterology-urology Devices
INTERLOCUTORY APPEAL	12.97	INTERLOCUTORY APPEAL FROM RULING OF PRESIDING OFFICER Hearing Procedures; Formal Evidentiary Public Hearing
INTERLOCUTORY APPEAL	17.18	INTERLOCUTORY APPEAL FROM RULING OF PRESIDING OFFICER Civil Money Penalties Hearings
INTERMITTENT	868.5955	INTERMITTENT MANDATORY VENTILATION ATTACHMENT Therapeutic Devices; Anesthesiology Devices
INTERMITTENT	890.1600	INTERMITTENT PRESSURE MEASUREMENT SYSTEM Physical Medicine Diagnostic Devices; Physical Medicine Devices
INTERNAL AGENCY REVIEW	10.75	INTERNAL AGENCY REVIEW OF DECISIONS General Administrative Procedures; Administrative Practices And Procedures
INTERNAL ANALGESIC/ANTIPYRETIC	201.322	OVER-THE COUNTER DRUG PRODUCTS CONTAINING INTERNAL ANALGESIC/ANTIPYRETIC ACTIVE INGREDIENTS Specific Labeling Requirements For Specific Drug Products; Labeling
INTERNAL USE	201.302	NOTICE TO MANUFACTURERS, PACKERS, AND DISTRIBUTORS OF DRUGS FOR INTERNAL USE WHICH CONTAIN MINERAL OIL Specific Labeling Requirements For Specific Drug Products; Labeling
INTERNAL USE	357.810	ACTIVE INGREDIENTS FOR DEODORANT DRUG PRODUCTS FOR INTERNAL USE Miscellaneous Internal Drug Products For Over-the-counter Human Use
INTERNAL USE	357.850	LABELING OF DEODORANT DRUG PRODUCTS FOR INTERNAL USE Miscellaneous Internal Drug Products For Over-the-counter Human Use
INTERNAL USE	878.4450	NONABSORBABLE GAUZE FOR INTERNAL USE Surgical Devices; General And Plastic Surgery Devices
INTERNATIONAL TRANSACTION	1313.33	CONTENTS OF AN INTERNATIONAL TRANSACTION DECLARATION Importation And Exportation Of Precursors And Essential Chemicals
INTERNATIONAL TRANSACTION	1313.34	DISTRIBUTION OF THE INTERNATIONAL TRANSACTION DECLARATION Importation And Exportation Of Precursors And Essential Chemicals
INTERNATIONAL TRANSACTIONS	1313.32	REQUIREMENT OF AUTHORIZATION FOR INTERNATIONAL TRANSACTIONS Importation And Exportation Of Precursors And Essential Chemicals
INTERNATIONAL TREATY	1308.46	CONTROL REQUIRED UNDER INTERNATIONAL TREATY Hearings; Schedules Of Controlled Substances
INTERPOSITIONAL IMPLANT	872.3970	INTERARTICULAR DISC PROSTHESIS (INTERPOSITIONAL IMPLANT) Prosthetic Devices; Dental Devices
INTERPRETATION	868.1900	DIAGNOSTIC PULMONARY-FUNCTION INTERPRETATION CALCULATOR Diagnostic Devices; Anesthesiology Devices
INTERSTATE	130.17	TEMPORARY PERMITS FOR INTERSTATE SHIPMENT OF EXPERIMENTAL PACKS OF FOOD VARYING FROM THE REQUIREMENTS OF DEFINITIONS AND STANDARDS OF IDENTITY General Provisions; Food Standards: General
INTERSTATE	1240.62	TURTLES INTRASTATE AND INTERSTATE REQUIREMENTS Specific Administrative Decisions Regarding Interstate Shipments; Control Of Communicable Diseases
INTERVERTEBRAL	888.3060	SPINAL INTERVERTEBRAL FIXATION ORTHOSIS Prosthetic Devices; Orthopedic Devices
INTRA-AGENCY MEMORANDA	20.62	INTER- OR INTRA-AGENCY MEMORANDA OR LETTERS Exemptions; Public Information
INTRA-AGENCY RECORDS	21.70	DISCLOSURE AND INTRA-AGENCY USE OF RECORDS IN PRIVACY ACT RECORD SYSTEMS; NO ACCOUNTING REQUIRED Disclosure Of Records In Privacy Act Record Systems To Persons Other Than The Subject Individual; Protection Of Privacy
INTRA-AORTIC BALLOON	870.3535	INTRO-AORTIC BALLOON AND CONTROL SYSTEM Cardiovascular Prosthetic Devices; Cardiovascular Devices
INTRA-VENOUS	880.5025	I.V. CONTAINER General Hospital And Personal Use Therapeutic Devices; General Hospital And Personal Use Devices
INTRA-VENOUS	880.5420	PRESSURE INFUSOR FOR AN I.V. BAG General Hospital And Personal Use Therapeutic Devices; General Hospital And Personal Use Devices

INTRACARDIAC	870.3470	INTRACARDIAC PATCH OR PLEDGET MADE OF POLYPROPYLENE, POLYETHYLENE TEREPHTHALATE, OR POLYTETRAFLUOROETHYLENE Cardiovascular Prosthetic Devices; Cardiovascular Devices
INTRACARDIAC	870.4430	CARDIOPULMONARY BYPASS INTRACARDIAC SUCTION CONTROL Cardiovascular Surgical Devices; Cardiovascular Devices
INTRACAVITARY	870.1270	INTRACAVITARY PHONO CATHETER SYSTEM Cardiovascular Diagnostic Devices; Cardiovascular Devices
INTRACEREBRAL	882.5840	IMPLANTED INTRACEREBRAL/SUBCORTICAL STIMULATOR FOR PAIN RELIEF Neurological Therapeutic Devices; Neurological Devices
INTRACRANIAL	882.1620	INTRACRANIAL PRESSURE MONITORING DEVICE Neurological Diagnostic Devices; Neurological Devices
INTRALUMINAL	870.4875	INTRALUMINAL ARTERY STRIPPER Cardiovascular Surgical Devices; Cardiovascular Devices
INTRAMAMMARY	526.88	AMOXICILLIN TRIHYDRATE FOR INTRAMAMMARY INFUSION Intramammary Dosage Forms
INTRAMAMMARY	529.365	CEPHAPIRIN SODIUM FOR INTRAMAMMARY INFUSION Certain Other Dosage Form New Animal Drugs
INTRAMAMMARY	526.464	CLOXACILLIN INTRAMAMMARY DOSAGE FORMS Intramammary Dosage Forms
INTRAMAMMARY	526.1130	HETACILLIN POTASSIUM FOR INTRAMAMMARY INFUSION Intramammary Dosage Forms
INTRAMAMMARY	526.1696	PENICILLIN INTRAMAMMARY DOSAGE FORMS Intramammary Dosage Forms
INTRAMAMMARY	526.1696b	PENICILLIN G PROCAINE-DIHYDROSTREPTOMYCIN IN SOYBEAN OIL FOR INTRAMAMMARY INFUSION (DRY COWS) Intramammary Dosage Forms
INTRAMAMMARY	526.1696c	PENICILLIN G PROCAINE-DIHYDROSTREPTOMYCIN SULFATE FOR INTRAMAMMARY INFUSION (DRY COWS) Intramammary Dosage Forms
INTRAMAMMARY	526.1696d	PENICILLIN G PROCAINE-NOVOBIOCIN FOR INTRAMAMMARY INFUSION Intramammary Dosage Forms
INTRAMEDULLARY	888.3020	INTRAMEDULLARY FIXATION ROD Prosthetic Devices; Orthopedic Devices
INTRANASAL SPLINT	874.4780	INTRANASAL SPLINT Surgical Devices; Ear, Nose And Throat Devices
INTRAOCULAR	886.3600	INTRAOCULAR LENS Prosthetic Devices; Ophthalmic Devices
INTRAOCULAR	886.4270	INTRAOCULAR GAS Surgical Devices; Ophthalmic Devices
INTRAOCULAR	886.4275	INTRAOCULAR FLUID Surgical Devices; Ophthalmic Devices
INTRAOCULAR	886.4280	INTRAOCULAR PRESSURE MEASURING DEVICE Surgical Devices; Ophthalmic Devices
INTRAOCULAR	886.4300	INTRAOCULAR LENS GUIDE Surgical Devices; Ophthalmic Devices
INTRAORAL	872.1810	INTRAORAL SOURCE X-RAY SYSTEM Diagnostic Devices; Dental Devices
INTRAORAL	872.4130	INTRAORAL DENTAL DRILL Surgical Devices; Dental Devices
INTRAORAL	872.4600	INTRAORAL LIGATURE AND WIRE LOCK Surgical Devices; Dental Devices
INTRAORAL	872.6890	INTRAORAL DENTAL WAX Miscellaneous Devices; Dental Devices
INTRAOSSEOUS	872.4880	INTRAOSSEOUS FIXATION SCREW OR WIRE Surgical Devices; Dental Devices
INTRASTATE	1240.62	TURTLES INTRASTATE AND INTERSTATE REQUIREMENTS Specific Administrative Decisions Regarding Interstate Shipments; Control Of Communicable Diseases
INTRAUTERINE	529.50	AMKACIN SULFATE INTRAUTERINE SOLUTION Certain Other Dosage Form New Animal Drugs
INTRAUTERINE	529.1044	GENTAMICIN SULFATE INTRAUTERINE SOLUTION Certain Other Dosage Form New Animal Drugs
INTRAUTERINE	884.2700	INTRAUTERINE PRESSURE MONITOR AND ACCESSORIES Obstetrical And Gynecological Monitoring Devices; Obstetrical And Gynecological Devices
INTRAUTERINE	884.5360	CONTRACEPTIVE INTRAUTERINE DEVICE (IUD) AND INTRODUCER Obstetrical And Gynecological Therapeutic Devices; Obstetrical And Gynecological Devices
INTRAVASCULAR	870.1200	DIAGNOSTIC INTRAVASCULAR CATHETER Cardiovascular Diagnostic Devices; Cardiovascular Devices
INTRAVASCULAR	870.3375	CARDIOVASCULAR INTRAVASCULAR FILTER Cardiovascular Prosthetic Devices; Cardiovascular Devices

INTRAVASCULAR	880.5200	INTRAVASCULAR CATHETER General Hospital And Personal Use Therapeutic Devices; General Hospital And Personal Use Devices
INTRAVASCULAR	880.5210	INTRAVASCULAR CATHETER SECUREMENT DEVICE General Hospital And Personal Use Therapeutic Devices; General Hospital And Personal Use Devices
INTRAVASCULAR	880.5440	INTRAVASCULAR ADMINISTRATION SET General Hospital And Personal Use Therapeutic Devices; General Hospital And Personal Use Devices
INTRAVASCULAR	880.5970	PERCUTANEOUS, IMPLANTED, LONG-TERM INTRAVASCULAR CATHETER General Hospital And Personal Use Therapeutic Devices; General Hospital And Personal Use Devices
INTRAVASCULAR	882.5150	INTRAVASCULAR OCCLUDING CATHETER Neurological Therapeutic Devices; Neurological Devices
INTRODUCER	870.1340	CATHETER INTRODUCER Cardiovascular Diagnostic Devices; Cardiovascular Devices
INTRODUCER	880.6920	SYRINGE NEEDLE INTRODUCER General Hospital And Personal Use Miscellaneous Devices; General Hospital And Personal Use Devices
INTRODUCER	884.5360	CONTRACEPTIVE INTRAUTERINE DEVICE (IUD) AND INTRODUCER Obstetrical And Gynecological Therapeutic Devices; Obstetrical And Gynecological Devices
INTRODUCER	884.5380	CONTRACEPTIVE TUBAL OCCLUSION DEVICE (TOD) AND INTRODUCER Obstetrical And Gynecological Therapeutic Devices; Obstetrical And Gynecological Devices
INTRODUCTION	868.5780	TUBE INTRODUCTION FORCEPS Therapeutic Devices; Anesthesiology Devices
INTRODUCTION	878.4200	INTRODUCTION/DRAINAGE CATHETER AND ACCESSORIES Surgical Devices; General And Plastic Surgery Devices
INVALIDITY, PATENT	314.52	NOTICE OF CERTIFICATION OF INVALIDITY OR NONINFRINGEMENT OF A PATENT Applications; Applications For FDA Approval To Market A New Drug
INVALIDITY, PATENT	314.95	NOTICE OF CERTIFICATION OF INVALIDITY OR NONINFRINGEMENT OF A PATENT Abbreviated Applications; Applications For FDA Approval To Market A New Drug
INVENTORIES	1304.04	MAINTENANCE OF RECORDS AND INVENTORIES General Information; Records And Reports Of Registrants
INVENTORY	1303.24	INVENTORY ALLOWANCE Individual Manufacturing Quotas; Quotas
INVENTORY	1304.11	INVENTORY REQUIREMENTS Inventory Requirements; Records And Reports Of Registrants
INVERT SUGAR	184.1859	INVERT SUGAR Listing Of Specific Substances Affirmed As GRAS; Direct Food Substances Affirmed As Generally Recognized As Safe
INVESTIGATION	2.10	EXAMINATION AND INVESTIGATION SAMPLES General Provisions; General Administrative Rulings And Decisions
INVESTIGATION	5.25	RESEARCH, INVESTIGATION, AND TESTING PROGRAMS AND HEALTH INFORMATION AND PROMOTION PROGRAMS General Redelegations Of Authority; Delegations Of Authority And Organization
INVESTIGATION	203.37	INVESTIGATION AND NOTIFICATION REQUIREMENTS Samples; Prescription Drug Marketing
INVESTIGATION	312.21	PHASES OF AN INVESTIGATION Investigational New Drug Application
INVESTIGATION	312.40	GENERAL REQUIREMENTS FOR USE OF AN INVESTIGATIONAL NEW DRUG IN A CLINICAL INVESTIGATION Investigational New Drug Application
INVESTIGATION	1404.311	INVESTIGATION AND REFERRAL Governmentwide Debarment And Suspension And Requirements For Drug-Free Workplace (Grants)
INVESTIGATIONAL DEVICE	5.413	APPROVAL, DISAPPROVAL, OR WITHDRAWAL OF APPROVAL OF APPLICATIONS AND ENTERING INTO AGREEMENTS FOR INVESTIGATIONAL DEVICE EXEMPTIONS Medical Devices And Radiological Health; Redelegations Of Authority; Delegations Of Authority And Organization
INVESTIGATIONAL DEVICE	812.36	TREATMENT USE OF AN INVESTIGATIONAL DEVICE Investigational Device Exemptions
INVESTIGATIONAL DEVICES	812.5	LABELING OF INVESTIGATIONAL DEVICES General Provisions; Investigational Device Exemptions
INVESTIGATIONAL DRUG	312.59	DISPOSITION OF UNUSED SUPPLY OF INVESTIGATIONAL DRUG Investigational New Drug Application
INVESTIGATIONAL DRUG	312.61	CONTROL OF THE INVESTIGATIONAL DRUG Investigational New Drug Application
INVESTIGATIONAL DRUGS	312.7	PROMOTION AND CHARGING FOR INVESTIGATIONAL DRUGS Investigational New Drug Application

INVESTIGATIONAL NEW DRUG	5.102	AUTHORITY TO APPROVE AND TO WITHDRAW APPROVAL OF A CHARGE FOR INVESTIGATIONAL NEW DRUGS Human Drugs; Redelegations Of Authority; Delegations Of Authority And Organization
INVESTIGATIONAL NEW DRUG	312.6	LABELING OF AN INVESTIGATIONAL NEW DRUG Investigational New Drug Application
INVESTIGATIONAL NEW DRUG	312.34	TREATMENT USE OF AN INVESTIGATIONAL NEW DRUG Investigational New Drug Application
INVESTIGATIONAL NEW DRUG	312.36	EMERGENCY USE OF AN INVESTIGATIONAL NEW DRUG Investigational New Drug Application
INVESTIGATIONAL NEW DRUG	312.40	GENERAL REQUIREMENTS FOR USE OF AN INVESTIGATIONAL NEW DRUG IN A CLINICAL INVESTIGATION Investigational New Drug Application
INVESTIGATIONAL NEW DRUG	320.31	APPLICABILITY OF REQUIREMENTS REGARDING AN INVESTIGATIONAL NEW DRUG APPLICATION Procedures For Determining The Bioavailability Of Drug Products; Bioavailability And Bioequivalence Requirements
INVESTIGATIONAL NEW DRUG	514.12	CONFIDENTIALITY OF DATA AND INFORMATION IN AN INVESTIGATIONAL NEW ANIMAL DRUG NOTICE General Provisions; New Animal Drug Applications
INVESTIGATIONAL NEW DRUG	601.50	CONFIDENTIALITY OF DATA AND INFORMATION IN AN INVESTIGATIONAL NEW DRUG NOTICE FOR A BIOLOGICAL PRODUCT Confidentiality Of Information; Licensing
INVESTIGATIONAL PLAN	812.25	INVESTIGATIONAL PLAN Application And Administrative Action; Investigational Device Exemptions
INVESTIGATIONAL USE	5.101	TERMINATION OF EXEMPTIONS FOR NEW DRUGS FOR INVESTIGATIONAL USE IN HUMAN BEINGS Human Drugs; Redelegations Of Authority; Delegations Of Authority And Organization
INVESTIGATIONAL USE	5.505	TERMINATION OF EXEMPTIONS FOR NEW DRUGS FOR INVESTIGATIONAL USE IN ANIMALS Animal Drugs; Redelegations Of Authority; Delegations Of Authority And Organization
INVESTIGATIONAL USE	71.37	EXEMPTION OF COLOR ADDITIVES FOR INVESTIGATIONAL USE Administrative Action On Petitions; Color Additive Petitions
INVESTIGATIONAL USE	170.17	EXEMPTION FOR INVESTIGATIONAL USE AND PROCEDURE FOR OBTAINING AUTHORIZATION TO MARKET EDIBLE PRODUCTS FORM EXPERIMENTAL ANIMALS General Provisions; Food Additives
INVESTIGATIONAL USE	312.160	DRUGS FOR INVESTIGATIONAL USE IN LABORATORY RESEARCH ANIMALS OR IN VITRO TESTS Investigational New Drug Application
INVESTIGATIONAL USE	511.1	NEW ANIMAL DRUGS FOR INVESTIGATIONAL USE EXEMPT FROM SECTION 512(A) OF THE ACT New Animal Drugs For Investigational Use
INVESTIGATIONAL USE	570.17	EXEMPTION FOR INVESTIGATIONAL USE AND PROCEDURE FOR OBTAINING AUTHORIZATION TO MARKET EDIBLE PRODUCTS FROM EXPERIMENTAL ANIMALS General Provisions; Food Additives
INVESTIGATIONS	312.56	REVIEW OF ONGOING INVESTIGATIONS Investigational New Drug Application
INVESTIGATIONS	314.54	PROCEDURE FOR SUBMISSION OF AN APPLICATION REQUIRING INVESTIGATIONS FOR APPROVAL OF A NEW INDICATION FOR/OTHER CHANGE FROM A LISTED DRUG Applications For FDA Approval To Market A New Drug
INVESTIGATIONS	812.27	REPORT OF PRIOR INVESTIGATIONS Application And Administrative Action; Investigational Device Exemptions
INVESTIGATIONS	812.46	MONITORING INVESTIGATIONS Responsibilities Of Sponsors; Investigational Device Exemptions
INVESTIGATIONS	1230.33	INVESTIGATIONS Regulations Under The Federal Caustic Poison Act
INVESTIGATIONS, CLINICAL	50.51	CLINICAL INVESTIGATIONS NOT INVOLVING GREATER THAN MINIMAL RISK Additional Safeguards For Children In Clinical Investigations; Protection Of Human Subjects
INVESTIGATIONS, CLINICAL	50.52	CLINICAL INVESTIGATIONS INVOLVING GREATER THAN MINIMAL RISK BUT PRESENTING THE PROSPECT OF DIRECT BENEFIT TO INDIVIDUAL SUBJECTS Additional Safeguards For Children In Clinical Investigations; Protection Of Human Subjects

INVESTIGATIONS, CLINICAL	50.53	CLINICAL INVESTIGATIONS INVOLVING GREATER THAN MINIMAL RISK AND NO PROSPECT OF DIRECT BENEFIT TO INDIVIDUAL SUBJECTS, BUT LIKELY TO YIELD GENERALIZABLE KNOWLEDGE ABOUT THE SUBJECTS' DISORDER OR CONDITION Additional Safeguards For Children In Clinical Investigations; Protection Of Human Subjects
INVESTIGATIONS, CLINICAL	50.54	CLINICAL INVESTIGATIONS NOT OTHERWISE APPROVABLE THAT PRESENT AN OPPORTUNITY TO UNDERSTAND, PREVENT, OR ALLEVIATE A SERIOUS PROBLEM AFFECTING THE HEALTH OR WELFARE OF CHILDREN Additional Safeguards For Children In Clinical Investigations; Protection Of Human Subjects
INVESTIGATOR	312.62	INVESTIGATOR RECORDKEEPING AND RECORD RETENTION Investigational New Drug Application
INVESTIGATOR	312.64	INVESTIGATOR REPORTS Investigational New Drug Application
INVESTIGATOR	312.70	DISQUALIFICATION OF A CLINICAL INVESTIGATOR Investigational New Drug Application
INVESTIGATOR	812.119	DISQUALIFICATION OF A CLINICAL INVESTIGATOR Responsibilities Of Investigators; Investigational Device Exemptions
INVESTIGATORS	312.53	SELECTING INVESTIGATORS AND MONITORS Investigational New Drug Application
INVESTIGATORS	312.55	INFORMING INVESTIGATORS Investigational New Drug Application
INVESTIGATORS	312.60	GENERAL RESPONSIBILITIES OF INVESTIGATORS Investigational New Drug Application
INVESTIGATOR'S	312.62	INSPECTION OF INVESTIGATOR'S RECORDS AND REPORTS Investigational New Drug Application
INVESTIGATORS	812.43	SELECTING INVESTIGATORS AND MONITORS Responsibilities Of Sponsors; Investigational Device Exemptions
INVESTIGATORS	812.45	INFORMING INVESTIGATORS Responsibilities Of Sponsors; Investigational Device Exemptions
INVESTIGATORS	812.100	GENERAL RESPONSIBILITIES OF INVESTIGATORS Responsibilities Of Investigators; Investigational Device Exemptions
INVESTIGATORS	812.110	SPECIFIC RESPONSIBILITIES OF INVESTIGATORS Responsibilities Of Investigators; Investigational Device Exemptions
INVOICE	1312.27	CONTENTS OF SPECIAL CONTROLLED SUBSTANCES INVOICE Exportation Of Controlled Substances; Importation And Exportation Of Controlled Substances
INVOICE	1312.28	DISTRIBUTION OF SPECIAL CONTROLLED SUBSTANCES INVOICE Exportation Of Controlled Substances; Importation And Exportation Of Controlled Substances
IODATE, CALCIUM	184.1206	CALCIUM IODATE Listing Of Specific Substances Affirmed As GRAS; Direct Food Substances Affirmed As Generally Recognized As Safe
IODATE, CUPROUS	184.1265	CUPROUS IODATE Listing Of Specific Substances Affirmed As GRAS; Direct Food Substances Affirmed As Generally Recognized As Safe
IODATE, POTASSIUM	184.1635	POTASSIUM IODATE Listing Of Specific Substances Affirmed As GRAS; Direct Food Substances Affirmed As Generally Recognized As Safe
IODIDE, DITHIAZANINE	520.763	DITHIAZANINE IODIDE ORAL DOSAGE FORMS Oral Dosage Form New Animal Drugs
IODIDE, POTASSIUM	172.375	POTASSIUM IODIDE Special Dietary And Nutritional Additives; Food Additives Permitted For Direct Addition To Food For Human Consumption
IODIDE, POTASSIUM	184.1634	POTASSIUM IODIDE Listing Of Specific Substances Affirmed As GRAS; Direct Food Substances Affirmed As Generally Recognized As Safe
IODIDE, POTASSIUM	582.5634	POTASSIUM IODIDE Nutrients And/or Dietary Supplements; Substances Generally Recognized As Safe
IODINATED CASEIN	520.1157	IODINATED CASEIN TABLETS Oral Dosage Form New Animal Drugs
IODINATED CASEIN	558.295	IODINATED CASEIN Specific New Animal Drugs For Use In Animal Feeds; New Animal Drugs For Use In Animal Feeds
IODINE	862.1640	PROTEIN-BOUND IODINE TEST SYSTEM Clinical Chemistry Test Systems; Clinical Chemistry And Clinical Toxicology Devices
IODIZED SALT	100.155	SALT AND IODIZED SALT Specific Administrative Rulings And Decisions; General
IODOCHLORHYDROXYQUIN	520.1158	IODOCHLORHYDROXYQUIN BOLUSES Oral Dosage Form New Animal Drugs

IRON	510.440	INJECTABLE IRON PREPARATIONS Requirements For Specific New Animal Drugs; New Animal Drugs
IRON	862.1410	IRON (NON-HEME) TEST SYSTEM Clinical Chemistry Test Systems; Clinical Chemistry And Clinical Toxicology Devices
IRON AMMONIUM CITRATE	172.430	IRON AMMONIUM CITRATE Anticaking Agents; Food Additives Permitted For Direct Addition To Food For Human Consumption
IRON AMMONIUM CITRATE	573.560	IRON AMMONIUM CITRATE Food Additive Listing; Food Additives Permitted In Feed And Drinking Water Of Animals
IRON-BINDING	862.1415	IRON-BINDING CAPACITY TEST SYSTEM Clinical Chemistry Test Systems; Clinical Chemistry And Clinical Toxicology Devices
IRON-CHOLINE CITRATE	172.370	IRON-CHOLINE CITRATE COMPLEX Special Dietary And Nutritional Additives; Food Additives Permitted For Direct Addition To Food For Human Consumption
IRON-CHOLINE CITRATE	573.580	IRON-CHOLINE CITRATE COMPLEX Food Additive Listing; Food Additives Permitted In Feed And Drinking Water Of Animals
IRON-CONTAINING	111.50	PACKAGING OF IRON-CONTAINING DIETARY SUPPLEMENTS Current Good Manufacturing Practice For Dietary Supplements
IRON DEXTRAN	520.1182	IRON DEXTRAN ORAL SUSPENSION Oral Dosage Form New Animal Drugs
IRON DEXTRAN	522.1182	IRON DEXTRAN COMPLEX INJECTION Implantation Or Injectable Dosage Form New Animal Drugs
IRON HYDROGENATED DEXTRAN	522.1183	IRON HYDROGENATED DEXTRAN INJECTION Implantation Or Injectable Dosage Form New Animal Drugs
IRON OXIDE	73.200	SYNTHETIC IRON OXIDE Foods; Listing Of Color Additives Exempt From Certification
IRON OXIDE	73.1200	SYNTHETIC IRON OXIDE Drugs; Listing Of Color Additives Exempt From Certification
IRON OXIDES	73.2250	IRON OXIDES Cosmetics; Listing Of Color Additives Exempt From Certification
IRON OXIDES	73.3125	IRON OXIDES Medical Devices; Listing Of Color Additives Exempt From Certification
IRON OXIDES	186.1374	IRON OXIDES Listing Of Specific Substances Affirmed As GRAS; Indirect Food Substances Affirmed As Generally Recognized As Safe
IRON REDUCED	582.5375	IRON REDUCED Nutrients And/or Dietary Supplements; Substances Generally Recognized As Safe
IRRADIATION	179.25	GENERAL PROVISIONS FOR FOOD IRRADIATION Radiation And Radiation Sources; Irradiation In The Production, Processing And Handling Of Food
IRRADIATION	179.45	PACKAGING MATERIALS FOR USE DURING THE IRRADIATION OF PREPACKAGED FOODS Packaging Materials For Irradiated Foods; Irradiation In The Production, Processing And Handling Of Food
IRREVERSIBLE MORBIDITY	314.510	APPROVAL BASED ON A SURROGATE ENDPOINT OR ON AN EFFECT ON A CLINICAL ENDPOINT OTHER THAN SURVIVAL OR IRREVERSIBLE MORBIDITY Applications For FDA Approval To Market A New Drug
IRREVERSIBLE MORBIDITY	601.41	APPROVAL BASED ON A SURROGATE ENDPOINT OR ON AN EFFECT ON A CLINICAL ENDPOINT OTHER THAN SURVIVAL OR IRREVERSIBLE MORBIDITY Accelerated Approval Of Biological Products; Licensing
IRRIGATING	880.6960	IRRIGATING SYRINGE General Hospital And Personal Use Miscellaneous Devices; General Hospital And Personal Use Devices
IRRIGATION, COLONIC	876.5220	COLONIC IRRIGATION SYSTEM Therapeutic Devices; Gastroenterology urology Devices
IRRIGATION DEVICE	886.4360	OCULAR SURGERY IRRIGATION DEVICE Surgical Devices; Ophthalmic Devices
IRRIGATION, ORAL	872.6510	ORAL IRRIGATION UNIT Miscellaneous Devices; Dental Devices
IRRIGATOR, NASAL	874.5550	POWERED NASAL IRRIGATOR Therapeutic Devices; Ear, Nose, And Throat Devices
IRRIGATOR, OSTOMY	876.5895	OSTOMY IRRIGATOR Therapeutic Devices; Gastroenterology-urology Devices
ISO-BUTANE	184.1165	N-BUTANE AND ISO-BUTANE Listing Of Specific Substances Affirmed As GRAS; Direct Food Substances Affirmed As Generally Recognized As Safe
ISOBUTYLENE	177.1420	ISOBUTYLENE POLYMERS Substances For Use As Basic Components Of Single And Repeated Use Food Contact Surfaces; Indirect Food Additives: Polymers

ISOBUTYLENE-BUTENE	177.1430	ISOBUTYLENE-BUTENE COPOLYMERS Substances For Use As Basic Components Of Single And Repeated Use Food Contact Surfaces; Indirect Food Additives: Polymers
ISOBUTYRATE	172.833	SUCROSE ACETATE ISOBUTYRATE (SAIB) Food Additives Permitted For Direct Addition To Food For Human Consumption
ISOCITRIC DEHYDROGENASE	862.1420	ISOCITRIC DEHYDROGENASE TEST SYSTEM Clinical Chemistry Test Systems; Clinical Chemistry And Clinical Toxicology Devices
ISOFLUPREDONE	524.1484b	NEOMYCIN SULFATE, ISOFLUPREDONE ACETATE, TETRACAINE HYDROCHLORIDE, AND MYRISTYL-GAMMA-PICOLINIUM CHLORIDE TOPICAL POWDER Ophthalmic And Topical Dosage Form New Animal Drugs
ISOFLUPREDONE	524.1484c	NEOMYCIN SULFATE, ISOFLUPREDONE ACETATE, TETRACAINE HYDROCHLORIDE OINTMENT Ophthalmic And Topical Dosage Form New Animal Drugs
ISOFLURANE	529.1186	ISOFLURANE Certain Other Dosage Form New Animal Drugs
ISOKINETIC	890.1925	ISOKINETIC TESTING AND EVALUATION SYSTEM Physical Medicine Diagnostic Devices; Physical Medicine Devices
ISOLATE, FISH PROTEIN	172.340	FISH PROTEIN ISOLATE Special Dietary And Nutritional Additives; Food Additives Permitted For Direct Addition To Food For Human Consumption
ISOLATED	876.5880	ISOLATED KIDNEY PERFUSION AND TRANSPORT SYSTEM AND ACCESSORIES Therapeutic Devices; Gastroenterology-urology Devices
ISOLATION	870.2600	SIGNAL ISOLATION SYSTEM Cardiovascular Monitoring Devices; Cardiovascular Devices
ISOLATION	870.2620	LINE ISOLATION MONITOR Cardiovascular Monitoring Devices; Cardiovascular Devices
ISOLATION	880.5450	PATIENT CARE REVERSE ISOLATION CHAMBER General Hospital And Personal Use Therapeutic Devices; General Hospital And Personal Use Devices
ISOLEUCINE	582.5381	ISOLEUCINE Nutrients And/or Dietary Supplements; Substances Generally Recognized As Safe
ISOMERASE	184.1372	INSOLUBLE GLUCOSE ISOMERASE ENZYME PREPARATIONS Listing Of Specific Substances Affirmed As GRAS; Direct Food Substances Affirmed As Generally Recognized As Safe
ISOMERASE	862.1570	PHOSPHOHEXOSE ISOMERASE TEST SYSTEM Clinical Chemistry Test Systems; Clinical Chemistry And Clinical Toxicology Devices
ISOMERASE	862.1720	TRIOSE PHOSPHATE ISOMERASE TEST SYSTEM Clinical Chemistry Test Systems; Clinical Chemistry And Clinical Toxicology Devices
ISOPARAFFINIC	172.882	SYNTHETIC ISOPARAFFINIC PETROLEUM HYDROCARBONS Multipurpose Additives; Food Additives Permitted For Direct Addition To Food For Human Consumption
ISOPARAFFINIC	178.3530	ISOPARAFFINIC PETROLEUM HYDROCARBONS, SYNTHETIC Certain Adjuvants And Production Aids; Indirect Food Additives: Adjuvants, Production Aids, And Sanitizers
ISOPHTHALATE	177.1240	1,4-CYCLOHEXYLENE DIMETHYLENE TEREPHTHALATE AND 1,4-CYCLOHEXYLENE DIMETHYLENE ISOPHTHALATE COPOLYMER Substances For Use As Basic Components Of Single And Repeated Use Food Contact Surfaces; Indirect Food Additives: Polymers
ISOPHTHALATE	177.1345	ETHYLENE/1,3-PHENYLENE OXYETHYLENE ISOPHTHALATE/TEREPHTHALATE COPOLYMER Substances For Use As Basic Components Of Single And Repeated Use Food Contact Surfaces; Indirect Food Additives: Polymers
ISOPROPAMIDE	520.1920	PROCHLORPERAZINE, ISOPROPAMIDE SUSTAINED RELEASE CAPSULES Oral Dosage Form New Animal Drugs
ISOPROPAMIDE	520.1921	PROCHLORPERAZINE, ISOPROPAMIDE, WITH NEOMYCIN SUSTAINED-RELEASE CAPSULES Oral Dosage Form New Animal Drugs
ISOPROPAMIDE	522.1920	PROCHLORPERAZINE, ISOPROPAMIDE FOR INJECTION Implantation Or Injectable Dosage Form New Animal Drugs
ISOPROPYL ALCOHOL	173.240	ISOPROPYL ALCOHOL Solvents, Lubricants, Release Agents And Related Substances; Secondary Direct Food Additives Permitted In Food For Human Consumption
ISOPROPYL CITRATE	184.1386	ISOPROPYL CITRATE Listing Of Specific Substances Affirmed As GRAS; Direct Food Substances Affirmed As Generally Recognized As Safe
ISOPROPYL CITRATE	582.6386	ISOPROPYL CITRATE Sequestrants; Substances Generally Recognized As Safe

ISOPROPYLIDENE	189.300	HYDROGENATED 4,4'-ISOPROPYLIDENE-DIPHENOLPHOSPHITE ESTER RESINS Substances Prohibited From Use In Human Food
ISOPROPYLIDENEDIPHENOL	175.380	XYLENE-FORMALDEHYDE RESINS CONDENSED WITH 4,4'-ISOPROPYLIDENEDIPHENOL-EPICHLOROHYDRIN EPOXY RESINS Substances For Use As Components Of Coatings; Indirect Food Additives
ISOPROPYLIDENEDIPHENOL	177.1440	4,4'-ISOPROPYLIDENEDIPHENOL-EPICHLOROHYDRIN RESINS MINIMUM MOLECULAR WEIGHT 10,000 Substances For Use As Basic Components Of Single And Repeated Use Food Contact Surfaces; Indirect Food Additives: Polymers
ISOPROPYLIDENEDIPHENOL	177.2280	4,4'-ISOPROPYLIDENEDIPHENOL-EPICHLOROHYDRIN THERMOSETTING EPOXY RESINS Substances For Use Only As Components Of Articles Intended For Repeated Use; Indirect Food Additives: Polymers
ISOPROTERENOL	201.305	ISOPROTERENOL INHALATION PREPARATIONS (PRESSURIZED AEROSOLS, NEBULIZERS, POWDERS) FOR HUMAN USE; WARNINGS Specific Labeling Requirements For Specific Drug Products; Labeling
ISOTOPE	886.1670	OPHTHALMIC ISOTOPE UPTAKE PROBE Diagnostic Devices; Ophthalmic Devices
ISSUANCE	5.33	ISSUANCE OF REPORTS OF MINOR VIOLATIONS General Redelegations Of Authority; Delegations Of Authority And Organization
ISSUANCE	5.34	ISSUANCE OF NOTICES RELATING TO PROPOSALS AND ORDERS FOR DEBARMENT AND DENIAL OF AN APPLICATION TO TERMINATE DEBARMENT General Redelegations Of Authority; Delegations Of Authority And Organization
ISSUANCE	5.100	ISSUANCE OF NOTICE IMPLEMENTING THE PROVISIONS OF THE DRUG AMENDMENTS OF 1962 Human Drugs; Redelegations Of Authority; Delegations Of Authority And Organization
ISSUANCE	5.105	ISSUANCE OF NOTICES RELATING TO PROPOSALS TO REFUSE APPROVAL OR TO WITHDRAW APPROVAL OF NEW DRUG APPLICATIONS AND THEIR SUPPLEMENTS Human Drugs; Redelegations Of Authority; Delegations Of Authority And Organization
ISSUANCE	5.109	ISSUANCE OF WRITTEN NOTICES CONCERNING PATENT INFORMATION, CURRENT GOOD MANUFACTURING PRACTICES AND FALSE OR MISLEADING LABELING OF NEW DRUGS Human Drugs; Redelegations Of Authority; Delegations Of Authority And Organization
ISSUANCE	5.202	ISSUANCE OF NOTICES OF OPPORTUNITY FOR A HEARING ON PROPOSALS FOR DENIAL OF APPROVAL OF APPLICATIONS FOR LICENSES Biologics; Redelegations Of Authority; Delegations Of Authority And Organization
ISSUANCE	5.203	ISSUANCE AND REVOCATION OF LICENSES FOR THE PROPAGATION OR MANUFACTURE AND PREPARATION OF BIOLOGICAL PRODUCTS Biologics; Redelegations Of Authority; Delegations Of Authority And Organization
ISSUANCE	5.301	ISSUANCE OF INITIAL EMERGENCY PERMIT ORDERS AND NOTICES OF CONFIRMATION OF EFFECTIVE DATE OF FINAL REGULATIONS ON FOOD FOR HUMAN AND ANIMAL CONSUMPTION Foods And Cosmetics; Redelegations Of Authority; Delegations Of Authority And Organization
ISSUANCE	5.400	ISSUANCE OF FEDERAL REGISTER DOCUMENTS TO RECOGNIZE OR TO WITHDRAW RECOGNITION OF A STANDARD TO MEET PREMARKET SUBMISSION REQUIREMENTS Medical Devices And Radiological Health; Redelegations Of Authority; Delegations Of Authority And Organization
ISSUANCE	5.401	ISSUANCE OF FEDERAL REGISTER DOCUMENTS PERTAINING TO EXEMPTIONS FROM PREMARKET NOTIFICATION Medical Devices And Radiological Health; Redelegations Of Authority; Delegations Of Authority And Organization
ISSUANCE	5.500	ISSUANCE OF FEDERAL REGISTER DOCUMENTS PERTAINING TO THE DETERMINATION OF SAFE LEVELS, NOTICE OF NEED FOR DEVELOPMENT OF AN ANALYTICAL METHOD, AND PROHIBITION OF CERTAIN EXTRALABEL DRUG USE Animal Drugs; Redelegations Of Authority; Delegations Of Authority And Organization
ISSUANCE	5.502	ISSUANCE OF NOTICES, PROPOSALS AND ORDERS RELATING TO NEW ANIMAL DRUGS AND MEDICATED FEED MILL LICENSE APPLICATIONS Animal Drugs; Redelegations Of Authority; Delegations Of Authority And Organization
ISSUANCE	5.504	ISSUANCE OF WRITTEN NOTICES CONCERNING PATENT INFORMATION, CURRENT GOOD MANUFACTURING PRACTICES AND FALSE OR MISLEADING LABELING OF NEW ANIMAL DRUGS AND FEEDS BEARING OR CONTAINING NEW ANIMAL DRUGS Animal Drugs; Redelegations Of Authority; Delegations Of Authority And Organization

ISSUANCE	12.20	INITIATION OF A HEARING INVOLVING THE ISSUANCE, AMENDMENT, OR REVOCATION OF A REGULATION Initiation Of Proceedings; Formal Evidentiary Public Hearing
ISSUANCE	12.21	INITIATION OF A HEARING INVOLVING THE ISSUANCE, AMENDMENT, OR REVOCATION OF AN ORDER Initiation Of Proceedings; Formal Evidentiary Public Hearing
ISSUANCE	108.7	ISSUANCE OR DENIAL OF A PERMIT General Provisions; Emergency Permit Control
ISSUANCE	1240.20	ISSUANCE AND POSTING OF CERTIFICATES FOLLOWING INSPECTION Control Of Communicable Diseases
ISSUANCE, LABELING	211.125	LABELING ISSUANCE Packaging And Labeling Control; Current Good Manufacturing Practice For Finished Pharmaceuticals
ISSUANCE, LICENSE	601.4	ISSUANCE AND DENIAL OF LICENSE General Provisions; Licensing
ISSUANCE, LICENSES	601.20	BIOLOGICS LICENSES; ISSUANCE AND CONDITIONS Biologics Licensing; Licensing
ISSUANCE, PERMIT	108.7	ISSUANCE OR DENIAL OF PERMIT General Provisions; Emergency Permit Control
ISSUANCE, PERMIT	1312.13	ISSUANCE OF IMPORT PERMIT Importation Of Controlled Substances; Importation And Exportation Of Controlled Substances
ISSUANCE, PERMIT	1312.23	ISSUANCE OF EXPORT PERMIT Exportation Of Controlled Substances; Importation And Exportation Of Controlled Substances
ISSUANCE, PRESCRIPTIONS	1306.05	MANNER OF ISSUANCE OF PRESCRIPTIONS General Information; Prescriptions
ISSUE	71.22	DECEPTION AS A BASIS FOR REFUSING TO ISSUE REGULATIONS; DECEPTIVE USE OF A COLOR ADDITIVE FOR WHICH A REGULATION HAS ISSUED Administrative Action On Petitions; Color Additive Petitions
ISSUE	1306.03	PERSONS ENTITLED TO ISSUE PRESCRIPTIONS General Information; Prescriptions
IUD	884.5360	CONTRACEPTIVE INTRAUTERINE DEVICE (IUD) AND INTRODUCER Obstetrical And Gynecological Therapeutic Devices; Obstetrical And Gynecological Devices
IVERMECTIN	520.1192	IVERMECTIN PASTE Oral Dosage Form New Animal Drugs
IVERMECTIN	520.1193	IVERMECTIN TABLETS AND CHEWABLES Oral Dosage Form New Animal Drugs
IVERMECTIN	520.1194	IVERMECTIN DRENCH Oral Dosage Form New Animal Drugs
IVERMECTIN	520.1195	IVERMECTIN LIQUID Oral Dosage Form New Animal Drugs
IVERMECTIN	520.1196	IVERMECTIN AND PYRANTEL PAMOATE CHEWABLE TABLET Oral Dosage Form New Animal Drug Applications
IVERMECTIN	520.1197	IVERMECTIN SUSTAINED-RELEASE BOLUS Oral Dosage Form New Animal Drugs
IVERMECTIN	522.1192	IVERMECTIN INJECTION Implantation Or Injectable Dosage Form New Animal Drugs
IVERMECTIN	522.1193	IVERMECTIN AND CLORSULON INJECTION Implantation Or Injectable Dosage Form New Animal Drugs
IVERMECTIN	524.1193	IVERMECTIN POUR-ON Ophthalmic And Topical Dosage Form New Animal Drugs
IVERMECTIN	524.1195	IVERMECTIN OTIC SUSPENSION Ophthalmic And Topical Dosage Form New Animal Drugs
IVERMECTIN	556.344	IVERMECTIN Specific Tolerances For Residues Of New Animal Drugs; Tolerances For Residues Of New Animal Drugs In Food
IVERMECTIN	558.300	IVERMECTIN New Animal Drugs For Use In Animal Feeds

J

JACK CHEESE	133.153	MONTEREY CHEESE AND MONTEREY JACK CHEESE Requirements For Specific Standardized Cheese And Related Products; Cheeses And Related Cheese Products
JACK CHEESE	133.154	HIGH-MOISTURE JACK CHEESE Requirements For Specific Standardized Cheese And Related Products; Cheeses And Related Cheese Products
JAMS	150.160	FRUIT PRESERVES AND JAMS Requirements For Specific Standardized Fruit Butters, Jellies, Preserves; Fruit Butters, Jellies, Preserves, And Related Products
JAMS	150.161	ARTIFICIALLY SWEETENED FRUIT PRESERVES AND JAMS Requirements For Specific Standardized Fruit Butters, Jellies, Preserves; Fruit Butters, Jellies, Preserves, And Related Products
JAPAN WAX	186.1555	JAPAN WAX Indirect Food Substances Affirmed As Generally Recognized As Safe
JELLY	150.140	FRUIT JELLY Requirements For Specific Standardized Fruit Butters, Jellies, Preserves; Fruit Butters, Jellies, Preserves, And Related Products
JELLY	150.141	ARTIFICIALLY SWEETENED FRUIT JELLY Requirements For Specific Standardized Fruit Butters, Jellies, Preserves; Fruit Butters, Jellies, Preserves, And Related Products
JET INJECTOR	872.4465	GAS-POWERED JET INJECTOR Surgical Devices; Dental Devices
JET INJECTOR	872.4475	SPRING-POWERED JET INJECTOR Surgical Devices; Dental Devices
JET LAVAGE	880.5475	JET LAVAGE General Hospital And Personal Use Therapeutic Devices; General Hospital And Personal Use Devices
JET, WATER	876.4650	WATER JET RENAL STONE DISLODGER SYSTEM Surgical Devices; Gastroenterology-urology Devices
JOINT, ANKLE	888.3100	ANKLE JOINT METAL/COMPOSITE SEMI-CONSTRAINED CEMENTED PROSTHESIS Prosthetic Devices; Orthopedic Devices
JOINT, ANKLE	888.3110	ANKLE JOINT METAL/POLYMER SEMI-CONSTRAINED CEMENTED PROSTHESIS Prosthetic Devices; Orthopedic Devices
JOINT, ANKLE	888.3120	ANKLE JOINT METAL/POLYMER NON-CONSTRAINED CEMENTED PROSTHESIS Prosthetic Devices; Orthopedic Devices
JOINT COMMITTEE	26.73	JOINT COMMITTEE Framework Provisions; Mutual Recognition Of Pharmaceutical GMP Reports, Medical Device Quality System Audit Reports
JOINT COMMITTEE, SECTORAL	26.17	ROLE AND COMPOSITION OF THE JOINT SECTORAL COMMITTEE Specific Sector Provisions For Pharmaceutical GMPs; Mutual Recognition Of Pharmaceutical GMP Reports, Medical Device Quality System Audit Reports
JOINT COMMITTEE, SECTORAL	26.47	ROLE AND COMPOSITION OF THE JOINT SECTORAL COMMITTEE Specific Sector Provisions For Medical Devices; Mutual Recognition Of Pharmaceutical GMP Reports, Medical Device Quality System Audit Reports
JOINT, ELBOW	888.3150	ELBOW JOINT METAL/POLYMER CONSTRAINED CEMENTED PROSTHESIS Prosthetic Devices; Orthopedic Devices
JOINT, ELBOW	888.3160	ELBOW JOINT METAL/POLYMER SEMI-CONSTRAINED CEMENTED PROSTHESIS Prosthetic Devices; Orthopedic Devices
JOINT, ELBOW	888.3170	ELBOW JOINT RADIAL (HEMI-ELBOW) POLYMER PROSTHESIS Prosthetic Devices; Orthopedic Devices
JOINT, ELBOW	888.3180	ELBOW JOINT HUMERAL (HEMI-ELBOW) METALLIC UNCEMENTED PROSTHESIS Prosthetic Devices; Orthopedic Devices
JOINT FILINGS	1309.33	FILING OF APPLICATION; JOINT FILINGS Registration Of Manufacturers, Distributors, Importers And Exporters Of List I Chemicals
JOINT, FINGER	888.3200	FINGER JOINT METAL/METAL CONSTRAINED UNCEMENTED PROSTHESIS Prosthetic Devices; Orthopedic Devices
JOINT, FINGER	888.3210	FINGER JOINT METAL/METAL CONSTRAINED CEMENTED PROSTHESIS Prosthetic Devices; Orthopedic Devices
JOINT, FINGER	888.3220	FINGER JOINT METAL/POLYMER CONSTRAINED CEMENTED PROSTHESIS Prosthetic Devices; Orthopedic Devices
JOINT, FINGER	888.3230	FINGER JOINT POLYMER CONSTRAINED CEMENTED PROSTHESIS Prosthetic Devices; Orthopedic Devices

JOINT, HIP	888.3300	HIP JOINT METAL CONSTRAINED CEMENTED OR UNCEMENTED PROSTHESIS Prosthetic Devices; Orthopedic Devices
JOINT, HIP	888.3310	HIP JOINT METAL/POLYMER CONSTRAINED CEMENTED OR UNCEMENTED PROSTHESIS Prosthetic Devices; Orthopedic Devices
JOINT, HIP	888.3320	HIP JOINT METAL/METAL SEMI-CONSTRAINED, WITH A CEMENTED ACETABULAR COMPONENT, PROSTHESIS Prosthetic Devices; Orthopedic Devices
JOINT, HIP	888.3330	HIP JOINT METAL/METAL SEMI-CONSTRAINED, WITH AN UNCEMENTED ACETABULAR COMPONENT, PROSTHESIS Prosthetic Devices; Orthopedic Devices
JOINT, HIP	888.3340	HIP JOINT METAL/COMPOSITE SEMI-CONSTRAINED CEMENTED PROSTHESIS Prosthetic Devices; Orthopedic Devices
JOINT, HIP	888.3350	HIP JOINT METAL/POLYMER SEMI-CONSTRAINED CEMENTED PROSTHESIS Prosthetic Devices; Orthopedic Devices
JOINT, HIP	888.3353	HIP JOINT METAL/CERAMIC/POLYMER SEMI-CONSTRAINED CEMENTED OR NONPOROUS UNCEMENTED PROSTHESIS Prosthetic Devices; Orthopedic Devices
JOINT, HIP	888.3358	HIP JOINT METAL/POLYMER/METAL SEMI-CONSTRAINED POROUS-COATED UNCEMENTED PROSTHESIS Prosthetic Devices; Orthopedic Devices
JOINT, HIP	888.3360	HIP JOINT FEMORAL (HEMI-HIP) METALLIC CEMENTED OR UNCEMENTED PROSTHESIS Prosthetic Devices; Orthopedic Devices
JOINT, HIP	888.3370	HIP JOINT (HEMI-HIP) ACETABULAR METAL CEMENTED PROSTHESIS Prosthetic Devices; Orthopedic Devices
JOINT, HIP	888.3380	HIP JOINT FEMORAL (HEMI-HIP) TRUNION-BEARING METAL/POLYACETAL CEMENTED PROSTHESIS Prosthetic Devices; Orthopedic Devices
JOINT, HIP	888.3390	HIP JOINT FEMORAL (HEMI-HIP) METAL/POLYMER CEMENTED OR UNCEMENTED PROSTHESIS Prosthetic Devices; Orthopedic Devices
JOINT, HIP	888.3400	HIP JOINT FEMORAL (HEMI-HIP) METALLIC RESURFACING PROSTHESIS Prosthetic Devices; Orthopedic Devices
JOINT, HIP	888.3410	HIP JOINT METAL/POLYMER SEMI-CONSTRAINED RESURFACING CEMENTED PROSTHESIS Prosthetic Devices; Orthopedic Devices
JOINT, KNEE	888.3480	KNEE JOINT FEMOROTIBIAL METALLIC CONSTRAINED CEMENTED PROSTHESIS Prosthetic Devices; Orthopedic Devices
JOINT, KNEE	888.3490	KNEE JOINT FEMOROTIBIAL METAL/COMPOSITE NON-CONSTRAINED CEMENTED PROSTHESIS Prosthetic Devices; Orthopedic Devices
JOINT, KNEE	888.3500	KNEE JOINT FEMOROTIBIAL METAL/COMPOSITE SEMI-CONSTRAINED CEMENTED PROSTHESIS Prosthetic Devices; Orthopedic Devices
JOINT, KNEE	888.3510	KNEE JOINT FEMOROTIBIAL METAL/POLYMER CONSTRAINED CEMENTED PROSTHESIS Prosthetic Devices; Orthopedic Devices
JOINT, KNEE	888.3520	KNEE JOINT FEMOROTIBIAL METAL/POLYMER NON-CONSTRAINED CEMENTED PROSTHESIS Prosthetic Devices; Orthopedic Devices
JOINT, KNEE	888.3530	KNEE JOINT FEMOROTIBIAL METAL/POLYMER SEMI-CONSTRAINED CEMENTED PROSTHESIS Prosthetic Devices; Orthopedic Devices
JOINT, KNEE	888.3540	KNEE JOINT PATELLOFEMORAL POLYMER/METAL SEMI-CONSTRAINED CEMENTED PROSTHESIS Prosthetic Devices; Orthopedic Devices
JOINT, KNEE	888.3550	KNEE JOINT PATELLOFEMOROTIBIAL POLYMER/METAL/METAL CONSTRAINED CEMENTED PROSTHESIS Prosthetic Devices; Orthopedic Devices
JOINT, KNEE	888.3560	KNEE JOINT PATELLOFEMOROTIBIAL POLYMER/METAL/POLYMER SEMI-CONSTRAINED CEMENTED PROSTHESIS Prosthetic Devices; Orthopedic Devices
JOINT, KNEE	888.3570	KNEE JOINT FEMORAL (HEMI-KNEE) METALLIC UNCEMENTED PROSTHESIS Prosthetic Devices; Orthopedic Devices
JOINT, KNEE	888.3580	KNEE JOINT PATELLAR (HEMI-KNEE) METALLIC RESURFACING UNCEMENTED PROSTHESIS Prosthetic Devices; Orthopedic Devices
JOINT, KNEE	888.3590	KNEE JOINT TIBIAL (HEMI-KNEE) METALLIC RESURFACING UNCEMENTED PROSTHESIS Prosthetic Devices; Orthopedic Devices

JOINT SECTORAL COMMITTEE	26.17	ROLE AND COMPOSITION OF THE JOINT SECTORAL COMMITTEE Specific Sector Provisions For Pharmaceutical GMPs; Mutual Recognition Of Pharmaceutical GMP Reports, Medical Device Quality System Audit Reports
JOINT SECTORAL COMMITTEE	26.47	ROLE AND COMPOSITION OF THE JOINT SECTORAL COMMITTEE Specific Sector Provisions For Medical Devices; Mutual Recognition Of Pharmaceutical GMP Reports, Medical Device Quality System Audit Reports
JOINT, SHOULDER	888.3640	SHOULDER JOINT METAL/METAL OR METAL/POLYMER CONSTRAINED CEMENTED PROSTHESIS Prosthetic Devices; Orthopedic Devices
JOINT, SHOULDER	888.3650	SHOULDER JOINT METAL/POLYMER NON-CONSTRAINED CEMENTED PROSTHESIS Prosthetic Devices; Orthopedic Devices
JOINT, SHOULDER	888.3660	SHOULDER JOINT METAL/POLYMER SEMI-CONSTRAINED CEMENTED PROSTHESIS Prosthetic Devices; Orthopedic Devices
JOINT, SHOULDER	888.3670	SHOULDER JOINT METAL/POLYMER/METAL NONCONSTRAINED OR SEMI-CONSTRAINED POROUS-COATED UNCEMENTED PROSTHESIS Prosthetic Devices; Orthopedic Devices
JOINT, SHOULDER	888.3680	SHOULDER JOINT GLENOID (HEMI-SHOULDER) METALLIC CEMENTED PROSTHESIS Prosthetic Devices; Orthopedic Devices
JOINT, SHOULDER	888.3690	SHOULDER JOINT HUMERAL (HEMI-SHOULDER) METALLIC UNCEMENTED PROSTHESIS Prosthetic Devices; Orthopedic Devices
JOINT, TEMPOROMANDIBULAR	872.3940	TOTAL TEMPOROMANDIBULAR JOINT PROSTHESIS Prosthetic Devices; Dental Devices
JOINT, TOE	888.3720	TOE JOINT POLYMER CONSTRAINED PROSTHESIS Prosthetic Devices; Orthopedic Devices
JOINT, TOE	888.3730	TOE JOINT PHALANGEAL (HEMI-TOE) POLYMER PROSTHESIS Prosthetic Devices; Orthopedic Devices
JOINT, WRIST	888.3750	WRIST JOINT CARPAL LUNATE POLYMER PROSTHESIS Prosthetic Devices; Orthopedic Devices
JOINT, WRIST	888.3760	WRIST JOINT CARPAL SCAPHOID POLYMER PROSTHESIS Prosthetic Devices; Orthopedic Devices
JOINT, WRIST	888.3770	WRIST JOINT CARPAL TRAPEZIUM POLYMER PROSTHESIS Prosthetic Devices; Orthopedic Devices
JOINT, WRIST	888.3780	WRIST JOINT POLYMER CONSTRAINED PROSTHESIS Prosthetic Devices; Orthopedic Devices
JOINT, WRIST	888.3790	WRIST JOINT METAL CONSTRAINED CEMENTED PROSTHESIS Prosthetic Devices; Orthopedic Devices
JOINT, WRIST	888.3800	WRIST JOINT METAL/POLYMER SEMI-CONSTRAINED CEMENTED PROSTHESIS Prosthetic Devices; Orthopedic Devices
JOINT, WRIST	888.3810	WRIST JOINT ULNAR (HEMI-WRIST) POLYMER PROSTHESIS Prosthetic Devices; Orthopedic Devices
JUDICIAL FORFEITURE	1316.78	JUDICIAL FORFEITURE Seizure, Forfeiture, And Disposition Of Property; Administrative Functions, Practices, And Procedures
JUDICIAL FORFEITURE	1316.95	PETITION FOR EXPEDITED RELEASE OF A CONVEYANCE IN A JUDICIAL FORFEITURE ACTION Expedited Forfeiture Proceedings For Certain Property; Administrative Functions, Practices, And Procedures
JUDICIAL FORFEITURE	1316.96	RULING ON A PETITION FOR EXPEDITED RELEASE OF A CONVEYANCE IN A JUDICIAL FORFEITURE ACTION Expedited Forfeiture Proceedings For Certain Property; Administrative Functions, Practices, And Procedures
JUDICIAL FORFEITURE	1316.97	INITIATING JUDICIAL FORFEITURE PROCEEDING AGAINST A CONVEYANCE WITHIN 60 DAYS OF FILING A CLAIM AND COST BOND Expedited Forfeiture Proceedings For Certain Property; Administrative Functions, Practices, And Procedures
JUDICIAL FORFEITURE	1316.98	SUBSTITUTE RES BOND IN A JUDICIAL FORFEITURE ACTION AGAINST A CONVEYANCE Expedited Forfeiture Proceedings For Certain Property; Administrative Functions, Practices, And Procedures
JUDICIAL REVIEW	12.30	JUDICIAL REVIEW AFTER WAIVER OF HEARING ON A REGULATION Initiation Of Proceedings; Formal Evidentiary Public Hearing
JUDICIAL REVIEW	12.159	COPIES OF PETITIONS FOR JUDICIAL REVIEW Judicial Review; Formal Evidentiary Public Hearing
JUDICIAL REVIEW	16.120	JUDICIAL REVIEW Judicial Review; Regulatory Hearing Before The Food And Drug Administration
JUDICIAL REVIEW	17.51	JUDICIAL REVIEW Civil Money Penalties Hearings
JUDICIAL REVIEW	20.46	JUDICIAL REVIEW OF PROPOSED DISCLOSURE Procedures And Fees; Public Information

JUDICIAL REVIEW	314.235	JUDICIAL REVIEW Hearing Procedures For New Drugs; Applications For FDA Approval To Market A New Drug
JUDICIAL REVIEW	514.235	JUDICIAL REVIEW Judicial Review; New Animal Drug Applications
JUDICIAL REVIEW	515.40	JUDICIAL REVIEW Judicial Review; Medicated Feed Mill License
JUDICIAL REVIEW	1316.68	COPIES OF PETITIONS FOR JUDICIAL REVIEW Administrative Hearings; Administrative Functions, Practices, And Procedures
JUICE, FRUIT	73.250	FRUIT JUICE Foods; Listing Of Color Additives Exempt From Certification
JUICE, FRUIT/VEGETABLE	101.30	PERCENTAGE JUICE DECLARATION FOR FOOD PURPORTING TO BE BEVERAGES THAT CONTAIN FRUIT OR VEGETABLE JUICE Specific Food Labeling Requirements; Food Labeling
JUICE, FRUIT/VEGETABLE	102.33	BEVERAGES THAT CONTAIN FRUIT OR VEGETABLE JUICE Requirements For Specific Nonstandardized Foods; Common Or Usual Name For Nonstandardized Foods
JUICE, GRAPEFRUIT	146.132	GRAPEFRUIT JUICE Requirements For Specific Standardized Canned Fruit Juices And Beverages; Canned Fruit Juices
JUICE, LEMON	146.114	LEMON JUICE Requirements For Specific Standardized Canned Fruit Juices And Beverages; Canned Fruit Juices
JUICE, ORANGE	146.135	ORANGE JUICE Requirements For Specific Standardized Canned Fruit Juices And Beverages; Canned Fruit Juices
JUICE, ORANGE	146.137	FROZEN ORANGE JUICE Requirements For Specific Standardized Canned Fruit Juices And Beverages; Canned Fruit Juices
JUICE, ORANGE	146.140	PASTEURIZED ORANGE JUICE Requirements For Specific Standardized Canned Fruit Juices And Beverages; Canned Fruit Juices
JUICE, ORANGE	146.141	CANNED ORANGE JUICE Requirements For Specific Standardized Canned Fruit Juices And Beverages; Canned Fruit Juices
JUICE, ORANGE	146.145	ORANGE JUICE FROM CONCENTRATE Requirements For Specific Standardized Canned Fruit Juices And Beverages; Canned Fruit Juices
JUICE, ORANGE	146.146	FROZEN CONCENTRATED ORANGE JUICE Requirements For Specific Standardized Canned Fruit Juices And Beverages; Canned Fruit Juices
JUICE, ORANGE	146.148	REDUCED ACID FROZEN CONCENTRATED ORANGE JUICE Requirements For Specific Standardized Canned Fruit Juices And Beverages; Canned Fruit Juices
JUICE, ORANGE	146.150	CANNED CONCENTRATED ORANGE JUICE Requirements For Specific Standardized Canned Fruit Juices And Beverages; Canned Fruit Juices
JUICE, ORANGE	146.151	ORANGE JUICE FOR MANUFACTURING Requirements For Specific Standardized Canned Fruit Juices And Beverages; Canned Fruit Juices
JUICE, ORANGE	146.152	ORANGE JUICE WITH PRESERVATIVE Requirements For Specific Standardized Canned Fruit Juices And Beverages; Canned Fruit Juices
JUICE, ORANGE	146.153	CONCENTRATED ORANGE JUICE FOR MANUFACTURING Requirements For Specific Standardized Canned Fruit Juices And Beverages; Canned Fruit Juices
JUICE, ORANGE	146.154	CONCENTRATED ORANGE JUICE WITH PRESERVATIVE Requirements For Specific Standardized Canned Fruit Juices And Beverages; Canned Fruit Juices
JUICE, PINEAPPLE	146.185	CANNED PINEAPPLE JUICE Requirements For Specific Standardized Canned Fruit Juices And Beverages; Canned Fruit Juices
JUICE, PRUNE	146.187	CANNED PRUNE JUICE Requirements For Specific Standardized Canned Fruit Juices And Beverages; Canned Fruit Juices
JUICE, TOMATO	156.145	TOMATO JUICE Requirements For Specific Standardized Vegetable Juices; Vegetable Juices
JUICE, VEGETABLE	73.260	VEGETABLE JUICE Foods; Listing Of Color Additives Exempt From Certification
JUICE, VEGETABLE	101.30	PERCENTAGE JUICE DECLARATION FOR FOOD PURPORTING TO BE BEVERAGES THAT CONTAIN FRUIT OR VEGETABLE JUICE Specific Food Labeling Requirements; Food Labeling

JUICE, VEGETABLE	102.33	BEVERAGES THAT CONTAIN FRUIT OR VEGETABLE JUICE Requirements For Specific Nonstandardized Foods; Common Or Usual Name For Nonstandardized Foods
JURISDICTION, PRODUCT	5.700	AUTHORITY RELATING TO DETERMINATION OF PRODUCT PRIMARY JURISDICTION Product Designation; Redelegations Of Authority; Delegations Of Authority And Organization
JURISDICTION, PRODUCT	3.6	PRODUCT JURISDICTION OFFICER Product Jurisdiction

K

KAESE, CHEESE	133.127	COOK CHEESE, KOCH KAESE Requirements For Specific Standardized Cheese And Related Products; Cheeses And Related Cheese Products
KANAMYCIN	524.1200	KANAMYCIN OPHTHALMIC AND TOPICAL DOSAGE FORMS Ophthalmic And Topical Dosage Form New Animal Drugs
KANAMYCIN	862.3520	KANAMYCIN TEST SYSTEM Clinical Toxicology Test Systems; Clinical Chemistry And Clinical Toxicology Devices
KANAMYCIN SULFATE	520.1204	KANAMYCIN SULFATE, AMINOPENTAMIDE HYDROGEN SULFATE, PECTIN, BISMUTH SUBCARBONATE, ACTIVATED ATTAPULGITE ORAL SUSPENSION Oral Dosage Form New Animal Drugs
KANAMYCIN SULFATE	520.1205	KANAMYCIN SULFATE, PECTIN, BISMUTH SUBCARBONATE, ACTIVATED ATTAPULGITE TABLETS Oral Dosage Form New Animal Drugs
KANAMYCIN SULFATE	522.1204	KANAMYCIN SULFATE INJECTION Implantation Or Injectable Dosage Form New Animal Drugs
KANAMYCIN SULFATE	524.1204	KANAMYCIN SULFATE, CALCIUM AMPHOMYCIN, AND HYDROCORTISONE ACETATE Ophthalmic And Topical Dosage Form New Animal Drugs
KAOLIN	186.1256	CLAY (KAOLIN) Listing Of Specific Substances Affirmed As GRAS; Indirect Food Substances Affirmed As Generally Recognized As Safe
KARAYA	872.3400	KARAYA AND SODIUM BORATE WITH OR WITHOUT ACACIA DENTURE ADHESIVE Prosthetic Devices; Dental Devices
KARAYA	872.3450	ETHYLENE OXIDE HOMOPOLYMER AND/OR KARAYA DENTURE ADHESIVE Prosthetic Devices; Dental Devices
KARAYA GUM	184.1349	KARAYA GUM (STERCULIA GUM) Listing Of Specific Substances Affirmed As GRAS; Direct Food Substances Affirmed As Generally Recognized As Safe
KARAYA GUM	201.319	WATER-SOLUBLE GUMS, AND HYDROPHILIC MUCILLOIDS (INCLUDING GUAR GUM, KARAYA GUM, XANTHAN GUM) AS ACTIVE INGREDIENTS; WARNINGS AND DIRECTIONS Specific Labeling Requirements For Specific Drug Products; Labeling
KELP	172.365	KELP Special Dietary And Nutritional Additives; Food Additives Permitted For Direct Addition To Food For Human Consumption
KELP	201.319	WATER-SOLUBLE GUMS, AND HYDROPHILIC MUCILLOIDS (INCLUDING KELP) AS ACTIVE INGREDIENTS; WARNINGS AND DIRECTIONS Specific Labeling Requirements For Specific Drug Products; Labeling
KENTUCKY	808.67	KENTUCKY Listing Of Specific State And Local Exemptions; Exemption From Federal Preemption Of State & Local Medical Device Requirements
KERATOLYTIC	346.20	KERATOLYTIC ACTIVE INGREDIENTS Anorectal Drug Products For Over-the-counter Human Use
KERATOME	886.4370	KERATOME Surgical Devices; Ophthalmic Devices
KERATOPROSTHESIS	886.33040	KERATOPROSTHESIS Prosthetic Devices; Ophthalmic Devices
KERATOSCOPE	886.1350	KERATOSCOPE Diagnostic Devices; Ophthalmic Devices
KERNEL, PALM	172.861	COCOA BUTTER SUBSTITUTE FROM COCONUT OIL, PALM KERNEL OIL, OR BOTH OILS Food Additives Permitted For Direct Addition To Food For Human Consumption
KERNEL, TAMARIND SEED	176.350	TAMARIND SEED KERNEL POWDER Substances For Use Only As Components Of Paper And Paperboard; Indirect Food Additives
KETAMINE HYDROCHLORIDE	522.1222	KETAMINE HYDROCHLORIDE INJECTABLE DOSAGE FORMS Implantation Or Injectable Dosage Form New Animal Drugs
KETENE DIMERS	176.120	ALKYL KETENE DIMERS Substances For Use Only As Components Of Paper And Paperboard; Indirect Food Additives
KETOGENIC STEROIDS	862.1385	17-HYDROXYCORTICOSTEROIDS (17-KETOGENIC STEROIDS) TEST SYSTEM Clinical Chemistry Test Systems; Clinical Chemistry And Clinical Toxicology Device
KETONES	862.1435	KETONES (NONQUANTITATIVE) TEST SYSTEM Clinical Chemistry Test Systems; Clinical Chemistry And Clinical Toxicology Devices
KETOPROFEN	522.1225	KETOPROFEN SOLUTION Implantation Or Injectable Dosage Form New Animal Drugs

KETOSTEROIDS	862.1430	17-KETOSTEROIDS TEST SYSTEM Clinical Chemistry Test Systems; Clinical Chemistry And Clinical Toxicology Devices
KIDNEY	876.5880	ISOLATED KIDNEY PERFUSION AND TRANSPORT SYSTEM AND ACCESSORIES Therapeutic Devices; Gastroenterology-urology Devices
KIIL TYPE INSERT	876.5830	HEMODIALYZER WITH DISPOSABLE INSERT (KIIL TYPE) Therapeutic Devices; Gastroenterology-urology Devices
KINASE	862.1215	CREATINE PHOSPHOKINASE/CREATINE KINASE OR ISOENZYMES TEST SYSTEM Clinical Chemistry Test Systems; Clinical Chemistry And Clinical Toxicology Devices
KINASE	862.1650	PYRUVATE KINASE TEST SYSTEM Clinical Chemistry Test Systems; Clinical Chemistry And Clinical Toxicology Devices
KIT	864.2260	CHROMOSOME CULTURE KIT Cell And Tissue Culture Products; Hematology And Pathology Devices
KIT	864.9650	QUALITY CONTROL KIT FOR BLOOD BANKING REAGENTS Products Used In Establishments That Manufacture Blood And Blood Products; Hematology And Pathology Devices
KIT	866.2480	QUALITY CONTROL KIT FOR CULTURE MEDIA Microbiology Devices; Immunology And Microbiology Devices
KIT	868.1100	ARTERIAL BLOOD SAMPLING KIT Diagnostic Devices; Anesthesiology Devices
KIT	868.5140	ANESTHESIA CONDUCTION KIT Therapeutic Devices; Anesthesiology Devices
KIT	870.1350	CATHETER BALLOON REPAIR KIT Cardiovascular Diagnostic Devices; Cardiovascular Devices
KIT	872.3570	OTC DENTURE REPAIR KIT Prosthetic Devices; Dental Devices
KIT	872.3600	PARTIALLY FABRICATED DENTURE KIT Prosthetic Devices; Dental Devices
KIT	876.5210	ENEMA KIT Therapeutic Devices; Gastroenterology-urology Devices
KIT	878.3925	PLASTIC SURGERY KIT AND ACCESSORIES Prosthetic Devices; General And Plastic Surgery Devices
KIT	880.5740	SUCTION SNAKEBITE KIT General Hospital And Personal Use Therapeutic Devices; General Hospital And Personal Use Devices
KIT	880.5760	CHEMICAL COLD PACK SNAKEBITE KIT General Hospital And Personal Use Therapeutic Devices; General Hospital And Personal Use Devices
KIT	880.5960	LICE REMOVAL KIT General Hospital And Personal Use Therapeutic Devices; General Hospital And Personal Use Device
KITS	864.1860	IMMUNOHISTOCHEMISTRY REAGENTS AND KITS Biological Stains; Hematology And Pathology Devices
KNIFE	886.4230	OPHTHALMIC KNIFE TEST DRUM Surgical Devices; Ophthalmic Devices
KLEBSIELLA SPP.	866.3340	KLEBSIELLA SPP. SEROLOGICAL REAGENTS Serological Reagents; Immunology And Microbiology Devices
KLUYVEROMYCES LACTIS	184.1388	LACTASE ENZYME PREPARATION FROM KLUYVEROMYCES LACTIS Listing Of Specific Substances Affirmed As GRAS; Direct Food Substances Affirmed As Generally Recognized As Safe
KNEE JOINT	888.3480	KNEE JOINT FEMOROTIBIAL METALLIC CONSTRAINED CEMENTED PROSTHESIS Prosthetic Devices; Orthopedic Devices
KNEE JOINT	888.3490	KNEE JOINT FEMOROTIBIAL METAL/COMPOSITE NON-CONSTRAINED CEMENTED PROSTHESIS Prosthetic Devices; Orthopedic Devices
KNEE JOINT	888.3500	KNEE JOINT FEMOROTIBIAL METAL/COMPOSITE SEMI-CONSTRAINED CEMENTED PROSTHESIS Prosthetic Devices; Orthopedic Devices
KNEE JOINT	888.3510	KNEE JOINT FEMOROTIBIAL METAL/POLYMER CONSTRAINED CEMENTED PROSTHESIS Prosthetic Devices; Orthopedic Devices
KNEE JOINT	888.3520	KNEE JOINT FEMOROTIBIAL METAL/POLYMER NON-CONSTRAINED CEMENTED PROSTHESIS Prosthetic Devices; Orthopedic Devices
KNEE JOINT	888.3530	KNEE JOINT FEMOROTIBIAL METAL/POLYMER SEMI-CONSTRAINED CEMENTED PROSTHESIS Prosthetic Devices; Orthopedic Devices

KNEE JOINT	888.3540	KNEE JOINT PATELLOFEMORAL POLYMER/METAL SEMI-CONSTRAINED CEMENTED PROSTHESIS Prosthetic Devices; Orthopedic Devices
KNEE JOINT	888.3550	KNEE JOINT PATELLOFEMORAL POLYMER/METAL/METAL CONSTRAINED CEMENTED PROSTHESIS Prosthetic Devices; Orthopedic Devices
KNEE JOINT	888.3560	KNEE JOINT PATELLOFEMORAL POLYMER/METAL/POLYMER SEMI-CONSTRAINED CEMENTED PROSTHESIS Prosthetic Devices; Orthopedic Devices
KNEE JOINT	888.3570	KNEE JOINT FEMORAL (HEMI-KNEE) METALLIC UNCEMENTED PROSTHESIS Prosthetic Devices; Orthopedic Devices
KNEE JOINT	888.3580	KNEE JOINT PATELLAR (HEMI-KNEE) METALLIC RESURFACING UNCEMENTED PROSTHESIS Prosthetic Devices; Orthopedic Devices
KNEE JOINT	888.3590	KNEE JOINT TIBIAL (HEMI-KNEE) METALLIC RESURFACING UNCEMENTED PROSTHESIS Prosthetic Devices; Orthopedic Devices
KOCH KAESE	133.127	COOK CHEESE, KOCH KAESE Requirements For Specific Standardized Cheese And Related Products; Cheeses And Related Cheese Products

L

LABEL	1302.04	LOCATION AND SIZE OF SYMBOL ON LABEL AND LABELING Labeling And Packaging Requirements For Controlled Substances
LABEL	606.121	CONTAINER LABEL Finished Product Control; Current Good Manufacturing Practice For Blood And Blood Components
LABEL	610.60	CONTAINER LABEL Labeling Standards; General Biological Products Standards
LABEL	610.61	PACKAGE LABEL Labeling Standards; General Biological Products Standards
LABEL	610.62	PROPER NAME; PACKAGE LABEL; LEGIBLE TYPE Labeling Standards; General Biological Products Standards
LABEL	801.15	MEDICAL DEVICES; PROMINENCE OF REQUIRED LABEL STATEMENTS General Labeling Provisions; Labeling
LABEL DECLARATION	201.312	MAGNESIUM SULFATE HEPTAHYDRATE; LABEL DECLARATION ON DRUG PRODUCTS Specific Labeling Requirements For Specific Drug Products; Labeling
LABEL DECLARATION	501.103	PETITIONS REQUESTING EXEMPTIONS FROM OR SPECIAL REQUIREMENTS FOR LABEL DECLARATION OF INGREDIENTS Exemptions From Animal Food Labeling Requirements; Animal Food Labeling
LABEL DESIGNATIONS	130.11	LABEL DESIGNATIONS OF INGREDIENTS FOR STANDARDIZED FOODS Food Standards: General
LABEL INFORMATION	1.20	PRESENCE OF MANDATORY LABEL INFORMATION General Labeling Requirements; General Enforcement Regulations
LABEL INFORMATION	1230.40	REQUIRED LABEL INFORMATION Regulations Under The Federal Caustic Poison Act
LABEL STATEMENTS	1.23	PROCEDURES FOR REQUESTING VARIATIONS AND EXEMPTIONS FROM REQUIRED LABEL STATEMENTS General Labeling Requirements; General Enforcement Regulations
LABEL STATEMENTS	1.24	EXEMPTIONS FROM REQUIRED LABEL STATEMENTS General Labeling Requirements; General Enforcement Regulations
LABEL STATEMENTS	101.65	IMPLIED NUTRIENT CONTENT CLAIMS AND RELATED LABEL STATEMENTS Specific Requirements For Nutrient Content Claims; Food Labeling
LABEL STATEMENTS	105.66	LABEL STATEMENTS RELATING TO USEFULNESS IN REDUCING OR MAINTAINING BODY WEIGHT Label Statements; Foods For Special Dietary Use
LABEL STATEMENTS	201.15	DRUGS; PROMINENCE OF REQUIRED LABEL STATEMENTS General Labeling Provisions; Labeling
LABELED, STRAMONIUM	250.12	STRAMONIUM PREPARATIONS LABELED WITH DIRECTIONS FOR USE IN SELF- MEDICATION REGARDED AS MISBRANDED Drugs Regarded As Misbranded; Specific Requirements For Specific Human Drugs
LABELER CODE	607.35	NOTIFICATION OF REGISTRANT; BLOOD PRODUCT ESTABLISHMENT REGISTRATION NUMBER AND NDC LABELER CODE Procedures For Domestic Blood Product Establishments; Establishment Registratn+product Listing, Manufacturers Human Blood/b.products
LABELER CODES	510.600	NAMES, ADDRESSES, AND DRUG LABELER CODES OF SPONSORS OF APPROVED APPLICATIONS Sponsors Of Approved Applications; New Animal Drugs
LABELING	70.25	LABELING REQUIREMENTS FOR COLOR ADDITIVES (OTHER THAN HAIR DYES) Packaging And Labeling; Color Additives
LABELING	99.405	APPLICABILITY OF LABELING, ADULTERATION, AND MISBRANDING AUTHORITY Dissemination Of Information On Unapproved/New Uses For Marketed Drugs, Biologics, and Devices
LABELING	201.58	REQUESTS FOR WAIVER OF REQUIREMENT FOR ADEQUATE AND WELL-CONTROLLED STUDIES TO SUBSTANTIATE CERTAIN LABELING STATEMENTS Labeling Requirements For Prescription Drugs And/or Insulin; Labeling
LABELING	201.200	DISCLOSURE OF DRUG EFFICACY STUDY EVALUATIONS IN LABELING AND ADVERTISING Labeling Claims For Drugs In Drug Efficacy Studies; Labeling
LABELING	211.125	LABELING ISSUANCE Packaging And Labeling Control; Current Good Manufacturing Practice For Finished Pharmaceuticals
LABELING	211.130	PACKAGING AND LABELING OPERATIONS Packaging And Labeling Control; Current Good Manufacturing Practice For Finished Pharmaceuticals

LABELING, ACTIVE INGREDIENTS	333.160	LABELING OF PERMITTED COMBINATIONS OF ACTIVE INGREDIENTS First Aid Antibiotic Drug Products; Topical Antimicrobial Drug Products For Over-the-counter Human Use
LABELING, ACTIVE INGREDIENTS	346.52	LABELING OF PERMITTED COMBINATIONS OF ANORECTAL ACTIVE INGREDIENTS Anorectal Drug Products For Over-the-counter Human Use
LABELING, ACTIVE INGREDIENTS	349.79	LABELING OF PERMITTED COMBINATIONS OF ACTIVE INGREDIENTS Ophthalmic Drug Products For Over-the-counter Human Use
LABELING, ACTIVE INGREDIENTS	352.60	LABELING OF PERMITTED COMBINATIONS OF ACTIVE INGREDIENTS Sunscreen Drug Products For Over-the-counter Human Use
LABELING, ANALGESIC	348.50	LABELING OF EXTERNAL ANALGESIC DRUG PRODUCTS External Analgesic Drug Products For Over-the-counter Human Use
LABELING, ANIMAL DRUGS	500.51	LABELING OF ANIMAL DRUGS; MISBRANDING Animal Drug Labeling Requirements; General
LABELING, ANIMAL FEED	501.110	ANIMAL FEED LABELING; COLLECTIVE NAMES FOR FEED INGREDIENTS Exemptions From Animal Food Labeling Requirements; Animal Food Labeling
LABELING, ANIMAL FOOD	501.3	IDENTITY LABELING OF ANIMAL FOOD IN PACKAGE FORM General Provisions; Animal Food Labeling
LABELING, ANIMAL FOOD	501.8	LABELING OF ANIMAL FOOD WITH NUMBER OF SERVINGS General Provisions; Animal Food Labeling
LABELING, ANIMAL FOOD	501.17	ANIMAL FOOD LABELING WARNING STATEMENTS General Provisions; Animal Food Labeling
LABELING, ANIMAL FOOD	501.100	ANIMAL FOOD; EXEMPTIONS FROM LABELING Exemptions From Animal Food Labeling Requirements; Animal Food Labeling
LABELING, ANIMAL FOODS	501.22	ANIMAL FOODS; LABELING OF SPICES, FLAVORINGS, COLORINGS, AND CHEMICAL PRESERVATIVES Specific Animal Food Labeling Requirements; Animal Food Labeling
LABELING, ANORECTAL	346.50	LABELING OF ANORECTAL DRUG PRODUCTS Labeling; Anorectal Drug Products For Over-the-counter Human Use
LABELING, ANORECTAL	346.52	LABELING OF PERMITTED COMBINATIONS OF ANORECTAL ACTIVE INGREDIENTS Labeling; Anorectal Drug Products For Over-the-counter Human Use
LABELING, ANTACID	331.30	LABELING OF ANTACID PRODUCTS Antacid Products For Over-the-counter (OTC) Human Use
LABELING, ANTHELMINTHIC	357.250	LABELING OF ANTHELMINTHIC DRUG PRODUCTS Anthelmintic Drug Products; Miscellaneous Internal Drug Products For OTC Human Use
LABELING, ANTIBIOTIC	510.106	LABELING OF ANTIBIOTIC AND ANTIBIOTIC-CONTAINING DRUGS INTENDED FOR USE IN MILK-PRODUCING ANIMALS Specific Administrative Rulings And Decisions; New Animal Drugs
LABELING, ANTICARIES	355.50	LABELING OF ANTICARIES DRUG PRODUCTS Anticaries Drug Products For Over-the-counter Human Use
LABELING, ANTIEMETIC	336.50	LABELING OF ANTIEMETIC DRUG PRODUCTS Antiemetic Drug Products For Over-the-counter Human Use
LABELING, ANTIFLATULENT	332.30	LABELING OF ANTIFLATULENT PRODUCTS Antiflatulent Products For Over-the-counter Human Use
LABELING, ANTIFUNGAL	333.250	LABELING OF ANTIFUNGAL DRUG PRODUCTS Topical Antifungal Drug Products; Topical Antimicrobial Drug Products For OTC Human Use
LABELING, ANTIHISTAMINE	341.72	LABELING OF ANTIHISTAMINE DRUG PRODUCTS Cold, Cough, Allergy, Bronchodilator, And Antiasthmatic Drug Products For OTC Human Use
LABELING, ANTITUSSIVE	341.74	LABELING OF ANTIHISTAMINE DRUG PRODUCTS Cold, Cough, Allergy, Bronchodilator, And Antiasthmatic Drug Products For OTC Human Use
LABELING, ASTRINGENT	347.50	LABELING OF ASTRINGENT DRUG PRODUCTS Skin Protectant Drug Products For Over-the-counter Human Use
LABELING, BRONCHODILATOR	341.76	LABELING OF BRONCHODILATOR DRUG PRODUCTS Cold, Cough, Allergy, Bronchodilator, Antiasthmatic Drug Products OTC Human Use
LABELING, CHOLECYSTOKINETIC	357.250	LABELING OF CHOLECYSTOKINETIC DRUG PRODUCTS Cholecystokinetic Drug Products; Miscellaneous Internal Drug Products For OTC Human Use
LABELING, CONDOMS	801.435	USER LABELING FOR LATEX CONDOMS Labeling
LABELING, CORN/CALLUS	358.550	LABELING OF CORN AND CALLUS REMOVER DRUG PRODUCTS Corn And Callus Remover Drug Products; Miscellaneous External Drug Products For OTC Human Use
LABELING, CORTICOSTEROIDS	510.410	CORTICOSTEROIDS FOR ORAL, INJECTABLE, AND OPHTHALMIC USE IN ANIMALS; WARNINGS AND LABELING REQUIREMENTS Requirements For Specific New Animal Drugs; New Animal Drugs
LABELING, COSMETIC	701.30	INGREDIENT NAMES ESTABLISHED FOR COSMETIC INGREDIENT LABELING Labeling Of Specific Ingredients; Cosmetic Labeling

LABELING, COSMETIC	740.10	LABELING OF COSMETIC PRODUCTS FOR WHICH ADEQUATE SUBSTANTIATION OF SAFETY HAS NOT BEEN OBTAINED Warning Statements; Cosmetic Product Warning Statements
LABELING, DANDRUFF	358.750	LABELING OF DRUG PRODUCTS FOR THE CONTROL OF DANDRUFF, SEBORRHEIC DERMATITIS AND PSORIASIS Miscellaneous External Drug Products For Over-the-counter Human Use
LABELING, DEODORANT	357.850	LABELING OF DEODORANT DRUG PRODUCTS FOR INTERNAL USE Miscellaneous Internal Drug Products For Over-the-counter Human Use
LABELING, DEVICE	820.120	DEVICE LABELING Labeling And Packaging Control; Quality System Regulation
LABELING, DEVICES	801.150	MEDICAL DEVICES; PROCESSING, LABELING, OR REPACKING Other Exemptions; Labeling
LABELING, DEVICES	801.437	USER LABELING FOR DEVICES THAT CONTAIN NATURAL RUBBER Special Requirements For Specific Devices; Labeling
LABELING, DEVICES	812.5	LABELING OF INVESTIGATIONAL DEVICES General Provisions; Investigational Device Exemptions
LABELING, DIAGNOSTIC	809.10	LABELING FOR IN VITRO DIAGNOSTIC PRODUCTS In Vitro Diagnostic Products For Human Use
LABELING, DIETARY SUPPLEM	101.36	NUTRITION LABELING OF DIETARY SUPPLEMENTS Food Labeling
LABELING, DRUG	312.6	LABELING OF AN INVESTIGATIONAL NEW DRUG Investigational New Drug Application
LABELING, DRUG	500.55	EXEMPTION FROM CERTAIN DRUG-LABELING REQUIREMENTS Animal Drug Labeling Requirements; General
LABELING, DRUGS	5.109	ISSUANCE OF WRITTEN NOTICES CONCERNING PATENT INFORMATION, CURRENT GOOD MANUFACTURING PRACTICES AND FALSE OR MISLEADING LABELING OF NEW DRUGS Human Drugs; Redelegations Of Authority; Delegations Of Authority And Organization
LABELING, DRUGS	5.504	ISSUANCE OF WRITTEN NOTICES CONCERNING PATENT INFORMATION, CURRENT GOOD MANUFACTURING PRACTICES AND FALSE OR MISLEADING LABELING OF NEW ANIMAL DRUGS AND FEEDS BEARING OR CONTAINING NEW ANIMAL DRUGS Animal Drugs; Redelegations Of Authority; Delegations Of Authority And Organization
LABELING, DRUGS	201.150	DRUGS; PROCESSING, LABELING, OR REPACKING Other Exemptions; Labeling
LABELING, DRUGS	510.105	LABELING OF DRUGS FOR USE IN MILK-PRODUCING ANIMALS Specific Administrative Rulings And Decisions; New Animal Drugs
LABELING, EAR DRYING	344.52	LABELING OF EAR DRYING AID DRUG PRODUCTS Topical Otic Over-the-counter Products For Human Use
LABELING, EARWAX	344.50	LABELING OF EARWAX REMOVAL AID DRUG PRODUCTS Topical Otic Over-the-counter Products For Human Use
LABELING, ESTRADIOL	201.313	ESTRADIOL LABELING Specific Labeling Requirements For Specific Drug Products; Labeling
LABELING, EXPECTORANT	341.78	LABELING OF EXPECTORANT DRUG PRODUCTS Active Ingredients; Cold, Cough, Allergy, Bronchodilator And Anti-asthmatic Drugs Products
LABELING, EYEWASH	349.78	LABELING OF EYEWASH DRUG PRODUCTS Ophthalmic Drug Products For Over-the-counter Human Use
LABELING, FIRST AID	333.150	LABELING OF FIRST AID ANTIBIOTIC DRUG PRODUCTS First Aid Antibiotic Drug Products; Topical Antimicrobial Drug Products For Over-the-counter Human Use
LABELING, FOOD	101.3	IDENTITY LABELING OF FOOD IN PACKAGED FORM Food Labeling
LABELING, FOOD	101.17	FOOD LABELING WARNING, NOTICE, AND SAFE HANDLING STATEMENTS General Provisions; Food Labeling
LABELING, FOOD	101.108	TEMPORARY EXEMPTIONS FOR PURPOSES OF CONDUCTING AUTHORIZED FOOD LABELING EXPERIMENTS Exemptions From Food Labeling Requirements; Food Labeling
LABELING, FOOD	101.100	FOOD; EXEMPTIONS FROM LABELING Exemptions From Food Labeling Requirements; Food Labeling
LABELING, FOODS	101.22	FOODS; LABELING OF SPICES, FLAVORINGS, COLORINGS AND CHEMICAL PRESERVATIVES Specific Food Labeling Requirements; Food Labeling
LABELING, HABIT-FORMING DRUGS	329.10	LABELING REQUIREMENTS FOR HABIT-FORMING DRUGS Labeling; Habit-forming Drugs
LABELING, HEARING AID	801.420	HEARING AID DEVICES; PROFESSIONAL AND PATIENT LABELING Special Requirements For Specific Devices; Labeling
LABELING, HYPOGLYCEMIC	310.517	LABELING FOR ORAL HYPOGLYCEMIC DRUGS OF THE SULFONYLUREA CLASS Requirements For Specific New Drugs Or Devices; New Drugs
LABELING, IN VITRO	809.10	LABELING FOR IN VITRO DIAGNOSTIC PRODUCTS In Vitro Diagnostic Products For Human Use

LABELING, MARGARINE	166.40	LABELING OF MARGARINE General Provisions; Margarine
LABELING, MEDICAL DEVICES	801.150	MEDICAL DEVICES; PROCESSING, LABELING, OR REPACKING Other Exemptions; Labeling
LABELING, NASAL DECONGEST	341.80	LABELING NASAL DECONGESTANT DRUG PRODUCTS Cold, Cough, Allergy, Bronchodilator, And Antiasthmatic Drug Products For OTC Human Use
LABELING, NUTRITION	101.9	NUTRITION LABELING OF FOOD General Provisions; Food Labeling
LABELING, NUTRITION	101.10	NUTRITION LABELING OF RESTAURANT FOODS General Provisions; Food Labeling
LABELING, NUTRITION	101.36	NUTRITION LABELING OF DIETARY SUPPLEMENTS Food Labeling
LABELING, NUTRITION	101.42	NUTRITION LABELING OF RAW FRUIT, VEGETABLES AND FISH Food Labeling
LABELING, NUTRITION	101.45	GUIDELINES FOR THE VOLUNTARY NUTRITION LABELING OF RAW FRUIT, VEGETABLES, AND FISH Food Labeling
LABELING, NUTRITIONAL	101.43	SUBSTANTIAL COMPLIANCE OF FOOD RETAILERS WITH THE GUIDELINES FOR THE VOLUNTARY NUTRITIONAL LABELING OF RAW FRUIT, VEGETABLES AND FISH Food Labeling
LABELING, OPHTHALMIC	349.50	LABELING OF OPHTHALMIC DRUGS PRODUCTS Ophthalmic Drug Products For Over-the-counter Human Use
LABELING, OPHTHALMIC	349.55	LABELING OF OPHTHALMIC ASTRINGENT DRUG PRODUCTS Ophthalmic Drug Products For Over-the-counter Human Use
LABELING, OPHTHALMIC	349.60	LABELING OF OPHTHALMIC DEMULCENT DRUG PRODUCTS Ophthalmic Drug Products For Over-the-counter Human Use
LABELING, OPHTHALMIC	349.65	LABELING OF OPHTHALMIC EMOLLIENT DRUG PRODUCTS Ophthalmic Drug Products For Over-the-counter Human Use
LABELING, OPHTHALMIC	349.70	LABELING OF OPHTHALMIC HYPERTONICITY DRUG PRODUCTS Ophthalmic Drug Products For Over-the-counter Human Use
LABELING, OPHTHALMIC	349.75	LABELING OF OPHTHALMIC VASOCONSTRICTOR DRUG PRODUCTS Ophthalmic Drug Products For Over-the-counter Human Use
LABELING, OTC DRUG	201.66	FORMAT AND CONTENT REQUIREMENTS FOR OTC DRUG PRODUCT LABELING Labeling Requirements For Over-the-counter Drugs; Labeling
LABELING, OTC DRUG	341.70	LABELING OF OTC DRUG PRODUCTS CONTAINING INGREDIENTS THAT ARE USED FOR TREATING CONCURRENT SYMPTOMS Cold, Cough, Allergy, Bronchodilator, And Antiasthmatic Drug Products For OTC Human Use
LABELING, PEDICULICIDE	358.650	LABELING OF PEDICULICIDE DRUG PRODUCTS Miscellaneous External Drug Products For Over-the-counter Human Use
LABELING, PHENINDIONE	201.310	PHENINDIONE; LABELING OF DRUG PREPARATIONS INTENDED FOR USE BY MAN Specific Labeling Requirements For Specific Drug Products; Labeling
LABELING, PRESCRIPTION DRUGS	5.107	EXTENSIONS OR STAYS OF EFFECTIVE DATES FOR COMPLIANCE WITH CERTAIN LABELING REQUIREMENTS FOR HUMAN PRESCRIPTION DRUGS Human Drugs; Redelegations Of Authority; Delegations Of Authority And Organization
LABELING, PRESCRIPTION DRUGS	201.56	GENERAL REQUIREMENTS ON CONTENT AND FORMAT OF LABELING FOR HUMAN PRESCRIPTION DRUGS Labeling Requirements For Prescription Drugs And/or Insulin; Labeling
LABELING, PRESCRIPTION DRUGS	201.57	SPECIFIC REQUIREMENTS ON CONTENT AND FORMAT OF LABELING FOR HUMAN PRESCRIPTION DRUGS Labeling Requirements For Prescription Drugs And/or Insulin; Labeling
LABELING, PROFESSIONAL	331.80	PROFESSIONAL LABELING Antacid Products For Over-the-counter (OTC) Human Use
LABELING, PROFESSIONAL	332.31	PROFESSIONAL LABELING Antiflatulent Products For Over-the-counter Human Use
LABELING, PROFESSIONAL	333.280	LABELING PROFESSIONAL Topical Antifungal Drug Products; Topical Antimicrobial Drug Products For OTC Human Use
LABELING, PROFESSIONAL	336.00	PROFESSIONAL LABELING Antiemetic Drug Products For Over-the-counter Human Use
LABELING, PROFESSIONAL	341.90	PROFESSIONAL LABELING Cold, Cough, Allergy, Bronchodilator, Antiasthmatic Drug Products OTC Human Use
LABELING, PROFESSIONAL	343.80	PROFESSIONAL LABELING Internal Analgesic, Antipyretic And Antirheumatic Drug Products For OTC Human Use
LABELING, PROFESSIONAL	349.80	PROFESSIONAL LABELING Ophthalmic Drug Products For Over-the-counter Human Use

LABELING, PROFESSIONAL	355.60	PROFESSIONAL LABELING Anticaries Drug Products For Over-the-counter Human Use
LABELING, PROFESSIONAL	357.180	PROFESSIONAL LABELING Anthelmintic Drug Products; Miscellaneous Internal Drug Products For OTC Human Use
LABELING, PROFESSIONAL	357.280	PROFESSIONAL LABELING Cholecystokinetic Drug Products; Miscellaneous Internal Drug Products For OTC Human Use
LABELING, PROFESSIONAL	801.420	HEARING AID DEVICES; PROFESSIONAL AND PATIENT LABELING Special Requirements For Specific Devices; Labeling
LABELING, REMINDER	200.200	PRESCRIPTION DRUGS; REMINDER ADVERTISEMENTS AND REMINDER LABELING TO PROVIDE PRICE INFORMATION TO CONSUMERS Prescription Drug Consumer Price Listing; General
LABELING, SALICYLATES	201.314	LABELING OF DRUG PREPARATIONS CONTAINING SALICYLATES Specific Labeling Requirements For Specific Drug Products; Labeling
LABELING, SAMPLE UNITS	203.38	SAMPLE LOT OR CONTROL NUMBERS; LABELING OF SAMPLE UNITS Prescription Drug Marketing
LABELING, SLEEP-AID	338.50	LABELING OF NIGHTTIME SLEEP-AID DRUG PRODUCTS Nighttime Sleep-aid Drug Products For Over-the-counter Human Use
LABELING, SODIUM	201.64	SODIUM LABELING Labeling Requirements For Over-the-counter Drugs; Labeling
LABELING, STIMULANT	340.50	LABELING OF STIMULANT DRUG PRODUCTS Stimulant Drug Products For Over-the-counter Human Use
LABELING, SUNSCREEN	352.52	LABELING OF SUNSCREEN DRUG PRODUCTS Sunscreen Drug Products For Over-the-counter Human Use
LABELING, TAMPONS	801.430	USER LABELING FOR MENSTRUAL TAMPONS Special Requirements For Specific Devices; Labeling
LABELING, WART REMOVER	358.150	LABELING OF WART REMOVER DRUG PRODUCTS Wart Remover Drug Products; Miscellaneous External Drug Products For Over-the-counter Human Use
LABELING, WINTERGREEN OIL	201.303	LABELING OF DRUG PREPARATIONS CONTAINING SIGNIFICANT PROPORTIONS OF WINTERGREEN OIL Specific Labeling Requirements For Specific Drug Products; Labeling
LABELS	161.30	DECLARATION OF QUANTITY OF CONTENTS ON LABELS FOR CANNED OYSTERS General Provisions; Fish And Shellfish
LABORATORIES, CONSULTING	200.10	CONTRACT FACILITIES (INCLUDING CONSULTING LABORATORIES) UTILIZED AS EXTRAMURAL FACILITIES BY PHARMACEUTICAL MANUFACTURERS General Provisions; General
LABORATORY	58.49	LABORATORY OPERATION AREAS Facilities; Good Laboratory Practice For Nonclinical Laboratory Studies
LABORATORY ACCREDITATION	5.303	ESTABLISHING STANDARDS AND APPROVING ACCREDITING BODIES UNDER THE NATIONAL LABORATORY ACCREDITATION PROGRAM Foods And Cosmetics; Redelegations Of Authority; Delegations Of Authority And Organization
LABORATORY ANIMAL	579.22	IONIZING RADIATION FOR TREATMENT OF LABORATORY ANIMAL DIETS Radiation And Radiation Sources; Irradiation In The Production, Process & Handling Of Animal Feed And Pet Food
LABORATORY ANIMALS	211.173	LABORATORY ANIMALS Laboratory Controls; Current Good Manufacturing Practice For Finished Pharmaceuticals
LABORATORY ASSAYS	225.158	LABORATORY ASSAYS Product Quality Assurance; Current Good Manufacturing Practice For Medicated Feeds
LABORATORY CONTROLS	225.58	LABORATORY CONTROLS Product Quality Control; Current Good Manufacturing Practice For Medicated Feeds
LABORATORY CONTROLS	226.58	LABORATORY CONTROLS Product Quality Control; Current Good Manufacturing Practice For Type A Medicated Articles
LABORATORY CONTROLS	606.140	LABORATORY CONTROLS Laboratory Controls; Current Good Manufacturing Practice For Blood And Blood Components
LABORATORY EQUIPMENT	862.2050	GENERAL PURPOSE LABORATORY EQUIPMENT LABELED OR PROMOTED FOR A SPECIFIC MEDICINAL USE Clinical Laboratory Instruments; Clinical Chemistry And Clinical Toxicology Devices
LABORATORY RECORDS	211.194	LABORATORY RECORDS Records And Reports; Current Good Manufacturing Practice For Finished Pharmaceuticals
LABORATORY RESEARCH	312.160	NEW DRUGS FOR INVESTIGATIONAL USE IN LABORATORY RESEARCH ANIMALS OR IN VITRO TESTS Exemptions From Section 505(a); New Drugs For Investigational Use
LABORATORY STUDY	58.130	CONDUCT OF A NONCLINICAL LABORATORY STUDY Protocol For And Conduct Of A Non-clinical Laboratory Study; Good Laboratory Practice For Nonclinical Laboratory Studies

LABORATORY STUDY	58.185	REPORTING OF NONCLINICAL LABORATORY STUDY RESULTS Records And Reports; Good Laboratory Practice For Nonclinical Laboratory Studies
LABORATORY TESTS	640.67	LABORATORY TESTS Source Plasma; Additional Standards For Human Blood And Blood Products
LABWARE	884.6160	ASSISTED REPRODUCTION LABWARE Obstetrical And Gynecological Devices
LACTASE	184.1387	LACTASE ENZYME PREPARATION FROM CANDIDA PSEUDOTROPICALIS Listing Of Specific Substances Affirmed As GRAS; Direct Food Substances Affirmed As Generally Recognized As Safe
LACTASE	184.1388	LACTASE ENZYME PREPARATION FROM KLUYVEROMYCES LACTIS Listing Of Specific Substances Affirmed As GRAS; Direct Food Substances Affirmed As Generally Recognized As Safe
LACTATE	73.165	FERROUS LACTATE Listing Of Color Additives Exempt From Certification
LACTATE	184.1207	CALCIUM LACTATE Listing Of Specific Substances Affirmed As GRAS; Direct Food Substances Affirmed As Generally Recognized As Safe
LACTATE	184.1311	FERROUS LACTATE Listing Of Specific Substances Affirmed As GRAS; Direct Food Substances Affirmed As Generally Recognized As Safe
LACTATE	184.1639	POTASSIUM LACTATE Listing Of Specific Substances Affirmed As GRAS; Direct Food Substances Affirmed As Generally Recognized As Safe
LACTATE	184.1768	SODIUM LACTATE Listing Of Specific Substances Affirmed As GRAS; Direct Food Substances Affirmed As Generally Recognized As Safe
LACTATE	522.1698	PENTAZOCINE LACTATE INJECTION Implantation Or Injectable Dosage Form New Animal Drugs
LACTATE	582.1207	CALCIUM LACTATE General Purpose Food Additives; Substances Generally Recognized As Safe
LACTATE	582.5311	FERROUS LACTATE Nutrients And/or Dietary Supplements; Substances Generally Recognized As Safe
LACTATE DEHYDROGENASE	862.1440	LACTATE DEHYDROGENASE TEST SYSTEM Clinical Chemistry Test Systems; Clinical Chemistry And Clinical Toxicology Devices
LACTATE DEHYDROGENASE	862.1445	LACTATE DEHYDROGENASE ISOENZYMES TEST SYSTEM Clinical Chemistry Test Systems; Clinical Chemistry And Clinical Toxicology Devices
LACTIC ACID	184.1061	LACTIC ACID Listing Of Specific Substances Affirmed As GRAS; Direct Food Substances Affirmed As Generally Recognized As Safe
LACTIC ACID	582.1061	LACTIC ACID General Purpose Food Additives; Substances Generally Recognized As Safe
LACTIC ACID	862.1450	LACTIC ACID TEST SYSTEM Clinical Chemistry Test Systems; Clinical Chemistry And Clinical Toxicology Devices
LACTIC DEHYDROGENASE	866.5560	LACTIC DEHYDROGENASE IMMUNOLOGICAL TEST SYSTEM Immunological Test Systems; Immunology And Microbiology Devices
LACTIDE	878.4493	ABSORBABLE POLY(GLYCOLIDE/L-LACTIDE) SURGICAL SUTURE Surgical Devices; General And Plastic Surgery Devices
LACTO ESTERS	172.852	GLYCERYL-LACTO ESTERS OF FATTY ACIDS Multipurpose Additives; Food Additives Permitted For Direct Addition To Food For Human Consumption
LACTOBACILLUS	184.1924	UREASE ENZYME PREPARATION FROM LACTOBACILLUS FERMENTUM Listing Of Specific Substances Affirmed As GRAS; Direct Food Substances Affirmed As Generally Recognized As Safe
LACTOBIONATE, CALCIUM	172.720	CALCIUM LACTOBIONATE Other Specific Usage Additives; Food Additives Permitted For Direct Addition To Food For Human Consumption
LACTOCOCCUS	184.1985	AMINOPEPTIDASE ENZYME PREPARATION DERIVED FROM LACTOCOCCUS LACTIS Listing Of Specific Substances Affirmed As GRAS; Direct Food Substances Affirmed As Generally Recognized As Safe
LACTOFERRIN	866.5570	LACTOFERRIN IMMUNOLOGICAL TEST SYSTEM Immunological Test Systems; Immunology And Microbiology Devices
LACTOGEN	862.1585	HUMAN PLACENTAL LACTOGEN TEST SYSTEM Clinical Chemistry Test Systems; Clinical Chemistry And Clinical Toxicology Devices
LACTOGEN	862.1625	PROLACTIN (LACTOGEN) TEST SYSTEM Clinical Chemistry Test Systems; Clinical Chemistry And Clinical Toxicology Devices
LACTONE	184.1318	GLUCONO DELTA-LACTONE Listing Of Specific Substances Affirmed As GRAS; Direct Food Substances Affirmed As Generally Recognized As Safe

LACTOSE	168.122	LACTOSE Requirements For Specific Standardized Sweeteners And Table Sirups; Sweeteners And Table Sirups
LACTOSE	184.1979a	REDUCED LACTOSE WHEY Listing Of Specific Substances Affirmed As GRAS; Direct Food Substances Affirmed As Generally Recognized As Safe
LACTYLATE	172.844	CALCIUM STEAROYL-2-LACTYLATE Multipurpose Additives; Food Additives Permitted For Direct Addition To Food For Human Consumption
LACTYLATE	172.846	SODIUM STEAROYL LACTYLATE Multipurpose Additives; Food Additives Permitted For Direct Addition To Food For Human Consumption
LACTYLATED ESTERS	172.850	LACTYLATED FATTY ACID ESTERS OF GLYCEROL AND PROPYLENE GLYCOL Multipurpose Additives; Food Additives Permitted For Direct Addition To Food For Human Consumption
LACTYLIC ESTERS	172.848	LACTYLIC ESTERS OF FATTY ACIDS Multipurpose Additives; Food Additives Permitted For Direct Addition To Food For Human Consumption
LAIDLOMYCIN	558.305	LAIDLOMYCIN PROPRIONATE POTASSIUM New Animal Drugs For Use In Animal Feeds
LAKES (D&C)	82.1051	LAKES (D&C) Drugs And Cosmetics; Listing Of Certified Provisionally Listed Colors And Specifications
LAKES (FD&C)	82.51	LAKES (FD&C) Foods, Drugs, And Cosmetics; Listing Of Certified Provisionally Listed Colors And Specifications
LAKES (EXT. D&C)	82.2051	LAKES (EXT. D&C) Externally Applied Drugs And Cosmetics; Listing Of Certified Provisionally Listed Colors And Specifications
LAMB	880.5640	LAMB FEEDING NIPPLE General Hospital And Personal Use Therapeutic Devices; General Hospital And Personal Use Devices
LAMINARIA	884.4250	HYGROSCOPIC LAMINARIA CERVICAL DILATOR Obstetrical And Gynecological Surgical Devices; Obstetrical And Gynecological Devices
LAMINATE	177.1390	LAMINATE STRUCTURES FOR USE AT TEMPERATURES OF 250 F AND ABOVE Substances For Use As Basic Components Of Single And Repeated Use Food Contact Surfaces; Indirect Food Additives: Polymers
LAMINATE	177.1395	LAMINATE STRUCTURES FOR USE AT TEMPERATURES BETWEEN 120 F AND 250 F Substances For Use As Basic Components Of Single And Repeated Use Food Contact Surfaces; Indirect Food Additives: Polymers
LAMP, FLUORESCENT	866.2600	WOOD'S FLUORESCENT LAMP Microbiology Devices; Immunology And Microbiology Devices
LAMP, INFRARED	890.5500	INFRARED LAMP Physical Medicine Therapeutic Devices; Physical Medicine Devices
LAMP, SURGICAL	878.4580	SURGICAL LAMP Surgical Devices; General And Plastic Surgery Devices
LAMP, ULTRAVIOLET	878.4630	ULTRAVIOLET LAMP FOR DERMATOLOGICAL DISORDERS Surgical Devices; General And Plastic Surgery Devices
LAMP, ULTRAVIOLET	878.4635	ULTRAVIOLET LAMP FOR TANNING Surgical Devices; General And Plastic Surgery Devices
LAMPS, MERCURY	1040.30	HIGH-INTENSITY MERCURY VAPOR DISCHARGE LAMPS Performance Standards For Light-emitting Products
LAMPS, ULTRAVIOLET	1040.20	SUNLAMP PRODUCTS AND ULTRAVIOLET LAMPS INTENDED FOR USE IN SUNLAMP PRODUCTS Performance Standards For Light-emitting Products
LANGUAGE, SPANISH	201.16	DRUGS; SPANISH-LANGUAGE VERSION OF CERTAIN REQUIRED STATEMENTS General Labeling Provisions; Labeling
LANGUAGE, SPANISH	290.6	SPANISH-LANGUAGE VERSION OF REQUIRED WARNING General Provisions; Controlled Drugs
LANGUAGE, SPANISH	801.16	MEDICAL DEVICES; SPANISH-LANGUAGE VERSION OF CERTAIN REQUIRED STATEMENTS General Labeling Provisions; Labeling
LAPAROSCOPE	884.1720	GYNECOLOGIC LAPAROSCOPE AND ACCESSORIES Obstetrical And Gynecological Diagnostic Devices; Obstetrical And Gynecological Devices
LAPAROSCOPIC	884.1730	LAPAROSCOPIC INSUFFLATOR Obstetrical And Gynecological Diagnostic Devices; Obstetrical And Gynecological Devices

LARYNGEAL	874.3730	LARYNGEAL PROSTHESIS (TAUB DESIGN) Prosthetic Devices; Ear, Nose, And Throat Devices
LARYNGOLOGY	874.4490	ARGON LASER FOR OTOLOGY, RHINOLOGY, AND LARYNGOLOGY Surgical Devices; Ear, Nose, And Throat Devices
LARYNGOSCOPE	868.5530	FLEXIBLE LARYNGOSCOPE Therapeutic Devices; Anesthesiology Devices
LARYNGOSCOPE	868.5540	RIGID LARYNGOSCOPE Therapeutic Devices; Anaesthesiology Devices
LARYNGOSTROBOSCOPE	874.4750	LARYNGOSTROBOSCOPE Surgical Devices; Ear, Nose, And Throat Devices
LARYNGOTRACHEAL	868.5170	LARYNGOTRACHEAL TOPICAL ANESTHESIA APPLICATOR Therapeutic Devices; Anesthesiology Devices
LARYNX	874.3375	BATTERY-POWERED ARTIFICIAL LARYNX Prosthetic Devices; Ear, Nose, And Throat Devices
LASALOCID	556.347	LASALOCID Specific Tolerances For Residues Of New Animal Drugs; Tolerances For Residues Of New Animal Drugs In Food
LASALOCID	558.311	LASALOCID Specific New Animal Drugs For Use In Animal Feeds; New Animal Drugs For Use In Animal Feeds
LASER	872.1745	LASER FLUORESCENCE CARIES DETECTION DEVICE Diagnostic Devices; Dental Devices
LASER	874.4490	ARGON LASER FOR OTOLOGY, RHINOLOGY, AND LARYNGOLOGY Surgical Devices; Ear, Nose, And Throat Devices
LASER	874.4500	EAR, NOSE, AND THROAT MICROSURGICAL CARBON-DIOXIDE LASER Surgical Devices; Ear, Nose, And Throat Devices
LASER	878.4810	LASER SURGICAL INSTRUMENT FOR USE IN PLASTIC SURGERY AND IN DERMATOLOGY Surgical Devices; General And Plastic Surgery Devices
LASER	884.4550	GYNECOLOGIC SURGICAL LASER Obstetrical And Gynecological Surgical Devices; Obstetrical And Gynecological Devices
LASER	886.1360	VISUAL FIELD LASER INSTRUMENT Diagnostic Devices; Ophthalmic Devices
LASER	886.4390	OPHTHALMIC LASER Surgical Devices; Ophthalmic Devices
LASER	886.4392	ND:YAG LASER FOR POSTERIOR CAPSULOTOMY AND PERIPHERAL IRIDECTOMY Surgical Devices; Ophthalmic Devices
LASER	1040.10	LASER PRODUCTS Performance Standards For Light-emitting Products
LASER	1040.11	SPECIFIC PURPOSE LASER PRODUCTS Performance Standards For Light-emitting Products
LATEX	801.435	USER LABELING FOR LATEX CONDOMS Special Requirements For Specific Devices; Labeling
LAURYL ETHER	522.1846	POLYOXYETHYLENE (23) LAURYL ETHER BLOCKS Implantation Or Injectable Dosage Form New Animal Drugs
LAURYL SULFATE	172.822	SODIUM LAURYL SULFATE Multipurpose Additives; Food Additives Permitted For Direct Addition To Food For Human Consumption
LAURYL VINYL ETHER	177.1970	VINYL CHLORIDE-LAURYL VINYL ETHER COPOLYMERS Substances For Use As Basic Components Of Single And Repeated Use Food Contact Surfaces; Indirect Food Additives: Polymers
LAVAGE	880.5475	JET LAVAGE General Hospital And Personal Use Therapeutic Devices; General Hospital And Personal Use Devices
LAVATORIES	1250.90	TOILET AND LAVATORIES Sanitation Facilities And Conditions On Vessels; Interstate Conveyance Sanitation
LAVATORY	1250.38	TOILET AND LAVATORY FACILITIES FOR USE OF FOOD HANDLING EMPLOYEES Food Service Sanitation On Land And Air Conveyances; Interstate Conveyance Sanitation
LAVATORY	1250.50	TOILET AND LAVATORY FACILITIES Equipment And Operation Of Land And Air Conveyances; Interstate Conveyance Sanitation
LAW	1316.64	PROPOSED FINDINGS OF FACT AND CONCLUSIONS OF LAW Administrative Hearings; Administrative Functions, Practices, And Procedures
LAW ENFORCEMENT	10.40	PROMULGATION OF REGULATIONS FOR THE EFFICIENT ENFORCEMENT OF THE LAW General Administrative Procedures; Administrative Practices And Procedures
LAW ENFORCEMENT	20.64	RECORDS OR INFORMATION COMPILED FOR LAW ENFORCEMENT PURPOSES Exemptions; Public Information

LAW ENFORCEMENT	201.125	DRUGS FOR USE IN TEACHING, LAW ENFORCEMENT, RESEARCH, AND ANALYSIS Exemptions From Adequate Directions For Use; Labeling
LAW ENFORCEMENT	801.125	MEDICAL DEVICES FOR USE IN TEACHING, LAW ENFORCEMENT, RESEARCH, AND ANALYSIS Exemptions From Adequate Directions For Use; Labeling
LAW ENFORCEMENT	1301.24	EXEMPTION OF LAW ENFORCEMENT OFFICIALS Requirements For Registration; Registration Of Manufacturers, Distributors, And Dispensers Of Controlled Substances
LAW ENFORCEMENT OFFICIALS	1309.26	EXEMPTION OF LAW ENFORCEMENT OFFICIALS Registration Of Manufacturers, Distributors, Importers And Exporters Of List I Chemicals
LAW, STATE/FEDERAL	1307.02	APPLICATION OF STATE LAW AND OTHER FEDERAL LAW General Information; Miscellaneous
LAWS	1003.5	EFFECT OF REGULATIONS ON OTHER LAWS General Provisions; Notification Of Defects Or Failure To Comply
LAY USE	801.405	LABELING OF ARTICLES INTENDED FOR LAY USE IN THE REPAIRING AND/OR REFITTING OF DENTURES Special Requirements For Specific Devices; Labeling
LEAD	862.3550	LEAD TEST SYSTEM Clinical Toxicology Test Systems; Clinical Chemistry And Clinical Toxicology Devices
LEAD	870.2350	ELECTROCARDIOGRAPH LEAD SWITCHING ADAPTOR Cardiovascular Monitoring Devices; Cardiovascular Devices
LEAD	870.3620	PACEMAKER LEAD ADAPTOR Cardiovascular Prosthetic Devices; Cardiovascular Devices
LEAD	882.1410	ELECTROENCEPHALOGRAPH ELECTRODE/LEAD TESTER Neurological Diagnostic Devices; Neurological Devices
LEAD AGENCIES	25.44	LEAD AND COOPERATING AGENCIES Preparation Of Environmental Documents; Environmental Impact Considerations
LEAD ACETATE	73.2396	LEAD ACETATE Cosmetics; Listing Of Color Additives Exempt From Certification
LEAD-LINED	872.1850	LEAD-LINED POSITION INDICATOR Diagnostic Devices; Dental Devices
LEAD SOLDERS	189.240	LEAD SOLDERS Substances Prohibited From Use In Human Food
LEAD, TIN-COATED	189.301	TIN-COATED LEAD FOIL CAPSULES FOR WINE BOTTLES Substances Prohibited From Use In Human Foods
LEAKAGE CURRENT	870.2640	PORTABLE LEAKAGE CURRENT ALARM Cardiovascular Monitoring Devices; Cardiovascular Devices
LEAKAGE DEFECTS	800.20	PATIENT EXAMINATION GLOVES AND SURGEON'S GLOVES; SAMPLE PLANS AND TEST METHODS FOR LEAKAGE DEFECTS; ADULTERATION Requirements For Specific Medical Device; General
LEATHER MEAL	573.540	HYDROLYZED LEATHER MEAL Food Additive Listing; Food Additives Permitted In Feed And Drinking Water Of Animals
LEAVES, COCA	1304.32	REPORTS OF MANUFACTURERS IMPORTING COCA LEAVES Reports; Records And Reports Of Registrants
LECITHIN	172.814	HYDROXYLATED LECITHIN Multipurpose Additives; Food Additives Permitted For Direct Addition To Food For Human Consumption
LECITHIN	184.1063	ENZYME-MODIFIED LECITHIN Listing Of Specific Substances Affirmed As GRAS; Direct Food Substances Affirmed As Generally Recognized As Safe
LECITHIN	184.1400	LECITHIN Listing Of Specific Substances Affirmed As GRAS; Direct Food Substances Affirmed As Generally Recognized As Safe
LECITHIN	582.1400	LECITHIN General Purpose Food Additives; Substances Generally Recognized As Safe
LECITHIN	862.1455	LECITHIN/SPHINGOMYELIN RATIO IN AMNIOTIC FLUID TEST SYSTEM Clinical Chemistry Test Systems; Clinical Chemistry And Clinical Toxicology Devices
LECTINS	864.9550	LECTINS AND PROTECTINS Products Used In Establishments That Manufacture Blood And Blood Products; Hematology And Pathology Devices
LEG CRAMPS	310.546	DRUG PRODUCTS CONTAINING ACTIVE INGREDIENTS OFFERED OTC FOR THE TREATMENT AND/OR PREVENTION OF NOCTURNAL LEG MUSCLE CRAMPS Requirements For Specific New Drugs Or Devices; New Drugs
LEGAL BASIS	120.9	LEGAL BASIS Hazard Analysis And Critical Control Point (HACCP) Systems
LEGAL GUARDIANS	21.75	RIGHTS OF LEGAL GUARDIANS Disclosure Of Records In Privacy Act Record Systems To Persons Other Than The Subject Individual; Protection Of Privacy

LEGIBLE	610.62	PROPER NAME; PACKAGE LABEL; LEGIBLE TYPE Labeling Standards; General Biological Products Standards
LEGISLATION	5.606	ACCEPTANCE OF ASSISTANCE FROM STATE AND LOCAL AUTHORITIES FOR ENFORCEMENT OF RADIATION CONTROL LEGISLATION AND REGULATIONS Radiation Control; Redelegations Of Authority; Delegations Of Authority And Organization
LEMON JUICE	146.114	LEMON JUICE Requirements For Specific Standardized Canned Fruit Juices And Beverages; Canned Fruit Juices
LEMONADE	146.120	FROZEN CONCENTRATE FOR LEMONADE Requirements For Specific Standardized Canned Fruit Juices And Beverages; Canned Fruit Juices
LEMONADE	146.121	FROZEN CONCENTRATE FOR ARTIFICIALLY SWEETENED LEMONADE Requirements For Specific Standardized Canned Fruit Juices And Beverages; Canned Fruit Juices
LEMONADE	146.126	FROZEN CONCENTRATE FOR COLORED LEMONADE Requirements For Specific Standardized Canned Fruit Juices And Beverages; Canned Fruit Juices
LENGTH	26.35	LENGTH AND PURPOSE OF TRANSITION PERIOD Specific Sector Provisions For Medical Devices; Mutual Recognition Of Pharmaceutical GMP Reports, Medical Device Quality System Audit Reports
LENS	886.1375	BAGOLINI LENS Diagnostic Devices; Ophthalmic Devices
LENS	886.1380	DIAGNOSTIC CONDENSING LENS Diagnostic Devices; Ophthalmic Devices
LENS	886.1390	FLEXIBLE DIAGNOSTIC FRESNEL LENS Diagnostic Devices; Ophthalmic Devices
LENS	886.1395	DIAGNOSTIC HRUBY FUNDUS LENS Diagnostic Devices; Ophthalmic Devices
LENS	886.1400	MADDOX LENS Diagnostic Devices; Ophthalmic Devices
LENS	886.1405	OPHTHALMIC TRIAL LENS SET Diagnostic Devices; Ophthalmic Devices
LENS	886.1410	OPHTHALMIC TRIAL LENS CLIP Diagnostic Devices; Ophthalmic Devices
LENS	886.1415	OPHTHALMIC TRIAL LENS FRAME Diagnostic Devices; Ophthalmic Devices
LENS	886.1420	OPHTHALMIC LENS GAUGE Diagnostic Devices; Ophthalmic Devices
LENS	886.3600	INTRAOCULAR LENS Prosthetic Devices; Ophthalmic Devices
LENS	886.4300	INTRAOCULAR LENS GUIDE Surgical Devices; Ophthalmic Devices
LENS	886.5844	PRESCRIPTION SPECTACLE LENS Therapeutic Devices; Ophthalmic Devices
LENS, CONTACT	800.10	CONTACT LENS SOLUTION; STERILITY Requirements For Specific Medical Devices; General
LENS, CONTACT	800.12	CONTACT LENS SOLUTIONS AND TABLETS; TAMPER-RESISTANT PACKAGING Requirements For Specific Medical Devices; General
LENS, CONTACT	886.1385	POLYMETHYLMETHACRYLATE (PMMA) DIAGNOSTIC CONTACT LENS Diagnostic Devices; Ophthalmic Devices
LENS, CONTACT	886.1430	OPHTHALMIC CONTACT LENS RADIUS MEASURING DEVICE Diagnostic Devices; Ophthalmic Devices
LENS, CONTACT	886.5420	CONTACT LENS INSERTER/REMOVER Therapeutic Devices; Ophthalmic Devices
LENS, CONTACT	886.5916	RIGID GAS PERMEABLE CONTACT LENS Therapeutic Devices; Ophthalmic Devices
LENS, CONTACT	886.5918	RIGID GAS PERMEABLE CONTACT LENS CARE PRODUCTS Therapeutic Devices; Ophthalmic Devices
LENS, CONTACT	886.5925	SOFT (HYDROPHILIC) CONTACT LENS Therapeutic Devices; Ophthalmic Devices
LENS, CONTACT	886.5928	SOFT (HYDROPHILIC) CONTACT LENS CARE PRODUCTS Therapeutic Devices; Ophthalmic Devices
LENS MEASURING INSTRUMENT	886.1425	LENS MEASURING INSTRUMENT Diagnostic Devices; Ophthalmic Devices
LENS RADIUS MEASURING	886.1430	OPHTHALMIC CONTACT LENS RADIUS MEASURING DEVICE Diagnostic Devices; Ophthalmic Devices
LENSES, IMPACT-RESISTANT	801.410	USE OF IMPACT-RESISTANT LENSES IN EYEGLASSES AND SUNGLASSES Special Requirements For Specific Devices; Labeling
LEPTOSPIRA SPP.	866.3350	LEPTOSPIRA SPP. SEROLOGICAL REAGENTS Serological Reagents; Immunology And Microbiology Devices
LESION	882.4400	RADIOFREQUENCY LESION GENERATOR Neurological Surgical Devices; Neurological Devices

LESION	882.4725	RADIOFREQUENCY LESION PROBE Neurological Surgical Devices; Neurological Devices
LESION	882.5500	LESION TEMPERATURE MONITOR Neurological Therapeutic Devices; Neurological Devices
LESSER ACTIONS	56.120	LESSER ADMINISTRATIVE ACTIONS Administrative Action For Noncompliance; Institutional Review Boards
LETTER, APPROVABLE	314.110	APPROVABLE LETTER TO THE APPLICANT FDA Action On Applications; Applications For FDA Approval To Market A New Drug
LETTER, NOT APPROVABLE	314.120	NOT APPROVABLE LETTER TO THE APPLICANT FDA Action On Applications; Applications For FDA Approval To Market A New Drug
LETTER, DESIGNATION	3.8	LETTER OF DESIGNATION Product Jurisdiction
LETTER, DESIGNATION	3.9	EFFECT OF LETTER OF DESIGNATION Product Jurisdiction
LETTERS, INTER-AGENCY	20.62	INTER- OR INTRA-AGENCY MEMORANDA OR LETTERS Exemptions; Public Information
LETTERS, OPINION	170.6	OPINION LETTERS ON FOOD ADDITIVE STATUS General Provisions; Food Additives
LETTERS, OPINION	570.6	OPINION LETTERS ON FOOD ADDITIVE STATUS General Provisions; Food Additives
LEUCINE	582.5406	LEUCINE Nutrients And/or Dietary Supplements; Substances Generally Recognized As Safe
LEUCINE AMINOPEPTIDASE	862.1460	LEUCINE AMINOPEPTIDASE TEST SYSTEM Clinical Chemistry Test Systems; Clinical Chemistry And Clinical Toxicology Devices
LEUKAPHERESIS	606.110	PLATELETPHERESIS, LEUKAPHERESIS, AND PLASMAPHERESIS Production And Process Controls; Current Good Manufacturing Practice For Blood And Blood Components
LEUKOCYTE	864.7660	LEUKOCYTE ALKALINE PHOSPHATASE TEST Hematology Kits And Packages; Hematology And Pathology Devices
LEUKOCYTE	864.7675	LEUKOCYTE PEROXIDASE TEST Hematology Kits And Packages; Hematology And Pathology Devices
LEUKOTOME	882.4600	LEUKOTOME Neurological Surgical Devices; Neurological Devices
LEVAMISOLE	524.1240	LEVAMISOLE Ophthalmic And Topical Dosage Form New Animal Drugs
LEVAMISOLE HYDROCHLORIDE	520.1242	LEVAMISOLE HYDROCHLORIDE ORAL DOSAGE FORMS Oral Dosage Form New Animal Drugs
LEVAMISOLE HYDROCHLORIDE	556.350	LEVAMISOLE HYDROCHLORIDE Specific Tolerances For Residues Of New Animal Drugs; Tolerances For Residues Of New Animal Drugs In Food
LEVAMISOLE HYDROCHLORIDE	558.315	LEVAMISOLE HYDROCHLORIDE (EQUIVALENT) Specific New Animal Drugs For Use In Animal Feeds; New Animal Drugs For Use In Animal Feeds
LEVAMISOLE PHOSPHATE	522.1244	LEVAMISOLE PHOSPHATE INJECTION Implantation Or Injectable Dosage Form New Animal Drugs
LEVAMISOLE RESINATE	520.1242d	LEVAMISOLE RESINATE Oral Dosage Form New Animal Drugs
LEVAMISOLE RESINATE	520.1242g	LEVAMISOLE RESINATE AND FAMPHUR PASTE Oral Dosage Form New Animal Drugs
LEVEL, OZONE	801.415	MAXIMUM ACCEPTABLE LEVEL OF OZONE Special Requirements For Specific Devices; Labeling
LEVEL SENSING	870.4340	CARDIOPULMONARY BYPASS LEVEL SENSING MONITOR AND/OR CONTROL Cardiovascular Surgical Devices; Cardiovascular Devices
LEVELS, ACTION	109.4	ESTABLISHMENT OF TOLERANCES, REGULATORY LIMITS AND ACTION LEVELS General Provisions; Unavoidable Contaminants In Food For Human Consumption & Food-packaging Material
LEVELS, ACTION	509.4	ESTABLISHMENT OF TOLERANCES, REGULATORY LIMITS AND ACTION LEVELS General Provisions; Unavoidable Contaminants In Animal Food & Food-packaging Material
LEVELS, SAFE	5.500	ISSUANCE OF FEDERAL REGISTER DOCUMENTS PERTAINING TO THE DETERMINATION OF SAFE LEVELS, NOTICE OF NEED FOR DEVELOPMENT OF AN ANALYTICAL METHOD, PROHIBITION OF CERTAIN EXTRALABEL DRUG USE Animal Drugs; Redelegations Of Authority; Delegations Of Authority And Organization
LEVELS, SAFE	530.22	SAFE LEVELS AND ANALYTICAL METHODS FOR FOOD-PRODUCING ANIMALS Extralabel Drug Use In Animals
LEVELS, SAFE	530.23	PROCEDURES FOR SETTING AND ANNOUNCING SAFE LEVELS Extralabel Drug Use In Animals

LEVELS, SAFE	530.40	SAFE LEVELS AND AVAILABILITY OF ANALYTICAL METHODS Extralabel Drug Use In Animals
LICE REMOVAL	880.5960	LICE REMOVAL KIT General Hospital And Personal Use Therapeutic Devices; General Hospital And Personal Use Devices
LICENSE	310.4	BIOLOGICS; PRODUCTS SUBJECT TO LICENSE CONTROL General Provisions; New Drugs
LICENSE	510.4	BIOLOGICS; PRODUCTS SUBJECT TO LICENSE CONTROL General Provisions; New Animal Drugs
LICENSE	601.4	ISSUANCE AND DENIAL OF LICENSE General Provisions; Licensing
LICENSE	601.5	REVOCATION OF LICENSE General Provisions; Licensing
LICENSE	601.6	SUSPENSION OF LICENSE General Provisions; Licensing
LICENSE, FEED MILL	5.501	APPROVAL OF NEW ANIMAL DRUG APPLICATIONS, MEDICATED FEED MILL LICENSE APPLICATIONS AND THEIR SUPPLEMENTS Animal Drugs; Redelegations Of Authority; Delegations Of Authority And Organization
LICENSE, FEED MILL	5.502	ISSUANCE OF NOTICES, PROPOSALS, AND ORDERS RELATING TO NEW ANIMAL DRUGS AND MEDICATED FEED MILL LICENSE APPLICATIONS Animal Drugs; Redelegations Of Authority; Delegations Of Authority And Organization
LICENSE, FEED MILL	510.301	RECORDS AND REPORTS CONCERNING EXPERIENCE WITH ANIMAL FEEDS BEARING OR CONTAINING NEW ANIMAL DRUGS FOR WHICH AN APPROVED MEDICATED FEED MILL LICENSE APPLICATION IS IN EFFECT Records And Reports; New Animal Drugs
LICENSE, FEED MILL	515.10	MEDICATED FEED MILL LICENSE APPLICATIONS Applications; Medicated Feed Mill License
LICENSE, FEED MILL	515.11	SUPPLEMENTAL MEDICATED FEED MILL LICENSE APPLICATIONS Applications; Medicated Feed Mill License
LICENSE, FEED MILL	515.20	APPROVAL OF MEDICATED FEED MILL LICENSE APPLICATIONS Administrative Actions On Licenses; Medicated Feed Mill License
LICENSE, FEED MILL	515.21	REFUSAL TO APPROVE A MEDICATED FEED MILL LICENSE APPLICATION Administrative Actions On Licenses; Medicated Feed Mill License
LICENSE, FEED MILL	515.22	SUSPENSION AND/OR REVOCATION OF APPROVAL OF A MEDICATED FEED MILL LICENSE Administrative Actions On Licenses; Medicated Feed Mill License
LICENSE, FEED MILL	515.23	VOLUNTARY REVOCATION OF A MEDICATED FEED MILL LICENSE Administrative Actions On Licenses; Medicated Feed Mill License
LICENSE, FEED MILL	515.24	NOTICE OF REVOCATION OF A MEDICATED FEED MILL LICENSE Administrative Actions On Licenses; Medicated Feed Mill License
LICENSE, FEED MILL	515.25	REVOCATION OF ORDER REFUSING TO APPROVE A MEDICATED FEED MILL LICENSE APPLICATION OR SUSPENDING OR REVOKING A LICENSE Administrative Actions On Licenses; Medicated Feed Mill License
LICENSE, FEED MILL	558.4	REQUIREMENT OF A MEDICATED FEED MILL LICENSE General Provisions; New Animal Drugs For Use In Animal Feeds
LICENSED ESTABLISHMENT	601.22	PRODUCTS IN SHORT SUPPLY; INITIAL MANUFACTURING AT OTHER THAN LICENSED ESTABLISHMENT Biologics Licensing; Licensing
LICENSED MANUFACTURERS	600.14	REPORTING OF BIOLOGICAL PRODUCT DEVIATIONS BY LICENSED MANUFACTURERS Establishment Standards; Biological Products: General
LICENSED MANUFACTURERS	606.171	REPORTING OF PRODUCT DEVIATIONS BY LICENSED MANUFACTURERS, UNLICENSED REGISTERED BLOOD ESTABLISHMENTS, AND TRANSFUSION SERVICES Current Good Manufacturing Practice For Blood And Blood Components
LICENSED PRODUCTS	601.25	REVIEW PROCEDURES TO DETERMINE LICENSED BIOLOGICAL PRODUCTS ARE SAFE, EFFECTIVE NOT MISBRANDED UNDER PRESCRIBED, RECOMMENDED, OR SUGGESTED CONDITIONS OF USE Biologics Licensing; Licensing
LICENSED PRODUCTS	601.26	RECLASSIFICATION PROCEDURES DETERMINE LICENSED BIOLOGICAL PRODUCTS ARE SAFE, EFFECTIVE, NOT MISBRANDED UNDER PRESCRIBED, RECOMMENDED, SUGGESTED CONDITIONS OF USE Biologics Licensing; Licensing
LICENSED PRODUCTS	610.53	DATING PERIODS FOR LICENSED BIOLOGICAL PRODUCTS Dating Period Limitations; General Biological Products Standards
LICENSES	5.202	ISSUANCE OF NOTICES OF OPPORTUNITY FOR A HEARING ON PROPOSALS FOR DENIAL OF APPROVAL OF APPLICATIONS FOR LICENSES, SUSPENSION OF LICENSES, OR REVOCATION OF LICENSES Biologics; Redelegations Of Authority; Delegations Of Authority And Organization
LICENSES	5.203	ISSUANCE AND REVOCATION OF LICENSES FOR THE PROPAGATION OR MANUFACTURE AND PREPARATION OF BIOLOGICAL PRODUCTS Biologics; Redelegations Of Authority; Delegations Of Authority And Organization

LICENSES	510.305	MAINTENANCE OF COPIES OF APPROVED MEDICATED FEED MILL LICENSES TO MANUFACTURE ANIMAL FEED BEARING OR CONTAINING NEW ANIMAL DRUGS Records And Reports; New Animal Drugs
LICENSES	601.2	APPLICATIONS FOR BIOLOGICS LICENSES; PROCEDURES FOR FILING General Provisions; Licensing
LICENSES	601.20	BIOLOGICS LICENSES; ISSUANCE AND CONDITIONS Biologics Licensing; Licensing
LICENSES	601.51	CONFIDENTIALITY OF DATA AND INFORMATION IN APPLICATIONS FOR BIOLOGICS LICENSES Confidentiality Of Information; Licensing
LICENSES	601.9	LICENSES; REISSUANCE General Provisions; Licensing
LICENSING	205.4	WHOLESALE DRUG DISTRIBUTOR LICENSING REQUIREMENT Guidelines For State Licensing Of Wholesale Prescription Drug Distributors
LICENSURE	205.5	MINIMUM REQUIRED INFORMATION FOR LICENSURE Guidelines For State Licensing Of Wholesale Prescription Drug Distributors
LICORICE	184.1408	LICORICE AND LICORICE DERIVATIVES Listing Of Specific Substances Affirmed As GRAS; Direct Food Substances Affirmed As Generally Recognized As Safe
LIDOCAINE	522.1662b	OXYTETRACYCLINE HYDROCHLORIDE WITH LIDOCAINE INJECTION Implantation Or Injectable Dosage Form New Animal Drugs
LIDOCAINE	862.3555	LIDOCAINE TEST SYSTEM Clinical Toxicology Test Systems; Clinical Chemistry And Clinical Toxicology Devices
LIFE -SUPPORTING DEVICES	860.93	CLASSIFICATION OF IMPLANTS, LIFE-SUPPORTING OR LIFE - SUSTAINING DEVICES Classification; Medical Device Classification Procedures
LIFE -SUSTAINING DEVICES	860.93	CLASSIFICATION OF IMPLANTS, LIFE -SUPPORTING OR LIFE -SUSTAINING DEVICES Classification; Medical Device Classification Procedures
LIFE-THREATENING	312.84	RISK-BENEFIT ANALYSIS IN REVIEW OF MARKETING APPLICATIONS FOR DRUGS TO TREAT LIFE-THREATENING AND SEVERELY-DEBILITATING ILLNESSES Investigational New Drug Application
LIFT	880.5500	AC-POWERED PATIENT LIFT General Hospital And Personal Use Therapeutic Devices; General Hospital And Personal Use Devices
LIFT	880.5510	NON-AC-POWERED PATIENT LIFT General Hospital And Personal Use Therapeutic Devices; General Hospital And Personal Use Devices
LIGATOR	876.4400	HEMORRHOIDAL LIGATOR Surgical Devices; Gastroenterology-urology Devices
LIGATURE	872.4600	INTRAORAL LIGATURE AND WIRE LOCK Surgical Devices; Dental Devices
LIGHT	101.56	NUTRIENT CONTENT CLAIMS FOR "LIGHT" OR "LITE" Specific Requirements For Nutrient Content Claims; Food Labeling
LIGHT	179.41	PULSED LIGHT FOR THE TREATMENT OF FOOD Radiation And Radiation Sources; Irradiation In The Production, Processing And Handling Of Food
LIGHT	872.4620	FIBER OPTIC DENTAL LIGHT Surgical Devices; Dental Devices
LIGHT	872.4630	DENTAL OPERATING LIGHT Surgical Devices; Dental Devices
LIGHT	874.4350	EAR, NOSE AND THROAT FIBEROPTIC LIGHT SOURCE AND CARRIER Surgical Devices; Ear, Nose And Throat Devices
LIGHT	876.4020	FIBEROPTIC LIGHT URETERAL CATHETER Surgical Devices; Gastroenterology-urology Devices
LIGHT	880.6320	AC-POWERED MEDICAL EXAMINATION LIGHT General Hospital And Personal Use Miscellaneous Devices; General Hospital And Personal Use Devices
LIGHT	880.6350	BATTERY-POWERED MEDICAL EXAMINATION LIGHT General Hospital And Personal Use Miscellaneous Devices; General Hospital And Personal Use Devices
LIGHT BEAM	892.5780	LIGHT BEAM PATIENT POSITION INDICATOR Therapeutic Devices; Radiology Devices
LIGHT CHAIN SPECIFIC	866.5550	IMMUNOGLOBULIN (LIGHT CHAIN SPECIFIC) IMMUNOLOGICAL TEST SYSTEM Immunological Test Systems; Immunology And Microbiology Devices
LIGHT CREAM	131.155	LIGHT CREAM Requirements For Specific Standardized Milk And Cream; Milk And Cream
LIGHT CREAM	131.157	LIGHT WHIPPING CREAM Requirements For Specific Standardized Milk And Cream; Milk And Cream

LIGHT PETROLEUM	172.884	ODORLESS LIGHT PETROLEUM HYDROCARBONS Multipurpose Additives; Food Additives Permitted For Direct Addition To Food For Human Consumption
LIGHT PETROLEUM	178.3650	ODORLESS LIGHT PETROLEUM HYDROCARBONS Certain Adjuvants And Production Aids; Indirect Food Additives: Adjuvants, Production Aids, And Sanitizers
LIGHT PETROLEUM	573.740	ODORLESS LIGHT PETROLEUM HYDROCARBONS Food Additive Listing; Food Additives Permitted In Feed And Drinking Water Of Animals
LIGHT SOURCE	352.71	LIGHT SOURCE (SOLAR STIMULATOR) Sunscreen Drug Products For Over-the-counter Human Use
LIGHTING	211.44	LIGHTING Buildings And Facilities; Current Good Manufacturing Practice For Finished Pharmaceuticals
LIGNIN SULFONATES	573.600	LIGNIN SULFONATES Food Additive Listing; Food Additives Permitted In Feed And Drinking Water Of Animals
LIGNOSULFONATE	172.715	CALCIUM LIGNOSULFONATE Other Specific Usage Additives; Food Additives Permitted For Direct Addition To Food For Human Consumption
LIMB	870.5800	COMPRESSIBLE LIMB SLEEVE Cardiovascular Therapeutic Devices; Cardiovascular Devices
LIMB	890.3410	EXTERNAL LIMB ORTHOTIC COMPONENT Physical Medicine Prosthetic Devices; Physical Medicine Devices
LIMB	890.3420	EXTERNAL LIMB PROSTHETIC COMPONENT Physical Medicine Prosthetic Devices; Physical Medicine Devices
LIMB	890.3500	EXTERNAL ASSEMBLED LOWER LIMB PROSTHESIS Physical Medicine Prosthetic Devices; Physical Medicine Devices
LIMB	890.5575	POWERED EXTERNAL LIMB OVERLOAD WARNING DEVICE Physical Medicine Therapeutic Devices; Physical Medicine Devices
LIMBURGER CHEESE	133.152	LIMBURGER CHEESE Requirements For Specific Standardized Cheese And Related Products; Cheeses And Related Cheese Products
LIMESTONE	184.1409	GROUND LIMESTONE Listing Of Specific Substances Affirmed As GRAS; Direct Food Substances Affirmed As Generally Recognized As Safe
LIMITATION, CERTIFICATES	81.32	LIMITATION OF CERTIFICATES General Specs+restrictns Provisional Color Additives For Foods, Drugs, And Cosmetics
LIMITATION, PACKAGE SIZE	201.307	SODIUM PHOSPHATES, PACKAGE SIZE LIMITATION, WARNINGS AND DIRECTIONS FOR OTC SALE Labeling
LIMITATIONS	530.11	LIMITATIONS Extralabel Drug Use In Animals
LIMITATIONS, CERTIFICATES	80.32	LIMITATIONS OF CERTIFICATES Certification Procedures; Color Additive Certification
LIMITATIONS, EXEMPTIONS	20.80	APPLICABILITY OF LIMITATIONS ON EXEMPTIONS Limitations On Exemptions; Public Information
LIMITATIONS OF EXEMPTIONS	862.9	LIMITATIONS OF EXEMPTIONS FROM SECTION 510(K) OF THE FEDERAL FOOD, DRUG, AND COSMETIC ACT Clinical Chemistry And Clinical Toxicology Devices
LIMITATIONS OF EXEMPTIONS	864.9	LIMITATIONS OF EXEMPTIONS FROM SECTION 510(K) OF THE FEDERAL FOOD, DRUG, AND COSMETIC ACT Hematology And Pathology Devices
LIMITATIONS OF EXEMPTIONS	866.9	LIMITATIONS OF EXEMPTIONS FROM SECTION 510(K) OF THE FEDERAL FOOD, DRUG, AND COSMETIC ACT Immunology And Microbiology Devices
LIMITATIONS OF EXEMPTIONS	868.9	LIMITATIONS OF EXEMPTIONS FROM SECTION 510(K) OF THE FEDERAL FOOD, DRUG, AND COSMETIC ACT Anesthesiology Devices
LIMITATIONS OF EXEMPTIONS	870.9	LIMITATIONS OF EXEMPTIONS FROM SECTION 510(K) OF THE FEDERAL FOOD, DRUG, AND COSMETIC ACT Cardiovascular Devices
LIMITATIONS OF EXEMPTIONS	872.9	LIMITATIONS OF EXEMPTIONS FROM SECTION 510(K) OF THE FEDERAL FOOD, DRUG, AND COSMETIC ACT Dental Devices
LIMITATIONS OF EXEMPTIONS	874.9	LIMITATIONS OF EXEMPTIONS FROM SECTION 510(K) OF THE FEDERAL FOOD, DRUG, AND COSMETIC ACT Ear, Nose, And Throat Devices
LIMITATIONS OF EXEMPTIONS	876.9	LIMITATIONS OF EXEMPTIONS FROM SECTION 510(K) OF THE FEDERAL FOOD, DRUG, AND COSMETIC ACT Gastroenterology-urology Devices

LIPASE	184.1415	ANIMAL LIPASE Listing Of Specific Substances Affirmed As GRAS; Direct Food Substances Affirmed As Generally Recognized As Safe
LIPASE	184.1420	LIPASE ENZYME PREPARATION DERIVED FROM RHIZOPUS NIVEUS Listing Of Specific Substances Affirmed As GRAS; Direct Food Substances Affirmed As Generally Recognized As Safe
LIPASE	862.1465	LIPASE TEST SYSTEM Clinical Chemistry Test Systems; Clinical Chemistry And Clinical Toxicology Devices
LIPID	862.1470	LIPID (TOTAL) TEST SYSTEM Clinical Chemistry Test Systems; Clinical Chemistry And Clinical Toxicology Devices
LIPIDS, DIETARY	101.73	HEALTH CLAIMS: DIETARY LIPIDS AND CANCER Food Labeling
LIPOPLASTY	878.5040	SUCTION LIPOPLASTY SYSTEM Surgical Devices; General And Plastic Surgery Devices
LIPOPROTEIN	862.1475	LIPOPROTEIN TEST SYSTEM Clinical Chemistry Test Systems; Clinical Chemistry And Clinical Toxicology Devices
LIPOPROTEIN	866.5580	ALPHA-1 LIPOPROTEIN IMMUNOLOGICAL TEST SYSTEM Immunological Test Systems; Immunology And Microbiology Devices
LIPOPROTEIN	866.5590	LIPOPROTEIN X IMMUNOLOGICAL TEST SYSTEM Immunological Test Systems; Immunology And Microbiology Devices
LIPOPROTEIN	866.5600	LOW-DENSITY LIPOPROTEIN IMMUNOLOGICAL TEST SYSTEM Immunological Test Systems; Immunology And Microbiology Devices
LIQUID	520.447	CLINDAMYCIN HYDROCHLORIDE LIQUID Oral Dosage Form New Animal Drugs
LIQUID	520.530	CYTHIOATE ORAL LIQUID Oral Dosage Form New Animal Drugs
LIQUID	520.1195	IVERMECTIN LIQUID Oral Dosage Form New Animal Drugs
LIQUID	520.1806	PIPERAZINE MONOHYDROCHLORIDE LIQUID Oral Dosage Form New Animal Drugs
LIQUID	520.2089	ROXARSONE LIQUID Oral Dosage Form New Animal Drugs
LIQUID	20.2456	TIAMULIN LIQUID CONCENTRATE Oral Dosage Form New Animal Drugs
LIQUID	524.86	AMITRAZ LIQUID Ophthalmic And Topical Dosage Form New Animal Drugs
LIQUID	524.1742	N-(MERCAPTOMETHYL) PHTHALIMIDE S-(*O,O*-DIMETHYL PHOSPHORODITHIOATE) EMULSIFIABLE LIQUID Ophthalmic And Topical Dosage Form New Animal Drugs
LIQUID BANDAGE	880.5090	LIQUID BANDAGE General Hospital And Personal Use Therapeutic Devices; General Hospital And Personal Use Devices
LIQUID CHROMATOGRAPHY	862.2250	GAS LIQUID CHROMATOGRAPHY SYSTEM FOR CLINICAL USE Clinical Laboratory Instruments; Clinical Chemistry And Clinical Toxicology Devices
LIQUID CHROMATOGRAPHY	862.2260	HIGH PRESSURE LIQUID CHROMATOGRAPHY SYSTEM FOR CLINICAL USE Clinical Laboratory Instruments; Clinical Chemistry And Clinical Toxicology Devices
LIQUID CRYSTAL	880.2200	LIQUID CRYSTAL FOREHEAD TEMPERATURE STRIP General Hospital And Personal Use Monitoring Devices; General Hospital And Personal Use Devices
LIQUID CRYSTAL	880.6970	LIQUID CRYSTAL VEIN LOCATOR General Hospital And Personal Use Miscellaneous Devices; General Hospital And Personal Use Devices
LIQUID CRYSTAL	884.2982	LIQUID CRYSTAL THERMOGRAPHIC SYSTEM Obstetrical And Gynecological Monitoring Devices; Obstetrical And Gynecological Devices
LIQUID CRYSTALLINE	21.2620	LIQUID CRYSTALLINE TRYPSIN, PERU BALSAM, CASTOR OIL Ophthalmic And Topical Dosage Form New Animal Drugs
LIQUID EGGS	160.115	LIQUID EGGS Requirements For Specific Standardized Eggs And Egg Products; Eggs And Egg Products
LIQUID FEED	558.465	POLOXYALENE FREE-CHOICE LIQUID TYPE C FEED General Provisions; New Animal Drugs For Use In Animal Feeds
LIQUID FEEDS	558.5	NEW ANIMAL DRUG REQUIREMENTS FOR LIQUID TYPE B FEEDS General Provisions; New Animal Drugs For Use In Animal Feeds
LIQUID MEDICATION	880.6430	LIQUID MEDICATION DISPENSER General Hospital And Personal Use Miscellaneous Devices; General Hospital And Personal Use Devices
LIQUID OXYGEN	68.5655	PORTABLE LIQUID OXYGEN UNIT Therapeutic Devices; Anesthesiology Devices

LIQUID STERILIZANTS	880.6885	LIQUID CHEMICAL STERILIZANTS/HIGH-LEVEL DISINFECTANTS General Hospital And Personal Use Miscellaneous Devices; General Hospital And Personal Use Devices
LIQUOR, CHOCOLATE	163.111	CHOCOLATE LIQUOR Requirements For Specific Standardized Cacao Products; Cacao Products
LIST, BLOOD PRODUCT	607.20	WHO MUST REGISTER AND SUBMIT A BLOOD PRODUCT LIST Procedures For Domestic Blood Product Establishments; Establishment Registratn+product Listing, Manufacturers Human Blood/b.products
LIST BLOOD PRODUCTS	607.22	HOW AND WHERETO REGISTER ESTABLISHMENTS AND LIST BLOOD PRODUCTS Procedures For Domestic Blood Product Establishments; Establishment Registratn+product Listing, Manufacturers Human Blood/b.products
LIST OF COMMITTEES	14.100	LIST OF STANDING ADVISORY COMMITTEES Standing Advisory Committees; Public Hearing Before A Public Advisory Committee
LIST, DEVICE	807.20	WHO MUST REGISTER AND SUBMIT A DEVICE LIST Procedures For Device Establishments; Establishment Registration And Device Listing For Manufacturers And Initial Importers Of Devices
LIST DEVICES	807.22	HOW AND WHERE TO REGISTER ESTABLISHMENTS AND LIST DEVICES Procedures For Device Establishments; Establishment Registration And Device Listing For Manufacturers And Initial Importers Of Devices
LIST, DRUG	207.20	WHO MUST REGISTER AND SUBMIT A DRUG LIST Procedures For Domestic Drug Establishments; Registration Of Drug Producers And Listing Of Drugs In Commercial Distribution
LIST, DRUG PRODUCT	314.162	REMOVAL OF A DRUG PRODUCT FROM THE LIST Applications For FDA Approval To Market A New Drug
LIST DRUGS	207.22	HOW AND WHERE TO REGISTER AND LIST DRUGS Procedures For Domestic Drug Establishments; Registration Of Drug Producers And Listing Of Drugs In Commercial Distribution
LIST, HCT/P	1271.21	WHEN DO I REGISTER, SUBMIT AN HCT/P LIST, AND SUBMIT UPDATES? Procedures For Registration And Listing; Human Cells, Tissues, And Cellular And Tissue-Based Products
LIST, HCT/P	1271.22	HOW AND WHERE DO I REGISTER AND SUBMIT AN HCT/P LIST? Procedures For Registration And Listing; Human Cells, Tissues, And Cellular And Tissue-Based Products
LISTED	26.10	REGULATORY AUTHORITIES NOT LISTED AS CURRENTLY EQUIVALENT Specific Sector Provisions For Pharmaceutical GMPs; Mutual Recognition Of Pharmaceutical GMP Reports, Medical Device Quality System Audit Reports
LISTED	26.67	SUSPENSION OF LISTED CONFORMITY ASSESSMENT BODIES Framework Provisions; Mutual Recognition Of Pharmaceutical GMP Reports, Medical Device Quality System Audit Reports
LISTED	26.68	WITHDRAWAL OF LISTED CONFORMITY ASSESSMENT BODIES Framework Provisions; Mutual Recognition Of Pharmaceutical GMP Reports, Medical Device Quality System Audit Reports
LISTED CHEMICALS	1300.2	DEFINITIONS RELATING TO LISTED CHEMICALS Definitions
LISTED DRUG	314.54	PROCEDURE FOR SUBMISSION OF AN APPLICATION REQUIRING INVESTIGATIONS FOR APPROVAL OF A NEW INDICATION FOR, OR OTHER CHANGE FROM, A LISTED DRUG Applications For FDA Approval To Market A New Drug
LISTED DRUG	314.93	PETITION TO REQUEST A CHANGE FROM A LISTED DRUG Applications For FDA Approval To Market A New Drug
LISTED DRUG	314.161	DETERMINATION OF REASONS FOR VOLUNTARY WITHDRAWAL OF A LISTED DRUG Applications For FDA Approval To Market A New Drug
LISTED DRUG	314.122	SUBMITTING AN ABBREVIATED APPLICATION FOR A LISTED DRUG THAT IS NO LONGER MARKETED Applications For FDA Approval To Market A New Drug
LISTED DRUG	314.161	DETERMINATION OF REASONS FOR VOLUNTARY WITHDRAWAL OF A LISTED DRUG Applications For FDA Approval To Market A New Drug
LISTERIA SPP.	866.3355	LISTERIA SPP. SEROLOGICAL REAGENTS Serological Reagents; Immunology And Microbiology Devices
LISTING	26.36	LISTING OF CABS Specific Sector Provisions For Medical Devices; Mutual Recognition Of Pharmaceutical GMP Reports, Medical Device Quality System Audit Reports
LISTING	26.46	LISTING OF ADDITIONAL CABS Specific Sector Provisions For Medical Devices; Mutual Recognition Of Pharmaceutical GMP Reports, Medical Device Quality System Audit Reports
LISTING	26.66	DESIGNATION AND LISTING PROCEDURES Framework Provisions; Mutual Recognition Of Pharmaceutical GMP Reports, Medical Device Quality System Audit Reports
LISTING	70.19	FEES FOR LISTING General Provisions; Color Additives

LISTING	71.27	LISTING AND EXEMPTION FROM CERTIFICATION ON THE COMMISSIONER'S INITIATIVE Administrative Action On Petitions; Color Additive Petitions
LISTING	331.11	LISTING OF SPECIFIC ACTIVE INGREDIENTS Active Ingredients; Antacid Products For Over-the-counter (OTC) Human Use
LISTING	807.31	ADDITIONAL LISTING INFORMATION Procedures For Device Establishments; Establishment Registration And Device Listing For Manufacturers And Initial Importers Of Devices
LISTING, BLOOD PRODUCT	607.21	TIMES FOR ESTABLISHMENT REGISTRATION AND BLOOD PRODUCT LISTING Procedures For Domestic Blood Product Establishments; Establishment Registratn+product Listing, Manufacturers Human Blood/b.products
LISTING, BLOOD PRODUCT	607.25	INFORMATION REQUIRED FOR ESTABLISHMENT REGISTRATION AND BLOOD PRODUCT LISTING Procedures For Domestic Blood Product Establishments; Establishment Registratn+product Listing, Manufacturers Human Blood/b.products
LISTING, BLOOD PRODUCT	607.30	UPDATING BLOOD PRODUCT LISTING INFORMATION Procedures For Domestic Blood Product Establishments; Establishment Registratn+product Listing, Manufacturers Human Blood/b.products
LISTING, BLOOD PRODUCT	607.31	ADDITIONAL BLOOD PRODUCT LISTING INFORMATION Procedures For Domestic Blood Product Establishments; Establishment Registratn+product Listing, Manufacturers Human Blood/b.products
LISTING, BLOOD PRODUCT	607.40	ESTABLISHMENT REGISTRATION AND BLOOD PRODUCT LISTING REQUIREMENTS FOR FOREIGN BLOOD PRODUCT ESTABLISHMENTS Procedures For Foreign Blood Product Establishments; Establishment Registratn+product Listing, Manufacturers Human Blood/b.products
LISTING, BLOOD PRODUCTS	207.7	ESTABLISHMENT REGISTRATION AND PRODUCT LISTING FOR HUMAN BLOOD AND BLOOD PRODUCTS AND FOR MEDICAL DEVICES General; Registration Of Drug Producers And Listing Of Drugs In Commercial Distribution
LISTING, BLOOD PRODUCTS	607.7	ESTABLISHMENT REGISTRATION AND PRODUCT LISTING OF BLOOD BANKS AND OTHER FIRMS MANUFACTURING HUMAN BLOOD AND BLOOD PRODUCTS General Provisions; Establishment Registratn+product Listing, Manufacturers Human Blood/b.products
LISTING, DEVICE	20.116	DRUG AND DEVICE LISTING INFORMATION Availability Of Specific Categories Of Records; Public Information
LISTING, DEVICE	807.21	TIMES FOR ESTABLISHMENT REGISTRATION AND DEVICE LISTING Procedures For Device Establishments; Establishment Registration And Device Listing For Manufacturers And Initial Importers Of Devices
LISTING, DEVICE	807.25	INFORMATION REQUIRED OR REQUESTED FOR ESTABLISHMENT REGISTRATION AND DEVICE LISTING Procedures For Device Establishments; Establishment Registration And Device Listing For Manufacturers And Initial Importers Of Devices
LISTING, DEVICE	807.30	UPDATING DEVICE LISTING INFORMATION Procedures For Device Establishments; Establishment Registration And Device Listing For Manufacturers And Initial Importers Of Devices
LISTING, DEVICE	807.40	ESTABLISHMENT REGISTRATION AND DEVICE LISTING FOR FOREIGN ESTABLISHMENTS IMPORTING OR OFFERING FOR IMPORT DEVICES INTO THE UNITED STATES Registration Procedures For Foreign Device Establishments; Establishment Registration And Device Listing For Manufacturers And Initial Importers Of Devices
LISTING, DRUG	20.116	DRUG AND DEVICE LISTING INFORMATION Availability Of Specific Categories Of Records; Public Information
LISTING, DRUG	207.21	TIMES FOR REGISTRATION AND DRUG LISTING Procedures For Domestic Drug Establishments; Registration Of Drug Producers And Listing Of Drugs In Commercial Distribution
LISTING, DRUG	207.25	INFORMATION REQUIRED IN REGISTRATION AND DRUG LISTING Procedures For Domestic Drug Establishments; Registration Of Drug Producers And Listing Of Drugs In Commercial Distribution
LISTING, DRUG	207.30	UPDATING DRUG LISTING INFORMATION Procedures For Domestic Drug Establishments; Registration Of Drug Producers And Listing Of Drugs In Commercial Distribution
LISTING, DRUG	207.31	ADDITIONAL DRUG LISTING INFORMATION Procedures For Domestic Drug Establishments; Registration Of Drug Producers And Listing Of Drugs In Commercial Distribution
LISTING, DRUG	207.35	NOTIFICATION OF REGISTRANT; DRUG ESTABLISHMENT REGISTRATION NUMBER AND DRUG LISTING NUMBER Procedures For Domestic Drug Establishments; Registration Of Drug Producers And Listing Of Drugs In Commercial Distribution
LISTING, DRUG	207.40	ESTABLISHMENT REGISTRATION AND DRUG LISTING REQUIREMENTS FOR FOREIGN ESTABLISHMENTS Procedure For Foreign Drug Establishments; Registration Of Drug Producers And Listing Of Drugs In Commercial Distribution

LISTING, HCT/P	1271.25	WHAT INFORMATION IS REQUIRED FOR ESTABLISHMENT REGISTRATION AND HCT/P LISTING? Procedures For Registration And Listing; Human Cells, Tissues, And Cellular And Tissue-Based Products
LISTINGS, BLOOD PRODUCT	607.37	INSPECTION OF ESTABLISHMENT REGISTRATIONS AND BLOOD PRODUCT LISTINGS Procedures For Domestic Blood Product Establishments; Establishment Registratn+product Listing, Manufacturers Human Blood/b.products
LISTINGS, COLOR ADDITIVES	81.10	TERMINATION OF PROVISIONAL LISTINGS OF COLOR ADDITIVES General Specs+restrictns Provisional Color Additives For Foods, Drugs, And Cosmetics
LISTINGS, DEVICE	807.37	INSPECTION OF ESTABLISHMENT REGISTRATION AND DEVICE LISTINGS Procedures For Device Establishments; Establishment Registration And Device Listing For Manufacturers And Initial Importers Of Devices
LISTINGS, DRUG	207.37	INSPECTION OF REGISTRATIONS AND DRUG LISTINGS Procedures For Domestic Drug Establishments; Registration Of Drug Producers And Listing Of Drugs In Commercial Distribution
LISTINGS, HCT/P	1271.37	WILL ESTABLISHMENT REGISTRATIONS AND HCT/P LISTINGS BE AVAILABLE FOR INSPECTION, AND HOW DO I REQUEST INFORMATION? Procedures For Registration And Listing; Human Cells, Tissues, And Cellular And Tissue-Based Products
LISTS	20.119	LISTS OF NAMES AND ADDRESSES Availability Of Specific Categories Of Records; Public Information
LISTS	81.1	PROVISIONAL LISTS OF COLOR ADDITIVES General Specs+restrictns Provisional Color Additives For Foods, Drugs, And Cosmetics
LISTS, WITNESS	17.25	EXCHANGES OF WITNESS LISTS, WITNESS STATEMENTS, AND EXHIBITS Civil Money Penalties Hearings
LITE	101.56	NUTRIENT CONTENT CLAIMS FOR "LIGHT" OR "LITE" Specific Requirements For Nutrient Content Claims; Food Labeling
LITHIUM	862.3560	LITHIUM TEST SYSTEM Clinical Toxicology Test Systems; Clinical Chemistry And Clinical Toxicology Devices
LITHOTRIPTOR	876.4480	ELECTROHYDRAULIC LITHOTRIPTOR Surgical Devices; Gastroenterology-urology Devices
LITHOTRIPTOR	876.4500	MECHANICAL LITHOTRIPTOR Surgical Devices; Gastroenterology-urology Devices
LITHOTRIPTOR	876.599	EXTRACORPOREAL SHOCK WAVE LITHOTRIPTER Therapeutic Devices; Gastroenterology-urology Devices
LOCAL ANESTHETIC	346.10	LOCAL ANESTHETIC ACTIVE INGREDIENTS Anorectal Drug Products For Over-the-counter Human Use
LOCAL AUTHORITIES	5.606	ACCEPTANCE OF ASSISTANCE FROM STATE AND LOCAL AUTHORITIES FOR ENFORCEMENT OF RADIATION CONTROL LEGISLATION AND REGULATIONS Radiation Control; Redelegations Of Authority; Delegations Of Authority And Organization
LOCAL CONTROL	240.30	MEASURES IN THE EVENT OF INADEQUATE LOCAL CONTROL Control Of Communicable Diseases
LOCAL OFFICIALS	20.84	DISCLOSURE TO CONSULTANTS, ADVISORY COMMITTEES, STATE AND LOCAL GOVT. OFFICIALS AND OTHER SPECIAL GOVERNMENT EMPLOYEES Limitations On Exemptions; Public Information
LOCAL OFFICIALS	20.88	COMMUNICATIONS WITH STATE AND LOCAL GOVERNMENT OFFICIALS Limitations On Exemptions; Public Information
LOCAL REQUIREMENTS	100.1	PETITIONS REQUESTING EXEMPTION FROM PREEMPTION FOR STATE OR LOCAL REQUIREMENTS State And Local Requirements; General
LOCATING	864.5260	AUTOMATED CELL-LOCATING DEVICE Automated And Semi-automated Hematology Devices; Hematology And Pathology Devices
LOCATION	201.17	DRUGS; LOCATION OF EXPIRATION DATE General Labeling Provisions; Labeling
LOCATION	1302.04	LOCATION AND SIZE OF SYMBOL ON LABEL Labeling And Packaging Requirements For Controlled Substances
LOCATION, EQUIPMENT	211.63	EQUIPMENT DESIGN, SIZE, AND LOCATION Equipment And Facilities; Current Good Manufacturing Practice For Finished Pharmaceuticals
LOCATIONS	1304.25	RECORDS FOR TREATMENT PROGRAMS WHICH COMPOUND NARCOTICS FOR TREATMENT PROGRAMS AND OTHER LOCATIONS Continuing Records; Records And Reports Of Registrants
LOCATIONS, SEPARATE	1309.23	SEPARATE REGISTRATION FOR SEPARATE LOCATIONS Registration Of Manufacturers, Distributors, Importers And Exporters Of List I Chemicals
LOCATOR	880.6970	LIQUID CRYSTAL VEIN LOCATOR General Hospital And Personal Use Miscellaneous Devices; General Hospital And Personal Use Devices

LOCATOR	874.1820	SURGICAL NERVE STIMULATOR/LOCATOR Diagnostic Devices; Ear, Nose, And Throat Devices
LOCATOR	886.4400	ELECTRONIC METAL LOCATOR Surgical Devices; Ophthalmic Devices
LOCK	872.4600	INTRAORAL LIGATURE AND WIRE LOCK Surgical Devices; Dental Devices
LOCUST BEAN GUM	184.1343	LOCUST (CAROB) BEAN GUM Listing Of Specific Substances Affirmed As GRAS; Direct Food Substances Affirmed As Generally Recognized As Safe
LOCUST BEAN GUM	582.7343	LOCUST BEAN GUM Stabilizers; Substances Generally Recognized As Safe
LOG, EQUIPMENT	211.182	EQUIPMENT CLEANING AND USE LOG Records And Reports; Current Good Manufacturing Practice For Finished Pharmaceuticals
LOGWOOD EXTRACT	73.1410	LOGWOOD EXTRACT Drugs; Listing Of Color Additives Exempt From Certification
LONG CHAIN	178.3780	POLYHYDRIC ALCOHOL ESTERS OF LONG CHAIN MONOBASIC ACIDS Certain Adjuvants And Production Aids; Indirect Food Additives: Adjuvants, Production Aids, And Sanitizers
LONG-TERM CATHETER	880.5970	PERCUTANEOUS, IMPLANTED, LONG-TERM INTRAVASCULAR CATHETER General Hospital And Personal Use Therapeutic Devices; General Hospital And Personal Use Devices
LONG-TERM STUDIES	310.303	CONTINUATION OF LONG-TERM STUDIES, RECORDS, AND REPORTS ON CERTAIN DRUGS FOR WHICH NEW DRUG APPLICATIONS HAVE BEEN APPROVED Records And Reports; New Drugs
LONG-TERM SUPPORT	868.5610	MEMBRANE LUNG FOR LONG-TERM PULMONARY SUPPORT Therapeutic Devices; Anesthesiology Devices
LOOK BACK	610.46	"LOOK BACK" REQUIREMENTS Testing Requirements For Communicable Disease Agents; General Biological Products Standards
LOOK BACK	610.47	"LOOK BACK" NOTIFICATION REQUIREMENTS FOR TRANSFUSION SERVICES Testing Requirements For Communicable Disease Agents; General Biological Products Standards
LOST FORMS	1305.12	LOST AND STOLEN ORDER FORMS Order Forms
LOT	203.38	SAMPLE LOT OR CONTROL NUMBERS; LABELING OF SAMPLE UNITS Prescription Drug Marketing
LOT	610.1	TESTS PRIOR TO RELEASE REQUIRED FOR EACH LOT Release Requirements; General Biological Products Standards
LOTION	524.1443	MICONAZOLE NITRATE CREAM; MICONAZOLE NITRATE LOTION; MICONAZOLE NITRATE SPRAY Ophthalmic And Topical Dosage Form New Animal Drugs
LOUPE	886.4770	LOW-POWER BINOCULAR LOUPE Therapeutic Devices; Ophthalmic Devices
LOUPES	886.4770	OPHTHALMIC OPERATING SPECTACLES (LOUPES) Surgical Devices; Ophthalmic Devices
LOW-ACID FOODS	108.35	THERMAL PROCESSING OF LOW-ACID FOODS PACKAGED IN HERMETICALLY SEALED CONTAINERS Specific Requirements And Conditions For Exemption From Or Compliance With An Emergency Permit; Emergency Permit Control
LOW-ACID FOODS	500.23	THERMALLY PROCESSED LOW-ACID FOODS PACKAGED IN HERMETICALLY SEALED CONTAINERS Specific Administrative Rulings And Decisions; General
LOW-DENSITY LIPOPROTEIN	866.5600	LOW-DENSITY LIPOPROTEIN IMMUNOLOGICAL TEST SYSTEM Immunological Test Systems; Immunology And Microbiology Devices
LOW-FAT COCOA	163.114	LOW-FAT COCOA Requirements For Specific Standardized Cacao Products; Cacao Products
LOW-MOISTURE MOZZARELLA	133.156	LOW-MOISTURE MOZZARELLA AND SCAMORZA CHEESE Requirements For Specific Standardized Cheese And Related Products; Cheeses And Related Cheese Products
LOW-MOISTURE MOZZARELLA	133.158	LOW-MOISTURE PART-SKIM MOZZARELLA AND SCAMORZA CHEESE Requirements For Specific Standardized Cheese And Related Products; Cheeses And Related Cheese Products
LOW SODIUM	133.116	LOW SODIUM CHEDDAR CHEESE Requirements For Specific Standardized Cheese And Related Products; Cheeses And Related Cheese Products
LOW SODIUM	133.121	LOW SODIUM COLBY CHEESE Requirements For Specific Standardized Cheese And Related Products; Cheeses And Related Cheese Products
LOW-VISION	886.5540	LOW-VISION MAGNIFIER Therapeutic Devices; Ophthalmic Devices
LOW-VISION	886.5870	LOW-VISION TELESCOPE Therapeutic Devices; Ophthalmic Devices

LOWER LIMB	890.3500	EXTERNAL ASSEMBLED LOWER LIMB PROSTHESIS Physical Medicine Prosthetic Devices; Physical Medicine Devices
LOWFAT YOGURT	131.203	LOWFAT YOGURT Requirements For Specific Standardized Milk And Cream; Milk And Cream
LSD	862.3580	LYSERGIC ACID DIETHYLAMIDE (LSD) TEST SYSTEM Clinical Toxicology Test Systems; Clinical Chemistry And Clinical Toxicology Devices
LUBRICANT	880.6375	PATIENT LUBRICANT General Hospital And Personal Use Miscellaneous Devices; General Hospital And Personal Use Devices
LUBRICANT	884.5310	CONDOM WITH SPERMICIDAL LUBRICANT Obstetrical And Gynecological Therapeutic Devices; Obstetrical And Gynecological Devices
LUBRICANTS	178.3570	LUBRICANTS WITH INCIDENTAL FOOD CONTACT Certain Adjuvants And Production Aids; Indirect Food Additives: Adjuvants, Production Aids, And Sanitizers
LUBRICANTS	178.3910	SURFACE LUBRICANTS USED IN THE MANUFACTURE OF METALLIC ARTICLES Certain Adjuvants And Production Aids; Indirect Food Additives: Adjuvants, Production Aids, And Sanitizers
LUBRICATING	878.4480	ABSORBABLE POWDER FOR LUBRICATING A SURGEON'S GLOVE Surgical Devices; General And Plastic Surgery Devices
LUFENURON	520.1288	LUFENURON TABLETS Oral Dosage Form New Animal Drugs
LUFENURON	520.1289	LUFENURON SUSPENSION Oral Dosage Form New Animal Drugs
LUFENURON	522.1289	LUFENURON SUSPENSION Implantation Or Injectable Dosage Form New Animal Drugs
LUFENURON	522.1446	MILBEMYCIN OXIME/LUFENURON TABLETS Implantation Or Injectable Dosage Form New Animal Drugs
LUMEN	880.5570	HYPODERMIC SINGLE LUMEN NEEDLE General Hospital And Personal Use Therapeutic Devices; General Hospital And Personal Use Device
LUMINESCENT	73.2995	LUMINESCENT ZINC SULFIDE Cosmetics; Listing Of Color Additives Exempt From Certification
LUNATE	888.3750	WRIST JOINT CARPAL LUNATE POLYMER PROSTHESIS Prosthetic Devices; Orthopedic Devices
LUNG	868.5610	MEMBRANE LUNG FOR LONG-TERM PULMONARY SUPPORT Therapeutic Devices; Anesthesiology Devices
LUNG	870.4220	CARDIOPULMONARY BYPASS HEART-LUNG MACHINE CONSOLE Cardiovascular Surgical Devices; Cardiovascular Devices
LUNG WATER	868.2450	LUNG WATER MONITOR Monitoring Devices; Anesthesiology Devices
LUPROSTIOL	522.1290	LUPROSTIOL Implantation Or Injectable Dosage Form New Animal Drugs
LUPUS ERYTHEMATOSUS	866.5820	SYSTEMIC LUPUS ERYTHEMATOSUS IMMUNOLOGICAL TEST SYSTEM Immunological Test Systems; Immunology And Microbiology Devices
LUTEINIZING HORMONE	522.1820	PITUITARY LUTEINIZING HORMONE FOR INJECTION Implantation Or Injectable Dosage Form New Animal Drugs
LUTEINIZING HORMONE	862.1485	LUTEINIZING HORMONE TEST SYSTEM Clinical Chemistry Test Systems; Clinical Chemistry And Clinical Toxicology Devices
LYMPHOCYTE	864.8500	LYMPHOCYTE SEPARATION MEDIUM Hematology Reagents; Hematology And Pathology Devices
LYMPHOCYTIC	866.3360	LYMPHOCYTIC CHORIOMENINGITIS VIRUS SEROLOGICAL REAGENTS Serological Reagents; Immunology And Microbiology Devices
LYSERGIC ACID	862.3580	LYSERGIC ACID DIETHYLAMIDE (LSD) TEST SYSTEM Clinical Toxicology Test Systems; Clinical Chemistry And Clinical Toxicology Devices
LYSINE	582.5411	LYSINE Nutrients And/or Dietary Supplements; Substances Generally Recognized As Safe
LYSING	864.8540	RED CELL LYSING REAGENT Hematology Reagents; Hematology And Pathology Devices
LYSIS	864.7275	EUGLOBULIN LYSIS TIME TESTS Hematology Kits And Packages; Hematology And Pathology Devices
LYSODEIKTICUS	173.135	CATALASE DERIVED FROM MICROCOCCUS LYSODEIKTICUS Enzyme Preparations And Microorganisms; Secondary Direct Food Additives Permitted In Food For Human Consumption
LYSOZYME	862.1490	LYSOZYME (MURAMIDASE) TEST SYSTEM Clinical Chemistry Test Systems; Clinical Chemistry And Clinical Toxicology Devices

M

MACARONI	139.110	MACARONI PRODUCTS Requirements For Specific Standardized Macaroni And Noodle Products; Macaroni And Noodle Products
MACARONI	139.115	ENRICHED MACARONI PRODUCTS Requirements For Specific Standardized Macaroni And Noodle Products; Macaroni And Noodle Products
MACARONI	139.117	ENRICHED MACARONI PRODUCTS WITH FORTIFIED PROTEIN Requirements For Specific Standardized Macaroni And Noodle Products; Macaroni And Noodle Products
MACARONI	139.120	MILK MACARONI PRODUCTS Requirements For Specific Standardized Macaroni And Noodle Products; Macaroni And Noodle Products
MACARONI	139.121	NONFAT MILK MACARONI PRODUCTS Requirements For Specific Standardized Macaroni And Noodle Products; Macaroni And Noodle Products
MACARONI	139.122	ENRICHED NONFAT MILK MACARONI PRODUCTS Requirements For Specific Standardized Macaroni And Noodle Products; Macaroni And Noodle Products
MACARONI	139.125	VEGETABLE MACARONI PRODUCTS Requirements For Specific Standardized Macaroni And Noodle Products; Macaroni And Noodle Products
MACARONI	139.135	ENRICHED VEGETABLE MACARONI PRODUCTS Requirements For Specific Standardized Macaroni And Noodle Products; Macaroni And Noodle Products
MACARONI	139.138	WHOLE WHEAT MACARONI PRODUCTS Requirements For Specific Standardized Macaroni And Noodle Products; Macaroni And Noodle Products
MACARONI	139.140	WHEAT AND SOY MACARONI PRODUCTS Requirements For Specific Standardized Macaroni And Noodle Products; Macaroni And Noodle Products
MACHINE, GAS	868.5160	GAS MACHINE FOR ANESTHESIA OR ANALGESIA Therapeutic Devices; Anesthesiology Devices
MACHINE, HEART-LUNG	870.4220	CARDIOPULMONARY BYPASS HEART-LUNG MACHINE CONSOLE Cardiovascular Surgical Devices; Cardiovascular Devices
MACROGLOBULIN	866.5620	ALPHA-2-MACROGLOBULIN IMMUNOLOGICAL TEST SYSTEM Immunological Test Systems; Immunology And Microbiology Devices
MADDOX LENS	886.1400	MADDOX LENS Diagnostic Devices; Ophthalmic Devices
MADURAMICIN AMMONIUM	556.375	MADURAMICIN AMMONIUM Specific Tolerances For Residues Of New Animal Drugs In Food; Tolerances For Residues Of New Animal Drugs In Food
MADURAMICIN AMMONIUM	558.340	MADURAMICIN AMMONIUM Specific New Animal Drugs For Use In Animal Feeds; New Animal Drugs For Use In Animal Feeds
MAGNESIUM	862.1495	MAGNESIUM TEST SYSTEM Clinical Chemistry Test Systems; Clinical Chemistry And Clinical Toxicology Devices
MAGNESIUM CARBONATE	184.1425	MAGNESIUM CARBONATE Listing Of Specific Substances Affirmed As GRAS; Direct Food Substances Affirmed As Generally Recognized As Safe
MAGNESIUM CARBONATE	582.1425	MAGNESIUM CARBONATE General Purpose Food Additives; Substances Generally Recognized As Safe
MAGNESIUM CHLORIDE	184.1426	MAGNESIUM CHLORIDE Listing Of Specific Substances Affirmed As GRAS; Direct Food Substances Affirmed As Generally Recognized As Safe
MAGNESIUM HYDROXIDE	184.1428	MAGNESIUM HYDROXIDE Listing Of Specific Substances Affirmed As GRAS; Direct Food Substances Affirmed As Generally Recognized As Safe
MAGNESIUM HYDROXIDE	582.1428	MAGNESIUM HYDROXIDE General Purpose Food Additives; Substances Generally Recognized As Safe
MAGNESIUM OXIDE	184.1431	MAGNESIUM OXIDE Listing Of Specific Substances Affirmed As GRAS; Direct Food Substances Affirmed As Generally Recognized As Safe
MAGNESIUM OXIDE	582.1431	MAGNESIUM OXIDE General Purpose Food Additives; Substances Generally Recognized As Safe
MAGNESIUM OXIDE	582.5431	MAGNESIUM OXIDE Nutrients And/or Dietary Supplements; Substances Generally Recognized As Safe
MAGNESIUM PHOSPHATE	184.1434	MAGNESIUM PHOSPHATE Listing Of Specific Substances Affirmed As GRAS; Direct Food Substances Affirmed As Generally Recognized As Safe

MAGNESIUM PHOSPHATE	582.5434	MAGNESIUM PHOSPHATE Nutrients And/or Dietary Supplements; Substances Generally Recognized As Safe
MAGNESIUM SILICATE	182.2437	MAGNESIUM SILICATE Anticaking Agents; Substances Generally Recognized As Safe
MAGNESIUM SILICATE	582.2437	MAGNESIUM SILICATE Anticaking Agents; Substances Generally Recognized As Safe
MAGNESIUM STEARATE	184.1440	MAGNESIUM STEARATE Listing Of Specific Substances Affirmed As GRAS; Direct Food Substances Affirmed As Generally Recognized As Safe
MAGNESIUM SULFATE	184.1443	MAGNESIUM SULFATE Listing Of Specific Substances Affirmed As GRAS; Direct Food Substances Affirmed As Generally Recognized As Safe
MAGNESIUM SULFATE	201.312	MAGNESIUM SULFATE HEPTAHYDRATE; LABEL DECLARATION ON DRUG PRODUCTS Specific Labeling Requirements For Specific Drug Products; Labeling
MAGNESIUM SULFATE	522.380	CHLORAL HYDRATE, PENTOBARBITAL, AND MAGNESIUM SULFATE STERILE AQUEOUS SOLUTION Implantation Or Injectable Dosage Form New Animal Drugs
MAGNESIUM SULFATE	582.5443	MAGNESIUM SULFATE Nutrients And/or Dietary Supplements; Substances Generally Recognized As Safe
MAGNET	870.3690	PACEMAKER TEST MAGNET Cardiovascular Prosthetic Devices; Cardiovascular Devices
MAGNET	886.4440	AC-POWERED MAGNET Surgical Devices; Ophthalmic Devices
MAGNET	886.4445	PERMANENT MAGNET Surgical Devices; Ophthalmic Devices
MAGNETIC	870.2800	MEDICAL MAGNETIC TAPE RECORDER Cardiovascular Monitoring Devices; Cardiovascular Devices
MAGNETIC RESONANCE	892.1000	MAGNETIC RESONANCE DIAGNOSTIC DEVICE Diagnostic Devices; Radiology Devices
MAGNIFIER	886.5540	LOW-VISION MAGNIFIER Therapeutic Devices; Ophthalmic Devices
MAGNIFYING	886.5840	MAGNIFYING SPECTACLES Therapeutic Devices; Ophthalmic Devices
MAIL	203.30	SAMPLE DISTRIBUTION BY MAIL OR COMMON CARRIER Prescription Drug Marketing
MAIL	203.31	SAMPLE DISTRIBUTION BY MEANS OTHER THAN MAIL OR COMMON CARRIER Prescription Drug Marketing
MAILING	200.5	MAILING OF IMPORTANT INFORMATION ABOUT DRUGS General Provisions; General
MAILING	203.36	FULFILLMENT HOUSES, SHIPPING AND MAILING SERVICES Prescription Drug Marketing
MAIN DISHES/DINNERS	102.28	FOOD PACKAGED FOR USE IN THE PREPARATION OF "MAIN DISHES" OR "DINNERS" Requirements For Specific Nonstandardized Foods; Common Or Usual Name For Nonstandardized Foods
MAINE	808.69	MAINE Listing Of Specific State And Local Exemptions; Exemption From Federal Preemption Of State & Local Medical Device Requirements
MAINTAINING WEIGHT	105.66	LABEL STATEMENTS RELATING TO USEFULNESS IN REDUCING OR MAINTAINING BODY WEIGHT Label Statements; Foods For Special Dietary Use
MAINTENANCE	58.63	MAINTENANCE AND CALIBRATION OF EQUIPMENT Equipment; Good Laboratory Practice For Nonclinical Laboratory Studies
MAINTENANCE	211.58	MAINTENANCE Buildings And Facilities; Current Good Manufacturing Practice For Finished Pharmaceuticals
MAINTENANCE	211.67	EQUIPMENT CLEANING AND MAINTENANCE Equipment And Facilities; Current Good Manufacturing Practice For Finished Pharmaceuticals
MAINTENANCE	510.305	MAINTENANCE OF COPIES OF APPROVED MEDICATED FEED MILL LICENSES TO MANUFACTURE ANIMAL FEED BEARING OR CONTAINING NEW ANIMAL DRUGS Records And Reports; New Animal Drugs
MAINTENANCE	1250.30	CONSTRUCTION, MAINTENANCE AND USE OF PLACES WHERE FOOD IS PREPARED, SERVED, OR STORED Food Service Sanitation On Land And Air Conveyances, And Vessels; Interstate Conveyance Sanitation
MAINTENANCE, RECORDS	205.50	MINIMUM REQUIREMENTS FOR STORAGE AND HANDLING OF PRESCRIPTION DRUGS AND ESTABLISHMENT AND MAINTENANCE OF PRESCRIPTION DRUG DISTRIBUTION RECORDS Guidelines For State Licensing Of Wholesale Prescription Drug Distributors

MAINTENANCE, RECORDS	320.36	REQUIREMENTS FOR MAINTENANCE OF RECORDS OF BIOEQUIVALENCE TESTING Procedures For Determining The Bioavailability or Bioequivalence Of Drug Products; Bioavailability And Bioequivalence Requirements
MAINTENANCE, RECORDS	1304.04	MAINTENANCE OF RECORDS AND INVENTORIES General Information; Records And Reports Of Registrants
MAINTENANCE, RECORDS	1310.05	MAINTENANCE OF RECORDS Records and Reports Of Listed Chemicals And Certain Machines
MAINTENANCE TREATMENT	1304.24	RECORDS FOR MAINTENANCE TREATMENT PROGRAMS AND DETOXIFICATION TREATMENT PROGRAMS Continuing Records; Records And Reports Of Registrants
MAINTAINED, RECORDS	1002.30	RECORDS TO BE MAINTAINED BY MANUFACTURERS Manufacturers' Records; Records and Reports
MAJOR AGENCY ACTION	25.60	ENVIRONMENTAL EFFECTS ABROAD OF MAJOR AGENCY ACTIONS Other Requirements; Environmental Impact Considerations
MALARIA	310.547	DRUG PRODUCTS CONTAINING QUININE OFFERED OTC FOR THE TREATMENT AND/OR PREVENTION OF MALARIA Requirements For Specific New Drugs Or Devices; New Drugs
MALEATE	522.23	ACEPROMAZINE MALEATE INJECTION Implantation Or Injectable Dosage Form New Animal Drugs
MALEATE	522.2063	PYRILAMINE MALEATE INJECTION Implantation Or Injectable Dosage Form New Animal Drugs
MALEIC ACID	872.3490	CARBOXYMETHYLCELLULOSE SODIUM AND/OR POLYVINYLMETHYLETHER MALEIC ACID CALCIUM-SODIUM DOUBLE SALT DENTURE ADHESIVE Prosthetic Devices; Dental Devices
MALEIC ANHYDRIDE	177.1820	STYRENE-MALEIC ANHYDRIDE COPOLYMERS Substances For Use As Basic Components Of Single And Repeated Use Food Contact Surfaces; Indirect Food Additives: Polymers
MALEIC ANHYDRIDE	872.3500	POLYVINYLMETHYLETHER MALEIC ANHYDRIDE (PVM-MA), ACID COPOLYMER, AND CARBOXYMETHYLCELLULOSE SODIUM (NACMC) DENTURE ADHESIVE Prosthetic Devices; Dental Devices
MALES	876.5160	UROLOGICAL CLAMP FOR MALES Therapeutic Devices; Gastroenterology-urology Devices
MALIC ACID	184.1069	MALIC ACID Listing Of Specific Substances Affirmed As GRAS; Direct Food Substances Affirmed As Generally Recognized As Safe
MALIC ACID	582.1069	MALIC ACID General Purpose Food Additives; Substances Generally Recognized As Safe
MALIC DEHYDROGENASE	862.1500	MALIC DEHYDROGENASE TEST SYSTEM Clinical Chemistry Test Systems; Clinical Chemistry And Clinical Toxicology Devices
MALLEABLE	882.5225	IMPLANTED MALLEABLE CLIP Neurological Therapeutic Devices; Neurological Devices
MALT	184.1443a	MALT Listing Of Specific Substances Affirmed As GRAS; Direct Food Substances Affirmed As Generally Recognized As Safe
MALT EXTRACT/SYRUP	184.1445	MALT SYRUP (MALT EXTRACT) Listing Of Specific Substances Affirmed As GRAS; Direct Food Substances Affirmed As Generally Recognized As Safe
MALT SPROUT	172.590	YEAST-MALT SPROUT EXTRACT Flavoring Agents And Related Substances; Food Additives Permitted For Direct Addition To Food For Human Consumption
MALTODEXTRIN	184.1444	MALTODEXTRIN Listing Of Specific Substances Affirmed As GRAS; Direct Food Substances Affirmed As Generally Recognized As Safe
MAMMOGRAPHIC	892.1710	MAMMOGRAPHIC X-RAY SYSTEM Diagnostic Devices; Radiology Devices
MAMMOGRAPHY	5.1000	AUTHORITY TO ENSURE THAT MAMMOGRAPHY FACILITIES MEET QUALITY STANDARDS Mammography Facilities; Redelegations Of Authority; Delegations Of Authority And Organization
MANAGEMENT	58.31	TESTING FACILITY MANAGEMENT Organization And Personnel; Good Laboratory Practice For Nonclinical Laboratory Studies
MANAGEMENT	820.20	MANAGEMENT RESPONSIBILITY Quality System Regulation
MANAGEMENT	1403.20	STANDARDS FOR FINANCIAL MANAGEMENT Post-award Requirements; Uniform Administrative Requirements For Grants And Coop. Agreements To State & Local Gov.
MANDATORY INFORMATION	1.20	PRESENCE OF MANDATORY LABELING INFORMATION General Labeling Requirements; General Enforcement Regulations

MANDATORY PASTEURIZATION	1240.61	MANDATORY PASTEURIZATION FOR ALL MILK AND MILK PRODUCTS IN FINAL PACKAGE FORM INTENDED FOR DIRECT HUMAN CONSUMPTION Specific Administrative Decisions Regarding Interstate Shipments; Control Of Communicable Diseases
MANDATORY RECALL	810.13	MANDATORY RECALL ORDER Mandatory Medical Device Recall Procedures; Medical Device Recall Authority
MANDATORY RECALL	810.14	CEASE DISTRIBUTION AND NOTIFICATION OR MANDATORY RECALL STRATEGY Mandatory Medical Device Recall Procedures; Medical Device Recall Authority
MANDATORY RECALL	810.15	COMMUNICATIONS CONCERNING A CEASE DISTRIBUTION AND NOTIFICATION OR MANDATORY RECALL ORDER Mandatory Medical Device Recall Procedures; Medical Device Recall Authority
MANDATORY RECALL	810.16	CEASE DISTRIBUTION AND NOTIFICATION OR MANDATORY RECALL STATUS REPORTS Mandatory Medical Device Recall Procedures; Medical Device Recall Authority
MANDATORY RECALL	810.17	TERMINATION OF A CEASE DISTRIBUTION AND NOTIFICATION OR MANDATORY RECALL ORDER Mandatory Medical Device Recall Procedures; Medical Device Recall Authority
MANDATORY STATEMENTS	99.103	MANDATORY STATEMENTS AND INFORMATION Dissemination Of Information On Unapproved/New Uses For Marketed Drugs, Biologics And Devices
MANDATORY VENTILATION	868.5955	INTERMITTENT MANDATORY VENTILATION ATTACHMENT Therapeutic Devices; Anesthesiology Devices
MANDELATE, METHENAMINE	520.2280	SULFAMETHIZOLE AND METHENAMINE MANDELATE TABLETS Oral Dosage Form New Animal Drugs
MANDIBULAR	872.3960	MANDIBULAR CONDYLE PROSTHESIS Prosthetic Devices; Dental Devices
MANDIBULAR	874.3695	MANDIBULAR IMPLANT FACIAL PROSTHESIS Prosthetic Devices; Ear, Nose, And Throat Devices
MANGANESE CHLORIDE	184.1446	MANGANESE CHLORIDE Listing Of Specific Substances Affirmed As GRAS; Direct Food Substances Affirmed As Generally Recognized As Safe
MANGANESE CHLORIDE	582.5446	MANGANESE CHLORIDE Nutrients And/or Dietary Supplements; Substances Generally Recognized As Safe
MANGANESE CITRATE	184.1449	MANGANESE CITRATE Listing Of Specific Substances Affirmed As GRAS; Direct Food Substances Affirmed As Generally Recognized As Safe
MANGANESE CITRATE	582.5449	MANGANESE CITRATE Nutrients And/or Dietary Supplements; Substances Generally Recognized As Safe
MANGANESE GLUCONATE	184.1452	MANGANESE GLUCONATE Listing Of Specific Substances Affirmed As GRAS; Direct Food Substances Affirmed As Generally Recognized As Safe
MANGANESE GLUCONATE	582.5452	MANGANESE GLUCONATE Nutrients And/or Dietary Supplements; Substances Generally Recognized As Safe
MANGANESE GLYCEROPHOSPHATE	582.5455	MANGANESE GLYCEROPHOSPHATE Nutrients And/or Dietary Supplements; Substances Generally Recognized As Safe
MANGANESE HYPOPHOSPHITE	182.8458	MANGANESE HYPOPHOSPHITE Nutrients; Substances Generally Recognized As Safe
MANGANESE HYPOPHOSPHITE	582.5458	MANGANESE HYPOPHOSPHITE Nutrients And/or Dietary Supplements; Substances Generally Recognized As Safe
MANGANESE SULFATE	184.1461	MANGANESE SULFATE Listing Of Specific Substances Affirmed As GRAS; Direct Food Substances Affirmed As Generally Recognized As Safe
MANGANESE SULFATE	582.5461	MANGANESE SULFATE Nutrients And/or Dietary Supplements; Substances Generally Recognized As Safe
MANGANESE VIOLET	73.2775	MANGANESE VIOLET Cosmetics; Listing Of Color Additives Exempt From Certification
MANGANOUS OXIDE	182.5464	MANGANOUS OXIDE Dietary Supplements; Substances Generally Recognized As Safe
MANGANOUS OXIDE	582.5464	MANGANOUS OXIDE Nutrients And/or Dietary Supplements; Substances Generally Recognized As Safe
MANIFESTATIONS	11.50	SIGNATURE MANIFESTATIONS Electronic Records; Electronic Signatures
MANIFOLD	870.4290	CARDIOPULMONARY BYPASS ADAPTOR, STOPCOCK, MANIFOLD, OR FITTING Cardiovascular Surgical Devices; Cardiovascular Devices
MANNITOL	180.25	MANNITOL Specific Requirements For Certain Food Additives; Food Additives Permitted In Food Or In Contact With Food On Interim Basis Pending Study
MANNITOL	582.5470	MANNITOL Nutrients And/or Dietary Supplements; Substances Generally Recognized As Safe
MANOMETER	870.1140	VENOUS BLOOD PRESSURE MANOMETER Cardiovascular Diagnostic Devices; Cardiovascular Devices

MANOMETER	880.2500	SPINAL FLUID MANOMETER General Hospital And Personal Use Monitoring Devices; General Hospital And Personal Use Devices
MANUAL	864.6160	MANUAL BLOOD CELL COUNTING DEVICE Manual Hematology Devices; Hematology And Pathology Devices
MANUAL	866.2180	MANUAL COLONY COUNTER Microbiology Devices; Immunology And Microbiology Devices
MANUAL	868.1030	MANUAL ALGESIMETER Diagnostic Devices; Anesthesiology Devices
MANUAL	868.5915	MANUAL EMERGENCY VENTILATOR Therapeutic Devices; Anesthesiology Devices
MANUAL	872.6855	MANUAL TOOTHBRUSH Miscellaneous Devices; Dental Devices
MANUAL	874.4420	EAR, NOSE, AND THROAT MANUAL SURGICAL INSTRUMENT Surgical Devices; Ear, Nose, And Throat Devices
MANUAL	876.4730	MANUAL GASTROENTEROLOGY-UROLOGY SURGICAL INSTRUMENT AND ACCESSORIES Surgical Devices; Gastroenterology-urology Devices
MANUAL	878.4800	MANUAL SURGICAL INSTRUMENT FOR GENERAL USE Surgical Devices; General And Plastic Surgery Devices
MANUAL	878.4950	MANUAL OPERATING TABLE AND MANUAL OPERATING CHAIR AND ACCESSORIES Surgical Devices; General And Plastic Surgery Devices
MANUAL	880.5120	MANUAL ADJUSTABLE HOSPITAL BED General Hospital And Personal Use Therapeutic Devices; General Hospital And Personal Use Devices
MANUAL	880.6785	MANUAL PATIENT TRANSFER DEVICE General Hospital And Personal Use Miscellaneous Devices; General Hospital And Personal Use Devices
MANUAL	882.4300	MANUAL CRANIAL DRILLS, BURRS, TREPHINES, AND THEIR ACCESSORIES Neurological Surgical Devices; Neurological Devices
MANUAL	882.4840	MANUAL RONGEUR Neurological Surgical Devices; Neurological Devices
MANUAL	884.4520	OBSTETRIC-GYNECOLOGIC GENERAL MANUAL INSTRUMENT Obstetrical And Gynecological Surgical Devices; Obstetrical And Gynecological Devices
MANUAL	884.4530	OBSTETRIC-GYNECOLOGIC SPECIALIZED MANUAL INSTRUMENT Obstetrical And Gynecological Surgical Devices; Obstetrical And Gynecological Devices
MANUAL	886.1770	MANUAL REFRACTOR Diagnostic Devices; Ophthalmic Devices
MANUAL	886.4350	MANUAL OPHTHALMIC SURGICAL INSTRUMENT Surgical Devices; Ophthalmic Devices
MANUAL	888.4540	ORTHOPEDIC MANUAL SURGICAL INSTRUMENT Surgical Devices; Orthopedic Devices
MANUAL	888.5980	MANUAL CAST APPLICATION AND REMOVAL INSTRUMENT Surgical Devices; Orthopedic Devices
MANUALS, FDA	20.107	FOOD AND DRUG ADMINISTRATION MANUALS Availability Of Specific Categories Of Records; Public Information
MANUFACTURE	5.203	ISSUANCE AND REVOCATION OF LICENSES FOR THE PROPAGATION OR MANUFACTURE AND PREPARATION OF BIOLOGICAL PRODUCTS Biologics; Redelegations Of Authority; Delegations Of Authority And Organization
MANUFACTURE	176.150	CHELATING AGENTS USED IN THE MANUFACTURE OF PAPER AND PAPERBOARD Substances For Use Only As Components Of Paper And Paperboard; Indirect Food Additives
MANUFACTURE	176.210	DEFOAMING AGENTS USED IN THE MANUFACTURE OF PAPER AND PAPERBOARD Substances For Use Only As Components Of Paper And Paperboard; Indirect Food Additives
MANUFACTURE	178.3010	ADJUVANT SUBSTANCES USED IN THE MANUFACTURE OF FOAMED PLASTICS Certain Adjuvants And Production Aids; Indirect Food Additives: Adjuvants, Production Aids, And Sanitizers
MANUFACTURE	178.3910	SURFACE LUBRICANTS USED IN THE MANUFACTURE OF METALLIC ARTICLES Certain Adjuvants And Production Aids; Indirect Food Additives: Adjuvants, Production Aids, And Sanitizers
MANUFACTURE	181.22	CERTAIN SUBSTANCES EMPLOYED IN THE MANUFACTURE OF FOOD-PACKAGING MATERIALS Specific Prior-sanctioned Food Ingredients; Prior-sanctioned Food Ingredients
MANUFACTURE	181.30	SUBSTANCES USED IN THE MANUFACTURE OF PAPER AND PAPERBOARD PRODUCTS USED IN FOOD PACKAGING Specific Prior-sanctioned Food Ingredients; Prior-sanctioned Food Ingredients

MANUFACTURE	510.7	CONSIGNEES OF NEW ANIMAL DRUGS FOR USE IN THE MANUFACTURE OF ANIMAL FEED General Provisions; New Animal Drugs
MANUFACTURE	510.305	MAINTENANCE OF COPIES OF APPROVED MEDICATED FEED MILL LICENSES TO MANUFACTURE ANIMAL FEED BEARING OR CONTAINING NEW ANIMAL DRUGS Records And Reports; New Animal Drugs
MANUFACTURE	610.42	RESTRICTIONS ON USE FOR FURTHER MANUFACTURE OF MEDICAL DEVICES Testing Requirements For Communicable Disease Agents; General Biological Products Standards
MANUFACTURE	610.50	DATE OF MANUFACTURE Dating Period Limitations; General Biological Products Standards
MANUFACTURE	640.102	MANUFACTURE OF IMMUNE GLOBULIN (HUMAN) Immune Globulin (Human); Additional Standards For Human Blood And Blood Products
MANUFACTURE	680.2	MANUFACTURE OF ALLERGENIC PRODUCTS Allergenic Products; Additional Standards For Miscellaneous Products
MANUFACTURE	1307.13	INCIDENTAL MANUFACTURE OF CONTROLLED SUBSTANCES Special Exemptions For Manufacture And Distribution Of Controlled Substances; Miscellaneous
MANUFACTURE, BULK	1301.43	APPLICATION FOR BULK MANUFACTURE OF SCHEDULE I AND II SUBSTANCES Action On Applications For Registration: Revocation Or Suspension Of Registration; Registration Of Manufacturers, Distributors, And Dispensers Of Controlled Substances
MANUFACTURER	99.201	MANUFACTURERS SUBMISSION TO THE AGENCY Dissemination Of Information On Unapproved/New Uses For Marketed Drugs, Biologics And Devices
MANUFACTURER	101.5	FOOD, NAME AND PLACE OF BUSINESS OF MANUFACTURER, PACKER, OR DISTRIBUTOR General Provisions; Food Labeling
MANUFACTURER	201.1	DRUGS; NAME AND PLACE OF BUSINESS OF MANUFACTURER, PACKER OR DISTRIBUTOR General Labeling Provisions; Labeling
MANUFACTURER	501.5	ANIMAL FOOD; NAME AND PLACE OF BUSINESS OF MANUFACTURER, PACKER, OR DISTRIBUTOR General Provisions; Animal Food Labeling
MANUFACTURER	701.12	NAME AND PLACE OF BUSINESS OF MANUFACTURER, PACKER, OR DISTRIBUTOR Package Form; Cosmetic Labeling
MANUFACTURER	801.1	MEDICAL DEVICES; NAME AND PLACE OF BUSINESS OF MANUFACTURER, PACKER OR DISTRIBUTOR General Labeling Provisions; Labeling
MANUFACTURER	821.25	DEVICE TRACKING SYSTEM AND CONTENT REQUIREMENTS: MANUFACTURER REQUIREMENTS Medical Device Tracking Requirements
MANUFACTURER	895.22	SUBMISSION OF DATA AND INFORMATION BY THE MANUFACTURER, DISTRIBUTOR, OR IMPORTER Banned Devices
MANUFACTURER	1003.10	DISCOVERY OF DEFECT OR FAILURE OF COMPLIANCE BY MANUFACTURER; NOTICE REQUIREMENTS Discovery Of Defect Or Failure To Comply; Notification Of Defects Or Failure To Comply
MANUFACTURER	1003.20	NOTIFICATION BY THE MANUFACTURER TO THE SECRETARY Notification; Notification Of Defects Or Failure To Comply
MANUFACTURER	1003.21	NOTIFICATION BY THE MANUFACTURER TO AFFECTED PERSONS Notification; Notification Of Defects Or Failure To Comply
MANUFACTURER	1230.12	MANUFACTURER; DISTRIBUTOR Labeling; Regulations Under The Federal Caustic Poison Act
MANUFACTURERS	5.604	MANUFACTURERS REQUIREMENT TO PROVIDE DATA TO ULTIMATE PURCHASERS OF ELECTRONIC PRODUCTS Radiation Control; Redelegations Of Authority; Delegations Of Authority And Organization
MANUFACTURERS	5.605	DEALER AND DISTRIBUTOR DIRECTION TO PROVIDE DATA TO MANUFACTURERS OF ELECTRONIC PRODUCTS Radiation Control; Redelegations Of Authority; Delegations Of Authority And Organization
MANUFACTURERS	133.10	NOTICE TO MANUFACTURERS, PACKERS, AND DISTRIBUTORS OF PASTEURIZED BLENDED CHEESE, PASTEURIZED PROCESS CHEESE, CHEESE FOOD, CHEESE SPREAD, AND RELATED FOODS General Provisions; Cheeses And Related Cheese Products

MANUFACTURERS	200.10	CONTRACT FACILITIES (INCLUDING CONSULTING LABORATORIES) UTILIZED AS EXTRAMURAL FACILITIES BY PHARMACEUTICAL MANUFACTURERS General Provisions; General
MANUFACTURERS	201.300	NOTICE TO MANUFACTURERS, PACKERS, AND DISTRIBUTORS OF GLANDULAR PREPARATIONS Specific Labeling Requirements For Specific Drug Products; Labeling
MANUFACTURERS	201.301	NOTICE TO MANUFACTURERS, PACKERS, AND DISTRIBUTORS OF ESTROGENIC HORMONE PREPARATIONS Specific Labeling Requirements For Specific Drug Products; Labeling
MANUFACTURERS	201.302	NOTICE TO MANUFACTURERS, PACKERS, AND DISTRIBUTORS OF DRUGS FOR INTERNAL USE WHICH CONTAIN MINERAL OIL Specific Labeling Requirements For Specific Drug Products; Labeling
MANUFACTURERS	600.14	REPORTING OF BIOLOGICAL PRODUCT DEVIATIONS BY LICENSED MANUFACTURERS Establishment Standards; Biological Products: General
MANUFACTURERS	606.171	REPORTING OF PRODUCT DEVIATIONS BY LICENSED MANUFACTURERS, UNLICENSED REGISTERED BLOOD ESTABLISHMENTS, AND TRANSFUSION SERVICES Current Good Manufacturing Practice For Blood And Blood Components
MANUFACTURERS	610.44	USE OF REFERENCE PANELS BY MANUFACTURERS OF TEST KITS Testing Requirements For Communicable Disease Agents; General Biological Products Standards
MANUFACTURERS	803.10	GENERAL DESCRIPTION OF REPORTS REQUIRED FROM USER FACILITIES, IMPORTERS AND MANUFACTURERS Medical Device Reporting
MANUFACTURERS	803.50	INDIVIDUAL ADVERSE EVENT REPORTS; MANUFACTURERS Manufacturer Reporting Requirements; Medical Device Reporting
MANUFACTURERS	809.20	GENERAL REQUIREMENTS FOR MANUFACTURERS AND PRODUCERS OF IN VITRO DIAGNOSTIC PRODUCTS Requirements For Manufacturers And Producers; In Vitro Diagnostic Products For Human Use
MANUFACTURERS	821.30	TRACKING OBLIGATIONS OF PERSONS OTHER THAN DEVICE MANUFACTURERS: DISTRIBUTOR REQUIREMENTS Medical Device Tracking Requirements
MANUFACTURERS	1002.30	RECORDS TO BE MAINTAINED BY MANUFACTURERS Manufacturers' Records; Records and Reports
MANUFACTURERS	1002.51	EXEMPTIONS FOR MANUFACTURERS OF PRODUCTS INTENDED FOR THE U.S. GOVERNMENT Exemptions From Records And Reports Requirements; Records and Reports
MANUFACTURERS	1005.25	SERVICE OF PROCESS ON MANUFACTURERS Bonding And Compliance Procedures; Importation Of Electronic Products
MANUFACTURERS	1304.22	RECORDS FOR MANUFACTURERS, DISTRIBUTORS, DISPENSERS, RESEARCHERS, IMPORTERS AND EXPORTERS Continuing Records; Records And Reports Of Registrants
MANUFACTURERS	1304.31	REPORTS FROM MANUFACTURERS IMPORTING NARCOTIC RAW MATERIAL Reports; Records And Reports Of Registrants
MANUFACTURERS	1304.32	REPORTS OF MANUFACTURERS IMPORTING COCA LEAVES Reports; Records And Reports Of Registrants
MANUFACTURERS	1309.27	EXEMPTION OF CERTAIN MANUFACTURERS Registration Of Manufacturers, Distributors, Importers And Exporters Of List I Chemicals
MANUFACTURER'S AGENTS	5.802	MANUFACTURER'S RESIDENT IMPORT AGENTS Imports And Exports; Redelegations Of Authority; Delegations Of Authority And Organization
MANUFACTURERS, FOREIGN	803.58	FOREIGN MANUFACTURERS Manufacturer Reporting Requirements; Medical Device Reporting
MANUFACTURER'S OBLIGATION	1004.1	MANUFACTURER'S OBLIGATION TO REPAIR, REPLACE, OR REFUND COST OF ELECTRONIC PRODUCTS Repurchase, Repairs, Or Replacement Of Electronic Products
MANUFACTURING	20.115	PRODUCT CODES FOR MANUFACTURING OR SALES DATES Availability Of Specific Categories Of Records; Public Information
MANUFACTURING	108.12	MANUFACTURING, PROCESSING, OR PACKING WITHOUT A PERMIT, OR IN VIOLATION OF A PERMIT General Provisions; Emergency Permit Control
MANUFACTURING	109.15	USE OF POLYCHLORINATED BIPHENYLS (PCB'S) IN ESTABLISHMENTS MANUFACTURING FOOD-PACKAGING MATERIALS General Provisions; Unavoidable Contaminants In Food For Human Consumption & Food-packaging Material
MANUFACTURING	163.117	COCOA WITH DIOCTYL SODIUM SULFOSUCCINATE FOR MANUFACTURING Requirements For Specific Standardized Cacao Products; Cacao Products
MANUFACTURING	172.520	COCOA WITH DIOCTYL SODIUM SULFOSUCCINATE FOR MANUFACTURING Flavoring Agents And Related Substances; Food Additives Permitted For Direct Addition To Food For Human Consumption

MANUFACTURING	201.122	DRUGS FOR PROCESSING, REPACKING, OR MANUFACTURING Exemptions From Adequate Directions For Use; Labeling
MANUFACTURING	225.35	USE OF WORK AREAS, EQUIPMENT, AND STORAGE AREAS FOR OTHER MANUFACTURING AND STORAGE PURPOSES Construction And Maintenance Of Facilities And Equipment; Current Good Manufacturing Practice For Medicated Feeds
MANUFACTURING	509.15	USE OF POLYCHLORINATED BIPHENYLS (PCB'S) IN ESTABLISHMENTS MANUFACTURING FOOD-PACKAGING MATERIALS General Provisions; Unavoidable Contaminants In Animal Food And Food-packaging Material
MANUFACTURING	601.22	PRODUCTS IN SHORT SUPPLY; INITIAL MANUFACTURING AT OTHER THAN LICENSED LOCATION Biologics Licensing; Licensing
MANUFACTURING	801.122	MEDICAL DEVICES FOR PROCESSING, REPACKING, OR MANUFACTURING Exemptions From Adequate Directions For Use; Labeling
MANUFACTURING	1301.73	PHYSICAL SECURITY CONTROLS FOR NON-PRACTITIONERS; COMPOUNDERS FOR NARCOTIC TREATMENT PROGRAMS; MANUFACTURING AND COMPOUNDING AREAS Security Requirements; Registration Of Manufacturers, Distributors, And Dispensers Of Controlled Substances
MANUFACTURING, CHEESE	133.109	BRICK CHEESE FOR MANUFACTURING Requirements For Specific Standardized Cheese And Related Products; Cheeses And Related Cheese Products
MANUFACTURING, CHEESE	133.114	CHEDDAR CHEESE FOR MANUFACTURING Requirements For Specific Standardized Cheese And Related Products; Cheeses And Related Cheese Products
MANUFACTURING, CHEESE	133.119	COLBY CHEESE FOR MANUFACTURING Requirements For Specific Standardized Cheese And Related Products; Cheeses And Related Cheese Products
MANUFACTURING, CHEESE	133.137	WASHED CURD CHEESE FOR MANUFACTURING Requirements For Specific Standardized Cheese And Related Products; Cheeses And Related Cheese Products
MANUFACTURING, CHEESE	133.145	GRANULAR CHEESE FOR MANUFACTURING Requirements For Specific Standardized Cheese And Related Products; Cheeses And Related Cheese Products
MANUFACTURING, CHEESE	133.161	MUENSTER AND MUNSTER CHEESE FOR MANUFACTURING Requirements For Specific Standardized Cheese And Related Products; Cheeses And Related Cheese Products
MANUFACTURING, CHEESE	133.189	SKIM MILK CHEESE FOR MANUFACTURING Requirements For Specific Standardized Cheese And Related Products; Cheeses And Related Cheese Products
MANUFACTURING, CHEESE	133.196	SWISS CHEESE FOR MANUFACTURING Requirements For Specific Standardized Cheese And Related Products; Cheeses And Related Cheese Products
MANUFACTURING, O.J.	146.151	ORANGE JUICE FOR MANUFACTURING Requirements For Specific Standardized Canned Fruit Juices And Beverages; Canned Fruit Juices
MANUFACTURING, O.J.	146.153	CONCENTRATED ORANGE JUICE FOR MANUFACTURING Requirements For Specific Standardized Canned Fruit Juices And Beverages; Canned Fruit Juices
MANUFACTURING PRACTICE (CGMP)	110.5	CURRENT GOOD MANUFACTURING PRACTICE Current Good Manufacturing Practice In Manufacturing, Packing, Or Holding Human Food
MANUFACTURING PRACTICE (CGMP)	113.5	CURRENT GOOD MANUFACTURING PRACTICE Thermally Processed Low-acid Foods Packaged In Hermetically Sealed Containers
MANUFACTURING PRACTICE (CGMP)	114.5	CURRENT GOOD MANUFACTURING PRACTICE Acidified Foods
MANUFACTURING PRACTICE (CGMP)	120.5	CURRENT GOOD MANUFACTURING PRACTICE Hazard Analysis And Critical Control Point (HACCP) Systems
MANUFACTURING PRACTICE (CGMP)	123.5	CURRENT GOOD MANUFACTURING PRACTICE Fish and Fishery Products
MANUFACTURING PRACTICE (CGMP)	129.1	CURRENT GOOD MANUFACTURING PRACTICE Processing And Bottling Of Bottled Drinking Water
MANUFACTURING PRACTICE (CGMP)	210.1	STATUS OF CURRENT GOOD MANUFACTURING PRACTICE REGULATIONS Current Good Manufacturing Practice In Manufacturing, Processing, Packing, Or Holding Of Drugs; General
MANUFACTURING PRACTICE (CGMP)	210.2	APPLICABILITY OF CURRENT GOOD MANUFACTURING PRACTICE REGULATIONS Current Good Manufacturing Practice In Manufacturing, Processing, Packing, Or Holding Of Drugs; General
MANUFACTURING PRACTICE (CGMP)	225.1	CURRENT GOOD MANUFACTURING PRACTICE Current Good Manufacturing Practice For Medicated Feeds
MANUFACTURING PRACTICE (CGMP)	226.1	CURRENT GOOD MANUFACTURING PRACTICE Current Good Manufacturing Practice For Type A Medicated Articles

MANUFACTURING PRACTICES (CGMP)	5.109	ISSUANCE OF WRITTEN NOTICES CONCERNING PATENT INFORMATION, CURRENT GOOD MANUFACTURING PRACTICES AND FALSE OR MISLEADING LABELING OF NEW DRUGS Human Drugs; Redelegations Of Authority; Delegations Of Authority And Organization
MANUFACTURING PRACTICES (CGMP)	5.504	ISSUANCE OF WRITTEN NOTICES CONCERNING PATENT INFORMATION, CURRENT GOOD MANUFACTURING PRACTICES AND FALSE OR MISLEADING LABELING OF NEW ANIMAL DRUGS AND FEEDS BEARING OR CONTAINING NEW ANIMAL DRUGS Animal Drugs; Redelegations Of Authority; Delegations Of Authority And Organization
MANUFACTURING QUOTAS	1303.21	INDIVIDUAL MANUFACTURING QUOTAS Individual Manufacturing Quotas; Quotas
MANUFACTURING QUOTAS	1303.22	PROCEDURE FOR APPLYING FOR INDIVIDUAL MANUFACTURING QUOTAS Individual Manufacturing Quotas; Quotas
MANUFACTURING QUOTAS	1303.23	PROCEDURE FOR FIXING INDIVIDUAL MANUFACTURING QUOTAS Individual Manufacturing Quotas; Quotas
MANUFACTURING QUOTAS	1303.25	INCREASE IN INDIVIDUAL MANUFACTURING QUOTAS Individual Manufacturing Quotas; Quotas
MANUFACTURING QUOTAS	1303.26	REDUCTION IN INDIVIDUAL MANUFACTURING QUOTAS Individual Manufacturing Quotas; Quotas
MANUFACTURING RESPONSIBILITY	607.7	ESTABLISHMENT REGISTRATION AND PRODUCT LISTING OF BLOOD BANKS AND OTHER FIRMS MANUFACTURING HUMAN BLOOD AND BLOOD PRODUCTS General Provisions; Establishment Registratn+product Listing, Manufacturers Human Blood/b.products
MANUFACTURING RESPONSIBILITY	610.63	DIVIDED MANUFACTURING RESPONSIBILITY TO BE SHOWN Labeling Standards; General Biological Products Standards
MANUFACTURING RESPONSIBILITY	640.71	MANUFACTURING RESPONSIBILITY Source Plasma; Additional Standards For Human Blood And Blood Products
MAPLE	168.140	MAPLE SIRUP Requirements For Specific Standardized Sweeteners And Table Sirups; Sweeteners And Table Sirups
MARBOFLOXACIN	520.1310	MARBOFLOXACIN TABLETS Oral Dosage Form New Animal Drugs
MARGARINE	166.40	LABELING OF MARGARINE General Provisions; Margarine
MARGARINE	166.110	MARGARINE Requirements For Specific Standardized Margarine; Margarine
MARIGOLD, AZTEC	73.295	TAGETES (AZTEC MARIGOLD) MEAL AND EXTRACT Foods; Listing Of Color Additives Exempt From Certification
MARKER	500.86	MARKER RESIDUE AND TARGET TISSUE Regulation of Carcinogenic Compounds Used In Food-Producing Animals; General
MARKER	866.5065	HUMAN ALLOTYPIC MARKER IMMUNOLOGICAL TEST SYSTEM Immunological Test Systems; Immunology And Microbiology Devices
MARKER	878.4660	SKIN MARKER Surgical Devices; General And Plastic Surgery Devices
MARKER	886.4570	OPHTHALMIC SURGICAL MARKER Surgical Devices; Ophthalmic Devices
MARKET, AUTHORIZATION	170.17	EXEMPTION FOR INVESTIGATIONAL USE AND PROCEDURE FOR OBTAINING AUTHORIZATION TO MARKET EDIBLE PRODUCTS FROM EXPERIMENTAL ANIMALS General Provisions; Food Additives
MARKET, AUTHORIZATION	570.17	EXEMPTION FOR INVESTIGATIONAL USE AND PROCEDURE FOR OBTAINING AUTHORIZATION TO MARKET EDIBLE PRODUCTS FROM EXPERIMENTAL ANIMALS General Provisions; Food Additives
MARKET, REMOVED FROM	216.24	DRUG PRODUCTS WITHDRAWN OR REMOVED FROM THE MARKET FOR REASONS OF SAFETY OR EFFECTIVENESS Pharmacy Compounding
MARKETED PRESCRIPTION DRUGS	310.305	RECORDS AND REPORTS CONTAINING ADVERSE DRUG EXPERIENCES ON MARKETED PRESCRIPTION DRUGS FOR HUMAN USE WITHOUT APPROVED NEW DRUG APPLICATIONS Records And Reports; New Drugs
MARKETING	310.500	DIGOXIN PRODUCTS FOR ORAL USE; CONDITIONS FOR MARKETING Requirements For Specific New Drugs or Devices; New Drugs
MARKETING	330.13	CONDITIONS FOR MARKETING INGREDIENTS RECOMMENDED FOR OVER-THE-COUNTER (OTC) USE UNDER THE OTC DRUG REVIEW Administrative Procedures; Over-the-counter Human Drugs Generally Recognized As Safe And Effective & Not Misbranded
MARKETING APPLICATIONS	312.84	RISK-BENEFIT ANALYSIS IN REVIEW OF MARKETING APPLICATIONS FOR DRUGS TO TREAT LIFE-THREATENING AND SEVERELY-DEBILITATING ILLNESSES Investigational New Drug Application

MARKING, FILM	892.1640	RADIOGRAPHIC FILM MARKING SYSTEM Diagnostic Devices; Radiology Devices
MASK	868.5550	ANESTHETIC GAS MASK Therapeutic Devices; Anesthesiology Devices
MASK	868.5560	GAS MASK HEAD STRAP Therapeutic Devices; Anesthesiology Devices
MASK	868.5570	NONREBREATHING MASK Therapeutic Devices; Anesthesiology Devices
MASK	868.5580	OXYGEN MASK Therapeutic Devices; Anesthesiology Devices
MASK	868.5590	SCAVENGING MASK Therapeutic Devices; Anesthesiology Devices
MASK	868.5600	VENTURI MASK Therapeutic Devices; Anesthesiology Devices
MASKER	874.3400	TINNITUS MASKER Prosthetic Devices; Ear, Nose, And Throat Devices
MASS SPECTROMETER	862.2860	MASS SPECTROMETER FOR CLINICAL USE Clinical Laboratory Instruments; Clinical Chemistry And Clinical Toxicology Devices
MASSACHUSETTS	808.71	MASSACHUSETTS Listing Of Specific State And Local Exemptions; Exemption From Federal Preemption Of State & Local Medical Device Requirements
MASSAGER	890.5650	POWERED INFLATABLE TUBE MASSAGER Physical Medicine Therapeutic Devices; Physical Medicine Devices
MASSAGER	890.5660	THERAPEUTIC MASSAGER Physical Medicine Therapeutic Devices; Physical Medicine Devices
MASSAGING PICK	872.6650	MASSAGING PICK OR TIP FOR ORAL HYGIENE Miscellaneous Devices; Dental Devices
MASTER FILES	314.420	DRUG MASTER FILES Miscellaneous Provisions; Applications For FDA Approval To Market A New Drug
MASTER-FORMULA	226.102	MASTER-FORMULA AND BATCH-PRODUCTION RECORDS Records And Reports; Current Good Manufacturing Practice For Type A Medicated Articles
MASTER HEARING AID	874.3330	MASTER HEARING AID Prosthetic Devices; Ear, Nose, And Throat Devices
MASTER RECORD	225.102	MASTER RECORD FILE AND PRODUCTION RECORDS Records And Reports; Current Good Manufacturing Practice For Medicated Feeds
MASTER RECORD	820.181	DEVICE MASTER RECORD Records; Quality System Regulation
MASTER RECORDS	211.186	MASTER PRODUCTION AND CONTROL RECORDS Records And Reports; Current Good Manufacturing Practice For Finished Pharmaceuticals
MATCHING	1403.24	MATCHING OR COST SHARING Post-award Requirements; Uniform Administrative Requirements For Grants And Coop. Agreements To State & Local Gov.
MATERIAL	870.3710	PACEMAKER REPAIR OR REPLACEMENT MATERIAL Cardiovascular Prosthetic Devices; Cardiovascular Devices
MATERIAL	872.3310	COATING MATERIAL FOR RESIN FILLINGS Prosthetic Devices; Dental Devices
MATERIAL	874.3620	EAR, NOSE AND THROAT SYNTHETIC POLYMER MATERIAL Prosthetic Devices; Ear, Nose And Throat Devices
MATERIAL	878.3500	POLYTETRAFLUOROETHYLENE WITH CARBON FIBERS COMPOSITE IMPLANT MATERIAL Prosthetic Devices; General And Plastic Surgery Devices
MATERIAL	882.4500	CRANIOPLASTY MATERIAL FORMING INSTRUMENT Neurological Surgical Devices; Neurological Devices
MATERIAL FACTS	1.21	FAILURE TO REVEAL MATERIAL FACTS General Labeling Requirements; General Enforcement Regulations
MATERIAL, SOURCE	640.22	COLLECTION OF SOURCE MATERIAL Platelets; Additional Standards For Human Blood And Blood Products
MATERIAL, SOURCE	640.32	COLLECTION OF SOURCE MATERIAL Plasma; Additional Standards For Human Blood And Blood Products
MATERIAL, SOURCE	640.52	COLLECTION OF SOURCE MATERIAL Cryoprecipitate; Additional Standards For Human Blood And Blood Products
MATERIAL, SOURCE	660.32	COLLECTION OF SOURCE MATERIAL Reagent Red Blood Cells; Additional Standards For Diagnostic Substances For Laboratory Tests
MATERIAL, SOURCE	660.33	TESTING OF SOURCE MATERIAL Reagent Red Blood Cells; Additional Standards For Diagnostic Substances For Laboratory Tests
MATERIALS	14.33	COMPILATION OF MATERIALS FOR MEMBERS OF AN ADVISORY COMMITTEE Meeting Procedures; Public Hearing Before A Public Advisory Committee

MATERIALS	173.357	MATERIALS USED AS FIXING AGENTS IN THE IMMOBILIZATION OF ENZYME PREPARATIONS Specific Usage Additives; Secondary Direct Food Additives Permitted In Food For Human Consumption
MATERIALS	211.110	SAMPLING AND TESTING OF IN-PROCESS MATERIALS AND DRUG PRODUCTS Production And Process Controls; Current Good Manufacturing Practice For Finished Pharmaceuticals
MATERIALS	610.15	CONSTITUENT MATERIALS General Provisions; General Biological Products Standards
MATERIALS EXAMINATION	211.122	MATERIALS EXAMINATION AND USAGE CRITERIA Packaging And Labeling Control; Current Good Manufacturing Practice For Finished Pharmaceuticals
MATERIALS, FOOD-PACKAGING	109.15	USE OF POLYCHLORINATED BIPHENYLS (PCB'S) IN ESTABLISHMENTS MANUFACTURING FOOD-PACKAGING MATERIALS General Provisions; Unavoidable Contaminants In Food For Human Consumption & Food-packaging Material
MATERIALS, FOOD-PACKAGING	178.3130	ANTISTATIC AND/OR ANTIFOGGING AGENTS IN FOOD-PACKAGING MATERIALS Certain Adjuvants And Production Aids; Indirect Food Additives: Adjuvants, Production Aids, And Sanitizers
MATERIALS, FOOD-PACKAGING	181.22	CERTAIN SUBSTANCES EMPLOYED IN THE MANUFACTURE OF FOOD-PACKAGING MATERIALS Specific Prior-sanctioned Food Ingredients; Prior-sanctioned Food Ingredients
MATERIALS, FOOD-PACKAGING	509.15	USE OF POLYCHLORINATED BIPHENYLS (PCB'S) IN ESTABLISHMENTS MANUFACTURING FOOD-PACKAGING MATERIALS General Provisions; Unavoidable Contaminants In Animal Food And Food-packaging Material
MATERIALS, PACKAGING	179.45	PACKAGING MATERIALS FOR USE DURING THE IRRADIATION OF PREPACKAGED FOODS Packaging Materials For Irradiated Foods; Irradiation In The Production, Processing And Handling Of Food
MATERIALS, PACKAGING	570.13	INDIRECT FOOD ADDITIVES RESULTING FROM PACKAGING MATERIALS PRIOR SANCTIONED FOR ANIMAL FEED AND PET FOOD General Provisions; Food Additives
MATERIALS, PACKAGING	570.14	INDIRECT FOOD ADDITIVES RESULTING FROM PACKAGING MATERIALS FOR ANIMAL FEED AND PET FOOD General Provisions; Food Additives
MATERIALS, PROMOTIONAL	314.550	PROMOTIONAL MATERIALS Accelerated Approval Of New Drugs; Applications For FDA Approval To Market A New Drug
MATERIALS, PROMOTIONAL	601.45	PROMOTIONAL MATERIALS Accelerated Approval Of Biological Products; Licensing
MATRIX	175.390	ZINC-SILICON DIOXIDE MATRIX COATINGS Substances For Use As Components Of Coatings; Indirect Food Additives
MATTRESS	880.5150	NONPOWERED FLOTATION THERAPY MATTRESS General Hospital And Personal Use Therapeutic Devices; General Hospital And Personal Use Devices
MATTRESS	880.5550	ALTERNATING PRESSURE AIR FLOTATION MATTRESS General Hospital And Personal Use Therapeutic Devices; General Hospital And Personal Use Devices
MATTRESS	880.5560	TEMPERATURE REGULATED WATER MATTRESS General Hospital And Personal Use Therapeutic Devices; General Hospital And Personal Use Devices
MATTRESS	880.6190	MATTRESS COVER FOR MEDICAL PURPOSES General Hospital And Personal Use Miscellaneous Devices; General Hospital And Personal Use Devices
MAXIMUM LEVEL OZONE	801.415	MAXIMUM ACCEPTABLE LEVEL OF OZONE Special Requirements For Specific Devices; Labeling
MAXWELL SPOT	886.1435	MAXWELL SPOT Diagnostic Devices; Ophthalmic Devices
MAYONNAISE	169.140	MAYONNAISE Requirements For Specific Standardized Food Dressings And Flavorings; Food Dressings And Flavorings
MDR PROCEDURES	803.17	WRITTEN MDR PROCEDURES General Provisions; Medical Device Reporting
MEAL, ALGAE	73.185	HAEMATOCOCCUS ALGAE MEAL Foods; Listing Of Color Additives Exempt From Certification
MEAL, ALGAE	73.275	DRIED ALGAE MEAL Foods; Listing Of Color Additives Exempt From Certification
MEAL, CORN	137.250	WHITE CORN MEAL Requirements For Specific Standardized Cereal Flours And Related Products; Cereal Flours And Related Products

MEAL, CORN	137.255	BOLTED WHITE CORN MEAL Requirements For Specific Standardized Cereal Flours And Related Products; Cereal Flours And Related Products
MEAL, CORN	137.265	DEGERMINATED WHITE CORN MEAL Requirements For Specific Standardized Cereal Flours And Related Products; Cereal Flours And Related Products
MEAL, CORN	137.270	SELF-RISING WHITE CORN MEAL Requirements For Specific Standardized Cereal Flours And Related Products; Cereal Flours And Related Products
MEAL, CORN	137.275	YELLOW CORN MEAL Requirements For Specific Standardized Cereal Flours And Related Products; Cereal Flours And Related Products
MEAL, CORN	137.280	BOLTED YELLOW CORN MEAL Requirements For Specific Standardized Cereal Flours And Related Products; Cereal Flours And Related Products
MEAL, CORN	137.285	DEGERMINATED YELLOW CORN MEAL Requirements For Specific Standardized Cereal Flours And Related Products; Cereal Flours And Related Products
MEAL, CORN	137.290	SELF-RISING YELLOW CORN MEAL Requirements For Specific Standardized Cereal Flours And Related Products; Cereal Flours And Related Products
MEAL, COTTONSEED	573.140	AMMONIATED COTTONSEED MEAL Food Additive Listing; Food Additives Permitted In Feed And Drinking Water Of Animals
MEAL, CRAMBE	573.310	CRAMBE MEAL, HEAT TOASTED Food Additive Listing; Food Additives Permitted In Feed And Drinking Water Of Animals
MEAL, HAEMATOCOCCUS ALGAE	73.185	HAEMATOCOCCUS ALGAE MEAL Foods; Listing Of Color Additives Exempt From Certification
MEAL, LEATHER	573.540	HYDROLYZED LEATHER MEAL Food Additive Listing; Food Additives Permitted In Feed And Drinking Water Of Animals
MEAL, TAGETES	73.295	TAGETES (AZTEC MARIGOLD) MEAL AND EXTRACT Foods; Listing Of Color Additives Exempt From Certification
MEALS	137.260	ENRICHED CORN MEALS Requirements For Specific Standardized Cereal Flours And Related Products; Cereal Flours And Related Products
MEANING	801.4	MEANING OF “INTENDED USES” General Labeling Provisions; Labeling
MEASLES VIRUS	866.3520	RUBEOLA (MEASLES) VIRUS SEROLOGICAL REAGENTS Serological Reagents; Immunology And Microbiology Devices
MEASUREMENT	864.8165	CALIBRATOR FOR HEMOGLOBIN OR HEMATOCRIT MEASUREMENT Hematology Reagents; Hematology And Pathology Devices
MEASUREMENT	870.1130	NONINVASIVE BLOOD PRESSURE MEASUREMENT SYSTEM Cardiovascular Diagnostic Devices; Cardiovascular Devices
MEASUREMENT	876.1620	URODYNAMICS MEASUREMENT SYSTEM Diagnostic Devices; Gastroenterology-urology Devices
MEASUREMENT	882.1540	GALVANIC SKIN RESPONSE MEASUREMENT DEVICE Neurological Diagnostic Devices; Neurological Devices
MEASUREMENT	882.1550	NERVE CONDUCTION VELOCITY MEASUREMENT DEVICE Neurological Diagnostic Devices; Neurological Devices
MEASUREMENT	882.1560	SKIN POTENTIAL MEASUREMENT DEVICE Neurological Diagnostic Devices; Neurological Devices
MEASUREMENT	882.1570	POWERED DIRECT-CONTACT TEMPERATURE MEASUREMENT DEVICE Neurological Diagnostic Devices; Neurological Devices
MEASUREMENT	890.1600	INTERMITTENT PRESSURE MEASUREMENT SYSTEM Physical Medicine Diagnostic Devices; Physical Medicine Devices
MEASURER	872.1500	GINGIVAL FLUID MEASURER Diagnostic Devices; Dental Devices
MEASURES	1240.30	MEASURES IN THE EVENT OF INADEQUATE LOCAL CONTROL Control Of Communicable Diseases
MEASURING	820.72	INSPECTION, MEASURING, AND TEST EQUIPMENT Production And Process Controls; Quality System Regulation
MEASURING	864.5950	BLOOD VOLUME MEASURING DEVICE Automated And Semi-automated Hematology Devices; Hematology And Pathology Devices
MEASURING	864.6400	HEMATOCRIT MEASURING DEVICE Manual Hematology Devices; Hematology And Pathology Devices
MEASURING	876.1800	URINE FLOW OR VOLUME MEASURING SYSTEM Diagnostic Devices; Gastroenterology-urology Devices
MEASURING	886.1425	LENS MEASURING DEVICE Diagnostic Devices; Ophthalmic Devices
MEASURING	886.1430	OPHTHALMIC CONTACT LENS RADIUS MEASURING DEVICE Diagnostic Devices; Ophthalmic Devices

MEASURING	886.1450	CORNEAL RADIUS MEASURING DEVICE Diagnostic Devices; Ophthalmic Devices
MEASURING	886.1460	STEREOPSIS MEASURING INSTRUMENT Diagnostic Devices; Ophthalmic Devices
MEASURING	886.4280	INTRAOCULAR PRESSURE MEASURING DEVICE Surgical Devices; Ophthalmic Devices
MEASURING	890.1575	FORCE-MEASURING PLATFORM Physical Medicine Diagnostic Devices; Physical Medicine Devices
MEASURING	890.5360	MEASURING EXERCISE EQUIPMENT Physical Medicine Therapeutic Devices; Physical Medicine Devices
MEAT	5.302	DETENTION OF MEAT, POULTRY, EGGS AND RELATED PRODUCTS Foods And Cosmetics; Redelegations Of Authority; Delegations Of Authority And Organization
MEATS	133.125	COLD-PACK CHEESE FOOD WITH FRUITS, VEGETABLES, OR MEATS Requirements For Specific Standardized Cheese And Related Products; Cheeses And Related Cheese Products
MEATS	133.168	PASTEURIZED BLENDED CHEESE WITH FRUITS, VEGETABLES, OR MEAT Requirements For Specific Standardized Cheese And Related Products; Cheeses And Related Cheese Products
MEATS	133.170	PASTEURIZED PROCESS CHEESE WITH FRUITS, VEGETABLES, OR MEATS Requirements For Specific Standardized Cheese And Related Products; Cheeses And Related Cheese Products
MEATS	133.174	PASTEURIZED PROCESS CHEESE FOOD WITH FRUITS, VEGETABLES, OR MEATS Requirements For Specific Standardized Cheese And Related Products; Cheeses And Related Cheese Products
MEATS	133.176	PASTEURIZED CHEESE SPREAD WITH FRUITS, VEGETABLES OR MEATS Requirements For Specific Standardized Cheese And Related Products; Cheeses And Related Cheese Products
MEATS	133.180	PASTEURIZED PROCESS CHEESE SPREAD WITH FRUITS, VEGETABLES, OR MEATS Requirements For Specific Standardized Cheese And Related Products; Cheeses And Related Cheese Products
MEBENDAZOLE	520.1320	MEBENDAZOLE ORAL Oral Dosage Form New Animal Drugs
MEBENDAZOLE	520.1326	MEBENDAZOLE TRICHLORFON ORAL DOSAGE FORMS Oral Dosage Form New Animal Drugs
MEBENDAZOLE	520.1326a	MEBENDAZOLE AND TRICHLORFON POWDER Oral Dosage Form New Animal Drugs
MEBENDAZOLE	520.1326b	MEBENDAZOLE AND TRICHLORFON PASTE Oral Dosage Form New Animal Drugs
MECHANICAL CHAIR	890.3100	MECHANICAL CHAIR Physical Medicine Prosthetic Devices; Physical Medicine Devices
MECHANICAL CLEANER	872.3530	MECHANICAL DENTURE CLEANER Prosthetic Devices; Dental Devices
MECHANICAL EQUIPMENT	211.68	AUTOMATIC, MECHANICAL, AND ELECTRONIC EQUIPMENT Equipment And Facilities; Current Good Manufacturing Practice For Finished Pharmaceuticals
MECHANICAL/HYDRAULIC	876.5280	IMPLANTED MECHANICAL/HYDRAULIC URINARY CONTINENCE DEVICE Therapeutic Devices; Gastroenterology-urology Devices
MECHANICAL LITHOTRIPTOR	876.4500	MECHANICAL LITHOTRIPTOR Surgical Devices; Gastroenterology-urology Devices
MECHANICAL STIMULATOR	882.1880	EVOKED RESPONSE MECHANICAL STIMULATOR Neurological Diagnostic Devices; Neurological Devices
MECHANICAL TABLE	890.3750	MECHANICAL TABLE Physical Medicine Prosthetic Devices; Physical Medicine Devices
MECHANICAL WALKER	890.3825	MECHANICAL WALKER Physical Medicine Prosthetic Devices; Physical Medicine Devices
MECHANICAL WHEELCHAIR	890.3850	MECHANICAL WHEELCHAIR Physical Medicine Prosthetic Devices; Physical Medicine Devices
MECLOFENAMIC ACID	520.1330	MECLOFENAMIC ACID GRANULES Oral Dosage Form New Animal Drugs
MECLOFENAMIC ACID	520.1331	MECLOFENAMIC ACID TABLETS Oral Dosage Form New Animal Drugs
MEDETOMIDINE	522.1335	MEDETOMIDINE HYDROCHLORIDE FOR INJECTION Implantation Or Injectable Dosage Form New Animal Drugs
MEDIA	864.2360	MYCOPLASMA DETECTION MEDIA AND COMPONENTS Cell And Tissue Culture Products; Hematology And Pathology Devices
MEDIA	864.9600	POTENTIATING MEDIA FOR IN VITRO DIAGNOSTIC USE Products Used In Establishments That Manufacture Blood And Blood Products; Hematology And Pathology Devices
MEDIA	882.1275	ELECTROCONDUCTIVE MEDIA Neurological Diagnostic Devices; Neurological Devices

MEDIA	884.6180	REPRODUCTIVE MEDIA AND SUPPLEMENTS Assisted Reproduction Devices; Obstetrical And Gynecological Devices
MEDIA, CULTURE	864.2220	SYNTHETIC CELL AND TISSUE CULTURE MEDIA AND COMPONENTS Cell And Tissue Culture Products; Hematology And Pathology Devices
MEDIA, CULTURE	866.2450	SUPPLEMENT FOR CULTURE MEDIA Microbiology Devices; Immunology And Microbiology Devices
MEDIA, CULTURE	866.2480	QUALITY CONTROL KIT FOR CULTURE MEDIA Microbiology Devices; Immunology And Microbiology Devices
MEDIA, CULTURE	876.5885	TISSUE CULTURE MEDIA FOR HUMAN EX VIVO TISSUE AND CELL CULTURE PROCESSING APPLICATIONS Therapeutic Devices; Gastroenterology-urology Devices
MEDIA, ELECTRONIC	10.205	ELECTRONIC MEDIA COVERAGE OF PUBLIC ADMINISTRATIVE PROCEEDINGS Electronic Media Coverage Public Admin. Proceedings; Guideline Policy Procedures; Administrative Practices And Procedures
MEDIA, ELECTRONIC	10.206	PROCEDURES FOR ELECTRONIC MEDIA COVERAGE OF AGENCY PUBLIC ADMINISTRATIVE PROCEEDINGS Electronic Media Coverage Public Admin. Proceedings; Guideline Policy Procedures; Administrative Practices And Procedures
MEDIASTINOSCOPE	874.4720	MEDIASTINOSCOPE AND ACCESSORIES Surgical Devices; Ear, Nose And Throat Devices
MEDICAL	21.33	MEDICAL RECORDS Requirements For Specific Categories Of Records; Protection Of Privacy
MEDICAL	640.62	MEDICAL SUPERVISION Source Plasma; Additional Standards For Human Blood And Blood Products
MEDICAL	868.6885	MEDICAL GAS YOKE ASSEMBLY Miscellaneous; Anesthesiology Devices
MEDICAL	870.2450	MEDICAL CATHODE-RAY TUBE DISPLAY Cardiovascular Monitoring Devices; Cardiovascular Devices
MEDICAL	870.2800	MEDICAL MAGNETIC TAPE RECORDER Cardiovascular Monitoring Devices; Cardiovascular Devices
MEDICAL	880.5045	MEDICAL RECIRCULATING AIR CLEANER General Hospital And Personal Use Therapeutic Devices; General Hospital And Personal Use Devices
MEDICAL	880.5160	THERAPEUTIC MEDICAL BINDER General Hospital And Personal Use Therapeutic Devices; General Hospital And Personal Use Devices
MEDICAL	880.5240	MEDICAL ADHESIVE TAPE AND ADHESIVE BANDAGE General Hospital And Personal Use Therapeutic Devices; General Hospital And Personal Use Devices
MEDICAL	880.5300	MEDICAL ABSORBENT FIBER General Hospital And Personal Use Therapeutic Devices; General Hospital And Personal Use Devices
MEDICAL	880.5780	MEDICAL SUPPORT STOCKING General Hospital And Personal Use Therapeutic Devices; General Hospital And Personal Use Devices
MEDICAL	880.6060	MEDICAL DISPOSABLE BEDDING General Hospital And Personal Use Miscellaneous Devices; General Hospital And Personal Use Devices
MEDICAL	880.6140	MEDICAL CHAIR AND TABLE General Hospital And Personal Use Miscellaneous Devices; General Hospital And Personal Use Devices
MEDICAL	880.6150	ULTRASONIC CLEANER FOR MEDICAL INSTRUMENTS General Hospital And Personal Use Miscellaneous Devices; General Hospital And Personal Use Devices
MEDICAL	880.6190	MATTRESS COVER FOR MEDICAL PURPOSES General Hospital And Personal Use Miscellaneous Devices; General Hospital And Personal Use Devices
MEDICAL	880.6280	MEDICAL INSOLE General Hospital And Personal Use Miscellaneous Devices; General Hospital And Personal Use Devices
MEDICAL	880.6320	AC-POWERED MEDICAL EXAMINATION LIGHT General Hospital And Personal Use Miscellaneous Devices; General Hospital And Personal Use Devices
MEDICAL	880.6350	BATTERY-POWERED MEDICAL EXAMINATION LIGHT General Hospital And Personal Use Miscellaneous Devices; General Hospital And Personal Use Devices
MEDICAL	880.6500	MEDICAL ULTRAVIOLET AIR PURIFIER General Hospital And Personal Use Miscellaneous Devices; General Hospital And Personal Use Devices
MEDICAL	880.6710	MEDICAL ULTRAVIOLET WATER PURIFIER General Hospital And Personal Use Miscellaneous Devices; General Hospital And Personal Use Devices

MEDICAL	880.6820	MEDICAL DISPOSABLE SCISSORS General Hospital And Personal Use Miscellaneous Devices; General Hospital And Personal Use Devices
MEDICAL	892.5050	MEDICAL CHARGED PARTICLE RADIATION THERAPY SYSTEM Therapeutic Devices; Radiology Devices
MEDICAL	892.5300	MEDICAL NEUTRON RADIATION THERAPY SYSTEM Therapeutic Devices; Radiology Devices
MEDICAL	1000.50	RECOMMENDATION FOR THE USE OF SPECIFIC AREA GONAD SHIELDING ON PATIENTS DURING MEDICAL DIAGNOSTIC X-RAY PROCEDURES Radiation Protection Recommendations; General
MEDICAL CARE	203.11	APPLICATIONS FOR REIMPORTATION TO PROVIDE EMERGENCY MEDICAL CARE Reimportation; Prescription Drug Marketing
MEDICAL CARE SPACES	1250.84	WATER IN GALLEYS AND MEDICAL CARE SPACES Sanitation Facilities And Conditions On Vessels; Interstate Conveyance Sanitation
MEDICAL DEVICE	5.411	MEDICAL DEVICE RECALL AUTHORITY Medical Devices And Radiological Health; Redelegations Of Authority; Delegations Of Authority And Organization
MEDICAL DEVICE	5.412	TEMPORARY SUSPENSION OF A MEDICAL DEVICE APPLICATION Medical Devices And Radiological Health; Redelegations Of Authority; Delegations Of Authority And Organization
MEDICAL DEVICE	5.415	AUTHORITY RELATING TO MEDICAL DEVICE REPORTING PROCEDURES Medical Devices And Radiological Health; Redelegations Of Authority; Delegations Of Authority And Organization
MEDICAL DEVICE	5.416	MEDICAL DEVICE TRACKING Medical Devices And Radiological Health; Redelegations Of Authority; Delegations Of Authority And Organization
MEDICAL DEVICES	5.27	PATENT TERM EXTENSIONS FOR HUMAN DRUG PRODUCTS, MEDICAL DEVICES, AND FOOD AND COLOR ADDITIVES General Redelegations Of Authority; Delegations Of Authority And Organization
MEDICAL DEVICES	5.402	DETENTION OF ADULTERATED OR MISBRANDED MEDICAL DEVICES Medical Devices And Radiological Health; Redelegations Of Authority; Delegations Of Authority And Organization
MEDICAL DEVICES	5.403	AUTHORIZATION TO USE ALTERNATIVE EVIDENCE FOR DETERMINATION OF THE EFFECTIVENESS OF MEDICAL DEVICES Medical Devices And Radiological Health; Redelegations Of Authority; Delegations Of Authority And Organization
MEDICAL DEVICES	5.404	NOTIFICATION TO PETITIONERS OF DETERMINATIONS MADE ON PETITIONS FOR RECLASSIFICATION OF MEDICAL DEVICES Medical Devices And Radiological Health; Redelegations Of Authority; Delegations Of Authority And Organization
MEDICAL DEVICES	5.406	NOTIFICATION TO SPONSORS OF DEFICIENCIES IN PETITIONS FOR RECLASSIFICATION OF MEDICAL DEVICES Medical Devices And Radiological Health; Redelegations Of Authority; Delegations Of Authority And Organization
MEDICAL DEVICES	5.407	APPROVAL, DISAPPROVAL, OR WITHDRAWAL OF APPROVAL OF PRODUCT DEVELOPMENT PROTOCOLS AND APPLICATIONS FOR PREMARKET APPROVAL FOR MEDICAL DEVICES Medical Devices And Radiological Health; Redelegations Of Authority; Delegations Of Authority And Organization
MEDICAL DEVICES	5.409	DETERMINATIONS THAT MEDICAL DEVICES PRESENT UNREASONABLE RISK OF SUBSTANTIAL HARM Medical Devices And Radiological Health; Redelegations Of Authority; Delegations Of Authority And Organization
MEDICAL DEVICES	5.410	ORDERS TO REPAIR OR REPLACE, OR MAKE REFUNDS FOR, MEDICAL DEVICES Medical Devices And Radiological Health; Redelegations Of Authority; Delegations Of Authority And Organization
MEDICAL DEVICES	5.417	AUTHORITY PERTAINING TO ACCREDITATION FUNCTIONS FOR MEDICAL DEVICES Medical Devices And Radiological Health; Redelegations Of Authority; Delegations Of Authority And Organization
MEDICAL DEVICES	207.7	ESTABLISHMENT REGISTRATION AND PRODUCT LISTING FOR HUMAN BLOOD AND BLOOD PRODUCTS AND FOR MEDICAL DEVICES General; Registration Of Drug Producers And Listing Of Drugs In Commercial Distribution
MEDICAL DEVICES	610.42	RESTRICTIONS ON USE FOR FURTHER MANUFACTURE OF MEDICAL DEVICES Testing Requirements For Communicable Disease Agents; General Biological Products Standards
MEDICAL DEVICES	801.1	MEDICAL DEVICES; NAME AND PLACE OF BUSINESS OF MANUFACTURER, PACKER OR DISTRIBUTOR General Labeling Provisions; Labeling

MEDICAL DEVICES	801.5	MEDICAL DEVICES; ADEQUATE DIRECTIONS FOR USE General Labeling Provisions; Labeling
MEDICAL DEVICES	801.6	MEDICAL DEVICES; MISLEADING STATEMENTS General Labeling Provisions; Labeling
MEDICAL DEVICES	801.15	MEDICAL DEVICES; PROMINENCE OF REQUIRED LABEL STATEMENTS General Labeling Provisions; Labeling
MEDICAL DEVICES	801.16	MEDICAL DEVICES; SPANISH-LANGUAGE VERSION OF CERTAIN REQUIRED STATEMENTS General Labeling Provisions; Labeling
MEDICAL DEVICES	801.63	MEDICAL DEVICES; WARNING STATEMENTS FOR DEVICES CONTAINING OR MANUFACTURED WITH CHLOROFLUOROCARBONS Labeling
MEDICAL DEVICES	801.116	MEDICAL DEVICES HAVING COMMONLY KNOWN DIRECTIONS Exemptions From Adequate Directions For Use; Labeling
MEDICAL DEVICES	801.122	MEDICAL DEVICES FOR PROCESSING, REPACKING, OR MANUFACTURING Exemptions From Adequate Directions For Use; Labeling
MEDICAL DEVICES	801.125	MEDICAL DEVICES FOR USE IN TEACHING, LAW ENFORCEMENT, RESEARCH, AND ANALYSIS Exemptions From Adequate Directions For Use; Labeling
MEDICAL DEVICES	801.127	MEDICAL DEVICES; EXPIRATION OF EXEMPTION Exemptions From Adequate Directions For Use; Labeling
MEDICAL DEVICES	801.150	MEDICAL DEVICES; PROCESSING, LABELING, OR REPACKING Other Exemptions; Labeling
MEDICAL FILES	20.63	PERSONNEL, MEDICAL, AND SIMILAR FILES, DISCLOSURE OF WHICH CONSTITUTES A CLEARLY UNWARRANTED INVASION OF PERSONAL PRIVACY Exemptions; Public Information
MEDICAL IMAGE	892.2010	MEDICAL IMAGE STORAGE DEVICE Diagnostic Devices; Radiology Devices
MEDICAL IMAGE	892.2020	MEDICAL IMAGE COMMUNICATIONS DEVICE Diagnostic Devices; Radiology Devices
MEDICAL IMAGE	892.2030	MEDICAL IMAGE DIGITIZER Diagnostic Devices; Radiology Devices
MEDICAL IMAGE	892.2040	MEDICAL IMAGE HARD COPY DEVICE Diagnostic Devices; Radiology Devices
MEDICAL USE, EXEMPTIONS	1301.26	EXEMPTIONS FROM IMPORT OR EXPORT REQUIREMENTS FOR PERSONAL MEDICAL USE Registration Of Manufacturers, Distributors, And Dispensers Of Controlled Substances
MEDICATED FEED MILL	5.501	APPROVAL OF NEW ANIMAL DRUG APPLICATIONS, MEDICATED FEED MILL LICENSE APPLICATIONS AND THEIR SUPPLEMENTS Animal Drugs; Redelegations Of Authority; Delegations Of Authority And Organization
MEDICATED FEED MILL	5.502	ISSUANCE OF NOTICES, PROPOSALS, AND ORDERS RELATING TO NEW ANIMAL DRUGS AND MEDICATED FEED MILL LICENSE APPLICATIONS Animal Drugs; Redelegations Of Authority; Delegations Of Authority And Organization
MEDICATED FEED MILL	510.301	RECORDS AND REPORTS CONCERNING EXPERIENCE WITH ANIMAL FEEDS BEARING OR CONTAINING NEW ANIMAL DRUGS FOR WHICH AN APPROVED MEDICATED FEED MILL LICENSE APPLICATION IS IN EFFECT Records And Reports; New Animal Drugs
MEDICATED FEED MILL	510.305	MAINTENANCE OF COPIES OF APPROVED MEDICATED FEED MILL LICENSES TO MANUFACTURE ANIMAL FEED BEARING OR CONTAINING NEW ANIMAL DRUGS Records And Reports; New Animal Drugs
MEDICATED FEED MILL	515.10	MEDICATED FEED MILL LICENSE APPLICATIONS Applications; Medicated Feed Mill License
MEDICATED FEED MILL	515.11	SUPPLEMENTAL MEDICATED FEED MILL LICENSE APPLICATIONS Applications; Medicated Feed Mill License
MEDICATED FEED MILL	515.20	APPROVAL OF MEDICATED FEED MILL LICENSE APPLICATIONS Administrative Actions On Licenses; Medicated Feed Mill License
MEDICATED FEED MILL	515.21	REFUSAL TO APPROVE A MEDICATED FEED MILL LICENSE APPLICATION Administrative Actions On Licenses; Medicated Feed Mill License
MEDICATED FEED MILL	515.22	SUSPENSION AND/OR REVOCATION OF APPROVAL OF A MEDICATED FEED MILL LICENSE Administrative Actions On Licenses; Medicated Feed Mill License
MEDICATED FEED MILL	515.23	VOLUNTARY REVOCATION OF A MEDICATED FEED MILL LICENSE Administrative Actions On Licenses; Medicated Feed Mill License
MEDICATED FEED MILL	515.24	NOTICE OF REVOCATION OF A MEDICATED FEED MILL LICENSE Administrative Actions On Licenses; Medicated Feed Mill License
MEDICATED FEED MILL	515.25	REVOCATION OF ORDER REFUSING TO APPROVE A MEDICATED FEED MILL LICENSE APPLICATION OR SUSPENDING OR REVOKING A LICENSE Administrative Actions On Licenses; Medicated Feed Mill License

MEDICATED FEED MILL	558.4	REQUIREMENT OF A MEDICATED FEED MILL LICENSE General Provisions; New Animal Drugs For Use In Animal Feeds
MEDICATED MILK	520.2200b	SULFACHLORPYRIDAZINE MEDICATED MILK AND DRINKING WATER Oral Dosage Form New Animal Drugs
MEDICATION DISPENSER	880.6430	LIQUID MEDICATION DISPENSER General Hospital And Personal Use Miscellaneous Devices; General Hospital And Personal Use Devices
MEDICATION GUIDE	208.20	CONTENT AND FORMAT OF A MEDICATION GUIDE Medication Guides For Prescription Drug Products
MEDICATION GUIDE	208.24	DISTRIBUTION AND DISPENSING A MEDICATION GUIDE Medication Guides For Prescription Drug Products
MEDICINE, VETERINARY	510.112	ANTIBIOTICS USED IN VETERINARY MEDICINE AND FOR NONMEDICAL PURPOSES; REQUIRED DATA Specific Administrative Rulings And Decisions; New Animal Drugs
MEDIUM	864.8500	LYMPHOCYTE SEPARATION MEDIUM Hematology Reagents; Hematology And Pathology Devices
MEDIUM	866.2440	AUTOMATED MEDIUM DISPENSING AND STACKING DEVICE Microbiology Devices; Immunology And Microbiology Devices
MEDIUM ASIAGO CHEESE	133.103	ASIAGO MEDIUM CHEESE Requirements For Specific Standardized Cheese And Related Products; Cheeses And Related Cheese Products
MEDIUM, CULTURE	866.1700	CULTURE MEDIUM FOR ANTIMICROBIAL SUSCEPTIBILITY TESTS Diagnostic Devices; Immunology And Microbiology Devices
MEDIUM, CULTURE	866.2300	MULTIPURPOSE CULTURE MEDIUM Microbiology Devices; Immunology And Microbiology Devices
MEDIUM, CULTURE	866.2320	DIFFERENTIAL CULTURE MEDIUM Microbiology Devices; Immunology And Microbiology Devices
MEDIUM, CULTURE	866.2330	ENRICHED CULTURE MEDIUM Microbiology Devices; Immunology And Microbiology Devices
MEDIUM, CULTURE	866.2350	MICROBIOLOGICAL ASSAY CULTURE MEDIUM Microbiology Devices; Immunology And Microbiology Devices
MEDIUM, CULTURE	866.2360	SELECTIVE CULTURE MEDIUM Microbiology Devices; Immunology And Microbiology Devices
MEDIUM, CULTURE	866.2390	TRANSPORT CULTURE MEDIUM Microbiology Devices; Immunology And Microbiology Devices
MEDIUM, CULTURE	866.2410	CULTURE MEDIUM FOR PATHOGENIC NEISSERIA SPP. Microbiology Devices; Immunology And Microbiology Devices
MEETINGS	0.65	MEETINGS AND CORRESPONDENCE General Administrative Procedures; Administrative Practices And Procedures
MEETINGS	14.22	MEETINGS OF AN ADVISORY COMMITTEE Meeting Procedures; Public Hearing Before A Public Advisory Committee
MEETINGS	14.25	PORTIONS OF ADVISORY COMMITTEE MEETINGS Meeting Procedures; Public Hearing Before A Public Advisory Committee
MEETINGS	14.27	DETERMINATION TO CLOSE PORTIONS OF ADVISORY COMMITTEE MEETINGS Meeting Procedures; Public Hearing Before A Public Advisory Committee
MEETINGS	14.60	MINUTES AND REPORTS OF ADVISORY COMMITTEE MEETINGS Records Of Meetings And Hearings Before Advisory Committees; Public Hearing Before A Public Advisory Committee
MEETINGS	14.61	TRANSCRIPTS OF ADVISORY COMMITTEE MEETINGS Records Of Meetings And Hearings Before Advisory Committees; Public Hearing Before A Public Advisory Committee
MEETINGS	14.130	CONDUCT OF TEPRSSC MEETINGS; AVAILABILITY OF TEPRSSC RECORDS Technical Electronic Products Radiation Safety Standards Committee; Public Hearing Before A Public Advisory Committee
MEETINGS	312.47	MEETINGS Investigational New Drug Application
MEGESTROL	520.1341	MEGESTROL ACETATE TABLETS Oral Dosage Form New Animal Drugs
MEGLUMINE	520.503	DIATRIZOATE MEGLUMINE AND DIATRIZOATE SODIUM ORAL SOLUTION Oral Dosage Form New Animal Drugs
MEGLUMINE	520.970a	FLUNIXIN MEGLUMINE GRANULES Oral Dosage Form New Animal Drugs
MEGLUMINE	520.970b	FLUNIXIN MEGLUMINE PASTE Oral Dosage Form New Animal Drugs
MEGLUMINE	522.563	DIATRIZOATE MEGLUMINE AND DIATRIZOATE SODIUM INJECTION Implantation Or Injectable Dosage Form New Animal Drugs
MEGLUMINE	522.970	FLUNIXIN MEGLUMINE SOLUTION Implantation Or Injectable Dosage Form New Animal Drugs
MEGLUMINE	556.286	FLUNIXIN MEGLUMINE Tolerances For Residues Of New Animal Drugs In Food
MELAMINE	177.1460	MELAMINE-FORMALDEHYDE RESINS IN MOLDED ARTICLES Substances For Use As Basic Components Of Single And Repeated Use Food Contact Surfaces; Indirect Food Additives: Polymers

MELARSOMINE	522.1362	MELARSOMINE DIHYDROCHLORIDE FOR INJECTION Implantation Or Injectable Dosage Form New Animal Drugs
MELATONIN	522.1350	MELATONIN IMPLANT Implantation Or Injectable Dosage Form New Animal Drugs
MELENGESTROL	556.380	MELENGESTROL ACETATE Specific Tolerances For Residues Of New Animal Drugs; Tolerances For Residues Of New Animal Drugs In Food
MELENGESTROL	558.342	MELENGESTROL ACETATE Specific New Animal Drugs For Use In Animal Feeds; New Animal Drugs For Use In Animal Feeds
MELLORINE	135.130	MELLORINE Requirements For Specific Standardized Frozen Desserts; Frozen Desserts
MEMBERS	5.30	AUTHORITY TO SELECT TEMPORARY VOTING MEMBERS FOR ADVISORY COMMITTEES AND TO SIGN CONFLICT OF INTEREST WAIVERS General Redelegations Of Authority; Delegations Of Authority And Organization
MEMBERS	13.10	MEMBERS OF A BOARD General Provisions; Public Hearing Before A Public Board Of Inquiry
MEMBERS	14.33	COMPILATION OF MATERIALS FOR MEMBERS OF AN ADVISORY COMMITTEE Meeting Procedures; Public Hearing Before A Public Advisory Committee
MEMBERS	14.80	QUALIFICATIONS FOR MEMBERS OF STANDING POLICY AND TECHNICAL ADVISORY COMMITTEES Members Of Advisory Committees; Public Hearing Before A Public Advisory Committee
MEMBERS	14.82	NOMINATIONS OF VOTING MEMBERS OF STANDING ADVISORY COMMITTEES Members Of Advisory Committees; Public Hearing Before A Public Advisory Committee
MEMBERS	14.90	AD HOC ADVISORY COMMITTEE MEMBERS Members Of Advisory Committees; Public Hearing Before A Public Advisory Committee
MEMBERS	14.95	COMPENSATION OF ADVISORY COMMITTEE MEMBERS Members Of Advisory Committees; Public Hearing Before A Public Advisory Committee
MEMBERS, NONVOTING	14.84	NOMINATIONS AND SELECTION OF NONVOTING MEMBERS OF STANDING TECHNICAL ADVISORY COMMITTEES Members Of Advisory Committees; Public Hearing Before A Public Advisory Committee
MEMBERS, NONVOTING	14.86	RIGHTS AND RESPONSIBILITIES OF NONVOTING MEMBERS OF ADVISORY COMMITTEES Members Of Advisory Committees; Public Hearing Before A Public Advisory Committee
MEMBERSHIP	14.127	MEMBERSHIP OF TEPRSSC Technical Electronic Products Radiation Safety Standards Committee; Public Hearing Before A Public Advisory Committee
MEMBERSHIP	14.147	MEMBERSHIP OF A COLOR ADDITIVE ADVISORY COMMITTEE Color Additive Advisory Committees; Public Hearing Before A Public Advisory Committee
MEMBERSHIP	56.107	IRB MEMBERSHIP Organization And Personnel; Institutional Review Boards
MEMBRANE	874.3930	TYMPANOSTOMY TUBE WITH SEMI-PERMEABLE MEMBRANE Prosthetic Devices; Ear, Nose And Throat Devices
MEMBRANE LUNG	868.5610	MEMBRANE LUNG FOR LONG-TERM PULMONARY SUPPORT Therapeutic Devices; Anesthesiology Devices
MEMBRANES	177.2550	REVERSE OSMOSIS MEMBRANES Substances For Use Only As Components Of Articles Intended For Repeated Use; Indirect Food Additives: Polymers
MEMBRANES	177.2910	ULTRA-FILTRATION MEMBRANES Substances For Use Only As Components Of Articles Intended For Repeated Use; Indirect Food Additives: Polymers
MEMBRANES, ION-EXCHANGE	173.20	ION-EXCHANGE MEMBRANES Polymer Substances For Food Treatment; Secondary Direct Food Additives Permitted In Food For Human Consumption
MEMBRANES, ION-EXCHANGE	173.21	PEFLUORINATED ION-EXCHANGE MEMBRANES Secondary Direct Food Additives Permitted In Food For Human Consumption
MEMORANDA, INTER-AGENCY	20.62	INTER- OR INTRA-AGENCY MEMORANDA OR LETTERS Exemptions; Public Information
MENADIONE	573.620	MENADIONE DIMETHYLPYRIMIDINOL BISULFITE Food Additive Listing; Food Additives Permitted In Feed And Drinking Water Of Animals
MENADIONE	573.625	MENADIONE NICOTINAMIDE BISULFITE Food Additives Permitted In Feed And Drinking Water Of Animals
MENHADEN OIL	184.1472	MENHADEN OIL Listing Of Specific Substances Affirmed As GRAS; Direct Food Substances Affirmed As Generally Recognized As Safe

MENSTRUAL CUP	884.5400	MENSTRUAL CUP Obstetrical And Gynecological Therapeutic Devices; Obstetrical And Gynecological Device
MENSTRUAL PAD	884.5425	SCENTED OR SCENTED DEODORIZED MENSTRUAL PAD Obstetrical And Gynecological Therapeutic Devices; Obstetrical And Gynecological Device
MENSTRUAL PAD	884.5435	UNSCENTED MENSTRUAL PAD Obstetrical And Gynecological Therapeutic Devices; Obstetrical And Gynecological Device
MENSTRUAL TAMPON	884.5460	SCENTED OR SCENTED DEODORIZED MENSTRUAL TAMPON Obstetrical And Gynecological Therapeutic Devices; Obstetrical And Gynecological Device
MENSTRUAL TAMPON	884.5470	UNSCENTED MENSTRUAL TAMPON Obstetrical And Gynecological Therapeutic Devices; Obstetrical And Gynecological Device
MENSTRUAL TAMPONS	801.430	USER LABELING FOR MENSTRUAL TAMPONS Special Requirements For Specific Devices; Labeling
MEPIVACAINE	522.1372	MEPIVACAINE HYDROCHLORIDE INJECTION Implantation Or Injectable Dosage Form New Animal Drugs
MERCAPTOBENZOTHIAZOLE	524.1376	2-MERCAPTOBENZOTHIAZOLE SOLUTION Ophthalmic And Topical Dosage Form New Animal Drugs
MERCAPTOIMIDAZOLINE	189.250	MERCAPTOIMIDAZOLINE AND 2-MERCAPTOIMIDAZOLINE Substances Prohibited From Indirect Addition To Human Food Through Contact Surfaces; Substances Prohibited From Use In Human Food
MERCAPTOMETHYL	524.1742	N-(MERCAPTOMETHYL) PHTHALIMIDE S-(*O,O*-DIMETHYL PHOSPHORODITHIOATE) EMULSIFIABLE LIQUID Ophthalmic And Topical Dosage Form New Animal Drugs
MERCURY	700.13	USE OF MERCURY COMPOUNDS IN COSMETICS INCLUDING USE AS SKINBLEACHING AGENTS IN COSMETIC PREPARATIONS ALSO REGARDED AS DRUGS Requirements For Specific Cosmetic Products; General
MERCURY	862.3600	MERCURY TEST SYSTEM Clinical Toxicology Test Systems; Clinical Chemistry And Clinical Toxicology Devices
MERCURY	872.3080	MERCURY AND ALLOY DISPENSER Prosthetic Devices; Dental Devices
MERCURY	872.3700	DENTAL MERCURY Prosthetic Devices; Dental Devices
MERCURY	880.2920	CLINICAL MERCURY THERMOMETER General Hospital And Personal Use Monitoring Devices; General Hospital And Personal Use Devices
MERCURY VAPOR	1040.30	HIGH-INTENSITY MERCURY VAPOR DISCHARGE LAMPS Performance Standards For Light-emitting Products
MESH BAG	870.3650	PACEMAKER POLYMERIC MESH BAG Cardiovascular Prosthetic Devices; Cardiovascular Devices
MESH, SURGICAL	878.3300	SURGICAL MESH Prosthetic Devices; General And Plastic Surgery Devices
METABISULFITE	182.3637	POTASSIUM METABISULFITE Chemical Preservatives; Substances Generally Recognized As Safe
METABISULFITE	182.3766	SODIUM METABISULFITE Chemical Preservatives; Substances Generally Recognized As Safe
METABISULFITE	582.3637	POTASSIUM METABISULFITE Chemical Preservatives; Substances Generally Recognized As Safe
METABISULFITE	582.3766	SODIUM METABISULFITE Chemical Preservatives; Substances Generally Recognized As Safe
METAL, BASE	872.3170	BASE METAL ALLOY Prosthetic Devices; Dental Devices
METAL LOCATOR	886.4400	ELECTRONIC METAL LOCATOR Surgical Devices; Ophthalmic Devices
METAL, PRECIOUS	872.3060	GOLD BASED ALLOYS AND PRECIOUS METAL ALLOYS FOR CLINICAL USE Prosthetic Devices; Dental Devices
METALLIC	178.3910	SURFACE LUBRICANTS USED IN THE MANUFACTURE OF METALLIC ARTICLES Certain Adjuvants And Production Aids; Indirect Food Additives: Adjuvants, Production Aids, And Sanitizers
METALLIC	888.3030	SINGLE/MULTIPLE COMPONENT METALLIC BONE FIXATION APPLIANCES AND ACCESSORIES Prosthetic Devices; Orthopedic Devices
METALLIC	888.3040	SMOOTH OR THREADED METALLIC BONE FIXATION FASTENER Prosthetic Devices; Orthopedic Devices
METAPHOSPHATE	182.6769	SODIUM METAPHOSPHATE Sequestrants; Substances Generally Recognized As Safe
METAPHOSPHATE	582.6769	SODIUM METAPHOSPHATE Sequestrants; Substances Generally Recognized As Safe

METASILICATE	184.1769	SODIUM METASILICATE Listing Of Specific Substances Affirmed As GRAS; Direct Food Substances Affirmed As Generally Recognized As Safe
METER	868.1780	INSPIRATORY AIRWAY PRESSURE METER Diagnostic Devices; Anesthesiology Devices
METER	868.1860	PEAK-FLOW METER FOR SPIROMETRY Diagnostic Devices; Anesthesiology Devices
METHACRYLATE	73.3121	POLY(HYDROXYETHYL METHACRYLATE)-DYE COPOLYMERS Medical Devices; Listing Of Color Additives Exempt From Certification
METHACRYLATE	73.3127	VINYL ALCOHOL/METHYL METHACRYLATE-DYE REACTION PRODUCTS Listing Of Color Additives Exempt From Certification
METHACRYLATE	177.1030	ACRYLONITRILE/BUTADIENE/STYRENE/METHYL METHACRYLATE COPOLYMER Substances For Use As Basic Components Of Single And Repeated Use Food Contact Surfaces; Indirect Food Additives: Polymers
METHACRYLATE	177.1830	STYRENE-METHYL METHACRYLATE COPOLYMERS Substances For Use As Basic Components Of Single And Repeated Use Food Contact Surfaces; Indirect Food Additives: Polymers
METHACRYLATE	177.2000	VINYLIDENE CHLORIDE/METHYL ACRYLATE/METHYL METHACRYLATE POLYMERS Substances For Use As Basic Components Of Single And Repeated Use Food Contact Surfaces; Indirect Food Additives: Polymers
METHACRYLATE	177.2465	POLYMETHYLMETHACRYLATE/POLY(TRIMETHOXYSILYLPROPYL) METHACRYLATE COPOLYMERS Substances For Use Only As Components Of Articles Intended For Repeated Use; Indirect Food Additives: Polymers
METHACRYLATE	882.5030	METHYL METHACRYLATE FOR ANEURYSMORRHAPHY Neurological Therapeutic Devices; Neurological Devices
METHACRYLATE	882.5300	METHYL METHACRYLATE FOR CRANIOPLASTY Neurological Therapeutic Devices; Neurological Devices
METHACRYLIC ACID	172.775	METHACRYLIC ACID-DIVINYLBENZENE COPOLYMER Other Specific Usage Additives; Food Additives Permitted For Direct Addition To Food For Human Consumption
METHADONE	862.3620	METHADONE TEST SYSTEM Clinical Toxicology Test Systems; Clinical Chemistry And Clinical Toxicology Devices
METHAMPHETAMINE	250.101	AMPHETAMINE AND METHAMPHETAMINE INHALERS REGARDED AS PRESCRIPTION DRUGS New Drug Or Prescription Status Of Specific Drugs; Specific Requirements For Specific Human Drugs
METHAMPHETAMINE	862.3610	METHAMPHETAMINE TEST SYSTEM Clinical Toxicology Test Systems; Clinical Chemistry And Clinical Toxicology Devices
METHAQUALONE	862.3630	METHAQUALONE TEST SYSTEM Clinical Toxicology Test Systems; Clinical Chemistry And Clinical Toxicology Devices
METHENAMINE MANDELATE	520.2280	SULFAMETHIZOLE AND METHENAMINE MANDELATE TABLETS Oral Dosage Form New Animal Drugs
METHANESULFONATE	529.2503	TRICAINE METHANESULFONATE Certain Other Dosage Form New Animal Drugs
METHIONINE	172.372	N-ACETYL-L-METHIONINE Special Dietary And Nutritional Additives; Food Additives Permitted For Direct Addition To Food For Human Consumption
METHIONINE	582.5475	METHIONINE Nutrients And/or Dietary Supplements; Substances Generally Recognized As Safe
METHIONINE HYDROXY ANALOG	582.5477	METHIONINE HYDROXY ANALOG AND ITS CALCIUM SALTS Nutrients And/or Dietary Supplements; Substances Generally Recognized As Safe
METHIONINE SULFATE	172.399	ZINC METHIONINE SULFATE Special Dietary And Nutritional Additives; Food Additives Permitted For Direct Addition To Food For Human Consumption
METHOCARBAMOL	520.1380	METHOCARBAMOL TABLETS Oral Dosage Form New Animal Drugs
METHOCARBAMOL	522.1380	METHOCARBAMOL INJECTION Implantation Or Injectable Dosage Form New Animal Drugs
METHOD	1210.16	METHOD OF BACTERIAL COUNT Inspection And Testing; Regulations Under The Federal Import Milk Act
METHOD, ANALYTICAL	5.500	ISSUANCE OF FEDERAL REGISTER DOCUMENTS PERTAINING TO THE DETERMINATION OF SAFE LEVELS, NOTICE OF NEED FOR DEVELOPMENT OF AN ANALYTICAL METHOD, PROHIBITION OF CERTAIN EXTRALABEL DRUG USE Animal Drugs; Redelegations Of Authority; Delegations Of Authority And Organization

METHOD OF PAYMENT	1309.12	TIME AND METHOD OF PAYMENT; REFUND Registration Of Manufacturers, Distributors, Importers And Exporters Of List I Chemicals
METHOD, REGULATORY	500.88	REGULATORY METHOD Regulation of Carcinogenic Compounds Used In Food-Producing Animals; General
METHODOLOGY	114.90	METHODOLOGY Production And Process Controls; Acidified Foods
METHODS	2.19	METHODS OF ANALYSIS General Provisions; General Administrative Rulings And Decisions
METHODS	5.602	TESTING PROGRAMS AND METHODS OF CERTIFICATION AND IDENTIFICATION FOR ELECTRONIC PRODUCTS Radiation Control; Redelegations Of Authority; Delegations Of Authority And Organization
METHODS	130.12	GENERAL METHODS FOR WATER CAPACITY AND FILL OF CONTAINERS General Provisions; Food Standards: General
METHODS	133.5	METHODS OF ANALYSIS General Provisions; Cheeses And Related Cheese Products
METHODS	163.5	METHODS OF ANALYSIS General Provisions; Cacao Products
METHODS	610.9	EQUIVALENT METHODS AND PROCESSES General Provisions; General Biological Products Standards
METHODS	1210.15	PASTEURIZATION; EQUIPMENT AND METHODS Inspection And Testing; Regulations Under The Federal Import Milk Act
METHODS, ANALYTICAL	320.29	ANALYTICAL METHODS FOR AN IN VIVO BIOAVAILABILITY STUDY Procedures For Determining The Bioavailability Or Bioequivalence Of Drug Products; Bioavailability And Bioequivalence Requirements
METHODS, ANALYTICAL	530.22	SAFE LEVELS AND ANALYTICAL METHODS FOR FOOD-PRODUCING ANIMALS Extralabel Drug Use In Animals
METHODS, ANALYTICAL	530.24	PROCEDURE FOR ANNOUNCING ANALYTICAL METHODS FOR DRUG RESIDUE QUANTIFICATION Extralabel Drug Use In Animals
METHODS, ANALYTICAL	530.40	SAFE LEVELS AND AVAILABILITY OF ANALYTICAL METHODS Extralabel Drug Use In Animals
METHOPRENE	520.1390	S-METHOPRENE Oral Dosage Form New Animal Drugs
METHYL ACRYLATE	177.1340	ETHYLENE-METHYL ACRYLATE COPOLYMER RESINS Substances For Use As Basic Components Of Single And Repeated Use Food Contact Surfaces; Indirect Food Additives: Polymers
METHYL ACRYLATE	177.1480	NITRILE RUBBER MODIFIED ACRYLONITRILE-METHYL ACRYLATE COPOLYMERS Substances For Use As Basic Components Of Single And Repeated Use Food Contact Surfaces; Indirect Food Additives: Polymers
METHYL ACRYLATE	177.1990	VINYLIDENE CHLORIDE/METHYL ACRYLATE COPOLYMERS Substances For Use As Basic Components Of Single And Repeated Use Food Contact Surfaces; Indirect Food Additives: Polymers
METHYL ACRYLATE	177.2000	VINYLIDENE CHLORIDE/METHYL ACRYLATE/METHYL METHACRYLATE POLYMERS Substances For Use As Basic Components Of Single And Repeated Use Food Contact Surfaces; Indirect Food Additives: Polymers
METHYL ALCOHOL	173.250	METHYL ALCOHOL RESIDUES Solvents, Lubricants, Release Agents And Related Substances; Secondary Direct Food Additives Permitted In Food For Human Consumption
METHYL ESTERS	172.225	METHYL AND ETHYL ESTERS OF FATTY ACIDS PRODUCED FROM EDIBLE FATS AND OILS Coatings, Films And Related Substances; Food Additives Permitted For Direct Addition To Food For Human Consumption
METHYL ESTERS	573.640	METHYL ESTERS OF HIGHER FATTY ACIDS Food Additive Listing; Food Additives Permitted In Feed And Drinking Water Of Animals
METHYL ETHYL CELLULOSE	172.872	METHYL ETHYL CELLULOSE Multipurpose Additives; Food Additives Permitted For Direct Addition To Food For Human Consumption
METHYL GLUCOSIDE	172.816	METHYL GLUCOSIDE-COCONUT OIL ESTER Multipurpose Additives; Food Additives Permitted For Direct Addition To Food For Human Consumption
METHYL GLUCOSIDE	178.3600	METHYL GLUCOSIDE-COCONUT OIL ESTER Certain Adjuvants And Production Aids; Indirect Food Additives: Adjuvants, Production Aids, And Sanitizers
METHYL GLUCOSIDE	573.660	METHYL GLUCOSIDE-COCONUT OIL ESTER Food Additive Listing; Food Additives Permitted In Feed And Drinking Water Of Animals
METHYL METHACRYLATE	73.3127	VINYL ALCOHOL/METHYL METHACRYLATE-DYE REACTION PRODUCTS Listing Of Color Additives Exempt From Certification

METHYL METHACRYLATE	177.1030	ACRYLONITRILE/BUTADIENE/STYRENE/METHYL METHACRYLATE COPOLYMER Substances For Use As Basic Components Of Single And Repeated Use Food Contact Surfaces; Indirect Food Additives: Polymers
METHYL METHACRYLATE	177.1830	STYRENE-METHYL METHACRYLATE COPOLYMERS Substances For Use As Basic Components Of Single And Repeated Use Food Contact Surfaces; Indirect Food Additives: Polymers
METHYL METHACRYLATE	177.2000	VINYLIDENE CHLORIDE/METHYL ACRYLATE/METHYL METHACRYLATE POLYMERS Substances For Use As Basic Components Of Single And Repeated Use Food Contact Surfaces; Indirect Food Additives: Polymers
METHYL METHACRYLATE	882.5030	METHYL METHACRYLATE FOR ANEURYSMORRHAPHY Neurological Therapeutic Devices; Neurological Devices
METHYL METHACRYLATE	882.5300	METHYL METHACRYLATE FOR CRANIOPLASTY Neurological Therapeutic Devices; Neurological Devices
METHYL SULFATE	173.385	SODIUM METHYL SULFATE Specific Usage Additives; Secondary Direct Food Additives Permitted In Food For Human Consumption
METHYLCELLULOSE	172.874	HYDROXYPROPYL METHYLCELLULOSE Multipurpose Additives; Food Additives Permitted For Direct Addition To Food For Human Consumption
METHYLCELLULOSE	182.1480	METHYLCELLULOSE Multiple Purpose GRAS Food Substances; Substances Generally Recognized As Safe
METHYLCELLULOSE	201.319	WATER-SOLUBLE GUMS, HYDROPHILIC GUMS AND HYDROPHILIC MUCILLOIDS (INCLUDING METHYLCELLULOSE) AS ACTIVE INGREDIENTS; REQUIRED WARNINGS AND DIRECTIONS Specific Labeling Requirements For Specific Drug Products; Labeling
METHYLCELLULOSE	582.1480	METHYLCELLULOSE General Purpose Food Additives; Substances Generally Recognized As Safe
METHYLENE BLUE	500.27	METHYLENE BLUE-CONTAINING DRUGS FOR USE IN ANIMALS Specific Administrative Rulings And Decisions; General
METHYLENE CHLORIDE	173.255	METHYLENE CHLORIDE Solvents, Lubricants, Release Agents And Related Substances; Secondary Direct Food Additives Permitted In Food For Human Consumption
METHYLENE CHLORIDE	700.19	USE OF METHYLENE CHLORIDE AS AN INGREDIENT IN COSMETIC PRODUCTS Requirements For Specific Cosmetic Products; General
METHYLENE DISALICYLATE	520.154b	SOLUBLE BACITRACIN METHYLENE DISALICYLATE AND STREPTOMYCIN SULFATE ORAL POWDER Oral Dosage Form New Animal Drugs
METHYLENE DISALICYLATE	558.76	BACITRACIN METHYLENE DISALICYLATE Specific New Animal Drugs For Use In Animal Feeds; New Animal Drugs For Use In Animal Feeds
METHYLENEBIS	189.280	4,4'-METHYLENEBIS (2-CHLOROANALINE) Substances Prohibited From Indirect Addition To Human Food Thru Food Contact Surfaces; Substances Prohibited From Use In Human Food
METHYLMALONIC ACID	862.1509	METHYLMALONIC ACID TEST SYSTEM Clinical Chemistry Test Systems; Clinical Chemistry And Clinical Toxicology Devices
METHYLPARABEN	184.1490	METHYLPARABEN Listing Of Specific Substances Affirmed As GRAS; Direct Food Substances Affirmed As Generally Recognized As Safe
METHYLPARABEN	556.390	METHYLPARABEN Specific Tolerances For Residues Of New Animal Drugs; Tolerances For Residues Of New Animal Drugs In Food
METHYLPARABEN	582.3490	METHYLPARABEN Chemical Preservatives; Substances Generally Recognized As Safe
METHYLPREDNISOLONE	520.1408	METHYLPREDNISOLONE TABLETS Oral Dosage Form New Animal Drugs
METHYLPREDNISOLONE	520.1409	METHYLPREDNISOLONE, ASPIRIN TABLETS Oral Dosage Form New Animal Drugs
METHYLPREDNISOLONE	522.1410	STERILE METHYLPREDNISOLONE ACETATE SUSPENSION Implantation Or Injectable Dosage Form New Animal Drugs
METHYLPREDNISOLONE	556.400	METHYLPREDNISOLONE Specific Tolerances For Residues Of New Animal Drugs; Tolerances For Residues Of New Animal Drugs In Food
METHYLSTYRENE	177.1635	POLY(P-METHYLSTYRENE) AND RUBBER-MODIFIED POLY(P-METHYLSTYRENE) Substances For Use As Basic Components Of Single And Repeated Use Food Contact Surfaces; Indirect Food Additives: Polymers
METHYLSTYRENE	178.3610	α-METHYLSTYRENE-VINYLTOLUENE RESINS, HYDROGENATED Certain Adjuvants And Production Aids; Indirect Food Additives: Adjuvants, Production Aids, And Sanitizers

METOSERPATE	520.1422	METOSERPATE HYDROCHLORIDE Oral Dosage Form New Animal Drugs
METOSERPATE	556.410	METOSERPATE HYDROCHLORIDE Specific Tolerances For Residues Of New Animal Drugs; Tolerances For Residues Of New Animal Drugs In Food
METREURYNTER	884.5050	METREURYNTER-BALLOON ABORTION SYSTEM Obstetrical And Gynecological Therapeutic Devices; Obstetrical And Gynecological Devices
MIBOLERONE	520.1430	MIBOLERONE Oral Dosage Form New Animal Drugs
MIBOLERONE	558.348	MIBOLERONE Specific New Animal Drugs For Use In Animal Feeds; New Animal Drugs For Use In Animal Feeds
MICA	73.1496	MICA Drugs; Listing Of Color Additives Exempt From Certification
MICA	73.2496	MICA Cosmetics; Listing Of Color Additives Exempt From Certification
MICONAZOLE	524.1443	MICONAZOLE NITRATE CREAM; MICONAZOLE NITRATE LOTION; MICONAZOLE NITRATE SPRAY Ophthalmic And Topical Dosage Form New Animal Drugs
MICRO CHEMISTRY	862.2170	MICRO CHEMISTRY ANALYZER FOR CLINICAL USE Clinical Laboratory Instruments; Clinical Chemistry And Clinical Toxicology Devices
MICROBIAL ENZYMES	173.150	MILK-CLOTTING ENZYMES, MICROBIAL Enzyme Preparations And Microorganisms; Secondary Direct Food Additives Permitted In Food For Human Consumption
MICROBIAL GROWTH	866.2560	MICROBIAL GROWTH MONITOR Microbiology Devices; Immunology And Microbiology Devices
MICROBIOLOGICAL	211.113	CONTROL OF MICROBIOLOGICAL CONTAMINATION Production And Process Controls; Current Good Manufacturing Practice For Finished Pharmaceuticals
MICROBIOLOGICAL	866.2350	MICROBIOLOGICAL ASSAY CULTURE MEDIUM Microbiology Devices; Immunology And Microbiology Devices
MICROBIOLOGICAL	866.2540	MICROBIOLOGICAL INCUBATOR Microbiology Devices; Immunology And Microbiology Devices
MICROBIOLOGICAL	866.2900	MICROBIOLOGICAL SPECIMEN COLLECTION AND TRANSPORT DEVICE Microbiology Devices; Immunology And Microbiology Devices
MICROCAPSULES	172.230	MICROCAPSULES FOR FLAVORING SUBSTANCES Coatings, Films And Related Substances; Food Additives Permitted For Direct Addition To Food For Human Consumption
MICROCCOCUS LYSODEIKTICUS	173.135	CATALASE DERIVED FROM MICROCCOCUS LYSODEIKTICUS Enzyme Preparations And Microorganisms; Secondary Direct Food Additives Permitted In Food For Human Consumption
MICROGLOBULIN	866.5630	BETA-2-MICROGLOBULIN IMMUNOLOGICAL TEST SYSTEM Immunological Test Systems; Immunology And Microbiology Devices
MICROINJECTORS	884.6150	ASSISTED REPRODUCTION MICROMANIPULATORS AND MICROINJECTORS Assisted Reproduction Devices; Obstetrical And Gynecological Devices
MICROMANIPULATORS	884.6150	ASSISTED REPRODUCTION MICROMANIPULATORS AND MICROINJECTORS Assisted Reproduction Devices; Obstetrical And Gynecological Devices
MICROORGANISM	866.2660	MICROORGANISM DIFFERENTIATION AND IDENTIFICATION DEVICE Microbiology Devices; Immunology And Microbiology Devices
MICROORGANISMS	173.320	CHEMICALS FOR CONTROLLING MICROORGANISMS IN CANE-SUGAR AND BEET-SUGAR MILLS Specific Usage Additives; Secondary Direct Food Additives Permitted In Food For Human Consumption
MICROORGANISMS, SALMONELLA	500.35	ANIMAL FEEDS CONTAMINATED WITH SALMONELLA MICROORGANISMS Specific Administrative Rulings And Decisions; General
MICROPARTICULATED PROTEIN	184.1498	MICROPARTICULATED PROTEIN PRODUCT Listing Of Specific Substances Affirmed As GRAS; Direct Food Substances Affirmed As Generally Recognized As Safe
MICROPIPETTE	884.6140	ASSISTED REPRODUCTION MICROPIPETTE FABRICATION INSTRUMENTS Assisted Reproduction Devices; Obstetrical And Gynecological Devices
MICROPOROUS FILTERS	177.2250	FILTERS, MICROPOROUS POLYMERIC Substances For Use Only As Components Of Articles Intended For Repeated Use; Indirect Food Additives: Polymers
MICROSCOPE	878.4700	SURGICAL MICROSCOPE AND ACCESSORIES Surgical Devices; General And Plastic Surgery Devices
MICROSCOPES	864.3600	MICROSCOPES AND ACCESSORIES Pathology Instrumentation And Accessories; Hematology And Pathology Devices
MICROSCOPES	884.6190	ASSISTED REPRODUCTIVE MICROSCOPES AND MICROSCOPE ACCESSORIES Assisted Reproduction Devices; Obstetrical And Gynecological Devices
MICROSECTIONS	864.3400	DEVICE FOR SEALING MICROSECTIONS Pathology Instrumentation And Accessories; Hematology And Pathology Devices

MICROSEDIMENTATION	864.5350	MICROSEDIMENTATION CENTRIFUGE Automated And Semi-automated Hematology Devices; Hematology And Pathology Devices
MICROSPHERE	870.1360	TRACE MICROSPHERE Cardiovascular Diagnostic Devices; Cardiovascular Devices
MICROSURGICAL	874.4500	EAR, NOSE AND THROAT MICROSURGICAL CARBON-DIOXIDE LASER Surgical Devices; Ear, Nose And Throat Devices
MICROSURGICAL	882.4525	MICROSURGICAL INSTRUMENT Neurological Surgical Devices; Neurological Devices
MICROTITER	866.2500	MICROTITER DILUTING AND DISPENSING DEVICE Microbiology Devices; Immunology And Microbiology Devices
MICROTITRATOR	862.2680	MICROTITRATOR FOR CLINICAL USE Clinical Laboratory Instruments; Clinical Chemistry And Clinical Toxicology Devices
MICROTOOLS	884.6130	ASSISTED REPRODUCTION MICROTOOLS Assisted Reproduction Devices; Obstetrical And Gynecological Devices
MICROWAVE	179.30	RADIOFREQUENCY RADIATION FOR THE HEATING OF FOOD, INCLUDING MICROWAVE FREQUENCIES Radiation And Radiation Sources; Irradiation In The Production, Processing And Handling Of Food
MICROWAVE	890.5275	MICROWAVE DIATHERMY Physical Medicine Therapeutic Devices; Physical Medicine Devices
MICROWAVE	1030.10	MICROWAVE OVENS Performance Standards For Microwave And Radio Frequency Emitting Products
MIDDLE EAR	874.3430	MIDDLE EAR MOLD Prosthetic Devices; Ear, Nose And Throat Devices
MIGRATING	182.70	SUBSTANCES MIGRATING FROM COTTON AND COTTON FABRICS USED IN DRY FOOD PACKAGING General Provisions; Substances Generally Recognized As Safe
MIGRATING	182.90	SUBSTANCES MIGRATING TO FOOD FROM PAPER AND PAPERBOARD PRODUCTS General Provisions; Substances Generally Recognized As Safe
MILBEMYCIN OXIME	520.1445	MILBEMYCIN OXIME TABLETS Oral Dosage Form New Animal Drugs
MILBEMYCIN OXIME	520.1446	MILBEMYCIN OXIME/LUFENURON TABLETS Oral Dosage Form New Animal Drugs
MILBEMYCIN OXIME	524.1446	MILBEMYCIN OXIME SOLUTION Ophthalmic And Topical Dosage Form New Animal Drugs
MILITARY	1301.23	EXEMPTION OF CERTAIN MILITARY AND OTHER PERSONNEL Registration Of Manufacturers, Distributors, And Dispensers Of Controlled Substances
MILK	131.110	MILK Requirements For Specific Standardized Milk And Cream; Milk And Cream
MILK	1210.25	PERMITS FOR PASTEURIZED MILK OR CREAM Permit Control; Regulations Under The Federal Import Milk Act
MILK	1210.26	PERMITS FOR RAW MILK OR CREAM Permit Control; Regulations Under The Federal Import Milk Act
MILK	1240.61	MANDATORY PASTEURIZATION FOR ALL MILK AND MILK PRODUCTS IN FINAL PACKAGE FORM INTENDED FOR DIRECT HUMAN CONSUMPTION Specific Administrative Decisions Regarding Interstate Shipments; Control Of Communicable Diseases
MILK, ACIDIFIED	131.111	ACIDIFIED MILK Requirements For Specific Standardized Milk And Cream; Milk And Cream
MILK ACT	1210.27	PERMITS WAIVING CLAUSES 2 AND 5, SECTION 2 OF THE FEDERAL IMPORT MILK ACT Permit Control; Regulations Under The Federal Import Milk Act
MILK ACT	1210.28	PERMITS WAIVING CLAUSE 4, SECTION 2 OF THE FEDERAL IMPORT MILK ACT Permit Control; Regulations Under The Federal Import Milk Act
MILK, BLUE-MOLD CHEESE	133.184	ROQUEFORT, SHEEP'S MILK BLUE-MOLD, AND BLUE-MOLD CHEESE FROM SHEEP'S MILK Requirements For Specific Standardized Cheese And Related Products; Cheeses And Related Cheese Products
MILK BREAD	136.130	MILK BREAD, ROLLS, AND BUNS Requirements For Specific Standardized Bakery Products; Bakery Products
MILK, BREAST	866.5170	BREAST MILK IMMUNOLOGICAL TEST SYSTEM Immunological Test Systems; Immunology And Microbiology Devices
MILK CHOCOLATE	163.130	MILK CHOCOLATE Requirements For Specific Standardized Cacao Products; Cacao Products
MILK CHOCOLATE	163.140	SKIM MILK CHOCOLATE Requirements For Specific Standardized Cacao Products; Cacao Products
MILK CHOCOLATE	163.155	MILK CHOCOLATE AND VEGETABLE FAT COATING Requirements For Specific Standardized Cacao Products; Cacao Products

MILK-CLOTTING ENZYMES	173.150	MILK-CLOTTING ENZYMES, MICROBIAL Enzyme Preparations And Microorganisms; Secondary Direct Food Additives Permitted In Food For Human Consumption
MILK, CONCENTRATED	131.115	CONCENTRATED MILK Requirements For Specific Standardized Milk And Cream; Milk And Cream
MILK, CONDENSED SWEETENED	131.120	SWEETENED CONDENSED MILK Requirements For Specific Standardized Milk And Cream; Milk And Cream
MILK, CULTURED	131.112	CULTURED MILK Requirements For Specific Standardized Milk And Cream; Milk And Cream
MILK, DRY NONFAT	131.125	NONFAT DRY MILK Requirements For Specific Standardized Milk And Cream; Milk And Cream
MILK, DRY WHOLE	131.147	DRY WHOLE MILK Requirements For Specific Standardized Milk And Cream; Milk And Cream
MILK, EVAPORATED	131.130	EVAPORATED MILK Requirements For Specific Standardized Milk And Cream; Milk And Cream
MILK, FORTIFIED	131.127	NONFAT DRY MILK FORTIFIED WITH VITAMINS A AND D Requirements For Specific Standardized Milk And Cream; Milk And Cream
MILK, GOAT'S	135.115	GOAT'S MILK ICE CREAM Requirements For Specific Standardized Frozen Desserts; Frozen Desserts
MILK MACARONI	139.120	MILK MACARONI PRODUCTS Requirements For Specific Standardized Macaroni And Noodle Products; Macaroni And Noodle Products
MILK MACARONI	139.121	NONFAT MILK MACARONI PRODUCTS Requirements For Specific Standardized Macaroni And Noodle Products; Macaroni And Noodle Products
MILK MACARONI	139.122	ENRICHED NONFAT MILK MACARONI PRODUCTS Requirements For Specific Standardized Macaroni And Noodle Products; Macaroni And Noodle Products
MILK, MEDICATED	520.2200	SULFACHLORPYRIDAZINE MEDICATED MILK AND DRINKING WATER Oral Dosage Form New Animal Drugs
MILK-PRODUCING ANIMALS	510.105	LABELING OF DRUGS FOR USE IN MILK-PRODUCING ANIMALS Specific Administrative Rulings And Decisions; New Animal Drugs
MILK-PRODUCING ANIMALS	510.106	LABELING OF ANTIBIOTIC AND ANTIBIOTIC-CONTAINING DRUGS INTENDED FOR USE IN MILK-PRODUCING ANIMALS Specific Administrative Rulings And Decisions; New Animal Drugs
MILK, SHEEP'S	133.184	ROQUEFORT, SHEEP'S MILK BLUE-MOLD, AND BLUE-MOLD CHEESE FROM SHEEP'S MILK Requirements For Specific Standardized Cheese And Related Products; Cheeses And Related Cheese Products
MILL, FEED	5.501	APPROVAL OF NEW ANIMAL DRUG APPLICATIONS, MEDICATED FEED MILL LICENSE APPLICATIONS AND THEIR SUPPLEMENTS Animal Drugs; Redelegations Of Authority; Delegations Of Authority And Organization
MILL, FEED	5.502	ISSUANCE OF NOTICES, PROPOSALS, AND ORDERS RELATING TO NEW ANIMAL DRUGS AND MEDICATED FEED MILL LICENSE APPLICATIONS Animal Drugs; Redelegations Of Authority; Delegations Of Authority And Organization
MILL, FEED	510.301	RECORDS AND REPORTS CONCERNING EXPERIENCE WITH ANIMAL FEEDS BEARING OR CONTAINING NEW ANIMAL DRUGS FOR WHICH AN APPROVED MEDICATED FEED MILL LICENSE APPLICATION IS IN EFFECT Records And Reports; New Animal Drugs
MILL, FEED	510.305	MAINTENANCE OF COPIES OF APPROVED MEDICATED FEED MILL LICENSES TO MANUFACTURE ANIMAL FEED BEARING OR CONTAINING NEW ANIMAL DRUGS Records And Reports; New Animal Drugs
MILL, FEED	515.10	MEDICATED FEED MILL LICENSE APPLICATIONS Applications; Medicated Feed Mill License
MILL, FEED	515.11	SUPPLEMENTAL MEDICATED FEED MILL LICENSE APPLICATIONS Applications; Medicated Feed Mill License
MILL, FEED	515.20	APPROVAL OF MEDICATED FEED MILL LICENSE APPLICATIONS Administrative Actions On Licenses; Medicated Feed Mill License
MILL, FEED	515.21	REFUSAL TO APPROVE A MEDICATED FEED MILL LICENSE APPLICATION Administrative Actions On Licenses; Medicated Feed Mill License
MILL, FEED	515.22	SUSPENSION AND/OR REVOCATION OF APPROVAL OF A MEDICATED FEED MILL LICENSE Administrative Actions On Licenses; Medicated Feed Mill License
MILL, FEED	515.23	VOLUNTARY REVOCATION OF A MEDICATED FEED MILL LICENSE Administrative Actions On Licenses; Medicated Feed Mill License
MILL, FEED	515.24	NOTICE OF REVOCATION OF A MEDICATED FEED MILL LICENSE Administrative Actions On Licenses; Medicated Feed Mill License
MILL, FEED	515.25	REVOCATION OF ORDER REFUSING TO APPROVE A MEDICATED FEED MILL LICENSE APPLICATION OR SUSPENDING OR REVOKING A LICENSE Administrative Actions On Licenses; Medicated Feed Mill License

MILL, FEED	558.4	REQUIREMENT OF A MEDICATED FEED MILL LICENSE General Provisions; New Animal Drugs For Use In Animal Feeds
MILLS, SUGAR	173.320	CHEMICAL FOR CONTROLLING MICROORGANISMS IN CANE-SUGAR AND BEET-SUGAR MILLS Specific Usage Additives; Secondary Direct Food Additives Permitted In Food For Human Consumption
MINCED CLAMS	102.49	FRIED CLAMS MADE FROM MINCED CLAMS Requirements For Specific Nonstandardized Foods; Common Or Usual Name For Nonstandardized Foods
MINCED FISH	102.45	FISH STICKS OR PORTIONS MADE FROM MINCED FISH Requirements For Specific Nonstandardized Foods; Common Or Usual Name For Nonstandardized Foods
MINERAL OIL	172.878	WHITE MINERAL OIL Multipurpose Additives; Food Additives Permitted For Direct Addition To Food For Human Consumption
MINERAL OIL	178.3620	MINERAL OIL Certain Adjuvants And Production Aids; Indirect Food Additives: Adjuvants, Production Aids, And Sanitizers
MINERAL OIL	201.302	NOTICE TO MANUFACTURERS, PACKERS, AND DISTRIBUTORS OF DRUGS FOR INTERNAL USE WHICH CONTAIN MINERAL OIL Specific Labeling Requirements For Specific Drug Products; Labeling
MINERAL OIL	573.680	MINERAL OIL Food Additive Listing; Food Additives Permitted In Feed And Drinking Water Of Animals
MINERAL PROTEIN FEED	520.2380a	THIABENDAZOLE TOP DRESSING AND MINERAL PROTEIN FEED BLOCK Oral Dosage Form New Animal Drugs
MINERAL REINFORCED	177.2355	MINERAL REINFORCED NYLON RESINS Substances For Use Only As Components Of Articles Intended For Repeated Use; Indirect Food Additives: Polymers
MINERALS, REDUCED	184.1979a	REDUCED MINERALS WHEY Listing Of Specific Substances Affirmed As GRAS; Direct Food Substances Affirmed As Generally Recognized As Safe
MINERALS, TRACE	582.80	TRACE MINERALS ADDED TO ANIMAL FEEDS General Provisions; Substances Generally Recognized As Safe
MINIMAL RISK	50.51	CLINICAL INVESTIGATIONS NOT INVOLVING GREATER THAN MINIMAL RISK Additional Safeguards For Children In Clinical Investigations; Protection Of Human Subjects
MINIMAL RISK	50.52	CLINICAL INVESTIGATIONS INVOLVING GREATER THAN MINIMAL RISK BUT PRESENTING THE PROSPECT OF DIRECT BENEFIT TO INDIVIDUAL SUBJECTS Additional Safeguards For Children In Clinical Investigations; Protection Of Human Subjects
MINIMAL RISK	50.53	CLINICAL INVESTIGATIONS INVOLVING GREATER THAN MINIMAL RISK AND NO PROSPECT OF DIRECT BENEFIT TO INDIVIDUAL SUBJECTS, BUT LIKELY TO YIELD GENERALIZABLE KNOWLEDGE ABOUT THE SUBJECTS' DISORDER OR CONDITION Additional Safeguards For Children In Clinical Investigations; Protection Of Human Subjects
MINIMAL RISK	56.110	EXPEDITED REVIEW PROCEDURES FOR CERTAIN KINDS OF RESEARCH INVOLVING NO MORE THAN MINIMAL RISK, AND FOR MINOR CHANGES IN APPROVED RESEARCH IRB Functions And Operations; Institutional Review Boards
MINIMUM INFORMATION	205.5	MINIMUM REQUIRED INFORMATION FOR LICENSURE Guidelines For State Licensing Of Wholesale Prescription Drug Distributors
MINIMUM QUALIFICATIONS	205.6	MINIMUM QUALIFICATIONS Guidelines For State Licensing Of Wholesale Prescription Drug Distributors
MINIMUM REQUIREMENTS	205.50	MINIMUM REQUIREMENTS FOR STORAGE AND HANDLING OF PRESCRIPTION DRUGS AND ESTABLISHMENT AND MAINTENANCE OF PRESCRIPTION DRUG DISTRIBUTION RECORDS Guidelines For State Licensing Of Wholesale Prescription Drug Distributors
MINNESOTA	808.73	MINNESOTA Listing Of Specific State And Local Exemptions; Exemption From Federal Preemption Of State & Local Medical Device Requirements
MINOR CHANGES	56.110	EXPEDITED REVIEW PROCEDURES FOR CERTAIN KINDS OF RESEARCH INVOLVING NO MORE THAN MINIMAL RISK, AND FOR MINOR CHANGES IN APPROVED RESEARCH IRB Functions And Operations; Institutional Review Boards
MINOR VIOLATIONS	5.33	ISSUANCE OF REPORTS OF MINOR VIOLATIONS General Redelegations Of Authority; Delegations Of Authority And Organization
MINUTES	14.60	MINUTES AND REPORTS OF ADVISORY COMMITTEE MEETINGS Records Of Meetings And Hearings Before Advisory Committees; Public Hearing Before A Public Advisory Committee
MIRROR	886.1500	HEADBAND MIRROR Diagnostic Devices; Ophthalmic Devices

MISBRANDED	5.402	DETENTION OF ADULTERATED OR MISBRANDED MEDICAL DEVICES Medical Devices And Radiological Health; Redelegations Of Authority; Delegations Of Authority And Organization
MISBRANDED	250.12	STRAMONIUM PREPARATIONS LABELED WITH DIRECTIONS FOR USE IN SELF-MEDICATION REGARDED AS MISBRANDED Drugs Regarded As Misbranded; Specific Requirements For Specific Human Drugs
MISBRANDED	330.1	GENERAL CONDITIONS FOR GENERAL RECOGNITION AS SAFE, EFFECTIVE AND NOT MISBRANDED General Provisions; Over-the-counter Human Drugs Generally Recognized As Safe And Effective & Not Misbranded
MISBRANDED	330.10	PROCEDURES FOR CLASSIFYING OTC DRUGS AS GENERALLY RECOGNIZED AS SAFE AND EFFECTIVE AND NOT MISBRANDED, AND FOR ESTABLISHING MONOGRAPHS Administrative Procedures; Over-the-counter Human Drugs Generally Recognized As Safe And Effective & Not Misbranded
MISBRANDED	330.14	ADDITIONAL CRITERIA AND PROCEDURES FOR CLASSIFYING OTC DRUGS AS GENERALLY RECOGNIZED AS SAFE AND EFFECTIVE AND NOT MISBRANDED Administrative Procedures; Over-the-counter Human Drugs Generally Recognized As Safe And Effective & Not Misbranded
MISBRANDED	601.25	REVIEW PROCEDURES TO DETERMINE THAT LICENSED BIOLOGICAL PRODUCTS ARE SAFE, EFFECTIVE, NOT MISBRANDED UNDER PRESCRIBED, RECOMMENDED, OR SUGGESTED CONDITIONS OF USE Biologics Licensing; Licensing
MISBRANDED	601.26	RECLASSIFICATION PROCEDURES TO DETERMINE THAT LICENSED BIOLOGICAL PRODUCTS ARE SAFE, EFFECTIVE, NOT MISBRANDED UNDER PRESCRIBED, RECOMMENDED, OR SUGGESTED CONDITIONS OF USE Biologics Licensing; Licensing
MISBRANDING	99.405	APPLICABILITY OF LABELING, ADULTERATION, AND MISBRANDING AUTHORITY Dissemination Of Information On Unapproved/New Uses For Marketed Drugs, Biologics, And Devices
MISBRANDING	101.18	MISBRANDING OF FOOD General Provisions; Food Labeling
MISBRANDING	207.39	MISBRANDING BY REFERENCE TO REGISTRATION OR TO REGISTRATION NUMBER Procedures For Domestic Drug Establishments; Registration Of Drug Producers And Listing Of Drugs In Commercial Distribution
MISBRANDING	314.170	ADULTERATION AND MISBRANDING OF AN APPROVED DRUG FDA Action On Applications; Applications For FDA Approval To Market A New Drug
MISBRANDING	500.51	LABELING OF ANIMAL DRUGS; MISBRANDING Animal Drug Labeling Requirements; General
MISBRANDING	501.18	MISBRANDING OF ANIMAL FOOD General Provisions; Animal Food Labeling
MISBRANDING	607.39	MISBRANDING BY REFERENCE TO ESTABLISHMENT REGISTRATION OR TO REGISTRATION NUMBER Procedures For Domestic Blood Product Establishments; Establishment Registratn+product Listing, Manufacturers Human Blood/b.products
MISBRANDING	701.1	MISBRANDING General Provisions; Cosmetic Labeling
MISBRANDING	710.8	MISBRANDING BY REFERENCE TO REGISTRATION OR TO REGISTRATION NUMBER Voluntary Registration Of Cosmetic Product Establishments
MISBRANDING	720.9	MISBRANDING BY REFERENCE TO FILING OR TO STATEMENT NUMBER Voluntary Filing Of Cosmetic Product Ingredient Composition Statements
MISBRANDING	807.39	MISBRANDING BY REFERENCE TO ESTABLISHMENT REGISTRATION OR TO REGISTRATION NUMBER Procedures For Device Establishments; Establishment Registration And Device Listing For Manufacturers And Initial Importers Of Devices
MISBRANDING	807.97	MISBRANDING BY REFERENCE TO PREMARKET NOTIFICATION Premarket Notification Procedures; Establishment Registration And Device Listing For Manufacturers And Initial Importers Of Devices
MISLEADING CONTAINERS	100.100	MISLEADING CONTAINERS Misbranding For Reasons Other Than Labeling; General
MISLEADING LABELING	5.109	ISSUANCE OF WRITTEN NOTICES CONCERNING PATENT INFORMATION, CURRENT GOOD MANUFACTURING PRACTICES AND FALSE OR MISLEADING LABELING OF NEW DRUGS Human Drugs; Redelegations Of Authority; Delegations Of Authority And Organization

MISLEADING LABELING	5.504	ISSUANCE OF WRITTEN NOTICES CONCERNING PATENT INFORMATION, CURRENT GOOD MANUFACTURING PRACTICES AND FALSE OR MISLEADING LABELING OF NEW ANIMAL DRUGS AND FEEDS BEARING OR CONTAINING NEW ANIMAL DRUGS Animal Drugs; Redelegations Of Authority; Delegations Of Authority And Organization
MISLEADING STATEMENTS	201.6	DRUGS; MISLEADING STATEMENTS General Labeling Provisions; Labeling
MISLEADING STATEMENTS	801.6	MEDICAL DEVICES; MISLEADING STATEMENTS General Labeling Provisions; Labeling
MISSISSIPPI	808.74	MISSISSIPPI Listing Of Specific State And Local Exemptions; Exemption From Federal Preemption Of State & Local Medical Device Requirements
MITIGATION	1316.79	PETITIONS FOR REMISSION OR MITIGATION OF FORFEITURE Seizure, Forfeiture, And Disposition Of Property; Administrative Functions, Practices, And Procedures
MIXED DAIRY CHOCOLATES	163.145	MIXED DAIRY PRODUCT CHOCOLATES Requirements For Specific Standardized Cacao Products; Cacao Products
MIXED ENZYME PRODUCT	184.1027	MIXED CARBOHYDRASE AND PROTEASE ENZYME PRODUCT Listing Of Specific Substances Affirmed As GRAS; Direct Food Substances Affirmed As Generally Recognized As Safe
MIXED NUTS	164.110	MIXED NUTS Requirements For Specific Standardized Tree Nut And Peanut Products; Tree Nut And Peanut Products
MIXER	868.5330	BREATHING GAS MIXER Therapeutic Devices; Anesthesiology Devices
MIXER	888.4210	CEMENT MIXER FOR CLINICAL USE Surgical Devices; Orthopedic Devices
MIXING	864.9195	BLOOD MIXING DEVICES AND BLOOD WEIGHING DEVICES Products Used In Establishments That Manufacture Blood And Blood Products; Hematology And Pathology Devices
MIXTURE	58.113	MIXTURE OF ARTICLES WITH CARRIERS Test And Control Articles; Good Laboratory Practice For Nonclinical Laboratory Studies
MIXTURE	864.8625	HEMATOLOGY QUALITY CONTROL MIXTURE Hematology Reagents; Hematology And Pathology Devices
MIXTURES	73.1	DILUENTS IN COLOR ADDITIVE MIXTURES FOR FOOD USE EXEMPT FROM CERTIFICATION Foods; Listing Of Color Additives Exempt From Certification
MIXTURES	73.1001	DILUENTS IN COLOR ADDITIVE MIXTURES FOR DRUG USE EXEMPT FROM CERTIFICATION Drugs; Listing Of Color Additives Exempt From Certification
MIXTURES	80.35	COLOR ADDITIVE MIXTURES; CERTIFICATION AND EXEMPTION FROM CERTIFICATION Certification Procedures; Color Additive Certification
MIXTURES	82.6	CERTIFIABLE MIXTURES General Provisions; Listing Of Certified Provisionally Listed Colors And Specifications
MIXTURES	102.37	MIXTURES OF EDIBLE FAT OR OIL AND OLIVE OIL Requirements For Specific Nonstandardized Foods; Common Or Usual Name For Nonstandardized Foods
MOBILE	892.1720	MOBILE X-RAY SYSTEM Diagnostic Devices; Radiology Devices
MODIFICATION	10.19	WAIVER, SUSPENSION, OR MODIFICATION OF PROCEDURAL REQUIREMENTS General Provisions; Administrative Practices And Procedures
MODIFICATION	12.26	MODIFICATION OR REVOCATION OF REGULATION OR ORDER Initiation Of Proceedings; Formal Evidentiary Public Hearing
MODIFICATION	312.42	CLINICAL HOLDS AND REQUESTS FOR MODIFICATION Investigational New Drug Application
MODIFICATION	640.74	MODIFICATION OF SOURCE PLASMA Source Plasma; Additional Standards For Human Blood And Blood Products
MODIFICATION	1301.51	MODIFICATION IN REGISTRATION Registration Of Manufacturers, Distributors, And Dispensers Of Controlled Substances
MODIFICATION	1303.33	WAIVER OF MODIFICATION OF RULES Hearings; Quotas
MODIFICATION	1309.61	MODIFICATION IN REGISTRATION Registration Of Manufacturers, Distributors, Importers And Exporters Of List I Chemicals
MODIFICATION	1312.43	WAIVER OR MODIFICATION OF RULES Hearings; Importation And Exportation Of Controlled Substances
MODIFICATION	1313.53	WAIVER OR MODIFICATION OF RULES Importation And Exportation Of Precursors And Essential Chemicals

MODIFICATION INSTRUMENT	874.3540	PROSTHESIS MODIFICATION INSTRUMENT FOR OSSICULAR REPLACEMENT SURGERY Prosthetic Devices; Ear, Nose And Throat Devices
MODIFICATIONS	640.6	MODIFICATIONS OF WHOLE BLOOD Whole Blood; Additional Standards For Human Blood And Blood Products
MODIFICATIONS	640.17	MODIFICATIONS FOR SPECIFIC PRODUCTS Red Blood Cells; Additional Standards For Human Blood And Blood Products
MODIFICATIONS, TEST	331.21	TEST MODIFICATIONS Testing Procedures; Antacid Products For Over-the-counter (OTC) Human Use
MODIFICATIONS, TEST	352.77	TEST MODIFICATIONS Testing Procedures; Sunscreen Drug Products For OTC Human Use
MODIFIERS	178.3790	POLYMER MODIFIERS IN SEMIRIGID AND RIGID VINYL CHLORIDE PLASTICS Certain Adjuvants And Production Aids; Indirect Food Additives: Adjuvants, Production Aids, And Sanitizers
MOIST HEAT PACK	890.5730	MOIST HEAT PACK Physical Medicine Therapeutic Devices; Physical Medicine Devices
MOIST STEAM CABINET	890.5250	MOIST STEAM CABINET Physical Medicine Therapeutic Devices; Physical Medicine Devices
MOISTURE	133.154	HIGH-MOISTURE JACK CHEESE Requirements For Specific Standardized Cheese And Related Products; Cheeses And Related Cheese Products
MOISTURE	868.5375	HEAT AND MOISTURE CONDENSER (ARTIFICIAL NOSE) Therapeutic Devices; Anesthesiology Devices
MOLD	874.3430	MIDDLE EAR MOLD Prosthetic Devices; Ear, Nose And Throat Devices
MOLD CHEESE, BLUE	133.184	ROQUEFORT, SHEEP'S MILK BLUE-MOLD, AND BLUE-MOLD CHEESE FROM SHEEP'S MILK Requirements For Specific Standardized Cheese And Related Products; Cheeses And Related Cheese Products
MOLDED ARTICLES	177.1460	MELAMINE-FORMALDEHYDE RESINS IN MOLDED ARTICLES Substances For Use As Basic Components Of Single And Repeated Use Food Contact Surfaces; Indirect Food Additives: Polymers
MOLDED ARTICLES	177.1900	UREA-FORMALDEHYDE RESINS IN MOLDED ARTICLES Substances For Use As Basic Components Of Single And Repeated Use Food Contact Surfaces; Indirect Food Additives: Polymers
MOLDED ARTICLES	177.2410	PHENOLIC RESINS IN MOLDED ARTICLES Substances For Use Only As Components Of Articles Intended For Repeated Use; Indirect Food Additives: Polymers
MOLECULAR SIEVE RESINS	173.40	MOLECULAR SIEVE RESINS Polymer Substances For Food Treatment; Secondary Direct Food Additives Permitted In Food For Human Consumption
MOLECULAR WEIGHT	172.820	POLYETHYLENE GYLCOL (MEAN MOLECULAR WEIGHT 200-9,500) Multipurpose Additives; Food Additives Permitted For Direct Addition To Food For Human Consumption
MOLECULAR WEIGHT	177.1440	4,4'-ISOPROPYLIDENEDIPHENOL-EPICHLOROHYDRIN RESINS MINIMUM MOLECULAR WEIGHT 10,000 Substances For Use As Basic Components Of Single And Repeated Use Food Contact Surfaces; Indirect Food Additives: Polymers
MOLECULAR WEIGHT	178.3750	POLYETHYLENE GLYCOL (MEAN MOLECULAR WEIGHT 200-9,500) Certain Adjuvants And Production Aids; Indirect Food Additives: Adjuvants, Production Aids, And, Sanitizers
MOLLUSCAN	1240.60	MOLLUSCAN SHELLFISH Specific Administrative Decisions Regarding Interstate Shipments; Control Of Communicable Diseases
MOMETASONE	524.1044h	GENTAMICIN SULFATE, MOMETASONE FUROATE, CLOTRIMAZOLE OTIC SUSPENSION Ophthalmic And Topical Dosage Form New Animal Drugs
MONENSIN	520.1448	MONENSIN ORAL DOSAGE FORMS Oral Dosage Form New Animal Drugs
MONENSIN	556.420	MONENSIN Specific Tolerances For Residues Of New Animal Drugs; Tolerances For Residues Of New Animal Drugs In Food
MONENSIN	558.355	MONENSIN Specific New Animal Drugs For Use In Animal Feeds; New Animal Drugs For Use In Animal Feeds
MONITOR	866.2560	MICROBIAL GROWTH MONITOR Microbiology Devices; Immunology And Microbiology Devices
MONITOR	868.2025	ULTRASONIC AIR EMBOLISM MONITOR Monitoring Devices; Anesthesiology Devices
MONITOR	868.2375	BREATHING FREQUENCY MONITOR Monitoring Devices; Anesthesiology Devices
MONITOR	868.2450	LUNG WATER MONITOR Monitoring Devices; Anesthesiology Devices

MONITOR	868.2480	CUTANEOUS CARBON DIOXIDE (PCCO2) MONITOR Monitoring Devices; Anesthesiology Devices
MONITOR	868.2500	CUTANEOUS OXYGEN MONITOR Monitoring Devices; Anesthesiology Devices
MONITOR	868.2600	AIRWAY PRESSURE MONITOR Monitoring Devices; Anesthesiology Devices
MONITOR	870.2300	CARDIAC MONITOR (INCLUDING CARDIOTACHOMETER AND RATE ALARM) Cardiovascular Monitoring Devices; Cardiovascular Devices
MONITOR	870.2620	LINE ISOLATION MONITOR Cardiovascular Monitoring Devices; Cardiovascular Devices
MONITOR	870.4330	CARDIOPULMONARY BYPASS ON-LINE BLOOD GAS MONITOR Cardiovascular Surgical Devices; Cardiovascular Devices
MONITOR	870.4340	CARDIOPULMONARY BYPASS LEVEL SENSING MONITOR AND/OR CONTROL Cardiovascular Surgical Devices; Cardiovascular Devices
MONITOR	880.2400	BED-PATIENT MONITOR General Hospital And Personal Use Monitoring Devices; General Hospital And Personal Use Devices
MONITOR	880.2420	ELECTRONIC MONITOR FOR GRAVITY FLOW INFUSION SYSTEMS General Hospital And Personal Use Monitoring Devices; General Hospital And Personal Use Devices
MONITOR	880.2460	ELECTRICALLY POWERED SPINAL FLUID PRESSURE MONITOR General Hospital And Personal Use Monitoring Devices; General Hospital And Personal Use Devices
MONITOR	882.1610	ALPHA MONITOR Neurological Diagnostic Devices; Neurological Devices
MONITOR	882.5500	LESION TEMPERATURE MONITOR Neurological Therapeutic Devices; Neurological Devices
MONITOR	884.2600	FETAL CARDIAC MONITOR Obstetrical And Gynecological Monitoring Devices; Obstetrical And Gynecological Devices
MONITOR	884.2620	FETAL ENCEPHALOGRAPHIC MONITOR Obstetrical And Gynecological Monitoring Devices; Obstetrical And Gynecological Devices
MONITOR	884.2640	FETAL PHONOCARDIOGRAPHIC MONITOR Obstetrical And Gynecological Monitoring Devices; Obstetrical And Gynecological Devices
MONITOR	884.2660	FETAL ULTRASONIC MONITOR AND ACCESSORIES Obstetrical And Gynecological Monitoring Devices; Obstetrical And Gynecological Devices
MONITOR	884.2700	INTRAUTERINE PRESSURE MONITOR AND ACCESSORIES Obstetrical And Gynecological Monitoring Devices; Obstetrical And Gynecological Devices
MONITOR	884.2720	EXTERNAL UTERINE CONTRACTION MONITOR AND ACCESSORIES Obstetrical And Gynecological Monitoring Devices; Obstetrical And Gynecological Devices
MONITOR	884.2730	HOME UTERINE ACTIVITY MONITOR Obstetrical And Gynecological Monitoring Devices; Obstetrical And Gynecological Devices
MONITOR	886.1510	EYE MOVEMENT MONITOR Diagnostic Devices; Ophthalmic Devices
MONITOR	892.1540	NONFETAL ULTRASONIC MONITOR Diagnostic Devices; Radiology Devices
MONITORING	26.15	MONITORING CONTINUED EQUIVALENCE Specific Sector Provisions For Pharmaceutical GMPs; Mutual Recognition Of Pharmaceutical GMP Reports, Medical Device Quality System Audit Reports
MONITORING	26.45	MONITORING CONTINUED EQUIVALENCE Specific Sector Provisions For Medical Devices; Mutual Recognition Of Pharmaceutical GMP Reports, Medical Device Quality System Audit Reports
MONITORING	26.69	MONITORING OF CONFORMITY ASSESSMENT BODIES Framework Provisions; Mutual Recognition Of Pharmaceutical GMP Reports, Medical Device Quality System Audit Reports
MONITORING	312.87	ACTIVE MONITORING OF CONDUCT AND EVALUATION OF CLINICAL TRIALS Drugs Intended To Treat Life-threatening And Severely-debilitating Illnesses; Investigational New Drug Application
MONITORING	812.46	MONITORING INVESTIGATIONS Responsibilities Of Sponsors; Investigational Device Exemptions
MONITORING	868.1850	MONITORING SPIROMETER Diagnostic Devices; Anesthesiology Devices
MONITORING	876.1725	GASTROINTESTINAL MOTILITY MONITORING SYSTEM Diagnostic Devices; Gastroenterology-urology Devices
MONITORING	882.1620	INTRACRANIAL PRESSURE MONITORING DEVICE Neurological Diagnostic Devices; Neurological Devices

MONITORING	884.2740	PERINATAL MONITORING SYSTEM AND ACCESSORIES Obstetrical And Gynecological Monitoring Devices; Obstetrical And Gynecological Devices
MONITORING	1403.40	MONITORING AND REPORTING PROGRAM PERFORMANCE Reports, Records, Retention, And Enforcement; Uniform Administrative Requirements For Grants And Coop. Agreements To State And Local Gov.
MONITORS	312.53	SELECTING INVESTIGATORS AND MONITORS Investigational New Drug Application
MONITORS	812.43	SELECTING INVESTIGATORS AND MONITORS Responsibilities Of Sponsors; Investigational Device Exemptions
MONO- AND DIESTERS	172.856	PROPYLENE GLYCOL MONO- AND DIESTERS OF FATS AND FATTY ACIDS Multipurpose Additives; Food Additives Permitted For Direct Addition To Food For Human Consumption
MONO- AND DIGLYCERIDES	172.834	ETHOXYLATED MONO- AND DIGLYCERIDES Multipurpose Additives; Food Additives Permitted For Direct Addition To Food For Human Consumption
MONO- AND DIGLYCERIDES	184.1101	DIACETYL TARTARIC ACID ESTERS OF MONO- AND DIGLYCERIDES Listing Of Specific Substances Affirmed As GRAS; Direct Food Substances Affirmed As Generally Recognized As Safe
MONO- AND DIGLYCERIDES	184.1505	MONO- AND DIGLYCERIDES Listing Of Specific Substances Affirmed As GRAS; Direct Food Substances Affirmed As Generally Recognized As Safe
MONO- AND DIGLYCERIDES	184.1521	MONOSODIUM PHOSPHATE DERIVATIVES OF MONO- AND DIGLYCERIDES Listing Of Specific Substances Affirmed As GRAS; Direct Food Substances Affirmed As Generally Recognized As Safe
MONO- AND DIGLYCERIDES	582.4101	DIACETYL TARTARIC ACID ESTERS OF MONO- AND DIGLYCERIDES OF EDIBLE FATS OR OILS, OR EDIBLE FAT-FORMING FATTY ACIDS Emulsifying Agents; Substances Generally Recognized As Safe
MONO- AND DIGLYCERIDES	582.4505	MONO- AND DIGLYCERIDES OF EDIBLE FATS OR OILS, OR EDIBLE FAT-FORMING ACIDS Emulsifying Agents; Substances Generally Recognized As Safe
MONO- AND DIGLYCERIDES	582.4521	MONOSODIUM PHOSPHATE DERIVATIVES OF MONO- AND DIGLYCERIDES OF EDIBLE FATS OR OILS, OR EDIBLE FAT-FORMING FATTY ACIDS Emulsifying Agents; Substances Generally Recognized As Safe
MONO- AND DIMETHYL NAPHTHALENE	172.824	SODIUM MONO- AND DIMETHYL NAPHTHALENE SULFONATES Multipurpose Additives; Food Additives Permitted For Direct Addition To Food For Human Consumption
MONO- AND DIOLEATE	573.800	POLYETHYLENE GLYCOL (400) MONO- AND DIOLEATE Food Additive Listing; Food Additives Permitted In Feed And Drinking Water Of Animals
MONO- AND DIOLEATES	573.820	POLYOXYETHYLENE GLYCOL (400) MONO- AND DIOLEATES Food Additive Listing; Food Additives Permitted In Feed And Drinking Water Of Animals
MONOAMMONIUM GLUTAMATE	182.1500	MONOAMMONIUM GLUTAMATE Multiple Purpose GRAS Food Substances; Substances Generally Recognized As Safe
MONOAMMONIUM GLUTAMATE	582.1500	MONOAMMONIUM GLUTAMATE General Purpose Food Additives; Substances Generally Recognized As Safe
MONOBASIC	178.3780	POLYHYDRIC ALCOHOL ESTERS OF LONG CHAIN MONOBASIC ACIDS Certain Adjuvants And Production Aids; Indirect Food Additives: Adjuvants, Production Aids, And Sanitizers
MONOBASIC	182.6215	MONOBASIC CALCIUM PHOSPHATE Sequestrants; Substances Generally Recognized As Safe
MONOBASIC	184.1141	AMMONIUM PHOSPHATE, MONOBASIC Listing Of Specific Substances Affirmed As GRAS; Direct Food Substances Affirmed As Generally Recognized As Safe
MONOBASIC	582.6215	MONOBASIC CALCIUM PHOSPHATE Sequestrants; Substances Generally Recognized As Safe
MONOCHLOROACETIC ACID	189.155	MONOCHLOROACETIC ACID Substances Generally Prohibited From Direct Addition Or Use As Human Food; Substances Prohibited From Use In Human Food
MONOGLYCERIDE	172.832	MONOGLYCERIDE CITRATE Multipurpose Additives; Food Additives Permitted For Direct Addition To Food For Human Consumption
MONOGLYCERIDES	172.828	ACETYLATED MONOGLYCERIDES Multipurpose Additives; Food Additives Permitted For Direct Addition To Food For Human Consumption
MONOGLYCERIDES	172.830	SUCCINYLATED MONOGLYCERIDES Multipurpose Additives; Food Additives Permitted For Direct Addition To Food For Human Consumption
MONOGLYCERYL CITRATE	172.755	STEARYL MONOGLYCERIDYL CITRATE Other Specific Usage Additives; Food Additives Permitted For Direct Addition To Food For Human Consumption

MONOGRAPH	330.11	NDA DEVIATIONS FROM APPLICABLE MONOGRAPH Administrative Procedures; Over-the-counter Human Drugs Generally Recognized As Safe And Effective & Not Misbranded
MONOGRAPHS	330.10	PROCEDURES FOR CLASSIFYING OTC DRUGS AS GENERALLY RECOGNIZED AS SAFE AND EFFECTIVE AND NOT MISBRANDED, AND ESTABLISHING MONOGRAPHS Administrative Procedures; Over-the-counter Human Drugs Generally Recognized As Safe And Effective & Not Misbranded
MONOHYDRATE, DEXTROSE	168.111	DEXTROSE MONOHYDRATE Requirements For Specific Standardized Sweeteners And Table Sirups; Sweeteners And Table Sirups
MONOHYDROCHLORIDE	184.1272	L-CYSTEINE MONOHYDROCHLORIDE Listing Of Specific Substances Affirmed As GRAS; Direct Food Substances Affirmed As Generally Recognized As Safe
MONOISOPROPYL CITRATE	582.6511	MONOISOPROPYL CITRATE Sequestrants; Substances Generally Recognized As Safe
MONOLAURATE	178.3760	POLYETHYLENE GLYCOL (400) MONOLAURATE Certain Adjuvants And Production Aids; Indirect Food Additives: Adjuvants, Production Aids, And Sanitizers
MONOMER	864.7300	FIBRIN MONOMER PARACOAGULATION TEST Hematology Kits And Packages; Hematology And Pathology Devices
MONOMER	888.4220	CEMENT MONOMER VAPOR EVACUATOR Surgical Devices; Orthopedic Devices
MONONITRATE	184.5878	THIAMINE MONONITRATE Listing Of Specific Substances Affirmed As GRAS; Direct Food Substances Affirmed As Generally Recognized As Safe
MONONITRATE	582.5878	THIAMINE MONONITRATE Nutrients And/or Dietary Supplements; Substances Generally Recognized As Safe
MONONUCLEOSIS	866.5640	INFECTIOUS MONONUCLEOSIS IMMUNOLOGICAL TEST SYSTEM Immunological Test Systems; Immunology And Microbiology Devices
MONOOLEATE	173.75	SORBITAN MONOOLEATE Polymer Substances And Polymer Adjuvants For Food Treatment; Secondary Direct Food Additives Permitted In Food Human Consumption
MONOOLEATE	184.1323	GLYCERYL MONOOLEATE Listing Of Specific Substances Affirmed As GRAS; Direct Food Substances Affirmed As Generally Recognized As Safe
MONOPOTASSIUM GLUTAMATE	182.1516	MONOPOTASSIUM GLUTAMATE Multiple Purpose GRAS Food Substances; Substances Generally Recognized As Safe
MONOPOTASSIUM GLUTAMATE	582.1516	MONOPOTASSIUM GLUTAMATE General Purpose Food Additives; Substances Generally Recognized As Safe
MONOSODIUM PHOSPHATE	184.1521	MONOSODIUM PHOSPHATE DERIVATIVES OF MONO- AND DIGLYCERIDES Listing Of Specific Substances Affirmed As GRAS; Direct Food Substances Affirmed As Generally Recognized As Safe
MONOSODIUM PHOSPHATE	582.4521	MONOSODIUM PHOSPHATE DERIVATIVES OF MONO- AND DIGLYCERIDES OF EDIBLE FATS OR OILS, OR EDIBLE FAT-FORMING FATTY ACIDS Emulsifying Agents; Substances Generally Recognized As Safe
MONOSTEARATE	172.842	SORBITAN MONOSTEARATE Multipurpose Additives; Food Additives Permitted For Direct Addition To Food For Human Consumption
MONOSTEARATE	184.1324	GLYCERYL MONOSTEARATE Listing Of Specific Substances Affirmed As GRAS; Direct Food Substances Affirmed As Generally Recognized As Safe
MONOSTEARATE	582.1324	GLYCERYL MONOSTEARATE General Purpose Food Additives; Substances Generally Recognized As Safe
MONOXIDE, CARBON	177.1312	ETHYLENE-CARBON MONOXIDE COPOLYMERS Substances For Use As Basic Components Of Single And Repeated Use Food Contact Surfaces; Indirect Food Additives: Polymers
MONOXIDE, CARBON	862.3220	CARBON MONOXIDE TEST SYSTEM Clinical Toxicology Test Systems; Clinical Chemistry And Clinical Toxicology Devices
MONOXIDE, CARBON	868.1430	CARBON MONOXIDE ANALYZER Diagnostic Devices; Anesthesiology Devices
MONTAN WAX ACIDS	178.3770	POLYHYDRIC ALCOHOL ESTERS OF OXIDATIVELY REFINED (GERSTHOFEN PROCESS) MONTAN WAX ACIDS Certain Adjuvants And Production Aids; Indirect Food Additives: Adjuvants, Production Aids, And Sanitizers
MONTEREY JACK CHEESE	133.153	MONTEREY CHEESE AND MONTEREY JACK CHEESE Requirements For Specific Standardized Cheese And Related Products; Cheeses And Related Cheese Products
MORANTEL TARTRATE	520.1450	MORANTEL TARTRATE ORAL DOSAGE FORMS Oral Dosage Form New Animal Drugs

MORANTEL TARTRATE	520.1450c	MORANTEL TARTRATE SUSTAINED-RELEASE TRILAMINATE CYLINDER/SHEET Oral Dosage Form New Animal Drugs
MORANTEL TARTRATE	556.425	MORANTEL TARTRATE Specific Tolerances For Residues Of New Animal Drugs; Tolerances For Residues Of New Animal Drugs In Food
MORANTEL TARTRATE	558.360	MORANTEL TARTRATE Specific New Animal Drugs For Use In Animal Feeds; New Animal Drugs For Use In Animal Feed
MORBIDITY	314.510	APPROVAL BASED ON A SURROGATE ENDPOINT OR ON AN EFFECT ON A CLINICAL ENDPOINT OTHER THAN SURVIVAL OR IRREVERSIBLE MORBIDITY Applications For FDA Approval To Market A New Drug
MORBIDITY	601.41	APPROVAL BASED ON A SURROGATE ENDPOINT OR ON AN EFFECT ON A CLINICAL ENDPOINT OTHER THAN SURVIVAL OR IRREVERSIBLE MORBIDITY Accelerated Approval Of Biological Products; Licensing
MORE	101.54	NUTRIENT CONTENT CLAIMS FOR "GOOD SOURCE," "HIGH," "MORE," AND "HIGH POTENCY" Specific Requirements For Nutrient Content Claims; Food Labeling
MORPHINE	862.3640	MORPHINE TEST SYSTEM Clinical Toxicology Test Systems; Clinical Chemistry And Clinical Toxicology Devices
MORPHOLINE	172.235	MORPHOLINE Coatings, Films And Related Substances; Food Additives Permitted For Direct Addition To Food For Human Consumption
MORTIERELLA VINACEAE	173.145	ALPHA-GALACTOSIDASE DERIVED FROM MORTIERELLA VINACEAE VAR. RAFFINOSEUTILIZER Enzyme Preparations And Microorganisms; Secondary Direct Food Additives Permitted In Food For Human Consumption
MOTILITY	876.1725	GASTROINTESTINAL MOTILITY MONITORING SYSTEM Diagnostic Devices; Gastroenterology-urology Devices
MOTIONS	12.99	MOTIONS Hearing Procedures; Formal Evidentiary Public Hearing
MOTIONS	17.32	MOTIONS Civil Money Penalties Hearings
MOTOR	882.4360	ELECTRIC CRANIAL DRILL MOTOR Neurological Surgical Devices; Neurological Devices
MOTOR	882.4370	PNEUMATIC CRANIAL DRILL MOTOR Neurological Surgical Devices; Neurological Devices
MOTORIZED	890.3800	MOTORIZED THREE-WHEELED VEHICLE Physical Medicine Prosthetic Devices; Physical Medicine Devices
MOTORS	878.4820	SURGICAL INSTRUMENT MOTORS AND ACCESSORIES/ATTACHMENTS Surgical Devices; General And Plastic Surgery Devices
MOUNT, X-RAY TUBE	892.1770	DIAGNOSTIC X-RAY TUBE MOUNT Diagnostic Devices; Radiology Devices
MOUTHPIECE	868.5620	BREATHING MOUTHPIECE Therapeutic Devices; Anesthesiology Devices
MOXIDECTIN	520.1451	MOXIDECTIN Oral Dosage Form New Animal Drugs
MOXIDECTIN	520.1452	MOXIDECTIN GEL Oral Dosage Form New Animal Drugs
MOXIDECTIN	522.1451	MOXIDECTIN Implantation Or Injectable Dosage Form New Animal Drugs
MOXIDECTIN	524.1451	MOXIDECTIN Ophthalmic And Topical Dosage Form New Animal Drugs
MOXIDECTIN	556.426	MOXIDECTIN Specific Tolerances For Residues Of New Animal Drugs; Tolerances For Residues Of New Animal Drugs In Food
MOZZARELLA CHEESE	133.155	MOZZARELLA CHEESE AND SCAMORZA CHEESE Requirements For Specific Standardized Cheese And Related Products; Cheeses And Related Cheese Products
MOZZARELLA CHEESE	133.156	LOW-MOISTURE MOZZARELLA AND SCAMORZA CHEESE Requirements For Specific Standardized Cheese And Related Products; Cheeses And Related Cheese Products
MOZZARELLA CHEESE	133.157	PART-SKIM MOZZARELLA AND SCAMORZA CHEESE Requirements For Specific Standardized Cheese And Related Products; Cheeses And Related Cheese Products
MOZZARELLA CHEESE	133.158	LOW-MOISTURE PART-SKIM MOZZARELLA AND SCAMORZA CHEESE Requirements For Specific Standardized Cheese And Related Products; Cheeses And Related Cheese Products
MUCPOLYSACCHARIDES	862.1505	MUCPOLYSACCHARIDES (NONQUANTITATIVE) TEST SYSTEM Clinical Chemistry Test Systems; Clinical Chemistry And Clinical Toxicology Devices

MUCOR MIEHEI	173.140	ESTERASE-LIPASE DERIVED FROM MUCOR MIEHEI Enzyme Preparations And Microorganisms; Secondary Direct Food Additives Permitted In Food For Human Consumption
MUCUS	884.1040	VISCOMETER FOR CERVICAL MUCUS Obstetrical And Gynecological Diagnostic Devices; Obstetrical And Gynecological Devices
MUENSTER CHEESE	133.160	MUENSTER AND MUNSTER CHEESE Requirements For Specific Standardized Cheese And Related Products; Cheeses And Related Cheese Products
MUENSTER CHEESE	133.161	MUENSTER AND MUNSTER CHEESE FOR MANUFACTURING Requirements For Specific Standardized Cheese And Related Products; Cheeses And Related Cheese Products
MULTIPLE AUTOANTIBODIES	866.5660	MULTIPLE AUTOANTIBODIES IMMUNOLOGICAL TEST SYSTEM Immunological Test Systems; Immunology And Microbiology Devices
MULTIPLE-DOSE	320.27	GUIDELINES ON THE DESIGN OF A MULTIPLE-DOSE IN VIVO BIOAVAILABILITY STUDY Procedures For Determining The Bioavailability Or Bioequivalence Of Drug Products; Bioavailability And Bioequivalence Requirements
MULTIPURPOSE CULTURE MEDIUM	866.2300	MULTIPURPOSE CULTURE MEDIUM Microbiology Devices; Immunology And Microbiology Devices
MULTIPURPOSE SYSTEM	864.5425	MULTIPURPOSE SYSTEM FOR IN VITRO COAGULATION STUDIES Automated And Semi-automated Hematology Devices; Hematology And Pathology Devices
MUMPS VIRUS	866.3380	MUMPS VIRUS SEROLOGICAL REAGENTS Serological Reagents; Immunology And Microbiology Devices
MUNSTER CHEESE	133.160	MUENSTER AND MUNSTER CHEESE Requirements For Specific Standardized Cheese And Related Products; Cheeses And Related Cheese Products
MUNSTER CHEESE	133.161	MUENSTER AND MUNSTER CHEESE FOR MANUFACTURING Requirements For Specific Standardized Cheese And Related Products; Cheeses And Related Cheese Products
MUPIROCIN	524.1465	MUPIROCIN OINTMENT Ophthalmic And Topical Dosage Form New Animal Drugs
MURAMIDASE	862.1490	LYSOZYME (MURAMIDASE) TEST SYSTEM Clinical Chemistry Test Systems; Clinical Chemistry And Clinical Toxicology Devices
MUSCLE	866.5120	ANTISMOOTH MUSCLE ANTIBODY IMMUNOLOGICAL TEST SYSTEM Immunological Test Systems; Immunology And Microbiology Devices
MUSCLE CRAMPS	310.546	DRUG PRODUCTS CONTAINING ACTIVE INGREDIENTS OFFERED OTC FOR THE TREATMENT AND/OR PREVENTION OF NOCTURNAL LEG MUSCLE CRAMPS Requirements For Specific New Drugs Or Devices; New Drugs
MUSCLE STIMULATOR	884.5940	POWERED VAGINAL MUSCLE STIMULATOR FOR THERAPEUTIC USE Obstetrical And Gynecological Therapeutic Devices; Obstetrical And Gynecological Devices
MUSCLE STIMULATOR	890.1850	DIAGNOSTIC MUSCLE STIMULATOR Physical Medicine Diagnostic Devices; Physical Medicine Devices
MUSCLE STIMULATOR	890.5850	POWERED MUSCLE STIMULATOR Physical Therapeutic Diagnostic Devices; Physical Medicine Devices
MUSCLE STIMULATOR	890.5860	ULTRASOUND AND MUSCLE STIMULATOR Physical Therapeutic Diagnostic Devices; Physical Medicine Devices
MUSHROOMS	155.201	CANNED MUSHROOMS Requirements For Specific Standardized Canned Vegetables; Canned Vegetables
MYCOBACTERIUM TUBERCULOSIS	866.3370	MYCOBACTERIUM TUBERCULOSIS IMMUNOFLUORESCENT REAGENTS Serological Reagents; Immunology And Microbiology Devices
MYCOPLASMA	610.30	TEST FOR MYCOPLASMA Mycoplasma; General Biological Products Standards
MYCOPLASMA	864.2360	MYCOPLASMA DETECTION MEDIA AND COMPONENTS Cell And Tissue Culture Products; Hematology And Pathology Devices
MYCOPLASMA	866.3375	MYCOPLASMA SPP. SEROLOGICAL REAGENTS Serological Reagents; Immunology And Microbiology Devices
MYOGLOBIN	866.5680	MYOGLOBIN IMMUNOLOGICAL TEST SYSTEM Immunological Test Systems; Immunology And Microbiology Devices
MYRISTYL-GAMMA-PICOLINIUM	524.1484	NEOMYCIN SULFATE, ISOFLUPREDONE ACETATE, TETRACAINE HYDROCHLORIDE, AND MYRISTYL-GAMMA-PICOLINIUM CHLORIDE, TOPICAL POWDER Ophthalmic And Topical Dosage Form New Animal Drugs

N

N-ACETYL-L-METHIONINE	172.372	N-ACETYL-L-METHIONINE Special Dietary And Nutritional Additives; Food Additives Permitted For Direct Addition To Food For Human Consumption
N-BUTANE	184.1165	N-BUTANE AND ISO-BUTANE Listing Of Specific Substances Affirmed As GRAS; Direct Food Substances Affirmed As Generally Recognized As Safe
N-BUTYL CHLORIDE	520.260	N-BUTYL CHLORIDE CAPSULES Oral Dosage Form New Animal Drugs
NACMC	872.3500	POLYVINYLMETHYLETHER MALEIC ANHYDRIDE (PVM-MA), ACID COPOLYMER, AND CARBOXYMETHYLCELLULOSE SODIUM (NACMC) DENTURE ADHESIVE Prosthetic Devices; Dental Devices
NAILBITING	310.536	DRUG PRODUCTS CONTAINING ACTIVE INGREDIENTS OFFERED OVER-THE-COUNTER (OTC) FOR USE AS A NAILBITING OR THUMBSUCKING DETERRENT Requirements For Specific New Drugs Or Devices; New Drugs
NALORPHINE HYDROCHLORIDE	522.1452	NALORPHINE HYDROCHLORIDE INJECTION Implantation Or Injectable Dosage Form New Animal Drugs
NALOXONE HYDROCHLORIDE	522.1462	NALOXONE HYDROCHLORIDE INJECTION Implantation Or Injectable Dosage Form New Animal Drugs
NALTREXONE	522.1465	NALTREXONE HYDROCHLORIDE INJECTION Implantation Or Injectable Dosage Form New Animal Drugs
NAME	101.5	FOOD, NAME AND PLACE OF BUSINESS OF MANUFACTURER, PACKER, OR DISTRIBUTOR General Provisions; Food Labeling
NAME	201.1	DRUGS; NAME AND PLACE OF BUSINESS OF MANUFACTURER, PACKER OR DISTRIBUTOR General Labeling Provisions; Labeling
NAME	501.5	ANIMAL FOOD; NAME AND PLACE OF BUSINESS OF MANUFACTURER, PACKER, OR DISTRIBUTOR General Provisions; Animal Food Labeling
NAME	610.64	NAME OF SELLING AGENT OR DISTRIBUTOR Labeling Standards; General Biological Products Standards
NAME	701.12	NAME AND PLACE OF BUSINESS OF MANUFACTURER, PACKER, OR DISTRIBUTOR Package Form; Cosmetic Labeling
NAME	801.1	MEDICAL DEVICES; NAME AND PLACE OF BUSINESS OF MANUFACTURER, PACKER OR DISTRIBUTOR General Labeling Provisions; Labeling
NAME, COMPENDIAL	299.5	DRUGS; COMPENDIAL NAME General Provisions; Drugs; Official Names And Established Names
NAME, PROPER	610.62	PROPER NAME; PACKAGE LABEL; LEGIBLE TYPE Labeling Standards; General Biological Products Standards
NAMES	20.32	DISCLOSURE OF FDA EMPLOYEE NAMES General Policy; Public Information
NAMES	20.119	LISTS OF NAMES AND ADDRESSES Availability Of Specific Categories Of Records; Public Information
NAMES	510.600	NAMES, ADDRESSES, AND DRUG LABELER CODES OF SPONSORS OF APPROVED APPLICATIONS Sponsors Of Approved Applications; New Animal Drugs
NAMES, COLLECTIVE	501.110	ANIMAL FEED LABELING; COLLECTIVE NAMES FOR FEED INGREDIENTS Exemptions From Animal Food Labeling Requirements; Animal Food Labeling
NAMES, DRUGS	299.4	ESTABLISHED NAMES FOR DRUGS General Provisions, Drugs; Official Names And Established Names
NAMES, INGREDIENT	701.30	INGREDIENT NAMES ESTABLISHED FOR COSMETIC INGREDIENT LABELING Labeling Of Specific Ingredients; Cosmetic Labeling
NAPHTHA	172.250	PETROLEUM NAPHTHA Coatings, Films And Related Substances; Food Additives Permitted For Direct Addition To Food For Human Consumption
NAPHTHALENE SULPHONATES	172.824	SODIUM MONO- AND DIMETHYL NAPHTHALENE SULFONATES Multipurpose Additives; Food Additives Permitted For Direct Addition To Food For Human Consumption
NAPHTHALENEDIYLCARBONYL	177.1637	POLY(OXY-1,2-ETHANEDIYLOXYCARBONYL-2,6-NAPHTHALENEDIYLCARBONYL) RESINS Substances For Use As Basic Components Of Single And Repeated Use Food Contact Surfaces; Indirect Food Additives: Polymers
NAPHTHENATE, COPPER	524.463	COPPER NAPHTHENATE SOLUTION Ophthalmic And Topical Dosage Form New Animal Drugs
NAPROXEN	520.1468	NAPROXEN GRANULES Oral Dosage Form New Animal Drugs

NARASIN	556.428	NARASIN Specific Tolerances For Residues Of New Animal Drugs, Tolerances For Residues Of New Animal Drugs In Food
NARASIN	558.363	NARASIN Specific New Animal Drugs For Use In Animal Feeds; New Animal Drugs For Use In Animal Feeds
NARCOTIC	1304.31	REPORTS FROM MANUFACTURERS IMPORTING NARCOTIC RAW MATERIAL Reports; Records And Reports Of Registrants
NARCOTIC DRUGS	1306.07	ADMINISTERING OR DISPENSING OF NARCOTIC DRUGS General Information; Prescriptions
NARCOTIC TREATMENT	1301.72	PHYSICAL SECURITY CONTROLS FOR NON-PRACTITIONERS; NARCOTIC TREATMENT PROGRAMS AND COMPOUNDERS FOR NARCOTIC TREATMENT PROGRAMS; STORAGE AREAS Security Requirements; Registration Of Manufacturers, Distributors, And Dispensers Of Controlled Substances
NARCOTIC TREATMENT	1301.73	PHYSICAL SECURITY CONTROLS FOR NON-PRACTITIONERS; COMPOUNDERS FOR NARCOTIC TREATMENT PROGRAMS; MANUFACTURING AND COMPOUNDING AREAS Security Requirements; Registration Of Manufacturers, Distributors, And Dispensers Of Controlled Substances
NARCOTIC TREATMENT	1301.74	OTHER SECURITY CONTROLS FOR NON-PRACTITIONERS; NARCOTIC TREATMENT PROGRAMS AND COMPOUNDERS FOR NARCOTIC TREATMENT PROGRAMS Security Requirements; Registration Of Manufacturers, Distributors, And Dispensers Of Controlled Substances
NARCOTICS	1304.25	RECORDS FOR TREATMENT PROGRAMS WHICH COMPOUND NARCOTICS FOR TREATMENT PROGRAMS AND OTHER LOCATIONS Continuing Records; Records And Reports Of Registrants
NASAL CANNULA	868.5340	NASAL OXYGEN CANNULA Therapeutic Devices; Anesthesiology Devices
NASAL CATHETER	868.5350	NASAL OXYGEN CATHETER Therapeutic Devices; Anesthesiology Devices
NASAL DECONGESTANT	341.20	NASAL DECONGESTANT ACTIVE INGREDIENTS Cold, Cough, Allergy, Bronchodilator, And Antiasthmatic Drug Products For OTC Human Use
NASAL DECONGESTANT	341.80	LABELING NASAL DECONGESTANT DRUG PRODUCTS Cold, Cough, Allergy, Bronchodilator, And Antiasthmatic Drug Products For OTC Human Use
NASAL DILATOR	874.3900	NASAL DILATOR Prosthetic Devices; Ear, Nose, And Throat Devices
NASAL IRRIGATOR	874.5550	POWERED NASAL IRRIGATOR Therapeutic Devices; Ear, Nose, And Throat Devices
NASAL SPLINT	874.5800	EXTERNAL NASAL SPLINT Therapeutic Devices; Ear, Nose, And Throat Devices
NASOPHARYNGEAL	868.5100	NASOPHARYNGEAL AIRWAY Therapeutic Devices; Anesthesiology Devices
NASOPHARYNGEAL	874.4175	NASOPHARYNGEAL CATHETER Surgical Devices; Ear, Nose, And Throat Devices
NASOPHARYNGEAL	882.1340	NASOPHARYNGEAL ELECTRODE Neurological Diagnostic Devices; Neurological Devices
NASOPHARYNGOSCOPE	874.4760	NASOPHARYNGOSCOPE (FLEXIBLE OR RIGID) AND ACCESSORIES Surgical Devices; Ear, Nose, And Throat Devices
NATAMYCIN	172.155	NATAMYCIN (PIMARICIN) Food Preservatives; Food Additives Permitted For Direct Addition To Food For Human Consumption
NATIONAL ARCHIVES	21.31	RECORDS STORED BY THE NATIONAL ARCHIVES AND RECORDS ADMINISTRATION Requirements For Specific Categories Of Records; Protection Of Privacy
NATIONAL DRUG CODE	201.2	DRUGS AND DEVICES; NATIONAL DRUG CODE NUMBERS General Labeling Provisions; Labeling
NATIONAL DRUG CONTROL	1401.2	THE OFFICE OF NATIONAL DRUG CONTROL POLICY - ORGANIZATION AND FUNCTIONS Public Availability Of Information
NATIONAL LAB ACCREDITATION	5.303	ESTABLISHING STANDARDS AND APPROVING ACCREDITING BODIES UNDER THE NATIONAL LABORATORY ACCREDITATION PROGRAM Foods And Cosmetics; Redelegations Of Authority; Delegations Of Authority And Organization
NATIONAL TECH. INFO	20.50	AVAILABILITY OF RECORDS AT NATIONAL TECHNICAL INFORMATION SERVICE Procedures And Fees; Public Information
NATIVE AMERICAN CHURCH	1307.31	NATIVE AMERICAN CHURCH Special Exempt Persons; Miscellaneous

NATRIURETIC PEPTIDE	862.1117	B-TYPE NATRIURETIC PEPTIDE TEST SYSTEM Clinical Chemistry Test Systems; Clinical Chemistry And Clinical Toxicology Devices
NATURAL DEFECTS	110.110	NATURAL OR UNAVOIDABLE DEFECTS IN FOOD FOR HUMAN USE THAT PRESENT NO HEALTH HAZARD Defect Action Levels; Current Good Manufacturing Practice In Manufacturing, Packing, Or Holding Human Food
NATURAL EXTRACTIVES	182.20	ESSENTIAL OILS, OLEORESINS (SOLVENT-FREE), AND NATURAL EXTRACTIVES (INCLUDING DISTILLATES) General Provisions; Substances Generally Recognized As Safe
NATURAL EXTRACTIVES	182.40	NATURAL EXTRACTIVES (SOLVENT-FREE) USED IN CONJUNCTION WITH SPICES, SEASONINGS, AND FLAVORINGS General Provisions; Substances Generally Recognized As Safe
NATURAL EXTRACTIVES	582.20	ESSENTIAL OILS, OLEORESINS (SOLVENT-FREE), AND NATURAL EXTRACTIVES (INCLUDING DISTILLATES) General Provisions; Substances Generally Recognized As Safe
NATURAL EXTRACTIVES	582.40	NATURAL EXTRACTIVES (SOLVENT-FREE) USED IN CONJUNCTION WITH SPICES, SEASONINGS, AND FLAVORINGS General Provisions; Substances Generally Recognized As Safe
NATURAL EXTRACTS	182.50	CERTAIN OTHER SPICES, SEASONINGS, ESSENTIAL OILS, OLEORESINS, AND NATURAL EXTRACTS General Provisions; Substances Generally Recognized As Safe
NATURAL EXTRACTS	582.50	CERTAIN OTHER SPICES, SEASONINGS, ESSENTIAL OILS, OLEORESINS, AND NATURAL EXTRACTS General Provisions; Substances Generally Recognized As Safe
NATURAL FLAVORING	172.510	NATURAL FLAVORING SUBSTANCES AND NATURAL SUBSTANCES USED IN CONJUNCTION WITH FLAVORS Flavoring Agents And Related Substances; Food Additives Permitted For Direct Addition To Food For Human Consumption
NATURAL RUBBER	801.437	USER LABELING FOR DEVICES THAT CONTAIN NATURAL RUBBER Special Requirements For Specific Devices; Labeling
NATURAL SEASONINGS	182.10	SPICES AND OTHER NATURAL SEASONINGS AND FLAVORINGS General Provisions; Substances Generally Recognized As Safe
NATURAL SEASONINGS	582.10	SPICES AND OTHER NATURAL SEASONINGS AND FLAVORINGS General Provisions; Substances Generally Recognized As Safe
NATURAL SILK SUTURE	878.5030	NATURAL NONABSORBABLE SILK SURGICAL SUTURE Surgical Devices; General And Plastic Surgery Devices
NATURAL SUBSTANCES	582.30	NATURAL SUBSTANCES USED IN CONJUNCTION WITH SPICES AND OTHER NATURAL SEASONINGS AND FLAVORINGS General Provisions; Substances Generally Recognized As Safe
ND:YAG LASER	886.4392	ND:YAG LASER FOR POSTERIOR CAPSULOTOMY AND PERIPHERAL IRIDECTOMY Surgical Devices; Ophthalmic Devices
NDA	330.11	NDA DEVIATIONS FROM APPLICABLE MONOGRAPH Administrative Procedures; Over-the-counter Human Drugs Generally Recognized As Safe And Effective & Not Misbranded
NDC	201.2	DRUGS AND DEVICES; NATIONAL DRUG CODE NUMBERS General Labeling Provisions; Labeling
NDC	607.35	NOTIFICATION OF REGISTRANT; BLOOD PRODUCT ESTABLISHMENT REGISTRATION NUMBER AND NDC LABELER CODE Procedures For Domestic Blood Product Establishments; Establishment Registratn+product Listing, Manufacturers Human Blood/b.products
NDGA	189.165	NORDIHYDORGUAIARETIC ACID (NDGA) Substances Generally Prohibited From Direct Addition Or Use As Human Food; Substances Prohibited From Use In Human Food
NEARPOINT RULER	886.1790	NEARPOINT RULER Diagnostic Devices; Ophthalmic Devices
NEBRASKA	808.77	NEBRASKA Listing Of Specific State And Local Exemptions; Exemption From Federal Preemption Of State & Local Medical Device Requirements
NEBULIZER	868.5630	NEBULIZER Therapeutic Devices; Anesthesiology Devices
NEBULIZER	868.5640	MEDICINAL NONVENTILATORY NEBULIZER (ATOMIZER) Therapeutic Devices; Anesthesiology Devices
NEBULIZERS	201.305	ISOPROTERENOL INHALATION PREPARATIONS (PRESSURIZED AEROSOLS, NEBULIZERS, POWDERS) FOR HUMAN USE; WARNINGS Specific Labeling Requirements For Specific Drug Products; Labeling
NEED, PERMIT	108.5	DETERMINATION OF THE NEED FOR A PERMIT General Provisions; Emergency Permit Control
NEEDLE	868.5090	EMERGENCY AIRWAY NEEDLE Therapeutic Devices; Anesthesiology Devices
NEEDLE	868.5150	ANESTHESIA CONDUCTION NEEDLE Therapeutic Devices; Anesthesiology Devices

NEEDLE	872.4730	DENTAL INJECTING NEEDLE Surgical Devices; Dental Devices
NEEDLE	880.5570	HYPODERMIC SINGLE LUMEN NEEDLE General Hospital And Personal Use Therapeutic Devices; General Hospital And Personal Use Devices
NEEDLE	880.5580	ACUPUNCTURE NEEDLE General Hospital And Personal Use Devices
NEEDLE	880.6920	SYRINGE NEEDLE INTRODUCER General Hospital And Personal Use Miscellaneous Devices; General Hospital And Personal Use Devices
NEEDLE	882.1350	NEEDLE ELECTRODE Neurological Diagnostic Devices; Neurological Devices
NEEDLE	882.4650	NEUROSURGICAL SUTURE NEEDLE Neurological Surgical Devices; Neurological Devices
NEEDLE	890.1385	DIAGNOSTIC ELECTROMYOGRAPH NEEDLE ELECTRODE Physical Medicine Diagnostic Devices; Physical Medicine Devices
NEEDLE-TYPE EPILATOR	878.5350	NEEDLE-TYPE EPILATOR Therapeutic Devices; General And Plastic Surgery Devices
NEEDLES, REPRODUCTION	884.6100	ASSISTED REPRODUCTION NEEDLES Assisted Reproduction Devices; Obstetrical And Gynecological Devices
NEGATIVE	868.5935	EXTERNAL NEGATIVE PRESSURE VENTILATOR Therapeutic Devices; Anesthesiology Devices
NEISSERIA SPP.	866.2410	CULTURE MEDIUM FOR PATHOGENIC NEISSERIA SPP. Microbiology Devices; Immunology And Microbiology Devices
NEISSERIA SPP.	866.3390	NEISSERIA SPP. DIRECT SEROLOGICAL TEST REAGENTS Serological Reagents; Immunology And Microbiology Devices
NEOMYCIN	520.1921	PROCHLORPERAZINE, ISOPROPAMIDE, WITH NEOMYCIN SUSTAINED-RELEASE CAPSULES Oral Dosage Form New Animal Drugs
NEOMYCIN	524.1484h	NEOMYCIN, PENICILLIN, POLYMIXIN, HYDROCORTISONE SUSPENSION Ophthalmic And Topical Dosage Form New Animal Drugs
NEOMYCIN	524.1600a	NYSTATIN, NEOMYCIN, THIOSTREPTON, AND TRIAMCINOLONE ACETONIDE OINTMENT Ophthalmic And Topical Dosage Form New Animal Drugs
NEOMYCIN	524.1600b	NYSTATIN, NEOMYCIN, THIOSTREPTON, AND TRIAMCINOLONE ACETONIDE OPHTHALMIC OINTMENT Ophthalmic And Topical Dosage Form New Animal Drugs
NEOMYCIN	556.430	NEOMYCIN Specific Tolerances For Residues Of New Animal Drugs; Tolerances For Residues Of New Animal Drugs In Food
NEOMYCIN SULFATE	520.82b	AMINOPROPAZINE FUMARATE, NEOMYCIN SULFATE TABLETS Oral Dosage Form New Animal Drugs
NEOMYCIN SULFATE	520.1484	NEOMYCIN SULFATE SOLUBLE POWDER Oral Dosage Form New Animal Drug Applications
NEOMYCIN SULFATE	520.1485	NEOMYCIN SULFATE ORAL SOLUTION Oral Dosage Form New Animal Drugs
NEOMYCIN SULFATE	522.1484	NEOMYCIN SULFATE STERILE SOLUTION Implantation Or Injectable Dosage Form New Animal Drugs
NEOMYCIN SULFATE	524.154	BACITRACIN OR BACITRACIN ZINC-NEOMYCIN SULFATE-POLYMIXIN B SULFATE OPHTHALMIC OINTMENT Ophthalmic And Topical Dosage Form New Animal Drugs
NEOMYCIN SULFATE	524.155	BACITRACIN ZINC-POLYMIXIN B SULFATE-NEOMYCIN SULFATE-HYDROCORTISONE OR HYDROCORTISONE ACETATE OPHTHALMIC OINTMENT Ophthalmic And Topical Dosage Form New Animal Drugs
NEOMYCIN SULFATE	524.960	FLUMETHASONE, NEOMYCIN SULFATE, AND POLYMYXIN B SULFATE OPHTHALMIC SOLUTION Ophthalmic And Topical Dosage Form New Animal Drugs
NEOMYCIN SULFATE	524.981c	FLUOCINOLONE ACETONIDE, NEOMYCIN SULFATE CREAM Ophthalmic And Topical Dosage Form New Animal Drugs
NEOMYCIN SULFATE	524.1484	NEOMYCIN SULFATE OPHTHALMIC AND TOPICAL DOSAGE FORMS Ophthalmic And Topical Dosage Form New Animal Drugs
NEOMYCIN SULFATE	524.1880	PREDNISOLONE-NEOMYCIN SULFATE OPHTHALMIC OINTMENT Ophthalmic And Topical Dosage Form New Animal Drugs
NEOMYCIN SULFATE	524.1881b	PREDNISOLONE ACETATE-NEOMYCIN SULFATE STERILE SUSPENSION Ophthalmic And Topical Dosage Form New Animal Drugs
NEOMYCIN SULFATE	524.1883	PREDNISOLONE SODIUM PHOSPHATE-NEOMYCIN SULFATE OPHTHALMIC OINTMENT Ophthalmic And Topical Dosage Form New Animal Drugs
NEOMYCIN SULFATE	558.364	NEOMYCIN SULFATE New Animal Drugs For Use In Animal Feeds
NEON	868.1670	NEON GAS ANALYZER Diagnostic Devices; Anesthesiology Devices

NEONATAL	880.5270	NEONATAL EYE PAD General Hospital And Personal Use Therapeutic Devices; General Hospital And Personal Use Devices
NEONATAL	880.5400	NEONATAL INCUBATOR General Hospital And Personal Use Therapeutic Devices; General Hospital And Personal Use Devices
NEONATAL	880.5410	NEONATAL TRANSPORT INCUBATOR General Hospital And Personal Use Therapeutic Devices; General Hospital And Personal Use Devices
NEONATAL	880.5700	NEONATAL PHOTOTHERAPY UNIT General Hospital And Personal Use Therapeutic Devices; General Hospital And Personal Use Devices
NEONATE	862.1113	BILIRUBIN (TOTAL AND UNBOUND) IN THE NEONATE TEST SYSTEM Clinical Chemistry Test Systems; Clinical Chemistry And Clinical Toxicology Devices
NEOSTIGMINE	522.1503	NEOSTIGMINE METHYLSULFATE INJECTION Implantation Or Injectable Dosage Form New Animal Drugs
NEPA PLANNING	25.10	POLICIES AND NEPA PLANNING General Provisions; Environmental Impact Considerations
NEPHELOMETER	862.2700	NEPHELOMETER FOR CLINICAL USE Clinical Laboratory Instruments; Clinical Chemistry And Clinical Toxicology Devices
NEQUINATE	556.440	NEQUINATE Specific Tolerances For Residues Of New Animal Drugs; Tolerances For Residues Of New Animal Drugs In Food
NEQUINATE	558.365	NEQUINATE Specific New Animal Drugs For Use In Animal Foods; New Animal Drugs For Use In Animal Feeds
NERVE	868.2775	ELECTRICAL PERIPHERAL NERVE STIMULATOR Monitoring Devices; Anesthesiology Devices
NERVE	870.3850	CAROTID SINUS NERVE STIMULATOR Cardiovascular Prosthetic Devices; Cardiovascular Devices
NERVE	874.1820	SURGICAL NERVE STIMULATOR/LOCATOR Diagnostic Devices; Ear, Nose, And Throat Devices
NERVE	882.1550	NERVE CONDUCTION VELOCITY MEASUREMENT DEVICE Neurological Diagnostic Devices; Neurological Devices
NERVE	882.5275	NERVE CUFF Neurological Therapeutic Devices; Neurological Devices
NERVE	882.5830	IMPLANTED DIAPHRAGMATIC/PHRENIC NERVE STIMULATOR Neurological Therapeutic Devices; Neurological Devices
NERVE	882.5870	IMPLANTED PERIPHERAL NERVE STIMULATOR FOR PAIN RELIEF Neurological Therapeutic Devices; Neurological Devices
NERVE	882.5890	TRANSCUTANEOUS ELECTRICAL NERVE STIMULATOR FOR PAIN RELIEF Neurological Therapeutic Devices; Neurological Devices
NERVOUS SYSTEM	882.5550	CENTRAL NERVOUS SYSTEM FLUID SHUNT AND COMPONENTS Neurological Therapeutic Devices; Neurological Devices
NET QUANTITY	101.105	DECLARATION OF NET QUANTITY OF CONTENTS WHEN EXEMPT Exemptions From Food Labeling Requirements; Food Labeling
NET QUANTITY	201.51	DECLARATION OF NET QUANTITY OF CONTENTS Labeling Requirements For Prescription Drugs And/or Insulin; Labeling
NET QUANTITY	201.62	DECLARATION OF NET QUANTITY OF CONTENTS Labeling Requirements For Over-the-counter Drugs; Labeling
NET QUANTITY	501.105	DECLARATION OF NET QUANTITY OF CONTENTS WHEN EXEMPT Exemptions From Animal Food Labeling Requirements; Animal Food Labeling
NET QUANTITY	701.13	DECLARATION OF NET QUANTITY OF CONTENTS Package Form; Cosmetic Labeling
NET QUANTITY	801.62	DECLARATION OF NET QUANTITY OF CONTENTS Labeling Requirements For Over-the-counter Devices; Labeling
NEUFCHATEL CHEESE	133.162	NEUFCHATEL CHEESE Requirements For Specific Standardized Cheese And Related Products; Cheeses And Related Cheese Products
NEUFCHATEL CHEESE	133.178	PASTEURIZED NEUFCHATEL CHEESE SPREAD WITH OTHER FOODS Requirements For Specific Standardized Cheese And Related Products; Cheeses And Related Cheese Products
NEURAL TUBE	101.79	HEALTH CLAIMS: FOLATE AND NEURAL TUBE DEFECTS Food Labeling
NEUROLEPTIC	862.3645	NEUROLEPTIC DRUGS RADIORECEPTOR ASSAY TEST SYSTEM Clinical Toxicology Test Systems; Clinical Chemistry And Clinical Toxicology Devices
NEUROLOGICAL	882.1480	NEUROLOGICAL ENDOSCOPE Neurological Diagnostic Devices; Neurological Devices
NEUROMUSCULAR	882.5810	EXTERNAL FUNCTIONAL NEUROMUSCULAR STIMULATOR Neurological Therapeutic Devices; Neurological Devices
NEUROMUSCULAR	882.5860	IMPLANTED NEUROMUSCULAR STIMULATOR Neurological Therapeutic Devices; Neurological Devices

NEUROSURGERY	882.4800	SELF-RETAINING RETRACTOR FOR NEUROSURGERY Neurological Surgical Devices; Neurological Devices
NEUROSURGICAL	882.4125	NEUROSURGICAL CHAIR Neurological Surgical Devices; Neurological Devices
NEUROSURGICAL	882.4440	NEUROSURGICAL HEADRESTS Neurological Surgical Devices; Neurological Devices
NEUROSURGICAL	882.4460	NEUROSURGICAL HEAD HOLDER (SKULL CLAMP) Neurological Surgical Devices; Neurological Devices
NEUROSURGICAL	882.4535	NONPOWERED NEUROSURGICAL INSTRUMENT Neurological Surgical Devices; Neurological Devices
NEUROSURGICAL	882.4650	NEUROSURGICAL SUTURE NEEDLE Neurological Surgical Devices; Neurological Devices
NEUTRON	892.5300	MEDICAL NEUTRON RADIATION THERAPY SYSTEM Therapeutic Devices; Radiology Devices
NEW ANIMAL DRUG	510.455	NEW ANIMAL DRUG REQUIREMENTS REGARDING FREE-CHOICE ADMINISTRATION IN FEEDS Requirements For Specific New Animal Drugs; New Animal Drugs
NEW ANIMAL DRUG	514.12	CONFIDENTIALITY OF DATA AND INFORMATION IN AN INVESTIGATIONAL NEW ANIMAL DRUG NOTICE General Provisions; New Animal Drug Applications
NEW ANIMAL DRUG	558.5	NEW ANIMAL DRUG REQUIREMENTS FOR LIQUID TYPE B FEEDS General Provisions; New Animal Drugs For Use In Animal Feeds
NEW ANIMAL DRUG APPLICATION	514.11	CONFIDENTIALITY OF DATA AND INFORMATION IN A NEW ANIMAL DRUG APPLICATION FILE General Provisions; New Animal Drug Applications
NEW ANIMAL DRUG APPLICATIONS	5.501	APPROVAL OF NEW ANIMAL DRUG APPLICATIONS AND THEIR SUPPLEMENTS Animal Drugs; Redelegations Of Authority; Delegations Of Authority And Organization
NEW ANIMAL DRUG APPLICATIONS	5.503	SUBMISSION OF AND EFFECTIVE APPROVAL DATES FOR ABBREVIATED NEW ANIMAL DRUG APPLICATIONS Animal Drugs; Redelegations Of Authority; Delegations Of Authority And Organization
NEW ANIMAL DRUG APPLICATIONS	514.8	SUPPLEMENTAL NEW ANIMAL DRUG APPLICATIONS General Provisions; New Animal Drug Applications
NEW ANIMAL DRUGS	5.502	ISSUANCE OF NOTICES, PROPOSALS, AND ORDERS RELATING TO NEW ANIMAL DRUGS AND MEDICATED FEED MILL LICENSE APPLICATIONS Animal Drugs; Redelegations Of Authority; Delegations Of Authority And Organization
NEW ANIMAL DRUGS	5.504	ISSUANCE OF WRITTEN NOTICES CONCERNING PATENT INFORMATION, CURRENT GOOD MANUFACTURING PRACTICES AND FALSE OR MISLEADING LABELING OF NEW ANIMAL DRUGS AND FEEDS BEARING OR CONTAINING NEW ANIMAL DRUGS Animal Drugs; Redelegations Of Authority; Delegations Of Authority And Organization
NEW ANIMAL DRUGS	201.115	NEW DRUGS OR NEW ANIMAL DRUGS Exemptions From Adequate Directions For Use; Labeling
NEW ANIMAL DRUGS	510.7	CONSIGNEES OF NEW ANIMAL DRUGS FOR USE IN THE MANUFACTURE OF ANIMAL FEED General Provisions; New Animal Drugs
NEW ANIMAL DRUGS	510.300	RECORDS AND REPORTS CONCERNING EXPERIENCE WITH NEW ANIMAL DRUGS FOR WHICH AN APPROVED APPLICATION IS IN EFFECT Records And Reports; New Animal Drugs
NEW ANIMAL DRUGS	510.301	RECORDS AND REPORTS CONCERNING EXPERIENCE WITH ANIMAL FEEDS BEARING OR CONTAINING NEW ANIMAL DRUGS FOR WHICH AN APPROVED MEDICATED FEED MILL LICENSE IS IN EFFECT Records And Reports; New Animal Drugs
NEW ANIMAL DRUGS	510.305	MAINTENANCE OF COPIES OF APPROVED MEDICATED FEED MILL LICENSES TO MANUFACTURE ANIMAL FEED BEARING OR CONTAINING NEW ANIMAL DRUGS Records And Reports; New Animal Drugs
NEW ANIMAL DRUGS	510.515	ANIMAL FEEDS BEARING OR CONTAINING NEW ANIMAL DRUGS SUBJECT TO THE PROVISIONS OF SECTION 512(N) OF THE ACT Animal Use Exemptions From Certification And Labeling Requirements; New Animal Drugs
NEW ANIMAL DRUGS	511.1	NEW ANIMAL DRUGS FOR INVESTIGATIONAL USE EXEMPT FROM SECTION 512(A) New Animal Drugs For Investigational Use
NEW ANIMAL DRUGS	514.2	APPLICATIONS FOR ANIMAL FEEDS BEARING OR CONTAINING NEW ANIMAL DRUGS General Provisions; New Animal Drug Applications
NEW ANIMAL DRUGS	514.80	RECORDS AND REPORTS CONCERNING EXPERIENCE WITH APPROVED NEW ANIMAL DRUGS Administrative Actions On Applications; New Animal Drug Applications

NEW ANIMAL DRUGS	530.13	EXTRALABEL USE FROM COMPOUNDING OF APPROVED NEW ANIMAL AND APPROVED HUMAN DRUGS Extralabel Drug Use In Animals
NEW ANIMAL DRUGS	556.1	GENERAL CONSIDERATIONS; TOLERANCES FOR RESIDUES OF NEW ANIMAL DRUGS IN FOOD General Provisions; Tolerances For Residues Of New Animal Drugs In Food
NEW DEVICES	860.134	PROCEDURES FOR "NEW DEVICES" UNDER SECTION 513(F) OF THE ACT AND RECLASSIFICATION OF CERTAIN DEVICES Reclassification; Medical Device Classification Procedures
NEW DRUG	20.117	NEW DRUG INFORMATION Availability Of Specific Categories Of Records; Public Information
NEW DRUG	310.6	APPLICABILITY OF "NEW DRUG" OR SAFETY/EFFECTIVENESS FINDINGS IN DRUG EFFICACY STUDY IMPLEMENTATION NOTICES AND NOTICES OF OPPORTUNITY FOR HEARING TO IDENTICAL, RELATED, AND SIMILAR DRUG PRODUCTS General Provisions; New Drugs
NEW DRUG	310.100	NEW DRUG STATUS OPINIONS; STATEMENT OF POLICY Specific Administrative Rulings And Decisions; New Drugs
NEW DRUG	310.103	NEW DRUG SUBSTANCES INTENDED FOR HYPERSENSITIVITY TESTING Specific Administrative Rulings And Decisions; New Drugs
NEW DRUG	310.502	CERTAIN DRUGS ACCORDED NEW DRUG STATUS THROUGH RULEMAKING PROCEDURES Requirements For Specific New Drugs Or Devices; New Drugs
NEW DRUG	312.6	LABELING OF AN INVESTIGATIONAL NEW DRUG Investigational New Drug Application
NEW DRUG	312.34	TREATMENT USE OF AN INVESTIGATIONAL NEW DRUG Investigational New Drug Application
NEW DRUG	312.36	EMERGENCY USE OF AN INVESTIGATIONAL NEW DRUG Investigational New Drug Application
NEW DRUG	312.40	GENERAL REQUIREMENTS FOR USE OF AN INVESTIGATIONAL NEW DRUG IN A CLINICAL INVESTIGATION Investigational New Drug Application
NEW DRUG	314.108	NEW DRUG PRODUCT EXCLUSIVITY Applications For FDA Approval To Market A New Drug
NEW DRUG	601.50	CONFIDENTIALITY OF DATA AND INFORMATION IN AN INVESTIGATIONAL NEW DRUG NOTICE FOR A BIOLOGICAL PRODUCT Confidentiality Of Information; Licensing
NEW DRUG APPLICATION	314.101	FILING AN APPLICATION AND RECEIVING AN ABBREVIATED NEW DRUG APPLICATION Applications For FDA Approval To Market A New Drug
NEW DRUG APPLICATION	314.127	REFUSAL TO APPROVE AN ABBREVIATED NEW DRUG APPLICATION Applications For FDA Approval To Market A New Drug
NEW DRUG APPLICATION	314.151	WITHDRAWAL OF APPROVAL OF AN ABBREVIATED NEW DRUG APPLICATION UNDER SECTION 505(J)(5) Applications For FDA Approval To Market A New Drug
NEW DRUG APPLICATION	314.152	NOTICE OF WITHDRAWAL OF APPROVAL OF AN ABBREVIATED NEW DRUG APPLICATION UNDER SECTION 505(J)(5) FDA Action On Applications; Applications For FDA Approval To Market A New Drug
NEW DRUG APPLICATION	314.153	SUSPENSION OF APPROVAL OF AN ABBREVIATED NEW DRUG APPLICATION FDA Action On Applications; Applications For FDA Approval To Market A New Drug
NEW DRUG APPLICATION	320.31	APPLICABILITY OF REQUIREMENTS REGARDING AN "INVESTIGATIONAL NEW DRUG APPLICATION" Procedures For Determining The Bioavailability Or Bioequivalence Of Drug Products; Bioavailability And Bioequivalence Requirements
NEW DRUG APPLICATIONS	5.103	APPROVAL OF NEW DRUG APPLICATIONS AND THEIR SUPPLEMENTS Human Drugs; Redelegations Of Authority; Delegations Of Authority And Organization
NEW DRUG APPLICATIONS	5.105	ISSUANCE OF NOTICES RELATING TO PROPOSALS TO REFUSE APPROVAL OR TO WITHDRAW APPROVAL OF NEW DRUG APPLICATIONS AND THEIR SUPPLEMENTS Human Drugs; Redelegations Of Authority; Delegations Of Authority And Organization
NEW DRUG APPLICATIONS	5.106	SUBMISSION OF AND EFFECTIVE APPROVAL DATES FOR ABBREVIATED NEW DRUG APPLICATIONS AND CERTAIN NEW DRUG APPLICATIONS Human Drugs; Redelegations Of Authority; Delegations Of Authority And Organization
NEW DRUG APPLICATIONS	310.201	EXEMPTION FOR CERTAIN DRUGS LIMITED BY NEW DRUG APPLICATIONS TO PRESCRIPTION SALE New Drugs Exempted From Prescription-dispensing Requirements; New Drugs
NEW DRUG APPLICATIONS	310.303	CONTINUATION OF LONG-TERM STUDIES, RECORDS, AND REPORTS ON CERTAIN DRUGS FOR WHICH NEW DRUG APPLICATIONS HAVE BEEN APPROVED Records And Reports; New Drugs

NICOTINAMIDE	573.625	MENADIONE NICOTINAMIDE BISULFITE Food Additive Listing; Food Additives Permitted In Feed And Drinking Water Of Animals
NICOTINATE	172.310	ALUMINUM NICOTINATE Special Dietary And Nutritional Additive; Food Additives Permitted For Direct Addition To Food For Human Consumption
NIFURPIRINOL	529.1526	NIFURPIRINOL CAPSULES Certain Other Dosage Form New Animal Drugs
NIGHTTIME SLEEP-AID	338.10	NIGHTTIME SLEEP-AID ACTIVE INGREDIENTS Active Ingredients; Nighttime Sleep-aid Drug Products For Over-the-counter Human Use
NIGHTTIME SLEEP-AID	338.50	LABELING OF NIGHTTIME SLEEP-AID DRUG PRODUCTS Labeling; Nighttime Sleep-aid Drug Products For Over-the-counter Human Use
NIPPLE	880.5630	NIPPLE SHIELD General Hospital And Personal Use Therapeutic Devices; General Hospital And Personal Use Devices
NIPPLE	880.5640	LAMB FEEDING NIPPLE General Hospital And Personal Use Therapeutic Devices; General Hospital And Personal Use Devices
NISIN PREPARATION	184.1538	NISIN PREPARATION Listing Of Specific Substances Affirmed As GRAS; Direct Food Substances Affirmed As Generally Recognized As Safe
NITARSONE	558.369	NITARSONE Specific New Animal Drugs For Use In Animal Feeds; New Animal Drugs For Use In Animal Feeds
NITENPYRAM	520.1510	NITENPYRAM Oral Dosage Form New Animal Drugs
NITRATE, MICONAZOLE	524.1443	MICONAZOLE NITRATE CREAM; MICONAZOLE NITRATE LOTION; MICONAZOLE NITRATE SPRAY Ophthalmic And Topical Dosage Form New Animal Drugs
NITRATE, POTASSIUM	172.160	POTASSIUM NITRATE Food Preservatives; Food Additives Permitted For Direct Addition To Food For Human Consumption
NITRATE, POTASSIUM	181.33	SODIUM NITRATE AND POTASSIUM NITRATE Specific Prior-sanctioned Food Ingredients; Prior-sanctioned Food Ingredients
NITRATE, SODIUM	172.170	SODIUM NITRATE Food Preservatives; Food Additives Permitted For Direct Addition To Food For Human Consumption
NITRATE, SODIUM	176.320	SODIUM NITRATE-UREA COMPLEX Substances For Use Only As Components Of Paper And Paperboard; Indirect Food Additives
NITRATE, SODIUM	181.33	SODIUM NITRATE AND POTASSIUM NITRATE Specific Prior-sanctioned Food Ingredients; Prior-sanctioned Food Ingredients
NITRATES	170.60	NITRITES AND/OR NITRATES IN CURING PREMIXES Specific Administrative Rulings And Decisions; Food Additives
NITRIC OXIDE	868.2380	NITRIC OXIDE ANALYZER Therapeutic Devices; Anesthesiology Devices
NITRIC OXIDE	868.5165	NITRIC OXIDE ADMINISTRATION APPARATUS Therapeutic Devices; Anesthesiology Devices
NITRILE	177.1480	NITRILE RUBBER MODIFIED ACRYLONITRILE-METHYL ACRYLATE COPOLYMERS Substances For Use As Basic Components Of Single And Repeated Use Food Contact Surfaces; Indirect Food Additives: Polymers
NITRITE	862.1510	NITRITE (NONQUANTITATIVE) TEST SYSTEM Clinical Chemistry Test Systems; Clinical Chemistry And Clinical Toxicology Devices
NITRITE, POTASSIUM	181.34	SODIUM NITRITE AND POTASSIUM NITRITE Specific Prior-sanctioned Food Ingredients; Prior-sanctioned Food Ingredients
NITRITE, SODIUM	172.175	SODIUM NITRITE Food Preservatives; Food Additives Permitted For Direct Addition To Food For Human Consumption
NITRITE, SODIUM	172.177	SODIUM NITRITE USED IN PROCESSING SMOKED CHUB Food Preservatives; Food Additives Permitted For Direct Addition To Food For Human Consumption
NITRITE, SODIUM	181.34	SODIUM NITRITE AND POTASSIUM NITRITE Specific Prior-sanctioned Food Ingredients; Prior-sanctioned Food Ingredients
NITRITE, SODIUM	573.700	SODIUM NITRITE Food Additive Listing; Food Additives Permitted In Feed And Drinking Water Of Animals
NITRITES	170.60	NITRITES AND/OR NITRATES IN CURING PREMIXES Specific Administrative Rulings And Decisions; Food Additives
NITROFURAZONE	524.1580	NITROFURAZONE OPHTHALMIC AND TOPICAL DOSAGE FORMS Ophthalmic And Topical Dosage Form New Animal Drugs

NITROGEN	184.1540	NITROGEN Listing Of Specific Substances Affirmed As GRAS; Direct Food Substances Affirmed As Generally Recognized As Safe
NITROGEN	582.1540	NITROGEN General Purpose Food Additives; Substances Generally Recognized As Safe
NITROGEN	862.1515	NITROGEN (AMINO-NITROGEN) TEST SYSTEM Clinical Chemistry Test Systems; Clinical Chemistry And Clinical Toxicology Devices
NITROGEN	862.1770	UREA NITROGEN TEST SYSTEM Clinical Chemistry Test Systems; Clinical Chemistry And Clinical Toxicology Devices
NITROGEN	868.1690	NITROGEN GAS ANALYZER Diagnostic Devices; Anesthesiology Devices
NITROGEN DIOXIDE	868.2385	NITROGEN DIOXIDE ANALYZER Monitoring Devices; Anesthesiology Devices
NITROGLYCERIN	250.300	NITROGLYCERIN FOR HUMAN USE; PACKAGING AND WARNINGS Special Packaging Requirements; Specific Requirements For Specific Human Drugs
NITROMIDE	558.376	NITROMIDE AND SULFANITRAN Specific New Animal Drugs For Use In Animal Feeds; New Animal Drugs For Use In Animal Feeds
NITROTHIAZOLE	556.20	2-ACETYLAMINO-5-NITROTHIAZOLE Tolerances For Residues Of New Animal Drugs In Food
NITROUS OXIDE	184.1545	NITROUS OXIDE Listing Of Specific Substances Affirmed As GRAS; Direct Food Substances Affirmed As Generally Recognized As Safe
NITROUS OXIDE	868.1700	NITROUS OXIDE GAS ANALYZER Diagnostic Devices; Anesthesiology Devices
NO RESIDUE	500.84	OPERATIONAL DEFINITION OF "NO RESIDUE" Regulation Of Carcinogenic Compounds Used In Food-producing Animals; General
NOCTURNAL MUSCLE CRAMPS	310.546	DRUG PRODUCTS CONTAINING ACTIVE INGREDIENTS OFFERED OTC FOR THE TREATMENT AND/OR PREVENTION OF NOCTURNAL LEG MUSCLE CRAMPS Requirements For Specific New Drugs Or Devices; New Drugs
NOISE GENERATOR	874.1120	ELECTRONIC NOISE GENERATOR FOR AUDIOMETRIC TESTING Diagnostic Devices; Ear, Nose, And Throat Devices
NOMINATIONS	14.82	NOMINATIONS OF VOTING MEMBERS OF STANDING ADVISORY COMMITTEES Members Of Advisory Committees; Public Hearing Before A Public Advisory Committee
NOMINATIONS	14.84	NOMINATIONS AND SELECTION OF NONVOTING MEMBERS OF STANDING TECHNICAL ADVISORY COMMITTEES Members Of Advisory Committees; Public Hearing Before A Public Advisory Committee
NON-ANTIFLATULENT	332.15	COMBINATION WITH NON-ANTIFLATULENT ACTIVE INGREDIENTS Active Ingredients; Antiflatulent Products For Over-the-counter Human Use
NON-FEDERAL	1403.26	NON-FEDERAL AUDIT Post-Award Requirements; Uniform Administrative Requirements For Grants And Coop. Agreements To State & Local Gov
NON-NARCOTIC	1312.30	SCHEDULE III, IV, AND V NON-NARCOTIC CONTROLLED SUBSTANCES REQUIRING AN IMPORT AND EXPORT PERMIT Exportation Of Controlled Substances; Importation And Exportation Of Controlled Substances
NON-PRACTITIONERS	1301.72	PHYSICAL SECURITY CONTROLS FOR NON-PRACTITIONERS; NARCOTIC TREATMENT PROGRAMS AND COMPOUNDERS FOR NARCOTIC TREATMENT PROGRAMS; STORAGE AREAS Security Requirements; Registration Of Manufacturers, Distributors, And Dispensers Of Controlled Substances
NON-PRACTITIONERS	1301.73	PHYSICAL SECURITY CONTROLS FOR NON-PRACTITIONERS; COMPOUNDERS FOR NARCOTIC TREATMENT PROGRAMS; MANUFACTURING AND COMPOUNDING AREAS Security Requirements; Registration Of Manufacturers, Distributors, And Dispensers Of Controlled Substances
NON-PRACTITIONERS;	1301.74	OTHER SECURITY CONTROLS FOR NON-PRACTITIONERS; NARCOTIC TREATMENT PROGRAMS AND COMPOUNDERS FOR NARCOTIC TREATMENT PROGRAMS Security Requirements; Registration Of Manufacturers, Distributors, And Dispensers Of Controlled Substances
NONABSORBABLE SUTURE	878.5000	NONABSORBABLE POLY(ETHYLENE TEREPHTHALATE) SURGICAL SUTURE Surgical Devices; General And Plastic Surgery Devices
NONABSORBABLE SUTURE	878.5010	NONABSORBABLE POLYPROPYLENE SURGICAL SUTURE Surgical Devices; General And Plastic Surgery Devices
NONABSORBABLE SUTURE	878.5020	NONABSORBABLE POLYAMIDE SURGICAL SUTURE Surgical Devices; General And Plastic Surgery Devices

NONABSORBABLE SUTURE	878.5030	NATURAL NONABSORBABLE SILK SURGICAL SUTURE Surgical Devices; General And Plastic Surgery Devices
NONALTERABLE	882.5330	PREFORMED NONALTERABLE CRANIOPLASTY PLATE Neurological Therapeutic Devices; Neurological Devices
NONANTACID	331.15	COMBINATION WITH NONANTACID ACTIVE INGREDIENTS Active Ingredients; Antacid Products For Over-the-counter (OTC) Human Use
NONCLINICAL LAB STUDY	58.130	CONDUCT OF A NONCLINICAL LABORATORY STUDY Protocol For And Conduct Of A Nonclinical Laboratory Study; Good Laboratory Practice For Nonclinical Laboratory Studies
NONCLINICAL LAB STUDY	58.185	REPORTING OF NONCLINICAL LABORATORY STUDY RESULTS Records And Reports; Good Laboratory Practice For Nonclinical Laboratory Studies
NONCOMPLYING	1005.3	IMPORTATION OF NONCOMPLYING GOODS PROHIBITED Importation Of Electronic Products
NONCONFORMING	820.90	NONCONFORMING PRODUCT Quality System Regulation
NONCONTINUOUS VENTILATOR	868.5905	NONCONTINUOUS VENTILATOR (IPPB) Therapeutic Devices; Anesthesiology Devices
NONFAT MILK	131.125	NONFAT DRY MILK Requirements For Specific Standardized Milk And Cream; Milk And Cream
NONFAT MILK	131.127	NONFAT DRY MILK FORTIFIED WITH VITAMINS A AND D Requirements For Specific Standardized Milk And Cream; Milk And Cream
NONFAT MILK	139.121	NONFAT MILK MACARONI PRODUCTS Requirements For Specific Standardized Macaroni And Noodle Products; Macaroni And Noodle Products
NONFAT MILK	139.122	ENRICHED NONFAT MILK MACARONI PRODUCTS Requirements For Specific Standardized Macaroni And Noodle Products; Macaroni And Noodle Products
NONFAT YOGURT	131.206	NONFAT YOGURT Requirements For Specific Standardized Milk And Cream; Milk And Cream
NONFOOD ANIMALS	530.30	EXTRALABEL DRUG USE IN NONFOOD ANIMALS Extralabel Drug Use In Animals
NONHUMAN	864.9160	BLOOD GROUP SUBSTANCES OF NONHUMAN ORIGIN FOR IN VITRO DIAGNOSTIC USE Products Used In Establishments That Manufacture Blood And Blood Products; Hematology And Pathology Devices
NONIMPLANTED	876.5310	NONIMPLANTED, PERIPHERAL ELECTRICAL CONTINENCE DEVICE Therapeutic Devices; Gastroenterology-urology Devices
NONIMPLANTED	876.5320	NONIMPLANTED ELECTRICAL CONTINENCE DEVICE Therapeutic Devices; Gastroenterology-urology Devices
NONINFLATABLE SPLINT	878.3910	NONINFLATABLE EXTREMITY SPLINT Prosthetic Devices; General And Plastic Surgery Devices
NONINFRINGEMENT, PATENT	314.52	NOTICE OF CERTIFICATION OF INVALIDITY OR NONINFRINGEMENT OF A PATENT Applications For FDA Approval To Market A New Drug
NONINFRINGEMENT, PATENT	314.95	NOTICE OF CERTIFICATION OF INVALIDITY OR NONINFRINGEMENT OF A PATENT Applications For FDA Approval To Market A New Drug
NONINVASIVE	870.1130	NONINVASIVE BLOOD PRESSURE MEASUREMENT SYSTEM Cardiovascular Diagnostic Devices; Cardiovascular Devices
NONINVASIVE	870.5550	EXTERNAL TRANSCUTANEOUS CARDIAC PACEMAKER (NONINVASIVE) Cardiovascular Therapeutic Devices; Cardiovascular Devices
NONMEDICAL PURPOSES	510.112	ANTIBIOTICS USED IN VETERINARY MEDICINE AND FOR NONMEDICAL PURPOSES, REQUIRED DATA Specific Administrative Rulings And Decisions; New Animal Drugs
NONNARCOTIC	1308.21	APPLICATION FOR EXCLUSION OF A NONNARCOTIC SUBSTANCE Excluded Nonnarcotic Substances; Schedules Of Controlled Substances
NONNARCOTIC	1308.31	APPLICATION FOR EXEMPTION OF A NONNARCOTIC PRESCRIPTION PRODUCT Exempted Prescription Products; Schedules Of Controlled Substances
NONPARTIES	12.89	PARTICIPATION OF NONPARTIES Hearing Procedures; Formal Evidentiary Public Hearing
NONREBREATHING	868.5570	NONREBREATHING MASK Therapeutic Devices; Anesthesiology Devices
NONREBREATHING	868.5870	NONREBREATHING VALVE Therapeutic Devices; Anesthesiology Devices
NONRESORBABLE GAUZE	878.4014	NONRESORBABLE GAUZE SPONGE FOR EXTERNAL USE General And Plastic Surgery Devices
NONROLLER-TYPE	870.4360	NONROLLER-TYPE CARDIOPULMONARY BYPASS BLOOD PUMP Cardiovascular Surgical Devices; Cardiovascular Devices
NONSPECIFIC PROPERTIES	660.54	POTENCY TESTS, SPECIFICITY TESTS, TESTS FOR HETEROSPECIFIC ANTIBODIES, AND ADDITIONAL TESTS FOR NONSPECIFIC PROPERTIES Anti-human Globulin; Additional Standards For Diagnostic Substances For Laboratory Tests

NONSPECIFIC REQUESTS	20.48	NONSPECIFIC AND OVERLY BURDENSOME REQUESTS Procedures And Fees; Public Information
NONTRANSPARENT CONTAINERS	161.173	CANNED WET PACKED SHRIMP IN TRANSPARENT OR NONTRANSPARENT CONTAINERS Requirements For Specific Standardized Fish And Shellfish; Fish And Shellfish
NONTREPONEMAL REAGENTS	866.3820	TREPONEMA PALLIDUM NONTREPONEMAL TEST REAGENTS Serological Reagents; Immunology And Microbiology Devices
NONVENTILATORY	868.5640	MEDICINAL NONVENTILATORY NEBULIZER (ATOMIZER) Therapeutic Devices; Anesthesiology Devices
NONVOTING	14.84	NOMINATIONS AND SELECTION OF NONVOTING MEMBERS OF STANDING TECHNICAL ADVISORY COMMITTEES Members Of Advisory Committees; Public Hearing Before A Public Advisory Committee
NONVOTING	14.86	RIGHTS AND RESPONSIBILITIES OF NONVOTING MEMBERS OF ADVISORY COMMITTEES Members Of Advisory Committees; Public Hearing Before A Public Advisory Committee
NOODLE	139.150	NOODLE PRODUCTS Requirements For Specific Standardized Macaroni And Noodle Products; Macaroni And Noodle Products
NOODLE	139.155	ENRICHED NOODLE PRODUCTS Requirements For Specific Standardized Macaroni And Noodle Products; Macaroni And Noodle Products
NOODLE	139.160	VEGETABLE NOODLE PRODUCTS Requirements For Specific Standardized Macaroni And Noodle Products; Macaroni And Noodle Products
NOODLE	139.165	ENRICHED VEGETABLE NOODLE PRODUCTS Requirements For Specific Standardized Macaroni And Noodle Products; Macaroni And Noodle Products
NOODLE	139.180	WHEAT AND SOY NOODLE PRODUCTS Requirements For Specific Standardized Macaroni And Noodle Products; Macaroni And Noodle Products
NORDIHYDORGUAIARETIC ACID	189.165	NORDIHYDORGUAIARETIC ACID (NDGA) Substances Generally Prohibited From Direct Addition Or Use As Human Food; Substances Prohibited From Use In Human Food
NORGESTOMET	522.850	ESTRADIOL VALERATE AND NORGESTOMET IN COMBINATION Implantation Or Injectable Dosage Form New Animal Drugs
NORMAL PROPYL ALCOHOL	573.880	NORMAL PROPYL ALCOHOL Food Additive Listing; Food Additives Permitted In Feed And Drinking Water Of Animals
NOSE	874.3620	EAR, NOSE, AND THROAT SYNTHETIC POLYMER MATERIAL Prosthetic Devices; Ear, Nose, And Throat Devices
NOSE	874.4140	EAR, NOSE, AND THROAT BURR Surgical Devices; Ear, Nose, And Throat Devices
NOSE	874.4250	EAR, NOSE, AND THROAT ELECTRIC OR PNEUMATIC SURGICAL DRILL Surgical Devices; Ear, Nose, And Throat Devices
NOSE	874.4350	EAR, NOSE, AND THROAT FIBEROPTIC LIGHT SOURCE CARRIER Surgical Devices; Ear, Nose, And Throat Devices
NOSE	874.4420	EAR, NOSE, AND THROAT MANUAL SURGICAL INSTRUMENT Surgical Devices; Ear, Nose, And Throat Devices
NOSE	874.4500	EAR, NOSE, AND THROAT MICROSURGICAL CARBON DIOXIDE LASER Surgical Devices; Ear, Nose, And Throat Devices
NOSE	874.5220	EAR, NOSE, AND THROAT DRUG ADMINISTRATION DEVICE Therapeutic Devices; Ear, Nose, And Throat Devices
NOSE	874.5300	EAR, NOSE AND THROAT EXAMINATION AND TREATMENT UNIT Therapeutic Devices; Ear, Nose, And Throat Devices
NOSE, ARTIFICIAL	868.5375	HEAT AND MOISTURE CONDENSER (ARTIFICIAL NOSE) Therapeutic Devices; Anesthesiology Devices
NOSE CLIP	868.6225	NOSE CLIP Miscellaneous; Anesthesiology Devices
NOSE PROSTHESIS	878.3680	NOSE PROSTHESIS Prosthetic Devices; General And Plastic Surgery Devices
NOT APPROVABLE	314.120	NOT APPROVABLE LETTER TO THE APPLICANT FDA Action On Applications; Applications For FDA Approval To Market A New Drug
NOTATION	21.53	NOTATION AND DISCLOSURE OF DISPUTED RECORDS Procedures For Requests For Amendment Of Records; Protection Of Privacy
NOTICE	1.90	NOTICE OF SAMPLING Imports And Exports; General Enforcement Regulation
NOTICE	5.100	ISSUANCE OF NOTICE IMPLEMENTING THE PROVISIONS OF THE DRUG AMENDMENTS OF 1962 Human Drugs; Redelegations Of Authority; Delegations Of Authority And Organization

NOTICE	5.500	ISSUANCE OF FEDERAL REGISTER DOCUMENTS PERTAINING TO THE DETERMINATION OF SAFE LEVELS, NOTICE OF NEED FOR DEVELOPMENT OF AN ANALYTICAL METHOD, AND PROHIBITION OF CERTAIN EXTRALABEL DRUG USE Animal Drugs; Redelegations Of Authority; Delegations Of Authority And Organization
NOTICE	12.23	NOTICE OF FILING OF OBJECTIONS Initiation Of Proceedings; Formal Evidentiary Public Hearing
NOTICE	12.35	NOTICE OF HEARING; STAY OF ACTION Initiation Of Proceedings; Formal Evidentiary Public Hearing
NOTICE	12.45	NOTICE OF PARTICIPATION Appearance And Participation; Formal Evidentiary Public Hearing
NOTICE	12.95	OFFICIAL NOTICE Hearing Procedures; Formal Evidentiary Public Hearing
NOTICE	13.5	NOTICE OF A HEARING BEFORE A BOARD General Provisions; Public Hearing Before A Public Board Of Inquiry
NOTICE	14.20	NOTICE OF HEARING BEFORE AN ADVISORY COMMITTEE Meeting Procedures; Public Hearing Before A Public Advisory Committee
NOTICE	15.20	NOTICE OF A PUBLIC HEARING BEFORE THE COMMISSIONER Procedures For Public Hearing Before The Commissioner; Public Hearing Before The Commissioner
NOTICE	15.21	NOTICE OF PARTICIPATION; SCHEDULE FOR HEARING Procedures For Public Hearing Before The Commissioner; Public Hearing Before The Commissioner
NOTICE	17.13	NOTICE OF HEARING Civil Money Penalties Hearings
NOTICE	21.20	PROCEDURES FOR NOTICE OF FDA PRIVACY ACT RECORD SYSTEMS FDA Privacy Act Record Systems; Protection Of Privacy
NOTICE	21.74	PROVIDING NOTICE THAT A RECORD IS DISPUTED Disclosure Of Records In Privacy Act Record Systems To Persons Other ThanThe Subject Individual; Protection Of Privacy
NOTICE	58.204	NOTICE OF AND OPPORTUNITY FOR HEARING ON PROPOSED DISQUALIFICATION Disqualification Of Testing Facilities; Good Laboratory Practice For Nonclinical Laboratory Studies
NOTICE	60.42	NOTICE OF HEARING Due Diligence Petitions; Patent Term Restoration
NOTICE	71.2	NOTICE OF FILING OF PETITION General Provisions; Color Additive Petitions
NOTICE	101.17	FOOD LABELING WARNING, NOTICE, AND SAFE HANDLING STATEMENTS General Provisions; Food Labeling
NOTICE	133.10	NOTICE TO MANUFACTURERS, PACKERS, AND DISTRIBUTORS OF PASTEURIZED BLENDED CHEESE, PASTEURIZED PROCESS CHEESE, CHEESE FOOD, CHEESE SPREAD, AND RELATED FOODS General Provisions; Cheeses And Related Cheese Products
NOTICE	201.300	NOTICE TO MANUFACTURERS, PACKERS, AND DISTRIBUTORS OF GLANDULAR PREPARATIONS Specific Labeling Requirements For Specific Drug Products; Labeling
NOTICE	201.301	NOTICE TO MANUFACTURERS, PACKERS, AND DISTRIBUTORS OF ESTROGENIC HORMONE PREPARATIONS Specific Labeling Requirements For Specific Drug Products; Labeling
NOTICE	201.302	NOTICE TO MANUFACTURERS, PACKERS, AND DISTRIBUTORS OF DRUGS FOR INTERNAL USE WHICH CONTAIN MINERAL OIL Specific Labeling Requirements For Specific Drug Products; Labeling
NOTICE	314.52	NOTICE OF CERTIFICATION OF INVALIDITY OR NONINFRINGEMENT OF A PATENT Applications For FDA Approval To Market A New Drug
NOTICE	314.95	NOTICE OF CERTIFICATION OF INVALIDITY OR NONINFRINGEMENT OF A PATENT Applications For FDA Approval To Market A New Drug
NOTICE	314.152	NOTICE OF WITHDRAWAL OF APPROVAL OF AN APPLICATION OR ABBREVIATED APPLICATION FOR A NEW DRUG FDA Action On Applications; Applications For FDA Approval To Market A New Drug
NOTICE	314.200	NOTICE OF OPPORTUNITY FOR HEARING; NOTICE OF PARTICIPATION AND REQUEST FOR HEARING; GRANT OR DENIAL OF HEARING Hearing Procedures For New Drugs; Applications For FDA Approval To Market A New Drug
NOTICE	514.12	CONFIDENTIALITY OF DATA AND INFORMATION IN AN INVESTIGATIONAL NEW ANIMAL DRUG NOTICE General Provisions; New Animal Drug Applications
NOTICE	514.116	NOTICE OF WITHDRAWAL OF APPROVAL OF APPLICATION Administrative Actions On Applications; New Animal Drug Applications
NOTICE	514.200	CONTENTS OF NOTICE OF OPPORTUNITY FOR A HEARING Hearing Procedures; New Animal Drug Applications

NOTICE	515.24	NOTICE OF REVOCATION OF A MEDICATED FEED MILL LICENSE Administrative Actions On Licenses; Medicated Feed Mill License
NOTICE	601.50	CONFIDENTIALITY OF DATA AND INFORMATION IN AN INVESTIGATIONAL NEW DRUG NOTICE FOR A BIOLOGICAL PRODUCT Confidentiality Of Information; Licensing
NOTICE	810.18	PUBLIC NOTICE Mandatory Medical Device Recall Procedures; Medical Device Recall Authority
NOTICE	1003.10	DISCOVERY OF DEFECT OR FAILURE OF COMPLIANCE BY MANUFACTURER; NOTICE REQUIREMENTS Discovery Of Defect Or Failure To Comply; Notification Of Defects Or Failure To Comply
NOTICE	1005.10	NOTICE OF SAMPLING Inspection And Testing; Importation Of Electronic Products
NOTICE	1316.06	NOTICE OF INSPECTION Administrative Inspections; Administrative Functions, Practices, And Procedures
NOTICE	1316.32	NOTICE OF PROCEEDING; TIME AND PLACE Enforcement Proceedings; Administrative Functions, Practices, And Procedures
NOTICE	1316.48	NOTICE OF APPEARANCE Administrative Hearings; Administrative Functions, Practices, And Procedures
NOTICE	1316.97	NOTICE PROVISIONS Expedited Forfeiture Proceedings For Certain Property; Administrative Functions, Practices, And Procedures
NOTICE	1404.312	NOTICE OF PROPOSED DEBARMENT Debarment; Governmentwide Debarment And Suspension And Requirements For Drug-Free Workplace (Grants)
NOTICE	1404.411	NOTICE OF SUSPENSION Suspension; Governmentwide Debarment And Suspension And Requirements For Drug-Free Workplace (Grants)
NOTICE, ADVANCE	1312.32	SCHEDULES II, III, IV: ADVANCE NOTICE Transshipment And In-Transit Shipment Of Controlled Substances; Importation And Exportation Of Controlled Substances
NOTICE, ADVANCE	1313.15	WAIVER OF 15-DAY ADVANCE NOTICE FOR REGULAR IMPORTERS Importation Of Listed Chemicals; Importation And Exportation Of Precursors And Essential Chemicals
NOTICE, ADVANCE	1313.24	WAIVER OF 15-DAY ADVANCE NOTICE FOR CHEMICAL EXPORTERS Exportation Of Listed Chemicals; Importation And Exportation Of Precursors And Essential Chemicals
NOTICE, ADVANCE	1313.31	ADVANCE NOTICE OF IMPORTATION FOR TRANSSHIPMENT OR TRANSFER Transshipment, In-Transit Shipments And International Transactions Involving Listed Chemicals; Importation And Exportation Of Precursors And Essential Chemicals
NOTICES	5.34	ISSUANCE OF NOTICES RELATING TO PROPOSALS AND ORDERS FOR DEBARMENT AND DENIAL OF AN APPLICATION TO TERMINATE DEBARMENT General Redelegations Of Authority; Delegations Of Authority And Organization
NOTICES	5.104	RESPONSES TO DRUG ENFORCEMENT ADMINISTRATION TEMPORARY SCHEDULING NOTICES Human Drugs; Redelegations Of Authority; Delegations Of Authority And Organization
NOTICES	5.105	ISSUANCE OF NOTICES RELATING TO PROPOSALS TO REFUSE APPROVAL OR TO WITHDRAW APPROVAL OF NEW DRUG APPLICATIONS AND THEIR SUPPLEMENTS Human Drugs; Redelegations Of Authority; Delegations Of Authority And Organization
NOTICES	5.109	ISSUANCE OF WRITTEN NOTICES CONCERNING PATENT INFORMATION, CURRENT GOOD MANUFACTURING PRACTICES AND FALSE OR MISLEADING LABELING OF NEW DRUGS Human Drugs; Redelegations Of Authority; Delegations Of Authority And Organization
NOTICES	5.202	ISSUANCE OF NOTICES OF OPPORTUNITY FOR A HEARING ON PROPOSALS FOR DENIAL OF APPROVAL OF APPLICATIONS FOR LICENSES, SUSPENSION OR REVOCATION OF LICENSES AND CERTAIN NOTICES Biologics; Redelegations Of Authority; Delegations Of Authority And Organization
NOTICES	5.301	ISSUANCE OF INITIAL EMERGENCY PERMIT ORDERS AND NOTICES OF CONFIRMATION OF EFFECTIVE DATE OF FINAL REGULATIONS ON FOOD FOR HUMAN AND ANIMAL CONSUMPTION Foods And Cosmetics; Redelegations Of Authority; Delegations Of Authority And Organization
NOTICES	5.502	ISSUANCE OF NOTICES, PROPOSALS, AND ORDERS RELATING TO NEW ANIMAL DRUGS AND MEDICATED FEED MILL LICENSE APPLICATIONS Animal Drugs; Redelegations Of Authority; Delegations Of Authority And Organization

NOTICES	5.504	ISSUANCE OF WRITTEN NOTICES CONCERNING PATENT INFORMATION, CURRENT GOOD MANUFACTURING PRACTICES AND FALSE OR MISLEADING LABELING OF NEW ANIMAL DRUGS AND FEEDS BEARING OR CONTAINING NEW ANIMAL DRUGS Animal Drugs; Redelegations Of Authority; Delegations Of Authority And Organization
NOTICES	10.80	DISSEMINATION OF DRAFT FEDERAL REGISTER NOTICES AND REGULATIONS General Administrative Procedures; Administrative Practices And Procedures
NOTICES	310.6	APPLICABILITY OF "NEW DRUG" OR SAFETY/EFFECTIVENESS FINDINGS IN DRUG EFFICACY STUDY IMPLEMENTATION NOTICES AND NOTICES OF OPPORTUNITY FOR HEARING TO IDENTICAL, RELATED, AND SIMILAR DRUG PRODUCTS General Provisions; New Drugs
NOTICES	514.121	SERVICE OF NOTICES AND ORDERS Administrative Actions On Applications; New Animal Drug Applications
NOTICES	515.26	SERVICE OF NOTICES AND ORDERS Administrative Actions On Licenses; Medicated Feed Mill License
NOTIFICATION	1.101	NOTIFICATION AND RECORDKEEPING Imports And Exports; General Enforcement Regulations
NOTIFICATION	5.204	NOTIFICATION OF RELEASE FOR DISTRIBUTION OF BIOLOGICAL PRODUCTS Biologics; Redelegations Of Authority; Delegations Of Authority And Organization
NOTIFICATION	5.401	ISSUANCE OF FEDERAL REGISTER DOCUMENTS PERTAINING TO EXEMPTIONS FROM PREMARKET NOTIFICATION Medical Devices And Radiological Health; Redelegations Of Authority; Delegations Of Authority And Organization
NOTIFICATION	5.404	NOTIFICATION TO PETITIONERS OF DETERMINATIONS MADE ON PETITIONS FOR RECLASSIFICATION OF MEDICAL DEVICES Medical Devices And Radiological Health; Redelegations Of Authority; Delegations Of Authority And Organization
NOTIFICATION	5.406	NOTIFICATION TO SPONSORS OF DEFICIENCIES IN PETITIONS FOR RECLASSIFICATION OF MEDICAL DEVICES Medical Devices And Radiological Health; Redelegations Of Authority; Delegations Of Authority And Organization
NOTIFICATION	5.603	NOTIFICATION OF DEFECTS IN, AND REPAIR OR REPLACEMENT OF, ELECTRONIC PRODUCTS Radiation Control; Redelegations Of Authority; Delegations Of Authority And Organization
NOTIFICATION	7.50	PUBLIC NOTIFICATION OF RECALL Recalls - Guidance On Policy, Procedures & Industry Responsibilities; Enforcement Policy
NOTIFICATION	21.40	PROCEDURES FOR SUBMITTING REQUESTS FOR NOTIFICATION AND ACCESS Procedures For Notification And Access To Records In Privacy Act Record Systems; Protection Of Privacy
NOTIFICATION	203.37	INVESTIGATION AND NOTIFICATION REQUIREMENTS Prescription Drug Marketing
NOTIFICATION	207.35	NOTIFICATION OF REGISTRANT; DRUG ESTABLISHMENT REGISTRATION NUMBER AND DRUG LISTING NUMBER Procedures For Domestic Drug Establishments; Registration Of Drug Producers And Listing Of Drugs In Commercial Distribution
NOTIFICATION	607.35	NOTIFICATION OF REGISTRANT; BLOOD PRODUCT ESTABLISHMENT REGISTRATION NUMBER AND NDC LABELER CODE Procedures For Domestic Blood Product Establishments; Establishment Registratn+product Listing, Manufacturers Human Blood/b.products
NOTIFICATION	710.6	NOTIFICATION OF REGISTRANT; COSMETIC PRODUCT ESTABLISHMENT REGISTRATION NUMBER Voluntary Registration Of Cosmetic Product Establishments
NOTIFICATION	720.7	NOTIFICATION OF PERSON SUBMITTING COSMETIC PRODUCT INGREDIENT STATEMENT Voluntary Filing Of Cosmetic Product Ingredient Composition Statements
NOTIFICATION	807.35	NOTIFICATION OF REGISTRANT Procedures For Device Establishments; Establishment Registration And Device Listing For Manufacturers And Initial Importers Of Devices
NOTIFICATION	810.10	CEASE DISTRIBUTION AND NOTIFICATION ORDER Mandatory Medical Device Recall Procedures; Medical Device Recall Authority
NOTIFICATION	810.12	WRITTEN REQUEST FOR REVIEW OF CEASE DISTRIBUTION AND NOTIFICATION ORDER Mandatory Medical Device Recall Procedures; Medical Device Recall Authority
NOTIFICATION	810.14	CEASE DISTRIBUTION AND NOTIFICATION OR MANDATORY RECALL STRATEGY Mandatory Medical Device Recall Procedures; Medical Device Recall Authority

NOTIFICATION	810.15	COMMUNICATIONS CONCERNING A CEASE DISTRIBUTION AND NOTIFICATION OR MANDATORY RECALL ORDER Mandatory Medical Device Recall Procedures; Medical Device Recall Authority
NOTIFICATION	810.16	CEASE DISTRIBUTION AND NOTIFICATION OR MANDATORY RECALL STATUS REPORTS Mandatory Medical Device Recall Procedures; Medical Device Recall Authority
NOTIFICATION	810.17	TERMINATION OF A CEASE DISTRIBUTION AND NOTIFICATION OR MANDATORY RECALL ORDER Mandatory Medical Device Recall Procedures; Medical Device Recall Authority
NOTIFICATION	1002.3	NOTIFICATION TO USER OF PERFORMANCE AND TECHNICAL DATA General Provisions; Records And Reports
NOTIFICATION	1003.20	NOTIFICATION BY THE MANUFACTURER TO THE SECRETARY Notification; Notification Of Defects Or Failure To Comply
NOTIFICATION	1003.21	NOTIFICATION BY THE MANUFACTURER TO AFFECTED PERSONS Notification; Notification Of Defects Or Failure To Comply
NOTIFICATION, PREMARKET	190.6	REQUIREMENT FOR PREMARKET NOTIFICATION New Dietary Ingredient Notification; Dietary Supplements
NOTIFICATION, PREMARKET	807.81	WHEN A PREMARKET NOTIFICATION SUBMISSION IS REQUIRED Premarket Notification Procedures; Establishment Registration And Device Listing For Manufacturers And Initial Importers Of Devices
NOTIFICATION, PREMARKET	807.85	EXEMPTION FROM PREMARKET NOTIFICATION Premarket Notification Procedures; Establishment Registration And Device Listing For Manufacturers And Initial Importers Of Devices
NOTIFICATION, PREMARKET	807.87	INFORMATION REQUIRED IN A PREMARKET NOTIFICATION SUBMISSION Premarket Notification Procedures; Establishment Registration And Device Listing For Manufacturers And Initial Importers Of Devices
NOTIFICATION, PREMARKET	807.90	FORMAT OF A PREMARKET NOTIFICATION SUBMISSION Premarket Notification Procedures; Establishment Registration And Device Listing For Manufacturers And Initial Importers Of Devices
NOTIFICATION, PREMARKET	807.97	MISBRANDING BY REFERENCE TO PREMARKET NOTIFICATION Premarket Notification Procedures; Establishment Registration And Device Listing For Manufacturers And Initial Importers Of Devices
NOTIFICATION, PREMARKET	807.100	FDA ACTION ON A PREMARKET NOTIFICATION Premarket Notification Procedures; Establishment Registration And Device Listing For Manufacturers And Initial Importers Of Devices
NOTIFICATION REQUIREMENTS	107.240	NOTIFICATION REQUIREMENTS Infant Formula Recalls; Infant Formula
NOTIFICATION REQUIREMENTS	610.47	"LOOK BACK" NOTIFICATION REQUIREMENTS FOR TRANSFUSION SERVICES Testing Requirements For Communicable Disease Agents; General Biological Products Standards
NOTIFICATION REQUIREMENTS	1003.30	APPLICATION FOR EXEMPTION FROM NOTIFICATION REQUIREMENTS Exemptions From Notification Requirements; Notification Of Defects Or Failure To Comply
NOVOBIOCIN	520.2345f	TETRACYCLINE PHOSPHATE COMPLEX AND SODIUM NOVOBIOCIN CAPSULES Oral Dosage Form New Animal Drugs
NOVOBIOCIN	520.2345g	TETRACYCLINE HYDROCHLORIDE AND SODIUM NOVOBIOCIN TABLETS Oral Dosage Form New Animal Drugs
NOVOBIOCIN	520.2345h	TETRACYCLINE HYDROCHLORIDE, SODIUM NOVOBIOCIN, AND PREDNISOLONE TABLETS Oral Dosage Form New Animal Drugs
NOVOBIOCIN	526.1590	NOVOBIOCIN OIL SUSPENSION Intramammary Dosage Forms
NOVOBIOCIN	526.1696d	PENICILLIN G PROCAINE-NOVOBIOCIN FOR INTRAMAMMARY INFUSION Intramammary Dosage Forms
NOVOBIOCIN	556.460	NOVOBIOCIN Specific Tolerances For Residues Of New Animal Drugs; Tolerances For Residues Of New Animal Drugs In Food
NOVOBIOCIN	558.415	NOVOBIOCIN Specific New Animal Drugs For Use In Animal Feeds; New Animal Drugs For Use In Animal Feeds
NUCLEAR CALIBRATION	892.1400	NUCLEAR SEALED CALIBRATION SOURCE Diagnostic Devices; Radiology Devices
NUCLEAR ELECTROCARDIOGRAPH	892.1410	NUCLEAR ELECTROCARDIOGRAPH SYNCHRONIZER Diagnostic Devices; Radiology Devices
NUCLEAR PHANTOM	892.1370	NUCLEAR ANTHROPOMORPHIC PHANTOM Diagnostic Devices; Radiology Devices
NUCLEAR PHANTOM	892.1380	NUCLEAR FLOOD SOURCE PHANTOM Diagnostic Devices; Radiology Devices
NUCLEAR SCANNER	892.1300	NUCLEAR RECTILINEAR SCANNER Diagnostic Devices; Radiology Devices
NUCLEAR SCANNER	892.1330	NUCLEAR WHOLE BODY SCANNER Diagnostic Devices; Radiology Devices

NUCLEAR SCANNING BED	892.1350	NUCLEAR SCANNING BED Diagnostic Devices; Radiology Devices
NUCLEAR TOMOGRAPHY	892.1310	NUCLEAR TOMOGRAPHY SYSTEM Diagnostic Devices; Radiology Devices
NUCLEAR UPTAKE PROBE	892.1320	NUCLEAR UPTAKE PROBE Diagnostic Devices; Radiology Devices
NUCLEAR WHOLE BODY COUNTER	892.1130	NUCLEAR WHOLE BODY COUNTER Diagnostic Devices; Radiology Devices
NUCLEOTIDASE	862.1520	5'-NUCLEOTIDASE TEST SYSTEM Clinical Chemistry Test Systems; Clinical Chemistry And Clinical Toxicology Devices
NUMBER, CODE	1308.03	ADMINISTRATION CONTROLLED SUBSTANCES CODE NUMBER General Information; Schedules Of Controlled Substances
NUMBER, REGISTRATION	207.35	NOTIFICATION OF REGISTRANT; DRUG ESTABLISHMENT REGISTRATION NUMBER AND DRUG LISTING NUMBER Procedures For Domestic Drug Establishments; Registration Of Drug Producers And Listing Of Drugs In Commercial Distribution
NUMBER, REGISTRATION	207.39	MISBRANDING BY REFERENCE TO REGISTRATION OR TO REGISTRATION NUMBER Procedures For Domestic Drug Establishments; Registration Of Drug Producers And Listing Of Drugs In Commercial Distribution
NUMBER, REGISTRATION	607.35	NOTIFICATION OF REGISTRANT; BLOOD PRODUCT ESTABLISHMENT REGISTRATION NUMBER AND NDC LABELER CODE Procedures For Domestic Blood Product Establishments; Establishment Registratn+product Listing, Manufacturers Human Blood/b.products
NUMBER, REGISTRATION	607.39	MISBRANDING BY REFERENCE TO ESTABLISHMENT REGISTRATION OR TO REGISTRATION NUMBER Procedures For Domestic Blood Product Establishments; Establishment Registratn+product Listing, Manufacturers Human Blood/b.products
NUMBER, REGISTRATION	710.6	NOTIFICATION OF REGISTRANT; COSMETIC PRODUCT ESTABLISHMENT REGISTRATION NUMBER Voluntary Registration Of Cosmetic Product Establishments
NUMBER, REGISTRATION	710.8	MISBRANDING BY REFERENCE TO REGISTRATION OR TO REGISTRATION NUMBER Voluntary Registration Of Cosmetic Product Establishments
NUMBER, REGISTRATION	807.39	MISBRANDING BY REFERENCE TO ESTABLISHMENT REGISTRATION OR TO REGISTRATION NUMBER Procedures For Device Establishments; Establishment Registration And Device Listing For Manufacturers And Initial Importers Of Devices
NUMBER, REGISTRATION	1271.27	WILL FDA ASSIGN ME A REGISTRATION NUMBER? Procedures For Registration And Listing; Human Cells, Tissues, And Cellular And Tissue-Based Products
NUMBER OF SERVINGS	501.8	LABELING OF ANIMAL FOOD WITH NUMBER OF SERVINGS General Provisions; Animal Food Labeling
NUMBER, STATEMENT	720.9	MISBRANDING BY REFERENCE TO FILING OR TO STATEMENT NUMBER Voluntary Filing Of Cosmetic Product Ingredient Composition Statements
NUMBERS, CONTROL	201.18	DRUGS; SIGNIFICANCE OF CONTROL NUMBERS General Labeling Provisions; Labeling
NUMBERS, LOT/CONTROL	203.38	SAMPLE LOT OR CONTROL NUMBERS Labeling; Prescription Drug Marketing
NUMBERS, NDC	201.2	DRUGS AND DEVICES; NATIONAL DRUG CODE NUMBERS General Labeling Provisions; Labeling
NURSING	330.2	PREGNANCY NURSING WARNING General Provisions; Over-the-counter Human Drugs Generally Recognized As Safe And Effective & Not Misbranded
NUTRIENT CONTENT	5.300	FOOD STANDARDS, FOOD ADDITIVES, GRAS SUBSTANCES, COLOR ADDITIVES, NUTRIENT CONTENT CLAIMS, AND HEALTH CLAIMS Foods And Cosmetics; Redelegations Of Authority; Delegations Of Authority And Organization
NUTRIENT CONTENT	101.13	NUTRIENT CONTENT CLAIMS - GENERAL PRINCIPLES General Provisions; Food Labeling
NUTRIENT CONTENT	101.54	NUTRIENT CONTENT CLAIMS FOR "GOOD SOURCE," "HIGH," "MORE," AND "HIGH POTENCY" Specific Requirements For Nutrient Content Claims; Food Labeling
NUTRIENT CONTENT	101.56	NUTRIENT CONTENT CLAIMS FOR "LIGHT" OR "LITE" Specific Requirements For Nutrient Content Claims; Food Labeling
NUTRIENT CONTENT	101.60	NUTRIENT CONTENT CLAIMS FOR THE CALORIE CONTENT OF FOODS Specific Requirements For Nutrient Content Claims; Food Labeling
NUTRIENT CONTENT	101.61	NUTRIENT CONTENT CLAIMS FOR THE SODIUM CONTENT OF FOODS Specific Requirements For Nutrient Content Claims; Food Labeling
NUTRIENT CONTENT	101.62	NUTRIENT CONTENT CLAIMS FOR FAT, FATTY ACID, AND CHOLESTEROL CONTENT OF FOODS Specific Requirements For Nutrient Content Claims; Food Labeling

O

OBESITY	250.11	THYROID-CONTAINING DRUG PREPARATIONS INTENDED FOR TREATMENT OF OBESITY IN HUMANS Drugs Regarded As Misbranded; Specific Requirements For Specific Human Drugs
OBJECTIONS	12.22	FILING OBJECTIONS AND REQUESTS FOR A HEARING ON A REGULATION OR ORDER Initiation Of Proceedings; Formal Evidentiary Public Hearing
OBJECTIONS	12.23	NOTICE OF FILING OF OBJECTIONS Initiation Of Proceedings; Formal Evidentiary Public Hearing
OBJECTIONS	12.24	RULING ON OBJECTIONS AND REQUESTS FOR HEARING Initiation Of Proceedings; Formal Evidentiary Public Hearing
OBJECTIONS	71.30	PROCEDURE FOR FILING OBJECTIONS TO REGULATIONS Administrative Action On Petitions; Color Additive Petitions
OBJECTIONS	171.110	PROCEDURE FOR OBJECTIONS AND HEARINGS Administrative Actions On Applications; Food Additive Petitions
OBJECTIONS	571.110	PROCEDURE FOR OBJECTIONS AND HEARINGS Administrative Actions On Applications; Food Additive Petitions
OBJECTIONS	1316.60	OBJECTIONS; OFFER OF PROOF Administrative Hearings; Administrative Functions, Practices, And Procedures
OBLET	520.1242b	LEVAMISOLE HYDROCHLORIDE TABLET OR OBLET (BOLUS) Oral Dosage Form New Animal Drugs
OBLET	520.2260a	SULFAMETHAZINE OBLET, TABLET, AND BOLUS Oral Dosage Form New Animal Drugs
OBLIGATION	1004.1	MANUFACTURER'S OBLIGATION TO REPAIR, REPLACE, OR REFUND COST OF ELECTRONIC PRODUCTS Repurchase, Repairs, Or Replacement Of Electronic Products
OBLIGATIONS	26.62	GENERAL OBLIGATIONS Framework Provisions; Mutual Recognition Of Pharmaceutical GMP Reports, Medical Device Quality System Audit Reports
OBLIGATIONS	26.75	SUSPENSION OF RECOGNITION OBLIGATIONS Framework Provisions; Mutual Recognition Of Pharmaceutical GMP Reports, Medical Device Quality System Audit Reports
OBLIGATIONS	312.52	TRANSFER OF OBLIGATIONS TO A CONTRACT RESEARCH ORGANIZATION Investigational New Drug Application
OBLIGATIONS	821.30	TRACKING OBLIGATIONS OF PERSONS OTHER THAN DEVICE MANUFACTURERS: DISTRIBUTOR REQUIREMENTS Medical Device Tracking Requirements
OBSTETRIC	884.2050	OBSTETRIC DATA ANALYZER Obstetrical And Gynecological Monitoring Devices; Obstetrical And Gynecological Devices
OBSTETRIC	884.2225	OBSTETRIC-GYNECOLOGIC ULTRASONIC IMAGER Obstetrical And Gynecological Monitoring Devices; Obstetrical And Gynecological Devices
OBSTETRIC	884.2960	OBSTETRIC ULTRASONIC TRANSDUCER AND ACCESSORIES Obstetrical And Gynecological Monitoring Devices; Obstetrical And Gynecological Devices
OBSTETRIC	884.4400	OBSTETRIC FORCEPS Obstetrical And Gynecological Surgical Devices; Obstetrical And Gynecological Devices
OBSTETRIC	884.4500	OBSTETRIC FETAL DESTRUCTIVE INSTRUMENT Obstetrical And Gynecological Surgical Devices; Obstetrical And Gynecological Devices
OBSTETRIC	884.4520	OBSTETRIC-GYNECOLOGIC GENERAL MANUAL INSTRUMENT Obstetrical And Gynecological Surgical Devices; Obstetrical And Gynecological Devices
OBSTETRIC	884.4530	OBSTETRIC-GYNECOLOGIC SPECIALIZED MANUAL INSTRUMENT Obstetrical And Gynecological Surgical Devices; Obstetrical And Gynecological Devices
OBSTETRIC	884.4900	OBSTETRIC TABLE AND ACCESSORIES Obstetrical And Gynecological Surgical Devices; Obstetrical And Gynecological Devices
OBSTETRIC	884.5100	OBSTETRIC ANESTHESIA SET Obstetrical And Gynecological Therapeutic Devices; Obstetrical And Gynecological Devices
OBSTRUCTION	868.5115	DEVICE TO RELIEVE ACUTE UPPER AIRWAY OBSTRUCTION Therapeutic Devices; Anesthesiology Devices
OBTURATOR	868.5650	ESOPHAGEAL OBTURATOR Therapeutic Devices; Anaesthesiology Devices
OCCLUDER	870.1370	CATHETER TIP OCCLUDER Cardiovascular Diagnostic Devices; Cardiovascular Devices

OCCLUDING	882.5150	INTRAVASCULAR OCCLUDING CATHETER Neurological Therapeutic Devices; Neurological Devices
OCCLUSION	870.2890	VESSEL OCCLUSION TRANSDUCE Cardiovascular Monitoring Devices; Cardiovascular Devices
OCCLUSION	880.5950	UMBILICAL OCCLUSION DEVICE General Hospital And Personal Use Therapeutic Devices; General Hospital And Personal Use Devices
OCCLUSION	884.5380	CONTRACEPTIVE TUBAL OCCLUSION DEVICE (TOD) AND INTRODUCER Obstetrical And Gynecological Therapeutic Devices; Obstetrical And Gynecological Devices
OCCLUSIVE WOUND DRESSING	878.4020	OCCLUSIVE WOUND DRESSING Surgical Devices; General And Plastic Surgery Devices
OCCULT BLOOD	864.6550	OCCULT BLOOD TEST Manual Hematology Devices; Hematology And Pathology Devices
OCEAN VESSELS	1301.25	REGISTRATION REGARDING OCEAN VESSELS, COMMERCIAL AIRCRAFT, AND CERTAIN OTHER ENTITIES Registration Of Manufacturers, Distributors, And Dispensers Of Controlled Substances
OCTADECYLAMINE	200.11	USE OF OCTADECYLAMINE IN STEAM LINES OF DRUG ESTABLISHMENTS General Provisions; General
OCTAFLUOROCYCLOBUTANE	173.360	OCTAFLUOROCYCLOBUTANE Specific Usage Additives; Secondary Direct Food Additives Permitted In Food For Human Consumption
OCTANE	176.160	CHROMIUM (CR III) COMPLEX N-ETHYL-N-HEPTADECYLFLUORO-OCTANE SULFONYL GLYCINE Substances For Use Only As Components Of Paper And Paperboard; Indirect Food Additives
OCULAR	882.1790	OCULAR PLETHYSMOGRAPH Neurological Diagnostic Devices; Neurological Devices
OCULAR	886.1040	OCULAR ESTHESIOMETER Diagnostic Devices; Ophthalmic Devices
OCULAR	886.4360	OCULAR SURGERY IRRIGATION DEVICE Surgical Devices; Ophthalmic Devices
OCULAR	886.4610	OCULAR PRESSURE APPLICATOR Surgical Devices; Ophthalmic Devices
ODORLESS LIGHT PETROLEUM	172.884	ODORLESS LIGHT PETROLEUM HYDROCARBONS Multipurpose Additives; Food Additives Permitted For Direct Addition To Food For Human Consumption
ODORLESS LIGHT PETROLEUM	178.3650	ODORLESS LIGHT PETROLEUM HYDROCARBONS Certain Adjuvants And Production Aids; Indirect Food Additives: Adjuvants, Production Aids, And Sanitizers
ODORLESS LIGHT PETROLEUM	573.740	ODORLESS LIGHT PETROLEUM HYDROCARBONS Food Additive Listing; Food Additives Permitted In Feed And Drinking Water Of Animals
OFFER OF PROOF	1316.60	OBJECTIONS; OFFER OF PROOF Administrative Hearings; Administrative Functions, Practices, And Procedures
OFFICE NAT. DRUG CONTROL	1401.2	THE OFFICE OF NATIONAL DRUG CONTROL POLICY - ORGANIZATION AND FUNCTIONS Public Availability Of Information
OFFICER	3.6	PRODUCT JURISDICTION OFFICER Product Jurisdiction
OFFICER, PRESIDING	12.60	PRESIDING OFFICER Presiding Officer; Formal Evidentiary Public Hearing
OFFICER, PRESIDING	12.70	AUTHORITY OF PRESIDING OFFICER Presiding Officer; Formal Evidentiary Public Hearing
OFFICER, PRESIDING	12.75	DISQUALIFICATION OF PRESIDING OFFICER Presiding Officer; Formal Evidentiary Public Hearing
OFFICER, PRESIDING	12.78	UNAVAILABILITY OF PRESIDING OFFICER Presiding Officer; Formal Evidentiary Public Hearing
OFFICER, PRESIDING	12.97	INTERLOCUTORY APPEAL FROM RULING OF PRESIDING OFFICER Hearing Procedures; Formal Evidentiary Public Hearing
OFFICER, PRESIDING	16.42	PRESIDING OFFICER Commissioner And Presiding Officer; Regulatory Hearing Before The Food And Drug Administration
OFFICER, PRESIDING	16.44	COMMUNICATION TO PRESIDING OFFICER AND COMMISSIONER Commissioner And Presiding Officer; Regulatory Hearing Before The Food And Drug Administration
OFFICER, PRESIDING	17.18	INTERLOCUTORY APPEAL FROM RULING OF PRESIDING OFFICER Civil Money Penalties Hearings
OFFICER, PRESIDING	17.19	AUTHORITY OF THE PRESIDING OFFICER Civil Money Penalties Hearings
OFFICER, PRESIDING	1316.52	PRESIDING OFFICER Administrative Hearings; Administrative Functions, Practices, And Procedures

OFFICER, PRESIDING	1316.62	APPEAL FROM RULING OF PRESIDING OFFICER Administrative Hearings; Administrative Functions, Practices, And Procedures
OFFICERS	5.20	GENERAL REDELEGATIONS OF AUTHORITY FROM THE COMMISSIONER TO OTHER OFFICERS OF THE FDA General Redelegations Of Authority; Delegations Of Authority And Organization
OFFICERS	1316.72	OFFICERS WHO WILL MAKE SEIZURES Seizure, Forfeiture, And Disposition Of Property; Administrative Functions, Practices, And Procedures
OFFICES, PUBLIC INFORMATION	5.1110	FDA PUBLIC INFORMATION OFFICES Organization; Delegations Of Authority And Organization
OFFICIAL COMPENDIA	369.7	WARNINGS REQUIRED BY OFFICIAL COMPENDIA Definitions And Interpretations; Interpretative Statements Re: Warnings On Drugs And Devices For OTC Sale
OFFICIAL NOTICE	12.95	OFFICIAL NOTICE Hearing Procedures; Formal Evidentiary Public Hearing
OFFICIAL RECORDS	5.23	DISCLOSURE OF OFFICIAL RECORDS AND AUTHORIZATION OF TESTIMONY General Redelegations Of Authority; Delegations Of Authority And Organization
OFFICIAL RELEASE	610.2	REQUESTS FOR SAMPLES AND PROTOCOLS; OFFICIAL RELEASE Release Requirements; General Biological Products Standards
OFFICIAL RELEASE	660.6	SAMPLES; PROTOCOLS; OFFICIAL RELEASE Antibody To Hepatitis B Surface Antigen; Additional Standards For Diagnostic Substances For Laboratory Tests
OFFICIAL RELEASE	660.29	SAMPLES; PROTOCOLS; OFFICIAL RELEASE Blood Grouping Reagent; Additional Standards For Diagnostic Substances For Laboratory Tests
OFFICIAL RELEASE	660.46	SAMPLES; PROTOCOLS; OFFICIAL RELEASE Hepatitis B Surface Antigen; Additional Standards For Diagnostic Substances For Laboratory Tests
OFFICIAL TRANSCRIPT	12.98	OFFICIAL TRANSCRIPT Hearing Procedures; Formal Evidentiary Public Hearing
OFFICIAL TRANSCRIPT	1316.63	OFFICIAL TRANSCRIPT; INDEX; CORRECTIONS Administrative Hearings; Administrative Functions, Practices, And Procedures
OFFICIAL, RESPONSIBLE	25.45	RESPONSIBLE AGENCY OFFICIAL Preparation of Environmental Documents; Environmental Impact Considerations
OFFICIALS	5.35	OFFICIALS AUTHORIZED TO MAKE CERTIFICATION UNDER 5 U.S.C. 605(B) FOR ANY PROPOSED AND FINAL RULES General Redelegations Of Authority; Delegations Of Authority And Organization
OFFICIALS, FOREIGN GOVERNMENT	20.89	COMMUNICATIONS WITH FOREIGN GOVERNMENT OFFICIALS Limitations On Exemptions; Public Information
OFFICIALS, LAW ENFORCEMENT	1301.26	EXEMPTION OF LAW ENFORCEMENT OFFICIALS Registration Of Manufacturers, Distributors, And Dispensers Of Controlled Substances
OFFICIALS, LAW ENFORCEMENT	1309.26	EXEMPTION OF LAW ENFORCEMENT OFFICIALS Registration Of Manufacturers, Distributors, Importers And Exporters Of List I Chemicals
OFFICIALS, PUBLIC HEALTH	5.10	DELEGATIONS FROM THE SECRETARY, THE ASSISTANT SECRETARY FOR HEALTH, AND PUBLIC HEALTH SERVICE OFFICIALS Delegations Of Authority To The Commissioner Of Food And Drug; Delegations Of Authority And Organization
OFFICIALS, STATE/LOCAL GOVERNMENT	20.84	DISCLOSURE TO CONSULTANTS, ADVISORY COMMITTEES, STATE AND LOCAL GOVERNMENT OFFICIALS AND OTHER SPECIAL GOVERNMENT EMPLOYEES Limitations On Exemptions; Public Information
OFFICIALS, STATE/LOCAL GOVERNMENT	20.88	COMMUNICATIONS WITH STATE AND LOCAL GOVERNMENT OFFICIALS Limitations On Exemptions; Public Information
OFFSET	176.130	ANTI-OFFSET SUBSTANCES Substances For Use Only As Components Of Paper And Paperboard; Indirect Food Additives
OHIO	808.86	OHIO Listing Of Specific State And Local Exemptions; Exemption From Federal Preemption Of State & Local Medical Device Requirements
OIL	522.1696c	PENICILLIN G PROCAINE IN OIL Implantation Or Injectable Dosage Form New Animal Drugs
OIL	526.1590	NOVOBIOCIN OIL SUSPENSION Intramammary Dosage Forms
OIL	526.1696a	PENICILLIN G PROCAINE IN OIL Intramammary Dosage Forms
OIL, CARROT	73.300	CARROT OIL Foods; Listing Of Color Additives Exempt From Certification
OIL, CASTOR	172.876	CASTOR OIL Multipurpose Additives; Food Additives Permitted For Direct Addition To Food For Human Consumption

OIL, CASTOR	178.3280	CASTOR OIL, HYDROGENATED Certain Adjuvants And Production Aids; Indirect Food Additives: Adjuvants, Production Aids, And Sanitizers
OIL, CASTOR	524.2620	LIQUID CRYSTALLINE TRYPSIN, PERU BALSAM, CASTOR OIL Ophthalmic And Topical Dosage Form New Animal Drugs
OIL, COCONUT	172.816	METHYL GLUCOSIDE-COCONUT OIL ESTER Multipurpose Additives; Food Additives Permitted For Direct Addition To Food For Human Consumption
OIL, COCONUT	172.861	COCOA BUTTER SUBSTITUTE FROM COCONUT OIL, PALM KERNEL OIL, OR BOTH OILS Food Additives Permitted For Direct Addition To Food For Human Consumption
OIL, COCONUT	573.660	METHYL GLUCOSIDE-COCONUT OIL ESTER Food Additive Listing; Food Additives Permitted In Feed And Drinking Water Of Animals
OIL, CORN ENDOSPERM	73.315	CORN ENDOSPERM OIL Foods; Listing Of Color Additives Exempt From Certification
OIL, EDIBLE	102.37	MIXTURES OF EDIBLE FAT OR OIL AND OLIVE OIL Requirements For Specific Nonstandardized Foods; Common Or Usual Name For Nonstandardized Foods
OIL, FISH	186.1551	HYDROGENATED FISH OIL Listing Of Specific Substances Affirmed As GRAS; Indirect Food Substances Affirmed As Generally Recognized As Safe
OIL, MENHADEN	184.1472	MENHADEN OIL Listing Of Specific Substances Affirmed As GRAS; Direct Food Substances Affirmed As Generally Recognized As Safe
OIL, MINERAL	172.878	WHITE MINERAL OIL Multipurpose Additives; Food Additives Permitted For Direct Addition To Food For Human Consumption
OIL, MINERAL	178.3620	MINERAL OIL Certain Adjuvants And Production Aids; Indirect Food Additives: Adjuvants, Production Aids, And Sanitizers
OIL, MINERAL	201.302	NOTICE TO MANUFACTURERS, PACKERS, AND DISTRIBUTORS OF DRUGS FOR INTERNAL USE WHICH CONTAIN MINERAL OIL Specific Labeling Requirements For Specific Drug Products; Labeling
OIL, MINERAL	573.680	MINERAL OIL Food Additive Listing; Food Additives Permitted In Feed And Drinking Water Of Animals
OIL, OLIVE	102.37	MIXTURES OF EDIBLE FAT OR OIL AND OLIVE OIL Requirements For Specific Nonstandardized Foods; Common Or Usual Name For Nonstandardized Foods
OIL, PALM KERNEL	172.861	COCOA BUTTER SUBSTITUTE FROM COCONUT OIL, PALM KERNEL OIL, OR BOTH OILS Food Additives Permitted For Direct Addition To Food For Human Consumption
OIL, RAPESEED	184.1555	RAPESEED OIL Listing Of Specific Substances Affirmed As GRAS; Direct Food Substances Affirmed As Generally Recognized As Safe
OIL OF RUE	184.1699	OIL OF RUE Listing Of Specific Substances Affirmed As GRAS; Direct Food Substances Affirmed As Generally Recognized As Safe
OIL, SHEANUT	184.1702	SHEANUT OIL Listing Of Specific Substances Affirmed As GRAS; Direct Food Substances Affirmed As Generally Recognized As Safe
OIL, SOYBEAN	172.723	EPOXIDIZED SOYBEAN OIL Food Additive Permitted For Direct Addition To Food For Human Consumption
OIL, SOYBEAN	526.1696b	PENICILLIN G PROCAINE-DIHYDROSTREPTOMYCIN IN SOYBEAN OIL FOR INTRAMAMMARY INFUSION (DRY COWS) Intramammary Dosage Forms
OIL, SPERM	173.275	HYDROGENATED SPERM OIL Solvents, Lubricants, Release Agents And Related Substances; Secondary Direct Food Additives Permitted In Food For Human Consumption
OIL, TALL	172.862	OLEIC ACID DERIVED FROM TALL OIL FATTY ACIDS Multipurpose Additives; Food Additives Permitted For Direct Addition To Food For Human Consumption
OIL, TALL	186.1557	TALL OIL Listing Of Specific Substances Affirmed As GRAS; Indirect Food Substances Affirmed As Generally Recognized As Safe
OIL, VEGETABLE	180.30	BROMINATED VEGETABLE OIL Specific Requirements For Certain Food Additives; Food Additives Permitted In Food Or In Contact With Food On Interim Basis Pending Study
OIL, WINTERGREEN	201.303	LABELING OF DRUG PREPARATIONS CONTAINING SIGNIFICANT PROPORTIONS OF WINTERGREEN OIL Specific Labeling Requirements For Specific Drug Products; Labeling

OILS	172.861	COCOA BUTTER SUBSTITUTE FROM COCONUT OIL, PALM KERNEL OIL, OR BOTH OILS Food Additives Permitted For Direct Addition To Food For Human Consumption
OILS, DRYING	181.26	DRYING OILS AS COMPONENTS OF FINISHED RESINS Specific Prior-sanctioned Food Ingredients; Prior-sanctioned Food Ingredients
OILS, EDIBLE	172.225	METHYL ESTERS OF FATTY ACIDS PRODUCED FROM EDIBLE FATS AND OILS Coatings, Films And Related Substances; Food Additives Permitted For Direct Addition To Food For Human Consumption
OILS, EDIBLE	582.4101	DIACETYL TARTARIC ACID ESTERS OF MONO- AND DIGLYCERIDES OF EDIBLE FATS OR OILS, OR EDIBLE FAT-FORMING FATTY ACIDS Emulsifying Agents; Substances Generally Recognized As Safe
OILS, EDIBLE	582.4505	MONO- AND DIGLYCERIDES OF EDIBLE FATS OR OILS, OR EDIBLE FAT-FORMING ACIDS Emulsifying Agents; Substances Generally Recognized As Safe
OILS, EDIBLE	582.4521	MONOSODIUM PHOSPHATE DERIVATIVES OF MONO- AND DIGLYCERIDES OF EDIBLE FATS OR OILS, OR EDIBLE FAT-FORMING FATTY ACIDS Emulsifying Agents; Substances Generally Recognized As Safe
OILS, ESSENTIAL	182.20	ESSENTIAL OILS, OLEORESINS (SOLVENT-FREE), AND NATURAL EXTRACTIVES (INCLUDING DISTILLATES) General Provisions; Substances Generally Recognized As Safe
OILS, ESSENTIAL	182.50	CERTAIN OTHER SPICES, SEASONINGS, ESSENTIAL OILS, OLEORESINS, AND NATURAL EXTRACTS General Provisions; Substances Generally Recognized As Safe
OILS, ESSENTIAL	582.20	ESSENTIAL OILS, OLEORESINS (SOLVENT-FREE),AND NATURAL EXTRACTIVES INCLUDING DISTILLATES) General Provisions; Substances Generally Recognized As Safe
OILS, ESSENTIAL	582.50	CERTAIN OTHER SPICES, SEASONINGS, ESSENTIAL OILS, OLEORESINS, AND NATURAL EXTRACTS General Provisions; Substances Generally Recognized As Safe
OINTMENT	524.154	BACITRACIN OR BACITRACIN ZINC-NEOMYCIN POLYMIXIN-B SULFATE OPHTHALMIC OINTMENT Ophthalmic And Topical Dosage Form New Animal Drugs
OINTMENT	524.155	BACITRACIN ZINC-POLYMYXIN B SULFATE, NEOMYCIN SULFATE-HYDROCORTISONE OR HYDROCORTISONE ACETATE OPHTHALMIC OINTMENT Ophthalmic And Topical Dosage Form New Animal Drugs
OINTMENT	524.390a	CHLORAMPHENICOL OPHTHALMIC OINTMENT Ophthalmic And Topical Dosage Form New Animal Drugs
OINTMENT	524.390d	CHLORAMPHENICOL-PREDNISOLONE OPHTHALMIC OINTMENT Ophthalmic And Topical Dosage Form New Animal Drugs
OINTMENT	524.402	CHLORHEXIDINE DIACETATE OINTMENT Ophthalmic And Topical Dosage Form New Animal Drugs
OINTMENT	524.575	CYCLOSPORINE OPHTHALMIC OINTMENT Ophthalmic And Topical Dosage Form New Animal Drugs
OINTMENT	524.1044c	GENTAMICIN OPHTHALMIC OINTMENT Ophthalmic And Topical Dosage Form New Animal Drugs
OINTMENT	524.1044d	GENTAMICIN SULFATE, BETAMETHASONE VALERATE OINTMENT Ophthalmic And Topical Dosage Form New Animal Drugs
OINTMENT	524.1044g	GENTAMICIN SULFATE, BETAMETHASONE VALERATE, CLOTRIMASOLE OINTMENT Ophthalmic And Topical Dosage Form New Animal Drugs
OINTMENT	524.1200a	KANAMYCIN OPHTHALMIC OINTMENT Ophthalmic And Topical Dosage Form New Animal Drugs
OINTMENT	524.1465	MUPIROCIN OINTMENT Ophthalmic And Topical Dosage Form New Animal Drugs
OINTMENT	524.1484a	NEOMYCIN SULFATE OPHTHALMIC OINTMENT Ophthalmic And Topical Dosage Form New Animal Drugs
OINTMENT	524.1484c	NEOMYCIN SULFATE, ISOFLUPREDONE ACETATE, TETRACAINE HYDROCHLORIDE OINTMENT Ophthalmic And Topical Dosage Form New Animal Drugs
OINTMENT	524.1484d	NEOMYCIN SULFATE, HYDROCORTISONE ACETATE, TETRACAINE HYDROCHLORIDE EAR OINTMENT Ophthalmic And Topical Dosage Form New Animal Drugs
OINTMENT	524.1484i	NEOMYCIN SULFATE, HYDROCORTISONE ACETATE, STERILE OINTMENT Ophthalmic And Topical Dosage Form New Animal Drugs
OINTMENT	524.1580b	NITROFURAZONE OINTMENT Ophthalmic And Topical Dosage Form New Animal Drugs
OINTMENT	524.1580e	NITROFURAZONE OINTMENT WITH BUTACAINE SULFATE Ophthalmic And Topical Dosage Form New Animal Drugs
OINTMENT	524.1600a	NYSTATIN, NEOMYCIN, THIOSTREPTON, AND TRIAMCINOLONE ACETONIDE OINTMENT Ophthalmic And Topical Dosage Form New Animal Drugs

OINTMENT	524.1600b	NYSTATIN, NEOMYCIN, THIOSTREPTON, AND TRIAMCINOLONE ACETONIDE OPHTHALMIC OINTMENT Ophthalmic And Topical Dosage Form New Animal Drugs
OINTMENT	524.1662	OXYTETRACYCLINE HYDROCHLORIDE, POLYMYXIN B SULFATE OPHTHALMIC OINTMENT Ophthalmic And Topical Dosage Form New Animal Drugs
OINTMENT	524.1880	PREDNISOLONE-NEOMYCIN SULFATE OPHTHALMIC OINTMENT Ophthalmic And Topical Dosage Form New Animal Drugs
OINTMENT	524.1883	PREDNISOLONE SODIUM PHOSPHATE-NEOMYCIN SULFATE OPHTHALMIC OINTMENT Ophthalmic And Topical Dosage Form New Animal Drugs
OLD ASIAGO CHEESE	133.104	ASIAGO OLD CHEESE Requirements For Specific Standardized Cheese And Related Products; Cheeses And Related Cheese Products
OLD DEVICES	860.84	CLASSIFICATION PROCEDURES FOR "OLD DEVICES" Classification; Medical Device Classification Procedures
OLEANDOMYCIN	556.480	OLEANDOMYCIN Specific Tolerances For Residues Of New Animal Drugs; Tolerances For Residues Of New Animal Drugs In Food
OLEANDOMYCIN	558.435	OLEANDOMYCIN Specific New Animal Drugs For Use In Animal Feeds; New Animal Drugs For Use In Animal Feeds
OLEATE, BUTYL	172.270	SULFATED BUTYL OLEATE Food Additives Permitted For Direct Addition To Food For Human Consumption
OLEATE, SODIUM	186.1770	SODIUM OLEATE Listing Of Specific Substances Affirmed As GRAS; Indirect Food Substances Affirmed As Generally Recognized As Safe
OLEATE SODIUM	522.1610	OLEATE SODIUM SOLUTION Implantation Or Injectable Dosage Form New Animal Drugs
OLEFIN	177.1520	OLEFIN POLYMERS Substances For Use As Basic Components Of Single And Repeated Use Food Contact Surfaces; Indirect Food Additives: Polymers
OLEIC ACID	172.862	OLEIC ACID DERIVED FROM TALL OIL FATTY ACIDS Multipurpose Additives; Food Additives Permitted For Direct Addition To Food For Human Consumption
OLEORESIN, PAPRIKA	73.345	PAPRIKA OLEORESIN Foods; Listing Of Color Additives Exempt From Certification
OLEORESIN, TUMERIC	73.615	TUMERIC OLEORESIN Foods; Listing Of Color Additives Exempt From Certification
OLEORESINS	182.20	ESSENTIAL OILS, OLEORESINS (SOLVENT-FREE), AND NATURAL EXTRACTIVES (INCLUDING DISTILLATES) General Provisions; Substances Generally Recognized As Safe
OLEORESINS	182.50	CERTAIN OTHER SPICES, SEASONINGS, ESSENTIAL OILS, OLEORESINS, AND NATURAL EXTRACTS General Provisions; Substances Generally Recognized As Safe
OLEORESINS	582.20	ESSENTIAL OILS, OLEORESINS (SOLVENT-FREE), AND NATURAL EXTRACTIVES (INCLUDING DISTILLATES) General Provisions; Substances Generally Recognized As Safe
OLEORESINS	582.50	CERTAIN OTHER SPICES, SEASONINGS, ESSENTIAL OILS, OLEORESINS, AND NATURAL EXTRACTS General Provisions; Substances Generally Recognized As Safe
OLESTRA	172.867	OLESTRA Food Additives Permitted For Direct Addition To Food For Human Consumption
OLIVE OIL	102.37	MIXTURES OF EDIBLE FAT OR OIL AND OLIVE OIL Requirements For Specific Nonstandardized Foods; Common Or Usual Name For Nonstandardized Foods
OLYMPIA OYSTERS	161.136	OLYMPIA OYSTERS Requirements For Specific Standardized Fish And Shellfish; Fish And Shellfish
OMEPRAZOLE	520.1615	OMEPRAZOLE Oral Dosage Form New Animal Drugs
ON-LINE	870.4330	CARDIOPULMONARY BYPASS ON-LINE BLOOD GAS MONITOR Cardiovascular Surgical Devices; Cardiovascular Devices
ONCOMETER	862.2720	PLASMA ONCOMETER FOR CLINICAL USE Clinical Laboratory Instruments; Clinical Chemistry And Clinical Toxicology Devices
ONCOMETRY	862.1530	PLASMA ONCOMETRY TEST SYSTEM Clinical Chemistry Test Systems; Clinical Chemistry And Clinical Toxicology Devices
ONDCP	1401.2	THE OFFICE OF NATIONAL DRUG CONTROL POLICY - ORGANIZATION AND FUNCTIONS Public Availability Of Information
ONDCP	1402.3	INFORMATION IN THE CUSTODY OF ONDCP Mandatory Declassification Review

ONDCP	1404.505	ONDCP RESPONSIBILITIES Governmentwide Debarment And Suspension And Requirements For Drug-Free Workplace (Grants)
ONION RINGS	102.39	ONION RINGS MADE FROM DICED ONION Requirements For Specific Nonstandardized Foods; Common Or Usual Name For Nonstandardized Foods
OPEN SYSTEMS	11.30	CONTROLS FOR OPEN SYSTEMS Electronic Records; Electronic Signatures
OPERATING HEADLAMP	886.4335	OPERATING HEADLAMP Surgical Devices; Ophthalmic Devices
OPERATING LIGHT	872.4630	DENTAL OPERATING LIGHT Surgical Devices; Dental Devices
OPERATING PROCEDURES	58.81	STANDARD OPERATING PROCEDURES Testing Facilities Operation; Good Laboratory Practice For Nonclinical Laboratory Studies
OPERATING PROCEDURES	120.6	SANITATION STANDARD OPERATING PROCEDURES Hazard Analysis And Critical Control Point (HACCP) Systems
OPERATING PROCEDURES	606.100	STANDARD OPERATING PROCEDURES Production And Process Controls; Current Good Manufacturing Practice For Blood And Blood Components
OPERATING ROOM	878.5070	AIR-HANDLING APPARATUS FOR A SURGICAL OPERATING ROOM Therapeutic Devices; General And Plastic Surgery Devices
OPERATING TABLE/CHAIR	878.4950	MANUAL OPERATING TABLE AND ACCESSORIES AND MANUAL OPERATING CHAIR AND ACCESSORIES Surgical Devices; General And Plastic Surgery Devices
OPERATING TABLES/CHAIRS	878.4960	OPERATING TABLES AND ACCESSORIES AND OPERATING CHAIRS AND ACCESSORIES Surgical Devices; General And Plastic Surgery Devices
OPERATION	58.49	LABORATORY OPERATION AREAS Facilities; Good Laboratory Practice For Nonclinical Laboratory Studies
OPERATIONAL PERIOD	26.11	START OF OPERATIONAL PERIOD Specific Sector Provisions For Pharmaceutical GMPs; Mutual Recognition Of Pharmaceutical GMP Reports, Medical Device Quality System Audit Reports
OPERATIONAL PERIOD	26.40	START OF THE OPERATIONAL PERIOD Specific Sector Provisions For Medical Devices; Mutual Recognition Of Pharmaceutical GMP Reports, Medical Device Quality System Audit Reports
OPERATIONS	56.108	IRB FUNCTIONS & OPERATIONS IRB Functions And Operations; Institutional Review Boards
OPERATIONS	113.87	OPERATIONS IN THE THERMAL PROCESSING ROOM Production And Process Controls; Thermally Processed Low-acid Foods Packaged In Hermetically Sealed Containers
OPERATIONS	211.130	PACKAGING AND LABELING OPERATIONS Packaging And Labeling Control; Current Good Manufacturing Practice For Finished Pharmaceuticals
OPERATIONS, SANITARY	110.35	SANITARY OPERATIONS Buildings And Facilities; Current Good Manufacturing Practice In Manufacturing, Packing, Or Holding Human Food
OPERATIONS, SANITARY	129.37	SANITARY OPERATIONS Buildings And Facilities; Processing And Bottling Of Bottled Drinking Water
OPERATIVE UNIT, DENTAL	872.6640	DENTAL OPERATIVE UNIT AND ACCESSORIES Miscellaneous Devices; Dental Devices
OPHTHALMIC	510.410	CORTICOSTEROIDS FOR ORAL, INJECTABLE, AND OPHTHALMIC USE IN ANIMALS, WARNINGS AND LABELING REQUIREMENTS Requirements For Specific New Animal Drugs; New Animal Drugs
OPHTHALMIC	524.154	BACITRACIN OR BACITRACIN ZINC-NEOMYCIN SULFATE-POLYMIXIN B SULFATE OPHTHALMIC OINTMENT Ophthalmic And Topical Dosage Form New Animal Drugs
OPHTHALMIC	524.155	BACITRACIN ZINC-POLYMIXIN B SULFATE-NEOMYCIN SULFATE-HYDROCORTISONE OR HYDROCORTISONE ACETATE OPHTHALMIC OINTMENT Ophthalmic And Topical Dosage Form New Animal Drugs
OPHTHALMIC	524.390	CHLORAMPHENICOL OPHTHALMIC AND TOPICAL DOSAGE FORMS Ophthalmic And Topical Dosage Form New Animal Drugs
OPHTHALMIC	524.575	CYCLOSPORINE OPHTHALMIC OINTMENT Ophthalmic And Topical Dosage Form New Animal Drugs
OPHTHALMIC	524.660	DIMETHYL SULFOXIDE OPHTHALMIC AND TOPICAL DOSAGE FORMS Ophthalmic And Topical Dosage Form New Animal Drugs
OPHTHALMIC	524.960	FLUMETHASONE, NEOMYCIN SULFATE, AND POLYMYXIN B SULFATE OPHTHALMIC SOLUTIONS Ophthalmic And Topical Dosage Form New Animal Drugs
OPHTHALMIC	524.981	FLUOCINOLONE ACETONIDE OPHTHALMIC AND TOPICAL DOSAGE FORMS Ophthalmic And Topical Dosage Form New Animal Drugs
OPHTHALMIC	524.1044	GENTAMICIN SULFATE OPHTHALMIC AND TOPICAL DOSAGE FORMS Ophthalmic And Topical Dosage Form New Animal Drugs

OPHTHALMIC	524.1044a	GENTAMICIN OPHTHALMIC SOLUTION Ophthalmic And Topical Dosage Form New Animal Drugs
OPHTHALMIC	524.1044c	GENTAMICIN OPHTHALMIC OINTMENT Ophthalmic And Topical Dosage Form New Animal Drug
OPHTHALMIC	524.1200	KANAMYCIN OPHTHALMIC AND TOPICAL DOSAGE FORMS Ophthalmic And Topical Dosage Form New Animal Drugs
OPHTHALMIC	524.1200b	KANAMYCIN OPHTHALMIC AQUEOUS SOLUTION Ophthalmic And Topical Dosage Form New Animal Drugs
OPHTHALMIC	524.1484	NEOMYCIN SULFATE OPHTHALMIC AND TOPICAL DOSAGE FORMS Ophthalmic And Topical Dosage Form New Animal Drugs
OPHTHALMIC	524.1484e	NEOMYCIN SULFATE AND POLYMYXIN B SULFATE OPHTHALMIC SOLUTION Ophthalmic And Topical Dosage Form New Animal Drugs
OPHTHALMIC	524.1580	NITROFURAZONE OPHTHALMIC AND TOPICAL DOSAGE FORMS Ophthalmic And Topical Dosage Form New Animal Drugs
OPHTHALMIC	524.1600	NYSTATIN OPHTHALMIC AND TOPICAL DOSAGE FORMS Ophthalmic And Topical Dosage Form New Animal Drugs
OPHTHALMIC	524.1600b	NYSTATIN, NEOMYCIN, THIOSTREPTON, AND TRIAMCINOLONE ACETONIDE OPHTHALMIC OINTMENT Ophthalmic And Topical Dosage Form New Animal Drugs
OPHTHALMIC	524.1662	OXYTETRACYCLINE HYDROCHLORIDE OPHTHALMIC AND TOPICAL DOSAGE FORMS Ophthalmic And Topical Dosage Form New Animal Drugs
OPHTHALMIC	524.1662b	OXYTETRACYCLINE HYDROCHLORIDE, POLYMYXIN B SULFATE OPHTHALMIC OINTMENT Ophthalmic And Topical Dosage Form New Animal Drugs
OPHTHALMIC	524.1880	PREDNISOLONE-NEOMYCIN SULFATE OPHTHALMIC OINTMENT Ophthalmic And Topical Dosage Form New Animal Drugs
OPHTHALMIC	524.1881	PREDNISOLONE ACETATE OPHTHALMIC AND TOPICAL DOSAGE FORMS Ophthalmic And Topical Dosage Form New Animal Drugs
OPHTHALMIC	524.1883	PREDNISOLONE SODIUM PHOSPHATE-NEOMYCIN SULFATE OPHTHALMIC OINTMENT Ophthalmic And Topical Dosage Form New Animal Drugs
OPHTHALMIC	524.1982	PROPARACAINE HYDROCHLORIDE OPHTHALMIC SOLUTION Ophthalmic And Topical Dosage Form New Animal Drugs
OPHTHALMIC	886.1120	OPHTHALMIC CAMERA Diagnostic Devices; Ophthalmic Devices
OPHTHALMIC	886.1140	OPHTHALMIC CHAIR Diagnostic Devices; Ophthalmic Devices
OPHTHALMIC	886.1405	OPHTHALMIC TRIAL LENS SET Diagnostic Devices; Ophthalmic Devices
OPHTHALMIC	886.1410	OPHTHALMIC TRIAL LENS CLIP Diagnostic Devices; Ophthalmic Devices
OPHTHALMIC	886.1415	OPHTHALMIC TRIAL LENS FRAME Diagnostic Devices; Ophthalmic Devices
OPHTHALMIC	886.1420	OPHTHALMIC LENS GAUGE Diagnostic Devices; Ophthalmic Devices
OPHTHALMIC	886.1430	OPHTHALMIC CONTACT LENS RADIUS MEASURING DEVICE Diagnostic Devices; Ophthalmic Devices
OPHTHALMIC	886.1640	OPHTHALMIC PREAMPLIFIER Diagnostic Devices; Ophthalmic Devices
OPHTHALMIC	886.1650	OPHTHALMIC BAR PRISM Diagnostic Devices; Ophthalmic Devices
OPHTHALMIC	886.1655	OPHTHALMIC FRESNEL PRISM Diagnostic Devices; Ophthalmic Devices
OPHTHALMIC	886.1665	OPHTHALMIC ROTARY PRISM Diagnostic Devices; Ophthalmic Devices
OPHTHALMIC	886.1670	OPHTHALMIC ISOTOPE UPTAKE PROBE Diagnostic Devices; Ophthalmic Devices
OPHTHALMIC	886.1680	OPHTHALMIC PROJECTOR Diagnostic Devices; Ophthalmic Devices
OPHTHALMIC	886.1760	OPHTHALMIC REFRACTOMETER Diagnostic Devices; Ophthalmic Devices
OPHTHALMIC	886.1860	OPHTHALMIC INSTRUMENT STAND Diagnostic Devices; Ophthalmic Devices
OPHTHALMIC	886.3100	OPHTHALMIC TANTALUM CLIP Prosthetic Devices; Ophthalmic Devices
OPHTHALMIC	886.3130	OPHTHALMIC CONFORMER Prosthetic Devices; Ophthalmic Devices
OPHTHALMIC	886.4230	OPHTHALMIC KNIFE TEST DRUM Surgical Devices; Ophthalmic Devices
OPHTHALMIC	886.4250	OPHTHALMIC ELECTROLYSIS UNIT Surgical Devices; Ophthalmic Devices
OPHTHALMIC	886.4350	MANUAL OPHTHALMIC SURGICAL INSTRUMENT Surgical Devices; Ophthalmic Devices

OPHTHALMIC	886.4390	OPHTHALMIC LASER Surgical Devices; Ophthalmic Devices
OPHTHALMIC	886.4570	OPHTHALMIC SURGICAL MARKER Surgical Devices; Ophthalmic Devices
OPHTHALMIC	886.4690	OPHTHALMIC PHOTOCOAGULATOR Surgical Devices; Ophthalmic Devices
OPHTHALMIC	886.4750	OPHTHALMIC EYE SHIELD Surgical Devices; Ophthalmic Devices
OPHTHALMIC	886.4770	OPHTHALMIC OPERATING SPECTACLES (LOUPES) Surgical Devices; Ophthalmic Devices
OPHTHALMIC	886.4790	OPHTHALMIC SPONGE Surgical Devices; Ophthalmic Devices
OPHTHALMIC	886.4855	OPHTHALMIC INSTRUMENT TABLE Surgical Devices; Ophthalmic Devices
OPHTHALMIC	886.5100	OPHTHALMIC BETA RADIATION SOURCE Therapeutic Devices; Ophthalmic Devices
OPHTHALMIC	886.5800	OPHTHALMIC BAR READER Therapeutic Devices; Ophthalmic Devices
OPHTHALMIC	886.5810	OPHTHALMIC PRISM READER Therapeutic Devices; Ophthalmic Devices
OPHTHALMIC ASTRINGENT	349.10	OPHTHALMIC ASTRINGENT Active Ingredients; Ophthalmic Drug Products For Over-the-counter Human Use
OPHTHALMIC ASTRINGENT	349.55	LABELING OF OPHTHALMIC ASTRINGENT DRUG PRODUCTS Labeling; Ophthalmic Drug Products For Over-the-counter Human Use
OPHTHALMIC DEMULCENT	349.60	LABELING OF OPHTHALMIC DEMULCENT DRUG PRODUCTS Labeling; Ophthalmic Drug Products For Over-the-counter Human Use
OPHTHALMIC DEMULCENTS	349.12	OPHTHALMIC DEMULCENTS Active Ingredients; Ophthalmic Drug Products For Over-the-counter Human Use
OPHTHALMIC DRUG	349.50	LABELING OF OPHTHALMIC DRUG PRODUCTS Labeling; Ophthalmic Drug Products For Over-the-counter Human Use
OPHTHALMIC EMOLLIENT	349.65	LABELING OF OPHTHALMIC EMOLLIENT DRUG PRODUCTS Labeling; Ophthalmic Drug Products For Over-the-counter Human Use
OPHTHALMIC EMOLLIENTS	349.14	OPHTHALMIC EMOLLIENTS Active Ingredients; Ophthalmic Drug Products For Over-the-counter Human Use
OPHTHALMIC HYPERTONICITY	349.16	OPHTHALMIC HYPERTONICITY AGENT Active Ingredients; Ophthalmic Drug Products For Over-the-counter Human Use
OPHTHALMIC HYPERTONICITY	349.70	LABELING OF OPHTHALMIC HYPERTONICITY DRUG PRODUCTS Labeling; Ophthalmic Drug Products For Over-the-counter Human Use
OPHTHALMIC PREPARATIONS	200.50	OPHTHALMIC PREPARATIONS AND DISPENSERS Requirements For Specific Classes Of Drugs; General
OPHTHALMIC VASOCONSTRICTOR	349.75	LABELING OF OPHTHALMIC VASOCONSTRICTOR DRUG PRODUCTS Labeling; Ophthalmic Drug Products For Over-the-counter Human Use
OPHTHALMIC VASOCONSTRICTORS	349.18	OPHTHALMIC VASOCONSTRICTORS Active Ingredients; Ophthalmic Drug Products For Over-the-counter Human Use
OPHTHALMOSCOPE	886.1570	OPHTHALMOSCOPE Diagnostic Devices; Ophthalmic Devices
OPIATE	862.3650	OPIATE TEST SYSTEM Clinical Toxicology Test Systems; Clinical Chemistry And Clinical Toxicology Devices
OPINION LETTERS	170.6	OPINION LETTERS ON FOOD ADDITIVE STATUS General Provisions; Food Additives
OPINION LETTERS	570.6	OPINION LETTERS ON FOOD ADDITIVE STATUS General Provisions; Food Additives
OPINIONS	310.100	NEW DRUG STATUS OPINIONS; STATEMENT OF POLICY Specific Administrative Rulings And Decisions; New Drugs
OPINIONS, ADVISORY	10.85	ADVISORY OPINIONS General Administrative Procedures; Administrative Practices And Procedures
OPINIONS, ADVISORY	12.83	ADVISORY OPINIONS Hearing Procedures; Formal Evidentiary Public Hearing
OPINIONS, ADVISORY	808.5	ADVISORY OPINIONS General Provisions; Exemption From Federal Preemption Of State & Local Medical Device Requirements
OPPORTUNITIES	7.87	RECORDS RELATED TO OPPORTUNITIES FOR PRESENTATION OF VIEWS CONDUCTED BEFORE REPORT OF CRIMINAL VIOLATION Criminal Violations; Enforcement Policy
OPPORTUNITY	5.202	ISSUANCE OF NOTICES OF OPPORTUNITY FOR A HEARING ON PROPOSALS FOR DENIAL OF APPROVAL OF APPLICATIONS FOR LICENSES, SUSPENSION OF LICENSES, OR REVOCATION OF LICENSES AND CERTAIN NOTICES Biologics; Redelegations Of Authority; Delegations Of Authority And Organization
OPPORTUNITY	7.84	OPPORTUNITY FOR PRESENTATION OF VIEWS BEFORE REPORT OF CRIMINAL VIOLATION Criminal Violations; Enforcement Policy

OPPORTUNITY	10.50	PROMULGATION OF REGULATIONS AND ORDERS AFTER AN OPPORTUNITY FOR A FORMAL EVIDENTIARY PUBLIC HEARING General Administrative Procedures; Administrative Practices And Procedures
OPPORTUNITY	58.204	NOTICE OF AND OPPORTUNITY FOR HEARING ON PROPOSED DISQUALIFICATION Disqualification Of Testing Facilities; Good Laboratory Practice For Nonclinical Laboratory Studies
OPPORTUNITY	310.6	APPLICABILITY OF "NEW DRUG" OR SAFETY/EFFECTIVENESS FINDINGS IN DRUG EFFICACY STUDY IMPLEMENTATION NOTICES AND NOTICES OF OPPORTUNITY FOR HEARING TO IDENTICAL, RELATED, AND SIMILAR DRUG PRODUCTS General Provisions; New Drugs
OPPORTUNITY	314.200	NOTICE OF OPPORTUNITY FOR HEARING; NOTICE OF PARTICIPATION AND REQUEST FOR HEARING; GRANT OR DENIAL OF HEARING Hearing Procedures For New Drugs; Applications For FDA Approval To Market A New Drug
OPPORTUNITY	514.200	CONTENTS OF NOTICE OF OPPORTUNITY FOR A HEARING Hearing Procedures; New Animal Drug Applications
OPPORTUNITY	515.30	CONTENTS OF NOTICE OF OPPORTUNITY FOR A HEARING Hearing Procedures; Medicated Feed Mill License
OPPORTUNITY	1404.313	OPPORTUNITY TO CONTEST PROPOSED DEBARMENT Debarment; Governmentwide Debarment And Suspension And Requirements For Drug-Free Workplace (Grants)
OPPORTUNITY	1404.412	OPPORTUNITY TO CONTEST PROPOSED SUSPENSION Suspension; Governmentwide Debarment And Suspension And Requirements For Drug-Free Workplace (Grants)
OPTICAL	886.5915	OPTICAL VISION AID Therapeutic Devices; Ophthalmic Devices
OPTOKINETIC	886.1200	OPTOKINETIC DRUM Diagnostic Devices; Ophthalmic Devices
ORAL	520.45	ALBENDAZOLE ORAL DOSAGE FORMS Oral Dosage Form New Animal Drugs
ORAL	520.82	AMINOPROPAZINE FUMARATE ORAL DOSAGE FORMS Oral Dosage Form New Animal Drugs
ORAL	520.88	AMOXICILLIN ORAL DOSAGE FORMS Oral Dosage Form New Animal Drugs
ORAL	520.90	AMPICILLIN ORAL DOSAGE FORMS Oral Dosage Form New Animal Drugs
ORAL	520.100	AMPROLIUM ORAL DOSAGE FORMS Oral Dosage Form New Animal Drugs
ORAL	520.154	BACITRACIN ORAL DOSAGE FORMS Oral Dosage Form New Animal Drugs
ORAL	520.300	CAMBENDAZOLE ORAL DOSAGE FORMS Oral Dosage Form New Animal Drugs
ORAL	520.315	CEFADROXIL POWDER FOR ORAL SUSPENSION Oral Dosage Form New Animal Drugs
ORAL	520.390	CHLORAMPHENICOL ORAL DOSAGE FORMS Oral Dosage Form New Animal Drugs
ORAL	520.445	CHLORTETRACYCLINE ORAL DOSAGE FORMS Oral Dosage Form New Animal Drugs
ORAL	520.530	CYTHIOATE ORAL LIQUID Oral Dosage Form New Animal Drugs
ORAL	520.540	DEXAMETHASONE ORAL DOSAGE FORMS Oral Dosage Form New Animal Drugs
ORAL	520.563	DIATRIZOATE MEGLUMINE AND DIATRIZOATE SODIUM ORAL SOLUTION Oral Dosage Form New Animal Drugs
ORAL	520.620	DIETHYLCARBAMAZINE ORAL DOSAGE FORMS Oral Dosage Form New Animal Drugs
ORAL	520.622	DIETHYLCARBAMAZINE CITRATE ORAL DOSAGE FORMS Oral Dosage Form New Animal Drugs
ORAL	520.763	DITHIAZANINE IODIDE ORAL DOSAGE FORMS Oral Dosage Form New Animal Drugs
ORAL	520.813	ENROFLOXACIN ORAL SOLUTION Oral Dosage Form New Animal Drugs
ORAL	520.903	FEBANTEL ORAL DOSAGE FORMS Oral Dosage Form New Animal Drugs
ORAL	520.905	FENBENDAZOLE ORAL DOSAGE FORMS Oral Dosage Form New Animal Drugs
ORAL	520.970	FLUNIXIN ORAL DOSAGE FORMS Oral Dosage Form New Animal Drugs
ORAL	520.1010	FUROSEMIDE ORAL DOSAGE FORMS Oral Dosage Form New Animal Drugs
ORAL	520.1044	GENTAMICIN SULFATE ORAL DOSAGE FORMS Oral Dosage Form New Animal Drugs

ORAL	520.1120	HALOXON ORAL DOSAGE FORMS Oral Dosage Form New Animal Drugs
ORAL	520.1130	HETACILLIN ORAL DOSAGE FORMS Oral Dosage Form New Animal Drugs
ORAL	520.1182	IRON DEXTRAN ORAL SUSPENSION Oral Dosage Form New Animal Drugs
ORAL	520.1242	LEVAMISOLE HYDROCHLORIDE ORAL DOSAGE FORMS Oral Dosage Form New Animal Drugs
ORAL	520.1263	LINCOMYCIN HYDROCHLORIDE MONOHYDRATE ORAL DOSAGE FORMS Oral Dosage Form New Animal Drugs
ORAL	520.1320	MEBENDAZOLE ORAL Oral Dosage Form New Animal Drugs
ORAL	520.1326	MEBENDAZOLE AND TRICHLORFON ORAL DOSAGE FORMS Oral Dosage Form New Animal Drugs
ORAL	520.1448	MONENSIN ORAL DOSAGE FORMS Oral Dosage Form New Animal Drugs
ORAL	520.1450	MORANTEL TARTRATE ORAL DOSAGE FORMS Oral Dosage Form New Animal Drugs
ORAL	520.1485	NEOMYCIN SULFATE ORAL SOLUTION Oral Dosage Form New Animal Drugs
ORAL	520.1696	PENICILLIN ORAL DOSAGE FORMS Oral Dosage Form New Animal Drugs
ORAL	520.1720	PHENYLBUTAZONE ORAL DOSAGE FORMS Oral Dosage Form New Animal Drugs
ORAL	520.1802	PIPERAZINE-CARBON DISULFIDE COMPLEX ORAL DOSAGE FORMS Oral Dosage Form New Animal Drugs
ORAL	520.2122	SPECTINOMYCIN DIHYDROCHLORIDE ORAL SOLUTION Oral Dosage Form New Animal Drugs
ORAL	520.2123	SPECTINOMYCIN DIHYDROCHLORIDE PENTAHYDRATE ORAL DOSAGE FORMS Oral Dosage Form New Animal Drugs
ORAL	520.2150	STANOZOLOL ORAL DOSAGE FORMS Oral Dosage Form New Animal Drugs
ORAL	520.2158	STREPTOMYCIN/DIHYDROSTREPTOMYCIN ORAL DOSAGE FORMS Oral Dosage Form New Animal Drugs
ORAL	520.2160	STYRYLPYRIDINIUM, DIETHYLCARBAMAZINE ORAL DOSAGE FORMS Oral Dosage Form New Animal Drugs
ORAL	520.2200	SULFACHLORPYRIDAZINE ORAL DOSAGE FORMS Oral Dosage Form New Animal Drugs
ORAL	520.2220	SULFADIMETHOXINE ORAL DOSAGE FORMS Oral Dosage Form New Animal Drugs
ORAL	520.2260	SULFAMETHAZINE ORAL DOSAGE FORMS Oral Dosage Form New Animal Drugs
ORAL	520.2261	SULFAMETHAZINE SODIUM ORAL DOSAGE FORMS Oral Dosage Form New Animal Drugs
ORAL	520.2325	SULFAQUINOXALINE ORAL DOSAGE FORMS Oral Dosage Form New Animal Drugs
ORAL	520.2345	TETRACYCLINE ORAL DOSAGE FORMS Oral Dosage Form New Animal Drugs
ORAL	520.2380	THIABENDAZOLE ORAL DOSAGE FORMS Oral Dosage Form New Animal Drugs
ORAL	520.2380b	THIABENDAZOLE DRENCH OR ORAL PASTE Oral Dosage Form New Animal Drugs
ORAL	520.2473	TIOXIDAZOLE ORAL DOSAGE FORMS Oral Dosage Form New Animal Drugs
ORAL	520.2482	TRIAMCINOLONE ACETONIDE ORAL POWDER Oral Dosage Form New Animal Drugs
ORAL	520.2520	TRICHLORFON ORAL DOSAGE FORMS Oral Dosage Form New Animal Drugs
ORAL	520.2611	TRIMETHOPRIM AND SULFADIAZINE ORAL PASTE Oral Dosage Form New Animal Drugs
ORAL	520.2612	TRIMETHOPRIM AND SULFADIAZINE ORAL SUSPENSION Oral Dosage Form New Animal Drugs
ORAL ADMINISTRATION	310.537	DRUG PRODUCTS CONTAINING ACTIVE INGREDIENTS OFFERED OVER-THE-COUNTER (OTC) FOR ORAL ADMINISTRATION FOR THE TREATMENT OF FEVER BLISTERS AND COLD SORES Requirements For Specific New Drugs Or Devices; New Drugs
ORAL CAVITY	872.6030	ORAL CAVITY ABRASIVE POLISHING AGENT Miscellaneous Devices; Dental Devices
ORAL CONTRACEPTIVES	310.501	PATIENT PACKAGE INSERTS FOR ORAL CONTRACEPTIVES Requirements For Specific New Drugs Or Devices; New Drugs
ORAL CORTICOSTEROIDS	510.410	CORTICOSTEROIDS FOR ORAL, INJECTABLE, AND OPHTHALMIC USE IN ANIMALS; WARNINGS AND LABELING REQUIREMENTS Requirements For Specific New Animal Drugs; New Animal Drugs

ORAL DIGITALIS	201.317	DIGITALIS AND RELATED CARDIOTONIC DRUGS FOR HUMAN USE IN ORAL DOSAGE FORMS; REQUIRED WARNING Specific Labeling Requirements For Specific Drug Products; Labeling
ORAL DIGOXIN	310.500	DIGOXIN PRODUCTS FOR ORAL USE; CONDITIONS FOR MARKETING Requirements For Specific New Drugs Or Devices; New Drugs
ORAL DISCUSSIONS	20.104	SUMMARIES OF ORAL DISCUSSIONS Availability Of Specific Categories Of Records; Public Information
ORAL DOSAGE IMPRINTING	330.3	IMPRINTING OF SOLID ORAL DOSAGE FORM DRUG PRODUCTS Over-the-counter Human Drugs Generally Recognized As Safe And Effective & Not Misbranded
ORAL HEARINGS	12.90	CONDUCT AT ORAL HEARINGS OR CONFERENCES Hearing Procedures; Formal Evidentiary Public Hearing
ORAL HYGIENE	872.6650	MASSAGING PICK OR TIP OR ORAL HYGIENE Miscellaneous Devices; Dental Devices
ORAL HYPOGLYCEMIC DRUGS	310.517	LABELING FOR ORAL HYPOGLYCEMIC DRUGS OF THE SULFONYLUREA CLASS Requirements For Specific New Drugs Or Devices; New Drugs
ORAL INGESTION	201.306	POTASSIUM SALT PREPARATIONS INTENDED FOR ORAL INGESTION BY MAN Specific Labeling Requirements For Specific Drug Products; Labeling
ORAL INGESTION	328.50	PRINCIPAL DISPLAY PANEL OF ALL OTC DRUG PRODUCTS INTENDED FOR ORAL INGESTION THAT CONTAIN ALCOHOL Over-the-counter Drug Products Intended For Oral Ingestion That Contain Alcohol
ORAL INHALATION	200.51	AQUEOUS-BASED DRUG PRODUCTS FOR ORAL INHALATION Requirements For Specific Classes Of Drugs; General
ORAL INSECT REPELLANTS	310.529	DRUG PRODUCTS CONTAINING ACTIVE INGREDIENTS OFFERED OVER-THE-COUNTER (OTC) FOR ORAL USE AS INSECT REPELLENTS Requirements For Specific New Drugs Or Devices; New Drugs
ORAL IRRIGATION	872.6510	ORAL IRRIGATION UNIT Miscellaneous Devices; Dental Devices
ORAL TESTIMONY	12.87	PURPOSE; ORAL AND WRITTEN TESTIMONY; BURDEN OF PROOF Hearing Procedures; Formal Evidentiary Public Hearing
ORAL WOUND HEALING	310.534	DRUG PRODUCTS CONTAINING ACTIVE INGREDIENTS OFFERED OVER-THE-COUNTER (OTC) FOR HUMAN USE AS ORAL WOUND HEALING AGENTS Requirements For Specific New Drugs Or Devices; New Drugs
ORANGE 1, VAT	73.3112	C.I. VAT ORANGE 1 Medical Devices; Listing Of Color Additives Exempt From Certification
ORANGE B	74.250	ORANGE B Foods; Listing Of Color Additives Subject To Certification
ORANGE JUICE	146.135	ORANGE JUICE Requirements For Specific Standardized Canned Fruit Juices And Beverages; Canned Fruit Juices
ORANGE JUICE	146.137	FROZEN ORANGE JUICE Requirements For Specific Standardized Canned Fruit Juices And Beverages; Canned Fruit Juices
ORANGE JUICE	146.140	PASTEURIZED ORANGE JUICE Requirements For Specific Standardized Canned Fruit Juices And Beverages; Canned Fruit Juices
ORANGE JUICE	146.141	CANNED ORANGE JUICE Requirements For Specific Standardized Canned Fruit Juices And Beverages; Canned Fruit Juices
ORANGE JUICE	146.145	ORANGE JUICE FROM CONCENTRATE Requirements For Specific Standardized Canned Fruit Juices And Beverages; Canned Fruit Juices
ORANGE JUICE	146.146	FROZEN CONCENTRATED ORANGE JUICE Requirements For Specific Standardized Canned Fruit Juices And Beverages; Canned Fruit Juices
ORANGE JUICE	146.148	REDUCED ACID FROZEN CONCENTRATED ORANGE JUICE Requirements For Specific Standardized Canned Fruit Juices And Beverages; Canned Fruit Juices
ORANGE JUICE	146.150	CANNED CONCENTRATED ORANGE JUICE Requirements For Specific Standardized Canned Fruit Juices And Beverages; Canned Fruit Juices
ORANGE JUICE	146.151	ORANGE JUICE FOR MANUFACTURING Requirements For Specific Standardized Canned Fruit Juices And Beverages; Canned Fruit Juices
ORANGE JUICE	146.152	ORANGE JUICE WITH PRESERVATIVE Requirements For Specific Standardized Canned Fruit Juices And Beverages; Canned Fruit Juices
ORANGE JUICE	146.153	CONCENTRATED ORANGE JUICE FOR MANUFACTURING Requirements For Specific Standardized Canned Fruit Juices And Beverages; Canned Fruit Juices

ORANGE JUICE	146.154	CONCENTRATED ORANGE JUICE WITH PRESERVATIVE Requirements For Specific Standardized Canned Fruit Juices And Beverages; Canned Fruit Juices
ORANGE NO. 4	74.1254	D&C ORANGE NO. 4 Drugs; Listing Of Color Additives Subject To Certification
ORANGE NO. 4	74.2254	D&C ORANGE NO. 4 Cosmetics; Listing Of Color Additives Subject To Certification
ORANGE NO. 4	82.1254	D&C ORANGE NO. 4 Drugs And Cosmetics; Listing Of Certified Provisionally Listed Colors And Specifications
ORANGE NO. 5	74.1255	D&C ORANGE NO. 5 Drugs; Listing Of Color Additives Subject To Certification
ORANGE NO. 5	74.2255	D&C ORANGE NO. 5 Cosmetics; Listing Of Color Additives Subject To Certification
ORANGE NO. 5	82.1255	D&C ORANGE NO. 5 Drugs And Cosmetics; Listing Of Certified Provisionally Listed Colors And Specifications
ORANGE NO. 10	74.1260	D&C ORANGE NO. 10 Drugs; Listing Of Color Additives Subject To Certification
ORANGE NO. 10	74.2260	D&C ORANGE NO. 10 Cosmetics; Listing Of Color Additives Subject To Certification
ORANGE NO. 10	82.1260	D&C ORANGE NO. 10 Drugs And Cosmetics; Listing Of Certified Provisionally Listed Colors And Specifications
ORANGE NO. 11	74.1261	D&C ORANGE NO. 11 Drugs; Listing Of Color Additives Subject To Certification
ORANGE NO. 11	74.2261	D&C ORANGE NO. 11 Cosmetics; Listing Of Color Additives Subject To Certification
ORANGE NO. 11	82.1261	D&C ORANGE NO. 11 Drugs And Cosmetics; Listing Of Certified Provisionally Listed Colors And Specifications
ORBIFLOXACIN	520.1616	ORBIFLOXACIN PASTE Oral Dosage Form New Animal Drugs
ORBITAL IMPLANT	886.3340	EXTRAOCULAR ORBITAL IMPLANT Prosthetic Devices; Ophthalmic Devices
ORDER	12.21	INITIATION OF A HEARING INVOLVING THE ISSUANCE, AMENDMENT, OR REVOCATION OF AN ORDER Initiation Of Proceedings; Formal Evidentiary Public Hearing
ORDER	12.22	FILING OBJECTIONS AND REQUESTS FOR A HEARING ON A REGULATION OR ORDER Initiation Of Proceedings; Formal Evidentiary Public Hearing
ORDER	12.26	MODIFICATION OR REVOCATION OF REGULATION OR ORDER Initiation Of Proceedings; Formal Evidentiary Public Hearing
ORDER	12.38	EFFECTIVE DATE OF AN ORDER Initiation Of Proceedings; Formal Evidentiary Public Hearing
ORDER	17.28	PROTECTIVE ORDER Civil Money Penalties Hearings
ORDER	20.83	DISCLOSURE REQUIRED BY COURT ORDER Limitations On Exemptions; Public Information
ORDER	514.120	REVOCATION OF ORDER REFUSING TO APPROVE AN APPLICATION OR SUSPENDING OR WITHDRAWING APPROVAL OF AN APPLICATION Administrative Actions On Applications; New Animal Drug Applications
ORDER	810.10	CEASE DISTRIBUTION AND NOTIFICATION ORDER Mandatory Medical Device Recall Procedures; Medical Device Recall Authority
ORDER	810.12	WRITTEN REQUEST FOR REVIEW OF CEASE DISTRIBUTION AND NOTIFICATION ORDER Mandatory Medical Device Recall Procedures; Medical Device Recall Authority
ORDER	810.13	MANDATORY RECALL ORDER Mandatory Medical Device Recall Procedures; Medical Device Recall Authority
ORDER	1301.37	ORDER TO SHOW CAUSE Action On Application For Registration: Revocation Or Suspension Of Registration; Registration Of Manufacturers, Distributors, And Dispensers Of Controlled Substances
ORDER	1309.46	ORDER TO SHOW CAUSE Registration Of Manufacturers, Distributors, Importers And Exporters Of List I Chemicals
ORDER, FINAL	58.206	FINAL ORDER ON DISQUALIFICATION Disqualification Of Testing Facilities; Good Laboratory Practice For Nonclinical Laboratory Studies
ORDER, FINAL	1301.36	SUSPENSION OF REGISTRATION; SUSPENSION PENDING FINAL ORDER; EXTENSION OF REGISTRATION PENDING FINAL ORDER Action On Application For Registration: Revocation Or Suspension Of Registration; Registration Of Manufacturers, Distributors, And Dispensers Of Controlled Substances

ORDER, FINAL	1301.46	FINAL ORDER Hearings; Registration Of Manufacturers, Distributors, And Dispensers Of Controlled Substances
ORDER, FINAL	1303.37	FINAL ORDER Hearings; Quotas
ORDER, FINAL	1308.45	FINAL ORDER Hearings; Schedules Of Controlled Substances
ORDER, FINAL	1309.44	SUSPENSION OF REGISTRATION PENDING FINAL ORDER Registration Of Manufacturers, Distributors, Importers And Exporters Of List I Chemicals
ORDER, FINAL	1309.45	EXTENSION OF REGISTRATION PENDING FINAL ORDER Registration Of Manufacturers, Distributors, Importers And Exporters Of List I Chemicals
ORDER, FINAL	1312.47	FINAL ORDER Hearings; Importation And Exportation Of Controlled Substances
ORDER, FINAL	1313.57	FINAL ORDER Hearings; Importation And Exportation Of Precursors And Essential Chemicals
ORDER, FINAL	1316.67	FINAL ORDER Administrative Hearings; Administrative Functions, Practices, And Procedures
ORDER FORMS	1305.03	DISTRIBUTIONS REQUIRING ORDER FORMS Order Forms
ORDER FORMS	1305.04	PERSONS ENTITLED TO OBTAIN AND EXECUTE ORDER FORMS Order Forms
ORDER FORMS	1305.05	PROCEDURE FOR OBTAINING ORDER FORMS Order Forms
ORDER FORMS	1305.06	PROCEDURE FOR EXECUTING ORDER FORMS Order Forms
ORDER FORMS	1305.08	PERSONS ENTITLED TO FILL ORDER FORMS Order Forms
ORDER FORMS	1305.09	PROCEDURE FOR FILLING ORDER FORMS Order Forms
ORDER FORMS	1305.10	PROCEDURE FOR ENDORSING ORDER FORMS Order Forms
ORDER FORMS	1305.11	UNACCEPTED AND DEFECTIVE ORDER FORMS Order Forms
ORDER FORMS	1305.12	LOST AND STOLEN ORDER FORMS Order Forms
ORDER FORMS	1305.13	PRESERVATION OF ORDER FORMS Order Forms
ORDER FORMS	1305.14	RETURN OF UNUSED ORDER FORMS Order Forms
ORDER FORMS	1305.15	CANCELLATION AND VOIDING OF ORDER FORMS Order Forms
ORDER FORMS	1305.16	SPECIAL PROCEDURE FOR FILLING CERTAIN ORDER FORMS Order Forms
ORDER, RECALL	810.13	MANDATORY RECALL ORDER Mandatory Medical Device Recall Procedures; Medical Device Recall Authority
ORDER, RECALL	810.15	COMMUNICATIONS CONCERNING A CEASE DISTRIBUTION AND NOTIFICATION OR MANDATORY RECALL ORDER Mandatory Medical Device Recall Procedures; Medical Device Recall Authority
ORDER, RECALL	810.16	CEASE DISTRIBUTION AND NOTIFICATION OR MANDATORY RECALL ORDER STATUS REPORTS Mandatory Medical Device Recall Procedures; Medical Device Recall Authority
ORDER, RECALL	810.17	TERMINATION OF A CEASE DISTRIBUTION AND NOTIFICATION OR MANDATORY RECALL ORDER Mandatory Medical Device Recall Procedures; Medical Device Recall Authority
ORDERS	5.34	ISSUANCE OF NOTICES RELATING TO PROPOSALS AND ORDERS FOR DEBARMENT AND DENIAL OF AN APPLICATION TO TERMINATE DEBARMENT General Redelegations Of Authority; Delegations Of Authority And Organization
ORDERS	5.301	ISSUANCE OF INITIAL EMERGENCY PERMIT ORDERS AND NOTICES OF CONFIRMATION OF EFFECTIVE DATE OF FINAL REGULATIONS ON FOOD FOR HUMAN AND ANIMAL CONSUMPTION Foods And Cosmetics; Redelegations Of Authority; Delegations Of Authority And Organization
ORDERS	5.410	ORDERS TO REPAIR OR REPLACE, OR MAKE REFUNDS FOR, MEDICAL DEVICES Medical Devices And Radiological Health; Redelegations Of Authority; Delegations Of Authority And Organization
ORDERS	5.502	ISSUANCE OF NOTICES, PROPOSALS, AND ORDERS RELATING TO NEW ANIMAL DRUGS AND MEDICATED FEED MILL LICENSE APPLICATIONS Animal Drugs; Redelegations Of Authority; Delegations Of Authority And Organization

ORPHAN-DRUG	316.31	SCOPE OF ORPHAN-DRUG EXCLUSIVE APPROVAL Orphan-Drug Exclusive Approval; Orphan Drugs
ORPHAN DRUG	316.40	TREATMENT USE OF A DESIGNATED ORPHAN DRUG Open Protocols For Investigations; Orphan Drugs
ORPHAN DRUGS	316.36	INSUFFICIENT QUANTITIES OF ORPHAN DRUGS Orphan-Drug Exclusive Approval; Orphan Drugs
ORPHAN PRODUCTS	5.900	ORPHAN PRODUCTS Orphan Products; Redelegations Of Authority; Delegations Of Authority And Organization
ORTHODONTIC	872.5410	ORTHODONTIC APPLIANCE AND ACCESSORIES Therapeutic Devices; Dental Devices
ORTHODONTIC	872.5470	ORTHODONTIC PLASTIC BRACKET Therapeutic Devices; Dental Devices
ORTHODONTIC	872.5500	EXTRAORAL ORTHODONTIC HEADGEAR Therapeutic Devices; Dental Devices
ORTHOPEDIC	888.4540	ORTHOPEDIC MANUAL SURGICAL INSTRUMENT Surgical Devices; Orthopedic Devices
ORTHOPEDIC	888.5850	NONPOWERED ORTHOPEDIC TRACTION APPARATUS AND ACCESSORIES Surgical Devices; Orthopedic Devices
ORTHOSIS, CRANIAL	882.5970	CRANIAL ORTHOSIS Neurological Therapeutic Devices; Neurological Devices
ORTHOSIS, FIXATION	888.3050	SPINAL INTERLAMINAL FIXATION ORTHOSIS Prosthetic Devices; Orthopedic Devices
ORTHOSIS, FIXATION	888.3060	SPINAL INTERVERTEBRAL BODY FIXATION ORTHOSIS Prosthetic Devices; Orthopedic Devices
ORTHOSIS, LIMB	890.3474	LIMB ORTHOSIS Physical Medicine Prosthetic Devices; Physical Medicine Devices
ORTHOSIS, PNEUMATIC	890.3610	RIGID PNEUMATIC STRUCTURE ORTHOSIS Physical Medicine Prosthetic Devices; Physical Medicine Devices
ORTHOSIS, TRUNCAL	890.3490	TRUNCAL ORTHOSIS Physical Medicine Prosthetic Devices; Physical Medicine Devices
ORTHOTIC	890.3025	PROSTHETIC AND ORTHOTIC ACCESSORY Physical Medicine Prosthetic Devices; Physical Medicine Devices
ORTHOTIC	890.3410	EXTERNAL LIMB ORTHOTIC COMPONENT Physical Medicine Prosthetic Devices; Physical Medicine Devices
OSCILLOMETER	870.2675	OSCILLOMETER Cardiovascular Monitoring Devices; Cardiovascular Devices
OSMOLALITY	862.1540	OSMOLALITY TEST SYSTEM Clinical Chemistry Test Systems; Clinical Chemistry And Clinical Toxicology Devices
OSMOMETER	862.2730	OSMOMETER FOR CLINICAL USE Clinical Laboratory Instruments; Clinical Chemistry And Clinical Toxicology Devices
OSMOSIS, REVERSE	177.2550	REVERSE OSMOSIS MEMBRANES Substances For Use Only As Components Of Articles Intended For Repeated Use; Indirect Food Additives: Polymers
OSMOTIC FRAGILITY	864.6600	OSMOTIC FRAGILITY TEST Manual Hematology Devices; Hematology And Pathology Devices
OSSICULAR	874.3450	PARTIAL OSSICULAR REPLACEMENT PROSTHESIS Prosthetic Devices; Ear, Nose And Throat Devices
OSSICULAR	874.3495	TOTAL OSSICULAR REPLACEMENT PROSTHESIS Prosthetic Devices; Ear, Nose And Throat Devices
OSSICULAR	874.3540	PROSTHESIS MODIFICATION INSTRUMENT FOR OSSICULAR REPLACEMENT SURGERY Prosthetic Devices; Ear, Nose And Throat Devices
OSTEOPOROSIS	101.72	HEALTH CLAIMS: CALCIUM AND OSTEOPOROSIS Food Labeling
OSTOMY	876.5895	OSTOMY IRRIGATOR Therapeutic Devices; Gastroenterology-urology Devices
OSTOMY	876.5900	OSTOMY POUCH AND ACCESSORIES Therapeutic Devices; Gastroenterology-urology Devices
OTC	201.66	FORMAT AND CONTENT REQUIREMENTS FOR OVER-THE-COUNTER (OTC) DRUG PRODUCT LABELING Labeling Requirements For Over-the-counter Drugs; Labeling
OTC	211.132	TAMPER-EVIDENT PACKAGING REQUIREMENTS FOR OVER-THE-COUNTER (OTC) HUMAN DRUG PRODUCTS Packaging And Labeling Control; Current Good Manufacturing Practice For Finished Pharmaceuticals
OTC	310.519	DRUG PRODUCTS MARKETED AS OVER-THE-COUNTER (OTC) DAYTIME SEDATIVES Requirements For Specific New Drugs Or Devices; New Drugs
OTC	310.527	DRUG PRODUCTS CONTAINING ACTIVE INGREDIENTS OFFERED OVER-THE-COUNTER (OTC) FOR EXTERNAL USE A HAIR GROWERS OR FOR HAIR LOSS PREVENTION Requirements For Specific New Drugs Or Devices; New Drugs

OTC	310.528	DRUG PRODUCTS CONTAINING ACTIVE INGREDIENTS OFFERED OVER-THE-COUNTER (OTC) FOR USE AS AN APHRODISIAC Requirements For Specific New Drugs Or Devices; New Drugs
OTC	310.529	DRUG PRODUCTS CONTAINING ACTIVE INGREDIENTS OFFERED OVER-THE-COUNTER (OTC) FOR ORAL USE AS INSECT REPELLENTS Requirements For Specific New Drugs Or Devices; New Drugs
OTC	310.530	TOPICALLY APPLIED HORMONE-CONTAINING DRUG PRODUCTS FOR OVER-THE-COUNTER (OTC) HUMAN USE Requirements For Specific New Drugs Or Devices; New Drugs
OTC	310.531	DRUG PRODUCTS CONTAINING ACTIVE INGREDIENTS OFFERED OVER-THE-COUNTER (OTC) FOR THE TREATMENT OF BOILS Requirements For Specific New Drugs Or Devices; New Drugs
OTC	310.532	DRUG PRODUCTS CONTAINING ACTIVE INGREDIENTS OFFERED OVER-THE-COUNTER (OTC) TO RELIEVE THE SYMPTOMS OF BENIGN PROSTATIC HYPERTROPHY Requirements For Specific New Drugs Or Devices; New Drugs
OTC	310.533	DRUG PRODUCTS CONTAINING ACTIVE INGREDIENTS OFFERED OVER-THE-COUNTER (OTC) FOR HUMAN USE AS AN ANTICHOLINERGIC IN COUGH-COLD DRUG PRODUCTS Requirements For Specific New Drugs Or Devices; New Drugs
OTC	310.534	DRUG PRODUCTS CONTAINING ACTIVE INGREDIENTS OFFERED OVER-THE-COUNTER (OTC) FOR HUMAN USE AS ORAL WOUND HEALING AGENTS Requirements For Specific New Drugs Or Devices; New Drugs
OTC	310.536	DRUG PRODUCTS CONTAINING ACTIVE INGREDIENTS OFFERED OVER-THE-COUNTER (OTC) FOR USE AS A NAILBITING OR THUMBSUCKING DETERRENT Requirements For Specific New Drugs Or Devices; New Drugs
OTC	310.537	DRUG PRODUCTS CONTAINING ACTIVE INGREDIENTS OFFERED OVER-THE-COUNTER (OTC) FOR TREATMENT OF FEVER BLISTERS AND COLD SORES Requirements For Specific New Drugs Or Devices; New Drugs
OTC	310.538	DRUG PRODUCTS CONTAINING ACTIVE INGREDIENTS OFFERED OVER-THE-COUNTER (OTC) FOR USE FOR INGROWN TOENAIL RELIEF Requirements For Specific New Drugs Or Devices; New Drugs
OTC	310.540	DRUG PRODUCTS CONTAINING ACTIVE INGREDIENTS OFFERED OVER-THE-COUNTER (OTC) FOR USE AS STOMACH ACIDIFIERS Requirements For Specific New Drugs Or Devices; New Drugs
OTC	310.541	OVER-THE-COUNTER (OTC) DRUG PRODUCTS CONTAINING ACTIVE INGREDIENTS OFFERED FOR USE IN THE TREATMENT OF HYPOPHOSPHATEMIA Requirements For Specific New Drugs Or Devices; New Drugs
OTC	310.542	OVER-THE-COUNTER (OTC) DRUG PRODUCTS CONTAINING ACTIVE INGREDIENTS OFFERED FOR USE IN THE TREATMENT OF HYPERPHOSPHATEMIA Requirements For Specific New Drugs Or Devices; New Drugs
OTC	310.543	DRUG PRODUCTS CONTAINING ACTIVE INGREDIENTS OFFERED OVER-THE-COUNTER (OTC) FOR HUMAN USE IN EXOCRINE PANCREATIC INSUFFICIENCY Requirements For Specific New Drugs Or Devices; New Drugs
OTC	310.544	DRUG PRODUCTS CONTAINING ACTIVE INGREDIENTS OFFERED OVER-THE-COUNTER (OTC) FOR USE AS A SMOKING DETERRENT Requirements For Specific New Drugs Or Devices; New Drugs
OTC	310.545	DRUG PRODUCTS CONTAINING CERTAIN ACTIVE INGREDIENTS OFFERED OVER-THE-COUNTER (OTC) FOR CERTAIN USES Requirements For Specific New Drugs Or Devices; New Drugs
OTC	310.546	DRUG PRODUCTS CONTAINING ACTIVE INGREDIENTS OFFERED OVER-THE-COUNTER (OTC) FOR THE TREATMENT AND/OR PREVENTION OF NOCTURNAL LEG MUSCLE CRAMPS Requirements For Specific New Drugs Or Devices; New Drugs
OTC	310.547	DRUG PRODUCTS CONTAINING QUININE OFFERED OVER-THE-COUNTER (OTC) FOR THE TREATMENT AND/OR PREVENTION OF MALARIA Requirements For Specific New Drugs Or Devices; New Drugs
OTC	310.548	DRUG PRODUCTS CONTAINING COLLOIDAL SILVER INGREDIENTS OFFERED OVER-THE-COUNTER (OTC) FOR THE TREATMENT OF DISEASE Requirements For Specific New Drugs Or Devices; New Drugs
OTC	328.50	PRINCIPAL DISPLAY PANEL OF ALL OTC DRUG PRODUCTS INTENDED FOR ORAL INGESTION THAT CONTAIN ALCOHOL Over-the-counter Drug Products Intended For Oral Ingestion That Contain Alcohol
OTC	330.10	PROCEDURES FOR CLASSIFYING OTC DRUGS AS GENERALLY RECOGNIZED AS SAFE AND EFFECTIVE AND NOT MISBRANDED Administrative Procedures; Over-the-counter Human Drugs Generally Recognized As Safe And Effective & Not Misbranded

OTC	330.12	STATUS OF OVER-THE-COUNTER (OTC) DRUGS PREVIOUSLY REVIEWED UNDER THE DRUG EFFICACY STUDY (DESI) Administrative Procedures; Over-the-counter Human Drugs Generally Recognized As Safe And Effective & Not Misbranded
OTC	330.13	CONDITIONS FOR MARKETING INGREDIENTS RECOMMENDED FOR OVER-THE-COUNTER (OTC) USE UNDER THE OTC DRUG REVIEW Administrative Procedures; Over-the-counter Human Drugs Generally Recognized As Safe And Effective & Not Misbranded
OTC	341.70	LABELING OF OTC DRUG PRODUCTS CONTAINING INGREDIENTS THAT ARE USED FOR TREATING CONCURRENT SYMPTOMS Cold, Cough, Allergy, Bronchodilator, And Antiasthmatic Drug Products For OTC Human Use
OTC	809.40	RESTRICTIONS ON THE SALE, DISTRIBUTION AND USE OF OTC TEST SAMPLE COLLECTION SYSTEMS FOR DRUGS OF ABUSE TESTING Requirements For Manufacturers And Producers; In Vitro Diagnostic Products For Human Use
OTC	864.3260	OTC TEST SAMPLE COLLECTION SYSTEMS FOR DRUGS OF ABUSE TESTING Pathology Instrumentation And Accessories; Hematology And Pathology Devices
OTC	872.3520	OTC DENTURE CLEANSER Prosthetic Devices; Dental Devices
OTC	872.3540	OTC DENTURE CUSHION OR PAD Prosthetic Devices; Dental Devices
OTC	872.3560	OTC DENTURE RELINER Prosthetic Devices; Dental Devices
OTC	872.3570	OTC DENTURE REPAIR KIT Prosthetic Devices; Dental Devices
OTIC	524.981e	FLUOCINOLONE ACETONIDE, DIMETHYL SULFOXIDE OTIC SOLUTION Ophthalmic And Topical Dosage Form New Animal Drugs
OTIC	524.1044b	GENTAMICIN SULFATE, BETAMETHASONE VALERATE OTIC SOLUTION Ophthalmic And Topical Dosage Form New Animal Drugs
OTIC	524.1044h	GENTAMICIN SULFATE, MOMETASONE FUROATE, CLOTRIMAZOLE OTIC SUSPENSION Ophthalmic And Topical Dosage Form New Animal Drugs
OTIC	524.1195	IVERMECTIN OTIC SUSPENSION Ophthalmic And Topical Dosage Form New Animal Drugs
OTIC	524.1484k	NEOMYCIN SULFATE, PREDNISOLONE, TETRACAINE, AND SQUALANE TOPICAL-OTIC SUSPENSION Ophthalmic And Topical Dosage Form New Animal Drugs
OTOLOGY	874.4490	ARGON LASER FOR OTOLOGY, RHINOLOGY, AND LARYNGOLOGY Surgical Devices; Ear, Nose, And Throat Devices
OTOSCOPE	874.4770	OTOSCOPE Surgical Devices; Ear, Nose, And Throat Devices
OUCHTERLONY	866.4600	OUCHTERLONY AGAR PLATE Immunology Laboratory Equipment And Reagents; Immunology And Microbiology Devices
OUTSIDE THE U.S.	814.15	RESEARCH CONDUCTED OUTSIDE THE UNITED STATES General; Premarket Approval Of Medical Devices
OVENS, MICROWAVE	1030.10	MICROWAVE OVENS Performance Standards For Microwave And Radio Frequency Emitting Products
OVER-THE-COUNTER	201.21	DECLARATION OF PRESENCE OF PHENYLALANINE AS A COMPONENT OF ASPARTAME IN OVER-THE-COUNTER AND PRESCRIPTION DRUGS FOR HUMAN USE General Labeling Provisions; Labeling
OVER-THE-COUNTER	201.66	FORMAT AND CONTENT REQUIREMENTS FOR OVER-THE-COUNTER (OTC) DRUGS PRODUCT LABELING Labeling Requirements For Over-the-counter Drugs; Labeling
OVER-THE-COUNTER	201.307	SODIUM PHOSPHATES, PACKAGE SIZE LIMITATION, WARNINGS AND DIRECTIONS FOR OVER-THE-COUNTER SALE Specific Labeling Requirements For Specific Drug Products; Labeling
OVER-THE-COUNTER	201.308	IPECAC SYRUP; WARNINGS AND DIRECTIONS FOR USE FOR OVER-THE-COUNTER SALE Specific Labeling Requirements For Specific Drug Products; Labeling
OVER-THE-COUNTER	201.315	OVER-THE-COUNTER DRUGS FOR MINOR SORE THROATS; SUGGESTED WARNING Specific Labeling Requirements For Specific Drug Products; Labeling
OVER-THE-COUNTER	201.322	OVER-THE-COUNTER DRUG PRODUCTS CONTAINING INTERNAL ANALGESIC/ANTIPYRETIC ACTIVE INGREDIENTS; REQUIRED ALCOHOL WARNING Specific Labeling Requirements For Specific Drug Products; Labeling
OVER-THE-COUNTER	211.132	TAMPER-EVIDENT PACKAGING REQUIREMENTS FOR OVER-THE-COUNTER (OTC) HUMAN DRUG PRODUCTS Packaging And Labeling Control; Current Good Manufacturing Practice For Finished Pharmaceuticals

OVER-THE-COUNTER	310.519	DRUG PRODUCTS MARKETED AS OVER-THE-COUNTER (OTC) DAYTIME SEDATIVES Requirements For Specific New Drugs Or Devices; New Drugs
OVER-THE-COUNTER	310.527	DRUG PRODUCTS CONTAINING ACTIVE INGREDIENTS OFFERED OVER-THE-COUNTER (OTC) FOR EXTERNAL USE A HAIR GROWERS OR FOR HAIR LOSS PREVENTION Requirements For Specific New Drugs Or Devices; New Drugs
OVER-THE-COUNTER	310.528	DRUG PRODUCTS CONTAINING ACTIVE INGREDIENTS OFFERED OVER-THE-COUNTER (OTC) FOR USE AS AN APHRODISIAC Requirements For Specific New Drugs Or Devices; New Drugs
OVER-THE-COUNTER	310.529	DRUG PRODUCTS CONTAINING ACTIVE INGREDIENTS OFFERED OVER-THE-COUNTER (OTC) FOR ORAL USE AS INSECT REPELLENTS Requirements For Specific New Drugs Or Devices; New Drugs
OVER-THE-COUNTER	310.530	TOPICALLY APPLIED HORMONE-CONTAINING DRUG PRODUCTS FOR OVER-THE-COUNTER (OTC) FOR HUMAN USE Requirements For Specific New Drugs Or Devices; New Drugs
OVER-THE-COUNTER	310.531	DRUG PRODUCTS CONTAINING ACTIVE INGREDIENTS OFFERED OVER-THE-COUNTER (OTC) FOR THE TREATMENT OF BOILS Requirements For Specific New Drugs Or Devices; New Drugs
OVER-THE-COUNTER	310.532	DRUG PRODUCTS CONTAINING ACTIVE INGREDIENTS OFFERED OVER-THE-COUNTER (OTC) TO RELIEVE THE SYMPTOMS OF BENIGN PROSTATIC HYPERTROPHY Requirements For Specific New Drugs Or Devices; New Drugs
OVER-THE-COUNTER	310.533	DRUG PRODUCTS CONTAINING ACTIVE INGREDIENTS OFFERED OVER-THE-COUNTER (OTC) FOR HUMAN USE AS AN ANTICHOLINERGIC IN COUGH-COLD DRUG PRODUCTS Requirements For Specific New Drugs Or Devices; New Drugs
OVER-THE-COUNTER	310.534	DRUG PRODUCTS CONTAINING ACTIVE INGREDIENTS OFFERED OVER-THE-COUNTER (OTC) FOR HUMAN USE AS ORAL WOUND HEALING AGENTS Requirements For Specific New Drugs Or Devices; New Drugs
OVER-THE-COUNTER	310.536	DRUG PRODUCTS CONTAINING ACTIVE INGREDIENTS OFFERED OVER-THE-COUNTER (OTC) FOR USE AS A NAILBITING OR THUMBSUCKING DETERRENT Requirements For Specific New Drugs Or Devices; New Drugs
OVER-THE-COUNTER	310.537	DRUG PRODUCTS CONTAINING ACTIVE INGREDIENTS OFFERED OVER-THE-COUNTER (OTC) FOR TREATMENT OF FEVER BLISTERS AND COLD SORES Requirements For Specific New Drugs Or Devices; New Drugs
OVER-THE-COUNTER	310.538	DRUG PRODUCTS CONTAINING ACTIVE INGREDIENTS OFFERED OVER-THE-COUNTER (OTC) FOR USE FOR INGROWN TOENAIL RELIEF Requirements For Specific New Drugs Or Devices; New Drugs
OVER-THE-COUNTER	310.540	DRUG PRODUCTS CONTAINING ACTIVE INGREDIENTS OFFERED OVER-THE-COUNTER (OTC) FOR USE AS STOMACH ACIDIFIERS Requirements For Specific New Drugs Or Devices; New Drugs
OVER-THE-COUNTER	310.541	OVER-THE-COUNTER (OTC) DRUG PRODUCTS CONTAINING ACTIVE INGREDIENTS OFFERED FOR USE IN THE TREATMENT OF HYPOPHOSPHATEMIA Requirements For Specific New Drugs Or Devices; New Drugs
OVER-THE-COUNTER	310.542	OVER-THE-COUNTER (OTC) DRUG PRODUCTS CONTAINING ACTIVE INGREDIENTS OFFERED FOR USE IN THE TREATMENT OF HYPERPHOSPHATEMIA Requirements For Specific New Drugs Or Devices; New Drugs
OVER-THE-COUNTER	310.543	DRUG PRODUCTS CONTAINING ACTIVE INGREDIENTS OFFERED OVER-THE-COUNTER (OTC) FOR HUMAN USE IN EXOCRINE PANCREATIC INSUFFICIENCY Requirements For Specific New Drugs Or Devices; New Drugs
OVER-THE-COUNTER	310.544	DRUG PRODUCTS CONTAINING ACTIVE INGREDIENTS OFFERED OVER-THE-COUNTER (OTC) FOR USE AS A SMOKING DETERRENT Requirements For Specific New Drugs Or Devices; New Drugs
OVER-THE-COUNTER	310.545	DRUG PRODUCTS CONTAINING CERTAIN ACTIVE INGREDIENTS OFFERED OVER-THE-COUNTER (OTC) FOR CERTAIN USES Requirements For Specific New Drugs Or Devices; New Drugs
OVER-THE-COUNTER	310.546	DRUG PRODUCTS CONTAINING ACTIVE INGREDIENTS OFFERED OVER-THE-COUNTER (OTC) FOR THE TREATMENT AND/OR PREVENTION OF NOCTURNAL LEG MUSCLE CRAMPS Requirements For Specific New Drugs Or Devices; New Drugs
OVER-THE-COUNTER	310.547	DRUG PRODUCTS CONTAINING QUININE OFFERED OVER-THE-COUNTER (OTC) FOR THE TREATMENT AND/OR PREVENTION OF MALARIA Requirements For Specific New Drugs Or Devices; New Drugs
OVER-THE-COUNTER	310.548	DRUG PRODUCTS CONTAINING COLLOIDAL SILVER INGREDIENTS OFFERED OVER-THE-COUNTER (OTC) FOR THE TREATMENT OF DISEASE Requirements For Specific New Drugs Or Devices; New Drugs

OVER-THE-COUNTER	330.12	STATUS OF OVER-THE-COUNTER (OTC) DRUGS PREVIOUSLY REVIEWED UNDER THE DRUG EFFICACY STUDY (DESI) Administrative Procedures; Over-the-counter Human Drugs Generally Recognized As Safe And Effective & Not Misbranded
OVER-THE-COUNTER	330.13	CONDITIONS FOR MARKETING INGREDIENTS RECOMMENDED FOR OVER-THE-COUNTER (OTC) USE UNDER THE OTC DRUG REVIEW Administrative Procedures; Over-the-counter Human Drugs Generally Recognized As Safe And Effective & Not Misbranded
OVERLOAD	890.5575	POWERED EXTERNAL LIMB OVERLOAD WARNING DEVICE Physical Medicine Therapeutic Devices; Physical Medicine Devices
OWNERSHIP	314.72	CHANGE IN OWNERSHIP OF AN APPLICATION Applications; Applications For FDA Approval To Market A New Drug
OWNERSHIP	316.27	CHANGE IN OWNERSHIP OF ORPHAN-DRUG DESIGNATION Orphan Drugs
OX BILE EXTRACT	184.1560	OX BILE EXTRACT Listing Of Specific Substances Affirmed As GRAS; Direct Food Substances Affirmed As Generally Recognized As Safe
OXALATE	862.1542	OXALATE TEST SYSTEM Clinical Chemistry Test Systems; Clinical Chemistry And Clinical Toxicology Devices
OXFENDAZOLE	520.1628	OXFENDAZOLE POWDER AND PELLETS Oral Dosage Form New Animal Drugs
OXFENDAZOLE	520.1629	OXFENDAZOLE PASTE Oral Dosage Form New Animal Drugs
OXFENDAZOLE	520.1630	OXFENDAZOLE SUSPENSION Oral Dosage Form New Animal Drugs
OXFENDAZOLE	520.1631	OXFENDAZOLE AND TRICHLORFON PASTE Oral Dosage Form New Animal Drugs
OXFENDAZOLE	556.495	OXFENDAZOLE Specific Tolerances For Residues Of New; Tolerance For Residues Of New Animal Drugs In Food
OXIBENDAZOLE	520.623	DIETHYLCARBAMAZINE CITRATE, OXIBENDAZOLE CHEWABLE TABLETS Oral Dosage Form New Animal Drugs
OXIBENDAZOLE	520.1638	OXIBENDAZOLE PASTE Oral Dosage Form New Animal Drugs
OXIBENDAZOLE	520.1640	OXIBENDAZOLE SUSPENSION Oral Dosage Form New Animal Drugs
OXIDASE SCREENING	866.2420	OXIDASE SCREENING TEST FOR GONORRHEA Microbiology Devices; Immunology And Microbiology Devices
OXIDATIVELY REFINED	178.3770	POLYHYDRIC ALCOHOL DIESTERS OF OXIDATIVELY REFINED (GERSTHOFEN PROCESS) MONTAN WAX ACIDS Certain Adjuvants And Production Aids; Indirect Food Additives: Adjuvants, Production Aids, And Sanitizers
OXIDE	177.2460	POLY(2,6-DIMETHYL-1,4-PHENYLENE) OXIDE RESINS Substances For Use Only As Components Of Articles Intended For Repeated Use; Indirect Food Additives: Polymers
OXIDE, CALCIUM	184.1210	CALCIUM OXIDE Listing Of Specific Substances Affirmed As GRAS; Direct Food Substances Affirmed As Generally Recognized As Safe
OXIDE, CALCIUM	582.1210	CALCIUM OXIDE General Purpose Food Additives; Substances Generally Recognized As Safe
OXIDE, CALCIUM	582.5210	CALCIUM OXIDE Nutrients And/or Dietary Supplements; Substances Generally Recognized As Safe
OXIDE, CHROMIUM	73.1015	CHROMIUM-COBALT-ALUMINUM OXIDE Drugs; Listing Of Color Additives Exempt From Certification
OXIDE, CHROMIUM	73.1327	CHROMIUM OXIDE GREENS Drugs; Listing Of Color Additives Exempt From Certification
OXIDE, CHROMIUM	73.2327	CHROMIUM OXIDE GREENS Cosmetics; Listing Of Color Additives Exempt From Certification
OXIDE, CHROMIUM	73.3110a	CHROMIUM-COBALT-ALUMINUM OXIDE Medical Devices; Listing Of Color Additives Exempt From Certification
OXIDE, CHROMIUM	73.3111	CHROMIUM OXIDE GREENS Medical Devices; Listing Of Color Additives Exempt From Certification
OXIDE, ETHYLENE	172.770	ETHYLENE OXIDE POLYMER Other Specific Usage Additives; Food Additives Permitted For Direct Addition To Food For Human Consumption
OXIDE, ETHYLENE	172.808	COPOLYMER CONDENSATES OF ETHYLENE OXIDE AND PROPYLENE OXIDE Multipurpose Additives; Food Additives Permitted For Direct Addition To Food For Human Consumption
OXIDE, ETHYLENE	872.3410	ETHYLENE OXIDE HOMOPOLYMER AND/OR CARBOXYMETHYLCELLULOSE SODIUM DENTURE ADHESIVE Prosthetic Devices; Dental Devices
OXIDE, ETHYLENE	872.3450	ETHYLENE OXIDE HOMOPOLYMER AND/OR KARAYA DENTURE ADHESIVE Prosthetic Devices; Dental Devices

OXIDE, ETHYLENE	880.6100	ETHYLENE OXIDE GAS AERATOR CABINET General Hospital And Personal Use Miscellaneous Devices; General Hospital And Personal Use Devices
OXIDE, ETHYLENE	880.6860	ETHYLENE OXIDE GAS STERILIZER General Hospital And Personal Use Miscellaneous Devices; General Hospital And Personal Use Devices
OXIDE, FERRIC	186.1300	FERRIC OXIDE Listing Of Specific Substances Affirmed As GRAS; Indirect Food Substances Affirmed As Generally Recognized As Safe
OXIDE, FERRIC	522.940	COLLOIDAL FERRIC OXIDE INJECTION Implantation Or Injectable Dosage Form New Animal Drugs
OXIDE, IRON	73.200	SYNTHETIC IRON OXIDE Foods; Listing Of Color Additives Exempt From Certification
OXIDE, IRON	73.1200	SYNTHETIC IRON OXIDE Drugs; Listing Of Color Additives Exempt From Certification
OXIDE, MAGNESIUM	184.1431	MAGNESIUM OXIDE Listing Of Specific Substances Affirmed As GRAS; Direct Food Substances Affirmed As Generally Recognized As Safe
OXIDE, MAGNESIUM	582.1431	MAGNESIUM OXIDE General Purpose Food Additives; Substances Generally Recognized As Safe
OXIDE, MAGNESIUM	582.5431	MAGNESIUM OXIDE Nutrients And/or Dietary Supplements; Substances Generally Recognized As Safe
OXIDE, MANGANOUS	582.5464	MANGANOUS OXIDE Nutrients And/or Dietary Supplements; Substances Generally Recognized As Safe
OXIDE, NITRIC	868.2380	NITRIC OXIDE ANALYZER Monitoring Devices; Anesthesiology Devices
OXIDE, NITRIC	868.5165	NITRIC OXIDE ADMINISTRATION APPARATUS Therapeutic Devices; Anesthesiology Devices
OXIDE, NITROUS	184.1545	NITROUS OXIDE Listing Of Specific Substances Affirmed As GRAS; Direct Food Substances Affirmed As Generally Recognized As Safe
OXIDE, NITROUS	868.1700	NITROUS OXIDE GAS ANALYZER Diagnostic Devices; Anesthesiology Devices
OXIDE, ZINC	73.1991	OXIDE Drugs; Listing Of Color Additives Exempt From Certification
OXIDE, ZINC	73.2991	ZINC OXIDE Cosmetics; Listing Of Color Additives Exempt From Certification
OXIDE, ZINC	182.8991	ZINC OXIDE Nutrients; Substances Generally Recognized As Safe
OXIDE, ZINC	582.599	ZINC OXIDE Nutrients And/or Dietary Supplements; Substances Generally Recognized As Safe
OXIDES, IRON	73.2250	IRON OXIDES Cosmetics; Listing Of Color Additives Exempt From Certification
OXIDES, IRON	73.3125	IRON OXIDES Medical Devices; Listing Of Color Additives Exempt From Certification
OXIDES, IRON	186.1374	IRON OXIDES Listing Of Specific Substances Affirmed As GRAS; Indirect Food Substances Affirmed As Generally Recognized As Safe
OXIDIZED POLYETHYLENE	172.260	OXIDIZED POLYETHYLENE Coatings, Films And Related Substances; Food Additives Permitted For Direct Addition To Food For Human Consumption
OXIDIZED POLYETHYLENE	177.1620	POLYETHYLENE, OXIDIZED Substances For Use As Basic Components Of Single And Repeated Use Food Contact Surfaces; Indirect Food Additives: Polymers
OXIME, MILBEMYCIN	520.1445	MILBEMYCIN OXIME TABLETS Oral Dosage Form New Animal Drugs
OXIME, MILBEMYCIN	520.1446	MILBEMYCIN OXIME/LUFENURON TABLETS Oral Dosage Form New Animal Drugs
OXIME, MILBEMYCIN	524.1446	MILBEMYCIN OXIME SOLUTION Ophthalmic And Topical Dosage Form New Animal Drugs
OXIMETER	870.1230	FIBEROPTIC OXIMETER CATHETER Cardiovascular Diagnostic Devices; Cardiovascular Devices
OXIMETER	870.2700	OXIMETER Cardiovascular Monitoring Devices; Cardiovascular Devices
OXIMETER	870.2710	EAR OXIMETER Cardiovascular Monitoring Devices; Cardiovascular Devices
OXYCHLORIDE	73.1162	BISMUTH OXYCHLORIDE Drugs; Listing Of Color Additives Exempt From Certification
OXYCHLORIDE	73.2162	BISMUTH OXYCHLORIDE Cosmetics; Listing Of Color Additives Exempt From Certification
OXYETHYLENE	177.1345	ETHYLENE/1,3-PHENYLENE OXYETHYLENE ISOPHTHALATE/TEREPHTHALATE COPOLYMER Substances For Use As Basic Components Of Single And Repeated Use Food Contact Surfaces; Indirect Food Additives: Polymers

OXYGEN ANALYZER	868.1200	INDWELLING BLOOD OXYGEN PARTIAL PRESSURE (PO2) ANALYZER Diagnostic Devices; Anesthesiology Device
OXYGEN ANALYZER	868.1720	OXYGEN GAS ANALYZER Diagnostic Devices; Anesthesiology Devices
OXYGEN CANNULA	868.5340	NASAL OXYGEN CANNULA Therapeutic Devices; Anesthesiology Devices
OXYGEN CATHETER	868.5350	NASAL OXYGEN CATHETER Therapeutic Devices; Anesthesiology Devices
OXYGEN CHAMBER	878.5650	TOPICAL OXYGEN CHAMBER FOR EXTREMITIES Therapeutic Devices; General And Plastic Surgery Devices
OXYGEN GENERATOR	868.5440	PORTABLE OXYGEN GENERATOR Therapeutic Devices; Anesthesiology Devices
OXYGEN MASK	868.5580	OXYGEN MASK Therapeutic Devices; Anesthesiology Devices
OXYGEN MONITOR	868.2500	CUTANEOUS OXYGEN MONITOR Monitoring Devices; Anesthesiology Devices
OXYGEN TENT	868.5700	NONPOWERED OXYGEN TENT Therapeutic Devices; Anesthesiology Devices
OXYGEN TENT	868.5710	ELECTRICALLY POWERED OXYGEN TENT Therapeutic Devices; Anesthesiology Devices
OXYGEN UNIT	868.5655	PORTABLE LIQUID OXYGEN UNIT Therapeutic Devices; Anesthesiology Devices
OXYGEN UPTAKE	868.1730	OXYGEN UPTAKE COMPUTER Diagnostic Devices; Anesthesiology Devices
OXYGENATOR	870.4350	CARDIOPULMONARY BYPASS OXYGENATOR Cardiovascular Surgical Devices; Cardiovascular Devices
OXYHEMOGLOBIN	868.1120	INDWELLING BLOOD OXYHEMOGLOBIN CONCENTRATION ANALYZER Diagnostic Devices; Anesthesiology Devices
OXYMORPHONE	522.1642	OXYMORPHONE HYDROCHLORIDE INJECTION Implantation Or Injectable Dosage Form New Animal Drugs
OXYSTEARIN	172.818	OXYSTEARIN Multipurpose Additives; Food Additives Permitted For Direct Addition To Food For Human Consumption
OXYTETRACYCLINE	520.1660	OXYTETRACYCLINE Oral Dosage Form New Animal Drugs
OXYTETRACYCLINE	522.1660	OXYTETRACYCLINE INJECTION Implantation Or Injectable Dosage Form New Animal Drugs
OXYTETRACYCLINE	556.500	OXYTETRACYCLINE Specific Tolerances For Residues Of New Animal Drugs; Tolerances For Residues Of New Animal Drugs In Food
OXYTETRACYCLINE	558.450	OXYTETRACYCLINE Specific New Animal Drugs For Use In Animal Feeds; New Animal Drugs For Use In Animal Feeds
OXYTETRACYCLINE HCL	520.1660	OXYTETRACYCLINE HYDROCHLORIDE CAPSULES Oral Dosage Form New Animal Drugs
OXYTETRACYCLINE HCL	522.1662	OXYTETRACYCLINE HYDROCHLORIDE IMPLANTATION OR INJECTABLE DOSAGE FORMS Implantation Or Injectable Dosage Form New Animal Drugs
OXYTETRACYCLINE HCL	524.1662	OXYTETRACYCLINE HYDROCHLORIDE OPHTHALMIC AND TOPICAL DOSAGE FORMS Ophthalmic And Topical Dosage Form New Animal Drugs
OXYTOCIN	522.1680	OXYTOCIN INJECTION Implantation Or Injectable Dosage Form New Animal Drugs
OYSTERS	161.30	DECLARATION OF QUANTITY OF CONTENTS ON LABELS FOR CANNED OYSTERS General Provisions; Fish And Shellfish
OYSTERS	161.130	OYSTERS Requirements For Specific Standardized Fish And Shellfish; Fish And Shellfish
OYSTERS	161.136	OLYMPIA OYSTERS Requirements For Specific Standardized Fish And Shellfish; Fish And Shellfish
OYSTERS	161.145	CANNED OYSTERS Requirements For Specific Standardized Fish And Shellfish; Fish And Shellfish
OZONE	173.368	OZONE Specific Usage Additives; Secondary Direct Food Additives Permitted In Food For Human Consumption
OZONE	184.1563	OZONE Listing Of Specific Substances Affirmed As GRAS; Direct Food Substances Affirmed As Generally Recognized As Safe
OZONE	801.415	MAXIMUM ACCEPTABLE LEVEL OF OZONE Special Requirements For Specific Devices; Labeling
OZONE-DEPLETING	201.320	WARNING STATEMENTS FOR DRUG PRODUCTS CONTAINING OR MANUFACTURED WITH CHLOROFLUOROCARBONS OR OTHER OZONE-DEPLETING SUBSTANCES Specific Labeling For Specific Drug Products; Labeling

OZONE-DEPLETING	801.63	MEDICAL DEVICES; WARNING STATEMENT FOR DEVICES CONTAINING OR MANUFACTURED WITH CHLOROFLUOROCARBONS OR OTHER CLASS I OZONE-DEPLETING SUBSTANCES Labeling Requirements For Over-the-counter Devices; Labeling
OZONE-DEPLETING	801.433	WARNING STATEMENTS FOR PRESCRIPTION AND RESTRICTED DEVICE PRODUCTS CONTAINING CHLOROFLUOROCARBONS OR OTHER OZONE-DEPLETING SUBSTANCES Specific Requirements For Specific Devices; Labeling

P

P-4000	189.175	P-4000 Substances Generally Prohibited From Direct Addition Or Use As Human Food; Substances Prohibited From Use In Human Food
PACEMAKER	870.1750	EXTERNAL PROGRAMMABLE PACEMAKER PULSE GENERATOR Cardiovascular Diagnostic Devices; Cardiovascular Devices
PACEMAKER	870.3600	EXTERNAL PACEMAKER PULSE GENERATOR Cardiovascular Prosthetic Devices; Cardiovascular Devices
PACEMAKER	870.3610	IMPLANTABLE PACEMAKER PULSE GENERATOR Cardiovascular Prosthetic Devices; Cardiovascular Devices
PACEMAKER	870.3620	PACEMAKER LEAD ADAPTOR Cardiovascular Prosthetic Devices; Cardiovascular Devices
PACEMAKER	870.3630	PACEMAKER GENERATOR FUNCTION ANALYZER Cardiovascular Prosthetic Devices; Cardiovascular Devices
PACEMAKER	870.3640	INDIRECT PACEMAKER GENERATOR FUNCTION ANALYZER Cardiovascular Prosthetic Devices; Cardiovascular Devices
PACEMAKER	870.3650	PACEMAKER POLYMERIC MESH BAG Cardiovascular Prosthetic Devices; Cardiovascular Devices
PACEMAKER	870.3670	PACEMAKER CHARGER Cardiovascular Prosthetic Devices; Cardiovascular Devices
PACEMAKER	870.3680	CARDIOVASCULAR PERMANENT OR TEMPORARY PACEMAKER ELECTRODE Cardiovascular Prosthetic Devices; Cardiovascular Devices
PACEMAKER	870.3690	PACEMAKER TEST MAGNET Cardiovascular Prosthetic Devices; Cardiovascular Devices
PACEMAKER	870.3700	PACEMAKER PROGRAMMERS Cardiovascular Prosthetic Devices; Cardiovascular Devices
PACEMAKER	870.3710	PACEMAKER REPAIR OR REPLACEMENT MATERIAL Cardiovascular Prosthetic Devices; Cardiovascular Devices
PACEMAKER	870.3720	PACEMAKER ELECTRODE FUNCTION TESTER Cardiovascular Prosthetic Devices; Cardiovascular Devices
PACEMAKER	870.3730	PACEMAKER SERVICE TOOLS Cardiovascular Prosthetic Devices; Cardiovascular Devices
PACEMAKER	870.5550	EXTERNAL TRANSCUTANEOUS CARDIAC PACEMAKER (NONINVASIVE) Cardiovascular Therapeutic Devices; Cardiovascular Devices
PACIFIC SALMON	161.170	CANNED PACIFIC SALMON Requirements For Specific Standardized Fish And Shellfish; Fish And Shellfish
PACIFIC WHITING	102.46	PACIFIC WHITING Requirements For Specific Nonstandardized Foods; Common Or Usual Name For Nonstandardized Foods
PACK, COLD	880.5760	CHEMICAL COLD PACK SNAKEBITE KIT General Hospital And Personal Use Therapeutic Devices; General Hospital And Personal Use Devices
PACK, COLD	890.5700	COLD PACK Physical Medicine Therapeutic Devices; Physical Medicine Devices
PACK, COLD	890.5710	HOT OR COLD DISPOSABLE PACK Physical Medicine Therapeutic Devices; Physical Medicine Devices
PACK, COLD	890.5720	WATER CIRCULATING HOT OR COLD PACK Physical Medicine Therapeutic Devices; Physical Medicine Devices
PACK, HEAT	890.5730	MOIST HEAT PACK Physical Medicine Therapeutic Devices; Physical Medicine Devices
PACK, WET	161.173	CANNED WET PACK SHRIMP IN TRANSPARENT OR NONTRANSPARENT CONTAINERS Requirements For Specific Standardized Fish And Shellfish; Fish And Shellfish
PACKAGE	101.1	PRINCIPAL DISPLAY PANEL OF PACKAGE FORM FOOD General Provisions; Food Labeling
PACKAGE	101.2	INFORMATION PANEL OF PACKAGE FORM FOOD General Provisions; Food Labeling
PACKAGE	501.1	PRINCIPAL DISPLAY PANEL OF PACKAGE FORM ANIMAL FOOD General Provisions; Animal Food Labeling
PACKAGE	501.2	INFORMATION PANEL OF PACKAGE FOR ANIMAL FOOD General Provisions; Animal Food Labeling
PACKAGE	501.3	IDENTITY LABELING OF ANIMAL FOOD IN PACKAGE FORM General Provisions; Animal Food Labeling
PACKAGE, FINAL	1240.61	MANDATORY PASTEURIZATION FOR ALL MILK AND MILK PRODUCTS IN FINAL PACKAGE FORM INTENDED FOR DIRECT HUMAN CONSUMPTION Specific Administrative Decisions Regarding Interstate Shipments; Control Of Communicable Diseases
PACKAGE INSERTS	310.501	PATIENT PACKAGE INSERTS FOR ORAL CONTRACEPTIVES Requirements For Specific New Drugs Or Devices; New Drugs
PACKAGE INSERTS	310.515	PATIENT PACKAGE INSERTS FOR ESTROGENS Requirements For Specific New Drugs Or Devices; New Drugs

PACKAGE INSERTS	357.152	PACKAGE INSERTS FOR ANTHELMINTIC DRUG PRODUCTS Anthelmintic Drug Products; Miscellaneous Internal Drug Products For Over-the-counter Human Use
PACKAGE LABEL	610.61	PACKAGE LABEL Labeling Standards; General Biological Products Standards
PACKAGE LABEL	610.62	PROPER NAME; PACKAGE LABEL; LEGIBLE TYPE Labeling Standards; General Biological Products Standards
PACKAGE SIZE	201.307	SODIUM PHOSPHATES, PACKAGE SIZE LIMITATION, WARNINGS AND DIRECTIONS FOR OVER-THE-COUNTER SALE Labeling
PACKAGED FOOD	179.21	SOURCES OF RADIATION USED FOR INSPECTION OF FOOD, FOR INSPECTION OF PACKAGED FOOD, AND FOR CONTROLLING FOOD PROCESSING Radiation And Radiation Sources; Irradiation In The Production, Processing And Handling Of Food
PACKAGED FORM	101.3	IDENTITY LABELING OF FOODS IN PACKAGED FORM Food Labeling
PACKAGED, HERMETICALLY	108.35	THERMAL PROCESSING OF LOW-ACID FOODS PACKAGED IN HERMETICALLY SEALED CONTAINERS Specific Requirements And Conditions For Exemption From Or Compliance With An Emergency Permit; Emergency Permit Control
PACKAGED, HERMETICALLY	500.23	THERMALLY PROCESSED LOW-ACID FOODS PACKAGED IN HERMETICALLY SEALED CONTAINERS Specific Administrative Rulings And Decisions; General
PACKAGING	70.20	PACKAGING REQUIREMENTS FOR STRAIGHT COLORS (OTHER THAN HAIR DYES) Packaging And Labeling; Color Additives
PACKAGING	109.15	USE OF POLYCHLORINATED BIPHENYLS (PCB'S) IN ESTABLISHMENTS MANUFACTURING FOOD-PACKAGING MATERIALS General Provisions; Unavoidable Contaminants In Food For Human Consumption & Food-packaging Material
PACKAGING	111.50	PACKAGING OF IRON-CONTAINING DIETARY SUPPLEMENTS Current Good Manufacturing Practice For Dietary Supplements
PACKAGING	179.45	PACKAGING MATERIALS FOR USE DURING THE IRRADIATION OF PREPACKAGED FOODS Packaging Materials For Irradiated Foods; Irradiation In The Production, Processing And Handling Of Food
PACKAGING	181.22	CERTAIN SUBSTANCES EMPLOYED IN THE MANUFACTURE OF FOOD-PACKAGING MATERIALS Specific Prior-sanctioned Food Ingredients; Prior-sanctioned Food Ingredients
PACKAGING	181.30	SUBSTANCES USED IN THE MANUFACTURE OF PAPER AND PAPERBOARD PRODUCTS USED IN FOOD PACKAGING Specific Prior-sanctioned Food Ingredients; Prior-sanctioned Food Ingredients
PACKAGING	182.70	SUBSTANCES MIGRATING FROM COTTON AND COTTON FABRICS USED IN DRY FOOD PACKAGING General Provisions; Substances Generally Recognized As Safe
PACKAGING	211.130	PACKAGING AND LABELING OPERATIONS Packaging And Labeling Control; Current Good Manufacturing Practice For Finished Pharmaceuticals
PACKAGING	226.80	PACKAGING AND LABELING Packaging And Labeling; Current Good Manufacturing Practice For Type A Medicated Articles
PACKAGING	355.20	PACKAGING CONDITIONS Anticaries Drug Products For Over-the-counter Human Use
PACKAGING	509.15	USE OF POLYCHLORINATED BIPHENYLS (PCB'S) IN ESTABLISHMENTS MANUFACTURED FOOD-PACKAGING MATERIALS General Provisions; Unavoidable Contaminants In Animal Food And Food-packaging Material
PACKAGING	570.13	INDIRECT FOOD ADDITIVES RESULTING FROM PACKAGING MATERIALS PRIOR SANCTIONED FOR ANIMAL FEED AND PET FOOD General Provisions; Food Additives
PACKAGING	570.14	INDIRECT FOOD ADDITIVES RESULTING FROM PACKAGING MATERIALS FOR ANIMAL FEED AND PET FOOD General Provisions; Food Additives
PACKAGING	800.12	CONTACT LENS SOLUTIONS AND TABLETS; TAMPER-RESISTANT PACKAGING Requirements For Specific Medical Devices; General
PACKAGING	820.130	DEVICE PACKAGING Labeling And Packaging Control; Quality System Regulation
PACKAGING	1302.07	LABELING AND PACKAGING REQUIREMENTS FOR IMPORTED AND EXPORTED SUBSTANCES Labeling And Packaging Requirements For Controlled Substances

PACKAGING, TAMPER-EVIDENT	211.132	TAMPER-EVIDENT PACKAGING REQUIREMENTS FOR OVER-THE-COUNTER (OTC) HUMAN DRUG PRODUCTS Packaging And Labeling Control; Current Good Manufacturing Practice For Finished Pharmaceuticals
PACKAGING, TAMPER-RESISTANT	700.25	TAMPER-RESISTANT PACKAGING REQUIREMENTS FOR COSMETIC PRODUCTS Requirements For Specific Cosmetic Products; General
PACKER	101.5	FOOD; NAME AND PLACE OF BUSINESS OF MANUFACTURER, PACKER, OR DISTRIBUTOR. General Provisions; Food Labeling
PACKER	201.5	DRUGS; NAME AND PLACE OF BUSINESS OF MANUFACTURER, PACKER, OR DISTRIBUTOR General Labeling Provisions; Labeling
PACKER	501.5	ANIMAL FOOD; NAME AND PLACE OF BUSINESS OF MANUFACTURER, PACKER, OR DISTRIBUTOR General Provisions; Animal Food Labeling
PACKER	701.12	NAME AND PLACE OF BUSINESS OF MANUFACTURER, PACKER, OR DISTRIBUTOR Package Form; Cosmetic Labeling
PACKER	801.1	MEDICAL DEVICES; NAME AND PLACE OF BUSINESS OF MANUFACTURER, PACKER OR DISTRIBUTOR General Labeling Provisions; Labeling
PACKERS	133.10	NOTICE TO MANUFACTURERS, PACKERS, AND DISTRIBUTORS OF PASTEURIZED BLENDED CHEESE, PASTEURIZED PROCESS CHEESE, CHEESE FOOD, CHEESE SPREAD, AND RELATED FOODS General Provisions; Cheeses And Related Cheese Products
PACKERS	201.300	NOTICE TO MANUFACTURERS, PACKERS, AND DISTRIBUTORS OF GLANDULAR PREPARATIONS Specific Labeling Requirements For Specific Drug Products; Labeling
PACKERS	201.301	NOTICE TO MANUFACTURERS, PACKERS, AND DISTRIBUTORS OF ESTROGENIC HORMONE PREPARATIONS Specific Labeling Requirements For Specific Drug Products; Labeling
PACKERS	201.302	NOTICE TO MANUFACTURERS, PACKERS, AND DISTRIBUTORS OF DRUGS FOR INTERNAL USE WHICH CONTAIN MINERAL OIL Specific Labeling Requirements For Specific Drug Products; Labeling
PACKING	108.12	MANUFACTURING, PROCESSING, OR PACKING WITHOUT A PERMIT, OR IN VIOLATION OF A PERMIT General Provisions; Emergency Permit Control
PACKS OF FOOD	130.17	TEMPORARY PERMITS FOR INTERSTATE SHIPMENT OF EXPERIMENTAL PACKS OF FOOD VARYING FROM THE REQUIREMENTS OF DEFINITIONS AND STANDARDS OF IDENTITY General Provisions; Food Standards: General
PAD, DENTURE	872.3540	OTC DENTURE CUSHION OR PAD Prosthetic Devices; Dental Devices
PAD, EYE	878.4440	EYE PAD Surgical Devices; General And Plastic Surgery Devices
PAD, EYE	880.5270	NEONATAL EYE PAD General Hospital And Personal Use Therapeutic Devices; General Hospital And Personal Use Devices
PAD, HEATING	890.5740	POWERED HEATING PAD Physical Medicine Therapeutic Devices; Physical Medicine Devices
PAD, MENSTRUAL	884.5425	SCENTED OR SCENTED DEODORIZED MENSTRUAL PAD Obstetrical And Gynecological Therapeutic Devices; Obstetrical And Gynecological Device
PAD, MENSTRUAL	884.5435	UNSCENTED MENSTRUAL PAD Obstetrical And Gynecological Therapeutic Devices; Obstetrical And Gynecological Device
PADDIE	882.4700	COTTONOID PADDIE Neurological Surgical Devices; Neurological Devices
PADDLES	870.5300	DC-DEFIBRILLATOR (INCLUDING PADDLES) Cardiovascular Therapeutic Devices; Cardiovascular Devices
PADS	890.3790	CANE, CRUTCH, AND WALKER TIPS AND PADS Physical Medicine Prosthetic Devices; Physical Medicine Devices
PAIN RELIEF	882.5840	IMPLANTED INTRACEREBRAL/SUBCORTICAL STIMULATOR FOR PAIN RELIEF Neurological Therapeutic Devices; Neurological Devices
PAIN RELIEF	882.5870	IMPLANTED PERIPHERAL NERVE STIMULATOR FOR PAIN RELIEF Neurological Therapeutic Devices; Neurological Devices
PAIN RELIEF	882.5880	IMPLANTED SPINAL CORD STIMULATOR FOR PAIN RELIEF Neurological Therapeutic Devices; Neurological Devices
PAIN RELIEF	882.5890	TRANSCUTANEOUS ELECTRICAL NERVE STIMULATOR FOR PAIN RELIEF Neurological Therapeutic Devices; Neurological Devices
PALM KERNEL OIL	172.861	COCOA BUTTER SUBSTITUTE FROM COCONUT OIL, PALM KERNEL OIL, OR BOTH OILS Food Additives Permitted For Direct Addition To Food For Human Consumption

PALMITATE	182.3149	ASCORBYL PALMITATE Chemical Preservatives; Substances Generally Recognized As Safe
PALMITATE	186.1771	SODIUM PALMITATE Listing Of Specific Substances Affirmed As GRAS; Indirect Food Substances Affirmed As Generally Recognized As Safe
PALMITATE	582.3149	ASCORBYL PALMITATE Chemical Preservatives; Substances Generally Recognized As Safe
PALMITATE	582.5936	VITAMIN A PALMITATE Nutrients And/or Dietary Supplements; Substances Generally Recognized As Safe
PALMITIC ACIDS	178.3450	ESTERS OF STEARIC AND PALMITIC ACIDS Certain Adjuvants And Production Aids; Indirect Food Additives: Adjuvants, Production Aids, And Sanitizers
PALMITOSTEARATE	184.1329	GLYCERYL PALMITOSTEARATE Listing Of Specific Substances Affirmed As GRAS; Direct Food Substances Affirmed As Generally Recognized As Safe
PAMOATE	520.1196	IVERMECTIN AND PYRANTEL PAMOATE CHEWABLE TABLET Oral Dosage Form New Animal Drugs
PAMOATE	520.1871	PRAZIQUANTEL/PYRANTEL PAMOATE TABLETS Oral Dosage Form New Animal Drugs
PAMOATE	520.1872	PRAZIQUANTEL, PYRANTEL PAMOATE, AND FEBANTEL TABLETS Oral Dosage Form New Animal Drugs
PAMOATE	520.2041	PYRANTEL PAMOATE CHEWABLE TABLETS Oral Dosage Form New Animal Drugs
PAMOATE	520.2042	PYRANTEL PAMOATE TABLETS Oral Dosage Form New Animal Drugs
PAMOATE	520.2043	PYRANTEL PAMOATE SUSPENSION Oral Dosage Form New Animal Drugs
PAMOATE	520.2044	PYRANTEL PAMOATE PASTE Oral Dosage Form New Animal Drugs
PANCREATIC INSUFFICIENCY	310.543	DRUG PRODUCTS CONTAINING ACTIVE INGREDIENTS OFFERED OTC FOR HUMAN USE IN EXOCRINE PANCREATIC INSUFFICIENCY Requirements For Specific New Drugs Or Devices; New Drugs
PANCREATIN	184.1583	PANCREATIN Listing Of Specific Substances Affirmed As GRAS; Direct Food Substances Affirmed As Generally Recognized As Safe
PANEL, DISPLAY	101.1	PRINCIPAL DISPLAY PANEL OF PACKAGE FORM FOOD General Provisions; Food Labeling
PANEL, DISPLAY	201.60	PRINCIPAL DISPLAY PANEL Labeling Requirements For Over-the-counter Drugs; Labeling
PANEL, DISPLAY	328.50	PRINCIPAL DISPLAY PANEL OF ALL OTC DRUG PRODUCTS INTENDED FOR ORAL INGESTION THAT CONTAIN ALCOHOL Over-the-counter Drug Products Intended For Oral Ingestion That Contain Alcohol
PANEL, DISPLAY	352.50	PRINCIPAL DISPLAY PANEL OF ALL SUNSCREEN DRUG PRODUCTS Sunscreen Drug Products For Over-the-counter Human Use
PANEL, DISPLAY	355.55	PRINCIPAL DISPLAY PANEL OF ALL FLUORIDE RINSE DRUG PRODUCTS Anticaries Drug Products For Over-the-counter Human Use
PANEL, DISPLAY	501.1	PRINCIPAL DISPLAY PANEL OF PACKAGE FORM ANIMAL FOOD General Provisions; Animal Food Labeling
PANEL, DISPLAY	701.10	PRINCIPAL DISPLAY PANEL Package Form; Cosmetic Labeling
PANEL, DISPLAY	801.60	PRINCIPAL DISPLAY PANEL Labeling Requirements For Over-the-counter Devices; Labeling
PANEL, INFORMATION	1.2	INFORMATION PANEL OF PACKAGE FOR ANIMAL FOOD General Provisions; Animal Food Labeling
PANEL, INFORMATION	101.2	INFORMATION PANEL OF PACKAGE FORM FOOD General Provisions; Food Labeling
PANEL, REFERENCE	660.3	REFERENCE PANEL Antibody To Hepatitis B Surface Antigen; Additional Standards For Diagnostic Substances For Laboratory Tests
PANELS	860.125	CONSULTATION WITH PANELS Reclassification; Medical Device Classification Procedures
PANELS, REFERENCE	610.44	USE OF REFERENCE PANELS BY MANUFACTURERS OF TEST KITS Testing Requirements For Communicable Disease Agents; General Biological Products Standards
PANTOGRAPH	872.3730	PANTOGRAPH Prosthetic Devices; Dental Devices
PANTOTHENAMIDE	172.335	D-PANTOTHENAMIDE Special Dietary And Nutritional Additives; Food Additives Permitted For Direct Addition To Food For Human Consumption
PANTOTHENATE	172.330	CALCIUM PANTOTHENATE, CALCIUM CHLORIDE DOUBLE SALT Special Dietary And Nutritional Additives; Food Additives Permitted For Direct Addition To Food For Human Consumption

PANTOTHENATE	184.1212	CALCIUM PANTOTHENATE Listing Of Specific Substances Affirmed As GRAS; Direct Food Substances Affirmed As Generally Recognized As Safe
PANTOTHENATE	582.1212	CALCIUM PANTOTHENATE Nutrients And/or Dietary Supplements; Substances Generally Recognized As Safe
PANTOTHENATE	582.5772	SODIUM PANTOTHENATE Nutrients And/or Dietary Supplements; Substances Generally Recognized As Safe
PANTOTHENYL ALCOHOL	582.5580	D-PANTOTHENYL ALCOHOL Nutrients And/or Dietary Supplements; Substances Generally Recognized As Safe
PAPAIN	184.1585	PAPAIN Listing Of Specific Substances Affirmed As GRAS; Direct Food Substances Affirmed As Generally Recognized As Safe
PAPAIN	582.1585	PAPAIN General Purpose Food Additives; Substances Generally Recognized As Safe
PAPER	176.150	CHELATING AGENTS USED IN THE MANUFACTURE OF PAPER AND PAPERBOARD Substances For Use Only As Components Of Paper And Paperboard; Indirect Food Additives
PAPER	176.170	COMPONENTS OF PAPER AND PAPERBOARD IN CONTACT WITH AQUEOUS AND FATTY FOODS Substances For Use Only As Components Of Paper And Paperboard; Indirect Food Additives
PAPER	176.180	COMPONENTS OF PAPER AND PAPERBOARD IN CONTACT WITH DRY FOOD Substances For Use Only As Components Of Paper And Paperboard; Indirect Food Additives
PAPER	176.210	DEFOAMING AGENTS USED IN THE MANUFACTURE OF PAPER AND PAPERBOARD Substances For Use Only As Components Of Paper And Paperboard; Indirect Food Additives
PAPER	181.30	SUBSTANCES USED IN THE MANUFACTURE OF PAPER AND PAPERBOARD PRODUCTS USED IN FOOD PACKAGING Specific Prior-sanctioned Food Ingredients; Prior-sanctioned Food Ingredients
PAPER	182.90	SUBSTANCES MIGRATING TO FOOD FROM PAPER AND PAPERBOARD PRODUCTS General Provisions; Substances Generally Recognized As Safe
PAPER	870.2810	PAPER CHART RECORDER Cardiovascular Monitoring Devices; Cardiovascular Devices
PAPER, ARTICULATION	872.6140	ARTICULATION PAPER Miscellaneous Devices; Dental Devices
PAPER POINT	872.3830	ENDODONTIC PAPER POINT Prosthetic Devices; Dental Devices
PAPERBOARD	176.150	CHELATING AGENTS USED IN THE MANUFACTURE OF PAPER AND PAPERBOARD Substances For Use Only As Components Of Paper And Paperboard; Indirect Food Additives
PAPERBOARD	176.170	COMPONENTS OF PAPER AND PAPERBOARD IN CONTACT WITH AQUEOUS AND FATTY FOODS Substances For Use Only As Components Of Paper And Paperboard; Indirect Food Additives
PAPERBOARD	176.180	COMPONENTS OF PAPER AND PAPERBOARD IN CONTACT WITH DRY FOOD Substances For Use Only As Components Of Paper And Paperboard; Indirect Food Additives
PAPERBOARD	176.210	DEFOAMING AGENTS USED IN THE MANUFACTURE OF PAPER AND PAPERBOARD Substances For Use Only As Components Of Paper And Paperboard; Indirect Food Additives
PAPERBOARD	181.30	SUBSTANCES USED IN THE MANUFACTURE OF PAPER AND PAPERBOARD PRODUCTS USED IN FOOD PACKAGING Specific Prior-sanctioned Food Ingredients; Prior-sanctioned Food Ingredients
PAPERBOARD	182.90	SUBSTANCES MIGRATING TO FOOD FROM PAPER AND PAPERBOARD PRODUCTS General Provisions; Substances Generally Recognized As Safe
PAPERS	17.31	FORM, FILING, AND SERVICE OF PAPERS Civil Money Penalties Hearings
PAPRIKA	73.340	PAPRIKA Foods; Listing Of Color Additives Exempt From Certification
PAPRIKA	73.345	PAPRIKA OLEORESIN Foods; Listing Of Color Additives Exempt From Certification
PARACOAGULATION	864.7300	FIBRIN MONOMER PARACOAGULATION TEST Hematology Kits And Packages; Hematology And Pathology Devices
PARAFFIN	172.275	SYNTHETIC PARAFFIN AND SUCCINIC DERIVATIVES Coatings, Films And Related Substances; Food Additives Permitted For Direct Addition To Food For Human Consumption

PARAFFIN	175.250	PARAFFIN (SYNTHETIC) Substances For Use As Components Of Coatings; Indirect Food Additives
PARAFFIN	890.5110	PARAFFIN BATH Physical Medicine Therapeutic Devices; Physical Medicine Devices
PARAINFLUENZA VIRUS	866.3400	PARAINFLUENZA VIRUS SEROLOGICAL REAGENTS Serological Reagents; Immunology And Microbiology Devices
PARATHYROID	862.1545	PARATHYROID HORMONE TEST SYSTEM Clinical Chemistry Test Systems; Clinical Chemistry And Clinical Toxicology Devices
PARENTERAL DRUG	310.509	PARENTERAL DRUG PRODUCTS IN PLASTIC CONTAINERS Requirements For Specific New Drugs Or Devices; New Drugs
PARENTERAL NUTRITION	201.323	ALUMINUM IN LARGE AND SMALL VOLUME PARENTERALS USED IN TOTAL PARENTERAL NUTRITION Specific Labeling Requirements For Specific Drug Products; Labeling
PARENTS	50.55	REQUIREMENTS FOR PERMISSION BY PARENTS OR GUARDIANS AND FOR ASSENT BY CHILDREN Additional Safeguards For Children In Clinical Investigations; Protection Of Human Subjects
PARMESAN CHEESE	133.165	PARMESAN AND REGGIANO CHEESE Requirements For Specific Standardized Cheese And Related Products; Cheeses And Related Cheese Products
PART-SKIM CHEESE	133.157	PART-SKIM MOZZARELLA AND SCAMORZA CHEESE Requirements For Specific Standardized Cheese And Related Products; Cheeses And Related Cheese Products
PART-SKIM CHEESE	133.158	LOW-MOISTURE PART-SKIM MOZZARELLA AND SCAMORZA CHEESE Requirements For Specific Standardized Cheese And Related Products; Cheeses And Related Cheese Products
PART-SKIM CHEESES	133.188	SEMISOFT PART-SKIM CHEESES Requirements For Specific Standardized Cheese And Related Products; Cheeses And Related Cheese Products
PART-SKIM CHEESES	133.191	PART-SKIM SPICED CHEESES Requirements For Specific Standardized Cheese And Related Products; Cheeses And Related Cheese Products
PARTE, EX	10.55	SEPARATION OF FUNCTIONS; EX PARTE COMMUNICATIONS General Administrative Procedures; Administrative Practices And Procedures
PARTE, EX	13.15	SEPARATION OF FUNCTIONS; EX PARTE COMMUNICATIONS; ADMINISTRATIVE SUPPORT General Provisions; Public Hearing Before A Public Board Of Inquiry
PARTE, EX	17.20	EX PARTE CONTACTS Civil Money Penalties Hearings
PARTE, EX	1316.51	CONDUCT OF HEARING AND PARTIES; EX PARTE COMMUNICATIONS Administrative Hearings; Administrative Functions, Practices, And Procedures
PARTIAL	864.7925	PARTIAL THROMBOPLASTIN TIME TESTS Hematology Kits And Packages; Hematology And Pathology Devices
PARTIAL	874.3450	PARTIAL OSSICULAR REPLACEMENT PROSTHESIS Prosthetic Devices; Ear, Nose And Throat Devices
PARTIAL DISCLOSURE	20.22	PARTIAL DISCLOSURE OF RECORDS General Policy; Public Information
PARTIAL FILLING	1306.13	PARTIAL FILLING OF PRESCRIPTIONS Controlled Substances Listed In Schedule II; Prescriptions
PARTIAL FILLING	1306.23	PARTIAL FILLING OF PRESCRIPTIONS Controlled Substances Listed In Schedules III, IV, And V; Prescriptions
PARTIAL PHOSPHORIC ACID	175.260	PARTIAL PHOSPHORIC ACID ESTERS OF POLYESTER RESINS Substances For Use As Components Of Coatings; Indirect Food Additives
PARTIAL PRESSURE	868.1150	INDWELLING BLOOD CARBON DIOXIDE PARTIAL PRESSURE (PCO2) ANALYZER Diagnostic Devices; Anesthesiology Devices
PARTIAL PRESSURE	868.1200	INDWELLING BLOOD OXYGEN PARTIAL PRESSURE (PO2) ANALYZER Diagnostic Devices; Anesthesiology Devices
PARTICIPANTS	12.85	DISCLOSURE OF DATA AND INFORMATION BY THE PARTICIPANTS Hearing Procedures; Formal Evidentiary Public Hearing
PARTICIPANTS	13.25	DISCLOSURE OF DATA AND INFORMATION BY THE PARTICIPANTS Hearing Procedures; Public Hearing Before A Public Board Of Inquiry
PARTICIPANTS'	1404.510	PARTICIPANTS' RESPONSIBILITIES Governmentwide Debarment And Suspension And Requirements For Drug-Free Workplace (Grants)
PARTICIPATE	12.82	PETITION TO PARTICIPATE IN FORMA PAUPERIS Hearing Procedures; Formal Evidentiary Public Hearing
PARTICIPATION	12.45	NOTICE OF PARTICIPATION Appearance And Participation; Formal Evidentiary Public Hearing
PARTICIPATION	12.50	ADVICE ON PUBLIC PARTICIPATION IN HEARINGS Appearance And Participation; Formal Evidentiary Public Hearing
PARTICIPATION	12.89	PARTICIPATION OF NONPARTIES Hearing Procedures; Formal Evidentiary Public Hearing

PARTICIPATION	15.21	NOTICE OF PARTICIPATION; SCHEDULE FOR HEARING Procedures For Public Hearing Before The Commissioner; Public Hearing Before The Commissioner
PARTICIPATION	26.7	PARTICIPATION IN THE EQUIVALENCE ASSESSMENT AND DETERMINATION Specific Sector Pharmaceutical GMPs; Mutual Recognition Of Pharmaceutical GMP Reports, Medical Device Quality System Audit Reports
PARTICIPATION	314.200	NOTICE OF OPPORTUNITY FOR HEARING; NOTICE OF PARTICIPATION AND REQUEST FOR HEARING; GRANT OR DENIAL OF HEARING Hearing Procedures For New Drugs; Applications For FDA Approval To Market A New Drug
PARTICLE COLLECTOR	874.4800	BONE PARTICLE COLLECTOR Surgical Devices; Ear, Nose, And Throat Devices
PARTIES	17.15	PARTIES TO THE HEARING Civil Money Penalties Hearings
PARTIES	1316.51	CONDUCT OF HEARING AND PARTIES; EX PARTE COMMUNICATIONS Administrative Hearings; Administrative Functions, Practices, And Procedures
PASSWORDS	11.300	CONTROLS FOR IDENTIFICATION CODES/PASSWORDS Electronic Records; Electronic Signatures
PASTE	520.45b	ALBENDAZOLE PASTE Oral Dosage Form New Animal Drugs
PASTE	520.300c	CAMBENDAZOLE PASTE Oral Dosage Form New Animal Drugs
PASTE	520.903a	FEBANTEL PASTE Oral Dosage Form New Animal Drugs
PASTE	520.903d	FEBANTEL-PRAZIQUANTEL PASTE Oral Dosage Form New Animal Drugs
PASTE	520.905c	FENBENDAZOLE PASTE Oral Dosage Form New Animal Drugs
PASTE	520.970	FLUNIXIN MEGLUMINE PASTE Oral Dosage Form New Animal Drugs
PASTE	520.1192	IVERMECTIN PASTE Oral Dosage Form New Animal Drugs
PASTE	520.1242g	LEVAMISOLE RESINATE AND FAMPHUR PASTE Oral Dosage Form New Animal Drugs
PASTE	520.1326b	MEBENDAZOLE AND TRICHLORFON PASTE Oral Dosage Form New Animal Drugs
PASTE	520.1629	OXFENDAZOLE PASTE Oral Dosage Form New Animal Drugs
PASTE	520.1631	OXFENDAZOLE AND TRICHLORFON PASTE Oral Dosage Form New Animal Drugs
PASTE	520.1638	OXIBENDAZOLE PASTE Oral Dosage Form New Animal Drugs
PASTE	520.1720c	PHENYLBUTAZONE PASTE Oral Dosage Form New Animal Drugs
PASTE	520.2044	PYRANTEL PAMOATE PASTE Oral Dosage Form New Animal Drugs
PASTE	520.2380b	THIABENDAZOLE DRENCH OR ORAL PASTE Oral Dosage Form New Animal Drugs
PASTE	520.2473b	TIOXIDAZOLE PASTE Oral Dosage Form New Animal Drugs
PASTE	520.2611	TRIMETHOPRIM AND SULFADIAZINE ORAL PASTE Oral Dosage Form New Animal Drugs
PASTEURIZATION	1210.15	PASTEURIZATION; EQUIPMENT AND METHODS Inspection And Testing; Regulations Under The Federal Import Milk Act
PASTEURIZATION	1240.61	MANDATORY PASTEURIZATION FOR ALL MILK AND MILK PRODUCTS IN FINAL PACKAGE FORM INTENDED FOR DIRECT HUMAN CONSUMPTION Specific Administrative Decisions Regarding Interstate Shipments; Control Of Communicable Diseases
PASTEURIZED	1210.25	PERMITS FOR PASTEURIZED MILK OR CREAM Permit Control; Regulations Under The Federal Import Milk Act
PASTEURIZED CHEESE	133.10	NOTICE TO MANUFACTURERS, PACKERS, AND DISTRIBUTORS OF PASTEURIZED BLENDED CHEESE, PASTEURIZED PROCESS CHEESE, CHEESE FOOD, CHEESE SPREAD, AND RELATED FOODS General Provisions; Cheeses And Related Cheese Products
PASTEURIZED CHEESE	133.167	PASTEURIZED BLENDED CHEESE Requirements For Specific Standardized Cheese And Related Products; Cheeses And Related Cheese Products
PASTEURIZED CHEESE	133.168	PASTEURIZED BLENDED CHEESE WITH FRUITS, VEGETABLES, OR MEATS Requirements For Specific Standardized Cheese And Related Products; Cheeses And Related Cheese Products
PASTEURIZED CHEESE	133.169	PASTEURIZED PROCESS CHEESE Requirements For Specific Standardized Cheese And Related Products; Cheeses And Related Cheese Products

PASTEURIZED CHEESE	133.170	PASTEURIZED PROCESS CHEESE WITH FRUITS, VEGETABLES, OR MEATS Requirements For Specific Standardized Cheese And Related Products; Cheeses And Related Cheese Products
PASTEURIZED CHEESE	133.171	PASTEURIZED PROCESS PIMENTO CHEESE Requirements For Specific Standardized Cheese And Related Products; Cheeses And Related Cheese Products
PASTEURIZED CHEESE	133.173	PASTEURIZED PROCESS CHEESE FOOD Requirements For Specific Standardized Cheese And Related Products; Cheeses And Related Cheese Products
PASTEURIZED CHEESE	133.174	PASTEURIZED PROCESS CHEESE FOOD WITH FRUITS, VEGETABLES, OR MEATS Requirements For Specific Standardized Cheese And Related Products; Cheeses And Related Cheese Products
PASTEURIZED CHEESE	133.175	PASTEURIZED CHEESE SPREAD Requirements For Specific Standardized Cheese And Related Products; Cheeses And Related Cheese Products
PASTEURIZED CHEESE	133.176	PASTEURIZED CHEESE SPREAD WITH FRUITS, VEGETABLES OR MEATS Requirements For Specific Standardized Cheese And Related Products; Cheeses And Related Cheese Products
PASTEURIZED CHEESE	133.178	PASTEURIZED NEUFCHATEL CHEESE SPREAD WITH OTHER FOODS Requirements For Specific Standardized Cheese And Related Products; Cheeses And Related Cheese Products
PASTEURIZED CHEESE	133.179	PASTEURIZED PROCESS CHEESE SPREAD Requirements For Specific Standardized Cheese And Related Products; Cheeses And Related Cheese Products
PASTEURIZED CHEESE	133.180	PASTEURIZED PROCESS CHEESE SPREAD WITH FRUITS, VEGETABLES, OR MEATS Requirements For Specific Standardized Cheese And Related Products; Cheeses And Related Cheese Products
PASTEURIZED ORANGE JUICE	146.140	PASTEURIZED ORANGE JUICE Requirements For Specific Standardized Canned Fruit Juices And Beverages; Canned Fruit Juices
PATCH	870.3470	INTRACARDIAC PATCH OR PLEDGET MADE OF POLYPROPYLENE, POLYETHYLENE TEREPHTHALATE, OR POLYTETRAFLUOROETHYLENE Cardiovascular Prosthetic Devices; Cardiovascular Devices
PATELLAR	888.3580	KNEE JOINT PATELLAR (HEMI-KNEE) METALLIC RESURFACING UNCEMENTED PROSTHESIS Prosthetic Devices; Orthopedic Devices
PATELLOFEMORAL	888.3540	KNEE JOINT PATELLOFEMORAL POLYMER/METAL SEMI-CONSTRAINED CEMENTED PROSTHESIS Prosthetic Devices; Orthopedic Devices
PATELLOFEMOROTIBIAL	888.3550	KNEE JOINT PATELLOFEMOROTIBIAL POLYMER/METAL/METAL CONSTRAINED CEMENTED PROSTHESIS Prosthetic Devices; Orthopedic Devices
PATELLOFEMOROTIBIAL	888.3560	KNEE JOINT PATELLOFEMOROTIBIAL POLYMER/METAL/POLYMER SEMI-CONSTRAINED CEMENTED PROSTHESIS Prosthetic Devices; Orthopedic Devices
PATENT	314.52	NOTICE OF CERTIFICATION OF INVALIDITY OR NONINFRINGEMENT OF A PATENT Applications For FDA Approval To Market A New Drug
PATENT	314.53	SUBMISSION OF PATENT INFORMATION Applications For FDA Approval To Market A New Drug
PATENT	314.95	NOTICE OF CERTIFICATION OF INVALIDITY OR NONINFRINGEMENT OF A PATENT Applications For FDA Approval To Market A New Drug
PATENT EXTENSIONS	5.27	PATENT TERM EXTENSIONS FOR HUMAN DRUG PRODUCTS, MEDICAL DEVICES, AND FOOD AND COLOR ADDITIVES; AND AUTHORITY TO PERFORM DUE DILIGENCE DETERMINATIONS AND INFORMAL HEARINGS General Redelegations Of Authority; Delegations Of Authority And Organization
PATENT INFORMATION	5.109	ISSUANCE OF WRITTEN NOTICES CONCERNING PATENT INFORMATION, CURRENT GOOD MANUFACTURING PRACTICES AND FALSE OR MISLEADING LABELING OF NEW DRUGS Human Drugs; Redelegations Of Authority; Delegations Of Authority And Organization
PATENT INFORMATION	5.504	ISSUANCE OF WRITTEN NOTICES CONCERNING PATENT INFORMATION, CURRENT GOOD MANUFACTURING PRACTICES AND FALSE OR MISLEADING LABELING OF NEW ANIMAL DRUGS AND FEEDS BEARING OR CONTAINING NEW ANIMAL DRUGS Animal Drugs; Redelegations Of Authority; Delegations Of Authority And Organization
PATHOGENIC NEISSERIA	866.2410	CULTURE MEDIUM FOR PATHOGENIC NEISSERIA SPP. Microbiology Devices; Immunology And Microbiology Devices
PATIENT CARE	870.5050	PATIENT CARE SUCTION APPARATUS Cardiovascular Therapeutic Devices; Cardiovascular Devices

PATIENT CARE	880.5450	PATIENT CARE REVERSE ISOLATION CHAMBER General Hospital And Personal Use Therapeutic Devices; General Hospital And Personal Use Device
PATIENT CRADLE	892.1830	RADIOLOGIC PATIENT CRADLE Diagnostic Devices; Radiology Devices
PATIENT EXAM GLOVE	880.6250	PATIENT EXAMINATION GLOVE General Hospital And Personal Use Miscellaneous Devices; General Hospital And Personal Use Devices
PATIENT EXAM GLOVES	800.20	PATIENT EXAMINATION GLOVES AND SURGEON'S GLOVES; SAMPLE PLANS AND TEST METHOD FOR LEAKAGE DEFECTS; ADULTERATION Requirements For Specific Medical Devices; General
PATIENT INSERTS	310.501	PATIENT PACKAGE INSERTS FOR ORAL CONTRACEPTIVES Requirements For Specific New Drugs Or Devices; New Drugs
PATIENT INSERTS	310.515	PATIENT PACKAGE INSERTS FOR ESTROGENS Requirements For Specific New Drugs Or Devices; New Drugs
PATIENT LABELING	801.420	HEARING AID DEVICES; PROFESSIONAL AND PATIENT LABELING Special Requirements For Specific Devices; Labeling
PATIENT LIFT	880.5500	AC-POWERED PATIENT LIFT General Hospital And Personal Use Therapeutic Devices; General Hospital And Personal Use Devices
PATIENT LIFT	880.5510	NON-AC-POWERED PATIENT LIFT General Hospital And Personal Use Therapeutic Devices; General Hospital And Personal Use Devices
PATIENT LUBRICANT	880.6375	PATIENT LUBRICANT General Hospital And Personal Use Miscellaneous Devices; General Hospital And Personal Use Devices
PATIENT MONITOR	880.2400	BED-PATIENT MONITOR General Hospital And Personal Use Monitoring Devices; General Hospital And Personal Use Devices
PATIENT POSITION	868.6820	PATIENT POSITION SUPPORT Miscellaneous; Anesthesiology Devices
PATIENT POSITION	892.5780	LIGHT BEAM PATIENT POSITION INDICATOR Therapeutic Devices; Radiology Devices
PATIENT ROTATION	890.5180	MANUAL PATIENT ROTATION BED Physical Medicine Therapeutic Devices; Physical Medicine Devices
PATIENT ROTATION	890.5225	POWERED PATIENT ROTATION BED Physical Medicine Therapeutic Devices; Physical Medicine Devices
PATIENT SAFETY	312.88	SAFEGUARDS FOR PATIENT SAFETY Drugs Intended To Treat Life-threatening And Severely-debilitating Illnesses; Investigational New Drug Application
PATIENT SCALE	880.2700	STAND-ON PATIENT SCALE General Hospital And Personal Use Monitoring Devices; General Hospital And Personal Use Devices
PATIENT SCALE	880.2720	PATIENT SCALE General Hospital And Personal Use Monitoring Devices; General Hospital And Personal Use Devices
PATIENT SUCTION APPARATUS	878.4680	NONPOWERED, SINGLE PATIENT, PORTABLE SUCTION APPARATUS Surgical Devices; General And Plastic Surgery Devices
PATIENT SUPPORT	892.5770	POWERED RADIATION THERAPY PATIENT SUPPORT ASSEMBLY Therapeutic Devices; Radiology Devices
PATIENT TRANSDUCER	870.2900	PATIENT TRANSDUCER AND ELECTRODE CABLE (INCLUDING CONNECTOR) Cardiovascular Monitoring Devices; Cardiovascular Devices
PATIENT TRANSFER	880.6775	POWERED PATIENT TRANSFER DEVICE General Hospital And Personal Use Miscellaneous Devices; General Hospital And Personal Use Devices
PATIENT TRANSFER	880.6785	MANUAL PATIENT TRANSFER DEVICE General Hospital And Personal Use Miscellaneous Devices; General Hospital And Personal Use Devices
PATIENT TRANSPORT	890.5150	POWERED PATIENT TRANSPORT Physical Medicine Therapeutic Devices; Physical Medicine Devices
PATIENTS	1000.50	RECOMMENDATION FOR THE USE OF SPECIFIC AREA GONAD SHIELDING ON PATIENTS DURING MEDICAL DIAGNOSTIC X-RAY PROCEDURES Radiation Protection Recommendations; General
PATTERN	892.1420	RADIONUCLIDE TEST PATTERN PHANTOM Diagnostic Devices; Radiology Devices
PAUPERIS	12.82	PETITION TO PARTICIPATE IN FORMA PAUPERIS Hearing Procedures; Formal Evidentiary Public Hearing
PAYMENT	1.91	PAYMENT FOR SAMPLES Imports And Exports; General Enforcement Regulations
PAYMENT	1005.11	PAYMENT FOR SAMPLES Inspection And Testing; Importation Of Electronic Products
PAYMENT	1309.12	TIME AND METHOD OF PAYMENT; REFUND Registration Of Manufacturers, Distributors, Importers And Exporters Of List I Chemicals

PAYMENT	1403.21	PAYMENT Post-Award Requirements; Uniform Administrative Requirements For Grants And Coop. Agreements To State & Local Gov
PAYMENTS	1404.615	GROUNDS FOR SUSPENSION OF PAYMENTS, SUSPENSION OR TERMINATION OF GRANTS, OR SUSPENSION OR DEBARMENT Governmentwide Debarment And Suspension And Requirements For Drug-Free Workplace (Grants)
PCB'S	109.15	USE OF POLYCHLORINATED BIPHENYLS (PCB'S) IN ESTABLISHMENTS MANUFACTURING FOOD-PACKAGING MATERIALS General Provisions; Unavoidable Contaminants In Food For Human Consumption & Food-packaging Material
PCB'S	109.30	TEMPORARY TOLERANCES FOR POLYCHLORINATED BIPHENYLS (PCB'S) Tolerances For Unavoidable Poisonous Or Deleterious Substances; Unavoidable Contaminants In Food For Human Consumption & Food-packaging Material
PCB'S	500.45	USE OF POLYCHLORINATED BIPHENYLS (PCB'S) IN THE PRODUCTION, HANDLING AND STORAGE OF ANIMAL FEED Specific Administrative Rulings And Decisions; General
PCB'S	509.15	USE OF POLYCHLORINATED BIPHENYLS (PCB'S) IN ESTABLISHMENTS MANUFACTURED FOOD-PACKAGING MATERIALS General Provisions; Unavoidable Contaminants In Animal Food And Food-packaging Material
PCB'S	509.30	TEMPORARY TOLERANCES FOR POLYCHLORINATED BIPHENYLS (PCB'S) Tolerances For Unavoidable Poisonous Or Deleterious Substances; Unavoidable Contaminants In Animal Food And Food-packaging Material
PDP	814.19	PRODUCT DEVELOPMENT PROTOCOL (PDP) General; Premarket Approval Of Medical Devices
PEACHES	145.170	CANNED PEACHES Requirements For Specific Standardized Canned Fruits; Canned Fruits
PEACHES	145.171	ARTIFICIALLY SWEETENED CANNED PEACHES Requirements For Specific Standardized Canned Fruits; Canned Fruits
PEAK-FLOW METER	868.1860	PEAK-FLOW METER FOR SPIROMETRY Diagnostic Devices; Anesthesiology Devices
PEANUT BUTTER	164.150	PEANUT BUTTER Requirements For Specific Standardized Tree Nut And Peanut Products; Tree Nut And Peanut Products
PEANUT SPREADS	102.23	PEANUT SPREADS Requirements For Specific Nonstandardized Foods; Common Or Usual Name For Nonstandardized Foods
PEARS	145.175	CANNED PEARS Requirements For Specific Standardized Canned Fruits; Canned Fruits
PEARS	145.176	ARTIFICIALLY SWEETENED CANNED PEARS Requirements For Specific Standardized Canned Fruits; Canned Fruits
PEAS	155.170	CANNED PEAS Requirements For Specific Standardized Canned Vegetables; Canned Vegetables
PEAS	155.172	CANNED DRY PEAS Requirements For Specific Standardized Canned Vegetables; Canned Vegetables
PEAS	158.170	FROZEN PEAS Requirements For Specific Standardized Frozen Vegetables; Frozen Vegetables
PECTIN	520.1204	KANAMYCIN SULFATE, AMINOPENTAMIDE HYDROGEN SULFATE, PECTIN, BISMUTH SUBCARBONATE, ACTIVATED ATTAPULGITE ORAL SUSPENSION Oral Dosage Form New Animal Drugs
PECTIN	520.1205	KANAMYCIN SULFATE, PECTIN, BISMUTH SUBCARBONATE, ACTIVATED ATTAPULGITE TABLETS Oral Dosage Form New Animal Drugs
PECINATE	582.1775	SODIUM PECTINATE General Purpose Food Additives; Substances Generally Recognized As Safe
PECTINS	184.1588	PECTINS Listing Of Specific Substances Affirmed As GRAS; Direct Food Substances Affirmed As Generally Recognized As Safe
PEDIATRIC	314.55	PEDIATRIC USE INFORMATION Applications For FDA Approval To Market A New Drug
PEDIATRIC	880.5140	PEDIATRIC HOSPITAL BED General Hospital And Personal Use Therapeutic Devices; General Hospital And Personal Use Devices
PEDIATRIC	880.5680	PEDIATRIC POSITION HOLDER General Hospital And Personal Use Therapeutic Devices; General Hospital And Personal Use Devices
PEDIATRIC STUDIES	601.27	PEDIATRIC STUDIES Biologics Licensing; Licensing
PEDIATRIC STUDIES	201.23	REQUIRED PEDIATRIC STUDIES General Labeling Provisions; Labeling
PEDIATRIC STUDIES	601.37	ANNUAL REPORTS FOR POSTMARKETING PEDIATRIC STUDIES Biologics Licensing; Licensing

PEDICLE SCREW	888.3070	PEDICLE SCREW SPINAL SYSTEM Prosthetic Devices; Orthopedic Devices
PEDICULICIDE	358.610	PEDICULICIDE ACTIVE INGREDIENTS Miscellaneous External Drug Products For Over-the-counter Human Use
PEDICULICIDE	358.650	LABELING OF PEDICULICIDE DRUG PRODUCTS Miscellaneous External Drug Products For Over-the-counter Human Use
PEELING	173.315	CHEMICALS USED IN WASHING OR TO ASSIST IN THE PEELING OF FRUITS AND VEGETABLES Specific Usage Additives; Secondary Direct Food Additives Permitted In Food For Human Consumption
PEFLUORINATED ION-EXCHANGE	173.21	PEFLUORINATED ION-EXCHANGE MEMBRANES Secondary Direct Food Additives Permitted In Food For Human Consumption
PELLETS	520.300b	CAMBENDAZOLE PELLETS Oral Dosage Form New Animal Drugs
PELLETS	520.1628	OXFENDAZOLE POWDER AND PELLETS Oral Dosage Form New Animal Drugs
PELLETS	520.2045	PYRANTEL TARTRATE POWDER; PYRANTEL TARTRATE PELLETS Oral Dosage Form New Animal Drugs
PENALTIES	17.34	DETERMINING THE AMOUNT OF PENALTIES AND ASSESSMENTS Civil Money Penalties Hearings
PENALTIES	205.8	VIOLATIONS AND PENALTIES Guidelines For State Licensing Of Wholesale Prescription Drug Distributors
PENALTIES	1230.49	PENALTIES Regulations Under The Federal Caustic Poison Act
PENICILLIN	520.1690	PENICILLIN ORAL DOSAGE FORMS Oral Dosage Form New Animal Drug Applications
PENICILLIN	524.1484h	NEOMYCIN, PENICILLIN, POLYMIXIN, HYDROCORTISONE SUSPENSION Ophthalmic And Topical Dosage Form New Animal Drugs
PENICILLIN	526.1696	PENICILLIN INTRAMAMMARY DOSAGE FORMS Intramammary Dosage Forms
PENICILLIN	556.510	PENICILLIN Specific Tolerances For Residues Of New Animal Drugs; Tolerances For Residues Of New Animal Drugs In Food
PENICILLIN	558.155	CHLORTETRACYCLINE, SULFATHIAZOLE, PENICILLIN Specific New Animal Drugs For Use In Animal Feeds; New Animal Drugs For Use In Animal Feeds
PENICILLIN	558.460	PENICILLIN Specific New Animal Drugs For Use In Animal Feeds; New Animal Drugs For Use In Animal Feeds
PENICILLIN CONTAMINATION	211.176	PENICILLIN CONTAMINATION Laboratory Controls; Current Good Manufacturing Practice For Finished Pharmaceuticals
PENICILLIN G PROCAINE	522.1696	PENICILLIN G PROCAINE IMPLANTATION AND INJECTABLE DOSAGE FORMS Implantation Or Injectable Dosage Form New Animal Drugs
PENICILLIN G PROCAINE	522.1696c	PENICILLIN G PROCAINE IN OIL Implantation Or Injectable Dosage Form New Animal Drugs
PENICILLIN G PROCAINE	526.1696a	PENICILLIN G PROCAINE IN OIL Intramammary Dosage Forms
PENICILLIN, PROCAINE	558.145	CHLORTETRACYCLINE, PROCAINE PENICILLIN, AND SULFAMETHAZINE Specific New Animal Drugs For Use In Animal Feeds; New Animal Drugs For Use In Animal Feeds
PENILE IMPLANT	876.3350	PENILE INFLATABLE IMPLANT Prosthetic Devices; Gastroenterology-urology Devices
PENILE IMPLANT	876.3630	PENILE RIGIDITY IMPLANT Prosthetic Devices; Gastroenterology-urology Devices
PENNSYLVANIA	808.88	PENNSYLVANIA Listing Of Specific State And Local Exemptions; Exemption From Federal Preemption Of State & Local Medical Device Requirements
PENTACHLOROPHENATE	178.3900	SODIUM PENTACHLOROPHENATE Certain Adjuvants And Production Aids; Indirect Food Additives; Adjuvants, Production Aids, And Sanitizers
PENTAERYTHRITOL	178.3690	PENTAERYTHRITOL ADIPATE-STEARATE Certain Adjuvants And Production Aids; Indirect Food Additives: Adjuvants, Production Aids, And Sanitizers
PENTAZOCINE	522.1698	PENTAZOCINE LACTATE INJECTION Implantation Or Injectable Dosage Form New Animal Drugs
PENTOBARBITAL	522.380	CHLORAL HYDRATE, PENTOBARBITAL, AND MAGNESIUM SULFATE STERILE AQUEOUS SOLUTION Implantation Or Injectable Dosage Form New Animal Drugs
PENTOBARBITAL	522.1704	SODIUM PENTOBARBITAL INJECTION Implantation Or Injectable Dosage Form New Animal Drugs
PENTOBARBITAL	522.2444b	SODIUM THIOPENTAL, SODIUM PENTOBARBITAL FOR INJECTION Implantation Or Injectable Dosage Form New Animal Drugs

PEPTIDE	862.1117	B-TYPE NATRIURETIC PEPTIDE TEST SYSTEM Clinical Chemistry Test Systems; Clinical Chemistry And Clinical Toxicology Devices
PEPTIDES	862.1135	C-PEPTIDES OF PROINSULIN TEST SYSTEM Clinical Chemistry Test Systems; Clinical Chemistry And Clinical Toxicology Devices
PEPTONES	184.1553	PEPTONES Listing Of Specific Substances Affirmed As GRAS; Direct Food Substances Affirmed As Generally Recognized As Safe
PERCENT OF INGREDIENTS	331.20	DETERMINATION OF PERCENT CONTRIBUTION OF ACTIVE INGREDIENTS Testing Procedures; Antacid Products For Over-the-counter (OTC) Human Use
PERCHA	872.3850	GUTTA PERCHA Prosthetic Devices; Dental Devices
PERCUSSOR	868.5665	POWERED PERCUSSOR Therapeutic Devices; Anesthesiology Devices
PERCUSSOR	882.1700	PERCUSSOR Neurological Diagnostic Devices; Neurological Devices
PERCUTANEOUS	870.1250	PERCUTANEOUS CATHETER Cardiovascular Diagnostic Devices; Cardiovascular Devices
PERCUTANEOUS	870.1310	VESSEL DILATOR FOR PERCUTANEOUS CATHETERIZATION Cardiovascular Diagnostic Devices; Cardiovascular Devices
PERFLUOROCARBON	177.1550	PERFLUOROCARBON RESINS Substances For Use As Basic Components Of Single And Repeated Use Food Contact Surfaces; Indirect Food Additives: Polymers
PERFLUOROCARBON	177.2400	PERFLUOROCARBON CURED ELASTOMERS Substances For Use Only As Components Of Articles Intended For Repeated Use; Indirect Food Additives: Polymers
PERFLUOROHEXANE	173.342	CHLOROFLUOROCARBON 113 AND PERFLUOROHEXANE Specific Usage Additives; Secondary Direct Food Additives Permitted In Food For Human Consumption
PERFORMANCE	1403.40	MONITORING AND REPORTING PROGRAM PERFORMANCE Reports, Records, Retention, And Enforcement; Uniform Administrative Requirements For Grants And Coop. Agreements To State And Local Gov.
PERFORMANCE DATA	1002.3	NOTIFICATION TO USER OF PERFORMANCE AND TECHNICAL DATA General Provisions; Records And Reports
PERFORMANCE STANDARD	860.132	PROCEDURES WHEN THE COMMISSIONER INITIATES A PERFORMANCE STANDARD OR PREMARKET APPROVAL PROCEEDING UNDER SECTION 514(B) OR 515(B) Reclassification; Medical Device Classification Procedures
PERFORMANCE STANDARD	898.12	PERFORMANCE STANDARD Performance Standards For Electrode Lead Wires And Patient Cables
PERFORMANCE STANDARDS	5.600	VARIANCES FROM PERFORMANCE STANDARDS FOR ELECTRONIC PRODUCTS Radiation Control; Redelegations Of Authority; Delegations Of Authority And Organization
PERFORMANCE STANDARDS	5.601	EXEMPTION OF ELECTRONIC PRODUCTS FROM PERFORMANCE STANDARDS AND PROHIBITED ACTS Radiation Control; Redelegations Of Authority; Delegations Of Authority And Organization
PERFUSION	876.5880	ISOLATED KIDNEY PERFUSION AND TRANSPORT SYSTEM AND ACCESSORIES Therapeutic Devices; Gastroenterology-urology Devices
PERIMETER	886.1605	PERIMETER Diagnostic Devices; Ophthalmic Devices
PERINATAL	884.2740	PERINATAL MONITORING SYSTEM AND ACCESSORIES Obstetrical And Gynecological Monitoring Devices; Obstetrical And Gynecological Devices
PERINEAL	884.5390	PERINEAL HEATER Obstetrical And Gynecological Therapeutic Devices; Obstetrical And Gynecological Devices
PERINEOMETER	884.1425	PERINEOMETER Obstetrical And Gynecological Diagnostic Devices; Obstetrical And Gynecological Devices
PERIOD, OPERATIONAL	26.11	START OF OPERATIONAL PERIOD Specific Sector Provisions For Pharmaceutical GMPs; Mutual Recognition Of Pharmaceutical GMP Reports, Medical Device Quality System Audit Reports
PERIOD, OPERATIONAL	26.40	START OF THE OPERATIONAL PERIOD Specific Sector Provisions For Medical Devices; Mutual Recognition Of Pharmaceutical GMP Reports, Medical Device Quality System Audit Reports
PERIOD, REVIEW	60.20	FDA ACTION ON REGULATORY REVIEW PERIOD DETERMINATION Regulatory Review Period Determinations; Patent Term Restoration
PERIOD, REVIEW	60.22	REGULATORY REVIEW PERIOD DETERMINATIONS Regulatory Review Period Determinations; Patent Term Restoration

PERIOD, REVIEW	60.24	REVISION OF REGULATORY REVIEW PERIOD DETERMINATIONS Regulatory Review Period Determinations; Patent Term Restoration
PERIOD, REVIEW	60.26	FINAL ACTION ON REGULATORY REVIEW PERIOD DETERMINATIONS Regulatory Review Period Determinations; Patent Term Restoration
PERIOD, TRANSITION	26.5	LENGTH OF TRANSITION PERIOD Specific Sector Provisions For Pharmaceutical GMPs; Mutual Recognition Of Pharmaceutical GMP Reports, Medical Device Quality System Audit Reports
PERIOD, TRANSITION	26.35	LENGTH AND PURPOSE OF TRANSITION PERIOD Specific Sector Provisions For Medical Devices; Mutual Recognition Of Pharmaceutical GMP Reports, Medical Device Quality System Audit Reports
PERIOD, TRANSITION	26.38	OTHER TRANSITION PERIOD ACTIVITIES Specific Sector Provisions For Medical Devices; Mutual Recognition Of Pharmaceutical GMP Reports, Medical Device Quality System Audit Reports
PERIODATE	573.240	CALCIUM PERIODATE Food Additive Listing; Food Additives Permitted In Feed And Drinking Water Of Animals
PERIODS, DATING	610.53	DATING PERIODS FOR LICENSED BIOLOGICAL PRODUCTS Dating Period Limitations; General Biological Products Standards
PERIODS, REVIEW	60.28	TIME FRAME FOR DETERMINING REGULATORY REVIEW PERIODS Regulatory Review Period Determinations; Patent Term Restoration
PERIPHERAL	868.2775	ELECTRICAL PERIPHERAL NERVE STIMULATOR Monitoring Devices; Anesthesiology Devices
PERIPHERAL	882.5870	IMPLANTED PERIPHERAL NERVE STIMULATOR FOR PAIN RELIEF Neurological Therapeutic Devices; Neurological Devices
PERISHABLES	1250.27	STORAGE OF PERISHABLES Food Service Sanitation On Land And Air Conveyances, And Vessels; Interstate Conveyance Sanitation
PERITONEAL	876.5630	PERITONEAL DIALYSIS SYSTEM AND ACCESSORIES Therapeutic Devices; Gastroenterology-urology Devices
PERITONEO-VENOUS	876.5955	PERITONEO-VENOUS SHUNT Therapeutic Devices; Gastroenterology-urology Devices
PERMANENT DISQUALIFICATION	19.55	PERMANENT DISQUALIFICATION OF FORMER EMPLOYEES Disqualification Conditions; Standards Of Conduct And Conflicts Of Interest
PERMANENT ELECTRODE	870.3680	CARDIOVASCULAR PERMANENT OR TEMPORARY PACEMAKER ELECTRODE Cardiovascular Prosthetic Devices; Cardiovascular Devices
PERMANENT-RESIDENT AGENT	316.22	PERMANENT-RESIDENT AGENT FOR FOREIGN SPONSOR Orphan Drugs
PERMANGANATE	250.108	POTASSIUM PERMANGANATE PREPARATIONS AS PRESCRIPTION DRUGS New Drug Or Prescription Status Of Specific Drugs; Specific Requirements For Specific Human Drugs
PERMEABLE, GAS	886.5916	RIGID GAS PERMEABLE CONTACT LENS Therapeutic Devices; Ophthalmic Devices
PERMEABLE, GAS	886.5918	RIGID GAS PERMEABLE CONTACT LENS CARE PRODUCTS Therapeutic Devices; Ophthalmic Devices
PERMEABILITY	876.5860	HIGH PERMEABILITY HEMODIALYSIS SYSTEM Therapeutic Devices; Gastroenterology-urology Devices
PERMISSIBLE	610.17	PERMISSIBLE COMBINATIONS General Provisions; General Biological Products Standards
PERMISSION	50.55	REQUIREMENTS FOR PERMISSION BY PARENTS OR GUARDIANS AND FOR ASSENT BY CHILDREN Additional Safeguards For Children In Clinical Investigations; Protection Of Human Subjects
PERMIT	5.301	ISSUANCE OF INITIAL EMERGENCY PERMIT ORDERS AND NOTICES OF CONFIRMATION OF EFFECTIVE DATE OF FINAL REGULATIONS ON FOOD FOR HUMAN AND ANIMAL CONSUMPTION Foods And Cosmetics; Redelegations Of Authority, Delegations Of Authority And Organization
PERMIT	108.5	DETERMINATION OF THE NEED FOR A PERMIT General Provisions; Emergency Permit Control
PERMIT	108.6	REVOCATION OF DETERMINATION OF NEED FOR PERMIT General Provisions; Emergency Permit Control
PERMIT	108.7	ISSUANCE OR DENIAL OF PERMIT General Provisions; Emergency Permit Control
PERMIT	108.10	SUSPENSION AND REINSTATEMENT OF PERMIT General Provisions; Emergency Permit Control
PERMIT	108.12	MANUFACTURING, PROCESSING, OR PACKING WITHOUT A PERMIT OR IN VIOLATION OF A PERMIT General Provisions; Emergency Permit Control
PERMIT	500.24	EMERGENCY PERMIT CONTROL Specific Administrative Rulings And Decisions; General
PERMIT	1210.20	APPLICATION FOR PERMIT Permit Control; Regulations Under The Federal Import Milk Act
PERMIT	1210.21	PERMIT NUMBER Permit Control; Regulations Under The Federal Import Milk Act

PERMIT	1210.30	HEARING PROCEDURE FOR PERMIT DENIAL, SUSPENSION, AND REVOCATION Regulations Under The Federal Import Milk Act
PERMIT	1312.16	CANCELLATION OF PERMIT; EXPIRATION DATE Importation Of Controlled Substances; Importation And Exportation Of Controlled Substances
PERMIT, EXPORT	1312.22	APPLICATION FOR EXPORT PERMIT Exportation Of Controlled Substances; Importation And Exportation Of Controlled Substances
PERMIT, EXPORT	1312.23	ISSUANCE OF EXPORT PERMIT Exportation Of Controlled Substances; Importation And Exportation Of Controlled Substances
PERMIT, EXPORT	1312.24	DISTRIBUTION OF COPIES OF EXPORT PERMIT Exportation Of Controlled Substances; Importation And Exportation Of Controlled Substances
PERMIT, EXPORT	1312.30	SCHEDULE III, IV, AND V NON-NARCOTIC CONTROLLED SUBSTANCES REQUIRING AN IMPORT AND EXPORT PERMIT Exportation Of Controlled Substances; Importation And Exportation Of Controlled Substances
PERMIT, IMPORT	1312.12	APPLICATION FOR IMPORT PERMIT Importation Of Controlled Substances; Importation And Exportation Of Controlled Substances
PERMIT, IMPORT	1312.13	ISSUANCE OF IMPORT PERMIT Importation Of Controlled Substances; Importation And Exportation Of Controlled Substances
PERMIT, IMPORT	1312.14	DISTRIBUTION OF COPIES OF IMPORT PERMIT Importation Of Controlled Substances; Importation And Exportation Of Controlled Substances
PERMIT, IMPORT	1312.30	SCHEDULE III, IV, AND V NON-NARCOTIC CONTROLLED SUBSTANCES REQUIRING AN IMPORT AND EXPORT PERMIT Exportation Of Controlled Substances; Importation And Exportation Of Controlled Substances
PERMITS	130.17	TEMPORARY PERMITS FOR INTERSTATE SHIPMENT OF EXPERIMENTAL PACKS OF FOOD VARYING FROM THE REQUIREMENTS OF DEFINITIONS AND STANDARDS OF IDENTITY General Provisions; Food Standards: General
PERMITS	1210.23	PERMITS GRANTED ON CERTIFICATE Permit Control; Regulations Under The Federal Import Milk Act
PERMITS	1210.24	TEMPORARY PERMITS Permit Control; Regulations Under The Federal Import Milk Act
PERMITS	1210.25	PERMITS FOR PASTEURIZED MILK OR CREAM Permit Control; Regulations Under The Federal Import Milk Act
PERMITS	1210.26	PERMITS FOR RAW MILK OR CREAM Permit Control; Regulations Under The Federal Import Milk Act
PERMITS	1210.27	PERMITS WAIVING CLAUSES 2 AND 5, SECTION 2 OF THE FEDERAL IMPORT MILK ACT Permit Control; Regulations Under The Federal Import Milk Act
PERMITS	1210.28	PERMITS WAIVING CLAUSE 4, SECTION 2 OF THE FEDERAL IMPORT MILK ACT Permit Control; Regulations Under The Federal Import Milk Act
PERMITTED COMBINATIONS	333.120	PERMITTED COMBINATIONS OF ACTIVE INGREDIENTS First Aid Antibiotic Drug Products; Topical Antimicrobial Drug Products For Over-the-counter Human Use
PERMITTED COMBINATIONS	333.160	LABELING OF PERMITTED COMBINATIONS OF ACTIVE INGREDIENTS First Aid Antibiotic Drug Products; Topical Antimicrobial Drug Products For Over-the-counter Human Use
PERMITTED COMBINATIONS	333.320	PERMITTED COMBINATIONS OF ACTIVE INGREDIENTS Topical Acne Drug Products; Topical Antimicrobial Drug Products For Over-the-counter Human Use
PERMITTED COMBINATIONS	343.22	PERMITTED COMBINATIONS OF ACTIVE INGREDIENTS FOR CARDIOVASCULAR-RHEUMATOLOGIC USE Internal Analgesic, Antipyretic, And Antirheumatic Drug Products For OTC Human Use
PERMITTED COMBINATIONS	346.22	PERMITTED COMBINATIONS OF ANORECTAL ACTIVE INGREDIENTS Anorectal Drug Products For Over-the-counter Human Use
PERMITTED COMBINATIONS	346.52	LABELING OF PERMITTED COMBINATIONS OF ANORECTAL ACTIVE INGREDIENTS Anorectal Drug Products For Over-the-counter Human Use
PERMITTED COMBINATIONS	349.30	PERMITTED COMBINATIONS OF ACTIVE INGREDIENTS Ophthalmic Drug Products For Over-the-counter Human Use
PERMITTED COMBINATIONS	349.79	LABELING OF PERMITTED COMBINATIONS OF ACTIVE INGREDIENTS Ophthalmic Drug Products For Over-the-counter Human Use
PERMITTED COMBINATIONS	352.20	PERMITTED COMBINATIONS OF ACTIVE INGREDIENTS Sunscreen Drug Products For Over-the-counter Human Use

PERMITTED COMBINATIONS	352.60	LABELING OF PERMITTED COMBINATIONS OF ACTIVE INGREDIENTS Sunscreen Drug Products For Over-the-counter Human Use
PERMITTED COMBINATIONS	358.720	PERMITTED COMBINATIONS OF ACTIVE INGREDIENTS Miscellaneous External Drug Products For Over-the-counter Human Use
PERMITTED EXTRALABEL USE	530.20	CONDITIONS FOR PERMITTED EXTRALABEL ANIMAL AND HUMAN DRUG USE IN FOOD-PRODUCING ANIMALS Extralabel Drug Use In Animals
PERNICIOUS ANEMIA	250.201	PREPARATIONS FOR THE TREATMENT OF PERNICIOUS ANEMIA Requirements For Drugs And Foods; Specific Requirements For Specific Human Drugs
PEROXIDASE	864.7675	LEUKOCYTE PEROXIDASE Hematology Kits And Packages; Hematology And Pathology Devices
PEROXIDE, BENZYL	184.1157	BENZYL PEROXIDE Listing Of Specific Substances Affirmed As Gras; Direct Food Substances Affirmed As Generally Recognized As Safe
PEROXIDE, HYDROGEN	178.1005	HYDROGEN PEROXIDE SOLUTION Substances Utilized To Control The Growth Of Microorganisms; Indirect Food Additives: Adjuvants, Production Aids, And Sanitizers
PEROXIDE, HYDROGEN	184.1366	HYDROGEN PEROXIDE Listing Of Specific Substances Affirmed As Gras; Direct Food Substances Affirmed As Generally Recognized As Safe
PEROXIDE, HYDROGEN	582.1366	HYDROGEN PEROXIDE General Purpose Food Additives; Substances Generally Recognized As Safe
PEROXIDES, ACETONE	172.802	ACETONE PEROXIDES Multipurpose Additives; Food Additives Permitted For Direct Addition To Food For Human Consumption
PEROXYACIDS	173.370	PEROXYACIDS Specific Usage Additives; Secondary Direct Food Additives Permitted In Food For Human Consumption
PERMANENT	886.4445	PERMANENT MAGNET Surgical Devices; Ophthalmic Devices
PERSON, NOTIFICATION	720.7	NOTIFICATION OF PERSON SUBMITTING COSMETIC PRODUCT INGREDIENT STATEMENT AND COSMETIC RAW MATERIAL COMPOSITION STATEMENT Voluntary Filing Of Cosmetic Product Ingredient Composition Statements
PERSONAL PRIVACY	20.63	PERSONNEL, MEDICAL, AND SIMILAR FILES, DISCLOSURE OF WHICH CONSTITUTES A CLEARLY UNWARRANTED INVASION OF PERSONAL PRIVACY Exemptions; Public Information
PERSONNEL	58.29	PERSONNEL Organization And Personnel; Good Laboratory Practice For Nonclinical Laboratory Studies
PERSONNEL	110.10	PERSONNEL General Provisions; Current Good Manufacturing Practice In Manufacturing, Packing, Or Holding Human Food
PERSONNEL	113.10	PERSONNEL General Provisions; Thermally Processed Low-acid Foods Packaged In Hermetically Sealed Containers
PERSONNEL	114.10	PERSONNEL General Provisions; Acidified Foods
PERSONNEL	205.7	PERSONNEL Guidelines For State Licensing Of Wholesale Prescription Drug Distributors
PERSONNEL	225.10	PERSONNEL General Provisions; Current Good Manufacturing Practice For Medicated Feeds
PERSONNEL	226.10	PERSONNEL General Provisions; Current Good Manufacturing Practice For Type A Medicated Articles
PERSONNEL	600.10	PERSONNEL Establishment Standards; Biological Products: General
PERSONNEL	606.20	PERSONNEL Organization And Personnel; Current Good Manufacturing Practice For Blood And Blood Components
PERSONNEL	820.25	PERSONNEL Quality System Requirements; Quality System Regulation
PERSONNEL	892.6500	PERSONNEL PROTECTIVE SHIELD Miscellaneous Devices; Radiology Devices
PERSONNEL	1301.23	EXEMPTION OF CERTAIN MILITARY AND OTHER PERSONNEL Registration Of Manufacturers, Distributors, And Dispensers Of Controlled Substances
PERSONNEL FILES	20.63	PERSONNEL, MEDICAL, AND SIMILAR FILES, DISCLOSURE OF WHICH CONSTITUTES A CLEARLY UNWARRANTED INVASION OF PERSONAL PRIVACY Exemptions; Public Information

PERSONNEL QUALIFICATIONS	211.25	PERSONNEL QUALIFICATIONS Organization And Personnel; Current Good Manufacturing Practice For Finished Pharmaceuticals
PERSONNEL RECORDS	21.32	PERSONNEL RECORDS Requirements For Specific Categories Of Records; Protection Of Privacy
PERSONNEL RESPONSIBILITIES	211.28	PERSONNEL RESPONSIBILITIES Organization And Personnel; Current Good Manufacturing Practice For Finished Pharmaceuticals
PERSONS	14.31	CONSULTATION BY AN ADVISORY COMMITTEE WITH OTHER PERSONS Meeting Procedures; Public Hearing Before A Public Advisory Committee
PERSONS	21.72	INDIVIDUAL CONSENT TO DISCLOSURE OF RECORDS TO OTHER PERSONS Disclosure Of Records In Privacy Act Record Systems To Persons Other ThanThe Subject Individual; Protection Of Privacy
PERSONS	821.30	TRACKING OBLIGATIONS OF PERSONS OTHER THAN DEVICE MANUFACTURERS: DISTRIBUTOR REQUIREMENTS Medical Device Tracking Requirements
PERSONS	1301.11	PERSONS REQUIRED TO REGISTER Registration Of Manufacturers, Distributors, And Dispensers Of Controlled Substances
PERSONS	1304.03	PERSONS REQUIRED TO KEEP RECORDS AND FILE REPORTS General Information; Records And Reports Of Registrants
PERSONS	1305.04	PERSONS ENTITLED TO OBTAIN AND EXECUTE ORDER FORMS Order Forms
PERSONS	1305.08	PERSONS ENTITLED TO FILL ORDER FORMS Order Forms
PERSONS	1306.03	PERSONS ENTITLED TO ISSUE PRESCRIPTIONS General Information; Prescriptions
PERSONS	1306.06	PERSONS ENTITLED TO FILL PRESCRIPTIONS General Information; Prescriptions
PERSONS	1309.21	PERSONS REQUIRED TO REGISTER Registration Of Manufacturers, Distributors, Importers And Exporters Of List I Chemicals
PERSONS	1404.205	INELIGIBLE PERSONS Governmentwide Debarment And Suspension And Requirements For Drug-Free Workplace (Grants)
PERU BALSAM	524.2620	LIQUID CRYSTALLINE TRYPSIN, PERU BALSAM, CASTOR OIL Ophthalmic And Topical Dosage Form New Animal Drugs
PERYLENE-DIONE	73.3120	16,17-DIMETHOXYDINAPHTHO [1,2,3-CD:3',2',1'-1M] PERYLENE-5,10-DIONE Medical Devices; Listing Of Color Additives Exempt From Certification
PESSARY	884.3575	VAGINAL PESSARY Obstetrical And Gynecological Prosthetic Devices; Obstetrical And Gynecological Devices
PESTICIDE	172.710	ADJUVANTS FOR PESTICIDE USE DILUTIONS Other Specific Usage Additives; Food Additives Permitted For Direct Addition To Food For Human Consumption
PESTICIDE	182.99	ADJUVANTS FOR PESTICIDE CHEMICALS General Provisions; Substances Generally Recognized As Safe
PESTICIDE	570.19	PESTICIDE CHEMICALS IN PROCESSED FOODS General Provisions; Food Additives
PESTICIDE	582.99	ADJUVANTS FOR PESTICIDE CHEMICALS General Provisions; Substances Generally Recognized As Safe
PESTICIDES	170.19	PESTICIDE CHEMICALS IN PROCESSED FOODS General Provisions; Food Additives
PET FOOD	570.13	INDIRECT FOOD ADDITIVES RESULTING FROM PACKAGING MATERIALS PRIOR SANCTIONED FOR ANIMAL FEED AND PET FOOD General Provisions; Food Additives
PET FOOD	570.14	INDIRECT FOOD ADDITIVES RESULTING FROM PACKAGING MATERIALS FOR ANIMAL FEED AND PET FOOD General Provisions; Food Additives
PETITION	10.30	CITIZEN PETITION General Administrative Procedures; Administrative Practices And Procedures
PETITION	12.82	PETITION TO PARTICIPATE IN FORMA PAUPERIS Hearing Procedures; Formal Evidentiary Public Hearing
PETITION	60.32	APPLICANT RESPONSE TO PETITION Due Diligence Petitions; Patent Term Restoration
PETITION	71.2	NOTICE OF FILING OF PETITION General Provisions; Color Additive Petitions
PETITION	71.18	PETITION FOR EXEMPTION FROM CERTIFICATION General Provisions; Color Additive Petitions
PETITION	171.6	AMENDMENT OF PETITION General Provisions; Food Additive Petitions
PETITION	171.7	WITHDRAWAL OF PETITION WITHOUT PREJUDICE General Provisions; Food Additive Petitions

PETITION	171.100	REGULATION BASED ON PETITION Administrative Actions On Applications; Food Additive Petitions
PETITION	314.93	PETITION TO REQUEST A CHANGE FROM A LISTED DRUG Applications For FDA Approval To Market A New Drug
PETITION	314.122	SUBMITTING AN ABBREVIATED APPLICATION FOR, OR A 505(J)(2)(C) PETITION THAT RELIES ON A LISTED DRUG THAT IS NO LONGER MARKETED Applications For FDA Approval To Market A New Drug
PETITION	571.6	AMENDMENT OF PETITION General Provisions; Food Additive Petitions
PETITION	571.7	WITHDRAWAL OF PETITION WITHOUT PREJUDICE General Provisions; Food Additive Petitions
PETITION	571.100	REGULATION BASED ON PETITION Administrative Actions On Applications; Food Additive Petitions
PETITION	860.123	RECLASSIFICATION PETITION: CONTENT AND FORM Reclassification; Medical Device Classification Procedures
PETITION	1316.92	PETITION FOR EXPEDITED RELEASE IN AN ADMINISTRATIVE FORFEITURE ACTION Administrative Functions, Practices, And Procedures
PETITION	1316.93	RULING ON PETITION FOR EXPEDITED RELEASE IN AN ADMINISTRATIVE FORFEITURE ACTION Administrative Functions, Practices, And Procedures
PETITION	1316.95	PETITION FOR EXPEDITED RELEASE OF A CONVEYANCE IN A JUDICIAL FORFEITURE ACTION Administrative Functions, Practices, And Procedures
PETITION	1316.96	RULING ON A PETITION FOR EXPEDITED RELEASE OF A CONVEYANCE IN A JUDICIAL FORFEITURE ACTION Expedited Forfeiture Proceedings For Certain Property; Administrative Functions, Practices, And Procedures
PETITIONERS	5.404	NOTIFICATION TO PETITIONERS OF DETERMINATIONS MADE ON PETITIONS FOR RECLASSIFICATION OF MEDICAL DEVICES Medical Devices And Radiological Health; Redelegations Of Authority; Delegations Of Authority And Organization
PETITIONS	5.29	PETITIONS UNDER PART 10 General Redelegations Of Authority; Delegations Of Authority And Organization
PETITIONS	5.404	NOTIFICATION TO PETITIONERS OF DETERMINATIONS MADE ON PETITIONS FOR RECLASSIFICATION OF MEDICAL DEVICES Medical Devices And Radiological Health; Redelegations Of Authority; Delegations Of Authority And Organization
PETITIONS	5.406	NOTIFICATION TO SPONSORS OF DEFICIENCIES IN PETITIONS FOR RECLASSIFICATION OF MEDICAL DEVICES Medical Devices And Radiological Health; Redelegations Of Authority; Delegations Of Authority And Organization
PETITIONS	12.159	COPIES OF PETITIONS FOR JUDICIAL REVIEW Judicial Review; Formal Evidentiary Public Hearing
PETITIONS	60.30	FILING, FORMAT, AND CONTENT OF PETITIONS Due Diligence Petitions; Patent Term Restoration
PETITIONS	60.34	FDA ACTION ON PETITIONS Due Diligence Petitions; Patent Term Restoration
PETITIONS	71.1	PETITIONS General Provisions; Color Additive Petitions
PETITIONS	71.6	EXTENSION OF TIME FOR STUDYING PETITIONS; SUBSTANTIVE AMENDMENTS; WITHDRAWAL OF PETITIONS WITHOUT PREJUDICE General Provisions; Color Additive Petitions
PETITIONS	71.15	CONFIDENTIALITY OF DATA AND INFORMATION IN COLOR ADDITIVE PETITIONS General Provisions; Color Additive Petitions
PETITIONS	100.1	PETITIONS REQUESTING EXEMPTION FROM PREEMPTION FOR STATE OR LOCAL REQUIREMENTS State And Local Requirements; General
PETITIONS	101.69	PETITIONS FOR NUTRIENT CONTENT CLAIMS Specific Requirements For Nutrient Content Claims; Food Labeling
PETITIONS	101.70	PETITIONS FOR HEALTH CLAIMS Specific Requirements For Health Claims; Food Labeling
PETITIONS	102.19	PETITIONS General Provisions; Common Or Usual Name For Nonstandardized Foods
PETITIONS	171.1	PETITIONS General Provisions; Food Additive Petitions
PETITIONS	501.103	PETITIONS REQUESTING EXEMPTIONS FROM OR SPECIAL REQUIREMENTS FOR LABEL DECLARATION OF INGREDIENTS Exemptions From Animal Food Labeling Requirements; Animal Food Labeling
PETITIONS	502.19	PETITIONS Common Or Usual Names For Nonstandardized Animal Foods

PETITIONS	509.5	PETITIONS General Provisions; Unavoidable Contaminants In Animal Food And Food-packaging Material
PETITIONS	571.1	PETITIONS General Provisions; Food Additive Petitions
PETITIONS	1316.68	COPIES OF PETITIONS FOR JUDICIAL REVIEW Administrative Hearings; Administrative Functions, Practices, And Procedures
PETITIONS	1316.79	PETITIONS FOR REMISSION OR MITIGATION OF FORFEITURE Seizure, Forfeiture, And Disposition Of Property; Administrative Functions, Practices, And Procedures
PETITIONS	1316.80	TIME FOR FILING PETITIONS Seizure, Forfeiture, And Disposition Of Property; Administrative Functions, Practices, And Procedures
PETITIONS	1316.81	OF PETITIONS Seizure, Forfeiture, And Disposition Of Property; Administrative Functions, Practices, And Procedures
PETROLATUM	172.880	PETROLATUM Multipurpose Additives; Food Additives Permitted For Direct Addition To Food For Human Consumption
PETROLATUM	178.3700	PETROLATUM Certain Adjuvants And Production Aids; Indirect Food Additives: Adjuvants, Production Aids, And Sanitizers
PETROLATUM	573.720	PETROLATUM Food Additive Listing; Food Additives Permitted In Feed And Drinking Water Of Animals
PETROLEUM	172.250	PETROLEUM NAPHTHA Coatings, Films And Related Substances; Food Additives Permitted For Direct Addition To Food For Human Consumption
PETROLEUM	172.882	SYNTHETIC ISOPARAFFINIC PETROLEUM HYDROCARBONS Multipurpose Additives; Food Additives Permitted For Direct Addition To Food For Human Consumption
PETROLEUM	172.884	ODORLESS LIGHT PETROLEUM HYDROCARBONS Multipurpose Additives; Food Additives Permitted For Direct Addition To Food For Human Consumption
PETROLEUM	178.3530	ISOPARAFFINIC PETROLEUM HYDROCARBONS, SYNTHETIC Certain Adjuvants And Production Aids; Indirect Food Additives: Adjuvants, Production Aids, And Sanitizers
PETROLEUM	178.3650	ODORLESS LIGHT PETROLEUM HYDROCARBONS Certain Adjuvants And Production Aids; Indirect Food Additives: Adjuvants, Production Aids, And Sanitizers
PETROLEUM	573.740	ODORLESS LIGHT PETROLEUM HYDROCARBONS Food Additive listing; Food Additives Permitted In Feed And Drinking Water Of Animals
PETROLEUM WAX	172.886	PETROLEUM WAX Multipurpose Additives; Food Additives Permitted For Direct Addition To Food For Human Consumption
PETROLEUM WAX	172.888	SYNTHETIC PETROLEUM WAX Multipurpose Additives; Food Additives Permitted For Direct Addition To Food For Human Consumption
PETROLEUM WAX	178.3710	PETROLEUM WAX Certain Adjuvants And Production Aids; Indirect Food Additives: Adjuvants, Production Aids, And Sanitizers
PETROLEUM WAX	178.3720	PETROLEUM WAX, SYNTHETIC Certain Adjuvants And Production Aids; Indirect Food Additives: Adjuvants, Production Aids, And Sanitizers
PH	862.1120	BLOOD GASES (PCO2PO2) AND BLOOD pH TEST SYSTEM Clinical Chemistry Test Systems; Clinical Chemistry And Clinical Toxicology Devices
PH	862.1550	URINARY pH (NONQUANTITATIVE) TEST SYSTEM Clinical Chemistry Test Systems; Clinical Chemistry And Clinical Toxicology Devices
PH	868.1170	INDWELLING BLOOD HYDROGEN ION CONCENTRATION (pH) ANALYZER Diagnostic Devices; Anesthesiology Devices
PH	876.1400	STOMACH pH ELECTRODE Diagnostic Devices; Gastroenterology-urology Devices
PHACOFRAGMENTATION	886.4670	PHACOFRAGMENTATION SYSTEM Surgical Devices; Ophthalmic Devices
PHAFFIA YEAST	73.355	PHAFFIA YEAST Foods; Listing Of Color Additives Exempt From Certification
PHALANGEAL	888.3730	TOE JOINT PHALANGEAL (HEMI-TOE) POLYMER PROSTHESIS Prosthetic Devices; Orthopedic Devices
PHANTOM	892.1370	NUCLEAR ANTHROPOMORPHIC PHANTOM Diagnostic Devices; Radiology Devices

PHANTOM	892.1380	NUCLEAR FLOOD SOURCE PHANTOM Diagnostic Devices; Radiology Devices
PHANTOM	892.1420	RADIONUCLIDE TEST PATTERN PHANTOM Diagnostic Devices; Radiology Devices
PHANTOM	892.1950	RADIOGRAPHIC ANTHROPOMORPHIC PHANTOM Diagnostic Devices; Radiology Devices
PHARMACEUTICAL MANUFACTURERS	200.10	CONTRACT FACILITIES (INCLUDING CONSULTING LABORATORIES) UTILIZED AS EXTRAMURAL FACILITIES BY PHARMACEUTICAL MANUFACTURERS General Provisions; General
PHARMACIES	1306.25	TRANSFER BETWEEN PHARMACIES OF PRESCRIPTION INFORMATION FOR SCHEDULES III, IV, AND V CONTROLLED SUBSTANCES FOR REFILL PURPOSES Controlled Substances Listed In Schedules III, IV And V; Prescriptions
PHARMACISTS	200.7	SUPPLYING PHARMACISTS WITH INDICATIONS AND DOSAGE INFORMATION General Provisions; General
PHARMACOLOGICAL EFFECT	320.28	CORRELATION OF BIOAVAILABILITY WITH AN ACUTE PHARMACOLOGICAL EFFECT OR CLINICAL EVIDENCE Procedures For Determining The Bioavailability Or Bioequivalence Of Drug Products; Bioavailability And Bioequivalence Requirements
PHASE 4 STUDIES	312.85	PHASE 4 STUDIES Drugs Intended To Treat Life-threatening And Severely-debilitating Illnesses; Investigational New Drug Application
PHASES OF INVESTIGATION	312.21	PHASES OF AN INVESTIGATION Investigational New Drug Application
PHENACETIN	201.309	ACETOPHENETIDIN (PHENACETIN)-CONTAINING PREPARATIONS; NECESSARY WARNING STATEMENT Specific Labeling Requirements For Specific Drug Products; Labeling
PHENINDIONE	201.310	PHENINDIONE; LABELING OF DRUG PREPARATIONS INTENDED FOR USE BY MAN Specific Labeling Requirements For Specific Drug Products; Labeling
PHENOBARBITAL	862.3660	PHENOBARBITAL TEST SYSTEM Clinical Toxicology Test Systems; Clinical Chemistry And Clinical Toxicology Devices
PHENOL	172.150	4-HYDROXYMETHYL-2,6-DI-TERT-BUTYL-PHENOL Food Preservatives; Food Additives Permitted For Direct Addition To Food For Human Consumption
PHENOL	178.2550	4-HYDROXYMETHYL-2,6-DI-TERT-BUTYL-PHENOL Antioxidants And Stabilizers; Indirect Food Additives: Adjuvants, Production Aids, And Sanitizers
PHENOLIC	177.2410	PHENOLIC RESINS IN MOLDED ARTICLES Substances For Use Only As Components Of Articles Intended For Repeated Use; Indirect Food Additives: Polymers
PHENOTHIAZINE	520.1802c	PIPERAZINE-CARBON DISULFIDE COMPLEX WITH PHENOTHIAZINE SUSPENSION Oral Dosage Form New Animal Drugs
PHENOTHIAZINE	520.2520	TRICHLORFON, PHENOTHIAZINE, AND PIPERAZINE DIHYDROCHLORIDE POWDER Oral Dosage Form New Animal Drugs
PHENOTHIAZINE	862.3670	PHENOTHIAZINE TEST SYSTEM Clinical Toxicology Test Systems; Clinical Chemistry And Clinical Toxicology Devices
PHENYLALANINE	201.21	DECLARATION OF PRESENCE OF PHENYLALANINE AS A COMPONENT OF ASPARTAME IN OVER-THE-COUNTER AND PRESCRIPTION DRUGS FOR HUMAN USE General Labeling Provisions; Labeling
PHENYLALANINE	582.5590	PHENYLALANINE Nutrients And/or Dietary Supplements; Substances Generally Recognized As Safe
PHENYLALANINE	862.1555	PHENYLALANINE TEST SYSTEM Clinical Chemistry Test Systems; Clinical Chemistry And Clinical Toxicology Devices
PHENYLAMINOANTHRAQUINONE	73.3100	1,4-BIS[4-(2-METHACRYLOXYETHYL)PHENYLAMINO]ANTHRAQUINONE COPOLYMERS Medical Devices; Listing Of Color Additives Exempt From Certification
PHENYLBUTAZONE	520.1720	PHENYLBUTAZONE ORAL DOSAGE FORMS Oral Dosage Form New Animal Drugs
PHENYLBUTAZONE	522.1720	PHENYLBUTAZONE INJECTION Implantation Or Injectable Dosage Form New Animal Drugs
PHENYLENE OXIDE	177.2460	POLY(2,6-DIMETHYL-1,4-PHENYLENE) OXIDE RESINS Substances For Use Only As Components Of Articles Intended For Repeated Use; Indirect Food Additives: Polymers
PHENYLENETEREPHTHALAMIDE	177.1632	POLY (PHENYLENETEREPHTHALAMIDE) RESINS Substances For Use As Basic Components Of Single And Repeated Use Food Contact Surfaces; Indirect Food Additives: Polymers

PHENYLKETONES	862.1560	URINARY PHENYLKETONES TEST SYSTEM Clinical Chemistry Test Systems; Clinical Chemistry And Clinical Toxicology Devices
PHLEBOGRAPH	870.2750	IMPEDANCE PHLEBOGRAPH Cardiovascular Monitoring Devices; Cardiovascular Devices
PHONOCARDIOGRAPH	870.2390	PHONOCARDIOGRAPH Cardiovascular Monitoring Devices; Cardiovascular Devices
PHONOCARDIOGRAPHIC	884.2640	FETAL PHONOCARDIOGRAPHIC MONITOR Obstetrical And Gynecological Monitoring Devices; Obstetrical And Gynecological Devices
PHONOCATHETER	870.1270	INTRACAVITARY PHONOCATHETER SYSTEM Cardiovascular Diagnostic Devices; Cardiovascular Devices
PHOSPHATASE, ACID	862.1020	ACID PHOSPHATASE (TOTAL OR PROSTATIC) TEST SYSTEM Clinical Chemistry Test Systems; Clinical Chemistry And Clinical Toxicology Devices
PHOSPHATASE, ALKALINE	862.1050	ALKALINE PHOSPHATASE OR ISOENZYMES TEST SYSTEM Clinical Chemistry Test Systems; Clinical Chemistry And Clinical Toxicology Devices
PHOSPHATASE, ALKALINE	864.7660	LEUKOCYTE ALKALINE PHOSPHATASE TEST Hematology Kits And Packages; Hematology And Pathology Devices
PHOSPHATE, ALUMINUM	582.1781	ALUMINUM PHOSPHATE General Purpose Food Additives; Substances Generally Recognized As Safe
PHOSPHATE, AMMONIUM	184.1141	AMMONIUM PHOSPHATE, DIBASIC Listing Of Specific Substances Affirmed As GRAS; Direct Food Substances Affirmed As Generally Recognized As Safe
PHOSPHATE, AMMONIUM	184.1141a	AMMONIUM PHOSPHATE, MONOBASIC Listing Of Specific Substances Affirmed As GRAS; Direct Food Substances Affirmed As Generally Recognized As Safe
PHOSPHATE, AMMONIUM	582.1141	AMMONIUM PHOSPHATE General Purpose Food Additives; Substances Generally Recognized As Safe
PHOSPHATE, CALCIUM	182.1217	CALCIUM PHOSPHATE Multiple Purpose GRAS Food Substances; Substances Generally Recognized As Safe
PHOSPHATE, CALCIUM	182.6215	MONOBASIC CALCIUM PHOSPHATE Sequestrants; Substances Generally Recognized As Safe
PHOSPHATE, CALCIUM	182.8217	CALCIUM PHOSPHATE Nutrients; Substances Generally Recognized As Safe
PHOSPHATE, CALCIUM	582.1217	CALCIUM PHOSPHATE General Purpose Food Additives; Substances Generally Recognized As Safe
PHOSPHATE, CALCIUM	582.5217	CALCIUM PHOSPHATE Nutrients And/or Dietary Supplements; Substances Generally Recognized As Safe
PHOSPHATE, CALCIUM	582.6215	MONOBASIC CALCIUM PHOSPHATE Sequestrants; Substances Generally Recognized As Safe
PHOSPHATE, DIAMMONIUM	573.320	DIAMMONIUM PHOSPHATE Food Additive Listing; Food Additives Permitted In Feed And Drinking Water Of Animals
PHOSPHATE, DIPOTASSIUM	182.6285	DIPOTASSIUM PHOSPHATE Sequestrants; Substances Generally Recognized As Safe
PHOSPHATE, DIPOTASSIUM	582.6285	DIPOTASSIUM PHOSPHATE Sequestrants; Substances Generally Recognized As Safe
PHOSPHATE, DISODIUM	182.6290	DISODIUM PHOSPHATE Sequestrants; Substances Generally Recognized As Safe
PHOSPHATE, DISODIUM	582.6290	DISODIUM PHOSPHATE Sequestrants; Substances Generally Recognized As Safe
PHOSPHATE, FERRIC	184.1301	FERRIC PHOSPHATE Listing Of Specific Substances Affirmed As GRAS; Direct Food Substances Affirmed As Generally Recognized As Safe
PHOSPHATE, FERRIC	582.5301	FERRIC PHOSPHATE Nutrients And/or Dietary Supplements; Substances Generally Recognized As Safe
PHOSPHATE, GLUCOSE-6	864.7360	ERYTHROCYTIC GLUCOSE-6-PHOSPHATE DEHYDROGENASE ASSAY Hematology Kits And Packages; Hematology And Pathology Devices
PHOSPHATE, MAGNESIUM	184.1434	MAGNESIUM PHOSPHATE Listing Of Specific Substances Affirmed As GRAS; Direct Food Substances Affirmed As Generally Recognized As Safe
PHOSPHATE, MAGNESIUM	582.5434	MAGNESIUM PHOSPHATE Nutrients And/or Dietary Supplements; Substances Generally Recognized As Safe
PHOSPHATE, MONOSODIUM	582.4521	MONOSODIUM PHOSPHATE DERIVATIVES OF MONO- AND DIGLYCERIDES OF EDIBLE FATS OR OILS, OR EDIBLE FAT-FORMING FATTY ACIDS Emulsifying Agents; Substances Generally Recognized As Safe
PHOSPHATE, RIBOFLAVIN	582.5697	RIBOFLAVIN-5-PHOSPHATE Nutrients And/or Dietary Supplements; Substances Generally Recognized As Safe
PHOSPHATE, SODIUM	182.1778	SODIUM PHOSPHATE Multiple Purpose GRAS Food Substances; Substances Generally Recognized As Safe

PHOSPHATE, SODIUM	182.6778	SODIUM PHOSPHATE Sequestrants; Substances Generally Recognized As Safe
PHOSPHATE, SODIUM	182.8778	SODIUM PHOSPHATE Nutrients; Substances Generally Recognized As Safe
PHOSPHATE, SODIUM	582.1778	SODIUM PHOSPHATE General Purpose Food Additives; Substances Generally Recognized As Safe
PHOSPHATE, SODIUM	582.6778	SODIUM PHOSPHATE Sequestrants; Substances Generally Recognized As Safe
PHOSPHATE, SODIUM	582.5778	SODIUM PHOSPHATE Nutrients And/or Dietary Supplements; Substances Generally Recognized As Safe
PHOSPHATE, SODIUM ACID	182.6085	SODIUM ACID PHOSPHATE Sequestrants; Substances Generally Recognized As Safe
PHOSPHATE, SODIUM ACID	582.6085	SODIUM ACID PHOSPHATE Sequestrants; Substances Generally Recognized As Safe
PHOSPHATE, SODIUM ALUMINUM	182.1781	SODIUM ALUMINUM PHOSPHATE Multiple Purpose GRAS Food Substances; Substances Generally Recognized As Safe
PHOSPHATE, TRICALCIUM	872.3930	TRICALCIUM PHOSPHATE GRANULES FOR DENTAL BONE REPAIR Prosthetic Devices; Dental Devices
PHOSPHATE, TRIOSE	862.1720	TRIOSE PHOSPHATE ISOMERASE TEST SYSTEM Clinical Chemistry Test Systems; Clinical Chemistry And Clinical Toxicology Devices
PHOSPHATED FLOUR	137.175	PHOSPHATED FLOUR Requirements For Specific Standardized Cereal Flours And Related Products; Cereal Flours And Related Products
PHOSPHATES, SODIUM	201.307	SODIUM PHOSPHATES, PACKAGE SIZE LIMITATION, WARNINGS AND DIRECTIONS FOR OVER-THE-COUNTER SALE Specific Labeling Requirements For Specific Drug Products; Labeling
PHOSPHOGLUCONATE	862.1565	6-PHOSPHOGLUCONATE DEHYDROGENASE TEST SYSTEM Clinical Chemistry Test Systems; Clinical Chemistry And Clinical Toxicology Devices
PHOSPHOHEXOSE	862.1570	PHOSPHOHEXOSE ISOMERASE TEST SYSTEM Clinical Chemistry Test Systems; Clinical Chemistry And Clinical Toxicology Devices
PHOSPHOKINASE	862.1215	CREATINE PHOSPHOKINASE/CREATINE KINASE OR ISOENZYMES TEST SYSTEM Clinical Chemistry Test Systems; Clinical Chemistry And Clinical Toxicology Devices
PHOSPHOLIPID	862.1575	PHOSPHOLIPID TEST SYSTEM Clinical Chemistry Test Systems; Clinical Chemistry And Clinical Toxicology Devices
PHOSPHORIC ACID	175.260	PARTIAL PHOSPHORIC ACID ESTERS OF POLYESTER RESINS Substances For Use As Components Of Coatings; Indirect Food Additives
PHOSPHORIC ACID	182.1073	PHOSPHORIC ACID Multiple Purpose GRAS Food Substances; Substances Generally Recognized As Safe
PHOSPHORIC ACID	582.1073	PHOSPHORIC ACID General Purpose Food Additives; Substances Generally Recognized As Safe
PHOSPHORODITHIOATE	524.1742	N-(MERCAPTOMETHYL) PHTHALIMIDE S-(*O,O*-DIMETHYL PHOSPHORODITHIOATE) EMULSIFIABLE LIQUID Ophthalmic And Topical Dosage Form New Animal Drugs
PHOSPHOTRANSFERASE	173.170	AMINOGLYCOSIDE 3'-PHOSPHOTRANSFERASE II Secondary Direct Food Additives Permitted In Food For Human Consumption
PHOSPHOTRANSFERASE	573.130	AMINOGLYCOSIDE 3'-PHOSPHOTRANSFERASE II Food Additives Permitted In Feed And Drinking Water Of Animals
PHOSPHORUS	862.1580	PHOSPHORUS (INORGANIC) TEST SYSTEM Clinical Chemistry Test Systems; Clinical Chemistry And Clinical Toxicology Devices
PHOTIC	882.1890	EVOKED RESPONSE PHOTIC STIMULATOR Neurological Diagnostic Devices; Neurological Devices
PHOTOCOAGULATOR	886.4690	OPHTHALMIC PHOTOCOAGULATOR Surgical Devices; Ophthalmic Devices
PHOTOELECTRIC	870.2780	HYDRAULIC, PNEUMATIC, OR PHOTOELECTRIC PLETHYSMOGRAPHS Cardiovascular Monitoring Devices; Cardiovascular Devices
PHOTOFLUOROGRAPHIC	892.1730	PHOTOFLUOROGRAPHIC X-RAY SYSTEM Diagnostic Devices; Radiology Devices
PHOTOMETER	862.2300	COLORIMETER, PHOTOMETER, OR SPECTROPHOTOMETER FOR CLINICAL USE Clinical Laboratory Instruments; Clinical Chemistry And Clinical Toxicology Devices
PHOTOMETER	862.2540	FLAME EMISSION PHOTOMETER FOR CLINICAL USE Clinical Laboratory Instruments; Clinical Chemistry And Toxicology Devices
PHOTOMETRIC	862.2160	DISCRETE PHOTOMETRIC CHEMISTRY ANALYZER FOR CLINICAL USE Clinical Laboratory Instruments; Clinical Chemistry And Clinical Toxicology Devices

PHOTOSTIMULATOR	886.1630	AC-POWERED PHOTOSTIMULATOR Diagnostic Devices; Ophthalmic Devices
PHOTOTHERAPY	880.5700	NEONATAL PHOTOTHERAPY UNIT General Hospital And Personal Use Therapeutic Devices; General Hospital And Personal Use Devices
PHRENIC NERVE	882.5830	IMPLANTED DIAPHRAGMATIC/PHRENIC NERVE STIMULATOR Neurological Therapeutic Devices; Neurological Devices
PHTHALATE	177.1630	POLYETHYLENE PHTHALATE POLYMERS Substances For Use As Basic Components Of Single And Repeated Use Food Contact Surfaces; Indirect Food Additives: Polymers
PHTHALIMIDE	524.1742	N-(MERCAPTOMETHYL) PHTHALIMIDE S-(*O,O*-DIMETHYL PHOSPHORODITHIOATE) EMULSIFIABLE LIQUID Ophthalmic And Topical Dosage Form New Animal Drugs
PHTHALOCYANINATO	74.3045	[PHTHALOCYANINATO(2-)] COPPER Listing Of Color Additives Subject To Certification
PHTHALOCYANINE	73.3124	PHTHALOCYANINE GREEN Medical Devices; Listing Of Color Additives Exempt From Certification
PHYSICAL ESTABLISHMENT	600.11	PHYSICAL ESTABLISHMENT, EQUIPMENT, ANIMALS, AND CARE Establishment Standards; Biological Products: General
PHYSICAL SECURITY	1301.72	PHYSICAL SECURITY CONTROLS FOR NON-PRACTITIONERS; NARCOTIC TREATMENT PROGRAMS AND COMPOUNDERS FOR NARCOTIC TREATMENT PROGRAMS; STORAGE AREAS Security Requirements; Registration Of Manufacturers, Distributors, And Dispensers Of Controlled Substances
PHYSICAL SECURITY	1301.73	PHYSICAL SECURITY CONTROLS FOR NON-PRACTITIONERS; COMPOUNDERS FOR NARCOTIC TREATMENT PROGRAMS; MANUFACTURING AND COMPOUNDING AREAS Security Requirements; Registration Of Manufacturers, Distributors, And Dispensers Of Controlled Substances
PHYSICAL SECURITY	1301.75	PHYSICAL SECURITY CONTROLS FOR PRACTITIONERS Security Requirements; Registration Of Manufacturers, Distributors, And Dispensers Of Controlled Substances
PHYSICAL THERAPY	890.5880	MULTI-FUNCTION PHYSICAL THERAPY TABLE Physical Medicine Therapeutic Devices; Physical Medicine Devices
PHYSICIANS	20.112	VOLUNTARY DRUG EXPERIENCE REPORTS SUBMITTED BY PHYSICIANS AND HOSPITALS Availability Of Specific Categories Of Records; Public Information
PHYSIOLOGICAL	870.2910	RADIOFREQUENCY PHYSIOLOGICAL SIGNAL TRANSMITTER AND RECEIVER Cardiovascular Monitoring Devices; Cardiovascular Devices
PHYSIOLOGICAL	882.1835	PHYSIOLOGICAL SIGNAL AMPLIFIER Neurological Diagnostic Devices; Neurological Devices
PHYSIOLOGICAL	882.1845	PHYSIOLOGICAL SIGNAL CONDITIONER Neurological Diagnostic Devices; Neurological Devices
PHYTATE	582.6219	CALCIUM PHYTATE Sequestrants; Substances Generally Recognized As Safe
PICHIA PASTORIS	573.750	PICHIA PASTORIS DRIED YEAST Food Additives Permitted In Feed And Drinking Water Of Animals
PICK, MASSAGING	872.6650	MASSAGING PICK OR TIP OR ORAL HYGIENE Miscellaneous Devices; Dental Devices
PICOLINIUM	524.1484	NEOMYCIN SULFATE, ISOFLUPREDONE ACETATE, TETRACAINE HYDROCHLORIDE, AND MYRISTYL-GAMMA-PICOLINIUM CHLORIDE, TOPICAL POWDER Ophthalmic And Topical Dosage Form New Animal Drugs
PICTURE ARCHIVING	892.2050	PICTURE ARCHIVING AND COMMUNICATIONS SYSTEM Diagnostic Devices; Radiology Devices
PIE	152.126	FROZEN CHERRY PIE Requirements For Specific Standardized Fruit Pies; Fruit Pies
PIER	1240.86	PROTECTION OF PIER WATER SYSTEM Source And Use Of Potable Water; Control Of Communicable Diseases
PIG PUMP	520.1044b	GENTAMICIN SULFATE PIG PUMP ORAL SOLUTION Oral Dosage Form New Animal Drugs
PIGMENT	178.3725	PIGMENT DISPERSANTS Certain Adjuvants And Production Aids; Indirect Food Additives: Adjuvants, Production Aids, And Sanitizers
PIMARICIN	172.155	NATAMYCIN (PIMARICIN) Food Preservatives; Food Additives Permitted For Direct Addition To Food For Human Consumption
PIMENTO	133.171	PASTEURIZED PROCESS PIMENTO CHEESE Requirements For Specific Standardized Cheese And Related Products; Cheeses And Related Cheese Products
PIN	872.3740	RETENTIVE AND SPLINTING PIN Prosthetic Devices; Dental Devices
PINEAPPLE	145.180	CANNED PINEAPPLE Requirements For Specific Standardized Canned Fruits; Canned Fruits

PINEAPPLE	145.181	ARTIFICIALLY SWEETENED CANNED PINEAPPLE Requirements For Specific Standardized Canned Fruits; Canned Fruits
PINEAPPLE	146.185	PINEAPPLE JUICE Requirements For Specific Standardized Canned Fruit Juices And Beverages; Canned Fruit Juices
PINWHEEL	882.1750	PINWHEEL Neurological Diagnostic Devices; Neurological Devices
PIPERAZINE	520.1802	PIPERAZINE-CARBON DISULFIDE COMPLEX ORAL DOSAGE FORMS Oral Dosage Form New Animal Drugs
PIPERAZINE	520.1807	PIPERAZINE Oral Dosage Form New Animal Drugs
PIPERAZINE	556.513	PIPERAZINE Tolerances For Residues Of New Animal Drugs In Food
PIPERAZINE CITRATE	520.763c	DITHIAZANINE IODIDE AND PIPERAZINE CITRATE SUSPENSION Oral Dosage Form New Animal Drugs
PIPERAZINE CITRATE	520.1803	PIPERAZINE CITRATE CAPSULES Oral Dosage Form New Animal Drugs
PIPERAZINE CITRATE	520.2380d	THIABENDAZOLE, PIPERAZINE CITRATE SUSPENSION Oral Dosage Form New Animal Drugs
PIPERAZINE DIHYDROCHLORIDE	520.1242	LEVAMISOLE HYDROCHLORIDE AND PIPERAZINE DIHYDROCHLORIDE Oral Dosage Form New Animal Drugs
PIPERAZINE DIHYDROCHLORIDE	520.2520g	TRICHLORFON, PHENOTHIAZINE, AND PIPERAZINE DIHYDROCHLORIDE POWDER Oral Dosage Form New Animal Drugs
PIPERAZINE MONOHYDROCHLORIDE	520.1806	PIPERAZINE MONOHYDROCHLORIDE LIQUID Oral Dosage Form New Animal Drugs
PIPERAZINE PHOSPHATE	520.1804	PIPERAZINE PHOSPHATE CAPSULES Oral Dosage Form New Animal Drugs
PIPERAZINE PHOSPHATE	520.1805	PIPERAZINE PHOSPHATE WITH THENIUM CLOSYLATE TABLETS Oral Dosage Form New Animal Drugs
PIPERAZINE PHOSPHATE	520.2380f	THIABENDAZOLE, PIPERAZINE PHOSPHATE POWDER Oral Dosage Form New Animal Drugs
PIPERONYL BUTOXIDE	178.3730	PIPERONYL BUTOXIDE AND PYRETHRINS AS COMPONENTS OF BAGS Certain Adjuvants And Production Aids; Indirect Food Additives: Adjuvants, Production Aids, And Sanitizers
PIPETTING	862.2750	PIPETTING AND DILUTING SYSTEM FOR CLINICAL USE Clinical Laboratory Instruments; Clinical Chemistry And Clinical Toxicology Devices
PIRLIMYCIN	556.515	PIRLIMYCIN Tolerances For Residues Of New Animal Drugs In Food
PIRLIMYCIN HYDROCHLORIDE	526.1810	PIRLIMYCIN HYDROCHLORIDE Intramammary Dosage Forms
PISTON SYRINGE	880.5860	PISTON SYRINGE General Hospital And Personal Use Therapeutic Devices; General Hospital And Personal Use Devices
PIT SEALANT	872.3765	PIT AND FISSURE SEALANT AND CONDITIONER Prosthetic Devices; Dental Devices
PITUITARY LUTEINIZING	522.1820	PITUITARY LUTEINIZING HORMONE FOR INJECTION Implantation Or Injectable Dosage Form New Animal Drugs
PLACE	12.91	TIME AND PLACE OF PREHEARING CONFERENCE Hearing Procedures; Formal Evidentiary Public Hearing
PLACE	1301.16	TIME AND PLACE OF HEARING Hearings; Registration Of Manufacturers, Distributors, And Dispensers Of Controlled Substances
PLACE	1303.36	TIME AND PLACE OF HEARING Hearings; Quotas
PLACE	1309.55	TIME AND PLACE OF HEARING Registration Of Manufacturers, Distributors, Importers And Exporters Of List I Chemicals
PLACE	1312.46	TIME AND PLACE OF HEARING Hearings; Importation And Exportation Of Controlled Substances
PLACE	1316.32	NOTICE OF PROCEEDING; TIME AND PLACE Enforcement Proceedings; Administrative Functions, Practices, And Procedures
PLACE	1316.53	TIME AND PLACE OF HEARING Administrative Hearings; Administrative Functions, Practices, And Procedures
PLACE OF BUSINESS	101.5	FOOD, NAME AND PLACE OF BUSINESS OF MANUFACTURER, PACKER, OR DISTRIBUTOR General Provisions; Food Labeling
PLACE OF BUSINESS	201.1	DRUGS; NAME AND PLACE OF BUSINESS OF MANUFACTURER, PACKER OR DISTRIBUTOR General Labeling Provisions; Labeling
PLACE OF BUSINESS	501.5	ANIMAL FOOD; NAME AND PLACE OF BUSINESS OF MANUFACTURER, PACKER, OR DISTRIBUTOR General Provisions; Animal Food Labeling

PLACE OF BUSINESS	701.12	NAME AND PLACE OF BUSINESS OF MANUFACTURER, PACKER, OR DISTRIBUTOR Package Form; Cosmetic Labeling
PLACE OF BUSINESS	801.1	MEDICAL DEVICES; NAME AND PLACE OF BUSINESS OF MANUFACTURER, PACKER OR DISTRIBUTOR General Labeling Provisions; Labeling
PLACEMENT	1230.10	PLACEMENT Labeling; Regulations Under The Federal Caustic Poison Act
PLACENTAL LACTOGEN	862.1585	HUMAN PLACENTAL LACTOGEN TEST SYSTEM Clinical Chemistry Test Systems; Clinical Chemistry And Clinical Toxicology Devices
PLAN	120.8	HAZARD ANALYSIS AND CRITICAL CONTROL POINT (HACCP) PLAN Hazard Analysis And Critical Control Point (HACCP) Systems
PLAN	123.6	HAZARD ANALYSIS AND HAZARD ANALYSIS CRITICAL CONTROL POINT (HACCP) PLAN Fish And Fishery Products
PLAN	812.25	INVESTIGATIONAL PLAN Application And Administrative Action; Investigational Device Exemptions
PLANNED STUDIES	99.203	REQUEST TO EXTEND THE TIME FOR COMPLETING PLANNED STUDIES Dissemination Of Information On Unapproved/New Uses For Marketed Drugs, Biologics, And Devices
PLANNED STUDIES	99.303	EXTENSION OF TIME FOR COMPLETING PLANNED STUDIES Dissemination Of Information On Unapproved/New Uses For Marketed Drugs, Biologics, And Devices
PLANNING, NEPA	25.10	POLICIES AND NEPA PLANNING General Provisions; Environmental Impact Considerations
PLANS	1004.2	PLANS FOR THE REPAIR OF ELECTRONIC PRODUCTS Repurchase, Repairs, Or Replacement Of Electronic Products
PLANS	1004.3	PLANS FOR THE REPLACEMENT OF ELECTRONIC PRODUCTS Repurchase, Repairs, Or Replacement Of Electronic Products
PLANS	1004.4	PLANS FOR REFUNDING THE COST OF ELECTRONIC PRODUCTS Repurchase, Repairs, Or Replacement Of Electronic Products
PLANS	1004.6	APPROVAL OF PLANS Repurchase, Repairs, Or Replacement Of Electronic Products
PLANS	1250.41	SUBMITTAL OF CONSTRUCTION PLANS Equipment And Operation Of Land And Air Conveyances; Interstate Conveyance Sanitation
PLANS	1250.62	SUBMITTAL OF CONSTRUCTION PLANS Servicing Areas For Land And Air Conveyances; Interstate Conveyance Sanitation
PLANS	1403.11	STATE PLANS Pre-Award Requirements; Uniform Administrative Requirements For Grants And Coop. Agreements To State & Local Gov.
PLANT	110.20	PLANT AND GROUNDS Buildings And Facilities; Current Good Manufacturing Practice In Manufacturing, Packing, Or Holding Human Food
PLANT	129.20	PLANT CONSTRUCTION AND DESIGN Buildings And Facilities; Processing And Bottling Of Bottled Drinking Water
PLANT MATERIAL	1308.35	EXEMPTION OF CERTAIN CANNABIS PLANT MATERIAL, AND PRODUCTS MADE THEREFROM, THAT CONTAIN TETRAHYDROCANNABINOIDS Schedules Of Controlled Substances
PLANTAGO SEED	201.319	WATER-SOLUBLE GUMS, HYDROPHILIC GUMS AND HYDROPHILIC MUCILLOIDS (INCLUDING PLANTAGO SEED (PSYLLIUM)) AS ACTIVE INGREDIENTS; REQUIRED WARNINGS AND DIRECTIONS Specific Labeling Requirements For Specific Drug Products; Labeling
PLANTS	1210.14	SANITARY INSPECTION OF PLANTS Inspection and Testing; Regulations Under The Federal Import Milk Act
PLASMA	640.30	PLASMA Plasma; Additional Standards For Human Blood And Blood Products
PLASMA	640.60	SOURCE PLASMA Source Plasma; Additional Standards For Human Blood And Blood Products
PLASMA	640.64	COLLECTION OF BLOOD FOR SOURCE PLASMA Source Plasma; Additional Standards For Human Blood And Blood Products
PLASMA	640.74	MODIFICATION OF SOURCE PLASMA Source Plasma; Additional Standards For Human Blood And Blood Products
PLASMA	866.2160	COAGULASE PLASMA Microbiology Devices; Immunology And Microbiology Devices
PLASMA	866.5700	WHOLE HUMAN PLASMA OR SERUM IMMUNOLOGICAL TEST SYSTEM Immunological Test Systems; Immunology And Microbiology Devices
PLASMA ONCOMETER	862.2720	PLASMA ONCOMETER FOR CLINICAL USE Clinical Laboratory Instruments; Clinical Chemistry And Clinical Toxicology Devices
PLASMA ONCOMETRY	862.1530	PLASMA ONCOMETRY TEST SYSTEM Clinical Chemistry Test Systems; Clinical Chemistry And Clinical Toxicology Devices

PLASMA PROTEIN FRACTION	640.90	PLASMA PROTEIN FRACTION (HUMAN) Plasma Protein Fraction (Human); Additional Standards For Human Blood And Blood Products
PLASMA VISCOMETER	862.2920	PLASMA VISCOMETER FOR CLINICAL USE Clinical Laboratory Instruments; Clinical Chemistry And Clinical Toxicology Devices
PLASMA WARMING DEVICE	864.9205	BLOOD AND PLASMA WARMING DEVICE Products Used In Establishments That Manufacture Blood And Blood Products; Hematology And Pathology Devices
PLASMAPHERESIS	606.110	PLATELETPHERESIS, LEUKAPHERESIS, AND PLASMAPHERESIS Production And Process Controls; Current Good Manufacturing Practice For Blood And Blood Components
PLASMAPHERESIS	640.65	PLASMAPHERESIS Source Plasma; Additional Standards For Human Blood And Blood Products
PLASMINOGEN	866.5715	PLASMINOGEN IMMUNOLOGICAL TEST SYSTEM Immunological Test Systems; Immunology And Microbiology Devices
PLASTIC BRACKET	872.5470	ORTHODONTIC PLASTIC BRACKET Therapeutic Devices; Dental Devices
PLASTIC CONTAINERS	310.509	PARENTERAL DRUG PRODUCTS IN PLASTIC CONTAINERS Requirements For Specific New Drugs Or Devices; New Drugs
PLASTIC SURGERY	878.3925	PLASTIC SURGERY KIT AND ACCESSORIES Prosthetic Devices; General And Plastic Surgery Devices
PLASTIC SURGERY	878.4810	LASER SURGICAL INSTRUMENT FOR USE IN PLASTIC SURGERY AND IN DERMATOLOGY Surgical Devices; General And Plastic Surgery Devices
PLASTIC TOOTH	872.3590	PREFORMED PLASTIC DENTURE TOOTH Prosthetic Devices; Dental Devices
PLASTICIZERS	178.3740	PLASTICIZERS IN POLYMERIC SUBSTANCES Certain Adjuvants And Production Aids; Indirect Food Additives: Adjuvants, Production Aids, And Sanitizers
PLASTICIZERS	181.27	PLASTICIZERS Specific Prior-sanctioned Food Ingredients; Prior-sanctioned Food Ingredients
PLASTICS	178.2650	ORGANOTIN STABILIZERS IN VINYL CHLORIDE PLASTICS Antioxidants And Stabilizers; Indirect Food Additives: Adjuvants, Production Aids, And Sanitizers
PLASTICS	178.3010	ADJUVANT SUBSTANCES USED IN THE MANUFACTURE OF FOAMED PLASTICS Certain Adjuvants And Production Aids; Indirect Food Additives: Adjuvants, Production Aids, And Sanitizers
PLASTICS	178.3790	POLYMER MODIFIERS IN SEMIRIGID AND RIGID VINYL CHLORIDE PLASTICS Certain Adjuvants And Production Aids; Indirect Food Additives: Adjuvants, Production Aids, And Sanitizers
PLATE	866.4600	OUCHTERLONY AGAR PLATE Immunology Laboratory Equipment And Reagents; Immunology And Microbiology Devices
PLATE	866.4800	RADIAL IMMUNODIFFUSION PLATE Immunology Laboratory Equipment And Reagents; Immunology And Microbiology Devices
PLATE	872.4760	BONE PLATE Surgical Devices; Dental Devices
PLATE	872.6200	BASE PLATE SHELLAC Miscellaneous Devices; Dental Devices
PLATE	882.4030	SKULL PLATE ANVIL Neurological Surgical Devices; Neurological Devices
PLATE	882.5320	PREFORMED ALTERABLE CRANIOPLASTY PLATE Neurological Therapeutic Devices; Neurological Devices
PLATE	882.5330	PREFORMED NONALTERABLE CRANIOPLASTY PLATE Neurological Therapeutic Devices; Neurological Devices
PLATE	882.5360	CRANIOPLASTY PLATE FASTENER Neurological Therapeutic Devices; Neurological Devices
PLATE ILLUMINATOR	886.1160	COLOR VISION PLATE ILLUMINATOR Diagnostic Devices; Ophthalmic Devices
PLATELET	864.5700	AUTOMATED PLATELET AGGREGATION SYSTEM Automated And Semi-automated Hematology Devices; Hematology And Pathology Devices
PLATELET	864.6650	PLATELET ADHESION TEST Manual Hematology Devices; Hematology And Pathology Devices
PLATELET	864.6675	PLATELET AGGREGOMETER Manual Hematology Devices; Hematology And Pathology Devices
PLATELET	864.7695	PLATELET FACTOR 4 RADIOIMMUNOASSAY Hematology Kits And Packages; Hematology And Pathology Devices
PLATELET	864.8175	CALIBRATOR FOR PLATELET COUNTING Hematology Reagents; Hematology And Pathology Devices

PLATELET	864.9575	ENVIRONMENTAL CHAMBER FOR STORAGE OF PLATELET CONCENTRATE Products Used In Establishments That Manufacture Blood And Blood Products; Hematology And Pathology Devices
PLATELETPHERESIS	606.110	PLATELETPHERESIS, LEUKAPHERESIS, AND PLASMAPHERESIS Production And Process Controls; Current Good Manufacturing Practice For Blood And Blood Components
PLATELETS	640.20	PLATELETS Platelets; Additional Standards For Human Blood And Blood Products
PLATFORM	890.1575	FORCE-MEASURING PLATFORM Physical Medicine Diagnostic Devices; Physical Medicine Devices
PLATFORM SCALE	890.3940	WHEELCHAIR PLATFORM SCALE Physical Medicine Prosthetic Devices; Physical Medicine Devices
PLEDGET	870.3470	INTRACARDIAC PATCH OR PLEDGET MADE OF POLYPROPYLENE, POLYETHYLENE TEREPHTHALATE, OR POLYTETRAFLUOROETHYLENE Cardiovascular Prosthetic Devices; Cardiovascular Devices
PLETHYSMOGRAPH	868.1750	PRESSURE PLETHYSMOGRAPH Diagnostic Devices; Anesthesiology Devices
PLETHYSMOGRAPH	868.1760	VOLUME PLETHYSMOGRAPH Diagnostic Devices; Anesthesiology Devices
PLETHYSMOGRAPH	870.2770	IMPEDANCE PLETHYSMOGRAPH Cardiovascular Monitoring Devices; Cardiovascular Devices
PLETHYSMOGRAPH	882.1790	OCULAR PLETHYSMOGRAPH Neurological Diagnostic Devices; Neurological Devices
PLETHYSMOGRAPHS	870.2780	HYDRAULIC, PNEUMATIC, OR PHOTOELECTRIC PLETHYSMOGRAPHS Cardiovascular Monitoring Devices; Cardiovascular Devices
PLINTH	890.3520	PLINTH Physical Medicine Prosthetic Devices; Physical Medicine Devices
PLOSS	868.1965	SWITCHING VALVE (PLOSS) Diagnostic Devices; Anesthesiology Devices
PLUMBING	211.48	PLUMBING Buildings And Facilities; Current Good Manufacturing Practice For Finished Pharmaceuticals
PLUMS	145.185	CANNED PLUMS Requirements For Specific Standardized Canned Fruits; Canned Fruits
PMA	814.9	CONFIDENTIALITY OF DATA AND INFORMATION IN A PREMARKET APPROVAL APPLICATION (PMA) FILE General; Premarket Approval Of Medical Devices
PMA	814.37	PMA AMENDMENTS AND RESUBMITTED PMA'S Premarket Approval Application (PMA); Premarket Approval Of Medical Devices
PMA	814.39	PMA SUPPLEMENTS Premarket Approval Application (PMA); Premarket Approval Of Medical Devices
PMA	814.40	TIME FRAMES FOR REVIEWING A PMA FDA Action On A PMA; Premarket Approval Of Medical Devices
PMA	814.42	FILING A PMA FDA Action On A PMA; Premarket Approval Of Medical Devices
PMA	814.44	PROCEDURES FOR REVIEW OF A PMA FDA Action On A PMA; Premarket Approval Of Medical Devices
PMA	814.45	DENIAL OF APPROVAL OF A PMA FDA Action On A PMA; Premarket Approval Of Medical Devices
PMA	814.46	WITHDRAWAL OF APPROVAL OF A PMA FDA Action On A PMA; Premarket Approval Of Medical Devices
PMA	814.47	TEMPORARY SUSPENSION OF APPROVAL OF A PMA FDA Action On A PMA; Premarket Approval Of Medical Devices
PMA'S, RESUBMITTED	814.37	PMA AMENDMENTS AND RESUBMITTED PMA'S Premarket Approval Application (PMA); Premarket Approval Of Medical Devices
PMMA	886.1385	POLYMETHYLMETHACRYLATE (PMMA) DIAGNOSTIC CONTACT LENS Diagnostic Devices; Ophthalmic Devices
PMMA	888.3027	POLYMETHYLMETHACRYLATE (PMMA) BONE CEMENT Prosthetic Devices; Orthopedic Devices
PNEUMATIC	870.2780	HYDRAULIC, PNEUMATIC, OR PHOTOELECTRIC PLETHYSMOGRAPHS Cardiovascular Monitoring Devices; Cardiovascular Devices
PNEUMATIC	874.4250	EAR, NOSE, AND THROAT ELECTRIC OR PNEUMATIC SURGICAL DRILL Surgical Devices; Ear, Nose, And Throat Devices
PNEUMATIC	878.5910	PNEUMATIC TOURNIQUET Therapeutic Devices; General And Plastic Surgery Devices
PNEUMATIC	882.4370	PNEUMATIC CRANIAL DRILL MOTOR Neurological Surgical Devices; Neurological Devices
PNEUMATIC	890.3610	RIGID PNEUMATIC STRUCTURE ORTHOSIS Physical Medicine Prosthetic Devices; Physical Medicine Devices
PNEUMOENCEPHALOGRAPHIC	892.1820	PNEUMOENCEPHALOGRAPHIC CHAIR Diagnostic Devices; Radiology Devices
PNEUMONITIS	866.5500	HYPERSENSITIVITY PNEUMONITIS IMMUNOLOGICAL TEST SYSTEM Immunological Test Systems; Immunology And Microbiology Devices

PNEUMOTACHOMETER	868.2550	PNEUMOTACHOMETER Monitoring Devices; Anesthesiology Devices
POINT, PAPER	872.3830	ENDODONTIC PAPER POINT Prosthetic Devices; Dental Devices
POINT, SILVER	872.3840	ENDODONTIC SILVER POINT Prosthetic Devices; Dental Devices
POISON	1230.13	LABELING OF "POISON" Labeling; Regulations Under The Federal Caustic Poison Act
POISON ACT	2.110	DEFINITION OF AMMONIA UNDER FEDERAL CAUSTIC POISON ACT Caustic Poisons; General Administrative Rulings And Decisions
POISONOUS SUBSTANCES	2.25	GRAIN SEED TREATED WITH POISONOUS SUBSTANCES; COLOR IDENTIFICATION TO PREVENT ADULTERATION OF HUMAN AND ANIMAL FOOD Human And Animal Foods; General Administrative Rulings And Decisions
POISONOUS SUBSTANCES	109.6	ADDED POISONOUS OR DELETERIOUS SUBSTANCES General Provisions; Unavoidable Contaminants In Food For Human Consumption & Food-packaging Material
POISONOUS SUBSTANCES	509.6	ADDED POISONOUS OR DELETERIOUS SUBSTANCES General Provisions; Unavoidable Contaminants In Animal Food And Food-packaging Material
POLICIES	25.5	POLICIES AND NEPA PLANNING General Provisions; Environmental Impact Considerations
POLICIES	203.34	POLICIES AND PROCEDURES, ADMINISTRATIVE SYSTEMS Prescription Drug Marketing
POLICY	7.40	RECALL POLICY Recalls - Guidance On Policy, Procedures & Industry Responsibilities; Enforcement Policy
POLICY	14.80	QUALIFICATIONS FOR MEMBERS OF STANDING POLICY AND TECHNICAL ADVISORY COMMITTEES Members Of Advisory Committees; Public Hearing Before A Public Advisory Committee
POLICY	20.20	POLICY ON DISCLOSURE OF FDA RECORDS General Policy; Public Information
POLICY	21.10	POLICY CONCERNING RECORDS ABOUT INDIVIDUALS General Provisions; Protection Of Privacy
POLICY	21.60	POLICY Exemptions; Protection Of Privacy
POLICY	310.100	NEW DRUG STATUS OPINIONS; STATEMENT OF POLICY Specific Administrative Rulings And Decisions; New Drugs
POLICY	369.4	WARNINGS SUGGESTED FOR DRUGS BY FORMAL OR INFORMAL STATEMENTS OF POLICY Definitions And Interpretations; Interpretative Statements Re: Warnings On Drugs And Devices For OTC Sale
POLICY	861.5	STATEMENT OF POLICY General; Procedures For Performance Standards Development
POLICY	1401.2	THE OFFICE OF NATIONAL DRUG CONTROL POLICY - ORGANIZATION AND FUNCTIONS Public Availability Of Information
POLICY	1404.115	POLICY Governmentwide Debarment And Suspension And Requirements For Drug Free Workplace (Grants)
POLIOVIRUS	866.3405	POLIOVIRUS SEROLOGICAL REAGENTS Serological Reagents; Immunology And Microbiology Devices
POLISHING AGENT	872.6030	ORAL CAVITY ABRASIVE POLISHING AGENT Miscellaneous Devices; Dental Devices
POLOXALENE	520.1840	POLOXALENE Oral Dosage Form New Animal Drugs
POLOXALENE	558.464	POLOXALENE Specific New Animal Drugs For Use In Animal Feeds; New Animal Drugs For Use In Animal Feeds
POLOXALENE	558.465	POLOXALENE FREE-CHOICE LIQUID TYPE C FEED Specific New Animal Drugs For Use In Animal Feeds; New Animal Drugs For Use In Animal Feeds
POLOXALENE	573.760	POLOXALENE Food Additive Listing; Food Additives Permitted In Feed And Drinking Water Of Animals
POLY-1-BUTENE	177.1570	POLY-1-BUTENE RESINS AND BUTENE/ETHYLENE COPOLYMERS Substances For Use As Basic Components Of Single And Repeated Use Food Contact Surfaces; Indirect Food Additives: Polymers
POLY(ARYLETHERKETONE)	177.2415	POLY(ARYLETHERKETONE) RESINS Substances For Use Only As Components Of Articles Intended For Repeated Use; Indirect Food Additive Polymers

POLY(DIMETHYL-PHENYLENE) OXIDE	177.2460	POLY(2,6-DIMETHYL-1,4-PHENYLENE) OXIDE RESINS Substances For Use Only As Components Of Articles Intended For Repeated Use; Indirect Food Additives: Polymers
POLY(ETHYLENE-TEREPHTHALATE)	878.5000	NONABSORBABLE POLY(ETHYLENETEREPHTHALATE) SURGICAL SUTURE Surgical Devices; General And Plastic Surgery Devices
POLY(GLYCOLIDE/L-LACTIDE)	878.4493	ABSORBABLE POLY(GLYCOLIDE/L-LACTIDE) SURGICAL SUTURE Surgical Devices; General And Plastic Surgery Devices
POLY(HYDROXYETHYL-METHACRYLATE)	73.3121	POLY(HYDROXYETHYL METHACRYLATE)-DYE COPOLYMERS Medical Devices; Listing Of Color Additives Exempt From Certification
POLY(METHYLSTYRENE)	177.1635	POLY(P-METHYLSTYRENE) AND RUBBER-MODIFIED POLY(P-METHYLSTYRENE) Substances For Use As Basic Components Of Single And Repeated Use Food Contact Surfaces; Indirect Food Additives: Polymers
POLY(OXY-ETHANEDIYLOXYCARBONYL)	177.1637	POLY(OXY-1,2-ETHANEDIYLOXYCARBONYL-2,6-NAPHTHALENEDIYLCARBONYL) RESINS Substances For Use As Basic Components Of Single And Repeated Use Food Contact Surfaces; Indirect Food Additives: Polymers
POLY-PENTAAZA	176.250	POLY-1,4,7,10,13-PENTAAZA-15-HYDROXYHEXADECANE Substances For Use Only As Components Of Paper And Paperboard; Indirect Food Additives
POLY(PHENYL-ENETEREPHTHALAMIDE)	177.1632	POLY (PHENYLENETEREPHTHALAMIDE) RESINS Substances For Use As Basic Components Of Single And Repeated Use Food Contact Surfaces; Indirect Food Additives: Polymers
POLY(PHENYL-ENETEREPHTHALATE)	177.1632	POLY(PHENYLENETEREPHTHALAMIDE) RESINS Substances For Use As Basic Components Of Single And Repeated Use Food Contact Surfaces; Indirect Food Additives: Polymers
POLY(TETRAMETHYLENE TEREPHTHALATE)	177.1660	POLY(TETRAMETHYLENE TEREPHTHALATE) Substances For Use As Basic Components Of Single And Repeated Use Food Contact Surfaces; Indirect Food Additives: Polymers
POLY (TRIMETHOXYSILYLPROPYL)	177.2465	POLYMETHYLMETHACRYLATE/POLY (TRIMETHOXYSILYLPROPYL) METHACRYLATE COPOLYMERS Substances For Use Only As Components Of Articles Intended For Repeated Use; Indirect Food Additives: Polymers
POLY(VINYL FLUORIDE)	175.270	POLY(VINYL FLUORIDE) RESINS Substances For Use As Components Of Coatings; Indirect Food Additives
POLY(VINYLPYRIDINE)	573.870	POLY(2-VINYLPYRIDINE-CO-STYRENE) Food Additive Listing; Food Additives Permitted In Feed And Drinking Water Of Animals
POLYACRYLAMIDE	172.255	POLYACRYLAMIDE Coatings, Films And Related Substances; Food Additives Permitted For Direct Addition To Food For Human Consumption
POLYACRYLAMIDE	173.10	MODIFIED POLYACRYLAMIDE RESIN Polymer Substances For Food Treatment; Secondary Direct Food Additives Permitted In Food For Human Consumption
POLYACRYLAMIDE	872.3420	CARBOXYMETHYLCELLULOSE SODIUM AND CATIONIC POLYACRYLAMIDE POLYMER DENTURE ADHESIVE Prosthetic Devices; Dental Devices
POLYACRYLAMIDE	872.3480	POLYACRYLAMIDE POLYMER DENTURE ADHESIVE Prosthetic Devices; Dental Devices
POLYACRYLATE	173.73	SODIUM POLYACRYLATE Polymer Substances For Food Treatment; Secondary Direct Food Additives Permitted In Food For Human Consumption
POLYACRYLATE	177.1211	CROSS-LINKED POLYACRYLATE COPOLYMERS Substances For Use As Basic Components Of Single And Repeated Use Food Contact Surfaces; Indirect Food Additives: Polymers
POLYAMIDE	878.5020	NONABSORBABLE POLYAMIDE SURGICAL SUTURE Surgical Devices; General And Plastic Surgery Devices
POLYAMIDE-IMIDE	177.2450	POLYAMIDE-IMIDE RESINS Substances For Use Only As Components Of Articles Intended For Repeated Use; Indirect Food Additives: Polymers
POLYARYLATE	177.1555	POLYARYLATE RESINS Substances For Use As Basic Components Of Single And Repeated Use Food Contact Surfaces; Indirect Food Additives: Polymers
POLYARYLETHERKETONE	177.1556	POLYARYLETHERKETONE RESINS Substances For Use As Basic Components Of Single And Repeated Use Food Contact Surfaces; Indirect Food Additives: Polymers
POLYARYLSULFONE	177.1560	POLYARYLSULFONE RESINS Substances For Use As Basic Components Of Single And Repeated Use Food Contact Surfaces; Indirect Food Additives: Polymers
POLYCARBONATE	175.365	VINYLIDENE CHLORIDE COPOLYMER COATINGS FOR POLYCARBONATE FILM Substances For Use As Components Of Coatings; Indirect Food Additives

POLYCARBONATE	177.1580	POLYCARBONATE RESINS Substances For Use As Basic Components Of Single And Repeated Use Food Contact Surfaces; Indirect Food Additives: Polymers
POLYCARBOPHIL	201.319	WATER-SOLUBLE GUMS, HYDROPHILIC GUMS AND HYDROPHILIC MUCILLOIDS (INCLUDING CALCIUM POLYCARBOPHIL, AND POLYCARBOPHIL) AS ACTIVE INGREDIENTS; REQUIRED WARNINGS AND DIRECTIONS Specific Labeling Requirements For Specific Drug Products; Labeling
POLYCHLORINATED BIPHENYLS	109.15	USE OF POLYCHLORINATED BIPHENYLS (PCB'S) IN ESTABLISHMENTS MANUFACTURING FOOD-PACKAGING MATERIALS General Provisions; Unavoidable Contaminants In Food For Human Consumption & Food-packaging Material
POLYCHLORINATED BIPHENYLS	109.30	TOLERANCES FOR POLYCHLORINATED BIPHENYLS (PCB'S) Tolerances For Unavoidable Poisonous Or Deleterious Substances; Unavoidable Contaminants In Food For Human Consumption & Food-packaging Material
POLYCHLORINATED BIPHENYLS	500.45	USE OF POLYCHLORINATED BIPHENYLS (PCB'S) IN THE PRODUCTION, HANDLING AND STORAGE OF ANIMAL FEED Specific Administrative Rulings And Decisions; General
POLYCHLORINATED BIPHENYLS	509.15	USE OF POLYCHLORINATED BIPHENYLS (PCB'S) IN ESTABLISHMENTS MANUFACTURING FOOD-PACKAGING MATERIALS General Provisions; Unavoidable Contaminants In Animal Food And Food-packaging Material
POLYCHLORINATED BIPHENYLS	509.30	TEMPORARY TOLERANCES FOR POLYCHLORINATED BIPHENYLS (PCB'S) Tolerances For Unavoidable Poisonous Or Deleterious Substances; Unavoidable Contaminants In Animal Food And Food-packaging Material
POLYDEXTROSE	172.84	POLYDEXTROSE Multipurpose Additives; Food Additives Permitted For Direct Addition To Food For Human Consumption
POLYEPOXY	177.1650	POLYSULFIDE POLYMER-POLYEPOXY RESINS Substances For Use As Basic Components Of Single And Repeated Use Food Contact Surfaces; Indirect Food Additives: Polymers
POLYESTER	175.260	PARTIAL PHOSPHORIC ACID ESTERS OF POLYESTER RESINS Substances For Use As Components Of Coatings; Indirect Food Additives
POLYESTER	177.1590	POLYESTER ELASTOMERS Substances For Use As Basic Components Of Single And Repeated Use Food Contact Surfaces; Indirect Food Additives: Polymers
POLYESTER	177.2420	POLYESTER RESINS, CROSS-LINKED Substances For Use Only As Components Of Articles Intended For Repeated Use; Indirect Food Additives: Polymers
POLYESTERCARBONATE	177.1585	POLYESTERCARBONATE RESINS Substances For Use As Basic Components Of Single And Repeated Use Food Contact Surfaces; Indirect Food Additives: Polymers
POLYETHER	177.2430	POLYETHER RESINS, CHLORINATED Substances For Use Only As Components Or Articles Intended For Repeated Use; Indirect Food Additives: Polymers
POLYETHERIMIDE	177.1595	POLYETHERIMIDE RESIN Substances For Use As Basic Components Of Single And Repeated Use Food Contact Surfaces; Indirect Food Additives: Polymers
POLYETHERSULFONE	177.2440	POLYETHERSULFONE RESINS Substances For Use Only As Components Of Articles Intended For Repeated Use; Indirect Food Additives: Polymers
POLYETHYLENE	172.260	OXIDIZED POLYETHYLENE Coatings, Films And Related Substances; Food Additives Permitted For Direct Addition To Food For Human Consumption
POLYETHYLENE	177.1600	POLYETHYLENE RESINS, CARBOXYL MODIFIED Substances For Use As Basic Components Of Single And Repeated Use Food Contact Surfaces; Indirect Food Additives: Polymers
POLYETHYLENE	177.1610	POLYETHYLENE, CHLORINATED Substances For Use As Basic Components Of Single And Repeated Use Food Contact Surfaces; Indirect Food Additives: Polymers
POLYETHYLENE	177.1615	POLYETHYLENE, FLUORINATED Substances For Use As Basic Components Of Single And Repeated Use Food Contact Surfaces; Indirect Food Additives: Polymers
POLYETHYLENE	177.1620	POLYETHYLENE, OXIDIZED Substances For Use As Basic Components Of Single And Repeated Use Food Contact Surfaces; Indirect Food Additives: Polymers
POLYETHYLENE	573.780	POLYETHYLENE Food Additive Listing; Food Additives Permitted In Feed And Drinking Water Of Animals
POLYETHYLENE GLYCOL	172.820	POLYETHYLENE GLYCOL (MEAN MOLECULAR WEIGHT 200-9,500) Multipurpose Additives; Food Additives Permitted For Direct Addition To Food For Human Consumption

POLYETHYLENE GLYCOL	178.3750	POLYETHYLENE GLYCOL (MEAN MOLECULAR WEIGHT 200-9,500) Certain Adjuvants And Production Aids; Indirect Food Additives: Adjuvants, Production Aids, And Sanitizers
POLYETHYLENE GLYCOL	178.3760	POLYETHYLENE GLYCOL (400) MONOLAURATE Certain Adjuvants And Production Aids; Indirect Food Additives: Adjuvants, Production Aids, And Sanitizers
POLYETHYLENE GLYCOL	573.800	POLYETHYLENE GLYCOL (400) MONO- AND DIOLEATES Food Additive Listing; Food Additives Permitted In Feed And Drinking Water Of Animals
POLYETHYLENE GLYCOL	573.820	POLYETHYLENE GLYCOL (400) MONO- AND DIOLEATES Food Additive Listing; Food Additives Permitted In Feed And Drinking Water Of Animals
POLYETHYLENE PHTHALATE	177.1630	POLYETHYLENE PHTHALATE POLYMERS Substances For Use As Basic Components Of Single And Repeated Use Food Contact Surfaces; Indirect Food Additives: Polymers
POLYETHYLENE TEREPHTHALATE	870.3470	INTRACARDIAC PATCH OR PLEDGET MADE OF POLYPROPYLENE, POLYETHYLENE TEREPHTHALATE, OR POLYTETRAFLUOROETHYLENE Cardiovascular Prosthetic Devices; Cardiovascular Devices
POLYGLYCEROL	172.854	POLYGLYCEROL ESTERS OF FATTY ACIDS Multipurpose Additives; Food Additives Permitted For Direct Addition To Food For Human Consumption
POLYHYDRIC ALCOHOL	178.3770	POLYHYDRIC ALCOHOL ESTERS OF OXIDATIVELY REFINED (GERSTHOFEN PROCESS) MONTAN WAX ACIDS Certain Adjuvants And Production Aids; Indirect Food Additives: Adjuvants, Production Aids, And Sanitizers
POLYHYDRIC ALCOHOL	178.3780	POLYHYDRIC ALCOHOL ESTERS OF LONG CHAIN MONOBASIC ACIDS Certain Adjuvants And Production Aids; Indirect Food Additives: Adjuvants, Production Aids, And Sanitizers
POLYMALEIC ACID	173.45	POLYMALEIC ACID AND ITS SODIUM SALT Polymer Substances For Food Treatment; Secondary Direct Food Additives Permitted In Food For Human Consumption
POLYMER	172.770	ETHYLENE OXIDE POLYMER Other Specific Usage Additives; Food Additives Permitted For Direct Addition To Food For Human Consumption
POLYMER	177.1650	POLYSULFIDE POLYMER-POLYEPOXY RESINS Substances For Use As Basic Components Of Single And Repeated Use Food Contact Surfaces; Indirect Food Additives: Polymers
POLYMER	177.2210	ETHYLENE POLYMER, CHLOROSULFONATED Substances For Use Only As Components Of Articles Intended For Repeated Use; Indirect Food Additives: Polymers
POLYMER	872.3420	CARBOXYMETHYLCELLULOSE SODIUM AND CATIONIC POLYACRYLAMIDE POLYMER DENTURE ADHESIVE Prosthetic Devices; Dental Devices
POLYMER	872.3480	POLYACRYLAMIDE POLYMER DENTURE ADHESIVE Prosthetic Devices; Dental Devices
POLYMER	874.3620	EAR, NOSE, AND THROAT SYNTHETIC POLYMER MATERIAL Prosthetic Devices; Ear, Nose, And Throat Devices
POLYMER MODIFIERS	178.3790	POLYMER MODIFIERS IN SEMIRIGID AND RIGID VINYL CHLORIDE PLASTICS Certain Adjuvants And Production Aids; Indirect Food Additives: Adjuvants, Production Aids, And Sanitizers
POLYMERIC	178.3740	PLASTICIZERS IN POLYMERIC SUBSTANCES Certain Adjuvants And Production Aids; Indirect Food Additives: Adjuvants, Production Aids, And Sanitizers
POLYMERIC COATINGS	175.300	RESINOUS AND POLYMERIC COATINGS Substances For Use As Components Of Coatings; Indirect Food Additives
POLYMERIC COATINGS	175.320	RESINOUS AND POLYMERIC COATINGS FOR POLYOLEFIN FILMS Substances For Use As Components Of Coatings; Indirect Food Additives
POLYMERIC FILTERS	177.2250	FILTERS, MICROPOROUS POLYMERIC Substances For Use Only As Components Of Articles Intended For Repeated Use; Indirect Food Additives: Polymers
POLYMERIC MESH BAG	870.3650	PACEMAKER POLYMERIC MESH BAG Cardiovascular Prosthetic Devices; Cardiovascular Devices
POLYMERIZATION	872.6070	ULTRAVIOLET ACTIVATOR FOR POLYMERIZATION Miscellaneous Devices; Dental Devices
POLYMERS	177.1420	ISOBUTYLENE POLYMERS Substances For Use As Basic Components Of Single And Repeated Use Food Contact Surfaces; Indirect Food Additives: Polymers
POLYMERS	177.1520	OLEFIN POLYMERS Substances For Use As Basic Components Of Single And Repeated Use Food Contact Surfaces; Indirect Food Additives: Polymers
POLYMERS	177.1630	POLYETHYLENE PHTHALATE POLYMERS Substances For Use As Basic Components Of Single And Repeated Use Food Contact Surfaces; Indirect Food Additives: Polymers

POLYMERS	177.1810	STYRENE BLOCK POLYMERS Substances For Use As Basic Components Of Single And Repeated Use Food Contact Surfaces; Indirect Food Additives: Polymers
POLYMERS	177.2000	VINYLIDENE CHLORIDE/METHYL ACRYLATE/METHYL METHACRYLATE POLYMERS Substances For Use As Basic Components Of Single And Repeated Use Food Contact Surfaces; Indirect Food Additives: Polymers
POLYMERS	178.2010	ANTIOXIDANTS AND/OR STABILIZERS FOR POLYMERS Antioxidants And Stabilizers; Indirect Food Additives: Adjuvants, Production Aids, And Sanitizers
POLYMERS	178.3295	CLARIFYING AGENTS FOR POLYMERS Certain Adjuvants And Production Aids; Indirect Food Additives: Adjuvants, Production Aids, And Sanitizers
POLYMERS	178.3297	COLORANTS FOR POLYMERS Certain Adjuvants And Production Aids; Indirect Food Additives: Adjuvants, Production Aids, And Sanitizers
POLYMETHYLMETHACRYLATE	177.2465	POLYMETHYLMETHACRYLATE/POLY (TRIMETHOXYSILYLPROPYL) METHACRYLATE COPOLYMERS Substances For Use Only As Components Of Articles Intended For Repeated Use; Indirect Food Additives: Polymers
POLYMETHYLMETHACRYLATE	886.1385	POLYMETHYLMETHACRYLATE (PMMA) DIAGNOSTIC CONTACT LENS Diagnostic Devices; Ophthalmic Devices
POLYMETHYLMETHACRYLATE	888.3027	POLYMETHYLMETHACRYLATE (PMMA) BONE CEMENT Prosthetic Devices; Orthopedic Devices
POLYMIXIN	524.1484h	NEOMYCIN, PENICILLIN, POLYMIXIN, HYDROCORTISONE SUSPENSION Ophthalmic And Topical Dosage Form New Animal Drugs
POLYMIXIN B SULFATE	524.154	BACITRACIN OR BACITRACIN ZINC-NEOMYCIN SULFATE-POLYMIXIN B SULFATE OPHTHALMIC OINTMENT Ophthalmic And Topical Dosage Form New Animal Drugs
POLYMIXIN B SULFATE	524.155	BACITRACIN ZINC-POLYMIXIN B SULFATE-NEOMYCIN SULFATE-HYDROCORTISONE OR HYDROCORTISONE ACETATE OPHTHALMIC OINTMENT Ophthalmic And Topical Dosage Form New Animal Drugs
POLYMYXIN B SULFATE	524.960	FLUMETHASONE, NEOMYCIN SULFATE, AND POLYMYXIN B SULFATE OPHTHALMIC SOLUTIONS Ophthalmic And Topical Dosage Form New Animal Drugs
POLYMYXIN B SULFATE	524.1484e	NEOMYCIN SULFATE AND POLYMYXIN B SULFATE OPHTHALMIC SOLUTION Ophthalmic And Topical Dosage Form New Animal Drugs
POLYMYXIN B SULFATE	524.1662b	OXYTETRACYCLINE HYDROCHLORIDE, POLYMYXIN B SULFATE OPHTHALMIC OINTMENT Ophthalmic And Topical Dosage Form New Animal Drugs
POLYOLEFIN	175.320	RESINOUS AND POLYMERIC COATINGS FOR POLYOLEFIN FILMS Substances For Use As Components Of Coatings; Indirect Food Additives
POLYOXYETHYLENE	573.820	POLYOXYETHYLENE GLYCOL (400) MONO- AND DIOLEATES Food Additive Listing; Food Additives Permitted In Feed And Drinking Water Of Animals
POLYOXYETHYLENE	520.1846	POLYOXYETHYLENE (23) LAURYL ETHER BLOCKS Oral Dosage Form New Animal Drugs
POLYOXYMETHYLENE	177.2470	POLYOXYMETHYLENE COPOLYMER Substances For Use Only As Components Of Articles Intended For Repeated Use; Indirect Food Additives: Polymers
POLYOXYMETHYLENE	177.2480	POLYOXYMETHYLENE HOMOPOLYMER Substances For Use Only As Components Of Articles Intended For Repeated Use; Indirect Food Additives: Polymers
POLYPHENYLENE	177.2490	POLYPHENYLENE SULFIDE RESINS Substances For Use Only As Components Of Articles Intended For Repeated Use; Indirect Food Additives: Polymers
POLYPHENYLENE	177.2500	POLYPHENYLENE SULFONE RESINS Substances For Use Only As Components Of Articles Intended For Repeated Use; Indirect Food Additives: Polymers
POLYPROPYLENE	870.3470	INTRACARDIAC PATCH OR PLEDGET MADE OF POLYPROPYLENE, POLYETHYLENE TEREPHTHALATE, OR POLYTETRAFLUOROETHYLENE Cardiovascular Prosthetic Devices; Cardiovascular Devices
POLYPROPYLENE	878.5010	NONABSORBABLE POLYPROPYLENE SURGICAL SUTURE Surgical Devices; General And Plastic Surgery Devices
POLYSORBATE 60	172.836	POLYSORBATE 60 Multipurpose Additives; Food Additives Permitted For Direct Addition To Food For Human Consumption
POLYSORBATE	60 573.840	POLYSORBATE 60 Food Additive Listing; Food Additives Permitted In Feed And Drinking Water Of Animals
POLYSORBATE 65	172.838	POLYSORBATE 65 Multipurpose Additives; Food Additives Permitted For Direct Addition To Food For Human Consumption

POLYSORBATE 80	172.623	CARRAGEENAN WITH POLYSORBATE 80 Gums, Chewing Gum Bases And Related Substances; Food Additives Permitted For Direct Addition To Food For Human Consumption
POLYSORBATE 80	172.840	POLYSORBATE 80 Multipurpose Additives; Food Additives Permitted For Direct Addition To Food For Human Consumption
POLYSORBATE 80	573.860	POLYSORBATE 80 Food Additive Listing; Food Additives Permitted In Feed And Drinking Water Of Animals
POLYSTYRENE	177.1640	POLYSTYRENE AND RUBBER-MODIFIED POLYSTYRENE Substances For Use As Basic Components Of Single And Repeated Use Food Contact Surfaces; Indirect Food Additives: Polymers
POLYSULFATED GLYCOSAMINOGLYCAN	522.1850	POLYSULFATED GLYCOSAMINOGLYCAN Implantation Or Injectable Dosage Form New Animal Drugs
POLYSULFIDE	177.1650	POLYSULFIDE POLYMER-POLYEPOXY RESINS Substances For Use As Basic Components Of Single And Repeated Use Food Contact Surfaces; Indirect Food Additives: Polymers
POLYSULFONE	177.1655	POLYSULFONE RESINS Substances For Use As Basic Components Of Single And Repeated Use Food Contact Surfaces; Indirect Food Additives: Polymers
POLYTETRAFLUOROETHYLENE	870.3470	INTRACARDIAC PATCH OR PLEDGET MADE OF POLYPROPYLENE, POLYETHYLENE TEREPHTHALATE, OR POLYTETRAFLUOROETHYLENE Cardiovascular Prosthetic Devices; Cardiovascular Devices
POLYTETRAFLUOROETHYLENE	872.3680	POLYTETRAFLUOROETHYLENE (PTFE) VITREOUS CARBON MATERIALS Prosthetic Devices; Dental Devices
POLYTETRAFLUOROETHYLENE	878.3500	POLYTETRAFLUOROETHYLENE WITH CARBON FIBERS COMPOSITE IMPLANT MATERIAL Prosthetic Devices; General And Plastic Surgery Devices
POLYTETRAFLUOROETHYLENE	878.4520	POLYTETRAFLUOROETHYLENE INJECTABLE Surgical Devices; General And Plastic Surgery Devices
POLYTETRAFLUOROETHYLENE	878.5035	NONABSORBABLE EXPANDED POLYTETRAFLUOROETHYLENE SURGICAL SUTURE Surgical Devices; General And Plastic Surgery Devices
POLYURETHANE	177.1680	POLYURETHANE RESINS Substances For Use As Basic Components Of Single And Repeated Use Food Contact Surfaces; Indirect Food Additives: Polymers
POLYVINYL	177.1670	POLYVINYL ALCOHOL FILM Substances For Use As Basic Components Of Single And Repeated Use Food Contact Surfaces; Indirect Food Additives: Polymers
POLYVINYLIDENE	177.2510	POLYVINYLIDENE FLUORIDE RESINS Substances For Use Only As Components Of Articles Intended For Repeated Use; Indirect Food Additives: Polymers
POLYVINYLMETHYLETHER	872.3490	CARBOXYMETHYLCELLULOSE SODIUM AND/OR POLYVINYLMETHYLETHER MALEIC ACID CALCIUM-SODIUM DOUBLE SALT DENTURE ADHESIVE Prosthetic Devices; Dental Devices
POLYVINYLMETHYLETHER	872.3500	POLYVINYLMETHYLETHER MALEIC ANHYDRIDE (PVM-MA), ACID COPOLYMER, AND CARBOXYMETHYLCELLULOSE SODIUM (NACMC) DENTURE ADHESIVE Prosthetic Devices; Dental Devices
POLYVINYLPOLYPYRROLIDONE	173.50	POLYVINYLPOLYPYRROLIDONE Polymer Substances For Food Treatment; Secondary Direct Food Additives Permitted In Food For Human Consumption
POLYVINYLPYRROLIDONE	173.55	POLYVINYLPYRROLIDONE Polymer Substances For Food Treatment; Secondary Direct Food Additives Permitted In Food For Human Consumption
PONAZURIL	520.1855	PONAZURIL Oral Dosage Form New Animal Drugs
POOLS	1250.89	SWIMMING POOLS Sanitation Facilities And Conditions On Vessels; Interstate Conveyance Sanitation
PORCELAIN POWDER	872.6660	PORCELAIN POWDER FOR CLINICAL USE Miscellaneous Devices; Dental Devices
PORCELAIN TOOTH	872.3920	PORCELAIN TOOTH Prosthetic Devices; Dental Devices
POROUS-COATED PROSTHESIS	888.3358	HIP JOINT METAL/POLYMER/METAL SEMI-CONSTRAINED POROUS-COATED UNCEMENTED PROSTHESIS Prosthetic Devices; Orthopedic Devices
PORPHOBILINOGEN	862.1590	PORPHOBILINOGEN TEST SYSTEM Clinical Chemistry Test Systems; Clinical Chemistry And Clinical Toxicology Devices
PORPHYRINS	862.1595	PORPHYRINS TEST SYSTEM Clinical Chemistry Test Systems; Clinical Chemistry And Clinical Toxicology Devices

PORT	880.5965	SUBCUTANEOUS, IMPLANTED, INTRAVASCULAR INFUSION PORT AND CATHETER General Hospital And Personal Use Therapeutic Devices; General Hospital And Personal Use Devices
PORTABLE	868.5440	PORTABLE OXYGEN GENERATOR Therapeutic Devices; Anesthesiology Devices
PORTABLE	868.5655	PORTABLE LIQUID OXYGEN UNIT Therapeutic Devices; Anesthesiology Devices
PORTABLE	868.6250	PORTABLE AIR COMPRESSOR Miscellaneous; Anesthesiology Devices
PORTABLE	870.2640	PORTABLE LEAKAGE CURRENT ALARM Cardiovascular Monitoring Devices; Cardiovascular Devices
PORTABLE	878.4680	NONPOWERED, SINGLE PATIENT, PORTABLE SUCTION APPARATUS Surgical Devices; General And Plastic Surgery Devices
PORTIONS	14.25	PORTIONS OF ADVISORY COMMITTEE MEETINGS Meeting Procedures; Public Hearing Before A Public Advisory Committee
PORTIONS	14.27	DETERMINATION TO CLOSE PORTIONS OF ADVISORY COMMITTEE MEETINGS Meeting Procedures; Public Hearing Before A Public Advisory Committee
POSITION HOLDER	880.5680	PEDIATRIC POSITION HOLDER General Hospital And Personal Use Therapeutic Devices; General Hospital And Personal Use Devices
POSITION INDICATING	872.1840	DENTAL X-RAY POSITION INDICATING DEVICE Diagnostic Devices; Dental Devices
POSITION INDICATOR	872.1850	LEAD-LINED POSITION INDICATOR Diagnostic Devices; Dental Devices
POSITION INDICATOR	892.5780	LIGHT BEAM PATIENT POSITION INDICATOR Therapeutic Devices; Radiology Devices
POSITION SUPPORT	868.6820	PATIENT POSITION SUPPORT Miscellaneous; Anesthesiology Devices
POSITIONER	872.5525	PREFORMED TOOTH POSITIONER Therapeutic Devices; Dental Devices
POSITIONING	890.3110	ELECTRIC POSITIONING CHAIR Physical Medicine Prosthetic Devices; Physical Medicine Devices
POSITIVE END	868.5965	POSITIVE END EXPIRATORY PRESSURE BREATHING ATTACHMENT Therapeutic Devices; Anesthesiology Devices
POSITRON CAMERA	892.1110	POSITRON CAMERA Diagnostic Devices; Radiology Devices
POST	872.3820	ROOT CANAL POST Prosthetic Devices; Dental Devices
POSTAPPROVAL INSPECTION	26.13	TRANSMISSION OF POSTAPPROVAL INSPECTION REPORTS Specific Sector Provisions For Pharmaceutical GMPs; Mutual Recognition Of Pharmaceutical GMP Reports, Medical Device Quality System Audit Reports
POSTAPPROVAL REQUIREMENTS	814.82	POSTAPPROVAL REQUIREMENTS Postapproval Requirements; Premarket Approval Of Medical Devices
POSTAPPROVAL REQUIREMENTS	814.126	POSTAPPROVAL REQUIREMENTS AND REPORTS Humanitarian Use Devices; Premarket Approval Of Medical Devices
POSTHEARING BRIEFS	17.43	POSTHEARING BRIEFS Civil Money Penalties Hearings
POSTING	1240.20	ISSUANCE AND POSTING OF CERTIFICATES FOLLOWING INSPECTION Control Of Communicable Diseases
POSTING	1316.94	POSTING OF SUBSTITUTE RES IN AN ADMINISTRATIVE FORFEITURE ACTION Expedited Forfeiture Proceedings For Certain Property; Administrative Functions, Practices, And Procedures
POSTMARKET SURVEILLANCE	5.414	POSTMARKET SURVEILLANCE Medical Devices And Radiological Health; Redelegations Of Authority; Delegations Of Authority And Organization
POSTMARKET VIGILANCE	26.50	ALERT SYSTEM AND EXCHANGE OF POSTMARKET VIGILANCE REPORTS Mutual Recognition Of Pharmaceutical GMP Reports, Medical Device Quality System Audit Reports
POSTMARKETING	314.80	POSTMARKETING REPORTING OF ADVERSE DRUG EXPERIENCES Applications; Applications For FDA Approval To Market A New Drug
POSTMARKETING	314.81	OTHER POSTMARKETING REPORTS Applications; Applications For FDA Approval To Market A New Drug
POSTMARKETING	314.98	POSTMARKETING REPORTS Abbreviated Applications; Applications For FDA Approval To Market A New Drug
POSTMARKETING	314.540	POSTMARKETING SAFETY REPORTING Applications For FDA Approval To Market A New Drug
POSTMARKETING	600.80	POSTMARKETING REPORTING OF ADVERSE EXPERIENCES Biological Products: General
POSTMARKETING	601.28	ANNUAL REPORTS OF POSTMARKETING PEDIATRIC STUDIES Biologics Licensing; Licensing

POSTMARKETING	601.44	POSTMARKETING SAFETY REPORTING Accelerated Approval Of Biological Products; Licensing
POSTMARKETING	601.70	ANNUAL PROGRESS REPORTS OF POSTMARKETING STUDIES Postmarketing Studies; Licensing
POSTURE	868.5365	POSTURE CHAIR FOR CARDIAC OR PULMONARY TREATMENT Therapeutic Devices; Anesthesiology Devices
POTABLE	1250.82	POTABLE WATER SYSTEMS Sanitation Facilities And Conditions On Vessels; Interstate Conveyance Sanitation
POTASSIUM	73.1125	POTASSIUM SODIUM COPPER CHLOROPHYLLIN (CHLOROPHYLLIN-COPPER COMPLEX) Drugs; Listing Of Color Additives Exempt From Certification
POTASSIUM	73.2125	POTASSIUM SODIUM COPPER CHLOROPHYLLIN (CHLOROPHYLLIN-COPPER COMPLEX) Cosmetics; Listing Of Color Additives Exempt From Certification
POTASSIUM	172.725	GIBBERELLIC ACID AND ITS POTASSIUM SALT Other Specific Usage Additives; Food Additives Permitted For Direct Addition To Food For Human Consumption
POTASSIUM	172.800	ACESULFAME POTASSIUM Multipurpose Additives; Food Additives Permitted For Direct Addition To Food For Human Consumption
POTASSIUM	201.306	POTASSIUM SALT PREPARATIONS INTENDED OR ORAL INGESTION BY MAN Specific Labeling Requirements For Specific Drug Products; Labeling
POTASSIUM	526.1130	HETACILLIN POTASSIUM FOR INTRAMAMMARY INFUSION Intramammary Dosage Forms
POTASSIUM	558.305	LAIDLOMYCIN PROPRIONATE POTASSIUM New Animal Drugs For Use In Animal Feeds
POTASSIUM	862.1600	POTASSIUM TEST SYSTEM Clinical Chemistry Test Systems; Clinical Chemistry And Clinical Toxicology Devices
POTASSIUM ACID TARTRATE	184.1077	POTASSIUM ACID TARTRATE Listing Of Specific Substances Affirmed As GRAS; Direct Food Substances Affirmed As Generally Recognized As Safe
POTASSIUM ACID TARTRATE	582.1077	POTASSIUM ACID TARTRATE General Purpose Food Additives; Substances Generally Recognized As Safe
POTASSIUM ALGINATE	184.1610	POTASSIUM ALGINATE Listing Of Specific Substances Affirmed As GRAS; Direct Food Substances Affirmed As Generally Recognized As Safe
POTASSIUM ALGINATE	582.7610	POTASSIUM ALGINATE Stabilizers; Substances Generally Recognized As Safe
POTASSIUM BICARBONATE	184.1613	POTASSIUM BICARBONATE Listing Of Specific Substances Affirmed As GRAS; Direct Food Substances Affirmed As Generally Recognized As Safe
POTASSIUM BICARBONATE	582.1613	POTASSIUM BICARBONATE General Purpose Food Additives; Substances Generally Recognized As Safe
POTASSIUM BISULFITE	182.3616	POTASSIUM BISULFITE Chemical Preservatives; Substances Generally Recognized As Safe
POTASSIUM BISULFITE	582.3616	POTASSIUM BISULFITE Chemical Preservatives; Substances Generally Recognized As Safe
POTASSIUM BROMATE	172.730	POTASSIUM BROMATE Other Specific Usage Additives; Food Additives Permitted For Direct Addition To Food For Human Consumption
POTASSIUM CARBONATE	184.1619	POTASSIUM CARBONATE Listing Of Specific Substances Affirmed As GRAS; Direct Food Substances Affirmed As Generally Recognized As Safe
POTASSIUM CARBONATE	582.1619	POTASSIUM CARBONATE General Purpose Food Additives; Substances Generally Recognized As Safe
POTASSIUM CHLORIDE	184.1622	POTASSIUM CHLORIDE Listing Of Specific Substances Affirmed As GRAS; Direct Food Substances Affirmed As Generally Recognized As Safe
POTASSIUM CHLORIDE	582.5622	POTASSIUM CHLORIDE Nutrients And/or Dietary Supplements; Substances Generally Recognized As Safe
POTASSIUM CITRATE	184.1625	POTASSIUM CITRATE Listing Of Specific Substances Affirmed As GRAS; Direct Food Substances Affirmed As Generally Recognized As Safe
POTASSIUM CITRATE	582.1625	POTASSIUM CITRATE General Purpose Food Additives; Substances Generally Recognized As Safe
POTASSIUM CITRATE	582.6625	POTASSIUM CITRATE Sequestrants; Substances Generally Recognized As Safe
POTASSIUM GLYCEROPHOSPHATE	582.5628	POTASSIUM GLYCEROPHOSPHATE Nutrients And/or Dietary Supplements; Substances Generally Recognized As Safe
POTASSIUM HYDROXIDE	184.1631	POTASSIUM HYDROXIDE Listing Of Specific Substances Affirmed As GRAS; Direct Food Substances Affirmed As Generally Recognized As Safe

POTASSIUM HYDROXIDE	582.1631	POTASSIUM HYDROXIDE General Purpose Food Additives; Substances Generally Recognized As Safe
POTASSIUM IODATE	184.1635	POTASSIUM IODATE Listing Of Specific Substances Affirmed As GRAS; Direct Food Substances Affirmed As Generally Recognized As Safe
POTASSIUM IODIDE	172.375	POTASSIUM IODIDE Special Dietary And Nutritional Additives; Food Additives Permitted For Direct Addition To Food For Human Consumption
POTASSIUM IODIDE	184.1634	POTASSIUM IODIDE Listing Of Specific Substances Affirmed As GRAS; Direct Food Substances Affirmed As Generally Recognized As Safe
POTASSIUM IODIDE	582.5634	POTASSIUM IODIDE Nutrients And/or Dietary Supplements; Substances Generally Recognized As Safe
POTASSIUM LACTATE	184.1639	POTASSIUM LACTATE Listing Of Specific Substances Affirmed As GRAS; Direct Food Substances Affirmed As Generally Recognized As Safe
POTASSIUM METABISULFITE	182.3637	POTASSIUM METABISULFITE Chemical Preservatives; Substances Generally Recognized As Safe
POTASSIUM METABISULFITE	582.3637	POTASSIUM METABISULFITE Chemical Preservatives; Substances Generally Recognized As Safe
POTASSIUM NITRATE	172.160	POTASSIUM NITRATE Food Preservatives; Food Additives Permitted For Direct Addition To Food For Human Consumption
POTASSIUM NITRATE	181.33	SODIUM NITRATE AND POTASSIUM NITRATE Specific Prior-sanctioned Food Ingredients; Prior-sanctioned Food Ingredients
POTASSIUM NITRITE	181.34	SODIUM NITRITE AND POTASSIUM NITRITE Specific Prior-sanctioned Food Ingredients; Prior-sanctioned Food Ingredients
POTASSIUM PERMANGANATE	250.108	POTASSIUM PERMANGANATE PREPARATIONS AS PRESCRIPTION DRUGS New Drug Or Prescription Status Of Specific Drugs; Specific Requirements For Specific Human Drugs
POTASSIUM SORBATE	182.3640	POTASSIUM SORBATE Chemical Preservatives; Substances Generally Recognized As Safe
POTASSIUM SORBATE	582.3640	POTASSIUM SORBATE Chemical Preservatives; Substances Generally Recognized As Safe
POTASSIUM SULFATE	182.1129	ALUMINUM POTASSIUM SULFATE Multiple Purpose GRAS Food Substances; Substances Generally Recognized As Safe
POTASSIUM SULFATE	184.1643	POTASSIUM SULFATE Listing Of Specific Substances Affirmed As GRAS; Direct Food Substances Affirmed As Generally Recognized As Safe
POTASSIUM SULFATE	582.1129	ALUMINUM POTASSIUM SULFATE General Purpose Food Additives; Substances Generally Recognized As Safe
POTASSIUM SULFATE	582.1643	POTASSIUM SULFATE General Purpose Food Additives; Substances Generally Recognized As Safe
POTASSIUM TARTRATE	184.1804	SODIUM POTASSIUM TARTRATE Listing Of Specific Substances Affirmed As GRAS; Direct Food Substances Affirmed As Generally Recognized As Safe
POTASSIUM TARTRATE	582.1804	SODIUM POTASSIUM TARTRATE General Purpose Food Additives; Substances Generally Recognized As Safe
POTASSIUM TARTRATE	582.6804	SODIUM POTASSIUM TARTRATE Sequestrants; Substances Generally Recognized As Safe
POTATO CHIPS	102.41	POTATO CHIPS MADE FROM DRIED POTATOES Requirements For Specific Nonstandardized Foods; Common Or Usual Name For Nonstandardized Foods
POTENCY	101.54	NUTRIENT CONTENT CLAIMS FOR "GOOD SOURCE," "HIGH," "MORE," AND "HIGH POTENCY" Specific Requirements For Nutrient Content Claims; Food Labeling
POTENCY	610.10	POTENCY General Provisions; General Biological Products Standards
POTENCY	610.21	LIMITS OF POTENCY Standard Preparations And Limits Of Potency; General Biological Products Standards
POTENCY	640.50	QUALITY CONTROL TEST FOR POTENCY Cryoprecipitate; Additional Standards For Human Blood And Blood Products
POTENCY	640.104	POTENCY Immune Globulin (Human); Additional Standards For Human Blood And Blood Products
POTENCY	660.4	POTENCY TEST Antibody To Hepatitis B Surface Antigen; Additional Standards For Diagnostic Substances For Laboratory Tests
POTENCY	660.22	POTENCY REQUIREMENTS WITH REFERENCE PREPARATIONS Blood Grouping Reagent; Additional Standards For Diagnostic Substances For Laboratory Tests

POWDER	520.2087	ROXARSONE SOLUBLE POWDER Oral Dosage Form New Animal Drugs
POWDER	520.2123b	SPECTINOMYCIN DIHYDROCHLORIDE PENTAHYDRATE SOLUBLE POWDER Oral Dosage Form New Animal Drugs
POWDER	520.2220a	SULFADIMETHOXINE ORAL SOLUTION AND SOLUBLE POWDER Oral Dosage Form New Animal Drugs
POWDER	520.2261b	SULFAMETHAZINE SODIUM SOLUBLE POWDER Oral Dosage Form New Animal Drugs
POWDER	520.2345d	TETRACYCLINE HYDROCHLORIDE SOLUBLE POWDER Oral Dosage Form New Animal Drugs
POWDER	520.2380f	THIABENDAZOLE, PIPERAZINE PHOSPHATE POWDER Oral Dosage Form New Animal Drugs
POWDER	520.2455	TIAMULIN SOLUBLE POWDER Oral Dosage Form New Animal Drugs
POWDER	520.2482	TRIAMCINOLONE ACETONIDE ORAL POWDER Oral Dosage Form New Animal Drugs
POWDER	520.2520g	TRICHLORFON, PHENOTHIAZINE, AND PIPERAZINE DIHYDROCHLORIDE POWDER Oral Dosage Form New Animal Drugs
POWDER	520.2613	TRIMETHOPRIM AND SULFADIAZINE POWDER Oral Dosage Form New Animal Drugs
POWDER	522.313	CEFTIOFUR SODIUM POWDER FOR INJECTION Implantation Or Injectable Dosage Form New Animal Drugs
POWDER	522.468	COLISTIMETHATE SODIUM POWDER FOR INJECTION Implantation Or Injectable Dosage Form New Animal Drugs
POWDER	522.1085	GUAIFENESIN STERILE POWDER Implantation Or Injectable Dosage Form New Animal Drugs
POWDER	522.1155	IMIDOCARB DIPROPIONATE STERILE POWDER Implantation Or Injectable Dosage Form New Animal Drugs
POWDER	524.1005	FURAZOLIDONE AEROSOL POWDER Ophthalmic And Topical Dosage Form New Animal Drugs
POWDER	524.1484	NEOMYCIN SULFATE, ISOFLUPREDONE ACETATE, TETRACAINE HYDROCHLORIDE, AND MYRISTYL-GAMMA-PICOLINIUM CHLORIDE, TOPICAL POWDER Ophthalmic And Topical Dosage Form New Animal Drugs
POWDER	524.1580c	NITROFURAZONE SOLUBLE POWDER Ophthalmic And Topical Dosage Form New Animal Drugs
POWDER	529.2464	TICARCILLIN POWDER Certain Other Dosage Form New Animal Drugs
POWDER, ABSORBABLE	878.4480	ABSORBABLE POWDER FOR LUBRICATING A SURGEON'S GLOVE Surgical Devices; General And Plastic Surgery Devices
POWDER, ALUMINUM	73.1645	ALUMINUM POWDER Drugs; Listing Of Color Additives Exempt From Certification
POWDER, ALUMINUM	73.2645	ALUMINUM POWDER Cosmetics; Listing Of Color Additives Exempt From Certification
POWDER, BEET	73.40	DEHYDRATED BEETS (BEET POWDER) Foods; Listing Of Color Additives Exempt From Certification
POWDER, BRONZE	73.1646	BRONZE POWDER Drugs; Listing Of Color Additives Exempt From Certification
POWDER, BRONZE	73.2646	BRONZE POWDER Cosmetics; Listing Of Color Additives Exempt From Certification
POWDER, COPPER	73.1647	COPPER POWDER Drugs; Listing Of Color Additives Exempt From Certification
POWDER, COPPER	73.2647	COPPER POWDER Cosmetics; Listing Of Color Additives Exempt From Certification
POWDER, PORCELAIN	872.6660	PORCELAIN POWDER FOR CLINICAL USE Miscellaneous Devices; Dental Devices
POWDER, TAMARIND SEED	176.350	TAMARIND SEED KERNEL POWDER Substances For Use Only As Components Of Paper And Paperboard; Indirect Food Additives
POWDER, TEST	866.1640	ANTIMICROBIAL SUSCEPTIBILITY TEST POWDER Diagnostic Devices; Immunology And Microbiology Devices
POWDER, VANILLA	169.170	VANILLA POWDER Requirements For Specific Standardized Food Dressings And Flavorings; Food Dressings And Flavorings
POWDER, VANILLA-VANILLIN	169.182	VANILLA-VANILLIN POWDER Requirements For Specific Standardized Food Dressings And Flavorings; Food Dressings And Flavorings
POWDERS	201.305	ISOPROTERENOL INHALATION PREPARATIONS (PRESSURIZED AEROSOLS, NEBULIZERS, POWDERS) FOR HUMAN USE; WARNINGS Specific Labeling Requirements For Specific Drug Products; Labeling
POWER OF ATTORNEY	1305.07	POWER OF ATTORNEY Order Forms

PRACTICES	10.10	SUMMARIES OF ADMINISTRATIVE PRACTICES AND PROCEDURES General Provisions; Administrative Practices And Procedures
PRACTICES	10.115	GOOD GUIDANCE PRACTICES General Administrative Procedures; Administrative Practices And Procedures
PRACTICES	812.7	PROHIBITION OF PROMOTION AND OTHER PRACTICES General Provisions; Investigational Device Exemptions
PRACTITIONER	1307.11	DISTRIBUTION BY DISPENSER TO ANOTHER PRACTITIONER Special Exemptions For Manufacture And Distribution Of Controlled Substances; Miscellaneous
PRACTITIONERS	1301.22	EXEMPTION OF AGENTS AND EMPLOYEES; AFFILIATED PRACTITIONERS Registration Of Manufacturers, Distributors, And Dispensers Of Controlled Substances
PRACTITIONERS	1301.75	PHYSICAL SECURITY CONTROLS FOR PRACTITIONERS Security Requirements; Registration Of Manufacturers, Distributors, And Dispensers Of Controlled Substances
PRACTITIONERS	1301.76	OTHER SECURITY CONTROLS FOR PRACTITIONERS Security Requirements; Registration Of Manufacturers, Distributors, And Dispensers Of Controlled Substances
PRALIDOXIME	522.1862	STERILE PRALIDOXIME CHLORIDE Implantation Or Injectable Dosage Form New Animal Drugs
PRAZIQUANTEL	520.903d	FEBENTAL-PRAZIQUANTEL PASTE Oral Dosage Form New Animal Drugs
PRAZIQUANTEL	520.1870	PRAZIQUANTEL TABLETS Oral Dosage Form New Animal Drugs
PRAZIQUANTEL	520.1871	PRAZIQUANTEL/PYRANTEL PAMOATE TABLETS Oral Dosage Form New Animal Drugs
PRAZIQUANTEL	520.1872	PRAZIQUANTEL, PYRANTEL PAMOATE AND FEBENTAL TABLETS Oral Dosage Form New Animal Drugs
PRAZIQUANTEL	522.1870	PRAZIQUANTEL INJECTABLE SOLUTION Implantation Or Injectable Dosage Form New Animal Drugs
PREALBUMIN	866.5060	PREALBUMIN IMMUNOLOGICAL TEST SYSTEM Immunological Test Systems; Immunology And Microbiology Devices
PREAMPLIFIER	886.1640	OPHTHALMIC PREAMPLIFIER Diagnostic Devices; Ophthalmic Devices
PREAPPROVAL INSPECTION	26.14	TRANSMISSION OF PREAPPROVAL INSPECTION REPORTS Specific Sector Provisions For Pharmaceutical GMPs; Mutual Recognition Of Pharmaceutical GMP Reports, Medical Device Quality System Audit Reports
PREBYPASS	870.4280	CARDIOPULMONARY PREBYPASS FILTER Cardiovascular Surgical Devices; Cardiovascular Devices
PRECISION ATTACHMENT	872.3165	PRECISION ATTACHMENT Prosthetic Devices; Dental Devices
PRECURSORS, IMMEDIATE	1308.47	CONTROL OF IMMEDIATE PRECURSORS Hearings; Schedules Of Controlled Substances
PREDICTIVE	868.1890	PREDICTIVE PULMONARY-FUNCTION VALUE CALCULATOR Diagnostic Devices; Anesthesiology Devices
PREDNISOLONE	520.1880	PREDNISOLONE TABLETS Oral Dosage Form New Animal Drugs
PREDNISOLONE	520.2345h	TETRACYCLINE HYDROCHLORIDE, SODIUM NOVOBIOCIN, AND PREDNISOLONE TABLETS Oral Dosage Form New Animal Drugs
PREDNISOLONE	520.2604	TRIMEPRAZINE TARTRATE AND PREDNISOLONE TABLETS Oral Dosage Form New Animal Drugs
PREDNISOLONE	520.2605	TRIMEPRAZINE TARTRATE AND PREDNISOLONE CAPSULES Oral Dosage Form New Animal Drugs
PREDNISOLONE	522.1883	PREDNISOLONE SODIUM PHOSPHATE INJECTION, STERILE Implantation Or Injectable Dosage Form New Animal Drugs
PREDNISOLONE	522.1884	PREDNISOLONE SODIUM SUCCINATE INJECTION Implantation Or Injectable Dosage Form New Animal Drugs
PREDNISOLONE	522.1885	PREDNISOLONE TERTIARY BUTYLACETATE SUSPENSION Implantation Or Injectable Dosage Form New Animal Drugs
PREDNISOLONE	524.390d	CHLORAMPHENICOL-PREDNISOLONE OPHTHALMIC OINTMENT Ophthalmic And Topical Dosage Form New Animal Drugs
PREDNISOLONE	524.1484k	NEOMYCIN SULFATE, PREDNISOLONE, TETRACAINE, AND SQUALANE TOPICAL-OTIC SUSPENSION Ophthalmic And Topical Dosage Form New Animal Drugs
PREDNISOLONE	524.1880	PREDNISOLONE-NEOMYCIN SULFATE OPHTHALMIC OINTMENT Ophthalmic And Topical Dosage Form New Animal Drugs
PREDNISOLONE	524.1883	PREDNISOLONE SODIUM PHOSPHATE-NEOMYCIN SULFATE OPHTHALMIC OINTMENT Ophthalmic And Topical Dosage Form New Animal Drugs
PREDNISOLONE	556.520	PREDNISOLONE Specific Tolerances For Residues Of New Animal Drugs; Tolerances For Residues Of New Animal Drugs In Food

PREDNISOLONE ACETATE	522.1881	STERILE PREDNISOLONE ACETATE AQUEOUS SUSPENSION Implantation Or Injectable Dosage Form New Animal Drugs
PREDNISOLONE ACETATE	524.1484f	NEOMYCIN SULFATE, PREDNISOLONE ACETATE, TETRACAINE HYDROCHLORIDE EARDROPS Ophthalmic And Topical Dosage Form New Animal Drugs
PREDNISOLONE ACETATE	524.1881	PREDNISOLONE ACETATE OPHTHALMIC AND TOPICAL DOSAGE FORMS Ophthalmic And Topical Dosage Form New Animal Drugs
PREDNISONE	522.1890	STERILE PREDNISONE SUSPENSION Implantation Or Injectable Dosage Form New Animal Drugs
PREDNISONE	556.530	PREDNISONE Specific Tolerances For Residues Of New Animal Drugs; Tolerances For Residues Of New Animal Drugs In Food
PREFORMED	872.3130	PREFORMED ANCHOR Prosthetic Devices; Dental Devices
PREFORMED	872.3285	PREFORMED CLASP Prosthetic Devices; Dental Devices
PREFORMED	872.3330	PREFORMED CROWN Prosthetic Devices; Dental Devices
PREFORMED	872.3360	PREFORMED CUSP Prosthetic Devices; Dental Devices
PREFORMED	872.3580	PREFORMED GOLD DENTURE TOOTH Prosthetic Devices; Dental Devices
PREFORMED	872.3590	PREFORMED PLASTIC DENTURE TOOTH Prosthetic Devices; Dental Devices
PREFORMED	872.5525	PREFORMED TOOTH POSITIONER Therapeutic Devices; Dental Devices
PREFORMED	872.6880	PREFORMED IMPRESSION TRAY Miscellaneous Devices; Dental Devices
PREFORMED	882.5320	PREFORMED ALTERABLE CRANIOPLASTY PLATE Neurological Therapeutic Devices; Neurological Devices
PREFORMED	882.5330	PREFORMED NONALTERABLE CRANIOPLASTY PLATE Neurological Therapeutic Devices; Neurological Devices
PREFORMED	882.5900	PREFORMED CRANIOSYNOSTOSIS STRIP Neurological Therapeutic Devices; Neurological Devices
PREGNANCY	201.63	PREGNANCY/BREAST-FEEDING WARNING Labeling Requirements For Over-the-counter Drugs; Labeling
PREGNANCY	330.2	PREGNANCY-NURSING WARNING General Provisions; Over-the-counter Human Drugs Generally Recognized As Safe And Effective & Not Misbranded
PREGNANCY	862.1270	ESTROGENS (TOTAL, IN PREGNANCY) TEST SYSTEM Clinical Chemistry Test Systems; Clinical Chemistry And Clinical Toxicology Devices
PREGNANEDIOL	862.1605	PREGNANEDIOL TEST SYSTEM Clinical Chemistry Test Systems; Clinical Chemistry And Clinical Toxicology Devices
PREGNANETRIOL	862.1610	PREGNANETRIOL TEST SYSTEM Clinical Chemistry Test Systems; Clinical Chemistry And Clinical Toxicology Devices
PREGNENOLONE	862.1615	PREGNENOLONE TEST SYSTEM Clinical Chemistry Test Systems; Clinical Chemistry And Clinical Toxicology Devices
PREHEARING CONFERENCE	12.91	TIME AND PLACE OF PREHEARING CONFERENCE Hearing Procedures; Formal Evidentiary Public Hearing
PREHEARING CONFERENCE	12.92	PREHEARING CONFERENCE PROCEDURE Hearing Procedures; Formal Evidentiary Public Hearing
PREHEARING CONFERENCE	1316.54	PREHEARING CONFERENCE Administrative Hearings; Administrative Functions, Practices, And Procedures
PREHEARING CONFERENCE	1316.57	SUBMISSION OF DOCUMENTARY EVIDENCE AND AFFIDAVITS AND IDENTIFICATION OF WITNESSES SUBSEQUENT TO PREHEARING CONFERENCE Administrative Hearings; Administrative Functions, Practices, And Procedures
PREHEARING CONFERENCES	17.21	PREHEARING CONFERENCES Civil Money Penalties Hearings
PREHEARING RULING	1316.55	PREHEARING RULING Administrative Hearings; Administrative Functions, Practices, And Procedures
PREJUDICE	71.6	EXTENSION OF TIME FOR STUDYING PETITIONS; SUBSTANTIVE AMENDMENTS; WITHDRAWAL OF PETITIONS WITHOUT PREJUDICE General Provisions; Color Additive Petitions
PREJUDICE	171.7	WITHDRAWAL OF PETITION WITHOUT PREJUDICE General Provisions; Food Additive Petitions
PREJUDICE	514.7	WITHDRAWAL OF APPLICATIONS WITHOUT PREJUDICE General Provisions; New Animal Drug Applications
PREJUDICE	571.7	WITHDRAWAL OF PETITION WITHOUT PREJUDICE General Provisions; Food Additive Petitions

PREMARKET APPROVAL	5.407	APPROVAL, DISAPPROVAL, OR WITHDRAWAL OF APPROVAL OF PRODUCT DEVELOPMENT PROTOCOLS AND APPLICATIONS FOR PREMARKET APPROVAL FOR MEDICAL DEVICES Medical Devices And Radiological Health; Redelegations Of Authority; Delegations Of Authority And Organization
PREMARKET APPROVAL	5.408	DETERMINATIONS CONCERNING THE TYPE OF VALID SCIENTIFIC EVIDENCE SUBMITTED IN A PREMARKET APPROVAL APPLICATION Medical Devices And Radiological Health; Redelegations Of Authority; Delegations Of Authority And Organization
PREMARKET APPROVAL	5.701	PREMARKET APPROVAL OF A PRODUCT THAT IS, OR CONTAINS A BIOLOGIC, A DEVICE, OR A DRUG Product Designation; Redelegations Of Authority; Delegations Of Authority And Organization
PREMARKET APPROVAL	814.9	CONFIDENTIALITY OF DATA AND INFORMATION IN A PREMARKET APPROVAL APPLICATION (PMA) FILE General; Premarket Approval Of Medical Devices
PREMARKET APPROVAL	860.132	PROCEDURES WHEN THE COMMISSIONER INITIATES A PERFORMANCE STANDARD OR PREMARKET APPROVAL PROCEEDING UNDER SECTION 514(B) OR 515(B) Reclassification; Medical Device Classification Procedures
PREMARKET APPROVAL	862.3	EFFECTIVE DATES OF REQUIREMENT FOR PREMARKET APPROVAL Clinical Chemistry And Clinical Toxicology Devices
PREMARKET APPROVAL	864.3	EFFECTIVE DATES OF REQUIREMENT FOR PREMARKET APPROVAL Hematology And Pathology Devices
PREMARKET APPROVAL	866.3	EFFECTIVE DATES OF REQUIREMENT FOR PREMARKET APPROVAL Immunology And Microbiology Devices
PREMARKET APPROVAL	868.3	EFFECTIVE DATES OF REQUIREMENT FOR PREMARKET APPROVAL Anesthesiology Devices
PREMARKET APPROVAL	870.3	EFFECTIVE DATES OF REQUIREMENT FOR PREMARKET APPROVAL Cardiovascular Devices
PREMARKET APPROVAL	872.3	EFFECTIVE DATES OF REQUIREMENT FOR PREMARKET APPROVAL Dental Devices
PREMARKET APPROVAL	874.3	EFFECTIVE DATES OF REQUIREMENT FOR PREMARKET APPROVAL Ear, Nose, And Throat Devices
PREMARKET APPROVAL	876.3	EFFECTIVE DATES OF REQUIREMENT FOR PREMARKET APPROVAL Gastroenterology-urology Devices
PREMARKET APPROVAL	878.3	EFFECTIVE DATES OF REQUIREMENT FOR PREMARKET APPROVAL General And Plastic Surgery Devices
PREMARKET APPROVAL	880.3	EFFECTIVE DATES OF REQUIREMENT FOR PREMARKET APPROVAL General Hospital And Personal Use Devices
PREMARKET APPROVAL	882.3	EFFECTIVE DATES OF REQUIREMENT FOR PREMARKET APPROVAL Neurological Devices
PREMARKET APPROVAL	884.3	EFFECTIVE DATES OF REQUIREMENT FOR PREMARKET APPROVAL Obstetrical And Gynecological Devices
PREMARKET APPROVAL	886.3	EFFECTIVE DATES OF REQUIREMENT FOR PREMARKET APPROVAL Ophthalmic Devices
PREMARKET APPROVAL	888.3	EFFECTIVE DATES OF REQUIREMENT FOR PREMARKET APPROVAL Orthopedic Devices
PREMARKET APPROVAL	890.3	EFFECTIVE DATES OF REQUIREMENT FOR PREMARKET APPROVAL Physical Medicine Devices
PREMARKET APPROVAL	892.3	EFFECTIVE DATES OF REQUIREMENTS FOR PREMARKET APPROVAL Radiology Devices
PREMARKET NOTIFICATION	5.401	ISSUANCE OF FEDERAL REGISTER DOCUMENTS PERTAINING TO EXEMPTIONS FROM PREMARKET NOTIFICATION Medical Devices And Radiological Health; Redelegations Of Authority; Delegations Of Authority And Organization
PREMARKET NOTIFICATION	190.6	REQUIREMENT FOR PREMARKET NOTIFICATION New Dietary Ingredient Notification; Dietary Supplements
PREMARKET NOTIFICATION	807.81	WHEN A PREMARKET NOTIFICATION SUBMISSION IS REQUIRED Premarket Notification Procedures; Establishment Registration And Device Listing For Manufacturers And Initial Importers Of Devices
PREMARKET NOTIFICATION	807.85	EXEMPTION FROM PREMARKET NOTIFICATION Premarket Notification Procedures; Establishment Registration And Device Listing For Manufacturers And Initial Importers Of Devices
PREMARKET NOTIFICATION	807.87	INFORMATION REQUIRED IN A PREMARKET NOTIFICATION SUBMISSION Premarket Notification Procedures; Establishment Registration And Device Listing For Manufacturers And Initial Importers Of Devices
PREMARKET NOTIFICATION	807.90	FORMAT OF A PREMARKET NOTIFICATION SUBMISSION Premarket Notification Procedures; Establishment Registration And Device Listing For Manufacturers And Initial Importers Of Devices
PREMARKET NOTIFICATION	807.97	MISBRANDING BY REFERENCE TO PREMARKET NOTIFICATION Premarket Notification Procedures; Establishment Registration And Device Listing For Manufacturers And Initial Importers Of Devices

PREMARKET NOTIFICATION	807.100	FDA ACTION ON A PREMARKET NOTIFICATION Premarket Notification Procedures; Establishment Registration And Device Listing For Manufacturers And Initial Importers Of Devices
PREMARKET SUBMISSION	5.400	ISSUANCE OF FEDERAL REGISTER DOCUMENTS TO RECOGNIZE OR TO WITHDRAW RECOGNITION OF A STANDARD TO MEET PREMARKET SUBMISSION REQUIREMENTS Medical Devices And Radiological Health; Redelegations Of Authority; Delegations Of Authority And Organization
PREMIXES	170.60	NITRITES AND/OR NITRATES IN CURING PREMIXES Specific Administrative Rulings And Decisions; Food Additives
PREPACKAGED FOODS	179.45	PACKAGING MATERIALS FOR USE DURING THE IRRADIATION OF PREPACKAGED FOODS Packaging Materials For Irradiated Foods; Irradiation In The Production, Processing And Handling Of Food
PREPARATION	25.20	ACTIONS REQUIRING PREPARATION OF AN ENVIRONMENTAL ASSESSMENT Agency Actions Requiring Environmental Consideration; Environmental Impact Considerations
PREPARATION	25.22	ACTIONS REQUIRING PREPARATION OF AN ENVIRONMENTAL IMPACT STATEMENT Agency Actions Requiring Environmental Consideration; Environmental Impact Considerations
PREPARATION, BIOLOGICAL PRODUCTS	5.203	ISSUANCE AND REVOCATION OF LICENSES FOR THE PROPAGATION OR MANUFACTURE AND PREPARATION OF BIOLOGICAL PRODUCTS Biologics; Redelegations Of Authority; Delegations Of Authority And Organization
PREPARATION, PRODUCT	113.81	PRODUCT PREPARATION Production And Process Controls; Thermally Processed Low-acid Foods Packaged In Hermetically Sealed Containers
PREPARATION, STANDARD	640.55	U.S. STANDARD PREPARATION Cryoprecipitate; Additional Standards For Human Blood And Blood Products
PREPARATIONS, CHEMICAL	1308.23	EXEMPTION OF CERTAIN CHEMICAL PREPARATIONS; APPLICATION Exempt Chemical Preparations; Schedules Of Controlled Substances
PREPARATIONS, CHEMICAL	1308.24	EXEMPTION CHEMICAL PREPARATIONS Exempt Chemical Preparations; Schedules Of Controlled Substances
PREPARATIONS, ENZYME	173.357	MATERIALS USED AS FIXING AGENTS IN THE IMMOBILIZATION OF ENZYME PREPARATIONS Specific Usage Additives; Secondary Direct Food Additives Permitted In Food For Human Consumption
PREPARATIONS, REFERENCE	660.52	REFERENCE PREPARATIONS Anti-human Globulin; Additional Standards For Diagnostic Substances For Laboratory Tests
PREPARED, FOOD	1250.30	CONSTRUCTION, MAINTENANCE AND USE OF PLACES WHERE FOOD IS PREPARED, SERVED, OR STORED Food Service Sanitation On Land And Air Conveyances, And Vessels; Interstate Conveyance Sanitation
PREPROGRAMMED	870.1435	SINGLE-FUNCTION PREPROGRAMMED DIAGNOSTIC COMPUTER Cardiovascular Diagnostic Devices; Cardiovascular Devices
PRESCRIBED	601.25	REVIEW PROCEDURES TO DETERMINE THAT LICENSED BIOLOGICAL PRODUCTS ARE SAFE, EFFECTIVE NOT MISBRANDED UNDER PRESCRIBED, RECOMMENDED, OR SUGGESTED CONDITIONS OF USE Biologics Licensing; Licensing
PRESCRIBED	601.26	RECLASSIFICATION PROCEDURES TO DETERMINE THAT LICENSED BIOLOGICAL PRODUCTS ARE SAFE, EFFECTIVE, AND NOT MISBRANDED UNDER PRESCRIBED, RECOMMENDED, OR SUGGESTED CONDITIONS OF USE Biologics Licensing; Licensing
PRESCRIPTION	1306.04	PURPOSE OF ISSUE OF PRESCRIPTION General Information; Prescriptions
PRESCRIPTION	1306.11	REQUIREMENT OF PRESCRIPTION Controlled Substances Listed In Schedule II; Prescriptions
PRESCRIPTION	1306.21	REQUIREMENT OF PRESCRIPTION Controlled Substances Listed In Schedules III, IV, And V; Prescriptions
PRESCRIPTION	1306.25	TRANSFER BETWEEN PHARMACIES OF PRESCRIPTION INFORMATION FOR SCHEDULES III, IV, AND V CONTROLLED SUBSTANCES FOR REFILL PURPOSES Controlled Substances Listed In Schedules III, IV, And V; Prescriptions
PRESCRIPTION	1306.26	DISPENSING WITHOUT PRESCRIPTION Controlled Substances Listed In Schedules III, IV, And V; Prescriptions
PRESCRIPTION CHEMICALS	201.120	PRESCRIPTION CHEMICALS AND OTHER PRESCRIPTION COMPONENTS Exemptions From Adequate Directions For Use; Labeling
PRESCRIPTION DEVICE	801.433	WARNING STATEMENTS FOR PRESCRIPTION AND RESTRICTED DEVICE PRODUCTS CONTAINING CHLOROFLUOROCARBONS OR OTHER OZONE-DEPLETING SUBSTANCES Specific Requirements For Specific Devices; Labeling

PRESCRIPTION DEVICES	801.109	PRESCRIPTION DEVICES Exemptions From Adequate Directions For Use; Labeling
PRESCRIPTION DEVICES	801.110	RETAIL EXEMPTIONS FOR PRESCRIPTION DEVICES Exemptions From Adequate Directions For Use; Labeling
PRESCRIPTION DRUG	5.108	AUTHORITY RELATING TO WAIVERS OR REDUCTIONS OF PRESCRIPTION DRUG USER FEES Human Drugs; Redelegations Of Authority; Delegations Of Authority And Organization
PRESCRIPTION DRUG	202.1	PRESCRIPTION DRUG ADVERTISEMENTS Prescription Drug Advertising
PRESCRIPTION DRUG	250.100	AMYL NITRITE INHALANT AS A PRESCRIPTION DRUG FOR HUMAN USE New Drug Or Prescription Status Of Specific Drugs; Specific Requirements For Specific Human Drugs
PRESCRIPTION DRUG	310.305	RECORDS AND REPORTS CONTAINING ADVERSE DRUG EXPERIENCES ON MARKETED PRESCRIPTION DRUGS FOR HUMAN USE WITHOUT APPROVED NEW DRUG APPLICATIONS Records And Reports; New Drugs
PRESCRIPTION DRUGS	5.107	EXTENSIONS OR STAYS OF EFFECTIVE DATES FOR COMPLIANCE WITH CERTAIN LABELING REQUIREMENTS FOR HUMAN PRESCRIPTION DRUGS Human Drugs; Redelegations Of Authority; Delegations Of Authority And Organization
PRESCRIPTION DRUGS	14.160	ESTABLISHMENT OF STANDING TECHNICAL ADVISORY COMMITTEES FOR HUMAN PRESCRIPTION DRUGS Advisory Committees For Human Prescription Drugs; Public Hearing Before A Public Advisory Committee
PRESCRIPTION DRUGS	200.200	PRESCRIPTION DRUGS; REMINDER ADVERTISEMENTS AND REMINDER LABELING TO PROVIDE PRICE INFORMATION TO CONSUMERS Prescription Drug Consumer Price Listing; General
PRESCRIPTION DRUGS	201.21	DECLARATION OF PRESENCE OF PHENYLALANINE AS A COMPONENT OF ASPARTAME IN OVER-THE-COUNTER AND PRESCRIPTION DRUGS FOR HUMAN USE General Labeling Provisions; Labeling
PRESCRIPTION DRUGS	201.22	PRESCRIPTION DRUGS CONTAINING SULFITES; REQUIRED WARNING STATEMENTS General Labeling Provisions; Labeling
PRESCRIPTION DRUGS	201.56	GENERAL REQUIREMENTS ON CONTENT AND FORMAT OF LABELING FOR HUMAN PRESCRIPTION DRUGS Labeling Requirements For Prescription Drugs And/or Insulin; Labeling
PRESCRIPTION DRUGS	201.57	SPECIFIC REQUIREMENTS ON CONTENT AND FORMAT OF LABELING FOR HUMAN PRESCRIPTION DRUGS Labeling Requirements For Prescription Drugs And/or Insulin; Labeling
PRESCRIPTION DRUGS	201.100	PRESCRIPTION DRUGS FOR HUMAN USE Exemptions From Adequate Directions For Use; Labeling
PRESCRIPTION DRUGS	203.50	REQUIREMENTS FOR WHOLESALE DISTRIBUTION OF PRESCRIPTION DRUGS Wholesale Distribution; Prescription Drug Marketing
PRESCRIPTION DRUGS	205.50	MINIMUM REQUIREMENTS FOR THE STORAGE AND HANDLING OF PRESCRIPTION DRUGS Guidelines For State Licensing Of Wholesale Prescription Drug Distributors
PRESCRIPTION DRUGS	250.101	AMPHETAMINE AND METHAMPHETAMINE INHALERS REGARDED AS PRESCRIPTION DRUGS New Drug Or Prescription Status Of Specific Drugs; Specific Requirements For Specific Human Drugs
PRESCRIPTION DRUGS	250.105	GELSEMIUM-CONTAINING PREPARATIONS REGARDED AS PRESCRIPTION DRUGS New Drug Or Prescription Status Of Specific Drugs; Specific Requirements For Specific Human Drugs
PRESCRIPTION DRUGS	250.108	POTASSIUM PERMANGANATE PREPARATIONS AS PRESCRIPTION DRUGS New Drug Or Prescription Status Of Specific Drugs; Specific Requirements For Specific Human Drugs
PRESCRIPTION DRUGS	300.50	FIXED-COMBINATION PRESCRIPTION DRUGS FOR HUMANS Combination Drugs; General
PRESCRIPTION DRUGS	310.305	RECORDS AND REPORTS CONTAINING ADVERSE DRUG EXPERIENCES ON MARKETED PRESCRIPTION DRUGS FOR HUMAN USE WITHOUT APPROVED NEW DRUG APPLICATIONS Records And Reports; New Drugs
PRESCRIPTION DRUGS	1309.28	EXEMPTION OF DISTRIBUTORS OF REGULATED PRESCRIPTION DRUG PRODUCTS Registration Of Manufacturers, Distributors, Importers And Exporters Of List I Chemicals
PRESCRIPTION-DISPENSING	369.3	WARNINGS REQUIRED ON DRUGS EXEMPTED FROM PRESCRIPTION-DISPENSING REQUIREMENTS OF SECTION 503(B)(1)(C) Definitions And Interpretations; Interpretative Statements Re: Warnings On Drugs And Devices For OTC Sale

PRESCRIPTION-EXEMPTION	310.200	PRESCRIPTION-EXEMPTION PROCEDURE New Drugs Exempted From Prescription-dispensing Requirements; New Drugs
PRESCRIPTION LENS	886.5844	PRESCRIPTION SPECTACLE LENS Therapeutic Devices; Ophthalmic Devices
PRESCRIPTION PRODUCT	1308.31	APPLICATION FOR EXEMPTION OF A NONNARCOTIC PRESCRIPTION PRODUCT Exempted Prescription Products; Schedules Of Controlled Substances
PRESCRIPTION PRODUCTS	1308.32	EXEMPTED PRESCRIPTION PRODUCTS Exempted Prescription Products; Schedules Of Controlled Substances
PRESCRIPTION REQUIREMENTS	290.2	EXEMPTION FROM PRESCRIPTION REQUIREMENTS General Provisions; Controlled Drugs
PRESCRIPTION REQUIREMENTS	329.20	EXEMPTION OF CERTAIN HABIT-FORMING DRUGS FROM PRESCRIPTION REQUIREMENTS Exemptions; Habit-forming Drugs
PRESCRIPTION SALE	310.201	EXEMPTION FOR CERTAIN DRUGS LIMITED BY NEW DRUG APPLICATIONS TO PRESCRIPTION SALE New Drugs Exempted From Prescription-dispensing Requirements; New Drugs
PRESCRIPTIONS	1306.03	PERSONS ENTITLED TO ISSUE PRESCRIPTIONS General Information; Prescriptions
PRESCRIPTIONS	1306.05	MANNER OF ISSUANCE OF PRESCRIPTIONS General Information; Prescriptions
PRESCRIPTIONS	1306.06	PERSONS ENTITLED TO FILL PRESCRIPTIONS General Information; Prescriptions
PRESCRIPTIONS	1306.12	REFILLING PRESCRIPTIONS Controlled Substances Listed In Schedule II; Prescriptions
PRESCRIPTIONS	1306.13	PARTIAL FILLING OF PRESCRIPTIONS Controlled Substances Listed In Schedule II; Prescriptions
PRESCRIPTIONS	1306.14	LABELING OF SUBSTANCES AND FILLING OF PRESCRIPTIONS Controlled Substances Listed In Schedule II; Prescriptions
PRESCRIPTIONS	1306.22	REFILLING OF PRESCRIPTIONS Controlled Substances Listed In Schedules III, IV, And V; Prescriptions
PRESCRIPTIONS	1306.23	PARTIAL FILLING OF PRESCRIPTIONS Controlled Substances Listed In Schedules III, IV, And V; Prescriptions
PRESCRIPTIONS	1306.24	LABELING OF SUBSTANCES AND FILLING OF PRESCRIPTIONS Controlled Substances Listed In Schedules III, IV, And V; Prescriptions
PRESENTATION	7.84	OPPORTUNITY FOR PRESENTATION OF VIEWS BEFORE REPORT OF CRIMINAL VIOLATION Criminal Violations; Enforcement Policy
PRESENTATION	7.85	CONDUCT OF A PRESENTATION OF VIEWS BEFORE REPORT OF CRIMINAL VIOLATION Criminal Violations; Enforcement Policy
PRESENTATION	7.87	RECORDS RELATED TO OPPORTUNITIES FOR PRESENTATION OF VIEWS CONDUCTED BEFORE REPORT OF CRIMINAL VIOLATION Criminal Violations; Enforcement Policy
PRESERVATION	26.74	PRESERVATION OF REGULATORY AUTHORITY Framework Provisions; Mutual Recognition Of Pharmaceutical GMP Reports, Medical Device Quality System Audit Reports
PRESERVATION	1002.31	PRESERVATION AND INSPECTION OF RECORDS Manufacturers' Records; Records and Reports
PRESERVATION	1305.13	PRESERVATION OF ORDER FORMS Order Forms
PRESERVATIVE	146.152	ORANGE JUICE WITH PRESERVATIVE Requirements For Specific Standardized Canned Fruit Juices And Beverages; Canned Fruit Juices
PRESERVATIVE	146.154	CONCENTRATED ORANGE JUICE WITH PRESERVATIVE Requirements For Specific Standardized Canned Fruit Juices And Beverages; Canned Fruit Juices
PRESERVATIVES	101.22	FOODS; LABELING OF SPICES, FLAVORINGS, COLORINGS AND CHEMICAL PRESERVATIVES Specific Food Labeling Requirements; Food Labeling
PRESERVATIVES	178.3800	PRESERVATIVES FOR WOOD Certain Adjuvants And Production Aids; Indirect Food Additives: Adjuvants, Production Aids, And Sanitizers
PRESERVATIVES	501.22	ANIMAL FOODS; LABELING OF SPICES, FLAVORINGS, COLORINGS, AND CHEMICAL PRESERVATIVES Specific Animal Food Labeling Requirements; Animal Food Labeling
PRESERVED FIGS	145.134	CANNED PRESERVED FIGS Requirements For Specific Standardized Canned Fruits; Canned Fruits
PRESERVES	150.160	FRUIT PRESERVES AND JAMS Requirements For Specific Standardized Fruit Butters, Jellies, Preserves; Fruit Butters, Jellies, Preserves, And Related Products
PRESERVES	150.161	ARTIFICIALLY SWEETENED FRUIT PRESERVES AND JAMS Requirements For Specific Standardized Fruit Butters, Jellies, Preserves; Fruit Butters, Jellies, Preserves, And Related Products

PRESIDING OFFICER	12.60	PRESIDING OFFICER Presiding Officer; Formal Evidentiary Public Hearing
PRESIDING OFFICER	12.70	AUTHORITY OF PRESIDING OFFICER Presiding Officer; Formal Evidentiary Public Hearing
PRESIDING OFFICER	12.75	DISQUALIFICATION OF PRESIDING OFFICER Presiding Officer; Formal Evidentiary Public Hearing
PRESIDING OFFICER	12.78	UNAVAILABILITY OF PRESIDING OFFICER Presiding Officer; Formal Evidentiary Public Hearing
PRESIDING OFFICER	12.97	INTERLOCUTORY APPEAL FROM RULING OF PRESIDING OFFICER Hearing Procedures; Formal Evidentiary Public Hearing
PRESIDING OFFICER	16.42	PRESIDING OFFICER Commissioner And Presiding Officer; Regulatory Hearing Before The Food And Drug Administration
PRESIDING OFFICER	16.44	COMMUNICATION TO PRESIDING OFFICER AND COMMISSIONER Commissioner And Presiding Officer; Regulatory Hearing Before The Food And Drug Administration
PRESIDING OFFICER	17.18	INTERLOCUTORY APPEAL FROM RULING OF PRESIDING OFFICER Civil Money Penalties Hearings
PRESIDING OFFICER	17.19	AUTHORITY OF THE PRESIDING OFFICER Civil Money Penalties Hearing
PRESIDING OFFICER	1316.52	PRESIDING OFFICER Administrative Hearings; Administrative Functions, Practices, And Procedures
PRESIDING OFFICER	1316.62	APPEAL FROM RULING OF PRESIDING OFFICER Administrative Hearings; Administrative Functions, Practices, And Procedures
PRESSURE	862.2260	HIGH PRESSURE LIQUID CHROMATOGRAPHY SYSTEM FOR CLINICAL USE Clinical Laboratory Instruments; Clinical Chemistry And Clinical Toxicology Devices
PRESSURE	868.1150	INDWELLING BLOOD CARBON DIOXIDE PARTIAL PRESSURE (PCO2) ANALYZER Diagnostic Devices; Anesthesiology Devices
PRESSURE	868.1200	INDWELLING BLOOD OXYGEN PARTIAL PRESSURE (PO2) ANALYZER Diagnostic Devices; Anesthesiology Devices
PRESSURE	868.1750	PRESSURE PLETHYSMOGRAPH Diagnostic Devices; Anesthesiology Devices
PRESSURE	868.1780	INSPIRATORY AIRWAY PRESSURE METER Diagnostic Devices; Anesthesiology Devices
PRESSURE	868.2600	AIRWAY PRESSURE MONITOR Monitoring Devices; Anesthesiology Devices
PRESSURE	868.2610	GAS PRESSURE GAUGE Monitoring Devices; Anesthesiology Devices
PRESSURE	868.2620	GAS PRESSURE CALIBRATOR Monitoring Devices; Anesthesiology Devices
PRESSURE	868.2700	PRESSURE REGULATOR Monitoring Devices; Anesthesiology Devices
PRESSURE	868.2875	DIFFERENTIAL PRESSURE TRANSDUCER Monitoring Devices; Anesthesiology Devices
PRESSURE	868.2900	GAS PRESSURE TRANSDUCER Monitoring Devices; Anesthesiology Devices
PRESSURE	868.5860	PRESSURE TUBING AND ACCESSORIES Therapeutic Devices; Anesthesiology Devices
PRESSURE	868.5935	EXTERNAL NEGATIVE PRESSURE VENTILATOR Therapeutic Devices; Anesthesiology Devices
PRESSURE	868.5965	POSITIVE END EXPIRATORY PRESSURE BREATHING ATTACHMENT Therapeutic Devices; Anesthesiology Devices
PRESSURE	870.1100	BLOOD PRESSURE ALARM Cardiovascular Diagnostic Devices; Cardiovascular Devices
PRESSURE	870.1110	BLOOD PRESSURE COMPUTER Cardiovascular Diagnostic Devices; Cardiovascular Devices
PRESSURE	870.1120	BLOOD PRESSURE CUFF Cardiovascular Diagnostic Devices; Cardiovascular Devices
PRESSURE	870.1130	NONINVASIVE BLOOD PRESSURE MEASUREMENT SYSTEM Cardiovascular Diagnostic Devices; Cardiovascular Devices
PRESSURE	870.1140	VENOUS BLOOD PRESSURE MANOMETER Cardiovascular Diagnostic Devices; Cardiovascular Devices
PRESSURE	870.2850	EXTRAVASCULAR BLOOD PRESSURE TRANSDUCER Cardiovascular Monitoring Devices; Cardiovascular Devices
PRESSURE	870.2870	CATHETER TIP PRESSURE TRANSDUCER Cardiovascular Monitoring Devices; Cardiovascular Devices
PRESSURE	870.4310	CARDIOPULMONARY BYPASS CORONARY PRESSURE GAUGE Cardiovascular Surgical Devices; Cardiovascular Devices
PRESSURE	880.2460	ELECTRICALLY POWERED SPINAL FLUID PRESSURE MONITOR General Hospital And Personal Use Monitoring Devices; General Hospital And Personal Use Devices

PRESSURE	880.5420	PRESSURE INFUSOR FOR AN I. V. BAG General Hospital And Personal Use Therapeutic Devices; General Hospital And Personal Use Devices
PRESSURE	880.5550	ALTERNATING PRESSURE AIR FLOTATION MATTRESS General Hospital And Personal Use Therapeutic Devices; General Hospital And Personal Use Devices
PRESSURE	880.6450	SKIN PRESSURE PROTECTORS General Hospital And Personal Use Miscellaneous Devices; General Hospital And Personal Use Devices
PRESSURE	882.1620	INTRACRANIAL PRESSURE MONITORING DEVICE Neurological Diagnostic Devices; Neurological Devices
PRESSURE	884.2700	INTRAUTERINE PRESSURE MONITOR AND ACCESSORIES Obstetrical And Gynecological Monitoring Devices; Obstetrical And Gynecological Devices
PRESSURE	886.4280	INTRAOCULAR PRESSURE MEASURING DEVICE Surgical Devices; Ophthalmic Devices
PRESSURE	886.4610	OCULAR PRESSURE APPLICATOR Surgical Devices; Ophthalmic Devices
PRESSURE	890.1600	INTERMITTENT PRESSURE MEASUREMENT SYSTEM Physical Medicine Diagnostic Devices; Physical Medicine Devices
PRESSURE	890.1615	MINIATURE PRESSURE TRANSDUCER Physical Medicine Diagnostic Devices; Physical Medicine Devices
PRESSURE	890.5765	PRESSURE-APPLYING DEVICE Physical Medicine Therapeutic Devices; Physical Medicine Devices
PRESSURE-SENSITIVE	175.125	PRESSURE-SENSITIVE ADHESIVES Substances For Use Only As Components Of Adhesives; Indirect Food Additives
PRESSURIZED	201.305	ISOPROTERENOL INHALATION PREPARATIONS (PRESSURIZED AEROSOLS, NEBULIZERS, POWDERS) FOR HUMAN USE; WARNINGS Specific Labeling Requirements For Specific Drug Products; Labeling
PRESUBMISSION REVIEW	20.44	PRESUBMISSION REVIEW OF REQUEST FOR CONFIDENTIALITY OF VOLUNTARILY SUBMITTED DATA OR INFORMATION Procedures And Fees; Public Information
PREVENT ADULTERATION	2.25	GRAIN SEED TREATED WITH POISONOUS SUBSTANCES; COLOR IDENTIFICATION TO PREVENT ADULTERATION OF HUMAN AND ANIMAL FOOD Human and Animal Foods; General Administrative Rulings And Decisions
PREVENTION	310.546	DRUG PRODUCTS CONTAINING ACTIVE INGREDIENTS OFFERED OTC FOR THE TREATMENT AND/OR PREVENTION OF NOCTURNAL LEG MUSCLE CRAMPS Requirements For Specific New Drugs Or Devices; New Drugs
PREVENTION	310.547	DRUG PRODUCTS CONTAINING QUININE OFFERED OTC FOR THE TREATMENT AND/OR PREVENTION OF MALARIA Requirements For Specific New Drugs Or Devices; New Drugs
PREVENTION	310.548	DRUG PRODUCTS CONTAINING COLLOIDAL SILVER INGREDIENTS OFFERED OTC FOR THE TREATMENT AND/OR PREVENTION OF DISEASE Requirements For Specific New Drugs Or Devices; New Drugs
PREVENTIVE ACTION	820.100	CORRECTIVE AND PREVENTIVE ACTION Quality System Regulation
PRICE INFORMATION	200.200	PRESCRIPTION DRUGS; REMINDER ADVERTISEMENTS AND REMINDER LABELING TO PROVIDE PRICE INFORMATION TO CONSUMERS Prescription Drug Consumer Price Listing; General
PRIMARY JURISDICTION	5.700	AUTHORITY RELATING TO DETERMINATION OF PRODUCT PRIMARY JURISDICTION Product Designation; Redelegations Of Authority; Delegations Of Authority And Organization
PRIMARY SOURCE	20.49	REFERRAL TO PRIMARY SOURCE OF RECORDS Procedures And Fees; Public Information
PRIMIDONE	520.1900	PRIMIDONE TABLETS Oral Dosage Form New Animal Drugs
PRIMIDONE	862.3680	PRIMIDONE TEST SYSTEM Clinical Toxicology Test Systems; Clinical Chemistry And Clinical Toxicology Devices
PRINCIPAL DISPLAY PANEL	101.1	PRINCIPAL DISPLAY PANEL OF PACKAGE FORM FOOD General Provisions; Food Labeling
PRINCIPAL DISPLAY PANEL	201.60	PRINCIPAL DISPLAY PANEL Labeling Requirements For Over-the-counter Drugs; Labeling
PRINCIPAL DISPLAY PANEL	328.50	PRINCIPAL DISPLAY PANEL OF ALL OTC DRUG PRODUCTS INTENDED FOR ORAL INGESTION THAT CONTAIN ALCOHOL Over-the-counter Drug Products Intended For Oral Ingestion That Contain Alcohol
PRINCIPAL DISPLAY PANEL	352.50	PRINCIPAL DISPLAY PANEL OF ALL SUNSCREEN Sunscreen Drug Products For Over-the-counter Human Use
PRINCIPAL DISPLAY PANEL	355.55	PRINCIPAL DISPLAY PANEL OF ALL FLUORIDE RINSE DRUG PRODUCTS Anticaries Drug Products For Over-the-counter Human Use

PRINCIPAL DISPLAY PANEL	501.1	PRINCIPAL DISPLAY PANEL OF PACKAGE FORM ANIMAL FOOD General Provisions; Animal Food Labeling
PRINCIPAL DISPLAY PANEL	701.10	PRINCIPAL DISPLAY PANEL Package Form; Cosmetic Labeling
PRINCIPAL DISPLAY PANEL	801.60	PRINCIPAL DISPLAY PANEL Labeling Requirements For Over-the-counter Devices; Labeling
PRINCIPLES	101.13	NUTRIENT CONTENT CLAIMS - GENERAL PRINCIPLES Food Labeling
PRINCIPLES	102.5	GENERAL PRINCIPLES General Provisions; Common Or Usual Name For Nonstandardized Foods
PRINCIPLES	104.5	GENERAL PRINCIPLES Nutritional Quality Guidelines For Foods
PRINCIPLES	170.20	GENERAL PRINCIPLES FOR EVALUATING THE SAFETY OF FOOD ADDITIVES Food Additive Safety; Food Additives
PRINCIPLES	312.22	GENERAL PRINCIPLES OF THE IND SUBMISSION Investigational New Drug Application
PRINCIPLES	502.5	GENERAL PRINCIPLES Common Or Usual Names For Nonstandardized Animal Foods
PRINCIPLES	570.20	GENERAL PRINCIPLES FOR EVALUATING THE SAFETY OF FOOD ADDITIVES Food Additive Safety; Food Additives
PRIOR INVESTIGATIONS	812.27	REPORT OF PRIOR INVESTIGATIONS Investigational Device Exemptions
PRIOR SANCTIONED	570.13	INDIRECT FOOD ADDITIVES RESULTING FROM PACKAGING MATERIALS PRIOR SANCTIONED FOR ANIMAL FEED AND PET FOOD General Provisions; Food Additives
PRIOR SANCTIONS	181.5	PRIOR SANCTIONS General Provisions; Prior-sanctioned Food Ingredients
PRISM	886.1650	OPHTHALMIC BAR PRISM Diagnostic Devices; Ophthalmic Devices
PRISM	886.1655	OPHTHALMIC FRESNEL PRISM Diagnostic Devices; Ophthalmic Devices
PRISM	886.1660	GONIOSCOPIC PRISM Diagnostic Devices; Ophthalmic Devices
PRISM	886.1665	OPHTHALMIC ROTARY PRISM Diagnostic Devices; Ophthalmic Devices
PRISM READER	886.5810	OPHTHALMIC PRISM READER Therapeutic Devices; Ophthalmic Devices
PRIVACY ACT	21.20	PROCEDURES FOR NOTICE OF FDA PRIVACY ACT RECORD SYSTEMS FDA Privacy Act Record Systems; Protection Of Privacy
PRIVACY ACT	21.70	DISCLOSURE AND INTRA-AGENCY USE OF RECORDS IN PRIVACY ACT RECORD SYSTEMS; NO ACCOUNTING REQUIRED Disclosure Of Records In Privacy Act Record Systems To Persons Other Than The Subject Individual; Protection Of Privacy
PRIVACY ACT	21.71	DISCLOSURE OF RECORDS IN PRIVACY ACT RECORD SYSTEMS; ACCOUNTING REQUIRED Disclosure Of Records In Privacy Act Record Systems To Persons Other ThanThe Subject Individual; Protection Of Privacy
PRIVACY ACT	21.73	ACCURACY, COMPLETENESS, TIMELINESS, AND RELEVANCE OF RECORDS DISCLOSED FROM PRIVACY ACT RECORD SYSTEMS Disclosure Of Records In Privacy Act Record Systems To Persons Other ThanThe Subject Individual; Protection Of Privacy
PRIVACY, PERSONAL	20.63	PERSONNEL, MEDICAL, AND SIMILAR FILES, DISCLOSURE OF WHICH CONSTITUTES A CLEARLY UNWARRANTED INVASION OF PERSONAL PRIVACY Exemptions; Public Information
PRIVATE CONTRACTOR	20.51	USE OF PRIVATE CONTRACTOR FOR COPYING Procedures And Fees; Public Information
PRIVILEGED INFORMATION	20.61	TRADE SECRETS AND COMMERCIAL OR FINANCIAL INFORMATION WHICH IS PRIVILEGED OR CONFIDENTIAL Exemptions; Public Information
PROBABLE CAUSE	1316.10	ADMINISTRATIVE PROBABLE CAUSE Administrative Inspections; Administrative Functions, Practices, And Procedures
PROBE	870.1220	ELECTRODE RECORDING CATHETER OR ELECTRODE RECORDING PROBE Cardiovascular Diagnostic Devices; Cardiovascular Devices
PROBE	870.1915	THERMODILUTION PROBE Cardiovascular Diagnostic Devices; Cardiovascular Devices
PROBE	870.2120	EXTRAVASCULAR BLOOD FLOW PROBE Cardiovascular Monitoring Devices; Cardiovascular Devices
PROBE	882.4725	RADIOFREQUENCY LESION PROBE Neurological Surgical Devices; Neurological Devices
PROBE	886.1670	OPHTHALMIC ISOTOPE UPTAKE PROBE Diagnostic Devices; Ophthalmic Devices
PROBE	892.1320	NUCLEAR UPTAKE PROBE Diagnostic Devices; Radiology Devices

PROCAINE	522.1696	PENICILLIN G PROCAINE IMPLANTATION AND INJECTABLE DOSAGE FORMS Implantation Or Injectable Dosage Form New Animal Drugs
PROCAINE	526.1696a	PENICILLIN G PROCAINE IN OIL Intramammary Dosage Forms
PROCAINE	526.1696b	PENICILLIN G PROCAINE-DIHYDROSTREPTOMYCIN IN SOYBEAN OIL FOR INTRAMAMMARY INFUSION (DRY COWS) Intramammary Dosage Forms
PROCAINE	526.1696c	PENICILLIN G PROCAINE-DIHYDROSTREPTOMYCIN SULFATE FOR INTRAMAMMARY INFUSION (DRY COWS) Intramammary Dosage Forms
PROCAINE	526.1696d	PENICILLIN G PROCAINE-NOVOBIOCIN FOR INTRAMAMMARY INFUSION Intramammary Dosage Forms
PROCAINE	558.145	CHLORTETRACYCLINE, PROCAINE PENICILLIN, AND SULFAMETHAZINE Specific New Animal Drugs For Use In Animal Feeds; New Animal Drugs For Use In Animal Feeds
PROCEDURAL REQUIREMENTS	10.19	WAIVER, SUSPENSION, OR MODIFICATION OF PROCEDURAL REQUIREMENTS General Provisions; Administrative Practices And Procedures
PROCEDURE	12.92	PREHEARING CONFERENCE PROCEDURE Hearing Procedures; Formal Evidentiary Public Hearing
PROCEDURE	16.60	HEARING PROCEDURE Procedure For Regulatory Hearing; Regulatory Hearing Before The Food And Drug Administration
PROCEDURE	56.110	EXPEDITED REVIEW PROCEDURE FOR CERTAIN KINDS OF RESEARCH INVOLVING NO MORE THAN MINIMAL RISK, AND FOR MINOR CHANGES IN APPROVED RESEARCH IRB Functions And Operations; Institutional Review Boards
PROCEDURE	71.30	PROCEDURE FOR FILING OBJECTIONS TO REGULATIONS Administrative Action On Petitions; Color Additive Petitions
PROCEDURE	130.5	PROCEDURE FOR ESTABLISHING A FOOD STANDARD General Provisions; Food Standards: General
PROCEDURE	170.17	EXEMPTION FOR INVESTIGATIONAL USE AND PROCEDURE FOR OBTAINING AUTHORIZATION TO MARKET EDIBLE PRODUCTS FROM EXPERIMENTAL ANIMALS General Provisions; Food Additives
PROCEDURE	171.110	PROCEDURE FOR OBJECTIONS AND HEARINGS Administrative Actions On Applications; Food Additive Petitions
PROCEDURE	171.130	PROCEDURE FOR AMENDING AND REPEALING TOLERANCES OR EXEMPTIONS FROM TOLERANCES Administrative Actions On Applications; Food Additive Petitions
PROCEDURE	310.200	PRESCRIPTION-EXEMPTION PROCEDURE New Drugs Exempted From Prescription-dispensing Requirements; New Drugs
PROCEDURE	314.54	PROCEDURE FOR SUBMISSION OF AN APPLICATION REQUIRING INVESTIGATIONS FOR APPROVAL OF A NEW INDICATION FOR, OR OTHER CHANGE FROM, A LISTED DRUG Applications For FDA Approval To Market A New Drug
PROCEDURE	314.201	PROCEDURE FOR HEARINGS Hearing Procedures For New Drugs; Applications For FDA Approval To Market A New Drug
PROCEDURE	514.201	PROCEDURE FOR HEARINGS Hearing Procedures; New Animal Drug Applications
PROCEDURE	530.23	PROCEDURE FOR SETTING AND ANNOUNCING SAFE LEVELS Extralabel Drug Use In Animals
PROCEDURE	530.24	PROCEDURE FOR ANNOUNCING ANALYTICAL METHODS FOR DRUG RESIDUE QUANTIFICATION Extralabel Drug Use In Animals
PROCEDURE	570.17	EXEMPTION FOR INVESTIGATIONAL USE AND PROCEDURE FOR OBTAINING AUTHORIZATION TO MARKET EDIBLE PRODUCTS FROM EXPERIMENTAL ANIMALS General Provisions; Food Additives
PROCEDURE	571.110	PROCEDURE FOR OBJECTIONS AND HEARINGS Administrative Actions On Applications; Food Additive Petitions
PROCEDURE	571.130	PROCEDURE FOR AMENDING AND REPEALING TOLERANCES OR EXEMPTIONS FROM TOLERANCES Administrative Actions On Applications; Food Additive Petitions
PROCEDURE	601.7	PROCEDURE FOR HEARINGS General Provisions; Licensing
PROCEDURE	1210.30	HEARING PROCEDURE FOR PERMIT DENIAL, SUSPENSION, AND REVOCATION Regulations Under The Federal Import Milk Act
PROCEDURE	1303.22	PROCEDURE FOR APPLYING FOR INDIVIDUAL MANUFACTURING QUOTAS Individual Manufacturing Quotas; Quotas
PROCEDURE	1303.23	PROCEDURE FOR FIXING INDIVIDUAL MANUFACTURING QUOTAS Individual Manufacturing Quotas; Quotas

PROCEDURE	1305.05	PROCEDURE FOR OBTAINING ORDER FORMS Order Forms
PROCEDURE	1305.06	PROCEDURE FOR EXECUTING ORDER FORMS Order Forms
PROCEDURE	1305.09	PROCEDURE FOR FILLING ORDER FORMS Order Forms
PROCEDURE	1305.10	PROCEDURE FOR ENDORSING ORDER FORMS Order Forms
PROCEDURE	1305.16	SPECIAL PROCEDURE FOR FILLING CERTAIN ORDER FORMS Order Forms
PROCEDURE	1307.21	PROCEDURE FOR DISPOSING OF CONTROLLED SUBSTANCES Disposal Of Controlled Substances; Miscellaneous
PROCEDURE	1402.5	APPEAL PROCEDURE Mandatory Declassification Review
PROCEDURES	1.23	PROCEDURES FOR REQUESTING VARIATIONS AND EXEMPTIONS FROM REQUIRED LABEL STATEMENTS General Labeling Requirements; General Enforcement Regulations
PROCEDURES	3.5	PROCEDURES FOR IDENTIFYING THE DESIGNATED AGENCY COMPONENT Product Jurisdiction
PROCEDURES	5.415	AUTHORITY RELATING TO MEDICAL DEVICE REPORTING PROCEDURES Medical Devices And Radiological Health; Redelegations Of Authority; Delegations Of Authority And Organization
PROCEDURES	10.10	SUMMARIES OF ADMINISTRATIVE PRACTICES AND PROCEDURES General Provisions; Administrative Practices And Procedures
PROCEDURES	10.206	PROCEDURES FOR ELECTRONIC MEDIA COVERAGE OF AGENCY PUBLIC ADMINISTRATIVE PROCEEDINGS Electronic Media Coverage Public Admin Proceedings; Guideline On Policy Procedures; Administrative Practices And Procedures
PROCEDURES	14.125	PROCEDURES OF TEPRSSC Technical Electronic Products Radiation Safety Standards Committee; Public Hearing Before A Public Advisory Committee
PROCEDURES	14.145	PROCEDURES OF A COLOR ADDITIVE ADVISORY COMMITTEE Color Additive Advisory Committees; Public Hearing Before A Public Advisory Committee
PROCEDURES	21.20	PROCEDURES FOR NOTICE OF FDA PRIVACY ACT RECORD SYSTEMS FDA Privacy Act Record Systems; Protection Of Privacy
PROCEDURES	21.40	PROCEDURES FOR SUBMITTING REQUESTS FOR NOTIFICATION AND ACCESS Procedures For Notification And Access To Records In Privacy Act Record Systems; Protection Of Privacy
PROCEDURES	21.50	PROCEDURES FOR SUBMITTING REQUESTS FOR AMENDMENT OF RECORDS Procedures For Requests For Amendment Of Records; Protection Of Privacy
PROCEDURES	25.15	GENERAL PROCEDURES Agency Actions Requiring Environmental Consideration; Environmental Impact Considerations
PROCEDURES	26.66	DESIGNATION AND LISTING PROCEDURES Framework Provisions; Mutual Recognition Of Pharmaceutical GMP Reports, Medical Device Quality System Audit Reports
PROCEDURES	58.81	STANDARD OPERATING PROCEDURES Testing Facilities Operation; Good Laboratory Practice For Nonclinical Laboratory Studies
PROCEDURES	60.44	HEARING PROCEDURES Due Diligence Hearings; Patent Term Restoration
PROCEDURES	106.1	STATUS AND APPLICABILITY OF THE QUALITY CONTROL PROCEDURES REGULATION General Provisions; Infant Formula Quality Control Procedures
PROCEDURES	110.40	EQUIPMENT AND PROCEDURES Equipment; Current Good Manufacturing Practice In Manufacturing, Packing, Or Holding Human Food
PROCEDURES	113.40	EQUIPMENT AND PROCEDURES Equipment; Thermally Processed Low-acid Foods Packaged In Hermetically Sealed Containers
PROCEDURES	123.11	SANITATION CONTROL PROCEDURES Fish And Fishery Products
PROCEDURES	129.40	EQUIPMENT AND PROCEDURES Equipment And Facilities; Processing And Bottling Of Bottled Drinking Water
PROCEDURES	203.34	POLICIES AND PROCEDURES, ADMINISTRATIVE SYSTEMS Samples; Prescription Drug Marketing
PROCEDURES	211.100	WRITTEN PROCEDURES; DEVIATIONS Production And Process Controls; Current Good Manufacturing Practice For Finished Pharmaceuticals

PROCEDURES	211.142	DISTRIBUTION PROCEDURES Holding And Distribution; Current Good Manufacturing Practice For Finished Pharmaceuticals
PROCEDURES	211.150	WAREHOUSING PROCEDURES Holding And Distribution; Current Good Manufacturing Practice For Finished Pharmaceuticals
PROCEDURES	225.65	EQUIPMENT CLEANOUT PROCEDURES Product Quality Control; Current Good Manufacturing Practice For Medicated Feeds
PROCEDURES	225.165	EQUIPMENT CLEANOUT PROCEDURES Product Quality Assurance; Current Good Manufacturing Practice For Medicated Feeds
PROCEDURES	226.40	PRODUCTION AND CONTROL PROCEDURES Product Quality Control; Current Good Manufacturing Practice For Type A Medicated Articles
PROCEDURES	314.71	PROCEDURES FOR SUBMISSION OF A SUPPLEMENT TO AN APPROVED APPLICATION Applications; Applications For FDA Approval To Market A New Drug
PROCEDURES	314.530	WITHDRAWAL PROCEDURES Applications For FDA Approval To Market A New Drug
PROCEDURES	320.32	PROCEDURES FOR ESTABLISHING OR AMENDING A BIOEQUIVALENCE REQUIREMENT Procedures For Determining The Bioavailability Or Bioequivalence Of Drug Products; Bioavailability And Bioequivalence Requirements
PROCEDURES	330.10	PROCEDURES FOR CLASSIFYING OTC DRUGS AS GENERALLY RECOGNIZED AS SAFE AND EFFECTIVE AND NOT MISBRANDED, AND ESTABLISHING MONOGRAPHS Administrative Procedures; Over-the-counter Human Drugs Generally Recognized As Safe And Effective & Not Misbranded
PROCEDURES	330.14	ADDITIONAL CRITERIA AND PROCEDURES FOR CLASSIFYING OTC DRUGS AS GENERALLY RECOGNIZED AS SAFE AND EFFECTIVE AND NOT MISBRANDED Administrative Procedures; Over-the-counter Human Drugs Generally Recognized As Safe And Effective & Not Misbranded
PROCEDURES	352.72	GENERAL TESTING PROCEDURES Sunscreen Drug Products For Over-the-counter Human Use
PROCEDURES	355.70	TESTING PROCEDURES FOR FLUORIDE DENTIFRICE PRODUCTS Anticaries Drug Products For Over-the-counter Human Use
PROCEDURES	515.31	PROCEDURES FOR HEARINGS Hearing Procedures; Medicated Feed Mill License
PROCEDURES	601.2	APPLICATIONS FOR BIOLOGICS LICENSES; PROCEDURES FOR FILING General Provisions; Licensing
PROCEDURES	601.25	REVIEW PROCEDURES TO DETERMINE LICENSED BIOLOGICAL PRODUCTS ARE SAFE, EFFECTIVE, AND NOT MISBRANDED UNDER PRESCRIBED, RECOMMENDED, OR SUGGESTED CONDITIONS OF USE Biologics Licensing; Licensing
PROCEDURES	601.26	RECLASSIFICATION PROCEDURES TO DETERMINE THAT LICENSED BIOLOGICAL PRODUCTS ARE SAFE, EFFECTIVE, AND NOT MISBRANDED UNDER PRESCRIBED, RECOMMENDED, OR SUGGESTED CONDITIONS OF USE Biologics Licensing; Licensing
PROCEDURES	601.43	WITHDRAWAL PROCEDURES Accelerated Approval Of Biological Products; Licensing
PROCEDURES	606.100	STANDARD OPERATING PROCEDURES Production And Process Controls; Current Good Manufacturing Practice For Blood And Blood Components
PROCEDURES	606.165	DISTRIBUTION AND RECEIPT; PROCEDURES AND RECORDS Records And Reports; Current Good Manufacturing Practice For Blood And Blood Components
PROCEDURES	640.120	ALTERNATIVE PROCEDURES Alternative Procedures; Additional Standards For Human Blood And Blood Products
PROCEDURES	660.53	CONTROLS FOR SEROLOGICAL PROCEDURES Anti-human Globulin; Additional Standards For Diagnostic Substances For Laboratory Tests
PROCEDURES	803.17	WRITTEN MDR PROCEDURES General Provisions; Medical Device Reporting
PROCEDURES	808.25	PROCEDURES FOR PROCESSING AN APPLICATION Exemption Procedures; Exemption From Federal Preemption Of State & Local Medical Device Requirements
PROCEDURES	814.44	PROCEDURES FOR REVIEW OF A PMA FDA Action On A PMA; Premarket Approval Of Medical Devices
PROCEDURES	814.116	PROCEDURES FOR REVIEW OF AN HDE Humanitarian Use Devices; Premarket Approval Of Medical Devices
PROCEDURES	860.84	CLASSIFICATION PROCEDURES FOR “OLD DEVICES” Classification; Medical Device Classification Procedures

PROCEDURES	860.130	GENERAL PROCEDURES UNDER SECTION 513(E) Reclassification; Medical Device Classification Procedures
PROCEDURES	860.132	PROCEDURES WHEN THE COMMISSIONER INITIATES A PERFORMANCE STANDARD OR PREMARKET APPROVAL PROCEEDING UNDER SECTION 514(B) OR 515(B) Reclassification; Medical Device Classification Procedures
PROCEDURES	860.134	PROCEDURES FOR "NEW DEVICES" UNDER SECTION 513(F) AND RECLASSIFICATION OF CERTAIN DEVICES Reclassification; Medical Device Classification Procedures
PROCEDURES	860.136	PROCEDURES FOR TRANSITIONAL PRODUCTS UNDER SECTION 520(L) OF THE ACT Reclassification; Medical Device Classification Procedures
PROCEDURES	895.21	PROCEDURES FOR BANNING A DEVICE General Provisions; Banned Devices
PROCEDURES	1010.13	SPECIAL TESTING PROCEDURES Performance Standards For Electronic Products: General
PROCEDURES	1270.9	WRITTEN PROCEDURES Human Tissue Intended For Transplantation
PROCEDURES	1301.17	SPECIAL PROCEDURES FOR CERTAIN APPLICATIONS Registration; Registration Of Manufacturers, Distributors, And Dispensers Of Controlled Substances
PROCEDURES	1301.90	EMPLOYEE SCREENING PROCEDURES Employee Screening - Non-practitioners; Registration Of Manufacturers, Distributors, And Dispensers Of Controlled Substances
PROCEDURES	1404.310	PROCEDURES Debarment; Governmentwide Debarment And Suspension And Requirements For Drug-Free Workplace (Grants)
PROCEDURES	1404.410	PROCEDURES Suspension; Governmentwide Debarment And Suspension And Requirements For Drug-Free Workplace (Grants)
PROCEDURES	1404.630	CERTIFICATION REQUIREMENTS AND PROCEDURES Drug-free Workplace Requirements; Governmentwide Debarment And Suspension And Requirements For Drug-Free Workplace (Grants)
PROCEEDING	1316.31	AUTHORITY FOR ENFORCEMENT PROCEEDING Enforcement Proceedings; Administrative Functions, Practices, And Procedures
PROCEEDING	1316.32	NOTICE OF PROCEEDING; TIME AND PLACE Enforcement Proceedings; Administrative Functions, Practices, And Procedures
PROCEEDING	1316.33	CONDUCT OF PROCEEDING Enforcement Proceedings; Administrative Functions, Practices, And Procedures
PROCEEDING	1316.34	RECORDS OF PROCEEDING Enforcement Proceedings; Administrative Functions, Practices, And Procedures
PROCEEDING	1316.97	INITIATING JUDICIAL FORFEITURE PROCEEDING AGAINST A CONVEYANCE WITHIN 60 DAYS OF FILING A CLAIM AND COST BOND Expedited Forfeiture Proceedings For Certain Property; Administrative Functions, Practices, And Procedures
PROCEEDINGS	10.25	INITIATION OF ADMINISTRATIVE PROCEEDINGS General Administrative Procedures; Administrative Practices And Procedures
PROCEEDINGS	10.205	ELECTRONIC MEDIA COVERAGE OF PUBLIC ADMINISTRATIVE PROCEEDINGS Electronic Media Coverage Public Admin Proceedings; Guideline Policy Procedures; Administrative Practices And Procedures
PROCEEDINGS	10.206	PROCEDURES FOR ELECTRONIC MEDIA COVERAGE OF AGENCY PUBLIC ADMINISTRATIVE PROCEEDINGS Electronic Media Coverage Public Admin Proceedings; Guideline Policy Procedures; Administrative Practices
PROCEEDINGS	13.30	PROCEEDINGS OF A BOARD Hearing Procedures; Public Hearing Before A Public Board Of Inquiry
PROCEEDINGS	14.5	PURPOSE OF PROCEEDINGS BEFORE AN ADVISORY COMMITTEE General Provisions; Public Hearing Before A Public Advisory Committee
PROCEEDINGS	20.86	DISCLOSURE IN ADMINISTRATIVE OR COURT PROCEEDINGS Limitations On Exemptions; Public Information
PROCEEDINGS	1308.43	INITIATION OF PROCEEDINGS FOR RULEMAKING Hearings; Schedules Of Controlled Substances
PROCESS	1005.25	SERVICE OF PROCESS ON MANUFACTURERS Bonding And Compliance Procedures; Importation Of Electronic Products
PROCESS	1401.2	EXPEDITED PROCESS Public Availability Of Information
PROCESS CHEESE	133.10	NOTICE TO MANUFACTURERS, PACKERS, AND DISTRIBUTORS OF PASTEURIZED BLENDED CHEESE, PASTEURIZED PROCESS CHEESE, CHEESE FOOD, CHEESE SPREAD, AND RELATED FOODS General Provisions; Cheeses And Related Cheese Products
PROCESS CHEESE	133.169	PASTEURIZED PROCESS CHEESE Requirements For Specific Standardized Cheese And Related Products; Cheeses And Related Cheese Products

PROCESS CHEESE	133.170	PASTEURIZED PROCESS CHEESE WITH FRUITS, VEGETABLES, OR MEATS Requirements For Specific Standardized Cheese And Related Products; Cheeses And Related Cheese Products
PROCESS CHEESE	133.171	PASTEURIZED PROCESS PIMENTO CHEESE Requirements For Specific Standardized Cheese And Related Products; Cheeses And Related Cheese Products
PROCESS CHEESE	133.173	PASTEURIZED PROCESS CHEESE FOOD Requirements For Specific Standardized Cheese And Related Products; Cheeses And Related Cheese Products
PROCESS CHEESE	133.174	PASTEURIZED PROCESS CHEESE FOOD WITH FRUITS, VEGETABLES, OR MEATS Requirements For Specific Standardized Cheese And Related Products; Cheeses And Related Cheese Products
PROCESS CHEESE	133.179	PASTEURIZED PROCESS CHEESE SPREAD Requirements For Specific Standardized Cheese And Related Products; Cheeses And Related Cheese Products
PROCESS CHEESE	133.180	PASTEURIZED PROCESS CHEESE SPREAD WITH FRUITS, VEGETABLES, OR MEATS Requirements For Specific Standardized Cheese And Related Products; Cheeses And Related Cheese Products
PROCESS CONTROLS	120.24	PROCESS CONTROLS Pathogen Reduction; Hazard Analysis And Critical Control Point (HACCP) Systems
PROCESS CONTROLS	123.16	PROCESS CONTROLS Fish And Fishery Products
PROCESS CONTROLS	820.70	PRODUCTION AND PROCESS CONTROLS Production And Process Controls; Quality System Regulation
PROCESS VALIDATION	820.75	PROCESS VALIDATION Production And Process Controls; Quality System Regulation
PROCESS VERIFICATION	120.25	PROCESS VERIFICATION FOR CERTAIN PROCESSORS Pathogen Reduction; Hazard Analysis And Critical Control Point (HACCP) Systems
PROCESSED FOODS	170.19	PESTICIDE CHEMICALS IN PROCESSED FOODS General Provisions; Food Additives
PROCESSED FOODS	570.19	PESTICIDE CHEMICALS IN PROCESSED FOODS General Provisions; Food Additives
PROCESSED, THERMALLY	500.23	THERMALLY PROCESSED LOW-ACID FOODS PACKAGED IN HERMETICALLY SEALED CONTAINERS Specific Administrative Rulings And Decisions; General
PROCESSES	110.80	PROCESSES AND CONTROLS Production And Process Controls; Current Good Manufacturing Practice In Manufacturing, Packing, Or Holding Human Food
PROCESSES	113.83	ESTABLISHING SCHEDULED PROCESSES Production And Process Controls; Thermally Processed Low-acid Foods Packaged In Hermetically Sealed Containers
PROCESSES	114.80	PROCESSES AND CONTROLS Production And Process Controls; Acidified Foods
PROCESSES	114.83	ESTABLISHING SCHEDULED PROCESSES Production And Process Controls; Acidified Foods
PROCESSES	114.89	DEVIATIONS FROM SCHEDULED PROCESSES Production And Process Controls; Acidified Foods
PROCESSES	129.80	PROCESSES AND CONTROLS Production And Process Controls; Processing And Bottling Of Bottled Drinking Water
PROCESSES	610.9	EQUIVALENT METHODS AND PROCESSES General Provisions; General Biological Products Standards
PROCESSING	5.304	APPROVAL OF SCHOOLS PROVIDING FOOD-PROCESSING INSTRUCTION Foods And Cosmetics; Redelegations Of Authority; Delegations Of Authority And Organization
PROCESSING	21.41	PROCESSING OF REQUESTS Procedures For Notification Of And Access To Records In Privacy Act Record Systems; Protection Of Privacy
PROCESSING	108.12	MANUFACTURING, PROCESSING, OR PACKING WITHOUT A PERMIT, OR IN VIOLATION OF A PERMIT General Provisions; Emergency Permit Control
PROCESSING	108.35	THERMAL PROCESSING OF LOW-ACID FOODS PACKAGED IN HERMETICALLY SEALED CONTAINERS Specific Requirements And Conditions For Exemption From Or Compliance With An Emergency Permit; Emergency Permit Control
PROCESSING	113.87	OPERATIONS IN THE THERMAL PROCESSING ROOM Production and Process Controls; Thermally Processed Low-acid Foods Packaged In Hermetically Sealed Containers
PROCESSING	113.89	DEVIATIONS IN PROCESSING, VENTING, OR CONTROL OF CRITICAL FACTORS Production And Process Controls; Thermally Processed Low-acid Foods Packaged In Hermetically Sealed Containers

PROCESSING	113.100	PROCESSING AND PRODUCTION RECORDS Records And Reports; Thermally Processed Low-acid Foods Packaged In Hermetically Sealed Containers
PROCESSING	179.21	SOURCES OF RADIATION USED FOR INSPECTION OF FOOD, FOR INSPECTION OF PACKAGED FOOD, AND FOR CONTROLLING FOOD PROCESSING Radiation And Radiation Sources; Irradiation In The Production, Processing And Handling Of Food
PROCESSING	179.39	ULTRAVIOLET RADIATION FOR THE PROCESSING AND TREATMENT OF FOOD Radiation And Radiation Sources; Irradiation In The Production, Processing And Handling Of Food
PROCESSING	201.122	DRUGS FOR PROCESSING, REPACKING, OR MANUFACTURING Exemptions From Adequate Directions For Use; Labeling
PROCESSING	201.150	DRUGS; PROCESSING, LABELING, OR REPACKING Other Exemptions; Labeling
PROCESSING	640.16	PROCESSING Red Blood Cells; Additional Standards For Human Blood And Blood Products
PROCESSING	640.24	PROCESSING Platelets; Additional Standards For Human Blood And Blood Products
PROCESSING	640.34	PROCESSING Plasma; Additional Standards For Human Blood And Blood Products
PROCESSING	640.54	PROCESSING Cryoprecipitate; Additional Standards For Human Blood And Blood Products
PROCESSING	640.68	PROCESSING Source Plasma; Additional Standards For Human Blood And Blood Products
PROCESSING	640.81	PROCESSING Albumin (Human); Additional Standards For Human Blood And Blood Products
PROCESSING	640.91	PROCESSING Plasma Protein Fraction (Human); Additional Standards For Human Blood And Blood Products
PROCESSING	660.21	PROCESSING Blood Grouping Reagent; Additional Standards For Diagnostic Substances For Laboratory Tests
PROCESSING	660.34	PROCESSING Reagent Red Blood Cells; Additional Standards For Diagnostic Substances For Laboratory Tests
PROCESSING	660.41	PROCESSING Hepatitis B Surface Antigen; Additional Standards For Diagnostic Substances For Laboratory Tests
PROCESSING	660.51	PROCESSING Anti-human Globulin; Additional Standards For Diagnostic Substances For Laboratory Tests
PROCESSING	801.122	MEDICAL DEVICES FOR PROCESSING, REPACKING, OR MANUFACTURING Exemptions From Adequate Directions For Use; Labeling
PROCESSING	801.150	MEDICAL DEVICES; PROCESSING, LABELING, OR REPACKING Other Exemptions; Labeling
PROCESSING	808.25	PROCEDURES FOR PROCESSING AN APPLICATION Exemption Procedures; Exemption From Federal Preemption Of State & Local Medical Device Requirements
PROCESSING	862.2100	CALCULATOR/DATA PROCESSING MODULE FOR CLINICAL USE Clinical Laboratory Instruments; Clinical Chemistry And Clinical Toxicology Devices
PROCESSING	864.3010	TISSUE PROCESSING EQUIPMENT Pathology Instrumentation And Accessories; Hematology And Pathology Devices
PROCESSING	864.9100	EMPTY CONTAINER FOR THE COLLECTION AND PROCESSING OF BLOOD AND BLOOD COMPONENTS Products Used In Establishments That Manufacture Blood And Blood Products; Hematology And Pathology Devices
PROCESSING	864.9145	PROCESSING SYSTEM FOR FROZEN BLOOD Products Used In Establishments That Manufacture Blood And Blood Products; Hematology And Pathology Devices
PROCESSOR, FILM	892.1900	AUTOMATIC RADIOGRAPHIC FILM PROCESSOR Diagnostic Devices; Radiology Devices
PROCESSOR, TISSUE	864.3875	AUTOMATED TISSUE PROCESSOR Pathology Instrumentation And Accessories; Hematology And Pathology Devices
PROCESSORS	120.25	PROCESS VERIFICATION FOR CERTAIN PROCESSORS Pathogen Reduction; Hazard Analysis And Critical Control Point (HACCP) Systems
PROCHLORPERAZINE	520.1920	PROCHLORPERAZINE, ISOPROPAMIDE SUSTAINED RELEASE CAPSULES Oral Dosage Form New Animal Drugs
PROCHLORPERAZINE	520.1921	PROCHLORPERAZINE, ISOPROPAMIDE, WITH NEOMYCIN SUSTAINED-RELEASE CAPSULES Oral Dosage Form New Animal Drugs
PROCHLORPERAZINE	522.1920	PROCHLORPERAZINE, ISOPROPAMIDE FOR INJECTION Implantation Or Injectable Dosage Form New Animal Drugs

PROCONVERTIN	864.7735	PROTHROMBIN-PROCONVERTIN TEST AND THROMBOTEST Hematology Kits And Packages; Hematology And Pathology Devices
PROCUREMENT	1303.12	PROCUREMENT QUOTAS Aggregate Production And Procurement Quotas; Quotas
PROCUREMENT	1403.36	PROCUREMENT Changes, Property, and Subawards; Uniform Administrative Requirements For Grants And Coop. Agreements To State & Local Gov.
PRODUCERS	809.20	GENERAL REQUIREMENTS FOR MANUFACTURERS AND PRODUCERS OF IN VITRO DIAGNOSTIC PRODUCTS Requirements For Manufacturers And Producers; In Vitro Diagnostic Products For Human Use
PRODUCT, BIOLOGICAL	600.14	REPORTING OF BIOLOGICAL PRODUCT DEVIATIONS BY LICENSED MANUFACTURERS Establishment Standards; Biological Products: General
PRODUCT, BIOLOGICAL	601.50	CONFIDENTIALITY OF DATA AND INFORMATION IN AN INVESTIGATIONAL NEW DRUG NOTICE FOR A BIOLOGICAL PRODUCT Confidentiality Of Information; Licensing
PRODUCT, BLOOD	607.20	WHO MUST REGISTER AND SUBMIT A BLOOD PRODUCT LIST Procedures For Domestic Blood Product Establishments; Establishment Registratn+product Listing, Manufacturers Human Blood/b.products
PRODUCT, BLOOD	607.21	TIMES FOR ESTABLISHMENT REGISTRATION AND BLOOD PRODUCT LISTING Procedures For Domestic Blood Product Establishments; Establishment Registratn+product Listing, Manufacturers Human Blood/b.products
PRODUCT, BLOOD	607.25	INFORMATION REQUIRED FOR ESTABLISHMENT REGISTRATION AND BLOOD PRODUCT LISTING Procedures For Domestic Blood Product Establishments; Establishment Registratn+product Listing, Manufacturers Human Blood/b.products
PRODUCT, BLOOD	607.30	UPDATING BLOOD PRODUCT LISTING INFORMATION Procedures For Domestic Blood Product Establishments; Establishment Registratn+product Listing, Manufacturers Human Blood/b.products
PRODUCT, BLOOD	607.31	ADDITIONAL BLOOD PRODUCT LISTING INFORMATION Procedures For Domestic Blood Product Establishments; Establishment Registratn+product Listing, Manufacturers Human Blood/b.products
PRODUCT, BLOOD	607.35	NOTIFICATION OF REGISTRANT; BLOOD PRODUCT ESTABLISHMENT REGISTRATION NUMBER AND NDC LABELER CODE Procedures For Domestic Blood Product Establishments; Establishment Registratn+product Listing, Manufacturers Human Blood/b.products
PRODUCT, BLOOD	607.37	INSPECTION OF ESTABLISHMENT REGISTRATIONS AND BLOOD PRODUCT LISTINGS Procedures For Domestic Blood Product Establishments; Establishment Registratn+product Listing, Manufacturers Human Blood/b.products
PRODUCT, BLOOD	607.40	ESTABLISHMENT REGISTRATION AND BLOOD PRODUCT LISTING REQUIREMENTS FOR FOREIGN BLOOD PRODUCT ESTABLISHMENTS Procedures For Foreign Blood Product Establishments; Establishment Registratn+product Listing, Manufacturers Human Blood/b.products
PRODUCT, BLOOD	607.65	EXEMPTIONS FOR BLOOD PRODUCT ESTABLISHMENTS Exemptions; Establishment Registratn+product Listing, Manufacturers Human Blood/ b.products
PRODUCT CODES	20.115	PRODUCT CODES FOR MANUFACTURING OR SALES DATES Availability Of Specific Categories Of Records; Public Information
PRODUCT, COSMETIC	710.6	NOTIFICATION OF REGISTRANT; COSMETIC PRODUCT ESTABLISHMENT REGISTRATION NUMBER Voluntary Registration Of Cosmetic Product Establishments
PRODUCT, COSMETIC	720.7	NOTIFICATION OF PERSON SUBMITTING COSMETIC PRODUCT INGREDIENT STATEMENT Voluntary Filing Of Cosmetic Product Ingredient Composition Statements
PRODUCT COVERAGE	26.4	PRODUCT COVERAGE Specific Sector Provisions For Pharmaceutical GMPs; Mutual Recognition Of Pharmaceutical GMP Reports, Medical Device Quality System Audit Reports
PRODUCT COVERAGE	26.33	PRODUCT COVERAGE Specific Sector Provisions For Medical Devices; Mutual Recognition Of Pharmaceutical GMP Reports, Medical Device Quality System Audit Reports
PRODUCT DEFECT	20.113	VOLUNTARY PRODUCT DEFECT REPORTS Availability Of Specific Categories Of Records; Public Information
PRODUCT DEVELOPMENT	5.407	APPROVAL, DISAPPROVAL, OR WITHDRAWAL OF APPROVAL OF PRODUCT DEVELOPMENT PROTOCOLS AND APPLICATIONS FOR PREMARKET APPROVAL FOR MEDICAL DEVICES Medical Devices And Radiological Health; Redelegations Of Authority; Delegations Of Authority And Organization
PRODUCT DEVELOPMENT	814.19	PRODUCT DEVELOPMENT PROTOCOL (PDP) General; Premarket Approval Of Medical Devices

PRODUCT DEVIATIONS	606.171	REPORTING OF PRODUCT DEVIATIONS BY MANUFACTURERS, UNLICENSED BLOOD ESTABLISHMENTS, AND TRANSFUSION SERVICES Records And Reports; Current Good Manufacturing Practices For Blood And Blood Components
PRODUCT, DRUG	201.66	FORMAT AND CONTENT REQUIREMENTS FOR OTC DRUG PRODUCT LABELING Labeling Requirements For Over-the-counter Drugs; Labeling
PRODUCT, DRUG	211.82	RECEIPT AND STORAGE OF UNTESTED COMPONENTS, DRUG PRODUCT CONTAINERS, AND CLOSURES Control Of Components And Drug Product Containers And Closures; Current Good Manufacturing Practice For Finished Pharmaceuticals
PRODUCT, DRUG	211.84	TESTING AND APPROVAL OR REJECTION OF COMPONENTS, DRUG PRODUCT CONTAINERS, AND CLOSURES Control Of Components And Drug Product Containers And Closures; Current Good Manufacturing Practice For Finished Pharmaceuticals
PRODUCT, DRUG	211.86	USE OF APPROVED COMPONENTS, DRUG PRODUCT CONTAINERS, AND CLOSURES Control Of Components And Drug Product Containers And Closures; Current Good Manufacturing Practice For Finished Pharmaceuticals
PRODUCT, DRUG	211.87	RETESTING OF APPROVED COMPONENTS, DRUG PRODUCT CONTAINERS, AND CLOSURES Control Of Components And Drug Product Containers And Closures; Current Good Manufacturing Practice For Finished Pharmaceuticals
PRODUCT, DRUG	211.89	REJECTED COMPONENTS, DRUG PRODUCT CONTAINERS, AND CLOSURES Control Of Components And Drug Product Containers And Closures; Current Good Manufacturing Practice For Finished Pharmaceuticals
PRODUCT, DRUG	211.94	DRUG PRODUCT CONTAINERS AND CLOSURES Control Of Components And Drug Product Containers And Closures; Current Good Manufacturing Practice For Finished Pharmaceuticals
PRODUCT, DRUG	211.134	DRUG PRODUCT INSPECTION Packaging And Labeling Control; Current Good Manufacturing Practice For Finished Pharmaceuticals
PRODUCT, DRUG	211.184	COMPONENT, DRUG PRODUCT CONTAINER, CLOSURE, AND LABELING RECORDS Records And Reports; Current Good Manufacturing Practice For Finished Pharmaceuticals
PRODUCT, DRUG	211.208	DRUG PRODUCT SALVAGING Returned And Salvaged Drug Products; Current Good Manufacturing Practice For Finished Pharmaceuticals
PRODUCT, DRUG	314.108	NEW DRUG PRODUCT EXCLUSIVITY Applications For FDA Approval To Market A New Drug
PRODUCT, DRUG	314.162	REMOVAL OF A DRUG PRODUCT FROM THE LIST Applications For FDA Approval To Market A New Drug
PRODUCT, ELECTRONIC	14.120	ESTABLISHMENT OF THE TECHNICAL ELECTRONIC PRODUCT RADIATION SAFETY STANDARD COMMITTEE (TEPRSSC) Technical Electronic Products Radiation Safety Standards Committee; Public Hearing Before A Public Advisory Committee
PRODUCT, ELECTRONIC	1003.2	DEFECT IN AN ELECTRONIC PRODUCT General Provisions; Notification Of Defects Or Failure To Comply
PRODUCT EVALUATION	26.42	EXCHANGE AND ENDORSEMENT OF PRODUCT EVALUATION REPORTS Specific Sector Provisions For Medical Devices; Mutual Recognition Of Pharmaceutical GMP Reports, Medical Device Quality System Audit Reports
PRODUCT EVALUATION	26.44	TRANSMISSION OF PRODUCT EVALUATION REPORTS Specific Sector Provisions For Medical Devices; Mutual Recognition Of Pharmaceutical GMP Reports, Medical Device Quality System Audit Reports
PRODUCT, FINAL	640.82	TESTS ON FINAL PRODUCT Albumin (Human); Additional Standards For Human Blood And Blood Products
PRODUCT, FINAL	640.92	TESTS ON FINAL PRODUCT Plasma Protein Fraction (Human); Additional Standards For Human Blood And Blood Products
PRODUCT, FINAL	640.103	THE FINAL PRODUCT Immune Globulin (Human); Additional Standards For Human Blood And Blood Products
PRODUCT, FINISHED	106.30	FINISHED PRODUCT EVALUATION Quality Control Procedures For Assuring Nutrient Content Of Infant Formulas; Infant Formula Quality Control Procedures
PRODUCT JURISDICTION	3.6	PRODUCT JURISDICTION OFFICER Product Jurisdiction
PRODUCT LISTING	207.7	ESTABLISHMENT REGISTRATION AND PRODUCT LISTING FOR HUMAN BLOOD AND BLOOD PRODUCTS AND FOR MEDICAL DEVICES General; Registration Of Drug Producers And Listing Of Drugs In Commercial Distribution
PRODUCT, NONCONFORMING	820.90	NONCONFORMING PRODUCT Quality System Regulation

PRODUCT PREPARATION	113.81	PRODUCT PREPARATION Production And Process Controls; Thermally Processed Low-acid Foods Packaged In Hermetically Sealed Containers
PRODUCT, PRESCRIPTION	1308.31	APPLICATION FOR EXEMPTION OF A NONNARCOTIC PRESCRIPTION PRODUCT Exempted Prescription Products; Schedules Of Controlled Substances
PRODUCT REPORTS	1002.10	ABBREVIATED REPORTS Required Manufacturers' Reports For Listed Electronic Products; Records And Reports
PRODUCTION	211.111	TIME LIMITATIONS ON PRODUCTION Production And Process Controls; Current Good Manufacturing Practice For Finished Pharmaceuticals
PRODUCTION	500.45	USE OF POLYCHLORINATED BIPHENYLS (PCB'S) IN THE PRODUCTION, HANDLING AND STORAGE OF ANIMAL FEED Specific Administrative Rulings And Decisions; General
PRODUCTION, BATCH	226.302	MASTER-FORMULA AND BATCH-PRODUCTION RECORDS Records And Reports; Current Good Manufacturing Practice For Type A Medicated Articles
PRODUCTION CONTROLS	820.70	PRODUCTION AND PROCESS CONTROLS Production And Process Controls; Quality System Regulation
PRODUCTION PROCEDURES	226.40	PRODUCTION AND CONTROL PROCEDURES Product Quality Control; Current Good Manufacturing Practice For Type A Medicated Articles
PRODUCTION QUOTAS	1303.11	AGGREGATE PRODUCTION QUOTAS Aggregate Production And Procurement Quotas; Quotas
PRODUCTION QUOTAS	1303.13	ADJUSTMENT OF AGGREGATE PRODUCTION QUOTAS Aggregate Production And Procurement Quotas; Quotas
PRODUCTION RECORD	211.192	PRODUCTION RECORD REVIEW Records And Reports; Current Good Manufacturing Practice For Finished Pharmaceuticals
PRODUCTION OF RECORDS	20.2	PRODUCTION OF RECORDS BY FDA EMPLOYEES Official Testimony And Information; Public Information
PRODUCTION RECORDS	113.100	PROCESSING AND PRODUCTION RECORDS Records And Reports; Thermally Processed Low-acid Foods Packaged In Hermetically Sealed Containers
PRODUCTION RECORDS	211.186	MASTER PRODUCTION AND CONTROL RECORDS Records And Reports; Current Good Manufacturing Practice For Finished Pharmaceuticals
PRODUCTION RECORDS	211.188	BATCH PRODUCTION AND CONTROL RECORDS Records And Reports; Current Good Manufacturing Practice For Finished Pharmaceuticals
PRODUCTION RECORDS	225.102	MASTER RECORD FILE AND PRODUCTION RECORDS Records And Reports; Current Good Manufacturing Practice For Medicated Feeds
PRODUCTION RECORDS	225.202	FORMULA, PRODUCTION, AND DISTRIBUTION RECORDS Records; Current Good Manufacturing Practice For Medicated Feeds
PRODUCTS, ALLERGENIC	680.1	ALLERGENIC PRODUCTS Additional Standards For Miscellaneous Products
PRODUCTS, ALLERGENIC	680.2	MANUFACTURE OF ALLERGENIC PRODUCTS Additional Standards For Miscellaneous Products
PRODUCTS, ANTACID	331.30	LABELING OF ANTACID PRODUCTS Antacid Products For Over-the-counter (OTC) Human Use
PRODUCTS, ANTIFLATULENT	332.30	LABELING OF ANTIFLATULENT PRODUCTS Antiflatulent Products For Over-the-counter Human Use
PRODUCTS, BIOLOGICAL	5.203	ISSUANCE AND REVOCATION OF LICENSES FOR THE PROPAGATION OR MANUFACTURE AND PREPARATION OF BIOLOGICAL PRODUCTS Biologics; Redelegations Of Authority; Delegations Of Authority And Organization
PRODUCTS, BIOLOGICAL	5.204	NOTIFICATION OF RELEASE FOR DISTRIBUTION OF BIOLOGICAL PRODUCTS Biologics; Redelegations Of Authority; Delegations Of Authority And Organization
PRODUCTS, BIOLOGICAL	601.25	REVIEW PROCEDURES TO DETERMINE THAT LICENSED BIOLOGICAL PRODUCTS ARE SAFE, EFFECTIVE, NOT MISBRANDED UNDER PRESCRIBED, RECOMMENDED, OR SUGGESTED CONDITIONS OF USE Biologics Licensing; Licensing
PRODUCTS, BIOLOGICAL	601.26	RECLASSIFICATION PROCEDURES TO DETERMINE THAT LICENSED BIOLOGICAL PRODUCTS ARE SAFE, EFFECTIVE, NOT MISBRANDED UNDER PRESCRIBED, RECOMMENDED, OR SUGGESTED CONDITIONS OF USE Biologics Licensing; Licensing
PRODUCTS, BIOLOGICAL	601.50	CONFIDENTIALITY OF DATA AND INFORMATION IN AN INVESTIGATIONAL NEW DRUG NOTICE FOR A BIOLOGICAL PRODUCT Confidentiality Of Information; Licensing
PRODUCTS, BIOLOGICAL	610.53	DATING PERIODS FOR LICENSED BIOLOGICAL PRODUCTS Dating Period Limitations; General Biological Products Standards

PRODUCTS, BLOOD	207.7	ESTABLISHMENT REGISTRATION AND PRODUCT LISTING FOR HUMAN BLOOD AND BLOOD PRODUCTS AND FOR MEDICAL DEVICES General; Registration Of Drug Producers And Listing Of Drugs In Commercial Distribution
PRODUCTS, BLOOD	607.7	ESTABLISHMENT REGISTRATION AND PRODUCT LISTING OF BLOOD BANKS AND OTHER FIRMS MANUFACTURING HUMAN BLOOD AND BLOOD PRODUCTS General Provisions; Establishment Registratn+product Listing, Manufacturers Human Blood/b.products
PRODUCTS, BLOOD	607.22	HOW AND WHERE TO REGISTER ESTABLISHMENTS AND LIST BLOOD PRODUCTS Procedures For Domestic Blood Product Establishments; Establishment Registratn+product Listing, Manufacturers Human Blood/b.products
PRODUCTS, CANNABIS PLANT	1308.35	EXEMPTION OF CERTAIN CANNABIS PLANT MATERIAL, AND PRODUCTS MADE THEREFROM, THAT CONTAIN TETRAHYDROCANNABINOIDS Schedules Of Controlled Substances
PRODUCTS, CELLULAR	1271.22	HOW AND WHERE DO I REGISTER AND SUBMIT AN HCT/P (HUMAN CELLS/TISSUES/PRODUCTS) LIST? Procedures For Registration And Listing; Human Cells, Tissues, And Cellular And Tissue-Based Products
PRODUCTS, COSMETIC	250.250	HEXACHLOROPHENE, AS A COMPONENT OF DRUG AND COSMETIC PRODUCTS Requirements For Drugs And Cosmetics; Specific Requirements For Specific Human Drugs
PRODUCTS, COSMETIC	700.14	USE OF VINYL CHLORIDE AS AN INGREDIENT, INCLUDING PROPELLANT OF COSMETIC AEROSOL PRODUCTS Requirements For Specific Cosmetic Products; General
PRODUCTS, COSMETIC	700.15	USE OF CERTAIN HALOGENATED SALICYLANILIDES AS INGREDIENTS IN COSMETIC PRODUCTS Requirements For Specific Cosmetic Products; General
PRODUCTS, COSMETIC	700.16	USE OF AEROSOL COSMETIC PRODUCTS CONTAINING ZIRCONIUM Requirements For Specific Cosmetic Products; General
PRODUCTS, COSMETIC	700.18	USE OF CHLOROFORM AS AN INGREDIENT IN COSMETIC PRODUCTS Requirements For Specific Cosmetic Products; General
PRODUCTS, COSMETIC	700.19	USE OF METHYLENE CHLORIDE AS AN INGREDIENT IN COSMETIC PRODUCTS Requirements For Specific Cosmetic Products; General
PRODUCTS, COSMETIC	700.25	TAMPER-RESISTANT PACKAGING REQUIREMENTS FOR COSMETIC PRODUCTS Requirements For Specific Cosmetic Products; General
PRODUCTS, COSMETIC	720.4	INFORMATION REQUESTED ABOUT COSMETIC PRODUCTS Voluntary Filing Of Cosmetic Product Ingredient Composition Statements
PRODUCTS, COSMETIC	740.10	LABELING OF COSMETIC PRODUCTS FOR WHICH ADEQUATE SUBSTANTIATION OF SAFETY HAS NOT BEEN OBTAINED Warning Statements; Cosmetic Product Warning Statements
PRODUCTS, DEVELOPMENT	601.21	PRODUCTS UNDER DEVELOPMENT Biologics Licensing; Licensing
PRODUCTS, DIAGNOSTIC	201.119	IN VITRO DIAGNOSTIC PRODUCTS Exemptions From Adequate Directions For Use; Labeling
PRODUCTS, DIAGNOSTIC	801.119	IN VITRO DIAGNOSTIC PRODUCTS Exemptions From Adequate Directions For Use; Labeling
PRODUCTS, DIAGNOSTIC	809.10	LABELING FOR IN VITRO DIAGNOSTIC PRODUCTS Labeling; In Vitro Diagnostic Products For Human Use
PRODUCTS, DIAGNOSTIC	809.20	GENERAL REQUIREMENTS FOR MANUFACTURERS AND PRODUCERS OF IN VITRO DIAGNOSTIC PRODUCTS Requirements For Manufacturers And Producers; In Vitro Diagnostic Products For Human Use
PRODUCTS, DEVICE	801.43	WARNING STATEMENTS FOR PRESCRIPTION AND RESTRICTED DEVICE PRODUCTS CONTAINING OR MANUFACTURED WITH CHLOROFLUOROCARBONS Specific Requirements For Specific Devices; Labeling
PRODUCTS, DRUG	5.27	PATENT TERM EXTENSIONS FOR HUMAN DRUG PRODUCTS, MEDICAL DEVICES, AND FOOD AND COLOR ADDITIVES; AND AUTHORITY TO PERFORM DUE DILIGENCE DETERMINATIONS AND INFORMAL HEARINGS General Redelegations Of Authority; Delegations Of Authority And Organization
PRODUCTS, DRUG	200.51	AQUEOUS-BASED DRUG PRODUCTS FOR ORAL INHALATION Requirements For Specific Classes Of Drugs; General
PRODUCTS, DRUG	201.312	MAGNESIUM SULFATE HEPTAHYDRATE; LABEL DECLARATION ON DRUG PRODUCTS Specific Labeling Requirements For Specific Drug Products; Labeling
PRODUCTS, DRUG	201.320	WARNING STATEMENTS FOR DRUG PRODUCTS CONTAINING OR MANUFACTURED WITH CHLOROFLUOROCARBONS Specific Labeling Requirements For Specific Drug Products; Labeling

PRODUCTS, DRUG	201.322	OVER-THE-COUNTER DRUG PRODUCTS CONTAINING INTERNAL ANALGESIC/ANTIPYRETIC ACTIVE INGREDIENTS; REQUIRED ALCOHOL WARNING Specific Labeling Requirements For Specific Drug Products; Labeling
PRODUCTS, DRUG	211.110	SAMPLING AND TESTING OF IN-PROCESS MATERIALS AND DRUG PRODUCTS Equipment; Current Good Manufacturing Practice For Finished Pharmaceuticals
PRODUCTS, DRUG	211.132	TAMPER-EVIDENT PACKAGING REQUIREMENTS FOR OTC HUMAN DRUG PRODUCTS Current Good Manufacturing Practice For Finished Pharmaceuticals
PRODUCTS, DRUG	211.204	RETURNED DRUG PRODUCTS Returned And Salvaged Drug Products; Current Good Manufacturing Practice For Finished Pharmaceuticals
PRODUCTS, DRUG	216.24	DRUG PRODUCTS WITHDRAWN OR REMOVED FROM THE MARKET FOR REASONS OF SAFETY OR EFFECTIVENESS Pharmacy Compounding
PRODUCTS, DRUG	250.250	HEXACHLOROPHENE, AS A COMPONENT OF DRUG AND COSMETIC PRODUCTS Requirements For Drugs And Cosmetics; Specific Requirements For Specific Human Drugs
PRODUCTS, DRUG	310.6	APPLICABILITY OF "NEW DRUG" OR SAFETY/EFFECTIVENESS FINDINGS IN DRUG EFFICACY STUDY IMPLEMENTATION NOTICES AND NOTICES OF OPPORTUNITY FOR HEARING TO IDENTICAL, RELATED, AND SIMILAR DRUG PRODUCTS General Provisions; New Drugs
PRODUCTS, DRUG	310.509	PARENTERAL DRUG PRODUCTS IN PLASTIC CONTAINERS Requirements For Specific New Drugs Or Devices; New Drugs
PRODUCTS, DRUG	310.518	DRUG PRODUCTS CONTAINING IRON OR IRON SALTS Requirements For Specific New Drugs Or Devices; New Drugs
PRODUCTS, DRUG	310.519	DRUG PRODUCTS MARKETED AS OTC DAYTIME SEDATIVES Requirements For Specific New Drugs Or Devices; New Drugs
PRODUCTS, DRUG	310.527	DRUG PRODUCTS CONTAINING ACTIVE INGREDIENTS OFFERED OTC FOR EXTERNAL USE A HAIR GROWERS OR FOR HAIR LOSS PREVENTION Requirements For Specific New Drugs Or Devices; New Drugs
PRODUCTS, DRUG	310.528	DRUG PRODUCTS CONTAINING ACTIVE INGREDIENTS OFFERED OTC FOR USE AS AN APHRODISIAC Requirements For Specific New Drugs Or Devices; New Drugs
PRODUCTS, DRUG	310.529	DRUG PRODUCTS CONTAINING ACTIVE INGREDIENTS OFFERED OTC FOR ORAL USE AS INSECT REPELLENTS Requirements For Specific New Drugs Or Devices; New Drugs
PRODUCTS, DRUG	310.530	TOPICALLY APPLIED HORMONE-CONTAINING DRUG PRODUCTS FOR OTC HUMAN USE Requirements For Specific New Drugs Or Devices; New Drugs
PRODUCTS, DRUG	310.531	DRUG PRODUCTS CONTAINING ACTIVE INGREDIENTS OFFERED OTC FOR THE TREATMENT OF BOILS Requirements For Specific New Drugs Or Devices; New Drugs
PRODUCTS, DRUG	310.532	DRUG PRODUCTS CONTAINING ACTIVE INGREDIENTS OFFERED OTC TO RELIEVE THE SYMPTOMS OF BENIGN PROSTATIC HYPERTROPHY Requirements For Specific New Drugs Or Devices; New Drugs
PRODUCTS, DRUG	310.533	DRUG PRODUCTS CONTAINING ACTIVE INGREDIENTS OFFERED OTC FOR HUMAN USE AS AN ANTICHOLINERGIC IN COUGH-COLD DRUG PRODUCTS Requirements For Specific New Drugs Or Devices; New Drugs
PRODUCTS, DRUG	310.534	DRUG PRODUCTS CONTAINING ACTIVE INGREDIENTS OFFERED OTC FOR HUMAN USE AS ORAL WOUND HEALING AGENTS Requirements For Specific New Drugs Or Devices; New Drugs
PRODUCTS, DRUG	310.536	DRUG PRODUCTS CONTAINING ACTIVE INGREDIENTS OFFERED OTC FOR USE AS A NAILBITING OR THUMBSUCKING DETERRENT Requirements For Specific New Drugs Or Devices; New Drugs
PRODUCTS, DRUG	310.537	DRUG PRODUCTS CONTAINING ACTIVE INGREDIENTS OFFERED OTC FOR TREATMENT OF FEVER BLISTERS AND COLD SORES Requirements For Specific New Drugs Or Devices; New Drugs
PRODUCTS, DRUG	310.538	DRUG PRODUCTS CONTAINING ACTIVE INGREDIENTS OFFERED OTC FOR USE FOR INGROWN TOENAIL RELIEF Requirements For Specific New Drugs Or Devices; New Drugs
PRODUCTS, DRUG	310.540	DRUG PRODUCTS CONTAINING ACTIVE INGREDIENTS OFFERED OTC FOR USE AS STOMACH ACIDIFIERS Requirements For Specific New Drugs Or Devices; New Drugs
PRODUCTS, DRUG	310.541	OTC DRUG PRODUCTS CONTAINING ACTIVE INGREDIENTS OFFERED FOR USE IN THE TREATMENT OF HYPOPHOSPHATEMIA Requirements For Specific New Drugs Or Devices; New Drugs
PRODUCTS, DRUG	310.542	OTC DRUG PRODUCTS CONTAINING ACTIVE INGREDIENTS OFFERED FOR USE IN THE TREATMENT OF HYPERPHOSPHATEMIA Requirements For Specific New Drugs Or Devices; New Drugs

PRODUCTS, DRUG	310.543	DRUG PRODUCTS CONTAINING ACTIVE INGREDIENTS OFFERED OTC FOR HUMAN USE IN EXOCRINE PANCREATIC INSUFFICIENCY Requirements For Specific New Drugs Or Devices; New Drugs
PRODUCTS, DRUG	310.544	DRUG PRODUCTS CONTAINING ACTIVE INGREDIENTS OFFERED OTC FOR USE AS A SMOKING DETERRENT Requirements For Specific New Drugs Or Devices; New Drugs
PRODUCTS, DRUG	310.545	DRUG PRODUCTS CONTAINING CERTAIN ACTIVE INGREDIENTS OFFERED OTC FOR CERTAIN USES Requirements For Specific New Drugs Or Devices; New Drugs
PRODUCTS, DRUG	310.546	DRUG PRODUCTS CONTAINING ACTIVE INGREDIENTS OFFERED OTC FOR THE TREATMENT AND/OR PREVENTION OF NOCTURNAL LEG MUSCLE CRAMPS Requirements For Specific New Drugs Or Devices; New Drugs
PRODUCTS, DRUG	310.547	DRUG PRODUCTS CONTAINING QUININE OFFERED OTC FOR THE TREATMENT AND/OR PREVENTION OF MALARIA Requirements For Specific New Drugs Or Devices; New Drugs
PRODUCTS, DRUG	310.548	DRUG PRODUCTS CONTAINING COLLOIDAL SILVER INGREDIENTS OFFERED OTC FOR THE TREATMENT AND/OR PREVENTION OF DISEASE Requirements For Specific New Drugs Or Devices; New Drugs
PRODUCTS, DRUG	314.92	DRUG PRODUCTS FOR WHICH ABBREVIATED APPLICATIONS MAY BE SUBMITTED Applications For FDA Approval To Market A New Drug
PRODUCTS, DRUG	328.50	PRINCIPAL DISPLAY PANEL OF ALL OTC DRUG PRODUCTS INTENDED FOR ORAL INGESTION THAT CONTAIN ALCOHOL Over-the-counter Drug Products Intended For Oral Ingestion That Contain Alcohol
PRODUCTS, DRUG	330.3	IMPRINTING OF SOLID ORAL DOSAGE FORM DRUG PRODUCTS Over-the-counter Human Drugs Generally Recognized As Safe And Effective & Not Misbranded
PRODUCTS, DRUG	333.150	LABELING OF FIRST AID ANTIBIOTIC DRUG PRODUCTS Topical Antimicrobial Drug Products For Over-the-counter Human Use
PRODUCTS, DRUG	333.250	LABELING OF ANTIFUNGAL DRUG PRODUCTS Topical Antimicrobial Drug Products For Over-the-counter Human Use
PRODUCTS, DRUG	333.350	LABELING OF ACNE DRUG PRODUCTS Topical Antimicrobial Drug Products For Over-the-counter Human Use
PRODUCTS, DRUG	336.50	LABELING OF ANTIEMETIC DRUG PRODUCTS Antiemetic Drug Products For Over-the-counter Human Use
PRODUCTS, DRUG	338.50	LABELING OF NIGHTTIME SLEEP-AID DRUG PRODUCTS Nighttime Sleep-aid Drug Products For Over-the-counter Human Use
PRODUCTS, DRUG	340.50	LABELING OF STIMULANT DRUG PRODUCTS Stimulant Drug Products For Over-the-counter Human Use
PRODUCTS, DRUG	341.70	LABELING OF OTC DRUG PRODUCTS CONTAINING INGREDIENTS USED FOR TREATING CONCURRENT SYMPTOMS Cold, Cough, Allergy, Bronchodilator, And Antiasthmatic Drug Products For OTC Human Use
PRODUCTS, DRUG	341.72	LABELING OF ANTIHISTAMINE DRUG PRODUCTS Cold, Cough, Allergy, Bronchodilator, And Antiasthmatic Drug Products For OTC Human Use
PRODUCTS, DRUG	341.74	LABELING OF ANTITUSSIVE DRUG PRODUCTS Cold, Cough, Allergy, Bronchodilator, And Antiasthmatic Drug Products For OTC Human Use
PRODUCTS, DRUG	341.76	LABELING OF BRONCHODILATOR DRUG PRODUCTS Cold, Cough, Allergy, Bronchodilator, And Antiasthmatic Drug Products For OTC Human Use
PRODUCTS, DRUG	341.78	LABELING OF EXPECTORANT DRUG PRODUCTS Cold, Cough, Allergy, Bronchodilator, And Antiasthmatic Drug Products For OTC Human Use
PRODUCTS, DRUG	341.80	LABELING OF NASAL DECONGESTANT DRUG PRODUCTS Cold, Cough, Allergy, Bronchodilator, And Antiasthmatic Drug Products For OTC Human Use
PRODUCTS, DRUG	344.50	LABELING OF EARWAX REMOVAL AID DRUG PRODUCTS Topical Otic Over-the-counter Products For Human Use
PRODUCTS, DRUG	344.52	LABELING OF EAR DRYING AID DRUG PRODUCTS Topical Otic Over-the-counter Products For Human Use
PRODUCTS, DRUG	346.52	LABELING OF ANORECTAL DRUG PRODUCTS Anorectal Drug Products For Over-the-counter Human Use
PRODUCTS, DRUG	347.50	LABELING OF ASTRINGENT DRUG PRODUCTS Skin Protectant Drug Products For Over-the-counter Human Use
PRODUCTS, DRUG	348.50	LABELING OF EXTERNAL ANALGESIC DRUG PRODUCTS External Analgesic Drug Products For Over-the-counter Human Use
PRODUCTS, DRUG	349.50	LABELING OF OPHTHALMIC DRUG PRODUCTS Ophthalmic Drug Products For Over-the-counter Human Use
PRODUCTS, DRUG	349.55	LABELING OF OPHTHALMIC ASTRINGENT DRUG PRODUCTS Ophthalmic Drug Products For Over-the-counter Human Use

PRODUCTS, DRUG	349.60	LABELING OF OPHTHALMIC DEMULCENT DRUG PRODUCTS Ophthalmic Drug Products For Over-the-counter Human Use
PRODUCTS, DRUG	349.65	LABELING OF OPHTHALMIC EMOLLIENT DRUG PRODUCTS Ophthalmic Drug Products For Over-the-counter Human Use
PRODUCTS, DRUG	349.70	LABELING OF OPHTHALMIC HYPERTONICITY DRUG PRODUCTS Ophthalmic Drug Products For Over-the-counter Human Use
PRODUCTS, DRUG	349.75	LABELING OF OPHTHALMIC VASOCONSTRICTOR DRUG PRODUCTS Ophthalmic Drug Products For Over-the-counter Human Use
PRODUCTS, DRUG	349.78	LABELING OF OPHTHALMIC EYEWASH DRUG PRODUCTS Ophthalmic Drug Products For Over-the-counter Human Use
PRODUCTS, DRUG	352.50	PRINCIPAL DISPLAY PANEL OF ALL SUNSCREEN DRUG PRODUCTS Sunscreen Drug Products For Over-the-counter Human Use
PRODUCTS, DRUG	352.52	LABELING OF SUNSCREEN DRUG PRODUCTS Sunscreen Drug Products For Over-the-counter Human Use
PRODUCTS, DRUG	355.50	LABELING OF ANTICARIES DRUG PRODUCTS Anticaries Drug Products For Over-the-counter Human Use
PRODUCTS, DRUG	355.55	PRINCIPAL DISPLAY PANEL OF ALL FLUORIDE RINSE DRUG PRODUCTS Anticaries Drug Products For Over-the-counter Human Use
PRODUCTS, DRUG	355.70	TESTING PROCEDURES FOR FLUORIDE DENTIFRICE DRUG PRODUCTS Anticaries Drug Products For Over-the-counter Human Use
PRODUCTS, DRUG	357.110	LABELING OF ANTHELMINTIC DRUG PRODUCTS Anthelmintic Drug Products; Miscellaneous Internal Drug Products For OTC Human Use
PRODUCTS, DRUG	357.152	PACKAGE INSERTS FOR ANTHELMINTIC DRUG PRODUCTS Anthelmintic Drug Products; Miscellaneous Internal Drug Products For OTC Human Use
PRODUCTS, DRUG	357.250	LABELING OF CHOLECYSTOKINETIC DRUG PRODUCTS Cholecystokinetic Drug Products; Miscellaneous Internal Drug Products For OTC Human Use
PRODUCTS, DRUG	357.810	ACTIVE INGREDIENTS FOR DEODORANT DRUG PRODUCTS FOR INTERNAL USE Deodorant Drug Products For Internal Use; Miscellaneous Internal Drug Products For OTC Human Use
PRODUCTS, DRUG	357.850	LABELING OF DEODORANT DRUG PRODUCTS FOR INTERNAL USE Deodorant Drug Products For Internal Use; Miscellaneous Internal Drug Products For OTC Human Use
PRODUCTS, DRUG	358.150	LABELING OF WART REMOVER DRUG PRODUCTS Miscellaneous External Drug Products For Over-the-counter Human Use
PRODUCTS, DRUG	358.550	LABELING OF CORN AND CALLUS REMOVER DRUG PRODUCTS Miscellaneous External Drug Products For Over-the-counter Human Use
PRODUCTS, DRUG	358.650	LABELING OF PEDICULICIDE DRUG PRODUCTS Miscellaneous External Drug Products For Over-the-counter Human Use
PRODUCTS, DRUG	358.750	LABELING OF DRUG PRODUCTS FOR THE CONTROL OF DANDRUFF, SEBORRHEIC DERMATITIS Miscellaneous External Drug Products For Over-the-counter Human Use
PRODUCTS, DRUG	1309.28	EXEMPTION OF DISTRIBUTORS OF REGULATED PRESCRIPTION DRUG PRODUCTS Registration Of Manufacturers, Distributors, Importers And Exporters Of List I Chemicals
PRODUCTS, DRUG	1310.11	REINSTATEMENT OF THE EXEMPTION OF DRUG PRODUCTS DISTRIBUTED UNDER THE FOOD DRUG AND COSMETIC ACT Records And Reports Of Listed Chemicals And Certain Machines
PRODUCTS, DRUG	1310.14	EXEMPTION OF DRUG PRODUCTS CONTAINING EPHEDRINE AND ANOTHER ACTIVE MEDICINAL INGREDIENT Records And Reports Of Listed Chemicals And Certain Machines
PRODUCTS, DRUG	1310.15	EXEMPT DRUG PRODUCTS CONTAINING EPHEDRINE AND ANOTHER ACTIVE MEDICINAL INGREDIENT Records And Reports Of Listed Chemicals And Certain Machines
PRODUCTS, ELECTRONIC	5.600	VARIANCES FROM PERFORMANCE STANDARDS FOR ELECTRONIC PRODUCTS Radiation Control; Redelegations Of Authority; Delegations Of Authority And Organization
PRODUCTS, ELECTRONIC	5.601	EXEMPTION OF ELECTRONIC PRODUCTS FROM PERFORMANCE STANDARDS AND PROHIBITED ACTS Radiation Control; Redelegations Of Authority; Delegations Of Authority And Organization
PRODUCTS, ELECTRONIC	5.602	TESTING PROGRAMS AND METHODS OF CERTIFICATION AND IDENTIFICATION FOR ELECTRONIC PRODUCTS Radiation Control; Redelegations Of Authority; Delegations Of Authority And Organization
PRODUCTS, ELECTRONIC	5.603	NOTIFICATION OF DEFECTS IN, AND REPAIR OR REPLACEMENT OF, ELECTRONIC PRODUCTS Radiation Control; Redelegations Of Authority; Delegations Of Authority And Organization

PRODUCTS, NOODLE	139.155	ENRICHED NOODLE PRODUCTS Requirements For Specific Standardized Macaroni And Noodle Products; Macaroni And Noodle Products
PRODUCTS, NOODLE	139.160	VEGETABLE NOODLE PRODUCTS Requirements For Specific Standardized Macaroni And Noodle Products; Macaroni And Noodle Products
PRODUCTS, NOODLE	139.165	ENRICHED VEGETABLE NOODLE PRODUCTS Requirements For Specific Standardized Macaroni And Noodle Products; Macaroni And Noodle Products
PRODUCTS, NOODLE	139.180	WHEAT AND SOY NOODLE PRODUCTS Requirements For Specific Standardized Macaroni And Noodle Products; Macaroni And Noodle Products
PRODUCTS, ORPHAN	5.900	ORPHAN PRODUCTS Orphan Products; Redelegations Of Authority; Delegations Of Authority And Organization
PRODUCTS, PRESCRIPTION	1308.32	EXEMPTED PRESCRIPTION PRODUCTS Exempted Prescription Products; Schedules Of Controlled Substances
PRODUCTS, SHORT SUPPLY	601.22	PRODUCTS IN SHORT SUPPLY; INITIAL MANUFACTURING AT OTHER THAN LICENSED ESTABLISHMENT Biologics Licensing; Licensing
PRODUCTS, TISSUE-BASED	1271.22	HOW AND WHERE DO I REGISTER AND SUBMIT AN HCT/P (HUMAN CELLS/TISSUES/PRODUCTS) LIST? Procedures For Registration And Listing; Human Cells, Tissues, And Cellular And Tissue-Based Products
PRODUCTS, TRANSITIONAL	860.136	PROCEDURES FOR TRANSITIONAL PRODUCTS UNDER SECTION 520(L) OF THE ACT Reclassification; Medical Device Classification Procedures
PROFESSIONAL LABELING	331.80	PROFESSIONAL LABELING Antacid Products For Over-the-counter (OTC) Human Use
PROFESSIONAL LABELING	332.31	PROFESSIONAL LABELING Antiflatulent Products For Over-the-counter Human Use
PROFESSIONAL LABELING	333.280	LABELING PROFESSIONAL Topical Antifungal Drug Products; Topical Antimicrobial Drug Products For OTC Human Use
PROFESSIONAL LABELING	336.80	PROFESSIONAL LABELING Antiemetic Drug Products For Over-the-counter Human Use
PROFESSIONAL LABELING	341.90	PROFESSIONAL LABELING Cold, Cough, Allergy, Bronchodilator, Antiasthmatic Drug Products OTC Human Use
PROFESSIONAL LABELING	343.80	PROFESSIONAL LABELING Internal Analgesic, Antipyretic And Antirheumatic Drug Products For OTC Human Use
PROFESSIONAL LABELING	349.80	PROFESSIONAL LABELING Ophthalmic Drug Products For Over-the-counter Human Use
PROFESSIONAL LABELING	355.60	PROFESSIONAL LABELING Anticaries Drug Products For Over-the-counter Human Use
PROFESSIONAL LABELING	357.180	PROFESSIONAL LABELING Anthelmintic Drug Products; Miscellaneous Internal Drug Products For OTC Human Use
PROFESSIONAL LABELING	357.280	PROFESSIONAL LABELING Cholecystokinetic Drug Products; Miscellaneous Internal Drug Products For OTC Human Use
PROFESSIONAL LABELING	801.420	HEARING AID DEVICES; PROFESSIONAL AND PATIENT LABELING Special Requirements For Specific Devices; Labeling
PROGESTERONE	522.1940	PROGESTERONE AND ESTRADIOL BENZOATE IN COMBINATION Implantation Or Injectable Dosage Form New Animal Drugs
PROGESTERONE	556.540	PROGESTERONE Specific Tolerances For Residues Of New Animal Drugs; Tolerances For Residues Of New Animal Drugs In Food
PROGESTERONE	862.1620	PROGESTERONE TEST SYSTEM Clinical Chemistry Test Systems; Clinical Chemistry And Clinical Toxicology Devices
PROGRAM AUTHORITIES	5.201	REDELEGATION OF THE CENTER FOR BIOLOGICS EVALUATION AND RESEARCH DIRECTOR'S PROGRAM AUTHORITIES Biologics; Redelegations Of Authority; Delegations Of Authority And Organization
PROGRAM INCOME	1403.25	PROGRAM INCOME Post-award Requirements; Uniform Administrative Requirements For Grants And Coop. Agreements To State And Local Gov.
PROGRAM PERFORMANCE	1403.40	MONITORING AND REPORTING PROGRAM PERFORMANCE Post-award Requirements; Uniform Administrative Requirements For Grants And Coop. Agreements To State And Local Gov.
PROGRAMMABLE	870.1425	PROGRAMMABLE DIAGNOSTIC COMPUTER Cardiovascular Diagnostic Devices; Cardiovascular Devices

PROGRAMMABLE	870.1750	EXTERNAL PROGRAMMABLE PACEMAKER PULSE GENERATOR Cardiovascular Diagnostic Devices; Cardiovascular Devices
PROGRAMMER	892.1870	RADIOGRAPHIC FILM/CASSETTE CHANGER PROGRAMMER Diagnostic Devices; Radiology Devices
PROGRAMMERS	870.3700	PACEMAKER PROGRAMMERS Cardiovascular Prosthetic Devices; Cardiovascular Devices
PROGRAMS, TESTING	5.25	RESEARCH, INVESTIGATION, AND TESTING PROGRAMS AND HEALTH INFORMATION AND PROMOTION PROGRAMS General Redelegations Of Authority; Delegations Of Authority And Organization
PROGRAMS, TESTING	5.602	TESTING PROGRAMS AND METHODS OF CERTIFICATION AND IDENTIFICATION FOR ELECTRONIC PRODUCTS Radiation Control; Redelegations Of Authority; Delegations Of Authority And Organization
PROGRAMS, TREATMENT	1301.72	PHYSICAL SECURITY CONTROLS FOR NON-PRACTITIONERS; NARCOTIC TREATMENT PROGRAMS AND COMPOUNDERS FOR NARCOTIC TREATMENT PROGRAMS; STORAGE AREAS Security Requirements; Registration Of Manufacturers, Distributors, And Dispensers Of Controlled Substances
PROGRAMS, TREATMENT	1301.73	PHYSICAL SECURITY CONTROLS FOR NON-PRACTITIONERS; COMPOUNDERS FOR NARCOTIC TREATMENT PROGRAMS; MANUFACTURING AND COMPOUNDING AREAS Security Requirements; Registration Of Manufacturers, Distributors, And Dispensers Of Controlled Substances
PROGRAMS, TREATMENT	1301.74	OTHER SECURITY CONTROLS FOR NON-PRACTITIONERS; NARCOTIC TREATMENT PROGRAMS AND COMPOUNDERS FOR NARCOTIC TREATMENT PROGRAMS Security Requirements; Registration Of Manufacturers, Distributors, And Dispensers Of Controlled Substances
PROGRAMS, TREATMENT	1304.24	RECORDS FOR MAINTENANCE TREATMENT PROGRAMS AND DETOXIFICATION TREATMENT PROGRAMS Continuing Records; Records And Reports Of Registrants
PROGRAMS, TREATMENT	1304.25	RECORDS FOR TREATMENT PROGRAMS WHICH COMPOUND NARCOTICS FOR TREATMENT PROGRAMS AND OTHER LOCATIONS Continuing Records; Records And Reports Of Registrants
PROGRESS REPORTS	601.70	ANNUAL PROGRESS REPORTS OF POSTMARKETING STUDIES Postmarketing Studies; Licensing
PROHIBITED	5.601	EXEMPTION OF ELECTRONIC PRODUCTS FROM PERFORMANCE STANDARDS AND PROHIBITED ACTS Radiation Control; Redelegations Of Authority; Delegations Of Authority And Organization
PROHIBITED	189.1	SUBSTANCES PROHIBITED FROM USE IN HUMAN FOOD General Provisions; Substances Prohibited From Use In Human Food
PROHIBITED	530.41	DRUGS PROHIBITED FOR EXTRALABEL USE IN ANIMALS Extralabel Drug Use In Animals
PROHIBITED	589.1	SUBSTANCES PROHIBITED FROM USE IN ANIMAL FOOD OR FEED Substances Prohibited From Use In Animal Food Or Feed
PROHIBITED	589.2000	ANIMAL PROTEINS PROHIBITED IN RUMINANT FEED Substances Prohibited From Use In Animal Food Or Feed
PROHIBITED	1005.3	IMPORTATION OF NONCOMPLYING GOODS PROHIBITED Importation Of Electronic Products
PROHIBITED	1312.29	DOMESTIC RELEASE PROHIBITED Exportation Of Controlled Substances; Importation And Exportation Of Controlled Substances
PROHIBITING	530.25	ORDERS PROHIBITING EXTRALABEL USES FOR DRUGS IN FOOD-PRODUCING ANIMALS Extralabel Drug Use In Animals
PROHIBITION	5.500	ISSUANCE OF FEDERAL REGISTER DOCUMENTS PERTAINING TO THE DETERMINATION OF SAFE LEVELS, NOTICE OF NEED FOR DEVELOPMENT OF AN ANALYTICAL METHOD, AND PROHIBITION OF CERTAIN EXTRALABEL DRUG USE Animal Drugs; Redelegations Of Authority; Delegations Of Authority And Organization
PROHIBITION	20.29	PROHIBITION ON WITHDRAWAL OF RECORDS FROM FDA FILES General Policy; Public Information
PROHIBITION	812.7	PROHIBITION OF PROMOTION AND OTHER PRACTICES General Provisions; Investigational Device Exemptions
PROHIBITIONS	530.21	PROHIBITIONS FOR FOOD-PRODUCING ANIMALS Extralabel Drug Use In Animals
PROINSULIN	862.1135	C-PEPTIDES OF PROINSULIN TEST SYSTEM Clinical Chemistry Test Systems; Clinical Chemistry And Clinical Toxicology Devices
PROJECTOR, OPHTHALMIC	886.1680	OPHTHALMIC PROJECTOR Diagnostic Devices; Ophthalmic Devices

PROLACTIN	862.1625	PROLACTIN (LACTOGEN) TEST SYSTEM Clinical Chemistry Test Systems; Clinical Chemistry And Clinical Toxicology Devices
PROLINE	582.5650	PROLINE Nutrients And/or Dietary Supplements; Substances Generally Recognized As Safe
PROMAZINE	520.1962	PROMAZINE HYDROCHLORIDE Oral Dosage Form New Animal Drugs
PROMAZINE	522.1222	KETAMINE HYDROCHLORIDE WITH PROMAZINE HYDROCHLORIDE AND AMINOPENTAMIDE HYDROGEN SULFATE INJECTION Implantation Or Injectable Dosage Form New Animal Drugs
PROMAZINE	522.1962	PROMAZINE HYDROCHLORIDE INJECTION Implantation Or Injectable Dosage Form New Animal Drugs
PROMINENCE	101.15	FOOD; PROMINENCE OF REQUIRED STATEMENTS General Provisions; Food Labeling
PROMINENCE	201.15	DRUGS; PROMINENCE OF REQUIRED LABEL STATEMENTS General Labeling Provisions; Labeling
PROMINENCE	501.15	ANIMAL FOOD; PROMINENCE OF REQUIRED STATEMENTS General Provisions; Animal Food Labeling
PROMINENCE	801.15	MEDICAL DEVICES; PROMINENCE OF REQUIRED LABEL STATEMENTS General Labeling Provisions; Labeling
PROMOTION	5.25	RESEARCH, INVESTIGATION, AND TESTING PROGRAMS AND HEALTH INFORMATION AND PROMOTION PROGRAMS General Redelegations Of Authority; Delegations Of Authority And Organization
PROMOTION	312.7	PROMOTION AND CHARGING FOR INVESTIGATIONAL DRUGS Investigational New Drug Application
PROMOTION	530.4	ADVERTISING AND PROMOTION Extralabel Drug Use In Animals
PROMOTION	812.7	PROHIBITION OF PROMOTION AND OTHER PRACTICES General Provisions; Investigational Device Exemptions
PROMOTIONAL MATERIALS	314.550	PROMOTIONAL MATERIALS Applications For FDA Approval To Market A New Drug
PROMOTIONAL MATERIALS	601.45	PROMOTIONAL MATERIALS Accelerated Approval Of Biological Products; Licensing
PROMULGATION, REGULATIONS	10.40	PROMULGATION OF REGULATIONS FOR THE EFFICIENT ENFORCEMENT OF THE LAW General Administrative Procedures; Administrative Practices And Procedures
PROMULGATION, REGULATIONS	10.50	PROMULGATION OF REGULATIONS AND ORDERS AFTER AN OPPORTUNITY FOR A FORMAL EVIDENTIARY PUBLIC HEARING General Administrative Procedures; Administrative Practices And Procedures
PROOF	1310.07	PROOF OF IDENTITY Records and Reports Of Listed Chemicals And Certain Machines
PROOF	1316.60	OBJECTIONS; OFFER OF PROOF Administrative Hearings; Administrative Functions, Practices, And Procedures
PROOF, BURDEN OF	12.87	PURPOSE; ORAL AND WRITTEN TESTIMONY; BURDEN OF PROOF Hearing Procedures; Formal Evidentiary Hearing
PROOF, BURDEN OF	17.33	THE HEARING AND BURDEN OF PROOF Civil Money Penalties Hearings
PROOF, BURDEN OF	1301.44	BURDEN OF PROOF Hearings; Registration Of Manufacturers, Distributors, And Dispensers Of Controlled Substances
PROOF, BURDEN OF	1303.36	BURDEN OF PROOF Hearings; Quotas
PROOF, BURDEN OF	1309.54	BURDEN OF PROOF Hearings; Registration Of Manufacturers, Distributors, Importers And Exporters Of List I Chemicals
PROOF, BURDEN OF	1312.45	BURDEN OF PROOF Hearings; Importation And Exportation Of Controlled Substances
PROOF, BURDEN OF	1313.55	BURDEN OF PROOF Hearings; Importation And Exportation Of Precursors And Essential Chemicals
PROOF, BURDEN OF	1316.56	BURDEN OF PROOF Administrative Hearings; Administrative Functions, Practices, And Procedures
PROPAGATION	5.203	ISSUANCE AND REVOCATION OF LICENSES FOR THE PROPAGATION OR MANUFACTURE AND PREPARATION OF BIOLOGICAL PRODUCTS Biologics; Redelegations Of Authority; Delegations Of Authority And Organization
PROPANE	184.1655	PROPANE Listing Of Specific Substances Affirmed As GRAS; Direct Food Substances Affirmed As Generally Recognized As Safe
PROPANE	582.1655	PROPANE General Purpose Food Additives; Substances Generally Recognized As Safe
PROPARACAINE	524.1982	PROPARACAINE HYDROCHLORIDE OPHTHALMIC SOLUTION Ophthalmic And Topical Dosage Form New Animal Drugs
PROPELLANT	700.14	USE OF VINYL CHLORIDE AS AN INGREDIENT, INCLUDING PROPELLANT OF COSMETIC AEROSOL PRODUCTS Requirements For Specific Cosmetic Products; General

PROPELLANTS	2.125	USE OF CHLOROFLUOROCARBON PROPELLANTS IN SELF-PRESSURIZED CONTAINERS Specific Provisions; General Administrative Rulings And Decisions
PROPELLANTS	189.191	CHLOROFLUOROCARBON PROPELLANTS Substances Generally Prohibited From Direct Addition Or Use As Human Food; Substances Prohibited From Use In Human Food
PROPELLANTS	300.100	CHLOROFLUOROCARBON PROPELLANTS Substances Generally Prohibited From Drugs; General
PROPELLANTS	700.23	CHLOROFLUOROCARBON PROPELLANTS Requirements For Specific Cosmetic Products; General
PROPELLANTS	801.417	CHLOROFLUOROCARBON PROPELLANTS Special Requirements For Specific Devices; Labeling
PROPER NAME	610.62	PROPER NAME; PACKAGE LABEL; LEGIBLE TYPE Labeling Standards; General Biological Products Standards
PROPERIDIN	866.5320	PROPERIDIN FACTOR B IMMUNOLOGICAL TEST SYSTEM Immunological Test Systems; Immunology And Microbiology Devices
PROPERTY	1403.31	REAL PROPERTY Changes, Property, and Subawards; Uniform Administrative Requirements For Grants And Coop. Agreements To State & Local Gov
PROPHYLAXIS CUP	872.6290	PROPHYLAXIS CUP Miscellaneous Devices; Dental Devices
PROPIONATE, CALCIUM	184.1221	CALCIUM PROPIONATE Listing Of Specific Substances Affirmed As GRAS; Direct Food Substances Affirmed As Generally Recognized As Safe
PROPIONATE, CALCIUM	582.3221	CALCIUM PROPIONATE Chemical Preservatives; Substances Generally Recognized As Safe
PROPIONATE, SODIUM	184.1784	SODIUM PROPIONATE Listing Of Specific Substances Affirmed As GRAS; Direct Food Substances Affirmed As Generally Recognized As Safe
PROPIONATE, SODIUM	582.3784	SODIUM PROPIONATE Chemical Preservatives; Substances Generally Recognized As Safe
PROPIONIC ACID	184.1081	PROPIONIC ACID Listing Of Specific Substances Affirmed As GRAS; Direct Food Substances Affirmed As Generally Recognized As Safe
PROPIONIC ACID	582.3081	PROPIONIC ACID Chemical Preservatives; Substances Generally Recognized As Safe
PROPIOPROMAZINE	522.2002	PROPIOPROMAZINE HYDROCHLORIDE INJECTION Implantation Or Injectable Dosage Form New Animal Drugs
PROPOFOL	522.2005	PROPOFOL INJECTION Implantation Or Injectable Dosage Form New Animal Drugs
PROPOSALS	5.34	ISSUANCE OF NOTICES RELATING TO PROPOSALS AND ORDERS FOR DEBARMENT AND DENIAL OF AN APPLICATION TO TERMINATE DEBARMENT General Redelegations Of Authority; Delegations Of Authority And Organization
PROPOSALS	5.105	ISSUANCE OF NOTICES RELATING TO PROPOSALS TO REFUSE APPROVAL OR TO WITHDRAW APPROVAL OF NEW DRUG APPLICATIONS AND THEIR SUPPLEMENTS Human Drugs; Redelegations Of Authority; Delegations Of Authority And Organization
PROPOSALS	5.202	ISSUANCE OF NOTICES OF OPPORTUNITY FOR A HEARING ON PROPOSALS FOR DENIAL OF APPROVAL OF APPLICATIONS FOR LICENSES, SUSPENSION OF LICENSES, OR REVOCATION OF LICENSES Biologics; Redelegations Of Authority; Delegations Of Authority And Organization
PROPOSALS	5.502	ISSUANCE OF NOTICES, PROPOSALS, AND ORDERS RELATING TO NEW ANIMAL DRUGS AND MEDICATED FEED MILL LICENSE APPLICATIONS Animal Drugs; Redelegations Of Authority; Delegations Of Authority And Organization
PROPOSALS, SETTLEMENT	10.110	SETTLEMENT PROPOSALS General Administrative Procedures; Administrative Practices And Procedures
PROPOXYPHENE	862.3700	PROPOXYPHENE TEST SYSTEM Clinical Toxicology Test Systems; Clinical Chemistry And Clinical Toxicology Devices
PROPYL ALCOHOL	573.880	NORMAL PROPYL ALCOHOL Food Additive Listing; Food Additives Permitted In Feed And Drinking Water Of Animals
PROPYL GALLATE	184.1660	PROPYL GALLATE Listing Of Specific Substances Affirmed As GRAS; Direct Food Substances Affirmed As Generally Recognized As Safe
PROPYL GALLATE	582.3660	PROPYL GALLATE Chemical Preservatives; Substances Generally Recognized As Safe
PROPYLENE	177.1980	VINYL CHLORIDE-PROPYLENE COPOLYMERS Substances For Use As Basic Components Of Single And Repeated Use Food Contact Surfaces; Indirect Food Additives: Polymers

PROPYLENE GLYCOL	172.765	SUCCISTEARIN (STEAROYL PROPYLENE GLYCOL HYDROGEN SUCCINATE) Other Specific Usage Additives; Food Additives Permitted For Direct Addition To Food For Human Consumption
PROPYLENE GLYCOL	172.850	LACTYLATED FATTY ACID ESTERS OF GLYCEROL AND PROPYLENE GLYCOL Multipurpose Additives; Food Additives Permitted For Direct Addition To Food For Human Consumption
PROPYLENE GLYCOL	172.856	PROPYLENE GLYCOL MONO- AND DIESTERS OF FATS AND FATTY ACIDS Multipurpose Additives; Food Additives Permitted For Direct Addition To Food For Human Consumption
PROPYLENE GLYCOL	172.858	PROPYLENE GLYCOL ALGINATE Multipurpose Additives; Food Additives Permitted For Direct Addition To Food For Human Consumption
PROPYLENE GLYCOL	184.1666	PROPYLENE GLYCOL Listing Of Specific Substances Affirmed As GRAS; Direct Food Substances Affirmed As Generally Recognized As Safe
PROPYLENE GLYCOL	500.50	PROPYLENE GLYCOL IN OR ON CAT FOOD Specific Administrative Rulings and Decisions; General
PROPYLENE GLYCOL	582.1666	PROPYLENE GLYCOL General Purpose Food Additives; Substances Generally Recognized As Safe
PROPYLENE GLYCOL	582.4666	PROPYLENE GLYCOL Emulsifying Agents; Substances Generally Recognized As Safe
PROPYLENE GLYCOL	589.1001	PROPYLENE GLYCOL IN OR ON CAT FOOD Substances Prohibited From Use In Animal Food Or Feed
PROPYLENE OXIDE	172.808	COPOLYMER CONDENSATES OF ETHYLENE OXIDE AND PROPYLENE OXIDE Multipurpose Additives; Food Additives Permitted For Direct Addition To Food For Human Consumption
PROPYLPARABEN	184.1670	PROPYLPARABEN Listing Of Specific Substances Affirmed As GRAS; Direct Food Substances Affirmed As Generally Recognized As Safe
PROPYLPARABEN	556.550	PROPYLPARABEN Specific Tolerances For Residues Of New Animal Drugs; Tolerances For Residues Of New Animal Drugs In Food
PROPYLPARABEN	582.3670	PROPYLPARABEN Chemical Preservatives; Substances Generally Recognized As Safe
PROSECUTION	1210.31	HEARING BEFORE PROSECUTION Regulations Under The Federal Import Milk Act
PROSECUTION	1316.24	EXEMPTION FROM PROSECUTION FOR RESEARCHERS Protection Of Researchers And Research Subjects; Administrative Functions, Practices, And Procedures
PROSTALENE	522.2012	PROSTALENE SOLUTION Implantation Or Injectable Dosage Form New Animal Drugs
PROSTATIC	862.1020	ACID PHOSPHATASE (TOTAL OR PROSTATIC) TEST SYSTEM Clinical Chemistry Test Systems; Clinical Chemistry And Clinical Toxicology Devices
PROSTATIC HYPERTROPHY	310.532	DRUG PRODUCTS CONTAINING ACTIVE INGREDIENTS OFFERED OVER-THE-COUNTER (OTC) TO RELIEVE THE SYMPTOMS OF BENIGN PROSTATIC HYPERTROPHY Requirements For Specific New Drugs Or Devices; New Drugs
PROSTHESIS	870.3450	VASCULAR GRAFT PROSTHESIS Cardiovascular Prosthetic Devices; Cardiovascular Devices
PROSTHESIS ADHESIVE	878.3750	EXTERNAL PROSTHESIS ADHESIVE Prosthetic Devices; General And Plastic Surgery Devices
PROSTHESIS, ANKLE	888.3100	ANKLE JOINT METAL/COMPOSITE SEMI-CONSTRAINED CEMENTED PROSTHESIS Prosthetic Devices; Orthopedic Devices
PROSTHESIS, ANKLE	888.3110	ANKLE JOINT METAL/POLYMER SEMI-CONSTRAINED CEMENTED PROSTHESIS Prosthetic Devices; Orthopedic Devices
PROSTHESIS, ANKLE	888.3120	ANKLE JOINT METAL/POLYMER NON-CONSTRAINED CEMENTED PROSTHESIS Prosthetic Devices; Orthopedic Devices
PROSTHESIS, BREAST	878.3530	SILICONE INFLATABLE BREAST PROSTHESIS Prosthetic Devices; General And Plastic Surgery Devices
PROSTHESIS, BREAST	878.3540	SILICONE GEL-FILLED BREAST PROSTHESIS Prosthetic Devices; General And Plastic Surgery Devices
PROSTHESIS, CHIN	878.3550	CHIN PROSTHESIS Prosthetic Devices; General And Plastic Surgery Devices
PROSTHESIS, EAR	878.3590	EAR PROSTHESIS Prosthetic Devices; General And Plastic Surgery Devices
PROSTHESIS, ELBOW	888.3150	ELBOW JOINT METAL/POLYMER CONSTRAINED CEMENTED PROSTHESIS Prosthetic Devices; Orthopedic Devices

PROSTHESIS, ELBOW	888.3160	ELBOW JOINT METAL/POLYMER SEMI-CONSTRAINED CEMENTED PROSTHESIS Prosthetic Devices; Orthopedic Devices
PROSTHESIS, ELBOW	888.3170	ELBOW JOINT RADIAL (HEMI-ELBOW) POLYMER PROSTHESIS Prosthetic Devices; Orthopedic Devices
PROSTHESIS, ELBOW	888.3180	ELBOW JOINT HUMERAL (HEMI-ELBOW) METALLIC UNCEMENTED PROSTHESIS Prosthetic Devices; Orthopedic Devices
PROSTHESIS, ESOPHAGEAL	878.3610	ESOPHAGEAL PROSTHESIS Prosthetic Devices; General And Plastic Surgery Devices
PROSTHESIS, FALLOPIAN	884.3650	FALLOPIAN TUBE PROSTHESIS Obstetrical And Gynecological Prosthetic Devices; Obstetrical And Gynecological Devices
PROSTHESIS, FINGER	888.3200	FINGER JOINT METAL/METAL CONSTRAINED UNCEMENTED PROSTHESIS Prosthetic Devices; Orthopedic Devices
PROSTHESIS, FINGER	888.3210	FINGER JOINT METAL/METAL CONSTRAINED CEMENTED PROSTHESIS Prosthetic Devices; Orthopedic Devices
PROSTHESIS, FINGER	888.3220	FINGER JOINT METAL/POLYMER CONSTRAINED CEMENTED PROSTHESIS Prosthetic Devices; Orthopedic Devices
PROSTHESIS, FINGER	888.3230	FINGER JOINT POLYMER CONSTRAINED CEMENTED PROSTHESIS Prosthetic Devices; Orthopedic Devices
PROSTHESIS, GLENOID FOSSA	872.3950	GLENOID FOSSA PROSTHESIS Prosthetic Devices; Dental Devices
PROSTHESIS, HIP	888.3300	HIP JOINT METAL CONSTRAINED CEMENTED OR UNCEMENTED PROSTHESIS Prosthetic Devices; Orthopedic Devices
PROSTHESIS, HIP	888.3310	HIP JOINT METAL/POLYMER CONSTRAINED CEMENTED OR UNCEMENTED PROSTHESIS Prosthetic Devices; Orthopedic Devices
PROSTHESIS, HIP	888.3320	HIP JOINT METAL/METAL SEMI-CONSTRAINED, WITH A CEMENTED ACETABULAR COMPONENT, PROSTHESIS Prosthetic Devices; Orthopedic Devices
PROSTHESIS, HIP	888.3330	HIP JOINT METAL/METAL SEMI-CONSTRAINED, WITH AN UNCEMENTED ACETABULAR COMPONENT, PROSTHESIS Prosthetic Devices; Orthopedic Devices
PROSTHESIS, HIP	888.3340	HIP JOINT METAL/COMPOSITE SEMI-CONSTRAINED CEMENTED PROSTHESIS Prosthetic Devices; Orthopedic Devices
PROSTHESIS, HIP	888.3350	HIP JOINT METAL/POLYMER SEMI-CONSTRAINED CEMENTED PROSTHESIS Prosthetic Devices; Orthopedic Devices
PROSTHESIS, HIP	888.3353	HIP JOINT METAL/CERAMIC/POLYMER SEMI-CONSTRAINED CEMENTED OR NONPOROUS UNCEMENTED PROSTHESIS Prosthetic Devices; Orthopedic Devices
PROSTHESIS, HIP	888.3358	HIP JOINT METAL/POLYMER/METAL SEMI-CONSTRAINED POROUS-COATED UNCEMENTED PROSTHESIS Prosthetic Devices; Orthopedic Devices
PROSTHESIS, HIP	888.3360	HIP JOINT FEMORAL (HEMI-HIP) METALLIC CEMENTED OR UNCEMENTED PROSTHESIS Prosthetic Devices; Orthopedic Devices
PROSTHESIS, HIP	888.3370	HIP JOINT (HEMI-HIP) ACETABULAR METAL CEMENTED PROSTHESIS Prosthetic Devices; Orthopedic Devices
PROSTHESIS, HIP	888.3380	HIP JOINT FEMORAL (HEMI-HIP) TRUNION-BEARING METAL/POLYACETAL CEMENTED PROSTHESIS Prosthetic Devices; Orthopedic Devices
PROSTHESIS, HIP	888.3390	HIP JOINT FEMORAL (HEMI-HIP) METAL/POLYMER CEMENTED OR UNCEMENTED PROSTHESIS Prosthetic Devices; Orthopedic Devices
PROSTHESIS, HIP	888.3400	HIP JOINT FEMORAL (HEMI-HIP) METALLIC RESURFACING PROSTHESIS Prosthetic Devices; Orthopedic Devices
PROSTHESIS, HIP	888.3410	HIP JOINT METAL/POLYMER SEMI-CONSTRAINED RESURFACING CEMENTED PROSTHESIS Prosthetic Devices; Orthopedic Devices
PROSTHESIS, INTERARTICULAR DISC	872.3970	INTERARTICULAR DISC PROSTHESIS (INTERPOSITIONAL IMPLANT) Prosthetic Devices; Dental Devices
PROSTHESIS, KNEE	888.3480	KNEE JOINT FEMOROTIBIAL METALLIC CONSTRAINED CEMENTED PROSTHESIS Prosthetic Devices; Orthopedic Devices
PROSTHESIS, KNEE	888.3490	KNEE JOINT FEMOROTIBIAL METAL/COMPOSITE NON-CONSTRAINED CEMENTED PROSTHESIS Prosthetic Devices; Orthopedic Devices
PROSTHESIS, KNEE	888.3500	KNEE JOINT FEMOROTIBIAL METAL/COMPOSITE SEMI-CONSTRAINED CEMENTED PROSTHESIS Prosthetic Devices; Orthopedic Devices

PROSTHESIS, KNEE	888.3510	KNEE JOINT FEMOROTIBIAL METAL/POLYMER CONSTRAINED CEMENTED PROSTHESIS Prosthetic Devices; Orthopedic Devices
PROSTHESIS, KNEE	888.3520	KNEE JOINT FEMOROTIBIAL METAL/POLYMER NON-CONSTRAINED CEMENTED PROSTHESIS Prosthetic Devices; Orthopedic Devices
PROSTHESIS, KNEE	888.3530	KNEE JOINT FEMOROTIBIAL METAL/POLYMER SEMI-CONSTRAINED CEMENTED PROSTHESIS Prosthetic Devices; Orthopedic Devices
PROSTHESIS, KNEE	888.3540	KNEE JOINT PATELLOFEMORAL POLYMER/METAL SEMI-CONSTRAINED CEMENTED PROSTHESIS Prosthetic Devices; Orthopedic Devices
PROSTHESIS, KNEE	888.3550	KNEE JOINT PATELLOFEMOROTIBIAL POLYMER/METAL/METAL CONSTRAINED CEMENTED PROSTHESIS Prosthetic Devices; Orthopedic Devices
PROSTHESIS, KNEE	888.3560	KNEE JOINT PATELLOFEMOROTIBIAL POLYMER/METAL/POLYMER SEMI-CONSTRAINED CEMENTED PROSTHESIS Prosthetic Devices; Orthopedic Devices
PROSTHESIS, KNEE	888.3570	KNEE JOINT FEMORAL (HEMI-KNEE) METALLIC UNCEMENTED PROSTHESIS Prosthetic Devices; Orthopedic Devices
PROSTHESIS, KNEE	888.3580	KNEE JOINT PATELLAR (HEMI-KNEE) METALLIC RESURFACING UNCEMENTED PROSTHESIS Prosthetic Devices; Orthopedic Devices
PROSTHESIS, KNEE	888.3590	KNEE JOINT TIBIAL (HEMI-KNEE) METALLIC RESURFACING UNCEMENTED PROSTHESIS Prosthetic Devices; Orthopedic Devices
PROSTHESIS, LARYNGEAL	874.3730	LARYNGEAL PROSTHESIS (TAUB DESIGN) Prosthetic Devices; Ear, Nose, And Throat Devices
PROSTHESIS, LOWER LIMB	890.3500	EXTERNAL ASSEMBLED LOWER LIMB PROSTHESIS Physical Medicine Prosthetic Devices; Physical Medicine Devices
PROSTHESIS, MANDIBULAR	872.3960	MANDIBULAR CONDYLE PROSTHESIS Prosthetic Devices; Dental Devices
PROSTHESIS, MANDIBULAR	874.3695	MANDIBULAR IMPLANT FACIAL PROSTHESIS Prosthetic Devices; Ear, Nose, And Throat Devices
PROSTHESIS, NOSE	878.3680	NOSE PROSTHESIS Prosthetic Devices; General And Plastic Surgery Devices
PROSTHESIS, OSSICULAR	874.3450	PARTIAL OSSICULAR REPLACEMENT PROSTHESIS Prosthetic Devices; Ear, Nose, And Throat Devices
PROSTHESIS, OSSICULAR	874.3495	TOTAL OSSICULAR REPLACEMENT PROSTHESIS Prosthetic Devices; Ear, Nose, And Throat Devices
PROSTHESIS, OSSICULAR	874.3540	PROSTHESIS MODIFICATION INSTRUMENT FOR OSSICULAR REPLACEMENT SURGERY Prosthetic Devices; Ear, Nose, And Throat Devices
PROSTHESIS, RESTORATION	878.3800	EXTERNAL AESTHETIC RESTORATION PROSTHESIS Prosthetic Devices; General And Plastic Surgery Devices
PROSTHESIS, SHOULDER	888.3640	SHOULDER JOINT METAL/METAL OR METAL/POLYMER CONSTRAINED CEMENTED PROSTHESIS Prosthetic Devices; Orthopedic Devices
PROSTHESIS, SHOULDER	888.3650	SHOULDER JOINT METAL/POLYMER NON-CONSTRAINED CEMENTED PROSTHESIS Prosthetic Devices; Orthopedic Devices
PROSTHESIS, SHOULDER	888.3660	SHOULDER JOINT METAL/POLYMER SEMI-CONSTRAINED CEMENTED PROSTHESIS Prosthetic Devices; Orthopedic Devices
PROSTHESIS, SHOULDER	888.3670	SHOULDER JOINT METAL/POLYMER/METAL NONCONSTRAINED OR SEMI-CONSTRAINED POROUS-COATED UNCEMENTED PROSTHESIS Prosthetic Devices; Orthopedic Devices
PROSTHESIS, SHOULDER	888.3680	SHOULDER JOINT GLENOID (HEMI-SHOULDER) METALLIC CEMENTED PROSTHESIS Prosthetic Devices; Orthopedic Devices
PROSTHESIS, SHOULDER	888.3690	SHOULDER JOINT HUMERAL (HEMI-SHOULDER) METALLIC UNCEMENTED PROSTHESIS Prosthetic Devices; Orthopedic Devices
PROSTHESIS, TEMPOROMANDIBULAR	872.3940	TOTAL TEMPOROMANDIBULAR JOINT PROSTHESIS Prosthetic Devices; Dental Devices
PROSTHESIS, TENDON	888.3025	PASSIVE TENDON PROSTHESIS Prosthetic Devices; Orthopedic Devices
PROSTHESIS, TESTICULAR	876.3750	TESTICULAR PROSTHESIS Prosthetic Devices; Gastroenterology-urology Devices
PROSTHESIS, TOE	888.3720	TOE JOINT POLYMER CONSTRAINED PROSTHESIS Prosthetic Devices; Orthopedic Devices
PROSTHESIS, TOE	888.3730	TOE JOINT PHALANGEAL (HEMI-TOE) POLYMER PROSTHESIS Prosthetic Devices; Orthopedic Devices

PROSTHESIS, TRACHEAL	878.3720	TRACHEAL PROSTHESIS Prosthetic Devices; General And Plastic Surgery Devices
PROSTHESIS, WRIST	888.3750	WRIST JOINT CARPAL LUNATE POLYMER PROSTHESIS Prosthetic Devices; Orthopedic Devices
PROSTHESIS, WRIST	888.3760	WRIST JOINT CARPAL SCAPHOID POLYMER PROSTHESIS Prosthetic Devices; Orthopedic Devices
PROSTHESIS, WRIST	888.3770	WRIST JOINT CARPAL TRAPEZIUM POLYMER PROSTHESIS Prosthetic Devices; Orthopedic Devices
PROSTHESIS, WRIST	888.3780	WRIST JOINT POLYMER CONSTRAINED PROSTHESIS Prosthetic Devices; Orthopedic Devices
PROSTHESIS, WRIST	888.3790	WRIST JOINT METAL CONSTRAINED CEMENTED PROSTHESIS Prosthetic Devices; Orthopedic Devices
PROSTHESIS, WRIST	888.3800	WRIST JOINT METAL/POLYMER SEMI-CONSTRAINED CEMENTED PROSTHESIS Prosthetic Devices; Orthopedic Devices
PROSTHESIS, WRIST	888.3810	WRIST JOINT ULNAR (HEMI-WRIST) POLYMER PROSTHESIS Prosthetic Devices; Orthopedic Devices
PROSTHETIC	870.3935	PROSTHETIC HEART VALVE HOLDER Cardiovascular Prosthetic Devices; Cardiovascular Devices
PROSTHETIC	870.3945	PROSTHETIC HEART VALVE SIZER Cardiovascular Prosthetic Devices; Cardiovascular Devices
PROSTHETIC	890.3025	PROSTHETIC AND ORTHOTIC ACCESSORY Physical Medicine Prosthetic Devices; Physical Medicine Devices
PROSTHETIC	890.3420	EXTERNAL LIMB PROSTHETIC COMPONENT Physical Medicine Prosthetic Devices; Physical Medicine Devices
PROSTHETIC	895.101	PROSTHETIC HAIR FIBERS Listing Of Banned Devices; Banned Devices
PROTEASE	184.1027	MIXED CARBOHYDRASE AND PROTEASE ENZYME PRODUCT Listing Of Specific Substances Affirmed As GRAS; Direct Food Substances Affirmed As Generally Recognized As Safe
PROTEASE	184.1150	BACTERIALLY DERIVED PROTEASE ENZYME PREPARATION Listing Of Specific Substances Affirmed As GRAS; Direct Food Substances Affirmed As Generally Recognized As Safe
PROTECTANT	346.14	PROTECTANT ACTIVE INGREDIENTS Anorectal Drug Products For Over-the-counter Human Use
PROTECTINS	864.9550	LECTINS AND PROTECTINS Products Used In Establishments That Manufacture Blood And Blood Products; Hematology And Pathology Devices
PROTECTION	1240.86	PROTECTION OF PIER WATER SYSTEM Source And Use Of Potable Water; Control Of Communicable Diseases
PROTECTIVE GARMENT	876.5920	PROTECTIVE GARMENT FOR INCONTINENCE Therapeutic Devices; Gastroenterology-urology Devices
PROTECTIVE ORDER	17.28	PROTECTIVE ORDER Civil Money Penalties Hearings
PROTECTIVE RESTRAINT	880.6760	PROTECTIVE RESTRAINT General Hospital And Personal Use Miscellaneous Devices; General Hospital And Personal Use Devices
PROTECTIVE SHIELD	892.6500	PERSONNEL PROTECTIVE SHIELD Miscellaneous Devices; Radiology Devices
PROTECTOR	868.5820	DENTAL PROTECTOR Therapeutic Devices; Anesthesiology Devices
PROTECTOR	872.6670	SILICATE PROTECTOR Miscellaneous Devices; Dental Devices
PROTECTORS	880.6450	SKIN PRESSURE PROTECTORS General Hospital And Personal Use Miscellaneous Devices; General Hospital And Personal Use Devices
PROTEIN	862.1630	PROTEIN (FRACTIONATION) TEST SYSTEM Clinical Chemistry Test Systems; Clinical Chemistry And Clinical Toxicology Devices
PROTEIN	862.1635	TOTAL PROTEIN TEST SYSTEM Clinical Chemistry Test Systems; Clinical Chemistry And Clinical Toxicology Devices
PROTEIN-BOUND IODINE	862.1640	PROTEIN-BOUND IODINE TEST SYSTEM Clinical Chemistry Test Systems; Clinical Chemistry And Clinical Toxicology Devices
PROTEIN, C-REACTIVE	866.5270	C-REACTIVE PROTEIN IMMUNOLOGICAL TEST SYSTEM Immunological Test Systems; Immunology And Microbiology Devices
PROTEIN FEED BLOCK	520.2380	THIABENDAZOLE TOP DRESSING AND MINERAL PROTEIN FEED BLOCK Oral Dosage Form New Animal Drugs
PROTEIN, FISH	172.340	FISH PROTEIN ISOLATE Special Dietary And Nutritional Additives; Food Additives Permitted For Direct Addition To Food For Human Consumption

PROTEIN, FISH	172.385	WHOLE FISH PROTEIN CONCENTRATE Special Dietary And Nutritional Additives; Food Additives Permitted For Direct Addition To Food For Human Consumption
PROTEIN, FORTIFIED	139.117	ENRICHED MACARONI PRODUCTS WITH FORTIFIED PROTEIN Requirements For Specific Standardized Macaroni And Noodle Products; Macaroni And Noodle Products
PROTEIN HYDROLYSATE	573.200	CONDENSED ANIMAL PROTEIN HYDROLYSATE Food Additive Listing; Food Additives Permitted In Feed And Drinking Water Of Animals
PROTEIN HYDROLYSATES	102.22	PROTEIN HYDROLYSATES Requirements For Specific Nonstandardized Foods; Common Or Usual Name For Nonstandardized Foods
PROTEIN, PLASMA	640.90	PLASMA PROTEIN FRACTION (HUMAN) Plasma Protein Fraction (Human); Additional Standards For Human Blood And Blood Products
PROTEIN PRODUCT	184.1498	MICROPARTICULATED PROTEIN PRODUCT Listing Of Specific Substances Affirmed As GRAS; Direct Food Substances Affirmed As Generally Recognized As Safe
PROTEIN, RETINOL-BINDING	866.5765	RETINOL-BINDING PROTEIN IMMUNOLOGICAL TEST SYSTEM Immunological Test Systems; Immunology And Microbiology Devices
PROTEIN, SOY	101.82	HEALTH CLAIMS: SOY PROTEIN AND RISK OF CORONARY HEART DISEASE Food Labeling
PROTEIN, URINARY	862.1645	URINARY PROTEIN OR ALBUMIN (NONQUANTITATIVE) TEST SYSTEM Clinical Chemistry Test Systems; Clinical Chemistry And Clinical Toxicology Devices
PROTEIN, WHEY	184.1979	WHEY PROTEIN CONCENTRATE Listing Of Specific Substances Affirmed As GRAS; Direct Food Substances Affirmed As Generally Recognized As Safe
PROTEIN, YEAST	172.325	BAKERS YEAST PROTEIN Special Dietary And Nutritional Additives; Food Additives Permitted For Direct Addition To Food For Human Consumption
PROTEINS, BENCE-JONES	866.5150	BENCE-JONES PROTEINS IMMUNOLOGICAL TEST SYSTEM Immunological Test Systems; Immunology And Microbiology Devices
PROTEUS SPP.	866.3410	PROTEUS SPP. (WEIL-FELIX) SEROLOGICAL REAGENTS Serological Reagents; Immunology And Microbiology Devices
PROTHROMBIN	864.7720	PROTHROMBIN CONSUMPTION TEST Hematology Kits And Packages; Hematology And Pathology Devices
PROTHROMBIN	864.7735	PROTHROMBIN-PROCONVERTIN TEST AND THROMBOTEST Hematology Kits And Packages; Hematology And Pathology Devices
PROTHROMBIN	864.7750	PROTHROMBIN TIME TEST Hematology Kits And Packages; Hematology And Pathology Devices
PROTHROMBIN	866.5735	PROTHROMBIN IMMUNOLOGICAL TEST SYSTEM Immunological Test Systems; Immunology And Microbiology Devices
PROTOCOL	58.120	PROTOCOL Protocol For And Conduct Of A Non-clinical Laboratory Study; Good Laboratory Practice For Nonclinical Laboratory Studies
PROTOCOL	312.30	PROTOCOL AMENDMENTS Investigational New Drug Application
PROTOCOL	814.19	PRODUCT DEVELOPMENT PROTOCOL (PDP) General; Premarket Approval Of Medical Devices
PROTOCOLS	5.407	APPROVAL, DISAPPROVAL, OR WITHDRAWAL OF APPROVAL OF PRODUCT DEVELOPMENT PROTOCOLS AND APPLICATIONS FOR PREMARKET APPROVAL FOR MEDICAL DEVICES Medical Devices And Radiological Health; Redelegations Of Authority; Delegations Of Authority And Organization
PROTOCOLS	312.83	TREATMENT PROTOCOLS Investigational New Drug Application
PROTOCOLS	320.30	INQUIRIES REGARDING BIOAVAILABILITY AND BIOEQUIVALENCE REQUIREMENTS AND REVIEW OF PROTOCOLS BY THE FDA Procedures For Determining The Bioavailability Or Bioequivalence Of Drug Products; Bioavailability And Bioequivalence Requirements
PROTOCOLS	610.2	REQUESTS FOR SAMPLES AND PROTOCOLS; OFFICIAL RELEASE Release Requirements; General Biological Products Standards
PROTOCOLS	660.6	SAMPLES; PROTOCOLS; OFFICIAL RELEASE Antibody To Hepatitis B Surface Antigen; Additional Standards For Diagnostic Substances For Laboratory Tests
PROTOCOLS	660.36	SAMPLES AND PROTOCOLS Reagent Red Blood Cells; Additional Standards For Diagnostic Substances For Laboratory Tests
PROTOCOLS	660.46	SAMPLES; PROTOCOLS; OFFICIAL RELEASE Hepatitis B Surface Antigen; Additional Standards For Diagnostic Substances For Laboratory Tests

PROTOCOLS	1301.18	RESEARCH PROTOCOLS Registration Of Manufacturers, Distributors, And Dispensers Of Controlled Substances
PROTRACTOR	888.4600	PROTRACTOR FOR CLINICAL USE Surgical Devices; Orthopedic Devices
PROVISIONAL LISTINGS	81.10	TERMINATION OF PROVISIONAL LISTINGS OF COLOR ADDITIVES General Specs+restrictns For Provisional Color Additives For Foods, Drugs, And Cosmetics
PROVISIONAL LISTS	81.1	PROVISIONAL LISTS OF COLOR ADDITIVES General Specs+restrictns For Provisional Color Additives For Foods, Drugs, And Cosmetics
PROVISIONS, FINAL	26.81	FINAL PROVISIONS Framework Provisions; Mutual Recognition Of Pharmaceutical GMP Reports, Medical Device Quality System Audit Reports
PROVOLONE CHEESE	133.181	PROVOLONE CHEESE Requirements For Specific Standardized Cheese And Related Products; Cheeses And Related Cheese Products
PRUNE JUICE	146.187	CANNED PRUNE JUICE Requirements For Specific Standardized Canned Fruit Juices And Beverages; Canned Fruit Juices
PRUNES	145.190	CANNED PRUNES Requirements For Specific Standardized Canned Fruits; Canned Fruits
PRUSSIATE, YELLOW	172.490	YELLOW PRUSSIATE OF SODA Anticaking Agents; Food Additives Permitted For Direct Addition To Food For Human Consumption
PRUSSIATE, YELLOW	573.1020	YELLOW PRUSSIATE OF SODA Food Additive Listing; Food Additives Permitted In Feed And Drinking Water Of Animals
PSEUDOMONAS SPP.	866.3415	PSEUDOMONAS SPP. SEROLOGICAL REAGENTS Serological Reagents; Immunology And Microbiology Devices
PSITTACINE BIRDS	1240.62	PSITTACINE BIRDS Specific Administrative Decisions Regarding Interstate Shipments; Control Of Communicable Diseases
PSORIASIS	358.710	ACTIVE INGREDIENTS FOR THE CONTROL OF DANDRUFF, SEBORRHEIC DERMATITIS, OR PSORIASIS Miscellaneous External Drug Products For Over-the-counter Human Use
PSORIASIS	358.750	LABELING OF DRUG PRODUCTS FOR THE CONTROL OF DANDRUFF, SEBORRHEIC DERMATITIS, OR PSORIASIS Miscellaneous External Drug Products For Over-the-counter Human Use
PSYLLIUM	201.319	WATER-SOLUBLE GUMS, HYDROPHILIC GUMS AND HYDROPHILIC MUCILLOIDS (INCLUDING PLANTAGO SEED (PSYLLIUM)) AS ACTIVE INGREDIENTS; REQUIRED WARNINGS AND DIRECTIONS Specific Labeling Requirements For Specific Drug Products; Labeling
PTFE	872.3680	POLYTETRAFLUOROETHYLENE (PTFE) VITREOUS CARBON MATERIALS Prosthetic Devices; Dental Devices
PTOSIS CRUTCH	886.5600	PTOSIS CRUTCH Therapeutic Devices; Ophthalmic Devices
PUBLIC ADMIN PROCEEDINGS	10.205	ELECTRONIC MEDIA COVERAGE OF PUBLIC ADMINISTRATIVE PROCEEDINGS Electronic Media Coverage Public Admin Proceedings; Guideline Policy Procedures; Administrative Practices And Procedures
PUBLIC ADMIN PROCEEDINGS	10.206	PROCEDURES FOR ELECTRONIC MEDIA COVERAGE OF AGENCY PUBLIC ADMINISTRATIVE PROCEEDINGS Electronic Media Coverage Public Admin Proceedings; Guideline Policy Procedures; Administrative Practices And Procedures
PUBLIC AVAILABILITY	803.9	PUBLIC AVAILABILITY OF REPORTS General Provisions; Medical Device Reporting
PUBLIC AVAILABILITY	806.40	PUBLIC AVAILABILITY OF REPORTS Medical Devices; Reports Of Corrections And Removals
PUBLIC CALENDAR	10.100	PUBLIC CALENDAR General Administrative Procedures; Administrative Practices And Procedures
PUBLIC, DISCLOSED TO	20.81	DATA AND INFORMATION PREVIOUSLY DISCLOSED TO THE PUBLIC Limitations On Exemptions; Public Information
PUBLIC DISCLOSURE	10.20	SUBMISSION OF DOCUMENTS TO DOCKETS MANAGEMENT BRANCH; COMPUTATION OF TIME; AVAILABILITY FOR PUBLIC DISCLOSURE General Administrative Procedures; Administrative Practices And Procedures
PUBLIC DISCLOSURE	56.122	PUBLIC DISCLOSURE OF INFORMATION REGARDING REVOCATION Administrative Actions For Noncompliance; Institutional Review Boards
PUBLIC DISCLOSURE	58.213	PUBLIC DISCLOSURE OF INFORMATION REGARDING DISQUALIFICATION Disqualification Of Testing Facilities; Good Laboratory Practice For Nonclinical Laboratory Studies
PUBLIC DISCLOSURE	312.130	AVAILABILITY FOR PUBLIC DISCLOSURE OF DATA AND INFORMATION IN AN IND Investigational New Drug Application

PUBLIC DISCLOSURE	314.430	AVAILABILITY FOR PUBLIC DISCLOSURE OF DATA AND INFORMATION IN AN APPLICATION Miscellaneous Provisions; Applications For FDA Approval To Market A New Drug
PUBLIC DISCLOSURE	316.52	AVAILABILITY FOR PUBLIC DISCLOSURE OF DATA AND INFORMATION IN REQUESTS AND APPLICATIONS Orphan Drugs
PUBLIC INQUIRIES	14.65	PUBLIC INQUIRIES AND REQUESTS FOR ADVISORY COMMITTEE RECORDS Records Of Meetings And Hearings Before Advisory Committees; Public Hearing Before A Public Advisory Committee
PUBLIC HEALTH	2.5	IMMINENT HAZARD TO THE PUBLIC HEALTH General Provisions; General Administrative Rulings And Decisions
PUBLIC HEALTH	25.16	PUBLIC HEALTH AND SAFETY EMERGENCIES Environmental Impact Considerations
PUBLIC HEARING	10.50	PROMULGATION OF REGULATIONS AND ORDERS AFTER AN OPPORTUNITY FOR A FORMAL EVIDENTIARY PUBLIC HEARING General Administrative Procedures; Administrative Practices And Procedures
PUBLIC HEARING	14.70	ADMINISTRATIVE RECORD OF A PUBLIC HEARING BEFORE AN ADVISORY COMMITTEE Records Of Meetings And Hearings Before Advisory Committees; Public Hearing Before A Public Advisory Committee
PUBLIC HEARING	15.20	NOTICE OF A PUBLIC HEARING BEFORE THE COMMISSIONER Procedures For Public Hearing Before The Commissioner; Public Hearing Before The Commissioner
PUBLIC HEARING	15.30	CONDUCT OF A PUBLIC HEARING BEFORE THE COMMISSIONER Procedures For Public Hearing Before The Commissioner; Public Hearing Before The Commissioner
PUBLIC INFORMATION OFFICES	5.1110	FDA PUBLIC INFORMATION OFFICES Organization; Delegations Of Authority And Organization
PUBLIC NOTICE	810.18	PUBLIC NOTICE Mandatory Medical Device Recall Procedures; Medical Device Recall Authority
PUBLIC NOTIFICATION	7.50	PUBLIC NOTIFICATION OF RECALL Recalls - Guidance On Policy, Procedures & Industry Responsibilities; Enforcement Policy
PUBLIC PARTICIPATION	12.50	ADVICE ON PUBLIC PARTICIPATION IN HEARINGS Appearance And Participation; Formal Evidentiary Public Hearing
PUBLICATION	71.20	PUBLICATION OF REGULATION Administrative Action On Petitions; Color Additive Petitions
PUBLICATION	316.28	PUBLICATION OF ORPHAN-DRUG DESIGNATIONS Orphan Drugs
PUBLICATION	601.8	PUBLICATION OF REVOCATION General Provisions; Licensing
PUBLICATION	1230.37	PUBLICATION Regulations Under The Federal Caustic Poison Act
PULMONARY	868.5365	POSTURE CHAIR FOR CARDIAC OR PULMONARY TREATMENT Therapeutic Devices; Anesthesiology Devices
PULMONARY	868.5610	MEMBRANE LUNG FOR LONG-TERM PULMONARY SUPPORT Therapeutic Devices; Anesthesiology Devices
PULMONARY-FUNCTION	868.1880	PULMONARY-FUNCTION DATA CALCULATOR Diagnostic Devices; Anesthesiology Devices
PULMONARY-FUNCTION	868.1890	PREDICTIVE PULMONARY-FUNCTION VALUE CALCULATOR Diagnostic Devices; Anesthesiology Devices
PULMONARY-FUNCTION	868.1900	DIAGNOSTIC PULMONARY-FUNCTION INTERPRETATION CALCULATOR Diagnostic Devices; Anesthesiology Devices
PULP	176.260	PULP FROM RECLAIMED FIBER Substances For Use Only As Components Of Paper And Paperboard; Indirect Food Additives
PULP	186.1673	PULP Listing Of Specific Substances Affirmed As GRAS; Indirect Food Substances Affirmed As Generally Recognized As Safe
PULP TESTER	872.1720	PULP TESTER Diagnostic Devices; Dental Devices
PULP TESTERS	872.1730	ELECTRODE GEL FOR PULP TESTERS Diagnostic Devices; Dental Devices
PULSATILE FLOW	870.4320	CARDIOPULMONARY BYPASS PULSATILE FLOW GENERATOR Cardiovascular Surgical Devices; Cardiovascular Devices
PULSATING DEVICE	870.5225	EXTERNAL COUNTER-PULSATING DEVICE Cardiovascular Therapeutic Devices; Cardiovascular Devices
PULSE GENERATOR	870.1750	EXTERNAL PROGRAMMABLE PACEMAKER PULSE GENERATOR Cardiovascular Diagnostic Devices; Cardiovascular Devices
PULSE GENERATOR	870.3600	EXTERNAL PACEMAKER PULSE GENERATOR Cardiovascular Prosthetic Devices; Cardiovascular Devices
PULSE GENERATOR	870.3610	IMPLANTABLE PACEMAKER PULSE GENERATOR Cardiovascular Prosthetic Devices; Cardiovascular Devices

PULSED LIGHT	179.39	PULSED LIGHT FOR THE TREATMENT OF FOOD Radiation And Radiation Sources; Irradiation In The Production, Processing And Handling Of Food
PUMP	520.1044	GENTAMICIN SULFATE PIG PUMP ORAL SOLUTION Oral Dosage Form New Animal Drugs
PUMP	870.1800	WITHDRAWAL-INFUSION PUMP Cardiovascular Diagnostic Devices; Cardiovascular Devices
PUMP	870.4360	NONROLLER-TYPE CARDIOPULMONARY BYPASS BLOOD PUMP Cardiovascular Surgical Devices; Cardiovascular Devices
PUMP	870.4370	ROLLER-TYPE CARDIOPULMONARY BYPASS BLOOD PUMP Cardiovascular Surgical Devices; Cardiovascular Devices
PUMP	870.4380	CARDIOPULMONARY BYPASS PUMP SPEED CONTROL Cardiovascular Surgical Devices; Cardiovascular Devices
PUMP	870.4390	CARDIOPULMONARY BYPASS PUMP TUBING Cardiovascular Surgical Devices; Cardiovascular Devices
PUMP	878.4780	POWERED SUCTION PUMP Surgical Devices; General And Plastic Surgery Devices
PUMP	880.5725	INFUSION PUMP General Hospital And Personal Use Therapeutic Devices; General Hospital And Personal Use Devices
PUMP	884.5150	NONPOWERED BREAST PUMP Obstetrical And Gynecological Devices
PUMP	884.5160	POWERED BREAST PUMP Obstetrical And Gynecological Devices
PUNCH	882.4750	SKULL PUNCH Neurological Surgical Devices; Neurological Devices
PUPILLOGRAPH	886.1690	PUPILLOGRAPH Diagnostic Devices; Ophthalmic Devices
PUPILLOMETER	886.1700	PUPILLOMETER Diagnostic Devices; Ophthalmic Devices
PURCHASERS	5.604	MANUFACTURERS REQUIREMENT TO PROVIDE DATA TO ULTIMATE PURCHASERS OF ELECTRONIC PRODUCTS Radiation Control; Redelegations Of Authority; Delegations Of Authority And Organization
PURCHASERS	1003.22	COPIES OF COMMUNICATIONS SENT TO PURCHASERS, DEALERS OR DISTRIBUTORS Notification; Notification Of Defects Or Failure To Comply
PURCHASING	820.50	PURCHASING CONTROLS Quality System Regulation
PURIFICATION, WATER	876.5665	WATER PURIFICATION SYSTEM FOR HEMODIALYSIS Therapeutic Devices; Gastroenterology-urology Devices
PURIFICATION, WATER	884.6170	ASSISTED REPRODUCTION WATER AND WATER PURIFICATION SYSTEMS Assisted Reproduction Devices; Obstetrical And Gynecological Devices
PURIFIER, AIR	880.6500	MEDICAL ULTRAVIOLET AIR PURIFIER General Hospital And Personal Use Miscellaneous Devices; General Hospital And Personal Use Devices
PURIFIER, WATER	880.6710	MEDICAL ULTRAVIOLET WATER PURIFIER General Hospital And Personal Use Miscellaneous Devices; General Hospital And Personal Use Devices
PURITY	610.13	PURITY General Provisions; General Biological Products Standards
PURPOSES, NONMEDICAL	510.112	ANTIBIOTICS USED IN VETERINARY MEDICINE AND FOR NONMEDICAL PURPOSES; REQUIRED DATA Specific Administrative Rulings And Decisions; New Animal Drugs
PVM-MA	872.3500	POLYVINYLMETHYLETHER MALEIC ANHYDRIDE (PVM-MA), ACID COPOLYMER, AND CARBOXYMETHYLCELLULOSE SODIUM (NACMC) DENTURE ADHESIVE Prosthetic Devices; Dental Devices
PYRANTEL PAMOATE	520.1196	IVERMECTIN AND PYRANTEL PAMOATE CHEWABLE TABLET Oral Dosage Form New Animal Drugs
PYRANTEL PAMOATE	520.1871	PRAZIQUANTEL/PYRANTEL PAMOATE TABLETS Oral Dosage Form New Animal Drugs
PYRANTEL PAMOATE	520.1872	PRAZIQUANTEL, PYRANTEL PAMOATE AND FEBENTAL TABLETS Oral Dosage Form New Animal Drugs
PYRANTEL PAMOATE	520.2041	PYRANTEL PAMOATE CHEWABLE TABLETS Oral Dosage Form New Animal Drugs
PYRANTEL PAMOATE	520.2042	PYRANTEL PAMOATE TABLETS Oral Dosage Form New Animal Drugs
PYRANTEL PAMOATE	520.2043	PYRANTEL PAMOATE SUSPENSION Oral Dosage Form New Animal Drugs
PYRANTEL PAMOATE	520.2044	PYRANTEL PAMOATE PASTE Oral Dosage Form New Animal Drugs
PYRANTEL TARTRATE	520.2045	PYRANTEL TARTRATE POWDER; PYRANTEL TARTRATE PELLETS Oral Dosage Form New Animal Drugs

PYRANTEL TARTRATE	556.560	PYRANTEL TARTRATE Specific Tolerances For Residues Of New Animal Drugs; Tolerances For Residues Of New Animal Drugs In Food
PYRANTEL TARTRATE	558.485	PYRANTEL TARTRATE Specific New Animal Drugs For Use In Animal Feeds; New Animal Drugs For Use In Animal Feeds
PYRETHRINS	178.3730	PIPERONYL BUTOXIDE AND PYRETHRINS AS COMPONENTS OF BAGS Certain Adjuvants And Production Aids; Indirect Food Additives: Adjuvants, Production Aids, And Sanitizers
PYRAZOL-3-ONE	73.3122	4-[(2,4-DIMETHYLPHENYL)AZO]-2,4-DIHYDRO-5-METHYL-2-PHENYL-3H-PYRAZOL-3-ONE Drugs; Listing Of Color Additives Exempt From Certification
PYRIDOXINE HYDROCHLORIDE	184.1676	PYRIDOXINE HYDROCHLORIDE Listing Of Specific Substances Affirmed As GRAS; Direct Food Substances Affirmed As Generally Recognized As Safe
PYRIDOXINE HYDROCHLORIDE	582.5676	PYRIDOXINE HYDROCHLORIDE Nutrients And/or Dietary Supplements; Substances Generally Recognized As Safe
PYRILAMINE MALEATE	522.2063	PYRILAMINE MALEATE INJECTION Implantation Or Injectable Dosage Form New Animal Drugs
PYROGALLOL	73.1375	PYROGALLOL Drugs; Listing Of Color Additives Exempt From Certification
PYROPHOSPHATE	182.1087	SODIUM ACID PYROPHOSPHATE Multiple Purpose GRAS Food Substances; Substances Generally Recognized As Safe
PYROPHOSPHATE	182.6787	SODIUM PYROPHOSPHATE Sequestrants; Substances Generally Recognized As Safe
PYROPHOSPHATE	182.6789	TETRA SODIUM PYROPHOSPHATE Sequestrants; Substances Generally Recognized As Safe
PYROPHOSPHATE	182.8223	CALCIUM PYROPHOSPHATE Nutrients; Substances Generally Recognized As Safe
PYROPHOSPHATE	184.1304	FERRIC PYROPHOSPHATE Listing Of Specific Substances Affirmed As GRAS; Direct Food Substances Affirmed As Generally Recognized As Safe
PYROPHOSPHATE	582.1087	SODIUM ACID PYROPHOSPHATE General Purpose Food Additives; Substances Generally Recognized As Safe
PYROPHOSPHATE	582.5223	CALCIUM PYROPHOSPHATE Nutrients And/or Dietary Supplements; Substances Generally Recognized As Safe
PYROPHOSPHATE	582.5304	FERRIC PYROPHOSPHATE Nutrients And/or Dietary Supplements; Substances Generally Recognized As Safe
PYROPHOSPHATE	582.5306	FERRIC SODIUM PYROPHOSPHATE Nutrients And/or Dietary Supplements; Substances Generally Recognized As Safe
PYROPHOSPHATE	582.6787	SODIUM PYROPHOSPHATE Sequestrants; Substances Generally Recognized As Safe
PYROPHOSPHATE	582.6789	TETRA SODIUM PYROPHOSPHATE Sequestrants; Substances Generally Recognized As Safe
PYROPHYLLITE	73.1400	PYROPHYLLITE Drugs; Listing Of Color Additives Exempt From Certification
PYROPHYLLITE	73.2400	PYROPHYLLITE Cosmetics; Listing Of Color Additives Exempt From Certification
PYROPHYLLITE	573.900	PYROPHYLLITE Food Additive Listing; Food Additives Permitted In Feed And Drinking Water Of Animals
PYRUVATE	862.1650	PYRUVATE KINASE TEST SYSTEM Clinical Chemistry Test Systems; Clinical Chemistry And Clinical Toxicology Devices
PYRUVIC ACID	862.1655	PYRUVIC ACID TEST SYSTEM Clinical Chemistry Test Systems; Clinical Chemistry And Clinical Toxicology Devices

Q

QUALIFICATIONS	14.80	QUALIFICATIONS FOR MEMBERS OF STANDING POLICY AND TECHNICAL ADVISORY COMMITTEES Members Of Advisory Committees; Public Hearing Before A Public Advisory Committee
QUALIFICATIONS	205.6	MINIMUM QUALIFICATIONS Guidelines For State Licensing Of Wholesale Prescription Drug Distributors
QUALIFICATIONS	211.25	PERSONNEL QUALIFICATIONS Organization And Personnel; Current Good Manufacturing Practice For Finished Pharmaceuticals
QUALITY ASPECTS	26.19	INFORMATION RELATING TO QUALITY ASPECTS Specific Sector Provisions For Pharmaceutical GMPs; Mutual Recognition Of Pharmaceutical GMP Reports, Medical Device Quality System Audit Reports
QUALITY ASSURANCE	20.114	DATA AND INFORMATION SUBMITTED PURSUANT TO COOPERATIVE QUALITY ASSURANCE AGREEMENTS Availability Of Specific Categories Of Records; Public Information
QUALITY ASSURANCE	58.35	QUALITY ASSURANCE UNIT Organization And Personnel; Good Laboratory Practice For Nonclinical Laboratory Studies
QUALITY ASSURANCE	892.1940	RADIOLOGIC QUALITY ASSURANCE INSTRUMENT Diagnostic Devices; Radiology Devices
QUALITY ASSURANCE	1000.55	RECOMMENDATION FOR QUALITY ASSURANCE PROGRAMS IN DIAGNOSTIC RADIOLOGY FACILITIES Radiation Protection Recommendations; General
QUALITY CONTROL	106.1	STATUS AND APPLICABILITY OF THE QUALITY CONTROL PROCEDURES REGULATION General Provisions; Infant Formula Quality Control Procedures
QUALITY CONTROL	211.22	RESPONSIBILITIES OF QUALITY CONTROL UNIT Organization And Personnel; Current Good Manufacturing Practice For Finished Pharmaceutical
QUALITY CONTROL	640.56	QUALITY CONTROL TEST FOR POTENCY Cryoprecipitate; Additional Standards For Human Blood And Blood Products
QUALITY CONTROL	862.1660	QUALITY CONTROL MATERIAL TEST SYSTEM Clinical Chemistry Test Systems; Clinical Chemistry And Clinical Toxicology Devices
QUALITY CONTROL	864.8625	HEMATOLOGY QUALITY CONTROL MIXTURE Hematology Reagents; Hematology And Pathology Devices
QUALITY CONTROL	864.9650	QUALITY CONTROL KIT FOR BLOOD BANKING REAGENTS Products Used In Establishments That Manufacture Blood And Blood Products; Hematology And Pathology Devices
QUALITY CONTROL	866.2480	QUALITY CONTROL KIT FOR CULTURE MEDIA Microbiology Devices; Immunology And Microbiology Devices
QUALITY STANDARDS	5.1000	AUTHORITY TO ENSURE THAT MAMMOGRAPHY FACILITIES MEET QUALITY STANDARDS Mammography Facilities; Redelegations Of Authority; Delegations Of Authority And Organization
QUALITY STANDARDS	900.12	QUALITY STANDARDS Quality Standards And Certification; Mammography
QUALITY STANDARDS	900.18	ALTERNATIVE REQUIREMENTS FOR SECTION 900.12 QUALITY STANDARDS Quality Standards And Certification; Mammography
QUALITY, SUBSTANDARD	130.14	GENERAL STATEMENTS OF SUBSTANDARD QUALITY AND SUBSTANDARD FILL OF CONTAINERS General Provisions; Food Standards: General
QUALITY SYSTEM	26.41	EXCHANGE AND ENDORSEMENT OF QUALITY SYSTEM EVALUATION REPORTS Specific Sector Provisions For Medical Devices; Mutual Recognition Of Pharmaceutical GMP Reports, Medical Device Quality System Audit Reports
QUALITY SYSTEM	26.43	TRANSMISSION OF QUALITY SYSTEM EVALUATION REPORTS Specific Sector Provisions For Medical Devices; Mutual Recognition Of Pharmaceutical GMP Reports, Medical Device Quality System Audit Reports
QUALITY SYSTEM	820.5	QUALITY SYSTEM Quality System Regulation
QUANTIFICATION	530.24	PROCEDURE FOR ANNOUNCING ANALYTICAL METHODS FOR DRUG RESIDUE QUANTIFICATION Extralabel Drug Use In Animals
QUANTITIES, INSUFFICIENT	316.36	INSUFFICIENT QUANTITIES OF ORPHAN DRUGS Orphan Drugs
QUANTITY	161.30	DECLARATION OF QUANTITY OF CONTENTS ON LABELS FOR CANNED OYSTERS General Provisions; Fish And Shellfish

QUANTITY, NET	101.105	DECLARATION OF NET QUANTITY OF CONTENTS WHEN EXEMPT Exemptions From Food Labeling Requirements; Food Labeling
QUANTITY; NET	201.51	DECLARATION OF NET QUANTITY OF CONTENTS Labeling Requirements For Prescription Drugs And/or Insulin; Labeling
QUANTITY, NET	201.62	DECLARATION OF NET QUANTITY OF CONTENTS Labeling Requirements For Over-the-counter Drugs; Labeling
QUANTITY, NET	501.105	DECLARATION OF NET QUANTITY OF CONTENTS WHEN EXEMPT Exemptions From Animal Food Labeling Requirements; Animal Food Labeling
QUANTITY, NET	701.13	DECLARATION OF NET QUANTITY OF CONTENTS Package Form; Cosmetic Labeling
QUANTITY, NET	801.62	DECLARATION OF NET QUANTITY OF CONTENTS Labeling Requirements For Over-the-counter Devices; Labeling
QUATERNARY AMMONIUM CHLORIDE	172.165	QUATERNARY AMMONIUM CHLORIDE COMBINATION Food Preservatives; Food Additives Permitted For Direct Addition To Food For Human Consumption
QUININE	172.575	QUININE Flavoring Agents And Related Substances; Food Additives Permitted For Direct Addition To Food For Human Consumption
QUININE	310.547	DRUG PRODUCTS CONTAINING QUININE OFFERED OVER-THE-COUNTER FOR THE TREATMENT AND/OR PREVENTION OF MALARIA Requirements For Specific New Drugs Or Devices; New Drugs
QUININE	862.3750	QUININE TEST SYSTEM Clinical Toxicology Test Systems; Clinical Chemistry And Clinical Toxicology Devices
QUOTA	1303.27	ABANDONMENT OF QUOTA Individual Manufacturing Quotas; Quotas
QUOTAS, MANUFACTURING	1303.21	INDIVIDUAL MANUFACTURING QUOTAS Individual Manufacturing Quotas; Quotas
QUOTAS, MANUFACTURING	1303.22	PROCEDURE FOR APPLYING FOR INDIVIDUAL MANUFACTURING QUOTAS Individual Manufacturing Quotas; Quotas
QUOTAS, MANUFACTURING	1303.23	PROCEDURE FOR FIXING INDIVIDUAL MANUFACTURING QUOTAS Individual Manufacturing Quotas; Quotas
QUOTAS, MANUFACTURING	1303.25	INCREASE IN INDIVIDUAL MANUFACTURING QUOTAS Individual Manufacturing Quotas; Quotas
QUOTAS, MANUFACTURING	1303.26	REDUCTION IN INDIVIDUAL MANUFACTURING QUOTAS Individual Manufacturing Quotas; Quotas
QUOTAS, PROCUREMENT	1303.12	PROCUREMENT QUOTAS Aggregate Production And Procurement Quotas; Quotas
QUOTAS, PRODUCTION	1303.11	AGGREGATE PRODUCTION QUOTAS Aggregate Production And Procurement Quotas; Quotas
QUOTAS, PRODUCTION	1303.13	ADJUSTMENT OF AGGREGATE PRODUCTION QUOTAS Aggregate Production And Procurement Quotas; Quotas

R

RABIESVIRUS	866.3460	RABIESVIRUS IMMUNOFLUORESCENT REAGENTS Serological Reagents; Immunology And Microbiology Devices
RACK	882.4215	CLIP RACK Neurological Surgical Devices; Neurological Devices
RACK	886.1750	SKIASCOPIC RACK Diagnostic Devices; Ophthalmic Devices
RACTOPAMINE	556.570	RACTOPAMINE Tolerances For Residues Of New Animal Drugs In Food
RACTOPAMINE	558.500	RACTOPAMINE New Animal Drugs For Use In Animal Feeds
RADIAL	888.3170	ELBOW JOINT RADIAL (HEMI-ELBOW) POLYMER PROSTHESIS Prosthetic Devices; Orthopedic Devices
RADIAL IMMUNODIFFUSION	866.4800	RADIAL IMMUNODIFFUSION PLATE Immunology Laboratory Equipment And Reagents; Immunology And Microbiology Devices
RADIANT WARMER	880.5130	INFANT RADIANT WARMER General Hospital And Personal Use Therapeutic Devices; General Hospital And Personal Use Devices
RADIATION	179.21	SOURCES OF RADIATION USED FOR INSPECTION OF FOOD, FOR INSPECTION OF PACKAGED FOOD, AND FOR CONTROLLING FOOD PROCESSING Radiation And Radiation Sources; Irradiation In The Production, Processing And Handling Of Food
RADIATION	179.26	IONIZING RADIATION FOR THE TREATMENT OF FOOD Radiation And Radiation Sources; Irradiation In The Production, Processing And Handling Of Food
RADIATION	179.30	RADIOFREQUENCY RADIATION FOR THE HEATING OF FOOD, INCLUDING MICROWAVE FREQUENCIES Radiation And Radiation Sources; Irradiation In The Production, Processing And Handling Of Food
RADIATION	179.39	ULTRAVIOLET RADIATION FOR THE PROCESSING AND TREATMENT OF FOOD Radiation And Radiation Sources; Irradiation In The Production, Processing And Handling Of Food
RADIATION	579.22	IONIZING RADIATION FOR TREATMENT OF ANIMAL DIETS Radiation And Radiation Sources; Irradiation In The Production, Process & Handling Of Animal Feed And Pet Food
RADIATION	579.40	IONIZING RADIATION FOR THE TREATMENT OF POULTRY FEED AND POULTRY FEED INGREDIENTS Irradiation In The Production, Processing And Handling Of Animal Feed And Pet Food
RADIATION	886.5100	OPHTHALMIC BETA RADIATION SOURCE Therapeutic Devices; Ophthalmic Devices
RADIATION	1002.20	REPORTING OF ACCIDENTAL RADIATION OCCURRENCES Manufacturers' Reports On Accidental Radiation Occurrences; Records And Reports
RADIATION CONTROL	5.606	ACCEPTANCE OF ASSISTANCE FROM STATE AND LOCAL AUTHORITIES FOR ENFORCEMENT OF RADIATION CONTROL LEGISLATION AND REGULATIONS Radiation Control; Redelegations Of Authority, Delegations Of Authority And Organization
RADIATION CONTROL	1000.15	EXAMPLES OF ELECTRONIC PRODUCTS SUBJECT TO THE RADIATION CONTROL FOR HEALTH AND SAFETY ACT OF 1968 Statements Of Policy And Interpretation, General
RADIATION SAFETY	14.120	ESTABLISHMENT OF THE TECHNICAL ELECTRONIC PRODUCT RADIATION SAFETY STANDARDS COMMITTEE (TEPRSSC) Technical Electronic Products Radiation Safety Standards Committee; Public Hearing Before A Public Advisory Committee
RADIATION THERAPY	892.5050	MEDICAL CHARGED-PARTICLE RADIATION THERAPY SYSTEM Therapeutic Devices; Radiology Devices
RADIATION THERAPY	892.5300	MEDICAL NEUTRON RADIATION THERAPY SYSTEM Therapeutic Devices; Radiology Devices
RADIATION THERAPY	892.5710	RADIATION THERAPY BEAM-SHAPING BLOCK Therapeutic Devices; Radiology Devices
RADIATION THERAPY	892.5750	RADIONUCLIDE RADIATION THERAPY SYSTEM Therapeutic Devices; Radiology Devices
RADIATION THERAPY	892.5770	POWERED RADIATION THERAPY PATIENT SUPPORT ASSEMBLY Therapeutic Devices; Radiology Devices
RADIATION THERAPY	892.5840	RADIATION THERAPY SIMULATION SYSTEM Therapeutic Devices; Radiology Devices

RADIATION THERAPY	892.5900	X-RAY RADIATION THERAPY SYSTEM Therapeutic Devices; Radiology Devices
RADIOACTIVE DRUGS	201.129	DRUGS; EXEMPTION FOR RADIOACTIVE DRUGS FOR RESEARCH USE Exemptions From Adequate Directions For Use; Labeling
RADIOACTIVE DRUGS	310.503	REQUIREMENTS REGARDING CERTAIN RADIOACTIVE DRUGS Requirements For Specific New Drugs Or Devices; New Drugs
RADIOALLERGOSORBENT	866.5750	RADIOALLERGOSORBENT (RAST) IMMUNOLOGICAL TEST SYSTEM Immunological Test Systems; Immunology And Microbiology Devices
RADIOCHROMATOGRAM	862.2400	DENSITOMETER/SCANNER (INTEGRATING, REFLECTANCE, TLC, OR RADIOCHROMATOGRAM) FOR CLINICAL USE Clinical Laboratory Instruments; Clinical Chemistry And Clinical Toxicology Devices
RADIOFREQUENCY	179.30	RADIOFREQUENCY RADIATION FOR THE HEATING OF FOOD, INCLUDING MICROWAVE FREQUENCIES Radiation And Radiation Sources; Irradiation In The Production, Processing And Handling Of Food
RADIOFREQUENCY	870.2910	RADIOFREQUENCY PHYSIOLOGICAL SIGNAL TRANSMITTER AND RECEIVER Cardiovascular Monitoring Devices; Cardiovascular Devices
RADIOFREQUENCY	882.4400	RADIOFREQUENCY LESION GENERATOR Neurological Surgical Devices; Neurological Devices
RADIOFREQUENCY	882.4725	RADIOFREQUENCY LESION PROBE Neurological Surgical Devices; Neurological Devices
RADIOFREQUENCY	886.1400	RADIOFREQUENCY ELECTROSURGICAL CAUTERY APPARATUS Surgical Devices; Ophthalmic Devices
RADIOGRAPHIC	892.1640	RADIOGRAPHIC FILM MARKING SYSTEM Diagnostic Devices; Radiology Devices
RADIOGRAPHIC	892.1840	RADIOGRAPHIC FILM Diagnostic Devices; Radiology Devices
RADIOGRAPHIC	892.1850	RADIOGRAPHIC FILM CASSETTE Diagnostic Devices; Radiology Devices
RADIOGRAPHIC	892.1860	RADIOGRAPHIC FILM/CASSETTE CHANGER Diagnostic Devices; Radiology Devices
RADIOGRAPHIC	892.1870	RADIOGRAPHIC FILM/CASSETTE CHANGER PROGRAMMER Diagnostic Devices; Radiology Devices
RADIOGRAPHIC	892.1880	WALL-MOUNTED RADIOGRAPHIC CASSETTE HOLDER Diagnostic Devices; Radiology Devices
RADIOGRAPHIC	892.1890	RADIOGRAPHIC FILM ILLUMINATOR Diagnostic Devices; Radiology Devices
RADIOGRAPHIC	892.1900	AUTOMATIC RADIOGRAPHIC FILM PROCESSOR Diagnostic Devices; Radiology Devices
RADIOGRAPHIC	892.1910	RADIOGRAPHIC GRID Diagnostic Devices; Radiology Devices
RADIOGRAPHIC	892.1920	RADIOGRAPHIC HEAD HOLDER Diagnostic Devices; Radiology Devices
RADIOGRAPHIC	892.1950	RADIOGRAPHIC ANTHROPOMORPHIC PHANTOM Diagnostic Devices; Radiology Devices
RADIOGRAPHIC	892.1960	RADIOGRAPHIC INTENSIFYING SCREEN Diagnostic Devices; Radiology Devices
RADIOGRAPHIC	1020.31	RADIOGRAPHIC EQUIPMENT Performance Standards For Ionizing Radiation Emitting Products
RADIOIMMUNOASSAY	864.7695	PLATELET FACTOR 4 RADIOIMMUNOASSAY Hematology Kits And Packages; Hematology And Pathology Devices
RADIOLOGIC	892.1830	RADIOLOGIC PATIENT CRADLE Diagnostic Devices; Radiology Devices
RADIOLOGIC	892.1940	RADIOLOGIC QUALITY ASSURANCE INSTRUMENT Diagnostic Devices; Radiology Devices
RADIOLOGIC	892.1980	RADIOLOGIC TABLE Diagnostic Devices; Radiology Devices
RADIOLOGY	1000.55	RECOMMENDATION FOR QUALITY ASSURANCE PROGRAMS IN DIAGNOSTIC RADIOLOGY FACILITIES Radiation Protection Recommendations; General
RADIONUCLIDE	892.1360	RADIONUCLIDE DOSE CALIBRATOR Diagnostic Devices; Radiology Devices
RADIONUCLIDE	892.1390	RADIONUCLIDE REBREATHING SYSTEM Diagnostic Devices; Radiology Devices
RADIONUCLIDE	892.1420	RADIONUCLIDE TEST PATTERN PHANTOM Diagnostic Devices; Radiology Devices
RADIONUCLIDE	892.5650	MANUAL RADIONUCLIDE APPLICATOR SYSTEM Therapeutic Devices; Radiology Devices
RADIONUCLIDE	892.5700	REMOTE CONTROLLED RADIONUCLIDE APPLICATOR SYSTEM Therapeutic Devices; Radiology Devices
RADIONUCLIDE	892.5730	RADIONUCLIDE BRACHYTHERAPY SOURCE Therapeutic Devices; Radiology Devices
RADIONUCLIDE	892.5740	RADIONUCLIDE TELETHERAPY SOURCE Therapeutic Devices; Radiology Devices

RADIONUCLIDE	892.5750	RADIONUCLIDE RADIATION THERAPY SYSTEM Therapeutic Devices; Radiology Devices
RADIORECEPTOR	862.3645	NEUROLEPTIC DRUGS RADIORECEPTOR ASSAY TEST SYSTEM Clinical Toxicology Test Systems; Clinical Chemistry And Clinical Toxicology Devices
RADIUS MEASURING	886.1430	OPHTHALMIC CONTACT LENS RADIUS MEASURING DEVICE Diagnostic Devices; Ophthalmic Devices
RADIUS MEASURING	886.1450	CORNEAL RADIUS MEASURING DEVICE Diagnostic Devices; Ophthalmic Devices
RAILROAD	1250.45	FOOD HANDLING FACILITIES ON RAILROAD CONVEYANCES Equipment And Operation Of Land And Air Conveyances; Interstate Conveyance Sanitation
RAILROAD	1250.51	RAILROAD CONVEYANCES; DISCHARGE OF WASTES Equipment And Operation Of Land And Air Conveyances; Interstate Conveyance Sanitation
RAISIN BREAD	136.160	RAISIN BREAD, ROLLS, AND BUNS Requirements For Specific Standardized Bakery Products; Bakery Products
RAPESEED OIL	184.1555	RAPESEED OIL Listing Of Specific Substances Affirmed As GRAS; Direct Food Substances Affirmed As Generally Recognized As Safe
RAST	866.5750	RADIOALLERGOSORBENT (RAST) IMMUNOLOGICAL TEST SYSTEM Immunological Test Systems; Immunology And Microbiology Devices
RATE ALARM	870.2300	CARDIAC MONITOR (INCLUDING CARDIOTACHOMETER AND RATE ALARM) Cardiovascular Monitoring Devices; Cardiovascular Devices
RATE, SEDIMENTATION	864.5800	AUTOMATED SEDIMENTATION RATE DEVICE Automated And Semi-automated Hematology Devices; Hematology And Pathology Devices
RATE, SEDIMENTATION	864.6700	ERYTHROCYTE SEDIMENTATION RATE TEST Manual Hematology Devices; Hematology And Pathology Devices
RAW FRUIT	101.42	NUTRITION LABELING OF RAW FRUIT, VEGETABLES AND FISH Food Labeling
RAW FRUIT	101.43	SUBSTANTIAL COMPLIANCE OF FOOD RETAILERS WITH THE GUIDELINES FOR THE VOLUNTARY NUTRITION LABELING OF RAW FRUIT, VEGETABLES, AND FISH Food Labeling
RAW FRUIT	101.44	IDENTIFICATION OF THE 20 MOST FREQUENTLY CONSUMED RAW FRUIT, VEGETABLES AND FISH IN THE UNITED STATES Food Labeling
RAW FRUIT	101.45	GUIDELINES FOR THE VOLUNTARY NUTRITION LABELING OF RAW FRUIT, VEGETABLES, AND FISH Food Labeling
RAW MATERIAL	1304.31	REPORTS FROM MANUFACTURERS IMPORTING NARCOTIC RAW MATERIAL Reports; Records And Reports Of Registrants
RAW MILK	1210.26	PERMITS FOR RAW MILK OR CREAM Permit Control; Regulations Under The Federal Import Milk Act
RAW SHRIMP	161.175	FROZEN RAW BREADED SHRIMP Requirements For Specific Standardized Fish And Shellfish; Fish And Shellfish
RAW SHRIMP	161.176	FROZEN RAW LIGHTLY BREADED SHRIMP Requirements For Specific Standardized Fish And Shellfish; Fish And Shellfish
REACCREDITATION	900.15	APPEALS OF ADVERSE ACCREDITATION OR REACCREDITATION DECISIONS THAT PRECLUDE CERTIFICATION OR RECERTIFICATION Quality Standards And Certification; Mammography
REACTION	606.170	ADVERSE REACTION FILE Records And Reports; Current Good Manufacturing Practice For Blood And Blood Components
REACTIONS	640.73	REPORTING OF FATAL DONOR REACTIONS Source Plasma; Additional Standards For Human Blood And Blood Products
READER, BAR	886.5800	OPHTHALMIC BAR READER Therapeutic Devices; Ophthalmic Devices
READER, PRISM	886.5810	OPHTHALMIC PRISM READER Therapeutic Devices; Ophthalmic Devices
READER, ZONE	866.2850	AUTOMATED ZONE READER Microbiology Devices; Immunology And Microbiology Devices
READING SYSTEM	886.5820	CLOSED-CIRCUIT TELEVISION READING SYSTEM Therapeutic Devices; Ophthalmic Devices
REAGENT	660.20	BLOOD GROUPING REAGENT Blood Grouping Reagent; Additional Standards For Diagnostic Substances For Laboratory Tests
REAGENT	660.30	REAGENT RED BLOOD CELLS Reagent Red Blood Cells; Additional Standards For Diagnostic Substances For Laboratory Tests
REAGENT	864.4010	GENERAL PURPOSE REAGENT Specimen Preparation Reagents; Hematology And Pathology Devices

REAGENT	864.4020	ANALYTE SPECIFIC REAGENT Specimen Preparation Reagents; Hematology And Pathology Devices
REAGENT	864.8100	BOTHROPS ATROX REAGENT Hematology Reagents; Hematology And Pathology Devices
REAGENT	864.8540	RED CELL LYSING REAGENT Hematology Reagents; Hematology And Pathology Devices
REAGENT	864.8950	RUSSELL VIPER VENOM REAGENT Hematology Reagents; Hematology And Pathology Devices
REAGENT	866.4100	COMPLEMENT REAGENT Immunology Laboratory Equipment And Reagents; Immunology And Microbiology Devices
REAGENTS	58.83	REAGENTS AND SOLUTIONS Testing Facilities Operation; Good Laboratory Practice For Nonclinical Laboratory Studies
REAGENTS	606.65	SUPPLIES AND REAGENTS Equipment; Current Good Manufacturing Practice For Blood And Blood Components
REAGENTS	809.30	RESTRICTIONS ON THE SALE, DISTRIBUTION AND USE OF ANALYTE SPECIFIC REAGENTS Requirements For Manufacturers And Producers; In Vitro Diagnostic Products For Human Use
REAGENTS	864.1860	IMMUNOHISTOCHEMISTRY REAGENTS AND KITS Biological Stains; Hematology And Pathology Devices
REAGENTS	864.9225	CELL-FREEZING APPARATUS AND REAGENTS FOR IN VITRO DIAGNOSTIC USE Products Used In Establishments That Manufacture Blood And Blood Products; Hematology And Pathology Devices
REAGENTS	864.9650	QUALITY CONTROL KIT FOR BLOOD BANKING REAGENTS Products Used In Establishments That Manufacture Blood And Blood Products; Hematology And Pathology Devices
REAGENTS	866.3010	ACINETOBACTER CALCOACETICUS SEROLOGICAL REAGENTS Serological Reagents; Immunology And Microbiology Devices
REAGENTS	866.3020	ADENOVIRUS SEROLOGICAL REAGENTS Serological Reagents; Immunology And Microbiology Devices
REAGENTS	866.3035	ARIZONA SPP. SEROLOGICAL REAGENTS Serological Reagents; Immunology And Microbiology Devices
REAGENTS	866.3040	APERGILLUS SPP. SEROLOGICAL REAGENTS Serological Reagents; Immunology And Microbiology Devices
REAGENTS	866.3060	BLASTOMYCES DERMATITIDIS SEROLOGICAL REAGENTS Serological Reagents; Immunology And Microbiology Devices
REAGENTS	866.3065	BORDETELLA SPP. SEROLOGICAL REAGENTS Serological Reagents; Immunology And Microbiology Devices
REAGENTS	866.3085	BRUCELLA SPP. SEROLOGICAL REAGENTS Serological Reagents; Immunology And Microbiology Devices
REAGENTS	866.3110	CAMPYLOBACTER FETUS SEROLOGICAL REAGENTS Serological Reagents; Immunology And Microbiology Devices
REAGENTS	866.3120	CHLAMYDIA SEROLOGICAL REAGENTS Serological Reagents; Immunology And Microbiology Devices
REAGENTS	866.3125	CITROBACTER SPP. SEROLOGICAL REAGENTS Serological Reagents; Immunology And Microbiology Devices
REAGENTS	866.3135	COCCIDIOIDES IMMITIS SEROLOGICAL REAGENTS Serological Reagents; Immunology And Microbiology Devices
REAGENTS	866.3140	CORYNEBACTERIUM SPP. SEROLOGICAL REAGENTS Serological Reagents; Immunology And Microbiology Devices
REAGENTS	866.3145	COXSACKIEVIRUS SEROLOGICAL REAGENTS Serological Reagents; Immunology And Microbiology Devices
REAGENTS	866.3165	CRYPTOCOCCUS NEOFORMANS SEROLOGICAL REAGENTS Serological Reagents; Immunology And Microbiology Devices
REAGENTS	866.3175	CYTOMEGALOVIRUS SEROLOGICAL REAGENTS Serological Reagents; Immunology And Microbiology Devices
REAGENTS	866.3200	ECHINOCOCCUS SPP. SEROLOGICAL REAGENTS Serological Reagents; Immunology And Microbiology Devices
REAGENTS	866.3205	ECHOVIRUS SEROLOGICAL REAGENTS Serological Reagents; Immunology And Microbiology Devices
REAGENTS	866.3220	ENTAMOEBA HISTOLYTICA SEROLOGICAL REAGENTS Serological Reagents; Immunology And Microbiology Devices
REAGENTS	866.3235	EPSTEIN-BARR VIRUS SEROLOGICAL REAGENTS Serological Reagents; Immunology And Microbiology Devices
REAGENTS	866.3240	EQUINE ENCEPHALOMYELITIS VIRUS SEROLOGICAL REAGENTS Serological Reagents; Immunology And Microbiology Devices
REAGENTS	866.3250	ERYSIPELOTHRIX RHUSIOPATHIAE SEROLOGICAL REAGENTS Serological Reagents; Immunology And Microbiology Devices
REAGENTS	866.3255	ESCHERICHIA COLI SEROLOGICAL REAGENTS Serological Reagents; Immunology And Microbiology Devices

REAGENTS	866.3270	FLAVOBACTERIUM SPP. SEROLOGICAL REAGENTS
		Serological Reagents; Immunology And Microbiology Devices
REAGENTS	866.3280	FRANCISELLA TULARENSIS SEROLOGICAL REAGENTS
		Serological Reagents; Immunology And Microbiology Devices
REAGENTS	866.3300	HAEMOPHILUS SPP. SEROLOGICAL REAGENTS
		Serological Reagents; Immunology And Microbiology Devices
REAGENTS	866.3305	HERPES SIMPLEX VIRUS SEROLOGICAL REAGENTS
		Serological Reagents; Immunology And Microbiology Devices
REAGENTS	866.3320	HISTOPLASMA CAPSULATUM SEROLOGICAL REAGENTS
		Serological Reagents; Immunology And Microbiology Devices
REAGENTS	866.3330	INFLUENZA VIRUS SEROLOGICAL REAGENTS
		Serological Reagents; Immunology And Microbiology Devices
REAGENTS	866.3340	KLEBSIELLA SPP. SEROLOGICAL REAGENTS
		Serological Reagents; Immunology And Microbiology Devices
REAGENTS	866.3350	LEPTOSPIRA SPP. SEROLOGICAL REAGENTS
		Serological Reagents; Immunology And Microbiology Devices
REAGENTS	866.3355	LISTERIA SPP. SEROLOGICAL REAGENTS
		Serological Reagents; Immunology And Microbiology Devices
REAGENTS	866.3360	LYMPHOCYTIC CHORIOMENINGITIS VIRUS SEROLOGICAL REAGENTS
		Serological Reagents; Immunology And Microbiology Devices
REAGENTS	866.3370	MYCOBACTERIUM TUBERCULOSIS IMMUNOFLUORESCENT REAGENTS
		Serological Reagents; Immunology And Microbiology Devices
REAGENTS	866.3375	MYCOPLASMA SPP. SEROLOGICAL REAGENTS
		Serological Reagents; Immunology And Microbiology Devices
REAGENTS	866.3380	MUMPS VIRUS SEROLOGICAL REAGENTS
		Serological Reagents; Immunology And Microbiology Devices
REAGENTS	866.3390	NEISSERIA SPP. DIRECT SEROLOGICAL TEST REAGENTS
		Serological Reagents; Immunology And Microbiology Devices
REAGENTS	866.3400	PARAINFLUENZA VIRUS SEROLOGICAL REAGENTS
		Serological Reagents; Immunology And Microbiology Devices
REAGENTS	866.3405	POLIOVIRUS SEROLOGICAL REAGENTS
		Serological Reagents; Immunology And Microbiology Devices
REAGENTS	866.3410	PROTEUS SPP. (WEIL-FELIX) SEROLOGICAL REAGENTS
		Serological Reagents; Immunology And Microbiology Devices
REAGENTS	866.3415	PSEUDOMONAS SPP. SEROLOGICAL REAGENTS
		Serological Reagents; Immunology And Microbiology Devices
REAGENTS	866.3460	RABIESVIRUS IMMUNOFLUORESCENT REAGENTS
		Serological Reagents; Immunology And Microbiology Devices
REAGENTS	866.3470	REOVIRUS SEROLOGICAL REAGENTS
		Serological Reagents; Immunology And Microbiology Devices
REAGENTS	866.3480	RESPIRATORY SYNCYTIAL VIRUS SEROLOGICAL REAGENTS
		Serological Reagents; Immunology And Microbiology Devices
REAGENTS	866.3490	RHINOVIRUS SEROLOGICAL REAGENTS
		Serological Reagents; Immunology And Microbiology Devices
REAGENTS	866.3500	RICKETTSIA SEROLOGICAL REAGENTS
		Serological Reagents; Immunology And Microbiology Devices
REAGENTS	866.3510	RUBELLA VIRUS SEROLOGICAL REAGENTS
		Serological Reagents; Immunology And Microbiology Devices
REAGENTS	866.3520	RUBEOLA (MEASLES) VIRUS SEROLOGICAL REAGENTS
		Serological Reagents; Immunology And Microbiology Devices
REAGENTS	866.3550	SALMONELLA SPP. SEROLOGICAL REAGENTS
		Serological Reagents; Immunology And Microbiology Devices
REAGENTS	866.3600	SCHISTOSOMA SPP. SEROLOGICAL REAGENTS
		Serological Reagents; Immunology And Microbiology Devices
REAGENTS	866.3630	SERRATIA SPP. SEROLOGICAL REAGENTS
		Serological Reagents; Immunology And Microbiology Devices
REAGENTS	866.3660	SHIGELLA SPP. SEROLOGICAL REAGENTS
		Serological Reagents; Immunology And Microbiology Devices
REAGENTS	866.3680	SPOROTHRIX SCHENCKII SEROLOGICAL REAGENTS
		Serological Reagents; Immunology And Microbiology Devices
REAGENTS	866.3700	STAPHYLOCOCCUS AUREUS SEROLOGICAL REAGENTS
		Serological Reagents; Immunology And Microbiology Devices
REAGENTS	866.3720	STREPTOCOCCUS SPP. EXOENZYME REAGENTS
		Serological Reagents; Immunology And Microbiology Devices
REAGENTS	866.3740	STREPTOCOCCUS SPP. SEROLOGICAL REAGENTS
		Serological Reagents; Immunology And Microbiology Devices
REAGENTS	866.3780	TOXOPLASMA GONDII SEROLOGICAL REAGENTS
		Serological Reagents; Immunology And Microbiology Devices
REAGENTS	866.3820	TREPONEMA PALLIDUM NONTREPONEMAL TEST REAGENTS
		Serological Reagents; Immunology And Microbiology Devices
REAGENTS	866.3830	TREPONEMA PALLIDUM TREPONEMAL TEST REAGENTS
		Serological Reagents; Immunology And Microbiology Devices
REAGENTS	866.3850	TRICHINELLA SPIRALIS SEROLOGICAL REAGENTS
		Serological Reagents; Immunology And Microbiology Devices

REAGENTS	866.3870	TRYPANOSOMA SPP. SEROLOGICAL REAGENTS Serological Reagents; Immunology And Microbiology Devices
REAGENTS	866.3900	VARICELLA-ZOSTER VIRUS SEROLOGICAL REAGENTS Serological Reagents; Immunology And Microbiology Devices
REAGENTS	866.3930	VIBRIO CHOLERAE SEROLOGICAL REAGENTS Serological Reagents; Immunology And Microbiology Devices
REAL PROPERTY	1403.31	REAL PROPERTY Changes, Property, and Subawards; Uniform Administrative Requirements For Grants And Coop. Agreements To State & Local Gov.
REASONS	514.110	REASONS FOR REFUSING TO FILE APPLICATIONS Administrative Actions On Applications; New Animal Drug Applications
REBASING	872.3760	DENTURE RELINING, REPAIRING, OR REBASING RESIN Prosthetic Devices; Dental Devices
REBREATHING	868.5675	REBREATHING DEVICE Therapeutic Devices; Anesthesiology Devices
REBREATHING	892.1390	RADIONUCLIDE REBREATHING SYSTEM Diagnostic Devices; Radiology Devices
RECALL	5.411	MEDICAL DEVICE RECALL AUTHORITY Medical Devices And Radiological Health; Redelegations Of Authority; Delegations Of Authority And Organization
RECALL	7.40	RECALL POLICY Recalls - Guidance on Policy, Procedures & Industry Responsibilities; Enforcement Policy
RECALL	7.41	HEALTH HAZARD EVALUATION AND RECALL CLASSIFICATION Recalls - Guidance on Policy, Procedures & Industry Responsibilities; Enforcement Policy
RECALL	7.42	RECALL STRATEGY Recalls - Guidance on Policy, Procedures & Industry Responsibilities; Enforcement Policy
RECALL	7.45	FOOD AND DRUG ADMINISTRATION-REQUESTED RECALL Recalls - Guidance on Policy, Procedures & Industry Responsibilities; Enforcement Policy
RECALL	7.46	FIRM-INITIATED RECALL Recalls - Guidance on Policy, Procedures & Industry Responsibilities; Enforcement Policy
RECALL	7.49	RECALL COMMUNICATIONS Recalls - Guidance on Policy, Procedures & Industry Responsibilities; Enforcement Policy
RECALL	7.50	PUBLIC NOTIFICATION OF RECALL Recalls - Guidance on Policy, Procedures & Industry Responsibilities; Enforcement Policy
RECALL	7.53	RECALL STATUS REPORTS Recalls - Guidance on Policy, Procedures & Industry Responsibilities; Enforcement Policy
RECALL	7.55	TERMINATION OF A RECALL Recalls - Guidance on Policy, Procedures & Industry Responsibilities; Enforcement Policy
RECALL	107.200	FDA-REQUIRED RECALL Infant Formula Recalls; Infant Formula
RECALL	107.230	ELEMENTS OF AN INFANT FORMULA RECALL Infant Formula Recalls; Infant Formula
RECALL	107.250	TERMINATION OF AN INFANT FORMULA RECALL Infant Formula Recalls; Infant Formula
RECALL	107.260	REVISION OF AN INFANT FORMULA RECALL Infant Formula Recalls; Infant Formula
RECALL	1270.43	RETENTION, RECALL AND DESTRUCTION OF HUMAN TISSUE Human Tissue Intended For Transplantation
RECALL, MANDATORY	810.13	MANDATORY RECALL ORDER Mandatory Medical Device Recall Procedures; Medical Device Recall Authority
RECALL, MANDATORY	810.14	CEASE DISTRIBUTION AND NOTIFICATION OR MANDATORY RECALL STRATEGY Mandatory Medical Device Recall Procedures; Medical Device Recall Authority
RECALL, MANDATORY	810.15	COMMUNICATIONS CONCERNING A CEASE DISTRIBUTION AND NOTIFICATION OR MANDATORY RECALL ORDER Mandatory Medical Device Recall Procedures; Medical Device Recall Authority
RECALL, MANDATORY	810.16	CEASE DISTRIBUTION AND NOTIFICATION OR MANDATORY RECALL ORDER STATUS REPORTS Mandatory Medical Device Recall Procedures; Medical Device Recall Authority
RECALL, MANDATORY	810.17	TERMINATION OF A CEASE DISTRIBUTION AND NOTIFICATION OR MANDATORY RECALL ORDER Mandatory Medical Device Recall Procedures; Medical Device Recall Authority
RECALLS	107.220	SCOPE AND EFFECT OF INFANT FORMULA RECALLS Infant Formula Recalls; Infant Formula

RECOGNITION, FDA	316.34	FDA RECOGNITION OF EXCLUSIVE APPROVAL Orphan Drugs
RECOGNIZED AS SAFE	5.300	FOOD STANDARDS, FOOD ADDITIVES, GENERALLY RECOGNIZED AS SAFE (GRAS) SUBSTANCES, COLOR ADDITIVES, NUTRIENT CONTENT CLAIMS, AND HEALTH CLAIMS Foods And Cosmetics; Redelegations Of Authority; Delegations Of Authority And Organization
RECOGNIZED AS SAFE	170.30	ELIGIBILITY FOR CLASSIFICATION AS GENERALLY RECOGNIZED AS SAFE (GRAS) Food Additive Safety; Food Additives
RECOGNIZED AS SAFE	170.35	AFFIRMATION OF GENERALLY RECOGNIZED AS SAFE (GRAS) STATUS Food Additive Safety; Food Additives
RECOGNIZED AS SAFE	182.1	SUBSTANCES THAT ARE GENERALLY RECOGNIZED AS SAFE General Provisions; Substances Generally Recognized As Safe
RECOGNIZED AS SAFE	184.1	SUBSTANCES ADDED DIRECTLY TO HUMAN FOOD AFFIRMED AS GENERALLY RECOGNIZED AS SAFE (GRAS) General Provisions; Direct Food Substances Affirmed As Generally Recognized As Safe
RECOGNIZED AS SAFE	186.1	SUBSTANCES ADDED INDIRECTLY TO HUMAN FOOD AFFIRMED AS GENERALLY RECOGNIZED AS SAFE (GRAS) General Provisions; Indirect Food Substances Affirmed As Generally Recognized As Safe
RECOGNIZED AS SAFE	330.10	PROCEDURES FOR CLASSIFYING OTC DRUGS AS GENERALLY RECOGNIZED AS SAFE AND EFFECTIVE AND NOT MISBRANDED Administrative Procedures; Over-the-counter Human Drugs Generally Recognized As Safe And Effective & Not Misbranded
RECOGNIZED AS SAFE	330.14	ADDITIONAL CRITERIA AND PROCEDURES FOR CLASSIFYING OTC DRUGS AS GENERALLY RECOGNIZED AS SAFE AND EFFECTIVE AND NOT MISBRANDED Administrative Procedures; Over-the-counter Human Drugs Generally Recognized As Safe And Effective & Not Misbranded
RECOGNIZED AS SAFE	570.30	ELIGIBILITY FOR CLASSIFICATION AS GENERALLY RECOGNIZED AS SAFE (GRAS) Food Additive Safety; Food Additives
RECOGNIZED AS SAFE	570.35	AFFIRMATION OF GENERALLY RECOGNIZED AS SAFE (GRAS) STATUS Food Additive Safety; Food Additives
RECOGNIZED AS SAFE	582.1	SUBSTANCES THAT ARE GENERALLY RECOGNIZED AS SAFE General Provisions; Substances Generally Recognized As Safe
RECOMMENDATION	1000.50	RECOMMENDATION FOR THE USE OF SPECIFIC AREA GONAD SHIELDING ON PATIENTS DURING MEDICAL DIAGNOSTIC X-RAY PROCEDURES Radiation Protection Recommendations; General
RECOMMENDATION	1000.55	RECOMMENDATION FOR QUALITY ASSURANCE PROGRAMS IN DIAGNOSTIC RADIOLOGY FACILITIES Radiation Protection Recommendations; General
RECOMMENDATION	1000.60	RECOMMENDATION ON ADMINISTRATIVELY REQUIRED DENTAL X-RAY EXAMINATIONS Radiation Protection Recommendations; General
RECOMMENDATIONS	10.90	FDA REGULATIONS, RECOMMENDATIONS, AND AGREEMENTS General Administrative Procedures; Administrative Practices And Procedures
RECOMMENDATIONS	14.174	ADVICE AND RECOMMENDATIONS IN WRITING Advisory Committees For Human Prescription Drugs; Public Hearing Before A Public Advisory Committee
RECOMMENDATIONS, WRITTEN	316.10	CONTENT AND FORMAT OF A REQUEST FOR WRITTEN RECOMMENDATIONS Orphan Drugs
RECOMMENDATIONS, WRITTEN	316.12	PROVIDING WRITTEN RECOMMENDATIONS Orphan Drugs
RECOMMENDATIONS, WRITTEN	316.14	REFUSAL TO PROVIDE WRITTEN RECOMMENDATIONS Orphan Drugs
RECOMMENDED	330.13	CONDITIONS FOR MARKETING INGREDIENTS RECOMMENDED FOR OVER-THE-COUNTER (OTC) USE UNDER THE OTC DRUG REVIEW Administrative Procedures; Over-the-counter Human Drugs Generally Recognized As Safe And Effective & Not Misbranded
RECOMMENDED USE	601.25	REVIEW PROCEDURES TO DETERMINE THAT LICENSED BIOLOGICAL PRODUCTS ARE SAFE, EFFECTIVE, NOT MISBRANDED UNDER PRESCRIBED, RECOMMENDED, OR SUGGESTED CONDITIONS OF USE Biologics Licensing; Licensing
RECOMMENDED USE	601.26	RECLASSIFICATION PROCEDURES TO DETERMINE THAT LICENSED BIOLOGICAL PRODUCTS ARE SAFE, EFFECTIVE, NOT MISBRANDED UNDER PRESCRIBED, RECOMMENDED, OR SUGGESTED CONDITIONS OF USE Biologics Licensing; Licensing
RECOMMENDED WARNING	369.20	DRUGS; RECOMMENDED WARNING AND CAUTION STATEMENTS Warning And Caution Statement For Drugs; Interpretative Statements Re: Warnings On Drugs And Devices For OTC Sale

RECONDITION	1.95	APPLICATION OF AUTHORIZATION TO RELABEL AND RECONDITION Imports And Exports; General Enforcement Regulations
RECONDITION	1.96	GRANTING OF AUTHORIZATION TO RELABEL AND RECONDITION Imports And Exports; General Enforcement Regulations
RECONDITIONING	1.99	COSTS CHARGEABLE IN CONNECTION WITH RELABELING AND RECONDITIONING INADMISSABLE IMPORTS Imports And Exports; General Enforcement Regulations
RECONSIDERATION	10.33	ADMINISTRATIVE RECONSIDERATION OF ACTION General Administrative Procedures; Administrative Practices And Procedures
RECONSIDERATION	12.139	RECONSIDERATION AND STAY OF ACTION Initial And Final Decisions; Formal Evidentiary Public Hearing
RECONSIDERATION	16.119	RECONSIDERATION AND STAY OF ACTION Reconsideration And Stay; Regulatory Hearing Before The Food And Drug Administration
RECORD	11.70	SIGNATURE/RECORD LINKING Electronic Records; Electronic Signatures
RECORD	12.100	ADMINISTRATIVE RECORD OF A HEARING Administrative Record; Formal Evidentiary Public Hearing
RECORD	12.105	EXAMINATION OF RECORD Administrative Record; Formal Evidentiary Public Hearing
RECORD	13.40	ADMINISTRATIVE RECORD OF ABOARD Records Of A Hearing Before A Board; Public Hearing Before A Public Board Of Inquiry
RECORD	13.45	EXAMINATION OF RECORD Records Of A Hearing Before A Board; Public Hearing Before A Public Board Of Inquiry
RECORD	13.50	RECORD FOR ADMINISTRATIVE DECISION Records Of A Hearing Before A Board; Public Hearing Before A Public Board Of Inquiry
RECORD	14.70	ADMINISTRATIVE RECORD OF A PUBLIC HEARING BEFORE AN ADVISORY COMMITTEE Records Of Meetings And Hearings Before Advisory Committees; Public Hearing Before A Public Advisory Committee
RECORD	14.75	EXAMINATION OF ADMINISTRATIVE RECORD AND OTHER ADVISORY COMMITTEE RECORDS Records Of Meetings And Hearings Before Advisory Committees; Public Hearing Before A Public Advisory Committee
RECORD	15.40	ADMINISTRATIVE RECORD Records Of A Public Hearing Before The Commissioner; Public Hearing Before The Commissioner
RECORD	15.45	EXAMINATION OF ADMINISTRATIVE RECORD Records Of A Public Hearing Before The Commissioner; Public Hearing Before The Commissioner
RECORD	16.80	ADMINISTRATIVE RECORD OF A REGULATORY HEARING Administrative Record And Decision; Regulatory Hearing Before The Food And Drug Administration
RECORD	16.85	EXAMINATION OF ADMINISTRATIVE RECORD Administrative Record And Decision; Regulatory Hearing Before The Food And Drug Administration
RECORD	16.95	ADMINISTRATIVE DECISION AND RECORD FOR DECISION Administrative Record And Decision; Regulatory Hearing Before The Food And Drug Administration
RECORD	17.41	THE ADMINISTRATIVE RECORD Civil Money Penalties Hearings
RECORD	21.74	PROVIDING NOTICE THAT A RECORD IS DISPUTED Disclosure Of Records In Privacy Act Record Systems To Persons Other ThanThe Subject Individual; Protection Of Privacy
RECORD	1316.46	INSPECTION OF RECORD Administrative Hearings; Administrative Functions, Practices, And Procedures
RECORD	1316.65	REPORT AND RECORD Administrative Hearings; Administrative Functions, Practices, And Procedures
RECORD, DEVICE HISTORY	820.184	DEVICE HISTORY RECORD Records; Quality System Regulation
RECORD, MASTER	225.102	MASTER RECORD FILE AND PRODUCTION RECORDS Records And Reports; Current Good Manufacturing Practice For Medicated Feeds
RECORD, MASTER	820.181	DEVICE MASTER RECORD Records; Quality System Regulation
RECORD, QUALITY SYSTEM	820.186	QUALITY SYSTEM RECORD Records; Quality System Regulation
RECORD, PRODUCTION	211.192	PRODUCTION RECORD REVIEW Records And Reports; Current Good Manufacturing Practice For Finished Pharmaceuticals
RECORD RETENTION	54.6	RECORDKEEPING AND RECORD RETENTION Financial Disclosure By Clinical Investigators

RECORD RETENTION	312.57	RECORDKEEPING AND RECORD RETENTION Responsibilities Of Sponsors And Investigators; Investigational New Drug Application
RECORD RETENTION	312.62	INVESTIGATOR RECORDKEEPING AND RECORD RETENTION Responsibilities Of Sponsors And Investigators; Investigational New Drug Application
RECORD SYSTEMS	21.20	PROCEDURES FOR NOTICE OF FDA PRIVACY ACT RECORD SYSTEMS FDA Privacy Act Record Systems; Protection Of Privacy
RECORD SYSTEMS	21.70	DISCLOSURE AND INTRA-AGENCY USE OF RECORDS IN PRIVACY ACT RECORD SYSTEMS; NO ACCOUNTING REQUIRED Disclosure Of Records In Privacy Act Record Systems To Persons Other ThanThe Subject Individual; Protection Of Privacy
RECORD SYSTEMS	21.71	DISCLOSURE OF RECORDS IN PRIVACY ACT RECORD SYSTEMS; ACCOUNTING REQUIRED Disclosure Of Records In Privacy Act Record Systems To Persons Other ThanThe Subject Individual; Protection Of Privacy
RECORD SYSTEMS	21.73	ACCURACY, COMPLETENESS, TIMELINESS, AND RELEVANCE OF RECORDS DISCLOSED FROM PRIVACY ACT RECORD SYSTEMS Disclosure Of Records In Privacy Act Record Systems To Persons Other ThanThe Subject Individual; Protection Of Privacy
RECORDER, CHART	870.2810	PAPER CHART RECORDER Cardiovascular Monitoring Devices; Cardiovascular Devices
RECORDER, TAPE	870.2800	MEDICAL MAGNETIC TAPE RECORDER Cardiovascular Monitoring Devices; Cardiovascular Devices
RECORDING	870.1220	ELECTRODE RECORDING CATHETER OR ELECTRODE RECORDING PROBE Cardiovascular Diagnostic Devices; Cardiovascular Devices
RECORDKEEPING	1.101	NOTIFICATION AND RECORDKEEPING Imports And Exports; General Enforcement Regulations
RECORDKEEPING	54.6	RECORDKEEPING AND RECORD RETENTION Financial Disclosure By Clinical Investigators
RECORDKEEPING	99.501	RECORDKEEPING AND REPORTS Dissemination Of Information On Unapproved/New Uses For Marketed Drugs, Biologics, and Devices
RECORDKEEPING	203.36	FULFILLMENT HOUSES, SHIPPING AND MAILING SERVICES, COMARKETING AGREEMENTS, AND THIRD-PARTY RECORDKEEPING Samples; Prescription Drug Marketing
RECORDKEEPING	312.57	RECORDKEEPING AND RECORD RETENTION Responsibilities Of Sponsors And Investigators; Investigational New Drug Application
RECORDKEEPING	312.62	INVESTIGATOR RECORDKEEPING AND RECORD RETENTION Responsibilities Of Sponsors And Investigators; Investigational New Drug Application
RECORDS	5.23	DISCLOSURE OF OFFICIAL RECORDS AND AUTHORIZATION OF TESTIMONY General Redelegations Of Authority; Delegations Of Authority And Organization
RECORDS	7.87	RECORDS RELATED TO OPPORTUNITIES FOR PRESENTATION OF VIEWS CONDUCTED BEFORE REPORT OF CRIMINAL VIOLATION Criminal Violations; Enforcement Policy
RECORDS	14.65	PUBLIC INQUIRIES AND REQUESTS FOR ADVISORY COMMITTEE RECORDS Records Of Meetings And Hearings Before Advisory Committees; Public Hearing Before A Public Advisory Committee
RECORDS	14.75	EXAMINATION OF ADMINISTRATIVE RECORD AND OTHER ADVISORY COMMITTEE RECORDS Records Of Meetings And Hearings Before Advisory Committees; Public Hearing Before A Public Advisory Committee
RECORDS	14.130	CONDUCT OF TEPRSSC MEETINGS; AVAILABILITY OF TEPRSSC RECORDS Technical Electronic Products Radiation Safety Standards Committee; Public Hearing Before A Public Advisory Committee
RECORDS	20.2	PRODUCTION OF RECORDS BY FDA EMPLOYEES Official Testimony And Information; Public Information
RECORDS	20.21	UNIFORM ACCESS TO RECORDS General Policy; Public Information
RECORDS	20.22	PARTIAL DISCLOSURE OF RECORDS General Policy; Public Information
RECORDS	20.23	REQUEST FOR EXISTING RECORDS General Policy; Public Information
RECORDS	20.24	PREPARATION OF NEW RECORDS General Policy; Public Information
RECORDS	20.26	INDEXES OF CERTAIN RECORDS General Policy; Public Information
RECORDS	20.27	SUBMISSION OF RECORDS MARKED AS CONFIDENTIAL General Policy; Public Information
RECORDS	20.29	PROHIBITION ON WITHDRAWAL OF RECORDS FROM FDA FILES General Policy; Public Information

RECORDS	20.40	FILING A REQUEST FOR RECORDS Procedures And Fees; Public Information
RECORDS	20.47	DENIAL OF A REQUEST FOR RECORDS Procedures And Fees; Public Information
RECORDS	20.49	REFERRAL TO PRIMARY SOURCE OF RECORDS Procedures And Fees; Public Information
RECORDS	20.50	AVAILABILITY OF RECORDS AT NATIONAL TECHNICAL INFORMATION SERVICE Procedures And Fees; Public Information
RECORDS	20.64	RECORDS OR INFORMATION COMPILED FOR LAW ENFORCEMENT PURPOSES Exemptions; Public Information
RECORDS	20.101	ADMINISTRATIVE ENFORCEMENT RECORDS Availability Of Specific Categories Of Records; Public Information
RECORDS	20.102	COURT ENFORCEMENT RECORDS Availability Of Specific Categories Of Records; Public Information
RECORDS	20.118	ADVISORY COMMITTEE RECORDS Availability Of Specific Categories Of Records; Public Information
RECORDS	21.10	POLICY CONCERNING RECORDS ABOUT INDIVIDUALS General Provisions; Protection Of Privacy
RECORDS	21.30	RECORDS OF CONTRACTORS Requirements For Specific Categories Of Records; Protection Of Privacy
RECORDS	21.31	RECORDS STORED BY THE NATIONAL ARCHIVES AND RECORDS ADMINISTRATION Requirements For Specific Categories Of Records; Protection Of Privacy
RECORDS	21.43	ACCESS TO REQUESTED RECORDS Procedures For Notification Of And Access To Records In Privacy Act Record Systems; Protection Of Privacy
RECORDS	21.50	PROCEDURES FOR SUBMITTING REQUESTS FOR AMENDMENT OF RECORDS Procedures For Requests For Amendment Of Records; Protection Of Privacy
RECORDS	21.51	RESPONSES TO REQUESTS FOR AMENDMENT OF RECORDS Procedures For Requests For Amendment Of Records; Protection Of Privacy
RECORDS	21.52	ADMINISTRATIVE APPEALS OF REFUSALS TO AMEND RECORDS Procedures For Requests For Amendment Of Records; Protection Of Privacy
RECORDS	21.53	NOTATION AND DISCLOSURE OF DISPUTED RECORDS Procedures For Requests For Amendment Of Records; Protection Of Privacy
RECORDS	21.54	AMENDED OR DISPUTED RECORDS RECEIVED FROM OTHER AGENCIES Procedures For Requests For Amendment Of Records; Protection Of Privacy
RECORDS	21.65	ACCESS TO RECORDS IN EXEMPT SYSTEMS Exemptions; Protection Of Privacy
RECORDS	21.70	DISCLOSURE AND INTRA-AGENCY USE OF RECORDS IN PRIVACY ACT RECORD SYSTEMS; NO ACCOUNTING REQUIRED Disclosure Of Records In Privacy Act Record Systems To Persons Other Than The Subject Individual; Protection Of Privacy
RECORDS	21.71	DISCLOSURE OF RECORDS IN PRIVACY ACT RECORD SYSTEMS; ACCOUNTING REQUIRED Disclosure Of Records In Privacy Act Record Systems To Persons Other Than The Subject Individual; Protection Of Privacy
RECORDS	21.72	INDIVIDUAL CONSENT TO DISCLOSURE OF RECORDS TO OTHER PERSONS Disclosure Of Records In Privacy Act Record Systems To Persons Other Than The Subject Individual; Protection Of Privacy
RECORDS	21.73	ACCURACY, COMPLETENESS, TIMELINESS, AND RELEVANCE OF RECORDS DISCLOSED FROM PRIVACY ACT RECORD SYSTEMS Disclosure Of Records In Privacy Act Record Systems To Persons Other Than The Subject Individual; Protection Of Privacy
RECORDS	25.43	RECORDS OF DECISION Preparation Of Environmental Documents; Environmental Impact Considerations
RECORDS	58.190	STORAGE AND RETRIEVAL OF RECORDS AND DATA Records And Reports; Good Laboratory Practice For Nonclinical Laboratory Studies
RECORDS	58.195	RETENTION OF RECORDS Records And Reports; Good Laboratory Practice For Nonclinical Laboratory Studies
RECORDS	106.100	RECORDS Records And Reports; Infant Formula Quality Control Procedures
RECORDS	107.280	RECORDS RETENTION Infant Formula Recalls; Infant Formula
RECORDS	113.100	PROCESSING AND PRODUCTION RECORDS Records And Reports; Thermally Processed Low-acid Foods Packaged In Hermetically Sealed Containers
RECORDS	114.100	RECORDS Records And Reports; Acidified Foods
RECORDS	120.12	RECORDS Hazard Analysis And Critical Control Point (HACCP) Systems

RECORDS	123.9	RECORDS Fish And Fishery Products
RECORDS	203.60	REQUEST AND RECEIPT FORMS, REPORTS AND RECORDS Prescription Drug Marketing
RECORDS	310.303	CONTINUATION OF LONG-TERM STUDIES, RECORDS, AND REPORTS ON CERTAIN DRUGS FOR WHICH NEW DRUG APPLICATIONS HAVE BEEN APPROVED Records And Reports; New Drugs
RECORDS	310.305	RECORDS AND REPORTS CONTAINING ADVERSE DRUG EXPERIENCES ON MARKETED PRESCRIPTION DRUGS FOR HUMAN USE WITHOUT APPROVED NEW DRUG APPLICATIONS Records And Reports; New Drugs
RECORDS	312.58	INSPECTION OF SPONSOR'S RECORDS AND REPORTS Investigational New Drug Application
RECORDS	312.68	INSPECTION OF INVESTIGATOR'S RECORDS AND REPORTS Investigational New Drug Application
RECORDS	510.300	RECORDS AND REPORTS CONCERNING EXPERIENCE WITH NEW ANIMAL DRUGS FOR WHICH AN APPROVED APPLICATION IS IN EFFECT Records And Reports; New Animal Drugs
RECORDS	510.301	RECORDS AND REPORTS CONCERNING EXPERIENCE WITH ANIMAL FEEDS BEARING OR CONTAINING NEW ANIMAL DRUGS FOR WHICH AN APPROVED MEDICATED FEED MILL APPLICATION IS IN EFFECT Records And Reports; New Animal Drugs
RECORDS	514.80	RECORDS AND REPORTS CONCERNING EXPERIENCE WITH APPROVED NEW ANIMAL DRUGS Administrative Actions On Applications; New Animal Drug Applications
RECORDS	600.12	RECORDS Establishment Standards; Biological Products: General
RECORDS	606.160	RECORDS Records And Reports; Current Good Manufacturing Practice For Blood And Blood Components
RECORDS	640.72	RECORDS Source Plasma; Additional Standards For Human Blood And Blood Products
RECORDS	803.18	FILES AND DISTRIBUTOR RECORDS General Provisions; Medical Device Reporting
RECORDS	806.20	RECORDS OF CORRECTIONS AND REMOVALS NOT REQUIRED TO BE REPORTED Medical Devices; Reports Of Corrections And Removals
RECORDS	806.30	FDA ACCESS TO RECORDS Medical Devices; Reports Of Corrections And Removals
RECORDS	812.140	RECORDS Records And Reports; Investigational Device Exemptions
RECORDS	821.60	RETENTION OF RECORDS Medical Device Tracking Requirements
RECORDS	1002.30	RECORDS TO BE MAINTAINED BY MANUFACTURERS Manufacturers' Records; Records and Reports
RECORDS	1002.31	PRESERVATION AND INSPECTION OF RECORDS Manufacturers' Records; Records and Reports
RECORDS	1002.40	RECORDS TO BE OBTAINED BY DEALERS AND DISTRIBUTORS Dealer And Distributor Records; Records and Reports
RECORDS	1002.41	DISPOSITION OF RECORDS OBTAINED BY DEALERS AND DISTRIBUTORS Dealer And Distributor Records; Records and Reports
RECORDS	1002.42	CONFIDENTIALITY OF RECORDS FURNISHED BY DEALERS AND DISTRIBUTORS Dealer And Distributor Records; Records and Reports
RECORDS	1270.33	RECORDS; GENERAL REQUIREMENTS Human Tissue Intended For Transplantation
RECORDS	1270.35	SPECIFIC RECORDS Human Tissue Intended For Transplantation
RECORDS	1304.03	PERSONS REQUIRED TO KEEP RECORDS AND FILE REPORTS General Information; Records And Reports Of Registrants
RECORDS	1304.04	MAINTENANCE OF RECORDS AND INVENTORIES General Information; Records And Reports Of Registrants
RECORDS	1304.21	GENERAL REQUIREMENTS FOR CONTINUING RECORDS Continuing Records; Records And Reports Of Registrants
RECORDS	1304.22	RECORDS FOR MANUFACTURERS, DISTRIBUTORS, DISPENSERS, RESEARCHERS, IMPORTERS, AND EXPORTERS Continuing Records; Records And Reports Of Registrants
RECORDS	1304.23	RECORDS FOR CHEMICAL ANALYSTS Continuing Records; Records And Reports Of Registrants
RECORDS	1304.24	RECORDS FOR MAINTENANCE TREATMENT PROGRAMS AND DETOXIFICATION TREATMENT PROGRAMS Continuing Records; Records And Reports Of Registrants

RECORDS	1304.25	RECORDS FOR TREATMENT PROGRAMS WHICH COMPOUND NARCOTICS FOR TREATMENT PROGRAMS AND OTHER LOCATIONS Continuing Records; Records And Reports Of Registrants
RECORDS	1310.03	PERSONS REQUIRED TO KEEP RECORDS AND FILE REPORTS Records and Reports Of Listed Chemicals And Certain Machines
RECORDS	1310.04	MAINTENANCE OF RECORDS Records and Reports Of Listed Chemicals And Certain Machines
RECORDS	1310.06	CONTENT OF RECORDS AND REPORTS Records and Reports Of Listed Chemicals And Certain Machines
RECORDS	1312.26	RECORDS REQUIRED OF EXPORTER Exportation Of Controlled Substances; Importation And Exportation Of Controlled Substances
RECORDS	1316.34	RECORDS OF PROCEEDING Enforcement Proceedings; Administrative Functions, Practices, And Procedures
RECORDS	1401.5	HOW TO REQUEST RECORDS Public Availability Of Information
RECORDS	1403.42	RETENTION AND ACCESS REQUIREMENTS FOR RECORDS Reports, Records, Retention, And Enforcement; Uniform Administrative Requirements For Grants And Coop. Agreements To State And Local Gov.
RECORDS, BATCH PRODUCTION	211.188	BATCH PRODUCTION AND CONTROL RECORDS Records And Reports; Current Good Manufacturing Practice For Finished Pharmaceuticals
RECORDS, BATCH PRODUCTION	226.102	MASTER-FORMULA AND BATCH-PRODUCTION RECORDS Records And Reports; Current Good Manufacturing Practice For Type A Medicated Articles
RECORDS, DISTRIBUTION	80.39	RECORDS OF DISTRIBUTION Certification Procedures; Color Additive Certification
RECORDS, DISTRIBUTION	205.50	MINIMUM REQUIREMENTS FOR STORAGE AND HANDLING OF PRESCRIPTION DRUGS AND ESTABLISHMENT AND MAINTENANCE OF PRESCRIPTION DRUG DISTRIBUTION RECORDS Guidelines For State Licensing Of Wholesale Prescription Drug Distributors
RECORDS, DISTRIBUTION	211.196	DISTRIBUTION RECORDS Records And Reports; Current Good Manufacturing Practice For Finished Pharmaceuticals
RECORDS, DISTRIBUTION	225.110	DISTRIBUTION RECORDS Records And Reports; Current Good Manufacturing Practice For Medicated Feeds
RECORDS, DISTRIBUTION	225.202	FORMULA, PRODUCTION, AND DISTRIBUTION RECORDS Records; Current Good Manufacturing Practice For Medicated Feeds
RECORDS, DISTRIBUTION	226.110	DISTRIBUTION RECORDS Records And Reports; Current Good Manufacturing Practice For Type A Medicated Articles
RECORDS, DISTRIBUTION	606.165	DISTRIBUTION AND RECEIPT; PROCEDURES AND RECORDS Records And Reports; Current Good Manufacturing Practice For Blood And Blood Components
RECORDS, FDA	20.3	CERTIFICATION AND AUTHENTICATION OF FDA RECORDS Official Testimony And Information; Public Information
RECORDS, FDA	20.20	POLICY ON DISCLOSURE OF FDA RECORDS General Policy; Public Information
RECORDS, FDA	20.31	RETENTION SCHEDULE OF REQUESTS FOR FDA RECORDS General Policy; Public Information
RECORDS, IRB	56.115	IRB RECORDS Records And Reports; Institutional Review Boards
RECORDS, LABELING	211.184	COMPONENT, DRUG PRODUCT CONTAINER, CLOSURE, AND LABELING RECORDS Records And Reports; Current Good Manufacturing Practice For Finished Pharmaceuticals
RECORDS, LABORATORY	211.194	LABORATORY RECORDS Records And Reports; Current Good Manufacturing Practice For Finished Pharmaceuticals
RECORDS, MEDICAL	21.33	MEDICAL RECORDS Requirements For Specific Categories Of Records; Protection Of Privacy
RECORDS, PERSONNEL	21.32	PERSONNEL RECORDS Requirements For Specific Categories Of Records; Protection Of Privacy
RECORDS, PRODUCTION	211.186	MASTER PRODUCTION AND CONTROL RECORDS Records And Reports; Current Good Manufacturing Practice For Finished Pharmaceuticals
RECORDS, PRODUCTION	225.102	MASTER RECORD FILE AND PRODUCTION RECORDS Records And Reports; Current Good Manufacturing Practice For Medicated Feeds
RECORDS, VETERINARY	530.5	VETERINARY RECORDS Extralabel Drug Use In Animals
RECTAL DILATOR	876.5450	RECTAL DILATOR Therapeutic Devices; Gastroenterology-urology Devices
RECTILINEAR SCANNER	892.1300	NUCLEAR RECTILINEAR SCANNER Diagnostic Devices; Radiology Devices

RED ALGAE	184.1121	RED ALGAE Listing Of Specific Substances Affirmed As GRAS; Direct Food Substances Affirmed As Generally Recognized As Safe
RED BLOOD CELL	864.7100	RED BLOOD CELL ENZYME ASSAY Hematology Kits And Packages; Hematology And Pathology Devices
RED BLOOD CELLS	640.10	RED BLOOD CELLS Red Blood Cells; Additional Standards For Human Blood And Blood Products
RED BLOOD CELLS	660.30	REAGENT RED BLOOD CELLS Reagent Red Blood Cells; Additional Standards For Diagnostic Substances For Laboratory Tests
RED CELL	864.5300	RED CELL INDICES DEVICE Automated And Semi-automated Hematology Devices; Hematology And Pathology Devices
RED CELL	864.8185	CALIBRATOR FOR RED CELL AND WHITE CELL COUNTING Hematology Reagents; Hematology And Pathology Devices
RED CELL	864.8540	RED CELL LYSING REAGENT Hematology Reagents; Hematology And Pathology Devices
RED NO. 2	74.302	CITRUS RED NO. 2 Foods; Listing Of Color Additives Subject To Certification
RED NO. 3	74.303	FD&C RED NO. 3 Foods; Listing Of Color Additives Subject To Certification
RED NO. 3	74.1303	FD&C RED NO. 3 Drugs; Listing Of Color Additives Subject To Certification
RED NO. 4	74.1304	FD&C RED NO. 4 Drugs; Listing Of Color Additives Subject To Certification
RED NO. 4	74.2304	FD&C RED NO. 4 Cosmetics; Listing Of Color Additives Subject To Certification
RED NO. 4	82.304	FD&C RED NO. 4 Foods, Drugs, And Cosmetics; Listing Of Certified Provisionally Listed Colors And Specifications
RED NO. 6	74.1306	D&C RED NO. 6 Drugs; Listing Of Color Additives Subject To Certification
RED NO. 6	74.2306	D&C RED NO. 6 Cosmetics; Listing Of Color Additives Subject To Certification
RED NO. 6	82.1306	D&C RED NO. 6 Drugs And Cosmetics; Listing Of Certified Provisionally Listed Colors And Specifications
RED NO. 7	74.1307	D&C RED NO. 7 Drugs; Listing Of Color Additives Subject To Certification
RED NO. 7	74.2307	D&C RED NO. 7 Cosmetics; Listing Of Color Additives Subject To Certification
RED NO. 7	82.1307	D&C RED NO. 7 Drugs And Cosmetics; Listing Of Certified Provisionally Listed Colors And Specifications
RED NO. 17	74.1317	D&C RED NO. 17 Drugs; Listing Of Color Additives Subject To Certification
RED NO. 17	74.2317	D&C RED NO. 17 Cosmetics; Listing Of Color Additives Subject To Certification
RED NO. 17	74.3230	D&C RED NO. 17 Medical Devices; Listing Of Color Additives Subject To Certification
RED NO. 17	82.1317	D&C RED NO. 17 Drugs And Cosmetics; Listing Of Certified Provisionally Listed Colors And Specifications
RED NO. 21	74.1321	D&C RED NO. 21 Drugs; Listing Of Color Additives Subject To Certification
RED NO. 21	74.2321	D&C RED NO. 21 Cosmetics; Listing Of Color Additives Subject To Certification
RED NO. 21	82.1321	D&C RED NO. 21 Drugs And Cosmetics; Listing Of Certified Provisionally Listed Colors And Specifications
RED NO. 22	74.1322	D&C RED NO. 22 Drugs; Listing Of Color Additives Subject To Certification
RED NO. 22	74.2322	D&C RED NO. 22 Cosmetics; Listing Of Color Additives Subject To Certification
RED NO. 22	82.1322	D&C RED NO. 22 Drugs And Cosmetics; Listing Of Certified Provisionally Listed Colors And Specifications
RED NO. 27	74.1327	D&C RED NO. 27 Drugs; Listing Of Color Additives Subject To Certification
RED NO. 27	74.2327	D&C RED NO. 27 Cosmetics; Listing Of Color Additives Subject To Certification
RED NO. 27	82.1327	D&C RED NO. 27 Drugs And Cosmetics; Listing Of Certified Provisionally Listed Colors And Specifications

RED NO. 28	74.1328	D&C RED NO. 28 Drugs; Listing Of Color Additives Subject To Certification
RED NO. 28	74.2328	D&C RED NO. 28 Cosmetics; Listing Of Color Additives Subject To Certification
RED NO. 28	82.1328	D&C RED NO. 28 Drugs And Cosmetics; Listing Of Certified Provisionally Listed Colors And Specifications
RED NO. 30	74.1330	D&C RED NO. 30 Drugs; Listing Of Color Additives Subject To Certification
RED NO. 30	74.2330	D&C RED NO. 30 Cosmetics; Listing Of Color Additives Subject To Certification
RED NO. 30	82.1330	D&C RED NO. 30 Drugs And Cosmetics; Listing Of Certified Provisionally Listed Colors And Specifications
RED NO. 31	74.1331	D&C RED NO. 31 Drugs; Listing Of Color Additives Subject To Certification
RED NO. 31	74.2331	D&C RED NO. 31 Cosmetics; Listing Of Color Additives Subject To Certification
RED NO. 31	82.1331	D&C RED NO. 31 Drugs And Cosmetics; Listing Of Certified Provisionally Listed Colors And Specifications
RED NO. 33	74.1333	D&C RED NO. 33 Drugs; Listing Of Color Additives Subject To Certification
RED NO. 33	74.2333	D&C RED NO. 33 Cosmetics; Listing Of Color Additives Subject To Certification
RED NO. 33	82.1333	D&C RED NO. 33 Drugs And Cosmetics; Listing Of Certified Provisionally Listed Colors And Specifications
RED NO. 34	74.1334	D&C RED NO. 34 Drugs; Listing Of Color Additives Subject To Certification
RED NO. 34	74.2334	D&C RED NO. 34 Cosmetics; Listing Of Color Additives Subject To Certification
RED NO. 34	82.1334	D&C RED NO. 34 Drugs And Cosmetics; Listing Of Certified Provisionally Listed Colors And Specifications
RED NO. 36	74.1336	D&C RED NO. 36 Drugs; Listing Of Color Additives Subject To Certification
RED NO. 36	74.2336	D&C RED NO. 36 Cosmetics; Listing Of Color Additives Subject To Certification
RED NO. 36	82.1336	D&C RED NO. 36 Drugs And Cosmetics; Listing Of Certified Provisionally Listed Colors And Specifications
RED NO. 39	74.1339	D&C RED NO. 39 Drugs; Listing Of Color Additives Subject To Certification
RED NO. 40	74.340	FD&C RED NO. 40 Foods; Listing Of Color Additives Subject To Certification
RED NO. 40	74.1340	FD&C RED NO. 40 Drugs; Listing Of Color Additives Subject To Certification
RED NO. 40	74.2340	FD&C RED NO. 40 Cosmetics; Listing Of Color Additives Subject To Certification
REDELEGATION	5.201	REDELEGATION OF THE CENTER FOR BIOLOGICS EVALUATION AND RESEARCH DIRECTOR'S PROGRAM AUTHORITIES Biologics, Redelegations Of Authority; Delegations Of Authority And Organization
REDELEGATIONS	5.20	GENERAL REDELEGATIONS OF AUTHORITY FROM THE COMMISSIONER TO OTHER OFFICERS OF THE FOOD AND DRUG ADMINISTRATION General Redelegations Of Authority; Delegations Of Authority And Organization
REDUCED ACID	146.148	REDUCED ACID FROZEN CONCENTRATED ORANGE JUICE Requirements For Specific Standardized Canned Fruit Juices And Beverages; Canned Fruit Juices
REDUCED, IRON	582.5375	IRON REDUCED Nutrients And/or Dietary Supplements; Substances Generally Recognized As Safe
REDUCED LACTOSE	184.1979a	REDUCED LACTOSE WHEY Listing Of Specific Substances Affirmed As GRAS; Direct Food Substances Affirmed As Generally Recognized As Safe
REDUCED MINERALS	184.1979b	REDUCED MINERALS WHEY Listing Of Specific Substances Affirmed As GRAS; Direct Food Substances Affirmed As Generally Recognized As Safe
REDUCING WEIGHT	105.66	LABEL STATEMENTS RELATING TO USEFULNESS IN REDUCING OR MAINTAINING BODY WEIGHT Label Statements; Foods For Special Dietary Use
REDUCTASE, GLUTATHIONE	864.7375	GLUTATHIONE REDUCTASE ASSAY Hematology Kits And Packages; Hematology And Pathology Devices
REDUCTION	1303.26	REDUCTION IN INDIVIDUAL MANUFACTURING QUOTAS Individual Manufacturing Quotas; Quotas

REFRACTOR	886.1770	MANUAL REFRACTOR Diagnostic Devices; Ophthalmic Devices
REFRIGERATION	115.50	REFRIGERATION OF SHELL EGGS HELD FOR RETAIL DISTRIBUTION Shell Eggs
REFRIGERATION	1250.34	REFRIGERATION EQUIPMENT Food Service Sanitation On Land And Air Conveyances, And Vessels; Interstate Conveyance Sanitation
REFRIGERATOR	864.9700	BLOOD STORAGE REFRIGERATOR AND BLOOD STORAGE FREEZER Products Used In Establishments That Manufacture Blood And Blood Products; Hematology And Pathology Devices
REFUND	1004.1	MANUFACTURER'S OBLIGATION TO REPAIR, REPLACE, OR REFUND COST OF ELECTRONIC PRODUCTS Repurchase, Repairs, Or Replacement Of Electronic Products
REFUND	1309.12	TIME AND METHOD OF PAYMENT; REFUND Registration Of Manufacturers, Distributors, Importers And Exporters Of List I Chemicals
REFUNDING	1004.4	PLANS FOR REFUNDING THE COST OF ELECTRONIC PRODUCTS Repurchase, Repairs, Or Replacement Of Electronic Products
REFUNDS	5.410	ORDERS TO REPAIR OR REPLACE, OR MAKE REFUNDS FOR, MEDICAL DEVICES Medical Devices And Radiological Health; Redelegations Of Authority; Delegations Of Authority And Organization
REFUSAL	1.94	HEARING ON REFUSAL OF ADMISSION Imports And Exports; General Enforcement Regulations
REFUSAL	314.125	REFUSAL TO APPROVE AN APPLICATION Applications For FDA Approval To Market A New Drug
REFUSAL	314.127	REFUSAL TO APPROVE AN ABBREVIATED NEW DRUG APPLICATION Applications For FDA Approval To Market A New Drug
REFUSAL	316.14	REFUSAL TO PROVIDE WRITTEN RECOMMENDATIONS Orphan Drugs
REFUSAL	316.25	REFUSAL TO GRANT ORPHAN DRUG DESIGNATION Orphan Drugs
REFUSAL	514.111	REFUSAL TO APPROVE AN APPLICATION Administrative Actions On Applications; New Animal Drug Applications
REFUSAL	515.21	REFUSAL TO APPROVE A MEDICATED FEED MILL LICENSE APPLICATION Administrative Actions On Licenses; Medicated Feed Mill License
REFUSAL	1316.12	REFUSAL TO ALLOW INSPECTION WITH AN ADMINISTRATIVE WARRANT Administrative Inspections; Administrative Functions, Practices, And Procedures
REFUSALS	21.52	ADMINISTRATIVE APPEALS OF REFUSALS TO AMEND RECORDS Procedures For Requests For Amendment Of Records; Protection Of Privacy
REFUSE	5.105	ISSUANCE OF NOTICES RELATING TO PROPOSALS TO REFUSE APPROVAL OR TO WITHDRAW APPROVAL OF NEW DRUG APPLICATIONS AND THEIR SUPPLEMENTS Human Drugs; Redelegations Of Authority; Delegations Of Authority And Organization
REFUSE	80.34	AUTHORITY TO REFUSE CERTIFICATION SERVICE Certification Procedures; Color Additive Certification
REFUSE	211.50	SEWAGE AND REFUSE Buildings And Facilities; Current Good Manufacturing Practice For Finished Pharmaceuticals
REFUSED	314.160	APPROVAL OF AN APPLICATION FOR WHICH APPROVAL WAS PREVIOUSLY REFUSED, SUSPENDED, OR WITHDRAWN FDA Action On Applications; Applications For FDA Approval To Market A New Drug
REFUSING	71.22	DECEPTION AS A BASIS FOR REFUSING TO ISSUE REGULATIONS; DECEPTIVE USE OF A COLOR ADDITIVE FOR WHICH A REGULATION HAS ISSUED Administrative Action On Petitions; Color Additive Petitions
REFUSING	514.110	REASONS FOR REFUSING TO FILE APPLICATIONS Administrative Actions On Applications; New Animal Drug Applications
REFUSING	514.120	REVOCATION OF ORDER REFUSING TO APPROVE AN APPLICATION OR SUSPENDING OR WITHDRAWING APPROVAL OF AN APPLICATION Administrative Actions On Applications; New Animal Drug Applications
REGENERATED	876.5600	SORBENT REGENERATED DIALYSATE DELIVERY SYSTEM FOR HEMODIALYSIS Therapeutic Devices; Gastroenterology-urology Devices
REGGIANO CHEESE	133.165	PARMESAN AND REGGIANO CHEESE Requirements For Specific Standardized Cheese And Related Products; Cheeses And Related Cheese Products
REGISTER	207.20	WHO MUST REGISTER AND SUBMIT A DRUG LIST Procedures For Domestic Drug Establishments; Registration Of Drug Producers And Listing Of Drugs In Commercial Distribution
REGISTER	207.22	HOW AND WHERE TO REGISTER AND LIST DRUGS Procedures For Domestic Drug Establishments; Registration Of Drug Producers And Listing Of Drugs In Commercial Distribution

REGISTER	607.20	WHO MUST REGISTER AND SUBMIT A BLOOD PRODUCT LIST Procedures For Domestic Blood Product Establishments; Establishment Registratn+product Listing, Manufacturers Human Blood/b.products
REGISTER	607.22	HOW AND WHERE TO REGISTER ESTABLISHMENTS AND LIST BLOOD PRODUCTS Procedures For Domestic Blood Product Establishments; Establishment Registratn+product Listing, Manufacturers Human Blood/b.products
REGISTER	710.1	WHO SHOULD REGISTER Voluntary Registration Of Cosmetic Product Establishments
REGISTER	710.3	HOW AND WHERE TO REGISTER Voluntary Registration Of Cosmetic Product Establishments
REGISTER	807.20	WHO MUST REGISTER AND SUBMIT A DEVICE LIST Procedures For Device Establishments; Establishment Registration And Device Listing For Manufacturers And Initial Importers Of Devices
REGISTER	807.22	HOW AND WHERE TO REGISTER ESTABLISHMENTS AND LIST DEVICES Procedures For Device Establishments; Establishment Registration And Device Listing For Manufacturers And Initial Importers Of Devices
REGISTER	1271.21	WHEN DO I REGISTER, SUBMIT AN HCT/P LIST, AND SUBMIT UPDATES? Procedures For Registration And Listing; Human Cells, Tissues, And Cellular And Tissue-Based Products
REGISTER	1271.22	HOW AND WHERE DO I REGISTER AND SUBMIT AN HCT/P LIST? Procedures For Registration And Listing; Human Cells, Tissues, And Cellular And Tissue-Based Products
REGISTER	1301.11	PERSONS REQUIRED TO REGISTER Registration Of Manufacturers, Distributors, And Dispensers Of Controlled Substances
REGISTER	1309.21	PERSONS REQUIRED TO REGISTER Registration Of Manufacturers, Distributors, Importers And Exporters Of List I Chemicals
REGISTERED	606.171	REPORTING OF PRODUCT DEVIATIONS BY LICENSED MANUFACTURERS, UNLICENSED REGISTERED BLOOD ESTABLISHMENTS, AND TRANSFUSION SERVICES Records And Reports; Current Good Manufacturing Practices For Blood And Blood Components
REGISTRANT	207.35	NOTIFICATION OF REGISTRANT; DRUG ESTABLISHMENT REGISTRATION NUMBER AND DRUG LISTING NUMBER Procedures For Domestic Drug Establishments; Registration Of Drug Producers And Listing Of Drugs In Commercial Distribution
REGISTRANT	607.35	NOTIFICATION OF REGISTRANT; BLOOD PRODUCT ESTABLISHMENT REGISTRATION NUMBER AND NDC LABELER CODE Procedures For Domestic Blood Product Establishments; Establishment Registratn+product Listing, Manufacturers Human Blood/b.products
REGISTRANT	710.6	NOTIFICATION OF REGISTRANT; COSMETIC PRODUCT ESTABLISHMENT REGISTRATION NUMBER Voluntary Registration Of Cosmetic Product Establishments
REGISTRANT	807.35	NOTIFICATION OF REGISTRANT Procedures For Device Establishments; Establishment Registration And Device Listing For Manufacturers And Initial Importers Of Devices
REGISTRANTS	1309.25	EXEMPTION OF CERTAIN CONTROLLED SUBSTANCE REGISTRANTS Registration Of Manufacturers, Distributors, Importers And Exporters Of List I Chemicals
REGISTRATION	207.7	ESTABLISHMENT REGISTRATION AND PRODUCT LISTING FOR HUMAN BLOOD AND BLOOD PRODUCTS AND FOR MEDICAL DEVICES General; Registration Of Drug Producers And Listing Of Drugs In Commercial Distribution
REGISTRATION	207.21	TIMES FOR REGISTRATION AND DRUG LISTING Procedures For Domestic Drug Establishments; Registration Of Drug Producers And Listing Of Drugs In Commercial Distribution
REGISTRATION	207.25	INFORMATION REQUIRED IN REGISTRATION AND DRUG LISTING Procedures For Domestic Drug Establishments; Registration Of Drug Producers And Listing Of Drugs In Commercial Distribution
REGISTRATION	207.26	AMENDMENTS TO REGISTRATION Procedures For Domestic Drug Establishments; Registration Of Drug Producers And Listing Of Drugs In Commercial Distribution
REGISTRATION	207.39	MISBRANDING BY REFERENCE TO REGISTRATION OR TO REGISTRATION NUMBER Procedures For Domestic Drug Establishments; Registration Of Drug Producers And Listing Of Drugs In Commercial Distribution
REGISTRATION	607.7	ESTABLISHMENT REGISTRATION AND PRODUCT LISTING OF BLOOD BANKS AND OTHER FIRMS MANUFACTURING HUMAN BLOOD AND BLOOD PRODUCTS General Provisions; Establishment Registratn+product Listing, Manufacturers Human Blood/b.products

REGISTRATION	607.21	TIMES FOR ESTABLISHMENT REGISTRATION AND BLOOD PRODUCT LISTING Procedures For Domestic Blood Product Establishments; Establishment Registratn+product Listing, Manufacturers Human Blood/b.products
REGISTRATION	607.25	INFORMATION REQUIRED FOR ESTABLISHMENT REGISTRATION AND BLOOD PRODUCT LISTING Procedures For Domestic Blood Product Establishments; Establishment Registratn+product Listing, Manufacturers Human Blood/b.products
REGISTRATION	607.26	AMENDMENTS TO ESTABLISHMENT REGISTRATION Procedures For Domestic Blood Product Establishments; Establishment Registratn+product Listing, Manufacturers Human Blood/b.products
REGISTRATION	607.39	MISBRANDING BY REFERENCE TO ESTABLISHMENT REGISTRATION OR TO REGISTRATION NUMBER Procedures For Domestic Blood Product Establishments; Establishment Registratn+product Listing, Manufacturers Human Blood/b.products
REGISTRATION	607.40	ESTABLISHMENT REGISTRATION AND BLOOD PRODUCT LISTING REQUIREMENTS FOR FOREIGN BLOOD PRODUCT ESTABLISHMENTS Procedures For Foreign Blood Product Establishments; Establishment Registratn+product Listing, Manufacturers Human Blood/b.products
REGISTRATION	710.2	TIME FOR REGISTRATION Voluntary Registration Of Cosmetic Product Establishments
REGISTRATION	710.5	AMENDMENTS TO REGISTRATION Voluntary Registration Of Cosmetic Product Establishments
REGISTRATION	710.8	MISBRANDING BY REFERENCE TO REGISTRATION OR TO REGISTRATION NUMBER Voluntary Registration Of Cosmetic Product Establishments
REGISTRATION	807.21	TIMES FOR ESTABLISHMENT REGISTRATION AND DEVICE LISTING Procedures For Device Establishments; Establishment Registration And Device Listing For Manufacturers And Initial Importers Of Devices
REGISTRATION	807.25	INFORMATION REQUIRED OR REQUESTED FOR ESTABLISHMENT REGISTRATION AND DEVICE LISTING Procedures For Device Establishments; Establishment Registration And Device Listing For Manufacturers And Initial Importers Of Devices
REGISTRATION	807.26	AMENDMENTS TO ESTABLISHMENT REGISTRATION Procedures For Device Establishments; Establishment Registration And Device Listing For Manufacturers And Initial Importers Of Devices
REGISTRATION	807.37	INSPECTION OF ESTABLISHMENT REGISTRATION AND DEVICE LISTINGS Procedures For Device Establishments; Establishment Registration And Device Listing For Manufacturers And Initial Importers Of Devices
REGISTRATION	807.39	MISBRANDING BY REFERENCE TO ESTABLISHMENT REGISTRATION OR TO REGISTRATION NUMBER Procedures For Device Establishments; Establishment Registration And Device Listing For Manufacturers And Initial Importers Of Devices
REGISTRATION	807.40	ESTABLISHMENT REGISTRATION AND DEVICE LISTING FOR FOREIGN ESTABLISHMENTS IMPORTING OR OFFERING FOR IMPORT DEVICES INTO THE UNITED STATES Registration Procedures For Foreign Device Establishments; Establishment Registration And Device Listing For Manufacturers And Initial Importers Of Devices
REGISTRATION	1271.25	WHAT INFORMATION IS REQUIRED FOR ESTABLISHMENT REGISTRATION AND HCT/P LISTING? Procedures For Registration And Listing; Human Cells, Tissues, And Cellular And Tissue-Based Products
REGISTRATION	1271.26	WHEN MUST I AMEND MY ESTABLISHMENT REGISTRATION? Procedures For Registration And Listing; Human Cells, Tissues, And Cellular And Tissue-Based Products
REGISTRATION	1301.13	APPLICATION FOR REGISTRATION; TIME FOR APPLICATION FOR REGISTRATION; REGISTRATION FOR INDEPENDENT ACTIVITIES Registration Of Manufacturers, Distributors, And Dispensers Of Controlled Substances
REGISTRATION	1301.25	REGISTRATION OF OCEAN VESSELS, AIRCRAFT, AND OTHER ENTITIES Registration Of Manufacturers, Distributors, And Dispensers Of Controlled Substances
REGISTRATION	1301.35	CERTIFICATE OF REGISTRATION; DENIAL OF REGISTRATION Registration Of Manufacturers, Distributors, And Dispensers Of Controlled Substances
REGISTRATION	1301.36	SUSPENSION OR REVOCATION OF REGISTRATION; SUSPENSION OF REGISTRATION PENDING FINAL ORDER; EXTENSION OF REGISTRATION PENDING FINAL ORDER Registration Of Manufacturers, Distributors, And Dispensers Of Controlled Substances
REGISTRATION	1301.51	MODIFICATION IN REGISTRATION Registration Of Manufacturers, Distributors, And Dispensers Of Controlled Substances

REGISTRATION	1301.52	TERMINATION OF REGISTRATION; TRANSFER OF REGISTRATION; DISTRIBUTION UPON DISCONTINUANCE OF BUSINESS Registration Of Manufacturers, Distributors, And Dispensers Of Controlled Substances
REGISTRATION	1309.22	SEPARATE REGISTRATION FOR INDEPENDENT ACTIVITIES Registration Of Manufacturers, Distributors, Importers And Exporters Of List I Chemicals
REGISTRATION	1309.23	SEPARATE REGISTRATION FOR SEPARATE LOCATIONS Registration Of Manufacturers, Distributors, Importers And Exporters Of List I Chemicals
REGISTRATION	1309.29	WAIVER OF REGISTRATION REQUIREMENT FOR CERTAIN ACTIVITIES Registration Of Manufacturers, Distributors, Importers And Exporters Of List I Chemicals
REGISTRATION	1309.31	TIME FOR APPLICATION FOR REGISTRATION; EXPIRATION DATE Registration Of Manufacturers, Distributors, Importers And Exporters Of List I Chemicals
REGISTRATION	1309.42	CERTIFICATE OF REGISTRATION; DENIAL OF REGISTRATION Registration Of Manufacturers, Distributors, Importers And Exporters Of List I Chemicals
REGISTRATION	1309.43	SUSPENSION OR REVOCATION OF REGISTRATION Registration Of Manufacturers, Distributors, Importers And Exporters Of List I Chemicals
REGISTRATION	1309.44	SUSPENSION OF REGISTRATION PENDING FINAL ORDER Registration Of Manufacturers, Distributors, Importers And Exporters Of List I Chemicals
REGISTRATION	1309.45	EXTENSION OF REGISTRATION PENDING FINAL ORDER Registration Of Manufacturers, Distributors, Importers And Exporters Of List I Chemicals
REGISTRATION	1309.61	MODIFICATION IN REGISTRATION Registration Of Manufacturers, Distributors, Importers And Exporters Of List I Chemicals
REGISTRATION	1309.62	TERMINATION OF REGISTRATION Registration Of Manufacturers, Distributors, Importers And Exporters Of List I Chemicals
REGISTRATION	1309.63	TRANSFER OF REGISTRATION Registration Of Manufacturers, Distributors, Importers And Exporters Of List I Chemicals
REGISTRATION	1310.09	TEMPORARY EXEMPTION FROM REGISTRATION Records And Reports Of Listed Chemicals And Certain Machines
REGISTRATION NUMBER	207.35	NOTIFICATION OF REGISTRANT; DRUG ESTABLISHMENT REGISTRATION NUMBER AND DRUG LISTING NUMBER Procedures For Domestic Drug Establishments; Registration Of Drug Producers And Listing Of Drugs In Commercial Distribution
REGISTRATION NUMBER	207.39	MISBRANDING BY REFERENCE TO REGISTRATION OR TO REGISTRATION NUMBER Procedures For Domestic Drug Establishments; Registration Of Drug Producers And Listing Of Drugs In Commercial Distribution
REGISTRATION NUMBER	607.35	NOTIFICATION OF REGISTRANT; BLOOD PRODUCT ESTABLISHMENT REGISTRATION NUMBER AND NDC LABELER CODE Procedures For Domestic Blood Product Establishments; Establishment Registratn+product Listing, Manufacturers Human Blood/b.products
REGISTRATION NUMBER	607.39	MISBRANDING BY REFERENCE TO ESTABLISHMENT REGISTRATION OR TO REGISTRATION NUMBER Procedures For Domestic Blood Product Establishments; Establishment Registratn+product Listing, Manufacturers Human Blood/b.products
REGISTRATION NUMBER	710.6	NOTIFICATION OF REGISTRANT; COSMETIC PRODUCT ESTABLISHMENT REGISTRATION NUMBER Voluntary Registration Of Cosmetic Product Establishments
REGISTRATION NUMBER	710.8	MISBRANDING BY REFERENCE TO REGISTRATION OR TO REGISTRATION NUMBER Voluntary Registration Of Cosmetic Product Establishments
REGISTRATION NUMBER	807.39	MISBRANDING BY REFERENCE TO ESTABLISHMENT REGISTRATION OR TO REGISTRATION NUMBER Procedures For Device Establishments; Establishment Registration And Device Listing For Manufacturers And Initial Importers Of Devices
REGISTRATION NUMBER	1271.27	WILL FDA ASSIGN ME A REGISTRATION NUMBER? Procedures For Registration And Listing; Human Cells, Tissues, And Cellular And Tissue-Based Products
REGISTRATIONS	207.37	INSPECTION OF REGISTRATIONS AND DRUG LISTINGS Procedures For Domestic Drug Establishments; Registration Of Drug Producers And Listing Of Drugs In Commercial Distribution

REGISTRATIONS	607.37	INSPECTION OF ESTABLISHMENT REGISTRATIONS AND BLOOD PRODUCT LISTINGS Procedures For Domestic Blood Product Establishments; Establishment Registratn+product Listing, Manufacturers Human Blood/b.products
REGISTRATIONS	710.7	INSPECTION OF REGISTRATIONS Voluntary Registration Of Cosmetic Product Establishments
REGISTRATIONS	1271.37	WILL ESTABLISHMENT REGISTRATIONS AND HCT/P LISTINGS BE AVAILABLE FOR INSPECTION, AND HOW DO I REQUEST INFORMATION? Procedures For Registration And Listing; Human Cells, Tissues, And Cellular And Tissue-Based Products
REGISTRATIONS	1301.12	SEPARATE REGISTRATIONS FOR SEPARATE LOCATIONS Registration Of Manufacturers, Distributors, And Dispensers Of Controlled Substances
REGULATED	880.5560	TEMPERATURE REGULATED WATER MATTRESS General Hospital And Personal Use Therapeutic Devices; General Hospital And Personal Use Devices
REGULATING	870.5900	THERMAL REGULATING SYSTEM Cardiovascular Therapeutic Devices; Cardiovascular Devices
REGULATION	12.20	INITIATION OF A HEARING INVOLVING THE ISSUANCE, AMENDMENT, OR REVOCATION OF A REGULATION Initiation Of Proceedings; Formal Evidentiary Public Hearing
REGULATION	12.22	FILING OBJECTIONS AND REQUESTS FOR A HEARING ON A REGULATION OR ORDER Initiation Of Proceedings; Formal Evidentiary Public Hearing
REGULATION	12.26	MODIFICATION OR REVOCATION OF REGULATION OR ORDER Initiation Of Proceedings; Formal Evidentiary Public Hearing
REGULATION	12.30	JUDICIAL REVIEW AFTER WAIVER OF HEARING ON A REGULATION Initiation Of Proceedings; Formal Evidentiary Public Hearing
REGULATION	12.37	EFFECTIVE DATE OF A REGULATION Initiation Of Proceedings; Formal Evidentiary Public Hearing
REGULATION	16.24	REGULATORY HEARING REQUIRED BY THE ACT OR A REGULATION Initiation Of Proceedings; Regulatory Hearing Before The Food And Drug Administration
REGULATION	71.20	PUBLICATION OF REGULATION Administrative Action On Petitions; Color Additive Petitions
REGULATION	71.22	DECEPTION AS A BASIS FOR REFUSING TO ISSUE REGULATIONS; DECEPTIVE USE OF A COLOR ADDITIVE FOR WHICH A REGULATION HAS ISSUED Administrative Action On Petitions; Color Additive Petitions
REGULATION	106.1	STATUS AND APPLICABILITY OF THE QUALITY CONTROL PROCEDURES REGULATION General Provisions; Infant Formula Quality Control Procedures
REGULATION	170.15	ADOPTION OF REGULATION ON INITIATIVE OF COMMISSIONER General Provisions; Food Additives
REGULATION	170.39	THRESHOLD OF REGULATION FOR SUBSTANCES USED IN FOOD-CONTACT ARTICLES Food Additive Safety; Food Additives
REGULATION	171.8	THRESHOLD OF REGULATION FOR SUBSTANCES USED IN FOOD-CONTACT ARTICLES General Provisions; Food Additive Petitions
REGULATION	171.100	REGULATION BASED ON PETITION Administrative Actions On Applications; Food Additive Petitions
REGULATION	171.102	EFFECTIVE DATE OF REGULATION Administrative Actions On Applications; Food Additive Petitions
REGULATION	174.6	THRESHOLD OF REGULATION FOR SUBSTANCES USED IN FOOD-CONTACT ARTICLES Indirect Food Additives: General
REGULATION	570.15	ADOPTION OF REGULATION ON INITIATIVE OF COMMISSIONER General Provisions; Food Additives
REGULATION	571.100	REGULATION BASED ON PETITION Administrative Actions On Applications; Food Additive Petitions
REGULATION	571.102	EFFECTIVE DATE OF REGULATION Administrative Actions On Applications; Food Additive Petitions
REGULATION	862.2	REGULATION OF CALIBRATORS General Provisions; Clinical Chemistry And Clinical Toxicology Devices
REGULATIONS	5.606	ACCEPTANCE OF ASSISTANCE FROM STATE AND LOCAL AUTHORITIES FOR ENFORCEMENT OF RADIATION CONTROL LEGISLATION AND REGULATIONS Radiation Control; Redelegations Of Authority; Delegations Of Authority And Organization
REGULATIONS	5.301	ISSUANCE OF INITIAL EMERGENCY PERMIT ORDERS AND NOTICES OF CONFIRMATION OF EFFECTIVE DATE OF FINAL REGULATIONS ON FOOD FOR HUMAN AND ANIMAL CONSUMPTION Foods And Cosmetics; Redelegations Of Authority; Delegations Of Authority And Organization

REGULATIONS	10.40	PROMULGATION OF REGULATIONS FOR THE EFFICIENT ENFORCEMENT OF THE LAW General Administrative Procedures; Administrative Practices And Procedures
REGULATIONS	10.50	PROMULGATION OF REGULATIONS AND ORDERS AFTER AN OPPORTUNITY FOR A FORMAL EVIDENTIARY PUBLIC HEARING General Administrative Procedures; Administrative Practices And Procedures
REGULATIONS	10.80	DISSEMINATION OF DRAFT FEDERAL REGISTER NOTICES AND REGULATIONS General Administrative Procedures; Administrative Practices And Procedures
REGULATIONS	10.90	FDA REGULATIONS, GUIDELINES, RECOMMENDATIONS, AND AGREEMENTS General Administrative Procedures; Administrative Practices And Procedures
REGULATIONS	19.5	REFERENCE TO DEPARTMENT REGULATIONS General Provisions; Standards Of Conduct And Conflicts Of Interest
REGULATIONS	20.25	RETROACTIVE APPLICATION OF REGULATIONS General Policy; Public Information
REGULATIONS	20.100	APPLICABILITY; CROSS-REFERENCE TO OTHER REGULATIONS Availability Of Specific Categories Of Records; Public Information
REGULATIONS	71.22	DECEPTION AS A BASIS FOR REFUSING TO ISSUE REGULATIONS; DECEPTIVE USE OF A COLOR ADDITIVE FOR WHICH A REGULATION HAS ISSUED Administrative Action On Petitions; Color Additive Petitions
REGULATIONS	71.30	PROCEDURE FOR FILING OBJECTIONS TO REGULATIONS Administrative Action On Petitions; Color Additive Petitions
REGULATIONS	100.2	STATE ENFORCEMENT OF FEDERAL REGULATIONS State And Local Requirements; General
REGULATIONS	210.1	STATUS OF CURRENT GOOD MANUFACTURING PRACTICE REGULATIONS Current Good Manufacturing Practice In Manufacturing, Processing, Packing, Or Holding Of Drugs; General
REGULATIONS	210.2	APPLICABILITY OF CURRENT GOOD MANUFACTURING PRACTICE REGULATIONS Current Good Manufacturing Practice In Manufacturing, Processing, Packing, Or Holding Of Drugs; General
REGULATIONS	369.21	DRUGS; WARNING AND CAUTION STATEMENTS REQUIRED BY REGULATIONS Warning And Caution Statements For Drugs; Interpretative Statements Re: Warnings On Drugs And Devices For OTC Sale
REGULATIONS	579.12	INCORPORATION OF REGULATIONS IN PART 179 General Provisions; Irradiation In The Production, Process & Handling Of Animal Feed And Pet Food
REGULATIONS	1003.5	EFFECT OF REGULATIONS ON OTHER LAWS General Provisions; Notification Of Defects Or Failure To Comply
REGULATIONS	1271.20	IF MY HCT/P'S DO NOT MEET THE CRITERIA IN SECTION 1271.10, AND I DO NOT QUALIFY FOR ANY OF THE EXCEPTIONS IN SECTION 1271.15, WHAT REGULATIONS APPLY? General Provisions; Human Cells, Tissues, And Cellular And Tissue-Based Products
REGULATIONS	1307.03	EXCEPTIONS TO REGULATIONS General Information; Miscellaneous
REGULATOR	868.2700	PRESSURE REGULATOR Monitoring Devices; Anesthesiology Devices
REGULATORY AUTHORITIES	26.10	REGULATORY AUTHORITIES NOT LISTED AS CURRENTLY EQUIVALENT Specific Sector Provisions For Pharmaceutical GMPs; Mutual Recognition Of Pharmaceutical GMP Reports, Medical Device Quality System Audit Reports
REGULATORY AUTHORITIES	26.34	REGULATORY AUTHORITIES Specific Sector Provisions For Medical Devices; Mutual Recognition Of Pharmaceutical GMP Reports, Medical Device Quality System Audit Reports
REGULATORY AUTHORITY	26.74	PRESERVATION OF REGULATORY AUTHORITY Framework Provisions; Mutual Recognition Of Pharmaceutical GMP Reports, Medical Device Quality System Audit Reports
REGULATORY COLLABORATION	26.18	REGULATORY COLLABORATION Specific Sector Provisions For Pharmaceutical GMPs; Mutual Recognition Of Pharmaceutical GMP Reports, Medical Device Quality System Audit Reports
REGULATORY COOPERATION	26.49	REGULATORY COOPERATION Specific Sector Provisions For Medical Devices; Mutual Recognition Of Pharmaceutical GMP Reports, Medical Device Quality System Audit Reports
REGULATORY HEARING	16.22	INITIATION OF REGULATORY HEARING Initiation Of Proceedings; Regulatory Hearing Before The Food And Drug Administration
REGULATORY HEARING	16.24	REGULATORY HEARING REQUIRED BY THE ACT OR A REGULATION Initiation Of Proceedings; Regulatory Hearing Before The Food And Drug Administration
REGULATORY HEARING	16.80	ADMINISTRATIVE RECORD OF A REGULATORY HEARING Administrative Record And Decision; Regulatory Hearing Before The Food And Drug Administration

REGULATORY HEARING	810.11	REGULATORY HEARING Mandatory Medical Device Recall Procedures; Medical Device Recall Authority
REGULATORY LIMITS	109.4	ESTABLISHMENT OF TOLERANCES, REGULATORY LIMITS, AND ACTION LEVELS General Provisions; Unavoidable Contaminants In Food For Human Consumption & Food-packaging Material
REGULATORY LIMITS	509.4	ESTABLISHMENT OF TOLERANCES, REGULATORY LIMITS, AND ACTION LEVELS General Provisions; Unavoidable Contaminants In Animal Food And Food-packaging Material
REGULATORY METHOD	500.88	REGULATORY METHOD Regulation Of Carcinogenic Compounds Used In Food-producing Animals; General
REGULATORY RESEARCH	312.86	FOCUSED FDA REGULATORY RESEARCH Drugs Intended To Treat Life-threatening And Severely-debilitating Illnesses; Investigational New Drug Application
REGULATORY REVIEW	60.20	FDA ACTION ON REGULATORY REVIEW PERIOD DETERMINATION Regulatory Review Period Determinations; Patent Term Restoration
REGULATORY REVIEW	60.22	REGULATORY REVIEW PERIOD DETERMINATIONS Regulatory Review Period Determinations; Patent Term Restoration
REGULATORY REVIEW	60.24	REVISION OF REGULATORY REVIEW PERIOD DETERMINATIONS Regulatory Review Period Determinations; Patent Term Restoration
REGULATORY REVIEW	60.26	FINAL ACTION ON REGULATORY REVIEW PERIOD DETERMINATIONS Regulatory Review Period Determinations; Patent Term Restoration
REGULATORY REVIEW	60.28	TIME FRAME FOR DETERMINING REGULATORY REVIEW PERIODS Regulatory Review Period Determinations; Patent Term Restoration
REIMPORTATION	203.10	RESTRICTIONS ON REIMPORTATION Prescription Drug Marketing
REIMPORTATION	203.11	APPLICATIONS FOR REIMPORTATION TO PROVIDE EMERGENCY MEDICAL CARE Prescription Drug Marketing
REINFORCED	177.2355	MINERAL REINFORCED NYLON RESINS Substances For Use Only As Components Of Articles Intended For Repeated Use; Indirect Food Additives: Polymers
REINFORCED	178.3850	REINFORCED WAX Certain Adjuvants And Production Aids; Indirect Food Additives: Adjuvants, Production Aids, And Sanitizers
REINSTATEMENT	56.123	REINSTATEMENT OF AN IRB OR AN INSTITUTION Administrative Actions For Noncompliance; Institutional Review Boards
REINSTATEMENT	58.219	REINSTATEMENT OF A DISQUALIFIED TESTING FACILITY Disqualification Of Testing Facilities; Good Laboratory Practice For Nonclinical Laboratory Studies
REINSTATEMENT	108.10	SUSPENSION AND REINSTATEMENT OF PERMIT General Provisions; Emergency Permit Control
REINSTATEMENT	1310.11	REINSTATEMENT OF THE EXEMPTION FOR DRUG PRODUCTS DISTRIBUTED UNDER THE FOOD, DRUG AND COSMETIC ACT Records And Reports Of Listed Chemicals And Certain Machines
REISSUANCE	601.9	LICENSES; REISSUANCE General Provisions; Licensing
REJECTED	211.89	REJECTED COMPONENTS, DRUG PRODUCT CONTAINERS, AND CLOSURES Control Of Components And Drug Product Containers And Closures; Current Good Manufacturing Practice For Finished Pharmaceuticals
REJECTED	1230.47	REJECTED CONTAINERS Imports; Regulations Under The Federal Caustic Poison Act
REJECTION	211.84	TESTING AND APPROVAL OR REJECTION OF COMPONENTS, DRUG PRODUCT CONTAINERS, AND CLOSURES Control Of Components And Drug Product Containers And Closures; Current Good Manufacturing Practice For Finished Pharmaceuticals
RELABEL	1.95	APPLICATION FOR AUTHORIZATION TO RELABEL AND RECONDITION Imports And Exports; General Enforcement Regulations
RELABEL	1.90	GRANTING OF AUTHORIZATION TO RELABEL AND RECONDITION Imports And Exports; General Enforcement Regulations
RELABELING	1.00	COSTS CHARGEABLE IN CONNECTION WITH RELABELING AND RECONDITIONING INADMISSABLE IMPORTS Imports And Exports; General Enforcement Regulations
RELABELING	1230.48	RELABELING OF CONTAINERS Imports; Regulations Under The Federal Caustic Poison Act
RELEASE	5.204	NOTIFICATION OF RELEASE FOR DISTRIBUTION OF BIOLOGICAL PRODUCTS Biologics; Redelegations Of Authority; Delegations Of Authority And Organization
RELEASE	211.165	TESTING AND RELEASE FOR DISTRIBUTION Laboratory Controls; Current Good Manufacturing Practice For Finished Pharmaceuticals
RELEASE	610.1	TESTS PRIOR TO RELEASE REQUIRED FOR EACH LOT Release Requirements; General Biological Products Standards

RELEASE	610.2	REQUESTS FOR SAMPLES AND PROTOCOLS; OFFICIAL RELEASE Release Requirements; General Biological Products Standards
RELEASE	660.6	SAMPLES; PROTOCOLS; OFFICIAL RELEASE Antibody To Hepatitis B Surface Antigen; Additional Standards For Diagnostic Substances For Laboratory Tests
RELEASE	660.46	SAMPLES; PROTOCOLS; OFFICIAL RELEASE Hepatitis B Surface Antigen; Additional Standards For Diagnostic Substances For Laboratory Tests
RELEASE	1230.45	NO VIOLATION; RELEASE Imports; Regulations Under The Federal Caustic Poison Act
RELEASE AGENTS	178.3860	RELEASE AGENTS Certain Adjuvants And Production Aids; Indirect Food Additives: Adjuvants, Production Aids, And Sanitizers
RELEASE AGENTS	181.28	RELEASE AGENTS Specific Prior-sanctioned Food Ingredients; Prior-sanctioned Food Ingredients
RELEASE ASSAY	864.7040	ADENOSINE TRIPHOSPHATE RELEASE ASSAY Hematology Kits And Packages; Hematology And Pathology Devices
RELEASE, DOMESTIC	1312.29	DOMESTIC RELEASE PROHIBITED Exportation Of Controlled Substances; Importation And Exportation Of Controlled Substances
RELEASE, DRUG	343.90	DISSOLUTION AND DRUG RELEASE TESTING Internal Analgesic, Antipyretic And Antirheumatic Drug Products for OTC Human Use
RELEASE, EXPEDITED	1316.92	PETITION FOR EXPEDITED RELEASE IN AN ADMINISTRATIVE FORFEITURE ACTION Expedited Forfeiture Proceedings For Certain Property; Administrative Functions, Practices, And Procedures
RELEASE, EXPEDITED	1316.93	RULING ON PETITION FOR EXPEDITED RELEASE IN AN ADMINISTRATIVE FORFEITURE ACTION Expedited Forfeiture Proceedings For Certain Property; Administrative Functions, Practices, And Procedures
RELEASE, EXPEDITED	1316.95	PETITION FOR EXPEDITED RELEASE OF A CONVEYANCE IN A JUDICIAL FORFEITURE ACTION Expedited Forfeiture Proceedings For Certain Property; Administrative Functions, Practices, And Procedures
RELEASE, EXPEDITED	1316.96	RULING ON A PETITION FOR EXPEDITED RELEASE OF A CONVEYANCE IN A JUDICIAL FORFEITURE ACTION Expedited Forfeiture Proceedings For Certain Property; Administrative Functions, Practices, And Procedures
RELEVANCE	21.73	ACCURACY, COMPLETENESS, TIMELINESS, AND RELEVANCE OF RECORDS DISCLOSED FROM PRIVACY ACT RECORD SYSTEMS Disclosure Of Records In Privacy Act Record Systems To Persons Other Than The Subject Individual; Protection Of Privacy
RELIEF, PAIN	882.5840	IMPLANTED INTRACEREBRAL/SUBCORTICAL STIMULATOR FOR PAIN RELIEF Neurological Therapeutic Devices; Neurological Devices
RELIEF, PAIN	882.5870	IMPLANTED PERIPHERAL NERVE STIMULATOR FOR PAIN RELIEF Neurological Therapeutic Devices; Neurological Devices
RELIEF, PAIN	882.5880	IMPLANTED SPINAL CORD STIMULATOR FOR PAIN RELIEF Neurological Therapeutic Devices; Neurological Devices
RELIEF, PAIN	882.5890	TRANSCUTANEOUS ELECTRICAL NERVE STIMULATOR FOR PAIN RELIEF Neurological Therapeutic Devices; Neurological Devices
RELINER	872.3560	OTC DENTURE RELINER Prosthetic Devices; Dental Devices
RELINING	872.3760	DENTURE RELINING, REPAIRING, OR REBASING RESIN Prosthetic Devices; Dental Devices
REMEDIES, ADMINISTRATIVE	10.45	COURT REVIEW OF FINAL ADMINISTRATIVE ACTION; EXHAUSTION OF ADMINISTRATIVE REMEDIES General Administrative Procedures; Administrative Practices And Procedures
REMEDIES, ADMINISTRATIVE	14.7	ADMINISTRATIVE REMEDIES General Provisions; Public Hearing Before A Public Advisory Committee
REMINDER LABELING	200.200	PRESCRIPTION DRUGS; REMINDER ADVERTISEMENTS AND REMINDER LABELING TO PROVIDE PRICE INFORMATION TO CONSUMERS Prescription Drug Consumer Price Listing; General
REMISSION OF FORFEITURE	1316.79	PETITIONS FOR REMISSION OR MITIGATION OF FORFEITURE Seizure, Forfeiture, And Disposition Of Property; Administrative Functions, Practices, And Procedures
REMOTE CONTROLLED	892.5700	REMOTE CONTROLLED RADIONUCLIDE APPLICATOR SYSTEM Therapeutic Devices; Radiology Devices
REMOVABLE SKIN CLIP	878.4320	REMOVABLE SKIN CLIP Surgical Devices; General And Plastic Surgery Devices
REMOVABLE SKIN STAPLE	878.4760	REMOVABLE SKIN STAPLE Surgical Devices; General And Plastic Surgery Devices
REMOVAL, CAST	888.5960	CAST REMOVAL INSTRUMENT Surgical Devices; Orthopedic Devices

REMOVAL, CAST	888.5980	MANUAL CAST APPLICATION AND REMOVAL INSTRUMENT Surgical Devices; Orthopedic Devices
REMOVAL, CLIP	882.4200	CLIP REMOVAL INSTRUMENT Neurological Surgical Devices; Neurological Devices
REMOVAL, DRUG PRODUCT	314.162	REMOVAL OF A DRUG PRODUCT FROM THE LIST Applications For FDA Approval To Market A New Drug
REMOVAL, EARWAX	344.10	EARWAX REMOVAL AID ACTIVE INGREDIENT Active Ingredients; Topical Otic Drug Products For Over-the-counter Human Use
REMOVAL, EARWAX	344.50	LABELING OF EARWAX REMOVAL AID DRUG PRODUCTS Labeling; Topical Otic Drug Products For Over-the-counter Human Use
REMOVAL, LICE	880.5960	LICE REMOVAL KIT General Hospital And Personal Use Therapeutic Devices; General Hospital And Personal Use Devices
REMOVALS	107.210	FIRM-INITIATED PRODUCT REMOVALS Infant Formula Recalls; Infant Formula
REMOVALS	806.10	REPORTS OF CORRECTIONS AND REMOVALS Medical Devices; Reports Of Corrections And Removals
REMOVALS	806.20	RECORDS OF CORRECTIONS AND REMOVALS NOT REQUIRED TO BE REPORTED Medical Devices; Reports Of Corrections And Removals
REMOVED	216.24	DRUG PRODUCTS WITHDRAWN OR REMOVED FROM THE MARKET FOR REASONS OF SAFETY OR EFFECTIVENESS Compounded Drug Products; Pharmacy Compounding
REMOVER, CORN/CALLUS	358.510	CORN AND CALLUS REMOVER ACTIVE INGREDIENTS Corn And Callus Remover Drug Products; Miscellaneous External Drug Products For OTC Human Use
REMOVER, CORN/CALLUS	358.550	LABELING OF CORN AND CALLUS REMOVER DRUG PRODUCTS Corn And Callus Remover Drug Products; Miscellaneous External Drug Products For OTC Human Use
REMOVER, LENS	886.5420	CONTACT LENS INSERTER/REMOVER Therapeutic Devices; Ophthalmic Devices
REMOVER, WART	358.110	WART REMOVER ACTIVE INGREDIENTS Wart Remover Drug Products; Miscellaneous External Drug Products For OTC Human Use
REMOVER, WART	358.150	LABELING OF WART REMOVER DRUG PRODUCTS Wart Remover Drug Products; Miscellaneous External Drug Products For OTC Human Use
RENAL STONE	876.4650	WATER JET RENAL STONE DISLODGER SYSTEM Surgical Devices; Gastroenterology-urology Devices
RENEWAL	14.40	ESTABLISHMENT AND RENEWAL OF ADVISORY COMMITTEES Establishment Of Advisory Committees; Public Hearing Before A Public Advisory Committee
RENIN	862.1085	ANGIOTENSIN I AND RENIN TEST SYSTEM Clinical Chemistry Test Systems; Clinical Chemistry And Clinical Toxicology Devices
RENNET	184.1685	RENNET (ANIMAL-DERIVED) AND CHYMOSIN PREPARATION (FERMENTATION-DERIVED) Listing Of Specific Substances Affirmed As GRAS; Direct Food Substances Affirmed As Generally Recognized As Safe
RENNET	582.1685	RENNET General Purpose Food Additives; Substances Generally Recognized As Safe
REOVIRUS	866.3470	REOVIRUS SEROLOGICAL REAGENTS Serological Reagents; Immunology And Microbiology Devices
REPACKING	201.122	DRUGS FOR PROCESSING, REPACKING, OR MANUFACTURING Exemptions From Adequate Directions For Use; Labeling
REPACKING	201.150	DRUGS; PROCESSING, LABELING, OR REPACKING Other Exemptions; Labeling
REPACKING	801.122	MEDICAL DEVICES FOR PROCESSING, REPACKING, OR MANUFACTURING Exemptions From Adequate Directions For Use; Labeling
REPACKING	801.150	MEDICAL DEVICES; PROCESSING, LABELING, OR REPACKING Other Exemptions; Labeling
REPAIR	5.410	ORDERS TO REPAIR OR REPLACE, OR MAKE REFUNDS FOR, MEDICAL DEVICES Medical Devices And Radiological Health; Redelegations Of Authority; Delegations Of Authority And Organization
REPAIR	5.603	NOTIFICATION OF DEFECTS IN, AND REPAIR OR REPLACEMENT OF, ELECTRONIC PRODUCTS Radiation Control; Redelegations Of Authority; Delegations Of Authority And Organization
REPAIR	870.1350	CATHETER BALLOON REPAIR KIT Cardiovascular Diagnostic Devices; Cardiovascular Devices
REPAIR	870.3710	PACEMAKER REPAIR OR REPLACEMENT MATERIAL Cardiovascular Prosthetic Devices; Cardiovascular Devices

REPAIR	872.3570	OTC DENTURE REPAIR KIT Prosthetic Devices; Dental Devices
REPAIR	872.3930	TRICALCIUM PHOSPHATE GRANULES FOR DENTAL BONE REPAIR Prosthetic Devices; Dental Devices
REPAIR	1004.1	MANUFACTURER'S OBLIGATION TO REPAIR, REPLACE, OR REFUND COST OF ELECTRONIC PRODUCTS Repurchase, Repairs, Or Replacement Of Electronic Products
REPAIR	1004.2	PLANS FOR THE REPAIR OF ELECTRONIC PRODUCTS Repurchase, Repairs, Or Replacement Of Electronic Products
REPAIRING	801.405	LABELING OF ARTICLES INTENDED FOR LAY USE IN THE REPAIRING AND/OR REFITTING OF DENTURES Special Requirements For Specific Devices; Labeling
REPAIRING RESIN	872.3760	DENTURE RELINING, REPAIRING, OR REBASING RESIN Prosthetic Devices; Dental Devices
REPEALING TOLERANCES	171.130	PROCEDURE FOR AMENDING AND REPEALING TOLERANCES OR EXEMPTIONS FROM TOLERANCES Administrative Actions On Applications; Food Additive Petitions
REPEALING TOLERANCES	571.130	PROCEDURE FOR AMENDING AND REPEALING TOLERANCES OR EXEMPTIONS FROM TOLERANCES Administrative Actions On Applications; Food Additive Petitions
REPEATED USE	177.2600	RUBBER ARTICLES INTENDED FOR REPEATED USE Substances For Use Only As Components Of Articles Intended For Repeated Use; Indirect Food Additives: Polymers
REPELLENTS, INSECT	310.529	DRUG PRODUCTS CONTAINING ACTIVE INGREDIENTS OFFERED OVER-THE-COUNTER (OTC) FOR ORAL USE AS INSECT REPELLENTS Requirements For Specific New Drugs Or Devices; New Drugs
REPLACE	5.410	ORDERS TO REPAIR OR REPLACE, OR MAKE REFUNDS FOR, MEDICAL DEVICES Medical Devices And Radiological Health; Redelegations Of Authority; Delegations Of Authority And Organization
REPLACE	1004.1	MANUFACTURER'S OBLIGATION TO REPAIR, REPLACE, OR REFUND COST OF ELECTRONIC PRODUCTS Repurchase, Repairs, Or Replacement Of Electronic Products
REPLACEMENT	5.603	NOTIFICATION OF DEFECTS IN, AND REPAIR OR REPLACEMENT OF, ELECTRONIC PRODUCTS Radiation Control; Redelegations Of Authority; Delegations Of Authority And Organization
REPLACEMENT	1004.3	PLANS FOR THE REPLACEMENT OF ELECTRONIC PRODUCTS Repurchase, Repairs, Or Replacement Of Electronic Products
REPLACEMENT MATERIAL	870.3710	PACEMAKER REPAIR OR REPLACEMENT MATERIAL Cardiovascular Prosthetic Devices; Cardiovascular Devices
REPLACEMENT, OSSICULAR	874.3450	PARTIAL OSSICULAR REPLACEMENT PROSTHESIS Prosthetic Devices; Ear, Nose, And Throat Devices
REPLACEMENT, OSSICULAR	874.3495	TOTAL OSSICULAR REPLACEMENT PROSTHESIS Prosthetic Devices; Ear, Nose, And Throat Devices
REPLACEMENT, OSSICULAR	874.3540	PROSTHESIS MODIFICATION INSTRUMENT FOR OSSICULAR REPLACEMENT SURGERY Prosthetic Devices; Ear, Nose, And Throat Devices
REPLACEMENT VALVE	870.3925	REPLACEMENT HEART VALVE Cardiovascular Prosthetic Devices; Cardiovascular Devices
REPORT	7.84	OPPORTUNITY FOR PRESENTATION OF VIEWS BEFORE REPORT OF CRIMINAL VIOLATION Criminal Violations; Enforcement Policy
REPORT	7.85	CONDUCT OF A PRESENTATION OF VIEWS BEFORE REPORT OF CRIMINAL VIOLATION Criminal Violations; Enforcement Policy
REPORT	7.87	RECORDS RELATED TO OPPORTUNITIES FOR PRESENTATION OF VIEWS CONDUCTED BEFORE REPORT OF CRIMINAL VIOLATION Criminal Violations; Enforcement Policy
REPORT	19.21	DUTY TO REPORT VIOLATIONS Reporting Of Violations; Standards Of Conduct And Conflicts Of Interest
REPORT	803.20	HOW TO REPORT Requirements For Individual Adverse Event Reports; Medical Device Reporting
REPORT	803.32	INDIVIDUAL ADVERSE EVENT REPORT DATA ELEMENTS User Facility Reporting Requirements; Medical Device Reporting
REPORT	803.42	INDIVIDUAL ADVERSE EVENT REPORT DATA ELEMENTS Importer Reporting Requirement; Medical Device Reporting
REPORT	803.52	INDIVIDUAL ADVERSE EVENT REPORT DATA ELEMENTS Manufacturer Reporting Requirements; Medical Device Reporting
REPORT	812.27	REPORT OF PRIOR INVESTIGATIONS Application And Administrative Action; Investigational Device Exemptions
REPORT	1240.45	REPORT OF DISEASE Control Of Communicable Diseases

REPORTS	20.112	VOLUNTARY DRUG EXPERIENCE REPORTS SUBMITTED BY PHYSICIANS AND HOSPITALS Availability Of Specific Categories Of Records; Public Information
REPORTS	20.113	VOLUNTARY PRODUCT DEFECT REPORTS Availability Of Specific Categories Of Records; Public Information
REPORTS	26.12	NATURE OF RECOGNITION OF INSPECTION REPORTS Specific Sector Provisions For Pharmaceutical GMPs; Mutual Recognition Of Pharmaceutical GMP Reports, Medical Device Quality System Audit Reports
REPORTS	26.13	TRANSMISSION OF POSTAPPROVAL INSPECTION REPORTS Specific Sector Provisions For Pharmaceutical GMPs; Mutual Recognition Of Pharmaceutical GMP Reports, Medical Device Quality System Audit Reports
REPORTS	26.14	TRANSMISSION OF PREAPPROVAL INSPECTION REPORTS Specific Sector Provisions For Pharmaceutical GMPs; Mutual Recognition Of Pharmaceutical GMP Reports, Medical Device Quality System Audit Reports
REPORTS	26.41	EXCHANGE AND ENDORSEMENT OF QUALITY SYSTEM EVALUATION REPORTS Specific Sector Provisions For Medical Devices; Mutual Recognition Of Pharmaceutical GMP Reports, Medical Device Quality System Audit Reports
REPORTS	26.42	EXCHANGE AND ENDORSEMENT OF PRODUCT EVALUATION REPORTS Specific Sector Provisions For Medical Devices; Mutual Recognition Of Pharmaceutical GMP Reports, Medical Device Quality System Audit Reports
REPORTS	26.43	TRANSMISSION OF QUALITY SYSTEM EVALUATION REPORTS Specific Sector Provisions For Medical Devices; Mutual Recognition Of Pharmaceutical GMP Reports, Medical Device Quality System Audit Reports
REPORTS	26.44	TRANSMISSION OF PRODUCT EVALUATION REPORTS Specific Sector Provisions For Medical Devices; Mutual Recognition Of Pharmaceutical GMP Reports, Medical Device Quality System Audit Reports
REPORTS	26.50	ALERT SYSTEM AND EXCHANGE OF POSTMARKET VIGILANCE REPORTS Specific Sector Provisions For Medical Devices; Mutual Recognition Of Pharmaceutical GMP Reports, Medical Device Quality System Audit Reports
REPORTS	99.501	RECORDKEEPING AND REPORTS Dissemination Of Information On Unapproved/New Uses For Marketed Drugs, Biologics, and Devices
REPORTS	203.60	REQUEST AND RECEIPT FORMS, REPORTS, AND RECORDS Prescription Drug Marketing
REPORTS	310.303	CONTINUATION OF LONG-TERM STUDIES, RECORDS, AND REPORTS ON CERTAIN DRUGS FOR WHICH NEW DRUG APPLICATIONS HAVE BEEN APPROVED Records And Reports; New Drugs
REPORTS	310.305	RECORDS AND REPORTS CONTAINING ADVERSE DRUG EXPERIENCES ON MARKETED PRESCRIPTION DRUGS FOR HUMAN USE WITHOUT APPROVED NEW DRUG APPLICATIONS Records And Reports; New Drugs
REPORTS	312.32	IND SAFETY REPORTS Investigational New Drug Application
REPORTS	312.58	INSPECTION OF SPONSOR'S RECORDS AND REPORTS Investigational New Drug Application
REPORTS	312.64	INVESTIGATOR REPORTS Investigational New Drug Application
REPORTS	312.68	INSPECTION OF INVESTIGATOR'S RECORDS AND REPORTS Investigational New Drug Application
REPORTS	314.81	OTHER POSTMARKETING REPORTS Applications; Applications For FDA Approval To Market A New Drug
REPORTS	314.98	POSTMARKETING REPORTS Abbreviated Applications; Applications For FDA Approval To Market A New Drug
REPORTS	510.300	RECORDS AND REPORTS CONCERNING EXPERIENCE WITH NEW ANIMAL DRUGS FOR WHICH AN APPROVED APPLICATION IS IN EFFECT Records And Reports; New Animal Drugs
REPORTS	510.301	RECORDS AND REPORTS CONCERNING EXPERIENCE WITH ANIMAL FEEDS BEARING OR CONTAINING NEW ANIMAL DRUGS FOR WHICH AN APPROVED MEDICATED FEED MILL LICENSE APPLICATION IS IN EFFECT Records And Reports; New Animal Drugs
REPORTS	514.80	RECORDS AND REPORTS CONCERNING EXPERIENCE WITH APPROVED NEW ANIMAL DRUGS Administrative Actions On Applications; New Animal Drug Applications
REPORTS	600.81	DISTRIBUTION REPORTS Biological Products: General
REPORTS	803.9	PUBLIC AVAILABILITY OF REPORTS General Provisions; Medical Device Reporting
REPORTS	803.10	GENERAL DESCRIPTION OF REPORTS REQUIRED FROM USER FACILITIES, IMPORTERS, AND MANUFACTURERS General Provisions; Medical Device Reporting
REPORTS	803.12	WHERE TO SUBMIT REPORTS General Provisions; Medical Device Reporting

REPORTS	803.53	FIVE-DAY REPORTS Manufacturer Reporting Requirements; Medical Device Reporting
REPORTS	803.55	BASELINE REPORTS Manufacturer Reporting Requirements; Medical Device Reporting
REPORTS	803.56	SUPPLEMENTAL REPORTS Manufacturer Reporting Requirements; Medical Device Reporting
REPORTS	806.10	REPORTS OF CORRECTIONS AND REMOVALS Medical Devices; Reports Of Corrections And Removals
REPORTS	806.40	PUBLIC AVAILABILITY OF REPORTS Medical Devices; Reports Of Corrections And Removals
REPORTS	810.16	CEASE DISTRIBUTION AND NOTIFICATION OR MANDATORY RECALL ORDER STATUS REPORTS Mandatory Medical Device Recall Procedures; Medical Device Recall Authority
REPORTS	812.150	REPORTS Records And Reports; Investigational Device Exemptions
REPORTS	814.84	REPORTS Postapproval Requirements; Premarket Approval Of Medical Devices
REPORTS	814.126	POSTAPPROVAL REQUIREMENTS AND REPORTS Humanitarian Use Devices; Premarket Approval Of Medical Devices
REPORTS	1002.7	PRODUCT REPORTS Required Manufacturers' Reports For Listed Electronic Products; Records And Reports
REPORTS	1002.10	SUBMISSION OF DATA AND REPORTS General Provisions; Records And Reports
REPORTS	1002.11	SUPPLEMENTAL REPORTS Required Manufacturers' Reports For Listed Electronic Products; Records And Reports
REPORTS	1002.12	ABBREVIATED REPORTS Required Manufacturers' Reports For Listed Electronic Products; Records And Reports
REPORTS	1304.03	PERSONS REQUIRED TO KEEP RECORDS AND FILE REPORTS General Information; Records And Reports Of Registrants
REPORTS	1304.31	REPORTS FROM MANUFACTURERS IMPORTING NARCOTIC RAW MATERIAL Reports; Records And Reports Of Registrants
REPORTS	1304.32	REPORTS OF MANUFACTURERS IMPORTING COCA LEAVES Reports; Records And Reports Of Registrants
REPORTS	1304.33	REPORTS TO ARCOS Reports; Records And Reports Of Registrants
REPORTS	1310.03	PERSONS REQUIRED TO KEEP RECORDS AND FILE REPORTS Records and Reports Of Listed Chemicals And Certain Machines
REPORTS	1310.05	REPORTS Records and Reports Of Listed Chemicals And Certain Machines
REPORTS	1310.06	CONTENT OF RECORDS AND REPORTS Records and Reports Of Listed Chemicals And Certain Machines
REPORTS, ADVERSE EVENT	803.30	INDIVIDUAL ADVERSE EVENT REPORTS; USER FACILITIES User Facility Reporting Requirements; Medical Device Reporting
REPORTS, ADVERSE EVENT	803.50	INDIVIDUAL ADVERSE EVENT REPORTS; MANUFACTURERS Manufacturer Reporting Requirements; Medical Device Reporting
REPORTS, ANNUAL	312.33	ANNUAL REPORTS Investigational New Drug Application
REPORTS, ANNUAL	316.30	ANNUAL REPORTS OF HOLDER OF ORPHAN-DRUG DESIGNATION Orphan Drugs
REPORTS, ANNUAL	601.28	ANNUAL REPORTS OF POSTMARKETING PEDIATRIC STUDIES Biologics Licensing; Licensing
REPORTS, ANNUAL	601.37	ANNUAL REPORTS OF POSTMARKETING PEDIATRIC STUDIES Biologics Licensing; Licensing
REPORTS, ANNUAL	601.70	ANNUAL PROGRESS REPORTS OF POSTMARKETING STUDIES Postmarketing Studies; Licensing
REPORTS, ANNUAL	803.33	ANNUAL REPORTS User Facility Reporting Requirements; Medical Device Reporting
REPORTS, ANNUAL	1002.13	ANNUAL REPORTS Required Manufacturers' Reports For Listed Electronic Products; Records And Reports
REPOSITORY CORTICOTROPIN	522.480	REPOSITORY CORTICOTROPIN INJECTION Implantation Or Injectable Dosage Form New Animal Drugs
REPRESENTATION	10.105	REPRESENTATION BY AN ORGANIZATION General Administrative Procedures; Administrative Practices And Procedures
REPRESENTATION	1316.50	APPEARANCE; REPRESENTATION; AUTHORIZATION Administrative Hearings; Administrative Functions, Practices, And Procedures
REPROCESSING	211.115	REPROCESSING Production And Process Controls; Current Good Manufacturing Practice For Finished Pharmaceuticals

REPROCESSING	820.115	REPROCESSING OF DEVICES OR COMPONENTS Production And Process Controls; Good Manufacturing Practice For Medical Devices
REPRODUCTION, ASSISTED	884.6100	ASSISTED REPRODUCTION NEEDLES Assisted Reproduction Devices; Obstetrical And Gynecological Devices
REPRODUCTION, ASSISTED	884.6110	ASSISTED REPRODUCTION CATHETERS Assisted Reproduction Devices; Obstetrical And Gynecological Devices
REPRODUCTION, ASSISTED	884.6120	ASSISTED REPRODUCTION ACCESSORIES Assisted Reproduction Devices; Obstetrical And Gynecological Devices
REPRODUCTION, ASSISTED	884.6130	ASSISTED REPRODUCTION MICROTOOLS Assisted Reproduction Devices; Obstetrical And Gynecological Devices
REPRODUCTION, ASSISTED	884.6140	ASSISTED REPRODUCTION MICROPIPETTE FABRICATION INSTRUMENTS Assisted Reproduction Devices; Obstetrical And Gynecological Devices
REPRODUCTION , ASSISTED	884.6150	ASSISTED REPRODUCTION MICROMANIPULATORS AND MICROINJECTORS Assisted Reproduction Devices; Obstetrical And Gynecological Devices
REPRODUCTION, ASSISTED	884.6160	ASSISTED REPRODUCTION LABWARE Assisted Reproduction Devices; Obstetrical And Gynecological Devices
REPRODUCTION, ASSISTED	884.6170	ASSISTED REPRODUCTION WATER AND WATER PURIFICATION SYSTEMS Assisted Reproduction Devices; Obstetrical And Gynecological Devices
REPRODUCTIVE, ASSISTED	884.6190	ASSISTED REPRODUCTIVE MICROSCOPES AND MICROSCOPE ACCESSORIES Assisted Reproduction Devices; Obstetrical And Gynecological Devices
REPRODUCTIVE MEDIA	884.6180	REPRODUCTIVE MEDIA AND SUPPLEMENTS Assisted Reproduction Devices; Obstetrical And Gynecological Devices
REQUEST	3.7	REQUEST FOR DESIGNATION Product Jurisdiction
REQUEST	20.23	REQUEST FOR EXISTING RECORDS General Policy; Public Information
REQUEST	20.40	FILING A REQUEST FOR RECORDS Procedures And Fees; Public Information
REQUEST	20.44	PRESUBMISSION REVIEW OF REQUEST FOR CONFIDENTIALITY OF VOLUNTARILY SUBMITTED DATA OR INFORMATION Procedures And Fees; Public Information
REQUEST	20.47	DENIAL OF A REQUEST FOR RECORDS Procedures And Fees; Public Information
REQUEST	20.52	REQUEST FOR REVIEW WITHOUT COPYING Procedures And Fees; Public Information
REQUEST	70.55	REQUEST FOR SCIENTIFIC STUDIES Safety Evaluation; Color Additives
REQUEST	80.21	REQUEST FOR CERTIFICATION Certification Procedures; Color Additive Certification
REQUEST	99.203	REQUEST TO EXTEND THE TIME FOR COMPLETING PLANNED STUDIES Dissemination Of Information On Unapproved/New Uses For Marketed Drugs, Biologics, And Devices
REQUEST	314.93	PETITION TO REQUEST A CHANGE FROM A LISTED DRUG Abbreviated Applications; Applications For FDA Approval To Market A New Drug
REQUEST	316.10	CONTENT AND FORMAT OF A REQUEST FOR WRITTEN RECOMMENDATIONS Orphan Drugs
REQUEST	316.20	CONTENT AND FORMAT OF A REQUEST FOR ORPHAN-DRUG DESIGNATION Orphan Drugs
REQUEST	810.12	WRITTEN REQUEST FOR REVIEW OF CEASE DISTRIBUTION AND NOTIFICATION ORDER Mandatory Medical Device Recall Procedures; Medical Device Recall Authority
REQUEST	1401.5	HOW TO REQUEST RECORDS Public Availability Of Information
REQUEST FORMS	203.60	REQUEST AND RECEIPT FORMS, REPORTS AND RECORDS Prescription Drug Marketing
REQUEST FOR HEARING	12.32	REQUEST FOR ALTERNATIVE FORM OF HEARING Initiation Of Proceedings; Formal Evidentiary Public Hearing
REQUEST FOR HEARING	60.40	REQUEST FOR HEARING Due Diligence Hearings; Patent Term Restoration
REQUEST FOR HEARING	314.200	NOTICE OF OPPORTUNITY FOR HEARING; NOTICE OF PARTICIPATION AND REQUEST FOR HEARING; GRANT OR DENIAL OF HEARING Hearing Procedures For New Drugs; Applications For FDA Approval To Market A New Drug
REQUEST FOR HEARING	1301.43	REQUEST FOR HEARING OR APPEARANCE; WAIVER Hearings; Registration Of Manufacturers, Distributors, And Dispensers Of Controlled Substances
REQUEST FOR HEARING	1303.34	REQUEST FOR HEARING OR APPEARANCE; WAIVER Hearings; Quotas
REQUEST FOR HEARING	1308.44	REQUEST FOR HEARING OR APPEARANCE; WAIVER Hearings; Schedules Of Controlled Substances

REQUEST FOR HEARING	1309.53	REQUEST FOR HEARING OR APPEARANCE; WAIVER Hearings; Registration Of Manufacturers, Distributors, Importers And Exporters Of List I Chemicals
REQUEST FOR HEARING	1312.44	REQUEST FOR HEARING OR APPEARANCE; WAIVER Hearings; Importation And Exportation Of Controlled Substances
REQUEST FOR HEARING	1313.54	REQUEST FOR HEARING Hearings; Importation And Exportation Of Precursors And Essential Chemicals
REQUEST FOR HEARING	1316.47	REQUEST FOR HEARING Administrative Hearings; Administrative Functions, Practices, And Procedures
REQUESTED INFORMATION	710.4	INFORMATION REQUESTED Voluntary Registration Of Cosmetic Product Establishments
REQUESTED INFORMATION	720.4	INFORMATION REQUESTED ABOUT COSMETIC PRODUCTS Voluntary Filing Of Cosmetic Product Ingredient Composition Statements
REQUESTED INFORMATION	807.25	INFORMATION REQUIRED OR REQUESTED FOR ESTABLISHMENT REGISTRATION AND DEVICE LISTING Procedures For Device Establishments; Establishment Registration And Device Listing For Manufacturers And Initial Importers Of Devices
REQUESTED RECORDS	21.43	ACCESS TO REQUESTED RECORDS Procedures For Notification Of And Access To Records In Privacy Act Record Systems; Protection Of Privacy
REQUESTERS	1401.12	FEES TO BE CHARGED - CATEGORIES OF REQUESTERS Public Availability Of Information
REQUESTING	1.23	PROCEDURES FOR REQUESTING VARIATIONS AND EXEMPTIONS FROM REQUIRED LABEL STATEMENTS General Labeling Requirements; General Enforcement Regulations
REQUESTING	501.103	PETITIONS REQUESTING EXEMPTIONS FROM OR SPECIAL REQUIREMENTS FOR LABEL DECLARATION OF INGREDIENTS Exemptions From Animal Food Labeling Requirements; Animal Food Labeling
REQUESTS	12.22	FILING OBJECTIONS AND REQUESTS FOR A HEARING ON A REGULATION OR ORDER Initiation Of Proceedings; Formal Evidentiary Public Hearing
REQUESTS	12.24	RULING ON OBJECTIONS AND REQUESTS FOR HEARING Initiation Of Proceedings; Formal Evidentiary Public Hearing
REQUESTS	14.65	PUBLIC INQUIRIES AND REQUESTS FOR ADVISORY COMMITTEE RECORDS Records Of Meetings And Hearings Before Advisory Committees; Public Hearing Before A Public Advisory Committee
REQUESTS	20.31	RETENTION SCHEDULE OF REQUESTS FOR FDA RECORDS General Policy; Public Information
REQUESTS	20.48	NONSPECIFIC AND OVERLY BURDENSOME REQUESTS Procedures And Fees; Public Information
REQUESTS	21.40	PROCEDURES FOR SUBMITTING REQUESTS FOR NOTIFICATION AND ACCESS Procedures For Notification Of And Access To Records In Privacy Act Record Systems; Protection Of Privacy
REQUESTS	21.41	PROCESSING OF REQUESTS Procedures For Notification Of And Access To Records In Privacy Act Record Systems; Protection Of Privacy
REQUESTS	21.42	RESPONSES TO REQUESTS Procedures For Notification Of And Access To Records In Privacy Act Record Systems; Protection Of Privacy
REQUESTS	21.50	PROCEDURES FOR SUBMITTING REQUESTS FOR AMENDMENT OF RECORDS Procedures For Requests For Amendment Of Records; Protection Of Privacy
REQUESTS	21.51	RESPONSES TO REQUESTS FOR AMENDMENT OF RECORDS Procedures For Requests For Amendment Of Records; Protection Of Privacy
REQUESTS	80.22	SAMPLES TO ACCOMPANY REQUESTS FOR CERTIFICATION Certification Procedures; Color Additive Certification
REQUESTS	201.58	REQUESTS FOR WAIVER OF REQUIREMENT FOR ADEQUATE AND WELL-CONTROLLED STUDIES TO SUBSTANTIATE CERTAIN LABELING STATEMENTS Labeling Requirements For Prescription Drugs And/or Insulin; Labeling
REQUESTS	203.35	STANDING REQUESTS Samples; Prescription Drug Marketing
REQUESTS	312.42	CLINICAL HOLDS AND REQUESTS FOR MODIFICATION Investigational New Drug Application
REQUESTS	316.23	TIMING OF REQUESTS FOR ORPHAN-DRUG DESIGNATION; DESIGNATION OF ALREADY APPROVED DRUGS Orphan Drugs
REQUESTS	316.52	AVAILABILITY FOR PUBLIC DISCLOSURE OF DATA AND INFORMATION IN REQUESTS AND APPLICATIONS Orphan Drugs
REQUESTS	610.2	REQUESTS FOR SAMPLES AND PROTOCOLS; OFFICIAL RELEASE Release Requirements; General Biological Products Standards

REQUESTS	803.15	REQUESTS FOR ADDITIONAL INFORMATION Medical Device Reporting
REQUIRED	1.23	PROCEDURES FOR REQUESTING VARIATIONS AND EXEMPTIONS FROM REQUIRED LABEL STATEMENTS General Labeling Requirements; General Enforcement Regulations
REQUIRED	1.24	EXEMPTIONS FROM REQUIRED LABEL STATEMENTS General Labeling Requirements; General Enforcement Regulations
REQUIRED	16.24	REGULATORY HEARING REQUIRED BY THE ACT OR A REGULATION Initiation Of Proceedings; Regulatory Hearing Before The Food And Drug Administration
REQUIRED	20.83	DISCLOSURE REQUIRED BY COURT ORDER Limitations On Exemptions; Public Information
REQUIRED	21.70	DISCLOSURE AND INTRA-AGENCY USE OF RECORDS IN PRIVACY ACT RECORD SYSTEMS; NO ACCOUNTING REQUIRED Disclosure Of Records In Privacy Act Record Systems To Persons Other ThanThe Subject Individual; Protection Of Privacy
REQUIRED	21.71	DISCLOSURE OF RECORDS IN PRIVACY ACT RECORD SYSTEMS; ACCOUNTING REQUIRED Disclosure Of Records In Privacy Act Record Systems To Persons Other ThanThe Subject Individual; Protection Of Privacy
REQUIRED	56.103	CIRCUMSTANCES IN WHICH IRB REVIEW IS REQUIRED General Provisions; Institutional Review Boards
REQUIRED	205.5	MINIMUM REQUIRED INFORMATION FOR LICENSURE Guidelines For State Licensing Of Wholesale Prescription Drug Distributors
REQUIRED	206.10	CODE IMPRINT REQUIRED Imprinting Of Solid Oral Dosage Form Drug Products For Human Use
REQUIRED	207.25	INFORMATION REQUIRED IN REGISTRATION AND DRUG LISTING Procedures For Domestic Drug Establishments; Registration Of Drug Producers And Listing Of Drugs In Commercial Distribution
REQUIRED	510.112	ANTIBIOTICS USED IN VETERINARY MEDICINE AND FOR NONMEDICAL PURPOSES; REQUIRED DATA Specific Administrative Rulings And Decisions; New Animal Drugs
REQUIRED	607.25	INFORMATION REQUIRED FOR ESTABLISHMENT REGISTRATION AND BLOOD PRODUCT LISTING Procedures For Domestic Blood Product Establishments; Establishment Registratn+product Listing, Manufacturers Human Blood/b.products
REQUIRED	803.10	GENERAL DESCRIPTION OF REPORTS REQUIRED FROM USER FACILITIES, IMPORTERS, AND MANUFACTURERS Medical Device Reporting
REQUIRED	807.81	WHEN A PREMARKET NOTIFICATION SUBMISSION IS REQUIRED Premarket Notification Procedures; Establishment Registration And Device Listing For Manufacturers And Initial Importers Of Devices
REQUIRED	807.25	INFORMATION REQUIRED OR REQUESTED FOR ESTABLISHMENT REGISTRATION AND DEVICE LISTING Procedures For Device Establishments; Establishment Registration And Device Listing For Manufacturers And Initial Importers Of Devices
REQUIRED	807.87	INFORMATION REQUIRED IN A PREMARKET NOTIFICATION SUBMISSION Premarket Notification Procedures; Establishment Registration And Device Listing For Manufacturers And Initial Importers Of Devices
REQUIRED	1230.40	REQUIRED LABEL INFORMATION Imports; Regulations Under The Federal Caustic Poison Act
REQUIRED	1301.11	PERSONS REQUIRED TO REGISTER Registration Of Manufacturers, Distributors, And Dispensers Of Controlled Substances
REQUIRED	1302.03	SYMBOL REQUIRED; EXCEPTIONS Labeling And Packaging Requirements For Controlled Substances
REQUIRED	1304.03	PERSONS REQUIRED TO KEEP RECORDS AND FILE REPORTS General Information; Records And Reports Of Registrants
REQUIRED	1308.46	CONTROL REQUIRED UNDER INTERNATIONAL TREATY Hearings; Schedules Of Controlled Substances
REQUIRED	1309.21	PERSONS REQUIRED TO REGISTER Registration Of Manufacturers, Distributors, Importers And Exporters Of List I Chemicals
REQUIRED	1310.03	PERSONS REQUIRED TO KEEP RECORDS AND FILE REPORTS Records and Reports Of Listed Chemicals And Certain Machines
REQUIRED	1312.26	RECORDS REQUIRED OF EXPORTER Exportation Of Controlled Substances; Importation And Exportation Of Controlled Substances
REQUIRED PEDIATRIC STUDIES	201.23	REQUIRED PEDIATRIC STUDIES General Labeling Provisions; Labeling
REQUIRED RECALL	107.200	FOOD AND DRUG ADMINISTRATION REQUIRED RECALL Infant Formula Recalls; Infant Formula
REQUIRED STATEMENTS	101.15	FOOD; PROMINENCE OF REQUIRED STATEMENTS General Provisions; Food Labeling

REQUIRED STATEMENTS	201.15	DRUGS; PROMINENCE OF REQUIRED LABEL STATEMENTS General Labeling Provisions; Labeling
REQUIRED STATEMENTS	201.16	DRUGS; SPANISH-LANGUAGE VERSION OF CERTAIN REQUIRED STATEMENTS General Labeling Provisions; Labeling
REQUIRED STATEMENTS	501.15	ANIMAL FOOD; PROMINENCE OF REQUIRED STATEMENTS General Provisions; Animal Food Labeling
REQUIRED STATEMENTS	801.15	MEDICAL DEVICES; PROMINENCE OF REQUIRED LABEL STATEMENTS General Labeling Provisions; Labeling
REQUIRED STATEMENTS	801.16	MEDICAL DEVICES; SPANISH-LANGUAGE VERSION OF CERTAIN REQUIRED STATEMENTS General Labeling Provisions; Labeling
REQUIRED STUDIES	201.23	REQUIRED PEDIATRIC STUDIES General Labeling Provisions; Labeling
REQUIRED TESTS	610.1	TESTS PRIOR TO RELEASE REQUIRED FOR EACH LOT Release Requirements; General Biological Products Standards
REQUIRED WARNING	201.22	PRESCRIPTION DRUGS CONTAINING SULFITES; REQUIRED WARNING STATEMENTS General Labeling Provisions; Labeling
REQUIRED WARNING	201.316	DRUGS WITH THYROID HORMONE ACTIVITY FOR HUMAN USE; REQUIRED WARNING Specific Labeling Requirements For Specific Drug Products; Labeling
REQUIRED WARNING	201.317	DIGITALIS AND RELATED CARDIOTONIC DRUGS FOR HUMAN USE IN ORAL DOSAGE FORMS; REQUIRED WARNING Specific Labeling Requirements For Specific Drug Products; Labeling
REQUIRED WARNING	201.322	OVER-THE COUNTER DRUG PRODUCTS CONTAINING INTERNAL ANALGESIC/ANTIPYRETIC ACTIVE INGREDIENTS; REQUIRED ALCOHOL WARNING Specific Labeling Requirements For Specific Drug Products; Labeling
REQUIRED WARNING	290.5	DRUGS; STATEMENT OF REQUIRED WARNING General Provisions; Controlled Drugs
REQUIRED WARNING	290.6	SPANISH-LANGUAGE VERSION OF REQUIRED WARNING General Provisions; Controlled Drugs
REQUIRED WARNING	369.21	DRUGS; WARNING AND CAUTION STATEMENTS REQUIRED BY REGULATIONS Warning And Caution Statement For Drugs; Interpretative Statements Re: Warnings On Drugs And Devices For OTC Sale
REQUIRED, WARNING	369.22	DRUGS; WARNING AND CAUTION STATEMENTS SPECIFICALLY REQUIRED BY LAW Warning And Caution Statement For Drugs; Interpretative Statements Re: Warnings On Drugs And Devices For OTC Sale
REQUIRED WARNINGS	201.319	WATER-SOLUBLE GUMS, HYDROPHILIC GUMS AND HYDROPHILIC MUCILLOIDS (INCLUDING AGAR, CARRAGEENAN, KELP, PSYLLIUM, XANTHAN GUM) AS ACTIVE INGREDIENTS; REQUIRED WARNINGS AND DIRECTIONS Specific Labeling Requirements For Specific Drug Products; Labeling
REQUIRED WARNINGS	369.7	WARNINGS REQUIRED BY OFFICIAL COMPENDIA Definitions And Interpretations; Interpretative Statements Re Warnings On Drugs And Devices For OTC Sale
REQUIRED WORDING	1230.11	REQUIRED WORDING Labeling; Regulations Under The Federal Caustic Poison Act
REQUIREMENT	5.604	MANUFACTURERS REQUIREMENT TO PROVIDE DATA TO ULTIMATE PURCHASERS OF ELECTRONIC PRODUCTS Radiation Control, Redelegations Of Authority; Delegations Of Authority And Organization
REQUIREMENT	56.104	EXEMPTIONS FROM IRB REQUIREMENT General Provisions; Institutional Review Boards
REQUIREMENT	56.105	WAIVER OF IRB REQUIREMENT General Provisions; Institutional Review Boards
REQUIREMENT	99.205	APPLICATION FOR EXEMPTION FROM THE REQUIREMENT TO FILE A SUPPLEMENTAL APPLICATION Dissemination Of Information On Unapproved/New Uses For Marketed Drugs, Biologics, and Devices
REQUIREMENT	99.305	EXEMPTION FROM THE REQUIREMENT TO FILE A SUPPLEMENTAL APPLICATION Dissemination Of Information On Unapproved/New Uses For Marketed Drugs, Biologics, and Devices
REQUIREMENT	190.6	REQUIREMENT FOR PREMARKET NOTIFICATION New Dietary Ingredient Notification; Dietary Supplements
REQUIREMENT	205.4	WHOLESALE DRUG DISTRIBUTOR LICENSING REQUIREMENT Guidelines For State Licensing Of Wholesale Prescription Drug Distributors
REQUIREMENT	312.20	REQUIREMENT FOR AN IND Investigational New Drug Application

REQUIREMENT	320.32	PROCEDURES FOR ESTABLISHING OR AMENDING A BIOEQUIVALENCE REQUIREMENT Procedures For Determining The Bioavailability Or Bioequivalence Of Drug Products; Bioavailability And Bioequivalence Requirements
REQUIREMENT	558.4	REQUIREMENT OF A MEDICATED FEED MILL LICENSE General Provisions; New Animal Drugs For Use In Animal Feeds
REQUIREMENT	803.13	ENGLISH REPORTING REQUIREMENT Medical Device Reporting
REQUIREMENT	862.3	EFFECTIVE DATES OF REQUIREMENT FOR PREMARKET APPROVAL Clinical Chemistry And Clinical Toxicology Devices
REQUIREMENT	864.3	EFFECTIVE DATES OF REQUIREMENT FOR PREMARKET APPROVAL Hematology And Pathology Devices
REQUIREMENT	866.3	EFFECTIVE DATES OF REQUIREMENT FOR PREMARKET APPROVAL Immunology And Microbiology Devices
REQUIREMENT	868.3	EFFECTIVE DATES OF REQUIREMENT FOR PREMARKET APPROVAL Anesthesiology Devices
REQUIREMENT	870.3	EFFECTIVE DATES OF REQUIREMENT FOR PREMARKET APPROVAL Cardiovascular Devices
REQUIREMENT	872.3	EFFECTIVE DATES OF REQUIREMENT FOR PREMARKET APPROVAL Dental Devices
REQUIREMENT	874.3	EFFECTIVE DATES OF REQUIREMENT FOR PREMARKET APPROVAL Ear, Nose, And Throat Devices
REQUIREMENT	876.3	EFFECTIVE DATES OF REQUIREMENT FOR PREMARKET APPROVAL Gastroenterology-urology Devices
REQUIREMENT	878.3	EFFECTIVE DATES OF REQUIREMENT FOR PREMARKET APPROVAL General And Plastic Surgery Devices
REQUIREMENT	880.3	EFFECTIVE DATES OF REQUIREMENT FOR PREMARKET APPROVAL General Hospital And Personal Use Devices
REQUIREMENT	882.3	EFFECTIVE DATES OF REQUIREMENT FOR PREMARKET APPROVAL Neurological Devices
REQUIREMENT	884.3	EFFECTIVE DATES OF REQUIREMENT FOR PREMARKET APPROVAL Obstetrical And Gynecological Devices
REQUIREMENT	886.3	EFFECTIVE DATES OF REQUIREMENT FOR PREMARKET APPROVAL Ophthalmic Devices
REQUIREMENT	888.3	EFFECTIVE DATES OF REQUIREMENT FOR PREMARKET APPROVAL Orthopedic Devices
REQUIREMENT	890.3	EFFECTIVE DATES OF REQUIREMENT FOR PREMARKET APPROVAL Physical Medicine Devices
REQUIREMENT	892.3	EFFECTIVE DATES OF REQUIREMENTS FOR PREMARKET APPROVAL Radiology Devices
REQUIREMENT	1306.11	REQUIREMENT OF PRESCRIPTION Controlled Substances Listed In Schedule II; Prescriptions
REQUIREMENT	1306.21	REQUIREMENT OF PRESCRIPTION Controlled Substances Listed In Schedules III, IV, And V; Prescriptions
REQUIREMENT	1309.29	WAIVER OF REGISTRATION REQUIREMENT FOR CERTAIN ACTIVITIES Registration Of Manufacturers, Distributors, Importers And Exporters Of List I Chemicals
REQUIREMENT	1312.11	REQUIREMENT OF AUTHORIZATION TO IMPORT Importation Of Controlled Substances; Importation And Exportation Of Controlled Substances
REQUIREMENT	1312.21	REQUIREMENT OF AUTHORIZATION TO EXPORT Exportation Of Controlled Substances; Importation And Exportation Of Controlled Substances
REQUIREMENT	1313.12	REQUIREMENT OF AUTHORIZATION TO IMPORT Importation Of Listed Chemicals; Importation And Exportation Of Precursors And Essential Chemicals
REQUIREMENT	1313.21	REQUIREMENT OF AUTHORIZATION TO EXPORT Exportation Of Listed Chemicals; Importation And Exportation Of Precursors And Essential Chemicals
REQUIREMENT	1313.32	REQUIREMENT OF AUTHORIZATION FOR INTERNATIONAL TRANSACTIONS Importation And Exportation Of Precursors And Essential Chemicals
REQUIREMENT	1316.07	REQUIREMENT FOR ADMINISTRATIVE INSPECTION WARRANT Administrative Inspections; Administrative Functions, Practices, And Procedures
REQUIREMENTS	5.107	EXTENSIONS OR STAYS OF EFFECTIVE DATES FOR COMPLIANCE WITH CERTAIN LABELING REQUIREMENTS FOR HUMAN PRESCRIPTION DRUGS Human Drugs; Redelegations Of Authority; Delegations Of Authority And Organization
REQUIREMENTS	5.400	ISSUANCE OF FEDERAL REGISTER DOCUMENTS TO RECOGNIZE OR TO WITHDRAW RECOGNITION OF A STANDARD TO MEET PREMARKET SUBMISSION REQUIREMENTS Medical Devices And Radiological Health; Redelegations Of Authority; Delegations Of Authority And Organization

REQUIREMENTS	10.19	WAIVER, SUSPENSION, OR MODIFICATION OF PROCEDURAL REQUIREMENTS General Provisions; Administrative Practices And Procedures
REQUIREMENTS	11.100	GENERAL REQUIREMENTS Electronic Records; Electronic Signatures
REQUIREMENTS	50.20	GENERAL REQUIREMENTS FOR INFORMED CONSENT Informed Consent Of Human Subjects; Protection Of Human Subjects
REQUIREMENTS	50.23	EXCEPTION FROM GENERAL REQUIREMENTS Informed Consent Of Human Subjects; Protection Of Human Subjects
REQUIREMENTS	50.24	EXEMPTION FROM INFORMED CONSENT REQUIREMENTS FOR EMERGENCY RESEARCH Informed Consent Of Human Subjects; Protection Of Human Subjects
REQUIREMENTS	50.55	REQUIREMENTS FOR PERMISSION BY PARENTS OR GUARDIANS AND FOR ASSENT BY CHILDREN Additional Safeguards For Children In Clinical Investigations; Protection Of Human Subjects
REQUIREMENTS	54.4	CERTIFICATION AND DISCLOSURE REQUIREMENTS Financial Disclosure By Clinical Investigators
REQUIREMENTS	70.20	PACKAGING REQUIREMENTS FOR STRAIGHT COLORS (OTHER THAN HAIR DYES) Packaging And Labeling; Color Additives
REQUIREMENTS	70.25	LABELING REQUIREMENTS FOR COLOR ADDITIVES (OTHER THAN HAIR DYES) Packaging And Labeling; Color Additives
REQUIREMENTS	100.1	PETITIONS REQUESTING EXEMPTION FROM PREEMPTION FOR STATE OR LOCAL REQUIREMENTS State And Local Requirements; General
REQUIREMENTS	101.14	HEALTH CLAIMS: GENERAL REQUIREMENTS General Provisions; Food Labeling
REQUIREMENTS	107.240	NOTIFICATION REQUIREMENTS Infant Formula Recalls; Infant Formula
REQUIREMENTS	108.19	ESTABLISHMENT OF REQUIREMENTS FOR EXEMPTION FROM SECTION 404 OF THE ACT General Provisions; Emergency Permit Control
REQUIREMENTS	120.14	APPLICATION OF REQUIREMENTS TO IMPORTED PRODUCTS Hazard Analysis And Critical Control Point (HACCP) Systems
REQUIREMENTS	123.12	SPECIAL REQUIREMENTS FOR IMPORTED PRODUCTS Fish And Fishery Products
REQUIREMENTS	130.10	REQUIREMENTS FOR FOODS NAMED BY USE OF A NUTRIENT CONTENT CLAIM AND A STANDARDIZED TERM General Provisions; Food Standards: General
REQUIREMENTS	130.17	TEMPORARY PERMITS FOR INTERSTATE SHIPMENT OF EXPERIMENTAL PACKS OF FOOD VARYING FROM THE REQUIREMENTS OF DEFINITIONS AND STANDARDS OF IDENTITY General Provisions; Food Standards: General
REQUIREMENTS	201.56	GENERAL REQUIREMENTS ON CONTENT AND FORMAT OF LABELING FOR HUMAN PRESCRIPTION DRUGS Labeling Requirements For Prescription Drugs And/or Insulin; Labeling
REQUIREMENTS	201.57	SPECIFIC REQUIREMENTS ON CONTENT AND FORMAT OF LABELING FOR HUMAN PRESCRIPTION DRUGS Labeling Requirements For Prescription Drugs And/or Insulin; Labeling
REQUIREMENTS	201.66	FORMAT AND CONTENT REQUIREMENTS FOR OTC DRUG PRODUCT LABELING Labeling Requirements For Over-the-counter Drugs; Labeling
REQUIREMENTS	203.32	DRUG SAMPLE STORAGE AND HANDLING REQUIREMENTS Prescription Drug Marketing
REQUIREMENTS	203.37	INVESTIGATION AND NOTIFICATION REQUIREMENTS Prescription Drug Marketing
REQUIREMENTS	203.50	REQUIREMENTS FOR WHOLESALE DISTRIBUTION OF PRESCRIPTION DRUGS Prescription Drug Marketing
REQUIREMENTS	205.50	MINIMUM REQUIREMENTS FOR STORAGE AND HANDLING OF PRESCRIPTION DRUGS AND ESTABLISHMENT AND MAINTENANCE OF PRESCRIPTION DRUG DISTRIBUTION RECORDS Guidelines For State Licensing Of Wholesale Prescription Drug Distributors
REQUIREMENTS	207.40	ESTABLISHMENT REGISTRATION AND DRUG LISTING REQUIREMENTS FOR FOREIGN ESTABLISHMENTS Procedure For Foreign Drug Establishments; Registration Of Drug Producers And Listing Of Drugs In Commercial Distribution
REQUIREMENTS	211.80	GENERAL REQUIREMENTS Control Of Components And Drug Product Containers And Closures; Current Good Manufacturing Practice For Finished Pharmaceuticals

REQUIREMENTS	606.120	LABELING, GENERAL REQUIREMENTS Finished Product Control; Current Good Manufacturing Practice For Blood And Blood Components
REQUIREMENTS	607.40	ESTABLISHMENT REGISTRATION AND BLOOD PRODUCT LISTING REQUIREMENTS FOR FOREIGN BLOOD PRODUCT ESTABLISHMENTS Procedures For Foreign Blood Product Establishments; Establishment Registratn+product Listing, Manufacturers Human Blood/b.products
REQUIREMENTS	610.40	TEST REQUIREMENTS Testing Requirements For Communicable Disease Agents; General Biological Products Standards
REQUIREMENTS	610.46	"LOOK BACK" REQUIREMENTS Testing Requirements For Communicable Disease Agents; General Biological Products Standards
REQUIREMENTS	610.47	"LOOK BACK" NOTIFICATION REQUIREMENTS FOR TRANSFUSION SERVICES Testing Requirements For Communicable Disease Agents; General Biological Products Standards
REQUIREMENTS	640.2	GENERAL REQUIREMENTS Whole Blood; Additional Standards For Human Blood And Blood Products
REQUIREMENTS	640.11	GENERAL REQUIREMENTS Red Blood Cells; Additional Standards For Human Blood And Blood Products
REQUIREMENTS	640.25	GENERAL REQUIREMENTS Platelets; Additional Standards For Human Blood And Blood Products
REQUIREMENTS	640.69	GENERAL REQUIREMENTS Source Plasma; Additional Standards For Human Blood And Blood Products
REQUIREMENTS	640.83	GENERAL REQUIREMENTS Albumin (Human); Additional Standards For Human Blood And Blood Products
REQUIREMENTS	640.93	GENERAL REQUIREMENTS Plasma Protein Fraction (Human); Additional Standards For Human Blood And Blood Products
REQUIREMENTS	640.101	GENERAL REQUIREMENTS Immune Globulin (Human); Additional Standards For Human Blood And Blood Products
REQUIREMENTS	660.2	GENERAL REQUIREMENTS Antibody To Hepatitis B Surface Antigen; Additional Standards For Diagnostic Substances For Laboratory Tests
REQUIREMENTS	660.22	POTENCY REQUIREMENTS WITH REFERENCE PREPARATIONS Blood Grouping Reagent; Additional Standards For Diagnostic Substances For Laboratory Tests
REQUIREMENTS	700.25	TAMPER-RESISTANT PACKAGING REQUIREMENTS FOR COSMETIC PRODUCTS Requirements For Specific Cosmetic Products; General
REQUIREMENTS	701.2	FORM OF STATING LABELING REQUIREMENTS General Provisions; Cosmetic Labeling
REQUIREMENTS	701.9	EXEMPTIONS FROM LABELING REQUIREMENTS General Provisions; Cosmetic Labeling
REQUIREMENTS	803.19	EXEMPTIONS OF VARIANCES, AND ALTERNATIVE REPORTING REQUIREMENTS Medical Device Reporting
REQUIREMENTS	803.40	INDIVIDUAL ADVERSE EVENT REPORTING REQUIREMENTS; IMPORTERS Importer Reporting Requirement; Medical Device Reporting
REQUIREMENTS	809.20	GENERAL REQUIREMENTS FOR MANUFACTURERS AND PRODUCERS OF IN VITRO DIAGNOSTIC PRODUCTS Requirements For Manufacturers And Producers; In Vitro Diagnostic Products For Human Use
REQUIREMENTS	812.18	IMPORT AND EXPORT REQUIREMENTS General Provisions; Investigational Device Exemptions
REQUIREMENTS	814.82	POSTAPPROVAL REQUIREMENTS Postapproval Requirements; Premarket Approval Of Medical Devices
REQUIREMENTS	814.124	INSTITUTIONAL REVIEW BOARD REQUIREMENTS Humanitarian Use Devices; Premarket Approval Of Medical Devices
REQUIREMENTS	814.126	POSTAPPROVAL REQUIREMENTS AND REPORTS Humanitarian Use Devices; Premarket Approval Of Medical Devices
REQUIREMENTS	820.180	GENERAL REQUIREMENTS Records; Quality System Regulation
REQUIREMENTS	821.25	DEVICE TRACKING SYSTEM AND CONTENT REQUIREMENTS; MANUFACTURER REQUIREMENTS Medical Device Tracking Requirements
REQUIREMENTS	821.30	TRACKING OBLIGATIONS OF PERSONS OTHER THAN DEVICE MANUFACTURERS: DISTRIBUTOR REQUIREMENTS Medical Device Tracking Requirements
REQUIREMENTS	900.13	REQUIREMENTS FOR CERTIFICATION Quality Standards And Certification; Mammography
REQUIREMENTS	900.18	ALTERNATIVE REQUIREMENTS FOR SECTION 900.12 QUALITY STANDARDS Quality Standards And Certification; Mammography

REQUIREMENTS	1003.10	DISCOVERY OF DEFECT OR FAILURE OF COMPLIANCE BY MANUFACTURER; NOTICE REQUIREMENTS Discovery Of Defect Or Failure To Comply; Notification Of Defects Or Failure To Comply
REQUIREMENTS	1003.30	APPLICATION FOR EXEMPTION FROM NOTIFICATION REQUIREMENTS Exemptions From Notification Requirements; Notification Of Defects Or Failure To Comply
REQUIREMENTS	1240.62	TURTLES INTRASTATE AND INTERSTATE REQUIREMENTS Specific Administrative Decisions Regarding Interstate Shipments; Control Of Communicable Diseases
REQUIREMENTS	1240.80	GENERAL REQUIREMENTS FOR WATER FOR DRINKING AND CULINARY PURPOSES Source And Use Of Potable Water; Control Of Communicable Diseases
REQUIREMENTS	1250.22	GENERAL REQUIREMENTS Food Service Sanitation On Land And Air Conveyances, And Vessels; Interstate Conveyance Sanitation
REQUIREMENTS	1250.26	SPECIAL FOOD REQUIREMENTS Food Service Sanitation On Land And Air Conveyances, And Vessels; Interstate Conveyance Sanitation
REQUIREMENTS	1250.63	GENERAL REQUIREMENTS Servicing Areas For Land And Air Conveyances; Interstate Conveyance Sanitation
REQUIREMENTS	1270.33	RECORDS, GENERAL REQUIREMENTS Procedures and Records; Human Tissue Intended For Transplantation
REQUIREMENTS	1271.15	ARE THERE ANY EXCEPTIONS FROM THE REQUIREMENTS OF THIS PART? General Provisions; Human Cells, Tissues, And Cellular And Tissue-Based Products
REQUIREMENTS	1301.26	EXEMPTIONS FROM IMPORT OR EXPORT REQUIREMENTS FOR PERSONAL MEDICAL USE Registration Of Manufacturers, Distributors, And Dispensers Of Controlled Substances
REQUIREMENTS	1301.71	SECURITY REQUIREMENTS GENERALLY Registration Of Manufacturers, Distributors, And Dispensers Of Controlled Substances
REQUIREMENTS	1302.05	EFFECTIVE DATES OF LABELING REQUIREMENTS Labeling And Packaging Requirements For Controlled Substances
REQUIREMENTS	1302.07	LABELING AND PACKAGING REQUIREMENTS FOR IMPORTED AND EXPORTED SUBSTANCES Labeling And Packaging Requirements For Controlled Substances
REQUIREMENTS	1304.11	INVENTORY REQUIREMENTS Inventory Requirements; Records And Reports Of Registrants
REQUIREMENTS	1304.21	GENERAL REQUIREMENTS FOR CONTINUING RECORDS Continuing Records; Records And Reports Of Registrants
REQUIREMENTS	1309.71	GENERAL SECURITY REQUIREMENTS Registration Of Manufacturers, Distributors, Importers And Exporters Of List I Chemicals
REQUIREMENTS	1316.76	REQUIREMENTS AS TO CLAIM AND BOND Seizure, Forfeiture, And Disposition Of Property; Administrative Functions, Practices, And Procedures
REQUIREMENTS	1403.42	RETENTION AND ACCESS REQUIREMENTS FOR RECORDS Reports, Records, Retention, And Enforcement; Uniform Administrative Requirements For Grants And Coop. Agreements To State And Local Gov.
REQUIREMENTS	1404.630	CERTIFICATION REQUIREMENTS AND PROCEDURES Governmentwide Debarment And Suspension And Requirements For Drug-Free Workplace (Grants)
REQUIRING	25.23	ACTIONS REQUIRING THE PREPARATION OF AN ENVIRONMENTAL IMPACT STATEMENT Agency Actions Requiring Environmental Consideration; Environmental Impact Considerations
REQUIRING	1312.30	SCHEDULE III, IV, AND V NON-NARCOTIC CONTROLLED SUBSTANCES REQUIRING AN IMPORT AND EXPORT PERMIT Exportation Of Controlled Substances; Importation And Exportation Of Controlled Substances
RES, SUBSTITUTE	1316.94	POSTING OF SUBSTITUTE RES IN AN ADMINISTRATIVE FORFEITURE ACTION Expedited Forfeiture Proceedings For Certain Property; Administrative Functions, Practices, And Procedures
RES, SUBSTITUTE	1316.98	SUBSTITUTE RES BOND IN A JUDICIAL FORFEITURE ACTION AGAINST A CONVEYANCE Expedited Forfeiture Proceedings For Certain Property; Administrative Functions, Practices, And Procedures
RESEARCH	5.25	RESEARCH, INVESTIGATION, AND TESTING PROGRAMS AND HEALTH INFORMATION AND PROMOTION PROGRAMS General Redelegations Of Authority; Delegations Of Authority And Organization

RESEARCH	5.201	REDELEGATION OF THE CENTER FOR BIOLOGICS EVALUATION AND RESEARCH DIRECTOR'S PROGRAM AUTHORITIES Biologics; Redelegations Of Authority; Delegations Of Authority And Organization
RESEARCH	20.105	TESTING AND RESEARCH CONDUCTED BY OR WITH FUNDS PROVIDED BY THE FDA Availability Of Specific Categories Of Records; Public Information
RESEARCH	56.109	IRB REVIEW OF RESEARCH IRB Functions And Operations; Institutional Review Boards
RESEARCH	56.110	EXPEDITED REVIEW PROCEDURES FOR CERTAIN KINDS OF RESEARCH INVOLVING NO MORE THAN MINIMAL RISK, AND FOR MINOR CHANGES IN APPROVED RESEARCH IRB Functions And Operations; Institutional Review Boards
RESEARCH	56.111	CRITERIA FOR IRB APPROVAL OF RESEARCH IRB Functions And Operations; Institutional Review Boards
RESEARCH	56.113	SUSPENSION OR TERMINATION OF IRB APPROVAL OF RESEARCH IRB Functions And Operations; Institutional Review Boards
RESEARCH	56.114	COOPERATIVE RESEARCH IRB Functions And Operations; Institutional Review Boards
RESEARCH	201.125	DRUGS FOR USE IN TEACHING, LAW ENFORCEMENT, RESEARCH, AND ANALYSIS Exemptions From Adequate Directions For Use; Labeling
RESEARCH	201.129	DRUGS; EXEMPTION FOR RADIOACTIVE DRUGS FOR RESEARCH USE Exemptions From Adequate Directions For Use; Labeling
RESEARCH	312.86	FOCUSED FDA REGULATORY RESEARCH Drugs Intended To Treat Life-threatening And Severely-debilitating Illnesses; Investigational New Drug Application
RESEARCH	801.125	MEDICAL DEVICES FOR USE IN TEACHING, LAW ENFORCEMENT, RESEARCH, AND ANALYSIS Exemptions From Adequate Directions For Use; Labeling
RESEARCH	814.15	RESEARCH CONDUCTED OUTSIDE THE UNITED STATES General; Premarket Approval Of Medical Devices
RESEARCH	1301.18	RESEARCH PROTOCOLS Registration Of Manufacturers, Distributors, And Dispensers Of Controlled Substances
RESEARCH	1301.32	ACTION ON APPLICATIONS FOR RESEARCH IN SCHEDULE I SUBSTANCES Registration Of Manufacturers, Distributors, And Dispensers Of Controlled Substances
RESEARCH ANIMALS	312.160	DRUGS FOR INVESTIGATIONAL USE IN LABORATORY RESEARCH ANIMALS OR IN VITRO TESTS Drugs For Investigational Use In Laboratory Research Animals Or In Vitro Tests; Investigational New Drug Application
RESEARCH, EMERGENCY	50.24	EXEMPTION FROM INFORMED CONSENT REQUIREMENTS FOR EMERGENCY RESEARCH Informed Consent Of Human Subjects; Protection Of Human Subjects
RESEARCH, EMERGENCY	312.54	EMERGENCY RESEARCH UNDER SECTION 50.24 OF THIS CHAPTER Investigational New Drug Application
RESEARCH, EMERGENCY	812.47	EMERGENCY RESEARCH UNDER SECTION 50.24 OF THIS CHAPTER Investigational Drug Exemptions
RESEARCH ORGANIZATION	312.52	TRANSFER OF OBLIGATIONS TO A CONTRACT RESEARCH ORGANIZATION Investigational New Drug Application
RESEARCH PROTOCOLS	1301.18	RESEARCH PROTOCOLS Registration Of Manufacturers, Distributors, And Dispensers Of Controlled Substances
RESEARCH SUBJECTS	1316.23	CONFIDENTIALITY OF IDENTITY OF RESEARCH SUBJECTS Protection Of Researchers And Research Subjects; Administrative Functions, Practices, And Procedures
RESEARCHERS	1304.22	RECORDS FOR MANUFACTURERS, DISTRIBUTORS, DISPENSERS, RESEARCHERS, IMPORTERS AND EXPORTERS Continuing Records; Records And Reports Of Registrants
RESEARCHERS	1316.24	EXEMPTION FROM PROSECUTION FOR RESEARCHERS Protection Of Researchers And Research Subjects; Administrative Functions, Practices, And Procedures
RESERVATION OF AUTHORITY	5.11	RESERVATION OF AUTHORITY Delegations Of Authority To The Commissioner Of Food And Drug; Delegations Of Authority And Organization
RESERVE SAMPLES	211.170	RESERVE SAMPLES Laboratory Controls; Current Good Manufacturing Practice For Finished Pharmaceuticals
RESERVOIR	870.4400	CARDIOPULMONARY BYPASS BLOOD RESERVOIR Cardiovascular Surgical Devices; Cardiovascular Devices
RESERVOIR BAG	868.5320	RESERVOIR BAG Therapeutic Devices; Anesthesiology Devices
RESIDENT AGENT	316.22	PERMANENT RESIDENT AGENT FOR FOREIGN SPONSOR Orphan Drugs

RESIDENT AGENTS	5.802	MANUFACTURER'S RESIDENT IMPORT AGENTS Imports And Exports; Redelegations Of Authority; Delegations Of Authority And Organization
RESIDUE	500.84	OPERATIONAL DEFINITION OF "NO RESIDUE" Regulation Of Carcinogenic Compounds Used In Food-producing Animals; General
RESIDUE	500.86	MARKER RESIDUE AND TARGET TISSUE Regulation Of Carcinogenic Compounds Used In Food-producing Animals; General
RESIDUE, DRUG	530.24	PROCEDURE FOR ANNOUNCING ANALYTICAL METHODS FOR DRUG RESIDUE QUANTIFICATION Extralabel Drug Use In Animals
RESIDUES, ALCOHOL	173.250	METHYL ALCOHOL RESIDUES Solvents, Lubricants, Release Agents And Related Substances; Secondary Direct Food Additives Permitted In Food For Human Consumption
RESIDUES, DRUGS	556.1	GENERAL CONSIDERATIONS; TOLERANCES FOR RESIDUES OF NEW ANIMAL DRUGS IN FOOD General Provisions; Tolerances For Residues Of New Animal Drugs In Food
RESIN	172.215	COUMARONE-INDENE RESIN Coatings, Films And Related Substances; Food Additives Permitted For Direct Addition To Food For Human Consumption
RESIN	172.280	TERPENE RESIN Coatings, Films And Related Substances; Food Additives Permitted For Direct Addition To Food For Human Consumption
RESIN	173.5	ACRYLATE-ACRYLAMIDE RESINS Polymer Substances For Food Treatment; Secondary Direct Food Additives Permitted In Food For Human Consumption
RESIN	173.10	MODIFIED POLYACRYLAMIDE RESIN Polymer Substances For Food Treatment; Secondary Direct Food Additives Permitted In Food For Human Consumption
RESIN	173.70	CHLOROMETHYLATED AMINATED STYRENE-DIVINYLBENZENE RESIN Polymer Substances For Food Treatment; Secondary Direct Food Additives Permitted In Food For Human Consumption
RESIN	177.1595	POLYETHERIMIDE RESIN Substances For Use As Basic Components Of Single And Repeated Use Food Contact Surfaces; Indirect Food Additives: Polymers
RESIN	573.120	ACRYLAMIDE-ACRYLIC ACID RESIN Food Additive Listing; Food Additives Permitted In Feed And Drinking Water Of Animals
RESIN	872.3140	RESIN APPLICATOR Prosthetic Devices; Dental Devices
RESIN	872.3200	RESIN TOOTH BONDING AGENT Prosthetic Devices; Dental Devices
RESIN	872.3300	HYDROPHILIC RESIN COATING FOR DENTURES Prosthetic Devices; Dental Devices
RESIN	872.3310	COATING MATERIAL FOR RESIN FILLINGS Prosthetic Devices; Dental Devices
RESIN	872.3670	RESIN IMPRESSION TRAY MATERIAL Prosthetic Devices; Dental Devices
RESIN	872.3690	TOOTH SHADE RESIN MATERIAL Prosthetic Devices; Dental Devices
RESIN	872.3750	BRACKET ADHESIVE RESIN AND TOOTH CONDITIONER Prosthetic Devices; Dental Devices
RESIN	872.3760	DENTURE RELINING, REPAIRING, OR REBASING RESIN Prosthetic Devices; Dental Devices
RESIN	872.3770	TEMPORARY CROWN AND BRIDGE RESIN Prosthetic Devices; Dental Devices
RESIN	872.3820	ROOT CANAL FILLING RESIN Prosthetic Devices; Dental Devices
RESIN-BONDED	177.2260	FILTERS, RESIN-BONDED Substances For Use Only As Components Of Articles Intended For Repeated Use; Indirect Food Additives: Polymers
RESINOUS COATINGS	175.300	RESINOUS AND POLYMERIC COATINGS Substances For Use As Components Of Coatings; Indirect Food Additives
RESINOUS COATINGS	175.320	RESINOUS AND POLYMERIC COATINGS FOR POLYOLEFIN FILMS Substances For Use As Components Of Coatings; Indirect Food Additives
RESINS	175.260	PARTIAL PHOSPHORIC ACID ESTERS OF POLYESTER RESINS Substances For Use As Components Of Coatings; Indirect Food Additives
RESINS	175.270	POLY(VINYL FLUORIDE) RESINS Substances For Use As Components Of Coatings; Indirect Food Additives
RESINS	175.380	XYLENE-FORMALDEHYDE RESINS CONDENSED WITH 4,4'-ISOPROPYLIDENEDIPHENOL-EPICHLOROHYDRIN EPOXY RESINS Substances For Use As Components Of Coatings; Indirect Food Additives
RESINS	176.110	ACRYLAMIDE-ACRYLIC ACID RESINS Substances For Use Only As Components Of Paper And Paperboard; Indirect Food Additives

RESINS	177.1330	IONOMERIC RESINS Substances For Use As Basic Components Of Single And Repeated Use Food Contact Surfaces; Indirect Food Additives: Polymers
RESINS	177.1340	ETHYLENE-METHYL ACRYLATE COPOLYMER RESINS Substances For Use As Basic Components Of Single And Repeated Use Food Contact Surfaces; Indirect Food Additives: Polymers
RESINS	177.1380	FLUOROCARBON RESINS Substances For Use As Basic Components Of Single And Repeated Use Food Contact Surfaces; Indirect Food Additives: Polymers
RESINS	177.1440	4,4'-ISOPROPYLIDENEDIPHENOL-EPICHLOROHYDRIN RESINS MINIMUM MOLECULAR WEIGHT 10,000 Substances For Use As Basic Components Of Single And Repeated Use Food Contact Surfaces; Indirect Food Additives: Polymers
RESINS	177.1460	MELAMINE-FORMALDEHYDE RESINS IN MOLDED ARTICLES Substances For Use As Basic Components Of Single And Repeated Use Food Contact Surfaces; Indirect Food Additives: Polymers
RESINS	177.1500	NYLON RESINS Substances For Use As Basic Components Of Single And Repeated Use Food Contact Surfaces; Indirect Food Additives: Polymers
RESINS	177.1550	PERFLUOROCARBON RESINS Substances For Use As Basic Components Of Single And Repeated Use Food Contact Surfaces; Indirect Food Additives: Polymers
RESINS	177.1555	POLYARYLATE RESINS Substances For Use As Basic Components Of Single And Repeated Use Food Contact Surfaces; Indirect Food Additives: Polymers
RESINS	177.1556	POLYARYLETHERKETONE RESINS Substances For Use As Basic Components Of Single And Repeated Use Food Contact Surfaces; Indirect Food Additives: Polymers
RESINS	177.1560	POLYARYLSULFONE RESINS Substances For Use As Basic Components Of Single And Repeated Use Food Contact Surfaces; Indirect Food Additives: Polymers
RESINS	177.1570	POLY-1-BUTENE RESINS AND BUTENE/ETHYLENE COPOLYMERS Substances For Use As Basic Components Of Single And Repeated Use Food Contact Surfaces; Indirect Food Additives: Polymers
RESINS	177.1580	POLYCARBONATE RESINS Substances For Use As Basic Components Of Single And Repeated Use Food Contact Surfaces; Indirect Food Additives: Polymers
RESINS	177.1585	POLYESTERCARBONATE RESINS Substances For Use As Basic Components Of Single And Repeated Use Food Contact Surfaces; Indirect Food Additives: Polymers
RESINS	177.1600	POLYETHYLENE RESINS, CARBOXYL MODIFIED Substances For Use As Basic Components Of Single And Repeated Use Food Contact Surfaces; Indirect Food Additives: Polymers
RESINS	177.1632	POLY(PHENYLENETEREPHTHALAMIDE) RESINS Substances For Use As Basic Components Of Single And Repeated Use Food Contact Surfaces; Indirect Food Additives: Polymers
RESINS	177.1637	POLY(OXY-1,2-ETHANEDIYLOXYCARBONYL-2,6-NAPHTHALENEDIYLCARBONYL) RESINS Substances For Use As Basic Components Of Single And Repeated Use Food Contact Surfaces; Indirect Food Additives: Polymers
RESINS	177.1650	POLYSULFIDE POLYMER-POLYEPOXY RESINS Substances For Use As Basic Components Of Single And Repeated Use Food Contact Surfaces; Indirect Food Additives: Polymers
RESINS	177.1655	POLYSULFONE RESINS Substances For Use As Basic Components Of Single And Repeated Use Food Contact Surfaces; Indirect Food Additives: Polymers
RESINS	177.1680	POLYURETHANE RESINS Substances For Use As Basic Components Of Single And Repeated Use Food Contact Surfaces; Indirect Food Additives: Polymers
RESINS	177.1900	UREA FORMALDEHYDE RESINS IN MOLDED ARTICLES Substances For Use As Basic Components Of Single And Repeated Use Food Contact Surfaces; Indirect Food Additives: Polymers
RESINS	177.2280	4,4'-ISOPROPYLIDENEDIPHENOL-EPICHLOROHYDRIN THERMOSETTING EPOXY RESINS Substances For Use Only As Components Of Articles Intended For Repeated Use; Indirect Food Additives: Polymers
RESINS	177.2355	MINERAL REINFORCED NYLON RESINS Substances For Use Only As Components Of Articles Intended For Repeated Use; Indirect Food Additives: Polymers
RESINS	177.2410	PHENOLIC RESINS IN MOLDED ARTICLES Substances For Use Only As Components Of Articles Intended For Repeated Use; Indirect Food Additives: Polymers
RESINS	177.2415	POLY(ARYLETHERKETONE) RESINS Indirect Food Additive Polymers

RESINS	177.2420	POLYESTER RESINS, CROSS-LINKED Substances For Use Only As Components Of Articles Intended For Repeated Use; Indirect Food Additives: Polymers
RESINS	177.2430	POLYETHER RESINS, CHLORINATED Substances For Use Only As Components Of Articles Intended For Repeated Use; Indirect Food Additives: Polymers
RESINS	177.2440	POLYETHERSULFONE RESINS Substances For Use Only As Components Of Articles Intended For Repeated Use; Indirect Food Additives: Polymers
RESINS	177.2450	POLYAMIDE-IMIDE RESINS Substances For Use Only As Components Of Articles Intended For Repeated Use; Indirect Food Additives: Polymers
RESINS	177.2460	POLY(2,6-DIMETHYL-1,4-PHENYLENE) OXIDE RESINS Substances For Use Only As Components Of Articles Intended For Repeated Use; Indirect Food Additives: Polymers
RESINS	177.2490	POLYPHENYLENE SULFIDE RESINS Substances For Use Only As Components Of Articles Intended For Repeated Use; Indirect Food Additives: Polymers
RESINS	177.2500	POLYPHENYLENE SULFONE RESINS Substances For Use Only As Components Of Articles Intended For Repeated Use; Indirect Food Additives: Polymers
RESINS	177.2510	POLYVINYLIDENE FLUORIDE RESINS Substances For Use Only As Components Of Articles Intended For Repeated Use; Indirect Food Additives: Polymers
RESINS	177.2710	STYRENE-DIVINYLBENZENE RESINS, CROSS-LINKED Substances For Use Only As Components Of Articles Intended For Repeated Use; Indirect Food Additives: Polymers
RESINS	178.3610	α-METHYLSTYRENE-VINYLTOLUENE RESINS, HYDROGENATED Certain Adjuvants And Production Aids; Indirect Food Additives: Adjuvants, Production Aids, And Sanitizers
RESINS	178.3930	TERPENE RESINS Certain Adjuvants And Production Aids; Indirect Food Additives: Adjuvants, Production Aids, And Sanitizers
RESINS	181.26	DRYING OILS AS COMPONENTS OF FINISHED RESINS Specific Prior-sanctioned Food Ingredients; Prior-sanctioned Food Ingredients
RESINS	181.32	ACRYLONITRILE POLYMERS AND RESINS Specific Prior-sanctioned Food Ingredients; Prior-sanctioned Food Ingredients
RESINS	189.300	HYDROGENATED 4,4'-ISOPROPYLIDENE-DIPHENOLPHOSPHITE ESTER RESINS Substances Prohibited From Indirect Addition To Human Food Through Food-contact Surfaces; Substances Prohibited From Use In Human Food
RESINS, ION-EXCHANGE	173.25	ION-EXCHANGE RESINS Polymer Substances For Food Treatment; Secondary Direct Food Additives Permitted In Food For Human Consumption
RESINS, SIEVE	173.40	MOLECULAR SIEVE RESINS Polymer Substances For Food Treatment; Secondary Direct Food Additives Permitted In Food For Human Consumption
RESISTANT, TAMPER	700.25	TAMPER-RESISTANT PACKAGING REQUIREMENTS FOR COSMETIC PRODUCTS Requirements For Specific Cosmetic Products; General
RESISTANT, TAMPER	800.12	CONTACT LENS SOLUTIONS AND TABLETS; TAMPER-RESISTANT PACKAGING Requirements For Specific Medical Devices; General
RESISTANT, WATER	352.76	DETERMINATION IF PRODUCT IS WATER RESISTANT OR VERY WATER RESISTANT Sunscreen Drug Products For Over-the-counter Human Use
RESONANCE, MAGNETIC	892.1000	MAGNETIC RESONANCE DIAGNOSTIC DEVICE Diagnostic Devices; Radiology Devices
RESOLUTION	312.48	DISPUTE RESOLUTION Investigational New Drug Application
RESOLUTION	314.103	DISPUTE RESOLUTION FDA Action On Applications; Applications For FDA Approval To Market A New Drug
RESPIRATOR SYNCHRONIZER	892.1970	RADIOGRAPHIC ECG/RESPIRATOR SYNCHRONIZER Diagnostic Devices; Radiology Devices
RESPIRATORY	866.3480	RESPIRATORY SYNCYTIAL VIRUS SEROLOGICAL REAGENTS Serological Reagents; Immunology And Microbiology Devices
RESPIRATORY	868.5450	RESPIRATORY GAS HUMIDIFIER Therapeutic Devices; Anesthesiology Devices
RESPONSE	60.32	APPLICANT RESPONSE TO PETITION Due Diligence Petitions; Patent Term Restoration
RESPONSE	882.1540	GALVANIC SKIN RESPONSE MEASUREMENT DEVICE Neurological Diagnostic Devices; Neurological Devices
RESPONSE	882.1870	EVOKED RESPONSE ELECTRICAL STIMULATOR Neurological Diagnostic Devices; Neurological Devices

RESPONSE	882.1880	EVOKED RESPONSE MECHANICAL STIMULATOR Neurological Diagnostic Devices; Neurological Devices
RESPONSE	882.1890	EVOKED RESPONSE PHOTIC STIMULATOR Neurological Diagnostic Devices; Neurological Devices
RESPONSE	882.1900	EVOKED RESPONSE AUDITORY STIMULATOR Neurological Diagnostic Devices; Neurological Devices
RESPONSE	1401.7	PROMPT RESPONSE Public Availability Of Information
RESPONSES	21.42	RESPONSES TO REQUESTS Procedures For Notification Of And Access To Records In Privacy Act Record Systems; Protection Of Privacy
RESPONSES	21.51	RESPONSES TO REQUESTS FOR AMENDMENT OF RECORDS Procedures For Requests For Amendment Of Records; Protection Of Privacy
RESPONSIBLE	25.45	RESPONSIBLE AGENCY OFFICIAL Preparation of Environmental Documents; Environmental Impact Considerations
RESPONSIBILITIES	14.86	RIGHTS AND RESPONSIBILITIES OF NONVOTING MEMBERS OF ADVISORY COMMITTEES Members Of Advisory Committees; Public Hearing Before A Public Advisory Committee
RESPONSIBILITIES	211.22	RESPONSIBILITIES OF QUALITY CONTROL UNIT Organization And Personnel; Current Good Manufacturing Practice For Finished Pharmaceuticals
RESPONSIBILITIES	211.28	PERSONNEL RESPONSIBILITIES Organization And Personnel; Current Good Manufacturing Practice For Finished Pharmaceuticals
RESPONSIBILITIES	312.50	GENERAL RESPONSIBILITIES OF SPONSORS Responsibilities Of Sponsors And Investigators; Investigational New Drug Application
RESPONSIBILITIES	312.60	GENERAL RESPONSIBILITIES OF INVESTIGATORS Responsibilities Of Sponsors And Investigators; Investigational New Drug Application
RESPONSIBILITIES	314.99	OTHER RESPONSIBILITIES OF AN APPLICANT OF AN ABBREVIATED APPLICATION Abbreviated Applications; Applications For FDA Approval To Market A New Drug
RESPONSIBILITIES	812.40	GENERAL RESPONSIBILITIES OF SPONSORS Responsibilities Of Sponsors; Investigational Device Exemptions
RESPONSIBILITIES	812.100	GENERAL RESPONSIBILITIES OF INVESTIGATORS Responsibilities Of Investigators; Investigational Device Exemptions
RESPONSIBILITIES	812.110	SPECIFIC RESPONSIBILITIES OF INVESTIGATORS Responsibilities Of Investigators; Investigational Device Exemptions
RESPONSIBILITIES	1309.72	FELONY CONVICTION; EMPLOYER RESPONSIBILITIES Registration Of Manufacturers, Distributors, Importers And Exporters Of List I Chemicals
RESPONSIBILITIES	1404.500	GSA RESPONSIBILITIES Governmentwide Debarment And Suspension And Requirements For Drug-Free Workplace (Grants)
RESPONSIBILITIES	1404.505	ONDCP RESPONSIBILITIES Governmentwide Debarment And Suspension And Requirements For Drug-Free Workplace (Grants)
RESPONSIBILITIES	1404.510	PARTICIPANTS' RESPONSIBILITIES Governmentwide Debarment And Suspension And Requirements For Drug-Free Workplace (Grants)
RESPONSIBILITY	1230.16	RESPONSIBILITY FOR LABELING DIRECTIONS FOR TREATMENT Labeling; Regulations Under The Federal Caustic Poison Act
RESPONSIBILITY	1402.2	RESPONSIBILITY Mandatory Declassification Review
RESPONSIBILITY, EMPLOYEE	1301.91	EMPLOYEE RESPONSIBILITY TO REPORT DRUG DIVERSION Employee Screening - Non-practitioners; Registration Of Manufacturers, Distributors, And Dispensers Of Controlled Substances
RESPONSIBILITY, EMPLOYEE	1309.73	EMPLOYEE RESPONSIBILITY TO REPORT DIVERSION Registration Of Manufacturers, Distributors, Importers And Exporters Of List I Chemicals
RESPONSIBILITY, MANAGEMENT	820.20	MANAGEMENT RESPONSIBILITY Quality System Regulation
RESPONSIBILITY, MANUFACTURING	610.63	DIVIDED MANUFACTURING RESPONSIBILITY TO BE SHOWN Labeling Standards; General Biological Products Standards
RESPONSIBILITY, MANUFACTURING	640.71	MANUFACTURING RESPONSIBILITY Source Plasma; Additional Standards For Human Blood And Blood Products
RESTAURANT FOODS	101.10	NUTRITION LABELING OF RESTAURANT FOODS General Provisions; Food Labeling
RESTORATION, AESTHETIC	878.3800	EXTERNAL AESTHETIC RESTORATION PROSTHESIS Prosthetic Devices; General And Plastic Surgery Devices

RESTRAINT	880.6760	PROTECTIVE RESTRAINT General Hospital And Personal Use Miscellaneous Devices; General Hospital And Personal Use Devices
RESTRICTED DEVICE	801.433	WARNING STATEMENTS FOR PRESCRIPTION AND RESTRICTED DEVICE PRODUCTS CONTAINING OR MANUFACTURED WITH CHLOROFLUOROCARBONS Labeling
RESTRICTIONS	70.5	GENERAL RESTRICTIONS ON USE OF COLOR ADDITIVES General Provisions; Color Additives
RESTRICTIONS	203.10	RESTRICTIONS ON REIMPORTATION Prescription Drug Marketing
RESTRICTIONS	203.20	SALES RESTRICTIONS Prescription Drug Marketing
RESTRICTIONS	314.520	APPROVAL WITH RESTRICTIONS TO ASSURE SAFE USE Accelerated Approval Of New Drugs; Applications For FDA Approval To Market A New Drug
RESTRICTIONS	601.42	APPROVAL WITH RESTRICTIONS TO ASSURE SAFE USE Accelerated Approval Of Biological Products; Licensing
RESTRICTIONS	610.42	RESTRICTIONS ON USE FOR FURTHER MANUFACTURE OF MEDICAL DEVICES Testing Requirements For Communicable Disease Agents; General Biological Products Standards
RESTRICTIONS	809.30	RESTRICTIONS ON THE SALE, DISTRIBUTION AND USE OF ANALYTE SPECIFIC REAGENTS Requirements For Manufacturers And Producers; In Vitro Diagnostic Products For Human Use
RESTRICTIONS	809.40	RESTRICTIONS ON THE SALE, DISTRIBUTION AND USE OF OTC TEST SAMPLE COLLECTION SYSTEMS FOR DRUGS OF ABUSE TESTING Requirements For Manufacturers And Producers; In Vitro Diagnostic Products For Human Use
RESTRICTIONS	1313.25	FOREIGN IMPORT RESTRICTIONS Exportation Of Listed Chemicals; Importation And Exportation Of Precursors And Essential Chemicals
RESTRICTIONS	1404.225	FAILURE TO ADHERE TO RESTRICTIONS Governmentwide Debarment And Suspension And Requirements For Drug-Free Workplace (Grants)
RESUBMITTED	814.37	PMA AMENDMENTS AND RESUBMITTED PMA'S Premarket Approval Application (PMA); Premarket Approval Of Medical Devices
RESUBMITTED	814.106	HDE AMENDMENTS AND RESUBMITTED HDE'S Humanitarian Use Devices; Premarket Approval Of Medical Devices
RESULTS	58.185	REPORTING OF NONCLINICAL LABORATORY STUDY RESULTS Records And Reports; Good Laboratory Practice For Nonclinical Laboratory Studies
RESURFACING	888.5	RESURFACING TECHNIQUE General Provisions; Orthopedic Devices
RESURFACING PROSTHESIS	888.3400	HIP JOINT FEMORAL (HEMI-HIP) METALLIC RESURFACING PROSTHESIS Prosthetic Devices; Orthopedic Devices
RESURFACING PROSTHESIS	888.3410	HIP JOINT METAL/POLYMER SEMI-CONSTRAINED RESURFACING CEMENTED PROSTHESIS Prosthetic Devices; Orthopedic Devices
RESURFACING PROSTHESIS	888.3580	KNEE JOINT PATELLAR (HEMI-KNEE) METALLIC RESURFACING UNCEMENTED PROSTHESIS Prosthetic Devices; Orthopedic Devices
RESURFACING PROSTHESIS	888.3590	KNEE JOINT TIBIAL (HEMI-KNEE) METALLIC RESURFACING UNCEMENTED PROSTHESIS Prosthetic Devices; Orthopedic Devices
RESUSCITATION	880.6080	CARDIOPULMONARY RESUSCITATION BOARD General Hospital And Personal Use Miscellaneous Devices; General Hospital And Personal Use Devices
RETAIL DISTRIBUTION	115.50	REFRIGERATION OF SHELL EGGS HELD FOR RETAIL DISTRIBUTION Shell Eggs
RETAIL EXEMPTIONS	801.110	RETAIL EXEMPTIONS FOR PRESCRIPTION DEVICES Exemptions From Adequate Directions For Use; Labeling
RETAILERS, FOOD	101.43	SUBSTANTIAL COMPLIANCE OF FOOD RETAILERS WITH THE GUIDELINES FOR THE VOLUNTARY NUTRITION LABELING OF RAW FRUIT, VEGETABLES, AND FISH Food Labeling
RETENTION, HUMAN TISSUE	1270.43	RETENTION, RECALL, AND DESTRUCTION OF HUMAN TISSUE Human Tissue Intended For Transplantation
RETENTION, RECORD	54.6	RECORDKEEPING AND RECORD RETENTION Financial Disclosure By Clinical Investigators
RETENTION, RECORD	312.57	RECORDKEEPING AND RECORD RETENTION Responsibilities Of Sponsors And Investigators; Investigational New Drug Application

RETENTION, RECORD	312.62	INVESTIGATOR RECORDKEEPING AND RECORD RETENTION Responsibilities Of Sponsors And Investigators; Investigational New Drug Application
RETENTION, RECORDS	58.195	RETENTION OF RECORDS Records And Reports; Good Laboratory Practice For Nonclinical Laboratory Studies
RETENTION, RECORDS	107.280	RECORDS RETENTION Infant Formula Recalls; Infant Formula
RETENTION, RECORDS	821.60	RETENTION OF RECORDS Medical Device Tracking Requirements
RETENTION, RECORDS	1403.42	RETENTION AND ACCESS REQUIREMENTS FOR RECORDS Reports, Records, Retention, And Enforcement; Uniform Administrative Requirements For Grants And Coop. Agreements To State And Local Gov.
RETENTION, SAMPLES	320.38	RETENTION OF BIOAVAILABILITY SAMPLES Procedures For Determining The Bioavailability Or Bioequivalence Of Drug Products; Bioavailability And Bioequivalence Requirements
RETENTION, SAMPLES	320.63	RETENTION OF BIOEQUIVALENCE SAMPLES Procedures For Determining The Bioavailability Or Bioequivalence Of Drug Products; Bioavailability And Bioequivalence Requirements
RETENTION, SAMPLES	600.13	RETENTION SAMPLES Establishment Standards; Biological Products: General
RETENTION SCHEDULE	20.31	RETENTION SCHEDULE OF REQUESTS FOR FDA RECORDS General Policy; Public Information
RETENTION, SUTURE	878.4930	SUTURE RETENTION DEVICE Surgical Devices; General And Plastic Surgery Devices
RETENTIVE PIN	872.3740	RETENTIVE AND SPLINTING PIN Prosthetic Devices, Dental Devices
RETESTING	211.87	RETESTING OF APPROVED COMPONENTS, DRUG PRODUCT CONTAINERS, AND CLOSURES Control Of Components And Drug Product Containers And Closures; Current Good Manufacturing Practice For Finished Pharmaceuticals
RETINOL-BINDING PROTEIN	866.5765	RETINOL-BINDING PROTEIN IMMUNOLOGICAL TEST SYSTEM Immunological Test Systems; Immunology And Microbiology Devices
RETINOSCOPE	886.1780	RETINOSCOPE Diagnostic Devices; Ophthalmic Devices
RETRACTOR	876.4530	GASTROENTEROLOGY-UROLOGY FIBEROPTIC RETRACTOR Surgical Devices; Gastroenterology-urology Devices
RETRACTOR	882.4800	SELF-RETAINING RETRACTOR FOR NEUROSURGERY Neurological Surgical Devices; Neurological Devices
RETRIEVAL	58.190	STORAGE AND RETRIEVAL OF RECORDS AND DATA Records And Reports; Good Laboratory Practice For Nonclinical Laboratory Studies
RETROACTIVE REGULATIONS	20.25	RETROACTIVE APPLICATION OF REGULATIONS General Policy; Public Information
RETURN	1305.14	RETURN OF UNUSED ORDER FORMS Order Forms
RETURN	870.4420	CARDIOPULMONARY BYPASS CARDIOTOMY RETURN SUCKER Cardiovascular Surgical Devices; Cardiovascular Devices
RETURNED DRUG	211.204	RETURNED DRUG PRODUCTS Returned And Salvaged Drug Products; Current Good Manufacturing Practice For Finished Pharmaceuticals
RETURNS	203.23	RETURNS Prescription Drug Marketing
REVERSE ISOLATION	880.5450	PATIENT CARE REVERSE ISOLATION CHAMBER General Hospital And Personal Use Therapeutic Devices; General Hospital And Personal Use Devices
REVERSE OSMOSIS	177.2550	REVERSE OSMOSIS MEMBRANES Substances For Use Only As Components Of Articles Intended For Repeated Use; Indirect Food Additives; Polymers
REVIEW	10.45	COURT REVIEW OF FINAL ADMINISTRATIVE ACTION; EXHAUSTION OF ADMINISTRATIVE REMEDIES General Administrative Procedures; Administrative Practices And Procedures
REVIEW	10.75	INTERNAL AGENCY REVIEW OF DECISIONS General Administrative Procedures; Administrative Practices And Procedures
REVIEW	12.125	APPEAL FROM OR REVIEW OF INITIAL DECISION Initial And Final Decisions; Formal Evidentiary Public Hearing
REVIEW	12.130	DECISION BY COMMISSIONER ON APPEAL OR REVIEW OF INITIAL DECISION Initial And Final Decisions; Formal Evidentiary Public Hearing
REVIEW	20.44	PRESUBMISSION REVIEW OF REQUEST FOR CONFIDENTIALITY OF VOLUNTARILY SUBMITTED DATA OR INFORMATION Procedures And Fees; Public Information
REVIEW	20.52	REQUEST FOR REVIEW WITHOUT COPYING Procedures And Fees; Public Information
REVIEW	56.109	IRB REVIEW OF RESEARCH IRB Functions And Operations; Institutional Review Boards

REVIEW	56.110	EXPEDITED REVIEW PROCEDURES FOR CERTAIN KINDS OF RESEARCH INVOLVING NO MORE THAN MINIMAL RISK, AND FOR MINOR CHANGES IN APPROVED RESEARCH IRB Functions And Operations; Institutional Review Boards
REVIEW	56.112	REVIEW BY INSTITUTION IRB Functions And Operations; Institutional Review Boards
REVIEW	130.6	REVIEW OF CODEX ALIMENTARIUS FOOD STANDARDS General Provisions; Food Standards: General
REVIEW	211.192	PRODUCTION RECORD REVIEW Records And Reports; Current Good Manufacturing Practice For Finished Pharmaceutical
REVIEW	312.56	REVIEW OF ONGOING INVESTIGATIONS Investigational New Drug Application
REVIEW	320.30	INQUIRIES REGARDING BIOAVAILABILITY AND BIOEQUIVALENCE REQUIREMENTS AND REVIEW OF PROTOCOLS BY THE FDA Procedures For Determining The Bioavailability Or Bioequivalence Of Drug Products; Bioavailability And Bioequivalence Requirements
REVIEW	330.13	CONDITIONS FOR MARKETING INGREDIENTS RECOMMENDED FOR OVER-THE-COUNTER (OTC) USE UNDER THE OTC DRUG REVIEW Administrative Procedures; Over-the-counter Human Drugs Generally Recognized As Safe And Effective & Not Misbranded
REVIEW	601.25	REVIEW PROCEDURES TO DETERMINE THAT LICENSED BIOLOGICAL PRODUCTS ARE SAFE, EFFECTIVE, AND NOT MISBRANDED UNDER PRESCRIBED, RECOMMENDED, OR SUGGESTED CONDITIONS OF USE Biologics Licensing; Licensing
REVIEW	810.12	WRITTEN REQUEST FOR REVIEW OF CEASE DISTRIBUTION AND NOTIFICATION ORDER Mandatory Medical Device Recall Procedures; Medical Device Recall Authority
REVIEW	814.44	PROCEDURES FOR REVIEW OF A PMA FDA Action On A PMA; Premarket Approval Of Medical Devices
REVIEW	814.116	PROCEDURES FOR REVIEW OF AN HDE Humanitarian Use Devices; Premarket Approval Of Medical Devices
REVIEW, ADMINISTRATIVE	1301.31	ADMINISTRATIVE REVIEW GENERALLY Registration Of Manufacturers, Distributors, And Dispensers Of Controlled Substances
REVIEW, ADMINISTRATIVE	1309.41	ADMINISTRATIVE REVIEW GENERALLY Registration Of Manufacturers, Distributors, Importers And Exporters Of List I Chemicals
REVIEW, IRB	56.103	CIRCUMSTANCES IN WHICH IRB REVIEW IS REQUIRED General Provisions; Institutional Review Boards
REVIEW, IRB	312.66	ASSURANCE OF IRB REVIEW Investigational New Drug Application
REVIEW, IRB	812.64	IRB'S CONTINUING REVIEW IRB Review And Approval; Investigational Device Exemptions
REVIEW, JUDICIAL	12.30	JUDICIAL REVIEW AFTER WAIVER OF HEARING ON A REGULATION Initiation Of Proceedings; Formal Evidentiary Public Hearing
REVIEW, JUDICIAL	12.159	COPIES OF PETITIONS FOR JUDICIAL REVIEW Judicial Review; Formal Evidentiary Public Hearing
REVIEW, JUDICIAL	16.120	JUDICIAL REVIEW Judicial Review; Regulatory Hearing Before The Food And Drug Administration
REVIEW, JUDICIAL	17.51	JUDICIAL REVIEW Civil Money Penalties Hearings
REVIEW, JUDICIAL	20.46	JUDICIAL REVIEW OF PROPOSED DISCLOSURE Procedures And Fees; Public Information
REVIEW, JUDICIAL	314.235	JUDICIAL REVIEW Hearing Procedures For New Drugs; Applications For FDA Approval To Market A New Drug
REVIEW, JUDICIAL	514.235	JUDICIAL REVIEW Judicial Review; New Animal Drug Applications
REVIEW, JUDICIAL	515.40	JUDICIAL REVIEW Judicial Review; Medicated Feed Mill License
REVIEW, JUDICIAL	1316.68	COPIES OF PETITIONS FOR JUDICIAL REVIEW Administrative Hearings; Administrative Functions, Practices, And Procedures
REVIEW BOARD	19.10	FDA CONFLICT OF INTEREST REVIEW BOARD General Provisions; Standards Of Conduct And Conflicts Of Interest
REVIEW BOARD	814.124	INSTITUTIONAL REVIEW BOARD REQUIREMENTS Humanitarian Use Devices; Premarket Approval Of Medical Devices
REVIEW PERIOD	60.20	FDA ACTION ON REGULATORY REVIEW PERIOD DETERMINATION Regulatory Review Period Determinations; Patent Term Restoration
REVIEW PERIOD	60.22	REGULATORY REVIEW PERIOD DETERMINATIONS Regulatory Review Period Determinations; Patent Term Restoration
REVIEW PERIOD	60.24	REVISION OF REGULATORY REVIEW PERIOD DETERMINATIONS Regulatory Review Period Determinations; Patent Term Restoration

REVIEW PERIOD	60.26	FINAL ACTION ON REGULATORY REVIEW PERIOD DETERMINATIONS Regulatory Review Period Determinations; Patent Term Restoration
REVIEW PERIOD	60.28	TIME FRAME FOR DETERMINING REGULATORY REVIEW PERIODS Regulatory Review Period Determinations; Patent Term Restoration
REVIEW TIME	3.10	STAY OF REVIEW TIME Product Jurisdiction
REVIEWED	330.12	STATUS OF OVER-THE-COUNTER (OTC) DRUGS PREVIOUSLY REVIEWED UNDER THE DRUG EFFICACY STUDY (DESI) Administrative Procedures; Over-the-counter Human Drugs Generally Recognized As Safe And Effective & Not Misbranded
REVIEWING	314.100	TIME FRAMES FOR REVIEWING APPLICATIONS AND ABBREVIATED APPLICATIONS FDA Action On Applications; Applications For FDA Approval To Market A New Drug
REVIEWING	814.40	TIME FRAMES FOR REVIEWING A PMA FDA Action On A PMA; Premarket Approval Of Medical Device
REVIEWING	814.114	TIMEFRAMES FOR REVIEWING AN HDE Humanitarian Use Devices; Premarket Approval Of Medical Devices
REVISION	60.24	REVISION OF REGULATORY REVIEW PERIOD DETERMINATIONS Regulatory Review Period Determinations; Patent Term Restoration
REVISION	107.260	REVISION OF AN INFANT FORMULA RECALL Infant Formula Recalls; Infant Formula
REVOCATION	5.202	ISSUANCE OF NOTICES OF OPPORTUNITY FOR A HEARING ON PROPOSALS FOR DENIAL OF APPROVAL OF APPLICATIONS FOR LICENSES, SUSPENSION OF LICENSES, OR REVOCATION OF LICENSES Biologics; Redelegations Of Authority; Delegations Of Authority And Organization
REVOCATION	5.203	ISSUANCE AND REVOCATION OF LICENSES FOR THE PROPAGATION OR MANUFACTURE AND PREPARATION OF BIOLOGICAL PRODUCTS Biologics; Redelegations Of Authority; Delegations Of Authority And Organization
REVOCATION	12.20	INITIATION OF A HEARING INVOLVING THE ISSUANCE, AMENDMENT, OR REVOCATION OF A REGULATION Initiation Of Proceedings; Formal Evidentiary Public Hearing
REVOCATION	12.21	INITIATION OF A HEARING INVOLVING THE ISSUANCE, AMENDMENT, OR REVOCATION OF AN ORDER Initiation Of Proceedings; Formal Evidentiary Public Hearing
REVOCATION	12.26	MODIFICATION OR REVOCATION OF REGULATION OR ORDER Initiation Of Proceedings; Formal Evidentiary Public Hearing
REVOCATION	56.122	PUBLIC DISCLOSURE OF INFORMATION REGARDING REVOCATION Administrative Actions For Noncompliance; Institutional Review Boards
REVOCATION	71.26	REVOCATION OF EXEMPTION FROM CERTIFICATION Administrative Action On Petitions; Color Additive Petitions
REVOCATION	108.6	REVOCATION OF DETERMINATION OF NEED FOR PERMIT General Provisions; Emergency Permit Control
REVOCATION	316.29	REVOCATION OF ORPHAN-DRUG DESIGNATION Orphan Drugs
REVOCATION	514.120	REVOCATION OF ORDER REFUSING TO APPROVE AN APPLICATION OR SUSPENDING OR WITHDRAWING APPROVAL OF AN APPLICATION Administrative Actions On Applications; New Animal Drug Applications
REVOCATION	515.22	SUSPENSION AND/OR REVOCATION OF APPROVAL OF A MEDICATED FEED MILL LICENSE Administrative Actions On Licenses; Medicated Feed Mill License
REVOCATION	515.23	VOLUNTARY REVOCATION OF A MEDICATED FEED MILL LICENSE Administrative Actions On Licenses; Medicated Feed Mill License
REVOCATION	515.24	NOTICE OF REVOCATION OF A MEDICATED FEED MILL LICENSE Administrative Actions On Licenses; Medicated Feed Mill License
REVOCATION	515.25	REVOCATION OF ORDER REFUSING TO APPROVE A MEDICATED FEED MILL LICENSE APPLICATION OR SUSPENDING OR REVOKING A LICENSE Administrative Actions On Licenses; Medicated Feed Mill License
REVOCATION	601.5	REVOCATION OF LICENSE General Provisions; Licensing
REVOCATION	601.8	PUBLICATION OF REVOCATION General Provisions; Licensing
REVOCATION	808.35	REVOCATION OF AN EXEMPTION Exemption Procedures; Exemption From Federal Preemption Of State & Local Medical Device Requirements
REVOCATION	861.34	AMENDMENT OR REVOCATION OF A STANDARD Procedures For Performance Standards Development And Publication; Procedures For Performance Standards Development
REVOCATION	900.13	REVOCATION OF ACCREDITATION AND OF ACCREDITATION BODY APPROVAL Quality Standards And Certification; Mammography
REVOCATION	900.14	SUSPENSION OR REVOCATION OF CERTIFICATES Quality Standards And Certification; Mammography

REVOCATION	1210.30	HEARING PROCEDURE FOR PERMIT DENIAL, SUSPENSION, AND REVOCATION Hearings; Regulations Under The Federal Import Milk Act
REVOCATION	1301.36	SUSPENSION OR REVOCATION OF REGISTRATION Registration Of Manufacturers, Distributors, And Dispensers Of Controlled Substances
REVOCATION	1309.43	SUSPENSION OR REVOCATION OF REGISTRATION Registration Of Manufacturers, Distributors, Importers And Exporters Of List I Chemicals
REWARD	203.70	APPLICATION FOR A REWARD Prescription Drug Marketing
RHEOENCEPHALOGRAPH	882.1825	RHEOENCEPHALOGRAPH Neurological Diagnostic Devices; Neurological Devices
RHEUMATOID	866.5775	RHEUMATOID FACTOR IMMUNOLOGICAL TEST SYSTEM Immunological Test Systems; Immunology And Microbiology Devices
RHEUMATOLOGIC	343.13	RHEUMATOLOGIC ACTIVE INGREDIENTS Internal Analgesic, Antipyretic And Antirheumatic Drug Products For OTC Human Use
RHEUMATOLOGIC	343.22	PERMITTED COMBINATIONS OF ACTIVE INGREDIENTS FOR CARDIOVASCULAR-RHEUMATOLOGIC USE Internal Analgesic, Antipyretic And Antirheumatic Drug Products For OTC Human Use
RHINOANEMOMETER	868.1800	RHINOANEMOMETER Diagnostic Devices; Anesthesiology Devices
RHINOLOGY	874.4490	ARGON LASER FOR OTOLOGY, RHINOLOGY, AND LARYNGOLOGY Surgical Devices; Ear, Nose, And Throat Devices
RHINOVIRUS	866.3490	RHINOVIRUS SEROLOGICAL REAGENTS Serological Reagents; Immunology And Microbiology Devices
RHIZOPUS NIVEUS	173.110	AMYLOGLUCOSIDASE DERIVED FROM RHIZOPUS NIVEUS Enzyme Preparations And Microorganisms; Secondary Direct Food Additives Permitted In Food For Human Consumption
RHIZOPUS NIVEUS	184.1420	LIPASE ENZYME PREPARATION DERIVED FROM RHIZOPUS NIVEUS Listing Of Specific Substances Affirmed As GRAS; Direct Food Substances Affirmed As Generally Recognized As Safe
RHIZOPUS ORYZAE	173.130	CARBOHYDRASE DERIVED FROM RHIZOPUS ORYZAE Enzyme Preparations And Microorganisms; Secondary Direct Food Additives Permitted In Food For Human Consumption
RHODE ISLAND	808.89	RHODE ISLAND Listing Of Specific State And Local Exemptions; Exemption From Federal Preemption Of State & Local Medical Device Requirements
RIBDAM	876.4560	RIBDAM Surgical Devices; Gastroenterology-urology Devices
RIBOFLAVIN	73.450	RIBOFLAVIN Foods; Listing Of Color Additives Exempt From Certification
RIBOFLAVIN	184.1695	RIBOFLAVIN Listing Of Specific Substances Affirmed As GRAS; Direct Food Substances Affirmed As Generally Recognized As Safe
RIBOFLAVIN	582.5695	RIBOFLAVIN Nutrients And/or Dietary Supplements; Substances Generally Recognized As Safe
RIBOFLAVIN-5-PHOSPHATE	184.1697	RIBOFLAVIN-5'-PHOSPHATE (SODIUM) Listing Of Specific Substances Affirmed As GRAS; Direct Food Substances Affirmed As Generally Recognized As Safe
RIBOFLAVIN-5-PHOSPHATE	582.5697	RIBOFLAVIN-5-PHOSPHATE Nutrients And/or Dietary Supplements; Substances Generally Recognized As Safe
RICE	137.350	ENRICHED RICE Requirements For Specific Standardized Cereal Flours And Related Products; Cereal Flours And Related Products
RICE BRAN WAX	172.890	RICE BRAN WAX Multipurpose Additives; Food Additives Permitted For Direct Addition To Food For Human Consumption
RICE HULLS	573.160	AMMONIATED RICE HULLS Food Additive Listing; Food Additives Permitted In Feed And Drinking Water Of Animals
RICKETTSIA	866.3500	RICKETTSIA SEROLOGICAL REAGENTS Serological Reagents; Immunology And Microbiology Devices
RIGHT	16.62	RIGHT TO COUNSEL Procedures Before A Regulatory Hearing; Regulatory Hearing Before The FDA
RIGHTS	14.86	RIGHTS AND RESPONSIBILITIES OF NONVOTING MEMBERS OF ADVISORY COMMITTEES Members Of Advisory Committees; Public Hearing Before A Public Advisory Committee

RIGHTS	21.75	RIGHTS OF LEGAL GUARDIANS Disclosure Of Records In Privacy Act Record Systems To Persons Other Than The Subject Individual; Protection Of Privacy
RIGID CONTACT LENS	886.5916	RIGID GAS PERMEABLE CONTACT LENS Therapeutic Devices; Ophthalmic Devices
RIGID CONTACT LENS	886.5918	RIGID GAS PERMEABLE CONTACT LENS CARE PRODUCTS Therapeutic Devices; Ophthalmic Devices
RIGID CONTAINERS	164.120	SHELLED NUTS IN RIGID OR SEMIRIGID CONTAINERS Requirements For Specific Standardized Tree Nut And Peanut Products; Tree Nut And Peanut Products
RIGID LARYNGOSCOPE	868.5540	RIGID LARYNGOSCOPE Therapeutic Devices; Anesthesiology Devices
RIGID ORTHOSIS	890.3610	RIGID PNEUMATIC STRUCTURE ORTHOSIS Physical Medicine Prosthetic Devices; Physical Medicine Devices
RIGID PLASTICS	177.1010	ACRYLIC AND MODIFIED ACRYLIC PLASTICS, SEMIRIGID AND RIGID Substances For Use As Basic Components Of Single And Repeated Use Food Contact Surfaces; Indirect Food Additives: Polymers
RIGID PLASTICS	178.3790	POLYMER MODIFIERS IN SEMIRIGID AND RIGID VINYL CHLORIDE PLASTICS Certain Adjuvants And Production Aids; Indirect Food Additives: Adjuvants, Production Aids, And Sanitizers
RIGIDITY ANALYZER	882.1020	RIGIDITY ANALYZER Neurological Diagnostic Devices; Neurological Devices
RIGIDITY IMPLANT	876.3630	PENILE RIGIDITY IMPLANT Prosthetic Devices; Gastroenterology-urology Devices
RING	870.3800	ANNULOPLASTY RING Cardiovascular Prosthetic Devices; Cardiovascular Devices
RING	872.5550	TEETHING RING Therapeutic Devices; Dental Devices
RING	880.6200	RING CUTTER General Hospital And Personal Use Miscellaneous Devices; General Hospital And Personal Use Devices
RINSE, FLUORIDE	355.55	PRINCIPAL DISPLAY PANEL OF ALL FLUORIDE RINSE DRUG PRODUCTS Anticaries Drug Products For Over-the-counter Human Use
RIPENED CHEESES	133.182	SOFT RIPENED CHEESES Requirements For Specific Standardized Cheese And Related Products; Cheeses And Related Cheese Products
RISK-BENEFIT	312.84	RISK-BENEFIT ANALYSIS IN REVIEW OF MARKETING APPLICATIONS FOR DRUGS TO TREAT LIFE-THREATENING AND SEVERELY-DEBILITATING ILLNESSES Drugs Intended To Treat Life-threatening And Severely-debilitating Illnesses; Investigational New Drug Application
RISK, CANCER	740.18	COAL TAR HAIR DYES POSING A RISK OF CANCER Warning Statements; Cosmetic Product Warning Statements
RISK, CORONARY	101.75	HEALTH CLAIMS: DIETARY SATURATED FAT AND CHOLESTEROL AND RISK OF CORONARY HEART DISEASE Food Labeling
RISK, CORONARY	101.77	HEALTH CLAIMS: FRUITS, VEGETABLES, AND GRAIN PRODUCTS THAT CONTAIN FIBER, PARTICULARLY SOLUBLE FIBER, AND RISK OF CORONARY HEART DISEASE Food Labeling
RISK, CORONARY	101.81	HEALTH CLAIMS: SOLUBLE FIBER FROM CERTAIN FOODS AND RISK OF CORONARY HEART DISEASE Food Labeling
RISK, CORONARY	101.82	HEALTH CLAIMS: SOY PROTEIN AND RISK OF CORONARY HEART DISEASE Food Labeling
RISK, CORONARY	101.83	HEALTH CLAIMS: PLANT STEROL/STANOL ESTERS AND RISK OF CORONARY HEART DISEASE Food Labeling
RISK, MINIMAL	50.51	CLINICAL INVESTIGATIONS NOT INVOLVING GREATER THAN MINIMAL RISK Additional Safeguards For Children In Clinical Investigations; Protection Of Human Subjects
RISK, MINIMAL	50.52	CLINICAL INVESTIGATIONS INVOLVING GREATER THAN MINIMAL RISK BUT PRESENTING THE PROSPECT OF DIRECT BENEFIT TO INDIVIDUAL SUBJECTS Additional Safeguards For Children In Clinical Investigations; Protection Of Human Subjects
RISK, MINIMAL	50.53	CLINICAL INVESTIGATIONS INVOLVING GREATER THAN MINIMAL RISK AND NO PROSPECT OF DIRECT BENEFIT TO INDIVIDUAL SUBJECTS, BUT LIKELY TO YIELD GENERALIZABLE KNOWLEDGE ABOUT THE SUBJECTS' DISORDER OR CONDITION Additional Safeguards For Children In Clinical Investigations; Protection Of Human Subjects

RISK, MINIMAL	56.110	EXPEDITED REVIEW PROCEDURES FOR CERTAIN KINDS OF RESEARCH INVOLVING NO MORE THAN MINIMAL RISK, AND FOR MINOR CHANGES IN APPROVED RESEARCH IRB Functions And Operations; Institutional Review Boards
RISK, SIGNIFICANT	812.66	SIGNIFICANT RISK DEVICE DETERMINATIONS IRB Review And Approval; Investigational Device Exemptions
RISK, UNREASONABLE	5.409	DETERMINATIONS THAT MEDICAL DEVICES PRESENT UNREASONABLE RISK OF SUBSTANTIAL HARM Medical Devices And Radiological Health; Redelegations Of Authority; Delegations Of Authority And Organization
ROBENIDINE HYDROCHLORIDE	556.580	ROBENIDINE HYDROCHLORIDE Specific Tolerances For Residues Of New Animal Drugs; Tolerances For Residues Of New Animal Drugs In Food
ROBENIDINE HYDROCHLORIDE	558.515	ROBENIDINE HYDROCHLORIDE Specific New Animal Drugs For Use In Animal Feeds; New Animal Drugs For Use In Animal Feeds
ROCKET IMMUNOELECTROPHORESIS	866.4830	ROCKET IMMUNOELECTROPHORESIS EQUIPMENT Immunology Laboratory Equipment And Reagents; Immunology And Microbiology Devices
ROCKING BED	868.5180	ROCKING BED Therapeutic Devices; Anesthesiology Devices
ROD	876.4270	COLOSTOMY ROD Surgical Devices; Gastroenterology-urology Devices
ROD	888.3020	INTRAMEDULLARY FIXATION ROD Prosthetic Devices; Orthopedic Devices
RODENT	1250.96	RODENT CONTROL Sanitation Facilities And Conditions On Vessels; Interstate Conveyance Sanitation
ROLLER-TYPE	870.4370	ROLLER-TYPE CARDIOPULMONARY BYPASS BLOOD PUMP Cardiovascular Surgical Devices; Cardiovascular Devices
ROLLS	136.110	BREAD, ROLLS, AND BUNS Requirements For Specific Standardized Bakery Products; Bakery Products
ROLLS	136.115	ENRICHED BREAD, ROLLS, AND BUNS Requirements For Specific Standardized Bakery Products; Bakery Products
ROLLS	136.130	MILK BREAD, ROLLS, AND BUNS Requirements For Specific Standardized Bakery Products; Bakery Products
ROLLS	136.160	RAISIN BREAD, ROLLS, AND BUNS Requirements For Specific Standardized Bakery Products; Bakery Products
ROLLS	136.180	WHOLE WHEAT BREAD, ROLLS, AND BUNS Requirements For Specific Standardized Bakery Products; Bakery Products
ROMANO CHEESE	133.183	ROMANO CHEESE Requirements For Specific Standardized Cheese And Related Products; Cheeses And Related Cheese Products
RONGEUR	882.4840	MANUAL RONGEUR Neurological Surgical Devices; Neurological Devices
RONGEUR	882.4845	POWERED RONGEUR Neurological Surgical Devices; Neurological Devices
ROOT CANAL	872.3810	ROOT CANAL POST Prosthetic Devices; Dental Devices
ROOT CANAL	872.3820	ROOT CANAL FILLING RESIN Prosthetic Devices; Dental Devices
ROQUEFORT CHEESE	133.184	ROQUEFORT, SHEEP'S MILK BLUE-MOLD, AND BLUE-MOLD CHEESE FROM SHEEP'S MILK Requirements For Specific Standardized Cheese And Related Products; Cheeses And Related Cheese Products
ROSIN	172.735	GLYCEROL ESTER OF WOOD ROSIN Other Specific Usage Additives; Food Additives Permitted For Direct Addition To Food For Human Consumption
ROSINS	178.3870	ROSINS AND ROSIN DERIVATIVES Certain Adjuvants And Production Aids; Indirect Food Additives: Adjuvants, Production Aids, And Sanitizers
ROTARY PRISM	886.1665	OPHTHALMIC ROTARY PRISM Diagnostic Devices; Ophthalmic Devices
ROTARY SCALER	872.4840	ROTARY SCALER Surgical Devices; Dental Devices
ROTATING	870.5925	AUTOMATIC ROTATING TOURNIQUET Cardiovascular Therapeutic Devices; Cardiovascular Devices
ROTATION	890.5180	MANUAL PATIENT ROTATION BED Physical Medicine Therapeutic Devices; Physical Medicine Devices
ROTATION	890.5225	POWERED PATIENT ROTATION BED Physical Medicine Therapeutic Devices; Physical Medicine Devices
ROXARSONE	520.2087	ROXARSONE SOLUBLE POWDER Oral Dosage Form New Animal Drugs Not Subject To Certification

ROXARSONE	520.2088	ROXARSONE TABLETS Oral Dosage Form New Animal Drugs Not Subject To Certification
ROXARSONE	520.2089	ROXARSONE LIQUID Oral Dosage Form New Animal Drugs
ROXARSONE	558.530	ROXARSONE Specific New Animal Drugs For Use In Animal Feeds; New Animal Drugs For Use In Animal Feeds
RUBBER	177.1480	NITRILE RUBBER MODIFIED ACRYLONITRILE-METHYL ACRYLATE COPOLYMERS Substances For Use As Basic Components Of Single And Repeated Use Food Contact Surfaces; Indirect Food Additives: Polymers
RUBBER	177.2600	RUBBER ARTICLES INTENDED FOR REPEATED USE Substances For Use Only As Components Of Articles Intended For Repeated Use; Indirect Food Additives: Polymers
RUBBER	801.437	USER LABELING FOR DEVICES THAT CONTAIN NATURAL RUBBER Labeling
RUBBER DAM	872.6300	RUBBER DAM AND ACCESSORIES Miscellaneous Devices; Dental Devices
RUBBER-MODIFIED	177.1635	POLY(P-METHYLSTYRENE) AND RUBBER-MODIFIED POLY(P-METHYLSTYRENE) Substances For Use As Basic Components Of Single And Repeated Use Food Contact Surfaces; Indirect Food Additives: Polymers
RUBBER-MODIFIED	177.1640	POLYSTYRENE AND RUBBER-MODIFIED POLYSTYRENE Substances For Use As Basic Components Of Single And Repeated Use Food Contact Surfaces; Indirect Food Additives: Polymers
RUBELLA VIRUS	866.3510	RUBELLA VIRUS SEROLOGICAL REAGENTS Serological Reagents; Immunology And Microbiology Devices
RUBEOLA VIRUS	866.3520	RUBEOLA (MEASLES) VIRUS SEROLOGICAL REAGENTS Serological Reagents; Immunology And Microbiology Devices
RUE	184.1698	RUE Listing Of Specific Substances Affirmed As GRAS; Direct Food Substances Affirmed As Generally Recognized As Safe
RUE, OIL	184.1699	OIL OF RUE Listing Of Specific Substances Affirmed As GRAS; Direct Food Substances Affirmed As Generally Recognized As Safe
RULEMAKING	310.502	CERTAIN DRUGS ACCORDED NEW DRUG STATUS THROUGH RULEMAKING PROCEDURES Requirements For Specific New Drugs Or Devices; New Drugs
RULEMAKING	1308.43	INITIATION OF PROCEEDINGS FOR RULEMAKING Hearings; Schedules Of Controlled Substances
RULER	886.1790	NEARPOINT RULER Diagnostic Devices; Ophthalmic Devices
RULES	5.35	OFFICIALS AUTHORIZED TO MAKE CERTIFICATION UNDER 5 U.S.C. 605(B) FOR ANY PROPOSED AND FINAL RULES General Redelegations Of Authority; Delegations Of Authority And Organization
RULES	14.39	ADDITIONAL RULES FOR A PARTICULAR ADVISORY COMMITTEE Meeting Procedures; Public Hearing Before A Public Advisory Committee
RULES	1303.33	WAIVER OF MODIFICATION OF RULES Hearings; Quotas
RULES	1312.43	WAIVER OR MODIFICATION OF RULES Hearings; Importation And Exportation Of Controlled Substances
RULES	1313.53	WAIVER OR MODIFICATION OF RULES Importation And Exportation Of Precursors And Essential Chemicals
RULES	1316.44	WAIVER OR MODIFICATION OF RULES Administrative Hearings; Administrative Functions, Practices, And Procedures
RULING	12.24	RULING ON OBJECTIONS AND REQUESTS FOR HEARING Initiation Of Proceedings; Formal Evidentiary Public Hearing
RULING	12.07	INTERLOCUTORY APPEAL FROM RULING OF PRESIDING OFFICER Hearing Procedures; Formal Evidentiary Public Hearing
RULING	17.18	INTERLOCUTORY APPEAL FROM RULING OF PRESIDING OFFICER Civil Money Penalties Hearings
RULING	1316.55	PREHEARING RULING Administrative Hearings; Administrative Functions, Practices, And Procedures
RULING	1316.62	APPEAL FROM RULING OF PRESIDING OFFICER Administrative Hearings; Administrative Functions, Practices, And Procedures
RULING	1316.93	RULING ON PETITION FOR EXPEDITED RELEASE IN AN ADMINISTRATIVE FORFEITURE ACTION Expedited Forfeiture Proceedings For Certain Property; Administrative Functions, Practices, And Procedures
RULING	1316.96	RULING ON A PETITION FOR EXPEDITED RELEASE OF A CONVEYANCE IN A JUDICIAL FORFEITURE ACTION Expedited Forfeiture Proceedings For Certain Property; Administrative Functions, Practices, And Procedures

RULINGS	1316.61	EXCEPTIONS TO RULINGS Administrative Hearings; Administrative Functions, Practices, And Procedures
RUMINANT	589.2000	ANIMAL PROTEINS PROHIBITED IN RUMINANT FEED Substances Prohibited From Use In Animal Food Or Feed
RUSSELL VIPER	864.8950	RUSSELL VIPER VENOM REAGENT Hematology Reagents; Hematology And Pathology Devices

S

S, COMPOUND	862.1185	COMPOUND S (11-DEOXYCORTISOL) TEST SYSTEM Clinical Chemistry Test Systems; Clinical Chemistry And Clinical Toxicology Devices
S, FACTOR	866.5330	FACTOR XIII, A, S, IMMUNOLOGICAL TEST SYSTEM Immunological Test Systems; Immunology And Microbiology Devices
SACCHARIN	180.37	SACCHARIN, AMMONIUM SACCHARIN, CALCIUM SACCHARIN, AND SODIUM SACCHARIN Specific Requirements For Certain Food Additives; Food Additives Permitted In Food Or In Contact With Food On Interim Basis Pending Study
SACCHAROMYCES	866.5785	ANTI-SACCHAROMYCES CEREVISIAE ANTIBODY (ASCA) TEST SYSTEMS Immunological Test Systems; Immunology And Microbiology Devices
SACCULOTOMY	874.3760	SACCULOTOMY TACK (CODY TACK) Prosthetic Devices; Ear, Nose, And Throat Devices
SAFE	330.1	GENERAL CONDITIONS FOR GENERAL RECOGNITION AS SAFE, EFFECTIVE AND NOT MISBRANDED General Provisions; Over-the-counter Human Drugs Generally Recognized As Safe And Effective & Not Misbranded
SAFE	330.10	PROCEDURES FOR CLASSIFYING OTC DRUGS AS GENERALLY RECOGNIZED AS SAFE AND EFFECTIVE AND NOT MISBRANDED, AND FOR ESTABLISHING MONOGRAPHS Administrative Procedures; Over-the-counter Human Drugs Generally Recognized As Safe And Effective & Not Misbranded
SAFE	601.25	REVIEW PROCEDURES TO DETERMINE THAT LICENSED BIOLOGICAL PRODUCTS ARE SAFE, EFFECTIVE, AND NOT MISBRANDED UNDER PRESCRIBED, RECOMMENDED, OR SUGGESTED CONDITIONS OF USE Biologics Licensing; Licensing
SAFE	601.26	RECLASSIFICATION PROCEDURES TO DETERMINE THAT LICENSED BIOLOGICAL PRODUCTS ARE SAFE, EFFECTIVE, AND NOT MISBRANDED UNDER PRESCRIBED, RECOMMENDED, OR SUGGESTED CONDITIONS OF USE Biologics Licensing; Licensing
SAFE, GENERALLY RECOGNIZED	5.300	FOOD STANDARDS, FOOD ADDITIVES, GENERALLY RECOGNIZED AS SAFE (GRAS) SUBSTANCES, COLOR ADDITIVES, NUTRIENT CONTENT CLAIMS, AND HEALTH CLAIMS Foods And Cosmetics; Redelegations Of Authority; Delegations Of Authority And Organization
SAFE, GENERALLY RECOGNIZED	170.30	ELIGIBILITY FOR CLASSIFICATION AS GENERALLY RECOGNIZED AS SAFE (GRAS) Food Additive Safety; Food Additives
SAFE, GENERALLY RECOGNIZED	170.35	AFFIRMATION OF GENERALLY RECOGNIZED AS SAFE (GRAS) STATUS Food Additive Safety; Food Additives
SAFE, GENERALLY RECOGNIZED	182.1	SUBSTANCES THAT ARE GENERALLY RECOGNIZED AS SAFE General Provisions; Substances Generally Recognized As Safe
SAFE, GENERALLY RECOGNIZED	184.1	SUBSTANCES ADDED DIRECTLY TO HUMAN FOOD AFFIRMED AS GENERALLY RECOGNIZED AS SAFE (GRAS) General Provisions; Direct Food Substances Affirmed As Generally Recognized As Safe
SAFE, GENERALLY RECOGNIZED	186.1	SUBSTANCES ADDED INDIRECTLY TO HUMAN FOOD AFFIRMED AS GENERALLY RECOGNIZED AS SAFE (GRAS) General Provisions; Indirect Food Substances Affirmed As Generally Recognized As Safe
SAFE, GENERALLY RECOGNIZED	330.10	PROCEDURES FOR CLASSIFYING OTC DRUGS AS GENERALLY RECOGNIZED AS SAFE AND EFFECTIVE AND NOT MISBRANDED Administrative Procedures; Over-the-counter Human Drugs Generally Recognized As Safe And Effective & Not Misbranded
SAFE, GENERALLY RECOGNIZED	330.14	ADDITIONAL CRITERIA AND PROCEDURES FOR CLASSIFYING OTC DRUGS AS GENERALLY RECOGNIZED AS SAFE AND EFFECTIVE AND NOT MISBRANDED Administrative Procedures; Over-the-counter Human Drugs Generally Recognized As Safe And Effective & Not Misbranded
SAFE, GENERALLY RECOGNIZED	570.30	ELIGIBILITY FOR CLASSIFICATION AS GENERALLY RECOGNIZED AS SAFE (GRAS) Food Additive Safety; Food Additives
SAFE, GENERALLY RECOGNIZED	570.35	AFFIRMATION OF GENERALLY RECOGNIZED AS SAFE (GRAS) STATUS Food Additive Safety; Food Additives
SAFE, GENERALLY RECOGNIZED	582.1	SUBSTANCES THAT ARE GENERALLY RECOGNIZED AS SAFE General Provisions; Substances Generally Recognized As Safe
SAFE HANDLING	101.17	FOOD LABELING WARNING, NOTICE, AND SAFE HANDLING STATEMENTS General Provisions; Food Labeling

SAFE LEVELS	5.500	ISSUANCE OF FEDERAL REGISTER DOCUMENTS PERTAINING TO THE DETERMINATION OF SAFE LEVELS, NOTICE OF NEED FOR DEVELOPMENT OF AN ANALYTICAL METHOD, AND PROHIBITION OF CERTAIN EXTRALABEL DRUG USE Animal Drugs; Redelegations Of Authority; Delegations Of Authority And Organization
SAFE LEVELS	530.22	SAFE LEVELS AND ANALYTICAL METHODS FOR FOOD-PRODUCING ANIMALS Extralabel Drug Use In Animals
SAFE LEVELS	530.23	PROCEDURES FOR SETTING AND ANNOUNCING SAFE LEVELS Extralabel Drug Use In Animals
SAFE LEVELS	530.40	SAFE LEVELS AND AVAILABILITY OF ANALYTICAL METHODS Extralabel Drug Use In Animals
SAFE USE	314.520	APPROVAL WITH RESTRICTIONS TO ASSURE SAFE USE Accelerated Approval Of New Drugs; Applications For FDA Approval To Market A New Drug
SAFE USE	601.42	APPROVAL WITH RESTRICTIONS TO ASSURE SAFE USE Accelerated Approval Of Biological Products; Licensing
SAFEGUARD CLAUSE	26.21	SAFEGUARD CLAUSE Specific Sector Provisions For Pharmaceutical GMPs; Mutual Recognition Of Pharmaceutical GMP Reports, Medical Device Quality System Audit Reports
SAFEGUARDS	312.88	SAFEGUARDS FOR PATIENT SAFETY Drugs Intended To Treat Life-threatening And Severely-debilitating Illnesses; Investigational New Drug Application
SAFER VACCINES	5.200	FUNCTIONS PERTAINING TO SAFER VACCINES Biologics; Redelegations Of Authority; Delegations Of Authority And Organization
SAFETY	14.120	ESTABLISHMENT OF THE TECHNICAL ELECTRONIC PRODUCT RADIATION SAFETY STANDARD COMMITTEE (TEPRSSC) Technical Electronic Products Radiation Safety Standards Committee; Public Hearing Before A Public Advisory Committee
SAFETY	25.16	PUBLIC HEALTH AND SAFETY EMERGENCIES Environmental Impact Considerations
SAFETY	70.40	SAFETY FACTORS TO BE CONSIDERED Safety Evaluation; Color Additives
SAFETY	70.42	CRITERIA FOR EVALUATING THE SAFETY OF COLOR ADDITIVES Safety Evaluation; Color Additives
SAFETY	170.20	GENERAL PRINCIPLES FOR EVALUATING THE SAFETY OF FOOD ADDITIVES Food Additive Safety; Food Additives
SAFETY	170.22	SAFETY FACTORS TO BE CONSIDERED Food Additive Safety; Food Additives
SAFETY	216.24	DRUG PRODUCTS WITHDRAWN OR REMOVED FROM THE MARKET FOR REASONS OF SAFETY OR EFFECTIVENESS Pharmacy Compounding
SAFETY	310.6	APPLICABILITY OF "NEW DRUG" OR SAFETY OR EFFECTIVENESS FINDINGS IN DRUG EFFICACY STUDY IMPLEMENTATION NOTICES AND NOTICES OF OPPORTUNITY FOR HEARING TO IDENTICAL, RELATED, AND SIMILAR DRUG PRODUCTS General Provisions; New Drugs
SAFETY	312.88	SAFEGUARDS FOR PATIENT SAFETY Drugs Intended To Treat Life-threatening And Severely-debilitating Illnesses; Investigational New Drug Application
SAFETY	315.3	GENERAL FACTORS RELEVANT TO SAFETY AND EFFECTIVENESS Diagnostic Radiopharmaceuticals
SAFETY	315.6	EVALUATION OF SAFETY Diagnostic Radiopharmaceuticals
SAFETY	570.20	GENERAL PRINCIPLES FOR EVALUATING THE SAFETY OF FOOD ADDITIVES Food Additive Safety; Food Additives
SAFETY	601.32	GENERAL FACTORS RELEVANT TO SAFETY AND EFFECTIVENESS Diagnostic Radiopharmaceuticals; Licensing
SAFETY	601.35	EVALUATION OF SAFETY Diagnostic Radiopharmaceuticals; Licensing
SAFETY	610.11	GENERAL SAFETY General Provisions; General Biological Products Standards
SAFETY	610.11a	INACTIVATED INFLUENZA VACCINE, GENERAL SAFETY TEST General Provisions; General Biological Products Standards
SAFETY	740.10	LABELING OF COSMETIC PRODUCTS FOR WHICH ADEQUATE SUBSTANTIATION OF SAFETY HAS NOT BEEN OBTAINED Warning Statements; Cosmetic Product Warning Statements
SAFETY	860.7	DETERMINATION OF SAFETY AND EFFECTIVENESS General; Medical Device Classification Procedures
SAFETY	1000.15	EXAMPLES OF ELECTRONIC PRODUCTS SUBJECT TO THE RADIATION CONTROL FOR HEALTH AND SAFETY ACT OF 1968 Statements Of Policy And Interpretation; General

SAFETY REPORTING	314.540	POSTMARKETING SAFETY REPORTING Applications For FDA Approval To Market A New Drug
SAFETY REPORTING	601.44	POSTMARKETING SAFETY REPORTING Accelerated Approval Of Biological Products; Licensing
SAFETY REPORTS	312.32	IND SAFETY REPORTS Investigational New Drug Application
SAFFRON	73.500	SAFFRON Foods; Listing Of Color Additives Exempt From Certification
SAFROLE	189.180	SAFROLE Substances Generally Prohibited From Direct Addition Or Use As Human Food; Substances Prohibited From Use In Human Food
SAFROLE-FREE EXTRACT	172.580	SAFROLE-FREE EXTRACT OF SASSAFRAS Flavoring Agents And Related Substances; Food Additives Permitted For Direct Addition To Food For Human Consumption
SAGO CHEESE	133.186	SAP SAGO CHEESE Requirements For Specific Standardized Cheese And Related Products; Cheeses And Related Cheese Products
SAIB	172.833	SUCROSE ACETATE ISOBUTYRATE (SAIB) Food Additives Permitted For Direct Addition To Food For Human Consumption
SALAD DRESSING	169.150	SALAD DRESSING Requirements For Specific Standardized Food Dressings And Flavorings; Food Dressings And Flavorings
SALE	801.421	HEARING AID DEVICES; CONDITIONS FOR SALE Special Requirements For Specific Devices; Labeling
SALE	809.30	RESTRICTIONS ON THE SALE, DISTRIBUTION AND USE OF ANALYTE SPECIFIC REAGENTS Requirements For Manufacturers And Producers; In Vitro Diagnostic Products For Human Use
SALE	809.40	RESTRICTIONS ON THE SALE, DISTRIBUTION AND USE OF OTC TEST SAMPLE COLLECTION SYSTEMS FOR DRUGS OF ABUSE TESTING Requirements For Manufacturers And Producers; In Vitro Diagnostic Products For Human Use
SALE, OTC	201.307	SODIUM PHOSPHATES, PACKAGE SIZE LIMITATION, WARNINGS AND DIRECTIONS FOR OVER-THE-COUNTER SALE Specific Labeling Requirements For Specific Drug Products; Labeling
SALE, OTC	201.308	IPECAC SYRUP; WARNINGS AND DIRECTIONS FOR USE FOR OVER-THE-COUNTER SALE Specific Labeling Requirements For Specific Drug Products; Labeling
SALE, PRESCRIPTION	310.201	EXEMPTION FOR CERTAIN DRUGS LIMITED BY NEW DRUG APPLICATIONS TO PRESCRIPTION SALE New Drugs Exempted From Prescription-dispensing Requirements; New Drugs
SALES	20.115	PRODUCT CODES FOR MANUFACTURING OR SALES DATES Availability Of Specific Categories Of Records; Public Information
SALES RESTRICTIONS	203.20	SALES RESTRICTIONS Prescription Drug Marketing
SALICYLANILIDES	700.15	USE OF CERTAIN HALOGENATED SALICYLANILIDES AS INGREDIENTS IN COSMETIC PRODUCTS Requirements For Specific Cosmetic Products; General
SALICYLATE	862.3830	SALICYLATE TEST SYSTEM Clinical Toxicology Test Systems; Clinical Chemistry And Clinical Toxicology Devices
SALICYLATES	201.314	LABELING OF DRUG PREPARATIONS CONTAINING SALICYLATES Specific Labeling Requirements For Specific Drug Products; Labeling
SALICYLIC ACID	520.2090	SALICYLIC ACID Certain Other Dosage Form New Animal Drugs
SALICYLIC ACID	556.590	SALICYLIC ACID Specific Tolerances For Residues Of New Animal Drugs; Tolerances For Residues Of New Animal Drugs In Food
SALINOMYCIN	558.550	SALINOMYCIN Specific New Animal Drugs For Use In Animal Feeds; New Animal Drugs For Use In Animal Feeds
SALIVA	872.6050	SALIVA ABSORBER Miscellaneous Devices; Dental Devices
SALMON, PACIFIC	161.170	CANNED PACIFIC SALMON Requirements For Specific Standardized Fish And Shellfish; Fish And Shellfish
SALMONELLA	500.35	ANIMAL FEEDS CONTAMINATED WITH SALMONELLA MICROORGANISMS Specific Administrative Rulings And Decisions; General
SALMONELLA	866.3550	SALMONELLA SPP. SEROLOGICAL REAGENTS Serological Reagents; Immunology And Microbiology Devices
SALT	100.155	SALT AND IODIZED SALT Specific Administrative Rulings And Decisions; General
SALT, POTASSIUM	201.306	POTASSIUM SALT PREPARATIONS INTENDED FOR ORAL INGESTION BY MAN Specific Labeling Requirements For Specific Drug Products; Labeling

SALT SOLUTIONS, BALANCED	864.2875	BALANCED SALT SOLUTIONS OR FORMULATIONS Cell And Tissue Culture Products; Hematology And Pathology Devices
SALTS OF FATTY ACIDS	172.863	SALTS OF FATTY ACIDS Multipurpose Additives; Food Additives Permitted For Direct Addition To Food For Human Consumption
SALTS OF FATTY ACIDS	573.914	SALTS OF VOLATILE FATTY ACIDS Food Additive Listing; Food Additives Permitted In Feed And Drinking Water Of Animals
SALVAGING	211.208	DRUG PRODUCT SALVAGING Returned And Salvaged Drug Products; Current Good Manufacturing Practice For Finished Pharmaceuticals
SAMPLE	203.32	DRUG SAMPLE STORAGE AND HANDLING REQUIREMENTS Samples; Prescription Drug Marketing
SAMPLE	203.33	DRUG SAMPLE FORMS Samples; Prescription Drug Marketing
SAMPLE	203.38	SAMPLE LOT OR CONTROL NUMBERS; LABELING OF SAMPLE UNITS Samples; Prescription Drug Marketing
SAMPLE	809.40	RESTRICTIONS ON THE SALE, DISTRIBUTION AND USE OF OTC TEST SAMPLE COLLECTION SYSTEMS FOR DRUGS OF ABUSE TESTING Requirements For Manufacturers And Producers; In Vitro Diagnostic Products For Human Use
SAMPLE	862.2310	CLINICAL SAMPLE CONCENTRATOR Clinical Laboratory Instruments; Clinical Chemistry And Clinical Toxicology Devices
SAMPLE	864.3260	OTC TEST SAMPLE COLLECTION SYSTEMS FOR DRUGS OF ABUSE TESTING Pathology Instrumentation And Accessories; Hematology And Pathology Devices
SAMPLE	1210.17	AUTHORITY TO SAMPLE AND INSPECT Regulations Under The Federal Import Milk Act
SAMPLE DISTRIBUTION	203.30	SAMPLE DISTRIBUTION BY MAIL OR COMMON CARRIER Samples; Prescription Drug Marketing
SAMPLE DISTRIBUTION	203.31	SAMPLE DISTRIBUTION BY MEANS OTHER THAN MAIL OR COMMON CARRIER Samples; Prescription Drug Marketing
SAMPLER	884.1550	AMNIOTIC FLUID SAMPLER (AMNIOCENTESIS TRAY) Obstetrical And Gynecological Diagnostic Devices; Obstetrical And Gynecological Devices
SAMPLER	884.1560	FETAL BLOOD SAMPLER Obstetrical And Gynecological Diagnostic Devices; Obstetrical And Gynecological Devices
SAMPLES	2.10	EXAMINATION AND INVESTIGATION SAMPLES General Provisions; General Administrative Rulings And Decisions
SAMPLES	71.4	SAMPLES, ADDITIONAL INFORMATION General Provisions; Color Additive Petitions
SAMPLES	80.22	SAMPLES TO ACCOMPANY REQUESTS FOR CERTIFICATION Certification Procedures; Color Additive Certification
SAMPLES	203.39	DONATION OF DRUG SAMPLES TO CHARITABLE INSTITUTIONS Samples; Prescription Drug Marketing
SAMPLES	211.170	RESERVE SAMPLES Laboratory Controls; Current Good Manufacturing Practice For Finished Pharmaceuticals
SAMPLES	600.13	RETENTION SAMPLES Establishment Standards; Biological Products: General
SAMPLES	601.33	FOREIGN ESTABLISHMENTS AND PRODUCTS: SAMPLES FOR EACH IMPORTATION Biologics Licensing; Licensing
SAMPLES	610.2	REQUESTS FOR SAMPLES AND PROTOCOLS; OFFICIAL RELEASE Release Requirements; General Biological Products Standards
SAMPLES	660.6	SAMPLES; PROTOCOLS; OFFICIAL RELEASE Antibody To Hepatitis B Surface Antigen; Additional Standards For Diagnostic Substances For Laboratory Tests
SAMPLES	660.36	SAMPLES AND PROTOCOLS Reagent Red Blood Cells; Additional Standards For Diagnostic Substances For Laboratory Tests
SAMPLES	660.46	SAMPLES; PROTOCOLS; OFFICIAL RELEASE Hepatitis B Surface Antigen; Additional Standards For Diagnostic Substances For Laboratory Tests
SAMPLES	1230.30	COLLECTION OF SAMPLES Administrative Procedures; Regulations Under The Federal Caustic Poison Act
SAMPLES	1230.31	WHERE SAMPLES MAY BE COLLECTED Administrative Procedures; Regulations Under The Federal Caustic Poison Act
SAMPLES	1230.32	ANALYZING OF SAMPLES Administrative Procedures; Regulations Under The Federal Caustic Poison Act
SAMPLES	1230.44	SAMPLES Imports; Regulations Under The Federal Caustic Poison Act

SAMPLES, PAYMENT	1.91	PAYMENT FOR SAMPLES Imports And Exports; General Enforcement Regulations
SAMPLES, PAYMENT	1005.11	PAYMENT FOR SAMPLES Inspection And Testing; Importation Of Electronic Products
SAMPLES, RETENTION	320.38	RETENTION OF BIOAVAILABILITY SAMPLES Procedures For Determining The Bioavailability Or Bioequivalence Of Drug Products; Bioavailability And Bioequivalence Requirements
SAMPLES, RETENTION	320.63	RETENTION OF BIOEQUIVALENCE SAMPLES Procedures For Determining The Bioavailability Or Bioequivalence Of Drug Products; Bioavailability And Bioequivalence Requirements
SAMPLING	1.90	NOTICE OF SAMPLING Imports And Exports; General Enforcement Regulations
SAMPLING	211.110	SAMPLING AND TESTING OF IN-PROCESS MATERIALS AND DRUG PRODUCTS Production And Process Controls; Current Good Manufacturing Practice For Finished Pharmaceuticals
SAMPLING	1005.10	NOTICE OF SAMPLING Inspection And Testing; Importation Of Electronic Products
SAMPLING, BLOOD	868.1100	ARTERIAL BLOOD SAMPLING KIT Diagnostic Devices; Anesthesiology Devices
SAMSOE CHEESE	133.185	SAMSOE CHEESE Requirements For Specific Standardized Cheese And Related Products; Cheeses And Related Cheese Products
SANCTIONED	570.13	INDIRECT FOOD ADDITIVES RESULTING FROM PACKAGING MATERIALS PRIOR SANCTIONED FOR ANIMAL FEED AND PET FOOD General Provisions; Food Additives
SANCTIONS	17.35	SANCTIONS Civil Money Penalties Hearings
SANCTIONS	181.5	PRIOR SANCTIONS General Provisions; Prior-sanctioned Food Ingredients
SANCTIONS	1404.635	REPORTING OF AND EMPLOYEE SANCTIONS FOR CONVICTIONS OF CRIMINAL DRUG OFFENSES Governmentwide Debarment And Suspension And Requirements For Drug-Free Workplace (Grants)
SANITARY FACILITIES	129.35	SANITARY FACILITIES Buildings And Facilities; Processing And Bottling Of Bottled Drinking Water
SANITARY INSPECTION	1210.11	SANITARY INSPECTION OF DIARY FARMS Regulations Under The Federal Import Milk Act
SANITARY INSPECTION	1210.14	SANITARY INSPECTION OF PLANTS Regulations Under The Federal Import Milk Act
SANITARY OPERATIONS	110.35	SANITARY OPERATIONS Buildings And Facilities; Current Good Manufacturing Practice In Manufacturing, Packing, Or Holding Human Food
SANITARY OPERATIONS	129.37	SANITARY OPERATIONS Buildings And Facilities; Processing And Bottling Of Bottled Drinking Water
SANITATION	120.6	SANITATION STANDARD OPERATING PROCEDURES Hazard Analysis And Critical Control Point (HACCP) Systems
SANITATION	123.11	SANITATION CONTROL PROCEDURES Fish And Fishery Products
SANITATION	211.56	SANITATION Buildings And Facilities; Current Good Manufacturing Practice For Finished Pharmaceuticals
SANITATION	1240.95	SANITATION OF WATER BOATS Source And Use Of Potable Water; Control Of Communicable Diseases
SANITIZING	178.1010	SANITIZING SOLUTIONS Substances Utilized To Control The Growth Of Microorganisms; Indirect Food Additives: Adjuvants, Production Aids, And Sanitizers
SAP SAGO CHEESE	133.186	SAP SAGO CHEESE Requirements For Specific Standardized Cheese And Related Products; Cheeses And Related Cheese Products
SASSAFRAS	172.580	SAFROLE-FREE EXTRACT OF SASSAFRAS Flavoring Agents And Related Substances; Food Additives Permitted For Direct Addition To Food For Human Consumption
SATURATED FAT	101.75	HEALTH CLAIMS: DIETARY SATURATED FAT AND CHOLESTEROL AND RISK OF CORONARY HEART DISEASE Food Labeling
SAVINGS	1.31	PACKAGE SIZE SAVINGS General Labeling Requirements; General Enforcement Regulations
SCALE	880.2700	STAND-ON PATIENT SCALE General Hospital And Personal Use Monitoring Devices; General Hospital And Personal Use Devices
SCALE	880.2720	PATIENT SCALE General Hospital And Personal Use Monitoring Devices; General Hospital And Personal Use Devices

SCALE	880.2740	SURGICAL SPONGE SCALE General Hospital And Personal Use Monitoring Devices; General Hospital And Personal Use Devices
SCALE	890.3940	WHEELCHAIR PLATFORM SCALE Physical Medicine Prosthetic Devices; Physical Medicine Devices
SCALER	872.4840	ROTARY SCALER Surgical Devices; Dental Devices
SCALER	872.4850	ULTRASONIC SCALER Surgical Devices; Dental Devices
SCALP CLIP	882.4150	SCALP CLIP Neurological Surgical Devices; Neurological Devices
SCALP ELECTRODE	884.2675	FETAL SCALP CIRCULAR (SPIRAL) ELECTRODE AND APPLICATOR Obstetrical And Gynecological Monitoring Devices; Obstetrical And Gynecological Devices
SCALP ELECTRODE	884.2685	FETAL SCALP CLIP ELECTRODE AND APPLICATOR Obstetrical And Gynecological Monitoring Devices; Obstetrical And Gynecological Devices
SCAMORZA CHEESE	133.155	MOZZARELLA CHEESE AND SCAMORZA CHEESE Requirements For Specific Standardized Cheese And Related Products; Cheeses And Related Cheese Products
SCAMORZA CHEESE	133.156	LOW-MOISTURE MOZZARELLA AND SCAMORZA CHEESE Requirements For Specific Standardized Cheese And Related Products; Cheeses And Related Cheese Products
SCAMORZA CHEESE	133.157	PART-SKIM MOZZARELLA AND SCAMORZA CHEESE Requirements For Specific Standardized Cheese And Related Products; Cheeses And Related Cheese Products
SCAMORZA CHEESE	133.158	LOW-MOISTURE PART-SKIM MOZZARELLA AND SCAMORZA CHEESE Requirements For Specific Standardized Cheese And Related Products; Cheeses And Related Cheese Products
SCANNER	882.1925	ULTRASONIC SCANNER CALIBRATION TEST BLOCK Neurological Diagnostic Devices; Neurological Devices
SCANNER	892.1220	FLUORESCENT SCANNER Diagnostic Devices; Radiology Devices
SCANNER, NUCLEAR BODY	892.1330	NUCLEAR WHOLE BODY SCANNER Diagnostic Devices; Radiology Devices
SCANNER, RECTILINEAR	892.1300	NUCLEAR RECTILINEAR SCANNER Diagnostic Devices; Radiology Devices
SCANNING BED, NUCLEAR	892.1350	NUCLEAR SCANNING BED Diagnostic Devices; Radiology Devices
SCAPHOID	888.3760	WRIST JOINT CARPAL SCAPHOID POLYMER PROSTHESIS Prosthetic Devices; Orthopedic Devices
SCAVENGING	868.5430	GAS-SCAVENGING APPARATUS Therapeutic Devices; Anesthesiology Devices
SCAVENGING MASK	868.5590	SCAVENGING MASK Therapeutic Devices; Anesthesiology Devices
SCHEDULE	15.21	NOTICE OF PARTICIPATION; SCHEDULE FOR HEARING Procedures For Public Hearing Before The Commissioner; Public Hearing Before The Commissioner
SCHEDULE I	1301.32	ACTION ON APPLICATIONS FOR RESEARCH IN SCHEDULE I SUBSTANCES Action On Application For Registration: Revocation Or Suspension Of Registration Registration Of Manufacturers, Distributors, And Dispensers Of Controlled Substances
SCHEDULE I	1308.11	SCHEDULE I Schedules; Schedules Of Controlled Substances
SCHEDULE I	1312.31	SCHEDULE I: APPLICATION FOR PRIOR WRITTEN APPROVAL Transshipment And In-transit Shipment Of Controlled Substances; Importation And Exportation Of Controlled Substances
SCHEDULE I AND II	1301.33	APPLICATION FOR BULK MANUFACTURE OF SCHEDULE I AND II SUBSTANCES Action On Application For Registration: Revocation Or Suspension Of Registration; Registration Of Manufacturers, Distributors, And Dispensers Of Controlled Substances
SCHEDULE I AND II	1301.34	APPLICATION FOR IMPORTATION OF SCHEDULE I AND II SUBSTANCES Action On Application For Registration: Revocation Or Suspension Of Registration; Registration Of Manufacturers, Distributors, And Dispensers Of Controlled Substances
SCHEDULE II	1308.12	SCHEDULE II Schedules; Schedules Of Controlled Substances
SCHEDULE III	1308.13	SCHEDULE III Schedules; Schedules Of Controlled Substances
SCHEDULE III, IV, V	1312.30	SCHEDULE III, IV, AND V NON-NARCOTIC CONTROLLED SUBSTANCES REQUIRING AN IMPORT AND EXPORT PERMIT Exportation Of Controlled Substances; Importation And Exportation Of Controlled Substances

SCHEDULE IV	1308.14	SCHEDULE IV Schedules; Schedules Of Controlled Substances
SCHEDULE V	1308.15	SCHEDULE V Schedules; Schedules Of Controlled Substances
SCHEDULED PROCESSES	113.83	ESTABLISHING SCHEDULED PROCESSES Production And Process Controls; Thermally Processed Low-acid Foods Packaged In Hermetically Sealed Containers
SCHEDULED PROCESSES	114.83	ESTABLISHING SCHEDULED PROCESSES Production And Process Controls; Acidified Foods
SCHEDULED PROCESSES	114.89	DEVIATIONS FROM SCHEDULED PROCESSES Production And Process Controls; Acidified Foods
SCHEDULES II, III, IV	1312.32	SCHEDULES II, III, IV: ADVANCE NOTICE Transshipment And In-Transit Shipment Of Controlled Substances; Importation And Exportation Of Controlled Substances
SCHEDULES III, IV, V	1306.26	TRANSFER BETWEEN PHARMACIES OF PRESCRIPTION INFORMATION FOR SCHEDULES III, IV, AND V CONTROLLED SUBSTANCES FOR REFILL PURPOSES Controlled Substances Listed In Schedules III, IV, And V; Prescriptions
SCHEDULING	5.104	RESPONSES TO DRUG ENFORCEMENT ADMINISTRATION TEMPORARY SCHEDULING NOTICES Human Drugs; Redelegations Of Authority; Delegations Of Authority And Organization
SCHEDULING, EMERGENCY	1308.49	EMERGENCY SCHEDULING Hearings; Schedules Of Controlled Substances
SCHIRMER STRIP	886.1800	SCHIRMER STRIP Diagnostic Devices; Ophthalmic Devices
SCHISTOSOMA SPP.	866.3600	SCHISTOSOMA SPP. SEROLOGICAL REAGENTS Serological Reagents; Immunology And Microbiology Devices
SCHOOLS	5.304	APPROVAL OF SCHOOLS PROVIDING FOOD-PROCESSING INSTRUCTION Foods And Cosmetics; Redelegations Of Authority; Delegations Of Authority And Organization
SCIENTIFIC EVIDENCE	5.408	DETERMINATIONS CONCERNING THE TYPE OF VALID SCIENTIFIC EVIDENCE SUBMITTED IN A PREMARKET APPROVAL APPLICATION Medical Devices And Radiological Health; Redelegations Of Authority; Delegations Of Authority And Organization
SCIENTIFIC STUDIES	70.55	REQUEST FOR SCIENTIFIC STUDIES Safety Evaluation; Color Additives
SCINTILLATION CAMERA	892.1100	SCINTILLATION (GAMMA) CAMERA Diagnostic Devices; Radiology Devices
SCISSORS	880.6820	MEDICAL DISPOSABLE SCISSORS General Hospital And Personal Use Miscellaneous Devices; General Hospital And Personal Use Devices
SCLERAL BUCKLING	886.3300	ABSORBABLE IMPLANT (SCLERAL BUCKLING METHOD) Prosthetic Devices; Ophthalmic Devices
SCLERAL SHELL	886.3800	SCLERAL SHELL Prosthetic Devices; Ophthalmic Devices
SCORING	1210.18	SCORING Regulations Under The Federal Import Milk Act
SCREEN	886.1810	TANGENT SCREEN (CAMPIMETER) Diagnostic Devices; Ophthalmic Devices
SCREEN	892.1960	RADIOGRAPHIC INTENSIFYING SCREEN Diagnostic Devices; Radiology Devices
SCREENING, EMPLOYEE	1301.90	EMPLOYEE SCREENING PROCEDURES Employee Screening - Non-practitioners; Registration Of Manufacturers, Distributors, And Dispensers Of Controlled Substances
SCREENING, GONORRHEA	866.2420	OXIDASE SCREENING TEST FOR GONORRHEA Microbiology Devices; Immunology And Microbiology Devices
SCREW	872.4880	INTRAOSSEOUS FIXATION SCREW OR WIRE Surgical Devices; Dental Devices
SCREW	888.3070	PEDICLE SCREW SPINAL SYSTEM Prosthetic Devices; Orthopedic Devices
SCREWDRIVER	882.4900	SKULLPLATE SCREWDRIVER Neurological Surgical Devices; Neurological Devices
SCROTAL SUPPORT	880.5820	THERAPEUTIC SCROTAL SUPPORT General Hospital And Personal Use Therapeutic Devices; General Hospital And Personal Use Devices
SEAFOOD COCKTAILS	102.54	SEAFOOD COCKTAILS Requirements For Specific Nonstandardized Foods; Common Or Usual Name For Nonstandardized Foods
SEAL, DEPARTMENT	5.22	CERTIFICATION OF TRUE COPIES AND USE OF DEPARTMENT SEAL General Redelegations Of Authority; Delegations Of Authority And Organization
SEALANT	872.3765	PIT AND FISSURE SEALANT AND CONDITIONER Prosthetic Devices; Dental Devices

SEALED	892.1400	NUCLEAR SEALED CALIBRATION SOURCE Diagnostic Devices; Radiology Devices
SEALED, HERMETICALLY	108.35	THERMAL PROCESSING OF LOW-ACID FOODS PACKAGED IN HERMETICALLY SEALED CONTAINERS Specific Requirements And Conditions For Exemption From Or Compliance With An Emergency Permit; Emergency Permit Control
SEALED, HERMETICALLY	500.23	THERMALLY PROCESSED LOW-ACID FOODS PACKAGED IN HERMETICALLY SEALED CONTAINERS Specific Administrative Rulings And Decisions; General
SEALING	1302.06	SEALING OF CONTROLLED SUBSTANCES Labeling And Packaging Requirements For Controlled Substances
SEALING	864.3400	DEVICE FOR SEALING MICROSECTIONS Pathology Instrumentation And Accessories; Hematology And Pathology Devices
SEALING	864.9750	HEAT-SEALING DEVICE Products Used In Establishments That Manufacture Blood And Blood Products; Hematology And Pathology Devices
SEALING GASKETS	177.1210	CLOSURES WITH SEALING GASKETS FOR FOOD CONTAINERS Substances For Use As Basic Components Of Single And Repeated Use Food Contact Surfaces; Indirect Food Additives: Polymers
SEASONINGS	182.10	SPICES AND OTHER NATURAL SEASONINGS AND FLAVORINGS General Provisions; Substances Generally Recognized As Safe
SEASONINGS	182.40	NATURAL EXTRACTIVES (SOLVENT-FREE) USED IN CONJUNCTION WITH SPICES, SEASONINGS, AND FLAVORINGS General Provisions; Substances Generally Recognized As Safe
SEASONINGS	182.50	CERTAIN OTHER SPICES, SEASONINGS, ESSENTIAL OILS, OLEORESINS, AND NATURAL EXTRACTS General Provisions; Substances Generally Recognized As Safe
SEASONINGS	582.10	SPICES AND OTHER NATURAL SEASONINGS AND FLAVORINGS General Provisions; Substances Generally Recognized As Safe
SEASONINGS	582.30	NATURAL SUBSTANCES USED IN CONJUNCTION WITH SPICES AND OTHER NATURAL SEASONINGS AND FLAVORINGS General Provisions; Substances Generally Recognized As Safe
SEASONINGS	582.40	NATURAL EXTRACTIVES (SOLVENT-FREE) USED IN CONJUNCTION WITH SPICES, SEASONINGS, AND FLAVORINGS General Provisions; Substances Generally Recognized As Safe
SEASONINGS	582.50	CERTAIN OTHER SPICES, SEASONINGS, ESSENTIAL OILS, OLEORESINS, AND NATURAL EXTRACTS General Provisions; Substances Generally Recognized As Safe
SEBORRHEIC DERMATITIS	358.710	ACTIVE INGREDIENTS FOR THE CONTROL OF DANDRUFF, SEBORRHEIC DERMATITIS, OR PSORIASIS Miscellaneous External Drug Products For Over-the-counter Human Use
SEBORRHEIC DERMATITIS	358.750	LABELING OF DRUG PRODUCTS FOR THE CONTROL OF DANDRUFF, SEBORRHEIC DERMATITIS, OR PSORIASIS Miscellaneous External Drug Products For Over-the-counter Human Use
SECONDHAND CONTAINERS	2.35	USE OF SECONDHAND CONTAINERS FOR THE SHIPMENT OR STORAGE OF FOOD AND ANIMAL FEED Human and Animal Foods; General Administrative Rulings And Decisions
SECRETARY	5.10	DELEGATIONS FROM THE SECRETARY OF HEALTH AND HUMAN SERVICES TO THE COMMISSIONER OF FOOD AND DRUGS Delegations Of Authority To The Commissioner Of Food And Drugs; Delegations Of Authority And Organization
SECRETORY	866.5380	FREE SECRETORY COMPONENT IMMUNOLOGICAL TEST SYSTEM Immunological Test Systems; Immunology And Microbiology Devices
SECRETS, TRADE	20.53	INDEXING TRADE SECRETS AND CONFIDENTIAL COMMERCIAL OR FINANCIAL INFORMATION Procedures And Fees; Public Information
SECRETS, TRADE	20.61	TRADE SECRETS AND COMMERCIAL OR FINANCIAL INFORMATION WHICH IS PRIVILEGED OR CONFIDENTIAL Exemptions; Public Information
SECTORAL COMMITTEE, JOINT	26.17	ROLE AND COMPOSITION OF THE JOINT SECTORAL COMMITTEE Specific Sector Provisions For Pharmaceutical GMPs; Mutual Recognition Of Pharmaceutical GMP Reports, Medical Device Quality System Audit Reports
SECTORAL COMMITTEE, JOINT	26.47	ROLE AND COMPOSITION OF THE JOINT SECTORAL COMMITTEE Specific Sector Provisions For Medical Devices; Mutual Recognition Of Pharmaceutical GMP Reports, Medical Device Quality System Audit Reports
SECTORAL CONTACT	26.72	SECTORAL CONTACT POINTS Framework Provisions; Mutual Recognition Of Pharmaceutical GMP Reports, Medical Device Quality System Audit Reports
SECUREMENT, CATHETER	880.5210	INTRAVASCULAR CATHETER SECUREMENT DEVICE General Hospital And Personal Use Therapeutic Devices; General Hospital And Personal Use Devices

SECURITY CONTROLS	1301.72	PHYSICAL SECURITY CONTROLS FOR NON-PRACTITIONERS; NARCOTIC TREATMENT PROGRAMS AND COMPOUNDERS FOR NARCOTIC TREATMENT PROGRAMS; STORAGE AREAS Security Requirements; Registration Of Manufacturers, Distributors, And Dispensers Of Controlled Substances
SECURITY CONTROLS	1301.73	PHYSICAL SECURITY CONTROLS FOR NON-PRACTITIONERS; COMPOUNDERS FOR NARCOTIC TREATMENT PROGRAMS; MANUFACTURING AND COMPOUNDING AREAS Security Requirements; Registration Of Manufacturers, Distributors, And Dispensers Of Controlled Substances
SECURITY CONTROLS	1301.74	OTHER SECURITY CONTROLS FOR NON-PRACTITIONERS; NARCOTIC TREATMENT PROGRAMS AND COMPOUNDERS FOR NARCOTIC TREATMENT PROGRAMS Security Requirements; Registration Of Manufacturers, Distributors, And Dispensers Of Controlled Substances
SECURITY CONTROLS	1301.75	PHYSICAL SECURITY CONTROLS FOR PRACTITIONERS Security Requirements; Registration Of Manufacturers, Distributors, And Dispensers Of Controlled Substances
SECURITY CONTROLS	1301.76	OTHER SECURITY CONTROLS FOR PRACTITIONERS Security Requirements; Registration Of Manufacturers, Distributors, And Dispensers Of Controlled Substances
SECURITY CONTROLS	1301.77	SECURITY CONTROLS FOR FREIGHT FORWARDING FACILITIES Security Requirements; Registration Of Manufacturers, Distributors, And Dispensers Of Controlled Substances
SECURITY REQUIREMENTS	1301.71	SECURITY REQUIREMENTS GENERALLY Security Requirements; Registration Of Manufacturers, Distributors, And Dispensers Of Controlled Substances
SECURITY REQUIREMENTS	1309.71	GENERAL SECURITY REQUIREMENTS Registration Of Manufacturers, Distributors, Importers And Exporters Of List I Chemicals
SEDATIVES, DAYTIME	310.519	DRUG PRODUCTS MARKETED AS OVER-THE-COUNTER (OTC) DAYTIME SEDATIVES Requirements For Specific New Drugs Or Devices; New Drugs
SEDIMENTATION RATE	864.5800	AUTOMATED SEDIMENTATION RATE DEVICE Automated And Semi-automated Hematology Devices; Hematology And Pathology Devices
SEDIMENTATION RATE	864.6700	ERYTHROCYTE SEDIMENTATION RATE TEST Manual Hematology Devices; Hematology And Pathology Devices
SEED, GRAIN	2.25	GRAIN SEED TREATED WITH POISONOUS SUBSTANCES; COLOR IDENTIFICATION TO PREVENT ADULTERATION OF HUMAN AND ANIMAL FOOD Human and Animal Foods; General Administrative Rulings And Decisions
SEED, TAMARIND	176.350	TAMARIND SEED KERNEL POWDER Substances For Use Only As Components Of Paper And Paperboard; Indirect Food Additives
SEEDLESS GRAPES	145.140	CANNED SEEDLESS GRAPES Requirements For Specific Standardized Canned Fruits; Canned Fruits
SEGMENTS FOR TESTING	640.15	SEGMENTS FOR TESTING Red Blood Cells; Additional Standards For Human Blood And Blood Products
SEIZURES	1316.72	OFFICERS WHO WILL MAKE SEIZURES Seizure, Forfeiture, And Disposition Of Property; Administrative Functions, Practices, And Procedures
SELAMECTIN	524.2008	SELAMECTIN Ophthalmic And Topical Dosage Form New Animal Drugs
SELECTING	312.53	SELECTING INVESTIGATORS AND MONITORS Investigational New Drug Application
SELECTING	812.43	SELECTING INVESTIGATORS AND MONITORS Responsibilities Of Sponsors; Investigational Device Exemptions
SELECTION	14.84	NOMINATIONS AND SELECTION OF NONVOTING MEMBERS OF STANDING TECHNICAL ADVISORY COMMITTEES Members Of Advisory Committees; Public Hearing Before A Public Advisory Committee
SELECTIVE	866.2360	SELECTIVE CULTURE MEDIUM Microbiology Devices; Immunology And Microbiology Devices
SELEGILINE	520.2098	SELEGILINE HYDROCHLORIDE TABLETS Oral Dosage Form New Animal Drugs
SELENIUM	520.2100	SELENIUM, VITAMIN E CAPSULES Oral Dosage Form New Animal Drugs
SELENIUM	522.2100	SELENIUM, VITAMIN E INJECTION Implantation Or Injectable Dosage Form New Animal Drugs
SELENIUM	524.2101	SELENIUM DISULFIDE SUSPENSION Ophthalmic And Topical Dosage Form New Animal Drugs

SELENIUM	573.920	SELENIUM Food Additive Listing; Food Additives Permitted In Feed And Drinking Water Of Animals
SELF-MEDICATION	250.12	STRAMONIUM PREPARATIONS LABELED WITH DIRECTIONS FOR USE IN SELF-MEDICATION REGARDED AS MISBRANDED Drugs Regarded As Misbranded; Specific Requirements For Specific Human Drugs
SELF-PRESSURIZED	2.125	USE OF CHLOROFLUOROCARBON PROPELLANTS IN SELF-PRESSURIZED CONTAINERS Specific Provisions; General Administrative Rulings And Decisions
SELF-PRESSURIZED	740.11	COSMETICS IN SELF-PRESSURIZED CONTAINERS Warning Statements; Cosmetic Product Warning Statements
SELF-RETAINING	882.4800	SELF-RETAINING RETRACTOR FOR NEUROSURGERY Neurological Surgical Devices; Neurological Devices
SELF-RISING CORN MEAL	137.270	SELF-RISING WHITE CORN MEAL Requirements For Specific Standardized Cereal Flours And Related Products; Cereal Flours And Related Products
SELF-RISING CORN MEAL	137.290	SELF-RISING YELLOW CORN MEAL Requirements For Specific Standardized Cereal Flours And Related Products; Cereal Flours And Related Products
SELF-RISING FLOUR	137.180	SELF-RISING FLOUR Requirements For Specific Standardized Cereal Flours And Related Products; Cereal Flours And Related Products
SELF-RISING FLOUR	137.185	ENRICHED SELF-RISING FLOUR Requirements For Specific Standardized Cereal Flours And Related Products; Cereal Flours And Related Products
SEMDURAMICIN	556.597	SEMDURAMICIN Tolerances For Residues Of New Animal Drugs In Food
SEMDURAMICIN	558.555	SEMDURAMICIN New Animal Drugs For Use In Animal Feeds
SEMIPERMEABLE MEMBRANE	874.3930	TYMPANOSTOMY TUBE WITH SEMIPERMEABLE MEMBRANE Prosthetic Devices; Ear, Nose And Throat Devices
SEMINAL FLUID (SPERM)	866.5800	SEMINAL FLUID (SPERM) IMMUNOLOGICAL TEST SYSTEM Immunological Test Systems; Immunology And Microbiology Devices
SEMIRIGID CONTAINERS	164.120	SHELLED NUTS IN RIGID OR SEMIRIGID CONTAINERS Requirements For Specific Standardized Tree Nut And Peanut Products; Tree Nut And Peanut Products
SEMIRIGID PLASTICS	177.1010	ACRYLIC AND MODIFIED ACRYLIC PLASTICS, SEMIRIGID AND RIGID Substances For Use As Basic Components Of Single And Repeated Use Food Contact Surfaces; Indirect Food Additives: Polymers
SEMIRIGID PLASTICS	178.3790	POLYMER MODIFIERS IN SEMIRIGID AND RIGID VINYL CHLORIDE PLASTICS Certain Adjuvants And Production Aids; Indirect Food Additives: Adjuvants, Production Aids, And Sanitizers
SEMISOFT CHEESES	133.187	SEMISOFT CHEESES Requirements For Specific Standardized Cheese And Related Products; Cheeses And Related Cheese Products
SEMISOFT CHEESES	133.188	SEMISOFT PART-SKIM CHEESES Requirements For Specific Standardized Cheese And Related Products; Cheeses And Related Cheese Products
SEMOLINA	137.320	SEMOLINA Requirements For Specific Standardized Cereal Flours And Related Products; Cereal Flours And Related Products
SENSING	870.4340	CARDIOPULMONARY BYPASS LEVEL SENSING MONITOR AND/OR CONTROL Cardiovascular Surgical Devices; Cardiovascular Devices
SENSITIVITY INDEX	874.1070	SHORT INCREMENT SENSITIVITY INDEX (SISI) ADAPTER Diagnostic Devices; Ear, Nose, And Throat Devices
SENSOR	870.4410	CARDIOPULMONARY BYPASS IN-LINE BLOOD GAS SENSOR Cardiovascular Surgical Devices; Cardiovascular Devices
SEPARATE REGISTRATION	1309.22	SEPARATE REGISTRATION FOR INDEPENDENT ACTIVITIES Registration Of Manufacturers, Distributors, Importers And Exporters Of List I Chemicals
SEPARATE REGISTRATION	1309.23	SEPARATE REGISTRATION FOR SEPARATE LOCATIONS Registration Of Manufacturers, Distributors, Importers And Exporters Of List I Chemicals
SEPARATE REGISTRATIONS	1301.12	SEPARATE REGISTRATIONS FOR SEPARATE LOCATIONS Registration Of Manufacturers, Distributors, And Dispensers Of Controlled Substances
SEPARATION OF FUNCTIONS	10.55	SEPARATION OF FUNCTIONS; EX PARTE COMMUNICATIONS General Administrative Procedures; Administrative Practices And Procedures
SEPARATION OF FUNCTIONS	13.15	SEPARATION OF FUNCTIONS; EX PARTE COMMUNICATIONS; ADMINISTRATIVE SUPPORT General Provisions; Public Hearing Before A Public Board Of Inquiry
SEPARATION MATERIAL	862.2230	CHROMATOGRAPHIC SEPARATION MATERIAL FOR CLINICAL USE Clinical Laboratory Instruments; Clinical Chemistry And Clinical Toxicology Devices

SEPARATION MEDIUM	864.8500	LYMPHOCYTE SEPARATION MEDIUM Hematology Reagents; Hematology And Pathology Devices
SEPARATOR	864.9245	AUTOMATED BLOOD CELL SEPARATOR Products Used In Establishments That Manufacture Blood And Blood Products; Hematology And Pathology Devices
SEPTOSTOMY	870.5175	SEPTOSTOMY CATHETER Cardiovascular Therapeutic Devices; Cardiovascular Devices
SERA	864.2800	ANIMAL AND HUMAN SERA Cell And Tissue Culture Products; Hematology And Pathology Devices
SERINE	582.5701	SERINE Nutrients And/or Dietary Supplements; Substances Generally Recognized As Safe
SEROLOGICAL	660.53	CONTROLS FOR SEROLOGICAL PROCEDURES Anti-human Globulin; Additional Standards For Diagnostic Substances For Laboratory Tests
SEROLOGICAL	866.3010	ACINETOBACTER CALCOACETICUS SEROLOGICAL REAGENTS Serological Reagents; Immunology And Microbiology Devices
SEROLOGICAL	866.3020	ADENOVIRUS SEROLOGICAL REAGENTS Serological Reagents; Immunology And Microbiology Devices
SEROLOGICAL	866.3035	ARIZONA SPP. SEROLOGICAL REAGENTS Serological Reagents; Immunology And Microbiology Devices
SEROLOGICAL	866.3040	ASPERGILLUS SPP. SEROLOGICAL REAGENTS Serological Reagents; Immunology And Microbiology Devices
SEROLOGICAL	866.3060	BLASTOMYCES DERMATITIDIS SEROLOGICAL REAGENTS Serological Reagents; Immunology And Microbiology Devices
SEROLOGICAL	866.3065	BORDETELLA SPP. SEROLOGICAL REAGENTS Serological Reagents; Immunology And Microbiology Devices
SEROLOGICAL	866.3085	BRUCELLA SPP. SEROLOGICAL REAGENTS Serological Reagents; Immunology And Microbiology Devices
SEROLOGICAL	866.3110	CAMPYLOBACTER FETUS SEROLOGICAL REAGENTS Serological Reagents; Immunology And Microbiology Devices
SEROLOGICAL	866.3120	CHLAMYDIA SEROLOGICAL REAGENTS Serological Reagents; Immunology And Microbiology Devices
SEROLOGICAL	866.3125	CITROBACTER SPP. SEROLOGICAL REAGENTS Serological Reagents; Immunology And Microbiology Devices
SEROLOGICAL	866.3135	COCCIDIOIDES IMMITIS SEROLOGICAL REAGENTS Serological Reagents; Immunology And Microbiology Devices
SEROLOGICAL	866.3140	CORYNEBACTERIUM SPP. SEROLOGICAL REAGENTS Serological Reagents; Immunology And Microbiology Devices
SEROLOGICAL	866.3145	COXSACKIEVIRUS SEROLOGICAL REAGENTS Serological Reagents; Immunology And Microbiology Devices
SEROLOGICAL	866.3165	CRYPTOCOCCUS NEOFORMANS SEROLOGICAL REAGENTS Serological Reagents; Immunology And Microbiology Devices
SEROLOGICAL	866.3175	CYTOMEGALOVIRUS SEROLOGICAL REAGENTS Serological Reagents; Immunology And Microbiology Devices
SEROLOGICAL	866.3200	ECHINOCOCCUS SPP. SEROLOGICAL REAGENTS Serological Reagents; Immunology And Microbiology Devices
SEROLOGICAL	866.3205	ECHOVIRUS SEROLOGICAL REAGENTS Serological Reagents; Immunology And Microbiology Devices
SEROLOGICAL	866.3220	ENTAMOEBA HISTOLYTICA SEROLOGICAL REAGENTS Serological Reagents; Immunology And Microbiology Devices
SEROLOGICAL	866.3235	EPSTEIN-BARR VIRUS SEROLOGICAL REAGENTS Serological Reagents; Immunology And Microbiology Devices
SEROLOGICAL	866.3240	EQUINE ENCEPHALOMYELITIS VIRUS SEROLOGICAL REAGENTS Serological Reagents; Immunology And Microbiology Devices
SEROLOGICAL	866.3250	ERYSIPELOTHRIX RHUSIOPATHIAE SEROLOGICAL REAGENTS Serological Reagents; Immunology And Microbiology Devices
SEROLOGICAL	866.3255	ESCHERICHIA COLI SEROLOGICAL REAGENTS Serological Reagents; Immunology And Microbiology Devices
SEROLOGICAL	866.3270	FLAVOBACTERIUM SPP. SEROLOGICAL REAGENTS Serological Reagents; Immunology And Microbiology Devices
SEROLOGICAL	866.3280	FRANCISELLA TULARENSIS SEROLOGICAL REAGENTS Serological Reagents; Immunology And Microbiology Devices
SEROLOGICAL	866.3300	HAEMOPHILUS SPP. SEROLOGICAL REAGENTS Serological Reagents; Immunology And Microbiology Devices
SEROLOGICAL	866.3305	HERPES SIMPLEX VIRUS SEROLOGICAL REAGENTS Serological Reagents; Immunology And Microbiology Devices
SEROLOGICAL	866.3320	HISTOPLASMA CAPSULATUM SEROLOGICAL REAGENTS Serological Reagents; Immunology And Microbiology Devices
SEROLOGICAL	866.3330	INFLUENZA VIRUS SEROLOGICAL REAGENTS Serological Reagents; Immunology And Microbiology Devices
SEROLOGICAL	866.3340	KLEBSIELLA SPP. SEROLOGICAL REAGENTS Serological Reagents; Immunology And Microbiology Devices
SEROLOGICAL	866.3350	LEPTOSPIRA SPP. SEROLOGICAL REAGENTS Serological Reagents; Immunology And Microbiology Devices

SEROLOGICAL	866.3355	LISTERIA SPP. SEROLOGICAL REAGENTS Serological Reagents; Immunology And Microbiology Devices
SEROLOGICAL	866.3360	LYMPHOCYTIC CHORIOMENINGITIS VIRUS SEROLOGICAL REAGENTS Serological Reagents; Immunology And Microbiology Devices
SEROLOGICAL	866.3375	MYCOPLASMA SPP. SEROLOGIAL REAGENTS Serological Reagents; Immunology And Microbiology Devices
SEROLOGICAL	866.3380	MUMPS VIRUS SEROLOGICAL REAGENTS Serological Reagents; Immunology And Microbiology Devices
SEROLOGICAL	866.3390	NEISSERIA SPP. DIRECT SEROLOGICAL TEST REAGENTS Serological Reagents; Immunology And Microbiology Devices
SEROLOGICAL	866.3400	PARAINFLUENZA VIRUS SEROLOGICAL REAGENTS Serological Reagents; Immunology And Microbiology Devices
SEROLOGICAL	866.3405	POLIOVIRUS SEROLOGICAL REAGENTS Serological Reagents; Immunology And Microbiology Devices
SEROLOGICAL	866.3410	PROTEUS SPP. (WEIL-FELIX) SEROLOGICAL REAGENTS Serological Reagents; Immunology And Microbiology Devices
SEROLOGICAL	866.3415	PSEUDOMONAS SPP. SEROLOGICAL REAGENTS Serological Reagents; Immunology And Microbiology Devices
SEROLOGICAL	866.3470	REOVIRUS SEROLOGICAL REAGENTS Serological Reagents; Immunology And Microbiology Devices
SEROLOGICAL	866.3480	RESPIRATORY SYNCYTIAL VIRUS SEROLOGICAL REAGENTS Serological Reagents; Immunology And Microbiology Devices
SEROLOGICAL	866.3490	RHINOVIRUS SEROLOGICAL REAGENTS Serological Reagents; Immunology And Microbiology Devices
SEROLOGICAL	866.3500	RICKETTSIA SEROLOGICAL REAGENTS Serological Reagents; Immunology And Microbiology Devices
SEROLOGICAL	866.3510	RUBELLA VIRUS SEROLOGICAL REAGENTS Serological Reagents; Immunology And Microbiology Devices
SEROLOGICAL	866.3520	RUBEOLA (MEASLES) VIRUS SEROLOGICAL REAGENTS Serological Reagents; Immunology And Microbiology Devices
SEROLOGICAL	866.3550	SALMONELLA SPP. SEROLOGICAL REAGENTS Serological Reagents; Immunology And Microbiology Devices
SEROLOGICAL	866.3600	SCHISTOSOMA SPP. SEROLOGICAL REAGENTS Serological Reagents; Immunology And Microbiology Devices
SEROLOGICAL	866.3630	SERRATIA SPP. SEROLOGICAL REAGENTS Serological Reagents; Immunology And Microbiology Devices
SEROLOGICAL	866.3660	SHIGELLA SPP. SEROLOGICAL REAGENTS Serological Reagents; Immunology And Microbiology Devices
SEROLOGICAL	866.3680	SPOROTHRIX SCHENCKII SEROLOGICAL REAGENTS Serological Reagents; Immunology And Microbiology Devices
SEROLOGICAL	866.3700	STAPHYLOCOCCUS AUREUS SEROLOGICAL REAGENTS Serological Reagents; Immunology And Microbiology Devices
SEROLOGICAL	866.3740	STREPTOCOCCUS SPP. SEROLOGICAL REAGENTS Serological Reagents; Immunology And Microbiology Devices
SEROLOGICAL	866.3780	TOXOPLASMA GONDII SEROLOGICAL REAGENTS Serological Reagents; Immunology And Microbiology Devices
SEROLOGICAL	866.3850	TRICHINELLA SPIRALIS SEROLOGICAL REAGENTS Serological Reagents; Immunology And Microbiology Devices
SEROLOGICAL	866.3870	TRYPANOSOMA SPP. SEROLOGICAL REAGENTS Serological Reagents; Immunology And Microbiology Devices
SEROLOGICAL	866.3900	VARICELLA-ZOSTER VIRUS SEROLOGICAL REAGENTS Serological Reagents; Immunology And Microbiology Devices
SEROLOGICAL	866.3930	VIBRIO CHOLERAE SEROLOGICAL REAGENTS Serological Reagents; Immunology And Microbiology Devices
SEROTONIN	862.1390	5-HYDROXYINDOLE ACETIC ACID/SEROTONIN TEST SYSTEM Clinical Chemistry Test Systems; Clinical Chemistry And Clinical Toxicology Devices
SERRATIA SPP.	866.3630	SERRATIA SPP. SEROLOGICAL REAGENTS Serological Reagents; Immunology And Microbiology Devices
SERUM	866.5700	WHOLE HUMAN PLASMA OR SERUM IMMUNOLOGICAL TEST SYSTEM Immunological Test Systems; Immunology And Microbiology Devices
SERUM GONADOTROPIN	522.1079	SERUM GONADOTROPIN AND CHORIONIC GONADOTROPIN Implantation Or Injectable Dosage Form New Animal Drugs
SERUMS	610.16	TOTAL SOLIDS IN SERUMS General Provisions; General Biological Products Standards
SERVICE	5.26	SERVICE FELLOWSHIPS General Redelegations Of Authority; Delegations Of Authority And Organization
SERVICE	12.80	FILING AND SERVICE OF SUBMISSIONS Hearing Procedures; Formal Evidentiary Public Hearing
SERVICE	17.7	SERVICE OF COMPLAINT Civil Money Penalties Hearings
SERVICE	17.31	FORM, FILING AND SERVICE OF PAPERS Civil Money Penalties Hearings

SERVICE	514.121	SERVICE OF NOTICES AND ORDERS Administrative Actions On Applications; New Animal Drug Applications
SERVICE	810.4	SERVICE OF ORDERS Medical Device Recall Authority
SERVICE	814.17	SERVICE OF ORDERS General; Premarket Approval Of Medical Devices
SERVICE	1005.25	SERVICE OF PROCESS ON MANUFACTURERS Bonding And Compliance Procedures; Importation Of Electronic Products
SERVICE, CERTIFICATION	80.34	AUTHORITY TO REFUSE CERTIFICATION SERVICE Certification Procedures; Color Additive Certification
SERVICE, GOVERNMENT	19.6	CODE OF ETHICS FOR GOVERNMENT SERVICE General Provisions; Standards Of Conduct And Conflicts Of Interest
SERVICE, INFORMATION	20.50	AVAILABILITY OF RECORDS AT NATIONAL TECHNICAL INFORMATION SERVICE Procedures And Fees; Public Information
SERVICES	515.26	SERVICES OF NOTICES AND ORDERS Administrative Actions On Licenses; Medicated Feed Mill License
SERVICES, CERTIFICATION	80.10	FEES FOR CERTIFICATION SERVICES General Procedures; Color Additive Certification
SERVICES, TRANSFUSION	610.47	" LOOK BACK" NOTIFICATION REQUIREMENTS FOR TRANSFUSION SERVICES Testing Requirements For Communicable Disease Agents; General Biological Products Standards
SERVICING	820.200	SERVICING Servicing; Quality System Regulation
SERVINGS, NUMBER OF	501.8	LABELING OF ANIMAL FOOD WITH NUMBER OF SERVINGS General Provisions; Animal Food Labeling
SESQUICARBONATE	184.1792	SODIUM SESQUICARBONATE Listing Of Specific Substances Affirmed As GRAS; Direct Food Substances Affirmed As Generally Recognized As Safe
SESQUICARBONATE	582.1792	SODIUM SESQUICARBONATE General Purpose Food Additives; Substances Generally Recognized As Safe
SET, CALIBRATION	874.1080	AUDIOMETER CALIBRATION SET Diagnostic Devices; Ear, Nose, And Throat Devices
SET, INTRAVASCULAR ADMINISTRATION	880.5440	INTRAVASCULAR ADMINISTRATION SET General Hospital And Personal Use Therapeutic Devices; General Hospital And Personal Use Devices
SET, TRANSFER	864.9875	TRANSFER SET Products Used In Establishments That Manufacture Blood And Blood Products; Hematology And Pathology Devices
SETTLEMENT	1404.315	SETTLEMENT AND VOLUNTARY EXCLUSION Governmentwide Debarment And Suspension And Requirements For Drug-Free Workplace (Grants)
SETTLEMENT PROPOSALS	10.110	SETTLEMENT PROPOSALS General Administrative Procedures; Administrative Practices And Procedures
SEVERELY-DEBILITATING	312.84	RISK-BENEFIT ANALYSIS IN REVIEW OF MARKETING APPLICATIONS FOR DRUGS TO TREAT LIFE-THREATENING AND SEVERELY-DEBILITATING ILLNESSES Investigational New Drug Application
SEVOFLURANE	529.2150	SEVOFLURANE Certain Other Dosage Form New Animal Drugs
SEWAGE	211.50	SEWAGE AND REFUSE Buildings And Facilities; Current Good Manufacturing Practice For Finished Pharmaceuticals
SGOT	862.1100	ASPARTATE AMINO TRANSFERASE (AST/SGOT) TEST SYSTEM Clinical Chemistry Test Systems; Clinical Chemistry And Clinical Toxicology Devices
SGPT	862.1030	ALANINE AMINO TRANSFERASE (ALT/SGPT) TEST SYSTEM Clinical Chemistry Test Systems; Clinical Chemistry And Clinical Toxicology Devices
SHADE, TOOTH	872.3690	TOOTH SHADE RESIN MATERIAL Prosthetic Devices; Dental Devices
SHEANUT OIL	184.1702	SHEANUT OIL Listing Of Specific Substances Affirmed As GRAS; Direct Food Substances Affirmed As Generally Recognized As Safe
SHEATH	884.5320	GLANS SHEATH Obstetrical And Gynecological Therapeutic Devices; Obstetrical And Gynecological Devices
SHEEP	500.65	EPINEPHRINE INJECTION FOR EMERGENCY TREATMENT OF ANAPHYLACTOID SHOCK IN CATTLE, HORSES, SHEEP, AND SWINE Requirements For Specific Animal Drugs; General

SHEEP'S MILK	133.184	ROQUEFORT, SHEEP'S MILK BLUE-MOLD, AND BLUE-MOLD CHEESE FROM SHEEP'S MILK Requirements For Specific Standardized Cheese And Related Products; Cheeses And Related Cheese Products
SHEET	520.1450c	MORANTEL TARTRATE SUSTAINED-RELEASE TRILAMINATE CYLINDER/SHEET Oral Dosage Form New Animal Drugs
SHEET	880.5180	BURN SHEET General Hospital And Personal Use Therapeutic Devices; General Hospital And Personal Use Devices
SHELL EGGS	115.50	REFRIGERATION OF SHELL EGGS HELD FOR RETAIL DISTRIBUTION Shell Eggs
SHELLAC	872.6200	BASE PLATE SHELLAC Miscellaneous Devices; Dental Devices
SHELLED NUTS	164.120	SHELLED NUTS IN RIGID OR SEMIRIGID CONTAINERS Requirements For Specific Standardized Tree Nut And Peanut Products; Tree Nut And Peanut Products
SHELLFISH	1240.60	MOLLUSCAN SHELLFISH Specific Administrative Decisions Regarding Interstate Shipments; Control Of Communicable Diseases
SHERBET	135.140	SHERBET Requirements For Specific Standardized Frozen Desserts; Frozen Desserts
SHIELD	880.5630	NIPPLE SHIELD General Hospital And Personal Use Therapeutic Devices; General Hospital And Personal Use Devices
SHIELD	886.4750	OPHTHALMIC EYE SHIELD Surgical Devices; Ophthalmic Devices
SHIELD	892.6500	PERSONNEL PROTECTIVE SHIELD Miscellaneous Devices; Radiology Devices
SHIELDING	1000.50	RECOMMENDATION FOR THE USE OF SPECIFIC AREA GONAD SHIELDING ON PATIENTS DURING MEDICAL DIAGNOSTIC X-RAY PROCEDURES Radiation Protection Recommendations; General
SHIGELLA SPP.	866.3660	SHIGELLA SPP. SEROLOGICAL REAGENTS Serological Reagents; Immunology And Microbiology Devices
SHIPMENT	2.35	USE OF SECONDHAND CONTAINERS FOR THE SHIPMENT OR STORAGE OF FOOD AND ANIMAL FEED Human and Animal Foods; General Administrative Rulings And Decisions
SHIPMENT	130.17	TEMPORARY PERMITS FOR INTERSTATE SHIPMENT OF EXPERIMENTAL PACKS OF FOOD VARYING FROM THE REQUIREMENTS OF DEFINITIONS AND STANDARDS OF IDENTITY General Provisions; Food Standards: General
SHIPMENT	600.15	TEMPERATURES DURING SHIPMENT Establishment Standards; Biological Products: General
SHIPMENTS	1312.15	SHIPMENTS IN GREATER OR LESS AMOUNT THAN AUTHORIZED Importation Of Controlled Substances; Importation And Exportation Of Controlled Substances
SHIPMENTS	1313.41	SUSPENSION OF SHIPMENTS Importation And Exportation Of Precursors And Essential Chemicals
SHIPPED	640.76	PRODUCTS STORED OR SHIPPED AT UNACCEPTABLE TEMPERATURES Source Plasma; Additional Standards For Human Blood And Blood Products
SHIPPING	203.36	FULFILLMENT HOUSES, SHIPPING AND MAILING SERVICES Prescription Drug Marketing
SHOCK, ANAPHYLACTOID	500.65	EPINEPHRINE INJECTION FOR EMERGENCY TREATMENT OF ANAPHYLACTOID SHOCK IN CATTLE, HORSES, SHEEP, AND SWINE Requirements For Specific Animal Drugs; General
SHOCK WAVE	876.5990	EXTRACORPOREAL SHOCK WAVE LITHOTRIPTER Therapeutic Devices; Gastroenterology-urology Devices
SHORT SUPPLY	601.22	PRODUCTS IN SHORT SUPPLY; INITIAL MANUFACTURING AT OTHER THAN LICENSED LOCATION Biologics Licensing; Licensing
SHORTWAVE	890.5290	SHORTWAVE DIATHERMY Physical Medicine Therapeutic Devices; Physical Medicine Devices
SHOULDER JOINT	888.3640	SHOULDER JOINT METAL/METAL OR METAL/POLYMER CONSTRAINED CEMENTED PROSTHESIS Prosthetic Devices; Orthopedic Devices
SHOULDER JOINT	888.3650	SHOULDER JOINT METAL/POLYMER NON-CONSTRAINED CEMENTED PROSTHESIS Prosthetic Devices; Orthopedic Devices
SHOULDER JOINT	888.3660	SHOULDER JOINT METAL/POLYMER SEMI-CONSTRAINED CEMENTED PROSTHESIS Prosthetic Devices; Orthopedic Devices
SHOULDER JOINT	888.3670	SHOULDER JOINT METAL/POLYMER/METAL NONCONSTRAINED OR SEMI-CONSTRAINED POROUS-COATED UNCEMENTED PROSTHESIS Prosthetic Devices; Orthopedic Devices

SHOULDER JOINT	888.3680	SHOULDER JOINT GLENOID (HEMI-SHOULDER) METALLIC CEMENTED PROSTHESIS Prosthetic Devices; Orthopedic Devices
SHOULDER JOINT	888.3690	SHOULDER JOINT HUMERAL (HEMI-SHOULDER) METALLIC UNCEMENTED PROSTHESIS Prosthetic Devices; Orthopedic Devices
SHOW CAUSE	1301.37	ORDER TO SHOW CAUSE Action On Application For Registration: Revocation Or Suspension Of Registration; Registration Of Manufacturers, Distributors, And Dispensers Of Controlled Substances
SHOW CAUSE	1309.46	ORDER TO SHOW CAUSE Registration Of Manufacturers, Distributors, Importers And Exporters Of List I Chemicals
SHRIMP	102.55	NONSTANDARDIZED BREADED COMPOSITE SHRIMP UNITS Requirements For Specific Nonstandardized Foods; Common Or Usual Name For Nonstandardized Foods
SHRIMP	161.173	CANNED WET PACKED SHRIMP IN TRANSPARENT OR NONTRANSPARENT CONTAINERS Requirements For Specific Standardized Fish And Shellfish; Fish And Shellfish
SHRIMP	161.175	FROZEN RAW BREADED SHRIMP Requirements For Specific Standardized Fish And Shellfish; Fish And Shellfish
SHRIMP	161.176	FROZEN RAW LIGHTLY BREADED SHRIMP Requirements For Specific Standardized Fish And Shellfish; Fish And Shellfish
SHUNT	874.3820	ENDOLYMPHATIC SHUNT Prosthetic Devices; Ear, Nose And Throat Devices
SHUNT	874.3850	ENDOLYMPHATIC SHUNT TUBE WITH VALVE Prosthetic Devices; Ear, Nose And Throat Devices
SHUNT	876.5955	PERITONEO-VENOUS SHUNT Therapeutic Devices; Gastroenterology-urology Devices
SHUNT	882.4545	SHUNT SYSTEM IMPLANTATION INSTRUMENT Neurological Surgical Devices; Neurological Devices
SHUNT	882.5550	CENTRAL NERVOUS SYSTEM FLUID SHUNT AND COMPONENTS Neurological Therapeutic Devices; Neurological Devices
SHUNT	886.3920	AQUEOUS SHUNT Prosthetic Devices; Ophthalmic Devices
SICKLE CELL	864.7825	SICKLE CELL TEST Hematology Kits And Packages; Hematology And Pathology Devices
SICILIANO CHEESE	133.111	CACIOCAVALLO SICILIANO CHEESE Requirements For Specific Standardized Cheese And Related Products; Cheeses And Related Cheese Products
SIEVE RESINS	173.40	MOLECULAR SIEVE RESINS Polymer Substances For Food Treatment; Secondary Direct Food Additives Permitted In Food For Human Consumption
SIGNAL	870.2050	BIOPOTENTIAL AMPLIFIER AND SIGNAL CONDITIONER Cardiovascular Monitoring Devices; Cardiovascular Devices
SIGNAL	870.2060	TRANSDUCER SIGNAL AMPLIFIER AND SIGNAL CONDITIONER Cardiovascular Monitoring Devices; Cardiovascular Devices
SIGNAL	870.2600	SIGNAL ISOLATION SYSTEM Cardiovascular Monitoring Devices; Cardiovascular Devices
SIGNAL	870.2910	RADIOFREQUENCY PHYSIOLOGICAL SIGNAL TRANSMITTER AND RECEIVER Cardiovascular Monitoring Devices; Cardiovascular Devices
SIGNAL	882.1420	ELECTROENCEPHALOGRAM (EEG) SIGNAL SPECTRUM ANALYZER Neurological Diagnostic Devices; Neurological Devices
SIGNAL	882.1430	ELECTROENCEPHALOGRAPH TEST SIGNAL GENERATOR Neurological Diagnostic Devices; Neurological Devices
SIGNAL	882.1835	PHYSIOLOGICAL SIGNAL AMPLIFIER Neurological Diagnostic Devices; Neurological Devices
SIGNAL	882.1845	PHYSIOLOGICAL SIGNAL CONDITIONER Neurological Diagnostic Devices; Neurological Devices
SIGNATURE	11.50	SIGNATURE MANIFESTATIONS Electronic Records; Electronic Signatures
SIGNATURE	1309.32	APPLICATION FORMS; CONTENTS; SIGNATURE Registration Of Manufacturers, Distributors, Importers And Exporters Of List I Chemicals
SIGNATURE, ELECTRONIC	11.200	ELECTRONIC SIGNATURE COMPONENTS AND CONTROLS Electronic Records; Electronic Signatures
SIGNATURE LINKING	11.70	SIGNATURE/RECORD LINKING Electronic Records; Electronic Signatures
SIGNIFICANT DECISIONS	10.70	DOCUMENTATION OF SIGNIFICANT DECISIONS IN ADMINISTRATIVE FILE General Administrative Procedures; Administrative Practices And Procedures
SIGNIFICANT IMPACT	25.41	FINDING OF NO SIGNIFICANT IMPACT Preparation Of Environmental Documents; Environmental Impact Considerations

SIGNIFICANT IMPACT	25.51	ENVIRONMENTAL ASSESSMENTS AND FINDINGS OF NO SIGNIFICANT IMPACT Public Participation And Notification Of Environmental Documents; Environmental Impact Considerations
SIGNIFICANT RISK	812.66	SIGNIFICANT RISK DEVICE DETERMINATIONS IRB Review And Approval; Investigational Device Exemptions
SILICA	182.1711	SILICA AEROGEL Multiple Purpose GRAS Food Substances; Substances Generally Recognized As Safe
SILICA	582.1711	SILICA AEROGEL General Purpose Food Additives; Substances Generally Recognized As Safe
SILICAS	584.700	HYDROPHOBIC SILICAS Food Substances Affirmed As Generally Recognized As Safe In Feed And Drinking Water
SILICATE	872.6670	SILICATE PROTECTOR Miscellaneous Devices; Dental Devices
SILICATE, ALUMINUM CALCIUM	182.2122	ALUMINUM CALCIUM SILICATE Anticaking Agents; Substances Generally Recognized As Safe
SILICATE, ALUMINUM CALCIUM	582.2122	ALUMINUM CALCIUM SILICATE Anticaking Agents; Substances Generally Recognized As Safe
SILICATE, CALCIUM	172.410	CALCIUM SILICATE Anticaking Agents; Food Additives Permitted For Direct Addition To Food For Human Consumption
SILICATE, CALCIUM	182.2227	CALCIUM SILICATE Anticaking Agents; Substances Generally Recognized As Safe
SILICATE, CALCIUM	573.260	CALCIUM SILICATE Food Additive Listing; Food Additives Permitted In Feed And Drinking Water Of Animals
SILICATE, CALCIUM	582.2227	CALCIUM SILICATE Anticaking Agents; Substances Generally Recognized As Safe
SILICATE, MAGNESIUM	182.2437	MAGNESIUM SILICATE Anticaking Agents; Substances Generally Recognized As Safe
SILICATE, MAGNESIUM	582.2437	MAGNESIUM SILICATE Anticaking Agents; Substances Generally Recognized As Safe
SILICATE, TRICALCIUM	182.2906	TRICALCIUM SILICATE Anticaking Agents; Substances Generally Recognized As Safe
SILICATE, TRICALCIUM	582.2906	TRICALCIUM SILICATE Anticaking Agents; Substances Generally Recognized As Safe
SILICON DIOXIDE	172.480	SILICON DIOXIDE Anticaking Agents; Food Additives Permitted For Direct Addition To Food For Human Consumption
SILICON DIOXIDE	175.390	ZINC-SILICON DIOXIDE MATRIX COATINGS Substances For Use As Components Of Coatings; Indirect Food Additives
SILICON DIOXIDE	573.940	SILICON DIOXIDE Food Additive Listing; Food Additives Permitted In Feed And Drinking Water Of Animals
SILICONE PROSTHESIS	878.3530	SILICONE INFLATABLE BREAST PROSTHESIS Prosthetic Devices; General And Plastic Surgery Devices
SILICONE PROSTHESIS	878.3540	SILICONE GEL-FILLED BREAST PROSTHESIS Prosthetic Devices; General And Plastic Surgery Devices
SILK, CORN	184.1262	CORN SILK AND CORN SILK EXTRACT Listing Of Specific Substances Affirmed As GRAS; Direct Food Substances Affirmed As Generally Recognized As Safe
SILK SUTURE	878.5030	NATURAL NONABSORBABLE SILK SURGICAL SUTURE Surgical Devices; General And Plastic Surgery Devices
SILVER	73.2500	SILVER Cosmetics; Listing Of Color Additives Exempt From Certification
SILVER	310.548	DRUG PRODUCTS CONTAINING COLLOIDAL SILVER INGREDIENTS OR SILVER SALTS OFFERED OTC FOR THE TREATMENT AND/OR PREVENTION OF DISEASE Requirements For Specific New Drugs Or Devices; New Drugs
SILVER SULFADIAZINE	524.802	ENROFLOXACIN, SILVER SULFADIAZINE SOLUTION Ophthalmic And Topical Dosage Form New Animal Drugs
SIMPLE DRILLS	882.4310	POWERED SIMPLE CRANIAL DRILLS, BURRS, TREPHINES, AND THEIR ACCESSORIES Neurological Surgical Devices; Neurological Device
SIMULATAN	886.1840	SIMULATAN (INCLUDING CROSSED CYLINDER) Diagnostic Devices; Ophthalmic Devices
SIMULATION	892.5840	RADIATION THERAPY SIMULATION SYSTEM Therapeutic Devices; Radiology Devices
SINGLE-DOSE	320.26	GUIDELINES ON THE DESIGN OF A SINGLE-DOSE IN VIVO BIOAVAILABILITY STUDY Procedures For Determining The Bioavailability Or Bioequivalence Of Drug Products; Bioavailability And Bioequivalence Requirements

SINGLE-FUNCTION	870.1435	SINGLE-FUNCTION PREPROGRAMMED DIAGNOSTIC COMPUTER Cardiovascular Diagnostic Devices; Cardiovascular Devices
SINGLE LUMEN	880.5570	HYPODERMIC SINGLE LUMEN NEEDLE General Hospital And Personal Use Therapeutic Devices; General Hospital And Personal Use Devices
SINUS, CAROTID	870.3850	CAROTID SINUS NERVE STIMULATOR Cardiovascular Prosthetic Devices; Cardiovascular Devices
SIRUP	520.1263	LINCOMYCIN HYDROCHLORIDE MONOHYDRATE TABLETS AND SIRUP Oral Dosage Form New Animal Drugs
SIRUP, CANE	168.130	CANE SIRUP Requirements For Specific Standardized Sweeteners And Table Sirups; Sweeteners And Table Sirups
SIRUP, GLUCOSE	168.120	GLUCOSE SIRUP Requirements For Specific Standardized Sweeteners And Table Sirups; Sweeteners And Table Sirups
SIRUP, GLUCOSE	168.121	DRIED GLUCOSE SIRUP Requirements For Specific Standardized Sweeteners And Table Sirups; Sweeteners And Table Sirups
SIRUP, MAPLE	168.140	MAPLE SIRUP Requirements For Specific Standardized Sweeteners And Table Sirups; Sweeteners And Table Sirups
SIRUP, SORGHUM	168.160	SORGHUM SIRUP Requirements For Specific Standardized Sweeteners And Table Sirups; Sweeteners And Table Sirups
SIRUP, TABLE	168.180	TABLE SIRUP Requirements For Specific Standardized Sweeteners And Table Sirups; Sweeteners And Table Sirups
SISI	874.1070	SHORT INCREMENT SENSITIVITY INDEX (SISI) ADAPTER Diagnostic Devices; Ear, Nose, And Throat Devices
SITUATION, EMERGENCY	290.10	DEFINITION OF EMERGENCY SITUATION General Provisions; Controlled Drugs
SITZ BATH	890.5125	NONPOWERED SITZ BATH Physical Medicine Therapeutic Devices; Physical Medicine Devices
SIZE, EQUIPMENT	211.63	EQUIPMENT DESIGN, SIZE, AND LOCATION Equipment And Facilities; Current Good Manufacturing Practice For Finished Pharmaceuticals
SIZE, PACKAGE	201.307	SODIUM PHOSPHATES, PACKAGE SIZE LIMITATION, WARNINGS AND DIRECTIONS FOR OVER-THE-COUNTER SALE Specific Labeling Requirements For Specific Drug Products; Labeling
SIZE, SYMBOL	1302.04	LOCATION AND SIZE OF SYMBOL ON LABEL Labeling And Packaging Requirements For Controlled Substances
SIZER	870.3945	PROSTHETIC HEART VALVE SIZER Cardiovascular Prosthetic Devices; Cardiovascular Devices
SKIM CHEESE	133.157	PART-SKIM MOZZARELLA AND SCAMORZA CHEESE Requirements For Specific Standardized Cheese And Related Products; Cheeses And Related Cheese Products
SKIM CHEESE	133.158	LOW-MOISTURE PART-SKIM MOZZARELLA AND SCAMORZA CHEESE Requirements For Specific Standardized Cheese And Related Products; Cheeses And Related Cheese Products
SKIM CHEESES	133.188	SEMISOFT PART-SKIM CHEESES Requirements For Specific Standardized Cheese And Related Products; Cheeses And Related Cheese Products
SKIM CHEESES	133.191	PART-SKIM SPICED CHEESES Requirements For Specific Standardized Cheese And Related Products; Cheeses And Related Cheese Products
SKIM MILK CHEESE	133.189	SKIM MILK CHEESE FOR MANUFACTURING Requirements For Specific Standardized Cheese And Related Products; Cheeses And Related Cheese Products
SKIM MILK CHOCOLATE	163.140	SKIM MILK CHOCOLATE Requirements For Specific Standardized Cacao Products; Cacao Products
SKIN CLIP	878.4320	REMOVABLE SKIN CLIP Surgical Devices; General And Plastic Surgery Devices
SKIN DEGREASER	878.4730	SURGICAL SKIN DEGREASER OR ADHESIVE TAPE SOLVENT Surgical Devices; General And Plastic Surgery Devices
SKIN, GRAPE	73.170	GRAPE SKIN EXTRACT (ENOCIANINA) Foods; Listing Of Color Additives Exempt From Certification
SKIN MARKER	878.4660	SKIN MARKER Surgical Devices; General And Plastic Surgery Devices
SKIN POTENTIAL	882.1560	SKIN POTENTIAL MEASUREMENT DEVICE Neurological Diagnostic Devices; Neurological Devices
SKIN PRESSURE	880.6450	SKIN PRESSURE PROTECTORS General Hospital And Personal Use Miscellaneous Devices; General Hospital And Personal Use Devices

SKIN RESPONSE	882.1540	GALVANIC SKIN RESPONSE MEASUREMENT DEVICE Neurological Diagnostic Devices; Neurological Devices
SKIN STAPLE	878.4760	REMOVABLE SKIN STAPLE Surgical Devices; General And Plastic Surgery Devices
SKINBLEACHING	700.13	USE OF MERCURY COMPOUNDS IN COSMETICS INCLUDING USE AS SKINBLEACHING AGENTS IN COSMETIC PREPARATIONS ALSO REGARDED AS DRUGS Requirements For Specific Cosmetic Products; General
SKULL	882.4030	SKULL PLATE ANVIL Neurological Surgical Devices; Neurological Devices
SKULL	882.4460	NEUROSURGICAL HEAD HOLDER (SKULL CLAMP) Neurological Surgical Devices; Neurological Devices
SKULL	882.4750	SKULL PUNCH Neurological Surgical Devices; Neurological Devices
SKULL	882.5960	SKULL TONGS FOR TRACTION Neurological Therapeutic Devices; Neurological Devices
SKULLPLATE	882.4900	SKULLPLATE SCREWDRIVER Neurological Surgical Devices; Neurological Devices
SLCG	862.1187	CONJUGATED SULFOLITHOCHOLIC ACID (SLCG) TEST SYSTEM Clinical Chemistry Test Systems; Clinical Chemistry And Clinical Toxicology Devices
SLEEP-AID	338.10	NIGHTTIME SLEEP-AID ACTIVE INGREDIENTS Active Ingredients; Nighttime Sleep-aid Drug Products For Over-the-counter Human Use
SLEEP-AID	338.50	LABELING OF NIGHTTIME SLEEP-AID DRUG PRODUCTS Labeling; Nighttime Sleep-aid Drug Products For Over-the-counter Human Use
SLEEVE	870.5800	COMPRESSIBLE LIMB SLEEVE Cardiovascular Therapeutic Devices; Cardiovascular Devices
SLIDE SPINNER	864.5850	AUTOMATED SLIDE SPINNER Automated And Semi-automated Hematology Devices; Hematology And Pathology Devices
SLIDE STAINER	864.3800	AUTOMATED SLIDE STAINER Pathology Instrumentation And Accessories; Hematology And Pathology Devices
SLIMICIDES	176.300	SLIMICIDES Substances For Use Only As Components Of Paper And Paperboard; Indirect Food Additives
SLING	890.3640	ARM SLING Physical Medicine Prosthetic Devices; Physical Medicine Devices
SLITLAMP	886.1850	AC-POWERED SLITLAMP BIOMICROSCOPE Diagnostic Devices; Ophthalmic Devices
SMOKED CHUB	172.177	SODIUM NITRITE USED IN PROCESSING SMOKED CHUB Food Preservatives; Food Additives Permitted For Direct Addition To Food For Human Consumption
SMOKING DETERRENT	310.544	DRUG PRODUCTS CONTAINING ACTIVE INGREDIENTS OFFERED OTC FOR USE AS A SMOKING DETERRENT Requirements For Specific New Drugs Or Devices; New Drugs
SNAKEBITE KIT	880.5740	SUCTION SNAKEBITE KIT General Hospital And Personal Use Therapeutic Devices; General Hospital And Personal Use Devices
SNAKEBITE KIT	880.5760	CHEMICAL COLD PACK SNAKEBITE KIT General Hospital And Personal Use Therapeutic Devices; General Hospital And Personal Use Devices
SOAKED CHEESE	133.136	WASHED CURD AND SOAKED CURD CHEESE Requirements For Specific Standardized Cheese And Related Products; Cheeses And Related Cheese Products
SOAP	701.20	DETERGENT SUBSTANCES, OTHER THAN SOAP, INTENDED FOR USE IN CLEANSING THE BODY Labeling Of Specific Ingredients; Cosmetic Labeling
SODA, YELLOW PRUSSIATE	172.490	YELLOW PRUSSIATE OF SODA Anticaking Agents; Food Additives Permitted For Direct Addition To Food For Human Consumption
SODA, YELLOW PRUSSIATE	573.1020	YELLOW PRUSSIATE OF SODA Food Additive Listing; Food Additives Permitted In Feed And Drinking Water Of Animals
SODIUM	101.61	NUTRIENT CONTENT CLAIMS FOR THE SODIUM CONTENT OF FOODS Specific Requirements For Nutrient Content Claims; Food Labeling
SODIUM	101.74	HEALTH CLAIMS: SODIUM AND HYPERTENSION Specific Requirements For Health Claims; Food Labeling
SODIUM	133.116	LOW SODIUM CHEDDAR CHEESE Requirements For Specific Standardized Cheese And Related Products; Cheeses And Related Cheese Products
SODIUM	133.121	LOW SODIUM COLBY CHEESE Requirements For Specific Standardized Cheese And Related Products; Cheeses And Related Cheese Products

SODIUM	184.1697	RIBOFLAVIN-5'-PHOSPHATE (SODIUM) Listing Of Specific Substances Affirmed As GRAS; Direct Food Substances Affirmed As Generally Recognized As Safe
SODIUM	201.64	SODIUM LABELING Labeling Requirements For Over-the-counter Drugs; Labeling
SODIUM	862.1665	SODIUM TEST SYSTEM Clinical Chemistry Test Systems; Clinical Chemistry And Clinical Toxicology Devices
SODIUM ACETATE	184.1721	SODIUM ACETATE Listing Of Specific Substances Affirmed As GRAS; Direct Food Substances Affirmed As Generally Recognized As Safe
SODIUM ACETATE	582.1721	SODIUM ACETATE General Purpose Food Additives; Substances Generally Recognized As Safe
SODIUM, ACETAZOLAMIDE	520.44	ACETAZOLAMIDE SODIUM SOLUBLE POWDER Oral Dosage Form New Animal Drugs
SODIUM ACETAZOLAMIDE	522.44	STERILE SODIUM ACETAZOLAMIDE Implantation Or Injectable Dosage Form New Animal Drugs
SODIUM ACID PHOSPHATE	182.6085	SODIUM ACID PHOSPHATE Sequestrants; Substances Generally Recognized As Safe
SODIUM ACID PHOSPHATE	582.6085	SODIUM ACID PHOSPHATE Sequestrants; Substances Generally Recognized As Safe
SODIUM ACID PYROPHOSPHATE	182.1087	SODIUM ACID PYROPHOSPHATE Multiple Purpose GRAS Food Substances; Substances Generally Recognized As Safe
SODIUM ACID PYROPHOSPHATE	582.1087	SODIUM ACID PYROPHOSPHATE General Purpose Food Additives; Substances Generally Recognized As Safe
SODIUM ALGINATE	184.1724	SODIUM ALGINATE Listing Of Specific Substances Affirmed As GRAS; Direct Food Substances Affirmed As Generally Recognized As Safe
SODIUM ALGINATE	582.7724	SODIUM ALGINATE Stabilizers; Substances Generally Recognized As Safe
SODIUM ALUMINOSILICATE	182.2727	SODIUM ALUMINOSILICATE Anticaking Agents; Substances Generally Recognized As Safe
SODIUM ALUMINOSILICATE	182.2729	SODIUM CALCIUM ALUMINOSILICATE, HYDRATED Anticaking Agents; Substances Generally Recognized As Safe
SODIUM ALUMINOSILICATE	582.2727	SODIUM ALUMINOSILICATE Anticaking Agents; Substances Generally Recognized As Safe
SODIUM ALUMINOSILICATE	582.2729	HYDRATED SODIUM CALCIUM ALUMINOSILICATE Anticaking Agents; Substances Generally Recognized As Safe
SODIUM ALUMINUM PHOSPHATE	182.1781	SODIUM ALUMINUM PHOSPHATE Multiple Purpose GRAS Food Substances; Substances Generally Recognized As Safe
SODIUM ALUMINUM PHOSPHATE	582.1781	SODIUM ALUMINUM PHOSPHATE General Purpose Food Additives; Substances Generally Recognized As Safe
SODIUM, ARSANILATE	558.60	ARSANILATE SODIUM Specific New Animal Drugs For Use In Animal Feeds; New Animal Drugs For Use In Animal Feeds
SODIUM, ARSENAMIDE	522.144	ARSENAMIDE SODIUM AQUEOUS INJECTION Implantation Or Injectable Dosage Form New Animal Drugs
SODIUM ASCORBATE	182.3731	SODIUM ASCORBATE Chemical Preservatives; Substances Generally Recognized As Safe
SODIUM ASCORBATE	582.3731	SODIUM ASCORBATE Chemical Preservatives; Substances Generally Recognized As Safe
SODIUM BENZOATE	184.1733	SODIUM BENZOATE Listing Of Specific Substances Affirmed As GRAS; Direct Food Substances Affirmed As Generally Recognized As Safe
SODIUM BENZOATE	582.3733	SODIUM BENZOATE Chemical Preservatives; Substances Generally Recognized As Safe
SODIUM BICARBONATE	184.1736	SODIUM BICARBONATE Listing Of Specific Substances Affirmed As GRAS; Direct Food Substances Affirmed As Generally Recognized As Safe
SODIUM BICARBONATE	582.1736	SODIUM BICARBONATE General Purpose Food Additives; Substances Generally Recognized As Safe
SODIUM BISULFITE	182.3739	SODIUM BISULFITE Chemical Preservatives; Substances Generally Recognized As Safe
SODIUM BISULFITE	582.3739	SODIUM BISULFITE Chemical Preservatives; Substances Generally Recognized As Safe
SODIUM BORATE	872.3400	KARAYA AND SODIUM BORATE WITH OR WITHOUT ACACIA DENTURE ADHESIVE Prosthetic Devices; Dental Devices
SODIUM CARBONATE	184.1742	SODIUM CARBONATE Listing Of Specific Substances Affirmed As GRAS; Direct Food Substances Affirmed As Generally Recognized As Safe

SODIUM CARBONATE	582.1742	SODIUM CARBONATE General Purpose Food Additives; Substances Generally Recognized As Safe
SODIUM CARBOXYMETHYL-CELLULOSE	182.1745	SODIUM CARBOXYMETHYLCELLULOSE Multiple Purpose GRAS Food Substances; Substances Generally Recognized As Safe
SODIUM CARBOXYMETHYL-CELLULOSE	582.1745	SODIUM CARBOXYMETHYLCELLULOSE General Purpose Food Additives; Substances Generally Recognized As Safe
SODIUM CASEINATE	182.1748	SODIUM CASEINATE Multiple Purpose GRAS Food Substances; Substances Generally Recognized As Safe
SODIUM CASEINATE	582.1748	SODIUM CASEINATE General Purpose Food Additives; Substances Generally Recognized As Safe
SODIUM, CEFTIOFUR	522.313	CEFTIOFUR SODIUM POWDER FOR INJECTION Implantation Or Injectable Dosage Form New Animal Drugs
SODIUM CHLORITE	173.325	ACIDIFIED SODIUM CHLORITE SOLUTIONS Specific Usage Additives; Secondary Direct Food Additives Permitted In Food For Human Consumption
SODIUM CHLORITE	186.1750	SODIUM CHLORITE Listing Of Specific Substances Affirmed As GRAS; Indirect Food Substances Affirmed As Generally Recognized As Safe
SODIUM CITRATE	184.1751	SODIUM CITRATE Listing Of Specific Substances Affirmed As GRAS; Direct Food Substances Affirmed As Generally Recognized As Safe
SODIUM CITRATE	582.1751	SODIUM CITRATE General Purpose Food Additives; Substances Generally Recognized As Safe
SODIUM CITRATE	582.6751	SODIUM CITRATE Sequestrants; Substances Generally Recognized As Safe
SODIUM, CLOPROSTENOL	522.460	CLOPROSTENOL SODIUM Implantation Or Injectable Dosage Form New Animal Drugs
SODIUM, COLISTIMETHATE	522.468	COLISTIMETHATE SODIUM POWDER FOR INJECTION Implantation Or Injectable Dosage Form New Animal Drugs
SODIUM COPPER CHLOROPHYLLIN	73.1125	POTASSIUM SODIUM COPPER CHLOROPHYLLIN (CHLOROPHYLLIN-COPPER COMPLEX) Drugs; Listing Of Color Additives Exempt From Certification
SODIUM COPPER CHLOROPHYLLIN	73.2125	POTASSIUM SODIUM COPPER CHLOROPHYLLIN (CHLOROPHYLLIN-COPPER COMPLEX) Cosmetics; Listing Of Color Additives Exempt From Certification
SODIUM DIACETATE	184.1754	SODIUM DIACETATE Listing Of Specific Substances Affirmed As GRAS; Direct Food Substances Affirmed As Generally Recognized As Safe
SODIUM DIACETATE	582.6754	SODIUM DIACETATE Sequestrants; Substances Generally Recognized As Safe
SODIUM, DIATRIZOATE	520.563	DIATRIZOATE MEGLUMINE AND DIATRIZOATE SODIUM ORAL SOLUTION Oral Dosage Form New Animal Drugs
SODIUM, DICLOXACILLIN	520.608	DICLOXACILLIN SODIUM MONOHYDRATE CAPSULES Oral Dosage Form New Animal Drugs
SODIUM, FLUPROSTENOL	522.995	FLUPROSTENOL SODIUM INJECTION Implantation Or Injectable Dosage Form New Animal Drugs
SODIUM FORMATE	186.1756	SODIUM FORMATE Listing Of Specific Substances Affirmed As GRAS; Indirect Food Substances Affirmed As Generally Recognized As Safe
SODIUM GLUCONATE	182.6757	SODIUM GLUCONATE Sequestrants; Substances Generally Recognized As Safe
SODIUM GLUCONATE	582.6757	SODIUM GLUCONATE Sequestrants; Substances Generally Recognized As Safe
SODIUM HEXAMETAPHOSPHATE	182.6760	SODIUM HEXAMETAPHOSPHATE Sequestrants; Substances Generally Recognized As Safe
SODIUM HEXAMETAPHOSPHATE	582.6760	SODIUM HEXAMETAPHOSPHATE Sequestrants; Substances Generally Recognized As Safe
SODIUM, HYALURONATE	522.1145	HYALURONATE SODIUM INJECTION Implantation Or Injectable Dosage Form New Animal Drugs
SODIUM HYDROXIDE	184.1763	SODIUM HYDROXIDE Listing Of Specific Substances Affirmed As GRAS; Direct Food Substances Affirmed As Generally Recognized As Safe
SODIUM HYDROXIDE	582.1763	SODIUM HYDROXIDE General Purpose Food Additives; Substances Generally Recognized As Safe
SODIUM HYPOPHOSPHITE	184.1764	SODIUM HYPOPHOSPHITE Listing Of Specific Substances Affirmed As GRAS; Direct Food Substances Affirmed As Generally Recognized As Safe
SODIUM LACTATE	184.1768	SODIUM LACTATE Listing Of Specific Substances Affirmed As GRAS; Direct Food Substances Affirmed As Generally Recognized As Safe

SODIUM LAURYL SULFATE	172.822	SODIUM LAURYL SULFATE Multipurpose Additives; Food Additives Permitted For Direct Addition To Food For Human Consumption
SODIUM LIOTHYRONINE	520.1284	SODIUM LIOTHYRONINE TABLETS Oral Dosage Form New Animal Drugs
SODIUM METABISULFITE	182.3766	SODIUM METABISULFITE Chemical Preservatives; Substances Generally Recognized As Safe
SODIUM METABISULFITE	582.3766	SODIUM METABISULFITE Chemical Preservatives; Substances Generally Recognized As Safe
SODIUM METAPHOSPHATE	182.6769	SODIUM METAPHOSPHATE Sequestrants; Substances Generally Recognized As Safe
SODIUM METAPHOSPHATE	582.6769	SODIUM METAPHOSPHATE Sequestrants; Substances Generally Recognized As Safe
SODIUM METASILICATE	184.1769a	SODIUM METASILICATE Listing Of Specific Substances Affirmed As GRAS; Direct Food Substances Affirmed As Generally Recognized As Safe
SODIUM METHYL SULFATE	173.385	SODIUM METHYL SULFATE Specific Usage Additives; Secondary Direct Food Additives Permitted In Food For Human Consumption
SODIUM NAPHTHALENE SULFONATES	172.824	SODIUM MONO- AND DIMETHYL NAPHTHALENE SULFONATES Multipurpose Additives; Food Additives Permitted For Direct Addition To Food For Human Consumption
SODIUM NITRATE	172.170	SODIUM NITRATE Food Preservatives; Food Additives Permitted For Direct Addition To Food For Human Consumption
SODIUM NITRATE	176.320	SODIUM NITRATE-UREA COMPLEX Substances For Use Only As Components Of Paper And Paperboard; Indirect Food Additives
SODIUM NITRATE	181.33	SODIUM NITRATE AND POTASSIUM NITRATE Specific Prior-sanctioned Food Ingredients; Prior-sanctioned Food Ingredients
SODIUM NITRITE	172.175	SODIUM NITRITE Food Preservatives; Food Additives Permitted For Direct Addition To Food For Human Consumption
SODIUM NITRITE	172.177	SODIUM NITRITE USED IN PROCESSING SMOKED CHUB Food Preservatives; Food Additives Permitted For Direct Addition To Food For Human Consumption
SODIUM NITRITE	181.34	SODIUM NITRITE AND POTASSIUM NITRITE Specific Prior-sanctioned Food Ingredients; Prior-sanctioned Food Ingredients
SODIUM NITRITE	573.700	SODIUM NITRITE Food Additive Listing; Food Additives Permitted In Feed And Drinking Water Of Animals
SODIUM NOVOBIOCIN	520.2345f	TETRACYCLINE PHOSPHATE COMPLEX AND SODIUM NOVOBIOCIN CAPSULES Oral Dosage Form New Animal Drugs
SODIUM NOVOBIOCIN	520.2345g	TETRACYCLINE HYDROCHLORIDE AND SODIUM NOVOBIOCIN TABLETS Oral Dosage Form New Animal Drugs
SODIUM NOVOBIOCIN	520.2345h	TETRACYCLINE HYDROCHLORIDE, SODIUM NOVOBIOCIN, AND PREDNISOLONE TABLETS Oral Dosage Form New Animal Drugs
SODIUM OLEATE	186.1770	SODIUM OLEATE Listing Of Specific Substances Affirmed As GRAS; Indirect Food Substances Affirmed As Generally Recognized As Safe
SODIUM, OLEATE	522.1010	OLEATE SODIUM SOLUTION Implantation Or Injectable Dosage Form New Animal Drugs
SODIUM PALMITATE	186.1771	SODIUM PALMITATE Listing Of Specific Substances Affirmed As GRAS; Indirect Food Substances Affirmed As Generally Recognized As Safe
SODIUM PANTOTHENATE	582.5772	SODIUM PANTOTHENATE Nutrients And/or Dietary Supplements; Substances Generally Recognized As Safe
SODIUM PECTINATE	582.1775	SODIUM PECTINATE General Purpose Food Additives; Substances Generally Recognized As Safe
SODIUM PENTACHLOROPHENATE	178.3900	SODIUM PENTACHLOROPHENATE Certain Adjuvants And Production Aids; Indirect Food Additives: Adjuvants, Production Aids, And Sanitizers
SODIUM PENTOBARBITAL	522.1704	SODIUM PENTOBARBITAL INJECTION Implantation Or Injectable Dosage Form New Animal Drugs
SODIUM PENTOBARBITAL	522.2444b	SODIUM THIOPENTAL, SODIUM PENTOBARBITAL FOR INJECTION Implantation Or Injectable Dosage Form New Animal Drugs
SODIUM PHOSPHATE	182.1778	SODIUM PHOSPHATE Multiple Purpose GRAS Food Substances; Substances Generally Recognized As Safe
SODIUM PHOSPHATE	182.6778	SODIUM PHOSPHATE Sequestrants; Substances Generally Recognized As Safe

SODIUM PHOSPHATE	182.8778	SODIUM PHOSPHATE Nutrients; Substances Generally Recognized As Safe
SODIUM PHOSPHATE	582.1778	SODIUM PHOSPHATE General Purpose Food Additives; Substances Generally Recognized As Safe
SODIUM PHOSPHATE	582.5778	SODIUM PHOSPHATE Nutrients And/or Dietary Supplements; Substances Generally Recognized As Safe
SODIUM PHOSPHATE	582.6778	SODIUM PHOSPHATE Sequestrants; Substances Generally Recognized As Safe
SODIUM PHOSPHATES	201.307	SODIUM PHOSPHATES, PACKAGE SIZE LIMITATION, WARNINGS AND DIRECTIONS FOR OVER-THE-COUNTER SALE Labeling
SODIUM POLYACRYLATE	173.73	SODIUM POLYACRYLATE Polymer Substances And Polymer Adjuvants; Secondary Direct Food Additives Permitted In Food For Human Consumption
SODIUM, PREDNISOLONE	522.1883	PREDNISOLONE SODIUM PHOSPHATE INJECTION, STERILE Implantation Or Injectable Dosage Form New Animal Drugs
SODIUM PROPIONATE	184.1784	SODIUM PROPIONATE Listing Of Specific Substances Affirmed As GRAS; Direct Food Substances Affirmed As Generally Recognized As Safe
SODIUM PROPIONATE	582.3784	SODIUM PROPIONATE Chemical Preservatives; Substances Generally Recognized As Safe
SODIUM PYROPHOSPHATE	182.6787	SODIUM PYROPHOSPHATE Sequestrants; Substances Generally Recognized As Safe
SODIUM PYROPHOSPHATE	182.6789	TETRA SODIUM PYROPHOSPHATE Sequestrants; Substances Generally Recognized As Safe
SODIUM PYROPHOSPHATE	582.5306	FERRIC SODIUM PYROPHOSPHATE Nutrients And/or Dietary Supplements; Substances Generally Recognized As Safe
SODIUM PYROPHOSPHATE	582.6787	SODIUM PYROPHOSPHATE Sequestrants; Substances Generally Recognized As Safe
SODIUM PYROPHOSPHATE	582.6789	TETRA SODIUM PYROPHOSPHATE Sequestrants; Substances Generally Recognized As Safe
SODIUM SACCHARIN	180.37	SACCHARIN, AMMONIUM SACCHARIN, CALCIUM SACCHARIN, AND SODIUM SACCHARIN Specific Requirements For Certain Food Additives; Food Additives Permitted In Food Or In Contact With Food On Interim Basis Pending Study
SODIUM SESQUICARBONATE	184.1792	SODIUM SESQUICARBONATE Listing Of Specific Substances Affirmed As GRAS; Direct Food Substances Affirmed As Generally Recognized As Safe
SODIUM SESQUICARBONATE	582.1792	SODIUM SESQUICARBONATE General Purpose Food Additives; Substances Generally Recognized As Safe
SODIUM SORBATE	182.3795	SODIUM SORBATE Chemical Preservatives; Substances Generally Recognized As Safe
SODIUM SORBATE	582.3795	SODIUM SORBATE Chemical Preservatives; Substances Generally Recognized As Safe
SODIUM STEARATE	573.280	FEED-GRADE CALCIUM STEARATE AND SODIUM STEARATE Food Additive Listing; Food Additives Permitted In Feed And Drinking Water Of Animals
SODIUM STEAROYL-LACTYLATE	172.846	SODIUM STEAROYL-2-LACTYLATE Multipurpose Additives; Food Additives Permitted For Direct Addition To Food For Human Consumption
SODIUM STEARYL FUMARATE	172.826	SODIUM STEARYL FUMARATE Multipurpose Additives; Food Additives Permitted For Direct Addition To Food For Human Consumption
SODIUM, SULFABROMOMETHAZINE	520.2170	SULFABROMOMETHAZINE SODIUM BOLUSES Oral Dosage Form New Animal Drugs
SODIUM, SULFABROMOMETHAZINE	556.620	SULFABROMOMETHAZINE SODIUM Specific Tolerances For Residues Of New Animal Drugs; Tolerances For Residues Of New Animal Drugs In Food
SODIUM SULFACHLOROPYRAZINE	520.2184	SODIUM SULFACHLOROPYRAZINE MONOHYDRATE Oral Dosage Form New Animal Drugs
SODIUM SULFACHLOROPYRAZINE	556.625	SODIUM SULFACHLOROPYRAZINE MONOHYDRATE Specific Tolerances For Residues Of New Animal Drugs; Tolerances For Residues Of New Animal Drugs In Food
SODIUM, SULFAMETHAZINE	520.2261	SULFAMETHAZINE SODIUM ORAL DOSAGE FORMS Oral Dosage Form New Animal Drugs
SODIUM SULFATE	182.1131	ALUMINUM SODIUM SULFATE Multiple Purpose GRAS Food Substances; Substances Generally Recognized As Safe
SODIUM SULFATE	186.1797	SODIUM SULFATE Listing Of Specific Substances Affirmed As GRAS; Indirect Food Substances Affirmed As Generally Recognized As Safe
SODIUM SULFATE	582.1131	ALUMINUM SODIUM SULFATE General Purpose Food Additives; Substances Generally Recognized As Safe

SODIUM SULFITE	182.3798	SODIUM SULFITE Chemical Preservatives; Substances Generally Recognized As Safe
SODIUM SULFITE	582.3798	SODIUM SULFITE Chemical Preservatives; Substances Generally Recognized As Safe
SODIUM SULFOSUCCINATE	163.117	COCOA WITH DIOCTYL SODIUM SULFOSUCCINATE FOR MANUFACTURING Requirements For Specific Standardized Cacao Products; Cacao Products
SODIUM SULFOSUCCINATE	172.520	COCOA WITH DIOCTYL SODIUM SULFOSUCCINATE FOR MANUFACTURING Flavoring Agents And Related Substances; Food Additives Permitted For Direct Addition To Food For Human Consumption
SODIUM SULFOSUCCINATE	172.810	DIOCTYL SODIUM SULFOSUCCINATE Multipurpose Additives; Food Additives Permitted For Direct Addition To Food For Human Consumption
SODIUM TARTRATE	184.1801	SODIUM TARTRATE Listing Of Specific Substances Affirmed As GRAS; Direct Food Substances Affirmed As Generally Recognized As Safe
SODIUM TARTRATE	184.1804	SODIUM POTASSIUM TARTRATE Listing Of Specific Substances Affirmed As GRAS; Direct Food Substances Affirmed As Generally Recognized As Safe
SODIUM TARTRATE	582.1804	SODIUM POTASSIUM TARTRATE General Purpose Food Additives; Substances Generally Recognized As Safe
SODIUM TARTRATE	582.6801	SODIUM TARTRATE Sequestrants; Substances Generally Recognized As Safe
SODIUM TARTRATE	582.6804	SODIUM POTASSIUM TARTRATE Sequestrants; Substances Generally Recognized As Safe
SODIUM, THIALBARBITONE	522.2404	THIALBARBITONE SODIUM FOR INJECTION Implantation Or Injectable Dosage Form New Animal Drugs
SODIUM THIAMYLAL	522.2424	SODIUM THIAMYLAL FOR INJECTION Implantation Or Injectable Dosage Form New Animal Drugs
SODIUM THIOPENTAL	522.2444	SODIUM THIOPENTAL IMPLANTATION OR INJECTABLE DOSAGE FORMS Implantation Or Injectable Dosage Form New Animal Drugs
SODIUM THIOSULFATE	184.1807	SODIUM THIOSULFATE Listing Of Specific Substances Affirmed As GRAS; Direct Food Substances Affirmed As Generally Recognized As Safe
SODIUM THIOSULFATE	582.6807	SODIUM THIOSULFATE Sequestrants; Substances Generally Recognized As Safe
SODIUM TRIPOLYPHOSPHATE	182.1810	SODIUM TRIPOLYPHOSPHATE Multiple Purpose GRAS Food Substances; Substances Generally Recognized As Safe
SODIUM TRIPOLYPHOSPHATE	182.6810	SODIUM TRIPOLYPHOSPHATE Sequestrants; Substances Generally Recognized As Safe
SODIUM TRIPOLYPHOSPHATE	582.1810	SODIUM TRIPOLYPHOSPHATE General Purpose Food Additives; Substances Generally Recognized As Safe
SODIUM TRIPOLYPHOSPHATE	582.6810	SODIUM TRIPOLYPHOSPHATE Sequestrants; Substances Generally Recognized As Safe
SOFT CHEESE	133.102	ASIAGO FRESH AND ASIAGO SOFT CHEESE Requirements For Specific Standardized Cheese And Related Products; Cheeses And Related Cheese Products
SOFT CHEESES	133.182	SOFT RIPENED CHEESES Requirements For Specific Standardized Cheese And Related Products; Cheeses And Related Cheese Products
SOFT CONTACT LENS	886.5925	SOFT (HYDROPHILIC) CONTACT LENS Therapeutic Devices; Ophthalmic Devices
SOFT CONTACT LENS	886.5928	SOFT (HYDROPHILIC) CONTACT LENS CARE PRODUCTS Therapeutic Devices; Ophthalmic Devices
SOLAR STIMULATOR	352.71	LIGHT SOURCE (SOLAR STIMULATOR) Sunscreen Drug Products For Over-the-counter Human Use
SOLDERS, LEAD	189.240	LEAD SOLDERS Substances Prohibited From Use In Human Food
SOLID ORAL DOSAGE	330.3	IMPRINTING OF SOLID ORAL DOSAGE FORM DRUG PRODUCTS Over-the-counter Human Drugs Generally Recognized As Safe And Effective & Not Misbranded
SOLIDS	610.16	TOTAL SOLIDS IN SERUMS General Provisions; General Biological Products Standards
SOLUBLE FIBER	101.81	HEALTH CLAIMS: SOLUBLE FIBER FROM CERTAIN FOODS AND RISK OF CORONARY HEART DISEASE Food Labeling
SOLUBLE POWDER	520.44	ACETAZOLAMIDE SODIUM SOLUBLE POWDER Oral Dosage Form New Animal Drugs
SOLUBLE POWDER	520.88d	AMOXICILLIN TRIHYDRATE SOLUBLE POWDER Oral Dosage Form New Animal Drugs
SOLUBLE POWDER	520.90e	AMPICILLIN TRIHYDRATE SOLUBLE POWDER Oral Dosage Form New Animal Drugs
SOLUBLE POWDER	520.110	APRAMYCIN SULFATE SOLUBLE POWDER Oral Dosage Form New Animal Drugs

SOLUBLE POWDER	520.154c	BACITRACIN ZINC SOLUBLE POWDER Oral Dosage Form New Animal Drugs
SOLUBLE POWDER	520.445a	CHLORTETRACYCLINE BISULFATE/SULFAMETHAZINE BISULFATE SOLUBLE POWDER Oral Dosage Form New Animal Drugs
SOLUBLE POWDER	520.1044c	GENTAMICIN SULFATE SOLUBLE POWDER Oral Dosage Form New Animal Drugs
SOLUBLE POWDER	520.1263b	LINCOMYCIN HYDROCHLORIDE MONOHYDRATE AND SPECTINOMYCIN SULFATE TETRAHYDRATE SOLUBLE POWDER Oral Dosage Form New Animal Drugs
SOLUBLE POWDER	520.1263c	LINCOMYCIN HYDROCHLORIDE SOLUBLE POWDER Oral Dosage Form New Animal Drugs
SOLUBLE POWDER	520.1660d	OXYTETRACYCLINE HYDROCHLORIDE SOLUBLE POWDER Oral Dosage Form New Animal Drugs
SOLUBLE POWDER	520.2087	ROXARSONE SOLUBLE POWDER Oral Dosage Form New Animal Drugs
SOLUBLE POWDER	520.2123b	SPECTINOMYCIN DIHYDROCHLORIDE PENTAHYDRATE SOLUBLE POWDER Oral Dosage Form New Animal Drugs
SOLUBLE POWDER	520.2220a	SULFAMETHOXINE ORAL SOLUTION AND SOLUBLE POWDER Oral Dosage Form New Animal Drugs
SOLUBLE POWDER	520.2261b	SULFAMETHAZINE SODIUM SOLUBLE POWDER Oral Dosage Form New Animal Drugs
SOLUBLE POWDER	520.2345d	TETRACYCLINE HYDROCHLORIDE SOLUBLE POWDER Oral Dosage Form New Animal Drugs
SOLUBLE POWDER	520.2455	TIAMULIN SOLUBLE POWDER Oral Dosage Form New Animal Drugs
SOLUBLE POWDER	524.1484	NEOMYCIN SULFATE SOLUBLE POWDER Ophthalmic And Topical Dosage Form New Animal Drugs
SOLUBLE POWDER	524.1580c	NITROFURAZONE SOLUBLE POWDER Ophthalmic And Topical Dosage Form New Animal Drugs
SOLUTION	520.48	ALTRENOGEST SOLUTION Oral Dosage Form New Animal Drugs
SOLUTION	520.563	DIATRIZOATE MEGLUMINE AND DIATRIZOATE SODIUM ORAL SOLUTION Oral Dosage Form New Animal Drugs
SOLUTION	520.813	ENROFLOXACIN ORAL SOLUTION Oral Dosage Form New Animal Drugs
SOLUTION	520.1044a	GENTAMICIN SULFATE ORAL SOLUTION Oral Dosage Form New Animal Drugs
SOLUTION	520.1044b	GENTAMICIN SULFATE PIG PUMP ORAL SOLUTION Oral Dosage Form New Animal Drugs
SOLUTION	520.1485	NEOMYCIN SULFATE ORAL SOLUTION Oral Dosage Form New Animal Drugs
SOLUTION	520.1696c	GENTAMICIN SULFATE ORAL SOLUTION Oral Dosage Form New Animal Drugs
SOLUTION	520.2122	PENICILLIN V POTASSIUM FOR ORAL SOLUTION Oral Dosage Form New Animal Drugs
SOLUTION	520.2185	STREPTOMYCIN SULFATE ORAL SOLUTION Oral Dosage Form New Animal Drugs
SOLUTION	520.2220a	SULFAMETHOXINE ORAL SOLUTION AND SOLUBLE POWDER Oral Dosage Form New Animal Drugs
SOLUTION	520.2261a	SULFAMETHAZINE SODIUM DRINKING WATER SOLUTION Oral Dosage Form New Animal Drugs
SOLUTION	522.82	AMINOPROPAZINE FUMARATE STERILE SOLUTION INJECTION Implantation Or Injectable Dosage Form New Animal Drugs
SOLUTION	522.380	CHLORAL HYDRATE, PENTOBARBITAL, AND MAGNESIUM SULFATE STERILE AQUEOUS SOLUTION Implantation Or Injectable Dosage Form New Animal Drugs
SOLUTION	522.690	DINOPROST TROMETHAMINE STERILE SOLUTION Implantation Or Injectable Dosage Form New Animal Drugs
SOLUTION	522.812	ENROFLOXACIN SOLUTION Implantation Or Injectable Dosage Form New Animal Drugs
SOLUTION	522.900	EUTHANASIA SOLUTION Implantation Or Injectable Dosage Form New Animal Drugs
SOLUTION	522.914	FENPROSTALENE SOLUTION Implantation Or Injectable Dosage Form New Animal Drugs
SOLUTION	522.960c	FLUMETHASONE SOLUTION Implantation Or Injectable Dosage Form New Animal Drugs
SOLUTION	522.970	FLUNIXIN MEGLUMINE SOLUTION Implantation Or Injectable Dosage Form New Animal Drugs
SOLUTION	522.1020	GELATIN SOLUTION Implantation Or Injectable Dosage Form New Animal Drugs
SOLUTION	522.1156	IMIDOCARB DIPROPIONATE SOLUTION Implantation Or Injectable Dosage Form New Animal Drugs

SOLUTION	522.1225	KETOPROFEN SOLUTION Implantation Or Injectable Dosage Form New Animal Drugs
SOLUTION	522.1290	LUPROSTIOL STERILE SOLUTION Implantation Or Injectable Dosage Form New Animal Drugs
SOLUTION	522.1484	NEOMYCIN SULFATE STERILE SOLUTION Implantation Or Injectable Dosage Form New Animal Drugs
SOLUTION	522.1610	OLEATE SODIUM SOLUTION Implantation Or Injectable Dosage Form New Animal Drugs
SOLUTION	522.1870	PRAZIQUANTEL INJECTABLE SOLUTION Implantation Or Injectable Dosage Form New Animal Drugs
SOLUTION	522.2012	PROSTALENE SOLUTION Implantation Or Injectable Dosage Form New Animal Drugs
SOLUTION	522.2121	SPECTINOMYCIN SULFATE SOLUTION Implantation Or Injectable Dosage Form New Animal Drugs
SOLUTION	522.2260	SULFAMETHAZINE INJECTABLE SOLUTION Implantation Or Injectable Dosage Form New Animal Drugs
SOLUTION	524.390b	CHLORAMPENICL OPHTHALMIC SOLUTION Ophthalmic And Topical Dosage Form New Animal Drugs
SOLUTION	524.463	COPPER NAPHTHENATE SOLUTION Ophthalmic And Topical Dosage Form New Animal Drugs
SOLUTION	524.660a	DIMETHYL SULFOXIDE SOLUTION Ophthalmic And Topical Dosage Form New Animal Drugs
SOLUTION	524.981b	FLUOCINOLONE ACETONIDE SOLUTION Ophthalmic And Topical Dosage Form New Animal Drugs
SOLUTION	524.981d	FLUOCINOLONE ACETONIDE, DIMETHYL SULFOXIDE SOLUTION Ophthalmic And Topical Dosage Form New Animal Drugs
SOLUTION	524.981e	FLUOCINOLONE ACETONIDE, DIMETHYL SULFOXIDE OTIC SOLUTION Ophthalmic And Topical Dosage Form New Animal Drugs
SOLUTION	524.1044a	GENTAMICIN OPHTHALMIC SOLUTION Ophthalmic And Topical Dosage Form New Animal Drugs
SOLUTION	524.1044b	GENTAMICIN SULFATE, BETAMETHASONE VALERATE OTIC SOLUTION Ophthalmic And Topical Dosage Form New Animal Drugs
SOLUTION	524.1200b	KANAMYCIN OPHTHALMIC AQUEOUS SOLUTION Ophthalmic And Topical Dosage Form New Animal Drugs
SOLUTION	524.1376	2-MERCAPTOBENZOTHIAZOLE SOLUTION Ophthalmic And Topical Dosage Form New Animal Drugs
SOLUTION	524.1446	MILBEMYCIN OXIME SOLUTION Ophthalmic And Topical Dosage Form New Animal Drugs
SOLUTION	524.1484e	NEOMYCIN SULFATE AND POLYMYXIN B SULFATE OPHTHALMIC SOLUTION Ophthalmic And Topical Dosage Form New Animal Drugs
SOLUTION	524.1484g	NEOMYCIN SULFATE-THIABENDAZOLE-DEXAMETHASONE SOLUTION Ophthalmic And Topical Dosage Form New Animal Drugs
SOLUTION	524.1982	PROPARACAINE HYDROCHLORIDE OPHTHALMIC SOLUTION Ophthalmic And Topical Dosage Form New Animal Drugs
SOLUTION	529.50	AMIKACIN SULFATE INTRAUTERINE SOLUTION Certain Other Dosage Form New Animal Drugs
SOLUTION	529.1030	FORMALIN SOLUTION Certain Other Dosage Form New Animal Drugs
SOLUTION	529.1044a	GENTAMICIN SULFATE INTRAUTERINE SOLUTION Certain Other Dosage Form New Animal Drugs
SOLUTION	529.1044b	GENTAMICIN SULFATE SOLUTION Certain Other Dosage Form New Animal Drugs
SOLUTION, ENZYME	864.9400	STABILIZED ENZYME SOLUTION Products Used In Establishments That Manufacture Blood And Blood Products; Hematology And Pathology Devices
SOLUTION, DYE/CHEMICAL	864.1850	DYE AND CHEMICAL SOLUTION STAINS Biological Stains; Hematology And Pathology Devices
SOLUTION, HYDROGEN PEROXIDE	178.1005	HYDROGEN PEROXIDE SOLUTION Substances Utilized To Control The Growth Of Microorganisms; Indirect Food Additives: Adjuvants, Production Aids, And Sanitizers
SOLUTION, SPECIFIC GRAVITY	864.9320	COPPER SULFATE SOLUTION FOR SPECIFIC GRAVITY DETERMINATIONS Products Used In Establishments That Manufacture Blood And Blood Products; Hematology And Pathology Devices
SOLUTIONS	58.83	REAGENTS AND SOLUTIONS Testing Facilities Operation; Good Laboratory Practice For Nonclinical Laboratory Studies
SOLUTIONS, BALANCED SALT	864.2875	BALANCED SALT SOLUTIONS OR FORMULATIONS Cell And Tissue Culture Products; Hematology And Pathology Devices
SOLUTIONS, CONTACT LENS	800.10	CONTACT LENS SOLUTIONS; STERILITY Requirements For Specific Medical Devices; General
SOLUTIONS, CONTACT LENS	800.12	CONTACT LENS SOLUTIONS AND TABLETS; TAMPER-RESISTANT PACKAGING Requirements For Specific Medical Devices; General

SOLUTIONS, OPHTHALMIC	524.960	FLUMETHASONE, NEOMYCIN SULFATE, AND POLYMYXIN B SULFATE OPHTHALMIC SOLUTIONS Ophthalmic And Topical Dosage Form New Animal Drugs
SOLUTIONS, SANITIZING	178.1010	SANITIZING SOLUTIONS Substances Utilized To Control The Growth Of Microorganisms; Indirect Food Additives: Adjuvants, Production Aids, And Sanitizers
SOLVENT	878.4730	SURGICAL SKIN DEGREASER OR ADHESIVE TAPE SOLVENT Surgical Devices; General And Plastic Surgery Devices
SOLVENT EXTRACTION	173.280	SOLVENT EXTRACTION PROCESS FOR CITRIC ACID Solvents, Lubricants, Release Agents And Related Substances; Secondary Direct Food Additives Permitted In Food For Human Consumption
SOLVENT-FREE EXTRACTIVES	182.40	NATURAL EXTRACTIVES (SOLVENT-FREE) USED IN CONJUNCTION WITH SPICES, SEASONINGS, AND FLAVORINGS General Provisions; Substances Generally Recognized As Safe
SOLVENT-FREE EXTRACTIVES	582.40	NATURAL EXTRACTIVES (SOLVENT-FREE) USED IN CONJUNCTION WITH SPICES, SEASONINGS, AND FLAVORINGS General Provisions; Substances Generally Recognized As Safe
SOLVENT-FREE OLEORESINS	182.20	ESSENTIAL OILS, OLEORESINS (SOLVENT-FREE), AND NATURAL EXTRACTIVES (INCLUDING DISTILLATES) General Provisions; Substances Generally Recognized As Safe
SOLVENT-FREE OLEORESINS	582.20	ESSENTIAL OILS, OLEORESINS (SOLVENT-FREE), AND NATURAL EXTRACTIVES (INCLUDING DISTILLATES) General Provisions; Substances Generally Recognized As Safe
SOMETRIBOVE	522.2112	STERILE SOMETRIBOVE ZINC SUSPENSION Implantation Or Injectable Dosage Form New Animal Drugs
SONIC	888.4580	SONIC SURGICAL INSTRUMENT AND ACCESSORIES/ATTACHMENTS Surgical Devices; Orthopedic Devices
SORBATE, CALCIUM	182.3225	CALCIUM SORBATE Chemical Preservatives; Substances Generally Recognized As Safe
SORBATE, CALCIUM	582.3225	CALCIUM SORBATE Chemical Preservatives; Substances Generally Recognized As Safe
SORBATE, POTASSIUM	182.3640	POTASSIUM SORBATE Chemical Preservatives; Substances Generally Recognized As Safe
SORBATE, POTASSIUM	582.3640	POTASSIUM SORBATE Chemical Preservatives; Substances Generally Recognized As Safe
SORBATE, SODIUM	182.3795	SODIUM SORBATE Chemical Preservatives; Substances Generally Recognized As Safe
SORBATE, SODIUM	582.3795	SODIUM SORBATE Chemical Preservatives; Substances Generally Recognized As Safe
SORBENT	876.5600	SORBENT REGENERATED DIALYSATE DELIVERY SYSTEM FOR HEMODIALYSIS Therapeutic Devices; Gastroenterology-urology Devices
SORBENT	876.5870	SORBENT HEMOPERFUSION SYSTEM Therapeutic Devices; Gastroenterology-urology Devices
SORBIC ACID	182.3089	SORBIC ACID Chemical Preservatives; Substances Generally Recognized As Safe
SORBIC ACID	582.3089	SORBIC ACID Chemical Preservatives; Substances Generally Recognized As Safe
SORBITAN MONOOLEATE	173.75	SORBITAN MONOOLEATE Polymer Substances And Polymer Adjuvants For Food Treatment; Secondary Direct Food Additives Permitted In Food Human Consumption
SORBITAN MONOSTEARATE	172.842	SORBITAN MONOSTEARATE Multipurpose Additives; Food Additives Permitted For Direct Addition To Food For Human Consumption
SORBITAN MONOSTEARATE	573.960	SORBITAN MONOSTEARATE Food Additive Listing; Food Additives Permitted In Feed And Drinking Water Of Animals
SORBITOL	184.1835	SORBITOL Listing Of Specific Substances Affirmed As GRAS; Direct Food Substances Affirmed As Generally Recognized As Safe
SORBITOL	582.5835	SORBITOL Nutrients And/or Dietary Supplements; Substances Generally Recognized As Safe
SORBITOL	862.1670	SORBITOL DEHYDROGENASE TEST SYSTEM Clinical Chemistry Test Systems; Clinical Chemistry And Clinical Toxicology Devices
SORBOSE	186.1839	SORBOSE Listing Of Specific Substances Affirmed As GRAS; Indirect Food Substances Affirmed As Generally Recognized As Safe
SORE THROATS	201.315	OVER-THE-COUNTER DRUGS FOR MINOR SORE THROATS; SUGGESTED WARNING Specific Labeling Requirements For Specific Drug Products; Labeling
SORES, COLD	310	537 DRUG PRODUCTS CONTAINING ACTIVE INGREDIENTS OFFERED OTC FOR TREATMENT OF FEVER BLISTERS AND COLD SORES Requirements For Specific New Drugs Or Devices; New Drugs

SORGHUM	168.160	SORGHUM SIRUP Requirements For Specific Standardized Sweeteners And Table Syrups; Sweeteners And Table Sirups
SOUND	870.2860	HEART SOUND TRANSDUCER Cardiovascular Monitoring Devices; Cardiovascular Devices
SOUND	876.4590	INTERLOCKING URETHRAL SOUND Surgical Devices; Gastroenterology-urology Devices
SOUR CREAM	131.160	SOUR CREAM Requirements For Specific Standardized Milk And Cream; Milk And Cream
SOUR CREAM	131.162	ACIDIFIED SOUR CREAM Requirements For Specific Standardized Milk And Cream; Milk And Cream
SOURCE	640.22	COLLECTION OF SOURCE MATERIAL Platelets; Additional Standards For Human Blood And Blood Products
SOURCE	640.32	COLLECTION OF SOURCE MATERIAL Plasma; Additional Standards For Human Blood And Blood Products
SOURCE	640.52	COLLECTION OF SOURCE MATERIAL Cryoprecipitate; Additional Standards For Human Blood And Blood Products
SOURCE	640.60	SOURCE PLASMA Source Plasma; Additional Standards For Human Blood And Blood Products
SOURCE	640.64	COLLECTION OF BLOOD FOR SOURCE PLASMA Source Plasma; Additional Standards For Human Blood And Blood Products
SOURCE	640.74	MODIFICATION OF SOURCE PLASMA Source Plasma; Additional Standards For Human Blood And Blood Products
SOURCE	660.32	COLLECTION OF SOURCE MATERIAL Reagent Red Blood Cells; Additional Standards For Diagnostic Substances For Laboratory Tests
SOURCE	660.33	TESTING OF SOURCE MATERIAL Reagent Red Blood Cells; Additional Standards For Diagnostic Substances For Laboratory Tests
SOURCE	872.1800	EXTRAORAL SOURCE X-RAY SYSTEM Diagnostic devices; Dental Devices
SOURCE	872.1810	INTRAORAL SOURCE X-RAY SYSTEM Diagnostic Devices; Dental Devices
SOURCE	886.5100	OPHTHALMIC BETA RADIATION SOURCE Therapeutic Devices; Ophthalmic Devices
SOURCE	892.1380	NUCLEAR FLOOD SOURCE PHANTOM Diagnostic Devices; Radiology Devices
SOURCE	892.1400	NUCLEAR SEALED CALIBRATION SOURCE Diagnostic Devices; Radiology Devices
SOURCE	892.5730	RADIONUCLIDE BRACHYTHERAPY SOURCE Therapeutic Devices; Radiology Devices
SOURCE	892.5740	RADIONUCLIDE TELETHERAPY SOURCE Therapeutic Devices; Radiology Devices
SOURCE	1250.28	SOURCE AND HANDLING OF ICE Food Service Sanitation On Land And Air Conveyances; Interstate Conveyance Sanitation
SOURCE CONTROLS	123.28	SOURCE CONTROLS Raw Molluscan Shellfish; Fish And Fishery Products
SOURCE, GOOD	101.54	NUTRIENT CONTENT CLAIMS FOR "GOOD SOURCE," "HIGH," "MORE," AND "HIGH POTENCY" Specific Requirements For Nutrient Content Claims; Food Labeling
SOURCE IDENTIFICATION	1250.25	SOURCE IDENTIFICATION AND INSPECTION OF FOOD AND DRINK Food Service Sanitation On Land And Air Conveyances, Interstate Conveyance Sanitation
SOURCE, LIGHT	352.71	LIGHT SOURCE (SOLAR STIMULATOR) Sunscreen Drug Products For Over-the-counter Human Use
SOURCE, LIGHT	874.4350	EAR, NOSE, AND THROAT FIBEROPTIC LIGHT SOURCE AND CARRIER Surgical Devices; Ear, Nose, And Throat Devices
SOURCE OF RECORDS	20.49	REFERRAL TO PRIMARY SOURCE OF RECORDS Procedures And Fees; Public Information
SOURCES OF INFORMATION	1301.93	SOURCES OF INFORMATION FOR EMPLOYEE CHECKS Employee Screening - Non-practitioners; Registration Of Manufacturers, Distributors, And Dispensers Of Controlled Substances
SOURCES OF RADIATION	179.21	SOURCES OF RADIATION USED FOR INSPECTION OF FOOD, FOR INSPECTION OF PACKAGED FOOD, AND FOR CONTROLLING FOOD PROCESSING Radiation And Radiation Sources; Irradiation In The Production, Processing And Handling Of Food
SOY	139.140	WHEAT AND SOY MACARONI PRODUCTS Requirements For Specific Standardized Macaroni And Noodle Products; Macaroni And Noodle Products
SOY	139.180	WHEAT AND SOY NOODLE PRODUCTS Requirements For Specific Standardized Macaroni And Noodle Products; Macaroni And Noodle Products

SOY PROTEIN	101.82	HEALTH CLAIMS: SOY PROTEIN AND RISK OF CORONARY HEART DISEASE Specific Requirements For Health Claims; Food Labeling
SOYBEAN OIL	172.723	EPOXIDIZED SOYBEAN OIL Food Additives Permitted For Direct Addition To Food For Human Consumption
SOYBEAN OIL	526.1696b	PENICILLIN G PROCAINE-DIHYDROSTREPTOMYCIN IN SOYBEAN OIL FOR INTRAMAMMARY INFUSION (DRY COWS) Intramammary Dosage Forms
SPANISH-LANGUAGE	201.16	DRUGS; SPANISH-LANGUAGE VERSION OF CERTAIN REQUIRED STATEMENTS General Labeling Provisions; Labeling
SPANISH-LANGUAGE	290.6	SPANISH-LANGUAGE VERSION OF REQUIRED WARNING General Provisions; Controlled Drugs
SPANISH-LANGUAGE	801.16	MEDICAL DEVICES; SPANISH-LANGUAGE VERSION OF CERTAIN REQUIRED STATEMENTS General Labeling Provisions; Labeling
SPECIFIC GRAVITY	864.9320	COPPER SULFATE SOLUTION FOR SPECIFIC GRAVITY DETERMINATIONS Products Used In Establishments That Manufacture Blood And Blood Products; Hematology And Pathology Devices
SPECIFIC GUARANTY	1230.21	SPECIFIC GUARANTY Regulations Under The Federal Caustic Poison Act
SPECIFICATIONS, COLORS	82.5	GENERAL SPECIFICATIONS FOR STRAIGHT COLORS General Provisions; Listing Of Certified Provisionally Listed Colors And Specifications
SPECIFICATIONS, NUTRIENT	107.100	NUTRIENT SPECIFICATIONS Nutrient Requirements; Infant Formula
SPECIFICITY	660.5	SPECIFICITY Antibody To Hepatitis B Surface Antigen; Additional Standards For Diagnostic Substances For Laboratory Tests
SPECIFICITY	660.26	SPECIFICITY TESTS Blood Grouping Reagent; Additional Standards For Diagnostic Substances For Laboratory Tests
SPECIFICITY	660.44	SPECIFICITY Hepatitis B Surface Antigen; Additional Standards For Diagnostic Substances For Laboratory Tests
SPECIFICITY	660.54	POTENCY TESTS, SPECIFICITY TESTS, TESTS FOR HETEROSPECIFIC ANTIBODIES, AND ADDITIONAL TESTS FOR NONSPECIFIC PROPERTIES Anti-human Globulin; Additional Standards For Diagnostic Substances For Laboratory Tests
SPECIMEN	58.51	SPECIMEN AND DATA STORAGE FACILITIES Facilities; Good Laboratory Practice For Nonclinical Laboratory Studies
SPECIMEN	862.1675	BLOOD SPECIMEN COLLECTION DEVICE Clinical Chemistry Test Systems; Clinical Chemistry And Clinical Toxicology Devices
SPECIMEN	864.3250	SPECIMEN TRANSPORT AND STORAGE CONTAINER Pathology Instrumentation And Accessories; Hematology And Pathology Devices
SPECIMEN	866.2900	MICROBIOLOGICAL SPECIMEN COLLECTION AND TRANSPORT DEVICE Microbiology Devices; Immunology And Microbiology Devices
SPECTACLE DISSOCIATION	886.1910	SPECTACLE DISSOCIATION TEST SYSTEM Diagnostic Devices; Ophthalmic Devices
SPECTACLE FRAME	886.5842	SPECTACLE FRAME Therapeutic Devices; Ophthalmic Devices
SPECTACLE LENS	886.5844	PRESCRIPTION SPECTACLE LENS Therapeutic Devices; Ophthalmic Devices
SPECTACLES, MAGNIFYING	886.5840	MAGNIFYING SPECTACLES Therapeutic Devices; Ophthalmic Devices
SPECTACLES, OPERATING	886.4770	OPHTHALMIC OPERATING SPECTACLES (LOUPES) Surgical Devices; Ophthalmic Devices
SPECTINOMYCIN	556.600	SPECTINOMYCIN Specific Tolerances For Residues Of New Animal Drugs; Tolerances For Residues Of New Animal Drugs In Food
SPECTINOMYCIN DIHYDROCHLORIDE	520.2122	SPECTINOMYCIN DIHYDROCHLORIDE ORAL SOLUTION Oral Dosage Form New Animal Drugs
SPECTINOMYCIN DIHYDROCHLORIDE	520.2123	SPECTINOMYCIN DIHYDROCHLORIDE PENTAHYDRATE ORAL DOSAGE FORMS Oral Dosage Form New Animal Drugs
SPECTINOMYCIN DIHYDROCHLORIDE	522.2120	SPECTINOMYCIN DIHYDROCHLORIDE INJECTION Implantation Or Injectable Dosage Form New Animal Drugs
SPECTINOMYCIN SULFATE	520.1263	LINCOMYCIN HYDROCHLORIDE MONOHYDRATE AND SPECTINOMYCIN SULFATE TETRAHYDRATE SOLUBLE POWDER Oral Dosage Form New Animal Drugs
SPECTINOMYCIN SULFATE	522.2121	SPECTINOMYCIN SULFATE SOLUTION Implantation Or Injectable Dosage Form New Animal Drugs
SPECTROMETER	862.2860	MASS SPECTROMETER FOR CLINICAL USE Clinical Laboratory Instruments; Clinical Chemistry And Clinical Toxicology Devices

SPECTROPHOTOMETER	862.2300	COLORIMETER, PHOTOMETER, OR SPECTROPHOTOMETER FOR CLINICAL USE Clinical Laboratory Instruments; Clinical Chemistry And Clinical Toxicology Devices
SPECTROPHOTOMETER	862.2850	ATOMIC ABSORPTION SPECTROPHOTOMETER FOR CLINICAL USE Clinical Laboratory Instruments; Clinical Chemistry And Clinical Toxicology Devices
SPECTRUM	882.1420	ELECTROENCEPHALOGRAM (EEG) SIGNAL SPECTRUM ANALYZER Neurological Diagnostic Devices; Neurological Devices
SPECULUM	878.1800	SPECULUM AND ACCESSORIES Diagnostic Devices; General And Plastic Surgery Devices
SPEED CONTROL	870.4380	CARDIOPULMONARY BYPASS PUMP SPEED CONTROL Cardiovascular Surgical Devices; Cardiovascular Devices
SPERM	866.5800	SEMINAL FLUID (SPERM) IMMUNOLOGICAL TEST SYSTEM Immunological Test Systems; Immunology And Microbiology Devices
SPERM OIL	173.275	HYDROGENATED SPERM OIL Solvents, Lubricants, Release Agents And Related Substances; Secondary Direct Food Additives Permitted In Food For Human Consumption
SPERMICIDAL	884.5310	CONDOM WITH SPERMICIDAL LUBRICANT Obstetrical And Gynecological Therapeutic Devices; Obstetrical And Gynecological Devices
SPF	352.73	DETERMINATION OF SPF VALUE Sunscreen Drug Products For Over-the-counter Human Use
SPHERE, EYE	886.3320	EYE SPHERE IMPLANT Prosthetic Devices; Ophthalmic Devices
SPHINGOMYELIN	862.1455	LECITHIN/SPHINGOMYELIN RATIO IN AMNIOTIC FLUID TEST SYSTEM Clinical Chemistry Test Systems; Clinical Chemistry And Clinical Toxicology Devices
SPICED CHEESES	133.190	SPICED CHEESES Requirements For Specific Standardized Cheese And Related Products; Cheeses And Related Cheese Products
SPICED CHEESES	133.191	PART-SKIM SPICED CHEESES Requirements For Specific Standardized Cheese And Related Products; Cheeses And Related Cheese Products
SPICED CHEESES	133.193	SPICED, FLAVORED STANDARDIZED CHEESES Requirements For Specific Standardized Cheese And Related Products; Cheeses And Related Cheese Products
SPICES	101.22	FOODS; LABELING OF SPICES, FLAVORINGS, COLORINGS AND CHEMICAL PRESERVATIVES Specific Food Labeling Requirements; Food Labeling
SPICES	182.10	SPICES AND OTHER NATURAL SEASONINGS AND FLAVORINGS General Provisions; Substances Generally Recognized As Safe
SPICES	182.40	NATURAL EXTRACTIVES (SOLVENT-FREE) USED IN CONJUNCTION WITH SPICES, SEASONINGS, AND FLAVORINGS General Provisions; Substances Generally Recognized As Safe
SPICES	182.50	CERTAIN OTHER SPICES, SEASONINGS, ESSENTIAL OILS, OLEORESINS, AND NATURAL EXTRACTS General Provisions; Substances Generally Recognized As Safe
SPICES	501.22	ANIMAL FOODS; LABELING OF SPICES, FLAVORINGS, COLORINGS, AND CHEMICAL PRESERVATIVES Specific Animal Food Labeling Requirements; Animal Food Labeling
SPICES	582.10	SPICES AND OTHER NATURAL SEASONINGS AND FLAVORINGS General Provisions; Substances Generally Recognized As Safe
SPICES	582.30	NATURAL SUBSTANCES USED IN CONJUNCTION WITH SPICES AND OTHER NATURAL SEASONINGS AND FLAVORINGS General Provisions; Substances Generally Recognized As Safe
SPICES	582.40	NATURAL EXTRACTIVES (SOLVENT FREE) USED IN CONJUNCTION WITH SPICES, SEASONINGS, AND FLAVORINGS General Provisions; Substances Generally Recognized As Safe
SPICES	582.50	CERTAIN OTHER SPICES, SEASONINGS, ESSENTIAL OILS, OLEORESINS, AND NATURAL EXTRACTS General Provisions; Substances Generally Recognized As Safe
SPINAL CORD	882.5850	IMPLANTED SPINAL CORD STIMULATOR FOR BLADDER EVACUATION Neurological Therapeutic Devices; Neurological Devices
SPINAL CORD	882.5880	IMPLANTED SPINAL CORD STIMULATOR FOR PAIN RELIEF Neurological Therapeutic Devices; Neurological Devices
SPINAL FLUID	866.5860	TOTAL SPINAL FLUID IMMUNOLOGICAL TEST SYSTEM Immunological Test Systems; Immunology And Microbiology Devices
SPINAL FLUID	880.2460	ELECTRICALLY POWERED SPINAL FLUID PRESSURE MONITOR General Hospital And Personal Use Monitoring Devices; General Hospital And Personal Use Devices
SPINAL FLUID	880.2500	SPINAL FLUID MANOMETER General Hospital And Personal Use Monitoring Devices; General Hospital And Personal Use Devices
SPINAL ORTHOSIS	888.3050	SPINAL INTERLAMINAL FIXATION ORTHOSIS Prosthetic Devices; Orthopedic Devices

SPINAL ORTHOSIS	888.3060	SPINAL INTERVERTEBRAL BODY FIXATION ORTHOSIS Prosthetic Devices; Orthopedic Devices
SPINAL SYSTEM	888.3070	PEDICLE SCREW SPINAL SYSTEM Prosthetic Devices; Orthopedic Devices
SPINNER	864.5850	AUTOMATED SLIDE SPINNER Automated And Semi-automated Hematology Devices; Hematology And Pathology Devices
SPIRAL ELECTRODE	884.2675	FETAL SCALP CIRCULAR (SPIRAL) ELECTRODE AND APPLICATOR Obstetrical And Gynecological Monitoring Devices; Obstetrical And Gynecological Devices
SPIROMETER	868.1840	DIAGNOSTIC SPIROMETER Diagnostic Devices; Anesthesiology Devices
SPIROMETER	868.1850	MONITORING SPIROMETER Diagnostic Devices; Anesthesiology Devices
SPIROMETER	868.5690	INCENTIVE SPIROMETER Therapeutic Devices; Anesthesiology Devices
SPIROMETRY	868.1860	PEAK-FLOW METER FOR SPIROMETRY Diagnostic Devices; Anesthesiology Devices
SPLINT	872.3890	ENDODONTIC STABILIZING SPLINT Prosthetic Devices; Dental Devices
SPLINT	874.4780	INTRANASAL SPLINT Surgical Devices; Ear, Nose, And Throat Devices
SPLINT	874.5800	EXTERNAL NASAL SPLINT Therapeutic Devices; Ear, Nose, And Throat Devices
SPLINT	878.3900	INFLATABLE EXTREMITY SPLINT Prosthetic Devices; General And Plastic Surgery Devices
SPLINT	878.3910	NONINFLATABLE EXTREMITY SPLINT Prosthetic Devices; General And Plastic Surgery Devices
SPLINT	890.3665	CONGENITAL HIP DISLOCATION ABDUCTION SPLINT Physical Medicine Prosthetic Devices; Physical Medicine Devices
SPLINT	890.3675	DENIS BROWN SPLINT Physical Medicine Prosthetic Devices; Physical Medicine Devices
SPLINTING PIN	872.3740	RETENTIVE AND SPLINTING PIN Prosthetic Devices; Dental Devices
SPONGE, NONRESORBABLE	878.4014	NONRESORBABLE GAUZE/SPONGE FOR EXTERNAL USE Surgical Devices; General And Plastic Surgery Devices
SPONGE, OPHTHALMIC	886.4790	OPHTHALMIC SPONGE Surgical Devices; Ophthalmic Devices
SPONGE SCALE	880.2740	SURGICAL SPONGE SCALE General Hospital And Personal Use Monitoring Devices; General Hospital And Personal Use Devices
SPONGE, VAGINAL	529.1003	FLUROGESTONE ACETATE-IMPREGNATED VAGINAL SPONGE Certain Other Dosage Form New Animal Drugs
SPONSOR	58.217	SUSPENSION OR TERMINATION OF A TESTING FACILITY BY A SPONSOR Disqualification Of Testing Facilities; Good Laboratory Practice For Nonclinical Laboratory Studies
SPONSOR, FOREIGN	316.22	PERMANENT-RESIDENT AGENT FOR FOREIGN SPONSOR Orphan Drugs
SPONSORS	5.406	NOTIFICATION TO SPONSORS OF DEFICIENCIES IN PETITIONS FOR RECLASSIFICATION OF MEDICAL DEVICES Medical Devices And Radiological Health; Redelegations Of Authority; Delegations Of Authority And Organization
SPONSORS	312.50	GENERAL RESPONSIBILITIES OF SPONSORS Responsibilities Of Sponsors And Investigators; Investigational New Drug Application
SPONSORS	510.600	NAMES, ADDRESSES, AND DRUG LABELER CODES OF SPONSORS OF APPROVED APPLICATIONS Sponsors Of Approved Applications; New Animal Drugs
SPONSORS	812.40	GENERAL RESPONSIBILITIES OF SPONSORS Responsibilities Of Sponsors; Investigational Device Exemptions
SPONSOR'S RECORDS	312.58	INSPECTION OF SPONSOR'S RECORDS AND REPORTS Responsibilities Of Sponsors And Investigators; Investigational New Drug Application
SPOROTHRIX SCHENCKII	866.3680	SPOROTHRIX SCHENCKII SEROLOGICAL REAGENTS Serological Reagents; Immunology And Microbiology Devices
SPOT	892.1620	CINE OR SPOT FLUOROGRAPHIC X-RAY CAMERA Diagnostic Devices; Radiology Devices
SPOT-FILM	892.1670	SPOT-FILM DEVICE Diagnostic Devices; Radiology Devices
SPOT, MAXWELL	886.1435	MAXWELL SPOT Diagnostic Devices; Ophthalmic Devices
SPRAY	524.1044e	GENTAMICIN SULFATE SPRAY Ophthalmic And Topical Dosage Form New Animal Drugs

SPRAY	524.1044f	GENTAMICIN SULFATE, BETAMETHASONE VALERATE TOPICAL SPRAY Ophthalmic And Topical Dosage Form New Animal Drugs
SPRAY	524.1443	MICONAZOLE NITRATE CREAM; MICONAZOLE NITRATE LOTION; MICONAZOLE NITRATE SPRAY Ophthalmic And Topical Dosage Form New Animal Drugs
SPRAY	524.1662	OXYTETRACYCLINE HYDROCHLORIDE AND HYDROCORTISONE SPRAY Ophthalmic And Topical Dosage Form New Animal Drugs
SPRAYS	740.12	FEMININE DEODORANT SPRAYS Warning Statements; Cosmetic Product Warning Statements
SPREAD, CHEESE	133.10	NOTICE TO MANUFACTURERS, PACKERS, AND DISTRIBUTORS OF PASTEURIZED BLENDED CHEESE, PASTEURIZED PROCESS CHEESE, CHEESE FOOD, CHEESE SPREAD, AND RELATED FOODS General Provisions; Cheeses And Related Cheese Products
SPREAD, CHEESE	133.175	PASTEURIZED CHEESE SPREAD Requirements For Specific Standardized Cheese And Related Products; Cheeses And Related Cheese Products
SPREAD, CHEESE	133.176	PASTEURIZED CHEESE SPREAD WITH FRUITS, VEGETABLES OR MEATS Requirements For Specific Standardized Cheese And Related Products; Cheeses And Related Cheese Products
SPREAD, CHEESE	133.178	PASTEURIZED NEUFCHATEL CHEESE SPREAD WITH OTHER FOODS Requirements For Specific Standardized Cheese And Related Products; Cheeses And Related Cheese Products
SPREAD, CHEESE	133.179	PASTEURIZED PROCESS CHEESE SPREAD Requirements For Specific Standardized Cheese And Related Products; Cheeses And Related Cheese Products
SPREAD, CHEESE	133.180	PASTEURIZED PROCESS CHEESE SPREAD WITH FRUITS, VEGETABLES, OR MEATS Requirements For Specific Standardized Cheese And Related Products; Cheeses And Related Cheese Products
SPREADER	868.5760	CUFF SPREADER Therapeutic Devices; Anesthesiology Devices
SPREADS, PEANUT	102.23	PEANUT SPREADS Requirements For Specific Nonstandardized Foods; Common Or Usual Name For Nonstandardized Foods
SPRING-POWERED	872.4475	SPRING-POWERED JET INJECTOR Surgical Devices; Dental Devices
SPROUT	172.590	YEAST-MALT SPROUT EXTRACT Flavoring Agents And Related Substances; Food Additives Permitted For Direct Addition To Food For Human Consumption
SQUALANE	524.1484k	NEOMYCIN SULFATE, PREDNISOLONE, TETRACAINE, AND SQUALANE TOPICAL-OTIC SUSPENSION Ophthalmic And Topical Dosage Form New Animal Drugs
STABILITY TESTING	211.166	STABILITY TESTING Laboratory Controls; Current Good Manufacturing Practice For Finished Pharmaceuticals
STABILIZED	864.9400	STABILIZED ENZYME SOLUTION Products Used In Establishments That Manufacture Blood And Blood Products; Hematology And Pathology Devices
STABILIZER	880.6980	VEIN STABILIZER General Hospital And Personal Use Miscellaneous Devices; General Hospital And Personal Use Devices
STABILIZERS	178.2010	ANTIOXIDANTS AND/OR STABILIZERS FOR POLYMERS Antioxidants And Stabilizers; Indirect Food Additives: Adjuvants, Production Aids, And Sanitizers
STABILIZERS	178.2650	ORGANOTIN STABILIZERS IN VINYL CHLORIDE PLASTICS Antioxidants And Stabilizers; Indirect Food Additives: Adjuvants, Production Aids, And Sanitizers
STABILIZERS	181.29	STABILIZERS Specific Prior-sanctioned Food Ingredients; Prior-sanctioned Food Ingredients
STACKING DEVICE	866.2410	AUTOMATED MEDIUM DISPENSING AND STACKING DEVICE Microbiology Devices; Immunology And Microbiology Devices
STAINER	864.3800	AUTOMATED SLIDE STAINER Pathology Instrumentation And Accessories; Hematology And Pathology Devices
STAINLESS STEEL	872.3350	GOLD OR STAINLESS STEEL CUSP Prosthetic Devices; Dental Devices
STAINLESS STEEL	878.4495	STAINLESS STEEL SUTURE Surgical Devices; General And Plastic Surgery Devices
STAINS	864.1850	DYE AND CHEMICAL SOLUTION STAINS Biological Stains; Hematology And Pathology Devices
STAIR-CLIMBING	890.3890	STAIR-CLIMBING WHEELCHAIR Physical Medicine Prosthetic Devices; Physical Medicine Devices
STAND	880.6990	INFUSION STAND General Hospital And Personal Use Miscellaneous Devices; General Hospital And Personal Use Devices

STAND	886.1860	OPHTHALMIC INSTRUMENT STAND Diagnostic Devices; Ophthalmic Devices
STAND-ON SCALE	880.2700	STAND-ON PATIENT SCALE General Hospital And Personal Use Monitoring Devices; General Hospital And Personal Use Devices
STANDARD	5.400	ISSUANCE OF FEDERAL REGISTER DOCUMENTS TO RECOGNIZE OR TO WITHDRAW RECOGNITION OF A STANDARD TO MEET PREMARKET SUBMISSION REQUIREMENTS Medical Devices And Radiological Health; Redelegations Of Authority; Delegations Of Authority And Organization
STANDARD	60.36	STANDARD OF DUE DILIGENCE Due Diligence Petitions; Patent Term Restoration
STANDARD	130.5	PROCEDURE FOR ESTABLISHING A FOOD STANDARD General Provisions; Food Standards: General
STANDARD	860.132	PROCEDURES WHEN THE COMMISSIONER INITIATES A PERFORMANCE STANDARD OR PREMARKET APPROVAL PROCEEDING UNDER SECTION 514(B) OR 515(B) OF THE ACT Reclassification; Medical Device Classification Procedures
STANDARD	861.24	EXISTING STANDARD AS A PROPOSED STANDARD Procedures For Performance Standards Development And Publication; Procedures For Performance Standards Development
STANDARD	861.34	AMENDMENT OR REVOCATION OF A STANDARD Procedures For Performance Standards Development And Publication; Procedures For Performance Standards Development
STANDARD	898.12	PERFORMANCE STANDARD Performance Standards For Electrode Lead Wires And Patient Cables
STANDARD OPERATING PROCEDURES	58.81	STANDARD OPERATING PROCEDURES Testing Facilities Operation; Good Laboratory Practice For Nonclinical Laboratory Studies
STANDARD OPERATING PROCEDURES	120.6	SANITATION STANDARD OPERATING PROCEDURES Hazard Analysis And Critical Control Point (HACCP) Systems
STANDARD OPERATING PROCEDURES	606.100	STANDARD OPERATING PROCEDURES Production And Process Controls; Current Good Manufacturing Practice For Blood And Blood Components
STANDARD PREPARATION	640.55	U.S. STANDARD PREPARATION Cryoprecipitate; Additional Standards For Human Blood And Blood Products
STANDARD PREPARATIONS	610.20	STANDARD PREPARATIONS Standard Preparations And Limits Of Potency; General Biological Products Standards
STANDARD-SETTING	10.95	PARTICIPATION IN OUTSIDE STANDARD-SETTING ACTIVITIES General Administrative Procedures; Administrative Practices And Procedures
STANDARD SUNSCREEN	352.70	STANDARD SUNSCREEN Testing Procedures; Sunscreen Drug Products For Over-the-counter Human Use
STANDARDIZED CHEESES	133.193	SPICED, FLAVORED STANDARDIZED CHEESES Requirements For Specific Standardized Cheese And Related Products; Cheeses And Related Cheese Products
STANDARDIZED FOOD	130.9	SULFITES IN STANDARDIZED FOOD Food Standards: General
STANDARDIZED FOODS	70.10	COLOR ADDITIVES IN STANDARDIZED FOODS AND NEW DRUGS General Provisions; Color Additives
STANDARDIZED FOODS	130.11	LABEL DESIGNATIONS OF INGREDIENTS FOR STANDARDIZED FOODS Food Standards: General
STANDARDIZED FOODS	170.10	FOOD ADDITIVES IN STANDARDIZED FOODS General Provisions; Food Additives
STANDARDIZED TERM	130.10	REQUIREMENTS FOR FOODS NAMED BY USE OF A NUTRIENT CONTENT CLAIM AND A STANDARDIZED TERM Food Standards: General
STANDARDS	5.300	FOOD STANDARDS, FOOD ADDITIVES, GENERALLY RECOGNIZED AS SAFE (GRAS) SUBSTANCES, COLOR ADDITIVES, NUTRIENT CONTENT CLAIMS, AND HEALTH CLAIMS Foods And Cosmetics; Redelegations Of Authority; Delegations Of Authority And Organization
STANDARDS	5.303	ESTABLISHING STANDARDS AND APPROVING ACCREDITING BODIES UNDER THE NATIONAL LABORATORY ACCREDITATION PROGRAM Foods And Cosmetics; Redelegations Of Authority; Delegations Of Authority And Organization
STANDARDS	5.1000	AUTHORITY TO ENSURE THAT MAMMOGRAPHY FACILITIES MEET QUALITY STANDARDS Mammography Facilities; Redelegations Of Authority; Delegations Of Authority And Organization
STANDARDS	14.120	ESTABLISHMENT OF THE TECHNICAL ELECTRONIC PRODUCT RADIATION SAFETY STANDARDS COMMITTEE (TEPRSSC) Technical Electronic Products Radiation Safety Standards Committee; Public Hearing Before A Public Advisory Committee

STANDARDS	130.6	REVIEW OF CODEX ALIMENTARIUS FOOD STANDARDS General Provisions; Food Standards: General
STANDARDS	130.8	CONFORMITY TO DEFINITIONS AND STANDARDS OF IDENTITY General Provisions; Food Standards: General
STANDARDS	130.17	TEMPORARY PERMITS FOR INTERSTATE SHIPMENT OF EXPERIMENTAL PACKS OF FOOD VARYING FROM THE REQUIREMENTS OF DEFINITIONS AND STANDARDS OF IDENTITY General Provisions; Food Standards: General
STANDARDS	130.20	FOOD ADDITIVES PROPOSED FOR USE IN FOODS FOR WHICH DEFINITIONS AND STANDARDS OF IDENTITY ARE ESTABLISHED Food Additives In Standardized Foods; Food Standards: General
STANDARDS	861.7	CONTENTS OF STANDARDS General; Procedures For Performance Standards Development
STANDARDS	861.20	SUMMARY OF STANDARDS DEVELOPMENT PROCESS Procedures For Performance Standards Development And Publication; Procedures For Performance Standards Development
STANDARDS	861.30	DEVELOPMENT OF STANDARDS Procedures For Performance Standards Development And Publication; Procedures For Performance Standards Development
STANDARDS	861.38	STANDARDS ADVISORY COMMITTEES Procedures For Performance Standards Development And Publication; Procedures For Performance Standards Development
STANDARDS	900.4	STANDARDS FOR ACCREDITATION BODIES Accreditation; Mammography
STANDARDS	900.22	STANDARDS FOR CERTIFICATION AGENCIES States As Certifiers; Mammography
STANDARDS	1403.20	STANDARDS FOR FINANCIAL MANAGEMENT Post-award Requirements; Uniform Administrative Requirements For Grants And Coop. Agreements To State & Local Gov.
STANDARDS, PERFORMANCE	5.600	VARIANCES FROM PERFORMANCE STANDARDS FOR ELECTRONIC PRODUCTS Radiation Control; Redelegations Of Authority; Delegations Of Authority And Organization
STANDARDS, PERFORMANCE	5.601	EXEMPTION OF ELECTRONIC PRODUCTS FROM PERFORMANCE STANDARDS AND PROHIBITED ACTS Radiation Control; Redelegations Of Authority; Delegations Of Authority And Organization
STANDARDS, QUALITY	900.12	QUALITY STANDARDS Quality Standards And Certification; Mammography
STANDARDS, QUALITY	900.18	ALTERNATIVE REQUIREMENTS FOR ß 900.12 QUALITY STANDARDS Quality Standards And Certification; Mammography
STANDING COMMITTEES	14.80	QUALIFICATIONS FOR MEMBERS OF STANDING POLICY AND TECHNICAL ADVISORY COMMITTEES Members Of Advisory Committees; Public Hearing Before A Public Advisory Committee
STANDING COMMITTEES	14.84	NOMINATIONS AND SELECTION OF NONVOTING MEMBERS OF STANDING TECHNICAL ADVISORY COMMITTEES Members Of Advisory Committees; Public Hearing Before A Public Advisory Committee
STANDING COMMITTEES	14.100	LIST OF STANDING ADVISORY COMMITTEES Standing Advisory Committees; Public Hearing Before A Public Advisory Committee
STANDING COMMITTEES	14.160	ESTABLISHMENT OF STANDING TECHNICAL ADVISORY COMMITTEES FOR HUMAN PRESCRIPTION DRUGS Advisory Committees For Human Prescription Drugs; Public Hearing Before A Public Advisory Committee
STANDING REQUESTS	203.35	STANDING REQUESTS Samples; Prescription Drug Marketing
STANDUP WHEELCHAIR	890.3900	STANDUP WHEELCHAIR Physical Medicine Prosthetic Devices; Physical Medicine Devices
STANNOUS CHLORIDE	172.180	STANNOUS CHLORIDE Food Preservatives; Food Additives Permitted For Direct Addition To Food For Human Consumption
STANNOUS CHLORIDE	184.1845	STANNOUS CHLORIDE (ANHYDROUS AND DIHYDRATED) Listing Of Specific Substances Affirmed As GRAS; Direct Food Substances Affirmed As Generally Recognized As Safe
STANNOUS CHLORIDE	582.3845	STANNOUS CHLORIDE Chemical Preservatives; Substances Generally Recognized As Safe
STANOL ESTERS	101.83	HEALTH CLAIMS: PLANT STEROL/STANOL ESTERS AND RISK OF CORONARY HEART DISEASE Specific Requirements For Health Claims; Food Labeling
STANOZOLOL	520.2150	STANOZOLOL ORAL DOSAGE FORMS Oral Dosage Form New Animal Drugs
STANOZOLOL	522.2150	STANOZOLOL STERILE SUSPENSION Implantation Or Injectable Dosage Form New Animal Drugs

STAPHYLOCOCCAL TYPING	866.2050	STAPHYLOCOCCAL TYPING BACTERIOPHAGE Microbiology Devices; Immunology And Microbiology Devices
STAPHYLOCOCCUS AUREUS	866.3700	STAPHYLOCOCCUS AUREUS SEROLOGICAL REAGENTS Serological Reagents; Immunology And Microbiology Devices
STAPLE	878.4750	IMPLANTABLE STAPLE Surgical Devices; General And Plastic Surgery Devices
STAPLE	878.4760	REMOVABLE SKIN STAPLE Surgical Devices; General And Plastic Surgery Devices
STARCH	172.892	FOOD STARCH-MODIFIED Multipurpose Additives; Food Additives Permitted For Direct Addition To Food For Human Consumption
STARCH	178.3520	INDUSTRIAL STARCH-MODIFIED Certain Adjuvants And Production Aids; Indirect Food Additives: Adjuvants, Production Aids, And Sanitizers
STARTER DISTILLATE	184.1848	STARTER DISTILLATE Listing Of Specific Substances Affirmed As GRAS; Direct Food Substances Affirmed As Generally Recognized As Safe
STATE AUTHORITIES	5.606	ACCEPTANCE OF ASSISTANCE FROM STATE AND LOCAL AUTHORITIES FOR ENFORCEMENT OF RADIATION CONTROL LEGISLATION AND REGULATIONS Radiation Control; Redelegations Of Authority; Delegations Of Authority And Organization
STATE ENFORCEMENT	100.2	STATE ENFORCEMENT OF FEDERAL REGULATIONS State And Local Requirements; General
STATE LAW	1307.02	APPLICATION OF STATE LAW AND OTHER FEDERAL LAW General Information; Miscellaneous
STATE OFFICIALS	20.84	DISCLOSURE TO CONSULTANTS, ADVISORY COMMITTEES, STATE AND LOCAL GOVERNMENT OFFICIALS AND OTHER SPECIAL GOVERNMENT EMPLOYEES Limitations On Exemptions; Public Information
STATE OFFICIALS	20.88	COMMUNICATIONS WITH STATE AND LOCAL GOVERNMENT OFFICIALS Limitations On Exemptions; Public Information
STATE PLANS	1403.11	STATE PLANS Pre-Award Requirements; Uniform Administrative Requirements For Grants And Coop. Agreements To State & Local Gov.
STATE REQUIREMENTS	100.1	PETITIONS REQUESTING EXEMPTION FROM PREEMPTION FOR STATE OR LOCAL REQUIREMENTS State And Local Requirements; General
STATEMENT	720.6	AMENDMENTS TO STATEMENT Voluntary Filing Of Cosmetic Product Ingredient Composition Statements
STATEMENT OF DOSAGE	201.55	STATEMENT OF DOSAGE Labeling Requirements For Prescription Drugs And/or Insulin; Labeling
STATEMENT, 510(K)	807.93	CONTENT AND FORMAT OF A 510(K) STATEMENT Registration Procedures For Foreign Device Establishments; Establishment Registration And Device Listing For Manufacturers And Initial Importers Of Devices
STATEMENT OF IDENTITY	201.50	STATEMENT OF IDENTITY Labeling Requirements For Prescription Drugs And/or Insulin; Labeling
STATEMENT OF IDENTITY	201.61	STATEMENT OF IDENTITY Labeling Requirements For Over-the-counter Drugs; Labeling
STATEMENT OF IDENTITY	801.61	STATEMENT OF IDENTITY Labeling Requirements For Over-the-counter Devices; Labeling
STATEMENT, IMPACT	25.22	ACTIONS REQUIRING PREPARATION OF AN ENVIRONMENTAL IMPACT STATEMENT Agency Actions Requiring Environmental Consideration; Environmental Impact Considerations
STATEMENT, INGREDIENT	720.7	NOTIFICATION OF PERSON SUBMITTING COSMETIC PRODUCT INGREDIENT STATEMENT Voluntary Filing Of Cosmetic Product Ingredient Composition Statements
STATEMENT OF INGREDIENTS	201.10	DRUGS; STATEMENT OF INGREDIENTS General Labeling Provisions; Labeling
STATEMENT NUMBER	720.9	MISBRANDING BY REFERENCE TO FILING OR TO STATEMENT NUMBER Voluntary Filing Of Cosmetic Product Ingredient Composition Statements
STATEMENT OF POLICY	310.100	NEW DRUG STATUS OPINIONS; STATEMENT OF POLICY Specific Administrative Rulings And Decisions; New Drugs
STATEMENT OF POLICY	861.5	STATEMENT OF POLICY General; Procedures For Performance Standards Development
STATEMENT, WARNING	201.309	ACETOPHENETIDIN (PHENACETIN)-CONTAINING PREPARATIONS; NECESSARY WARNING STATEMENT Specific Labeling Requirements For Specific Drug Products; Labeling
STATEMENT, WARNING	290.5	DRUGS; STATEMENT OF REQUIRED WARNING General Provisions; Controlled Drugs
STATEMENTS	101.93	CERTAIN TYPES OF STATEMENTS FOR DIETARY SUPPLEMENTS Specific Requirements For Descriptive Claims That Are Neither Nutrient Content Nor Health Claims; Food Labeling

STATEMENTS	514.15	UNTRUE STATEMENTS IN APPLICATIONS General Provisions; New Animal Drug Applications
STATEMENTS	720.8	CONFIDENTIALITY OF STATEMENTS Voluntary Filing Of Cosmetic Product Ingredient Composition Statements
STATEMENTS, IMPACT	25.42	ENVIRONMENTAL IMPACT STATEMENTS Preparation Of Environmental Documents; Environmental Impact Considerations
STATEMENTS, IMPACT	25.52	ENVIRONMENTAL IMPACT STATEMENTS Public Participation and Notification Of Environmental Documents; Environmental Impact Considerations
STATEMENTS, LABEL	1.23	PROCEDURES FOR REQUESTING VARIATIONS AND EXEMPTIONS FROM REQUIRED LABEL STATEMENTS General Labeling Requirements; General Enforcement Regulations
STATEMENTS, LABEL	1.24	EXEMPTIONS FROM REQUIRED LABEL STATEMENTS General Labeling Requirements; General Enforcement Regulations
STATEMENTS, LABEL	101.65	IMPLIED NUTRIENT CONTENT CLAIMS AND RELATED LABEL STATEMENTS Specific Requirements For Nutrient Content Claims; Food Labeling
STATEMENTS, LABEL	105.66	LABEL STATEMENTS RELATING TO USEFULNESS IN REDUCING OR MAINTAINING BODY WEIGHT Label Statements; Foods For Special Dietary Use
STATEMENTS, LABEL	201.15	DRUGS; PROMINENCE OF REQUIRED LABEL STATEMENTS General Labeling Provisions; Labeling
STATEMENTS, LABEL	801.15	MEDICAL DEVICES; PROMINENCE OF REQUIRED LABEL STATEMENTS General Labeling Provisions; Labeling
STATEMENTS, LABELING	101.17	FOOD LABELING WARNING, NOTICE, AND SAFE HANDLING STATEMENTS General Provisions; Food Labeling
STATEMENTS, LABELING	201.58	REQUESTS FOR WAIVER OF REQUIREMENT FOR ADEQUATE AND WELL-CONTROLLED STUDIES TO SUBSTANTIATE CERTAIN LABELING STATEMENTS Labeling Requirements For Prescription Drugs And/or Insulin; Labeling
STATEMENTS, MANDATORY	99.103	MANDATORY STATEMENTS AND INFORMATION Dissemination Of Information On Unapproved/New Uses For Marketed Drugs, Biologics, And Devices
STATEMENTS, MISLEADING	201.6	DRUGS; MISLEADING STATEMENTS General Labeling Provisions; Labeling
STATEMENTS, MISLEADING	801.6	MEDICAL DEVICES; MISLEADING STATEMENTS General Labeling Provisions; Labeling
STATEMENTS OF POLICY	369.4	WARNINGS SUGGESTED FOR DRUGS BY FORMAL OR INFORMAL STATEMENTS OF POLICY Definitions And Interpretations; Interpretative Statements Re: Warnings On Drugs And Devices For OTC Sale
STATEMENTS, QUALITY	130.14	GENERAL STATEMENTS OF SUBSTANDARD QUALITY AND SUBSTANDARD FILL OF CONTAINER General Provisions; Food Standards: General
STATEMENTS, REQUIRED	101.15	FOOD; PROMINENCE OF REQUIRED STATEMENTS General Provisions; Food Labeling
STATEMENTS, REQUIRED	201.16	DRUGS; SPANISH-LANGUAGE VERSION OF CERTAIN REQUIRED STATEMENTS General Labeling Provisions; Labeling
STATEMENTS, REQUIRED	501.15	ANIMAL FOOD; PROMINENCE OF REQUIRED STATEMENTS General Provisions; Animal Food Labeling
STATEMENTS, REQUIRED	801.16	MEDICAL DEVICES; SPANISH-LANGUAGE VERSION OF CERTAIN REQUIRED STATEMENTS General Labeling Provisions; Labeling
STATEMENTS, WARNING	101.17	FOOD LABELING WARNING AND NOTICE STATEMENTS General Provisions; Food Labeling
STATEMENTS, WARNING	201.22	PRESCRIPTION DRUGS CONTAINING SULFITES; REQUIRED WARNING STATEMENTS General Labeling Provisions; Labeling
STATEMENTS, WARNING	201.320	WARNING STATEMENTS FOR DRUG PRODUCTS CONTAINING OR MANUFACTURED WITH CHLOROFLUOROCARBONS Labeling
STATEMENTS, WARNING	369.8	WARNING STATEMENTS IN RELATION TO CONDITIONS FOR USE Definitions And Interpretations; Interpretative Statements Re: Warnings On Drugs And Devices For OTC Sale
STATEMENTS, WARNING	369.10	CONSPICUOUSNESS OF WARNING STATEMENTS Definitions And Interpretations; Interpretative Statements Re: Warnings On Drugs And Devices For OTC Sale
STATEMENTS, WARNING	369.20	DRUGS; RECOMMENDED WARNING AND CAUTION STATEMENTS Warning And Caution Statement For Drugs; Interpretative Statements Re: Warnings On Drugs And Devices For OTC Sale
STATEMENTS, WARNING	369.21	DRUGS; WARNING AND CAUTION STATEMENTS REQUIRED BY REGULATIONS Warning And Caution Statement For Drugs; Interpretative Statements Re: Warnings On Drugs And Devices For OTC Sale

STATEMENTS, WARNING	369.22	DRUGS; WARNING AND CAUTION STATEMENTS SPECIFICALLY REQUIRED BY LAW Warning And Caution Statement For Drugs; Interpretative Statements Re: Warnings On Drugs And Devices For OTC Sale
STATEMENTS, WARNING	501.17	ANIMAL FOOD LABELING WARNING STATEMENTS General Provisions; Animal Food Labeling
STATEMENTS, WARNING	740.1	ESTABLISHMENT OF WARNING STATEMENTS General; Cosmetic Product Warning Statements
STATEMENTS, WARNING	740.2	CONSPICUOUSNESS OF WARNING STATEMENTS General; Cosmetic Product Warning Statements
STATEMENTS, WARNING	801.63	MEDICAL DEVICES; WARNING STATEMENTS FOR DEVICES CONTAINING OR MANUFACTURED WITH CHLOROFLUOROCARBONS Labeling
STATEMENTS, WARNING	801.433	WARNING STATEMENTS FOR PRESCRIPTION AND RESTRICTED DEVICE PRODUCTS CONTAINING OR MANUFACTURED WITH CHLOROFLUOROCARBONS Labeling
STATEMENTS, WARNINGS	369.4	WARNINGS SUGGESTED FOR DRUGS BY FORMAL OR INFORMAL STATEMENTS OF POLICY Definitions And Interpretations; Interpretative Statements Re: Warnings On Drugs And Devices For OTC Sale
STATEMENTS, WITNESS	17.25	EXCHANGE OF WITNESS LISTS, WITNESS STATEMENTS, AND EXHIBITS Civil Money Penalties Hearings
STATIONARY	892.1680	STATIONARY X-RAY SYSTEM Diagnostic Devices; Radiology Devices
STATISTICAL TECHNIQUES	820.250	STATISTICAL TECHNIQUES Statistical Techniques; Quality System Regulation
STATUS	106.1	STATUS AND APPLICABILITY OF THE QUALITY CONTROL PROCEDURES REGULATION General Provisions; Infant Formula Quality Control Procedures
STATUS	210.1	STATUS OF CURRENT GOOD MANUFACTURING PRACTICE REGULATIONS Current Good Manufacturing Practice In Manufacturing, Processing, Packing, Or Holding Of Drugs; General
STATUS	312.45	INACTIVE STATUS Investigational New Drug Application
STATUS	330.12	STATUS OF OVER-THE-COUNTER (OTC) DRUGS PREVIOUSLY REVIEWED UNDER THE DRUG EFFICACY STUDY (DESI) Administrative Procedures; Over-the-counter Human Drugs Generally Recognized As Safe And Effective & Not Misbranded
STATUS	610.19	STATUS OF SPECIFIC PRODUCTS; GROUP A STREPTOCOCCUS General Provisions; General Biological Products Standards
STATUS	820.86	ACCEPTANCE STATUS Acceptance Activities; Quality System Regulation
STATUS, FOOD ADDITIVE	170.6	OPINION LETTERS ON FOOD ADDITIVE STATUS General Provisions; Food Additives
STATUS, FOOD ADDITIVE	170.38	DETERMINATION OF FOOD ADDITIVE STATUS Food Additive Safety; Food Additives
STATUS, FOOD ADDITIVE	570.6	OPINION LETTERS ON FOOD ADDITIVE STATUS General Provisions; Food Additives
STATUS, FOOD ADDITIVE	570.38	DETERMINATION OF FOOD ADDITIVE STATUS Food Additive Safety; Food Additives
STATUS, GRAS	170.35	AFFIRMATION OF GENERALLY RECOGNIZED AS SAFE (GRAS) STATUS Food Additive Safety; Food Additives
STATUS, GRAS	570.35	AFFIRMATION OF GENERALLY RECOGNIZED AS SAFE (GRAS) STATUS Food Additive Safety; Food Additives
STATUS, HUD	814.102	DESIGNATION OF HUD STATUS Humanitarian Use Devices; Premarket Approval Of Medical Devices
STATUS, NEW DRUG	310.100	NEW DRUG STATUS OPINIONS; STATEMENT OF POLICY Specific Administrative Rulings And Decisions; New Drugs
STATUS, NEW DRUG	310.502	CERTAIN DRUGS ACCORDED NEW DRUG STATUS THROUGH RULEMAKING PROCEDURES Requirements For Specific New Drugs Or Devices; New Drugs
STATUS, ORPHAN-DRUG	316.21	VERIFICATION OF ORPHAN-DRUG STATUS Orphan Drugs
STATUS, RECALL	7.53	RECALL STATUS REPORTS Recalls - Guidance On Policy, Procedures & Industry Responsibilities; Enforcement Policy
STATUS, RECALL ORDER	810.16	CEASE DISTRIBUTION AND NOTIFICATION OR MANDATORY RECALL ORDER STATUS REPORTS Mandatory Medical Device Recall Procedures; Medical Device Recall Authority
STAY OF ACTION	10.35	ADMINISTRATIVE STAY OF ACTION General Administrative Procedures; Administrative Practices And Procedures
STAY OF ACTION	12.35	NOTICE OF HEARING; STAY OF ACTION Initiation Of Proceedings; Formal Evidentiary Public Hearing

STAY OF ACTION	12.139	RECONSIDERATION AND STAY OF ACTION Initial And Final Decisions; Formal Evidentiary Public Hearing
STAY OF ACTION	16.119	RECONSIDERATION AND STAY OF ACTION Reconsideration And Stay; Regulatory Hearing Before The Food And Drug Administration
STAY OF REVIEW TIME	3.10	STAY OF REVIEW TIME Product Jurisdiction
STAYS	5.107	EXTENSIONS OR STAYS OF EFFECTIVE DATES FOR COMPLIANCE WITH CERTAIN LABELING REQUIREMENTS FOR HUMAN PRESCRIPTION DRUGS Human Drugs; Redelegations Of Authority; Delegations Of Authority And Organization
STEAM CABINET	890.5250	MOIST STEAM CABINET Physical Medicine Therapeutic Devices; Physical Medicine Devices
STEAM LINES	200.11	USE OF OCTADECYLAMINE IN STEAM LINES OF DRUG ESTABLISHMENTS General Provisions; General
STEAM STERILIZER	880.6880	STEAM STERILIZER General Hospital And Personal Use Miscellaneous Devices; General Hospital And Personal Use Devices
STEARATE	178.3690	PENTAERYTHRITOL ADIPATE-STEARATE Certain Adjuvants And Production Aids; Indirect Food Additives: Adjuvants, Production Aids, And Sanitizers
STEARATE, CALCIUM	184.1229	CALCIUM STEARATE Listing Of Specific Substances Affirmed As GRAS; Direct Food Substances Affirmed As Generally Recognized As Safe
STEARATE, CALCIUM	573.280	FEED-GRADE CALCIUM STEARATE AND SODIUM STEARATE Food Additive Listing; Food Additives Permitted In Feed And Drinking Water Of Animals
STEARATE, MAGENSIUM	184.1440	MAGNESIUM STEARATE Listing Of Specific Substances Affirmed As GRAS; Direct Food Substances Affirmed As Generally Recognized As Safe
STEARATE, SODIUM	573.280	FEED-GRADE CALCIUM STEARATE AND SODIUM STEARATE Food Additive Listing; Food Additives Permitted In Feed And Drinking Water Of Animals
STEARATE, ZINC	182.8994	ZINC STEARATE Nutrients; Substances Generally Recognized As Safe
STEARATE, ZINC	582.5994	ZINC STEARATE Nutrients And/or Dietary Supplements; Substances Generally Recognized As Safe
STEARIC ACID	178.3450	ESTERS OF STEARIC AND PALMITIC ACIDS Certain Adjuvants And Production Aids; Indirect Food Additives: Adjuvants, Production Aids, And Sanitizers
STEARIC ACID	184.1090	STEARIC ACID Listing Of Specific Substances Affirmed As GRAS; Direct Food Substances Affirmed As Generally Recognized As Safe
STEAROTHERMOPHILUS	184.1012	α-AMYLASE ENZYME PREPARATION FROM BACILLUS STEAROTHERMOPHILUS Listing Of Specific Substances Affirmed As GRAS; Direct Food Substances Affirmed As Generally Recognized As Safe
STEAROYL PROPYLENE GLYCOL	172.765	SUCCISTEARIN (STEAROYL PROPYLENE GLYCOL HYDROGEN SUCCINATE) Other Specific Usage Additives; Food Additives Permitted For Direct Addition To Food For Human Consumption
STEAROYL-LACTYLATE	172.844	CALCIUM STEAROYL-2-LACTYLATE Multipurpose Additives; Food Additives Permitted For Direct Addition To Food For Human Consumption
STEAROYL LACTYLATE	172.846	SODIUM STEAROYL LACTYLATE Multipurpose Additives; Food Additives Permitted For Direct Addition To Food For Human Consumption
STEARYL CITRATE	172.755	STEARYL MONOGLYCERIDYL CITRATE Other Specific Usage Additives; Food Additives Permitted For Direct Addition To Food For Human Consumption
STEARYL CITRATE	184.1851	STEARYL CITRATE Listing Of Specific Substances Affirmed As GRAS; Direct Food Substances Affirmed As Generally Recognized As Safe
STEARYL CITRATE	582.6851	STEARYL CITRATE Sequestrants; Substances Generally Recognized As Safe
STEARYL FUMARATE	172.826	SODIUM STEARYL FUMARATE Multipurpose Additives; Food Additives Permitted For Direct Addition To Food For Human Consumption
STEEL	178.3300	CORROSION INHIBITORS USED FOR STEEL OR TINPLATE Certain Adjuvants And Production Aids; Indirect Food Additives: Adjuvants, Production Aids, And Sanitizers
STEEL	872.3350	GOLD OR STAINLESS STEEL CUSP Prosthetic Devices; Dental Devices
STEEL	878.4495	STAINLESS STEEL SUTURE Surgical Devices; General And Plastic Surgery Devices

STEERABLE	870.1280	STEERABLE CATHETER Cardiovascular Diagnostic Devices; Cardiovascular Devices
STEERABLE	870.1290	STEERABLE CATHETER CONTROL SYSTEM Cardiovascular Diagnostic Devices; Cardiovascular Devices
STENT	876.4620	URETERAL STENT Surgical Devices; Gastroenterology-urology Devices
STENT	884.3900	VAGINAL STENT Obstetrical And Gynecological Prosthetic Devices; Obstetrical And Gynecological Devices
STERCULIA GUM	184.1349	KARAYA GUM (STERCULIA GUM) Listing Of Specific Substances Affirmed As GRAS; Direct Food Substances Affirmed As Generally Recognized As Safe
STERCULIA GUM	582.7349	STERCULIA GUM Stabilizers; Substances Generally Recognized As Safe
STEREOPSIS	886.1460	STEREOPSIS MEASURING INSTRUMENT Diagnostic Devices; Ophthalmic Devices
STEREOSCOPE	886.1870	STEREOSCOPE Diagnostic Devices; Ophthalmic Devices
STEREOSCOPIC	886.1880	FUSION AND STEREOSCOPIC TARGET Diagnostic Devices; Ophthalmic Devices
STEREOTAXIC	882.4560	STEREOTAXIC INSTRUMENT Neurological Surgical Devices; Neurological Devices
STERILIZANTS	880.6885	LIQUID CHEMICAL STERILIZANTS/HIGH-LEVEL DISINFECTANTS General Hospital And Personal Use Miscellaneous Devices; General Hospital And Personal Use Devices
STERILE	522.44	STERILE SODIUM ACETAZOLAMIDE Implantation Or Injectable Dosage Form New Animal Drugs
STERILE	522.82	AMINOPROPAZINE FUMARATE STERILE SOLUTION INJECTION Implantation Or Injectable Dosage Form New Animal Drugs
STERILE	522.90a	AMPICILLIN TRIHYDRATE STERILE SUSPENSION Implantation Or Injectable Dosage Form New Animal Drugs
STERILE	522.90b	AMPICILLIN TRIHYDRATE FOR STERILE SUSPENSION Implantation Or Injectable Dosage Form New Animal Drugs
STERILE	522.314	CEFTIOFUR HYDROCHLORIDE STERILE SUSPENSION Implantation Or Injectable Dosage Form New Animal Drugs
STERILE	522.380	CHLORAL HYDRATE, PENTOBARBITAL, AND MAGNESIUM SULFATE STERILE AQUEOUS SOLUTION Implantation Or Injectable Dosage Form New Animal Drugs
STERILE	522.690	DINOPROST TROMETHAMINE STERILE SOLUTION Implantation Or Injectable Dosage Form New Animal Drugs
STERILE	522.1085	GUAIFENESIN STERILE POWDER Implantation Or Injectable Dosage Form New Animal Drugs
STERILE	522.1155	IMIDOCARB DIPROPIONATE STERILE POWDER Implantation Or Injectable Dosage Form New Animal Drugs
STERILE	522.1290	LUPROSTIOL STERILE SOLUTION Implantation Or Injectable Dosage Form New Animal Drugs
STERILE	522.1410	STERILE METHYLPREDNISOLONE ACETATE SUSPENSION Implantation Or Injectable Dosage Form New Animal Drugs
STERILE	522.1484	NEOMYCIN SULFATE STERILE SOLUTION Implantation Or Injectable Dosage Form New Animal Drugs
STERILE	522.1696a	PENICILLIN G BENZATHINE AND PENICILLIN G PROCAINE STERILE SUSPENSION Implantation Or Injectable Dosage Form New Animal Drugs
STERILE	522.1862	STERILE PRALIDOXIME CHLORIDE Implantation Or Injectable Dosage Form New Animal Drugs
STERILE	522.1881	STERILE PREDNISOLONE ACETATE AQUEOUS SUSPENSION Implantation Or Injectable Dosage Form New Animal Drugs
STERILE	522.1883	PREDNISOLONE SODIUM PHOSPHATE INJECTION, STERILE Implantation Or Injectable Dosage Form New Animal Drugs
STERILE	522.1890	STERILE PREDNISONE SUSPENSION Implantation Or Injectable Dosage Form New Animal Drugs
STERILE	522.2112	STERILE SOMETRIBOVE ZINC SUSPENSION Implantation Or Injectable Dosage Form New Animal Drugs
STERILE	522.2150	STANOZOLOL STERILE SUSPENSION Implantation Or Injectable Dosage Form New Animal Drugs
STERILE	522.2483	STERILE TRIAMCINOLONE ACETONIDE SUSPENSION Implantation Or Injectable Dosage Form New Animal Drugs
STERILE	522.2610	TRIMETHOPRIM AND SULFADIAZINE STERILE SUSPENSION Implantation Or Injectable Dosage Form New Animal Drugs
STERILE	524.1484	NEOMYCIN SULFATE, HYDROCORTISONE ACETATE, STERILE OINTMENT Ophthalmic And Topical Dosage Form New Animal Drugs
STERILE	524.1881b	PREDNISOLONE ACETATE-NEOMYCIN SULFATE STERILE SUSPENSION Ophthalmic And Topical Dosage Form New Animal Drugs

STERILITY	610.12	STERILITY General Provisions; General Biological Products Standards
STERILITY	800.10	CONTACT LENS SOLUTION; STERILITY Requirements For Specific Medical Devices; General
STERILIZATION	880.2800	STERILIZATION PROCESS INDICATOR General Hospital And Personal Use Monitoring Devices; General Hospital And Personal Use Devices
STERILIZATION	880.6850	STERILIZATION WRAP General Hospital And Personal Use Miscellaneous Devices; General Hospital And Personal Use Devices
STERILIZER, BOILING WATER	872.6710	BOILING WATER STERILIZER Miscellaneous Devices; Dental Devices
STERILIZER, DRY HEAT	872.6730	ENDODONTIC DRY HEAT STERILIZER Miscellaneous Devices; Dental Devices
STERILIZER, DRY-HEAT	880.6870	DRY-HEAT STERILIZER General Hospital And Personal Use Miscellaneous Devices; General Hospital And Personal Use Devices
STERILIZER, GAS	880.6860	ETHYLENE OXIDE GAS STERILIZER General Hospital And Personal Use Miscellaneous Devices; General Hospital And Personal Use Devices
STERILIZER, STEAM	880.6880	STEAM STERILIZER General Hospital And Personal Use Miscellaneous Devices; General Hospital And Personal Use Devices
STERILIZER, TONOMETER	886.1940	TONOMETER STERILIZER Diagnostic Devices; Ophthalmic Devices
STEROID, ANABOLIC	1308.25	EXCLUSION OF A VETERINARY ANABOLIC STEROID IMPLANT PRODUCT; APPLICATION Schedules Of Controlled Substances
STEROID, ANABOLIC	1308.26	EXCLUDED VETERINARY ANABOLIC STEROID IMPLANT PRODUCTS Schedules Of Controlled Substances
STEROID, ANABOLIC	1308.33	EXEMPTION OF CERTAIN ANABOLIC STEROID PRODUCTS; APPLICATION Schedules Of Controlled Substances
STEROID, ANABOLIC	1308.34	EXEMPT ANABOLIC STEROID PRODUCTS Schedules Of Controlled Substances
STEROIDS	862.1385	17-HYDROXYCORTICOSTEROIDS (17-KETOGENIC STEROIDS) TEST SYSTEM Clinical Chemistry Test Systems; Clinical Chemistry And Clinical Toxicology Devices
STEROL ESTERS	101.83	HEALTH CLAIMS: PLANT STEROL/STANOL ESTERS AND RISK OF CORONARY HEART DISEASE Specific Requirements For Health Claims; Food Labeling
STETHOSCOPE	868.1910	ESOPHAGEAL STETHOSCOPE Diagnostic Devices; Anesthesiology Devices
STETHOSCOPE	868.1920	ESOPHAGEAL STETHOSCOPE WITH ELECTRICAL CONDUCTORS Diagnostic Devices; Anesthesiology Devices
STETHOSCOPE	868.1930	STETHOSCOPE HEAD Diagnostic Devices; Anesthesiology Devices
STETHOSCOPE	870.1875	STETHOSCOPE Cardiovascular Diagnostic Devices; Cardiovascular Devices
STETHOSCOPE	884.2900	FETAL STETHOSCOPE Obstetrical And Gynecological Monitoring Devices; Obstetrical And Gynecological Devices
STIMULANT	340.10	STIMULANT ACTIVE INGREDIENT Active Ingredient; Stimulant Drug Products For Over-the-counter Human Use
STIMULANT	340.50	LABELING OF STIMULANT DRUG PRODUCTS Labeling; Stimulant Drug Products For Over-the-counter Human Use
STIMULATOR	352.71	LIGHT SOURCE (SOLAR STIMULATOR) Sunscreen Drug Products For Over-the-counter Human Use
STIMULATOR	868.2775	ELECTRICAL PERIPHERAL NERVE STIMULATOR Monitoring Devices; Anesthesiology Devices
STIMULATOR	870.3850	CAROTID SINUS NERVE STIMULATOR Cardiovascular Prosthetic Devices; Cardiovascular Devices
STIMULATOR	874.1800	AIR OR WATER CALORIC STIMULATOR Diagnostic Devices; Ear, Nose, And Throat Devices
STIMULATOR	874.1820	SURGICAL NERVE STIMULATOR/LOCATOR Diagnostic Devices; Ear, Nose, And Throat Devices
STIMULATOR	882.1870	EVOKED RESPONSE ELECTRICAL STIMULATOR Neurological Diagnostic Devices; Neurological Devices
STIMULATOR	882.1880	EVOKED RESPONSE MECHANICAL STIMULATOR Neurological Diagnostic Devices; Neurological Devices
STIMULATOR	882.1890	EVOKED RESPONSE PHOTIC STIMULATOR Neurological Diagnostic Devices; Neurological Devices
STIMULATOR	882.1900	EVOKED RESPONSE AUDITORY STIMULATOR Neurological Diagnostic Devices; Neurological Devices

STIMULATOR	882.5800	CRANIAL ELECTROTHERAPHY STIMULATOR Neurological Therapeutic Devices; Neurological Devices
STIMULATOR	882.5810	EXTERNAL FUNCTIONAL NEUROMUSCULAR STIMULATOR Neurological Therapeutic Devices; Neurological Devices
STIMULATOR	882.5820	IMPLANTED CEREBELLAR STIMULATOR Neurological Therapeutic Devices; Neurological Devices
STIMULATOR	882.5830	IMPLANTED DIAPHRAGMATIC/PHRENIC NERVE STIMULATOR Neurological Therapeutic Devices; Neurological Devices
STIMULATOR	882.5840	IMPLANTED INTRACEREBRAL/SUBCORTICAL STIMULATOR FOR PAIN RELIEF Neurological Therapeutic Devices; Neurological Devices
STIMULATOR	882.5850	IMPLANTED SPINAL CORD STIMULATOR FOR BLADDER EVACUATION Neurological Therapeutic Devices; Neurological Devices
STIMULATOR	882.5860	IMPLANTED NEUROMUSCULAR STIMULATOR Neurological Therapeutic Devices; Neurological Devices
STIMULATOR	882.5870	IMPLANTED PERIPHERAL NERVE STIMULATOR FOR PAIN RELIEF Neurological Therapeutic Devices; Neurological Devices
STIMULATOR	882.5880	IMPLANTED SPINAL CORD STIMULATOR FOR PAIN RELIEF Neurological Therapeutic Devices; Neurological Devices
STIMULATOR	882.5890	TRANSCUTANEOUS ELECTRICAL NERVE STIMULATOR FOR PAIN RELIEF Neurological Therapeutic Devices; Neurological Devices
STIMULATOR	884.5940	POWERED VAGINAL MUSCLE STIMULATOR FOR THERAPEUTIC USE Obstetrical And Gynecological Therapeutic Devices; Obstetrical And Gynecological Devices
STIMULATOR	890.1850	DIAGNOSTIC MUSCLE STIMULATOR Physical Medicine Diagnostic Devices; Physical Medicine Devices
STIMULATOR	890.5850	POWERED MUSCLE STIMULATOR Physical Therapeutic Diagnostic Devices; Physical Medicine Devices
STIMULATOR	890.5860	ULTRASOUND AND MUSCLE STIMULATOR Physical Therapeutic Diagnostic Devices; Physical Medicine Devices
STIRRED CURD CHEESE	133.144	GRANULAR AND STIRRED CURD CHEESE Requirements For Specific Standardized Cheese And Related Products; Cheeses And Related Cheese Products
STOCKING	880.5780	MEDICAL SUPPORT STOCKING General Hospital And Personal Use Therapeutic Devices; General Hospital And Personal Use Devices
STOLEN ORDER FORMS	1305.12	LOST AND STOLEN ORDER FORMS Order Forms
STOMACH	876.1400	STOMACH pH ELECTRODE Diagnostic Devices; Gastroenterology-urology Devices
STOMACH ACIDIFIERS	310.540	DRUG PRODUCTS CONTAINING ACTIVE INGREDIENTS OFFERED OVER-THE-COUNTER (OTC) FOR USE AS STOMACH ACIDIFIERS Requirements For Specific New Drugs Or Devices; New Drugs
STONE	876.4650	WATER JET RENAL STONE DISLODGER SYSTEM Surgical Devices; Gastroenterology-urology Devices
STONE	876.4680	URETERAL STONE DISLODGER Surgical Devices; Gastroenterology-urology Devices
STONES	862.1780	URINARY CALCULI (STONES) TEST SYSTEM Clinical Chemistry Test Systems; Clinical Chemistry And Clinical Toxicology Devices
STOOL	868.6700	ANESTHESIA STOOL Miscellaneous; Anesthesiology Devices
STOPCOCK	870.4290	CARDIOPULMONARY BYPASS ADAPTOR, STOPCOCK, MANIFOLD, OR FITTING Cardiovascular Surgical Devices; Cardiovascular Devices
STORAGE	2.35	USE OF SECONDHAND CONTAINERS FOR THE SHIPMENT AND STORAGE OF FOOD AND ANIMAL FEED Human and Animal Foods; General Administrative Rulings And Decisions
STORAGE	58.51	SPECIMEN AND DATA STORAGE FACILITIES Facilities; Good Laboratory Practice For Nonclinical Laboratory Studies
STORAGE	58.190	STORAGE AND RETRIEVAL OF RECORDS AND DATA Records And Reports; Good Laboratory Practice For Nonclinical Laboratory Studies
STORAGE	203.32	DRUG SAMPLE STORAGE AND HANDLING REQUIREMENTS Samples; Prescription Drug Marketing
STORAGE	205.50	MINIMUM REQUIREMENTS FOR THE STORAGE AND HANDLING OF PRESCRIPTION DRUGS AND FOR ESTABLISHMENT AND MAINTENANCE OF PRESCRIPTION DRUG DISTRIBUTION RECORDS Guidelines For State Licensing Of Wholesale Prescription Drug Distributors
STORAGE	211.82	RECEIPT AND STORAGE OF UNTESTED COMPONENTS, DRUG PRODUCT CONTAINERS, AND CLOSURES Control Of Components And Drug Product Containers And Closures; Current Good Manufacturing Practice For Finished Pharmaceuticals
STORAGE	500.45	USE OF POLYCHLORINATED BIPHENYLS (PCB'S) IN THE PRODUCTION, HANDLING, AND STORAGE OF ANIMAL FEED Specific Administrative Rulings And Decisions; General

STORAGE	820.150	STORAGE Handling, Storage, Distribution, And Installation; Quality System Regulation
STORAGE	864.3250	SPECIMEN TRANSPORT AND STORAGE CONTAINER Pathology Instrumentation And Accessories; Hematology And Pathology Devices
STORAGE	864.9575	ENVIRONMENTAL CHAMBER FOR STORAGE OF PLATELET CONCENTRATE Products Used In Establishments That Manufacture Blood And Blood Products; Hematology And Pathology Devices
STORAGE	864.9700	BLOOD STORAGE REFRIGERATOR AND BLOOD STORAGE FREEZER Products Used In Establishments That Manufacture Blood And Blood Products; Hematology And Pathology Devices
STORAGE	892.2010	MEDICAL IMAGE STORAGE DEVICE Diagnostic Devices; Radiology Devices
STORAGE	1250.27	STORAGE OF PERISHABLES Food Service Sanitation On Land And Air Conveyances, And Vessels; Interstate Conveyance Sanitation
STORAGE	1250.83	STORAGE OF WATER PRIOR TO TREATMENT Sanitation Facilities And Conditions On Vessels; Interstate Conveyance Sanitation
STORAGE AREAS	225.35	USE OF WORK AREAS, EQUIPMENT, AND STORAGE AREAS FOR OTHER MANUFACTURING AND STORAGE PURPOSE Construction And Maintenance Of Facilities And Equipment; Current Good Manufacturing Practice For Medicated Feeds
STORAGE AREAS	225.135	WORK AND STORAGE AREAS Facilities And Equipment; Current Good Manufacturing Practice For Medicated Feeds
STORAGE AREAS	1301.72	PHYSICAL SECURITY CONTROLS FOR NON-PRACTITIONERS; NARCOTIC TREATMENT PROGRAMS AND COMPOUNDERS FOR NARCOTIC TREATMENT PROGRAMS; STORAGE AREAS Security Requirements; Registration Of Manufacturers, Distributors, And Dispensers Of Controlled Substances
STORED	640.76	PRODUCTS STORED OR SHIPPED AT UNACCEPTABLE TEMPERATURES Source Plasma; Additional Standards For Human Blood And Blood Products
STORED	1250.30	CONSTRUCTION, MAINTENANCE AND USE OF PLACES WHERE FOOD IS PREPARED, SERVED, OR STORED Food Service Sanitation On Land And Air Conveyances, And Vessels; Interstate Conveyance Sanitation
STRAIGHT COLORS	70.20	PACKAGING REQUIREMENTS FOR STRAIGHT COLORS (OTHER THAN HAIR DYES) Packaging And Labeling; Color Additives
STRAIGHT COLORS	82.5	GENERAL SPECIFICATIONS FOR STRAIGHT COLORS General Provisions; Listing Of Certified Provisionally Listed Colors And Specifications
STRAMONIUM	250.12	STRAMONIUM PREPARATIONS LABELED WITH DIRECTIONS FOR USE IN SELF-MEDICATION REGARDED AS MISBRANDED Drugs Regarded As Misbranded; Specific Requirements For Specific Human Drugs
STRAP	868.5560	GAS MASK HEAD STRAP Therapeutic Devices; Anesthesiology Devices
STRATEGY, RECALL	7.42	RECALL STRATEGY Recalls - Guidance On Policy, Procedures & Industry Responsibilities; Enforcement Policy
STRATEGY, RECALL	810.14	CEASE DISTRIBUTION AND NOTIFICATION OR MANDATORY RECALL STRATEGY Mandatory Medical Device Recall Procedures; Medical Device Recall Authority
STREPTOCOCCUS	610.19	STATUS OF SPECIFIC PRODUCTS; GROUP A STREPTOCOCCUS General Provisions; General Biological Products Standards
STREPTOCOCCUS	866.3720	STREPTOCOCCUS SPP. EXOENZYME REAGENTS Serological Reagents; Immunology And Microbiology Devices
STREPTOCOCCUS	866.3740	STREPTOCOCCUS SPP. SEROLOGICAL REAGENTS Serological Reagents; Immunology And Microbiology Devices
STREPTOMYCIN	520.154b	SOLUBLE BACITRACIN METHYLENE DISALICYLATE AND STREPTOMYCIN SULFATE ORAL POWDER Oral Dosage Form New Animal Drugs
STREPTOMYCIN	520.2158	STREPTOMYCIN/DIHYDROSTREPTOMYCIN ORAL DOSAGE FORMS Oral Dosage Form New Animal Drugs
STREPTOMYCIN	556.610	STREPTOMYCIN Specific Tolerances For Residues Of New Animal Drugs; Tolerances For Residues Of New Animal Drugs In Food
STRETCHER	880.6900	HAND-CARRIED STRETCHER General Hospital And Personal Use Miscellaneous Devices; General Hospital And Personal Use Devices
STRETCHER	880.6910	WHEELED STRETCHER General Hospital And Personal Use Miscellaneous Devices; General Hospital And Personal Use Devices
STRETCHER	890.3690	POWERED WHEELED STRETCHER Physical Medicine Prosthetic Devices; Physical Medicine Devices

STRIP	880.2200	LIQUID CRYSTAL FOREHEAD TEMPERATURE STRIP General Hospital And Personal Use Monitoring Devices; General Hospital And Personal Use Devices
STRIP	882.5900	PREFORMED CRANIOSYNOSTOSIS STRIP Neurological Therapeutic Devices; Neurological Devices
STRIP	886.1800	SCHIRMER STRIP Diagnostic Devices; Ophthalmic Devices
STRIPPABLE FOOD COATINGS	175.230	HOT-MELT STRIPPABLE FOOD COATINGS Substances For Use As Components Of Coatings; Indirect Food Additives
STRIPPER	870.4875	INTRALUMINAL ARTERY STRIPPER Cardiovascular Surgical Devices; Cardiovascular Devices
STRIPPER	870.4885	EXTERNAL VEIN STRIPPER Cardiovascular Surgical Devices; Cardiovascular Devices
STRUCTURE, FIELD	5.115	FIELD STRUCTURE Organization; Delegations Of Authority And Organization
STRUCTURE ORTHOSIS	890.3610	RIGID PNEUMATIC STRUCTURE ORTHOSIS Physical Medicine Prosthetic Devices; Physical Medicine Devices
STUDIES	20.106	STUDIES AND REPORTS PREPARED BY OR WITH FUNDS PROVIDED BY THE FDA Availability Of Specific Categories Of Records; Public Information
STUDIES	58.10	APPLICABILITY TO STUDIES PERFORMED UNDER GRANTS AND CONTRACTS General Provisions; Good Laboratory Practice For Nonclinical Laboratory Studies
STUDIES	70.55	REQUEST FOR SCIENTIFIC STUDIES Safety Evaluation; Color Additives
STUDIES	99.203	REQUEST TO EXTEND THE TIME FOR COMPLETING PLANNED STUDIES Dissemination Of Information On Unapproved/New Uses For Marketed Drugs, Biologics, And Devices
STUDIES	99.303	EXTENSION OF TIME FOR COMPLETING PLANNED STUDIES Dissemination Of Information On Unapproved/New Uses For Marketed Drugs, Biologics, And Devices
STUDIES	310.303	CONTINUATION OF LONG-TERM STUDIES, RECORDS, AND REPORTS ON CERTAIN DRUGS FOR WHICH NEW DRUG APPLICATIONS HAVE BEEN APPROVED Records And Reports; New Drugs
STUDIES	601.70	ANNUAL PROGRESS REPORTS OF POSTMARKETING STUDIES Postmarketing Studies; Licensing
STUDIES	864.5425	MULTIPURPOSE SYSTEM FOR IN VITRO COAGULATION STUDIES Automated And Semi-automated Hematology Devices; Hematology And Pathology Devices
STUDIES, FOREIGN	312.110	FOREIGN CLINICAL STUDIES NOT CONDUCTED UNDER AN IND Miscellaneous; New Drugs For Investigational Use
STUDIES, PEDIATRIC	201.23	REQUIRED PEDIATRIC STUDIES Labeling
STUDIES, PEDIATRIC	601.27	PEDIATRIC STUDIES Biologics Licensing; Licensing
STUDIES, PEDIATRIC	601.37	ANNUAL REPORTS OF POSTMARKETING PEDIATRIC STUDIES Biologics Licensing; Licensing
STUDIES, PHASE 4	312.85	PHASE 4 STUDIES Drugs Intended To Treat Life-threatening And Severely-debilitating Illnesses; Investigational New Drug Application
STUDIES, WELL CONTROLLED	201.58	REQUESTS FOR WAIVER OF REQUIREMENT FOR ADEQUATE AND WELL-CONTROLLED STUDIES TO SUBSTANTIATE CERTAIN LABELING STATEMENTS Labeling Requirements For Prescription Drugs And/or Insulin; Labeling
STUDIES, WELL CONTROLLED	314.126	ADEQUATE AND WELL-CONTROLLED STUDIES FDA Action On Applications; Applications For FDA Approval To Market A New Drug
STUDIES, WELL-CONTROLLED	514.117	ADEQUATE AND WELL-CONTROLLED STUDIES Administrative Actions On Applications; New Animal Drug Applications
STUDY	58.33	STUDY DIRECTOR Organization And Personnel; Good Laboratory Practice For Nonclinical Laboratory Studies
STUDY	58.130	CONDUCT OF A NONCLINICAL LABORATORY STUDY Protocol For And Conduct Of A Nonclinical Laboratory Study; Good Laboratory Practice For Nonclinical Laboratory Studies
STUDY	58.185	REPORTING OF NONCLINICAL LABORATORY STUDY RESULTS Records And Reports; Good Laboratory Practice For Nonclinical Laboratory Studies
STUDY	320.25	GUIDELINES FOR THE CONDUCT OF AN IN VIVO BIOAVAILABILITY STUDY Procedures For Determining The Bioavailability Or Bioequivalence Of Drug Products; Bioavailability And Bioequivalence Requirements
STUDY	320.26	GUIDELINES ON THE DESIGN OF A SINGLE-DOSE IN VIVO BIOAVAILABILITY STUDY Procedures For Determining The Bioavailability Or Bioequivalence Of Drug Products; Bioavailability And Bioequivalence Requirements

STUDY	320.27	GUIDELINES ON THE DESIGN OF A MULTIPLE-DOSE IN VIVO BIOAVAILABILITY STUDY Procedures For Determining The Bioavailability Or Bioequivalence Of Drug Products; Bioavailability And Bioequivalence Requirements
STUDY	320.29	ANALYTICAL METHODS FOR AN IN VIVO BIOAVAILABILITY STUDY Procedures For Determining The Bioavailability Or Bioequivalence Of Drug Products; Bioavailability And Bioequivalence Requirements
STUDY, DRUG EFFICACY	201.200	DISCLOSURE OF DRUG EFFICACY STUDY EVALUATIONS IN LABELING AND ADVERTISING Labeling Claims For Drugs In Drug Efficacy Studies; Labeling
STUDY, DRUG EFFICACY	310.6	APPLICABILITY OF “NEW DRUG” OR SAFETY OR EFFECTIVENESS FINDINGS IN DRUG EFFICACY STUDY IMPLEMENTATION NOTICES AND NOTICES OF OPPORTUNITY FOR HEARING TO IDENTICAL, RELATED, AND SIMILAR DRUG PRODUCTS General Provisions; New Drugs
STUDY, DRUG EFFICACY	330.12	STATUS OF OVER-THE-COUNTER (OTC) DRUGS PREVIOUSLY REVIEWED UNDER THE DRUG EFFICACY STUDY (DESI) Administrative Procedures; Over-the-counter Human Drugs Generally Recognized As Safe And Effective & Not Misbranded
STYLET	868.5790	TRACHEAL TUBE STYLET Therapeutic Devices; Anesthesiology Devices
STYLET	870.1380	CATHETER STYLET Cardiovascular Diagnostic Devices; Cardiovascular Devices
STYRENE	173.70	CHLOROMETHYLATED AMINATED STYRENE-DIVINYLBENZENE RESIN Polymer Substances For Food Treatment; Secondary Direct Food Additives Permitted In Food For Human Consumption
STYRENE	177.1020	ACRYLONITRILE/BUTADIENE/STYRENE COPOLYMER Substances For Use As Basic Components Of Single And Repeated Use Food Contact Surfaces; Indirect Food Additives: Polymers
STYRENE	177.1030	ACRYLONITRILE/BUTADIENE/STYRENE/METHYL METHACRYLATE COPOLYMER Substances For Use As Basic Components Of Single And Repeated Use Food Contact Surfaces; Indirect Food Additives: Polymers
STYRENE	177.1040	ACRYLONITRILE/STYRENE COPOLYMER Substances For Use As Basic Components Of Single And Repeated Use Food Contact Surfaces; Indirect Food Additives: Polymers
STYRENE	177.1050	ACRYLONITRILE/STYRENE COPOLYMER MODIFIED WITH BUTADIENE/STYRENE ELASTOMER Substances For Use As Basic Components Of Single And Repeated Use Food Contact Surfaces; Indirect Food Additives: Polymers
STYRENE	177.1810	STYRENE BLOCK POLYMERS Substances For Use As Basic Components Of Single And Repeated Use Food Contact Surfaces; Indirect Food Additives: Polymers
STYRENE	177.1820	STYRENE-MALEIC ANHYDRIDE COPOLYMERS Substances For Use As Basic Components Of Single And Repeated Use Food Contact Surfaces; Indirect Food Additives: Polymers
STYRENE	177.1830	STYRENE-METHYL METHACRYLATE COPOLYMERS Substances For Use As Basic Components Of Single And Repeated Use Food Contact Surfaces; Indirect Food Additives: Polymers
STYRENE	177.2710	STYRENE-DIVINYLBENZENE RESINS, CROSS-LINKED Substances For Use Only As Components Of Articles Intended For Repeated Use; Indirect Food Additives: Polymers
STYRENE	573.870	POLY(2-VINYLPYRIDINE-CO-STYRENE) Food Additive Listing; Food Additives Permitted In Feed And Drinking Water Of Animals
STYRYLPYRIDINIUM	520.2160	STYRYLPYRIDINIUM, DIETHYLCARBAMAZINE ORAL DOSAGE FORMS Oral Dosage Form New Animal Drugs
SUBAWARDS	1403.35	SUBAWARDS TO DEBARRED AND SUSPENDED PARTIES Post-award Requirements; Uniform Administrative Requirements For Grants And Coop. Agreements To State & Local Gov.
SUBGRANT	1403.12	SPECIAL GRANT OR SUBGRANT CONDITIONS FOR “HIGH-RISK” GRANTEES Pre-Award Requirements; Uniform Administrative Requirements For Grants And Coop. Agreements To State & Local Gov.
SUBGRANTS	1403.37	SUBGRANTS Post-Award Requirements; Uniform Administrative Requirements For Grants And Coop. Agreements To State & Local Gov.
SUBJECTS	50.52	CLINICAL INVESTIGATIONS INVOLVING GREATER THAN MINIMAL RISK BUT PRESENTING THE PROSPECT OF DIRECT BENEFIT TO INDIVIDUAL SUBJECTS Additional Safeguards For Children In Clinical Investigations; Protection Of Human Subjects

SUBMISSIONS	15.25	WRITTEN SUBMISSIONS Procedures For Public Hearing Before The Commissioner; Public Hearing Before The Commissioner
SUBMISSIONS	316.4	ADDRESS FOR SUBMISSIONS Orphan Drugs
SUBMISSIONS	312.35	SUBMISSIONS FOR TREATMENT USE Investigational New Drug Application
SUBMIT	207.20	WHO MUST REGISTER AND SUBMIT A DRUG LIST Procedures For Domestic Drug Establishments; Registration Of Drug Producers And Listing Of Drugs In Commercial Distribution
SUBMIT	607.20	WHO MUST REGISTER AND SUBMIT A BLOOD PRODUCT LIST Procedures For Domestic Blood Product Establishments; Establishment Registratn+product Listing, Manufacturers Human Blood/b.products
SUBMIT	803.12	WHERE TO SUBMIT REPORTS General Provisions; Medical Device Reporting
SUBMIT	807.20	WHO MUST REGISTER AND SUBMIT A DEVICE LIST Procedures For Device Establishments; Establishment Registration And Device Listing For Manufacturers And Initial Importers Of Devices
SUBMITTAL	1250.41	SUBMITTAL OF CONSTRUCTION PLANS Equipment And Operation Of Land And Air Conveyances; Interstate Conveyance Sanitation
SUBMITTAL	1250.62	SUBMITTAL OF CONSTRUCTION PLANS Servicing Areas For Land And Air Conveyances; Interstate Conveyance Sanitation
SUBMITTED, VOLUNTARILY	20.44	PRESUBMISSION REVIEW OF REQUEST FOR CONFIDENTIALITY OF VOLUNTARILY SUBMITTED DATA OR INFORMATION Procedures And Fees; Public Information
SUBMITTED, VOLUNTARILY	20.111	DATA AND INFORMATION SUBMITTED VOLUNTARILY TO THE FDA Availability Of Specific Categories Of Records; Public Information
SUBMITTING	21.40	PROCEDURES FOR SUBMITTING REQUESTS FOR NOTIFICATION AND ACCESS Procedures For Notification Of And Access To Records In Privacy Act Record Systems; Protection Of Privacy
SUBMITTING	21.50	PROCEDURES FOR SUBMITTING REQUESTS FOR AMENDMENT OF RECORDS Procedures For Requests For Amendment Of Records; Protection Of Privacy
SUBMITTING	314.122	SUBMITTING AN ABBREVIATED APPLICATION FOR A LISTED DRUG THAT IS NO LONGER MARKETED Applications For FDA Approval To Market A New Drug
SUBMITTING	720.7	NOTIFICATION OF PERSON SUBMITTING COSMETIC PRODUCT INGREDIENT STATEMENT Voluntary Filing Of Cosmetic Product Ingredient Composition Statements
SUBPERIOSTEAL	872.3645	SUBPERIOSTEAL IMPLANT MATERIAL Prosthetic Devices; Dental Devices
SUBPOENAS	17.27	HEARING SUBPOENAS Civil Money Penalties Hearings
SUBSTANCE, CONTROLLED	1309.25	EXEMPTION OF CERTAIN CONTROLLED SUBSTANCE REGISTRANTS Registration Of Manufacturers, Distributors, Importers And Exporters Of List I Chemicals
SUBSTANCE, NONNARCOTIC	1308.21	APPLICATION FOR EXCLUSION OF A NONNARCOTIC SUBSTANCE Excluded Nonnarcotic Substances; Schedules Of Controlled Substances
SUBSTANCES	70.11	RELATED SUBSTANCES General Provisions; Color Additives
SUBSTANCES	176.130	ANTI OFFSET SUBSTANCES Substances For Use Only As Components Of Paper And Paperboard; Indirect Food Additives
SUBSTANCES	178.3010	ADJUVANT SUBSTANCES USED IN THE MANUFACTURE OF FOAMED PLASTICS Certain Adjuvants And Production Aids; Indirect Food Additives: Adjuvants, Production Aids, And Sanitizers
SUBSTANCES	178.3740	PLASTICIZERS IN POLYMERIC SUBSTANCES Certain Adjuvants And Production Aids; Indirect Food Additives: Adjuvants, Production Aids, And Sanitizers
SUBSTANCES	181.22	CERTAIN SUBSTANCES EMPLOYED IN THE MANUFACTURE OF FOOD-PACKAGING MATERIALS Specific Prior-sanctioned Food Ingredients; Prior-sanctioned Food Ingredients
SUBSTANCES	181.30	SUBSTANCES USED IN THE MANUFACTURE OF PAPER AND PAPERBOARD PRODUCTS USED IN FOOD PACKAGING Specific Prior-sanctioned Food Ingredients; Prior-sanctioned Food Ingredients
SUBSTANCES	182.70	SUBSTANCES MIGRATING FROM COTTON AND COTTON FABRICS USED IN DRY FOOD PACKAGING General Provisions; Substances Generally Recognized As Safe
SUBSTANCES	182.90	SUBSTANCES MIGRATING TO FOOD FROM PAPER AND PAPERBOARD PRODUCTS General Provisions; Substances Generally Recognized As Safe

SUBSTANCES	310.103	NEW DRUG SUBSTANCES INTENDED FOR HYPERSENSITIVITY TESTING Specific Administrative Rulings And Decisions; New Drugs
SUBSTANCES	329.1	HABIT-FORMING DRUGS WHICH ARE CHEMICAL DERIVATIVES OF SUBSTANCES SPECIFIED IN SECTION 502(D) OF THE FEDERAL FOOD, DRUG, AND COSMETIC ACT Derivatives Designated As Habit Forming; Habit-forming Drugs
SUBSTANCES	701.20	DETERGENT SUBSTANCES, OTHER THAN SOAP, INTENDED FOR USE IN CLEANSING THE BODY Labeling Of Specific Ingredients; Cosmetic Labeling
SUBSTANCES	864.9160	BLOOD GROUP SUBSTANCES OF NONHUMAN ORIGIN FOR IN VITRO DIAGNOSTIC USE Products Used in Establishments That Manufacture Blood And Blood Products; Hematology And Pathology Devices
SUBSTANCES	1301.32	ACTION ON APPLICATIONS FOR RESEARCH IN SCHEDULE I SUBSTANCES Actions On Application For Registration: Revocation Or Suspension Of Registration; Registration Of Manufacturers, Distributors, And Dispensers Of Controlled Substances
SUBSTANCES	1301.33	APPLICATION FOR BULK MANUFACTURE OF SCHEDULE I AND II SUBSTANCES Action On Application For Registration: Revocation Or Suspension Of Registration; Registration Of Manufacturers, Distributors, And Dispensers Of Controlled Substances
SUBSTANCES	1301.34	APPLICATION FOR IMPORTATION OF SCHEDULE I AND II SUBSTANCES Action On Application For Registration: Revocation Or Suspension Of Registration; Registration Of Manufacturers, Distributors, And Dispensers Of Controlled Substances
SUBSTANCES	1306.14	LABELING OF SUBSTANCES AND FILLING OF PRESCRIPTIONS Controlled Substances Listed In Schedule II; Prescriptions
SUBSTANCES	1306.24	LABELING OF SUBSTANCES AND FILLING OF PRESCRIPTIONS Controlled Substances Listed In Schedules III, IV, And V; Prescriptions
SUBSTANCES, CONTROLLED	290.1	CONTROLLED SUBSTANCES General Provisions; Controlled Drugs
SUBSTANCES, CONTROLLED	312.69	HANDLING OF CONTROLLED SUBSTANCES Responsibilities Of Sponsors And Investigators; Investigational New Drug Application
SUBSTANCES, CONTROLLED	1300.01	DEFINITIONS RELATING TO CONTROLLED SUBSTANCES Definitions
SUBSTANCES, CONTROLLED	1302.06	SEALING OF CONTROLLED SUBSTANCES Labeling And Packaging Requirements For Controlled Substances
SUBSTANCES, CONTROLLED	1306.26	TRANSFER BETWEEN PHARMACIES OF PRESCRIPTION INFORMATION FOR SCHEDULES III, IV, AND V CONTROLLED SUBSTANCES FOR REFILL PURPOSES Controlled Substances Listed In Schedules III, IV, And V; Prescriptions
SUBSTANCES, CONTROLLED	1307.13	INCIDENTAL MANUFACTURE OF CONTROLLED SUBSTANCES Special Exemptions For Manufacture And Distribution Of Controlled Substances; Miscellaneous
SUBSTANCES, CONTROLLED	1307.21	PROCEDURE FOR DISPOSING OF CONTROLLED SUBSTANCES Disposal Of Controlled Substances; Miscellaneous
SUBSTANCES, CONTROLLED	1307.22	DISPOSAL OF CONTROLLED SUBSTANCES BY THE ADMINISTRATION Disposal Of Controlled Substances; Miscellaneous
SUBSTANCES, CONTROLLED	1308.03	ADMINISTRATION CONTROLLED SUBSTANCES CODE NUMBER General Information; Schedules Of Controlled Substances
SUBSTANCES, CONTROLLED	1312.27	CONTENTS OF SPECIAL CONTROLLED SUBSTANCES INVOICE Exportation Of Controlled Substances; Importation And Exportation Of Controlled Substances
SUBSTANCES, CONTROLLED	1312.28	DISTRIBUTION OF SPECIAL CONTROLLED SUBSTANCES INVOICE Exportation Of Controlled Substances; Importation And Exportation Of Controlled Substances
SUBSTANCES, CONTROLLED	1312.30	SCHEDULE III, IV, AND V NON-NARCOTIC CONTROLLED SUBSTANCES REQUIRING AN IMPORT AND EXPORT PERMIT Exportation Of Controlled Substances; Importation And Exportation Of Controlled Substances
SUBSTANCES, EXCLUDED	1308.22	EXCLUDED SUBSTANCES Excluded Nonnarcotic Substances; Schedules Of Controlled Substances
SUBSTANCES, FLAVORING	172.230	MICROCAPSULES FOR FLAVORING SUBSTANCES Coatings, Films And Related Substances; Food Additives Permitted For Direct Addition To Food For Human Consumption
SUBSTANCES, FLAVORING	172.510	NATURAL FLAVORING SUBSTANCES AND NATURAL SUBSTANCES USED IN CONJUNCTION WITH FLAVORS Flavoring Agents And Related Substances; Food Additives Permitted For Direct Addition To Food For Human Consumption
SUBSTANCES, FLAVORING	172.515	SYNTHETIC FLAVORING SUBSTANCES AND ADJUVANTS Favoring Agents And Related Substances; Food Additives Permitted For Direct Addition To Food For Human Consumption

SUBSTANCES, FLAVORING	182.60	SYNTHETIC FLAVORING SUBSTANCES AND ADJUVANTS General Provisions; Substances Generally Recognized As Safe
SUBSTANCES, FLAVORING	582.60	SYNTHETIC FLAVORING SUBSTANCES AND ADJUVANTS General Provisions; Substances Generally Recognized As Safe
SUBSTANCES, FOOD-CONTACT	170.39	THRESHOLD OF REGULATION FOR SUBSTANCES USED IN FOOD-CONTACT ARTICLES Food Additive Safety; Food Additives
SUBSTANCES, FOOD-CONTACT	171.8	THRESHOLD OF REGULATION FOR SUBSTANCES USED IN FOOD-CONTACT ARTICLES Food Additive Petitions
SUBSTANCES, FOOD-CONTACT	174.6	THRESHOLD OF REGULATION FOR SUBSTANCES USED IN FOOD-CONTACT ARTICLES Indirect Food Additives: General
SUBSTANCES, GRAS	5.300	FOOD STANDARDS, FOOD ADDITIVES, GENERALLY RECOGNIZED AS SAFE (GRAS) SUBSTANCES, COLOR ADDITIVES, NUTRIENT CONTENT CLAIMS, AND HEALTH CLAIMS Foods And Cosmetics; Redelegations Of Authority; Delegations Of Authority And Organization
SUBSTANCES, GRAS	182.1	SUBSTANCES THAT ARE GENERALLY RECOGNIZED AS SAFE General Provisions; Substances Generally Recognized As Safe
SUBSTANCES, GRAS	184.1	SUBSTANCES ADDED DIRECTLY TO HUMAN FOOD AFFIRMED AS GENERALLY RECOGNIZED AS SAFE (GRAS) General Provisions; Direct Food Substances Affirmed As Generally Recognized As Safe
SUBSTANCES, GRAS	186.1	SUBSTANCES ADDED INDIRECTLY TO HUMAN FOOD AFFIRMED AS GENERALLY RECOGNIZED AS SAFE (GRAS) General Provisions; Indirect Food Substances Affirmed As Generally Recognized As Safe
SUBSTANCES, GRAS	582.1	SUBSTANCES THAT ARE GENERALLY RECOGNIZED AS SAFE General Provisions; Substances Generally Recognized As Safe
SUBSTANCES, IMPORT/EXPORT	1302.07	LABELING AND PACKAGING REQUIREMENTS FOR IMPORTED AND EXPORTED SUBSTANCES Labeling And Packaging Requirements For Controlled Substances
SUBSTANCES, NATURAL	582.30	NATURAL SUBSTANCES USED IN CONJUNCTION WITH SPICES AND OTHER NATURAL SEASONINGS AND FLAVORINGS General Provisions; Substances Generally Recognized As Safe
SUBSTANCES, POISONOUS	2.25	GRAIN SEED TREATED WITH POISONOUS SUBSTANCES; COLOR IDENTIFICATION TO PREVENT ADULTERATION OF HUMAN AND ANIMAL FOOD Human And Animal Foods; General Administrative Rulings And Decision
SUBSTANCES, POISONOUS	109.6	ADDED POISONOUS OR DELETERIOUS SUBSTANCES General Provisions; Unavoidable Contaminants In Food For Human Consumption & Food-packaging Material
SUBSTANCES PROHIBITED	189.1	SUBSTANCES PROHIBITED FROM USE IN HUMAN FOOD General Provisions; Substances Prohibited From Use In Human Food
SUBSTANCES PROHIBITED	589.1	SUBSTANCES PROHIBITED FROM USE IN ANIMAL FOOD OR FEED General Provisions; Substances Prohibited From Use In Animal Food Or Feed
SUBSTANDARD	130.14	GENERAL STATEMENTS OF SUBSTANDARD QUALITY AND SUBSTANDARD FILL OF CONTAINER General Provisions; Food Standards: General
SUBSTANTIAL COMPLIANCE	101.43	SUBSTANTIAL COMPLIANCE OF FOOD RETAILERS WITH THE GUIDELINES FOR THE VOLUNTARY NUTRITION LABELING OF RAW FRUIT, VEGETABLES, AND FISH Food Labeling
SUBSTANTIAL HARM	5.409	DETERMINATIONS THAT MEDICAL DEVICES PRESENT UNREASONABLE RISK OF SUBSTANTIAL HARM Medical Devices And Radiological Health; Redelegations Of Authority; Delegations Of Authority And Organization
SUBSTANTIATION	740.10	LABELING OF COSMETIC PRODUCTS FOR WHICH ADEQUATE SUBSTANTIATION OF SAFETY HAS NOT BEEN OBTAINED Warning Statements; Cosmetic Product Warning Statements
SUBSTANTIVE AMENDMENTS	71.6	EXTENSION OF TIME FOR STUDYING PETITIONS; SUBSTANTIVE AMENDMENTS; WITHDRAWAL OF PETITIONS WITHOUT PREJUDICE General Provisions; Color Additive Petitions
SUBSTITUTE, COCOA BUTTER	172.861	COCOA BUTTER SUBSTITUTE FROM COCONUT OIL, PALM KERNEL OIL, OR BOTH OILS Food Additives Permitted For Direct Addition To Food For Human Consumption
SUBSTITUTE, COCOA BUTTER	184.1259	COCOA BUTTER SUBSTITUTE Listing Of Specific Substances Affirmed As GRAS; Direct Food Substances Affirmed As Generally Recognized As Safe
SUBSTITUTE, DURA	882.5910	DURA SUBSTITUTE Neurological Therapeutic Devices; Neurological Devices

SUBSTITUTE RES	1316.94	POSTING OF SUBSTITUTE RES IN AN ADMINISTRATIVE FORFEITURE ACTION Expedited Forfeiture Proceedings For Certain Property; Administrative Functions, Practices, And Procedures
SUBSTITUTE RES	1316.98	SUBSTITUTE RES BOND IN A JUDICIAL FORFEITURE ACTION AGAINST A CONVEYANCE Expedited Forfeiture Proceedings For Certain Property; Administrative Functions, Practices, And Procedures
SUCCINATE	172.765	SUCCISTEARIN (STEAROYL PROPYLENE GLYCOL HYDROGEN SUCCINATE) Other Specific Usage Additives; Food Additives Permitted For Direct Addition To Food For Human Consumption
SUCCINATE	520.784	DOXYLAMINE SUCCINATE TABLETS Oral Dosage Form New Animal Drugs
SUCCINATE	522.784	DOXYLAMINE SUCCINATE INJECTION Implantation Or Injectable Dosage Form New Animal Drugs
SUCCINIC ACID	184.1091	SUCCINIC ACID Listing Of Specific Substances Affirmed As GRAS; Direct Food Substances Affirmed As Generally Recognized As Safe
SUCCINIC ACID	582.1091	SUCCINIC ACID General Purpose Food Additives; Substances Generally Recognized As Safe
SUCCINIC DERIVATIVES	172.275	SYNTHETIC PARAFFIN AND SUCCINIC DERIVATIVES Coatings, Films And Related Substances; Food Additives Permitted For Direct Addition To Food For Human Consumption
SUCCINYLATED MONOGLYCERIDES	172.830	SUCCINYLATED MONOGLYCERIDES Multipurpose Additives; Food Additives Permitted For Direct Addition To Food For Human Consumption
SUCCISTEARIN	172.765	SUCCISTEARIN (STEAROYL PROPYLENE GLYCOL HYDROGEN SUCCINATE) Other Specific Usage Additives; Food Additives Permitted For Direct Addition To Food For Human Consumption
SUCKER	870.4420	CARDIOPULMONARY BYPASS CARDIOTOMY RETURN SUCKER Cardiovascular Surgical Devices; Cardiovascular Devices
SUCRALOSE	172.831	SUCRALOSE Multipurpose Additives; Food Additives Permitted For Direct Addition To Food For Human Consumption
SUCROSE	172.859	SUCROSE FATTY ACID ESTERS Multipurpose Additives; Food Additives Permitted For Direct Addition To Food For Human Consumption
SUCROSE	184.1854	SUCROSE Listing Of Specific Substances Affirmed As GRAS; Direct Food Substances Affirmed As Generally Recognized As Safe
SUCROSE ACETATE ISOBUTYRATE	172.833	SUCROSE ACETATE ISOBUTYRATE (SAIB) Multipurpose Additives; Food Additives Permitted For Direct Addition To Food For Human Consumption
SUCTION	868.6810	TRACHEOBRONCHIAL SUCTION CATHETER Miscellaneous; Anesthesiology Devices
SUCTION	870.4270	CARDIOPULMONARY BYPASS CARDIOTOMY SUCTION LINE BLOOD FILTER Cardiovascular Surgical Devices; Cardiovascular Devices
SUCTION	870.4430	CARDIOPULMONARY BYPASS INTRACARDIAC SUCTION CONTROL Cardiovascular Surgical Devices; Cardiovascular Devices
SUCTION	870.5050	PATIENT CARE SUCTION APPARATUS Cardiovascular Therapeutic Devices; Cardiovascular Devices
SUCTION	874.5350	SUCTION ANTICHOKE DEVICE Therapeutic Devices; Ear, Nose, And Throat Devices
SUCTION	878.4680	NON-POWERED, SINGLE PATIENT, PORTABLE SUCTION APPARATUS Surgical Devices; General And Plastic Surgery Devices
SUCTION	878.5040	SUCTION LIPOPLASTY SYSTEM Surgical Devices; General And Plastic Surgery Devices
SUCTION	880.5740	SUCTION SNAKEBITE KIT General Hospital And Personal Use Therapeutic Devices; General Hospital And Personal Use Devices
SUCTION	880.6740	VACUUM-POWERED BODY FLUID SUCTION APPARATUS General Hospital And Personal Use Miscellaneous Devices; General Hospital And Personal Use Devices
SUCTION	884.1175	ENDOMETRIAL SUCTION CURETTE AND ACCESSORIES Obstetrical And Gynecological Diagnostic Devices; Obstetrical And Gynecological Devices
SUCTION PUMP	878.4780	POWERED SUCTION PUMP Surgical Devices; General And Plastic Surgery Devices
SUGAR ALCOHOLS	101.80	HEALTH CLAIMS: DIETARY SUGAR ALCOHOLS AND DENTAL CARIES Food Labeling
SUGAR BEET EXTRACT	172.585	SUGAR BEET EXTRACT FLAVOR BASE Flavoring Agents And Related Substances; Food Additives Permitted For Direct Addition To Food For Human Consumption

SUGAR, CANE/BEET	173.320	CHEMICALS FOR CONTROLLING MICROORGANISMS IN CANE-SUGAR AND BEET-SUGAR MILLS Specific Usage Additives; Secondary Direct Food Additives Permitted In Food For Human Consumption
SUGAR, CORN	184.1857	CORN SUGAR Listing Of Specific Substances Affirmed As GRAS; Direct Food Substances Affirmed As Generally Recognized As Safe
SUGAR, INVERT	184.1859	INVERT SUGAR Listing Of Specific Substances Affirmed As GRAS; Direct Food Substances Affirmed As Generally Recognized As Safe
SUGGESTIONS	1402.7	SUGGESTIONS AND COMPLAINTS Mandatory Declassification Review
SUITABILITY	640.3	SUITABILITY OF DONOR Whole Blood; Additional Standards For Human Blood And Blood Products
SUITABILITY	640.12	SUITABILITY OF DONOR Red Blood Cells; Additional Standards For Human Blood And Blood Products
SUITABILITY	640.21	SUITABILITY OF DONORS Platelets; Additional Standards For Human Blood And Blood Products
SUITABILITY	640.31	SUITABILITY OF DONORS Plasma; Additional Standards For Human Blood And Blood Products
SUITABILITY	640.51	SUITABILITY OF DONORS Cryoprecipitate; Additional Standards For Human Blood And Blood Products
SUITABILITY	640.63	SUITABILITY OF DONOR Source Plasma; Additional Standards For Human Blood And Blood Products
SUITABILITY	660.31	SUITABILITY OF THE DONOR Reagent Red Blood Cells; Additional Standards For Diagnostic Substances For Laboratory Tests
SUITABILITY	1270.21	DETERMINATION OF DONOR SUITABILITY FOR HUMAN TISSUE INTENDED FOR TRANSPLANTATION Human Tissue Intended For Transplantation
SULFABROMOMETHAZINE	520.2170	SULFABROMOMETHAZINE SODIUM BOLUSES Oral Dosage Form New Animal Drugs
SULFABROMOMETHAZINE	556.620	SULFABROMOMETHAZINE SODIUM Specific Tolerances For Residues Of New Animal Drugs; Tolerances For Residues Of New Animal Drugs In Food
SULFACHLOROPYRAZINE	520.2184	SODIUM SULFACHLOROPYRAZINE MONOHYDRATE Oral Dosage Form New Animal Drugs
SULFACHLOROPYRAZINE	556.625	SODIUM SULFACHLOROPYRAZINE MONOHYDRATE Specific Tolerances For Residues Of New Animal Drugs; Tolerances For Residues Of New Animal Drugs In Food
SULFACHLORPYRIDAZINE	520.2200	SULFACHLORPYRIDAZINE ORAL DOSAGE FORMS Oral Dosage Form New Animal Drugs
SULFACHLORPYRIDAZINE	522.2200	SULFACHLORPYRIDAZINE Implantation Or Injectable Dosage Form New Animal Drugs
SULFACHLORPYRIDAZINE	556.630	SULFACHLORPYRIDAZINE Specific Tolerances For Residues Of New Animal Drugs; Tolerances For Residues Of New Animal Drugs In Food
SULFADIAZINE	520.2610	TRIMETHOPRIM AND SULFADIAZINE TABLETS Oral Dosage Form New Animal Drugs
SULFADIAZINE	520.2611	TRIMETHOPRIM AND SULFADIAZINE ORAL PASTE Oral Dosage Form New Animal Drugs
SULFADIAZINE	520.2612	TRIMETHOPRIM AND SULFADIAZINE ORAL SUSPENSION Oral Dosage Form New Animal Drugs
SULFADIAZINE	520.2613	TRIMETHOPRIM AND SULFADIAZINE POWDER Oral Dosage Form New Animal Drugs
SULFADIAZINE	522.2610	TRIMETHOPRIM AND SULFADIAZINE STERILE SUSPENSION Implantation Or Injectable Dosage Form New Animal Drugs
SULFADIAZINE	524.802	ENROFLOXACIN, SILVER SULFADIAZINE EMULSION Ophthalmic And Topical Dosage Form New Animal Drugs
SULFADIMETHOXINE	520.2220	SULFADIMETHOXINE ORAL DOSAGE FORMS Oral Dosage Form New Animal Drugs
SULFADIMETHOXINE	522.2220	SULFADIMETHOXINE INJECTION Implantation Or Injectable Dosage Form New Animal Drugs
SULFADIMETHOXINE	556.640	SULFADIMETHOXINE Specific Tolerances For Residues Of New Animal Drugs; Tolerances For Residues Of New Animal Drugs In Food
SULFADIMETHOXINE	558.575	SULFADIMETHOXINE, ORMETOPRIM Specific New Animal Drugs For Use In Animal Feeds; New Animal Drugs For Use In Animal Feeds
SULFAETHOXYPYRIDAZINE	520.2240	SULFAETHOXYPYRIDAZINE Oral Dosage Form New Animal Drugs
SULFAETHOXYPYRIDAZINE	522.2240	SULFAETHOXYPYRIDAZINE Implantation Or Injectable Dosage Form New Animal Drugs

SULFAETHOXYPYRIDAZINE	556.650	SULFAETHOXYPYRIDAZINE Specific Tolerances For Residues Of New Animal Drugs; Tolerances For Residues Of New Animal Drugs In Food
SULFAETHOXYPYRIDAZINE	558.579	SULFAETHOXYPYRIDAZINE Specific New Animal Drugs For Use In Animal Feeds; New Animal Drugs For Use In Animal Feeds
SULFAMERAZINE	556.660	SULFAMERAZINE Specific Tolerances For Residues Of New Animal Drugs; Tolerances For Residues Of New Animal Drugs In Food
SULFAMERAZINE	558.582	SULFAMERAZINE Specific New Animal Drugs For Use In Animal Feeds; New Animal Drugs For Use In Animal Feeds
SULFAMETHAZINE	520.445a	CHLORTETRACYCLINE BISULFATE/SULFAMETHAZINE BISULFATE SOLUBLE POWDER Oral Dosage Form New Animal Drugs
SULFAMETHAZINE	520.2260	SULFAMETHAZINE ORAL DOSAGE FORMS Oral Dosage Form New Animal Drugs
SULFAMETHAZINE	520.2261	SULFAMETHAZINE SODIUM ORAL DOSAGE FORMS Oral Dosage Form New Animal Drugs
SULFAMETHAZINE	522.2260	SULFAMETHAZINE INJECTABLE SOLUTION Implantation Or Injectable Dosage Form New Animal Drugs
SULFAMETHAZINE	556.670	SULFAMETHAZINE Specific Tolerances For Residues Of New Animal Drugs; Tolerances For Residues Of New Animal Drugs In Food
SULFAMETHAZINE	558.140	CHLORTETRACYCLINE AND SULFAMETHAZINE Specific New Animal Drugs For Use In Animal Feeds; New Animal Drugs For Use In Animal Feeds
SULFAMETHAZINE	558.145	CHLORTETRACYCLINE, PROCAINE PENICILLIN, AND SULFAMETHAZINE Specific New Animal Drugs For Use In Animal Feeds; New Animal Drugs For Use In Animal Feeds
SULFAMETHAZINE	558.630	TYLOSIN AND SULFAMETHAZINE Specific New Animal Drugs For Use In Animal Feeds; New Animal Drugs For Use In Animal Feeds
SULFAMETHIZOLE	520.2280	SULFAMETHIZOLE AND METHENAMINE MANDELATE TABLETS Oral Dosage Form New Animal Drugs
SULFAMIC ACID	186.1093	SULFAMIC ACID Listing Of Specific Substances Affirmed As GRAS; Indirect Food Substances Affirmed As Generally Recognized As Safe
SULFANITRAN	520.2320	SULFANITRAN AND AKLOMIDE IN COMBINATION Oral Dosage Form New Animal Drugs
SULFANITRAN	556.680	SULFANITRAN Specific Tolerances For Residues Of New Animal Drugs; Tolerances For Residues Of New Animal Drugs In Food
SULFANITRAN	558.376	NITROMIDE AND SULFANITRAN Specific New Animal Drugs For Use In Animal Feeds; New Animal Drugs For Use In Animal Feeds
SULFAQUINOXALINE	520.2325	SULFAQUINOXALINE ORAL DOSAGE FORMS Oral Dosage Form New Animal Drugs
SULFAQUINOXALINE	556.685	SULFAQUINOXALINE Specific Tolerances For Residues Of New Animal Drugs; Tolerances For Residues Of New Animal Drugs In Food
SULFAQUINOXOLINE	558.586	SULFAQUINOXOLINE Specific New Animal Drugs For Use In Animal Feeds; New Animal Drugs For Use In Animal Feeds
SULFATE, ALUMINUM	182.1125	ALUMINUM SULFATE Multiple Purpose GRAS Food Substances; Substances Generally Recognized As Safe
SULFATE, ALUMINUM	182.1127	ALUMINUM AMMONIUM SULFATE Multiple Purpose GRAS Food Substances; Substances Generally Recognized As Safe
SULFATE, ALUMINUM	182.1129	ALUMINUM POTASSIUM SULFATE Multiple Purpose GRAS Food Substances; Substances Generally Recognized As Safe
SULFATE, ALUMINUM	182.1131	ALUMINUM SODIUM SULFATE Multiple Purpose GRAS Food Substances; Substances Generally Recognized As Safe
SULFATE, ALUMINUM	582.1125	ALUMINUM SULFATE General Purpose Food Additives; Substances Generally Recognized As Safe
SULFATE, ALUMINUM	582.1127	ALUMINUM AMMONIUM SULFATE General Purpose Food Additives; Substances Generally Recognized As Safe
SULFATE, ALUMINUM	582.1129	ALUMINUM POTASSIUM SULFATE General Purpose Food Additives; Substances Generally Recognized As Safe
SULFATE, ALUMINUM	582.1131	ALUMINUM SODIUM SULFATE General Purpose Food Additives; Substances Generally Recognized As Safe

SULFATE, AMMONIUM	184.1143	AMMONIUM SULFATE Listing Of Specific Substances Affirmed As GRAS; Direct Food Substances Affirmed As Generally Recognized As Safe
SULFATE, AMMONIUM	582.1143	AMMONIUM SULFATE General Purpose Food Additives; Substances Generally Recognized As Safe
SULFATE, CALCIUM	184.1230	CALCIUM SULFATE Listing Of Specific Substances Affirmed As GRAS; Direct Food Substances Affirmed As Generally Recognized As Safe
SULFATE, CALCIUM	582.5230	CALCIUM SULFATE Nutrients And/or Dietary Supplements; Substances Generally Recognized As Safe
SULFATE, COPPER	184.1261	COPPER SULFATE Listing Of Specific Substances Affirmed As GRAS; Direct Food Substances Affirmed As Generally Recognized As Safe
SULFATE, COPPER	864.9320	COPPER SULFATE SOLUTION FOR SPECIFIC GRAVITY DETERMINATIONS Products Used In Establishments That Manufacture Blood And Blood Products; Hematology And Pathology Devices
SULFATE, FERRIC	184.1307	FERRIC SULFATE Listing Of Specific Substances Affirmed As GRAS; Direct Food Substances Affirmed As Generally Recognized As Safe
SULFATE, FERROUS	184.1315	FERROUS SULFATE Listing Of Specific Substances Affirmed As GRAS; Direct Food Substances Affirmed As Generally Recognized As Safe
SULFATE, FERROUS	582.5315	FERROUS SULFATE Nutrients And/or Dietary Supplements; Substances Generally Recognized As Safe
SULFATE, MAGNESIUM	184.1443	MAGNESIUM SULFATE Listing Of Specific Substances Affirmed As GRAS; Direct Food Substances Affirmed As Generally Recognized As Safe
SULFATE, MAGNESIUM	201.312	MAGNESIUM SULFATE HEPTAHYDRATE; LABEL DECLARATION ON DRUG PRODUCTS Specific Labeling Requirements For Specific Drug Products; Labeling
SULFATE, MAGNESIUM	582.5443	MAGNESIUM SULFATE Nutrients And/or Dietary Supplements; Substances Generally Recognized As Safe
SULFATE, MANGANESE	184.1461	MANGANESE SULFATE Listing Of Specific Substances Affirmed As GRAS; Direct Food Substances Affirmed As Generally Recognized As Safe
SULFATE, MANGANESE	582.5461	MANGANESE SULFATE Nutrients And/or Dietary Supplements; Substances Generally Recognized As Safe
SULFATE, POTASSIUM	184.1643	POTASSIUM SULFATE Listing Of Specific Substances Affirmed As GRAS; Direct Food Substances Affirmed As Generally Recognized As Safe
SULFATE, POTASSIUM	582.1643	POTASSIUM SULFATE General Purpose Food Additives; Substances Generally Recognized As Safe
SULFATE, SODIUM LAURYL	172.822	SODIUM LAURYL SULFATE Multipurpose Additives; Food Additives Permitted For Direct Addition To Food For Human Consumption
SULFATE, SODIUM METHYL	173.385	SODIUM METHYL SULFATE Specific Usage Additives; Secondary Direct Food Additives Permitted In Food For Human Consumption
SULFATE, ZINC	182.8997	ZINC SULFATE Nutrients; Substances Generally Recognized As Safe
SULFATE, ZINC	582.5997	ZINC SULFATE Nutrients And/or Dietary Supplements; Substances Generally Recognized As Safe
SULFATE, ZINC METHIONINE	172.399	ZINC METHIONINE SULFATE Special Dietary And Nutritional Additives; Food Additives Permitted For Direct Addition To Food For Human Consumption
SULFATED BUTYL OLEATE	172.270	SULFATED BUTYL OLEATE Coatings, Films And Related Substances; Food Additives Permitted For Direct Addition To Food For Human Consumption
SULFATHIAZOLE	556.690	SULFATHIAZOLE Specific Tolerances For Residues Of New Animal Drugs; Tolerances For Residues Of New Animal Drugs In Food
SULFATHIAZOLE	558.155	CHLORTETRACYCLINE, SULFATHIAZOLE, PENICILLIN Specific New Animal Drugs For Use In Animal Feeds; New Animal Drugs For Use In Animal Feeds
SULFHEMOGLOBIN	864.7490	SULFHEMOGLOBIN ASSAY Hematology Kits And Packages; Hematology And Pathology Devices
SULFIDE	73.2995	LUMINESCENT ZINC SULFIDE Cosmetics; Listing Of Color Additives Exempt From Certification
SULFIDE	177.2490	POLYPHENYLENE SULFIDE RESINS Substances For Use Only As Components Of Articles Intended For Repeated Use; Indirect Food Additives: Polymers
SULFIDE DETECTION	872.1870	SULFIDE DETECTION DEVICE Diagnostic Devices; Dental Devices

SULFISOXAZOLE	520.2330	SULFISOXAZOLE TABLETS Oral Dosage Form New Animal Drugs
SULFITE	182.3798	SODIUM SULFITE Chemical Preservatives; Substances Generally Recognized As Safe
SULFITE	582.3798	SODIUM SULFITE Chemical Preservatives; Substances Generally Recognized As Safe
SULFITES	130.9	SULFITES IN STANDARDIZED FOOD General Provisions; Food Standards: General
SULFITES	201.22	PRESCRIPTION DRUGS CONTAINING SULFITES; REQUIRED WARNING STATEMENTS General Labeling Provisions; Labeling
SULFOLITHOCHOLIC ACID	862.1187	CONJUGATED SULFOLITHOCHOLIC ACID (SLCG) TEST SYSTEM Clinical Chemistry Test Systems; Clinical Chemistry And Clinical Toxicology Devices
SULFOMYXIN	522.2340	SULFOMYXIN Implantation Or Injectable Dosage Form New Animal Drugs
SULFOMYXIN	556.700	SULFOMYXIN Specific Tolerances For Residues Of New Animal Drugs; Tolerances For Residues Of New Animal Drugs In Food
SULFONAMIDE	558.15	ANTIBIOTIC, NITROFURAN, AND SULFONAMIDE DRUGS IN THE FEED OF ANIMALS General Provisions; New Animal Drugs For Use In Animal Feeds
SULFONAMIDE	862.3850	SULFONAMIDE TEST SYSTEM Clinical Toxicology Test Systems; Clinical Chemistry And Clinical Toxicology Devices
SULFONATES	172.824	SODIUM MONO- AND DIMETHYL NAPHTHALENE SULFONATES Multipurpose Additives; Food Additives Permitted For Direct Addition To Food For Human Consumption
SULFONATES	573.600	LIGNIN SULFONATES Food Additive Listing; Food Additives Permitted In Feed And Drinking Water Of Animals
SULFONE	177.2500	POLYPHENYLENE SULFONE RESINS Substances For Use Only As Components Of Articles Intended For Repeated Use; Indirect Food Additives: Polymers
SULFONIC ACID	173.395	TRIFLUOROMETHANE SULFONIC ACID Specific Usage Additives; Secondary Direct Food Additives Permitted In Food For Human Consumption
SULFONYL GLYCINE	176.160	CHROMIUM (CR III) OF COMPLEX OF N-ETHYL-N-HEPTADECYLFLUORO-OCTANE SULFONYL GLYCINE Substances For Use Only As Components Of Paper And Paperboard; Indirect Food Additives
SULFONYLUREA	310.517	LABELING FOR ORAL HYPOGLYCEMIC DRUGS OF THE SULFONYLUREA CLASS Requirements For Specific New Drugs Or Devices; New Drugs
SULFOSUCCINATE	163.117	COCOA WITH DIOCTYL SODIUM SULFOSUCCINATE FOR MANUFACTURING Requirements For Specific Standardized Cacao Products; Cacao Products
SULFOSUCCINATE	172.520	COCOA WITH DIOCTYL SODIUM SULFOSUCCINATE FOR MANUFACTURING Flavoring Agents And Related Substances; Food Additives Permitted For Direct Addition To Food For Human Consumption
SULFOSUCCINATE	172.810	DIOCTYL SODIUM SULFOSUCCINATE Multipurpose Additives; Food Additives Permitted For Direct Addition To Food For Human Consumption
SULFOXIDE, DIMETHYL	524.660	DIMETHYL SULFOXIDE OPHTHALMIC AND TOPICAL DOSAGE FORMS Ophthalmic And Topical Dosage Form New Animal Drugs
SULFOXIDE, DIMETHYL	524.981d	FLUOCINOLONE ACETONIDE, DIMETHYL SULFOXIDE SOLUTION Ophthalmic And Topical Dosage Form New Animal Drugs
SULFOXIDE, DIMETHYL	524.981e	FLUOCINOLONE ACETONIDE, DIMETHYL SULFOXIDE OTIC SOLUTION Ophthalmic And Topical Dosage Form New Animal Drugs
SULFUR DIOXIDE	182.3862	SULFUR DIOXIDE Chemical Preservatives; Substances Generally Recognized As Safe
SULFUR DIOXIDE	582.3862	SULFUR DIOXIDE Chemical Preservatives; Substances Generally Recognized As Safe
SULFURIC ACID	184.1095	SULFURIC ACID Listing Of Specific Substances Affirmed As GRAS; Direct Food Substances Affirmed As Generally Recognized As Safe
SULFURIC ACID	582.1095	SULFURIC ACID General Purpose Food Additives; Substances Generally Recognized As Safe
SUMMARIES	10.10	SUMMARIES OF ADMINISTRATIVE PRACTICES AND PROCEDURES General Provisions; Administrative Practices And Procedures
SUMMARIES	20.104	SUMMARIES OF ORAL DISCUSSIONS Availability Of Specific Categories Of Records; Public Information
SUMMARY	861.20	SUMMARY OF STANDARDS DEVELOPMENT PROCESS Procedures For Performance Standards Development And Publication; Procedures For Performance Standards Development

SUMMARY	1316.58	SUMMARY OF TESTIMONY; AFFIDAVITS Administrative Hearings; Administrative Functions, Practices, And Procedures
SUMMARY DECISION	16.26	DENIAL OF HEARING AND SUMMARY DECISION Initiation Of Proceedings; Regulatory Hearing Before The Food And Drug Administration
SUMMARY DECISIONS	12.93	SUMMARY DECISIONS Hearing Procedures; Formal Evidentiary Public Hearing
SUMMARY DECISIONS	17.17	SUMMARY DECISIONS Civil Money Penalties Hearings
SUMMARY, 510(K)	807.92	CONTENT AND FORMAT OF A 510(K) SUMMARY Premarket Notification Procedures; Establishment Registration And Device Listing For Manufacturers And Initial Importers Of Devices
SUNGLASSES	801.410	USE OF IMPACT-RESISTANT LENSES IN EYEGLASSES AND SUNGLASSES Special Requirements For Specific Devices; Labeling
SUNGLASSES	886.5850	SUNGLASSES (NONPRESCRIPTION) Therapeutic Devices; Ophthalmic Devices
SUNLAMP	1040.20	SUNLAMP PRODUCTS AND ULTRAVIOLET LAMPS INTENDED FOR USE IN SUNLAMP PRODUCTS Performance Standards For Light-emitting Products
SUNSCREEN	352.10	SUNSCREEN ACTIVE INGREDIENTS Sunscreen Drug Products For Over-the-counter Human Use
SUNSCREEN	352.50	PRINCIPAL DISPLAY PANEL OF ALL SUNSCREEN DRUG PRODUCTS Sunscreen Drug Products For Over-the-counter Human Use
SUNSCREEN	352.52	LABELING OF SUNSCREEN DRUG PRODUCTS Sunscreen Drug Products For Over-the-counter Human Use
SUNSCREEN	352.70	STANDARD SUNSCREEN Sunscreen Drug Products For Over-the-counter Human Use
SUNSCREEN	700.35	COSMETICS CONTAINING SUNSCREEN INGREDIENTS Requirements For Specific Cosmetic Products; General
SUNTANNING PREPARATIONS	740.19	SUNTANNING PREPARATIONS Cosmetic Product Warning Statements
SUPERVISION	640.62	MEDICAL SUPERVISION Source Plasma; Additional Standards For Human Blood And Blood Products
SUPPLEMENT	314.71	PROCEDURES FOR SUBMISSION OF A SUPPLEMENT TO AN APPROVED APPLICATION Applications; Applications For FDA Approval To Market A New Drug
SUPPLEMENT	866.2450	SUPPLEMENT FOR CULTURE MEDIA Microbiology Devices; Immunology And Microbiology Devices
SUPPLEMENTAL	25.34	DRAFT, FINAL, AND SUPPLEMENTAL ENVIRONMENTAL IMPACT STATEMENTS Preparation Of Environmental Documents; Environmental Impact Considerations
SUPPLEMENTAL APPLICATION	99.205	APPLICATION FOR EXEMPTION FROM THE REQUIREMENT TO FILE A SUPPLEMENTAL APPLICATION Dissemination Of Information On Unapproved/New Uses For Marketed Drugs, Biologics, and Devices
SUPPLEMENTAL APPLICATION	99.305	EXEMPTION FROM THE REQUIREMENT TO FILE A SUPPLEMENTAL APPLICATION Dissemination Of Information On Unapproved/New Uses For Marketed Drugs, Biologics, And Devices
SUPPLEMENTAL APPLICATIONS	514.8	SUPPLEMENTAL NEW ANIMAL DRUG APPLICATIONS General Provisions; New Animal Drug Applications
SUPPLEMENTAL APPLICATIONS	514.106	APPROVAL OF SUPPLEMENTAL APPLICATIONS Administrative Actions On Applications; New Animal Drug Applications
SUPPLEMENTAL APPLICATIONS	515.11	SUPPLEMENTAL MEDICATED FEED MILL LICENSE APPLICATIONS Applications; Medicated Feed Mill License
SUPPLEMENTAL APPLICATIONS	812.35	SUPPLEMENTAL APPLICATIONS Application And Administrative Action; Investigational Device Exemptions
SUPPLEMENTAL REPORTS	803.56	SUPPLEMENTAL REPORTS Manufacturer Reporting Requirements; Medical Device Reporting
SUPPLEMENTAL REPORTS	1002.11	SUPPLEMENTAL REPORTS Required Manufacturers' Reports For Listed Electronic Products; Records And Reports
SUPPLEMENTS, APPLICATION	314.70	SUPPLEMENTS AND OTHER CHANGES TO AN APPROVED APPLICATION Applications; Applications For FDA Approval To Market A New Drug
SUPPLEMENTS, APPLICATION	314.97	SUPPLEMENTS AND OTHER CHANGES TO AN APPROVED ABBREVIATED APPLICATION Abbreviated Applications; Applications For FDA Approval To Market A New Drug
SUPPLEMENTS, DIETARY	101.36	NUTRITION LABELING OF DIETARY SUPPLEMENTS Specific Nutrition Labeling Requirements And Guidelines; Food Labeling
SUPPLEMENTS, DIETARY	101.93	CERTAIN TYPES OF STATEMENTS FOR DIETARY SUPPLEMENTS Specific Requirements For Descriptive Claims That Are Neither Nutrient Content Nor Health Claims; Food Labeling
SUPPLEMENTS, DIETARY	111.50	PACKAGING OF IRON-CONTAINING DIETARY SUPPLEMENTS Current Good Manufacturing Practice For Dietary Supplements

SUPPLEMENTS, MEDIA	884.6180	REPRODUCTIVE MEDIA AND SUPPLEMENTS Assisted Reproduction Devices; Obstetrical And Gynecological Devices
SUPPLEMENTS, NADA	5.501	APPROVAL OF NEW ANIMAL DRUG APPLICATIONS, MEDICATED FEED MILL LICENSE APPLICATIONS, AND THEIR SUPPLEMENTS Animal Drugs; Redelegations Of Authority; Delegations Of Authority And Organization
SUPPLEMENTS, NDA	5.103	APPROVAL OF NEW DRUG APPLICATIONS AND THEIR SUPPLEMENTS Human Drugs; Redelegations Of Authority; Delegations Of Authority And Organization
SUPPLEMENTS, NDA	5.105	ISSUANCE OF NOTICES RELATING TO PROPOSALS TO REFUSE APPROVAL OR TO WITHDRAW APPROVAL OF NEW DRUG APPLICATIONS AND THEIR SUPPLEMENTS Human Drugs; Redelegations Of Authority; Delegations Of Authority And Organization
SUPPLEMENTS, PMA	814.39	PMA SUPPLEMENTS Premarket Approval Application (PMA); Premarket Approval Of Medical Devices
SUPPLIER	1307.12	DISTRIBUTION TO SUPPLIER Special Exemptions For Manufacture And Distribution Of Controlled Substances; Miscellaneous
SUPPLIES	606.65	SUPPLIES AND REAGENTS Equipment; Current Good Manufacturing Practice For Blood And Blood Components
SUPPLIES	864.2240	CELL AND TISSUE CULTURE SUPPLIES AND EQUIPMENT Cell And Tissue Culture Products; Hematology And Pathology Devices
SUPPLIES	864.9050	BLOOD BANK SUPPLIES Products Used In Establishments That Manufacture Blood And Blood Products; Hematology And Pathology Devices
SUPPLIES	1403.33	SUPPLIES Post-award Requirements; Uniform Administrative Requirements For Grants And Coop. Agreements To State & Local Gov.
SUPPLY	601.22	PRODUCTS IN SHORT SUPPLY; INITIAL MANUFACTURING AT OTHER THAN LICENSED ESTABLISHMENT Biologics Licensing; Licensing
SUPPLY	312.59	DISPOSITION OF UNUSED SUPPLY OF INVESTIGATIONAL DRUG Responsibilities Of Sponsors And Investigators; Investigational New Drug Application
SUPPLY FACILITIES	58.45	ANIMAL SUPPLY FACILITIES Facilities; Good Laboratory Practice For Nonclinical Laboratory Studies
SUPPORT	13.15	SEPARATION OF FUNCTIONS; EX PARTE COMMUNICATIONS; ADMINISTRATIVE SUPPORT General Provisions; Public Hearing Before A Public Board Of Inquiry
SUPPORT	868.5280	BREATHING TUBE SUPPORT Therapeutic Devices; Anesthesiology Devices
SUPPORT	868.5610	MEMBRANE LUNG FOR LONG-TERM PULMONARY SUPPORT Therapeutic Devices; Anesthesiology Devices
SUPPORT	868.6820	PATIENT POSITION SUPPORT Miscellaneous; Anesthesiology Devices
SUPPORT	876.5970	HERNIA SUPPORT Therapeutic Devices; Gastroenterology-urology Devices
SUPPORT	880.5780	MEDICAL SUPPORT STOCKING General Hospital And Personal Use Therapeutic Devices; General Hospital And Personal Use Devices
SUPPORT	880.5820	THERAPEUTIC SCROTAL SUPPORT General Hospital And Personal Use Therapeutic Devices; General Hospital And Personal Use Devices
SUPPORT	892.5770	POWERED RADIATION THERAPY PATIENT SUPPORT ASSEMBLY Therapeutic Devices; Radiology Devices
SUPPORT GEL	866.4900	SUPPORT GEL Immunology Laboratory Equipment And Reagents; Immunology And Microbiology Devices
SUPRAPUBIC CATHETER	876.5090	SUPRAPUBIC UROLOGICAL CATHETER AND ACCESSORIES Therapeutic Devices; Gastroenterology-urology Devices
SURFACE-ACTIVE	178.3400	EMULSIFIERS AND/OR SURFACE-ACTIVE AGENTS Certain Adjuvants And Production Aids; Indirect Food Additives: Adjuvants, Production Aids, And Sanitizers
SURFACE ANTIGEN	660.1	ANTIBODY TO HEPATITIS B SURFACE ANTIGEN Antibody To Hepatitis B Surface Antigen; Additional Standards For Diagnostic Substances For Laboratory Tests
SURFACE ANTIGEN	660.40	HEPATITIS B SURFACE ANTIGEN Hepatitis B Surface Antigen; Additional Standards For Diagnostic Substances For Laboratory Tests
SURFACE ELECTRODE	870.2370	ELECTROCARDIOGRAPH SURFACE ELECTRODE TESTER Cardiovascular Monitoring Devices; Cardiovascular Devices

SURFACE LUBRICANTS	178.3910	SURFACE LUBRICANTS USED IN THE MANUFACTURE OF METALLIC ARTICLES Certain Adjuvants And Production Aids; Indirect Food Additives: Adjuvants, Production Aids, And Sanitizers
SURGEON'S GLOVE	878.4460	SURGEON'S GLOVE Surgical Devices; General And Plastic Surgery Devices
SURGEON'S GLOVE	878.4480	ABSORBABLE POWDER FOR LUBRICATING A SURGEON'S GLOVE Surgical Devices; General And Plastic Surgery Devices
SURGEON'S GLOVES	800.20	PATIENT EXAMINATION GLOVES AND SURGEON'S GLOVES; SAMPLE PLANS AND TEST METHOD FOR LEAKAGE DEFECTS; ADULTERATION Requirements For Specific Medical Devices; General
SURGEON'S GLOVING CREAM	878.4470	SURGEON'S GLOVING CREAM Surgical Devices; General And Plastic Surgery Devices
SURGERY	874.3540	PROSTHESIS MODIFICATION INSTRUMENT FOR OSSICULAR REPLACEMENT SURGERY Prosthetic Devices; Ear, Nose And Throat Devices
SURGERY	886.4360	OCULAR SURGERY IRRIGATION DEVICE Surgical Devices; Ophthalmic Devices
SURGERY, PLASTIC	878.3925	PLASTIC SURGERY KIT AND ACCESSORIES Prosthetic Devices; General And Plastic Surgery Devices
SURGERY, PLASTIC	878.4810	LASER SURGICAL INSTRUMENT FOR USE IN PLASTIC SURGERY AND IN DERMATOLOGY Surgical Devices; General And Plastic Surgery Devices
SURGICAL APPAREL	878.4040	SURGICAL APPAREL Surgical Devices; General And Plastic Surgery Devices
SURGICAL CAMERA	878.4160	SURGICAL CAMERA AND ACCESSORIES Surgical Devices; General And Plastic Surgery Devices
SURGICAL, CRYOGENIC	882.4250	CRYOGENIC SURGICAL DEVICE Neurological Surgical Devices; Neurological Devices
SURGICAL DEGREASER	878.4730	SURGICAL SKIN DEGREASER OR ADHESIVE TAPE SOLVENT Surgical Devices; General And Plastic Surgery Devices
SURGICAL DILATOR	870.4475	SURGICAL VESSEL DILATOR Cardiovascular Surgical Devices; Cardiovascular Devices
SURGICAL DRAPE	878.4370	SURGICAL DRAPE AND DRAPE ACCESSORIES Surgical Devices; General And Plastic Surgery Devices
SURGICAL DRILL	874.1820	SURGICAL NERVE STIMULATOR/LOCATOR Diagnostic Devices; Ear, Nose, And Throat Devices
SURGICAL DRILL	874.4250	EAR, NOSE, AND THROAT ELECTRIC OR PNEUMATIC SURGICAL DRILL Surgical Devices; Ear, Nose, And Throat Devices
SURGICAL INSTRUMENT	874.4420	EAR, NOSE, AND THROAT MANUAL SURGICAL INSTRUMENT Surgical Devices; Ear, Nose, And Throat Devices
SURGICAL INSTRUMENT	876.4730	MANUAL GASTROENTEROLOGY-UROLOGY SURGICAL INSTRUMENT AND ACCESSORIES Surgical Devices; Gastroenterology-urology Devices
SURGICAL INSTRUMENT	878.4800	MANUAL SURGICAL INSTRUMENT FOR GENERAL USE Surgical Devices; General And Plastic Surgery Devices
SURGICAL INSTRUMENT	878.4810	LASER SURGICAL INSTRUMENT FOR USE IN PLASTIC SURGERY AND IN DERMATOLOGY Surgical Devices; General And Plastic Surgery Devices
SURGICAL INSTRUMENT	886.4350	MANUAL OPHTHALMIC SURGICAL INSTRUMENT Surgical Devices; Ophthalmic Devices
SURGICAL INSTRUMENT	888.4540	ORTHOPEDIC MANUAL SURGICAL INSTRUMENT Surgical Devices; Orthopedic Devices
SURGICAL INSTRUMENT	888.4580	SONIC SURGICAL INSTRUMENT AND ACCESSORIES/ATTACHMENTS Surgical Devices; Orthopedic Devices
SURGICAL INSTRUMENT MOTORS	878.4820	SURGICAL INSTRUMENT MOTORS AND ACCESSORIES/ATTACHMENTS Surgical Devices; General And Plastic Surgery Devices
SURGICAL INSTRUMENTS	870.4500	CARDIOVASCULAR SURGICAL INSTRUMENTS Cardiovascular Surgical Devices; Cardiovascular Devices
SURGICAL LAMP	878.4580	SURGICAL LAMP Surgical Devices; General And Plastic Surgery Devices
SURGICAL LASER	884.4550	GYNECOLOGIC SURGICAL LASER Obstetrical And Gynecological Surgical Devices; Obstetrical And Gynecological Devices
SURGICAL MARKER	886.4570	OPHTHALMIC SURGICAL MARKER Surgical Devices; Ophthalmic Devices
SURGICAL MESH	878.3300	SURGICAL MESH Prosthetic Devices; General And Plastic Surgery Devices
SURGICAL MICROSCOPE	878.4700	SURGICAL MICROSCOPE AND ACCESSORIES Surgical Devices; General And Plastic Surgery Devices
SURGICAL OP. ROOM	878.5070	AIR-HANDLING APPARATUS FOR A SURGICAL OPERATING ROOM Therapeutic Devices; General And Plastic Surgery Devices

SURGICAL SCALE	880.2740	SURGICAL SPONGE SCALE General Hospital And Personal Use Monitoring Devices; General Hospital And Personal Use Devices
SURGICAL SUTURE	878.4493	ABSORBABLE POLY(GLYCOLIDE/L-LACTIDE) SURGICAL SUTURE Surgical Devices; General And Plastic Surgery Devices
SURGICAL SUTURE	878.4830	ABSORBABLE SURGICAL GUT SUTURE Surgical Devices; General And Plastic Surgery Devices
SURGICAL SUTURE	878.5000	NONABSORBABLE POLY(ETHYLENE TEREPHTHALATE) SURGICAL SUTURE Surgical Devices; General And Plastic Surgery Devices
SURGICAL SUTURE	878.5010	NONABSORBABLE POLYPROPYLENE SURGICAL SUTURE Surgical Devices; General And Plastic Surgery Devices
SURGICAL SUTURE	878.5020	NONABSORBABLE POLYAMIDE SURGICAL SUTURE Surgical Devices; General And Plastic Surgery Devices
SURGICAL SUTURE	878.5030	NATURAL NONABSORBABLE SILK SURGICAL SUTURE Surgical Devices; General And Plastic Surgery Devices
SURGICAL SUTURE	878.5035	NONABSORBABLE EXPANDED POLYTETRAFLUOROETHYLENE SURGICAL SUTURE Surgical Devices; General And Plastic Surgery Devices
SURROGATE ENDPOINT	314.510	APPROVAL BASED ON A SURROGATE ENDPOINT OR ON AN EFFECT ON A CLINICAL ENDPOINT OTHER THAN SURVIVAL OR IRREVERSIBLE MORBIDITY Applications For FDA Approval To Market A New Drug
SURROGATE ENDPOINT	601.41	APPROVAL BASED ON A SURROGATE ENDPOINT OR ON AN EFFECT ON A CLINICAL ENDPOINT OTHER THAN SURVIVAL OR IRREVERSIBLE MORBIDITY Accelerated Approval Of Biological Products; Licensing
SURVEILLANCE, POSTMARKET	5.414	POSTMARKET SURVEILLANCE Medical Devices And Radiological Health; Redelegations Of Authority; Delegations Of Authority And Organization
SURVIVAL	314.510	APPROVAL BASED ON A SURROGATE ENDPOINT OR ON AN EFFECT ON A CLINICAL ENDPOINT OTHER THAN SURVIVAL OR IRREVERSIBLE MORBIDITY Applications For FDA Approval To Market A New Drug
SURVIVAL	601.41	APPROVAL BASED ON A SURROGATE ENDPOINT OR ON AN EFFECT ON A CLINICAL ENDPOINT OTHER THAN SURVIVAL OR IRREVERSIBLE MORBIDITY Accelerated Approval Of Biological Products; Licensing
SUSCEPTIBILITY TEST	866.1620	ANTIMICROBIAL SUSCEPTIBILITY TEST DISC Diagnostic Devices; Immunology And Microbiology Devices
SUSCEPTIBILITY TEST	866.1640	ANTIMICROBIAL SUSCEPTIBILITY TEST POWDER Diagnostic Devices; Immunology And Microbiology Devices
SUSCEPTIBILITY TESTS	866.1700	CULTURE MEDIUM FOR ANTIMICROBIAL SUSCEPTIBILITY TESTS Diagnostic Devices; Immunology And Microbiology Devices
SUSPENDED	314.160	APPROVAL OF AN APPLICATION FOR WHICH APPROVAL WAS PREVIOUSLY REFUSED, SUSPENDED, OR WITHDRAWN FDA Action On Applications; Applications For FDA Approval To Market A New Drug
SUSPENDED	1403.35	SUBAWARDS TO DEBARRED AND SUSPENDED PARTIES Post-award Requirements; Uniform Administrative Requirements For Grants And Coop. Agreements To State & Local Gov.
SUSPENDING	514.120	REVOCATION OF ORDER REFUSING TO APPROVE AN APPLICATION OR SUSPENDING OR WITHDRAWING APPROVAL OF AN APPLICATION Administrative Actions On Applications; New Animal Drug Applications
SUSPENDING	515.25	REVOCATION OF ORDER REFUSING TO APPROVE A MEDICATED FEED MILL LICENSE APPLICATION OR SUSPENDING OR REVOKING A LICENSE Administrative Actions On Licenses; Medicated Feed Mill License
SUSPENSION	10.19	WAIVER, SUSPENSION, OR MODIFICATION OF PROCEDURAL REQUIREMENTS General Provisions; Administrative Practices And Procedures
SUSPENSION	26.16	SUSPENSION Specific Sector Provisions For Pharmaceutical GMPs; Mutual Recognition Of Pharmaceutical GMP Reports, Medical Device Quality System Audit Reports
SUSPENSION	26.67	SUSPENSION OF LISTED CONFORMITY ASSESSMENT BODIES Framework Provisions; Mutual Recognition Of Pharmaceutical GMP Reports, Medical Device Quality System Audit Reports
SUSPENSION	108.10	SUSPENSION AND REINSTATEMENT OF PERMIT General Provisions; Emergency Permit Control
SUSPENSION	520.45a	ALBENDAZOLE SUSPENSION Oral Dosage Form New Animal Drugs
SUSPENSION	520.88b	AMOXICILLIN TRIHYDRATE FOR ORAL SUSPENSION Oral Dosage Form New Animal Drugs
SUSPENSION	520.88c	AMOXICILLIN TRIHYDRATE ORAL SUSPENSION Oral Dosage Form New Animal Drugs
SUSPENSION	520.88h	AMOXICILLIN TRIHYDRATE AND CLAVULANATE POTASSIUM FOR ORAL SUSPENSION Oral Dosage Form New Animal Drugs
SUSPENSION	520.90d	AMPICILLIN TRIHYDRATE FOR ORAL SUSPENSION Oral Dosage Form New Animal Drugs

SUSPENSION	520.300	CAMBENDAZOLE SUSPENSION Oral Dosage Form New Animal Drugs
SUSPENSION	520.315	CEFADROXIL POWDER FOR ORAL SUSPENSION Oral Dosage Form New Animal Drugs
SUSPENSION	520.390c	CHLORAMPHENICOL PALMITATE ORAL SUSPENSION Oral Dosage Form New Animal Drugs
SUSPENSION	520.763c	DITHIAZANINE IODIDE AND PIPERAZINE CITRATE SUSPENSION Oral Dosage Form New Animal Drugs
SUSPENSION	520.903	FEBANTEL SUSPENSION Oral Dosage Form New Animal Drugs
SUSPENSION	520.905a	FENBENDAZOLE SUSPENSION Oral Dosage Form New Animal Drugs
SUSPENSION	520.1130b	HETACILLIN POTASSIUM ORAL SUSPENSION Oral Dosage Form New Animal Drugs
SUSPENSION	520.1182	IRON DEXTRAN ORAL SUSPENSION Oral Dosage Form New Animal Drugs
SUSPENSION	520.1204	KANAMYCIN SULFATE, AMINOPENTAMIDE HYDROGEN SULFATE, PECTIN, BISMUTH SUBCARBONATE, ACTIVATED ATTAPULGITE ORAL SUSPENSION Oral Dosage Form New Animal Drugs
SUSPENSION	520.1289	LUFENURON SUSPENSION Oral Dosage Form New Animal Drugs
SUSPENSION	520.1630	OXFENDAZOLE SUSPENSION Oral Dosage Form New Animal Drugs
SUSPENSION	520.1640	OXIBENDAZOLE SUSPENSION Oral Dosage Form New Animal Drugs
SUSPENSION	520.1802a	PIPERAZINE-CARBON DISULFIDE COMPLEX SUSPENSION Oral Dosage Form New Animal Drugs
SUSPENSION	520.1802c	PIPERAZINE-CARBON DISULFIDE COMPLEX WITH PHENOTHIAZINE SUSPENSION Oral Dosage Form New Animal Drugs
SUSPENSION	520.2043	PYRANTEL PAMOATE SUSPENSION Oral Dosage Form New Animal Drugs
SUSPENSION	520.2158c	DIHYDROSTREPTOMYCIN ORAL SUSPENSION Oral Dosage Form New Animal Drugs
SUSPENSION	520.2220c	SULFADIMETHOXINE ORAL SUSPENSION Oral Dosage Form New Animal Drugs
SUSPENSION	520.2380d	THIABENDAZOLE, PIPERAZINE CITRATE SUSPENSION Oral Dosage Form New Animal Drugs
SUSPENSION	520.2612	TRIMETHOPRIM AND SULFADIAZINE ORAL SUSPENSION Oral Dosage Form New Animal Drugs
SUSPENSION	522.88	STERILE AMOXICILLIN TRIHYDRATE FOR SUSPENSION Implantation Or Injectable Dosage Form New Animal Drugs
SUSPENSION	522.90a	AMPICILLIN TRIHYDRATE STERILE SUSPENSION Implantation Or Injectable Dosage Form New Animal Drugs
SUSPENSION	522.90b	AMPICILLIN TRIHYDRATE FOR STERILE SUSPENSION Implantation Or Injectable Dosage Form New Animal Drugs
SUSPENSION	522.161	BETAMETHASONE ACETATE AND BETAMETHASONE DISODIUM PHOSPHATE AQUEOUS SUSPENSION Implantation Or Injectable Dosage Form New Anima/Drugs
SUSPENSION	522.163	BETAMETHASONE DIPROPIONATE AND BETAMETHASONE SODIUM PHOSPHATE AQUEOUS SUSPENSION Implantation Or Injectable Dosage Form New Animal Drugs
SUSPENSION	522.314	CEFTIOFUR HYDROCHLORIDE STERILE SUSPENSION Implantation Or Injectable Dosage Form New Animal Drugs
SUSPENSION	522.542	DEXAMETHASONE-21-ISONICOTINATE SUSPENSION Implantation Or Injectable Dosage Form New Animal Drugs
SUSPENSION	522.960	FLUMETHASONE SUSPENSION Implantation Or Injectable Dosage Form New Animal Drugs
SUSPENSION	522.1081	CHORIONIC GONADOTROPIN FOR INJECTION; CHORIONIC GONADOTROPIN SUSPENSION Implantation Or Injectable Dosage Form New Animal Drugs
SUSPENSION	522.1410	STERILE METHYLPREDNISOLONE ACETATE SUSPENSION Implantation Or Injectable Dosage Form New Animal Drugs
SUSPENSION	522.1696a	PENICILLIN G BENZATHINE AND PENICILLIN G PROCAINE STERILE SUSPENSION Implantation Or Injectable Dosage Form New Animal Drugs
SUSPENSION	522.1696b	PENICILLIN G PROCAINE AQUEOUS SUSPENSION Implantation Or Injectable Dosage Form New Animal Drugs
SUSPENSION	522.1881	STERILE PREDNISOLONE ACETATE AQUEOUS SUSPENSION Implantation Or Injectable Dosage Form New Animal Drugs
SUSPENSION	522.1885	PREDNISOLONE TERTIARY BUTYLACETATE SUSPENSION Implantation Or Injectable Dosage Form New Animal Drugs
SUSPENSION	522.1890	STERILE PREDNISONE SUSPENSION Implantation Or Injectable Dosage Form New Animal Drugs

SUSPENSION	522.2112	STERILE SOMETRIBOVE ZINC SUSPENSION Implantation Or Injectable Dosage Form New Animal Drugs
SUSPENSION	522.2150	STANOZOLOL STERILE SUSPENSION Implantation Or Injectable Dosage Form New Animal Drugs
SUSPENSION	522.2483	STERILE TRIAMCINOLONE ACETONIDE SUSPENSION Implantation Or Injectable Dosage Form New Animal Drugs
SUSPENSION	522.2610	TRIMETHOPRIM AND SULFADIAZINE STERILE SUSPENSION Implantation Or Injectable Dosage Form New Animal Drugs
SUSPENSION	524.1044h	GENTAMICIN SULFATE, MOMETASONE FUROATE, CLOTRIMAZOLE OTIC SUSPENSION Ophthalmic And Topical Dosage Form New Animal Drugs
SUSPENSION	524.1195	IVERMECTIN OTIC SUSPENSION Ophthalmic And Topical Dosage Form New Animal Drugs
SUSPENSION	524.1484h	NEOMYCIN, PENICILLIN, POLYMIXIN, HYDROCORTISONE SUSPENSION Ophthalmic And Topical Dosage Form New Animal Drugs
SUSPENSION	524.1484k	NEOMYCIN SULFATE, PREDNISOLONE, TETRACAINE, AND SQUALANE TOPICAL-OTIC SUSPENSION Ophthalmic And Topical Dosage Form New Animal Drugs
SUSPENSION	524.1881b	PREDNISOLONE ACETATE-NEOMYCIN SULFATE STERILE SUSPENSION Ophthalmic And Topical Dosage Form New Animal Drugs
SUSPENSION	524.2101	SELENIUM DISULFIDE SUSPENSION Ophthalmic And Topical Dosage Form New Animal Drugs
SUSPENSION	526.1590	NOVOBIOCIN OIL SUSPENSION Intramammary Dosage Forms
SUSPENSION	529.400	CHLORHEXIDINE TABLETS AND SUSPENSION Certain Other Dosage Form New Animal Drugs
SUSPENSION	1404.200	DEBARMENT OR SUSPENSION Governmentwide Debarment And Suspension And Requirements For Drug-Free Workplace (Grants)
SUSPENSION	1404.405	CAUSES FOR SUSPENSION Governmentwide Debarment And Suspension And Requirements For Drug-Free Workplace (Grants)
SUSPENSION	1404.411	NOTICE OF SUSPENSION Governmentwide Debarment And Suspension And Requirements For Drug-Free Workplace (Grants)
SUSPENSION	1404.412	OPPORTUNITY TO CONTEST SUSPENSION Governmentwide Debarment And Suspension And Requirements For Drug-Free Workplace (Grants)
SUSPENSION	1404.415	PERIOD OF SUSPENSION Governmentwide Debarment And Suspension And Requirements For Drug-Free Workplace (Grants)
SUSPENSION	1404.420	SCOPE OF SUSPENSION Governmentwide Debarment And Suspension And Requirements For Drug-Free Workplace (Grants)
SUSPENSION	1404.615	GROUNDS FOR SUSPENSION OF PAYMENTS, SUSPENSION OR TERMINATION OF GRANTS, OR SUSPENSION OR DEBARMENT Governmentwide Debarment And Suspension And Requirements For Drug-Free Workplace (Grants)
SUSPENSION OF APPLICATION	5.412	TEMPORARY SUSPENSION OF A MEDICAL DEVICE APPLICATION Medical Devices And Radiological Health; Redelegations Of Authority; Delegations Of Authority And Organization
SUSPENSION OF APPROVAL	56.113	SUSPENSION OR TERMINATION OF IRB APPROVAL OF RESEARCH IRB Functions And Operations; Institutional Review Boards
SUSPENSION OF APPROVAL	314.153	SUSPENSION OF APPROVAL OF AN ABBREVIATED NEW DRUG APPLICATION Applications For FDA Approval To Market A New Drug
SUSPENSION OF APPROVAL	515.22	SUSPENSION AND/OR REVOCATION OF APPROVAL OF A MEDICATED FEED MILL LICENSE Administrative Actions On Licenses; Medicated Feed Mill License
SUSPENSION OF APPROVAL	814.47	TEMPORARY SUSPENSION OF APPROVAL OF A PMA FDA Action On A PMA; Premarket Approval Of Medical Devices
SUSPENSION OF APPROVAL	814.120	TEMPORARY SUSPENSION OF APPROVAL OF A HDE Humanitarian Use Devices; Premarket Approval Of Medical Devices
SUSPENSION OF CERTIFICATES	900.14	SUSPENSION OR REVOCATION OF CERTIFICATES Quality Standards And Certification; Mammography
SUSPENSION OF LICENSE	601.6	SUSPENSION OF LICENSE General Provisions; Licensing
SUSPENSION OF LICENSES	5.202	ISSUANCE OF NOTICES OF OPPORTUNITY FOR A HEARING ON PROPOSALS FOR DENIAL OF APPROVAL OF APPLICATIONS FOR LICENSES, SUSPENSION OF LICENSES, OR REVOCATION OF LICENSES Biologics; Redelegations Of Authority; Delegations Of Authority And Organization
SUSPENSION OF OBLIGATIONS	26.75	SUSPENSION OF RECOGNITION OBLIGATIONS Framework Provisions; Mutual Recognition Of Pharmaceutical GMP Reports, Medical Device Quality System Audit Reports

SUSPENSION, PERMIT	1210.30	HEARING PROCEDURE FOR PERMIT DENIAL, SUSPENSION, AND REVOCATION Hearings; Regulations Under The Federal Import Milk Act
SUSPENSION OF REGISTRATION	1301.36	SUSPENSION OR REVOCATION OF REGISTRATION; SUSPENSION OF REGISTRATION PENDING FINAL ORDER Action On Application For Registration; Registration Of Manufacturers, Distributors, And Dispensers Of Controlled Substances
SUSPENSION OF REGISTRATION	1309.43	SUSPENSION OR REVOCATION OF REGISTRATION Registration Of Manufacturers, Distributors, Importers And Exporters Of List I Chemicals
SUSPENSION OF REGISTRATION	1309.44	SUSPENSION OF REGISTRATION PENDING FINAL ORDER Registration Of Manufacturers, Distributors, Importers And Exporters Of List I Chemicals
SUSPENSION OF SHIPMENTS	1313.41	SUSPENSION OF SHIPMENTS Importation And Exportation Of Precursors And Essential Chemicals
SUSPENSION, TESTING FACILITY	58.217	SUSPENSION OR TERMINATION OF A TESTING FACILITY BY A SPONSOR Disqualification Of Testing Facilities; Good Laboratory Practice For Nonclinical Laboratory Studies
SUSTAINED-RELEASE	520.1197	IVERMECTIN SUSTAINED-RELEASE BOLUS Oral Dosage Form New Animal Drugs
SUSTAINED-RELEASE	520.1450c	MORANTEL TARTRATE SUSTAINED-RELEASE TRILAMINATE CYLINDER/SHEET Oral Dosage Form New Animal Drugs
SUSTAINED RELEASE	520.1920	PROCHLORPERAZINE, ISOPROPAMIDE SUSTAINED RELEASE CAPSULES Oral Dosage Form New Animal Drugs
SUSTAINED-RELEASE	520.1921	PROCHLORPERAZINE, ISOPROPAMIDE, WITH NEOMYCIN SUSTAINED-RELEASE CAPSULES Oral Dosage Form New Animal Drugs
SUSTAINED-RELEASE	520.2260b	SULFAMETHAZINE SUSTAINED-RELEASE BOLUSES Oral Dosage Form New Animal Drugs
SUSTAINED-RELEASE	520.2260c	SULFAMETHAZINE SUSTAINED-RELEASE TABLETS Oral Dosage Form New Animal Drugs
SUTURE	882.4650	NEUROSURGICAL SUTURE NEEDLE Neurological Surgical Devices; Neurological Devices
SUTURE	878.4493	ABSORBABLE POLY(GLYCOLIDE/L-LACTIDE) SURGICAL SUTURE Surgical Devices; General And Plastic Surgery Devices
SUTURE	878.4495	STAINLESS STEEL SUTURE Surgical Devices; General And Plastic Surgery Devices
SUTURE	878.4830	ABSORBABLE SURGICAL GUT SUTURE Surgical Devices; General And Plastic Surgery Devices
SUTURE	878.5000	NONABSORBABLE POLY(ETHYLENE TEREPHTHALATE) SURGICAL SUTURE Surgical Devices; General And Plastic Surgery Devices
SUTURE	878.5010	NONABSORBABLE POLYPROPYLENE SURGICAL SUTURE Surgical Devices; General And Plastic Surgery Devices
SUTURE	878.5020	NONABSORBABLE POLYAMIDE SURGICAL SUTURE Surgical Devices; General And Plastic Surgery Devices
SUTURE	878.5030	NATURAL NONABSORBABLE SILK SURGICAL SUTURE Surgical Devices; General And Plastic Surgery Devices
SUTURE	878.5035	NONABSORBABLE EXPANDED POLYTETRAFLUOROETHYLENE SURGICAL SUTURE Surgical Devices; General And Plastic Surgery Devices
SUTURE RETENTION	878.4930	SUTURE RETENTION DEVICE Surgical Devices; General And Plastic Surgery Devices
SWEET CHOCOLATE	163.123	SWEET CHOCOLATE Requirements For Specific Standardized Cacao Products; Cacao Products
SWEET CHOCOLATE	163.153	SWEET CHOCOLATE AND VEGETABLE FAT COATING Requirements For Specific Standardized Cacao Products; Cacao Products
SWEET COCOA	163.150	SWEET COCOA AND VEGETABLE FAT COATING Requirements For Specific Standardized Cacao Products; Cacao Products
SWEETENED, ARTIFICIALLY	145.116	ARTIFICIALLY SWEETENED CANNED APRICOTS Requirements For Specific Standardized Canned Fruits; Canned Fruits
SWEETENED, ARTIFICIALLY	145.126	ARTIFICIALLY SWEETENED CANNED CHERRIES Requirements For Specific Standardized Canned Fruits; Canned Fruits
SWEETENED, ARTIFICIALLY	145.131	ARTIFICIALLY SWEETENED CANNED FIGS Requirements For Specific Standardized Canned Fruits; Canned Fruits
SWEETENED, ARTIFICIALLY	145.136	ARTIFICIALLY SWEETENED CANNED FRUIT COCKTAIL Requirements For Specific Standardized Canned Fruits; Canned Fruits
SWEETENED, ARTIFICIALLY	145.171	ARTIFICIALLY SWEETENED CANNED PEACHES Requirements For Specific Standardized Canned Fruits; Canned Fruits
SWEETENED, ARTIFICIALLY	145.176	ARTIFICIALLY SWEETENED CANNED PEARS Requirements For Specific Standardized Canned Fruits; Canned Fruits
SWEETENED, ARTIFICIALLY	145.181	ARTIFICIALLY SWEETENED CANNED PINEAPPLE Requirements For Specific Standardized Canned Fruits; Canned Fruits

SYNTHETIC PETROLEUM WAX	172.888	SYNTHETIC PETROLEUM WAX Multipurpose Additives; Food Additives Permitted For Direct Addition To Food For Human Consumption
SYNTHETIC PETROLEUM WAX	178.3720	PETROLEUM WAX, SYNTHETIC Certain Adjuvants And Production Aids; Indirect Food Additives: Adjuvants, Production Aids, And Sanitizers
SYNTHETIC POLYMER MATERIAL	874.3620	EAR, NOSE, AND THROAT SYNTHETIC POLYMER MATERIAL Prosthetic Devices; Ear, Nose, And Throat Devices
SYRINGE	870.1650	ANGIOGRAPHIC INJECTOR AND SYRINGE Cardiovascular Diagnostic Devices; Cardiovascular Devices
SYRINGE	870.1670	SYRINGE ACTUATOR FOR AN INJECTOR Cardiovascular Diagnostic Devices; Cardiovascular Devices
SYRINGE	872.6770	CARTRIDGE SYRINGE Miscellaneous Devices; Dental Devices
SYRINGE	880.5860	PISTON SYRINGE General Hospital And Personal Use Therapeutic Devices; General Hospital And Personal Use Devices
SYRINGE	880.6920	SYRINGE NEEDLE INTRODUCER General Hospital And Personal Use Miscellaneous Devices; General Hospital And Personal Use Devices
SYRINGE	880.6960	IRRIGATING SYRINGE General Hospital And Personal Use Miscellaneous Devices; General Hospital And Personal Use Devices
SYRUP, CLENBUTEROL	520.4512	CLENBUTEROL SYRUP Oral Dosage Form New Animal Drugs
SYRUP, CORN	184.1865	CORN SYRUP Listing Of Specific Substances Affirmed As GRAS; Direct Food Substances Affirmed As Generally Recognized As Safe
SYRUP, CORN	184.1866	HIGH FRUCTOSE CORN SYRUP Listing Of Specific Substances Affirmed As GRAS; Direct Food Substances Affirmed As Generally Recognized As Safe
SYRUP, CORN	573.530	HYDROGENATED CORN SYRUP Food Additive Listing; Food Additives Permitted In Feed And Drinking Water Of Animals
SYRUP, DIETHYLCARBAMAZINE	520.622	DIETHYLCARBAMAZINE CITRATE SYRUP Oral Dosage Form New Animal Drugs
SYRUP, IPECAC	201.308	IPECAC SYRUP; WARNINGS AND DIRECTIONS FOR USE FOR OVER-THE-COUNTER SALE Specific Labeling Requirements For Specific Drug Products; Labeling
SYRUP, MALT	184.1445	MALT SYRUP (MALT EXTRACT) Listing Of Specific Substances Affirmed As GRAS; Direct Food Substances Affirmed As Generally Recognized As Safe
SYSTEM	864.5425	MULTIPURPOSE SYSTEM FOR IN VITRO COAGULATION STUDIES Automated And Semi-automated Hematology Devices; Hematology And Pathology Devices
SYSTEM	864.5620	AUTOMATED HEMOGLOBIN SYSTEM Automated And Semi-automated Hematology Devices; Hematology And Pathology Devices
SYSTEM	864.5700	AUTOMATED PLATELET AGGREGATION SYSTEM Automated And Semi-automated Hematology Devices; Hematology And Pathology Devices
SYSTEM	864.7340	FIBRINOGEN DETERMINATION SYSTEM Hematology Kits And Packages; Hematology And Pathology Devices
SYSTEM	864.7440	ELECTROPHORETIC HEMOGLOBIN ANALYSIS SYSTEM Hematology Kits And Packages; Hematology And Pathology Devices
SYSTEM	864.9125	VACUUM-ASSISTED BLOOD COLLECTION SYSTEM Products Used In Establishments That Manufacture Blood And Blood Products; Hematology And Pathology Devices
SYSTEM	864.9145	PROCESSING SYSTEM FOR FROZEN BLOOD Products Used In Establishments That Manufacture Blood And Blood Products; Hematology And Pathology Devices
SYSTEM	864.9175	AUTOMATED BLOOD GROUPING AND ANTIBODY TEST SYSTEM Products Used In Establishments That Manufacture Blood And Blood Products; Hematology And Pathology Devices
SYSTEM	868.5270	BREATHING SYSTEM HEATER Therapeutic Devices; Anesthesiology Devices
SYSTEM	870.1130	NONINVASIVE BLOOD PRESSURE MEASUREMENT SYSTEM Cardiovascular Diagnostic Devices; Cardiovascular Devices
SYSTEM	870.1270	INTRACAVITARY PHONOCATHETER SYSTEM Cardiovascular Diagnostic Devices; Cardiovascular Devices
SYSTEM	870.1290	STEERABLE CATHETER CONTROL SYSTEM Cardiovascular Diagnostic Devices; Cardiovascular Devices
SYSTEM	870.2600	SIGNAL ISOLATION SYSTEM Cardiovascular Monitoring Devices; Cardiovascular Devices

SYSTEM	870.3535	INTRA-AORTIC BALLOON AND CONTROL SYSTEM Cardiovascular Prosthetic Devices; Cardiovascular Devices
SYSTEM	870.5900	THERMAL REGULATING SYSTEM Cardiovascular Therapeutic Devices; Cardiovascular Devices
SYSTEM	874.3310	HEARING AID CALIBRATOR AND ANALYSIS SYSTEM Prosthetic Devices; Ear, Nose, And Throat Devices
SYSTEM	876.1620	URODYNAMICS MEASUREMENT SYSTEM Diagnostic Devices; Gastroenterology-urology Devices
SYSTEM	876.1725	GASTROINTESTINAL MOTILITY MONITORING SYSTEM Diagnostic Devices; Gastroenterology-urology Devices
SYSTEM	876.1800	URINE FLOW OR VOLUME MEASURING SYSTEM Diagnostic Devices; Gastroenterology-urology Devices
SYSTEM	876.4650	WATER JET RENAL STONE DISLODGER SYSTEM Surgical Devices; Gastroenterology-urology Devices
SYSTEM	876.5220	COLONIC IRRIGATION SYSTEM Therapeutic Devices; Gastroenterology-urology Devices
SYSTEM	876.5600	SORBENT REGENERATED DIALYSATE DELIVERY SYSTEM FOR HEMODIALYSIS Therapeutic Devices; Gastroenterology-urology Devices
SYSTEM	876.5630	PERITONEAL DIALYSIS SYSTEM AND ACCESSORIES Therapeutic Devices; Gastroenterology-urology Devices
SYSTEM	876.5665	WATER PURIFICATION SYSTEM FOR HEMODIALYSIS Therapeutic Devices; Gastroenterology-urology Devices
SYSTEM	876.5820	HEMODIALYSIS SYSTEM AND ACCESSORIES Therapeutic Devices; Gastroenterology-urology Devices
SYSTEM	876.5860	HIGH PERMEABILITY HEMODIALYSIS SYSTEM Therapeutic Devices; Gastroenterology-urology Devices
SYSTEM	876.5870	SORBENT HEMOPERFUSION SYSTEM Therapeutic Devices; Gastroenterology-urology Devices
SYSTEM	876.5880	ISOLATED KIDNEY PERFUSION AND TRANSPORT SYSTEM AND ACCESSORIES Therapeutic Devices; Gastroenterology-urology Devices
SYSTEM	882.1855	ELECTROENCEPHALOGRAM (EEG) TELEMETRY SYSTEM Neurological Diagnostic Devices; Neurological Devices
SYSTEM	882.4545	SHUNT SYSTEM IMPLANTATION INSTRUMENT Neurological Surgical Devices; Neurological Devices
SYSTEM	884.2980	TELETHERMOGRAPHIC SYSTEM Obstetrical And Gynecological Monitoring; Obstetrical And Gynecological Devices
SYSTEM	884.2982	LIQUID CRYSTAL THERMOGRAPHIC SYSTEM Obstetrical And Gynecological Monitoring; Obstetrical And Gynecological Devices
SYSTEM	886.5820	CLOSED-CIRCUIT TELEVISION READING SYSTEM Therapeutic Devices; Ophthalmic Devices
SYSTEM, ALERT	26.20	ALERT SYSTEM Specific Sector Provisions For Pharmaceutical GMPs; Mutual Recognition Of Pharmaceutical GMP Reports, Medical Device Quality System Audit Reports
SYSTEM, ALERT	26.50	ALERT SYSTEM AND EXCHANGE OF POSTMARKET VIGILANCE REPORTS Specific Sector Provisions For Medical Devices; Mutual Recognition Of Pharmaceutical GMP Reports, Medical Device Quality System Audit Reports
SYSTEM, CENTRAL NERVOUS	882.5550	CENTRAL NERVOUS SYSTEM FLUID SHUNT AND COMPONENTS Neurological Therapeutic Devices; Neurological Devices
SYSTEM, COMMUNICATION	890.3700	NONPOWERED COMMUNICATION SYSTEM Physical Medicine Prosthetic Devices; Physical Medicine Devices
SYSTEM, COMMUNICATION	890.3710	POWERED COMMUNICATION SYSTEM Physical Medicine Prosthetic Devices; Physical Medicine Devices
SYSTEM, DEVICE TRACKING	821.25	DEVICE TRACKING SYSTEM AND CONTENT REQUIREMENTS; MANUFACTURER REQUIREMENTS Medical Device Tracking Requirements
SYSTEM, IMAGING	876.1300	INGESTIBLE TELEMETRIC GASTROINTESTINAL CAPSULE IMAGING SYSTEM Diagnostic Devices; Gastroenterology-urology Devices
SYSTEM, IMAGING	892.1550	ULTRASONIC PULSED DOPPLER IMAGING SYSTEM Diagnostic Devices; Radiology Devices
SYSTEM, IMAGING	892.1560	ULTRASONIC PULSED ECHO IMAGING SYSTEM Diagnostic Devices; Radiology Devices
SYSTEM, IMAGING	892.1630	ELECTROSTATIC X-RAY IMAGING SYSTEM Diagnostic Devices; Radiology Devices
SYSTEM, QUALITY	26.41	EXCHANGE AND ENDORSEMENT OF QUALITY SYSTEM EVALUATION REPORTS Specific Sector Provisions For Medical Devices; Mutual Recognition Of Pharmaceutical GMP Reports, Medical Device Quality System Audit Reports
SYSTEM, QUALITY	26.43	TRANSMISSION OF QUALITY SYSTEM EVALUATION REPORTS Specific Sector Provisions For Medical Devices; Mutual Recognition Of Pharmaceutical GMP Reports, Medical Device Quality System Audit Reports
SYSTEM, QUALITY	820.5	QUALITY SYSTEM General Provisions; Quality System Regulation

SYSTEM, QUALITY	820.186	QUALITY SYSTEM RECORD Records; Quality System Regulation
SYSTEM, WATER	1240.86	PROTECTION OF PIER WATER SYSTEM Source And Use Of Potable Water; Control Of Communicable Diseases
SYSTEMIC LUPUS	866.5820	SYSTEMIC LUPUS ERYTHEMATOSUS IMMUNOLOGICAL TEST SYSTEM Immunological Test Systems; Immunology And Microbiology Devices
SYSTEMS	21.21	CHANGES IN SYSTEMS AND NEW SYSTEMS FDA Privacy Act Record Systems; Protection Of Privacy
SYSTEMS	21.61	EXEMPT SYSTEMS Exemptions; Protection Of Privacy
SYSTEMS	21.65	ACCESS TO RECORDS IN EXEMPT SYSTEMS Exemptions; Protection Of Privacy
SYSTEMS	203.34	POLICIES AND PROCEDURES; ADMINISTRATIVE SYSTEMS Samples; Prescription Drug Marketing
SYSTEMS	864.9300	AUTOMATED COOMBS TEST SYSTEMS Products Used In Establishments That Manufacture Blood And Blood Products; Hematology And Pathology Devices
SYSTEMS	880.2420	ELECTRONIC MONITOR FOR GRAVITY FLOW INFUSION SYSTEMS General Hospital And Personal Use Monitoring Devices; General Hospital And Personal Use Devices
SYSTEMS	888.3070	PEDICLE SCREW SPINAL SYSTEMS Prosthetic Devices; Orthopedic Devices
SYSTEMS, CLOSED	11.10	CONTROLS FOR CLOSED SYSTEMS Electronic Records; Electronic Signatures
SYSTEMS, COLLECTION	809.40	RESTRICTIONS ON THE SALE, DISTRIBUTION AND USE OF OTC TEST SAMPLE COLLECTION SYSTEMS FOR DRUGS OF ABUSE TESTING Requirements For Manufacturers And Producers; In Vitro Diagnostic Products For Human Use
SYSTEMS, COLLECTION	864.3260	OTC TEST SAMPLE COLLECTION SYSTEMS FOR DRUGS OF ABUSE TESTING Pathology Instrumentation And Accessories; Hematology And Pathology Devices
SYSTEMS, OPEN	11.30	CONTROLS FOR OPEN SYSTEMS Electronic Records; Electronic Signatures
SYSTEMS, RECORD	21.20	PROCEDURES FOR NOTICE OF FDA PRIVACY ACT RECORD SYSTEMS FDA Privacy Act Record Systems; Protection Of Privacy
SYSTEMS, RECORD	21.70	DISCLOSURE AND INTRA-AGENCY USE OF RECORDS IN PRIVACY ACT RECORD SYSTEMS; NO ACCOUNTING REQUIRED Disclosure Of Records In Privacy Act Record Systems To Persons Other Than The Subject Individual; Protection Of Privacy
SYSTEMS, RECORD	21.71	DISCLOSURE OF RECORDS IN PRIVACY ACT RECORD SYSTEMS; ACCOUNTING REQUIRED Disclosure Of Records In Privacy Act Record Systems To Persons Other Than The Subject Individual; Protection Of Privacy
SYSTEMS, RECORD	21.73	ACCURACY, COMPLETENESS, TIMELINESS, AND RELEVANCE OF RECORDS DISCLOSED FROM PRIVACY ACT RECORD SYSTEMS Disclosure Of Records In Privacy Act Record Systems To Persons Other Than The Subject Individual; Protection Of Privacy
SYSTEMS, WATER	1250.42	WATER SYSTEMS; CONSTANT TEMPERATURE BOTTLES Equipment And Operation Of Land And Air Conveyances; Interstate Conveyance Sanitation
SYSTEMS, WATER	1250.82	POTABLE WATER SYSTEMS Sanitation Facilities And Conditions On Vessels; Interstate Conveyance Sanitation

T

TABLE, ANESTHETIC	868.6100	ANESTHETIC CABINET, TABLE, OR TRAY Miscellaneous; Anesthesiology Devices
TABLE, INSTRUMENT	886.4855	OPHTHALMIC INSTRUMENT TABLE Surgical Devices; Ophthalmic Devices
TABLE, MECHANICAL	890.3750	MECHANICAL TABLE Physical Medicine Prosthetic Devices; Physical Medicine Devices
TABLE, MEDICAL	880.6140	MEDICAL CHAIR AND TABLE General Hospital And Personal Use Miscellaneous Devices; General Hospital And Personal Use Devices
TABLE, OBSTETRIC	884.4900	OBSTETRIC TABLE AND ACCESSORIES Obstetrical And Gynecological Surgical Devices; Obstetrical And Gynecological Devices
TABLE, OPERATING	878.4950	MANUAL OPERATING TABLE AND ACCESSORIES AND MANUAL OPERATING CHAIR AND ACCESSORIES Surgical Devices; General And Plastic Surgery Devices
TABLE, PHYSICAL THERAPY	890.5880	MULTI-FUNCTION PHYSICAL THERAPY TABLE Physical Medicine Therapeutic Devices; Physical Medicine Devices
TABLE, POWERED	890.3760	POWERED TABLE Physical Medicine Prosthetic Devices; Physical Medicine Devices
TABLE, RADIOLOGIC	892.1980	RADIOLOGIC TABLE Diagnostic Devices; Radiology Devices
TABLE SIRUP	168.180	TABLE SIRUP Requirements For Specific Standardized Sweeteners And Table Syrups; Sweeteners And Table Sirups
TABLE, UROLOGICAL	876.4890	UROLOGICAL TABLE AND ACCESSORIES Surgical Devices; Gastroenterology-urology Devices
TABLES, OPERATING	878.4960	OPERATING TABLES AND ACCESSORIES AND OPERATING CHAIRS AND ACCESSORIES Surgical Devices; General And Plastic Surgery Devices
TABLET	520.1196	IVERMECTIN AND PYRANTEL PAMOATE CHEWABLE TABLET Oral Dosage Form New Animal Drugs
TABLET	520.1242b	LEVAMISOLE HYDROCHLORIDE TABLET OR OBLET (BOLUS) Oral Dosage Form New Animal Drugs
TABLET	520.2260a	SULFAMETHAZINE OBLET, TABLET, AND BOLUS Oral Dosage Form New Animal Drugs
TABLETS	520.23	ACEPROMAZINE MALEATE TABLETS Oral Dosage Form New Animal Drugs
TABLETS	520.82a	AMINOPROPAZINE FUMARATE TABLETS Oral Dosage Form New Animal Drugs
TABLETS	520.82b	AMINOPROPAZINE FUMARATE, NEOMYCIN SULFATE TABLETS Oral Dosage Form New Animal Drugs
TABLETS	520.88a	AMOXICILLIN TRIHYDRATE FILM-COATED TABLETS Oral Dosage Form New Animal Drugs
TABLETS	520.88f	AMOXICILLIN TRIHYDRATE TABLETS Oral Dosage Form New Animal Drugs
TABLETS	520.88g	AMOXICILLIN TRIHYDRATE AND CLAVULANATE POTASSIUM FILM-COATED TABLETS Oral Dosage Form New Animal Drugs
TABLETS	520.90b	AMPICILLIN TRIHYDRATE TABLETS Oral Dosage Form New Animal Drugs
TABLETS	520.246	BUTORPHANOL TARTRATE TABLETS Oral Dosage Form New Animal Drugs
TABLETS	520.310	CARAMIPHEN ETHANEDISULFONATE AND AMMONIUM CHLORIDE TABLETS Oral Dosage Form New Animal Drugs
TABLETS	520.312	CARNIDAZOLE TABLETS Oral Dosage Form New Animal Drugs
TABLETS	520.314	CEFADROXIL TABLETS Oral Dosage Form New Animal Drugs
TABLETS	520.390a	CHLORAMPHENICOL TABLETS Oral Dosage Form New Animal Drugs
TABLETS	520.420	CHLOROTHIAZIDE TABLETS AND BOLUSES Oral Dosage Form New Animal Drugs
TABLETS	520.434	CHLORPHENESIN CARBAMATE TABLETS Oral Dosage Form New Animal Drugs
TABLETS	520.445c	CHLORTETRACYCLINE TABLETS AND BOLUSES Oral Dosage Form New Animal Drugs
TABLETS	520.455	CLOMIPRAMINE HYDROCHLORIDE TABLETS Oral Dosage Form New Animal Drugs
TABLETS	520.531	CYTHIOATE TABLETS Oral Dosage Form New Animal Drugs

TABLETS	520.540b	DEXAMETHASONE TABLETS AND BOLUSES Oral Dosage Form New Animal Drugs
TABLETS	520.540c	DEXAMETHASONE CHEWABLE TABLETS Oral Dosage Form New Animal Drugs
TABLETS	520.581	DICHLOROPHENE TABLETS Oral Dosage Form New Animal Drugs
TABLETS	520.622a	DIETHYLCARBAMAZINE CITRATE TABLETS Oral Dosage Form New Animal Drugs
TABLETS	520.622c	DIETHYLCARBAMAZINE CITRATE CHEWABLE TABLETS Oral Dosage Form New Animal Drugs
TABLETS	520.623	DIETHYLCARBAMAZINE CITRATE, OXIBENDAZOLE CHEWABLE TABLETS Oral Dosage Form New Animal Drugs
TABLETS	520.763a	DITHIAZANINE IODIDE TABLETS Oral Dosage Form New Animal Drugs
TABLETS	520.784	DOXYLAMINE SUCCINATE TABLETS Oral Dosage Form New Animal Drugs
TABLETS	520.804	ENALAPRIL TABLETS Oral Dosage Form New Animal Drugs
TABLETS	520.812	ENROFLOXACIN TABLETS Oral Dosage Form New Animal Drugs
TABLETS	520.816	EPSIPRANTEL TABLETS Oral Dosage Form New Animal Drugs
TABLETS	520.863	ETHYLISOBUTRAZINE HYDROCHLORIDE TABLETS Oral Dosage Form New Animal Drugs
TABLETS	520.960	FLUMETHASONE TABLETS Oral Dosage Form New Animal Drugs
TABLETS	520.1130c	HETACILLIN POTASSIUM TABLETS Oral Dosage Form New Animal Drugs
TABLETS	520.1157	IODINATED CASEIN TABLETS Oral Dosage Form New Animal Drugs
TABLETS	520.1193	IVERMECTIN TABLETS AND CHEWABLES Oral Dosage Form New Animal Drugs
TABLETS	520.1205	KANAMYCIN SULFATE, PECTIN, BISMUTH SUBCARBONATE, ACTIVATED ATTAPULGITE TABLETS Oral Dosage Form New Animal Drugs
TABLETS	520.1242e	LEVAMISOLE HYDROCHLORIDE EFFERVESCENT TABLETS Oral Dosage Form New Animal Drugs
TABLETS	520.1263a	LINCOMYCIN HYDROCHLORIDE MONOHYDRATE TABLETS AND SIRUP Oral Dosage Form New Animal Drugs
TABLETS	520.1284	SODIUM LIOTHYRONINE TABLETS Oral Dosage Form New Animal Drugs
TABLETS	520.1288	LUFENURON TABLETS Oral Dosage Form New Animal Drugs
TABLETS	520.1310	MARBOFLOXACIN TABLETS Oral Dosage Form New Animal Drugs
TABLETS	520.1331	MECLOFENAMIC ACID TABLETS Oral Dosage Form New Animal Drugs
TABLETS	520.1341	MEGESTROL ACETATE TABLETS Oral Dosage Form New Animal Drugs
TABLETS	520.1380	METHOCARBAMOL TABLETS Oral Dosage Form New Animal Drugs
TABLETS	520.1408	METHYLPREDNISOLONE TABLETS Oral Dosage Form New Animal Drugs
TABLETS	520.1409	METHYLPREDNISOLONE, ASPIRIN TABLETS Oral Dosage Form New Animal Drugs
TABLETS	520.1446	MILBEMYCIN OXIME/LUFENURON TABLETS Oral Dosage Form New Animal Drugs
TABLETS	520.1616	ORBIFLOXACIN TABLETS Oral Dosage Form New Animal Drugs
TABLETS	520.1660c	OXYTETRACYCLINE HYDROCHLORIDE TABLETS/BOLUSES Oral Dosage Form New Animal Drugs
TABLETS	520.1696d	PENICILLIN V POTASSIUM TABLETS Oral Dosage Form New Animal Drugs
TABLETS	520.1720a	PHENYLBUTAZONE TABLETS AND BOLUSES Oral Dosage Form New Animal Drugs
TABLETS	520.1805	PIPERAZINE PHOSPHATE WITH THENIUM CLOSYLATE TABLETS Oral Dosage Form New Animal Drugs
TABLETS	520.1870	PRAZIQUANTEL TABLETS Oral Dosage Form New Animal Drugs
TABLETS	520.1871	PRAZIQUANTEL/PYRANTEL PAMOATE TABLETS Oral Dosage Form New Animal Drugs
TABLETS	520.1872	PRAZIQUANTEL, PYRANTEL PAMOATE AND FEBANTEL TABLETS Oral Dosage Form New Animal Drugs

TABLETS	520.1880	PREDNISOLONE TABLETS Oral Dosage Form New Animal Drugs
TABLETS	520.1900	PRIMIDONE TABLETS Oral Dosage Form New Animal Drugs
TABLETS	520.2041	PYRANTEL PAMOATE CHEWABLE TABLETS Oral Dosage Form New Animal Drugs
TABLETS	520.2042	PYRANTEL PAMOATE TABLETS Oral Dosage Form New Animal Drugs
TABLETS	520.2088	ROXARSONE TABLETS Oral Dosage Form New Animal Drugs
TABLETS	520.2098	SELEGILINE HYDROCHLORIDE TABLETS Oral Dosage Form New Animal Drugs
TABLETS	520.2123	SPECTINOMYCIN DIHYDROCHLORIDE PENTAHYDRATE TABLETS Oral Dosage Form New Animal Drugs
TABLETS	520.2150a	STANOZOLOL TABLETS Oral Dosage Form New Animal Drugs
TABLETS	520.2150b	STANOZOLOL CHEWABLE TABLETS Oral Dosage Form New Animal Drugs
TABLETS	520.2158b	DIHYDROSTREPTOMYCIN TABLETS Oral Dosage Form New Animal Drugs
TABLETS	520.2200c	SULFACHLORPYRIDAZINE TABLETS Oral Dosage Form New Animal Drugs
TABLETS	520.2220b	SULFADIMETHOXINE TABLETS AND BOLUSES Oral Dosage Form New Animal Drugs
TABLETS	520.2220d	SULFADIMETHOXINE-ORMETOPRIM TABLETS Oral Dosage Form New Animal Drugs
TABLETS	520.2240b	SULFAETHOXYPYRIDAZINE TABLETS Oral Dosage Form New Animal Drugs
TABLETS	520.2260c	SULFAMETHAZINE SUSTAINED-RELEASE TABLETS Oral Dosage Form New Animal Drugs
TABLETS	520.2280	SULFAMETHIZOLE AND METHENAMINE MANDELATE TABLETS Oral Dosage Form New Animal Drugs
TABLETS	520.2330	SULFISOXAZOLE TABLETS Oral Dosage Form New Animal Drugs
TABLETS	520.2345b	TETRACYCLINE TABLETS Oral Dosage Form New Animal Drugs
TABLETS	520.2345g	TETRACYCLINE HYDROCHLORIDE AND SODIUM NOVOBIOCIN TABLETS Oral Dosage Form New Animal Drugs
TABLETS	520.2345h	TETRACYCLINE HYDROCHLORIDE, SODIUM NOVOBIOCIN, AND PREDNISOLONE TABLETS Oral Dosage Form New Animal Drugs
TABLETS	520.2362	THENIUM CLOSYLATE TABLETS Oral Dosage Form New Animal Drugs
TABLETS	520.2481	TRIAMCINOLONE ACETONIDE TABLETS Oral Dosage Form New Animal Drugs
TABLETS	520.2582	TRIFLUPROMAZINE HYDROCHLORIDE TABLETS Oral Dosage Form New Animal Drugs
TABLETS	520.2604	TRIMEPRAZINE TARTRATE AND PREDNISOLONE TABLETS Oral Dosage Form New Animal Drugs
TABLETS	520.2610	TRIMETHOPRIM AND SULFADIAZINE TABLETS Oral Dosage Form New Animal Drugs
TABLETS	529.400	CHLORHEXIDINE TABLETS AND SUSPENSION Certain Other Dosage Form New Animal Drugs
TABLETS	800.12	CONTACT LENS SOLUTIONS AND TABLETS; TAMPER RESISTANT PACKAGING Requirements For Specific Medical Devices; General
TACK (CODY)	874.3760	SACCULOTOMY TACK (CODY TACK) Prosthetic Devices; Ear, Nose And Throat Devices
TAG	1210.22	FORM OF TAG Permit Control; Regulations Under The Federal Import Milk Act
TAGETES	73.295	TAGETES (AZTEC MARIGOLD) MEAL AND EXTRACT Foods; Listing Of Color Additives Exempt From Certification
TALC	73.1550	TALC Drugs; Listing Of Color Additives Exempt From Certification
TALL OIL	172.862	OLEIC ACID DERIVED FROM TALL OIL FATTY ACIDS Multipurpose Additives; Food Additives Permitted For Direct Addition To Food For Human Consumption
TALL OIL	186.1557	TALL OIL Listing Of Specific Substances Affirmed As GRAS; Indirect Food Substances Affirmed As Generally Recognized As Safe
TAMARIND	176.350	TAMARIND SEED KERNEL POWDER Substances For Use Only As Components Of Paper And Paperboard; Indirect Food Additives

TAMPER-EVIDENT	211.132	TAMPER-EVIDENT PACKAGING REQUIREMENTS FOR OVER-THE-COUNTER HUMAN DRUG PRODUCTS Packaging And Labeling Control; Current Good Manufacturing Practice For Finished Pharmaceuticals
TAMPER-RESISTANT	700.25	TAMPER-RESISTANT PACKAGING REQUIREMENTS FOR COSMETIC PRODUCTS Requirements For Specific Cosmetic Products; General
TAMPER-RESISTANT	800.12	CONTACT LENS SOLUTIONS AND TABLETS; TAMPER-RESISTANT PACKAGING Requirements For Specific Medical Devices; General
TAMPON	884.5460	SCENTED OR SCENTED DEODORIZED MENSTRUAL TAMPON Obstetrical And Gynecological Therapeutic Devices; Obstetrical And Gynecological Device
TAMPON	884.5470	UNSCENTED MENSTRUAL TAMPON Obstetrical And Gynecological Therapeutic Devices; Obstetrical And Gynecological Device
TAMPONS	801.430	USER LABELING FOR MENSTRUAL TAMPONS Special Requirements For Specific Devices; Labeling
TANGENT SCREEN	886.1810	TANGENT SCREEN (CAMPIMETER) Diagnostic Devices; Ophthalmic Devices
TANNIC ACID	184.1097	TANNIC ACID Listing Of Specific Substances Affirmed As GRAS; Direct Food Substances Affirmed As Generally Recognized As Safe
TANNIC ACID	201.304	TANNIC ACID AND BARIUM ENEMA PREPARATIONS Specific Labeling Requirements For Specific Drug Products; Labeling
TANNING LAMP	878.4635	ULTRAVIOLET LAMP FOR TANNING Surgical Devices; General And Plastic Surgery Devices
TANTALUM CLIP	886.3100	OPHTHALMIC TANTALUM CLIP Prosthetic Devices; Ophthalmic Devices
TAPE, ADHESIVE	880.5240	MEDICAL ADHESIVE TAPE AND ADHESIVE BANDAGE General Hospital And Personal Use Therapeutic Devices; General Hospital And Personal Use Devices
TAPE, NYSTAGMUS	886.1905	NYSTAGMUS TAPE Diagnostic Devices; Ophthalmic Devices
TAPE RECORDER	870.2800	MEDICAL MAGNETIC TAPE RECORDER Cardiovascular Monitoring Devices; Cardiovascular Devices
TAPE SOLVENT	878.4730	SURGICAL SKIN DEGREASER OR ADHESIVE TAPE SOLVENT Surgical Devices; General And Plastic Surgery Devices
TAR, COAL	740.18	COAL TAR HAIR DYES POSING A RISK OF CANCER Warning Statements; Cosmetic Product Warning Statements
TARGET	886.1880	FUSION AND STEREOSCOPIC TARGET Diagnostic Devices; Ophthalmic Devices
TARGET TISSUE	500.86	MARKER RESIDUE AND TARGET TISSUE Regulation of Carcinogenic Compounds Used In Food-Producing Animals; General
TARTARIC ACID	184.1099	TARTARIC ACID Listing Of Specific Substances Affirmed As GRAS; Direct Food Substances Affirmed As Generally Recognized As Safe
TARTARIC ACID	184.1101	DIACETYL TARTARIC ACID ESTERS OF MONO- AND DIGLYCERIDES Listing Of Specific Substances Affirmed As GRAS; Direct Food Substances Affirmed As Generally Recognized As Safe
TARTARIC ACID	582.1099	TARTARIC ACID General Purpose Food Additives; Substances Generally Recognized As Safe
TARTARIC ACID	582.4101	DIACETYL TARTARIC ACID ESTERS OF MONO- AND DIGLYCERIDES OF EDIBLE FATS OR OILS, OR EDIBLE FAT-FORMING FATTY ACIDS Emulsifying Agents; Substances Generally Recognized As Safe
TARTARIC ACID	582.6099	TARTARIC ACID Sequestrants; Substances Generally Recognized As Safe
TARTRATE, POTASSIUM ACID	184.1077	POTASSIUM ACID TARTRATE Listing Of Specific Substances Affirmed As GRAS; Direct Food Substances Affirmed As Generally Recognized As Safe
TARTRATE, POTASSIUM ACID	582.1077	POTASSIUM ACID TARTRATE General Purpose Food Additives; Substances Generally Recognized As Safe
TARTRATE, SODIUM	184.1801	SODIUM TARTRATE Listing Of Specific Substances Affirmed As GRAS; Direct Food Substances Affirmed As Generally Recognized As Safe
TARTRATE, SODIUM	582.6801	SODIUM TARTRATE Sequestrants; Substances Generally Recognized As Safe
TARTRATE, SODIUM POTASSIUM	184.1804	SODIUM POTASSIUM TARTRATE Listing Of Specific Substances Affirmed As GRAS; Direct Food Substances Affirmed As Generally Recognized As Safe
TARTRATE, SODIUM POTASSIUM	582.6804	SODIUM POTASSIUM TARTRATE Sequestrants; Substances Generally Recognized As Safe
TARTRATE, SODIUM POTASSIUM	582.1804	SODIUM POTASSIUM TARTRATE General Purpose Food Additives; Substances Generally Recognized As Safe

TAUB DESIGN PROSTHESIS	874.3730	LARYNGEAL PROSTHESIS (TAUB DESIGN) Prosthetic Devices; Ear, Nose, And Throat Devices
TAURINE	573.980	TAURINE Food Additive Listing; Food Additives Permitted In Feed And Drinking Water Of Animals
TBHQ	172.185	TBHQ Food Preservatives; Food Additives Permitted For Direct Addition To Food For Human Consumption
TEACHING	201.125	DRUGS FOR USE IN TEACHING, LAW ENFORCEMENT, RESEARCH, AND ANALYSIS Exemptions From Adequate Directions For Use; Labeling
TEACHING	801.125	DEVICES FOR USE IN TEACHING, LAW ENFORCEMENT, RESEARCH, AND ANALYSIS Exemptions From Adequate Directions For Use; Labeling
TECHNICAL	14.80	QUALIFICATIONS FOR MEMBERS OF STANDING POLICY AND TECHNICAL ADVISORY COMMITTEES Members Of Advisory Committees; Public Hearing Before A Public Advisory Committee
TECHNICAL	14.84	NOMINATIONS AND SELECTION OF NONVOTING MEMBERS OF STANDING TECHNICAL ADVISORY COMMITTEES Members Of Advisory Committees; Public Hearing Before A Public Advisory Committee
TECHNICAL	14.120	ESTABLISHMENT OF THE TECHNICAL ELECTRONIC PRODUCT RADIATION SAFETY STANDARD COMMITTEE (TEPRSSC) Technical Electronic Products Radiation Safety Standards Committee; Public Hearing Before A Public Advisory Committee
TECHNICAL	14.160	ESTABLISHMENT OF STANDING TECHNICAL ADVISORY COMMITTEES FOR HUMAN PRESCRIPTION DRUGS Advisory Committees For Human Prescription Drugs; Public Hearing Before A Public Advisory Committee
TECHNICAL	20.50	AVAILABILITY OF RECORDS AT NATIONAL TECHNICAL INFORMATION SERVICE Procedures And Fees; Public Information
TECHNICAL DATA	1002.3	NOTIFICATION TO USER OF PERFORMANCE AND TECHNICAL DATA General Provisions; Records and Reports
TECHNOLOGY TRANSFER	5.24	AUTHORITY RELATING TO TECHNOLOGY TRANSFER General Redelegations Of Authority; Delegations Of Authority And Organization
TEE DRAIN	868.5995	TEE DRAIN (WATER TRAP) Therapeutic Devices; Anesthesiology Devices
TEETH BLEACHING	872.6475	HEAT SOURCE FOR BLEACHING TEETH Miscellaneous Devices; Dental Devices
TEETHING RING	872.5550	TEETHING RING Therapeutic Devices; Dental Devices
TELEMETRIC	876.1300	INGESTIBLE TELEMETRIC GASTROINTESTINAL CAPSULE IMAGING SYSTEM Diagnostic Devices; Gastroenterology-urology Devices
TELEMETRY	882.1855	ELECTROENCEPHALOGRAM (EEG) TELEMETRY SYSTEM Neurological Diagnostic Devices; Neurological Devices
TELEPHONE	870.2920	TELEPHONE ELECTROCARDIOGRAPH TRANSMITTER AND RECEIVER Cardiovascular Monitoring Devices; Cardiovascular Devices
TELETHERAPY	892.5740	RADIONUCLIDE TELETHERAPY SOURCE Therapeutic Devices; Radiology Devices
TELETHERMOGRAPHIC SYSTEM	884.2980	TELETHERMOGRAPHIC SYSTEM Obstetrical And Gynecological Monitoring, Obstetrical And Gynecological Devices
TELEVISION READING SYSTEM	886.5820	CLOSED-CIRCUIT TELEVISION READING SYSTEM Therapeutic Devices; Ophthalmic Devices
TELEVISION RECEIVERS	1020.10	TELEVISION RECEIVERS Performance Standards For Ionizing Radiation Emitting Products
TEMPERATURE	870.4250	CARDIOPULMONARY BYPASS TEMPERATURE CONTROLLER Cardiovascular Surgical Devices; Cardiovascular Devices
TEMPERATURE	880.2200	LIQUID CRYSTAL FOREHEAD TEMPERATURE STRIP General Hospital And Personal Use Monitoring Devices; General Hospital And Personal Use Devices
TEMPERATURE	880.5560	TEMPERATURE REGULATED WATER MATTRESS General Hospital And Personal Use Therapeutic Devices; General Hospital And Personal Use Devices
TEMPERATURE	882.1570	POWERED DIRECT-CONTACT TEMPERATURE MEASUREMENT DEVICE Neurological Diagnostic Devices; Neurological Devices
TEMPERATURE	882.5500	LESION TEMPERATURE MONITOR Neurological Therapeutic Devices; Neurological Devices
TEMPERATURE, CONSTANT	1250.42	WATER SYSTEMS; CONSTANT TEMPERATURE BOTTLES Equipment And Operation Of Land And Air Conveyances; Interstate Conveyance Sanitation

TEMPERATURE, CONSTANT	1250.85	DRINKING FOUNTAINS AND COOLERS; ICE; CONSTANT TEMPERATURE BOTTLES Sanitation Facilities And Conditions On Vessels; Interstate Conveyance Sanitation
TEMPERATURES	177.1390	LAMINATE STRUCTURES FOR USE AT TEMPERATURES OF 250 F AND ABOVE Substances For Use As Basic Components Of Single And Repeated Use Food Contact Surfaces; Indirect Food Additives: Polymers
TEMPERATURES	177.1395	LAMINATE STRUCTURES FOR USE AT TEMPERATURES BETWEEN 120 F AND 250 F Substances For Use As Basic Components Of Single And Repeated Use Food Contact Surfaces; Indirect Food Additives: Polymers
TEMPERATURES	600.15	TEMPERATURES DURING SHIPMENT Establishment Standards; Biological Products: General
TEMPERATURES	640.76	PRODUCTS STORED OR SHIPPED AT UNACCEPTABLE TEMPERATURES Source Plasma; Additional Standards For Human Blood And Blood Products
TEMPLATE	888.4800	TEMPLATE FOR CLINICAL USE Surgical Devices; Orthopedic Devices
TEMPORARY CROWN	872.3770	TEMPORARY CROWN AND BRIDGE RESIN Prosthetic Devices; Dental Devices
TEMPORARY DISQUALIFICATION	19.45	TEMPORARY DISQUALIFICATION OF FORMER EMPLOYEES Disqualification Conditions; Standards Of Conduct And Conflicts Of Interest
TEMPORARY ELECTRODE	870.3680	CARDIOVASCULAR PERMANENT OR TEMPORARY PACEMAKER ELECTRODE Cardiovascular Prosthetic Devices; Cardiovascular Devices
TEMPORARY EXEMPTION	1310.09	TEMPORARY EXEMPTION FROM REGISTRATION Records And Reports Of Listed Chemicals And Certain Machines
TEMPORARY EXEMPTIONS	101.108	TEMPORARY EXEMPTIONS FOR PURPOSES OF CONDUCTING AUTHORIZED FOOD LABELING EXPERIMENTS Exemptions From Food Labeling Requirements; Food Labeling
TEMPORARY MEMBERS	5.30	AUTHORITY TO SELECT TEMPORARY VOTING MEMBERS OF ADVISORY COMMITTEES AND AUTHORITY TO SIGN CONFLICT OF INTEREST WAIVERS General Redelegations Of Authority; Delegations Of Authority And Organization
TEMPORARY PERMITS	130.17	TEMPORARY PERMITS FOR INTERSTATE SHIPMENT OF EXPERIMENTAL PACKS OF FOOD VARYING FROM THE REQUIREMENTS OF DEFINITIONS AND STANDARDS OF IDENTITY General Provisions; Food Standards: General
TEMPORARY PERMITS	1210.24	TEMPORARY PERMITS Permit Control; Regulations Under The Federal Import Milk Act
TEMPORARY SCHEDULING	5.104	RESPONSES TO DRUG ENFORCEMENT ADMINISTRATION TEMPORARY SCHEDULING NOTICES Human Drugs; Redelegations Of Authority; Delegations Of Authority And Organization
TEMPORARY SUSPENSION	5.412	TEMPORARY SUSPENSION OF A MEDICAL DEVICE APPLICATION Medical Devices And Radiological Health; Redelegations Of Authority; Delegations Of Authority And Organization
TEMPORARY SUSPENSION	814.47	TEMPORARY SUSPENSION OF APPROVAL OF A PMA FDA Action On A PMA; Premarket Approval Of Medical Devices
TEMPORARY SUSPENSION	814.120	TEMPORARY SUSPENSION OF APPROVAL OF A HDE Humanitarian Use Devices; Premarket Approval Of Medical Devices
TEMPORARY TOLERANCES	509.30	TEMPORARY TOLERANCES FOR POLYCHLORINATED BIPHENYLS (PCB'S) Tolerances For Unavoidable Poisonous Or Deleterious Substances; Unavoidable Contaminants In Animal Food And Food-packaging Material
TEMPOROMANDIBULAR	872.3940	TOTAL TEMPOROMANDIBULAR JOINT PROSTHESIS Prosthetic Devices; Dental Devices
TENDON	888.3025	PASSIVE TENDON PROSTHESIS Prosthetic Devices; Orthopedic Devices
TENT	868.5700	NONPOWERED OXYGEN TENT Therapeutic Devices; Anesthesiology Devices
TENT	868.5710	ELECTRICALLY POWERED OXYGEN TENT Therapeutic Devices; Anesthesiology Devices
TEPRSSC	14.120	ESTABLISHMENT OF THE TECHNICAL ELECTRONIC PRODUCT RADIATION SAFETY STANDARD COMMITTEE (TEPRSSC) Technical Electronic Products Radiation Safety Standards Committee; Public Hearing Before A Public Advisory Committee
TEPRSSC	14.122	FUNCTIONS OF TEPRSSC Technical Electronic Products Radiation Safety Standards Committee; Public Hearing Before A Public Advisory Committee
TEPRSSC	14.125	PROCEDURES OF TEPRSSC Technical Electronic Products Radiation Safety Standards Committee; Public Hearing Before A Public Advisory Committee
TEPRSSC	14.127	MEMBERSHIP OF TEPRSSC Technical Electronic Products Radiation Safety Standards Committee; Public Hearing Before A Public Advisory Committee

TEPRSSC	14.130	CONDUCT OF TEPRSSC MEETINGS; AVAILABILITY OF TEPRSSC RECORDS Technical Electronic Products Radiation Safety Standards Committee; Public Hearing Before A Public Advisory Committee
TEREPHTHALATE	177.1240	1,4-CYCLOHEXYLENE DIMETHYLENE TEREPHTHALATE AND 1,4-CYCLOHEXYLENE DIMETHYLENE ISOPHTHALATE COPOLYMER Substances For Use As Basic Components Of Single And Repeated Use Food Contact Surfaces; Indirect Food Additives: Polymers
TEREPHTHALATE	177.1315	ETHYLENE-1,4-CYCLOHEXYLENE DIMETHYLENE TEREPHTHALATE COPOLYMERS Substances For Use As Basic Components Of Single And Repeated Use Food Contact Surfaces; Indirect Food Additives: Polymers
TEREPHTHALATE	177.1345	ETHYLENE/1,3-PHENYLENE OXYETHYLENE ISOPHTHALATE/TEREPHTHALATE COPOLYMER Substances For Use As Basic Components Of Single And Repeated Use Food Contact Surfaces; Indirect Food Additives: Polymers
TEREPHTHALATE	177.1660	POLY(TETRAMETHYLENE TEREPHTHALATE) Substances For Use As Basic Components Of Single And Repeated Use Food Contact Surfaces; Indirect Food Additives: Polymers
TEREPHTHALATE	870.3470	INTRACARDIAC PATCH OR PLEDGET MADE OF POLYPROPYLENE, POLYETHYLENE TEREPHTHALATE, OR POLYTETRAFLUOROETHYLENE Cardiovascular Prosthetic Devices; Cardiovascular Devices
TEREPHTHALATE	878.5000	NONABSORBABLE POLY(ETHYLENE TEREPHTHALATE) SURGICAL SUTURE Surgical Devices; General And Plastic Surgery Devices
TERMINATE	5.34	ISSUANCE OF NOTICES RELATING TO PROPOSALS AND ORDERS FOR DEBARMENT AND DENIAL OF AN APPLICATION TO TERMINATE DEBARMENT General Redelegations Of Authority; Delegations Of Authority And Organization
TERMINATION	5.101	TERMINATION OF EXEMPTIONS FOR NEW DRUGS FOR INVESTIGATIONAL USE IN HUMAN BEINGS Human Drugs; Redelegations Of Authority; Delegations Of Authority And Organization
TERMINATION	5.505	TERMINATION OF EXEMPTIONS FOR NEW DRUGS FOR INVESTIGATIONAL USE IN ANIMALS Animal Drugs; Redelegations Of Authority; Delegations Of Authority And Organization
TERMINATION	7.55	TERMINATION OF A RECALL Recalls - Guidance On Policy, Procedures & Industry Responsibilities; Enforcement Policy
TERMINATION	14.55	TERMINATION OF ADVISORY COMMITTEES Establishment Of Advisory Committees; Public Hearing Before A Public Advisory Committee
TERMINATION	26.80	ENTRY INTO FORCE, AMENDMENT, AND TERMINATION Framework Provisions; Mutual Recognition Of Pharmaceutical GMP Reports, Medical Device Quality System Audit Reports
TERMINATION	56.113	SUSPENSION OR TERMINATION OF IRB APPROVAL OF RESEARCH IRB Functions And Operations; Institutional Review Boards
TERMINATION	58.217	SUSPENSION OR TERMINATION OF A TESTING FACILITY BY A SPONSOR Disqualification Of Testing Facilities; Good Laboratory Practice For Nonclinical Laboratory Studies
TERMINATION	81.10	TERMINATION OF PROVISIONAL LISTINGS OF COLOR ADDITIVES General Specs+restrictns Provisional Color Additives For Foods, Drugs, And Cosmetics
TERMINATION	99.403	TERMINATION OF APPROVALS OF APPLICATIONS FOR EXEMPTION Dissemination Of Information On Unapproved/New Uses For Marketed Drugs, Biologics, And Devices
TERMINATION	107.250	TERMINATION OF AN INFANT FORMULA RECALL Infant Formula Recalls; Infant Formula
TERMINATION	312.44	TERMINATION Administrative Actions; Investigational New Drug Application
TERMINATION	314.560	TERMINATION OF REQUIREMENTS Accelerated Approval Of New Drugs; Applications For FDA Approval To Market A New Drug
TERMINATION	601.46	TERMINATION REQUIREMENTS Accelerated Approval Of Biological Products; Licensing
TERMINATION	810.17	TERMINATION OF A CEASE DISTRIBUTION AND NOTIFICATION OR MANDATORY RECALL ORDER Mandatory Medical Device Recall Procedures; Medical Device Recall Authority
TERMINATION	1301.52	TERMINATION OF REGISTRATION; TRANSFER OF REGISTRATION; DISTRIBUTION UPON DISCONTINUANCE OF BUSINESS Registration Of Manufacturers, Distributors, And Dispensers Of Controlled Substances
TERMINATION	1309.62	TERMINATION OF REGISTRATION Registration Of Manufacturers, Distributors, Importers And Exporters Of List I Chemicals

TERMINATION	1403.44	TERMINATION FOR CONVENIENCE Post-award Requirements; Uniform Administrative Requirements For Grants And Coop. Agreements To State & Local Gov.
TERMINATION	1404.615	GROUNDS FOR SUSPENSION OF PAYMENTS, SUSPENSION OR TERMINATION OF GRANTS, OR SUSPENSION OR DEBARMENT Governmentwide Debarment And Suspension And Requirements For Drug-Free Workplace (Grants)
TERMINOLOGY	25.5	TERMINOLOGY General Provisions; Environmental Impact Considerations
TERMS AND CONDITIONS	107.50	TERMS AND CONDITIONS Exempt Infant Formulas; Infant Formula
TERPENE	172.280	TERPENE RESIN Coatings, Films And Related Substances; Food Additives Permitted For Direct Addition To Food For Human Consumption
TERPENE	178.3930	TERPENE RESINS Certain Adjuvants And Production Aids; Indirect Food Additives: Adjuvants, Production Aids, And Sanitizers
TERRITORIAL APPLICATION	26.79	TERRITORIAL APPLICATION Framework Provisions; Mutual Recognition Of Pharmaceutical GMP Reports, Medical Device Quality System Audit Reports
TEST	58.47	FACILITIES FOR HANDLING TEST AND CONTROL ARTICLES Facilities; Good Laboratory Practice For Nonclinical Laboratory Studies
TEST	58.105	TEST AND CONTROL ARTICLE CHARACTERIZATION Test And Control Articles; Good Laboratory Practice For Nonclinical Laboratory Studies
TEST	58.107	TEST AND CONTROL ARTICLE HANDLING Test And Control Articles; Good Laboratory Practice For Nonclinical Laboratory Studies
TEST	331.21	TEST MODIFICATIONS Testing Procedures; Antacid Products For Over-the-counter (OTC) Human Use
TEST	352.77	TEST MODIFICATIONS Testing Procedures; Sunscreen Drug Products For Over-the-counter Human Use
TEST	610.11a	INACTIVATED INFLUENZA VACCINE, GENERAL SAFETY TEST General Provisions; General Biological Products Standards
TEST	610.30	TEST FOR MYCOPLASMA Mycoplasma; General Biological Products Standards
TEST	610.40	TEST REQUIREMENTS Testing Requirements For Communicable Disease Agents; General Biological Products Standards
TEST	640.56	QUALITY CONTROL TEST FOR POTENCY Cryoprecipitate; Additional Standards For Human Blood And Blood Products
TEST	660.4	POTENCY TEST Antibody To Hepatitis B Surface Antigen; Additional Standards For Diagnostic Substances For Laboratory Tests
TEST	660.43	POTENCY TEST Hepatitis B Surface Antigen; Additional Standards For Diagnostic Substances For Laboratory Tests
TEST	864.6550	OCCULT BLOOD TEST Manual Hematology Devices; Hematology And Pathology Devices
TEST	864.6600	OSMOTIC FRAGILITY TEST Manual Hematology Devices; Hematology And Pathology Devices
TEST	864.6650	PLATELET ADHESION TEST Manual Hematology Devices; Hematology And Pathology Devices
TEST	864.6700	ERYTHROCYTE SEDIMENTATION RATE TEST Manual Hematology Devices; Hematology And Pathology Devices
TEST	864.7290	FACTOR DEFICIENCY TEST Hematology Kits And Packages; Hematology And Pathology Devices
TEST	864.7300	FIBRIN MONOMER PARACOAGULATION TEST Hematology Kits And Packages; Hematology And Pathology Devices
TEST	864.7660	LEUKOCYTE ALKALINE PHOSPHATASE TEST Hematology Kits And Packages; Hematology And Pathology Devices
TEST	864.7675	LEUKOCYTE PEROXIDASE TEST Hematology Kits And Packages; Hematology And Pathology Devices
TEST	864.7720	PROTHROMBIN CONSUMPTION TEST Hematology Kits And Packages; Hematology And Pathology Devices
TEST	864.7735	PROTHROMBIN-PROCONVERTIN TEST AND THROMBOTEST Hematology Kits And Packages; Hematology And Pathology Devices
TEST	864.7750	PROTHROMBIN TIME TEST Hematology Kits And Packages; Hematology And Pathology Devices
TEST	864.7825	SICKLE CELL TEST Hematology Kits And Packages; Hematology And Pathology Devices
TEST	864.7875	THROMBIN TIME TEST Hematology Kits And Packages; Hematology And Pathology Devices

TEST	864.7900	THROMBOPLASTIN GENERATION TEST Hematology Kits And Packages; Hematology And Pathology Devices
TEST	866.1620	ANTIMICROBIAL SUSCEPTIBILITY TEST DISC Diagnostic Devices; Immunology And Microbiology Devices
TEST	866.1640	ANTIMICROBIAL SUSCEPTIBILITY TEST POWDER Diagnostic Devices; Immunology And Microbiology Devices
TEST	866.2420	OXIDASE SCREENING TEST FOR GONORRHEA Microbiology Devices; Immunology And Microbiology Devices
TEST	866.3290	GONOCOCCAL ANTIBODY TEST (GAT) Serological Reagents; Immunology And Microbiology Devices
TEST	870.3690	PACEMAKER TEST MAGNET Cardiovascular Prosthetic Devices; Cardiovascular Devices
TEST	882.1430	ELECTROENCEPHALOGRAPH TEST SIGNAL GENERATOR Neurological Diagnostic Devices; Neurological Devices
TEST	882.1925	ULTRASONIC SCANNER CALIBRATION TEST BLOCK Neurological Diagnostic Devices; Neurological Devices
TEST	1210.13	TUBERCULIN TEST Inspection And Testing; Regulations Under The Federal Import Milk Act
TEST DRUM	886.4230	OPHTHALMIC KNIFE TEST DRUM Surgical Devices; Ophthalmic Devices
TEST EQUIPMENT	820.72	INSPECTION, MEASURING, AND TEST EQUIPMENT Production And Process Controls; Quality System Regulation
TEST KITS	610.44	USE OF REFERENCE PANELS BY MANUFACTURERS OF TEST KITS Testing Requirements For Communicable Disease Agents; General Biological Products Standards
TEST METHOD	800.20	PATIENT EXAMINATION GLOVES AND SURGEON'S GLOVES; SAMPLE PLANS AND TEST METHOD FOR LEAKAGE DEFECTS; ADULTERATION Requirements For Specific Medical Devices; General
TEST PATTERN	892.1420	RADIONUCLIDE TEST PATTERN PHANTOM Diagnostic Devices; Radiology Devices
TEST PROCEDURES	1010.13	TEST PROCEDURES Performance Standards For Electronic Products: General
TEST REAGENTS	866.3390	NEISSERIA SPP. DIRECT SEROLOGICAL TEST REAGENTS Serological Reagents; Immunology And Microbiology Devices
TEST REAGENTS	866.3820	TREPONEMA PALLIDUM NONTREPONEMAL TEST REAGENTS Serological Reagents; Immunology And Microbiology Devices
TEST REAGENTS	866.3830	TREPONEMA PALLIDUM TREPONEMAL TEST REAGENTS Serological Reagents; Immunology And Microbiology Devices
TEST SAMPLE	809.40	RESTRICTIONS ON THE SALE, DISTRIBUTION AND USE OF OTC TEST SAMPLE COLLECTION SYSTEMS FOR DRUGS OF ABUSE TESTING Requirements For Manufacturers And Producers; In Vitro Diagnostic Products For Human Use
TEST SAMPLE	864.3260	OTC TEST SAMPLE COLLECTION SYSTEMS FOR DRUGS OF ABUSE TESTING Pathology Instrumentation And Accessories; Hematology And Pathology Devices
TEST SYSTEM	862.1020	ACID PHOSPHATASE (TOTAL OR PROSTATIC) TEST SYSTEM Clinical Chemistry Test Systems; Clinical Chemistry And Clinical Toxicology Devices
TEST SYSTEM	862.1025	ADRENOCORTICOTROPIC HORMONE (ACTH) TEST SYSTEM Clinical Chemistry Test Systems; Clinical Chemistry And Clinical Toxicology Devices
TEST SYSTEM	862.1030	ALANINE AMINO TRANSFERASE (ALT/SGPT) TEST SYSTEM Clinical Chemistry Test Systems; Clinical Chemistry And Clinical Toxicology Devices
TEST SYSTEM	862.1035	ALBUMIN TEST SYSTEM Clinical Chemistry Test Systems; Clinical Chemistry And Clinical Toxicology Devices
TEST SYSTEM	862.1040	ALDOLASE TEST SYSTEM Clinical Chemistry Test Systems; Clinical Chemistry And Clinical Toxicology Devices
TEST SYSTEM	862.1050	ALKALINE PHOSPHATASE OR ISOENZYMES TEST SYSTEM Clinical Chemistry Test Systems; Clinical Chemistry And Clinical Toxicology Devices
TEST SYSTEM	862.1060	DELTA-AMINOLEVULINIC ACID TEST SYSTEM Clinical Chemistry Test Systems; Clinical Chemistry And Clinical Toxicology Devices
TEST SYSTEM	862.1065	AMMONIA TEST SYSTEM Clinical Chemistry Test Systems; Clinical Chemistry And Clinical Toxicology Devices
TEST SYSTEM	862.1070	AMYLASE TEST SYSTEM Clinical Chemistry Test Systems; Clinical Chemistry And Clinical Toxicology Devices
TEST SYSTEM	862.1075	ANDROSTENEDIONE TEST SYSTEM Clinical Chemistry Test Systems; Clinical Chemistry And Clinical Toxicology Devices

TEST SYSTEM	862.1080	ANDROSTERONE TEST SYSTEM Clinical Chemistry Test Systems; Clinical Chemistry And Clinical Toxicology Devices
TEST SYSTEM	862.1085	ANGIOTENSIN I AND RENIN TEST SYSTEM Clinical Chemistry Test Systems; Clinical Chemistry And Clinical Toxicology Devices
TEST SYSTEM	862.1090	ANGIOTENSIN CONVERTING ENZYME (A.C.E.) TEST SYSTEM Clinical Chemistry Test Systems; Clinical Chemistry And Clinical Toxicology Devices
TEST SYSTEM	862.1095	ASCORBIC ACID TEST SYSTEM Clinical Chemistry Test Systems; Clinical Chemistry And Clinical Toxicology Devices
TEST SYSTEM	862.1110	BILIRUBIN (TOTAL OR DIRECT) TEST SYSTEM Clinical Chemistry Test Systems; Clinical Chemistry And Clinical Toxicology Devices
TEST SYSTEM	862.1113	BILIRUBIN (TOTAL AND UNBOUND) IN THE NEONATE TEST SYSTEM Clinical Chemistry Test Systems; Clinical Chemistry And Clinical Toxicology Devices
TEST SYSTEM	862.1115	URINARY BILIRUBIN AND ITS CONJUGATES (NONQUANTITATIVE) TEST SYSTEM Clinical Chemistry Test Systems; Clinical Chemistry And Clinical Toxicology Devices
TEST SYSTEM	862.1117	B-TYPE NATRIURETIC PEPTIDE TEST SYSTEM Clinical Chemistry Test Systems; Clinical Chemistry And Clinical Toxicology Devices
TEST SYSTEM	862.1118	BIOTINIDASE TEST SYSTEM Clinical Chemistry Test Systems; Clinical Chemistry And Clinical Toxicology Devices
TEST SYSTEM	862.1120	BLOOD GASES AND BLOOD pH TEST SYSTEM Clinical Chemistry Test Systems; Clinical Chemistry And Clinical Toxicology Devices
TEST SYSTEM	862.1130	BLOOD VOLUME TEST SYSTEM Clinical Chemistry Test Systems; Clinical Chemistry And Clinical Toxicology Devices
TEST SYSTEM	862.1135	C-PEPTIDES OF PROINSULIN TEST SYSTEM Clinical Chemistry Test Systems; Clinical Chemistry And Clinical Toxicology Devices
TEST SYSTEM	862.1140	CALCITONIN TEST SYSTEM Clinical Chemistry Test Systems; Clinical Chemistry And Clinical Toxicology Devices
TEST SYSTEM	862.1145	CALCIUM TEST SYSTEM Clinical Chemistry Test Systems; Clinical Chemistry And Clinical Toxicology Devices
TEST SYSTEM	862.1155	HUMAN CHORIONIC GONADOTROPIN (HCG) TEST SYSTEM Clinical Chemistry Test Systems; Clinical Chemistry And Clinical Toxicology Devices
TEST SYSTEM	862.1160	BICARBONATE/CARBON DIOXIDE TEST SYSTEM Clinical Chemistry Test Systems; Clinical Chemistry And Clinical Toxicology Devices
TEST SYSTEM	862.1165	CATECHOLAMINES (TOTAL) TEST SYSTEM Clinical Chemistry Test Systems; Clinical Chemistry And Clinical Toxicology Devices
TEST SYSTEM	862.1170	CHLORIDE TEST SYSTEM Clinical Chemistry Test Systems; Clinical Chemistry And Clinical Toxicology Devices
TEST SYSTEM	862.1175	CHOLESTEROL (TOTAL) TEST SYSTEM Clinical Chemistry Test Systems; Clinical Chemistry And Clinical Toxicology Devices
TEST SYSTEM	862.1177	CHOLYLGLYCINE TEST SYSTEM Clinical Chemistry Test Systems; Clinical Chemistry And Clinical Toxicology Devices
TEST SYSTEM	862.1180	CHYMOTRYPSIN TEST SYSTEM Clinical Chemistry Test Systems; Clinical Chemistry And Clinical Toxicology Devices
TEST SYSTEM	862.1185	COMPOUND S (11-DEOXYCORTISOL) TEST SYSTEM Clinical Chemistry Test Systems; Clinical Chemistry And Clinical Toxicology Devices
TEST SYSTEM	862.1187	CONJUGATED SULFOLITHOCHOLIC ACID (SLCG) TEST SYSTEM Clinical Chemistry Test Systems; Clinical Chemistry And Clinical Toxicology Devices
TEST SYSTEM	862.1190	COPPER TEST SYSTEM Clinical Chemistry Test Systems; Clinical Chemistry And Clinical Toxicology Devices

TEST SYSTEM	862.1195	CORTICOIDS TEST SYSTEM Clinical Chemistry Test Systems; Clinical Chemistry And Clinical Toxicology Devices
TEST SYSTEM	862.1200	CORTICOSTERONE TEST SYSTEM Clinical Chemistry Test Systems; Clinical Chemistry And Clinical Toxicology Devices
TEST SYSTEM	862.1205	CORTISOL (HYDROCORTISONE AND HYDROXYCORTICOSTERONE) TEST SYSTEM Clinical Chemistry Test Systems; Clinical Chemistry And Clinical Toxicology Devices
TEST SYSTEM	862.1210	CREATINE TEST SYSTEM Clinical Chemistry Test Systems; Clinical Chemistry And Clinical Toxicology Devices
TEST SYSTEM	862.1215	CREATINE PHOSPHOKINASE/CREATINE KINASE OR ISOENZYMES TEST SYSTEM Clinical Chemistry Test Systems; Clinical Chemistry And Clinical Toxicology Devices
TEST SYSTEM	862.1225	CREATININE TEST SYSTEM Clinical Chemistry Test Systems; Clinical Chemistry And Clinical Toxicology Devices
TEST SYSTEM	862.1230	CYCLIC AMP TEST SYSTEM Clinical Chemistry Test Systems; Clinical Chemistry And Clinical Toxicology Devices
TEST SYSTEM	862.1240	CYSTINE TEST SYSTEM Clinical Chemistry Test Systems; Clinical Chemistry And Clinical Toxicology Devices
TEST SYSTEM	862.1250	DESOXYCORTICOSTERONE TEST SYSTEM Clinical Chemistry Test Systems; Clinical Chemistry And Clinical Toxicology Devices
TEST SYSTEM	862.1255	2,3-DIPHOSPHOGLYCERIC ACID TEST SYSTEM Clinical Chemistry Test Systems; Clinical Chemistry And Clinical Toxicology Devices
TEST SYSTEM	862.1260	ESTRADIOL TEST SYSTEM Clinical Chemistry Test Systems; Clinical Chemistry And Clinical Toxicology Devices
TEST SYSTEM	862.1265	ESTRIOL TEST SYSTEM Clinical Chemistry Test Systems; Clinical Chemistry And Clinical Toxicology Devices
TEST SYSTEM	862.1270	ESTROGENS (TOTAL, IN PREGNANCY) TEST SYSTEM Clinical Chemistry Test Systems; Clinical Chemistry And Clinical Toxicology Devices
TEST SYSTEM	862.1275	ESTROGENS (TOTAL, NONPREGNANCY) TEST SYSTEM Clinical Chemistry Test Systems; Clinical Chemistry And Clinical Toxicology Devices
TEST SYSTEM	862.1280	ESTRONE TEST SYSTEM Clinical Chemistry Test Systems; Clinical Chemistry And Clinical Toxicology Devices
TEST SYSTEM	862.1285	ETIOCHOLANOLONE TEST SYSTEM Clinical Chemistry Test Systems; Clinical Chemistry And Clinical Toxicology Devices
TEST SYSTEM	862.1290	FATTY ACIDS TEST SYSTEM Clinical Chemistry Test Systems; Clinical Chemistry And Clinical Toxicology Devices
TEST SYSTEM	862.1295	FOLIC ACID TEST SYSTEM Clinical Chemistry Test Systems; Clinical Chemistry And Clinical Toxicology Devices
TEST SYSTEM	862.1300	FOLLICLE-STIMULATING HORMONE TEST SYSTEM Clinical Chemistry Test Systems; Clinical Chemistry And Clinical Toxicology Devices
TEST SYSTEM	862.1305	FORMIMINOGLUTAMIC ACID (FIGLU) TEST SYSTEM Clinical Chemistry Test Systems; Clinical Chemistry And Clinical Toxicology Devices
TEST SYSTEM	862.1310	GALACTOSE TEST SYSTEM Clinical Chemistry Test Systems; Clinical Chemistry And Clinical Toxicology Devices
TEST SYSTEM	862.1315	GALACTOSE-1-PHOSPHATE URIDYL TRANSFERASE TEST SYSTEM Clinical Chemistry Test Systems; Clinical Chemistry And Clinical Toxicology Devices
TEST SYSTEM	862.1320	GASTRIC ACIDITY TEST SYSTEM Clinical Chemistry Test Systems; Clinical Chemistry And Clinical Toxicology Devices
TEST SYSTEM	862.1325	GASTRIN TEST SYSTEM Clinical Chemistry Test Systems; Clinical Chemistry And Clinical Toxicology Devices

TEST SYSTEM	862.1330	GLOBULIN TEST SYSTEM Clinical Chemistry Test Systems; Clinical Chemistry And Clinical Toxicology Devices
TEST SYSTEM	862.1335	GLUCAGON TEST SYSTEM Clinical Chemistry Test Systems; Clinical Chemistry And Clinical Toxicology Devices
TEST SYSTEM	862.1340	URINARY GLUCOSE (NONQUANTITATIVE) TEST SYSTEM Clinical Chemistry Test Systems; Clinical Chemistry And Clinical Toxicology Devices
TEST SYSTEM	862.1345	GLUCOSE TEST SYSTEM Clinical Chemistry Test Systems; Clinical Chemistry And Clinical Toxicology Devices
TEST SYSTEM	862.1360	GAMMA-GLUTAMYL TRANSPEPTIDASE AND ISOENZYMES TEST SYSTEM Clinical Chemistry Test Systems; Clinical Chemistry And Clinical Toxicology Devices
TEST SYSTEM	862.1365	GLUTATHIONE TEST SYSTEM Clinical Chemistry Test Systems; Clinical Chemistry And Clinical Toxicology Devices
TEST SYSTEM	862.1370	HUMAN GROWTH HORMONE TEST SYSTEM Clinical Chemistry Test Systems; Clinical Chemistry And Clinical Toxicology Devices
TEST SYSTEM	862.1375	HISTIDINE TEST SYSTEM Clinical Chemistry Test Systems; Clinical Chemistry And Clinical Toxicology Devices
TEST SYSTEM	862.1377	URINARY HOMOCYSTINE (NONQUANTITATIVE) TEST SYSTEM Clinical Chemistry Test Systems; Clinical Chemistry And Clinical Toxicology Devices
TEST SYSTEM	862.1380	HYDROXYBUTYRIC DEHYDROGENASE TEST SYSTEM Clinical Chemistry Test Systems; Clinical Chemistry And Clinical Toxicology Devices
TEST SYSTEM	862.1385	17-HYDROXYCORTICOSTEROIDS (17-KETOGENIC STEROIDS) TEST SYSTEM Clinical Chemistry Test Systems; Clinical Chemistry And Clinical Toxicology Devices
TEST SYSTEM	862.1390	5-HYDROXYINDOLE ACETIC ACID/SEROTONIN TEST SYSTEM Clinical Chemistry Test Systems; Clinical Chemistry And Clinical Toxicology Devices
TEST SYSTEM	862.1395	17-HYDROXYPROGESTERONE TEST SYSTEM Clinical Chemistry Test Systems; Clinical Chemistry And Clinical Toxicology Devices
TEST SYSTEM	862.1400	HYDROXPROLINE TEST SYSTEM Clinical Chemistry Test Systems; Clinical Chemistry And Clinical Toxicology Devices
TEST SYSTEM	862.1405	IMMUNOREACTIVE INSULIN TEST SYSTEM Clinical Chemistry Test Systems; Clinical Chemistry And Clinical Toxicology Devices
TEST SYSTEM	862.1410	IRON (NON-HEME) TEST SYSTEM Clinical Chemistry Test Systems; Clinical Chemistry And Clinical Toxicology Devices
TEST SYSTEM	862.1415	IRON-BINDING CAPACITY TEST SYSTEM Clinical Chemistry Test Systems; Clinical Chemistry And Clinical Toxicology Device
TEST SYSTEM	862.1420	ISOCITRIC DEHYDROGENASE TEST SYSTEM Clinical Chemistry Test Systems; Clinical Chemistry And Clinical Toxicology Devices
TEST SYSTEM	862.1430	17-KETOSTEROIDS TEST SYSTEM Clinical Chemistry Test Systems; Clinical Chemistry And Clinical Toxicology Devices
TEST SYSTEM	862.1435	KETONES (NONQUANTITATIVE) TEST SYSTEM Clinical Chemistry Test Systems; Clinical Chemistry And Clinical Toxicology Devices
TEST SYSTEM	862.1440	LACTATE DEHYDROGENASE TEST SYSTEM Clinical Chemistry Test Systems; Clinical Chemistry And Clinical Toxicology Devices
TEST SYSTEM	862.1445	LACTATE DEHYDROGENASE ISOENZYMES TEST SYSTEM Clinical Chemistry Test Systems; Clinical Chemistry And Clinical Toxicology Devices
TEST SYSTEM	862.1450	LACTIC ACID TEST SYSTEM Clinical Chemistry Test Systems; Clinical Chemistry And Clinical Toxicology Devices
TEST SYSTEM	862.1455	LECITHIN/SPHINGOMYELIN RATIO IN AMNIOTIC FLUID TEST SYSTEM Clinical Chemistry Test Systems; Clinical Chemistry And Clinical Toxicology Devices
TEST SYSTEM	862.1460	LEUCINE AMINOPEPTIDASE TEST SYSTEM Clinical Chemistry Test Systems; Clinical Chemistry And Clinical Toxicology Devices

TEST SYSTEM	862.1465	LIPASE TEST SYSTEM Clinical Chemistry Test Systems; Clinical Chemistry And Clinical Toxicology Devices
TEST SYSTEM	862.1470	LIPID (TOTAL) TEST SYSTEM Clinical Chemistry Test Systems; Clinical Chemistry And Clinical Toxicology Devices
TEST SYSTEM	862.1475	LIPOPROTEIN TEST SYSTEM Clinical Chemistry Test Systems; Clinical Chemistry And Clinical Toxicology Devices
TEST SYSTEM	862.1485	LUTEINIZING HORMONE TEST SYSTEM Clinical Chemistry Test Systems; Clinical Chemistry And Clinical Toxicology Devices
TEST SYSTEM	862.1490	LYSOZYME (MURAMIDASE) TEST SYSTEM Clinical Chemistry Test Systems; Clinical Chemistry And Clinical Toxicology Devices
TEST SYSTEM	862.1495	MAGNESIUM TEST SYSTEM Clinical Chemistry Test Systems; Clinical Chemistry And Clinical Toxicology Devices
TEST SYSTEM	862.1500	MALIC DEHYDROGENASE TEST SYSTEM Clinical Chemistry Test Systems; Clinical Chemistry And Clinical Toxicology Devices
TEST SYSTEM	862.1505	MUCOPOLYSACCHARIDES (NONQUANTITATIVE) TEST SYSTEM Clinical Chemistry Test Systems; Clinical Chemistry And Clinical Toxicology Devices
TEST SYSTEM	862.1509	METHYLMALONIC ACID TEST SYSTEM Clinical Chemistry Test Systems; Clinical Chemistry And Clinical Toxicology Devices
TEST SYSTEM	862.1510	NITRITE (NONQUANTITATIVE) TEST SYSTEM Clinical Chemistry Test Systems; Clinical Chemistry And Clinical Toxicology Devices
TEST SYSTEM	862.1515	NITROGEN (AMINO-NITROGEN) TEST SYSTEM Clinical Chemistry Test Systems; Clinical Chemistry And Clinical Toxicology Devices
TEST SYSTEM	862.1520	5'-NUCLEOTIDASE TEST SYSTEM Clinical Chemistry Test Systems; Clinical Chemistry And Clinical Toxicology Devices
TEST SYSTEM	862.1530	PLASMA ONCOMETRY TEST SYSTEM Clinical Chemistry Test Systems; Clinical Chemistry And Clinical Toxicology Devices
TEST SYSTEM	862.1535	ORNITHINE CARBAMYL TRANSFERASE TEST SYSTEM Clinical Chemistry Test Systems; Clinical Chemistry And Clinical Toxicology Devices
TEST SYSTEM	862.1540	OSMOLALITY TEST SYSTEM Clinical Chemistry Test Systems; Clinical Chemistry And Clinical Toxicology Devices
TEST SYSTEM	862.1542	OXALATE TEST SYSTEM Clinical Chemistry Test Systems; Clinical Chemistry And Clinical Toxicology Devices
TEST SYSTEM	862.1545	PARATHYROID HORMONE TEST SYSTEM Clinical Chemistry Test Systems; Clinical Chemistry And Clinical Toxicology Devices
TEST SYSTEM	862.1550	URINARY pH (NONQUANTITATIVE) TEST SYSTEM Clinical Chemistry Test Systems; Clinical Chemistry And Clinical Toxicology Devices
TEST SYSTEM	862.1555	PHENYLALANINE TEST SYSTEM Clinical Chemistry Test Systems; Clinical Chemistry And Clinical Toxicology Devices
TEST SYSTEM	862.1560	URINARY PHENYLKETONES TEST SYSTEM Clinical Chemistry Test Systems; Clinical Chemistry And Clinical Toxicology Devices
TEST SYSTEM	862.1565	6-PHOSPHOGLUCONATE DEHYDROGENASE TEST SYSTEM Clinical Chemistry Test Systems; Clinical Chemistry And Clinical Toxicology Devices
TEST SYSTEM	862.1570	PHOSPHOHEXOSE ISOMERASE TEST SYSTEM Clinical Chemistry Test Systems; Clinical Chemistry And Clinical Toxicology Devices
TEST SYSTEM	862.1575	PHOSPHOLIPID TEST SYSTEM Clinical Chemistry Test Systems; Clinical Chemistry And Clinical Toxicology Devices
TEST SYSTEM	862.1580	PHOSPHORUS (INORGANIC) TEST SYSTEM Clinical Chemistry Test Systems; Clinical Chemistry And Clinical Toxicology Devices

TEST SYSTEM	862.1585	HUMAN PLACENTAL LACTOGEN TEST SYSTEM Clinical Chemistry Test Systems; Clinical Chemistry And Clinical Toxicology Devices
TEST SYSTEM	862.1590	PORPHOBILINOGEN TEST SYSTEM Clinical Chemistry Test Systems; Clinical Chemistry And Clinical Toxicology Devices
TEST SYSTEM	862.1595	PORPHYRINS TEST SYSTEM Clinical Chemistry Test Systems; Clinical Chemistry And Clinical Toxicology Devices
TEST SYSTEM	862.1600	POTASSIUM TEST SYSTEM Clinical Chemistry Test Systems; Clinical Chemistry And Clinical Toxicology Devices
TEST SYSTEM	862.1605	PREGNANEDIOL TEST SYSTEM Clinical Chemistry Test Systems; Clinical Chemistry And Clinical Toxicology Devices
TEST SYSTEM	862.1610	PREGNANETRIOL TEST SYSTEM Clinical Chemistry Test Systems; Clinical Chemistry And Clinical Toxicology Devices
TEST SYSTEM	862.1615	PREGNENOLONE TEST SYSTEM Clinical Chemistry Test Systems; Clinical Chemistry And Clinical Toxicology Devices
TEST SYSTEM	862.1620	PROGESTERONE TEST SYSTEM Clinical Chemistry Test Systems; Clinical Chemistry And Clinical Toxicology Devices
TEST SYSTEM	862.1625	PROLACTIN (LACTOGEN) TEST SYSTEM Clinical Chemistry Test Systems; Clinical Chemistry And Clinical Toxicology Devices
TEST SYSTEM	862.1630	PROTEIN (FRACTIONATION) TEST SYSTEM Clinical Chemistry Test Systems; Clinical Chemistry And Clinical Toxicology Devices
TEST SYSTEM	862.1635	TOTAL PROTEIN TEST SYSTEM Clinical Chemistry Test Systems; Clinical Chemistry And Clinical Toxicology Devices
TEST SYSTEM	862.1640	PROTEIN-BOUND IODINE TEST SYSTEM Clinical Chemistry Test Systems; Clinical Chemistry And Clinical Toxicology Devices
TEST SYSTEM	862.1645	URINARY PROTEIN OR ALBUMIN (NON-QUANTITATIVE) TEST SYSTEM Clinical Chemistry Test Systems; Clinical Chemistry And Clinical Toxicology Devices
TEST SYSTEM	862.1650	PYRUVATE KINASE TEST SYSTEM Clinical Chemistry Test Systems; Clinical Chemistry And Clinical Toxicology Devices
TEST SYSTEM	862.1655	PYRUVIC ACID TEST SYSTEM Clinical Chemistry Test Systems; Clinical Chemistry And Clinical Toxicology Devices
TEST SYSTEM	862.1665	SODIUM TEST SYSTEM Clinical Chemistry Test Systems; Clinical Chemistry And Clinical Toxicology Devices
TEST SYSTEM	862.1670	SORBITOL DEHYDROGENASE TEST SYSTEM Clinical Chemistry Test Systems; Clinical Chemistry And Clinical Toxicology Devices
TEST SYSTEM	862.1680	TESTOSTERONE TEST SYSTEM Clinical Chemistry Test Systems; Clinical Chemistry And Clinical Toxicology Devices
TEST SYSTEM	862.1685	THYROXINE-BINDING GLOBULIN TEST SYSTEM Clinical Chemistry Test Systems; Clinical Chemistry And Clinical Toxicology Devices
TEST SYSTEM	862.1690	THYROID-STIMULATING HORMONE TEST SYSTEM Clinical Chemistry Test Systems; Clinical Chemistry And Clinical Toxicology Devices
TEST SYSTEM	862.1695	FREE THYROXINE TEST SYSTEM Clinical Chemistry Test Systems; Clinical Chemistry And Clinical Toxicology Devices
TEST SYSTEM	862.1700	TOTAL THYROXINE TEST SYSTEM Clinical Chemistry Test Systems; Clinical Chemistry And Clinical Toxicology Devices
TEST SYSTEM	862.1705	TRIGLYCERIDE TEST SYSTEM Clinical Chemistry Test Systems; Clinical Chemistry And Clinical Toxicology Devices
TEST SYSTEM	862.1710	TOTAL TRIIODOTHYRONINE TEST SYSTEM Clinical Chemistry Test Systems; Clinical Chemistry And Clinical Toxicology Devices

TEST SYSTEM	862.1715	TRIIODOTHYRONINE UPTAKE TEST SYSTEM Clinical Chemistry Test Systems; Clinical Chemistry And Clinical Toxicology Devices
TEST SYSTEM	862.1720	TRIOSE PHOSPHATE ISOMERASE TEST SYSTEM Clinical Chemistry Test Systems; Clinical Chemistry And Clinical Toxicology Devices
TEST SYSTEM	862.1725	TRYPSIN TEST SYSTEM Clinical Chemistry Test Systems; Clinical Chemistry And Clinical Toxicology Devices
TEST SYSTEM	862.1730	FREE TYROSINE TEST SYSTEM Clinical Chemistry Test Systems; Clinical Chemistry And Clinical Toxicology Devices
TEST SYSTEM	862.1770	UREA NITROGEN TEST SYSTEM Clinical Chemistry Test Systems; Clinical Chemistry And Clinical Toxicology Devices
TEST SYSTEM	862.1775	URIC ACID TEST SYSTEM Clinical Chemistry Test Systems; Clinical Chemistry And Toxicology Devices
TEST SYSTEM	862.1780	URINARY CALCULI (STONES) TEST SYSTEM Clinical Chemistry Test Systems; Clinical Chemistry And Clinical Toxicology Devices
TEST SYSTEM	862.1785	URINARY UROBILINOGEN (NONQUANTITATIVE) TEST SYSTEM Clinical Chemistry Test Systems; Clinical Chemistry And Clinical Toxicology Devices
TEST SYSTEM	862.1790	UROPORPHYRIN TEST SYSTEM Clinical Chemistry Test Systems; Clinical Chemistry And Clinical Toxicology Devices
TEST SYSTEM	862.1795	VANILMANDELIC ACID TEST SYSTEM Clinical Chemistry Test Systems; Clinical Chemistry And Clinical Toxicology Devices
TEST SYSTEM	862.1805	VITAMIN A TEST SYSTEM Clinical Chemistry Test Systems; Clinical Chemistry And Clinical Toxicology Devices
TEST SYSTEM	862.1810	VITAMIN B12 TEST SYSTEM Clinical Chemistry Test Systems; Clinical Chemistry And Clinical Toxicology Devices
TEST SYSTEM	862.1815	VITAMIN E TEST SYSTEM Clinical Chemistry Test Systems; Clinical Chemistry And Clinical Toxicology Devices
TEST SYSTEM	862.1820	XYLOSE TEST SYSTEM Clinical Chemistry Test Systems; Clinical Chemistry And Clinical Toxicology Devices
TEST SYSTEM	862.1825	VITAMIN D TEST SYSTEM Clinical Chemistry Test Systems; Clinical Chemistry And Clinical Toxicology Devices
TEST SYSTEM	862.3030	ACETAMINOPHEN TEST SYSTEM Clinical Toxicology Test Systems; Clinical Chemistry And Clinical Toxicology Devices
TEST SYSTEM	862.3035	AMIKACIN TEST SYSTEM Clinical Toxicology Test Systems; Clinical Chemistry And Clinical Toxicology Devices
TEST SYSTEM	862.3040	ALCOHOL TEST SYSTEM Clinical Toxicology Test Systems; Clinical Chemistry And Clinical Toxicology Devices
TEST SYSTEM	862.3050	BREATH-ALCOHOL TEST SYSTEM Clinical Toxicology Test Systems; Clinical Chemistry And Clinical Toxicology Devices
TEST SYSTEM	862.3100	AMPHETAMINE TEST SYSTEM Clinical Toxicology Test Systems; Clinical Chemistry And Clinical Toxicology Devices
TEST SYSTEM	862.3110	ANTIMONY TEST SYSTEM Clinical Toxicology Test Systems; Clinical Chemistry And Clinical Toxicology Devices
TEST SYSTEM	862.3120	ARSENIC TEST SYSTEM Clinical Toxicology Test Systems; Clinical Chemistry And Clinical Toxicology Devices
TEST SYSTEM	862.3150	BARBITURATE TEST SYSTEM Clinical Toxicology Test Systems; Clinical Chemistry And Clinical Toxicology Devices
TEST SYSTEM	862.3170	BENZODIAZEPAM TEST SYSTEM Clinical Toxicology Test Systems; Clinical Chemistry And Clinical Toxicology Devices
TEST SYSTEM	862.3220	CARBON MONOXIDE TEST SYSTEM Clinical Toxicology Test Systems; Clinical Chemistry And Clinical Toxicology Devices

TEST SYSTEM	862.3240	CHOLINESTERASE TEST SYSTEM Clinical Toxicology Test Systems; Clinical Chemistry And Clinical Toxicology Devices
TEST SYSTEM	862.3250	COCAINE AND COCAINE METABOLITE TEST SYSTEM Clinical Toxicology Test Systems; Clinical Chemistry And Clinical Toxicology Devices
TEST SYSTEM	862.3270	CODEINE TEST SYSTEM Clinical Toxicology Test Systems; Clinical Chemistry And Clinical Toxicology Devices
TEST SYSTEM	862.3300	DIGITOXIN TEST SYSTEM Clinical Toxicology Test Systems; Clinical Chemistry And Clinical Toxicology Devices
TEST SYSTEM	862.3320	DIGOXIN TEST SYSTEM Clinical Toxicology Test Systems; Clinical Chemistry And Clinical Toxicology Devices
TEST SYSTEM	862.3350	DIPHENYLHYDANTOIN TEST SYSTEM Clinical Toxicology Test Systems; Clinical Chemistry And Clinical Toxicology Devices
TEST SYSTEM	862.3380	ETHOSUXIMIDE TEST SYSTEM Clinical Toxicology Test Systems; Clinical Chemistry And Toxicology Devices
TEST SYSTEM	862.3450	GENTAMICIN TEST SYSTEM Clinical Toxicology Test Systems; Clinical Chemistry And Toxicology Devices
TEST SYSTEM	862.3520	KANAMYCIN TEST SYSTEM Clinical Toxicology Test Systems; Clinical Chemistry And Clinical Toxicology Devices
TEST SYSTEM	862.3550	LEAD TEST SYSTEM Clinical Toxicology Test Systems; Clinical Chemistry And Clinical Toxicology Devices
TEST SYSTEM	862.3555	LIDOCAINE TEST SYSTEM Clinical Toxicology Test Systems; Clinical Chemistry And Clinical Toxicology Devices
TEST SYSTEM	862.3560	LITHIUM TEST SYSTEM Clinical Toxicology Test Systems; Clinical Chemistry And Clinical Toxicology Devices
TEST SYSTEM	862.3580	LYSERGIC ACID DIETHYLAMIDE (LSD) TEST SYSTEM Clinical Toxicology Test Systems; Clinical Chemistry And Clinical Toxicology Devices
TEST SYSTEM	862.3600	MERCURY TEST SYSTEM Clinical Toxicology Test Systems; Clinical Chemistry And Clinical Toxicology Devices
TEST SYSTEM	862.3610	METHAMPHETAMINE TEST SYSTEM Clinical Toxicology Test Systems; Clinical Chemistry And Clinical Toxicology Devices
TEST SYSTEM	862.3620	METHADONE TEST SYSTEM Clinical Toxicology Test Systems; Clinical Chemistry And Clinical Toxicology Devices
TEST SYSTEM	862.3630	METHAQUALONE TEST SYSTEM Clinical Toxicology Test Systems; Clinical Chemistry And Clinical Toxicology Devices
TEST SYSTEM	862.3640	MORPHINE TEST SYSTEM Clinical Toxicology Test Systems; Clinical Chemistry And Clinical Toxicology Devices
TEST SYSTEM	862.3645	NEUROLEPTIC DRUGS RADIORECEPTOR ASSAY TEST SYSTEM Clinical Toxicology Test Systems; Clinical Chemistry And Clinical Toxicology Devices
TEST SYSTEM	862.3650	OPIATE TEST SYSTEM Clinical Toxicology Test Systems; Clinical Chemistry And Clinical Toxicology Devices
TEST SYSTEM	862.3660	PHENOBARBITAL TEST SYSTEM Clinical Toxicology Test Systems; Clinical Chemistry And Clinical Toxicology Devices
TEST SYSTEM	862.3670	PHENOTHIAZINE TEST SYSTEM Clinical Toxicology Test Systems; Clinical Chemistry And Clinical Toxicology Devices
TEST SYSTEM	862.3680	PRIMIDONE TEST SYSTEM Clinical Toxicology Test Systems; Clinical Chemistry And Clinical Toxicology Devices
TEST SYSTEM	862.3700	PROPOXYPHENE TEST SYSTEM Clinical Toxicology Test Systems; Clinical Chemistry And Clinical Toxicology Devices
TEST SYSTEM	862.3750	QUININE TEST SYSTEM Clinical Toxicology Test Systems; Clinical Chemistry And Clinical Toxicology Devices

TEST SYSTEM	862.3830	SALICYLATE TEST SYSTEM Clinical Toxicology Test Systems; Clinical Chemistry And Clinical Toxicology Devices
TEST SYSTEM	862.3850	SULFONAMIDE TEST SYSTEM Clinical Toxicology Test Systems; Clinical Chemistry And Clinical Toxicology Devices
TEST SYSTEM	862.3870	CANNABINOID TEST SYSTEM Clinical Toxicology Test Systems; Clinical Chemistry And Clinical Toxicology Devices
TEST SYSTEM	862.3880	THEOPHYLLINE TEST SYSTEM Clinical Toxicology Test Systems; Clinical Chemistry And Clinical Toxicology Devices
TEST SYSTEM	862.3900	TOBRAMYCIN TEST SYSTEM Clinical Toxicology Test Systems; Clinical Chemistry And Clinical Toxicology Devices
TEST SYSTEM	862.3910	TRICYCLIC ANTIDEPRESSANT DRUGS TEST SYSTEM Clinical Toxicology Test Systems; Clinical Chemistry And Clinical Toxicology Devices
TEST SYSTEM	862.3950	VANCOMYCIN TEST SYSTEM Clinical Toxicology Test Systems; Clinical Chemistry And Clinical Toxicology Devices
TEST SYSTEM	864.9175	AUTOMATED BLOOD GROUPING AND ANTIBODY TEST SYSTEM Products Used In Establishments That Manufacture Blood And Blood Products; Hematology And Pathology Devices
TEST SYSTEM	866.5040	ALBUMIN IMMUNOLOGICAL TEST SYSTEM Immunological Test Systems; Immunology And Microbiology Devices
TEST SYSTEM	866.5060	PREALBUMIN IMMUNOLOGICAL TEST SYSTEM Immunological Test Systems; Immunology And Microbiology Devices
TEST SYSTEM	866.5065	HUMAN ALLOTYPIC MARKER IMMUNOLOGICAL TEST SYSTEM Immunological Test Systems; Immunology And Microbiology Devices
TEST SYSTEM	866.5080	ALPHA-1-ANTICHYMOTRYPSIN IMMUNOLOGICAL TEST SYSTEM Immunological Test Systems; Immunology And Microbiology Devices
TEST SYSTEM	866.5090	ANTIMITOCHONDRIAL ANTIBODY IMMUNOLOGICAL TEST SYSTEM Immunological Test Systems; Immunology And Microbiology Devices
TEST SYSTEM	866.5100	ANTINUCLEAR ANTIBODY IMMUNOLOGICAL TEST SYSTEM Immunological Test Systems; Immunology And Microbiology Devices
TEST SYSTEM	866.5110	ANTIPARIETAL ANTIBODY IMMUNOLOGICAL TEST SYSTEM Immunological Test Systems; Immunology And Microbiology Devices
TEST SYSTEM	866.5120	ANTISMOOTH MUSCLE ANTIBODY IMMUNOLOGICAL TEST SYSTEM Immunological Test Systems; Immunology And Microbiology Devices
TEST SYSTEM	866.5130	ALPHA-1-ANTITRYPSIN IMMUNOLOGICAL TEST SYSTEM Immunological Test Systems; Immunology And Microbiology Devices
TEST SYSTEM	866.5150	BENCE-JONES PROTEINS IMMUNOLOGICAL TEST SYSTEM Immunological Test Systems; Immunology And Microbiology Devices
TEST SYSTEM	866.5160	BETA-GLOBULIN IMMUNOLOGICAL TEST SYSTEM Immunological Test Systems; Immunology And Microbiology Devices
TEST SYSTEM	866.5170	BREAST MILK IMMUNOLOGICAL TEST SYSTEM Immunological Test Systems; Immunology And Microbiology Devices
TEST SYSTEM	866.5200	CARBONIC ANHYDRASE B AND C IMMUNOLOGICAL TEST SYSTEM Immunological Test Systems; Immunology And Microbiology Devices
TEST SYSTEM	866.5210	CERULOPLASMIN IMMUNOLOGICAL TEST SYSTEM Immunological Test Systems; Immunology And Microbiology Devices
TEST SYSTEM	866.5220	COHN FRACTION II IMMUNOLOGICAL TEST SYSTEM Immunological Test Systems; Immunology And Microbiology Devices
TEST SYSTEM	866.5230	COLOSTRUM IMMUNOLOGICAL TEST SYSTEM Immunological Test Systems; Immunology And Microbiology Devices
TEST SYSTEM	866.5240	COMPLEMENT COMPONENTS IMMUNOLOGICAL TEST SYSTEM Immunological Test Systems; Immunology And Microbiology Devices
TEST SYSTEM	866.5250	COMPLEMENT C1 INHIBITOR (INACTIVATOR) IMMUNOLOGICAL TEST SYSTEM Immunological Test Systems; Immunology And Microbiology Devices
TEST SYSTEM	866.5260	COMPLEMENT C3B INACTIVATOR IMMUNOLOGICAL TEST SYSTEM Immunological Test Systems; Immunology And Microbiology Devices
TEST SYSTEM	866.5270	C-REACTIVE PROTEIN IMMUNOLOGICAL TEST SYSTEM Immunological Test Systems; Immunology And Microbiology Devices
TEST SYSTEM	866.5320	PROPERIDIN FACTOR B IMMUNOLOGICAL TEST SYSTEM Immunological Test Systems; Immunology And Microbiology Devices
TEST SYSTEM	866.5330	FACTOR XIII, A, S, IMMUNOLOGICAL TEST SYSTEM Immunological Test Systems; Immunology And Microbiology Devices
TEST SYSTEM	866.5340	FERRITIN IMMUNOLOGICAL TEST SYSTEM Immunological Test Systems; Immunology And Microbiology Devices
TEST SYSTEM	866.5350	FIBRINOPEPTIDE A IMMUNOLOGICAL TEST SYSTEM Immunological Test Systems; Immunology And Microbiology Devices

TEST SYSTEM	866.5360	COHN FRACTION IV IMMUNOLOGICAL TEST SYSTEM Immunological Test Systems; Immunology And Microbiology Devices
TEST SYSTEM	866.5370	COHN FRACTION V IMMUNOLOGICAL TEST SYSTEM Immunological Test Systems; Immunology And Microbiology Devices
TEST SYSTEM	866.5380	FREE SECRETORY COMPONENT IMMUNOLOGICAL TEST SYSTEM Immunological Test Systems; Immunology And Microbiology Devices
TEST SYSTEM	866.5400	ALPHA-GLOBULIN IMMUNOLOGICAL TEST SYSTEM Immunological Test Systems; Immunology And Microbiology Devices
TEST SYSTEM	866.5420	ALPHA-1-GLYCOPROTEINS IMMUNOLOGICAL TEST SYSTEM Immunological Test Systems; Immunology And Microbiology Devices
TEST SYSTEM	866.5425	ALPHA-2-GLYCOPROTEINS IMMUNOLOGICAL TEST SYSTEM Immunological Test Systems; Immunology And Microbiology Devices
TEST SYSTEM	866.5430	BETA-2-GLYCOPROTEIN I IMMUNOLOGICAL TEST SYSTEM Immunological Test Systems; Immunology And Microbiology Devices
TEST SYSTEM	866.5440	BETA-2-GLYCOPROTEIN III IMMUNOLOGICAL TEST SYSTEM Immunological Test Systems; Immunology And Microbiology Devices
TEST SYSTEM	866.5460	HAPTOGLOBIN IMMUNOLOGICAL TEST SYSTEM Immunological Test Systems; Immunology And Microbiology Devices
TEST SYSTEM	866.5470	HEMOGLOBIN IMMUNOLOGICAL TEST SYSTEM Immunological Test Systems; Immunology And Microbiology Devices
TEST SYSTEM	866.5490	HEMOPEXIN IMMUNOLOGICAL TEST SYSTEM Immunological Test Systems; Immunology And Microbiology Devices
TEST SYSTEM	866.5500	HYPERSENSITIVITY PNEUMONITIS IMMUNOLOGICAL TEST SYSTEM Immunological Test Systems; Immunology And Microbiology Devices
TEST SYSTEM	866.5510	IMMUNOGLOBULINS A, G, M, D, AND E IMMUNOLOGICAL TEST SYSTEM Immunological Test Systems; Immunology And Microbiology Devices
TEST SYSTEM	866.5520	IMMUNOGLOBULIN G (FAB FRAGMENT SPECIFIC) IMMUNOLOGICAL TEST SYSTEM Immunological Test Systems; Immunology And Microbiology Devices
TEST SYSTEM	866.5530	IMMUNOGLOBULIN G (FC FRAGMENT SPECIFIC) IMMUNOLOGICAL TEST SYSTEM Immunological Test Systems; Immunology And Microbiology Devices
TEST SYSTEM	866.5540	IMMUNOGLOBULIN G (FD FRAGMENT SPECIFIC) IMMUNOLOGICAL TEST SYSTEM Immunological Test Systems; Immunology And Microbiology Devices
TEST SYSTEM	866.5550	IMMUNOGLOBULIN (LIGHT CHAIN SPECIFIC) IMMUNOLOGICAL TEST SYSTEM Immunological Test Systems; Immunology And Microbiology Devices
TEST SYSTEM	866.5560	LACTIC DEHYDROGENASE IMMUNOLOGICAL TEST SYSTEM Immunological Test Systems; Immunology And Microbiology Devices
TEST SYSTEM	866.5570	LACTOFERRIN IMMUNOLOGICAL TEST SYSTEM Immunological Test Systems; Immunology And Microbiology Devices
TEST SYSTEM	866.5580	ALPHA-1-LIPOPROTEIN IMMUNOLOGICAL TEST SYSTEM Immunological Test Systems; Immunology And Microbiology Devices
TEST SYSTEM	866.5590	LIPOPROTEIN X IMMUNOLOGICAL TEST SYSTEM Immunological Test Systems; Immunology And Microbiology Devices
TEST SYSTEM	866.5600	LOW-DENSITY LIPOPROTEIN IMMUNOLOGICAL TEST SYSTEM Immunological Test Systems; Immunology And Microbiology Devices
TEST SYSTEM	866.5620	ALPHA-2-MACROGLOBULIN IMMUNOLOGICAL TEST SYSTEM Immunological Test Systems; Immunology And Microbiology Devices
TEST SYSTEM	866.5630	BETA-2-MICROGLOBULIN IMMUNOLOGICAL TEST SYSTEM Immunological Test Systems; Immunology And Microbiology Devices
TEST SYSTEM	866.5640	INFECTIOUS MONONUCLEOSIS IMMUNOLOGICAL TEST SYSTEM Immunological Test Systems; Immunology And Microbiology Devices
TEST SYSTEM	866.5660	MULTIPLE AUTOANTIBODIES IMMUNOLOGICAL TEST SYSTEM Immunological Test Systems; Immunology And Microbiology Devices
TEST SYSTEM	866.5680	MYOGLOBIN IMMUNOLOGICAL TEST SYSTEM Immunological Test Systems; Immunology And Microbiology Devices
TEST SYSTEM	866.5700	WHOLE HUMAN PLASMA OR SERUM IMMUNOLOGICAL TEST SYSTEM Immunological Test Systems; Immunology And Microbiology Devices
TEST SYSTEM	866.5715	PLASMINOGEN IMMUNOLOGICAL TEST SYSTEM Immunological Test Systems; Immunology And Microbiology Devices
TEST SYSTEM	866.5735	PROTHROMBIN IMMUNOLOGICAL TEST SYSTEM Immunological Test Systems; Immunology And Microbiology Devices
TEST SYSTEM	866.5750	RADIOALLERGOSORBENT (RAST) IMMUNOLOGICAL TEST SYSTEM Immunological Test Systems; Immunology And Microbiology Devices
TEST SYSTEM	866.5765	RETINOL-BINDING PROTEIN IMMUNOLOGICAL TEST SYSTEM Immunological Test Systems; Immunology And Microbiology Devices
TEST SYSTEM	866.5775	RHEUMATOID FACTOR IMMUNOLOGICAL TEST SYSTEM Immunological Test Systems; Immunology And Microbiology Devices
TEST SYSTEM	866.5800	SEMINAL FLUID (SPERM) IMMUNOLOGICAL TEST SYSTEM Immunological Test Systems; Immunology And Microbiology Devices
TEST SYSTEM	866.5820	SYSTEMIC LUPUS ERYTHEMATOSUS IMMUNOLOGICAL TEST SYSTEM Immunological Test Systems; Immunology And Microbiology Devices

TEST SYSTEM	866.5860	TOTAL SPINAL FLUID IMMUNOLOGICAL TEST SYSTEM Immunological Test Systems; Immunology And Microbiology Devices
TEST SYSTEM	866.5870	THYROID AUTOANTIBODY IMMUNOLOGICAL TEST SYSTEM Immunological Test Systems; Immunology And Microbiology Devices
TEST SYSTEM	866.5880	TRANSFERRIN IMMUNOLOGICAL TEST SYSTEM Immunological Test Systems; Immunology And Microbiology Devices
TEST SYSTEM	866.5890	INTER-ALPHA TRYPSIN INHIBITOR IMMUNOLOGICAL TEST SYSTEM Immunological Test Systems; Immunology And Microbiology Devices
TEST SYSTEM	866.6010	TUMOR ASSOCIATED ANTIGEN IMMUNOLOGICAL TEST SYSTEM Tumor Associated Antigen Immunological Test Systems; Immunology And Microbiology Devices
TEST SYSTEM	886.1910	SPECTACLE DISSOCIATION TEST SYSTEM Diagnostic Devices; Ophthalmic Devices
TEST SYSTEMS	864.9300	AUTOMATED COOMBS TEST SYSTEMS Products Used In Establishments That Manufacture Blood And Blood Products; Hematology And Pathology Devices
TEST SYSTEMS	866.5785	ANTI-SACCHAROMYCES CEREVISIAE ANTIBODY (ASCA) TEST SYSTEMS Immunological Test Systems; Immunology And Microbiology Devices
TESTER	866.1070	COLOR VISION TESTER Diagnostic Devices; Ophthalmic Devices
TESTER	870.2370	ELECTROCARDIOGRAPH SURFACE ELECTRODE TESTER Cardiovascular Monitoring Devices; Cardiovascular Devices
TESTER	870.3720	PACEMAKER ELECTRODE FUNCTION TESTER Cardiovascular Prosthetic Devices; Cardiovascular Devices
TESTER	870.5325	DEFIBRILLATOR TESTER Cardiovascular Therapeutic Devices; Cardiovascular Devices
TESTER	872.1720	PULP TESTER Diagnostic Devices; Dental Devices
TESTER	874.1090	AUDITORY IMPEDANCE TESTER Diagnostic Devices; Ear, Nose, And Throat Devices
TESTER	882.1410	ELECTROENCEPHALOGRAPH ELECTRODE/LEAD TESTER Neurological Diagnostic Devices; Neurological Devices
TESTERS	72.1730	ELECTRODE GEL FOR PULP TESTERS Diagnostic Devices; Dental Devices
TESTICULAR	876.3750	TESTICULAR PROSTHESIS Prosthetic Devices; Gastroenterology-urology Devices
TESTIMONY	5.23	DISCLOSURE OF OFFICIAL RECORDS AND AUTHORIZATION OF TESTIMONY General Redelegations Of Authority; Delegations Of Authority And Organization
TESTIMONY	12.87	PURPOSE; ORAL AND WRITTEN TESTIMONY; BURDEN OF PROOF Hearing Procedures; Formal Evidentiary Public Hearing
TESTIMONY	20.1	TESTIMONY BY FOOD AND DRUG ADMINISTRATION EMPLOYEES Official Testimony And Information; Public Information
TESTIMONY	1316.58	SUMMARY OF TESTIMONY; AFFIDAVITS Administrative Hearings; Administrative Functions, Practices, And Procedures
TESTING	20.105	TESTING AND RESEARCH CONDUCTED BY OR WITH FUNDS PROVIDED BY THE FDA Availability Of Specific Categories Of Records; Public Information
TESTING	211.84	TESTING AND APPROVAL OR REJECTION OF COMPONENTS, DRUG PRODUCT CONTAINERS, AND CLOSURES Control Of Components And Drug Product Containers And Closures; Current Good Manufacturing Practice For Finished Pharmaceuticals
TESTING	211.110	SAMPLING AND TESTING OF IN-PROCESS MATERIALS AND DRUG PRODUCTS Production And Process Controls; Current Good Manufacturing Practice For Finished Pharmaceuticals
TESTING	211.165	TESTING AND RELEASE FOR DISTRIBUTION Laboratory Controls; Current Good Manufacturing Practice For Finished Pharmaceuticals
TESTING	211.166	STABILITY TESTING Laboratory Controls; Current Good Manufacturing Practice For Finished Pharmaceuticals
TESTING	211.167	SPECIAL TESTING REQUIREMENTS Laboratory Controls; Current Good Manufacturing Practice For Finished Pharmaceuticals
TESTING	310.103	NEW DRUG SUBSTANCES INTENDED FOR HYPERSENSITIVITY TESTING Specific Administrative Rulings And Decisions; New Drugs
TESTING	320.34	REQUIREMENTS FOR BATCH TESTING AND CERTIFICATION BY THE FDA Procedures For Determining The Bioavailability Or Bioequivalence Of Drug Products; Bioavailability And Bioequivalence Requirements
TESTING	320.35	REQUIREMENTS FOR IN VITRO TESTING OF EACH BATCH Procedures For Determining The Bioavailability Or Bioequivalence Of Drug Products; Bioavailability And Bioequivalence Requirements

TESTING	320.36	REQUIREMENTS FOR MAINTENANCE OF RECORDS OF BIOEQUIVALENCE TESTING Procedures For Determining The Bioavailability Or Bioequivalence Of Drug Products; Bioavailability And Bioequivalence Requirements
TESTING	343.90	DISSOLUTION AND DRUG RELEASE TESTING Internal Analgesic, Antipyretic And Antirheumatic Drug Products For OTC Human Use
TESTING	606.151	COMPATIBILITY TESTING Laboratory Controls; Current Good Manufacturing Practice For Blood And Blood Components
TESTING	640.5	TESTING THE BLOOD Whole Blood; Additional Standards For Human Blood And Blood Products
TESTING	640.14	TESTING THE BLOOD Red Blood Cells; Additional Standards For Human Blood And Blood Products
TESTING	640.15	SEGMENTS FOR TESTING Red Blood Cells; Additional Standards For Human Blood And Blood Products
TESTING	640.23	TESTING THE BLOOD Platelets; Additional Standards For Human Blood And Blood Products
TESTING	640.33	TESTING THE BLOOD Plasma; Additional Standards For Human Blood And Blood Products
TESTING	640.53	TESTING THE BLOOD Cryoprecipitate; Additional Standards For Human Blood And Blood Products
TESTING	660.33	TESTING OF SOURCE MATERIAL Reagent Red Blood Cells; Additional Standards For Diagnostic Substances For Laboratory Tests
TESTING	809.40	RESTRICTIONS ON THE SALE, DISTRIBUTION AND USE OF OTC TEST SAMPLE COLLECTION SYSTEMS FOR DRUGS OF ABUSE TESTING Requirements For Manufacturers And Producers; In Vitro Diagnostic Products For Human Use
TESTING	864.3260	OTC TEST SAMPLE COLLECTION SYSTEMS FOR DRUGS OF ABUSE TESTING Pathology Instrumentation And Accessories; Hematology And Pathology Devices
TESTING	874.1060	ACOUSTIC CHAMBER FOR AUDIOMETRIC TESTING Diagnostic Devices; Ear, Nose, And Throat Devices
TESTING	874.1100	EARPHONE CUSHION FOR AUDIOMETRIC TESTING Diagnostic Devices; Ear, Nose, And Throat Devices
TESTING	874.1120	ELECTRONIC NOISE GENERATOR FOR AUDIOMETRIC TESTING Diagnostic Devices; Ear, Nose, And Throat Devices
TESTING	890.1925	ISOKINETIC TESTING AND EVALUATION SYSTEM Physical Medicine Diagnostic Devices; Physical Medicine Devices
TESTING FACILITY	58.15	INSPECTION OF A TESTING FACILITY General Provisions; Good Laboratory Practice For Nonclinical Laboratory Studies
TESTING FACILITY	58.31	TESTING FACILITY MANAGEMENT Organization And Personnel; Good Laboratory Practice For Nonclinical Laboratory Studies
TESTING FACILITY	58.217	SUSPENSION OR TERMINATION OF A TESTING FACILITY BY A SPONSOR Disqualification Of Testing Facilities; Good Laboratory Practice For Nonclinical Laboratory Studies
TESTING FACILITY	58.219	REINSTATEMENT OF A DISQUALIFIED TESTING FACILITY Disqualification Of Testing Facilities; Good Laboratory Practice For Nonclinical Laboratory Studies
TESTING PROCEDURES	352.72	GENERAL TESTING PROCEDURES Sunscreen Drug Products For Over-the-counter Human Use
TESTING PROCEDURES	355.70	TESTING PROCEDURES FOR FLUORIDE DENTIFRICE DRUG PRODUCTS Anticaries Drug Products For Over-the-counter Human Use
TESTING PROGRAMS	5.25	RESEARCH, INVESTIGATION, AND TESTING PROGRAMS AND HEALTH INFORMATION AND PROMOTION PROGRAMS General Redelegations Of Authority; Delegations Of Authority And Organization
TESTING PROGRAMS	5.602	TESTING PROGRAMS AND METHODS OF CERTIFICATION AND IDENTIFICATION FOR ELECTRONIC PRODUCTS Radiation Control; Redelegations Of Authority; Delegations Of Authority And Organization
TESTOSTERONE	522.842	ESTRADIOL BENZOATE AND TESTOSTERONE PROPIONATE IN COMBINATION Implantation Or Injectable Dosage Form New Animal Drugs
TESTOSTERONE	556.710	TESTOSTERONE PROPIONATE Specific Tolerances For Residues Of New Animal Drugs; Tolerances For Residues Of New Animal Drugs In Food
TESTOSTERONE	862.1680	TESTOSTERONE TEST SYSTEM Clinical Chemistry Test Systems; Clinical Chemistry And Clinical Toxicology Devices
TESTS	312.160	DRUGS FOR INVESTIGATIONAL USE IN LABORATORY RESEARCH ANIMALS OR IN VITRO TESTS Investigational New Drug Application

TESTS	610.1	TESTS PRIOR TO RELEASE REQUIRED FOR EACH LOT Release Requirements; General Biological Products Standards
TESTS	640.67	LABORATORY TESTS Source Plasma; Additional Standards For Human Blood And Blood Products
TESTS	640.82	TESTS ON FINAL PRODUCT Albumin (Human); Additional Standards For Human Blood And Blood Products
TESTS	640.92	TESTS ON FINAL PRODUCT Plasma Protein Fraction (Human); Additional Standards For Human Blood And Blood Products
TESTS	660.25	POTENCY TESTS WITHOUT REFERENCE PREPARATIONS Blood Grouping Reagent; Additional Standards For Diagnostic Substances For Laboratory Tests
TESTS	660.27	SPECIFICITY TESTS AND AVIDITY TESTS Blood Grouping Reagent; Additional Standards For Diagnostic Substances For Laboratory Tests
TESTS	660.54	POTENCY TESTS, SPECIFICITY TESTS, TESTS FOR HETEROSPECIFIC ANTIBODIES, AND ADDITIONAL TESTS FOR NONSPECIFIC PROPERTIES Anti-human Globulin; Additional Standards For Diagnostic Substances For Laboratory Tests
TESTS	680.3	TESTS Additional Standards For Miscellaneous Products
TESTS	864.7140	ACTIVATED WHOLE BLOOD CLOTTING TIME TESTS Hematology Kits And Packages; Hematology And Pathology Devices
TESTS	864.7275	EUGLOBULIN LYSIS TIME TESTS Hematology Kits And Packages; Hematology And Pathology Devices
TESTS	864.7925	PARTIAL THROMBOPLASTIN TIME TESTS Hematology Kits And Packages; Hematology And Pathology Devices
TESTS	866.1700	CULTURE MEDIUM FOR ANTIMICROBIAL SUSCEPTIBILITY TESTS Diagnostic Devices; Immunology And Microbiology Devices
TETRA SODIUM PYROPHOSPHATE	182.6789	TETRA SODIUM PYROPHOSPHATE Sequestrants; Substances Generally Recognized As Safe
TETRA SODIUM PYROPHOSPHATE	582.6789	TETRA SODIUM PYROPHOSPHATE Sequestrants; Substances Generally Recognized As Safe
TETRACAINE	524.1484k	NEOMYCIN SULFATE, PREDNISOLONE, TETRACAINE, AND SQUALANE TOPICAL-OTIC SUSPENSION Ophthalmic And Topical Dosage Form New Animal Drugs
TETRACAINE HYDROCHLORIDE	524.1484b	NEOMYCIN SULFATE, ISOFLUPREDONE ACETATE, TETRACAINE HYDROCHLORIDE, AND MYRISTYL-GAMMA-PICOLINIUM CHLORIDE TOPICAL POWDER Ophthalmic And Topical Dosage Form New Animal Drugs
TETRACAINE HYDROCHLORIDE	524.1484c	NEOMYCIN SULFATE, ISOFLUPREDONE ACETATE, TETRACAINE HYDROCHLORIDE OINTMENT Ophthalmic And Topical Dosage Form New Animal Drugs
TETRACAINE HYDROCHLORIDE	524.1484d	NEOMYCIN SULFATE, HYDROCORTISONE ACETATE, TETRACAINE HYDROCHLORIDE EAR OINTMENT Ophthalmic And Topical Dosage Form New Animal Drugs
TETRACAINE HYDROCHLORIDE	524.1484f	NEOMYCIN SULFATE, PREDNISOLONE ACETATE, TETRACAINE HYDROCHLORIDE EARDROPS Ophthalmic And Topical Dosage Form New Animal Drugs
TETRACYCLINE	520.2345	TETRACYCLINE ORAL DOSAGE FORMS Oral Dosage Form New Animal Drugs
TETRACYCLINE	556.720	TETRACYCLINE Specific Tolerances For Residues Of New Animal Drugs; Tolerances For Residues Of New Animal Drugs In Food
TETRAETHYLENE GLYCOL	178.3940	TETRAETHYLENE GLYCOL DI-(2-ETHYLHEXOATE) Certain Adjuvants And Production Aids; Indirect Food Additives: Adjuvants, Production Aids, And Sanitizers
TETRAHYDROCANNABINOIDS	1308.35	EXEMPTION OF CERTAIN CANNABIS PLANT MATERIAL, AND PRODUCTS MADE THEREFROM, THAT CONTAIN TETRAHYDROCANNABINOIDS Schedules Of Controlled Substances
TETRAHYDROFURAN	178.3950	TETRAHYDROFURAN Certain Adjuvants And Production Aids; Indirect Food Additives: Adjuvants, Production Aids, And Sanitizers
TETRAHYDROTHIADIAZINE	176.230	3,5-DIMETHYL-1,3,5,*2H*-TETRAHYDROTHIADIAZINE-2-THIONE Substances For Use Only As Components Of Paper And Paperboard; Indirect Food Additives
TETRAMETHYLENE	177.1660	POLY(TETRAMETHYLENE TEREPHTHALATE) Substances For Use As Basic Components Of Single And Repeated Use Food Contact Surfaces; Indirect Food Additives: Polymers
TEXAS	808.93	TEXAS Listing Of Specific State And Local Exemptions; Exemption From Federal Preemption Of State & Local Medical Device Requirements

TEXTILES	177.2800	TEXTILES AND TEXTILE FIBERS Substances For Use Only As Components Of Articles Intended For Repeated Use; Indirect Food Additives: Polymers
TEXTRYLS	177.1850	TEXTRYLS Substances For Use As Basic Components Of Single And Repeated Use Food Contact Surfaces; Indirect Food Additives: Polymers
THBP	172.190	THBP Food Preservatives; Food Additives Permitted For Direct Addition To Food For Human Consumption
THENIUM CLOSYLATE	520.1805	PIPERAZINE PHOSPHATE WITH THENIUM CLOSYLATE TABLETS Oral Dosage Form New Animal Drugs
THENIUM CLOSYLATE	520.2362	THENIUM CLOSYLATE TABLETS Oral Dosage Form New Animal Drugs
THEOPHYLLINE	862.3880	THEOPHYLLINE TEST SYSTEM Clinical Toxicology Test Systems; Clinical Chemistry And Clinical Toxicology Devices
THERAPEUTIC	868.5460	THERAPEUTIC HUMIDIFIER FOR HOME USE Therapeutic Devices; Anesthesiology Devices
THERAPEUTIC	880.5160	THERAPEUTIC MEDICAL BINDER General Hospital And Personal Use Therapeutic Devices; General Hospital And Personal Use Devices
THERAPEUTIC	880.5820	THERAPEUTIC SCROTAL SUPPORT General Hospital And Personal Use Therapeutic Devices; General Hospital And Personal Use Devices
THERAPEUTIC	892.5930	THERAPEUTIC X-RAY TUBE HOUSING ASSEMBLY Therapeutic Devices; Radiology Devices
THERAPY, ELECTROCONVULSIVE	882.5940	ELECTROCONVULSIVE THERAPY DEVICE Neurological Therapeutic Devices; Neurological Devices
THERAPY MATTRESS	880.5150	NONPOWERED FLOTATION THERAPY MATTRESS General Hospital And Personal Use Therapeutic Devices; General Hospital And Personal Use Devices
THERAPY, PHYSICAL	890.5880	MULTI-FUNCTION PHYSICAL THERAPY TABLE Physical Medicine Therapeutic Devices; Physical Medicine Devices
THERAPY, RADIATION	892.5050	MEDICAL CHARGED-PARTICLE RADIATION THERAPY SYSTEM Therapeutic Devices; Radiology Devices
THERAPY, RADIATION	892.5300	MEDICAL NEUTRON RADIATION THERAPY SYSTEM Therapeutic Devices; Radiology Devices
THERAPY, RADIATION	892.5710	RADIATION THERAPY BEAM-SHAPING BLOCK Therapeutic Devices; Radiology Devices
THERAPY, RADIATION	892.5750	RADIONUCLIDE RADIATION THERAPY SYSTEM Therapeutic Devices; Radiology Devices
THERAPY, RADIATION	892.5770	POWERED RADIATION THERAPY PATIENT SUPPORT ASSEMBLY Therapeutic Devices; Radiology Devices
THERAPY, RADIATION	892.5840	RADIATION THERAPY SIMULATION SYSTEM Therapeutic Devices; Radiology Devices
THERAPY, RADIATION	892.5900	X-RAY RADIATION THERAPY SYSTEM Therapeutic Devices; Radiology Devices
THERAPY, ULTRASONIC	1050.10	ULTRASONIC THERAPY PRODUCTS Performance Standards For Sonic, Infrasonic, And Ultrasonic Radiation-emitting Products
THERMAL PROCESSING	108.35	THERMAL PROCESSING OF LOW-ACID ANIMAL FOODS PACKAGED IN HERMETICALLY SEALED CONTAINERS Specific Requirements And Conditions For Exemption From Or Compliance With An Emergency Permit; Emergency Permit Control
THERMAL PROCESSING	113.87	OPERATIONS IN THE THERMAL PROCESSING ROOM Production And Process Controls; Thermally Processed Low-acid Foods Packaged In Hermetically Sealed Containers
THERMAL REGULATING	870.5900	THERMAL REGULATING SYSTEM Cardiovascular Therapeutic Devices; Cardiovascular Devices
THERMALLY PROCESSED	500.23	THERMALLY PROCESSED LOW-ACID FOODS PACKAGED IN HERMETICALLY SEALED CONTAINERS Specific Administrative Rulings And Decisions; General
THERMODILUTION	870.1915	THERMODILUTION PROBE Cardiovascular Diagnostic Devices; Cardiovascular Devices
THERMOGRAPHIC SYSTEM	884.2982	LIQUID CRYSTAL THERMOGRAPHIC SYSTEM Obstetrical And Gynecological Monitoring; Obstetrical And Gynecological Devices
THERMOMETER	880.2900	CLINICAL COLOR CHANGE THERMOMETER General Hospital And Personal Use Monitoring Devices; General Hospital And Personal Use Devices
THERMOMETER	880.2910	CLINICAL ELECTRONIC THERMOMETER General Hospital And Personal Use Monitoring Devices; General Hospital And Personal Use Devices

THERMOMETER	880.2920	CLINICAL MERCURY THERMOMETER General Hospital And Personal Use Monitoring Devices; General Hospital And Personal Use Devices
THERMOSETTING RESINS	177.2280	4,4'-ISOPROPYLIDENEDIPHENOL-EPICHLOROHYDRIN THERMOSETTING EPOXY RESINS Substances For Use Only As Components Of Articles Intended For Repeated Use; Indirect Food Additives: Polymers
THIABENDAZOLE	520.2380	THIABENDAZOLE ORAL DOSAGE FORMS Oral Dosage Form New Animal Drugs
THIABENDAZOLE	524.1484g	NEOMYCIN SULFATE-THIABENDAZOLE-DEXAMETHASONE SOLUTION Ophthalmic And Topical Dosage Form New Animal Drugs
THIABENDAZOLE	556.730	THIABENDAZOLE Specific Tolerances For Residues Of New Animal Drugs; Tolerances For Residues Of New Animal Drugs In Food
THIABENDAZOLE	558.615	THIABENDAZOLE Specific New Animal Drugs For Use In Animal Feeds; New Animal Drugs For Use In Animal Feeds
THIALBARBITONE SODIUM	522.2404	THIALBARBITONE SODIUM FOR INJECTION Implantation Or Injectable Dosage Form New Animal Drugs
THIAMINE HYDROCHLORIDE	184.1875	THIAMINE HYDROCHLORIDE Listing Of Specific Substances Affirmed As GRAS; Direct Food Substances Affirmed As Generally Recognized As Safe
THIAMINE HYDROCHLORIDE	582.5875	THIAMINE HYDROCHLORIDE Nutrients And/or Dietary Supplements; Substances Generally Recognized As Safe
THIAMINE MONONITRATE	184.1878	THIAMINE MONONITRATE Listing Of Specific Substances Affirmed As GRAS; Direct Food Substances Affirmed As Generally Recognized As Safe
THIAMINE MONONITRATE	582.5878	THIAMINE MONONITRATE Nutrients And/or Dietary Supplements; Substances Generally Recognized As Safe
THIAMYLAL	522.2424	SODIUM THIAMYLAL FOR INJECTION Implantation Or Injectable Dosage Form New Animal Drugs
THIN-LAYER CHROMATOGRAPHY	862.2270	THIN-LAYER LIQUID CHROMATOGRAPHY SYSTEM FOR CLINICAL USE Clinical Laboratory Instruments; Clinical Chemistry And Clinical Toxicology Devices
THIOCYANATE	558.248	ERYTHROMYCIN THIOCYANATE Specific New Animal Drugs For Use In Animal Feeds; New Animal Drugs For Use In Animal Feeds
THIODIPROPIONATE	182.3280	DILAURYL THIODIPROPIONATE Chemical Preservatives; Substances Generally Recognized As Safe
THIODIPROPIONATE	582.3280	DILAURYL THIODIPROPIONATE Chemical Preservatives; Substances Generally Recognized As Safe
THIODIPROPIONIC ACID	182.3109	THIODIPROPIONIC ACID Chemical Preservatives; Substances Generally Recognized As Safe
THIODIPROPIONIC ACID	582.3109	THIODIPROPIONIC ACID Chemical Preservatives; Substances Generally Recognized As Safe
THIONE	176.230	3,5-DIMETHYL-1,3,5,*2H*-TETRAHYDROTHIADIAZINE-2-THIONE Substances For Use Only As Components Of Paper And Paperboard; Indirect Food Additives
THIOPENTAL	522.2444	SODIUM THIOPENTAL IMPLANTATION OR INJECTABLE DOSAGE FORMS Implantation Or Injectable Dosage Form New Animal Drugs
THIOSTREPTON	524.1600a	NYSTATIN, NEOMYCIN, THIOSTREPTON, AND TRIAMCINOLONE ACETONIDE OINTMENT Ophthalmic And Topical Dosage Form New Animal Drugs
THIOSTREPTON	524.1600b	NYSTATIN, NEOMYCIN, THIOSTREPTON, AND TRIAMCINOLONE ACETONIDE OPHTHALMIC OINTMENT Ophthalmic And Topical Dosage Form New Animal Drugs
THIOSULFATE	184.1807	SODIUM THIOSULFATE Listing Of Specific Substances Affirmed As GRAS; Direct Food Substances Affirmed As Generally Recognized As Safe
THIOSULFATE	582.6807	SODIUM THIOSULFATE Sequestrants; Substances Generally Recognized As Safe
THIOUREA	189.190	THIOUREA Substances Generally Prohibited From Direct Addition Or Use As Human Food; Substances Prohibited From Use In Human Food
THIRD-PARTY	203.36	FULFILLMENT HOUSES, SHIPPING AND MAILING SERVICES, COMARKETING AGREEMENTS, AND THIRD-PARTY RECORDKEEPING Samples; Prescription Drug Marketing
THORPE TUBE	868.2320	UNCOMPENSATED THORPE TUBE FLOWMETER Monitoring Devices; Anesthesiology Devices
THORPE TUBE	868.2340	COMPENSATED THORPE TUBE FLOWMETER Monitoring Devices; Anesthesiology Devices
THREE-WHEELED VEHICLE	890.3800	MOTORIZED THREE-WHEELED VEHICLE Physical Medicine Prosthetic Devices; Physical Medicine Devices
THREONINE	582.5881	THREONINE Nutrients And/or Dietary Supplements; Substances Generally Recognized As Safe

THRESHOLD	170.39	THRESHOLD OF REGULATION FOR SUBSTANCES USED IN FOOD-CONTACT ARTICLES Food Additive Safety; Food Additives
THRESHOLD	171.8	THRESHOLD OF REGULATIONS FOR SUBSTANCES USED IN FOOD-CONTACT ARTICLES General Provisions; Food Additive Petitions
THRESHOLD	174.6	THRESHOLD OF REGULATION FOR SUBSTANCES USED IN FOOD-CONTACT ARTICLES Indirect Food Additives: General
THROAT	874.3620	EAR, NOSE, AND THROAT SYNTHETIC POLYMER MATERIAL Prosthetic Devices; Ear, Nose, And Throat Devices
THROAT	874.4140	EAR, NOSE, AND THROAT BUR Surgical Devices; Ear, Nose, And Throat Devices
THROAT	874.4250	EAR, NOSE, AND THROAT ELECTRIC OR PNEUMATIC SURGICAL DRILL Surgical Devices; Ear, Nose, And Throat Devices
THROAT	874.4350	EAR, NOSE, AND THROAT FIBEROPTIC LIGHT SOURCE AND CARRIER Surgical Devices; Ear, Nose, And Throat Devices
THROAT	874.4420	EAR, NOSE, AND THROAT MANUAL SURGICAL INSTRUMENT Surgical Devices; Ear, Nose, And Throat Devices
THROAT	874.4500	EAR, NOSE, AND THROAT MICROSURGICAL CARBON DIOXIDE LASER Surgical Devices; Ear, Nose, And Throat Devices
THROAT	874.5220	EAR, NOSE, AND THROAT DRUG ADMINISTRATION DEVICE Therapeutic Devices; Ear, Nose, And Throat Devices
THROAT	874.5300	EAR, NOSE, AND THROAT EXAMINATION AND TREATMENT UNIT Therapeutic Devices; Ear, Nose, And Throat Devices
THROATS, SORE	201.315	OVER-THE-COUNTER DRUGS FOR MINOR SORE THROATS; SUGGESTED WARNING Specific Labeling Requirements For Specific Drug Products; Labeling
THROMBIN	864.7875	THROMBIN TIME TEST Hematology Kits And Packages; Hematology And Pathology Devices
THROMBOPLASTIN	864.7900	THROMBOPLASTIN GENERATION TEST Hematology Kits And Packages; Hematology And Pathology Devices
THROMBOPLASTIN	864.7925	PARTIAL THROMBOPLASTIN TIME TESTS Hematology Kits And Packages; Hematology And Pathology Devices
THROMBOTEST	864.7735	PROTHROMBIN-PROCONVERTIN TEST AND THROMBOTEST Hematology Kits And Packages; Hematology And Pathology Devices
THUMBSUCKING	310.536	DRUG PRODUCTS CONTAINING ACTIVE INGREDIENTS OFFERED OTC FOR USE AS A NAILBITING OR THUMBSUCKING DETERRENT Requirements For Specific New Drugs Or Devices; New Drugs
THYROID AUTOANTIBODY	866.5870	THYROID AUTOANTIBODY IMMUNOLOGICAL TEST SYSTEM Immunological Test Systems; Immunology And Microbiology Devices
THYROID DRUGS	250.11	THYROID-CONTAINING DRUG PREPARATIONS INTENDED FOR TREATMENT OF OBESITY IN HUMANS Drugs Regarded As Misbranded; Specific Requirements For Specific Human Drugs
THYROID HORMONES	201.316	DRUGS WITH THYROID HORMONE ACTIVITY FOR HUMAN USE; REQUIRED WARNING Specific Labeling Requirements For Specific Drug Products; Labeling
THYROID-STIMULATING	862.1690	THYROID-STIMULATING HORMONE TEST SYSTEM Clinical Chemistry Test Systems; Clinical Chemistry And Clinical Toxicology Devices
THYROXINE	862.1695	FREE THYROXINE TEST SYSTEM Clinical Chemistry Test Systems; Clinical Chemistry And Clinical Toxicology Devices
THYROXINE	862.1700	TOTAL THYROXINE TEST SYSTEM Clinical Chemistry Test Systems; Clinical Chemistry And Clinical Toxicology Devices
THYROXINE-BINDING	862.1685	THYROXINE-BINDING GLOBULIN TEST SYSTEM Clinical Chemistry Test Systems; Clinical Chemistry And Clinical Toxicology Devices
TIAMULIN	520.2455	TIAMULIN SOLUBLE POWDER Oral Dosage Form New Animal Drugs
TIAMULIN	520.2456	TIAMULIN LIQUID CONCENTRATE Oral Dosage Form New Animal Drugs
TIAMULIN	556.738	TIAMULIN Specific Tolerances For Residues Of New Animal Drugs; Tolerances For Residues Of New Animal Drugs In Food
TIAMULIN	558.600	TIAMULIN Specific New Animal Drugs For Use In Animal Feeds; New Animal Drugs For Use In Animal Feeds
TIBIAL	888.3590	KNEE JOINT TIBIAL (HEMI-KNEE) METALLIC RESURFACING UNCEMENTED PROSTHESIS Prosthetic Devices; Orthopedic Devices
TICARCILLIN	529.2464	TICARCILLIN POWDER Certain Other Dosage Form New Animal Drugs

TILETAMINE HYDROCHLORIDE	522.2470	TILETAMINE HYDROCHLORIDE AND ZOLAZEPAM HYDROCHLORIDE FOR INJECTION Implantation Or Injectable Dosage Form New Animal Drugs
TILMICOSIN	556.735	TILMICOSIN Specific Tolerances For Residues Of New Animal Drugs; Tolerances For Residues Of New Animal Drugs In Food
TILMICOSIN	558.618	TILMICOSIN Specific New Animal Drugs For Use In Animal Feeds; New Animal Drugs For Use In Animal Feeds
TILMICOSIN PHOSPHATE	522.2471	TILMICOSIN PHOSPHATE INJECTION Implantation Or Injectable Dosage Form New Animal Drugs
TIME	864.6100	BLEEDING TIME DEVICE Manual Hematology Devices; Hematology And Pathology Devices
TIME	864.7140	ACTIVATED WHOLE BLOOD CLOTTING TIME TESTS Hematology Kits And Packages; Hematology And Pathology Devices
TIME	864.7275	EUGLOBULIN LYSIS TIME TESTS Hematology Kits And Packages; Hematology And Pathology Devices
TIME	864.7750	PROTHROMBIN TIME TEST Hematology Kits And Packages; Hematology And Pathology Devices
TIME	864.7875	THROMBIN TIME TEST Hematology Kits And Packages; Hematology And Pathology Devices
TIME	864.7925	PARTIAL THROMBOPLASTIN TIME TESTS Hematology Kits And Packages; Hematology And Pathology Devices
TIME FOR APPLICATION	1301.13	APPLICATION FOR REGISTRATION; TIME FOR APPLICATION FOR REGISTRATION; EXPIRATION DATE Applications For Registration; Registration Of Manufacturers, Distributors, And Dispensers Of Controlled Substances
TIME FOR APPLICATION	1309.31	TIME FOR APPLICATION FOR REGISTRATION; EXPIRATION DATE Registration Of Manufacturers, Distributors, Importers And Exporters Of List I Chemicals
TIME FOR APPLICATION	1329.31	TIME FOR APPLICATION FOR REGISTRATION; EXPIRATION DATE Registration Of Manufacturers, Distributors, Importers And Exporters Of List I Chemicals
TIME, COMPUTATION	10.20	SUBMISSION OF DOCUMENTS TO DOCKETS MANAGEMENT BRANCH; COMPUTATION OF TIME; AVAILABILITY FOR PUBLIC DISCLOSURE General Administrative Procedures; Administrative Practices And Procedures
TIME, COMPUTATION	17.30	COMPUTATION OF TIME Civil Money Penalties Hearings
TIME, COMPUTATION	810.3	COMPUTATION OF TIME Medical Device Recall Authority
TIME, EXTEND	99.203	REQUEST TO EXTEND THE TIME FOR COMPLETING PLANNED STUDIES Dissemination Of Information On Unapproved/New Uses For Marketed Drugs, Biologics, And Devices
TIME, EXTENSION	71.6	EXTENSION OF TIME FOR STUDYING PETITIONS; SUBSTANTIVE AMENDMENTS; WITHDRAWAL OF PETITIONS WITHOUT PREJUDICE General Provisions; Color Additive Petitions
TIME, EXTENSION	99.303	EXTENSION OF TIME FOR COMPLETING PLANNED STUDIES Dissemination Of Information On Unapproved/New Uses For Marketed Drugs, Biologics, And Devices
TIME, EXTENSION	1401.8	EXTENSION OF TIME Public Availability Of Information
TIME FOR FILING	1316.80	TIME FOR FILING PETITIONS Seizure, Forfeiture, And Disposition Of Property; Administrative Functions, Practices, And Procedures
TIME FRAME	60.28	TIME FRAME FOR DETERMINING REGULATORY REVIEW PERIODS Regulatory Review Period Determinations; Patent Term Restoration
TIME FRAMES	314.100	TIME FRAMES FOR REVIEWING APPLICATIONS AND ABBREVIATED APPLICATIONS FDA Action On Applications; Applications For FDA Approval To Market A New Drug
TIME FRAMES	814.40	TIME FRAMES FOR REVIEWING A PMA FDA Action On A PMA; Premarket Approval Of Medical Devices
TIME OF HEARING	1301.45	TIME AND PLACE OF HEARING Hearings; Registration Of Manufacturers, Distributors, And Dispensers Of Controlled Substances
TIME OF HEARING	1303.36	TIME AND PLACE OF HEARING Hearings; Quotas
TIME OF HEARING	1309.55	TIME AND PLACE OF HEARING Registration Of Manufacturers, Distributors, Importers And Exporters Of List I Chemicals
TIME OF HEARING	1312.46	TIME AND PLACE OF HEARING Hearings; Importation And Exportation Of Controlled Substances
TIME OF HEARING	1313.56	TIME AND PLACE OF HEARING Hearings; Importation And Exportation Of Precursors And Essential Chemicals

TIME OF HEARING	1316.53	TIME AND PLACE OF HEARING Administrative Hearings; Administrative Functions, Practices, And Procedures
TIME OF INSPECTION	600.21	TIME OF INSPECTION Establishment Inspection; Biological Products: General
TIME LIMITATIONS	20.41	TIME LIMITATIONS Procedures And Fees; Public Information
TIME LIMITATIONS	211.111	TIME LIMITATIONS ON PRODUCTION Production And Process Controls; Current Good Manufacturing Practice For Finished Pharmaceuticals
TIME OF PAYMENT	1309.12	TIME AND METHOD OF PAYMENT; REFUND Registration Of Manufacturers, Distributors, Importers And Exporters Of List I Chemicals
TIME OF PREHEARING	12.91	TIME AND PLACE OF PREHEARING CONFERENCE Hearing Procedures; Formal Evidentiary Public Hearing
TIME OF PROCEEDING	1316.32	NOTICE OF PROCEEDING; TIME AND PLACE Enforcement Proceedings; Administrative Functions, Practices, And Procedures
TIME FOR REGISTRATION	710.2	TIME FOR REGISTRATION Voluntary Registration Of Cosmetic Product Establishments
TIME, REVIEW	3.10	STAY OF REVIEW TIME Product Jurisdiction
TIMED-RELEASE DRUGS	500.26	TIMED-RELEASE DOSAGE FORM DRUGS Specific Administrative Rulings And Decisions; General
TIMELINESS	21.73	ACCURACY, COMPLETENESS, TIMELINESS, AND RELEVANCE OF RECORDS DISCLOSED FROM PRIVACY ACT RECORD SYSTEMS Disclosure Of Records In Privacy Act Record Systems To Persons Other ThanThe Subject Individual; Protection Of Privacy
TIMER	880.2930	APGAR TIMER General Hospital And Personal Use Monitoring Devices; General Hospital And Personal Use Devices
TIMES FOR FILING	720.2	TIMES FOR FILING Voluntary Filing Of Cosmetic Product Ingredient Composition Statements
TIMES, REGISTRATION	207.21	TIMES FOR REGISTRATION AND DRUG LISTING Procedures For Domestic Drug Establishments; Registration Of Drug Producers And Listing Of Drugs In Commercial Distribution
TIMES, REGISTRATION	607.21	TIMES FOR ESTABLISHMENT REGISTRATION AND BLOOD PRODUCT LISTING Procedures For Domestic Blood Product Establishments; Establishment Registratn+product Listing, Manufacturers Human Blood/b.products
TIMES, REGISTRATION	807.21	TIMES FOR ESTABLISHMENT REGISTRATION AND DEVICE LISTING Procedures For Device Establishments; Establishment Registration And Device Listing For Manufacturers And Initial Importers Of Devices
TIMING	316.23	TIMING OF REQUESTS FOR ORPHAN DRUG DESIGNATION, DESIGNATION OF ALREADY APPROVED DRUGS Orphan Drugs
TIN-COATED	189.301	TIN-COATED LEAD FOIL CAPSULES FOR WINE BOTTLES Substances Prohibited From Use In Human Foods
TINNITUS	874.3400	TINNITUS MASKER Prosthetic Devices; Ear, Nose, And Throat Devices
TINPLATE	178.3300	CORROSION INHIBITORS USED FOR STEEL OR TINPLATE Certain Adjuvants And Production Aids; Indirect Food Additives: Adjuvants, Production Aids, And Sanitizers
TIOXIDAZOLE	520.2473	TIOXIDAZOLE ORAL DOSAGE FORMS Oral Dosage Form New Animal Drugs
TIP	870.1370	CATHETER TIP OCCLUDER Cardiovascular Diagnostic Devices; Cardiovascular Devices
TIP	870.2870	CATHETER TIP PRESSURE TRANSDUCER Cardiovascular Monitoring Devices; Cardiovascular Devices
TIP	872.6650	MASSAGING PICK OR TIP FOR ORAL HYGIENE Miscellaneous Devices; Dental Devices
TIPS	890.3790	CANE, CRUTCH, AND WALKER TIPS AND PADS Physical Medicine Prosthetic Devices; Physical Medicine Devices
TIPPED	880.6025	ABSORBENT TIPPED APPLICATOR General Hospital And Personal Use Miscellaneous Devices; General Hospital And Personal Use Devices
TISSUE-BASED PRODUCTS	1271.22	HOW AND WHERE DO I REGISTER AND SUBMIT AN HCT/P (HUMAN CELLS/TISSUES/PRODUCTS) LIST? Procedures For Registration And Listing; Human Cells, Tissues, And Cellular And Tissue-Based Products
TISSUE CULTURE	864.2220	SYNTHETIC CELL AND TISSUE CULTURE MEDIA AND COMPONENTS Cell And Tissue Culture Products; Hematology And Pathology Devices
TISSUE CULTURE	864.2240	CELL AND TISSUE CULTURE SUPPLIES AND EQUIPMENT Cell And Tissue Culture Products; Hematology And Pathology Devices

TISSUE CULTURE	876.5885	TISSUE CULTURE MEDIA FOR HUMAN EX VIVO TISSUE AND CELL CULTURE PROCESSING APPLICATIONS Therapeutic Devices; Gastroenterology-urology Devices
TISSUE, HUMAN	1270.21	DETERMINATION OF DONOR SUITABILITY FOR HUMAN TISSUE INTENDED FOR TRANSPLANTATION Human Tissue Intended For Transplantation
TISSUE, HUMAN	1270.43	RETENTION, RECALL, AND DESTRUCTION OF HUMAN TISSUE Human Tissue Intended For Transplantation
TISSUE PROCESSING	864.3010	TISSUE PROCESSING EQUIPMENT Pathology Instrumentation And Accessories; Hematology And Pathology Devices
TISSUE PROCESSOR	864.3875	AUTOMATED TISSUE PROCESSOR Pathology Instrumentation And Accessories; Hematology And Pathology Devices
TISSUE, TARGET	500.86	MARKER RESIDUE AND TARGET TISSUE Regulation Of Carcinogenic Compounds Used In Food-producing Animals; General
TISSUES, HUMAN	1271.22	HOW AND WHERE DO I REGISTER AND SUBMIT AN HCT/P (HUMAN CELLS/TISSUES/PRODUCTS) LIST? Procedures For Registration And Listing; Human Cells, Tissues, And Cellular And Tissue-Based Products
TITANIUM DIOXIDE	73.575	TITANIUM DIOXIDE Foods; Listing Of Color Additives Exempt From Certification
TITANIUM DIOXIDE	73.1575	TITANIUM DIOXIDE Drugs; Listing Of Color Additives Exempt From Certification
TITANIUM DIOXIDE	73.2575	TITANIUM DIOXIDE Cosmetics; Listing Of Color Additives Exempt From Certification
TITANIUM DIOXIDE	73.3126	TITANIUM DIOXIDE Medical Devices; Listing Of Color Additives Exempt From Certification
TLC	862.2400	DENSITOMETER/SCANNER (INTEGRATING, REFLECTANCE, TLC, OR RADIOCHROMATOGRAM) FOR CLINICAL USE Clinical Laboratory Instruments; Clinical Chemistry And Clinical Toxicology Devices
TOASTED COTTONSEED FLOUR	73.140	TOASTED PARTIALLY DEFATTED COOKED COTTONSEED FLOUR Foods; Listing Of Color Additives Exempt From Certification
TOASTED CRAMBE MEAL	573.310	CRAMBE MEAL, HEAT TOASTED Food Additive Listing; Food Additives Permitted In Feed And Drinking Water Of Animals
TOBRAMYCIN	862.3900	TOBRAMYCIN TEST SYSTEM Clinical Toxicology Test Systems; Clinical Chemistry And Clinical Toxicology Devices
TOCOPHEROL	182.8892	α-TOCOPHEROL ACETATE Nutrients; Substances Generally Recognized As Safe
TOCOPHEROL	582.5892	α-TOCOPHEROL ACETATE Nutrients And/or Dietary Supplements; Substances Generally Recognized As Safe
TOCOPHEROLS	182.3890	TOCOPHEROLS Chemical Preservatives; Substances Generally Recognized As Safe
TOCOPHEROLS	182.8890	TOCOPHEROLS Nutrients; Substances Generally Recognized As Safe
TOCOPHEROLS	184.1890	α-TOCOPHEROLS Listing Of Specific Substances Affirmed As GRAS; Direct Food Substances Affirmed As Generally Recognized As Safe
TOCOPHEROLS	582.3890	TOCOPHEROLS Chemical Preservatives; Substances Generally Recognized As Safe
TOCOPHEROLS	582.5890	TOCOPHEROLS Nutrients And/or Dietary Supplements; Substances Generally Recognized As Safe
TOD	884.5380	CONTRACEPTIVE TUBAL OCCLUSION DEVICE (TOD) AND INTRODUCER Obstetrical And Gynecological Therapeutic Devices; Obstetrical And Gynecological Devices
TOE JOINT	888.3720	TOE JOINT POLYMER CONSTRAINED PROSTHESIS Prosthetic Devices; Orthopedic Devices
TOE JOINT	888.3730	TOE JOINT PHALANGEAL (HEMI-TOE) POLYMER PROSTHESIS Prosthetic Devices; Orthopedic Devices
TOENAIL	310.538	DRUG PRODUCTS CONTAINING ACTIVE INGREDIENTS OFFERED OTC FOR USE FOR INGROWN TOENAIL RELIEF Requirements For Specific New Drugs Or Devices; New Drugs
TOILET ARTICLES	1250.44	DRINKING UTENSILS AND TOILET ARTICLES Equipment And Operation Of Land And Air Conveyances; Interstate Conveyance Sanitation
TOILET FACILITIES	211.52	WASHING AND TOILET FACILITIES Buildings And Facilities; Current Good Manufacturing Practice For Finished Pharmaceuticals
TOILET FACILITIES	1250.38	TOILET AND LAVATORY FACILITIES FOR USE OF FOOD-HANDLING EMPLOYEES Food Service Sanitation On Land And Air Conveyances, And Vessels; Interstate Conveyance Sanitation

TOILET FACILITIES	1250.50	TOILET AND LAVATORY FACILITIES Equipment And Operation Of Land And Air Conveyances; Interstate Conveyance Sanitation
TOILETS	1250.90	TOILETS AND LAVATORIES Sanitation Facilities And Conditions On Vessels; Interstate Conveyance Sanitation
TOLAZOLINE	522.2474	TOLAZOLINE HYDROCHLORIDE INJECTION Implantation Or Injectable Dosage Form New Animal Drugs
TOLERANCES	109.4	ESTABLISHMENT OF TOLERANCES, REGULATORY LIMITS AND ACTION LEVELS General Provisions; Unavoidable Contaminants In Food For Human Consumption & Food-packaging Material
TOLERANCES	109.30	TOLERANCES FOR POLYCHLORINATED BIPHENYLS (PCB'S) Tolerances For Unavoidable Poisonous Or Deleterious Substances; Unavoidable Contaminants In Food For Human Consumption & Food-packaging Material
TOLERANCES	170.18	TOLERANCES FOR RELATED FOOD ADDITIVES General Provisions; Food Additives
TOLERANCES	171.130	PROCEDURE FOR AMENDING AND REPEALING TOLERANCES OR EXEMPTIONS FROM TOLERANCES Administrative Actions On Applications; Food Additive Petitions
TOLERANCES	509.4	ESTABLISHMENT OF TOLERANCES, REGULATORY LIMITS AND ACTION LEVELS General Provisions; Unavoidable Contaminants In Animal Food And Food-packaging Material
TOLERANCES	509.30	TEMPORARY TOLERANCES FOR POLYCHLORINATED BIPHENYLS (PCB'S) Tolerances For Unavoidable Poisonous Or Deleterious Substances; Unavoidable Contaminants In Animal Food And Food-packaging Material
TOLERANCES	556.1	GENERAL CONSIDERATIONS; TOLERANCES FOR RESIDUES OF NEW ANIMAL DRUGS IN FOOD General Provisions; Tolerances For Residues Of New Animal Drugs In Food
TOLERANCES	570.18	TOLERANCES FOR RELATED FOOD ADDITIVES General Provisions; Food Additives
TOLERANCES	571.130	PROCEDURE FOR AMENDING AND REPEALING TOLERANCES OR EXEMPTIONS FROM TOLERANCES Administrative Actions On Applications; Food Additive Petitions
TOLNAFTATE	524.2350	TOLNAFTATE CREAM Ophthalmic And Topical Dosage Form New Animal Drugs
TOLUENE	520.580	DICHLOROPHENE AND TOLUENE CAPSULES Oral Dosage Form New Animal Drugs
TOMATO	155.191	TOMATO CONCENTRATES Requirements For Specific Standardized Canned Vegetables; Canned Vegetables
TOMATO	156.145	TOMATO JUICE Requirements For Specific Standardized Vegetable Juices; Vegetable Juices
TOMATOES	155.190	CANNED TOMATOES Requirements For Specific Standardized Canned Vegetables; Canned Vegetables
TOMOGRAPHIC	892.1740	TOMOGRAPHIC X-RAY SYSTEM Diagnostic Devices; Radiology Devices
TOMOGRAPHY	892.1200	EMISSION COMPUTED TOMOGRAPHY SYSTEM Diagnostic Devices; Radiology Devices
TOMOGRAPHY	892.1310	NUCLEAR TOMOGRAPHY SYSTEM Diagnostic Devices; Radiology Devices
TOMOGRAPHY	892.1750	COMPUTED TOMOGRAPHY X-RAY SYSTEM Diagnostic Devices; Radiology Devices
TOMOGRAPHY	1020.33	COMPUTED TOMOGRAPHY (CT) EQUIPMENT Performance Standards For Ionizing Radiation Emitting Products
TONE/TONER/TONIC	500.52	USE OF TERMS SUCH AS "TONIC," "TONE," "TONER" OR "CONDITIONER" IN THE LABELING OF PREPARATIONS INTENDED FOR USE IN OR ON ANIMALS Animal Drug Labeling Requirements; General
TONGS	874.5370	TONGS ANTICHOKE DEVICE Therapeutic Devices; Ear, Nose And Throat Devices
TONGS	882.5960	SKULL TONGS FOR TRACTION Neurological Therapeutic Devices; Neurological Devices
TONGUE	880.6230	TONGUE DEPRESSOR General Hospital And Personal Use Miscellaneous Devices; General Hospital And Personal Use Devices
TONOMETER	886.1930	TONOMETER AND ACCESSORIES Diagnostic Devices; Ophthalmic Devices
TONOMETER	886.1940	TONOMETER STERILIZER Diagnostic Devices; Ophthalmic Devices
TOOLS	870.3730	PACEMAKER SERVICE TOOLS Cardiovascular Prosthetic Devices; Cardiovascular Devices
TOOTH	872.3200	RESIN TOOTH BONDING AGENT Prosthetic Devices; Dental Devices
TOOTH	872.3580	PREFORMED GOLD DENTURE TOOTH Prosthetic Devices; Dental Devices

TOOTH	872.3590	PREFORMED PLASTIC DENTURE TOOTH Prosthetic Devices; Dental Devices
TOOTH	872.3690	TOOTH SHADE RESIN MATERIAL Prosthetic Devices; Dental Devices
TOOTH	872.3750	BRACKET ADHESIVE RESIN AND TOOTH CONDITIONER Prosthetic Devices; Dental Devices
TOOTH	872.3900	POSTERIOR ARTIFICIAL TOOTH WITH A METAL INSERT Prosthetic Devices; Dental Devices
TOOTH	872.3910	BACKING AND FACING FOR AN ARTIFICIAL TOOTH Prosthetic Devices; Dental Devices
TOOTH	872.3920	PORCELAIN TOOTH Prosthetic Devices; Dental Devices
TOOTH	872.5525	PREFORMED TOOTH POSITIONER Therapeutic Devices; Dental Devices
TOOTHBRUSH	872.6855	MANUAL TOOTHBRUSH Miscellaneous Devices; Dental Devices
TOOTHBRUSH	872.6865	POWERED TOOTHBRUSH Miscellaneous Devices; Dental Devices
TOP DRESSING	520.2380	THIABENDAZOLE TOP DRESSING AND MINERAL PROTEIN FEED BLOCK Oral Dosage Form New Animal Drugs
TOPICAL	524.390	CHLORAMPHENICOL OPHTHALMIC AND TOPICAL DOSAGE FORMS Ophthalmic And Topical Dosage Form New Animal Drugs
TOPICAL	524.660	DIMETHYL SULFOXIDE OPHTHALMIC AND TOPICAL DOSAGE FORMS Ophthalmic And Topical Dosage Form New Animal Drugs
TOPICAL	524.981	FLUOCINOLONE ACETONIDE OPHTHALMIC AND TOPICAL DOSAGE FORMS Ophthalmic And Topical Dosage Form New Animal Drugs
TOPICAL	524.1044	GENTAMICIN SULFATE OPHTHALMIC AND TOPICAL DOSAGE FORMS Ophthalmic And Topical Dosage Form New Animal Drugs
TOPICAL	524.1200	KANAMYCIN OPHTHALMIC AND TOPICAL DOSAGE FORMS Ophthalmic And Topical Dosage Form New Animal Drugs
TOPICAL	524.1484	NEOMYCIN SULFATE OPHTHALMIC AND TOPICAL DOSAGE FORMS Ophthalmic And Topical Dosage Form New Animal Drugs
TOPICAL	524.1580	NITROFURAZONE OPHTHALMIC AND TOPICAL DOSAGE FORMS Ophthalmic And Topical Dosage Form New Animal Drugs
TOPICAL	524.1600	NYSTATIN OPHTHALMIC AND TOPICAL DOSAGE FORMS Ophthalmic And Topical Dosage Form New Animal Drugs
TOPICAL	524.1662	OXYTETRACYCLINE HYDROCHLORIDE OPHTHALMIC AND TOPICAL DOSAGE FORMS Ophthalmic And Topical Dosage Form New Animal Drugs
TOPICAL	524.1881	PREDNISOLONE ACETATE OPHTHALMIC AND TOPICAL DOSAGE FORMS Ophthalmic And Topical Dosage Form New Animal Drugs
TOPICAL ANESTHESIA	868.5170	LARYNGOTRACHEAL TOPICAL ANESTHESIA APPLICATOR Therapeutic Devices; Anesthesiology Devices
TOPICAL-OTIC	524.1484	NEOMYCIN SULFATE, PREDNISOLONE, TETRACAINE, AND SQUALANE TOPICAL-OTIC SUSPENSION Ophthalmic And Topical Dosage Form New Animal Drugs
TOPICAL OXYGEN	878.5650	TOPICAL OXYGEN CHAMBER FOR EXTREMITIES Therapeutic Devices; General And Plastic Surgery Devices
TOPICAL POWDER	524.1484b	NEOMYCIN SULFATE, ISOFLUPREDONE ACETATE, TETRACAINE HYDROCHLORIDE, AND MYRISTYL-GAMMA-PICOLINIUM CHLORIDE, TOPICAL POWDER Ophthalmic And Topical Dosage Form New Animal Drugs
TOPICALLY APPLIED	310.530	TOPICALLY APPLIED HORMONE-CONTAINING DRUG PRODUCTS FOR OVER-THE-COUNTER (OTC) HUMAN USE Requirements For Specific New Drugs Or Devices; New Drugs
TOTAL SOLIDS	610.16	TOTAL SOLIDS IN SERUMS General Provisions; General Biological Products Standards
TOTAL SPINAL FLUID	866.5860	TOTAL SPINAL FLUID IMMUNOLOGICAL TEST SYSTEM Immunological Test Systems; Immunology And Microbiology Devices
TOURNIQUET	870.5925	AUTOMATIC ROTATING TOURNIQUET Cardiovascular Therapeutic Devices; Cardiovascular Devices
TOURNIQUET	878.5900	NONPNEUMATIC TOURNIQUET Therapeutic Devices; General And Plastic Surgery Devices
TOURNIQUET	878.5910	PNEUMATIC TOURNIQUET Therapeutic Devices; General And Plastic Surgery Devices
TOXICOLOGY	862.3200	CLINICAL TOXICOLOGY CALIBRATOR Clinical Toxicology Test Systems; Clinical Chemistry And Clinical Toxicology Devices
TOXICOLOGY	862.3280	CLINICAL TOXICOLOGY CONTROL MATERIAL Clinical Toxicology Test Systems; Clinical Chemistry And Clinical Toxicology Devices
TOXOPLASMA GONDII	866.3780	TOXOPLASMA GONDII SEROLOGICAL REAGENTS Serological Reagents; Immunology And Microbiology Devices

TOYNBEE	874.1925	TOYNBEE DIAGNOSTIC TUBE Diagnostic Devices; Ear, Nose, And Throat Devices
TRACE MICROSPHERE	870.1360	TRACE MICROSPHERE Cardiovascular Diagnostic Devices; Cardiovascular Devices
TRACE MINERALS	582.80	TRACE MINERALS ADDED TO ANIMAL FEEDS General Provisions; Substances Generally Recognized As Safe
TRACEABILITY	820.65	TRACEABILITY Identification And Traceability; Quality System Regulation
TRACHEAL	868.5730	TRACHEAL TUBE Therapeutic Devices; Anesthesiology Devices
TRACHEAL	868.5750	INFLATABLE TRACHEAL TUBE CUFF Therapeutic Devices; Anesthesiology Devices
TRACHEAL	868.5770	TRACHEAL TUBE FIXATION DEVICE Therapeutic Devices; Anesthesiology Devices
TRACHEAL	868.5790	TRACHEAL TUBE STYLET Therapeutic Devices; Anesthesiology Devices
TRACHEAL	868.5795	TRACHEAL TUBE CLEANING BRUSH Therapeutic Devices; Anesthesiology Devices
TRACHEAL	878.3720	TRACHEAL PROSTHESIS Prosthetic Devices; General And Plastic Surgery Devices
TRACHEAL/BRONCHIAL	868.5740	TRACHEAL/BRONCHIAL DIFFERENTIAL VENTILATION TUBE Therapeutic Devices; Anesthesiology Devices
TRACHEOBRONCHIAL	868.6810	TRACHEOBRONCHIAL SUCTION CATHETER Miscellaneous; Anesthesiology Devices
TRACHEOSTOMY	868.5800	TRACHEOSTOMY TUBE AND TUBE CUFF Therapeutic Devices; Anesthesiology Devices
TRACKING	5.416	MEDICAL DEVICE TRACKING Medical Devices And Radiological Health; Redelegations Of Authority; Delegations Of Authority And Organization
TRACKING	821.30	TRACKING OBLIGATIONS OF PERSONS OTHER THAN DEVICE MANUFACTURERS: DISTRIBUTOR REQUIREMENTS Medical Device Tracking Requirements
TRACKING, DEVICE	821.25	DEVICE TRACKING SYSTEM AND CONTENT REQUIREMENTS; MANUFACTURER REQUIREMENTS Medical Device Tracking Requirements
TRACKING, DEVICES	821.20	DEVICES SUBJECT TO TRACKING Medical Device Tracking Requirements
TRACTION	882.5960	SKULL TONGS FOR TRACTION Neurological Therapeutic Devices; Neurological Devices
TRACTION	888.5850	NONPOWERED ORTHOPEDIC TRACTION APPARATUS AND ACCESSORIES Surgical Devices; Orthopedic Devices
TRACTION	888.5890	NONINVASIVE TRACTION COMPONENT Surgical Devices; Orthopedic Devices
TRACTION	890.5900	POWERED TRACTION EQUIPMENT Physical Medicine Therapeutic Devices; Physical Medicine Devices
TRACTION	890.5925	TRACTION ACCESSORY Physical Medicine Therapeutic Devices; Physical Medicine Devices
TRADE SECRETS	20.53	INDEXING TRADE SECRETS AND CONFIDENTIAL COMMERCIAL OR FINANCIAL INFORMATION Procedures And Fees; Public Information
TRADE SECRETS	20.61	TRADE SECRETS AND COMMERCIAL OR FINANCIAL INFORMATION WHICH IS PRIVILEGED OR CONFIDENTIAL Exemptions; Public Information
TRAGACANTH	184.1351	GUM TRAGACANTH Listing Of Specific Substances Affirmed As GRAS; Direct Food Substances Affirmed As Generally Recognized As Safe
TRAGACANTH	201.319	WATER-SOLUBLE GUMS, AND HYDROPHILIC MUCILLOIDS (INCLUDING GUAR GUM, KARAYA GUM, TRAGACANTH, XANTHAN GUM) AS ACTIVE INGREDIENTS; REQUIRED WARNINGS AND DIRECTIONS Specific Labeling Requirements For Specific Drug Products; Labeling
TRAGACANTH	582.7351	GUM TRAGACANTH Stabilizers; Substances Generally Recognized As Safe
TRAINER	874.3320	GROUP HEARING AID OR GROUP AUDITORY TRAINER Prosthetic Devices; Ear, Nose, And Throat Devices
TRAINING	120.13	TRAINING Hazard Analysis And Critical Control Point (HACCP) Systems
TRAINING	123.10	TRAINING Fish And Fishery Products
TRANSABDOMINAL	884.1600	TRANSABDOMINAL AMNIOSCOPE (FETOSCOPE) AND ACCESSORIES Obstetrical And Gynecological Diagnostic Devices; Obstetrical And Gynecological Devices
TRANSACTION, INTERNATIONAL	1313.33	CONTENTS OF AN INTERNATIONAL TRANSACTION DECLARATION Importation And Exportation Of Precursors And Essential Chemicals

TRANSACTION, INTERNATIONAL	1313.34	DISTRIBUTION OF THE INTERNATIONAL TRANSACTION DECLARATION Importation And Exportation Of Precursors And Essential Chemicals
TRANSACTIONS, COVERED	404.220	CONTINUATION OF COVERED TRANSACTIONS Governmentwide Debarment And Suspension And Requirements For Drug-Free Workplace
TRANSACTIONS, EXCLUDED	1310.08	EXCLUDED TRANSACTIONS Records And Reports Of Listed Chemicals And Certain Machines
TRANSACTIONS, INTERNATIONAL	1313.32	REQUIREMENT OF AUTHORIZATION FOR INTERNATIONAL TRANSACTIONS Importation And Exportation Of Precursors And Essential Chemicals
TRANSCERVICAL	884.1660	TRANSCERVICAL ENDOSCOPE (AMNIOSCOPE) AND ACCESSORIES Obstetrical And Gynecological Diagnostic Devices; Obstetrical And Gynecological Devices
TRANSCRIPT	12.98	OFFICIAL TRANSCRIPT Hearing Procedures; Formal Evidentiary Public Hearing
TRANSCRIPT	1316.63	OFFICIAL TRANSCRIPT; INDEX; CORRECTIONS Administrative Hearings; Administrative Functions, Practices, And Procedures
TRANSCRIPTS	14.61	TRANSCRIPTS OF ADVISORY COMMITTEE MEETINGS Records Of Meetings And Hearings Before Advisory Committees; Public Hearing Before A Public Advisory Committee
TRANSCUTANEOUS	870.5550	EXTERNAL TRANSCUTANEOUS CARDIAC PACEMAKER (NONINVASIVE) Cardiovascular Therapeutic Devices; Cardiovascular Devices
TRANSCUTANEOUS	882.5890	TRANSCUTANEOUS ELECTRICAL NERVE STIMULATOR FOR PAIN RELIEF Neurological Therapeutic Devices; Neurological Devices
TRANSDUCER	868.2875	DIFFERENTIAL PRESSURE TRANSDUCER Monitoring Devices; Anesthesiology Devices
TRANSDUCER	868.2885	GAS FLOW TRANSDUCER Monitoring Devices; Anesthesiology Devices
TRANSDUCER	868.2900	GAS PRESSURE TRANSDUCER Monitoring Devices; Anesthesiology Devices
TRANSDUCER	870.2060	TRANSDUCER SIGNAL AMPLIFIER AND SIGNAL CONDITIONER Cardiovascular Monitoring Devices; Cardiovascular Devices
TRANSDUCER	870.2840	APEX CARDIOGRAPHIC TRANSDUCER Cardiovascular Monitoring Devices; Cardiovascular Devices
TRANSDUCER	870.2850	EXTRAVASCULAR BLOOD PRESSURE TRANSDUCER Cardiovascular Monitoring Devices; Cardiovascular Devices
TRANSDUCER	870.2860	HEART SOUND TRANSDUCER Cardiovascular Monitoring Devices; Cardiovascular Devices
TRANSDUCER	870.2870	CATHETER TIP PRESSURE TRANSDUCER Cardiovascular Monitoring Devices; Cardiovascular Devices
TRANSDUCER	870.2880	ULTRASONIC TRANSDUCER Cardiovascular Monitoring Devices; Cardiovascular Devices
TRANSDUCER	870.2890	VESSEL OCCLUSION TRANSDUCER Cardiovascular Monitoring Devices; Cardiovascular Devices
TRANSDUCER	870.2900	PATIENT TRANSDUCER AND ELECTRODE CABLE (INCLUDING CONNECTOR) Cardiovascular Monitoring Devices; Cardiovascular Devices
TRANSDUCER	882.1950	TREMOR TRANSDUCER Neurological Diagnostic Devices; Neurological Devices
TRANSDUCER	884.2960	OBSTETRIC ULTRASONIC TRANSDUCER AND ACCESSORIES Obstetrical And Gynecological Monitoring Devices; Obstetrical And Gynecological Devices
TRANSDUCER	890.1615	MINIATURE PRESSURE TRANSDUCER Physical Medicine Diagnostic Devices; Physical Medicine Devices
TRANSDUCER	892.1570	DIAGNOSTIC ULTRASONIC TRANSDUCER Diagnostic Devices; Radiology Devices
TRANSFER	864.9875	TRANSFER SET Products Used In Establishments That Manufacture Blood And Blood Products; Hematology And Pathology Devices
TRANSFER	880.6775	POWERED PATIENT TRANSFER DEVICE General Hospital And Personal Use Miscellaneous Devices; General Hospital And Personal Use Devices
TRANSFER	880.6785	MANUAL PATIENT TRANSFER DEVICE General Hospital And Personal Use Miscellaneous Devices; General Hospital And Personal Use Devices
TRANSFER	1306.25	TRANSFER BETWEEN PHARMACIES OF PRESCRIPTION INFORMATION FOR SCHEDULES 111, IV, AND V CONTROLLED SUBSTANCES FOR REFILL PURPOSES Controlled Substances Listed In Schedules III, IV And V; Prescriptions
TRANSFER	1313.31	ADVANCE NOTICE OF IMPORTATION FOR TRANSSHIPMENT OR TRANSFER Transshipment, In-Transit Shipments And International Transactions Involving Listed Chemicals; Importation And Exportation Of Precursors And Essential Chemicals
TRANSFER OF OBLIGATIONS	312.52	TRANSFER OF OBLIGATIONS TO A CONTRACT RESEARCH ORGANIZATION Investigational New Drug Application

TRANSFER OF REGISTRATION	1301.52	TERMINATION OF REGISTRATION; TRANSFER OF REGISTRATION; DISTRIBUTION UPON DISCONTINUANCE OF BUSINESS Registration Of Manufacturers, Distributors, And Dispensers, Of Controlled Substances
TRANSFER OF REGISTRATION	1309.63	TRANSFER OF REGISTRATION Registration Of Manufacturers, Distributors, Importers And Exporters Of List I Chemicals
TRANSFER, TECHNOLOGY	5.24	AUTHORITY RELATING TO TECHNOLOGY TRANSFER General Redelegations Of Authority; Delegations Of Authority And Organization
TRANSFERASE	862.1100	ASPARTATE AMINO TRANSFERASE (AST/SGOT) TEST SYSTEM Clinical Chemistry Test Systems; Clinical Chemistry And Clinical Toxicology Devices
TRANSFERASE	862.1315	GALACTOSE-1-PHOSPHATE URIDYL TRANSFERASE TEST SYSTEM Clinical Chemistry Test Systems; Clinical Chemistry And Clinical Toxicology Devices
TRANSFERASE	862.1535	ORNITHINE CARBAMYL TRANSFERASE TEST SYSTEM Clinical Chemistry Test Systems; Clinical Chemistry And Clinical Toxicology Devices
TRANSFERRIN	866.5880	TRANSFERRIN IMMUNOLOGICAL TEST SYSTEM Immunological Test Systems; Immunology And Microbiology Devices
TRANSFUSION SERVICES	606.171	REPORTING OF PRODUCT DEVIATIONS BY MANUFACTURERS, UNLICENSED BLOOD ESTABLISHMENTS, AND TRANSFUSION SERVICES Records And Reports; Current Good Manufacturing Practices For Blood And Blood Components
TRANSFUSION SERVICES	610.47	"LOOK BACK" NOTIFICATION REQUIREMENTS FOR TRANSFUSION SERVICES Testing Requirements For Communicable Disease Agents; General Biological Products Standards
TRANSILLUMINATOR	886.1945	TRANSILLUMINATOR Diagnostic Devices; Ophthalmic Devices
TRANSILLUMINATOR	892.1970	TRANSILLUMINATOR FOR BREAST EVALUATION Diagnostic Devices; Radiology Devices
TRANSITION ACTIVITIES	26.8	OTHER TRANSITION ACTIVITIES Specific Sector Provisions For Pharmaceutical GMPs; Mutual Recognition Of Pharmaceutical GMP Reports, Medical Device Quality System Audit Reports
TRANSITION PERIOD	26.5	LENGTH OF TRANSITION PERIOD Specific Sector Provisions For Pharmaceutical GMPs; Mutual Recognition Of Pharmaceutical GMP Reports, Medical Device Quality System Audit Reports
TRANSITION PERIOD	26.35	LENGTH AND PURPOSE OF TRANSITION PERIOD Specific Sector Provisions For Medical Devices; Mutual Recognition Of Pharmaceutical GMP Reports, Medical Device Quality System Audit Reports
TRANSITION PERIOD	26.38	OTHER TRANSITION PERIOD ACTIVITIES Specific Sector Provisions For Medical Devices; Mutual Recognition Of Pharmaceutical GMP Reports, Medical Device Quality System Audit Reports
TRANSITIONAL ARRANGEMENTS	26.64	TRANSITIONAL ARRANGEMENTS Framework Provisions; Mutual Recognition Of Pharmaceutical GMP Reports, Medical Device Quality System Audit Reports
TRANSITIONAL PRODUCTS	860.136	PROCEDURES FOR TRANSITIONAL PRODUCTS UNDER SECTION 520(L) OF THE ACT Reclassification; Medical Device Classification Procedures
TRANSMITTER	870.2910	RADIOFREQUENCY PHYSIOLOGICAL SIGNAL TRANSMITTER AND RECEIVER Cardiovascular Monitoring Devices; Cardiovascular Devices
TRANSMITTER	870.2920	TELEPHONE ELECTROCARDIOGRAPH TRANSMITTER AND RECEIVER Cardiovascular Monitoring Devices; Cardiovascular Devices
TRANSPARENT CONTAINERS	161.173	CANNED WET PACKED SHRIMP IN TRANSPARENT OR NONTRANSPARENT CONTAINERS Requirements For Specific Standardized Fish And Shellfish; Fish And Shellfish
TRANSPEPTIDASE	862.1360	GAMMA-GLUTAMYL TRANSPEPTIDASE AND ISOENZYMES TEST SYSTEM Clinical Chemistry Test Systems; Clinical Chemistry And Clinical Toxicology Devices
TRANSPLANTATION	1270.21	DETERMINATION OF DONOR SUITABILITY FOR HUMAN TISSUE INTENDED FOR TRANSPLANTATION Human Tissue Intended For Transplantation
TRANSPORT	864.3250	SPECIMEN TRANSPORT AND STORAGE CONTAINER Pathology Instrumentation And Accessories; Hematology And Pathology Devices
TRANSPORT	866.2390	TRANSPORT CULTURE MEDIUM Microbiology Devices; Immunology And Microbiology Devices
TRANSPORT	866.2900	MICROBIOLOGICAL SPECIMEN COLLECTION AND TRANSPORT DEVICE Microbiology Devices; Immunology And Microbiology Devices
TRANSPORT	876.5880	ISOLATED KIDNEY PERFUSION AND TRANSPORT SYSTEM AND ACCESSORIES Therapeutic Devices; Gastroenterology-urology Devices
TRANSPORT	880.5410	NEONATAL TRANSPORT INCUBATOR General Hospital And Personal Use Therapeutic Devices; General Hospital And Personal Use Devices

TRANSPORT	890.5150	POWERED PATIENT TRANSPORT Physical Medicine Therapeutic Devices; Physical Medicine Devices
TRANSSHIPMENT	1313.31	ADVANCE NOTICE OF IMPORTATION FOR TRANSSHIPMENT OR TRANSFER Transshipment, In-Transit Shipments And International Transactions Involving Listed Chemicals; Importation And Exportation Of Precursors And Essential Chemicals
TRAP, WATER	868.5995	TEE DRAIN (WATER TRAP) Therapeutic Devices; Anesthesiology Devices
TRAPEZIUM	888.3770	WRIST JOINT CARPAL TRAPEZIUM POLYMER PROSTHESIS Prosthetic Devices; Orthopedic Devices
TRAY, AMNIOCENTESIS	884.1550	AMNIOTIC FLUID SAMPLER (AMNIOCENTESIS TRAY) Obstetrical And Gynecological Diagnostic Devices; Obstetrical And Gynecological Devices
TRAY, ANESTHETIC	868.6100	ANESTHETIC CABINET, TABLE, OR TRAY Miscellaneous; Anesthesiology Devices
TRAY, FLUORIDE	872.6870	DISPOSABLE FLUORIDE TRAY Miscellaneous Devices; Dental Devices
TRAY, IMPRESSION	872.3670	RESIN IMPRESSION TRAY MATERIAL Prosthetic Devices; Dental Devices
TRAY, IMPRESSION	872.6880	PREFORMED IMPRESSION TRAY Miscellaneous Devices; Dental Devices
TREASURY	17.54	DEPOSIT IN THE TREASURY OF THE UNITED STATES Civil Money Penalties Hearings
TREATED GRAIN	2.25	GRAIN SEED TREATED WITH POISONOUS SUBSTANCES; COLOR IDENTIFICATION TO PREVENT ADULTERATION OF HUMAN AND ANIMAL FOOD Human And Animal Foods; General Administrative Rulings And Decisions
TREATMENT	80.37	TREATMENT OF BATCH PENDING CERTIFICATION Certification Procedures; Color Additive Certification
TREATMENT	80.38	TREATMENT OF BATCH AFTER CERTIFICATION Certification Procedures; Color Additive Certification
TREATMENT	179.26	IONIZING RADIATION FOR THE TREATMENT OF FOOD Radiation And Radiation Sources; Irradiation In The Production, Processing And Handling Of Food
TREATMENT	179.39	ULTRAVIOLET RADIATION FOR THE PROCESSING AND TREATMENT OF FOOD Radiation And Radiation Sources; Irradiation In The Production, Processing And Handling Of Food
TREATMENT	179.41	PULSED LIGHT FOR THE TREATMENT OF FOOD Radiation And Radiation Sources; Irradiation In The Production, Processing And Handling Of Food
TREATMENT	579.22	IONIZING RADIATION FOR TREATMENT OF ANIMAL DIETS Radiation And Radiation Sources; Irradiation In The Production, Process & Handling Of Animal Feed And Pet Food
TREATMENT	579.40	IONIZING RADIATION FOR THE TREATMENT OF POULTRY FEED AND POULTRY FEED INGREDIENTS Radiation And Radiation Sources; Irradiation In The Production, Process & Handling Of Animal Feed And Pet Food
TREATMENT	868.5365	POSTURE CHAIR FOR CARDIAC OR PULMONARY TREATMENT Therapeutic Devices; Anesthesiology Devices
TREATMENT	1240.90	APPROVAL OF TREATMENT ABOARD CONVEYANCES Source And Use Of Potable Water; Control Of Communicable Diseases
TREATMENT	1250.83	STORAGE OF WATER PRIOR TO TREATMENT Sanitation Facilities And Conditions On Vessels; Interstate Conveyance Sanitation
TREATMENT, DIRECTIONS	1230.14	DIRECTIONS FOR TREATMENT Labeling; Regulations Under The Federal Caustic Poison Act
TREATMENT, DIRECTIONS	1230.15	RESPONSIBILITY FOR LABELING DIRECTIONS FOR TREATMENT Labeling; Regulations Under The Federal Caustic Poison Act
TREATMENT, DIRECTIONS	1230.10	EXEMPTION FROM LABELING DIRECTIONS FOR TREATMENT Labeling; Regulations Under The Federal Caustic Poison Act
TREATMENT, EMERGENCY	500.65	EPINEPHRINE INJECTION FOR EMERGENCY TREATMENT OF ANAPHYLACTOID SHOCK IN CATTLE, HORSES, SHEEP, AND SWINE Requirements For Specific Animal Drugs, General
TREATMENT, NARCOTIC	1301.72	PHYSICAL SECURITY CONTROLS FOR NON-PRACTITIONERS; NARCOTIC TREATMENT PROGRAMS AND COMPOUNDERS FOR NARCOTIC TREATMENT PROGRAMS; STORAGE AREAS Security Requirements; Registration Of Manufacturers, Distributors, And Dispensers Of Controlled Substances
TREATMENT, NARCOTIC	1301.73	PHYSICAL SECURITY CONTROLS FOR NON-PRACTITIONERS; COMPOUNDERS FOR NARCOTIC TREATMENT PROGRAMS; MANUFACTURING AND COMPOUNDING AREAS Security Requirements; Registration Of Manufacturers, Distributors, And Dispensers Of Controlled Substances

TREATMENT, NARCOTIC	1301.74	OTHER SECURITY CONTROLS FOR NON-PRACTITIONERS; NARCOTIC TREATMENT PROGRAMS AND COMPOUNDERS FOR NARCOTIC TREATMENT PROGRAMS Security Requirements; Registration Of Manufacturers, Distributors, And Dispensers Of Controlled Substances
TREATMENT PROGRAMS	1304.24	RECORDS FOR MAINTENANCE TREATMENT PROGRAMS AND DETOXIFICATION TREATMENT PROGRAMS Continuing Records; Records And Reports Of Registrants
TREATMENT PROGRAMS	1304.25	RECORDS FOR TREATMENT PROGRAMS WHICH COMPOUND NARCOTICS FOR TREATMENT PROGRAMS AND OTHER LOCATIONS Continuing Records; Records And Reports Of Registrants
TREATMENT PROTOCOLS	312.83	TREATMENT PROTOCOLS Investigational New Drug Application
TREATMENT UNIT, ENT	874.5300	EAR, NOSE, AND THROAT EXAMINATION AND TREATMENT UNIT Therapeutic Devices; Ear, Nose, And Throat Devices
TREATMENT USE	312.34	TREATMENT USE OF AN INVESTIGATIONAL NEW DRUG Investigational New Drug Application
TREATMENT USE	312.35	SUBMISSIONS FOR TREATMENT USE Investigational New Drug Application
TREATMENT USE	316.40	TREATMENT USE OF A DESIGNATED ORPHAN DRUG Orphan Drugs
TREATY, INTERNATIONAL	1308.46	CONTROL REQUIRED UNDER INTERNATIONAL TREATY Hearings; Schedules Of Controlled Substances
TREMOR	882.1950	TREMOR TRANSDUCER Neurological Diagnostic Devices; Neurological Devices
TRENBOLONE	556.739	TRENBOLONE Specific Tolerances For Residues Of New Animal Drugs; Tolerances For Residues Of New Animal Drugs In Food
TRENBOLONE ACETATE	522.2476	TRENBOLONE ACETATE Implantation Or Injectable Dosage Form New Animal Drugs
TRENBOLONE ACETATE	522.2477	TRENBOLONE ACETATE AND ESTRADIOL Implantation Or Injectable Dosage Form New Animal Drugs
TRENBOLONE ACETATE	522.2478	TRENBOLONE ACETATE AND ESTRADIOL BENZOATE Implantation Or Injectable Dosage Form New Animal Drugs
TREPHINES	882.4300	MANUAL CRANIAL DRILLS, BURRS, TREPHINES, AND THEIR ACCESSORIES Neurological Surgical Devices; Neurological Devices
TREPHINES	882.4305	POWERED COMPOUND CRANIAL DRILLS, BURRS, TREPHINES, AND THEIR ACCESSORIES Neurological Surgical Devices; Neurological Devices
TREPHINES	882.4310	POWERED SIMPLE CRANIAL DRILLS, BURRS, TREPHINES, AND THEIR ACCESSORIES Neurological Surgical Devices; Neurological Devices
TREPONEMA PALLIDUM	866.3820	TREPONEMA PALLIDUM NONTREPONEMAL TEST REAGENTS Serological Reagents; Immunology And Microbiology Devices
TREPONEMA PALLIDUM	866.3830	TREPONEMA PALLIDUM TREPONEMAL TEST REAGENTS Serological Reagents; Immunology And Microbiology Devices
TRI-(ACETOXYSTEARATE)	178.3505	GLYCERYL TRI-(12-ACETOXYSTEARATE) Certain Adjuvants And Production Aids; Indirect Food Additives: Adjuvants, Production Aids, And Sanitizers
TRIACETIN	184.1901	TRIACETIN Listing Of Specific Substances Affirmed As GRAS; Direct Food Substances Affirmed As Generally Recognized As Safe
TRIACETIN	582.1901	TRIACETIN General Purpose Food Additives; Substances Generally Recognized As Safe
TRIAL LENS	886.1405	OPHTHALMIC TRIAL LENS SET Diagnostic Devices; Ophthalmic Devices
TRIAL LENS	886.1410	OPHTHALMIC TRIAL LENS CLIP Diagnostic Devices; Ophthalmic Devices
TRIAL LENS	886.1415	OPHTHALMIC TRIAL LENS FRAME Diagnostic Devices; Ophthalmic Devices
TRIALS, CLINICAL	312.87	ACTIVE MONITORING OF CONDUCT AND EVALUATION OF CLINICAL TRIALS Investigational New Drug Application
TRIAMCINOLONE ACETONIDE	520.2481	TRIAMCINOLONE ACETONIDE TABLETS Oral Dosage Form New Animal Drugs
TRIAMCINOLONE ACETONIDE	520.2482	TRIAMCINOLONE ACETONIDE ORAL POWDER Oral Dosage Form New Animal Drugs
TRIAMCINOLONE ACETONIDE	522.2483	STERILE TRIAMCINOLONE ACETONIDE SUSPENSION Implantation Or Injectable Dosage Form New Animal Drugs
TRIAMCINOLONE ACETONIDE	524.1600a	NYSTATIN, NEOMYCIN, THIOSTREPTON, AND TRIAMCINOLONE ACETONIDE OINTMENT Ophthalmic And Topical Dosage Form New Animal Drugs
TRIAMCINOLONE ACETONIDE	524.1600b	NYSTATIN, NEOMYCIN, THIOSTREPTON, AND TRIAMCINOLONE ACETONIDE OPHTHALMIC OINTMENT Ophthalmic And Topical Dosage Form New Animal Drugs

TRIAMCINOLONE ACETONIDE	524.2481	TRIAMCINOLONE ACETONIDE CREAM Ophthalmic And Topical Dosage Form New Animal Drugs
TRIBUTYRIN	184.1903	TRIBUTYRIN Listing Of Specific Substances Affirmed As GRAS; Direct Food Substances Affirmed As Generally Recognized As Safe
TRICAINE METHANESULFONATE	529.2503	TRICAINE METHANESULFONATE Certain Other Dosage Form New Animal Drugs
TRICALCIUM PHOSPHATE	872.3930	TRICALCIUM PHOSPHATE GRANULES FOR DENTAL BONE REPAIR Prosthetic Devices; Dental Devices
TRICALCIUM SILICATE	182.2906	TRICALCIUM SILICATE Anticaking Agents; Substances Generally Recognized As Safe
TRICALCIUM SILICATE	582.2906	TRICALCIUM SILICATE Anticaking Agents; Substances Generally Recognized As Safe
TRICHINELLA SPIRALIS	866.3850	TRICHINELLA SPIRALIS SEROLOGICAL REAGENTS Serological Reagents; Immunology And Microbiology Devices
TRICHLORFON	520.1326	MEBENDAZOLE AND TRICHLORFON ORAL DOSAGE FORMS Oral Dosage Form New Animal Drugs
TRICHLORFON	520.1631	OXFENDAZOLE AND TRICHLORFON PASTE Oral Dosage Form New Animal Drugs
TRICHLORFON	520.2380e	THIABENDAZOLE WITH TRICHLORFON Oral Dosage Form New Animal Drugs
TRICHLORFON	520.2520	TRICHLORFON ORAL DOSAGE FORMS Oral Dosage Form New Animal Drugs
TRICHLOROETHYLENE	173.290	TRICHLOROETHYLENE Solvents, Lubricants, Release Agents And Related Substances; Secondary Direct Food Additives Permitted In Food For Human Consumption
TRICHODERMA	184.1250	CELLULASE ENZYME PREPARATION DERIVED FROM TRICHODERMA LONGIBRANCHIATUM Listing Of Specific Substances Affirmed As GRAS; Direct Food Substances Affirmed As Generally Recognized As Safe
TRICYCLIC ANTIDEPRESSANT	862.3910	TRICYCLIC ANTIDEPRESSANT DRUGS TEST SYSTEM Clinical Toxicology Test Systems; Clinical Chemistry And Clinical Toxicology Devices
TRIETHYL CITRATE	184.1911	TRIETHYL CITRATE Listing Of Specific Substances Affirmed As GRAS; Direct Food Substances Affirmed As Generally Recognized As Safe
TRIFLUOROMETHANE	173.395	TRIFLUOROMETHANE SULFONIC ACID Specific Usage Additives; Secondary Direct Food Additives Permitted In Food For Human Consumption
TRIFLUPROMAZINE	520.2582	TRIFLUPROMAZINE HYDROCHLORIDE TABLETS Oral Dosage Form New Animal Drugs
TRIFLUPROMAZINE	522.2582	TRIFLUPROMAZINE HYDROCHLORIDE INJECTION Implantation Or Injectable Dosage Form New Animal Drugs
TRIGLYCERIDE	862.1705	TRIGLYCERIDE TEST SYSTEM Clinical Chemistry Test Systems; Clinical Chemistry And Clinical Toxicology Devices
TRIIODOTHYRONINE	862.1710	TOTAL TRIIODOTHYRONINE TEST SYSTEM Clinical Chemistry Test Systems; Clinical Chemistry And Clinical Toxicology Devices
TRIIODOTHYRONINE	862.1715	TRIIODOTHYRONINE UPTAKE TEST SYSTEM Clinical Chemistry Test Systems; Clinical Chemistry And Clinical Toxicology Devices
TRILAMINATE CYLINDER/SHEET	520.1450c	MORANTEL TARTRATE SUSTAINED-RELEASE TRILAMINATE CYLINDER/SHEET Oral Dosage Form New Animal Drugs
TRIMEPRAZINE TARTRATE	520.2604	TRIMEPRAZINE TARTRATE AND PREDNISOLONE TABLETS Oral Dosage Form New Animal Drugs
TRIMEPRAZINE TARTRATE	520.2605	TRIMEPRAZINE TARTRATE AND PREDNISOLONE CAPSULES Oral Dosage Form New Animal Drugs
TRIMETHOPRIM	520.2610	TRIMETHOPRIM AND SULFADIAZINE TABLETS Oral Dosage Form New Animal Drugs
TRIMETHOPRIM	520.2611	TRIMETHOPRIM AND SULFADIAZINE ORAL PASTE Oral Dosage Form New Animal Drugs
TRIMETHOPRIM	520.2612	TRIMETHOPRIM AND SULFADIAZINE ORAL SUSPENSION Oral Dosage Form New Animal Drugs
TRIMETHOPRIM	520.2613	TRIMETHOPRIM AND SULFADIAZINE POWDER Oral Dosage Form New Animal Drugs
TRIMETHOPRIM	522.2610	TRIMETHOPRIM AND SULFADIAZINE STERILE SUSPENSION Implantation Or Injectable Dosage Form New Animal Drugs
TRIOSE PHOSPHATE	862.1720	TRIOSE PHOSPHATE ISOMERASE TEST SYSTEM Clinical Chemistry Test Systems; Clinical Chemistry And Clinical Toxicology Devices
TRIPELENNAMINE	522.2615	TRIPELENNAMINE HYDROCHLORIDE INJECTION Implantation Or Injectable Dosage Form New Animal Drugs

TRIPELENNAMINE	556.741	TRIPELENNAMINE Tolerances For Residues Of New Animal Drugs In Food
TRIPHOSPHATE	864.7040	ADENOSINE TRIPHOSPHATE RELEASE ASSAY Hematology Kits And Packages; Hematology And Pathology Devices
TRIPOLYPHOSPHATE	182.1810	SODIUM TRIPOLYPHOSPHATE Multiple Purpose GRAS Food Substances; Substances Generally Recognized As Safe
TRIPOLYPHOSPHATE	182.6810	SODIUM TRIPOLYPHOSPHATE Sequestrants; Substances Generally Recognized As Safe
TRIPOLYPHOSPHATE	582.1810	SODIUM TRIPOLYPHOSPHATE General Purpose Food Additives; Substances Generally Recognized As Safe
TRIPOLYPHOSPHATE	582.6810	SODIUM TRIPOLYPHOSPHATE Sequestrants; Substances Generally Recognized As Safe
TRISTEARATE	172.811	GLYCERYL TRISTEARATE Multipurpose Additives; Food Additives Permitted For Direct Addition To Food For Human Consumption
TROCAR	870.1390	TROCAR Cardiovascular Diagnostic Devices; Cardiovascular Devices
TROMETHAMINE	522.690	DINOPROST TROMETHAMINE STERILE SOLUTION Implantation Or Injectable Dosage Form New Animal Drugs
TRUE COPIES	5.22	CERTIFICATION OF TRUE COPIES AND USE OF DEPARTMENT SEAL General Redelegations Of Authority; Delegations Of Authority And Organization
TRUNCAL ORTHOSIS	890.3490	TRUNCAL ORTHOSIS Physical Medicine Prosthetic Devices; Physical Medicine Devices
TRUNION-BEARING	888.3380	HIP JOINT FEMORAL (HEMI-HIP) TRUNION-BEARING METAL/POLYACETAL CEMENTED PROSTHESIS Prosthetic Devices; Orthopedic Devices
TRYPANOSOMA SPP.	866.3870	TRYPANOSOMA SPP. SEROLOGICAL REAGENTS Serological Reagents; Immunology And Microbiology Devices
TRYPSIN	184.1914	TRYPSIN Listing Of Specific Substances Affirmed As GRAS; Direct Food Substances Affirmed As Generally Recognized As Safe
TRYPSIN	524.2620	LIQUID CRYSTALLINE TRYPSIN, PERU BALSAM, CASTOR OIL Ophthalmic And Topical Dosage Form New Animal Drugs
TRYPSIN INHIBITOR	866.5890	INTER-ALPHA TRYPSIN INHIBITOR IMMUNOLOGICAL TEST SYSTEM Immunological Test Systems; Immunology And Microbiology Devices
TRYPTOPHANE	582.5915	TRYPTOPHANE Nutrients And/or Dietary Supplements; Substances Generally Recognized As Safe
TUBAL OCCLUSION	884.5380	CONTRACEPTIVE TUBAL OCCLUSION DEVICE (TOD) AND INTRODUCER Obstetrical And Gynecological Therapeutic Devices; Obstetrical And Gynecological Devices
TUBE	101.79	HEALTH CLAIMS: FOLATE AND NEURAL TUBE DEFECTS Food Labeling
TUBE	864.6150	CAPILLARY BLOOD COLLECTION TUBE Manual Hematology Devices; Hematology And Pathology Devices
TUBE	868.2320	UNCOMPENSATED THORPE TUBE FLOWMETER Monitoring Devices; Anesthesiology Devices
TUBE	868.2340	COMPENSATED THORPE TUBE FLOWMETER Monitoring Devices; Anesthesiology Devices
TUBE	868.5280	BREATHING TUBE SUPPORT Therapeutic Devices; Anesthesiology Devices
TUBE	868.5720	BRONCHIAL TUBE Therapeutic Devices; Anesthesiology Devices
TUBE	868.5730	TRACHEAL TUBE Therapeutic Devices; Anesthesiology Devices
TUBE	868.5740	TRACHEAL/BRONCHIAL DIFFERENTIAL VENTILATION TUBE Therapeutic Devices; Anesthesiology Devices
TUBE	868.5750	INFLATABLE TRACHEAL TUBE CUFF Therapeutic Devices; Anesthesiology Devices
TUBE	868.5770	TRACHEAL TUBE FIXATION DEVICE Therapeutic Devices; Anesthesiology Devices
TUBE	868.5780	TUBE INTRODUCTION FORCEPS Therapeutic Devices; Anesthesiology Devices
TUBE	868.5790	TRACHEAL TUBE STYLET Therapeutic Devices; Anesthesiology Devices
TUBE	868.5800	TRACHEOSTOMY TUBE AND TUBE CUFF Therapeutic Devices; Anesthesiology Devices
TUBE	870.2450	MEDICAL CATHODE-RAY TUBE DISPLAY Cardiovascular Monitoring Devices; Cardiovascular Devices
TUBE	872.6570	IMPRESSION TUBE Miscellaneous Devices; Dental Devices
TUBE	874.1925	TOYNBEE DIAGNOSTIC TUBE Diagnostic Devices; Ear, Nose, And Throat Devices

TUBE	874.3850	ENDOLYMPHATIC SHUNT TUBE WITH VALVE Prosthetic Devices; Ear, Nose And Throat Devices
TUBE	874.3880	TYMPANOSTOMY TUBE Prosthetic Devices; Ear, Nose And Throat Devices
TUBE	874.3930	TYMPANOSTOMY TUBE WITH SEMI-PERMEABLE MEMBRANE Prosthetic Devices; Ear, Nose And Throat Devices
TUBE	876.5980	GASTROINTESTINAL TUBE AND ACCESSORIES Therapeutic Devices; Gastroenterology-urology Devices
TUBE	884.3650	FALLOPIAN TUBE PROSTHESIS Obstetrical And Gynecological Prosthetic Devices; Obstetrical And Gynecological Devices
TUBE	888.4230	CEMENT VENTILATION TUBE Surgical Devices; Orthopedic Devices
TUBE	890.5650	POWERED INFLATABLE TUBE MASSAGER Physical Medicine Therapeutic Devices; Physical Medicine Devices
TUBE, X-RAY	892.1760	DIAGNOSTIC X-RAY TUBE HOUSING ASSEMBLY Diagnostic Devices; Radiology Devices
TUBE, X-RAY	892.1770	DIAGNOSTIC X-RAY TUBE MOUNT Diagnostic Devices; Radiology Devices
TUBE, X-RAY	892.5930	THERAPEUTIC X-RAY TUBE HOUSING ASSEMBLY Therapeutic Devices; Radiology Devices
TUBES	1020.20	COLD-CATHODE GAS DISCHARGE TUBES Performance Standards For Ionizing Radiation Emitting Products
TUBERCULIN	1210.13	TUBERCULIN TEST Inspection And Testing; Regulations Under The Federal Import Milk Act
TUBERCULOSIS	866.3370	MYCOBACTERIUM TUBERCULOSIS IMMUNOFLUORESCENT REAGENTS Serological Reagents; Immunology And Microbiology Devices
TUBING	868.5860	PRESSURE TUBING AND ACCESSORIES Therapeutic Devices; Anesthesiology Devices
TUBING	868.5975	VENTILATOR TUBING Therapeutic Devices; Anesthesiology Devices
TUBING	870.4210	CARDIOPULMONARY BYPASS VASCULAR CATHETER, CANNULA, OR TUBING Cardiovascular Surgical Devices; Cardiovascular Devices
TUBING	870.4390	CARDIOPULMONARY BYPASS PUMP TUBING Cardiovascular Surgical Devices; Cardiovascular Devices
TUMOR ASSOCIATED	866.6010	TUMOR ASSOCIATED ANTIGEN IMMUNOLOGICAL TEST SYSTEM Tumor Associated Antigen Immunological Test Systems; Immunology And Microbiology Devices
TUNA	161.190	CANNED TUNA Requirements For Specific Standardized Fish And Shellfish; Fish And Shellfish
TUNING FORK	882.1525	TUNING FORK Neurological Diagnostic Devices; Neurological Devices
TURBOT	102.57	GREENLAND TURBOT Requirements For Specific Nonstandardized Foods; Common Or Usual Name For Nonstandardized Foods
TURMERIC	73.600	TURMERIC Foods; Listing Of Color Additives Exempt From Certification
TURMERIC OLEORESIN	73.615	TURMERIC OLEORESIN Foods; Listing Of Color Additives Exempt From Certification
TURTLES	1240.62	TURTLES INTRASTATE AND INTERSTATE REQUIREMENTS Specific Administrative Decisions Regarding Interstate Shipments; Control Of Communicable Diseases
TWEEZER-TYPE EPILATOR	878.5360	TWEEZER-TYPE EPILATOR Therapeutic Devices; General And Plastic Surgery Devices
TWO-POINT	882.1200	TWO-POINT DISCRIMINATOR Neurological Diagnostic Devices; Neurological Devices
TYLOSIN	520.2640	TYLOSIN Oral Dosage Form New Animal Drugs
TYLOSIN	522.2640	TYLOSIN INJECTABLE DOSAGE FORMS Implantation Or Injectable Dosage Form New Animal Drugs
TYLOSIN	556.740	TYLOSIN Specific Tolerances For Residues Of New Animal Drugs; Tolerances For Residues Of New Animal Drugs In Food
TYLOSIN	558.625	TYLOSIN Specific New Animal Drugs For Use In Animal Feeds; New Animal Drugs For Use In Animal Feeds
TYLOSIN	558.630	TYLOSIN AND SULFAMETHAZINE Specific New Animal Drugs For Use In Animal Feeds; New Animal Drugs For Use In Animal Feeds
TYMPANOSTOMY	874.3880	TYMPANOSTOMY TUBE Prosthetic Devices; Ear, Nose And Throat Devices
TYMPANOSTOMY	874.3930	TYMPANOSTOMY TUBE WITH SEMI-PERMEABLE MEMBRANE Prosthetic Devices; Ear, Nose And Throat Devices

TYPE	610.62	PROPER NAME; PACKAGE LABEL; LEGIBLE TYPE Labeling Standards; General Biological Products Standard
TYPE B FEEDS	558.5	NEW ANIMAL DRUG REQUIREMENTS FOR LIQUID TYPE B FEEDS General Provisions; New Animal Drugs For Use In Animal Feeds
TYPE C FEED	558.465	POLOXALENE FREE-CHOICE LIQUID TYPE C FEED Specific New Animal Drugs For Use In Animal Feeds; New Animal Drugs For Use In Animal Feeds
TYPING, STAPHYLOCOCAL	866.2050	STAPHYLOCOCCAL TYPING BACTERIOPHAGE Microbiology Devices; Immunology And Microbiology Devices
TYROSINE	582.5920	TYROSINE Nutrients And/or Dietary Supplements; Substances Generally Recognized As Safe

U

ULNAR	888.3810	WRIST JOINT ULNAR (HEMI-WRIST) POLYMER PROSTHESIS Prosthetic Devices; Orthopedic Devices
ULTRA-FILTRATION	177.2910	ULTRA-FILTRATION MEMBRANES Substances For Use Only As Components Of Articles Intended For Repeated Use; Indirect Food Additives: Polymers
ULTRAMARINE BLUE	73.50	ULTRAMARINE BLUE Foods; Listing Of Color Additives Exempt From Certification
ULTRAMARINES	73.2725	ULTRAMARINES Cosmetics; Listing Of Color Additives Exempt From Certification
ULTRASONIC CLEANER	880.6150	ULTRASONIC CLEANER FOR MEDICAL INSTRUMENTS General Hospital And Personal Use Miscellaneous Devices; General Hospital And Personal Use Devices
ULTRASONIC DIATHERMY	890.5300	ULTRASONIC DIATHERMY Physical Medicine Therapeutic Devices; Physical Medicine Devices
ULTRASONIC IMAGER	884.2225	OBSTETRIC-GYNECOLOGIC ULTRASONIC IMAGER Obstetrical And Gynecological Monitoring Devices; Obstetrical And Gynecological Devices
ULTRASONIC IMAGING	892.1550	ULTRASONIC PULSED DOPPLER IMAGING SYSTEM Diagnostic Devices; Radiology Devices
ULTRASONIC IMAGING	892.1560	ULTRASONIC PULSED ECHO IMAGING SYSTEM Diagnostic Devices; Radiology Devices
ULTRASONIC MONITOR	868.2025	ULTRASONIC AIR EMBOLISM MONITOR Monitoring Devices; Anesthesiology Devices
ULTRASONIC MONITOR	884.2660	FETAL ULTRASONIC MONITOR AND ACCESSORIES Obstetrical And Gynecological Monitoring Devices; Obstetrical And Gynecological Devices
ULTRASONIC MONITOR	892.1540	NONFETAL ULTRASONIC MONITOR Diagnostic Devices; Radiology Devices
ULTRASONIC SCALER	872.4850	ULTRASONIC SCALER Surgical Devices; Dental Devices
ULTRASONIC SCANNER	882.1925	ULTRASONIC SCANNER CALIBRATION TEST BLOCK Neurological Diagnostic Devices; Neurological Devices
ULTRASONIC THERAPY	1050.10	ULTRASONIC THERAPY PRODUCTS Performance Standards For Sonic, Infrasonic, And Ultrasonic Radiation-emitting Products
ULTRASONIC TRANSDUCER	870.2880	ULTRASONIC TRANSDUCER Cardiovascular Monitoring Devices; Cardiovascular Devices
ULTRASONIC TRANSDUCER	884.2960	OBSTETRIC ULTRASONIC TRANSDUCER AND ACCESSORIES Obstetrical And Gynecological Monitoring Devices; Obstetrical And Gynecological Devices
ULTRASONIC TRANSDUCER	892.1570	DIAGNOSTIC ULTRASONIC TRANSDUCER Diagnostic Devices; Radiology Devices
ULTRAVIOLET ACTIVATOR	872.6070	ULTRAVIOLET ACTIVATOR FOR POLYMERIZATION Miscellaneous Devices; Dental Devices
ULTRAVIOLET DETECTOR	872.6350	ULTRAVIOLET DETECTOR Miscellaneous Devices; Dental Devices
ULTRAVIOLET LAMP	878.4630	ULTRAVIOLET LAMP FOR DERMATOLOGIC DISORDERS Surgical Devices; General And Plastic Surgery Devices
ULTRAVIOLET LAMP	878.4635	ULTRAVIOLET LAMP FOR TANNING Surgical Devices; General And Plastic Surgery Devices
ULTRAVIOLET LAMPS	1040.20	SUNLAMP PRODUCTS AND ULTRAVIOLET LAMPS INTENDED FOR USE IN SUNLAMP PRODUCTS Performance Standards For Light-emitting Products
ULTRAVIOLET PURIFIER	880.6500	MEDICAL ULTRAVIOLET AIR PURIFIER General Hospital And Personal Use Miscellaneous Devices; General Hospital And Personal Use Devices
ULTRAVIOLET PURIFIER	880.6710	MEDICAL ULTRAVIOLET WATER PURIFIER General Hospital And Personal Use Miscellaneous Devices; General Hospital And Personal Use Devices
ULTRAVIOLET RADIATION	179.39	ULTRAVIOLET RADIATION FOR THE PROCESSING AND TREATMENT OF FOOD Radiation And Radiation Sources; Irradiation In The Production, Processing And Handling Of Food
UMBILICAL	880.5950	UMBILICAL OCCLUSION DEVICE General Hospital And Personal Use Therapeutic Devices; General Hospital And Personal Use Devices
UNACCEPTABLE	640.76	PRODUCTS STORED OR SHIPPED AT UNACCEPTABLE TEMPERATURES Source Plasma; Additional Standards For Human Blood And Blood Products
UNACCEPTED	1305.11	UNACCEPTED AND DEFECTIVE ORDER FORMS Order Forms

UNAPPROVED APPLICATION	314.60	AMENDMENTS TO AN UNAPPROVED APPLICATION Applications; Applications For FDA Approval To Market A New Drug
UNAPPROVED APPLICATION	314.65	WITHDRAWAL BY THE APPLICANT OF AN UNAPPROVED APPLICATION Applications; Applications For FDA Approval To Market A New Drug
UNAPPROVED APPLICATION	314.96	AMENDMENTS TO AN UNAPPROVED ABBREVIATED APPLICATION Abbreviated Applications; Applications For FDA Approval To Market A New Drug
UNAPPROVED DRUGS	5.801	EXPORT OF UNAPPROVED DRUGS Imports And Exports; Redelegations Of Authority; Delegations Of Authority And Organization
UNAVAILABILITY	12.78	UNAVAILABILITY OF PRESIDING OFFICER Presiding Officer; Formal Evidentiary Public Hearing
UNAVOIDABILITY	109.7	UNAVOIDABILITY General Provisions; Unavoidable Contaminants In Food For Human Consumption & Food-packaging Material
UNAVOIDABILITY	509.7	UNAVOIDABILITY General Provisions; Unavoidable Contaminants In Animal Food And Food-packaging Material
UNCERTAIN	20.45	SITUATIONS IN WHICH CONFIDENTIALITY IS UNCERTAIN Procedures And Fees; Public Information
UNCOMPENSATED	868.2320	UNCOMPENSATED THORPE TUBE FLOWMETER Monitoring Devices; Anesthesiology Devices
UNDECYLENATE	522.204	BOLDENONE UNDECYLENATE INJECTION Implantation Or Injectable Dosage Form New Animal Drugs
UNIFORM ACCESS	20.21	UNIFORM ACCESS TO RECORDS General Policy; Public Information
UNIT	868.5655	PORTABLE LIQUID OXYGEN UNIT Therapeutic Devices; Anesthesiology Devices
UNIT	870.4300	CARDIOPULMONARY BYPASS GAS CONTROL UNIT Cardiovascular Surgical Devices; Cardiovascular Devices
UNIT	876.4300	ENDOSCOPIC ELECTROSURGICAL UNIT AND ACCESSORIES Surgical Devices; Gastroenterology-urology Devices
UNIT	880.5700	NEONATAL PHOTOTHERAPY UNIT General Hospital And Personal Use Therapeutic Devices; General Hospital And Personal Use Devices
UNIT	890.5940	CHILLING UNIT Physical Medicine Therapeutic Devices; Physical Medicine Devices
UNIT	890.5950	POWERED HEATING UNIT Physical Medicine Therapeutic Devices; Physical Medicine Devices
UNIT, QUALITY	58.35	QUALITY ASSURANCE UNIT Organization And Personnel; Good Laboratory Practice For Nonclinical Laboratory Studies
UNIT, QUALITY	211.22	RESPONSIBILITIES OF QUALITY CONTROL UNIT Organization And Personnel; Current Good Manufacturing Practice For Finished Pharmaceuticals
UNITS, SAMPLE	203.38	SAMPLE LOT OR CONTROL NUMBERS; LABELING OF SAMPLE UNITS Prescription Drug Marketing
UNITED STATES	17.54	DEPOSIT IN THE TREASURY OF THE UNITED STATES Civil Money Penalties Hearings
UNITED STATES	101.44	IDENTIFICATION OF THE 20 MOST FREQUENTLY CONSUMED RAW FRUIT, VEGETABLES, AND FISH IN THE UNITED STATES Food Labeling
UNITED STATES	807.40	ESTABLISHMENT REGISTRATION AND DEVICE LISTING FOR FOREIGN ESTABLISHMENTS IMPORTING OR OFFERING FOR IMPORT DEVICES INTO THE UNITED STATES Registration Procedures For Foreign Device Establishments; Establishment Registration And Device Listing For Manufacturers And Initial Importers Of Devices
UNITED STATES	814.15	RESEARCH CONDUCTED OUTSIDE THE UNITED STATES General; Premarket Approval Of Medical Devices
UNITED STATES	1010.5	EXEMPTIONS FOR PRODUCTS INTENDED FOR UNITED STATES GOVERNMENT USE Performance Standards For Electronic Products: General
UNLICENSED ESTABLISHMENTS	606.171	REPORTING OF PRODUCT DEVIATIONS BY MANUFACTURERS, UNLICENSED BLOOD ESTABLISHMENTS, AND TRANSFUSION SERVICES Records And Reports; Current Good Manufacturing Practices For Blood And Blood Components
UNREASONABLE RISK	5.409	DETERMINATIONS THAT MEDICAL DEVICES PRESENT UNREASONABLE RISK OF SUBSTANTIAL HARM Medical Devices And Radiological Health; Redelegations Of Authority; Delegations Of Authority And Organization
UNTESTED COMPONENTS	211.82	RECEIPT AND STORAGE OF UNTESTED COMPONENTS, DRUG PRODUCT CONTAINERS, AND CLOSURES Control Of Components And Drug Product Containers And Closures; Current Good Manufacturing Practice For Finished Pharmaceuticals

UNTRUE STATEMENTS	514.15	UNTRUE STATEMENTS IN APPLICATIONS General Provisions; New Animal Drug Applications
UNUSED SUPPLY	312.59	DISPOSITION OF UNUSED SUPPLY OF INVESTIGATIONAL DRUG Investigational New Drug Application
UPDATING	207.30	UPDATING DRUG LISTING INFORMATION Procedures For Domestic Drug Establishments; Registration Of Drug Producers And Listing Of Drugs In Commercial Distribution
UPDATING	607.30	UPDATING BLOOD PRODUCT LISTING INFORMATION Procedures For Domestic Blood Product Establishments; Establishment Registratn+product Listing, Manufacturers Human Blood/b.products
UPDATING	807.30	UPDATING DEVICE LISTING INFORMATION Procedures For Device Establishments; Establishment Registration And Device Listing For Manufacturers And Initial Importers Of Devices
UPTAKE	862.1715	TRIIODOTHYRONINE UPTAKE TEST SYSTEM Clinical Chemistry Test Systems; Clinical Chemistry And Clinical Toxicology Devices
UPTAKE	868.1730	OXYGEN UPTAKE COMPUTER Diagnostic Devices; Anesthesiology Devices
UPTAKE	886.1670	OPHTHALMIC ISOTOPE UPTAKE PROBE Diagnostic Devices; Ophthalmic Devices
UPTAKE	892.1320	NUCLEAR UPTAKE PROBE Diagnostic Devices; Radiology Devices
UREA	176.320	SODIUM NITRATE-UREA COMPLEX Substances For Use Only As Components Of Paper And Paperboard; Indirect Food Additives: Paper And Paperboard Components
UREA	184.1923	UREA Listing Of Specific Substances Affirmed As GRAS; Direct Food Substances Affirmed As Generally Recognized As Safe
UREA-FORMALDEHYDE	177.1900	UREA-FORMALDEHYDE RESINS IN MOLDED ARTICLES Substances For Use As Basic Components Of Single And Repeated Use Food Contact Surfaces; Indirect Food Additives: Polymers
UREA NITROGEN	862.1770	UREA NITROGEN TEST SYSTEM Clinical Chemistry Test Systems; Clinical Chemistry And Clinical Toxicology Devices
UREASE	184.1924	UREASE ENZYME PREPARATION FROM LACTOBACILLUS FERMENTUM Listing Of Specific Substances Affirmed As GRAS; Direct Food Substances Affirmed As Generally Recognized As Safe
URETERAL	876.4020	FIBEROPTIC LIGHT URETERAL CATHETER Surgical Devices; Gastroenterology-urology Devices
URETERAL	876.4620	URETERAL STENT Surgical Devices; Gastroenterology-urology Devices
URETERAL	876.4680	URETERAL STONE DISLODGER Surgical Devices; Gastroenterology-urology Devices
URETERAL	876.5470	URETERAL DILATOR Therapeutic Devices; Gastroenterology-urology Devices
URETHRAL	876.4590	INTERLOCKING URETHRAL SOUND Surgical Devices; Gastroenterology-urology Devices
URETHRAL	876.5520	URETHRAL DILATOR Therapeutic Devices; Gastroenterology-urology Devices
URETHROTOME	876.4770	URETHROTOME Surgical Devices; Gastroenterology-urology Devices
URIC ACID	862.1775	URIC ACID TEST SYSTEM Clinical Chemistry Test Systems; Clinical Chemistry And Toxicology Devices
URIDYL TRANSFERASE	862.1315	GALACTOSE-1-PHOSPHATE URIDYL TRANSFERASE TEST SYSTEM Clinical Chemistry Test Systems; Clinical Chemistry And Clinical Toxicology Devices
URINALYSIS	862.2900	AUTOMATED URINALYSIS SYSTEM Clinical Laboratory Instruments; Clinical Chemistry And Clinical Toxicology Devices
URINARY BILIRUBIN	862.1115	URINARY BILIRUBIN AND ITS CONJUGATES (NONQUANTITATIVE) TEST SYSTEM Clinical Chemistry Test Systems; Clinical Chemistry And Clinical Toxicology Devices
URINARY CALCULI	862.1780	URINARY CALCULI (STONES) TEST SYSTEM Clinical Chemistry Test Systems; Clinical Chemistry And Clinical Toxicology Devices
URINARY CONTINENCE	876.5270	IMPLANTED ELECTRICAL URINARY CONTINENCE DEVICE Therapeutic Devices; Gastroenterology-urology Devices
URINARY CONTINENCE	876.5280	IMPLANTED MECHANICAL/HYDRAULIC URINARY CONTINENCE DEVICE Therapeutic Devices; Gastroenterology-urology Devices
URINARY HOMOCYSTINE	862.1377	URINARY HOMOCYSTINE (NONQUANTITATIVE) TEST SYSTEM Clinical Chemistry Test Systems; Clinical Chemistry And Clinical Toxicology Devices

URINARY pH	862.1550	URINARY pH (NONQUANTITATIVE) TEST SYSTEM Clinical Chemistry Test Systems; Clinical Chemistry And Clinical Toxicology Devices
URINARY PHENYLKETONES	862.1560	URINARY PHENYLKETONES TEST SYSTEM Clinical Chemistry Test Systems; Clinical Chemistry And Clinical Toxicology Devices
URINARY UROBILINOGEN	862.1785	URINARY UROBILINOGEN (NONQUANTITATIVE) TEST SYSTEM Clinical Chemistry Test Systems; Clinical Chemistry And Clinical Toxicology Devices
URINE	876.1800	URINE FLOW OR VOLUME MEASURING SYSTEM Diagnostic Devices; Gastroenterology-urology Devices
URINE	876.5250	URINE COLLECTOR AND ACCESSORIES Therapeutic Devices; Gastroenterology-urology Devices
UROBILINOGEN	862.1785	URINARY UROBILINOGEN (NONQUANTITATIVE) TEST SYSTEM Clinical Chemistry Test Systems; Clinical Chemistry And Clinical Toxicology Devices
URODYNAMICS	876.1620	URODYNAMICS MEASUREMENT SYSTEM Diagnostic Devices; Gastroenterology-urology Devices
UROLOGICAL	876.4890	UROLOGICAL TABLE AND ACCESSORIES Surgical Devices; Gastroenterology-urology Devices
UROLOGICAL	876.5090	SUPRAPUBIC UROLOGICAL CATHETER AND ACCESSORIES Therapeutic Devices; Gastroenterology-urology Devices
UROLOGICAL	876.5130	UROLOGICAL CATHETER AND ACCESSORIES Therapeutic Devices; Gastroenterology-urology Devices
UROLOGICAL	876.5160	UROLOGICAL CLAMP FOR MALES Therapeutic Devices; Gastroenterology-urology Devices
UROLOGY	876.1075	GASTROENTEROLOGY-UROLOGY BIOPSY INSTRUMENT Diagnostic Devices; Gastroenterology-urology Devices
UROLOGY	876.4370	GASTROENTEROLOGY-UROLOGY EVACUATOR Surgical Devices; Gastroenterology-urology Devices
UROLOGY	876.4530	GASTROENTEROLOGY-UROLOGY FIBEROPTIC RETRACTOR Surgical Devices; Gastroenterology-urology Devices
UROLOGY	876.4730	MANUAL GASTROENTEROLOGY-UROLOGY SURGICAL INSTRUMENT AND ACCESSORIES Surgical Devices; Gastroenterology-urology Devices
UROPORPHYRIN	862.1790	UROPORPHYRIN TEST SYSTEM Clinical Chemistry Test Systems; Clinical Chemistry And Clinical Toxicology Devices
U.S. GOVERNMENT	1002.51	EXEMPTIONS FOR MANUFACTURERS OF PRODUCTS INTENDED FOR THE U.S. GOVERNMENT Exemptions From Records And Reports Requirements; Records and Reports
U.S. STANDARD	640.55	U.S. STANDARD PREPARATION Cryoprecipitate; Additional Standards For Human Blood And Blood Products
USAGE CRITERIA	211.122	MATERIALS EXAMINATION AND USAGE CRITERIA Packaging And Labeling Control; Current Good Manufacturing Practice For Finished Pharmaceuticals
USE, CONDITIONS	369.8	WARNING STATEMENTS IN RELATION TO CONDITIONS FOR USE Definitions And Interpretations; Interpretative Statements Re: Warnings On Drugs And Devices For OTC Sale
USE, DIRECTIONS	107.20	DIRECTIONS FOR USE Labeling; Infant Formula
USE, DIRECTIONS	201.3	DRUGS; ADEQUATE DIRECTIONS FOR USE Labeling
USE, DIRECTIONS	250.12	STRAMONIUM PREPARATIONS LABELED WITH DIRECTIONS FOR USE IN SELF-MEDICATION REGARDED AS MISBRANDED Drugs Regarded As Misbranded; Specific Requirements For Specific Human Drugs
USE, DIRECTIONS	801.5	MEDICAL DEVICES; ADEQUATE DIRECTIONS FOR USE General Labeling Provisions; Labeling
USE, EMERGENCY	312.36	EMERGENCY USE OF AN INVESTIGATIONAL NEW DRUG Investigational New Drug Application
USE, EXTRALABEL	5.500	ISSUANCE OF FEDERAL REGISTER DOCUMENTS PERTAINING TO THE DETERMINATION OF SAFE LEVELS, NOTICE OF NEED FOR DEVELOPMENT OF AN ANALYTICAL METHOD, AND PROHIBITION OF CERTAIN EXTRALABEL DRUG USE Animal Drugs; Redelegations Of Authority; Delegations Of Authority And Organization
USE, EXTRALABEL	530.10	PROVISION PERMITTING EXTRALABEL USE OF ANIMAL DRUGS Extralabel Drug Use In Animals
USE, EXTRALABEL	530.13	EXTRALABEL USE FROM COMPOUNDING OF APPROVED NEW ANIMAL AND HUMAN DRUGS Extralabel Drug Use In Animals
USE, EXTRALABEL	530.20	CONDITIONS FOR PERMITTED EXTRALABEL ANIMAL AND HUMAN DRUG USE IN FOOD-PRODUCTS Extralabel Drug Use In Animals

USE, EXTRALABEL	530.30	EXTRALABEL DRUG USE IN NONFOOD ANIMALS Extralabel Drug Use In Animals
USE, EXTRALABEL	530.41	DRUGS PROHIBITED FOR EXTRALABEL USE IN ANIMALS Extralabel Drug Use In Animals
USE, HUMAN	110.110	NATURAL OR UNAVOIDABLE DEFECTS IN FOOD FOR HUMAN USE THAT PRESENT NO HEALTH HAZARD Defect Action Levels; Current Good Manufacturing Practice In Manufacturing, Packing, Or Holding Human Food
USE, HUMAN	201.20	DECLARATION OF PRESENCE OF FD&C YELLOW NO. 5 AND/OR FD&C YELLOW NO. 6 IN CERTAIN DRUGS FOR HUMAN USE General Labeling Provisions; Labeling
USE, HUMAN	201.21	DECLARATION OF PRESENCE OF PHENYLALANINE AS A COMPONENT OF ASPARTAME IN OVER-THE-COUNTER AND PRESCRIPTION DRUGS FOR HUMAN USE General Labeling Provisions; Labeling
USE, HUMAN	201.100	PRESCRIPTION DRUGS FOR HUMAN USE Exemptions From Adequate Directions For Use; Labeling
USE, HUMAN	201.305	ISOPROTERENOL INHALATION PREPARATIONS (PRESSURIZED AEROSOLS, NEBULIZERS, POWDERS) FOR HUMAN USE; WARNINGS Specific Labeling Requirements For Specific Drug Products; Labeling
USE, HUMAN	201.316	DRUGS WITH THYROID HORMONE ACTIVITY FOR HUMAN USE; REQUIRED WARNING Specific Labeling Requirements For Specific Drug Products; Labeling
USE, HUMAN	201.317	DIGITALIS AND RELATED CARDIOTONIC DRUGS FOR HUMAN USE IN ORAL DOSAGE FORMS; REQUIRED WARNING Specific Labeling Requirements For Specific Drug Products; Labeling
USE, HUMAN	250.100	AMYL NITRITE INHALANT AS A PRESCRIPTION DRUG FOR HUMAN USE New Drug Or Prescription Status Of Specific Drugs; Specific Requirements For Specific Human Drugs
USE, HUMAN	250.102	DRUG PREPARATIONS INTENDED FOR HUMAN USE CONTAINING CERTAIN "CORONARY VASODILATORS" New Drug Or Prescription Status Of Specific Drugs; Specific Requirements For Specific Human Drugs
USE, HUMAN	310.305	RECORDS AND REPORTS CONTAINING ADVERSE DRUG EXPERIENCES ON MARKETED PRESCRIPTION DRUGS FOR HUMAN USE WITHOUT APPROVED NEW DRUG APPLICATIONS Records And Reports; New Drugs
USE, HUMAN	310.530	TOPICALLY APPLIED HORMONE-CONTAINING DRUG PRODUCTS FOR OVER-THE-COUNTER HUMAN USE Requirements For Specific New Drugs Or Devices; New Drugs
USE, HUMAN	310.533	DRUG PRODUCTS CONTAINING ACTIVE INGREDIENTS OFFERED OTC FOR HUMAN USE AS AN ANTICHOLINERGIC IN COUGH-COLD DRUG PRODUCTS Requirements For Specific New Drugs Or Devices; New Drugs
USE, HUMAN	310.534	DRUG PRODUCTS CONTAINING ACTIVE INGREDIENTS OFFERED OTC FOR HUMAN USE AS ORAL WOUND HEALING AGENTS Requirements For Specific New Drugs Or Devices; New Drugs
USE, HUMAN	310.543	DRUG PRODUCTS CONTAINING ACTIVE INGREDIENTS OFFERED OTC FOR HUMAN USE IN EXOCRINE PANCREATIC INSUFFICIENCY Requirements For Specific New Drugs Or Devices; New Drugs
USE, INDICATIONS FOR	814.110	NEW INDICATIONS FOR USE Humanitarian Use Devices; Premarket Approval Of Medical Devices
USE, INTERNAL	357.810	ACTIVE INGREDIENTS FOR DEODORANT DRUG PRODUCTS FOR INTERNAL USE Deodorant Drug Products For Internal Use; Miscellaneous Internal Drug Products For Over-the-counter Human Use
USE, INTERNAL	357.850	LABELING OF DEODORANT DRUG PRODUCTS FOR INTERNAL USE Deodorant Drug Products For Internal Use; Miscellaneous Internal Drug Products For Over-the-counter Human Use
USE, INVESTIGATIONAL	5.101	TERMINATION OF EXEMPTIONS FOR NEW DRUGS FOR INVESTIGATIONAL USE IN HUMAN BEINGS Human Drugs; Redelegations Of Authority; Delegations Of Authority And Organization
USE, INVESTIGATIONAL	5.505	TERMINATION OF EXEMPTIONS FOR NEW DRUGS FOR INVESTIGATIONAL USE IN ANIMALS Animal Drugs; Redelegations Of Authority; Delegations Of Authority And Organization
USE, INVESTIGATIONAL	312.160	DRUGS FOR INVESTIGATIONAL USE IN LABORATORY RESEARCH ANIMALS OR IN VITRO TESTS Drugs For Investigational Use In Laboratory Research Animals Or In Vitro Tests; Investigational New Drug Application
USE LOG	211.182	EQUIPMENT CLEANING AND USE LOG Records And Reports; Current Good Manufacturing Practice For Finished Pharmaceuticals

USE, PEDIATRIC	314.55	PEDIATRIC USE INFORMATION Applications For FDA Approval To Market A New Drug
USE OF REFERENCE PANELS	610.44	USE OF REFERENCE PANELS BY MANUFACTURERS OF TEST KITS Testing Requirements For Communicable Disease Agents; General Biological Products Standards
USE, REPEATED	177.2600	RUBBER ARTICLES INTENDED FOR REPEATED USE Substances For Use Only As Components Of Articles Intended For Repeated Use; Indirect Food Additives: Polymers
USE, RESEARCH	201.125	DRUGS FOR USE IN TEACHING, LAW ENFORCEMENT, RESEARCH, AND ANALYSIS Exemptions From Adequate Directions For Use; Labeling
USE, RESEARCH	201.129	DRUGS; EXEMPTION FOR RADIOACTIVE DRUGS FOR RESEARCH USE Exemptions From Adequate Directions For Use; Labeling
USE, RESEARCH	801.125	MEDICAL DEVICES FOR USE IN TEACHING, LAW ENFORCEMENT, RESEARCH, AND ANALYSIS Exemptions From Adequate Directions For Use; Labeling
USE, RESTRICTIONS ON	610.42	RESTRICTIONS ON USE FOR FURTHER MANUFACTURE OF MEDICAL DEVICES Testing Requirements For Communicable Disease Agents; General Biological Products Standards
USE, SAFE	314.520	APPROVAL WITH RESTRICTIONS TO ASSURE SAFE USE Accelerated Approval Of New Drugs; Applications For FDA Approval To Market A New Drug
USE, SAFE	601.42	APPROVAL WITH RESTRICTIONS TO ASSURE SAFE USE Accelerated Approval Of Biological Products; Licensing
USE OF TERMS	500.52	USE OF TERMS SUCH AS "TONIC," "TONE," "TONER," OR "CONDITIONER" IN THE LABELING OF PREPARATIONS INTENDED FOR USE IN OR ON ANIMALS Animal Drug Labeling Requirements; General
USE, TREATMENT	312.34	TREATMENT USE OF AN INVESTIGATIONAL NEW DRUG Investigational New Drug Application
USE, TREATMENT	312.35	SUBMISSIONS FOR TREATMENT USE Investigational New Drug Application
USE, TREATMENT	316.40	TREATMENT USE OF A DESIGNATED ORPHAN DRUG Orphan Drugs
USE, TREATMENT	812.36	TREATMENT USE OF AN INVESTIGATIONAL DEVICE Investigational Device Exemptions
USER FACILITIES	803.10	GENERAL DESCRIPTION OF REPORTS REQUIRED FROM USER FACILITIES, IMPORTERS, AND MANUFACTURERS Medical Device Reporting
USER FACILITIES	803.30	INDIVIDUAL ADVERSE EVENT REPORTS; USER FACILITIES Medical Device Reporting
USER LABELING	801.430	USER LABELING FOR MENSTRUAL TAMPONS Special Requirements For Specific Devices; Labeling
USER LABELING	801.435	USER LABELING FOR LATEX CONDOMS Special Requirements For Specific Devices; Labeling
USER LABELING	801.437	USER LABELING FOR DEVICES THAT CONTAIN NATURAL RUBBER Special Requirements For Specific Devices; Labeling
USER NOTIFICATION	1002.3	NOTIFICATION TO USER OF PERFORMANCE AND TECHNICAL DATA General Provisions; Records and Reports
USES, EXTRALABEL	530.25	ORDERS PROHIBITING EXTRALABEL USES FOR DRUGS IN FOOD-PRODUCING ANIMALS Extralabel Drug Use In Animals
USES, INTENDED	201.128	MEANING OF "INTENDED USES" Exemptions From Adequate Directions For Use; Labeling
USES, INTENDED	801.4	MEANING OF "INTENDED USES" General Labeling Provisions; Labeling
UTENSILS	110.40	EQUIPMENT AND UTENSILS Equipment; Current Good Manufacturing Practice In Manufacturing, Packing, Or Holding Human Food
UTENSILS	1250.44	DRINKING UTENSILS AND TOILET ARTICLES Equipment And Operation Of Land And Air Conveyances; Interstate Conveyance Sanitation
UTERINE MONITOR	884.2720	EXTERNAL UTERINE CONTRACTION MONITOR AND ACCESSORIES Obstetrical And Gynecological Monitoring Devices; Obstetrical And Gynecological Devices
UTERINE MONITOR	884.2730	HOME UTERINE ACTIVITY MONITOR Obstetrical And Gynecological Monitoring Devices; Obstetrical And Gynecological Devices
UTEROTUBAL	884.1300	UTEROTUBAL CARBON DIOXIDE INSUFFLATOR AND ACCESSORIES Obstetrical And Gynecological Diagnostic Devices; Obstetrical And Gynecological Devices

V

VACCINE, INFLUENZA	610.11	INACTIVATED INFLUENZA VACCINE, GENERAL SAFETY TEST General Provisions; General Biological Products Standards
VACCINES	5.200	FUNCTIONS PERTAINING TO SAFER VACCINES Biologics; Redelegations Of Authority; Delegations Of Authority And Organization
VACUUM ABORTION	884.5070	VACUUM ABORTION SYSTEM Obstetrical And Gynecological Therapeutic Devices; Obstetrical And Gynecological Devices
VACUUM-ASSISTED	864.9125	VACUUM-ASSISTED BLOOD COLLECTION SYSTEM Products Used In Establishments That Manufacture Blood And Blood Products; Hematology And Pathology Devices
VACUUM EXTRACTOR	884.4340	FETAL VACUUM EXTRACTOR Obstetrical And Gynecological Surgical Devices; Obstetrical And Gynecological Devices
VACUUM-POWERED	880.6740	VACUUM-POWERED BODY FLUID SUCTION APPARATUS General Hospital And Personal Use Miscellaneous Devices; General Hospital And Personal Use Devices
VAGINAL DOUCHE	884.5900	THERAPEUTIC VAGINAL DOUCHE APPARATUS Obstetrical And Gynecological Therapeutic Devices; Obstetrical And Gynecological Devices
VAGINAL INSUFFLATOR	884.5920	VAGINAL INSUFFLATOR Obstetrical And Gynecological Therapeutic Devices; Obstetrical And Gynecological Devices
VAGINAL PESSARY	884.3575	VAGINAL PESSARY Obstetrical And Gynecological Prosthetic Devices; Obstetrical And Gynecological Devices
VAGINAL SPONGE	529.1003	FLUROGESTONE ACETATE-IMPREGNATED VAGINAL SPONGE Certain Other Dosage Form New Animal Drugs
VAGINAL STENT	884.3900	VAGINAL STENT Obstetrical And Gynecological Prosthetic Devices; Obstetrical And Gynecological Devices
VAGINAL STIMULATOR	884.5940	POWERED VAGINAL MUSCLE STIMULATOR FOR THERAPEUTIC USE Obstetrical And Gynecological Therapeutic Devices; Obstetrical And Gynecological Devices
VALIDATION	120.11	VERIFICATION AND VALIDATION Hazard Analysis And Critical Control Point (HACCP) Systems
VALIDATION	820.75	PROCESS VALIDATION Production And Process Controls; Quality System Regulation
VALINE	582.5925	VALINE Nutrients And/or Dietary Supplements; Substances Generally Recognized As Safe
VALVE	868.1965	SWITCHING VALVE (PLOSS) Diagnostic Devices; Anesthesiology Devices
VALVE	868.5870	NONREBREATHING VALVE Therapeutic Devices; Anesthesiology Devices
VALVE	870.3925	REPLACEMENT HEART VALVE Cardiovascular Prosthetic Devices; Cardiovascular Devices
VALVE	870.3935	PROSTHETIC HEART VALVE HOLDER Cardiovascular Prosthetic Devices; Cardiovascular Devices
VALVE	870.3945	PROSTHETIC HEART VALVE SIZER Cardiovascular Prosthetic Devices; Cardiovascular Devices
VALVE	874.3850	ENDOLYMPHATIC SHUNT TUBE WITH VALVE Prosthetic Devices; Ear, Nose And Throat Devices
VANCOMYCIN	862.3950	VANCOMYCIN TEST SYSTEM Clinical Toxicology Test Systems; Clinical Chemistry And Clinical Toxicology Devices
VANILLA	169.175	VANILLA EXTRACT Requirements For Specific Standardized Food Dressings And Flavorings; Food Dressings And Flavorings
VANILLA	169.176	CONCENTRATED VANILLA EXTRACT Requirements For Specific Standardized Food Dressings And Flavorings; Food Dressings And Flavorings
VANILLA	169.177	VANILLA FLAVORING Requirements For Specific Standardized Food Dressings And Flavorings; Food Dressings And Flavorings
VANILLA	169.178	CONCENTRATED VANILLA FLAVORING Requirements For Specific Standardized Food Dressings And Flavorings; Food Dressings And Flavorings
VANILLA	169.179	VANILLA POWDER Requirements For Specific Standardized Food Dressings And Flavorings; Food Dressings And Flavorings

VANILLA-VANILLIN	169.180	VANILLA-VANILLIN EXTRACT Requirements For Specific Standardized Food Dressings And Flavorings; Food Dressings And Flavorings
VANILLA-VANILLIN	169.181	VANILLA-VANILLIN FLAVORING Requirements For Specific Standardized Food Dressings And Flavorings; Food Dressings And Flavorings
VANILLA-VANILLIN	169.182	VANILLA-VANILLIN POWDER Requirements For Specific Standardized Food Dressings And Flavorings; Food Dressings And Flavorings
VANILMANDELIC ACID	862.1795	VANILMANDELIC ACID TEST SYSTEM Clinical Chemistry Test Systems; Clinical Chemistry And Clinical Toxicology Devices
VAPOR EVACUATOR	888.4220	CEMENT MONOMER VAPOR EVACUATOR Surgical Devices; Orthopedic Devices
VAPOR, MERCURY	1040.30	HIGH-INTENSITY MERCURY VAPOR DISCHARGE LAMPS Performance Standards For Light-emitting Products
VAPOR, WATER	868.1975	WATER VAPOR ANALYZER Diagnostic Devices; Anesthesiology Devices
VAPORIZER	868.5880	ANESTHETIC VAPORIZER Therapeutic Devices; Anesthesiology Devices
VARIANCES	5.600	VARIANCES FROM PERFORMANCE STANDARDS FOR ELECTRONIC PRODUCTS Radiation Control; Redelegations Of Authority; Delegations Of Authority And Organization
VARIANCES	803.19	EXEMPTIONS OF VARIANCES, AND ALTERNATIVE REPORTING REQUIREMENTS Medical Device Reporting
VARIANCES	821.2	EXEMPTIONS AND VARIANCES Medical Device Tracking Requirements
VARIANCES	898.14	EXEMPTIONS AND VARIANCES Performance Standards For Electrode Lead Wires And Patient Cables
VARIANCES	1010.4	VARIANCES Performance Standards For Electronic Products: General
VARIATIONS	1.23	PROCEDURES FOR REQUESTING VARIATIONS AND EXEMPTIONS FROM REQUIRED LABEL STATEMENTS General Labeling Requirements; General Enforcement Regulations
VARICELLA-ZOSTER VIRUS	866.3900	VARICELLA-ZOSTER VIRUS SEROLOGICAL REAGENTS Serological Reagents; Immunology And Microbiology Devices
VARNISH	872.3260	CAVITY VARNISH Prosthetic Devices; Dental Devices
VASCULAR	870.3250	VASCULAR CLIP Cardiovascular Prosthetic Devices; Cardiovascular Devices
VASCULAR	870.3450	VASCULAR GRAFT PROSTHESIS Cardiovascular Prosthetic Devices; Cardiovascular Devices
VASCULAR	870.4210	CARDIOPULMONARY BYPASS VASCULAR CATHETER, CANNULA, OR TUBING Cardiovascular Surgical Devices; Cardiovascular Devices
VASCULAR	870.4450	VASCULAR CLAMP Cardiovascular Surgical Devices; Cardiovascular Devices
VASOCONSTRICTOR	346.12	VASOCONSTRICTOR ACTIVE INGREDIENTS Anorectal Drug Products For Over-the-counter Human Use
VASOCONSTRICTOR	349.75	LABELING OF OPHTHALMIC VASOCONSTRICTOR DRUG PRODUCTS Labeling; Ophthalmic Drug Products For Over-the-counter Human Use
VASOCONSTRICTORS	349.18	OPHTHALMIC VASOCONSTRICTORS Active Ingredients; Ophthalmic Drug Products For Over-the-counter Human Use
VASODILATORS	250.102	DRUG PREPARATIONS INTENDED FOR HUMAN USE CONTAINING CERTAIN "CORONARY VASODILATORS" New Drug Or Prescription Status Of Specific Drugs; Specific Requirements For Specific Human Drugs
VAT ORANGE 1	73.3112	C.I. VAT ORANGE 1 Medical Devices; Listing Of Color Additives Exempt From Certification
VECTORCARDIOGRAPH	870.2400	VECTORCARDIOGRAPH Cardiovascular Monitoring Devices; Cardiovascular Devices
VEGETABLE	139.125	VEGETABLE MACARONI PRODUCTS Requirements For Specific Standardized Macaroni And Noodle Products; Macaroni And Noodle Products
VEGETABLE	139.135	ENRICHED VEGETABLE MACARONI PRODUCTS Requirements For Specific Standardized Macaroni And Noodle Products; Macaroni And Noodle Products
VEGETABLE	139.160	VEGETABLE NOODLE PRODUCTS Requirements For Specific Standardized Macaroni And Noodle Products; Macaroni And Noodle Products
VEGETABLE	139.165	ENRICHED VEGETABLE NOODLE PRODUCTS Requirements For Specific Standardized Macaroni And Noodle Products; Macaroni And Noodle Products

VEGETABLE FAT	163.150	SWEET COCOA AND VEGETABLE FAT COATING Requirements For Specific Standardized Cacao Products; Cacao Products
VEGETABLE FAT	163.153	SWEET CHOCOLATE AND VEGETABLE FAT COATING Requirements For Specific Standardized Cacao Products; Cacao Products
VEGETABLE FAT	163.155	MILK CHOCOLATE AND VEGETABLE FAT COATING Requirements For Specific Standardized Cacao Products; Cacao Products
VEGETABLE JUICE	73.260	VEGETABLE JUICE Foods; Listing Of Color Additives Exempt From Certification
VEGETABLE JUICE	102.33	BEVERAGES THAT CONTAIN FRUIT OR VEGETABLE JUICE Requirements For Specific Nonstandardized Foods; Common Or Usual Name For Nonstandardized Foods
VEGETABLE OIL	180.30	BROMINATED VEGETABLE OIL Specific Requirements For Certain Food Additives; Food Additives Permitted In Food Or In Contact With Food On Interim Basis Pending Study
VEGETABLES	101.42	NUTRITION LABELING OF RAW FRUIT, VEGETABLES AND FISH Food Labeling
VEGETABLES	101.43	SUBSTANTIAL COMPLIANCE OF FOOD RETAILERS WITH THE GUIDELINES FOR THE VOLUNTARY NUTRITION LABELING OF RAW FRUIT, VEGETABLES, AND FISH Food Labeling
VEGETABLES	101.44	IDENTIFICATION OF THE 20 MOST FREQUENTLY CONSUMED RAW FRUIT, VEGETABLES, AND FISH IN THE UNITED STATES Food Labeling
VEGETABLES	101.45	GUIDELINES FOR THE VOLUNTARY NUTRITION LABELING OF RAW FRUIT, VEGETABLES, AND FISH Food Labeling
VEGETABLES	101.76	HEALTH CLAIMS: FIBER-CONTAINING GRAIN PRODUCTS, FRUITS AND VEGETABLES AND CANCER Food Labeling
VEGETABLES	107.77	HEALTH CLAIMS: FRUIT, VEGETABLES, AND GRAIN PRODUCTS THAT CONTAIN FIBER, PARTICULARLY SOLUBLE FIBER, AND RISK OF CORONARY HEART DISEASE Food Labeling
VEGETABLES	101.78	HEALTH CLAIMS: FRUITS AND VEGETABLES AND CANCER Food Labeling
VEGETABLES	133.125	COLD-PACK CHEESE FOOD WITH FRUITS, VEGETABLES, OR MEATS Requirements For Specific Standardized Cheese And Related Products; Cheeses And Related Cheese Products
VEGETABLES	133.168	PASTEURIZED BLENDED CHEESE WITH FRUITS, VEGETABLES, OR MEATS Requirements For Specific Standardized Cheese And Related Products; Cheeses And Related Cheese Products
VEGETABLES	133.170	PASTEURIZED PROCESS CHEESE WITH FRUITS, VEGETABLES, OR MEATS Requirements For Specific Standardized Cheese And Related Products; Cheeses And Related Cheese Products
VEGETABLES	133.174	PASTEURIZED PROCESS CHEESE FOOD WITH FRUITS, VEGETABLES, OR MEATS Requirements For Specific Standardized Cheese And Related Products; Cheeses And Related Cheese Products
VEGETABLES	133.176	PASTEURIZED CHEESE SPREAD WITH FRUITS, VEGETABLES OR MEATS Requirements For Specific Standardized Cheese And Related Products; Cheeses And Related Cheese Products
VEGETABLES	133.180	PASTEURIZED PROCESS CHEESE SPREAD WITH FRUITS, VEGETABLES, OR MEATS Requirements For Specific Standardized Cheese And Related Products; Cheeses And Related Cheese Products
VEGETABLES	155.200	CERTAIN OTHER CANNED VEGETABLES Requirements For Specific Standardized Canned Vegetables; Canned Vegetables
VEGETABLES	173.315	CHEMICALS USED IN WASHING OR TO ASSIST IN THE PEELING OF FRUITS AND VEGETABLES Specific Usage Additives; Secondary Direct Food Additives Permitted In Food For Human Consumption
VEIN	870.4885	EXTERNAL VEIN STRIPPER Cardiovascular Surgical Devices; Cardiovascular Devices
VEIN	880.6970	LIQUID CRYSTAL VEIN LOCATOR General Hospital And Personal Use Miscellaneous Devices; General Hospital And Personal Use Devices
VEIN	880.6980	VEIN STABILIZER General Hospital And Personal Use Miscellaneous Devices; General Hospital And Personal Use Devices
VELOCITY, CONDUCTION	882.1550	NERVE CONDUCTION VELOCITY MEASUREMENT DEVICE Neurological Diagnostic Devices; Neurological Devices
VENA CAVA	870.3260	VENA CAVA CLIP Cardiovascular Prosthetic Devices; Cardiovascular Devices

VENOM, RUSSELL VIPER	864.8950	RUSSELL VIPER VENOM REAGENT Hematology Reagents; Hematology And Pathology Devices
VENOUS	870.1140	VENOUS BLOOD PRESSURE MANOMETER Cardiovascular Diagnostic Devices; Cardiovascular Devices
VENOUS	876.5955	PERITONEO-VENOUS SHUNT Therapeutic Devices; Gastroenterology-urology Devices
VENTILATION	211.46	VENTILATION, AIR FILTRATION, AIR HEATING AND COOLING Buildings And Facilities; Current Good Manufacturing Practice For Finished Pharmaceuticals
VENTILATION	868.5740	TRACHEAL/BRONCHIAL DIFFERENTIAL VENTILATION TUBE Therapeutic Devices; Anesthesiology Devices
VENTILATION	868.5955	INTERMITTENT MANDATORY VENTILATION ATTACHMENT Therapeutic Devices; Anesthesiology Devices
VENTILATION	888.4230	CEMENT VENTILATION TUBE Surgical Devices; Orthopedic Devices
VENTILATOR	868.5895	CONTINUOUS VENTILATOR Therapeutic Devices; Anesthesiology Devices
VENTILATOR	868.5905	NONCONTINUOUS VENTILATOR (IPPB) Therapeutic Devices; Anesthesiology Devices
VENTILATOR	868.5915	MANUAL EMERGENCY VENTILATOR Therapeutic Devices; Anesthesiology Devices
VENTILATOR	868.5925	POWERED EMERGENCY VENTILATOR Therapeutic Devices; Anesthesiology Devices
VENTILATOR	868.5935	EXTERNAL NEGATIVE PRESSURE VENTILATOR Therapeutic Devices; Anesthesiology Devices
VENTILATOR	868.5975	VENTILATOR TUBING Therapeutic Devices; Anesthesiology Devices
VENTING	113.89	DEVIATIONS IN PROCESSING, VENTING, OR CONTROL OF CRITICAL FACTORS Production And Process Controls; Thermally Processed Low-acid Foods Packaged In Hermetically Sealed Containers
VENTRICULAR	870.3545	VENTRICULAR BYPASS (ASSIST) DEVICE Cardiovascular Prosthetic Devices; Cardiovascular Devices
VENTRICULAR	882.4060	VENTRICULAR CANNULA Neurological Surgical Devices; Neurological Devices
VENTRICULAR	882.4100	VENTRICULAR CATHETER Neurological Surgical Devices; Neurological Devices
VENTURI MASK	868.5600	VENTURI MASK Therapeutic Devices; Anesthesiology Devices
VERIFICATION	21.44	VERIFICATION OF IDENTITY Procedures For Notification Of And Access To Records In Privacy Act Record Systems; Protection Of Privacy
VERIFICATION	120.11	VERIFICATION AND VALIDATION Hazard Analysis And Critical Control Point (HACCP) Systems
VERIFICATION	120.25	PROCESS VERIFICATION FOR CERTAIN PROCESSORS Pathogen Reduction; Hazard Analysis And Critical Control Point (HACCP) Systems
VERIFICATION	123.8	VERIFICATION Fish And Fishery Products
VERIFICATION	316.21	VERIFICATION OF ORPHAN-DRUG STATUS Orphan Drugs
VERSION, SPANISH	290.6	SPANISH-LANGUAGE VERSION OF REQUIRED WARNING General Provisions; Controlled Drugs
VERSION, SPANISH	801.16	MEDICAL DEVICES; SPANISH-LANGUAGE VERSION OF CERTAIN REQUIRED STATEMENTS General Labeling Provisions; Labeling
VERXITE	573.1000	VERXITE Food Additive Listing; Food Additives Permitted In Feed And Drinking Water Of Animals
VESSEL	868.1575	GAS COLLECTION VESSEL Diagnostic Devices; Anesthesiology Devices
VESSEL	870.1310	VESSEL DILATOR FOR PERCUTANEOUS CATHETERIZATION Cardiovascular Diagnostic Devices; Cardiovascular Devices
VESSEL	870.2890	VESSEL OCCLUSION TRANSDUCER Cardiovascular Monitoring Devices; Cardiovascular Devices
VESSEL	870.4475	SURGICAL VESSEL DILATOR Cardiovascular Surgical Devices; Cardiovascular Devices
VESSELS, OCEAN	1301.25	REGISTRATION REGARDING OCEAN VESSELS, AIRCRAFT, AND OTHER ENTITIES Registration Of Manufacturers, Distributors, And Dispensers Of Controlled Substances
VETERINARY DRUGS	201.105	VETERINARY DRUGS Exemptions From Adequate Directions For Use; Labeling
VETERINARY FEED	558.6	VETERINARY FEED DIRECTIVE DRUGS New Animal Drugs For Use In Animal Feeds

VETERINARY IMPLANT	1308.25	EXCLUSION OF A VETERINARY ANABOLIC STEROID IMPLANT PRODUCT; APPLICATION Schedules Of Controlled Substances
VETERINARY IMPLANT	1308.26	EXCLUDED VETERINARY ANABOLIC STEROID IMPLANT PRODUCTS Schedules Of Controlled Substances
VETERINARY MEDICINE	510.112	ANTIBIOTICS USED IN VETERINARY MEDICINE AND FOR NONMEDICAL PURPOSES; REQUIRED DATA Specific Administrative Rulings And Decisions; New Animal Drugs
VETERINARY RECORDS	530.5	VETERINARY RECORDS Extralabel Drug Use In Animals
VIBRATOR	884.5960	GENITAL VIBRATOR FOR THERAPEUTIC USE Obstetrical And Gynecological Therapeutic Devices; Obstetrical And Gynecological Devices
VIBRATOR	890.5975	THERAPEUTIC VIBRATOR Physical Medicine Therapeutic Devices; Physical Medicine Devices
VIBRATORY	884.4270	VIBRATORY CERVICAL DILATORS Obstetrical And Gynecological Surgical Devices; Obstetrical And Gynecological Devices
VIBRIO CHOLERAE	866.3930	VIBRIO CHOLERAE SEROLOGICAL REAGENTS Serological Reagents; Immunology And Microbiology Devices
VIBROCARDIOGRAPH	870.2310	APEX CARDIOGRAPH (VIBROCARDIOGRAPH) Cardiovascular Monitoring Devices; Cardiovascular Devices
VIEW BOX	864.9185	BLOOD GROUPING VIEW BOX Products Used In Establishments That Manufacture Blood And Blood Products; Hematology And Pathology Devices
VIGILANCE, POSTMARKET	26.50	ALERT SYSTEM AND EXCHANGE OF POSTMARKET VIGILANCE REPORTS Specific Sector Provisions For Medical Devices; Mutual Recognition Of Pharmaceutical GMP Reports, Medical Device Quality System Audit Reports
VINYL ACETATE	175.350	VINYL ACETATE/CROTONIC ACID COPOLYMER Substances For Use As Components Of Coatings; Indirect Food Additives
VINYL ACETATE	177.1350	ETHYLENE-VINYL ACETATE COPOLYMERS Substances For Use As Basic Components Of Single And Repeated Use Food Contact Surfaces; Indirect Food Additives: Polymers
VINYL ACETATE	177.1360	ETHYLENE-VINYL ACETATE-VINYL ALCOHOL COPOLYMERS Substances For Use As Basic Components Of Single And Repeated Use Food Contact Surfaces; Indirect Food Additives: Polymers
VINYL ALCOHOL	73.3127	VINYL ALCOHOL/METHYL METHACRYLATE-DYE REACTION PRODUCTS Listing Of Color Additives Exempt From Certification
VINYL ALCOHOL	177.1360	ETHYLENE-VINYL ACETATE-VINYL ALCOHOL COPOLYMERS Substances For Use As Basic Components Of Single And Repeated Use Food Contact Surfaces; Indirect Food Additives: Polymers
VINYL CHLORIDE	177.1950	VINYL CHLORIDE-ETHYLENE COPOLYMERS Substances For Use As Basic Components Of Single And Repeated Use Food Contact Surfaces; Indirect Food Additives: Polymers
VINYL CHLORIDE	177.1960	VINYL CHLORIDE-HEXENE-1 COPOLYMERS Substances For Use As Basic Components Of Single And Repeated Use Food Contact Surfaces; Indirect Food Additives: Polymers
VINYL CHLORIDE	177.1970	VINYL CHLORIDE-LAURYL VINYL ETHER COPOLYMERS Substances For Use As Basic Components Of Single And Repeated Use Food Contact Surfaces; Indirect Food Additives: Polymers
VINYL CHLORIDE	177.1980	VINYL CHLORIDE-PROPYLENE COPOLYMERS Substances For Use As Basic Components Of Single And Repeated Use Food Contact Surfaces; Indirect Food Additives: Polymers
VINYL CHLORIDE	178.2650	ORGANOTIN STABILIZERS IN VINYL CHLORIDE PLASTICS Antioxidants And Stabilizers; Indirect Food Additives: Adjuvants, Production Aids, And Sanitizers
VINYL CHLORIDE	178.3790	POLYMER MODIFIERS IN SEMIRIGID AND RIGID VINYL CHLORIDE PLASTICS Certain Adjuvants And Production Aids; Indirect Food Additives: Adjuvants, Production Aids, And Sanitizers
VINYL CHLORIDE	700.14	USE OF VINYL CHLORIDE AS AN INGREDIENT, INCLUDING PROPELLANT OF COSMETIC AEROSOL PRODUCTS Requirements For Specific Cosmetic Products; General
VINYL FLUORIDE	175.270	POLY(VINYL FLUORIDE) RESINS Substances For Use As Components Of Coatings; Indirect Food Additives
VINYLIDENE CHLORIDE	175.360	VINYLIDENE CHLORIDE COPOLYMER COATINGS FOR NYLON FILM Substances For Use As Components Of Coatings; Indirect Food Additives
VINYLIDENE CHLORIDE	175.365	VINYLIDENE CHLORIDE COPOLYMER COATINGS FOR POLYCARBONATE FILM Substances For Use As Components Of Coatings; Indirect Food Additives
VINYLIDENE CHLORIDE	177.1990	VINYLIDENE CHLORIDE/METHYL ACRYLATE COPOLYMERS Substances For Use As Basic Components Of Single And Repeated Use Food Contact Surfaces; Indirect Food Additives: Polymers

VINYLIDENE CHLORIDE	177.2000	VINYLIDENE CHLORIDE/METHYL ACRYLATE/METHYL METHACRYLATE POLYMERS Substances For Use As Basic Components Of Single And Repeated Use Food Contact Surfaces; Indirect Food Additives: Polymers
VINYLPYRIDINE	573.870	POLY(2-VINYLPYRIDINE-CO-STYRENE) Food Additive Listing; Food Additives Permitted In Feed And Drinking Water Of Animals
VINYLTOLUENE	178.3610	α-METHYLSTYRENE-VINYLTOLUENE RESINS, HYDROGENATED Certain Adjuvants And Production Aids; Indirect Food Additives: Adjuvants, Production Aids, And Sanitizers
VIOLATION	7.84	OPPORTUNITY FOR PRESENTATION OF VIEWS BEFORE REPORT OF CRIMINAL VIOLATION Criminal Violations; Enforcement Policy
VIOLATION	7.85	CONDUCT OF A PRESENTATION OF VIEWS BEFORE REPORT OF CRIMINAL VIOLATION Criminal Violations; Enforcement Policy
VIOLATION	7.87	RECORDS RELATED TO OPPORTUNITIES FOR PRESENTATION OF VIEWS CONDUCTED BEFORE REPORT OF CRIMINAL VIOLATION Criminal Violations; Enforcement Policy
VIOLATION	108.12	MANUFACTURING, PROCESSING, OR PACKING WITHOUT A PERMIT, OR IN VIOLATION OF A PERMIT General Provisions; Emergency Permit Control
VIOLATION	1230.45	NO VIOLATION; RELEASE Imports; Regulations Under The Federal Caustic Poison Act
VIOLATION	1230.46	VIOLATION Imports; Regulations Under The Federal Caustic Poison Act
VIOLATION	1404.620	EFFECT OF VIOLATION Governmentwide Debarment And Suspension And Requirements For Drug-Free Workplace (Grants)
VIOLATIONS	5.33	ISSUANCE OF REPORTS OF MINOR VIOLATIONS General Redelegations Of Authority; Delegations Of Authority And Organization
VIOLATIONS	19.21	DUTY TO REPORT VIOLATIONS Reporting Of Violations; Standards Of Conduct And Conflicts Of Interest
VIOLATIONS	205.8	VIOLATIONS AND PENALTIES Guidelines For State Licensing Of Wholesale Prescription Drug Distributors
VIOLET NO. 2	74.1602	D&C VIOLET NO. 2 Drugs; Listing Of Color Additives Subject To Certification
VIOLET NO. 2	74.2602	D&C VIOLET NO. 2 Cosmetics; Listing Of Color Additives Subject To Certification
VIOLET NO. 2	74.2602a	Ext. D&C VIOLET NO. 2 Cosmetics; Listing Of Color Additives Subject To Certification
VIOLET NO. 2	74.3602	D&C VIOLET NO. 2 Medical Devices; Listing Of Color Additives Subject To Certification
VIOLET NO. 2	82.1602	D&C VIOLET NO. 2 Drugs And Cosmetics; Listing Of Certified Provisionally Listed Colors And Specifications
VIOLET, GENTIAN	500.29	GENTIAN VIOLET FOR USE IN ANIMAL FEEDS Specific Administrative Rulings And Decisions; General
VIOLET, GENTIAN	500.30	GENTIAN VIOLET FOR ANIMAL DRUG USE Specific Administrative Rulings And Decisions; General
VIOLET, GENTIAN	589.1000	GENTIAN VIOLET Substances Prohibited From Use In Animal Food Or Feed
VIOLET, MANGANESE	73.2775	MANGANESE VIOLET Cosmetics; Listing Of Color Additives Exempt From Certification
VIPER VENOM	864.8950	RUSSELL VIPER VENOM REAGENT Hematology Reagents; Hematology And Pathology Devices
VIRGINIAMYCIN	556.750	VIRGINIAMYCIN Specific Tolerances For Residues Of New Animal Drugs; Tolerances For Residues Of New Animal Drugs In Food
VIRGINIAMYCIN	558.635	VIRGINIAMYCIN Specific New Animal Drugs For Use In Animal Feeds; New Animal Drugs For Use In Animal Feeds
VIRUS, CHORIOMENINGITIS	866.3360	LYMPHOCYTIC CHORIOMENINGITIS VIRUS SEROLOGICAL REAGENTS Serological Reagents; Immunology And Microbiology Devices
VIRUS, ENCEPHALOMYELITIS	866.3240	EQUINE ENCEPHALOMYELITIS VIRUS SEROLOGICAL REAGENTS Serological Reagents; Immunology And Microbiology Devices
VIRUS, EPSTEIN-BARR	866.3235	EPSTEIN-BARR VIRUS SEROLOGICAL REAGENTS Serological Reagents; Immunology And Microbiology Devices
VIRUS, HERPES SIMPLEX	866.3305	HERPES SIMPLEX VIRUS SEROLOGICAL REAGENTS Serological Reagents; Immunology And Microbiology Devices
VIRUS, INFLUENZA	866.3330	INFLUENZA VIRUS SEROLOGICAL REAGENTS Serological Reagents; Immunology And Microbiology Devices
VIRUS, MUMPS	866.3380	MUMPS VIRUS SEROLOGICAL REAGENTS Serological Reagents; Immunology And Microbiology Devices

VIRUS, PARAINFLUENZA	866.3400	PARAINFLUENZA VIRUS SEROLOGICAL REAGENTS Serological Reagents; Immunology And Microbiology Devices
VIRUS, POLIO	866.3405	POLIOVIRUS SEROLOGICAL REAGENTS Serological Reagents; Immunology And Microbiology Devices
VIRUS, RABIES	866.3460	RABIESVIRUS SEROLOGICAL REAGENTS Serological Reagents; Immunology And Microbiology Devices
VIRUS, REO	866.3470	REOVIRUS SEROLOGICAL REAGENTS Serological Reagents; Immunology And Microbiology Devices
VIRUS, RESPIRATORY SYNCYTIAL	866.3480	RESPIRATORY SYNCYTIAL VIRUS SEROLOGICAL REAGENTS Serological Reagents; Immunology And Microbiology Devices
VIRUS, RHINO	866.3490	RHINOVIRUS SEROLOGICAL REAGENTS Serological Reagents; Immunology And Microbiology Devices
VIRUS, RUBELLA	866.3510	RUBELLA VIRUS SEROLOGICAL REAGENTS Serological Reagents; Immunology And Microbiology Devices
VIRUS, RUBEOLA	866.3520	RUBEOLA (MEASLES) VIRUS SEROLOGICAL REAGENTS Serological Reagents; Immunology And Microbiology Devices
VIRUS, VARICELLA-ZOSTER	866.3900	VARICELLA-ZOSTER VIRUS SEROLOGICAL REAGENTS Serological Reagents; Immunology And Microbiology Devices
VISCOMETER	884.1040	VISCOMETER FOR CERVICAL MUCUS Obstetrical And Gynecological Diagnostic Devices; Obstetrical And Gynecological Devices
VISCOMETER, PLASMA	862.2920	PLASMA VISCOMETER FOR CLINICAL USE Clinical Laboratory Instruments; Clinical Chemistry And Clinical Toxicology Devices
VISION AID	886.5900	ELECTRONIC VISION AID Therapeutic Devices; Ophthalmic Devices
VISION AID	886.5910	IMAGE INTENSIFICATION VISION AID Therapeutic Devices; Ophthalmic Devices
VISION AID	886.5915	OPTICAL VISION AID Therapeutic Devices; Ophthalmic Devices
VISION, COLOR	886.1070	COLOR VISION TESTER Diagnostic Devices; Ophthalmic Devices
VISION, COLOR	886.1160	COLOR VISION PLATE ILLUMINATOR Diagnostic Devices; Ophthalmic Devices
VISION, LOW	886.5540	LOW-VISION MAGNIFIER Therapeutic Devices; Ophthalmic Devices
VISION, LOW	886.5870	LOW-VISION TELESCOPE Therapeutic Devices; Ophthalmic Devices
VISUAL ACUITY	886.1150	VISUAL ACUITY CHART Diagnostic Devices; Ophthalmic Devices
VISUAL FIELD	886.1360	VISUAL FIELD LASER INSTRUMENT Diagnostic Devices; Ophthalmic Devices
VITAMIN A	131.127	NONFAT DRY MILK FORTIFIED WITH VITAMINS A AND D Requirements For Specific Standardized Milk And Cream; Milk And Cream
VITAMIN A	184.1930	VITAMIN A Listing Of Specific Substances Affirmed As GRAS; Direct Food Substances Affirmed As Generally Recognized As Safe
VITAMIN A	582.5930	VITAMIN A Nutrients And/or Dietary Supplements; Substances Generally Recognized As Safe
VITAMIN A	862.1805	VITAMIN A TEST SYSTEM Clinical Chemistry Test Systems; Clinical Chemistry And Clinical Toxicology Devices
VITAMIN A ACETATE	582.5933	VITAMIN A ACETATE Nutrients And/or Dietary Supplements; Substances Generally Recognized As Safe
VITAMIN A PALMITATE	582.5936	VITAMIN A PALMITATE Nutrients And/or Dietary Supplements; Substances Generally Recognized As Safe
VITAMIN B12	184.1945	VITAMIN B12 Listing Of Specific Substances Affirmed As GRAS; Direct Food Substances Affirmed As Generally Recognized As Safe
VITAMIN B12	582.5945	VITAMIN B12 Nutrients And/or Dietary Supplements; Substances Generally Recognized As Safe
VITAMIN B12	862.1810	VITAMIN B12 TEST SYSTEM Clinical Chemistry Test Systems; Clinical Chemistry And Clinical Toxicology Devices
VITAMIN D	131.127	NONFAT DRY MILK FORTIFIED WITH VITAMINS A AND D Requirements For Specific Standardized Milk And Cream; Milk And Cream
VITAMIN D	184.1950	VITAMIN D Listing Of Specific Substances Affirmed As GRAS; Direct Food Substances Affirmed As Generally Recognized As Safe
VITAMIN D	862.1825	VITAMIN D TEST SYSTEM Clinical Chemistry Test Systems; Clinical Chemistry And Clinical Toxicology Devices
VITAMIN D2	582.5950	VITAMIN D2 Nutrients And/or Dietary Supplements; Substances Generally Recognized As Safe

VITAMIN D3	582.5953	VITAMIN D3 Nutrients And/or Dietary Supplements; Substances Generally Recognized As Safe
VITAMIN E	520.2100	SELENIUM, VITAMIN E CAPSULES Oral Dosage Form New Animal Drugs
VITAMIN E	522.2100	SELENIUM, VITAMIN E INJECTION Implantation Or Injectable Dosage Form New Animal Drugs
VITAMIN E	862.1815	VITAMIN E TEST SYSTEM Clinical Chemistry Test Systems; Clinical Chemistry And Clinical Toxicology Devices
VITREOUS ASPIRATION	886.4150	VITREOUS ASPIRATION AND CUTTING INSTRUMENT Surgical Devices; Ophthalmic Devices
VITREOUS CARBON	872.3680	POLYTETRAFLUOROETHYLENE (PTFE) VITREOUS CARBON MATERIALS Prosthetic Devices; Dental Devices
VITRO, IN	201.119	IN VITRO DIAGNOSTIC PRODUCTS Exemptions From Adequate Directions For Use; Labeling
VITRO, IN	312.160	DRUGS FOR INVESTIGATIONAL USE IN LABORATORY RESEARCH ANIMALS OR IN VITRO TESTS Investigational New Drug Application
VITRO, IN	320.35	REQUIREMENTS FOR IN VITRO TESTING OF EACH BATCH Procedures For Determining The Bioavailability Or Bioequivalence Of Drug Products; Bioavailability And Bioequivalence Requirements
VITRO, IN	801.119	IN VITRO DIAGNOSTIC PRODUCTS Exemptions From Adequate Directions For Use; Labeling
VITRO, IN	809.10	LABELING FOR IN VITRO DIAGNOSTIC PRODUCTS Labeling; In Vitro Diagnostic Products For Human Use
VITRO, IN	809.20	GENERAL REQUIREMENTS FOR MANUFACTURERS AND PRODUCERS OF IN VITRO DIAGNOSTIC PRODUCTS Requirements For Manufacturers And Producers; In Vitro Diagnostic Products For Human Use
VITRO, IN	864.5425	MULTIPURPOSE SYSTEM FOR IN VITRO COAGULATION STUDIES Automated And Semi-automated Hematology Devices; Hematology And Pathology Devices
VITRO, IN	864.9160	BLOOD GROUP SUBSTANCES OF NONHUMAN ORIGIN FOR IN VITRO DIAGNOSTIC USE Products Used In Establishments That Manufacture Blood And Blood Products; Hematology And Pathology Devices
VITRO, IN	864.9225	CELL-FREEZING APPARATUS AND REAGENTS FOR IN VITRO DIAGNOSTIC USE Products Used In Establishments That Manufacture Blood And Blood Products; Hematology And Pathology Devices
VITRO, IN	864.9275	BLOOD BANK CENTRIFUGE FOR IN VITRO DIAGNOSTIC USE Products Used In Establishments That Manufacture Blood And Blood Products; Hematology And Pathology Devices
VITRO, IN	864.9600	POTENTIATING MEDIA FOR IN VITRO DIAGNOSTIC USE Products Used In Establishments That Manufacture Blood And Blood Products; Hematology And Pathology Devices
VIVO, EX	876.5885	TISSUE CULTURE MEDIA FOR HUMAN EX VIVO TISSUE AND CELL CULTURE PROCESSING APPLICATIONS Therapeutic Devices; Gastroenterology-urology Devices
VIVO, IN	320.21	REQUIREMENTS FOR SUBMISSION OF IN VIVO BIOAVAILABILITY AND BIOEQUIVALENCE DATA Procedures For Determining The Bioavailability Or Bioequivalence Of Drug Products; Bioavailability And Bioequivalence Requirements
VIVO, IN	320.22	CRITERIA FOR WAIVER OF EVIDENCE OF IN VIVO BIOAVAILABILITY OR BIOEQUIVALENCE Procedures For Determining The Bioavailability Or Bioequivalence Of Drug Products; Bioavailability And Bioequivalence Requirements
VIVO, IN	320.23	BASIS FOR DEMONSTRATING IN VIVO BIOAVAILABILITY OR BIOEQUIVALENCE Procedures For Determining The Bioavailability Or Bioequivalence Of Drug Products; Bioavailability And Bioequivalence Requirements
VIVO, IN	320.25	GUIDELINES FOR THE CONDUCT OF AN IN VIVO BIOAVAILABILITY STUDY Procedures For Determining The Bioavailability Or Bioequivalence Of Drug Products; Bioavailability And Bioequivalence Requirements
VIVO, IN	320.26	GUIDELINES ON THE DESIGN OF A SINGLE-DOSE IN VIVO BIOAVAILABILITY STUDY Procedures For Determining The Bioavailability Or Bioequivalence Of Drug Products; Bioavailability And Bioequivalence Requirements
VIVO, IN	320.27	GUIDELINES ON THE DESIGN OF A MULTIPLE-DOSE IN VIVO BIOAVAILABILITY STUDY Procedures For Determining The Bioavailability Or Bioequivalence Of Drug Products; Bioavailability And Bioequivalence Requirements

VIVO, IN	320.29	ANALYTICAL METHODS FOR AN IN VIVO BIOAVAILABILITY STUDY Procedures For Determining The Bioavailability Or Bioequivalence Of Drug Products; Bioavailability And Bioequivalence Requirements
VOIDING OF FORMS	1305.15	CANCELLATION AND VOIDING OF ORDER FORMS Order Forms
VOLATILE FATTY ACIDS	573.914	SALTS OF VOLATILE FATTY ACIDS Food Additive Listing; Food Additives Permitted In Feed And Drinking Water Of Animals
VOLTAGE	892.1700	DIAGNOSTIC X-RAY HIGH VOLTAGE GENERATOR Diagnostic Devices; Radiology Devices
VOLUME	201.323	ALUMINUM IN LARGE AND SMALL VOLUME PARENTERALS USED IN TOTAL PARENTERAL NUTRITION Specific Labeling Requirements For Specific Drug Products; Labeling
VOLUME	864.5950	BLOOD VOLUME MEASURING DEVICE Automated And Semi-automated Hematology Devices; Hematology And Pathology Devices
VOLUME	868.1760	VOLUME PLETHYSMOGRAPH Diagnostic Devices; Anesthesiology Devices
VOLUME	868.1870	GAS VOLUME CALIBRATOR Diagnostic Devices; Anesthesiology Devices
VOLUME	876.1800	URINE FLOW OR VOLUME MEASURING SYSTEM Diagnostic Devices; Gastroenterology-urology Devices
VOLUNTARILY SUBMITTED	20.44	PRESUBMISSION REVIEW OF REQUEST FOR CONFIDENTIALITY OF VOLUNTARILY SUBMITTED DATA OR INFORMATION Procedures And Fees; Public Information
VOLUNTARILY SUBMITTED	20.111	DATA AND INFORMATION SUBMITTED VOLUNTARILY TO THE FDA Availability Of Specific Categories Of Records; Public Information
VOLUNTARY EXCLUSION	1404.210	VOLUNTARY EXCLUSION Governmentwide Debarment And Suspension And Requirements For Drug-Free Workplace (Grants)
VOLUNTARY EXCLUSION	1404.315	SETTLEMENT AND VOLUNTARY EXCLUSION Governmentwide Debarment And Suspension And Requirements For Drug-Free Workplace (Grants)
VOLUNTARY GUIDELINES	101.43	SUBSTANTIAL COMPLIANCE OF FOOD RETAILERS WITH THE GUIDELINES FOR THE VOLUNTARY NUTRITION LABELING OF RAW FRUIT, VEGETABLES, AND FISH Food Labeling
VOLUNTARY GUIDELINES	101.45	GUIDELINES FOR THE VOLUNTARY NUTRITION LABELING OF RAW FRUIT, VEGETABLES, AND FISH Food Labeling
VOLUNTARY REPORTS	20.112	VOLUNTARY DRUG EXPERIENCE REPORTS SUBMITTED BY PHYSICIANS AND HOSPITALS Availability Of Specific Categories Of Records; Public Information
VOLUNTARY REPORTS	20.113	VOLUNTARY PRODUCT DEFECT REPORTS Availability Of Specific Categories Of Records; Public Information
VOLUNTARY REVOCATION	515.23	VOLUNTARY REVOCATION OF A MEDICATED FEED MILL LICENSE Administrative Actions On Licenses; Medicated Feed Mill License
VOLUNTARY WITHDRAWAL	314.161	DETERMINATION OF REASONS FOR VOLUNTARY WITHDRAWAL OF A LISTED DRUG Applications For FDA Approval To Market A New Drug
VOTING MEMBERS	5.30	AUTHORITY TO SELECT TEMPORARY VOTING MEMBERS OF ADVISORY COMMITTEES AND AUTHORITY TO SIGN CONFLICT OF INTEREST WAIVERS General Redelegations Of Authority; Delegations Of Authority And Organization
VOTING MEMBERS	14.82	NOMINATIONS OF VOTING MEMBERS OF STANDING ADVISORY COMMITTEES Members Of Advisory Committees; Public Hearing Before A Public Advisory Committee

W

WAIVER	10.19	WAIVER, SUSPENSION, OR MODIFICATION OF PROCEDURAL REQUIREMENTS General Provisions; Administrative Practices And Procedures
WAIVER	12.30	JUDICIAL REVIEW AFTER WAIVER OF HEARING ON A REGULATION Initiation Of Proceedings; Formal Evidentiary Public Hearing
WAIVER	20.43	WAIVER OR REDUCTION OF FEES Procedures And Fees; Public Information
WAIVER	56.105	WAIVER OF IRB REQUIREMENT General Provisions; Institutional Review Boards
WAIVER	201.58	REQUESTS FOR WAIVER OF REQUIREMENT FOR ADEQUATE AND WELL-CONTROLLED STUDIES TO SUBSTANTIATE CERTAIN LABELING STATEMENTS Labeling Requirements For Prescription Drugs And/or Insulin; Labeling
WAIVER	320.22	CRITERIA FOR WAIVER OF EVIDENCE OF IN VIVO BIOAVAILABILITY OR BIOEQUIVALENCE Procedures For Determining The Bioavailability Or Bioequivalence Of Drug Products; Bioavailability And Bioequivalence Requirements
WAIVER	500.90	WAIVER OF REQUIREMENTS Regulation of Carcinogenic Compounds Used In Food-Producing Animals; General
WAIVER	1301.43	REQUEST FOR HEARING OR APPEARANCE; WAIVER Hearings; Registration Of Manufacturers, Distributors, And Dispensers Of Controlled Substances
WAIVER	1309.29	WAIVER OF REGISTRATION REQUIREMENT FOR CERTAIN ACTIVITIES Registration Of Manufacturers, Distributors, Importers And Exporters Of List I Chemicals
WAIVER	1303.33	WAIVER OF MODIFICATION OF RULES Hearings; Quotas
WAIVER	1303.34	REQUEST FOR HEARING OR APPEARANCE; WAIVER Hearings; Quotas
WAIVER	1308.45	REQUEST FOR HEARING OR APPEARANCE; WAIVER Hearings; Schedules Of Controlled Substances
WAIVER	1309.53	REQUEST FOR HEARING OR APPEARANCE; WAIVER Registration Of Manufacturers, Distributors, Importers And Exporters Of List I Chemicals
WAIVER	1312.43	WAIVER OR MODIFICATION OF RULES Hearings; Importation And Exportation Of Controlled Substances
WAIVER	1312.44	REQUEST FOR HEARING OR APPEARANCE; WAIVER Hearings; Importation And Exportation Of Controlled Substances
WAIVER	1313.15	WAIVER OF 15-DAY ADVANCE NOTICE FOR REGULAR IMPORTERS Importation Of Listed Chemicals; Importation And Exportation Of Precursors And Essential Chemicals
WAIVER	1313.24	WAIVER OF 15-DAY ADVANCE NOTICE FOR CHEMICAL EXPORTERS Exportation Of Listed Chemicals; Importation And Exportation Of Precursors And Essential Chemicals
WAIVER	1313.53	WAIVER OR MODIFICATION OF RULES Importation And Exportation Of Precursors And Essential Chemicals
WAIVER	1316.44	WAIVER OR MODIFICATION OF RULES Administrative Hearings; Administrative Functions, Practices, And Procedures
WAIVER	1316.49	WAIVER OF HEARING Administrative Hearings; Administrative Functions, Practices, And Procedures
WAIVER	1401.13	WAIVER OR REDUCTION OF FEES Public Availability Of Information
WAIVERS	5.30	AUTHORITY TO SELECT TEMPORARY VOTING MEMBERS OF ADVISORY COMMITTEES AND AUTHORITY TO SIGN CONFLICT OF INTEREST WAIVERS General Redelegations Of Authority; Delegations Of Authority And Organization
WAIVERS	5.108	AUTHORITY RELATING TO WAIVERS OR REDUCTIONS OF PRESCRIPTION DRUG USER FEES Human Drugs; Redelegations Of Authority; Delegations Of Authority And Organization
WAIVERS	312.10	WAIVERS Investigational New Drug Application
WAIVERS	314.90	WAIVERS Applications; Applications For FDA Approval To Market A New Drug
WAIVERS	600.90	WAIVERS Reporting Of Adverse Experiences; Biological Products: General
WAIVERS	812.10	WAIVERS General Provisions; Investigational Device Exemptions
WALKER	890.3825	MECHANICAL WALKER Physical Medicine Prosthetic Devices; Physical Medicine Devices

WALKER TIPS	890.3790	CANE, CRUTCH, AND WALKER TIPS AND PADS Physical Medicine Prosthetic Devices; Physical Medicine Devices
WARDS	50.56	WARDS Additional Safeguards For Children In Clinical Investigations; Protection Of Human Subjects
WAREHOUSING	110.93	WAREHOUSING AND DISTRIBUTION Production And Process Controls; Current GMP In Manufacturing, Packing, Or Holding Human Food
WAREHOUSING	211.142	WAREHOUSING PROCEDURES Holding And Distribution; Current Good Manufacturing Practice For Finished Pharmaceuticals
WARMER	872.6100	ANESTHETIC WARMER Miscellaneous Devices; Dental Devices
WARMER	880.5130	INFANT RADIANT WARMER General Hospital And Personal Use Therapeutic Devices; General Hospital And Personal Use Devices
WARMING DEVICE	864.9205	BLOOD AND PLASMA WARMING DEVICE Products Used In Establishments That Manufacture Blood And Blood Products; Hematology And Pathology Devices
WARNING	101.17	FOOD LABELING WARNING, NOTICE, AND SAFE HANDLING STATEMENTS General Provisions; Food Labeling
WARNING	201.22	PRESCRIPTION DRUGS CONTAINING SULFITES; REQUIRED WARNING STATEMENTS General Labeling Provisions; Labeling
WARNING	201.309	ACETOPHENETIDIN (PHENACETIN)-CONTAINING PREPARATIONS; NECESSARY WARNING STATEMENT Specific Labeling Requirements For Specific Drug Products; Labeling
WARNING	201.315	OVER-THE-COUNTER DRUGS FOR MINOR SORE THROATS; SUGGESTED WARNING Specific Labeling Requirements For Specific Drug Products; Labeling
WARNING	201.316	DRUGS WITH THYROID HORMONE ACTIVITY FOR HUMAN USE; REQUIRED WARNING Specific Labeling Requirements For Specific Drug Products; Labeling
WARNING	201.317	DIGITALIS AND RELATED CARDIOTONIC DRUGS FOR HUMAN USE IN ORAL DOSAGE FORMS; REQUIRED WARNING Specific Labeling Requirements For Specific Drug Products; Labeling
WARNING	201.320	WARNING STATEMENTS FOR DRUG PRODUCTS CONTAINING OR MANUFACTURED WITH CHLOROFLUOROCARBONS Specific Labeling Requirements For Specific Drug Products; Labeling
WARNING	201.322	OVER-THE COUNTER DRUG PRODUCTS CONTAINING INTERNAL ANALGESIC/ANTIPYRETIC ACTIVE INGREDIENTS; REQUIRED ALCOHOL WARNING Specific Labeling Requirements For Specific Drug Products; Labeling
WARNING	290.5	DRUGS; STATEMENT OF REQUIRED WARNING General Provisions; Controlled Drugs
WARNING	369.8	WARNING STATEMENTS IN RELATION TO CONDITIONS FOR USE Definitions And Interpretations; Interpretative Statements Re: Warnings On Drugs And Devices For OTC Sale
WARNING	369.10	CONSPICUOUSNESS OF WARNING STATEMENTS Definitions And Interpretations; Interpretative Statements Re: Warnings On Drugs And Devices For OTC Sale
WARNING	369.20	DRUGS; RECOMMENDED WARNING AND CAUTION STATEMENTS Warning And Caution Statements For Drugs; Interpretative Statements Re: Warnings On Drugs And Devices For OTC Sale
WARNING	369.21	DRUGS; WARNING AND CAUTION STATEMENTS REQUIRED BY REGULATIONS Warning And Caution Statements For Drugs; Interpretative Statements Re: Warnings On Drugs And Devices For OTC Sale
WARNING	369.22	DRUGS; WARNING AND CAUTION STATEMENTS SPECIFICALLY REQUIRED BY LAW Warning And Caution Statements For Drugs; Interpretative Statements Re: Warnings On Drugs And Devices For OTC Sale
WARNING	501.17	ANIMAL FOOD LABELING WARNING STATEMENTS General Provisions; Animal Food Labeling
WARNING	740.1	ESTABLISHMENT OF WARNING STATEMENTS General; Cosmetic Product Warning Statements
WARNING	740.2	CONSPICUOUSNESS OF WARNING STATEMENTS General; Cosmetic Product Warning Statements
WARNING	801.63	MEDICAL DEVICES; WARNING STATEMENTS FOR DEVICES CONTAINING OR MANUFACTURED WITH CHLOROFLUOROCARBONS Labeling Requirements For Over-the-counter Devices; Labeling

WARNING	801.433	WARNING STATEMENTS FOR PRESCRIPTION AND RESTRICTED DEVICE PRODUCTS CONTAINING OR MANUFACTURED WITH CHLOROFLUOROCARBONS Specific Requirements For Specific Devices; Labeling
WARNING, LIMB OVERLOAD	890.5575	POWERED EXTERNAL LIMB OVERLOAD WARNING DEVICE Physical Medicine Therapeutic Devices; Physical Medicine Devices
WARNING, PREGNANCY/ BREAST-FEEDING	201.63	PREGNANCY/BREAST-FEEDING WARNING Labeling Requirements For Over-the-counter Drugs; Labeling
WARNING, PREGNANCY-NURSING	330.2	PREGNANCY-NURSING WARNING General Provisions; Over-the-counter Human Drugs Generally Recognized As Safe And Effective & Not Misbranded
WARNING, SPANISH	290.6	SPANISH-LANGUAGE VERSION OF REQUIRED WARNING General Provisions; Controlled Drugs
WARNINGS	201.305	ISOPROTERENOL INHALATION PREPARATIONS (PRESSURIZED AEROSOLS, NEBULIZERS, POWDERS) FOR HUMAN USE; WARNINGS Specific Labeling Requirements For Specific Drug Products; Labeling
WARNINGS	201.307	SODIUM PHOSPHATES; PACKAGE SIZE LIMITATION, WARNINGS AND DIRECTIONS FOR OVER-THE-COUNTER SALE Specific Labeling Requirements For Specific Drug Products; Labeling
WARNINGS	201.308	IPECAC SYRUP; WARNINGS AND DIRECTIONS FOR USE FOR OVER-THE-COUNTER SALE Specific Labeling Requirements For Specific Drug Products; Labeling
WARNINGS	201.319	WATER-SOLUBLE GUMS, HYDROPHILIC GUMS AND HYDROPHILIC MUCILLOIDS AS ACTIVE INGREDIENTS; REQUIRED WARNINGS AND DIRECTIONS Specific Labeling Requirements For Specific Drug Products; Labeling
WARNINGS	369.3	WARNINGS REQUIRED ON DRUGS EXEMPTED FROM PRESCRIPTION-DISPENSING REQUIREMENTS OF SECTION 503(B)(1)(C) Definitions And Interpretations; Interpretative Statements Re: Warnings On Drugs And Devices For OTC Sale
WARNINGS	369.4	WARNINGS SUGGESTED FOR DRUGS BY FORMAL OR INFORMAL STATEMENTS OF POLICY Definitions And Interpretations; Interpretative Statements Re: Warnings On Drugs And Devices For OTC Sale
WARNINGS	369.7	WARNINGS REQUIRED BY OFFICIAL COMPENDIA Definitions And Interpretations; Interpretative Statements Re: Warnings On Drugs And Devices For OTC Sale
WARNINGS	369.9	GENERAL WARNINGS RE ACCIDENTAL INGESTION BY CHILDREN Definitions And Interpretations; Interpretative Statements Re: Warnings On Drugs And Devices For OTC Sale
WARNINGS	510.410	CORTICOSTEROIDS FOR ORAL, INJECTABLE, AND OPHTHALMIC USE IN ANIMALS; WARNINGS AND LABELING REQUIREMENTS Requirements For Specific New Animal Drugs; New Animal Drugs
WARRANT	1316.07	REQUIREMENT FOR ADMINISTRATIVE INSPECTION WARRANT; EXCEPTIONS Administrative Inspections; Administrative Functions, Practices, And Procedures
WARRANT	1316.09	APPLICATION FOR ADMINISTRATIVE INSPECTION WARRANT Administrative Inspections; Administrative Functions, Practices, And Procedures
WARRANT	1316.12	REFUSAL TO ALLOW INSPECTION WITH AN ADMINISTRATIVE WARRANT Administrative Inspections; Administrative Functions, Practices, And Procedures
WARRANTS	1316.11	EXECUTION OF WARRANTS Administrative Inspections; Administrative Functions, Practices, And Procedures
WART REMOVER	358.110	WART REMOVER ACTIVE INGREDIENTS Wart Remover Drug Products; Miscellaneous External Drug Products For Over-the-counter Human Use
WART REMOVER	358.150	LABELING OF WART REMOVER DRUG PRODUCTS Wart Remover Drug Products; Miscellaneous External Drug Products For Over-the-counter Human Use
WASH WATER	1250.87	WASH WATER Sanitation Facilities And Conditions On Vessels; Interstate Conveyance Sanitation
WASHED CURD CHEESE	133.136	WASHED CURD AND SOAKED CURD CHEESE Requirements For Specific Standardized Cheese And Related Products; Cheeses And Related Cheese Products
WASHED CURD CHEESE	133.137	WASHED CURD CHEESE FOR MANUFACTURING Requirements For Specific Standardized Cheese And Related Products; Cheeses And Related Cheese Products
WASHER	884.1185	ENDOMETRIAL WASHER Obstetrical And Gynecological Diagnostic Devices; Obstetrical And Gynecological Devices
WASHERS	880.6800	WASHERS FOR BODY WASTE RECEPTACLES General Hospital And Personal Use Miscellaneous Devices; General Hospital And Personal Use Devices

WASHING, CHEMICAL	173.315	CHEMICALS USED IN WASHING OR TO ASSIST IN THE PEELING OF FRUITS AND VEGETABLES Specific Usage Additives; Secondary Direct Food Additives Permitted In Food For Human Consumption
WASHING FACILITIES	211.52	WASHING AND TOILET FACILITIES Buildings And Facilities; Current Good Manufacturing Practice For Finished Pharmaceuticals
WASHINGTON	808.97	WASHINGTON Listing Of Specific State And Local Exemptions; Exemption From Federal Preemption Of State & Local Medical Device Requirements
WASTE	880.6730	BODY WASTE RECEPTACLE General Hospital And Personal Use Miscellaneous Devices; General Hospital And Personal Use Devices
WASTE	880.6800	WASHERS FOR BODY WASTE RECEPTACLES General Hospital And Personal Use Miscellaneous Devices; General Hospital And Personal Use Devices
WASTES	1250.51	RAILROAD CONVEYANCES; DISCHARGE OF WASTES Equipment And Operation Of Land And Air Conveyances; Interstate Conveyance Sanitation
WASTES	1250.52	DISCHARGE OF WASTES ON HIGHWAY CONVEYANCES Equipment And Operation Of Land And Air Conveyances; Interstate Conveyance Sanitation
WASTES	1250.53	DISCHARGE OF WASTES ON AIR CONVEYANCES Equipment And Operation Of Land And Air Conveyances; Interstate Conveyance Sanitation
WASTES	1250.70	DISPOSAL OF HUMAN WASTES Servicing Areas For Land And Air Conveyances; Interstate Conveyance Sanitation
WASTES	1250.93	DISCHARGE OF WASTES Sanitation Facilities And Conditions On Vessels; Interstate Conveyance Sanitation
WATER	1250.83	STORAGE OF WATER PRIOR TO TREATMENT Sanitation Facilities And Conditions On Vessels; Interstate Conveyance Sanitation
WATER	1250.84	WATER IN GALLEYS AND MEDICAL CARE SPACES Sanitation Facilities And Conditions On Vessels; Interstate Conveyance Sanitation
WATER	1250.86	WATER FOR MAKING ICE Sanitation Facilities And Conditions On Vessels; Interstate Conveyance Sanitation
WATER	1250.87	WASH WATER Sanitation Facilities And Conditions On Vessels; Interstate Conveyance Sanitation
WATER BOATS	1240.95	SANITATION OF WATER BOATS Source And Use Of Potable Water; Control Of Communicable Diseases
WATER, BOILER	173.310	BOILER WATER ADDITIVES Specific Usage Additives; Secondary Direct Food Additives Permitted In Food For Human Consumption
WATER BOTTLE	880.6085	HOT/COLD WATER BOTTLE General Hospital And Personal Use Miscellaneous Devices; General Hospital And Personal Use Devices
WATER, BOTTLED	165.110	BOTTLED WATER Requirements For Specific Standardized Beverages; Beverages
WATER CALORIC STIMULATOR	874.1800	AIR OR WATER CALORIC STIMULATOR Diagnostic Devices; Ear, Nose, And Throat Devices
WATER CAPACITY	130.12	GENERAL METHODS FOR WATER CAPACITY AND FILL OF CONTAINERS General Provisions; Food Standards: General
WATER CIRCULATING	890.5720	WATER CIRCULATING HOT OR COLD PACK Physical Medicine Therapeutic Devices; Physical Medicine Devices
WATER, DRINKING	520.100	AMPROLIUM DRINKING WATER Oral Dosage Form New Animal Drugs
WATER, DRINKING	520.1242	LEVAMISOLE HYDROCHLORIDE DRENCH AND DRINKING WATER Oral Dosage Form New Animal Drugs
WATER, DRINKING	520.2200b	SULFACHLORPYRIDAZINE MEDICATED MILK AND DRINKING WATER Oral Dosage Form New Animal Drugs
WATER, DRINKING	520.2240a	SULFAETHOXYPYRIDAZINE DRINKING WATER Oral Dosage Form New Animal Drugs
WATER, DRINKING	520.2261a	SULFAMETHAZINE SODIUM DRINKING WATER SOLUTION Oral Dosage Form New Animal Drugs
WATER, DRINKING	520.2325	SULFAQUINOXALINE DRINKING WATER Oral Dosage Form New Animal Drugs
WATER, DRINKING/CULINARY	1240.80	GENERAL REQUIREMENTS FOR WATER FOR DRINKING AND CULINARY PURPOSES Source And Use Of Potable Water; Control Of Communicable Diseases
WATER ICES	135.160	WATER ICES Requirements For Specific Standardized Frozen Desserts; Frozen Desserts
WATER-INSOLUBLE	177.1400	HYDROXYETHYL CELLULOSE FILM, WATER-INSOLUBLE Substances For Use As Basic Components Of Single And Repeated Use Food Contact Surfaces; Indirect Food Additives: Polymers

WATER JET	876.4650	WATER JET RENAL STONE DISLODGER SYSTEM Surgical Devices; Gastroenterology-urology Devices
WATER, LUNG	868.2450	LUNG WATER MONITOR Monitoring Devices; Anesthesiology Devices
WATER MATTRESS	880.5560	TEMPERATURE REGULATED WATER MATTRESS General Hospital And Personal Use Therapeutic Devices; General Hospital And Personal Use Devices
WATER PURIFICATION	876.5665	WATER PURIFICATION SYSTEM FOR HEMODIALYSIS Therapeutic Devices; Gastroenterology-urology Devices
WATER PURIFICATION	884.6170	ASSISTED REPRODUCTION WATER AND WATER PURIFICATION SYSTEMS Assisted Reproduction Devices; Obstetrical And Gynecological Devices
WATER PURIFIER	880.6710	MEDICAL ULTRAVIOLET WATER PURIFIER General Hospital And Personal Use Miscellaneous Devices; General Hospital And Personal Use Devices
WATER RESISTANT	352.76	DETERMINATION IF PRODUCT IS WATER RESISTANT OR VERY WATER RESISTANT Sunscreen Drug Products For Over-the-counter Human Use
WATER-SOLUBLE GUMS	201.319	WATER-SOLUBLE GUMS, HYDROPHILIC GUMS AND HYDROPHILIC MUCILLOIDS AS ACTIVE INGREDIENTS; REQUIRED WARNINGS AND DIRECTIONS Specific Labeling Requirements For Specific Drug Products; Labeling
WATER STERILIZER	872.6710	BOILING WATER STERILIZER Miscellaneous Devices; Dental Devices
WATER SYSTEM	1240.86	PROTECTION OF PIER WATER SYSTEM Source And Use Of Potable Water; Control Of Communicable Diseases
WATER SYSTEMS	1250.42	WATER SYSTEMS; CONSTANT TEMPERATURE BOTTLES Equipment And Operation Of Land And Air Conveyances; Interstate Conveyance Sanitation
WATER SYSTEMS	1250.82	POTABLE WATER SYSTEMS Sanitation Facilities And Conditions On Vessels; Interstate Conveyance Sanitation
WATER TRAP	868.5995	TEE DRAIN (WATER TRAP) Therapeutic Devices; Anesthesiology Devices
WATER VAPOR	868.1975	WATER VAPOR ANALYZER Diagnostic Devices; Anesthesiology Devices
WATERING EQUIPMENT	1250.67	WATERING EQUIPMENT Servicing Areas For Land And Air Conveyances; Interstate Conveyance Sanitation
WATERING POINTS	1240.83	APPROVAL OF WATERING POINTS Source And Use Of Potable Water; Control Of Communicable Diseases
WAVE, SHOCK	876.5990	EXTRACORPOREAL SHOCK WAVE LITHOTRIPTER Therapeutic Devices; Gastroenterology-urology Devices
WAX	178.3850	REINFORCED WAX Certain Adjuvants And Production Aids; Indirect Food Additives: Adjuvants, Production Aids, And Sanitizers
WAX BEANS	155.120	CANNED GREEN BEANS AND CANNED WAX BEANS Requirements For Specific Standardized Canned Vegetables; Canned Vegetables
WAX, BRAN	172.890	RICE BRAN WAX Multipurpose Additives; Food Additives Permitted For Direct Addition To Food For Human Consumption
WAX, CANDELILLA	184.1976	CANDELILLA WAX Listing Of Specific Substances Affirmed As GRAS; Direct Food Substances Affirmed As Generally Recognized As Safe
WAX, CARNAUBA	184.1978	CARNAUBA WAX Listing Of Specific Substances Affirmed As GRAS; Direct Food Substances Affirmed As Generally Recognized As Safe
WAX, CARNAUBA	582.1978	CARNAUBA WAX General Purpose Food Additives; Substances Generally Recognized As Safe
WAX, DENTAL	872.6890	INTRAORAL DENTAL WAX Miscellaneous Devices; Dental Devices
WAX, JAPAN	186.1555	JAPAN WAX Listing Of Specific Substances Affirmed As GRAS; Indirect Food Substances Affirmed As Generally Recognized As Safe
WAX, MONTAN	178.3770	POLYHYDRIC ALCOHOL ESTERS OF OXIDATIVELY REFINED (GERSTHOFEN PROCESS) MONTAN WAX ACIDS Certain Adjuvants And Production Aids; Indirect Food Additives: Adjuvants, Production Aids, And Sanitizers
WAX, PETROLEUM	172.886	PETROLEUM WAX Multipurpose Additives; Food Additives Permitted For Direct Addition To Food For Human Consumption
WAX, PETROLEUM	172.888	SYNTHETIC PETROLEUM WAX Multipurpose Additives; Food Additives Permitted For Direct Addition To Food For Human Consumption
WAX, PETROLEUM	178.3710	PETROLEUM WAX Certain Adjuvants And Production Aids; Indirect Food Additives: Adjuvants, Production Aids, And Sanitizers

WAX, PETROLEUM	178.3720	PETROLEUM WAX, SYNTHETIC Certain Adjuvants And Production Aids; Indirect Food Additives: Adjuvants, Production Aids, And Sanitizers
WEIGHING	864.9195	BLOOD MIXING DEVICES AND BLOOD WEIGHING DEVICES Products Used In Establishments That Manufacture Blood And Blood Products; Hematology And Pathology Devices
WEIGHT, BODY	105.66	LABEL STATEMENTS RELATING TO USEFULNESS IN REDUCING OR MAINTAINING BODY WEIGHT Label Statements; Foods For Special Dietary Use
WEIGHT, MOLECULAR	172.820	POLYETHYLENE GLYCOL (MEAN MOLECULAR WEIGHT 200-9,500) Multipurpose Additives; Food Additives Permitted For Direct Addition To Food For Human Consumption
WEIGHT, MOLECULAR	177.1440	4,4'-ISOPROPYLIDENEDIPHENOL-EPICHLOROHYDRIN RESINS MINIMUM MOLECULAR WEIGHT 10,000 Substances For Use As Basic Components Of Single And Repeated Use Food Contact Surfaces; Indirect Food Additives: Polymers
WEIGHT, MOLECULAR	178.3750	POLYETHYLENE GLYCOL (MEAN MOLECULAR WEIGHT 200-9,500) Certain Adjuvants And Production Aids; Indirect Food Additives: Adjuvants, Production Aids, And Sanitizers
WEIL-FELIX	866.3410	PROTEUS SPP. (WEIL-FELIX) SEROLOGICAL REAGENTS Serological Reagents; Immunology And Microbiology Devices
WELFARE	50.54	CLINICAL INVESTIGATIONS NOT OTHERWISE APPROVABLE THAT PRESENT AN OPPORTUNITY TO UNDERSTAND, PREVENT, OR ALLEVIATE A SERIOUS PROBLEM AFFECTING THE HEALTH OR WELFARE OF CHILDREN Additional Safeguards For Children In Clinical Investigations; Protection Of Human Subjects
WELL-CONTROLLED STUDIES	201.58	REQUESTS FOR WAIVER OF REQUIREMENT FOR ADEQUATE AND WELL-CONTROLLED STUDIES TO SUBSTANTIATE CERTAIN LABELING STATEMENTS Labeling Requirements For Prescription Drugs And/or Insulin; Labeling
WELL-CONTROLLED STUDIES	314.126	ADEQUATE AND WELL-CONTROLLED STUDIES FDA Action On Applications; Applications For FDA Approval To Market A New Drug
WELL-CONTROLLED STUDIES	514.117	ADEQUATE AND WELL-CONTROLLED STUDIES Administrative Actions On Applications; New Animal Drug Applications
WEST VIRGINIA	808.98	WEST VIRGINIA Listing Of Specific State And Local Exemptions; Exemption From Federal Preemption Of State & Local Medical Device Requirements
WET PACK SHRIMP	161.173	CANNED WET PACK SHRIMP IN TRANSPARENT OR NONTRANSPARENT CONTAINERS Requirements For Specific Standardized Fish And Shellfish; Fish And Shellfish
WHEAT	136.180	WHOLE WHEAT BREAD, ROLLS, AND BUNS Requirements For Specific Standardized Bakery Products; Bakery Products
WHEAT	137.190	CRACKED WHEAT Requirements For Specific Standardized Cereal Flours And Related Products; Cereal Flours And Related Products
WHEAT	137.195	CRUSHED WHEAT Requirements For Specific Standardized Cereal Flours And Related Products; Cereal Flours And Related Products
WHEAT	137.200	WHOLE WHEAT FLOUR Requirements For Specific Standardized Cereal Flours And Related Products; Cereal Flours And Related Products
WHEAT	137.205	BROMATED WHOLE WHEAT FLOUR Requirements For Specific Standardized Cereal Flours And Related Products; Cereal Flours And Related Products
WHEAT	139.138	WHOLE WHEAT MACARONI PRODUCTS Requirements For Specific Standardized Macaroni And Noodle Products; Macaroni And Noodle Products
WHEAT	139.140	WHEAT AND SOY MACARONI PRODUCTS Requirements For Specific Standardized Macaroni And Noodle Products; Macaroni And Noodle Products
WHEAT	139.180	WHEAT AND SOY NOODLE PRODUCTS Requirements For Specific Standardized Macaroni And Noodle Products; Macaroni And Noodle Products
WHEAT GLUTEN	184.1322	WHEAT GLUTEN Listing Of Specific Substances Affirmed As GRAS; Direct Food Substances Affirmed As Generally Recognized As Safe
WHEELCHAIR	890.3850	MECHANICAL WHEELCHAIR Physical Medicine Prosthetic Devices; Physical Medicine Devices
WHEELCHAIR	890.3860	POWERED WHEELCHAIR Physical Medicine Prosthetic Devices; Physical Medicine Devices
WHEELCHAIR	890.3880	SPECIAL GRADE WHEELCHAIR Physical Medicine Prosthetic Devices; Physical Medicine Devices
WHEELCHAIR	890.3890	STAIR-CLIMBING WHEELCHAIR Physical Medicine Prosthetic Devices; Physical Medicine Devices

WHEELCHAIR	890.3900	STANDUP WHEELCHAIR Physical Medicine Prosthetic Devices; Physical Medicine Devices
WHEELCHAIR	890.3910	WHEELCHAIR ACCESSORY Physical Medicine Prosthetic Devices; Physical Medicine Devices
WHEELCHAIR	890.3920	WHEELCHAIR COMPONENT Physical Medicine Prosthetic Devices; Physical Medicine Devices
WHEELCHAIR	890.3930	WHEELCHAIR ELEVATOR Physical Medicine Prosthetic Devices; Physical Medicine Devices
WHEELCHAIR	890.3940	WHEELCHAIR PLATFORM SCALE Physical Medicine Prosthetic Devices; Physical Medicine Devices
WHEELED	890.3800	MOTORIZED THREE-WHEELED VEHICLE Physical Medicine Prosthetic Devices; Physical Medicine Devices
WHEELED STRETCHER	880.6910	WHEELED STRETCHER General Hospital And Personal Use Miscellaneous Devices; General Hospital And Personal Use Devices
WHEELED STRETCHER	890.3690	POWERED WHEELED STRETCHER Physical Medicine Prosthetic Devices; Physical Medicine Devices
WHEY	184.1979	WHEY Listing Of Specific Substances Affirmed As GRAS; Direct Food Substances Affirmed As Generally Recognized As Safe
WHEY	184.1979a	REDUCED LACTOSE WHEY Listing Of Specific Substances Affirmed As GRAS; Direct Food Substances Affirmed As Generally Recognized As Safe
WHEY	184.1979b	REDUCED MINERALS WHEY Listing Of Specific Substances Affirmed As GRAS; Direct Food Substances Affirmed As Generally Recognized As Safe
WHEY	573.450	FERMENTED AMMONIATED CONDENSED WHEY Food Additive Listing; Food Additives Permitted In Feed And Drinking Water Of Animals
WHIPPED CREAM	131.25	WHIPPED CREAM PRODUCTS CONTAINING FLAVORING OR SWEETENING General Provisions; Milk And Cream
WHIPPING CREAM	131.157	LIGHT WHIPPING CREAM Requirements For Specific Standardized Milk And Cream; Milk And Cream
WHITE CELL	864.8185	CALIBRATOR FOR RED CELL AND WHITE CELL COUNTING Hematology Reagents; Hematology And Pathology Devices
WHITE CORN	137.211	WHITE CORN FLOUR Requirements For Specific Standardized Cereal Flours And Related Products; Cereal Flours And Related Products
WHITE CORN	137.250	WHITE CORN MEAL Requirements For Specific Standardized Cereal Flours And Related Products; Cereal Flours And Related Products
WHITE CORN	137.255	BOLTED WHITE CORN MEAL Requirements For Specific Standardized Cereal Flours And Related Products; Cereal Flours And Related Products
WHITE CORN	137.265	DEGERMINATED WHITE CORN MEAL Requirements For Specific Standardized Cereal Flours And Related Products; Cereal Flours And Related Products
WHITE CORN	137.270	SELF-RISING WHITE CORN MEAL Requirements For Specific Standardized Cereal Flours And Related Products; Cereal Flours And Related Products
WHITE MINERAL OIL	172.878	WHITE MINERAL OIL Multipurpose Additives; Food Additives Permitted For Direct Addition To Food For Human Consumption
WHITES, EGG	160.140	EGG WHITES Requirements For Specific Standardized Eggs And Egg Products; Eggs And Egg Products
WHITES, EGG	160.145	DRIED EGG WHITES Requirements For Specific Standardized Eggs And Egg Products; Eggs And Egg Products
WHITES, EGG	160.150	FROZEN EGG WHITES Requirements For Specific Standardized Eggs And Egg Products; Eggs And Egg Products
WHITING	102.46	PACIFIC WHITING Requirements For Specific Nonstandardized Foods; Common Or Usual Name For Nonstandardized Foods
WHOLE	12.28	DENIAL OF HEARING IN WHOLE OR IN PART Initiation Of Proceedings; Formal Evidentiary Public Hearing
WHOLE BLOOD	640.1	WHOLE BLOOD Whole Blood; Additional Standards For Human Blood And Blood Products
WHOLE BLOOD	640.6	MODIFICATIONS OF WHOLE BLOOD Whole Blood; Additional Standards For Human Blood And Blood Products
WHOLE BLOOD	864.7140	ACTIVATED WHOLE BLOOD CLOTTING TIME TESTS Hematology Kits And Packages; Hematology And Pathology Devices

WHOLE BLOOD	864.7500	WHOLE BLOOD HEMOGLOBIN ASSAYS Hematology Kits And Packages; Hematology And Pathology Devices
WHOLE BODY	892.1130	NUCLEAR WHOLE BODY COUNTER Diagnostic Devices; Radiology Devices
WHOLE BODY	892.1330	NUCLEAR WHOLE BODY SCANNER Diagnostic Devices; Radiology Devices
WHOLE MILK	131.147	DRY WHOLE MILK Requirements For Specific Standardized Milk And Cream; Milk And Cream
WHOLE DURUM FLOUR	137.225	WHOLE DURUM FLOUR Requirements For Specific Standardized Cereal Flours And Related Products; Cereal Flours And Related Products
WHOLE FISH PROTEIN	172.385	WHOLE FISH PROTEIN CONCENTRATE Special Dietary And Nutritional Additives; Food Additives Permitted For Direct Addition To Food For Human Consumption
WHOLE PLASMA	866.5700	WHOLE HUMAN PLASMA OR SERUM IMMUNOLOGICAL TEST SYSTEM Immunological Test Systems; Immunology And Microbiology Devices
WHOLE WHEAT	136.180	WHOLE WHEAT BREAD, ROLLS, AND BUNS Requirements For Specific Standardized Bakery Products; Bakery Products
WHOLE WHEAT	137.200	WHOLE WHEAT FLOUR Requirements For Specific Standardized Cereal Flours And Related Products; Cereal Flours And Related Products
WHOLE WHEAT	137.205	BROMATED WHOLE WHEAT FLOUR Requirements For Specific Standardized Cereal Flours And Related Products; Cereal Flours And Related Products
WHOLE WHEAT	139.138	WHOLE WHEAT MACARONI PRODUCTS Requirements For Specific Standardized Macaroni And Noodle Products; Macaroni And Noodle Products
WHOLESALE DISTRIBUTION	203.50	REQUIREMENTS FOR WHOLESALE DISTRIBUTION OF PRESCRIPTION DRUGS Prescription Drug Marketing
WHOLESALE DISTRIBUTOR	205.4	WHOLESALE DRUG DISTRIBUTOR LICENSING REQUIREMENT Guidelines For State Licensing Of Wholesale Prescription Drug Distributors
WINE BOTTLES	189.301	TIN-COATED LEAD FOIL CAPSULES FOR WINE BOTTLES Substances Prohibited From Use In Human Foods
WINTERGREEN OIL	201.303	LABELING OF DRUG PREPARATIONS CONTAINING SIGNIFICANT PROPORTIONS OF WINTERGREEN OIL Specific Labeling Requirements For Specific Drug Products; Labeling
WIRE	870.1330	CATHETER GUIDE WIRE Cardiovascular Diagnostic Devices; Cardiovascular Devices
WIRE	872.4600	INTRAORAL LIGATURE AND WIRE LOCK Surgical Devices; Dental Devices
WIRE	872.4880	INTRAOSSEOUS FIXATION SCREW OR WIRE Surgical Devices; Dental Devices
WITHDRAW APPROVAL	5.102	AUTHORITY TO APPROVE AND WITHDRAW APPROVAL OF A CHARGE FOR INVESTIGATIONAL NEW DRUGS Human Drugs; Redelegations Of Authority; Delegations Of Authority And Organization
WITHDRAW APPROVAL	5.105	ISSUANCE OF NOTICES RELATING TO PROPOSALS TO REFUSE APPROVAL OR TO WITHDRAW APPROVAL OF NEW DRUG APPLICATIONS AND THEIR SUPPLEMENTS Human Drugs; Redelegations Of Authority; Delegations Of Authority And Organization
WITHDRAW RECOGNITION	5.400	ISSUANCE OF FEDERAL REGISTER DOCUMENTS TO RECOGNIZE OR TO WITHDRAW RECOGNITION OF A STANDARD TO MEET PREMARKET SUBMISSION REQUIREMENTS Medical Devices And Radiological Health; Redelegations Of Authority; Delegations Of Authority And Organization
WITHDRAWAL	870.1800	WITHDRAWAL-INFUSION PUMP Cardiovascular Diagnostic Devices; Cardiovascular Devices
WITHDRAWAL OF APPLICATION	314.65	WITHDRAWAL BY THE APPLICANT OF AN UNAPPROVED APPLICATION Applications; Applications For FDA Approval To Market A New Drug
WITHDRAWAL OF APPLICATIONS	514.7	WITHDRAWAL OF APPLICATIONS WITHOUT PREJUDICE General Provisions; New Animal Drug Applications
WITHDRAWAL OF APPLICATIONS	1301.16	AMENDMENTS TO AND WITHDRAWAL OF APPLICATIONS Registration Of Manufacturers, Distributors, And Dispensers Of Controlled Substances
WITHDRAWAL OF APPROVAL	5.407	APPROVAL, DISAPPROVAL, OR WITHDRAWAL OF APPROVAL OF PRODUCT DEVELOPMENT PROTOCOLS AND APPLICATIONS FOR PREMARKET APPROVAL FOR MEDICAL DEVICES Medical Devices And Radiological Health; Redelegations Of Authority; Delegations Of Authority And Organization

WITHDRAWAL OF APPROVAL	5.413	APPROVAL, DISAPPROVAL, WITHDRAWAL OF APPROVAL OF APPLICATIONS AND ENTERING INTO AGREEMENTS FOR INVESTIGATIONAL DEVICE EXEMPTIONS Medical Devices And Radiological Health; Redelegations Of Authority; Delegations Of Authority And Organization
WITHDRAWAL OF APPROVAL	314.150	WITHDRAWAL OF APPROVAL OF AN APPLICATION OR ABBREVIATED APPLICATION Applications For FDA Approval To Market A New Drug
WITHDRAWAL OF APPROVAL	314.151	WITHDRAWAL OF APPROVAL OF AN ABBREVIATED NEW DRUG APPLICATION UNDER SECTION 505(J)(5) OF THE ACT Applications For FDA Approval To Market A New Drug
WITHDRAWAL OF APPROVAL	314.152	NOTICE OF WITHDRAWAL OF APPROVAL OF AN APPLICATION OR ABBREVIATED APPLICATION FOR A NEW DRUG FDA Action On Applications; Applications For FDA Approval To Market A New Drug
WITHDRAWAL OF APPROVAL	514.115	WITHDRAWAL OF APPROVAL OF APPLICATIONS Administrative Actions On Applications; New Animal Drug Applications
WITHDRAWAL OF APPROVAL	514.116	NOTICE OF WITHDRAWAL OF APPROVAL OF APPLICATION Administrative Actions On Applications; New Animal Drug Applications
WITHDRAWAL OF APPROVAL	814.46	WITHDRAWAL OF APPROVAL OF A PMA FDA Action On A PMA; Premarket Approval Of Medical Devices
WITHDRAWAL OF APPROVAL	814.118	DENIAL OF APPROVAL OR WITHDRAWAL OF APPROVAL OF AN HDE Humanitarian Use Devices; Premarket Approval Of Medical Devices
WITHDRAWAL OF APPROVAL	900.6	WITHDRAWAL OF APPROVAL Accreditation; Mammography
WITHDRAWAL OF APPROVAL	900.24	WITHDRAWAL OF APPROVAL States As Certifiers; Mammography
WITHDRAWAL, CONFORMITY ASSESSMENT	26.68	WITHDRAWAL OF LISTED CONFORMITY ASSESSMENT BODIES Framework Provisions; Mutual Recognition Of Pharmaceutical GMP Reports, Medical Device Quality System Audit Reports
WITHDRAWAL, IND	312.38	WITHDRAWAL OF AN IND Investigational New Drug Application
WITHDRAWAL, LISTED DRUG	314.161	DETERMINATION OF REASONS FOR VOLUNTARY WITHDRAWAL OF A LISTED DRUG Applications For FDA Approval To Market A New Drug
WITHDRAWAL OF PETITION	171.7	WITHDRAWAL OF PETITION WITHOUT PREJUDICE General Provisions; Food Additive Petitions
WITHDRAWAL OF PETITION	571.7	WITHDRAWAL OF PETITION WITHOUT PREJUDICE General Provisions; Food Additive Petitions
WITHDRAWAL OF PETITIONS	71.6	EXTENSION OF TIME FOR STUDYING PETITIONS; SUBSTANTIVE AMENDMENTS; WITHDRAWAL OF PETITIONS WITHOUT PREJUDICE General Provisions; Color Additive Petitions
WITHDRAWAL OF RECORDS	20.29	PROHIBITION ON WITHDRAWAL OF RECORDS FROM FDA FILES General Policy; Public Information
WITHDRAWAL PROCEDURES	314.530	WITHDRAWAL PROCEDURES Accelerated Approval Of New Drugs; Applications For FDA Approval To Market A New Drug
WITHDRAWAL PROCEDURES	601.43	WITHDRAWAL PROCEDURES Accelerated Approval Of Biological Products; Licensing
WITHDRAWALS OF APPLICATIONS	1309.36	AMENDMENTS TO AND WITHDRAWALS OF APPLICATIONS Registration Of Manufacturers, Distributors, Importers And Exporters Of List I Chemicals
WITHDRAWING APPROVAL	514.120	REVOCATION OF ORDER REFUSING TO APPROVE AN APPLICATION OR SUSPENDING OR WITHDRAWING APPROVAL OF AN APPLICATION Administrative Actions On Applications; New Animal Drug Applications
WITHDRAWN	216.24	DRUG PRODUCTS WITHDRAWN OR REMOVED FROM THE MARKET FOR REASONS OF SAFETY OR EFFECTIVENESS Pharmacy Compounding
WITHDRAWN	314.160	APPROVAL OF AN APPLICATION OR ABBREVIATED APPLICATION FOR WHICH APPROVAL WAS PREVIOUSLY REFUSED, SUSPENDED, OR WITHDRAWN FDA Action On Applications; Applications For FDA Approval To Market A New Drug
WITNESS	17.25	EXCHANGE OF WITNESS LISTS, WITNESS STATEMENTS, AND EXHIBITS Civil Money Penalties Hearings
WITNESSES	17.37	WITNESSES Civil Money Penalties Hearings
WITNESSES	1316.57	SUBMISSION OF DOCUMENTARY EVIDENCE AND AFFIDAVITS AND IDENTIFICATION OF WITNESSES SUBSEQUENT TO PREHEARING CONFERENCE Administrative Hearings; Administrative Functions, Practices, And Procedures
WOOD	178.3800	PRESERVATIVES FOR WOOD Certain Adjuvants And Production Aids; Indirect Food Additives: Adjuvants, Production Aids, And Sanitizers

WOOD ROSIN	172.735	GLYCEROL ESTER OF WOOD ROSIN Other Specific Usage Additives; Food Additives Permitted For Direct Addition To Food For Human Consumption
WOOD'S FLUORESCENT LAMP	866.2600	WOOD'S FLUORESCENT LAMP Microbiology Devices; Immunology And Microbiology Devices
WORK AREAS	225.35	USE OF WORK AREAS, EQUIPMENT, AND STORAGE AREAS FOR OTHER MANUFACTURING AND STORAGE PURPOSES Construction And Maintenance Of Facilities And Equipment; Current Good Manufacturing Practice For Medicated Feeds
WORK AREAS	225.135	WORK AND STORAGE AREAS Facilities And Equipment; Current Good Manufacturing Practice For Medicated Feeds
WOUND DRESSING	878.4018	HYDROPHILIC WOUND DRESSING Surgical Devices; General And Plastic Surgery Devices
WOUND DRESSING	878.4020	OCCLUSIVE WOUND DRESSING Surgical Devices; General And Plastic Surgery Devices
WOUND DRESSING	878.4022	HYDROGEL WOUND DRESSING AND BURN DRESSING Surgical Devices; General And Plastic Surgery Devices
WOUND HEALING	310.534	DRUG PRODUCTS CONTAINING ACTIVE INGREDIENTS OFFERED OTC FOR HUMAN USE AS ORAL WOUND HEALING AGENTS Requirements For Specific New Drugs Or Devices; New Drugs
WRAP	880.6850	STERILIZATION WRAP General Hospital And Personal Use Miscellaneous Devices; General Hospital And Personal Use Devices
WRIST JOINT	888.3750	WRIST JOINT CARPAL LUNATE POLYMER PROSTHESIS Prosthetic Devices; Orthopedic Devices
WRIST JOINT	888.3760	WRIST JOINT CARPAL SCAPHOID POLYMER PROSTHESIS Prosthetic Devices; Orthopedic Devices
WRIST JOINT	888.3770	WRIST JOINT CARPAL TRAPEZIUM POLYMER PROSTHESIS Prosthetic Devices; Orthopedic Devices
WRIST JOINT	888.3780	WRIST JOINT POLYMER CONSTRAINED PROSTHESIS Prosthetic Devices; Orthopedic Devices
WRIST JOINT	888.3790	WRIST JOINT METAL CONSTRAINED CEMENTED PROSTHESIS Prosthetic Devices; Orthopedic Devices
WRIST JOINT	888.3800	WRIST JOINT METAL/POLYMER SEMI-CONSTRAINED CEMENTED PROSTHESIS Prosthetic Devices; Orthopedic Devices
WRIST JOINT	888.3810	WRIST JOINT ULNAR (HEMI-WRIST) POLYMER PROSTHESIS Prosthetic Devices; Orthopedic Devices
WRITING	14.174	ADVICE AND RECOMMENDATIONS IN WRITING Advisory Committees For Human Prescription Drugs; Public Hearing Before A Public Advisory Committee
WRITTEN APPROVAL	1312.31	SCHEDULE I: APPLICATION FOR PRIOR WRITTEN APPROVAL Transshipment And In-transit Shipment Of Controlled Substances; Importation And Exportation Of Controlled Substances
WRITTEN NOTICES	5.109	ISSUANCE OF WRITTEN NOTICES CONCERNING PATENT INFORMATION, CURRENT GOOD MANUFACTURING PRACTICES AND FALSE OR MISLEADING LABELING OF NEW DRUGS Human Drugs; Redelegations Of Authority; Delegations Of Authority And Organization
WRITTEN NOTICES	5.504	ISSUANCE OF WRITTEN NOTICES CONCERNING PATENT INFORMATION, CURRENT GOOD MANUFACTURING PRACTICES AND FALSE OR MISLEADING LABELING OF NEW ANIMAL DRUGS AND FEEDS BEARING OR CONTAINING NEW ANIMAL DRUGS Animal Drugs; Redelegations Of Authority; Delegations Of Authority And Organization
WRITTEN PROCEDURES	211.100	WRITTEN PROCEDURES; DEVIATIONS Production And Process Controls; Current Good Manufacturing Practice For Finished Pharmaceuticals
WRITTEN PROCEDURES	803.17	WRITTEN MDR PROCEDURES General Provisions; Medical Device Reporting
WRITTEN PROCEDURES	1270.9	WRITTEN PROCEDURES Human Tissue Intended For Transplantation
WRITTEN RECOMMENDATIONS	316.10	CONTENT AND FORMAT OF A REQUEST FOR WRITTEN RECOMMENDATIONS Orphan Drugs
WRITTEN RECOMMENDATIONS	316.12	PROVIDING WRITTEN RECOMMENDATIONS Orphan Drugs
WRITTEN RECOMMENDATIONS	316.14	REFUSAL TO PROVIDE WRITTEN RECOMMENDATIONS Orphan Drugs
WRITTEN REQUEST	810.12	WRITTEN REQUEST FOR REVIEW OF CEASE DISTRIBUTION AND NOTIFICATION ORDER Mandatory Medical Device Recall Procedures; Medical Device Recall Authority

X

X, LIPOPROTEIN	866.5590	LIPOPROTEIN X IMMUNOLOGICAL TEST SYSTEM Immunological Test Systems; Immunology And Microbiology Devices
X-RAY	872.1800	EXTRAORAL SOURCE X-RAY SYSTEM Diagnostic devices; Dental Devices
X-RAY	872.1810	INTRAORAL SOURCE X-RAY SYSTEM Diagnostic Devices; Dental Devices
X-RAY	872.1820	DENTAL X-RAY EXPOSURE ALIGNMENT DEVICE Diagnostic Devices; Dental Devices
X-RAY	872.1840	DENTAL X-RAY POSITION INDICATING DEVICE Diagnostic Devices; Dental Devices
X-RAY	872.1905	DENTAL X-RAY FILM HOLDER Diagnostic Devices; Dental Devices
X-RAY	892.1600	ANGIOGRAPHIC X-RAY SYSTEM Diagnostic Devices; Radiology Devices
X-RAY	892.1610	DIAGNOSTIC X-RAY BEAM-LIMITING DEVICE Diagnostic Devices; Radiology Devices
X-RAY	892.1620	CINE OR SPOT FLUOROGRAPHIC X-RAY CAMERA Diagnostic Devices; Radiology Devices
X-RAY	892.1630	ELECTROSTATIC X-RAY IMAGING SYSTEM Diagnostic Devices; Radiology Devices
X-RAY	892.1650	IMAGE-INTENSIFIED FLUOROSCOPIC X-RAY SYSTEM Diagnostic Devices; Radiology Devices
X-RAY	892.1660	NON-IMAGE-INTENSIFIED FLUOROSCOPIC X-RAY SYSTEM Diagnostic Devices; Radiology Devices
X-RAY	892.1680	STATIONARY X-RAY SYSTEM Diagnostic Devices; Radiology Devices
X-RAY	892.1700	DIAGNOSTIC X-RAY HIGH VOLTAGE GENERATOR Diagnostic Devices; Radiology Devices
X-RAY	892.1710	MAMMOGRAPHIC X-RAY SYSTEM Diagnostic Devices; Radiology Devices
X-RAY	892.1720	MOBILE X-RAY SYSTEM Diagnostic Devices; Radiology Devices
X-RAY	892.1730	PHOTOFLUOROGRAPHIC X-RAY SYSTEM Diagnostic Devices; Radiology Devices
X-RAY	892.1740	TOMOGRAPHIC X-RAY SYSTEM Diagnostic Devices; Radiology Devices
X-RAY	892.1750	COMPUTED TOMOGRAPHY X-RAY SYSTEM Diagnostic Devices; Radiology Devices
X-RAY	892.1760	DIAGNOSTIC X-RAY TUBE HOUSING ASSEMBLY Diagnostic Devices; Radiology Devices
X-RAY	892.1770	DIAGNOSTIC X-RAY TUBE MOUNT Diagnostic Devices; Radiology Devices
X-RAY	892.5900	X-RAY RADIATION THERAPY SYSTEM Therapeutic Devices; Radiology Devices
X-RAY	892.5930	THERAPEUTIC X-RAY TUBE HOUSING ASSEMBLY Therapeutic Devices; Radiology Devices
X-RAY	1000.50	RECOMMENDATION FOR THE USE OF SPECIFIC AREA GONAD SHIELDING ON PATIENTS DURING MEDICAL DIAGNOSTIC X-RAY PROCEDURES Radiation Protection Recommendations; General
X-RAY	1000.60	RECOMMENDATIONS ON ADMINISTRATIVELY REQUIRED DENTAL X-RAY EXAMINATIONS Radiation Protection Recommendations; General
X-RAY	1020.30	DIAGNOSTIC X-RAY SYSTEMS AND THEIR MAJOR COMPONENTS Performance Standards For Ionizing Radiation Emitting Products
X-RAY	1020.40	CABINET X-RAY SYSTEM Performance Standards For Ionizing Radiation Emitting Products
XANTHAN GUM	172.695	XANTHAN GUM Gums, Chewing Gum Bases And Related Substances; Food Additives Permitted For Direct Addition To Food For Human Consumption
XANTHAN GUM	201.319	WATER-SOLUBLE GUMS, AND HYDROPHILIC MUCILLOIDS (INCLUDING GUAR GUM, KARAYA GUM, XANTHAN GUM) AS ACTIVE INGREDIENTS; REQUIRED WARNINGS AND DIRECTIONS Specific Labeling Requirements For Specific Drug Products; Labeling
XANTHAN GUM	573.1010	XANTHAN GUM Food Additive Listing; Food Additives Permitted In Feed And Drinking Water Of Animals
XANTHATE	573.300	CHOLINE XANTHATE Food Additive Listing; Food Additives Permitted In Feed And Drinking Water Of Animals

XYLAZINE HYDROCHLORIDE	522.2662	XYLAZINE HYDROCHLORIDE INJECTION Implantation Or Injectable Dosage Form New Animal Drugs
XYLENE	175.380	XYLENE-FORMALDEHYDE RESINS CONDENSED WITH 4,4'-ISOPROPYLIDENEDIPHENOL-EPICHLOROHYDRIN EPOXY RESINS Substances For Use As Components Of Coatings; Indirect Food Additives
XYLITOL	172.395	XYLITOL Special Dietary And Nutritional Additives; Food Additives Permitted For Direct Addition To Food For Human Consumption
XYLOSE	862.1820	XYLOSE TEST SYSTEM Clinical Chemistry Test Systems; Clinical Chemistry And Clinical Toxicology Devices

Y

YAG	886.4392	ND:YAG LASER FOR POSTERIOR CAPSULOTOMY AND PERIPHERAL IRIDECTOMY Surgical Devices; Ophthalmic Devices
YEAST	73.355	PHAFFIA YEAST Foods; Listing Of Color Additives Exempt From Certification
YEAST	172.325	BAKERS YEAST PROTEIN Special Dietary And Nutritional Additives; Food Additives Permitted For Direct Addition To Food For Human Consumption
YEAST	172.590	YEAST-MALT SPROUT EXTRACT Flavoring Agents And Related Substances; Food Additives Permitted For Direct Addition To Food For Human Consumption
YEAST	172.898	BAKERS YEAST GLYCAN Multipurpose Additives; Food Additives Permitted For Direct Addition To Food For Human Consumption
YEAST	184.1983	BAKERS YEAST EXTRACT Listing Of Specific Substances Affirmed As GRAS; Direct Food Substances Affirmed As Generally Recognized As Safe
YEAST	573.750	PICHIA PASTORIS DRIED YEAST Food Additives Permitted In Feed And Drinking Water Of Animals
YEASTS	172.896	DRIED YEASTS Multipurpose Additives; Food Additives Permitted For Direct Addition To Food For Human Consumption
YELLOW CORN	137.215	YELLOW CORN FLOUR Requirements For Specific Standardized Cereal Flours And Related Products; Cereal Flours And Related Products
YELLOW CORN	137.275	YELLOW CORN MEAL Requirements For Specific Standardized Cereal Flours And Related Products; Cereal Flours And Related Products
YELLOW CORN	137.280	BOLTED YELLOW CORN MEAL Requirements For Specific Standardized Cereal Flours And Related Products; Cereal Flours And Related Products
YELLOW CORN	137.285	DEGERMINATED YELLOW CORN MEAL Requirements For Specific Standardized Cereal Flours And Related Products; Cereal Flours And Related Products
YELLOW CORN	137.290	SELF-RISING YELLOW CORN MEAL Requirements For Specific Standardized Cereal Flours And Related Products; Cereal Flours And Related Products
YELLOW NO. 5	74.705	FD&C YELLOW NO. 5 Foods; Listing Of Color Additives Subject To Certification
YELLOW NO. 5	74.1705	FD&C YELLOW NO. 5 Drugs; Listing Of Color Additives Subject To Certification
YELLOW NO. 5	82.705	FD&C YELLOW NO. 5 Foods, Drugs, And Cosmetics; Listing Of Certified Provisionally Listed Colors And Specifications
YELLOW NO. 5/6	201.20	DECLARATION OF PRESENCE OF FD&C YELLOW NO. 5 OR FD&C YELLOW NO. 6 IN CERTAIN DRUGS FOR HUMAN USE General Labeling Provisions; Labeling
YELLOW NO. 6	74.706	FD&C YELLOW NO. 6 Foods; Listing Of Color Additives Subject To Certification
YELLOW NO. 6	74.1706	FD&C YELLOW NO. 6 Drugs; Listing Of Color Additives Requiring Certification
YELLOW NO. 6	74.2706	FD&C YELLOW NO. 6 Cosmetics; Listing Of Color Additives Requiring Certification
YELLOW NO. 6	82.706	FD&C YELLOW NO. 6 Foods, Drugs, And Cosmetics; Listing Of Certified Provisionally Listed Colors And Specifications
YELLOW NO. 7	74.1707	D&C YELLOW NO. 7 Drugs; Listing Of Color Additives Subject To Certification
YELLOW NO. 7	74.1707a	EXT. D&C YELLOW NO. 7 Drugs; Listing Of Color Additives Subject To Certification
YELLOW NO. 7	74.2707	D&C YELLOW NO. 7 Cosmetics; Listing Of Color Additives Subject To Certification
YELLOW NO. 7	74.2707a	EXT. D&C YELLOW NO. 7 Cosmetics; Listing Of Color Additives Subject To Certification
YELLOW NO. 7	82.1707	D&C YELLOW NO. 7 Drugs And Cosmetics; Listing Of Certified Provisionally Listed Colors And Specifications
YELLOW NO. 7	82.2707a	EXT. D&C YELLOW NO. 7 Externally Applied Drugs And Cosmetics; Listing Of Certified Provisionally Listed Colors And Specifications

YELLOW NO. 8	74.1708	D&C YELLOW NO. 8 Drugs; Listing Of Color Additives Subject To Certification
YELLOW NO. 8	74.2708	D&C YELLOW NO. 8 Cosmetics; Listing Of Color Additives Subject To Certification
YELLOW NO. 8	82.1708	D&C YELLOW NO. 8 Drugs And Cosmetics; Listing Of Certified Provisionally Listed Colors And Specifications
YELLOW NO. 10	74.1710	D&C YELLOW NO. 10 Drugs; Listing Of Color Additives Subject To Certification
YELLOW NO. 10	74.2710	D&C YELLOW NO. 10 Cosmetics; Listing Of Color Additives Subject To Certification
YELLOW NO. 10	74.3710	D&C YELLOW NO. 10 Medical Devices; Listing Of Color Additives Subject To Certification
YELLOW NO. 10	82.1710	D&C YELLOW NO. 10 Drugs And Cosmetics; Listing Of Certified Provisionally Listed Colors And Specifications
YELLOW NO. 11	74.1711	D&C YELLOW NO. 11 Drugs; Listing Of Color Additives Subject To Certification
YELLOW NO. 11	74.2711	D&C YELLOW NO. 11 Cosmetics; Listing Of Color Additives Subject To Certification
YELLOW PRUSSIATE	172.490	YELLOW PRUSSIATE OF SODA Anticaking Agents; Food Additives Permitted For Direct Addition To Food For Human Consumption
YELLOW PRUSSIATE	573.1020	YELLOW PRUSSIATE OF SODA Food Additive Listing; Food Additives Permitted In Feed And Drinking Water Of Animals
YIELD	211.103	CALCULATION OF YIELD Production And Process Controls; Current Good Manufacturing Practice For Finished Pharmaceuticals
YOGURT	131.200	YOGURT Requirements For Specific Standardized Milk And Cream; Milk And Cream
YOGURT	131.203	LOWFAT YOGURT Requirements For Specific Standardized Milk And Cream; Milk And Cream
YOGURT	131.206	NONFAT YOGURT Requirements For Specific Standardized Milk And Cream; Milk And Cream
YOHIMBINE	522.2670	YOHIMBINE INJECTION Implantation Or Injectable Dosage Form New Animal Drugs
YOKE	868.6885	MEDICAL GAS YOKE ASSEMBLY Miscellaneous; Anesthesiology Devices
YOLKS	160.180	EGG YOLKS Requirements For Specific Standardized Eggs And Egg Products; Eggs And Egg Products
YOLKS	160.185	DRIED EGG YOLKS Requirements For Specific Standardized Eggs And Egg Products; Eggs And Egg Products
YOLKS	160.190	FROZEN EGG YOLKS Requirements For Specific Standardized Eggs And Egg Products; Eggs And Egg Products

Z

ZEIN	184.1984	ZEIN Listing Of Specific Substances Affirmed As GRAS; Direct Food Substances Affirmed As Generally Recognized As Safe
ZERANOL	522.2680	ZERANOL Implantation Or Injectable Dosage Form New Animal Drugs
ZERANOL	556.760	ZERANOL Specific Tolerances For Residues Of New Animal Drugs; Tolerances For Residues Of New Animal Drugs In Food
ZINC	175.390	ZINC-SILICON DIOXIDE MATRIX COATINGS Substances For Use As Components Of Coatings; Indirect Food Additives
ZINC, BACITRACIN	524.154	BACITRACIN OR BACITRACIN ZINC-NEOMYCIN SULFATE-POLYMIXIN B SULFATE OPHTHALMIC OINTMENT Ophthalmic And Topical Dosage Form New Animal Drugs
ZINC, BACITRACIN	524.155	BACITRACIN ZINC-POLYMIXIN-B SULFATE-NEOMYCIN SULFATE-HYDROCORTISONE OR HYDROCORTISONE ACETATE OPHTHALMIC OINTMENT Ophthalmic And Topical Dosage Form New Animal Drugs
ZINC, BACITRACIN	558.78	BACITRACIN ZINC Specific New Animal Drugs For Use In Animal Feeds; New Animal Drugs For Use In Animal Feeds
ZINC CHLORIDE	182.8985	ZINC CHLORIDE Nutrients; Substances Generally Recognized As Safe
ZINC CHLORIDE	582.5985	ZINC CHLORIDE Nutrients And/or Dietary Supplements; Substances Generally Recognized As Safe
ZINC GLUCONATE	182.8988	ZINC GLUCONATE Nutrients; Substances Generally Recognized As Safe
ZINC GLUCONATE	582.5988	ZINC GLUCONATE Nutrients And/or Dietary Supplements; Substances Generally Recognized As Safe
ZINC METHIONATE SULFATE	172.399	ZINC METHIONINE SULFATE Special Dietary And Nutritional Additives; Food Additives Permitted For Direct Addition To Food For Human Consumption
ZINC OXIDE	73.1991	ZINC OXIDE Drugs; Listing Of Color Additives Exempt From Certification
ZINC OXIDE	73.2991	ZINC OXIDE Cosmetics; Listing Of Color Additives Exempt From Certification
ZINC OXIDE	182.8991	ZINC OXIDE Nutrients; Substances Generally Recognized As Safe
ZINC OXIDE	582.5991	ZINC OXIDE Nutrients And/or Dietary Supplements; Substances Generally Recognized As Safe
ZINC, SOMETRIBOVE	522.2112	STERILE SOMETRIBOVE ZINC SUSPENSION Implantation Or Injectable Dosage Form New Animal Drugs
ZINC STEARATE	182.8994	ZINC STEARATE Nutrients; Substances Generally Recognized As Safe
ZINC STEARATE	582.5994	ZINC STEARATE Nutrients And/or Dietary Supplements; Substances Generally Recognized As Safe
ZINC SULFATE	182.8997	SULFATE Nutrients; Substances Generally Recognized As Safe
ZINC SULFATE	582.5997	ZINC SULFATE Nutrients And/or Dietary Supplements; Substances Generally Recognized As Safe
ZINC SULFIDE	73.2995	LUMINESCENT ZINC SULFIDE Cosmetics; Listing Of Color Additives Exempt From Certification
ZIRCONIUM	700.16	USE OF AEROSOL COSMETIC PRODUCTS CONTAINING ZIRCONIUM Requirements For Specific Cosmetic Products; General
ZOALENE	556.770	ZOALENE Specific Tolerances For Residues Of New Animal Drugs; Tolerances For Residues Of New Animal Drugs In Food
ZOALENE	558.080	ZOALENE Specific New Animal Drugs For Use In Animal Feeds; New Animal Drugs For Use In Animal Feeds
ZOLAZEPAM	522.2470	TILETAMINE HYDROCHLORIDE AND ZOLAZEPAM HYDROCHLORIDE FOR INJECTION Implantation Or Injectable Dosage Form New Animal Drugs
ZONE READER	866.2850	AUTOMATED ZONE READER Microbiology Devices; Immunology And Microbiology Devices
ZOSTER	866.3900	VARICELLA-ZOSTER VIRUS SEROLOGICAL REAGENTS Serological Reagents; Immunology And Microbiology Devices